MEDICAL PHYSIOLOGY

PRINCIPLES FOR CLINICAL MEDICINE

FIFTH EDITION

BMA

MEDICAL PHYSIOLOGY

PRINCIPLES FOR CLINICAL MEDICINE

FIFTH EDITION

EDITED BY

RODNEY A. RHOADES, PHD

Professor Emeritus
Department of Cellular and Integrative Physiology
Indiana University School of Medicine
Indianapolis, Indiana

DAVID R. BELL, PHD

Associate Professor
Department of Cellular and Integrative Physiology
Indiana University School of Medicine–Fort Wayne
Fort Wayne, Indiana

 Wolters Kluwer

Philadelphia • Baltimore • New York • London
Buenos Aires • Hong Kong • Sydney • Tokyo

Acquisitions Editor: Crystal Taylor
Development Editor: Andrea Vosburgh
Editorial Coordinator: Annette Ferran
Marketing Manager: Michael McMahon
Production Project Manager: Bridgett Dougherty
Design Coordinator: Stephan Druding
Art Manager: Jennifer Clements
Manufacturing Coordinator: Margie Orzech
Prepress Vendor: SPi Global

5th edition

Library of Congress Cataloging-in-Publication Data
Names: Rhoades, Rodney, editor. | Bell, David R., 1952- editor.
Title: Medical physiology : principles for clinical medicine / edited by Rodney A. Rhoades, David R. Bell.
Other titles: Medical physiology (Rhoades)
Description: Fifth edition. | Philadelphia : Wolters Kluwer, [2018] | Includes index.
Identifiers: LCCN 2017007248 | ISBN 9781496310460
Subjects: | MESH: Physiological Phenomena
Classification: LCC QP34.5 | NLM QT 104 | DDC 612—dc23 LC record available at https://lccn.loc.gov/2017007248

PREFACE

Human physiology is the science that explains how cells, tissues, and organs interact and function as an integrated system. The fifth edition of *Medical Physiology: Principles for Clinical Medicine* provides the latest information of how the body's different systems work to allow it to cope with changes in its internal and external environment. In so doing, it addresses how the body maintains optimal health and ensures survival. Although the emphasis of the fifth edition is on normal physiology, discussion of pathophysiology is also undertaken to show how altered functions are involved in disease processes. We have also created a new type of chapter essay for this edition that highlights how physiology integrates with others medical sciences, such as those involving pharmacology, biochemistry, genetics, and clinical diagnostics. This enrichment of basic physiology reinforces fundamental physiologic principles while also demonstrating how basic concepts in physiology connect to clinical medicine.

Our mission for this edition has developed out of our decades of teaching and mentoring medical students and the feedback we have received from them. We have created a multicomponent learning resource to address three questions that medical students have told us most concern them in their academic training; these are, *"What should I know?"*, *"How do I know that I know it well?"*, and *"How does what I learn fit into medicine?"* Each component of this textbook presents a learning opportunity with these questions in mind. We have attempted to maximize those opportunities to the fullest, while providing a clear, accurate, and up-to-date introduction to medical physiology.

▶ AUDIENCE AND FUNCTION

This book, like the previous edition, is written for medical students but will be useful to dental, graduate nursing, and veterinary students as well. This is neither an encyclopedic textbook nor is it intended to be a condensed, descriptive review. Rather, the fifth edition focuses on the basic key physiologic principles necessary to understand human function and its fundamental context in clinical medicine. An additional important objective for the fifth edition is to demonstrate to the student that physiology is key to understanding other medical sciences such as pharmacology and pathophysiology. Although the book is written primarily with the student in mind, the fifth edition will also be a clear, concise, and helpful reference for physicians and other health care professionals.

All chapters were substantially revised, updated, and edited to achieve unity of voice and to be as concise and lucid as possible. In the fifth edition, each chapter is written to eliminate minutia and minimize the compilation of isolated facts. The chapters are written by medical school faculty members who are experts in their field and who have had decades of experience teaching physiology. They have focused their material on what is important for medical students to know. We have purposefully avoided discussion of research laboratory methods and/or historical material. Although such issues are important in other contexts, most medical students are too busy to be burdened by such information and prefer to focus on the essentials. We have also avoided topics that are as yet unsettled, while recognizing that new research constantly provides fresh insights and sometimes challenges old ideas.

▶ CONTENT AND ORGANIZATION

This book begins with a discussion of basic physiological concepts, such as homeostasis and cell signaling, in Chapter 1. Chapter 2 covers the cell membrane, membrane transport, and the cell membrane potential. Most of the remaining chapters discuss the different organ systems: nervous (Chapters 3–7), muscle (Chapter 8), cardiovascular (Chapters 11–17), respiratory (Chapters 18–21), renal (Chapters 22–23), gastrointestinal (Chapters 25–26), endocrine (Chapters 30–35), and reproductive physiology (Chapters 36–38). Special chapters on the blood (Chapter 9) and immunology (Chapter 10) are included. The immunology chapter emphasizes physiological applications of immunology. Chapters on acid–base regulation (Chapter 24), temperature regulation (Chapter 28), and exercise (Chapter 29) discuss these complex, integrated functions. The order of presentation of topics follows that of most United States medical school courses in physiology. After the first two chapters, the other chapters may be read pretty much as stand-alone units, and some chapters may be skipped if the subjects are taught in other courses (e.g., neurobiology or immunology).

▶ CHANGES FOR THE FIFTH EDITION

For the fifth edition, new expert contributing authors have been brought on board to update and rewrite major sections of the text and their associated ancillary learning tools. The entire Neurophysiology, Gastrointestinal, and Renal physiology sections of the text, as well as the chapters on Blood and Immunology, have been revised by these new authors.

In addition, in response to requests from students and instructors from around the globe, we have moved a portion of our formative assessment tools, formally only available online, into the print copy of the fifth edition of *Medical Physiology: Principles for Clinical Medicine*. Annotated, multiple choice review questions, in which explanations for the right and wrong answers are provided, as well as similar, annotated *Clinical Application Exercises* have been moved into the print copy of the new edition. The print copy, however, also retains links to additional online review questions, clinical application exercises, and advanced clinical problem-solving exercises.

Finally, we have replaced the Bench to Bedside essays found in the previous edition with new *Integrated Medical Sciences* essays. This new feature for the fifth edition is designed to highlight for the novice student the connections between physiology and the other basic medical sciences they are studying, such as pharmacology, pathology, and introductory clinical medicine.

▶ KEY FEATURES AND LEARNING TOOLS

As with previous editions, the fifth edition includes numerous ancillary student-learning tools as complements to the basic text.

Active Learning Objectives

Medical students deal with huge amounts of information from their courses and an ever-expanding electronic medium. This often makes the task of determining what they really need to know difficult for them. Students benefit greatly if they have a set of clear objectives placed in front of them *before* trying to sift through their learning materials. Our active learning objectives direct the student to apply the concepts and processes contained in the chapter, rather than memorize facts. Basic descriptors or table of content-type lists at the start of a chapter do not tell students what they should really know. Simply naming items in a chapter that the student has not yet learned does not help students cope with what they should do with material once it is mastered. *Active Learning Objectives* at the start of each chapter in this text are purposely designed to indicate what a student should be able to do with chapter material once it has been mastered. Medical students need to have understanding of the workings of physiology to become competent problem solvers. Toward that end, our *active learning objectives* direct students to explain, predict, and postulate rather than simply describe, define, and recite.

Annotated Chapter Review Questions, Clinical Application Exercises, and Advanced Clinical Problem-Solving Exercises

Providing students with formative assessment tools is essential if they are to determine what they know, whether they know it well, and what they do not know or do not know well. The fifth edition of *Medical Physiology: Principles for Clinical Medicine* provides the student with a multitiered approach to formative self-assessment. This edition provides student with over 350 USMLE-type multiple choice questions that are organized by chapter content and keyed with explanations for both the right and all the wrong answers. In this manner, students are able to gauge their understanding of the physiology they are learning at the moment. It is very frustrating for students to study hard, get a practice question wrong, and then be clueless as to both why another choice was right and theirs and the remaining choices wrong. We have found that the inability to recognize wrong choices as such is a leading cause of poor student

performance. By providing students with explanation for the wrong as well as the right answers with their chapter practice questions, we allow them to identify gaps in their understanding while at the same time avoiding the frustration that arises when they are given review questions without explanations for the answers. Just as important, when given a complete explanation for the choices in a question, students can better determine whether they know material well.

In the fifth edition, we take this type of formative assessment to additional, more complex levels in order to hone student problem-solving skills. We provide two *Clinical Application Exercises* with each of the 38 chapters in the book, with at least one of these placed in the print copy. These exercises are small clinical vignettes purposely circumscribed to the chapter content only. Multiple questions are asked of the student based on the vignette and the chapter content. Explanations are provided for the answers to all the questions. This type of "story problem" method of developing problem-solving skills differs from the typical case study format sometimes used in medical education. In the latter, it is not uncommon to include materials in the case for which the student has not yet learned or to which they have not yet been introduced. This can be counterproductive in that it can confound and misdirect students as they try to focus on the solution to the problem at hand. They can encounter difficulties with solving the case study because they have yet to be taught certain clinical information and complexity, rather than because their understanding of what they are currently learning is lacking. The latter is most important to them when trying to master physiology. Our *Clinical Application Exercises* are focused. They are designed to help students better learn how to apply the physiology they are learning at the moment to real clinically relevant problems. As such, this type of problem-solving exercise further helps the student identify what physiology they understand well and what they do not. In addition, this type of exercise helps students immediately see the clinical relevance of the physiology they are learning; it shows them where the physiology fits into clinical medicine.

Lastly, the online resources that accompany this book contain 38 *Advanced Clinical Problem-Solving Exercises*. These are longer, more involved clinical scenarios, each with multiple questions asked and explanations for the answers provided. These advanced exercises draw on a student's understanding of multiple disciplines within physiology as well as other biomedical fields. These exercises are more like true clinical case studies. They allow a student to evaluate their ability to integrate multiple disciplines within physiology and apply their collective understanding toward the answers to a multifaceted clinical problem. These advanced exercises elevate a student's problem-solving skills and further enhance their ability to determine what they do and do not know well. The exercises further illustrate for the student how the study of physiology fits into clinical medicine.

Clinical Focus Essays and Integrated Medical Sciences Essays

For the fifth edition, the *Clinical Focus* boxes of each chapter have again been updated. These short essays deal with pathophysiology, physiopharmacology, and clinical correlates of physiology, including those involved with basic therapeutics and clinical evaluation tools. A new feature, the *Integrated Medical Sciences* essay, has been added to each chapter in the fifth edition. These essays address a growing trend in medical student education of integrating the whole of medical sciences, basic and clinical, within given units of instruction. It is now common for students to be taught pathology, pathophysiology, pharmacology, and introductory medicine simultaneously, so as to create a fuller understanding of a particular disease or medical condition. The *Integrated Medical Sciences Essay* contained in each chapter is directed at connecting the physiology within a chapter to another type of medical science. Together both types of chapter-based essays gives students additional insight into the connection between physiology and the understanding of human disease and its treatment.

▶ ADDITIONAL EDUCATIONAL FEATURES

The fifth edition incorporates many features designed to facilitate learning and guide the student along his or her study of physiology. In-print features included in the fifth edition are as follows.

- **Illustrations and Tables.** The text again contains abundant full-color figures and flow diagrams. Review tables are also provided as useful summaries of material explained in more detail in the text. The illustrations in the text often show interrelationships between different variables or components of a system. These color illustrations are more than just visually appealing. As first employed in the fourth edition, the fifth edition of the text continues the use of color art as an instructive tool. Rather than applying color arbitrarily, color itself is used with purpose and delivers meaning. Graphs, diagrams, and flow charts, for example, incorporate a coordinated scheme; red is used to indicate stimulatory, augmented, or increased effects, whereas blue connotes inhibitory, impaired, or decreased effects.

  A coordinated color scheme is likewise used throughout to depict transport systems. This key, in which membrane pores are blue, primary active transporters are red, facilitated transporters are purple, cell chemical receptors are green, cotransporters are orange, and voltage-gated transporters are yellow, adds a level of instructiveness to the figures not seen in other physiology textbooks. By differentiating these elements integral to the workings of physiology by their function, the fifth edition artwork reinforces their purpose to teach students, rather than merely representing. These beautiful full-color conceptual diagrams guide students to an understanding of the general underpinnings of physiology. Figures work with text to provide meaningful, comprehensible content.

- **Bulleted Chapter Summaries.** These bulleted statements provide a concise summative description of the chapter and provide a good review checklist of the chapter.

- **Key Concept Subheadings.** Secondary chapter subheadings are depicted in bold in the text and written as active concept statements designed to convey the key point(s) of a given section. Unlike typical textbook subheadings that simply title a section, these are given in active full sentence form. For example, instead of heading a section "Edema," the heading instead becomes "Edema impairs diffusional transport across capillaries." In this way, the key idea in a section is immediately obvious. When taken together in a chapter, these statement subheadings give the student another means of chapter review.

- **Boldfacing.** Key terms are boldfaced upon their first appearance in a chapter. These terms are explained in the text and defined in the glossary for quick reference.

- **Abbreviations and Normal Values.** An appendix of common abbreviations in physiology and a table of normal blood, plasma, or serum values are included inside the book covers for convenient access. All abbreviations are defined when first used in the text, but the table of abbreviations in the appendix serves as a useful quick access of abbreviations commonly used in physiology and medicine. Normal values for blood are also embedded in the text, but the table on the inside front and back covers provides a more complete and easily accessible reference.

- **Index.** A comprehensive index allows the student to easily look up material in the text.

- **Glossary.** A glossary of all boldfaced terms in the text is included for quick access to definition of terms. Students will appreciate the book's inclusion of such a helpful and useful tool.

Readability and Design

The text is a pleasure to read, and topics are developed logically. Difficult concepts are explained clearly, in a unified voice, and supported with plentiful illustrations. Minutiae and esoteric topics are avoided. The fifth edition interior design not only makes navigating the text easier but also draws the reader in with immense visual appeal and strategic use of color. Likewise, the design highlights the pedagogical features, making them easier to find and use.

Ancillary Package

Still more features round out the colossal ancillary package online at www.thePoint.com/Rhoades. These bonus offerings provide ample opportunities for self-assessment, additional reading on tangential topics, and animated versions of the artwork to further elucidate the more complex concepts. Look for this icon thePoint® appearing throughout the text indicating associated online features.

- **Additional Formative Assessments.** In addition to the formative assessment tools now placed in the print copy of the textbook, additional chapter review questions and clinical application exercises are available online along with advanced clinical problem solving exercises. As with the print copy of the textbook, all questions are analytical in nature and test the student's ability to apply physiological principles to solving problems rather than test basic fact-based recall. They contain explanations for right and wrong answers. These chapter-based questions were written by the author of the corresponding chapter and not contracted out to a question-writing service.

- **Suggested Reading.** A short list of recent review articles, monographs, book chapters, classic papers, or Web sites where students can obtain additional information associated with each chapter is provided online.

- **Animations.** The fifth edition contains online animations illustrating difficult physiology concepts.

- **Image Bank for Instructors.** An image bank containing all of the figures in the book, in both pdf and jpeg formats, is available for download from our Web site at thePoint®.

- **Instructor Test Bank.** Also, for the fifth edition, the extensive test bank written by subject matter experts has been updated and expanded for instructors using this textbook in their course.

In closing, we would like to add that the discipline of physiology is changing. In the past 20 years, there has been an overemphasis of genetics placed into our understanding of "what makes us tick." This has resulted in the belief that genes control human biology, which has had a major impact on medical physiology and the way diseases are treated. This has recently changed with the emergence of the new science of epigenetics, which means regulation above the gene. Epigenetics can be defined as a mechanism by which genes can be switched off and on, but the genes themselves and their genetic code are not altered. This means protein synthesis and other cellular functions can be controlled above the level of the gene. New research shows that lifestyle and environmental signals can modify and regulate gene activity and shows such things as exercise, nutrition, stress, trauma, emotions, attitude, social engagement, and toxins can modify gene function without altering the genetic code. More striking is that these epigenetic modifications can be passed onto the next generation. Epigenetics is upending our understanding of physiologic regulation and has opened up a new frontier in human physiology and the future of medicine. Physiology is moving from structure–function relationships to epigenetics-induced functional relationships. This has resulted in new research that shows only approximately 15% of chronic diseases (e.g., hypertension, heart disease, asthma, obesity, diabetes, osteoporosis, arthritis, cancer, etc.) are specifically linked to genetics. The remaining 85% are factors due to lifestyle choices (e.g., diet, exercise, physical stress, emotional stress) and to environmental factors (toxins, smoking, pesticides, substance abuse, etc.). The fifth edition of *Medical Physiology: Principles of Clinical Medicine* provides all students with a strong foundation to be built upon by this new and exciting field of epigenetics.

We would like to express our deepest thanks and appreciation to all of the contributing authors. Without their expertise and cooperation, this fifth edition would have not been possible. We also wish to express our appreciation to all of our students and colleagues who have provided helpful comments and criticisms during the revision of this book. We would also like to give thanks for a job well done to our editorial staff for their guidance and assistance in significantly improving each edition of this book. A very special thanks goes to our Developmental Editor, Kelly Horvath, who was a delight to work with and whose patience and editorial talents were essential to the completion of the fifth edition of this book. We are indebted as well to our artist, Jennifer Clements. Finally, we would like to thank Crystal Taylor, our Acquisitions Editor at Wolters Kluwer, for her support, vision, and commitment to this book. We are indebted to her administrative talents and her managing of the staff and material resources for this project.

Last we would like to thank our wives, Pamela Bell and Judy Rhoades, for their love, patience, support, and understanding of our need to devote a great deal of personal time and energy to the development of this book.

Rodney A. Rhoades, PhD
David R. Bell, PhD

CONTRIBUTORS

David R. Bell, PhD
Associate Professor
Department of Cellular and Integrative Physiology
Indiana University School of Medicine–Fort Wayne
Fort Wayne, Indiana

Bonnie L. Blazer-Yost, PhD
Professor
Departments of Biology, Integrative and Cellular
    Physiology, and Anatomy and Cell Biology
Indiana University–Purdue University
Indianapolis, Indiana

Robert V. Considine, PhD
Associate Professor
Departments of Medicine and Physiology
Indiana University School of Medicine
Indianapolis, Indiana

Jeffrey S. Elmendorf, PhD
Associate Professor of Cellular and Integrative Physiology
Department of Physiology
Indiana University School of Medicine
Indianapolis, Indiana

Jennelle Durnett Richardson, PhD
Assistant Professor
Department of Clinical Pharmacology and Toxicology
Indiana University School of Medicine
Indianapolis, Indiana

Rodney A. Rhoades, PhD
Professor Emeritus
Department of Cellular and Integrative Physiology
Indiana University School of Medicine
Indianapolis, Indiana

Denise Slayback-Barry, PhD
Lecturer
Department of Biology
Indiana University–Purdue University Indianapolis
Indianapolis, Indiana

Robert Sweazey, PhD
Associate Professor of Anatomy and Cell Biology
Indiana University School of Medicine–Fort Wayne
Fort Wayne, Indiana

Frank A. Witzmann, PhD
Professor
Department of Cellular and Integrative Physiology
Indiana University School of Medicine
Indianapolis, Indiana

Robert W. Yost, PhD
Senior Lecturer
Department of Biology
Indiana University–Purdue University
Indianapolis, Indiana

CONTENTS

Preface v

Contributors ix

PART I — CELLULAR PHYSIOLOGY — 1

CHAPTER 1 — Medical Physiology: An Overview — 1
Scope of Medical Physiology 1
Future Direction of Medical Physiology 2

CHAPTER 2 — Cell Signaling, Membrane Transport, and Membrane Potential — 5
Basis of Physiologic Regulation 5
Plasma Membrane Structure 8
Solute Transport Mechanisms 9
Water Movement across the Plasma Membrane 18
Resting Membrane Potential 20
Communication and Signaling Modes 22
Molecular Basis of Cellular Signaling 23
Second Messengers 27

PART II — NEUROMUSCULAR PHYSIOLOGY — 34

CHAPTER 3 — Action Potential, Synaptic Transmission, and Nerve Function — 34
The Nervous System 34
Action Potentials 37
Synaptic Transmission 43
Neurotransmission 45

CHAPTER 4 — Sensory Physiology — 55
Sensory Systems 55
Somatosensory System 60
Visual System 62
Auditory System 69
Vestibular System 75
Gustatory and Olfactory Systems 78

CHAPTER 5 — Motor System — 86
Skeleton as Framework for Movement 86
Muscle Function and Body Movement 86
Nervous System Components for the Control of Movement 87
Spinal Cord in the Control of Movement 90
Supraspinal Influences on Motor Control 93
Cerebral Cortex Role in Motor Control 95
Basal Ganglia and Motor Control 98
Cerebellum in the Control of Movement 100

CHAPTER 6 — Autonomic Nervous System — 105
Anatomy of the Autonomic Nervous System 105
Neurotransmitters of the Autonomic Nervous System 106

The Parasympathetic Nervous System   110
Sympathetic Nervous System   112
Autonomic Integration   115

CHAPTER 7   Integrative Functions of the Central Nervous System   124

Hypothalamus   124
Brain Electrical Activity   132
Functional Components of the Forebrain   134
Higher Cognitive Skills   138

CHAPTER 8   Skeletal and Smooth Muscle   144

Skeletal Muscle   144
Motor Neurons and Excitation—Contraction Coupling in Skeletal Muscle   146
Mechanics of Skeletal Muscle Contraction   152
Skeletal Muscle Metabolism and Fiber Types   160
Muscle Plasticity, Epigenetics, and Endocrine Muscle   161
Smooth Muscle   161

PART III   BLOOD AND IMMUNOLOGY   173

CHAPTER 9   Blood Composition and Function   173

Blood Functions   173
Whole Blood   174
Soluble Components of Blood and their Tests   174
Formed Elements of Blood and Common Diagnostic Tests   177
Red Blood Cells   180
White Blood Cells   183
Platelet Formation   185
Blood Cell Formation   185
Blood Clotting   187

CHAPTER 10   Immunology, Organ Interaction, and Homeostasis   195

Immune System Components   195
Immune System Activation   196
Immune Detection System   198
Immune System Defenses   198
Cell-Mediated and Humoral Responses   201
Acute and Chronic Inflammation   208
Chronic Inflammation   210
Anti-inflammatory Drugs   211
Organ Transplantation and Immunology   212
Immunologic Disorders   213
Neuroendoimmunology   215

PART IV   CARDIOVASCULAR PHYSIOLOGY   221

CHAPTER 11   Overview of the Cardiovascular System and Hemodynamics   221

Functional Organization   222
Physics of Blood Containment and Movement   223
Physical Dynamics of Blood Flow   227
Distribution of Pressure, Flow, Velocity, and Blood Volume   233

CHAPTER 12   Electrical Activity of the Heart   237

Electrophysiology of Cardiac Muscle   237

Pathophysiology of Abnormal Generation of Cardiac Action Potentials   242
The Electrocardiogram   244

CHAPTER 13 Cardiac Muscle Mechanics and the Cardiac Pump   260
Cardiac Excitation–Contraction Coupling   260
The Cardiac Cycle   263
Determinants of Myocardial Performance   265
Determinants of Myocardial Oxygen Demand and Clinical Evaluation of
Cardiac Performance   270
Cardiac Output   272
The Measurement of Cardiac Output   274
Imaging Techniques for Measuring Cardiac Structures, Volumes,
Blood Flow, and Cardiac Output   275

CHAPTER 14 The Systemic Circulation   282
Determinants of Arterial Pressures   282
Arterial Pressure Measurement   284
Peripheral and Central Blood Volume   287
Coupling of Vascular and Cardiac Function   288

CHAPTER 15 Microcirculation and Lymphatic System   296
Structure and Function of the Microcirculation   296
The Lymphatic System   298
Solute Exchange between the Vasculature and Tissues   299
Water Exchange between the Vasculature and Interstitium   301
Regulation of Microvascular Resistance   304

CHAPTER 16 Special Circulations   315
Coronary Circulation   315
Cerebral Circulation   317
Circulation of the Small Intestine   320
Hepatic Circulation   322
Skeletal Muscle Circulation   323
Cutaneous Circulation   324
Fetal and Placental Circulations   326

CHAPTER 17 Control Mechanisms in Cardiovascular Function   334
Autonomic Neural Control of the Cardiovascular System   334
Hormonal Control of the Cardiovascular System   340
Circulatory Shock   344

PART V RESPIRATORY PHYSIOLOGY   352

CHAPTER 18 Ventilation and the Mechanics of Breathing   352
Lung Structural and Functional Relationships   353
Pulmonary Pressures and Airflow during Breathing   354
Spirometry and Lung Volumes   359
Minute Ventilation   362
Elastic Properties of Lung and Chest Wall   366
Airway Resistance and the Work of Breathing   373

CHAPTER 19 Gas Transfer and Transport   383
Gas Diffusion and Uptake   383
Diffusing Capacity   385
Gas Transport by the Blood   386
Respiratory Causes of Hypoxemia   389

CHAPTER 20   Pulmonary Circulation and Ventilation/Perfusion   398
Functional Organization   398
Hemodynamic Features   399
Fluid Exchange in Pulmonary Capillaries   403
Blood Flow Distribution in the Lungs   404
Shunts and Venous Admixture   407

CHAPTER 21   Control of Ventilation   412
Neural and Voluntary Control of Breathing   412
Neural Reflexes in The Control of Breathing   415
Physiologic Responses to Altered Oxygen and Carbon Dioxide   418
Control of Breathing during Sleep   421
Control of Breathing in Unusual Environments   423

PART VI   RENAL PHYSIOLOGY AND BODY FLUIDS   430

CHAPTER 22   Kidney Function   430
Overview of Renal Function   430
Nephron: Functional Unit of The Kidney   431
Renal Blood Flow   434
Glomerular Filtration   435
Glomerular Hemodynamic Forces   437
Tubular Reabsorption   439
Tubule Secretion   444
Urinary Concentration Mechanisms   446
Renal Clearance and Assessing Glomerular Function   451
Micturition   455

CHAPTER 23   Regulation of Fluid and Electrolyte Balance   460
Fluid Compartments of the Body   460
Fluid Balance   464
Disturbances in Fluid–Electrolyte Balance   467
Sodium Balance   468
Potassium Balance   475
Calcium Balance   477
Magnesium Balance   478
Phosphate Balance   479

CHAPTER 24   Acid–Base Homeostasis   485
Basic Principles of Acid–Base Interaction   485
Metabolic Production of Acids   487
Integration of the Body's Buffering Systems   488
Regulation of Intracellular pH   496
Physiologic Disturbances of Acid–Base Balance   496

PART VII   GASTROINTESTINAL PHYSIOLOGY   508

CHAPTER 25   Gastrointestinal System Functions   508
Functional Overview of Digestive System   508
Salivary Secretion   509
Gastric Secretion   511
Pancreatic Secretion   514

Biliary Secretion 518
Intestinal Secretion 521
Carbohydrate Digestion and Absorption 522
Lipid Digestion and Absorption 525
Protein Digestion and Absorption 528
Vitamin Absorption 531
Electrolyte and Mineral Absorption 533
Water Absorption 536

CHAPTER 26 **Liver Functions and Immune Surveillance**   **541**
Liver Structure and Function 541
Drug Metabolism in the Liver 544
Energy Metabolism in the Liver 545
Protein and Amino Acid Metabolism in the Liver 549
Liver as a Nutrient Storage Organ 550
Endocrine Functions of the Liver 552
Liver and Immune Responses 553

CHAPTER 27 **Motility and Gastrointestinal Regulation**   **557**
Organization of the Digestive System 557
Gastrointestinal System Motility 560
Esophageal and Gastric Motility 562
Small Intestinal Motility 564
Large Intestinal Motility 565
Smooth Muscle Contraction 569
Neural Control of Gut Motility and Digestive Function 571
Synaptic Transmission in the Enteric Nervous System 574
Enteric Motor Neurons 576

PART VIII   TEMPERATURE REGULATION AND EXERCISE PHYSIOLOGY   **587**

CHAPTER 28 **Regulation of Body Temperature**   **587**
Body Temperature and Heat Transfer 587
Balance between Heat Production and Heat Loss 590
Metabolic Rate and Heat Production at Rest 591
Heat Dissipation 594
Thermoregulatory Control 598
Thermoregulatory Responses during Exercise 601
Heat Acclimatization 603
Responses to Cold 604
Clinical Aspects of Thermoregulation 606

CHAPTER 29 **Exercise Physiology**   **614**
Oxygen Uptake and Exercise 614
Cardiovascular Responses to Exercise 615
Respiratory Responses to Exercise 619
Skeletal Muscle and Bone Responses to Exercise 621
Obesity, Aging, and Immune Responses to Exercise 623

PART IX   ENDOCRINE PHYSIOLOGY   **628**

CHAPTER 30 **Endocrine Control Mechanisms**   **628**
General Endocrine Concepts 628
Chemical Nature of Hormones 631

Measurement of Circulating Hormones   633
Mechanisms of Hormone Action   637

CHAPTER 31   Hypothalamus and the Pituitary Gland   643
Hypothalamic–Pituitary Axis   643
Posterior Pituitary Hormones   645
Anterior Pituitary Hormones   646

CHAPTER 32   Thyroid Gland   660
Thyroid Hormone Synthesis, Secretion, and Metabolism   660
Thyroid Hormone Effects on the Body   664
Abnormalities of Thyroid Function in Adults   667

CHAPTER 33   Adrenal Gland   673
Adrenal Cortex Synthesizes and Secretes Steroid Hormones   673
Adrenal Medulla Catecholamines   683

CHAPTER 34   Endocrine Pancreas   689
Islets of Langerhans   689
Mechanisms of Islet Hormone Synthesis and Secretion   690
Insulin and Glucagon Action   694
Diabetes Mellitus   698

CHAPTER 35   Endocrine Regulation of Calcium,
Phosphate, and Bone Homeostasis   705
Overview of Calcium and Phosphate in the Body   705
Calcium and Phosphate Metabolism   707
Plasma Calcium and Phosphate Regulation   709
Bone Dysfunction   713

PART X   REPRODUCTIVE PHYSIOLOGY   718

CHAPTER 36   Male Reproductive System   718
Endocrine Glands of the Male Reproductive System   718
Testicular Function and Regulation   718
Spermatogenesis   724
Endocrine Function of the Testis   726
Androgen Action and Male Development   728
Male Reproductive Disorders   731

CHAPTER 37   Female Reproductive System   736
Hormonal Regulation of the Female Reproductive System   736
Female Reproductive Organs   737
Ovarian Cycle   739
Menstrual Cycle   744
Infertility   749

CHAPTER 38   Fertilization, Pregnancy, and Fetal Development   753
Fertilization and Implantation   753
Placental Nutrient Uptake, Waste Elimination, and Gas Exchange   756
Hormones Required for a Successful Pregnancy   757
Postpartum Lactation   762
Puberty Onset   763
Sexual Development   765

Appendix A: Common Abbreviations in Physiology   773

Appendix B: Normal Blood, Plasma, or Serum Values   776

Glossary   779

Index   835

the**Point**® *Visit* http://thepoint.lww.com/rhoades5e *for additional chapter review Q&A, Clinical Application Exercises, animations, and more!*

1

Medical Physiology: An Overview

Human physiology is the science that explains how cells, tissues, and organs interact to allow the body to function while coping with changes in the internal and external environment. Accordingly, an important facet of physiology is to examine how the body's different systems are integrated to maintain optimal health and our survival.

▶ SCOPE OF MEDICAL PHYSIOLOGY

This book is designed to examine the connection between the basic sciences, human function, and health. As such, the basic concepts discussed in this book highlight the relationship between physiology and foundational principles for clinical medicine. Often, interns, residents, and practicing physicians become so focused on medical diagnoses, bodily traumas, and the treatment of diseases that they lose sight of the most important part of medicine—that is, *living* and *thriving*. Long before any disease leads to dysfunction and threatens a person's life, the human body is faced with difficulties and constraints to its survival that are dictated by natural chemical, physical, and biological laws. The body expends enormous energy to maintain normal function and to survive. In short, the body encounters many challenges to simply *stay alive*.

Organ systems are designed to regulate the body's internal environment.

The human body is made up of more than 65 trillion cells. That is correct—over 65 *trillion*, not 65 billion—that are organized into specialized tissues and organs that regulate our internal environment in a manner compatible to sustain cell function and survival. Organ systems are responsible for regulating many essential cellular processes to maintain the physical and chemical conditions of the extracellular fluid within a narrow range. These include water volume and osmolality, essential electrolyte concentrations, metabolic substrates, oxygen and carbon dioxide gas concentrations, pH, and temperature, to name a few. The ability to maintain a relative consistency in the chemical and physical environment surrounding the cells of our body, in the face of a variable external environment, is called **homeostasis**.

The ability to control our internal environment against the very intense challenges to our survival is part of the scientific essence of physiology. For example, the human body can be viewed as a warm, wet organism that has to survive in a cool, dry, and harsh world. The energy our body needs to keep all cellular processes running is derived from the oxidation of basic organic molecules. However, this essential metabolic process consumes enormous amounts of oxygen and, in the process, dumps enough carbon dioxide into the extracellular fluid to potentially quickly lower the blood pH from 7.4 to 1. Therefore, the body must contend constantly with the threat of dehydration, hypothermia, and acidosis. The metabolic fuel required for energy-producing oxidation reactions in our body is needed continually by the cells. Some organs, particularly the brain, can only use glucose as a source of energy for this purpose. Moreover, the nutrient sources that originate from the foods in our diet comprise complex polymer molecules of highly different chemical natures. These complex sources require energy and coordinated processing in the body in order to be rendered into monomeric substrates that can be taken up and utilized by the cells. The oxygen needed to combine with these substrates to produce energy for cellular functions moves into cells by simple diffusion, which itself does not require expenditure of energy by the cells. However, considerable effort is required for the lungs to take up enough oxygen to meet the consumption required by the body. Furthermore, the energy-saving process of diffusion will not work unless oxygen is brought to within 100 μM of all cells in the body. Without some sort of mechanism to bring oxygen from the external environment into our body and then transport it to within 100 μM of our cells, we could not survive. Similarly, carbon dioxide cannot simply percolate out of the body from wherever it is produced. It must be neutralized, transported away from cells, and expelled from the body via the lungs. The transport of nutrients, oxygen, carbon dioxide, and waste products to and from cells is carried out by the circulatory system. In tandem with the circulatory system is the respiratory system, also with a mechanical component, that takes oxygen from the atmosphere and transfers it to the circulatory system and expels acid-producing CO_2 as well. It is little wonder that should either our cardiovascular or pulmonary systems become dysfunctional, major complications can occur in a manner of minutes, even death.

Finally, normal function of the body and its survival as a whole require communication between our internal and external environment. Our survival is, therefore, very much dependent on such communication in support of the regulation of basic physiologic conditions within our body.

The neuroendocrine system provides a communication network to tissues and organs.

In the form of a central and peripheral nervous system, we exploit the existence of electrolytes in our body fluids to create voltages and currents that can be used to very rapidly trigger and transmit information about our external and internal environment in the form of electrical signals. These nerve signals, in turn, synthesize and deliver chemical messengers to specialized tissue and different organs. Not only does our body use such a neurological sensing and signaling system to regulate functions throughout our body, but also it exploits the existence of the blood transport system afforded by the cardiovascular system to transmit chemical signal and regulatory molecules from one organ system to another. This type of internal long-distance electrochemical information system dovetails with neurogenic mechanisms in what is collectively called the **neuroendocrine system**.

This system is essential in regulating growth, development, tissue repair, and defense. The neuroendocrine system is also involved in the maintenance of all cellular metabolic pathways, storage of metabolic substrates, electrolyte composition of the extracellular fluid, maintenance of bone structure, sexual maturation, reproduction, and birth.

In addition to long-distance electrochemical transmission, nerve signals are conducted through simple reflex circuits of the nervous system. These simple reflexes send neural signals to various systems that coordinate the activity of the different systems so that the whole body can work efficiently. For example, these simple reflexes are involved in activating muscles in our body to generate the force and motion needed for the survival advantage of physical mobility, the pumping of blood by the heart through our circulatory system, the mechanical movement of air in and out of our lungs, and the processing of complex foods by our digestive tract.

Nerve transmission and simple reflexes that are distributed across the entire system are also important in protective functions. For example, the withdrawal reflex (e.g., withdrawing the hand from a hot stove), cough reflex, stretch reflex, and scratch reflex are all part of the body's protective defense system. These simple reflexes set in motion a process to ensure that the body's tissues and organs are protected from exposure to heat, light, pressure, and toxic chemicals.

Another example of how these simple reflexes act as part of the body's defenses is by turning a contractile skeletal muscle into an endocrine organ. When skeletal muscle contracts, cytokines and peptides, called **myokines**, are released and work in an endocrine-like fashion. The myokines exert a local effect on muscle metabolism and are involved in muscle repair and hypertrophy (increased muscle mass). These myokines are also involved in other tissue regeneration, repair, and immunomodulation. In fact, immunoregulation was the focus of early myokine research. One of the first myokines to be identified and found to be secreted into the bloodstream in response to muscle contractions was **interleukin-6 (IL-6)**. IL-6 is a cytokine that is involved in the body's inflammatory response to fighting infection and repairing wounds. The secretion of IL-6 increases in exponential fashion proportional to the length of exercise and muscle mass. In practical

terms, the exercise-induced IL-6 secretion assumes physiological importance in the protection against certain types of chronic diseases.

Another myokine, brain-derived neurotrophic factor (BDNF), is also activated with exercise. Although BDNF is produced by contracting muscle, it is not released into the circulation but rather stays within skeletal tissue to enhance mitochondrial oxidation of fat. Paradoxically, BDNF activation through exercise also increases in the brain. BDNF is structurally related to growth factors and exerts its effects on neuronal development, growth, maintenance, and repair. BDNF secretion is also involved as a key component in the hypothalamic pathway that controls body mass and energy homeostasis.

To date, exercise-induced BDNF is the only mechanism known to stimulate growth of new nerve tissue. Moreover, long-term aerobic exercise has been shown to improve brain function and causes a significant enlargement in the hippocampus and cortex in the human brain. These exercise-induced changes in neural function have major practical implications in maintaining brain health and the treatment of neurological disorders. For example, dementia is a neurological disorder in which the brain loses its plasticity (i.e., loss of cellular function, circuit repair, and nerve restoration), particularly in the hippocampus.

Another structural component of the body's communication network that is involved in the body's defense system is the **autonomic nervous system**. For example, the vagus nerve, which receives and sends neural reflex signals to many organs, plays an important role in cardiac and gastrointestinal functions as well as inflammation. Stimulation of the vagus nerve inhibits the synthesis of key inflammatory mediators, which are known to exacerbate the conditions of rheumatoid arthritis and other autoimmune diseases. The use of bioelectrical stimulation at specific sites in the autonomic nervous system holds the promise of treating autoimmune diseases physiologically without the use of drugs.

Lastly, the neuroendocrine system as well as other organ systems will not work well, if at all, unless the chemical composition and volume of the aqueous environment surrounding all our cells is well regulated. The renal system has this primary responsibility in the body. Together the renal, neural, endocrine, cardiovascular, and pulmonary systems work in an integrated fashion to make valuable survival functions in our body possible.

▶ FUTURE DIRECTION OF MEDICAL PHYSIOLOGY

Physiology is playing an ever-important role in creating the framework for the maintenance of optimal health and the body's defenses. Moreover, new discoveries in physiology are continuing to provide insight into our understanding of "what makes us tick," as well as charting new directions for future medical therapies and interventions. Many of these discoveries are due to the fact that physiology applies new knowledge from other disciplines (e.g., molecular biology, genetics, immunology, biophysics, and bioengineering) to address important questions about the function and maintenance of human systems and their regulation.

Physiology is the bridge to the new science of epigenetics.

Many of the new discoveries regarding human function and survival are a result of the convergence of several of these disciplines with physiology providing a bridge between molecular/cellular events and organ function. In the past, there has been an overemphasis of genetic determination that has resulted in a scientific belief that the genes control human biology. For a long time, medical scientists thought human function and survival were controlled primarily by the body's DNA blueprint. This belief has had a major impact on the way medical scientists thought about human function. However, DNA is not the real business end of physiology, because it does not carry out many of the activities at the cellular and organ level that are required to keep the body functional and healthy. Those activities are performed mainly by proteins that get expressed by the cell's DNA, and it is these proteins that cause muscles to contract, that power the brain, that turn consumed food into nutrients that can be absorbed, and that transport oxygen throughout the circulatory system. The role of the DNA is to carry the code for all these different proteins. There is no doubt that the genetic code is a starting point and is certainly necessary. But it does not explain many of the functional variations at the tissue and organ levels or the body's ability to survive.

If only the DNA mattered, then identical twins would always be functionally identical, but this is not the case. A number of studies demonstrate that identical twins who are genetically identical show phenotypical and other differences. For instance, in one case of identical twins living together in the same family and being exposed to essentially the same environment, one twin is thin and the other overweight, one tall and the other not, one diabetic and the other not, and one an alcoholic and the other not.

Phenotypical differences between twins become even more striking when they live apart. One could argue that these phenotypic differences seen in identical twins are due to DNA mutation. However, new research shows that this is very seldom the case with twins, yet their lives are different. If nothing happened to the DNA blueprint in these individuals, then the question is, why are the differences occurring? The answer comes from an emerging field in biology called **epigenetics** that is having a major impact on human physiology. Epigenetics involves regulation at a level above the gene and shows that many of the body's functions, capabilities, and personality are not fixed at birth. Epigenetics is upending our understanding of how life is controlled. Epigenetics studies how lifestyle and environmental signals modify and regulate gene activity. The discipline is providing a link between the external environment, genetics, and human function and shows that such things as exercise, nutrition, stress, emotions, trauma, substance abuse, and social engagement can modify gene function without altering the genes themselves or their genetic code. Even more striking

is the fact that epigenetic modifications can be passed on to the next generation.

Epigenetics can be defined as a mechanism by which genes can be switched off and on, but the genes themselves and their genetic code are not altered. Epigenetics controls how the genes in the cell's DNA are used. There are regions on the outer structure of the genes, called the *epigenome*, that act like switches that can turn the cell's DNA on and off. Epigenetics is now recognized as the mechanism during development that determines whether cells, which have the same genetic code, become liver, brain, or muscle cells during cell differentiation.

Finally, epigenetics has opened a new frontier in human physiology and medicine. For example, new research shows that ~15% of the chronic diseases (e.g., obesity, diabetes, hypertension, heart attacks, strokes, and certain types of cancer) are specifically linked to genetics. The remaining factors (85%) are due to lifestyle choices (e.g., diet, exercise, emotional stress) and environmental factors (e.g., toxic agents, smoking, and substance abuse).

A paradigm shift is occurring in human physiology.

Physiologists are no longer seeing the body as just a machine in which descriptive biology is attached to structure–function events. In the last 10 years, physiologists have begun identifying specific cellular mechanisms and finding the missing links between nature and nurture. They can connect specific changes in lifestyles that alter whole body function, sometimes forever. Huge areas of physiology are being influenced by epigenetics and are filled with remarkable intrigue and complexity. This has caused a paradigm shift in physiology. Physiology has moved from structure–function relationships to epigenetics-induced functional relationships.

As a result of the shift, important discoveries are being made in neurophysiology and neuroscience, especially in neurological disorders such as dementia and Alzheimer's disease. Epigenetic-induced functional changes are also leading to new discoveries in obesity, an epidemic that has spread worldwide in the last 15 years. In the United States, approximately two out of every three adults are overweight. Obesity is linked to a wide range of health problems, including cardiovascular diseases and type 2 diabetes. In addition to overeating, the two studies cited in *Cell* and *Nature* provides evidence that a parent's diet can directly influence the epigenetic modification that predisposes the offspring to obesity and diabetes.

Another field in which epigenetic-induced functional changes are playing a significant role is in the field of aging. Aging can be defined as "the progressive decline of organ function that eventually results in disease and mortality." Because identical twins with the same genetic makeup can age differently, the questions for future research will be to determine how epigenetics changes with age. Equally important will be to determine how epigenetics can have both a positive and a negative effect on aging. For example, a sedentary lifestyle with poor eating habits, high stress, smoking, and excessive drinking can cause epigenetic-induced effects that accelerate the aging process, whereas a healthy lifestyle that includes daily exercise, good nutrition, smoking avoidance, and less stress can cause epigenetic-induced changes that keep the body younger.

The paradigm shift also has altered our thinking in unexpected frontiers of human function. In the past, medical scientists have been seeing human health and survival from the level of the gene alone. Epigenetic-induced physiologic changes are making it clear that individuals are no longer doomed by their genes. In practical terms, research indicates that about 20% of an individual's health, diseases, and how he or she ages is due to genetics, and the remaining 80% is dependent on epigenetic-induced alterations in human function via changes in lifestyle and environment.

For example, many of the diseases that cause early death (high blood pressure, obesity, diabetes, heart attacks, strokes, cancer) are classified as 80/20: 20% due to genetics and 80% related to lifestyle.

In summary, physiology has provided remarkable insight into how cells, tissues, and organs interact and how they are regulated. The interaction between physiology and the new science of epigenetics is showing that human health and survival are no longer determined by the limits set by our genes. Moreover, recent medical advances show the extraordinary influence individuals have over the control of their health, quality of life, and how they age. Lastly, the paradigm shift in epigenetic-induced functional changes is providing new directions for the future treatment of human diseases.

2 Cell Signaling, Membrane Transport, and Membrane Potential

Active Learning Objectives

Upon mastering the material in this chapter, you should be able to:

- Compare and contrast negative and positive feedback and explain the importance of these processes to homeostasis.
- Explain the difference between steady state and equilibrium, including the role of energy expenditure in these concepts.
- List the types of molecules that constitute the plasma membrane and explain how they are assembled to form a selectively permeable barrier.
- Describe how the plasma membrane maintains an internal environment that differs significantly from the extracellular fluid.
- Contrast how voltage-gated channels and ligand-gated channels are opened.
- Compare and contrast carrier-mediated transport systems with channels.
- Describe primary active transport and explain how secondary active transport is different.

- Describe the properties of epithelial cells that are necessary to produce directional movement of solutes and water.
- Outline the mechanisms that many cells use to regulate their volume when exposed to osmotic stress.
- List the key components of the Goldman equation and explain why this equation gives the value of the membrane potential.
- Explain why the resting membrane potential of most cells is close to the Nernst potential for K^+.
- Compare and contrast autocrine, paracrine, and endocrine signaling in the control of cell function.
- Describe how second messengers both regulate and amplify signal transduction.
- Explain the major differences between intracellular signal transduction by G-protein–coupled receptors and tyrosine kinase receptors.
- Describe how intracellular calcium concentration is regulated and used in intracellular signal transduction.

Cellular Physiology

The scope of physiology ranges from the functions of individual molecules and cells to the interaction of our bodies with the external world. Understanding how the different cell types that constitute tissues are controlled, how they interact both within and with other tissues, and how they adapt to changing conditions is central to the study of physiology. To maintain health, conditions in the body must be optimized through closely regulated processes that require efficient communication between cells and tissues.

Cells are delineated by their plasma membrane, a barrier that separates the cytosol from the extracellular fluid (ECF). The plasma membrane keeps ions, metabolites, and cell proteins needed for normal cell function from leaking out, allows specific ions and molecules to enter, and blocks entry of factors not needed by the cell. To function in coordination with the rest of the organism, cells send and receive information that is first processed by specific plasma membrane proteins.

This chapter discusses topics related to regulation of cellular homeostasis and communication between cells and tissues. Specific topics include the internal environment, types of membrane transport mechanisms for ions and other solutes, steady state and equilibrium, intercellular and intracellular communication, negative and positive feedback, feedforward control, and intracellular signal transduction cascades.

▶ BASIS OF PHYSIOLOGIC REGULATION

Our bodies are made up of incredibly complex and delicate materials, and we are constantly subjected to all kinds of disturbances, yet we keep going for a lifetime. It is clear

that conditions and processes in the body must be closely controlled and regulated—that is, kept within appropriate values. Below, we consider, in broad terms, physiologic regulation in the body.

Stable internal environment is essential for normal cell function.

The 19th-century French physiologist Claude Bernard was the first to formulate the concept of the internal environment (*milieu intérieur*). He pointed out that an external environment surrounds multicellular organisms (air or water) and a liquid internal environment (ECF) surrounds the cells that make up the organism. Cells are not directly exposed to the external world but, rather, interact with it through their surrounding environment, which is continuously renewed by the circulating blood.

For optimal cell, tissue, and organ function in animals, several facets of the internal environment must be maintained within narrow limits. These include but are not limited to (1) oxygen and carbon dioxide tensions; (2) concentrations of glucose and other metabolites; (3) osmotic pressure; (4) concentrations of hydrogen, potassium, calcium, and magnesium ions; and (5) temperature. Departures from optimal conditions may result in dysfunction, disease, or death. Bernard stated, "Stability of the internal environment is the primary condition for a free and independent existence." He recognized that an animal's independence from changing external conditions is related to its capacity to maintain a relatively constant internal environment. A good example is the ability of warm-blooded animals to live

5

in different climates. Over a wide range of external temperatures, core temperature in mammals is maintained constant by both physiologic and behavioral mechanisms. This stability offers great flexibility and has an obvious survival value.

Homeostasis is the maintenance of steady states in the body by coordinated physiologic mechanisms.

The key to maintaining the stability of the body's internal environment is the masterful coordination of important regulatory mechanisms in the body. The renowned physiologist Walter B. Cannon captured the spirit of the body's capacity for self-regulation by defining the term **homeostasis** as the maintenance of steady states in the body by coordinated physiologic mechanisms.

Understanding the concept of homeostasis is important for understanding and analyzing normal and pathologic conditions in the body. To function optimally under a variety of conditions, the body must sense departures from normal and then be able to activate mechanisms for restoring physiologic conditions to normal. Deviations from normal conditions may vary between too high and too low, so mechanisms exist for opposing changes in either direction.

Homeostatic regulation of a physiologic variable often involves several cooperating mechanisms activated at the same time or in succession. The more important a variable, the more numerous and complicated are the mechanisms that operate to keep it at the desired value. When the body is unable to restore physiologic variables, then disease or death can result. The ability to maintain homeostatic mechanisms varies over a person's lifetime, with some homeostatic mechanisms not being fully developed at birth and others declining with age. For example, a newborn infant cannot concentrate urine as well as an adult and is, therefore, less able to tolerate water deprivation. Older adults are less able to tolerate stresses, such as exercise or changing weather, than are younger adults.

The term *homeostasis* traditionally refers to the ECF that bathes our tissues—but it can also be applied to conditions within cells. In fact, the ultimate goal of maintaining a constant internal environment is to promote intracellular homeostasis, and toward this end, conditions in the cytosol of cells are closely regulated.

Negative feedback promotes stability, and feedforward control anticipates change.

Feedback is a flow of information along a closed loop. The components of a simple negative-feedback control system include a regulated variable, sensor, controller, and effector (Fig. 2.1). Each component controls the next component. Various disturbances may arise within or outside the system and cause undesired changes in the regulated variable. With **negative feedback**, a regulated variable is sensed, information is fed back to the controller, and the effector acts to oppose change (hence the term *negative*).

A familiar example of a negative-feedback control system is the thermostatic control of room temperature. Room temperature (regulated variable) is subjected to disturbances. For example, room temperature falls on a cold

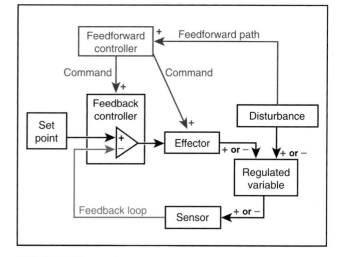

Figure 2.1 **Elements of negative-feedback and feedforward control systems.** In a negative-feedback control system, information flows along a closed loop. The regulated variable is sensed, and information about its level is provided to a feedback controller, which compares it with a desired value (set point). If there is a difference, an error signal is generated, which drives the effector to bring the regulated variable closer to the desired value. A feedforward controller generates commands without directly sensing the regulated variable, although it may sense a disturbance. Feedforward controllers often operate through feedback controllers.

day. A thermometer (sensor) in the thermostat (controller) detects the room temperature. The thermostat is set for a certain temperature (set point). The controller compares the actual temperature (feedback signal) with the set point temperature, and an error signal is generated if the room temperature falls below the set temperature. The error signal activates the furnace (effector). The resulting change in room temperature is monitored, and when the temperature rises sufficiently, the furnace is turned off. Such a negative-feedback system allows some fluctuation in room temperature, but the components act together to maintain the set temperature. Effective communication between the sensor and effector is important in keeping these oscillations to a minimum.

Similar negative-feedback systems exist to maintain homeostasis in the body. For example, the maintenance of water and salts in the body is referred to as **osmoregulation** or fluid balance. During exercise, loss of water from sweating results in an increased concentration of salts in the blood and tissue fluids, which is sensed by the cells in the brain (see Chapter 23). The brain responds by telling the kidneys to reduce secretion of water and also by increasing the sensation of thirst. Together, the reduction in water loss in the kidneys and increased water intake return the blood and tissue fluids to the correct osmotic concentration. This negative-feedback system allows for minor fluctuations in water and salt concentrations in the body but rapidly compensates for disturbances to restore acceptable osmotic conditions.

Feedforward control is another strategy for regulating body systems, particularly when a change with time is desired. In this case, a command signal is generated, which specifies the target or goal. The moment-to-moment operation of the controller is "open loop"; that is, the regulated variable itself is not sensed. Feedforward control mechanisms often sense a disturbance and can, therefore, take corrective action that anticipates change. For example, heart rate and breathing increase even before a person has begun to exercise.

Feedforward control usually acts in combination with negative-feedback systems. One example is picking up a pencil. The movements of the arm, hand, and fingers are directed by the cerebral cortex (feedforward controller); the movements are smooth, and forces are appropriate only in part because of the feedback of visual information and sensory information from receptors in the joints and muscles. Another example of this combination occurs during exercise. Respiratory and cardiovascular adjustments closely match muscular activity, so that arterial blood oxygen and carbon dioxide tensions (the partial pressure of a gas in a liquid) hardly change during all but exhausting exercise (see Chapter 21). Importantly, control system function can adapt over time. Past experience and learning can change the control system's output so that it behaves more efficiently or appropriately.

Although homeostatic control mechanisms usually act for the good of the body, they are sometimes deficient, inappropriate, or excessive. Many diseases, such as cancer, diabetes, and hypertension, develop because of defects in control mechanisms. Formation of a scar is an example of an important homeostatic mechanism for healing wounds, but in many chronic diseases, such as pulmonary fibrosis, hepatic cirrhosis, and renal interstitial disease, scar formation goes awry and becomes excessive.

Positive feedback promotes a change in one direction.

With **positive feedback**, a variable is sensed and action is taken to reinforce a change of the variable. The term positive refers to the response being in the same direction, leading to a cumulative or amplified effect. Positive feedback does not lead to stability or regulation, but to the opposite—a progressive change in one direction. One example of positive feedback is the sensation of needing to urinate. As the bladder fills, mechanosensors in the bladder are stimulated, and the smooth muscle in the bladder wall begins to contract (see Chapter 23). As the bladder continues to fill and become more distended, the contractions increase and the need to urinate becomes more urgent. Responding to the need to urinate results in a sensation of immediate relief upon emptying the bladder, and this is positive feedback. Another example of positive feedback occurs during the follicular phase of the menstrual cycle. The female sex hormone estrogen stimulates the release of luteinizing hormone, which in turn causes further estrogen synthesis by the ovaries. This positive feedback culminates in ovulation (see Chapter 37). Positive feedback, if unchecked, can lead to a vicious cycle and dangerous situations. For example, a heart may be so weakened by disease that it cannot provide adequate blood flow to the muscle tissue of the heart. This leads to a further reduction in cardiac pumping ability, even less coronary blood flow, and further deterioration of cardiac function. The physician's task sometimes is to disrupt detrimental cyclical positive-feedback loops.

Steady state and equilibrium are both stable conditions, but energy is required to maintain a steady state.

Simplistically, the whole body can be divided into two major compartments: intracellular fluid and ECF, which are separated by cell plasma membranes. The fluid component of the body constitutes about 60% of the total body weight. The intracellular fluid compartment constitutes about two thirds of the body's water and contains potassium, other ions, and proteins. The ECF compartment constitutes one third of the body's water (~20% body weight) and consists of all the body fluids outside of cells, including the interstitial fluid that bathes the cells, lymph, blood plasma, and specialized fluids such as cerebrospinal fluid. It is primarily a sodium chloride ($NaCl$) and sodium carbonate ($NaHCO_3$) solution that can be divided into three subcompartments: the interstitial fluid (lymph and plasma); plasma that circulates as the extracellular component of blood; and transcellular fluid, which is a set of fluids that are outside of normal compartments, such as cerebrospinal fluid, digestive fluids, and mucus.

When two compartments are in **equilibrium**, *opposing forces are balanced*, and there is no net transfer of a particular substance or energy from one compartment to the other. Equilibrium occurs if sufficient time for exchange has been allowed and if no physical or chemical driving force would favor net movement in one direction or the other. For example, osmotic equilibrium between cells and ECF is normally present in the body because of the high water permeability of most cell membranes. An equilibrium condition, if undisturbed, remains stable. No energy expenditure is required to maintain an equilibrium state.

Equilibrium and steady state are sometimes confused with each other. A **steady state** is simply a condition that does not change with time. It indicates that the amount or concentration of a substance in a compartment is constant. In a steady state, there is no net gain or net loss of a substance in a compartment. Steady state and equilibrium both suggest stable conditions, but a steady state does not necessarily indicate an equilibrium condition, and energy expenditure may be required to maintain a steady state. For example, in most body cells, there is a steady state for Na^+ ions; the amounts of Na^+ entering and leaving cells per unit time are equal. But intracellular and extracellular Na^+ ion concentrations are far from equilibrium. Extracellular $[Na^+]$ is much higher than intracellular $[Na^+]$, and Na^+ tends to move into cells down concentration and electrical gradients. The cell continuously uses metabolic energy to pump Na^+ out of the cell to maintain the cell in a steady state with respect to Na^+ ions. In living systems, conditions are often displaced from equilibrium by the constant expenditure of metabolic energy.

Figure 2.2 illustrates the distinctions between steady state and equilibrium. In Figure 2.2A, the fluid level in the sink is constant (a steady state) because the rates of inflow and outflow are equal. If we were to increase the rate of

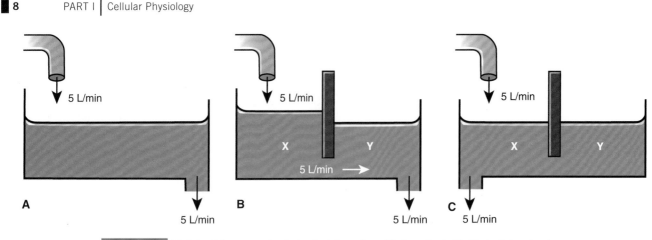

Figure 2.2 **Models of the concepts of steady state and equilibrium.** Parts **(A–C)** depict a steady state. In **(C)**, compartments X and Y are in equilibrium.

inflow (open the tap), the fluid level would rise, and with time, a new steady state might be established at a higher level. In Figure 2.2B, the fluids in compartments X and Y are not in equilibrium (the fluid levels are different), but the system as a whole and each compartment are in a steady state, because inputs and outputs are equal. In Figure 2.2C, the system is in a steady state and compartments X and Y are in equilibrium. Note that the term *steady state* can apply to a single or several compartments; the term *equilibrium* describes the relation between at least two adjacent compartments that can exchange matter or energy with each other.

▶ PLASMA MEMBRANE STRUCTURE

The first theory of membrane structure proposed that cells were surrounded by a double layer of lipid molecules, a **lipid bilayer**. However, this theory, based on the knowledge that lipid molecules form bilayers with low permeability to water-soluble molecules, did not explain the selective movement of certain water-soluble compounds, such as glucose and amino acids, across the plasma membrane. In 1972, Singer and Nicolson proposed the **fluid mosaic model** of the plasma membrane, which described the organization and interaction of proteins with the lipid bilayer (Fig. 2.3). With minor modifications, this model is still accepted as the correct picture of the structure of the plasma membrane.

Plasma membrane consists of different types of membrane lipids with different functions.

Lipids found in cell membranes can be classified into two broad groups: **phospholipids,** which contain fatty acids as part of the molecule and **cholesterol**, which does not have a fatty acid in its structure.

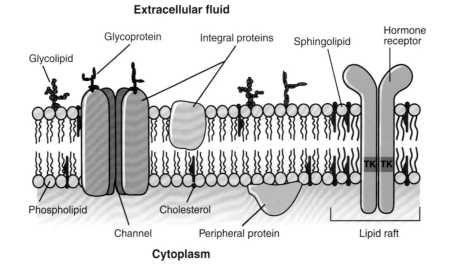

Figure 2.3 **The fluid mosaic model of the plasma membrane.** Lipids are arranged in a bilayer. Cholesterol provides rigidity to the bilayer. Integral proteins are embedded in the bilayer and often span it. Some membrane-spanning proteins form pores and channels. In some specialized cases, transmembrane pores on adjacent cells fuse together to form gap junctions that facilitate communication between the two cells. Other membrane-spanning proteins are receptors. Peripheral proteins do not penetrate the bilayer. Lipid rafts form stable microdomains composed of sphingolipids and cholesterol.

Phospholipids are the most abundant complex lipids found in cell membranes. They are amphipathic molecules formed by two fatty acids (normally, one saturated and one unsaturated) and one phosphoric acid group substituted on the backbone of a glycerol or sphingosine molecule. This arrangement produces a hydrophobic area formed by the two fatty acids and a polar hydrophilic head. When phospholipids are arranged in a bilayer, the polar heads are on the outside and the hydrophobic fatty acids on the inside (see Fig. 2.3). It is difficult for water-soluble molecules and ions to pass directly through the hydrophobic interior of the lipid bilayer.

The phospholipids, with a backbone of sphingosine (a long amino alcohol), are usually called *sphingolipids* and are present in all plasma membranes in small amounts. They are especially abundant in brain and nerve cells. Ceramide is a lipid second messenger that is generated from the sphingolipid sphingomyelin.

Glycolipids are lipid molecules that contain sugars and sugar derivatives (instead of phosphoric acid) in the polar head. They are located mainly in the outer half of the lipid bilayer, with the sugar molecules facing the ECF. Proteins can associate with the plasma membrane by linkage to the extracellular sugar moiety of glycolipids.

Cholesterol is an important component of mammalian plasma membranes. The proportion of cholesterol in plasma membranes varies from 10% to 50% of total lipids. Cholesterol has a rigid structure that stabilizes the cell membrane and reduces the natural mobility of lipids and proteins to move in the membrane. Some cell functions, such as the response of immune system cells to the presence of an antigen, depend on the ability of membrane proteins to move in the plane of the membrane to bind the antigen. A decrease in membrane fluidity resulting from an increase in cholesterol will impair these functions.

Aggregates of sphingolipids and cholesterol can form stable microdomains termed **lipid rafts** that diffuse laterally in the phospholipid bilayer. The protein **caveolin** is present in a subset of lipid rafts (termed **caveolae**), causing the raft to form a cavelike structure. It is believed that one function of both noncaveolar and caveolar lipid rafts is to facilitate interactions between specific proteins by selectively including (or excluding) these proteins from the raft microdomain. For example, lipid rafts can mediate the assembly of membrane receptors and intracellular signaling proteins as well as the sorting of plasma membrane proteins for internalization.

Proteins are integrally and peripherally associated with the plasma membrane.

Proteins are the second major component of the plasma membrane, present in about equal proportion by weight with the lipids. Two different types of proteins are associated with the plasma membrane. **Integral proteins** are embedded in the lipid bilayer, and many span it completely. The polypeptide chain of these proteins may cross the lipid bilayer once or may make multiple passes across it. The membrane-spanning segments usually contain amino acids with nonpolar side chains and are arranged in an ordered α-helical conformation. **Peripheral proteins** do not penetrate the lipid bilayer. They are in contact with the outer side of only one of the lipid layers—either the layer facing the cytoplasm or the layer facing the ECF (see Fig. 2.3). Many membrane proteins have carbohydrate molecules (sugar molecules) attached to the part of the protein that is exposed to the ECF and are termed **glycoproteins**. Some of the integral membrane proteins can move in the plane of the membrane, like small boats floating in the "sea" formed by the lipid bilayer. Other membrane proteins are anchored to the cytoskeleton inside the cell or to proteins of the extracellular matrix.

The proteins in the plasma membrane play a variety of roles. Many peripheral membrane proteins are enzymes, and many membrane-spanning integral proteins are carriers or channels for the movement of water-soluble molecules and ions into and out of the cell. **Gap junctions** are specialized protein channels, made of the protein **connexin**, that facilitate direct cell to cell communication. Six connexins assemble in the plasma membrane of a cell to form a half channel called a **connexon**. Two connexons aligned between two neighboring cells then join end to end to form an intercellular channel between the plasma membranes of adjacent cells. Gap junctions allow the flow of ions and small molecules between the cytosol of neighboring cells, thereby providing rapid transmission of electrical signals between cells in the heart, smooth muscle cells, and some nerve cells. Gap junctions are thought to play a role in the control of cell growth and differentiation by allowing adjacent cells to share a common intracellular environment. Often when a cell is injured, gap junctions close, isolating a damaged cell from its neighbors.

Membrane proteins also have a structural role, for example, maintaining the biconcave shape of the erythrocyte. Finally, some membrane proteins serve as highly specific receptors on the outside of the cell membrane to which extracellular molecules, such as hormones, can bind. If the receptor is a membrane-spanning protein, it provides a mechanism for converting an extracellular signal into an intracellular response.

▶ SOLUTE TRANSPORT MECHANISMS

All cells must import oxygen, sugars, amino acids, and small ions and export carbon dioxide, metabolic wastes, and secretions. At the same time, specialized cells require mechanisms to transport molecules such as enzymes, hormones, and neurotransmitters. The movement of large molecules is carried out by endocytosis and exocytosis: the transfer of substances into or out of the cell by vesicle formation and vesicle fusion with the plasma membrane. Cells also have mechanisms for the rapid movement of ions and solute molecules across the plasma membrane. These mechanisms are of two general types: **passive transport**, which requires no direct expenditure of metabolic energy, and **active transport**, which uses metabolic energy to move solutes across the plasma membrane.

Import of extracellular materials occurs through phagocytosis and endocytosis.

Phagocytosis is the ingestion of large particles or microorganisms, usually occurring only in specialized cells such as macrophages (Fig. 2.4). An important function of macrophages is to remove invading bacteria from the body. The phagocytic vesicle (1 to 2 μm in diameter) is almost as large as the phagocytic cell itself. Phagocytosis requires a specific stimulus. It occurs only after the extracellular particle has bound to the extracellular surface. The particle is then enveloped by expansion of the cell membrane around it.

Endocytosis is a general term for the process in which a region of the plasma membrane is pinched off to form an endocytic vesicle inside the cell. During vesicle formation, some fluid, dissolved solutes, and particulate material from the extracellular medium are trapped inside the vesicle and internalized by the cell. Endocytosis produces much smaller endocytic vesicles (0.1 to 0.2 μm in diameter) than phagocytosis. It occurs in almost all cells and is termed a constitutive process, because it occurs continually and specific stimuli are not required. In further contrast to phagocytosis, endocytosis originates with the formation of depressions in the cell membrane. The depressions pinch off within a few minutes after forming and give rise to endocytic vesicles inside the cell.

Two types of endocytosis can be distinguished (see Fig. 2.4). **Fluid-phase endocytosis** is the nonspecific uptake of the ECF and all its dissolved solutes. The material is trapped inside the endocytic vesicle as it is pinched off inside the cell. The amount of extracellular material internalized by this process is directly proportional to its concentration in the extracellular solution. **Receptor-mediated endocytosis** is a more efficient process, which uses receptors on the cell surface to bind specific molecules. These receptors accumulate at specific depressions known as **coated pits**, so named because the cytosolic surface of the membrane at this site is covered with a coat of several proteins. The coated pits pinch off continually to form endocytic vesicles, providing the cell with a mechanism for rapid internalization of a large amount of a specific molecule without the need to endocytose large volumes of ECF. The receptors also increase the uptake of molecules present at low concentrations outside the cell. Receptor-mediated endocytosis is the mechanism by which cells take up a variety of important molecules, including hormones, growth factors, and serum transport proteins such as the iron carrier **transferrin**. Foreign substances, such as diphtheria toxin and certain viruses, also enter cells by this pathway.

Export of macromolecules occurs through exocytosis.

Many cells synthesize important macromolecules that are destined for **exocytosis** or export from the cell. These molecules are synthesized in the endoplasmic reticulum, modified in the Golgi apparatus, and packed inside transport vesicles. The vesicles move to the cell surface, fuse with the cell membrane, and release their contents outside the cell (see Fig. 2.4).

There are two exocytic pathways—constitutive and regulated. The continuous secretion of mucus by **goblet cells** in the small intestine is an example of the *constitutive pathway* of exocytosis that is present in all cells. In other cells, macromolecules are stored inside the cell in secretory vesicles. These

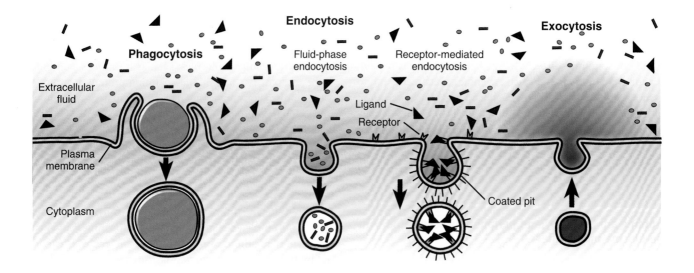

Figure 2.4 **The transport of macromolecules across the plasma membrane by the formation of vesicles.** Particulate matter in the extracellular fluid (ECF) is engulfed and internalized by phagocytosis. During fluid-phase endocytosis, ECF and dissolved macromolecules enter the cell in endocytic vesicles that pinch off at depressions in the plasma membrane. Receptor-mediated endocytosis uses membrane receptors at coated pits to bind and internalize specific solutes (ligands). Exocytosis is the release of macromolecules destined for export from the cell. These are packed inside secretory vesicles that fuse with the plasma membrane and release their contents outside the cell.

vesicles fuse with the cell membrane and release their contents only when a specific extracellular stimulus arrives at the cell membrane. This process, termed the *regulated pathway*, is responsible for the rapid "on-demand" secretion of many specific hormones, neurotransmitters, and digestive enzymes.

Uncharged solutes cross the plasma membrane by passive diffusion.

Any solute will tend to uniformly occupy the entire space available to it. This movement, known as **diffusion**, is a result of the spontaneous Brownian (random) movement that all molecules experience. A drop of ink placed in a glass of water will diffuse and slowly color all the water. The net result of diffusion is the movement of substances from regions of high concentration to regions of low concentration. Diffusion is an effective way for substances to move short distances.

The speed with which the diffusion of a solute in water occurs depends on the difference of concentration, the size of the molecules, and the possible interactions of the diffusible substance with water. These different factors appear in **Fick's law**, which describes the diffusion of any solute in water. In its simplest formulation, Fick's law can be written as:

$$J = DA(C_1 - C_2)/\Delta X \tag{1}$$

where J is the flow of solute from region 1 to region 2 in the solution; D is the diffusion coefficient of the solute, which is determined by factors such as solute molecular size and interactions of the solute with water; A is the cross-sectional area through which the flow of solute is measured; C is the concentration of the solute at regions 1 and 2; and ΔX is the distance between regions 1 and 2. Sometimes, J is expressed in units of amount of substance per unit area per unit time, for example, $mol/cm^2/h$, and is also referred to as the solute **flux**.

The principal force driving the passive diffusion of an uncharged solute across the plasma membrane is the difference of concentration between the inside and the outside of the cell. In the case of an electrically charged solute, such as an ion, diffusion is also driven by the membrane potential, which is the electrical gradient across the membrane. Movement of charged solutes and the membrane potential will be discussed in greater detail later in this chapter.

Diffusion across a membrane has no preferential direction; it can occur from the outside of the cell toward the inside or from the inside of the cell toward the outside. For any substance, it is possible to measure the **permeability coefficient (P)**, which gives the speed of the diffusion across a unit area of plasma membrane for a defined driving force. Fick's law for the diffusion of an uncharged solute across a membrane can be written as

$$J = PA(C_1 - C_2) \tag{2}$$

which is similar to equation 1. P includes the membrane thickness, the diffusion coefficient of the solute within the membrane, and the solubility of the solute in the membrane. Dissolved gases such as oxygen and carbon dioxide have high permeability coefficients and diffuse rapidly across the plasma membrane. As a result, gas exchange in the lungs is very effective. Diffusion across the plasma membrane implies

that the diffusing solute enter the lipid bilayer to cross it; thus, the solute's solubility in a lipid solvent (e.g., olive oil or chloroform) compared with its solubility in water is important in determining its permeability coefficient.

A substance's solubility in oil compared with its solubility in water is its **partition coefficient**. Lipophilic (lipid-soluble) substances, such as gases, steroid hormones, and anesthetic drugs, which mix well with the lipids in the plasma membrane, have high partition coefficients and, as a result, high permeability coefficients; they tend to cross the plasma membrane easily. Hydrophilic (water-soluble) substances, such as ions and sugars, do not interact well with the lipid component of the membrane, have low partition coefficients and low permeability coefficients, and diffuse across the membrane more slowly.

Solutes such as oxygen readily diffuse across the lipid part of the plasma membrane by simple diffusion. Thus, the relationship between the rate of movement and the difference in concentration between the two sides of the membrane is linear (Fig. 2.5). The larger the difference in concentration $(C_1 - C_2)$, the greater the amount of substance crossing the membrane per unit time.

Integral membrane proteins facilitate diffusion of solutes across the plasma membrane.

For many solutes of physiologic importance, such as ions, sugars, and amino acids, the rate of transport across the plasma membrane is much faster than expected for simple diffusion through a lipid bilayer. Furthermore, the relationship between transport rate and concentration difference of these hydrophilic

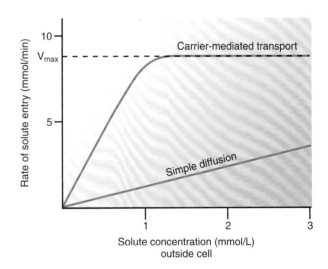

Figure 2.5 Solute transport across a plasma membrane by simple or facilitated diffusion. In simple diffusion, the rate of solute entry increases linearly with extracellular concentration of the solute. Assuming no change in intracellular concentration, increasing the extracellular concentration increases the gradient that drives solute entry. In facilitated diffusion, the rate of transport is much faster, and increases linearly as the extracellular solute concentration increases. The increase in transport is limited by the availability of channels and carriers. Once all are occupied by solute, further increases in extracellular concentration have no effect on the rate of transport. A maximum rate of transport (V_{max}) is achieved that cannot be exceeded.

substances follows a curve that reaches a plateau (see Fig. 2.5). Membrane transport with these characteristics is often called **facilitated diffusion** or **carrier-mediated diffusion**, because an integral membrane protein facilitates the movement of a solute through the membrane. Integral membrane proteins can form pores, channels, or carriers, each of which facilitates the transport of specific molecules across the membrane.

There are a limited number of pores, channels, and carriers in any cell membrane; thus, increasing the concentration of the solute initially uses the existing "spare" pores, channels, or carriers to transport the solute at a higher rate than by simple diffusion. As the concentration of the solute increases further and more solute molecules associate with the pore, channel, or carrier, the transport system eventually reaches saturation, when all the pores, channels, and carriers are involved in translocating molecules of solute. At this point, additional increases in solute concentration do not increase the rate of solute transport (see Fig. 2.5).

The types of integral membrane protein transport mechanisms considered here can transport a solute along its concentration gradient only, as in simple diffusion. Net movement stops when the concentration of the solute has the same value on both sides of the membrane. At this point, with reference to equation 2, $C_1 = C_2$ and the value of J is 0. The transport systems function until the solute concentrations have equilibrated. However, equilibrium is attained much faster than with simple diffusion.

Membrane pores

A pore provides a conduit through the lipid bilayer that is always open to both sides of the membrane. **Aquaporins** in the plasma membranes of specific kidney and gastrointestinal tract cells permit the rapid movement of water. Within the **nuclear pore complex**, which regulates movement of molecules into and out of the nucleus, is an aqueous pore that only allows the passive movement of molecules smaller than 45 kDa and excludes molecules larger than 62 kDa. The **mitochondrial permeability transition pore** and **mitochondrial voltage-dependent anion channel** (**VDAC**), which cross the inner and outer mitochondrial membranes, promote mitochondrial failure when formed, resulting in the generation of **reactive oxygen species** and cell death.

Gated channels

Small ions, such as Na^+, K^+, Cl^-, and Ca^{2+}, cross the plasma membrane faster than would be expected based on their partition coefficients in the lipid bilayer. The electrical charge of an ion makes it difficult for the ion to move across the lipid bilayer. The excitation of nerves, the contraction of muscle, the beating of the heart, and many other physiologic events are possible because of the ability of small ions to enter or leave the cell rapidly. This movement occurs through selective ion channels.

Ion channels are composed of several polypeptide subunits that span the plasma membrane and contain a gate that determines if the channel is open or closed. Specific stimuli cause a conformational change in the protein subunits to open the gate, creating an aqueous channel through which the ions can move (Fig. 2.6). In this way, ions do not have to enter the lipid bilayer to cross the membrane; they are always in an aqueous medium.

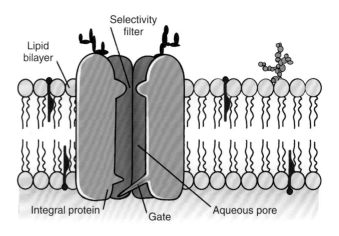

Figure 2.6 **An ion channel.** The polypeptide subunits of integral proteins that span the plasma membrane provide an aqueous pore through which ions can cross the membrane. Different types of gating mechanisms are used to open and close ion channels that are often selective for a specific ion.

When the channels are open, the ions diffuse rapidly from one side of the membrane to the other down the concentration gradient. Specific interactions between the ions and the sides of the channel produce an extremely rapid rate of ion movement; in fact, ion channels permit a much faster rate of solute transport (about 10^8 ions/s) than the carrier-mediated systems discussed below. Ion channels have a selectivity filter, which regulates the transport of certain classes of ions such as anions or cations or specific ions such as Na^+, K^+, Ca^{2+}, and Cl^- (see Fig. 2.6).

In general, ion channels exist either fully open or completely closed, and they open and close very rapidly. The frequency with which a channel opens is variable, and the time the channel remains open (usually a few milliseconds) is also variable. The overall rate of ion transport across a membrane can be controlled by changing the frequency of a channel opening or by changing the time a channel remains open.

Most ion channels usually open in response to a specific stimulus. Ion channels can be classified according to their gating mechanisms, the signals that make them open or close. There are voltage-gated channels and ligand-gated channels. Some ion channels are more like membrane pores in that they are always open; these ion transport proteins are referred to as *nongated channels*.

Voltage-gated ion channels open when the membrane potential changes beyond a certain threshold value. Channels of this type are involved in conducting the excitation signal along nerve axons and include sodium and potassium channels (see Chapter 3). Voltage-gated ion channels are found in many cell types. It is thought that some charged amino acids located in a membrane-spanning α-helical segment of the channel protein are sensitive to the transmembrane potential. Changes in the membrane potential cause these amino acids to move and induce a conformational change of the protein that opens the way for the ions.

Ligand-gated ion channels cannot open unless they first bind to a specific agonist. The opening of the gate is produced by a conformational change in the protein induced by the ligand binding. The ligand can be a neurotransmitter arriving from the extracellular medium. It can also be an intracellular second messenger, produced in response to some cell activity or hormone that reaches the ion channel from the inside of the cell. The nicotinic acetylcholine receptor channel found in the postsynaptic neuromuscular junction (see Chapters 3 and 8) is a ligand-gated ion channel that is opened by an extracellular ligand (acetylcholine). Examples of ion channels gated by intracellular messengers also abound in nature. This type of gating mechanism allows the channel to open or close in response to events that occur at other locations in the cell. For example, a sodium channel gated by intracellular cyclic guanosine monophosphate (cGMP) is located in the rod cells of the retina and opens in the presence of cGMP (see Chapter 4). The generalized structure of one subunit of an ion channel gated by cyclic nucleotides is shown in Figure 2.7. There are six membrane-spanning regions, and a cyclic nucleotide-binding site is exposed to the cytosol. The functional protein is a tetramer of four identical subunits. Other cell membranes have potassium channels that open when the intracellular concentration of calcium ions increases. Several known channels respond to inositol 1,4,5-trisphosphate, the activated part of G proteins, or adenosine triphosphate (ATP). The epithelial chloride channel that is mutated in cystic fibrosis is normally gated by ATP.

Carrier-mediated transport moves a range of ions and organic solutes passively across membranes.

In contrast to pores and ion channels, integral membrane proteins that form carriers provide a conduit through the membrane that is never open to both sides of the membrane at the same time. This is due to the presence of two gates (Fig. 2.8). During carrier-mediated transport, binding of the solute to one side of the carrier induces a conformational change in the protein, which closes one gate and opens the second gate, allowing the solute to pass through the membrane. As with pores and channels, carriers function until the solute concentrations have equilibrated.

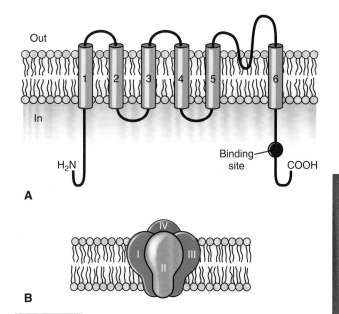

A

B

Figure 2.7 **Structure of a cyclic nucleotide-gated ion channel.** **(A)** The secondary structure of a single subunit has six membrane-spanning regions and a binding site for cyclic nucleotides on the cytosolic side of the membrane. **(B)** Four identical subunits (I–IV) assemble together to form a functional channel that provides a hydrophilic pathway across the plasma membrane.

Carrier-mediated transport systems have several characteristics:

- They allow the transport of polar (hydrophilic) molecules at rates much higher than that expected from the partition coefficient of these molecules.

- They eventually reach saturation at high substrate concentration (see Fig. 2.5).

- They have structural specificity, meaning each carrier system recognizes and binds specific chemical structures (a carrier for D-glucose will not bind or transport L-glucose).

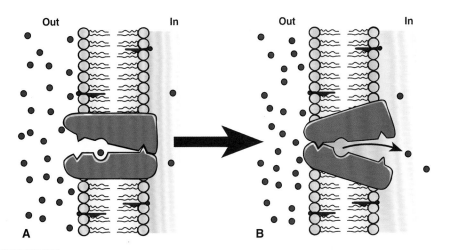

A   **B**

Figure 2.8 **The role of a carrier protein in facilitated diffusion of solute molecules across a plasma membrane.** In this example, solute transport into the cell is driven by the high solute concentration outside compared with inside. **(A)** Binding of extracellular solute to the membrane-spanning integral protein triggers a change in conformation that exposes the bound solute to the interior of the cell. **(B)** Bound solute readily dissociates from the carrier because of the low intracellular concentration of solute. The release of solute allows the carrier to revert to its original conformation **(A)** to begin the cycle again.

- They show competitive inhibition by molecules with similar chemical structure. For example, carrier-mediated transport of D-glucose occurs at a slower rate when molecules of D-galactose are also present. This is because galactose, structurally similar to glucose, competes with glucose for the available glucose carrier proteins.

A specific example of carrier-mediated transport is the movement of glucose from the blood to the interior of cells. Most mammalian cells use blood glucose as a major source of cellular energy, and glucose is transported into cells down its concentration gradient. The transport process in many cells, such as erythrocytes and the cells of fat, liver, and muscle tissues, involves a plasma membrane protein called *GLUT1* (glucose transporter-1). The erythrocyte GLUT1 has an affinity for D-glucose that is about 2,000-fold greater than the affinity for L-glucose. It is an integral membrane protein that contains 12 membrane-spanning α-helical segments.

Carrier-mediated transport, like simple diffusion, does not have a directional preference. It functions equally well bringing its specific solutes into or out of the cell, depending on the concentration gradient. Net movement by carrier-mediated transport ceases once the concentrations inside and outside the cell become equal.

The **anion exchange protein** (**AE1**), the predominant integral protein in the mammalian erythrocyte membrane, provides a good example of the reversibility of transporter action. AE1 is folded into at least 12 transmembrane a helices and normally permits the one-for-one exchange of Cl^- and HCO_3^- ions across the plasma membrane. The direction of ion movement is dependent only on the concentration gradients of the transported ions. AE1 has an important role in transporting CO_2 from the tissues to the lungs. The erythrocytes in systemic capillaries pick up CO_2 from tissues and convert it to HCO_3^-, which exits the cells via AE1. When the erythrocytes enter pulmonary capillaries, the AE1 allows plasma HCO_3^- to enter erythrocytes, where it is converted back to CO_2 for expiration by the lungs (see Chapter 21).

Active transport systems move solutes against gradients.

All the passive transport mechanisms tend to bring the cell into equilibrium with the ECF. Cells must oppose these equilibrating systems and preserve intracellular concentrations of solutes, in particular ions that are compatible with life.

Primary active transport

Integral membrane proteins that directly use metabolic energy to transport ions against a gradient of concentration or electrical potential are known as **ion pumps**. The direct use of metabolic energy to carry out transport defines a **primary active transport mechanism**. The source of metabolic energy is ATP synthesized by mitochondria, and the different ion pumps hydrolyze ATP to ADP using the energy stored in the third phosphate bond to carry out transport. Ion pumps also are called **ATPases**, because of the ability to hydrolyze ATP.

The most abundant ion pump in higher organisms is the sodium–potassium pump or **Na$^+$/K$^+$-ATPase**. It is found in the plasma membrane of practically every eukaryotic cell and is responsible for maintaining the low sodium and high potassium concentrations in the cytoplasm by transporting sodium out of the cell and potassium ions in. The sodium–potassium pump is an integral membrane protein consisting of two subunits. The α subunit has 10 transmembrane segments and is the catalytic subunit that mediates active transport. The smaller β subunit has one transmembrane segment and is essential for the proper assembly and membrane targeting of the pump. The Na$^+$/K$^+$-ATPase is known as a **P-type ATPase** because the protein is phosphorylated during the transport cycle (Fig. 2.9). The pump counterbalances the tendency of sodium ions to enter the cell passively and the tendency of potassium ions to leave passively. It maintains a high intracellular potassium concentration, which is necessary for protein synthesis. It also plays a role in the resting membrane potential by maintaining ion gradients. The sodium–potassium pump can be inhibited either by metabolic poisons that stop the synthesis and supply of ATP or by specific pump blockers, such as **digoxin**, a **cardiac glycoside** used to treat a variety of cardiac conditions.

As the Na$^+$/K$^+$-ATPase specifically moves sodium and potassium ions against their concentration or electrical potential, a number of other pumps move specific substrates across membranes utilizing the energy released by ATP hydrolysis.

- **Calcium pumps** are P-type ATPases located in the plasma membrane and the membrane of intracellular organelles. Plasma membrane Ca^{2+}-ATPases pump calcium out of the cell. Calcium pumps in the membrane of the endoplasmic reticulum and in the sarcoplasmic reticulum membrane within muscle cells (termed *SERCAs* for sarcoplasmic and endoplasmic reticulum calcium ATPases) pump calcium into the lumen of these organelles. The organelles store calcium and, as a result, help maintain a low cytosolic concentration of this ion.

- The **H$^+$/K$^+$-ATPase** is a P-type ATPase present in the luminal membrane of the parietal cells in the oxyntic (acid-secreting) glands of the stomach. By pumping protons into the lumen of the stomach in exchange for potassium ions, this pump maintains the low pH in the stomach that is necessary for proper digestion. It is also found in the colon and in the collecting ducts of the kidney. Its role in the kidney is to secrete H$^+$ ions into the urine, when blood pH falls, and to reabsorb K$^+$ ions (see Chapter 24).

- **Proton pumps** or **H$^+$-ATPases** are found in the membranes of the lysosomes and the Golgi apparatus. They pump protons from the cytosol into these organelles, keeping the inside of the organelles more acidic than the rest of the cell. These pumps are classified as **V-type ATPases** because they were first discovered in intracellular vacuolar structures, but are also present in plasma membranes. The secretion of protons by the V-type ATPase in **osteoclasts** helps to solubilize the bone mineral and creates an acidic environment for bone breakdown by enzymes. The proton pump in the kidney is present in the same cells as the H$^+$/K$^+$-ATPase and helps to secrete H$^+$ ions into the urine when blood pH falls.

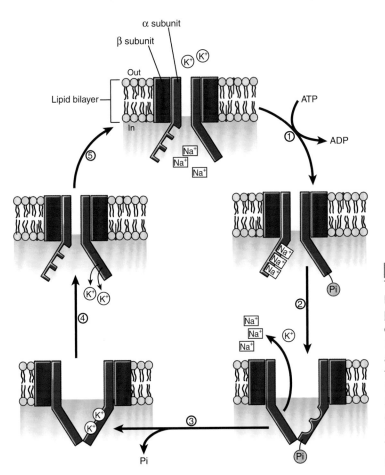

Figure 2.9 **Function of the sodium–potassium pump.** The pump is composed of two large α subunits that hydrolyze ATP and transport the ions. The two smaller β subunits are molecular chaperones that facilitate the correct integration of the α subunits into the membrane. In step 1, three intracellular Na^+ bind to the α subunit, and ATP is hydrolyzed to ADP. Phosphorylation (Pi) of the α subunit results in a conformational change, exposing the Na^+ to the extracellular space (step 2). In step 3, the Na^+ diffuses away and two K^+ bind, resulting in dephosphorylation of the α subunit. Dephosphorylation returns the α subunit to an intracellular conformation. The K^+ diffuses away, and ATP is rebound to start the cycle over again (step 6).

- **ATP-binding cassette (ABC) transporters** are a super-family of transporters composed of two transmembrane domains and two cytosolic nucleotide-binding domains. The transmembrane domains recognize specific solutes and transport them across the membrane using a number of different mechanisms, including conformational change. The nucleotide-binding domain, or ABC domain, has a highly conserved sequence. ABC transporters are involved in a number of cellular processes, including nutrient uptake, cholesterol and lipid trafficking, resistance to cytotoxic drugs and antibiotics, cellular immune response, and stem cell biology.

- **ABCA1**, a member of the ABC subfamily A, has an important role in effluxing cholesterol, phospholipids, and other metabolites out of cells. ABCA1 transfers lipids and cholesterol to lipid-poor **high-density lipoproteins (HDLs)**. ABCA1 is a unique ABC transporter because it is also a receptor, binding the lipid-poor HDL to facilitate the loading of the cholesterol that the transporter is moving out of the cell.

- ABC subfamily C transporters play a crucial role in the development of **multidrug resistance (MDR)**. There are a number of different transporters encoded by multiple MDR genes. The MDR1 transporter is widely distributed in the liver, brain, lung, kidney, pancreas, and small intestine and transports a wide range of antibiotics, antivirals, and chemotherapeutic drugs out of the cell. **MDR-associated protein transporters** are a related class of ABCC transporters that also interfere with antibiotic and chemotherapy. The cystic fibrosis transmembrane conductance regulator (ABCC7) is another member of this family.

- **Organic anion transporting polypeptides (OATPs)** are members of the solute carrier family and are highly expressed in the liver, kidney, and brain. OATPs transport anionic and cationic chemicals, steroid, and peptide backbones generally into cells. **Thyroxine**, **bile acids**, and **bilirubin** are important solutes transported by OATPs. These transporters also import agents such 3-hydroxy-3-methylglutaryl-CoA reductase inhibitors (statins), angiotensin-converting enzyme inhibitors, angiotensin receptor II antagonists, and cardiac glycosides into cells.

- **F-type ATPases** are located in the inner mitochondrial membrane. This type of proton pump normally functions in reverse. Instead of using the energy stored in ATP molecules to pump protons, its principal function is to synthesize ATP by using the energy stored in a gradient of protons that is crossing the inner mitochondrial membrane down its concentration gradient. The proton gradient is generated by the respiratory chain.

Secondary active transport

The net effect of ion pumps is maintenance of the various environments needed for the proper functioning of organelles, cells, and organs. Metabolic energy is expended by the pumps to create and maintain the differences in ion concentrations. Besides the importance of local ion concentrations for cell function, differences in concentrations represent stored energy. An ion releases potential energy when it moves down an electrochemical gradient, and this energy can be used to perform work. Cells have developed several carrier mechanisms to transport one solute against its concentration gradient by using the energy stored in the favorable gradient of another solute. In mammals, most of these mechanisms use sodium as the driver solute and use the energy of the sodium gradient to carry out the "uphill" transport of another important solute (Fig. 2.10). Because the sodium gradient is maintained by the action of the Na$^+$/K$^+$-ATPase, the function of these transport systems depends on the function of the Na$^+$/K$^+$-ATPase. Thus, they are called **secondary active transport mechanisms** because they depend on the supply of energy to the sodium–potassium pump. Disabling the pump with metabolic inhibitors or pharmacologic blockers causes these transport systems to stop when the sodium gradient has been dissipated.

Similar to passive carrier-mediated systems, secondary active transport systems are integral membrane proteins; they have specificity for the solute they transport and show saturation kinetics and competitive inhibition. They differ, however, in two respects. First, they cannot function in the absence of the driver ion, the ion that moves along its electrochemical gradient and supplies energy. Second, they transport the solute against its own concentration or electrochemical gradient. Functionally, the different secondary active transport systems can be classified into two groups: **symport** (cotransport) systems, in which the solute being transported moves in the same direction as the sodium ion, and **antiport** (exchange) systems, in which the sodium ion and the solute move in opposite directions.

Examples of symport mechanisms are the sodium-coupled sugar transport system and the several sodium-coupled amino acid transport systems found in the small intestine and the renal tubule. The symport systems allow efficient

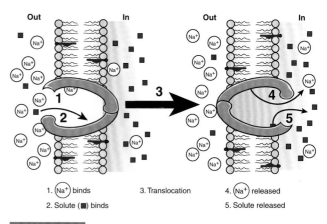

1. Na$^+$ binds   3. Translocation   4. Na$^+$ released
2. Solute (■) binds   5. Solute released

Figure 2.10 **Mechanism of secondary active transport.** A solute is moved against its concentration gradient by coupling it to Na$^+$ moving down a favorable gradient. Binding of extracellular Na$^+$ to the carrier protein (step 1) may increase the affinity of binding sites for solute, so that solute also can bind to the carrier (step 2), even though its extracellular concentration is low. A conformational change in the carrier protein (step 3) exposes the binding sites to the cytosol, where Na$^+$ readily dissociates because of the low intracellular Na$^+$ concentration (step 4). The release of Na$^+$ decreases the affinity of the carrier for solute and forces the release of the solute inside the cell (step 5), where solute concentration is already high. The free carrier then reverts to the conformation required for step 1, and the cycle begins again.

absorption of nutrients even when the nutrients are present at low concentrations. The sodium-dependent glucose transporter-1 (SGLT1) in the human intestine contains 664 amino acids in a single polypeptide chain with 14 membrane-spanning segments (Fig. 2.11). One complete cycle or turnover of

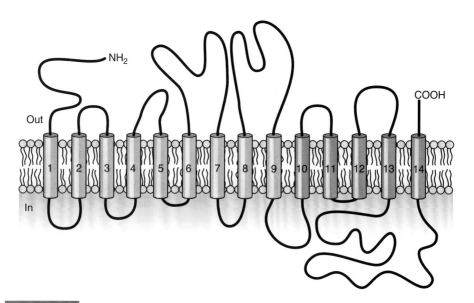

Figure 2.11 **A model of the secondary structure of the Na$^+$–glucose cotransporter protein (SGLT1) in the human intestine.** The polypeptide chain of 664 amino acids passes back and forth across the membrane 14 times. Each membrane-spanning segment consists of 21 amino acids arranged in an α-helical conformation. Both the NH$_2$ and the COOH ends are located on the extracellular side of the plasma membrane. In the functional protein, the membrane-spanning segments are clustered together to provide a hydrophilic pathway across the plasma membrane. The N-terminal portion of the protein, including helices 1 to 9, is required to couple Na$^+$ binding to glucose transport. The five helices (10–14) at the C terminus form the transport pathway for glucose.

a single SGLT1 protein, illustrated in Figure 2.10, can occur 1,000 times/s at 37°C. In reality, this cycle probably involves a coordinated trapping–release cycle and/or tilt of membrane-spanning segments rather than the simplistic view presented in Figure 2.10. Another example of a symport system is the family of sodium-coupled phosphate transporters (termed *NaPi*, types I and II) in the intestine and renal proximal tubule. These transporters have six to eight membrane-spanning segments and contain 460 to 690 amino acids. Sodium-coupled chloride transporters in the kidney are targets for inhibition by specific diuretics. The Na^+–Cl^- cotransporter in the distal tubule, known as *NCC*, is inhibited by thiazide diuretics, and the Na^+–K^+–$2Cl^-$ cotransporter in the ascending limb of the loop of Henle, referred to as *NKCC*, is inhibited by bumetanide.

The most important examples of antiporters are the Na^+/H^+ exchange and Na^+/Ca^{2+} exchange systems, found mainly in the plasma membrane of many cells. The first uses the sodium gradient to remove protons from the cell, controlling the intracellular pH and counterbalancing the production of protons in metabolic reactions. It is an **electroneutral system** because there is no net movement of charge. One Na^+ enters the cell for each H^+ that leaves. The second antiporter removes calcium from the cell and, together with the different calcium pumps, helps maintain a low cytosolic calcium concentration. It is an **electrogenic system** because there is a net movement of charge with three Na^+ entering the cell and one Ca^{2+} ion leaving in each cycle (Clinical Focus 2.1).

The structures of the symport and antiport protein transporters that have been characterized (see Fig. 2.11) share a common property with ion channels (see Fig. 2.8) and equilibrating carriers, namely, the presence of multiple membrane-spanning segments within the polypeptide chain. This supports the concept that, regardless of the mechanism, the membrane-spanning regions of a transport protein form a hydrophilic pathway for rapid transport of ions and solutes across the hydrophobic interior of the membrane lipid bilayer.

Transcellular transport

Epithelial cells occur in layers or sheets that allow the directional movement of solutes not only across the plasma membrane but also from one side of the cell layer to the other. Such regulated movement is achieved because the plasma membranes of epithelial cells have two distinct regions with different morphologies and different transport systems. These regions are the **apical membrane**, facing the lumen, and the **basolateral membrane**, facing the blood supply (Fig. 2.12). The specialized or polarized organization of the cells is maintained by the presence of **tight junctions** at the areas of contact between adjacent cells. Tight junctions prevent proteins on the apical membrane from migrating to the basolateral membrane and those on the basolateral membrane from migrating to the apical membrane. Thus, the entry and

Cellular Physiology

CLINICAL FOCUS | 2.1

Hexose Malabsorption in the Intestine

Malabsorption of hexoses in the intestine can be the indirect result of a number of circumstances, such as an increase in intestinal motility or defects in digestion because of pancreatic insufficiency. Although less common, malabsorption may be a direct result of a specific defect in hexose transport. Regardless of the cause, the symptoms are common and include diarrhea, abdominal pain, and gas. The challenge is to identify the cause so proper treatment can be applied. Some infants develop a copious watery diarrhea when fed milk that contains glucose or galactose or the disaccharides lactose and sucrose. The latter are degraded to glucose, galactose, and fructose by enzymes in the intestine. The dehydration can begin during the first day of life and can lead to rapid death if not corrected. Fortunately, the symptoms disappear when a carbohydrate-free formula fortified with fructose is used instead of milk. This condition is a rare inherited disease known as **glucose–galactose malabsorption (GGM)**, and about 200 severe cases have been reported worldwide. At least 10% of the general population has glucose or lactose intolerance, however, and it is possible that these people may have milder forms of the disease. A specific defect in absorption of glucose and galactose can be demonstrated by tolerance tests in which oral administration of these monosaccharides produces little or no increase in plasma glucose or galactose. The primary defect lies in the Na^+–glucose cotransporter protein

(SGLT, Fig. 2.11), located in the apical plasma membrane of intestinal epithelial cells (Fig. 2.12). Glucose and galactose have very similar structures, and both are substrates for transport by SGLT. Fructose transport is not affected by a defect in SGLT because a specific fructose transporter named **GLUT5** is present in the apical membrane. Human SGLT was cloned in 1989, and almost 30 different mutations have been identified in GGM patients. Many of the mutations produce premature cessation of SGLT protein synthesis or disrupt the trafficking of mature SGLT to the apical plasma membrane. In a few cases, the SGLT reaches the apical membrane but is no longer capable of glucose transport. The result in all cases is that functional SGLT proteins are not present in the apical membrane so glucose and galactose remain in the lumen of the intestine. As these solutes accumulate in the lumen, the osmolality of the fluids increases and retards absorption of water, leading to diarrhea and severe water loss from the body. Identification of specific changes in defective SGLT proteins in patients has provided clues about the specific amino acids that are essential for the normal function of SGLT. At the same time, advances in molecular biology have allowed a better understanding of the genetic defect at the cellular level and how this leads to the clinical symptoms. GGM is an example of how information from a disease can further understanding of physiology and vice versa. ■

Apical (luminal) side

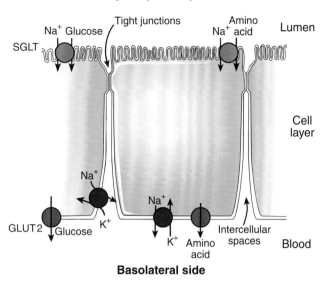

Basolateral side

Figure 2.12 **The localization of transport systems to different regions of the plasma membrane in epithelial cells of the small intestine.** In a polarized cell, the entry and exit of solutes such as glucose, amino acids, and Na$^+$ occur at opposite sides of the cell. Active entry of glucose and amino acids is restricted to the apical membrane, and exit requires equilibrating carriers located only in the basolateral membrane. For example, glucose enters on sodium-dependent glucose transporter (SGLT) and exits on glucose transporter-2 (GLUT2). Na$^+$ that enters via the apical symporters is pumped out by the Na$^+$/K$^+$-ATPase on the basolateral membrane. The result is a net movement of solutes from the luminal side of the cell to the basolateral side, ensuring efficient absorption of glucose, amino acids, and Na$^+$ from the intestinal lumen.

exit steps for solutes can be localized to opposite sides of the cell. This is the key to **transcellular transport** across epithelial cells.

An example is the absorption of glucose in the small intestine. Glucose enters the intestinal epithelial cells by active transport using the electrogenic Na$^+$–glucose cotransporter system (SGLT) in the apical membrane. This increases the intracellular glucose concentration above the blood glucose concentration, and the glucose molecules move passively out of the cell and into the blood via an equilibrating carrier mechanism (GLUT2) in the basolateral membrane (see Fig. 2.12). The intestinal GLUT2, like the erythrocyte GLUT1, is a sodium-independent transporter that moves glucose down its concentration gradient. Unlike GLUT1, the GLUT2 transporter can accept other sugars such as galactose and fructose that are also absorbed in the intestine. The Na$^+$/K$^+$-ATPase that is located in the basolateral membrane pumps out the sodium ions that enter the cell with the glucose molecules on SGLT. The polarized organization of the epithelial cells and the integrated functions of the plasma membrane transporters form the basis by which cells accomplish transcellular movement of both glucose and sodium ions.

► WATER MOVEMENT ACROSS THE PLASMA MEMBRANE

Water can move rapidly in and out of cells, but the partition coefficient of water into lipids is low; therefore, the permeability of the membrane lipid bilayer for water is also low. Specific membrane proteins that function as water channels explain the rapid movement of water across the plasma membrane. These water channels are small (30 kDa), integral membrane proteins known as **aquaporins**. Of the thirteen known mammalian aquaporins, eight are expressed in the kidney, where water movement across the plasma membrane is particularly rapid.

In the kidney, aquaporin-2 (AQP2) is abundant in the collecting duct and is the target of the hormone **arginine vasopressin**, also known as *antidiuretic hormone*. This hormone increases water transport in the collecting duct in part by stimulating the recruitment of AQP2 proteins into the apical plasma membrane. AQP2 has a critical role in inherited and acquired disorders of water reabsorption by the kidney (see Chapter 22).

Water movement across the plasma membrane is driven by differences in osmotic pressure.

The spontaneous movement of water across a membrane driven by a gradient of water concentration is the process known as **osmosis**. The water moves from an area of high concentration of water to an area of low concentration. Concentration is defined by the number of particles per unit of volume; thus, a solution with a high concentration of solutes has a low concentration of water, and vice versa. Osmosis can be viewed as the movement of water from a solution of high water concentration (low concentration of solute) toward a solution with a lower concentration of water (high solute concentration). Osmosis is a passive transport mechanism that tends to equalize the total solute concentrations of the solutions on both sides of every membrane.

If a cell that is in osmotic equilibrium is transferred to a more dilute solution, water will enter the cell, the cell volume will increase, and the solute concentration of the cytoplasm will be reduced. If the cell is transferred to a more concentrated solution, water will leave the cell, the cell volume will decrease, and the solute concentration of the cytoplasm will increase.

The driving force for the movement of water across the plasma membrane is the difference in water concentration between the two sides of the membrane. For historical reasons, this driving force is not called the chemical gradient of water but the difference in osmotic pressure. The **osmotic pressure** of a solution is defined as the pressure necessary to stop the net movement of water across a selectively permeable membrane that separates the solution from pure water. When a membrane separates two solutions of different osmotic pressure, water will move from the solution with low osmotic pressure (high water and low solute concentrations) to the solution of high osmotic pressure (low water and high solute concentrations). In this context, the term *selectively permeable* means that the membrane is permeable to water but not solutes. In reality, most biologic membranes contain membrane transport proteins that permit solute movement.

The osmotic pressure of a solution depends on the *number* of particles dissolved in it, the total concentration of all solutes, regardless of the type of solutes present. Many solutes, such as salts, acids, and bases, dissociate in water, so the number of particles is greater than the molar concentration. For example, NaCl dissociates in water to give Na^+ and Cl^-, so one molecule of NaCl will produce two osmotically active particles. In the case of $CaCl_2$, there are three particles per molecule. The equation giving the osmotic pressure of a solution is

$$\pi = nRTC \qquad (3)$$

where π is the osmotic pressure of the solution, n is the number of particles produced by the dissociation of one molecule of solute (2 for NaCl, 3 for $CaCl_2$), R is the universal gas constant (0.0821 L·atm/mol·K), T is the absolute temperature, and C is the concentration of the solute in mol/L. Osmotic pressure can be expressed in atmospheres (atm). Solutions with the same osmotic pressure are called **isosmotic**. A solution is **hyperosmotic** with respect to another solution if it has a higher osmotic pressure and **hyposmotic** if it has a lower osmotic pressure.

Equation 3, called the *van't Hoff equation*, is valid only when applied to very dilute solutions, in which the particles of solutes are so far away from each other that no interactions occur between them. Generally, this is not the case at physiologic concentrations. Interactions between dissolved particles, mainly between ions, cause the solution to behave as if the concentration of particles is less than the theoretical value (nC). A correction coefficient, called the *osmotic coefficient* (Φ) of the solute, needs to be introduced in the equation. Therefore, the osmotic pressure of a solution can be written more accurately as

$$\pi = nRT\Phi C \qquad (4)$$

The osmotic coefficient varies with the specific solute and its concentration. It has values between 0 and 1. For example, the osmotic coefficient of NaCl is 1.00 in an infinitely dilute solution but changes to 0.93 at the physiologic concentration of 0.15 mol/L.

At any given T, because R is constant, equation 4 shows that the osmotic pressure of a solution is directly proportional to the term $n\Phi C$. This term is known as the **osmolality** or *osmotic concentration* of a solution and is expressed in Osm/kg H_2O. Most physiologic solutions such as blood plasma contain many different solutes, and each contributes to the total osmolality of the solution. The osmolality of a solution containing a complex mixture of solutes is usually measured by freezing point depression. The freezing point of an aqueous solution of solutes is lower than that of pure water and depends on the total number of solute particles. Compared with pure water, which freezes at 0°C, a solution with an osmolality of 1 Osm/kg H_2O will freeze at −1.86°C. The ease with which osmolality can be measured has led to the wide use of this parameter for comparing the osmotic pressure of different solutions. The osmotic pressures of

physiologic solutions are not trivial. Consider blood plasma, for example, which usually has an osmolality of 0.28 Osm/kg H_2O, determined by freezing point depression. Equation 4 shows that the osmotic pressure of plasma at 37°C is 7.1 atm, about seven times greater than the atmospheric pressure.

Many cells can regulate their volume.

Cell volume changes can occur in response to changes in the osmolality of ECF in both normal and pathophysiologic situations. Accumulation of solutes can also produce volume changes by increasing the intracellular osmolality. Many cells can correct these volume changes. Volume regulation is particularly important in the brain where cell swelling can have serious consequences because expansion is strictly limited by the rigid skull.

Tonicity

A solution's osmolality is determined by the total concentration of all the solutes present. In contrast, the solution's **tonicity** is determined by the concentrations of only those solutes that do not enter ("penetrate") the cell. Tonicity determines cell volume, as illustrated in the following examples. Na^+ behaves as a nonpenetrating solute because it is pumped out of cells by the Na^+/K^+-ATPase at the same rate that it enters. A solution of NaCl at 0.2 Osm/kg H_2O is hypo-osmotic compared with cell cytosol at 0.3 Osm/kg H_2O. The NaCl solution is also **hypotonic** because cells will accumulate water and swell when placed in this solution. A solution containing a mixture of NaCl (0.3 Osm/kg H_2O) and urea (0.1 Osm/kg H_2O) has a total osmolality of 0.4 Osm/kg H_2O and will be hyperosmotic compared with cell cytosol. The solution is **isotonic**, however, because it produces no permanent change in cell volume. The reason is that cells shrink initially as a result of loss of water, but urea is a penetrating solute that rapidly enters the cells. Urea entry increases the intracellular osmolality, so water also enters and increases the volume. Entry of water ceases when the urea concentration is the same inside and outside the cells. At this point, the total osmolality both inside and outside the cells will be 0.4 Osm/kg H_2O and the cell volume will be restored to normal. By extension, it can be seen that normal blood plasma is an isotonic solution because Na^+ is the predominant plasma solute and is nonpenetrating. This stabilizes cell volume while other plasma solutes (glucose, amino acids, phosphate, urea, etc.) enter and leave the cells as needed.

Volume regulation mechanisms

When cell volume increases because of extracellular hypotonicity, the response of many cells is rapid activation of transport mechanisms that tend to decrease the cell volume. Different cells use different **regulatory volume decrease (RVD) mechanisms** to move solutes out of the cell and decrease the number of particles in the cytosol, causing water to leave the cell. Because cells have high intracellular concentrations of potassium, many RVD mechanisms involve an increased efflux of K^+, either by stimulating the opening of potassium channels or by activating symport mechanisms for KCl. Other cells activate the efflux of amino acids, such as taurine or proline. The net result is a decrease in intracellular solute content and a reduction of cell volume close to its original value.

When placed in a **hypertonic** solution, cells rapidly lose water and their volume decreases. In many cells, a decreased volume triggers **regulatory volume increase (RVI) mechanisms**, which increase the number of intracellular particles, bringing water back into the cells. Because Na^+ is the main extracellular ion, many RVI mechanisms involve an influx of sodium into the cell. Na^+–Cl^- symport, Na^+–K^+–$2Cl^-$ symport, and Na^+/H^+ antiport are some of the mechanisms activated to increase the intracellular concentration of Na^+ and increase the cell volume toward its original value.

Mechanisms based on an increased Na^+ influx are effective for only a short time because, eventually, the sodium pump will increase its activity and reduce intracellular Na^+ to its normal value. Cells that regularly encounter hypertonic ECFs have developed additional mechanisms for maintaining normal volume. These cells can synthesize specific organic solutes, enabling them to increase intracellular osmolality for a long time and avoiding altering the concentrations of ions they must maintain within a narrow range of values. The organic solutes are usually small molecules that do not interfere with normal cell function when they accumulate inside the cell. For example, cells of the medulla of the mammalian kidney can increase the level of the enzyme aldose reductase when subjected to elevated extracellular osmolality. This enzyme converts glucose to an osmotically active solute, sorbitol. Brain cells can synthesize and store inositol. Synthesis of sorbitol and inositol represent different answers to the problem of increasing the total intracellular osmolality, allowing normal cell volume to be maintained in the presence of hypertonic ECF.

Oral rehydration therapy is driven by solute transport.

Oral administration of rehydration solutions has dramatically reduced the mortality resulting from cholera and other diseases that involve excessive losses of water and solutes from the gastrointestinal tract. The main ingredients of rehydration solutions are glucose, NaCl, and water. The glucose and Na^+ ions are reabsorbed by SGLT1 and other transporters in the epithelial cells lining the lumen of the small intestine (see Fig. 2.12). Deposition of these solutes on the basolateral side of the epithelial cells increases the osmolality in that region compared with the intestinal lumen and drives the osmotic absorption of water. Absorption of glucose, and the obligatory increases in absorption of NaCl and water, helps to compensate for excessive diarrheal losses of salt and water.

▶ RESTING MEMBRANE POTENTIAL

The different passive and active transport systems are coordinated in a living cell to maintain intracellular ions and other solutes at concentrations compatible with life. Consequently, the cell does not equilibrate with the ECF but rather exists in a **steady state** with the extracellular solution. For example, intracellular Na^+ concentration (10 mmol/L in a muscle cell) is much lower than extracellular Na^+ concentration

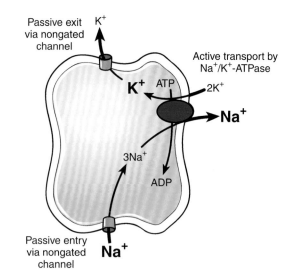

Figure 2.13 **The concept of a steady state.** Na^+ enters a cell through nongated Na^+ channels, moving passively down the electrochemical gradient. The rate of Na^+ entry is matched by the rate of active transport of Na^+ out of the cell via the Na^+/K^+-ATPase. The intracellular concentration of Na^+ remains low and constant. Similarly, the rate of passive K^+ exit through nongated K^+ channels is matched by the rate of active transport of K^+ into the cell via the pump. The intracellular K^+ concentration remains high and constant. During each cycle of the ATPase, two K^+ are exchanged for three Na^+, and one molecule of adenosine triphosphate (ATP) is hydrolyzed. *Boldfaced* and *lightfaced* fonts indicate high and low ion concentrations, respectively.

(140 mmol/L), so Na^+ enters the cell by passive transport through nongated (always open) Na^+ channels. The rate of Na^+ entry is matched, however, by the rate of active transport of Na^+ out of the cell via the sodium–potassium pump (Fig. 2.13). The net result is that intracellular Na^+ is maintained constant and at a low level, even though Na^+ continually enters and leaves the cell. The reverse is true for K^+, which is maintained at a high concentration inside the cell relative to the outside. The passive exit of K^+ through nongated K^+ channels is matched by active entry via the pump (see Fig. 2.13). Maintenance of this steady state with ion concentrations inside the cell different from those outside the cell is the basis for the difference in electrical potential across the plasma membrane or the **resting membrane potential**.

Ion movement is driven by the electrochemical potential.

If there are no differences in temperature or hydrostatic pressure between the two sides of a plasma membrane, two forces drive the movement of ions and other solutes across the membrane. One force results from the difference in the concentration of a substance between the inside and the outside of the cell and the tendency of every substance to move from areas of high concentration to areas of low concentration. The other force results from the difference in electrical potential between the two sides of the membrane and applies only to ions and other electrically charged solutes. When a difference in electrical potential exists, positive ions tend to move toward the negative side, whereas negative ions tend to move toward the positive side.

The sum of these two driving forces is called the *gradient* (or difference) of **electrochemical potential** across the membrane for a specific solute. It measures the tendency of that solute to cross the membrane. The expression of this force is given by

$$\Delta\mu = RT \ln C_i/C_o + zF(E_i - E_o) \tag{5}$$

where μ represents the electrochemical potential ($\Delta\mu$ is the difference in electrochemical potential between two sides of the membrane); C_i and C_o are the concentrations of the solute inside and outside the cell, respectively; E_i is the electrical potential inside the cell measured with respect to the electrical potential outside the cell (E_o); R is the universal gas constant (2 cal/mol·K); T is the absolute temperature (K); z is the valence of the ion; and F is the Faraday constant (23 cal/mV·mol). By inserting these units in equation 5 and simplifying, the electrochemical potential will be expressed in cal/mol, which is the unit of energy. If the solute is not an ion and has no electrical charge, then z = 0 and the last term of the equation becomes zero. In this case, the electrochemical potential is defined only by the different concentrations of the uncharged solute, called the **chemical potential**. The driving force for solute transport becomes solely the difference in chemical potential.

Net ion movement is zero at the equilibrium potential.

Net movement of an ion into or out of a cell continues as long as the driving force exists. Net movement stops and equilibrium is reached only when the driving force of electrochemical potential across the membrane becomes zero. The condition of equilibrium for any permeable ion will be $\Delta\mu = 0$. Substituting this condition into equation 5, we obtain

$$0 = RT \ln\left(\frac{C_i}{C_o}\right) + zF(E_i - E_o)$$

$$E_i - E_o = -\frac{RT}{zF} \ln\left(\frac{C_i}{C_o}\right) \tag{6}$$

$$E_i - E_o = -\frac{RT}{zF} \ln\left(\frac{C_o}{C_i}\right)$$

Equation 6, known as the **Nernst equation**, gives the value of the electrical potential difference ($E_i - E_o$) necessary for a specific ion to be at equilibrium. This value is known as the **Nernst equilibrium potential** for that particular ion, and it is expressed in millivolts (mV). At the equilibrium potential, the tendency of an ion to move in one direction because of the difference in concentrations is exactly balanced by the tendency to move in the opposite direction because of the difference in electrical potential. At this point, the ion will be in equilibrium and there will be no net movement. By converting to $\log_{10}$ and assuming a physiologic temperature of 37°C and a value of +1 for z (for Na^+ or K^+), the Nernst equation can be expressed as

$$E_i - E_o = 61 \text{LOG}_{10}(C_o/C_i) \tag{7}$$

Because Na^+ and K^+ (and other ions) are present at different concentrations inside and outside a cell, it follows from equation 7 that the equilibrium potential will be different for each ion.

Resting membrane potential is determined by the passive movement of several ions.

The resting membrane potential is the electrical potential difference across the plasma membrane of a normal living cell in its unstimulated state. It can be measured directly by the insertion of a microelectrode into the cell with a reference electrode in the ECF. The resting membrane potential is determined by those ions that can cross the membrane and are prevented from attaining equilibrium by active transport systems. Potassium, sodium, and chloride ions can cross the membranes of every living cell, and each of these ions contributes to the resting membrane potential. By contrast, the permeability of the membrane of most cells to divalent ions is so low that it can be ignored in this context.

The **Goldman equation** gives the value of the membrane potential (in mV) when all the permeable ions are accounted for:

$$E_i - E_o = \frac{RT}{F} \ln\left(\frac{P_K[K^+]_o + P_{Na}[Na^+]_o + P_{Cl}[Cl^-]_i}{P_K[K^+]_i + P_{Na}[Na^+]_i + P_{Cl}[Cl^-]_o}\right) \tag{8}$$

where P_K, P_{Na}, and P_{Cl} represent the permeability of the membrane to potassium, sodium, and chloride ions, respectively, and brackets indicate the concentration of the ion inside (i) and outside (o) the cell. If a specific cell is not permeable to one of these ions, the contribution of the impermeable ion to the membrane potential will be zero. For a cell that is permeable to an ion other than the three considered in the Goldman equation, that ion will contribute to the membrane potential and must be included in equation 8.

It can be seen from equation 8 that the contribution of any ion to the membrane potential is determined by the membrane's permeability to that particular ion. The higher the permeability of the membrane to one ion relative to the others, the more that ion will contribute to the membrane potential. The plasma membranes of most living cells are much more permeable to potassium ions than to any other ion. Making the assumption that P_{Na} and P_{Cl} are zero relative to P_K, equation 8 can be simplified to

$$E_i - E_o = \frac{RT}{F} \ln\left(\frac{P_K[K^+]_o}{P_K[K^+]_i}\right)$$

$$E_i - E_o = \frac{RT}{F} \ln\left(\frac{[K^+]_o}{[K^+]_i}\right) \tag{9}$$

which is the Nernst equation for the equilibrium potential for K^+ (see equation 6). This illustrates two important points:

- In most cells, the resting membrane potential is close to the equilibrium potential for K^+.
- The resting membrane potential of most cells is dominated by K^+ because the plasma membrane is more permeable to this ion compared with the others.

Cellular Physiology

As a typical example, the K^+ concentrations outside and inside a muscle cell are 3.5 and 155 mmol/L, respectively. Substituting these values in equation 7 gives a calculated equilibrium potential for K^+ of −100 mV, negative inside the cell relative to the outside. Measurement of the resting membrane potential in a muscle cell yields a value of −90 mV (negative inside). This value is close to, although not the same as, the equilibrium potential for K^+.

The reason the resting membrane potential in the muscle cell is less negative than the equilibrium potential for K^+ is as follows. Under physiologic conditions, there is passive entry of Na^+ ions. This entry of positively charged ions has a small but significant effect on the negative potential inside the cell. Assuming intracellular Na^+ to be 10 mmol/L and extracellular Na^+ to be 140 mmol/L, the Nernst equation gives a value of +70 mV for the Na^+ equilibrium potential (positive inside the cell). This is far from the resting membrane potential of −90 mV. Na^+ makes only a small contribution to the resting membrane potential because membrane permeability to Na^+ is low compared with that of K^+.

The contribution of Cl^- ions need not be considered because the resting membrane potential in the muscle cell is the same as the equilibrium potential for Cl^-. Therefore, there is no net movement of chloride ions.

In most cells, as shown above using a muscle cell as an example, the equilibrium potentials of K^+ and Na^+ are different from the resting membrane potential, which indicates that neither K^+ ions nor Na^+ ions are at equilibrium. Consequently, these ions continue to cross the plasma membrane via specific nongated channels, and these passive ion movements are *directly* responsible for the resting membrane potential.

The Na^+/K^+-ATPase is important *indirectly* for maintaining the resting membrane potential because it sets up the gradients of K^+ and Na^+ that drive passive K^+ exit and Na^+ entry. During each cycle of the pump, two K^+ ions are moved into the cell in exchange for three Na^+, which are moved out (see Fig. 2.13). Because of the unequal exchange mechanism, the pump's activity contributes slightly (about −5 mV) to the negative potential inside the cell.

▶ COMMUNICATION AND SIGNALING MODES

The human body has several means of transmitting information between cells. These mechanisms include direct communication between adjacent cells, autocrine and paracrine signaling, and the release of neurotransmitters and hormones produced by nerves and endocrine cells.

Cells communicate locally by paracrine and autocrine signaling.

Cells may signal to each other via the local release of chemical substances. In **paracrine signaling**, a chemical is liberated from a cell and diffuses a short distance through the ECF to act on nearby cells. Paracrine-signaling factors affect only the immediate environment and bind with high specificity to cell receptors on the plasma membrane of the receiving cell. They are also rapidly destroyed by extracellular enzymes or bound to extracellular matrix, thus preventing their widespread diffusion. **Nitric oxide (NO)** is an example of a paracrine-signaling molecule because it has an intrinsically short half-life and thus can affect cells located directly next to the NO-producing cell. NO has major roles in mediating vascular smooth muscle tone, facilitating neurotransmission in the central nervous system (CNS) and modulating immune responses (see Chapters 15 and 25).

In contrast, during **autocrine signaling**, the cell releases a chemical messenger into the ECF that binds to a receptor on the surface of the cell that secreted it. Eicosanoids (e.g., prostaglandins) are examples of signaling molecules that can act in an autocrine manner. These molecules act as local hormones to influence a variety of physiologic processes such as uterine smooth muscle contraction during pregnancy.

Nervous system provides for rapid and targeted communication.

The CNS includes the brain and spinal cord, which links the CNS to the peripheral nervous system (PNS), which is composed of nerves or bundles of neurons. Together, the CNS and the PNS integrate and coordinate a vast number of sensory processes and motor responses. The basic functions of the nervous system are to acquire sensory input from both the internal and external environment, integrate the input, and then activate a response to the stimuli. Sensory input to the nervous system can occur in many forms, such as taste, sound, blood pH, hormones, balance or orientation, pressure, or temperature, and these inputs are converted to signals that are sent to the brain or spinal cord. In the sensory centers of the brain and spinal cord, the input signals are rapidly integrated, and then a response is generated. The response is generally a motor output and is a signal that is transmitted to the organs and tissues, where it is converted into an action such as a change in heart rate, sensation of thirst, release of hormones, or a physical movement. The nervous system is also organized for discrete activities; it has an enormous number of "private lines" for sending messages from one distinct locus to another. The conduction of information along nerves occurs via electrical signals, called *action potentials*, and signal transmission between nerves or between nerves and effector structures takes place at a **synapse**. Synaptic transmission is mediated by the release of specific chemicals or **neurotransmitters** from the nerve terminals. Innervated cells have receptors in their cell membranes that selectively bind neurotransmitters. Chapter 3 discusses the actions of various neurotransmitters and how they are synthesized and degraded. Chapters 4 to 6 discuss the role of the nervous system in coordinating and controlling body functions.

Endocrine system provides for slower and more diffuse communication.

The endocrine system produces hormones in response to a variety of stimuli that are instrumental in establishing and maintaining homeostasis in the body. In contrast to the rapid, directed effects resulting from neuronal stimulation, responses to hormones are much slower (seconds to hours) in onset, and the effects often last longer. Hormones are secreted from endocrine glands and tissues and are broadcast to all parts of the body by the bloodstream. A particular cell can only respond to a hormone if it possesses the appropriate hormone receptor. Hormone effects may be focused. For example, arginine vasopressin specifically increases the water permeability of kidney collecting duct cells but does not alter the water permeability of other cells. Hormone effects can also be diffuse, influencing practically every cell in the body. For example, thyroxine has a general stimulatory effect on metabolism. Hormones play a critical role in controlling such body functions as growth, metabolism, and reproduction.

Cells that are not traditional endocrine cells produce a special category of chemical messengers called **tissue growth factors**. These growth factors are protein molecules that influence cell division, differentiation, and cell survival. They may exert effects in an autocrine, paracrine, or endocrine fashion. Many growth factors have been identified. **Nerve growth factor** enhances nerve cell development and stimulates the growth of axons. **Epidermal growth factor (EGF)** stimulates the growth of epithelial cells in the skin and other organs. **Platelet-derived growth factor** stimulates the proliferation of vascular smooth muscle and endothelial cells. **Insulin-like growth factors** stimulate the proliferation of a wide variety of cells and mediate many of the effects of growth hormone. Growth factors appear to be important in the development of multicellular organisms and in the regeneration and repair of damaged tissues.

Nervous and endocrine control systems overlap.

The distinction between nervous and endocrine control systems is not always clear. This is because the nervous system exerts control over endocrine gland function, most if not all endocrine glands are innervated by the PNS, and these nerves can directly control the endocrine function of the gland. In addition, the innervation of endocrine tissues can also regulate blood flow within the gland, which can impact the distribution and thus function of the hormone. On the other hand, hormones can affect the CNS to alter behavior and mood. Adding to this highly integrated relationship is the presence of specialized nerve cells, called **neuroendocrine**, or **neurosecretory cells**, which directly convert a neural signal into a hormonal signal. These cells thus directly convert electrical energy into chemical energy, and activation of a neurosecretory cell results in hormone secretion. Examples are the hypothalamic neurons, which liberate releasing factors that control secretion by the anterior pituitary gland, and the hypothalamic neurons, which secrete arginine vasopressin and oxytocin into the circulation. In addition, many proven or potential neurotransmitters found in nerve terminals are also well-known hormones, including arginine vasopressin, cholecystokinin, enkephalins, norepinephrine, secretin, and vasoactive intestinal peptide. Therefore, it is sometimes difficult to classify a particular molecule as either a hormone or a neurotransmitter.

▶ MOLECULAR BASIS OF CELLULAR SIGNALING

The study of intercellular communication has led to the identification of many complex signaling systems that are used by the body to network and coordinate functions. A general outline for a **signal cascade** is as follows: Signaling is initiated by binding of a first messenger to its appropriate ligand-binding site on the outer surface domain of its relevant membrane receptor. This results in activation of the receptor; the receptor may adopt a new conformation, form aggregates (multimerize), and/or become phosphorylated or dephosphorylated. These changes often result in association of adapter signaling molecules that couple the activated receptor to downstream molecules that transduce and amplify the signal through the cell by activating specific effector molecules and generating a second messenger. The outcome of the signal transduction cascade is a physiologic response, such as secretion, movement, growth, division, or death. It is important to remember these physiologic responses are the collective result of a multitude of signaling messengers that transmit signals to the cells in various tissues.

Plasma membrane receptors activate signal transduction pathways.

Cellular receptors are divided into two general types: *cell-surface receptors* and *intracellular receptors*. Three general classes of cell-surface receptors have been identified: *G protein–coupled receptors (GPCRs)*, *ion channel–linked receptors*, and *enzyme-linked receptors*. Intracellular receptors include steroid and thyroid hormone receptors and are discussed in a later section in this chapter. Cell-surface receptors are often found in **lipid rafts** that can compartmentalize and organize assembly of signaling complexes.

G protein–coupled receptors transmit signals through trimeric G proteins.

With more than 1,000 members, **G protein–coupled receptors (GPCRs)** are the largest family of cell-surface receptors. These receptors regulate their effector targets through the intermediary activity of a separate membrane-bound adapter protein complex called the *trimeric guanosine triphosphate (GTP)-binding regulatory protein* or *trimeric G protein* (Fig. 2.14). GPCRs mediate cellular responses to numerous types of first messenger signaling molecules, and many first messenger ligands can activate several different GPCRs.

GPCRs have a ligand-binding extracellular domain, separated by a seven-pass transmembrane-spanning region from the cytosolic regulatory domain at the other end of the molecule, where the receptor interacts with the membrane-bound G protein. Binding of ligand or hormone to the extracellular domain results in a conformational change in the receptor that is transmitted to the cytosolic regulatory domain. This conformational change allows an interaction of the ligand-bound, activated receptor with a trimeric G protein associated with the inner leaflet of the plasma membrane. The interaction between the ligand-bound, activated receptor and the G protein, in turn, activates the G protein, which dissociates from the receptor and transmits the signal to its effector enzyme or ion channel.

The trimeric G proteins are named for their requirement for **guanosine triphosphate (GTP)** binding and hydrolysis. Heterotrimeric G proteins are composed of three distinct subunits tethered to the plasma membrane. The G

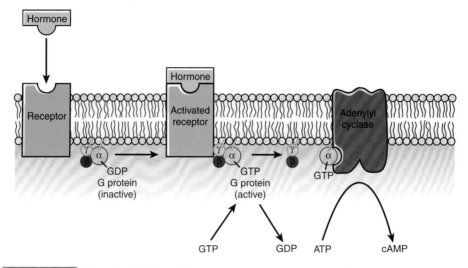

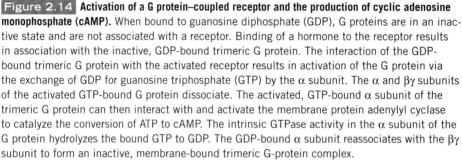

Figure 2.14 Activation of a G protein–coupled receptor and the production of cyclic adenosine monophosphate (cAMP). When bound to guanosine diphosphate (GDP), G proteins are in an inactive state and are not associated with a receptor. Binding of a hormone to the receptor results in association with the inactive, GDP-bound trimeric G protein. The interaction of the GDP-bound trimeric G protein with the activated receptor results in activation of the G protein via the exchange of GDP for guanosine triphosphate (GTP) by the α subunit. The α and βγ subunits of the activated GTP-bound G protein dissociate. The activated, GTP-bound α subunit of the trimeric G protein can then interact with and activate the membrane protein adenylyl cyclase to catalyze the conversion of ATP to cAMP. The intrinsic GTPase activity in the α subunit of the G protein hydrolyzes the bound GTP to GDP. The GDP-bound α subunit reassociates with the βγ subunit to form an inactive, membrane-bound trimeric G-protein complex.

protein subunits are the α subunit, which binds and hydrolyzes GTP, and the β and γ subunits, which form a stable, tight noncovalent-linked βγ dimer. When the α subunit binds **guanosine diphosphate (GDP)**, it associates with the βγ subunits to form a trimeric complex that can interact with the cytoplasmic domain of the GPCR. The conformational change that occurs upon ligand binding causes the GDP-bound trimeric (αβγ complex) G protein to associate with the ligand-bound receptor. The association of the GDP-bound trimeric complex with the GPCR activates the exchange of GDP for GTP. Displacement of GDP by GTP is favored in cells because GTP is in higher concentration. The displacement of GDP by GTP causes the α subunit to dissociate from the receptor and from the βγ subunits of the G protein. This exposes an effector-binding site on the α subunit, which then associates with an effector enzyme (e.g., AC or phospholipase C [PLC]) to result in the generation of second messengers (e.g., cAMP or IP$_3$ and DAG). The hydrolysis of GTP to GDP by the α subunit results in the reassociation of the α and βγ subunits, which are then ready to repeat the cycle.

Two major effector molecules regulated by G-protein subunits are **adenylyl cyclase (AC)** and PLC. The association of an activated Gα subunit with AC can either stimulate or inhibit the production of cAMP. Association of an α$_s$ subunit (s for stimulatory) promotes the activation of AC and production of cAMP. The association of an α$_i$ (i for inhibitory) subunit promotes the inhibition of AC and a decrease in cAMP. Thus, bidirectional regulation of AC is achieved by coupling different classes of cell-surface receptors to the enzyme by either G$_s$ or G$_i$ (Fig. 2.15).

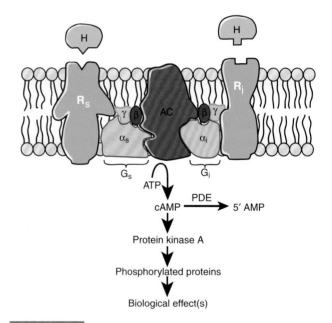

Figure 2.15 Stimulatory and inhibitory coupling of G proteins to adenylyl cyclase. Stimulatory (G$_s$) and inhibitory (G$_i$) G proteins couple hormone binding to the receptor with either activation or inhibition of adenylyl cyclase (AC). Each G protein is a trimer consisting of Gα, Gβ, and Gγ subunits. The Gα subunits in G$_s$ and G$_i$ are distinct and provide the specificity for either activation or inhibition of AC. Hormones that bind "stimulatory" receptors (R$_s$) are coupled to AC through stimulatory G proteins (G$_s$). Conversely, hormones inhibit AC by binding "inhibitory" receptors (R$_i$) coupled to AC through inhibitory G proteins (G$_i$). Intracellular levels of cyclic adenosine monophosphate (cAMP) are modulated by the activity of phosphodiesterase (PDE), which converts cAMP to 5'AMP and turns off the signaling pathway by reducing the level of cAMP.

In addition to α_s and α_i subunits, other isoforms of G-protein subunits have been described. For example, α_q activates PLC, resulting in the production of the second messengers, DAG, and inositol trisphosphate. Another Gα subunit, α_T or **transducin**, is expressed in photoreceptor tissues and has an important role in signaling in light-sensing rod cells in the retina by activation of the effector *cGMP phosphodiesterase (PDE)*, which degrades cGMP to 5′GMP (see Chapter 4). G-protein subunits are expressed in different combinations in different tissues, which contributes to both the specificity of the transduced signal and the second messenger produced.

Ion channel–linked receptors mediate cell signaling by regulating the intracellular concentration of specific ions.

Ion channels may be opened or closed by changing the membrane potential or by the binding of ligands to membrane receptors. In some cases, the receptor and ion channel are the same molecule such as at the neuromuscular junction, where the neurotransmitter acetylcholine binds to a muscle membrane nicotinic cholinergic receptor. In other cases, the receptor and ion channel are linked via a G protein, second messengers, and other downstream effector molecules, as in the muscarinic cholinergic receptor on cells innervated by parasympathetic postganglionic nerve fibers. Ion channels are also directly activated by cGMP or cAMP produced by receptor activation. This mode of ion channel control is predominantly found in the sensory tissues for sight, smell, and hearing and in the smooth muscle surrounding blood vessels. The opening or closing of ion channels plays a key role in signaling between electrically excitable cells, such as nerve and muscle.

Tyrosine kinase receptors signal through adapter proteins to activate the mitogen-activated protein kinase pathway.

Many hormones and growth factors (**mitogens**) signal their target cells by binding to receptors that have tyrosine kinase activity, resulting in phosphorylation of tyrosine residues in the receptor and other target proteins. Tyrosine kinase receptors either have an intrinsic tyrosine kinase within the cytoplasmic region of the receptor (Fig. 2.16) or, when activated, associate with a cytoplasmic tyrosine kinase.

Structurally, **tyrosine kinase receptors** consist of a hormone-binding region that is exposed to the extracellular

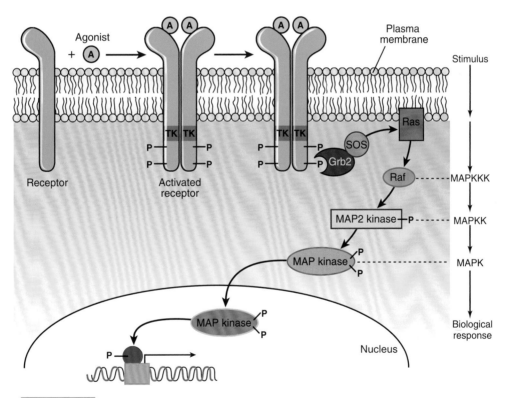

Figure 2.16 **A signaling pathway for tyrosine kinase receptors.** Binding of agonist to the tyrosine kinase receptor (TK) causes dimerization, activation of the intrinsic tyrosine kinase activity, and phosphorylation of the receptor subunits. The phosphotyrosine residues serve as docking sites for intracellular proteins, such as Grb2, which recruits Son of Sevenless (SOS), a guanine nucleotide exchange factor, to the receptor complex. SOS interacts with and modulates the activity of Ras by promoting the exchange of GDP for GTP. Ras-GTP (active form) activates the serine/threonine kinase Raf, initiating a phosphorylation cascade that results in the activation of mitogen-activated protein kinase (MAPK). MAPK translocates to the nucleus and phosphorylates transcription factors to modulate gene transcription. The right side of the figure illustrates the hierarchical organization of the MAPK signaling cascade. The generic names in this pathway are shown aligned to specific members of a typical tyrosine kinase pathway. Proteins with P attached represent phosphorylation at either tyrosine or serine/threonine residues.

space, a transmembrane region, and a cytoplasmic tail domain. Examples of **ligands** for these receptors include the hormones insulin or growth factors such as epidermal-, fibroblast-, and platelet-derived growth factors. The signaling cascades generated by the activation of tyrosine kinase receptors can result in the transcription of genes involved in growth, cellular differentiation, and movements (crawling or shape change).

The tyrosine kinase signaling pathway begins with the agonist binding to the extracellular portion of the receptor (see Fig. 2.16), which causes two of the agonist-bound receptors to associate (**dimerization**), and in turn, activating the built-in or associated tyrosine kinase. The activated tyrosine kinase then phosphorylates tyrosine residues in the other subunit (cross-phosphorylation) of the dimer to fully activate the receptor and create docking sites for additional signaling molecules or adapter proteins that have a specific sequence called an **SH2 domain**. The SH2-containing adapter proteins may be serine/threonine protein kinases, phosphatases, or other bridging proteins that transmit the signal from an activated receptor to many signaling pathways, resulting in a cellular response. A notable difference in signaling pathways activated by tyrosine kinase receptors is that they do not generate second messengers such as cAMP or cGMP.

One other signaling pathway associated with activated tyrosine kinase receptors results in activation of monomeric GTPase monomeric. Members of the **ras** family of *monomeric G proteins* are activated by many tyrosine kinase receptor growth factor agonists and, in turn, activate an intracellular signaling cascade that involves the phosphorylation and activation of protein kinases called *mitogen-activated protein kinases (MAPKs)*. In this pathway, the activated MAPK translocates to the nucleus, where it activates transcription of a cohort of genes needed for proliferation and survival or cell death (Clinical Focus 2.2).

Hormone receptors bind specific hormones to initiate cell signaling in the cells.

Hormone receptors reside either on the cell surface and bind **peptide hormones**, or inside the cell where they bind **steroid hormones**. *Peptide hormone* receptors are usually GPCRs and effect signaling by generation of second messengers such as cAMP and IP_3 and by the release of calcium from its storage compartments. Steroid hormones bind either to soluble receptors located in the cytosol or nucleus or to receptors already bound to the promoter elements of target genes. Cytoplasmic or nuclear steroid hormone receptors include the sex hormone receptors (androgens, estrogen,

---

CLINICAL FOCUS | 2.2

Tyrosine Kinase Inhibitors for Chronic Myeloid Leukemia

Cancer can result from defects in critical signaling molecules that regulate cell properties such as proliferation, differentiation, and survival. Normal cellular regulatory proteins or **protooncogenes** may become altered by mutation or abnormally expressed during cancer development. **Oncogenes**, the altered proteins that arise from protooncogenes, are in many cases signal transduction proteins that normally function in the regulation of cellular proliferation. Examples of signaling molecules that can become oncogenic span the entire signal transduction pathway and include ligands (e.g., growth factors), receptors, adapter and effector molecules, and transcription factors.

There are many examples of how normal cellular proteins can be converted into oncoproteins. One occurs in **chronic myeloid leukemia (CML)**. This disease is characterized by increased and unregulated clonal proliferation of myeloid cells in the bone marrow. CML results from an inherited chromosomal abnormality that involves a reciprocal translocation or exchange of genetic material between chromosomes 9 and 22 and was the first malignancy to be linked to a genetic abnormality. The translocation is referred to as the *Philadelphia chromosome* and results in the fusion of the *bcr* gene with part of the cellular *abl* (c-*abl*) gene. The c-*abl* gene encodes a protein tyrosine kinase. This abnormal Bcr–Abl fusion protein has unregulated tyrosine kinase activity, and through SH2 and SH3 binding domains, the mutant protein binds to and phosphorylates the interleukin-3β(c) receptor. This receptor is linked to control of cell proliferation, and the

expression of the unregulated Bcr–Abl protein activates signaling pathways that speed up cell division. The BCR–ABL protein also inhibits DNA repair, causing genomic instability and making the cell more susceptible to developing further genetic abnormalities.

The chromosomal translocation that results in the formation of the Bcr–Abl oncoprotein occurs during the development of hematopoietic stem cells, and the observance of a shorter Philadelphia 22 chromosome is diagnostic of this cancer. The translocation results initially in a CML that is characterized by a progressive leukocytosis (increase in number of circulating white blood cells) and the presence of circulating immature blast cells. However, other secondary mutations may spontaneously occur within the mutant stem cell and can lead to acute leukemia, a rapidly progressing disease that is often fatal.

Historically, CML was treated with chemotherapy, interferon administration, and bone marrow transplantation. With the understanding of the molecules and signaling pathways that result in this devastating cancer, targeted therapeutic strategies to attenuate the disease have been developed. Imatinib mesylate was the first tyrosine kinase inhibitor (developed in 2001) that could reduce the signaling activity of Bcr–Abl. Additional, more potent tyrosine kinase inhibitors have since been developed activities has been developed. These drugs can induce complete remission of CML and greatly improve the quality of life and lifespan of the patient. ■

and progesterone), glucocorticoid receptors (cortisol), and mineralocorticoid receptors (aldosterone). Examples of DNA-bound steroid hormone receptors include vitamin A, vitamin D, retinoid, and thyroid hormone receptors.

Generally, steroid hormone receptors have four recognized domains. The N-terminal *variable domain* is a region with little similarity between receptors. A centrally located *DNA-binding domain* consists of two globular motifs where zinc is coordinated with cysteine residues (**zinc finger**). This domain controls the target gene that will be activated and may also have sites for phosphorylation by protein kinases involved in modifying the transcriptional activity of the receptor. Between the central DNA-binding and the C-terminal hormone-binding domains is located a *hinge domain*, which controls the movement of the receptor to the nucleus. The carboxyl-terminal *hormone-binding and dimerization domain* binds the hormone and then allows the receptor to dimerize, a necessary step for binding to DNA. Steroid hormones bound to their receptor move to the nucleus, where the complex binds to the promoter region of a hormone-responsive gene. The targeted DNA sequence in the promoter is called a **hormone response element** (**HRE**). Binding of the hormone receptor complex to the HRE can either activate or repress transcription, resulting in newly synthesized proteins and/or enzymes that will affect cellular metabolism. Steroid hormone signaling is shown in Figure 2.17. In contrast to

steroid hormones, thyroid hormones, retinoic acid, vitamin A, and vitamin D bind to receptors that are already associated with the DNA response elements of target genes. The unoccupied receptors are inactive until the hormone binds, and they serve as repressors in the absence of hormone. Thyroid hormone receptor action is discussed in Chapter 32.

▶ SECOND MESSENGERS

The concept of second messengers and their central role in signaling began with Earl Sutherland, Jr., who was awarded the Nobel Prize in 1971 "for his discoveries concerning the mechanisms of action of hormones." Sutherland discovered cAMP and showed it was a critical intermediate in cellular responses to hormones. **Second messengers** transmit and amplify signals from receptors to downstream target molecules inside the cell. There are three general types of second messengers: hydrophilic, water-soluble messengers, such as IP_3, cAMP, cGMP, or Ca^{2+}; hydrophobic water-insoluble, lipid messengers, which are generally associated with lipid-rich membranes such as DAG and phosphatidylinositols (e.g., PIP_3); and gases, such as NO, CO, and reactive oxygen species (ROS), which can diffuse both through the cytosol as well as across cell membranes. A critical feature of second messengers is that they are able to be *rapidly synthesized and degraded* by cellular enzymes, rapidly sequestered in a membrane-bound organelle or vesicle or have a restricted distribution within the cell. It is the rapid appearance and disappearance that allow second messengers to amplify and then terminate signaling reactions, allowing fine-tuning of the response. For example, when a cell receptor is only briefly stimulated with a ligand, the generation of a second messenger will terminate rapidly. Conversely, when a large amount of ligand persists to stimulate a receptor, the increased levels of second messenger in the cell will be sustained for a longer period of time before termination. Each cell in the body is programmed to respond to specific combinations of first and second messengers, and these messengers elicit distinct physiologic response in different cell types. For example, the neurotransmitter acetylcholine can cause heart muscle to relax, skeletal muscle to contract, and secretory cells to secrete.

cAMP is the predominant second messenger in all cells.

Many peptide hormones and catecholamines bind specific GPCRs to produce an almost immediate increase in the intracellular concentration of cAMP. For these ligands, the receptor is coupled to a stimulatory G protein ($G\alpha_s$), which activates AC, a large transmembrane protein that converts intracellular ATP to cAMP. An increase in cAMP activates **cAMP-dependent protein kinase** (also called **protein kinase A** or **PKA**) and also directly activates some calcium channels.

Some hormones act to decrease cAMP synthesis by binding receptors that are coupled to an inhibitory ($G\alpha_i$) G protein. The intracellular signal provided by cAMP is rapidly terminated by its hydrolysis to 5′AMP by a family of

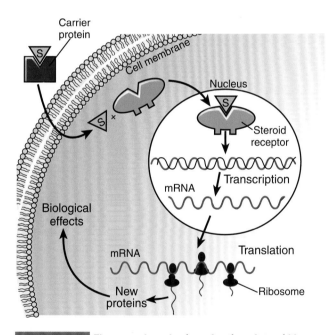

Figure 2.17 **The general mechanism of action of steroid hormones.** Steroid hormones (S) are lipid soluble and pass through the plasma membrane, where they bind to a cognate receptor in the cytoplasm. The steroid hormone receptor complex then moves to the nucleus and binds to a HRE in the promoter of specific hormone-responsive genes. Binding of the steroid hormone–receptor complex to the response element initiates transcription of the gene to form mRNA. The mRNA moves to the cytoplasm, where it is translated into a protein that participates in a cellular response. Thyroid hormones act by a similar mechanism, although their receptors are already bound to an HRE, repressing gene expression.

enzymes known as **phosphodiesterases** (**PDEs**), which can be activated by high levels of cyclic nucleotides or by other signal transduction processes.

Protein kinase A mediates the signaling effects of cAMP.

cAMP activates PKA, which, in turn, phosphorylates cellular proteins, ion channels, and transcription factors, altering their activity or function to achieve a cellular response. PKA is a tetramer consisting of two catalytic and two regulatory subunits in the inactive state. When cAMP in the cell increases, two molecules of cAMP bind to each of the regulatory subunits, causing them to dissociate from the catalytic subunits, activating PKA to (Fig. 2.18).

In some cell types, cAMP can directly bind to and alter the activity of ion channels. Cyclic nucleotide–gated ion channels may be regulated by either cAMP or cGMP and are especially important in the olfactory and visual systems. Odorant receptors are coupled to G proteins and, when stimulated by a specific odorant, activate AC to generate cAMP. The cAMP then binds a cAMP-gated ion channel that opens to allow calcium (Ca^{2+}) into the cell causing membrane "depolarization" (influx of positive ions) as part of the sensing of the odor.

cGMP and NO are important second messengers in smooth muscle and sensory cells.

The second messenger cGMP is generated by the enzyme guanylyl cyclase (GC). Protein kinase G (PKG) is the main target of cGMP, but it can also directly activate ion channels and pumps that modulate cytoplasmic Ca^{2+} levels in smooth muscle (see Chapters 8 and 15) and sensory tissue (see Chapter 6). There are two forms of GC, a soluble, cytoplasmic form and a membrane localized form. Soluble GC is a heterodimeric protein that contains two heme (an organic compound consisting of iron bound to a heterocyclic ring called *porphyrin*) prosthetic groups. Soluble GC is a target for the paracrine-signaling molecule nitric oxide (NO), which binds to the heme prosthetic groups and activates GC leading to the production of cGMP. NO produced by endothelial cells diffuses into smooth muscle cells to increase cGMP, activating PKG and ion channels, which reduce cytoplasmic Ca^{2+} concentrations, resulting in smooth muscle relaxation. Degradation of cGMP is mediated by a PDE activated by high levels of cGMP. The very short half-life of NO also serves to terminate the signaling pathway. NO was initially called **endothelial-derived relaxing factor** (**EDRF**). Research showing that EDRF was actually the gas NO resulted in a Nobel Prize in 1998 awarded to Robert Furchgott, Louis Ignarro, and Ferid Murad. NO is produced by the enzyme nitric oxide synthase (NOS) in a reaction that converts L-arginine to L-citrulline.

The transmembrane form of GC is a receptor for atrial natriuretic (ANP) peptide produced by cardiomyocytes in response to increased blood volume. Binding of ANP to transmembrane GC in the kidney increases cGMP, which stimulates Na^+ excretion to reduce blood volume (see Chapter 23).

Diacylglycerol and inositol trisphosphate are derived from lipid in the plasma membrane

Some GPCRs are coupled to the effector enzyme, **phospholipase C** (**PLC**), which is localized to the inner leaflet of the plasma membrane. Similar to other GPCRs, binding of a ligand or an agonist to the receptor results in activation of the associated G protein, usually $G\alpha_q$ (or G_q). Depending on the isoform of the G protein associated with the receptor, either the α or the $\beta\gamma$ subunit may stimulate PLC, resulting in hydrolysis of the membrane phospholipid PIP_2 into **DAG** and **IP_3**. Both DAG and IP_3 serve as second messengers in the cell (Fig. 2.19).

In its second messenger role, DAG accumulates in the plasma membrane and activates the membrane-bound calcium- and lipid-sensitive enzyme **protein kinase C** (**PKC**). When activated, PKC phosphorylates specific proteins in the cell to produce appropriate physiologic effects such as cell proliferation. Tumor-promoting **phorbol esters** that mimic the structure of DAG can bypass receptors to directly activate PKC, resulting in cellular proliferation.

IP_3 promotes the release of calcium ions into the cytoplasm by activation of endoplasmic or sarcoplasmic reticulum IP_3-gated calcium release channels (see Chapter 8). The concentration of free calcium ions in the cytoplasm of most cells is in the range of 10^{-7} M and may abruptly increase

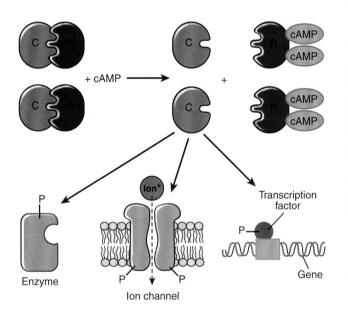

Figure 2.18 Activation and targets of protein kinase A. Inactive protein kinase A consists of two regulatory subunits complexed with two catalytic subunits. Activation of adenylyl cyclase results in increased cytosolic levels of cAMP. Two molecules of cAMP bind to each of the regulatory subunits, leading to the release of the active catalytic subunits. These subunits can then phosphorylate target enzymes, ion channels, or transcription factors, resulting in a cellular response. R, regulatory subunit; C, catalytic subunit; P, phosphate group.

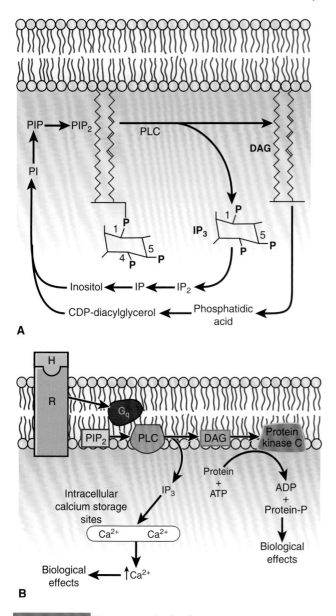

Figure 2.19 **The phosphatidylinositol second messenger system. (A)** The successive phosphorylation of phosphatidylinositol (PI) leads to the generation of phosphatidylinositol 4,5-bisphosphate (PIP_2). Phospholipase C (PLC) catalyzes the breakdown of PIP_2 to inositol trisphosphate (IP_3) and 1,2-DAG, which are used for signaling and can be recycled to generate phosphatidylinositol. **(B)** Binding of hormone (H) to a G protein–coupled receptor (R) releases G_q, a G protein that couples the receptor to PLC. Activated PLC cleaves PIP_2 to IP_3 and DAG. IP_3 interacts with calcium release channels in the endoplasmic reticulum, to release calcium to the cytoplasm. Increased intracellular calcium activates calcium-dependent enzymes. An accumulation of DAG in the plasma membrane activates the calcium- and phospholipid-dependent enzyme protein kinase C and phosphorylation of its downstream targets. Protein-P, phosphorylated protein.

1,000 times or more. The resulting increase in free cytoplasmic calcium synergizes with the action of DAG in the activation of some forms of PKC and may also activate many other calcium-dependent processes.

To terminate DAG and IP_3 signaling, the molecules are rapidly removed from the cytoplasm. IP_3 is dephosphorylated to inositol, which can be reused for phosphoinositide synthesis. DAG is converted to phosphatidic acid by the addition of a phosphate group to carbon number 3. Phosphatidic acid can then be used for the resynthesis of membrane inositol phospholipids (see Fig. 2.19). On removal of the IP_3 signal, calcium is quickly pumped back into its storage sites, restoring cytoplasmic calcium concentrations to low prestimulus levels.

Ceramide, another lipid second messenger, is generated from sphingomyelin through the action of plasma membrane–associated sphingomyelinase. Activation of sphingomyelinase occurs through binding of **cytokines** that mediate immune and inflammatory responses (e.g., **tumor necrosis factor [TNF]** and **interleukin-1**) to their receptors. These activated receptors then couple to sphingomyelinase, leading to its activation and generation of ceramide and subsequent activation of the MAPK pathway.

Cells use calcium as a second messenger by keeping resting intracellular calcium levels low.

The level of cytosolic calcium in an unstimulated cell is about 10,000 times lower than in the ECF (10^{-7} M vs. 10^{-3} M). This large calcium gradient is maintained by the limited permeability of the plasma membrane to calcium, by calcium transporters in the plasma membrane that extrude calcium, by calcium pumps in the membranes of intracellular organelles that store calcium, and by cytoplasmic and organellar proteins that bind calcium to buffer its free cytoplasmic concentration. Several plasma membrane ion channels serve to increase cytosolic calcium levels. Either these ion channels are voltage gated and open when the plasma membrane depolarizes or they may be controlled by PKA or PKC phosphorylation.

The endoplasmic reticulum has two other main types of ion channels that release calcium into the cytoplasm when activated. The small water-soluble molecule IP_3 activates the IP_3-gated **calcium release channel** in the membrane of the endoplasmic or sarcoplasmic (a specialized type of endoplasmic reticulum in smooth and striated muscle) reticulum. The activated channel opens to allow calcium to flow down a concentration gradient into the cytoplasm. The **ryanodine receptor** is structurally similar to the IP_3-gated channels and is located in the sarcoplasmic reticulum of muscle cells and neurons. In cardiac and skeletal muscle, ryanodine receptors release calcium to trigger muscle contraction when an action potential invades the transverse tubule system of these cells. Both types of channels are regulated by positive feedback, in which the released cytosolic calcium can bind to the receptor to enhance further calcium release. This form of positive feedback is referred to as **calcium-induced calcium release** and causes the calcium to be released suddenly in a spike, followed by a wavelike flow of the ion throughout the cytoplasm (see Chapters 8 and 13).

Increasing cytosolic free calcium activates many different signaling pathways and leads to numerous physiologic events, such as muscle contraction, neurotransmitter secretion, and cytoskeletal polymerization. Calcium acts as a second messenger in one of two ways:

- It binds directly to an effector target such as PKC to promote in its activation.
- It binds to an intermediary cytosolic calcium-binding protein such as calmodulin.

Calmodulin is a small protein (16 kDa) with four binding sites for calcium. The binding of calcium to calmodulin causes a dramatic conformational change and increases the affinity of the molecule for its effectors (Fig. 2.20). Calcium–calmodulin complexes bind to and activate a variety of cellular proteins, including protein kinases that are important in physiologic processes, such as smooth muscle contraction (myosin light-chain kinase; see Chapter 8) and hormone synthesis (aldosterone synthesis; see Chapter 33).

Two mechanisms operate to terminate calcium action: IP_3 is dephosphorylated by cellular phosphatases to inactivate it. In addition, the plasma membrane, endoplasmic reticulum, sarcoplasmic reticulum, and mitochondrial membranes all have ATP-driven calcium pumps that transport the free calcium out of the cytosol to the extracellular space or into an intracellular organelle. Lowering cytosolic calcium concentrations shifts the equilibrium to release calcium from calmodulin, which then dissociates from the various proteins that were activated, and the cell returns to its basal state.

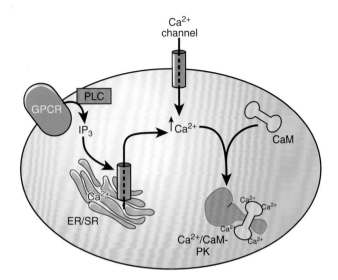

Figure 2.20 **The role of calcium in intracellular signaling and activation of calcium–calmodulin-dependent protein kinases.** Membrane-bound ion channels allow the entry of calcium from the extracellular space or release from internal stores (e.g., endoplasmic reticulum, sarcoplasmic reticulum in muscle cells, and mitochondria). Calcium can also be released from intracellular stores via the G-protein–mediated activation of phospholipase C (PLC) and the generation of inositol trisphosphate (IP_3). IP_3 releases calcium from the endoplasmic or sarcoplasmic reticulum in muscle cells by activating calcium ion channels. When intracellular calcium rises, four calcium ions complex with calmodulin protein (CaM) to induce a conformational change. Ca^{2+}/CaM can then bind to target proteins including Ca^{2+}/CaM-PKs, which then phosphorylate other substrates.

INTEGRATED MEDICAL SCIENCES

Phosphodiesterase, Angina, Pulmonary Hypertension, and Erectile Dysfunction—What is the Link?

A phosphodiesterase (PDE) is an enzyme that hydrolyzes a phosphodiester bond. Cyclic nucleotide PDEs are important in the clinical setting as they control the cellular levels of the second messengers, cyclic adenosine and guanosine monophosphate (cAMP and cGMP), and the signal transduction pathways modulated by these molecules. There are many cyclic nucleotide PDEs, which are classified according to sequence, regulation, substrate specificity, and tissue distribution. The tissue-specific expression of PDEs presents an opportunity to target a specific PDE with an inhibitory or activating drug.

Therapeutic agents for angina pectoris (severe chest pain resulting from insufficient blood supply to cardiovascular tissues) include the administration of nitrates, a commonly used agent that reduces myocardial oxygen demand. Nitrates act as an exogenous source of nitric oxide (NO), which stimulates soluble GCs to increase cGMP, transducing a signal that promotes relaxation of vascular smooth muscle in arteries and veins. The salutary effect of nitrates in treating myocardial ischemia is to dilate veins, which allows blood to translocate from inside the ventricles into the peripheral tissues. This reduces stretch and strain on the heart, which reduces myocardial oxygen demand. A common side effect of nitrates is **tachyphylaxis**, or reduced responsiveness to a chronically used drug. The search for new drugs to treat angina pectoris and other similar cardiovascular diseases led to the discovery of sildenafil, which is now marketed under the trade name Viagra. Sildenafil is a fairly selective inhibitor of PDE5, and its administration enhances cGMP levels in vascular smooth muscle cells, leading to vasodilation. Unfortunately, the relatively short half-life thwarted the usefulness of this drug as a practical treatment for chronic angina. In addition, several side effects were noted during clinical trials including the ability of sildenafil to augment the vasodilatory effects of nitrates. One other interesting, common side effect noted was penile erection, and subsequent clinical trials validated the use of this drug as an effective therapeutic agent for **erectile dysfunction (ED)**.

There are many clinical causes of ED, including psychological conditions such as depression. Common clinical conditions associated with ED include vascular disease; diabetes; neurologic conditions such as spinal cord injury, multiple sclerosis, and Parkinson disease; and numerous inflammatory conditions. During sexual stimulation, the penile cavernosal arteries relax and dilate, allowing increased blood flow. This increase in blood volume and compression of the trabecular muscle result in collapse and obstruction of venous outflow to produce a rigid erection. For an erection to occur, NO activates soluble guanylate cyclase causing increased synthesis of cGMP. Cellular levels of cGMP reflect a balance of activities between NO production by NO synthase and degradation of cGMP by cyclic PDE. Thus, the use of a transient inhibitor of PDE5, the main PDE in the cav-ernosal arteries and trabecular muscle, provides a rational, temporary vasodilation in those tissues.

Following its wide use as a therapeutic drug for ED, another application for sildenafil was discovered: the treatment of pulmonary hypertension under the trade name Revatio. Pulmonary hypertension results from high blood pressure in the pulmonary circulation. It is a highly progressive disease with a poor prognosis due to the ensuing right heart dysfunction and is often fatal. The usefulness of Revatio is based on the findings that in animal models of pulmonary hypertension, the levels of PDE5 increase in the pulmonary aorta and other arteries of the lung, leading to decreased cGMP and increased tone in this vessel. Thus, administration of sildenafil has a beneficial effect by increasing cGMP and relaxation. As more is learned about PDEs, it is likely that additional uses for sildenafil and other compounds that target cyclic PDEs will be discovered. ■

Chapter Summary

- Homeostasis is the maintenance of steady states in the body by coordinated physiologic mechanisms.
- Negative and positive feedbacks are used to modulate the body's responses to changes in the environment.
- Steady state and equilibrium are distinct conditions. Steady state is a condition that does not change over time, whereas equilibrium represents a balance between opposing forces.
- Macromolecules cross the plasma membrane by endocytosis and exocytosis.
- Passive movement of a solute across a membrane dissipates the gradient (driving force) and reaches an equilibrium at which point there is no net movement of solute.
- Simple diffusion is the passage of lipid-soluble solutes across the plasma membrane by diffusion through the lipid bilayer.
- Facilitated diffusion is the passage of water-soluble solutes and ions through a hydrophilic pathway created by a membrane-spanning integral protein.
- Facilitated diffusion of small ions is mediated by specific pores and ion channel proteins.
- Active transport uses a metabolic energy source to move solutes against gradients, and the process prevents a state of equilibrium.
- Polarized organization of epithelial cells ensures directional movement of solutes and water across the epithelial layer.

- Water crosses plasma membranes rapidly via channel proteins termed aquaporins. Water movement is a passive process driven by differences in osmotic pressure.
- Cells regulate their volume by moving solutes in or out to drive osmotic entry or exit of water, respectively.
- The driving force for ion transport is the sum of the electrical and chemical gradients, known as the gradient of electrochemical potential across the membrane.
- The resting membrane potential is determined by the passive movements of several ions through nongated channels, which are always open. It is described most accurately by the Goldman equation, which takes into account the differences in membrane permeability of different ions. In a muscle cell, for example, the membrane permeability to Na^+ is low compared with K^+ and the resting membrane potential is a result primarily of passive exit of K^+.
- Cellular communication is essential to integrate and coordinate the systems of the body so they can participate in different functions.
- A hallmark of cellular signaling is that it is regulated with a variety of mechanisms to both activate and terminate signal transduction.
- Activators of signal transduction pathways include ions, gases, small peptides, protein hormones, metabolites, and steroids.
- Receptors are the receivers of signaling molecules; they are located either on the plasma membrane or within the cell.
- Second messengers are important for amplification and flow of the signal received by plasma membrane receptors.

Chapter Review Questions

1. If a region or compartment is in a steady state with respect to a particular substance, then:

   A. The amount of the substance in the compartment is increasing.
   B. The amount of the substance in the compartment is decreasing.
   C. The amount of the substance in the compartment does not change with respect to time.
   D. There is no movement into or out of the compartment.
   E. The compartment must be in equilibrium with its surroundings.

The correct answer is C. In a steady state, the amount or concentration of a substance in a compartment does not change with respect to time. Although there may be considerable movements into and out of the compartment, there is no net gain or loss. Steady states in the body often do not represent an equilibrium condition, but they are displaced from equilibrium by the constant expenditure of metabolic energy.

2. Parathyroid hormone (PTH) acts on osteoblasts in bone and tubular cells in the kidneys. It binds to the PTH receptor to cause the release of calcium from bone to the body, a process called osteolysis. If you treated an animal with PTH, and measure an increase in cAMP levels in bone tissue, what kind of receptor would you predict that PTH binds to?

   A. Steroid
   B. Tyrosine kinase
   C. Nuclear
   D. G protein coupled
   E. Orphan

The correct answer is D. Generation of cAMP in cells occurs in response to activation of adenylyl cyclase, an effector that is coupled to GPCRs. The other answers represent different intracellular signaling mechanisms that do not require generation of cAMP.

3. A single cell within a culture of freshly isolated cardiac muscle cells is injected with a fluorescent dye that cannot cross cell membranes. Within minutes, several adjacent cells become fluorescent. The most likely explanation for this observation is the presence of:

   A. Ryanodine receptors.
   B. IP3 receptors.
   C. Transverse tubules.
   D. Desmosomes.
   E. Gap junctions.

The correct answer is E. Cardiac muscle cells have many gap junctions that allow the rapid transmission of electrical activity and the coordination of heart muscle contraction. Gap junctions are pores composed of paired connexons that allow the passage of ions, nucleotides, and other small molecules between cells.

4. Treatment of intestinal epithelial cells with the mitochondrial toxin 2,4 dinitrophenol results in loss of ATP synthesis. Transport of glucose from the apical (luminal) side of the cell to basolateral side of the intestinal cell will be reduced because:

   A. GLUT2 will not be able to transport glucose across apical membrane.
   B. Na^+ gradient across apical membrane to inside of cell will be dissipated.
   C. SLGT on apical membrane will not be able to transport Na^+ out of the cell.
   D. Na^+/K^+-ATPase on basolateral membrane will not be able to transport glucose out of cell.
   E. K^+ gradient across basolateral membrane to inside of cell will increase.

The correct answer is B. Loss of ATP will stop Na^+/K^+-ATPase on the basolateral membrane from expelling Na^+ out of the cell; thus, the Na^+ gradient across apical membrane will be dissipated and SLGT will not be able to cotransport Na^+ and glucose. GLUT2 transports glucose across the basolateral membrane and SLGT transports Na^+ into the cell, not out of the cell. Na^+/K^+-ATPase does not transport glucose and loss of Na^+/K^+-ATPase will decrease the K^+ gradient across the basolateral membrane.

Clinical Application Exercises 2.1

COLLAPSING FOOTBALL PLAYER

A 16-year-old boy arrives in a Texas hospital emergency room in a semiconscious state. He was transported there after collapsing during football practice. The temperature that August afternoon had reached 102°C and the team had been practicing for nearly 4 hours before the young man collapsed. On the way to the hospital, an IV was inserted, and he received intravenous fluids. He is complaining of dizziness when he sits up, cramping, and nausea and had vomited before being transported to the hospital. Upon examination, the boy is found to be in excellent general physical condition, although his heart rate and temperature are elevated and he is unable to give a urine sample. He claims that he only had a few sips of water during practice and just suddenly blacked out. After oral rehydration for several hours, the young man is released.

QUESTIONS

1. What disorder is consistent with this patient's symptoms?

2. How should this be treated to return the patient to a homeostatic condition?

ANSWERS

1. Symptoms of this patient are consistent with exercise-induced dehydration, a serious and potentially life-threatening condition in which the volume of water within the body compartments is insufficient for normal functions. Many athletes do not adequately maintain hydration during exercise and consequently experience the adverse effects of dehydration, which reduces the ability to tolerate prolonged exercise.

2. Dehydration is a condition that occurs when the loss of body fluids and water is greater than the amount that is taken into the body. The amount of water lost from the body by sweating depends on both the environment and the body fluid composition before exercise. The early effects of dehydration are to increase heart rate, and this impairs heat transfer from muscle contraction to skin where it is dissipated for cooling. Sweating is the body's way of getting rid of heat that is produced by muscle exercise. Sweat glands produce a fluid that is derived from the interstitial spaces and capillaries in the skin and is similar to plasma in that it is composed principally of water and sodium ions. As sweat evaporates, it dissipates the excess heat to cool the body; the dryer the air, the quicker evaporation and cooling occur to help maintain homeostasis. The water lost by sweating comes from the fluid compartments of the body, thus there is a net increase in the concentration of the electrolytes (ions) in the body fluid creating hypertonicity (having a higher concentration of electrolytes inside the fluid compartments than outside). If the water lost due to exercise and sweating is not replaced, dehydration can occur. The solution to the problem is to drink plenty of fluids before and during exercise. Without adequate water replacement, the water and electrolyte imbalance can lead to heat stroke and even death. Rarely is there a necessity to replace lost sodium, although many sports drinks contain both sodium and potassium. The reason for this is that ingested water, which is rapidly absorbed through the gut, results in adequate salt replacement by absorption from the contents of the digestive tract.

3 Action Potential, Synaptic Transmission, and Nerve Function

Active Learning Objectives

Upon mastering the material in this chapter, you should be able to:

- Explain how the organization of the nervous system supports afferent and efferent function.
- Describe how the access of components from the blood to the brain is restricted.
- Explain the specialized functions of the different cell types of the nervous system.
- Explain the mechanism by which components are transported between the soma and neuronal terminals.
- Relate ion channel function to membrane potentials.
- Explain how movement of ions through ion channels can produce an action potential.

- Explain why an action potential is unidirectional and how it propagates without decrement.
- Describe how myelination and diameter can affect axonal conduction velocity.
- Describe how electrical and chemical transmission differ.
- Explain how the specializations of the synapse contribute to synaptic transmission.
- Compare ionotropic and metabotropic receptors in terms of cell signaling.
- Describe different synaptic transmitters in terms of their signaling, function, and role in disease.

▶ THE NERVOUS SYSTEM

The nervous system is comprised of cells in the brain and spinal cord and extends throughout the body via a system of nerves and **ganglia** (clusters of neurons and support cells located outside of the brain and spinal cord). Together the brain and spinal cord are referred to as the **central nervous system** (**CNS**), while the nerves and ganglia throughout the rest of the body comprise the **peripheral nervous system** (**PNS**). The PNS has both **afferent** and **efferent** functions. The afferent function consists of collecting information from sensory and organ systems and transmitting that information to the CNS for processing. The efferent function consists of communicating with organ systems to maintain or adjust function as well as controlling motor function. The PNS also can be divided into the motor system (efferent), sensory systems (afferent), and autonomic nervous system (both efferent and afferent). These will each be covered in subsequent chapters.

Access to the central nervous system is restricted by the blood–brain barrier.

Access of molecules into the CNS from the blood is limited by the **blood–brain barrier** (**BBB**). The BBB is the result of a layer of capillary endothelial cells connected by tight junctions and surrounded by astrocytes cells (Fig. 3.1). Together, these effectively restrict transepithelial permeability and serve to protect the brain from infection and toxins. Molecules such as water, CO_2, O_2, amino acids, and glucose have essentially free access across these cells via carrier-mediated processes and active transport. Lipophilic molecules can also freely cross the BBB. In addition to the BBB, there is also a comparable blood–cerebrospinal fluid (CSF) barrier where the epithelial cells of the choroid plexus form tight junctions and restrict access from the blood into the CSF.

The primary cell types in the nervous system, neurons and glia, differ in function and morphology.

The primary cell types found in the nervous system are **neurons** and **glia** (Fig. 3.2). Neurons mainly function to store, communicate, and integrate information. Glia support this function and can be divided into multiple cell groups based on morphology and function. **Microglia** are found in the CNS and are related in origin and function to macrophages. They serve an immune function in that they phagocytose damaged cells, invade microorganisms, and secrete immune mediators. **Schwann cells** are found in the PNS and form **myelin**, which wraps around neuronal processes (axons) in a manner that contributes to faster signal conduction along the nerve. **Oligodendrocytes** perform a similar function in the CNS. This myelin sheath can be disrupted, as happens with multiple sclerosis, resulting in abnormalities in signal conduction. The myelin surrounding the axons is responsible for CNS **white matter**: the portions of the CNS that histologically appear white rather than gray. The **gray matter** is composed of cell bodies and nonmyelin surrounded processes. **Astrocytes** are star-shaped cells in the CNS that extend processes to blood vessels and neurons. The processes along blood vessels contribute to the blood–brain barrier. Astrocytic processes along neurons support neuronal function by providing metabolic support, regulating blood flow, helping to maintain appropriate extracellular concentrations of certain ions such as

Peripheral capillary

Brain capillary

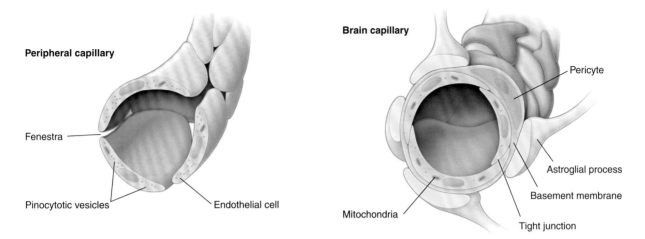

Fenestra

Pinocytotic vesicles

Endothelial cell

Mitochondria

Pericyte

Astroglial process

Basement membrane

Tight junction

Figure 3.1 **The blood–brain barrier restricts access from the capillary into the brain.** In the periphery, capillary endothelial cells have gaps (termed *fenestrae*) between them and use intra-cellular pinocytotic vesicles to facilitate the transcapillary transport of fluid and soluble molecules. In contrast, CNS vessels are sealed by tight junctions between the endothelial cells. The cells have fewer pinocytotic vesicles and are surrounded by pericytes and astroglial processes. In addition, capillary endothelial cells in the CNS have more mitochondria than those in systemic vessels; these mitochondria may reflect the energy requirements necessary for CNS endothelial cells to transport certain molecules into the CNS and transport other molecules out of the CNS.

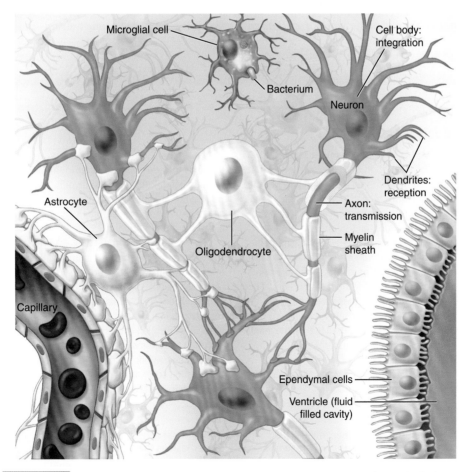

Microglial cell

Cell body: integration

Bacterium

Neuron

Dendrites: reception

Axon: transmission

Myelin sheath

Astrocyte

Oligodendrocyte

Capillary

Ependymal cells

Ventricle (fluid filled cavity)

Figure 3.2 **Glial cell types in the CNS.** Microglia serve an immune function in the CNS similar to that of macrophages in the PNS. Oligodendrocytes sheath central neuronal axons with myelin, which is accomplished by Schwann cells in the PNS. Astrocytes have many functions within the CNS including providing neuronal support and contributing to the blood–brain barrier. Ependymal cells contribute to a blood–CSF barrier.

potassium, and recycling signaling molecules such as glutamate and GABA from the extracellular space back to the neuron. **Ependymal cells** line the ventricles of the brain and central canal of the spinal cord and contribute to the blood–CSF barrier. Glial cells are more numerous than neurons and can proliferate particularly in response to injury or infection. This ability to proliferate can also go awry and result in gliomas: tumors derived from glial cells.

Neurons are less numerous than glial cells and, like glia, can be classified into different types based on morphology and function (Fig. 3.3). All neurons have a **soma** (cell body) containing the nucleus and primary organelles such as the endoplasmic reticulum and Golgi apparatus. From the soma arise processes. In the case of **unipolar** cells, there is only one process that extends from the soma, whereas **bipolar cells** have two processes and **multipolar** cells have multiple processes. In all of these, one process is the **axon**. The portion of the axon that extends from the cell body, the **axon hillock**, is generally a thickened area. This narrows into the initial segment of the axon. Many axons are wrapped in a **myelin sheath** by Schwann cells (PNS) or oligodendrocytes (CNS). These myelin sheaths are made of lipids and proteins and are wound around the axon at regular intervals. In between myelin sheaths are the **nodes of Ranvier**: short segments of the axon that are unmyelinated. The myelin sheath contributes to faster signal conduction along the axon. At the end of the axon is a specialized segment referred to as the **presynaptic terminal** or **bouton**. The axon functions to transmit information in the form of an action potential and to transport materials to the presynaptic terminal. The axon has an efferent function in that it releases signaling molecules to transmit information. The other process that extends from the soma is the **dendrite**. The neuron may have one process sharing dendritic and axonal function (unipolar), one process that is a dendrite (bipolar), or multiple dendritic processes (multipolar). In the case of multipolar neurons, each dendritic branch can have branches off of it producing extensive arborization. Dendrites serve an afferent function in that they gather information and transmit it to the soma. They form **postsynaptic terminals** receiving information released from the axon's presynaptic terminal. Dendrites can also receive information from glia, immune cells, and other cells. The dendrite–soma–axon forms the basic structure of the neuron. The specialized region in which molecules are released from the axon and interact with dendrites is referred to as the **synapse**. A simplified schematic of a synapse consists of a presynaptic terminal at the end of an axon adjacent to a postsynaptic terminal at the end of a dendrite. The 20 nm extracellular space between the terminals is the **synaptic cleft**. The presynaptic terminal contains vesicles that release chemical messengers into the synaptic cleft where they can then interact with receptors that are located on the postsynaptic terminal. Synaptic transmission is discussed below.

CLINICAL FOCUS | 3.1

Multiple Sclerosis

Multiple sclerosis (MS) is an autoimmune disease that is characterized by demyelination of axons in the CNS. It affects more women than men with onset in young to middle adulthood. Most cases of MS proceed in a relapsing–remitting pattern where an individual experiences symptoms for a few days followed by a few months that are symptom free. Some individuals do not fully recover between episodes but demonstrate continued deterioration such that progressive attacks are more severe. A small percentage of individuals experience a progressive disease without remission. The etiology of the disease is not well understood but may be influenced by exposure to the Epstein-Barr virus and vitamin D deficiency. Common symptoms experienced in the early phases of the disease include visual disturbances, tingling sensations, and motor disturbances.

The demyelination that occurs during MS is due to production of antibodies that target oligodendroctyes. This leads to their inability to fully myelinate axons and can result in damage to axonal processes. Other immune cells also infiltrate the damaged area along with astrocytes, which can result in scarring around the damaged axon. During the remission phase, remyelination can occur. However, over time the ability to effectively remyelinate the axon decreases.

The decreased myelination impacts the conduction velocity along that axon. Normally the electrical impulse undergoes saltatory conduction (i.e., jumping from one node of Ranvier to the next). This results in fast conduction and a signal that does not undergo diminution as it progresses. Because of disruption to the myelin, the ability of the axon to propagate the electrical signal is disrupted. Either a stronger signal is now needed to reach its destination or the signal simply is not conducted.

The symptoms an individual experiences will depend on the location of the myelin disruption. The disease commonly affects myelination of optic nerves, spinal tracts, and brain regions. Lesions in the optic nerves can result in blurred and double vision. Demyelination of axons forming the motor tracts from the spinal cord to the brain can result in muscle weakness, uncontrolled movement, and eventual paralysis. Other common symptoms include paresthesia (tingling sensation), reduced sensations, impaired bladder function, and cognitive dysfunction such as memory loss and impaired attention.

Diagnosis of MS is based on symptom patterns along with magnetic resonance imaging (MRI) and nerve conduction tests. Using MRI, lesions in the central white matter can be identified. However, it is not uncommon to identify white matter lesions that are not associated with symptoms. Nerve conduction tests are used to determine conduction velocity after stimulation of peripheral nerves. Treatment of MS includes using strong anti-inflammatory medication during acute exacerbations and immunomodulators to decrease the frequency and progression of future exacerbations. ■

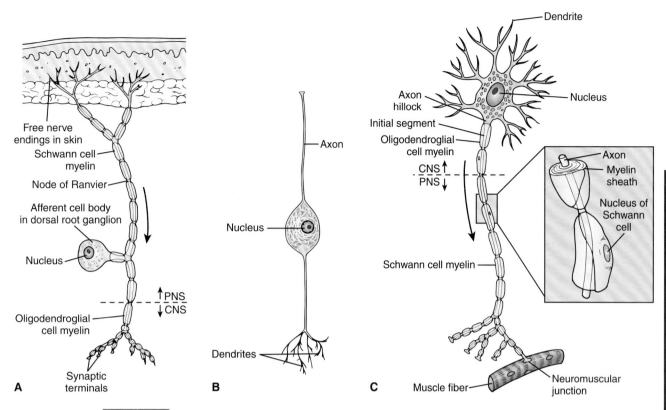

Figure 3.3 **Morphological classification of neurons. (A)** Unipolar and pseudounipolar neurons have a single process emanating from the soma. Generally, pseudounipolar neurons serve an afferent sensory function: receiving information from the periphery and transmitting that information to the CNS. **(B)** Bipolar neurons have a single axon and a single dendrite emanating from the soma. These are relatively rare but are utilized in some regions such as the retina and other special senses. **(C)** Multipolar neurons have a single axon and many dendrites emanating from the soma. These are the most common neuronal type and are found in the brain, spinal cord, and autonomic ganglia.

Neuronal transport is mediated by cytoskeletal components and occurs to and from the soma.

Like other cells, the nucleus and protein synthesis machinery is located in the neuronal soma. Because the synapse can be as far as 80 cm from the soma, there must be a mechanism of transport to move proteins to the terminals. As occurs in other cells, proteins are synthesized from mRNA by the ribosomes on the rough endoplasmic reticulum and then packaged into vesicles by the Golgi apparatus (Fig. 3.4). Many proteins are constitutively expressed, whereas the expression of others are induced after the activation of specific transcription factors. Once packaged, the proteins are ready for transport to the terminals via the neuronal cytoskeleton. The neuronal cytoskeleton consists of neurofilaments, microfilaments, and microtubules. **Neurofilaments** are the intermediate-sized component of the cytoskeleton and provide structural rigidity to the axon. **Microfilaments** are smaller in diameter than neurofilaments and are involved in extension of dendrites and axons during development, structural support, and organelle transport. They are composed of actin, which provides a track for its contractile partner myosin and enables cell migration/neuronal process extension. **Microtubules** are the largest in diameter and play a crucial role in transport of organelles and other material from the soma to the processes. This is aided by associated proteins, such as kinesin and dynein, which interact with the microtubule and the

organelle to move them in different directions. The transport of cellular materials can occur in an anterograde or retrograde fashion. **Anterograde transport** occurs when organelles and other material are transported from the soma to the neuronal processes. Anterograde transport can be either slow at a rate of ~1 mm/d or fast at a rate of ~400 mm/d. Fast anterograde transport occurs for organelles, vesicles, and membrane glycoproteins. One of the organelles that undergoes transport to the processes is mitochondria. As in other cells, mitochondria are responsible for energy production by providing the cell with ATP. They also serve to maintain calcium homeostasis, can generate reactive oxygen species, and are the site of monoamine oxidase, which is involved in the degradation of monoamine neurotransmitters. In these ways, mitochondria transported to the axon contribute to neuronal signaling. **Retrograde transport**, about 200 mm/d, occurs when material is transported from the process to the soma in order to, for example, interact with nuclear receptors or to be degraded by lysosomes.

▶ ACTION POTENTIALS

Just as there needs to be a mechanism for transporting cellular materials between the soma and neuronal processes, neurons also need a mechanism by which a signal is propagated from the dendritic nerve ending to the presynaptic terminal

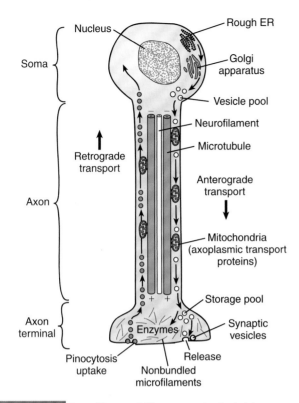

Figure 3.4 **Axonal transport.** The neuronal cytoskeleton provides rigidity to the axon and is involved in the transport of components to and from the soma. Organelles, vesicles, and other material moves along microtubule from the soma to the axonal terminal in an anterograde fashion. Material can also be transported in a retrograde fashion to the soma. ER, endoplasmic reticulum.

of the axon where neurotransmitters are stored and from which they are released. This propagation of signal is accomplished electrically through opening of voltage-gated ion channels, depolarization of neuronal membranes, and subsequent production of action potentials.

A neuron's resting membrane potential is heavily influenced by the electrochemical gradient of potassium ions.

As discussed in Chapter 2, channels that are permeable to specific ions reside within the plasma membrane. Some of these channels are sensitive to voltage and open and close at different membrane potentials. The **resting membrane potential** of the neuron is −70 mV (Fig. 3.5). This is heavily influenced by potassium's **equilibrium potential** (i.e., the concentration at which there is a balance between the forces due to the concentration gradient for the ion and the electrical gradient across the membrane), which can be calculated by the **Nernst equation** (see Chapter 2). At rest, there is a higher concentration of potassium ions inside the cell and a higher concentration of sodium ions outside the cell. The potassium concentration is ~140 mM intracellularly compared with 4 mM extracellularly, whereas sodium is 14 mM intracellularly and 140 mM extracellularly. Certain potassium channel subtypes are open at rest allowing potassium to leak out and thus establishing a negative charge differential across the plasma membrane. The **Na⁺/K⁺-ATPase** continually pumps K⁺ back into the cell thus maintaining the high intracellular K⁺ concentration and its concentration

gradient. The maintenance of the K⁺ and Na⁺ concentration gradients are critical for the production of action potentials.

Changes to the permeability of potassium and sodium ions can result in hyperpolarization or depolarization of the membrane.

Although the resting membrane potential for a neuron is −70 mV, this membrane potential can change due to the influx or efflux of ions, particularly sodium and potassium ions. There are a number of subtypes of sodium and potassium channels each with its own set of characteristics. Some of these channels open/close in response to binding of neurotransmitters to receptors, whereas others respond to voltage changes. The subtypes also differ in the length of time they remain open. Additionally, some of the channels inactivate meaning that there is a period of time in which the channel cannot be opened.

When ion channels open such that there is an influx of positive ions resulting in a less negative potential difference across the membrane, the membrane is said to **depolarize**. The primary ion responsible for neuronal depolarization is Na⁺ and it can enter the cell through a number of different types of sodium channels. Ligands can bind receptors to open sodium channels such as occurs when acetylcholine binds to nicotinic receptors. Other sodium channels have voltage sensors and open in response to voltage changes. The influx of positive ions leads to an **excitatory postsynaptic potential** (EPSP) (Fig. 3.6). The channel remains open for a finite period of time depending on its time constant τ. The increase in the membrane potential is also bounded in space by the space constant λ: when an ion channel opens, the ions enter and distribute along the membrane decreasing in magnitude with increasing distance from the channel. This spread of the signal over short segments of the membrane is known as **electrotonic conduction** or **passive conduction**. If there were only one ion channel subtype on the cell membrane, the membrane potential reached could be determined by the Nernst potential for that ion. Clearly this is not the case and the membrane potential ultimately reached at any portion of the membrane depends on additional factors. Because the opening of ion channels changes the permeability of the membrane to that ion, the number and types of channels open are of critical importance. The **Goldman equation** allows for the calculation of the membrane potential by including the permeability of K⁺, Na⁺, and Cl⁻, the ions that are the major contributors to membrane potential. Additionally, the opening of nearby channels can lead to a **temporal summation** or a **spatial summation** of EPSPs. Once all of the channels return to their unstimulated states, the membrane potential returns to its resting potential.

Although the opening of sodium channels can produce a more positive membrane potential, the opening of potassium channels will lead to the efflux of potassium ions. Some of the potassium channels respond to the binding of ligands to receptors, such as the binding of opioids to the opioid receptors, while other potassium channels are voltage sensitive. When potassium leaves the cell, the membrane potential **hyperpolarizes** (becomes more negative) thus creating an **inhibitory postsynaptic potential** (IPSP). The production of IPSPs serves to dampen a neuron's excitability. The summation of the membrane's EPSPs and IPSPs will determine the final membrane potential.

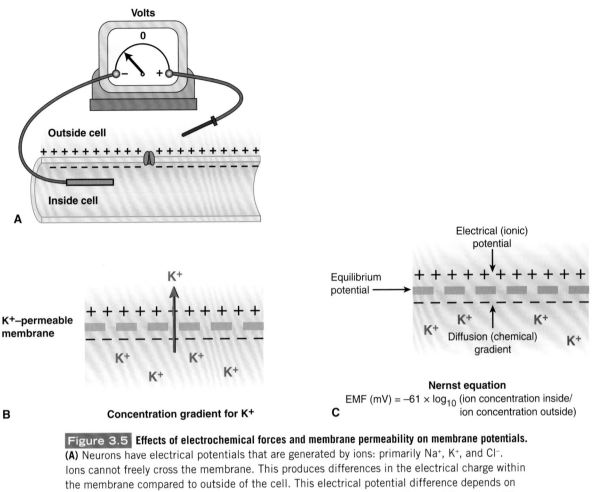

Nernst equation
EMF (mV) = −61 × log$_{10}$ (ion concentration inside/
ion concentration outside)

Figure 3.5 **Effects of electrochemical forces and membrane permeability on membrane potentials.**
(A) Neurons have electrical potentials that are generated by ions: primarily Na$^+$, K$^+$, and Cl$^-$.
Ions cannot freely cross the membrane. This produces differences in the electrical charge within
the membrane compared to outside of the cell. This electrical potential difference depends on
the charge of the ions on either side of the membrane, their concentration, and the permeability of the membrane to those ions. This voltage across the membrane represents the potential
difference and can be measured with a microelectrode. The opening of channels permeable to
a particular ion, such as K$^+$, results in the flow of that ion through the channel and a change
in the membrane potential. **(B)** The flow of an ion through an open channel will depend on the
concentration and electrical gradients across the membrane. Potassium ions have a higher
concentration within the cell, which produces a concentration gradient and leads to the efflux
of potassium out of the cell. However, because of the lack of permeability of the membrane to
anions, the inside of the membrane is negatively charged, which disfavors the movement of K$^+$
ions out of the cell. **(C)** The equilibrium potential is the membrane potential at which these two
forces are in balance and there is no net movement of the ion through open channels. The equilibrium potential can be calculated using the Nernst equation.

Action potential has different phases depending on the permeability of the sodium and potassium channels.

If through temporal or spatial summation, the EPSPs reach
a particular membrane potential referred to as the **threshold
potential** (generally around −55 mV) at the axon hillock, a
cascade of opening of voltage-gated sodium channels and
subsequent **action potential** can result. Once the threshold potential is reached, enough Na$^+$ channels open that the
membrane potential of the cell approaches the Nernst potential for Na$^+$, which is typically around +50 mV. Because K$^+$
channels open to repolarize the cell membrane through the
efflux of positive potassium ions, the membrane potential
generally peaks around +30 mV.

Action potentials have a characteristic shape that can
be divided into specific phases based on the **conductance** of

Na$^+$ and K$^+$ ions. Conductance is the flow of ions across the
membrane and is increased with opening of ion channels. It is
the inverse of **resistance**, which is greatest when channels are
closed and permeability is decreased. The membrane potential
at rest is typically around −70 mV reflecting the permeability
of the cell to K$^+$ and the nearness to its equilibrium potential of −90 mV as calculated by the Nernst equation. As the
cell is stimulated, local EPSPs are produced by the opening of
channels that allow the influx of Na$^+$ ions, which results in an
increase in the membrane potential to more positive values. If
the opening of Na$^+$ channels is timed appropriately or the Na$^+$
channels are close to one another, the EPSPs can summate to
reach the threshold potential of approximately −55 mV.

If, and only if, the membrane potential reaches the threshold, an action potential will be initiated (Fig. 3.7). In this way,

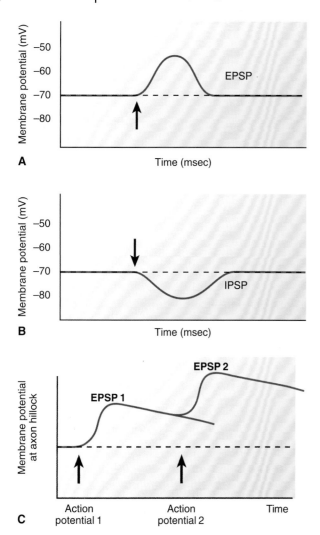

Figure 3.6 Excitatory and inhibitory postsynaptic potentials. **(A)** Inflow of positive ions at a synapse *(arrow)* makes the postsynaptic membrane potential less negative and increases the likelihood of the postsynaptic cell producing an action potential. This is an excitatory postsynaptic potential (EPSP). **(B)** Inflow of negative ions or outflow of positive ones makes the postsynaptic membrane potential more negative and decreases the likelihood of the cell producing an action potential. This is an inhibitory postsynaptic potential (IPSP). **(C)** If the time constant is long enough, some of the depolarization from the first EPSP is still present when the second occurs, and the individual depolarizations can summate.

the generation of an action potential is an **all-or-none phenomenon**. Unlike the electrotonic potential, the action potential is a **propagated potential**, which means that it regenerates and is capable of moving long distances along the nerve process. The density and characteristics of voltage-gated sodium channels are responsible for this phenomenon. These are found at high density along the axon hillock near the soma of the cell, which is sometimes referred to as the "trigger zone" because of its ability to generate action potentials. These channels are activated (open) at the threshold potential. Because of the density of the voltage-gated sodium channels along the axon, there is a relatively large influx of sodium across the membrane although the actual quantity of ions moving is very small: <0.01% of the cell's sodium pool. This leads to

the spreading of positive ions to the adjacent membrane and subsequent achievement of its threshold potential leading to opening of voltage-gated sodium channels in these nearby regions and the propagation of the action potential along the axon. As sodium ions enter, the action potential enters the **depolarization** phase, which is composed of a **rising phase**, in which the membrane is approaching a potential of 0 mV, and an **overshoot** phase, in which the membrane potential rises above 0 mV. The membrane potential approaches the equilibrium potential for sodium (+60 mV) but reaches a maximum of only about +30 mV due to inactivation of these channels and opening of potassium channels.

The **inactivation** of the sodium channel is phenomenon by which the channel is inaccessible to Na+ ions even though the membrane potential is such that the channel should be activated. It is accomplished by the closing of a "gate" on the intracellular side of the channel that prevents the passage of ions through the channel. The voltage-gated sodium channel exists in three states: rested, active, and inactivated. In the **resting state** the channel is closed but capable of being opened. In the **active state**, the membrane potential has risen enough such that a conformational change occurs that removes the "gate" on the extracellular side of the channel that was preventing ionic passage. The active state is rapidly followed by the **inactive state** during which time the channel cannot be opened. The inactive state only lasts for a few milliseconds before the channel returns to the resting state. Until this return, another action potential cannot be generated because inactivated sodium channels are impermeable to Na+ ions. This timeframe is known as the **absolute refractory period**.

At the peak of the action potential, voltage-gated potassium channels open and a relatively large efflux of K+ occurs. Because the majority of the sodium channels are inactivated and thus there is no movement of Na+ ions, this results in a rapid **repolarization**: a decrease in the membrane potential returning towards the resting membrane potential. Certain potassium channels remain open even after the potential returns to resting values. This leads to **afterhyperpolarization**, which is a decrease in the membrane potential past the resting potential. This is short lived since these channels rapidly close leading to a return to the resting membrane potential.

During the repolarization phase, more of the sodium channels are returning to the rested state and are again capable of producing an action potential. However, because there is a proportion of these channels that are still inactivated and due to the open K+ channels, it would take a stronger-than-normal stimulus to produce another action potential. This timeframe is the **relative refractory period**.

The absolute and relative refractory periods are important in the propagation of action potentials along the axon in a unidirectional manner (Fig. 3.8). As one membrane segment depolarizes to produce an action potential, the entering sodium ions spread to adjacent regions on either side of the depolarized membrane. In the forward direction, this changes the membrane potential of the adjacent segment, which results in the opening of Na+ channels and production of another action potential. In this manner, there is a sequential depolarization of adjacent segments of the axon

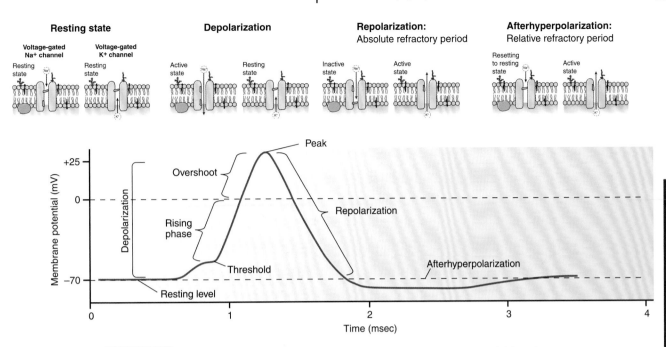

Figure 3.7 **Phases of the action potential. Resting:** When the membrane potential is at its resting level, both Na+ and K+ channels are closed. **Depolarized:** During depolarization, Na+ channels open allowing the influx of Na+. K+ channels remain closed. **Repolarization:** As the membrane potential becomes more positive, certain voltage-gated K+ channels open allowing the efflux of K+ and returning the membrane potential towards resting levels. After opening, Na+ channels inactivate. During the period of inactivation of Na+ channels, another action potential cannot be generated and the membrane is in an absolute refractory period. **Afterhyperpolarization:** As the membrane potential reaches the resting membrane potential, K+ channels remain open leading to a membrane potential that is more negative than the resting membrane potential. During this time the population of Na+ channels are returning to their resting closed state. While some of the Na+ channels remain inactivated, it will take greater stimulation to produce an action potential and the membrane is in a relative refractory period. Once the K+ channels close, the membrane returns to the resting membrane potential.

producing action potentials down the axon like a wave. Because all action potentials reach the same amplitude, the action potential propagates without decrement. In the adjacent membrane region that has already produced an action potential, many of the Na+ channels are inactivated and thus in the absolute or relative refractory period, which decreases the opportunity to produce another action potential and propagate the signal backwards.

With all action potentials reaching the same peak, the strength of a signal cannot be encoded by the action potential amplitude. Instead, strong signals increase the frequency of action potentials. This is accomplished by decreasing the time that the membrane is in the resting state.

Conduction velocity is influenced by myelination and fiber diameter.

The speed at which action potentials propagate along the axon is determined by the characteristics of the axon itself specifically the degree of myelination and the axonal diameter. Each axon can be surrounded by many myelin sheaths. The myelin sheaths wrap around an axon at regular intervals and act as insulators. This means that there are few sodium channels and very little current flow across the myelinated membrane. The myelin sheaths are interspersed by unmyelinated regions at the nodes of Ranvier. These unmyelinated regions have a high density of sodium channels. The myelin insulation allows the charge to "jump" from one node to the next by reducing the "leak" of current out of the membrane (Fig. 3.9). This is known as **"saltatory conduction"** and can be up to 50 times faster than propagation without myelination. Importantly, the current does not diminish as it is shunted from one node to the next. The larger the diameter of the axon, the faster the conduction due to decreased internal resistance and greater spread of the sodium ions to the adjacent membrane. Axons can be up to 20 μm in diameter. In contrast, the smallest diameter axons (0.4 μm) are unmyelinated. The speed of conduction is slower in unmyelinated fibers because the action potential is traveling continuously along the axon without jumping over various segments. Unlike the myelinated fibers, sodium channels are distributed evenly along the unmyelinated axon.

Axons in the PNS are bundled together and enveloped by a fibrous coat, known as the **epineurium**, to form nerves. There are three main types of nerve fibers: A, B, and C. These are distinguished based on myelination, diameter size, and function. The A fibers are large (2 to 20 μm in diameter), myelinated fibers that are involved in touch, proprioception, motor function and some pain transmission. This class is

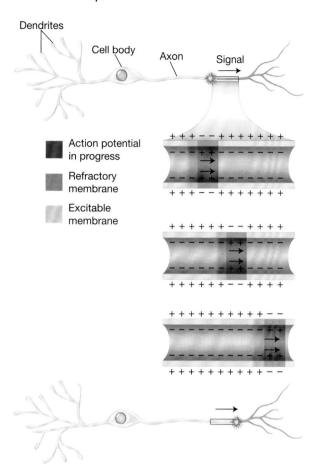

Dendrites

Cell body

Axon

Signal

■ Action potential in progress

■ Refractory membrane

□ Excitable membrane

Figure 3.8 **Unidirectional propagation of action potentials.** During the action potential, there is spread of sodium to adjacent segments of the depolarized membrane. Sodium channels that have just been involved in an action potential are inactivated for a few milliseconds and cannot be re-opened during that time resulting in the membrane being in a refractory period. Sodium channels on the membrane on the side of the action potential that has not yet generated an action potential, are in the rested state and can open and generate an action potential resulting in an excitable membrane. In this way, the action potential propagates unidirectionally towards the axon.

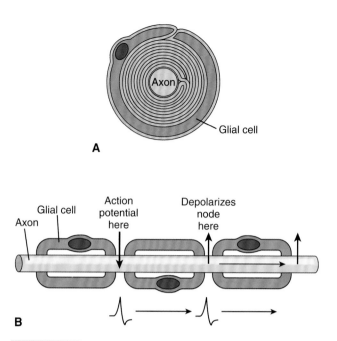

Axon

Glial cell

A

Glial cell

Axon

Action potential here

Depolarizes node here

B

Figure 3.9 **Myelination allows for saltatory conduction. (A)** Glial cells, oligodendrocytes in the CNS and Schwann cells in the PNS, wrap certain axons in overlapping layers of myelin, which serves to insulate the axon and limit ion flow from the underlying membrane. **(B)** Interspersed at regular intervals between myelinated segments are unmyelinated regions, nodes of Ranvier, where action potentials are generated due to the high density of sodium channels. The depolarization spreads through the insulated internodal region to the next node where another action potential is produced. Myelination results in a faster propagation of the action potential along the axon.

further divided based on the diameter size and function into α, β, γ, and δ subtypes (Table 3.1). The autonomic preganglionic fiber, which is the portion of the autonomic fiber that precedes the soma, is classified as a B fiber and is smaller in diameter than A fibers and only lightly myelinated. The portion of the sympathetic fibers that extends from the soma into the periphery is classified as a C fiber and is small in diameter and unmyelinated. Other C fibers carry information about pain and temperature. Unmyelinated fibers are particularly

TABLE 3.1 Nerve Fiber Classification				
Fiber Type	**Fiber Diameter (μm)**	**Myelination**	**Conduction Velocity (m/sec)**	**Function**
Aα	12–20	Yes	80–120	Proprioception, motor
Aβ	5–12	Yes	40–70	Touch, pressure
Aγ	3–6	Yes	15–40	Muscle spindles
Aδ	2–5	Yes	8–30	Pain, temperature
B	<3	Lightly	3–15	Preganglionic, autonomic
C: Dorsal Root	0.4–1.2	No	0.5–2.3	Pain, temperature
C: Sympathetic	0.3–1.3	No	0.7–2.3	Postganglionic

sensitive to local anesthetics, which block sodium channels. Thus, injection of a local anesthetic is used to block the transmission of information from a painful stimulus. At higher concentrations, local anesthetics can affect myelinated fibers as well. Larger fibers are more sensitive to compression, which can then cause loss of sensitivity to stimuli carried by those fibers.

▶ SYNAPTIC TRANSMISSION

The function of action potentials is to communicate information throughout the nervous system by transmitting the information from one portion of the nerve to another. Once the action potential reaches the nerve terminal, the signal can be communicated to another neuron or cell through a process known as synaptic transmission.

Neurons communicate both by electrical and chemical synapses.

Synaptic transmission is accomplished either at **electrical synapses** or chemical synapses. Electrical communication occurs via **gap junctions**, which occur when the terminals from two neurons are in close apposition to one another. Ions can flow directly from one neuron to the other through special channels. In this way, depolarization can spread from one neuron to another without any intermediaries such as neurotransmitters. Gap junctions are found in the embryonic nervous system but rarely occur in adult mammalian neurons (although they are present in adult smooth and cardiac muscle cells).

   Chemical synapses consist of the presynaptic terminal, synaptic cleft, and the postsynaptic terminal. The depolarization of the presynaptic terminal results in the release of chemical messengers, **neurotransmitters**, into the synaptic cleft. These communicate the signal to the postsynaptic cell by interacting with specific proteins, **receptors**, on the postsynaptic terminal. The activation of these receptors alters cellular function by opening channels, activating intracellular messengers, and directly or indirectly affecting transcription.

Upon depolarization of the presynaptic membrane, vesicles mobilize to the membrane, dock, and release neurotransmitters.

Neurotransmitters are stored in vesicles. These vesicles contain membrane proteins that are important for anchoring the vesicle within the cytoplasm, releasing the vesicle, and allowing its mobilization and docking at the synaptic terminal where the vesicle can release its contents into the synaptic cleft. Vesicles and associated proteins are synthesized in the soma and transported via axoplasmic transport. Some of the vesicles contain neuropeptides as signaling molecules that are also synthesized in the soma. Others contain neurotransmitters that can be synthesized within the cytosol of the presynaptic terminal or even within the vesicle itself. The neurotransmitter can be recycled and repackaged into the vesicles either by reuptake back into the presynaptic terminal or via degradation in the synaptic cleft and the reuptake of precursors into the presynaptic terminal.

The arrival of the action potential at the nerve terminal leads to a multistep process resulting in the release of neurotransmitters into the synaptic cleft (Fig. 3.10). When the membrane of the presynaptic terminal depolarizes, voltage-gated calcium channels open allowing the influx of calcium. At rest, the intracellular free calcium concentration is around 100 nM, whereas the extracellular calcium concentration is 1.25 mM. Upon depolarization, the intracellular calcium concentration increases to ~1 μM. This increase in intracellular calcium results in activation of various kinases, which phosphorylate a class of proteins called **synapsins**. Synapsins serve to anchor neurotransmitter-containing vesicles within the cytoplasm. Their phosphorylation leads to their release from cytoskeletal proteins and mobilization of vesicles to the **active zone** of the synaptic membrane, which is an area containing calcium channels and proteins, such as syntaxin, which specialize in vesicular release. The proteins on the plasma membrane interact with specific vesicular membrane proteins, such as synaptotagmin and synaptobrevin. These vesicular and axonal membrane proteins are called **SNARES**. The interaction of the vesicular and axonal membrane SNARES enables the vesicle to "dock" (i.e., line up appropriately for the release of its neurotransmitter content into the synaptic cleft). Once docked, the vesicular membrane fuses with the synaptic membrane and releases the vesicular contents into the synaptic cleft.

   Presynaptic fibers are generally associated with a predominant neurotransmitter but other chemical messengers can also be present in the presynaptic bouton. Neurotransmitters are small molecules that interact with postsynaptic receptors to produce a fast cellular response. These are generally found in small vesicles that are located near the active zone. Neuropeptides are contained in larger vesicles that are located more diffusely throughout the presynaptic terminal. Rather than producing a fast postsynaptic response, neuropeptides tend to have a modulatory effect on the response to neurotransmitters. In some cases, neuromodulators can be colocalized within the same vesicle as the neurotransmitter. Other signaling molecules can undergo nonvesicular release from the presynaptic terminal. These are generally synthesized "on-demand" in response to the influx of calcium. Some examples of these are prostaglandins, cannabinoids, and nitric oxide. Other molecules signaling through receptors include hormones, such as the steroid hormones, and growth factors, such as neurotrophins.

   Once the neurotransmitter has been released into the synaptic cleft, it binds a **receptor**, which is a protein that interacts with ligands to produce a cellular effect without changing the ligand. Receptors can be located on the cell surface or intracellularly. Neurotransmitters can also act at receptors on the presynaptic terminal, sometimes referred to as **autoreceptors**, to modulate further neurotransmitter release. Autoreceptors may act to open certain potassium channels resulting in more positive ions escaping and a hyperpolarization of the membrane and/or to decrease activity of calcium channels resulting in a decrease in vesicular mobilization and neurotransmitter release. In this way, release of neurotransmitter can negatively modulate further

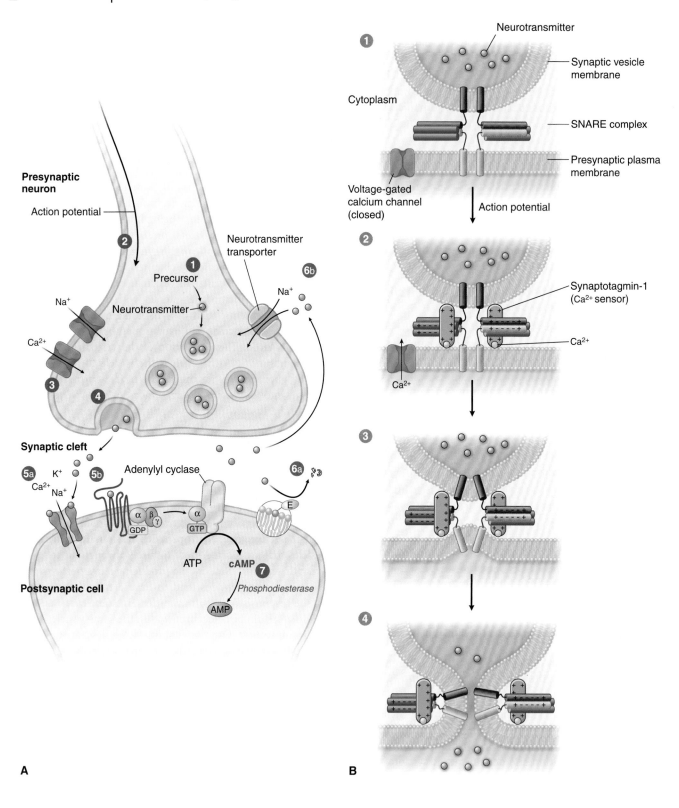

A

B

release from the same fiber. Receptors will be discussed in depth below.

Neurotransmitter actions can be terminated via diffusion, degradation, or cellular uptake.

In order to reach receptors, neurotransmitters diffuse away from the release site. They will be at the highest concentration within the synaptic cleft and often interact with receptors located in the postsynaptic density. The distance from which neurotransmitters diffuse away from the presynaptic terminal depends in part on the mechanisms by which that neurotransmitter is removed from the extracellular space, such as degradation or uptake into nearby cells. Some neurotransmitters, such as acetylcholine, are degraded by extracellular enzymes that are abundant in the synaptic cleft. Other neurotransmitters, such as norepinephrine, are taken up into the presynaptic terminal and either degraded by intracellular enzymes or recycled into vesicles for subsequent release. Neurotransmitters can also be taken up into glia or other cells where they undergo degradation or recycling back to neurons. This occurs with the amino acid neurotransmitter glutamate (Fig. 3.11). The further away from the release site the neurotransmitter travels, the lower its concentration will be at subsequent receptors. Some neurotransmitters have endocrine effects (i.e., the concentration can remain high enough as it diffuses from its release site that it can still activate distant receptors). These can be referred to as **neurohormones**.

▶ NEUROTRANSMISSION

Much of neuronal cellular communication is accomplished by the actions of a handful of small molecule neurotransmitters. These neurotransmitters are stored in vesicles and released in a calcium-dependent fashion from the presynaptic terminal. Once in the synaptic cleft, the neurotransmitters interact with receptors on the postsynaptic terminal and, in certain cases, also with autoreceptors on the presynaptic terminal that function to reduce further neurotransmitter release. The receptors with which the neurotransmitter interacts are specific for that particular neurotransmitter and are generally named for that neurotransmitter (e.g., dopamine interacts with dopamine receptors). There are often many subtypes of receptors with which a particular neurotransmitter can interact.

Neurotransmitters commonly act at ionotropic or metabotropic receptors to produce an effect.

In general, the small molecule neurotransmitters of the central nervous system interact with ligand-gated (**ionotropic**) or G protein–coupled (**metabotropic**) receptors. The ligand-gated receptors possess an internal pore that connects the extracellular and intracellular environments. These receptors respond to binding of the neurotransmitter by assuming a conformation in which the pore is open and permeable to particular ions thus allowing the flow of those ions into or out of the cell. Receptors that form pores that are permeable to sodium, such as certain glutamate receptors, will result in a depolarization of the membrane and an **excitatory** effect. Those that form pores that are permeable to chloride ions, such as certain GABA receptors, will result in hyperpolarization of the membrane and an **inhibitory** effect (Fig. 3.12). The ligand-gated receptor is able to respond quickly, within a few milliseconds, to the binding of the neurotransmitter and thus mediates **fast synaptic transmission**.

G protein–coupled receptors (GPCRs) respond more slowly to interaction with a neurotransmitter because of the initiation of a signaling cascade within the cell. Chapter 2

Figure 3.10 **Steps in synaptic transmission. (A)** Synaptic transmission can be divided into a series of steps that couple electrical depolarization of the presynaptic neuron to chemical signaling between the presynaptic and postsynaptic cells. (*1*) A neuron synthesizes neurotransmitters from precursors and stores the transmitter in vesicles. (*2*) An action potential traveling down the neuron depolarizes the presynaptic nerve terminal. (*3*) Membrane depolarization activates voltage-dependent Ca^{2+} channels, allowing Ca^{2+} entry into the presynaptic nerve terminal. (*4*) The increased cytosolic Ca^{2+} enables vesicle fusion with the plasma membrane of the presynaptic neuron, with subsequent release of neurotransmitter into the synaptic cleft. (*5*) Neurotransmitter diffuses across the synaptic cleft and binds to postsynaptic receptors. (*5a*) Neurotransmitter binding to ionotropic receptors causes channel opening and changes the permeability of the postsynaptic membrane to ions. This may also result in a change in the postsynaptic membrane potential. (*5b*) Neurotransmitter binding to metabotropic receptors on the postsynaptic cell activates intracellular signaling cascades; the example shows G protein activation leading to the formation of cAMP by adenylyl cyclase. In turn, such a signaling cascade can activate other ion-selective channels (not shown). (*6*) Signal termination is accomplished by removal of transmitter from the synaptic cleft. (*6a*) Transmitter can be degraded by enzymes (*E*) in the synaptic cleft. (*6b*) Alternatively, transmitter can be recycled into the presynaptic cell by reuptake transporters. (*7*) Signal termination can also be accomplished by enzymes (such as phosphodiesterase) that degrade postsynaptic intracellular signaling molecules (such as cAMP). **(B)** (*1*) Synaptic vesicles are tethered close to the plasma membrane of the presynaptic neuron by several protein–protein interactions. The most important of these interactions involve SNARE (soluble *N*-ethylmaleimide-sensitive factor attachment protein receptor) proteins present in both the vesicle membrane and the plasma membrane. The SNARE proteins include synaptobrevin (red), syntaxin-1 (yellow), and SNAP-25 (green). Voltage-gated Ca^{2+} channels are located in the plasma membrane in close proximity to these SNARE complexes; this facilitates the sensing of Ca^{2+} entry by Ca^{2+}-binding proteins (synaptotagmin-1, in blue) localized to the presynaptic plasma membrane and/or the synaptic vesicle membrane. (*2–4*) Voltage-gated calcium channels open in response to an action potential, allowing entry of extracellular Ca^{2+} into the cell. The increase in intracellular Ca^{2+} triggers binding of synaptotagmin-1 to the SNARE complex and fusion of the vesicle membrane with the plasma membrane, releasing neurotransmitter molecules into the synaptic cleft. Several additional proteins may also be involved in the regulation of synaptic vesicle fusion (*not shown*).

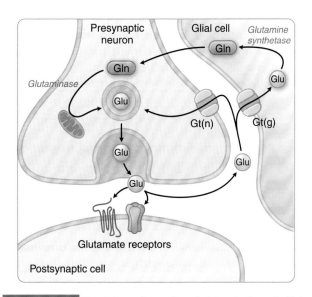

Figure 3.11 **Reuptake and recycling of glutamate through glial cells.** Glutamate diffuses out of the synaptic cleft where it is transported into the presynaptic terminal by the neuronal glutamate transporter [Gt(n)] or into glial cells by the glial glutamate transporter [Gt(g)]. In the glial cell it can be converted to glutamine (Gln), which is then transferred to the neuron and converted back to glutamate via the mitochondria-associated glutaminase.

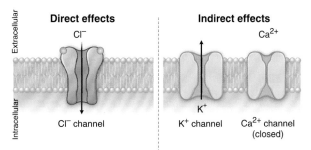

A Effects of inhibitory neurotransmitters

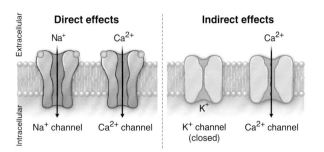

B Effects of excitatory neurotransmitters

Figure 3.12 **Receptor activation can produce inhibitory or excitatory effects. (A)** Inhibitory neurotransmitters *hyperpolarize* membranes by inducing a net outward current, by promoting either an influx of anions (e.g., opening a Cl⁻ channel) or an efflux of cations (e.g., opening a K⁺ channel). **(B)** Excitatory neurotransmitters *depolarize* membranes by inducing a net inward current, either by enhancing inward current (e.g., opening a Na⁺ or Ca²⁺ channel) or by reducing outward current (e.g., closing a K⁺ channel). Potassium channel closure, independent of changes in the resting membrane potential, also increases the resting membrane resistance and renders the cell more responsive to excitatory postsynaptic currents.

describes the function of GPCRs in detail. Briefly, once activated by the receptor, the alpha subunit of the G protein will bind to GTP (replacing GDP), which allows the dissociation of the alpha and beta/gamma subunits. The alpha and the beta/gamma subunits can independently interact with various effector molecules. The hydrolysis of GTP to GDP results in the reassociation of the subunits and the termination of their effects. There are three main types of GPCRs categorized by the action of its alpha subunit (Fig. 3.13).

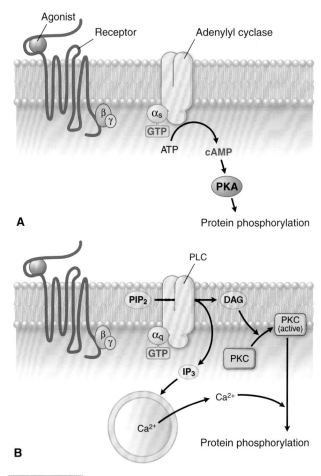

Figure 3.13 **Activation of adenylyl cyclase (AC) and phospholipase C (PLC) by G proteins.** G proteins can interact with several different types of effector molecules. The subtype of Gα protein that is activated often determines which effector the G protein will activate. Two of the most common Gα subunits are Gα_s and Gα_q, which stimulate adenylyl cyclase and phospholipase C, respectively. **(A)** When stimulated by Gα_s, adenylyl cyclase converts ATP to cyclic AMP (cAMP). cAMP then activates protein kinase A (PKA), which phosphorylates a number of specific cytosolic proteins. **(B)** When stimulate by Gα_q, phospholipase C (PLC) cleaves the membrane phospholipid phosphatidylinositol-4,5-bisphosphate (PIP₂) into diacylglycerol (DAG) and inositol-1, 4, 5-triphosphate (IP₃). DAG diffuses in the membrane to activate protein kinase C (PKC), which then phosphorylates specific cellular proteins. IP₃ stimulates release of Ca²⁺ from the endoplasmic reticulum into the cytosol. Calcium release also stimulates protein phosphorylation events that lead to changes in protein activation. Although not shown, the Bγ subunits of G proteins can also affect certain cellular signal transduction cascades.

The alpha subunit of the G_s type of GPCRs activates adenylyl cyclase to produce cAMP, which can then stimulate protein kinase A and result in protein phosphorylation. The alpha subunit of the G_i type of GPCRs inhibits the activity of adenylyl cyclase thus decreasing intracellular cAMP levels. At the same time, the beta/gamma subunits of G_i can lead to the opening of certain potassium channels, resulting in flow of potassium out of the cell and hyperpolarization of the membrane, and the closing of certain calcium channels, resulting in a reduction of calcium available for the mobilization of neurotransmitter vesicles and a decreased release of neurotransmitters from the presynaptic terminal. The alpha subunit of the G_q type of GPCRs activates phospholipase C, which catalyzes the hydrolysis of phosphatidylinositol 4, 5-bisphophate (PIP_2) to inositol triphosphate (IP$_3$) and diacylglycerol (DAG). These molecules activate separate signaling pathways. IP$_3$ diffuses through the cytoplasm to the endoplasmic reticulum where it induces the release of calcium. DAG activates protein kinase C (PKC) leading to phosphorylation of proteins.

It is not uncommon for a particular neurotransmitter to interact with receptors that have both ionotropic and metabotropic subtypes. The exact function of the neurotransmitter will be determined by its site of release as well as the action and distribution of its specific receptors.

Classical neurotransmitters are small molecules.

In general, a given presynaptic terminal will synthesize and store one type of neurotransmitter in its vesicles. Often this presynaptic terminal will also store large molecule

CLINICAL FOCUS | 3.2

Addiction

Many people have pleasurable experiences associated with food, alcohol, coffee, shopping, etc. and can disengage from these activities when these experiences are having negative consequences on the individual such as causing health or financial issues. Addiction occurs when an individual feels a compulsion to engage in intensely pleasurable behaviors despite negative consequences. These euphoric experiences can result from drugs, alcohol, food, shopping, sex, gambling, etc. The rewarding aspect of these behaviors involve activation of certain brain circuitry. The activation of these pathways produce reinforcing effects and a compulsion for the individual to continue activating these pathways. It is not well understood why one individual is more susceptible to addiction than another but seems to involve genetics, personality, and environmental factors.

Drugs can produce physical dependence and tolerance without producing addiction. Physical dependence is the body's adaptation to continued exposure to the drug. This adaptation can result in tolerance, which is a decreased responsiveness to the same dose of drug. In order to reach the same euphoric experience, the individual must increase the dose of the drug. Because the body has now adapted to the presence of the drug, sudden discontinuation of the drug will lead to a response that is decreased relative to the normal response in the absence of the drug. This can produce withdrawal symptoms. Withdrawal symptoms often are opposite of what is experienced with the drug: if the drug produces a calming effect and slowed heart rate, withdrawal from the drug may produce irritation and an increased heart rate. Many drugs produce physical dependence and in order to avoid withdrawal, patients are tapered off of the drug rather than abruptly discontinuing it.

Addiction is less common than physical dependence. Its development is associated with activation of a dopaminergic brain pathway often referred to as the "reward pathway." Dopaminergic cells in the ventral tegmental area (VTA) of the brainstem send axonal processes to another brain region, the nucleus accumbens. Dopamine is released in the nucleus accumbens and acts on dopamine receptors. Activation of this pathway is reinforcing in that it motivates behaviors that will lead to further activation of the pathway. Drugs of abuse activate this pathway. Some, such as cocaine, enhance the activation of the pathway by blocking the reuptake of dopamine and thereby increasing its concentration and time in the synaptic cleft to produce greater stimulation of dopamine receptors in the nucleus accumbens. Others, such as amphetamine, stimulate release of dopamine from the presynaptic terminal in addition to blocking the reuptake of dopamine.

Dopaminergic neurons in the VTA are inhibited by GABAergic fibers. These release GABA, which acts to decrease the excitability of the dopaminergic fiber, thus decreasing the release of dopamine into the nucleus accumbens and the activation of dopamine receptors. Some drugs, such as opioids, act at receptors on the GABAergic fibers to remove their inhibition of the dopaminergic fibers. This disinhibition results in the activation of these fibers, the release of dopamine in the nucleus accumbens, and activation of dopamine receptors.

In order to activate the dopamine reward pathway, drugs must cross the BBB and access the CNS. Drugs that are more lipophilic, will more readily cross the BBB and can provide greater stimulation. Thus, the lipophilicity of a drug can impact its abuse potential. Heroin is a more lipophilic form of morphine. It is metabolized in the CNS to morphine where it acts at opioid receptors in the VTA to disinhibit the dopaminergic projections into the nucleus accumbens. It is more addictive than morphine because it is better able to cross the BBB thus providing a faster and greater stimulation of the reward pathway. In general, short-acting lipophilic drugs have a greater addictive potential than longer-acting drugs that are more hydrophilic because they can achieve a greater concentration at the target receptor. ■

neuromodulators that have been synthesized in the cell body and transported to the presynaptic terminal. These are usually neuropeptides and serve to modulate the postsynaptic actions of the primary neurotransmitter. Classically, neurotransmitters consist of small molecules including amine molecules and certain amino acids.

Acetylcholine

Acetylcholine (ACh) was one of the first neurotransmitters discovered and played an important role in identifying chemical transmission as a primary means of neuronal communication. In the PNS, acetylcholine is involved in producing muscle contraction as well as mediating the responses of the parasympathetic nervous system, both of which will be discussed in depth in later chapters. Acetylcholine is also a transmitter in the CNS where it is involved in arousal and attention. Acetylcholine is synthesized within the presynaptic terminal from acetyl CoA and choline by the enzyme choline acetyltransferase. Once synthesized in the cytosol, ACh is transported into vesicles via the vesicle-associated transporter (VAT). Upon release into the synaptic cleft, it is degraded by acetylcholinesterase back to choline and acetate. Choline is then recycled back into the presynaptic terminal to be used again in the synthesis of ACh.

The receptors with which ACh interacts are referred to as **cholinergic receptors**. There are two main classes of cholinergic receptors: nicotinic and muscarinic receptors. These are named for substances that selectively activate these receptors. **Nicotinic receptors** have subtypes that are located at the neuromuscular junction, autonomic ganglia, and central nervous system. Nicotinic receptors are ligand-gated ion channels that are permeable to sodium and calcium. The influx of sodium results in local membrane depolarization and can lead to the activation of voltage-gated sodium channels and the production of an action potential. At the neuromuscular junction, this can result in muscle contraction. Excess stimulation of these receptors can also result in a depolarizing blockade in which there is a sustained contraction of the muscle. This will be discussed in depth in Chapter 6.

Muscarinic receptors have multiple subtypes, M_1–M_5, and are all GPCRs. M_1, M_3, and M_5 receptors interact with G_q proteins and thus lead to activation of phospholipase C and the eventual increase in intracellular calcium from intracellular stores along with activation of protein kinase C. M_2 and M_4 receptors interact with G_i proteins leading to decreased protein kinase A activation, hyperpolarization of the membrane due to the activation of certain potassium channels, and reduction in vesicular mobilization due to the closing of certain calcium channels. In addition to actions in the brain, muscarinic receptors are also the primary receptor of the parasympathetic nervous system.

Agents that block cholinergic transmission in the CNS are used clinically in the treatment of motion sickness and in restoring the DA–ACh balance lost in the initial stages of Parkinson's disease. Agents that enhance cholinergic transmission in the brain by inhibiting acetylcholinesterase are currently approved for use in Alzheimer's disease and may delay cognitive impairment in the initial stages of the disease.

Monoamines

The monoamines consist of catecholamines and serotonin. The catecholamines (dopamine, norepinephrine, and epinephrine) are all synthesized from a common precursor, tyrosine, and share a common synthetic pathway. All of the monoamines are metabolized by monoamine oxidase (MAO), which is located in the mitochondria of both the presynaptic and postsynaptic terminals. Inhibitors of monoamine oxidase can result in elevated levels of the monoamines. When a monoamine oxidase inhibitor is combined with another drug that increases the levels of a particular monoamine, life-threatening situations can result: hypertensive crisis with norepinephrine and serotonin syndrome with serotonin.

Norepinephrine and Epinephrine

Norepinephrine (NE) and epinephrine, sometimes referred to as adrenalin, both act at adrenergic receptors. Norepinephrine is synthesized within the vesicle from dopamine. Dopamine (DA) is transported into the vesicle via the vesicular monoamine transporter (VMAT). If the enzyme dopamine β hydroxylase is present in the vesicle, it will convert DA to NE. Norepinephrine can be further converted to epinephrine if the enzyme phenylethanolamine-*N*-methyltransferase is present. Most epinephrine is secreted from chromaffin cells of the adrenal medulla and circulates peripherally in response to activation of the sympathetic nervous system. Similarly, NE has a primary role in the sympathetic nervous system but it has a larger role in the CNS than does epinephrine. In the CNS, NE is involved in arousal, blood pressure regulation, control of mood, pain modulation, and appetite suppression. Once released, NE can interact with adrenergic receptors located on the postsynaptic and presynaptic terminals. It can then be taken back up into the presynaptic terminal by the norepinephrine transporter. Once in the cytosol, it can be reloaded into the vesicle via VMAT or be metabolized via MAO. It can also be degraded in nonneuronal cells by catechol-*O*-methyltransferase (COMT).

There are two main receptor subtypes with which norepinephrine and epinephrine interact: alpha and beta adrenergic receptors. Alpha adrenergic receptors have two subtypes. The α_1 receptor activates the G_s protein, while the α_2 receptor activates the G_i protein. The α_2 receptor can be found presynaptically where it can function to decrease further release of NE. There are three beta adrenergic receptors (β_1, β_2, and β_3), which all activate G_s proteins.

Adrenergic receptors in the CNS are involved in alertness, pain transmission, and central control of sympathetic tone. Agents that result in the activation of these receptors in the CNS are used clinically in the treatment of attention deficit hyperactivity disorder, narcolepsy, hypertension, depression, pain, and appetite suppression. Certain stimulant drugs, like amphetamine and methamphetamine, also produce effects through activation of adrenergic receptors.

Dopamine

Dopamine was once thought to simply be the precursor for norepinephrine but now it is recognized as a neurotransmitter in its own right. It is involved in control of movement, reward and addiction, and development of schizophrenia. Similar to norepinephrine and epinephrine, DA is released into the synaptic cleft where it can interact with postsynaptic and presynaptic receptors. It is transported back into the presynaptic terminal by the dopamine transporter where it can be re-loaded into vesicles via transport by VMAT. It can also be broken down within the terminal by MAO. Like other catecholamines, it can also be taken up by nonneuronal cells and metabolized by COMT.

Dopamine receptors are classified D_1–D_5 and are all GPCRs. D_1 and D_5 receptors activate the G_s protein, whereas D_2, D_3, and D_4 receptors activate G_i proteins. Activation of dopaminergic receptors is used clinically in the treatment of Parkinson's disease. Inhibition of these receptors is used clinically in the treatment of schizophrenia and psychosis. Many addictive drugs and behaviors involve activation of the dopaminergic "reward pathway."

Serotonin

Serotonin, or 5-hydroxytryptamine (5-HT), is a monoamine but not a catecholamine. In the brain, it is involved in mood control, appetite control, and nausea. It is synthesized from tryptophan rather than from tyrosine. Once synthesized in the cytosol, it is transported into the vesicle via VMAT. Upon release into the synaptic cleft, it interacts with postsynaptic and presynaptic receptors. Its action is terminated via reuptake into the presynaptic terminal by the serotonin transporter. Like the other monoamines, serotonin can be degraded within the presynaptic terminal by MAO or recycled into vesicles by VMAT.

Serotonin receptors represent ionotropic and metabotropic receptor classes. The 5-HT_1 receptors are linked to G_i protein, 5-HT_2 receptors are linked to G_q proteins, 5-HT_3 receptors are ligand-gated and permeable to sodium, and 5-HT_{4-7} receptors are linked to G_s proteins. Clinically, serotonin reuptake inhibitors are used to treat depression by enhancing the levels of serotonin in the synaptic cleft. Agents that activate selective serotonin receptors are used clinically to treat migraines and anxiety. Agents that block serotonin receptors in the chemoreceptor trigger zone are used in the treatment of nausea and vomiting. Certain drugs, such as LSD, produce hallucinations through activation of specific 5-HT receptors.

Glutamate

Glutamate is the principle excitatory neurotransmitter of the CNS. It plays a role in learning and memory, pain transmission, epilepsy, and some forms of neurotoxicity. Glutamate in nerve terminals is converted from α-ketoglutarate, from the Krebs cycle, and stored in vesicles. It then undergoes calcium-dependent vesicular release into the synaptic cleft where it can interact pre- and postsynaptically with its receptors. Its action is terminated by uptake via an excitatory amino acid transporter back into the presynaptic terminal or into glial cells where it is converted to glutamine by the enzyme glutamine synthase. Glutamine can then leave the glial cell and enter the presynaptic terminal via the glutamine transporter. It is converted in the presynaptic terminal back to glutamate and then taken up into the vesicle via a vesicular glutamate transporter.

Glutamate receptors are named for particular ligands that activate them. The *N*-methyl-ᴅ-aspartate (NMDA), α-amino-3-hydroxy-5-methyl-4-isoxazole propionic acid (AMPA), and kainate (KA) receptors are ionotropic receptors. AMPA and KA receptors are permeable to Na^+ and K^+, whereas NMDA receptors are also permeable to Ca^{2+}. The Ca^{2+} permeability is thought to play a role in the toxicity associated with excess glutamate in the synaptic cleft. There are also metabotropic receptors linked to either the G_q or G_i protein.

There are not very many clinical therapies that directly target glutamate receptors. Inhibitors of NMDA receptors may be of some benefit in treating epilepsy and are used to delay cognitive impairment in patients with Alzheimer's disease. Ketamine inhibits the NMDA receptor by blocking the pore and is used clinically as a general anesthetic and an analgesic. Ketamine is associated with hallucinations. Another blocker of the NMDA receptor pore, the illicit drug PCP, is also a hallucinogen.

GABA

The amino acid γ-aminobutyric acid (GABA) is the primary inhibitory neurotransmitter of the CNS. It functions as a CNS depressant. Disruption of GABA signaling may be involved in anxiety and epilepsy. GABA is synthesized from glutamate by glutamic acid decarboxylase (GAD). It is stored in vesicles and undergoes calcium-dependent vesicular release. GABA is taken into the presynaptic terminal and glial cells by a GABA transporter. It can then be recycled into vesicles or metabolized by GABA transaminase.

GABA interacts with two receptor subtypes. $GABA_A$ receptors are ligand-gated ion channels that are permeable to chloride ions. When the pore opens, these negatively charged ions enter the cell and lead to hyperpolarization thus decreasing the excitability of the cell. $GABA_B$ receptors are G_i proteins that lead to hyperpolarization of cells via the opening of certain potassium channels and decreased vesicular release via inhibition of presynaptic voltage-gated calcium channels.

Activation of GABA receptors has a number of clinical uses. Drugs that activate $GABA_A$ receptors are utilized to treat anxiety, insomnia, and epilepsy and to produce anesthesia. Drugs that activate $GABA_B$ receptors are utilized as muscle relaxants. Alcohol also activates the $GABA_A$ receptor resulting in general CNS depression.

Histamine

Histamine plays a major role in the immune response. It also functions as a minor neurotransmitter in the CNS. There are four identified histamine receptors, H_1–H_4, which are all GPCRs. The H_1 receptors are linked to G_q proteins. Inhibition of these receptors produce drowsiness, which is a side effect of antihistamines that cross the blood–brain barrier. These inhibitors of the H_1 receptors have therapeutic utility in over-the-counter sleep aids. In the periphery, H_1 receptor antagonists ("antihistamines") are mainly used in treating allergic responses and in symptoms of colds and flu. They are also used to reduce itch. H_2 receptor antagonists are used to reduce acid secretions in the gut.

Purines

ATP is another small molecule that can act in a neurotransmitter role. It is stored in vesicles and often found in terminals that also contain norepinephrine. ATP modulates the actions of NE through interacting with P_2 purine receptors. P_{2x} receptors are ionotropic, whereas P_{2Y} receptors are metabotropic. Adenosine is produced from the hydrolysis of ATP and can also act as a neuromodulator. It differs from ATP in that it is not stored in vesicles. It acts at distinct metabotropic receptors: P_1 receptors with subtypes A_{1-3}. Effects of caffeine are mediated in part by the inhibition of certain adenosine receptors.

Neuropeptides often serve a neuromodulatory role.

There are a number of neuropeptides that play a role in cellular signaling in the CNS. Many of these are also active in the GI tract and endocrine glands. Neuropeptide precursors are synthesized in the cell body where they are packaged into vesicles by the Golgi apparatus. The precursors are cleaved into smaller peptides by proteases within the vesicle as they are transported to the terminal. Generally, the neuropeptides are co-localized within the same presynaptic terminal with vesicles containing classical neurotransmitters. A strong stimulus can release both the neurotransmitter and neuropeptide into the synaptic cleft. The neuropeptide can modulate the function of the neurotransmitter through activation of its presynaptic and postsynaptic receptors, usually GPCRs. Neuropeptides diffuse out of the synaptic cleft and are degraded by extracellular proteases. Some of the neuropeptide receptors have become primary drug targets.

Opioids

Opioids are a class of compounds that can produce morphine-like effects. These effects are mediated through activation of opioid receptors. The endogenous compounds that activate these receptors are the neuropeptides endorphin, enkephalin, and dynorphin. There are three opioid receptor subtypes: μ, κ, and δ. All three activate G_i proteins resulting in decreased cAMP levels, opening of certain K^+ channels, and closing of certain Ca^{2+} channels. Thus, opioids decrease synaptic transmission by hyperpolarizing the cell and decreasing calcium-mediated vesicular mobilization. This is useful therapeutically in treating pain. However, activation of opioid receptors in other regions of the brain can produce respiratory depression, which is a particular concern with opioid overdose. Opioids can also produce euphoria and addiction through activation of receptors in the "reward pathway" of the brain. The analgesic effects can be separated from the addictive potential through appropriate prescribing practices.

Nonclassical neurotransmitters can be synthesized and released "on demand."

There are a number of other mediators of synaptic transmission that do not undergo classical vesicular storage and release yet are still formed in a calcium-dependent manner. Adenosine, discussed above, is an example of this. Once formed, these nonclassical neurotransmitters diffuse out of the presynaptic terminal to interact with pre- and postsynaptic receptors.

Eicosanoids

There are a number of putative synaptic transmitters that are formed from arachidonic acid. Phospholipase A2 liberates arachidonic acid from the cell membrane. Arachidonic acid can then act as a substrate for a number of enzymes. The presence of the enzyme cyclooxygenase will result in the production of prostaglandins and thromboxane, whereas the presence of the enzyme lipoxygenase will result in the production of leukotrienes. These lipid mediators are not stored in vesicles but will diffuse across the membrane to interact with pre- and postsynaptic receptors. The leukotrienes are implicated in inflammatory responses involved in asthma. The prostanoids play a role in modulating inflammatory responses, pain, and temperature. Certain anti-inflammatory drugs, such as aspirin, act by inhibiting the activation of cyclooxygenase.

Cannabinoids

Endocannabinoids are other lipid mediators derived from arachidonic acid derivatives. These compounds are formed in response to increased intracellular calcium. The fate of arachidonic acid is dependent on the particular enzymes that are present in the cell thus determining which, if any, arachidonic acid derivatives will be formed. Endocannabinoids are not stored in vesicles but instead diffuse across the membrane where they can act pre- or postsynaptically with cannabinoid receptors. These are the same receptors at which delta-9-tetrahydrocannabinol (THC), one of the active constituents of marijuana, acts. There are two receptor subtypes: CB_1 and CB_2. Both are linked to G_i proteins. CB_2 is mostly found in the periphery, whereas CB_1 receptors are located in the CNS. Endocannabinoid signaling may play roles in pain modulation and appetite. Compounds that activate these receptors have been explored therapeutically as analgesics and for the treatment of obesity.

INTEGRATED MEDICAL SCIENCES

Parkinson's Disease

Parkinson's disease is neurodegenerative disease characterized by the loss of dopaminergic neurons in the brain, particularly in the basal ganglia, and subsequent motor deficiencies. Individuals experience slowed movement, rigidity, tremor, and postural instability. Nonmotor symptoms, such as depression and constipation, may also accompany the disease.

Because of the loss of dopamine-synthesizing neurons in the brain, a logical treatment strategy is to increase dopamine in the brain. However, there are limitations to this approach. If dopamine is administered to a patient by an oral or other systemic route, it will not reach the central nervous system because of two mechanisms. Dopamine, along with other catecholamines, are degraded by COMT and MAO. The liver contains high concentrations of both of these enzymes. As dopamine circulates in the bloodstream and passes through the liver, much of it will be degraded. Any dopamine that remains, will not get into the brain because dopamine does not readily cross the BBB.

In order to increase the concentration of dopamine in the brain, a different strategy is employed. In the body's synthesis of dopamine, tyrosine is converted to dopa, which is then converted to dopamine by the enzyme dopa decarboxylase. One of the stereoisomers of dopa, levodopa or L-dopa, is transported across the BBB and thus has access to the brain where it can be converted to dopamine. Administration of levodopa is a therapeutic strategy in the treatment of Parkinson's disease. In order to prevent levodopa's conversion peripherally, it is administered with carbidopa, an inhibitor of dopa decarboxylase that will not cross the BBB. Another concern with levodopa is that, like dopamine, it is also a substrate for COMT and MAO and is degraded as it passes through the liver. Thus, entacapone, which inhibits COMT, can be administered along with levodopa/carbidopa to prevent the metabolism of levodopa. Inhibitors of monoamine oxidase such as rasagiline can also decrease metabolism of levodopa. With the addition of these drugs, levodopa is able to access the brain where it is then converted to dopamine and serves to increase the low levels of dopamine resulting from loss of dopaminergic neurons. Other strategies in treating Parkinson's disease include using drugs that directly activate dopamine receptors, such as bromocriptine and pramipexole. ■

Neuromuscular Physiology

Chapter Summary

- The nervous system is divided into the central, consisting of the brain and spinal cord, and peripheral, consisting of ganglia and nerves, nervous systems.
- The blood–brain barrier limits access of molecules into the CNS and is formed by a layer of capillary endothelial cells connected by tight junctions and surrounded by astrocytes cells.
- The primary cells in the nervous system are neurons and glia. Glial cells function as neuronal support cells. Schwann cells and oligodendrocytes insulate axons and increase the speed of conduction. Microglia serve an immune function. Astrocytes help form the blood–brain barrier, provide metabolic support to neurons, regulate blood flow, are involved in reuptake and recycling of some synaptic transmitters, and help regulate ionic concentrations in the extracellular fluid.
- Neurons function in storage, communication and integration of information. This is accomplished through dendrites, which receive and transmit information to the soma, and axons, which propagate action potentials and communicate via the release of synaptic transmitters.
- The neuronal cytoskeleton is involved in transporting materials both from the soma to the neuronal processes (anterograde transport) and from the processes to the soma (retrograde transport).
- Ion channels can be ligand-gated or voltage-gated. They are selective for particular ions. Once opened, the flow of ions into or out of the cell is determined by both electrical and concentration gradients for that ion and can be calculated by the Nernst equation.
- When ion channels open and allow the influx or efflux of ions, there is an excitatory or inhibitory postsynaptic potential. These summate both temporally and spatially to influence the net effect on membrane potential.
- Once the membrane potential depolarizes to a threshold potential, a cascade of opening of voltage-gated sodium channels occurs resulting in further depolarization and the production of an action potential.
- The return of the membrane potential to the resting membrane potential is accomplished by the opening of potassium channels and the inactivation of sodium channels.
- Action potentials are self-propagating, unidirectional, and can travel the length of the axon.
- The conduction velocity of action potentials is influenced by the diameter of the nerve fibers and their myelination. Fibers that are unmyelinated and small, C fibers, have the slowest conduction velocity and carry information about pain and temperature. Large myelinated fibers, Aα fibers, have the fastest conduction velocity and carry information about proprioception and motor function.
- The synapse consists of specialized segments of neurons used for chemical communication between cells consisting of the presynaptic terminal, the synaptic cleft, and the postsynaptic terminals of dendrites.
- Most neurotransmitters and neuropeptides are contained in vesicles that are located within the presynaptic terminal.
- The arrival of an action potential at the presynaptic terminal results in the opening of voltage-gated calcium channels, which results in calcium-dependent mobilization of vesicles to the membrane where the synaptic transmitter is released into the synaptic cleft to interact with receptors.
- Some synaptic transmitters are synthesized "on demand" in a calcium-dependent manner but are not stored in vesicles.
- Neurotransmitters can interact with receptors that are located on presynaptic or postsynaptic terminals. Generally, interactions with presynaptic receptors serve as negative modulators of release. Most neuronal receptors can be classified as ionotropic or metabotropic.

Chapter Review Questions

1. How would hypokalemia (low extracellular potassium levels) affect neuronal excitability?

   A. Decreased excitability due to a decreased resting membrane potential
   B. Decreased excitability due to an increased resting membrane potential
   C. Increased excitability due to a decreased resting membrane potential
   D. Increased excitability due to an increased resting membrane potential
   E. It would not affect neuronal excitability

The correct answer is A. Neuronal resting membrane potentials are near the potassium equilibrium potential and there is a much higher concentration of K^+ inside the cell compared with outside the cell. If extracellular K^+ levels decrease, this will result in K^+ leaving the cell to re-establish an equilibrium potential and balance electrochemical forces. When K^+ leaves the cell, the resting membrane potential will decrease making it harder to produce an action potential.

2. Metyrosine is a drug that will inhibit the conversion of tyrosine to dopamine. What other effects will it have?

   A. Activation of the reward pathway
   B. Decreased production of epinephrine
   C. Decreased production of serotonin
   D. Development of schizophrenia
   E. Increased metabolism of norepinephrine

The correct answer is B. Tyrosine is the rate-limiting step in the synthesis of all catecholamines (dopamine, norepinephrine, and epinephrine). Inhibiting the conversion of tyrosine to dopamine will decrease all catecholamines. C is incorrect because, while it is a monoamine, serotonin is not a catecholamine. A and D are incorrect because dopamine is involved in both the reward pathway and schizophrenia. E is incorrect because this will not affect either COMT or MAO, which are the enzymes involved in the degradation of norepinephrine.

3. Activation of opioid receptors leads to the inhibition of voltage-gated calcium channels. What effect will the inhibition of voltage-gated calcium channels have on cellular signaling?

   A. Decrease postsynaptic receptor activation
   B. Decrease the postsynaptic membrane potential
   C. Increase neurotransmitter release
   D. Increase the postsynaptic membrane potential
   E. Inhibit the generation of action potentials

The correct answer is A. Voltage-gated calcium channels are involved in mobilization of vesicles to the plasma membrane enabling them to dock and release neurotransmitters. By inhibiting these calcium channels, there will be less neurotransmitter available to activate postsynaptic receptors.

4. Increasing the refractory period would have what effect on action potentials?

   A. Decrease the amplitude of action potentials
   B. Decrease the frequency of action potentials
   C. Increase the amplitude of action potentials
   D. Increase the frequency of action potentials
   E. Increase Na^+ conductance

The correct answer is B. The refractory period is the time when sodium channels are inactivated (absolute refractory) or converting back to the rested state (relative refractory). During the absolute refractory period, the sodium channels cannot be activated and an action potential cannot be generated. During the relative refractory period, a larger stimulus is required to generate an action potential. The amplitude of the action potential does not change but there will be a decrease in the frequency of action potential production.

Clinical Application Exercises 3.1

EPISODIC ATAXIA

A previously healthy 30-year-old man notices difficulty completing his usual exercise routine over the last 2 weeks because of excessive fatigue during the workout. He consults his primary physician, and the physical examination reveals lymphadenopathy (enlarged lymph nodes) in the neck, axilla, and groin. Further evaluation by computed tomographic scanning of the chest and abdomen shows enlargement of thoracic and abdominal lymph nodes. He is referred to an oncologist, and a diagnosis of lymphoma is made. He is started on chemotherapy consisting of weekly intravenous infusions of the combination chemotherapy treatment known as *MOPP* (mechlorethamine, Oncovin, procarbazine, and prednisone). He experienced the expected nausea and general fatigue typical of this type treatment but tolerated these side effects

relatively well. A few days after the third round of treatment he begins to notice numbness in the toes of both feet. He also notices a burning, stinging sensation in the same areas. These abnormal sensations persist and over the next 2 weeks migrate from the toes up to the level of the ankles. At his next follow-up appointment with the oncologist, the previously enlarged lymph nodes have returned to normal size. Examination of the neurological function of his feet and legs shows that he cannot feel the touch of a ball of cotton or a vibrating tuning fork on his toes. The oncologist tells him that the foot symptoms are due to "transportation troubles in the nerves" from the chemotherapy, and that the symptoms will get better over the next few months.

QUESTIONS

1. Which of the chemotherapy medications is the most likely cause of the abnormal sensations?

2. Why did these abnormal sensations start in the toes?

3. What is the pathogenesis (mechanism) of the nerve dysfunction?

4. Why did the symptoms begin several weeks after the treatment began?

ANSWERS

1. Vincristine (Oncovin) is the medication in the combination of chemotherapy drugs that causes the nerve dysfunction. This medication is in the vinca alkaloids, which are named for the family of plants they are derived from.

2. The patient's symptoms fit in the category of peripheral neuropathy. This group of disorders can be caused by abnormal function of the nerve cell body, the axon, or the myelin sheath. Abnormalities of axon function are the most common cause of peripheral neuropathy. In this type of nerve disease,

the longest axons show the effects of the abnormal function first; therefore, the abnormalities appear first in the toes and distal feet. If the condition worsens, the abnormal sensations spread from the toes to the mid-foot and potentially up to the ankle or mid-shin. The fingers may also show similar abnormalities. Neuropathies like our patient's are said to be "length dependent," because the longest axons show the abnormalities first.

3. Vincristine (Oncovin) impairs transport along the axons by interfering with the assembly of tubulin into microtubules. Microtubules provide the framework for fast axoplasmic transport from the nucleus to the nerve terminal and also for retrograde transport from the nerve terminal back to the nucleus.

4. The nerve dysfunction effects of the vincristine (Oncovin) are cumulative and usually begin only after several cycles of treatment. If the medication is able to be discontinued, nerve function tends to improve over subsequent months as axoplasmic transport is restored and the axons become able to maintain their structure normally.

thePoint® *Visit* http://thepoint.lww.com/rhoades5e *for additional chapter review Q&A, Clinical Application Exercises, animations, and more!*

Active Learning Objectives

Upon mastering the material in this chapter, you should be able to:

- Explain how the many different forms of environmental energy are transformed by sensory receptors into the same type of electrical signal in sensory nerves.
- Explain why sensory receptors have specificity to sense one type of environmental energy over another.
- Explain why sensory receptors that respond to a selective type of environmental energy can respond to a different type, yet the sensation produced is the same as if the receptor responded to the energy for which it was designed to detect.
- Postulate the value of adaptation of sensory receptors in relation to the selective type of sensation the receptor is designed to detect.
- Explain why action potential frequency is the mode of information transmission in sensory systems and how changes in that frequency are related to graded electrical potential changes in sensory receptors.

- Explain how the processes of adaptation and accommodation permit the processing of a wide-range of stimulus intensities.
- Distinguish between rod and cone vision and point out the special features of each type.
- Describe the organization of the retina, relating each layer to its specific and integrated function in the process of vision.
- Explain, in terms of biochemical events, the process of light and dark adaptation in the retina.
- Describe the processing of sound by different parts of the auditory system, the mechanism of auditory transduction and how the frequency of sound is represented along the basilar membrane.
- Explain the roles of the different components of the vestibular system in detecting rotational and linear acceleration.
- Explain the sensory processes in the related senses of smell and taste and the receptors and mechanisms that underlie chemoreception.

▶ SENSORY SYSTEMS

The survival of any organism depends on having adequate information about the external environment as well as information about the state of internal bodily processes and functions. Information about our external and internal environments is gathered by the sensory branch of the nervous system. This chapter discusses the functions of the sensory receptors, the organs that permit us to gather this information. The discussion emphasizes somatic sensations, that is, those dealing with the external aspect of the body. This chapter does not specifically address the visceral sensations that come from internal organs.

Sensory systems transform physical and chemical signals from the external and internal environments into information in the form of nerve action potentials.

Events in our external and internal worlds must first be translated into signals that our nervous systems can process. Despite the wide range of types of information to be sensed and acted on, a small set of common principles underlies all sensory processes.

Furthermore, although the human body contains a very large number of different sensory receptors, these receptors have many functional features in common. The process of sensation essentially involves sampling small amounts of energy from the environment by sensory receptors and using it to control the generation of trains of action potentials in sensory nerves (Fig. 4.1). The pattern of sensory action potentials, along with the specific nature of the sensory receptor and its neural pathways in the brain, provides an internal representation of a specific component of the external world.

The wide variety of specialized sensory functions is a result of structural and physiologic adaptations that fit a particular receptor for its role in the organism. Finally, the process of sensation is a portion of the more complex process of perception, in which sensory information is integrated with previously learned information and other sensory inputs, enabling us to make judgments about the quality, intensity, and relevance of what is being sensed.

Sensory Stimuli

A factor in the environment that produces an effective response in a sensory receptor is called a **stimulus**. Typical stimuli include electromagnetic quantities, such as radiant heat or light; mechanical quantities, such as pressure, sound waves, or vibrations; and chemical qualities, such as acidity and molecular shape and size. Regardless of the sensation experienced, all sensory systems transmit information to the brain with the four common attributes of intensity, modality (e.g., temperature, pressure, and vision), location, and duration.

A fundamental property of sensory receptors is their ability to respond to different intensities of stimulation with a representative output. *Intensity* is a measure of the energy content available to interact with a sensory receptor. Most individuals are familiar with sensory intensity in phenomena such as the brightness of light, the "loudness" of sound, or the force of physical contact of an object with the body.

Sensory receptors are capable of detecting the modality of a stimulus. **Sensory modality** is the term used to describe the *kind* of sensation present. Common sensory modalities include taste, smell, touch, sight, and hearing (the traditional five senses) as well as more complex sensations,

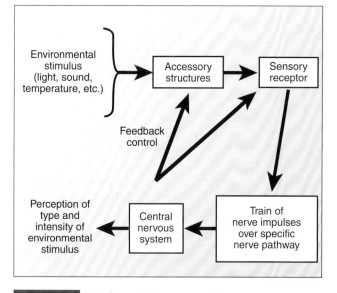

Figure 4.1 **A basic model for the translation of an environmental stimulus into a perception.** Although the details vary with each type of sensory modality, the overall process is similar.

such as slipperiness or wetness. Sensory modality is often used to classify receptors. For example, **photoreceptors** that serve a visual function sense light. **Chemoreceptors** detect chemical signals and serve the senses of taste and smell. **Mechanoreceptors** sense physical deformation and serve such sensory modalities as touch, hearing, and the amount of stress in a tendon or muscle. Heat, or its relative lack, is detected by **thermoreceptors**. Many sensory modalities are a combination of simpler sensations. For example, the sensation of wetness is actually composed of sensations of pressure and temperature. You can demonstrate this phenomenon yourself by placing your hand in a plastic bag and then immersing it in cold water. Even though your hand will remain dry inside the bag, the perception you sense will be that of wetness.

Although sensory receptors are often classified by the modality to which they respond, other sensory receptors are classified instead by their "vantage point" in the body. For example, exteroceptors detect stimuli from outside the body, interoceptors detect internal stimuli, and proprioceptors (*proprio* is Latin for "one's own") provide information about the positions of joints, muscle activity, and the orientation of the body in space. Nociceptors (pain receptors) detect noxious agents, both internally and externally.

It is often difficult to communicate a precise definition of a sensory modality because of the subjective perception or affect that accompanies it. This property has to do with the psychologic feeling attached to the stimulus. For example, the sensation of cold or touch can be pleasurable at certain intensities but rise to an impression of discomfort or extreme pain at very high intensities. Stimulus affect, however, is more complex than a combination of intensity with modality. Previous experience and learning play a role in determining the affect of a sensory perception. The affect of identical visual stimuli, for example, can provoke pleasure, anger, fear, or terror depending on the experiences of those perceiving the stimulus and the memories associated with it.

Sensory receptors

Most sensory receptors are optimized to respond preferentially to a single kind of environmental stimulus. The usual stimulus for the eye is light and that for the ear is sound. This specificity is a result of several features that match a receptor to its preferred stimulus. The usual and appropriate stimulus for a type of sensory receptor is called its **adequate stimulus**. This is the stimulus for which the receptor has the lowest **threshold**, that is, the lowest stimulus intensity that can be reliably detected. Threshold, however, can be difficult to quantify because it can vary over time, and it can be altered by either the presence of interfering stimuli or the action of accessory structures. These structures are not considered part of the sensory receptor itself. However, in many cases, such as the lens of the eye or the structures of the outer and middle ear, these structures either enhance the specific sensitivity of the receptor or exclude unwanted stimuli. Often these accessory structures are part of a control system that adjusts the sensitivity according to the information being received (see Fig. 4.1).

Most receptors will respond to additional types of stimuli other than the adequate stimulus. However, the threshold for the inappropriate stimuli is much higher. For example, applying pressure to the eye will cause one to "see lights" although light itself is the energy type for which the sensory receptors of the eye have the lowest threshold. Finally, almost all receptors can be stimulated electrically to produce sensations that mimic the one usually associated with that receptor.

Another feature of sensory system specificity resides in the central nervous system (CNS). The CNS pathway over which sensory information travels also plays a role in determining the nature of the perception. Information arriving by way of the optic nerve, for example, is always perceived as light and never as sound. This is known as the concept of the labeled line. Details of all neural sensory pathways are beyond the scope of this text.

Sensory transduction changes environmental energy into sensory nerve action potentials.

The key physiologic function of a sensory receptor is to translate nonelectrical forms of environmental energy into electrical events that can be transmitted and processed by the nervous system. These electrical events are nerve action potentials, which are the fundamental units of information in the nervous system. A device that translates one form of energy signal into another is called a **transducer**. Sensory receptors in the body, therefore, can be thought of as biologic transducers. A typical sequence of electrical events in the sensory transduction process is shown in Figure 4.2.

Generator potentials

Figure 4.2 depicts a typical type of sensory receptor system. This example is a representation of a mechanoreceptor, that is, a receptor that translates energy in the form of a physical stress or strain (i.e., pressure, displacement, stretch, etc.) into an electrical signal that will eventually produce action potentials in sensory nerves. In the example in Figure 4.2, the tip of the receptor has been deflected (strained) and held

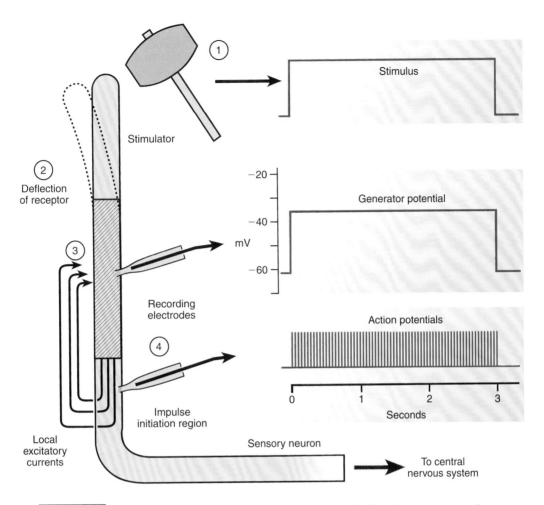

Figure 4.2 **The relation between an applied stimulus and the production of sensory nerve action potentials.** (See text for details.)

constant (1, 2). This deflection deforms the cell membrane of the receptor, causing a portion of it to become more permeable to positive ions (*shaded region*, 3). The resulting influx of positive charge (mostly from influx of sodium ions) then leads to a localized depolarization, called the **generator potential**. The generator potential creates a current through the axon membrane called a **local excitatory current**. It provides the link between the formation of the generator potential and the excitation of the nerve fiber membrane.

The production of the generator potential is of critical importance in the transduction process because it is the step in which information related to stimulus intensity and duration is transduced. The strength (intensity) of the stimulus applied (in Fig. 4.2, the amount of deflection) determines the size of the generator potential depolarization. Varying the intensity of the stimulation will correspondingly vary the generator potential, although the changes are not always directly proportional to the intensity. This is called a **graded response**, in contrast to the all-or-none response of an action potential, and causes a similar gradation of the strength of the local excitatory currents.

As generator potential depolarizing currents spread down the neuron, they reach a region of the receptor membrane called the **impulse initiation region** (4). This region is the location of the formation of action potentials in the neuron

and constitutes the next important link in the sensory process. In this region, action potentials are produced at a frequency related to the strength of the current caused by the generator potential (and, thus, related to the original stimulus intensity). This region is, therefore, sometimes called the *coding region*, because here the generator potential and flowing current caused by the initial stimulus to the receptor are translated into a train of action potentials of a frequency proportional (but not always in a linear fashion) to the size of the generator potential. The train of action potentials in the sensory nerve fiber is the electrical signal that is sent to the CNS. In complex sensory organs that contain a great many individual receptors, the generator potential may arise from several sources within the organ. Thus, the train of action potentials formed results from multiple inputs to the sensory receptor. In these instances, the generator potential is often called a **receptor potential**.

Sensory receptor subspecialization

The mechanoreceptor depicted in Figure 4.2 is an example of a sensory receptor subtype called a *long receptor*. In such receptors, the formation of the receptor and action potentials occurs in the same cell, the neuron. This type of sensory receptor-to-action potential arrangement is best suited for transmission of sensory information over long distances. In these cases, the initial physical stimulus creates a local receptor potential, which triggers action potentials that are then transmitted all the way to the CNS. The sense of touch and temperature at the tips of our fingers or toes are examples of sensory transduction by long receptors.

In contrast, other types of sensory transduction in the body occur through *short receptors*. Short receptors are highly specialized cells that, like long receptors, also convert a nonelectrical form of energy into a receptor potential within them. However, in short receptors, the receptor generator potential and neuron action potentials occur in two different cells. In short receptors, the local potential event is coupled to the release of neurotransmitter stored in the receptor cell. This transmitter is then released onto synapses of a postsynaptic primary afferent neuron in which it produces another localized generator potential. It is this second generator potential that triggers trains of action potentials in the afferent neuron that are then sent to the CNS. Short receptors produce either depolarizing or hyperpolarizing generator potentials. In some cases, the same receptor can produce depolarizing and hyperpolarizing generator potentials depending on the stimulus they receive. Positive generator potentials increase the rate of transmitter release, whereas negative potentials decrease this rate. The photoreceptors of the eye and the hair receptors of the cochlea of the inner ear are examples of short sensory receptors (see below).

Sensory nerve impulse frequency is modulated by the magnitude and duration of the generator potential.

Figure 4.3 shows how the magnitude and duration of a generator potential affect subsequent action potential formation in a sensory nerve. This example is typical of what might be seen in a long receptor-type sensory nerve. The threshold (*black line* in the figure) is a critical level of depolarization. Membrane potential changes below this level are caused by the local excitatory currents and vary in proportion to them,

whereas the membrane activity above the threshold level consists of locally produced **action potentials**. The lower trace shows a series of different stimuli applied to the receptor, and the upper trace shows the resulting electrical events in the impulse initiation region. No stimulus is given at A, and the membrane voltage is at the resting potential. At B, a small stimulus is applied, producing a generator potential too small to bring the impulse initiation region to threshold; no action potential activity results, and the stimulus causing this level of potential generation would, therefore, not be sensed at all. In contrast, at C, a brief stimulus of greater intensity is given, and the resulting generator potential displacement is of sufficient amplitude to trigger a single action potential. As in all excitable and all-or-none nerve membranes, the action potential is immediately followed by repolarization, often to a level that transiently hyperpolarizes the membrane potential because of temporarily high potassium conductance. Because the brief stimulus has been removed by this time, no further action potentials are produced. A longer stimulus of the same intensity (D) produces repetitive action potentials because, as the membrane repolarizes from the action potential, local excitatory currents are still flowing. They bring the repolarized membrane to threshold at a rate proportional to their strength. During this time interval, the fast sodium channels of the membrane are being reset, and another action potential is triggered as soon as the membrane potential reaches threshold. As long as the stimulus is maintained, this process will repeat itself at a rate determined by the stimulus intensity. If the intensity of the stimulus is increased (E), the local excitatory currents will be stronger and the threshold will be reached more rapidly. This will result in a reduction of the time between each action potential and, as a consequence, a higher action potential frequency. This change in action potential frequency is critical in communicating the intensity of the stimulus to the CNS.

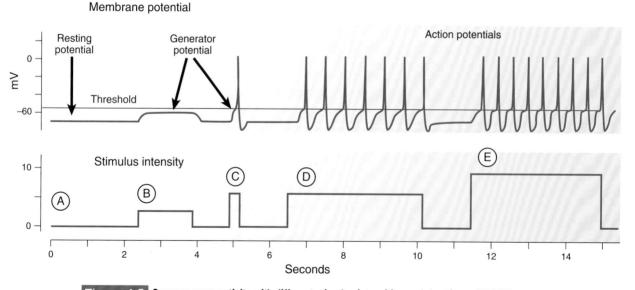

Figure 4.3 **Sensory nerve activity with different stimulus intensities and durations. (A)** With no stimulus, the membrane is at rest. **(B)** A subthreshold stimulus produces a generator potential too small to cause membrane excitation. **(C)** A brief, but intense, stimulus can cause a single action potential. **(D)** Maintaining this stimulus leads to a train of action potentials. **(E)** Increasing the stimulus intensity leads to an increase in the action potential firing rate.

Sustained sensory receptor stimulation can result in diminished action potential generation over time.

In the presence of a constant stimulus, only some sensory receptors maintain the magnitude of their initial generator potential. In most sensory receptors, the magnitude of the generator potential decays over time, even when the original sensory stimulus does not change. This phenomenon is called *adaptation*. Figure 4.4A shows the output from a receptor in which there is no adaptation. As long as the stimulus is maintained, there is a steady rate of action potential firing. Figure 4.4B shows slow adaptation; the generator potential amplitude declines slowly, and the interval between the action potentials increases correspondingly. Figure 4.4C demonstrates rapid adaptation; the action potential frequency falls rapidly and then maintains a constant slow rate that does not show further adaptation. Responses in which there is little or no adaptation are called *tonic*, whereas those in which significant adaptation occurs are called *phasic*. In some cases, tonic receptors may be called *intensity receptors* and phasic receptors may be called *velocity receptors*. Many receptors—*muscle spindles*, for example—show a combination of responses. On application of a stimulus, a rapidly adapting phasic response is followed by a steady tonic response. Both of these responses may be graded by the intensity of the stimulus. As a receptor adapts, the sensory input to the CNS is reduced, and the sensation is perceived as less intense.

The phenomenon of adaptation is important in preventing "sensory overload" and allows less important or unchanging environmental stimuli to be partially ignored. When a change occurs, however, the phasic response will occur again, and the sensory input will become temporarily more noticeable. Rapidly adapting receptors are also important in sensory systems that must sense the rate of change of a stimulus, especially when the intensity of the stimulus can vary over a range that would overload a tonic receptor.

Receptor adaptation can occur at several places in the transduction process. As mentioned above, adaptation of the generator potential itself can produce adaptation of the overall sensory response. In some cases, however, the receptor's sensitivity is changed by the action of accessory structures, as in the constriction of the pupil of the eye in the presence of bright light (an example of feedback-controlled adaptation). In the sensory cells of the eye, light-controlled changes in the amounts of the visual pigments also can change the basic sensitivity of the receptors and produce adaptation. Finally, the phenomenon of **accommodation** in the impulse initiation region of the sensory nerve fiber can slow the rate of action potential production even though the generator potential may show no change. Accommodation refers to a gradual increase in threshold caused by prolonged nerve depolarization and results from the inactivation of sodium channels.

Perception of sensory information requires encoding and decoding electrical signals sent to the central nervous system.

Environmental stimuli that have been partially processed by a sensory receptor must be conveyed to the CNS in such a way that the complete range of the intensity of the stimulus is preserved. After the acquisition of sensory stimuli, the process of perception involves the subsequent encoding and transmission of the sensory signal to the CNS. Further processing, or decoding, yields biologically useful information.

Compression

The first step in the encoding process is **compression**. Even when the receptor sensitivity is modified by accessory structures and adaptation, the range of input intensities is large, as shown in Figure 4.5. At the left is a 100-fold range in the intensity of a stimulus. At the right is an intensity scale that results from events in the sensory receptor. In most receptors, the magnitude of the generator potential is not exactly proportional to the stimulus intensity; it increases less and less as the stimulus intensity increases. Furthermore, the frequency of the action potentials produced in the impulse initiation region

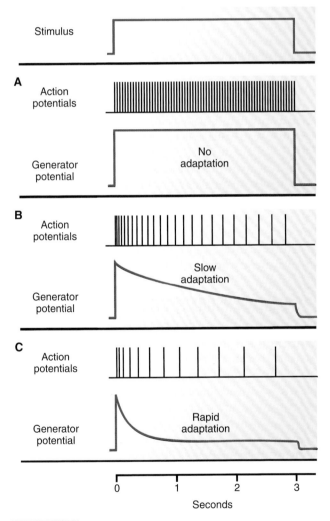

Figure 4.4 **Adaptation.** Adaptation in a sensory receptor is often related to a decline in the generator potential with time. **(A)** The generator potential is maintained without decline, and the action potential frequency remains constant. **(B)** A slow decline in the generator potential is associated with slow adaptation. **(C)** In a rapidly adapting receptor, the generator potential declines rapidly.

Neuromuscular Physiology

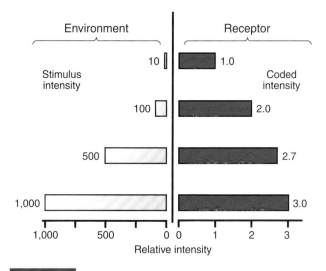

Figure 4.5 **Compression in sensory process.** By a variety of means, a wide range of input intensities is coded into a much narrower range of responses that can be represented by variations in action potential frequency.

is also not proportional to the strength of the local excitatory currents. Recall from Chapter 3 that there is an upper limit to the number of action potentials that can be generated in a neuron per second because each action potential has a finite refractory period. All these factors are responsible for the process of compression. Because of compression, changes in the intensity of an already large stimulus cause a smaller change in action potential frequency than the same change would cause if the change occurred from an initially small stimulus. As a result, the hundredfold variation in the stimulus is compressed into a threefold range after the receptor has processed the stimulus. Some information is necessarily lost in this process, although integrative processes in the CNS can restore the information or compensate for its absence. Physiologic evidence for compression is based on the observed nonlinear (logarithmic or power function) relation between the actual intensity of a stimulus and its perceived intensity.

Transmission

Sensory information initiated at the receptor level must be transmitted to the CNS in order to be perceived. The encoding processes in the receptors provide some basis for this transfer by producing a series of action potentials related to the stimulus intensity. A special process is necessary for the complete transfer, however, because of the nature of the conduction of action potentials. Recall that action potentials are transmitted down nerve axons without decrement; their duration and amplitude do not change during conduction and, therefore, the parameters of electrical potential amplitude and duration are lost as a means of transmitting different forms of information to the CNS. It follows, then, that the only information that can be conveyed by a single action potential is its presence or absence.

Relationships between and among action potentials, however, can convey large amounts of information, and this is the system found in the sensory transmission process. To review, in physiologic sensory systems, an input signal

provided by some physical quantity is continuously measured and converted into an electrical signal (the generator potential), whose amplitude is proportional to the input signal. This signal controls the frequency of action potential generation in the impulse initiation region of a sensory nerve fiber. Because action potentials are of a constant height and duration, the amplitude information of the original input signal is now in the form of a frequency of action potential generation. These trains of action potentials are sent along a nerve pathway to some distant point in the CNS, where they produce an electrical voltage in postsynaptic neurons that is proportional to the frequency of the arriving pulses. This voltage is a replica of the original input voltage. Further processing can produce a graphic record of the input data, which is accompanied by processing and interpretation in the CNS.

Perception

The interpretation of encoded and transmitted information into a perception requires several other factors. For instance, the interpretation of sensory input by the CNS depends on the neural pathway it takes to the brain. For this reason, for example, all information arriving on the optic nerves is interpreted as light, even though the signal may have arisen as a result of pressure applied to the eyeball. The localization of a cutaneous sensation to a particular part of the body also depends on the particular pathway it takes to the CNS. Often a sensation (usually pain) arising in a visceral structure (e.g., heart, gallbladder) is perceived as coming from a portion of the body surface, because developmentally related nerve fibers come from these anatomically different regions and converge on the same spinal neurons. Such a sensation is called referred pain.

The remainder of this chapter deals with specific sensory receptors, including those of both the somatosensory system (i.e., touch, temperature, and pain) and the special senses (i.e., sight, hearing, taste, and smell).

▶ SOMATOSENSORY SYSTEM

The somatosensory system is a diverse sensory system comprising receptors and processing centers that produce sensory modalities, such as touch (tactile sensation), proprioception (body position), temperature, and pain (nociception). These sensory receptors cover the skin and epithelia, skeletal muscles, bones and joints, and internal organs. The system responds to a number of diverse stimuli using the following receptors: mechanoreceptors (movement/pressure), chemoreceptors, thermoreceptors, and nociceptors. Signals from these receptors pass via sensory nerves through tracts in the spinal cord and into the brain. Neural information is processed primarily in the somatosensory area in the parietal lobe of the cerebral cortex (see Chapter 7 for details).

Skin receptors provide sensory information from the body surface.

The skin has a rich supply of sensory receptors that provide cutaneous sensations, such as touch (light and deep pressure), temperature (warmth and coldness), and pain. More complicated modalities include itch, tickle, wetness, etc. Cutaneous receptors can be mapped over the skin by using highly localized stimuli of heat, cold, pressure, or vibration. Some areas require a high degree of spatial localization (e.g., fingertips, lips) and, accordingly, have a high density of specific sensory receptors. These areas are correspondingly well represented in the somatosensory area of the cerebral cortex (see Chapter 7).

Touch

Several types of sensory receptors are involved in the sensation of touch (Fig. 4.6). Three of these receptors (Merkel disk, Meissner corpuscles, and Pacinian corpuscles) are distributed on various areas of the skin but are concentrated in regions void of hair (e.g., palm of the hand). The nerve endings of these receptors respond to pressure and texture; therefore, they are classified as **mechanoreceptors**. All mechanoreceptor types are stimulated by deformation of the nerve ending. When these receptors are deformed, either due to pressure or release of pressure, they cause receptor potentials to be generated by opening pressure-sensitive cation ion channels in the axon membrane, which allows an influx of positive ions resulting in the eventual formation of an action potential in the sensory neuron.

The *Merkel disks* are receptors (located in the lowest layer of the epidermis) that exhibit slow adaptation and also respond to steady pressure. These receptors provide information about an object's qualities like edges or its curves.

The *Meissner corpuscle* is a cutaneous touch receptor that mediates the sensation of light touch. These receptors are distributed on various areas of the skin but are concentrated in areas sensitive to light touch, such as fingertips, lips, and nipples. The Meissner corpuscle is a capsule surrounding the ending of a sensory nerve (or nerves) that wind

between stacks of flattened Schwann cells in the capsule interior. These Schwann "cushions" dissipate deformation of the nerve endings caused by sustained points of pressure on the skin. This makes these receptors rapidly adaptive.

The *Pacinian corpuscle* is a very highly adaptive receptor that is responsible for sensitivity to pressure. These receptors comprise nerve endings surrounded by multiple gel-like envelopes within capsules that act collectively as a mechanical filter. When the capsule is deformed, for example, by a point of pressure, the underlying nerve ending is deformed and generates an electrical potential. However, the multiple gel envelopes almost immediately equalize pressure surrounding the nerve, thus removing any local deformation. Consequently, the local receptor potential ceases to be generated. The nerve ending only deforms again to generate an electrical potential when the initial deforming pressure is suddenly released. Thus, the Pacinian corpuscle best detects the beginning and ending of a rapidly changing local pressure on the skin. This makes these receptors especially suited for detecting fast-changing stimuli such as vibrations.

Thermoreceptors are sensory nerve endings that code for absolute and relative temperatures.

Humans can discriminate a wide range of innocuous and noxious skin temperatures that are sensed by temperature receptors (thermoreceptors). Thermoreceptors belong to the transient receptor potential (TRP) family of cation channels with each thermally sensitive TRP responsive to a different temperature range. Cooling or warming the skin opens these TRP channels allowing inward flow of cations to depolarize the cell. Information about the thermal stimulus is relayed to the CNS by two groups of low conduction velocity temperature sensitive fibers that are either thinly myelinated fibers (cold) or nonmyelinated (warm). The density of temperature

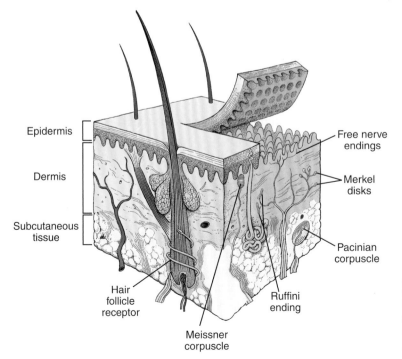

Figure 4.6 **Tactile receptors in the skin.** Cutaneous receptors providing a sense of touch, pressure, stretch, and vibration are the Merkel disk, Meissner corpuscle, Pacinian corpuscle, Ruffini corpuscle, and the hair follicle receptor.

receptors differs at various places on the body surface. They are present in much lower numbers than cutaneous mechanoreceptors, and there are many more cold receptors than warm receptors. The adequate stimulus for a warm thermoreceptor is warming, which results in a concomitant increase in rate of firing of its action potentials. Conversely, cooling results in a decrease in the action potential discharge rate of the warm receptor. Cold-sensitive thermoreceptors give rise to the sensation of cooling or coldness. Their rate of firing increases with cooling and decreases during warming.

The perception of temperature stimuli is closely related to the properties of the receptors. The phasic component of the response is apparent in our adaptation to sudden immersion in, for example, a warm bath. The sensation of warmth, apparent at first, soon fades away, and a less intense impression of the steady temperature may remain. Moving to somewhat cooler water produces an immediate sensation of cold that soon fades away. Over an intermediate temperature range (the "comfort zone"), there is no appreciable temperature sensation. This range is approximately 30°C to 36°C for a small area of skin; the range is narrower when the whole body is exposed. Outside this range, steady temperature sensation depends on the ambient (skin) temperature. Skin temperatures in the range of 20°C up to about 45°C are sensed as innocuous. Skin temperatures >45°C or <17°C elicit warm and cold pain respectively, but this sensation arises from thermosensitive TRP pain receptors, not innocuous thermal receptors. Cold pain can be activated under certain abnormal conditions, such as with peripheral neuropathy or from chemotherapy. Severely diabetic patients who suffer from peripheral neuropathy and patients who undergo chemotherapy can exhibit a cold-induced pain response when they touch a cool object. Both result from damage to the receptors. At very high skin temperatures (>45°C), there is a sensation of *paradoxical cold*, caused by nonspecific activation of a part of the innocuous cold receptor population.

Temperature perception is subject to considerable processing by higher centers. Although the perceived sensations reflect the activity of specific receptors, the phasic component of temperature perception may take several minutes to be completed, whereas the adaptation of the receptors is complete within seconds.

Nociceptors are free nerve endings that trigger the pain sensation in the brain.

Pain is sensed by a population of specific receptors called nociceptors. These receptors can detect mechanical, thermal, or chemical changes and are found in the skin, on internal organ surfaces, and on joint surfaces. The transduction mechanism for nociception is not completely understood, but TRP channels have been implicated for some types of nociception such as temperature-evoked pain responses. Nociceptors typically have a high threshold for mechanical, chemical, and thermal stimuli (or a combination) of intensity sufficient to cause tissue destruction, but these receptors can be sensitized by substances associated with tissue damage and inflammation so that when a previously innocuous stimulus is applied to a damaged area of the skin it is now perceived as painful.

All pain receptors are free nerve endings, and the concentration of these receptors varies throughout the body, mostly on the skin and less so in the internal organs. In the skin, these are the free endings of thin myelinated and nonmyelinated fibers with low conduction velocities. The skin has many more points at which pain can be elicited than it has mechanically or thermally sensitive sites. Because of the high threshold of pain receptors (compared with that of other cutaneous receptors), we are usually unaware of their existence.

Superficial pain may often have two components: an immediate, sharp, and highly localizable *initial pain* and, after a latency of about 1 second, a longer lasting and more diffuse *delayed pain*. These two submodalities appear to be mediated by different nerve fiber endings and are served by unique pathways in the CNS. In addition to their normally high thresholds, both cutaneous and deep pain receptors show little adaptation, a fact that is unpleasant but biologically necessary. Deep and visceral pain appears to be sensed by similar nerve endings, which may also be stimulated by local metabolic conditions such as an electrolyte imbalance leading to muscle cramps. **Referred pain** is a term to describe the pain phenomenon that is perceived at a site adjacent to the site of injury. Sometimes this pain is also referred to as reflective pain. One of the best examples of referred pain is during an episode of ischemia brought on by a myocardial infarction (heart attack) in which pain is frequently experienced in the neck, left arm, shoulders, and back rather than in the chest, the original site of injury.

▶ VISUAL SYSTEM

All sensory systems are important, but, in primates, the visual system is one of the most important. The visual system is the part of the CNS that enables mammals to process visual details. The eyes are organs that detect light and convert it to electrochemical impulses. Complex neural pathways exist that connect the eye, via the **optic nerve**, to the visual cortex and other areas of the brain, which converts these impulses into visual images.

Eyes comprise three layers of specialized tissue.

The eye is a fluid-filled, spherical organ enclosed by a three-layered structure of specialized tissue (Fig. 4.7). The outermost layer consists of a tough layer of connective tissue, called the *sclera*, and accounts for most of the eye's mechanical strength. The **extraocular muscles** are attached to the sclera and control movement of the eye in the socket. The anterior portion of the sclera consists of the transparent cornea through which light rays pass into the interior of the eye. The middle layer of the eye is known as the *choroid layer*, which is highly pigmented and highly vascularized. The *iris* is housed in the middle layer and is a circular smooth muscle structure that forms the *pupil*. The pupil is the neurally controlled aperture that controls the amount of light admitted to the interior of the eye. The iris also gives the eye

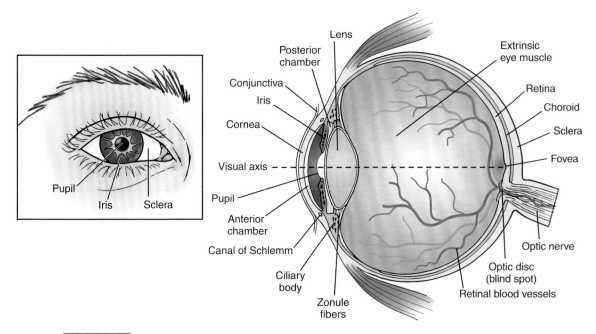

Figure 4.7 **The major parts of the human eye.** This is a view showing the relative positions of its optical and structural parts.

its characteristic color. The transparent *lens* is located just behind the iris and is suspended from the *ciliary body* by strands of fibers called *zonule fibers*. It is important to note that the lens is not part of any of the three tissue layers. The iris separates the space between the cornea and the lens into *anterior* and *posterior chambers*. The anterior chamber is filled with a clear watery liquid, called aqueous humor. The aqueous humor carries nutrients to the cornea and lens, both of which lack a vascular supply, and is continuously secreted by ciliary epithelial cells located behind the iris. As aqueous fluid accumulates, it is drained through the *canal of Schlemm* and into the venous circulation. If the drainage of aqueous humor is impaired, pressure builds up in the anterior chamber, and internal structures are compressed, leading to optic nerve damage. This condition is known as **glaucoma** and can cause blindness. The posterior chamber lies behind the iris and contains vitreous humor, a clear gelatinous liquid that helps to maintain the spherical shape of the eye.

The innermost layer of the eye is actually a bilayer. The outer of these two layers is a one-cell sheet called the pigment epithelium, which supports and helps maintain the inner *retina*, which extends from the optic disk all the way to the edge of the pupil. The retina contains the photosensitive cells called *rods* and *cones*, and its cells function in the early stage of image processing. Light information from the rods and cones is relayed via the optic nerve to the cerebral cortex and other parts of the brain. Slightly off to the nasal side of the retina is the optic disc, where the optic nerve leaves the retina. There are no photoreceptor cells here, resulting in a *blind spot* in the field of vision.

Eye structures modify incoming light rays to focus an image on the retina.

Light is defined as electromagnetic radiation composed of packets of energy called photons that travel in the form of waves. The distance between two waves is known as wavelengths. Light represents a small part of the electromagnetic spectrum, and the photoreceptors of the eye are sensitive only to wavelengths between 770 nm (red) and 380 nm (violet). The familiar colors of the spectrum all lay between these limits. A wide range of intensities (amplitude or height of the wave), from a single photon to the direct light of the sun, exists in nature. Light rays travel in a straight line in a given medium. Light rays are refracted (bent) as they pass between different media (e.g., air → liquid, or air → glass), and the degree light is bent gives rise to the various refractive indices.

Focusing

Light waves entering the eye must be focused to a common point, called a *focal point*, which in the eye is the *fovea* of the retina. The distance from the lens to the focal point is the *focal length*, which is determined by the length of an individual's eyeball (Fig. 4.8). In the eye, the focal length remains constant so incoming light waves must be refracted by the structures of the eye so they converge onto the fovea. When the refractive power of the eye structures matches the length of the eyeball and the image is focused on the fovea (emmetropia), the image is sharp. Most of the refraction of light waves is accomplished by the eye's convex surfaces the cornea and the lens. The most powerful refractory component of the eye is the curved cornea, which accounts for roughly two-thirds of the eyes refractive power. If the curvature of the cornea is not symmetric, the light waves are scattered with some being focused in front of the fovea and others behind it. This condition is called **astigmatism** and is treated with lens having multiple focal points. The refractive power of the cornea is constant so refraction of light by the lens is important for our ability to adjust our vision between distant and near objects. Light rays from distant objects are relatively parallel when they reach the eye, while lights rays from near objects are more divergent and must undergo greater refraction to focus them on the retina (see Fig. 4.8). The increased refraction

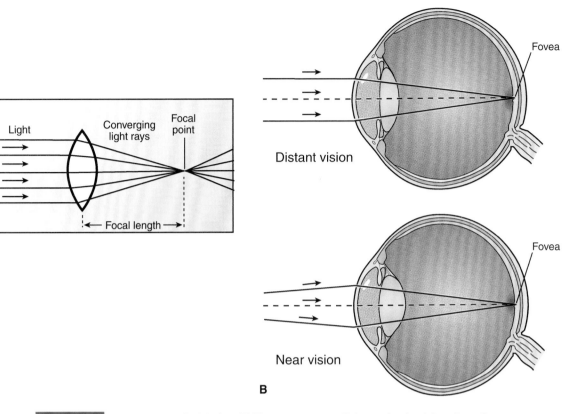

Figure 4.8 **The eye as an optical device. (A)** The eye converges light to a focal point on the retina by a convex lens. **(B)** During fixation, the center of the image falls on the fovea. For distant vision, the lenses are flattened, and rays from a distant object are brought to a sharp focus. For near vision, the lens curvature increases with accommodation, and rays from a nearby object are focused.

needed to focus the near object is supplied by the lens. The muscles of the ciliary body change the refractive strength of the lens by changing the curvature of the lens. For distant objects, the muscles are relaxed, the lens is flat and weakly refractive. For near objects, the ciliary muscles contract and the curvature and thickness of the lens are increased resulting in increased refraction that focuses the near object on the fovea. This ability to adjust the strength of the lens for close vision is called *accommodation*. With age, the lens loses its elasticity, and the point of vision moves farther away. This condition is called *presbyopia*, and supplemental refractive power, in the form of external lenses (reading glasses), is required for distinct near vision. Another problem is the loss of transparency in the lens, a condition known as *cataracts*. Normally, the elastic fibers in the lens are completely transparent to visible light. These fibers occasionally become opaque, and light rays have difficulty passing through the opaque lens. Cataracts can occur at different ages, but their prevalence increases with age. Cataracts are treated by surgical removal of the defective lens. An artificial lens may be implanted in its place, or eyeglasses may be used to replace the refractive power of the lens.

Errors of refraction, which are common, lead to blurred images because the images are focused in front of or behind the fovea (Fig. 4.9). They can be corrected with external lenses (eyeglasses or contact lenses). Farsightedness, or *hyperopia*, is caused by the eyeball being physically too short to focus on distant objects. The natural accommodation mechanism

may compensate for distance vision, but near objects will be focused behind the retina. The use of a positive (converging) lens corrects this error. If the eyeball is too long, nearsightedness, or *myopia*, results. In effect, the converging power of the eye is too great; close vision is clear, but the eye cannot focus on distant objects. Concave lenses can correct this defect.

Light regulation

The iris, which has both sympathetic and parasympathetic innervation, controls the diameter of the pupil (Fig. 4.10). It is capable of a 30-fold change in area and in the amount of light admitted to the eye. This change is under complex reflex control, and bright light entering just one eye will cause the appropriate constriction response in both eyes. As with a camera, when the pupil is constricted, less light enters, but the image is focused more sharply because the more poorly focused peripheral rays are cut off.

Phototransduction by retinal cells converts light energy into neural electrical–chemical signals.

The major function of the eye is to focus light waves from the external environment and create an image of the visual world on the retina, which serves much the same function as film in a camera. The retina is an extension of the CNS and is a multilayered structure containing the photoreceptor cells and a complex web of several types of nerve cells (Fig. 4.11). Light striking the photosensitive cells (rods and cons) initiates a cascade of chemical and electrical events that ultimately

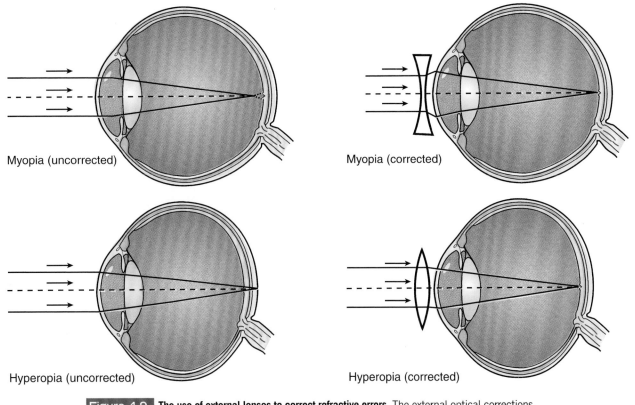

Figure 4.9 **The use of external lenses to correct refractive errors.** The external optical corrections change the effective focal length of the natural optical components. For myopia (nearsightedness), vision is corrected with concave lenses, which divert light before it reaches the eye. For hyperopia (farsightedness), vision is corrected with convex lenses, which converge light before it reaches the eye.

trigger nerve impulses, which are sent to various visual centers in the brain through the fibers of the optic nerve.

The retina consists of a stack of four main neuronal layers: pigment epithelium, the photoreceptor layer, the neural network layer, and, finally, the ganglion cell layer. The pigment layer consists of cells with high melanin content and functions to sharpen an image by preventing the scattering of stray light. People with albinism lack this pigment layer and have blurred vision that cannot be corrected effectively with external glasses.

The photoreceptor layer captures the visual image on the rods and cones. The rods and cones are packed tightly side by side, with a density of many thousand per square millimeter.

The photoreceptors are divided into two classes. The cones are responsible for **photopic** (daytime) vision, which is in color (chromatic), and the rods are responsible for **scotopic** (nighttime) vision, which is not in color. Their functions are basically similar, although they have important structural and biochemical differences.

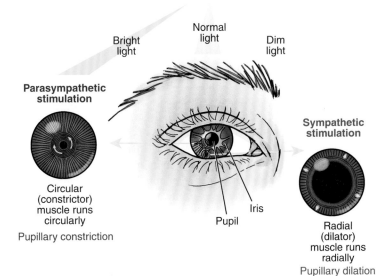

Figure 4.10 **Control of pupillary diameter.** The pupil controls the amount of light entering the eye by variable contraction of the iris muscles to admit more or less light. The iris muscles run circular (constriction) and radial (dilation).

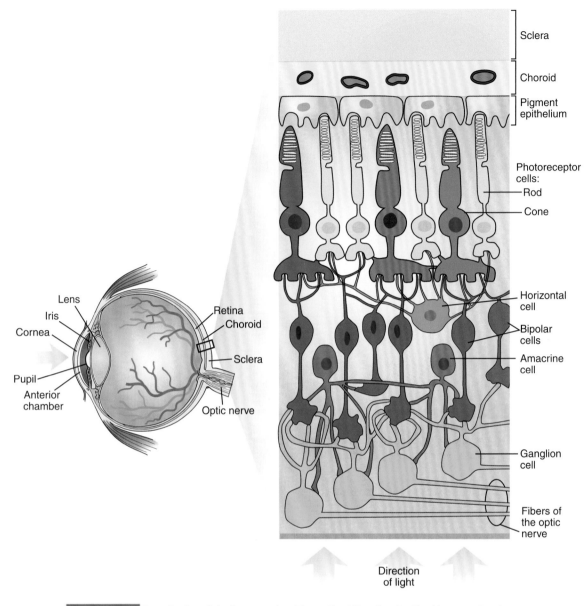

Figure 4.11 **Organization of the human retina.** The path of the visual retinal layers extends from the rods and cones (photoreceptor cells) to the ganglion cells. The light-sensitive end of the photoreceptor cells faces the choroid layer and away from the incoming light. The amacrine and horizontal cells are involved in retinal processing of visual input.

Cone cells have an *outer segment* that tapers to a point (Fig. 4.12). Three different photopigments are associated with cone cells. The pigments differ in the wavelength of light that optimally excites them. The peak spectral sensitivity for the *red-sensitive pigment* is 560 nm; for the *green-sensitive pigment*, it is about 530 nm; and for the *blue-sensitive pigment*, it is about 420 nm. The corresponding photoreceptors are called *red*, *green*, and *blue cones*, respectively. At wavelengths away from the optimum, the pigments still absorb light but with reduced sensitivity. Because of the interplay between light intensity and wavelength, a retina with only one class of cones would not be able to detect colors unambiguously. The presence of two of the three pigments in each cone removes this uncertainty. Color-blind individuals, who have a genetic lack of one or more of the pigments or who lack an associated transduction mechanism, cannot distinguish between the affected colors. Loss

of a single color system produces *dichromatic vision*, and lack of two of the systems causes *monochromatic vision*. If all three are lacking, vision is monochromatic and depends only on the rods.

A **rod cell** is long, slender, and cylindrical, and it is larger than a cone cell (see Fig. 4.12). Its outer segment contains numerous photoreceptor disks composed of cellular membrane in which the molecules of the photopigment **rhodopsin** are embedded. The lamellae near the tip are regularly shed and replaced with new membrane synthesized at the opposite end of the outer segment. The *inner segment*, connected to the *outer segment* by a modified cilium, contains the cell nucleus, many mitochondria that provide energy for the phototransduction process, and other cell organelles. At the base of the cell is a synaptic body that makes contact with one or more bipolar nerve cells and liberates a transmitter substance in response to changing light levels.

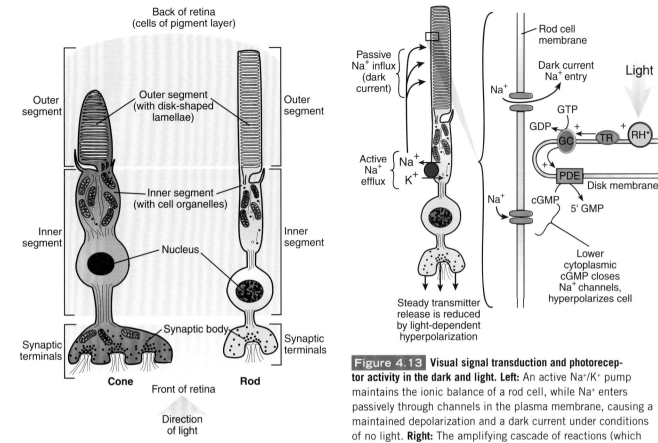

Figure 4.12 **Structure of photoreceptors of the human retina.** The rod and cones of the eye's photoreceptors contain three segments. The outer segment contains stacked flattened membrane disks that contain an abundance of light-absorbing photopigments. The inner segment houses the metabolic machinery, and the synaptic terminal stores and releases neurotransmitters.

Figure 4.13 **Visual signal transduction and photoreceptor activity in the dark and light. Left:** An active Na^+/K^+ pump maintains the ionic balance of a rod cell, while Na^+ enters passively through channels in the plasma membrane, causing a maintained depolarization and a dark current under conditions of no light. **Right:** The amplifying cascade of reactions (which take place in the disk membrane of a photoreceptor) allows a single activated rhodopsin molecule to control the hydrolysis of 500,000 cyclic guanosine monophosphate (cGMP) molecules. (See text for details of the reaction sequence.) In the presence of light, the reactions lead to a depletion of cGMP, resulting in the closing of cell membrane Na^+ channels and the production of a hyperpolarizing generator potential. The release of neurotransmitter decreases during stimulation by light. RH*, activated rhodopsin; TR, transducin; GC, guanylyl cyclase; PDE, phosphodiesterase; GTP, guanosine triphosphate; GDP, guanosine diphosphate.

The visual pigments of the photoreceptor cells convert light to a nerve signal. This process is best understood in the case of the rod cells. In the dark, the pigment rhodopsin (or *visual purple*) consists of a light-trapping chromophore (the part of a molecule responsible for its color) called **scotopsin**, which is chemically conjugated with **11-*cis*-retinal**, the aldehyde form of vitamin A1. When struck by light, rhodopsin undergoes a series of rapid chemical transitions, with the final intermediate form metarhodopsin II providing the critical link between this reaction series and the electrical response. The end products of the light-induced transformation are the original scotopsin and an all-*trans* form of retinal, now dissociated from each other. Under conditions of both light and dark, the all-*trans* form of retinal is isomerized back to the 11-*cis* form, and the rhodopsin is reconstituted. All of these reactions take place in the highly folded membranes comprising the outer segment of the rod cell.

The cellular process of visual signal transduction is shown in Figure 4.13. The coupling of the light-induced reactions and the electrical response involves the activation of **transducin**, a G protein; the associated exchange of guanosine triphosphate for guanosine diphosphate activates a phosphodiesterase. This, in turn, catalyzes the breakdown of cyclic guanosine monophosphate (cGMP) to 5′-GMP. When cellular cGMP levels are high (as in the dark), membrane sodium channels are kept open, and the cell is relatively depolarized. Under these conditions, there is a tonic release of neurotransmitter from the synaptic body of the rod cell. A decrease in the level of cGMP as a result of light-induced reactions causes the cell to close its sodium channels and hyperpolarize, thus reducing the release of neurotransmitter.

This change is the signal that is further processed by the nerve cells of the retina to form the final response in the optic nerve. An active sodium pump maintains the cellular concentration at proper levels. A large amplification of the light response takes place during the coupling steps. One activated rhodopsin molecule will activate ~ 500 transducins, each of which activates the hydrolysis of several thousand cGMP molecules. Under proper conditions, a rod cell can respond to a single photon striking the outer segment. The processes in cone cells are similar, although there are three different opsins (with different spectral sensitivities), and the specific transduction mechanism is also different. In addition, the overall sensitivity of the transduction process is lower.

In the light, much rhodopsin is in its unconjugated form, and the sensitivity of the rod cell is relatively low. During the process of *dark adaptation*, which takes about 40 minutes to complete, the stores of rhodopsin are gradually built up, with a consequent increase in sensitivity (by as much as 25,000 times). Cone cells adapt more quickly than rods, but their final sensitivity is much lower. The reverse process, *light adaptation*, takes about 5 minutes.

Neural network layer

Bipolar cells, **horizontal cells**, and **amacrine cells** comprise the *neural network layer* of the eye (see Fig. 4.11). These cells together are responsible for considerable initial processing of visual information. Because the distances between neurons here are so small, most cellular communication involves the electrotonic spread of cell potentials, rather than propagated action potentials. Light stimulation of the photoreceptors produces hyperpolarization that is transmitted to the bipolar cells. Some of these cells respond with a depolarization that is excitatory to the ganglion cells, whereas other cells respond with a hyperpolarization that is inhibitory. The horizontal cells also receive input from rod and cone cells but spread information laterally, causing inhibition of the bipolar cells on which they synapse. Another important aspect of retinal processing is **lateral inhibition**. A strongly stimulated receptor cell can inhibit, via lateral inhibitory pathways, the response of neighboring cells that are less well illuminated. This has the effect of increasing the apparent contrast at the edge of an image. Amacrine cells also send information laterally but synapse on ganglion cells.

Ganglion cell layer

In the *ganglion cell layer* (see Fig. 4.11), the results of retinal processing are finally integrated by the *retinal ganglion cells*, whose axons form the optic nerve. These cells are tonically active, sending action potentials into the optic nerve at an average rate of five per second, even when unstimulated. Input from other cells converging on the ganglion cells modifies this rate up or down.

Many kinds of information regarding color, brightness, contrast, and so on are passed along the optic nerve. The output of individual photoreceptor cells converges on the ganglion cells. In keeping with their role in visual acuity, relatively few cone cells converge on a ganglion cell, especially in the fovea, where the ratio is nearly 1:1. Rod cells, however, are highly convergent, with as many as 300 rods converging on a single ganglion cell. Although this mechanism reduces the sharpness of an image, it allows for a great increase in light sensitivity.

Signals from the retina are modified and separated before reaching the thalamus and visual cortex.

The retina, unlike a camera, does not simply send a picture to the brain. The retina spatially encodes (compresses) the image to fit the limited capacity of the optic nerve. Encoding

is necessary because there are 100 times more photoreceptor cells than ganglion cells. The encoding is carried out by bipolar and ganglion cells. Once the image is spatially encoded, the signal is sent out the optic nerve (via the axons of the ganglion cells) through the optic chiasm to the *lateral geniculate nucleus (LGN)*, which is in the thalamus.

Image crossover

Information from the right and left visual fields is transmitted to opposite sides of the brain. The optic nerves, each carrying about 1 million fibers from each retina, enter the rear of the orbit and pass to the underside of the brain to the **optic chiasma** (Fig. 4.14), where about half of the fibers from each eye "cross over" to the other side. Fibers from the temporal side of the retina do not cross the midline but travel in the optic tract on the same side of the brain. Fibers originating from the nasal side of the retina cross the optic chiasma and travel in the optic tract to the opposite side of the brain.

The first stop in the brain where information is modified from the visual pathways is the LGN, where the divided output information is separated and relayed by fiber bundles known as optic radiations (geniculocalcarine tract) to different zones of the *visual cortex* in the occipital lobe of the brain (see Fig. 4.14). Mechanisms in the visual cortex detect and integrate visual information, such as shape, contrast, line, and intensity, into a coherent visual perception.

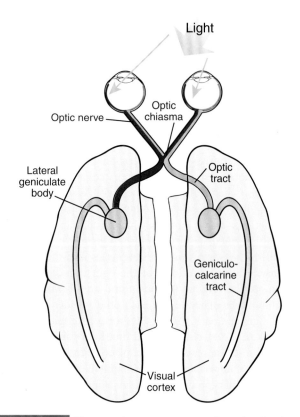

Figure 4.14 The central nervous system pathway for visual information. Fibers from the right visual field will stimulate the left half of each retina, and nerve impulses will be transmitted to the left hemisphere.

Information from the optic nerves is also sent to the **suprachiasmatic nucleus** of the hypothalamus, where it participates in the regulation of circadian rhythms; to the *pretectal nuclei*, which are concerned with the control of visual fixation and pupillary reflexes; and to the **superior colliculus**, which coordinates simultaneous bilateral eye movements, such as tracking and convergence.

Depth perception

Depth perception is the visual ability to see the world in three dimensions ("3D") and arises from binocular depth cues that require input from both eyes. Depth perception is lost when one eye is damaged. By using two images of the same scene obtained from slightly different angles, it is possible to triangulate the distance to an object. For example, if the object is far away, the disparity of the image falling on both retinas will be small. If the object is closer, the disparity of that image will be large. This binocular oculomotor cue provides a high degree of accuracy for depth/perception.

Visual reflexes are partially under central nervous system control.

Like a modern camera, the eye has an autofocus for light and an autofocus for distance as well as other reflex mechanisms. These reflexes are mediated, in part, by CNS control of the eye, and they help to adapt and protect the eye from injury.

Pupillary reflex

The *pupillary light reflex* controls the diameter of the pupil in response to the intensity of light. Light shone in one eye elicits a pupillary light reflex that causes both pupils to contract. Greater light intensity causes both pupils to become smaller and allow less light in. In contrast, lower intensity causes both pupils to become larger, allowing more light in. The pupillary reflex pathway begins with retinal ganglion cells, which convey information from the photoreceptors via their axons in the optic nerve. The optic nerve is responsible for the afferent limb of the pupillary reflex (i.e., it senses the incoming light). Light level information is processed in the brainstem and adjusting signals are relayed by parasympathetic fibers that travel in the oculomotor nerve, which is the efferent limb of the pupillary reflex (i.e., it causes the muscles to constrict the pupil).

The pupillary light reflex performs several important functions: (1) it regulates the intensity of light that falls on the retina, thereby assisting in adaption to various levels of light and darkness; (2) it aids in retinal sensitivity to light; and (3) it protects the eye from retinal damage from overexposure to light. Aging affects the pupillary light reflex, which becomes most apparent in going from a light to dark room. With aging, it takes longer to adjust to the darkness.

Corneal reflex

The corneal reflex is an involuntary blinking of the eyelids elicited by stimulation of the cornea with a foreign object or bright light. Just touching one cornea causes the reflex blinking of both eyes. The corneal reflex protects the eye from foreign material and bright lights. The reflex is mediated through the ophthalmic branch of the trigeminal nerve (fifth cranial nerve) that senses the stimulus on the cornea. The seventh cranial nerve (facial nerve) initiates the motor response.

This reflex is absent in newborns and is diminished with use of contacts lenses. The corneal reflex is often part of a neurologic examination, especially when a patient is evaluated for a coma.

Accommodation reflex

Accommodation is the process whereby the eye changes its optical power to maintain a clear vision on an object as its distance changes. This reflex action is in response to focusing on a near object and then looking at a distant object (and vice versa). When someone accommodates to a near object, three things happen. First, the eyes converge, which is accomplished by simultaneous activation of extraocular eye muscles (medial recti), to bring the fovea in line with the object. Second, the ciliary muscles contract increasing the curvature of the lenses and producing more refraction. This shortens the focal length so that the image lands on both retinas. Third, the pupil constricts in order to prevent divergent light rays from hitting the periphery of the retina and resulting in blurred vision. Pupillary constriction also improves the depth of vision. All three things happen automatically as part of the accommodation reflex. Accommodation is another visual reflex that diminishes with age. At about 60 years of age, most people will have noticed a decrease in their ability to focus on close objects.

▶ AUDITORY SYSTEM

The sensory systems of primates are astonishing achievements in design and efficiency. In many ways, the ear stands out among the sensory organs. The ear is the organ that not only receives sound but also plays a major role in the sense of balance and body position. The ear is part of the auditory system and consists of three components: the outer, middle, and inner ear (Fig. 4.15). The outer ear collects sound, and the middle ear amplifies the sound pressure before transmitting it to the fluid-filled inner ear. The inner ear houses two separate sensory systems: the auditory system, which contains the **cochlea** whose receptors convert sound waves into nerve impulses, and the vestibular system, which is involved in balance and special position.

Sound is an oscillating pressure wave composed of frequencies that are transmitted through different media.

Sound is an oscillating pressure wave traveling through a medium (i.e., air, liquid, solid). This process involves both compression and rarefaction (Fig. 4.16). The distance between the compression peaks is called the wavelength of sound and is inversely related to the frequency. Hearing in humans is normally limited to frequencies between 20 and 20,000 Hz (1 Hz = 1 hertz = 1 cycle/s). Bats, however, can

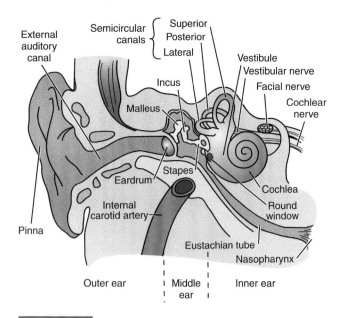

Figure 4.15 **Basic structure of the human ear.** Each ear consists of three basic parts: the external, middle, and the inner ear. The structures of the middle and inner ear are encased in the temporal bone of the skull. The external and middle portions of the ear transmit airborne sound waves to the fluid-filled inner ear, amplifying sound waves in the process.

perceive a wider range of frequencies (2 to 110 kHz). Pitch or tone of sound is determined by frequency. *Sinusoidal sound waves* (those that have regularly repeating oscillations) contain all of their energy at one frequency and are perceived as pure tones. *Complex sound waves*, such as those in speech or music, consist of the addition of several simpler waveforms of different frequencies and amplitudes. Intensity or loudness depends on the amplitude of the sound wave and is expressed as a decibel (dB) scale:

$$\text{Decibels(dB)} = 20 \log P/P_0$$

where P is sound pressure and P_0 is a reference level pressure (i.e., threshold for normal hearing at the best frequency). The threshold stimulus is set at 0 dB ($\log P/P_0 = 0$). For a sound that is 10 times greater than the reference, the expression becomes

$$dB = 20 \log (0.020 / 0.002) = 20$$

Thus, any two sounds having a 10-fold difference in intensity have a decibel difference of 20. A 100-fold difference would mean a 40-dB difference, and a 1,000-fold difference would mean a 60-dB difference. Common sounds range from a faint whisper (20 dB) to a jet taking off (140 dB).

External ear captures and amplifies sound.

An overall view of the human ear is shown in Figure 4.15. The auricle or *pinna*, the visible portion of the outer ear, collects sound waves and channels them down the external auditory canal. The external auditory canal extends inward through the temporal bone and its inner end is sealed by the *tympanic membrane* (eardrum), a thin, oval, slightly conical, flexible membrane. An incoming pressure wave traveling down the external auditory canal causes the eardrum to vibrate back and forth in step with the compressions and rarefactions of the sound wave. This is the first mechanical step in the transduction of sound. The overall acoustic effect of the outer ear structures is to produce an amplification of 10 to 15 dB in the frequency range broadly centered around 3,000 Hz.

Middle ear mechanically converts tympanic membrane vibrations to fluid waves in the inner ear.

The middle ear is an air-filled cavity containing three tiny bones that couple vibration of the eardrum into waves in the fluid and membranes of the inner ear. The hollow space of the middle ear is called the *tympanic cavity*. The eustachian tube connects the tympanic cavity to the pharynx, and the tube opens briefly during swallowing, allowing equalization of the pressures on either side of the eardrum. During rapid external pressure changes (such as during takeoff or descent in an airplane), the unequal forces displace the eardrum. Such physical deformation may cause discomfort or pain

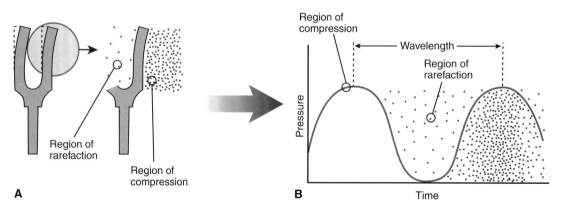

Figure 4.16 **Sound wave formations. (A)** Sound waves generated from a tuning fork cause molecules ahead of the advancing arm to be compressed and the molecules behind the arm to be rarified. **(B)** Sound waves are propagated as sinusoidal, alternating regions of compression and rarefaction of air molecules. The wavelength of a sinusoidal wave is the spatial period between two peak compression waves.

and, by restricting the motion of the tympanic membrane, may impair hearing. Blockages of the eustachian tube or fluid accumulation in the middle ear (as a result of an infection) can also lead to difficulties with hearing.

Bridging the gap between the tympanic membrane and the inner ear is a chain of three small bones, the ossicles (Fig. 4.17). The *malleus* (Latin for "hammer") is attached to the eardrum in such a way that the back-and-forth movement of the eardrum causes a rocking movement of the malleus. The *incus* (Latin for "anvil") connects the head of the malleus to the third bone, the *stapes* (Latin for "stirrup"). This last bone, through its oval footplate, connects to the oval window of the inner ear and is anchored there by an annular ligament.

Four separate suspensory ligaments hold the ossicles in position in the middle ear cavity. The superior and lateral ligaments lie roughly in the plane of the ossicular chain and anchor the head and shaft of the malleus. The anterior ligament attaches the head of the malleus to the anterior wall of the middle ear cavity, and the posterior ligament runs from the head of the incus to the posterior wall of the cavity. The suspensory ligaments allow the ossicles sufficient freedom to transmit the vibrations of the tympanic membrane to the oval window. This mechanism is especially important because, although the eardrum is suspended in air, the oval

window seals off a fluid-filled chamber. When sound waves in air strike liquid, most of the energy (~99%) is reflected off the liquid and lost. The middle ear allows the impedance matching of sound traveling in air to acoustic waves traveling in a system of fluids and membranes in the inner ear. Two principles of mechanical advantage are used in the impedance matching, the hydraulic principle and the lever principle. The shape of the articulated ossicular chain is lever-like and has a lever ratio of about 1.3:1.0, producing a slight gain in force. In addition, the vibrating portion of the tympanic membrane is coupled to the smaller area of the oval window (~a 17:1 ratio). These conditions result in a pressure gain of around 25 dB, largely compensating for the energy loss going from an air to fluid environment. Although the efficiency depends on the frequency, approximately 60% of the sound energy that strikes the eardrum is transmitted to the oval window.

The middle ear is also able to substantially dampen sound conduction when faced with very loud sounds by noise-induced reflex contraction of the middle-ear muscles. These small muscles are attached to the ossicular chain and help hold the bones in position and modify their function (see Fig. 4.17). The *tensor tympani muscle* inserts on the malleus (near the center of the eardrum), passes diagonally through the middle ear cavity, and enters the *tensor canal*, in which it is anchored. Contraction of this muscle limits the vibration amplitude of the eardrum and makes sound transmission less efficient. The *stapedius muscle* attaches to the stapes near its connection to the incus and runs posteriorly to the mastoid bone. Its contraction changes the axis of oscillation of the ossicular chain and causes dissipation of excess movement before it reaches the oval window. These muscles are activated by a reflex (simultaneously in both ears) and contract in response to moderate and loud sounds. They act to reduce the transmission of sound to the inner ear, thereby protecting its delicate structures. This is called the **acoustic reflex**.

Because this reflex requires up to 150 milliseconds to operate (depending on the loudness of the stimulus), it cannot provide protection from sharp or sudden bursts of sound.

The process of sound transmission can bypass the ossicular chain entirely. If a vibrating object, such as a tuning fork, is placed against a bone of the skull (typically the mastoid), the vibrations are transmitted mechanically to the fluid of the inner ear, where the normal processes act to complete the hearing process. Bone conduction is used as a means of diagnosing hearing disorders that may arise because of lesions in the ossicular chain. Some hearing aids employ bone conduction to overcome such deficits.

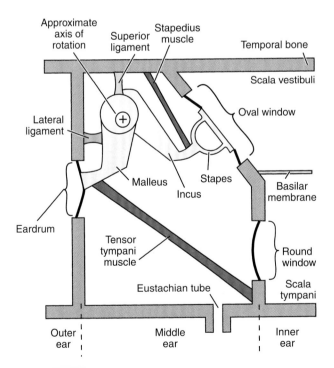

Figure 4.17 **Model of the middle ear.** Vibrations from the eardrum are transmitted by the lever system formed by the ossicular chain to the oval window of the scala vestibuli. The anterior and posterior ligaments, part of the suspensory system for the ossicles, are not shown. The combination of the four suspensory ligaments produces a virtual pivot point (marked by a *cross*); its position varies with the frequency and intensity of the sound. The stapedius and tensor tympani muscles modify the lever function of the ossicular chain.

Inner ear transduces sound.

The inner ear is a tortuous system of fluid-filled channels located in the temporal bone, the cochlea, which contains the receptor cells for the auditory system, and the semicircular canals and otolithic organs, which contain the receptor cells for the vestibular system. The inner ear is where the actual process of sound transduction takes place.

Cochlea

The cochlea defines the overall structure of the auditory transducer. It propagates the mechanical signals as waves in fluid and membranes and transduces them to nerve impulses, which are transmitted to the brain.

The auditory structures are located in the cochlea (Fig. 4.18), part of a cavity in the temporal bone called the *bony labyrinth*. The cochlea (Latin for "snail") is a fluid-filled spiral tube that arises from a cavity called the *vestibule*, with which the organs of balance also communicate. It is partitioned longitudinally into three divisions (canals) called the *scala vestibuli* (into which the oval window opens), the *scala tympani* (sealed off from the middle ear by the *round window*), and the membranous *scala media* (in which the sensory cells are located). Arising from the bony center axis of the spiral (the *modiolus*) is a winding shelf called the osseous spiral lamina. Opposite to it on the outer wall of the spiral is the *spiral ligament*, and connecting these two structures and forming the floor of the scala media is a highly flexible connective tissue sheet, the **basilar membrane**, which runs for almost the entire length of the cochlea. The hair cells, which are the actual sensory receptors, are located on the upper surface of the basilar membrane (see Fig. 4.18B). They are called "hair cells" because each has a bundle of hair-like cilia at the end that project away from the basilar membrane.

The Reissner membrane forms the roof of the scala media separating it from the scala vestibuli (see Fig. 4.18). The scala vestibuli communicates with the scala tympani at the apical (distal) end of the cochlea via the helicotrema, a small opening where a portion of the basilar membrane is missing. The scala vestibuli and scala tympani are filled with **perilymph**, a fluid high in sodium and low in potassium. The scala media contains **endolymph**, a fluid high in potassium and low in sodium. The endolymph is secreted by the **stria vascularis**, a layer of fibrous vascular tissue along the outer wall of the scala media. Because the cochlea is filled with incompressible fluid and is encased in hard bone, pressure changes caused by the in-and-out motion at the oval window (driven by the stapes) are relieved by an out-and-in motion of the flexible round window membrane.

Sensory structures

The **organ of Corti** is formed by structures located on the upper surface of the basilar membrane and runs the length of the scala media (see Fig. 4.18B). It contains one row of some 3,000 *inner hair cells*. The arch of Corti and other specialized supporting cells separate the inner hair cells from the three or four rows of *outer hair cells* (about 12,000) located on the stria vascularis side. The rows of inner and outer hair cells are inclined slightly toward each other and are covered by the *tectorial membrane*.

Nerve fibers from cell bodies located in the *spiral ganglia* form radial bundles on their way to synapse with the inner hair cells. Each nerve fiber makes synaptic connection with only one hair cell, but each hair cell is served by 8 to 30 fibers. Although the inner hair cells comprise only 20% of the hair cell population, they receive 95% of the afferent

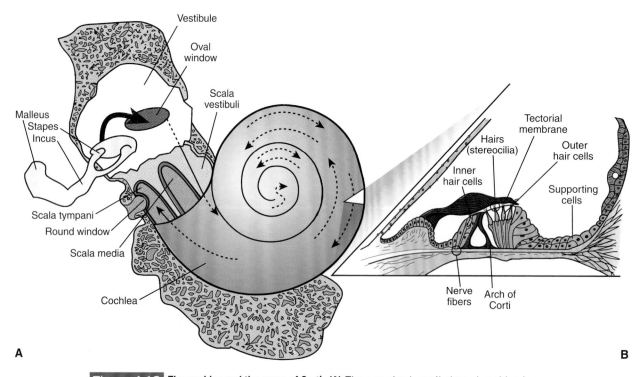

A

B

Figure 4.18 **The cochlea and the organ of Corti. (A)** The pea-sized, snail-shaped cochlea is a coiled system and is the hearing portion of the inner ear. **(B)** An enlargement of a cross-section of the organ of Corti, showing the relationships among the hair cells and the membranes. Hair cells in the organ of Corti transduce fluid movement into neural signals.

fibers. In contrast, many outer hair cells are each served by a single external spiral ganglion nerve fiber. The collected afferent fibers are bundled in the *cochlear nerve*, which exits the inner ear via the *internal auditory meatus*. Many efferent fibers also innervate the cochlea. They serve to enhance pitch discrimination and the ability to distinguish sounds in the presence of noise. Evidence suggests that efferent fibers to the outer hair cells cause them to shorten (contract), altering the mechanical properties of the cochlea.

Hair Cells

The hair cells of the inner and the outer rows are similar anatomically. Both sets extend upward into the scala media toward the tectorial membrane (see Fig. 4.18B). Extensions of the outer hair cells actually touch the tectorial membrane, whereas those of the inner hair cells appear to stop just short of contact. The hair cells make synaptic contact with afferent neurons using glutamate as their neurotransmitter.

At the apical end of each inner hair cell is a projecting bundle of about 50 **stereocilia**, rod-like structures packed in parallel, slightly curved rows. Minute strands link the free ends of the stereocilia together, so the bundle tends to move as a unit. The height of the individual stereocilia increases toward the outer edge of the cell (toward the stria vascularis), giving a sloping appearance to the bundle. Along the cochlea, the inner hair cells remain constant in size, whereas their stereocilia increase in height from about 4 μm at the basal end to 7 μm at the apical end. The outer hair cells are more elongated than the inner cells, and their size and height of stereocilia increase along the cochlea from base to apex. Each stereocilium contains cross-linked and closely packed actin filaments and, near the tip, a cation-selective transduction channel.

The process of mechanical transduction in hair cells is shown in Figure 4.19. When a hair bundle is deflected slightly (the threshold is <0.5 nm) toward the stria vascularis, minute mechanical forces open the transduction channels, and cations enter the cells. The resulting **depolarization**, roughly proportional to the deflection, causes the opening of voltage-gated calcium channels in the hair cell basolateral membrane. As in other synaptic systems, the entry of calcium ions results in migration and fusion of the synaptic vesicles with the cell membrane and the release of transmitter molecules, which generates action potentials in the afferent nerve. Approximately 15% of the transduction channels are open in the absence of any deflection, and bending in the direction of the modiolus of the cochlea results in hyperpolarization, thereby increasing the range of stereocilia movement that can be sensed.

The response time of hair cells is remarkable; they can detect repetitive motions of up to 100,000 times per second. They can, therefore, provide information throughout the course of a single cycle of a sound wave. Such rapid response is also necessary for the accurate localization of sound sources. When a sound comes from directly in front of a listener, the waves arrive simultaneously at both ears. If the sound originates off to one side, it reaches one ear sooner than the other and is slightly more intense at the nearer ear. The difference in arrival time is on the order of tenths of a millisecond, and the rapid response of the hair cells allows them to provide temporal input to the auditory cortex. The timing and intensity information are processed by groups of cells in the central auditory pathways to produce an accurate perception of the location of the sound source.

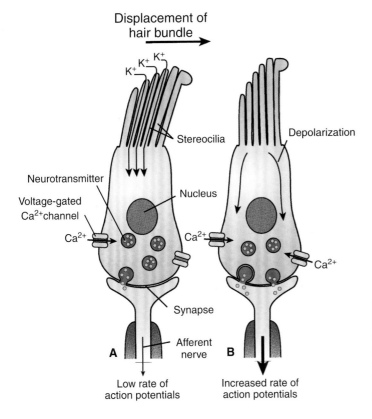

Figure 4.19 Role of stereocilia in sound transduction in the hair cells of the ear. (A) Deflection of the stereocilia opens apical K+ channels. **(B)** K+ enters the hair cells causing depolarization. The resulting depolarization opens the voltage-gated Ca2+ channels at the basal end of the cell. This causes the release of the neurotransmitter, thereby exciting the afferent nerve.

Neuromuscular Physiology

Organ of Corti

The actual transduction of sound requires an interaction among the tectorial membrane, the hair cells, and the basilar membrane. When a sound wave is transmitted to the oval window by the ossicular chain, a pressure wave travels up the scala vestibuli and down the scala tympani (Fig. 4.20). The canals of the cochlea, being encased in bone, are not deformed, and movements of the round window allow the small volume change needed for the transmission of the pressure wave. Resulting currents in the cochlear fluids produce an undulating distortion in the basilar membrane. Because the stiffness and width of the membrane vary with its length (it is wider and less stiff at the apex than at the base), the membrane deformation takes the form of a "traveling wave," which has its maximal amplitude at a position along the membrane corresponding to the particular frequency of the sound wave (Fig. 4.21). Low-frequency sounds cause a maximal displacement of the membrane near its apical end (near the helicotrema), whereas high-frequency sounds produce their maximal effect at the basal end (near the oval

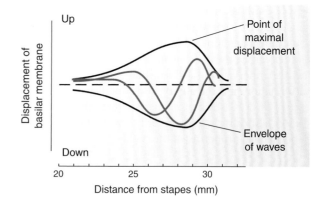

Figure 4.21 **Membrane localization of different frequencies.** An illustration of a traveling sound wave being displaced along the basilar membrane at two instants. Over time, the peak excursions of many such waves form an envelope of displacement with a maximal value at about 28 mm from the stapes (lower portion). At this position, its stimulating effect on the hair cells will be most intense. Low frequency (<800 Hz) produces maximal effects at the apex of the basilar membrane.

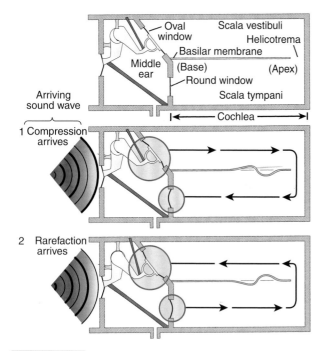

Figure 4.20 **The mechanics of the cochlea, showing the action of the structures responsible for pitch discrimination (with only the basilar membrane of the organ of Corti shown).** When the compression phase of a sound wave arrives at the eardrum, the ossicles transmit it to the oval window, which is pushed inward. A pressure wave travels up the scala vestibuli and (via the helicotrema) down the scala tympani. To relieve the pressure, the round window membrane bulges outward. Associated with the pressure waves are small eddy currents that cause a traveling wave of displacement to move along the basilar membrane from base to apex. The arrival of the next rarefaction phase reverses these processes. The frequency of the sound wave, interacting with the differences in the mass, width, and stiffness of the basilar membrane along its length, determines the characteristic position at which the membrane displacement is maximal.

window). As the basilar membrane moves up, the stereocilia of the hair cells are displaced against the tectorial membrane and subjected to lateral shearing forces that stimulate the cells.

Because of the tuning effect of the basilar membrane, only hair cells located at a particular place along the membrane are maximally stimulated by a given frequency (pitch). This localization is the essence of the **place theory** of pitch discrimination, and the mapping of specific tones (pitches) to specific areas of the basilar membrane is called **tonotopic organization**. As the signals from the cochlea ascend through the complex pathways of the auditory system in the brain, the tonotopic organization of the neural elements is at least partially preserved, and pitch can be spatially localized throughout the system.

The sense of pitch is further sharpened by the resonant characteristics of the different-length stereocilia along the length of the cochlea and by the frequency–response selectivity of neurons in the auditory pathway. The cochlea acts as both a transducer for sound waves and a frequency analyzer that sorts out the different pitches so that they can be separately distinguished. In the midrange of hearing (around 1,000 Hz), the human auditory system can sense a difference in frequency of as little as 3 Hz. The tonotopic organization of the basilar membrane has facilitated the invention of prosthetic devices whose aim is to provide some replacement of auditory function to people suffering from deafness that arises from severe malfunction of the middle or inner ear.

Central Auditory Pathways

Nerve fibers from the cochlea enter the spiral ganglion. From there, fibers are sent to the *dorsal* and *ventral cochlear nuclei*. The complex pathway that finally ends at the **auditory cortex** in the superior portion of the **temporal lobe** of the brain involves several sets of synapses and considerable crossing over and intermediate processing. As with the eye, there is a spatial correlation between cells in the sensory organ and specific locations in the primary auditory cortex. In this case, the representation is called a *tonotopic map*, with different pitches being represented by different locations, even though the firing rates of the cells no longer correspond to the frequency of sound originally presented to the inner ear.

Pitch discrimination

Pitch represents the perceived frequency of sound and is one of the major auditory attributes to musical tones. Pitch is determined by frequency (cycles per second or hertz) and corresponds closely to the repetition rate of sound waves. Although pitch discrimination (the ability to distinguish between various frequencies) is based on quantified frequencies, it is not an objective physical property but is completely subjective and is based on perceived sound. The basilar membrane is involved in determining pitch. Recall that different regions of the basilar membrane vibrate maximally at different frequencies (i.e., each frequency exhibits a peak vibration at different regions of the membrane). For example, higher frequencies vibrate best at the narrow end nearest the oval window, whereas low frequencies vibrate optimally at the wide end of the membrane. Sound waves themselves do not have pitch; distinguishing the perceived tones requires the work of the human brain. Each region of the basilar membrane is linked to a specific region of the auditory cortex, and tones are perceived in different regions of the primary auditory cortex. Tone deafness is the lack of relative pitch, or the inability to discriminate between musical notes. However, tone-deaf individuals can fully interpret the pitch of human speech.

Loudness

Loudness is another auditory attribute and depends on the amplitude of vibration. Sound waves originating from louder sounds cause the eardrum to vibrate more vigorously. Recall that a greater tympanic membrane deflection is converted into greater amplitude of basilar membrane movement. The CNS interprets this membrane oscillation and hair bending as a louder sound. Although the auditory system is sensitive and can detect very faint sounds, loud sounds can be damaging to the ear. The protective reflexes of the middle ear cannot sufficiently attenuate very loud sounds. Violent vibrations of the basilar membrane that come from sources, such as a jet engine or a siren for example, can permanently damage hair cells leading to noise-induced hearing loss.

Hearing loss results from a mechanical or a neural problem.

Hearing loss or *deafness* refers to conditions in which individuals are unable to detect or perceive some frequencies of sound. Deafness can be temporary or permanent (partial or complete). Hearing loss is the second most common physical disability in North America, affecting approximately 10% of the population. Hearing impairments are categorized by their types—*conductive deafness* and *sensorineural deafness*. A conductive deafness is an impairment resulting from dysfunction in any of the mechanisms that normally conducts sound waves through the outer ear, the eardrum, or the bones of the middle ear. Common causes include the buildup of earwax that blocks the ear canal, a ruptured eardrum, a middle ear infection, and restriction of ossicular movements of bony structures of the middle ear.

A sensorineural hearing impairment occurs as a result of (1) inner ear dysfunction, especially of the cochlea, where sound vibrations are converted into neural signals, or (2) any part of the brain that subsequently processes these signals. The vast majority of sensorineural deafness is associated with damage to the hair cells of the organ of Corti in the cochlea. This dysfunction may be present from birth from genetic or developmental abnormalities or arise through trauma or disease. Diseases causing deafness include measles, meningitis, mumps, and venereal disease. Some medications can also cause irreversible damage to the hair cells. The most important group is the aminoglycosides (primarily gentamicin). Other medications include anti-inflammatory (nonsteroidal) and diuretic drugs. Narcotic painkillers, in particular Vicodin and OxyContin, are another source of medication that causes permanent hearing loss.

One of the most common causes of partial hearing loss is sensory presbycusis, the progressive loss of the ability to hear high-frequency sounds with increasing age as the hair cells near the base of the cochlea "wear out" over time.

▶ VESTIBULAR SYSTEM

Animals need feedback information to keep track of which end is up—that is, how they are oriented to their surrounding environment. The eyes help to keep track of movement with other objects, but the sensory system that also contributes to our balance and sense of spatial orientation is the vestibular system. This system provides dominant input about movement and equilibrioception. The vestibular apparatus sends signals primarily to the neural structures that control the eye movements and keep us upright.

Vestibular system comprises the semicircular canal and the otoliths.

Each ear contains three **semicircular canals** and two **otolithic organs**, the *utricle* and the *saccule* (Fig. 4.22). These structures are located in the bony labyrinth of the temporal bone. As with hearing, the basic sensing elements are hair cells. The semicircular canals are arranged three dimensionally in planes that are at approximately right angles to each other and are called the *horizontal* (lateral), *anterior* (superior), and the *posterior canals* (inferior). The anterior and posterior canals may be collectively called *vertical semicircular canals*. The three pairs of canals work in a push–pull fashion—when one canal is stimulated, its corresponding partner on the other side is inhibited and vice versa. The push–pull systems allow

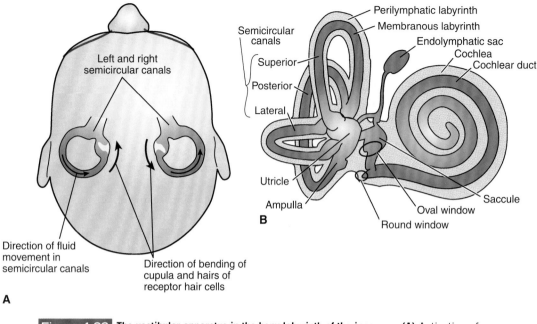

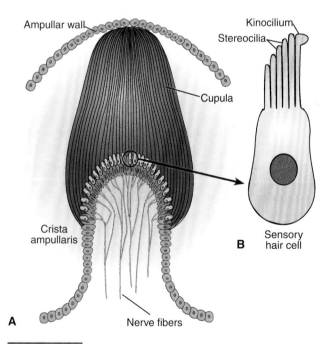

Figure 4.22 **The vestibular apparatus in the bony labyrinth of the inner ear. (A)** Activation of hair cells in the semicircular canals. **(B)** The semicircular canals sense rotary acceleration and motion, whereas the utricle and saccule sense linear acceleration and static position.

the vestibular apparatus to sense all directions of rotation. For example, the right horizontal canal gets stimulated during head rotations to the right, and the left horizontal canal gets stimulated by head rotations to the left. Vertical canals are coupled in a crossed fashion, which allows for stimulations for the anterior canal to also be inhibitory for the contralateral posterior canal and vice versa.

Movement of fluid within the horizontal canal corresponds to rotation of the head around a vertical axis (i.e., the neck). The anterior and posterior canals detect rotations of the head in the sagittal plane (i.e., nodding) and in the frontal plane, as when doing a cartwheel. Both anterior and posterior canals are oriented at ~ 45° between the frontal and sagittal planes. Near its junction with the utricle, each canal has a swollen portion called the *ampulla*. Each ampulla contains the sensory structure for that semicircular canal called the *crista ampullaris*. The crista ampullaris contains the vestibular receptor cells, which are hair cells similar to those in the auditory system, and supporting cells. The crista ampullaris is encapsulated in the *cupula*, a gelatinous mass (Fig. 4.23). Each vestibular hair cell contains up to a 100 stereocilia and a single longer kinocilium. Rotation of the head causes movement of the endolymph fluid in the canals pushing the cupula, which in turn displaces the hair cell stereocilia. Mechanical movement of the vestibular hair cell stereocilia is transduced into an electrical signal in a manner similar to that described above for auditory hair cells. The responses of the stereocilia to movement are polarized with movements of the stereocilia toward the kinocilium depolarizing the hair cell while movement in the opposite direction hyperpolarizes the hair cell. The hair cells synapse with terminal endings of afferent neurons whose axons form the *vestibular nerve*. Depolarization increases

the release of neurotransmitters in the hair cells, bringing about a concomitant increased rate of firing in the afferent fibers. Conversely, hyperpolarization inhibits neurotransmitter release from the hair cells and, in turn, decreases the frequency of action potentials in the afferent nerve fibers.

Figure 4.23 **The sensory structure of the semicircular canals. (A)** The crista ampullaris contains the hair (receptor) cells, and the whole structure is deflected by motion of the endolymph. **(B)** An individual hair cell.

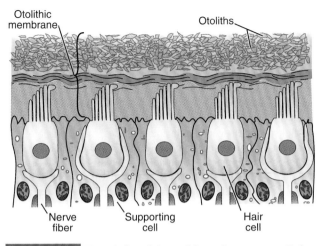

Figure 4.24 The relation of the otoliths to the sensory cells in the macula of the utricle and saccule. The gravity-driven movement of the otoliths stimulates the hair cells.

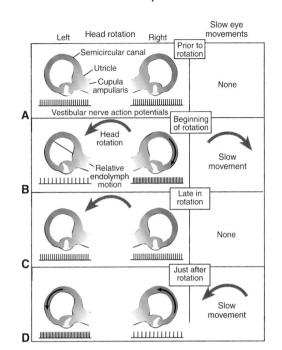

Figure 4.25 The role of the semicircular canals in sensing rotary acceleration. This sensation is linked to compensatory eye movements by the vestibulo-ocular reflex. Only the horizontal canals are considered here. This pair of canals is shown as if one were looking down through the top of a head looking toward the top of the page. Within the ampulla of each canal is the cupula, an extension of the crista ampullaris, the structure that senses motion in the endolymph fluid in the canal. Below each canal is the action potential train recorded from the vestibular nerve. **(A)** The head is still, and equal nerve activity is seen on both sides. There are no associated eye movements (right column). **(B)** The head has begun to rotate to the left. The inertia of the endolymph causes it to lag behind the movement, producing a fluid current that stimulates the cupulae (*arrows* show the direction of the relative movements). Because the two canals are mirror images, the neural effects are opposite on each side (the cupulae are bent in relatively opposite directions). The reflex action causes the eyes to move slowly to the right, opposite to the direction of rotation (right column). They then snap back and begin the slow movement again as rotation continues. The fast movement is called *rotatory nystagmus*. **(C)** As rotation continues, the endolymph "catches up" with the canal because of fluid friction and viscosity, and there is no relative movement to deflect the cupulae. Equal neural output comes from both sides, and the eye movements cease. **(D)** When the rotation stops, the inertia of the endolymph causes a current in the same direction as the preceding rotation, and the cupulae are again deflected, this time in a manner opposite to that shown in **(B)**. The slow eye movements now occur in the same direction as the former rotation.

Otolithic organs

Whereas the semicircular canals respond to rotation, the otolithic organs sense linear accelerations. The utricle and the saccule, are sac-like structures located in the vestibule, the bony chamber located between the semicircular canals and the cochlea. Calcium carbonate crystals, the **otoliths**, are suspended on a viscous gel layer over the hair cells and are heavier than their surroundings (Fig. 4.24). During linear acceleration, the otoliths get displaced, which, in turn, deflects the hair cell stereocilia producing a sensory signal. Most of the utricular signals elicit eye movements, whereas most of the saccular signals go to muscles that control our posture.

Vestibulo-ocular reflex

The **vestibulo-ocular reflex** (VOR) is a reflexive eye movement that stabilizes images on the retina during head movement (Fig. 4.25). This reflex produces an eye movement in the direction opposite to the head movement, thus preserving the image on the center of the visual field. The reflex is initiated when the rotation of the head is detected, which, in turn, triggers an inhibitory signal to the extraocular muscles on one side and an excitatory signal to the muscle on the other side. This reflex results in a compensatory movement of the eyes. The VOR has both rotational and translational inputs. When the head rotates about any axis (horizontal, vertical, or torsional), distant images are stabilized by rotating the eyes about the same axis, but in the opposite direction. Because slight head movements are happening all of the time, the VOR is very important for stabilizing vision. Patients whose VOR is impaired find it extremely difficult to read, because they cannot stabilize the eyes during small head movements. The VOR is driven by signals from the vestibular apparatus in the inner ear. The semicircular canals detect head rotation and drive the rotational VOR. The otoliths detect head translation and drive the translational VOR.

Vestibular dysfunction may result in immobilization.

The vestibular system is exquisitely designed to maintain equilibrium. However, there are instances wherein the system malfunctions and patients become immobilized. One such condition is vertigo, which results in dizziness or spinning movement when the patient is stationary. Vertigo is

CLINICAL FOCUS | 4.1

Cochlear Implants

Disorders of hearing are broadly divided into the following categories: conductive hearing loss, related to structures of the outer and middle ear; sensorineural hearing loss ("nerve deafness"), dealing with the mechanisms of the cochlea and peripheral nerves; and central hearing loss, concerning processes that lie in higher portions of the central nervous system. Damage to the cochlea, especially to the hair cells of the organ of Corti, produces sensorineural hearing loss by several means. Prolonged exposure to loud occupational or recreational noises can lead to hair cell damage, including mechanical disruption of the stereocilia. Such damage is localized along the basilar membrane at a position related to the pitch of the sound that produced it. Antibiotics such as streptomycin and certain diuretics can cause rapid and irreversible damage to hair cells similar to that caused by noise, but it occurs over a broad range of frequencies. Diseases such as meningitis, especially in children, can also lead to sensorineural hearing loss. In carefully selected patients, the use of a cochlear implant can restore some function to the profoundly deaf. The device consists of an external microphone, amplifier, and speech processor coupled by a plug-and-socket connection, magnetic induction, or a radiofrequency link to a receiver implanted under the skin over the mastoid bone. Stimulating wires then lead to the cochlea. A multielectrode intracochlear implant (with up to 22 active elements spaced along it) can be inserted into the basal turn of the scala tympani. The linear spatial arrangement of the electrodes takes advantage of the tonotopic organization of the cochlea, and some pitch (frequency) discrimination is possible. The external processor separates the speech signal into several frequency bands that contain the most critical speech information, and the multielectrode assembly presents the separated signals to the appropriate locations along the cochlea. In some devices, the signals are presented in rapid sequence, rather than simultaneously, to minimize interference between adjacent areas.

When implanted successfully, such a device can restore much of the ability to understand speech. Considerable training of the patient and fine-tuning of the speech processor are necessary. The degree of restoration of function ranges from recognition of critical environmental sounds to the ability to converse over a telephone. Cochlear implants are most successful in adults who became deaf after having learned to speak and hear naturally. Success in children depends critically on their age and linguistic ability; currently, implants are approved for use in children as young as 12 months of age.

Infrequent problems with infection, device failure, and natural growth of the auditory structures may limit the usefulness of cochlear implants for some patients. In certain cases, psychologic and social considerations may discourage the advisability of using auditory prosthetic devices in general. From a technical standpoint, however, continual refinements in the design of implantable devices and the processing circuit are extending the range of subjects who may benefit from cochlear implants. Research directed at external stimulation of higher auditory structures may eventually lead to even more effective treatments of profound hearing loss. ■

---

often associated with nausea and vomiting as well as difficulties with standing or walking. If the otolithic organs are stimulated rhythmically, as by the motion of a ship or automobile, distressing symptoms of **motion sickness** (dizziness and nausea) may result. Over time, these symptoms diminish and disappear. The underlying mechanisms causing vertigo and motion sickness are as yet poorly understood.

▶ GUSTATORY AND OLFACTORY SYSTEMS

Unlike the photoreceptors of the eye and the mechanoreceptors of the ear, the receptors of taste and smell are chemoreceptors. The sensation of taste and smell are two other sensory mechanisms that provide specific information about the external environment. In lower animals, the mechanisms of taste (gustation) and smell (olfaction) play a major role in finding food, seeking prey, finding directions, bonding with offspring and mates, and avoiding danger/predators. In the case of humans, most of these neural signals are associated with food, fragrance, and odors of our surrounding environment. Although our taste and smell are less sensitive than those of other species, millions of dollars are spent on additives to make our food taste better and on deodorants and perfumes to make

us more desirable/attractive and sociable. Chemoreceptors are also found in areas of the body not normally associated with taste or smell. For example, cells expressing bitter taste receptors are found in the nose, airway, and digestive tract. When receptors in the nose are stimulated by irritants, protective reflexes like apnea, coughing, and sneezing are elicited to keep the irritant from entering the lungs.

Taste buds house the receptor cells for gustatory sensations.

Taste is important for determining whether food in the mouth is a dangerous substance or safe to consume. The receptors for taste are packaged primarily in **taste buds** that are distributed across the upper alimentary canal with the majority of human taste buds located in the oral cavity (~5,000). The largest numbers are located on the dorsal portion of the tongue with smaller populations found in the epithelium of the soft palate, upper esophagus, epiglottis, and upper larynx. Taste buds are onion-shaped structures containing 50 to 100 elongated cells. Taste receptor cells are modified epithelial cells that extend from the basal lamina to the epithelial surface where their apical microvilli extend into an opening in the epithelium, the **taste pore**, to sample chemical compounds that are dissolved

CLINICAL FOCUS | 4.2

Vertigo

A common medical complaint is dizziness. This symptom may be a result of several factors, such as cerebral ischemia ("feeling faint"), reactions to medication, disturbances in gait, or disturbances in the function of the vestibular apparatus and its central nervous system (CNS) connections. Such disturbances can produce the phenomenon of vertigo, which may be defined as the illusion of motion (usually rotation) when no motion is actually occurring. Vertigo is often accompanied by autonomic nervous system symptoms of nausea, vomiting, sweating, and pallor. The body uses three integrated systems to establish its place in space: the vestibular system, which senses position and rotation of the head; the visual system, which provides spatial information about the external environment; and the somatosensory system, which provides information from joint, skin, and muscle receptors about limb position. Several forms of vertigo can arise from disturbances in these systems. Physiologic vertigo can result when there is discordant input from the three systems. Seasickness results from the unaccustomed repetitive motion of a ship (sensed via the vestibular system). Rapidly changing visual fields can cause visually induced motion sickness, and space sickness is associated with multiple-input disturbances. Central positional vertigo can arise from lesions in cranial nerve VIII (as may be associated with multiple sclerosis or some tumors), vertebrovascular insufficiency (especially in older adults), or from the impingement of vascular loops on neural structures. It is commonly present with other CNS symptoms. Peripheral vertigo arises from disturbances in the vestibular apparatus itself. The problem may be either unilateral or bilateral. Causes include trauma, physical defects in the labyrinthine system, and pathologic syndromes such as Ménière disease. As in the cochlea, aging produces considerable hair cell loss in the cristae and maculae of the vestibular system. Rotational or caloric stimulation can be used as an indicator of the degree of vestibular function. The most common form of peripheral vertigo is benign paroxysmal positional vertigo (BPPV). This is a severe vertigo, with incidence increasing with age. Episodes appear rapidly and are limited in duration (from minutes to days). They are usually brought on by assuming a particular position of the head such as one might do when painting a ceiling. BPPV is thought to be a result of the presence of canaliths, debris in the lumen of one of the semicircular canals. The offending particles are usually clumps of otoconia (otoliths) that have been shed from the maculae of the saccule and utricle, whose passages are connected to the semicircular canals. These clumps act as gravity-driven pistons in the canals, and their movement causes the endolymph to flow, producing the sensation of

rotary motion. Because they are in the lowest position, the posterior canals are the most frequently affected. In addition to the rotating sensation, this input gives rise, via the vestibular system, to a pattern of nystagmus (eye movements) appropriate to the spurious input.

The specific site of the problem can be determined by using the *Dix–Hallpike maneuver*, which is a series of physical maneuvers (changes in head and body position). By observing the resulting pattern of nystagmus and reported symptoms, the location of the defect can be deduced. Another set of maneuvers known as the *canalith repositioning procedure of Epley* can cause gravity to collect the loose canaliths and deposit them away from the lumen of the semicircular canal. This procedure is highly effective in cases of true BPPV, with a cure rate of up to 85% on the first attempt and nearly 100% on a subsequent attempt. Patients can be taught to perform the procedure on themselves if the problem returns.

Ménière disease is a syndrome of uncertain (but peripheral) origin associated with vertigo. Its cause(s) and precipitating factors are not well understood. Typical associated findings include fluctuating hearing loss and tinnitus (ringing in the ears). Episodes involve increased fluid pressure in the labyrinthine system, and symptoms may decrease in response to salt restriction and diuretics. Other cases of peripheral vertigo may be caused by trauma (usually unilateral) or by toxins or drugs (such as some antibiotics); this type is often bilateral.

Central and peripheral vertigo often may be differentiated on the basis of their specific symptoms. Peripheral vertigo is more severe, and its nystagmus shows a delay (latency) in appearing after a position change. Its nystagmus fatigues and can be reduced by visual fixation. Position-sensitive and of finite duration, the condition usually involves a horizontal orientation. Central vertigo, usually less severe, shows a vertically oriented nystagmus without latency and fatigability; it is not suppressed by visual fixation and may be of long duration.

Treatment of vertigo, beyond that mentioned above, can involve bed rest and vestibular inhibiting drugs (such as some antihistamines). However, these treatments are not always effective and may delay the natural compensation that can be aided by physical motion such as walking (unpleasant as that may be). In severe cases that require surgical intervention (labyrinthectomy), patients can often achieve a workable position sense via the other sensory inputs involved in maintaining equilibrium. Some activities, such as underwater swimming, must be avoided by those with an impaired sense of orientation, because false cues may lead to moving in inappropriate directions and increase the risk of drowning. ■

in saliva (Fig. 4.26). Only food that is dissolved in saliva can attach to taste receptor cells and evoke a neural sensation of taste. Taste buds contain four different types of cells. Type I cells act like glial cells providing support and maintaining the extracellular environment within the taste bud. Type I cells do

not make synaptic contact with afferent nerve fibers. Type II cells contain, in separate subpopulations, receptors for sweet, bitter, and umami. Although this cell type is in close apposition with numerous afferent fibers, they do not make traditional synaptic contacts with these fibers. Instead, these cells exhibit

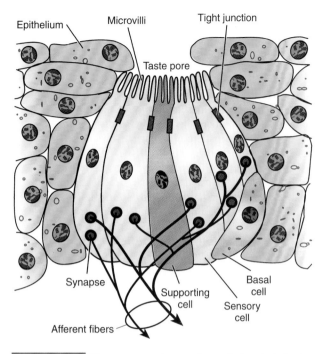

Figure 4.26 **The sensory and supporting cells in a taste bud.** Taste buds consist of taste cells surrounded by supporting epithelial cells and basal cells. The afferent nerve synapses with the basal areas of the sensory cells.

nonvesicular release of ATP that activates purinergic receptors on afferent nerve fibers and adjacent taste cells. Type III cells contain receptors for sour and form conventional synapses with afferent fibers. This cell type releases several transmitters that interact with sensory nerve endings and other cells in the taste bud. The fourth cell type, the basal cell, acts as a precursor cell able to differentiate into new taste cells. The sensory cells typically have a lifespan of 10 days. They are continually replenished by new sensory cells formed from the basal cells. When a sensory cell is replaced by a maturing basal cell, the old synaptic connections are broken, and new ones must be formed.

Stimulating the taste cells alters the cell's ion channels producing a depolarizing potential. Similar to other sensory receptors, a depolarizing potential leads to neurotransmitter release. The neurotransmitter, in turn, triggers action potentials in the afferent nerve fibers that are sent to the solitary nucleus in the brainstem via the seventh, ninth, and tenth cranial nerves. From the solitary nucleus, information is distributed to cell groups that are involved in feeding and digestive-related functions like salivary and gastrointestinal secretions, and swallowing. Taste information is also relayed, via the thalamus, to cortical structures responsible for the appreciation of taste quality and intensity as well as to cortical areas where multisensory integration of taste, smell, tactile, and temperature cues occurs to produce the experience we call flavor.

Gustatory chemoreception distinguishes five primary taste categories.

Among thousands of different taste sensations, humans can discriminate between five specific tastes received by the gustatory receptors. These are salty, sweet, bitter, sour, and umami,

which means "savory" or "meaty" in Japanese. There are also two "accessory qualities" of taste sensation, alkaline (soapy) and metallic. Recently, evidence has accumulated for another taste sensation—fatty. Scientists have identified several potential receptors for free fatty acids. These receptors are located on taste receptor cells and their activation results in an intracellular cascade similar to that observed following receptor activation by sweet, bitter, and umami compounds. Although the reception of different taste qualities has been traditionally represented as occupying specific parts of the tongue (i.e., the receptors for sweetness just behind the tip of the tongue, sour receptors predominately along the sides, salt at the tip, and bitter across the rear), this "tongue map" theory has been refuted. Rather, all areas respond to all five tastes, although there are differences in regional sensitivities to these tastes.

Gustatory transduction is mediated by multiple receptors and intracellular mechanisms.

Multiple receptors and intracellular mechanisms are responsible for converting the chemical compounds in food into the experience we call taste. Some mechanisms involve simple ion channels in the taste receptor membrane, while others activate intracellular second messenger cascades.

Salt receptors

Salt plays a critical role in ion and water homeostasis in animals, especially mammals. NaCl is especially important in the mammalian kidney as an osmotically active molecule that facilitates passive reuptake of water. Because of this, salt elicits a pleasant taste in most mammals, which potentially can lead to excess salt intake in the human diet. The simplest receptor found in the mouth is the NaCl receptor. Na^+ ions enter the taste cell directly through apical cation channels to depolarize the cell. This sodium channel is known as the epithelial sodium channel, or ENaC, and is composed of three subunits. The ENaC channel can be blocked by the drug amiloride in many mammals. However, the action of amiloride is much less pronounced on the salt receptors in humans leading to the speculation that there are additional membrane receptors or ion channels responsible for salt taste.

Sour receptors

Sour taste signals the presence of acidic foods (H^+ ions in solution). The citric taste of lemon is a good example of our experience with the distinctive sour taste. Sour taste can be mildly pleasant in small quantities and is linked to salt flavor. However, in larger quantities, it becomes more of an unpleasant taste. This is a protective mechanism and can signal an overripe fruit, rotten meat, or other spoiled foods that can be toxic to the body because of bacterial contamination. Also, sour taste signals acid (H^+ ions) to the brain, which can warn of serious tissue damage.

Although not completely understood, several mechanisms have been proposed for sour transduction. One mechanism is a simple proton channel that allows hydrogen ions to flow directly into the receptor cell to acidify the cytosol, while another mechanism is the simple diffusion of acids through the plasma membrane into the cell where they dissociate to acidify the cytosol. The acidification of the cytosol opens proton-sensitive sodium channels, the influx of sodium ions producing membrane depolarization, the opening of a voltage-gated Ca^+ channels, and the release of neurotransmitters from the receptor cell.

Receptors for bitter, sweet, and umami

Two families of taste **G protein–coupled receptors** (GPCR) have been identified, type 1 and type 2 taste receptors (T1R and T2R). The T1R family mediates sweet and umami taste, while the T2R mediates bitter taste.

The bitter taste is almost completely unpleasant in humans. This is because many nitrogenous organic compounds that have a pharmacologic effect on humans taste bitter. These include caffeine, the stimulant in coffee, and nicotine, the addictive compound in cigarettes. Almost all poisonous/toxic plants taste bitter; the bitter taste is a protective mechanism, and the reaction is a last-line warning system before the compound is swallowed to cause injury or death. However, humans have evolved a very sophisticated sense for bitter substances and can distinguish between many different compounds. As a result, they have overcome their innate aversion to certain bitter tastes, as evidenced by the wide consumption of caffeinated drinks enjoyed around the world.

Cells that detect bitter flavors have a large number of receptors, each of which responds to a different bitter flavor. This mechanism allows for many different classes of bitter compounds, which can be chemically very different, to be detected. TheT2R bitter taste receptors are GPCRs that are coupled to the G-protein **gustducin**, a G-protein similar to the visual G-protein, transducin. When a bitter compound binds to a T2R, gustducin is released. Gustducin activates phospholipase C, which stimulates the synthesis of 1,4,5-inositol triphosphate (IP_3) leading to Ca^{2+} release from intracellular stores. The increased intracellular calcium activates monovalent cation channels leading to membrane depolarization, opening of ATP permeable membrane channels and ATP release into the synaptic cleft. ATP in turn activates the afferent nerve endings in the taste bud.

Sweet taste signals the presence of readily usable food. Humans have evolved to seek out the highest caloric intake, because our ancient ancestors were not always sure when the next meal would be available. Carbohydrates have a very high caloric count (therefore, much energy) and, as a result, have been sought out by humans throughout history. Unfortunately, this evolutionary pattern has led to a craving of sweets. Carbohydrates are used as direct energy (sugars) and as stored energy (glycogen).

Like bitter taste T2Rs, the T1Rs for sweet taste transduction involve GPCRs. Three different T1Rs have been identified, T1R1, T1R2, and T1R3. The receptor for sweet compounds is the heterodimer T1R2 + T1R3. Binding of natural sugars, such as cane sugar, to this receptor, activates the GPCR and produces ATP release in manner similar to that described for bitter receptors. Artificial sweeteners, including saccharin, sucralose, and aspartame produce the experience of sweet by binding to the same sweet receptor, but at different binding domains

The umami taste signals the presence of meat, thus encouraging the intake of peptides and proteins. The signal is triggered by amino acids (specifically glutamine), which are used to build such things as muscle, organs, transport molecules (e.g., hemoglobin), and cellular enzymes. Amino acids are critical for the human body to function; therefore, it is important to have a steady supply. Hence, the umami taste signals a pleasurable response to a desirable, nutritionally rich source of protein. The umami receptor is a T1R1 + T1R3 heterodimer GPCR. Glutamate binds to the GPCR and activates the same second-messenger pathways used for transduction of bitter and sweet compounds. Umami receptors also respond to monosodium glutamate (commonly known as *MSG*), which is used as a food additive and is very popular in Asia, especially Japanese dishes.

Taste information is relayed to the brainstem by cranial nerves VII, IX, and X.

These nerves terminate in the solitary nucleus, a nucleus that also processes visceral sensory information and is in involved in visceral-related functions. From the solitary nucleus, information is relayed to cell groups that are involved in feeding and digestive-related functions like salivary and gastrointestinal secretions and swallowing. Taste information is also relayed, via the thalamus, to cortical structures responsible for the appreciation of taste quality and intensity as well as to cortical areas where multisensory integration of taste, smell, tactile, and temperature cues occurs to produce the experience we call flavor.

Olfactory apparatus serves the sense of smell.

The olfactory system has a much broader function than does the gustatory system. It not only participates in the selection and enjoyment of food but also is involved in detecting smell from the surrounding environment (e.g., fragrance of flowers, other people, and dangerous odors that can be harmful to the body). Olfaction compliments the gustatory system and is especially important for the appreciation of flavor. For example, savoring the aroma of a glass of red wine often surpasses the gustatory system. Many patients who complain about the loss of taste often have an olfactory disorder.

Humans, compared with some mammals, have a relative poor sense of smell. However, the human olfactory system is still quite extraordinary insofar as the nose contains >5 million olfactory receptors that can differentiate thousands of different odorants, even some that differ by a single molecular component. The olfactory system detects odorants that are inhaled through the nose, where they contact the olfactory epithelium, which contains various olfactory receptors. The receptor organ for olfaction, the olfactory epithelium, is located in the roof of the nasal cavity. Normally, there is little airflow in this region of the nasal tract, but sniffing serves to direct air upward, increasing the likelihood of an odor being detected. Another route that volatile compounds in food can reach the olfactory epithelium is retronasally from the oral cavity. In contrast to the taste sensory cells, the olfactory receptor cells are bipolar neurons and, as such, are *primary receptors*. These cells are interspersed among supporting cells that bind the cells together at their sensory ends and basal

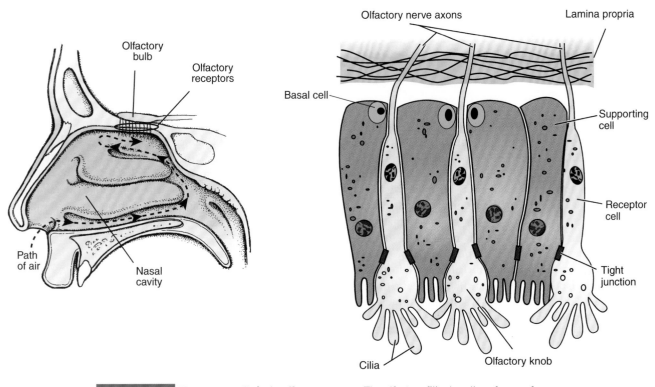

Figure 4.27 **The sensory cells in the olfactory mucosa.** The olfactory filia, bundles of axons forming the olfactory nerve, represent axons of olfactory receptor cells that terminate in the olfactory bulb.

cells (Fig. 4.27). Like taste receptor cells, olfactory receptor neurons continuously turnover with new receptor cells arising from basal cells. The epithelium is covered by a watery mucus through which odors diffuse to reach the olfactory receptors that are located on the cilia of the olfactory receptor neurons. Rather than binding specific ligands like most receptors, olfactory receptors have a binding affinity for a range of odorant molecules. This allows the olfactory system to discriminate among a wide range of different odor molecules. Olfactory thresholds vary widely from substance to substance; the threshold concentration for the detection of ethyl ether (used as a general anesthetic) is around 5.8 mg/L air, whereas that for methyl mercaptan (the odor of garlic) is ~0.5 ng/L. This represents a 10 million–fold difference in sensitivity.

The detection of an odor occurs when an odor diffuses into the mucus overlying the receptors and is transported to the receptors on the olfactory receptor dendrites by odorant binding proteins. The odors bind to olfactory receptors, which are GPCRs. This binding causes the production of cAMP, which binds to, and opens, cyclic nucleotide gated ion channels in the ciliary membrane allowing sodium and calcium ions to enter the cell. This inward flow depolarizes the olfactory receptor cell to produce a generator potential, which in turn causes action potentials in the afferent fiber. The frequency of the action potentials is dependent on the concentration of the odorant. The action potentials in the axons of the olfactory receptor neurons travel along the olfactory nerves (afferent nerves) through small foramina in the bony cribriform plate above the nose to terminate

in the **olfactory bulbs** located at the base of the forebrain. The transiting of the very thin olfactory axons through the cribriform plate makes this region particularly sensitive to traumatic injury that severs the olfactory receptor neuron axons leading to olfactory loss. The olfactory bulbs serve as a relay station that transmits information from the nose to other brain areas and consist of a complex multilayer neural architecture. As they enter the olfactory bulb, the olfactory nerve axon terminals cluster in a ball-like neural junction known as the *olfactory glomeruli*. Complex processing of odor information occurs within the glomerulus with each glomerulus receiving input primarily from olfactory receptor neurons that express similar olfactory receptors. Thus, the glomeruli serve to categorize the sense of smell. Glomeruli are also permeated by dendrites from neurons termed *mitral and tufted cells*, which, in turn, relay information to the olfactory cortex located at the base of the temporal lobe. The primary olfactory cortex further processes olfactory information to determine odor quality and familiarity and is associated with the learning and remembering of odors. Projections from primary olfactory cortex converge on forebrain association cortices associated with the processing of flavor as well limbic areas associated with emotions and memory, which may explain the close association of odors with these functions.

Sneezing

The olfactory mucosa also contains receptors and sensory fibers sensitive to irritants and certain odorous substances, such as peppermint and chlorine, that play a role in the initiation of reflex responses (e.g., sneezing) that result from irritation of the nasal tract. Sneezing typically occurs when an irritant passes through the nasal hairs to reach the nasal mucosa. This triggers the release of histamine, which, in turn, irritates the nerve cells in the nose, resulting in an electrical signal being sent to the brain via the trigeminal nerve to initiate the sneeze reflex. The brain then activates the pharyngeal, tracheal, and chest muscles to expel a large volume of air from the lungs through the nose and mouth. Sneezing is also triggered by sinus infection and allergies.

INTEGRATED MEDICAL SCIENCES

Age-Related Macular Degeneration

The leading cause of impaired vision in older adults is a condition known as age-related macular degeneration (AMD). Approximately 30% of persons older than 75 years have some form of this condition, which leads to decreased sharpness of central vision, especially in bright light. It is a progressive condition whose causes are obscure and for which no cure is available. Despite this, a number of current and emerging treatments can slow the progress of the disease and provide some improvement in function.

Approximately 10% to 15% of AMD cases are of the *exudative*, or "wet" type. It arises from a proliferation of blood vessels in the choroid layer of the retina behind the macula lutea. The macula lutea, which includes the fovea, is a small area of densely packed photoreceptor cells directly at the fixation point of straight-ahead vision. This area is responsible for our sharpest vision, and its mechanical disruption by the proliferating blood vessels leads to distortion of vision and progressive loss of photoreceptors. The process does not affect areas of the retina peripheral to the macula, so some peripheral vision is preserved. The remaining AMD cases are of the "dry" type, with changes in the retinal pigment epithelium and "drusen" (see below). This latter form poses less immediate threat to vision, and high levels of zinc and antioxidants can slow its progress.

The retina of an affected eye is characterized by the presence of drusen, which are localized deposits of cellular debris whose accumulation further distorts the shape of the retina. An important diagnostic tool, and one used by patients to follow the course of the disease, is the *Amsler grid*, a grid of black lines forming a horizontal and vertical meshwork. The lines, when observed by one eye at a time, will appear "wavy" when the retina is distorted. Thus, the patient may detect sudden changes in visual acuity and seek therapy.

The causes of the condition are still obscure. There is a strong suggestion of a hereditary predisposition, and it is most common among white persons from North America and Europe. The primary biologic agent associated with the process of choroidal neovascularization is a molecule known as vascular endothelial growth factor (VEGF). This factor is responsible for the growth of blood vessels (angiogenesis) at many sites in the body. Some evidence also implicates an inflammatory process as an etiologic factor. The disease is also associated with cardiovascular illness and the risk factors that it entails. Smoking also is an aggravating factor. Although there are no natural animal models for a human-like AMD, several transgenic mouse models can duplicate features of the disease and are being used in its study. It has recently been found that mice with a congenital lack of photoreceptors can be made sensitive to light by the retinal implantation of stem cell–derived embryonic photoreceptor (rod) cells. This appears to be a promising avenue for eventual treatments of a number of types of blindness including AMD.

For the early stages of AMD, there is little specific treatment available. Some help has been obtained by dietary supplementation with high-dose antioxidant vitamins (such as C, E, and β carotene) and zinc. Lutein may provide some benefit, and the effects of corticosteroids are being investigated. For advanced wet AMD, there are several avenues of therapy, and others are being developed. One of the first such approaches was *laser coagulation* of the invading blood vessels. This is a thermal process, and it damages the overlying retinal tissue, producing blind spots. This limits its use to areas outside the macula. A more recent technique is *verteporfin photodynamic laser treatment*; in this method, a dye is injected into the retinal circulation. When a low-intensity laser pulse activates it, the dye releases free radicals that cause occlusion in the target vessel without damage to the retina. Because of the role of VEGF in the neovascularization process, use of an isoform-specific VEGF inhibitor called *pegaptanib sodium* has achieved some success. This drug is well tolerated, and its effectiveness has been established. However, it must be delivered by a long series of intraocular injections. The U.S. Food and Drug Administration has recently approved the drug *ranibizumab*, which is a recombinant humanized monoclonal antibody fragment that targets all of the isoforms of VEGF. In patients with the wet form of AMD, this drug (trade name Lucentis) has been shown to improve vision. Finally, a surgical approach involves the implantation of a miniature telescope in the anterior chamber of the eye in the place of the biologic lens. This optical system spreads the light usually delivered to the fovea (5° of the visual field) to a much larger and usually undamaged surface of the more peripheral retina (a 55° field). Such a device, medically suited to only a subgroup of AMD patients, requires relearning of the visual process but can significantly increase the amount of useful vision.

AMD is a serious detriment to the quality of life in older persons and to those who care for them. Current research is aimed at a better understanding of the cause and progression of the disease and at finding specific therapies that are both effective and practical to apply. ■

Neuromuscular Physiology

Chapter Summary

- Sensory transduction takes place in a series of steps, starting with stimuli from the external or internal environment and ending with neural processing in the central nervous system.
- The structure of sensory organs optimizes their response to the preferred types of stimuli.
- A stimulus gives rise to a generator potential, which, in turn, causes action potentials to be produced in the associated sensory nerve.
- The speeds of adaptation of particular sensory receptors are related to their biologic roles.
- Specific sensory receptors for a variety of types of tactile stimulation are located in the skin.
- Somatic pain is associated with the body surface and the musculature; visceral pain is associated with the internal organs.
- The sensory function of the eyeball is determined by structures that form and adjust images and by structures that transform images into neural signals.
- The retina contains a number of layers and several cell types, each with a specific role in the process of visual transduction.
- The rod cells in the retina have a high sensitivity to light but produce less distinct images without color, whereas the cones provide sharp color vision with less sensitivity to light.
- The visual transduction process requires many steps, beginning with the absorption of light and ending with an electrical response.

- The outer ear receives sound waves and passes them to the middle ear. They are transmitted by the bones of the middle ear and passed to the inner ear, where the process of sound transduction takes place.
- The transmission of sound through the middle ear greatly increases the efficiency of its detection, whereas its protective mechanisms guard the inner ear from damage caused by extremely loud sounds. Disturbances in this transmission process can lead to hearing impairments.
- Sound vibrations enter the cochlea through the oval window and travel along the basilar membrane, where their energy is transformed into neural signals in the organ of Corti.
- Displacements of the basilar membrane cause deformation of the hair cells, the ultimate transducers of sound. Different sites along the basilar membrane are sensitive to different frequencies.
- The vestibular apparatus senses the position of the head and its movements by detecting small deflections of its sensory structures.
- Taste is mediated by sensory epithelial cells in the taste buds. There are five fundamental taste sensations: sweet, sour, salty, bitter, and umami. Several receptors and mechanisms are responsible for taste transduction.
- Smell is detected by nerve cells in the olfactory mucosa. Thousands of different odors can be detected and distinguished.

Chapter Review Questions

1. While recording the responses of a mechanoreceptor to stimulation of the skin, an investigator observes an increase in the number of action potentials. This increase usually signifies

   A. increased intensity of a stimulus.
   B. cessation of a stimulus.
   C. adaptation of the receptor.
   D. a constant and maintained stimulus.
   E. an increase in the action potential conduction velocity.

The correct answer is A. Receptors code an increase in stimulus intensity by an increase in action potential frequency. A larger stimulus produces a larger depolarization of the receptor membrane (i.e., a larger generator potential), which in turn is translated into a larger number of action potentials at impulse initiation region of the receptor. Cessation of a stimulus would produce a rapid decrease in action potential frequency, and adaptation of the receptor would also decrease action potential frequency. A constant and maintained stimulus would result in a steady response or a decrease due to adaptation. The action potential velocity is determined by the membrane properties of the nerve and would not be affected.

2. The transduction of sweet compounds is the result of:

   A. receptor cell intracellular osmotic changes caused by diffusion of sweet molecules into the taste pore.
   B. binding of sweet compounds to sweet specific ion channels in the apical membranes of taste receptor cells.
   C. sweet compounds binding to T1R G-protein–coupled receptors in the apical membranes of taste receptor cells.

   D. sweet molecules binding to T2R G-protein–coupled receptors in the basolateral membrane of taste receptor cells.
   E. sweet molecules entering the taste receptor cell through sweet channels in the apical membranes of taste receptor cells.

The correct answer is C. Sweet compounds bind to the T1R family of GPCRs taste receptors that are located on the microvilli of taste bud cells. Diffusion of sweet taste compounds into the mucus-filled taste pore is necessary for sweet taste transduction, but it does not produce significant osmotic changes inside the taste receptor cells. Sweet receptors are not coupled directly to ion channels. The T2R family of taste receptors are responsible for the transduction of bitter compounds and they are located on the microvilli. There are no sweet membrane channels in taste cells, but the entry of H+ ions through membrane proton channels is a mechanism for sour taste transduction.

3. As a person enters middle age, reading text at distances less than arm's length becomes more difficult. This difficulty is most likely due to:

   A. Age-related loss of in the transparency of the lens
   B. Age-related distortion of the cornea
   C. Age-related loss of receptor cells in the foveal region of the retina
   D. Age-related shortening of the eyeball
   E. Age-related changes in the elasticity of the lens

The correct answer is E. The person is suffering from presbyopia, the age-related inability to focus on close objects. Due to age-related changes in the elasticity of the lens the ability to accommodate decreases and the lens shape cannot be sufficiently curved so the near object is focused behind the retina. A loss of lens transparency (cataracts), corneal distortion (astigmatism) or receptor cells in the region of the fovea (age-related macular degeneration) would impair both near and far vision. As a person ages only minor changes occur in the shape of the eyeball.

4. Due to long-term exposure to a loud noise, the hair cells on the basilar membrane in the inner ear were damaged. The site of the damage was in the region of the basilar membrane nearest to the oval window. The damage will result in:

   A. diminished sensitivity to all frequencies of sound
   B. loss of hearing of high frequencies
   C. loss of hearing of low frequencies
   D. an exaggerated sensitivity to loud sounds
   E. complete deafness in the affected ear

The correct answer is B. The basilar membrane contains a tonotopic map with high frequencies represented at the base of the basilar membrane near the oval window and low frequencies represented at its apex near the helicotrema. Long-term exposure to excessive noise at a particular frequency results in damage to the area of the basilar membrane responsible for coding that frequency.

5. On a moonlit night, human vision is monochromatic and less acute than vision during daytime. This is because:

   A. objects are being illuminated by monochromatic light and there is no opportunity for color to be produced.
   B. the cone cells of the retina, while more closely packed than rod cell, have lower sensitivity to light of all colors.
   C. light rays of low intensity do not carry information as to color.
   D. retinal photoreceptor cells that have become dark adapted can no longer respond to carrying wavelengths of light.
   E. at low light levels, the lens cannot accommodate to sharpen vision.

Correct answer is B. The cone cells are responsible for color vision and are densely packed in the centrally located fovea, the region of the retina where the image is most focused. However, cone cells cease to function if light levels are too low. In such cases, the single-pigment rod cells that have greater sensitivity, but are located farther away from the fovea and show greater convergence, provide monochromatic but more diffuse vision. The color composition of light does not depend on its intensity, and dark adaptation does not change the spectral sensitivity of the receptors. While focusing mechanisms may be less effective with low light, they still function.

Clinical Application Exercises 4.1

A 43-year-old business executive who recently returned from a flight overseas complains that he had difficulty "popping" his ears as the plane descend and that he experienced significant pain during the episode. After landing, the pain gradually disappeared, but sounds are muffled and he has a sensation of fullness in the ears. His attempts to pop his ears continue to be unsuccessful. Otoscopic examination of his ears reveals tight and slightly distended eardrums with effusion, but no signs of infection, and

the patient's temperature is normal. The patient had a difficult time hearing a finger snap adjacent to either ear. In a Rinne test bone conduction was superior to air conduction in both ears.

1. What does the tight distended ear drum and effusion in this patient suggest?
2. Why did the patient experience a reduction in his ability to hear?
3. What does the inability to "pop" the ears suggest?

ANSWERS

1. The tight distended eardrum suggests that there is an increase in pressure within the middle ear, perhaps due in part to his inability to equalize the external and middle ear pressures ("the pop" during the plane's descent). The effusion suggests an accumulation of fluid in the normally air-filled cavity is also contributing to the increased pressure.
2. The patient decreased ability to hear a sound reflects a conductive hearing loss that is a result of increased pressure in the middle ear cavity. The increased pressure results in a less flexible tympanic membrane and consequently less energy is transmitted to the ossicles in response to an incoming sound. The increased pressure and fluid can also interfere with the

movement of the ossicles further reducing energy transfer through the middle ear. The Rinne test is a relatively simple test that can quickly screen for the presence of conductive hearing loss.

3. The inability to "pop," or equalize the pressure between the outer and middle ear, indicates that the eustachian tube is blocked. The eustachian tube interconnects the middle ear cavity and the pharynx and is the route by which pressures in the outer and middle ear are equalized.

*the*Point® *Visit* http://thepoint.lww.com/rhoades5e *for additional chapter review Q&A, Clinical Application Exercises, animations, and more!*

Neuromuscular Physiology

Active Learning Objectives

Upon mastering the material in this chapter, you should be able to:

- Explain the role of α motor neurons in skeletal muscle control.
- Explain how the actions of the muscle spindle and Golgi tendon organ regulate muscle action.
- Compare the anatomy of motor neuron subgroups in relation to the muscle groups controlled.
- Explain the three classic spinal reflexes and what they reveal about spinal cord circuitry.
- Explain how the brainstem nuclei and their associated descending spinal cord tracts influence motor neuron function.
- Explain the role of the cerebral cortex, the corticospinal tract, and the basal ganglia in the control of movement.
- Explain the influence of the cerebellum in motor control.

The finger movements of a neurosurgeon manipulating microsurgical instruments while repairing a cerebral aneurysm and the eye–hand–body coordination of a professional basketball player making a rimless three-point shot are two examples of the motor control system operating at high skill levels. The coordinated contraction of the hip flexors and ankle extensors to clear a slight pavement irregularity encountered during walking is a familiar example of the motor control system working at a seemingly automatic level. Conversely, the stiff-legged stride of a patient who experienced a stroke and the swaying walk of an intoxicated person are examples of perturbed motor control.

Although our understanding of the physiology of the motor system is still far from complete, a significant fund of knowledge exists. This chapter proceeds through the major components of the motor system, beginning with the skeleton and ending with the brain.

▶ SKELETON AS FRAMEWORK FOR MOVEMENT

Bones are the body's framework as well as a system of levers. They are the elements that move. The way adjacent bones articulate determines the motion and range of movement at a joint. Ligaments hold the bones together across the joint. Movements are described on the basis of the anatomic planes through which the skeleton moves and the physical structure of the joint. Most joints move in only one plane, but some permit movement in multiple anatomic reference planes (Fig. 5.1).

Hinge joints, such as the elbow, are uniaxial, permitting movements in the sagittal plane. The wrist is an example of a biaxial joint. The shoulder is a multiaxial joint; movement can occur in oblique planes as well as the three major planes of that joint. **Flexion** and extension describe movements in the sagittal plane. Flexion movements decrease the angle between the moving body segments. **Extension** describes movement in the opposite direction. **Abduction** moves the body part away from the midline, whereas **adduction** moves the body part toward the midline.

▶ MUSCLE FUNCTION AND BODY MOVEMENT

Muscles span joints and are attached at two or more points to the bony levers of the skeleton. The muscles provide the power that moves the body's levers. Muscles are described in terms of their origin and insertion attachment sites. The *origin* tends to be the more fixed, less mobile location, whereas the *insertion* refers to the skeletal site that is more mobile. Movement occurs when a muscle generates force on its attachment sites and undergoes shortening. This type of action is termed an **isotonic** or **concentric contraction**. Another form of muscular action is a controlled lengthening while still generating force. This is an **eccentric contraction**. A muscle may also generate force but hold its attachment sites static, as in **isometric contraction**.

Because muscle contraction can produce movement in only one direction, at least two muscles opposing each other at a joint are needed to achieve motion in more than one direction. When a muscle produces movement by shortening, it is an **agonist**. The prime mover is the muscle that contributes most to the movement. Muscles that oppose the action of the prime mover are **antagonists**. The quadriceps and hamstring muscles are examples of agonist–antagonist pairs in knee extension and flexion. During both simple and light-load skilled movements, the antagonist is relaxed. Contraction of the agonist with concomitant relaxation of the antagonist occurs by the nervous system function of **reciprocal inhibition**. Cocontraction of agonist and antagonist occurs during movements that require precise control.

A muscle functions as a **synergist** if it contracts at the same time as the agonist while cooperating in producing the movement. Synergistic action can aid in the following: producing a movement (e.g., the activity of both the flexor carpi ulnaris and extensor carpi ulnaris is used in producing ulnar deviation of the wrist), eliminating unwanted movements (e.g., the activity of wrist extensors prevents flexion of the wrist when the finger flexors contract in closing the hand), or stabilizing proximal joints (e.g., isometric contractions of muscles of the forearm, upper arm, shoulder, and trunk accompany a forceful grip of the hand).

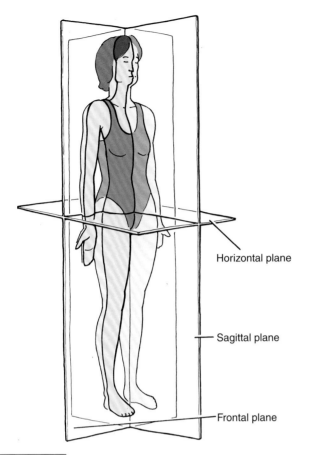

Figure 5.1 **Anatomic reference planes.** The figure is shown in the standard anatomic position with the associated primary reference planes.

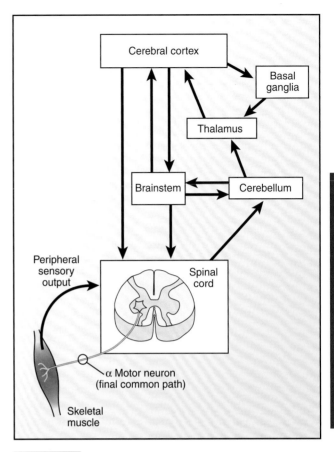

Figure 5.2 **Motor control system.** α Motor neurons are the final common path for motor control. Peripheral sensory input and spinal cord tracts that descend from the brainstem and the cerebral cortex influence the motor neurons. The cerebellum and basal ganglia contribute to motor control by modifying the brainstem and the cortical activity.

Neuromuscular Physiology

► NERVOUS SYSTEM COMPONENTS FOR THE CONTROL OF MOVEMENT

Here, we identify the components of the nervous system that are predominantly involved in the control of motor function and discuss the probable roles for each of them. It is important to appreciate that even the simplest reflex or voluntary movement requires the interaction of multiple levels of the nervous system (Fig. 5.2).

Motor neurons are the final common path for motor control.

The α motor neurons are referred to as final common pathway neurons because they are the route by which the central nervous system (CNS) controls the skeletal muscles. These motor neurons located in the ventral horns of the spinal gray matter and brainstem cranial nerve nuclei are influenced both by local reflex circuitry and by pathways that descend from the brainstem and cerebral cortex. The brainstem-derived pathways include the rubrospinal, vestibulospinal, and reticulospinal tracts; the cortical pathways are the corticospinal and corticobulbar tracts. Although some of the cortically derived axons terminate directly on motor neurons, most of the axons of the cortical-derived and brainstem-derived tracts terminate on interneurons, which then influence motor neuron function. The outputs of the

basal ganglia and cerebellum, two additional motor-related areas, provide fine-tuning of the cortical and brainstem influences on motor neuron functions.

Motor neurons segregate into two major categories, α (alpha) and γ (gamma). **α Motor neurons** innervate the extrafusal muscle fibers, which are responsible for force generation. **γ Motor neurons** innervate the intrafusal muscle fibers, which are components of the muscle spindle. γ Motor neurons are discussed further in the next section. An α motor neuron controls several muscle fibers, 10 to 1,000, depending on the muscle. The term **motor unit** describes an α motor neuron, its axon, the branches of the axon, the neuromuscular junction synapses at the distal end of each axon branch, and all of the extrafusal muscle fibers innervated by that motor neuron (Fig. 5.3). When a motor neuron is activated, all of its muscle fibers are also activated.

α Motor neurons can be separated into two general populations according to their cell body size and axon diameter. The larger cells have a high threshold to synaptic stimulation, have fast action potential conduction velocities, and are active in high-effort force generation. They innervate fast-twitch, high-force but fatigable muscle fibers. The smaller motor neurons have lower thresholds to synaptic stimulation,

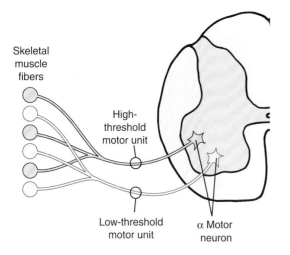

Skeletal
muscle
fibers

High-
threshold
motor unit

Low-threshold
motor unit

α Motor
neuron

Figure 5.3 **Motor unit structure.** A motor unit consists of an α motor neuron and a group of extrafusal muscle fibers it innervates. Functional characteristics, such as activation threshold, twitch speed, twitch force, and resistance to fatigue, are determined by the motor neuron. Low-threshold and high-threshold motor units are shown.

conduct action potentials at slightly slower velocities, and innervate slow-twitch, low-force, fatigue-resistant muscle fibers (see Chapter 8). The muscle fibers of a motor unit are homogeneous, either fast-twitch or slow-twitch. The twitch property of a muscle fiber is determined by the motor neuron, presumably by trophic substances released at the neuromuscular junction.

The organization into different motor unit types has important functional consequences for the production of smooth, coordinated contractions. The smallest neurons have the lowest threshold and are, therefore, activated first when synaptic activity is low. These motor units produce sustainable, relatively low-force tonic contractions in slow-twitch,

fatigue-resistant muscle fibers. If additional force is required, synaptic drive from higher centers increases the action potential firing rate of the initially activated motor neurons and then activates additional motor units of the same type. If yet higher force levels are needed, the larger motor neurons are recruited, but their contribution is less sustained as a result of their tendency for fatigability. This orderly process of motor unit recruitment obeys what is called the **size principle**—the smaller motor neurons are activated first. A logical corollary of this arrangement is that muscles concerned with endurance, such as antigravity muscles, contain predominantly slow-twitch muscle fibers in accordance with their function of continuous postural support. Muscles that contain predominantly fast-twitch fibers, including many physiologic flexors, are capable of producing high-force but less sustainable contractions.

Afferent muscle innervation provides feedback for motor control.

The muscles, joints, and ligaments are innervated with sensory receptors that inform the CNS about body position and muscle activity. Skeletal muscles contain muscle spindles, **Golgi tendon organs** (GTOs), free nerve endings, and some pacinian corpuscles. Joints contain Ruffini endings and pacinian corpuscles, joint capsules contain nerve endings, and ligaments contain Golgi tendon–like organs. Together, these are the proprioceptors, providing sensation from the deep somatic structures. The output of these receptors provides feedback that is necessary for the control of movements.

Muscle spindles provide information about the muscle length and the velocity at which the muscle is being stretched. GTOs provide information about the force being generated. Spindles are located in the mass of the muscle, in parallel with the extrafusal muscle fibers. GTOs are located at the junction of the muscle and its tendons, in series with the muscle fibers (Fig. 5.4).

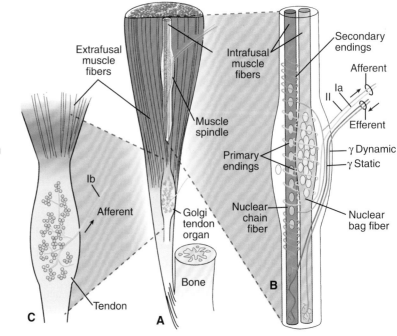

Figure 5.4 **Muscle spindle and Golgi tendon organ (GTO) structure. (A)** Muscle spindles are arranged parallel to extrafusal muscle fibers; GTOs are in series. **(B)** This enlarged spindle shows the following: nuclear bag and nuclear chain types of intrafusal fibers; afferent innervation by Ia axons, which provide primary endings to both types of fibers; type II axons, which have secondary endings mainly on chain fibers; and motor innervation by the two types of γ motor axons: static and dynamic. **(C)** An enlarged GTO. The sensory receptor endings interdigitate with the collagen fibers of the tendon. The axon is type Ib.

Muscle spindles

Muscle spindles are sensory receptors found in almost all of the skeletal muscles. They occur in greatest density in small muscles serving fine movements, such as those of the hand, and in the deep muscles of the neck. Muscle spindles function primarily to detect changes in muscle length and the rate of change in muscle length. They convey this information to the CNS via sensory neurons, and the brain processes this information to help determine position. The response of muscle spindles to changes in length also play a key role in regulating muscle contraction, by activating motor neurons via the stretch reflex to resist muscle stretch.

The muscle spindle, named for its long fusiform shape, is attached at both ends to extrafusal muscle fibers. Within the spindle's expanded middle portion is a fluid-filled capsule containing about seven specialized striated muscle fibers entwined by sensory nerve terminals. These intrafusal muscle fibers have contractile filaments at both ends. The noncontractile midportion contains the cell nuclei (see Fig. 5.4B). γ Motor neurons innervate the contractile elements. There are two types of intrafusal fibers: **nuclear bag fibers**, named for the large number of nuclei packed into the midportion, and **nuclear chain fibers**, in which the nuclei are arranged in a longitudinal row. There are about twice as many nuclear chain fibers as nuclear bag fibers per spindle. The nuclear bag type fibers are further classified as *bag₁* and *bag₂*, based on whether they respond best in the dynamic or static phase of muscle stretch, respectively.

Sensory axons surround both the noncontractile midportion and paracentral region of the contractile ends of the intrafusal fiber. The sensory axons are categorized as primary (type Ia) and secondary (type II). The axons of both types are myelinated. Type Ia axons are larger in diameter (12 to 20 μm) than type II axons (6 to 12 μm) and have faster conduction velocities. Type Ia axons have spiral endings that wrap around the middle of the intrafusal muscle fiber (see Fig. 5.4B). Both nuclear bag and nuclear chain fibers are innervated by type Ia axons. Type II axons innervate mainly nuclear chain fibers and have nerve endings that are located along the contractile components on either side of the type Ia spiral ending. The nerve endings of both primary and secondary sensory axons of the muscle spindles respond to stretch by generating action potentials that convey information to the CNS about changes in the muscle length and the velocity of length change (Fig. 5.5). The primary endings temporarily cease generating action potentials during the release of a muscle stretch (Fig. 5.6).

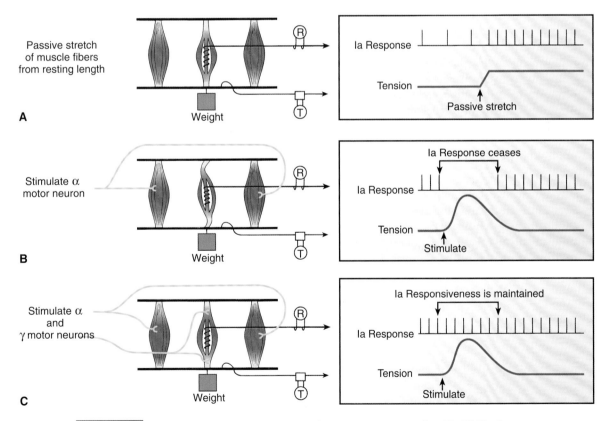

Figure 5.5 **Action potentials (R) from type Ia endings versus muscle tension (T). (A)** The Ia sensory endings from the muscle spindles discharge at a slow rate when the muscle is at its resting length and show an increased firing rate when the muscle is stretched. **(B)** α Motor neuron activation shortens the muscle and releases tension on the muscle spindle. Ia activity ceases temporarily during the tension release. **(C)** Concurrent α and γ motor neuron activation, as occurs in normal, voluntary muscle contraction, shortens the muscle spindle along with the extrafusal fibers, maintaining the spindle's responsiveness to the stretch.

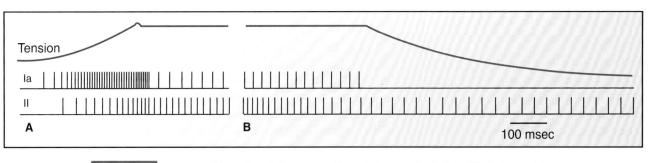

Figure 5.6 **Response of types Ia and II sensory endings during muscle stretch. (A)** During rapid stretch, type Ia endings show a greater firing rate increase, whereas type II endings show only a modest increase. **(B)** With the release of the stretch, Ia endings cease firing, whereas firing of type II endings slows. Ia endings report both the velocity and the length of muscle stretch; type II endings report the length.

Golgi tendon organs

GTOs are 1-mm-long, slender receptors encapsulated within the tendons of the skeletal muscles (see Fig. 5.4A, C). The distal pole of a GTO is anchored in collagen fibers of the tendon. The proximal pole is attached to the ends of the extrafusal muscle fibers. This arrangement places the GTO in series with the extrafusal muscle fibers such that contractions of the muscle stretch the GTO.

A large-diameter, myelinated type Ib afferent axon arises from each GTO. These axons are slightly smaller in diameter than the type Ia variety, which innervate the muscle spindle. Muscle contraction stretches the GTO and generates action potentials in type Ib axons. The GTO output provides information to the CNS about the force of the muscle contraction.

Information entering the spinal cord via type Ia and Ib axons is directed to many targets, including the spinal interneurons that give rise to the **spinocerebellar tracts**. These tracts convey information to the cerebellum about the status of muscle length and tension.

Motor neurons adjust spindle output during muscle contraction.

α Motor neurons innervate the extrafusal muscle fibers, and γ motor neurons innervate the intrafusal fibers. Cells bodies of both α and γ motor neurons reside in the ventral horns of the spinal cord and in nuclei of the cranial motor nerves. Nearly one third of all motor nerve axons are destined for intrafusal muscle fibers. This high number reflects the complex role of the spindles in motor system control. Intrafusal muscle fibers likewise constitute a significant portion of the total number of muscle cells, yet they contribute little or nothing to the total force generated when the muscle contracts. Rather, the contractions of intrafusal fibers play a modulating role in spindle output as they alter the length and, thereby, the sensitivity of the muscle spindles.

Even when the muscle is at rest, the muscle spindles are slightly stretched, and type Ia afferent nerves exhibit a slow discharge of action potentials. Contraction of the muscle increases the firing rate in type Ib axons from GTOs, whereas type Ia axons temporarily cease or reduce firing because the shortening of the surrounding extrafusal fibers unloads the intrafusal muscle fibers. If a load on the spindle were reinstituted, the Ia nerve endings would resume their sensitivity

to stretch. The role of the γ motor neurons is to "reload" the spindle during muscle contraction by activating the contractile elements of the intrafusal fibers. This is accomplished by coordinated activation of the α and γ motor neurons during muscle contraction (see Fig. 5.5).

The γ motor neurons and the intrafusal fibers they innervate are traditionally referred to as the **fusimotor system**. Axons of the γ neurons terminate distal to the sensory endings on the striated poles of the spindle's muscle fibers (see Fig. 5.4B).

γ Motor neurons are designated *dynamic* and *static*. This functional distinction is based on findings that dynamic γ motor neurons are associated with dynamic nuclear bag fibers and stimulation of dynamic γ neurons enhances the response of type Ia sensory axons to stretch but only during the dynamic (muscle length–changing) phase of a muscle stretch. During the static phase of the stretch (muscle length increase maintained), stimulation of the static γ motor neurons, which are associated with static nuclear bag fibers and nuclear chain fibers, enhances the response of type Ia and II sensory axons. These differences allow for largely independent control of the nuclear bag and nuclear chain fibers in the spindle and suggest that the motor system has the ability to monitor muscle length more precisely in some muscles and the speed of contraction in others.

▶ SPINAL CORD IN THE CONTROL OF MOVEMENT

Muscles interact extensively in the maintenance of posture and the production of coordinated movement. The circuitry of the spinal cord automatically controls much of this interaction. Sensory feedback from muscles reaches motor neurons of related muscles and, to a lesser degree, of more distant muscles. In addition to activating local circuits, muscles and joints transmit sensory information up the spinal cord to higher centers. This information is processed and can be relayed back to influence spinal cord circuits.

Spinal motor anatomy correlates with function.

The cell bodies of the spinal motor neurons are grouped into pools in the spinal cord ventral horns. A pool consists of the motor neurons that serve a particular muscle. The number of motor neurons that control a muscle varies in direct proportion to the delicacy of control required. The motor neurons are organized so that those innervating the axial muscles are grouped medially and those innervating the limbs are located laterally in the ventral horn. The lateral limb motor neuron areas are further organized so that neurons innervating flexors are positioned more dorsal and extensors more ventral in the ventral horn. A motor neuron pool may extend over several spinal segments, and the axons then emerge in the ventral nerve roots of two or even three adjacent spinal levels. A physiologic advantage to such an arrangement is that injury to a single nerve root, as might occur by herniation of an intervertebral disk, will not completely paralyze a muscle.

Between the spinal cord's dorsal and ventral horns lies the intermediate zone, which contains an extensive network of interneurons that interconnect motor neuron pools. Some interneurons make connections in their own cord segment; others have longer axon projections that travel in the white matter to terminate in other segments of the spinal cord. These longer interneurons, termed propriospinal cells, carry information that aids coordinated movement. The importance of spinal cord interneurons is clear, insofar as they comprise most of the neurons in the spinal cord as well as being the origin of most of the synapses on motor neurons.

Spinal cord mediates reflex activity.

The spinal cord contains neural circuitry to generate **reflexes**, stereotypical actions produced in response to a stimulus. One function of a reflex is to generate a rapid response. A familiar example is the rapid, involuntary withdrawal of a hand after touching a dangerously hot object well before the heat or pain is perceived. This type of reflex protects the organism before higher CNS levels identify the problem. Some reflexes are simple, whereas others are much more complex. Even the simplest requires coordinated action in which the agonist muscle contracts while the antagonist relaxes. The functional unit of a reflex consists of a sensor, an afferent pathway, an integrating center, an efferent pathway, and an effector. The sensory receptors for spinal reflexes are the proprioceptors and cutaneous receptors. Impulses initiated in these receptors travel along afferent nerves to the spinal cord, where interneurons and motor neurons constitute the integrating center. The motor neurons are the efferent pathway to the effector organs, the skeletal muscles. The responsiveness of such a functional unit can be modulated by higher motor centers acting through descending pathways to facilitate or inhibit its activation.

Study of the three types of spinal reflexes—the myotatic, the inverse myotatic, and the flexor withdrawal—provides a basis for understanding the general mechanism of reflexes.

Myotatic (muscle stretch) reflex

Stretching or elongating a muscle—such as when the patellar tendon is tapped with a reflex hammer or when a quick change in posture is made—causes it to contract within a short duration. The duration between the onset of a stimulus and the response, the **latency period**, is on the order of 30 ms

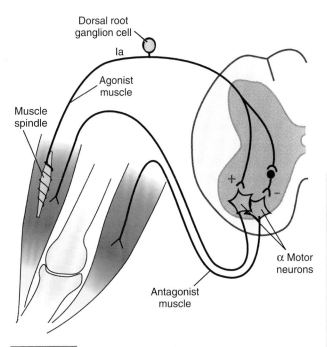

Figure 5.7 **Myotatic reflex circuitry.** Ia afferent axons from the muscle spindle make excitatory monosynaptic contact with homonymous motor neurons and with inhibitory interneurons that synapse on motor neurons of antagonist muscles. The plus sign indicates excitation; the minus sign indicates inhibition.

for a knee-jerk reflex in a human. This response, called the **myotatic** or **muscle stretch reflex**, is a result of monosynaptic circuitry, in which an afferent sensory neuron synapses directly on the efferent motor neuron (Fig. 5.7). The stretch activates muscle spindles and type Ia axons from the spindle carry action potentials to the spinal cord, where they synapse directly on motor neurons of the same (homonymous) muscle that was stretched and on motor neurons of synergistic muscles. Activation of these motor neurons causes the muscles to contract.

Collateral branches of type Ia axons also synapse on interneurons, whose action then inhibits motor neurons of antagonist muscles (see Fig. 5.7). This synaptic pattern, called *reciprocal inhibition*, serves to coordinate muscles of opposing function around a joint. Secondary (type II) spindle afferent fibers also synapse with homonymous motor neurons, providing excitatory input through both monosynaptic and polysynaptic pathways.

The myotatic reflex has two components: a *phasic part*, exemplified by the quick limb movement after tendon tapping with the reflex hammer, and a more sustained *tonic part*, thought to be important for posture maintenance. These two components blend together, but either one may predominate, depending on whether other synaptic activity, such as from cutaneous afferent neurons or pathways descending from higher centers, influences the motor response. Primary spindle afferent fibers probably mediate the tendon jerk, with secondary afferent fibers contributing mainly to the tonic phase of the reflex. The myotatic reflex performs many functions. At the most general level, it produces rapid corrections of motor output in the moment-to-moment control of movement. It also forms the basis for postural reflexes, which maintain body position despite a varying range of loads on the body.

Inverse myotatic reflex

Active contraction of a muscle also causes reflex inhibition of the contraction. This response is called the **inverse myotatic reflex** because it produces an effect that is opposite to that of the myotatic reflex. Active muscle contraction stimulates GTOs, producing action potentials in the type Ib afferent axons. Those axons synapse on inhibitory interneurons that inhibit motor neurons that innervated the muscles served by the activated GTOs and on excitatory interneurons that excite motor neurons of antagonist muscles (Fig. 5.8). The function of the inverse myotatic reflex appears to be a tension feedback system that can adjust the strength of contraction during sustained activity.

The inverse myotatic reflex, like the myotatic reflex, has a more potent influence on the physiologic extensor muscles than on the flexor muscles, suggesting that the two reflexes act together to maintain optimal responses in the antigravity muscles during postural adjustments. Another hypothesis about the conjoint function is that both of these reflexes contribute to the smooth generation of tension in muscle by regulating muscle stiffness.

Flexor withdrawal reflex

Cutaneous stimulation—such as touch, pressure, heat, cold, or tissue damage—can elicit a **flexor withdrawal reflex**. This reflex consists of a contraction of flexors and a relaxation of extensors in the stimulated limb. The action may be accompanied by a contraction of the extensors on the contralateral

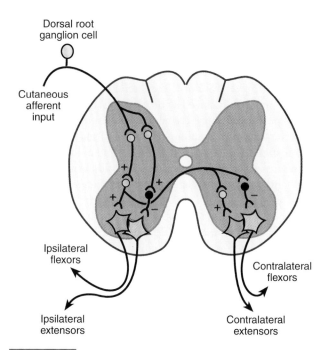

Figure 5.9 **Flexor withdrawal reflex circuitry.** Stimulation of cutaneous afferents activates ipsilateral flexor muscles via excitatory interneurons. Ipsilateral extensor motor neurons are inhibited. Contralateral extensor motor neuron activation provides postural support for the limb that is flexing.

side. The axons of cutaneous sensory receptors synapse on interneurons in the dorsal horn. Those interneurons act ipsilaterally to excite the motor neurons of flexor muscles and inhibit those of extensor muscles. Collaterals of interneurons cross the midline to excite contralateral extensor motor neurons and inhibit flexors (Fig. 5.9).

There are two types of flexor withdrawal reflexes: those that result from innocuous stimuli and those that result from potentially injurious stimulation. The first type produces a localized flexor response accompanied by slight or no limb withdrawal; the second type produces widespread flexor contraction throughout the limb and abrupt withdrawal. The function of the first type of reflex is less obvious but may be a general mechanism for adjusting the movement of a body part when an obstacle is detected by cutaneous sensory input. The function of the second type is protection of the individual. The endangered part is rapidly removed, and postural support of the opposite side is strengthened to support the body.

Collectively, these reflexes provide for stability and postural support (the myotatic and inverse myotatic) and mobility (flexor withdrawal). The reflexes provide a foundation of automatic responses on which more complicated voluntary movements are built.

Spinal cord injury alters voluntary and reflex motor activity.

When the spinal cord of a human or other mammal is severely injured, voluntary and reflex movements are immediately lost caudal to the level of injury. This acute impairment of function is called **spinal shock**. The loss of voluntary

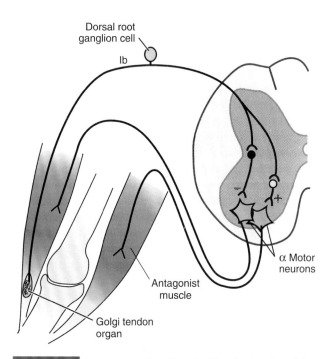

Figure 5.8 **Inverse myotatic reflex circuitry.** Contraction of the agonist muscle activates the Golgi tendon organ and Ib afferents that synapse on interneurons that inhibit agonist motor neurons and excite the motor neurons of the antagonist muscle.

motor control is termed **plegia**, and the loss of reflexes is termed **areflexia**. Spinal shock may last from days to months, depending on the severity of cord injury. Reflexes caudal to the spinal cord damage tend to return, as may some degree of voluntary control depending on the extent of spinal cord damage. As recovery proceeds, myotatic reflexes become hyperactive, as demonstrated by an excessively vigorous response to tapping the muscle tendon with a reflex hammer. A single tap, or limb repositioning that produces a change in the muscle length, may also provoke *clonus*, a condition characterized by repetitive contraction and relaxation of a muscle in an oscillating fashion every second or so. Flexor withdrawal reflexes may also reappear and be provoked by lesser stimuli than normally would be required. The acute loss and eventual overactivity of each of these reflexes results from the lack of influence of the neural tracts that descend from higher motor control centers to the motor neurons and associated interneuron pools.

▶ SUPRASPINAL INFLUENCES ON MOTOR CONTROL

Descending signals from the brainstem and cortex influence the rate of motor neuron firing and the recruitment of additional motor neurons to increase the force of muscle contraction. The brainstem contains the neural circuitry for initiating locomotion and for controlling posture. The maintenance of posture requires coordinated activity of both axial and limb muscles in response to input from proprioceptors and spatial position sensors such as the vestibular apparatus in the inner ear. Cerebral cortex input through the corticospinal system is necessary for the control of fine individual movements of the distal limbs and digits. Each higher level of the nervous system acts on lower levels to produce appropriate, more refined movements.

Brainstem is the origin of descending tracts that influence posture and movement.

Three brainstem nuclear groups give rise to descending motor tracts that influence spinal cord motor neurons and their associated interneurons. These are the **vestibular nuclear complex**, the **reticular formation**, and the **red nucleus** (Fig. 5.10). The other major descending influence on the motor neurons is the corticospinal tract, the only volitional control pathway in the motor system. This tract is discussed later. In most cases, the descending pathways act through synaptic connections on interneurons. The connection is less commonly made directly with motor neurons.

Rubrospinal tract

The red nucleus of the mesencephalon receives major input from both the cerebellum and the cerebral cortical motor areas. Output via the rubrospinal tract is directed predominantly to contralateral spinal motor neurons that are involved with movements of the upper limbs. The axons of the rubrospinal tract are located in the lateral spinal white matter. Rubrospinal action enhances the function of motor

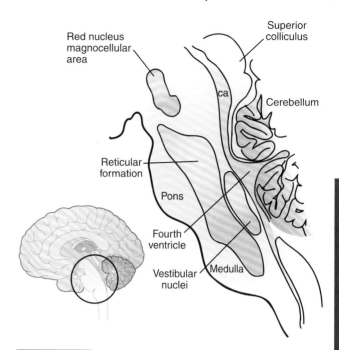

Figure 5.10 **Brainstem nuclei of descending motor pathways.** The magnocellular portion of the red nucleus is the origin of the rubrospinal tract. The lateral vestibular nucleus is the source of the vestibulospinal tract. The reticular formation is the source of two tracts: one from the pontine portion and one from the medulla. Structures illustrated are from the monkey. ca, cerebral aqueduct.

neurons innervating upper limb flexor muscles while inhibiting extensors. This tract may also influence γ motor neuron function. In higher mammals like humans, the corticospinal tract supersedes most of the function of the rubrospinal tract. However, when lesions of corticospinal tract occur, the rubrospinal tract can partially compensate for the loss of corticospinal tract inputs to upper limb motor neurons.

Vestibulospinal tract

The principal functions of the vestibular system are to activate neck, trunk, and limb muscles to maintain posture in response to movements of the head and to maintain visual fixation of the eyes on a target as the head moves through space. Both of these functions are controlled by pathways that originate from the vestibular complex located in the pons and medulla. The vestibular complex contains four nuclei (*superior*, *lateral*, *medial*, and *inferior vestibular nuclei*) that receive information about the position of the head in space from the vestibular apparatus of the inner ear and proprioceptive information from the spinal cord (see Chapter 4). The vestibular nuclei also are reciprocally connected to the cerebellum and adjacent reticular formation from which it receives sensory- and motor-related inputs. The vestibular nuclei modify both muscle tone and initiate automatic adjustments in posture through two descending pathways, the lateral and medial vestibulospinal tracts.

The *lateral vestibulospinal tract* originates primarily from cells in the lateral vestibular nucleus. The axons from these cells travel in the ipsilateral ventrolateral spinal cord white matter to terminate at all levels of the spinal cord where they excite interneurons and α and γ extensor motor neurons that innervate truncal and proximal limb muscles. These extensor motor neurons and their musculature are important for maintaining posture and modulating posture-related reflexes that help stabilize the body's position against the forces of gravity. As these axons descend, they give off collateral branches at multiple spinal cord levels, which ensure proper coordination of postural reflexes across multiple levels. Lesions in the brainstem secondary to stroke or trauma may abnormally enhance the influence of the vestibulospinal tract and produce dramatic clinical manifestations.

The *medial vestibulospinal tract* arises primarily from the medial vestibulospinal nucleus and descends bilaterally to terminate on motor neuron in the cervical spinal cord that controls neck extensor and flexor muscles. The function of this pathway is to reflexively activate neck muscles in response to changes in head position.

Reticulospinal tracts

The reticular formation is a complicated network of neurons located in the central gray matter core of the brainstem. Many neurons within this network have wide ranging connections that modulate neuronal activity at several CNS levels. Within this network are also discrete circuits that control a diverse set of functions such as sleep, autonomic functions, and eye movements. Within the medial regions of the caudal reticular formation are groups of large neurons that are involved in somatic motor control of both cranial nerve and spinal cord motor neurons. These neurons receive afferent inputs from the many areas including the spinal cord, vestibular nuclei, cerebellum, lateral hypothalamus, **globus pallidus (GP)**, midbrain, and sensorimotor cortex.

Two descending tracts important in the control of spinal lower motor neurons arise from medial reticular formation cells. These pathways mostly influence motor neurons that innervate truncal and limb extensor muscles. Through their influence on gamma motor neurons, these pathways modulate muscle tone and help make anticipatory adjustments in posture during movement. The **medial (pontine) reticulospinal tract** arises from pontine reticular nuclei and descends bilaterally with an ipsilateral preponderance in the anterior spinal cord white matter. This pathway relays excitatory action potentials to interneurons that influence α and γ motor neuron pools. The **medullary reticulospinal tract** arises from the reticular formation in the medulla and descends ipsilateral in the spinal cord white matter adjacent to the anterior horn. This pathway has an inhibitory influence on interneurons that modulate extensor motor neurons.

Terminations of the brainstem motor tracts correlate with their functions.

The rubrospinal tract terminates mostly on interneurons in the lateral spinal intermediate zone, but it also has some monosynaptic connections directly on motor neurons to muscles of the extremities. This tract supplements the corticospinal tract for independent movements of the upper extremities.

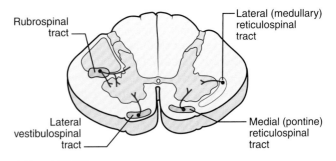

Figure 5.11 **Brainstem motor control tracts.** The vestibulospinal and reticulospinal tracts influence motor neurons that control axial and proximal limb muscles. The rubrospinal tract influences motor neurons controlling distal limb muscles.

The vestibulospinal and reticulospinal tracts terminate in the ventromedial part of the intermediate zone, an area in the gray matter containing propriospinal interneurons (Fig. 5.11). There are also some direct connections with motor neurons of the neck and back muscles and the proximal limb muscles. These tracts are the main CNS pathways for maintaining posture and head position during movement.

In accordance with their medial or lateral distributions to spinal motor neurons, the reticulospinal and vestibulospinal tracts are thought to be most important for the control of axial and proximal limb muscles, whereas the rubrospinal (and corticospinal) tracts are most important for the control of distal limb muscles, particularly the flexors

Sensory and motor systems work together to control posture.

The maintenance of an upright posture in humans requires active muscular resistance against gravity. For movement to occur, the initial posture must be altered by flexing some body parts against gravity. Balance must be maintained during movement, which is achieved by postural reflexes initiated by several key sensory systems. Vision, the vestibular system, and the somatosensory system are important for postural reflexes.

Somatosensory input provides information about the position and movement of one part of the body with respect to others. The vestibular system provides information about the position and movement of the head and neck with respect to the external world. Vision provides both types of information as well as information about objects in the external world. Visual and vestibular reflexes interact to produce coordinated head and eye movements associated with a shift in gaze. Vestibular reflexes and somatosensory neck reflexes interact to produce reflex changes in limb muscle activity. The quickest of these compensations occurs at about twice the latency of the monosynaptic myotatic reflex. These response types are termed **long-loop reflexes**. The extra time reflects the action of other neurons at different anatomic levels of the nervous system.

CLINICAL FOCUS | 5.1

Decerebrate Rigidity

A patient with a history of poorly controlled hypertension is brought to the emergency department because of sudden collapse and subsequent unresponsiveness. A neurologic examination performed about 30 minutes after onset shows no response to verbal stimuli. Spontaneous movements of the limbs are absent. A painful stimulus, compression of the soft tissue of the supraorbital ridge, causes immediate extension of the neck and all of the limbs. This posture relaxes within a few seconds after the stimulation is stopped. After the patient is stabilized medically, he undergoes a magnetic resonance imaging (MRI) study of the brain. The study demonstrates a large area of hemorrhage bilaterally in the upper portion of the brainstem.

The posture this patient demonstrated in response to the noxious stimulus is termed **decerebrate rigidity**. Its occurrence is associated with lesions of the mesencephalon that eliminate the influence of higher brainstem and cortical centers. The abnormal posture is a result of extreme antigravity extensor muscle activation by the unopposed action of the reticulospinal and vestibulospinal tracts. A model of this condition can be produced in experimental animals by a surgical lesion located between the mesencephalon and pons. ■

▶ CEREBRAL CORTEX ROLE IN MOTOR CONTROL

The cerebral cortical areas concerned with motor function exert the highest level of motor control with influence exerted over both involuntary and voluntary movements. The cortex modifies sensory-evoked involuntary movements like brainstem- and spinal cord–related postural adjustments by continuous modulation of brainstem descending motor pathways and spinal cord reflex pathways. The role of this modulation is readily apparent when it is damaged (see Clinical Focus 5.1). Cortical control of skilled voluntary movements, most of which involve the distal extremities, is accomplished through connections with cranial nerve motor nuclei and spinal cord interneurons and motor neurons that control skilled movement. In addition, these skilled movements are performed on a background of ongoing postural adjustments, which the cortex accomplishes by simultaneously activating brainstem descending motor pathways.

Several distinct cortical areas participate in voluntary movement.

Cortical areas associated with voluntary motor control are located in the frontal lobe rostral to the central sulcus and include **primary motor cortex (M1)**, Brodmann area 4 (Fig. 5.12), Brodmann area 6, which lies just rostral to M1, and the frontal eye fields, Brodmann area 8, which control conjugate horizontal movements of the eyes. Brodmann area 6 contains the *supplementary motor area (SMA)* located on the medial surface of the frontal lobe and *premotor area* located on the lateral surface. Other cortical areas that contribute to descending control of voluntary movements are the postcentral gyrus, areas 1, 2, and 3 and areas 5 and 7 of the parietal lobe. Each of these motor areas contributes fibers to the corticospinal tract, the principal efferent motor pathway from the cortex. Motor-related cortex receives information from cortical and subcortical areas that process sensory information, from cortical areas that underlie the motivation to move, and from subcortical structures that modulate motor activity like the cerebellum and basal ganglia.

Primary motor cortex

The primary motor cortex, Brodmann area 4, is located in the precentral gyrus laterally and anterior paracentral lobule medially. Like the organization of primary somatosensory cortex, the body is represented as somatotopic maps (Fig. 5.13). Those parts of the body that perform fine

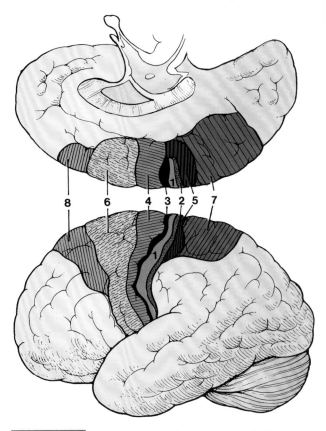

Figure 5.12 **Brodmann cytoarchitectural map of the human cerebral cortex.** Area *4* is the primary motor cortex; area *6* is the premotor cortex on the lateral and supplementary motor area on the medial aspect of the hemisphere. Area *8* influences voluntary conjugate eye movements. Areas *1, 2, 3, 5,* and *7* have sensory functions but also contribute axons to the corticospinal tract. The central sulcus divides areas *4* and *3*.

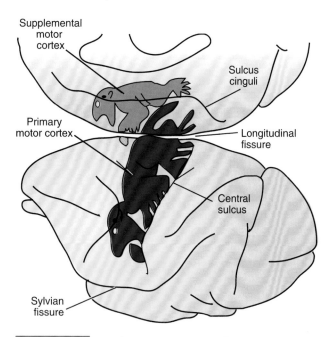

Supplemental
motor
cortex

Sulcus
cinguli

Primary
motor cortex

Longitudinal
fissure

Central
sulcus

Sylvian
fissure

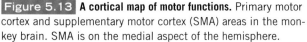

Figure 5.13 **A cortical map of motor functions.** Primary motor cortex and supplementary motor cortex (SMA) areas in the monkey brain. SMA is on the medial aspect of the hemisphere.

movements, such as the digits and the facial muscles, are controlled by a greater number of neurons that occupy more cortical territory than the neurons for the body parts only capable of gross movements. However, studies have shown that the somatotopic map, especially in areas representing the distal extremities, is not just a simple one-to-one representation. Stimulation at a specific location often activates multiple muscles over several spinal segments and muscles are often represented at more than one cortical location.

Neurons in M1 have the capability to encode the control of muscle force, muscle length, joint movement, and position. Low-level electrical stimulation of surgically exposed M1 produces twitch-like contraction of a few muscles or, less commonly, a single muscle. Movements elicited from M1 have the lowest stimulation thresholds and are the most discrete of any movements elicited by stimulation of motor cortical areas. Studies have shown that stronger stimulation of greater duration can elicit coordinated movements of multiple muscle groups indicating that M1 is important for controlling the number of muscles activated for a particular movement as well the trajectory and force of movements.

Stimulation of M1 body areas produces contralateral movement, whereas stimulation of cortical areas where the head is represented may produce bilateral motor responses. Destruction of any part of the primary motor cortex leads to immediate paralysis of the muscles controlled by that area. In humans, some function may return days to months later, but the movements lack the fine degree of muscle control of the normal state. For example, after a lesion in the arm area of M1, the use of the hand recovers, but the capacity for discrete finger movements does not.

M1 receives somatosensory input, both cutaneous and proprioceptive, as well as motor-related inputs from the cerebellum and basal ganglia via the thalamus. Other afferent

projections come from the contralateral motor cortex and many other ipsilateral cortical areas. There are many axons between the precentral (motor) and postcentral (somatosensory) gyri and many connections with visual cortical areas. Because M1 receives continuous sensory feedback regarding the performance of a movement, the cortical motor neurons can alter ongoing motor activity in response to peripheral sensory feedback. For example, cells innervating a particular muscle may respond to cutaneous stimuli originating in the area of skin that moves when that muscle is active, and they may respond to proprioceptive stimulation from the muscle to which they are related. The primary motor cortex also has the capability to control the flow of somatosensory information to motor control centers by way of efferent fibers from the primary motor cortex that terminate in brain areas that contribute to ascending somatic sensory pathways.

The importance of close coupling of sensory and motor functions is demonstrated by two cortically controlled reflexes that were originally described in experimental animals as being important for maintaining normal body support during locomotion—the placing and hopping reactions. The **placing reaction** can be demonstrated in a cat by holding it so that its limbs hang freely. Contact of any part of the animal's foot with the edge of a table provokes immediate placement of the foot on the table surface. The **hopping reaction** is demonstrated by holding an animal so that it stands on one leg. If the body is moved forward, backward, or to the side, the leg hops in the direction of the movement so that the foot is kept directly under the shoulder or hip, stabilizing the body position. Lesions of the contralateral precentral or postcentral gyrus abolish placing. Hopping is abolished by a contralateral lesion of the precentral gyrus.

Supplementary motor area

The SMA (**supplementary motor area**) portion of Brodmann area 6 is located on the medial surface of the hemispheres, above the cingulate sulcus and rostral to the leg area of the primary motor cortex (see Fig. 5.13). SMA receives input from other motor cortical areas and the prefrontal cortex, is reciprocally connected with M1, and projects directly to the reticular formation and spinal cord. Experimental lesions in M1 eliminate the ability of SMA stimulation to produce movements. SMA contains a somatotopic map of the body although it is less precisely organized than that of M1. Electrical stimulation of SMA produces movements, but a higher stimulus strength is required than for M1 and the movements produced are more complex and frequently bilateral. Evidence suggests that SMA is important for the movements we decide to make volitionally. The importance of SMA for this type of motor programming was demonstrated in a series of experiments that examined functional activation of motor cortical activities using functional magnetic resonance. When a subject makes a series of random finger movements, activity is localized to the hand area of M1. When the subject plans and makes a movement of the fingers, activity is observed in both M1 and SMA, and when finger movements are mentally planned, but not undertaken, activity is only observed in SMA. Additionally, lesions in SMA reduce the number of self-initiated movements but have little effect on movements evoked by external sensory cues.

Premotor cortex

The premotor cortex occupies an area on the lateral surface of the frontal lobe rostral to M1. Like SMA, there is a somatotopic representation of the body and higher stimulus intensities are required to elicit movements. Premotor cortex receives inputs from somatosensory areas of the parietal cortex and projects to M1, reticular formation motor-related nuclei, and spinal cord areas associated with control of proximal limb and axial muscles. The significant sensory input from cortical areas associated with somatosensory and visual movement processing and outputs to motor groups associated with postural adjustments suggests that premotor cortex is important for guiding movements in response to sensory stimuli. Studies have shown that premotor cortex neurons play an important role in conditional motor tasks where a particular movement, like reaching in a particular direction, is made in response to a learned cue such as a visual stimulus. In these cases, premotor neurons are active following the presentation of the visual cue and before the movement is made. Lesions in this area impair the ability to carry out conditional motor tasks. This and other evidence suggests that the premotor cortex is important for encoding movement intention and the selection of movements based on environmental sensory inputs.

Primary somatosensory cortex

The primary somatosensory cortex (S1) (Brodmann areas 1, 2, and 3) lies on the postcentral gyrus (see Fig. 5.13) and has a role in movement. Electrical stimulation here can produce movement, but thresholds are two to three times higher than in M1. S1 is reciprocally interconnected with M1 in a somatotopic pattern—for example, arm areas of sensory cortex project to arm areas of motor cortex. S1 Efferent fibers travel in the corticospinal tract and terminate in the dorsal horn areas of the spinal cord. These terminations are thought for modifying sensory inputs related to movements.

Superior parietal lobe

The superior parietal lobe (Brodmann areas 5 and 7) also has important motor functions. This area processes somatosensory and visual information related to the location of objects in the visual field. In addition to contributing a small number of fibers to the corticospinal tract, the superior parietal lobe is well connected to the motor areas in the frontal lobe, especially SMA and the frontal eye field Brodmann area 8. Studies in animals and humans suggest this area is important for the use of complex sensory information in the production of movement and for relaying visual information related to conjugate eye movements.

Corticospinal tract is the primary efferent path from the cortex.

The corticospinal tract originates from motor areas of the cerebral cortex with the largest contribution from M1 (about one third of the fibers). The corticospinal tract is also referred to as the **pyramidal tract** because it traverses the medullary pyramids on its way to the spinal cord (Fig. 5.14). All other descending motor tracts emanating from the brainstem are generally grouped together as the extrapyramidal system. In

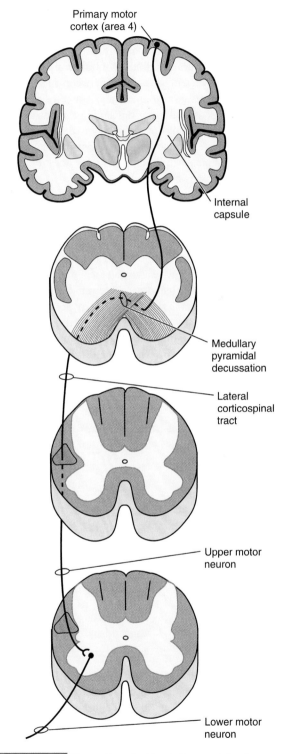

Figure 5.14 **The corticospinal tract.** Axons arising from cortical motor areas descend through the internal capsule, decussate in the medulla, descend in the lateral funiculus of the spinal cord as the lateral corticospinal tract, and terminate on motor neurons and interneurons in the ventral horn areas of the spinal cord. Note the designations of upper and lower motor neurons.

primates, 10% to 20% of corticospinal fibers end directly on motor neurons; the others end on interneurons associated with motor neurons.

From the cerebral cortex, the corticospinal tract axons descend through the brain along a path located between the basal ganglia and the thalamus, known as the **internal capsule**. They then continue along the ventral brainstem as the **cerebral peduncles** and through the pyramids of the medulla. Most of the corticospinal axons cross the midline in the medullary pyramids and descend in the contralateral spinal cord. Thus, the motor cortex in each hemisphere controls the muscles on the contralateral side of the body. After crossing in the medulla, the corticospinal axons descend in the dorsal lateral white matter of the spinal cord as the *lateral corticospinal tract* and terminate in lateral motor pools that control distal muscles of the limbs. A smaller group of axons does not cross in the medulla and descends as the *ventral (anterior) corticospinal tract* in the anterior spinal white matter. These axons terminate in the motor neuron pools and adjacent intermediate zone interneurons that control the axial and proximal musculature. Damage to the corticospinal tract above the level of the decussation or lateral corticospinal tract at the level of the spinal cord produces a characteristic type of impairment in motor control. Initially, muscles are weak and flaccid, but following a period of recovery, spastic paralysis, hypertonia, hyperreflexia, and pathologic reflexes (Babinski sign) are observed.

In addition to the direct corticospinal tract, there are other indirect pathways by which cortical fibers influence motor function. Some cortical efferent fibers project to the reticular formation and then to the spinal cord via the reticulospinal tract; others project to the red nucleus and then to the spinal cord via the rubrospinal tract. A third group of axons, termed corticonuclear or corticobulbar, terminate on brainstem interneurons and cranial nerve motor neurons for control of head musculature.

▶ BASAL GANGLIA AND MOTOR CONTROL

The **basal ganglia** are a group of subcortical nuclei located primarily in the base of the forebrain, with some in the diencephalon and upper brainstem. The striatum, globus pallidus, subthalamic nucleus, and substantia nigra comprise the basal ganglia. Input is derived primarily from motor areas of the cerebral cortex, and output is directed back, via a relay in the thalamus, to cortical areas concerned with movement. Basal ganglia action influences the entire motor system and plays a role in the preparation and execution of coordinated movements.

The forebrain (telencephalic) components of the basal ganglia consist of the **striatum**, which is made up of the **caudate nucleus** and the **putamen**, and the **globus pallidus (GP)**. The GP has two subdivisions: the *external segment (GPe)*, adjacent to the medial aspect of the putamen, and the *internal segment (GPi)*, medial to the GPe. The other main nuclei of the basal ganglia are the **subthalamic nucleus** in the diencephalon and the **substantia nigra** in the mesencephalon.

INTEGRATED MEDICAL SCIENCES

Spasticity

Patients who develop lesions, such as an infarction in the internal capsule of the cerebrum or spinal cord trauma affecting the anterior two thirds of the cord, develop a disorder of motor control that is termed **spasticity**. Clinically, this is characterized by (1) weakness, slowness, and clumsiness of movement in body parts below the level of the lesion; (2) abnormally increased muscle stiffness to passive lengthening; and (3) abnormally overactive muscle stretch reflexes in those limbs. The neurophysiology of spasticity is not fully understood, but the spinal interneurons involved with motor control become disinhibited and overactive as a result of impaired inhibitory influence from the upper brainstem and cortex. Although the increased muscle tone is partly compensatory for the impaired corticospinal tract input, it may become so exaggerated that the affected limbs become even less functional as a result of the exaggerated muscle stiffness.

Several medications are available to treat spasticity. These medications target either the CNS pathways associated with motor control or peripherally on muscles themselves. Dantrolene is a medication that directly targets the skeletal muscle dissociating the excitation–contraction coupling by interfering with Ca^{++} release from the sarcoplasmic reticulum (see Chapter 8). Baclofen is a medication that is an analog of the CNS inhibitory neurotransmitter γ-aminobutyric acid (GABA). When given orally or intrathecally using an implanted baclofen pump, it binds to the GABAB receptor and inhibits motor activity at the spinal cord level. Injection of precisely diluted amounts of the toxin produced by the anaerobic bacteria *Clostridium botulinum* into the abnormal muscles is another method to reduce their overactivity. The toxin impairs release of the neurotransmitter acetylcholine from the motor neuron at the synapse with muscle cells. Proper dosing partially weakens the muscle and reduces the abnormally increased muscle stiffness. Gaining an understanding of the physiology of synaptic vesicle release and discovering the role of the botulinum toxins in disturbing this process have allowed treatment with these agents to be made available to patients with conditions that formerly were difficult to treat. ∎

Basal ganglia are extensively interconnected.

Although the circuitry of the basal ganglia appears complex at first glance, it can be simplified into input, output, and internal pathways (Fig. 5.15). Input from the cerebral cortex is directed to the striatum and the subthalamic nucleus. The predominant nerve cell type in the striatum is termed the medium spiny neuron, based on its cell body size and dendritic structure. This type of neuron receives input from all of the cerebral cortex except for the primary visual and auditory areas. The input is roughly somatotopic and is excitatory via glutamate-containing neurons. The caudate receives most of its input from prefrontal, premotor, and supplementary motor cortex, whereas the putamen receives most of the cortical input from primary sensorimotor cortical areas. The subthalamus receives excitatory glutaminergic input from cortical areas concerned with motor function, including eye movement.

Basal ganglia output is from the GPi and a segment of the substantia nigra. This output is inhibitory using γ-aminobutyric acid (GABA) as the neurotransmitter. The GPi output is directed to ventrolateral and ventral anterior nuclei of the thalamus, which feed back to the cortical motor areas. Output from the GPi is also directed to a region in the upper brainstem termed the midbrain extrapyramidal area. This area then projects to the neurons of the reticulospinal tract. The substantia nigra output arises from the **pars reticulata (SNr)**, which is histologically similar to the GPi. The output is directed to the same areas that receive GPi terminations as well as to the superior colliculus of the mesencephalon, which is involved in eye movement control.

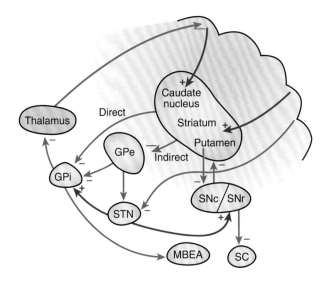

Figure 5.15 **Basal ganglia nuclei and circuitry.** The neuronal circuit of cerebral cortex to striatum to globus pallidus interna (GPi) to thalamus and back to the cortex is the main pathway for basal ganglia influence on motor control. Note the direct and indirect pathways of the striatum, GPi, globus pallidus externa (GPe), and subthalamic nucleus (SNT). GPi output also flows to the midbrain extrapyramidal area (MBEA). The substantia nigra pars reticulata (SNr) to superior colliculus (SC) pathway is important in eye movements. Excitatory pathways are shown in *red*, and inhibitory pathways are shown in *blue*. SNc, substantia nigra pars compacta.

The GPe, the subthalamic nucleus, and the pars compacta region of the substantia nigra (SNc) form important internal pathways within the basal ganglia. The GPe receives GABAergic inhibitory inputs from the striatum. The output of the GPe is inhibitory via GABA release and is directed to both the GPi and the subthalamic nucleus. The subthalamic nucleus output is excitatory and is directed to the GPi and the SNr. The striatum–GPe–subthalamic nucleus–GPi circuit has been termed the **indirect pathway**. Activation of this pathway inhibits the excitatory drive of the thalamus on cortical motor neurons resulting in decreased motor activity. In contrast, activation of the **direct pathway**, in which output from the striatum directly inhibits GPi, increases thalamic excitatory outputs and increases motor activity (see Fig. 5.15). The SNc output modulates cortical influence on the striatum via dopamine-releasing neurons, which are excitatory to the direct pathway and inhibitory to the indirect pathway.

Functions of the basal ganglia are partially revealed by disease.

Basal ganglia diseases produce profound motor dysfunction in humans. The disorders can result in reduced motor activity, **hypokinesis**, or abnormally enhanced activity, **hyperkinesis**. Two well-known neurologic conditions that show histologic abnormality in basal ganglia structures, **Parkinson's disease (PD)** and **Huntington's disease**, illustrate the effects of basal ganglia dysfunction. Patients with PD show a general slowness of initiation of movement and paucity of movement when in motion. The latter takes the form of reduced arm swing and lack of truncal swagger when walking. These patients also have a resting tremor of the hands, described as "pill rolling." The tremor stops when the hand goes into active motion. At autopsy, patients with PD show a severe loss of dopamine-containing neurons in the SNc region. Patients with Huntington's disease have uncontrollable, quick, brief movements of individual limbs. These movements are similar to what a normal individual might show when flicking a fly off a hand or when quickly reaching up to scratch an itchy nose. At autopsy, a severe loss of inhibitory striatal neurons is found.

The function of the basal ganglia in normal individuals appears to be modulation of cortical output. One theory is that the primary action is to allow desired motions to proceed and to inhibit undesirable movements. Neuronal activity is increased in the appropriate areas of the basal ganglia prior to the actual execution of movement. The basal ganglia act as a brake on undesirable motion by inhibitory output from the GPi back to the thalamus and, ultimately, to the motor cortex. The loss of dopamine-releasing neurons in PD is thought to enhance this braking effect by increasing the activity of the indirect basal ganglia pathway. This effect ultimately increases the inhibitory output of the GPi because the subthalamus is disinhibited. Decreased motor action is the final result. Hyperkinetic disorders such as Huntington's disease are thought to result from decreased GPi output resulting from loss of inhibitory influence of the direct pathway.

Stereotactic Neurosurgery and Deep Brain Stimulation for Parkinson's Disease

Parkinson's disease (PD) is a degenerative central nervous system disorder producing a generalized slowness of movement and resting tremor of the hands. Loss of dopamine-producing neurons in the substantia nigra pars compacta is the underlying pathology of the condition. Treatment with medications that stimulate an increased production of dopamine by the surviving substantia nigra neurons has revolutionized the management of Parkinson's disease. Unfortunately, the benefit of the medications tends to lessen after 5 to 10 years of treatment. After prolonged treatment, patients develop medication-related motor complications and motor fluctuations characterized by dyskinesias and the sudden loss of medication effectiveness producing significant decreases in mobility. Improved knowledge of basal ganglia circuitry has enabled neurosurgeons to develop surgical procedures to ameliorate some of the effects of the advancing disease.

Degeneration of the dopamine-releasing cells of the substantia nigra decreases inhibitory output of the striatum. The effect of this decreased inhibition on the direct pathway is less inhibition of GPi. The effect of decreased inhibitory drive on the indirect pathway results is increased activity in GPi. Enhanced activity of the GPi increases its inhibitory influence on the thalamus and results in decreased excitatory drive back to the cerebral cortex and hence decreased muscle activity.

The use of stereotactic neurosurgery to treat Parkinson's disease is generally employed after a patient no longer shows good symptomatic control with medications. Stereotactic neurosurgery is a technique in which a small probe is precisely placed into a target within the brain. Magnetic resonance imaging (MRI) of the brain defines the three-dimensional location of the desired target. The surgical probe is introduced into the brain through a small hole made in the skull and is guided to the target by the surgeon by using the MRI coordinates.

In the past, stereotactic surgery was directed at the GPi, which was destroyed by heating or passing direct electrical current through the probe. This destruction of GPi reduces the inhibitory outflow of the GPi and movement improves.

More recently, GPi lesioning has been replaced by the implantation of neurostimulators whose electrodes are implanted in either the GPi or a second target, the subthalamic nucleus. Although the exact mechanism involved is not known, high-frequency electrical stimulation of the GPi or subthalamic nucleus paradoxically reduces their output and, thereby, improves function of the thalamocortical pathway for motor control. Although DBS has proven to be an effective treatment for PD, like medical management of this disorder, it is only a symptomatic treatment and does not slow disease progression. ■

In addition to the basal ganglia motor loops described above, the basal ganglia also have circuits that are involved in modulating emotional and intellectual activities of the cortex and disorders in these functions frequently accompany the motor signs observed in basal ganglia disorders (Clinical Focus 5.2).

▶ CEREBELLUM IN THE CONTROL OF MOVEMENT

The **cerebellum**, or "little brain," lies caudal to the occipital lobe and is attached to the posterior aspect of the brainstem through three-paired fiber tracts: *inferior, middle,* and *superior* **cerebellar peduncles**. The cerebellum receives motor, sensory, and cognitive information from peripheral sensory receptors, the brainstem, and the cerebral cortex. Most of this information reaches the cerebellum through the inferior and middle cerebellar peduncles. The efferent projections of the cerebellum arise from three pairs of intrinsic nuclei located deep inside the cerebellum: the **fastigial**, interposed (or **interpositus**), and **dentate**. In some classification schemes, the interposed nucleus is further divided into the *emboliform* and *globose* nuclei. Cerebellar efferents are directed mainly to other motor control areas of the CNS and mostly exit the cerebellum in the superior cerebellar peduncle.

Structural divisions of the cerebellum correlate with function.

The cerebellar surface is arranged in multiple, parallel, longitudinal folds termed folia. Two deep transverse fissures divide the cerebellum into three main morphologic components—*anterior, posterior,* and *flocculonodular lobes* (Fig. 5.16). The cerebellum can also be divided in the sagittal plane into three functional subdivisions—the *vestibulocerebellum, spinocerebellum,* and *cerebrocerebellum*. These divisions appear in sequence during evolution. The lateral cerebellar hemispheres increase in size along with expansion of the cerebral cortex. The three divisions have similar intrinsic circuitry; thus, the function of each depends on the nature of the inputs it receives and output nucleus to which it projects.

The **vestibulocerebellum** is composed of the flocculonodular lobe. It receives input from the vestibular system about the position of the head in space and information concerning eye movements and visual processing from brainstem nuclei. Output goes either directly or indirectly, through a relay in the fastigial nucleus, back to vestibular and reticular formation nuclei in the brainstem The vestibulocerebellum functions to control equilibrium by influencing brainstem descending motor pathways associated with axial and proximal musculature and eye movements through actions on brainstem cranial nerve nuclei. Damage to the vestibulocerebellum is associated with falling, gait problems, tremor of the head or trunk, and deficiencies in pursuit eye movements.

The medially placed **spinocerebellum** consists of the midline vermis plus the medial portion of the lateral hemispheres, called the intermediate or paravermal zone (see Fig. 5.16). There is a somatotopic representation of the body in the spinocerebellum with the legs represented anteriorly

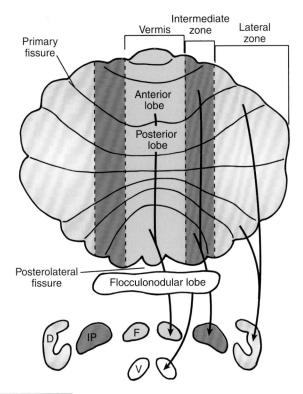

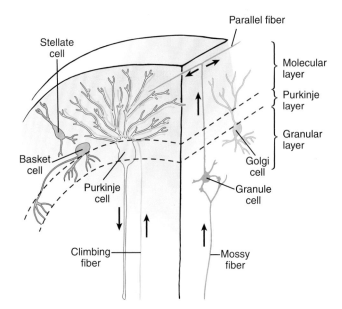

the cerebral cortex, relayed through the middle cerebellar peduncles from nuclei in the pons. The cortical areas that are prominent in motor control are the sources for most of this input. Output is directed to the dentate nuclei and, ultimately, to the motor and premotor areas of the cerebral cortex via the ventrolateral thalamus. The cerebrocerebellum is thought to participate in the planning, control, and timing of learned skilled movements, especially those that involve the distal extremities.

Intrinsic circuitry of the cerebellum is regulated by Purkinje cells.

The cerebellar cortex is composed of five types of neurons arranged into three layers (Fig. 5.17). The *molecular layer* is the outermost and consists mostly of axons and dendrites plus two types of interneurons: stellate cells and basket cells. The *Purkinje cell layer* contains the Purkinje cells, whose dendrites reach upward into the molecular layer in a fanlike array. The Purkinje cells are the efferent neurons of the cerebellar cortex. Their action is inhibitory, with GABA being the neurotransmitter. Deep to the Purkinje cells is the granular layer, containing Golgi cells and small local circuit neurons, the granule cells. The granule cells are numerous; there are more granule cells in the cerebellum than neurons in the entire cerebral cortex!

Afferent axons to the cerebellar cortex are of two types: **mossy fibers** and **climbing fibers**. Mossy fibers arise from spinal cord and brainstem neurons, including those of the pons that receive input from the cerebral cortex. Mossy fibers make multicontact excitatory glutaminergic synapses on granule cells. The axons of the granule cells then ascend to the molecular layer and bifurcate, forming the parallel fibers that travel perpendicular to and synapse with the dendrites of hundreds

Figure 5.16 **The structure of the cerebellum.** The three lobes are shown: anterior, posterior, and flocculonodular. The functional divisions are demarcated by color. The vestibulocerebellum (*white*) is the flocculonodular lobe, which projects to the vestibular (V) nuclei. The spinocerebellum includes the vermis (*green*) and the intermediate zone (*violet*), which project to the fastigial (F) and interposed (IP) nuclei, respectively. The cerebrocerebellum (*pink*) projects to the dentate nuclei (D).

and the head posteriorly. The trunk is mapped to the vermis and the limbs more laterally in the intermediate zone. Spinocerebellar pathways carrying somatosensory information terminate in the vermis and intermediate zones in somatotopic arrangements. The auditory, visual, and vestibular systems, and sensorimotor cortex, through a relay in the pons, also project to this portion of the cerebellum. Output from the vermis is directed to the fastigial nuclei, which project through the inferior cerebellar peduncle to the vestibular nuclei, reticular formation of the pons and medulla, and upper levels of the spinal cord. Output from the intermediate zones goes to the interposed nuclei and then to the red nucleus, and ventrolateral nucleus of the thalamus, which is connected motor cortical areas. The vermal component of the spinocerebellum mainly controls axial musculature important for posture and locomotion. The intermediate zone of the spinocerebellum is thought to be important for controlling ongoing limb movements. The intermediate zone receives information regarding an intended movement from cortical motor areas and sensory feedback information about the actual movements from ascending spinocerebellar pathways. It then compares intention to actual movements and sends correcting signals back to motor cortex.

The **cerebrocerebellum** occupies the lateral aspects of the cerebellar hemispheres. Input comes exclusively from

Figure 5.17 **Cerebellar circuitry.** The cell types and action potential pathways are shown. Mossy fibers are afferent from the spinal cord and the cerebral cortex. Climbing fibers are afferent from the inferior olive nucleus in the medulla and synapse directly on the Purkinje cells. The Purkinje cells are the efferent pathways of the cerebellum.

of Purkinje cells. Mossy fibers discharge at high tonic rates, 50 to 100 Hz, which increase further during voluntary movement. When mossy fiber input is of sufficient strength to bring a Purkinje cell to threshold, a single action potential results.

Climbing fibers arise from the inferior olive, a nucleus in the medulla. Each climbing fiber synapses directly on the dendrites of a Purkinje cell and exerts a strong excitatory influence. One action potential in a climbing fiber produces a burst of action potentials in the Purkinje cell, called a *complex spike*. Climbing fibers also synapse with basket, Golgi, and stellate interneurons, which then make inhibitory contact with adjacent Purkinje cells. This circuitry allows a climbing fiber to produce excitation in one Purkinje cell and inhibition in adjacent ones.

Mossy and climbing fibers also give off excitatory collateral axons to the deep cerebellar nuclei before reaching the cerebellar cortex. The cerebellar cortical output (Purkinje cell efferents) is inhibitory to the cerebellar and vestibular nuclei, but the final output of the cerebellum by those nuclei is predominantly excitatory.

Cerebellar lesions reveal functions of the cerebellum.

Lesions of the cerebellum produce impairment in the coordinated action of agonists, antagonists, and synergists. This impairment is clinically known as **ataxia**. The control of limb, axial, and cranial muscles may be impaired depending on the site of the cerebellar lesion. As noted above, lesions of the vestibulocerebellum or medial aspects of the spinocerebellum are manifest as problems with balance, truncal ataxia, and eye movements. The swaying, legs spaced far apart, walk of an intoxicated individual is a vivid example of truncal ataxia. More laterally placed cerebellar lesions produce limb ataxias that manifest as lower limb ataxia or the coarse jerking motions of an arm and hand during reaching for an object instead of the expected, smooth actions. This jerking-type motion is referred to as **action tremor**.

Cerebellar lesions can also produce a reduction in muscle tone, called **hypotonia**, which manifests as a decrease in the level of resistance to passive joint movement present in a normal relaxed person. Myotatic reflexes produced by tapping a tendon with a reflex hammer reverberate for several cycles (pendular reflexes) because of impaired damping from the reduced muscle tone. The hypotonia is likely a result of impaired processing of cerebellar afferent action potentials from the muscle spindles and GTOs.

Other common deficits associated with lateral lesions are the inability to perform rapid alternating movements, garbled speech, and a decrease in motor learning. In addition to its participation in controlling posture and movements, research has suggested that the cerebellum has a role in many other cognitive and affective behaviors and is associated with deficits in language.

Chapter Summary

- The contraction of skeletal muscle produces movement by acting on the skeleton.
- Motor neurons activate the skeletal muscles.
- Sensory feedback from muscles is important for precise control of contraction.
- The output of sensory receptors, such as the muscle spindle, can be adjusted.

- The spinal cord is the source of reflexes that are important in the initiation and control of movement.
- Spinal cord function is influenced by higher centers in the brainstem.
- The highest level of motor control comes from the cerebral cortex and this control is exerted via the corticospinal tract.
- The basal ganglia and the cerebellum provide feedback to the motor control areas of the cerebral cortex and brainstem.

Chapter Review Questions

1. Which type of motor unit is of prime importance in generating the muscle power necessary for the maintenance of posture?

   A. Low threshold, fatigue resistant
   B. High threshold, fatigable
   C. Intrafusal, gamma controlled
   D. High threshold, high force
   E. Extrafusal, gamma controlled

The correct answer is A. The maintenance of posture requires continuous muscle action. The low-threshold, for activation, fatigue-resistant motor units are the type active in postural control. Intrafusal muscle fibers do not contribute to force generation.

2. An elderly man suffers a stroke that damages the vestibulocerebellum. What type of deficit in motor function would you expect to see?

   A. Paralysis accompanied by spasticity
   B. Falling and gait problems
   C. Hyperreflexia
   D. Muscle rigidity
   E. Difficulty learning new skilled movements

The correct answer is B. The vestibulocerebellum works with the vestibular system to help control equilibrium. Damage to the vestibulocerebellum may also produce a truncal tremor and problems with smooth pursuit eye movements. Paralysis accompanied by spasticity is associated with damage to the corticospinal tract. Muscle rigidity and hyperreflexia are most frequently observed following damage to upper motor neurons. Muscle rigidity is also seen in some diseases of the basal ganglia. Difficulty learning new skilled movements is associated with damage to the more laterally located cerebrocerebellum.

3. A disease that produces decreased inhibitory input to the internal segment of globus pallidus should have what effect on the motor areas of the cerebral cortex?

   A. Increased excitatory feedback directly to the cortex
   B. No effect
   C. Decreased excitatory output from the thalamus to the cortex
   D. Increased excitatory output from the putamen to the cortex
   E. Increased excitatory output from the thalamus to the cortex

The correct answer is C. Decreased inhibitory input to the GPi from the putamen would enhance inhibitory output from the GPi to the thalamus. Thalamic output excites the cortex hence increased inhibition by GPI would decrease excitatory output from the thalamus back to the cortex and decreased motor cortical activity. There is no direct projection to the cortex from GPi or the putamen so there could be no direct effects.

4. A patient suffers a stroke that damages the primary motor cortex. An MRI would reveal damage in what part of the brain?

   A. Superior parietal lobe
   B. Superior temporal lobe
   C. Precentral gyrus
   D. Postcentral gyrus
   E. Medial hemisphere rostral to the paracentral lobule

The correct answer is C. The primary motor cortex is located along the precentral gyrus. The superior parietal lobe is association cortex responsible for processing somatosensory information and integrating visual and spatial inputs. The superior temporal lobe is associated with language reception. The postcentral gyrus is the location of primary somatosensory cortex. The medial hemisphere rostral to the paracentral lobule is the location of supplementary motor cortex.

Clinical Application Exercises 5.1

UNSTEADY CARPENTER

A 55-year-old carpenter presents for evaluation of unsteadiness of stance and difficulty walking. The problem began about 1 year ago when he noted a feeling of imbalance when attempting to put his trousers on while standing. He had to start doing this sitting on the side of the bed instead. A few months later, he also noted an unsteady feeling when standing or maneuvering in tight crowds. Over the past year, the difficulties have mildly worsened. His health has been good otherwise. There is no family history of similar problems.

Examination shows normal strength of lower extremity muscles. Muscle stretch reflexes (myotatic) are normal, as is sensory functions of the legs. When the patient is asked to perform a coordination test in which he has to run the heel of one foot down the shin of the opposite leg, he has difficulty keeping either heel accurately tracking along the shin and tends to lose his balance. During standing, he places his feet wider apart than normal, and when asked to move the feet closer together, he becomes notably unsteady. He cannot satisfactorily perform a "walk the line" test like administered by police to check for intoxication. Upper extremity coordination as assessed by having him touch the end of either index finger to the tip of his nose is normal. The patient denies drinking alcohol prior to the appointment today.

QUESTIONS

1. A lesion in which portion of the nervous system would best account for the pattern of dysfunction demonstrated on the neurological examination.

2. What does the fact that the patient's problems involve lower extremity coordination but spare coordination in the upper limbs say about the location of the lesion?

ANSWERS

1. The most likely site of damage in this patient is the cerebellum. Problems in coordination are often due to dysfunction of the cerebellum, particularly when the neurological examination does not show abnormalities in sensation or muscle strength in the involved limbs.

2. The patient demonstrates coordination deficits in the lower extremities in the form of problems smoothly running the heel of one foot along the shin of the opposite leg, so-called heel-to-shin dystaxia. The difficulty he demonstrates when attempting the "walking the line" test is also a manifestation of the type of neurological dysfunction found in this type of patient. Selective atrophy of the anterior midline components of the spinocerebellum is observed at autopsy in patients with predominantly lower extremity coordination deficits like these. Magnetic resonance imaging of the brain can also demonstrate this selective atrophy. Remember that there is a somatotopic map of the body on the surface of the cerebellum and the legs are represented anteriorly. The term "anterior cerebellar syndrome" is used to define the combination of clinical and anatomical deficits observed in this patient.

thePoint® Visit http://thepoint.lww.com/rhoades5e *for additional chapter review Q&A, Clinical Application Exercises, animations, and more!*

6 Autonomic Nervous System

Active Learning Objectives

Upon mastering the material in this chapter, you should be able to:

- Describe the autonomic nervous system efferent pathways from the CNS to effector organs and explain how these differ from the pathway of a motor neuron to the neuromuscular junction.
- Explain the difference between the parasympathetic and sympathetic nervous systems in terms of preganglionic neuronal location, ganglia location, and primary neurotransmitters utilized.
- Describe what is meant by cotransmitters in the autonomic nervous system.
- Explain nonadrenergic, noncholinergic transmission and give an example.
- Describe the function of the parasympathetic and sympathetic nervous systems.

- Explain the role of cholinergic and adrenergic receptors in autonomic function.
- Explain the importance of tonic activation of the parasympathetic and sympathetic nervous systems.
- Explain how presynaptic inhibition influences autonomic function.
- Describe autonomic integration in the eye.
- Explain what is meant by epinephrine reversal.
- Explain the mechanism of organophosphate poisoning and its physiological consequences and treatment.
- Describe how stimulants can produce life-threatening effects.
- Explain the "cheese syndrome" associated with monoamine oxidase.
- Describe the physiological impact and treatment of pheochromocytoma.

The nervous system can be divided into the **central nervous system** (**CNS**) composed of the brain and spinal cord and the **peripheral nervous system** (**PNS**) composed of nerves and ganglia localized outside of the CNS. The **autonomic nervous system** (**ANS**) is one of the components of the PNS and functions as a conduit between the CNS and peripheral organs. Information regarding the subconscious functioning of peripheral organs is transmitted to homeostatic control centers in the hypothalamus and brainstem. Autonomic regulation of these organs is modulated by input from the hypothalamus and brainstem based on these internal stimuli as well as external stimuli that are interpreted by other regions of the brain such as the limbic system and cerebral cortex. In this way, the ANS maintains homeostasis and stimulates necessary alterations in organ function in response to stimuli.

▶ ANATOMY OF THE AUTONOMIC NERVOUS SYSTEM

The two main divisions of the ANS are the **parasympathetic nervous system** (**PSNS**) and the **sympathetic nervous system** (**SNS**). The PSNS can be thought of as the "rest and digest" system that is active in controlling normal organ functioning. The SNS, on the other hand, is the "fight-or-flight" system that modulates organ functioning during times of arousal, stress, and danger. The **enteric nervous system** in the gut is often considered a third division of the ANS. The enteric nervous system regulates the functioning of the gastrointestinal system. It will be discussed separately from the PSNS and SNS in Chapter 27. The PSNS and SNS are both two-neuron pathways, but they differ in the location of these neurons.

Autonomic nervous system consists of a two-neuron efferent pathway.

The CNS communicates with muscles via the somatic motor system, which consists of a cell body in the spinal cord with an axon terminating on muscles at a specialized ending called the **neuromuscular junction** (Fig. 6.1). Unlike this one-neuron pathway, the ANS consists of a two-neuron efferent pathway. The first neuron is located in the brainstem, regions of the sacral spinal cord, or the intermediolateral gray area of the spinal cord. Its axon projects to a second neuron located outside of the CNS. The cell bodies of these second neurons are clustered together with supporting cells in a **ganglion**. The axon from the second neuron projects to the effector organ. The ANS is oriented from the ganglia such that the first neuron is called the **preganglionic neuron**, its axon is the **preganglionic fiber** or **preganglionic axon**, the second neuron is the ganglionic cell or **postganglionic neuron**, and its axon is the **postganglionic fiber** or **postganglionic axon**. The preganglionic axon is a lightly myelinated, slow conducting B fiber, whereas the postganglionic axon is an unmyelinated C fiber that is of smaller diameter and is slower conducting than the preganglionic axon. Unlike an axon from a motor neuron that terminates in a synaptic bouton at the muscle, the postganglionic axons consist of multiple regions specialized for neurotransmitter release, called **varicosities**, that are aligned along the axon in a manner similar to beads on a string. Rather than releasing neurotransmitter from a presynaptic terminal juxtaposed to receptors on postsynaptic neuronal dendrites as occurs in the CNS, neurotransmitters from varicosities are released into extracellular fluid where they diffuse short distances to effector cells and receptors. In some organs, these effector cells can then transmit this information electrically to another cell via gap junctions. The varicosities enable a stimulated fiber to simultaneously

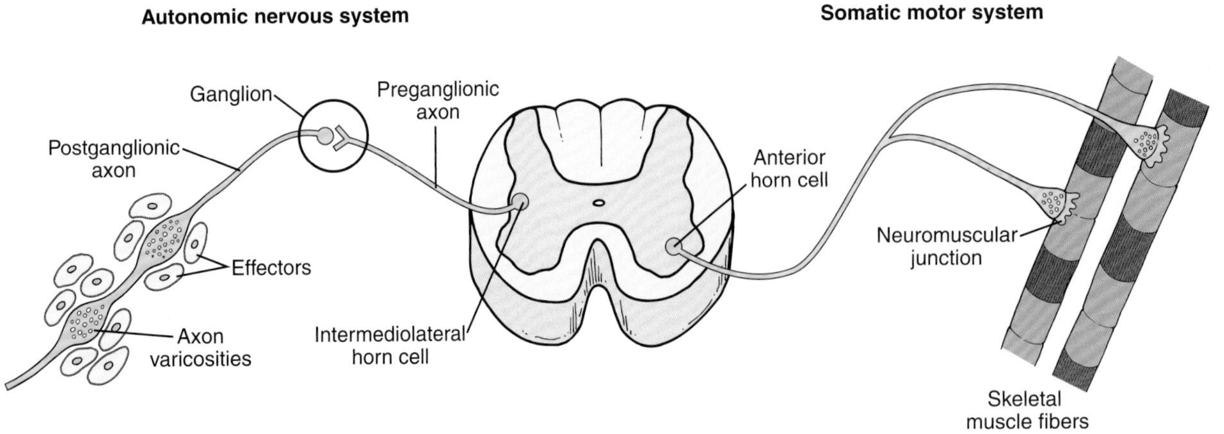

Autonomic nervous system   **Somatic motor system**

Figure 6.1 **Comparison of the autonomic nervous system (ANS) with somatic motor system.** The ANS utilizes a two-neuron pathway and releases neurotransmitters at varicosities. The somatic motor system utilizes a one-neuron pathway and releases neurotransmitters from localized nerve endings.

release neurotransmitter in multiple areas of the effector organ resulting in a greater coordinated impact on the organ.

PSNS and SNS differ in the location of their preganglionic neurons and ganglia.

There are differences in the location of the preganglionic and postganglionic neurons for the PSNS and SNS (Fig. 6.2). The preganglionic neurons of the PSNS are located in the brainstem and the sacral (lowest) portion of the spinal cord. Together, these regions are referred to as the **craniosacral** regions. The PSNS ganglia are located in or near the effector organs. This results in a long preganglionic fiber and a short postganglionic fiber. The brainstem PSNS preganglionic fibers are associated with a number of **cranial nerves** ([CN] Table 6.1).

The preganglionic neurons of the SNS are located in the middle regions of the spinal cord: the thoracic and lumbar portions (see Fig. 6.2). Together, these are referred to as the **thoracolumbar** region. Most SNS ganglia are **paravertebral**, located alongside the spinal cord in a **sympathetic "chain."** Because of the proximity to the spinal cord, the SNS preganglionic fibers are short and the postganglionic fibers are long. There are, however, a few exceptions to this generality in which the SNS preganglionic fiber is much longer. The celiac, super mesenteric, and inferior mesenteric ganglia are **collateral (prevertebral) ganglia** located beyond the paravertebral ganglia in the abdomen and pelvis. Another exception is the adrenal medulla, which is an end organ directly innervated by sympathetic *preganglionic* axons in a manner analogous to sympathetic ganglia but lacking postganglionic neurons. Thus, the preganglionic axons projecting to the adrenal medulla and collateral ganglia are longer than the typically short SNS preganglionic axons.

▶ NEUROTRANSMITTERS OF THE AUTONOMIC NERVOUS SYSTEM

The autonomic nervous system consists primarily of cholinergic and adrenergic fibers. The PSNS is fully a cholinergic system, whereas the SNS utilizes acetylcholine, norepinephrine, and epinephrine. The primary neurotransmitter of the postganglionic SNS fiber depends on the target organ.

All preganglionic axons release acetylcholine.

All neurons sending projections outside of the CNS (motor, PSNS, and SNS) are cholinergic neurons that release acetylcholine (ACh) from their axons (Fig. 6.3). Acetylcholine released from motor neurons interacts with nicotinic receptors at the neuromuscular junction (NMJ), whereas ACh from preganglionic fibers interacts with nicotinic receptors on the postsynaptic membrane of the sympathetic and parasympathetic postganglionic neurons. The nicotinic receptors at the NMJ are a different subtype (N_m) than those at the ganglia (N_n), but both open cation-selective ion channels when activated by ACh or another agonist. This allows the influx of sodium ions (Na^+) and the depolarization of the membrane, which can ultimately result in a muscle contraction at the NMJ or depolarization of postganglionic fibers. This depolarization can generate an action potential that is conducted by nerve fibers and can result in neurotransmitter release.

Parasympathetic postganglionic neurons release acetylcholine, whereas sympathetic postganglionic neurons primarily release norepinephrine.

In the PSNS, all postganglionic neurons are cholinergic synthesizing and releasing ACh from the postganglionic axon onto the effector organ where it can interact with muscarinic receptors. Acetylcholine is synthesized from choline, which is transported into the presynaptic terminal, and acetyl CoA, from mitochondria (Fig. 6.4). Acetylcholine is synthesized in the cytosol and then transported into the synaptic vesicle via the vesicle-associated transporter (VAT). Upon depolarization of the membrane from an action potential and opening of voltage-gated calcium channels, the synaptic vesicles are released from cytosolic anchor proteins, migrate to fuse with the plasma membrane, and release ACh where it can then act on pre- and postsynaptic cholinergic receptors. After release, ACh is rapidly hydrolyzed to choline and acetate by the enzyme acetylcholinesterase located on the postsynaptic membrane. Choline can then be transported back into the presynaptic terminal where it can be recycled to produce more ACh.

The SNS primarily utilizes norepinephrine (NE), sometimes referred to as noradrenaline, as the signaling molecule

Sympathetic Division

Parasympathetic Division

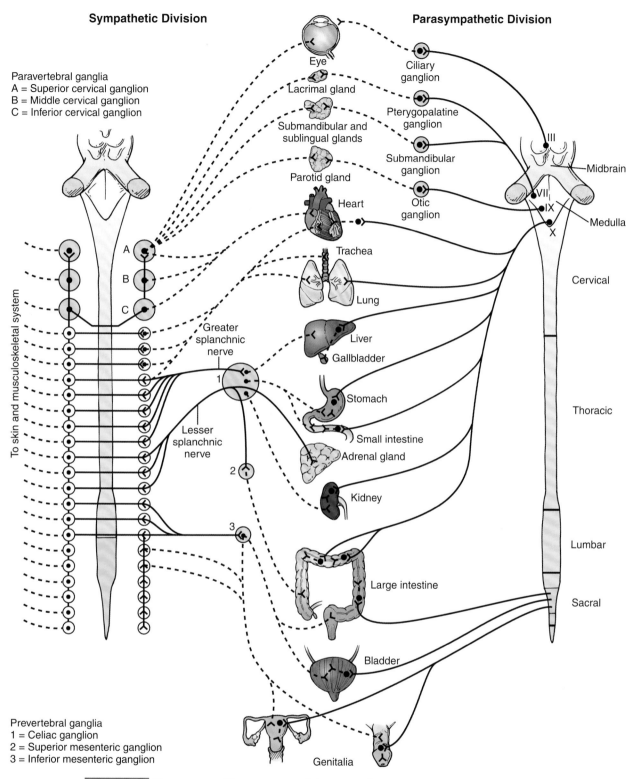

Neuromuscular Physiology

Figure 6.2 **The organ-specific arrangement of the autonomic nervous system.** Preganglionic axons are indicated by *solid lines* and postganglionic axons by *dashed lines*. Sympathetic axons destined for the skin and musculoskeletal system are shown on the left side of the spinal cord. Note the named paravertebral and prevertebral ganglia.

released from postganglionic axons. Norepinephrine is a catecholamine that is synthesized from the common catecholamine precursor tyrosine (see Fig. 6.4). Tyrosine is transported into the presynaptic terminal and converted to L-dopa and then dopamine. Dopamine is transported into synaptic

vesicles by the vesicular monoamine transporter (VMAT). If the vesicle contains the enzyme dopamine-β-hydroxylase, dopamine will be converted to NE. Excess dopamine in the cytosol can be metabolized by monoamine oxidase (MAO), an enzyme located on the outer membrane of mitochondria.

TABLE 6.1 Origins and Innervations of PSNS Fibers

Location of Preganglionic Neuron	Preganglionic Axons Location	Preganglionic Axon Termination	Postganglionic Effector Organ	Major Effects
Edinger-Westphal nucleus	Oculomotor (CN III)	Ciliary ganglion	Eye	Miosis (pupillary constriction)
				Accommodation
Superior salivatory nuclei	Facial (CN VII)	Pterygopalatine ganglion	Lacrimal gland	Tearing
			Nasal and palatal mucosa	Increased secretions
		Submandibular ganglion	Submandibular and sublingual glands	Salivation
Inferior salivatory nuclei	Glossopharyngeal (CN IX)	Otic ganglion	Parotid gland	Salivation
Nucleus ambiguus	Vagus (CN X)	Cardiac plexus	Heart	Decreased heart rate, Decreased contraction force
Dorsal motor nuclei		Pulmonary plexus	Lungs	Bronchoconstriction
				Increased secretions in lungs
		Myenteric plexus	GI tract	Increased motility of stomach
		Submucosal plexus		Increased secretions in stomach
Sacral spinal cord (S2, S3, S4)	Pelvic nerves	Myenteric plexus	Lower large intestine	Increased motility
		Submucosal plexus		Increased secretions
		Hypogastric plexus	Bladder	Urination
			Reproductive organs	Erection

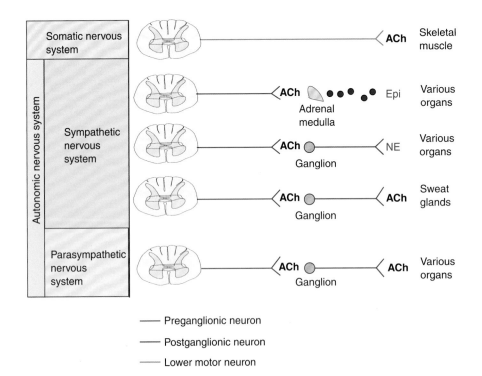

Figure 6.3 **Neurotransmitters synthesized and released from somatic and autonomic nervous systems.** Acetylcholine (ACh) is released from all axons exiting the spinal cord. Postganglionic axons of the parasympathetic nervous system also release ACh. Postganglionic axons of the sympathetic nervous system primarily release NE except for those innervating sweat glands, which release ACh. Sympathetic activation of the adrenal medulla by preganglionic fibers releases epinephrine directly into the bloodstream.

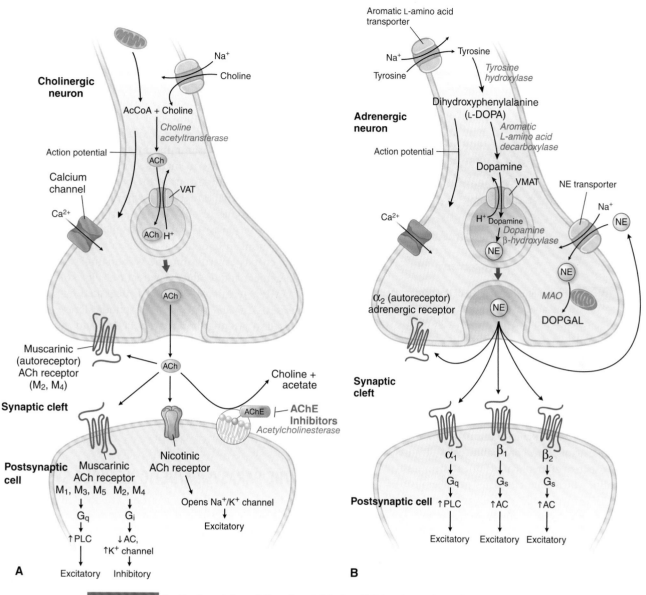

Figure 6.4 **Synthesis and degradation of acetylcholine (ACh) and norepinephrine (NE).**
(A) Acetylcholine is synthesized from choline and acetyl CoA, transported into vesicles, released in a calcium-dependent manner, and degraded in the synaptic cleft by acetylcholinesterase. It can interact with muscarinic or nicotinic receptors (shown in the figure on the same membrane but generally located on different postsynaptic cells). **(B)** Tyramine is the precursor for all catecholamines. It is converted to dopamine in the cytosol, which is then transported into vesicles. Once there, it can be converted to NE if the enzyme dopamine-β-hydroxylase is present. Norepinephrine is released in a calcium-dependent manner. It is removed via uptake by the norepinephrine transporter, or it can diffuse away, be taken up by other cell types, and degraded by catechol-*O*-methyltransferase (not shown). In the cytosol, it can be recycled into the vesicle or degraded by monoamine oxidase. After release, NE can interact with α or β adrenergic receptors (shown in the figure on the same membrane but generally located on different postsynaptic cells).

Upon release, NE interacts with presynaptic and postsynaptic adrenergic receptors. Unlike ACh, which is degraded in the extracellular fluid, NE either diffuses away where it can then be degraded by catechol-*O*-methyltransferase (COMT) or is taken back into the presynaptic terminal by the norepinephrine transporter where it can either be degraded by MAO or recycled to the vesicle by VMAT.

There are two instances in which neurotransmitters other than NE are utilized by the SNS at effector organs (see Fig. 6.3). The first is the sweat glands where sympathetic

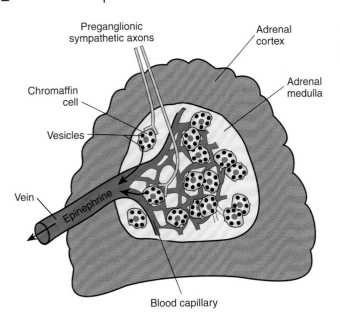

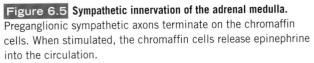

Figure 6.5 **Sympathetic innervation of the adrenal medulla.** Preganglionic sympathetic axons terminate on the chromaffin cells. When stimulated, the chromaffin cells release epinephrine into the circulation.

postganglionic axons synthesize and release ACh, which then interacts with muscarinic receptors to increase sweating. The second involves the adrenal medulla. The adrenal medulla receives input from the *preganglionic* sympathetic axon and acts as the sympathetic ganglion. Instead of a postganglionic axon, the chromaffin cells of adrenal medulla directly secrete epinephrine (Epi), sometimes referred to as adrenalin, which then circulates peripherally as a hormone (Fig. 6.5). Epinephrine is synthesized from NE by the enzyme phenylethanolamine *N*-methyltransferase. Approximately 20% of the neurotransmitter released by the adrenal medulla is NE with Epi accounting for the other 80%.

Many autonomic postganglionic fibers contain cotransmitters along with acetylcholine and norepinephrine.

In addition to the primary neurotransmitter in the autonomic fibers, autonomic postganglionic fibers can contain cotransmitters that modulate the function of the primary neurotransmitter. Often these cotransmitters are packaged into large dense core vesicles in the cell body and transported inside the vesicle to the synapse. There are also instances of the cotransmitters colocalizing with the primary neurotransmitter in the smaller clear vesicles that are located closer to the synaptic membrane. The vesicles that are further from the membrane require more stimulation to produce vesicular mobilization and neurotransmitter release. In this way, the action of a primary neurotransmitter can be modulated during times of greater activity and thus affect the duration or extent of the postsynaptic response. There are a number of cotransmitters that have been identified including ATP, adenosine, neuropeptide Y, vasoactive intestinal peptide, calcitonin gene-related peptide, and substance P.

Some autonomic fibers contain neither acetylcholine nor norepinephrine but instead undergo nonadrenergic noncholinergic (NANC) transmission.

Although most ANS fibers are associated with acetylcholine, norepinephrine, or epinephrine, there are ANS fibers that communicate using other signaling molecules such as nitric oxide. Nitric oxide is found in certain nonadrenergic noncholinergic (NANC) fibers of the parasympathetic nervous system such as those that innervate the corpus cavernosum of the penis. Nitric oxide is a gas that is formed presynaptically from arginine by the enzyme nitric oxide synthase in response to increased cytosolic calcium. Once synthesized, it diffuses out of the presynaptic terminal and into the postsynaptic terminal where it activates guanylyl cyclase to produce cyclic GMP (cGMP). Cyclic GMP activates protein kinase G, which phosphorylates certain proteins resulting in smooth muscle relaxation and vasodilation. In the corpus cavernosum, the vasodilation leads to penile engorgement and produces an erection (Fig. 6.6). The action of cGMP is terminated when it is degraded by phosphodiesterases. Certain types of phosphodiesterase inhibitors are used to prolong the action of cGMP and thus can treat disorders such as erectile dysfunction. Examples of these medications include sildenafil (brand name Viagra), tadalafil (Cialis), and vardenafil (Levitra). The enteric nervous system also utilizes nitric oxide as well as other compounds such as vasointestinal peptide and substance P in NANC fibers.

▶ THE PARASYMPATHETIC NERVOUS SYSTEM

The PSNS preganglionic neurons are located in specific cranial nerve nuclei (collection of neurons) in the brainstem (see Table 6.1 and Fig. 6.2). The preganglionic axons travel along that cranial nerve to terminate at parasympathetic ganglia. These may be true ganglia (i.e., a collection of cells), located a short distance from the organ that the postganglionic axon innervates as is seen with CN III, CN VII, and CN IX. The "ganglia" of CN X (vagus nerve) and pelvic nerves, however, are not true ganglia but are rather nerve plexuses, branching networks of intersecting nerves and blood vessels, located in or very near the effector organ itself.

PSNS is involved in normal homeostatic functions.

Many organs receive moderate PSNS input under normal conditions when an individual is in a nonstressed and safe environment. This regulation can be modulated by either increasing or decreasing the PSNS input. As much as 75% of parasympathetic activity may be mediated through the vagus nerve. Activation of the PSNS will result in miosis (pupillary contraction), accommodation of the lens, lacrimation (tearing), salivation, bradycardia, decreased force of myocardial contraction, bronchoconstriction and increased bronchial secretions, increased digestive secretions and GI motility, and urination (Table 6.2).

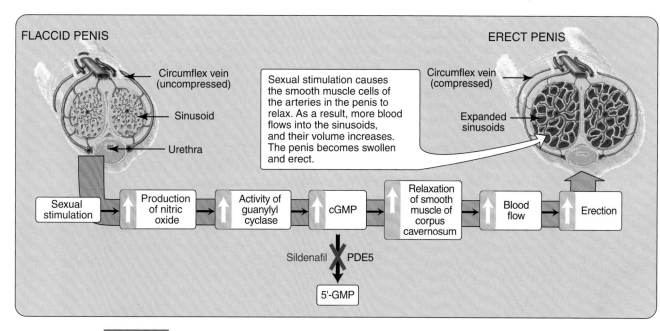

FLACCID PENIS

ERECT PENIS

Circumflex vein (uncompressed)

Sinusoid

Urethra

Sexual stimulation causes the smooth muscle cells of the arteries in the penis to relax. As a result, more blood flows into the sinusoids, and their volume increases. The penis becomes swollen and erect.

Circumflex vein (compressed)

Expanded sinusoids

Sexual stimulation → ↑ Production of nitric oxide → ↑ Activity of guanylyl cyclase → ↑ cGMP → ↑ Relaxation of smooth muscle of corpus cavernosum → ↑ Blood flow → ↑ Erection

Sildenafil ✗ PDE5

5'-GMP

Figure 6.6 **Nonadrenergic noncholinergic (NANC) parasympathetic nervous system (PSNS) mediation of penile erection.** Release of nitric oxide from NANC PSNS postganglionic fibers acts in the corpus cavernosum to produce smooth muscle relaxation allowing engorgement from increased blood flow and subsequent erection. Phosphodiesterase type 5 (PDE5) converts cGMP to 5'-GMP. This can be inhibited with selective PDE5 inhibitors such as sildenafil.

TABLE 6.2 Responses of Effectors to Parasympathetic and Sympathetic Stimulation

Region	Effector	Parasympathetic (Muscarinic Receptors)	Sympathetic (Adrenergic Receptors)
Eye and face	Pupil	Constriction: via circular muscle contraction (miosis)	Dilation (α_1): via radial muscle contraction (mydriasis)
	Ciliary muscle	Contraction	Relaxation (β_2)
	Müller muscle	None	Contraction (α_1)
	Lacrimal gland	Secretion	None
	Nasal glands	Secretion	Inhibition (α_1)
	Salivary glands	Secretion	Amylase secretion (β)
Skin	Sweat glands	None	Secretion (cholinergic muscarinic receptors)
	Piloerector muscles	None	Contraction (α_1)
Blood vessels	Skin (general)	None	Constriction (α)
	Facial and neck muscles	Dilation (?)*	Constriction (α)
			Dilation (?)*
	Skeletal muscle	None	Dilation (β_2)
			Constriction (α)
	Viscera	None	Constriction (α_1)
Heart	SA Node	Decrease heart rate	Increase heart rate ($\beta_1 > \beta_2$)
	AV Node	Decrease conduction velocity	Increase conduction velocity ($\beta_1 > \beta_2$)
	Atria	Decrease contractility	Increase contractility ($\beta_1 > \beta_2$)
	Ventricle	Little effect	Increase contractility ($\beta_1 > \beta_2$)

(Continued)

Neuromuscular Physiology

TABLE 6.2	Responses of Effectors to Parasympathetic and Sympathetic Stimulation (*Continued*)		
Region	**Effector**	**Parasympathetic (Muscarinic Receptors)**	**Sympathetic (Adrenergic Receptors)**
Lungs	Bronchioles	Constriction	Dilation (β_2)
	Glands	Secretion	Decrease (α_1)
			Increase (β_2)
Gastrointestinal tract	Wall muscles	Contraction	Relaxation (α, β_2)
	Sphincters	Relaxation	Contraction (α_1)
	Glands	Stimulation	Inhibition (α_2)
Metabolic functions	Liver	None	Glycogenolysis and gluconeogenesis (α_1, β_2)
	Pancreas (insulin)	Increased secretion	Decreased secretion (α_2)
	Adipose Cells	None	Lipolysis (β_3)
	Kidney	None	Renin release (β_1)
	Adrenal medulla	None	Secretion of epinephrine (cholinergic nicotinic)
Urinary system	Ureter	Relaxation	Contraction (α_1)
	Detrusor	Contraction	Relaxation ($\beta_3 > \beta_2$)[†]
	Sphincter	Relaxation	Contraction (α_1)
Reproductive system	Uterus	Variable	Contraction (α_1)
			Relaxation (β_2)
	Genitalia	Erection (ACh and NO)	Ejaculation/vaginal contraction (α)
Autonomic nerve endings	Autoreceptor	Presynaptic inhibition of ACh release	Presynaptic inhibition of NE release (α_2)
	Heteroreceptor	Presynaptic inhibition of NE release	Presynaptic inhibition of ACh release (α_2)

ACh, acetylcholine; NO, nitric oxide.

[*]The status of vasodilator activity to the face (as shown in blushing and/or response to cold exposure) is still controversial.

[†]The contribution of β_2 vs. β_3 to detrusor relaxation is controversial.

PSNS effects are mediated by cholinergic receptors.

For many organs, there is a 1:1 relationship between the PSNS preganglionic axon and postganglionic axon such that the effect of preganglionic activation is very discrete at the effector organ. Upon activation, the postganglionic axons of the PSNS release ACh, which activates muscarinic receptors in the effector organ. The muscarinic receptor family consists of five subtypes linked to two different signaling pathways. As Figure 3.13 shows, the M_1, M_3, and M_5 receptors activate the G_q G protein. This results in the hydrolysis of phosphatidylinositol bisphosphate (PIP$_2$) to diacylglycerol (DAG) and inositol triphosphate (IP$_3$). Diacylglycerol can activate protein kinase C resulting in phosphorylation of numerous proteins. Inositol triphosphate can release calcium from the endoplasmic reticulum, which can result in cellular contraction. Other cellular signaling molecules are also activated and ultimately modulate cellular function. The M_2 and M_4 muscarinic subtypes activate the G_i G protein. This results in decreased adenylyl cyclase activity and subsequent decreased cAMP levels. Additionally, activation of the G_i G protein can open certain potassium channels leading to membrane hyperpolarization and close certain presynaptic calcium channels ultimately decreasing vesicular release. Some M_2 receptors are located presynaptically and serve as autoreceptors producing negative feedback to decrease further release of ACh. The primary muscarinic receptors in autonomic effector organs are M_2 in the heart, which can result in *decreased* heart rate and force of contraction, and M_3 in smooth muscle and glands, which can result in *increased* secretions and muscle contraction.

▶ SYMPATHETIC NERVOUS SYSTEM

The SNS preganglionic axons exit the spinal cord along with axons from motor neurons via the ventral root (Fig. 6.7). These myelinated fibers separate from the motor neuron axons to form the white rami communicantes of the thoracic and lumbar nerves and terminate on the corresponding sympathetic ganglion. These SNS preganglionic axons

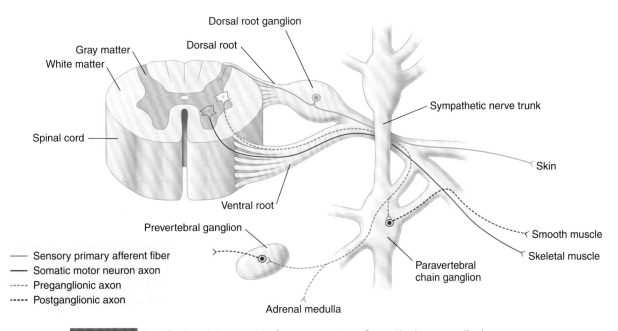

Figure 6.7 **Organization of the sympathetic nervous system.** Sympathetic preganglionic axons (*dashed blue line*) exit the spinal thoracic and lumbar segments of the spinal cord together with somatic motor neuron axon (*red line*) via the ventral root and project to paravertebral and prevertebral ganglia. Sympathetic postganglionic axons (*dashed purple line*) exit the ganglia and project to effector organs. The adrenal medulla receives preganglionic input and has no postganglionic axons. Sensory information is received by neurons with cell bodies in the dorsal root ganglion and axons projecting to both the periphery and spinal cord (*solid green line*).

can send additional projections to nearby ganglia along the sympathetic trunk. This results in a broader, more divergent effect of SNS preganglionic neuronal activation in contrast to the more discrete and direct effect of PSNS preganglionic neuronal activation. Also in contrast to the PSNS, it is the SNS *postganglionic* fibers that travel in nerve bundles to the effector organ.

SNS mediates responses in times of arousal, stress, and danger.

Like the PSNS, some organs receive tonic activation through the SNS allowing mediation by either increasing or decreasing the level of SNS activity. Activation of the SNS prepares a person to fight or flee through an increase in oxygen and energy delivery to key organs (Table 6.3).

SNS effects are mediated mostly by adrenergic receptors.

Upon activation of the SNS, many organs receive input from preganglionic neurons at multiple levels of the spinal cord due to the divergence of preganglionic axons along the sympathetic chain. It has been estimated that one sympathetic preganglionic neuron affects twenty postganglionic neurons. The effects of NE at the effector organs are relatively short lived due to the reuptake of NE by the NET and diffusion from the synapse. The effects of epinephrine, however, persist much longer. Epinephrine is released from the adrenal medulla after activation by SNS preganglionic axons and acts

TABLE 6.3 Effects of Sympathetic Activation on the Fight or Flight Response	
Organ Effect	**Utility for the Fight or Flight Response**
Increased heart rate and force of contraction	Increase oxygen delivery to key organs and muscles
Bronchodilation	Increase oxygen supply
Mydriasis (pupillary dilation)	Increase light entering eye
Vasoconstriction	Decreases blood flow if injured
	Redistribute blood flow to key organs
Vasodilation in skeletal muscle	Increase energy delivery for increased energy consumption
Relaxation of the gut and bladder	Decrease energy delivery to organs not utilized during the danger
Sweating	Dissipation of heat generated from increased muscle activity
Increased gluconeogenesis and glycogenolysis in liver	Increase energy supplies for delivery to key organs
Decreased insulin secretion	Maintain energy supplies for delivery to key organs
Lipolysis in adipose tissue	Use fatty acids to provide energy for skeletal muscles

as a hormone circulating peripherally and activating multiple organs.

With the exception of the sweat glands, which utilize muscarinic cholinergic receptors, stimulation of the SNS results in activation of adrenergic receptors at effector organs. The primary adrenergic subtypes are α_1, α_2 and β_1, β_2, and β_3. The α_1 receptors interact with G_q G proteins to ultimately increase protein kinase C activity and intracellular calcium levels (see Fig. 3.13). The α_1 receptors on vascular smooth muscle mediate vasoconstriction and maintenance of adequate blood pressure. The α_2 receptors interact with G_i G proteins and function presynaptically to inhibit further sympathetic output. The α_2 receptors also act postsynaptically to inhibit insulin output. The β_1, β_2, and β_3 receptors all interact with the G_s G protein. The β_1 receptors function in the heart to increase heart rate, the force of contraction, and conduction velocity. Activation of β_2 receptors produce bronchodilation, vasodilation in skeletal muscle, and increased plasma glucose levels through actions in the liver. The β_3 receptors are involved in relaxation of the bladder and lipolysis in adipose tissue (Clinical Focus 6.1).

CLINICAL FOCUS | 6.1

Indirect Sympathomimetics

An **indirect sympathomimetic** is a compound that is transported into the cell via the NET, via the DAT, and sometimes by SERT. It is then transported into the synaptic vesicle via the vesicular monoamine transporter (VMAT), as shown in the figure. Once in the vesicle, the indirect sympathomimetic displaces NE or DA from the vesicle. This increases NE or DA in the cytosol where it can be metabolized by monoamine oxidase. Because of the large cytosolic concentration, the NET or DAT will operate in reverse transporting the monoamine *into* the synaptic cleft thus producing a calcium-independent neurotransmitter release.

Methamphetamine is a stimulant drug of abuse that is an example of an indirect sympathomimetic. In addition to its indirect sympathomimetic activity, it has other actions such as inhibition of MAO, which can also increase adrenergic or dopaminergic activation. Increased release of NE in the periphery mimics certain functions of the SNS. Increased release of DA can activate the dopamine "reward" pathway between the ventral tegmental area and nucleus accumbens in the brain that is involved in the development of addiction. The activation of peripheral and central adrenergic receptors by methamphetamine can result in cardiac arrhythmias, myocardial infarction, stroke, and seizures, whereas activation of the reward pathway contributes to the addictive potential of methamphetamine.

Cocaine is another stimulant drug that activates both the dopamine reward pathway and adrenergic receptors, but it does so through a different mechanism. Cocaine inhibits reuptake of the monoamines at their individual transporters: the norepinephrine transporter (NET), dopamine transporter (DAT), and serotonin transporter (SERT). It also acts as a local anesthetic through inhibition of voltage-gated sodium channels. Similar to methamphetamine, elevated NE by cocaine can produce cardiac arrhythmias, myocardial infarction, stroke, and seizures. Overdoses of stimulants can be treated with benzodiazepines, which act at the $GABA_A$ receptor to produce CNS depression and counteract the CNS effects of stimulants.

Pseudoephedrine, brand name Sudafed, is also an indirect sympathomimetic. It produces decongestion through the

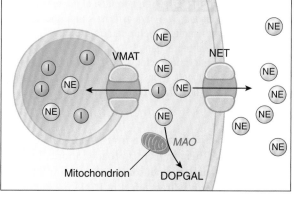

Indirect sympathomimetics act to increase release of norepinephrine. Indirect sympathomimetics (I) gain access to the cytosol via transport by the norepinephrine transporter (NET). They are then transported into vesicles via the vesicular monoamine transporter (VMAT). Inside the vesicle, these displace the stored NE resulting in an increase in cytosolic NE. Some of the NE is degraded by monoamine oxidase (MAO) and some is transported out of the cell via the reversal of the NET. Thus, NE is released independent of an action potential.

vasoconstriction of the nasal mucosa by increasing the release of NE and subsequent activation of the α_1 adrenergic receptor. At therapeutic doses, pseudoephedrine produces less cardiac and abuse concerns than accompany methamphetamine. Pseudoephedrine can be chemically converted to methamphetamine and for this reason is an over-the-counter drug that is highly regulated.

Pseudoephedrine contains a warning that it should not be taken by a person who has recently received a monoamine oxidase inhibitor (MAOI). The MAOIs are antidepressant drugs that are primarily used to treat depression that is refractory to other treatments. By inhibiting MAO, these drugs increase the concentration of norepinephrine, dopamine, and serotonin in the presynaptic terminal. This becomes dangerous in the presence of indirect sympathomimetics. Normally, a

Neuromuscular Physiology

portion of the NE that pseudoephedrine displaced from the vesicle into the cytosol is metabolized by MAO resulting in a smaller concentration of NE that is released into the synaptic cleft. When MAO is inhibited, there is a larger concentration of NE released and the subsequent activation of α_1 receptors on the vasculature and β_1 receptors in the heart can result in a life-threatening increase in blood pressure.

This dangerously high concentration of NE and hypertensive crisis can also occur when foods containing tyramine are ingested when MAO is inhibited. Tyramine is another indirect sympathomimetic that is normally degraded by intestinal and hepatic MAO. When MAO is inhibited, tyramine is able to enter the circulation and affect sympathetic nerve terminals and lead to increased NE release and potential hypertensive crisis. Tyramine is found in aged cheese, smoked fish, cured meats, and some beers. Because of this, the tyramine–MAOI effect is sometimes referred to as the "cheese effect." ■

▶ AUTONOMIC INTEGRATION

The autonomic nervous system is always active. As the "rest-and-digest" system, many organs receive input from the PSNS during "normal" activity. These include the eyes, heart, lungs, gut, and bladder. Parasympathetic activity can be increased when needed or decreased during times of increased SNS activation. Some organs are maintained by SNS basal activity. The blood vessels do not receive innervation from the PSNS, and thus, systemic vascular resistance is maintained by SNS input.

Presynaptic inhibition can modulate functioning of the ANS.

In addition to acting as autoreceptors to decrease further release of its own neurotransmitter, the M_2 and α_2 receptors can also function as **heteroreceptors** decreasing the release of other neurotransmitters. Input from the PSNS or SNS can be decreased through actions of one system on the presynaptic terminals of the other system's postganglionic axons. Release of ACh, for example, at postganglionic PSNS varicosities can activate M_2 receptors on SNS postganglionic fibers and *decrease* the release of NE from these fibers. Conversely, activation of the SNS can result in the release of NE from postganglionic varicosities and the activation of α_2 receptors on PSNS postganglionic fibers resulting in *decreased* ACh release from these fibers. In this way, an increase in the activity of one system can directly decrease the activity of the other system.

Eye receiving input from the PSNS and SNS is an example of autonomic integration.

In most organs, modulation of function occurs by integration of signals from the PSNS and SNS. The eye is an example of such integration. Pupillary constriction and dilation occur through the contraction of two different types of muscles. The circular muscles, also known as sphincter or constrictor muscles, form concentric rings around the pupil (Fig. 6.8). Activation of the PSNS results in the contraction of these muscles and pupillary constriction (**miosis**). The radial muscles radiate out perpendicularly from the circular muscles. Contraction of the radial muscles in response to SNS activation results in pupillary dilation (**mydriasis**). Certain drugs of abuse affect these systems. One of the signs

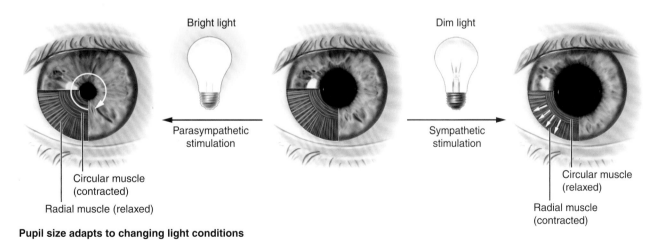

Bright light

Parasympathetic stimulation

Dim light

Sympathetic stimulation

Circular muscle (contracted)

Radial muscle (relaxed)

Circular muscle (relaxed)

Radial muscle (contracted)

Pupil size adapts to changing light conditions

Figure 6.8 **Autonomic regulation of pupil light adaptation.** The iris contains two muscles that are involved in modulation of pupillary diameter. The circular muscles surround the pupil in concentric rings, whereas the radial muscles are perpendicular to the circular muscles. Activation of the parasympathetic nervous system in response to bright light results in contraction of the circular muscles and pupillary constriction. Activation of the sympathetic nervous system in response to dim light results in contraction of the radial muscles and pupillary dilation.

of opioid use is pinpoint pupils. This occurs due to opioids relieving inhibition on parasympathetic fibers innervating the iris thus producing release of ACh, activation of muscarinic receptors, and contraction of circular muscles. Conversely, amphetamine and cocaine increase the activation of adrenergic receptors in the iris resulting in contraction of radial muscles and pupillary dilation.

Both the PSNS and the SNS also contribute to intraocular pressure. Activation of adrenergic receptors increases the secretion of aqueous humor from the ciliary body epithelium into the anterior chamber of the eye (Fig. 6.9). The aqueous humor exits the anterior chamber through the canal of Schlemm after flowing through the trabecular meshwork. Contraction of the ciliary muscle by muscarinic receptor activation opens the trabecular meshwork and allows more aqueous humor to flow out thus decreasing intraocular pressure. Elevated intraocular pressure can lead to glaucoma with damage to the optic nerve and potential vision loss. Open-angle glaucoma is associated with increased secretion of aqueous humor and/or decreased draining. Treatment is aimed at decreasing secretion of aqueous humor through targeting adrenergic receptors and increasing outflow by targeting muscarinic receptors or other receptors involved in contracting the ciliary muscle to open the trabecular meshwork.

Vasculature receives input only from the SNS, which contributes to the maintenance of appropriate blood pressure.

Unlike the eye, the vasculature receives input only from the SNS: there are no PSNS fibers that terminate on the vasculature. Thus, the SNS is responsible for maintaining

systemic vascular resistance, which, along with cardiac output, determines blood pressure. Blood pressure can be influenced by many factors, and the SNS must respond quickly and appropriately. When a person stands up, gravitational force leads to pooling of blood at the feet due to increased venous volume. In order to counteract this and continue systemic blood flow to all organs, there must be a concomitant increase in blood pressure. This increase is critical in maintaining adequate blood supply to the brain and is accomplished primarily through activation of SNS fibers that terminate on vascular smooth muscle cells and release norepinephrine to interact with α_1 adrenergic receptors. This activates the G_q G protein leading to increased intracellular calcium levels enabling the interaction of actin and myosin and vasoconstriction of the blood vessel thus counteracting the venous dilation. The lack of appropriate blood pressure modulation can result in orthostatic hypotension.

While there is no PSNS innervation of the vasculature, there are muscarinic receptors on endothelial cells that can be activated by ACh released by other cells, such as platelets and macrophages. Activation of these muscarinic receptors leads to the synthesis of nitric oxide (Fig. 6.10). Nitric oxide diffuses into the smooth muscle cell where it activates guanylyl cyclase and ultimately results in vasodilation. These muscarinic receptors can also be activated by exogenous compounds such as those found in certain mushrooms like those of the *Clitocybe* species. This can lead to "fast" mushroom poisoning characterized by activation of muscarinic receptors on endothelial cells and in the sweat glands and other organs resulting in hypotension, sweating, diarrhea, vomiting, bradycardia, bronchoconstriction, salivation, and visual disturbances.

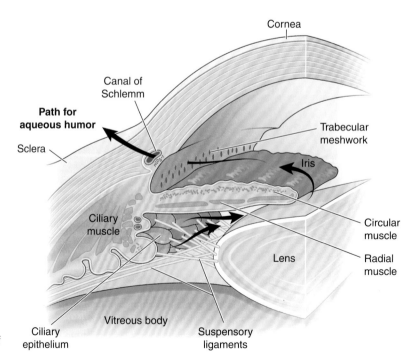

Figure 6.9 **Autonomic regulation of aqueous humor and intraocular pressure.** Aqueous humor is produced by the ciliary body epithelium in response to sympathetic nervous system activation and secreted into the anterior chamber of the eye. Activation of the parasympathetic nervous system results in the contraction of ciliary muscle and opening of the canal of Schlemm, which allows the outflow of the aqueous humor from the anterior chamber and the lowering of intraocular pressure. Activation of the SNS will relax the ciliary muscle decreasing outflow through the canal of Schlemm.

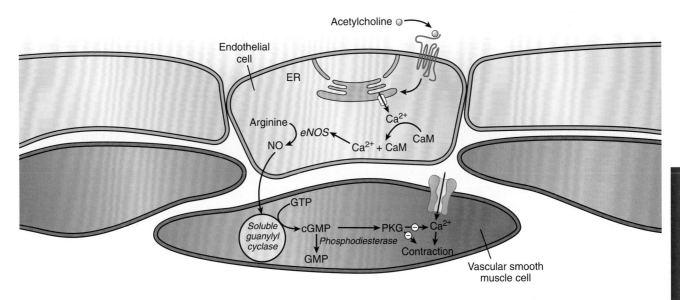

Figure 6.10 **Nitric oxide-induced vascular smooth muscle relaxation.** Acetylcholine can interact with receptors on endothelial cells to stimulate the synthesis of nitric oxide (NO). Nitric oxide diffuses into the vascular smooth muscle cell where it activates guanylyl cyclase to produce cGMP, which increases activity of protein kinase G (PKG). Protein kinase G phosphorylates calcium channels and actin/myosin proteins to decrease contraction, which results in vasodilation.

Epinephrine can produce both vasoconstriction and vasodilation.

Another role of the SNS is providing appropriate blood supply to skeletal muscle. This is particularly important during the "fight-or-flight" response. Unlike norepinephrine, which mostly acts locally in the area in which it was released from the postganglionic fiber, epinephrine circulates throughout the periphery as a peripheral hormone. As such, it interacts with α_1 receptors on vascular smooth muscle cells to produce vasoconstriction. This is part of the rationale for the use of "EpiPens" to counteract anaphylactic shock. During anaphylaxis, histamine produces vasodilation and bronchoconstriction through the activation of histamine receptors. Epinephrine counters this through actions on α_1 receptors to produce vasoconstriction and on β_2 receptors to produce bronchodilation.

Epinephrine also interacts with β_2 receptors on blood vessels in skeletal muscle to produce *vasodilation* thus facilitating the delivery of oxygen, glucose, and nutrients to meet increased metabolic demands. Despite this vasodilation, the net effect observed with systemic epinephrine is an *increase* in blood pressure because the β_2 vasodilation in skeletal muscle is masked by the much greater α_1 vasoconstriction. However, if the α_1 receptors are blocked, a decrease in blood pressure can be measured. This phenomenon

is referred to as "**epinephrine reversal**" because initially an increase in blood pressure is observed but this increase can be "reversed" by blocking α_1 receptors resulting in a decrease in blood pressure. This reversal does not occur with norepinephrine because it has little effect on β_2 receptors (Clinical Focus 6.2).

Receptors and pathways of the autonomic nervous system provide targets for therapeutics.

Because of the breadth of functions of the ANS, there are many medications that target the different receptors of the autonomic nervous system. In some cases, **agonists** are utilized to mimic the action of ACh or NE at its receptors. At other times, **antagonists** are utilized to block the function of ACh or NE at its receptor and decrease overall activity. Common medical conditions involving the ANS and their treatments are listed in Table 6.4. It is also important to remember that while activating or blocking part of the autonomic nervous system can produce therapeutically beneficial effects, it can also result in predictable adverse effects. In fact, many drugs for which "dry mouth" is an associated adverse effect produce some degree of muscarinic receptor blockade. Overall, drugs targeting functions modulated by the autonomic nervous system are highly effective and useful.

Neuromuscular Physiology

CLINICAL FOCUS | 6.2

Pheochromocytoma

Pheochromocytoma is a rare tumor of the chromaffin cells of the adrenal medulla. These are the same cells that receive preganglionic sympathetic input and release epinephrine, along with some NE, in the systemic circulation. In a person with pheochromocytoma, there is an overproduction of epinephrine and NE leading to high concentrations of catecholamines in the blood. This results in increased blood pressure, increased heart rate, and generalized sweating. Additionally, headaches are common possibly due to effects on the vasculature. Most tumors are removed surgically; however, drugs can be used to manage symptoms prior to surgery.

One approach to reduce symptoms is to block adrenergic receptors. Phenoxybenzamine irreversibly blocks both α_1 and α_2 receptors. Blocking α_1 receptors will decrease vasoconstriction and relieve the hypertension. However, the drop in blood pressure can activate the baroreceptor reflex (see Chapter 17) leading to *increased* NE release into the heart and *increases* in heart rate due to activation of the β_1 receptor. Once phenoxybenzamine begins lowering the blood pressure, a β-adrenergic receptor blocker can be added to block the baroreceptor-induced adrenergic stimulation of the heart. If the β-blocker is started before phenoxybenzamine, blood pressure can *increase* due to blockade of β_2 receptors in skeletal muscle: activation of β_2 receptors by epinephrine normally results in vasodilation but this effect is masked by the interaction of epinephrine with α_1 receptors and the overwhelming vasoconstriction that ensues. Because the β_2 receptor–induced vasodilation acts to mitigate the α_1 receptors, removal of β_2 vasodilation by the β-blocker can further *increase* blood pressure. A second approach to reduce symptoms of pheochromocytoma is the use of calcium channel blockers, which produce vasodilation and also decrease the stimulation of the heart. A third approach is the use of metyrosine. Metyrosine blocks the enzyme tyrosine hydroxylase, which converts tyrosine to L-dopa. This is the rate-limiting step in the synthesis of catecholamines. This results in the reduction of catecholamines throughout the body, which also produces a number of adverse effects including sedation, depression, and movement dysfunction. While not ideal, these approaches can provide short-term management of symptoms until the tumor can be removed. ■

TABLE 6.4　Common Treatments for Disorders Involving the ANS

		Indication	Example Generic	Example Brand Name
ACh Receptor Agonists:	Muscarinic Receptor	Glaucoma	Pilocarpine	Salagen
		Sjögren's syndrome		
		Urinary retention	Bethanechol	Urecholine
		Bronchoprovocation testing	Methacholine	Provocholine
	Nicotinic Receptor	Smoking cessation	Varenicline (partial agonist)	Chantix
ACh Receptor Antagonists:	Muscarinic Receptor	COPD	Ipratropium	Atrovent
		Motion sickness	Scopolamine	Transderm-Scop
		Overactive bladder	Tolterodine	Detrol
			Darifenacin	Enablex
		Parkinson's disease	Trihexyphenidyl	Artane
	Nicotinic Receptor: (Depolarizing)	Surgical paralysis	Succinylcholine	Anectine
	Nicotinic Receptor: (Nondepolarizing)	Surgical paralysis	Rocuronium	Zemuron

TABLE 6.4 Common Treatments for Disorders Involving the ANS (*Continued*)

		Indication	Example Generic	Example Brand Name
NE Receptor Agonists:	Nonselective agonist	Anaphylaxis	Epinephrine	EpiPen
	Indirect Sympathomimetic	Decongestant	Pseudoephedrine	Sudafed
		Narcolepsy	Amphetamine	Adderall, Vyvanse
		ADHD	Amphetamine	Adderall, Vyvanse
			Methylphenidate	Ritalin
	α_1 Receptor	Decongestant	Phenylephrine	Sudafed-PE
	α_2 Receptor	Hypertension	Clonidine	Catapres
		Glaucoma	Brimonidine	Agan P
	β_1	Cardiogenic shock	Dobutamine	Dobutrex
	β_2	Asthma rescue therapy	Albuterol	Proventil HFA
		Asthma prophylaxis	Salmeterol	Serevent Diskus
		Premature labor	Terbutaline	Terbulin
	β_3	Overactive bladder	Mirabegron	Myrbetriq
Blockers of NE Transporter:		ADHD	Atomoxetine	Strattera
NE Receptor Antagonists:	α_1	Hypertension	Terazosin	Hytrin
		Benign prostatic hyperplasia	Tamsulosin	Flomax
	β_1	Hypertension	Metoprolol	Lopressor
		Arrhythmias		
		Angina		
		Congestive heart failure		
		Glaucoma	Timolol	Timoptic

INTEGRATED MEDICAL SCIENCES

Organophosphate Poisoning

Acetylcholinesterase (AChE), the enzyme that hydrolyzes ACh, can be inhibited by different types of compounds. Some of these, such as the therapeutic drugs pyridostigmine (Mestinon) and donepezil (Aricept), inhibit the enzyme *reversibly*. This will allow the buildup of ACh in the synaptic cleft or at the neuromuscular junction (NMJ), thus increasing the signaling by ACh. However, when the ACh reaches high enough concentrations, it will outcompete the AChE inhibitor preventing toxic levels of ACh from accumulating. In this manner, these compounds are useful in the treatment of myasthenia gravis (pyridostigmine) and some cognitive symptoms of Alzheimer's disease (donepezil).

Other inhibitors of AChE interact with the enzyme *irreversibly*. This results in ACh accumulating and reaching toxic concentrations. At the NMJ, this can produce a **depolarizing blockade** of the nicotinic receptor. A depolarizing blockade occurs initially when the nicotinic receptor, a ligand-gated ion channel allowing the influx of sodium ions, remains open due to the continued binding of ACh and later due to the desensitization of the nicotinic receptor such that it can no longer be activated. The open channel of the nicotinic receptor prevents the repolarization of the membrane, which is critical for the reactivation of voltage-gated sodium channels that are involved in propagating action potentials necessary for muscle contraction. Thus, the muscle cannot undergo the normal contraction–relaxation cycle and is paralyzed. Irreversible inhibitors of AChE are often organophosphates. One use of organophosphates is as pesticides. In the United

(*Continued*)

States, the pesticides used are selective for insects but will affect humans at very high doses. Thus, organophosphate poisoning has been reported in farm workers or others who experience accidental high exposure to these pesticides. Another use of organophosphates is in chemical weapons such as the nerve gas sarin. Many of these compounds are deadly in very small amounts and act very quickly. Exposure to such compounds was a concern during the Gulf Wars.

Upon exposure to organophosphates, a person can experience stimulation of muscarinic receptors throughout the body due to the accumulation of ACh along with muscle weakness and paralysis due to the depolarizing blockade at the NMJ. Peripheral activation of muscarinic receptors will mimic parasympathetic nervous system activity along with activation of the sympathetic muscarinic receptors at the sweat glands. Thus, one experiences symptoms represented by the mnemonic DUMBBELSS: diarrhea, urination, meiosis, bradycardia, bronchoconstriction, emesis (vomiting), lacrimation (tearing), salivation, and sweating. In the CNS, activation of muscarinic receptors can lead to generalized seizures along with decreased respiratory drive through inhibition in the respiratory center in the brainstem.

Organophosphate poisoning can be treated with two drugs. One of these, pralidoxime, is a strong nucleophile that can remove the organophosphate from the AChE. This allows for the regeneration of the enzyme such that it can begin metabolizing the toxic levels of ACh. Pralidoxime cannot cross the blood–brain barrier. Because of this, it is effective at treating the paralysis and the stimulation of PSNS effector organs but it is ineffective in treating the CNS toxicity. For this, another drug is necessary. Atropine is an antagonist at muscarinic receptors and will block the access of ACh to muscarinic receptors. It is effective both peripherally and centrally and thus will block the overstimulation of all muscarinic receptors; however, it does not interact with nicotinic receptors and will not reverse the paralysis. Thus, both pralidoxime and atropine are necessary for the treatment of organophosphate poisoning. Autoinjectors containing both of these compounds are used by the military to protect against chemical weapon attacks. However, organophosphate poisoning must be treated quickly due to "aging" in which stronger bonds are formed between the organophosphate and AChE that are essentially unbreakable by pralidoxime. Aging occurs at different rates depending on the organophosphate but can occur in as little as 10 minutes. ■

Chapter Summary

- The autonomic nervous system consists of a two-neuron efferent pathway.
- The preganglionic neurons of the parasympathetic nervous system are located in the craniosacral regions of the spinal cord/brainstem and project to ganglia that are located in or near the effector organ.
- The preganglionic neurons of the sympathetic nervous system are located in the thoracolumbar region of the spinal cord and mostly project to ganglia that are located near the spinal cord in the sympathetic chain.
- The adrenal medulla is innervated by a sympathetic preganglionic axon and secretes epinephrine into the peripheral circulation.
- All preganglionic axons release acetylcholine.
- Parasympathetic postganglionic neurons release acetylcholine, whereas sympathetic postganglionic neurons primarily release norepinephrine with acetylcholine release from sweat glands.
- Many autonomic postganglionic fibers contain cotransmitters along with acetylcholine and norepinephrine.
- Some autonomic fibers contain neither acetylcholine nor norepinephrine but instead undergo nonadrenergic noncholinergic (NANC) transmission.
- The parasympathetic nervous system mediates normal functioning under nonstressed conditions, whereas the sympathetic nervous system mediates functioning during arousal, stress, and danger.
- The PSNS effects are mediated by cholinergic receptors, whereas the SNS effects are mostly mediated by adrenergic receptors.
- Both the PSNS and SNS exhibit tonic activity at different organs.
- Presynaptic inhibition can modulate functioning of the ANS by decreasing further output.
- The eye receives input from the PSNS and SNS and is an example of autonomic integration.
- The vasculature receives input only from the SNS, which is responsible for maintaining appropriate blood pressure.
- Epinephrine can produce vasoconstriction through actions at the α_1 receptor and vasodilation through actions at the β_2 receptor.
- There are many therapeutic drugs that target receptors involved in autonomic function.
- Stimulant drugs of abuse produce dangerous stimulation of peripheral adrenergic receptors.
- The combination of an indirect sympathomimetic and monoamine oxidase inhibitor can result in a hypertensive crisis.
- Pheochromocytoma is a tumor resulting in overproduction of catecholamines and can be treated by blocking their receptors or synthesis.
- Organophosphate poisoning can occur from exposure to pesticides or certain chemical weapons and will irreversibly block the breakdown of acetylcholine.

Chapter Review Questions

1. A 6-year-old boy is brought to the emergency department by his mother. The boy was playing in the yard while she was gardening and ingested some berries. He presents with dilated pupils, dry skin, elevated temperature, and apparent disorientation. His symptoms are a result of a chemical in the berries interfering with the normal functioning of the autonomic nervous system. Inhibition of which receptor or enzyme is consistent with the boy's presentation?

   A. Acetylcholinesterase
   B. α-Adrenergic receptor
   C. β-Adrenergic receptor
   D. Muscarinic cholinergic receptor
   E. Nicotinic cholinergic receptor

The correct answer is D. The muscarinic receptor antagonist, atropine, is found in the plant *Atropa belladonna*, also known as "deadly nightshade." This can produce symptoms described as "hot as a hare, blind as a bat, dry as a bone, red as a beet, and mad as a wet hen." The boy is dry due to the lack of sweating and salivating. The lack of sweating also contributes to an increased temperature. The blockade of muscarinic receptors in the eye results in pupillary dilation and blurred vision. Blocking muscarinic receptors in the CNS can result in confusion and delirium.

2. A 65-year-old male patient is being treated for erectile dysfunction with a drug that prolongs the actions of nitric oxide. Which of the following would be a predictable adverse effect?

   A. Constipation
   B. Sweating
   C. Muscle paralysis
   D. Pupillary dilation
   E. Vasodilation

The correct answer is E. Nitric oxide relaxes smooth muscle in the corpus cavernosum of the penis allowing an increased blood flow and subsequent erection. Phosphodiesterase inhibitors are used in the treatment of erectile dysfunction to decrease the breakdown of nitric oxide increasing the smooth muscle relaxation and erection. Nitric oxide also produces vasodilation. By decreasing the breakdown of nitric oxide, the phosphodiesterase inhibitors can lead to hypotension particularly in individuals taking nitrates for treatment of the cardiovascular disorder angina pectoris.

3. A 65-year-old male has been using propranolol, a nonselective β-adrenergic receptor antagonist, for the treatment of hypertension. Today he experienced an allergic reaction and visited the emergency department since he was having trouble breathing. He was administered epinephrine. Which

of the following would be a predictable effect of epinephrine administration in this patient?

A. Bronchodilation
B. Shock (resulting from low blood pressure)
C. Stroke (resulting from high blood pressure)
D. Tachycardia (increased heart rate)
E. Dry mouth

The correct answer is C. Epinephrine usually produces bronchodilation through activation of β_2 receptors, increased heart rate through activation of β_1 receptors, and vasoconstriction through the activation of α_1 receptors. However, the current patient is receiving propranolol, which will block the access of epinephrine to β receptors, which could prevent epinephrine from producing bronchodilation and increasing heart rate. Thus, answers A and D are incorrect.

The allergic reaction experienced by this patient is mediated by histamine release, which produces bronchoconstriction and vasodilation leading to a decrease in blood pressure, which can result in shock. Epinephrine is usually an effective treatment since it will produce bronchodilation, through activation of β_2 receptors, and vasoconstriction, through activation of α_1 receptors, resulting in a normalization of blood pressure. The α_1-mediated vasoconstriction is usually mitigated through activation of β_2 receptors, which produce vasodilation and thus prevent dangerously high blood pressures from occurring. In the absence of β_2 receptor stimulation, which can occur due to blocking of the receptors by a β receptor antagonist such as propranolol, there can be a large increase in blood pressure resulting in a hypertensive crisis, which can produce a stroke.

Clinical Application Exercises 6.1

A 13-year-old boy is staying a few days with his grandparents for a summer visit. His grandparents promise him a banana split if he will clean and straighten their garden workshed. During the cleanup, the boy accidentally knocked over a half-filled glass bottle of liquid, which falls and breaks on the cement floor of the shed. Thinking of being responsible, he calmly goes into the house for some paper towels and proceeds to clean up the spill in the workshed. He does not wear gloves or any kind of eye and face protection during the cleanup because the need for such protection does not cross his mind.

Within minutes after finishing the cleanup, the boy starts to salivate uncontrollably, and his eyes burn and tear up. His lower bowel churns, and he feels nauseated and dizzy. His muscles start to twitch. It is difficult for him to inhale, and he wheezes with each breath. He tries to walk to the house to get his grandparents, but he has great difficulty walking, and, along the way, he starts to urinate involuntarily. He collapses at the doorstep of his grandparents' home, where they immediately discover him and, realizing he is in severe physical distress, decide to immediately drive him to their local hospital emergency department, where he is immediately admitted and attended. Upon brief questioning by the attending ED physician, it is discovered that the shed the boy was cleaning contained fertilizers and a few herbicides and pesticides. The grandparents could not remember exactly which chemicals were present in the shed, but they do recall that a neighbor once said that one of the bottles contained the same chemical they heard was used to treat head lice.

QUESTIONS

1. Is there a single internal physiological factor that could account for the 13-year-old patient's presentation of both skeletal muscle and systemic autonomic effects following his exposure to the chemical he was trying to clean up? If so, what would cause exposure of the boy to that factor?
2. What chemical is used to treat head lice that would also create the symptoms exhibited by the patient in this case? To what class of chemical does this agent belong and what is its mechanism of action? Since the subject in this case did not ingest the liquid in the bottle he broke, how did the chemical(s) in the bottle enter his body?
3. How does exposure to excessive acetylcholine cause skeletal muscle twitching followed by muscle weakness?
4. Based on what you know thus far from this case presentation, how should the organophosphate poisoning in this boy be managed?

ANSWERS

1. The skeletal muscle twitching, followed by muscle weaknesses, including difficulty inhaling, is characteristic of overexposure of striated muscle to acetylcholine. Acetylcholine is also the parasympathetic neurotransmitter responsible for lacrimation, salivation, and bronchiole constriction. Acetylcholine stimulates GI acid secretion and intestinal motility and causes contraction of the urinary bladder, resulting in micturition. The boy's symptoms in this case could all be caused by excessive amounts of acetylcholine in the region of nicotinic receptors (primarily motor endplates) and autonomic muscarinic receptors (primarily parasympathetic postganglionic fiber receptors). Excessive concentrations of acetylcholine at synapses can result from excessive release from appropriate nerve endings or impaired chemical inactivation of acetylcholine by acetylcholinesterases in the synapses.

2. Malathion is a common pesticide used as a garden or agricultural insecticide to control mosquito populations or plant pests. It is also found in some treatments of head lice. Malathion belongs to a class of drugs called organophosphates, which are acetylcholinesterase inhibitors. Therefore, exposure to this agent mimics the effects of excess

acetylcholine at end organs including striated muscle and several end organs whose function is modulated by the parasympathetic nervous system.

Malathion is highly lipid soluble, and individuals can receive toxic concentrations of this agent by absorption through the skin, inhalation, or absorption through mucous membranes including the mouth and eyes. Malathion belongs to a subclass of organophosphates called irreversible cholinesterase inhibitors. Once bound to the cholinesterase, they become inactivated only extremely slowly. This factor enhances the potential toxicity of these agents.

3. Acetylcholine activates nicotinic receptors in skeletal muscle cells, which are ligand-gated sodium channels. Activation of these channels leads to a local depolarization of the membrane, which spreads electrotonically until they activate fast voltage-gated sodium channels in the vicinity. Activation of these channels causes the formation of action potentials in the muscle fiber and muscle contraction. Inhibition of acetylcholinesterases causes random activation of muscle fibers throughout the body, resulting in muscle twitching and fasciculations. However, continual activation of nicotinic receptors results in general muscle fiber depolarization and a shift in the resting membrane potential of the muscle fiber to more positive values. In many cases, this partial depolarization at rest locks the fast sodium channels in their inactive state, rendering them unable to trigger an action potential. A muscle fiber in this condition cannot contract, and, when this effect is spread to large groups of fibers, an overall muscle weakness ensues.

4. There are three primary goals to treatment of a patient poisoned by organophosphate exposure—decontaminate the patient, support cardiovascular and respiratory function, and restore normal acetylcholine balance. Skin of an exposed patient should be washed and the eyes flushed. The patient should be intubated and mechanically ventilated if unconscious. Atropine, a muscarinic-receptor antagonist, should be used to counteract all the parasympathomimetic actions of the organophosphate. In addition, pralidoxime should be used immediately in order to break the binding of the organophosphate with the cholinesterase. This is especially important with malathion because the bond of this agent with the enzyme "ages" over time, making it more and more difficult to dislodge the organophosphate from the enzyme.

thePoint® *Visit* http://thepoint.lww.com/rhoades5e *for additional chapter review Q&A, Clinical Application Exercises, animations, and more!*

7 Integrative Functions of the Central Nervous System

Active Learning Objectives

Upon mastering the material in this chapter, you should be able to:

- Explain how the hypothalamus interfaces among the endocrine, autonomic, and limbic systems.
- Explain the difference between a biologic rhythm and a cycle.
- Explain the role of the brainstem reticular formation in consciousness and arousal.
- Describe the stages of sleep.
- Describe the functional organization of the forebrain.
- Describe the functional components of the limbic system.

- Explain the brain's reward system.
- Explain the components of the limbic system involved in mood disorders and schizophrenia.
- Explain the concept of long-term potentiation and how it relates to learning.
- Describe the areas of the brain that involve memory and learning.
- Describe how the Broca and Wernicke areas serve as functional components of language.

The **central nervous system (CNS)** detects and processes sensory information from both the internal and external environment. The motor system directs information to skeletal muscle for locomotion and fine movement. Both sensory and motor functions do not occur in isolation but are highly coordinated. The CNS can be grouped according to integrative function and comprises the following:

I. Brainstem
   A. Medulla
   B. Pons
   C. Midbrain
II. Cerebellum
III. Forebrain
   A. Diencephalon
   B. Hypothalamus
   C. Thalamus
   D. Cerebrum
      1. Basal nuclei
      2. Cerebral cortex

The brain areas involved in the coordinated activities of human behavior is part of the central integrative system. An overview of these components is shown in Figure 7.1 with their respective integrative functions shown in Table 7.1. Functional components of the central integrative system can be classified into several broad categories. *Motivational systems*, for example, are those responsible for the human drive to satisfy basic needs such as hunger and thirst. Another is the group of systems that carry out functions such as *learning* and *memory*. These systems endow us with the ability to acquire and store new information. Still another category involves systems responsible for *communication*, specifically the form of the written and spoken language. Many of these functions depend on input from the **brainstem** regarding both internal and external environmental stimuli; the brainstem is also critical to the efferent control mediated by higher centers.

This chapter describes the brain areas responsible for these integrated functions and also examines how the brain collects and integrates information from its two hemispheres, specifically how similar functions are located on one side of the brain.

▶ HYPOTHALAMUS

The **hypothalamus** is a portion of the brain that contains a number of small nuclei with a variety of functions. The hypothalamus is located just below the thalamus and just above the brainstem. It forms the inferior part of the **diencephalon**. The diencephalon includes the hypothalamus, thalamus, and subthalamus (see Fig. 7.1). The rostral border of the hypothalamus is at the optic chiasm, and its caudal border is at the mammillary body. The hypothalamus is connected to the endocrine system via the pituitary and serves as the major regulator of endocrine function. On the basal surface of the hypothalamus, exiting the median eminence, the pituitary stalk contains the **hypothalamohypophyseal portal blood vessels** (Fig. 7.2). Parvocellular neurons within specific nuclei of the hypothalamus secrete *hypothalamic-releasing hormones*, also designated as **releasing factors**, into the portal vessels. Releasing factors then regulate the secretion of most hormones of the endocrine system. The releasing factors are transported to the anterior pituitary, where they stimulate secretion of trophic hormones to other glands of the endocrine system (see Chapter 31). The pituitary stalk also contains the axons of magnocellular neurons whose cell bodies are located in the supraoptic and paraventricular hypothalamic nuclei. These axons form the **hypothalamohypophyseal tract** within the pituitary stalk and represent the efferent limbs of neuroendocrine reflexes that lead to the secretion of the hormones **arginine vasopressin** and **oxytocin** into the blood. These hormones are made in the magnocellular neurons and released by their axon terminals next to the blood vessels within the posterior pituitary.

In the adult human, the hypothalamus is approximately the size of an almond. Despite its small size, it serves as the major headquarters for coordinating many functions. The hypothalamus is responsible for coordinating autonomic

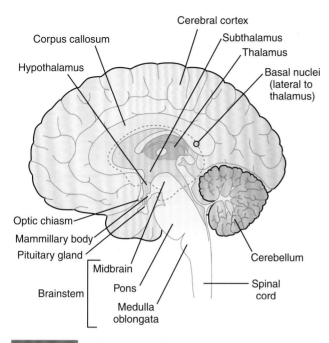

Figure 7.1 **A midsagittal section through the brain showing the most prominent components.** The cerebrum (*dotted line*) is the largest portion of the human brain.

TABLE 7.1	Major Integrative Functions of the Brain	
Component	**Major Functions**
Cerebral cortex	Sensory perception
	Voluntary movement
	Language
	Personality
	Memory/decision making
Basal nuclei	Muscle tone
	Coordination of slow/sustained movements
	Inhibition of random movement
Thalamus	Relay station for all synaptic input
	Awareness of sensations
	Motor control
Hypothalamus	Homeostatic regulation for temperature, thirst, food intake, fluid balance, and biological rhythms
	Relay station between nervous and endocrine systems
	Sex drive and sexual behavior
	Emotions and behavior patterns
	Sleep–wake cycle
Cerebellum	Balance
	Muscle tone
	Coordination of voluntary muscle activity
Brainstem	Origin of most peripheral cranial nerves
	Control centers for cardiovascular, respiratory, and digestive regulation
	Regulation of muscle reflexes for posture and equilibrium
	Some control of sleep–wake cycle
	Integration of all synaptic input from spinal cord
	Activation of cerebral cortex

reflexes of the brainstem and the spinal cord. It also activates the endocrine and somatic motor systems when responding to signals generated either within the hypothalamus or brainstem or in higher centers, such as the **limbic system**, where emotion and motivation are generated. The hypothalamus accomplishes this integration by virtue of its unique location at the interface between the limbic system and the endocrine and autonomic nervous systems.

As a major regulator of homeostasis, the hypothalamus receives input about the internal environment of the body via signals in the blood and from visceral afferents relaying through the brainstem. In most of the brain, capillary endothelial cells are connected by tight junctions that prevent substances in the blood from entering the brain. These tight junctions are part of the blood–brain barrier. The **blood–brain barrier** is missing in several small regions of the brain, called **circumventricular organs**, which are adjacent to the fluid-filled ventricular spaces. Important circumventricular organs are in and near the hypothalamus. Capillaries in these regions, similar to those in other organs, are fenestrated ("leaky"), allowing the cells of hypothalamic nuclei to sample the composition of the blood. Neurons in the hypothalamus then initiate the mechanisms necessary to maintain levels of constituents at a given set point, fixed within narrow limits by specific hypothalamic nuclei. Homeostatic functions regulated by the hypothalamus include body temperature, water and electrolyte balance, blood glucose levels, and energy balance.

Hypothalamus consists of distinct nuclei that interface between the endocrine, autonomic, and limbic systems.

The nuclei of the hypothalamus have ill-defined boundaries, despite their customary depiction (see Fig. 7.2). Many are named according to their anatomic location (e.g., anterior hypothalamic nuclei, ventromedial nucleus) or for the structures they lay next to (e.g., the periventricular nucleus surrounds the third ventricle, the **suprachiasmatic nucleus [SCN]** lies above the optic chiasm).

The hypothalamus receives afferent inputs from all levels of the CNS and makes reciprocal connections with the limbic system via fiber tracts in the **fornix**, stria terminalis, and ventral amygdalofugal pathway. The hypothalamus also makes extensive reciprocal connections with the brainstem, including the reticular formation and the medullary centers of cardiovascular, respiratory, and gastrointestinal regulation. Many of these connections travel within the

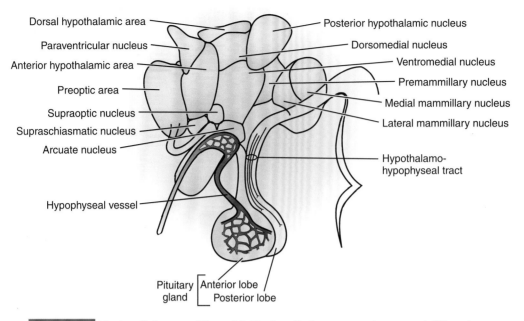

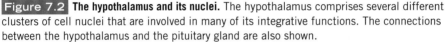

Figure 7.2 **The hypothalamus and its nuclei.** The hypothalamus comprises several different clusters of cell nuclei that are involved in many of its integrative functions. The connections between the hypothalamus and the pituitary gland are also shown.

medial forebrain bundle, which also connects the brainstem with the cerebral cortex.

Several major connections of the hypothalamus are one-way rather than reciprocal. One of these, the **mammillothalamic tract**, carries information from the mammillary bodies of the hypothalamus to the anterior nucleus of the **thalamus**, from where information is relayed to limbic regions of the cerebral cortex. A second one-way pathway carries visual information from the retina to the SCN of the hypothalamus via the optic nerve. Through this retinal input, the light cues of the day–night cycle entrain or synchronize the "biologic clock" of the brain to the external clock. The hypothalamus also projects directly to the spinal cord to activate sympathetic and parasympathetic preganglionic neurons (see Chapter 6).

Hypothalamus regulates energy balance by integrating metabolism and eating behavior.

Energy balance is the homeostasis of energy in living systems. It is measured by the following equation: energy intake = energy expenditure + storage. *Energy intake* is dependent on diet, which is mainly regulated by hunger and calories consumed. *Energy expenditure* is the sum of internal heat production and external work produced. Internal heat production is, in turn, the **basal metabolic rate (BMR)** and the thermal effect of food. External work is determined by physical activity. The hypothalamus plays a key role in regulating the homeostatic energy balance by controlling both energy intake and energy expenditure.

The regulation of eating behavior by the hypothalamus is part of a complex pathway that regulates food intake and energy expenditure in the face of changes in nutritional state. In general, the hypothalamus regulates caloric intake, use, and storage in a manner that tends to maintain the body weight in adulthood. The presumptive set point around which it

attempts to stabilize body weight is remarkably constant but can be altered by changes in physical activity, composition of the diet, emotional states, stress, aging, pregnancy, and pharmacologic agents such as atypical antipsychotics.

The melanocortin system in the **arcuate nucleus** (ARC) of the hypothalamus is a critical CNS component in the regulation of energy homeostasis (Fig. 7.3). Within ARC, **proopiomelanocortin (POMC)** neurons act as **anorexigenic** (causing loss of appetite) neurons, their stimulation leading to decreased food intake and increased energy expenditure. The decrease in feeding by POMC is mediated by alpha-melanocyte–stimulating hormone, a cleaved product of POMC, which simulates melanocortin receptors. Studies have shown that mutations in melanocortin receptor 4 are associated with early-onset obesity and mutations of this receptor occur in some patients with severe obesity. A second group of neurons in the ARC are **orexigenic** (stimulating appetite) and contain **neuropeptide Y (NPY)** and **agouti-related protein (AgRP)**. Stimulation of NPY/AgRP neurons causes increased food intake, decreased energy expenditure, and inhibition of POMC neurons.

Both POMC and NPY/AgRP neurons express receptors for a wide variety of energy-related hormones and sample levels of these peptides through nearby circumventricular organs. Some, such as **leptin** and **insulin**, act on a longtime course to maintain body weight and blood glucose levels, and others, such as **cholecystokinin (CCK)** and ghrelin, act on a shorter time course to regulate initiation or cessation of a meal. Additionally, these neurons receive feeding-related sensory signals from the external environment as well as from the viscera, the latter being relayed via the vagus nerve and brainstem nuclei.

Outputs from ARC that mediate alterations in feeding or energy expenditure include projections to the lateral

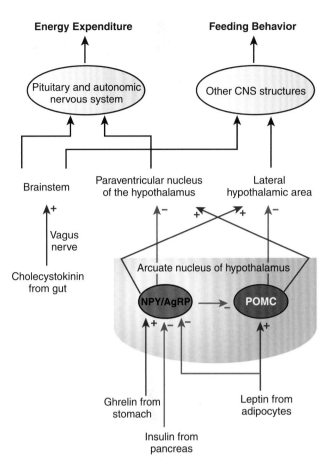

Energy Expenditure   **Feeding Behavior**

Pituitary and autonomic nervous system   Other CNS structures

Brainstem   Paraventricular nucleus of the hypothalamus   Lateral hypothalamic area

Vagus nerve

Cholecystokinin from gut

Arcuate nucleus of hypothalamus

NPY/AgRP   POMC

Ghrelin from stomach   Leptin from adipocytes

Insulin from pancreas

Figure 7.3 **Control of energy expenditure and feeding behavior.** Cell groups within the arcuate nucleus (ARC) integrate peripheral orexigenic and anorexigenic signals. Activation of neuropeptide Y (NPY)/agouti-related protein (AgRP) neurons within the ARC leads to a decrease in energy expenditure acting via the paraventricular nucleus of the hypothalamus (PVH), the pituitary, and the autonomic nervous system and an increase in feeding behavior acting via the lateral hypothalamic area (LH) and several other central nervous system (CNS) sites. Activation of proopiomelanocortin (POMC) neurons within the ARC increases energy expenditure and decreases feeding behavior acting via the same areas. Additional peripheral signals activate the brainstem via the vagus nerve; the brainstem then bypasses the ARC to act directly on the PVH and LH. Many central pathways (*not shown*) involved in feeding behavior and/or energy expenditure also project to the ARC, PVH, and LH.

hypothalamus, dorsomedial nucleus, ventromedial nucleus, and paraventricular nucleus, as well as a variety of structures in the brainstem and forebrain cortical and limbic systems involved with feeding and motivated behaviors. Direct output to parasympathetic and sympathetic preganglionic neurons in the brainstem and spinal cord derives from the paraventricular nucleus and the ARC. The paraventricular nucleus engages the endocrine system through its regulation of the pituitary gland.

A key player in the regulation of body weight is the hormone leptin, which is released by white fat cells (adipocytes). As fat stores increase, plasma leptin levels increase; conversely, as fat stores are depleted, leptin levels decrease. Leptin inhibits the NPY/AgRP cells and stimulates the POMC cells in the ARC (see Fig. 7.3). Physiologic responses to low leptin levels (starvation) are initiated by the hypothalamus to increase food intake, decrease energy expenditure, decrease reproductive function, decrease body temperature, and increase parasympathetic activity. Physiologic responses to high leptin levels (obesity) are initiated by the hypothalamus to decrease food intake, increase energy expenditure, and increase sympathetic activity.

In addition to long-term regulation of body weight, the hypothalamus also regulates eating behavior more acutely. Factors that limit the amount of food ingested during a single feeding episode originate in the gastrointestinal tract and influence the hypothalamic regulatory centers. These include the following: sensory signals carried by the vagus nerve that signify stomach filling; chemical signals giving rise to the sensation of satiety, including absorbed nutrients (glucose, certain amino acids, and fatty acids); and gastrointestinal hormones, especially CCK, produced by endocrine cells in the gut wall, and ghrelin, produced by the epithelia of the stomach and small intestine. CCK is a satiety signal mediated via the vagus, whereas ghrelin stimulates food intake by activating the NPY/AgRP neurons in the ARC. The success of bariatric gastric bypass surgery as a treatment for morbid obesity is thought in part to be a result of the decreased secretion of ghrelin.

The complex regulation of energy balance and the evolutionary pressure to conserve energy conspire to make weight loss a difficult proposition. The ready availability of highly palatable foods and the decrease in exercise levels required to procure them are thought to underlie the increased incidence of obesity in the population. Drugs that target increasing energy expenditure by activating the sympathetic nervous system have unacceptable side effects. Drug development has focused on blocking components of the central modulatory system that influences appetite (see Integrated Medical Sciences). One target has been the **endocannabinoids**. These compounds are the endogenous ligands for the cannabinoid receptors that mediate the effects of the psychoactive component of marijuana. Endocannabinoid receptors play an important role in energy homeostasis and are expressed in brain areas like the ARC, in several forebrain regions associated with motivated behavior, and in peripheral tissues associated with energy regulation like the liver and pancreas. Although the role of endocannabinoids in energy homeostasis is not completely understood, studies suggest that they modulate neuronal and tissue responses at multiple sites associated with energy homeostasis.

Sexual drive and sexual behavior are controlled by the hypothalamus.

The anterior and preoptic hypothalamic areas are sites for regulating gonadotropic hormone secretion and sexual behavior. Neurons in the medial preoptic area (MPOA) secrete **gonadotropin-releasing hormone (GnRH)**, beginning at puberty, in response to signals that are not understood. These neurons contain receptors for gonadal steroid hormones, testosterone, and/or estradiol, which regulate GnRH secretion in either a cyclic (female) or a continual (male) pattern following the onset of puberty.

INTEGRATED MEDICAL SCIENCES

Central Nervous System: Therapeutic Targets for Antiobesity Medications

Marijuana (*Cannabis sativa*) has been a cultivated crop with known psychoactive properties for more than 4,000 years, but its active constituent, Δ-9 tetrahydrocannabinol, was not isolated until 1964. A high-affinity binding site in the brain was discovered in 1988, and the first cannabinoid receptor (CB_1) was cloned in 1991. Soon thereafter, the first endogenous cannabinoid, anandamide, was discovered. These findings set the stage for intensive efforts to develop pharmacological agonists and antagonists that target the endogenous pathways. A major motivation for this effort was the hyperphagia and carbohydrate craving associated with marijuana intoxication, with the implication that the endocannabinoids are involved in the regulation of feeding behavior. Researchers hoped that by targeting this system, they could increase appetite in cancer patients and to decrease appetite in obese patients.

Early studies of the selective CB_1 receptor blocker, rimonabant, found peripheral and central effects that influenced metabolism, decreased food intake, weight, triglyceride levels, low-density lipoprotein levels, C-reactive protein levels, and insulin resistance while increasing high-density lipoprotein levels. Despite early success and approval as a weight loss drug in Europe, rimonabant had to be withdrawn from the market because of a serious gastrointestinal and adverse psychiatric side effects.

Another target for antiobesity drugs is the brain's serotonin system, which plays a role in energy and glucose homeostasis. Hypothalamic anorexigenic POMC neurons express serotonergic receptors and treatments that reduce central serotonergic signaling can produce weight gain. Lorcaserin, a centrally acting serotonin 5-HT_{2c} receptor agonist, reduces body weight primarily by reducing food intake. Animal evidence suggests that lorcaserin works by activating anorexigenic POMC neurons.

Melanocortin receptors are also being investigated as potential targets for treating obesity. Melanocortin receptor 4 (Mc4) is expressed on neurons in the feeding pathways and defects in Mc4 cause of a form autosomal dominant obesity. Mc4 receptor agonists have been shown to induce weight loss in animal studies and early-phase clinical trials, but the agents have yet to make it to market. ■

At a critical period in fetal development, circulating testosterone secreted by the testes of a male fetus changes the characteristics of cells in the MPOA that are destined later in life to secrete GnRH. These cells, which would secrete GnRH cyclically at puberty had they not been exposed to androgens prenatally, are transformed into cells that secrete GnRH continually at a homeostatically regulated level. As a result, males exhibit a steady-state secretion rate for gonadotropic hormones and, consequently, for testosterone (see Chapter 36).

In the absence of androgens in fetal blood during development, the MPOA remains unchanged, so that at puberty, the GnRH-secreting cells begin to secrete in a cyclic pattern. This pattern is reinforced and synchronized throughout female reproductive life by the cyclic feedback of ovarian steroids, estradiol and progesterone, on secretion of GnRH by the hypothalamus during the menstrual cycle (see Chapter 37).

Steroid levels during prenatal and postnatal development are known to mediate differentiation of sexually dimorphic regions of the brain of most vertebrate species. Sexually dimorphic brain anatomy, behavior, and susceptibility to neurologic and psychiatric illness are evident in humans. However, with the exception of the GnRH-secreting cells, it has been difficult to definitively show a steroid dependency for sexually dimorphic differentiation in the human brain.

In addition to secretion of GnRH, the MPOA has other roles in male sexual behavior. The MPOA is indirectly connected with all sensory systems and through reciprocal connections can modify sensory processing biasing it to prefer sexually relevant sensory cues. Outputs from the MPOA terminate in hypothalamic and brainstem nuclei that regulate autonomic and somatic behaviors related to male sexual responses. The paraventricular nucleus also contributes to male sexual behavior and its cells release oxytocin during sexual arousal, which has a powerful pro-erectile effect.

The ventromedial nucleus of the hypothalamus (VMH) has been implicated in the central neural control of female sexual behavior regulating fertility and sexual receptivity. Neurons in the VMH express high levels of estrogen receptors and injections of estrogen into the VMH in ovariectomized animals restore sexual receptivity.

Hypothalamus contains the "master biological clock" that controls rhythms and cycles.

Many physiologic events or processes repeat themselves with periodicity. When an event is repeated daily in the body, it is *rhythmic*, and if the event follows a monthly pattern, it is *cyclic*. Daily rhythms are referred to as **circadian rhythms**. Circadian rhythms are linked to the light–dark cycle. If the activity occurs during the day, the circadian rhythm is referred to as a **diurnal rhythm**. *Nocturnal* is another type of daily rhythm, and the behavior is characterized by activity during the night. Nocturnal rhythms are most prevalent in other vertebrates. An example of a diurnal rhythm is body temperature, which peaks during the day and falls at night. An example of the monthly cycle is the female menstrual cycle, in which a series of physiologic events is repeated every 28 days.

Circadian rhythms are endogenous (within the body) because they persist even in the absence of time cues, such as day–night cycles for light and dark periods, lunar cycles for monthly rhythms, or changes in temperature and day length for seasonal change. Accordingly, most organisms, including humans, are said to possess an endogenous timekeeper, a so-called **biologic clock** that times the body's regulated functions.

The machinery of the biological clock resides within the cell. Nearly every mammalian tissue examined has been shown to have an intrinsic molecular circadian clock that utilizes transcriptional–translational feedback loops involving several genes termed *clock genes*. Some examples of clock genes are *CLOCK* and *BMAL1*, which encode proteins in the positive arm of the loop, and the *PERIOD* (*PER*) and *CRYTOCHROME* (*CRY*) families that are part of the negative arm.

Most homeostatically regulated functions exhibit peaks and valleys of activity that recur approximately daily. The hypothalamus plays a critical role in regulating all of these biologic rhythms serving as the "master clock" for synchronization of these daily rhythms. Within the hypothalamus, the SCN has a special role. Its cells have the properties of an oscillator whose spontaneous firing patterns change dramatically during a day–night cycle. This diurnal cycle of activity is maintained *in vitro* and is an internal property of SCN cells that employ the same molecular clocks seen in other tissues. The SCN has multiple cellular clocks that interact with other hypothalamic and nonhypothalamic nuclei to produce a functional network that controls the circadian rhythms of the body. The SCN is also key to entrainment of the body's endogenous circadian rhythms to the world's external day–night clock. The SCN receives light information from special nonimage forming photoreceptors containing melanopsin located in the ganglion cell layer of the retina. This information travels to the SCN in the **retinohypothalamic tract** of the optic nerve and is important for resetting the brain's internal biological clock to match the external clock.

Figure 7.4 illustrates some of the circadian rhythms of the body. One of the most vivid is alertness, which peaks in the afternoon and is lowest in the hours preceding and following sleep. Another, body temperature, ranges ~1°C (~2°F) throughout the day, with the low point occurring during sleep. Plasma levels of growth hormone increase greatly during sleep, in keeping with this hormone's metabolic role as a glucose-sparing agent during the nocturnal fast. Cortisol, on the other hand, has its highest daily plasma level prior to arising in the morning. The mechanism by which the SCN can regulate diverse functions is partly related to its control of the production of melatonin by the pineal gland. Melatonin levels increase with decreasing light as night ensues.

Other homeostatically regulated functions exhibit diurnal patterns as well. When they are all in synchrony, they function harmoniously and impart a feeling of well-being. When there is a disruption in rhythmic pattern, such as by sleep deprivation or when passing too rapidly through several time zones, the period required for reentrainment of the SCN to the new day–night pattern is characterized by a

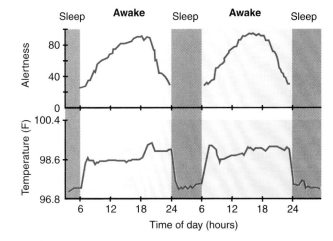

Figure 7.4 Circadian rhythms in some homeostatically regulated functions during two 24-hour periods. Alertness is measured on an arbitrary scale between sleep and most alert. Note that alertness and body temperature parallel each other.

feeling of malaise and physiologic distress. This is commonly experienced as "jet lag" in travelers crossing several time zones or by workers changing from day shift to night shift or from night shift to day shift. In such cases, the hypothalamus requires time to "reset its clock" before the regular rhythms are restored and a feeling of well-being ensues. The SCN uses the new pattern of light–darkness, as perceived in the retina, to entrain its firing rate to a pattern consistent with the external world. Resetting the clock may be facilitated by the judicious use of exogenous melatonin and by altering exposure to light.

In addition to entrainment of the biologic clock by light cycles, recent evidence suggests that the homeostatic mechanisms related to energy balance are intimately associated with the sleep–wake cycle. The link is a set of neurons in the posterolateral hypothalamus that express **hypocretin** (also known as *orexin*). These neurons are sensitive to nutritional status, stimulate appetite, and project to arousal areas of the brain. The SCN projects to the hypocretin-containing neurons and also receives input from the ARC. A functional deficit in hypocretin was found to be the cause of **narcolepsy** in animal models of this disease. Narcolepsy is characterized by sudden short episodes of sleep, cataplexy (a sudden episode of muscle weakness leading to collapse), excessive daytime sleepiness, vivid dreams, and sleep paralysis during sleep onset or upon waking.

Reticular formation governs the regulation of muscle tone, pain, and the sleep–awake pattern.

The brainstem contains anatomic groupings of cell bodies clearly identified as the nuclei of cranial sensory and motor nerves or as relay cells of ascending sensory or descending motor systems. The remaining cell groups of the brainstem, located in the central core, constitute a diffuse-appearing system of neurons with widely branching axons, known as the **reticular formation**. The reticular formation constitutes the core of the brainstem and runs through the midbrain, pons, and medulla. A unique characteristic of neurons of the reticular formation is their widespread system of axon collaterals, which make extensive synaptic contacts and, in some cases, travel over long distances terminating in multiple areas of the CNS. The reticular formation contains small neural networks with varied functions including the regulation of arousal, the sleep–wake cycle, modulation of pain, regulation of muscle tone, and control of autonomic functions like heart rate and respiration.

Ascending reticular activating system mediates consciousness and arousal.

The ascending reticular activating system connects to areas in the cortex, thalamus, and hypothalamus and is important for cortical arousal and the maintenance of consciousness. Sensory neurons bring peripheral sensory information such as touch, hearing, or vision to the CNS via specific pathways that ascend and synapse with specific nuclei of the thalamus, which, in turn, innervate primary sensory areas of the cerebral cortex. Each modality has, in addition, a nonspecific form of sensory transmission in that axons of the ascending fibers send collateral branches to cells of the reticular formation (Fig. 7.5). The latter, in turn, send their axons to the cerebral cortex or to the intralaminar nuclei of the thalamus, which innervate wide areas of the cerebral cortex and limbic system. In the cerebral cortex and limbic system, the influence of the nonspecific projections from the reticular formation is arousal of the organism. This series of connections modulates forebrain neurons and is termed the **ascending reticular activating system**. So important is the ascending reticular activating system

to the state of arousal that a malfunction in the reticular formation, particularly the rostral portion, can lead to a loss of consciousness and coma.

Monoaminergic reticular formation cell groups

Monoaminergic neurons release monoamine neurotransmitters derived from aromatic amino acids (e.g., tryptophan, tyrosine, and phenylalanine). Monoamine neurotransmitters include *histamine*, *catecholamines* (e.g., dopamine, norepinephrine [NE], and epinephrine), and *tryptamines* (melatonin and serotonin). Most monoaminergic cells are located in the reticular formation and axons arising from these cell groups innervate all parts of the CNS via widespread, divergent pathways. The cortex, limbic system, and basal ganglia are richly innervated by catecholaminergic (noradrenergic and dopaminergic) and serotonergic nerve terminals emanating from brainstem nuclei that contain relatively few cell bodies compared with their extensive terminal projections. From neurochemical manipulation of monoaminergic neurons in the limbic system, it is apparent that they play a major role in determining emotional state.

Dopamine pathways

Dopaminergic neurons are located in three major pathways originating from cell groups in either the midbrain (the substantia nigra and ventral tegmental area) or the hypothalamus (Fig. 7.6). The **nigrostriatal system** consists of neurons with cell bodies in the substantia nigra (pars compacta) and terminals in the striatum (caudate and putamen) located in the basal ganglia. This dopaminergic pathway is essential for maintaining normal muscle tone and modulating voluntary movements (see Chapter 5). The **tuberoinfundibular system** of dopaminergic neurons is located entirely within the hypothalamus, with cell bodies in the arcuate nucleus and periventricular nuclei and terminals in the median eminence on the ventral surface of the hypothalamus. The tuberoinfundibular system is responsible for the secretion of hypothalamic releasing factors into a portal system that carries them

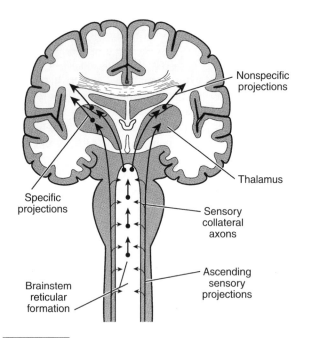

Figure 7.5 **The brainstem reticular formation and reticular activating system.** Ascending sensory tracts send axon collateral fibers to the reticular formation. Neurons in the reticular formation give rise to fibers that synapse in the intralaminar nuclei of the thalamus and the cerebral cortex. From the intralaminar nuclei, thalamic projections influence widespread areas of the cerebral cortex and limbic system.

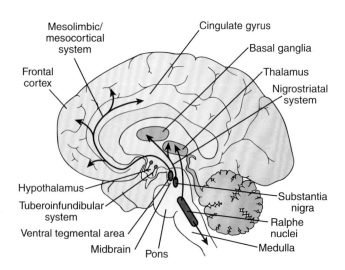

Figure 7.6 **The origins and projections of the three major dopaminergic systems and the origins and projections of the serotonergic system.**

through the pituitary stalk into the anterior pituitary lobe (see Chapter 31). The dopaminergic neurons inhibit prolactin release from the pituitary.

A major dopaminergic pathway in the brain is the *mesolimbic pathway*, part of the **mesolimbic/mesocortical system**. This pathway begins in the ventral tegmental area of the midbrain of the brainstem and innervates the limbic system, hippocampus, and the medial prefrontal cortex. It functions in modulating behavioral responses, especially motivation and drive, to stimuli through the neurotransmitter dopamine (see Fig. 7.6). The ventral tegmental nucleus is also part of the brain's reward system. Drugs that increase dopaminergic transmission, such as cocaine, which inhibits dopamine reuptake, and amphetamine, which promotes dopamine release and inhibits its reuptake, lead to repeated administration and abuse in part because they stimulate dopamine transmission in the brain's reward system. The mesolimbic dopaminergic system is also one site of action of **neuroleptic** drugs, which are used to treat schizophrenia (discussed later) and other psychotic conditions.

The *mesocortical pathway* is another dopaminergic pathway of the mesolimbic/mesocortical system that connects the ventral tegmentum to the cerebral cortex, particularly the frontal lobes. This pathway is essential to the normal cognitive function of the prefrontal cortex. It is thought to be involved in motivation and emotional response and, like the mesolimbic system, its dysfunction has been implicated in schizophrenia

Noradrenaline pathways

Noradrenergic neurons (containing NE) are located in cell groups in the medulla and pons (Fig. 7.7). The medullary cell groups project to the spinal cord where they influence cardiovascular regulation and other autonomic functions. Cell groups in the pons include the lateral system, which innervates

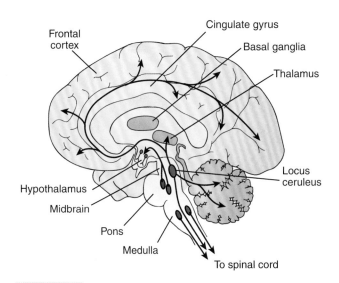

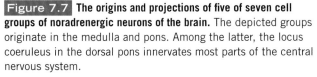

Figure 7.7 **The origins and projections of five of seven cell groups of noradrenergic neurons of the brain.** The depicted groups originate in the medulla and pons. Among the latter, the locus coeruleus in the dorsal pons innervates most parts of the central nervous system.

the basal forebrain and hypothalamus, and the locus ceruleus, which sends efferent fibers to nearly all parts of the CNS.

Noradrenergic neurons innervate all parts of the limbic system and the cerebral cortex, where they play a major role in setting **mood** (sustained emotional state) and **affect** (the emotion itself; e.g., euphoria, depression, anxiety). Drugs that alter noradrenergic transmission have profound effects on mood and affect. For example, reserpine, which depletes brain NE, induces a state of depression. Drugs that enhance NE availability, such as monoamine oxidase inhibitors (MAOIs) and inhibitors of reuptake, reverse this depression. Amphetamines and cocaine have effects on boosting noradrenergic transmission similar to those described for dopaminergic transmission; they inhibit reuptake and/or promote the release of NE. Increased noradrenergic transmission results in an elevation of mood, which further contributes to the potential for abusing such drugs, despite the depression that follows when drug levels fall. Some of the unwanted consequences of cocaine or amphetamine-like drugs reflect the increased noradrenergic transmission, in both the periphery and the CNS. This can result in a hypertensive crisis, myocardial infarction, or stroke, in addition to marked swings in affect, starting with euphoria and ending with profound depression.

Serotonin pathways

Serotonergic neurons (neurons containing serotonin) also innervate most parts of the CNS. Most serotonergic neurons are located in the *raphe nuclei,* groups of cells located adjacent to the midline of the brainstem (see Fig. 7.6). Serotonin functions in mood, feeding, memory, sleep, and cognitive skills and decreased serotonin levels are linked to affective disorders like depression (discussed later). Drugs that increase serotonin transmission are effective antidepressant agents. Serotonin also regulates heart rate and breathing, and low levels of serotonin have been linked to **sudden infant death syndrome (SIDS)** or *crib death*. SIDS is marked by the sudden death of an infant that is unexpected by history and cannot be explained following a thorough forensic autopsy. An abnormality in serotonin signaling is not the only link; the frequency of SIDS appears to be a strong function of the infant's sex, age, birth weight, sleeping position, and ethnicity as well as the education and socioeconomic status of the parents.

Consciousness and arousal neurochemical systems

Reticular formation monoaminergic cell groups are not the only neurochemical systems that modulate cortical activity. The *tuberomammillary* nucleus of the lateral hypothalamus contains *histaminergic* neurons (cells using histamine as a neurotransmitter). The histaminergic projections from the tuberomammillary nucleus to the cerebral cortex increase cortical activation and alertness. These neurons fire at a high level during wakefulness and are important in the maintenance of wakefulness. Oral antihistamines, which reduce allergy symptoms through their peripheral action, also block central histamine activity resulting in a common side effect of these drugs, drowsiness.

The *cholinergic system* (cells containing acetylcholine) is a nonmonoaminergic system that modulates cortical activity.

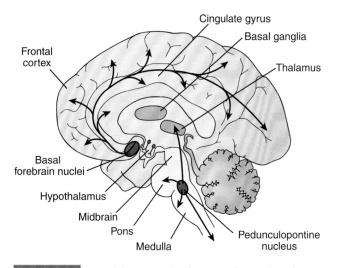

Figure 7.8 **The origins and projections of major cholinergic neurons.** Cholinergic neurons in the basal forebrain nuclei innervate all regions of the cerebral cortex. Cholinergic neurons in the brainstem's pedunculopontine nucleus provide a major input to the thalamus and also innervate the brainstem and spinal cord. Cholinergic interneurons are found in the basal ganglia. Not shown are peripherally projecting neurons, the somatic motor neurons, and autonomic preganglionic neurons, which also are cholinergic.

Concentrations of cholinergic neurons in the brainstem reticular formation *pedunculopontine nucleus* project to the thalamus, spinal cord, and other regions of the brainstem (Fig. 7.8). These cholinergic neurons influence the state of arousal by activating thalamocortical neurons involved in wakefulness and altering the activity of thalamic nuclei that are part of the sleep system.

The basal forebrain region contains prominent populations of cholinergic neurons that project to the hippocampus and to all regions of the cerebral cortex (see Fig. 7.8). These cholinergic neurons are known generically as **basal forebrain nuclei** and include the septal nuclei, the nucleus basalis, and the nucleus accumbens. Cholinergic projections from the nucleus basalis terminate throughout the cerebral cortex and contribute to wakefulness as well as REM sleep. Cortical cholinergic connections are also thought to be important for selective attention and projections to the hippocampus are involved in memory.

▶ BRAIN ELECTRICAL ACTIVITY

The influence of the ascending reticular activating system on the brain's activity can be monitored via **electroencephalography (EEG)**. EEG is a sensitive recording device for picking up the electrical activity of the brain's surface through electrodes placed on designated sites on the scalp. This noninvasive tool measures simultaneously, via multiple leads, the electrical activity of the major areas of the cerebral cortex. It is also the best diagnostic tool available for detecting abnormalities in electrical activity, such as in epilepsy, and for diagnosing sleep disorders.

The detected electrical activity reflects the extracellular recording of the myriad postsynaptic potentials in cortical

neurons underlying the electrode. The summated electrical potentials recorded from moment to moment in each lead are influenced greatly by the input of sensory information from the thalamus via specific and nonspecific projections to the cortical cells, as well as inputs that course laterally between different regions of the cortex.

EEG waves have characteristic patterns corresponding to state of arousal.

EEG frequency usually ranges from <1 to about 30 Hz, and amplitude, or height of the wave, usually ranges from 20 to 100 μV. Because the waves are a summation of activity in a complex network of neuronal processes, they are highly variable. However, during various states of consciousness, EEG waves have certain characteristic patterns. At the highest state of alertness, when sensory input is greatest, the waves are of high frequency and low amplitude, as many units discharge asynchronously. At the opposite end of the alertness scale, when sensory input is at its lowest in deep sleep, a synchronized EEG has the characteristics of low frequency and high amplitude. An absence of EEG activity is indicative of brain death.

EEG wave patterns are classified according to their frequency (Fig. 7.9). **Alpha waves**, a rhythm ranging from

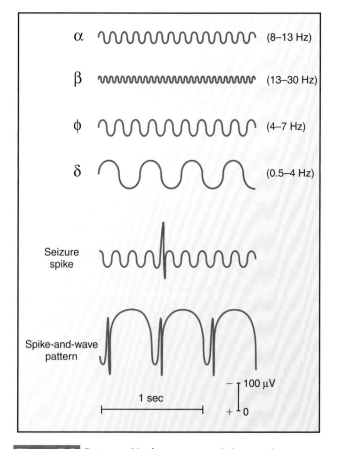

Figure 7.9 **Patterns of brain waves recorded on an electroencephalogram.** Wave patterns are designated alpha, beta, theta, or delta waves, based on frequency and relative amplitude. In epilepsy, abnormal spikes and large summated waves appear, as many neurons are activated simultaneously.

8 to 13 Hz, are observed when the person is awake but relaxed with the eyes closed. When the eyes are open, the added visual input to the cortex imparts a faster rhythm to the EEG, ranging from 13 to 30 Hz and designated **beta waves**. The slowest waves recorded occur during sleep: **theta waves** at 4 to 7 Hz and **delta waves** at 0.5 to 4 Hz, in deepest sleep.

Abnormal wave patterns are seen in **epilepsy**, a neurologic disorder of the brain characterized by spontaneous discharges of highly synchronized electrical activity, resulting in abnormalities ranging from momentary lapses of attention, to seizures of varying severity, to loss of consciousness. The characteristic waveform signifying seizure activity is the appearance of spikes or sharp peaks, as abnormally large numbers of units fire simultaneously. Examples of spike activity occurring singly and in a spike-and-wave pattern are shown in Figure 7.9.

Sleep stages are defined by the EEG.

Sleep follows a circadian rhythm and is a naturally recurring altered state of consciousness, in which sensory and motor activity are suspended, as well as nearly all voluntary muscle activity. During sleep, a heightened anabolic state exists, accentuating growth and rejuvenating the immune, nervous, and muscular systems. Sleep is regulated by the body's biologic clock. The biologic clock for sleep works in tandem with adenosine, a neurotransmitter that inhibits most of the functional processes associated with wakefulness. Adenosine rises over the course of the day, and high levels induce drowsiness. The hormone melatonin is released during drowsy sleep and causes a concomitant gradual decrease in core temperature.

Electrical activity associated with sleep was first discovered in 1937 using EEG. The EEG recorded during sleep reveals a persistently changing pattern of wave amplitudes and frequencies, indicating that the brain remains continually active even in the deepest stages of sleep. The EEG pattern recorded during sleep varies in a cyclic fashion that repeats approximately every 90 minutes, starting from the time of falling asleep to awakening 7 to 8 hours later. Sleep stages and other characteristics of sleep are assessed in a specialized sleep laboratory. Measurements include EEG of brain waves, *electrooculography* of eye movements, and *electromyography* of skeletal muscle activity. Sleep is divided into two broad categories: **rapid eye movement (REM)** and **nonrapid eye movement (NREM) sleep**. Each type has a distinct set of functional, neurologic, and physiologic features. NREM is further divided into three substages: N1, N2, and N3. N3 is also called *delta sleep* or slow-wave sleep (SWS). Sleep proceeds in cycles of REM and NREM, the order normally being awake → N1 → N2 → N3 → N2 → REM. There is a greater amount of deep sleep early in the night, whereas the proportion of REM sleep increases later in the night and just before natural awakening. In humans, each cycle lasts on the average 90 to 110 minutes. NREM sleep stages are characterized as follows:

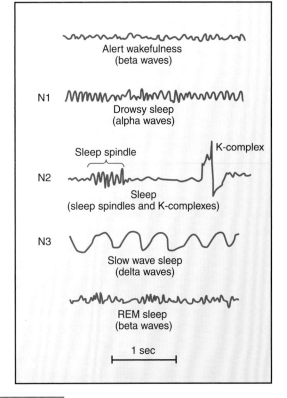

Figure 7.10 **The brain wave patterns during a normal sleep cycle.** Sleep cycles are divided into REM (rapid eye movement) and NREM (nonrapid eye movement) sleep. NREM is further divided into three stages: N1, N2, and N3.

- Stage N1 is referred to as *somnolence* or *drowsy sleep stage*. Brain waves in stage N1 transition from the beta waves of wakefulness to alpha waves then theta waves (Fig. 7.10). During this transition, sudden twitches, also known as myoclonus, are often observed. Also, muscle tone and most conscious awareness are lost during N1.

- During stage N2, muscular activity decreases and conscious awareness of the external environment completely disappears. The brain waves continue to slow and sleep spindles and K-complexes characteristic of this stage appear (Fig. 7.10). This stage occupies 45% to 55% of total sleep in adults.

- Stage N3 is characterized as *deep* or SWS with the presence of delta waves ranging from 0.4 to 4 Hz. During this stage physiological indicators such as breathing rate, heart rate, and blood pressure are all at their lowest levels. This is the stage in which night terrors, bed wetting (nocturnal enuresis), sleepwalking, and sleeptalking (somniloquy) most frequently occur.

REM sleep

REM sleep is characterized by REM as well as rapid, low-voltage EEG and accounts for 20% to 25% of total sleep time in most adults. This is the stage in which most memorable dreaming occurs. REM sleep is also known as paradoxical sleep, because of the seeming contradictions in its characteristics. First, the EEG exhibits unsynchronized, high-frequency, low-amplitude waves (i.e., a beta rhythm), which is more typical of the awake state than sleep (see Fig. 7.10), yet the subject is difficult to arouse. Second, the autonomic nervous system is in a state of excitation; blood pressure and heart rate are increased and breathing is irregular. In males, autonomic excitation in REM sleep includes penile erection. This reflex is used in diagnosing impotence, to determine whether erectile failure is based on a neurologic or a vascular defect (in which case, erection does not accompany REM sleep). When subjects are awakened during an REM period, they usually report dreaming. Accordingly, it is customary to consider REM sleep as dream sleep. Another curious characteristic of REM sleep is that most voluntary muscles are temporarily paralyzed. Two exceptions, in addition to the muscles of respiration, include the extraocular muscles, which contract rhythmically to produce the REMs, and the muscles of the middle ear, which protect the inner ear (see Chapter 4). Muscle paralysis is caused by an active inhibition of motor neurons mediated by a group of neurons located close to the locus coeruleus in the brainstem. Many of us have experienced this muscle paralysis on waking from a bad dream, feeling momentarily incapable of running from danger. In certain sleep disorders in which skeletal muscle contraction is not temporarily paralyzed in REM sleep, subjects act out dream sequences with disturbing results, with no conscious awareness of this happening.

Sleep in humans varies with developmental stage. Newborns sleep ~16 hours per day, of which about 50% is spent in REM sleep. Normal adults sleep 7 to 8 hours per day, of which about 25% is spent in REM sleep. The percentage of REM sleep declines further with age, together with a loss of the ability to achieve stage 3. Sleep debt is not getting enough sleep. Sleep deprivation can lead to a large sleep debt, which can cause emotional, mental, and physical problems. Sleep deprivation leads to increased irritability, cognitive impairment, memory loss, muscle tremors, and daytime yawning. Sleep deprivation has been shown to be a risk factor for weight gain, hypertension, type 2 diabetes, and impaired immune system. There are many reasons for poor sleep that lead to sleep deprivation. Pain, drugs, and stress can be the culprit. Sleep apnea, narcolepsy, insomnia, restless leg syndrome, and peripheral neuropathy are also causes of poor sleep.

Abnormal brain waves indicate epilepsy.

Seizure is defined as a transient symptom of abnormal excessive electrical activity in the brain. This abnormal activity is readily apparent in the EEG and different types of seizures have characteristic patterns that can aid in seizure classification. Several different etiologies can cause seizures including metabolic imbalances, high fever, drug abuse, genetic abnormalities, and brain infections or injury. **Epilepsy** is a chronic condition where an individual has recurrent seizures. The abnormal electrical activity of a seizure results from abnormal neuronal physiology that produces neuronal hyperexcitabilty and hypersynchrony. This abnormal pathology occurs in groups of neurons and represents an imbalance between excitatory and inhibitory ionic currents in the neuronal networks of the brain. Many of the drugs used to treat epilepsy target the neuronal ion channels responsible

for these currents in an attempt to bring the networks back into balance. Although these drugs may control the seizures of epilepsy, they do not cure the disease and patients must often take these drugs their entire lives.

The outward effect of epileptic seizures is manifested in three different ways: (1) dramatic wild thrashing movements (tonic–clonic seizure), (2) mild loss of awareness, or (3) convulsions. Seizure is not always accompanied by convulsions. Sometimes, a patient may lose consciousness and "slump" to the ground or they may only temporarily stare into space. After a seizure, while the brain is recovering, there is often a transient loss of memory.

Generalized and partial seizures

Epilepsy is not a single disease having many clinical manifestations and underlying pathologies. Seizures are generally divided into two major types, *generalized* and *partial*, based on where the seizures start and how they then spread.

Generalized seizures usually start in both hemispheres or at a subcortical location like the thalamus. The tonic-clonic seizure (formerly known as *grand mal seizures*) is an example of a generalized seizure that affects the entire brain. In the *tonic phase*, the person quickly loses consciousness and the skeletal muscles suddenly become tense, causing extremities to be pulled inward toward the body.

During the *clonic phase*, muscles contract and relax rapidly, causing convulsions. The convulsions can range from exaggerated twitches to violent shaking. The eyes typically roll back or close, and the tongue often suffers bruises sustained from jaw contractions. Incontinence sometimes occurs. Confusion and complete amnesia are usually experienced upon regaining consciousness. Due to physical and nervous exhaustion, a long sleep period usually follows a tonic–clonic seizure.

Absence seizures (formerly known as *petit mal seizures*) are another type of generalized seizure. In this seizure, the person appears to stare into space with episodes lasting up to 10 seconds. Partial seizures (formerly known as *focal seizures*) begin in a localized area of the brain and the disturbances produced reflect the function of that brain area. For example, a disturbance may include a twitching of part of the body, such as a limb, or a deceptive or illusory sensation. Partial seizures are further classified as (1) simple partial (with no interruption to consciousness) and (2) complex partial (interrupts consciousness). Most often, partial seizures result from a localized lesion or functional abnormality. Examples include scar tissue that adheres to adjacent brain tissue, a brain tumor that compresses a discrete area of the brain, or dysfunctional local circuitry. Although starting at a specific locality, in some cases, partial seizures can spread to encompass larger brain areas.

▶ FUNCTIONAL COMPONENTS OF THE FOREBRAIN

The forebrain is the rostral most portion of the brain consisting of diencephalon and cerebrum. The forebrain is responsible for controlling body temperature, reproductive

Left hemisphere   **Right hemisphere**

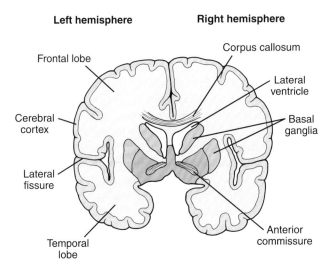

Figure 7.11 The cerebral hemispheres and some deep structures in a coronal section through the rostral forebrain. The corpus callosum is the major commissure that interconnects the right and left hemispheres. The anterior commissure connects rostral components of the right and left temporal lobes. The cortex is an outer rim of gray matter (neuronal cell bodies and dendrites); deep to the cortex is white matter (axonal projections) and then subcortical gray matter.

functions, eating, sleeping, emotional affect, and cognitive functions.

The cerebrum consists of the **cerebral cortex** and the subcortical structures rostral to the diencephalon. The cortex, a few-millimeters thick outer shell of the cerebrum, has a rich, multilayered array of neurons and their processes form columns perpendicular to the surface. The axons of cortical neurons give rise to descending fiber tracts and intrahemispheric and interhemispheric fiber tracts, which, together with ascending axons coursing toward the cortex, make up the prominent white matter underlying the outer cortical gray matter. A deep sagittal fissure divides the cortex into a right and left hemisphere, each of which receives sensory input from and sends its motor output to the opposite side of the body. A set of **commissures** containing axonal fibers interconnects the two hemispheres, so that processed neural information from one side of the forebrain is transmitted to the opposite hemisphere. The largest of these commissures is the **corpus callosum**, which interconnects a major portion of the cerebral hemispheres (Fig. 7.11). Among the subcortical structures located in the cerebrum are the amygdala and hippocampus, components of the limbic system, which regulates emotional response, and the basal ganglia (caudate, putamen, and globus pallidus), which are essential for coordinating motor activity (see Chapter 5) and thought processes.

Cerebral cortex comprises three functional areas: sensory, motor, and association areas.

In the human brain, the surface of the cerebral cortex is highly convoluted, with gyri (singular, **gyrus**) and sulci (singular, **sulcus**), which are akin to hills and valleys, respectively. Deep sulci are also called *fissures*. Two deep fissures form prominent landmarks on the surface of the cortex. The **central**

sulcus divides the **frontal lobe** from the **parietal lobe**, and the **lateral sulcus** (sylvian fissure) divides the **parietal lobe** from the **temporal lobe** (Fig. 7.12). The **occipital lobe** has less prominent sulci separating it from the parietal and temporal lobes.

Topographically, the cerebral cortex is divided into areas of specialized functions, including the primary sensory areas for vision (occipital cortex), hearing (temporal cortex), somatic sensation (postcentral gyrus), and motor functions (precentral gyrus) (see Chapters 4 and 5). Sensory and motor functions are controlled by cortical structures in the contralateral hemisphere (see Chapters 4 and 5). Particular cognitive functions or components of these functions may also be lateralized to one side of the brain.

As shown in Figure 7.12, these well-defined areas comprise only a small fraction of the surface of the cerebral cortex. The majority of the remaining cortical area is known as **association cortex**, where the processing of neural information is performed at the highest levels of which the organism is capable. Among vertebrates, the human cortex contains the most extensive association areas. The association areas are also sites of long-term memory, and they control such

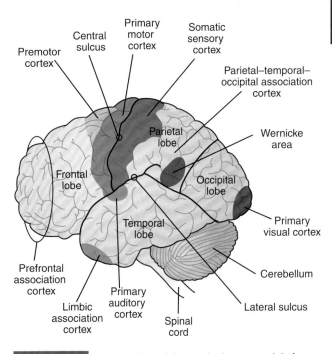

Figure 7.12 The four lobes of the cerebral cortex and their respective functional areas. The central sulcus and lateral sulcus (sylvian fissure) are prominent landmarks used in defining the cortical lobes. The primary auditory cortex is responsible for hearing. The limbic association cortex is the region for motivation, emotions, and memory. The prefrontal association cortex is responsible for voluntary activity, decision making, and personality traits. The premotor cortex is responsible for coordinating complex motor movements. The primary motor cortex is responsible for voluntary movements. The somatosensory cortex is responsible for sensation and proprioception. The parietal–temporal–occipital association cortex integrates all sensory input and is important in language. The Wernicke area is responsible for speech comprehension.

human functions as language acquisition, speech, musical ability, mathematical ability, complex motor skills, abstract thought, symbolic thought, and other cognitive functions.

Association areas interconnect and integrate information from the primary sensory and motor areas via intrahemispheric connections. The **parietal–temporal–occipital association cortex** integrates neural information contributed by visual, auditory, and somatic sensory experiences. The **prefrontal association cortex** is that part of the frontal lobe other than the specific motor and speech regions. The prefrontal area governs the following: **attention**, the monitoring of the external and internal milieu; **intention**, the shaping of behavior in accordance with internal motivation and external context; and **execution**, the orchestration of complex sensorimotor sequences in a seamless path toward a goal. The selection and execution of cognitively driven sensorimotor sequences depend on integration with the basal ganglia in the same way that motor sequences do (see Chapter 5). The prefrontal cortex is responsible for executive function (making decisions) and, by virtue of its connections with the limbic system, coordinates emotionally motivated behaviors. The prefrontal cortex receives neural input from the other association areas and regulates motivated behaviors by direct input to the premotor cortex, which serves as the association area of the motor cortex.

Cortical and subcortical structures are part of the architectural design of the limbic system.

The limbic system comprises specific areas of the cortex and subcortical structures interconnected via circuitous pathways (Fig. 7.13). Originally, the limbic system was considered to be restricted to a ring of cortical structures surrounding the corpus callosum, including the **cingulate gyrus**, parahippocampal gyrus, and **hippocampus**, together with the fiber tracts that interconnect them with the diencephalic components of the limbic system, the hypothalamus, and the anterior thalamus. Current descriptions of the limbic system also include the **amygdala** (deep in the temporal lobe); the **nucleus accumbens** (the limbic portion of the basal ganglia), a region of the brain involved in pleasure and reward; the **septal nuclei** (at the base of the forebrain); the **prefrontal cortex** (anterior and inferior components of the frontal lobe); and the habenula (in the diencephalon).

The limbic system supports a variety of functions, including memory, behavior, emotions, and olfaction. The limbic system operates by influencing the automatic nervous system and the endocrine system. It plays a role in sexual arousal and the emotional "high" derived from recreational drugs. Many of the psychiatric disorders, including bipolar disorder, depression, schizophrenia, and dementia, involve malfunctions in the limbic system.

Circuitous loops of fiber tracts interconnect the limbic structures. The main circuit links the hippocampus to the mammillary body of the hypothalamus by way of the fornix, the hypothalamus to the anterior thalamic nuclei via the mammillothalamic tract, and the anterior thalamus to the cingulate gyrus by widespread, anterior thalamic projections. To complete the circuit, the cingulate gyrus connects with the hippocampus. Other structures of the limbic system form smaller loops within this major circuit, forming the basis for a wide range of emotional behaviors. Of particular importance is the amygdala, which interconnects with the hypothalamus, thalamus, and basal forebrain region via the stria terminalis and ventral amygdalofugal pathway and is responsible for fear conditioning and vigilance in learning and memory paradigms.

The fornix also connects the hippocampus to the base of the forebrain, where the septal nuclei and nucleus accumbens reside. The prefrontal cortex and other areas of association cortex provide the limbic system with information based on previous learning and currently perceived needs. Inputs from the brainstem provide visceral and somatic sensory signals, including tactile, pressure, pain, and temperature information from the skin and sexual organs and pain information from the visceral organs.

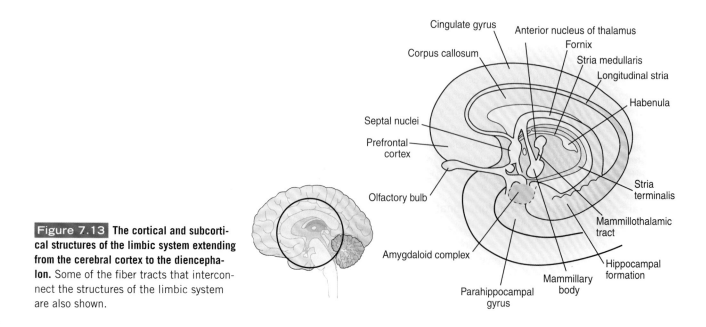

Figure 7.13 The cortical and subcortical structures of the limbic system extending from the cerebral cortex to the diencephalon. Some of the fiber tracts that interconnect the structures of the limbic system are also shown.

Brain's reward system

Experimental studies beginning early in the last century demonstrated that stimulating the limbic system or creating lesions in various parts of the limbic system can alter emotional states. Most of our knowledge comes from animal studies, but emotional feelings are reported by humans when limbic structures are stimulated during brain surgery or when limbic structures are activated during epileptic seizures.

Electrical stimulation of various sites in the limbic system produces either pleasurable (rewarding) or unpleasant (aversive) feelings. To study these findings, researchers use electrodes implanted in the brains of animals. When electrodes are implanted in structures presumed to generate rewarding feelings and the animals are allowed to deliver current to the electrodes by pressing a bar, repeated and prolonged self-stimulation is seen. Other needs—such as food, water, and sleep—are neglected. The sites that provoke the highest rates of electrical self-stimulation are in the ventral limbic areas, including the septal nuclei and nucleus accumbens.

The limbic system is one component of an extensive brain reward system. The reward system is responsible for increasing the likelihood of behaviors that have a positive outcome (positive reinforcement) and are pleasurable. Extensive studies have shown that activation of the mesolimbic dopamine system involving connections between the ventral tegmental nucleus, nucleus accumbens, and prefrontal cortex plays a major role in mediating reward. The nucleus accumbens is thought to be one site of action of addictive drugs, including opiates, alcohol, nicotine, cocaine, and amphetamine.

Aggressive behavior

A fight or flight response, including the autonomic components (see Chapter 6) and postures of rage and aggression characteristic of fighting behavior, can be elicited by electrical stimulation of sites in the hypothalamus and amygdala. If the frontal cortical connections to the limbic system are severed, rage postures and aggressiveness become permanent, illustrating the importance of the higher centers in restraining aggression and, presumably, in invoking it at appropriate times. By contrast, bilateral removal of the amygdala results in a placid animal that cannot be provoked.

Sexual arousal

The biologic basis of human sexual activity is poorly understood because of its complexity and because findings derived from nonhuman animal studies cannot be extrapolated. The major reason for this limitation is that the cerebral cortex, uniquely developed in the human brain, plays a more important role in governing human sexual activity than the instinctive or olfactory-driven behaviors in nonhuman primates and lower mammalian species. Nevertheless, several parallels in human and nonhuman sexual activities exist, indicating that hormones and the limbic system, in general, coordinate sex drive and mating behavior, with higher centers exerting more or less overriding influences.

Copulation in mammals is under parasympathetic control and is coordinated by reflexes of the sacral spinal cord, including male penile erection and ejaculation reflexes and engorgement of female erectile tissues as well as the muscular spasms of the orgasmic response. Copulatory behaviors and postures can be elicited in animals by stimulating parts of the hypothalamus, olfactory system, and limbic areas.

Human sexual desire and arousal are largely mediated through the limbic system. The amygdala contributes to the emotional aspects of sex and stimulation produces sexual feelings. The ventral striatum and septal nuclei are involved in the pleasure and rewarding aspects of sex. The orbitofrontal cortex, located at the base of the frontal lobe, has been implicated in mate selection, and frontal and parietal cortical areas play a role in sexual excitement. Ablation studies have shown that sexual behavior requires an intact connection of the limbic system with the frontal cortex. Additional important determinants of human sexual activity are the higher cortical functions of learning and memory, which serve to either reinforce or suppress the signals that initiate sexual responding, including the sexual reflexes coordinated by the sacral spinal cord.

Limbic system malfunction leads to psychiatric disorders.

The major psychiatric disorders, including affective disorders and schizophrenia, are disabling diseases with a genetic predisposition and no known cure. The functional basis for these disorders remains obscure. In particular, the role that environmental influences play on individuals with a genetic predisposition to developing a disorder is also unclear. Altered states of the brain's monoaminergic systems have been a major focus as possible underlying factors, based on extensive human studies in which neurochemical imbalances in catecholamines, acetylcholine (ACh), and serotonin have been observed. Another reason for focusing on the monoaminergic systems is that the most effective drugs used in treating psychiatric disorders are agents that alter monoaminergic transmission.

Affective disorders

Illnesses marked by abnormal mood regulation and associated signs and symptoms are called **affective disorders** or **mood disorders**. The word "mood" describes an overall emotional tone that provides the background to one's internal life. Mood disorders are broadly classified as **major depression**, which can be so profound as to provoke suicide, and **bipolar disorders** (formerly known as *manic–depressive disorder*), in which periods of profound depression are followed by periods of mania, in a cyclic pattern. Neurophysiologic studies indicate that depressed patients show decreased use of brain NE. In manic periods, NE transmission increases. Whether in depression or in mania, all patients seem to have decreased brain serotonergic transmission, suggesting that serotonin may exert an underlying permissive role in abnormal mood swings, in contrast with NE, whose transmission, in a sense, titrates the mood from highest to lowest extremes. The most effective treatments for depression are antidepressant drugs such as monoamine oxidase inhibitors (MAOIs), tricyclics, and **selective serotonin reuptake inhibitors (SSRIs)** that inhibit monoamine reuptake or increase levels of monoamines in the synaptic cleft. A therapeutic response to monoamine-based treatments ensues only after treatment is repeated over time. Similarly, when treatment stops, symptoms may not reappear for several weeks. This time lag in treatment response is presumably a result of alterations in the long-term regulation of receptor and second messenger systems in relevant regions of the brain. With electroconvulsive therapy, these pharmacological treatments have in common the ability to alter noradrenergic and serotonergic transmission serving the limbic system. These observations led to the most common theory of depression, the monoamine deficiency hypothesis.

The potential to understand depression—and other complex mental illness—has dramatically increased with the advent of both molecular genetics and imaging technologies. An important finding was that a genetic variation in the serotonin reuptake transporter predicts susceptibility to depression subsequent to a number of stressors. Functional imaging studies have suggested the possible involvement of several brain structures in depression. A meta-analysis suggests that the volume of anterior cingulate cortex and adjacent medial prefrontal cortex is reduced in individuals at risk of familial major depression and deep brain stimulation in this region has produced clinical benefits.

The need for new technologies to explain disease mechanisms is highlighted by the other side of the affective spectrum. The most effective long-term treatment for mania is lithium, although antipsychotic (neuroleptic) drugs, which block dopamine receptors, are effective in the acute treatment of mania. The therapeutic actions of lithium remain unknown, but the drug has an important action on a receptor-mediated second messenger system. Lithium interferes with regeneration of phosphatidylinositol in neuronal membranes by blocking the hydrolysis of inositol-1-phosphate. Depletion of phosphatidylinositol in the membrane renders it incapable of responding to receptors that use this second messenger system. Without knowing the specific neuroanatomic underpinnings of the disease, it is difficult to ascertain how these mechanisms might relate to amelioration of symptoms.

Schizophrenia: uncoupling of thought and emotional processes

Schizophrenia is the collective name for a group of psychotic disorders that vary greatly in symptoms among individuals. It most commonly manifests as delusions (fixed false beliefs) or hallucinations (perceptions in the absence of a stimulus) that are coupled to disorganized thinking and speech. Delusions are usually persecutory in nature and tend to follow themes of being harassed, cheated on, conspired against, drugged, or poisoned. Hallucination (most commonly hearing voices) is often paranoid in nature and frequently tied to social dysfunctions. The onset of symptoms typically occurs in young adulthood with ~2% reporting lifetime prevalence.

No neurologic tests for schizophrenia currently exist, and the diagnosis is based exclusively on the patient's self-reporting experiences and observed behavior.

Although the exact cause of schizophrenia is not known, increased dopamine activity in the mesolimbic pathway of the brain (see Fig. 7.6) is commonly found in patients with schizophrenia. The mainstay of treatment is antipsychotic medication that primarily works on suppressing dopamine activity. Older antipsychotic drugs (**neuroleptics** or typical antipsychotics) achieve their therapeutic effect by blocking dopamine D_2 receptors. Current research is focused on finding a subtype of dopamine receptor that mediates mesocortical/mesolimbic dopaminergic transmission but does not affect the nigrostriatal system, which controls motor function (see Fig. 7.6). So far, neuroleptic drugs that block one pathway almost always block the other as well, leading to unwanted neurologic side effects, including abnormal involuntary movements after long-term treatment or Parkinsonism in the short term. Similarly, some patients with Parkinson's disease who receive L-DOPA to augment dopaminergic transmission in the nigrostriatal pathway must be taken off the medication because they develop hallucinations.

Newer, so-called **atypical antipsychotics** have relatively low affinity for dopamine receptors and their clinical effectiveness as a schizophrenia treatment does not correlate with D_2 antagonism. Instead, this class of drug antagonizes serotonin receptors suggesting that dopamine may not be the primary culprit in the disease. Evidence is accumulating that hypofunctioning of glutamate signaling through the N-methyl-D-aspartate (NMDA) receptor (see Chapter 3) may also contribute to schizophrenia. The idea is that glutamate may normally activate pathways that inhibit dopaminergic pathways; and therefore, hypofunction of NMDA receptor signaling would increase dopamine signaling.

▶ HIGHER COGNITIVE SKILLS

Cognitive science is the interdisciplinary study of how information (e.g., concerning perception, language, speech, reasoning, and emotion) is represented and transformed in the brain. Multiple research disciplines (neuroscience, psychology, artificial intelligence, and learning science) are brought together to investigate neural circuitry and modular brain organization in an effort to analyze information from low-level learning and decision-making mechanisms to high-level logic and planning.

Cerebral cortex and the limbic system provide the architectural components for learning and memory systems.

Learning involves synthesizing different types of information in order to acquire (1) new knowledge, (2) behavior, (3) skills, (4) values, and (5) preferences. Memory allows the brain to store information for later retrieval and consists of both long-term (i.e., days, months, years) and short-term (i.e., minutes, hours) storage. Memory and learning are inextricably linked because part of the learning process involves the assimilation of new information and its commitment to memory. The most likely sites of learning in the human brain are the large association areas of the cerebral cortex, in coordination with subcortical structures deep in the temporal lobe, including the hippocampus and amygdala. The association areas draw on sensory information from sensory areas of the cortex and on emotional feelings transmitted via the limbic system. This information is integrated with previously learned skills and stored memory, which presumably also reside in the association areas.

The learning process itself is poorly understood, but it can be studied experimentally at the synaptic level in isolated slices of mammalian brain or in more simple invertebrate nervous systems. Synapses subjected to repeated presynaptic neuronal stimulation show changes in the excitability of postsynaptic neurons. These changes include the facilitation of neuronal firing, altered patterns of neurotransmitter release, second messenger formation, and, in intact organisms, evidence that learning occurred. The phenomenon of increased excitability and altered chemical state on repeated synaptic stimulation is known as **long-term potentiation**, a persistence beyond the cessation of electrical stimulation, as is expected of learning and memory. Ca^{2+} entry through activation of NMDA receptors is critical to the development of long-term potentiation. An early event in long-term potentiation is a series of protein phosphorylations induced by receptor-activated second messengers leading to activation of a host of intracellular proteins and altered excitability. In addition to biochemical changes in synaptic efficacy associated with learning at the cellular level, structural alterations occur. The number of connections between sets of neurons increases as a result of experience. The cellular and structural changes that accompany the formation of long-term memories reflect changes in gene expression within the neuron. Inhibition of transcription or translation blocks the formation of long-term, but not short-term, memories in model systems.

Much of our knowledge about human memory formation and retrieval is based on studies of patients in whom stroke, brain injury, or surgery resulted in memory disorders. Such knowledge is then examined in more rigorous experiments in nonhuman primates capable of cognitive functions or using brain imaging such as functional MRI. From these combined approaches, we know that memories are not stored in a single place in the brain but are distributed across many areas. The prefrontal cortex is essential for coordinating the formation of memory, starting from a learning experience in the cerebral cortex, then processing the information, and communicating it to the subcortical limbic structures. The prefrontal cortex receives sensory input from the parietal, occipital, and temporal lobes and emotional input from the limbic system. Drawing on skills such as language and mathematical ability, the prefrontal cortex integrates these inputs in light of previously acquired learning. The prefrontal cortex can thus be considered the site of working memory, where new experiences are processed, as opposed to sites that consolidate the memory and store it.

The processed information is then transmitted to the hippocampus, where it is consolidated over several hours into a more permanent form that is stored in, and can be retrieved from, the association cortices.

Long-term and short-term memory comprise the brain's memory system.

Memory can be divided into that which can be recalled for only a brief period (seconds to minutes) and that which can be recalled for weeks to years. Newly acquired learning experiences can be readily recalled for only a few minutes or more using **short-term memory**. An example of short-term memory is looking up a telephone number, repeating it mentally until you finish dialing the number, then promptly forgetting it as you focus your attention on starting the conversation. Short-term memory is a product of working memory; the decision to process information further for permanent storage is based on judgment as to its importance or on whether it is associated with a significant event or emotional state. An active process involving the hippocampus must be employed to make a memory more permanent.

The conversion of short-term to **long-term memory** is facilitated by repetition, by adding more than one sensory modality to learn the new experience (e.g., writing down a newly acquired fact at the same time one hears it spoken) and, even more effective, by tying the experience (through the limbic system) to a strong, meaningful emotional context. The role of the hippocampus in consolidating the memory is reinforced by its participation in generating the emotional state with which the new experience is associated.

Declarative memory (sometimes referred to as *explicit memory*) is one of the two subtypes of long-term memory systems. Declarative memory refers to the memory system that consciously recalls facts and events (e.g., birthdays, state capitals, musical event, and personal events). The counterpart to declarative memory is known **nondeclarative** or **procedural memory** (sometimes referred to as *implicit memory*) and refers to unconscious memories of how to do things (e.g., from tying shoes to driving a car). These memories are automatically retrieved and occur without the need for conscious control.

Memory loss

Memory loss can be partial or total. Unfortunately, memory loss is considered normal when it comes with aging. Sudden memory loss can result from brain trauma, strokes, cardiac arrest, meningitis, and epilepsy. Trauma to the brain is not the only cause of sudden memory loss. It may appear as a side effect from chemotherapy in which cytotoxic drugs are used to treat cancer or from drugs that are prescribed for lowering blood cholesterol. Sudden memory loss can be permanent or temporary. When it is caused by other medical conditions, such as Alzheimer's disease or Parkinson's disease, memory loss is gradual and tends to be permanent.

Cholinergic innervations play an important role in memory, with ACh as the major neurotransmitter in cognitive function, learning, and memory. Loss of cholinergic function is associated with **dementia**, an impairment

Alzheimer's Disease

Alzheimer's disease (AD) is the most common cause of dementia in older adults. The cause of the disease still is unknown, and there is no cure. In 2016, an estimated 5.4 million people in the United States suffered from AD. Although the disease usually begins after the age of 65 years and the risk of AD goes up with age, it is important to note that AD is not a normal part of aging. The aging of the baby boom generation has made AD one of the fastest growing diseases. Estimates indicate that by the year 2050, more than 13 million people in the United States will suffer from AD.

Cognitive deficits are the primary symptoms of AD. Early on, there is mild memory impairment that is most notable in the patient's inability to consolidate short-term memories to long-term memories. As the disease progresses, memory problems increase with long-term memories now involved, and difficulties with language are generally observed, including word-finding problems and decreased verbal fluency. Many patients also exhibit difficulty with visuospatial tasks. Personality changes are common, and patients become disoriented as the memory problems worsen. A progressive deterioration of function follows and, at late stages, the patient is bedridden, nearly mute, unresponsive, and incontinent. A definitive diagnosis of AD is not possible until autopsy, but the constellation of symptoms and disease progression allow a reasonably certain diagnosis.

Gross pathology consistent with AD is mild-to-severe cortical atrophy (depending on the age of onset and the age at death). Microscopic pathology indicates two classic signs of the disease even at the earliest stages: the presence of senile plaques (SPs) and neurofibrillary tangles (NFTs). As the disease progresses, synaptic and neuronal loss or atrophy and an increase in SPs and NFTs occur.

Although many neurotransmitter systems are implicated in AD, the most consistent pathology is the loss or atrophy of cholinergic neurons in the basal forebrain. Medications that ameliorate the cognitive symptoms of AD are cholinergic function enhancers. These observations emphasize the importance of cholinergic systems in cognitive function. ■

---

of memory, memory loss, abstract thinking, and judgment (see Clinical Focus Box 7.1).

Language and speech are coordinated in specific areas within the association cortex.

The ability to communicate by language, verbally and in writing, is one of the most difficult cognitive functions to study because only humans are capable of these skills. Thus, much of our knowledge of language processing in the brain has been inferred from clinical data by studying patients with aphasias—disturbances in producing or understanding the meaning of words—following brain injury, surgery, or other damage to the cerebral cortex.

Two areas appear to play an important role in language and speech: the **Wernicke area**, in the upper temporal lobe, and the **Broca area**, in the frontal lobe (Fig. 7.14). Both of these areas are located in the association cortex, adjacent to cortical areas that are essential in language communication. The Wernicke area is in the parietal–temporal–occipital association cortex, a major association area for processing sensory information from the somatic sensory, visual, and auditory cortices. The Broca area is in the prefrontal association cortex, adjacent to the portion of the motor cortex that regulates movement of the muscles of the mouth, tongue, and throat (i.e., the structures used in the mechanical production of speech). A fiber tract, the arcuate fasciculus, connects the Wernicke area with the Broca area to coordinate aspects of understanding and executing speech and language skills.

Clinical evidence indicates that the Wernicke area is essential for the comprehension, recognition, and construction of words and language, whereas the Broca area is essential for the mechanical production of speech. Patients with a defect in the Broca area show evidence of comprehending a spoken or written word but are not able to say the word. In contrast, patients with damage in the Wernicke area can produce speech, but the words they put together have little meaning.

Language is a highly lateralized function of the brain residing in the left hemisphere (see Clinical Focus Box 7.2). This dominance is observed in left-handed as well as right-handed people. Moreover, it is language that is lateralized, not the reception or production of speech. Thus, native signers (people who use sign language) who have been deaf since birth still show left-hemisphere language function.

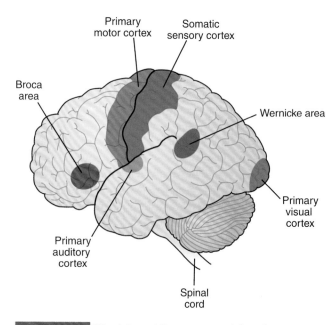

Figure 7.14 Wernicke and Broca areas and the primary motor, visual, auditory, and somatic sensory cortices.

The Split Brain

Patients with life-threatening, intractable epileptic seizures were treated in the past by surgical commissurotomy or cutting of the corpus callosum (see Fig. 7.11). This procedure effectively cut off most of the neuronal communication between the left and right hemispheres and vastly improved patient status because seizure activity no longer spread back and forth between the hemispheres.

There was a remarkable absence of overt signs of disability following commissurotomy; patients retained their original motor and sensory functions, learning and memory, personality, talents, emotional responding, and so on. This outcome was not unexpected because each hemisphere has bilateral representation of most known functions; moreover, those ascending (sensory) and descending (motor) neuronal systems that crossed to the opposite side were known to do so at levels lower than the corpus callosum.

Notwithstanding this appearance of normalcy, following commissurotomy, patients were shown to be impaired to the extent that one hemisphere literally did not know what the other was doing. It was further shown that each hemisphere processes neuronal information differently from the other and that some cerebral functions are confined exclusively to one hemisphere.

In an interesting series of studies by Nobel laureate Roger Sperry and colleagues, these patients with a so-called split brain were subjected to psychophysiological testing in which each disconnected hemisphere was examined independently. Their findings confirmed what was already known: sensory and motor functions are controlled by cortical structures in the contralateral hemisphere. For example, visual signals from the left visual field were perceived in the right occipital lobe, and there were contralateral controls for auditory, somatic sensory, and motor functions. (Note that the olfactory system is an exception, as odorant chemicals applied to one nostril are perceived in the olfactory lobe on the same side.) However, the scientists were surprised to find that language ability was controlled almost exclusively by the left hemisphere. Thus, if

an object was presented to the left brain via any of the sensory systems, the subject could readily identify it by the spoken word. However, if the object was presented to the right hemisphere, the subject could not find words to identify it. This was not due to an inability of the right hemisphere to perceive the object, as the subject could easily identify it among other choices by nonverbal means, such as feeling it while blindfolded. From these and other tests, it became clear that the right hemisphere was mute; it could not produce language.

In accordance with these findings, anatomic studies show that areas in the temporal lobe concerned with language ability, including the Wernicke area, are anatomically larger in the left hemisphere than in the right in a majority of humans, and this is seen even prenatally. Corroborative evidence of language ability in the left hemisphere is shown in persons who have had a stroke, where aphasias are most severe if the damage is on the left side of the brain. Analysis of people who are deaf who communicated by sign language prior to a stroke has shown that sign language is also a left-hemisphere function. These patients show the same kinds of grammatical and syntactical errors in their signing following a left-hemisphere stroke as do speakers.

In addition to language ability, the left hemisphere excels in mathematical ability, symbolic thinking, and sequential logic. The right hemisphere, on the other hand, excels in visuospatial ability, such as three-dimensional constructions with blocks and drawing maps, and in musical sense, artistic sense, and other higher functions that computers seem less capable of emulating. The right brain exhibits some ability in language and calculation, but at the level of children aged 5 to 7. It has been postulated that both sides of the brain are capable of all of these functions in early childhood, but the larger size of the language area in the left temporal lobe favors development of that side during language acquisition, resulting in nearly total specialization for language on the left side for the rest of one's life. ■

Chapter Summary

- The central nervous system receives, integrates, and acts on sensory stimuli from the internal and external environment.
- The hypothalamus is the interface between the endocrine, autonomic, and limbic systems.
- The hypothalamus regulates energy balance by integrating metabolism and eating behavior, it controls the gonads and sexual activity, and it contains the biological clock.
- The brainstem reticular formation is critical for arousal and sleep and contains several monoaminergic systems that innervate the forebrain.
- The ascending reticular activating system mediates consciousness and arousal.
- An electroencephalogram records electrical activity at the brain's surface in waves that have characteristic patterns depending on the state of arousal.
- The forebrain contains the thalamus, hypothalamus, and the cerebrum and processes sensory information, executes behavior, makes decisions, and is capable of learning.

- The cerebral cortex comprises three parts: sensory, motor, and association areas.
- The limbic system contains cortical and subcortical structures and is the seat of emotions, containing a reward system, mediating aggression, and coordinating sexual activity as well as being implicated in psychiatric disorders.
- Affective disorders are marked by abnormal mood regulation.
- Schizophrenia involves disordered thought processes, delusions, and hallucinations.
- Memory and learning require the cerebral cortex and limbic system.
- Declarative and nondeclarative memory involve different central nervous system structures.
- Cholinergic innervation is associated with cognitive function.
- Language and speech are coordinated in specific areas of the association cortex.

Chapter Review Questions

1. A patient's wife complains that several times during the last few weeks, her husband struck her as he flailed around violently during sleep. The husband indicates that when he wakes up during one of these sessions, he has been dreaming. What is the likely cause of his problem?

   A. Increased muscle tone during stage N3 sleep
   B. Increased drive to the motor cortex during REM sleep
   C. Lack of behavioral inhibition by the prefrontal cortex during sleep
   D. Lack of abolished muscle tone during REM sleep
   E. Abnormal functioning of the amygdala during paradoxical sleep

The correct answer is D. Dreams are associated with REM sleep. Normally, a person does not "act out" his or her dream because all of the motor neurons to muscles other than those for respiration, the middle ear, and the extraocular eye muscles are inhibited, abolishing muscle tone. If this inhibition does not occur, a person exhibits marked and often dangerous movement during dreaming. Muscle tone is reduced but not abolished during N3 stage sleep; however, movements are not an issue, presumably because dreaming does not occur. (Note, however, that sleepwalking occurs in slow-wave sleep.) Increased activity in motor areas of the cortex (or other areas) during REM sleep normally would not cause movement because motor neurons are inhibited.

2. A scientist develops a reagent that allows identification of leptin-sensing neurons in the CNS. The reagent is a fluorescent compound that binds to the plasma membrane of cells that sense leptin. Where would application of this reagent to sections of the brain result in fluorescent staining?

   A. Arcuate nucleus of the hypothalamus
   B. Mammillary nuclei of the hypothalamus
   C. Paraventricular nucleus of the hypothalamus
   D. Preoptic nucleus of the hypothalamus
   E. Ventromedial nucleus of the thalamus

The correct answer is A. Leptin is released into the blood from the body's adipocytes. Circulating leptin levels are sensed by neurons in the arcuate nucleus, which are located in an area that does not possess a blood–brain barrier. While other regions of the hypothalamus also lack a blood–brain barrier, the other hypothalamic nuclei listed do not contain leptin-sensing cells.

3. Classical language disorders include Broca aphasia, where people can understand language but cannot produce it, and Wernicke aphasia, where a person cannot understand language and produces only nonsensical language. However, damage to regions of the left frontal and parietal lobes outside Broca and Wernicke areas also produce language deficits. Where would an individual who presents poststroke with an ability to understand language but is able to produce only nonsensical language most likely have suffered damage?

   A. Thalamocortical tract
   B. Reticular activating system
   C. Prefrontal lobe
   D. Fornix
   E. Arcuate fasciculus

The correct answer is E. The arcuate fasciculus is the fiber bundle connecting the Broca and Wernicke areas. Damage to the arcuate fasciculus disconnects the Wernicke area, which is responsible for recognition and construction of words and language, from Broca area, which is responsible for language production. The fornix connects the hippocampus with the hypothalamus and basal fore-brain. The thalamocortical tract connects the thalamus with the cortex and the reticular activating system connects the brainstem with the thalamus and cortex. Although both are important for modulating cortical activity, damage does not produce the deficit seen in this individual. The prefrontal lobe is not a fiber bundle.

4. A blindfolded subject is asked to verbally identify a common object presented to her left hand. She is not allowed to touch the object with her right hand. Which of the following structures must be intact for her to complete the task?

   A. Primary somatic cortex on the left side of her brain
   B. Primary visual cortex on the right side of her brain
   C. Fornix
   D. Corpus callosum
   E. Hippocampus

The correct answer is D. Somatic sensory information from the left hand would be perceived in the right cortex, which does not generate language. To verbally explain what the object is, the information must cross to the left hemisphere. This crossing occurs through the corpus callosum. The fornix and hippocampus would be involved in storing memories about particular items, not retrieving the memory. Neither the primary somatic sensory cortex on the left side nor the visual cortex on either side plays a role in identifying an object placed in the left hand by tactile cues.

5. Which region of the central nervous system is thought to be the "master clock" for controlling various biological rhythms?

   A. Pineal body
   B. Reticular formation
   C. Forebrain
   D. Hypothalamus
   E. Pituitary gland

The correct answer is D. Biological rhythms (monthly, cir-cadian) are controlled by the suprachiasmatic nucleus in the hypothalamus. The pituitary is sometimes referred to as the "master" endocrine gland. It secretes nine hormones that regulate homeostasis. The pineal gland secretes serotonin and melatonin, two hormones that affect wake–sleep patterns. However, they are not the master clock for the control of the body's biological rhythms. The reticular formation modulates alertness and sleep but does not control biological rhythms. The forebrain controls body temperature, reproductive functions, eating, sleeping, and some emotions.

Clinical Application Exercises 7.1

STROKE

A 67-year-old man was taken to see his physician by his wife. For the preceding 2 days, the patient's wife had noticed that he did not seem to make sense when he spoke. She also indicated that he seemed a little disoriented and did not respond appropriately to her questions. He has no obvious motor or somatic sensory deficits. On examination, the physician concludes that the man had a stroke in a region of one of his cerebral hemispheres. As part of the diagnosis, the physician tests the man's visual fields and notices a decreased awareness of stimuli presented to one visual field.

QUESTIONS

1. Which side of the brain most likely suffered the stroke?
2. Which regions of the hemisphere suffered the stroke?
3. What information from the case history gives the answers to Questions 1 and 2?
4. Which visual field is affected by the stroke?

ANSWERS

1. The stroke occurred on the left side of the brain.

2. The stroke involved the superior posterior temporal lobe encompassing the Wernicke area and the occipital lobe encompassing the primary visual cortex.

3. Language deficits indicate involvement of the left hemisphere. The fluent but nonsensical speech indicates involvement of the Wernicke area. The visual field deficit indicates a loss in the visual cortex. The lack of motor or somatic sensory defi-cits excludes the posterior frontal and anterior parietal lobes.

4. The right visual field would be affected, because visual fields are represented in the contralateral hemispheres.

thePoint® *Visit* http://thepoint.lww.com/rhoades5e *for additional chapter review Q&A, Clinical Application Exercises, animations, and more!*

Neuromuscular Physiology

Active Learning Objectives

Upon mastering the material in this chapter, you should be able to:

- Explain why impulses from motor neurons are required for contraction of skeletal muscle and how these impulses are linked to the generation of a muscle contraction.
- For each specific step in excitation–contraction coupling, explain how failure of that step prevents skeletal muscle contraction.
- Explain why skeletal muscle contraction is either "full on" or "full off."
- Explain how force generation in muscle is enhanced by rapid repeated stimulation of the muscle and describe the value of this mechanism in terms of specific mechanical task muscles perform.
- Explain how lack of oxygen to skeletal muscle locks it into a rigor state.

- Distinguish between isotonic and isometric conditions and contractions and give examples of each.
- Explain the molecular basis of why the production of isometric force depends on the length to which muscle is stretched prior to its activation.
- Show how the isometric length–tension curve sets the limit for all isotonic contractions in skeletal muscle.
- Contrast the important steps in the regulation of smooth versus skeletal muscle contraction by calcium.
- Explain why the contraction of a smooth muscle cell can be graded, whereas that of a skeletal muscle fiber cannot.
- Explain the mechanism by which smooth muscle reduces its energy requirements during long-duration, sustained contractions.

Muscles are organs composed of cells that possess the ability to contract and thus generate force and movement. Muscle is commonly classified into subtypes based on anatomical location and appearance; the muscle associated primarily with the bones of the skeleton is designated **skeletal muscle**; that comprising the heart is called **cardiac muscle** and that found in the alimentary tract, the genitourinary system, blood vessels, bronchi, and the eye is called **smooth muscle** because of its unremarkable or "smooth" appearance when viewed microscopically. These anatomical classifications, however, are of little use for distinguishing the physiology of these structural subtypes of muscle. The mechanics of muscle contraction and relaxation as well as the regulation of the proteins that produce contraction are markedly different among the three histologic muscle types. This chapter explains the cellular physiology and mechanics of skeletal and smooth muscle. The physiology and mechanics of cardiac muscle are discussed in Chapters 12 and 13.

► SKELETAL MUSCLE

Skeletal muscle is primarily associated with the bones of the skeleton. However, this type of muscle also makes up the muscles of the diaphragm, parts of the esophagus, and the muscles that move the eye. Muscles of the skeleton are responsible for the large and forceful movements involved in walking, running, and lifting heavy objects. Large skeletal muscle groups, such as those in the lower back, buttocks, and thigh, are also responsible for stabilizing the body's position against gravity, which prevents teetering. Smaller skeletal muscle groups are used for the fine, delicate movements of the hands, such as those used in writing, playing the violin,

and manipulating tiny objects. Finally, a secondary function of skeletal muscle is the production of heat. Heat given off in shivering or during general exercise is an important source of heat generation in the body (see Chapters 28 and 29).

Protein filaments provide the architecture and contractile machinery of skeletal muscle.

All skeletal muscles have a striated (striped) appearance when viewed with a light or an electron microscope. This appearance comes from the highly organized arrangement of protein filaments within each muscle cell. These filaments make up the structure of the contractile mechanism and participate in the chemical processes that are responsible for converting chemical energy into mechanical energy and thus mechanical function of muscle.

A whole skeletal muscle is composed of numerous muscle cells, also called **muscle fibers**, which can measure to 100 μm in diameter and many centimeters long. Most skeletal muscle generates the ATP needed to supply energy for contraction through oxidative metabolism and therefore have an abundant supply of **mitochondria**.

Thick and thin filaments

The process of muscle contraction requires an elaborate subcellular structure of contractile proteins that couple cyclic biochemical interactions into physical movements. These contractile proteins are polymers that are described generically as thick filaments and thin filaments. The fundamental unit of a thick filament is **myosin** (molecular weight, ~500,000), a complex protein molecule with several distinct regions (Fig. 8.1).

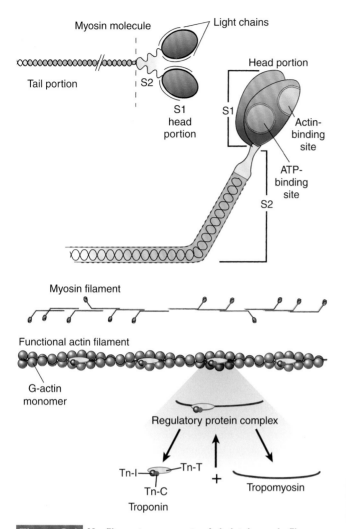

Figure 8.1 **Myofilament components of skeletal muscle fibers.** The myosin molecular structure is assembled into thick filaments. The S2 region of the molecule is flexible, whereas the tail region is stiff. Actin-binding sites are located on the globular myosin heads, which also contain light chains with ATPase activity. Thin (actin) filaments of skeletal muscle are formed from actin monomers (G-actin) into helical F-actin strands, which associate with a troponin–tropomyosin complex to form the functional actin filament. ATP, adenosine triphosphate; Tn-C, troponin-C; Tn-T, troponin-T; Tn-I, troponin-I; G-actin, globular actin; F-actin, filamentous actin.

The molecule consists of a long, straight portion (the "tail" portion) composed of light meromyosin. The remainder of the molecule, heavy meromyosin (HMM), consists of a protein chain, the *S2 region* (or *subfragment 2*), that serves as a flexible link between the tail and a globular head portion, called the *S1 region* (or *subfragment 1*). This region contains a binding site for another protein filament with the muscle cell called **actin** (which comprises the bulk of the thin filament; see Fig. 8.1) and an ATP-binding site that is involved in the supply of energy for the actual process of contraction. The S1 region is responsible for the enzymatic and chemical activity that results in actin and myosin interaction that generates muscle contraction.

Two low molecular weight peptide chains, called "light chains," are loosely attached to the S1 region; one called the essential light chain stabilizes myosin, whereas the other, called the **regulatory light chain**, is phosphorylated during muscle activity and serves to modulate muscle function. Functional myosin molecules are paired; their tail and S2 regions are wound about each other along their lengths and the two heads lie adjacent to each other. The tail regions, formed from close packing of the individual myosin dimers into thick filaments, are aligned so that they are tail to tail in the center of the thick filament and extend outward from the center in both directions, creating a bare zone (i.e., no heads protruding) in the middle of the filament (see Fig. 8.1).

Each actin filament (thin filament) is composed of two helically entwined strands, or chains, of repeating subunits of the globular protein, G-actin (molecular weight, 41,700). This chain is called *F-actin* (or *filamentous actin*; see Fig. 8.1). In the groove formed down the length of the helix, there is an end-to-end series of fibrous protein molecules (molecular weight, 50,000) called **tropomyosin**. Near one end of each tropomyosin molecule is a protein complex called **troponin**, composed of troponin-C (Tn-C), troponin-T (Tn-T), and troponin-I (Tn-I). The tropomyosin: troponin complexes are involved in regulating the interaction of actin with myosin heads that are responsible for muscle contraction and relaxation as described later in this chapter.

Interdigitating regions of thick and thin filaments in structural units called sarcomeres form the contractile unit of a muscle fiber.

A single muscle fiber (cell) is divided lengthwise into several hundred to several thousand parallel tubular structures called **myofibrils**. Each myofibril has alternating light and dark bands that repeat at regular intervals down the length of their long axis. These bands arise from the way actin and myosin interdigitate within the myofibril and give the fiber as a whole a striated (striped) appearance (Fig. 8.2). For ultrastructural identification, the bands are named as follows: **A band**—bounded by dark bands of the myofibril due to overlap of actin and myosin; **H zone**—a narrow, lighter-colored region in the center of the A band; **M line**—a prominent feature found at the center of the H zone in many skeletal muscles; **I band**—a low-density region between the A bands; and the **Z line**—a dark structure (also called a Z disk because the myofibril is a three-dimensional structure) that crosses the I band. The filaments of the I band attach to the Z line and extend in both directions into the adjacent A bands. The I band contains only thin (actin) filaments, whereas the A bands contain the thick (myosin) filaments and the overlapping projections of the thin filaments from the I band. The fundamental repeating unit of the myofibril bands is called a **sarcomere** and is defined as the space between (and including) two successive Z lines. Every myofibril contains numerous sarcomeres arranged in series along its long axis.

Neuromuscular Physiology

Neuromuscular Toxins in Nature: From Physiological Exploration to Clinical Pharmacology

The ability of skeletal muscle to provide force and movement to an organism has obvious survival value in the animal kingdom. Many animals and even plants have evolved the ability to manufacture toxins that can inactivate skeletal muscle as a means of warding off predation by another species or immobilizing a species as prey. Discovery of natural compounds that alter key physiological processes in the human body often leads to their use as tools to reveal the cellular mechanisms of those processes. It is not uncommon to then take this one step further and combine an understanding of the toxin's structure and its physiological actions to create related compounds that can be used clinically to manipulate physiological phenomena for a medicinal benefit to a human patient.

The legendary arrow poisons, collectively called *curare*, used by indigenous peoples of South and Central America to paralyze and kill animals for food, are an example of the transition in medicine of natural discovery, to revealed physiology, to clinical pharmacotherapeutics. First brought to Europe near the end of the 16th century, curare is a collection of alkaloids collected and prepared from plants of the *Strychnos* genus native to South and Central America. In the middle of the 19th century, experiments on the actions of curare preparations on animals revealed that the agents induced flaccid paralysis of skeletal muscle, including the diaphragm needed for pulmonary ventilation, while not affecting other life processes like the pumping of the heart. Claude Bernard demonstrated that the mechanism of action involved interference of the action of the motor neurons to activate the muscle rather than affecting the muscle itself thus demonstrating the principle of neuromuscular transmission in the genesis of muscle contraction. Over the next century, many investigations revealed that the curare alkaloids acted on the motor endplate, not the nerve itself, by competitively binding to the acetylcholine-binding sites of the nicotinic receptor/depolarizing ion channel of the skeletal muscle cell membrane. When bound by the alkaloids, acetylcholine released from the muscle's motor nerve terminals cannot open the nicotinic channels. This then renders the muscle incapable of generating depolarizing endplate potentials large enough to trigger action potentials, which are responsible for initiating skeletal muscle contraction. This accounts for the observation of induction of flaccid paralysis in animals exposed to curare. From the 1930s on, this collective knowledge was used to develop semisynthetic and synthetic substitutes for curare that could be used medically as a means to induce muscle relaxation in patients. Today, a handful of such competitive agents are available for therapeutic applications and are collectively called neuromuscular blocking agents. These agents are widely used to induce temporary flaccid paralysis in patients as a means of facilitating extensive surgical procedures and reducing the level of general anesthesia that would otherwise be needed to induce muscle relaxation during surgery. In this manner, levels of anesthesia that might otherwise suppress cardiopulmonary function can be avoided. ■

► MOTOR NEURONS AND EXCITATION–CONTRACTION COUPLING IN SKELETAL MUSCLE

Chemically, actin and myosin have a natural affinity for one another, and when added together in solution, the myosin head will bind tightly to sites on the G-actin monomer subunits of F-actin. The binding link between these two filaments is called a **crossbridge**. This binding is prevented from occurring spontaneously in the myofilaments of skeletal muscle by the presence of tropomyosin on the thin filament, which sterically interferes with the ability of myosin to access its actin-binding site. This is the state of skeletal muscle at rest (i.e., when it is not contracting) and is what is responsible for the flaccid property of resting skeletal muscle.

As described above, actin and myosin are arranged in a highly organized, interdigitating, structure within the sarcomere. In this arrangement, it is easy to see that it is possible for actin filaments to slide longitudinally between the myosin filaments (see Fig. 8.2). Consequently, if actin and myosin were allowed to temporarily bind to one another and the actin then somehow pulled into the A band of the sarcomere, then released at its new position and allowed to reattach to another myosin-binding site, and subsequently allowed to repeat the entire cycle in a "hand-over-hand"–like manner, the entire sarcomere would generate force and shorten. Extrapolated to all the sarcomeres in all the muscle fibers within a muscle, the muscle itself would then generate force and shorten as well, that is, it would contract.

The postulate that actin and myosin can interact in a controlled manner as just described is called the sliding filament hypothesis and it has long since been established as the molecular mechanism of muscle contraction. The activation and cycling of actin and myosin crossbridge formation during a muscle contraction is called the **crossbridge cycle**.

If skeletal muscle as a tissue is to be able to generate force and movement in any structure in the body, the crossbridge cycle must be activated and deactivated in a controlled manner. As it turns out, skeletal muscle cells cannot be activated on their own. All such cells require activation by neurotransmitters released from motor neuron axons that terminate at specialized regions of the skeletal muscle cell membrane. This process of activation of the skeletal muscle crossbridge cycle is called **excitation–contraction coupling**.

Skeletal muscle is considered an "excitable" tissue, which means that its membrane potential is linked to activation of that cell's function (i.e., contraction). Depolarization of

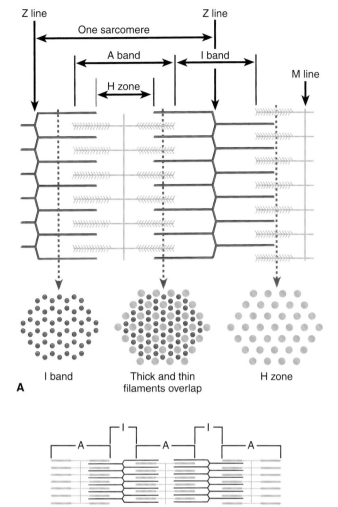

A

I band   Thick and thin filaments overlap   H zone

B   Increased overlap and sarcomere shortening

Figure 8.2 **Molecular organization of the sarcomere. (A)** The arrangement of the elements in a sarcomere viewed longitudinally and as cross sections through selected regions of the sarcomere. Note the overlap of myofilaments at different parts of the sarcomere. **(B)** Depiction of optimum myofilament overlap.

the skeletal muscle cell membrane of sufficient magnitude can lead to the generation of muscle cell action potentials. These action potentials trigger an increase in intracellular Ca^{2+} concentration within the muscle fiber, which activates the crossbridge cycle (see below for details). *Skeletal muscle cells, however, cannot generate these action potentials by themselves.* Instead, they must receive a depolarizing signal from motor neurons that innervate the cell. Consequently, all skeletal muscles are externally controlled; they cannot contract without a signal from the somatic nervous system. This system is analogous to the light in a table lamp. The light cannot be turned on unless an electric cord connects the lamp to a source of electrical current. Similarly, skeletal muscle cells cannot contract if the motor neurons innervating them are severed or transmission of the chemical signal to the underlying fiber is blocked. In such situations, the muscle group innervated by the motor neuron is flaccidly paralyzed.

Electrochemical events at the neuromuscular junction link synaptic transmission to muscle contraction.

The sequence of events that leads to contraction of skeletal muscle starts with action potentials that travel down somatic motor nerves originating in the central nervous system (CNS). Upon reaching a muscle cell, the axon of a motor neuron typically branches into several terminals, each of which constitutes the *presynaptic* portion of a special synapse called the **neuromuscular junction** (NMJ), also called the **myoneural junction** or **motor endplate** (Fig. 8.3). The terminals, which are all covered by a Schwann cell, lie in grooves or "gullies" in the surface of the muscle cell membrane. The signal from motor neuron to muscle cell is transmitted in the form of a chemical transmitter (neurotransmitter) released from nerve to muscle across the NMJ. This type of synapse has a close association between the membranes of the nerve and muscle. It possesses many of the features of a nerve–nerve synapse in the CNS. It is somewhat simpler, however, because it is only excitatory and accepts no feedback from the postsynaptic cell.

Within the axoplasm of the motor nerve terminals are located numerous membrane-enclosed vesicles containing **acetylcholine** (**ACh**). The postsynaptic portion of the NMJ or **endplate membrane** is formed into postjunctional folds that contain **nicotinic ACh receptors** (see Chapter 3). When action potentials reach the terminal of a motor axon, the resulting depolarization of the terminal causes membrane channels to open, allowing external calcium ions to enter the axon. This causes the axoplasmic vesicles of ACh to migrate to the inner surface of the axon membrane, where they fuse with the membrane and release their contents. The vesicles are all approximately the same size and all release about the same amount, or **quantum**, of neurotransmitter. ACh released from the motor neuron terminal then diffuses across the NMJ and binds to the ACh receptors in the motor endplate. When two ACh molecules are bound to a receptor, it undergoes a conformational change, opening a channel in the receptor that allows the relatively free passage of sodium and potassium ions down their respective electrochemical gradients. The opening of the channels depends only on the presence of neurotransmitter and not on membrane voltage. Sodium and potassium permeability changes occur simultaneously, and both ions share the same membrane channels, but the change in electrochemical gradient for sodium into the cell is much greater than that for potassium out of the cell. The result is a net inward positive current that depolarizes the postsynaptic membrane. This positive voltage change is called the **endplate potential** (see Fig. 8.3).

Endplate depolarization is *graded*, that is, its amplitude will vary with the number of receptors bound with ACh. The number of receptors bound is proportional to the local concentration of ACh in the NMJ, which, in turn, is a function both of the ACh released by the motor neuron and the amount of ACh degraded by **acetylcholinesterase** (**AChE**) in the NMJ. ACh binds loosely to its receptor. It can detach and diffuse away and become hydrolyzed into choline and acetate by AChE thus terminating its function as a transmitter molecule. At low action potential frequencies of the motor

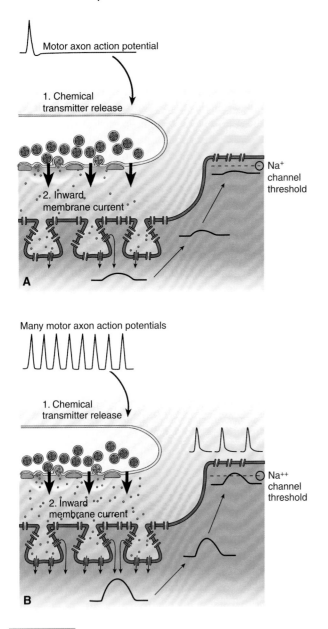

Figure 8.3 **Structural features and electrical activity of the neuromuscular junction.** Acetylcholine released from membrane vesicles is released from the axon terminal and diffuses to the underlying membrane where it binds with nicotinic receptors concentrated in invaginations of the sarcolemma. **(A)** An action potential from a motor neuron releases a small amount of acetylcholine into the myoneural junction. This causes a small depolarization in the motor endplate that is transmitted electrotonically with decrement through the myoplasm. In this instance, the amount of depolarization reaching the first available voltage-gated sodium channel in the cell membrane is below threshold for opening the channel, and no muscle action potential is formed. **(B)** With greater frequency of action potentials reaching the motor axon terminal, acetylcholine reaches a higher concentration in the myoneural junction. This creates a larger endplate potential than that seen with lesser motor neuron activation. Although this potential proceeds down the muscle fiber with decrement, the depolarization reaching the first sodium channels is greater than its threshold, and the muscle fiber is able to fire an action potential. Action potentials are self-reinforcing and travel down the muscle fiber without decrement in a manner similar to that seen in unmyelinated nerve fibers.

neuron, the amount of ACh entering the synaptic cleft will be low (see Fig. 8.3A) and reduced even further by the action of AChE. Consequently, the concentration of ACh receptors occupied will be very low and result in a small amplitude endplate depolarization. Endplate potentials are simple local depolarizations that are transmitted outward (leak) from underneath the motor endplate. They diffuse from their initial location with decrement, that is, their amplitude decays as they travel through the myoplasm.

The motor endplate does not contain voltage-gated sodium channels and cannot produce self-propagating action potentials. However, regions of the sarcolemma away from the motor endplates do contain voltage-gated sodium channels. If these channels can be brought to threshold, the muscle fiber can form self-propagating action potentials in a manner analogous to that in unmyelinated neurons. However, because endplate potentials move down the muscle fiber with decrement, an initial small endplate potential may be of insufficient amplitude by the time it reaches a sodium channel to bring that channel to its threshold for firing an action potential (see Fig. 8.3A). In contrast, large amounts of ACh released into the NMJ will cause a proportionally larger endplate depolarization (see Fig. 8.3B). Although this endplate potential will also spread out across the muscle membrane with decrement, enough depolarization will reach the region of voltage-gated sodium channels in the muscle membrane to cause them to open and fire an action potential. Once formed, the muscle action potential will then be self-propagating along the muscle cell membrane as with a nonmyelinated nerve fiber.

Neuromuscular transmission can be altered by toxins, drugs, and trauma.

Under normal circumstances, the endplate potential is much more than sufficient to produce a muscle action potential; this reserve, or safety factor, can help preserve function under abnormal conditions. However, neuromuscular transmission is subject to interference at several steps. Presynaptic blockade can occur if calcium does not enter the presynaptic terminal. Inhibition of choline uptake by the presynaptic terminal, such as that caused by the drug hemicholinium, results in the depletion of ACh in the neuron. **Botulinum toxin** interferes with SNARE proteins responsible for vesicular ACh release and results in flaccid paralysis of muscle exposed to the toxin, including the diaphragm, resulting in death.

Postsynaptic blockade can result from drugs that can bind to nicotinic receptors in the muscle membrane. Derivatives of **curare**, originally used as arrow poison in South America, bind tightly to ACh receptors without opening it thus reducing the endplate potential, which can result in flaccid muscle paralysis. Several classes of agents, such as **physostigmine (eserine)** and organophosphates (malathion and sarin gas), are potent inhibitors of AChE. At sufficient strength, they can result in excessive concentration of ACh at the NMJ, which forces the muscle cell to remain depolarized and unable to return from the inactive state to form additional action potentials. This is called depolarization blockade and can result in death.

CLINICAL FOCUS | 8.2

Treatment of Focal Dystonias with Botulinum Toxin

Focal dystonias are neuromuscular disorders characterized by involuntary and repetitive or sustained skeletal muscle contractions that cause twisting, turning, or squeezing movements in a body part. Abnormal postures and considerable pain, as well as physical impairment, often result. Usually, the abnormal contraction is limited to a small and specific region of muscles, hence the term focal (by itself). *Dystonia* means "faulty contraction." *Spasmodic torticollis* and *cervical dystonia* (involving neck and shoulder muscles), *blepharospasm* (eyelid muscles), *strabismus* and **nystagmus** (extraocular muscles), *spasmodic dysphonia* (vocal muscles), *hemifacial spasm* (facial muscles), *writer's cramp* (finger muscles in the forearm), achalasia (failure of the lower esophageal sphincter to open), and other esophageal dysmotilities are common dystonias. Such problems are neurologic, not psychiatric, in origin, and sufferers can have severe impairment of daily social and occupational activities.

Although the specific cause is located somewhere in the central nervous system, the exact nature of dystonia is unknown. A genetic predisposition to the disorder may exist in some cases. Centrally acting drugs are of limited effectiveness, and surgical denervation, which carries a significant risk of permanent and irreversible paralysis, may provide only temporary relief.

Botulinum toxin, which produces chemical denervation of motor neurons resulting in flaccid muscle paralysis, has shown potential benefit in the treatment of dystonias. Botulinum toxin is produced from the bacterium *Clostridium botulinum* and is one of the most potent natural toxins known; a lethal dose for a human adult is about 2 to 3 µg. Type A toxin, the complex form most often used therapeutically, is sold under the trade names Botox and Oculinum.

The toxin first binds to the cell membrane of presynaptic nerve terminals in skeletal muscles. The initial binding does not appear to produce flaccid paralysis until the toxin is actively transported into the cell, a process requiring more than an hour. Once inside the cell, the toxin disrupts calcium-mediated acetylcholine release by binding to docking proteins that link synaptic vesicles containing neurotransmitter to the nerve terminal membrane. This prevents action potentials from being able to release neurotransmitter into the neuromuscular junction. The transmission block at the neuromuscular junction of the affected neuron is irreversible. The nerve terminals begin to degenerate, and the denervated muscle fibers atrophy. Eventually, new nerve terminals sprout from the axons of affected nerves and make new synaptic contact with the chemically denervated muscle fibers. During the period of denervation, which may be several months, the patient usually experiences considerable relief of symptoms. The relief is temporary, however, and the treatment must be repeated when reinnervation has occurred.

Clinically, highly diluted toxin is injected into the individual muscles involved in the dystonia. Recently, Botox has been used to treat retractable esophageal dysmotility disorders that are refractory to other forms of treatment. In dystonias involving muscles in the extremities, injections of toxin are done in conjunction with electrical measurements of muscle activity (electromyography) to pinpoint the muscles involved. Patients typically begin to experience relief in a few days to a week. Depending on the specific disorder, relief may be dramatic and may last for several months or more. The abnormal contractions and associated pain are greatly reduced, speech can become clear again, eyes reopen and cease uncontrolled movements, and, often, normal activities can be resumed.

The principal adverse effect is a temporary weakness of the injected muscles. However, this property has been exploited for the cosmetic use of Botox for releasing muscle tension in the face and forehead, which then reduces the formation of wrinkles in the skin. Studies have shown that the toxin's activity is confined to the injected muscles, with no toxic effects noted elsewhere. Long-term effects of the treatment, if any, are unknown. ∎

However, in carefully controlled doses, AChE inhibitors can temporarily alleviate symptoms of **myasthenia gravis**, an autoimmune condition that results in a loss of postsynaptic ACh receptors with resultant small endplate potentials and muscular weakness. Partial inhibition of the enzymatic degradation of ACh allows it to remain in higher concentration in the NMJ and, thus, to compensate for the loss of receptor molecules.

Muscle action potentials release calcium from the sarcoplasmic reticulum to activate the crossbridge cycle.

The outer surface of a skeletal muscle fiber is surrounded by an electrically excitable cell membrane supported by an external meshwork of fine fibrous material. Together, these layers form the cell's surface coat, the **sarcolemma**. The sarcolemma generates and conducts action potentials much like those of nerve cells. Contained wholly within a skeletal muscle cell is another set of membranes called the **sarcoplasmic reticulum** (**SR**), which is a specialization of the endoplasmic reticulum (Fig. 8.4). The SR is specially adapted for the uptake, storage, and release of calcium ions. Calcium ATPase pumps in the SR allow it to concentrate calcium within the SR to concentrations more than a million times that in the cytoplasm of the myofiber at rest. This organelle plays a key role in controlling muscle contraction because the free calcium ion concentration within the muscle fiber is critical in controlling the processes of contraction and relaxation (see details below). Within each sarcomere, the SR consists of two distinct portions. The longitudinal element forms a system of hollow sheets and tubes that are closely associated with the myofibrils. The ends of the longitudinal

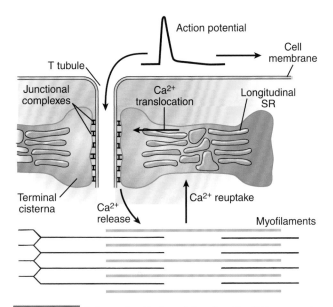

Figure 8.4 **Release of activator calcium from the SR by a muscle action potential.** Excitation–contraction coupling involves opening of ryanodine receptors by action potentials in the T-tube membrane with subsequence release of calcium stored in the SR into the sarcoplasm. This activates actin–myosin crossbridge cycling. The SR is replenished with calcium from the sarcoplasm by Ca^{2+} ATPase pumps in the SR membrane. (See text for details of the process.) T tubule, transverse tubule; SR, sarcoplasmic reticulum.

elements terminate in a system of **terminal cisternae** (or *lateral sacs*). These contain a protein, calsequestrin, that weakly binds calcium, and most of the stored calcium is located in this region.

Closely associated with both the terminal cisternae and the sarcolemma are the **transverse tubules (T tubules)**, which are inward extensions of the cell membrane whose interior is continuous with the extracellular space. Although they traverse the muscle fiber, T tubules do not open into its interior. The association of a T tubule and the two terminal cisternae at its sides is called a **triad**, a structure important in excitation–contraction coupling.

The large diameter of skeletal muscle cells places interior myofilaments out of range of the immediate influence of events at the cell surface, but the T tubules, SR, and their associated structures act as a specialized internal communication system that allows the surface action potential signal to penetrate to interior parts of the cell. This process begins in skeletal muscle with the electrical excitation of the surface membrane (see Fig. 8.4). An action potential sweeps rapidly down the length of the fiber and propagates down the T-tubule membrane. This propagation results in numerous action potentials traveling toward the center of the fiber, one in each T tubule.

At some point along the T tubule, the action potential reaches the region of a triad where the presence of the action potential is communicated to the terminal cisternae of the SR. Here, the T-tubule action potential affects specific protein molecules called **dihydropyridine receptors (DHPRs)**. These molecules, which are embedded in the T-tubule membrane in clusters of four, serve as voltage sensors that respond

to the T-tubule action potential. They are located in the region of the triad where the T-tubule and SR membranes are the closest together. Each cluster is located in close proximity to a specific channel protein called a **ryanodine receptor (RyR)**, which is embedded in the SR membrane. The RyR serves as a controllable channel through which calcium ions can move readily when it is in the open state. These are therefore sometimes termed **calcium release channels**. DHPR and RyR form a functional unit called a **junctional complex** (see Fig. 8.4).

When the muscle is at rest, the RyR is closed; when T-tubule depolarization reaches the DHPR, it causes the RyR to open and release calcium from the SR. This leads to rapid release of calcium ions from the terminal cisternae into the intracellular space surrounding the myofilaments.

State of total relaxation to maximum contraction in skeletal muscle is tightly linked to its intracellular calcium concentration.

In resting skeletal muscle, the free calcium ion concentration in the region of the myofilaments is low ($<10^{-7}$ M), and crossbridge formation is prevented by the troponin–tropomyosin complex of the thin myofilaments (Fig. 8.5). In this state, the muscle is thus relaxed. The trigger for activating the skeletal muscle crossbridge cycle is the rapid release of calcium from the SR, down its huge electrochemical gradient, upon stimulation by action potentials entering the triad. When RyR open in the SR in response to action potentials, intracellular calcium ion concentrations increase above resting levels and these ions begin to bind to the Tn-C subunit associated with each tropomyosin molecule. Through the action of Tn-I and Tn-T, calcium binding causes the tropomyosin molecule to

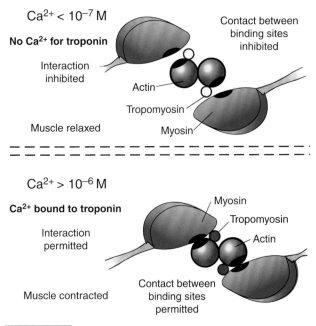

Figure 8.5 **The calcium switch for controlling skeletal muscle contraction.** Calcium ions, via the troponin–tropomyosin complex, control the blocking of the interaction between the myosin heads (the crossbridges) and the active site on the thin filaments. The thin filaments are seen in cross section.

change its position slightly, uncovering the myosin-binding sites on the actin filaments (see Fig. 8.5). The myosin is thus allowed to interact with actin, and the events of the crossbridge cycle begin. As long as calcium ions are bound to the Tn-C subunit, the crossbridge cycle can occur. The switching action of the calcium–troponin–tropomyosin complex in skeletal muscle is extended by the structure of the thin filaments, which allows one troponin molecule, via its tropomyosin connection, to control seven actin monomers. Because the calcium control in striated muscle is exercised through the thin filaments, it is termed **actin-linked regulation**.

The relationship between the relative force developed by actin–myosin binding and the calcium concentration in the region of the myofilaments is steep. At a calcium concentration of $<1 \times 10^{-7}$ M, the interaction between actin and myosin is negligible, whereas an increase in the calcium concentration to 1×10^{-5} M produces essentially full-force development. This process is saturable, so that further increases in calcium concentration lead to little increase in force. In skeletal muscle, an excess of calcium ions is usually present during activation, and the contractile system is nearly saturated. There is only one burst of calcium ion release for each action potential. If action potential generation in the muscle ceases, intracellular calcium concentration in the myofibrils rapidly decreases and the fiber returns to the relaxed state. This is because Ca^{2+} ATPases or **"calcium pumps"** in the SR membranes continuously take up free calcium ions from the myofilament space and actively pump them into the interior of the SR. In the absence of a stimulus for calcium release from the SR, the continuous activity of the calcium pumps reduces calcium in the region of the myofilaments to a low level ($<1 \times 10^{-7}$ M). Once returned into the SR, the resequestered calcium ions are moved along the longitudinal elements to storage sites in the terminal cisternae, and the system is ready to be activated again. This entire process takes place in a few tenths of a millisecond and may be repeated many times each second.

Cyclic interaction of actin and myosin is the molecular engine driving muscle contraction.

The process of contraction of a muscle cell involves a controlled cyclic interaction between the actin and myosin in a sequence of events called the **crossbridge cycle**. This cycle is shown in Figure 8.6. Briefly, this cycle involves attachment of thick-filament myosin heads to binding sites on actin monomers to form a crossbridge, incremental movement of the filaments to produce sarcomere shortening, detachment of the filaments, and then reattachment of myosin heads along the thin filaments and the repeat of the incremental movement, etc. This cycling of crossbridges results both in force generation and movement through the whole muscle fiber.

At rest, the myosin heads in the thick filaments are bound with **adenosine diphosphate** (**ADP**) and inorganic phosphate (P_i) and situated 90° to the axis of the thin filaments. Myosin does not bind to actin in this state because of the inhibitory action of tropomyosin on crossbridge attachments (see Fig. 8.6, step 1). When the inhibition

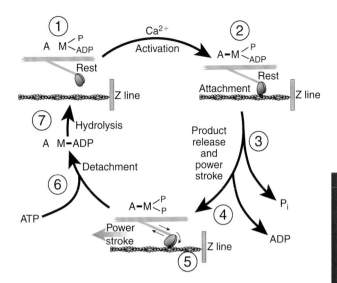

Figure 8.6 **Events of the crossbridge cycle in skeletal muscle.** (*1*) At rest, adenosine diphosphate (ADP) and inorganic phosphate are bound to the myosin head, which is in position to interact with actin. The interaction, however, is blocked allosterically by tropomyosin. (*2*) Inhibition of actin–myosin interaction is removed by binding of calcium to troponin-C; the myosin head binds to actin. The release of ADP and phosphate (*3*) changes the conformation of the myosin head from 90° to 45°, stretching the myosin S2 region (*4*). Recoil of the S2 region creates the power stroke (*5*). The still-attached crossbridge is now in the rigor state. Detachment (*6*) is possible when a new adenosine triphosphate (ATP) molecule binds to the myosin head and is subsequently hydrolyzed. Energy from ATP hydrolysis resets the myosin head from a 45° conformation back to its original 90° conformation (*7*), thereby returning the myosin and actin positions to their original resting state. These cyclic reactions can continue as long as the ATP supply remains and activation via Ca^{2+} maintained. (See text for further details.) A, actin; M, myosin; P_i, inorganic phosphate ion; –, chemical bond.

of tropomyosin on crossbridge attachments is removed by calcium released from the SR, actin and myosin bind strongly, and the crossbridges become firmly attached (see Fig. 8.6, step 2). When myosin is attached to actin, it loses its affinity for ADP and P_i, which then fall free into the cytoplasm (see Fig. 8.6, step 3). The release of ADP and P_i from myosin causes it to undergo a conformational change while the tip of the myosin head is still attached to actin so that the portion of the head attached to the flexible S2 region pitches outward by 45°, thereby stretching this flexible chain region in a manner similar to stretching a spring (see Fig. 8.6, step 4). This "spring," however, snaps back as soon as it is stretched, and because myosin is still attached to actin, the snapping back pulls the actin filaments past the myosin filaments in a movement called the **power stroke** (see Fig. 8.6, step 5).

Following this movement (which results in a relative filament displacement of around 10 nm) the actin–myosin binding is still strong and the crossbridge cannot detach. At this point in the cycle, the state of actin–myosin binding is termed a **rigor crossbridge** (see Fig. 8.6, step 5). If ATP is present, it binds to the rigor crossbridge at the myosin head, resulting in detachment of myosin from actin (see Fig. 8.6, step 6). The ATPase property of myosin then partially hydrolyzes ATP to ADP and P_i, which remain bound to the myosin head. The energy released from this hydrolysis is used to reset the conformation of the myosin head to its original 90° orientation (see Fig. 8.6, step 7) and the newly recharged myosin head can begin the cycle of attachment, power stroke, and detachment again (see Fig. 8.6, step 1). This cycle can repeat as long as the muscle is activated, a sufficient supply of ATP is available, and the physical limit to shortening has not been reached. If cellular energy stores are depleted, as happens after death, the crossbridges cannot detach because of the lack of ATP, and the cycle stops in an attached state (see Fig. 8.6, step 5). This produces an overall stiffness of the muscle, which is observed as the **rigor mortis** that sets in shortly after death.

Cellular structure of skeletal muscle transforms crossbridge cycling into mechanical motion.

The amounts of movement (~10 nM) and force (~4 pN) produced by a single crossbridge interaction are exceedingly small, and it is the organization of the myofilaments into sarcomeres that both allows useful work to be done and creates the chemically and mechanically cyclic nature of the crossbridge reactions. In a shortening muscle, the power stroke moves the two sets of filaments past one another, and the crossbridge that is formed following one "turn" of the cycle involves myosin attaching to a new actin monomer further down the thin filament.

The width of the A bands (thick-filament areas) in striated muscle remains constant, regardless of the length of the entire muscle fiber, whereas the width of the I bands (thin-filament–only areas) varies directly with the length of the fiber. The spacing between Z lines also depends directly on the length of the fiber. The lengths of the thin and thick myofilaments themselves remain constant despite changes in fiber length. The sliding filament theory proposes that changes in overall fiber length are directly associated with changes in the overlap between the thick and thin filaments. During contraction, this is accomplished by the interaction of the globular heads of the myosin molecules with binding sites on the actin filaments.

A muscle contains many thousands of sarcomeres placed end to end (in series). This arrangement has the effect of adding up all the small sarcomere length changes into a large overall shortening of the muscle. Similarly, the amount of force exerted by a single sarcomere is small (a few hundred micronewtons), but, again, there are thousands of sarcomeres side by side (in parallel), resulting in the production of considerable force.

▶ MECHANICS OF SKELETAL MUSCLE CONTRACTION

Muscle is a biologic motor; it produces physical work by transforming chemical energy into the generation of mechanical force and movement. In spite of the remarkable variety of controlled muscular movements that humans can make, the fundamental mechanical events of the contraction process can be described by a relatively small set of specially defined functions that result from particular muscle capabilities.

Contraction of large muscle groups results from temporal and spatial summation of single contractions in muscle units.

A single muscle action potential leads to a single brief contraction of the muscle called a **twitch** (Fig. 8.7). Though the contractile machinery (i.e., the crossbridge cycle mechanism) may be fully activated (or nearly so) during a twitch, some of this active contraction must be used first to stretch a series of elastic elements in the muscle fibers before full contractile force can be transmitted to the muscle fiber. Thus, the amount of force produced by a single twitch is relatively low. The duration of the action potential in a skeletal muscle fiber, however, is shorter (~5 ms) than the duration of a twitch (tens or hundreds of milliseconds). For this reason, the muscle fiber could be activated again long before the muscle has relaxed. Figure 8.7 also shows the result of stimulating a muscle that has not yet fully relaxed from a prior stimulus at a time well outside its electrical refractory period from that prior stimulus. In this situation, the force of the second stimulus is added to the remaining force from the first stimulus, and significant additional force is developed. These are called partially fused twitches. It is possible to fuse twitches so closely together that the force of the second stimulus is added to the first even before the first force has begun to decline. This results in even greater force enhancement in the muscle beyond that of a single twitch. Thus, in skeletal muscle, the interplay of the electrical refractory period and the duration of mechanical contraction allow individual

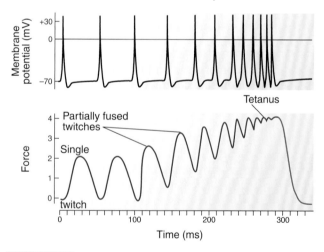

Figure 8.7 Twitch contraction, temporal summation, and fused tetanus. The first two contractions on the left are in response to a single action potential and are examples of a muscle twitch. When the interval between successive activations allows twitches to partially relax in between contractions, peak tension increases but oscillates. This is called *partial tetanus*. As the interval between successive stimuli steadily decreases, twitches fuse on top of one another, and a sustained, smooth higher force is created. This is called *tetanus*.

twitch contractions to be summed *temporally* to enhance the contractile force generated by a muscle fiber.

As an extension of this phenomenon, if many stimuli are given repeatedly and rapidly to the muscle fiber, the result is an amplified, *sustained* contraction called tetanus (see Fig. 8.7). When the contractions occur so close together that no fluctuations in force are observed, a **fused tetanus** is said to result. The repetition rate at which this occurs is the **tetanic fusion frequency**, typically 20 to 60 stimuli per second (which is considered beyond normal physiological stimuli), with the higher rates found in muscles that contract and relax rapidly. The amount of force produced in tetanus is typically several times that of a twitch, and the disparity between the two is expressed as the **tetanus/twitch ratio**. Figure 8.7 shows these effects in a special situation, in which the interval between successive stimuli is steadily reduced. Because it involves events that occur close together in time, a tetanus is a form of temporal summation.

Intracellular Ca^{2+} levels are different in muscle cells during fused tetanus than during partially fused twitches (partial tetanus). Action potentials do not vary significantly in muscle cells, and each action potential releases essentially the same amount of Ca^{2+} from the SR. This amount of calcium is sufficient to cause a twitch. During partial tetanus, the calcium increase with each action potential is very transient because SR Ca^{2+} ATPases rapidly pump the calcium back into the SR. Calcium levels peak and return to control levels repeatedly in this situation, and any increase, or summation, of twitch force is due to incomplete mechanical relaxation of a twitch before the muscle fiber is activated again. During complete, or fused, tetanus, the muscle fiber is activated at a frequency that adds calcium into the intracellular space faster than it can be pumped into the SR. In this case, total tension and intracellular calcium concentration are maximized, with the high calcium levels continually activating muscle contraction as long as the muscle is stimulated at a high enough rate. Although tetanic contractions can greatly increase force generation in skeletal muscle, this type of stimulation is poorly tolerated by the tissue and can damage the muscle cells. Therefore, full fused tetanus is rarely seen in skeletal muscle.

Muscles are organized into functional neuromuscular units that allow partial activation of a whole muscle.

Skeletal muscles are made up of many muscle fibers. These fibers require innervation from a motor neuron in order to be functional, but each fiber is not innervated by its own single neuron. Instead, a typical motor axon branches as it courses through the muscle, with each of its terminal branches innervating a single muscle fiber. In this manner, one neuron can innervate many muscle fibers. In some cases, a neuron may innervate just a few fibers, whereas in other cases, a single neuron, through its branching, can innervate tens of thousands of muscle fibers. A group of muscle fibers innervated by branches of the same motor neuron is called a **motor unit** (Fig. 8.8). Any large muscle group (e.g., the biceps, quadriceps, and gastrocnemius) contains many motor units. This arrangement allows for activation of only

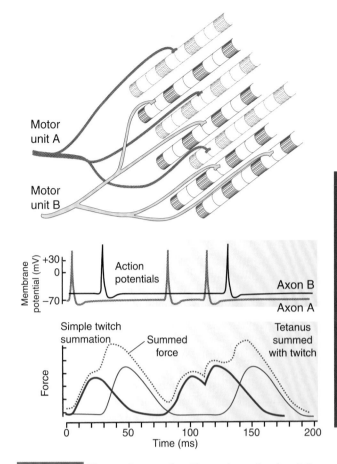

Figure 8.8 Motor unit summation. Two motor units, *A* and *B*, are shown above with their motor nerve action potentials and muscle twitches below. The summed force represents the total force generated in the whole muscle due to activation of both motor units. In the first example, fibers from motor units *A* and *B* are activated in close succession. Total force is the simple spatial summation of twitches from the two motor units; in the second, a brief tetanus in motor unit *A* (temporal summation) is spatially added to a twitch from motor unit *B* to create an even larger generation of force by the whole muscle.

a portion of the muscle at any one time and represents a *spatial* control of muscle contraction. The pattern of activation is determined by the CNS and distribution of the motor axons among the muscle fibers.

A motor unit functions as a single mechanical entity; all the fibers in the unit will contract together when an action potential travels from the CNS through the motor neuron to the muscle fibers contained within the unit. Independent contraction of only some of the fibers in a motor unit is impossible. Therefore, the motor unit is normally the smallest functional unit of a muscle. In muscles adapted for fine and precise control, only a few muscle fibers are associated with a given motor axon. In muscles in which high force is more important, a single motor axon controls many thousands of muscle fibers.

The number of motor units active at any one time determines the total force produced by a muscle; as more motor units are brought into play, the force increases. This is called **motor unit summation** or **spatial summation** and is illustrated in the lower tracing of Figure 8.8. This type of modulation of muscle function endows a large muscle group with the ability to deal with light loads with an economy of function. When light loads need to be moved or supported with the muscles of an arm, for example, the CNS activates only enough motor units to accomplish the task of moving such a load rather than modulating the contractile force of all the muscle fibers in all the muscle groups in the arm. This becomes particularly valuable for conservation of energy when a limb needs to support a weight that is far less than the maximum weight that could be supported by all the muscle groups in the limb. If the CNS were to simply stimulate all the muscle fibers in the arm (and, thus, all the motor units) to just hold pieces of paper, it would be expending far more energy than necessary to accomplish the task. Furthermore, there is no value in the CNS simply reducing the frequency of motor neuron action potential generation to all the arm muscles in order to reduce the total force generated by those muscles. Such a temporal reduction will only reduce the peak force generated by the muscle while producing oscillations in muscle contraction and relaxation. Instead of using either of these modes of muscle activation, the CNS causes fused contractions in reduced numbers of muscle fibers to produce a submaximal sustained contraction. It should be noted that holding objects against gravity is not quite this simple. A balance of activation of opposing muscle groups in a limb is primarily used to hold limbs steady against any opposing force, as sustained tetany is not tolerated well by muscle. During a sustained contraction involved with more complex movements, the CNS continually changes the pattern of activity, and the burden of contraction is shared among the motor units. This results in smooth contractions, with the force precisely controlled to either stabilize parts of the body or produce desired movements.

The mechanical environment of skeletal muscle modifies its contractile activity.

Mechanical factors external to the muscle influence the force, speed, and extent of shortening of contraction. These factors include the position to which a muscle is stretched prior to contracting and the load the muscle tries to move once contraction has begun. These relationships have been revealed by experiments in which mechanical conditions can be controlled to aid in the analysis of muscle contraction. Generally, these experimental arrangements represent "artificial" conditions that are better controlled and less complex than those encountered in real daily activities. Nevertheless, these types of analyses demonstrate how certain mechanical variables alter the contractile performance of muscle. Such relationships are not only applicable to understanding skeletal muscle as a mechanical engine but also represent critical components that control the mechanical performance of the heart as well. An understanding of such mechanical relationships is critical to understanding the function of the heart in health and disease in the clinical setting (see Chapter 13).

Isometric muscle contraction occurs when muscle contracts against a load that is too heavy to move.

If a muscle is attached on its ends to a permanent fixture so that it cannot move when activated, the muscle will express its contractile activity by developing force without shortening. This simplest type of contraction is termed an **isometric contraction** (meaning "same length") and is demonstrated when we try to push or pull an immovable object or an object whose mass is beyond our ability to move. In such situations, we feel our muscles contract, tense, and "harden," although our muscles do not actually shorten and no object is moved.

A representation of an isometric muscle contraction is shown in Figure 8.9 in which a dissected muscle group was fixed at a given length to a recording apparatus such that it could not shorten when stimulated. In this example, the muscle is stimulated only once but with sufficient strength to activate all its motor units. This produces a single twitch in which isometric force develops relatively rapidly followed by a subsequent slower isometric relaxation. The durations of both contraction time and relaxation time are related to the rate at which calcium ions can be delivered to and removed from the region of the crossbridges, the actual sites of force development. During an isometric contraction, the muscle consumes energy to fuel the processes that generate and maintain force, though no actual physical work is done on the external environment because no movement takes place.

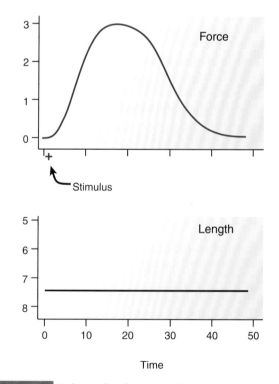

Figure 8.9 An isometric twitch contraction in skeletal muscle. In an experimental setting, the length of a muscle segment (preload) is fixed by attaching it to an immovable support and is thus held constant when the muscle is stimulated to contract. Upon delivery of a single maximum stimulus to the muscle, the muscle develops force over time and then relaxes. The muscle develops force but does not shorten. Therefore, the contraction is, by definition, isometric. (Force, length, and time units in the figure are arbitrary.)

Isotonic contractions in muscle result in movement of a load because the muscle generates a force greater than the load.

When conditions are arranged so that a muscle can generate a force larger than the load (weight) to which it is attached, the muscle will have the ability to shorten and thus move the mass. To accomplish load movement, the muscle first develops force enough to equal the weight of that mass to which it is attempting to move then begins to move the mass with whatever force-generating capability remains for that muscle contraction. This type of contraction is called **isotonic contraction** (meaning "same force"). In the simplest conditions, this constant force is the load a muscle moves. This load is also called an **afterload**, because its magnitude and presence are not apparent to the muscle until *after* it has begun to shorten.

An example of an isotonic contraction is shown in Figure 8.10, which shows the result of an experiment where a dissected muscle group is attached to a recording device on one end whereas a weight attached at the opposite end is less than the peak force capability of the muscle. When the muscle is stimulated, it will begin to develop force without shortening because it takes some time to build force and, initially, this force is less than that needed to lift the weight. This phase of the contraction is thus isometric (see Fig. 8.10, phase 1). After sufficient force has been generated, the muscle will begin to shorten and lift the load (see Fig. 8.10, phase 2). The contraction is now isotonic because the force exerted by the muscle is sufficient to match the weight the muscle is contracting against (Fig. 8.10, flat portion of the solid red line in the upper tracing). Beyond this point, developed force in the muscle needed to hold the weight is translated into muscle shortening (Fig. 8.10, solid line, lower tracing). As relaxation

begins (phase 3, Fig. 8.10), the muscle lengthens at constant force because it is still supporting the load. This phase of relaxation is isotonic, and the muscle is re-extended by the weight. When the muscle has been extended sufficiently to return to its original length, conditions again become isometric (phase 4, Fig. 8.10), and the remaining force in the muscle declines as it would in a purely isometric twitch. In almost all situations encountered in daily life, isotonic contraction is preceded by isometric force development; such contractions are called **mixed contractions** (isometric–isotonic–isometric).

The extent and velocity of shortening of a mixed contraction depend on the afterload the muscle moves. Figure 8.11 presents a series of three muscle contractions. In the figure, contraction A has the lowest afterload, the afterload in contraction B is twice that of A, and the afterload for contraction C is too heavy for the muscle to move at all. Note from this figure that the isotonic contraction with a heavy afterload cannot move that load as fast or as far as it can a lighter afterload. This unique relationship between afterload and the extent and velocity of shortening of isotonic contraction are described in detail below.

The mechanics of muscle contraction are altered by its initial passive stretch and its afterload.

Two important relationships describe the mechanics of muscle contraction:

The length–tension relationship—This represents the effect of changing the length to which resting muscle is stretched prior to contraction on the force generated by an *isometric* contraction. This passive precontraction or **resting length** is often called the **preload**.

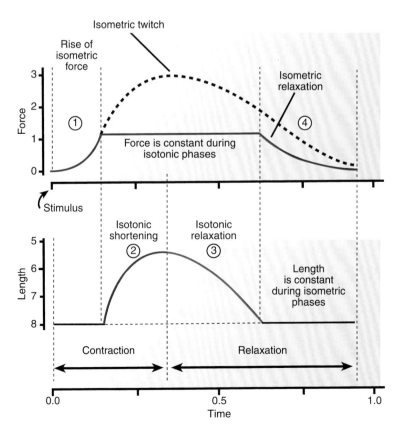

Figure 8.10 **A single isotonic contraction of skeletal muscle.** In an experimental setting, a segment of skeletal muscle is stretched to a given length and attached to a supported weight that can be moved. When this muscle segment is stimulated, the first part of the contraction will be isometric until sufficient force has developed to lift the weight. Once enough force has developed to support the weight, the contracting muscle can then move the weight by shortening. During shortening and isotonic relaxation, the force is constant (isotonic conditions), and during the final relaxation, conditions are again isometric because the muscle no longer lifts the weight. The *dotted line* in the force trace show the isometric twitch that would have resulted if the force had been too large (greater than three units) for the muscle to lift.

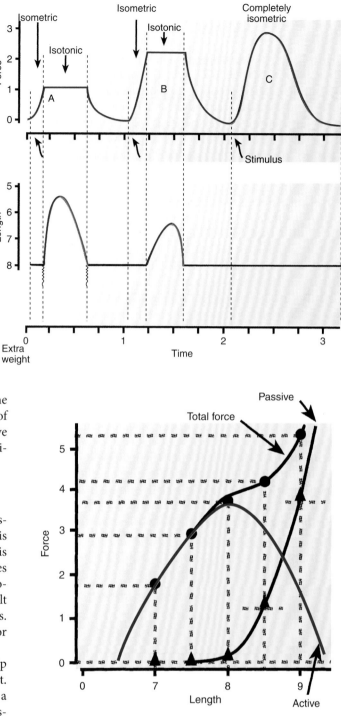

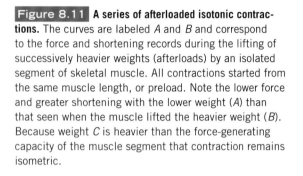

Figure 8.11 **A series of afterloaded isotonic contractions.** The curves are labeled *A* and *B* and correspond to the force and shortening records during the lifting of successively heavier weights (afterloads) by an isolated segment of skeletal muscle. All contractions started from the same muscle length, or preload. Note the lower force and greater shortening with the lower weight (*A*) than that seen when the muscle lifted the heavier weight (*B*). Because weight *C* is heavier than the force-generating capacity of the muscle segment that contraction remains isometric.

The force–velocity relationship—This represents the effect of *afterload* on the initial velocity of shortening of an isotonic contraction of muscle. Importantly, the curve defining this relationship is itself variable in that its position is altered by muscle preload.

The isometric length–tension relationship

Resting, noncontracting, skeletal muscle has nonlinear elastic properties when stretched to different lengths. When it is short, it is slack and will not resist passive extension. As it is made longer and longer, however, its resisting force increases more and more (Fig. 8.12, lower black line). This is analogous to the increasing stiffness, or resistance to stretch, felt when a rubber band is stretched to longer and longer lengths. This resisting force is called **passive force**, passive length, or **resting force** and is related to the preload on the muscle.

In an *isometric* muscle contraction, the relationship between *active* force and passive length is much different. Figure 8.12 also shows a graph of the force produced by a series of isometric twitches made over a range of initial passive muscle lengths (upper black line). The resulting tension (passive tension plus active isometric force) at each muscle length is plotted with the total peak force from each twitch related to its initial passive length (top black line in the figure). The difference between the total force and the passive force at any length is the active force (red line in the figure). *Note that the active length–tension curve is parabolic.* Consequently, isometric force has a maximum in relation to its preload, or initial resting length. The length corresponding to this maximum isometric force is called the optimum length (L_o). If the muscle is set to a shorter or longer length than L_o and then stimulated, it produces less force. At extremely short or long lengths, it produces no force at all.

Figure 8.12 **A length–tension curve for skeletal muscle.** Contractions are made at several resting lengths, and the resting (passive) and peak (total) forces for each twitch are plotted on a graph depicting force versus the length to which muscle was stretched prior to stimulation of contraction (preload). Total force in activated muscle is a sum of passive stretch (resistive) force plus the active force generated by activation of muscle crossbridges. Subtraction of the passive curve from the total force curve yields the active force curve (*red*). Note that the parabolic relationship between isometric force generation and initial stretch indicates that there is an optimum length (L_o) to which muscle can be stretched prior to contraction that will yield maximum isometric force generation in activated muscle. (Force and length are listed in arbitrary units.)

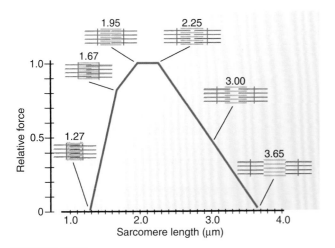

Figure 8.13 **Effect of filament overlap on force generation.** The force a muscle can produce depends on the amount of overlap between the thick and thin filaments, because this determines how many crossbridges can interact effectively. (See text for details.)

The parabolic length–tension relationship for isometric contraction in the whole muscle results from changes in actin and myosin overlap within the sarcomere as the muscle is stretched prior to contraction. The effect of actual sarcomere length on force generation is summarized in Figure 8.13. At lengths near L_O, where the isometric force that the muscle can generate is maximum, the number of actin-binding sites opposite the number of myosin heads available for contraction are also at a maximum. (Note in Fig. 8.13 that "maximum" isometric force is not a true "peak"; force does not vary with the degree of overlap between 1.95 and 2.25 μm. This occurs because further overlap of the bare zone along the thick filaments at the center of the A band, where no myosin heads are present, does not lead to an increase in the number of attached crossbridges.)

When the muscle is stretched beyond its normal resting length, decreased filament overlap occurs with increasing stretch. This reduces the amount of force that can be produced because less length of the thin filaments interdigitates with A-band thick filaments and fewer crossbridges are available for any subsequent contraction. At passive resting lengths shorter than L_O, additional geometric and physical factors play a role in myofilament interactions. Because muscle is a "telescoping" system, there is a physical limit to the degree to which resting muscle length can decrease. As thin myofilaments penetrate the A band from opposite sides, they begin to meet in the middle and interfere with each other (1.67 μm). At the extreme, further reduction in resting lengths are limited by the thick filaments of the A band being forced against the structure of the Z lines (1.27 μm). Note, however, if one were to start at this extreme short length and then stretch the muscle to various passive lengths up to and including L_O, any resulting isometric contraction would increase as the passive initial length of the muscle increases. This region of the isometric length–tension curve is sometimes called the *ascending limb* of the length–tension curve.

In this region, isometric force is positively proportional to initial resting length (preload).

It should be noted that, in the body, many skeletal muscles are confined by their attachments to a relatively short region of the isometric length–tension curve near L_O. Although this positions preload on the muscle to an area that can produce the best possible force generation in the muscle, it also means that skeletal muscle is essentially incapable of adjusting its force-generating capacity by repositioning itself at different locations on the length–tension curve prior to initiating contractions. However, the isometric length–tension relationships points to a potentially exploitable property of the sliding filament property of any muscle type. Modulation of active contraction is possible in muscle, hollow, organs like the heart (cardiac muscle) and blood vessels, bronchi, the alimentary tract, and reproductive organs (smooth muscle).

The isotonic force–velocity relationship

When all the motor units in the muscle are stimulated (i.e., no motor units left to recruit), a muscle's performance can still be affected by loading conditions. The extent and velocity of shortening of skeletal muscle during an isotonic contraction are negatively related to afterload but positively related to preload. Everyday experience shows that light loads can be lifted faster than heavy ones. A detailed analysis of this observation can be performed experimentally by arranging a muscle so that it can be presented with a series of afterloads (Fig. 8.14; also see Fig. 8.11). Measuring the **initial velocity**—the velocity during the earliest part of the shortening—at various afterloads shows that lighter loads are lifted more quickly and heavier loads more slowly. (Initial velocity is measured because the muscle soon begins to slow down upon moving the load.) When all the initial velocity measurements are related to each corresponding afterload lifted, an inverse relationship known as the *force–velocity curve* is obtained (see Fig. 8.14). This relationship shows that as the afterload is decreased, the velocity of shortening increases in a hyperbolic manner; that is, velocity increases more and more as the load is reduced. It also shows that if the applied load is greater than the maximal force capability of the muscle, no shortening will result, and the contraction will be isometric. This point is known as F_{max}.

Note from the force–velocity relationship that there is a theoretical velocity that the muscle could obtain if it does not have to move any afterload. This represents the *maximum velocity* of shortening capability of that muscle and is, therefore, known as V_{max}. In practice, a completely unloaded contraction is very difficult to arrange. Instead, V_{max} is obtained by mathematical extrapolation of the force–velocity curve. The value of V_{max} represents the maximal rate of crossbridge cycling for that muscle; it is directly related to the biochemistry of the actin–myosin ATPase activity in a particular muscle type and can be used to compare the properties of different muscles. Figure 8.14 shows force–velocity curves for a fully activated muscle. When measurements are made on a fully activated muscle, a force–velocity curve defines

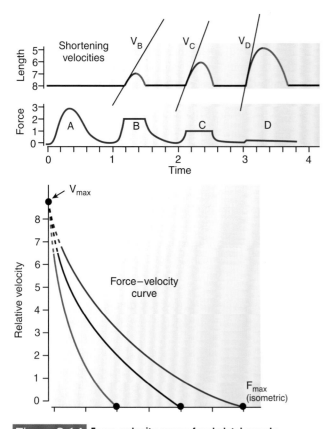

Figure 8.14 **Force–velocity curves for skeletal muscle.**
Contractions at four different afterloads (decreasing left to right)
are shown in the **top figure**. The initial shortening velocity (slope)
is measured (V_B, V_C, V_D) and the corresponding force and velocity
points plotted on the axes in the **bottom** graph (*black line*). Note
that the initial velocity of shortening increases when afterload
is decreased. The *black line* represents the force–velocity curve
of a muscle starting at a less than maximum preload. The *red
line* represents the relationship with the muscle starting at its
optimum preload, and the *blue line* shows the relationship at
the lowest preload in this example. Notice how preload affects
the velocity of shortening of any isotonic contraction and F_{max},
but neither it nor afterload affects V_{max} (maximal velocity of
shortening).

the upper limits of the muscle's isotonic capability. Factors
that modify muscle performance, such as fatigue or incomplete stimulation (e.g., fewer motor units activated), result
in operation *below* the limits defined by this force–velocity
curve.

Importantly, the force–velocity relationship is also
affected by the preload on the muscle prior to contraction
(see red and blue relationships in Fig. 8.14). The curve is
shifted to the right with increasing preload and to the left with
decreased preload, providing the muscle is not operating at
lengths longer than L_O. As revealed by crossbridge overlap
on the ascending limb of the length–tension relationship,
the effect of preload on the force–velocity relationship must
reflect the number of crossbridges available for an isotonic
contraction as well. Examination of Figure 8.14 reveals that
with increasing preload, a given afterload can be moved faster
than those starting at a lesser preload. The figure also shows

that at any given velocity, a larger afterload can be moved if
started at a higher preload (providing it does not exceed L_O)
than that which can be moved with a lesser preload. In short,
muscle performance is enhanced by bringing more crossbridges into play at the start of a contraction. Although this
feature of muscle behavior does not have direct physiologic
application to skeletal muscle *in situ*, which has a fixed initial
length through attachment to the skeleton, this relationship
becomes important in muscles in hollow organs were preload may be altered prior to a contraction (e.g., the heart and
blood vessels).

Figure 8.14 also reveals an important feature of the physiology of skeletal muscle with regard to crossbridge cycling.
Note that although preload alters the force–velocity position
for isotonic contractions, the V_{max} is not altered by preload
or any afterloads moved; force–velocity relationships starting at low preloads end up at the same V_{max} as those starting
with many more crossbridges available. Consequently, V_{max}
is said to be totally load independent in skeletal muscle. This
is because V_{max} reflects the intrinsic biochemical interaction
of actin and myosin during crossbridge cycling, and this
interaction for skeletal muscle actin and myosin cannot be
altered by any normal physiologic mechanism in the body;
modification of muscle performance at this cellular level is
not possible in skeletal muscle. However, as discussed later
in this chapter and in Chapter 13, such cell-based performance modification is available physiologically in cardiac
and smooth muscle.

Loading conditions affect the extent of skeletal muscle shortening.

The length–tension curve represents the effect of initial resting length, or preload, on the isometric contraction of skeletal muscle. At initial lengths less than and up to L_O, such a
curve represents the maximum force that can be generated at
any given preload. No shortening occurs in these situations,
and the curve depicts muscle contraction as if all its energy
was channeled into the development of force only. Points A,
B, C, and D in Figure 8.15, for example, are the maximum
force generation capability of a muscle at initial lengths (arbitrary units) of 8, 6.6, 5.6, and 5, respectively, with point A
in this diagram being representative of the isometric force-
generating capacity of the muscle at L_O.

Although the length–tension curve depicts isometric
contraction of muscle, it also represents the limit of shortening of any isotonic contraction of that muscle. Isotonic
contractions involve an initial isometric phase during
which force is generated up to the point it equals the load
the muscle is trying to move (i.e., the afterload). Once that
point is reached, the muscle then shortens while moving that
load (the isotonic phase). Such contractions are depicted in
Figure 8.15 by the solid red arrows that end at B, C, and D
in the diagram. Note that during an isotonic contraction, the
muscle will eventually reach a length where the maximum
isometric force capable by the muscle matches the afterload
the muscle is trying to move (see *dashed lines* in Fig. 8.15
that converge on point C). At this point, the muscle cannot

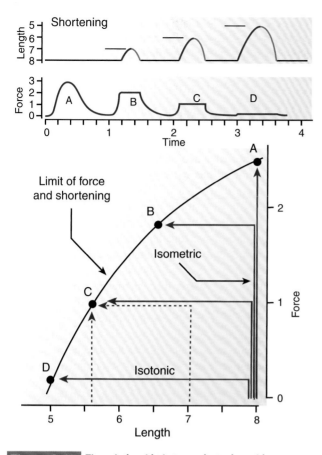

Figure 8.15 **The relationship between isotonic and isometric contractions.** The **top** graphs show the contractions from Figure 8.14, with different amounts of shortening. The **bottom** graph shows that, for contractions *B, C,* and *D,* the initial portion is isometric (the line moves upward at constant length) until the afterload force is reached. The muscle then shortens at the afterload force (the line moves to the left) until its length reaches a limit determined (at least approximately) by the isometric length–tension curve. The *dotted lines* show that the same final force/length point can be reached by several different approaches. Relaxation data, not shown on the graph, would trace out the same pathways in reverse.

shorten anymore; to do so would place the muscle at a length where its maximum isometric force-generating capability was less than the load it was trying to move—an impossibility. In other words, isotonic contractions of a muscle moving any afterload cannot move past a limit defined by the isometric length–tension curve for that muscle.

As Figure 8.15 shows, a lightly afterloaded muscle will shorten farther than one starting from the same initial length, or preload, but bearing a heavier load. Also note that if the muscle begins its shortening from a reduced initial length, its subsequent extent, or amount, of shortening will be reduced because the muscle starts closer to the limit imposed by the isometric–length tension curve. The following constraints apply to the extent of shortening of an isotonic contraction of skeletal muscle: (1) increasing preload up to L_o increases the extent of shortening (the distance the afterload is moved); (2) at any given preload, increasing the afterload decreases the extent of shortening; (3) at any given preload, the final

length of the muscle at the end of an isometric contraction is proportional to the afterload; and (4) the isometric length–tension curve sets the limit to the extent of shortening of any isotonic contraction of skeletal muscle.

The example above demonstrates that the distance a muscle can move an object is progressively reduced as the muscle tries to move increasingly heavy loads. One might postulate that skeletal muscle could move a heavy load as far as it could a light load if there was some way to change (enhance) the intrinsic contractile ability of the muscle cell itself. Such intrinsic enhancement would reveal itself as a shift of the ascending limb of the length–tension curve upward and to the left relative to its original position. Such a shift would move the limit to extent of shortening to the left, thereby letting the muscle move the heavy load as far as it could move the light load in its original state. Unfortunately, such intrinsic or cellular/molecular enhancement of the contractile mechanism is not possible under any normal physiologic conditions for skeletal muscle fibers. In essence, skeletal muscle shortening is constrained by its loading conditions; it *cannot* modify its cellular contractile capabilities independently of preload and afterload. Such is not the case with cardiac muscle, however, which can shift its length–tension relationship (see Chapter 13).

Skeletal attachments create lever systems that modify muscle action.

Anatomic location also places restrictions on muscle function by limiting the amount of shortening or determining the kinds of loads encountered. Skeletal muscle is generally attached to bone, and bones are attached to each other. The bones and muscles together constitute a **lever system** (Fig. 8.16). In most cases, the system works at a mechanical disadvantage with respect to the force exerted. For example, curling the forearm requires muscles that are attached near the fulcrum of the arm "lever" rather than at the more mechanically advantageous position near the hand. This means that in order to pull the forearm toward the upper arm while holding a weight in the hand, the muscle must exert a much greater force than the actual weight of the load being lifted (the muscle force is increased by the same ratio that the length change at the end of the extremity is increased). In the case of the human forearm, the biceps brachii, when moving a force applied to the hand, must exert a force at its insertion on the radius that is approximately seven times as great. However, the shortening capability of skeletal muscle by itself is rather limited. The benefit of the skeletal lever system is that it multiplies the distance over which an extremity can be moved. Consequently, it is possible to create large distances moved by the end of the limb with only small actual shortening of the muscle responsible for moving the limb. In this forearm example, the resulting movement of the hand is approximately seven times as far and seven times as rapid as the shortening of the muscle itself.

Finally, acting independently, a muscle can only shorten. Force must be used to relengthen the muscle, and this force must be provided externally. In some cases, gravity provides

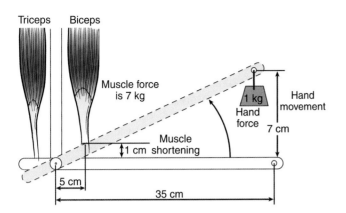

Figure 8.16 **Antagonistic pairs and the lever system of skeletal muscle.** Contraction of the biceps muscle lifts the lower arm (flexion) and elongates the triceps, whereas contraction of the triceps lowers the arm and hand (extension) and elongates the biceps. The bones of the lower arm are pivoted at the elbow joint (the fulcrum of the lever). The force of the biceps is applied through its tendon close to the fulcrum. The hand is seven times as far away from the elbow joint. Thus, the hand will move seven times as far (and fast) as the biceps shortens (lever ratio, 7:1), but the biceps will have to exert seven times as much force as the hand is supporting.

the external restoring force. However, in the body, muscles are often arranged in **antagonistic pairs** of **flexors** and **extensors**. In this manner, for example, a shortened biceps can be relengthened by the action of the triceps; the triceps, in turn, is relengthened by contraction of the biceps.

▶ SKELETAL MUSCLE METABOLISM AND FIBER TYPES

ATP derived from glucose metabolism is the primary energy-supplying mechanism for skeletal muscle contraction.

Cellular processes within skeletal muscle cells must supply biochemical energy to the contractile mechanism because contracting muscles perform work. Additional energy is required to pump the calcium ions involved in the control of contraction and for other cellular functions as well. In muscle cells, as in other cells, this energy ultimately comes from the universal high-energy compound, ATP, which is predominantly derived from oxidative metabolism. Although ATP is the immediate fuel for the contraction process, its concentration in the muscle cell is never high enough to sustain a long series of contractions. Most of the immediate energy supply is held in an "energy pool" of the compound creatine phosphate or phosphocreatine (PCr), which is in chemical equilibrium with ATP. After a molecule of ATP has been split and yielded its energy, the resulting ADP molecule is readily rephosphorylated to ATP by the high-energy phosphate group from a creatine phosphate molecule. The creatine phosphate pool is restored by ATP from the various cellular metabolic pathways. These reactions (of which the last two are the reverse of each other) can be summarized as follows:

ATP → ADP + P_i + Energy for contraction and other cell systems

ADP + PCr → ATP + Creatine (i.e., rephosphorylation of ATP)

ATP + Cr → ADP + PCr (restoration of PCr)

Because of the chemical equilibria involved, the concentration of PCr can fall to low levels before the ATP concentration shows a significant decline. It has been shown experimentally that when 90% of PCr has been used, the ATP concentration has fallen by only 10%. This situation results in a steady source of ATP for contraction that is maintained despite variations in energy supply and demand. Creatine phosphate is the most important storage form of high-energy phosphate; together with some other smaller sources, this energy reserve is sometimes called the **creatine phosphate pool**.

Glycolysis, an **anaerobic** pathway, and **oxidative phosphorylation**, an aerobic pathway, are the two major metabolic paths that supply energy to the energy-requiring reactions in the cell and to the mechanisms that replenish the creatine phosphate pool. Glucose for the glycolytic pathway may be derived from circulating blood glucose or from **glycogen**, which is the polymer storage form of glucose in skeletal muscle and liver cells. Glycolysis is rapid but extracts only a small fraction of the energy contained in the glucose molecule (2 ATPs per glucose molecule). The end product of anaerobic glycolysis is **lactic acid** or lactate. Under conditions of sufficient oxygen, this is converted to **pyruvic acid** or pyruvate, which enters the **Krebs cycle** where substrates are made available for oxidative phosphorylation (36 ATPs per glucose). **Glucose** is the preferred fuel for skeletal muscle contraction at higher levels of exercise. At maximal work levels, almost all the energy used is derived from glucose produced by glycogen breakdown in muscle tissue and from blood-borne glucose from dietary sources. Muscle has performance limitations based on its structure and energy-conversion processes; as such, its efficiency is much <100%, and it produces relatively large quantities of heat, which must be dealt with by the organism that it is serving (see Chapter 28).

Metabolic differences among muscle fibers affect their ability to sustain contraction.

Although the basic structural features of the sarcomeres and the thick–thin-filament interactions are essentially the same among skeletal muscles, the chemical reactions that supply the contractile system with energy vary. The enzymatic properties (i.e., the rate of ATP hydrolysis) of actomyosin ATPase also vary. A typical skeletal muscle usually contains a mixture of fiber types with different metabolic properties. However, in most muscles, a particular type predominates.

Muscle fibers are sometimes classified as oxidative (types I and IIA, or red fibers) or glycolytic (type IIB, or white fibers). Red fibers utilize oxidative metabolism for contraction and owe their color to the presence of **myoglobin**, which is a hemoglobin-like molecule that can bind, store, and release oxygen. It is abundant in muscle fibers that depend heavily on aerobic metabolism for their ATP supply, where it facilitates oxygen diffusion (and serves as a minor auxiliary oxygen source) in times of heavy demand. Such muscles generally perform heavy sustained tasks. The quadriceps and gluteal muscles that continually maintain posture while standing are examples of this type of muscle. Red muscle fibers are divided into **slow-twitch fibers** and **fast-twitch fibers** on the basis of their contraction speed. The differences in rates of contraction (shortening velocity or force development) arise from differences in actomyosin ATPase activity (i.e., in the basic crossbridge cycling rate).

White muscle fibers, which contain little myoglobin, are fast-twitch fibers that rely primarily on glycolytic metabolism. They contain significant amounts of stored glycogen, which can be broken down rapidly to provide a quick source of energy. They are responsible for quick, short muscle movements such as those involved with movement and blinking of the eyes. Although they contract rapidly and powerfully, they fatigue easily (i.e., their endurance is limited).

Fast muscles, both white and red, not only contract rapidly but also relax rapidly. Rapid relaxation requires a high rate of calcium pumping by the SR, which is abundant in these muscles. In such muscles, the energy used for calcium pumping can be as much as 30% of the total consumed. Fast muscles are supplied by large motor axons with high conduction velocities; this correlates with their ability to make quick and rapidly repeated contractions.

▶ MUSCLE PLASTICITY, EPIGENETICS, AND ENDOCRINE MUSCLE

Skeletal muscle structure and metabolism are not immutable, as has long been evidenced by the differing effects of aerobic exercise, isometric load exercise (e.g., weight lifting) and immobility on muscle mass, velocity of shortening, and metabolism (see Chapter 29 for details). The emerging science of epigenetics is revealing potential mechanisms by which external conditions from the environment, such as exercise, inactivity, proper nutrition, malnutrition, and exposures to medication or even environmental toxins, cannot just influence the body acutely but affect the genome through epigenetic mechanisms. Epigenetics is an evolving science that seeks to describe and explain how gene expression is changed independent of genotype yet can be heritable by subsequent generations. This provides a mechanism, for example, as to how environmental/nutritional conditions to which parents are exposed can be passed on to not just their immediate offspring, in the form of altered suppression or activation of their genes, but also possibly onto the next several generations as well. Insights into epigenetic effects on muscle cell growth, differentiation, and regeneration have beneficial potential for rehabilitative medicine and perhaps even set the stage for the ability to grow and replace muscle cells lost from trauma or disease.

Recent experimental studies have provided evidence for epigenetic modification of muscle type. For example, although muscle fiber phenotype (fast vs. slow) is determined during development by interactions with other elements in the developing body, such as motor neuron innervation type and thyroid hormone, it has been shown in cell culture that slow fibers only grow into slow fibers and fast fibers only into fast fibers. This cannot occur unless epigenetic changes occurred in the muscle precursor cells (i.e., stable and heritable changes in the muscle DNA of each type).

Other factors have been revealed recently to affect the epigenome of muscle. Obesity has been shown to induce epigenetic changes in a lipogenic muscle gene that alters the muscle cell to be smaller and thus better able to store fat. This muscle feature is a hallmark of type 2 diabetes, which has been correlated with obesity. Acetylation and deacetylation of histones are considered epigenetic modes of altering gene expression. Recently, HDAC inhibitors have been shown to promote regeneration of muscle cells while inhibiting fibrosis and fat deposition in dystrophic muscle. Finally, numerous recent studies have shown that the epigenetic marks of DNA methylation, histone acetylation, and miRNA are all altered by aerobic and isometric exercise, not just in skeletal muscle but also in other cell types such as leukocytes, brain cells, and adipose tissue.

Exercise is known to have beneficial effects on several tissues in the body besides skeletal muscle. Recent evidence that exercise creates epigenetic marks in tissues beyond just skeletal muscle suggests that muscle may not only be an epigenetic target, but that itself may send signals to other tissues to induce epigenetic changes in them as well. As introduced in Chapter 1, muscle cells are now believed to have an endocrine function via secretion of myokines. Myokines released from skeletal muscle have been shown to "cross-talk" with adipocytes and have been suggested to attenuate the insulin resistance and low-grade inflammation associated with type 2 diabetes (such attenuation is a known beneficial effect of regular exercise). In contrast, a lack of beneficial myokine secretion with aging of muscle has been suggested to link deterioration of that tissue to increased incidence of age-related neuromotor, neurocognitive, metabolic, and oncologic disorders, although evidence in support of this postulate is more correlative than causative at this time.

▶ SMOOTH MUSCLE

Smooth muscle is by far the most physiologically complicated and versatile muscle tissue in the body. Recall that skeletal muscle can only be activated intermittently and requires a signal from a neuron innervating it to do so. Furthermore, its relaxation is not active but rather the absence of contraction. In contrast, smooth muscle can both contract and relax actively on its own, as well as in response to neural signals, hormonal signals, chemical signals, or physical signals. Smooth muscle can exhibit twitch contraction, summed twitch contraction, or graded contractions, that is, smooth increments in contraction or relaxation rather than the "all-on-all-off" characteristic of skeletal muscle. In blood vessels, for example, vascular smooth muscle must be partially tonically active at all times (i.e., it must remain partially contracted continuously). This seemingly energy-intensive process is mitigated by the fact that smooth muscle, unlike skeletal muscle, contains special crossbridge cycling switching mechanisms that allow it to sustain high force at low levels of energy expenditure.

Smooth muscle outside the cardiovascular system is often labeled *visceral smooth muscle*. This muscle plays a key role in the functions of the digestive system, where it physically manipulates gut contents in complex ways during digestion and moves materials and fluids along the alimentary tract. Visceral smooth muscle in the gut, therefore, must be able to contract or relax phasicly in response to the many different types of neural, chemical, and physical signals involved in digestion. Some smooth muscles such as that in sphincters (circular bands of muscle that can stop flow in tubular organs) can remain contracted and closed for long periods but then relax and open transiently before contracting and closing again. Sphincters at the ends of the esophagus, the stomach, and rectum operate in this fashion (see Chapter 25). Smooth muscle in still other visceral organs remains relaxed most of the time but then can contract strongly in response to physiologic stimuli. Smooth muscles in the esophagus, urinary bladder, and gall bladder behave in this manner. As an ultimate case of this type of behavior, the smooth muscle of the uterus is relatively inactive during most of a woman's life but contracts and relaxes rapidly and powerfully over several hours during parturition.

Multiple smooth muscle contraction and relaxation patterns can be produced by neural, humoral, chemical, physical, and intrinsic stimuli.

Contractile activity in smooth muscle is initiated in markedly different ways compared to that for skeletal muscle. In the previous section, it was shown that skeletal muscle cannot contract without neural input to the muscle cells. Furthermore, the neural inputs to skeletal muscle fibers (via motor neurons) are often initiated from conscious thought and are, therefore, voluntary. Although smooth muscle is indeed innervated (predominantly by postganglionic sympathetic fibers), innervation is not necessary for smooth muscle contraction and modulation of smooth muscle contraction is totally involuntary. Rather than being the sole means of muscle activation, neurotransmitters from autonomic neurons are but one of many factors that modulate smooth muscle contraction. Contractile activity in smooth muscle is also modulated by numerous receptor-dependent and receptor-independent (direct-acting) contractile and relaxing agents.

Smooth muscle does have excitable tissue properties and can form action potential–driven contractions in response to signals from the ANS or other contractile agents. Smooth muscle can also summate contractions in response to a train or small burst of ANS-driven action potentials in a manner similar to temporal summation and tetanus in skeletal muscle. However, there are many other contractile agents that can mobilize calcium in smooth muscle and cause contraction without the formation of any action potentials. Furthermore, calcium levels and associated contraction of smooth muscle

can be *graded* in response to changes in its *resting* membrane potential. For example, stepwise incremental depolarizations of the resting membrane potential in smooth muscle cause proportional incremental increases in smooth muscle tone (contractile force); the contractions are graded and not linked to the formation of action potentials. Similar graded contraction or relaxation responses can be elicited in smooth muscle without any change in membrane potential at all. This can occur in response to changes in the concentration of many different constrictor and relaxing agents that act on smooth muscle directly or act via membrane receptors. Thus, whereas skeletal muscle is like a lamp with an on-and-off switch that will not work unless it is linked to a source of electricity, smooth muscle is like a light with a dimmer switch and its own source of excitation.

Altering smooth muscle contraction changes the internal dimensions of hollow organs.

Just a few basic tissue types accomplish a wide variety of smooth muscle tasks. The simplest smooth muscle arrangement is found in the arteries and veins of the circulatory system. Smooth muscle cells are oriented in the circumference of a vessel so that their shortening results in reducing the vessel's diameter. This reduction may range from a slight narrowing to a complete obstruction of the vessel lumen, depending on the physiologic needs of the body or organ. Tonic contraction there helps the vessels counteract the expanding force on the arteries due to exposure to high internal blood pressure. The blood pressure provides a force that would otherwise relengthen the cells in the vessel walls. In the smallest arterioles, the muscle layer may consist of single cells wrapped around the vessel, where it functions as an on/off spigot to control blood flow to very precise regions within an organ.

The circular arrangement of smooth muscle is also prominent in the airways of the lungs, where it regulates the resistance to airflow entering and exiting that organ. The circular smooth muscles making up sphincters, such as those in the gastrointestinal and urogenital tracts, have a special nerve supply and participate in complex reflex behavior. The intestinal tracts (small and large) contain two regions of smooth muscle cells within their walls. The outermost muscle layer, which is relatively thin, runs along the length of the intestine. The inner muscle layer, thicker and more powerful, has a circular arrangement. Coordinated alternating contractions and relaxations of these two layers mix and propel the contents of the intestine, although most of the motive power is provided by circular muscle. Control of the motility of the intestine through these smooth muscle layers is very complex and discussed in more detail in Chapter 25.

The most complex arrangement of smooth muscle is found in organs such as the urinary bladder and uterus. Numerous layers and orientations of muscle fibers are present, and the effect of their contraction is an overall reduction of the volume of these organs. The relengthening force, in the case of these hollow organs, is provided by the gradual accumulation of contents. In the urinary bladder, for example, the muscle is gradually stretched as the emptied organ fills again with urine from the kidneys.

Smooth muscle cells lack the ultrastructural organization of skeletal muscle fibers.

In contrast to skeletal muscle fibers, smooth muscle cells are small relative to the tissue they make up. Individual smooth muscle cells are 100 to 300 μm long and 5 to 10 μm in diameter. When isolated from the tissue, the cells are roughly cylindrical along most of their length and taper at the ends. Electron microscopy reveals that the cell margins contain many areas of small membrane invaginations, called **caveolae**, which are sites of specialized cell functions. There is no organized T-tubular system in smooth muscle, and its SR is not as abundant as that of skeletal muscle. The bulk of the cell interior is occupied by three types of myofilaments: thick, thin, and intermediate (Fig. 8.17). The *thin filaments* are similar to those of skeletal muscle but lack the troponin protein complex. The *thick filaments* are composed of myosin molecules, as in skeletal muscle, but the arrangement of the individual molecules into filaments is different; the filaments are called *side-polar*, with one crossbridge orientation on one side of the filament and the opposite orientation on the other. There is no central bare region without crossbridges. The **intermediate filaments** are so named because their diameter of 10 nm is between that of the thick and thin filaments. Intermediate filaments appear to have a cytoskeletal, rather than a contractile, function. Prominent throughout the cytoplasm are small, dark-staining areas called **dense bodies**. They are associated with the thin and intermediate filaments and are considered analogous to the Z lines of skeletal muscle. Dense bodies associated with the cell margins are often called **membrane-associated dense bodies**

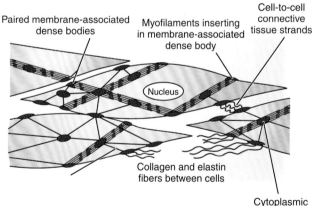

Neuromuscular Physiology

(or patches) or **focal adhesions**. They appear to serve as anchors for thin filaments and transmit the force of contraction to adjacent cells.

Smooth muscle lacks the regular sarcomere structure of skeletal muscle. There is some association among dense bodies down the length of a cell and a tendency of thick filaments to show a degree of lateral grouping. However, it appears that the lack of a strongly periodic arrangement of the contractile apparatus is an adaptation of smooth muscle associated with its ability to function over a wide range of lengths in its host organ and to develop high forces.

Cell-to-cell transmission of force and electrical activity

A proposed arrangement of the smooth muscle contractile and force transmission system is also shown in simple terms in Figure 8.17. Assemblies of myofilaments are anchored within the cell by the dense bodies and at the cell margins by the membrane-associated dense bodies. The contractile apparatus lies oblique to the long axis of the cell. Many of the membrane-associated dense bodies are opposite on one another in adjacent cells and may provide continuity of force transmission between the contractile apparatus in each cell. Collagen and elastin fibers run throughout the tissue. These fibers, along with reticular connective tissue, comprise the **connective tissue matrix** or stroma found in all smooth muscle tissues. It connects the cells and gives integrity to the whole tissue.

Smooth muscle cells are also coupled electrically, although this functional connectivity predominates in visceral, rather than vascular, smooth muscle. The structure most responsible for this coupling is the **gap junction**, which allows a low-resistance pathway for electrical currents to flow between cells. This facilitates the propagation of changes in membrane potential throughout tissue containing many smooth muscle cells.

Calcium required to contract smooth muscle is obtained from both extracellular and intracellular sources.

An elevation of intracellular calcium concentration to about 1 μM is an absolute requirement for smooth muscle contraction, and a reduction in smooth muscle calcium concentration below this level is considered a primary mechanism of smooth muscle relaxation. Figure 8.18 gives an overall picture of calcium regulation in smooth muscle. These processes may be grouped into those modulating *calcium entry*, intracellular *calcium liberation*, and *calcium exit* from the cell. Calcium enters the cell through several pathways, including voltage-gated 1,4 dihydropyridine calcium channels, numerous different ligand-gated channels, stretch-activated calcium channels, and a relatively small number of unregulated **"leak" channels**. Calcium leaks can permit the continual passive entry of small amounts of extracellular calcium in the cell, though this entry is negligible under normal conditions. Such leaks can affect cell calcium entry in pathological conditions such as that seen contributing to excessive smooth muscle contraction in arterial hypertension.

Figure 8.17 **A schematic representation of the contractile system and cell-to-cell connections in smooth muscle.** Note regions of association between thick and thin filaments that are anchored by the cytoplasmic and membrane-associated dense bodies. A network of intermediate filaments provides some spatial organization (see, especially, the *left side*). Several types of cell-to-cell mechanical connections are shown, including direct connections and connections to the extracellular connective tissue matrix. (Structures are not necessarily drawn to scale.)

Paired membrane-associated dense bodies · Myofilaments inserting in membrane-associated dense body · Cell-to-cell connective tissue strands · Nucleus · Collagen and elastin fibers between cells · Cytoplasmic dense body

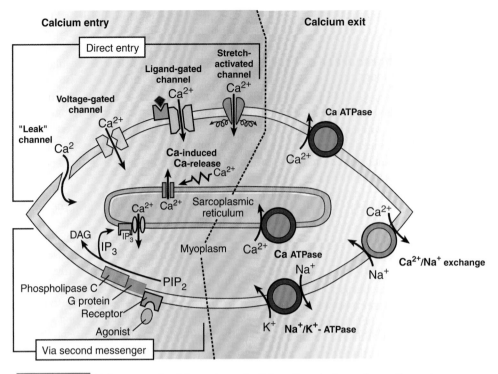

Figure 8.18 Major routes of calcium entry and exit from the cytoplasm of smooth muscle. Calcium enters the smooth muscle cell through voltage-gated, ligand-gated, stretch-activated, and "leak" calcium channels. Calcium can also be released into the cell from the sarcoplasmic reticulum (SR) by inositol 1,4,5-trisphosphate (IP$_3$) and Ca^{2+}-stimulated calcium release. The processes on the *left side* increase cytoplasmic calcium and promote contraction; those on the *right side* decrease internal calcium and promote relaxation. DAG, diacylglycerol; IP$_3$, inositol 1,4,5-trisphosphate; PIP$_2$, phosphatidylinositol 4,5-bisphosphate.

Membrane depolarization opens voltage-gated calcium channels in smooth muscle that operate in a similar fashion to the voltage-gated sodium channels in nerves and skeletal muscle. These channels have an activation gate that opens during membrane depolarization starting at approximately −55 to −40 mV, along with an inactivation gate that closes with membrane depolarization several milliseconds after opening of the activation gate. The opening and closing kinetics of these types of calcium channels are much slower compared to that seen in voltage-gated sodium channels. The kinetics of voltage-gated calcium channels provides a short window of opportunity through which calcium can enter the cell from the extracellular fluid through the channel, down an enormous electrochemical gradient. This calcium influx is a significant source of activator calcium for smooth muscle contraction and the rapid influx of positive charge from the calcium can also create an action potential. This action potential, however, is of smaller amplitude and has a slower upstroke than those caused by voltage-gated sodium channels in nerve and skeletal muscles.

Literally dozens of different ligand-gated calcium channels exist among the smooth muscles in the body, with each type designed to be activated by a specific agonist, or ligand. Many of these channels open to allow calcium entry into the cell and initiate contraction when a specific agonist binds to a specific receptor on the channel. Many hormones and neurotransmitters contract smooth muscle through this

mechanism. Activation of ligand-gated calcium channels generally does not generate action potentials in smooth muscle except at very high agonist concentrations. Contraction of smooth muscle via ligand-gated calcium channels without the generation of an action potential is called **pharmacomechanical coupling**. Some ligands for ligand-gated calcium channels operate by decreasing the open time of the channel or decreasing the amount of time the channel is open. These ligand channels decrease intracellular calcium concentration and cause the muscle to relax.

Within the cell, the major storage site of calcium is the SR. In some types of smooth muscles, its capacity is quite small and these tissues are strongly dependent on extracellular calcium for their function. Nevertheless, calcium from the SR can contribute to smooth muscle contraction. Calcium is released from the SR by at least two mechanisms, including inositol 1,4,5-trisphosphate (IP$_3$)-induced release and a **calcium-induced calcium release**. IP$_3$ formation is caused by activation of phospholipase C (PLC), by ligand-activated membrane receptors, which forms IP$_3$ and diacylglycerol (DAG) from phosphatidylinositol 4,5-bisphosphate (PIP$_2$) in the cell. In this latter mechanism, calcium that has entered the cell via a membrane channel causes additional calcium release from the SR, amplifying its activating effect.

Smooth muscle cells also have calcium channels that open whenever the smooth muscle cell is stretched. These channels are distinct entities and not simply voltage-gated or ligand-gated calcium channels that are opened by stretch (although there is some evidence to suggest that voltage-gated channels can be affected this way in some circumstances). Stretch-activated calcium channels play a key role in the regulation of vascular smooth muscle contraction because they are linked to the control of organ blood flow (see Chapter 15).

During a contraction, the cytoplasmic calcium level is significantly elevated. Following contraction, calcium leaves the myoplasm in two directions: a portion of it is returned to storage in the SR by a Ca^{2+}-ATPase pump in the SR membrane and the rest is ejected from the cell by two mechanisms. The more important of these is via a membrane Ca^{2+} ATPase pump located in the cell membrane. The second mechanism, also located in the plasma membrane, is sodium–calcium exchange, which exchanges the entry of three sodium ions to the extrusion of one calcium ion. This mechanism derives its energy from the large sodium gradient across the plasma membrane of the cell; thus, it depends critically on the operation of the cell membrane Na^+/K^+-ATPase.

Phosphorylation of myosin filament proteins is necessary for smooth muscle contraction.

Calcium ions are central to the control of contraction in smooth muscle, where they modulate the crossbridge cycle. However, the thin filaments of smooth muscle lack troponin, and the actin-linked regulation of skeletal muscle is not present in smooth muscle. Control of smooth muscle contraction relies instead on the thick filaments and is, therefore, called **myosin-linked regulation**. In actin-linked regulation, the contractile system is in a constant state of "inhibited readiness" and calcium ions remove the inhibition. In the case of smooth muscle myosin-linked regulation, the role of calcium is to cause *activation from* a resting state of the contractile system. This general scheme is shown in Figure 8.19. At rest, there is little cyclic interaction between the myosin and actin filaments because of a special feature of its myosin molecules. The S2 portion of each myosin molecule (the paired "head" portion) contains four protein *light chains*. Two are necessary for actin–myosin interaction (essential light chains, M.W. 16,000) and two others are critical requirements for smooth muscle contraction (regulatory light chains, M.W. 20,000) Regulatory light chains contain specific amino acid residues that can be

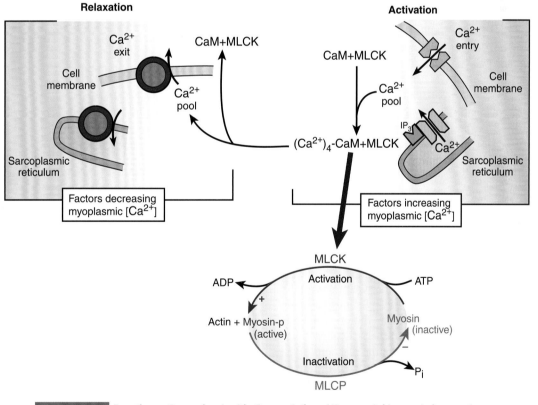

Figure 8.19 **Reaction pathways involved in the regulation of the crossbridge cycle in smooth muscle.** Activation (*upper right*) begins when cytoplasmic calcium levels are increased and calcium binds to calmodulin (CaM), activating the myosin light-chain kinase (MLCK). The kinase (*lower right*) catalyzes the phosphorylation of myosin, changing it to an active form (myosin-P, or Mp). The phosphorylated myosin can then participate in a mechanical crossbridge cycle much like that in skeletal muscle, although much slower. When calcium levels are reduced, calcium leaves CaM, the kinase is inactivated, and the myosin light-chain phosphatase (MLCP) dephosphorylates the myosin, making it inactive. The crossbridge cycle stops, and the muscle relaxes. A, actin; ADP, adenosine diphosphate; ATP, adenosine triphosphate; P_i, inorganic phosphate ion.

phosphorylated by ATP. The enzyme responsible for this **phosphorylation** is **myosin light-chain kinase (MLCK)**. When the regulatory light chains are phosphorylated, the myosin heads can interact in a cyclic fashion with actin in the crossbridge cycle much as in skeletal muscle. The cycle will not initiate if myosin is not phosphorylated. It is important to note that the ATP molecule that phosphorylates a myosin light chain is separate and distinct from the one consumed as an energy source by the mechanochemical reactions of the crossbridge cycle.

For myosin phosphorylation to occur, the MLCK must be activated. This activation is subject to control. Closely associated with the MLCK is **calmodulin (CaM)**, a smaller protein that binds calcium ions. When four calcium ions are bound, the CaM protein activates its associated MLCK and light-chain phosphorylation can proceed (see Fig. 8.19). It is this MLCK-activating step that is sensitive to the cytoplasmic calcium concentration. At levels below 10^{-7} M Ca^{2+}, no calcium is bound to CaM and no contraction can take place. When cytoplasmic calcium concentration is $>10^{-5}$ M, the binding sites on CaM are fully occupied, light-chain phosphorylation proceeds at maximal rate, and contraction occurs. Between these extreme limits, variations in the internal calcium concentration can cause corresponding gradations in the contractile force. This, then, is the biochemical mechanism responsible for the fact that smooth muscle contraction can be graded between total relaxation and total contraction in contrast to the all-or-none–type limitation to the activation of skeletal muscle contraction. The phosphorylation of myosin can be reversed by the enzyme **myosin light-chain phosphatase (MLCP)**. The activity of this phosphatase appears to be only partially regulated; that is, there is always some enzymatic activity, even while the muscle is contracting. During contraction, however, MLCK-catalyzed phosphorylation proceeds at a significantly higher rate, and phosphorylated myosin predominates. There is some evidence that during ligand-gated contraction, a DAG-PKC (protein kinase C)–dependent pathway results in the formation of the protein CPI-17, which inhibits MLCP, thereby favoring the phosphorylated myosin state (Fig. 8.20). This ligand-gated pathway also activates the RhoGTP/Rho kinase system, which activates CPI-17 and inhibits MLCP by phosphorylation of those proteins.

When the cytoplasmic calcium concentration falls, MLCK activity is reduced because the calcium dissociates from the CaM, and myosin dephosphorylation (catalyzed by the phosphatase) predominates. Because dephosphorylated myosin has a low affinity for actin, the reactions of the crossbridge cycle can no longer take place. Relaxation is, thus, brought about by mechanisms that lower cytoplasmic calcium concentrations, decrease MLCK activity, or increase MLCP activity. It also appears that actin itself can be made more or less available to interact with phosphorylated myosin and, thereby, control the extent of contraction of vascular smooth muscle. PKC, in particular, appears to activate a cascade of signal proteins within the smooth muscle cell to increase the availability of actin to then interact with phosphorylated myosin and promote vascular smooth muscle contraction.

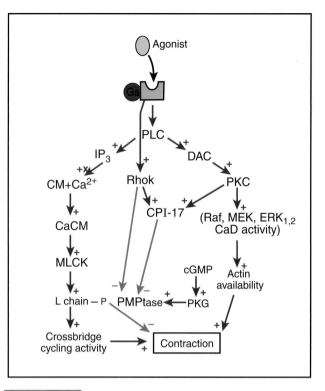

Figure 8.20 **Regulation of contraction and relaxation in vascular smooth muscle.** Myosin must be phosphorylated in vascular smooth muscle in order for crossbridge cycling to occur. The extent of myosin phosphorylation is believed to result from the net effect of phosphorylation by myosin light-chain kinase (MLCK) and dephosphorylation by myosin light-chain phosphatase (MPtase). Contraction of vascular smooth muscle occurs through activation of MLCK and inactivation of MPtase. Other proteins also regulate contraction/relaxation by making actin more or less available to participate in crossbridge cycling with phosphorylated myosin.

Latch state is a mechanism that reduces the energy cost of continual smooth muscle contraction.

Smooth muscle contractile activity cannot be divided clearly into twitch and tetanus, as in skeletal muscle. In some cases, smooth muscle makes rapid phasic contractions, followed by complete relaxation. In other cases, smooth muscle can maintain a low level of active tension for long periods without cyclic contraction and relaxation. This type of long-maintained contraction is called *tonus* (rather than tetanus) or a *tonic contraction*. This is typical of smooth muscle activated by ligand-gated calcium channels (hormonal, pharmacologic, or metabolic factors), whereas phasic activity is more closely associated with activation of voltage-gated calcium channels (i.e., by membrane depolarization). This extended maintenance of force capability is called the **latch state** (Fig. 8.21). The latch state is not a rigor state but rather a state of reduced rate of crossbridge cycling so that each remains in the attached state for a longer portion of its total cycle. This condition is characterized by low levels of ATP utilization and has a lower calcium concentration requirement. Thus, the latch state can be thought of as a means of reducing the energy requirements of long, sustained contractions over what might be expected based on energy usage

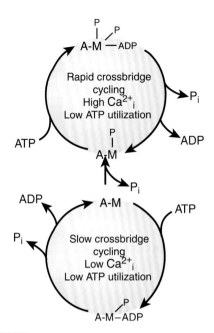

Neuromuscular Physiology

Figure 8.21 **Representation of the latch state of the crossbridge cycle.** When myosin is dephosphorylated while attached to actin, the crossbridge cycling slows because actin and myosin remain attached for longer durations than normal. This allows the muscle to maintain tonic contractions for long periods of time with a reduced energy requirement for adenosine triphosphate (ATP) and at a reduced intracellular calcium concentration. P, protein; P_i, inorganic phosphate; ADP, adenosine diphosphate; A-M, actin bound to myosin.

in short-duration contractions. Current understanding of the latch state suggests that it may occur when myosin is dephosphorylated while still attached to actin during the crossbridge cycle. In this sense, this dephosphorylation of myosin is different than that which appears to be involved in smooth muscle relaxation.

Mechanical activity in smooth muscle is adapted for its specialized physiologic roles.

The isometric length–tension relationship in smooth muscle is generally much broader than that for skeletal muscle with a less defined L_O. This allows smooth muscle to function at higher capacities over a significantly greater range of lengths. This is a beneficial property in that smooth muscle is not constrained by skeletal attachments but is instead often found lining hollow organs whose dimensions, and thus the stretch placed on the muscle therein, can vary over a large range.

Smooth muscle exhibits more resistance to stretch on the ascending limb of this relationship (Fig. 8.22A), which is mostly a result of the network of connective tissue that supports the smooth muscle cells and resists overextension.

The contraction and relaxation of smooth muscle is much slower than that of skeletal or cardiac muscle though it can maintain contraction far longer. The source of these differences lies largely in the chemistry of the interaction between the actin and myosin of smooth muscle. The inherent rate of the actomyosin ATPase correlates strongly with the velocity of shortening of the intact muscle. Most smooth muscles require several seconds (or even minutes) to develop maximal isometric force. A smooth muscle that contracts 100 times more slowly than a skeletal muscle will have an actomyosin ATPase that is 100 times as slow. The major source of this difference in rates is the myosin molecules; the actin found in smooth and skeletal muscles is rather similar. There is also a close association in smooth muscle between maximal shortening velocity and degree of myosin light-chain phosphorylation. The high economy of tension maintenance, typically 300 to 500 times greater than that in skeletal muscle of similar size, is vital to the physiologic function of smooth muscle. Compared with that in skeletal muscle, the crossbridge cycle in smooth muscle is hundreds

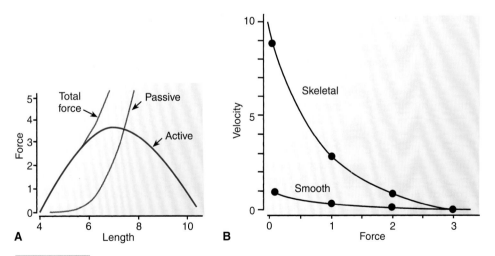

Figure 8.22 **Smooth muscle mechanical characteristics. (A)** Typical length–tension curves from skeletal and smooth muscle. Note the greater range of operating lengths for smooth muscle and the leftward shift of the passive (resting) tension curve compared to that seen in skeletal muscle. **(B)** Skeletal and smooth muscle force–velocity curves. Although the peak forces may be similar, the maximum shortening velocity of smooth muscle is typically 100 times lower than that of skeletal muscle. (Force and length units are arbitrary.)

of times slower, and, therefore, much more time is spent with the crossbridges in the attached phase of the cycle.

The force–velocity curve for smooth muscle (Fig. 8.22B) reflects the differences in crossbridge functions described previously. Although smooth muscle contains one third to one fifth as much myosin as skeletal muscle, the longer smooth muscle myofilaments and the slower crossbridge cycling rate allow it to produce as much force per unit of cross-sectional area as does skeletal muscle. Thus, the maximum values for smooth muscle on the force axis would be similar, whereas the maximum (and intermediate) velocity values are different. Furthermore, smooth muscle can have a set of force–velocity curves, each corresponding to a different level of myosin light-chain phosphorylation. This is often revealed by the particular maximum contractions and contraction velocities produced by different receptor-mediated contractile agonists in smooth muscle. That is, three different contractile agonists of the same concentration on the same smooth muscle can produce three different levels of maximum force generation and contraction velocities in that smooth muscle.

Smooth muscle has several modes of relaxation.

The central cause of smooth muscle relaxation is a reduction in the internal (cytoplasmic) calcium concentration, a process that is itself the result of several mechanisms. Electrical repolarization of the plasma membrane leads to a decrease in the influx of calcium ions, whereas the plasma membrane calcium pump and the sodium–calcium exchange mechanism (to a lesser extent) actively promote calcium efflux. Most important quantitatively is the uptake of calcium back into the SR. The net result of lowering the calcium concentration is a reduction in MLCK activity so that dephosphorylation of myosin can predominate over phosphorylation.

The cyclic nucleotides cAMP and cGMP are second messenger links between smooth muscle relaxation and activation of specific smooth muscle membrane receptors by chemical ligands that mediate relaxation of smooth muscle. For example, binding of catecholamines to beta-adrenergic receptors in vascular and bronchial smooth muscle activates **adenylyl cyclase**, which results in the formation of **cyclic adenosine monophosphate (cAMP)**. Cyclic AMP triggers a cascade of events within the smooth muscle that lead to relaxation. These include enhanced calcium uptake by the SR and decrease myosin light-chain phosphorylation.

Nitric oxide and nitrodilators such as nitroglycerin relax smooth muscle by activating guanyl cyclase and producing the second messenger **cGMP**. This leads to the activation of **cGMP-dependent protein kinase (PKG)**, which promotes the opening of calcium-activated potassium ion channels in the cell membrane, leading to hyperpolarization and subsequent relaxation. PKG also activates MLCP and blocks the activity of agonist-evoked **phospholipase C (PLC)**. This action reduces the liberation of stored calcium ions by IP$_3$.

INTEGRATED MEDICAL SCIENCES

Smooth Muscle: A Target for Medical Pharmaceutical Therapies

The highly diverse mechanisms used by smooth muscle to initiate contraction and relaxation allow it to perform very diverse tasks in the body. A lot of these tasks revolve around the location of smooth muscle in what can be called hollow organs (e.g., blood vessels, airways, the GI, and genitourinary systems and bladders). Smooth muscle can maintain constant tone, stay relaxed until needed to contract, and stay contracted until needing to relax and can both contract and relax in coordinated fashion to provide directional movement. In this manner, smooth muscle is employed to alter air flow in the lungs, blood flow to tissues as well as blood distribution and blood pressure in the body, and to function as sphincters and movement mechanisms to produce swallowing, movement, and evacuation in the GI tract as well as contraction of the bladder. These widespread organ-specific locations and functions, coupled with the multiple intracellular mechanisms and membrane receptors regulating contraction of smooth muscle, make this tissue a rich target for medical pharmacotherapy. For example, nitroglycerin is used in patients suffering an ischemic attack of the heart, which occurs when the heart's demand for oxygen exceeds its supply.

Nitroglycerin produces NO in blood vessels, which directly relaxes its smooth muscle through intracellular activation of guanylate cyclase and production of cGMP. This highly lipid-soluble and rapidly absorbed agent is taken to produce acute, significant dilation of arteries and veins, which causes redistribution of blood out of the pulmonary and cardiac area and into the periphery. This redistribution reduces and relieves stress on the heart and its need for oxygen thereby relieving the ischemic attack. Drugs that block the 1,4 DHP or L-type calcium channel have similar smooth muscle uses. These channels are predominantly found in the cardiovascular system. Like NO, they are a non–receptor-acting agent that binds to and inhibits the ability of the channel to allow calcium into smooth muscle. This causes a graded reduction in smooth muscle tone that then dilates arteries, reducing blood pressure and stress on the heart. These vascular actions of these agents are used to treat hypertension (high blood pressure) and to prevent myocardial ischemia. α1-Adrenergic receptor blockers are a class of membrane receptor selective agents that block the action of catecholamines, usually released from vascular sympathetic nerve endings, on vascular smooth muscle.

They are used to dilate arteries, primarily to reduce arterial blood pressure, and have wide use in the treatment of hypertension and myocardial ischemia. Cholinergic mimics and anticholinergic agents (blockers) have been used as pharmacotherapy for many GI and urinary bladder disorders. They can be used to decrease or increase motility in the GI tract, aid swallowing, and either reduce urinary retention or calm overactive bladder muscle.

As the student will start to surmise while progressing through this book, smooth muscle is an inviting target for pharmacological intervention in the treatment of organ system diseases and malfunctions. It will become equally apparent however that actual successful development of such therapies must be complicated by the fact that an agent used to ameliorate one malfunction in one organ system could spill its effects to another system at the same time and thus may cause serious undesired side effects. There are too many agonists and receptors in common with too many systems innervated by the ANS for this not to be a real possibility. Such ANS receptor elements are also found in the CNS and skeletal muscle system further enhancing the possibility of complex side effects. Pharmaceuticals for bladder or GI issues can spill over to other cholinergic processes such as tear formation and salivation. Even more complexity arises if these drugs have spillover effects on skeletal muscle and the CNS as well. Sometimes, route of administration of a drug used for smooth muscle effect can mitigate spill over side effects in organs. For example, β-adrenergic agonists (stimulators) relax bronchiole smooth muscle but also stimulate the heart, which could prove dangerous. However, when inhaled, these drugs dilate bronchiole smooth muscle without spilling into the general circulation. In this manner, they can be used to counteract dangerous bronchospasm in asthma without affecting the CV system. Considerations such as those mentioned above are the driving force for the need of much training of physicians so that they can properly utilize and understand the full medical consequences of using smooth muscle drugs. ■

Neuromuscular Physiology

Chapter Summary

- The contractile apparatus of skeletal muscle is made up of thick (myosin-containing) and thin (actin-containing) myofilaments arranged in interdigitating arrays called *sarcomeres*.
- Contraction is brought about by a cyclic enzymatic interaction between the myofilaments. The biochemical reactions of the crossbridge cycle require adenosine triphosphate (ATP) as a metabolic fuel and a ligand to detach myofilament attachments during the cycle. The power stroke of the myosin head provides the force to cause relative myofilament sliding and sarcomere shortening.
- The neuromuscular junction is a specialized chemical synapse that links motor nerve impulses to skeletal muscle activation.
- In skeletal muscle, membrane excitation is coupled to contraction by the action of calcium on the troponin regulatory complex on the thin filaments.
- Calcium ions are stored in the sarcoplasmic reticulum (SR) and released to the myofilaments upon depolarization of the T-tubule membranes. Binding of calcium ions to troponin allows the interaction of actin and myosin filaments in the crossbridge cycle. Relaxation occurs when calcium is pumped back into the SR.
- Muscle contraction is fueled by ATP derived primarily from the metabolism of carbohydrates and also by metabolism of amino acids and fatty acids.
- Contractile force in whole muscle is adjusted by varying the number of motor units stimulated and by the temporal pattern of stimulation. Brief, single contractions are twitches; summed contractions are tetanies.
- The length–tension curve relates initial, passive muscle length to isometric tension, and the force–velocity curve relates speed of shortening to the force exerted by the muscle while moving a load in an isotonic contraction.
- Skeletal attachments act as a lever system, increasing the effective range of muscle shortening while reducing the amount of force a limb can exert. These attachments also confine the muscle operation to near the optimal length on its length–tension curve.
- Metabolic and structure adaptations fit skeletal muscle for a variety of roles. Red muscle fibers are associated with aerobic metabolism, whereas white muscles operate anaerobically.
- Muscle metabolism and structure are plastic and may be altered epigenetically while also acting as an endocrine organ to affect other tissues in the body through epigenetic mechanisms.
- Smooth muscle is specialized for a wide range of tonic- and phasic-type contractions. It has an actin–myosin contractile system, although it lacks the regular ultrastructure of skeletal muscle.
- Regulation of smooth muscle contraction is under the control of calcium ions, which promote, via the enzyme myosin light-chain kinase, phosphorylation of the light chains on the myosin molecule. This enables it to interact with actin and participate in the crossbridge cycle.
- Smooth muscle cells are electrically coupled, and activation spreads from cell to cell throughout the tissue. Nerves may initiate contraction or modify its duration and extent.
- The latch state of smooth muscle is a condition of slow crossbridge cycling that conserves energy during tonic contractions.
- Relaxation in smooth muscle is associated with a lowering of cytoplasmic calcium and dephosphorylation of the myosin through the action of a phosphatase enzyme.

Chapter Review Questions

1. Anticholinesterases are a class of drug that inhibits the enzyme (acetylcholinesterases) that breaks down acetylcholine after its release into the neuromuscular junction. What would happen in a skeletal muscle motor unit in response to a train of motor nerve impulses if the muscle was exposed to an anticholinesterase?

   A. Acetylcholine concentration would be higher in motor endplate synapse compared to normal.
   B. Endplate potentials would be diminished compared to normal.
   C. It would require that the motor neuron be stimulated at a higher frequency than normal to produce normal amounts of acetylcholine at the motor endplate.
   D. The skeletal muscle fiber could not form action potentials.
   E. Release of acetylcholine from the motor neuron would be impaired compared to normal.

The correct answer is A. The concentration of acetylcholine in the myoneural junction is the net effect of acetylcholine being released into the junction from the motor neuron and decomposition of the acetylcholine by acetylcholinesterase in the junction. Anticholinesterases would block acetylcholine breakdown and, thus, increase the concentration of acetylcholine in the junction. The endplate potential is proportional to the number of nicotinic receptors occupied by acetylcholine and would likely be increased in the presence of an anticholinesterase. This would make it more likely that muscle action potentials could be formed. In the presence of anticholinesterases, less motor neuron stimulation (and, thus, less acetylcholine release) would be needed to maintain normal concentrations of acetylcholine in the vicinity of the motor endplate. Finally, anticholinesterases have no effect on the release of acetylcholine from the motor neuron terminal; these enzymes affect acetylcholine after it is released.

2. At L_o, a muscle group produces a maximum isometric twitch force of 10 g, whereas it produces a maximum isometric force of 7 g at 0.5 L_o. If the muscle length at L_o is 5 cm, which of the following is true?

   A. Starting a 0.5 L_o, the muscle cannot produce an isotonic contraction with a 6 g afterload.
   B. Starting a 0.5 L_o, the muscle can move an 8 g load to 0.4 L_o.
   C. An isotonic contraction of 10 g load can occur at 0.8 L_o.
   D. Starting at L_o, the muscle can move a 9 g load to 2.5 cm.
   E. Starting at L_o, the muscle can move a 7 g load to 2.5 cm.

The correct answer is E. Starting at an optimum preload, the muscle will shorten if it moves a load less than the maximum isometric force possible at its optimum preload until it reaches a length whose maximum isometric force-generating capacity is the same as the load being moved. The muscle cannot move beyond this point because, at shorter lengths, there will not be enough crossbridges to support the load being moved.

3. Blockade of calcium release from the SR of smooth muscle will:

   A. Prevent contraction of the muscle following activation of ligand-gated calcium channels
   B. Prevent action potential generation in the smooth muscle cells
   C. Not affect tonic contraction of the muscle stimulated by activation of ligand-gated membrane channels
   D. Not affect the contribution to contraction of the muscle as stimulated by IP_3
   E. Result in total relaxation of the smooth muscle

The correct answer is C. Activator calcium for contraction of smooth muscle is obtained from both the release of calcium from the SR and the entry of calcium from the extracellular fluid through one or more of the types of calcium channels in the muscle cell membrane. Opening of ligand-gated channels will contract the muscle tonically, first due to the influx of calcium through the channel and second by activating the latch state mechanism in the cell. Action potential generation in smooth muscle occurs primarily from calcium influx through membrane channels and not from release of calcium from the SR. IP_3 is a trigger for the release of calcium from the SR and contributes to smooth muscle contraction when it is produced in response to activation of membrane receptors by the appropriate ligand. Blockade of calcium release from the SR would not relax smooth muscle because of the availability of activator calcium from the extracellular fluid via membrane calcium channels.

Clinical Application Exercises 8.1

For the past month, a 48-year-old male office manager has noticed increasing muscle weakness about his shoulders, neck, and arms along with occasional trouble getting both eyes to focus on single objects (diplopia). While cleaning out his garage of garden fertilizers, malathion insecticide, and herbicides, he becomes easily fatigued, especially while reaching up and outward to clean shelving. He decides to get examined by his family physician. During his history and physical examination 2 weeks later, the patient reveals he's had no sudden falls or injuries at any time prior to appearance of his symptoms nor did he experience symptoms solely while cleaning his garage. He does complain that he is becoming more short of breath while doing physical activity. His physical revealed that he had normal reflexes but had trouble abducting or holding out his arms for any length of time. When he tries to keep his eyelids open for extended time, his left lid starts to droop. When the patient tries to smile, his smile resembles a grimace though the patient is not in any pain. Based on this history and examination, the patient's physician orders an electromyogram study of the patient's proximal arm muscles along with an acetylcholine receptor (AChR) antibody test. The myogram shows marked diminution of compound muscle action potentials within 12 minutes of stimulated electroshocks to the muscle of four stimuli per second (normal equals about 30 minutes). The anti-AChR antibody test indicates the presence of acetylcholine receptor antibodies in the patient's plasma. Based on the patient's physical and test results, the patient's physician concludes that the patient has myasthenia gravis and decides to forego an IV edrophonium test on the patient to further verify the diagnosis. He instead immediately prescribes therapy that includes intermittent use of pyridostigmine for the patient's symptoms.

QUESTIONS

1. Based on the patient's history and physical examination, postulate what type of potential causes were eliminated by the patient's physician and why.

2. Postulate what connection might exist between the patient's physical and clinical test results and his symptoms.

3. The physician determines that the patient has myasthenia gravis. What findings are consistent with this result?

ANSWERS

1. Skeletal muscle contraction requires motor nerve stimulation, release of acetylcholine transmitter at the NM junction, and excitation contraction coupling of muscle. There is no indication of traumatic injury to the CNS or peripheral nerves in this patient that would indicate a neurogenic cause of his symptoms. Similarly, this is not indicated by intact muscle reflexes as well as the fact that his symptoms are bilateral and involve more central muscles (eyelids, shoulders, neck). That

is not consistent with a specific anatomical injury. Malathion is an organophosphate anticholinesterase that can amplify acetylcholine effects at the NM junction rather than reduce them. Although in severe cases this can cause depolarization paralysis by locking the muscle sodium channel in the inactive state, this is an acute response that can result in death from paralysis of the diaphragm as well as widespread acute ANS effects. Thus, it is unlikely that the patient was exposed to toxic concentrations of malathion while cleaning his garage.

2. ACh receptor antibodies occur when acetylcholine receptors are attacked by the body's immune system. If such an attack should damage or destroy the nicotinic acetylcholine receptors in the NM junction, this would account for muscle weakness. This would also account for the patient's fatigue, inability to sustain muscle contractions (arm abduction, eye lid lift), and the fact that his muscles weaken when stimulated repetitively. When stimulated repetitively (as in the EMG test), the acetylcholine released from motor neurons diminishes over time in a process called presynaptic rundown with the effect of diminished activation of underlying skeletal muscle. However, in a normal patient, stimuli of two to four cycles per second produce a 10% to 15% reduction in muscle activation after about 30 minutes of stimulation. When this rundown phenomenon is coupled with lots of ACh receptors on the NM junction, rundown occurs much more rapidly.

3. Myasthenia gravis is a prototypical autoimmune disease that destroys nicotinic ACh receptors (via compliment fixation), stimulates internalization of the receptors, and otherwise destroys and reduces the NM surface folds. The result is markedly reduced ACh receptors available for any quanta of ACh released by motor neurons at the NM junction. Because muscle contraction depends on the number of nicotinic receptors bound to ACh, MG results in weak muscle contractions in response to normal motor neuron stimulation and exacerbates synaptic rundown. The patient's history, physical, and test results are consistent with the onset of myasthenia gravis, which generally appears suddenly without notable cause and affects muscle of the face and upper extremities and bulbar muscles.

thePoint® *Visit* http://thepoint.lww.com/rhoades5e *for additional chapter review Q&A, Clinical Application Exercises, animations, and more!*

9 Blood Composition and Function

Active Learning Objectives

Upon mastering the material in this chapter, you should be able to:

- Understand the four major functions of blood: transport, hemostasis, regulation of temperature and pH, and immunity.
- Describe the components of plasma and understand the main functions of each.
- Understand the types of formed elements and their unique roles in homeostasis.
- Explain how the blood lipid profile can be applied to assess cardiovascular health.
- Discuss the clinical usefulness of blood tests including the basic metabolic panel, complete metabolic panel, hematocrit, and complete blood cell count.

- Understand how the composition of red blood cells contributes to their function.
- Distinguish the different types of anemia and understand how these differences dictate the type of treatment protocols that would be used.
- Explain the function and clinical significance of erythropoietin as well as its involvement in blood doping.
- Understand how white blood cells form an integrated response to prevent and fight infections.
- Track the roles and components of the blood clotting phases, from immediate actions to wound healing.
- Explain the relevance of blood type in blood transfusions.

The viability and metabolism of the body's cells are dependent on adequate perfusion by the blood. In the systemic circulation, arterial blood transports components necessary for maintaining a relatively stable and constant cellular environment (e.g., osmolytes, nutrients, and O_2) and blood flowing through the veins transports CO_2 and other metabolic waste products to the lungs, kidneys, and liver for disposal. Because the heart is a dual pump, the blood flowing in the pulmonary arteries carries O_2-poor blood to the lungs for oxygenation, and the aorta carries the oxygenated blood to the systemic circulation (see Chapter 11). The components carried by the blood are delivered to individual cells as it passes through an extensive network of thin-walled capillaries. Blood also plays a role in maintaining body temperature and in fighting infection (see Chapter 10). Blood tests are common diagnostic tools to detect a wide variety of homeostatic imbalances.

▶ BLOOD FUNCTIONS

Blood is a dynamic and complex living connective tissue. The cellular and plasma components of blood work in concert to perform four major roles: transport substances, regulate hemostasis, maintain a stable internal environment by regulation of temperature and pH, and aid in resisting infection or disease via the immune system.

Transport

As the primary means of long-distance transport in the body, blood carries an abundance of important substances including electrolytes, amino acids, sugar, proteins, lipids, minerals,

hormones, and waste products. Depending upon the solubility of an individual substance, it can be transported as freely dissolved in solution (plasma), bound to a specific carrier protein (e.g., iron bound to transferrin), bound to a nonspecific carrier protein (e.g., hormones bound to albumin), or within blood cells. Control of the transport modality can be physiologically regulated. For example, about half of circulating calcium is in its free form (Ca^{2+}), whereas the other half is complexed to albumin or anions. In contrast, O_2 is preferentially transported within **red blood cells (RBCs)** bound to **hemoglobin (Hgb or Hb)**, and <1% is freely dissolved in the plasma.

Hemostasis

Complex and efficient mechanisms have evolved to prevent blood loss from a damaged vessel. The arrest of bleeding is called **hemostasis**. The failure to stop bleed after injury is called **hemorrhage** and can result in life-threatening blood loss if not controlled by the physiological mechanisms that constitute normal hemostasis.

Homeostasis

Homeostasis, as a physiological term, means maintaining a relatively steady state to create an optimal internal environment. The blood system plays a pivotal role in preserving homeostasis by maintaining pH and temperature. Plasma proteins form an immediately available buffer system to modulate the acid equivalents produced by most metabolic reactions. In addition, blood carries excess acid and base equivalents to organs such as the kidney and lungs for elimination. The rapidly circulating blood is an excellent conduit for transporting the heat generated by metabolic reactions and, therefore, plays a major role in thermoregulation by sequestering blood in the core as a result of vasoconstriction when ambient temperature is low or by dissipating heat by peripheral vasodilation when the environment is hot or the body has generated internal heat (e.g., during exercise).

Immunity

White blood cells (**WBCs**) are involved in the body's defense against infection and disease caused by pathogens and in the clearance of foreign antigens (see Chapter 10). Although the unbroken skin and mucous membranes act as barriers to the entry of infectious agents into the body, microbes can penetrate or bypass these frontline defenses. WBCs, working in conjunction with various proteins, monitor the blood to detect the presence of microorganisms and other foreign substances. In most cases, the blood's defense system is efficient enough to eliminate the pathogens or to prevent their spreading before they can cause substantial bodily harm.

▶ WHOLE BLOOD

The average adult has about 5 L (5.3 quarts) of whole blood; men have 5.0 to 6.0 L and women have 4.5 to 5.5 L. Approximately 55% of the total blood volume is liquid (**plasma**), and about 45% is formed elements (i.e., RBCs, WBCs, and **platelets**). Blood accounts for 6% to 8% of the body weight of a healthy adult.

Blood is a specialized connective tissue.

Blood is considered a type of connective tissue because the cells that give rise to the formed elements are derived from the same lineage as the cells that form bone, cartilage and the dermal layer of skin. Connective tissues are made of cells surrounded by matrix. In the case of blood, the matrix is the plasma, whereas the formed elements are RBCs, WBCs, and platelets. These are called the *formed elements* because only the WBCs are truly cells containing a nucleus. RBCs contain a nucleus during development in the red bone marrow but lose this when fully mature. Platelets are pieces of cells derived during the controlled breakage of a large cell called a **megakaryocyte**.

When fully oxygenated, whole blood is bright red, which results from the oxygenated iron in Hgb. Oxygen-poor blood (deoxygenated blood) has a lower level of oxygenated Hbg and is darker red. Oxygen-poor blood is carried by veins in the systemic circulation and by arteries in the pulmonary circulation. Systemic veins, when viewed through the skin, typically appear blue in color from the deflection of light when it penetrates the skin.

Medical terms related to whole blood often begin with **hem/o-** or **hemat/o-** from the Greek word *haima* for blood. For instance, **hemolysis** is the premature destruction of **erythrocytes**, and a **hematologist** is a blood specialist.

To obtain whole blood for laboratory analysis, an **anticoagulant** must be added to keep the blood from clotting. Clinically, there are not many assays in which whole blood is used; whole blood is so viscous (normal viscosity values are 3.5 to 5.5) and cloudy that it interferes with the chemical reactions in the test solutions. However, there is a continual effort to develop assays that overcome these problems, which would allow the use of whole blood for fast diagnosis in emergency situations.

▶ SOLUBLE COMPONENTS OF BLOOD AND THEIR TESTS

Because blood is the transport medium for many organic and inorganic substances in the body, analysis of the components of the plasma is important for detecting normal as well as pathological processes. Reference values such as those in Table 9.1 are well documented and can be useful in detecting diseases or other physiological changes that disrupt homeostasis.

TABLE 9.1 Basic Components of Plasma

Class	Substance	Reference Range
Cations	Sodium (Na^+)	136–145 mEq/L
	Potassium (K^+)	3.5–5.0 mEq/L
	Calcium (Ca^{2+})	4.2–5.2 mEq/L
	Magnesium (Mg^{2+})	1.5–2.0 mEq/L
	Iron (Fe^{3+})	50–170 µg/dL
	Copper (Cu^{2+})	70–155 µg/dL
	Hydrogen (H^+)	35–45 nmol/L
Anions	Chloride (Cl^-)	95–105 mEq/L
	Bicarbonate (HCO_3^-)	22–26 mEq/L
	Lactate$^-$	0.67–1.8 mEq/L
	Sulfate (SO_4^{2-})	0.9–1.1 mEq/L
	Phosphate ($HPO_4^{2-}/H_2PO_4^-$)	3.0–4.5 mg/dL
Proteins	Total	6–8 g/dL
	Albumin	3.5–5.5 g/dL
	Globulin	2.3–3.5 g/dL
Fats	Cholesterol	150–200 mg/dL
	Phospholipids	150–220 mg/dL
	Triglycerides	35–160 mg/dL
Carbohydrates	Glucose	70–110 mg/dL
Other examples	Alkaline phosphatase	20–140 U/L
	Alanine transaminase	8–20 U/L
	Aspartate transaminase	9–40 U/mL
	Bilirubin (total)	0.1–1.0 mg/dL
	Blood urea nitrogen	7–18 mg/dL
	Creatinine	0.6–1.2 mg/dL
	Ketones	0.2–2.0 mg/dL
	Vitamin A	0.15–0.6 µg/mL
	Vitamin B_{12}	200–800 pg/mL
	Vitamin C	0.4–1.5 mg/dL

Plasma becomes serum after the removal of clotting factors.

Plasma is about 93% H_2O, with the remaining 7% composed of dissolved or suspended solutes (6% organic substances and 1% inorganic substances). Plasma is obtained by collecting whole blood in a tube containing an anticoagulant that prevents blood clotting, such as ethylenediaminetetraacetic acid (EDTA), citrate, or heparin. Centrifugation is used to separate the blood into an upper plasma layer; a lower RBC layer; and a thin interface, the **buffy coat** that contains WBCs and platelets. When blood is allowed to clot or coagulate before centrifugation, the liquid remaining at the top of the tube is **serum**, which is now devoid of the soluble clotting factors that precipitated with the clot.

Plasma from a patient who has fasted overnight is a cloudy, pale, or grayish yellow liquid. If the blood is drawn shortly after a meal, it may appear milky, due to a high lipid or chylomicron content. Reddish samples are generally due to the presence of Hgb due to hemolysis of RBCs.

For many biochemical tests, plasma and serum can be used interchangeably. For some tests, only serum can be used because the clotting factors in plasma interfere with the assay. For coagulation tests, only plasma can be used because all clotting factors need to be present.

Plasma can be stored frozen below −20°C for future analysis, but it must be frozen within 6 to 8 hours after donation to preserve clotting factors. **Fresh frozen plasma** (**FFP**) can be stored in a blood bank for 1 to 7 years and can be used for therapeutic plasma exchange called **plasmapheresis**. FFP that is frozen after 6 and 24 hours of whole blood collection has a lower level of labile coagulation factors, such as factors V and VIII.

Analyzing blood samples reveals a patient's health status.

For most analyses, blood samples are drawn from a patient's arm vein (**venipuncture**). In newborns, blood is collected by a **heel stick**. Blood from an artery is collected mainly to measure arterial blood gases (i.e., O_2 and CO_2) and HCO_3^- to determine their levels related to lung function. Systemic arterial blood will be highest in the blood gases because it has not moved through the tissues where extraction of these gases takes place. For the same reason, arterial concentrations of drugs are higher than venous concentrations. For most other analyses, values from the arterial and venous sides are equal.

Blood is collected into evacuated collection tubes, designed to fill with a predetermined volume (typically 7 mL) of blood. The rubber stoppers are color-coded according to the additive present in the tube. Lavender-top tubes contain EDTA to bind Ca^{2+}, a critical component for blood clotting, and are used to produce plasma for hematology tests such as **complete blood counts** (**CBCs**) and blood typing. Light-blue tubes contain sodium citrate, an alternative anticoagulant, and are used for coagulation tests. Green tubes contain the anticoagulant heparin and are used to obtain plasma for a variety of clinical chemistry tests. Red tubes (serum separator tubes) contain clot activators and are used to produce serum.

Plasma contains ions, carbohydrates, lipid, and proteins.

Electrolytes are salts found in their dissociated form (ions) in the blood and are either negatively (**anions**) or positively (**cations**) charged. Ions are important for maintaining a normal osmotic balance in the plasma. In general, when the body needs to move water from one compartment to another, an electrolyte is transported under controlled conditions (often hormonally regulated) and water follows passively.

In plasma, Na^+ is the most abundant cation and Cl^- and HCO_3^- are the major anions. These three ions are the main osmotically active solutes in the extracellular fluid. Because electrolytes are the most easily transported substances between cellular compartments, their movement will directly impact osmotic water gain or loss. For example, a gain or loss of Na^+ will cause blood volume expansion or contraction, respectively, and can contribute to blood pressure. Other functions of electrolytes include their role in membrane excitability and pH regulation.

Like electrolytes, proteins also contain a net charge that is usually negative at physiological pH. Because of the principle of **electroneutrality** in blood plasma, the sum of positive charges equals the sum of negative charges.

Proteins are the most abundant organic components in the plasma. Almost 1,400 different *plasma proteins* accounting for 6 to 8 g of protein/dL of blood have been identified. The proteins function as enzymes, hormones, antibodies, and transporters as well as contribute to plasma osmolality and acid–base balance. Proteins in the blood form the **oncotic pressure**, also known as **colloidal osmotic pressure**. Because proteins do not diffuse across cell membranes, this component of the total osmotic pressure is relatively stable, whereas the osmotic pressure due to electrolytes is more changeable due to the ease of transport of these ions across cell membrane.

Albumins account for about 60% of the total plasma protein concentration, which makes these proteins a significant contributor the oncotic pressure. Albumin also serves as a nonspecific transport protein that noncovalently binds a variety of molecules including fatty acids, hormones, drugs, and other substances. Albumins are produced by the liver, and a low concentration in serum might indicate liver disease and/or malnutrition.

Globulins constitute about 36% of the total plasma proteins and are often subdivided into categories called **alpha** (**α**), **beta** (**β**), and **gamma** (**γ**) **globulins**. These designations are historical and arise from a now outdated technique of column chromatography, which separated the plasma membrane proteins by molecular weights. Because the categories separate proteins by size, each of the subdivisions represents multiple types of proteins. In general, the α and β globulins are proteins produced in the liver, whereas γ globulins are produced by cells of the immune system. The α and β globulin fractions contain specific transport proteins (e.g., metal-binding proteins, hormone-binding proteins, and lipid-binding proteins) and proteins involved in clot retraction. γ-Globulins are predominately antibodies produced by B lymphocytes (see Chapter 10).

Fibrinogens are large proteins synthesized in the liver that account for about 4% of the total plasma proteins. Fibrinogen is enzymatically cleaved by thrombin to yield the insoluble protein **fibrin**, which forms the fibrous, meshlike structure of a blood clot. *In vitro*, plasma depleted of fibrinogen will not coagulate. Higher fibrinogen levels are correlated with an increased risk of stroke.

The main **plasma lipids** are cholesterol, phospholipids, and triglycerides. Cholesterol is an important component of cell membranes as well as a precursor for steroid hormone synthesis. Phospholipids are used for building plasma membranes. Triglycerides are important for the transfer of food-derived energy into cells. Because of their hydrophobic nature, plasma lipids are transported together with proteins.

Dietary triglycerides and dietary cholesterol are assembled into chylomicrons by the intestinal mucosa before they are released into the lymph and ultimately into the circulation (see Chapter 25). Triglycerides and cholesterol from the liver are packaged as **very-low-density lipoproteins** (**VLDLs**) and **intermediate-density** or **low-density** (**LDL**) **lipoproteins** that are released into the plasma for the delivery of lipids including cholesterol to other tissues. LDLs are absorbed by peripheral cells in which the cholesterol content of the lipoproteins is removed and used by the cells. The liver and small intestine also form and release **high-density lipoprotein** (**HDL**), which binds serum cholesterol and delivers it back to the liver, where it is passed from the body via bile (see Chapter 26).

The major **plasma carbohydrate** is glucose. Glucose is a primary energy source for all the body's cells and the sole source of energy for cells such as nerves and cardiac muscle cells. Serum glucose level is tightly regulated within 70 to 110 mg/dL (4 to 8 mmol/L), except for the time immediately following a meal. Hence, blood glucose tests are standardized in regard to food intake (e.g., fasting test and 2-hour postprandial test). When quantifying blood glucose, the serum must be promptly separated from the RBCs because they can utilize glucose in the absence of insulin.

Blood lipid profile helps determine a patient's cardiovascular disease risk.

A standard **blood lipid profile** includes values for total cholesterol, LDL, HDL, and total triglycerides. These profiles give important information about a patient's risk of developing heart-related conditions, with high LDL, low HDL, and/or high triglycerides being positive indicators of health risk. With high LDL (the form that takes cholesterol to tissues), the risk of cholesterol plaque buildup in arterial walls (**atherosclerosis**) is increased. Low HDL levels indicate decreased ability to remove excess cholesterol. Many people with obesity, heart disease, and/or diabetes also have high triglyceride levels. Several ratios, such as the LDL/HDL ratio and the total cholesterol/HDL ratio, are also used as risk predictors.

Basic and complete metabolic panels are indicators of metabolic health.

The **basic metabolic panel** (**BMP**) includes the analysis of glucose, Ca^{2+}, Na^+, K^+, CO_2 (or HCO_3^-), Cl^-, blood urea nitrogen (**BUN**), and creatinine. The latter two are metabolites that are filtered out of plasma by the kidneys, so their increase in plasma might indicate kidney dysfunction. Abnormalities in an **electrolyte panel** (i.e., electrolytes and HCO_3^-) commonly portend problems with fluid balance, such as edema or hypertension.

The **complete metabolic panel** (**CMP**) includes the tests of the BMP and the following components: albumin;

total protein; bilirubin; and the liver enzymes alkaline phosphatase (ALP), alanine transaminase (ALT), and aspartate transaminase (AST). The latter four components are used to assess liver health and function. For instance, ALP is elevated in hepatitis, ALT is commonly increased with bile duct obstruction, and AST levels are used to monitor general liver damage. Bilirubin is a waste product of the liver, produced from RBC breakdown and recycling. Depending on its form (conjugated or unconjugated), bilirubin levels can be used to identify a problem that occurs before the liver (e.g., hemolytic anemia), a problem within the liver (e.g., cirrhosis), or a problem after the liver (e.g., bile duct blockage).

Abnormal serum protein patterns in electrophoresis reveal health problems.

Serum protein electrophoresis is a common method for separating blood proteins in a solid matrix (mostly cellular acetate) according to their size and charge (Fig. 9.1). Globular blood proteins (albumin, and α, β, and γ globulins) normally form five main peaks, or zones, on the matrix: albumin (59%), $α_1$-zone (4%), $α_2$-zone (7.5%), β-zone (12%), and γ-zone (17.5%). These zones might differ in size and/or pattern in a patient with certain types of anemias, during acute inflammation, or in the presence of an autoimmune disease. For instance, an increased $β_1$ peak is typical of iron deficiency anemia due to an increased level of the iron-binding protein transferrin. Another example is a largely increased γ-globulin peak observed in most patients with multiple myeloma due to the presence of proteins produced by abnormal plasma cells.

Immunologic assays detect and measure serum antigens and antibodies.

A wide variety of serum immunoassays are available in which a precipitation reaction occurs as a result of antibodies combining with a specific antigen. The detection and quantification of either antibodies or antigens are diagnostically valuable. For example, elevation of smooth muscle cell antibodies and antinuclear antibodies points toward autoimmune hepatitis. Elevation of gluten or antigliadin antibodies can be used to diagnose celiac disease. Abnormal antigens found on microorganisms can be detected by immunoassay and can often identify infected people before symptoms appear; alternatively, the presence of circulating antibodies to particular antigens can be used to determine if a recovered patient was infected with a particular microorganism.

In some cases, antigens interact with antibodies to produce large aggregates visible to the naked eye, for example, the instantaneously readable ABO-Rh blood typing test kits, in which the antigens on the surface of RBCs agglutinate with added antibodies. In most cases, however, the antigen–antibody complexes must be labeled for detection (e.g., dyes, fluorescent reagents, or radioactive isotopes).

The blood tests that are cited are meant to be neither an exclusive nor an exhaustive list. However, it is useful to remember that the blood is the conduit for moving both nutrients and waste products and is, therefore, where many of the aberrations of homeostasis will be expressed.

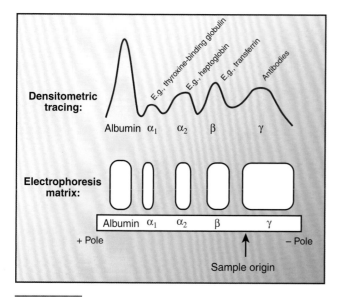

Figure 9.1 **Serum protein electrophoresis.** Globular serum proteins are electrophoretically separated on a matrix and stained. The intensity of the dye can be densitometrically quantified, and changes to the normal pattern are clinically relevant.

▶ FORMED ELEMENTS OF BLOOD AND COMMON DIAGNOSTIC TESTS

As noted above, the formed elements of blood are RBCs (**erythrocytes**), WBCs (**leukocytes**), and platelets (**thrombocytes**) (Fig. 9.2). The term **formed elements** is used instead of cells because only the WBCs are truly cells. Each microliter of blood contains 4 to 6 million RBCs, 4,500 to 10,000 WBCs, and 150,000 to 400,000 thrombocytes.

Blood is a viscous liquid.

The **viscosity** of blood, a measure of resistance to flow, is 3.5 to 5.5 times that of H_2O. The plasma viscosity is about 1.5 to 1.8 times that of H_2O. Blood viscosity increases or decreases as the total number of formed elements or plasma proteins changes. Pumping high-viscosity blood puts strain on the heart because a higher pressure is required to achieve tissue perfusion. High-viscosity blood also tends to coagulate more easily. In healthy people, a slight increase in the relative blood cell concentration and, hence, blood viscosity can occur (for instance, due to dehydration), but this is easily tolerated. However, in those with more viscous blood, such as the substantial increase in RBCs seen in patients with lung disease, the increase may lead to cardiovascular problems. Decreased blood viscosity is usually indicative of other serious health problems that cause a decrease in plasma proteins or formed elements. Because RBCs are the most abundant of the formed elements, a decrease in viscosity is often associated with anemias.

Hematocrits determine the amount of red blood cells.

The **hematocrit (Hct)** is the fraction of the total blood volume composed of RBCs and is a simple and important screening diagnostic procedure in the evaluation of hematologic disease. Hct can be determined by centrifugation of anticoagulated blood within small capillary tubes to separate blood cells from the plasma. A simple measurement of the RBC volume (length of RBCs in capillary tube) divided by the total length of the RBCs plus the plasma provides the percentage of RBCs (Fig. 9.3). Hct values have a relatively large normal range and vary somewhat between men and women (Table 9.2).

A decreased Hct indicates **anemia** that can arise from deficiencies in RBC production or premature RBC lysis. Immediately after hemorrhage, the Hct does not change because cells and plasma are lost in equal proportions.

Increased Hct values indicate **polycythemia** and may result from either an increase in the rate of production or a decrease in the rate of destruction of RBCs. For example, respiratory disease can cause polycythemia because the lack of O_2 due to the compromised gas exchange signals the synthesis of additional

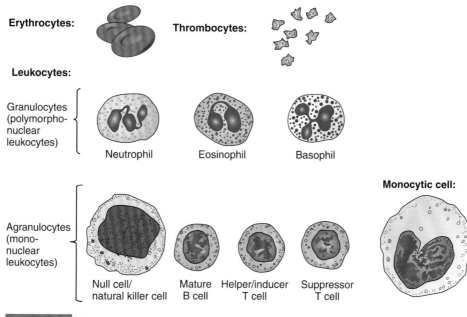

Figure 9.2 **Formed elements of blood.** This figure shows the size and appearance of cells and cell fragments (thrombocytes) found in the circulation.

RBCs. Dehydration, which decreases the H_2O content and thus the volume of plasma, also results in an increased Hct.

Complete blood counts determine the number and type of cells in the blood.

The complete blood count (CBC) includes the following tests, which are summarized in Table 9.2: RBC count, Hgb, Hct, and RBC indices; total WBC count and differential WBC count; and platelet count and mean platelet volume. The cell counts can be given as absolute values or as percentages. In Table 9.2, the reference values of a healthy adult are given both in commonly used and in standardized international (SI) units. The values vary slightly with age, sex, and physiologic condition (e.g., pregnancy and elevation above sea level).

In evaluating patients for hematologic diseases, it is important to determine the total number of circulating RBCs and the Hgb concentration in the blood. This information is used to determine whether the patient is anemic. The Hct can only be used to determine anemia when fluid status is taken into account. With a given RBC count, Hgb, and Hct, several other important **blood indices** can be derived.

The **mean corpuscular (or cell) volume (MCV)** is the index most often used, because it reflects the average volume of each RBC. It is calculated as follows:

$$MCV = \frac{Hematocrit}{RBC\ (cells/L)} \quad (1)$$

Cells of normal size are described as **normocytic**, cells with a low MCV are **microcytic**, and those with a high MCV are **macrocytic**.

The **mean corpuscular (or cell) hemoglobin (MCH)** value is an estimate of the average Hgb content of each RBC. It is derived as follows:

$$MCH = \frac{Blood\ hemoglobin\ (g/L)}{RBC\ (cells/L)} \quad (2)$$

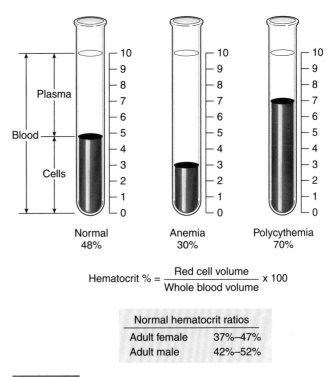

$$Hematocrit\ \% = \frac{Red\ cell\ volume}{Whole\ blood\ volume} \times 100$$

Normal hematocrit ratios	
Adult female	37%–47%
Adult male	42%–52%

Figure 9.3 **The hematocrit.** After centrifugation of whole blood, the relative amount of RBCs as a function of total blood volume is calculated as shown in the figure. Normal values differ slightly according to gender. Anemia that results from a loss of RBCs will result in a low hematocrit; various physiological or pathophysiological states that result in high RBC production will result in a high hematocrit (polycythemia).

MCH values usually rise or fall as the MCV is increased or decreased. The MCH is also often related to the **mean corpuscular (or cell) hemoglobin concentration (MCHC)** because the RBC count is usually related to the Hct. Exceptions to this rule yield important diagnostic clues.

Abbreviation	Test	Reference Range	Standardized International Reference	Significance (Examples)
RBC	Red blood cells			
	Men	4.3–5.9 × 10⁶/μL	4.3–5.9 × 10¹²/L	
	Women	3.5–5.5 × 10⁶/μL	3.5–5.5 × 10¹²/L	
Hgb or Hb	Hemoglobin			↓: anemia, severe bleeding
	Men	13.5–17.5 g/dL	2.09–2.71 mmol/L	↑: too many made, fluid loss, polycythemia
	Women	12–16 g/dL	1.86–2.48 mmol/L	
Hct or Ht	Hematocrit			
	Men	41%–53%	0.41–0.53	
	Women	36%–46%	0.36–0.46	
MCV	Mean cell volume	80–100 μm³	80–100 fL	↓: iron deficiency, thalassemia
MCH	Mean cell hemoglobin	25.4–34.6 pg/cell	0.39–0.54 fmol/cell	↑: B12 and folate deficiency (MCH variable)
MCHC	Mean cell hemoglobin concentration	31%–36% Hb/cell	4.81–5.58 mmol Hb/L	↓: deficient Hgb synthesis
				↑: spherocytosis

TABLE 9.2 Complete Blood Count

TABLE 9.2 Complete Blood Count (*Continued*)

Abbreviation	Test	Reference Range	Standardized International Reference	Significance (Examples)
RDW	RBC distribution width	11.7%–14.2%	0.12–0.14	↑: mixed population, immature cells
WBC	White blood cells	4,500–11,000/µL	$4.5–11 \times 10^{-3}$/L	↓: some medication, autoimmune diseases, bone marrow diseases, severe infections
Neutrophil	Neutrophils (54%–62%)	4,000–7,000/µL	$4–7 \times 10^{-3}$/L	
Lymph	Lymphocytes (25%–33%)	2,500–5,000/µL	$2.5–5 \times 10^{-3}$/L	↑: infection (abscess, meningitis, pneumonia, appendicitis, tonsillitis), inflammation, leukemia, stress, dead tissue (burns, heart attack, gangrene)
Mono	Monocytes (3%–7%)	100–1,000/µL	$0.1–1 \times 10^{-3}$/L	
Eos	Eosinophils (1%–3%)	0–500/µL	$0–0.5 \times 10^{-3}$/L	
Baso	Basophils (0%–1%)	0–100/µL	$0–0.1 \times 10^{-3}$/L	
Plt	Platelet count	$0.15–0.4 \times 10^{6}$/µL	$0.15–0.4 \times 10^{12}$/L	↓: not enough made, bleeding, systemic lupus erythematosus, pernicious anemia, hypersplenism, leukemia, chemotherapy
MPV	Mean platelet volume	7.5–11.5 µm³	7.5–11.5 fL	
				↑: too many made, young cells

CLINICAL FOCUS | 9.1

Blood Group Systems

The blood types A, B, AB, and O are based on the presence or absence of distinct antigens A or B on red blood cells. People with red blood cells that are covered with A or B molecules are said to have *type A* blood or *type B* blood, respectively. If both molecules, A and B, are present, the blood is *type AB*; if neither is present, it is called *type O*.

Knowing about a person's blood type is important for blood transfusions. Inappropriate combinations between the blood groups of the recipient and the donor can lead to potentially fatal agglutination, because the recipient's immune system will have antibodies against the donor's red blood cells. Agglutination of red blood cells may lead to improper blood circulation, release of hemoglobin (Hgb) that crystallizes, and eventually, kidney failure.

People have antibodies of the immunoglobulin (Ig) M class directed against the antigens that are not on their red blood cells. Therefore, someone with an A blood type will have anti-B antibodies; someone with B blood type will have anti-A antibodies and someone with O blood type will have antibodies to both A and B. Consequently, a person with type A blood can receive type A or O, not type B or AB blood. Correspondingly, a person with type B can receive type B or O blood. Those with AB blood can receive blood from anyone and are, therefore, **universal recipients**. Type O persons can receive blood from a type O person only but can donate to all groups and are, thus, called **universal donors**.

Blood type can further be useful for a person's identification, such as in forensic medicine and for the identification of family relationships in disputed parentage. The latter is possible because the blood type of the child is related to that of the parents. For instance, if the mother and the child have type O blood, the father's blood type must be A, B, or O. If the mother has type O blood, but the child presents with type B blood, the father's blood type must be B or AB.

In the United States, type O blood is most common, followed by types A, B, and AB, in this order. Because blood groups are genetically determined, the frequency of the ABO blood groups varies in populations throughout the world, even in different groups within a given country. Most likely, these differences will gradually disappear because of increased mobility and a greater societal acceptance of interracial marriages. In the United States, whites are currently characterized by a higher frequency of type A and a lower frequency of type B in comparison with African and Asian Americans, whereas Indians have a low frequency of both A and B types.

The blood type of a person is unrelated to his or her Hgb type because blood antigens are membrane factors, whereas Hgb is dissolved within the cell cytoplasm.

The other blood group system that is often grouped with the ABO system involves **Rhesus factors** (**Rh**), which are unique cell surface proteins. The Rh system is complex, involving three genes producing Rh antigens C, D, and E. Rh D, the most important, is found in the blood of 85% of people, who are classified as Rh+ (Rhesus positive). The remaining 15% are said to be Rh− (Rhesus negative).

Knowing about the Rh factor is important in pregnant women because a baby's life can be endangered if it inherits Rh+ blood from its father but the mother is Rh−. Late in pregnancy, or at parturition, the baby's blood may cross to the mother's system, where the erythrocytes are recognized as foreign and antibodies are formed against them. These antibodies pose a serious threat to any future Rh+ babies of hers in following pregnancies. To prevent

(Continued)

this, the mother receives anti-Rhesus D immunoglobulin around the 28th week of pregnancy and right after delivery. The Ig attaches to Rh+ cells from the baby in the mother's bloodstream and destroys them, preventing the triggers for the mother's immune system to produce its own anti-D antibody.

Similar problems of incompatibility between mother and child generally do not exist for the ABO blood groups because the antibodies formed belong to the IgM class and cannot cross the placenta. In rare cases, however, a type O mother with a fetus that is A, B, or AB can deliver a newborn with signs of immunologic defense reactions such as jaundice, anemia, and elevated bilirubin levels. Because of these complications, all pregnant women are advised to determine their ABO/Rh blood type during their first prenatal visit. ∎

The MCHC provides an index of the average Hgb content in the mass of circulating RBCs. It is calculated as follows:

$$\text{MCHC} = \frac{\text{Blood hemoglobin (g/L)}}{\text{Hematocrit}} = \frac{\text{MCH}}{\text{MCV}} \quad (3)$$

Low MCHC indicates deficient Hgb synthesis, and the cells are described as **hypochromic**. High MCHC values are rare because, normally, the Hgb concentration is close to the saturation point in RBCs.

The **red cell distribution width** (**RDW**) measures the degree of the average size dispersion of RBCs, thereby indicating the degree of **anisocytosis** (variation in RBC size).

The **oxygen-carrying capacity** (**OCC**) is the maximum amount of O_2 that can be carried by the Hgb contained in 1 dL (100 mL) of blood. Each gram of Hgb can combine with and transport 1.34 mL of O_2, which is equal to 20 ml of O_2 per dL under ideal conditions. The actual measured value will vary with the total Hgb, which is why anemia can lead to severe hypoxemia.

The **white blood cell count** (**WBCC**) measures the concentration of leukocytes in blood. A high WBCC is called **leukocytosis** and occurs in infection, allergy, systemic illness, inflammation, tissue injury, and leukemia. A low WBCC is called **leukopenia** and can be present in some viral infections, immunodeficiency states, and bone marrow failures. The WBCC is often used to monitor a patient's recovery from illness.

For diagnostic purposes, a **differential white blood count** (**diff**) will determine which type of WBC is affected. This information provides invaluable clues to the type of illness but also requires careful analysis. Unlike for RBCs, there are various storage pools for WBCs outside of blood, such as the spleen, lymph nodes, and lymphoid tissues. Therefore, changes in blood concentrations indicate changes in the storage pool equilibrium. Second, the number and percentage of WBCs vary somewhat with age. Newborns have a high WBCC for the first weeks after birth with a high percentage of **neutrophils**. In childhood, **lymphocytes** are predominant, whereas, at senescence, total WBCC might be decreased. Last, the CBC provides a **platelet count** (**PltC**), which is used as a starting point in the diagnosis of disorders of hemostasis. Decreased PltsC may be the result of bone marrow failure or of peripheral platelet destruction.

Blood smears detect blood parasites and other hematologic disorders.

A visual analysis of the blood cells in a **blood smear** can validate any cellular abnormalities or determine the presence of certain microorganisms. Light microscopic analysis allows the differentiation of WBCs according to their morphologic appearance and staining characteristics and the identification of immature cells, so-called **band cells**. A blood smear analysis will reveal the characteristic "sickle" shapes of RBCs, which occur in sickle cell anemia and changes in the appearance of the cytoplasm of RBCs infected with parasites such as those that cause malaria.

Blood smears are commonly stained with a polychrome stain such as **Wright-Giemsa** after fixing the cells with methanol. The orange **eosin** dye stains the basic components of the cell, such as the granules of eosinophils and the Hgb of RBCs. The **methylene blue** dye stains the acid components of the cell, such as the DNA and RNA. Many other stains are available.

A **relative WBC count** indicates which individual cell type is expanded relative to the other types of cells and aids in disease diagnosis. The numbers of particular types of WBCs will change in response to infection, allergic reactions, inflammation, and certain cancers such as leukemia. **Leukopenia** indicates an inadequate number of WBCs, whereas **leukocytosis** refers to high numbers of WBCs.

▶ RED BLOOD CELLS

RBCs are the most abundant of the formed elements and are the principle means of delivering O_2 to cells via the circulatory system. The cell's cytoplasm is rich in Hgb, an iron-containing biomolecule that readily binds with O_2. Mammalian mature RBCs are unique among those of vertebrates because they lack a nucleus and most organelles. RBCs are produced in the red bone marrow and circulate for about 100 to 120 days before being recycled by macrophages.

Erythrocytes consist mainly of hemoglobin, a unique pigment containing heme groups in which iron atoms bind to oxygen.

Mature RBCs have the shape of flexible biconcave disks, with a diameter of about 7 μm and a maximum thickness of 2.5 μm (Fig. 9.4). This form maximizes the amount of surface area available for gas exchange. Hgb within the RBCs is responsible for binding and transporting 99% of the O_2 need by the tissues. The other 1% is dissolved in the plasma.

Each RBC contains several hundred Hgb molecules of about 64,500 daltons consisting of four globin polypeptides and four heme groups, the iron-carrying portion. These polypeptide chains contain two α-globin molecules and two molecules of another type of globin chain (β, γ, δ, or ε). Four types of mature Hgb molecules, designated by their polypeptide composition, can be found in human RBCs. The most prevalent in adults is HbA (see Fig. 9.4) consisting of two

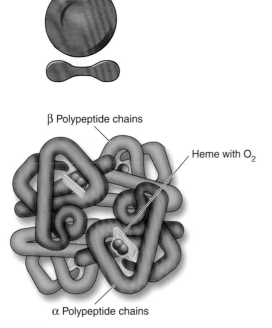

β Polypeptide chains

Heme with O_2

α Polypeptide chains

Figure 9.4 | **Structure of hemoglobin A.** Erythrocytes have the shape of biconcave disks, with a diameter of about 7 μm and a maximum thickness of 2.5 μm. They contain several hundreds of hemoglobin molecules, each consisting of four polypeptide chains (two α and two β), with each chain containing iron bound to its heme group. (Modified from McArdle WD, Katch FI, Katch VL. *Essentials of Exercise Physiology.* Baltimore, MA: Lippincott Williams & Wilkins, 2005.)

α and two β polypeptide chains ($\alpha_2\beta_2$). HbA_2, which makes up about 1.5% to 3% of total adult Hgb, has the subunit formula $\alpha_2\delta_2$. **Embryonic hemoglobin (EH)** is present in RBCs during early human development. It consists of two α chains and two ε chains ($\alpha_2\varepsilon_2$). The production of ε chains ceases at about the 3rd month of fetal development. **Fetal hemoglobin (FH)** (also known as **hemoglobin F** or $\alpha_2\gamma_2$) replaces EH and is the major Hgb component during intrauterine life. Both EH and FH have higher O_2-binding affinity to assure proper O_2 levels for the fetus within the uterine environment. FH levels in circulating RBCs decrease rapidly during infancy but are maintained at a concentration of ~0.5% in adults.

The production of each type of globin chain is controlled by individual structural genes with five different loci. Mutations can occur anywhere in the five gene loci resulting in the production of abnormal Hgb molecules leading to **hemoglobinopathies**. The most common of this spectrum of diseases is sickle cell anemia, which is caused by a point mutation (single amino acid substitution) in the β-globin chain. Another relatively well-known type of hemoglobinopathies is the thalassemias. These fall into two types, the α and β thalassemias named for the globin chain containing the mutation.

Each globin chain is bound to a heme group, which is a complex porphyrin ring containing a central iron atom (see Fig. 9.4). O_2 binds to the iron forming **oxyhemoglobin (HbO_2)**, the O_2-saturated form of Hgb, which transports O_2 from the lungs to tissues. When the O_2 is released to the tissues, HbO_2 becomes reduced hemoglobin (**deoxyhemoglobin**).

The reduced hemoglobin has an increased affinity for carbon dioxide, which reversibly binds the α and β chains to remove this waste product of metabolism from the tissues. However, only a minor portion of the CO_2 is removed in this way. In the tissues, the removal of CO_2 takes place via three mechanisms. Approximately 5% to 7% is bound to plasma proteins, 10% to 15% binds to Hgb, and the remainder is converted into H_2CO_3. CO_2 diffuses into RBCs and is converted to H_2CO_3 by carbonic anhydrase. It then freely dissociates into HCO_3^- and H^+. The H^+ ions are buffered by Hgb, and, in this way, erythrocytes also play a major role in maintaining blood pH. Hgb also acts as a buffer by ionizing the imidazole ring of histidines in the protein.

Carbon monoxide (CO) can rapidly replace O_2 in HbO_2, forming the almost irreversible binding, which accounts for the asphyxiating properties of CO. Nitrates and certain other chemicals oxidize the iron in Hgb from the ferrous (Fe^{+2}) to the ferric state (Fe^{+3}), resulting in the formation of **methemoglobin (metHb)**. The O_2 carried by MetHb is so tightly bound to ferric iron that it is not released to the tissues and is, therefore, useless in respiration.

Cyanosis, the dark-blue coloration of skin associated with **anoxia**, becomes evident when the concentration of reduced hemoglobin exceeds 5 g/dL. If the condition is caused only by a diminished O_2 supply it is reversible by the administration of O_2, but if it is caused by the accumulation of stabilized metHb, then administration of O_2 alone will not be effective.

Changes in erythrocyte morphology provide insights into specific blood disorders.

The RBC membrane is attached to a complex skeleton of fibrous intracellular proteins. This structural arrangement gives the cell stability and its unique shape, while providing the flexibility necessary to move through small, curved vessels such as capillary beds. Changes in RBC composition may lead to morphologic changes, which can provide valuable clinical information regarding the nature of the pathology (Fig. 9.5).

Changes in size and shape

Large variation in the size of RBCs is referred to as **anisocytosis**. Larger-than-normal RBCs are termed **macrocytes**; smaller-than-normal RBCs are referred to as **microcytes**. Irregularly shaped RBCs are referred to as **poikilocytes**. **Echinocytes**, or **Burr cells**, are spiked RBCs generated by alterations in the plasma environment. **Schistocytes** are fragments of RBCs damaged during blood flow through abnormal blood vessels or cardiac prostheses.

Changes in color

Abnormal Hgb content of RBCs may lead to changes in the staining pattern of cells observed on stained, dried films. Normal cells appear red-orange throughout, with a slight central pallor as a result of the cell shape (**normochromic**). **Hypochromic** cells appear pale with only a ring of deeply colored Hgb on the periphery. Other pathologic variations in RBC appearance include **spherocytes**, small, densely staining cells with a loss of biconcavity due to membrane abnormalities, and **target cells** (also known as *codocytes* or

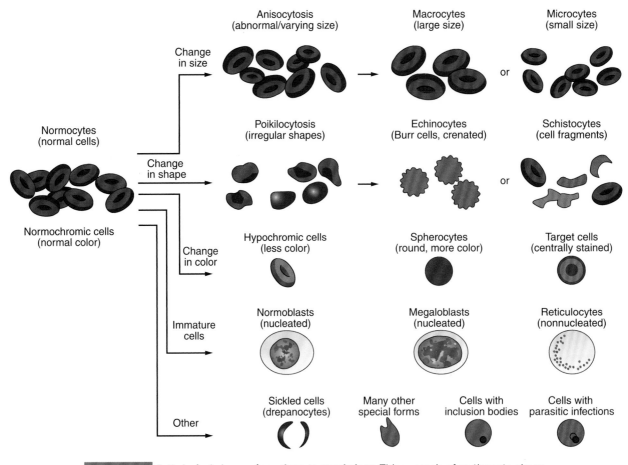

Figure 9.5 **Pathological changes in erythrocyte morphology.** This synopsis of erythrocyte abnormalities presents a variety of possible deviations that are helpful for the diagnosis of anemias and other diseases.

leptocytes), which are cells with a densely staining central area and a pale surrounding area. Target cells are observed in liver disease and after splenectomy.

Immaturity

Normal mature RBCs do not have cell nuclei, and, therefore, the presence of nucleated RBCs in peripheral blood is of diagnostic significance. An immature form of RBCs that still has its nucleus, the **normoblast**, is found predominately in the red bone marrow (see Fig. 9.8) but is seen in the blood in several types of anemias, especially when the marrow is actively responding to a demand for new RBCs. In seriously ill patients, the appearance of normoblasts in peripheral blood is a grave prognostic sign preceding death, often by several hours. Another nucleated RBC, the **megaloblast**, is unusually large due to delayed nuclear maturation and cell division. Megaloblasts are seen in peripheral blood in pernicious anemia and folic acid deficiency. **Reticulocytes** are immature cells that have shed their nucleus but still retain residual nuclear material that can be made visible with a specific vital stain. A small amount of these normal RBC precursors is found in the blood. However, an increase in the percentage of reticulocytes is an indicator that the level of **erythropoiesis** (RBC formation) in bone marrow has increased. For instance, if a patient is anemic or has had a

recent hemorrhage and the bone marrow is functioning properly, the reticulocyte count will be elevated.

Other changes

There is an abundance of other RBC abnormalities, which may result from hemoglobinopathies (e.g., **drepanocytes**, or sickled cells), liver diseases (e.g., **stomatocytes**, or mouth cells), and hematopoiesis outside the bone marrow cavity (e.g., **dacryocytes**, or teardrop cells). Cells may contain **protein inclusion bodies** from abnormal Hgb precipitation, lead poisoning, or abnormal erythropoiesis. Four species of the protist genus *Plasmodium* are able to infect human RBCs and cause malaria. Another protist genus, *Babesia*, causes malaria-like symptoms. Both *Plasmodium* and *Babesia* infections can be diagnosed by observing the characteristic appearance of the cytoplasm of RBCs from infected individuals.

Red blood cell components are recycled.

Following their release from the bone marrow, RBCs circulate in the blood for about 120 days before they become senescent and die. To maintain homeostasis under normal conditions, about 200 billion replacement RBCs are made daily, and this can be increased several-fold under conditions of RBC loss. Some of the old cells break up in the bloodstream, causing **hemolysis**, but the majority are engulfed by

TABLE 9.3 Iron Profile			
Name	Reference Range	Standardized International Reference	Description
Iron	50–170 µg/dL	9–30 µmol/L	Amount of iron bound to transferrin in blood
Ferritin	150–200 ng/mL	15–200 µg/L	Storage form of excess iron
Total iron-binding capacity (TIBC)	252–479 µg/dL	45–86 µmol/L	Amount of iron needed to bind to all transferrin
Unsaturated iron-binding capacity ([UIBC] TIBC—iron)	202–309 µg/dL	36–56 µmol/L	Transferrin not bound to iron
Transferrin (measured)	200–380 mg/dL	2–3.8 g/L	Transferrin not bound to iron
Transferrin saturation (iron/TIBC)	20%–50%	0.2–0.5	Percentage of transferrin with iron bound to it

macrophages of the liver, spleen, and bone marrow. This phagocytic process initiates the recycling of important components of the RBCs including iron and protein.

Hgb released from hemolyzed RBCs binds to proteins such as **haptoglobin** or **hemopexin**, which are cleared from the circulation by macrophages in the liver. Hgb from RBCs that are engulfed by macrophages is broken down into globin and heme within the phagocytic cells. The globin portion is catabolized by proteases into constituent amino acids that are reused for protein synthesis. Heme is broken down into free iron, and the components of the **porphyrin** ring surrounding the iron are first metabolized to **biliverdin**, a green substance that is further reduced to **bilirubin**. Bilirubin is then carried by albumin to the liver, where it is chemically conjugated to glucuronide before it is excreted in the bile.

Unconjugated and conjugated bilirubin is measured clinically to detect and monitor liver or gallbladder dysfunction. Blood samples with high bilirubin are said to be **icteric**, which is evident as a yellow-green tint and increased stickiness. Those with icteric blood develop jaundice. The yellow coloration that is characteristic of jaundice can be seen in the skin, particularly under the fingernails. The increase in bilirubin to a level that can be visually detected is a serious situation because a high level of bilirubin that is not bound to plasma proteins is neurotoxic.

Most of the iron in the body is involved in RBC production, so it is carefully reclaimed during RBC turnover. Free iron is toxic, so, in the blood, iron is transported bound to the protein **transferrin**. Cells that need iron possess membrane receptors to which transferrin binds and is then internalized. Inside the cell, the ferric iron is released and either incorporated into new heme or stored, predominately in the liver, bound to **ferritin**. Although iron recycling is efficient, small amounts are continuously lost and must be replenished by dietary intake. In women, iron loss is increased during menstrual bleeding. The majority of iron in the diet is derived from heme in meat (*organic iron*), but iron can also be provided by the absorption of *inorganic iron* (mainly Fe^{2+}, some Fe^{3+}) by intestinal epithelial cells.

Iron profile evaluates iron stores of the body.

To sufficiently differentiate between iron deficiency anemias and the anemia of inflammation as well as to diagnose iron overload, an **iron profile** is analyzed (Table 9.3). **Serum iron levels** ("iron" for short) are measured with a simple colorimetric method. However, the body's iron stores are better represented by *ferritin* levels, which are measured by radioimmunoassays or enzyme immunoassays. **Total iron-binding capacity (TIBC)** is the amount of iron needed to bind to all iron-binding proteins. Because *transferrin* represents the largest quantity of iron-binding proteins, TIBC is an indirect assessment of the amount of transferrin present. The **unsaturated iron-binding capacity (UIBC)** is the calculated amount of transferrin that is not occupied by iron. UIBC equals TIBC minus iron. Last, the **transferrin saturation** is calculated as the ratio of serum iron and TIBC values and is expressed as a percentage. Thus, it indicates the percentage of transferrin with iron bound to it, which is 20% to 40% in healthy people. In addition to serum analyses, bone marrow aspiration and biopsies are helpful to determine total body iron stores.

In iron deficiency, the iron level is low, but the TIBC is increased so that transferrin saturation becomes low. TIBC is also increased in pregnancy. In iron excess, the iron level will be high and the TIBC will be low or normal, resulting in an increase of transferrin saturation. To evaluate a patient's nutritional status or liver function, typically the protein transferrin will be monitored because it is produced in the liver. Anemias often present with characteristic changes in iron profiles.

▶ WHITE BLOOD CELLS

WBCs, or leukocytes, are delivered by the blood to sites of infection or tissue disruption, where they defend the body against infecting organisms, foreign compounds, and damaged tissue.

Leukocytes contain five diverse cell types and constitute part of the immune system.

The five main types of WBCs are **neutrophils**, **eosinophils**, **basophils**, **lymphocytes**, and **monocytes** (see Figs. 9.2 and 9.7). These cells are further divided into two classifications, **granulocytes** and **agranulocytes**, based on the presence and absence of intracellular granules that can be visualized with a light microscope. The neutrophils, eosinophils, and basophils are granular leukocytes, whereas the lymphocytes (i.e., B cells and T cells) and monocytes are agranular. Although monocytes and lymphocytes may also possess cytoplasmic granules, they are not as numerous or as distinct on a regularly stained blood smear. The nuclei of mature granular cells are multilobed, and these cell types are often called **polymorphonuclear leukocytes** (**PMNs**) as opposed to the monocytes and lymphocytes, which are termed **mononuclear leukocytes**.

Hematopoietic stem cells reside predominately in the red bone marrow and can be further differentiated into various lineages by a group of cytokines that are collectively called **colony-stimulating factors**. The first stage of differentiation is into two lineages giving rise to myeloid and lymphoid progenitor cells. Myeloid progenitor cells give rise to RBCs and all WBCs except the lymphocytes, which are part of the lymphoid lineage (Fig. 9.7).

Neutrophils defend against bacterial and fungal infection through phagocytosis.

Neutrophils are the most prevalent leukocyte in the peripheral blood of a healthy person (50% to 70% of all leukocytes). They are highly motile phagocytic cells and are the first defensive cell type to be recruited to a site of inflammation. The primary mission of neutrophils is to find bacteria or fungi and neutralize them by **phagocytosis**. This process can be described in four steps (Fig. 9.6). Defects in neutrophil function quickly lead to massive infection and, quite often, death.

Step 1: Recognition of foreign invader

When bacteria or their products are recognized and bound by circulating antibodies, the bacteria release chemotactic factors that attract neutrophils. Neutrophils recognize the bacteria as foreign by binding to the antibodies via Fc receptors. Bacteria can also interact with tissue cells, lymphocytes, or platelets that then release factors that attract and activate neutrophils.

Step 2: Invagination of cell membrane

At the site of infection, neutrophils engulf the invading pathogen by phagocytosis. Phagocytosis is facilitated when the bacteria are coated with host defense proteins known as **opsonins** (for details see Chapter 10).

Step 3: Phagosome formation

The generated phagocytic vacuole, or **phagosome**, fuses with intracellular granules containing digestive enzymes and **defensins**. The defensins are cationic proteins that kill the microorganisms by creating holes in their plasma membranes. Agents stored in neutrophil granules include lysozyme, a bacteriolytic enzyme, and myeloperoxidase, which reacts with hydrogen peroxide to generate potent, bacteria-killing oxidants. One of these oxidants is hypochlorous acid (HOCl), a chemical typically found in household bleach. Granules also contain proteases (e.g., collagenase).

Step 4: Killing of pathogens

An important step for the effective destruction of pathogens is the activation of the enzyme **nicotinamide adenine dinucleotide phosphate** (**NADPH**) oxidase. The oxidase is dormant in resting cells but activated by its interaction with a G protein and cytosolic molecules generated during phagocytosis. Enzyme activation leads to the catalytic production of **superoxide ion**, a toxic free radical, within the phagosome. The generation of superoxide and other potent reactive agents is collectively termed the **respiratory burst** or **oxidative burst**. The reactive agents kill bacteria directly or participate in secondary free radical reactions to generate other potent antimicrobial agents such as **hydrogen peroxide**.

The important role of the **NADPH oxidase** (**NOX**) for efficient host protection against invading pathogens becomes evident in **chronic granulomatous disease**, which is caused by the lack of phagocytic NADPH oxidase and is characterized by recurrent life-threatening bacterial and fungal infections.

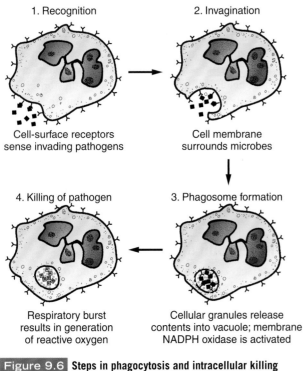

1. Recognition
Cell-surface receptors sense invading pathogens

2. Invagination
Cell membrane surrounds microbes

3. Phagosome formation
Cellular granules release contents into vacuole; membrane NADPH oxidase is activated

4. Killing of pathogen
Respiratory burst results in generation of reactive oxygen

Figure 9.6 **Steps in phagocytosis and intracellular killing by neutrophils.** Neutrophils are the first defensive cells to be recruited to a site of inflammation. NADPH, nicotinamide adenine dinucleotide phosphate.

Eosinophils are inflammatory cells that defend against parasitic infections.

In a healthy person, eosinophils are 2% to 4% of circulating WBCs and are readily identified by their characteristic appearance and the staining of their cytoplasm. As the name implies, *eosin*ophils take on a deep orange-red color during staining with eosin or other acidic dyes. Like neutrophils, eosinophils migrate in response to chemotactic signals and exhibit a metabolic burst when activated. Eosinophils participate in the defense against parasites such as roundworms. They are effective against these large invading organisms by releasing their granular contents that contain nitric oxide and cytotoxic enzymes. These WBCs increase several-fold in number in response to parasitic infections. Eosinophils are involved in allergic reactions because they are also activated by allergens. They also play a role in wound healing by releasing growth factors and chemicals that neutralize inflammatory mediators.

Basophils release histamine, causing the inflammation of allergic and antigen reactions.

The number of basophils in the blood smear of a healthy individual ranges from 0% to 2% of the WBCs. This makes basophils even rarer than eosinophils, and many differential counts show none. Basophils have multiple pleomorphic, deep-staining granules throughout their cytoplasm and are readily stained with basic dyes. Basophils are recruited to sites of cellular injury or inflammation. The granules contain **heparin** and **histamine**, which have anticoagulant and vasodilating properties, respectively. When stimulated, the granular contents are released from the cells and mediate increases in regional blood flow and the attraction of other WBCs, including eosinophils.

Mast cells are capable of synthesizing many of the same mediators as basophils. However, the two cell types appear to be distinct despite their close similarities. Basophils and mast cells must be activated to release their inflammatory mediators in a process called **degranulation**. In an allergic response, activation and degranulation occurs when antigens that are already tightly bound to basophils and mast cells bind to immunoglobulin E (IgE). Other direct or supportive stimuli for degranulation include mechanical injury, activated complement proteins, and drugs. When antigens are allergens such as pollen, the release of histamine and heparin can cause the characteristic symptoms of allergy.

Monocytes migrate from the blood stream and become macrophages.

Monocytes, together with lymphocytes, belong to the category of mononuclear leukocytes. Monocytes constitute about 2% to 8% of total WBCs and can be distinguished from lymphocytes based on their pale-blue or blue-gray cytoplasm when stained with a Wright stain. On activation, monocytes leave blood vessels, migrate into tissue, and fully differentiate into macrophages, which are large phagocytic cells. Macrophages contain granules with lytic enzymes and chemicals that are used to destroy ingested microbes, antigens, and other foreign substances. In addition to their role as phagocytes, macrophages secrete chemotactic signals for other WBCs and fibroblasts. Macrophages are

considered an integral part of the immune system because they serve as antigen-presenting cells to B cells and T cells (see Chapter 10).

Lymphocytes contain three cell types that participate in the immune system.

About 16% to 45% of leukocytes are lymphocytes composed of **B cells** and **T cells** and **natural killer (NK)** cells. In a stained light microscopic slide, the different types of lymphocytes cannot be distinguished from each other. Circulating lymphocytes possess a deeply stained nucleus that is large in relation to the remainder of the cell so that often only a small rim of cytoplasm appears around the nucleus (see Fig. 9.2). Some lymphocytes such as NK cells have a broader band of cytoplasm, closely resembling monocytes. Most of the lymphocytes are found outside the blood in lymphoid organs or connective tissue.

Most circulating lymphocytes are T cells or T lymphocytes (for *thymus-dependent lymphocytes*). They participate in cell-mediated immune defenses and are divided into several subtypes (e.g., helper T cells, cytotoxic T cells, and NK cells) (see Chapter 10). Some 20% to 30% of circulating lymphocytes are B cells. B cells are bone marrow–derived lymphocytes. When a B cell encounters its specific antigen (**clonal selection**), it becomes activated and proceeds to replicate itself (**clonal expansion**). Many of these clonal B cells will mature into **plasma cells**. Plasma cells are highly specialized cells with the ability to produce large amounts of target-specific antibody. They are larger and stain darker than naive B cells because of the large amount of protein being synthesized within.

▶ PLATELET FORMATION

Platelets are irregularly shaped, disklike fragments of their precursor cell, the megakaryocyte. They are one fourth to one third the size of erythrocytes (1.5 to 3.0 μm). As megakaryocytes develop, they undergo a process of controlled fragmentation that results in the release of over 1,000 platelets per cell. Several factors stimulate megakaryocytes to release platelets within the bone marrow sinusoids. This includes the hormone thrombopoietin, which is mainly generated by the liver and the kidneys and released in response to low numbers of circulating platelets. Platelets have no nucleus but possess important proteins, nucleotide, and other factors, which are stored in intracellular granules and secreted when platelets are activated during platelet aggregation in response to vascular injury.

▶ BLOOD CELL FORMATION

Blood cells must be continuously replenished. As already stated, RBCs survive in the circulation for about 120 days. Platelets have an average lifespan of 15 to 45 days, but many of them are immediately consumed as they participate in day-to-day hemostasis. Leukocytes have a variable lifespan. Some lymphocytes circulate for 1 year or longer after production. Neutrophils, constantly guarding body fluids and

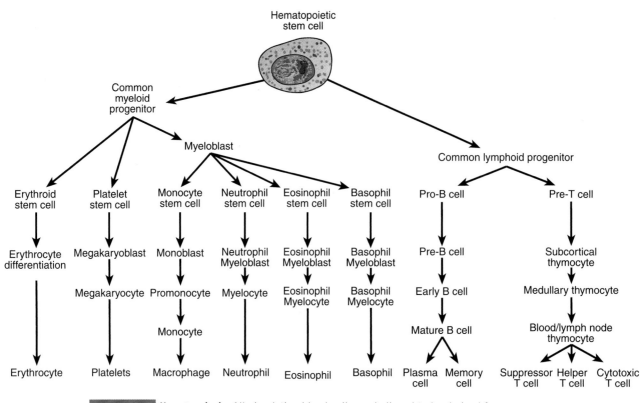

Hematopoietic stem cell

Figure 9.7 **Hematopoiesis.** All circulating blood cells are believed to be derived from a common bone marrow progenitor—the hematopoietic stem cell. Its differentiation along different lineages via increasingly more committed stem cells toward the final blood cells depends on the encountered conditions and hematopoietins.

tissues against infection, have a circulating half-life of hours to days.

Hematopoiesis takes place in bone marrow and lymphatic tissue.

Hematopoiesis, the process of blood cell generation, occurs in healthy adults in the red bone marrow and lymphatic tissues, such as the spleen, thymus, and lymph nodes (Fig. 9.7). During fetal development, hematopoietic cells are present in high levels in the liver, spleen, and blood. Shortly before birth, blood cell production gradually begins to shift to the marrow. By age 20 years, the marrow in the cavities of many long bones becomes inactive, and blood cell production mainly occurs in the marrow of iliac crest, sternum, pelvis, and ribs. In some disease states such as leukemia, liver and lymphatic tissue can resume their hematopoietic function in adulthood, which is called **extramedullary hematopoiesis**.

Mature blood cells originate from a multipotent stem cell.

Blood cell production begins with the proliferation of **multipotent stem cells** called **hematopoietic stem cells**. Depending on the stimulating factors, the progeny of hematopoietic stem cells can form any of the WBCs, RBCs

and platelets. Colony-stimulating factors are cytokines or hormones that control the lineage fate of cells as well as the speed of cellular production. This large family of factors can be released by various tissues in response to a variety of stimuli. Examples of the hematopoietic factors include **erythropoietin** (RBC production), **M-CSF** (monocyte production), **G-CSF** (granulocyte production), and **multi-CSF** (granulocyte, monocyte, and platelet production).

Erythropoiesis is regulated by the renal hormone erythropoietin.

Erythropoiesis is the process by which RBCs are produced. A major factor controlling RBC production is erythropoietin. This hormone is released by the kidneys in response to decreased O_2 delivery. The decreased O_2 can come from a variety of causes including hemorrhage; anemia; respiratory problems, such as **emphysema** or **chronic obstructive pulmonary disease** (**COPD**); or a sudden change in altitude (lower O_2 pressure). Erythropoietin regulates the differentiation of the uncommitted stem cells to enter the erythrocyte lineage, forming, in order, normoblasts (also called *erythroblasts*); reticulocytes; and, finally, mature erythrocytes, which enter the bloodstream (Fig. 9.8). A common symptom of patients with chronic kidney disease is anemia from the lack of erythropoietin. Genetically engineered human erythropoietin and some longer lasting analogs are used for the treatment of anemia due to chemotherapy or chronic kidney

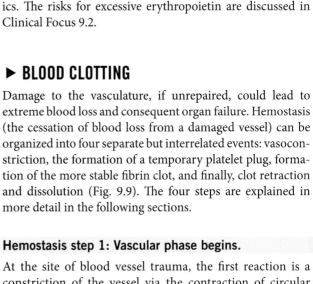

Figure 9.8 **Erythropoiesis.** Erythrocytes are the result of a process that begins with uncommitted stem cells and involves a series of differentiations in the bone marrow sinusoids until the final erythrocytes enter the bloodstream by diapedesis.

failure. The availability of the recombinant form combined with the extra O_2 carrying capacity of increased numbers of RBCs has made this a drug of abuse for endurance athletics. The risks for excessive erythropoietin are discussed in Clinical Focus 9.2.

▶ BLOOD CLOTTING

Damage to the vasculature, if unrepaired, could lead to extreme blood loss and consequent organ failure. Hemostasis (the cessation of blood loss from a damaged vessel) can be organized into four separate but interrelated events: vasoconstriction, the formation of a temporary platelet plug, formation of the more stable fibrin clot, and finally, clot retraction and dissolution (Fig. 9.9). The four steps are explained in more detail in the following sections.

Hemostasis step 1: Vascular phase begins.

At the site of blood vessel trauma, the first reaction is a constriction of the vessel via the contraction of circular smooth muscles surrounding the vasculature. This reaction is known as a **myogenic contraction**. At the site of injured tissue, there is also the release of potent chemical substances such as ADP, tissue factor, and prostacyclin that play roles in subsequent phases. The injured endothelial cells also release factors such as **endothelins** that,

---

CLINICAL FOCUS | 9.2

Blood Doping and Erythropoietin

Charging an athlete with "blood doping" is unfortunately a common newspaper headline during world-class endurance sports events. It refers to methods of enhancing an athlete's red blood cell count in advance of a competition to increase oxygen delivery capacity and to reduce muscle fatigue.

One blood doping method is autologous blood transfusion, in which the athlete's own red blood cells are stored and then transfused back in, a week or less before competition. This technique is not only illegal but also dangerous. An increase in hematocrit (Hct) results in increased blood viscosity, which is a major contributor for developing heart disease.

Another way to increase Hct is by increasing the concentration, production, or activity of the hormone erythropoietin. Erythropoietin is a glycoprotein produced primarily by peritubular cells in adult kidneys in response to low blood O_2. It promotes the proliferation and differentiation of erythrocyte precursors,

The human gene for erythropoietin was cloned in 1985, leading to the production of recombinant human erythropoietin (rhEPO) and genetically engineered darbepoetin, which has an increased lifetime compared to the recombinant hormone form. The large-scale production of synthetic erythropoietin made it available for the treatment of anemias. Currently, rhEPO and darbepoetin are approved in the United States to treat patients with severe anemias associated with chronic renal failure, AIDS, and chemotherapy. Usage in the treatment of severe anemias is limited in order

to balance the benefits of increased red cell production with the health risks associated with an elevated Hct, especially for those who are ill.

Within the first 4 years after the synthetic versions of erythropoietin became available, more than 17 high-performance athletes died from the consequences of drug-induced blood clots. The fatal dangers of excess erythropoietin include sudden death from stroke or heart failure or the development of antierythropoietin antibodies, which, ironically, cause the destruction of red blood cells.

During drug screening, increased Hct and elevated reticulocyte blood count indicate potential erythropoietin abuse. In this case, urine and blood tests will be done. Erythropoietin in urine can only be detected for a few days after injection. Erythropoietin in blood has a half-life of about 2 weeks. In addition, normal Hct levels show a wide range and can be substantially altered by dehydration or training at high altitudes. This creates a challenge for the correct testing time because the drug-induced increase in red cell mass can last up to a few months.

Currently, exogenous hormone can be distinguished from the body's normal hormone by electrophoresis, but this is valid only while the exogenous hormone is in the bloodstream. Because repeated pre-event testing for every athlete is not realistic, exposing illegal blood doping will remain a challenge. Additionally, drug advancements will require concomitant advancements in drug-detection procedures. ■

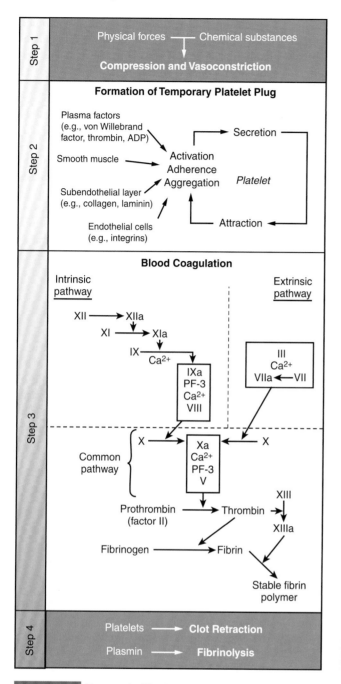

Step 1
Physical forces ——— Chemical substances
Compression and Vasoconstriction

Step 2
Formation of Temporary Platelet Plug

Plasma factors
(e.g., von Willebrand
factor, thrombin, ADP)
→ Secretion
Smooth muscle
→ Activation
Adherence
Aggregation *Platelet*
Subendothelial layer
(e.g., collagen, laminin)
Endothelial cells
(e.g., integrins)
← Attraction ←

Step 3
Blood Coagulation

Intrinsic pathway
Extrinsic pathway

XII → XIIa
XI → XIa
IX
Ca^{2+}
IXa
PF-3
Ca^{2+}
VIII

III
Ca^{2+}
VIIa ← VII

X → Xa ← X
Ca^{2+}
PF-3
V

Common pathway

XIII

Prothrombin → Thrombin →
(factor II)
XIIIa

Fibrinogen → Fibrin
Stable fibrin polymer

Step 4
Platelets ——→ **Clot Retraction**
Plasmin ——→ **Fibrinolysis**

Figure 9.9 **Hemostasis.** The blood's response to blood vessel injury can be viewed as four interrelated steps. Roman numerals refer to coagulation factors. ADP, adenosine diphosphate; PF-3, platelet factor 3.

in addition to stimulating smooth muscle contraction, act as growth factors in the stimulation of endothelial, muscle, and fibroblast proliferation. The endothelial cells at the site of injury become "sticky," enhancing the platelet reaction in the second phase.

Hemostasis step 2: Platelet plug forms.

The immediate goal of hemostasis is the fast production of a physical barrier that covers the opening in the blood vessel. The initial plug is composed of platelets (thrombocytes).

The exposure of collagen from the injured vascular wall, as well as the sticky surface formed by the endothelial cell reaction, cause the platelets to adhere to the vessels and become activated. Platelet activation causes a shape change and aggregation of the platelet to each other and the exposed collagen. The platelets contract and release their vesicle contents (**degranulation**) initiating both acute and long-term effects. ADP and the prostanoid **thromboxane A2**, released by the platelets together with the exposed collagen on the vessel wall stimulate further platelet aggregation and degranulation, ultimately forming the platelet plug. Other granule contents have additional roles. Serotonin and thromboxane A2 stimulate vascular contraction, released clotting factors and Ca^{2+} play roles in the subsequent clotting cascade, and **platelet-derived growth factor** (PDGF) promotes vessel repair. The positive feedback loop that promotes further platelet aggregation and degranulation is tempered by **prostacyclin**, a prostaglandin that is released from the damaged endothelial cells and inhibits platelet activation, and by circulating enzymes in the blood that break down ADP.

Importance of the platelet count for hemostasis

As in many cases, disease states can illustrate the importance of a physiological process. **Thrombocytopenia**, a low number of platelets, can be due to decreased production of platelets (e.g., in patients with serious bone marrow disease or undergoing chemotherapy), unavailability of platelets (e.g., platelets can be sequestered in an enlarged spleen), or accelerated destruction of platelets (e.g., coating of platelets with IgG and removal by phagocytes). **Immune thrombocytopenia purpura** is a common autoimmune disorder caused by autoantibodies to platelets. Purpura (meaning *bruising*) and multiple bruises all over the body as a result of minor bumps are one of the first characteristics of thrombocytopenia. This serves to illustrate that, under normal physiological conditions, the platelets plug the minor breaks in the vessels, particularly the small vessels, that are a part of daily life.

Hemostasis step 3: Thrombin catalyzes the conversion of fibrinogen into fibrin to form a stable clot.

The first two phases are not sufficient to close large breaks in blood vessels. A **blood clot** that can form on the top of a platelet plug is a network composed of the insoluble protein **fibrin**, which traps blood cells, platelets, and fluid. The fibrin is formed from the proteolytic degradation of fibrinogen, a normal blood protein, by the enzyme **thrombin**. The activation of thrombin and blood clot formation takes place according to a carefully orchestrated signal transduction cascade called **coagulation**.

The coagulation cascade is mediated by the sequential activation of a series of **coagulation factors**, proteins synthesized in the liver that circulate in the plasma in an inactive state. The coagulation factors are referred to by Roman numerals in a sequence based on the order in which they were discovered, not the order in which they occur in the cascade. The plasma coagulation factors and their common names are listed in Table 9.4.

TABLE 9.4 Factors of the Coagulation Cascade

Scientific Name	Common Names	Pathway
Factor I	Fibrinogen	Both
Factor II	Prothrombin	Both
Factor III	Tissue factor; tissue thromboplastin	Extrinsic
Factor IV	Calcium	Both
Factor V	Proaccelerin; labile factor; accelerator (Ac−) globulin	Both
Factor VI (Va)	Accelerin	
Factor VII	Proconvertin; serum prothrombin conversion accelerator (SPCA); cothromboplastin	Extrinsic
Factor VIII	Antihemophilic factor A; platelet cofactor 1; antihemophilic globulin (AHG)	Intrinsic
Factor IX	Christmas factor; platelet thromboplastin component (PTC); antihemophilic factor B	Intrinsic
Factor X	Prothrombinase; Stuart–Prower factor	Both
Factor XI	Plasma thromboplastin antecedent (PTA)	Intrinsic
Factor XII	Hageman factor; contact factor	Intrinsic
Factor XIII	Fibrin-stabilizing factor (FSF); protransglutaminase; fibrinoligase	Both

Two separate coagulation cascades, the **intrinsic coagulation pathway** and the **extrinsic coagulation pathway**, result in blood clotting in response to different initiation signals (see Fig. 9.9). The final steps in fibrin formation are common to both pathways. Many of the steps in all parts of the cascade require both phospholipids and Ca^{2+} as cofactors. Hence, sequestering Ca^{2+} by Ca^{2+} chelators such as EDTA is used as a means to inhibit blood coagulation.

The intrinsic pathway is so named because the necessary factors are contained within the blood. Exposed collagen from the vessel activates factor XII, which then activates factor XI, which, in turn, activates factor IX. Activated factor IX binds factor VII in the presence of Ca^{2+} to activate factor X, which is the beginning of the common pathway.

For the initiation of the extrinsic pathway, a factor extrinsic to blood but released from injured tissue, called **tissue thromboplastin (factor III)** or *tissue factor*, is required. Factor III, in the presence of Ca^{2+}, forms a complex with factor VII, and this enzyme complex activates factor X.

Any attempt to describe a distinct division of coagulation into the two separate pathways is an oversimplification. For instance, the factor VII/III complex of the extrinsic pathway can stimulate the formation of activated factor IX in the intrinsic pathway. Although the concept of independently acting intrinsic versus extrinsic coagulation pathways is incorrect, the pathways are still important to understand because they are monitored individually in the clinical coagulation tests: The activated partial thromboplastin time monitors the intrinsic pathway, and prothrombin time monitors the extrinsic pathway, as outlined below.

Common coagulation pathway

The final events leading to fibrin formation by either pathway result from the activation of the common pathway. Activated factor X, in the presence of platelet phospholipids and Ca^{2+}, activates the conversion of proenzyme prothrombin to thrombin. Thrombin is a proteolytic enzyme that catalyses the cleavage of fibrinogen to fibrin. A plasma enzyme, fibrin-stabilizing factor (factor XIII), catalyses the formation of covalent bonds between strands of polymerized fibrin, stabilizing and tightening the blood clot.

Thrombin is also a potent platelet and endothelial cell stimulus and enhances the participation of these cells in coagulation during hemostasis step 2. To a large extent, the interaction of coagulation factors occurs on the surfaces of platelets and endothelial cells. Although plasma can eventually clot in the absence of surface contact, localization and assembly of coagulation factors on cell surfaces amplifies reaction rates by several orders of magnitude.

Prothrombin time

The extrinsic system is evaluated by determination of the **prothrombin time** (**PT**). The PT, reported as time in seconds, represents how long a plasma sample takes to clot after a mixture of thromboplastin (factor III) and calcium chloride are added. A time longer than 11 to 13 seconds indicates a deficiency in prothrombin or other clotting factors that affect prothrombin.

PT is commonly used to measure the effectiveness of **coumarin**-type anticoagulant drugs. If a patient with a prolonged PT must have surgery, it is important that the PT be brought within a normal range before surgery.

Activated partial thromboplastin time

The test used to monitor the activity of the intrinsic system is the **activated partial thromboplastin time (aPTT)**. The aPTT measures the clotting time of plasma, from the activation of factor XII by a commercial biologic aPTT reagent through the formation of a fibrin clot. If a patient's value does not fall into the reference range of 25 to 38 seconds, additional tests are necessary to determine the exact cause of the coagulation problem.

The aPTT is used to monitor **heparin** therapy. Heparin is an injectable anticoagulant that prevents the formation and growth of clots and may be advised for patients with severe heart disease. All anticlotting therapies carry the risk of bleeding, and so appropriate heparin or coumarin dosing is critical.

The deficiency or deletion of any one factor of the cascade can lead to **hemophilia**. For instance, individuals deficient in factor VIII (antihemophilic factor) have hemophilia A, a severe X chromosome–linked condition with prolonged bleeding time on tissue injury as a result of delayed clotting. Hemophilia B is the result of factor IX deficiency.

Hemostasis step 4: Plasmin mediates fibrinolysis.

Clot retraction is a phenomenon that may occur within minutes or hours after clot formation. The clot draws together, which pulls the torn edges of the vessel closer together, reducing residual bleeding and stabilizing the injury site. The retraction requires the action of platelets, which contain actin and myosin. Clot retraction reduces the size of the injured area, making it easier for fibroblasts, smooth muscle cells, and endothelial cells to start wound healing.

Fibrinolysis is the process of breaking down the product of coagulation, a fibrin clot. The main enzyme in fibrinolysis is **plasmin**, which cleaves the fibrin mesh at various places, leading to the formation of circulating fragments that are cleared by other proteases or by the kidney and liver. Plasmin is a serine protease that circulates as the inactive proenzyme **plasminogen**. Plasminogen is converted to plasmin by a **tissue plasminogen activator** (TPA), which is released by activated endothelial cells.

Control of the clotting cascade

A number of endogenous anticoagulants exert control over the clotting cascade. Platelet function is strongly inhibited, for example, by the endothelial cell metabolite prostacyclin, which is generated from arachidonic acid during cellular activation. Thrombin bound to **thrombomodulin** on the surface of endothelial cells converts **protein C** to an active protease. Activated protein C and its cofactor, protein S, restrain further coagulation by proteolysis of factors Va and VIIIa. Furthermore, activated protein C augments fibrinolysis by blocking an inhibitor of TPA. Finally,

antithrombin III is a potent inhibitor of proteases involved in the coagulation cascade such as thrombin. The activity of antithrombin III is accelerated by small amounts of heparin, a mucopolysaccharide released by both basophils and endothelial cells.

Chemoattractants, mitogens, and growth factors

While the blood clot resolves, multiple factors participate in wound healing. Optimal wound healing requires the generation of new tissue cells as well as the recruitment of new blood vessels to nourish the repairing tissue. **Chemoattractants** attract smooth muscle cells, inflammatory cells, and fibroblasts to the wound, and **mitogens** induce them to proliferate. **Growth factors**, a subclass of cytokines, induce stem cells to differentiate. There are many different growth factors involved in this proliferative phase of wound repair, including **platelet-derived growth factor** (**PDGF**) and **epidermal growth factor** (**EGF**). They control the wound contraction and continuous remodeling of tissue and collagen over an extended period for ultimate healing.

An important event during wound healing is **angiogenesis**, the formation of new blood vessels. Platelets play an important role because they secrete factors that induce proliferation, migration, and differentiation of two of the major components of blood vessels: endothelial cells and smooth muscle cells. There are at least 20 angiogenic growth factors and 30 angiogenesis inhibitors found naturally in the body. These keep most blood vessels in a quiescent state but shift the balance toward formation of new vessels after injury.

A detailed understanding of the events that regulate angiogenesis has profound therapeutic potential. For example, exogenously applied *angiogenesis-inducing agents* may prove useful in accelerating the repair of tissue damaged by thrombi in the pulmonary, cerebral, or cardiac circulation. In addition, angiogenic factors may assist in the repair of lesions that normally repair slowly—or not at all—such as skin ulcers in patients who are bedridden or diabetic. On the other hand, *inhibition of angiogenesis* may prove particularly useful in the treatment of patients with cancer, because growing tumors require the recruitment of blood vessels to survive.

INTEGRATED MEDICAL SCIENCES

Anemias: Understanding the Physiology and the Importance of an Accurate Diagnosis

Anemia is a spectrum of diseases that are characterized by low O_2-carrying capacity of blood. The lack of O_2 delivery to tissues for ATP and energy production accounts for the common symptoms, which include weakness, shortness of breath, inability to maintain temperature, and chronic mental and physical fatigue.

The first line of treatment for anemia is often a transfusion of packed red cells. However, because the disease has multiple causes, effective and safe long-term treatment requires that the underlying pathology be diagnosed. Anemias can be divided into three general classifications with multiple subtypes within each classification. The three general causes of anemia are (1) insufficient RBCs, (2) decreases in hemoglobin production, and (3) abnormal hemoglobin.

1. Insufficient RBCs

Insufficient RBCs can arise from hemorrhage, hemolysis, or a decrease in RBC production. Hemorrhagic anemia can result from an acute blood loss or a more chronic loss such as a bleeding ulcer. This would be considered a secondary consequence of the injury or disease that causes the blood loss. In an otherwise healthy individual, curing the source of the blood loss will rapidly attenuate the anemia.

Hemolytic anemia means that the RBCs are breaking down prematurely. There are rare genetic defects that can result in red cell lysis. One of these is glucose-6-phosphate deficiency (G6PD), an X-linked recessive disorder, and another is sickle cell disease. Sickle cell disease is also classified as a hemoglobinopathy and will be considered as such in the third classification.

Certain infectious agents can also cause red blood cell lysis. These agents include many gram-positive bacteria (e.g., *Streptococcus*, *Enterococcus*, and *Staphylococcus*) and some parasites (e.g., *Plasmodium*, the malaria parasite). Hemolytic anemia can also be caused by autoimmune disorders due to autoantibodies or complement fragments deposited on red blood cells.

Finally, hemolytic anemia is a consequence of transfusion with mistyped blood during which the naturally occurring antibodies in the recipient's body recognize the transfused blood as having foreign antigens and mount an effective immune response to destroy the foreign RBCs.

A decrease in the production of RBCs produces aplastic anemia. This form of the disease can also have multiple causes including radiation, cytotoxic drugs (e.g., drugs used in cancer chemotherapy), environmental pollutants (e.g., benzene), or autoimmune disease. The cells that are most vulnerable to these insults are hematopoietic stem cells, so aplastic anemia will usually result in a decrease of not only RBCs but also WBCs and platelets.

2. Decreased Hemoglobin Production

The second major category of anemias comprises those that arise from decreases in hemoglobin production. The two most common subtypes are iron deficiency anemia and pernicious anemia. Iron is required for the production of mature, O_2-binding hemoglobin, and most of the iron in the body is contained in hemoglobin. Iron deficiency anemia can arise from a lack of dietary iron. Alternatively, and more difficult to treat, are the iron deficiencies that arise from an inadequate absorption of iron or the loss of iron stores. The inadequate absorption of iron can be due to intestinal diseases such as celiac disease or Crohn's disease. With the increasing prevalence of gastric bypass surgery and the accompanying malabsorptive syndrome, the risk of iron deficiency anemia is likely to increase in this patient population. Liver is the main storage organ for iron and, therefore, severe liver disease can cause iron deficiency. Iron deficiency anemia is considered the main global form of anemia.

Pernicious anemia is a deficiency in vitamin B_{12}, which is necessary to form erythrocyte maturation factor. A glycoprotein called *intrinsic factor* is produced by the stomach and aids in the uptake of vitamin B_{12}. Once absorbed, the vitamin B_{12} complexes with the intrinsic factor to form erythrocyte maturation factor, which is necessary for normal RBC production. Pernicious anemia can be caused by a dietary deficiency of vitamin B_{12} or by a defect in the intestinal transport process.

3. Abnormal Hemoglobins

The final major category of anemias includes the genetic diseases that result in the formation of abnormal hemoglobin, notably sickle cell disease and thalassemia. Sickle cell disease is caused by a two–amino acid mutation in the globin protein. Due to the conformational change induced by this mutation, deoxygenation causes a change in cellular morphology leading to stiff, crescent-shaped cells that can get trapped in small capillaries blocking blood flow and causing tissue anoxia, swelling, and pain. The sickled cells also lyse more easily. Interestingly, this was one of the first genetic diseases in which the actual mutation was discovered. Despite decades of research, there is still no cure and only moderately effective treatment.

There are two types of thalassemia, α and β, which have mutations in the α- and β-globin chains respectively. The mutations in thalassemia result in a decrease in the synthesis of the pertinent globin protein resulting in lowered levels of hemoglobin production.

Importance of Accurate Diagnosis

An elucidation of the pathophysiology is crucial for appropriate treatment of anemia. Perhaps the best illustration of this principle is provided by the example of the importance of distinguishing iron deficiency anemia from hemolytic anemia—both of which can occur in developing countries where a simple hematocrit would be the most available diagnostic test.

In hemolytic anemia, the RBCs are being lysed prematurely and, hence, the body is attempting to recycle the important components of the RBC breakdown including the iron of the heme moiety. Because iron is toxic, the liberated ion is bound to the protein transferrin for transport to the storage sites and is bound to ferritin during storage. During a hemolytic event, both of those proteins are likely to be saturated with iron.

Iron deficiency anemia is effectively treated with increases in dietary iron or iron supplementation. Misdiagnosis of hemolytic anemia as iron deficiency anemia could lead to treatment that would increase toxic (ionic) form of iron in the body because of a lack of iron-binding proteins. ∎

Chapter Summary

- Blood is a dynamic connective tissue with four major functions: to transport substances throughout the body; to protect the body from losing blood due to an injury; to maintain body homeostasis in regard to pH, water, heat, osmolality, and other factors; and to circulate cells and chemicals that defend the body against disease.
- Adult humans have ~5 L of whole blood, with about 45% of formed elements suspended in 55% plasma and solutes.
- Plasma is the liquid portion of blood. It contains 93% water and 7% of molecules such as proteins (e.g., albumin, globulins, fibrinogen, enzymes, and hormones), lipids (e.g., cholesterol and triglycerides), carbohydrates (glucose), electrolytes (e.g., Na^+, Cl^-, and HCO_3^-), and cellular waste (e.g., bilirubin, urea, and creatinine).
- Serum contains all components of plasma except for substances involved in blood clotting.
- Erythrocytes are anuclear, disk-shaped cells that deliver oxygen throughout the body via hemoglobin (Hgb). Adult Hgb is made of an α and β globin and a heme group containing an iron surrounded by a porphyrin ring that binds the O_2. Although changes in Hgb are rare, several hundred abnormal variants exist, producing clinical disorders such as sickle cell disease.
- Erythrocytes are characterized by number, shape, size, color, and maturity, providing valuable indicators for the diagnosis of anemias, hemoglobinopathies, infections, and other illnesses.
- Anemias can be caused by decreased red blood cells, decreased hemoglobin production, or abnormal hemoglobin. Treatment is dependent on the type of anemia.
- At the end of about a 4-month lifetime, old erythrocytes are engulfed by macrophages. Their hemoglobin, including iron, is recycled while generating the diagnostic waste product bilirubin. Excess bilirubin or an inability to process the bilirubin due to liver or gallbladder disease causes jaundice.
- Leukocytes are classified morphologically as granulocytes (eosinophils, basophils, and neutrophils) and agranulocytes (monocytes and lymphocytes).
- Leukocytes defend the body against infection using phagocytosis and various antimicrobial weapons, release mediators to control inflammation, and contribute to wound healing.
- Neutrophils are early inflammatory mediators, killing and engulfing microbes. Eosinophils support neutrophils and can

also attack multicellular parasites. Basophils support eosinophils and are also associated with asthma and allergies. T lymphocytes are the key players for cell-mediated immunity. B lymphocytes make antibodies that attack bacteria and toxins.
- Hematopoiesis is the development of circulating blood cells from the uncommitted hematopoietic stem cell of bone marrow. Immature cells differentiate along cell lineages into mature cells promoted by hematopoietins and other cytokines.
- Erythropoietin is a hormone produced by the kidney and in response to low blood O_2 and promotes erythropoiesis in bone marrow. Patients on dialysis often require erythropoietin intake to maintain a normal hematocrit (Hct). Some aerobic athletes abuse erythropoietin to illegally increase their Hct.
- Thrombocytes (platelets) are irregularly shaped, small, anuclear, cell-derived structures that, together with plasma proteins, control blood clotting and promote wound healing.
- Blood coagulation involves the fast formation of a weak platelet plug of aggregated platelets, which is expanded and stabilized into a more robust plug made of cells, platelets, and insoluble fibrin molecules. Endothelial cells, blood coagulation factors, Ca^{2+}, and mediators released by platelets, control the coagulation cascade. Thrombin is necessary for fibrin clot formation and plasmin for its dissolution.
- The basic metabolic panel reveals blood values for glucose, electrolytes, BUN, and creatinine; the CMP additionally gives values for albumin, total protein, ALP, ALT, AST, and bilirubin.
- The complete blood count includes WBC counts, red blood cell counts, platelet counts, hemoglobin (Hgb), hematocrit, red blood cell indices (mean cell volume, mean cell Hgb, and mean cell Hgb concentration), and mean platelet volume.
- Other routinely performed blood tests include blood viscosity, specific gravity, serum protein electrophoresis, erythrocyte sedimentation rate, coagulation tests (e.g., bleeding time, platelet count, prothrombin time, and activated partial prothrombin time), and iron profile (serum iron, ferritin, total iron-binding capacity, UIBC, transferrin, and transferrin saturation).
- For blood transfusions, donor and recipient blood must be compatible to avoid agglutination between erythrocyte-associated A, B, and Rh antigens and anti-A, anti-B, and anti-Rh antibodies.

Chapter Review Questions

1. Which of the following would be expected to contain relatively high numbers of functional hematopoietic progenitor cells?

   A. Adult circulating blood
   B. Adult spleen
   C. Adult thymus
   D. Circulating blood in children
   E. Umbilical cord blood

The correct answer is E. Umbilical cord blood, derived from the circulating blood of newborn infants, possesses high levels of hematopoietic progenitors. Levels of circulating progenitors rapidly decrease after birth. The spleens of adult humans function as a hematopoietic organ only in certain disease states, such as leukemia. The thymus is only involved in hematopoiesis prior to birth. Circulating blood cells in the adult or child are mature cells.

2. The clinician in charge hands you a patient's chart that contains results from about 50 blood tests. He asks you to have a look at the patient's granulocyte count. From the following list of cells, to which cells does the clinician refer?

   A. B cells
   B. Macrophages
   C. Monocytes
   D. Neutrophils
   E. T cells

The correct answer is D. The granulocyte group consists of neutrophils, eosinophils, and basophils, while monocytes/macrophages and lymphocytes (B cells and T cells) belong to the category of agranulocytes.

3. Polycythemia vera is a hereditary neoplastic bone marrow disorder characterized by abnormally high red blood cell production. The steady state concentration of a substance in serum can provide additional information to confirm the diagnosis of the patient with polycythemia vera. Its level is typically low. This is in contrast to patients with secondary polycythemia, which is caused by respiratory conditions like emphysema that stimulate erythrocyte production. Which of the following substances is most likely tested?

   A. Albumin
   B. Bilirubin
   C. Erythropoietin
   D. Haptoglobin
   E. Plasmin

The correct answer is C. Erythropoietin (EPO) is a hormone that stimulates bone marrow to produce erythrocytes. Its steady state condition in serum can provide useful information in the assessment of various anemic and polycythemic conditions. In polycythemia vera, EPO is low in response to the constant overproduction of erythrocytes due to mutated myeloblasts. On the other hand, any condition that leads to increased production of EPO (physiologically from hypoxia) will cause overproduction of erythrocytes and is defined as secondary polycythemia.

4. Which of the following patients is most likely to have a low total iron-binding capacity (TIBC) and normal transferrin saturation?

   A. A heavily menstruating woman
   B. A patient with chronic colon cancer leading to colonic bleeding
   C. A person with severe hookworm infestation
   D. A trauma victim, hematocrit 18, after being given large amounts of saline IV in the ER
   E. A woman nursing twins recently born by a complicated caesarian section

The correct answer is D. The low hematocrit of 18 (normal 36 to 50 depending on age and gender) indicates acute severe blood loss and concomitant loss of transferrin (low TIBC). To avoid hemorrhagic shock, the victim is treated with saline, which equally diluted all blood components. Since he probably lost both transferrin and serum iron, he might have a normal transferrin saturation, the ratio of serum iron and TIBC. On the other hand, all other patients are likely to suffer from iron deficiency anemia due to chronic blood loss or increased iron demand. Iron deficiency anemia presents with low serum iron and elevated transferrin (high TIBC), which increases when iron stores are low. Percent transferrin saturation is typically low due to the insufficient iron. Typical causes for iron deficiency anemia are blood loss due to heavy menstruation and chronic occult bleeding, often from the GI tract, such as present in the patient with colon cancer and the patient with parasites. Blood loss during a complicated caesarian section and increased iron requirement for lactation might cause anemia for the woman with the twins.

Clinical Application Exercises 9.1

ANEMIC BABY

A 24-year-old woman delivered her first child, a baby girl, in mid December. She felt well prepared by the nurse to take care of the baby at home, and breast-feeding was less of a problem than anticipated. The new family had a wonderful traditional Christmas at the farm where they live. In mid January, she visited her family doctor. He noted that the baby looked pale. He determined the blood type of the baby as AB+ and the blood type of the mother as A+. The hemoglobin (Hgb) content of the baby's blood sample was 11.1 mg/dL, and the mother's blood sample revealed Hgb content of 11.8 mg/dL. A follow-up visit was scheduled in early February.

At the February meeting, the baby looked very pale and bluish; the mother looked very tired and said that the baby cries a lot. The Hgb of the baby was now 9 mg/dL, and the mother's 12 mg/dL. A smear of the baby's blood revealed severe hypochromic microcytic erythrocytes with marked anisocytosis and poikilocytosis. The sample contained target cells and schistocytes. The pediatrician became concerned and was suspicious about the presence of a blood disorder, possibly thalassemia, a condition in which insufficient Hgb is formed. The result of a genetic analysis confirmed the diagnosis of beta-thalassemia major.

QUESTIONS

1. What can be concluded from the January Hgb values of the baby and the mother?

2. Explain the baby's red blood cell (RBC) values and morphology at the February visit.

3. The automated blood analysis performed in February resulted in an elevated white blood cell (WBC) count of the baby. Can you explain why?

ANSWERS

1. The baby's Hgb value of 11.1 mg/dL is at the low end of the normal range of a 1-month-old baby (11 to 15 mg/dL); the mother's Hgb value of 11.8 mg/dL is slightly low (12 to 16 mg/dL), but not unusual for a new mother. No interventions are necessary at this point. Since low iron is the most common reason for anemia of babies, it is necessary to closely monitor Hgb values in the near future. Blood type incompatibility between the mother and the child, which can lead to mild anemia in newborns, is unlikely since it occurs when the mother is type O and her baby is type A, B, or AB. Rhesus incompatibility is excluded.

2. The major form of beta-thalassemia leads to a significant deficiency in the beta chain production of Hgb so that there is little or no Hgb A present. As a consequence, Hgb A2 and Hgb F are elevated and for a short time substitute for the Hgb A loss. However, with decreasing Hgb F, starting soon after birth, the baby becomes hypoxic (pale), which stimulates erythropoietin production. Erythropoietin stimulates intramedullary, and eventually extramedullary, hematopoiesis.

Newly produced blood cells are small (microcytic) and contain only alpha chains (hypochromic), which may precipitate in the developing red blood cells (RBCs). This leads to target cells with dense center and pale rim, and generally to a cell population of unequal size (anisocytosis), abnormal shape (poikilocytosis), and cell breakage (schistocytes). Small-size cells lead to low hematocrit and low mean corpuscular volume (MCV). Inadequate Hgb production leads to decreased MCH, MCHC, and oxygen carrying capacity. Anisocytosis leads to increased RDW.

3. The initial automated blood cell count after the disease manifests often erroneously reveals an elevated WBC. This is due to the presence of nucleated RBCs (regenerative response by bone marrow), which are indistinguishable from leukocytes in hematology analyzers, regardless of technique. For this reason, the WBC count needs to be corrected for the number of nucleated RBCs, which is typically done when a differential blood cell count reveals more than 10 nucleated red cells per 100 white cells. After the correction, the WBC should be within the normal range in the baby.

thePoint® Visit http://thepoint.lww.com/rhoades5e *for additional chapter review Q&A, Clinical Application Exercises, animations, and more!*

Active Learning Objectives

Upon mastering the material in this chapter you should be able to:

- Explain immune system triggers and self-tolerance.
- Apply the roles of both noncellular and cellular components of innate immunity in maintaining body homeostasis.
- Explain how the adaptive immune system achieves its three main features: specificity, diversity, and memory.
- Describe the mechanisms for both exogenous and endogenous antigen presentation in cell-mediated immunity.
- Explain the roles of T-cell subtypes in adaptive immunity.
- Apply the clinical relevance of both innate and adaptive immunity.
- Identify three ways that the five classes of antibodies work to eliminate antigens.

- Recognize the signs and functions of both acute and chronic inflammation.
- Explain the clinical roles of pro- and anti-inflammatory cytokines.
- Explain the influences of the immune system upon organ transplantation.
- Recognize immune system disorders by immune system reactions.
- Apply the principle of immune surveillance in hematologic malignancies.
- Describe the interaction among the immune, neuronal, and hormonal systems in maintaining homeostasis.

You have learned about the neuromuscular system in the previous section. In later chapters, you will learn about other systems, such as the cardiovascular, respiratory, renal, and endocrine systems. By contrast, you will see in this chapter that the immune system is made up of a wide variety of disease-fighting cells found throughout the body in the plasma, lymph, tissues, and the various organs. The immune system is essential in our survival and the maintenance of optimal health. The immune system protects the body with a set of remarkable molecules called **antibodies** that can recognize an infinite range of foreign invaders known as **pathogens** (e.g., viruses, bacteria, parasites, toxins) and tags them for destruction by the other components of the immune system.

Immunology is the study of the functional process in which foreign matter (living and nonliving) is either destroyed or rendered harmless by the body's defense system. The functional role of the immune defense system is to (1) protect the body against infection and pathogens (bacteria, fungi, viruses, parasites, and other microbes), (2) neutralize and/or destroy foreign matter, and (3) function as an immune surveillance system (screening and neutralizing malignant cells).

Immunologic disorders (autoimmune diseases, hypersensitivities, and immune deficiency) result from malfunctions or inappropriate responses of the immune system. Immunology has contributed to great advances in medicine following Louis Pasteur's theory that microorganisms are the cause of infection and his role in the understanding that vaccination prevents infectious disease. The aim of this chapter is to present basic components of the immune system and their roles in organ function and homeostasis.

▶ IMMUNE SYSTEM COMPONENTS

The immune system defends the body from disease by identifying and killing pathogens, cells infected with a pathogen, and tumor cells. The immune system detects a wide variety of pathogenic agents, from bacteria to viruses to parasitic worms. In addition to identifying pathogens, it must also distinguish them from the body's own healthy cells and tissues. Detection is a complex, ongoing process because pathogens evolve rapidly, producing adaptations that help them avoid detection by the immune system and allow successful infection.

The defense system consists of layered specificity and sophistication.

The immune system protects the body against infection with layered defenses of increasing specificity. The surface barriers (e.g., skin and membrane secretions) are the first line of defense to prevent pathogens such as bacteria and viruses from entering the body. The skin and respiratory tract provide an important chemical barrier by secreting antimicrobial agents. Within the genitourinary and gastrointestinal tracts, commensal floras serve as biological barriers by competing with pathogenic bacteria for food and in some conditions changing their environment such as pH or available iron. The biologic barrier reduces the probability that pathogens will reach a critical mass to cause an illness.

If pathogens breach these barriers, the **innate immune system** provides an immediate nonspecific response. The innate response is triggered when specific molecules on the microbes are identified by pattern recognition receptors on immune cells. If pathogens successfully evade the innate response, the body engages a third layer of protection, the **adaptive immune system**, which is activated by the innate response. The cells of the adaptive response are specifically recruited to participate in the reactions, and this is why this immunity is also called *acquired immunity*. The adaptive system can change its response during an infection to improve recognition of the pathogen.

Thymus and bone marrow are linked to the adaptive immune system.

The thymus and bone marrow are called **primary** (or **central**) **lymphoid organs**. Lymphocytes, the specialized leukocytes involved in adaptive immunity, mature and become *immunocompetent* in the primary lymphoid organs. All lymphoid cells originate, and some complete the maturation process within the bone marrow, while others move to secondary lymphoid organs to complete the process (e.g., the eosinophil moves to the spleen to complete maturation). The exception is the pre-T cell, which leaves the bone marrow and undergoes complete maturation in the thymus before it is released into the body.

Lymph nodes, tonsils, mucosa-associated lymphoid tissues (MALT), and the spleen are considered **secondary** (or **peripheral**) **lymphoid organs**. These are the organs where mature immune cells participate in specific immune reactions. Figure 10.1 shows the primary and secondary lymphatic tissues.

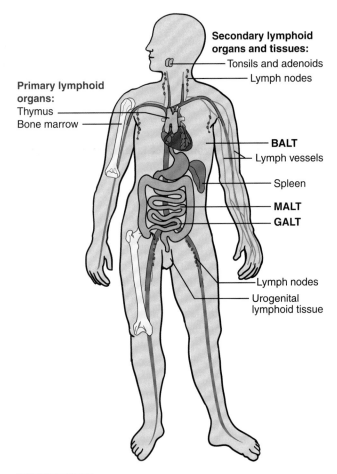

Primary lymphoid organs:
Thymus
Bone marrow

Secondary lymphoid organs and tissues:
Tonsils and adenoids
Lymph nodes

BALT
Lymph vessels
Spleen
MALT
GALT

Lymph nodes
Urogenital lymphoid tissue

Figure 10.1 Leukocytes are special cells of the adaptive immune system called lymphocytes. The thymus and bone marrow are primary sites of lymphocyte maturation. Lymph nodes, tonsils, the spleen, and lymphoid tissues of the lung, gut, and urogenital tract are secondary lymphoid organs. BALT, bronchus-associated lymphoid tissue; MALT, mucosa-associated lymphoid tissue; GALT, gut-associated lymphoid tissue.

Lymph nodes are the sites through which blood, lymph, and immune cells are filtered. These encapsulated organs are located throughout the body at junctions of **lymphatic vessels** and are optimized for interaction between professional **antigen-presenting cells** (**APCs**) and T and B lymphocytes. The movement of lymph through lymph vessels and lymph nodes is supported by skeletal muscle movement but is otherwise passive. There is no organ similar to the heart to pressurize the lymph flow.

During a bacterial infection, lymph nodes swell as a result of proliferation of immune cells. Palpation of lymph nodes is part of most orderly clinical examinations and gives an indication of the activity level of the immune system.

There is an abundance *of noncellular immune elements* in blood and lymph. They include, but are not limited to, antibodies, communication molecules (e.g., cytokines and chemokines), and complement and are presented later in the chapter. Additionally, the innate immune system encompasses a variety of fluids and other noncellular components with antimicrobial characteristics.

▶ IMMUNE SYSTEM ACTIVATION

The paramount function of the immune system is to recognize and destroy pathogens that enter the body. Pathogens contain **antigens**; some are surface antigens and some are antigens found within the pathogen. **Immunogens** are antigens that activate the immune system, whereas **haptens** are antigens that do not activate the immune system. However, haptens do become immunogenic when linked to a carrier molecule. Infection activates the immune system. More specifically, infection activates B and T lymphocytes through receptor recognition of immunogens.

Pathogens may enter the body through a cut in the skin (as in the case of the hepatitis B virus), through membranes of the respiratory tract (as in the case of the measles virus), or through membranes of the digestive tract (as in the case of *Salmonella* bacteria). Pathogens may also be transmitted by insect bites (as in the case of malaria) or by sexual encounter (as in the case of the human immunodeficiency virus).

In addition to pathogens, other organic and nonorganic foreign substances can be antigenic. Pesticides, cosmetics, and exhaust particles are examples of such substances that can activate the immune system.

Large complex molecules and proteins are the best activators of the immune system.

Proteins are by far the best immunogens. Their complex three-dimensional conformation, their electrical charges, and their ability to aggregate make them good candidates to be recognized by immune defenses. Polysaccharides, lipids, and nucleic acids by themselves induce either weak or no immune responses. But, similar to proteins, their immunogenicity increases with increasing complexity. For instance, polymers made of only one type of nucleic acid are poor immunogens, whereas polymers made of more than one base are usually well recognized. Lastly, size matters. Typically, only molecules with a molecular weight of 4,000 Da or above elicit an immune response.

Because proteins are immunologically the best recognized, they are used for many vaccines to improve the immunogenicity. For instance, the vaccine targeting the carbohydrate capsule of *Streptococcus pneumoniae* is conjugated to protein carriers to enhance its effectiveness.

The immune system is highly selective: Severe injury and necrosis activate immune defenses, whereas mild injury and apoptosis do not.

Immune activation is selective. For example, injury and necrosis will activate the immune system, but mild injury and apoptosis will not. The activation is coupled with other body defenses and reactions. When the body sustains injury, whether from physical force, cellular breakdown, or biologic infection, the decision for immune activation depends on the severity and type of the injury.

If the cells are not too severely injured, the injured and surrounding cells may adapt to the event by changing their size, number, and functions, without activating immune responses. For instance, if skeletal muscle cells are stressed within tolerable limits, they become larger and improve their functions. This is the principle for strength exercise. On the other hand, severe cellular stress can activate the immune system, for instance, in the case of airway cells that respond strongly to irritating smoke particles with inflammation.

If the damage to a cell is too great, the cell will die by **necrosis**. The immune system is activated when the cells collapse and their cellular contents, including lysosomal enzymes, leak into the surrounding tissue, causing further damage and starting an intense inflammatory response. As a result, chemokines (chemicals leaked from the damaged cells) attract various types of immune cells to the injury site, which further activate the immune response. An example is Duchenne muscular dystrophy. In this case, severe muscle fiber stress and necrosis as a result of genetically induced muscle weakness lead to immune responses that further promote the death of dystrophic muscle. Necrosis pulls the body away from homeostasis.

In contrast to necrotic cell death, cell death by **apoptosis** will not cause immune activation. Apoptotic cells die without bursting apart, and as a result, no damaging substances are released from the cells and an inflammatory response is avoided. Apoptosis is the body's nonpathologic process of removing cells. This is especially important for the role of the immune system in maintaining homeostasis and regulating tolerance. Every day, several million B and T lymphocytes are generated, and most are removed, apoptotically, by negative selection (or activation-induced cell death).

Complementary processes are involved in the activation of innate and adaptive immunity.

Exogenous pathogens, such as extracellular bacteria and intracellular bacteria, that are still identifiable from outside the cell are first attacked by the **innate immunity** response (Fig. 10.2). Phagocytic cells, such as neutrophils, eosinophils (in some cases), macrophages, and dendritic cells, might engulf the antigen or destroy it with enzymes and nitrogen and oxygen reactive species. The pathogen might directly

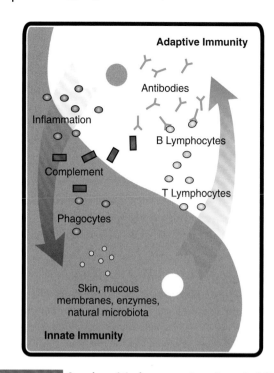

Figure 10.2 **Overview of the immune system.** As part of the innate immunity, the human body has natural measures to prevent entry of microbes. When these barriers are overcome, a foreign invader first encounters phagocytes. Complement may bind to the invader and facilitate its phagocytosis or lysis. Inflammation may develop. If the innate immune system cannot destroy the invader, the adaptive immune system is activated with T and B lymphocytes as its effectors. Although the innate and adaptive immune systems are characterized by contrasting functions and timing, they work closely together and rely on each other to succeed in removing invading pathogens.

activate **complement** proteins, another important component of the innate immune system. This activation can result in **lysis** of the organism and promote activation of the *inflammatory response*, or **inflammation**.

If some antigenic (immunogenic) parts of the microorganism are present on the plasma membrane of the host cell, antibodies can bind directly to the infected cell and activate complement (discussed later in detail).

If the pathogen evades the innate immune system and no antibodies are present, then the **adaptive immune system** response is triggered. Adaptive immunity is also activated by viruses, protozoan parasites, and intracellular bacteria that cannot be recognized from outside the cell (endogenous antigens). In this case, the antigen is degraded intracellularly into smaller pieces. The pieces are then incorporated into the host cells' plasma membranes and, together with **major histocompatibility complex (MHC) proteins**, are presented to T cells (discussed later in detail). These microorganisms are ultimately eliminated together with the infected host cell.

Although innate and adaptive immune systems are characterized by contrasting functions and timing, they work together in ways that obscure their differences. For instance, the initiation and adequate functioning of the innate system often depend on the presence of elements of the adaptive immune system such as small amounts of specific antibody in blood plasma. The reverse is true as well. Antibodies and other mediators of the adaptive immune system depend on elements that are typically associated with the innate immune system such as phagocytic macrophages and dendritic cells.

Only when working together can the innate and adaptive immune systems prevent the establishment and long-term survival of infectious agents. For this reason, Figure 10.2 is presented as an analogy of the Chinese yin and yang symbol, in which opposites intertwine and complement each other toward a greater whole.

▶ IMMUNE DETECTION SYSTEM

The task of the immune system is to screen millions of molecules and eliminate them when recognized as not being a normal part of the body. For that, it has to discriminate *self* from *nonself*. To destroy only pathogens while leaving the body's own healthy cells intact requires a complex system because microorganisms and human tissue are made of closely related material.

Self-tolerance, the action of the immune system to forgo attacking the body's own cells and proteins, is a multistep process that begins in the thymus during T-lymphocyte development. This process, called *central tolerance*, is discussed later in this chapter. Self-tolerance then continues with additional mechanisms outside of the thymus (called *peripheral tolerance*).

Tolerance.

In addition to self-antigen, a few important foreign antigens also do not activate the immune system (called *acquired* or *induced tolerance*). We generally tolerate food-derived foreign particles (although foods can induce allergic reactions). A mother does not reject the fetal molecules that are derived from both maternal and paternal genes. We also do not destroy the nonpathogenic microbiota in our gastrointestinal tract, even though they are foreign to the human body.

Any mistake in the process of self/nonself discrimination can have devastating consequences to the organism. When a body erroneously mounts an immune response against its own tissues, it can lead to an *autoimmune disease*.

▶ IMMUNE SYSTEM DEFENSES

The immune system employs three mechanisms to mount its defense against invaders: physical barriers, innate immunity, and adaptive immunity.

Surface barriers are the immune system's first line of defense.

As mentioned earlier, our immune system protects against infection with layered defenses of increasing specificity. The physical barrier (e.g., skin and membrane secretions) is the first line of defense to prevent pathogens, such as bacteria and viruses, from entering the body. Our skin is an effective anatomical and physiologic barrier against microorganisms. First, the cells of the epidermal layers are dry and

densely packed, making them an inhospitable environment to many bacteria. Furthermore, salty secretions from sweat glands and oily secretions from the sebaceous glands associated with hair follicles create a hyperosmotic and slightly acidic skin environment, which dehydrates bacteria and discourages those that prefer a neutral pH for colonization. Additionally, the continuous desquamation of skin cells eliminates bacteria adhering to epithelial cells. Lastly, many noninvasive, nonpathogenic commensal (part of normal microbiota) microorganisms on the skin prevent growth of harmful microorganisms in a process called *competitive exclusion*.

Acidity is also used in other places of the body as an antimicrobial tool. For example, natural microbiota alters the fluids of the vaginal and urinary tracts to an acidic pH below 4.5 so that yeast and other microorganisms cannot grow. Likewise, parietal cells of the stomach create highly acidic gastric juice below pH 3 to hinder microbial growth.

Other factors such as low oxygen tension and fever are said to contribute to the barrier defenses, but their physiologic impact is still under discussion. The role of fever is thought to have a direct negative effect on certain microorganisms and may also enhance the efficiency of **phagocytosis**.

Cellular secretions

Various types of cellular secretions destroy and eliminate pathogens. *Mucus* prevents microorganisms from adhering to epithelial cells and contains antibacterial components. The mucus of the gut blocks, inactivates, or destroys pathogens associated with food before they can enter the body. A thin layer of mucus covering the airway from the nose to the bronchioles traps inhaled viruses, bacteria, pollens, and other particles and facilitates their removal before they can damage the airway lining cells. The flow of *saliva* helps to wash away bacteria attached to food particles while attacking them with thiocyanates and lysozymes. Saliva can also contain antibodies that destroy oral bacteria. *Tears* and *nasal secretions* contain similar antibacterial components.

The list of chemical factors with antimicrobial characteristics is long. It includes **pepsin** in the stomach, **defensins** produced by immunologic cells, and **surfactant** of the lung, to name a few. **Interferons** are a group of proteins that are produced by cells following viral infection. Complement is a group of serum proteins that circulate in an inactive state and can be activated by a variety of specific and nonspecific immunologic mechanisms. Their actions, sometimes called the *humoral component* of the innate immune system, are discussed in more detail below. An important role of these chemical factors is to connect the three lines of immune system defenses.

Innate immunity is the second line of defense.

The ability of cells and tissues to respond to and to get rid of environmental challenges is an ancient evolutionary development that has persisted through vertebrate development as innate immunity. The zoologist Metchnikoff discovered that cells of a starfish could phagocytose invaders, which

TABLE 10.1	Characteristics of the Innate and Adaptive Immune Systems	
	Innate	**Adaptive**
System	Primitive, found in invertebrates	Evolved in early vertebrates
	Present at birth	Develops and changes throughout life
	Responses do not change over time	
Stimulation	Not required	Required
Specificity	Minimal	Highly specific
	Fixed	Self vs. nonself discrimination
	Microbial pattern recognition	Diverse antigen receptors mediate responses.
Response	Within minutes	Develops over days
	No change in quality and quantity over time	Improved by previous exposure
Memory	No	Yes
Soluble factors	Lysozyme, complement, acute-phase proteins, interferon, cytokines	Antibodies, cytokines (interleukins), interferon
Cells	Phagocytic leukocytes, natural killer cells	T cells, B cells

TABLE 10.2	Cellular Elements of the Innate Immune System	
Cells	**Main Function**	**Phagocytosis**
Neutrophils	Kill microorganisms intracellularly and extracellularly	+
Macrophages	Kill microorganisms intracellularly and extracellularly	+
	Present antigen	
Dendritic cells	Phagocytose pathogens	+
	Present antigen	
NK and LAK cells	Destroy virus-infected and tumor cells	−
Eosinophils	Secrete factors that kill certain parasites and worms	+
Mast cells	Release factors that increase blood flow and vascular permeability	−

NK, natural killer; LAK, lymphokine-activated killer.

means that this 600-million-year-old invertebrate possesses an innate immune system. Innate immunity is also called *nonspecific* or *natural immunity*. Table 10.1 contrasts its basic characteristics in humans with the characteristics of the adaptive immune system (also called the *acquired immune system*). The innate immune system:

- Is present at birth
- Persists throughout life
- Can be mobilized rapidly and acts quickly
- Attacks all antigens fairly equally because it recognizes patterns on pathogens
- Maintains the quantity and quality of the response (the response does not change over time)

Phagocytic leukocytes

As mentioned in Chapter 9, the leukocytes commonly found in the blood are neutrophils, lymphocytes, monocytes, eosinophils, and basophils. Most of these leukocytes participate in innate immune mechanisms, while lymphocytes are involved with the adaptive immune response. Innate leukocytes include phagocytes (neutrophils, macrophages, eosinophils, and dendritic cells) and nonphagocytic cells as outlined in Table 10.2. Phagocytic cells identify and eliminate pathogens, either by attacking larger pathogens through contact or by engulfing them. They can be activated by T-cell

contact (e.g., neutrophils), by T-cell cytokines (e.g., macrophages), or by contact with pathogens. For the latter, phagocytes can directly sense the pathogens through a group of transmembrane receptors, the so-called **toll-like receptors.**

On activation, phagocytes engulf the microorganism, particle, or cell debris. The engulfed matter is enclosed within vacuoles and enzymatically digested, after fusion with lysosomes. Some pathogens have been coated with **opsonins** in a process called opsonization to render them more attractive to phagocytosis. Examples of opsonins are immunoglobulin G (IgG) antibody and the C3b molecule of the complement system.

Neutrophils (one of the *polymorphonuclear leukocytes*) recognize chemicals produced by bacteria in a cut or scratch and migrate toward them. Once arrived, they ingest the bacteria and kill them. For killing, neutrophils use proteolytic enzymes and reactive oxygen and nitrogen species produced as part of the respiratory burst (see Chapter 9).

Macrophages are derived from circulating monocytes. Once monocytes migrate into tissue, they differentiate and become the larger, more powerful, phagocytic macrophages. Macrophages kill, like neutrophils, by using the respiratory burst and proteolytic enzymes. Macrophages secrete various cytokines that attract other leukocytes to the infection site and initiate the acute-phase inflammatory response. Finally, macrophages act as professional APCs to T cells and are hence an important bridge to the T-cell–mediated immunity of the adaptive immune response.

Macrophages can circulate in lymph vessels (*wandering, nonfixed macrophages),* or they can reside in connective tissue, in lymph nodules, along the digestive tract, in the lungs, in the spleen, and in other places (*mature, fixed macrophages*). Fixed macrophages are part of the **reticulo-endothelial system (RES)**, which, in addition to removing pathogens, also removes old cells and cellular debris from the bloodstream. Some of these macrophages have their own names. For instance, the macrophages along certain blood vessels in the liver are called **Kupffer cells**, whereas the macrophages of the joints are called *synovial A cells. Microglial cells* are macrophages located in the brain.

Eosinophils are best known as participants in allergic reactions, where they might detoxify some of the inflammation-inducing substances. But they are also primarily evolved to secrete factors that punch small holes in worms and other parasites, causing them to die. However, they can also be phagocytic when the situation permits.

Dendritic cells are, like macrophages, a critical link between the innate and adaptive immune systems. They exist in an immature form throughout the epithelium of the skin (e.g., **Langerhans cells**), the respiratory tract, and the gastrointestinal tract. After phagocytosis of pathogens, the cells mature and travel to regional lymph nodes, where they activate T cells, which then activate B cells to produce antibodies against the pathogen.

Nonphagocytic leukocytes

In addition to phagocytic cells, innate leukocytes also comprise nonphagocytic cells (NK and LAK cells and mast cells) as shown in Table 10.2. NK cells attack aberrant body cells such as virus-infected cells and malignant cells. They release the cytolytic protein **perforin**, which forms a pore in the plasma membrane of the target cell. Proteolytic enzymes, such as **granzyme**, are also part of the NK's cytoplasmic granules. When released, it enters the target cell and induces apoptosis. On exposure to lymphocyte secretions, such as interleukin-2 and interferon-γ, NK cells become **lymphokine-activated killer (LAK) cells**, which are even more efficient in killing than NK cells.

Mast cells are present in most tissues in the vicinity of blood vessels and contain many granules rich in **histamine** and **heparin**. They are especially prominent under coverings lining the body surfaces such as the skin, mouth, nose, lung mucosa, and digestive tract. Although best known for their roles in allergy and anaphylaxis of the adaptive immune system (see next section), mast cells play an important role in the innate system as well. Additionally, they are intimately involved in wound healing. They release factors that increase blood flow and vascular permeability, bringing components of immunity to the site of infection. In combination with IgE antibody from B cells, mast cells can also target parasites that are too large to be phagocytosed, such as intestinal worms.

Adaptive immunity is the third line of defense.

Adaptive immunity uses three important features in its method of attack: specificity, diversity, and memory. Microbes that escape the onslaught of cells and molecules of the innate immune system face attack by T cells, B cells, and B-cell products of the adaptive immune system, also called the *acquired immune system*. Table 10.1 lists its basic characteristics in comparison to those of the innate immune system. The adaptive immune system:

- Is a relatively recent evolutionary development and characteristic of jawed vertebrates
- Is activated by thousands of diverse antigens, which are presented as glycoproteins on the surface of bacteria, as coat proteins of viruses, as microbial toxins, or as membranes of infected cells
- Responds with the proliferation of cells and the generation of antibodies that specifically assault the invading pathogens
- Responds slowly, being fully activated about 4 days after the immunologic threat
- Is capable of immunologic memory, so that repeated exposure to the same infectious agent results in improved resistance against it

Specificity

The *specificity* of the adaptive immune system is created by **antigen recognition molecules**, which are synthesized prior to the exposure to antigen and, in B lymphocytes, can be modified during the immune response to make them even more specific to the antigen.

Several different types of recognition molecules participate in this anticipatory defense system: (1) specific receptors on T and B lymphocytes; (2) major histocompatibility complex (MHC) proteins (recognition molecules on cell surfaces); and (3) antibodies, which are the secreted form of B-cell receptors (BCRs). Though the general structure for the types of molecules is very similar (e.g., every antibody has a typical Y-shaped form), a small region of each individual molecule is different (e.g., for antibodies, the hypervariable region) and allows only the binding of a specific antigen. There are five major classes of antibodies: IgG, IgM, IgA, IgE, and IgD. Each class of antibody plays a unique role in immune defense and will be discussed later in the chapter.

An antigen might be recognized by several recognition molecules. To stay with the example of antibodies, most protein antigens have several **epitopes** (the part of the antigen that binds the antibody) and, hence, are recognized by different B cells, which release different antibodies to mount a **polyclonal antibody** response.

On the other hand, closely related antigens may share epitopes (**cross-reactivity**). For instance, antibodies that are induced by some microbial antigens cross-react with polysaccharide antigens found on red blood cells and are the basis for the ABO blood type system (see Chapter 9).

Diversity

The *diversity* of adaptive immune responses is based on a huge variety of antigen receptor configurations, essentially one receptor for each different antigen that might be encountered.

In a person, there are about 10^{18} different possible **T-cell receptors (TCRs)**, with each T cell expressing one type of receptor. The **B-cell receptor (BCR)** is a membrane-bound form of immunoglobulin. There are about 10^{14} possible

BCRs, again one type of immunoglobulin per B cell. The molecule diversity is mainly achieved by variable recombination of gene segments prior to exposure to antigen and, in the case of immunoglobulins, additionally by mutation of the molecules after exposure to antigen.

The recognition of an antigen by the lymphocyte with the best-fitting receptor occurs mainly in the local lymph node and induces the activation, proliferation, and differentiation of the responsive cell, a process known as **clonal selection**. Figure 10.3 shows the process for four B cells with different BCRs and one recognizing the antigen. Not evident in the figure is that optimal activation and proliferation of B cells (and, equally, T cells) occur only when costimulatory signals, secondary to TCR and BCR bindings, are present.

Clonal selection amplifies the number of T or B lymphocytes that are programmed to specifically respond to the inciting stimulus. In Figure 10.3, this means all resulting cells from proliferation have the same BCRs, ready to bind

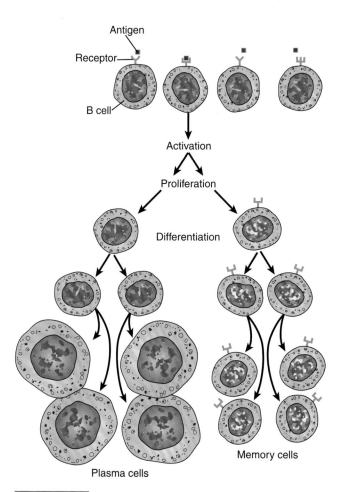

Figure 10.3 **Clonal selection of lymphocytes.** Only the clone of the lymphocyte that has the unique ability to recognize the antigen of interest proliferates and generates progenitor cells. These cells are specific to the inducing antigen but may have different functions. In the case of B cells, plasma cell clones produce antibodies, and memory cell clones enhance subsequent immune responses to the specific antigen. In the case of T cells, the clones become either effector cells or memory cells. Clonal selection occurs in secondary lymph organs such as the local lymph nodes.

more of the specific antigen. However, not all cells possess the exact identical functional characteristics. For instance, antigen binding to the BCRs induces their development into memory cells and into **plasma cells.** Plasma cells are much larger and are capable of producing and secreting antibodies. Initially, the plasma cells produce IgM antibodies and later can switch to produce IgG, IgA, or IgE antibodies when antibodies with different functional capabilities are needed. This maturation process is called *Ig isotype switching.*

Similarly, clonal proliferation of T cells can lead to the generation of more *antigen-specific T cells* and to the production of **effector T cells**, such as T helper and cytotoxic T cells, and memory T cells.

Memory

The *memory* of the adaptive immune system is based on the fact that some descendants in the expanded B-cell and T-cell clones function as **memory cells** (see Fig. 10.3). These cells mimic the reactive specificity of the original lymphocytes that responded to the antigen and accelerate the responsiveness of the immune system when the antigen is encountered again (**anamnestic response**) and are the basis for immunization via vaccinations.

▶ CELL-MEDIATED AND HUMORAL RESPONSES

The adaptive immune system has two major branches, cell-mediated and humoral immunity. As mentioned previously, though presented as distinct systems, no part of the immune system works separately. Rather, they all work in a cooperative fashion using cytokines and other means as a communication system.

B cells mediate the humoral immune response, whereas T cells regulate the cell-mediated immune response. The CD4 and CD8 ("CD" for cluster differentiation) T-cell subgroups of the cellular immunity branch of the adaptive immune system have different functions and are divided into additional subtypes according to their molecular appearance, cytokine sensitivity and production, and specific functions. These cells form, with each other and with other immune cells, a complex network of communication and immune response that is the basis for the efficiency, flexibility, and longevity of the adaptive immune system.

Figure 10.4 graphically summarizes important interactions between cells. The figure is organized according to the type of encountered antigen (exogenous or endogenous, left), to the sensing and responding processes (antigen recognition and presentation and immune response, top), and to the involvement of T cells and B cells (cellular and humoral immune responses, right).

Cell-mediated response involves activation of T cells and release of cytokines.

Cell-mediated responses in adaptive immunity do not involve antibodies or complement but, rather, activation of macrophages, NK cells, T lymphocytes, and the release of various cytokines in response to an antigen.

T cells continuously patrol the body and check the foreignness of antigen. T cells become activated when antigen

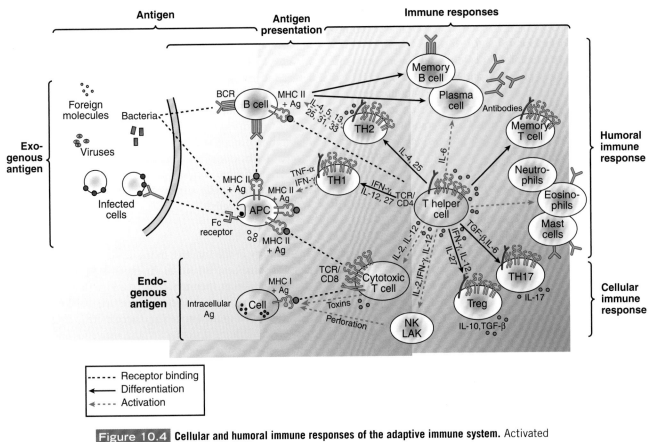

Figure 10.4 Cellular and humoral immune responses of the adaptive immune system. Activated T cells accomplish cellular immune responses, and B cells and antibodies mediate humoral immune responses. Exogenous antigen activates B cells by binding to the B-cell receptor (BCR) and T cells by binding to the T-cell receptor (TCR). However, TCR binding only occurs when antigen peptide is presented by antigen-presenting cells in association with major histocompatibility complex (MHC) proteins. The TCR is either associated with CD4 or CD8, depending on the T-cell type. Endogenous antigens are presented via MHC class I proteins to cytotoxic T cells, which destroy the host cell together with the intracellular pathogen. The immune response involves two paths, one using B cells (humoral immune response) and one using cytotoxic T cells (cellular immune response). Millions of different B and T cell types exist to recognize millions of different antigens. Ag, antigen; CD, cluster of differentiation; IFN, interferon; IL, interleukin; LAK, lymphokine-activated killer cell; NK, natural killer cell; TH, T helper cell; TNF, tumor necrosis factor; Treg, T regulatory cell.

binds to the specific TCR plus a costimulatory element. Antigen can only bind to the TCR when presented by APCs in combination with MHC proteins. The complex interaction between the T cell and the APC is called **immunologic synapse**. Figure 10.5 and Table 10.3 summarize the events at the synapse.

Major histocompatibility complex

The **major histocompatibility complex** (**MHC**) is a large genomic region or gene family found in most vertebrates. When activated, MHC genes produce MHC molecules that play an important role in the immune system.

In the case that the antigen stems from extracellular proteins or phagocytosed bacteria, the antigen is digested intracellularly within phagolysosomes and associated with **MHC class II molecules** on the surface of professional APCs, which then present the antigen to the TCR of CD4+ T helper cells (see Fig. 10.4, upper part). Only macrophages,

dendritic cells, and B cells can do so; hence, they are professional APCs.

Macrophages become APCs when they upregulate their MHC II molecules in response to infection and appropriate stimuli such as **lipopolysaccharide** (**LPS**) from gram-negative bacteria. *Dendritic cells* are mostly found in peripheral tissue, where they ingest, accumulate, and process antigens. *B cells* become activated by ligand binding to the BCR. Hence, they present the particular antigen to which the antibody that they express is directed. For other antigens, B cells are not very efficient APCs.

Almost every nucleated cell in the body can present antigen to the TCR of CD8+ cytotoxic T cells. In this case, the antigen is an intracellular pathogen, which is degraded in the cytosol and associated with **MHC class I molecules** (see Fig. 10.5, lower part).

Similarly, exposure to foreign antigen from a tissue graft triggers an immune reaction in the body. This occurs when

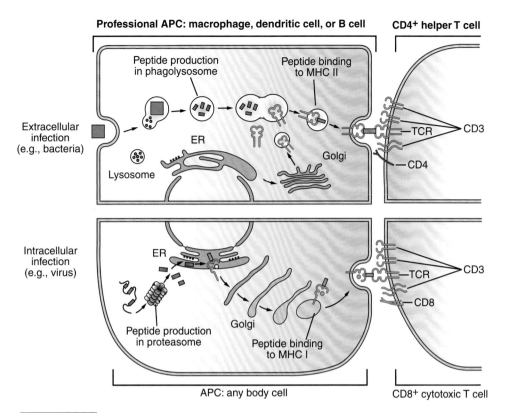

Figure 10.5 **Antigens bind to antigen-specific receptors.** An antigen cannot induce an immune response by itself. They first have to bind to an antigen-specific receptor to trigger a response. Extracellular antigens are phagocytosed, degraded within the phagolysosome, and associated with MHC class II molecules. The antigen is then presented on the surface of the professional antigen-presenting cell, where it binds to the T-cell receptor (TCR) of CD4+ T helper cells. Macrophages, dendritic cells, and B cells are such antigen-presenting cells. Intracellular antigen is digested within the proteasome, associated with MHC class I molecules, and presented on the surface of the cell to the TCR of CD8+ cytotoxic T cells. Almost every cell in the body can present antigen in this latter way. CD, cluster of differentiation; ER, endoplasmic reticulum; MHC, major histocompatibility complex.

the tissues of the donor and the recipient are not *histocompatible*, which explains the origin for the name "major histocompatibility complex."

T-Cell differentiation

T cells, or *T lymphocytes*, play a central role in cell-mediated immunity. They are distinguished from other lymphocytes, such as B cells and NK cells, by the presence of a special receptor on their cell membrane called T-cell receptors (TCRs). "T" stands for *thymus*, the principal organ responsible for the T cell's maturation.

There are more than 160 known clusters that coat the surface of leukocytes and many other cells. CD3 is found on the surface of all T cells as the signal transduction portion of the TCR. CD3 is made up of three dimers: two zeta proteins, a gamma and an epsilon protein, and an epsilon and a delta protein. CD4 and CD8 are two that are used to identify functionally critical T-cell subsets. To give two other examples, CD14 is an LPS-binding protein receptor on monocytes and macrophages, and CD69 is used as a marker of cell activation for T, B, NK cells, and macrophages.

When the cells coming from the bone marrow reach the thymus, T cells have neither CD4 nor CD8 (*double-negative T cells*), but they then start expressing both markers (*double-positive T cells*) while undergoing maturation in the thymus. During development in the thymus, double-positive cells differentiate into cells with either CD4 or CD8 receptors (single positives). This is a critical step because during this time they develop their ability to distinguish self from nonself peptides.

First, cells that do not bind MHC/antigen complexes within 3 to 4 days will die. The rest of the cells undergo **positive selection**. This means that T cells that proliferate bind antigen (self- or nonself) complexed with class I or II MHC proteins. This happens in the *cortex of the thymus*. At this time, cells that are specific to MHC II protein retain CD4, lose CD8, and become T helper cells. T cells that are specific to MHC I protein retain CD8, lose CD4, and become cytotoxic T cells.

After positive selection, cells undergo **negative selection** in the *medulla of the thymus*. During this process, cells that bind with *high affinity* to MHC/self-antigen complexes die by apoptosis. This is important because these cells would have later reacted with self-peptides and caused autoim-

TABLE 10.3	T-cell–Mediated Immunity Response According to the Source of Antigen		
	Intracellular Infection (Viruses, Some Bacteria)	**Extracellular Infection (Bacteria)**	**Extracellular Proteins: Antigen surfaces, Vaccines, and Toxins**
Location of antigen	Cytosol	Phagosomes	Endosomes
Antigen-presenting cell (APC)	Any cell	Professional: macrophages, dendritic cells, B cells	B cells
Location of APC	Anywhere	Lymphoid and connective tissue, body cavities, epithelium	Lymphoid tissues, blood
Molecules of display	MHC class I	MHC class II	MHC class II
Antigen recognition cells	CD8$^+$ CTL cells	CD4$^+$ T$_H$1 and CD4$^+$ T$_H$2 cells	CD4$^+$ T$_H$2 cells
Response	Release of cytotoxic effector molecules	Release of macrophage-activating effector molecules and molecules that support B-cell activation and differentiation	B-cell activation and CD4$^+$ T$_H$ cell activation
Effect	Death of infected cell	Killing of bacteria and parasites; supported activation of B cells	Secretion of antibody to eliminate bacteria and toxins

MHC, major histocompatibility complex; CD, cluster of differentiation; CTL, cytotoxic T cells; T$_H$1, T helper 1 cell; T$_H$2, T helper 2 cell.

mune diseases. Cells that bind with low affinity to MHC/self-antigen are allowed to leave the thymus. They will later only be activated by high-affinity binding to MHC/*foreign* antigen complex, the adequate signal for immune activation.

Killer, helper, and memory T cells

Killer T cells, helper T cells, and memory T cells can be distinguished based on their immunologic function. *Killer T cells*, or cytotoxic T cells (CD8$^+$), are lymphocytes that release lymphotoxins and are designed to kill cells that are infected with viruses or other pathogens.

Helper T cells (CD4$^+$) help determine which types of immune response the body will take to attack a particular pathogen. These cells have no cytotoxic activity and do not directly kill infected cells or clear pathogens. Instead, helper T cells control the immune response by directing other cells to perform these tasks. They function to regulate both the innate and adaptive immune responses. Helper T cells direct the immune response by secreting lymphokines. They stimulate proliferation of B cells and cytotoxic T cells, attract neutrophils, and activate macrophages. The differentiation of CD4$^+$ T helper cells into best-known subtypes T helper 1 and T helper 2 cells occurs after activation in the peripheral lymphoid system. Each subtype produces a distinct set of effector molecules.

For instance, *T helper 1 cells* release the macrophage-activating effector molecules interferon-γ and tumor necrosis factor-α (TNF-α). The resultant actions, called T helper 1 response, or *type 1 response*, support activities of macrophages and cytotoxic T cells of the cellular immune system.

On the other hand, *T helper 2 cells* produce effector molecules, such as interleukin-4, interleukin-5, interleukin-13, interleukin-25, interleukin-31, and interleukin-33, among numerous other cytokines. This T helper 2 response, or *type 2 response*, promotes the actions of B cells and hence the humoral immune system.

Research shows that the dual T helper cell model is far too simple. For instance, many helper T cells express cytokines from both profiles. They are often named *T helper zero* cells. Additionally, other immune cells express many of the specific T-cell cytokines as well. Several new models have been proposed that include T helper 17 cells, but as of today, there is no unanimously accepted approach.

About 10% of CD4$^+$ T cells are *T regulatory cells* (previously known as *suppressor T cells*). Several subtypes have been named (e.g., Treg, Tr1, and Th3 cells). All have in common the ability to suppress immune responses and, hence, are important in maintaining immune homeostasis. For instance, they are known to inhibit the production of cytotoxic T cells when they are no longer needed. On the other hand, it has been shown that T helper cells are also capable of "regulating" their own responses. Regulatory T cells are thought to be associated with induction of tolerance to microbiota at mucosal surfaces.

Lastly, *memory T cells* are lymphocytes that become "experienced" by having encountered an antigen during a prior infection or a previous vaccination, or a cancer cell. Memory T cells, when encountering an invader (pathogen) for the second time, mount a faster and stronger immune response than they did the first time. These cells are critical for the longevity of specific immunity. T helper cells are absolutely required for B-cell memory cells to develop and antibody class-switching events to occur.

Time delay in cell-mediated immunity

T cells and their products may exert their effects in concert with other effector cells, such as neutrophils, eosinophils, and mast cells. The secretion of T-cell factors that recruit and activate other cells takes time, and thus, the consequences of T-cell activation are not noticeable until 24 to 48 hours after antigen challenge.

An example illustrating this time delay is the *delayed-type hypersensitivity reaction* (see Immunologic Disorders) to purified protein derivative (PPD), a response used to assess prior exposure to the bacteria that cause tuberculosis.

Injected under the skin, PPD elicits the familiar inflammatory reaction characterized by local erythema and edema 1 to 2 days post injection.

Cell-mediated immune responses, although slow to develop, are potent and versatile. They provide the main defense against many pathogens. T cells are also responsible for the rejection of transplanted tissue grafts (see Organ Transplantation and Immunology) and the containment of the growth of neoplastic cells (see Immunologic Disorders). A deficiency in T-cell immunity, such as that associated with AIDS, predisposes the affected patient to a wide array of serious, life-threatening infections.

Humoral immunity is mediated by secreting antibodies.

Binding of antigen to the BCR activates B lymphocytes, which then start proliferating. Activated B cells can differentiate into plasma cells that secrete antibodies. If the activated B cell is in the presence of an activated T helper cell, the T helper cell releases cytokines that support B-cell activation (see Fig. 10.4). In the classical, somewhat outdated model, B-cell–activating cytokines, such as IL-4, are part of the T helper 2 response. Most new cells become plasma cells, which produce antibodies for 4 to 5 days, resulting in a high level of antibodies in plasma and other body fluids. These antibodies can bind specifically to the antigenic determinant that induced their secretion. When T helper cells are involved with the activation, other B-cell clones become long-lived *memory B cells* (see Figs. 10.3 and 10.4).

In contrast to the time-delayed response of cell-mediated immunity, antibodies can induce immediate responses to antigens and, thereby, provoke *immediate hypersensitivity reactions* (see Immunologic Disorders).

Antibody structure

Antibodies are also called **immunoglobulins**. The primary structure of an antibody is illustrated in Figure 10.6. Each antibody molecule consists of four polypeptide chains (two *heavy chains* and two *light chains*) held together as a Y-shaped molecule by one or more disulfide bridges.

There are two isotypes of the light chain, κ and λ. Heavy chains have five different isotypes (α, δ, ε, γ, and μ), which

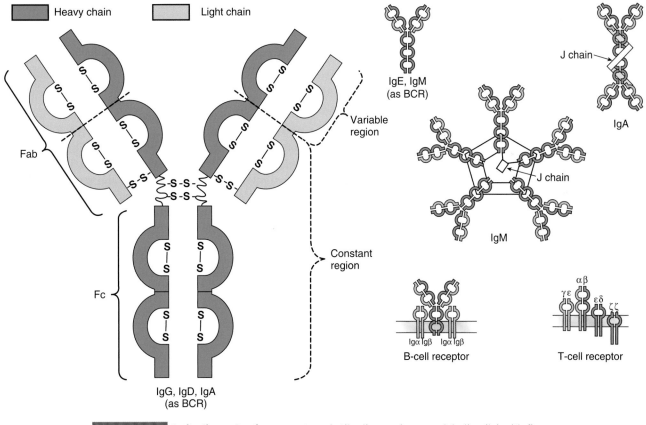

Figure 10.6 **Antibodies and antigen receptors.** Antibodies are immunoglobulins (Ig) with five isotypes (IgA, IgD, IgE, IgG, IgM). Not shown are the four IgG and two IgA subtypes. The basic unit of each isotype is a monomer that consists of two heavy chains and two light chains held together in a Y-configuration by disulfide bonds. The heavy chains are different for the various isotypes. Each end portion of the "Y" is called *Fab* because it is the *fragment* that contains the *antigen-binding* site. A second, *crystallizable fragment* is called *Fc*. When secreted from plasma cells, IgG, IgD, and IgE consist of one monomer, IgA of two (dimer), and IgM of five (pentamer). The B-cell receptor resembles a membrane-associated Ig monomer. The T-cell receptor contains a Fab-like structure.

constitute five different classes of antibody, each with different effector functions (see below). Each polypeptide chain possesses both a *constant region*, where the protein structure is highly conserved, and a *variable region*, where considerable amino acid sequence heterogeneity is found.

The amino terminal domains at each end of the forked portion of the "Y" of both the heavy and light chains are known as the *Fab regions* ("Fab" for fragment, antigen binding), which contain the antigen-binding regions. Antibodies are flexible in that the Fab arm can wave, bend, and rotate. This freedom of movement allows it to more easily conform to the shape of the antigen.

The carboxy terminal end of the heavy chain is termed the *Fc region* ("Fc" for fragment, crystallizable). Neutrophils, monocytes, and mast cells can recognize Fc regions via their Fc receptors, which facilitate effector mechanisms such as phagocytosis.

Classes of antibodies

The BCR is on the surface of a B cell and is comprised of an antibody coupled with a dimer (an Igα protein and Igβ protein). The Igα and Igβ dimer is associated with signal transduction, whereas the antibody is focused upon antigen binding. There are five classes (isotypes) of antibodies (IgA, IgD, IgE, IgG, and IgM), each class has specific biologic properties that deal with particular antigens. Figure 10.6 and Table 10.4 summarize the shape and functions of the five major classes of antibodies. Figure 10.6 additionally shows the similarities of the TCR and the BCR to immunoglobulin.

IgM is the dominant class of antibody secreted during a primary immune response (first encounter with the antigen). It has a half-life of about 5 days. The BCR on a naïve B cell (B cells that have not yet been activated by antigen) express IgM is a monomer. However, IgM, when secreted by plasma cells, consists of five "Y" units, held together by a *joining chain (J chain)*. The size of secretory IgM and its many antigen-binding sites provide the molecule with an excel-

lent capacity for agglutination of bacteria and blood cells. Although it has the potential to bind 10 antigens, in reality, it usually only binds five. Fixed macrophages of the RES system efficiently and quickly remove such agglutinated antigens.

IgD is found in plasma (in very small amounts) and on the surface of naïve B cells. It has a half-life of 3 days. An exact function is not known, but its ubiquitous presence in the animal kingdom suggests an important role. It has been postulated to be involved in the induction of immune tolerance. IgD was also found to activate basophils and mast cells. IgD serum concentration does increase during chronic infection but is not associated with any particular disease.

IgG is the major antibody produced in response to secondary and higher-order antigen encounters (subsequent encounters with antigens from which memory cells were created). Secondary immune responses are turned on faster and produce more antibodies that have a higher affinity to the antigen compared with primary responses. Hence, IgG is the most prevalent antibody in serum and is responsible for adaptive immunity to bacteria and other microorganisms. IgG exists in serum as a monomer. There are four IgG subclasses in humans (IgG1, IgG2, IgG3, and IgG4). IgG has a half-life of 23 days, the longest among the antibody classes. When bound to antigen, IgG can activate serum complement and cause opsonization. It can cross the placenta and is secreted into colostrum, protecting the fetus as well as the newborn from infection.

IgA usually exists as a polymer, when secreted from plasma cells, of the fundamental Y-shaped antibody unit. In most IgA molecules, a *joining chain (J chain)* holds together two antibody units (dimeric form). As the IgA passes through epithelial cells, an additional antigenic fragment, the *secretory piece*, is added. In this conformation, IgA is actively secreted into saliva, tears, colostrum, and mucus and hence is known as secretory immunoglobulin (sIgA). IgA is also found in serum, mainly as IgA1 isotype produced by bone

TABLE 10.4	Characteristics of Different Antibody Classes				
Characteristic	IgG	IgA	IgM	IgD	IgE
Molecular weight (×10⁻³ Da)	150	150, 400	900	180	190
Serum concentration (mg/dL)	600–1,500	85–300	50–400	<15	0.01–0.03
Serum concentration (% of Ig)	~76	~15	8	1	0.002
Half-life (days)	~23	~6	~5	~3	~3
Crosses placenta	+ (not IgG4)	–	–	–	–
Enters secretions	+	+ +	–	–	–
Agglutinates particles	+	+	+ + +	–	–
Allergic reactions	+	–	–	–	+ + + +
Complement fixation	+ (not IgG4)	–	+ +	–	–
Fc receptor binding to monocytes and neutrophils	+ +	–	+	–	–

Ig, immunoglobulin.

marrow B cells, while IgA2 subtype is present in secretions. As part of the BCR, IgA is found in the monomeric form. IgA has a half-life of about 6 days.

IgE is a monomeric antibody that is slightly larger than IgG but has a relatively short half-life of about 3 days. IgE avidly binds via its Fc region to cells, such as mast cells, basophils and eosinophils, which are involved in allergic reactions and antiparasitic immunity.

Antibody action

Antibodies act against antigens in three ways: they neutralize the antigen, opsonize the antigen, or stimulate complement fixation to assist phagocytes.

1. **Neutralization.** Antibodies can bind to antigens, forming easily recognizable antibody–antigen complexes, which are removed by phagocytosis. Antibodies can also immobilize and agglutinate infectious agents so that a virus cannot penetrate the host cell or a microbe cannot colonize mucosal tissue.

2. **Opsonization.** IgG antibodies bind to bacteria or virus-infected cells at the Fab region. That way, the pathogen is "tagged" or "opsonized" for destruction by free radicals and enzymes or phagocytosis. For phagocytosis, the Fc portion of the antibody binds to Fc receptors on phagocytes. Some complement components (e.g., C3b and C4b) can also act as opsonins.

3. **Complement fixation.** Complement is a group of at least nine distinct proteins that circulate in plasma in an inactive form, but involves about 25 proteins and protein fragments (Fig. 10.7). When the inactive proteins hydrolyze, they become activated. A cascade of events occurs when the first protein, C1, recognizes preformed IgM or IgG antibody–antigen complexes (*classical pathway*). A side entry to this pathway exists, in which complexes between bacterial mannose residues and the plasma protein mannose-binding lectin start the cascade (*lectin pathway*). C-reactive protein (CRP) can also provide a side entry into this pathway by directly activating C1. The cascade leads to the formation of C3 convertase through the activation of C4 and C2, and so the convertase has the form C4b2a.

A homologous variant of C3 convertase (C3bBb) is also formed in an *alternative pathway,* which is activated by small amounts of C3b that spontaneously hydrolyzed from C3. When C3b binds to human cells, it is quickly inactivated. However, C3b becomes protected when it binds to microbial cell surfaces and forms C3 convertase through factors D and B. The complex is stabilized by properdin (factor P). A side entry to this pathway exists, in which complexes between bacterial mannose residues and the plasma protein mannose-binding lectin start the cascade (*lectin pathway*). Compared to the classical pathway, the alternative pathway is faster, but requires specific microbial antigens for activation.

The C3 convertases of both pathways lead to the formation of complement C3b that activates C5 convertase, another critical enzyme in the cascade that forms C5b. In the classical pathway, C5 convertase is created when C3b joins

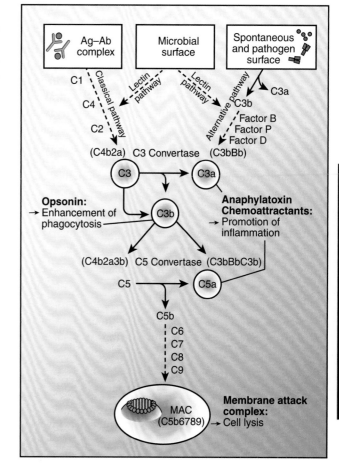

Figure 10.7 **The complement system attacks the surface of foreign cells.** Over 20 different complement (C) proteins supplement antibody activity in a cascade of events to eliminate pathogens. The goal is to produce opsonin C3b, anaphylatoxins, chemoattractants C4a, C3a and C5a, and the membrane attack complex (MAC). The MAC inserts into membranes of cells and causes their lysis. Antigen–antibody (Ag–Ab) complexes and other (e.g., C-reactive protein and lectin pathway components) activate the classical pathway. Spontaneously hydrolyzed C3 and pathogen surface molecules (e.g., lipopolysaccharides) activate the alternative pathway and other (e.g., lectin pathway components) activate the classical pathway. The lectin pathway is initiated in response to microbial surfaces (mannose). The classical and alternative pathways use structurally different but functionally identical enzymes C3 convertase and C5 convertase.

C4b2a (C4b2a3b) and in the alternative pathway when C3b joins C3bBb (C3bBbC3b). C5b and the four complement proteins, C6, C7, C8, and C9, form the **membrane attack complex** (**MAC**). The tubular MAC inserts into the membrane of the target cell because of its hydrophobic nature. This allows free passage of small molecules, ions, and water, resulting in the target cell's death.

Serum levels of C3 and CRP can be used to assess inflammation in a patient. Low serum levels of C3 indicate that C3 is being consumed (activated) in complement activation pathways, whereas an elevated level of CRP correlates with increased inflammation.

▶ ACUTE AND CHRONIC INFLAMMATION

Inflammation is the initial response of the body to infection or trauma. Although *microbial infection* is probably the most common cause of inflammation, many types of *tissue injury* can also evoke inflammatory reactions. These include mechanical injury, radiation, burns, frostbites, chemical irritants, and tissue necrosis resulting from lack of oxygen or nutrients. The body instigates a nonspecific cascade of physiologic processes in vascularized tissues involving elements of the innate immune system, with the goal of repairing cellular damage and restoring the tissue to its normal function.

- In a normal response, inflammation is self-initiating, temporally self-propagating, and self-terminating.

- Inflammation is a necessary response to tissue injury, and human life without inflammation is unthinkable.

- On the other hand, inflammation often overshoots in its reactions, which leads to a vicious circle of repeated injury and persistent inflammation.

- Inflammation is closely connected with all kinds of illnesses, so that anti-inflammatory therapy is at the heart of many treatments.

Acute inflammation is a short-term process and is characterized by five cardinal signs.

Inflammation is a protective mechanism that involves immune cells, blood vessels, and cellular mediators. Inflammation is a general response and therefore is part of the innate mechanism of immunity, as compared to adaptive immunity, which is specific for each pathogen. The purpose of inflammation is to (1) eliminate the initial cause of tissue injury, (2) clear out necrotic cells and remove damaged tissue caused from the initial insult, and (3) initiate tissue repair. The classical signs of acute inflammation follow a common pattern with the first sign consisting of redness, followed by heat, swelling, and pain.

With increasing knowledge of the complexity of inflammation, many models have been put forward to expand on these four signs to categorize the great variety of inflammatory responses. However, they are not yet commonly applied in clinics beyond the addition of a fifth sign, **functio laesa** (the loss of function), which is a consequence of tissue operating at conditions out of homeostasis and spending energy on repair processes.

Acute inflammation is a three-step process involving vasodilation and cell migration from blood to tissue.

Inflammation can be classified as either acute or chronic. *Acute inflammation* is the body's initial response and involves the movement of leukocytes, especially granulocytes from the blood into the damaged tissue. Prolonged inflammation is known as *chronic inflammation* and involves a shift in the type of cells present at the injured site. The progressive shift in cells types triggers the destruction and healing process.

At the onset of infection, burn, or other injuries, acute inflammation is initiated by cells already present in the injured tissue—mainly, resident macrophages, dendritic cells, Kupffer cells, and mast cells. These cells release inflammatory mediators that bring about functional changes to tissues. One of the immediate changes observed in acute inflammation is vasodilation and, therefore, increased permeability. Vasodilation increases blood flow to the injured area, sometimes up to 10-fold. Additionally, the vessel walls become leaky, on the one hand, as a result of the injury-related necrosis of endothelial cells and, on the other hand, as a result of chemically directed retraction of endothelial cells. Histamine, which is released by mast cells, basophils, eosinophils, and platelets, is a major chemical mediator of this process.

Water, salts, and small proteins, such as fibrinogen, exit from the plasma into the damaged area. Fibrinogen forms fibrin networks to trap microorganisms. The leaking fluid is called **exudate**, or *pus* in the case of an infected wound. If an infected area is further liquefied and shielded from surrounding tissue, it is called an *abscess. Blisters,* pools of lymph fluid, may form in response to burns, infections, or irritating agents.

Blood leukocytes are attracted and migrate into the inflamed tissue in a regulated process that evolves three steps: in the first step, called **margination**, neutrophils are recruited and form bridges with endothelial cells at the inflamed sites. These bridges are made of *selectins*, P-selectins and E-selectins on endothelial cells and L-selectins on neutrophils.

In the second step, called **rolling**, the neutrophils move along the endothelium like tumbleweed since the bridges are initially loose and form and detach continuously. Eventually, a more firm connection is built between *integrins* on neutrophils and *intracellular* and *vascular adhesion molecules* on the endothelial side.

In the third step, leukocyte extravasation, commonly called **diapedesis** or *transmigration*, neutrophils actively migrate through the blood vessel basement membrane into the tissue, with the aid of the firm connection but without damaging the endothelial cells. After neutrophils, *monocytes* are attracted and differentiate into *macrophages* when leaving the blood. Later, *lymphocytes* follow as well. The appearance and increase of the nongranular cells, monocytes, and lymphocytes mark the transition into chronic inflammation.

Inflammatory mediators develop and maintain the inflammatory response.

Inflammatory mediators are soluble molecules that act locally at the site of damage and coordinate the inflammatory responses.

Exogenous mediators (produced outside of the body) include **endotoxin**, which is a family of bacterial toxins associated with the LPS complex of gram-negative bacteria. Endotoxin is released from lysed bacteria and binds to receptors on monocytes and macrophages, thus activating them.

Endogenous mediators (produced within the body) are released from injured or activated cells at the inflammation site. Numerous substances are known to regulate inflammation, many of them with seemingly redundant functions. Table 10.5 lists a few of them. Some have previously been

TABLE 10.5 Endogenous Inflammatory Mediators

Type	Name	Released by	Some Actions
Chemical mediators	Histamine	Mast cells, basophils, eosinophils, leukocytes, platelets	Vasodilation, vascular permeability
	Lysosomal compounds	Neutrophils, macrophages	Vascular permeability, complement activation
	Eicosanoids: Prostaglandins Thromboxanes Leukotrienes	Many cells, neutrophils	Vasoactive properties, platelet aggregation, prolongation of edema
	Platelet-activating factor	Neutrophils, monocytes, mast cells, eosinophils	Vascular permeability, neutrophil migration, bronchoconstriction
	Serotonin	Mast cells, platelets	Vasoconstriction
	Cytokines	Lymphocytes, monocytes	Vasoactive and chemotactic properties
	Chemokines	Tissue cells, endothelial cells, leukocytes	Chemotaxis of inflammatory effector cells
Gases	Nitric oxide	Endothelial cells, macrophages	Vascular smooth muscle relaxation and vasodilation, microbe killing, inhibitor of platelet actions
Neuropeptides	Tachykinins, kinins	Mainly sensory neurons	Vasodilation, vascular permeability, smooth muscle contraction, mucus secretion, pain
Plasma factors	Complement (C5a, C3a, and others)	Enzymes from dying cells; antigen–antibody complexes; endotoxins; products of kinin, coagulation, and fibrinolytic system	Chemotaxis, degranulation of phagocytes, mast cells and platelets, cytolytic activity, opsonization of bacteria
	Kinins	Coagulation factor XII	Vascular permeability, mediators of pain, nonvascular smooth muscle contraction
	Coagulation factors	Coagulation factor XII	Conversion of fibrinogen to fibrin
	Fibrinolytic system		Plasmin lyses fibrin

mentioned as part of the innate immune system, demonstrating the tight interplay between the two immune responses.

Many mediators can interact with multiple receptors, often exerting diverse functions. For instance, histamine interacts with three receptors. The H_1 receptors mediate acute proinflammatory vascular effects, whereas activation of H_2 receptors results in anti-inflammatory actions. H_3 receptors are involved in the control of histamine release.

From a structural standpoint, inflammatory mediators belong to many different categories, such as proteins (e.g., complement, antibodies, and acute-phase reactants), lipids (e.g., prostaglandins and platelet-activating factor), amines (e.g., histamine), gases (e.g., nitric oxide), kinins (e.g., bradykinin), and neuropeptides (e.g., substance P).

Many *plasma-derived mediators* are present as precursors and need enzymatic cleavage to become active. For example, thrombin is necessary to transform soluble fibrinogen into insoluble fibrin for blood clots. *Cell-derived mediators* can be preformed (e.g., histamine) or synthesized as needed (e.g., prostaglandin).

Inflammatory cytokine profiles provide a biomarker for the severity of inflammation.

An important class of mediators is **cytokines**, which are cell products synthesized de novo in response to immune stimuli. They generally act over short distances and short time spans. They comprise **lymphokines** (cytokines made by lymphocytes), **monokines** (made by monocytes), **chemokines** (chemoattractants made by various cells), and **interleukins** ([ILs] made by one leukocyte and acting on other leukocytes).

Certain inflammatory diseases present with characteristic cytokine profiles; therefore, inhibition of proinflammatory cytokines may aid in the treatment of the diseases. The classical proinflammatory cytokine triad contains IL-1, IL-6, and TNF-α. The following presents the sources, major targets, and principal activities along with those of IL-8 and INF-γ, two additional clinically relevant factors.

Interleukin-1: IL-1 is produced by phagocytes, lymphocytes, endothelial cells, and others. It targets lymphocytes, macrophages, and endothelial cells. It activates cells, increases adhesion molecule expression, and induces fever and the release of acute-phase proteins.

Interleukin-6: IL-6 is produced by lymphocytes, macrophages, fibroblasts, and endothelial cells. It targets B cells and hepatocytes. In addition to B-cell differentiation, it promotes the production of acute-phase proteins and fever.

Tumor necrosis factor-α: TNF-α derives from macrophages, mast cells, lymphocytes, and endothelial cells. Its targets and effects are similar to those of IL-1. It additionally stimulates angiogenesis and has other functions outside of inflammation.

Interleukin-8: IL-8 has been renamed *CXCL8*, because it is a chemokine produced by macrophages, endothelial cells, and fibroblasts. It targets neutrophils, basophils, and T cells. Its major actions are chemotactic and angiogenic.

Interferon-γ: IFN-γ is produced by T cells and NK cells. It targets leukocytes, tissue cells, and T helper cells. In inflammation, it activates phagocytes and enhances leukocyte–endothelial adherence.

In response to IL-1, IL-6, and others, the liver produces **acute-phase blood proteins**. An important member is *C-reactive protein*, which is used to monitor the severity and progression of some cardiovascular diseases. Another example is *acute-phase serum amyloid A (A-SAA) proteins*. One of their functions is to recruit immune cells to the inflammation site. Blood concentration of A-SAA might rise up to 1,000-fold during inflammation; therefore, A-SAAs are used as markers in some autoimmune diseases.

Acute inflammation is a stereotypic, highly complex process that self-terminates.

Modern molecular biology superimposes many additional layers of complexity onto the presented model of acute inflammation as a stereotypic process of the vascular system. For instance, it has been shown that factors independent of the vasculature can trigger and modulate aspects of both inflammation and repair. *Vibration* and *hypoxia* can lead to histamine release of mast cells (degranulation), hence initiating inflammatory events. *Mechanical load* on tendon fibroblasts can modulate their response during inflammation, promoting their destruction at high loads and protecting them from inflammatory apoptosis at low loads. Furthermore, proinflammatory molecules can be upregulated without any concomitant invasion and stimulation of inflammatory cells.

Under normal circumstances, acute inflammation terminates itself. Chemical mediators disappear either because of their short half-life or because they are enzymatically inactivated (e.g., kininases inactivate kinins). Substrates are consumed, and the lymph flow carries the mediators away faster than they can be produced. *Anti-inflammatory cytokines* such as IL-4, IL-10, and transforming growth factor-β (TGF-β) induce repair of damaged tissue. Following inflammation, tissue that is capable of regeneration will be almost completely restored; otherwise, *scar formation* occurs.

▶ CHRONIC INFLAMMATION

Chronic inflammation is the failure to resolve inflammation and usually causes serious bodily harm. Chronic inflammation develops when neither agent nor host is strong enough to overwhelm the other (e.g., in the case of osteomyelitis, the infection of bone), when there is prolonged exposure to the toxic agent (e.g., silicosis from inhalation of crystalline silica dust), or as part of autoimmune disease processes (e.g., rheumatoid arthritis [RA]). It can also develop without being preceded by acute inflammation (e.g., in tuberculosis).

Chronic inflammation may last weeks, months, or years and leads to *chronic wounds,* which are composed of loosely arranged connective tissue (*granulation tissue*), infiltrated fibroblasts, and inflammatory cells. The vastly dominant inflammatory cell type is the *monocyte/macrophage*. It is the only cell type present in the case of chronic inflammation resulting from a nonantigenic agent such as a suture thread. In the case that the injurious agents are also antigenic, other cell types appear, such as lymphocytes, plasma cells, and eosinophils.

Chronic inflammation is characterized by ongoing tissue damage from reactive oxygen and nitrogen species and proteases that are secreted by inflammatory cells. Other products, such as arachidonic acid and proinflammatory cytokines, amplify as well as propagate the damage. This weakens the body and makes it even more susceptible to infection and further inflammation. It is a vicious cycle that leads to clinical symptoms typical of chronic inflammatory diseases, such as RA, atherosclerosis, or psoriasis.

Overwhelming inflammation combined with immune suppression can lead to *systemic inflammatory response syndrome*, which is called **sepsis**, or *septicemia*, when the inflammation is a result of infection. In severe cases, it might lead to organ failure and death.

Chronic inflammatory diseases can affect every body part and are usually indicated by adding the suffix "itis." For instance, *myocarditis* is inflammation of the heart, and *nephritis* is inflammation of the kidney. Some inflammatory conditions do not follow the conventional terminology. The most common one is *asthma*. Another example is *pneumonia*, which is more commonly used than "pneumonitis" to name the chronic inflammatory infection of the lung.

Chronic inflammation is both a symptom and a cause of disease.

The role of inflammation in diseases that were not traditionally categorized as inflammatory diseases has been viewed with considerable interest. In fact, inflammation is now recognized to be a critical pathologic component underlying many illnesses, ranging from Alzheimer's and Parkinson's diseases to diabetes and certain types of cancer (e.g., colon cancer).

Although the association between inflammation and disease has been recognized for a long time, inflammation was mainly considered a result, rather than a cause, of the illness. Studies clearly show that the destructive, self-promoting cycle between oxidative stress and resulting inflammation causes more oxidative stress and contributes to the development of chronic illnesses. This chain of events may initially occur at a low, asymptomatic level, which leads to the speculation that *subacute chronic inflammation* might be the cause of a number of illnesses.

▶ ANTI-INFLAMMATORY DRUGS

Corticosteroids are the most powerful anti-inflammatory drugs currently available. They inhibit the accumulation of neutrophils at sites of inflammation (see Neuroendoimmunology) but also have widespread effects on other inflammatory cells and processes. This is the reason for abundant side effects ranging from osteoporosis to disruption of the hypothalamic–hypophyseal axis.

Nonsteroidal anti-inflammatory drugs inhibit **cyclooxygenase** (**COX**), which is involved in prostaglandin production. There are at least two isoforms of this enzyme. *COX-1 is* found in platelets and catalyzes the production of thromboxane A2. COX-1 inhibition leads to diminished blood clotting, so COX-1 inhibitors are used as blood thinners. *COX-2* is expressed in blood vessels and macrophages in response to inflammation and leads to production of prostaglandin I2.

Drugs that specifically block COX-2 can decrease inflammatory pain. However, both COX-1 and COX-2 are present at many more sites, which is a reason for the abundant pharmacologic side effects of COX inhibitors.

Alternative therapies include a wide variety of natural substances that have antioxidative and anti-inflammatory features but often have poorly understood mechanisms of action. These include nutrients such as vitamin E, herbs such as chamomile, and supplements such as glucosamines, to name a few.

Because the side effects or uncertainties associated with many existing anti-inflammatory drugs often outweigh their benefits, a main focus of twenty-first century medical research is to develop novel anti-inflammatory strategies. Novel drugs generally aim to target specific pathways instead of general inflammatory processes. For instance, leukotriene modifiers, a relatively new class of anti-inflammatory medications for asthma patients, specifically block the products of arachidonic acid metabolism and can prevent an asthma episode before it even starts (Clinical Focus 10.1).

Blood and Immunology

CLINICAL FOCUS | 10.1

Allergies and Asthma

Epidemiologic data confirm what many parents suspect. Their children are more susceptible to allergies than they were when they were children. In the past 30 years, the prevalence of allergies has been greatly increasing in the United States and in the world, proportionally more so in more developed countries compared with developing countries and more so in urban areas compared with rural areas.

One hypothesis, often referred to as the hygiene model, proposes that the children who are not exposed to enough antigens or innocuous microorganisms from soil or water early in their childhood do not adequately develop their T helper 1 cell response. This leads to an imbalance toward T helper 2 cell responses, which are present from birth. Even though this hypothesis is disputed as being grossly oversimplified, it is undisputed that the various manifestations of allergic inflammation are the result of wrongly activated T helper 2 cells.

Asthma is one common allergic disorder. About 23 million people, including almost 7 million children, have asthma. It belongs to type I hypersensitivity disorders and is characterized by airways that narrow excessively in response to many provoking stimuli.

Typical allergens causing atopic (also called extrinsic or allergic) asthma are animal dander, dust mite excrement, grass pollen, and cockroach antigen. Strenuous exercise or worry might trigger nonatopic asthma.

The symptoms of both asthma types are similar and include wheezing, breathlessness, chest tightness, and coughing. They are caused by chronic inflammation of the airways, which leads to their swelling. Bronchoconstriction, excess production of mucus, and increased collagen deposition further exacerbate the restriction of the airflow to and from the lungs.

The development of asthmatic symptoms ultimately depends on the presence of T helper 2 cells, which secrete a characteristic cytokine repertoire. Of these, interleukin-4 activates B cells to produce IgE antibodies, which bind to mast cells, eosinophils, and other airway cells via receptors that are specific for the Fc portions of the antibodies. On subsequent encounters with the antigen, IgE cross-linking on mast cells and eosinophils occurs and leads to the cells' activation and their release of histamine and leukotrienes. These and other inflammatory factors recruit inflammatory cells including additional T helper 2 cells to the lung, resulting in a damaging immunologic positive feedback cycle. IL-4, together with IL-9, is also necessary for mast cell maturation, and along with IL-5, for eosinophil recruitment.

Several treatments are available that interfere with the different allergic processes. Antihistamine drugs for treatment of atopic asthma target H1 receptors or prevent the release of histamine from mast cells. Epinephrine and β2-selective agonists for treatment of systemic anaphylaxis modulate the sympathetic nervous system (see Neuroendoimmunology). Corticosteroids and leukotriene receptor antagonists aim at suppression of inflammation. Immunotherapy (also called hyposensitization or desensitization) involves subcutaneous injections that contain gradually increasing doses of an allergen extract. The immune system responds with gradually increasing levels of IgG antibodies, making fewer IgE antibodies, which seems to help a patient avoid dramatic reactions when the allergen is then encountered naturally. Patients of allergen immunotherapy are at risk of developing systemic anaphylaxis. Antibody titers should be monitored throughout the hyposensitization process to insure the patient is making fewer IgE antibodies and that the therapy is not increasing the levels of IgE. ■

► ORGAN TRANSPLANTATION AND IMMUNOLOGY

Both organs and tissues can be transplanted. Transplanted organs include heart, liver, lungs, pancreas, intestines, and thymus. Transplanted tissues include cornea, skin, heart valves, veins, bones, and tendons. The latter two are both referred to as *musculoskeletal grafts*. Worldwide, the kidney is the most common organ transplant, although musculoskeletal transplants (tissue) outnumber the kidney more than 10-fold.

Organ transplantation is indicated when irreversible damage has occurred and alternative treatments are not applicable. In the early 1950s, the first human kidney was transplanted to the donor's identical twin brother. Such transplantation is called an **isograft** because the donor and recipient are genetically identical. In 2014, ~17,000 kidneys were transplanted in the United States, most often as **allografts** between genetically different people.

Kidneys, lungs, and livers can come from living donors, because a person can live with one kidney or one lung and because liver tissue regenerates. Transplantations of heart, whole lung, pancreas, and cornea come from deceased donors. So-called nonvital transplantations refer to the replacement of body parts such as the hands, knees, uterus, trachea, and larynx.

Tissue that is transplanted in the same person from one place to another place is called an **autograft**. Skin and blood vessels are common tissues for such transplantation.

Although, technologically, transplantation is a viable therapeutic option, it is mainly limited by the immensely larger demand for transplantation organs compared with their availability. An attempt to overcome the donor organ shortage is **xenograft** transplantation, in which animal tissue is used to replace human organs.

Two other avenues for replacement of tissue are the development of *artificial organs* and the transplantation of *stem cells,* which may be able to regenerate full-functioning organs. The latter is the "holy grail" of organ replacement because the new tissue would not be recognized as foreign if the stem cells came from the recipient. However, as of today, these alternatives do not work nearly as well as allografted organs.

There are many issues to consider and challenges to overcome, involving ethical, social, and legal problems. To name just a few, xenotransplantation may introduce new diseases to humans and lead to disrespectful treatment of animals. The use of animal and artificial organs may raise problems of self-esteem as a result of the presence of nonhuman tissue in the body, and stem cell therapy may confer the risk of developing cancer (Clinical Focus 10.2).

Histocompatibility is the most important criterion for a match in organ donation.

The major reason for the donor organ shortage is that the blood and tissue of the donor and the blood and tissue of the recipient need to have similar immunogenic markers in

---

CLINICAL FOCUS | 10.2

Bone Marrow Transplantation

When a patient has a terminal bone marrow disease, such as leukemia or aplastic anemia, often the only possibility for a cure is bone marrow transplantation. In this procedure, healthy bone marrow cells are used to replace the patient's diseased hematopoietic system.

To identify a suitable donor, blood leukocytes of prospective donors, usually close relatives, are screened to determine whether their antigenic patterns match those of the patient. A transplant is successful only if the donor's human leukocyte antigens closely match those of the recipient. Because there are several antigenic determinants and each can be occupied by any one of several genes, there are thousands of possible combinations of leukocyte antigens. The odds that any person's cells will randomly match those of another are less than one in a million.

The antigenic composition of leukocytes in bone marrow and peripheral blood are identical, so analysis of blood leukocytes usually provides enough information to determine whether the transplanted cells will engraft successfully. Peripheral blood stem cells are currently being tested to see whether they can be used as effectively as bone marrow blood stem cells. This would minimize the pain and risks undertaken by donors.

If significantly different from the recipient's tissue type, transplanted leukocytes may be recognized as foreign by the patient's immune system and, therefore, be rejected, a phenomenon called host-versus-graft response. However, this often plays a minor role because the patient's immune system is immunocompromised and not able to react strongly.

The major problem of bone marrow transplantations is the fact that functional T cells in the graft recognize the host's tissue as foreign and mount an immune response. This is called **graft-versus-host disease (GVHD)**. The disease often begins with a skin rash, as transplanted lymphocytes invade the dermis, and may end in death as lymphocytes destroy every organ system in the marrow recipient.

There are several methods to decrease the morbidity of marrow transplantations. Immunosuppressive agents, including steroids, cyclosporine, and anti–T-cell antiserum, effectively decrease the immune function of the transplanted lymphocytes. Another useful approach involves the physical removal of T cells from bone marrow prior to transplantation.

T-cell–depleted bone marrow is much less capable than untreated marrow of causing acute GVHD. These techniques have substantially increased the potential pool of bone marrow donors for a given patient. ■

order to avoid rejection. For instance, virtually all kidneys with unmatched *ABO blood group* between donor and recipient will be rejected within minutes.

For successful transplantation, it is critical that the recipient does not have, or develop, antibodies against the donor's **human leukocyte antigen** (**HLA**). HLAs are the human antigens of the MHC complex, MHC classes I and II proteins. The MHC class I antigens are present on the surface of most body cells but were first discovered on leukocytes. HLA-A, HLA-B, and HLA-C are MHC class I antigens, and HLA-DP, HLA-DQ, and HLA-DR are MHC class II antigens, present on professional APCs. There are numerous alleles for each of the HLA antigens. For minimum requirements, transplantation doctors look at only six of the many HLA antigens (two A, two B, and two DRB1 antigens) for matching donors to patients. However, research has shown that a match in more HLA antigens and additional factors such as race, age, and sex improves the patient's chances for successful transplantation.

Although tissue typing decreases the risk of transplant rejection, the match between donor and recipient is never perfect (except for identical twins), and the reality of clinical practice requires organ transfer between less well-matched pairs. So, clinical symptoms resulting from immune attacks on the transplant are fairly common. Table 10.6 summarizes the three different types of graft rejections according to time elapsed after transplantation.

Immunosuppressive drugs greatly decrease the risk of rejection. There are numerous drugs available that inhibit different immune processes. Many of them downregulate the unwanted T-cell responses. Total destruction of T cells and other leukocytes by whole-body irradiation is used before bone marrow transplantation (see Chapter 9). However, the compromised immune system of a bone marrow transplant patient predisposes them to develop graft-versus-host disease (GVHD).

TABLE 10.6 Types of Graft Rejection

Type	Time	Reasons
Hyperacute	Minutes–hours	Preformed antidonor antibodies trigger type II hypersensitivity reactions (antibody-mediated destruction of cells).
Acute	Days–weeks	HLA incompatibility or inappropriate connection to blood supply triggers T-cell activation and type IV hypersensitivity reactions (CD4$^+$ T$_H$1 cells and activated macrophages); complicated by antibody-mediated rejection.
Chronic	Months–years	Unclear; cellular and humoral mechanisms involved; mediated by growth factors and repair mechanisms; recurrence of disease possible

HLA, human leukocyte antigen; CD4$^+$ T$_H$1, CD4$^+$ T helper type 1 cell.

▶ IMMUNOLOGIC DISORDERS

Table 10.7 presents a list of immune disorders, which is not intended to be comprehensive but to provide a framework that can be expanded in pathology courses. It is organized into hypersensitivity, immunodeficiency, and autoimmune disorders.

Many diseases involve mechanisms that fit into several of the categories presented in the table. One such example is **rheumatoid arthritis** (**RA**). An unknown antigen that stimulates antibody formation, characteristic of *type III hypersensitivity*, is thought to initiate RA. This leads to joint damage, and as a result, autoantigens are released, which perpetuate the disease in the typical fashion of an *autoimmune disease*. T cells are recruited and activate macrophages, which leads to *chronic inflammation*. Immune responses dominated by T helper 1 cells, characteristic of *type IV hypersensitivity* reactions, maintain the disease.

Immunological dysfunction is due to overreaction, failure to respond, and incorrect response.

As mentioned previously, under normal conditions, when infection resolves, the inflammatory response is actively terminated to prevent unnecessary damage to tissue. Failure to do so results in chronic inflammation and tissue damage. Accordingly, the immune system must be sensitive to pathogens and use sufficient power to defeat the pathogen, because even small numbers of residual microorganisms might soon endanger the body again. But this principle comes at the risk of overdoing it. Asthma, familial Mediterranean fever (recurrent episodes of peritonitis, pleuritis, and arthritis), and Crohn's disease (an inflammatory bowel disease) are examples in which chronic inflammation leads to immunologic disorders and tissue damage.

Overreaction

The most common immunologic disorders due to immune system overreaction are known as **allergies**, or **anaphylaxis** in the case of severe forms. Allergies are categorized into four types, hypersensitivity disorder type I to type IV (see Table 10.7). Depending on the sensitivity of the affected person, allergies can cause localized or systemic symptoms. *Localized reactions* are usually biphasic, with an immediate response mainly resulting from histamine (erythema, wheal, and flare) and a delayed response that may last several hours. *Systemic reactions* might be life threatening because of the sudden loss of blood pressure as a result of general vasodilation and due to airway obstruction as a result of smooth muscle contraction.

Immunodeficiency

Immunodeficiency (or **immune deficiency**) is a condition in which the immune system's ability to fight infections is compromised or absent. Many disorders fall into this category, although the number of patients with congenital immunodeficiency diseases is low.

The study of these diseases reveals important insights into cellular and molecular immunology. For instance, *severe combined immunodeficiency (SCID)*, commonly known as "bubble-boy disease," represents a group of congenital

TABLE 10.7 Immune Disorders

Names	Mechanisms	Examples (Common Names)
Hypersensitivity disorders: Damaging immune responses elicited by allergens		
Immediate hypersensitivity		
Type I (immediate hypersensitivity)	Allergens cause cross-linking of IgE bound to mast cells, basophils, and eosinophils, which release vasoactive mediators (e.g., histamine); symptoms within minutes.	• Atopic diseases: hives, asthma, hay fever • Food allergy • Systemic anaphylaxis
Type II (Ag–Ab cytotoxicity)	Cytotoxicity is mediated by antibody (primarily IgG) directed against epitopes on surface membrane of host cells; symptoms within hours.	• Newborn hemolytic anemia • Autoimmune hemolytic anemia • Blood transfusion reactions
Type III (Ag–Ab immune complex disease)	Circulating immune complexes (primarily IgG) escape phagocytosis and cause deposits in tissues or blood vessels leading to inflammation; symptoms within hours.	• Serum sickness • Glomerulonephritis • Farmer's lung
Delayed-type hypersensitivity		
Type IV (cell-mediated hypersensitivity)	Memory TH$_1$ cells cause cell-mediated immunity resulting in tissue damage by macrophages; symptoms within day(s).	• Contact dermatitis • Photoallergic dermatitis • Celiac disease
Immunodeficiency disorders: Caused by deficiencies (primary) or by illness in previously healthy person (secondary)		
B cell	Deficiency in antibody-mediated immunity	• Bruton agammaglobulinemia • IgA deficiency
T cell	Deficiency in cell-mediated immunity	• DiGeorge syndrome
Combined	Combined deficiency of cellular and humoral immunity	• SCIDs
Nonspecific	Mediated by deficient phagocytic and/or natural killer cells	• Leukocyte adhesion deficiency • Chronic granulomatous disease
Complement	Defects of individual components or control proteins	• C1, C4, C2, C3, C5–C9 deficiencies • Hereditary angioedema with lack of C1 inhibitor
Acquired	Human immunodeficiency virus (HIV) infects and kills CD$_4^+$ T helper cells, which leads to massive immunosuppression and increased risk of cancer.	• AIDS • Prolonged serious disorders such as cancers, kidney failure, and diabetes
Autoimmune disorders: Immune responses against the body's own tissue		
Antibody mediated	A specific antibody targets a particular antigen, which leads to its destruction and the signs of the disease.	• Autoimmune hemolytic anemia • Myasthenia gravis • Graves disease
Immune complex mediated	Antibodies complex with autoantigen into large molecules, which circulate around the body and cause destruction.	• Systemic lupus erythematosus • Rheumatoid arthritis
T cell mediated	T cells recognize autoantigen, which leads to tissue destruction without requiring the production of autoantibody.	• Multiple sclerosis • Type 1 diabetes • Hashimoto thyroiditis
Mediated by complement deficiency	Deficiencies in complement components often lead to or predispose to the development of autoimmunity.	• Systemic lupus erythematosus

Ag–Ab, antigen–antibody; CD, cluster of differentiation; Ig, immunoglobulin; SCIDs, severe combined immunodeficiencies; AIDS, acquired immunodeficiency syndrome; T$_H$1, T helper 1 cell.

disorders characterized by little or no immune response. Research showed that in the X-linked form of SCID, T cells are dysfunctional as a result of a mutation of the IL-2 receptor gamma chain. The immunologic consequences of missing IL-2 signaling are severe and may lead to the patient's death within the first year of life.

Autoimmunity

Autoimmunity is the failure of the body to recognize its own tissue as self, which allows an attack against its own cells and tissues. Diseases that result from such aberrant immune response are termed *autoimmune disease*. Examples of autoimmune diseases include type 1 diabetes mellitus, systemic lupus erythematosus (SLE), Graves disease, and RA. Genetic and environmental factors as well as gender contribute to the development of many of the autoimmune diseases. For example, ~75% of the 24 million Americans who suffer from autoimmune disease are women.

Although many contributing factors are involved, the trigger mechanism(s) for most autoimmune diseases is still poorly understood. For example, SLE is a chronic autoimmune disease that affects multiple organs. Its etiology is largely unknown, although it is clear that a genetic predisposition to its development exists, because an identical twin has 3 to 10 times increased risk for developing SLE. Hormones most likely play a role because 90% of patients with SLE are women. Lastly, it seems that environmental factors such as certain medications, stress, and other diseases can also lead to SLE.

Immune cells may become carcinogenic, leading to hematologic malignancies.

Not represented in Table 10.7 are *hematologic malignancies* because they typically form a separate category. Like any cell, cells of the immune system can grow out of control. **Multiple myeloma** is a typically incurable cancer of plasma cells. **Leukemia**, the abnormal proliferation of leukocytes, is the most common childhood cancer. **Lymphomas** encompass a diverse group of cancers that present as solid masses specific to the lymphatic system.

Immune **oncogenes**, genes that turn immune cells into tumor cells, can be induced spontaneously, but their persistence is increased in the presence of carcinogens or viruses. For example, the human T-cell leukemia virus type 1 can cause acute T-cell leukemia and lymphoma, and the Epstein-Barr virus of the herpes family is linked to Burkitt lymphoma and, possibly, Hodgkin lymphoma.

Tumor antigens can identify tumor cells and serve as potential candidates for cancer immunotherapy.

Some leukemias and lymphomas are easy to differentiate histologically from healthy tissue, whereas others are not. For instance, lymphoblasts in acute lymphoblastic leukemia look different than normal lymphocytes. However, the cells in chronic lymphocytic leukemia are hard to distinguish from normal cells, and the disease might resemble a physiologically high lymphocyte count. In this case, identification of specific cell markers, which are not present in healthy cells, aids the diagnosis. Hence, tumor immunology identifies **tumor-specific antigens**.

Some tumor antigens, such as overexpressed normal antigen, mutated proteins, or viral antigens, can be recognized by the immune system, in a process called **immune surveillance**, to develop antitumor immunity. This is one reason why people immunocompromised as a result of previous infections, other diseases, or age are more likely to develop cancers. For instance, although most people with Burkitt lymphoma remain healthy when infected, it is the most common childhood cancer in central Africa, where malaria and malnutrition contribute to immune weakness.

Nevertheless, most cancers occur in people with a normal immune system. The reason is that tumor cells develop mechanisms to evade or disturb appropriate immune responses. These include the loss or reduced expression of MHC proteins and the secretion of immunosuppressive cytokines such as TGF-β, to name two mechanisms.

In the rapidly advancing field of **cancer immunotherapy**, which aims to enhance the body's natural defenses against malignant tumors, we can utilize components of the anticancer immune response that have already developed in that patient. Active immunization with tumor antigens and cytokine or antitumor antibody therapies are examples of this type of treatment.

▶ NEUROENDOIMMUNOLOGY

Neuroendoimmunology is the study of the interaction between the nervous, endocrine, and immune systems. These systems exchange an extensive amount of information between the three systems. For example, they share a number of the same receptors and synthesize and secrete some of the same molecules, such as peptide hormones, neurotransmitters, and cytokines. Accordingly, it is not surprising that the strength of a person's immune response depends not only on the type of antigen or the person's age and genetic makeup but also on the person's lifestyle choices, emotional character, and ability to cope with stress.

Immune system is downregulated by neuronal and hormonal stress signals.

It is well known that the physiologic stress response downregulates the immune system. As a consequence, people who are stressed, for instance, from losing a relative or facing examinations, become sick more easily.

During an immune response, cytokines such as IL-1 stimulate the activity of paraventricular hypothalamic neurons to release corticotropin-releasing factor (CRF), which starts the physiologic *stress response* aimed at bringing a person back to physiologic homeostasis. In the pituitary gland, CRF stimulates the expression of pro-opiomelanocortin, which is converted to β-endorphin and adrenocorticotrophic hormone (ACTH). ACTH stimulates the release of corticosteroids from the adrenal gland. All these molecules can also bind directly to receptors on immune cells. The resulting actions are generally inhibitory to the immune system but can also be immunosupportive, especially during acute stress.

Activating the stress response via CRF also leads to activation of the *sympathetic nervous system* and the release of its neurotransmitter norepinephrine. Cytokines that are released by immune cells or by cells of the central nervous system (CNS), such as microglia and astrocytes, can also activate this autonomic path of the stress response.

A network of autonomic nerve fibers sends information from the CNS to the spleen and lymph nodes, where the fibers end in close proximity to T lymphocytes and macrophages. These immune cells, as well as B cells, express β-adrenergic receptors of the β-2 subtype, and their activation generally leads to inhibition of cell function. For instance, adrenergic activation of T cells decreases their expression of integrins, so that they cannot migrate out of the blood vessel into the tissue. This explains the high T-cell counts seen soon after a stressful event.

Furthermore, sympathetic nerve endings and the adrenal medulla release **endorphins** and **enkephalins**. These opioid peptides oppose the body's stress response and help maintain homeostasis during times of physical and psychological stress. The immune responses to opioids vary, depending on the subtype and concentration of the neuropeptide.

Melatonin, a hormone of the pineal gland, seems to regulate the opioid network and consequently the immune system. Melatonin therapy is under consideration to support the immune system of older people to avoid **immunosenescence**.

Neural mediators

Immune cells that are present in neuronal and other tissue produce mediators that can act on the brain.

Lymphokines and *thymokines* (produced by the thymus) carry the information flow from the peripheral immune system to the CNS. For instance, IL-1 is a potent mitogen for astroglial cells. In addition, it acts at hypothalamic cells causing fever. IL-2 promotes the division and maturation of oligodendrocytes, and IL-3 supports the survival of cholinergic neurons.

CD4+ T cells and macrophages manufacture *neuropeptides*, such as vasoactive intestinal peptide and pituitary adenylate cyclase–activating polypeptide, which were previously thought to be products solely of nonimmune cells.

Until recently, it was also believed that the brain was devoid of peripheral immune cells. It is now known that immune cells can pass the blood–brain barrier and enter the CNS. It seems that they are present to regulate physiologic neuronal events and potentially contribute to maintaining the brain's ability for cognitive functions. On the other hand, neuronal immune cells may trigger autoimmune disorders. For instance, autoimmune T cells that recognize myelin basic protein of myelin sheaths can be associated with the development of the demyelinating disease **multiple sclerosis**.

Hormonal mediators

The endocrine system and immune system work in concert to maintain homeostasis. Many hormones have been linked to immune functions. For instance, cells of the immune system also produce (e.g., monocytes and lymphocytes) and use (e.g., lymphocytes, monocytes, and NK cells) **thyroid-stimulating hormone**, best known as a regulator of metabolic functions. *Growth hormone, prolactin,* female and male *sex hormones,* and *leptin* are additional examples of hormones that unequivocally modulate the immune system.

In the typical fashion of hormones, they exert multiple actions in many target cells, depending on the hormone concentration, the target tissue, and the environment. This makes hormones optimal regulators of the body's homeostasis but usually causes unwanted side effects when used clinically.

Unpredictable side effects are one reason why use of *immune therapies* using hormones or neuropeptides is still limited. For example, the adrenal hormone epinephrine is given to reverse some effects of allergy. The field of psychoneuroimmunology builds on the fact that positive experiences and behavior may also boost the immune system. Thus, immunocompromised people are counseled to find ways for dealing with stress. However, many new applications are being developed, and targeted enhancement or suppression of immune cell functions will play a more important role in our approaches to help the body to maintain homeostasis.

INTEGRATED MEDICAL SCIENCES

Vitamin D Deficiency

Vitamin D is a fat-soluble vitamin that is best associated with the ultraviolet (UV) rays of the sun and calcium absorption. It has been suggested that exposure of the face, arms, legs, and back to the UV rays from the sun for, 5 to 30 minutes any time from 10 AM until 3 PM, at least twice each week would promote adequate vitamin D synthesis and maintain the protective properties of endogenous vitamin D. However, this exposure to UV rays comes with the potential health risk of developing melanoma. Individuals who do not get the adequate exposure to the UV rays of the sun, must consume exogenous vitamin D found in foods or taken as dietary supplements. Vitamin D is found naturally in foods such as milk, egg yolks, salmon, trout, and cod liver oil. Vitamin D is also available as a dietary supplement and is often added to foods in which it does not naturally occur. The vitamin D obtained from the diet is absorbed by the small intestine and is transported to the liver by vitamin D–binding protein (VDBP). Our body activates vitamin D in both the liver and the kidneys. It first becomes 25-hydroxyvitamin D (calcidiol) in the liver and later is converted to 1,25-dihydroxyvitamin D (calcitriol) in the kidneys. Vitamin D promotes absorption of calcium in the gastrointestinal tract and is best associated with bone growth and strength.

Vitamin D deficiency has long been associated with rickets in children, when bones become brittle and soft due to the lack of calcium, and osteoporosis in the elderly. There is, however, increasing evidence that suggests that vitamin D plays a significant role in immune function, both in regulating immune responses to infectious diseases and regulating the immune cells' interactions with the commensal, nonpathogenic microbiota of the digestive tract. It has been suggested that numerous cell types, including the majority of immune cells, can express the vitamin D receptor (VDR) and that vitamin D can influence the activation or differentiation of these cells *in vitro*. If this modulation is also found during immune responses in vivo, the presence of vitamin D could influence both the innate and adaptive immune systems. Dysregulation of immune responses can result in the susceptibility to infectious diseases and the development of various autoimmune conditions such as Crohn disease (CD), multiple sclerosis (MS), and systemic lupus erythematosus (SLE).

Crohn disease is an inflammatory bowel disease, typically affecting the ileum and/or the colon. The inflammation in the lining of the intestines obstructs normal functioning of the digestive system and leads to altered flow of digested foods and decreased absorption of nutrients and interferes with the body's first line of defense at the mucosa. The inflammatory response leads to collateral damage of healthy cells that compromises the integrity of the mucosa, and the cells that secrete antimicrobial peptides are damaged or lost. Symptoms usually develop over time and can be cyclic, coming with periods of disease exacerbations and remissions. There is no clear evidence that a specific type of food is a trigger or aggravator of the disease; however, some patients find that consuming a low-fat diet with limited dairy products and limited fiber helps control symptoms. It has also been suggested that patients with a low plasma vitamin D level may have an increased risk of developing CD; experiencing an exacerbation of CD symptoms; developing cancers, in particular colorectal cancer; and developing *Clostridium difficile* infection.

Vitamin D deficiency has also been associated with MS. MS is characterized by the demyelination of the neurons of the central nervous system (CNS). The demyelination presents early in the disease as weakness or tingling in arms and legs, loss of coordination, loss of mental focus, and/or loss of balance. MS is, typically, a disease with relapses and periods of remissions. It has long been reported that individuals who grow up in higher latitudes are more likely to develop MS grow up in higher latitudes are more likely to develop MS than those who grow up in the tropics. Individuals living in the tropics have increased exposure to the ultraviolet light from sunshine and subsequent production of vitamin D compared to those living at higher latitudes. Recent evidence suggests that vitamin D deficiency is a risk factor for both the development of and progression of MS.

Systemic lupus erythematosus is a chronic inflammatory disease that can affect various organs of the body such as joints, the skin, lungs, blood vessels, kidneys, and the heart. SLE patients also may experience relapses and remissions. SLE patients, typically, have low serum levels of vitamin D. This vitamin D insufficiency could be correlated with the photosensitivity associated with SLE, limiting SLE patients' exposure to the sun. Another contributing factor could be that many of the drugs used to treat SLE interfere with vitamin D metabolism and could also contribute to this deficiency. Some studies have suggested that vitamin D deficiency may be a risk factor for SLE or contribute to the progression of SLE, but the data are controversial at this point. These discrepancies could be due to polymorphisms in the VDR gene; however, this has not yet been clearly demonstrated. Also, yet to be determined is whether vitamin D insufficiency changes the course and prognosis of SLE. Correlating a specific environmental agent to the development and/or progression of SLE is difficult as there are genetic, hormonal, environmental, and immunological contributions to this autoimmune disease. Other complications associated with SLE studies are that patients may experience a mixture of symptoms and the comorbidity of SLE with other autoimmune diseases.

Chronic inflammation in response to infectious agents or due to autoimmune disease can also be associated with the development of cancer. At this time, current data suggest that vitamin D can be considered as a potential anti-inflammatory agent and an immunosuppressant and could be used to treat autoimmune diseases such as CD, MS, and SLE. It is also likely that vitamin D plays a role supporting the regulatory efforts at the mucosa, helping immune cells determine what microbes are commensal microbiota and what microbes are pathogenic and require removal. ■

Blood and Immunology

Chapter Summary

- The immune system encompasses the thymus, spleen, bone marrow, and lymph system. It additionally includes cells such as macrophages, dendritic cells, mast cells, and blood leukocytes (neutrophils, lymphocytes, monocytes, eosinophils, basophils) as well as noncellular elements that are part of every tissue throughout the body.
- Antigens are substances that bind to T- and B-cell receptors or antibodies. Haptens and immunogens are antigens, but only immunogens elicit an immune response. Proteins are by far the best immunogens.
- The immune system operates by distinguishing between self and foreign molecules inside our body and then tolerating normal body substances (including nonpathogenic microbiota) and eliminating any invasion by foreign substances causing harm.
- Immune system defenses include physical, chemical, and mechanical barriers to entry of pathogens.
- The immune system is divided into two systems: the innate system as the immediate response to common immunogens and the adaptive system, which changes throughout life and which has memory to respond to an encounter more strongly the second time compared with the first time.
- Innate and adaptive systems work together so that the immune response is adjusted to the type and severity of the threat.
- The innate leukocytes include phagocytes that encompass neutrophils, macrophages, and dendritic cells. They have receptors for pathogens and for complement-coated and antibody-coated antigens. They can engulf and destroy pathogens while influencing immune responses via immune mediators. Macrophages and dendritic cells can act as professional antigen-presenting cells to T cells. Even though eosinophils are involved in allergy and other immune responses, such as antiparasitic immunity, they can also be phagocytic.
- The innate leukocytes further include nonphagocytic cells such as natural killer cells that attack aberrant body cells, and mast cells that participate in the defense against pathogens and wound healing.
- Acute inflammation is the body's initial response to infectious microorganisms and injuries. It is characterized by the dilation and leaking of blood vessels and subsequent attraction of circulating leukocytes to the inflammatory site, where innate immune responses occur.
- Adaptive immunity is a delayed, but highly effective immune response against specific antigens. The adaptive immune system is divided into cell-mediated immunity, which targets infected cells and is managed by T cells, and humoral immunity, which deals with infectious agents in blood and tissues and is mediated by B cells and their antibodies.
- B cells and T cells have antigen-specific receptors, which bind directly to antigen in the case of B cells or antigen presented on major histocompatibility complex (MHC) proteins in the case of T cells.
- Stimulation by antigen plus other signals causes T lymphocytes to proliferate into clones of effector cells and memory cells.
- Cytotoxic CD8$^+$ T cells recognize and kill cells that are infected with intracellular pathogen presented on class I MHC protein. CD4$^+$ T helper 1 cells activate macrophages that present antigen on class II major histocompatibility complex (MHC) protein. CD4$^+$ T helper 2 cells activate B cells that present antigen on class II MHC protein.
- Activated B cells become antibody-producing plasma cells and can become memory cells, with the help of CD4$^+$T helper cells.
- Antibodies are antigen-specific binding proteins that neutralize and opsonize antigen and activate complement to promote various immune reactions. Antibodies occur in five isotypes with special functions, IgM, IgD, IgG, IgA, and IgE.
- Graft rejection is the major complication of organ transplantation and depends mainly on the allogeneic differences in the human leukocyte antigens between organ donor and recipient.
- Graft-versus-host disease is a complication of bone marrow transplants. Newly engrafted T cells can begin attacking the host cells when the patient is immunocompromised.
- Every acute illness evokes an immunologic response, and nearly every prolonged serious illness interferes with the immune system.
- Chronic inflammation leads to tissue damage and clinical symptoms. It is associated with the predisposition to a multitude of diseases.
- Immune system diseases are caused by abnormal or absent immunologic mechanisms, whether humoral, cell-mediated, or both.
- The actions of the immune system are closely interrelated with the actions of the endocrine and the neuronal systems to keep body homeostasis. Forthcoming information on these interactions will lead to novel treatments.

Chapter Review Questions

1. Penicillin is a small antibiotic molecule with a molecular weight of 300 Da. It binds to an enzyme in the wall of many bacteria, which leads to the weakening of the cell wall. As an unwanted side effect, penicillin can also bind to serum or tissue proteins, and the resulting complexes can cause allergic reactions. Which of the following describes the role of penicillin in the immune response?

   A. Antigen
   B. Hapten
   C. Immunogen
   D. Opsonin
   E. Pathogen

The correct answer is B. Penicillin is a hapten. Haptens are too small to be immunogenic on their own, but when attached to a macromolecule, usually a protein, they become quite immunogenic. Penicillin, alone, is not immunogenic. Penicillin is not acting as an opsonin, as opsonins attach to pathogens and facilitate phagocytosis. The pathogen is the disease-causing agent.

2. Herpes simplex viruses tend to infect cells of the skin or mucous membranes. Which of the following are the correct types of host cells, antigen-presenting molecules, and target cells that are involved in the immune response against the infection?

   A. Host Cells Type of Antigen-Presenting Molecules Target Cells
   B. Any cell MHC class I Cytotoxic T cell
   C. Any cell MHC class II T helper cell
   D. Any cell MHC class I T helper cell
   E. B cell MHC class II Cytotoxic T cell
   F. Dendritic cell MHC class II Cytotoxic T cell

The correct answer is A. Endogenous antigens, such as viruses replicating within a host cell, are digested by enzymes of the host cell and presented, coupled to MHC class I molecules, to activated CD8+ cytotoxic T cells, which destroy the infected cell. Class I MHC molecules are expressed by almost all host cells with a nucleus. In the case that the antigen stems from extracellular proteins or phagocytosed bacteria, the antigen is digested intracellularly within phagolysosomes and associated with MHC class II molecules, which then present the antigen to the TCR of CD4+ T helper cells. Macrophages, dendritic cells, and B cells are professional antigen-presenting cells and present to CD4+ T cells.

3. On a train ride, a 2nd-year medical student overhears part of the conversation between a couple who are sitting next to her. The man says "The bad news is that I need an allogeneic graft, but the good news is that I do not need to take immunosuppressive drugs." Which of the following transplant types are they discussing?

   A. Bone marrow
   B. Cartilage
   C. Heart
   D. Kidney
   E. Liver

The correct answer is B. An allogeneic graft is a transplant between genetically different people. The immune system of the recipient will recognize the slightly different antigen on the surface of the donor cells (human leukocyte antigens) as foreign and start an immune attack, leading to graft rejection. Immunosuppressive drugs minimize the rejection by suppressing both cell-mediated and antibody-mediated immunity. Hence, tissue rejection and immunosuppressive therapy rely on tissue vascularization. The only avascular tissue in the list of answers is cartilage. Unlike other connective tissues, cartilage does not contain blood vessels. Since there is no risk of an immune reaction, chondrocytes from cadaver cartilage can be allografted.

4. A 34-year-old male complains about recurring infections that truly bother him since he is preparing for the iron man triathlon. Blood is drawn for extensive analysis of the patient's health. It might also include a test for which of the following hormones that, when increased, might explain the increased susceptibility to infection?

   A. Adrenaline
   B. Cortisol
   C. Estrogen
   D. Growth hormone
   E. Insulin

The correct answer is B. Severe exercise induces stress to which the body responds with increased cortisol levels. Cortisol depresses a variety of immune responses. Although severe exercise induces the synthesis of adrenaline, there is no known correlation to increased susceptibility to infection. Estrogens tend to stimulate immune responses. Moreover, there is no indication why estrogen levels should be increased in this patient. Growth hormone does not depress immune responses. While stress alters insulin needs, stress does not raise insulin levels. In fact, insulin secretion is reduced during exercise, partly through the suppressive actions of cortisol.

5. A newborn has signs of birth defects including hearing loss and cardiovascular abnormalities that could be attributed to congenital rubella syndrome. A rubella antibody test is ordered and the results reveal the presence of specific IgM and IgG antibodies against the rubella virus in the newborn's cord blood. Which of the following can be concluded?

   A. The mother has passed IgM antibodies in utero to the child.
   B. The mother has passed both IgM and IgG in utero to the child.
   C. The newborn has no evidence of infection with rubella virus.
   D. The newborn was infected with rubella virus during pregnancy.
   E. This is a false-positive result since a newborn cannot have antibodies.

The correct answer is D. The presence of IgM antibodies against the rubella virus in a newborn indicates that the newborn was infected during pregnancy since the mother's IgM antibodies cannot pass to the baby through the umbilical cord and the present IgM has to be fetal antibody. Since congenital rubella syndrome occurs because of infection during the first trimester, or to a lesser extent the second trimester, the mother's system has most likely developed both IgM and IgG antibodies by week 24. IgG antibodies can be passed from the mother to the child in utero, so that the IgG of the newborn could be a mixture of both the maternal and the fetal IgG.

If the baby's blood would only contain IgG but no IgM, it would indicate that the mother has passed IgG, obtained by immunization, to the child, which may protect the child from rubella infection during the initial months of life. However, the presence of IgM antibodies in the baby indicates a recent infection with, not immunity to, the rubella virus. Although false-positive results are possible due to some cross-reactivity of test components, the reasoning that a newborn cannot have antibodies is a false statement.

Clinical Application Exercises 10.1

A MYSTERIOUS AUTOIMMUNE DISEASE

A Japanese male, age 31, was treated by his family practitioner for bacterial tonsillitis. His tonsils had been removed at age 11. Two weeks later, the patient complained of blood in his urine and some flank pain. A low-grade fever was present. A urine dipstick analysis showed mild proteinuria. Urine analysis revealed some casts and dysmorphic erythrocytes. The patient was treated for urinary tract infection but the symptoms remained. Based on elevated creatinine concentration, the decision for a kidney biopsy

was made. Immunofluorescence microscopy revealed prominent globular deposits of IgA in the mesangium of the glomeruli. The complement components C3, C5b-9, properdin, and various lectins were found. The diagnosis of IgA nephropathy was made. IgA nephropathy is a rare disorder of the immune system in which IgA is deposited in the kidney mesangium, slowly destroying it in about 50% of the cases.

QUESTIONS

1. What is the role of IgA in the immune system?
2. Discuss the rationale for immunosuppressants, dietary changes, and antihypertensive and diuretic medication as part of the treatment of a patient with IgA nephropathy.

3. To which complement pathways do the glomerular deposits C3, C5b-9, properdin, and lectins point?

ANSWERS

1. IgA is known as the secretory immunoglobulin and can make up 10% to 15% of the antibody found in a healthy person. It is secreted as dimers into fluids that have access to the body surfaces such as saliva, tears, and mucus of the gut and the lungs. IgA can directly bind pathogens preventing their colonization of mucosal tissues. It seems that the antibody in IgA nephropathy is derived from the mucosal immune system of the respiratory or gastrointestinal tract. The IgA deposited within the glomerular mesangium is predominantly J chain containing polymeric IgA1.
2. The treatments for a patient with IgA nephropathy depend very much on the severity of the disease symptoms that range from asymptomatic to severe kidney impairment. The risk/benefit ratio of the treatments has to be carefully considered. *Immunosuppressants* such as steroids address the aberration of the immune system in producing excess IgA and the inflammation of the kidney caused by irritating casts. Fish oil capsules, rich in omega-3-polyunsaturated fatty acids, have been prescribed to produce altered, less effective proinflam-

matory eicosanoids. Food and vitamins with antioxidant capabilities reduce inflammation as well.

Recommendations for additional *dietary changes* include a low-salt and low-potassium diet to control the high blood pressure that can occur as a consequence of kidney failure. Low-protein diets have been recommended to slow the progression of kidney damage.

To minimize any additional stress on the kidneys, *high blood pressure medication* and *diuretic therapy* might be warranted. If there is a risk for end-stage renal failure, dialysis and kidney transplantation need to be considered.

3. The presence of C3 together with C5b-9 and properdin in the glomerular casts points toward the activation of the complement cascade via the alternative pathway. Recent evidence indicates that deposits of mannose-binding protein and other lectins point toward the involvement of the lectin pathway in a significant number of cases.

the**Point** *Visit* http://thepoint.lww.com/rhoades5e *for additional chapter review Q&A, Clinical Application Exercises, animations, and more!*

11 Overview of the Cardiovascular System and Hemodynamics

Active Learning Objectives

Upon mastering the material in this chapter, you should be able to:

- Explain why the outputs of the left and right ventricle are interdependent.
- Explain how the parallel arrangement of arteries supplying systemic organs allows independent control of each organ's blood flow.
- Explain the mechanisms by which major physiological vasoconstrictors and vasodilators contract or relax arteries and veins.
- Explain how vascular compliance is altered by transmural pressure, vascular smooth muscle contraction, and aging.
- Explain why changes in blood volume alter volume in the venous side of the circulation more than on the arterial side of the circulation.
- Explain why standing causes blood to drain from regions above the heart and pool in the lower extremities.

- Predict the consequences of a distended vessel on vascular wall stress, the probability of vessel dissection and rupture, as well as the ability of a blood vessel to contract.
- Explain how changes in arterial pressure and vascular resistance alter blood flow and how blood flow and vascular resistance affect arterial pressure.
- Predict how a vasoactive agent will directly affect blood pressure or organ blood flow.
- Explain the hemodynamic consequences of polycythemia and dehydration on the ability of the heart to supply adequate blood flow to systemic organs.
- Predict and explain how pathological changes in cardiovascular anatomy and the physical composition of blood can create flow murmurs.

The cardiovascular system is commonly described as a fluid transport system that delivers substances to the tissues of the body while removing the by-products of tissue metabolism. This definition is technically correct. However, lost in such a bland, generic, definition is a sense of how absolutely essential the cardiovascular system is for *human survival*. Chemical energy that supports cell functions in the body is generated predominantly by oxidative metabolism. Transport of oxygen across the cell membrane to support this metabolism occurs by simple passive diffusion. Although this process has the advantage of saving the cell from expending energy to transport oxygen into it, transport by passive diffusion has significant limitations. Transporting anything by simple diffusion encounters the problem of being able to get the substance from one point to another "on time." The consumption of oxygen by any cell requires that oxygen must be transported across the cell membrane at a rate (flux) that is sufficient to match the rate at which it is consumed. Failure to do so will eventually stop oxidative metabolism and result in cell death. The key problem of transport by passive diffusion is that the time required for any molecule to diffuse from one point to another is markedly affected by the distance to be traveled between the points. For example, it takes ~5 seconds for an O_2 molecule to diffuse 100 μm. (This distance happens to be compatible with the size of our cells, their nearest source of O_2, and the rate of O_2 usage by cellular metabolism.) However, the physics of

random molecular motion, which drives passive diffusion, is such that each 10-fold increase in the distance that must be traveled increases the average time required to traverse that distance by a factor of 100. Therefore, should the diffusion distance in the example above increase from 100 to 1,000 μm (1 mm), the average time a molecule would take to traverse that distance would increase to 500 seconds. A further increase from 1 mm to 1 cm would increase the average time to 50,000 seconds, or almost 14 hours!

The time factor limitation associated with passive diffusion has important consequences for us as a living organism. First, the dependency of our cells to get oxygen by passive diffusion necessitates that we have some sort of transport system that can bring oxygen to within 100 μm of every cell. Although our pulmonary system gets oxygen from the atmosphere into our body, it is the ultimate function of a circulatory system to get that oxygen close enough to every cell so that it can enter by diffusion. Because of dependency on oxygen diffusion the survival of even the tiniest oxygen-consuming multicellular organisms requires the existence of a circulatory, or transport system, for oxygen. It is not surprising that when our cardiovascular system malfunctions, we as a whole malfunction, and if this system fails, our body "fails." The physics of passive diffusion places another constraint on our body. If for any reason a cell becomes separated by more than 100 μm from its nearest source of oxygen, as arranged by the existing cardiovascular system, that cell will receive insufficient O_2 to meet its metabolic needs. In such a state, the cell will malfunction and likely die. As shall be seen later in this text, tissue ischemia is a major clinical problem in the cardiovascular and other organ systems in the body.

Although the oxygen problem is reason enough for a large organism like ourselves to have an absolute requirement for a cardiovascular system, our oxidative metabolism creates another problem that this same system helps resolve. Carbon dioxide produced by oxidative metabolism can rapidly acidify the cell and its surroundings if it is not removed. Fortunately, CO_2 transport out of our cells can also utilize simple diffusion because a circulatory system to transport it away and carry it to the lungs is already in place. In addition to its essential role in delivering oxygen and removing carbon dioxide, the cardiovascular system is also exploited for other tasks in the body. For example, the body uses hormones for the control of important physiologic functions, and these hormones are transported from the site of their production to their target organs in the bloodstream by the cardiovascular system. In addition, blood serves as a reservoir for heat. The cardiovascular system plays an important role in the control of heat exchange between the body and external environment by controlling the amount of blood flowing through the skin, which is the site of heat transfer between our body and the external environment.

▶ FUNCTIONAL ORGANIZATION

Students in medical physiology should be familiar with the anatomic organization of the heart and blood vessels from their courses in gross anatomy and histology. A brief overview is contained in this section. The cardiovascular system is a fluid transport system for the movement of blood throughout the body. This is represented in simplistic form in Figure 11.1. Blood is driven through the cardiovascular system through the actions of a hollow muscular pump called the **heart**. The heart is a four-chambered muscular organ that contracts and relaxes in a regular repeating cycle to pump blood.

The heart is really two pumps connected in series. The left heart is composed of the **left atrium** and **left ventricle**, separated by the **mitral valve**. Contraction of the left ventricle is responsible for pumping blood to all systemic organs except the lungs. During contraction of the ventricles, blood exits the left ventricle through the **aortic valve** into a single tubular conduit called the **aorta**. The aorta is an **artery**, which, by definition, is any blood vessel that carries blood away from the heart and to the tissues of the body. The aorta branches into successively smaller arteries, many of which are given anatomic names. These arteries, in turn, branch into millions of smaller vessels of ~10 to 1,000 μm in external diameter, called **arterioles**. Arterioles, in turn, terminate into billions of **capillaries**, which are the main site of transport of water, gases, electrolytes, substrates, and waste products between the bloodstream and the extracellular fluid. Blood from the capillaries of all the systemic organs coalesces into thin-walled **venules**, which merge into **veins**. By definition,

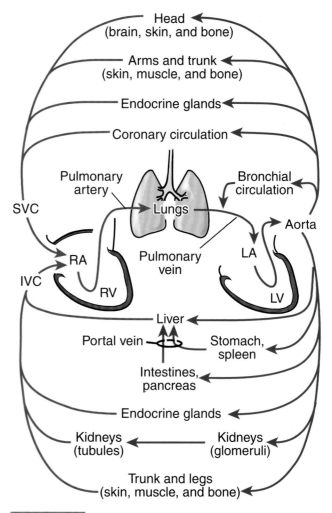

Figure 11.1 **A model of the cardiovascular system.** The right and left sides of the heart are aligned in series, as are the systemic circulation and the pulmonary circulation. In contrast, the circulations of the organs other than the lungs are in parallel. Each organ receives blood from the aorta and returns it to the vena cava. Exceptions to the series arrangements exist between the splanchnic veins and portal circulation of the liver as well as the glomerular and tubular capillary networks of the kidney. SVC, superior vena cava; IVC, inferior vena cava; RA, right atrium; RV, right ventricle; LA, left atrium; LV, left ventricle.

a vein is any blood vessel that returns blood from the tissues back into the heart. Small veins eventually merge to form two large single veins called the **superior vena cava** (**SVC**) and the **inferior vena cava** (**IVC**). The SVC collects blood from the brain, head, and upper extremities above the level of the heart, whereas the IVC collects blood from all regions below the level of the heart. Both of these large veins empty into the **right atrium**. The right atrium is the upper chamber of the right heart. It is separated from the **right ventricle** by the **tricuspid valve**. The right ventricle pumps blood through the **pulmonic valve** into the **pulmonary artery** and thence into the lungs. Blood exiting the lungs is returned to the **left atrium** where it passes through the **mitral valve** and into the left ventricle, completing the circulatory loop. As a whole, the period of time the heart spends in contraction and pumping blood out of the ventricles is called **systole**, and the time it spends in relaxation when these chamber fill with blood is called **diastole**.

The outputs of the right and left heart are interdependent because they are connected in series.

The right and left heart are said to be arranged *in series*, or in line, one after the other, right to left. For this simple reason, it cannot be possible for the flow output of the right and left heart to differ for anything more than a few seconds. To see how this must be, imagine a bucket of water interposed between two pumps. One pump removes water from the bucket and pumps it into the inflow of the second pump. The second pump takes this inflow and returns it into the bucket for the first pump to remove again. It is obvious that if the outputs of the two pumps are not matched identically, the level of water in the bucket will change; the bucket could be drained dry or could overflow. This is analogous to the situation in the pulmonary circulation that is interposed between the right and left heart. Because of the series arrangement between the right and left heart, an output of the left heart that exceeds that of the right by as little as 2% will drain the pulmonary circulation dry within minutes. Conversely, if the right heart output exceeded the left by a similar amount, the pulmonary circulation would overflow and the person would drown in his or her own body fluids. Clearly, neither of these situations arises in a healthy person. The implication of this is that some mechanism must function to closely match the outputs of the right and left heart. Such a mechanism exists at the level of the muscle cell and is based on the same principles of skeletal muscle cell mechanics you have read about in earlier chapters. Details of these principles as they apply to the heart are discussed in Chapter 13.

The series arrangement of the right and left heart also implies that malfunctions in the left heart will be transmitted back into the pulmonary circulation and the right heart. This in turn can cause the respiratory system to malfunction. Indeed, one of the first clinical signs of left heart failure is respiratory distress. Conversely, problems originating on the right side of the circulation affect the output of the left heart and imperil the blood supply to all systemic organs. For example, large blood clots, which can form in the major veins of the leg and abdomen following surgery can break away and slip through the right heart chambers and into the pulmonary artery where they eventually lodge, to form a **pulmonary embolus**. If such emboli block enough of the pulmonary circulation, the left heart may not be able to pump enough blood for the individual to survive.

Parallel arrangement of organ circulations permits independent control of blood flow in individual organs.

The metabolic demands of our muscles, digestive system, brain, and so on are different relative to one another and relative to their own resting values, depending on the activity in the organ at a given time. The arterial system delivers blood to organ systems that are arranged in a *parallel*, or side-by-side, network. Therefore, in most cases, blood flow into one organ system is not dependent on blood flow through another organ upstream. The parallel arterial distribution system of organ blood supply allows the adjustment of blood flow to an individual organ to meet its own needs without creating major disturbances in the blood supply to other organs. A notable exception to this arrangement, however, is seen in the portal circulation. Venous outflow from the intestines and other splanchnic organs drains into the liver through the **portal vein** before being emptied into the IVC. The liver obtains blood from the portal vein as well as its own arterial supply (see Fig. 11.1) and can be considered to be arranged in series with much of the splanchnic circulation.

Vascular smooth muscle actively controls the diameter of arteries and veins.

All blood vessels, except capillaries, have a similar basic structure. They are both lined with a single layer of epithelial cells called the **endothelium**. The media of these vessels contain circular layers of smooth muscle cells, whereas the outermost layer, called the **adventitia**, is composed of collagen and elastin fibers that add flexible structural integrity to arteries and veins. Because the smooth muscle within blood vessels is arranged in circular layers, contraction or relaxation of these muscles will reduce or widen, respectively, the lumen diameter of arteries and veins. Altering blood vessel caliber has a profound effect on organ blood flow and blood volume distribution in the cardiovascular system. These effects are discussed later in this chapter.

There are literally scores of normal physiologic, pathologic, and clinical pharmacologic agents that can alter the contraction and relaxation of arterial and venous smooth muscle. These take the form of direct physical forces and chemical agents, hormones, paracrine substances, and receptor-mediated hormonal and neurotransmitter agonists. For example, the contraction of arteries and veins is modified by transmitters released from sympathetic nerve endings that enter these vessels through their adventitial layer and act on specific receptors on the smooth muscle membrane. The endothelium, which stands at the interface between blood plasma and the rest of the vessel wall, is the source of important paracrine agents that have major, receptor-independent, direct effects on blood vessel contraction. A simplified, partial list of factors that contract or relax vascular smooth muscle is provided in Table 11.1.

▶ PHYSICS OF BLOOD CONTAINMENT AND MOVEMENT

The study of the physical variables related to the containment and movement of blood in the cardiovascular system is called **hemodynamics**. From an engineering standpoint, an accurate description of all the hemodynamic phenomena in the cardiovascular system is complex. Fortunately, the human body deals with these phenomena and their control in a considerably simplified manner. The cardiovascular system behaves much as if the heart were producing an average steady flow through a series of solid pipes, similar to the flow of water through a city's water distribution system. Thus, basic principles of fluid dynamics can be applied to the understanding of cardiovascular phenomena.

> **TABLE 11.1** A Simplified List of Direct (Non–Receptor-Mediated) and Smooth Muscle Membrane Receptor–Mediated Factors that Contract or Relax Vascular Smooth Muscle

Vasoconstrictor/Vasodilator	Mechanism of Action
Vasoconstrictors: Nonreceptor mediated	
Ca^{2+}	Enters cells through membrane channels or is released from the SR by IP_3; binds to calmodulin, which activates myosin light chain kinase to initiate crossbridge attachment and cycling
K^+	Membrane depolarization resulting in activation of L-type Ca^{2+} channels
O_2 (normoxia)	Maintenance of contraction
Membrane depolarization	Activation of L-type Ca^{2+} channels
Increased transmural pressure	Activation of stretch-activated plasma membrane calcium channels
Action potentials	Membrane depolarization resulting in the activation of L-type calcium channels
Vasoconstrictors: Receptor mediated	
Norepinephrine	Activation of α adrenoceptors to open receptor-operated plasma membrane calcium channels, activation of phospholipase C, formation of IP_3 and DAG, and release of calcium from the SR
Epinephrine	Same as norepinephrine
Acetylcholine	Activation of muscarinic M_2 receptors to open receptor-operated plasma membrane calcium channels, activation of phospholipase C, formation of IP_3 and DAG, and release of calcium from the SR; inhibits cAMP formation
Serotonin	Activation of $5\text{-}HT_2$ receptors to open receptor-operated plasma membrane calcium channels
Vasopressin	Activation of V receptors to open receptor-operated plasma membrane calcium channels
Angiotensin II	Activation of angiotensin II receptors to open receptor-operated plasma membrane channels and activate IP_3/DAG pathways
Endothelin	Activation of ET_A receptors with stimulation of phospholipase C, formation of IP_3, and release of calcium from the SR
ATP	Activation of P_{2x} receptor to open receptor-operated plasma membrane channels
Vasodilators: Direct	
Nitric oxide	Activation of guanylate cyclase, formation of cGMP, and calcium removal
Hyperpolarization	Inhibition of voltage-gated Ca^{2+} channels
Decreased transmural pressure	Inhibition of stretch-operated Ca^{2+} channels
Hypoxia	Decreased ATP supply and formation of adenosine. H_2S and/or CO formation
Hypercapnia	Likely from acidosis
Acidosis	Unknown
Hyperosmolarity	Unknown
CO	Stimulation of calcium activated K^+ channels
cAMP	Activation of cAMP-dependent kinase resulting in a phosphorylation cascade that reduces intracellular calcium concentration
cGMP	Activation of cGMP-dependent kinase resulting in a phosphorylation cascade that reduces intracellular calcium concentration
Hyperkalemia (low levels)	Membrane hyperpolarization resulting from activation of the Na^+/K^+ pump

TABLE 11.1	A Simplified List of Direct (Non–Receptor-Mediated) and Smooth Muscle Membrane Receptor–Mediated Factors that Contract or Relax Vascular Smooth Muscle (*Continued*)
Vasoconstrictor/Vasodilator	**Mechanism of Action**
	Vasodilators: Receptor mediated
Epinephrine	β_2 receptor–mediated activation of adenyl cyclase resulting in the formation of cAMP
Adenosine	A_1, A_{2A}, and A_{2B} receptor activation of ATP-dependent K^+ channels leading to membrane hyperpolarization and closure of voltage-gated calcium channels
Histamine	H_2 receptor activation of adenyl cyclase with formation of cAMP
PGI_2	Activation of adenyl cyclase with formation of cAMP

5-HT, 5-hydroxytryptamine; ATP, adenosine triphosphate; cAMP, cyclic adenosine monophosphate; cGMP, cyclic guanosine monophosphate; DAG, diacylglycerol; IP_3, inositol 1,4,5-triphosphate; PGI_2, prostacyclin; SR, sarcoplasmic reticulum.

Fluid cannot move through a system unless some energy is applied to it. In fluid dynamics, this energy is in the form of a difference in **pressure**, or pressure gradient, between two points in the system. Pressure is expressed as units of force, or weight, per unit area. A familiar example of this is in the pounds per square inch (psi) recommendation stamped on the side of tires. The psi indicates the pressure to which a tire should be inflated with air above atmospheric pressure. Inflating a tire to 32 psi signifies that 32 more pounds press against every square inch of the inner tire surface than against the outside of the tire.

The pressure exerted at any level within a column of fluid reflects the collective weight of all the fluid above that level as it is pulled down by the acceleration of gravity. It is defined as

$$P = \rho gh \qquad (1)$$

where P = pressure, ρ = the density of the fluid, g = the acceleration of gravity, and h = the height of the column of fluid above the layer where pressure is being measured. The force represented by pressure in a fluid system is often described as the force that is able to push a column of fluid in a tube straight up against gravity. In this way, the magnitude of the force resulting from fluid pressure can be measured by how high the column of fluid rises in the tube (Fig. 11.2A). In physiologic systems, pressure is expressed in this manner as centimeters H_2O, or the more practical mm Hg. Because mercury is much denser than water, it will not be pushed as far upward by typical pressures seen within the cardiovascular system and thus devices containing columns of mercury (or based on them) can be easily used to measure physiological pressures in the body.

In the human cardiovascular system, aortic pressure peaks during contraction of the heart to about 120 mm Hg and drops to about 80 mm Hg when the heart relaxes. The peak pressure is called the **systolic pressure,** whereas the minimum arterial pressure value during relaxation of the heart is called the **diastolic pressure**. Thus, if one end of a tube were to be inserted into the aorta with the other end connected to a column of mercury sitting perpendicular to the ground at the level of the heart, that column of mercury would rise 120 mm during systole and fall to 80 mm during diastole. In clinical practice, human arterial pressure is reported as systolic over diastolic pressure or, in this example, 120/80 mm Hg. (Our mean arterial pressure is not the arithmetic mean of systolic and diastolic pressure but is instead about 93 mm Hg, because the time the heart spends relaxing is longer than the time it spends contracting and ejecting blood into the aorta.)

The pressure outside the body or on the outside of any hollow structure within the body other than the intrapleural space is about the same as atmospheric pressure. Atmospheric pressure at sea level is usually about 760 mm Hg. In reality, therefore, our average blood pressure is 93 + 760 mm Hg, or 863 mm Hg. However, pressures within our body are never expressed in this technically exact manner. Instead, the effect of atmospheric pressure is simply ignored and taken as a *zero reference point*. The pressure reported in our systems then is really the difference of pressure within that system relative to atmospheric pressure.

The effect of gravity on pressure within blood vessels in our body is significant when we are standing, as shown in Figure 11.2B for veins. In a recumbent position, pressure in the veins is between 2 and 10 mm Hg. However, when one stands, the influence of gravity subtracts ~40 mm Hg of pressure from the vessels in the head and adds about 90 mm Hg of pressure to those in the feet. If human veins were rigid tubes, this would not have a profound effect on the circulation or on distribution of blood within sections of the cardiovascular system. However, veins are flexible structures. Therefore, when one stands, blood tends to pool in the veins of the lower extremities. This is responsible in part for the swelling and aching of the feet after standing for long periods of time. Standing in this manner can even result in such extensive pooling that fainting will occur from an inability of enough blood to return to the heart to be pumped to the brain.

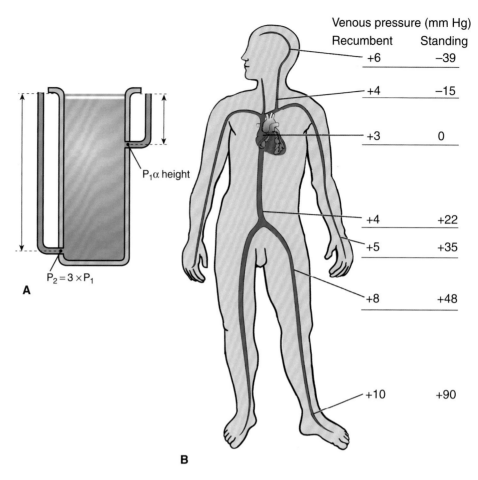

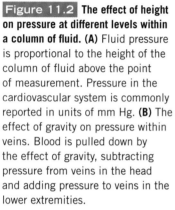

Figure 11.2 **The effect of height on pressure at different levels within a column of fluid.** **(A)** Fluid pressure is proportional to the height of the column of fluid above the point of measurement. Pressure in the cardiovascular system is commonly reported in units of mm Hg. **(B)** The effect of gravity on pressure within veins. Blood is pulled down by the effect of gravity, subtracting pressure from veins in the head and adding pressure to veins in the lower extremities.

Blood vessel volume is a function of vessel flexibility and the pressure difference across the vessel wall.

The volume of fluid within a container made of inflexible walls, such as a glass bottle, is the same no matter what the pressure difference is between the inside and the outside of the bottle. In such a container, the walls cannot move when any difference in pressure is applied across them. In contrast, the walls of arteries and especially veins are flexible. Consequently, the volume contained within them is a function of both the pressure difference across their wall, called the **transmural pressure**, and the degree of flexibility within the vascular wall. Transmural pressure, or P_T, is always defined as the difference in pressure *inside versus outside* a hollow structure. Large transmural pressures within flexible vessels create large intravascular volumes; small transmural pressures in stiff-walled vessels produce small intravascular volumes.

There are a couple of ways of depicting pressure/volume interrelationships in blood vessels. The volume of blood contained in the vessel for a given transmural pressure is called the **vascular capacitance** and is calculated as

$$\text{Capacitance} = \text{Volume}/P_T \qquad (2)$$

where volume and transmural pressure are generally given the units of mL and mm Hg, respectively. However, physiologists are interested in how volume changes in a distensible blood vessel for a given change in pressure. The change in volume for a given change in transmural

pressure is called the **compliance** and is given by the equation

$$C = \Delta V/\Delta P \qquad (3)$$

where the Δ signifies the before/after change of volume or pressure. (Note that although pressure outside the vessel is taken to be atmospheric pressure, or zero baseline, this pressure can become positive in certain pathologic conditions and thus reduce the distending effect of pressure within the blood vessel.) The compliance equation can be rearranged to yield two useful relationships relating changes in either volume or pressure as a function of the other in a blood vessel. For example, one can write $\Delta V = C \times \Delta P_T$. This equation indicates that the change in volume contained within a vascular segment will be great in a vessel with high compliance and a large change in intravascular pressure. One can also write $\Delta P_T = \Delta V/C$, which indicates that adding volume into a vascular segment will produce a large increase in pressure within the vessel if the volume added is large and the vascular compliance is low.

Both capacitance and compliance can be used as measures of the distensibility, or flexibility, of a blood vessel; highly distensible vessels have a higher capacitance and compliance than vessels with stiff walls, *if the vessels have the same initial dimensions.* For this reason, veins have a higher capacitance and compliance than arteries of similar size. However, the use of these variables to measure vessel flexibility fails when vessels of significantly different sizes are compared.

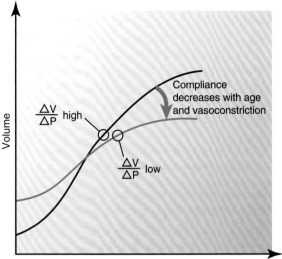

Figure 11.3 **The relationship between volume and transmural pressure in arteries.** Volume within an artery increases with an increase in transmural pressure because the arterial wall is flexible. An artery is stiffest at low and high transmural pressure, respectively. The slope of this arterial volume–pressure relationship at any point is a measure of arterial compliance. Arterial compliance also decreases with aging and with the level of active smooth muscle contraction in the artery.

For example, a large stiff-walled vessel may have a higher capacitance value than a tiny flexible vessel. For this reason, one should use the *percentage* increase in volume for a given increase in pressure as a means of comparing distensibility between vessels and segments of the vasculature of different sizes. This value is sometimes called *vascular distensibility.* As explained in later chapters, vascular capacitance, compliance, and distensibility are critical determinants of the performance of the heart, the stress and workload placed on the heart, and the amount of oxygen the heart must receive to function properly. All of these factors are important parameters in understanding the consequences of heart diseases and their treatments.

Even within a given artery or vein, capacitance, distensibility, and compliance are not constants in the cardiovascular system. For example, as transmural pressure increases, compliance decreases (Fig. 11.3). In addition, arterial compliance decreases with age at any given transmural pressure and is reduced by contraction of the smooth muscle within either arteries or veins. This active control over pressure/volume interrelationships in the vascular tree is an important component of the moment-to-moment control of blood volume distribution in the body and cardiac performance.

Blood vessels must overcome wall stress to be able to contract.

Any transmural pressure within an artery or a vein exerts a force on the vessel wall that would tend to rip the wall apart were it not for the opposing forces supplied by the muscle and connective tissue of the vessel wall, as shown in Figure 11.4. This force is called **tension** and is equal to the

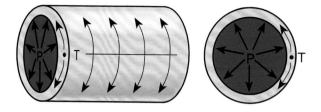

Figure 11.4 **Pressure and tension in a cylindrical blood vessel.** The tension is the force that would pull the vessel apart along an imaginary line along the length of the vessel. Tension in the vessel wall is related by the law of Laplace, as described in the text. P, pressure; T, tension.

product of the transmural pressure and the vessel radius. This relationship is called the **law of Laplace**. However, this relationship applies directly only to cylinders with thin walls. Blood vessel walls are sufficiently thick, so that in reality, this force is equal to a **wall stress** (the product of pressure [P] and the radius [r] divided by the wall thickness [w], or $S = P \times r/w$). There are many consequences of this relationship in distensible tubes. First, because tension and stress are related to vessel radius, small vessels are able to withstand higher pressures than vessels of larger diameters. For this reason, capillaries (inner diameter ~10 μm) can withstand relatively high intravascular pressure even though they are composed only of a single cell layer of endothelial cells. In arteries and veins, vessels with thick walls relative to their radius are able to withstand higher pressure than vessels with large r/w ratios because wall stress is lower in the former. Finally, tension and stress, not simply pressure, are the true forces that must be overcome to contract any hollow organ, such as a blood vessel or the heart. As is discussed later in the text, tension and stress are important determinants of energy requirements for the contraction of hollow organs.

▶ PHYSICAL DYNAMICS OF BLOOD FLOW

An understanding of the physical factors and laws that govern the movement of blood in the cardiovascular system is critically important to the overall understanding of cardiovascular functions. The amount of blood that flows through any segment of the cardiovascular system is a function of blood pressure, vascular geometry, and the dynamic fluid characteristics of blood. In any tube of a given diameter, the amount of flow through the tube is proportional to the *difference* in pressure between one end of the tube and the other (Fig. 11.5); doubling the pressure difference doubles the flow, whereas halving the difference halves the flow. Flow through a cylindrical tube is related to the length of the tube in an inverse proportion, for example, keeping all other determinants of flow unchanged, doubling only the length of a cylindrical tube reduces flow through the tube by half; tripling the length reduces flow to 1/3 that of the shorter tube and so forth. However, as shown in Figure 11.5, flow through a tube is profoundly affected by the tube radius in that flow is proportional to the radius[4]. Consequently, doubling the tube radius results in a 16-fold increase in flow. This radius effect

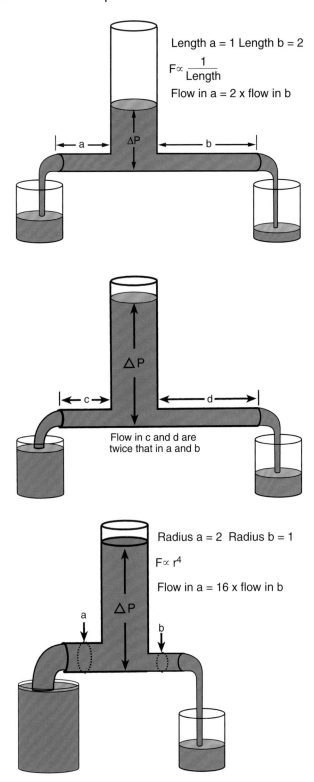

Length a = 1 Length b = 2

$$F \propto \frac{1}{Length}$$

Flow in a = 2 x flow in b

Flow in c and d are twice that in a and b

Radius a = 2 Radius b = 1

$$F \propto r^4$$

Flow in a = 16 x flow in b

Figure 11.5 **The influence of pressure difference, tube length, and tube radius on flow.** The pressure difference driving flow (ΔP) is the result of the height of the column of fluid above the openings of tubes a and b. Increasing the pressure *difference* increases flow; if pressure was the same at the entrance and exit of the tubes, there would be no flow. Flow is reduced in direct proportion to the length of the tube through which it flows. Because flow is also a function of the fourth power of the tube radius, small changes in radius have a marked effect on flow. F, flow; P, pressure.

has a profound influence therefore on blood flow through any tubular system.

Finally, fluid flow through a tube is affected by the viscosity, or the "thickness and stickiness," of the fluid. Thick, sticky fluids will not flow as easily as thinner, watery fluid. The units of viscosity are given in poise, named after Louis Poiseuille. Water has a viscosity of ~0.01 Poise or 1 centipoise (cP). Blood is a suspension of proteins and cells in a salt solution. As such, it is more viscous than water. Blood plasma has a viscosity of 1.7 cP whereas whole blood, which contains plasma and blood cells, has a viscosity of ~4 cP.

Relationships between pressure, fluid flow, and resistance are quantified by Poiseuille's law.

In the 1840s, Louis Poiseuille conducted experiments that resulted in a mathematical relationship to describe flow in a cylindrical tube. This has become known as **Poiseuille's law**. This law states that:

$$Q = (P_1 - P_2)\pi r^4 / 8\eta l \tag{4}$$

where Q = flow, $(P_1 - P_2)$ = the pressure difference between the beginning and the end of the tube, r = the tube radius, l = the tube length, η = viscosity, and π and 8 are constants of proportionality. This relationship can also be written as $(P_1 - P_2) = Q\ 8\eta l/\pi r^4$. In this form, the term $8\eta l/\pi r^4$ is the resistance to blood flow and is sometimes simply designated as R. Flow resistance is a measure of how easily the fluid can pass through a tube for any given pressure difference. In physiology, it is easier to calculate resistance as $(P_1 - P_2)/Q$ and to express it in units of mm Hg/mL/min. This value is then called a peripheral resistance unit (PRU).

Poiseuille's law gives two of the most fundamentally important relationships used to describe flow and pressure in the cardiovascular system, which are

$$Q = (P_1 - P_2)/R \text{ and } (P_1 - P_2) = Q \times R \tag{5}$$

These equations indicate that flow is proportional to the pressure difference between the entrance and exit points of the tube and inversely proportional to the resistance (i.e., as resistance increases, flow decreases). It also tells us that at any given flow, the pressure drop along any two points down the length of the tube is proportional to resistance. In the cardiovascular system, the final "end of the tube" is considered the right atrium. Pressure in the right atrium is about 2 mm Hg but considered sufficiently close to zero to be ignored. Therefore, $(P_1 - P_2)$ becomes simply P. Furthermore, for the purpose of understanding basic hemodynamics, P is usually taken as the mean pressure within an artery or vein. Consequently, Poiseuille's law reduces to

$$Q = P/R \text{ or } P = Q \times R \tag{6}$$

Applied to the whole cardiovascular system, this law indicates that arterial pressure is the product of the flow output of the heart, called the **cardiac output**, and the resistance to flow provided by all the blood vessels in the circulation. This resistance is termed the **total peripheral vascular resistance (TPR)** or less correctly just vascular resistance.

Strictly speaking, Poiseuille's law applies only to nonpulsatile flow of a homogenous fluid in uniform, rigid, nonbranching cylindrical tubes. Because none of these characteristics is met in the cardiovascular system, one might imagine that Poiseuille's law cannot be applied to blood flow. However, this law can be and is applied to the cardiovascular system. Note that this application is not used as a simplified expedient that ignores some of the complex realities of the physics of blood flow. Rather, it is used in the manner expressed in equation 6 because the cardiovascular system truly does behave as if the heart pumped blood at a steady flow, producing a mean arterial pressure as the result of a single TPR. Thus, equation 6 predicts that mean arterial pressure will increase if the cardiac output, TPR, or both increase. It also predicts that at a constant mean arterial pressure, blood flow through any portion of the vascular tree will increase if TPR decreases. These simple cause-and-effect relationships are the critical determinants of arterial pressure and organ blood flow in the cardiovascular system and will be described more fully in the context of cardiovascular control mechanisms in Chapter 17.

The series and parallel arrangement of blood vessels within an organ affects vascular resistance in the organ.

The cardiovascular system is composed of millions of vessels of different sizes. There are two simple rules used to determine how many vessels of different sizes combine to give a single resistance to flow. In a system composed of different-sized tubes arranged in series (i.e., sequentially, or end to end), the total resistance of that system is simply the sum of the individual resistances. Or

$$R_{total} = \Sigma R_{individual} \qquad (7)$$

For example, if one were to examine the resistance of a segment of a tapering artery where the proximal 1 cm of length had an R = 1 PRU, the next 1 cm had an R = 2 PRUs, and the 1 cm after that an R = 5 PRUs, the resistance across the entire 3-cm length of that artery would be 1 + 2 + 5 = 8 PRUs. Nevertheless, the arterial portion of the cardiovascular system is not a single long blood vessel; the aorta branches into parallel vessels that reach the billions once at the level of capillaries. The total resistance in a system of parallel tubes is given by

$$1/R_{total} = \Sigma \left(1/R_{individual}\right) \qquad (8)$$

That is, the reciprocal of the total resistance in vessels arranged in parallel is the sum of the reciprocals of the individual resistances. Using the arterial segments mentioned above but arranged in parallel, one would arrive at $1/R_{total}$ = 1/1 + 1/2 + 1/8 = 13/8 and R_{total} = 8/13, or ~0.62 PRUs. In this case, the total resistance of the system of the three arteries arranged in parallel is not only less than it would have been if they were arranged in series but is actually less than the resistance of any one segment in the circuit. In most cases, this is a *general* rule of thumb; adding similar-sized arteries in parallel reduces resistance, whereas losing similar-sized arteries in parallel raises vascular resistance. The first case is seen following long-term aerobic training such as distance running,

in which the arteriolar and capillary network increase their numbers in parallel, thereby reducing resistance to blood flow.

The effect of summed series and parallel elements in the cardiovascular system is a bit more complicated than can be described solely by equations 7 and 8. In the cardiovascular system, vessels get smaller as they proceed from arteries down to arterioles and capillaries, which tends to increase resistance to flow. However, the number of vessels arranged in parallel also increases dramatically in this direction, which tends to decrease resistance. The effect of these two phenomena on the relative resistances of individual sections of the vasculature depends on whether the number of vessels added in parallel can compensate for the resistance effects of adding vessels that individually have a high resistance. This interplay between series and parallel elements is a primary factor in creating the form of the pressure profile, or ΔP, along the arterial side of the circulation, as will be explained later in this chapter.

High blood flow velocity decreases lateral pressure while increasing shear stress on the arterial wall.

In addition to arterial blood flow, physiologists are often interested in how fast the bloodstream is flowing within the artery. This variable is called **flow velocity** and is usually expressed in cm/s. Fluid flow velocity is given simply by the volume of flow per second (cm^3/s) divided by the cross-sectional area of the system through which the fluid is flowing (Fig. 11.6A). If a flow of 200 mL/s in a tube is forced through another narrower tube, the flow must go through that smaller opening faster to maintain a volume flow at 200 mL/s. Conversely, if this fluid is allowed to expand into a much larger cross-sectional area, it can move slower and still deliver 200 mL/s. This relationship holds whether applied to a single tube or a composite cross-sectional area of many tubes arranged in parallel such as a cross-section through a portion of the vascular system. Thus, at a constant flow, a decrease in the cross-sectional area through which the flow is moving increases flow velocity, and an increase in area decreases flow velocity.

The blood pressure exerted against the walls of an artery and the energy required to move blood at a certain velocity are interrelated by the **Bernoulli principle**. The total energy of blood flow in a blood vessel is the sum of its potential energy (represented as pressure against the vascular wall) and its kinetic energy resulting from its velocity (kinetic energy = ½ mv^2, where m = mass and v = velocity). Because energy is conserved, the total of potential and kinetic energy at any given point in a system is constant. Consequently, any increase in one form of the energy has to come at the expense of the other. For example, as flow velocity increases, lateral pressure must decrease to keep the total energy of the system constant (see Fig. 11.6A). This principle is seen with flow in the aorta, where high velocity reduces lateral pressure relative to that measured directly facing the flow stream, which equals the total energy in the system. This phenomenon is exploited clinically to evaluate the severity of the hemodynamic consequences of a stenotic (narrowed) aortic valve. A catheter with two pressure sensors is placed in the heart such that one lies within the ventricle and the other just

Vascular Abnormalities Associated with Chronic Primary Arterial Hypertension

Chronic arterial hypertension is a cardiovascular disease in which the main manifestation is a consistent elevation of arterial pressure ≥140/90 mm Hg. Diagnosis of chronic arterial hypertension in a patient is established following documentation of elevated pressure in a clinical setting at several time points over a period of many weeks or months. It is now estimated that up to 1/3 of the population of the United States suffers from chronic arterial hypertension. Such a high prevalence of hypertension represents a serious national health issue as this condition is the leading risk factor in stroke and a major contributor to morbidity and mortality associated with heart failure, coronary artery disease, atherosclerosis, and renal insufficiency.

Arterial hypertension is largely asymptomatic until it results in end-stage organ disease, which is a major reason it often goes unrecognized and untreated by the patient. Left untreated, arterial hypertension becomes more severe over time, as the result of progressive arterial and renal abnormalities resulting from exposure to high arterial pressure. People with untreated arterial hypertension do not get better on their own and must stay on medication throughout their lives to keep their arterial pressure down. Keeping pressure down in hypertension has been clinically established to provide definitive benefits to the patient, whereas failure to control hypertension results in definitive harm.

Chronic arterial hypertension is classified broadly as either *primary* or *secondary hypertension*. Secondary hypertension indicates that the elevated blood pressure is a secondary result of some other primary disease. Hypertension is a secondary consequence of renal artery stenosis, renal parenchymal disease, primary aldosteronism, pheochromocytoma, aortic coarctation, and thyrotoxicosis. Although there are several mechanisms known to cause secondary hypertension, only 5% to 15% of all cases of chronic arterial hypertension fall into that category. The remaining individuals with hypertension have what is called *primary hypertension* (formerly called essential hypertension). The primary classification indicates that the hypertension is a primary condition in its own right of unknown cause. Currently, the etiology of primary hypertension is believed to be the result of a mosaic of initiating factors and regulatory dysfunctions with strong genetic predispositions involving both. Many of these appear to involve the kidney and salt balance.

Regardless of the mechanisms creating hypertension, chronic arterial hypertension is the result of elevated peripheral vascular resistance. For this reason, hypertension is considered a *vascular* disease involving the arterioles because these vessels are the major contributor to total peripheral vascular resistance. Arteries are not static vessels; they enlarge when chronically exposed to high blood flow, regress when flow is chronically low and importantly increase in thickness when exposed to high internal stress. In chronic hypertension, arteries have thickened vascular walls and narrowed lumens. This thickening is due to vascular smooth muscle cell hypertrophy and hyperplasia as well as increased water content and accumulation of fibrotic material in the arterial wall. These factors increase the resistance of arteries, even in the absence of all vascular smooth muscle contraction, and acts as a structural, geometric amplifier of any vasoconstrictor stimuli. As a result, all arterial smooth muscle contraction, from maximum dilation to maximum contraction, results in exaggerated increases in arterial resistance in arteries from hypertensive, compared with normotensive, patients. In addition, the arterial smooth muscle cells in hypertensive patients are hypersensitive to vasoconstrictor stimuli and hyposensitive to vasodilators. This too results in an exaggerated increase in vascular resistance with any vasoconstrictor stimuli and a reduced capacity of the arteries to counteract this exaggeration with any vasodilator stimuli. Lost production and bioavailability of arterial endothelial nitric oxide, which is a potent vasodilator and antimitogenic agent for smooth muscle, is also a consequence of hypertension and may be a contributing factor in the maintenance and inevitable progression of that disease. ■

---

across the narrowed aortic valve. A high-velocity jet of blood moving through the narrowed valve causes a significant drop in lateral pressure detected by the aortic sensor compared with the ventricular sensor; the hemodynamic severity of the stenosis is proportional to this pressure difference.

All flowing fluids exert a "rubbing" force against the inner wall of the cylinder in which they flow. This force is analogous to what one feels by rubbing the palms of the hands together and is called **shear stress**. Shear stress on the inner walls of arteries increases proportionally as the velocity of flow near the wall increases. This has several medical implications. The release of paracrine substances from the endothelium as well as other transport phenomena is stimulated by increased shear stress on the endothelial cells that line the arteries. These factors are instrumental in stimulating capillary growth and endothelial repair and may be the

link between the growth of new blood vessels in skeletal muscle and periodic increases in muscle blood flow during aerobic exercise training. In atherosclerosis, plaque formation is reduced in areas of high shear but increased in low shear areas of arteries (see Clinical Focus 11.2).

High blood flow velocity can create turbulent flow in arteries.

Flow velocity also affects the organization of fluid layers of blood in arteries. Normally, blood cells flow in a streamlined or bullet-shaped profile with the concentric layers of fluid flow in arteries as illustrated in Figure 11.6B. In three dimensions, this can be envisioned as a set of thin telescoping cylindrical layers projecting from the inner arterial wall out uniformly to the center of the vessel. Flow velocity is highest in the center of a streamline blood flow and lowest adjacent

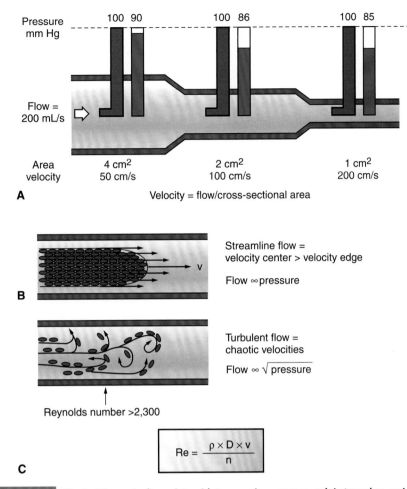

Figure 11.6 **Effect of flow velocity on lateral intravascular pressure, axial streaming, and turbulent flow. (A)** If any given flow of blood is forced through progressively smaller cross-sectional areas, the velocity of blood flow must increase. The Bernoulli principle states that increased flow velocity reduces the lateral pressure of the flow stream exerted against the wall of the vessel. **(B)** The distribution of red blood cells in a blood vessel depends on flow velocity. As flow velocity increases, red blood cells move toward the center of the blood vessel (axial streaming), where velocity is highest. Axial streaming of red blood cells creates a cell-free layer of plasma along the inner vessel wall. **(C)** At high flow velocity, the kinetic forces of flowing fluid overcome the viscous forces holding layers of fluid together, resulting in turbulent flow.

to the inner arterial wall. However, if total flow velocity becomes too high in an artery, the kinetic energy of the flow streams overcomes the tendency of the fluid layers to stick together from viscous forces. When this happens, the fluid layers break apart and become random and chaotic. This is a condition called **turbulence** (see Fig. 11.6C). Turbulence is a wasteful process that dissipates pressure energy in the cardiovascular system, which could otherwise be used to produce flow. The tendency to produce turbulent flow is expressed in a mathematical term called the **Reynolds number, Re**. The Reynolds number is a measure of the ratio of kinetic energy in the system (which will pull fluid layers apart) and the viscous component of the system (which holds the fluid layers together). It is represented by

$$Re = \rho Dv / \eta \qquad (9)$$

where ρ = fluid density, D = inner vessel diameter, v = flow velocity, and η = blood viscosity. Generally, a Re

$\geq$ 2,300 indicates that turbulence will occur in the fluid stream. Clearly, large-diameter vessels, high flow velocity, and low blood viscosity favor turbulence in the cardiovascular system.

Whereas laminar flow in arteries is silent, turbulent flow creates sounds. These sounds are clinically known as **murmurs** or **bruits**. Certain diseases, such as atherosclerosis and rheumatic fever, can scar the aortic or pulmonic valves, creating narrow openings and high flow velocities when blood is forced through them. A clinician can detect this problem by listening to the noise created by the resultant turbulence.

Blood viscosity is increased by red blood cells and low flow velocity.

The fluid dynamic properties of blood are more complicated than they are for a simple homogenous fluid such as water because blood is a suspension of proteins and cells in an aqueous medium. The study of the fluid dynamic properties

CLINICAL FOCUS | 11.2

Hemodynamic Localization of Atherosclerosis

Atherosclerosis is a chronic inflammatory response of the walls of large arteries that is initiated by injury to the endothelium. The exact cause of this injury is unknown. However, exposure to high serum lipid levels in the form of low-density lipoprotein (LDL) cholesterol and triglycerides is considered a prime factor, with contributions from hypertension, the chemicals in cigarette smoke, viruses, toxins, and homocysteinemia. With injury, the normal endothelium barrier becomes compromised. Leukocytes, primarily monocytes, adhere and infiltrate the arterial intima. These cells and the intima itself accumulate large quantities of lipoproteins, mainly LDL, which, when oxidized, further damages the artery, stimulates the production of damaging reactive oxygen species, and sets up a more aggressive local arterial inflammatory response. This results in mitotic activation of arterial smooth muscle cells, which migrate into the intima, as well as the activation of monocytes, which transform into macrophages that engulf lipid to become foam cells. These factors stimulate intimal thickening and smooth muscle cell growth with infiltrates the intima. These processes create an inwardly directed growth of the arterial wall, which encroaches on the arterial lumen and creates the appearance of fatty streaks on the inner arterial wall. Over time, this streak grows with the accumulation of extracellular matrix proteins, the development of a fibrous cap over the atheroma, and the development of a lipid-laden necrotic core-containing debris, foam cells, crystallized cholesterol, and calcium deposits. This creates what is often called an *atherosclerotic plaque*.

Atherosclerotic plaques usually only develop in a portion of the circumference of the arterial wall. The wall opposite the plaque can actively contract in response to vasoconstrictor stimuli, whereas the wall beneath the plaque becomes weakened, creating an arterial aneurysm that can rupture. Atherosclerotic plaques stimulate platelet aggregation and blood clot formation that can totally occlude an artery. Furthermore, the plaque is friable and can rupture, spilling debris into the arterial lumen, which further stimulate clot formation. Plaque rupture in the arteries of the heart is a primary cause of death from heart attack.

Branch points and curvatures alter flow velocity and shear stress at the arterial wall; these are decreased along the inner curvature of a flow stream and at the upstream edge of branch points. It is well known that plaque formation in the vascular system does not occur either uniformly or totally randomly in the vascular tree. Instead, it develops at branch points and bifurcations and along the inner curvature of arteries. All these areas contain regions of low flow velocity and low shear stress. The anatomic characteristics of the coronary arteries, the bifurcation of the common carotid arteries, and the entry to the renal arteries make these regions especially susceptible to the accumulation of atherosclerotic plaques and thus place the blood supply to the heart, the brain, and the kidneys at risk. ■

of blood is called **rheology**. The presence of proteins and cells in blood has two important hemodynamic consequences. First, blood viscosity increases exponentially as the blood hematocrit increases (Fig. 11.7A). Thus, as the hematocrit increases, the flow resistance against which the heart must pump increases, which substantially increases the cardiac workload. Although there are some pathologic instances in which a patient's hematocrit can increase because of overproduction of red blood cells (e.g., polycythemia vera), an abnormally increased hematocrit can come from the loss of blood plasma without a proportional loss of red blood cells. This can occur in severe dehydration, loss of plasma from severe burns, or inappropriate loss of water through the renal system as a consequence of kidney disease. Secondly, red blood cells tend to clump if blood flow velocity is sluggish (see Fig. 11.7B). This clumping raises blood viscosity, which can be a negative complicating factor in any condition that adversely diminishes the overall flow output of the heart, such as circulatory shock or heart failure.

Axial streaming of cells in flowing blood

Application of Poiseuille's law to fluid dynamics assumes that the fluid is homogeneous, that is, it is made of one element of uniform composition, such as water. Although all

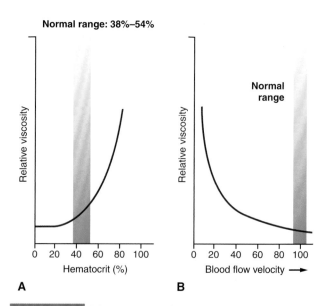

Figure 11.7 **Effect of hematocrit and flow velocity on blood viscosity. (A)** Increases in hematocrit above normal values produce a sharp increase in viscosity, causing marked increase in resistance to flow. **(B)** Blood viscosity in arteries increases dramatically whenever flow velocity slows to low levels. Thus, the resistance to flow is higher in a slow-moving arterial stream than in one that moves with high velocity.

suspensions are nonhomogeneous fluids, the properties of blood cells flowing in arteries are such that blood behaves hemodynamically as if it were a homogeneous fluid. Blood cells tend to compact densely in the center of the flow stream, leaving a thin layer of cell-free plasma against the vascular wall. This is called **axial streaming**. As such, blood flows more as a compact bulk fluid than as a mixed conglomeration of particles in suspension. For this reason, Poiseuille's law can be applied to the cardiovascular system as it was written. However, axial streaming of blood does tend to separate this bulk flow into two components of different viscosities. The fluid near the vessel wall is essentially cell-free plasma and has a viscosity of only 1.7 cP, as opposed to 4 cP for whole blood. In large vessels such as the aorta, this low-viscosity layer is only a small percentage of the average viscosity of the flow stream. Thus, for all practical purposes, the viscosity of the blood flowing through the entire aortic cross-section can be considered to be 4 cP. However, in small arterioles (<300 μm interior diameter) and capillaries, this thin layer becomes a greater percentage of the total volume contained within the vessel and thus contributes a greater percentage to the total viscosity of the blood traveling through those vessels. When fluid flows through these smaller vessels, fluid viscosity, as a whole, decreases. This is called the *Fahraeus-Lindqvist effect*, and it is responsible for reducing blood viscosity, and therefore, flow resistance when blood flows through extremely small vessels such as capillaries. This makes it easier for blood to flow through vessels that otherwise have extremely high resistances.

▶ DISTRIBUTION OF PRESSURE, FLOW, VELOCITY, AND BLOOD VOLUME

Meaningful insights into characteristics of the cardiovascular system can be obtained by examination of the distribution of flow, velocity, pressure, and volume within the system. For example, because veins are more compliant than arteries, one would expect that more of the total volume of blood in the cardiovascular system would reside in the venous rather than the arterial side of the circulation. This is precisely the case; about 2/3 of total blood volume is contained in veins relative to arteries and about 80% of that is contained in small veins. Less than 20% of total blood volume is contained in arteries and capillaries with the rest of the volume contained in the pulmonary circulation and chambers of the heart. Also, because cross-sectional area increases greatly from arteries to the arterioles and to the capillaries, the lowest blood flow velocity occurs through the capillary network (Fig. 11.8). This slow velocity through this exchange segment of the vascular system has the beneficial effect of allowing more time for the exchange of material between the cardiovascular system and the extracellular fluid.

The heart is an intermittent pump; it generates high pressure within the ventricles when it contracts during

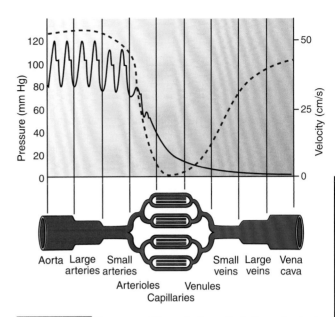

Figure 11.8 Pressure and flow velocity profile in the systemic circulation. The arterial portion of the circulation is characterized by high, pulsatile pressure and high flow velocity. This profile changes to one of low pressure and velocity without pulsatile character in the veins. The largest drop in mean arterial pressure occurs across the arteriolar segment of the circulation, indicating that this is the sight of highest vascular resistance in the cardiovascular system.

systole, which then drops to near zero during diastole. However, because arteries are compliant, some of the ejected blood into the arteries distends these vessels. During diastole, recoil of the arteries pushes blood forward against the downstream vascular resistance, generating a significant diastolic pressure. For this reason, diastolic pressure drops to only about 80 mm Hg in the aorta as compared with near zero in the ventricles.

Examination of the pressure profile across the cardiovascular system (see Fig. 11.8) shows that the largest drop of pressure occurs across the arterioles, indicating that this is the site of the greatest vascular resistance in the cardiovascular system. Although there are many more arterioles than arteries in the cardiovascular system (resistances in parallel), this large pressure drop indicates that their reduction in individual size dominates over the addition of parallel vessels. Similarly, although individual capillaries are very small (~10 μm), so many of these lie in parallel that resistance across the capillaries is actually lower than that across the arterioles; hence, the pressure drop across the capillary segment of the circulation is less than that across the arterioles.

Pressures within the arteries of the pulmonary circulation are not the same as those in the systemic circulation. Pulmonary arterial pressure is about 25/8 mm Hg. Because the outputs of the right and left heart are the same, the low pressure in the pulmonary circulation must indicate, according to Poiseuille's law, that vascular resistance is much lower in the pulmonary circulation than in all the organs combined that make up the systemic circulation.

INTEGRATED MEDICAL SCIENCES

Arteriogenesis

The growth of the human body from birth through puberty and to adulthood requires that the circulatory system needed to support the body as a whole grows with it. There are two ways in which the arterial system grows normally as the body grows. Arteries must grow larger and new arteries formed (**arteriogenesis**) and new capillaries must increase in numbers as well (**angiogenesis**). The process of arteriogenesis is currently the subject of intense multidisciplinary research. Insight into the mechanisms for stimulation of vascular growth hold the promise of development of methods of a circumventing damage to tissue below an occluded artery by stimulation of new collateral arterial growth through and around tissue downstream from the occlusion. Such new methods could circumvent current medical treatments (e.g., surgery) that are at best palliative by nature, as well as be a means to counteract abnormal growth of vessels that support tumor growth and create stenosis of arteries in disease states. Currently, the term angiogenesis is mainly applied to the development of collateral circulation around subacute, or slowly developing, arterial occlusions. In occlusive disease, only angiogenesis provides the means to fix flow deficits caused by an arterial occlusion. Angiogenesis requires both the enlargement of small, preexisting collateral arterial vessels and an increase in their length. Such a process must involve stimulation of arterial wall cell proliferation. Consequently, arteriogenesis is a process of active growth not passive distention of existing vessels.

Unlike angiogenesis, arteriogenesis is not initially stimulated by hypoxia. Instead, it is initiated by alterations in collateral vessel fluid dynamics. The pressure gradient between the upstream high pressure origin of the collateral above the occlusion and the downstream low pressure postocclusion zone initiates a marked increase in blood flow through the collateral vessel. It is now well established that the initiating stimulant for angiogenesis is the increased fluid shear stress on arterial endothelial cells that accompanies the increased blood flow. Because there is no direct cell-to-cell contact between the endothelial and smooth muscle layers in arteries, the signal from this mechanical force on the intima to underlying cells in the arterial wall has to be through a diffusible molecule. The current most likely candidate for this transmitter is NO through shear stress stimulation of endothelial NO synthase. Blockade of the NO pathway totally blocks angiogenesis. Sheer stress stimulates the endothelial calcium channel, TRPV4 (transient receptor potential cation channel) and because NO synthase is a calcium activated enzyme, it is likely this channel serves as the means by which shear stress activates NO production in the endothelium.

Although NO appears to be intimately responsible for the initiation of arteriogenesis, endothelium NO synthase is noted as primarily being involved in inducing arterial dilation rather than arterial wall proliferation. Therefore, other proliferative pathways must be involved in arteriogenesis. Arteriogenesis is known to be stimulated by inflammation. Circulating monocytes, either directly or by activating arterial wall macrophages, are known to stimulate arterial wall secretion of chemokines, growth factors, and proteases that can be involved in arterial wall growth and remodeling. Sheer stress stimulates monocyte chemoattractant protein (MCP-1) secretion by arterial smooth muscle cells and this factor seems to be involved in stimulation of cell growth in angiogenesis though its exact mechanism of action is not known. MCP-1 attracts monocytes in adventitia and activates indwelling macrophages. These in turn are thought to increase inflammatory factors such as TNFα and inducible NO synthase (iNOS, see Chapter 16). These findings indicate that the immune system in the body could play a key role in arteriogenesis. Recent studies have indicated that, in addition to macrophages, immune cells such as natural killer (NK) cells, T helper cells, and regulatory T lymphocytes all may have roles in arteriogenesis though their exact roles in this type of vascular growth remain to be fully elucidated. ■

Chapter Summary

- The cardiovascular system is a fluid transport system that delivers substances to the tissues of the body while removing the by-products of metabolism.
- The heart is composed of two pumps connected in series. The right heart pumps blood into the lungs. The left heart pumps blood through the rest of the body.
- Pressure is created within the atria and ventricles of the heart by contraction of the cardiac muscle. The directional opening of valves, which prevent back flow between chambers, ensures forward movement of blood through the heart.
- Arteries transport blood from the heart to the organs. Veins transport blood from the organs to the heart.
- Capillaries are the primary site of transport between blood and the extracellular fluid.
- Altering vascular smooth muscle contraction changes blood vessel radius and hemodynamic properties.
- Vascular resistance is inversely proportional to the internal radius of the blood vessel.
- The volume contained within any vascular segment is a function of transmural pressure and the compliance of the vascular wall.
- Transmural pressure in blood vessels produces wall tension and stress that must be overcome for the vessel to contract.
- Blood flow and pressure throughout the vascular system is created in accordance with the principles of Poiseuille's law; flow is proportional to the pressure gradient between points in the circulation but inversely proportional to vascular resistance.
- The velocity of blood flow in an artery affects lateral pressure and inner wall shear stress in arteries as well as transport across the capillary wall. It is also a factor in the creation of turbulence in the arterial flow stream.
- Flow resistance in arteries is influenced by hematocrit and blood flow velocity.
- The hemodynamic profile in the cardiovascular system is the result of the combined effects of all the relationships and laws governing the containment and movement of blood in the cardiovascular system.

Chapter Review Questions

1. Which of the following will exacerbate the pooling of blood that occurs in veins in the lower extremities of a patient when he or she assumes an upright position?

    A. Activation of α-adrenergic receptors on veins by the sympathetic nervous system
    B. Administration of the antianginal drug, nitroglycerin, which directly dilates veins more than arteries
    C. Administration of the emergency antihypertensive agent sodium nitroprusside, which directly dilates arteries more than veins
    D. Applying external compression to the lower extremities
    E. Placing the patient in supine position with his or her legs up

The correct answer is B. Blood preferentially pools in the veins versus the arteries under the influence of gravity because veins are more compliant than arteries. Contraction of vascular smooth muscle reduces vascular compliance, whereas relaxation of vascular smooth muscle increases compliance. For this reason, nitroglycerin will increase the amount of blood pooling in veins in a patient when he or she stands upright. This is a contributing factor in postural hypotension that often occurs in patients taking nitroglycerin for angina or heart failure. The enhancement of pooling temporarily causes a reduction in cardiac output severe enough to reduce arterial pressure and cause fainting. Drugs that predominantly relax arteries over veins do not cause much blood pooling upon standing and are generally not associated with postural hypotension. Anything that contracts or compresses veins helps prevent venous pooling. Indeed, contraction of veins by activation of the sympathetic nervous system is the body's natural mechanism for preventing postural hypotension. Support stockings, mechanical compression devices used on the legs after surgery, and the lower body compression suits worn by jet fighter pilots are ways in which application of external compression on the veins counteracts venous pooling in the lower extremities. Laying a patient horizontally with his or her legs up does take advantage of the effects of gravity on blood volume in the veins, except in this case gravity helps drain blood from the legs and send it to the heart.

2. A patient has a significantly reduced blood hematocrit as the result of a chronic bleeding ulcer. A heart murmur during systole can be heard with a stethoscope over the patient's chest. The likely cause of this murmur is:

    A. increased aortic turbulence due to decreased blood viscosity.
    B. creation of streamline flow in the aorta during systole.
    C. reduced resistance to blood flow in the arterioles.
    D. increased aortic turbulence due to increased blood viscosity.
    E. reduced flow velocity in the aorta during systole.

The correct answer is A. Blood viscosity is significantly related to blood hematocrit; blood with a low hematocrit has low viscosity. As reflected in the Reynolds number, turbulence is most likely to occur with low viscosity fluid at high flow velocity in vessels with large internal diameters. This corresponds to the situation in the aorta during systole in a patient with an abnormally low hematocrit. Streamline flow is silent and does not cause murmurs. Although the decreased hematocrit decreases resistance to flow through any segment of the arterial system, this effect in arterioles is unrelated to any murmur heard from the heart.

3. Which of the following describes the properties of an aortic segment exposed to a transmural pressure of 200 mm Hg compared to the same segment exposed to a transmural pressure of 100 mm Hg?

   A. Compliance will be reduced, and wall tension will be increased.

   B. Compliance will be increased, and wall tension will be reduced.
   C. Compliance and wall tension will be reduced.
   D. Compliance and wall tension will be increased.
   E. Neither compliance nor wall tension will change.

The correct answer is A. Re: Arteries are stiffer (reduced compliance) at very low and very high pressures. An aortic segment exposed to an internal net pressure of 200 mm Hg is distended to a place on its volume versus pressure curve where the aorta is stiffer, or less compliant. Tension = P × r is higher in cylindrical tubes like the aorta at high transmural pressures and large radii. In the segment exposed to 200 mm Hg versus 100 mm Hg not only is distending pressure higher but the aorta will be distended to a slightly larger radius and therefore that will also increase wall tension. None of the other choices to the question correctly state that an aorta under high pressure is both in a region of low compliance and exposed to high wall tension.

Clinical Application Exercises 11.1

A 15-year-old girl sees her family physician for a routine school athletic physical. Upon routine examination, her physician detects a reasonably loud systolic murmur when placing her stethoscope parasternally around the second and third intercostal spaces, but her physical exam is otherwise unremarkable. The patient mentions that she has no trouble with her endurance and does not feel her physical activities are limited. Her doctor orders a chest x-ray and an echocardiogram to investigate possible sources of the murmur. The echocardiogram reveals a small to moderate, irregular atrial septal defect (openings in the atrial septum between the left and the right atrium) with turbulent left-to-right shunting of blood through the septal defect, and a somewhat enlarged right atrium. This enlargement is consistent with findings from her chest x-ray, which also indicates slight dilation of the pulmonary arteries.

QUESTIONS

1. What is the source of the systolic murmur in this patient?
2. Why is this patient largely asymptomatic with respect to her physical endurance?
3. What is the cause of the patients enlarged right atrium and dilated pulmonary artery?

ANSWERS

1. Heart murmurs are indicative of turbulent flow conditions in the heart. The likely source of this murmur, therefore, is flow from the left to right atrium through the irregular septal defect that occurs because left atrial pressure is higher than right atrial pressure in the cardiovascular system.
2. Because left atrial pressure is higher than right atrial pressure, blood flows from left to right through the septal defect. Thus, there is a tendency for oxygenated blood to leak into the pulmonary circulation rather than deoxygenated blood mixing with the systemic blood supply. In this patient, the oxygen output to her systemic circulation is not significantly affected unless there is marked reduction in left ventricular output due to the shunt. This is unlikely given her physical presentation and the modest size of her septal defect.
3. Increased flow through the atrial septal defect increases volume on the right side of the heart, starting at the right atrium. This results in dilation of these structures and eventual hypertrophy as the atrial muscle and arterial tree respond to increased stress caused by chronic exposure to increased volume loading.

thePoint® *Visit* http://thepoint.lww.com/rhoades5e *for additional chapter review Q&A, Clinical Application Exercises, animations, and more!*

Active Learning Objectives

Upon mastering the material in this chapter, you should be able to:

- Contrast electromechanical coupling in cardiac muscle versus skeletal muscle and explain how this dictates differences in mechanical mechanisms of contraction in the two muscles.
- Explain how changes in membrane voltage-gated channels for sodium, potassium, and calcium create the five phases of atrial and ventricular muscle action potentials.
- Explain why it is not possible to create tetanic contraction in cardiac muscle.
- Predict how changes in ventricular conductances or plasma concentrations of sodium, calcium, and potassium will change the amplitude, duration, and refractory period of action potentials in ventricular muscle cells.
- Predict how changes in ion conductances or plasma concentration of electrolytes could enhance or inhibit the formation of ectopic foci in the myocardium.
- Explain how changes in plasma potassium concentration can lead to ventricular arrhythmias and cardiac arrest.
- Relate the changes in membrane conductances for sodium, calcium, and potassium in cardiac nodal tissue to their automaticity.
- Predict and explain how the intrinsic firing rate of the SA node is altered by (a) a change in the resting membrane potential and (b) the rate of decay of K+ conductance in phase 4 of the nodal potential.
- Explain the mechanism for the effects of acetylcholine and norepinephrine on sinoatrial node rhythmicity and how that relates to the effect of these neurotransmitters on heart rate.
- Predict ways in which the conduction of action potentials can be slowed through the myocardium.
- Explain how the normal ECG waveform is created by electrical events in the myocardium.
- Describe how the frontal and horizontal electrocardiogram lead systems can be used to determine the orientation of the heart, the direction of electrical activation of the heart, and changes in the muscle mass of atria and ventricles.
- Use electrocardiogram recordings to identify atrial and ventricular tachycardias, fibrillation, premature atrial contractions, premature ventricular contractions, and heart blocks.
- Use QRS vector analysis of electrocardiogram recordings to identify atrial and ventricular hypertrophy as well as abnormal myocardial conduction pathways.
- Identify myocardial ischemia, injury, infarction, and plasma electrolyte disturbances from electrocardiogram recordings.
- Predict changes in electrocardiograms caused by various cardiac drugs on the basis of their effect on ion conductances in the myocardium.

Although the heart is composed of striated muscle, it only shares a few functional similarities with skeletal muscle tissue. The basic mechanical–molecular cyclic interaction of actin and myosin is similar in cardiac and skeletal muscle. The links between action potential generation, calcium, and the initiation of contraction are similar on a general level, as are principles of muscle mechanics related to the effects of preload and afterload. However, beyond these few shared characteristics, the physiology of cardiac muscle is very different from that of skeletal muscle.

Skeletal muscle cannot contract without being connected to a motor neuron, which is analogous to a lamp that cannot be turned on if its cord is unplugged. Also, like the light bulb in the lamp, skeletal muscle fibers are either switched on (activated) or off (not activated). Finally, at any given preload and afterload, the force-generating capacity of skeletal muscle cannot be altered by any normal physiological mechanism. In contrast, cardiac muscle does not require innervation to be activated, and the intrinsic force generation capacity of each cardiac cell can be enhanced or depressed, much like a light on a dimmer switch. In this and the following chapter, the unique electrical and mechanical properties of cardiac muscle are addressed. This chapter focuses on the electrophysiology of cardiac muscle, whereas the mechanical properties of this muscle tissue are examined in Chapter 13.

► ELECTROPHYSIOLOGY OF CARDIAC MUSCLE

The heart is composed of muscle whose contraction is coupled to the generation of action potentials within its cells. However, cardiac muscle does not require action potentials from nerves to activate its own electrical activity (a fact that makes heart transplantation operations possible). Although both branches of the autonomic nervous system (ANS) innervate the heart, the ANS modulates cardiac function rather than initiates it. Furthermore, there are no anatomic or functional correlates of neuromuscular motor units in the heart. Therefore, the heart cannot recruit neuromuscular units to enhance its force-generating capacity.

Cardiac cells are electrically connected and can generate their own action potentials.

All myocardial cells are coupled electrically through gap junctions at points called *nexi* (Fig. 12.1). This allows the generation of an action potential in one myocardial cell to spread rapidly to all cells in the heart. Thus, electrically, the heart behaves as a **functional syncytium**, as if it was

237

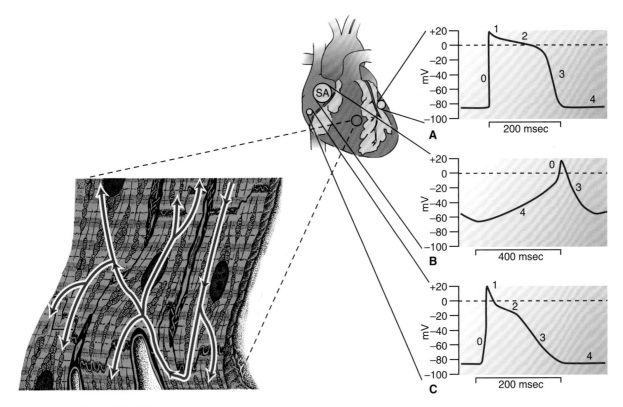

Figure 12.1 Cardiac action potentials (mV) recorded from ventricular (A), sinoatrial node (B), and atrial cells (C). Note the difference in the time scale of the sinoatrial cell. Numbers 0 to 4 refer to the phases of the action potential (see text). **Inset:** Diagrammatic representation of the nature of the functional syncytium in the myocardium. The colored line represents the path of excitation through the myocardium as it passes from cell to cell through gap junctions at cardiac nexi. SA, sinoatrial.

activated totally as one large cell. The advantage of this electrical connectivity between all cells is that it helps the heart contract as a large, coordinated mechanical unit for the purpose of pumping blood. The heart could not function as a pump if its millions of cells activated randomly. However, the syncytial character of the myocardium also means that the contractile force of the whole heart, unlike that of skeletal muscle groups, cannot be modulated by the recruitment of motor units. During systole, all cardiac cells are activated; there are no cells left to recruit. Although the syncytial character of the heart enables it to function better as a muscular pump, it also means that the activation of any cell in the heart can inadvertently activate the heart as a whole. This can have detrimental effects on the ability of the heart to pump blood, as is discussed later in this chapter.

Cardiac cells possess the unique property of **automaticity**; that is, they have the ability to generate their own action potentials without the need for chemical or electrical stimuli from other sources. Furthermore, some specialized cardiac cells display the property of **rhythmicity** or the ability to generate these potentials in a regular, repetitive manner. Unfortunately, however, in pathologic conditions, automaticity can occur in any myocardial cell as well as in the **Purkinje fibers**. This results in abnormal activation of the heart and its pumping mechanism.

Two major types of unique action potentials characterize electrical excitation of the heart.

The appearance of cardiac action potential is significantly different from the rapid spike characteristics seen in neurons or skeletal muscle fibers. There are two broad types of cardiac action potentials. Those characteristics of ventricular and atrial muscle, as well as of Purkinje fibers, are called "fast response" action potentials, whereas those observed in the sinoatrial (SA) node and atrioventricular (AV) node are called "slow response" action potentials (see Fig. 12.1). The fast response is divided into five phases (Fig. 12.2). The initial rapid depolarization of the cell membrane is designated phase 0. Phase 1 represents the subsequent partial repolarization of the membrane, which is followed by phase 2. Phase 2 is unique to cardiac muscle and is often called the plateau region of the action potential. Phase 3 is the rapid repolarization phase of the action potential, and phase 4 is the resting membrane potential.

Changes in plasma potassium concentration markedly alter the myocardial resting membrane potential.

Many ion channels contribute to the creation of the fast response (see Fig. 12.2). The resting membrane potential (phase 4) is primarily a K^+ diffusion potential, and thus, it is sensitive to changes in external K^+ concentration. High plasma

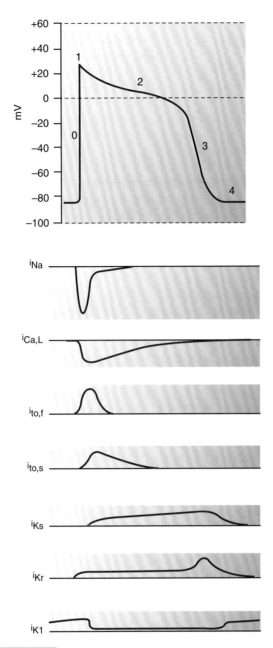

Figure 12.2 **Changes in cationic currents responsible for the formation of the fast response action potential in ventricular muscle cells.** I = current flow; downward deflections indicate current flow into the cell, whereas upward deflections indicate current flow out of the cell. The rise in action potential (phase 0) is caused by rapidly increasing Na$^+$ current carried by voltage-gated Na$^+$ channels. Na$^+$ current falls rapidly because voltage-gated Na$^+$ channels are self-inactivating by depolarization. K$^+$ current rises briefly, causing phase 1 ($i_{to,f}$ and $i_{to,s}$) and then falls precipitously as transient outward currents cease and I_{K1} channels close. Ca^{2+} channels are opened by depolarization and are responsible, along with closed I_{K1} channels, for phase 2. K$^+$ current begins to increase because of the delayed opening of I_{Kr} and I_{Ks} channels by depolarization and an increase in intracellular calcium concentration. These currents are responsible for rapid repolarization in phase 3. Once repolarization occurs, Na$^+$ and Ca^{2+} channels are returned to their resting state. Repolarization reopens I_{K1} channels and re-establishes phase 4.

K$^+$ concentrations (hyperkalemia) depolarize cardiac cells, whereas hypokalemia hyperpolarizes the tissue. Both of these conditions can adversely affect cardiac function, and for this reason, plasma K$^+$ levels are monitored carefully in a clinical setting. There is a small component of Na$^+$ influx to the cardiac resting membrane potential, making the resting membrane potential slightly more positive than the K$^+$ Nernst potential.

Cardiac cells have an intrinsic resting membrane potential buffer system that attenuates changes in membrane potential caused by changes in external K$^+$ concentration in the range of the normal plasma concentration of about 4.4 mM. Potassium conductance in myocardial cells is altered by the extracellular concentration of potassium surrounding the cells. Over the physiologic extremes of 2 to 7 mM, K$^+$ conductance increases when external K$^+$ concentration increases and decreases when external K$^+$ concentration decreases. Thus, the depolarizing effect of elevated external K$^+$ concentration is somewhat buffered by an increased K$^+$ efflux caused by a modest increase in K$^+$ conductance. However, this system cannot totally counteract the effect of external potassium on cardiac resting membrane potential. Hyperkalemia will always depolarize the cell at rest; it is simply that the magnitude of this depolarization is less than that predicted by the Nernst equation because of the effect of external K$^+$ on potassium conductance in cardiac myocytes.

Voltage-gated sodium and potassium channels initiate and terminate phase 0 of the fast response.

The five phases of the ventricular action potential results from overlapping time-dependent activation and inactivation of membrane channels for Na$^+$, K$^+$, and Ca^{2+} (Fig. 12.3; also see Fig. 12.2). The rapid upstroke in phase 0 occurs by a mechanism similar to that seen in nerve or skeletal muscle. Membrane depolarization up to and more positive than ~ −55 mV opens "activation, or m, gates" on voltage-sensitive Na$^+$ channels, allowing a rapid influx of Na$^+$ through the now open channel pore down its steep electrochemical gradient. This influx further depolarizes the membrane, opening more Na$^+$ channels, and so on, which causes a rapid, self-reinforcing increase in the depolarization of the cell. Opening of the cardiac Na$^+$ channel is self-limiting, however, in that the same depolarization that causes conformational changes in the channel protein to open the channel also triggers a conformation change in a second gate in the channel that closes it several milliseconds later (the h or "inactivation gate"). Past the peak of the cardiac action potential, more than 99% of the sodium channels are in the inactivated state. The inactivation gate will remain closed unless the cell membrane potential drops below about −50 mV. Like nerve and skeletal muscle cells, ventricular (and atrial) cells cannot fire a second action potential until the sodium channel is reset to its resting state by the cell returning to its normal resting membrane potential.

In nerve and skeletal muscle, a K$^+$ channel opens soon after the initial Na$^+$ event and the resulting facilitation of K$^+$ efflux through this channel rapidly repolarizes those cells. This does not occur in cardiac cells. In cardiac cells, the K$^+$

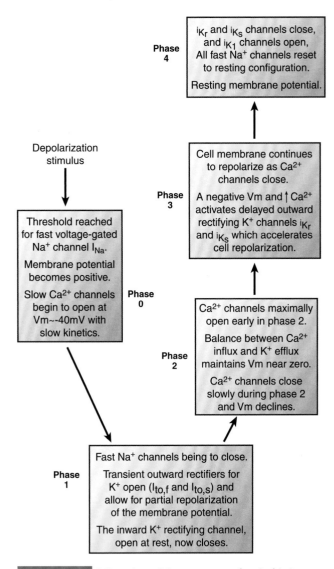

Phase 4
i_{Kr} and i_{Ks} channels close, and i_{K1} channels open, All fast Na$^+$ channels reset to resting configuration.
Resting membrane potential.

Depolarization stimulus

Phase 3
Cell membrane continues to repolarize as Ca^{2+} channels close.
A negative Vm and ↑ Ca^{2+} activates delayed outward rectifying K$^+$ channels i_{Kr} and i_{Ks} which accelerates cell repolarization.

Phase 0
Threshold reached for fast voltage-gated Na$^+$ channel I$_{Na}$.
Membrane potential becomes positive.
Slow Ca^{2+} channels begin to open at Vm~-40mV with slow kinetics.

Phase 2
Ca^{2+} channels maximally open early in phase 2.
Balance between Ca^{2+} influx and K$^+$ efflux maintains Vm near zero.
Ca^{2+} channels close slowly during phase 2 and Vm declines.

Phase 1
Fast Na$^+$ channels being to close.
Transient outward rectifiers for K$^+$ open (I$_{to,f}$ and I$_{to,s}$) and allow for partial repolarization of the membrane potential.
The inward K$^+$ rectifying channel, open at rest, now closes.

Figure 12.3 A flow chart of the events associated with the ventricular action potential. (See text for details.)

conductance (I_{K1}) largely responsible for the resting membrane potential is actually suppressed during depolarization, thus impairing any potential rapid repolarization of the tissue. There is some enhanced K$^+$ conductance at this time point in the cardiac action potential, but it is short lived, owing to the opening and then closing of two potassium channels that allow only K$^+$ to exit the cell. These channels are called **transient outward (to) K$^+$ channels** and are designated I$_{to,fast}$ and I$_{to,slow}$, owing to slightly different channel time kinetics. Coupled with the inactivation of the sodium channels in the membrane, the opening of these transient potassium channels causes the small, temporary repolarization of the cardiac cell that is designated as phase 1.

Cardiac cell refractory period is prolonged by opening of slow voltage-gated Ca^{2+} channels.

All cardiac cells contain **L-type Ca^{2+} channels** (also called 1,4 dihydropyridine calcium channels). These channels open and close with depolarization in a manner analogous to that seen for the Na$^+$ channel, except their opening and closing

kinetics are much slower. They also open at slightly less negative potentials than do the fast sodium channels (~ −40 to −50 mV). The electrochemical gradient for Ca^{2+} is enormous in cardiac cells, and the conductance of these channels for calcium is high. Once the calcium channels open, positive charge from Ca^{2+} rushes into the cell. During this phase of the action potential, hyperpolarizing effects of potassium efflux are diminished because of the inactivation of I$_{K1}$ channels by membrane depolarization. In phase 2, the influx of positive current from calcium approximately matches the remaining positive efflux carried by K$^+$ exiting through a few open K$^+$ channels in the membrane. This balance causes the membrane potential to remain relatively constant at a positive value and helps create phase 2 of the action potential. Eventually, the Ca$^+$ channels close, preventing further depolarizing currents from entering the cell, which aids in the eventual repolarization of the cell membrane.

The plateau phase of the ventricular action potential has a significant functional effect on cardiac muscle activation. Because of the plateau phase of the cardiac action potential, cardiac cells have a long refractory period, and a single contraction of cardiac muscle (twitch contraction) is completed before a second action potential can be generated. Thus, *cardiac cells cannot be tetanized*. This is a fortunate consequence in that tetany of cardiac muscle would obviously not be compatible with the function of the heart as a blood pump.

Repolarization of cardiac muscle cells involves activation of K$^+$ channels.

Outwardly rectifying potassium channels open later during phase 2 and during phase 3 of the cardiac muscle action potential. I$_{Ks}$ and I$_{Kr}$, which stand for slow and rapid outwardly rectifying potassium currents carried by these channels, intensify during phase 3 and bring about a rapid repolarization of the cell membrane. I$_{Ks}$ and I$_{Kr}$ channels are opened by depolarization and closed by membrane hyperpolarization starting at about −55 mV. The activation, or reopening, of I$_{K1}$ late in phase 3 is responsible for finishing the repolarization of the cell membrane and the reestablishment of the resting membrane potential. The increased intracellular calcium concentration from calcium influx during phase 2 appears to reactivate I$_{K1}$.

The sinoatrial node initiates and maintains the rhythm of electrical activation in the heart.

For the heart to function as an efficient pump, action potentials and subsequent myocardial contraction must be generated and spread through the myocardium in a regular, repetitive and organized manner. This will not occur if cardiac cells express their automaticity in a random, unpredictable fashion. Normally, prior to each contraction of the heart, cardiac electrical activity is initiated by a modified set of muscle cells on the posterior aspect of the right atrium at the junction of the superior and inferior vena cava, called the sinoatrial or SA node (Fig. 12.4). Once the SA node initiates an action potential, it travels through both atria at a rate of 0.1 to 1.0 m/s and coalesces at a second area of specialized

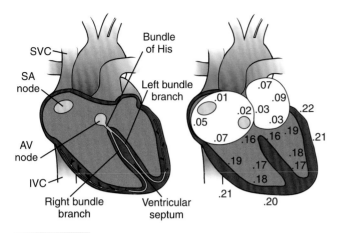

Figure 12.4 **The timing of the excitation of various areas of the heart (in seconds).** AV, atrioventricular; SA, sinoatrial; SVC, superior vena cava; IVC, inferior vena cava.

conduction tissue called the atrioventricular or AV node. This node lies at the junction between the atria and ventricles in the ventricular septum. Conduction through the AV node is slow (~0.05 m/s), which delays the movement of the cardiac action potential into the ventricles. This delay has the important effect of allowing more time for the ventricles to fill during diastole. The conduction of action potentials through the AV node shows directional preference; that is, action potentials travel more easily from the atria through the AV node toward the ventricles than in the opposite direction. The AV node is the only normal pathway by which action potentials from the atria travel into the ventricles because the connective tissue between the atria and ventricles and encircling the heart valves acts as an electrical insulator. Once the action potential emerges from the AV node, it enters the **bundle of His**, which splits into left and right bundle branches; these branches, in turn, give rise to Purkinje fibers, which line the entire endocardial surface of both ventricles. The ventricle itself is then activated in sequence from the septum/papillary muscle and endocardium to the epicardium and from the apex to the base of the heart. Action potentials travel so fast through Purkinje fibers (~4 m/s) that the underlying ventricular muscle is activated almost essentially simultaneously. Coupled with the syncytial nature of ventricular muscle cells, this ensures that the ventricular cells contract essentially *en masse* once activated, which allows for the effective pumping of blood out of the ventricles.

Unique recycling changes of ion conductances create automaticity and rhythmicity in cardiac nodal tissue.

Action potentials generated in the SA and AV nodes are smaller than those seen in cardiac muscle (Fig. 12.5). Nodal tissue does not contain fast voltage-gated Na^+ channels. The action potential is carried entirely by slow, L-type voltage-gated Ca^{2+} channels. Although these channels can carry small amounts of sodium current, the channel is conductive predominantly for calcium. The slow opening and closing of these channels with membrane depolarization create a phase 0 that is slower compared with that in ventricular cells. In addition, nodal action potentials start from more positive

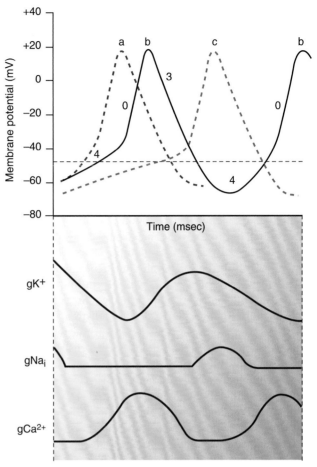

Figure 12.5 **Sinoatrial (SA) nodal potential as a function of time.** (*b*) Representation of a normal pacemaker potential. (*a*) Effect of norepinephrine. (*c*) Effect of acetylcholine. The light horizontal *dashed line* indicates threshold potential, which is set at about −48 mV in this example. The more rapidly rising phase 4 in the presence of norepinephrine (*a*) results from enhanced Na^+ permeability. The hyperpolarization and slower rise in phase 4 in the presence of acetylcholine (ACh) result from decreased Na^+ permeability and increased K^+ permeability, as a result of the opening of ACh-activated K^+ channels. For the purpose of illustration only, the cationic current changes responsible for the formation of the SA nodal potential are aligned temporally with the normal pacemaker potential in the figure. The pacemaker potential reaches threshold primarily through a slow decay of potassium conductance (gK^+) in the SA node. g, ionic conductance.

resting membrane potential (~ −65 mV), show no plateau, and exhibit a slow phase 3 as compared with that in cardiac muscle cells.

The most unique characteristic of action potentials in the SA and AV nodes is the spontaneous, progressive, and recycling depolarization that occurs in phase 4. This forms the basis for automaticity and rhythmicity in the SA node and cannot be considered a true "resting" potential. Shortly after phase 3, these cells experience a slight increase in Na^+ conductance (gNa^+_i), which results in a small inward depolarizing sodium current called $I_{Na,f}$ or "funny" sodium current. The depolarizing effects of this current tend to halt and reverse further hyperpolarization of the cell membrane (see Fig. 12.5). The sodium channel mediating this effect belongs to a class of cyclic nucleotide–gated channels called *hyperpolarization-activated cyclic nucleotide–gated* (more commonly known as *HCN*) *channels*. Defects in this channel have been associated with abnormal hyperpolarization of nodal tissue, which results in an abnormally slow or irregular heart rate. The $I_{Na,f}$ current, however, is not responsible by itself for the automaticity of nodal tissue. Throughout phase 4 of nodal action potentials, there is a progressive reduction in K⁺ conductance. This tends to progressively reduce K⁺ efflux from the cell, allowing it to slowly depolarize over time. This depolarization eventually begins to activate slow Ca⁺ channels, and the membrane becomes depolarized in a reinforcing manner that commonly results in an action potential. Depolarization later inactivates the Ca²⁺ channels, and the cell repolarizes as a result of K⁺ efflux. The decay of gK⁺ in phase 4 then occurs again, thus repeating the action potential cycle. The cycling decay of gK⁺ during phase 4 of the nodal action potential is the primary source of the automaticity of the SA node and its resulting rhythm.

The SA node possesses the highest intrinsic rate of spontaneous action potential generation of the specialized conduction tissues. The true intrinsic rate of the SA node is about 100 impulses/minute. However, the node rate is modified by innervation from the ANS. With a normal, intact ANS, the node's intrinsic rate is suppressed to about 70 to 80 impulses/min (see Chapter 17). This physiological resting rate is still greater than the intrinsic rates of the AV node (~50 impulses/min) or the Purkinje fibers (<20 impulses/min). Consequently, the SA node continually activates the heart before any other cardiac tissue can generate its own action potential. For this reason, the SA node is often called the heart's "pacemaker," because its rate of firing determines the **heart rate** or number of contractions per minute of the heart as a whole.

Both branches of the ANS affect phase 4 of the action potentials in the nodal tissue. ACh from the vagus nerve, which innervates the SA node, slows the heart rate. ACh enhances K⁺ conductance in nodal tissue and hyperpolarizes the resting membrane potential while also decreasing the slope in phase 4. Consequently, it takes longer for the cells to spontaneously reach a threshold and fewer action potentials are generated in a given period of time. Conversely, norepinephrine or activation of the sympathetic nervous system increases cation conductance in nodal tissue (primarily sodium) and increases the slope in phase 4 of the SA node thereby increasing the heart rate (see Fig. 12.5).

Although the AV node can generate its own repetitive action potentials, it does not do so unless the intrinsic activity of the SA node is severely suppressed. This can happen with excess activation of the vagus nerve to the SA node, administration of drugs that increase ACh at the SA node, or in certain disease states. When this occurs, the heart does not stop but instead becomes paced by the AV node. In this situation, the AV node is considered to be a **secondary pacemaker**. Although the heart beats rhythmically when paced by the AV node, it does so at a substantially reduced rate compared with that seen with the SA node as the heart's pacemaker. In extreme conditions, when both the SA and AV nodes are not functional, the Purkinje fibers will pace the heart. However,

the intrinsic Purkinje pacing is so slow that the flow output of the heart is insufficient for little more than keeping a person alive; normal activity is not possible. Patients in such conditions require pacing of the heart at a higher, controlled, rate by an implantable artificial pacemaker device.

Action potential conduction velocity through the myocardium is proportional to the amplitude and phase 0 upstroke of the cardiac action potential.

The conduction of action potentials through myocardial cells is affected by the characteristics of the action potentials themselves. Conduction velocity is increased when the amplitude of the action potential is increased. This increase in amplitude may occur as a result of factors that enhance Na⁺ and Ca²⁺ influx into the cell. Alternately, it can also occur by hyperpolarization of the resting membrane potential, which results in a larger amplitude action potential once the cell is activated. Hypokalemia and increased K⁺ conductance are two such mechanisms that can lead to hyperpolarization of myocardial muscle cells. Conduction of action potentials through the myocardium is also increased whenever the rate of depolarization in phase 0 is increased (increased dV/dt). This increase in depolarization amplitude and rate of rise will happen with any factor that increases the electrochemical gradient for Na⁺ in phase 0, such as cell membrane hyperpolarization or an increase in the ratio of extracellular to intracellular Na⁺ concentration. Conversely, partial depolarization of the cell membrane at rest will reduce the size and rate of rise of subsequent cardiac muscle action potentials, which in turn will slow the conduction of these action potentials from cell to cell through the myocardium. If myocardial cells become depolarized to the extent that their fast sodium channels cannot reset from the inactive state, then these cells will generate action potentials from the activation of the slow Ca²⁺ channels in the membrane instead. However, these action potentials will exhibit a slow rate of depolarization in phase 0 and small amplitude. As such, they will be conducted slowly through the myocardial muscle tissue.

Conduction specifically through the AV node is affected by normal physiologic phenomena. ACh from the vagus nerve decreases, whereas norepinephrine from sympathetic nerves enhances conduction velocity through the AV node. This helps the activation of the ventricles to match the pace set by the rate in the SA node. In addition, conduction through the AV node is sensitive to repetitive stimulation. Continuous stimulation at high rates results in an increased refractoriness of the nodal tissue. This helps prevent the ventricles from being driven at abnormally high rates, which can impinge on ventricular diastolic filling and hence the output of the whole heart.

▶ PATHOPHYSIOLOGY OF ABNORMAL GENERATION OF CARDIAC ACTION POTENTIALS

The efficient function of the heart as a blood pump capable of meeting any metabolic need of the body is totally dependent on generation of an appropriate, regular frequency of activation (i.e., beats per minute) and proper conduction of each individual impulse in normal sequence through the

Mechanism of Ventricular Reentry Tachycardias

Abnormal rhythms of the heart are characteristic of many pathologic conditions in the heart and by themselves can pose serious health risks to a patient. However, cardiac arrhythmias can be especially problematic when they cause the ventricles to be paced at extremely high rates. Starting at heart rates of about 180 beats/min and above, the time available for the filling of the heart during diastole becomes so compromised that even though the heart is beating many times per minute, its output is reduced as a result of decreased filling time. At this level, the faster the heart rate is, the lower is the output of the heart. It is not uncommon for individuals with ventricular tachycardia to have heart rates in excess of 250 beats/min.

In some cases, myocardial ischemia and injury can induce severe ventricular arrhythmias, called **reentry tachycardias**. Following myocardial infarction, dead zones of myocardium exist within the heart. These zones cannot conduct electrical impulses. However, because of the syncytial nature of the myocardium, electrical impulses simply flow around the dead zone and activate the remaining healthy myocardium. However, during conditions of ischemia and ischemic injury, ischemic portions of the myocardium will allow action potentials to proceed through them in one direction but not in the opposite direction. This is called *unidirectional blockade*. If action potentials are able to leak through this damaged area and emerge in the healthy myocardium after that area is past its refractory period, this "reentry" action potential will reactivate the healthy myocardium, sending another action potential generation through the injured tissue path. This

process will then repeat again and again in an endless circle. This condition is called a *circus rhythm* and, as impulses generated from this circus cycle radiate outward through the syncytium, they activate the heart as a whole, at the rate set by the circus rhythm. Such circuits thus become secondary pacemakers. They may involve a few myocardial cells or large portions of the myocardium and often end up pacing the heart at an abnormally high rate (>250 beats/min). The AV node, because it is naturally predisposed to unidirectional conduction, is a common site of such reentry problems.

Although the presence of a unidirectional conduction block in the myocardium is necessary for the creation of reentry tachycardias, it is not, by itself, sufficient to set up a circus rhythm. The area of myocardium that receives action potentials from the reentry circuit as they emerge from the injured tissue zone must be over their refractory period for the cycle to be initiated anew. Consequently, *any* condition that slows but does not block the conduction of action potentials through the myocardium or shortens the refractory period of healthy cardiac cells will increase the probability of creating ventricular reentry tachycardias. Furthermore, should these factors enhance automaticity in cardiac cells, there is a good chance that such tachycardias will be initiated. Unfortunately, myocardial ischemia, by causing partial membrane depolarization, brings cells closer to their activation threshold and creates action potentials of small amplitude, which are conducted slowly through the myocardium, thus enhancing the probability of creating reentry tachycardias. ∎

myocardium. Any deviation in the origin of an action potential from a site other than the SA node and any abnormal deviation in heart rate, its regularity, or its conduction path can result in poor mechanical performance of the heart as a blood pump. Some such deviations can be life threatening. In the following section, we will examine some of the electrophysiological mechanisms responsible for the generation of abnormal action potentials in the myocardium. Abnormal rhythm mechanisms will be included in the section of this chapter devoted to the electrocardiogram (ECG).

Abnormal pacemaker sites can appear anywhere in the heart in response to factors that cause partial depolarization or shortened refractory periods in cardiac cells.

Certain pathologic conditions, such as myocardial ischemia, scar tissue formation after myocardial infarction, or alterations of myocardial electrolyte handling, can result in action potentials being generated in areas of the myocardium other than the SA node. These areas are called **ectopic foci**. Because cardiac muscle is an electrical syncytium, any cell or group of cells that act as ectopic foci can activate the rest of the myocardium. This in turn will result in extra, abnormal, activations of myocardial contraction. In some cases, these ectopic foci fire only occasionally, but they can fire repetitively and

at high intrinsic rhythms in a manner that impairs the performance of the heart as a pump. Problems occur when such action potentials are generated in the myocardium in a random fashion or when the myocardium becomes paced by the ectopic foci at rates too high to allow proper filling of the ventricular chambers with blood (**ventricular tachycardias**).

Ischemic conditions predispose the heart to the formation of ectopic foci. Ischemia inhibits Na^+/K^+ pump activity, resulting in intracellular accumulation of Na^+ and Ca^{2+} and partial membrane depolarization. This places the resting membrane potential in the affected cells closer to their threshold for the activation of action potentials. In this state, small additional depolarizing stimuli, which would be insufficient to push normal cells to their threshold, will push the membrane potential of affected cells beyond their threshold, causing them to fire and activate the rest of the heart. Hyperkalemia, hypercalcemia, distention of the heart chambers (as occurs in congestive heart failure and hypertension), clinical treatment of chronic heart failure with cardiac glycosides (which partially inhibits the Na^+/K^+ pump), all can partially depolarize the resting membrane potential and predisposes affected cells to become ectopic foci. Any factor that reduces potassium conductance at rest or increases intracellular calcium concentration can also partially depolarize cardiac cells (e.g., catecholamines, see Chapter 13). Fatigue and emotional or physical stress predispose a person to the formation of ectopic foci because such states activate the sympathetic nervous system and release catecholamines onto the heart. The probability of the formation of ectopic foci in the heart is also increased by an individual's use of caffeine or nicotine, the former which increases intracellular calcium concentration in myocardial cells through multiple mechanisms. Besides partial depolarizing factors, agents that shorten the refractory period of cardiac cells allows affected cells increased time to be reactivated by any depolarizing stimulus and thus increase the probability of ectopic foci occurring in the heart. Catecholamines, which increase gK^+ in phase 3 of the action potential, and a class of drugs called calcium channel blockers, used to treat angina and hypertension, are examples of agents shortening the refractory period of myocardial cells.

In addition to the ion channels responsible for myocardial and nodal action potentials, heart muscle cells also contain a group of non–voltage-gated K^+ channels called inwardly rectifying (IR) K^+ channels, because they pass the inward current of K^+ more easily than the outward currents. However, this is simply an electrophysiologic classification in that there are never any inwardly directed potassium currents during the cardiac action potential (the cell membrane never reaches the potassium reversal potential of -90 to -100 mV). It is the outward current through these channels that is physiologically important. $I_{K, ATP}$ potassium channels are inactivated by intracellular adenosine triphosphate (ATP) and activated by adenosine diphosphate. They are thought to be one link between cardiac metabolism and membrane potential. These channels may be responsible for the reduced refractory period seen during myocardial ischemia. Other IR channels, such as I_K, ACh, and $I_{K, ado}$ (ACh, acetylcholine; ado, adenosine), are ligand-gated potassium channels that hyperpolarize the ventricular muscle cell. These channels may mediate cholinergic-mediated atrial arrhythmias (abnormally slow heart rates) as well as certain antiarrhythmic effects of adenosine, which is used to treat certain arrhythmic conditions.

Triggered activity

Triggered activity is a type of ventricular arrhythmia in which the abnormal beats are not generated *de novo* from ectopic foci. Instead, such activity follows a previous action potential or action potentials, which "triggers" an abnormal depolarization of ventricular muscle before the prior action potential is complete. This type of arrhythmia is most likely to occur in ventricular muscle cells and can result in individual **premature ventricular contractions** (**PVCs**) or, when repeated, more detrimental sustained tachycardias. Such conditions in the heart are difficult to terminate as one abnormal action potential triggers another and another, etc. Triggered activity is not self-activating, but it is thus considered self-sustaining.

Triggered activity is classified by the timing of the appearance of the abnormal depolarization. **Early afterdepolarizations** (**EADs**) are depolarizations of the muscle cells that occur late in phase 2 or early in phase 3 of the ventricular action potential. They are more likely to be induced by factors that prolong the action potential duration, such that slow calcium channels have time to recover and, thus, fire an additional low-amplitude action potential. Consequently,

bradycardias favor the formation of EADs as do factors that impair outward potassium current in phase 3 of the ventricular action potential. EADs are also more likely to occur with excessive stretch of the ventricular chambers, such as that which occurs with congestive heart failure.

Delayed afterdepolarizations (**DADs**) occur late in phase 3 or during phase 4 of the ventricular action potential. DADs are associated with increased intracellular calcium accumulation and concentration within the myocytes, which can thus depolarize the cell membrane to the threshold for an action potential. DAD formation is favored by tachycardias. Tachycardias increase the number of times per minute calcium channels open during phase 2 and thus can bring calcium into the cell faster than it can be removed, allowing its concentration to rise over time. Agents such as norepinephrine, which enhances calcium entry into myocardial cells, as well as cardiac glycosides, which reduce intracellular calcium removal, favor DAD formation.

▶ THE ELECTROCARDIOGRAM

When activated, the heart is a concentrated locus of time-varying electrical potentials in the body. When a portion of the myocardium becomes depolarized from an action potential, its polarity is temporarily reversed, becoming positive on the inside and negative on the outside relative to the neighboring inactivated tissue. When this reversal occurs, it temporarily creates two neighboring regions of opposite charge, or polarity, within the myocardium (Fig. 12.6). This difference in polarity between two locations is called a **dipole**. Electrical currents readily flow from one pole of a dipole to the other through any media between the poles that can conduct them. The intracellular and extracellular fluids in the body are largely composed of electrolyte solution, which is a good conductor of electricity. Thus, any dipole, formed at any time and in any direction within the myocardium, is transmitted through the body as a current between the ends of the dipole. This current then radiates outward through the body to the surface of the skin where it can be detected.

An electrocardiogram (ECG) is an amplified, timed recording of the electrical activity of the heart, as detected on the surface of the body. The recording gives a plot of voltage as a function of time. It results from the composite effect of all the different types of action potentials generated in the myocardium during activation and the resulting magnitude and orientation of the dipoles created by them. Although it is correct to say that the electrical activity in the heart is responsible for creating the ECG, the physician looks at this process in reverse; that is, the physician examines the ECG to create a picture of the electrical activity in the heart.

The ECG is one of the most useful diagnostic tools available in medicine. However, it is important to understand what information can and cannot be gained from the analysis of an ECG. The ECG can be used to detect abnormalities in heart rhythm and conduction, myocardial ischemia and infarction, plasma electrolyte imbalances, and effects of numerous drugs. One can also gain information from the ECG about the anatomic orientation of the heart, the size of the atria and ventricles, and the path taken by action potentials through the heart

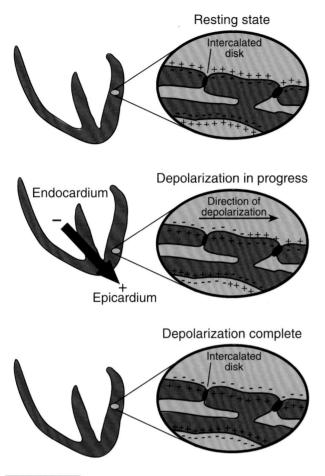

Resting state

Depolarization in progress

Depolarization complete

Figure 12.6 **Example of a cardiac dipole.** Partially depolarized myocardium creates a dipole. *Arrows* show the direction of the net dipole caused by depolarization of a portion of the myocardium. Dipoles are present only when a portion of the myocardium is in the process of depolarization or repolarization while other portions are not. They are not formed when the entire myocardium is depolarized or repolarized.

during normal or abnormal activation (e.g., the average direction of activation of the ventricles). The ECG, however, cannot give *direct* information about the contractile performance of the heart, which is equally important in the evaluation of myocardial status in a clinical setting. Other tools must be used for such an evaluation, and these will be discussed in later chapters.

The normal ECG depicts cyclic electrical activity and conduction in the atria and ventricles.

It is useful to recognize that certain well-known elements of the ECG represent key events in the electrical activation of the heart even though a detailed explanation of how all the myocardial action potentials during activation of the heart end up creating the common ECG is complex. The basic normal ECG is represented in Figure 12.7. Standard ECGs are run at a recording speed of 25 mm/s with 10 mm of vertical amplitude set to equal 10 mV. The first sign of electrical activity in the heart as revealed by the ECG is a small, rounded, upward (positive) deflection called a **P wave**. The P wave is caused by depolarization of the atria (not just the SA node). After a short interval, a complex, short-duration, high-amplitude, spike-like potential is observed. This

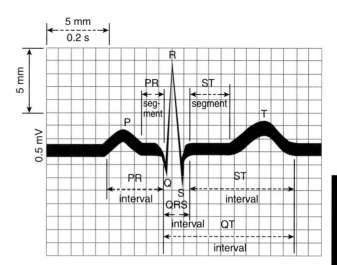

Figure 12.7 **The major waveforms and intervals associated with the normal electrocardiogram (ECG).** P, P wave; Q, Q wave. R, R wave; S, S wave; T, T wave.

potential is called the **QRS complex** and is caused by depolarization of the ventricles. By definition, within this complex, the first downward deflection after the P wave is called a **Q wave**, the next upward deflection is called an **R wave**, and the next subsequent downward deflection is called an **S wave**. However, depending on the location of the ECG recording on the body, the Q and S deflections may not be seen, and ventricular depolarization may appear only as an R wave on the ECG. Nevertheless, such a wave is still often called the QRS complex. Following the QRS complex, the entire ventricular mass is depolarized and thus there are no potential differences (dipoles) between areas of the myocardium. For this reason, no deflections are registered on the ECG, and the ECG is said to be **isoelectric** or at zero potential. Next, ventricular repolarization produces a broad wave of modest upward amplitude called a **T wave**. Repolarization of the atria is not seen in a typical ECG because it occurs during the same interval as the QRS complex and is lost in that signal.

Intervals between waves in the ECG are of physiologic and clinical importance. The **PR interval** is the time from the beginning of the P wave to the start of the QRS complex and represents the amount of time the action potential takes to travel from the SA node through the AV node. The PR interval typically lasts 0.12 to 0.20 seconds. Because most of this time represents the delay of the action potential conduction through the AV node, clinically significant inhibition of conduction through the AV node is often reflected as a lengthening of the PR interval. The **QRS interval** represents the interval of time that the action potential takes to travel from the end of the AV node through the ventricles (normally 0.06 to 0.1 seconds). The duration of the QRS complex is roughly equivalent to the duration of the P wave, despite the much greater muscle mass of the ventricles, because of the very rapid, synchronous excitation of the ventricles. The normal conduction pathway through the bundle branches, Purkinje fibers, and ventricular muscle is the most efficient and rapid mode of action potential travel. Therefore, any pathway other than this, which would thus be considered abnormal, will take longer and be reflected on the ECG as an abnormally long QRS interval.

The time between the initiation of the QRS complex and the end of the T wave is called the **QT interval**. If the ventricular action potential and QT interval are compared, the QRS complex corresponds to initial ventricular depolarization, the ST segment corresponds to the plateau phase, and the T wave corresponds to repolarization. The relationship between a single ventricular action potential and the events of the QT interval is approximate because the events of the QT interval represent the combined influence of all of the ventricular action potentials. Regardless, the QT interval measures the total duration of ventricular activation. The QT interval is altered by drugs or conditions that alter the rate of myocardial repolarization in phase 3 (e.g., altered K⁺ conductance), but it is also naturally inversely proportional to the heart rate. When corrected for existing heart rate, abnormal QT intervals indicate problems with repolarization of the ventricles. For example, if ventricular repolarization is

delayed, the QT interval is prolonged. Delayed repolarization is associated with the genesis of ventricular arrhythmias. Therefore, when corrected for the existing heart rate, the presence of a prolonged QT interval on the ECG is considered an ominous sign requiring medical attention.

Moment-to-moment orientation and magnitude of net dipoles in the heart determine the formation of the ECG.

The formation of the standard waveforms within the ECG can be explained as arising from the orientation and magnitude of the net, or collective average, of all the dipoles that are created throughout the heart during the electrical activation of the myocardium. To explain further, consider the voltage changes produced in which the body serves as a volume conductor and the heart generates a collection of changing dipoles (Fig. 12.8). In this example, an electrocardiographic

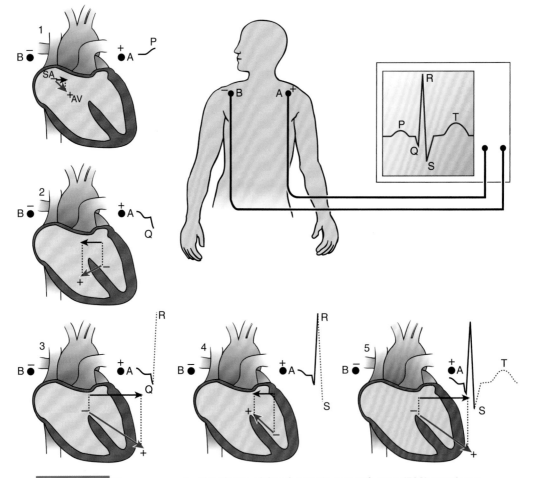

Figure 12.8 **The sequence of major dipoles giving rise to electrocardiogram (ECG) waveforms.** The *black arrows* are projected vectors that represent the magnitude and direction of a major dipole in the myocardium, which is depicted as a *red arrow* in the diagram. The magnitude is proportional to the mass of myocardium involved. The direction is determined by the orientation of depolarized and polarized regions of the myocardium. The vertical dashed lines project the vector onto the *A–B* coordinate (analogous to lead I); it is this component of the vector (represented here by the *black arrow*) that is sensed and recorded on the ECG. In panel 5, the last areas of the ventricles to depolarize are the first to repolarize (i.e., repolarization proceeds in a direction opposite to that of depolarization). The projection of the vector (*black arrow*) for repolarization points to the more positive electrode (*A*) as opposed to the less positive electrode (*B*) and so an upward deflection is recorded on this lead. AV, atrioventricular; SA, sinoatrial.

recorder is connected between points A and B such that point A is positive relative to point B. In this alignment, when the positive pole of a dipole is directed primarily more toward point A than point B, the ECG is deflected upward, and when it is primarily pointed toward point B, it is deflected downward.

In Figure 12.8, the red arrows show (in two dimensions, projected onto the frontal plane of the body) the direction of the *net* dipole resulting from the many individual dipoles present at that interval of time. The length of the red arrow is proportional to the magnitude (voltage) of the net dipole, which is related to the mass of myocardium from which it is generated. The black arrows show the magnitude of the dipole component that is projected onto or "parallel" to the imaginary line between points A and B (the points of recording electrodes). Importantly, it is this projected component that determines the amplitude and polarity of voltage that will be recorded on the ECG. As indicated in panel 1 of Figure 12.8, atrial excitation results from a wave of depolarization that originates in the SA node and spreads over the atria. The net dipole generated by this excitation has a magnitude proportional to the mass of the atrial muscle involved and a direction indicated by the red arrow. The head of the arrow points toward the positive end of the dipole, where the atrial muscle is not yet depolarized. The negative end of the dipole is located at the tail of the arrow, where depolarization has already occurred. Because point A is positive relative to point B on the recording electrodes, there will be an upward deflection of the ECG. The magnitude of this upward deflection depends on two factors: (1) the amount of tissue generating the dipole (with amplitude being proportional to the tissue mass involved) and (2) the orientation of the dipole relative to a parallel line connecting points A and B (i.e., black arrow projected in parallel onto the imaginary line connecting A and B. This latter relationship is demonstrated in Figure 12.9.

Imagine a net dipole in the atria muscle pointing directly along the line connecting point B to point A with its positive end pointing directly at point A. This depolarization will then create a positive deflection as described above. For the sake of example only, we shall assign this deflection an amplitude of +4 mm on the ECG recorder. Should this same depolarization, however, proceed directly from point A toward point B, with the positive end pointing directly at point B, a downward deflection of the same amplitude would result (i.e., a 4-mm deflection, but in the negative direction or, −4 mm). The amplitude of the deflection will thus vary in this example between −4 mm and +4 mm, depending on the angle of the net dipole relative to the line connecting A and B. Should, instead, the net dipole point toward A at a 45° angle, the deflection would be a positive 2 mm; if it points at 90° (perpendicular) to the line connecting A and B, it would not be pointing at either pole and no deflection would be recorded on the ECG. (Note, this is not the only way, however, in which the deflection will register zero. Once the atria are completely depolarized, no voltage difference, or dipole, will exist between A and B and thus the ECG would record zero deflection during that time).

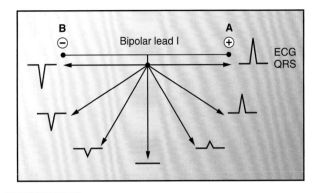

Figure 12.9 **A representation of how the orientation of a dipole in the ventricle relative to a bipolar lead affects the magnitude and polarity of a hypothetical deflection on the electrocardiogram (ECG).** Positive ends of the dipole pointed to the positive pole of the lead create positive deflections in the complex. Dipoles pointed directly to the positive pole, parallel to the lead, create the most positive deflection waves. Those pointed directly to the negative pole create the most negative deflection. Dipoles oriented perpendicularly to the lead have no component pointed to either pole of the lead and thus register no deflection of the ECG. QRS, QRS complex and interval.

Although the preceding discussion is an oversimplification, it presents the basic principles that create the pattern of the common ECG. For example, after the P wave, the ECG returns to its baseline or isoelectric level. During this time, the wave of depolarization moves through the AV node, the AV bundle, the bundle branches, and the Purkinje system. The dipoles created by the depolarization of these structures are too small to produce a deflection on the ECG. However, the depolarization of ventricular structures does create deflections on the ECG. The net dipole that results from the initial depolarization of the septum is shown in panel 2 of Figure 12.8. This depolarization is pointed toward point B and away from point A because the left side of the septum depolarizes before the right side. This orientation creates a small downward deflection produced on the ECG, called the Q wave although in many instances, the normal Q wave is so small that it is not apparent. Next, the wave of depolarization spreads via the Purkinje system across the inside surface of the free walls of the ventricles. Depolarization of free-wall ventricular muscle proceeds from the innermost layers of muscle (subendocardium) to the outermost layers (subepicardium). Because the muscle mass of the left ventricle is much greater than that of the right ventricle, the net dipole during this phase has the direction indicated in panel 3. The deflection of the ECG is upward because the dipole is directed at point A and it is large because of the great mass of tissue involved. This upward deflection is the R wave. The last portions of the ventricle to depolarize generate a net dipole with the direction shown in panel 4, and thus, the deflection on the ECG is downward. This final deflection is the S wave. The ECG tracing returns to baseline when all of the ventricular muscle becomes depolarized and all dipoles associated with ventricular depolarization disappear. The ST segment, or the period between the end of the S wave and the beginning of the T wave, is generally isoelectric. This indicates that no dipoles large enough to influence the ECG exist because all ventricular muscle is depolarized (the action potentials of all ventricular cells are in phase 2).

Repolarization, like depolarization, generates a dipole because the voltage of the depolarized area is different from that of the repolarized areas. The dipole associated with atrial repolarization does not appear as a separate deflection on the ECG because it generates a low voltage and because it is masked by the much larger QRS complex, which is present at the same time. Ventricular repolarization is not as orderly as ventricular depolarization. The duration of ventricular action potentials is longer in subendocardial myocardium than in subepicardial myocardium. The longer duration of subendocardial action potentials means that even though subendocardial cells were the first to depolarize, they are the last to repolarize. Because subepicardial cells repolarize first, the subepicardium is positive (outside) relative to the subendocardium; that is, the polarity of the net dipole of repolarization is the same as the polarity of the dipole of depolarization. This results in an upward deflection because, as in depolarization, point A is positive with respect to point B. This deflection is the T wave (see panel 5, Fig. 12.8). The T wave has a longer duration than the QRS complex because repolarization does not proceed as a synchronized, propagated wave. Instead, the timing of repolarization is a function of the properties of individual cells, such as the number of particular K$^+$ channels.

ECG evaluation is standardized by the use of a designated 12-lead system.

There are two broad classes of evaluations performed using the ECG. One of these involves the evaluation of abnormalities in the basic ECG form. Arrhythmias, conduction abnormalities, electrolyte disturbances, drug effects, and myocardial metabolic disorders such as ischemia can be detected as abnormal patterns from a single ECG electrode placed at most any location on the surface of the body. However, as explained in the previous section, the amplitude and, in some cases, the polarity of various waveforms within the ECG depend in part on the orientation of the net dipoles generated in the heart relative to the position of the electrode system measuring the ECG. Clearly then, the same electrical event in the heart will appear differently if it is viewed by two different electrode systems with different orientations relative to the event. For this reason, a standard system consisting of 12 specifically located "leads" or electrode connections is used in the clinical application of ECG recordings in patients. This system provides two benefits. First, it provides a standard framework for identifying patterns in the ECG, thus making the recognition of abnormalities easier and more consistent. Second, it enables an investigator to see the electrical activity in the heart from many different "views and angles" at any given point in time. Analysis of these different "views" is used to gain information about the orientation of the heart, the size of its chambers, and the general direction of activation in the myocardium during any interval.

Imaging cardiac electrical activity in the frontal plane uses six standardized frontal limb leads.

The first ECG lead system was developed by Willem Einthoven and was based on the idea that the heart sits at the center of a triangle in the frontal plane of the body, with vertices at the right shoulder, left shoulder, and pubic region (Fig. 12.10A). This triangle is called the **Einthoven triangle**, and it is the simplest geometry in which straight lines can completely encircle the heart in the frontal plane (i.e., 360°). In this system, the arms and the legs are considered extensions of the vertices, and ECG electrodes are placed on the right arm, left arm, and left leg, with the right leg serving as an electrical ground. This creates a series of connections, or leads, between each pair of electrodes. These leads are bipolar, meaning each has a positive and a negative pole. In lead I, the right arm is negative and the left arm is positive. In lead II, the right arm is negative and the left leg is positive. Finally, in lead III, the left arm is negative and the left leg positive. If one imagines the frontal plane of a person as being represented by a 360° circle, lead I has its positive pole placed at "3 o'clock" on the circle and is designated by convention as 0°. The positive pole of lead II is at 60°, and that for lead III is at 120°, as shown in Figure 12.10B. This lead system is often called the **standard frontal lead system**.

Another lead system is arranged in the frontal plane of the body and consists of three unipolar leads created by special connections between the same electrodes placed on the arms and legs for the creation of the standard frontal lead system. These leads are single positive poles relative to the center of the chest, which is considered zero potential. These leads are often called **augmented leads** because the ECG recording device amplifies the signal from these leads relative to those obtained in the bipolar leads (see Fig. 12.10B). The pole of one lead sits at the right shoulder at 210° and is called the *augmented right*, or *aVR*, lead; another sits at the left shoulder at 330° and is called *augmented left* or *aVL*. The final augmented lead is oriented at 90° in the frontal plane and is called the *augmented foot* or *aVF lead*. The augmented and standard limb leads give a two-dimensional picture of the electrical activity of the heart as it would be viewed, or projected, onto the frontal plane of the body.

Imaging cardiac electrical activity in the horizontal plane uses six standardized chest leads.

In a typical 12-lead ECG system, a series of six unipolar leads is arranged in a horizontal plane around the chest as shown in Figure 12.11. These are sometimes called **precordial or chest leads** and are designated V$_1$ to V$_6$, from just to the right of the sternum in the 4th intercostal space to the axial line in the 5th intercostal space. These leads give a two-dimensional picture of the electrical activity in the heart as it would be viewed from above or below a horizontal plane bisecting the heart. In addition, because these leads lie so close to the surface of the heart, individual precordial leads give detailed information about the electrical activity in the small, specific portions of the heart that lie beneath each electrode.

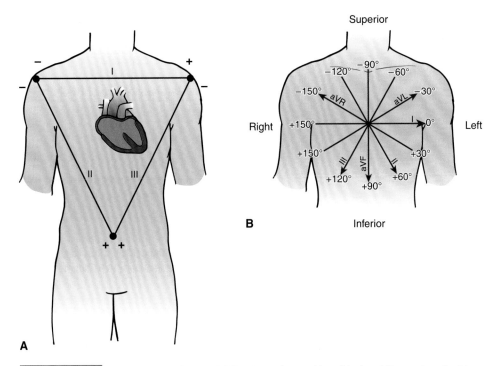

Figure 12.10 **The Einthoven triangle. (A)** The heart is considered to be at the center of a triangle, each corner of which serves as the location for electrodes that send signals as a positive and negative pole pair to the electrocardiogram recorder. The three resulting bipolar leads are designated I, II, and III. By convention, positive dipoles generated within the myocardium that are pointed in the direction of the positive pole of any of the leads create a positive deflection that is recorded from that lead; when pointed at the negative pole, a negative deflection is recorded. **(B)** The hexaxial reference system for the standard bipolar and augmented unipolar chest leads. The limb leads give information on cardiac dipole vectors in the frontal plane and are referenced by the angle, in degrees, in which the vector points as if viewed in two dimensions in the frontal plane. aVF, augmented foot lead; aVL, augmented left lead; aVR, augmented right lead.

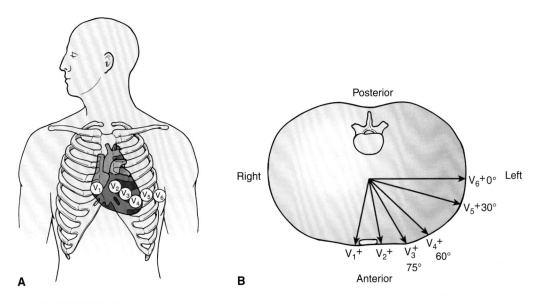

Figure 12.11 **Unipolar chest leads. (A)** V_1 is just to the right of the sternum in the 4th intercostal space. V_2 is just to the left of the sternum in the 4th intercostal space. V_4 is in the 5th interspace in the midclavicular line. V_3 is positioned midway between V_2 and V_4. V_5 is in the 5th intercostal space in the anterior axillary line. V_6 is in the 5th intercostal space in the midaxillary line. The reference, or zero, voltage for the unipolar chest leads is electronically combined out of the three limb leads. **(B)** The orientation of the unipolar chest leads in the horizontal plane. The chest leads give information on cardiac dipole vectors as if viewed in two dimensions in the horizontal plane.

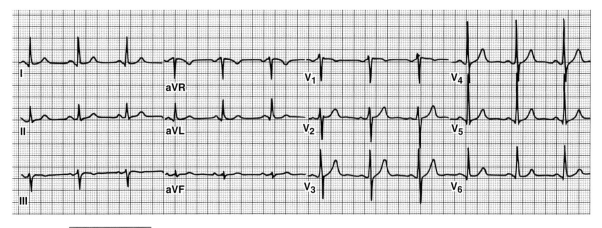

Figure 12.12 **The clinical standard 12-lead electrocardiogram.** Six limb leads and six chest leads are shown. Two dark horizontal lines (10 mm) are calibrated to be 1 mV. Dark vertical lines represent 0.2 seconds. aVF, augmented foot lead; aVL, augmented left lead; aVR, augmented right lead.

Typical tracings from the six frontal and six horizontal lead systems are shown in Figure 12.12. In general, clinical practice, all 12 leads are recorded in a patient at the same time. Because the chart speed for all ECG recordings is standardized at 25 mm/s, this affords an easy means of converting intervals between two points on the ECG into an estimate of heart rate in beats/minute (bpm). For example, R waves that are uniformly separated by 25 mm indicate that the heart is beating at 60 (bpm), 20 mm = 75 bpm, 15 mm = 100 bpm, 10 mm = 150 bpm, and 5 mm = 300 bpm.

Figure 12.12 shows that a P wave is always followed by a QRS complex of uniform shape and size. The PR interval is 0.16 seconds (normal 0.10 to 0.20 seconds). This measurement indicates that the conduction velocity of the action potential from the SA node to the ventricular muscle is normal. The average time between R waves (successive heart beats) is about 0.84 seconds, making the heart rate ~71 beats/min.

Information about the orientation of the heart, ventricular size, and conduction pathways can be obtained through the frontal QRS vector.

As explained above, changes in the magnitude and direction of any momentary cardiac dipole will cause changes in a given ECG lead, as determined by the orientation of the dipole relative to the orientation of the specific lead. Theoretically, by comparing the magnitude of any portion of an ECG simultaneously in several leads, one could work backward and determine what the orientation of the net dipole was in the heart at that moment that gave rise to those magnitudes in the different leads. In other words, by comparing, for example, the magnitude of positive deflections at one point of the ECG in leads I, II, and III at the same time, one could "triangulate in" and get an estimate of the direction of the net dipole in the heart at that precise moment. That direction could then be represented by a vector on the hexaxial reference system shown in Figure 12.10B.

Hypothetically, one could construct a set of vectors resulting from the net dipoles produced in the myocardium during every millisecond of cardiac activation. However, this is impractical and is not necessary to glean important clinical information from the ECG. Instead, what is done is to focus on a major cardiac event that is represented within the ECG, construct a general average vector for that event, and then plot its vector in the frontal or horizontal plane as desired. The most common event that is analyzed this way from the ECG is ventricular depolarization and its associated ECG QRS complex. In broad terms, the overall direction of depolarization through the myocardium is from the right atrium to the left atrium, down through the AV node, and on into the ventricles. The heart in most people sits in the chest at an angle with the apex pointing toward the lower left portions of the lungs. Thus, the long axis of the ventricles from base to apex is roughly a line pointed downward from the area of the right shoulder to the lower left side of the chest. Therefore, if one could "see" the *average* grand trek of depolarization through the ventricles projected in two dimensions onto a flat screen in the frontal plane of the body, it would appear as a vector proceeding from the upper right of the ventricles down toward the heart apex at an angle of ~ +60° on the hexaxial reference system. This average vector is called the **mean electrical QRS axis**, or mean QRS vector. Using the standard bipolar limb leads, the Einthoven triangle, and the hexaxial reference system for orientation, the *average* deflections of the QRS complexes giving rise to this mean vector would be most positive in lead II and least positive in lead III and in between these two magnitudes in lead I. (In fact, the axiom that I + III = II for average QRS amplitudes always applies and can be used as a handy check for proper connection of the frontal ECG leads.)

Calculating the average deflection for the mean QRS vector is simplified for the purpose of clinical analysis. Basically, a line along the isoelectric line of the ECG from a given lead (e.g., lead I) is drawn through the QRS complex, and the amplitude of the peak of the R wave is measured. The peak of any negative deflections in the QRS complex below the isoelectric line (usually the sum of the negative peaks of the Q and S waves) is subtracted from the R wave peak to give the average QRS amplitude in that lead. This value is then plotted in the appropriate polarity direction along the line on a hexaxial plot system that corresponds to the lead from which the ECG was measured (see Fig. 12.10). A perpendicular line is then dropped from that point. This same procedure is repeated for at least one other lead in the same plane (e.g., either lead II or lead III in this example). A line drawn from the center of the hexaxial system to the point of intersection of the perpendiculars represents the mean QRS vector in the heart at that moment in time.

An example of the application of mean QRS vector determination is shown in Figure 12.13, which uses the ECG tracings in Figure 12.12. For this example, the Einthoven triangle is used to calculate the vector. The net magnitude of the QRS complex in leads I, II, and III is measured and plotted on the appropriate axis. A perpendicular line is dropped from each of the plotted points and a vector drawn between the center of the triangle and the intersection of the perpendicular lines. This gives a mean QRS axis of +3°. Another simple

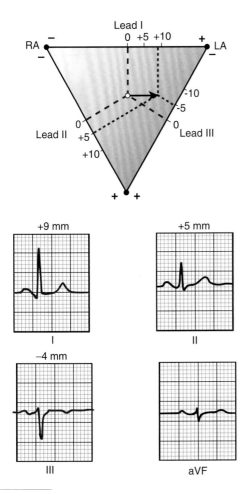

Figure 12.13 **An example of the determination of the mean QRS axis.** This axis can be estimated by using the Einthoven triangle and the net voltage of the QRS complex in any two of the bipolar limb leads. It can also be estimated by inspection of the six limb leads and select the lead most closely yielding a net voltage of zero (see text for details). Electrocardiogram tracings are from Figure 12.12 and are enlarged to facilitate viewing. aVF, augmented foot; LA, left arm; LL, left leg; RA, right arm.

way of estimating the mean QRS vector in this example is to simply examine the QRS complex in the six frontal limb leads and select the lead in which the average QRS complex deflection is closest to zero. As discussed earlier, when the cardiac dipole is perpendicular to a particular lead, the net deflection is zero. Once the net QRS deflection closest to zero is identified, it follows that the mean electrical axis is perpendicular to that lead, and the hexaxial reference system can be used to determine the angle of that axis. In Figure 12.12, the lead in which the net QRS deflection is closest to zero is lead aVF (enlarged in Fig. 12.13). Lead I is perpendicular to the axis of lead aVF, and because the QRS complex is upward in lead I, the mean electrical axis points to the left arm. Thus, the mean QRS vector is estimated to be about 0°, which is not far from the +3° of the actual calculated vector from the frontal bipolar leads.

The mean QRS vector is influenced by the position of the heart in the chest, the properties of the cardiac conduction system, the excitation and repolarization properties of the ventricular myocardium, and the size of the ventricles. Because the last three of these influences are most significant, the mean QRS electrical axis can provide valuable information about a variety of cardiac diseases. In healthy people, the mean QRS vector is oriented within the range of −10° to +110° on the hexaxial system. Sometimes, a rough measure is used to determine whether the QRS vector is normal in an individual by simply determining whether the amplitude of the QRS vector is positive in both lead I and lead II; if they both are, then the patient is considered to have an essentially normal QRS vector and no further vector analysis is required. A mean QRS vector outside these limits however indicates serious changes in either the path of depolarization in the ventricles or abnormal ventricular muscle mass, such as **left ventricular hypertrophy (LVH) or right ventricular hypertrophy** (RVH). Ventricular hypertrophy is a serious condition resulting from either ventricle being exposed chronically to high arterial pressure or outflow obstruction (e.g., valve stenosis). In LVH, the mean QRS vector shows what is called **left axis deviation**. In this situation, the mean QRS points to the upper left quadrant of the hexaxial projection between −30° and −90°. RVH is the approximate mirror image of this abnormal orientation with the mean QRS vector pointing between +110° and +150°.

Any ECG can detect abnormalities in the electrical activation and conduction in the heart.

Several types of ECG recordings are listed in Figure 12.14. Unless stated otherwise, one may assume these ECGs are from lead II. The ECG in Figure 12.14A shows **respiratory sinus arrhythmia**, which is an increase in the heart rate with inspiration and a decrease with expiration. The presence of a P wave before each QRS complex indicates that these beats originate in the SA node. Intervals between successive R waves of 1.08, 0.88, 0.88, 0.80, 0.66, and 0.66 seconds correspond to heart rates of 56, 68, 68, 75, 91, and 91 beats/min. The interval between the beginning of the P wave and the end of the T wave is uniform, and the change in the interval between beats is primarily accounted for by the variation in

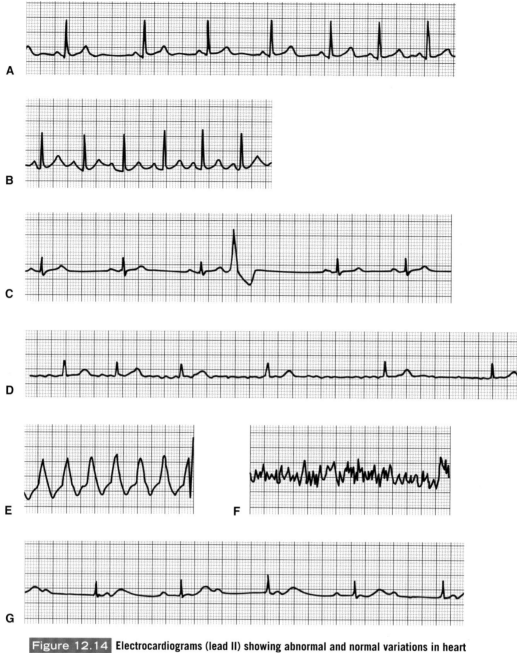

Figure 12.14 Electrocardiograms (lead II) showing abnormal and normal variations in heart rhythms. **(A)** Respiratory sinus arrhythmia. **(B)** Sinus tachycardia. **(C)** Premature ventricular complex. **(D)** Atrial fibrillation. **(E)** Ventricular tachycardia. **(F)** Ventricular fibrillation. **(G)** Complete atrioventricular block.

time between the end of the T wave and the beginning of the P wave. Although the heart rate changes, the interval during which electrical activation of the atria and ventricles occurs does not change nearly as much as the interval between beats, indicating that it is variations in the activation of the heart at the SA node that is altering rhythm and not variation in conduction through the heart. Respiratory sinus arrhythmia is caused by cyclic changes in sympathetic and parasympathetic neural activity to the SA node that accompanies respiration. It is observed in people with healthy hearts, although this particular example is somewhat dramatic. Respiratory sinus arrhythmia is usually accentuated in patients under anesthesia.

In the tracing in Figure 12.14B, the pattern of P, QRS, and T waves is normal but the interval between successive P, QRS, or T waves is much shorter than normal. This represents an SA node–driven increase in heart rate and is called a **sinus tachycardia**. Conversely, a slow SA node–driven heart rate is called a **sinus bradycardia** (image not shown). These patterns would be seen in a healthy person during acute exercise or in a well-conditioned athlete during rest, respectively.

Occasionally, the heart will be activated by the spontaneous generation of an action potential in one of the ventricular cells. Such an ectopic foci and its effect on the ECG are shown in Figure 12.14C. The first three QRS complexes in this panel are preceded by P waves; then, after the T wave of the third QRS complex, a QRS complex of increased voltage and longer duration occurs. This complex will result in a PVC, or premature ventricular contraction. The term PVC is used to designate the appearance of the firing of the ventricular ectopic foci on the ECG even though the ECG is the electrical, not mechanical, representation of the event. The premature complex in this example is not preceded by a P wave and is followed by a pause before the next normal P wave and QRS complex. In panel C, the ectopic focus is probably in the Purkinje system or ventricular muscle, where an aberrant pacemaker reached threshold before being depolarized by the normal wave of excitation. Once the ectopic focus triggers an action potential, the excitation is propagated over the ventricles. The abnormal pattern of excitation accounts for the greater voltage, change of mean electrical axis, and longer duration (inefficient conduction) of the QRS complex. Although the abnormal wave of excitation reached the AV node, retrograde conduction usually dies out in the AV node because the node preferentially conducts in the antegrade direction. The next normal atrial excitation (P wave) occurs but is hidden by the inverted T wave associated with the abnormal QRS complex. This normal wave of atrial excitation does not result in ventricular excitation because, when the impulse arrives at the AV node, a portion of the node is still refractory from excitation by the PVC. As a consequence, the next "scheduled" ventricular beat is missed. A prolonged interval following a premature ventricular beat is called a **compensatory pause**.

Ectopic pacemakers can also occur in the atria, where they are called **premature atrial contractions (PACs)**. Occasional PVCs or PACs are not uncommon in healthy people. However, the probability of their occurrence is increased by cigarette smoking, nicotine in any form, physical or emotional stress, caffeine, and fatigue. In some circumstances, atrial ectopic foci fire in a random, or chaotic, manner. This is shown in Figure 12.14D, which is an ECG from a patient with **atrial fibrillation**. In this condition, many small, random waves of depolarization occur in the atria such that actual, concentrated atrial contraction does not occur. In a patient with atrial fibrillation, the AV node conducts action potentials whenever a wave of atrial excitation happens to reach it and it is not refractory to activation. Thus, the ventricles are activated in a "hit or miss" fashion. The resulting ventricular rate is consequently highly irregular and can be detected in the patient as a highly irregular arterial pulse. Unless there are other abnormalities, conduction through the AV node and ventricles is normal, and the resulting QRS complex is normal. However, without a concentrated area of depolarization in the atria, the ECG shows QRS complexes that are not preceded by P waves. Atrial fibrillation is associated with numerous disease states, such as cardiomyopathy, pericarditis, hypertension, and hyperthyroidism, but it sometimes occurs in otherwise healthy people.

Sometimes, specific atrial ectopic foci can fire repetitively at a high rate, serving as an ectopic pacemaker that drives the ventricles. Such atrial ectopic foci can fire regularly at a rate of >200 times/min. This high rate of atrial activation can get through the AV node and activate the ventricles at the same rate. This is called **supraventricular tachycardia**. Such a condition may pace the heart at a rate too fast to allow adequate ventricular filling during diastole. Because pacing of the heart in this condition results from an ectopic focus in the atria, the resulting ECG may look similar to that of a person with sinus tachycardia.

An abnormally high rate of ventricular activation generated from ventricular ectopic foci is called **ventricular tachycardia** (Fig. 12.14E). Ventricular tachycardia results in a bizarre inefficient activation of the ventricles and extremely high heart rates that limit the period of ventricular filling. When filling is compromised, the flow output of the heart falls. This can result in syncope (dizziness/fainting) and can even lead to sudden death. Ventricular tachycardia is often a prelude to the life-threatening event of **ventricular fibrillation** (Fig. 12.14F). Ventricular fibrillation is characterized by the random, uncoordinated activation of millions of ventricular cells and results in no pumping of blood by the heart. Unless the heart can be converted through intervention to a regular rhythm, ventricular fibrillation will result in death.

In some situations, there is a disconnect between the electrical activation of the atria and the ventricles. This results from an impairment of the conduction of action potentials from the atria through the AV node and can result in a dissociation of P waves and QRS complexes in the ECG. In Figure 12.14G, both P waves and QRS complexes are present, but the timing of each is independent of the others. This situation is called **complete atrioventricular block** (sometimes called *third-degree block* or *complete heart block*) because the AV node fails to conduct impulses from the atria to the ventricles. Because the AV node is the only electrical connection between these areas, the pacemaker activities of the two muscles become entirely independent. In addition, a secondary pacemaker must take over the regular pacing of the heart. Because this usually involves the AV node or Purkinje systems that have low intrinsic rhythmicity, the heart will be paced at a lower-than-normal rate. In this example, the distance between P waves is about 0.8 seconds, giving an atrial rate of 75 beats/min. However, the distance between R waves averages 1.2 seconds, giving a ventricular rate of 50 beats/min. It is this rate that is responsible for pumping blood from the heart. In this example, the atrial pacemaker is probably in the SA node, and the ventricular pacemaker is probably in a lower portion of the AV node or bundle of His. Patients with complete atrioventricular block cannot resume normal activity without having an electronic pacemaker device permanently implanted in their chest to pace the heart at a faster rate.

AV block is not always complete. Sometimes, the PR interval is lengthened beyond normal limits (e.g., >0.2 seconds), but all atrial excitations are eventually conducted to the ventricles. This is **first-degree atrioventricular block**. When some, but not all, of the atrial excitations are conducted by the AV node, the block is called **second-degree atrioventricular block**. Second-degree heart block is often divided into two subtypes; type I (or Wenckebach) and type II (or Mobitz). In type I 2nd-degree heart block, the P–R interval gets successively longer with each heart beat until a P wave fails to be conducted to the ventricles, resulting in a pause of the heart. In this case, the R–R interval after the pause is less than the preceding R–R waves and more than the R–R interval after the pause. In type I 2nd-degree block, P–R intervals are normal, but occasionally, a P wave appears and is not conducted to the ventricles.

In some disease states, either the left or right branch of the bundle of His cannot transmit excitation. Depending on the branch affected, this type of heart block is called **right bundle branch block** or **left bundle branch block**. In these conditions, the portion of the heart with the blocked branch receives delayed activation from the unblocked side of the ventricles because the route of activation is less efficient. The resulting QRS complex becomes widened and will sometimes display a characteristic split in the R wave, creating a double-positive spiked appearance of the QRS complex. These twin spikes are designated as R and R[1].

Changes in electrically active ventricular mass are obtained by contrasting the ECGs from multiple leads.

In the healthy heart, the left ventricle is larger than the right and dominates the resulting QRS complex. Leads with positive poles most closely aligned on the left side of the body yield the most positive average QRS deflections (e.g., leads I, II, V_5, and V_6). Indeed, the only ECG lead that shows an average negative QRS deflection in a healthy person is aVR because the normal wave of activation of the ventricles proceeds away from this positive electrode. In addition, the ECGs from a healthy heart recorded from the precordial leads create a pattern of progressively greater average positive deflections from V_1 to V_6.

Chronic pulmonary hypertension or pulmonary emboli increase stress on the right ventricle each time it contracts. Over time, this creates **right ventricular hypertrophy**, or a larger than normal mass of the right ventricle. This mass pulls the net dipoles during ventricular activation to the right and can create mirror image deflections of the QRS complex in all the leads compared with that seen in the healthy heart. For example, lead aVR becomes positively deflected, V_6 downwardly deflected, V_1 net positively deflected, and so on. The recording in Figure 12.15 shows some effects of right ventricular enlargement on the ECG. The increased mass of right ventricular muscle results in R waves greater than S waves in lead V_1, which is the opposite of the pattern seen in the normal heart. This increased right ventricular mass also creates a large S waves in lead I and a larger than normal R waves in lead aVF. Finally, right ventricular hypertrophy causes a shift in the dipole of ventricular depolarization to the right (i.e., **right axis deviation**, QRS vector at +120° to +180°).

Other types of cardiac hypertrophy can be detected through the examination of simultaneous waveforms from different ECG leads. For example, Figure 12.16A shows an ECG from a patient with atrial hypertrophy. This is reflected by the P waves of lead III, which have a higher amplitude than normal because of the enlarged muscle mass of the atria. Figure 12.16B represents an ECG characteristic of left ventricular hypertrophy. In contrast to the ECG from a normal heart, there is a prominent S wave in V_1 and V_5 and the R waves are > +35 mm, which together are considered indicative of left ventricular hypertrophy. R waves > 11 mm in aVL

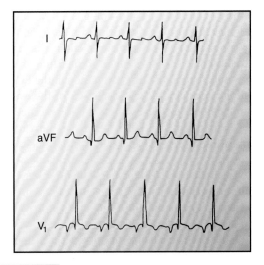

Figure 12.15 Effects of right ventricular hypertrophy on electrocardiogram recordings. Leads I, aVF, and V_1 of a patient are shown.

or 15 mm in lead I are also indicative of left ventricular hypertrophy. Left ventricular hypertrophy rotates the direction of the major dipole associated with ventricular depolarization more to the left than usual (i.e., **left axis deviation**, a negative QRS vector > −30°).

Metabolic conditions, drug effects, and myocardial injury can be detected by unique features of individual ECG waveforms.

Individual ECGs often contain tell-tale indicators of certain abnormal conditions. For example, plasma electrolyte disturbances create signature abnormalities in the ECG. Hyperkalemia alters potassium conductance in phase 3 of the cardiac action potential and thus affects repolarization patterns in the ventricles, resulting in a characteristic spiking or "mountain" characteristic of the T wave (Fig. 12.17). Hypokalemia also affects the myocardium in phase 3 and produces a characteristic flattened T wave followed by a secondary repolarization ECG wave, designated by the letter U. Both of these unique ECG waveforms are valuable indicators to clinicians because high plasma K^+ levels can induce fibrillation in the heart, whereas low K^+ can seriously suppress cardiac excitability and prevent cardiac activation.

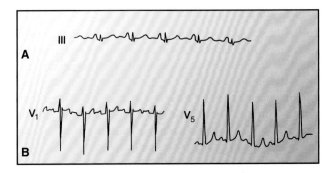

Figure 12.16 Effects of atrial and left ventricular hypertrophy on electrocardiogram recordings. **(A)** Large P waves (lead III) caused by atrial hypertrophy. **(B)** Altered QRS complex (leads V_1 and V_5) produced by left ventricular hypertrophy.

CLINICAL FOCUS | 12.2

Electrocardiogram Exercise Testing

Myocardial ischemia is one of the most dangerous cardiac conditions existing in patients with common forms of coronary artery disease, hypertension, heart failure, and arrhythmias. The electrocardiogram (ECG) can easily detect ischemia in a patient but only while the patient is connected to the ECG recorder. In the case of patients with coronary artery disease, the oxygen demand of the heart at rest may be within the ability of the compromised coronary artery system to deliver oxygen to the tissues (see Chapters 15 and 16 for more details). Thus, when the patient is at rest (e.g., at the doctor's office), evidence of myocardial ischemia may not appear on the ECG. However, should the oxygen demands of the heart increase from increased physical activity or emotional stress in the patient, the heart's diseased arterial system may not be able to meet the new, higher oxygen demand and the heart will become ischemic. This condition can result in heart attack and death.

Cardiac stress testing uses the ECG coupled with exercise as a means to examine the limits of oxygen delivery in the heart of patients with coronary artery disease. This testing has

many clinical modalities but all basically create increased oxygen demand in the heart through exercise while the patient's ECG and/or hemodynamic indicators are measured. Isometric stress (handgrip), dynamic (aerobic) stress (such as bicycle or treadmill ergometry), or combinations of the two are used to increase oxygen demand in the heart and in the body as a whole. The response of the patient as well as the response of the heart to this increased demand is monitored and analyzed. Extensive clinical guidelines for using and interpreting exercise stress testing have been developed and are continually updated by the American Heart Association and other professional cardiopulmonary or sports medicine groups.

Exercise testing has long been used for the diagnosis of the hemodynamic consequences, or severity, of obstructive coronary artery disease. Such obstructions do not often manifest themselves at rest but can be revealed by evidence of myocardial ischemia on the ECG and by other means during exercise stress testing. The use of exercise stress testing to reveal this "silent coronary artery disease" is invaluable to the clinician. ■

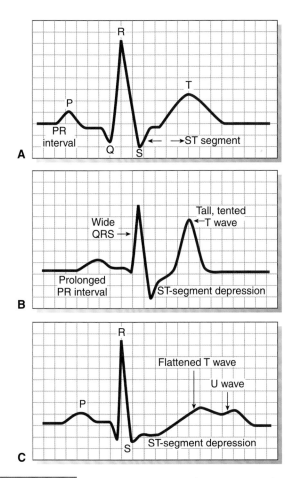

Figure 12.17 **The effects of hyperkalemia and hypokalemia on the electrocardiogram (ECG). (A)** A normal ECG tracing. **(B)** Effect of hyperkalemia. Notice the presence of the distinctive, tall, tented T wave. **(C)** Effect of hypokalemia with the appearance of a U wave.

Calcium is one of the triggers for increasing potassium conductance during phase 3. Abnormally low serum Ca^{2+} levels can thus delay the repolarization of the ventricles, and this is revealed on the ECG as an abnormally long QT interval, which can predispose the heart to EADs. Similarly, many antiarrhythmic drugs exert their effects by either increasing or decreasing the electrical refractory period in the ventricles. The toxic effects of such drugs can be detected by their effect on the QT interval of the ECG.

Finally, the clinically useful indication of myocardial ischemia, injury, and infarction (tissue death) can be detected by the ECG. One of the first signs of ischemia in the heart is an inversion of the T wave, as shown in Figure 12.18. Acute injury to the myocardium resulting from progressive ischemia results in an easily detected elevation of the ST segment. Infarction, or death of myocardial tissue, is recognized by the development of Q waves in front of these elevated ST segments. Old infarctions also present with abnormal Q waves in patients but without the elevated ST segment associate with acute ischemia. With myocardial ischemia, the cells in the ischemic region partially depolarize to a lower resting membrane potential because of a lowering of the potassium ion concentration gradient, although they are still capable of firing action potentials. As a consequence, a dipole is present during the TP interval in injured hearts because of the voltage difference between normal (polarized) and abnormal (partially polarized) tissue. However, no dipole is present during the ST interval because depolarization is uniform and complete in both injured and normal tissue (this is the plateau period of ventricular action potentials). Because the ECG is designed so that the TP interval is recorded as zero voltage, the true zero during the ST interval is recorded as a positive or negative deflection.

Figure 12.18 Electrocardiogram (ECG) changes associated with myocardial ischemia. The initial effects of ischemia create an inverted T wave in the ECG. The distinctive ST segment elevation in ischemic myocardium is associated with myocardial injury. The apparent zero baseline of the ECG before depolarization is below zero because of partial depolarization of the injured area. After depolarization (during the action potential plateau), all areas are depolarized and true zero is recorded. Because zero baseline is set arbitrarily (on the ECG recorder), a depressed diastolic baseline (TP segment) and an elevated ST segment cannot be distinguished. Regardless of the mechanism, this is referred to as an elevated ST segment. The ECG of a patient with acute myocardial infarction produces a Q wave in association with ST segment elevation. An abnormal Q wave without ST segment elevation (not shown) is indicative of an old myocardial infarction.

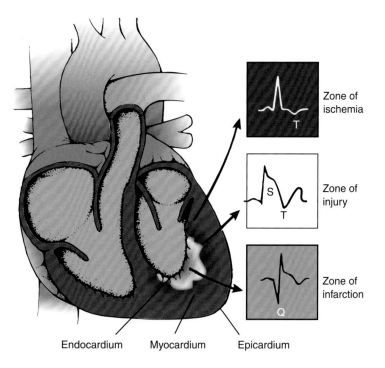

Zone of ischemia

Zone of injury

Zone of infarction

Endocardium   Myocardium   Epicardium

INTEGRATED MEDICAL SCIENCES

Sodium Channel Blockers for the Treatment of Cardiac Arrhythmias

Cardiac arrhythmias refer to any deviation from normal in the origin, conduction, or rhythm of the electrical activation of the heart. Such deviations in activation can result in erratic and insufficient mechanical activation of the heart such that blood flow output to peripheral organs is compromised. Some such arrhythmias can cause a patient to collapse and even die. The diagnosis of cardiac arrhythmias is aided greatly by detailed analyses using the electrocardiogram. However, management of many arrhythmias in patients is exceptionally challenging. Many drugs that modify cardiac electrical processes have been employed for the control and management of arrhythmias. Although these drugs can often be proarrhythmic and come with many other side effects, they are still used extensively as pharmacotherapy for many types of arrhythmias. These drugs have different, diverse, effects on electrical properties of cardiac cells. As an aid to organizing antiarrhythmic agents, these drugs have been collated into four categories (I, II, III, and IV) based on key electrophysiological properties common to the drugs in each category. Type I agents are called sodium channel antagonists because they impair the fast sodium channel kinetics in myocardial cells. Beta-adrenergic receptor antagonists (inhibitors) comprise the type II antiarrhythmics. Type III agents are primarily potassium channel inhibitors that extend the myocardial

cell refractory period. And type IV agents are comprised of the calcium channel blockers, verapamil and diltiazem as well as adenosine.

Type I agents slow phase 0 of the myocardial action potential and thereby slow conduction of action potentials through the myocardium. There are actually three subclasses of type I agents. Type I_A drugs show moderate phase 0 inhibition though they also tend to block K^+ channels and therefore extend cell refractory period. By impairing sodium channel operation, they tend to raise the effective action potential threshold in atrial and ventricular cells and thereby reduce cell excitability. These agents are most effective in ischemic tissue. Their salutatory effect is to convert unidirectional conduction blocks in damaged myocardium into bidirectional blocks thereby quashing reentry arrhythmias. Quinidine (a derivative of quinine), procainamide, and disopyramide belong to this subclass. Although these agents can be used to treat various atrial and ventricular arrhythmias as well as reentry tachycardias, they are so fraught with undesirable autonomic and other side effects that they are generally reserved for acute treatment of life-threatening arrhythmias. Quinidine, in particular, is so cardiotoxic that it has to be discontinued in roughly 50% of patients after just a single use. Class I_B agents have the smallest phase 0 sodium channel effect of the type I agents. Their primary beneficial effect is that they enhance K⁺ conductance in Purkinje fibers without altering the resting membrane potential. This enables these cells, which are often prone to form ectopic foci, to more easily counteract depolarizing stimuli that could otherwise create an ectopic foci. These agents gain access to the fast sodium channel in the active and inactive state so they are especially effective in rapidly cycling or partially depolarized tissue. For this reason, they reduce conduction velocity in ischemic tissue but not normal myocardium and are especially effective in reducing ectopic-based tachycardias. Lidocaine is the prototypical drug in this class, which are the agents of choice in the acute treatment of sustained ventricular tachycardia and the prevention of ventricular fibrillation. Agents in this class have anesthetic properties and cause CNS disturbances, which make them unsuitable for anything other than very short-term control of arrhythmias. Class I_c agents produce the most marked phase 0 depression in myocardial cells. Flecainide was one of the early drugs developed in this category, which also includes the agent propafenone. These agents however tend to be actually pro-arrhythmic in many instances and are therefore rarely used. Flecainide has actually been shown to increase mortality in arrhythmic patients.

The complex effects and side effects of Class I agents make them difficult to use in the long-term management of arrhythmias. Similar problems are associated with Class III agents as well. For this reason, current clinical management utilizes surgical and medical device modalities for the long-term treatment of arrhythmias. Fluoroscopic cardiac catheterization procedures and sophisticated electrophysiological recording analyses are now used to pinpoint areas of ectopic foci, which can then be physically destroyed by cryo- and radio ablation catheters, which can be inserted into the CV system and guided to the area of ectopy. Implantable, programmable, cardiac pacemaker/defibrillator units are now being used as a surgical solution to both abnormal bradyarrhythmias (pacemaker capability) as well as sudden emergent ventricular tachycardia or fibrillation (automatic fibrillation detection and electrical defibrillation). ■

Chapter Summary

- Cardiac cells do not require signals from nerve fibers to generate action potentials.
- Specialized nodal tissue cells exhibit the properties of automaticity and rhythmicity.
- Gap junctions at nexus between adjacent cells allow the heart to behave as a functional syncytium.
- Opening of voltage-gated sodium and calcium channels and the closing of voltage-gated potassium channels initiate action potentials in cardiac muscle cells.
- Action potentials in atrial and ventricular muscle cells have an extended depolarization plateau that creates an extended refractory period in the cardiac muscle cell.
- Cardiac muscle cells are repolarized following the depolarization phase by the closing of voltage-gated calcium channels and the opening of voltage-gated and ligand-gated potassium channels.
- A recycling decay and resetting of potassium conductance in the nodal cell membrane create recycling pacemaker potentials in the sinoatrial node.
- The sinoatrial node initiates electrical activity in the normal heart.
- Norepinephrine increases pacemaker activity and the speed of action potential conduction, whereas acetylcholine decreases pacemaker activity and the speed of action potential conduction.
- Electrical activity initiated at the sinoatrial node spreads preferentially in sequence across the atria, through the atrioventricular node, through the Purkinje system, and to ventricular muscle.
- The atrioventricular node delays the entry of action potentials into the ventricular system.
- Purkinje fibers transmit electrical activity rapidly into the inner layers of the ventricular myocardium.
- The conduction of electrical activity through the myocardium is a function of the amplitude of action potentials and the rate of rise of depolarization of action potentials in phase 0.
- An electrocardiogram is a recording of the time-varying voltage differences between repolarized and depolarized regions of the heart.
- The electrocardiogram provides clinically useful information about the rate, rhythm, pattern of depolarization, and mass of electrically active cardiac muscle.
- Changes in cardiac metabolism and plasma electrolytes as well as the effects of medicinal drugs on the electrical activity of the heart can be detected by an electrocardiogram.

Chapter Review Questions

1. The cardiac glycoside, digitalis, is a common agent used for the treatment of heart failure. This agent inhibits the Na^+/K^+ ATPase in all heart cells. Its therapeutic benefit arises from its effect in ventricular tissue where it causes intracellular calcium concentration to rise and enhances the contractile force that can be generated by the failing heart. However, the electrophysiological effects of a therapeutic concentration of this drug also alter Purkinje fibers in a manner that make them prone to the creation of ectopic foci and ventricular arrhythmias. The mechanism for these electrophysiological effects of a therapeutic level of digitalis most likely is because this drug will:

   A. Bring Purkinje fibers cells closer to their action potential threshold and reduce the action potential refractory period
   B. Bring Purkinje fibers closer to their action potential threshold and increase the action potential refractory period
   C. Bring Purkinje fibers farther from their action potential threshold and increase the action potential refractory period
   D. Block activation of sodium and calcium channels in Purkinje fibers
   E. Inactivate outwardly rectifying K^+ channels in Purkinje fibers

The correct answer is A. Inhibition of the Na^+/K^+ ATPase results in partial depolarization of the cell membrane thus bringing the myocardial cells closer to their threshold. Increases in intracellular calcium activate outwardly rectifying potassium channels in myocardial cells that terminate the depolarization phase of the action potential. This, therefore, shortens the cell refractory period. The effects of digitalis then on both the resting membrane potential and action potential refractory period increase the probability that Purkinje fibers can fire their own action potentials and do so at a high repetitive rate. Total blockade of sodium and calcium ion channels responsible for action potential formation would stop the heart. Therefore, this could not be a therapeutic effect of digitalis. Inhibiting outward rectifying potassium channels would extend, not shorten the refractory period of Purkinje fibers.

2. In a patient with left ventricular hypertrophy, which of the following ECG leads would register the most negative mean amplitude in the QRS complex?

   A. Lead I
   B. V_5
   C. Lead III
   D. Lead aVR
   E. aVF

The correct answer is C. In left ventricular hypertrophy, the increased left-sided muscle mass swings the mean QRS axis strongly to the left beyond the normal limit of –30°. At this angle, the mean QRS amplitude would be strongly positive in lead I. Mean QRS amplitude in aVR is negative in the normal heart; in left ventricular hypertrophy, it would be positive. Similarly, V_5 is in the area of the hypertrophy and would register a positive deflection during ventricular activation. Both leads III and aVF would show negative mean amplitudes in the QRS complex, but the angle relative to lead III would be greater away from the positive pole of that lead compared to the aVF positive poll. Therefore, lead III would show the greater negative mean amplitude.

3. Which of the following would be an indicator of an abnormal electrical condition in the heart?

   A. A mean frontal QRS vector at 67°
   B. PR interval of 180 msec
   C. A negative QRS deflection in lead III
   D. Presence of a T wave in lead II
   E. A positive average QRS deflection in lead aVR

The correct answer is E. A positive mean QRS deflection in lead aVR would imply that action potentials originated near the apex of the heart with the ventricles then being activated from the apex to the base, or in a retrograde fashion. This is the opposite of the typical path of activation. A mean QRS vector of + 67° and a PR interval of 180 msec are normal, and T waves are not abnormal occurrences in ECG leads. A normal heart could have a mean QRS vector between − 10° and + 30°. Such normal vectors would nevertheless produce a negative mean amplitude of the QRS complex in lead III.

4. Several antiarrhythmic drugs prolong phase 3 of the ventricular action potential. The clinical electrophysiological presentation of a patient on this drug would most likely resemble one with:

   A. High-amplitude spiked T waves on the ECG and hyperkalemia
   B. Prolonged QT intervals on the ECG and increased probability of forming early afterdepolarizations (EADs)
   C. Shortened PR intervals on the ECG and increased probability of forming early afterdepolarizations
   D. Split R waves on the ECG (i.e., an R followed by an R′ wave)
   E. Prolonged QRS intervals and second-degree heart block

The correct answer is B. Early afterdepolarizations (EADs) are action potentials that fire in cardiac cells (usually Purkinje fibers) during phase 3 of the action potential. Any factor that prolongs phase 3 of action potentials in cardiac cells makes those cells prone to the formation of EADs. Phase 3 of the cardiac action potentials represents the rapid repolarization phase of the action potential, which is determined, in part, by the duration of the QT interval on the ECG. Therefore, patients on drugs that prolong phase 3 of the cardiac action potential will often have prolonged QT intervals and an increased probability of forming EADs. Such drugs would not be expected to have any effect on plasma potassium concentration and would, therefore, create neither hyperkalemia in the patient nor its associated characteristic spiked T wave on the ECG. PR intervals occur before activation of the ventricular myocardium and reflect conduction time between activation of the atria and the start of activation of the ventricles; they are not indicative of any phase 3 phenomena in the ventricular myocardium. Split R waves are characteristic of bundle branch block and are a phenomenon distinct from EADs. Prolonged QRS intervals reflect the duration of conduction of action potentials through the ventricular myocardium and not the repolarization phase of the ventricular muscle cells.

Clinical Application Exercises 12.1

A 58-year-old woman arrived in the emergency department complaining of sudden onset of palpitations, light-headedness, and shortness of breath. These symptoms began ~2 hours previously. On examination, her blood pressure is 95/70 mm Hg, and her heart rate is essentially regular but averages 170 beats/min. An ECG demonstrates an abnormally shaped P wave, but the P–R interval is within normal limits. Her physical examination is otherwise unremarkable.

QUESTIONS

1. What is the likely cause of the patient's high heart rate?
2. Postulate why the patient was lighted, dizzy, and short of breath.
3. If you were to use a drug to manage this patient's electro-physiological situation, what cardiac action of that drug would be beneficial in alleviating the symptoms the patient is experiencing?

ANSWERS

1. The abnormal P wave on the ECG indicates that the atria are being activated from a location other than the SA node. For this site to be able to override the automaticity of the SA node, it must have an intrinsic rate higher than that of the SA node. The essentially normal P–R interval suggests that these ectopic foci are located near the AV node, which is allowing action potentials from these foci to reach the ventricles through this normal pathway. However, because the ectopic foci have a high intrinsic rate, the ventricular rate matches that of the ectopic foci. This type of arrhythmia is called supraventricular (i.e., above the ventricles) tachycardia.
2. When the ventricular rate is extremely rapid, there is little opportunity for ventricular filling to occur; despite the high heart rate, cardiac output falls in this setting (see Chapter 13 for more detailed explanation). This leads to hypotension and associated symptoms such as light-headedness and shortness of breath.
3. Drugs that can slow conduction through the AV node are useful in treating atrial fibrillation. These include beta-blockers and L-type calcium channel blockers. By slowing AV nodal conduction, these drugs reduce the rate of excitation of the ventricles or prevent all atrial activation from activating the ventricles as well. This will result in a slowing of the ventricular rate even though the atria may be tachycardic due to the ectopic foci there. At a slower ventricular rate, there is more time for filling, and the output of the heart is increased.

thePoint *Visit* http://thepoint.lww.com/rhoades5e *for additional chapter review Q&A, Clinical Application Exercises, animations, and more!*

Cardiovascular Physiology

Cardiovascular Physiology

Active Learning Objectives

Upon mastering the material in this chapter, you should be able to:

- Explain how altering calcium regulatory mechanism in myocardial cells can alter the inotropic state of the heart.
- Identify the correct chronological sequence of any set of hemodynamic variables associated with the cardiac cycle.
- Identify the mechanisms responsible for the four primary heart sounds and correctly relate abnormalities in these sounds to valve defects, changes in aortic pressure, or changes in ventricular filling.
- Explain how changes in myocardial contractility can counteract inhibitory effects of reduced preload or increased afterload on cardiac muscle contraction.
- Explain how changes in preload or afterload can counteract the inhibitory effects of reduced contractility on cardiac muscle contraction.
- Use left ventricular pressure volume loops to demonstrate how changes in left ventricular end-diastolic volume, aortic pressure, and contractility affect stroke volume.
- Explain the relationship of Starling's law of the heart to the mechanisms that match the output of the right and left ventricles.

- Explain how Starling's law of the heart helps maintain cardiac output when either afterload or heart rate is increased.
- Predict how changes in heart rate, aortic pressure, stroke volume, and contractility, either alone or in combination, alter myocardial oxygen demand.
- Identify physiologic and pathologic factors that alter cardiac output through changes in ventricular filling and explain the mechanisms responsible for this alteration.
- Identify pathologic factors that are associated with a negative inotropic state in the heart and explain known mechanisms for their effect.
- Explain how the reciprocal relationship between heart rate and stroke volume helps maintain cardiac output at high heart rates and at low heart rates.
- Explain the principles of the thermodilution technique for measurement of cardiac output.
- Explain the benefits and limitations associated with modern imaging techniques for determining ventricular volumes and cardiac output.

Clinical interest in cardiac electrophysiology stems from the fact that this activity is coupled to the contraction of the heart and movement of blood through the cardiovascular system. The metabolic demands of different organs of the body and the body as a whole vary widely in an individual, and the output of blood pumped by the heart must match those demands. Clearly, the output of the heart must be able to be increased when tissue metabolic demands increase. The heart can do so by changing its rate of activation (heart rate, or beats per minute) and by changing the contractile force it generates with each beat.

The heart is made of striated muscle and shares several contractile properties with that tissue. However, two modes of enhancing contractile force in skeletal muscle, tetanic contraction and motor unit recruitment, are unavailable to cardiac muscle and therefore cannot be used to enhance contractile performance of the heart. Cardiac muscle cannot form tetanic contractions because the extended action potential refractory period of atrial and ventricular muscle fibers is too long to allow for the force of individual twitch contractions to be summated (Fig. 13.1). Furthermore, because the heart is an electromechanical syncytium, it does not possess motor units that could otherwise be recruited to enhance force generation of the heart as a whole. As discussed in the previous chapter, all healthy cells of the heart are activated with each heart beat; there are no spare units to recruit.

In spite of its syncytial properties, cardiac muscle has the ability to enhance its force-generation capacity via two

mechanisms that are not available to skeletal muscle. Recall that skeletal muscle exhibits an active isometric length–tension relationship whereby increases in passive length, up to an optimum length, generate more isometric force. However, because skeletal muscles are attached to fixed points on the skeleton, moving passively along the ascending limb of the length–tension curve is not done by that tissue to adjust contractile force. The heart, in contrast, is a hollow, compliant, organ. It can, and does, have its force generation modified by moving along the ascending limb of the length–tension relationship. Finally, skeletal muscle fibers do not contain mechanisms whereby the intrinsic force-generation capacity of individual cells can be enhanced by normal physiological means. In contrast, the force-generation capacity of the heart muscle can be altered at the level of the individual cell.

► CARDIAC EXCITATION–CONTRACTION COUPLING

Coupling of electrical activity in the heart to myocardial contraction uses mechanisms similar to those previously discussed for skeletal muscle. Cardiac cells contain an extensive network of T tubules that extend longitudinally along myocardial fibers (in contrast to the radial arrangement in skeletal muscle). As in skeletal muscle, these carry action potentials deep into the cell, where they open dihydropyridine (DHP) calcium channels that are coupled to ryanodine-sensitive calcium release channels in the sarcoplasmic reticulum

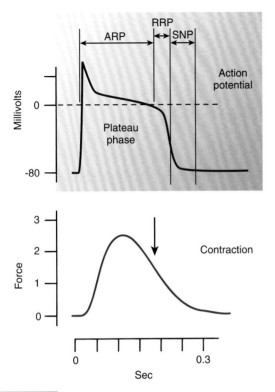

Figure 13.1 **A representation of a cardiac muscle action potential and isometric twitch.** The duration of the mechanical twitch in the muscle is completed before the muscle is over its refractory period to normal stimuli. Therefore, twitch contractions cannot be summed on one another and the muscle cannot be tetanized. ARP, absolute refractory period; RRP, relative refractory period; SNP, period of supranormal excitability.

(SR). The actin–myosin crossbridge cycling and its control through Ca^{2+}/troponin/tropomyosin interactions are essentially the same in both forms of striated muscle. Finally, both muscles use Ca^{2+} adenosine triphosphatase (ATPase) pumps in the SR to remove Ca^{2+} from the intracellular fluid while accumulating it into the SR against a concentration gradient. Ca^{2+} ATPase pumps and Na^+/Ca^{2+} in the cell membrane also reduce intracellular calcium levels and terminate contraction.

Force of contraction in cardiac muscle can be altered by changes in intracellular calcium concentration.

In contrast to skeletal muscle fibers, intracellular Ca^{2+} levels and force generation can be altered in individual myocardial cells by normal physiologic as well as abnormal conditions (Fig. 13.2). Physiologically, myocardial contraction is modified by altering the gating of the DHP Ca^{2+} channels in the cell membrane that are activated during phase 2 of the action potential. Calcium entering the cell through this channel, however, is not itself used to directly cause contraction. Only about 10% of this Ca^{2+} contributes to contraction of the muscle. Instead, increased intracellular Ca^{2+} through the channel serves as a trigger to release substantial stores of Ca^{2+} from within the SR (**calcium-induced calcium release**). In cardiac muscle, the DHP channel is linked to the ryanodine-sensitive calcium release channel in the SR membrane such that each

quantum of calcium admitted through the DHP channel causes a small, localized release of calcium from one ryanodine receptor channel in the SR. These localized intracellular bursts of calcium are called **calcium sparks**. The localized high concentration caused by a single calcium spark does not diffuse within the intracellular fluid to neighboring receptors in the SR at a high enough concentration to cause them to open. Instead, multiple calcium sparks are summated temporally and spatially within the myocardial cell to produce overall increases in intracellular Ca^{2+} concentration. In addition, high intracellular calcium concentrations stimulate the pumping of Ca^{2+} into the SR by the Ca^{2+} ATPase. This creates a larger pool of calcium within the SR that can be released with the next opening of the SR calcium release channels.

In general, anything that enhances calcium influx during the action potential in a cardiac myocyte results in a greater intracellular release of Ca^{2+} and thus a stronger subsequent contraction. This is the basis of the ability of cardiac muscle cells to alter their strength of contraction at the level of the cell. Furthermore, Ca^{2+} influx through the voltage-dependent channels in cardiac muscle cells can be modified by cellular regulatory mechanisms. For example, catecholamines (e.g., norepinephrine released from sympathetic nerve endings, or circulating epinephrine) activate β_1-adrenergic receptors on the myocardial cell membrane. These receptors are linked to adenylate cyclase within the myocardial cell by a stimulatory G protein (G_s). When catecholamine binds to the β_1 receptor, the adenylate cyclase is activated and converts adenosine triphosphate (ATP) to cyclic adenosine monophosphate (cAMP). In myocardial cells, cAMP binds to a cAMP-dependent protein kinase that phosphorylates the voltage-dependent Ca^{2+} channel. This phosphorylation increases the channel's probability of opening and increases the average time the channel remains open once activated. This enhances Ca^{2+} influx with every action potential, resulting in more intracellular calcium release and a stronger contraction in each myocardial cell. Extended to the whole heart, this increases the force-generating capacity of that organ and increases the output of blood with each contraction.

Modification of force generation in individual myocardial cells is independent of loading conditions on cardiac muscle.

The ability of myocardial cells to change the strength of contraction at the level of the cell represents a key additional mechanism by which the heart as a whole can adjust its contractile ability. Importantly, this modification of force-generating capacity can occur *independently* of any changes in force caused by alterations in preload or afterload on the heart. In cardiac muscle, this additional cellular factor is superimposed on any effects of preload or afterload on the heart. Thus, at *any* given loading condition, the heart exposed to an agent that enhances its ability to contract at the level of each cell can potentially generate more isometric force and move a greater load faster and farther compared with a heart not exposed to that agent. A modification of the contractile ability of muscle at the cellular level, independent of loading conditions, is said

Cardiovascular Physiology

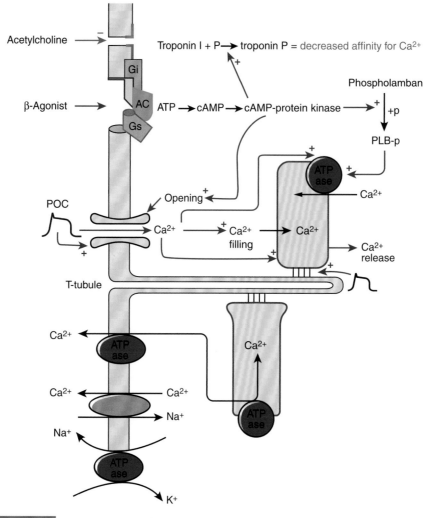

Figure 13.2 **Handling of calcium in a cardiac muscle cell.** Action potentials cause an influx of calcium through voltage-gated L-type 1,4 dihydropyridine (1,4 DHP) Ca^{2+} channels. Influx through one channel is closely linked to the release of calcium from one calcium-release channel in the sarcoplasmic reticulum (SR) (calcium-induced calcium release), causing a "calcium spark." Modulation of the influx of calcium through DHP channels through G protein–linked receptor and non–receptor-mediated events gives rise to the capability of modulation of the inotropic state of a single myocardial cell. Cyclic adenosine monophosphate (cAmp)-mediated mechanisms also speed relaxation during β-adrenergic receptor–mediated increases in cell contractility. Calcium pumps in the plasma and SR membranes as well as secondary active transport mechanisms for calcium in the plasma membrane return intracellular calcium concentration to low levels between action potential activations of the cell. AC, adenylate cyclase; ATPase, adenosine triphosphatase; Gi, inhibitory G protein; Gs, stimulatory G protein; P, phosphate; PLB-p, phosphorylated phospholamban; POC, potential-operated Ca^{2+} channel.

to be a modification of the **inotropic state** or **contractility** of the muscle. The ability to modify muscle performance by changing the inotropic state of individual cells is a hallmark of the physiologic control of myocardial contraction. (Note that the term "contractility" is often applied loosely in clinical settings to any factor that alters the contraction of the heart. This is a misapplication of the term. It is important for both physiologists and clinicians to distinguish alterations in the performance of the heart resulting from changes in inotropic state from those changes caused by alterations in loading conditions. In this text, the term "contractility" will be used to refer to the inotropic state of the heart.)

Factors that alter myocardial contractility

Many agents and conditions can alter the contractility, or inotropic state, of the heart (see Fig. 13.2 for reference). For example, β-receptor agonists are positive inotropic agents, whereas β-receptor antagonists are negative inotropic agents. Agents that block 1,4 DHP Ca^{2+} channels reduce myocardial contractility and thus exert a negative inotropic effect on the heart. Acetylcholine, released from parasympathetic nerve endings in the heart, is a negative inotropic agent. It inhibits adenyl cyclase through an inhibitory G protein and stimulates guanylate cyclase to produce cyclic guanosine monophosphate, which inhibits the opening of the Ca^{2+} channel.

The DHP channels require ATP for normal functioning. Therefore, myocardial ischemia, or lack of oxygen to the heart, results in an inhibition of the Ca^{2+} channel and is considered a negative inotropic condition. Also, acidosis (increased plasma H^+), which often occurs in ischemic conditions, inhibits myocardial contractility by interfering with the DHP channel.

Other non–receptor-mediated factors can alter intracellular calcium concentration and thus alter myocardial contractility. For example, Na^+/Ca^{2+} exchangers in the myocardial cell membrane remove Ca^{2+} from the interior of the cell through a secondary active transport mechanism linked to the influx of sodium moving down its electrochemical gradient. The reduction of the sodium gradient by the inhibition of Na^+/K^+ ATPase results in Ca^{2+} accumulation within the cell and a positive inotropic effect. This mechanism is responsible for the positive inotropic effects of cardiac glycosides such as digitalis and is the basis of their use to augment myocardial performance in heart failure. Methylxanthines, such as caffeine and theophylline (an asthma medication and component of tea), inhibit cAMP phosphodiesterase. These compounds therefore increase cAMP accumulation and induce a positive inotropic effect.

Activation of the sympathetic nervous system increases heart rate (a positive chronotropic effect) at the same time as it enhances myocardial contractility through β_1-receptor activation. Rapid heart rates require the rapid removal of calcium and inactivation of actin/myosin interactions so that relaxation can be accelerated and diastole shortened to accommodate the high heart rate. Generation of cAMP and protein kinase A by norepinephrine phosphorylates a regulatory protein in the SR called **phospholamban**. This protein normally inhibits the Ca^{2+} ATPase pumps in the SR. However, when it is phosphorylated, its inhibitory effects are removed and the pumping activity of the Ca^{2+} ATPase increases, thus removing calcium from the intracellular fluid. In addition, phosphorylation of troponin 1 by protein kinase A decreases its affinity for calcium, allowing the cellular actin–myosin crossbridges to break their attachments. The effects on calcium pumping in the SR and on calcium affinity accelerate the relaxation of myocardial cells.

▶ THE CARDIAC CYCLE

The cyclic contraction and relaxation of the myocardium sets into motion a sequence of events resulting in time-dependent changes in ventricular and aortic pressure, ventricular volumes, and flow into and out of the heart. A graphical representation of the variations in hemodynamic variables associated with the cardiac cycle is depicted in Figure 13.3. Such representations usually depict hemodynamic variables in the left heart and systemic circulation along with a tracing of the electrocardiogram (ECG) and heart sounds.

Ventricular systole consists of isovolumic contraction followed by a rapid, then reduced, blood ejection phase.

The peak of the R wave on the ECG is used to signify the start of systole, or the contraction phase of the cardiac

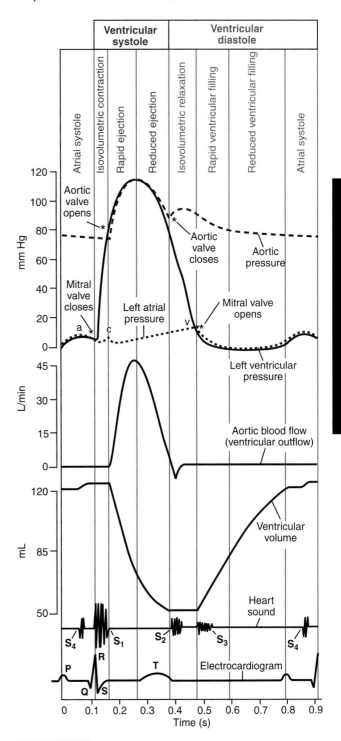

Figure 13.3 The timing of various hemodynamic and related events during the cardiac cycle. (See text for details.)

cycle. This contraction increases intraventricular pressure above end-diastolic levels, which are about 0 to 5 mm Hg in healthy adults. As soon as left ventricular pressure during systole exceeds that in the left atrium, the mitral valve closes. This closure is associated with a series of broad low-pitched sounds that are a result of the vibrations of blood and the chordae tendineae in the ventricles. This is called the **first heart sound**. (These sounds are not made by valve leaflets "slapping against one another.") The intensity of the first heart sound is proportional to the strength of myocardial contraction, and its evaluation is useful in clinical diagnosis.

As the ventricle contracts, there is a period of time when intraventricular pressure exceeds that in the left atrium but is still less than that in the aorta. In this state, all cardiac valves remain closed and the ventricle builds pressure without moving blood. This phase of systole is called **isovolumic contraction**. Once the left ventricular pressure exceeds that in the aorta, the aortic valve opens and blood leaves the ventricle, initiating the ejection phase of the cardiac cycle. During systole, a healthy heart ejects from 45% to about 67% of the volume of blood that was in the ventricles at the end of the previous diastole. Initially, with the opening of the aortic valve, ventricular volume decreases rapidly; about 70% of the volume to be ejected exits the ventricle during the first third of the ejection phase. This period is thus called the **rapid ejection phase**. During this phase, blood enters the vascular system faster than it can drain into the veins. Consequently, aortic pressure as well as intraventricular pressure rise during this phase. This causes a temporary increase in aortic blood volume, which distends the aortic wall and increases lateral aortic pressure. This pressure may actually slightly exceed left ventricular pressure, but the combined potential and kinetic energy during ejection is still greater in the ventricle than in the aorta, so flow proceeds outward. However, as some of the kinetic energy of outflow is converted into the potential energy that distends the aorta walls, the rate at which blood exits the ventricle begins to decline. This is called the **reduced ejection phase**.

During the reduced ejection phase, the T wave of the ECG begins and the aortic and ventricular pressures start to decline. When ventricular pressure eventually declines below total aortic pressure (lateral and kinetic components), the aortic valve closes. The sudden cessation of ventricular outflow caused by valve closure creates a **second heart sound**, whose intensity is proportional to the intensity of the valve closure. Intensity is increased whenever aortic or pulmonary arterial pressure is abnormally high and is taken as a clinical indicator of possible systemic or pulmonary hypertension, respectively. The closure of the aortic valve also causes a transient, sharp, small drop in aortic pressure, creating a notch in the aortic pressure profile, called the **incisura**. This event is used sometimes to demarcate the transition from systole to diastole in the cardiac cycle. However, systole is more correctly considered to be concluded when the T wave ends on the ECG.

The difference in the volume of blood in the ventricle at the end of diastole, or the **end-diastolic volume**, and the volume of blood in the ventricle at the end of systole, or the **end-systolic volume**, is called the **stroke volume**. Stroke volume represents the volume of blood ejected with each contraction of the ventricles (see Fig. 13.3). Note that the ventricles do not empty completely with each contraction. At the end of systole, a significant volume of blood remains in the ventricles and is called the **residual volume**. Residual volume is decreased by increases in myocardial contractility and heart rate and increased whenever the heart is weakened (e.g., heart failure) or when faced with increased outflow resistance (e.g., aortic valve stenosis). Thus, examination of end-systolic volume is useful clinically as an indicator of potential abnormal conditions affecting the heart.

Ventricular diastole consists of isovolumic relaxation followed by a rapid, then reduced, filling phase.

During the relaxation, or diastolic, phase of the cardiac cycle, volume and pressure changes proceed in the reverse of that seen for systole. Early in left ventricular relaxation, the aortic and mitral valves are closed, and the ventricle relaxes isovolumetrically. Thus, this phase of the cardiac cycle is called **isovolumic relaxation**. Once ventricular pressure falls below that of the left atrium, the mitral valve opens and the filling of the ventricle begins. Just prior to this occurrence, abrupt cessation of ventricular distention and the deceleration of blood create a faint third heart sound not normally heard in healthy people. However, the third heart sound is amplified in abnormally stiff or distended ventricles, such as that associated with heart failure, and its presence is therefore considered a serious sign of underlying cardiac abnormalities.

During diastole, the ventricles fill in a rapid filling phase, followed by a reduced filling phase (**rapid ventricular filling** and **reduced ventricular filling** phases). During rapid ventricular filling, ventricular pressure actually continues to decline because ventricular relaxation occurs more rapidly than does the filling of the ventricle. During the reduced filling phase, sometimes called diastasis, ventricular pressure starts to increase. The appearance of the P wave on the ECG and atrial contraction begin at the end of the reduced filling phase of the cardiac cycle. Atrial contraction produces a faint fourth heart sound that is amplified with A–V valve stenosis or with stiff ventricles. Diastole is considered to be concluded with the appearance of the next R wave on the ECG.

Abnormal conditions in heart valves are revealed by changes in venous pressure waveforms.

Occasionally, pressure waveforms in the jugular vein are depicted on cardiac cycle diagrams. When the right atrium contracts, a retrograde pressure pulse wave is sent backward into the jugular vein. This is called the **A wave**. Factors that impede the flow of blood from the atria to the ventricles, such as tricuspid valve stenosis, increase the amplitude of the A wave. A second venous pulse wave, called the **C wave**, is seen as an increase followed by a decrease in venous pulse pressure during the early phase of systole. The upslope of this wave is created by the bulging of the tricuspid valve into the right atrium during ventricular contraction, which sends a wave into the jugular vein. This wave is combined with a lateral transmission of the carotid systolic arterial pulse to the adjacent jugular vein. The subsequent decrease in pressure in the C wave is caused by the descent of the base of the heart and atrial stretch. Failure of the tricuspid valve to completely close during ventricular systole results in the propulsion of blood back into the atrium and vena cava and results in a high-amplitude C wave. The **V wave** of the venous pulse is seen as a gradual pressure increase during reduced ejection and isovolumic relaxation followed by a pressure decrease during the rapid-filling phase of the cycle. This wave is created first by continual peripheral venous return of blood to the atrium against a closed tricuspid valve followed by the sudden decrease in atrial distention caused by rapid ventricular filling. Tricuspid valve stenosis increases resistance to the filling of the right ventricle, which is indicated by an attenuation of the descending phase of the V wave.

Analysis of venous waveforms along with skilled attention to analysis of heart sounds was once used to provide clinical insights into cardiovascular disease involving cardiac valves. Such physical examination diagnosis is still valuable today in that, in the hands of a skilled physician, it provides a quick, inexpensive analysis of important aspects of the patient's heart that can signal dangerous conditions and their severity without having to wait for time-consuming technical instrument analyses to be scheduled, completed, and analyzed. Nevertheless, this type of indirect diagnosis is difficult and often benefits from additional high-tech analyses provided by several modern imaging techniques used to reveal valve condition and motion in the heart (as discussed later in this chapter).

▶ DETERMINANTS OF MYOCARDIAL PERFORMANCE

Evaluation of the mechanical properties of the heart as they relate to its function as a pump is important to physiologists and physicians. Myocardial performance is hindered in pathologic conditions such as ischemia and heart failure and evaluation of such conditions as they affect the heart as a pump are clinically important. In the previous section, it was shown that alteration of the inotropic state of individual myocardial cells is one means by which the contractile performance of the heart can be altered. However, as is the case with skeletal muscle, preload and afterload effects are also important determinants of the performance of cardiac muscle. These factors interact collectively with inotropic conditions to modify the overall contractile performance of the heart.

Preload, afterload, and inotropic state interact to alter contractile performance of cardiac muscle.

Four factors influence the contractile strength of cardiac muscle: (1) preload, or the initial length to which the muscle is stretched prior to contraction; (2) afterload, or all the forces against which cardiac muscle must contract to generate pressure and shorten; (3) contractility (inotropic state); and (4) the indirect inotropic effect of increased heart rate. This last effect is caused by the fact that high heart rates bring Ca^{2+} into the cell faster than it can be removed thereby increasing intracellular Ca^{2+} concentration and contractility. This helps the heart eject more blood with each contraction, thereby compensating for reduced filling time associated with high rates.

Cardiac muscle exhibits an active length–tension relationship similar to that of skeletal muscle (Fig. 13.4, red tracing). However, because the heart is a distensible hollow organ, its preload, or stretch just prior to contraction, could theoretically be altered by the degree to which it is filled at the end of diastole. Although cardiac muscle is not anchored to a fixed

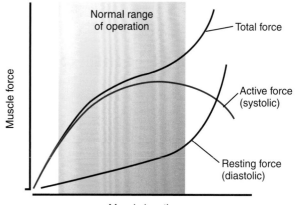

Figure 13.4 **The isometric length–tension curve for isolated cardiac muscle.** Cardiac muscle displays a parabolic active length–tension relationship similar to that of skeletal muscle but shows considerably more passive resistance to stretch at L_o. Functionally, this passive property constrains the heart to contraction along the ascending limb of the length–tension curve.

object in the body, it is intrinsically stiffer than skeletal muscle and exhibits significant passive resistance to stretch at a length corresponding to L_O (Fig. 13.4, lower black tracing). The functional consequence of this structural property is that cardiac muscle is *constrained to contract from lengths $\leq L_O$*; it does not get stretched past that length even with extreme filling. In practical terms, the heart is therefore constrained to operate on the ascending limb of the active length–tension curve. As a consequence, in the whole heart increased filling during diastole increases end-diastolic stretch (preload) and thus the potential isometric contractile force that can be generated in the heart once it begins to contract.

Effects of preload and afterload on cardiac contraction are enhanced or depressed by changes in cardiac inotropic state.

A key to understanding the contraction of cardiac muscle is to understand that *the effect of any given loading condition on the contractile force generated by the heart can be altered by the inotropic state of the heart.* At any given preload, isometric force generation of cardiac muscle is increased by positive inotropic stimuli and decreased by negative inotropic stimuli (Fig. 13.5). In other words, the ascending limb of the isometric length–tension relationship is shifted upward and to the left by a positive inotrope and downward and to the right by a negative inotrope.

In cardiac muscle, the isotonic force–velocity relationship is similar to that seen in skeletal muscle (Fig. 13.6A, black tracing); contractions with lighter loads proceed with greater velocity than those with heavier loads. Also, changes in preload have effects on the position of the isotonic force–velocity relationship that are similar to those seen in skeletal muscle. Decreasing preload reduces the velocity of shortening obtainable at any given afterload as well as the magnitude of the afterload the cardiac muscle can move at any given velocity (Fig. 13.6A, blue tracing vs. black tracing). Increasing preload, naturally, has the opposite effect on afterloaded contraction

Cardiomyopathies: Abnormalities of Heart Muscle

Heart disease takes many forms. Although some of these are related to problems with the valves or the electrical conduction system, many are a result of malfunctions of the cardiac muscle itself. These conditions, called *cardiomyopathies*, result in impaired heart function that may range from asymptomatic conditions to malfunctions causing sudden death.

There are several types of cardiomyopathy, and they have several causes. In *hypertrophic cardiomyopathy*, an enlargement of the ventricular muscle fibers occurs because of a chronic overload, such as that caused by systemic arterial hypertension or a stenotic aortic valve. Such muscle may fail because its high metabolic demands cannot be met, ischemia results, or fatal electrical arrhythmias develop. *Congestive* or *dilated cardiomyopathy* refers to cardiac muscle so weakened that it cannot pump strongly enough to empty the heart properly with each beat. This condition is common in patients with chronic congestive heart failure. In *restrictive cardiomyopathy*, the muscle becomes so stiffened and inextensible that the heart cannot fill properly between beats. Chronic poisoning with heavy metals, such as cobalt or lead, can produce *toxic cardiomyopathy*, and the skeletal muscle degeneration associated with muscular dystrophy is often accompanied by cardiomyopathy.

Cardiomyopathy arising from *viral myocarditis* is difficult to diagnose and may show no symptoms until death occurs.

The action of some enteroviruses (e.g., coxsackievirus B) may cause an autoimmune response that does the actual damage to the muscle. This damage may occur at the subcellular level by interfering with energy metabolism while producing little apparent structural disruption. Such conditions, which can usually only be diagnosed by direct muscle biopsy, are difficult to treat effectively, although spontaneous recovery can occur. In tropical regions, infection with a trypanosome (Chagas disease) can produce chronic cardiomyopathy. The tick-borne spirochete infection called Lyme disease can cause heart muscle damage and lead to heart block. Excessive and chronic consumption of alcohol can also cause cardiomyopathy that is often reversible if total abstinence is maintained.

Another important kind of cardiomyopathy arises from ischemia. An acute ischemic episode may be followed by a *stunned myocardium*, with reduced mechanical performance, whereas chronic ischemia can further reduce mechanical performance. Ischemic tissue has impaired calcium handling, which can lead to destructively high levels of internal calcium. These conditions can be improved by reestablishing an adequate oxygen supply (e.g., following clot dissolution or coronary bypass surgery), but even this treatment is risky because rapidly restoring the blood flow to ischemic tissue can lead to the production of oxygen radicals that cause significant cellular damage. ■

(Fig. 13.6A, red tracing vs. black tracing). However, the entire force–velocity relationship for any given preload can be shifted by the inotropic state of the heart. Positive inotropic influences shift the force–velocity curve at any given preload upward and to the right. In this manner, a positive inotropic agent allows cardiac muscle, at any given preload, to move a heavier load faster than that possible with normal myocardium at the same preload (Fig. 13.6B). Negative inotropic agents obviously have the opposite effect.

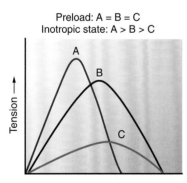

Preload: A = B = C
Inotropic state: A > B > C

Figure 13.5 **Three isometric twitch contractions of isolated cardiac muscle from identical preloads.** A = effect of a positive inotropic agent. Note the increased peak, and rate of tension development compared to a normal contraction (B). C = effect of a negative inotropic agent.

Unlike skeletal muscle, the theoretical maximum rate of crossbridge cycling, or V_{max}, in cardiac muscle is not constant. It is enhanced by positive inotropic stimuli and depressed by negative inotropic influences (note dashed extension of the tracings in Fig. 13.6B). Furthermore, V_{max} is not appreciably affected by loading conditions in cardiac muscle. Therefore, V_{max} can be thought of as a reflection of the inotropic state of the heart and indices that are related to V_{max} can be used to estimate the state of myocardial contractility (see below).

Figure 13.7 demonstrates how preload, afterload, and inotropic state interact to define myocardial contraction. As stated previously, positive inotropic stimuli shift the ascending limb of the isometric length–tension relationship upward and to the left, enabling cardiac muscle to generate more isometric force at any preload, whereas negative inotropic stimuli have the opposite effect. However, recall that the isometric length–tension relationship also represents the limit of how far muscle can shorten during an isotonic contraction against an afterload. Therefore, at any given combination of preload and afterload, a positive inotropic stimulus allows cardiac muscle to shorten farther, and/or shorten a given distance against a greater afterload, compared to that obtained by cardiac muscle in the normal inotropic state (note the extent of the horizontal blue, black, and red arrows in Fig. 13.7). Unfortunately, this effect of inotropic state can be detrimental to the heart as well. As can be seen in Figure 13.7, any negative inotropic stimuli will reduce the ability of the heart

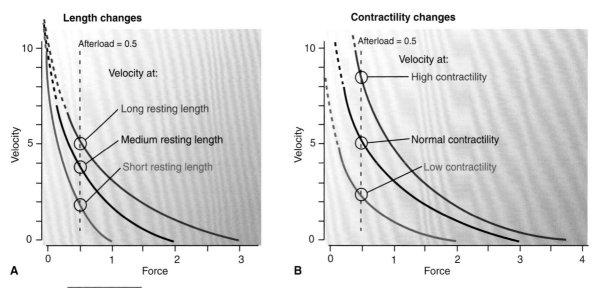

Figure 13.6 **Effect of length and contractility changes on the force–velocity curves of cardiac muscle. (A)** Decreased starting length (with constant contractility) produces lower velocities of shortening at a given afterload but does not affect the theoretical maximum velocity of shortening (V_{max}). Because of the presence of resting force (characteristic of heart muscle), it is impossible to make a direct measure of a zero-force contraction at each length. **(B)** Increased contractility produces increased velocity of shortening at a constant muscle length; V_{max} is proportional to the inotropic state.

to shorten with any given preload and afterload combination and/or reduce the load the heart can contract against for any specified amount of shortening.

Variables associated with contractile performance of the heart *in situ* are analogous to afterload, preload, and muscle shortening.

The relationships among preload, afterload, shortening, and contractility elucidated from studies of isolated cardiac muscle determine the output characteristics of the whole heart. To understand how loading conditions and contractility affect myocardial shortening and flow output, one needs to identify variables in the whole heart that are analogous to those in isolated muscle. For example, stroke volume, or

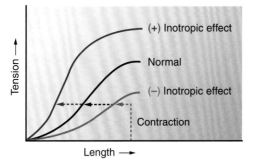

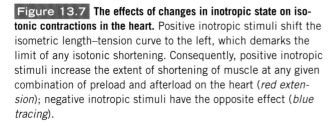

Figure 13.7 **The effects of changes in inotropic state on isotonic contractions in the heart.** Positive inotropic stimuli shift the isometric length–tension curve to the left, which demarks the limit of any isotonic shortening. Consequently, positive inotropic stimuli increase the extent of shortening of muscle at any given combination of preload and afterload on the heart (*red extension*); negative inotropic stimuli have the opposite effect (*blue tracing*).

the amount of blood ejected with each contraction of the heart, is analogous to muscle shortening. Similar analogs are used for preload and afterload. As the left ventricle fills with blood during diastole, it becomes stretched. The magnitude of this stretch is related to the blood volume in the heart at the end of diastole (**left ventricular end-diastolic volume, or LVEDV**). Therefore, LVEDV can be used as an indirect estimate of the preload in the left ventricle in the heart *in situ*. Because changes in end-diastolic volume result in changes in end-diastolic pressure, left ventricular end-diastolic pressure (LVEDP) is sometimes used as an indicator of preload on the heart. However, caution must be used in applying LVEDP as a measure of preload. Myocardial compliance can be reduced by myocardial ischemia, infarction, or hypertrophy. All of these will result in an increased end-diastolic pressure at any given LVEDV. Consequently, a person recording an increased LVEDP in this situation could assume, erroneously, that preload in the heart was increased when in fact it was not.

Pulmonary Wedge Pressure

Another hemodynamic measure of preload on the heart is the **pulmonary wedge pressure**. For this determination, a catheter with an inflatable balloon on its tip is inserted through a large systemic vein, through the right atria and ventricle and into the pulmonary artery until it encounters an artery smaller than the catheter. At this point, the balloon is inflated. The catheter now forms a quasi-lateral conduit between the pulmonary vessels downstream that empty into the left atrium and a pressure transducer on its other end outside the body. This pressure reflects changes in left atrial pressure, which is a determinant of left ventricular filling and thus preload. However, often, the pulmonary wedge pressure itself is used as a marker of impaired ventricular performance, such as in heart failure. In a failing heart, blood backs up into the pulmonary circulation to the extent it can markedly elevate the wedge pressure. Normal pulmonary wedge pressure is about 6 to 12 mm Hg with 18 mm Hg or above considered indicative of heart failure. The pulmonary wedge pressure is also elevated in pulmonary hypertension and is used to help determine the hemodynamic severity of mitral valve stenosis as well as a means of monitoring the effects of IV fluid infusions into a patient. In the latter case the infusion is adjusted to keep the wedge pressure between 12 and 14 mm Hg to prevent inducing pulmonary edema with the infusion.

Aortic pressure and wall stress are major determinants of afterload on the heart *in situ*.

In the left ventricle as a whole, afterload is equal to all the forces the muscle must overcome to eject a given volume of blood. The major contributor to afterload in the heart is ventricular or aortic pressure. However, it is more difficult to contract a distended heart against a given pressure than a normal heart, because cardiac wall tension and stress increase at any pressure when the average radius of the ventricular chamber increases. Therefore, tension (T = pressure × chamber radius) and wall stress (S = pressure × chamber radius/chamber thickness) are considered more accurate estimators of cardiac afterload. Other forces, such as those needed to overcome blood inertia, accelerate blood, or overcome the resistance of the valves, also contribute to afterload, especially in pathologic conditions.

The use of technologies to give better determinations of ventricular volumes, preload, and afterload in the heart encompasses a large portion of clinical cardiology. These techniques are not used in common physical examination but instead are used by qualified specialists in cardiology for the evaluation of myocardial performance as well as for the diagnosis and treatment of patients with serious heart conditions. A full description of these technologies is beyond the scope of this book.

Pressure–volume loops can be used to reveal the effects of loading conditions and inotropic state on the performance of the whole heart.

The effects of preload, afterload, and contractility on the stroke volume of the heart can be easily seen with a tracing of the changes in intraventricular pressure and volume during the cardiac cycle, called a **pressure–volume loop**, as shown in Figure 13.8. In this depiction of the cardiac cycle, the time for each portion of the cycle is, by convention, not indicated on the loop. The cardiac cycle starts at point A with the beginning of diastole. Filling of the ventricle proceeds along the line connecting point A to point B, where B represents LVEDV, which can be considered a measure of the preload on the heart. Isovolumic contraction proceeds from this point to point C, where ventricular pressure equals that in the aorta. At this point, the aortic valve opens and blood is then ejected from the ventricle against aortic pressure, which is the prime determinant of afterload

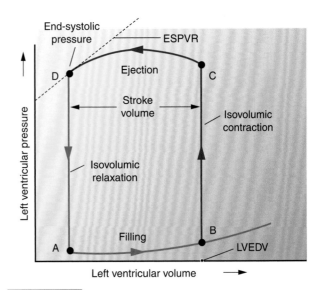

Figure 13.8 | **A pressure–volume loop representation of the cardiac cycle.** (See text for details.) LVEDV, left ventricular end-diastolic volume; ESPVR, end-systolic pressure–volume relationship.

in the normal heart. The volume in the ventricle during ejection proceeds from point C to point D, which represents the pressure and volume in the heart at the end of systole (LVESV). The volume change between LVEDV and LVESV is the stroke volume. At a given inotropic state, point D falls on a line that represents the most pressure the ventricle can generate at a given preload. This line is analogous to the isometric length–tension curve for isolated muscle, which sets the mechanical barrier for how far isotonic shortening in muscle can proceed. Finally, the heart undergoes isovolumic relaxation from point D back to point A where the diastole begins the cycle again.

Stroke volume is positively related to the level of the inotropic state of the heart.

The pressure and volume in the heart at the end of systole represent the point to which the heart has shortened such that the maximum force it can develop at that length just equals aortic afterload. The heart cannot shorten beyond this point because, at shorter circumferences, the number of crossbridges available would not enable the heart to generate enough force to move blood against existing ventricular stress. If one were to examine several cardiac cycles starting at various LVEDV (preloads) and proceeding against several different aortic pressures (afterloads), a linear plot of ventricular end-systolic pressure versus left ventricular end-systolic volume would emerge. This line is called the **end-systolic pressure–volume relationship** (**ESPVR**), and at a given inotropic state, it sets the limit to the extent of shortening cardiac muscle can produce against a given afterload. The position of this line is independent of loading conditions on the heart. *The ESPVR can only be changed by a change in the inotropic state of the heart;* it does not matter which combination of afterloaded and preloaded contractions was used to create the line. As a result, the ESPVR is considered to be an ideal indicator of the inotropic state of the heart.

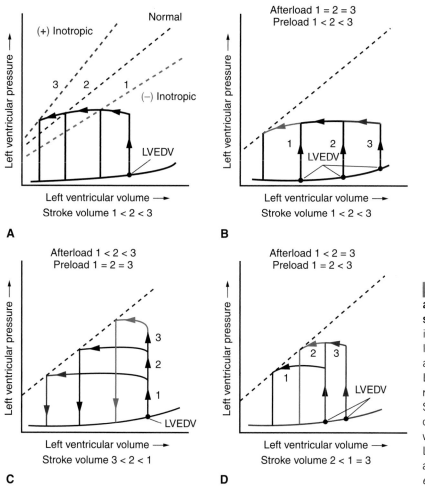

Figure 13.9 **Determinants of stroke volume as revealed by pressure–volume loop representations of the cardiac cycle.** **(A)** Effect of inotropic state on stroke volume at a given left ventricular end-diastolic volume (LVEDV) and arterial pressure. **(B)** Effect of changes in LVEDV on stroke volume at constant arterial pressure and inotropic state (the Frank-Starling relationship). **(C)** Immediate effect of changes in afterload on ventricular stroke volume at constant LVEDV. **(D)** Changes in LVEDV following an initial increase in afterload and during the next cycle. *Red, increased effect; Blue, decreased effect.*

Relative to a normal heart, a positive inotropic influence will rotate the ESPVR upward and to the left, whereas a negative inotropic influence will rotate it downward and to the right. The consequence of such a shift for the whole heart is that, *at any given afterload and preload, stroke volume is increased by a positive inotropic influence and decreased by a negative inotropic influence* (Fig. 13.9A, tracings 3 and 1, respectively).

Stroke volume is increased by an increase in preload or a decrease in afterload.

By themselves, loading conditions also affect the output of the heart. In Figure 13.9B, cardiac cycles 1, 2, and 3 occur with the same afterload and inotropic state but at different LVEDVs and thus preloads. As LVEDV increases, the initial muscle length from which myocardial contraction proceeds increases as well. Once contraction is initiated, the heart ejects blood until the pressure and volume remaining in the ventricle meets the ESPVR. As can be seen in the figure, greater initial LVEDV results in a greater stroke volume. This relationship was elucidated by Ernest Starling in the early 1900s and is known now as **Starling's law of the heart** (also known as *Starling law of the heart*, or heterometric regulation of stroke volume). Put simply, this law states, "*the more the heart fills, the more it pumps.*" The relationship between ventricular filling and the stroke output of the heart is important

in relation to the performance of the heart as a pump. As will be seen in subsequent sections, this relationship is involved in both physiologic and pathologic conditions affecting **cardiac output** (**CO**), which is the total flow output of the heart per minute.

Afterload, by itself, has a negative impact on stroke volume in the heart (see Fig. 13.9C). Each of the three cardiac cycles in this figure start from the same LVEDV and inotropic state but proceed against three different afterloads. Blood cannot be ejected from the heart until left ventricular pressure meets aortic pressure. When the ventricle contracts at higher afterloads (pressure), it will not shorten much before encountering the ESPVR. This results in a lower stroke volume compared with contractions proceeding against lower afterloads. It also indicates that pumping against elevated arterial pressure will have a detrimental effect on the output of the heart unless certain compensatory mechanisms are brought into play. One such mechanism is shown in Figure 13.9D. Cardiac cycle 1 in this figure represents a contraction at normal LVEDV against a normal arterial pressure. Loop 2 (blue tracing) represents the first contraction against an elevated arterial pressure from the same LVEDV and results in a reduced stroke volume. However, this reduced stroke volume means that the heart did not empty to its normal residual volume at the end of systole. In other words, the volume in the heart prior to its next filling is larger than normal.

Thus, when the normal diastolic inflow in the next cardiac cycle is added to this increased residual volume, an increased LVEDV will be created for the next contraction (contraction 3 on the figure, red tracing). By Starling's law, an augmented stroke volume will result. Heterometric regulation is thus a mechanism that enables the heart, within certain limits, to maintain a normal stroke volume against elevated arterial pressure.

Starling's law of the heart is also responsible for the remarkable balancing of the output of the right and left ventricles. For example, if the output of the right heart should suddenly exceed that of the left, inflow into the left ventricle would also increase, increasing LVEDV and thus subsequent output of the left ventricle; the same balance would occur if output from the right heart were suddenly decreased. In this manner, Starling's law automatically ensures that outputs of the right and left ventricle will always match.

In heart failure, the inability of individual cardiac cells to contract normally (a negative inotropic effect) can be somewhat compensated for by the fact that reduced ejection allows blood to pool in the left ventricle, increasing LVEDV and subsequently enhancing stroke volume. In fact, in heart failure, additional compensatory mechanisms in the body increase salt and water retention in the body as a whole in an effort to increase venous pressure and thus filling of the heart in diastole. This then increases stroke volume via Starling's law. Similarly, when the heart is paced at a slow rate, such as from the atrioventricular node or Purkinje fibers, the detrimental effect of this low rate on the cardiac output is somewhat attenuated by the fact that the ventricles have more time to fill and thus augment the stoke volume with each contraction according to Starling's law of the heart.

Although Starling's law applies qualitatively to the heart in any condition, the quantitative effect of this law is attenuated whenever compliance of the myocardium is decreased. This is because when the heart is stiff, little stretch of the myocardium is produced with a normal amount of ventricular filling. Consequently, subsequent stroke volumes become compromised. For example, myocardial hypertrophy, from chronic ventricular overload (common with chronic aortic valve stenosis or systemic hypertension) reduces ventricular compliance. Reduced ventricular compliance is especially problematic when it results from scar tissue formed in the heart following myocardial infarction, chronic inflammation, or infiltrative processes from infections.

Sometimes bleeding may occur between the ventricles and the pericardium, such as if a coronary artery is punctured during a cardiac catheterization procedure. This bleeding, called **cardiac tamponade**, compresses the ventricles, making it difficult for them to expand during diastole, thus reducing diastolic filling and subsequent stroke volumes. A similar situation can arise from scar tissue following **pericarditis**, in which inflamed pericardium eventually adheres to the epicardial tissue. This condition essentially wraps the heart in stiff tissue, which then impedes stretch of the heart chamber with ventricular filling.

▶ DETERMINANTS OF MYOCARDIAL OXYGEN DEMAND AND CLINICAL EVALUATION OF CARDIAC PERFORMANCE

The heart is only 0.3% of adult body weight, yet at rest, it uses ~7% of the body's total O_2 consumption. Due to the high O_2 requirements of the heart, this organ uses a wide variety of metabolic substrates. Two thirds of myocardial energy needs are derived from the metabolism of fatty acids because, on a molar basis, these compounds generate more ATP than sugars or amino acids. However, the metabolism of glucose supplies about 20% of the heart's energy needs, with the remaining needs supplied by the metabolism of lactate, pyruvate, amino acids, and certain ketones.

Any index that could predict cardiac O_2 demand under a variety of hemodynamic conditions would be invaluable to clinical medicine. Myocardial oxygen demand is one half of the important relationship between supply and demand of oxygen to the heart. Myocardial ischemia results whenever demand exceeds supply (see Chapters 15 and 16 for more details). For this reason alone, determination of oxygen demand by the heart is clinically valuable. However, a single index for determination of myocardial oxygen demand has yet to be found that is both clinically practical and universally applicable to all conditions that alter cardiac O_2 demands. Nevertheless, in an effort to develop such an index, much has been learned about determinants of oxygen consumption by the heart. This information has proved invaluable to the management of cardiovascular disease and acute cardiovascular crises in the clinical setting.

Increased afterload increases myocardial oxygen demand more than does increased preload, shortening, or inotropic state.

The energy output of a system is classically defined as the sum of heat production and work. Work, in turn, is defined as a force acting through a distance, or work = force × distance. As applied to work done by the left ventricle, work is given as the product of stroke volume and either arterial pressure, ventricular pressure, or wall stress during systole. It can also be determined by the area circumscribed by the pressure–volume loop depiction of the cardiac cycle.

In the heart, energy and, thus, oxygen must be used both to generate pressure and to pump blood (i.e., generate force and movement). In would seem that determination of cardiac work would be one way in which the oxygen demand of the heart could be determined. To the contrary, work alone is a poor indicator of the magnitude of oxygen demand by the heart. Consider that in relation to the mathematics of basic physics, there is no distinction between whether a given quantity of cardiac work results from a large stroke volume ejected against a low pressure or a small stroke volume ejected against a high pressure; the same quantitative value of work would emerge. However, in the heart, much more O_2 is required to generate force than to shorten. *Thus, more oxygen is needed to meet increased work resulting from an increase in cardiac stress than from an increase in stroke volume.* In other words, pumping against a high pressure is more oxygen costly, than pumping a high flow output, even if the total work is the same in both situations. Consequently, changes in cardiac work are poor quantitative indicators of myocardial oxygen demand.

Beyond the basal O_2 consumption needed to run membrane pumps and basic cellular activity, four major determinants of increased O_2 demand in the heart have been identified; these are (1) increased systolic pressure (or, more accurately, ventricular wall stress), (2) increased extent of muscle shortening (or stroke volume), (3) increased heart rate, and (4) positive inotropic stimuli.

Combinations of hemodynamic variables have been used as estimators of O_2 demand in the heart. The double product or systolic pressure × heart rate and the tension–time index (the area under the left ventricular systolic pressure curve) have been used as predictors of cardiac O_2 demand. Changes in these indices reflect changes in the oxygen needs of the heart caused by changes in heart rate and force generation, but they often fail in instances in which certain drugs, such as β-adrenergic agonists, produce positive inotropic and chronotropic (heart rate) effects on the heart at the same time that their vasodilator effects cause a drop in blood pressure. Indices based on the ESPVR can give accurate estimates of cardiac O_2 demand, but the determination of such relationships is currently impractical in a clinical setting.

Even without a good index available as a predictor of cardiac oxygen demand, it is useful to simply recognize factors proven to alter myocardial O_2 demand. For example, hypertension, or high blood pressure, greatly increases the stress load on the heart and thus increases O_2 demand. Similarly, exercise increases heart rate, myocardial contractility, and, in unconditioned people, blood pressure as well, all of which synergize to create greatly increase myocardial O_2 demand. In patients who have compromised coronary circulations resulting from atherosclerosis, it is thus helpful to realize that decreasing blood pressure and heart rate will lessen the possibility that cardiac oxygen demand will exceed cardiac oxygen supply.

Ejection fraction and hemodynamic evaluations are used as simple clinical indices of myocardial performance.

Clinical evaluation of the condition of the heart in patients is technologically demanding. Although many technologies and devices are available to evaluate the heart, their implementation requires specialized skills in cardiology. Nevertheless, there are some relatively simple means of estimating cardiac performance available to the clinician.

Clinicians can obtain a two-dimensional view of changes in the size of the ventricular chambers during the cardiac cycle through the use of ultrasound images employed in echocardiography. Ventricular chamber cross-sectional area during systole and diastole can be estimated from these real-time images and used to estimate stroke volume in the heart. The change in areas between diastole and systole is expressed as a percentage of the area during diastole and is called the **ejection fraction**. In normal hearts, the ejection fraction is 45% to 67%; values ≤40% indicate impaired performance.

Variables that reflect changes in the inotropic state of the heart have proven difficult to develop and are often used on the basis of what works reasonably well in the clinic. For example, the peak rate of rise of pressure in the left ventricle during isovolumic contraction (peak dP/dt) has been shown to reflect alterations in myocardial contractility and to be affected little by changes in preload or afterload. A peak dP/dt < 1,200 mm Hg/s indicates abnormally low myocardial contractility. Under certain conditions, inferences about the inotropic state of the heart can be made from the inspection of plots of stroke volume, CO, or stroke work versus LVEDV, LVEDP, or pulmonary wedge pressure. For example, the only way a patient's stroke volume can decrease or remain unchanged in the face of an increased ventricular preload is if the patient's heart is experiencing a negative inotropic influence. Conversely, increased stroke volume in the face of an unchanged or decreased preload indicates that the heart is under a positive inotropic influence.

There are certain physiologic and pathologic conditions that are known to change loading conditions and the inotropic state of the heart. For example, drugs that relax venous smooth muscle lower central venous and ultimately right atrial pressure, thereby reducing the filling of the heart and ventricular preload. Standing causes pooling of blood in the lower extremities and also decreases pressure in the right atrium with a similar effect unless the body corrects for this by activation of sympathetic nerves to veins (see Chapter 17). This occurs in normal healthy individuals. However, drugs used to treat certain cardiovascular diseases such as hypertension and myocardial ischemia can counteract this corrective reflex by directly relaxing vascular smooth muscle. Such an effect in veins causes a drop in preload when the patient taking such medications rises from a supine to standing position. Nitroglycerin in particular, which is used to treat angina, preferentially dilates veins over arteries. People taking these medications are susceptible to fainting on standing as a result of excess pooling of blood in the veins. This pooling causes a precipitous drop in right atrial pressure, reduced output of the heart, and transient low blood pressure (postural hypotension). Conversely, lying down and raising the feet increases right atrial pressure and stroke output of the heart by Starling mechanism.

Contraction of skeletal muscle squeezes muscle veins, increasing venous pressure and forcing blood back to the heart. This factor augments ventricular output during exercise by increasing preload. The lack of contraction, such as in someone standing at attention for exceeding long times, prevents this skeletal "squeezing" action and can thus lead to pooling of blood in the lower extremities and result in fainting.

Certain hormones, neurotransmitters, and drugs are known to affect the contractility of the heart. For example, any condition that activates the sympathetic nervous system will enhance myocardial contractility. Conditions that activate the vagus nerve to the heart or inhibit sympathetic activity can have a negative inotropic effect. Such conditions are discussed in later chapters. Myocarditis and myocardial ischemia can be assumed to be negative inotropic influences. Certain anesthetics, such as barbiturates, antiarrhythmic drugs, and Ca^{2+} antagonists, are known negative inotropic agents.

Heart Failure and Muscle Mechanics

Heart failure is a condition where the heart is unable to maintain sufficient output to meet the body's normal metabolic needs. In its mildest forms, it can be seen as an abnormally low ejection fraction without causing outward symptoms in the patient. In its most severe form, it relegates an individual unable to do the simplest physical activities with physical distress even present at rest. Once begun, heart failure usually becomes more severe over a period of months to years as the result of a progressive deterioration of ventricular muscle, remodeling of the ventricular walls and chambers, and sometimes ventricular wall stiffening with impairment of diastolic filling. The progression of pathologic factors that impair heart output, such as chronic hypertension, recurring myocardial ischemia or infarction, diabetes mellitus, etc., also contributes to the progression of heart failure. The term *congestive heart failure* refers to fluid congestion in the lungs that often accompanies heart failure. However, a person can have impaired ventricular performance without pulmonary congestion or fluid congestion independent of a malfunctioning heart. Although the pathogenic mechanisms that can lead to heart failure are many, complex, and incompletely understood at this time, the mechanical consequences of heart failure are more clearly understood and will be the focus of this discussion.

The most typical forms of heart failure involve impairment of systolic function, which can be seen graphically by a downward and rightward rotation of the ESPVR on a pressure volume loop diagram (see Fig. 13.9A). In congestive heart failure, retention of salt and water by the body and the reduced contractility of the heart failure combine to increase LVEDV. Although this helps increase stroke volume in the otherwise failing heart, the weakened heart cannot shorten as far and with normal velocity at normal afterloads. This

negative effect is much greater than any benefit imparted by a higher than normal LVEDV. Consequently, stroke output is diminished in the failing heart. However, the decreased slope of the ESPVR and the rather flattened Starling relationship (SV as a function of LVEDV; see Fig. 13.7) change the response of a failing heart to *changes* in afterload and preload in important ways compared to a normal heart. A simple comparison of pressure–volume loops in a failing heart versus a normal heart shows that a reduction in afterload produces a much greater increase in stroke volume in the failing heart than in the normal heart (see Fig. 13.9). Furthermore, the more flattened Starling relationship in the failing heart indicates that a reduction in LVEDV and/or LVEDP will have a smaller effect on reducing stroke volume in the failing heart than the normal heart (see Fig. 13.7). In short, the output of a failing heart is more sensitive to changes in afterload and less sensitive to changes in preload than a normal heart. This property is both a benefit and hindrance to the management of heart failure. It means that vasodilators will have an especially good benefit in patients in heart failure in that a reduction in arterial pressure will greatly augment stroke volume, whereas any venodilation effect to reduce stroke volume will be muted. The latter allows the clinician to reduce fluid volume in the patient, to relieve pulmonary and systemic congestion, without having a major impact on stroke output via Starling mechanism. However, it also means that a patient in heart failure is exquisitely sensitive to the negative effects of any increase in systemic arterial pressure and cannot get the benefit of increasing output by increasing preload without resulting in pulmonary and systemic congestion. This emphasizes the importance of the need to avoid factors that increase blood pressure and/or salt and water retention in the clinical management of patients with congestive heart failure. ■

▶ CARDIAC OUTPUT

Cardiac output (CO) is defined as the volume of blood ejected from the heart per unit time. The usual resting values for adults are 5 to 6 L/min, but this can rise to >25 L/min during heavy exercise. CO divided by body surface area is called the **cardiac index**. When it is necessary to normalize CO to compare outputs among people of different sizes, either cardiac index or CO divided by body weight can be used. CO is the product of heart rate (HR) and stroke volume (SV):

$$CO = SV \times HR \qquad (1)$$

Previously discussed factors that alter stroke volume as well as factors that alter HR combine to produce an overall effect on CO. These are outlined in Table 13.1. If HR remains constant, CO increases in proportion to stroke volume, and therefore, factors that increase stroke volume are also considered factors that can increase CO. If stroke volume

remains constant, CO increases in proportion to an HR up to about 180 beats/min. Therefore, within this limit, factors that increase HR also increase CO.

Cardiac output is maintained over a large range of heart rates through a reciprocal interaction between heart rate and stroke volume.

Heart rate can vary from <50 beats/min in a resting, physically fit person to >180 beats/min during maximal exercise. Changes in HR, however, do not necessarily cause proportional changes in CO because HR inversely affects stroke volume. As the HR increases, the duration of the cardiac cycle and diastole decreases thus reducing the time available for ventricular filling. This causes end-diastolic volume and thus stroke volume to fall. Events in the myocardium compensate to some degree for the decreased time available for filling. First, increases in HR reduce the duration of the action potential and, thus, the

TABLE 13.1 Factors Influencing Cardiac Output

Positive Effects

Increased end-diastolic fiber length (Starling's law, preload)

Increased ventricular end-diastolic volume or pressure

Decreased afterload (decrease aortic pressure or LV wall stress)

Increased contractility

Sympathetic stimulation via norepinephrine acting on β_1 receptors

Circulating epinephrine acting on β_1 receptors (minor)

Intrinsic changes in contractility in response to changes in heart rate

Positive inotropic drugs, for example, digitalis, β-receptor agonists, and phosphodiesterase inhibitors

Hypertrophy (i.e., from aerobic exercise training if filling is not impaired due to decreased compliance)

Negative Effects

Decreased end-diastolic fiber length

Decreased ventricular end-diastolic volume or pressure

Decreased ventricular compliance; hypertrophy, cardiac tamponade, scar tissue

Increased afterload (increased aortic pressure, ventricular systolic pressure, or LV wall stress)

Increased ventricular radius

Negative inotropic drugs; calcium-channel blockers, general anesthetics, β-blockers

Disease (coronary artery disease, myocarditis, cardiomyopathy, etc.)

Myocardial ischemia

Ventricular and supraventricular tachycardias

Irregular heart rate

Atrial fibrillation

Aortic valve stenosis

Mitral valve stenosis

---

the longer duration of diastole results in greater ventricular filling, increased end-diastolic fiber length, and thus an increased stroke volume. This somewhat compensates for the decreased HR. This balance works until the HR is <20 beats/min. At this point, additional increases in end-diastolic fiber length cannot augment stroke volume further because the maximum of the ventricular function curve has been reached. At HRs < 20 beats/min, CO falls in proportion to decreases in HR.

Changes in heart rate occur through the reciprocal activation of parasympathetic and sympathetic nerves to the heart.

Normal physiologic changes in HR are brought about by reciprocal changes in the activity of the parasympathetic (vagus) and sympathetic nerves to the sinoatrial (SA) node. Vagal effects predominate with regard to HR such that the HR is actually held down from what it would be if paced by the SA node without any autonomic influences present. Nevertheless, decreasing vagal activity and increasing sympathetic nerve activity to the heart bring about increases in HR. A decreased HR results from opposite changes in both branches of the autonomic nervous system.

The release of norepinephrine by sympathetic nerves not only increases the HR (see Chapter 12) but also dramatically increases the force of contraction through its positive inotropic effect. Furthermore, norepinephrine increases conduction velocity in the heart, resulting in a more efficient and rapid ejection of blood from the ventricles. These effects, summarized in Figure 13.10, maintain the stroke volume as the HR increases when the sympathetic nervous system is activated. The effect of HR on CO depends on the extent of

---

duration of systole, so the time available for diastolic filling decreases less than it would if the rate was changed through a change in the duration of diastole alone. Second, as discussed earlier, faster HRs are accompanied by an increase in the force of contraction, which tends to augment stroke volume. The increased contractility is sometimes called *treppe* or the *staircase phenomenon.* In reality, CO is not adversely affected by the decreased filling time during high HRs unless the rate exceeds ~180 beats/min. A person with abnormal tachycardia, however, (e.g., 250 to 300 beats/min caused by an ectopic ventricular pacemaker) has such reduced filling time that the patient usually has a reduction in CO despite the very high HR.

The reciprocal relationship between HR and the duration of diastole can also have a positive effect when HR is low. As the HR falls, the duration of ventricular diastole increases, and

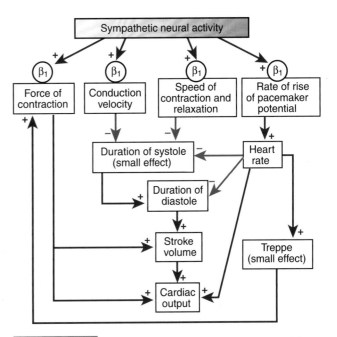

Figure 13.10 Effects of increased sympathetic neural activity on heart rate, stroke volume, and cardiac output (CO). Various effects of norepinephrine on the heart compensate for the decreased duration of diastole and hold stroke volume relatively constant, so that CO increases with increasing heart rate. *Red lines indicate stimulatory influences. Blue lines indicate inhibitory influences.*

concomitant changes in ventricular filling and contractility. CO increases proportionately over a broad range, when the HR increases physiologically as a result of an increase in sympathetic nervous system activity (as during exercise).

In summary, changing stroke volume and HR regulate CO. Stroke volume is influenced by the contractile force of the ventricular myocardium and by the force opposing ejection (the aortic pressure or afterload). Myocardial contractile force depends on ventricular end-diastolic fiber length (Starling's law) and myocardial contractility. Contractility is influenced by four major factors:

1. Norepinephrine released from cardiac sympathetic nerves and, to a much lesser extent, circulating norepinephrine and epinephrine released from the adrenal medulla
2. Certain hormones and drugs, including glucagon, isoproterenol, and digitalis (which increase contractility), and anesthetics (which decrease contractility)
3. Disease states, such as coronary artery disease, myocarditis, bacterial toxemia, and alterations in plasma electrolytes and acid–base balance
4. Intrinsic changes in contractility with changes in HR

▶ THE MEASUREMENT OF CARDIAC OUTPUT

The ability to measure CO is important for performing physiologic studies involving the heart and managing clinical problems in patients with heart disease or heart failure. In current clinical practice, CO in an individual is most often estimated as the product of HR, obtained by an ECG, and stroke volume, obtained by echocardiography. Alternatively, CO can be measured by measuring flow velocity and aortic cross-sectional area with Doppler ultrasound techniques. However, older techniques that were used to measure CO did so accurately by applying the principles of mass balance with indicator dilution techniques. This technique, in a slightly different iteration, is still employed today in experimental settings as a means for measuring oxygen consumption in an organ. An overview of the basic principles involved in determining CO and organ oxygen consumption by mass balance and indicator dilution techniques is summarized below.

Indicator–dilution techniques are applications of the principle of mass balance to determine cardiac output.

The use of mass balance to measure CO is best understood by considering the measurement of an unknown volume of liquid in a beaker. Dispersing a known quantity of dye throughout the liquid and then measuring the concentration of dye in a sample of liquid can determine the volume. Because mass is conserved, the quantity of dye (A) in the liquid is equal to the concentration of dye in the liquid (C) times the volume of liquid (V), $A = C \times V$. Because A is known and C can be measured, V can be calculated as $V = A/C$.

When the principle of mass balance is applied to CO, the goal is to measure the volume of blood flowing through the heart per unit of time. A known amount of dye or other indicator is injected, and the concentration of the dye or indicator is measured over time. This method for determining CO is called the **indicator dilution technique**. In this application, a known amount of indicator (A) is injected into the circulation, and the blood downstream is serially sampled after the indicator has had a chance to mix (Fig. 13.11). The indicator is usually injected on the venous side of the circulation (often into the right ventricle or pulmonary artery but, occasionally, directly into the left ventricle), and sampling is performed from a distal artery. The resulting concentration of

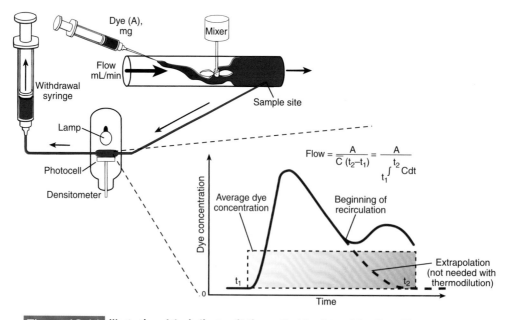

$$\text{Flow} = \frac{A}{\overline{C}\,(t_2 - t_1)} = \frac{A}{\int_{t_1}^{t_2} C\,dt}$$

Figure 13.11 **Illustration of the indicator dilution method for determining flow.** The volume per minute flowing in the tube equals the quantity of indicator (in this example, an injected dye (A) divided by the average dye concentration (C) at the sample site, multiplied by the time between the appearance (t_1) and disappearance (t_2) of the dye. The downslope of the dye concentration curve shows the effects of recirculation of the dye (*solid line*) and the semilogarithmic extrapolation of the downslope (*dashed line*) used to correct for recirculation.

indicator in the distal arterial blood (C) changes with time. First, the concentration rises as the portion of the indicator carried by the fastest-moving blood reaches the arterial sampling point. Concentration rises to a peak as the majority of indicator arrives and falls off as the indicator carried by the slower moving blood arrives. Before the last of the indicator arrives, the indicator carried by the blood flowing through the shortest pathways comes around again (recirculation). To correct for this recirculation, the downslope of the curve is assumed to be semilogarithmic, and the arterial value is extrapolated to zero indicator concentration. The average concentration of indicator can be determined by measuring the indicator concentration continuously from its first appearance (t_1) until its disappearance (t_2). The average concentration during that period (C_{ave}) is determined, and CO is calculated as

$$CO = A / [C_{ave}(t_1 - t_2)] \qquad (2)$$

Note the similarity between this equation and the one for calculating volume in a beaker. On the left is volume per minute (rather than volume, as in the beaker example). In the numerator on the right is the amount of indicator (A) and in the denominator is the average concentration over time (rather than absolute concentration, as in the beaker example). Concentration, volume, and amount appear in both examples, but time is present in the denominator on both sides in equation 2.

A better means of determining CO by the indicator dilution method involves the process of **thermodilution**. A Swan-Ganz catheter (a soft, flow-directed catheter with a balloon at the tip) is placed into a large vein and threaded through the right atrium and ventricle so that its tip lies in the pulmonary artery. The catheter is designed to allow a known amount of ice-cold saline solution to be injected into the right side of the heart via a side port in the catheter. This solution decreases the temperature of the surrounding blood. The magnitude of the decrease in temperature depends on the volume of blood that mixes with the solution, which depends on CO. A thermistor on the catheter tip (located downstream in the pulmonary artery) measures the fall in blood temperature. The CO can be determined using calculations similar to those described for the indicator dilution method. The advantage to this method is that there is no recirculation of the "indicator" and thus more accurate measurements of CO can be determined and done repeatedly over any interval of time. The thermodilution measure of cardiac output is considered the current gold standard for CO determination clinically; all other techniques are measured against it.

Another way in which the principle of mass balance is used to calculate CO takes advantage of the continuous entry of oxygen into the blood via the lungs. This measurement is based on **Fick's principle** and is the oldest means of accurately measuring CO. Although very accurate, measurement of CO by Fick's principle is very cumbersome and impractical for modern day clinical use. However, the principles espoused by this technique have instructional value to medical students.

To explain, in a steady state, the oxygen leaving the lungs (per unit time) via the pulmonary veins must equal the oxygen entering the lungs via the mixed venous blood returning from systemic organs plus that added into that blood at the pulmonary capillaries from air taken into the lungs during ventilation. When the body is at rest and in a steady state, without having recently been exposed to exercise or a meal, the amount of oxygen entering the blood through the lungs per minute is equal to the amount consumed by body metabolism per minute ($\dot{V}O_2$). Therefore, O_2 in blood leaving the lungs equals O_2 input via the pulmonary artery plus O_2 added by respiration.

The O_2 output via the pulmonary veins, in mL of O_2 per minute, is equal to the pulmonary vein O_2 content (vO_2, in mL of O_2 per mL of whole blood) multiplied by the CO (in this case, mL of whole blood per minute). Because O_2 is neither added nor subtracted from the blood as it passes from the pulmonary veins through the left heart to the systemic arteries, the arterial O_2 content (aO_2) multiplied by the CO is a reasonable measure of the O_2 output via pulmonary veins. Similarly, O_2 input via the pulmonary artery is equal to mixed venous blood oxygen input to the right heart and is given as mixed venous blood O_2 content (vO_2) multiplied by the CO. As indicated above, in the steady state, O_2 added by respiration is equal to oxygen consumption. Thus,

$$CO \times aO_2 = CO \times vO_2 + \dot{V}O_2 \qquad (3)$$

which rearranges to

$$CO = \dot{V}O_2 / (aO_2 - vO_2) \qquad (4)$$

Thus, CO can be calculated with measurement of systemic arterial blood oxygen content, pulmonary arterial (mixed venous) blood oxygen content, and total body oxygen consumption. This method measures CO accurately. However, the needed elements to the equation require cardiac catheterization and measurement of whole body oxygen consumption, which are impractical to use in the clinical setting. The relationship above, however, can be rearranged to yield a way to determine O_2 consumption in an organ or region of tissue ($\dot{V}O_2$) if total blood flow into the region can be measured along with the difference between the arterial oxygen content entering the region and that exiting it in its venous effluent (tissue $aO_2 - vO_2$, also called *oxygen extraction*). This determination is a more common application of Fick's principle in cardiovascular physiology.

▶ IMAGING TECHNIQUES FOR MEASURING CARDIAC STRUCTURES, VOLUMES, BLOOD FLOW, AND CARDIAC OUTPUT

A variety of clinical techniques employ imaging modalities as a means of measuring/estimating ventricular wall and chamber dimensions, the motion of the ventricles during the cardiac cycle and CO. Those that use time-dependent images of the heart to estimate the difference between end-diastolic and end-systolic volumes, which gives stroke volume, can be combined with HR to estimate CO.

Technetium-99 scans use tagging of radioactive technetium-99 to red blood cells or albumin as a means of measuring blood distribution. The radiation (gamma rays) emitted by the large pool(s) of blood in the cardiac chambers is measured using a specially designed gamma camera. The emitted radiation is proportional to the amount of technetium bound to the blood (easily determined by sampling the tagged blood) and the volume of blood in the heart. Using computerized analysis, the amount of radiation emitted by the left (or right) ventricle during various portions of the cardiac cycle can be determined (Fig. 13.12). Stroke volume is determined by measuring the difference in the amount of radiation measured at

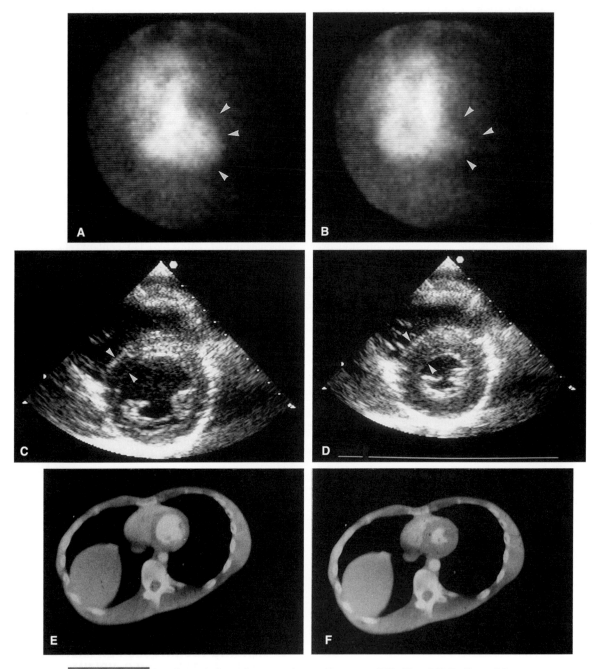

Figure 13.12 **Imaging techniques for measuring cardiac output (CO). (A and B)** Radionuclide angiograms. The *white arrowheads* in **(A)** show the boot-shaped left ventricle during cardiac diastole when it is maximally filled with radionuclide-labeled blood. In **(B)**, much of the apex appears to be missing (*white arrowheads*) because cardiac systole has caused the blood to be ejected as the intraventricular volume decreases. **(C and D)** Two-dimensional echocardiograms. In this cross-sectional view, the left ventricle appears as a ring. *White arrowheads* indicate wall thickness. In diastole **(C)**, the ventricle is large and the wall is thinned; during systole **(D)**, the wall thickens and the ventricular size decreases. **(E and F)** Ultrafast (cine) computed tomography. The ventricular size and the wall thickness can be assessed during diastole and systole, and the change in ventricular size can be used to calculate CO.

the end of diastole with that at the end of systole and multiplying this number by the HR to yield the CO.

Echocardiography is a noninvasive technique that uses ultrasound waves to produce one-dimensional and two-dimensional, real-time images of the heart. This is the same technology that is used to image the fetus during pregnancy. The one-dimensional mode of this technique, called **M-mode echocardiography** (where M stands for motion), sends a single beam of ultrasound across a single position to reveal a reflection of the thickness of ventricular walls and the chambers of the heart as well as the motion of the walls and valves during the cardiac cycle. The dimensions of the great vessels can also be imaged with this technique. Detection of abnormalities in ventricular wall motion is an important clinical sign of underlying muscle disease. For example, ischemic myocardium does not contract, and infarcted areas of the heart weaken and actually bulge outward, rather than contract inward, during systole. Such abnormalities are also imaged with two-dimensional echocardiography (**2D echocardiography**). In 2D echocardiography, the ultrasound probe rapidly scans back and forth to give a two-dimensional cross-sectional image of the heart. Simply changing the position of the probe can change the plane of this cross section and the resulting image. This method is used as a noninvasive means of estimating stroke volume and ejection fraction in the heart of a patient.

Doppler ultrasound is also used to detect blood flow across the valves of the heart and in blood vessels. The velocity of blood flow can be determined by measuring the Doppler shift (change in sound frequency) that occurs when the ultrasound wave is reflected off moving blood, much the same way a radar device measures the speed of a moving object. In cardiac evaluation, the velocity of blood flow is color-coded by computer. Turbulence and unusually high flow velocities, such as that which occurs when blood is ejected through a stenotic aortic valve, can be detected. The direction of flow can also be color-coded. This is useful in detecting regurgitation through valves such as in **mitral valve prolapse**.

In sum, echocardiography can be used to measure changes in ventricular chamber size (see Fig. 13.12C, D), aortic diameter, and aortic blood flow velocity occurring throughout the cardiac cycle. CO can also be estimated with this information in one of two ways. First, the change in ventricular volume occurring with each beat (stroke volume) can be determined and multiplied by the HR. Second, the average aortic blood flow velocity can be measured (just above or below the aortic valve) and multiplied by the measured aortic cross-sectional area to give aortic blood flow (which is nearly identical to CO). Ultrafast (cine) computed tomography and magnetic resonance imaging provide cross-sectional views of the heart during different phases of the cardiac cycle (see Fig. 13.12E, F). Stroke volume (and CO) can be calculated using the same principles described for radionuclide techniques or echocardiography. However, when ventricular volume changes are estimated from simply cross-sectional data, assumptions are made about ventricular geometry. The heart is not a sphere, and therefore, its volume cannot be estimated from a single dimension. Estimates using two or three axis approximations of an ellipse can be used with imaging techniques to give a better estimate of volume changes, but these assumptions can lead to errors in calculating CO. Nevertheless, such estimates of ventricular volumes and CO from imaging ventricular dimensions have proven useful in clinical applications.

INTEGRATED MEDICAL SCIENCES

Current Pharmacotherapy for the Management of Chronic Congestive Heart Failure

Although there are many factors that can lead to chronic congestive heart failure (CHF), the goals of any therapy for the condition are essentially the same: to (1) improve organ perfusion and (2) relieve pulmonary and systemic congestion caused by fluid accumulation in the patient. In Clinical Focus 13.2, it was shown that both these factors can be improved by exposing the patient to generalized vasodilation. Although this was part of early treatment strategies for CHF decades ago, it is now relegated primarily for short-term benefit in acute heart failure emergencies. Newer pharmacotherapies, based on better understanding of cellular malfunctions in heart failure, are now part of modern management of the syndrome. CHF can be thought of as a slow "shock" syndrome in which mechanisms the body normally uses acutely to support blood pressure and cardiac output, when those variable are impaired acutely,

instead become activated chronically (see Chapter 17). These compensations include support of blood pressure by activation of the sympathetic nervous system and the formation of a peptide, angiotensin II, which is a vasoconstrictor that also promotes salt and water retention by the kidneys. This latter factor supports cardiac output acutely through Starling mechanism. However, the cardiovascular system does not respond well to continual activation of these systems over months, which is what occurs in chronic heart failure. Chronic exposure to angiotensin II results in overproduction of oxygen radicals in the heart as well as blood vessels and is also mitogenic. Long-term exposure to catecholamines is especially damaging to the heart as well. In chronic heart failure, continual exposure to these agents results in accelerated apoptosis in the heart and malignant myocardial remodeling in which the heart, already weakened, becomes dilated and thus under even more stress. For this reason, current therapeutic recommendations are that patients be given drugs that block the actions of catecholamines and reduce angiotensin II in the circulation at the very first sign of heart failure (i.e., reduced ejection fraction) even if the patient has yet to show any symptoms of that cardiac impairment. Current recommendations are to place patients on β-adrenergic receptor blockers and a class of drugs called angiotensin-converting enzyme inhibitors (ACE inhibitors) to blunt the formation of angiotensin II in the body. The benefit of the latter is easy to visualize in that not only does this agent block the long-term pathological effects of sustained angiotensin II exposure but it also benefits the CHF patient by lowering vascular resistance and reducing water retention. The beneficial effects of a β-receptor antagonist (or "beta-blocker") are not apparent at first glance and were actually an unexpected finding during clinical trials examining the use of such agents for other purposes. Acutely, inhibition of cardiac beta receptors produces negative inotropic and chronotropic effects on the heart; these obviously, at first, would appear to be detrimental to patients with CHF. Indeed, within the first month of treatment with a beta-blocker, CHF patients may actually present with a slightly even more reduced ejection fraction. However, with sustained treatment over several months, their ejection fraction improves to a level higher than that seen when they first showed signs of failure. Current theory proposes that beta-blockers provide an "antiremodeling" effect on the failing heart, thus preventing and even reversing the malignant cardiac dilation seen with the disease. Beta-blockers

and ACE inhibitors have been proven to improve exercise tolerance and endurance and extend the lives of patients with CHF.

In spite of recent evidence of the benefit beta-blockers and ACE inhibitors in patients with CHF, it is generally difficult to manage the patient with more severe forms of CHF without the use of drugs to help remove water and salt from the body. Diuretics are a class of drugs that help the kidney excrete sodium and water (see Chapter 23) and are used extensively to aid patients with CHF, especially in terms of alleviating pulmonary and systemic edema. Finally, it would seem that finding a means to stimulate the failing heart might help patients with CHF by directly increasing its output in spite of its weakened or damaged tissue. To date however, there is only one agent used for this purpose, that being the cardiac glycoside digitalis. Digitalis has been used to treat "heart ailments" in medicine for over 200 years. It partially inhibits the Na^+/K^+ ATPase in myocardial cells, which secondarily impairs calcium extrusion from the cells by the Na^+/Ca^{2+} exchanger in the plasma membrane. This results in a positive inotropic effect on the heart and an improvement in its stroke output. However, digitalis also has detrimental electrophysiological properties on the heart and is a difficult drug to manage in the patient with CHF. Digitalis partially depolarizes the resting cell membrane potential, reduces the amplitude of myocardial action potentials and dV/dt, shortens the cell refractory period, and causes phase 4 in Purkinje fibers to drift toward threshold for firing an action potential. All of these factors taken together are a recipe for stimulating the formation of ectopic foci in the heart and creating favorable conditions for the creation of re-entry tachycardias and DADs. ■

Chapter Summary

- Tetanic contraction is prevented in cardiac muscle because the cardiac muscle action potential is longer than the duration of the contraction.
- Cardiac muscle operates at lengths along the ascending limb of the isometric length–tension curve.
- The velocity of shortening of cardiac muscle is inversely related to the force being exerted.
- Cardiac muscle contraction can be regulated by changes in contractility.
- The contractility of cardiac muscle is changed by inotropic interventions that include changes in the heart rate, the presence of circulating catecholamines, or sympathetic nerve stimulations.
- Calcium enters a cardiac muscle cell during the plateau of the action potential and promotes the release of internal calcium stores in the sarcoplasmic reticulum.
- Changes in cardiac muscle contractility are associated with changes in the amount of calcium released by calcium-induced calcium release mechanisms.
- Ventricular ejection is divided into rapid and reduced ejection phases.
- Stroke volume is the amount of blood ejected from the ventricles during one systole; it is the difference between ventricular end-diastolic and end-systolic volumes.

- The ventricles do not empty completely during systole, leaving a residual volume in the ventricle for the next filling cycle.
- Ventricular filling is divided into rapid and reduced filling phases.
- Heart sounds during the cardiac cycle are related to the opening and closing of valves in the heart.
- Venous pressure waves can detect abnormalities in atrioventricular valves.
- Cardiac output is the total flow output of the heart per minute and is the product of stroke volume times heart rate.
- Stroke volume is determined by end-diastolic fiber length, afterload, and contractility. Heart rate influences ventricular filling time and stroke volume so that changes in cardiac output caused by changes in heart rate are attenuated.
- The influence of the heart rate on cardiac output depends on simultaneous effects that enhance ventricular contractility at high rates.
- Cardiac energy demands are determined by collective contributions from ventricular wall stress, heart rate, stroke volume, and contractility.
- The energy cost of work in the heart is greater for work done to generate pressure than for work done to eject blood.
- Cardiac output can be measured by methods that rely on mass balance or cardiac imaging.

Chapter Review Questions

1. Which of the following pathological conditions would be most likely to create a prominent third heart sound during normal auscultation of the heart with a stethoscope?

    A. Congestive heart failure
    B. Systemic arterial hypertension
    C. Increased heart rate by 20 beats/min
    D. Decreased heart rate by 20 beats/min
    E. Arterial hypotension (i.e., lower-than-normal blood pressure)

The correct answer is A. The third heart sound is not normally heard, and its presence is a sign of underlying pathology. Its intensity is increased whenever the heart fills into a distended chamber or stiff ventricle. Ventricular distention is common in heart failure as blood backs up into the failing heart. High arterial pressure would increase the intensity of the second heart sound. Hypotension would not be expected to affect the third heart sound because the third sound occurs during diastole. Modest changes in heart rates are normal physiological adjustments and would not create an abnormal heart sound.

2. If a normal heart has a left ventricular end-diastolic volume (LVEDV) of 140 mL, a stroke volume of 70 mL, and a cardiac output of 4.9 L/min and pumps against a systolic arterial pressure of 120 mm Hg, which would be true of this same

heart if its LVEDV became 160 mL and its stroke volume 70 mL without a change in its cardiac output or mean systolic arterial pressure?

    A. Myocardial oxygen consumption would be increased because heart rate increased.
    B. Myocardial oxygen consumption would be decreased due to an increase in extent of shortening of the heart muscle.
    C. Myocardial oxygen consumption would be increased because the heart is now under a positive inotropic influence.
    D. Myocardial oxygen consumption would be increased due to an increase in systolic wall stress.
    E. Myocardial oxygen consumption would be decreased because passive diastolic stretch on the ventricle is decreased.

The correct answer is D. An increase in LVEDV without a change in stroke volume means the radius of the ventricular chamber is greater than normal throughout systole. Without any change in mean systolic pressure, the increased chamber radius increases ventricular wall stress. Wall stress is a major determinant of myocardial oxygen consumption. Filling did not increase in the ventricle, but rather just the volume about which filling occurred. Heart rate did not change in this example because CO = SV × HR and SV and CO did not change. A heart with increased LVEDV should increase SV if nothing else changed in the myocardium. The only way stroke volume could not change in this example when preload is increased is if the heart was under a negative inotropic influence. A negative inotropic condition would tend to reduce myocardial oxygen consumption, but this effect would likely be negated by the increased wall stress, which is the most significant variable in determining myocardial oxygen consumption in this instance.

3. Which of the following will promote an increase in the stroke volume of the heart?

   A. A reduction in venous tone
   B. A pneumothorax
   C. Dehydration
   D. General anesthetics
   E. Skeletal muscle contraction

The correct answer is E. Contraction of the skeletal muscle, such as that which occurs during walking or running, compresses veins and increases their intravascular pressure. This increased pressure translocates blood to the thoracic cavity (central circulation), increasing right atrial pressure, which, in turn, increases ventricular diastolic filling and preload. This increased preload results in an increased stroke volume by Starling mechanism. A reduction in venous tone (smooth muscle contraction in the veins) would have the opposite effect on right atrial pressure and filling and, thus, reduce stroke volume. Dehydration will reduce all body water volume, including that in the intravascular compartment. However, the primary effect of a loss of volume in the body on the vascular compartment is in the venous side of the circulation. Therefore, dehydration reduces venous and right atrial pressure. General anesthetics are negative inotropes and, by reducing myocardial contractility, will reduce stroke volume at any given preload or afterload.

4. Caffeine inhibits cyclic adenosine monophosphate phosphodiesterase and stimulates the release of calcium from the sarcoplasmic reticulum. The effect of caffeine on the heart through this effect would most likely be:

   A. an increase in myocardial contractility.
   B. a decrease in the rate of ventricular relaxation.
   C. a decrease in calcium influx into myocardial cells during systole.
   D. a reduction in the amount of calcium pumped into the sarcoplasmic reticulum during diastole.
   E. a decrease in myocardial contractility.

The correct answer is A. Both effects of calcium would increase the intracellular calcium concentration in the heart, thereby augmenting myocardial contractility. Cyclic adenosine monophosphate (cAMP) phosphodiesterase increases intracellular cAMP concentration by inhibiting its breakdown. cAMP increases the open time of L-type calcium channels and indirectly stimulates pumping of calcium into the sarcoplasmic reticulum by calcium ATPases. The latter effect helps speed myocardial relaxation during diastole.

5. All of the following would be consistent with the finding of heart failure in a patient except:

   A. elevated pulmonary wedge pressure (a measure of pressure in the left atrium).
   B. decreased stroke volume at normal arterial pressure.
   C. a left ventricular ejection fraction under 30%.
   D. an increased left ventricular max dP/dt (the maximum rate of change in ventricular pressure during systole).
   E. a left ventricular Starling relationship that is shifted downward and to the right compared to a normal heart.

The correct answer is D. A heart that is failing exhibits the characteristics of a heart under the influence of negative inotropic effects. The heart pumps either a normal stroke volume, but at a higher than normal filling pressure, or a reduced stroke volume in spite of a normal or even increased filling pressure. The latter is most common in heart failure beyond its earliest manifestations and is reflected in an abnormally high pulmonary wedge pressure. Another way of viewing this pressure is to consider that because the heart cannot produce a normal output, blood backs up into the pulmonary circuit, thereby raising pressure there. The ejection fraction of a normal heart is above 45% and up to 67%. A heart in failure has a flattened Starling relationship of stroke volume to preload; compared to a normal heart, a failing heart produces very little augmentation of stroke volume for a given increase in preload (ventricular filling).

Clinical Application Exercises 13.1

POSTINFARCTION LEFT VENTRICULAR FAILURE

A 44-year-old female patient with a 2-pack-a-day cigarette habit since age 18 suffers a major myocardial infarction in the anterior wall of her left ventricle from a blood clot in the proximal portion of her left anterior descending artery. The clot is removed through intracoronary administration of tPA during an emergency cardiac catheterization procedure, but the removal does not occur until 4 hours after the onset of the infarction. Twenty-four hours later, the patient is conscious with a heart rate of 100, BP of 110/85, occasional PVCs, dyspnea, tachypnea, and pulmonary rales. An echocardiogram of the patient reveals an abnormally large residual volume, increased left ventricular diastolic and systolic dimensions, an ejection fraction of 29%, and reduced peak systolic aortic flow velocity. During systole, there is asymmetric contraction of the left ventricle with bulging of the anterior wall. Chest x-ray evaluation reveals pulmonary congestion and edema formation.

QUESTIONS

1. What general diagnosis would you give this patient 24 hours after her myocardial infarction and what is its primary cause?

2. What explains the patient's echocardiographic findings?

3. What is the likely systolic left ventricular wall stress in this patient during the cardiac cycle as compared to someone without this magnitude of myocardial infarction?

4. General vasodilator therapy, which dilates both arteries and veins, will enhance stroke volume and relieve pulmonary congestion in this patient. What is the mechanical mechanism of this generalized vasodilator effect on the heart such that it can both increase stroke volume and reduce pulmonary congestion? How does this effect differ from that which would be seen in a normal healthy patient given this same therapy?

ANSWERS

1. The patient has all the symptoms of left ventricular congestive heart failure resulting from loss of a large amount of left ventricular contractile mass due to cell death and injury from the myocardial infarction. The infarction was likely exacerbated by the extended time the myocardium was ischemic followed by ischemia reperfusion injury during restoration of LAD patency by tPA.

2. Ischemic or infarcted myocardium is acontractile and cannot generate force during systole. The magnitude of this patient's infarction has obviously impaired myocardial contractile performance as evidenced by the low EF and low peak aortic flow velocity. As such, residual volume rises in this patient, and the diastolic volume of the heart increases. However, in spite of the increased preload, the heart is unable to sustain a normal stroke volume. This conclusion is consistent with the observation of a larger than normal end-systolic volume, abnormally low ejection fraction, reduced peak aortic flow velocity, and small pulse pressure. Infarcted myocardium is stiffer than normal myocardium but retains some compliance although it cannot contract. Therefore, the increase in ventricular pressure during systole causes the infracted portion of the myocardium to bulge during myocardial contraction.

3. Although the patient's blood pressure is slightly lower than average, the increased ventricular chamber size throughout the cardiac cycle places this heart in a condition of increased wall stress. This stress is especially high in the infracted, bulging region of the myocardium where the radius of curvature of the ventricular wall is greatly increased.

4. Stroke volume in the heart is determined by preload, afterload, and the position or slope of the left ventricular end-systolic pressure volume relationship (ESPVR). Preload, in part, is a function of venous filling pressure, and afterload is largely determined by peripheral vascular resistance and systolic arterial pressure. The ESPVR determines the final point to which the heart can contract at any given set of loading conditions and has a positive slope on a plot of left ventricular pressure versus volume. Contraction proceeds from any given afterload and preload until the PV point intersects the ESPVR. This determines the extent of shortening and the resulting stroke volume. Decreasing afterload positions any contraction at a level where the ESPVR is farther away from the start of contraction. Thus, extent of shortening can increase when afterload is decreased, and stroke volume will thus increase when afterload is decreased. As can be seen with any plot of the ESPVR in conjunction with a left ventricular pressure–volume loop, this enhancement of stroke volume is small in a heart with a steep ESPVR, such as a normal healthy heart. However, a heart in failure has an ESPVR with a decreased slope resulting from that heart's depressed contractile state. The result of this decreased slope is that the left end of the ESPVR is tilted up and to the left on the left ventricular pressure–volume relationship. When contractions from any loading condition proceed with this ESPVR in place, the contraction is able to proceed farther than that seen for the same loading conditions in a normal healthy heart. Thus, augmentation of stroke volume by decreasing afterload with an arterial dilator is greater in a failing heart than in a normal heart.

   Venodilation decreases venous filling pressure and tends to reduce stroke volume. However, venodilation allows for fluid to exit the central circulatory blood volume compartment thus reducing blood volume in the pulmonary circulation and in the heart. This helps reduce pulmonary congestion and ventricular wall stress, respectively. A normal heart has a steep Frank-Starling relationship, which means that small changes in preload have marked effects on stroke volume. Thus, venodilation in a healthy heart will likely cause a precipitous drop in cardiac output owing to this steep relationship. However, the Frank-Starling curve in a failing heart is very flat; rather large decreases in venous filling pressure (preload) can occur with minimal negative effect on stroke volume. This small effect to reduce stroke volume is offset by the concurrent arterial dilation afterload effect. Thus, implementation of generalized vasodilator therapy in a patient in heart failure augments stroke volume while at the same time reducing pulmonary congestion and myocardial wall stress.

thePoint* *Visit* http://thepoint.lww.com/rhoades5e *for additional chapter review Q&A, Clinical Application Exercises, animations, and more!*

Active Learning Objectives

Upon mastering the material in this chapter, you should be able to:
- Predict how changes in arterial compliance and stroke volume will change pulse pressure.
- Use analysis of the arterial pressure waveform to determine what changes in heart rate, stroke volume, and vascular resistance created the waveform.
- Explain how shifts between central and systemic venous volumes affect cardiac output.
- Explain how changes in central venous pressure cause changes in cardiac output and how changes in cardiac output cause changes in central venous pressure.

- Given the interdependence between cardiac output and central venous pressure, explain how equilibrium between the two variables arises in the cardiovascular system.
- Explain how and why the equilibrium point between cardiac output and central venous pressure is shifted by changes in blood volume, venous tone, arteriolar resistance, and the inotropic state of the heart both individually and collectively.

The aorta and other large-diameter arteries do not contribute significantly to total vascular resistance and therefore do not play a role in the regulation of organ blood flow or systemic arterial pressure. However, their elastic characteristics influence a variety of important cardiovascular variables. These include the arterial pressure waveform, cardiac work, and the effects of changes in vascular characteristics on cardiac output (CO).

▶ DETERMINANTS OF ARTERIAL PRESSURES

Four principal pressures of physiologic interest are contained within the arterial pressure waveform. These are the mean, systolic, diastolic, and pulse pressure. Clinically, arterial pressure is often reported as the ratio of systolic (peak) over diastolic pressure (lowest pressure), for example, as 120/80 or 110/75 mm Hg. Arterial pressure varies between these two values from heart beat to heart beat.

Mean arterial pressure is determined by cardiac output and systemic vascular resistance.

Mean arterial pressure, $\overline{P}_a$, is the integrated average arterial pressure determined from the arterial pressure waveform (Fig. 14.1). However, it is often approximated from the equation

$$\overline{P}_a = P_d + 1/3(P_s - P_d) \qquad (1)$$

where P_d is the diastolic pressure, P_s is the systolic pressure, and $P_s - P_d$ is the **pulse pressure**. Mean arterial pressure is closer to P_d, instead of the arithmetic average of P_s and P_d, because the duration of diastole is about twice as long as that of systole. In addition, because right atrial pressure is small and near zero (~2 mm Hg), mean arterial pressure can be approximated as equal to the product of cardiac output and total systemic vascular resistance (CO × SVR).

Changes in stroke volume and arterial compliance alter pulse pressure.

Pulse pressure is a function of stroke volume and arterial compliance. Figure 14.2A shows how aortic pressures change in response to a simple chance in aortic volume alone (i.e., when there is no concurrent change in aortic compliance, SVR, and heart rate). During rapid ejection, blood enters the aorta faster than it can run off, thus increasing aortic volume. This increased volume increases pressure in the aorta to a peak value in accordance with the interrelationship between aortic compliance, volume, and pressure in a hollow, flexible structure ($\Delta P = \Delta V/C$). During the reduced ejection phase of the cardiac cycle and diastole, blood exits the arteries faster than it enters. Volume in the aorta thus declines as does the pressure associated with it. If the amount of blood ejected into the aorta doubled, without a change in heart rate or SVR, this would double CO and, from Poiseuille's law, double mean arterial pressure. Changes in pressure resulting from increases and decreases in arterial blood volume would oscillate about this new mean. In the example in which ejection doubles, most of this volume would be ejected in the rapid ejection phase. Therefore, the amount ejected that exceeds runoff would also approximately double, and this would be reflected by a larger increase in the systolic pressure. Runoff would be little affected by this increased ejection, so diastolic pressure would fall at about the same *rate* as before but from this new, higher systolic pressure. Therefore, pressure would drop to a diastolic pressure that is considerably higher than before. Thus, an increase in mean aortic, systolic, diastolic, and pulse pressure results when stroke volume increases without any concurrent change in SVR, arterial compliance, and heart rate.

As explained in Chapter 11, arterial compliance is not constant but rather, in part, a function of transmural pressure in the arterial system; arterial compliance decreases as transmural pressure is increased (Fig. 14.2B). Because of this,

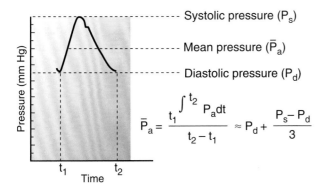

Figure 14.1 **Definition of mean arterial pressure.** Mean pressure is the area under the pressure curve divided by the time interval. This can be approximated as the diastolic pressure plus one third the pulse pressure.

a given change in aortic volume at a low initial volume, and thus transmural pressure, causes a relatively small change in arterial pressure with ejection. However, at a high initial transmural pressure, that same change in volume causes much larger changes in arterial pressure.

The general effect of a change in arterial compliance on arterial pressures is shown in Figure 14.2C. When arteries are stiff, they flex less during systole and recoil less during diastole. Therefore, for a given change in volume, decreased arterial compliance causes an increase in systolic pressure and a *decrease* in diastolic pressure. Note, if there is no change in arterial resistance or heart rate, mean arterial pressure will *not* change with a change in arterial compliance alone.

Interactions among stroke volume, heart rate, and systemic vascular resistance alter arterial pressures.

When CO changes in the face of a constant vascular resistance, mean arterial pressure is influenced according to the

formula $\bar{P}_a = CO \times SVR$. The influence of a change in CO on mean arterial pressure is independent of the cause of the change, whether it comes from heart rate or stroke volume. In contrast, the effect of a change in CO on pulse pressure greatly depends on whether and how stroke volume or heart rate change.

If an increase in heart rate is balanced by a proportional and opposite change in stroke volume, *mean* arterial pressure does not change because CO remains constant. However, the decrease in stroke volume that occurs in this situation results in a diminished pulse pressure. The reduced volume ejected during systole reduces systolic blood pressure, while the increased heart rate provides less time for pressure to diminish during diastole. Consequently, diastolic pressure increases. An increase in stroke volume accompanied by a decreased heart rate such that there is no change in CO likewise causes no change in mean arterial pressure. The increased stroke volume, however, produces a rise in pulse pressure. The increased volume ejected with each contraction causes a greater rise in systolic pressure. However, because heart rate is reduced, there is more time available for blood to run out of the arteries during diastole. Arterial volume, therefore, drops to a lower value than before, and diastolic pressure decreases as a result.

Another way to think about these events is depicted in Figure 14.3A. The first two pressure waves have systolic pressure of 120 mm Hg, a diastolic pressure of 80 mm Hg, and a mean arterial pressure of 93 mm Hg. Heart rate in this example is 72 beats/min. After the second beat, the heart rate is slowed to 60 beats/min, but stroke volume is increased sufficiently to maintain the same CO. The longer time interval between beats allows the diastolic pressure to fall to a new (lower) value of 70 mm Hg. The next systole, however, produces an increase in pulse pressure because of the ejection

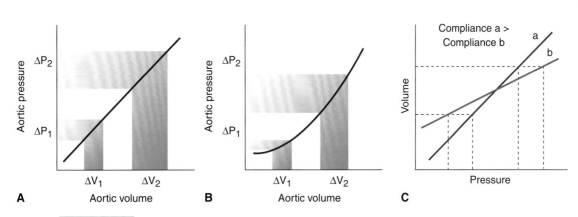

Figure 14.2 **Relationship between aortic volume and pressure. (A)** Theoretical condition where aortic compliance is independent of aortic volume. The change in volume (ΔV_1) causes the change in pressure (ΔP_1). A larger volume increment (ΔV_2), without changing heart rate or arterial compliance, produces a larger mean pressure and larger change in pressure (ΔP_2). **(B)** In the cardiovascular system, aortic compliance decreases as aortic volume and pressure increase. A change in volume (ΔV_1) is associated with a change in pressure (ΔP_1). The same change in volume at a higher initial volume (ΔV_2) causes a much larger change in pressure (ΔP_2) because of reduced compliance at the higher mean pressure. **(C)** With the same ΔV, ΔP increases with a decrease in compliance.

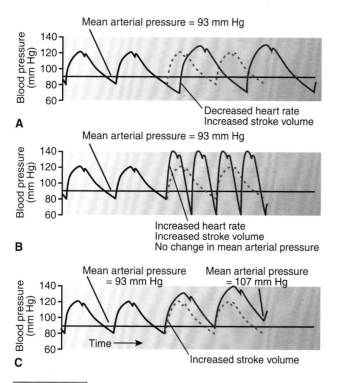

Figure 14.3 **Effects of changes in heart rate (HR), stroke volume, and systemic vascular resistance (SVR) on arterial pressure.** **(A)** Effect of increased stroke volume on arterial pressure with constant cardiac output (CO) and SVR. When cardiac output is held constant by lowering HR, there is no change in mean arterial pressure (93 mm Hg), but systolic pressure increases while diastolic pressure decreases. **(B)** Effect of increased HR *and* stroke volume with no change in mean arterial pressure because of decreased SVR. After the first two beats, stroke volume and HR are increased. Pulse pressure increases around an unchanged mean arterial pressure. Systolic pressure is higher due to higher ejection associated with the increased stroke volume, and diastolic pressure is lower than the control because the lower SVR allows more rapid runoff of blood out of the arterial system during diastole. **(C)** Effect of increased stroke volume, with constant HR and SVR. CO, mean arterial pressure, systolic pressure, diastolic pressure, and pulse pressure are all increased.

of a greater stroke volume, so systolic pressure rises to 130 mm Hg. The pressure then falls to the new (lower) diastolic pressure, and the cycle is repeated. Mean arterial pressure does not change because CO and SVR are constant, and the increased pulse pressure is distributed around the same mean arterial pressure as before. In some instances, mean arterial pressure may remain constant despite a change in CO because of an alteration in SVR. A good example of this is **dynamic exercise** (e.g., running or swimming). Dynamic exercise often produces little change in mean arterial pressure because the increase in CO is balanced by a decrease in SVR. The increase in CO is caused by increases in both heart rate and stroke volume. The elevated stroke volume results in a higher systolic pressure, whereas diastolic pressure is lower because the fall in SVR allows greater runoff from the aorta during diastole (Fig. 14.3B).

Figure 14.3C shows what happens if CO is increased by increasing stroke volume with no change in heart rate or SVR. The increased stroke volume occurs at the time of the next expected beat, and the diastolic pressure is as it was for previous beats, or 80 mm Hg. After a transition beat, the increased stroke volume results in an elevation in systolic pressure to 140 mm Hg. The decline of blood pressure during reduced ejection and diastole is not affected by the increased stroke volume alone, and thus, pressure falls normally during this phase, but from the higher systolic pressure caused by the increased stroke volume. This results in an increased final diastolic pressure (~90 mm Hg). Thus, in this new steady state, systolic, diastolic, and mean arterial pressures are all higher, the latter because CO increased while SVR remained unchanged. Also, in this instance, a further change in pulse pressure is created because the increase in mean arterial pressure (to 107 mm Hg) pushes oscillations in pressure to a less compliant region of the arterial system (see Fig. 14.2). Consequently, the increase in pulse pressure seen results from both higher stroke volume and decreased arterial compliance.

► ARTERIAL PRESSURE MEASUREMENT

Arterial blood pressure can be measured by direct or indirect (noninvasive) methods. In the laboratory and sometimes in clinical research or diagnostic settings, a cannula or pressure transducer-tipped catheter can be placed in an artery to measure pressure directly. During simple physical exams in clinical practice, however, blood pressure is usually measured indirectly.

Sphygmomanometry is a noninvasive measurement of blood pressure in humans.

Clinical measurement of blood pressure is almost always determined noninvasively by sphygmomanometry. In this method, a **sphygmomanometer** that registers pressure is connected to an inflatable cuff that is wrapped around the patient's arm and inflated so that the external pressure on the arm exceeds systolic blood pressure. When this occurs, blood flow into the arm is cut off, and no pulses can be detected distal to the cuff (Fig. 14.4). The external pressure in the cuff is measured by the height of a column of mercury in the manometer connected to the cuff or by means of a mechanical or digital electronic manometer that has been calibrated to a column of mercury. To measure blood pressure, the air in the cuff is first slowly released until blood can leak past the occlusion at the peak of systole. At this point, blood spurts past the point of partial occlusion at high velocity, resulting in turbulence. The vibrations associated with the turbulence are in the audible range and can be heard with a stethoscope placed over the brachial artery. These noises are known as **Korotkoff sounds**. The pressure corresponding to the first Korotkoff sound is the systolic pressure. As pressure in the cuff continues to fall, the brachial artery returns toward its normal shape and both the turbulence and Korotkoff sounds cease. The pressure at which the Korotkoff sounds cease is the diastolic pressure.

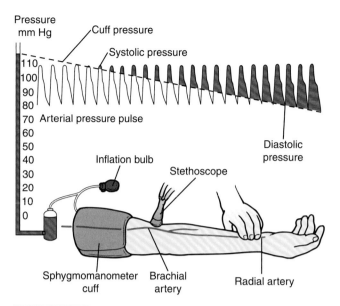

Figure 14.4 **Indirect measurement of arterial blood pressure by sphygmomanometry and radial artery auscultation.** To measure arterial blood pressure, an inflatable cuff connected to a pressure gauge calibrated against a mercury manometer is placed around the upper arm above the brachial artery. The cuff is initially inflated to a pressure ~20 mm Hg above the expected systolic pressure for the patient, and a stethoscope is placed lightly over the brachial region. Initially, there is no audible sound from the region. The cuff is deflated at a rate of 2 to 3 mm Hg/s, while the clinician listens for sounds created by pulsatile squirting of blood past the partially occluded artery. These sounds are called *Korotkoff sounds* and are characterized by five phases: Phase I is the level at which the first appearance of clear, repetitive, tapping sounds occur. This coincides approximately with the resumption of a palpable pulse. The pressure at which the first two tapping sounds occur is considered to be the peak systolic pressure. Phase II is the interval in which a swishing murmur is added to the initial tapping as the cuff is deflated. Phase III is characterized by distinct, crisp, and louder sounds as the artery opens more with cuff deflation. Phase IV is the start of a distinct muffled sound with a soft, blowing quality. Phase V is the final silence due to the resumption of full laminar flow in the brachial artery. The last sound heard before the onset of Phase V is considered to represent diastolic arterial pressure. If sounds persist from Phase IV to zero pressure, then the pressure recorded at the start of Phase IV is considered the diastolic blood pressure.

Although the cuff method has become a standard means of evaluating blood pressure in clinical settings, it is not without artifact errors. A cuff that is too narrow will give a falsely high pressure because the pressure in the cuff is not fully transmitted to the underlying artery. Ideally, cuff width should be ~1.5 times the diameter of the limb at the measurement site. Obesity may contribute to an inaccurate assessment if the cuff used is too small. In older adults (or those who have "stiff" or hard-to-compress blood vessels from other causes, such as arteriosclerosis), additional external pressure may be required to compress the blood vessels and stop the flow. This extra pressure can give a falsely high estimate of blood pressure.

Age, race, gender, diet, and body weight affect the range of arterial pressure seen in humans.

Although the range of blood pressures in the population as a whole is rather broad, changes in a given patient are of diagnostic importance. Normal arterial blood pressure in adults is often stated to be 120/80 mm Hg. However, these values were originally derived from insurance actuary data in acculturated Western societies after World War II. A more extensive examination of arterial pressures in the human population suggests that true normal human systolic pressure may be 100 to 110 mm Hg, with a corresponding diastolic pressure of 60 to 70 mm Hg.

In Western societies, arterial pressure is dependent on age. Systolic blood pressure rises throughout life, whereas diastolic blood pressure rises until the sixth decade of life, after which it stays relatively constant. Blood pressure is higher among African Americans than among Caucasian Americans, but it is lower in aboriginal Negroid populations than in Western Caucasian populations. Blood pressure is higher among men than among premenopausal women of similar age. Dietary fat and salt intake, as well as obesity, is associated with higher blood pressure. Other factors that affect blood pressure are excessive alcohol intake and psychosocial stress (which elevates pressure) as well as potassium and calcium intake and physical activity (which are associated with reductions in arterial pressure).

Cardiovascular Physiology

CLINICAL FOCUS | 14.1

Arteriosclerosis and Systolic Hypertension in Older Adults

Arteriosclerosis is broadly defined as any disease leading to the general hardening of the arteries. This includes reduced arterial compliance associated with the aging process, restenosis phenomena, hardening following transplantations of vascular segments, and even atherosclerosis. However, these conditions all differ significantly with respect to key immune, mitogenic, and lipid components to their pathology.

Typically, the term arteriosclerosis is used to describe the general progressive decrease in arterial compliance that accompanies the aging process (see Chapter 11). Even in the absence of true systemic arterial hypertension and atherosclerosis, people in their 70s have considerably stiffer arterial walls than those in their 20s. The process that leads to this arterial stiffening is thought to be related to the continual,

(Continued)

day-to-day exposure of arteries to oxygen radicals produced in the arterial wall as the by-product of natural metabolism. However, it likely also results from accumulated insults to the arteries from lifestyles, dietary factors, and pathological conditions known to stimulate excess radical production such as tobacco use, hypertension, and diabetes mellitus.

Arteries in young people owe their suppleness to the orderly arrangement of elastin and collagen in the arterial wall. Oxygen radicals attack the collagen and elastin matrix within the arteries, causing both strand breaks and rearrangement of elastic fibers into a more random configuration. This random configuration results in a stiffer artery. In the absence of any other pathologic factors, an older person with this type of arteriosclerosis exhibits an increased pulse pressure associated with an increased systolic pressure, lower diastolic pressure, and normal mean arterial pressure compared with a healthy younger person (healthy meaning with a blood pressure of 120/80 mm Hg). People with this condition have what is called **systolic hypertension** because only their systolic pressure is elevated.

Until somewhat recently, older patients with systolic hypertension were not treated with typical antihypertensive drug therapies for their condition. These people did not have the underlying vascular abnormalities associated with the increased arterial resistance of essential hypertension because their mean arterial pressure was not changed. Furthermore, their age led some to believe that there may be no long-term benefit to drug treatments, especially when such treatments were associated with serious side effects. However, systolic hypertension is no longer neglected in older adults and is instead treated with therapies designed to reduce arterial pressure. The reason for this is that systolic pressure is a key component of the wall stress that must be overcome for the heart to effectively contract. High systolic pressure leads to high stress that in turn increases oxygen demand by the heart. This places the heart at risk for ischemia and arrhythmias, whose incidence can be reduced if the systolic pressure is decreased. In addition, clinical trials have shown that the morbidity and mortality of older patients with systolic hypertension are improved with treatment to lower the systolic pressure, thus providing evidence-based support for antihypertensive therapies in older adults with systolic hypertension. ◼

CLINICAL FOCUS | 14.2

Effects of Age, Race, and Diet on Arterial Pressure

The common assignment of normal arterial pressure as 120/80 mm Hg is a convenient benchmark that does not reveal that arterial pressure varies within populations as a result of a variety of genetic and environmental factors. Human blood pressure values are a continuum. There is no clear dividing point within the human population that separates subpopulations with normal distributions around a "normal" mean arterial pressure from a different subpopulation distributed around a distinctly elevated arterial pressure. As stated in this chapter, the 120/80 mm Hg dividing point for normalcy has more to do with data sampling than with what is intrinsically supposed to be normal arterial pressure in healthy human beings.

In Western societies, mean arterial pressure and pulse pressure drift upward with age, even if the patient does not have clinically established classic arterial hypertension. However, members of primitive societies do not show this increase in pressure with age, have lower systolic and diastolic pressures as a group compared with members of Western societies, do not develop standard arteriosclerosis, and actually exhibit a slow decline in blood pressure with aging. An interesting observation in members of these primitive groups is that they consume potash as a seasoning, which contains high amounts of potassium. There is laboratory evidence that potassium may reduce wall stiffening although it has been speculated that chronic high NaCl consumption may lead to decreased compliance and impaired passive relaxation of arteries.

African Americans have a higher incidence and severity of arterial hypertension than do Caucasians of the same gender and similar age. In particular, their arterial pressure seems to be sensitive to salt in the diet. However, Negroid groups in primitive societies do not show this propensity to hypertension at all. An interesting hypothesis proposed to explain these dual effects in the same racial profile has pointed out that human populations that evolved in hot, dry environments have restricted access to water and salt. Those living in such environments possess traits that better enable them to aggressively extract and conserve salt and water from their diet and have a survival advantage over those who cannot conserve salt as well. However, placing such people in a Western society, where salt is a ubiquitous component of canned and processed foods, restaurant meals, etc., has been proposed to create a situation whereby too much salt is added into a body that is predisposed to conserve it aggressively. This then is thought to result in hypertension. It is believed also that such salt loading leads to a stiffening of the arterial walls as well as hypervolemia in the circulation. The influence of diet, rather than race, on hypertension is further supported by the observation that populations such as native Asians that consume a largely vegetarian diet have lower arterial pressures than the same populations raised and living in Western cultures, whereas native Japanese, who consume large amounts of salt in their diet (soy sauces, salted fish, etc.), have a high incidence of arterial hypertension and hypertension-related stroke. ◼

▶ PERIPHERAL AND CENTRAL BLOOD VOLUME

As discussed in Chapter 11, blood volume is distributed among the various segments of the circulatory system according to their basic geometry and compliance. Total blood volume in a 70-kg adult is 5.0 to 5.6 L. Approximately 80% of the total blood volume is located in the **systemic circulation** (i.e., the total volume minus the volume in the heart and lungs), and about 60% of the total blood volume (or 75% of the systemic blood volume) is located on the venous side of the circulation. The blood present in the small arteries and capillaries is only about 20% of the total blood volume. Because most of the systemic blood volume is in the veins, it is not surprising that changes in systemic blood volume primarily create changes in venous volume.

Large compliance in the veins allows them to accommodate high volumes with little change in pressure.

Systemic veins are ~20 times more compliant than systemic arteries; small changes in venous pressure are, therefore, associated with large changes in venous volume. If 500 mL of blood is infused into the circulation, about 80% (400 mL) locates in the systemic circulation. This increase in systemic blood volume raises what is termed the **mean circulatory filling pressure** (**MCP**) by a few mm Hg. MCP is the pressure that equilibrates into all segments of the vascular system when the heart is stopped. It is normally about 7 mm Hg and is a measure of how "full" the vascular system is or how "tightly" blood is contained within the vasculature as a result of vascular tone. A small rise in filling pressure, distributed throughout the systemic circulation, has a much larger effect on the volume of systemic veins than systemic arteries. Because of the much higher compliance of veins than arteries, 95% of the 400 mL from the example above (or 380 mL) will be distributed into the veins whereas only 5% (20 mL) will be found in arteries.

Central blood volume is useful for evaluating the effects of changes in blood volume on cardiac output.

Because of Starling's law of the heart, the filling of the heart is one of the key determinants of CO. In considering the role of distribution of blood volume in filling the heart, it is useful to divide the blood volume into central (or intrathoracic) and systemic (or extrathoracic) portions. The **central blood volume** includes the blood in the superior vena cava and the intrathoracic portions of the inferior vena cava, right atrium and ventricle, pulmonary circulation, and left atrium; this constitutes ~25% of the total blood volume. The central blood volume can be decreased or increased by shifts in blood to and from the extrathoracic blood volume. From a functional standpoint, the most important components of the **extrathoracic blood volume** are the *veins of the extremities and abdominal cavity*. Blood shifts readily between these veins and the vessels of the central blood volume. Blood in the central and extrathoracic arteries can be ignored in these shifts because of their low compliance. Furthermore, the extrathoracic blood volume in the neck and head is of little importance because there is far less blood in these regions and blood volume inside the cranium cannot change much because the skull is rigid. The volume of blood in the veins of the abdomen and extremities is about equal to the central blood volume; therefore, about half of the total blood volume is involved in shifts in distribution that affect the filling of the heart.

Central venous pressure is a qualitative measure of central blood volume.

Under normal conditions, changes in **central venous pressure** are a good reflection of central blood volume because the compliance of the intrathoracic vessels tends to be constant. Central venous pressure can be measured by placing the tip of a catheter in the right atrium. In general, the use of central venous pressure to assess changes in central blood volume depends on the assumption that the right heart is capable of pumping normally. In certain situations, however, the physiologic meaning of central venous pressure is changed. For example, if the tricuspid valve is incompetent (i.e., cannot close completely), right ventricular pressure is transmitted to the right atrium during ventricular systole, creating an abnormally high central venous pressure that is not primarily a result of increased central venous volume. Also, central venous pressure does not necessarily reflect left atrial or left ventricular filling pressure. Abnormalities in right or left heart function or in pulmonary vascular resistance can make it difficult to predict left atrial pressure from central venous pressure.

Unfortunately, measurements of the peripheral venous pressure, such as the pressure in an arm or leg vein, are subject to too many influences to be helpful in most clinical situations (e.g., partial occlusion caused by positioning or venous valves).

Cardiac output is sensitive to changes in central blood volume.

Consider what happens if blood is steadily infused into the inferior vena cava of a healthy person. As this occurs, the volume of blood returning to the chest—the venous return—is transiently greater than the volume leaving it, that is, the CO. This difference between the input and output of blood produces an increase in central blood volume. It will occur first in the right atrium, where the accompanying increase in pressure enhances right ventricular filling, end-diastolic fiber length (preload), and stroke volume. Increased flow into the lungs increases pulmonary blood volume and filling of the left atrium. The output of the left ventricle will increase according to Starling's law of the heart so that the output of the two ventricles exactly matches. CO will increase until it equals the sum of the previous venous return to the heart plus the infusion of new blood.

It can be surmised from the discussion above that central blood volume is altered by two events: changes in total blood volume and changes in the distribution of total blood volume between central and extrathoracic regions. An increase in total blood volume can occur as a result of an infusion of fluid, the retention of salt and water by the kidneys, or a shift in fluid from the interstitial space to plasma (see Chapter 15).

A decrease in blood volume can occur as a result of hemorrhage; fluid losses through sweat, vomiting, or diarrhea; or the transfer of fluid from plasma into the interstitial space. In the absence of compensatory events, changes in total blood volume result in proportional changes in both central and extrathoracic blood volume. For example, a moderate hemorrhage (10% of blood volume) with no distribution shift would cause a 10% decrease in central blood volume. The reduced central blood volume would, in the absence of compensatory events, lead to decreased filling of the ventricles and diminished stroke volume and CO.

Central blood volume can be altered by a shift in blood volume to or away from the extrathoracic circulation. Shifts in the distribution of blood volume occur for two reasons: a change in transmural pressure or a change in venous compliance. Changes in the transmural pressure of vessels in the chest or periphery can either enlarge (increased P_T) or diminish (decreased P_T) their size. Because total blood volume is finite, volume shifts in response to changes in transmural pressure in one region affect the volume of the other region. Imagine a long balloon filled with water. If it is slowly turned vertically on its long axis, the lower end of the balloon has the greatest transmural pressure because of the weight of the water pressing from above. Consequently, the lower end of the balloon will bulge and the upper end will shrink; volume increases in the lower end at the expense of a loss at the upper end.

The best physiologic example of a change in regional transmural pressure occurs when a person stands up. Standing increases the transmural pressure in the blood vessels of the legs because it creates a vertical column of blood between the heart and those vessels. The arterial and venous pressures at the ankles during standing can easily be increased almost 100 mm Hg higher than those in the individual in the recumbent position. The increased transmural pressure (outside pressure is still atmospheric) results in little distention of arteries because of their low compliance but results in considerable distention of veins because of their high compliance. In fact, ~550 mL of blood is needed to fill the stretched veins of the legs and feet when an average person stands up. Filling of the veins of the buttocks and pelvis also increases but to a lesser extent than the lower extremities, because the increase in transmural pressure is less. When a person stands, blood continues to be pumped by the heart at the same rate and stroke volume for one or two beats. However, much of the blood reaching the legs remains in the veins as they become passively stretched to their new size by the increased venous transmural pressure. This decreases the return of blood to the chest. Since CO exceeds venous return for a few beats, the central blood volume falls (as does the end-diastolic fiber length, stroke volume, and CO). Once the veins of the legs reach their new steady-state volume, the venous return again equals CO. Thus, the equality between venous return and CO is re-established even though the central blood volume is reduced by 550 mL. However, the new CO and venous return are decreased (relative to what they were before standing) because of the reduction in central blood volume. Without compensation, the decrease in systemic arterial pressure resulting from this decreased CO could cause a drop in brain blood flow and loss of consciousness. Compensatory mechanisms are then obviously required to maintain arterial pressure in the face of decreased CO (as well as blood flow to the brain). For example, upon standing, sympathetic nerves to peripheral veins are activated causing them to contract. The resulting decrease in venous compliance results in a redistribution of blood volume toward the central blood volume. The redistribution of blood toward the central blood volume helps to maintain ventricular filling and CO during standing. Therefore, it is not surprising that venoconstriction is one important compensatory mechanism for supporting CO following a drop in this output due to hemorrhage. Such compensatory mechanisms are discussed in detail in Chapters 16 and 17.

▶ COUPLING OF VASCULAR AND CARDIAC FUNCTION

Vascular volume, pressure, and resistance alter the output performance of the heart. However, because the cardiovascular system is a true "circulatory" system, it is equally true that the output of the heart alters vascular volumes and pressures. Therefore, equilibrium must exist between these two functional relationships. This section will explain how interactions between the cardiac and vascular systems combine to affect CO and cardiac work.

Compliant arteries reduce cardiac work.

One of the more important consequences of the elastic nature of large arteries is that it reduces cardiac work and myocardial oxygen demand. Consider the example in which the heart pumps blood for 4 seconds at a constant flow of 100 mL/s (6 L/min) into rigid arteries with a resistance of 16.67 peripheral resistance units (PRUs). This would then generate a constant pressure of 100 mm Hg, and cardiac work over the 4 seconds would be simply pressure (P) × total volume (V) or 100 mm Hg × 400 mL = 40,000 mm Hg mL. If the heart pumped intermittently and ejected blood at 100 mL/s into noncompliant arteries during the first half-second of the cycle only (i.e., 200 mL/s for 0.5 seconds), the same volume of 400 mL would be moved over the 4-second interval in this example. However, pressure would rise to 200 mm Hg during each ejection and drop to 0 mm Hg during relaxation. Although no work would be done during relaxation, work done during the contraction would be 80,000 mm Hg mL (200 mm Hg × 400 mL). In this situation, the oxygen demand on the heart would be substantially increased even though the output of the heart over 4 seconds was no better than that in the steady flow example. In contrast, if this same intermittent flow was ejected into arteries with infinite compliance (i.e., infinite ability to expand with an increase in volume), the pressure in the system would neither rise during systole nor fall during diastole and thus remain at an average of 100 mm Hg. The same 400 mL would be ejected over 4 seconds in this situation but, against a constant pressure of 100 mm Hg, the work done would again equal to 40,000 mm Hg mL.

Of course, arteries in the human body are neither totally rigid nor infinitely compliant. This example serves to demonstrate that decreased arterial compliance increases cardiac work and oxygen demand. By extension, myocardial oxygen demand will be increased by any factor that reduces arterial compliance, even if all other factors, such as arterial pressure, stroke volume, and heart rate, do not change. For this reason, because arterial compliance decreases as we age, the heart of an older person is confronted by increased oxygen demand compared to a younger person even if all other variables affecting cardiac oxygen demand are the same between the two individuals.

Increasing venous filling pressure increases cardiac output, but increasing cardiac output decreases venous pressure.

Central venous pressure is one of the key determinants of the filling of the right heart and, by extension of Starling's law of the heart, a key determinant of CO; increased pressure leads to increased diastolic filling and thus increased cardiac output. However, the heart is a pump set in a *circulatory* system. An increase in CO into the arterial segment of the circulation must come at the expense of volume and pressure in the venous side of the circulation. Yet, this would then reduce venous pressure and ventricular filling and CO according to Starling's law. This poses two questions: (1) how is CO maintained above or below the resting value and (2) what determines equilibrium between CO and central venous pressure when each variable can independently affect the other?

A demonstration of how CO alters central venous and arterial pressure is shown in Figure 14.5 A to D, where a theoretical circulation is depicted with a pump, a "venous" side of the circulation, and an "arterial" side of the circulation. For the sake of example, peripheral vascular resistance is set at 25 PRUs, and the veins are 24 times more compliant than the arteries. When the heart is stopped (CO equals zero), arterial and venous pressures are in equilibrium and dependent only on blood volume and the compliance of the vascular system. This value is called the **MCP** and is generally about 7 mm Hg. MCP is the value from which arterial pressure increases and venous pressure decreases once the heart starts to pump blood.

Using the example in Figure 14.5 with an MCP of 7 mm Hg, let us start the heart at an output of 1 L/min. The initial effect of this pumping by the heart will be to translocate blood from the venous side into the arterial side of the circulation. This will increase pressure in the arterial component of the circulation while it reduces pressure on the venous side. These changes in pressure will continue until enough pressure difference between the arteries and veins has been created to move blood at a rate of 1 L/min across the resistance vessels in the circulation. At that point, the rate of blood moving across the resistance and into the "venous" side of the circulation will match the rate of blood exiting the "venous" side of the circulation through the pump and into the "arterial side." By Poiseuille's law, this will occur when the pressure difference is 25 mm Hg. What needs to be determined next is how much of this 25 mm Hg is added to the arterial side and how much is subtracted from the venous side of the circulation. In this example, the veins are 24 times more compliant than the arteries. Stated another way, this means the arteries are 24 times stiffer than the veins. Consequently, for every 25 mm Hg of pressure difference between the arterial and venous sides of the circulation, 24 mm Hg is added to the MCP on the arterial side (increasing its pressure to 31 mm Hg) and 1 mm Hg is subtracted from the MCP on the venous side (reducing it to 6 mm Hg). Thus, for every 1 L/min of increase in CO, 24 mm Hg will be added to arterial pressure and 1 mm Hg subtracted from the venous pressure once the system reaches its steady state. In our example, at a CO of 2 L/min, arterial pressure becomes 55 mm Hg and venous pressure becomes 5 mm Hg (Fig. 14.5C); a CO of 4 L/min would yield an arterial pressure of 103 mm Hg and a central venous pressure of 3 mm Hg (Fig. 14.5D).

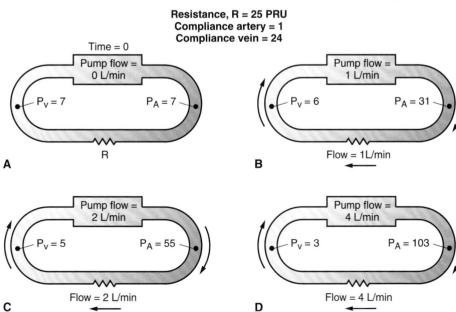

Figure 14.5 **A model of the effect of cardiac output (CO) on arterial and central venous pressures. (A)** At zero CO, all pressures equal the mean circulatory pressure. **(B)** Changes in pressures once CO is increased to 1 L/min. **(C)** Changes in pressures once CO is increased to 2 L/min. **(D)** Changes in pressures once CO is increased to 4 L/min. P_A, arterial pressure; P_V, central venous pressure; PRU, peripheral resistance unit; R, resistance.

Cardiovascular Physiology

Depictions of the interrelationships between venous pressure and cardiac output are used to predict how cardiac output and central venous pressure are altered by changes in vascular and cardiac variables.

Two graphical means can be used to depict the interrelationship between CO and central venous pressure. One relationship, called the **cardiac function curve**, plots CO as a function of central venous pressure (Fig. 14.6A). This is simply an extension of Starling's law of the heart and shows that CO will increase with an increase in central venous pressure. The cardiac function curve is characteristic of the heart itself in that only factors affecting the heart affect the position and shape of this curve. This curve can be produced even in a heart separated from the vasculature.

A second relationship, called the **vascular function curve**, shows how central venous pressure changes as a function of CO (Fig. 14.6B). First note that, contrary to mathematical convention, the independent variable of the vascular function curve (CO) is placed on the *y* axis and the dependent variable in this relationship, the venous pressure, is placed on the *x* axis. (For reasons to be seen below, this facilitates combining the cardiac function curve with the vascular function curve on the same plot.) The position and slope of the vascular function curve are affected only by factors associated with blood vessels, such as vascular resistance, vascular compliance, and blood volume. The vascular function curve is independent of characteristics of the heart and can be observed even if the heart in the circulation is replaced by an artificial pump. This curve shows that when CO is zero, venous pressure equals MCP and that venous pressure *decreases* as CO increases, until it results in venous collapse (at approximately −2 mm Hg), which limits further increases in CO.

Figure 14.7 A to F depicts vascular and cardiac function curves plotted together and demonstrates how equilibrium between CO and venous pressure is obtained in a variety of altered states. The cardiac function and vascular function curves both depict true functional relationships between cardiac output and venous pressure. Both of these must operate simultaneously in the intact cardiovascular system. Consequently, in any condition of the intact cardiovascular system, the value of CO and venous pressure must satisfy both the functional expression for the cardiac function curve and the vascular function curve at the same time. Although this could be solved mathematically, this can be seen more easily by reading the CO and venous pressure at the intersection of the cardiac function and venous function curve. Equilibrium for this system exists at the point of intersection between the two curves because it is at the intersection where one pair of CO and central venous pressure satisfies both the cardiac and vascular function curves (see Fig. 14.7A). If venous pressure were to suddenly increase, an increased CO would initially result. However, this elevated output would tend to reduce venous pressure, which would then reduce CO, and so on. The end result, after an initial perturbation in either venous pressure or CO, will be to return these two variables to the intersection of the cardiac function and vascular function curves. This is, therefore, the equilibrium point for the system.

Several factors influence the vascular and cardiac function curves. For example, increases in blood or plasma volume (hypervolemia) "fill" the cardiovascular system more and thus raise MCP. This does not significantly alter venous compliance but does shift the entire vascular function curve in parallel fashion to the right of the normal relationship (Fig. 14.7B). This shift will also be seen with increased venous tone (venous smooth muscle contraction), which "squeezes" the blood contained in the veins, thus raising their internal pressure. Conversely, hypovolemia or decreased venous tone has the opposite effect.

A change in arteriolar tone has a different effect on the vascular function curves than does a change in venous tone (Fig. 14.7C). Because little blood volume is contained in the arterioles, changes in arteriolar tone do not significantly affect MCP. However, reduced arteriolar resistance makes it easier for the heart to eject a given stroke volume with any venous filling pressure, whereas increased resistance has the opposite effect. In other words, with reduced systemic resistance, a smaller than normal ΔP is required to maintain flow through the peripheral arterial and into the veins at the same rate it is being pumped out of the veins and into the arteries (i.e., the CO). This diminishes the rise in arterial pressure and the drop in venous pressure at any cardiac output. By extension, therefore, at any central venous pressure, reducing arteriolar resistance augments CO and increasing such resistance impedes CO.

Changes in the inotropic state of the heart alter the cardiac, but not the vascular, function curve. By definition, a positive inotropic influence will allow the heart to produce a larger output at any given venous pressure, whereas a negative inotropic influence will have the opposite effect (Fig. 14.7D).

The net effect of changes in cardiac and venous function curves on the equilibrium values for CO and venous pressure is shown in Figure 14.7E, where point "A" represents

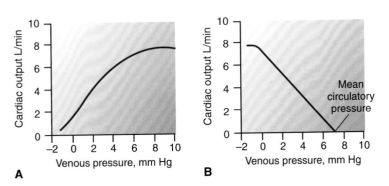

Figure 14.6 **(A) The cardiac function curve.** The curve is a representation of Starling's law of the heart. It is a function of characteristics of the heart only. **(B) The vascular function curve.** The relationship shows how venous pressure changes in response to a change in cardiac output (CO). It is a function of vascular variables only. The independent variable, CO, in this depiction is placed by convention on the *y* axis.

A

B

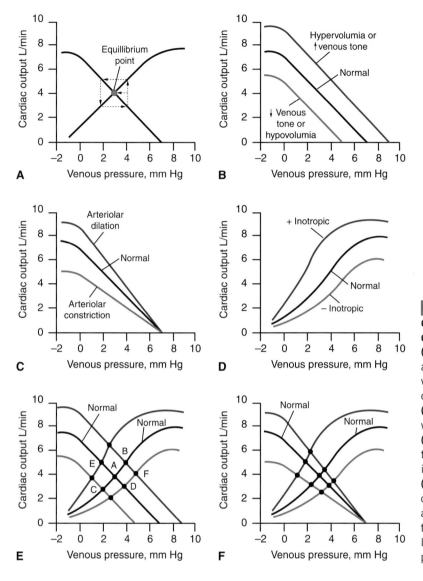

Figure 14.7 **Various equilibriums in cardiac output (CO) and venous pressure resulting from changes in cardiac function or vascular variables.** **(A)** The equilibrium point for concurrent cardiac and vascular function curves; only one set of values for cardiac output and venous pressure can satisfy both relationships at the same time. **(B)** Effect of changing either blood volume or venous tone on the vascular function curve. **(C)** Effect of changes in arteriolar resistance on the vascular function curve. **(D)** Effect of changes in inotropic state on the cardiac function curve. **(E)** Possible equilibrium points obtained by the conjunction of altered myocardial contractility and vascular function curves. **(F)** Effects of alteration in vascular resistance on cardiac and vascular function curves and the resulting equilibrium points for CO and venous pressure.

the normal equilibrium. Factors that increase blood volume such as an intravenous infusion of blood or fluid will shift the vascular function curve in parallel to the right, whereas loss of volume (i.e., hemorrhage, loss of body water, etc.) will shift that curve to the left. The new equilibrium points for these conditions, points "B" and "C," respectively, correctly show the interrelationship between CO and venous pressure that would be predicted by Starling's law of the heart; an increased CO is associated with an elevated venous (and hence filling) pressure, whereas a decreased CO is associated with a decreased venous pressure. Similarly, the equilibrium points associated with changes in the inotropic state of the heart, "D" and "E," also correspond with our understanding of the definition of contractility. A positive inotropic influence allows for an elevated CO at a lower-than-normal venous pressure, whereas a negative inotropic influence has the opposite effect. A positive inotrope, according to cardiac muscle mechanics, will allow the heart to empty more in spite of reduced preload. Stated another way, by Starling's law of the heart, a decreased filling pressure must decrease CO, and therefore, the only way the heart can produce a larger-than-normal output in spite of reduced filling is to enhance its intrinsic cellular contractile ability by inotropic mechanisms.

Figure 14.7E also serves to illustrate how the body can compensate for a failing heart. In acute heart failure, a patient's condition is likely best represented by point "D" on the graph, with reduced CO in spite of an elevated venous pressure because of the inability of the weakened heart to move blood from the veins to the arteries. However, with continued failure, the body's compensatory mechanisms stimulate the kidneys to retain salt and water, resulting in a hypervolemic shift in the vascular function curve to the right. This shift, through Starling mechanism, helps augment the CO of the failing myocardium, as shown by point "F" on the diagram. Again, stated another way, one way in which a heart in a negative inotropic state (as in heart failure) can increase its output is to exploit Starling's law and increase its preload.

Figure 14.7F demonstrates the effects of alterations in vascular resistance on the cardiac and vascular function curves. Elevated resistance impedes CO and also requires an increased drop in venous pressure to establish the elevated gradient across the elevated resistance that will be needed to produce peripheral flow that matches a given CO. Decreased resistance has the opposite effect. The precise location of the new equilibrium depends on the relative degree to which the vascular and cardiac function curves are affected by changes in resistance.

INTEGRATED MEDICAL SCIENCES

Atherosclerosis and Unexpected Consequences of Drug Therapies for Hyperlipoproteinemia

In Chapter 11, it was noted that hemodynamics influences the localization of the development of atherosclerosis in the arterial tree. However, it is not correct to conclude from that phenomenon that atherosclerotic development is a simple accumulation or "sludging" of cholesterol in favorable loci along the arterial system. In reality, atherosclerosis is a vessel wall injury/inflammation phenomenon involving infiltration of the arterial wall by immune system cells, transformation of arterial macrophages, and stimulation of cytokines and mitotic events leading to infiltrative growth into the vessel lumen and eventual cellular necrosis and calcification in the vascular wall. A fully developed atherosclerotic lesion in the arterial wall is called a **plaque**, and it is composed of a necrotic core of dead cells, fibrin, cholesterol crystals, and calcium covered with a fibrotic cap. Plaques are unstable lesions. They do not just encroach on the arterial lumen but can also rupture. When this occurs, the exposed plaque interior and related debris from the rupture activate a blood coagulation cascade resulting in a thrombus forming in the artery that can completely occlude the vessel lumen. Such occlusion in cerebral or coronary arteries and infarction downstream tissue in those organs, can occur, often resulting in death.

The initiating injury in the pathogenesis of atherosclerosis or factors that appear to initiate the process are currently the subject of significant research efforts, as are investigations into the mechanism of plaque rupture. High intraarterial pressure, diabetes mellitus, and tobacco smoking are thought to cause vessel wall and/or endothelial injury that may induce or at least accelerate the atherosclerotic process, especially in the context of increased vessel wall oxygen radical production associated with those diseases. Genetic factors (i.e., family history) also seem to be involved in the development of the disease in some individuals. Regardless of the nature of initial endothelial injury, such injury causes monocytes to enter the vessel intima where they release cytokines that stimulate smooth muscle movement into the subendothelium.

It has long been known that high plasma cholesterol and triglyceride levels (i.e., hyperlipidemia), and in particular high low-density lipoprotein cholesterol levels (LDL), are associated with significant atherosclerotic development. LDLs are responsible for cholesterol transport into cells and are now known to become oxidized in this process. These particles are thought to radicalize other organic molecules in the vessel wall and initiate a cascade of vessel wall injury if present in too high a concentration. Oxidized LDLs stimulate inflammatory cytokine production in the vessel wall and reduce wall NO production, which is a natural antiatherogenic molecule. The monocytes that were attracted into the intima transform into macrophages, which scavenge and ingest oxidized LDL. These cells undergo apoptosis as they collect lipids into the wall forming the necrotic core of a developing atherosclerotic lesion. Such cascading processes have been postulated to be key factors in the development of atherosclerosis.

Diets that are high in cholesterol are a known risk factor in the development of atherosclerosis, and efforts to limit fats and cholesterol in the diet have long been a mainstay in the clinical treatment and management of hyperlipidemia and hypercholesterolemia. However, dietary restriction presents difficult problems for the long-term management of hyperlipidemic conditions as patient compliance can be erratic or difficult to maintain over a long period of time. In such instances, pharmacotherapies are now available to help patients control lipids in their bloodstream. One of the more popular of these agents is the HMG-CoA reductase inhibitors, which are more generically known as **statins**. Statins inhibit the conversion of HMG-CoA to mevalonate, which is a very early step in cholesterol synthesis. This reduces intracellular cholesterol levels, primarily in the liver, which activates a protease that cleaves sterol regulatory element–binding proteins (SREBP). SREBP upregulates LDL receptor expression, which then increases endocytosis of LDL, thereby lowering serum LDL levels. This effect also stimulates the hepatic extraction of LDL precursors (very low–density lipoprotein remnants; VLDL remnants), which further decreases circulating LDL levels.

Statins have become key pharmacological agents for the treatment of hyperlipoproteinemias. However, they have also proven to be an example of the many instances in which drugs originally designed for one purpose produce side effects that could not be predicted based on known actions of the drug. Early use of statins was shown to unexpectedly produce various forms of myotoxicities in patients taking them, especially in skeletal muscle. These serious effects included various myopathies, muscle inflammation, myalgia and rhabdomyolysis, or breakdown of muscle fibers. This breakdown can cause severe pain. More importantly, myoglobin released from the muscle cell breakdown is converted to other products in the kidney, which results in renal damage as well.

The causes of myotoxicity associated with statin use are not fully understood at this time. However, HMG-CoA reductase is also needed to synthesize coenzyme Q10 (ubiquinone), which is found in the mitochondria of the heart, skeletal muscle, liver, and kidney. Coenzyme Q10 is a potent antioxidant, and it is speculated that statins may reduce the content of this factor in skeletal muscle thereby interfering with muscle energy processes in a manner that can cause myopathy. Although this may be a reasonable hypothesis, ubiquinone consumption as a dietary supplement has not proven to be effective in reducing statin myopathies and clinical trials to date have failed to show a connection between such induced myopathies and tissue coenzyme Q10 levels. ■

Chapter Summary

- Cardiac output and systemic vascular resistance determine mean arterial pressure.
- Stroke volume and arterial compliance are the main determinants of pulse pressure.
- Arterial compliance decreases as transmural arterial pressure increases.
- Changes in arterial pulse waveforms can be used to ascertain changes in hemodynamic variables that affect blood pressure.
- Shifts in blood volume between the periphery (extrathoracic blood volume) and the chest (central blood volume) influence preload and cardiac output.

- Central venous pressure and cardiac output are interrelated.
- The cardiac function curve depicts how cardiac output varies as a function of central venous pressure.
- The vascular function curve depicts how central venous pressure varies as a function of cardiac output.
- Cardiac and vascular function curves can predict new equilibriums for cardiac output and central venous pressure whenever cardiac or vascular properties are changed.

Chapter Review Questions

1. Mean arterial pressure changes if:

   A. Heart rate increases, with no changes in cardiac output or systemic vascular resistance
   B. Stroke volume changes, with no changes in heart rate or systemic vascular resistance
   C. Arterial compliance changes, with no changes in cardiac output or systemic vascular resistance
   D. Heart rate doubles, and systemic vascular resistance is halved, with no change in stroke volume
   E. Cardiac output doubles, and systemic vascular resistance is halved, with no change in heart rate

The correct answer is B. Mean arterial pressure is given by P = CO × TPR. An increase in stroke volume without a change in heart rate would increase cardiac output, and without an opposing change in vascular resistance, mean arterial pressure will then increase. Mean arterial pressure will not change if there is no change in both cardiac output and total vascular resistance. When there is no change in cardiac output and systemic vascular resistance, a change in arterial compliance will alter pulse pressure but not mean arterial pressure. Doubling heart rate without changing stroke volume will double cardiac output. If cardiac output is doubled while total vascular resistance is halved, the changes in the two components of mean arterial pressure will be in equal but opposite directions, and therefore, there will be no change in mean arterial pressure.

2. If the compliance of veins were equal to that of arteries, the change in central blood volume with standing would be:

   A. Less than normal
   B. Greater than normal
   C. The same as normal
   D. Less than normal but still more than the change in the arteries
   E. Less than normal and less than the change in the arteries

The correct answer is A. Blood pools in peripheral veins at the expense of the central circulation upon standing because veins

are about 20 times more compliant than arteries. Equal compliances in the arteries and veins would translocate blood into the lower extremities in the same proportion under the influence of gravity. If the compliance of veins were the same collectively as the arteries, there would be no such preferential pooling of blood into the peripheral venous system and, because the veins would have a lower-than-normal compliance, that pooling would be less than normal.

3. Dobutamine is a pure β1-adrenergic receptor agonist that is used to augment stroke volume in the heart following acute heart failure from myocardial infarction (heart attack). It does not affect smooth muscle tone in blood vessels. Which of the following effects best describes the effect of dobutamine on the cardiac and vascular function curves?

   A. It will shift the cardiac function curve downward and to the right without changing the position of the vascular function curve.
   B. It will shift the cardiac function curve upward and to the left while shifting the vascular function curve parallel and to the right.
   C. It will shift the cardiac function curve upward and to the left without changing the position of the vascular function curve.
   D. It will shift the cardiac function curve upward and to the left while decreasing mean circulatory filling pressure.
   E. It will shift the cardiac function curve upward and to the left while increasing mean circulatory filling pressure.

The correct answer is C. Dobutamine, by activating beta1-adrenergic receptors, is a positive inotrope. Inotropism is a characteristic of the heart. Dobutamine, as a positive inotrope, will shift the cardiac function curve upward and to the left. This drug does not alter smooth muscle tone in blood vessels and therefore would not have any direct effect on the position of the vascular function curve. Mean circulatory filling pressure is a function of the vascular system and blood volume when the heart is stopped, not when it is active under any inotropic condition.

4. Which of the following would be characteristic of an individual with compensated heart failure (i.e., one in which the body attempts to compensate for the consequences of reduced myocardial contractility of the failing heart by retaining fluid in the body and activating the sympathetic nervous system)?

   A. An increased mean circulatory pressure and reduced cardiac output

   B. A decreased central venous pressure and decreased cardiac output

   C. An increased mean circulatory pressure and increased cardiac output

   D. No change in central venous pressure and cardiac output

   E. A decreased central venous pressure and increased cardiac output

The correct answer is A. Heart failure is defined as a cardiac output that is too low for normal patient function. Therefore, almost always, a patient in heart failure, especially that bad enough to activate compensatory mechanisms in the body, will have a lower-than-normal cardiac output in spite of activation of the sympathetic nervous system, which attempts to stimulate the heart to compensate for its failing muscle. Furthermore, activation of hormonal and neurological systems in the body (see Chapter 17 for details) causes the patient to retain salt and water. This retention elevates mean circulatory and, thus, central venous pressure as a means to augment the stroke output of the failing heart via Starling mechanism.

5. A supine patient's blood pressure is measured in a clinical setting using the sphygmomanometric method with the measurement made in a patient's arm that is dangling over the edge of the table on which the patient is supine. A patient's blood pressure measured in this manner will:

   A. Underestimate the patient's true blood pressure

   B. Yield an accurate estimate of the patient's blood pressure

   C. Overestimate the patient's true blood pressure

   D. Prevent measurement of the patient's blood pressure

   E. Negate an ability to hear Korotkoff sounds in the brachial artery

The correct answer is C. Measuring blood pressure in an arm at a level below the level of the heart will not negate hearing Korotkoff sounds or obtaining a value for systolic and diastolic blood pressure. However, the effect of gravity will add pressure along any column of blood below the level of the heart in proportion to its distance below the level of the right atrium (i.e., along the arm). Therefore, the recording of the patient's arterial blood pressure will be higher than that obtained if the arm was in the correct position for measurement, that is, at the level of the right atrium.

Clinical Application Exercises 14.1

A 20-year-old male with no history of cardiopulmonary illness undergoes surgery to repair fractures to his right femur and tibia resulting from an auto accident. He is placed in a large full-leg cast with screws and pins to set the bones in their proper position. The cast and pins effectively immobilize the patient's leg. Three days postoperatively while lying in bed, he experiences sudden onset of chest discomfort and shortness of breath. His blood pressure is 100/75 mm Hg, and his heart rate is 105 beats/min. However, the ECG shows no changes suggestive of cardiac ischemia. There is swelling and tenderness in the left leg, which began about 3 days earlier. The attending physician performs a Wells bedside score to rule out the presence of a pulmonary embolism (needed score should be <4); however, the patient's score is 6. He orders a D-dimer ELISA test, which tests for cross-linked fibrin degradation products indicative of fibrinolysis of venous thromboembolism. High levels of fibrin degradation products are indicated, but because such test results can occur in the postoperative state, the physician decides to order a multidetector rotating CT scan of the patient's chest. This scan confirms the presence of a large pulmonary embolism involving the right pulmonary artery.

QUESTIONS

Note: For this set of questions, you will only be concerned with the broad immediate hemodynamic aspects of this patient's condition.

1. How are the patient's chest discomfort, shortness of breath, and arterial hypotension explained?

2. Is right ventricular pressure likely to be increased or decreased? Why?

3. Would intravenous infusion of additional fluids (such as blood or plasma) help the patient's arterial blood pressure?

ANSWERS

1. The patient's symptoms are caused by pulmonary embolism. In reality, the full effects of pulmonary embolism are complex and lead to systemic hypoxia through a cascade of alterations in the lungs. For the purpose of just the hemodynamic consequences of this condition, one should note that with a pulmonary embolism, a piece of blood clot located in a peripheral vein (in this case, a leg vein) breaks off and is carried through the right heart to a pulmonary artery where it lodges. Such emboli often arise from deep vein thrombosis following fractures and surgery to the limbs and leg bones. With a pulmonary embolism, blood flow from the pulmonary artery to the left heart is obstructed (i.e., pulmonary vascular resistance increases), resulting in elevated pulmonary arterial pressure. The sudden rise in pressure causes distention of the artery, which may contribute to the sensation of chest discomfort. Increased pulmonary arterial pressure (pulmonary hypertension) leads to right heart failure. Because left atrial (and left ventricular) filling is reduced (as a result of lack of blood flow from the lungs),

left-side cardiac output also falls. The fall in cardiac output causes a reflex increase in heart rate (see Chapter 17). The result is a combination of right- and left-side heart failure, producing the signs and symptoms seen in this patient.

2. The right ventricular pressure is likely to be increased because the blood clot in the pulmonary artery acts as a form of obstruction that raises the pulmonary artery resistance.

3. The problem here is increased afterload of the right ventricle caused by partial obstruction of the outflow tract. Because of this obstructed outflow, the diastolic volume of the right ventricle is already high. It is unlikely that infusing additional fluids into the veins will improve cardiac output because the extra filling of the right ventricle is unlikely to increase the force of contraction.

thePoint® *Visit* http://thepoint.lww.com/rhoades5e *for additional chapter review Q&A, Clinical Application Exercises, animations, and more!*

Active Learning Objectives

Upon mastering the material in this chapter, you should be able to:

- Predict how changes in the blood concentration, tissue concentration, and capillary permeability of a substance alter its transport by diffusion across the capillaries.
- Explain how changes in perfused capillary density affect transport of substances across capillaries.
- Predict whether a capillary will reabsorb or filter water based on changes in mean capillary hydrostatic pressure, plasma oncotic pressure, interstitial hydrostatic pressure, or interstitial oncotic pressure.
- Predict whether tissue edema will form based on changes in factors that affect capillary fluid filtration or lymph drainage of tissue.
- Explain how changes in precapillary and postcapillary resistances can increase or decrease fluid movement out of capillaries and how these changes may induce or attenuate edema formation.

- Identify and explain the changes in capillary filtration and lymph flow that are associated with, anaphylaxis, hypovolemic shock, diabetes mellitus, tissue inflammation, tissue traumas, and common disorders of plasma proteins.
- Explain myogenic and metabolic mechanisms responsible for autoregulation of blood flow.
- Explain the phenomena of active hyperemia, reactive hyperemia, and flow-mediated vasodilation.
- Explain the effects of the sympathetic nervous system on vascular resistance, venous pressure and capacitance, capillary transport, and capillary filtration in the microcirculation.
- Predict the effects of impairment of the endothelial NO system on arterial function.

▶ STRUCTURE AND FUNCTION OF THE MICROCIRCULATION

The microvasculature (Fig. 15.1) is considered to begin where the smallest arteries enter organs and end where the smallest veins, the venules, exit organs. In between are microscopic arteries, the arterioles, precapillary sphincters, and the capillaries. The arterioles divide into progressively smaller vessels so that each section of the tissue has its own specific microvessels. Depending on an animal's size, the largest arterioles have an inner diameter of 100 to 400 μm, and the largest venules have a diameter of 200 to 800 μm.

The microcirculation is the site of exchange of nutrients, water, gas, and small molecules between the plasma and the tissues. Under normal conditions, the capillaries do not allow exchange of peptides, proteins, and other large molecules between tissues and plasma. Virtually every cell in the body is in close contact with a capillary. The microcirculation regulates blood flow to individual organs, the distribution of blood flow within organs, diffusion distances between an organ's blood supply and tissues, as well as the capillary surface area available for exchange of materials between the plasma and tissues. In conjunction with cardiac output, it helps maintain arterial blood pressure by altering total peripheral vascular resistance and diastolic filling of the heart (see Chapters 11, 13, and 14).

Control of total peripheral vascular resistance is predominantly at the level of the arterial microvasculature.

VSM cells wrap around the arterioles at an ~90° angle to the long axis of the vessel. This arrangement is efficient because the tension developed by the VSM cell can then be almost

totally directed toward maintaining or changing vessel diameter against the pressure within the vessel. In most organs, arteriolar muscle cells operate at about half their maximal length. If the muscle cells fully relax, the diameter of the vessel can nearly double, which increases blood flow dramatically because flow increases as the fourth power of the vessel radius. When the muscle cells contract, the arterioles constrict; with intense stimulation, the arterioles can literally close for brief periods of time.

Normally, all microvessels, other than capillaries, are partially, tonically constricted. However, the magnitude of contraction can be variable from maximum relaxation (no contraction) to total closure of small arterioles and sphincters (maximum contraction). Tonic constriction of small arteries and arterioles maintains the relatively high vascular resistance in organs. Constriction results from the release of norepinephrine by the sympathetic nervous system, from a myogenic mechanism intrinsic to the smooth muscle (to be discussed later) and from other chemical and physical factors that are active at the level of the smooth muscle cell. The smallest arteries, combined with the arterioles of the microcirculation, constitute the resistance blood vessels; together, they regulate about 70% to 80% of the total vascular resistance with the remainder of the resistance about equally divided between the capillary beds and venules. For this reason, this portion of the microvasculature is the site of control of organ blood flow as well as arterial blood pressure in the whole organism.

Vessel radius of arterioles controls the major portion of vascular resistance and is determined by the transmural pressure gradient and wall tension, as expressed by the law of Laplace (see Chapter 11). Changes in wall tension developed

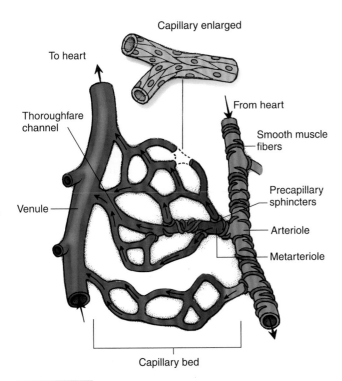

Figure 15.1 **A diagrammatic representation of the components of the microvasculature.** Arterioles control blood flow into a region of tissue and, along with precapillary sphincters, control the distribution of blood flow within the capillary network.

by the level of contraction of arteriolar smooth muscle cells directly alter vessel radius and, thus, vascular resistance. Most arterioles can dilate 60% to 100% from their resting diameter and can maintain a 40% to 50% constriction for long periods because of the latch state properties of VSM. Therefore, large decreases and increases in organ vascular resistance and blood flow are well within the capability of the microscopic blood vessels. For example, a 20-fold increase in blood flow can occur in contracting skeletal muscle during exercise because of intense dilation of arterioles in the muscle circulation, whereas blood flow in the same vasculature can be reduced to 20% to 30% of normal during reflex activation of sympathetic nerve activity. There is a constant balance between the regulation of vascular resistance to help maintain arterial pressure (see Chapter 17) and the regulation required to allow each tissue to receive sufficient blood flow to sustain its metabolism. If needed, the body will reduce blood flow to most organs, such as the skeletal muscles and splanchnic organs, in order to sustain arterial pressure to preserve flow to the heart and brain.

Exchange of water and materials between blood and tissues occurs across capillaries.

Capillaries, with inner diameters of about 4 to 8 μm, are the smallest vessels of the vascular system. Networks of capillaries arranged in parallel provide for most of the exchange area between blood plasma and tissue cells. A capillary is an endothelial tube surrounded by a basement membrane composed of dense connective tissue (Fig. 15.2). Capillaries in mammals do not have VSM cells and are unable to actively

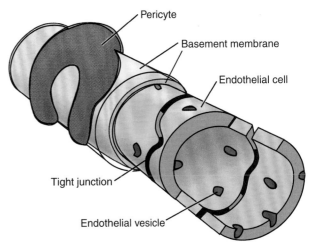

Figure 15.2 **The various layers of a mammalian capillary.** Adjacent endothelial cells are held together by tight junctions, which have occasional gaps. Water-soluble molecules pass through pores formed where tight junctions are imperfect. Vesicle formation and the diffusion of lipid-soluble molecules through endothelial cells provide other pathways for exchange.

change their inner diameter. Pericytes (Rouget cells), wrapped around the outside of the basement membrane, may be a primitive form of VSM cell and may add structural integrity to the capillary.

Although they are small in diameter and individually have a high vascular resistance, the parallel arrangement of many thousands of capillaries per cubic millimeter of tissue minimizes their collective resistance. Nevertheless, the capillary lumen is so small that red blood cells must fold into a shape resembling a parachute as they pass through and virtually fill the entire lumen. The small diameter of the capillary and the thin endothelial wall minimize the diffusion path for molecules from the capillary core to the tissue just outside the vessel. In fact, the diffusion path is so short that most gases and inorganic ions can pass through the capillary wall in <2 ms.

Passage of molecules occurs between and through capillary endothelial cells.

The exchange function of the capillary is intimately linked to the structure of its endothelial cells and basement membrane. Lipid-soluble molecules and gases such as oxygen and carbon dioxide readily pass through the lipid components of endothelial cell membranes. Water-soluble molecules, however, must diffuse through water-filled pathways formed in the capillary wall between adjacent endothelial cells. These pathways, known as **pores**, are not cylindrical holes but complex passageways formed by irregular tight junctions (see Fig. 15.2). The pores are partially filled with a matrix of small fibers of submicron dimensions, which acts partially to sieve the molecules approaching the water-filled pore. Most pores permit only molecules with a radius <3 to 6 nm to pass through the vessel wall. Thus, only water, inorganic ions, glucose, amino acids, and similar small, water-soluble solutes pass through the pores, whereas large molecules, such as serum albumin, globular proteins and blood cell components, are excluded.

The fiber matrix and the small spaces in the basement membrane, as well as those between endothelial cells, explain why the vessel wall behaves as if only about 1% of the total surface area were available for the exchange of water-soluble molecules. A limited number of large pores, or possibly defects, allow virtually any large molecule in blood plasma to pass through the capillary wall. Even though few large pores exist, there is enough that nearly all the serum albumin molecules leak out of the cardiovascular system each day.

The porosity of capillaries is not the same in all organs. The capillaries of the brain and spinal cord have virtually continuous tight junctions between adjacent endothelial cells; consequently, only the smallest water-soluble molecules pass through their capillary walls. Capillaries in cardiac and skeletal muscle also have relatively low porosity to water and small water-soluble molecules. In contrast, capillaries in the intestines, the liver, and the glomerulus of the kidney have capacities for large water transport. Those in the spleen and bone marrow have capillary pores so large that they allow for the passage of cellular elements between the blood and those organs.

An alternative pathway for water-soluble molecules through the capillary wall is via **endothelial vesicles** (see Fig. 15.2). Membrane-bound vesicles form on either side of the capillary wall by pinocytosis. Exocytosis occurs when the vesicle reaches the opposite side of the endothelial cell. The vesicles appear to migrate randomly between the luminal and abluminal sides of the endothelial cell. Even the largest molecules may cross the capillary wall in this way.

Venules collect blood from capillaries and act as a blood reservoir.

After the blood passes through the capillaries, it enters the venules, which are endothelial tubes usually surrounded by a monolayer of VSM cells. The venule vascular muscle cells are smaller in diameter and longer than those of arterioles, which may reflect that they do not need powerful muscles to oppose their internal pressure, which is much lower than those in arterioles. The smallest venules are unique because they are more permeable than capillaries to large and small molecules. This property appears to be related to frequent and large pores at tight junctions between adjacent venular endothelial cells. It is probable that much of the exchange of large water-soluble molecules occurs as the blood passes through small venules. The permeability of this venular microvasculature can be affected by local agents. Histamine, for example, increases venular permeability. This is part of the mechanism responsible for local tissue fluid accumulation in cutaneous tissue and mucous membranes in response to allergic reactions.

Venules are an important component of the blood reservoir system. At rest, approximately two thirds of the total blood volume is within the venous system with perhaps more than half of this volume within venules. Although the blood moves within this venous reservoir, it moves slowly, much like water in a reservoir behind a river dam. If venule radius is decreased, the volume of blood in this tissue reservoir can decrease up to 20 mL/kg of tissue and be translocated into the large veins. In this manner, the volume of blood readily available for circulation can increase by more than 1 L in a 70-kg person. Such a large change in the available blood volume can substantially improve the venous return of blood to the heart in response to a loss of blood volume due to hemorrhage or severe dehydration. Another example of this reservoir translocation occurs during blood donation procedures. The volume of blood typically removed from blood donors is about 500 mL, or about 10% of the total blood volume. However, this usually causes no ill effects because the venules and veins decrease their reservoir volume to restore the circulating blood volume.

▶ THE LYMPHATIC SYSTEM

Lymphatic vessels are microvessels that form an interconnected system of simple endothelial "drainage" tubes within tissues. They transport fluid, serum proteins, lipids, and even foreign substances from the interstitial spaces back to the circulation. Collecting lymphatic fluid from the organs is important because a volume of fluid equal to the plasma volume is filtered from the blood through the capillary pores and into the interstitium every day. The gastrointestinal tract, liver, skin, and lungs have the most extensive lymphatic systems; the latter helps prevent excess fluid accumulation in lung tissue, which would otherwise impair gas exchange. The central nervous system, in contrast, may not contain any lymph vessels.

The lymphatic system typically begins as blind-ended tubes, or **lymphatic bulbs**, which drain into the meshwork of interconnected lymphatic vessels. These vessels, like veins, contain one-way valves that open in the direction of larger vessels downstream (Fig. 15.3). The lymphatic vessels coalesce into increasingly more developed and larger collection vessels. These larger vessels in the tissue and the macroscopic lymphatic vessels outside the organs have contractile cells similar to VSM cells. These cells spontaneously contract, perhaps as a result of contractile endothelial cells, to generate pressure and thus flow toward the larger lymphatic vessels downstream. In addition, external compression of these lymphatic structures helps propel lymphatic fluid away from the tissues and toward the central circulation. Physical massage of body structures, such as the limbs, or organ movements, such as those that occur with intestinal motility during digestion or those associated with skeletal muscle contractions during movement, externally compress lymphatic vessels and move lymph away from the interstitium of the organ.

Lymphatic vessels mechanically collect fluid from tissue fluid between cells.

In all organ systems, more fluid is filtered than absorbed by the capillaries and plasma proteins diffuse into the interstitial spaces through the large pore system. By removing the fluid, the lymphatic vessels also collect the proteins and return them back into the bloodstream. This process is essential

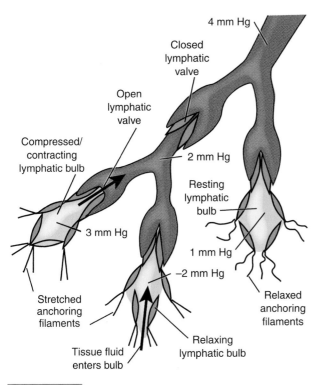

Figure 15.3 **Lymphatic vessels: basic structure and functions.** The contraction–relaxation cycle of lymphatic bulbs **(bottom)** is the fundamental process that removes excess water and plasma proteins from the interstitial spaces. Pressures along the lymphatics are generated by lymphatic vessel contractions and by organ movements, which cause alternately external compression and relaxation stimuli.

because the osmotic attraction of water by plasma proteins is the most important factor that retains water in the vascular system (see below).

The ability of lymphatic vessels to change diameter extrinsically or intrinsically is important for lymph formation and protein removal. The movement of fluid from tissue to the lymphatic vessel lumen is passive. The lymphatic vessels relax after a compression or contraction and allow fluid to enter. Once the interstitial fluid is in a lymphatic vessel, it is called **lymph**. In the smallest lymphatic vessels and, to some extent, in the larger lymphatic vessels, the endothelial cells are overlapped rather than fused together, as in blood capillaries. The overlapped portions of the cells are attached to anchoring filaments, which extend into the tissue (see Fig. 15.3). When stretched, anchoring filaments pull apart the free edges of the endothelial cells and create openings that allow tissue fluid and molecules carried in the fluid to enter. When lymphatic vessels are allowed to passively relax, the pressure in the lumen becomes slightly lower than in the interstitial space, and tissue fluid enters the lymphatic vessel. When the lymphatic bulb or vessel next actively contracts or is compressed, the overlapped cells are mechanically sealed to hold the lymph. The pressure developed inside the lymphatic vessel forces the lymph into the next downstream segment of the lymphatic system. Because the anchoring filaments are stretched during this process, the overlapped cells can again be parted during the relaxation of the lymphatic

vessel the way in which a stretched rubber band snaps back after the release of its stretch. This compression–relaxation cycle facilitates the uptake and flow of fluid from the interstitium into and down the lymphatic channels. The compression–relaxation cycle increases in frequency and vigor when excess water is in the lymph vessels, whether lymphatic smooth muscle cells or the contractile lymphatic endothelial cells control it. Conversely, less fluid in the lymphatic vessels allows the vessels to become quiet and pump less fluid. This simple regulatory system ensures that the fluid status of the organ's interstitial environment is appropriate.

The active and passive compression of lymphatic bulbs and vessels also provides the force needed to propel the lymph back to the venous side of the blood circulation. To maintain directional lymph flow, microscopic lymphatic bulbs and vessels as well as large lymphatic vessels have one-way valves (see Fig. 15.3). These valves allow lymph to flow only from the tissue toward the progressively larger lymphatic vessels and, finally, into large veins in the chest cavity.

Lymphatic pressures are only a few mm Hg in the bulbs and smallest lymphatic vessels and as high as 10 to 20 mm Hg during contractions of larger lymphatic vessels. This progression from lower to higher lymphatic pressures is possible because, as each lymphatic segment contracts, it develops a slightly higher pressure than in the next lymphatic vessel, and the lymphatic valve momentarily opens to allow lymph flow. When the activated lymphatic vessel relaxes, its pressure is again lower than that in the next vessel, and the lymphatic valve closes.

▶ SOLUTE EXCHANGE BETWEEN THE VASCULATURE AND TISSUES

The increasing numbers of vessels through successive branches of the arterial tree and on into the capillaries and venules dramatically increase the surface area of the microvasculature. The large surface area of the capillaries and smallest venules is important because most exchange of nutrients, wastes, and fluid occurs across these tiny vessels. Exchange between the interstitial fluid and capillaries occur via three processes: (1) diffusion, (2) filtration, and (3) vesicular transport.

Transport across capillaries is enhanced by increasing their collective surface area and reducing diffusion distances from capillaries to cells.

The spacing of microvessels in the tissues determines the distance molecules must diffuse from the blood to the interior of the tissue cells. In the example shown in Figure 15.4A, nutrients are supplied to a single cell by a single capillary. The density of dots at various locations represents the concentration of blood-borne molecules across the cell interior. Note that as molecules travel farther from the capillary, their concentration decreases substantially because the volume into which diffusion proceeds increases as the square of the distance. Furthermore, the time required for traversing a given distance by diffusion increases as an exponential function of distance (see Chapter 11). Consequently, if molecules are

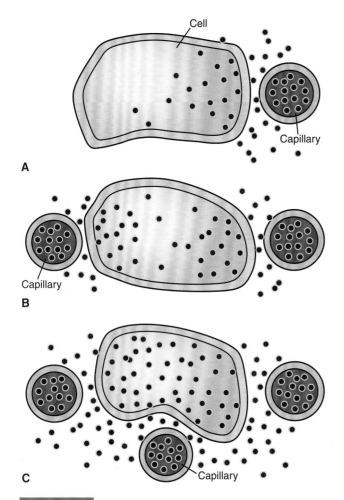

A

B

C

Figure 15.4 **The effect of the number of perfused capillaries on cell concentration of blood-borne molecules.** Molecules are represented by *dots* in the figure. **(A)** With one capillary, the left side of the cell has a low concentration. **(B)** The concentration can be substantially increased if a second capillary is perfused. **(C)** The perfusion of three capillaries around the cell increases concentrations of blood-borne molecules throughout the cell.

consumed by cells (e.g., oxygen), the concentration of substances surrounding them will decrease dramatically as the distance between the cell and its nearest capillary is increased.

Increasing the number of microvessels reduces diffusion distances from a given point inside a cell to the nearest capillary. Doing so minimizes the dilution of molecules within the cells caused by large diffusion distances. At any given moment during resting conditions, only about 40% to 60% of the capillaries are perfused by red blood cells in most organs. The capillaries not in use do contain blood, but it is not moving. As shown in Figure 15.4B and C, increasing the number of perfused capillaries with moving blood decreases diffusion distances and elevates cell concentrations of molecules derived from the blood. However, it is equally true that decreasing the number of perfused capillaries increases diffusion distances, decreases exchange, and reduces cell concentration of molecules, thereby threatening cell function and survival. This can result from severely constricting arterioles or by destroying existing capillaries in disease states such as diabetes mellitus.

Transport of a solute across capillaries is enhanced by increasing the capillary permeability and concentration gradient for the solute across the capillaries.

Diffusion is by far the most important means for moving solutes across capillary walls. Fick's law of diffusion (see Chapter 2) states that the rate of diffusion, or flux, of a solute between blood and tissue is represented by

$$J_s = PS(C_b - C_t) \tag{1}$$

where J_s is the net movement of solute (often expressed in mol/min per 100 g of tissue), PS is the **permeability surface area coefficient**, and C_b and C_t are the blood and tissue concentrations of the solute, respectively. The PS coefficient is directly related to the diffusion coefficient of the solute in the capillary wall and the vascular surface area available for exchange, whereas it is inversely related to the diffusion distance. The surface area and diffusion distance are determined, in part, by the number of microvessels with active blood flow as explained in the previous section. The diffusion coefficient is relatively constant, unless the capillaries are damaged, because it depends on the anatomic properties of the vessel wall (e.g., the size and abundance of pores) and the chemical nature of the material that is diffusing.

Concentrations of solutes in tissue and blood, as well as the number of perfused capillaries and thus diffusional surface area and distances, are all affected by physiological and pathophysiological conditions whether they occur acutely or chronically. Consequently, microvascular exchange can be altered dynamically. For example, perfused capillary density increases significantly when metabolic activity in a tissue increases as compared with the tissue at rest. This is seen in exercising muscle or in the intestinal circulation during digestion.

It is important to remember that diffusion rate for a molecule between two points depends on the *difference* between the high and low concentrations, not the specific concentrations at each point. For example, if the cell consumes a particular solute, the concentration in the cell will decrease, and for a constant concentration in blood plasma, the diffusion gradient will enlarge and thus increase the rate of diffusion into the cell, providing the tissue is supplied with sufficient blood. If the cell ceases to use as much of a given solute, the concentration in the cell will increase and the rate of diffusion consequently will decrease.

Increases in vascular permeability, surface area, or blood flow enhance the diffusion of small molecules from the blood.

As a result of diffusional losses and gains of molecules as blood passes through the tissues, the concentrations of various molecules in venous blood can be very different from those in arterial blood. The **extraction** (E), or **extraction ratio**, of material from blood perfusing a tissue can be calculated from the arterial (C_a) and venous (C_v) blood concentration as

$$E = (C_a - C_v)/C_a \tag{2}$$

If the blood loses material to the tissue, the value of E is positive and has a maximum value of 1 (e.g., if all material is removed from the arterial blood in which case $C_v = 0$). An E value of 0 indicates that no loss or gain occurred. A negative E value indicates that the tissue added material to the blood.

The total mass of material lost or gained by the blood can be calculated as

$$\text{Amount lost or gained} = E \times \dot{Q} \times C_a \qquad (3)$$

where E is extraction, $\dot{Q}$ is blood flow, and C_a is the arterial concentration. Although this equation is useful for calculating the total amount of material exchanged between tissue and blood, it does not allow a direct determination of how changes in vascular permeability and exchange surface area influence the extraction process. The extraction can be related to the permeability (P) and surface area (A) available for exchange as well as the blood flow, ($\dot{Q}$), by the equation:

$$E = 1 - e^{-(PA/\dot{Q})} \qquad (4)$$

where e is the base of the natural system of logarithms. This equation predicts that physiologically induced changes in the number of perfused capillaries, which alter surface area, and changes in blood flow are important determinants of exchange processes. A change that increases permeability, increases surface area, or decreases flow will increase extraction. Changes in the opposite direction decrease extraction. The inverse effect of blood flow on extraction occurs because, as flow increases, less time is available for exchange whereas a slowing of flow allows more time for exchange.

Ordinarily, the blood flow and total perfused surface area change in the same direction, although by different relative amounts. For example, surface area can generally, at most, double or be reduced by about half; however, blood flow can increase threefold to fivefold or more in some tissues, such as skeletal muscle during exercise, or decrease by about half in many organs without threatening viable tissue. The net effect is that extraction is rarely more than doubled or decreased by half relative to the resting value in most organs. This is still an important range because changes in extraction can compensate for reduced blood flow, or enhance exchange when blood flow is increased.

Diffusion-limited and flow-limited transport

Gases, small lipid-soluble molecules, water, simple sugars, and ions can diffuse so rapidly across capillaries that their transport from capillary to tissue or vice versa is not limited by their rate of diffusion. Their transport is limited only by their rate of delivery into the capillary network from blood perfusing the network. Therefore, such transport is called **flow-limited transport**. For a flow-limited transport substance, increasing blood flow increases the effective concentration of a substance in the capillary and thus accelerates its outward diffusion; that is, the faster materials are brought into the capillary by blood flow, relative to their diffusion out of the capillary, the higher will be their flux out of the capillaries and into the tissues.

In contrast to solutes characterized by flow-limited transport, larger lipophobic molecules, such as sucrose, polysaccharides, and proteins, have difficulty diffusing across the capillary membrane or through capillary pores. These substances may be delivered in large quantities into capillaries by blood flow but nevertheless exhibit low transcapillary flux. Their transport between the bloodstream and tissues is therefore limited by their rate of diffusion into or out of capillaries and not appreciably altered by their rate of delivery into the capillary network by blood flow. Such transport is called **diffusion-limited transport**.

In pathologic conditions, substances that otherwise exhibit flow-limited transport can become diffusion limited. This can occur when diffusion distances between capillaries and cells become too great to allow the rapid exchange of materials, as seen in the lungs when the transport of normally highly diffusible oxygen is impaired by infection or fluid accumulation in the pulmonary interstitium, which markedly increases diffusion distances. This phenomenon can also occur in any organ when the number of perfused capillaries is radically reduced (e.g., by low flow, blood clots, etc.), thereby reducing the capillary surface area available for diffusion while also increasing the distance between tissue cells and the nearest capillaries with blood flow.

▶ WATER EXCHANGE BETWEEN THE VASCULATURE AND INTERSTITIUM

To supply blood into the microvessels of tissues, the heart generates pressure and pumps blood into the arterial system. However, because the intercapillary hydrostatic pressure is greater than that in the interstitium, pressurized filtration of water occurs through pores in capillaries. This loss of water out of the cardiovascular system, if unchecked, could lead to severe hypovolemia and collapse of the cardiovascular system. However, plasma contains a significant amount of proteins (mostly albumin) that cannot cross the capillary membrane or pores. They are considered, therefore, "effective osmols" across the capillary membrane. Coupled with the fact that there is little protein in interstitial fluid, a net osmotic force favoring fluid flow into the capillaries exists. The osmotic pressure due to protein is often called a colloid oncotic, or just oncotic, pressure to indicate that the osmotic force is generated by the colligative properties of proteins. Intercapillary oncotic pressure is the major factor maintaining fluid retention in the cardiovascular system and thus is the key factor counteracting capillary fluid losses through capillary pores.

The interplay between net hydrostatic and oncotic forces determines the net direction of fluid exchange across the capillaries.

In organs other than the kidney, capillary hydrostatic pressure declines from the arterial to the venule end of the capillary. Typically, capillary hydrostatic pressure is ~40 mm Hg at its arteriolar end and ~15 mm Hg at the venule end. The average capillary pressure is different in each organ, ranging from about 15 mm Hg in intestinal villus capillaries to 55 mm Hg in the kidney glomerulus. The interstitial hydrostatic pressure ranges from slightly negative to 8 to

10 mm Hg. The oncotic pressure of plasma proteins is typically 18 to 25 mm Hg in mammals and is highly dependent on the plasma albumin concentration, which can change in abnormal conditions.

In the body as a whole, the net inward oncotic force does not balance the net outward hydrostatic pressure forces. Thus, there exists a slightly positive overall force for the filtration of fluid out of the capillaries. Most organs continuously form lymph, which supports the concept that capillary and venular filtration pressures generally are larger than absorption pressures. The balance of pressures is likely +1 to +2 mm Hg outward in most organs. However, this is not true for each capillary in the body all the time. Some capillaries can be net reabsorbers of water; others filter along their entire length, whereas still others can filter at their arteriolar end and reabsorb at their venule end, depending on existing hydrostatic pressure in both capillary locations. Based on directly measured capillary hydrostatic and plasma oncotic pressures, the entire length of the capillaries in skeletal muscle filter slightly all of the time, whereas the lower capillary pressures in the intestinal mucosa, brain, and pulmonary capillaries primarily favor absorption along the entire capillary length. Standing also causes high capillary hydrostatic pressures from gravitational effects on the blood in the arterial and venous vessels and results in excessive filtration in the lower extremities.

Effects of capillary oncotic and hydrostatic pressure on fluid flux are modified by these pressures in the interstitium.

A small amount of plasma protein enters the interstitial space; these proteins and, perhaps, native proteins of the space generate the tissue colloid osmotic pressure. This pressure of 2 to 5 mm Hg offsets part of the colloid osmotic pressure in the plasma. This is, in a sense, a filtration, or outward "drawing," pressure that opposes the blood colloid osmotic pressure. The hydrostatic pressure on the tissue side of the endothelial pores is the tissue hydrostatic pressure. The water volume in the interstitial space and tissue distensibility determine this pressure. The magnitude of tissue hydrostatic pressure in various tissues during resting conditions is a matter of debate. Tissue pressure is probably slightly below atmospheric pressure (negative) to slightly positive (about +3 mm Hg) during normal hydration of the interstitial space and becomes positive when that space contains excess water. Tissue hydrostatic pressure thus is a filtration force when negative and is an absorption force when positive. Tissue hydrostatic pressure can be increased by organ compression, such as in a skeletal muscle during contraction and decreased by severe dehydration.

If water is removed from the interstitial space, the hydrostatic pressure becomes negative (Fig. 15.5). As a general rule, about 500 to 1,000 mL of fluid can be withdrawn from the interstitial space of the entire body to help replace water losses resulting, for example, from excessive sweating, diarrhea, vomiting, or blood loss. In contrast, if a substantial amount of water is added to the interstitial space, the tissue hydrostatic pressure increases. However, a margin of

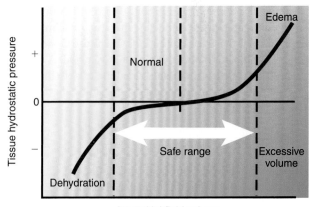

Figure 15.5 **Variations in tissue hydrostatic pressure as interstitial fluid volume is altered.** Under normal conditions, tissue pressure is slightly negative (subatmospheric), but an increase in volume can cause the pressure to become positive. If the interstitial fluid volume exceeds the "safe range," high tissue hydrostatic pressures and edema will be present. Tissue dehydration can cause negative tissue hydrostatic pressures.

safety exists over a wide range of tissue fluid volumes (see Fig. 15.5), which helps avoid excessive tissue hydration. The ability of tissues to allow substantial changes in interstitial volume with only small changes in pressure indicates that the interstitial space is distensible.

The Starling-Landis equation quantifies fluid flow across the capillaries.

At the end of the 19th century, the English physiologist Ernest Starling first postulated the role of hydrostatic and colloid osmotic pressures in determining fluid movement across capillaries. In the 1920s, the American physiologist Eugene Landis obtained experimental proof for Starling's hypothesis. The relationship is defined for a single capillary by the **Starling-Landis equation**:

$$J_V = K_h A \left[(P_c - P_t) - \sigma(COP_p - COP_t) \right] \qquad (5)$$

J_V is the net volume of fluid moving across the capillary wall per unit of time (μm^3/min). K_h is the **hydraulic conductivity for water**, which is the fluid permeability of the capillary wall. It can be thought of as the ease with which water crosses the capillary wall. K_h is expressed as μm^3/min/mm Hg (μm^2 of capillary surface area per minute per mm Hg pressure difference). The value of K_h increases up to fourfold from the arterial to the venous end of a typical capillary. A, in equation 5, is the vascular surface area, P_c is the capillary hydrostatic pressure, and P_t is the tissue hydrostatic pressure. COP_p and COP_t represent the plasma and tissue colloid osmotic pressures, respectively, and σ is the **reflection coefficient** for plasma proteins. The value of σ is 1 when molecules cannot cross the membrane (i.e., they are 100% "reflected") and 0 when molecules freely cross the membrane (i.e., they are not reflected at all). This coefficient is included because the microvascular wall is slightly permeable to plasma proteins, preventing the full expression of the two colloid osmotic pressures. Typical σ values for plasma proteins in the microvasculature exceed 0.9 in most organs other than the liver and spleen, which have capillaries that are permeable to plasma proteins. The reflection coefficient is normally relatively constant but can be decreased dramatically by hypoxia, inflammatory processes, and tissue injury. This leads to increased fluid filtration because the effective fluid-retaining power of the colloid osmotic pressure is reduced when the vessel wall becomes more permeable to plasma proteins.

The extrapolation of fluid filtration or absorption for a single capillary to fluid exchange in a whole tissue is difficult. Within organs, there are regional variations in microvascular pressures, possible filtration and absorption of fluid in vessels other than capillaries, and physiologically and pathologically induced variations in the available surface area for capillary exchange. Therefore, for whole organs, a measurement of total fluid movement relative to the mass of the tissue is used. To take into account the various hydraulic conductivities and total surface areas of all vessels involved, the volume (mL) of fluid moved per minute for a change of 1 mm Hg in net capillary filtration pressure for each 100 g of tissue is determined. This value is called the **capillary filtration coefficient (CFC, or K_f)**, even though it is likely that fluid exchange also occurs in venules. CFC values in tissues such as skeletal muscle and the small intestine are typically in the range of 0.025 to 0.16 mL/min/mm Hg per 100 g of tissue. The CFC replaces the combined hydraulic conductivity (K_h) and capillary surface area (A) variables in the Starling-Landis equation that apply for filtration across a single capillary.

The CFC can change if fluid permeability, the surface area (determined by the number of perfused microvessels), or both are altered. For example, during the intestinal absorption of products of digestion, especially lipids, both capillary fluid permeability and perfused surface area increase. Therefore, CFC increases dramatically. In contrast, during exercise, CFC increases in the skeletal muscle vasculature occur primarily through increased perfused capillary surface area with only small increases in fluid permeability.

The hydrostatic and colloid osmotic pressure differences across capillary walls, called the **Starling forces**, cause the movement of solutes along with the water into the interstitial spaces. However, most solutes transferred to the tissues move across capillary walls by simple diffusion, not by bulk flow of fluid.

Capillary hydrostatic pressure is altered by changes in precapillary and postcapillary resistance as well as arteriolar and venule blood pressure.

Normal physiologic processes in the body do not control plasma colloid osmotic pressure, tissue hydrostatic pressure, or tissue colloid osmotic pressure. Thus, manipulation of these parameters cannot be used to regulate filtration and reabsorption of fluid at the capillary. Such regulation is accomplished, therefore, by the adjustment of capillary hydrostatic pressure.

Capillary hydrostatic pressure (P_c) is influenced by four major variables: precapillary resistance (R_{pre}), postcapillary resistance (R_{post}), arterial blood pressure (P_a), and venous blood pressure (P_v). Precapillary and postcapillary resistances can be calculated from the pressure dissipated across the respective vascular regions divided by the total tissue blood flow ($\dot{Q}$), which is essentially equal for both regions:

$$R_{pre} = (P_a - P_c)/\dot{Q} \tag{6}$$

$$R_{post} = (P_c - P_v)/\dot{Q} \tag{7}$$

In most organ vasculatures, the precapillary resistance is three to six times higher than the postcapillary resistance. This has a substantial effect on capillary pressure. To demonstrate the effect of precapillary and postcapillary resistances on capillary pressure, we use the equations for the precapillary and postcapillary resistances to solve for blood flow:

$$\dot{Q} = (P_a - P_c)/R_{pre} = (P_c - P_v)/R_{post} \tag{8}$$

The two equations to the right of the flow term can be solved for capillary pressure:

$$P_c = \frac{(R_{post}/R_{pre})P_a + P_v}{1 + (R_{post}/R_{pre})} \tag{9}$$

Equation 9 indicates that the ratio of postcapillary to precapillary resistance, rather than the absolute magnitude of either resistance, determines the effect of arterial pressure (P_a) on capillary hydrostatic pressure. It also shows that venous pressure substantially influences capillary pressure and that the denominator influences both pressure effects. At a typical postcapillary to precapillary resistance ratio of 0.16:1, it is obvious that venous pressure has a larger effect on capillary hydrostatic pressure than does arteriolar pressure. Furthermore, with this ratio, the denominator of equation 9 will be 1.16, which means about 80% of a change in venous pressure will be reflected back to the capillaries.

As can be seen by equation 9, capillary hydrostatic pressure will increase with arteriolar vasodilation. This is a change that accompanies increased tissue metabolism. Furthermore, with increased tissue metabolism, the postcapillary to precapillary resistance ratio increases because precapillary resistance decreases more than postcapillary resistance. This change in the ratio will increase P_c. Because the balance of hydrostatic and colloid osmotic pressures is usually −2 to +2 mm Hg, a 10- to 15-mm Hg increase in capillary pressure during maximum vasodilation can cause a profound increase in filtration. The increased filtration associated with microvascular dilation is usually associated with a large increase in lymph production, which removes excess tissue fluid.

When sympathetic nervous system stimulation causes a substantial increase in precapillary resistance and a proportionately smaller increase in postcapillary resistance, the capillary pressure can decrease up to 15 mm Hg and, thereby, greatly increase the absorption of tissue fluid. This process is an important compensatory mechanism the body uses to combat the early stages of circulatory shock (see Chapter 17).

Cardiovascular Physiology

Edema impairs diffusional transport across the capillaries.

Changes in capillary hydrostatic pressure and plasma protein concentration can have a profound effect on filtration at the capillary. The hydrostatic pressure involved in transcapillary fluid exchange depends on how the microvasculature dissipates the prevailing arterial and venous pressures. As seen above, certain interplay of these variables can lead to substantial filtration across the capillaries. Although plasma protein concentration does not vary moment to moment, its circulating value in normal circumstances is determined largely by the rate of protein synthesis in the liver, where most of the plasma proteins are made. Disorders that either impair albumin synthesis or promote the loss of albumin result in reduced plasma protein concentration, a lowered plasma colloid osmotic pressure, and excessive fluid filtration at the capillaries.

Edema is a condition in which there is excessive accumulation of fluid in tissue spaces (i.e., the interstitium). Edema interferes with capillary transport by increasing diffusion distances between capillaries and tissue cells. Sometimes this accumulation can be substantial, and in addition to affecting capillary transport, this extreme can cause circulatory collapse if the edematous fluid was derived from a loss of plasma volume. For example, edema formation in the abdominal cavity (known as **ascites**) can allow large quantities of fluid to collect in the abdominal space.

Anything that causes excess fluid filtration at the capillaries or impairs fluid transport through the lymph channels can create edema (Fig. 15.6). One of the common causes of edema formation is a loss of albumin from the plasma.

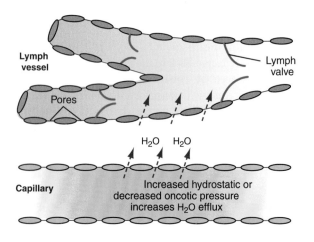

↑ Lymph flow:
↑ capillary hydrostatic pressure
↑ capillary surface area
↑ capillary permeability
↑ tissue metabolism
↑ muscle activity (massage)

↑ Edema formation:
↓ plasma protein concentration
arteriolar dilation
venous obstruction
lymphatic obstruction
histamine (↑ capillary permeability)
all factors that ↑ lymph flow except
 muscle massage

Figure 15.6 Determinants of water efflux from the capillary and into a lymph vessel. Factors that increase lymph flow or favor the formation of edema are listed in the figure.

Edema will occur when albumin concentrations drop below 2.5 g/100 mL. This is common in diseases in which the liver is unable to manufacture albumin or in kidney diseases in which albumin and other proteins are lost into the urine. Burns, by destroying capillary integrity, cause edema through increased capillary permeability, loss of albumin through damaged vessels, and inflammatory vasodilation. Hives, which are a form of localized edema associated with allergic reactions, result from an increase in capillary permeability, venule permeability, and arterial dilation, all caused by histamine release during the allergic response. Obstruction of veins, usually from blood clot formation after surgery, is another common cause of edema, as is the obstruction of lymph channels at lymph nodes by infections.

▶ REGULATION OF MICROVASCULAR RESISTANCE

The VSM cells around arterioles and venules respond to a wide variety of physical and chemical stimuli, which alter their diameter and resistance. The following sections consider the various physical and chemical conditions in tissues that affect the contractile state of VSM and, thus, affect resistance in the microvasculature.

Myogenic regulation causes arterioles to actively contract or relax in response to changes in intravascular pressure.

VSM can rapidly *actively* contract in response to being stretched and, conversely, can *actively* relax, when passively shortened. This process, called *myogenic regulation*, is activated in arterioles (and somewhat in venules) when microvascular pressure is increased or decreased. Myogenic mechanisms are extremely fast and appear to be able to adjust to the most rapid pressure changes. In fact, VSM may be able to contract or relax when the load on the muscle is increased or decreased, respectively, without any change in muscle length. These responses are known to persist for as long as the initial stimulus is present, unless vasoconstriction reduces blood flow to the extent that tissue becomes severely hypoxic and nonfunctional.

The cellular mechanisms responsible for myogenic regulation are not entirely understood, but several possibilities are likely involved. VSM contraction is sensitive to external calcium concentration. Adding calcium to the cytoplasm from the extracellular fluid through membrane channels would activate the smooth muscle cell and result in contraction. Conversely, limiting calcium entry through channels would allow calcium pumps to remove calcium ions from the cytoplasm and favor relaxation. VSM contains calcium ion-selective channels that open in response to increased membrane stretch or tension. These **stretch-activated calcium channels** are distinct from receptor-operated or voltage-gated calcium channels. These channels may contribute to myogenic adjustments of VSM contraction and thus vascular resistance whenever internal arterial pressure changes.

Another mechanism postulated for the myogenic response is that a nonspecific cation channel is opened in proportion to cell membrane stretch or tension. The entry of sodium ions through these open channels could depolarize the cell and lead to the opening of voltage-activated calcium channels, followed by calcium entry into the cell and contraction. During reduced stretch or tension, the nonspecific channels would close and allow hyperpolarization to occur. Most recently, evidence suggests that the modulation of the calcium sensitivity of the contractile apparatus in VSM is a key factor in the myogenic response. The sensitivity of the apparatus to calcium is modified by the relative activities of myosin light chain kinase and myosin light chain phosphatase in the smooth muscle and appears to suggest a role for latch state mechanisms in the myogenic response. Regardless of the precise mechanism involved, arteries contract and depolarize as their intravascular pressure is increased and relax and hyperpolarize as their intravascular pressure is decreased.

The myogenic response is particularly helpful in preventing tissue edema when venous pressure is elevated by more than 5 to 10 mm Hg above the typical resting values. The elevation of venous pressure results in an increase in capillary pressure, which would normally favor filtration and edema formation. However, some of this pressure is transmitted back into the arteriolar segment of the microcirculation. Myogenic arteriolar constriction in response to the stretch caused by this increase in pressure lowers the transmission of arterial pressure into the capillaries and thus minimizes the risk of edema (although this occurs at the expense of decreased blood flow). This mechanism is called the **venous–arteriolar response**.

Myogenic regulation may also be one factor that helps control organ blood flow in the face of changes in arterial pressure. This process, called **autoregulation**, or autoregulation of blood flow, is discussed in greater detail later in this chapter.

Organ blood flow is increased by increased tissue metabolism through local, nonneurogenic mechanisms.

In almost all organs, an increase in tissue metabolism is associated with increased blood flow and extraction of oxygen to meet the increased metabolic needs of the tissues. This phenomenon is called **active hyperemia**. In addition, a reduction in oxygen within the blood, hypoxia, and hypoxemia are associated with the dilation of the arterioles and increased blood flow (assuming neural reflexes to hypoxia are not activated). In both active hyperemia and local ischemia, the changes in arteriolar resistance are in the direction of increasing blood flow, thereby indirectly increasing oxygen delivery to the tissue. This helps tissues receive the oxygen necessary to meet their metabolic requirements.

The mechanisms behind the local regulation of the microvasculature in response to the metabolic needs of tissues involve many different types of cellular events. Decreased O_2, increased CO_2, increased H^+, and increased adenosine all accompany increased tissue metabolism and all are vasodilators, but none of these alone seems to be able to explain active hyperemia or the local vascular response to ischemia.

Oxygen is not stored in appreciable amounts in tissues, and the oxygen concentration will fall to nearly zero in about 1 minute if blood flow is stopped in any organ. Therefore, an increase in metabolic rate would decrease the tissue oxygen concentration and possibly directly signal the vascular muscle to relax by limiting the production of adenosine triphosphate (ATP) needed for the contraction of smooth muscle cells. Alternatively, the depletion of ATP could release the inhibitory effect of ATP on K^+ channels, which would result in hyperpolarization of the cell membrane, reduced activation of voltage-gated calcium channels, and vasodilation. Oxygen, or the lack thereof, has been shown to play a role at the molecular level in the operation of L-type calcium channels as well as the ATP-gated K^+ channels. However, no known effect of oxygen on these channels appears to adequately explain the effects of oxygen on smooth muscle contraction. Most recent evidence appears to point to H_2S metabolism in the mitochondria as the intercellular molecular link between hypoxia and vasodilation. H_2S metabolism in the mitochondria decreases during hypoxia, and thus, H_2S increases intracellularly. H_2S is a direct vasodilator at very low concentrations and it mimics the tissue effects of hypoxia in arteries as well as other oxygen-sensing tissues (i.e., peripheral neural chemoreceptors; see Chapter 17).

Although there are many possibilities to consider for cellular mediators of hypoxic vasodilation or active hyperemia in the vascular system, practical concerns exist as to cause and effect mechanisms. For example, many arterioles have a normal to slightly increased periarteriolar oxygen tension during skeletal muscle contractions because the increased delivery of oxygen through elevated blood flow offsets the increased use of oxygen by tissues immediately around the arteriole. Yet, the skeletal muscle arterioles remain dilated. Therefore, as long as blood flow is allowed to increase substantially, it is unlikely that oxygen availability at the arteriolar wall is a major factor in the sustained vasodilation that occurs during increased metabolism. In fact, it appears that the signal causing arteries to dilate when tissue metabolism increases may not exist in the arterial wall at all but, instead, may originate in the surrounding tissues. In addition, recent studies indicate that VSM cells are not particularly responsive to a broad range of oxygen tensions. Only unusually low or high oxygen tensions seem to be associated with direct changes in VSM force. However, either oxygen depletion from an organ's cells or an increased metabolic rate does cause the formation and release of free adenosine, Krebs cycle intermediates, and, in hypoxic conditions, lactic acid. It is possible then that these factors contribute to the phenomenon of active hyperemia. In addition to hypoxia itself, high concentration of either CO_2 or H^+ causes relaxation of the VSM. However, the levels of CO_2 and H^+ produced at the tissue, even with high metabolism, cannot dilate arterioles to the level seen during the active hyperemia. Furthermore, usually only transient increases in venous blood and interstitial tissue acidity occur if blood flow through an organ with increased metabolism is allowed to increase appropriately.

Of all the metabolite concentrations that change during both tissue metabolism and hypoxic conditions, adenosine is the best situated to account for arteriolar dilation and increased blood flow. Adenosine increases during both hypoxia and increased tissue metabolism, readily diffuses out of tissue cells to the adjacent microcirculation, and is a potent vasodilator. It appears to be an important mediator of active hyperemia in the heart and brain but is not believed to be the sole mediator of the ability of tissues to match their blood flow to their moment-to-moment metabolic needs. A summary of the metabolic regulation of blood flow is shown in Figure 15.7.

Arteriolar contractile state is affected by vasoactive chemicals released by endothelial cells.

Substances released by endothelial cells are important contributors to local vascular regulation. The most important of these is **nitric oxide (NO)** (formerly known as **endothelium-derived relaxing factor** or **EDRF**). NO is a potent vasodilator. It is formed by Ca^{2+}–calmodulin complex activation of endothelial **NO synthase (eNOS)** that acts on the amino acid L-arginine to produce L-citrulline and NO (Fig. 15.8). Endothelial NO synthase is constitutively expressed in endothelial cells and NO is

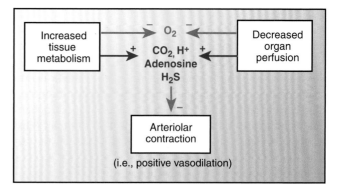

Figure 15.7 An overview of metabolic effects on arteriolar contraction. Decreased organ perfusion or increased tissue metabolism produce factors that dilate microcirculatory arteries and veins.

released continuously at rest from all arterial, microvascular, venular, and lymphatic endothelial cells. It causes the relaxation of VSM by activating guanylate cyclase within the muscle cells to increase the production of cyclic guanosine monophosphate (cGMP). cGMP activates various cGMP-dependent protein kinases that, in turn, activate calcium-lowering mechanisms in

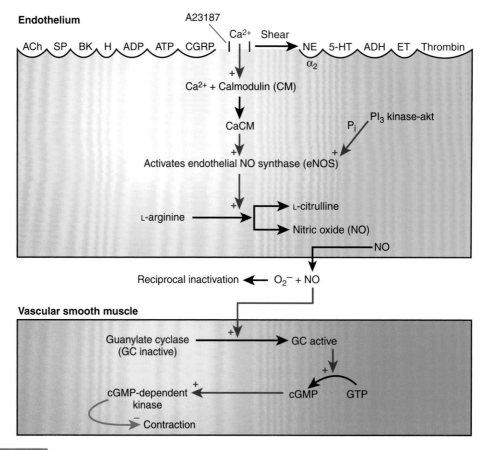

Figure 15.8 The mechanism of endothelium-dependent relaxation of vascular smooth muscle by nitric oxide (NO). Ach, acetylcholine; SP, substance P; BK, bradykinin; H, histamine; ADP, adenosine diphosphate; ATP, adenosine triphosphate; CGRP, calcitonin gene–related peptide; A23187, calcium ionophore; NE, norepinephrine; 5-HT, serotonin; ADH, antidiuretic hormone (vasopressin); ET, endothelin; +, stimulation; cGMP, cyclic guanosine monophosphate; GTP, guanosine triphosphate; eNOS, endothelial constitutive nitric oxide synthase. Phosphorylation of eNOS by an IP_3–akt pathway enhances eNOS sensitivity and NO generation.

the smooth muscle cell, thus inhibiting contraction. Numerous compounds, such as acetylcholine, histamine, and adenine nucleotides (ATP and ADP) as well as hypertonic conditions and hypoxia, cause the release of NO. Adenosine causes NO release from endothelial cells, although it also directly relaxes VSM cells through adenosine receptors.

Shear stress generated by blood moving past endothelial cells is an important mechanism for endothelial NO release and may represent the major, truly physiological, NO agonist in the cardiovascular system. Frictional forces between moving blood and the stationary endothelial cells distort the endothelial cells, opening special potassium channels and causing endothelial cell hyperpolarization. This increases calcium ion entry into the cell down the increased electrical gradient. The elevated cytosolic calcium ion concentration in the endothelium activates endothelial NOS to form more NO, and the blood vessels dilate in response.

This mechanism is used to coordinate vasodilation among various-sized arterioles and small arteries in the same vascular network. For example, an increase in metabolism in a small area of tissue will result in an initial local dilation of small arterioles and precapillary sphincters in that area. This dilation reduces arteriolar resistance, opens/fills additional capillary vessels, and thus increases blood flow into the tissue. However, because blood flow has increased, initial flow velocity and thus endothelial shear stress must increase in the larger arterioles upstream from the initial dilated network as well. This increased shear stress in upstream arteries causes their endothelial cells to release NO and relax their smooth muscle thereby allowing more flow to enter the downstream, open and dilated network. This process is called **flow-mediated vasodilation** and has been observed in cerebral, skeletal muscle, and small intestinal vasculatures. This process repeats further upstream in a phenomenon that has been called *ascending dilation*. As larger arterioles and small arteries control much more of the total vascular resistance than do small arterioles, the dilation of the larger resistance vessels enhances blood flow to meet the needs of the tissue downstream, which at this time has more open capillaries to receive and distribute flow.

Flow-mediated dilation is also seen in conduit arteries. However, the dilation of these vessels has no significant effect on vascular resistance and likely represents a negative feedback mechanism to control shear stress on the arterial endothelium; small dilation of larger arteries has no effect on blood flow per se but does reduce flow velocity and wall shear stress. This effect may be important because many transport processes across the endothelium as well as the interaction of blood cell elements with this vascular barrier are altered by changes in endothelial shear stress. Endothelial cells of arterioles also release vasodilatory prostaglandins and an uncharacterized hyperpolarizing factor when blood flow and shear stress are increased. However, NO appears to be the dominant vasodilator molecule for flow-dependent regulation.

In healthy individuals, basal release of NO exerts important bioregulatory effects in the cardiovascular system. Its basal vasodilatory contribution provides a major antihypertensive effect. Blood pressure increases rapidly by 75% or more if a person is given an agent that blocks endothelial NO production. NO also inhibits platelet aggregation and therefore serves as an anticoagulant. It inhibits neutrophil–endothelial interactions, is antimitogenic to VSM, and promotes the recovery of the endothelium from injury. As these processes are all involved in the pathogenesis of atherosclerosis, NO is considered an antiatherogenic agent.

NO is now known to be generated by NO synthase isoforms in other tissues and some of these have been implicated in cell communication and pathological conditions. Some neural cells contain a neuronal NOS (nNOS) and use NO as a signaling molecule. Smooth muscle and some immune cells contain an inducible form of NOS called iNOS. Inducible NOS is activated by endotoxins, LPS, and other infectious and inflammatory response molecules. Septic shock (see Chapter 17), especially that caused by abdominal infections (i.e., from a ruptured appendix, perforated ulcer, or, in females, from a reproductive organ infection) causes so much activation of iNOS in the VSM of arteries in splanchnic organs that blood pressure plummets rendering the body as a whole ischemic. Such septic shock, if not corrected rapidly, results in death.

Endothelial cells release a 21- amino acid peptide called **endothelin**, which is the most potent vasoconstrictor of blood vessels in the body. Although extremely small amounts are released under natural conditions, endothelin is thought to play a role in several vascular diseases such as hypertension and atherosclerosis. The vasoconstriction occurs because of a cascade of events beginning with phospholipase C activation and leading to activation of protein kinase C (PKC) (see Chapter 2). Two major types of endothelin receptors have been identified and others may exist. The constrictor function of endothelin is mediated by type B endothelin receptors. Type A endothelin receptors cause hyperplasia and hypertrophy of vascular muscle cells and the release of NO from endothelial cells. The precise function of endothelin in the normal vasculature is not clear; however, it is active during embryologic development. In knockout mice, the absence of the endothelin A receptor results in serious cardiac defects so that newborns are not viable. An absence of the type B receptor is associated with an enlarged colon, eventually leading to death.

In damaged heart tissue, such as in an infarct, cardiac endothelial cells increase endothelin production. The endothelin stimulates both VSM and cardiac muscle to contract more vigorously and induces the growth of surviving cardiac cells. However, excessive stimulation and hypertrophy of cells appear to contribute to heart failure, failure of contractility, and excessive enlargement of the heart. Part of the stimulation of endothelin production in the injured heart may be the cause of the damage per se. Also, increased formation of angiotensin II and norepinephrine during chronic heart disease stimulates endothelin production, probably at the gene expression level. The activation of PKC increases the expression of the c-*jun* protooncogene, which, in turn, activates the preproendothelin-1 gene.

Endothelin has also been implicated as a contributor to renal vascular failure, the pulmonary and systemic hypertension associated with insulin resistance, and the spasmodic contraction of cerebral blood vessels exposed to blood after a brain injury or a hemorrhagic stroke.

CLINICAL FOCUS | 15.1

Methods of Assessing Endothelial Function in Cardiovascular Disease

Hypertension, atherosclerosis, and diabetes mellitus all damage arterial endothelium and impair beneficial nitric oxide (NO)–mediated functions in those vessels. This damage and loss of NO invariably causes the vasculature to become prospasmodic, prothrombotic, and proatherogenic, which worsens the cardiovascular outcomes with these diseases. Assessment of the patient's condition and future risks in these diseases often focuses on the key variable associated with the condition: blood pressure in hypertension, high serum LDL to HDL ratios in atherosclerosis, and blood glucose levels in diabetes. More recent use of biochemical markers of NO biology, vascular inflammation, endothelial damage, adhesion molecules, and vascular repair have also been used to assess endothelial damage in cardiovascular disease. However, biochemical markers of NO or vascular disease are generally poor indicators of endothelial function, owing to the large variability in individual production and response to these biomarkers. From the perspective of the vascular system, what is of greater importance for the individual is what variables actually reflect the *functional state of the vasculature and its endothelium*. A patient could have modest hypertension, borderline plasma lipoprotein levels, and minimal glucose intolerance but could be at extreme risk for cardiovascular complications if these conditions were causing serious damage to the arterial endothelium. A physician could intervene earlier and more aggressively than he or she might otherwise if there were indications that these modest risk factors were causing more than modest disease in the vasculature.

The determination of impaired function of arterial endothelium in cardiovascular disease is relatively easy when assessments are made invasively on isolated arteries or in experimental animals. Indeed, these types of invasive studies have been the primary means by which the impairment of the endothelium NO system has been shown to be associated so extensively with cardiovascular disease. This type of invasive analysis, however, cannot be employed as a clinical test for patients. To be able to detect impaired endothelial function in patients as a regular clinical test, any technique employed needs to be noninvasive.

Recently, a noninvasive indicator for the assessment of vascular damage in cardiovascular disease, called flow-mediated vasodilation (FMD), has become increasingly used in clinical practice as a means of assessing the functional integrity of the vascular endothelium in various cardiovascular diseases. FMD is caused by shear stress–induced release of endothelial NO when arteries are exposed to high laminar

flow. It is known that the endothelial NO system is impaired in all forms of cardiovascular disease and that it may be an early sign of the damaging effects of a disease on the vascular system. FMD can be induced in patients by placing an inflatable blood pressure cuff around a patient's arm, pressurizing the cuff to stop blood flow for 5 minutes, and then releasing the cuff pressure rapidly to allow a reactive hyperemia to occur. The hyperemia increases blood flow and shear stress on the intima of the brachial artery. This shear induces an FMD that can be measured as a diameter change of the brachial artery using Doppler ultrasound. Poor dilation is taken to be a measure of impaired endothelial function in the patient. This method of FMD is superior in clinical applications to intra-arterial drug infusion dilation methods that were first used for assessment of endothelial function because the former, but not the latter, is noninvasive. Clinical studies continue to demonstrate correlations between impaired FMD and poor cardiovascular outcomes or cardiovascular disease. This likely reflects findings of numerous studies that have shown a strong correlation between endothelial dysfunction and increased risk of cardiovascular events, such as myocardial ischemia, infarction, and cardiac death. Several studies have shown that impaired FMD in the brachial artery is actually a good indicator of the severity of coronary artery disease in the patient. The relative success of this technique has spurred the development and investigation of additional FDM applications and additional noninvasive measures designed to assess endothelial function in patients. Most recently, FDM has been investigated as a tool to assess the functional state of penile cavernous arteries in erectile dysfunction. Newer, non-FMD vasodilation assessment methods are now being employed for clinical assessment of endothelial function. Digital pulse volume amplitude tonometry, where reactive hyperemia is measured by monitoring changes in the arterial pulse wave during the hyperemia, is one such hyperemia-based noninvasive method used to assess endothelial function in the microcirculation. Oscillometric ezFMD is a recently developed, sophisticated, noninvasive method for measuring endothelial function. This technique measures oscillations in arterial volume due to changes in blood pressure at different external inflatable cuff pressures. ezFMD is lower in patients with cardiovascular disease and is easier to perform and more reliable than standard FMD for assessment of endothelial function in CV disease. However, the future usefulness of this technique awaits further clinical validation. ■

Sympathetic nervous system regulates blood pressure and flow by constricting the microvessels.

Although the microvasculature uses local control mechanisms to adjust vascular resistance based on the physical and chemical environment of the tissue, the active tone of all arteries and veins is significantly affected by the sympathetic nervous system, which is tonically active to all such vessels. Sympathetic nerves release norepinephrine, which is a vasoconstrictor, onto the surface of smooth muscle cells in arteries and veins. Because sympathetic nerves form an extensive meshwork of axons over the exterior of the microvessels, all VSM cells are likely to receive norepinephrine tonically.

Furthermore, the diffusion path from nerve terminals to vascular smooth muscle is only a few microns. Thus, norepinephrine released from sympathetic nerves rapidly reaches the vascular muscle, where it activates membrane α_1-adrenergic receptors, resulting in vasoconstriction within 2 to 5 seconds.

Although changes in sympathetic nerve activity to blood vessels can markedly change organ blood flow, sympathetic neural effects on the vasculature are not used to control local tissue flow. Instead, they are used as part of the mechanisms used to modify total peripheral vascular resistance in order to control arterial blood pressure. As explained in more detail in Chapter 17, the arterial pressure is monitored moment to moment by neural reflex mechanisms that adjust the cardiac output and systemic vascular resistance via autonomic efferent nerves leading to the heart, arteries, and veins. This system is used to buffer moment-to-moment deviations in blood pressure from normal. This neural-based system allows for rapid adjustment of vascular resistance, which must occur quickly because rapid changes in body position or sudden exertion require immediate responses to maintain or increase arterial pressure. In the defense against

hypotension, the sympathetic nervous system routinely overrides local regulatory mechanisms in most organs, with the exception of the brain and the heart. In addition, although activation of sympathetic nerves to the skeletal muscle vasculature is intense during exercise, this vasoconstrictive action is overcome by massive local dilatory signals within skeletal muscle during such activity (see Chapter 26 for details).

Most organs control their blood flow via local autoregulation.

If the arterial blood pressure to an organ is decreased to the extent that blood flow is compromised, the arterioles of that organ respond by vasodilating. This resistance change counteracts the effects of reduced arterial pressure on organ blood flow and thereby returns it toward its original value in spite of the drop in perfusion pressure. If arterial pressure is elevated, flow is initially increased. However, with a minute or so, vascular resistance increases to the extent that blood flow is returned to, or close to, its original level. These responses of an organ system's vasculature to changes in perfusion pressure stand in stark contrast to what would

CLINICAL FOCUS | 15.2

Ischemia–Reperfusion Injury

Ischemia causes a condition of reduced oxygen availability in which the oxygen demand of the tissue is greater than its oxygen supply. This can arise from impaired tissue blood flow or stimulation of excessive demand relative to the blood flow a tissue is capable of receiving. In the heart, ischemia most often occurs acutely by the development of an obstruction in one of the major coronary arteries, usually from a blood clot forming in the area of an advanced or ruptured atherosclerotic plaque. Ischemia cannot be tolerated for long in myocardial tissue. Immediately upon the start of ischemia, the myocardial tissue affected becomes noncontractile. Within 1 hour, the oxygen-dependent cellular ionic and water balance mechanisms fail creating cellular and mitochondrial swelling with loss of membrane integrity. Calcium extrusion mechanisms fail, and the resultant increase in intracellular calcium activates enzymes that destroy myocardial tissue. If acute ischemia involves enough myocardial tissue, the initial loss of contractile ability may be enough to cause acute failure of the heart and circulatory collapse (cardiogenic shock; see Chapter 17). However, even with small ischemic episodes, the resulting tissue damage within hours after the initial insult may be enough to cause death from cardiac failure.

The best course of action to reduce ischemic damage in the heart would seem to be to restore myocardial blood flow as quickly as possible and/or change hemodynamic variables in such a way as to reduce myocardial oxygen demand. If the ischemia results from a blood clot in a major coronary artery, the physician can inject "clot-busting" drugs, such as streptokinase or tissue-type plasminogen activator (t-PA), through

a catheter into the major coronary artery affected by the clot. These drugs dissolve the clot and restore flow. However, in reality, restoration of blood flow is often associated with the induction of additional damage to the myocardium rather than preservation of tissue. This phenomenon is called *ischemia–reperfusion injury*. During the ischemic period, the arterioles downstream from the coronary arterial obstruction dilate through myogenic and metabolic mechanisms, lowering vascular resistance. On removal of the obstruction, this lowered resistance allows for a substantial reactive hyperemia, which floods the previously ischemic area with blood elements and oxygen. This hyperemia is associated with cell death from necrosis and apoptosis. The area also becomes infiltrated with neutrophils.

Tissue damage from ischemia–reperfusion injury results from the increased generation of oxygen radicals in the myocardium and endothelium as well as from infiltrating neutrophils and macrophages. Myocardial sources may result from the ischemic conversion of xanthine dehydrogenase to xanthine oxidase, which produces superoxide from xanthine. Damaged mitochondria likely also act as a source of oxygen radicals. The radicals also impair mitochondrial recovery and adenosine triphosphate generation, which leads to cell death. The neutrophil infiltration also sets up a local inflammatory response, which produces damaging cytokines and additional reactive oxygen species in the tissues as well as complement activation. Recent studies suggest that nitric oxide attenuates ischemia–reperfusion injury, perhaps by quenching superoxide produced in the condition. ∎

be observed in a rigid, nonliving tube where flow is directly proportional to perfusion pressure. The ability of an organ to maintain normal or near normal blood flow in the face of changes in the pressure driving flow is called *autoregulation of blood flow*, or simply **autoregulation** (Fig. 15.9). Autoregulation is a local control phenomenon involving responses of the arterial smooth muscle to local stimuli; it occurs independently of any nerve activity on the organ and can be observed in circulations with impaired functions of arterial endothelium. Autoregulation appears to be primarily related to metabolic and myogenic control mechanisms as well as to an increased release of NO if tissue oxygen availability decreases.

The efficacy of autoregulation is not 100% in all organs, and there are upper and lower pressure limits to its effectiveness as well. When arterial pressure in vessels to an

organ decreases, the vasodilation that ensues from autoregulation may bring organ blood flow back toward, but not fully to, its original value. Even if the blood flow might not be brought all the way back to its original level before the pressure dropped, autoregulation is still considered to have occurred; it is just that the efficacy of this autoregulation is <100%. The cerebral, cardiac, and renal circulations have the highest autoregulation efficacy. Skeletal muscle and intestinal vasculatures exhibit less well-developed autoregulation, whereas autoregulation is essentially absent in the circulation of the skin. Furthermore, in all organs, there are limits to the vasodilation of arteries that occurs during autoregulation; once vessels are maximally dilated, there is no capacity left to serve as an autoregulatory correction for further drops in pressure. Autoregulation capacity ceases at arterial pressures below 50 mm Hg in most organs because at that point, the small vessels that predominantly control resistance and blood flow are maximally dilated. In an analogous manner, the ability of arterioles to constrict against progressively higher pressures has a maximum. Eventually, the pressure pushing the vessel open is too much for the contractile force generated by the arterial smooth muscle. Pressures beyond this point tend then to increase organ blood flow and capillary hydrostatic pressure. Thus, there is an upper pressure limit as well to the autoregulation of blood flow in organs.

An example of autoregulation, based on data from the cerebral vasculature, is shown in Figure 15.9. Note that the arterioles continue to dilate at arterial pressures below 60 mm Hg, when blood flow begins to decrease significantly as arterial pressure is further lowered though they clearly cannot dilate sufficiently to maintain normal blood flow at those low arterial pressures. At greater-than-normal arterial pressures, the arterioles constrict via autoregulatory mechanisms. However, if the mean arterial pressure is elevated appreciably above 150 to 160 mm Hg, the vessel walls cannot maintain sufficient tension to oppose passive distention by the high arterial pressure. The result is excessive blood flow and high microvascular pressures, eventually leading to the rupture of small vessels, excess fluid filtration into the tissue, and edema. In fact, the reason that autoregulation regulates to counteract too much of an increase in blood flow to an organ likely represents a means to control against excessively high intracapillary pressure whenever blood pressure is increased.

A phenomenon related to autoregulation is **reactive hyperemia**. When blood flow to any organ is stopped or reduced by vascular compression for more than a few seconds, vascular resistance dramatically decreases. The absence of blood flow allows vasodilatory chemicals to accumulate while hypoxia occurs; the vessels also dilate as a result of decreased myogenic stimulation (low intravascular pressure). As soon as the vascular compression is removed, blood flow is dramatically increased for a few minutes because the vessels became dilated during the stoppage of flow. This phenomenon represents reactive hyperemia because it is a reaction to the previous period of ischemia.

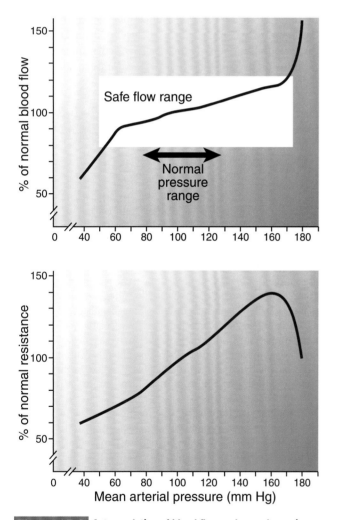

Figure 15.9 Autoregulation of blood flow and vascular resistance as mean arterial pressure is altered. The safe range for blood flow is about 80% to 125% of normal and usually occurs at arterial pressures of 60 to 160 mm Hg as a result of active adjustments of vascular resistance. At pressures above about 160 mm Hg, vascular resistance decreases because the pressure forces dilation to occur; at pressures below 60 mm Hg, the vessels are fully dilated, and resistance cannot be appreciably decreased further.

INTEGRATED MEDICAL SCIENCES

Diabetes Mellitus and Microvascular Disease: Prognosis and Current Treatments

Diabetes mellitus is a condition in which the body cannot regulate uptake and storage of glucose and lipids in tissues. Most commonly, this is due to either a lack of insulin production (type 1 or juvenile diabetes) or a primary inability of tissues to respond to insulin (type 2 or what is sometimes called adult-onset diabetes; see Chapter 34 for more details). This disease is reaching worldwide epidemic proportions and is the source of serious cardiovascular and other end organ morbidities. Both types of diabetes are associated with significant genetic, inheritance components. In addition, obesity is believed to play a causal role in type 2 diabetes in that it increases the requirement for insulin to the extent that even the high insulin concentrations provided by the pancreatic beta cells are insufficient. This overall condition is called insulin resistance.

More than 95% of persons with diabetes go through acute complications associated with periodic swings of hyperglycemia and hypoglycemia. Acute hyperglycemia leads to ketoacidosis and dehydration caused by excessive loss of water in the urine, which results from a high filtration of glucose at the glomerular capillaries such that the glucose entering the renal tubule exceeds its transport capacity. Consequently, this excess tubular fluid glucose generates osmotic retention of water in the tubules, which is subsequently lost out of the body by urination. Acute ketoacidosis and dehydration can lead to diabetic coma and death. The latter can occur with hypoglycemia as well, which can also acutely impair cognitive function and consciousness (see Chapter 34 for details).

Repeated exposure to hyperglycemic conditions brings about pathologies in the cardiovascular system involving both micro and macro blood vessels. Macrovascular complications center around inflammation-induced acceleration of atherosclerosis caused by tissue hyperglycemia. This accelerates peripheral, coronary, and cerebral vascular disease leading to increased incidence of ischemic disease, infarction in the heart, brain, and peripheral organs (i.e., heart attacks and strokes). Microvascular disease involves primarily the retina, kidneys, and nervous system. The microvasculature of the retina experiences microaneurysms, excessive angiogenesis, and edema in the retinal tissue caused by a leaky microvasculature after repeated exposures to hyperglycemic conditions. These conditions are associated with blindness, which is a common outcome of chronic diabetes. The glomerulus of the kidney becomes damaged in diabetes due to intrarenal hypertension and leakage of protein into the renal tubule. Intrarenal hypertension is linked to elevated levels of angiotensin II, and the exposure of the tubule to protein chronically leads to a glomerular inflammatory response. This along with intrarenal hypertension causes renal microvascular damage and glomerular fibrosis and scarring. These in turn can lead to end-stage renal disease in diabetics.

Periods of hyperglycemia over time cause reduced nitric oxide (NO) production by endothelial cells, increased reactivity of vascular smooth muscle to norepinephrine, and a reduced ability of microvessels to participate in tissue repair. The mechanism of many of the vascular abnormalities associated with diabetes appears to stem from the fact that hyperglycemia activates protein kinase C (PKC) in endothelial cells. PKC inhibits NO synthase, so NO formation is gradually suppressed. This leads to the loss of an important vasodilatory stimulus (NO) and results in vasoconstriction. PKC-beta has been shown to be activated by glucose and is thought to mediate microvascular leakage and angiogenesis. Hyperglycemia also increases activity in the polyol pathway resulting in increased sorbitol in tissues. This causes increases in extracellular osmolality, fluid accumulation, and pressure, which further damages tissues, including nerve axons. This latter effect may contribute to neuropathies, which are common in chronic diabetes.

Currently, treatment of type 1 diabetes is centered around aggressive control of plasma glucose levels with the goal of reducing exposure of the body to hyperglycemic conditions. Clinical trials have shown that when doing so, such control greatly reduces the microvascular consequences of hyperglycemia and thereby delays impaired vision and renal function. Other than insulin administration and strict adherence to diet regimens, there is little that pharmacotherapy can do in type 1 diabetes because of the loss of pancreatic beta cells. Type II diabetes though has many pharmacological options that can be used to help patients although such individuals also eventually lose beta cells over time and become increasingly dependent on insulin. However, other pharmacological agents are available to help with various pathological consequences of type 2 diabetes. Antihyperglycemic agents such as biguanides can help control hyperglycemic episodes by increasing tissue uptake of glucose and increasing islet cell responsiveness. In addition, insulin release can be enhanced by sulfonylureas and meglitinides. Both these agents bind (at different sites) to ATP-dependent K^+ channels on beta cell membranes, causing depolarization-mediated calcium entry into the cells, which stimulate insulin release. Glucose-stimulated insulin secretion can be effected by GLP-1 receptor analogues, which mimic natural incretins in the body. These agents also suppress appetite by increasing the feeling of satiety. Finally, alpha-glucosidase inhibitors help moderate the spikes in plasma glucose levels that occur after a meal by suppressing digestion of carbohydrates in the intestine and thus diminishing the amount of glucose absorbed during a meal. ∎

Cardiovascular Physiology

Chapter Summary

- The microcirculation controls the transport of water and substances between the tissues and blood.
- Transport of gases and lipid-soluble molecules occurs by diffusion across endothelial cells.
- Transport of water-soluble molecules occurs by diffusion through pores in adjacent endothelial cells.
- Transport of substances across a capillary by diffusion depends on the concentration gradient of the substance and the permeability of the capillary to the substance.
- Flow-limited transport of a substance is limited by the amount of blood flow that can be delivered to the tissues.
- Diffusion-limited transport of a substance is limited by the diffusion of the substance across the capillary membrane rather than by the amount of flow delivered to the tissues.
- Bulk filtration or absorption of water occurs across capillaries through pores in adjacent endothelial cells.
- Plasma hydrostatic and colloid osmotic pressures are the primary forces affecting fluid filtration and absorption across capillary walls.
- The ratio of postcapillary to precapillary resistance is a major determinant of capillary hydrostatic pressure.
- Lymphatic vessels collect excess water and protein molecules from the interstitial space between cells.
- Myogenic arteriolar regulation is a response to increased tension or stretch of the vessel wall muscle cells.
- By-products of metabolism cause the dilation of arterioles.
- Nitric oxide from the endothelium is a major local vasodilator of arterioles and controller of multiple vascular functions and properties.
- The axons of the sympathetic nervous system release norepinephrine, which constricts the arterioles and venules.
- Autoregulation of blood flow allows some organs to maintain nearly constant blood flow when arterial blood pressure is changed.

Chapter Review Questions

1. During digestion, intestinal smooth muscle motility as well as the energy requiring processes used for cellular transport of substrates across the intestinal mucosa are greatly increased. Oxygen transport in the intestine during digestion would therefore likely be characterized by:

   A. increased delivery into the microcirculation and increased transport across the capillary network into the intestinal tissues.
   B. decreased delivery into the microcirculation and decreased transport across the capillary network into the intestinal tissues.
   C. conversion of flow-limited transport of oxygen into diffusion-limited transport of oxygen due to low capillary blood oxygen concentration.
   D. decreased capillary surface area available for transport of oxygen.
   E. increased diffusion distances between the capillaries for the transport of oxygen.

The correct answer is A. Increased oxygen requiring activities by the intestine during digestion would induce an active hyperemia through dilation of arterioles. This would increase blood flow to the intestine and increase the number of open perfused capillaries. Therefore, oxygen delivery to the intestine would be increased, the capillary surface area available for diffusion would increase, and the distance between an intestinal cell and its nearest capillary would decrease. Thus, transport into and across the intestinal capillaries would increase and help meet the increased oxygen demand of that tissue during digestion. Diffusion parameters would be improved, not reduced, so that diffusion-limited transport would not occur.

2. Which of the following would limit local edema formation following traumatic injury to a lower limb?

   A. Capillary injury
   B. Formation of blood clots in the veins of the limb
   C. Sympathetic nerve–mediated arteriolar vasoconstriction
   D. Traumatic arterial dilation in the limb
   E. Rupture of lymphatic channels in the limb

The correct answer is C. Arteriolar vasoconstriction will decrease the postcapillary:precapillary ratio of resistance and therefore reduce hydrostatic pressure in the capillary. This reduces filtration of fluid at the capillary and can even cause the capillary to absorb water from the interstitium. Either of these would attenuate edema formation. Arteriolar dilation would have opposite effect. Capillary injury would disrupt capillary integrity and increase their hydraulic conductance, which would exacerbate edema formation. Blood clots in the large veins will increase back pressure into the capillaries that favor water filtration and edema formation. Rupture of lymph channels would remove the vascular component used to drain excess fluid from the interstitial spaces and thereby favor edema formation.

3. Chronic high arterial pressure severely damages the arterial endothelium. Compared to a normal arteriole, an arteriole in an individual with chronic arterial hypertension would:

   A. not constrict when intravascular pressure is increased.
   B. dilate when exogenous NO is applied to the vessel wall.
   C. constrict in response to norepinephrine.
   D. dilate less in response to increased blood flow (flow-mediated dilation.
   E. dilate when blood flow is reduced.

The correct answer is D. The endothelium is the key source of the potent vasodilator NO. NO production by the endothelium is increased by fluid shear stress resulting in arterial dilation (flow-mediated dilation). Arteries with impaired endothelium would display impaired flow-mediated dilation. Endothelial damage in hypertension would not affect myogenic constriction in response to an increase in intravascular pressure or metabolic dilation in response to low blood flow because these factors work at the level of the smooth muscle cell. Similarly, the receptors for norepinephrine that link the sympathetic nervous system to vasoconstriction are on the smooth muscle cells and not the endothelium.

4. Histamine is released in tissues in response to allergens and mediates characteristic changes in the microcirculation associated with an allergic reaction. Histamine is an arterial vasodilator that also affects the cytoskeleton of capillary endothelium such that the clefts between cells are opened. Which of the following would characterize the microcirculation in response to histamine?

    A. Capillary hydrostatic pressure and the capillary filtration coefficient are reduced.
    B. Capillary hydrostatic pressure is reduced, but the capillary filtration coefficient is increased.
    C. Capillary hydrostatic pressure is increased, but the capillary filtration coefficient is decreased.
    D. Capillary hydrostatic pressure and the capillary filtration coefficient are increased.
    E. Capillary hydrostatic pressure and the capillary filtration coefficient are unchanged.

The correct answer is D. Arteriolar dilation favors an increase in capillary hydrostatic pressure and an increase in blood flow into the capillary network. Venules also are dilated by histamine, which further augments blood flow, although it mitigates somewhat the increase in pressure caused by precapillary vasodilation. The capillary filtration coefficient is a measure of the hydraulic conductance across the capillary and is increased by anything that increases capillary permeability. The increase in filtration across capillary networks in response to an allergic reaction is often great enough to cause local edema in the tissue.

5. All of the following will favor tissue edema formation in splanchnic organs except:

    A. metabolic dilation of arterioles.
    B. activation of sympathetic nerves to the organs.
    C. chronic congestive heart failure.
    D. hypoalbuminemia.
    E. portal hypertension.

The correct answer is B. Arterial vasoconstriction to splenic organs will reduce intracapillary hydrostatic pressure and thus reduce the probability of edema formation. Dilation of arterioles in such organs will have the opposite effect, especially because organs of the digestive tract have such high capillary filtration coefficients. Hypoalbuminemia reduces plasma oncotic pressure everywhere in the body and favors capillary fluid filtration and edema formation. Portal hypertension increases capillary hydrostatic pressure in the small intestines and, in spite of a notable venous–arteriolar response in that organ, is almost always associated with intestinal edema formation and ascites.

Clinical Application Exercises 15.1

ANAPHYLACTIC SHOCK

A 5-year-old female child with no apparent medical problems is stung by a bee while playing in a park near her home. Her reaction to the sting is a greater-than-normal redness and itching at the site of the sting but is otherwise uneventful. During her regularly scheduled physical exam, her physician warns her mother that this child might be allergic to bee venom and should try to avoid being stung again. Unfortunately, 1 month later while playing in her backyard near a bed of flowers, she gets stung on her hand by a bee. Her mother cleans the sting with some rubbing alcohol and applies a bandage. However, the area of the sting begins to turn red and swells, developing a wheal of about a 3-cm radius around the sting within 2 minutes. The girl then rapidly develops severe difficulty breathing and is rushed to the emergency room. Examination in the ER reveals a swollen stiff abdomen, inflamed bronchi and evidence of significant airway restriction suggestive of angioedema, blood pressure of 80/50, and a heart rate of 120 with urticaria and angioedema over her arms, abdomen, and thighs.

QUESTIONS

1. What is the cause of the redness and swelling at the site of the bee sting on this girl's hand?
2. What is the likely predominant mediator of the redness and swelling at the sight of the bee sting?
3. What is the cause of the patient's low systolic, diastolic, and pulse pressures?

4. Treatment for a patient in anaphylactic shock includes an IM injection of epinephrine. This agent activates β-adrenergic receptors in the bronchi, which dilates airways. This helps counteract bronchial constriction caused by histamine and other anaphylactic agents during an anaphylaxis. How does this injection of epinephrine also help counteract the shock associated with anaphylaxis?

ANSWERS

1. The redness is caused by local cutaneous vasodilation. This contributes to the swelling, but the presence of a wheal in the area of the sting is indicative of local edema formation.

2. The girl is likely experiencing an IgE-mediated hypersensitivity reaction in response to bee venom, which serves as an antigen in the reaction. This reaction is characterized by extensive and severe degranulation of basophils with release of histamine. Histamine dilates arteries while increasing capillary and venule permeability. The vasodilation reddens the skin, whereas that effect in conjunction with increased capillary permeability causes enhanced local capillary filtration, protein exudation, and edema responsible for the wheal.

3. Low systolic and pulse pressure results from decreased stroke volume, increased arterial compliance, or both. Although histamine dilates arteries, compliance is a function of large arteries, and histamine acts predominantly at the microcirculatory level. The girl in this case exhibits sign of anaphylactic shock from her second bee sting. In this situation, excessive high levels of histamine are generated systemically and cause massive arterial dilation, resulting in a decrease in mean arterial and diastolic blood pressure. Furthermore, as in the case of local edema formation, systemic release of histamine causes massive capillary filtration and edema formation, especially in the abdomen where organ capillary filtration coefficients and surface area are naturally high. This effect is consistent with the presentation of a swollen stiff abdomen. The massive plasma efflux from the splanchnic area as well as other areas of the body (responsible for the angioedema) causes a large translocation of fluid out of the circulation, a drop in central venous filling pressure, and a decrease in stroke volume, which contributes to the low pulse pressure. The decreased stroke volume depresses cardiac output, which, with the decrease in vascular resistance, leads to shock.

4. Stimulation of β-adrenergic receptors in the heart increase contractility and thereby enhances stroke volume and cardiac output. Epinephrine activates both β- and α-adrenergic receptors on arteries. Because the latter predominate in all vascular beds except the brain and heart, this vascular effect increases vascular resistance and helps raise blood pressure.

thePoint® *Visit* http://thepoint.lww.com/rhoades5e *for additional chapter review Q&A, Clinical Application Exercises, animations, and more!*

Active Learning Objectives

Upon mastering the material in this chapter, you should be able to:

- Explain why coronary blood flow is greater during diastole than during systole.
- Explain why myocardial ischemia and infarction are more likely and potentially more severe in the endocardium than in the epicardium.
- Explain why the heart must depend solely on autoregulation of blood flow to maintain myocardial oxygen delivery in the face of decreased perfusion pressure.
- Predict how variables related to myocardial oxygen demand will affect blood flow to the heart including activation of the sympathetic nervous system.
- Explain why general sympathetic nervous system activation has no appreciable effect on blood flow to the brain.
- Explain how autoregulation of blood flow protects the blood–brain barrier.

- Explain the importance of cerebral hyperemia following exposure to CO_2 or H^+.
- Explain the mechanism responsible for adjusting the efficacy of autoregulation in the intestine and skeletal muscle on the basis of tissue metabolic demand.
- Explain the mechanisms responsible for active hyperemia in the intestine and skeletal muscle.
- Explain the benefit of the hepatic arterial buffer response.
- Explain the role of the sympathetic nervous system in the control of the cutaneous circulation.
- Explain gas exchange limitations and the fetus's susceptibility to ischemia in the context of the arrangement of the fetal circulation.
- Explain the advantages and disadvantages to the close connection between the fetal circulation and maternal blood chemistry.
- Explain the mechanisms responsible for the transition from the fetal cardiovascular system to that of a newborn.

The vascular system of every organ has special characteristics that are designed to meet the specific functions and specialized needs of that organ. In this chapter, the characteristics unique to the circulations of the heart, brain, small intestine, liver, skeletal muscle, and skin are described. Table 16.1 presents data on blood flow and oxygen use by these different organs and tissues. In addition, the anatomy and physiology of the fetal/placental circulation are presented along with the circulatory changes that occur at birth. The pulmonary and renal circulations are discussed in Chapters 19 and 22, respectively.

► CORONARY CIRCULATION

The coronary circulation provides blood flow to the heart, which consumes as much or more oxygen at rest as does the same mass of skeletal muscle during vigorous exercise (see Table 16.1). Furthermore, the heart can increase its flow four- to fivefold in order to provide for increased oxygen needs during exercise. This increase in available blood flow above that at rest constitutes what is called **coronary reserve**. The ability to increase blood flow to provide additional oxygen is imperative for the heart because even during resting conditions it extracts a near-maximum amount of oxygen from blood. Essentially, all capillaries are open and perfused in the heart during resting conditions, and its oxygen extraction at rest is greater than any other organ in the body. Coupled with the fact that the heart's ability to use anaerobic glycolysis is limited, the only way remaining to increase its tissue oxygen delivery is for it to increase its blood flow.

Cardiac blood flow occurs primarily during diastole because of inhibition of flow from cardiac contraction during systole.

An examination of the data in Table 16.1 reveals that, as a ratio of its blood flow at rest to its oxygen demand (O_2 consumption), the heart is the most *under*perfused organ in the body! This may seem surprising considering the importance of the heart to the well-being of the body as a whole. However, it reflects the fact that the contraction of the heart gets in the way of its own blood supply. Blood flow through the left ventricle decreases to a minimum during systole because the small intramuscular blood vessels are compressed and actually physically "sheared" closed by contraction and compressed by pressure generated in the muscle. Blood flow in the left coronary artery during cardiac systole is only 10% to 30% of that during diastole, and much of that represents flow to the epicardium where arteries are either outside cardiac muscle or shallowly imbedded there. The heart is perfused from the epicardial (outside) surface to the endocardial (inside) surface. The mechanical compression of systole has a more negative effect on the blood flow through the endocardial layers, where compressive forces are higher and microvascular pressures are lower, than it does in the epicardium. In heart diseases of all types, therefore, the subendocardial layers of the heart suffer more severe impairment and ischemia than do the epicardial layers.

In diastole, the heart musculature is relaxed and blood flow to the heart is not impeded by cardiac muscle effects (Fig. 16.1). Compared to its effect in the left ventricle, the compression effect of systole on blood flow is minimal in the

> **TABLE 16.1** Blood Flow and Oxygen Consumption of the Major Systemic Organs Estimated for a 70-kg Adult Man

Organ	Mass (kg)	Flow (mL/100 g/min)	Total Flow (mL/min)	Oxygen Use (mL/100 g/min)	Total Oxygen Use (mL/min)
Heart	0.5				
Rest		60–80	250	7.0–9.0	25–40
Exercise		200–300	1,000–1,200	25.0–40.0	65–85
Brain	1.4	50–60	750	4.0–5.0	50–60
Small intestine	3				
Rest		30–40	1,500	1.5–2.0	50–60
Absorption		45–70	2,200–2,600	2.5–3.5	80–110
Liver	1.8–2.0				
Total		100–300	1,400–1,500	13.0–14.0	180–200
Portal		70–90	1,100	5.0–7.0	
Hepatic artery		30–40	350	5.0–7.0	
Muscle	28				
Rest		2–6	750–1,000	0.2–0.4	60
Exercise		40–100	15,000–20,000	8.0–15.0	2,400
Skin	2.0–2.5				
Rest		1–3	200–500	0.1–0.2	2–4
Exercise		5–15	1,000–2,500		

right ventricle as a result of the lower pressures developed by that chamber. The heart's dependence on receiving most of its blood flow during diastole has no obvious deleterious effects on total coronary blood flow even during maximal exercise. However, in people with partially occluded coronary arteries (i.e., as in coronary artery disease), a simple increase in heart rate may decrease the time spent in diastole to the extent that any coronary blood flow increase needed to match increased myocardial demands is impaired.

Coronary blood flow is closely linked to cardiac oxygen demand.

Mean coronary blood flow increases significantly whenever myocardial oxygen demand is increased, as occurs with positive inotropic stimuli, increased systolic pressure and wall stress, or an increase in heart rate. Most of the small arteries and arterioles of the heart are surrounded by cardiac muscle cells and are exposed to molecules released by cardiac cells into the interstitial space. Many of these molecules cause dilation of the coronary arterioles and therefore could be responsible for coronary active hyperemia. However, no single mechanism adequately explains the dilation of coronary arterioles and small arteries when the metabolic rate of the heart is increased or when pathologic or experimental means are used to restrict blood flow. For example, the concentration of the potent vasodilator adenosine, which is derived from the breakdown of

adenosine triphosphate in cardiac cells, increases whenever cardiac metabolism is increased or blood flow to the heart is decreased. Blockade of the vasodilator actions of adenosine with theophylline, however, does not prevent coronary vasodilation when cardiac work is increased, blood flow is suppressed, or the arterial blood is depleted of oxygen. This implies that other vasodilatory stimuli may be involved in metabolic regulation of coronary blood flow. Vasodilatory prostaglandins, H^+, CO_2, nitric oxide (NO), and decreased availability of oxygen, as well as myogenic mechanisms, are capable of contributing to coronary vascular regulation. However, the exact nature of the link between coronary blood flow and cardiac metabolism is not fully understood at this time.

Direct and indirect actions of sympathetic nerves dilate coronary arteries.

Coronary arteries and arterioles are innervated by the sympathetic nervous system and contain predominantly β_2-adrenergic receptors (in fact coronary arterioles may contain only β_2-adrenergic receptors). The large-surface epicardial arteries also contain some α_1-adrenergic receptors. Activation of the sympathetic nervous system to the heart, such as during exercise, results in coronary vasodilation and a marked increase in coronary blood flow. This increase is primarily an active hyperemia in response to increased cardiac metabolism brought about by sympathetic

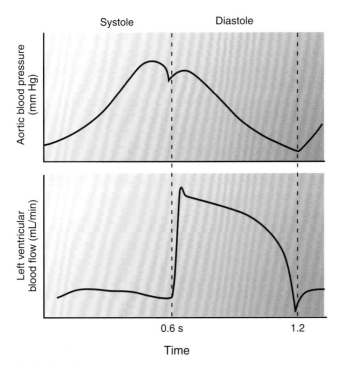

Systole Diastole

Figure 16.1 **Aortic blood pressure and left coronary blood flow during the cardiac cycle.** Note that left coronary artery blood flow decreases dramatically during the isovolumetric phase of systole, prior to opening of the aortic valve. Left coronary artery blood flow remains lower during systole than during diastole because of compression of the coronary blood vessels in the contracting myocardium. The left ventricle receives most of its arterial blood inflow during diastole.

nerve stimulation of heart rate and myocardial contractility. However, direct β-adrenergic–mediated coronary vasodilation also contributes to the hyperemia following sympathetic nerve activation.

It appears that the α_1-adrenergic receptors of the epicardial arteries may play a role in supporting endocardial perfusion during exercise. Intramural pressure in the heart during contraction is greater in the endocardium than in the epicardium. This creates a tendency for blood to be pushed backward away from the endocardium and toward the epicardium during systole. Adding some α-adrenergic constrictor influence to the epicardial vessels during heavy exercise helps minimize backflow and loss of blood from endocardial muscle.

Autoregulation of coronary blood flow has different limits in the endocardium versus the epicardium.

All of the heart's capillaries receive blood flow, even at normal heart rates and cardiac outputs. Thus, at rest, extraction of oxygen from the coronary circulation by the heart is essentially maximized. The only way the heart can maintain oxygen delivery in the face of a reduction in perfusion pressure is by local autoregulation of blood flow. The heart is a strong autoregulator of blood flow and can maintain near-normal flows over a wide range of perfusion pressures. However, higher pressure and compressive forces in the endocardium versus the epicardium make blood flow

more constrained during systole in the endocardium than in the outer layers of the heart. Consequently, the endocardial arterioles must dilate more during diastole to compensate for a more severe reduction of flow during systole. This means that the endocardium has less coronary reserve than does the epicardium. For this reason, the low-pressure limit for autoregulation in the endocardial layer is greater (i.e., at a higher intravascular pressure) than that in the epicardial layer. Endocardial arterial dilation reaches a maximum when arterial pressure drops to ~70 mm Hg, whereas maximum dilation in the epicardial arteries is not reached until pressure is ~40 mm Hg. For this reason, whenever blood flow to the heart is severely restricted, the endocardium is first to suffer damage, and the resulting area of myocardial injury and infarction is larger in the endocardium than in the epicardium.

▶ CEREBRAL CIRCULATION

The ultimate organ of life is the brain. Even the determination of death often depends on whether the brain is viable. The most common cause of brain injury results from some form of impaired brain blood flow. This may occur as the result of occlusion of vessels secondary to atherosclerotic processes and thrombosis, accidents to arteries in the neck or brain resulting in intercranial hemorrhage, or **aneurysms** that occur as a result of vessel wall tearing.

Brain blood flow remains essentially constant over a large range of arterial blood pressures.

The cerebral circulation shares many of the overarching physiologic characteristics of the coronary circulation. Like the heart, the brain has a high metabolic rate (see Table 16.1), extracts a large amount of oxygen from blood, and has a limited ability to use anaerobic glycolysis for metabolism. However, the blood vessels of the brain either are not, or at best sparsely, innervated by sympathetic nerves (e.g., in pial arteries) or are largely unresponsive to catecholamines. The brain, like the coronary vasculature, has an excellent ability to autoregulate blood flow with a range encompassing about 50 to 160 mm Hg. The vasculature of the brainstem exhibits the most precise autoregulation, with good but less precise regulation of blood flow in the cerebral cortex. This regional variation in autoregulatory ability has clinical implications because the region of the brain most likely to suffer at low arterial pressure is the cortex. Thus, hypotension will lead to unconsciousness before the automatic cardiovascular and ventilatory regulatory functions of the brainstem are compromised.

Both the cerebral arteries and cerebral arterioles appear to be involved in cerebral vascular autoregulation and other types of vascular responses. In fact, the arteries can change their resistance almost proportionately to the arterioles during autoregulation. This may occur in part because cerebral arteries exhibit myogenic vascular responses and because they are partially to fully embedded in the brain tissues. As such, they are likely influenced by the same vasoactive chemicals in the interstitial space as the arterioles. Although the brain vasculature exhibits myogenic vascular responses and may use this mechanism as a major contributor to autoregulation, a variety of mechanisms appear to be responsible for cerebral vascular autoregulation with no one variable as a clear, sole mediator. For example, when blood flow is maintained at normal levels, regardless of the arterial blood pressure, little extra adenosine, K^+, H^+, or other vasodilator metabolites are released, and brain tissue Po_2 remains relatively constant. However, increasing concentrations of any of these chemicals can cause significant vasodilation and increased blood flow.

Brain microvessels are uniquely sensitive to vasodilation by CO_2 and H^+.

Neural function is significantly impaired by acidosis. The cerebral circulation dilates markedly in response to elevations of CO_2 and H^+, and it is the most sensitive circulation in the body to these metabolites. Cerebral blood flow increases markedly when CO_2 levels increase in brain tissue or the cerebral spinal fluid. Cerebral hyperemia in response to hypercapnia, or acidosis, which is created by hypercapnia, is likely a means to wash these neurologically damaging agents out of the brain tissue. Both CO_2 and H^+ are formed when cerebral metabolism is increased by nerve action potentials, such as during normal brain

activation. In addition, interstitial K^+ is elevated when many action potentials are fired. The cause of dilation in response to both K^+ and CO_2 involves the formation of NO. However, the mechanism is not necessarily the typical endothelial formation of NO. The source of NO appears to be NO synthase in neurons as well as endothelial cells. Recent evidence indicates that the upper and lower limits of autoregulation in the brain are influenced by CO_2. Hypercapnia, which increases cerebral blood flow, shifts the upper limit of autoregulation at this higher flow to the left and the lower limit to the right. This narrows the range of pressure over which autoregulation is effective in hypercapnic states.

Reactions of cerebral blood flow to chemicals released by increased brain activity, such as CO_2, H^+, adenosine, and K^+, are part of the overall process of matching the brain's metabolic needs to the blood supply containing nutrients and oxygen. The 10% to 30% increase in blood flow in brain areas excited by peripheral nerve stimulation, mental activity, or visual activity may be related to these three substances released from active nerve cells. The cerebral vasculature also dilates when the oxygen content of arterial blood is reduced, but this effect is not as pronounced as that seen with an elevation of CO_2 in the vicinity of cerebral vessels.

CLINICAL FOCUS | 16.1

Exertional Angina

Coronary artery disease can produce many focal atherosclerotic lesions in the main coronary arteries. In spite of what may be extensive vascular disease, patients with this disease often do not present with any symptoms when they are at rest. However, once they increase their activity (e.g., walking upstairs, doing physical work, etc.), they suddenly exhibit shortness of breath and severe chest pain (angina pectoris). Angina is a sign that the myocardium is ischemic and injured. The angina induced by physical activity is often called "exertional angina," and it is the most common form of angina in patients with coronary artery disease.

The reason that patients with this condition are asymptomatic at rest, but show signs of myocardial ischemia upon exertion, is rooted in the flow autoregulation capacity of the myocardium and the strong link between coronary blood flow and tissue metabolism. When focal lesions within the large coronary arteries encroach on the lumen, they raise resistance to flow at that point. This increased resistance dissipates pressure downstream and would decrease flow in the artery as a whole if the downstream arterioles could not alter their resistance. However, the arterioles downstream are the site of autoregulation in the coronary circulation. When their internal pressure decreases, they respond by dilating in an effort to reduce resistance so that flow can be restored to its original value. Thus, any upstream increase in vascular resistance, such as that caused by a partial obstruction,

is compensated by a decrease in resistance in the vessels downstream from the obstruction. In this manner, the total resistance of the arterial circuit is returned to normal and, therefore, so is blood flow. The patient would obviously be unaware of this situation and exhibit no untoward symptoms. It has been estimated that a single focal obstruction in a major coronary must reduce lumen diameter by more than 90% before arteriolar diameter is maximized through autoregulatory mechanisms at rest. Thus, no resting ischemia and symptoms would occur until further reduction of the lumen beyond that point.

With coronary obstructions that are not severe enough to cause ischemia at rest, problems occur once the metabolic demands of the heart increase. The same arterioles that are dilated to compensate for upstream obstructions in the main coronary arteries are also the vessels that need to dilate to increase blood flow to the myocardium whenever activity of the heart is increased. If a portion of their dilating capacity is used to simply maintain resting flow, there may not be enough dilating capacity left (or **coronary reserve**) to augment blood flow to meet an increased oxygen demand by the heart. In that situation, the oxygen demand of the heart exceeds its oxygen supply, and ischemia, with the appearance of angina, occurs. This is why people with coronary artery disease can suffer from ischemia and angina upon exertion yet experience no ill effects at rest. ■

Cerebral vessels are insensitive to hormones and sympathetic nerve activity.

Catecholamines, as well as other circulating vasoconstrictor and vasodilator hormones, do not play much of a role in moment-to-moment regulation of cerebral blood flow. The blood–brain barrier effectively prevents constrictor and dilator agents in blood plasma from reaching the cerebral vascular smooth muscle, and the vessel wall contains enzymes that inactivate catecholamines, serotonin, and other neurological transmitters. Sympathetic nerves sparsely innervate the cerebral arteries and arterioles, if at all, and these vessels are essentially devoid of α-adrenergic or β-adrenergic receptors. Most of the adrenergic innervations appear to be located on the pial vessels rather than on the brain parenchyma. Stimulation of these nerves produces only mild vasoconstriction. If, however, sympathetic activity to the cerebral vasculature is permanently interrupted, the cerebral vasculature has a decreased ability to autoregulate blood flow at high arterial pressures, and the integrity of the blood–brain barrier is more easily disrupted. Therefore, some aspect of sympathetic nerve activity other than the routine regulation of vascular resistance is important for the maintenance of normal cerebral vascular function. This may relate to a neurogenic trophic factor that promotes the health of endothelial and smooth muscle cells in the cerebral microvessels.

The cerebral vasculature adapts to chronic high blood pressure.

In conditions of chronic hypertension, cerebral vascular resistance increases, thereby allowing cerebral blood flow and, presumably, capillary pressures to be normal. The adaptation of cerebral vessels to sustained hypertension lets them maintain vasoconstriction at arterial pressures that would overcome the contractile ability of a normal vasculature (Fig. 16.2).

The mechanisms that enable the cerebral vasculature to adjust the autoregulatory range upward appear to be hypertrophy of the vascular smooth muscle as well as a mechanical constraint to vasodilation as a result of more muscle tissue, more connective tissue, or both. The drawback to such adaptation is partial loss of the ability to dilate and regulate blood flow at low arterial pressures. This loss occurs because the changed passive structural properties of the resistance vessels restrict increases in vessel diameter at subnormal pressures and, in doing so, increase resistance. In fact, the lower pressure limit for autoregulation of blood flow can be almost as high as the normal mean arterial pressure (see Fig. 16.2). This can be problematic if the arterial blood pressure is rapidly lowered to normal in a person whose vasculature has adapted to hypertension. The person may faint from inadequate brain blood flow, even though the arterial pressure is in the normal range. Fortunately, a gradual reduction in arterial pressure, over weeks or months, returns autoregulation to a more normal pressure range.

Cerebral edema impairs blood flow to the brain.

The cranium encases the brain in a rigid bony structure. As such, should the brain begin to swell, intracranial pressure will increase dramatically. **Cerebral edema**, or an excessive

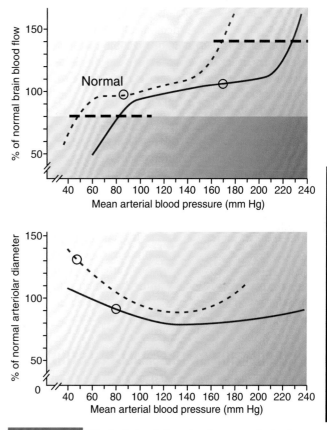

Figure 16.2 **Effects of chronic hypertension on cerebral autoregulation.** This condition is associated with a rightward shift in the arterial pressure range over which autoregulation of cerebral blood flow occurs (*upper panel*) because, for any given arterial pressure, resistance vessels of the brain have smaller-than-normal diameters (*lower panel*). As a consequence, people with hypertension can tolerate high arterial pressures that would otherwise cause vascular damage in healthy people. However, they risk reduced blood flow and brain hypoxia at low arterial pressures that are easily tolerated by healthy people. Effect of hypertension = solid line.

accumulation of fluid in the brain tissue, can greatly increase intracranial pressure. Such edema can be caused by a variety of traumatic or pathological conditions including infection, tumors, or trauma to the head that causes massive arteriolar dilation. In addition, frank bleeding into the brain tissue after a hemorrhagic stroke or trauma can increase intracranial pressure as well. In each case, as the intracranial pressure increases, the venules and veins are partially collapsed because their intravascular pressure is low. As these outflow vessels collapse, their resistance increases and capillary pressure rises (see Chapter 15). This increased capillary pressure favors increased filtration of fluid into the brain, which further raises the intracranial pressure. The end result is a positive feedback system in which intracranial pressure will become so high as to begin to compress small arterioles and decrease blood flow.

Excessive intracranial pressure is life threatening and a major clinical problem. Various means are used to try to reduce high intracranial pressure. For example, hypertonic mannitol can be infused into the circulation as a means of osmotically extracting water from an edematous brain. Sometimes opening of the skull and drainage of cerebrospinal fluid or hemorrhaged blood, if any, may be necessary. Hemorrhaged blood is particularly a problem because clotted blood contains denatured hemoglobin that destroys NO. This, in turn, leads to inappropriate vasoconstriction of the arterioles in the area of the hemorrhage, which further compromises blood supply to the brain.

If blood flow to the pons and medulla of the brain is decreased, tissue hypoxia activates sympathetic nervous system control centers in the brain. This results in massive sympathetic outflow to the organs of the body, resulting in severe vasoconstriction. This response is called the **Cushing reflex**. During this reflex, flow to the kidney may be so compromised as to prevent the formation of urine. The skin pales from removal of blood from that circulation, and ischemia can be produced in the intestine. However, this massive systemic vasoconstriction creates a marked elevation of mean arterial pressure (up to ~270 mm Hg), which helps open brain arterioles in the face of high external compression. Nevertheless, although blood flow may improve, the increase in arterial pressure elevates microvascular pressures, which worsens cerebral edema.

▶ CIRCULATION OF THE SMALL INTESTINE

The small intestine completes the digestion of food, which releases nutrients that are then absorbed into the circulation. At rest, the intestine receives about 20% of the cardiac output and uses about 20% of the body's oxygen consumption; both of these numbers nearly double after a large meal. Unless intestinal blood flow can increase during a large meal, food digestion and absorption simply do not occur.

The vasculature of the small intestine is elaborate. Small arteries and veins penetrate the muscular wall of the bowel and form a microvascular distribution system in the submucosa (Fig. 16.3). The muscle layers receive small arterioles from the submucosal vascular plexus; other small arterioles continue into individual vessels of the deep submucosa around glands and to the villi of the mucosa. Small arteries and larger arterioles preceding the separate muscle and submucosal–mucosal vasculatures control about 70% of the intestinal vascular resistance. The small arterioles of the muscle, submucosa, and mucosal layers can partially adjust blood flow locally to meet the needs of these small areas of tissue.

Autoregulation efficiency in the small intestine is modulated by intestinal oxygen consumption.

Compared with other major organs, the circulation of the small intestine has a poorly developed autoregulatory response in most conditions. As a result, blood flow usually declines in response to locally decreased arterial pressure because resistance does not adequately decrease to help counteract the effect of low pressure on blood flow. However, this is not true for all metabolic states of the intestine. Results of experiments

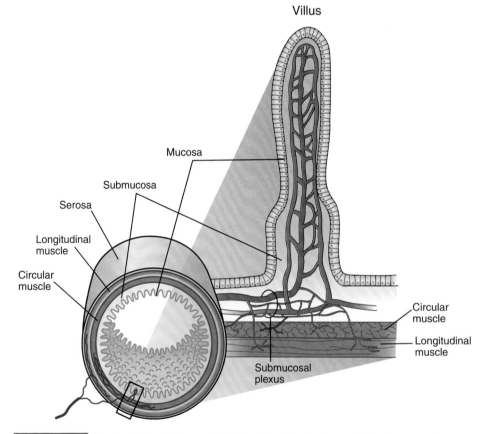

Figure 16.3 **The circulation of the small intestine.** The intestinal vasculature is unusual because branches from a common set of vessels located in the submucosa serve three different tissues, the muscle layers, submucosa, and mucosal layer. Small arteries and arterioles preceding the separate muscle, submucosal, and mucosal vasculatures regulate most of the intestinal vascular resistance.

on autoregulation in the small intestine indicate that the intestine is a poor autoregulator of blood flow when it is not occupied with digestion, and thus, its oxygen consumption is low, whereas it is a much better autoregulator during digestion, when its oxygen consumption is high. An understanding of this phenomenon is clear if one realizes that what the intestine needs during high metabolic demand is the oxygen in the blood more so than the blood itself. This can be accomplished by opening more capillaries, thereby increasing extraction of oxygen from the blood flow the intestine is receiving, rather than increasing total blood flow in order to increase total bulk oxygen delivery to the intestine. Applying this concept to the response of intestinal blood flow to a drop in arterial pressure, it follows that if perfusion pressure to the intestine were to drop, causing an initial drop in flow, the intestine could obtain the *oxygen* it needed to compensate for the drop in blood flow by simply taking the flow it gets and extracting more oxygen from it, rather than initiating an autoregulatory vasodilation to restore total flow. Indeed, this is exactly what happens as the intestinal permeability surface area coefficient increases, following a drop in perfusion pressure and blood flow, before frank vasodilation of the arterioles are used to correct the decrease in flow. Thus, when intestinal metabolism is low, a local decrease in oxygen delivery, brought on by an initial drop in blood pressure and blood flow, is compensated for first by dilating precapillary sphincters to open more capillary surface area for diffusion of oxygen. Should demand increase or total flow be compromised further, more capillaries will open until perfused capillary surface area is maximized. At that point, any further compensation for the effects of a drop in perfusion pressure and blood flow would have to be met by autoregulation of blood flow. Experimental evidence has validated this scenario repeatedly. In short, when oxygen delivery to the intestine is compromised, the intestine extracts more oxygen from its blood supply first and adjusts its total blood supply second. During digestion, all capillaries are open or nearly so in the intestine. While in this state, with oxygen extraction essentially maximized, any reduction in oxygen supply to the intestine caused by a temporary local drop in perfusion pressure and blood flow is then compensated for by strong local autoregulation.

The fact that the intestine seems to extract oxygen from blood before altering vascular resistance has a certain appeal for whole-body homeostatic mechanisms. Even though effects of changes in vascular resistance in the intestine on the body as a whole are lessened by the parallel arrangement of organ circulations in the body, the intestine and splanchnic circulations are still enormous. Manipulating extraction of oxygen before adjusting vascular resistance and flow might be a means of meeting the metabolic needs of the intestine while minimizing the effects of those needs on the blood supply to other organs.

Effects of elevated venous pressure on fluid filtration by intestinal capillaries.

The intestine has one of the highest capillary filtration coefficients among organ systems in the body. This could be potentially troublesome because the small intestine has a very large anatomical capillary surface area and it is not a good autoregulator of blood flow for most of the time during the day (i.e., when it is not occupied with digestion). These two factors make the intestine a prime source of fluid loss in the body whenever mean capillary hydrostatic pressure increases, such as what might occur whenever arterial or venous pressure increase.

Recall that, when arterial pressure is increased, strong autoregulation in an organ not only controls organ blood flow, but it also attenuates any increase in mean capillary hydrostatic pressure that would otherwise cause increased capillary fluid filtration. In a weak autoregulatory state, arteriolar constrictor responses to an increase in arterial pressure are weak and loss of fluid through the intestinal capillaries is likely. Furthermore, loss of fluid from the capillaries of the huge intestinal circulation could be significant in situations in which intestinal venous pressures rise. Such venous pressure rises can occur in liver disease or with portal vein obstructions because the intestinal veins are connected in series to the portal circulation of the liver. However, elevation of venous pressure in the intestinal circulation, which would increase capillary hydrostatic pressure, causes sustained myogenic arteriolar constriction that decreases capillary hydrostatic pressure. This is called the **venous–arteriolar response** and is a means by which major fluid loss from the capillaries is prevented in the intestine in the face of an elevation of intestinal venous pressure. Although the venous arteriolar response occurs in other organ systems, it appears to be strongest in the intestinal circulation.

High blood flow is required in the intestinal mucosal for absorption of nutrients.

The intestinal mucosa receives about 60% to 70% of the total intestinal blood flow. Blood flows of 70 to 100 mL/min/100 g in this specialized tissue are probable and much higher than the average blood flow for the total intestinal wall (see Table 16.1). This blood flow can exceed the resting blood flow in the heart and brain. The intestinal mucosa is composed of individual projections of tissue called *villi*. The interstitial space of the villi is mildly hyperosmotic (~400 mOsm/kg H_2O) at rest as a result of NaCl. During food absorption, the interstitial osmolality increases to 600 to 800 mOsm/kg H_2O near the villus tip, compared with 400 mOsm/kg H_2O near the villus base. The primary cause of high osmolalities in the villi appears to be greater absorption than removal of NaCl and nutrient molecules. There is also a possible countercurrent exchange process in which materials absorbed into the capillary blood diffuse from the venules into the incoming blood in the arterioles thus trapping them in the interstitium of the villus.

Lipid absorption causes a greater increase in intestinal blood flow and oxygen consumption, a condition known as **absorptive hyperemia**, than either carbohydrate or amino acid absorption. During absorption of all three classes of nutrients, the mucosa releases adenosine and CO_2 and oxygen is depleted, which may all stimulate mucosal hyperemia. The hyperosmotic lymph and venous blood that leave the villus to enter the submucosal tissues around the major resistance vessels are also major contributors to absorptive hyperemia. Hyperosmolality resulting from NaCl induces endothelial cells to release NO and dilate the major resistance arterioles in the submucosa. In contrast, hyperosmolality resulting from large organic molecules that do not enter endothelial cells does not cause appreciable increases in NO formation and produces much less of an increase in blood flow than equivalent hyperosmolality resulting from NaCl. These observations suggest that NaCl entering the endothelial cells is essential to induce NO formation though the mechanism of this effect is not clearly known at this time.

The active absorption of amino acids and carbohydrates and the metabolic processing of lipids into chylomicrons by mucosal epithelial cells place a major burden on the microvasculature of the small intestine. There is an extensive network of capillaries just below the villus epithelial cells. The villus capillaries are unusual in that portions of the cytoplasm are missing, so that the two opposing surfaces of the endothelial cell membranes appear to be fused. These areas of fusion, or closed fenestrae, are thought to facilitate the uptake of absorbed materials by capillaries. However, large molecules, such as plasma proteins, do not easily cross the fenestrated areas because the reflection coefficient for the intestinal vasculature is >0.9, about the same as in skeletal muscle and the heart.

Low capillary pressure in intestinal villi facilitates water absorption.

Although the mucosal layer of the small intestine has a high blood flow both at rest and during food absorption, the capillary blood pressure is usually 13 to 18 mm Hg and seldom higher than 20 mm Hg during food absorption. Therefore, plasma colloidal osmotic pressure is higher than capillary blood pressure, favoring the absorption of water brought into the villi. During lipid absorption, the plasma protein reflection coefficient for the overall intestinal vasculature is decreased from a normal value of more than 0.9 to about 0.7. It is assumed that most of the decrease in reflection coefficient occurs in the mucosal capillaries. This lowers the ability of plasma proteins to counteract capillary filtration, with the net result that fluid is added to the interstitial space. Eventually, this fluid must be removed. Not surprisingly, the highest rates of intestinal lymph formation normally occur during fat absorption.

Sympathetic nerve activity greatly decreases intestinal blood flow and volume.

The intestinal vasculature is richly innervated by sympathetic nerve fibers and contains predominantly α_1-adrenoceptors. Thus, major reductions in gastrointestinal blood flow and venous volume occur whenever sympathetic nerve activity is increased, such as during strenuous exercise or periods of pathologically low arterial blood pressure. In particular, venoconstriction in the intestine during hemorrhage helps to mobilize blood to the central circulation and helps compensates for the blood loss. Gastrointestinal blood flow is about 25% of the cardiac output at rest; a reduction in this blood flow, by heightened sympathetic activity, allows more vital functions to be supported with the available cardiac output.

However, in combination with a low arterial blood pressure (**hypotension**), gastrointestinal blood flow can be so drastically decreased by sympathetically mediated vasoconstriction that mucosal tissue damage can result. Nevertheless, this intestinal vasoconstriction can be seen as using the large size of the intestinal vasculature to the advantage of the body as a whole by providing a significant contribution of increased total peripheral vascular resistance to help counteract hypotension. In severe hypotension, perfusion to the heart and brain takes priority over that to other organs.

▶ HEPATIC CIRCULATION

The hepatic circulation perfuses the liver, which is one of the largest organs in the body. The liver is primarily an organ that maintains the organic chemical composition of the blood plasma. For example, all plasma proteins are produced by the liver, and the liver adds glucose from stored glycogen to the blood. The liver also removes damaged blood cells and bacteria and detoxifies many man-made or natural organic chemicals that have entered the body.

The hepatic circulation is perfused by gastrointestinal venous blood and hepatic arterial supply.

The human liver has a large blood flow, about 1.5 L/min or 25% of the resting cardiac output. It is perfused by both arterial blood through the hepatic artery and venous blood from the portal vein that has passed through the stomach, small intestine, pancreas, spleen, and portions of the large intestine. The venous blood arriving via the hepatic portal vein accounts for about 67% to 80% of the total liver blood flow (see Table 16.1). The remaining 20% to 33% of the total flow is through the hepatic artery. The majority of hepatic portal blood flow is determined by the flow through the stomach and small intestine.

The liver tissue efficiently extracts oxygen from the blood. About half of the oxygen used by the liver is derived from venous blood, even though the splanchnic organs have removed one third to one half of the available oxygen. The hepatic arterial circulation provides additional oxygen. The liver has a high metabolic rate and is a large organ; consequently, it has the largest oxygen consumption of all organs in a resting person (see Chapter 27). The liver vasculature is arranged into subunits that allow the arterial and portal blood to mix and provide nutrition for the liver cells (Fig. 16.4). Each subunit, called an *acinus*, is about 300 to 350 μm long and wide. In humans, usually three acini occur together. The core of each acinus is supplied by a single terminal portal venule; sinusoidal capillaries originate from this venule (see Fig. 16.4). The endothelial cells of the capillaries have fenestrated regions with discrete openings that facilitate exchange between the plasma and interstitial spaces. The capillaries do not have a basement membrane, which partially contributes to their high permeability. The terminal hepatic arteriole to each acinus is paired with the terminal portal venule at the acinus core, and blood from the arteriole and blood from the venule jointly perfuse the capillaries. The intermixing of the arterial and portal blood tends to be intermittent because the vascular smooth

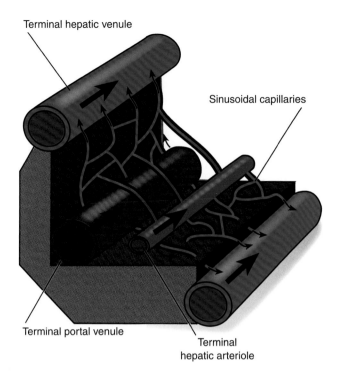

Figure 16.4 **Liver acinus microvascular anatomy.** A single liver acinus, the basic subunit of liver structure, is supplied by a terminal portal venule and a terminal hepatic arteriole. The mixture of portal venous and arterial blood occurs in the sinusoidal capillaries formed from the terminal portal venule. Usually, two terminal hepatic venules drain the sinusoidal capillaries at the external margins of each acinus.

muscle of the small arteriole alternately constricts and relaxes. This prevents arteriolar pressure from causing a sustained reversed flow in the sinusoidal capillaries, where pressures are 7 to 10 mm Hg. The best evidence for this is that hepatic artery and portal venous blood first mix at the level of the capillaries in each acinus. The sinusoidal capillaries are drained by the terminal hepatic venules at the outer margins of each acinus; usually, at least two hepatic venules drain each acinus.

Regulation of hepatic arterial and portal venous blood flow requires an interactive control system.

The regulation of portal venous and hepatic arterial blood flows is an interactive process: hepatic arterial flow increases and decreases reciprocally with the portal venous blood flow. This mechanism, known as the **hepatic arterial buffer response**, can compensate or buffer about 25% of the decrease or increase in portal blood flow. Exactly how this is accomplished is still under investigation, but vasodilatory metabolite accumulation, possibly adenosine, during decreased portal flow, as well as increased metabolite removal during elevated portal flow, is thought to influence the resistance of the hepatic arterioles.

One might suspect that during digestion, when gastrointestinal blood flow and, therefore, portal venous blood flow are increased, the gastrointestinal hormones in portal venous blood would influence hepatic vascular resistance. However, at concentrations in portal venous blood equivalent to those during digestion, none of the major hormones appears to influence hepatic blood flow. Therefore, the

increased hepatic blood flow during digestion would appear to be determined primarily by vascular responses of the gastrointestinal vasculatures.

The vascular resistances of the hepatic arterial and portal venous vasculatures are increased during sympathetic nerve activation, which also suppresses the hepatic arterial buffer mechanism. When the sympathetic nervous system is activated, about half the blood volume of the liver can be expelled into the general circulation. Because up to 15% of the total blood volume is in the liver, constriction of the hepatic vasculature can significantly increase the circulating blood volume during times of cardiovascular stress.

► SKELETAL MUSCLE CIRCULATION

The circulation of skeletal muscle makes up 30% to 40% of an adult's body weight and is the largest mass of tissue in the body. At rest, the skeletal muscle vasculature accounts for about 25% of systemic vascular resistance and individual muscles receive a low blood flow of about 2 to 6 mL/min/100 g. The dominant mechanism controlling skeletal muscle resistance at rest is the sympathetic nervous system; local regulatory mechanisms become more significant with high muscle metabolism (e.g., during exercise). Resting skeletal muscle has remarkably low oxygen consumption per 100 g of tissue, but its large mass makes its metabolic rate a major contributor to the total oxygen consumption in a resting person.

Skeletal muscle blood flow is varied by both sympathetic neural and local metabolic factors.

Skeletal muscle blood flow can increase 10- to 20-fold or more during the maximal vasodilation associated with high-performance aerobic exercise. During exercise, the effects of accumulation of local vasodilator agents in active skeletal muscle far exceed the effect of sympathetic neural vasoconstriction on vessels within the skeletal muscle. Under such circumstances, total muscle blood flow may increase to values three or more times higher than resting cardiac output. Cardiac output itself can increase fivefold during strenuous exercise, with most all of that increase due to an increase in skeletal muscle blood flow.

Skeletal muscle arteries contain both α- and β-adrenoceptors, but the former predominate. Thus, activation of the sympathetic nervous system causes vasoconstriction in skeletal muscle. Vascular resistance can easily double from resting values as a result of increased sympathetic nerve activity, with a resulting significant decrease in muscle blood flow. Such neurogenic vasoconstriction in skeletal muscle can override local metabolic demands of the tissue to such an extent that the muscle can become modestly ischemic even though metabolism in resting skeletal muscle is very low. Fortunately, skeletal muscle cells can survive long periods with minimal oxygen supply so this low blood flow does not result in marked cell damage or death. The effects of the sympathetic nervous system on the skeletal muscle circulation is a means to exploit this very large circulation for the benefit of the body as a whole rather than to control muscle blood flow. Activation of the sympathetic nervous system to the large mass that is skeletal muscle causes a large increase in total

Cardiovascular Physiology

systemic vascular resistance that is used to help maintain or limit a drop in arterial blood pressure when cardiac output is compromised (i.e., as in hemorrhage or dehydration). Like in the small intestine, this ability enables the heart and brain to be perfused in preference to other organs less critical to the acute survival of the individual. In addition, contraction of the skeletal muscle venules and veins forces blood from these vessels into the central circulation. This action helps counteract deficits in cardiac output that accompany losses of blood volume. In sum, the skeletal muscle vasculature can either place major demands on the cardiopulmonary system through massive vasodilation during exercise or respond as if expendable with intense vasoconstriction during a hypovolemic and/or hypotensive crisis.

Muscle blood flow is markedly affected by numerous local vasoactive agents.

As discussed in Chapter 15, many potential local regulatory mechanisms adjust blood flow to the metabolic needs of tissues. In skeletal muscle, as in the small intestine, autoregulation efficacy is dependent on local metabolism with increased efficacy associated with high metabolism. In addition, to ensure the best possible supply of nutrients, particularly oxygen, even mild exercise causes sufficient vasodilation to perfuse virtually all of the capillaries, rather than just 25% to 50% of them as occur at rest. Like the small intestine, skeletal muscle first recruits more capillaries to enhance oxygen extraction in times of increased oxygen need and adjusts blood flow if needed after extraction is essentially maximized.

The most dynamic local vascular regulatory phenomenon in skeletal muscle is the coupling of muscle blood flow to its activity (i.e., active hyperemia). In fast-twitch muscles, which primarily depend on anaerobic metabolism, the accumulation of hydrogen ions from lactic acid is potentially a major contributor to active hyperemia in that muscle type. In slow-twitch skeletal muscles, oxidative metabolic requirements can increase up to 20 times during heavy exercise. This level of muscle activity is associated with a very large muscle hyperemia. It is not hard to imagine that whatever causes metabolically linked vasodilation is in ample supply at high metabolic rates and/or possesses strong vasodilator activity. Increased CO_2, H^+, K^+, and adenosine as well as hypoxia are all vasodilatory to skeletal muscle arterioles, and such changes in these metabolites occur in exercising muscle. However, none of the observed changes of these metabolites alone explains muscle active hyperemia. Recently, the skeletal muscle vasodilation and hyperemia associated with exercise have been shown to be dependent upon NO, although the mechanism by which muscle activity is linked to NO in active hyperemia is not known.

It is difficult to account for large active hyperemia in skeletal muscle on the vasodilatory effect of low tissue oxygen tensions alone. During rhythmic muscle contractions, the blood flow during the relaxation phase can be high, and thus it is unlikely that the muscle becomes significantly hypoxic during submaximal aerobic exercise (i.e., high flow enhances oxygen delivery in the relaxation phase). Although the tissue oxygen content likely decreases as exercise intensity increases, the reduction does not compromise the

high aerobic metabolic rate except with the most demanding forms of exercise. Once the vasodilation and increased blood flow associated with exercise are established (in 1 to 2 minutes), the microvasculature is probably capable of maintaining ample oxygen for most workloads, perhaps up to 75% to 80% of maximum performance, because remarkably little additional lactic acid accumulates in the blood. Although H^+ is a vasodilator and lactic acid formation increases during hypoxia as well as with anaerobic metabolism, studies in humans and animals indicate that this acid is present only during the first several minutes of submaximal exercise.

The ability of skeletal muscle to meet its oxygen needs for sustained activity is not limitless. Near-maximum or maximum exercise exhausts the ability of the microvasculature to meet tissue oxygen needs, and hypoxic conditions rapidly develop, limiting the performance of the muscles. The burning sensation and muscle fatigue during maximum exercise, or at any time that muscle blood flow is inadequate to provide adequate oxygen, are partially a consequence of hypoxia. This type of burning sensation is particularly evident when a muscle must hold a weight in a steady position. In this situation, the contraction of the muscle compresses the microvessels, which severely reduces muscle blood flow. The combination of throttled flow with the energy demands of holding a weight against gravity creates marked hypoxia in muscle tissues.

▶ CUTANEOUS CIRCULATION

The structure of the skin vasculature differs according to location in the body. In all areas, an arcade of arterioles exists at the boundary of the dermis and the subcutaneous tissue over fatty tissues and skeletal muscles (Fig. 16.5). From this arteriolar arcade, arterioles ascend through the dermis into the superficial layers of the dermis, adjacent to the epidermal layers. These arterioles form a second network in the superficial dermal tissue and perfuse the extensive capillary loops that extend upward into the dermal papillae just beneath the epidermis. The dermal vasculature also provides the vessels that surround hair follicles, sebaceous glands, and sweat glands. The capillary loops are essentially perpendicular to the surface of the skin such that only the tips are in close proximity to the outermost layer of that tissue. All the capillaries from the superficial skin layers are drained by venules, which form a venous plexus in the superficial dermis and eventually drain into many large venules and small veins beneath the dermis. The vascular pattern just described is modified in the tissues of the hand, feet, ears, nose, and some areas of the face, in that direct vascular connections between arterioles and venules, known as **arteriovenous anastomoses**, occur primarily in the superficial dermal tissues (see Fig. 16.5). By contrast, relatively few arteriovenous anastomoses exist in the major portion of human skin over the limbs and torso. The anastomoses lead to a venous plexus that lies parallel to the skin surface and thus are positioned, when open, to direct blood into a large surface area of small blood vessels just underlying the surface of the skin.

Control of the cutaneous circulation is dominated by the sympathetic nervous system. At rest in a cool room, cutaneous vessels are significantly constricted by norepinephrine

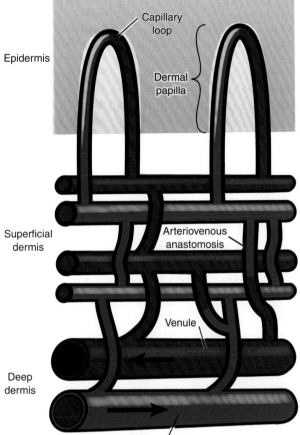

Figure 16.5 **Vascular organization in the skin.** The microcirculation of the skin is composed of a network of large arterioles and venules in the deep dermis, which send branches to the superficial network of smaller arterioles and venules. Arteriovenous anastomoses allow direct flow from arterioles into venules, which increase blood flow into a venous plexus when dilated. The capillary loops into the dermal papillae beneath the epidermis are supplied and drained by microvessels of the superficial dermal vasculature. Only the tips of the loops are in close contact with the surface of the skin.

from sympathetic nerves that innervate all areas of the circulation except the cutaneous capillaries. Innervation is especially dense on the arteriovenous anastomoses. Nerve blockade of the cutaneous circulation results in maximal vasodilation of its vessels because the vessels possess essentially no intrinsic active smooth muscle tone.

Adjustment of cutaneous blood flow is used for temperature regulation.

The skin is a large organ, representing 10% to 15% of total body mass, with a large external surface area that is positioned at the interface of the body with the external environment. The primary functions of the skin are protection of the body from the external environment and dissipation or conservation of heat as part of the mechanisms involved in body temperature regulation. If a great amount of heat must be conserved, the sympathetic nerves to the cutaneous circulation are activated, and all vessels there, especially

the anastomoses, constrict and restrict blood from entering the skin near the external environment (see Chapter 28 for details). If a great amount of heat must be dissipated, dilation of the arteriovenous anastomoses allows substantially increased skin blood flow to the venous plexuses, thereby increasing heat loss to the environment. This allows vasculatures of the hands and feet and, to a lesser extent, the face, neck, and ears to lose heat efficiently in a warm environment.

The skin has one of the lowest metabolic rates in the body and requires relatively little blood flow for purely nutritive functions. Consequently, despite its large mass, its resting metabolism does not place a major flow demand on the cardiovascular system. However, in warm climates, body temperature regulation requires that warm blood from the body core be carried to the external surface, where heat transfer to the environment can occur. Therefore, at typical indoor temperatures and during warm weather, skin blood flow is usually far in excess of the need for tissue nutrition. The reddish color of the skin during exercise in a warm environment reflects the large blood flow and dilation of skin arterioles and venules brought into play for dissipation of excess body heat (see Table 16.1).

The increase in the skin's blood flow when the body is exposed to heat probably occurs through two main mechanisms. First, an increase in body core temperature causes a reflex increase in the activity of *sympathetic cholinergic nerves*, which release acetylcholine. Acetylcholine release near sweat glands leads to the breakdown of a plasma protein (kininogen) to form bradykinin, a potent dilator of skin blood vessels, which also increases the release of NO as a major component of the dilatory mechanism. Second, simply increasing skin temperature will cause the blood vessels to dilate. This can result from heat applied to the skin from the external environment, heat from underlying active skeletal muscle, or increased blood temperature as it enters the skin.

Total skin blood flows of 5 to 8 L/min have been estimated in humans during vigorous exercise in a hot environment. During mild-to-moderate exercise in a warm environment, skin blood flow can equal or exceed blood flow to the skeletal muscles. Exercise tolerance can, therefore, be lower in a warm environment because the vascular resistance of the skin, combined with very low vascular resistance in muscle, is too low to maintain an appropriate arterial blood pressure, even at maximum cardiac output. One of the adaptations to exercise is an ability to increase blood flow in skin and dissipate more heat. In addition, aerobically trained humans are capable of higher sweat production rates than normal, which accelerates heat loss from the surface of the skin (see Chapter 28).

Most humans live in cool-to-cold regions, where body heat conservation is imperative. The sensation of cool or cold skin or a lowered body core temperature elicits a reflex increase in sympathetic nerve activity, which causes vasoconstriction of blood vessels in the skin, especially the arteriovenous anastamoses. Heat loss is minimized because the skin becomes a poorly perfused insulator, rather than a heat dissipater. As long as the skin temperature is higher than about 10°C to 13°C (50°F to 55°F), the neurally induced vasoconstriction is sustained. However, at lower tissue temperatures, the vascular smooth muscle cells progressively lose their contractile ability, and the vessels passively dilate to various extents. The reddish color of the hands, face, and ears on a cold day demonstrates increased blood flow and vasodilation as a result of low skin temperatures. To some extent, this cold-mediated vasodilation is useful because it lessens the chance of cold injury to exposed skin. However, if this process included most of the body surface, such as occurs when the body is submerged in cold water or inadequate clothing is worn hypothermia can result (see Chapters 28 and 29 for more detail).

▶ FETAL AND PLACENTAL CIRCULATIONS

The development of a human fetus depends on nutrient, gas, water, and waste exchange in the maternal and fetal portions of the placenta. The human fetal placenta is supplied by two **umbilical arteries**, which branch from the internal iliac arteries of the fetus. The fetal placenta is drained by a single **umbilical vein** (Fig. 16.6). The umbilical vein of the fetus returns oxygen and nutrients from the mother's body to the fetal cardiovascular system, and the umbilical arteries bring

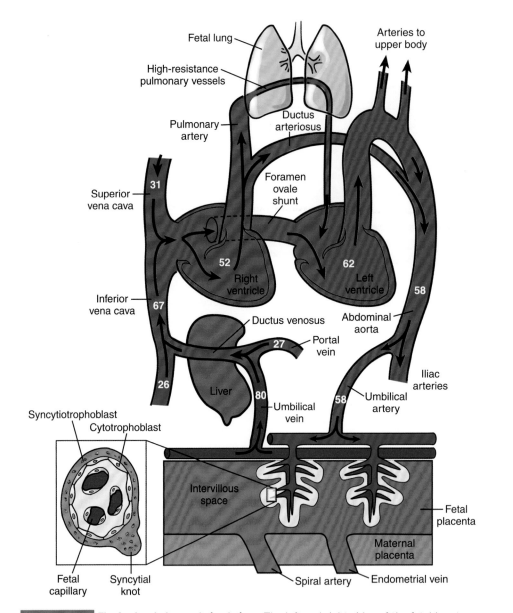

Figure 16.6 **The fetal and placental circulations.** The left and right sides of the fetal heart are separated solely for the purpose of illustration to emphasize the right-to-left shunt of blood through the open foramen ovale in the atrial septum and the right-to-left shunt through the ductus arteriosus. *Arrows* indicate the direction of blood flow. The numbers represent the percent saturation of blood hemoglobin with oxygen in the fetal circulation. Closure of the ductus venosus, foramen ovale, ductus arteriosus, and placental vessels at birth and the dilation of the pulmonary vasculature establish the adult circulation pattern. The insert is a cross-sectional view of a fetal placental villus, one of the branches of the treelike fetal vascular system in the placenta. The fetal capillaries provide incoming blood, and the sinusoidal capillaries act as the venous drainage. The villus is completely surrounded by the maternal blood, and the exchange of nutrients and wastes occurs across the fetal syncytiotrophoblast.

in blood laden with carbon dioxide and waste products from the fetus to be transferred to the mother's blood. Although many liters of oxygen and carbon dioxide, together with hundreds of grams of nutrients and wastes, are exchanged between the mother and fetus each day, the exchange of red blood cells or white blood cells is a rare event. This large chemical exchange without cellular exchange is possible because the fetal and maternal bloods are kept almost completely separate.

The fundamental anatomic and physiologic structure for fetal/maternal exchange is the placental villus. As the umbilical arteries enter the fetal placenta, they divide into many branches that penetrate the placenta toward the maternal system. These small arteries divide in a pattern similar to a fir tree, the placental villi being the small branches. The fetal capillaries bring in the fetal blood from the umbilical arteries, and then, blood leaves through sinusoidal capillaries to the umbilical venous system. Exchange occurs in the fetal capillaries and probably to some extent in the sinusoidal capillaries. The mother's vascular system forms a reservoir around the treelike structure such that her blood envelops the placental villi.

As shown in Figure 16.6, the outermost layer of the placental villus is the syncytiotrophoblast, where exchange by passive diffusion, facilitated diffusion, and active transport between fetus and mother occurs through fully differentiated epithelial cells. The underlying cytotrophoblast is composed of less differentiated cells, which can form additional syncytiotrophoblast cells as required. As cells of the syncytiotrophoblast die, they form syncytial knots, and eventually, these break off into the mother's blood system surrounding the fetal placental villi.

The placental vasculatures of both the fetus and the mother adapt to the size of the fetus as well as to the oxygen available within the maternal blood. For example, a minimal placental vascular anatomy will provide for a small fetus, but as the fetus develops and grows, a complex tree of placental vessels is essential to provide the surface area needed for the fetal–maternal exchange of gases, nutrients, and wastes. If the mother moves to a higher altitude, where less oxygen is available, the complexity of the placental vascular tree increases, compensating with additional areas for exchange. If this type of adaptation does not take place, the fetus may be underdeveloped or may die from a lack of oxygen.

During fetal development, the fetal tissues invade and cause partial degeneration of the maternal endometrial lining of the uterus. The result, after about 10 to 16 weeks of gestation, is an intervillous space between fetal placental villi that is filled with maternal blood. Instead of microvessels, there is a cavernous blood-filled space. The intervillous space is supplied by 100 to 200 **spiral arteries** of the maternal endometrium and is drained by the **endometrial veins**. During gestation, the spiral arteries enlarge in diameter and simultaneously lose their vascular smooth muscle layer. It is the arteries preceding them that actually regulate blood flow through the placenta. At the end of gestation, the total maternal blood flow to the intervillous space is ~600 to 1,000 mL/min, which represents about 15% to 25% of the resting cardiac output. In comparison, the fetal placenta has a blood flow of about 600 mL/min, which represents about 50% of the fetal cardiac output.

The exchange of materials across the syncytiotrophoblast layer follows the typical pattern for all cells. Gases, primarily oxygen and carbon dioxide, and nutrient lipids move by simple diffusion from the site of highest concentration to the site of lowest concentration. Small ions are moved predominantly by active transport processes. The GLUT1 transport protein passively transfers glucose, and amino acids require primarily facilitated diffusion through specific carrier proteins in the cell membranes, such as the system A transporter protein.

Large molecular weight peptides and proteins and many large, charged, water-soluble molecules used in pharmacologic treatments do not readily cross the placenta. Part of the transfer of large molecules probably occurs between the cells of the syncytiotrophoblast layer and by pinocytosis and exocytosis. Lipid-soluble molecules diffuse through the lipid bilayer of cell membranes. For example, lipid-soluble anesthetic agents in the mother's blood do enter and depress the fetus. As a consequence, anesthesia during pregnancy is somewhat risky for the fetus.

Placental exchange of oxygen and carbon dioxide is limited.

In spite of intimate contact between fetal and maternal circulations in the placenta, the combined structures of this interface still create a significant diffusion barrier for oxygen. Special fetal adaptations are required for oxygen exchange, because of the limitations of passive exchange across the placenta. The Po_2 of maternal arterial blood is about 80 to 100 mm Hg, whereas that of the incoming blood in the umbilical artery is about 20 to 25 mm Hg. This difference in oxygen tension provides a large driving force for exchange, but results in an increase in the Po_2 of fetal blood in the umbilical vein to only 30 to 35 mm Hg. Fortunately, **fetal hemoglobin** carries more oxygen at a low Po_2 than adult hemoglobin carries at a Po_2 two to three times higher. In addition, the concentration of hemoglobin in fetal blood is about 20% higher than in adult blood. The net result is that the fetus has sufficient oxygen to support its metabolism and growth but does so at low oxygen tensions, using the unique properties of fetal hemoglobin. After birth, when much more efficient oxygen exchange occurs in the lung, the newborn gradually replaces the red cells containing fetal hemoglobin with red cells containing adult hemoglobin.

Absence of lung ventilation in the fetus necessitates a bypass arrangement around the fetal pulmonary circulation.

After the umbilical vein leaves the fetal placenta, it passes through the abdominal wall at the future site of the umbilicus (navel). The umbilical vein enters the liver's portal venous circulation, although the bulk of the oxygenated venous blood passes directly through the liver in the **ductus venosus** (see Fig. 16.6). However, the liver receives blood with relatively high oxygen content. This content supports red blood cell development in the liver, which is the site of such development in the fetus, unlike the bone marrow in an adult. The low-oxygen-content venous blood from the lower body and the high-oxygen-content placental venous blood mix in the inferior vena cava. The oxygen content of the blood returning from the lower body is about twice that of venous blood returning from the upper body in the superior vena cava. The two streams of blood from the superior and inferior vena cava do not completely mix as they enter the right atrium. In addition, in the fetus, the lungs are collapsed and have a high vascular resistance. The net result is that oxygen-rich blood from the inferior vena cava is shunted from the right atrium into the left atrium through an opening in the atrial septum called the **foramen ovale**. Thus, the more oxygen-rich blood largely bypasses the fetal lungs. The upper-body blood generally enters the right ventricle as in the adult and is used to support growth of the lungs, which remain collapsed in the fetus. Perfusion of the collapsed (or noninflated) lungs of the fetus is minimal because the pulmonary vascular resistance is high in collapsed lungs. High fetal pulmonary vascular resistance may also result because the pulmonary vasculature has the unusual characteristic of vasoconstriction at low oxygen tensions. The minimal amount of venous blood returning from the lungs to the left atrium and the preferential passage of oxygenated venous blood into the left atrium through the foramen ovale allows blood in the left ventricle to have oxygen content about 20% higher than that in the right ventricle. This relatively high-oxygen-content blood supplies the coronary vasculature, the head, and the brain.

The right ventricle actually pumps at least twice as much blood as the left ventricle during fetal life. In fact, the infant at birth has a relatively more muscular right ventricular wall than the adult. The right ventricle pumps blood into the systemic arterial circulation via a shunt called the **ductus arteriosus**, which connects the pulmonary artery and aorta (see Fig. 16.6). For ductus arteriosus blood to enter the initial part of the descending aorta, the right ventricle must develop a higher pressure than the left ventricle, which is the exact opposite of the pattern in the adult. The blood in the descending aorta has less oxygen content than that in the left ventricle and ascending aorta because of the mixture of less well-oxygenated blood from the right ventricle. This difference is crucial because about two thirds of this blood must be used to perfuse the placenta and pick up additional oxygen. In this situation, a lack of oxygen content is useful.

Transition from fetal to neonatal life requires complex changes in the structure of the fetal circulation after birth.

After the newborn is delivered, the initial ventilatory movements cause the lungs to expand with air thereby reducing pulmonary vascular resistance substantially (along with pulmonary arterial pressure). The right ventricle now perfuses the lungs, and the circulation pattern in the newborn switches to that of an adult. The highly perfused, ventilated lungs immediately after birth allow a large amount of oxygen-rich blood to enter the left atrium and left ventricle and out into the aorta during systole. The increased oxygen tension in the aortic blood may provide the signal for closure of the ductus arteriosus, although suppression of vasodilator prostaglandins cannot be discounted. In any event, the ductus arteriosus constricts to virtual closure and, over time, becomes anatomically fused. Simultaneously, the increased oxygen to the peripheral tissues causes constriction in most body organs, and the sympathetic nervous system also stimulates the peripheral arterioles to constrict. The net result is that the left ventricle now pumps against a higher resistance. The combination of greater resistance and higher blood flow raises systemic arterial pressure and, in doing so, increases the mechanical load on the left ventricle, which leads to ventricle hypertrophy over time. In time, the reduced workload on the right ventricle, which now pumps against a much lower pressure, causes its larger mass to subside, as that of the left ventricle increases. A larger left ventricular mass versus that of the right ventricle is eventually complete and set in about 6 months after birth.

During all the processes just described, the open foramen ovale must be sealed after birth to prevent backflow of blood from the left to the right atrium. After birth, left atrial pressure increases from the returning blood from the lungs and exceeds right atrial pressure. This pressure difference passively pushes a tissue flap on the left side of the foramen ovale against the open atrial septum. In time, the tissues of the atrial septum fuse; however, an anatomic passage that is probably only passively sealed can be documented in some adults. The ductus venosus in the liver is open for several days after birth but gradually closes and is obliterated within 2 to 3 months.

After the fetus begins breathing, the fetal placental and umbilical vessels undergo progressive vasoconstriction to force placental blood into the fetal body, minimizing the possibility of fetal hemorrhage through the placental vessels. Vasoconstriction is related to physical trauma, increased oxygen availability, sensitivity of these vessels to circulating catecholamines, and less of a signal for vasodilator chemicals and prostaglandins in the fetal tissue. The intense vasoconstriction of the umbilical vessels helps collapse the ductus venous and increase systemic vascular resistance in the newborn because the placenta, which lies in parallel with the fetal cardiovascular system, is removed from the vascular circuitry. This also helps systemic, or left-sided, pressure exceed that in the right thus facilitating reversal of flow through the ductus arteriosus, exposing it to the vasoconstrictive effects of high oxygen tensions.

The final event of gestation is separation of the fetal and maternal placenta as a unit from the lining of the uterus. The separation process begins almost immediately after the fetus is expelled, but external delivery of the placenta can require up to 30 minutes. The separation occurs along the decidua spongiosa, a maternal structure, and requires that blood flow in the mother's spiral arteries be stopped. The cause of the placental separation may be mechanical, as the uterus surface area is greatly reduced by removal of the fetus and folds away from the uterine lining. Normally, about 500 to 600 mL of maternal blood is lost in the process of placental separation. However, as maternal blood volume increases 1,000 to 1,500 mL during gestation, this blood loss is not a significant concern.

CLINICAL FOCUS | 16.2

Atrial Septal Defects

The four-chamber configuration of the adult heart is necessary for the proper oxygenation of blood and the delivery of carbon dioxide to the lungs. The separation of the ventricles into the mammalian right and left chambers occurs early during normal fetal development of the ventricular septum. The atria in the fetus communicate through the foramen ovale. This is a normal configuration in the fetus, which provides a beneficial mechanism for diverting oxygenated blood from the inferior vena cava away from the lungs (which are nonfunctional in utero) and into the systemic circulation. In some people, the foramen ovale does not close after birth. This is called an atrial septal defect.

ASD is twice as prevalent in women as in men. The consequences of retaining this defect through childhood and on into adulthood vary. People with atrial septal defects (ASDs) can live well into their seventh or eighth decade, although life expectancy is not normal; mortality increases about 6% per year after age 40. The primary hemodynamic abnormality associated with ASD is shunting of blood from the left to the right atrium, which can cause right ventricular overload and mild pulmonary hypertension. Mixing of blood between the right and left heart can be minimal, in which case the individual can be largely asymptomatic, or more severe, resulting in slowed growth and reduced exercise endurance. In this case, young patients often exhibit increased endurance after surgical correction of ASD. More serious effects from ASD can arise about the fourth decade of life, as the patient accumulates additional cardiovascular diseases, such as stiffened ventricles from chronic arterial hypertension and coronary artery disease. In those cases, the communication between the atria allows for more serious right-side overload, resulting in symptomatic pulmonary hypertension and distended atria. The latter can cause tricuspid regurgitation and atrial fibrillation. Surgical correction is required to remedy ASD. This often involves simply inserting a catheter tipped with a collapsed patch through the ASD, opening the patch, pulling it against the septal opening, and thereby sealing the ASD. Tissue then grows over the patch, making the repair permanent. ■

INTEGRATED MEDICAL SCIENCES

The Evolution of Drug-Eluting Stents (DES) for the Treatment of Coronary Artery Disease

Coronary artery bypass grafts (CABGs) were one of the first solutions for the prevention of angina and heart attack caused by blockage of coronary arteries from atherosclerotic plaques. However, CABG procedures involve significant invasive surgery and risk for the patient. Furthermore, such graphs have proven to be susceptible to restenosis over time. Balloon angioplasty, when first developed, was intended as a less invasive, more risk-adverse treatment for alleviating the hemodynamic consequences of atherosclerotic plaques in the coronary arteries. This technique used a coronary arterial catheter fitted with an inflatable balloon on its tip that was inserted into the vascular system percutaneously and then guided into the coronary artery to the site of the plaque. Once opposite the plaque, the balloon was inflated and deflated repeatedly to "smash, flatten, and stretch" the plaque and artery, which was often sclerotic. This physical dilation effectively removed the narrowing at the site of the plaque. The problem with this procedure was that such repeated stretching and smashing of the lesion and arterial wall traumatized the artery, which would then trigger a local injury and inflammatory response. This inflammation would stimulate smooth muscle cell growth, resulting in intimal hyperplasia and significant restenosis of the artery.

An early solution to the restenosis problem was to outfit the balloon catheter with a cylindrical, flexible wire mesh that could be expanded to a wide diameter directly into the coronary arterial wall upon inflation of the balloon. This wire mesh tube, called a stent, provided support scaffolding in the arterial wall at the site of plaques and held the artery in an open position. Early arterial stents were made of bare metal and proved to be a nidus for thrombosis and focal inflammation. Laser polishing of the stent surface to an ultrasmooth finish was an early attempt at making the stent less procoagulant. However, implanting a bare metal mesh into the artery acted as an injury stimulus of its own, which in turn stimulated local tissue growth and restenosis.

For this reason, drug-eluting stents (DESs) impregnated with antiproliferative agents were developed to counteract the inflammatory tissue growth associated with earlier bare metal stents. DES is essentially a local drug-delivery system to help stop the process of restenosis and thrombosis in the stented arterial wall at the site of coronary plaques. The first DESs were coated with polymers containing either antimitogenic agents or antirejection agents. Some contained a reabsorbable polymer coating that reduced the risk of blood clot formation.

(Continued)

The first U.S. Food and Drug Administration–approved DESs contained sirolimus (an antirejection drug) or paclitaxel (an antimitotic) as the eluent. In healthy arteries, such as those of animals used to test the stents, antimitotic eluents proved beneficial by preventing intimal hyperplasia without affecting the endothelial growth and integrity surrounding the metal stent. However, in atherosclerotic arteries of humans, the function and health of the endothelium are severely impaired. In these vessels, antimitotic eluents impaired protective and antithrombotic endothelial growth over the metal stents leaving the stented region susceptible to blood clot formation. As DESs continued to increase in clinical use, it was discovered that although DESs were an improvement over bare metal stents, especially in terms of preventing neointimal proliferation and restenosis, evidence accumulated suggesting problematic "delayed reactions" to the stents resulting in thrombus formation in the artery long after the stent had be placed. This late stent thrombosis was often lethal. It is now understood that early polymers that contained the active antiproliferative drugs used to coat the DES stayed on the stent long after their drug is gone. The remaining polymer was shown to then act as a source of delayed reactions and inflammation leading to thrombosis. For this reason, the latest generation of DES designs has been developed to avoid this polymer-based reaction. This has led to the development of biodegradable polymer coatings on the stents, bioresorbable scaffolds, biocompatible polymers, and polymer-free DES. One of the latest developments are gene-eluting stents, which are intended to relieve restenosis and thrombosis by accelerating re-endothelialization around the stent scaffold, thereby reestablishing the natural antimitotic and antithrombotic properties of the endothelium of healthy arteries. ■

Chapter Summary

- The coronary and cerebral circulations are predominantly controlled by local metabolic factors.
- The coronary and cerebral circulations have high autoregulatory efficacy.
- Activation of the sympathetic nerves to the heart results in direct and indirect coronary arterial dilation.
- The blood vessels of the brain are largely unaffected by circulating hormones and vasoactive compounds because of the blood–brain barrier.
- Cerebral arteries are more sensitive to dilating effects of CO_2 and H^+ than are other systemic arteries.
- Blood flow to specific regions of the brain increase in response to increase neuronal activity and metabolism in those regions.
- Intestinal hyperemia during digestion is related to increased metabolism during digestion and nutrient absorption and is mediated in part by elevated sodium chloride concentration in the tissue and the release of nitric oxide.
- Autoregulation efficiency in the intestine and skeletal muscle is enhanced by increased tissue metabolism in those organs.
- The liver receives the portal venous blood from the gastrointestinal organs as its main blood supply, which is supplemented by hepatic arterial blood.

- Skeletal muscle tissue receives minimal blood flow at rest because of its limited oxygen requirements, but flow and oxygen use can increase markedly during intense muscle activity.
- The skin has a low oxygen requirement, but the high blood flow during warm temperatures or exercise allows for dissipation of a large amount of heat to the environment from the skin.
- Sympathetic neural vasoconstriction in the intestines, skeletal muscles, and skin help counteract hypotensive episodes.
- The fetus obtains nutrients and oxygen from the mother's blood supply using the combined maternal and fetal placental circulations.
- The circulation in the fetus bypasses the fetal lungs and obtains oxygen from placental exchange. The fetal circulation is designed to deliver blood with the highest-oxygen content to the developing brain, liver (for red blood cell generation), and upper extremities.
- Closures of the foramen ovale, ductus venous, and ductus arteriosus in the fetus, as well as collapse of the umbilical arteries and vein, are essential for the transformation of the fetal circulatory system into that of an air-breathing newborn.

Chapter Review Questions

1. Which of the following would be an expected response by the coronary vasculature to changes in blood or systemic hemodynamics in the body? (To simplify your answer, ignore any reflex actions that may accompany the following situations.)

   A. Increased blood flow when the heart workload is increased
   B. Decreased vascular resistance when coronary arterial blood pressure is increased
   C. Increased vascular resistance when mean arterial pressure is reduced from 90 to 70 mm Hg by hemorrhage
   D. Decreased blood flow when blood oxygen content is reduced
   E. Increased vascular resistance during aerobic exercise

The correct answer is A. Increased workload on the heart increases myocardial oxygen consumption, which results in coronary vasodilation and increased coronary blood flow (an active hyperemia response). Intra-arterial pressure increases in the coronary artery would cause a myogenic (autoregulatory) increase in resistance. A drop in arterial pressure from 90 to 70 mm Hg, if transferred to the coronary arteries, would not increase but rather decrease coronary vascular resistance due to strong autoregulation in the coronary circulation. Decreased blood oxygen content increases coronary blood flow. Exercise increases heart rate and contraction that increase myocardial metabolism, which in turn causes vasodilation.

2. In a generalized whole-body activation of the sympathetic nervous system, which of the following special circulations would show the least change in blood flow?

   A. Coronary
   B. Cerebral
   C. Small intestine
   D. Skeletal muscle
   E. Skin

The correct answer is B. The cerebral circulation is essentially unresponsive to sympathetic nerve stimulation. Flow in the skin, intestine, and skeletal muscle decreases markedly in response to activation of their sympathetic nerve supply. Coronary blood flow will increase upon activation of sympathetic nerves to the heart through direct and indirect mechanisms.

3. Which of the following will decrease cutaneous blood flow?

   A. Fever
   B. Histamine
   C. Chemical components of sweat
   D. Local subcutaneous anesthetics (i.e., procaine, Xylocaine, etc.)
   E. Exposure of the body to atmospheric temperatures of 35°F

The correct answer is E. Exposure of the body to cold temperatures results in sympathetic nerve–mediated cutaneous vasoconstriction and reduced blood flow. Histamine is an endothelium-dependent dilator, which relaxes skin vessels. Sweat contains the vasodilator bradykinin and facilitates increased cutaneous blood flow when body temperature increases. Neurogenic tone in the cutaneous circulation is higher than in any other organ system, and its abolition by anesthetics results in marked cutaneous vasodilation.

4. If arterial pressure downstream from a thrombus in the superior mesenteric artery is reduced by 20 mm Hg and the intestine is in the fasting state (i.e., not in the midst of digestion), which of the following would then best represent the condition of the intestinal circulation as compared to its condition during a fasting state?

   A. Decreased blood flow and decreased oxygen delivery to the intestine
   B. Increased blood flow and increased oxygen delivery to the intestine
   C. Decreased blood flow and normal oxygen delivery to the intestine
   D. Decreased blood flow and increased oxygen delivery to the intestine
   E. No change in blood flow or oxygen delivery to the intestine

The correct answer is C. The intestine is a poor autoregulator of blood flow in the fasted state, so flow decreases with a decrease in perfusion pressure. However, oxygen delivery to the intestine is maintained by increasing extraction of oxygen by the intestinal capillaries.

5. Myocardial ischemia and potential infarction following reduction of blood flow in the left anterior descending coronary artery generally occur in the endocardium before the epicardium and are generally more severe in deep layers of the heart than those near its surface. What is the primary reason for these phenomena?

   A. There are no β-adrenergic receptors on arterioles in the endocardium.
   B. The endocardium does not receive blood flow from the left anterior descending coronary artery.
   C. Inotropic stimulation of the endocardium is greater than that in the epicardium.
   D. The epicardium supplied by the left anterior descending coronary artery receives significant collateral blood flow from the right coronary artery.
   E. The lower limit of blood flow autoregulatory capacity in the endocardium is at a higher mean arterial pressure than that for the epicardium.

The correct answer is E. All coronary arteries and arterioles contain beta receptors. The endocardium receives blood flow from the left anterior descending artery (LAD) via septal and other penetrating arteries. There is no known difference between the epi- and endocardium in terms of their response to inotropic stimuli. In humans, there are essentially no collateral arterial connections between the LAD, circumflex, or right coronary arteries to supply blood to any myocardial layer. On the other hand, intramyocardial pressure from muscle contraction and transmission of ventricular blood pressure is greatest in the endocardium of the heart and diminishes considerably toward the epicardium. As a consequence, blood supply to the endocardium is more severely restricted for a longer time in the endocardium during systole, which necessitates a greater dilation of the endocardial arterioles during diastole to supply the oxygen demand of the myocardial tissue. Thus, during diastole, endocardial arterioles are more dilated than those in the epicardium, and their dilatory reserve is used up to a greater extent. The endocardium, therefore, will reach maximum dilation and thence progress into ischemia at a higher mean arterial pressure before the epicardium (usually commencing at ~70 mm Hg compared to ~40 mm Hg for the epicardium).

Clinical Application Exercises 16.1

A 55-year-old male 5′10″ tall weighing 220 lb with a history of high plasma cholesterol and moderate hypertension has come to his family physician complaining of an increasing occurrence of chest pains and "tightness in the middle of my chest" while doing basic yard work. However, he does not experience these pains whenever he is sitting quietly while watching TV or upon arising in the morning after a good night's rest. In the office, the patient's physical exam reveals a blood pressure of 155/93 and heart rate of 85 with no evidence of any heart murmurs or pulmonary congestion. Upon orders from his physician, the patient undergoes a treadmill stress test in which his ECG is monitored continually. The patient's heart rate increases rapidly to over 110 beats/min during the early portion of the test, but upon the first increase in

the slope of the treadmill, his ECG shows ST-segment depression, followed in rapid succession by ECG tracings showing ST-segment elevation. With this change in the ECG, the treadmill test is stopped.

The patient is then scheduled for a coronary angiography. Contrast dye imaging of the patient's coronary circulation during the angiography reveals a 5-mm long stenosis in the proximal portion of the LAD encroaching on 50% of that vessel's lumen. This stenosis is corrected by placement of a coronary stent in the stenotic region and expanding the stent until the artery assumes a normal internal diameter in that region. Six months after this procedure, the patient reports that he no longer has chest pains upon simple physical exertion.

QUESTIONS

1. What is the source of the patient's original chest pain?
2. What is the most likely effect of the patient's LAD stenosis on resting coronary blood flow? Could this effect be related to the lack of chest pain in the patient at rest?
3. What does ST-segment elevation signify in a patient?
4. Why did the patient experience chest pain during physical exertion but not during a resting state?

5. Nitroglycerin is a potent, fast-acting dilator of arteries and veins that is prescribed to patients to prevent exertional angina. However, the ability of nitroglycerin to relieve exertional angina in patients with coronary arterial steno-

sis is unrelated to its ability to directly dilate the coronary circulation and increase myocardial oxygen supply. What physiological mechanism is responsible for this lack of effect of nitroglycerin on the coronary circulation?

ANSWERS

1. The patient's chest pain results from myocardial ischemia, which can occur through any mechanism by which the myocardial oxygen supply cannot meet myocardial oxygen demand. This condition is called angina pectoris.

2. The stenosis will likely have no effect on the patient's resting coronary blood flow because the heart is an effective auto-regulator of blood flow. Any increase in coronary vascular resistance (and potential decrease in coronary blood flow) caused by the stenosis will be counteracted by autoregulatory dilation of downstream arterioles in the vascular bed supplied by the LAD. Thus, total coronary vascular resistance and blood flow will be restored. This is likely in the patient in this case because the patient did not complain of any chest pain while at rest. Such chest pain would have occurred if the stenosis was severe enough to reduce resting blood flow to the LAD region below the metabolic needs of that segment of myocardium. If metabolic demands of the heart can be met by its blood supply, no ischemia or chest pain will occur.

3. The changes in the ST segment during exercise are a cardinal sign of myocardial ischemia and injury.

4. During physical exertion, heart rate and myocardial contractility must increase in order to increase cardiac output to meet the demand for blood flow from exercising skeletal muscles. In addition, in unfit individuals, the increase in heart rate upon physical exertion is greater than in a physically fit individual and such individuals also experience an increase in blood pressure because their skeletal muscle vasodilation with exercise is poor. The increases in heart rate, blood pressure, and contractility all contribute to a great increase in oxygen demand in the heart. This demand can only be met by increasing myocardial blood flow because the heart cannot extract more oxygen from its arterial supply above that which occurs during resting conditions. The increase in myocardial blood flow following increased myocardial activity and

oxygen demand (active hyperemia) occurs normally as the result of metabolic dilation of myocardial arterioles. However, in the patient in this case, a portion of that arteriolar dilating capacity (**coronary reserve**) has been used up in order to compensate for the resistance of the upstream stenosis. Ischemia will occur if the remaining coronary reserve is not sufficient to produce the myocardial blood flow necessary to match myocardial oxygen supply with its increased exercise-induced demand. When this occurs, demand exceeds supply and ischemia with angina occurs.

5. Any increase in oxygen demand by the heart from any source is met by arteriolar dilation of the coronary circulation; thus, oxygen supply is matched with oxygen demand. In moderate coronary artery disease, autoregulation of blood flow in the heart compensates for the increased resistance caused by stenoses in the coronary arteries, and there is no ischemia at rest. However, anything that increases myocardial oxygen demand requires further dilation of coronary arterioles. If the arterioles are driven by this demand to use up their coronary reserve, they reach maximum dilation, that is, they cannot dilate any further. Thus, any more myocardial oxygen demand will exceed oxygen supply by blood flow; ischemia with angina will ensue. Although a potent vasodilator, nitroglycerin cannot relieve angina in a patient by dilating coronary arteries to increase blood supply because if the patient is at the point of experiencing angina, the arteries in the ischemic area are already maximally dilated. Nitroglycerin relieves ischemia by reducing myocardial oxygen demand. By dilating peripheral arteries and veins, nitroglycerin "unloads" the heart by allowing blood to redistribute to the veins and arteries and out of the cardiac chambers. This reduces chamber diameter throughout the cardiac cycle, which reduces myocardial wall stress (via the law of Laplace) and thus myocardial oxygen demand.

the**Point** *Visit* http://thepoint.lww.com/rhoades5e *for additional chapter review Q&A, Clinical Application Exercises, animations, and more!*

Control Mechanisms in Cardiovascular Function

Active Learning Objectives

Upon mastering the material in this chapter, you should be able to:

- Correctly identify changes in cardiovascular variables that can cause hypotension and explain the mechanism of their effect.
- Explain the mechanisms whereby hypotension leads to the activation of the baroreceptor reflex that corrects the hypotension.
- Explain the mechanisms that activate the chemoreceptor reflexes and the CNS ischemic response as they relate to support of blood pressure to support human survival.
- Explain why the baroreceptor reflex and not the chemoreceptor reflex or CNS ischemic reflex is responsible for buffering moment-to-moment changes in mean arterial pressure.
- Explain the neurohumoral mechanisms used in the body to counteract hypotension and how these synergize with neurogenic reflexes to control blood pressure.

- Explain why the baroreceptors do not set the mean arterial pressure in the body.
- Explain the mechanism of pressure diuresis and how this is involved with the kidney's role in setting the long-term level of mean arterial pressure.
- Explain how standing results in a decrease in cardiac output and blood pressure and what mechanisms are used by the body to restore normal blood pressure and cardiac output after standing.
- Explain the mechanisms responsible for progressive shock.
- Correctly identify the initiating causes and characteristic complications of shock caused by each of the following: hemorrhage, severe vomiting, sweating, diarrhea, decreased fluid and electrolyte intake, kidney damage, adrenal cortical destruction, severe burns, intestinal obstructions, general or spinal anesthesia, fever, emotional stress, anaphylaxis, and sepsis.

The mechanisms controlling the cardiovascular system involve individual and cooperative processes among neural, humoral, and local organ control mechanisms. Local vascular control mechanisms were discussed in Chapter 15. This chapter will focus on neural and humoral mechanisms that are primarily involved with the control of central blood volume and arterial pressure. Central blood volume plays a key role in determining cardiac output, whereas the ability of the body to maintain a relatively constant arterial blood pressure ensures proper regulation of organ blood flow.

Neural control of the cardiovascular system involves sympathetic and parasympathetic branches of the autonomic nervous system (ANS). Blood volume and arterial pressure are monitored by stretch receptors in the heart and arteries. Afferent nerve traffic from these receptors is integrated in the medulla oblongata with other afferent information to modulate activity in sympathetic and parasympathetic nerves that adjust heart rate, myocardial contraction, arterial resistance, and venous tone. In this way, cardiac output and systemic vascular resistance (SVR) are adjusted to maintain arterial pressure. This control is further refined by the action of hormones, such as **arginine vasopressin (AVP)** (also called **antidiuretic hormone** or **ADH**), **angiotensin II, aldosterone**, and **atrial natriuretic peptide (ANP),** that participate in the regulation of blood volume through their effects on salt and water balance.

Neural control of cardiac output and SVR plays a larger role in the rapid moment-to-moment regulation of arterial pressure, whereas hormones contribute to longer-term regulatory mechanisms. In some situations, factors other than blood volume and arterial pressure regulation strongly influence cardiovascular control mechanisms. These situations

include the fight-or-flight response, diving under water, thermoregulation, transferring from the supine to standing position, and exercise.

▶ AUTONOMIC NEURAL CONTROL OF THE CARDIOVASCULAR SYSTEM

Neural regulation of the cardiovascular system involves the postganglionic parasympathetic and sympathetic neurons, which are triggered by preganglionic neurons in the brain (parasympathetic) and spinal cord (sympathetic and parasympathetic). Afferent inputs influencing these neurons operate as sensors for arterial pressure and blood volume within locations in the cardiovascular system. They also sense conditions within other organs and the external environment.

Neurogenic control of the heart involves reciprocal activation of parasympathetic and sympathetic nerves.

Autonomic control of the heart and blood vessels was described in the section on the ANS earlier in this text. The heart is innervated by parasympathetic (vagus) and sympathetic nerve fibers (Fig. 17.1). Both types of fibers are tonically active; that is, they exhibit a steady stream of action potential firing at rest. Acetylcholine (ACh) released from parasympathetic fibers binds to muscarinic receptors of the sinoatrial (SA) and atrioventricular nodes as well as the specialized conducting tissues. Stimulation of parasympathetic fibers causes a slowing of the heart rate and a reduction in conduction velocity through the AV node. The ventricular muscle is only sparsely innervated by parasympathetic nerve fibers. Stimulation of these fibers has a negative, but small, inotropic effect. Some cardiac parasympathetic fibers end on

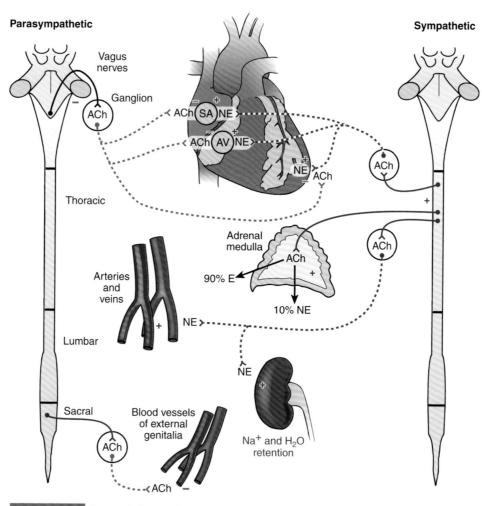

Figure 17.1 **Autonomic innervation of the cardiovascular system.** *Solid lines* represent pre-ganglionic fibers; *dashed lines* represent postganglionic fibers. *Red lines* represent sympathetic nerve pathways. *Blue lines* represent parasympathetic pathways. ACh, acetylcholine; NE, norepinephrine; E, epinephrine; SA, sinoatrial node; AV, atrioventricular node.

sympathetic nerves and inhibit the release of norepinephrine (NE) from sympathetic nerve fibers. Therefore, even in the presence of sympathetic nervous system activity, parasympathetic activation reduces cardiac contractility.

Sympathetic fibers to the heart release NE, which binds to β_1-adrenergic receptors in the SA node, the atrioventricular node, specialized conducting tissues, and cardiac muscle. Stimulation of these fibers cause increased heart rate, increased conduction velocity through the AV node, and a positive inotropic effect (contractility). Activity along the two divisions of the ANS changes in a reciprocal manner to create changes in heart rate. For example, an increase in heart rate is brought about by a simultaneous decrease in parasympathetic and an increase in sympathetic nerve activity to the heart. However, control of heart rate is dominated by parasympathetic effects. Activation of the parasympathetic system can slow the heart even when the sympathetic system is maximally activated. At submaximal sympathetic rates, activation of the vagus nerve can totally suppress the SA node and temporarily cause the heart to stop.

In contrast to the relationships controlling heart rate, control of cardiac contractility is dominated by sympathetic effects; the inotropic state is only minimally affected by vagal activity. Consequently, myocardial contractility is primarily modulated by the level of the activity in the sympathetic nerves to ventricular muscle.

Sympathetic fibers innervate arteries and veins of all the major systemic organs except the brain (see Fig. 17.1). These fibers tonically release NE, which binds to α_1-adrenergic and β_2-adrenergic receptors on blood vessels. However, because the arteries of all vascular beds except the heart and brain contain more α_1- than β_2-adrenergic receptors, activation of the sympathetic nerves to the systemic circulations causes vasoconstriction and an increase in vascular resistance. Circulating epinephrine, released from modified sympathetic nerve endings in the adrenal medulla, binds to α_1-adrenergic and β_2-adrenergic receptors of vascular smooth muscle cells, as well. However, the affinity of both β_1 and β_2 receptors for epinephrine is greater than that for NE. Therefore, at low circulating concentrations, epinephrine essentially activates only β receptors, with the effect of increasing cardiac output (chronotropic and inotropic effects) while *decreasing* SVR.

There is no known parasympathetic innervation of blood vessels in systemic organs with the exception of those of the external genitalia. Postganglionic parasympathetic fibers release ACh and nitric oxide (NO) to blood vessels in the external genitalia. ACh causes the further release of NO from endothelial cells, which results in vascular smooth muscle relaxation and vasodilation. These fibers mediate erection in males and engorgement of the female genitalia during sexual arousal.

Arterial effects of spinal cord injury

The steady train of sympathetic nerve activity, or tone, to blood vessels, the heart, and the adrenal medulla produces a background level of sympathetic vasoconstriction, cardiac stimulation, and adrenal catecholamine secretion in the body. All of these factors contribute to the maintenance of normal blood pressure. This tonic activity is generated by excitatory signals from the medulla oblongata. When the spinal cord is acutely transected and these excitatory signals can no longer reach sympathetic preganglionic fibers, their tonic firing is reduced and blood pressure falls. This effect is known as **spinal shock**. Humans have spinal reflexes of cardiovascular significance. For example, the stimulation of pain fibers entering the spinal cord below the level of a chronic spinal cord transection can cause reflex vasoconstriction and increased blood pressure.

Integrative functions of the medulla and functional cardiovascular centers

The medulla oblongata has three major cardiovascular functions: (1) generating tonic excitatory signals to spinal sympathetic preganglionic fibers, (2) integrating cardiovascular reflexes, and (3) integrating signals from supramedullary neural networks, circulating hormones, and drugs. Specific pools of neurons are responsible for elements of these functions. Neurons in the rostral ventrolateral nucleus (RVL) are normally active and provide tonic excitatory activity to the spinal cord. Specific pools of neurons within the RVL have actions on the heart and blood vessels. RVL neurons are critical in mediating reflex inhibition or activation of sympathetic nerves to the heart and blood vessels.

The cell bodies of cardiac preganglionic parasympathetic neurons are located in the nucleus ambiguus; the activity of these neurons is influenced by reflex input as well as input from respiratory neurons. Respiratory sinus arrhythmia (an oscillating momentary increase in heart rate with inspiration followed by a decreased rate with expiration) is primarily the result of the influence of medullary respiratory neurons that inhibit firing of preganglionic parasympathetic neurons during inspiration and excite these neurons during expiration.

Centers in the hypothalamus and medulla that mediate cardiovascular reflexes to control arterial blood pressure are not anatomically precise locales in those regions of the brain. For that reason, cardiovascular "centers" affecting the heart and blood vessels are often identified by their functional effects. Classically, three cardiovascular control centers have been so identified in the brain. These are (1) the vasomotor center, (2) the cardioaccelerating center, and (3) the cardioinhibitory center. The vasomotor is tonically active and subdivided into a *pressor area*, which causes vasoconstriction, and a *depressor area*, which causes vasodilation by inhibiting the pressor area. The cardioaccelerating center increases heart rate and myocardial contractility by activating sympathetic nerves to the heart. The cardioinhibitory center decreases heart rate and contractility by activating parasympathetic nerves to the heart.

The baroreceptor reflex buffers changes in mean arterial pressure by modulating cardiac output and total peripheral vascular resistance.

The most important reflex controls of the cardiovascular system originate in mechanoreceptors located in the aorta, carotid sinuses, atria, ventricles, and pulmonary vessels. These mechanoreceptors are sensitive to the stretch of the walls of these structures. The firing rate of nerves from these mechanoreceptors increases when the wall is stretched by increased transmural pressure. For this reason, mechanoreceptors in the aorta and carotid sinuses are called **baroreceptors**, or, sometimes, arterial baroreceptors or high-pressure receptors. Mechanoreceptors in the atria, ventricles, and pulmonary vessels primarily sense pressure changes brought about by changes in blood volume. Therefore, these receptors are referred to as volume receptors, low-pressure baroreceptors, or **cardiopulmonary baroreceptors**.

Increased pressure in the carotid sinus and aorta stretches **carotid sinus baroreceptors** and **aortic baroreceptors** and raises their firing rate. Nerve fibers from carotid sinus baroreceptors join the glossopharyngeal (cranial nerve IX) nerves and travel to the NTS. Nerve fibers from the aortic baroreceptors, located in the wall of the arch of the aorta, travel with the vagus (cranial nerve X) nerves to the NTS. The increased action potential traffic reaching the NTS leads to excitation of nucleus ambiguus neurons and inhibition of firing of RVL neurons. This results in increased parasympathetic and decreased sympathetic neural activity to the heart, resistance vessels (primarily arterioles), and veins (Fig. 17.2).

Stimulation of baroreceptor nerves stimulates the vasodepressor area and the cardioinhibitory area while depressing the cardiostimulatory area. The net effect of an increase in BP in the carotid sinus and/or aortic arch, therefore, is to reflexively decrease peripheral vascular resistance, venous tone, heart rate, and ventricular contraction. The combined effects on the veins and heart reduce cardiac output, which, when combined with the decreased SVR, reduces blood pressure back toward normal levels. This completes a negative feedback loop by which an increase in mean arterial pressure can be attenuated and returned to normal. Conversely, decreases in arterial pressure (and decreased stretch of the baroreceptors) increase sympathetic neural activity and decrease parasympathetic neural activity, resulting in increased heart rate, stroke volume, and SVR; this increases blood pressure toward the normal level. If the fall in mean arterial pressure is large, increased sympathetic neural activity to veins is added to the above responses, causing contraction of the venous smooth muscle and reducing venous compliance. Decreased venous compliance shifts blood toward the central circulation thereby increasing right atrial pressure, which increases stroke volume.

The neural reflex activation of the cardiovascular system in response to changes in mean arterial pressure in order to maintain that pressure within narrow limits is called the **baroreceptor reflex**. This reflex is extremely sensitive; the firing rate from nerves exiting the baroreceptors can sense changes in arterial pressure as little as 0.001 mm Hg. These receptors also respond rapidly to the rate of rise in arterial

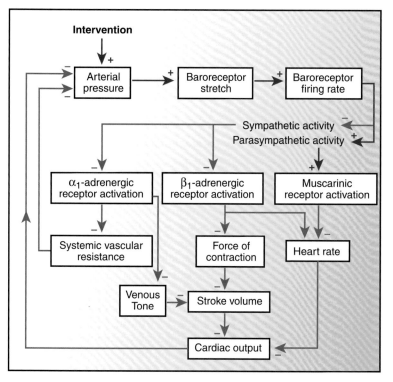

Figure 17.2 **Baroreceptor neural reflex responses to increased arterial pressure.** An intervention that elevates arterial pressure (either mean arterial pressure or pulse pressure), stretches the baroreceptors, and initiates the reflex. The resulting reduced systemic vascular resistance and cardiac output return arterial pressure toward the level existing before the intervention. Hormonal responses (not shown in the figure) involving adrenal epinephrine and the renin–angiotensin system are also involved in the reflex but more so as a response to hypotension rather than to hypertension. *Red (+) arrows* signify that an increase in the variable at the *arrow tail* leads to an increase in the variable at the *arrowhead* (and a decrease at the tail leads to a decrease at the head). *Blue (−) arrows* signify that a decrease in the variable at the *arrow tail* leads to a decrease in the variable at the *arrowhead*.

pressure and changes in pulse pressure; firing rate is greater in systole compared to diastole and greater early rather than later in systole. Increased pulse pressure will activate baroreceptor firing even in the absence of a change in mean arterial pressure.

The baroreceptor reflex also activates hormonal systems affecting blood pressure.

In addition to its effects on autonomic neural outflow to the heart and blood vessels, the baroreceptor reflex modifies the level of hormones that affect arterial pressure. Its most important influence is on the **renin–angiotensin–aldosterone system (RAAS)**. Decreased baroreceptor firing from decreased systemic arterial pressure results in increased sympathetic nerve activity to the kidneys, which, through activation of renal β_2 receptors, causes the kidneys to release renin. Renin converts a precursor peptide, formed in the liver, called angiotensinogen into the peptide angiotensin I. Angiotensin I is then enzymatically cleaved by **angiotensin-converting enzyme (ACE)** in the pulmonary endothelium to produce an active peptide called **angiotensin II**. In addition to neurogenic effects on renal release of renin, a reduction in arterial pressure in the renal artery, specifically in the afferent arteriole, also stimulates renin release. Finally, renin release is stimulated by a systemic decrease in plasma sodium concentration. Angiotensin II is a potent vasoconstrictor and stimulates the release of a steroid hormone called **aldosterone** from the adrenal gland, which causes the kidney to reabsorb salt and water. Activation of RAAS increases vascular resistance and blood volume, ultimately causing blood pressure to rise. The details of the renin–angiotensin–aldosterone system (RAAS) are discussed later in this chapter and in Chapter 23.

Information on the firing rate of the baroreceptors is also projected to the paraventricular nucleus of the hypothalamus, where the release of **arginine vasopressin (AVP;** also called **antidiuretic hormone** or **ADH)** by the posterior pituitary is controlled (see Chapter 23 for more detail). AVP release is increased by a decrease in circulating plasma volume, which decreases the firing rate or atrial low-pressure baroreceptors. AVP is a vasoconstrictor that also activates vasopressin receptors in the kidney to cause an increase in renal reabsorption of water, which in turn increases blood volume. An increase in arterial pressure causes decreased AVP release and increased excretion of water by the kidneys. Hormonal effects on salt and water balance and, ultimately, on cardiac output and blood pressure are powerful, but they occur more slowly (a timescale of many hours to days) than neurogenic effects, which work in seconds to minutes.

Baroreceptor activation is site-, pressure-, and time-dependent.

The effective range of the carotid sinus baroreceptor mechanism is ~40 mm Hg (when the receptor stops firing) to 180 mm Hg (when the firing rate reaches a maximum) (Fig. 17.3). Aortic baroreceptors initiate activation at about 70 mm Hg and reach maximum firing at a higher pressure.

An important property of the baroreceptor reflex is that it adapts during a period of 1 to 2 days to the prevailing mean arterial pressure. When the mean arterial pressure is suddenly raised, baroreceptor firing increases. If arterial pressure is held at the higher level, baroreceptor firing declines during the next few seconds. Firing rate then continues to decline more slowly until it returns to the original firing rate over the next 1 to 2 days. Consequently, if the mean arterial pressure is maintained at an elevated level, the tendency for

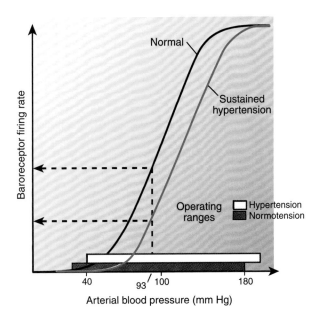

Figure 17.3 **Carotid sinus baroreceptor nerve firing rate and mean arterial pressure.** With normal conditions, a mean arterial pressure of 93 mm Hg is near the midrange of the firing rates for the nerves. Sustained hypertension causes the operating range to shift to the right, putting 93 mm Hg at the lower end of the firing range for the nerves. Aortic baroreceptors show similar relationships, except the point at which pressure activates the receptor and reaches maximum response is higher than that seen in carotid receptors (not shown).

the baroreceptors to initiate a decrease in cardiac output and SVR quickly disappears. This occurs, in part, because of a reduction in the rate of baroreceptor firing for a given mean arterial pressure (see Fig. 17.3). This is an example of **receptor adaptation**. A "resetting" of the reflex in the central nervous system (CNS) occurs as well.

The baroreceptor mechanism can be viewed as a "first line of defense" in the maintenance of normal blood pressure; it makes possible the rapid control of blood pressure needed with changes in posture or acute blood or water loss. However, this control mechanism does not provide for the long-term control of blood pressure.

Baroreceptor reflex in the preservation of cerebral and coronary blood flow during hypotension

By contributing to the control of mean arterial pressure, the baroreceptor reflex indirectly functions to help secure blood flow to two vital organs: the heart and brain. The increase in SVR caused by activation of the baroreceptor reflex whenever arterial pressure drops does not include sympathetic arterial constriction in the heart and brain. The combination of a minimal or no vasoconstrictor effect of sympathetic nerves on cerebral blood vessels and a robust autoregulatory response keep brain blood flow nearly normal despite modest decreases in arterial pressure (see Chapter 15). Activation of sympathetic nerves to the heart causes β_2-adrenergic receptor–mediated dilation of coronary arterioles and β_1-adrenergic receptor–mediated increases in cardiac muscle metabolism (see Chapter 16). The net effect is a marked

increase in coronary blood flow. In summary, when arterial pressure drops, the generalized vasoconstriction caused by the baroreceptor reflex restores blood pressure without vasoconstricting the brain and heart. This, coupled with strong local autoregulatory capacity, prevents, within limits, a reduction in blood flow in the heart and brain whenever blood pressure falls.

Cardiopulmonary baroreceptors sense central blood volume.

Cardiopulmonary baroreceptors are located in the cardiac atria, at the junction of the great veins and atria, in the ventricular myocardium, and in pulmonary vessels. Their nerve fibers run in the vagus nerve to the NTS, with projections to supramedullary areas as well. Unloading (i.e., decreasing the stretch) of the cardiopulmonary receptors by reducing central blood volume results in increased sympathetic nerve activity to the heart and blood vessels and decreased parasympathetic nerve activity to the heart. In addition, the cardiopulmonary reflex interacts with the baroreceptor reflex. Unloading of the cardiopulmonary receptors enhances the baroreceptor reflex, and loading the cardiopulmonary receptors, by increasing central blood volume, inhibits the baroreceptor reflex. Like the arterial baroreceptors, decreased stretch of the cardiopulmonary baroreceptors activates the RAAS and increases the release of AVP; this latter effect plays a role in regulating plasma volume whenever there are significant losses in plasma (~15%).

Chemoreceptors for P_{CO_2}, pH, and P_{O_2} affect mean arterial pressure.

The **carotid** and **aortic bodies** are specialized structures located in the areas of the carotid sinus and aortic arch that sense changes in blood O_2, CO_2, and pH. These structures are sometimes referred to as **chemoreceptors**. The carotid and aortic chemoreceptors are primarily involved with control of ventilation (see Chapter 21), but they also affect the cardiovascular system through neurogenic reflexes. Peripheral chemoreceptors send impulses to the NTS and increase firing rate when the P_{O_2} or pH of the arterial blood is low, the P_{CO_2} of arterial blood is increased, or the flow through the bodies is low or stopped. There are also central medullary chemoreceptors that increase their firing rate primarily in response to elevated arterial P_{CO_2}, which likely reflects a response to a decrease in pH in the brain.

The increased firing of both peripheral and central chemoreceptors leads to profound peripheral vasoconstriction that significantly elevates arterial pressure. If respiratory movements are voluntarily stopped, the vasoconstriction is more intense and a striking bradycardia and decreased cardiac output occur. This response pattern is typical of the diving response (discussed later). As in the case of the baroreceptor reflex, the coronary and cerebral circulations are not subject to sympathetic vasoconstrictor effects and instead exhibit vasodilation as a result of the combination of the direct effect of the abnormal blood gases and local metabolism.

In addition to its importance when arterial blood gases are abnormal, the chemoreceptor reflex is important in the cardiovascular response to severe hypotension. As blood pressure falls, blood flow through the carotid and aortic bodies decreases and chemoreceptor firing increases, probably because of changes in local P_{CO_2}, pH, and P_{O_2}. The chemoreceptor reflex, however, does not respond to a change in blood pressure itself until mean arterial pressure drops to about 80 mm Hg. Therefore, this reflex is not involved in maintenance of normal blood pressure on a moment-to-moment basis but rather serves as a secondary emergency reflex if blood pressure continues to fall in spite of activation of the baroreceptor reflex.

Pain and myocardial ischemia initiate cardiovascular reflexes.

Two reflex cardiovascular responses to pain occur. In the most common reflex, pain causes increased sympathetic activity to the heart and blood vessels, coupled with decreased parasympathetic activity to the heart. These events lead to increases in cardiac output, SVR, and mean arterial pressure. An example of this reaction is the **cold pressor response**, which is the elevated blood pressure that normally occurs from pain associated with placing an extremity in ice water. The increase in blood pressure produced by this challenge is exaggerated in several forms of hypertension.

A second type of response is produced by deep pain. The stimulation of deep pain fibers associated with crushing injuries, disruption of joints, testicular trauma, or distention of the abdominal organs results in diminished sympathetic activity and enhanced parasympathetic activity with decreased cardiac output, SVR, and blood pressure. This hypotensive response contributes to cardiovascular shock from severe trauma (see below).

Myocardial ischemia in the posterior and inferior myocardium causes reflex bradycardia and hypotension. The bradycardia results from increased parasympathetic tone. Dilation of systemic arterioles and veins in this situation is caused by withdrawal of sympathetic tone. This response mimics that following an injection of bradykinin, 5-hydroxytryptamine (serotonin), certain prostaglandins, or various other compounds into the coronary arteries supplying the posterior and inferior regions of the ventricles. This reflex is responsible for the bradycardia and hypotension that can occur in response to acute infarction of the posterior or inferior myocardium.

Higher-order ANS responses alter blood pressure and cardiac output.

The highest levels of organization in the ANS are the supramedullary networks of neurons with centers in the limbic cortex, amygdala, and hypothalamus. These supramedullary networks orchestrate cardiovascular responses to specific patterns of emotion and behavior by their projections to the ANS. Unlike the medulla, supramedullary networks do not contribute to the tonic maintenance of blood pressure, nor are they necessary for most cardiovascular reflexes. However, they do modulate reflex reactivity and can affect the behavior of the heart and systemic circulation.

Fear

On stimulation of certain areas in the hypothalamus, cats demonstrate a stereotypical rage response, with spitting, clawing, tail lashing, and back arching. This is accompanied by the autonomic fight-or-flight response described in Chapter 6. This reaction occurs naturally whenever the cat feels threatened and/or experiences fear. Cardiovascular responses include elevated heart rate and blood pressure. The initial behavioral pattern during the fight-or-flight response includes increased skeletal muscle tone and general alertness. There is increased sympathetic neural activity to blood vessels and the heart. The result of this cardiovascular response is an increase in cardiac output (by increasing both heart rate and stroke volume), SVR, and arterial pressure. There is some evidence that sympathetic cholinergic fibers to the muscle arteries elicit a neurogenic vasodilation in skeletal muscles in lower placental mammals like cats and dogs. However, whether these fibers exist in humans is questionable and it is believed that skeletal muscle vasodilation in the pre-exercise phase of fight-or-flight scenarios in humans is likely due to extraneuronal cholinergic mechanisms that stimulate NO release from the arterial endothelium. When the fight-or-flight response is consummated by actual fight or flight, arterioles in skeletal muscle dilate because of accumulation of local metabolites from the exercising muscles. This vasodilation may outweigh the sympathetic vasoconstriction in other organs, and SVR may actually fall. With a fall in SVR, mean arterial pressure returns toward normal despite the increase in cardiac output.

Emotional stress

Emotional situations often provoke the fight-or-flight response in humans, but it is usually not accompanied by muscle activity and vasodilation (e.g., medical students taking an examination). The massive vasodilation in skeletal muscle associated with exercise, which helps prevent an elevation of blood pressure upon activation of the sympathetic nervous system, is lost when the system is activated by emotional stress alone. For this reason, it has been postulated that repeated elevations in arterial pressure caused by dissociation of the cardiovascular component of the fight-or-flight response from the muscular exercise component are harmful.

Certain emotional experiences induce **vasovagal syncope** (fainting). Stimulation of specific areas of the cerebral cortex can lead to a sudden relaxation of skeletal muscles, depression of respiration, and loss of consciousness. The cardiovascular events accompanying these somatic changes include profound parasympathetic-induced bradycardia and withdrawal of resting sympathetic vasoconstrictor tone. There is a dramatic drop in heart rate, cardiac output, and SVR. The resultant decrease in mean arterial pressure results in unconsciousness (fainting) because of lowered cerebral blood flow. Vasovagal syncope appears in lower animals as the "playing dead" response typical of the opossum.

Exercise

Exercise causes activation of supramedullary neural networks that inhibit the activity of the baroreceptor reflex. The inhibition of medullary regions involved in the baroreceptor reflex is called *central command*. Central command results in withdrawal of parasympathetic tone to the heart, with a resulting increase in heart rate and cardiac output. The increased cardiac output supplies the added requirement for blood flow to exercising muscle. As exercise intensity increases, central command adds sympathetic tone that further increases heart rate and contractility. It also recruits sympathetic vasoconstriction that redistributes blood flow away from splanchnic organs and resting skeletal muscle to exercising muscle. In addition, afferent impulses from exercising skeletal muscle terminate in the RVL, where they further augment sympathetic tone. In spite of augmentation of sympathetic activity during exercise, the local metabolic vasodilator influences in exercising muscle overwhelm any enhanced sympathetic activity to its arteries, resulting in a net vasodilation and hyperemia in the exercising muscle. During exercise, blood flow of the skin is largely influenced by temperature regulation and dilates with an increase in body temperature as described in Chapters 16, 28, and 29.

Diving response

The **diving response** is best observed in seals and ducks, but it also occurs in humans. An experienced diver can exhibit intense slowing of the heart rate (parasympathetic) and peripheral vasoconstriction (sympathetic) of the extremities and splanchnic regions when his or her face is submerged in cold water. With breath holding during the dive, arterial Po_2 falls and pH falls as Pco_2 rises. This activates the chemoreceptor reflex, which reinforces the diving response. The arterioles of the brain and heart do not constrict, and therefore, they preferentially receive the cardiac output. This heart–brain circuit makes use of the oxygen stored in the blood that would normally be used by the other tissues, especially skeletal muscle. Once the diver surfaces, the heart rate and cardiac output increase substantially; vasodilation replaces peripheral vasoconstriction, restoring nutrient flow and washing out accumulated waste products.

Behavioral conditioning

Cardiovascular responses can be conditioned. Both classical and operant conditioning techniques have been used to raise and lower the blood pressure and heart rate of animals. Humans can also be taught to alter their heart rate and blood pressure using a variety of behavioral techniques, such as biofeedback. Behavioral conditioning of cardiovascular responses has significant clinical implications. Animal and human studies indicate that psychological stress can raise blood pressure, increase atherogenesis, and predispose the person to fatal cardiac arrhythmias. These effects are thought to result from activation of the fight-or-flight response in the absence of actual muscle activity. Other studies have shown beneficial effects of behavior patterns designed to introduce a sense of relaxation and well-being such as meditation. Some clinical regimens have employed these factors in the management of cardiovascular disease as an adjunct to pharmacotherapies.

Baroreceptor override

Supramedullary responses can override the baroreceptor reflex. For example, the fight-or-flight response causes the heart rate to rise above normal levels despite a simultaneous rise in arterial pressure. In such circumstances, the neurons connecting the hypothalamus to medullary areas inhibit the baroreceptor reflex and allow the corticohypothalamic response to predominate. Also, during exercise, input from supramedullary regions inhibits the baroreceptor reflex, promoting increased sympathetic tone and decreased parasympathetic tone despite an increase in arterial pressure. Moreover, the various cardiovascular response patterns do not necessarily occur in isolation, as previously described. Many response patterns interact, reflecting the extensive neural interconnections between all levels of the CNS and interaction with various elements of the local control systems. For example, the baroreceptor reflex interacts with thermoregulatory responses. Cutaneous arterial tone is totally dependent on the activity of sympathetic nerves to the arteries and veins in the skin; local, nonneurogenic, regulatory control is minor or absent in the cutaneous circulation. This circulation participates in body temperature regulation but also serve the baroreceptor reflex. At moderate levels of heat stress, the baroreceptor reflex can cause cutaneous arteriolar and venule constriction, to support arterial blood pressure, despite elevated core temperature. However, with severe heat stress, the baroreceptor reflex cannot overcome the neurogenic cutaneous vasodilation; as a result, arterial pressure regulation may fail.

▶ HORMONAL CONTROL OF THE CARDIOVASCULAR SYSTEM

As introduced earlier in this chapter, various hormones play a role in the control of the cardiovascular system. Important hormonal control mechanisms involved in cardiovascular homeostasis include epinephrine from the adrenal medulla, AVP (ADH) from the posterior pituitary gland, renin from the kidney, and ANP from the cardiac atrium.

Circulating epinephrine exerts different cardiovascular effects from those caused by sympathetic nerves.

When the sympathetic nervous system is activated, the adrenal medulla releases epinephrine (>90%) and NE (<10%) into the bloodstream. Changes in the circulating NE concentration from medullary release are small relative to changes in NE resulting from the direct release from nerve endings close to vascular smooth muscle and cardiac cells. Increased circulating epinephrine, however, is significant and contributes to skeletal muscle vasodilation during the fight-or-flight response and exercise. In these cases, epinephrine binds to β_2-adrenergic receptors of skeletal muscle arteriolar smooth muscle cells and causes vasorelaxation. In the heart, circulating epinephrine binds to cardiac cell β_1-adrenergic receptors and reinforces the effect of NE released from sympathetic nerve endings to increase myocardial contractility (and thus SV) and heart rate, which together increase cardiac output.

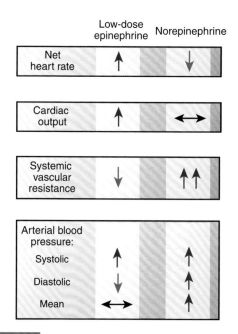

Figure 17.4 **A comparison of the effects of intravenous infusions of low concentrations of epinephrine and norepinephrine.**
↑, increase; ↓, decrease; ↔, no change; ↑↑, marked increase.

A comparison of the responses to infusions of low concentrations of epinephrine and NE illustrates not only the different effects of the two hormones but also the different reflex response each one elicits (Fig. 17.4). Epinephrine and NE have similar direct positive inotropic and chronotropic effects on the heart, but NE elicits a powerful baroreceptor reflex

because it causes significant systemic vasoconstriction that increases SVR and thus mean arterial pressure. The increased arterial pressure reflexively increases parasympathetic tone to the heart, which significantly decreases heart rate and masks some of the direct cardiac effects of NE. The net effect of the cardiovascular systems response to NE is to create little change in cardiac output while increasing SVR along with systolic, diastolic, and mean arterial pressure (i.e., a pressor response). In contrast, low concentrations of epinephrine cause vasodilation in skeletal muscle and splanchnic beds because it preferentially activates the β_2 over the α_1-adrenoceptors on the arterial system of those vascular beds. SVR may actually fall, but the cardiac output increases and mean arterial pressure does not rise. Consequently, the baroreceptor reflex is not elicited, parasympathetic tone to the heart is not increased, and the direct effects of epinephrine to increase heart rate and myocardial contractility are evident. At high concentrations, epinephrine binds to α_1-adrenergic receptors and causes peripheral vasoconstriction; this level of epinephrine is probably never reached except when it is administered as a drug.

Denervated organs, such as transplanted hearts, are hyperresponsive to circulating levels of epinephrine and NE. This increased sensitivity to neurotransmitters is referred to as **denervation hypersensitivity**. Several factors contribute to denervation hypersensitivity, including the absence of sympathetic nerve endings to take up circulating NE and epinephrine actively, leaving more transmitter available for binding to receptors. In addition, denervation results in upregulation of neurotransmitter receptors in target cells. This situation occurs in transplanted hearts, which are denervated because they

CLINICAL FOCUS | 17.1

Nitric Oxide in Cardiovascular Disease

The vascular endothelial nitric oxide (NO) system is responsible for some of the most important and beneficial functions involved with the maintenance and health of the cardiovascular system. Its intense vasodilatory effect antagonizes arterial spasm that might otherwise restrict organ blood flow while it also prevents significant resting hypertension. It has antithrombotic (antiplatelet aggregation) and antiatherogenic properties, attenuates ischemia–reperfusion injury, and suppresses smooth muscle growth after vascular injury while stimulating endothelial wound healing. In addition, it chemically quenches superoxide (O_2^-), which can cause extensive vascular damage.

It is reasonable to postulate that the endothelial NO system might be recruited by the cardiovascular system to counteract the effects of various cardiovascular diseases. Unfortunately, the endothelial NO system has been shown to be impaired in all forms of cardiovascular disease, including acute and chronic hypertension, hyperlipidemia and atherosclerosis, diabetes mellitus, ischemic injury, stroke, vascular transplantations, and heart failure.

As a result of an impaired NO system, people with cardiovascular disease are left with a cardiovascular system that

is prone to vascular spasm and hypertension, thrombosis, atherosclerosis, and stenosis from abnormal vascular wall growth following arterial injury. In addition, NO from the endothelium cannot abate the presence of oxygen radicals, or reactive oxygen species, in the vasculature. Such radicals, which arise from extravascular and intravascular sources, are overproduced in all cardiovascular disease. The enzyme NADP oxidase appears to be a prime source of the radicals. Exposure of vessels to reactive oxygen species damages the arterial endothelium and converts NO synthase to a form that makes oxygen radicals from L-arginine instead of NO.

NO itself seems to be able to attenuate ischemia–reperfusion injury, but widespread use of nitrates as substitutes for endogenous NO has not proven effective in treating many cardiovascular diseases. Instead, additional therapies are being directed to combat the effects of reactive oxygen species on the vasculature. Attenuation of these effects may cause the vascular endothelium to sustain less damage, thereby enhancing their ability to produce NO while at the same time improving the bioavailability of NO to the arterial smooth muscle by removing chemical species that quench any NO produced. ■

have to be removed from a donor. During exercise, circulating levels of NE and epinephrine increase. A transplanted heart cannot get direct inotropic benefit from sympathetic nerves to the heart; they have none. However, because of denervation hypersensitivity, transplanted hearts exhibit enhanced responsiveness to circulating catecholamines, which allow them to perform almost as well as normal hearts.

Renin–angiotensin–aldosterone system supports blood pressure and volume.

The control of total blood volume is involved in regulating arterial pressure. Because changes in total blood volume lead to changes in central blood volume, changes in total blood volume alter ventricular end-diastolic volume and cardiac output, which in turn can alter arterial pressure. Hormonal control of blood volume depends on hormones that regulate salt and water balance as well as red blood cell formation.

Reduced arterial pressure and reduced blood volume cause the release of **renin** from the kidneys. Renin release is mediated by the sympathetic nervous system and by the direct effect of lowered arterial pressure within the afferent arterioles of the kidneys. Renin is a proteolytic enzyme that catalyzes the conversion of angiotensinogen, a plasma protein, to angiotensin I, which, in turn, is converted to angiotensin II in the lung by **angiotensin-converting enzyme (ACE)** (Fig. 17.5). Renin is released during blood loss, even before blood pressure falls, and the resulting rise in plasma angiotensin II from renin release increases the SVR.

Angiotensin II elevates arterial pressure by the following actions: (1) it is a powerful direct arteriolar vasoconstrictor. In some circumstances, it is present in plasma in concentrations sufficient to increase SVR; (2) it potentiates the vasoconstrictive effects of the sympathetic nervous system on blood vessels by enhancing NE release from sympathetic nerve endings, reducing neuronal reuptake of NE, and sensitizing vascular smooth muscle to the actions of NE; (3) it causes the release of **aldosterone** from the adrenal cortex,

which in turn promotes sodium reabsorption in the distal convoluted tubules of the kidney; (4) it reduces sodium excretion by increasing sodium reabsorption by proximal tubules of the kidney; (5) it causes the release of AVP from the posterior pituitary gland, which both promotes water reabsorption from the collecting ducts of the kidney and directly causes systemic arterial vasoconstriction; and (6) it stimulates thirst to promote ingestion of water. Thus, the RAAS promotes elevation of arterial pressure by rapid effects to increase SVR and longer-term effects to promote increases in circulating blood volume through retention of sodium in the body (which indirectly leads to water retention), a reduction in water excretion, and a stimulation of water intake by stimulating thirst.

The RAAS is activated by pathophysiological states.

The RAAS is also involved as a common cause of secondary hypertension (i.e., an increased blood pressure that occurs secondary to a primary pathological condition). Renal artery stenosis, most commonly resulting from atherosclerotic plaques that encroach on the lumen of the main renal arteries, reduces afferent arteriolar pressure in the kidney. This activates the RAAS resulting in elevated renin and angiotensin II levels, which raise arterial pressure.

In patients with congestive heart failure, the reduced output of the failing heart activates the RAAS and increases plasma renin and angiotensin II concentrations. AII stimulates sodium and thus water retention in the body, which augments central venous pressure. This represents the body's attempt to compensate for a failing heart by augmenting ventricular preload to increase output via Starling's law of the heart. However, this action of the RAAS also creates pulmonary edema and congestion as well as an elevation of SVR. This latter effect impairs stroke output of the heart. More importantly, chronic activation of the RAAS stimulates malignant ventricular remodeling resulting in a dilated heart and increased ventricular wall stress. Over time, this leads

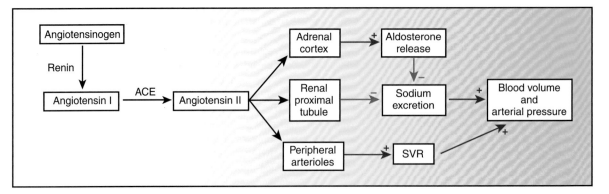

Figure 17.5 **The renin–angiotensin–aldosterone system (RAAS).** This system plays an important role in the regulation of arterial blood pressure and blood volume. Angiotensin II stimulates receptors in the adrenal cortex to produce aldosterone, the renal proximal tubule to promote sodium reabsorption, and arterial smooth muscle to cause vasoconstriction. *Red (+) arrows* signify that an increase in the variable at the *arrow tail* leads to an increase in the variable at the *arrowhead* (and a decrease at the tail leads to a decrease at the head). *Blue (−) arrows* signify that an increase in the variable at the *arrow tail* leads to a decrease in the variable at the *arrowhead* (and a decrease at the tail leads to an increase at the head). ACE, angiotensin-converting enzyme; SVR, systemic vascular resistance.

to a vicious cycle of progressive heart failure. Recognition of the role of RAAS in progressive heart failure has led to the successful use of drugs that inhibit this system as a means of treating all levels of heart failure.

The RAAS also plays a critical role when salt and water intake is reduced. Angiotensin II plays an important role in increasing SVR, as well as blood volume, in people on a low-salt diet. Consequently, if an ACE inhibitor is given to such people, blood pressure falls.

Arginine vasopressin primarily regulates blood volume.

AVP (also called antidiuretic hormone or ADH) is released by the posterior pituitary gland controlled by the hypo-thalamus. AVP is released by increased plasma osmolality, decreased baroreceptor and cardiopulmonary receptor firing (i.e., from a decrease in plasma volume), and various types of stress, such as physical injury or surgery. In addition, circu-lating angiotensin II stimulates AVP release and excess AVP is secreted by lung cancer cells. Although AVP is a vasocon-strictor, it is not ordinarily present in plasma in high enough concentrations to exert an effect on blood vessels. However, in special circumstances (e.g., severe hemorrhage), it prob-ably contributes to increased SVR in support of blood pres-sure. AVP exerts its major effect on the cardiovascular system by causing the retention of water by the kidneys, which is an important part of the neural/humoral mechanisms that regulate blood volume (see Chapter 23).

Stretch-activated release of atrial natriuretic peptide counteracts volume overload.

Atrial natriuretic peptide (ANP) is a 28-amino acid poly-peptide that is synthesized and stored in the atrial muscle cells and released into the bloodstream when the atria are stretched. When central blood volume and atrial stretch are increased, ANP secretion rises, leading to higher sodium excretion and a reduction in blood volume (see Chapter 23). It also inhibits renin release as well as aldosterone and AVP secretion. The role of ANP in controlling blood volume appears to be important in submammalian sea-dwelling animals (e.g., fish), and its role in the day-to-day control of volume and pressure in the cardiovascular system is ques-tionable. However, increased ANP (along with decreased aldosterone and AVP) may be partially responsible for the reduction in blood volume that occurs with prolonged bed rest. Furthermore, ANP is elevated in conditions associ-ated with abnormal retention of fluid in the body such as that which occurs in congestive heart failure (CHF). ANP is known to exist in two subtypes, ANP-A and ANP-B. A blood test for the B type has been developed and is currently used to assess the severity of CHF with ANP-B levels positively correlated with the severity of CHF.

Renal hypoxia stimulates red blood cell production.

The final step in blood volume regulation is production of erythrocytes. **Erythropoietin** is a hormone released by the kidneys in response to either hypoxia or reduced hematocrit (which reduces the total oxygen content in the blood). This hormone causes bone marrow to increase production of red blood cells, raising the total mass of circulating red cells. This can occur when the individual breathes air with a low Po_2 (i.e., at high altitude) for long periods of time or following hemorrhage.

A reduced hematocrit can result not just from frank loss of red blood cells but also from dilution of the existing number of red blood cells in the circulation. For example, an increase in circulating AVP and aldosterone that enhances salt and water retention lowers (dilutes) the hematocrit. The decrease in hematocrit stimulates erythropoietin release, which stimulates red blood cell synthesis and, therefore, bal-ances the increase in plasma volume with a larger red blood cell mass.

Short- and long-term blood pressure control involves different cardiovascular mechanisms.

Different mechanisms are responsible for the short-term and long-term control of blood pressure. Short-term control depends on activation of neurohumoral reflexes and systems as described earlier. A good example of short-term control of blood pressure occurs in the body's response to standing (Fig. 17.6). Upon standing, blood immediately pools in the peripheral veins. This pooling results in an initial decrease in cardiac output and blood pressure that activates neural mechanisms, primarily the baroreceptor reflex that rapidly restores cardiac output and mean arterial pressure. Standing also activates the low-pressure, volume-receptor control mechanisms involving AVP and ANP as well as the RAAS. These mechanisms are not critical in restoring pressure in the short term but can become important if the person is required to stand for long periods of time without moving. The activation of the RAAS on acute standing, however, can confound clinical measurements of plasma renin activity, leading to temporary increases in plasma renin levels that do not reflect the concentration of renin at rest. Therefore, such measurements are obtained with the patient in the supine position. In contrast to standing, contracting muscles in the lower limbs compresses veins and moves blood out of the periphery and into the central circulation thereby having a negative effect on peripheral venous pooling.

None of the neural and humoral mechanisms that are used by the body to control blood pressure in the short term seem to be involved in the long-term setting of mean arterial pressure. It appears that the kidney is responsible for setting the absolute level of mean arterial pressure, about which the neural/humoral mechanisms described above try to control on a moment-to-moment basis. The setting of mean arterial pressure and its long-term control depend on salt and water excretion by the kidneys. Although the excretion of salt and water by the kidneys is regulated by some of the neural and hormonal mechanisms mentioned earlier in this chapter, it is also regulated by arterial pressure. Increased arterial pressure results in increased excretion of salt and water. This phenom-enon is known as **pressure diuresis**. As long as mean arte-rial pressure is elevated, salt and water excretion will exceed the normal rate because of pressure diuresis. Importantly, pressure diuresis persists until it lowers blood volume and

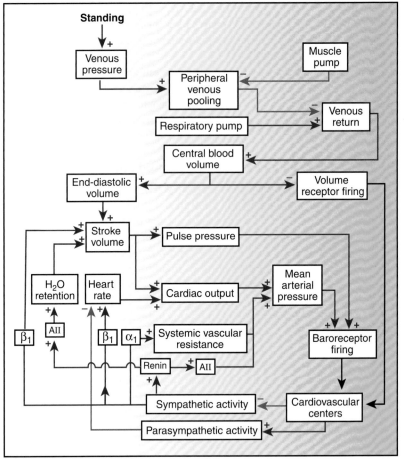

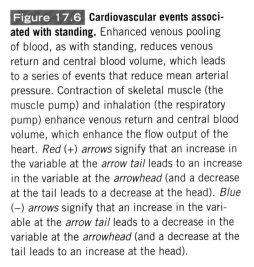

Figure 17.6 **Cardiovascular events associated with standing.** Enhanced venous pooling of blood, as with standing, reduces venous return and central blood volume, which leads to a series of events that reduce mean arterial pressure. Contraction of skeletal muscle (the muscle pump) and inhalation (the respiratory pump) enhance venous return and central blood volume, which enhance the flow output of the heart. *Red (+) arrows* signify that an increase in the variable at the *arrow tail* leads to an increase in the variable at the *arrowhead* (and a decrease at the tail leads to a decrease at the head). *Blue (–) arrows* signify that an increase in the variable at the *arrow tail* leads to a decrease in the variable at the *arrowhead* (and a decrease at the tail leads to an increase at the head).

cardiac output sufficiently to return mean arterial pressure to its original set level. A decrease in mean arterial pressure has the opposite effect on salt and water excretion; reduced pressure diuresis increases blood volume and cardiac output until mean arterial pressure is returned to its original set level.

Pressure diuresis is a slow but persistent mechanism for regulating arterial pressure. It will eventually return pressure to its original set point. In hypertensive patients, salt and water excretion are normal, but at a higher arterial pressure. If this were not the case, pressure diuresis would inexorably bring arterial pressure back to normal.

► CIRCULATORY SHOCK

Circulatory shock is a condition of generalized cardiovascular failure characterized by insufficient organ blood flow. It is also often accompanied by hypotension. This condition can result in deterioration of all tissues in the body and eventual death. In general, if shock is not corrected, it will result in impaired tissue oxygen delivery, generalized muscle weakness, renal failure, depressed mental function, or even unconsciousness. Decreased body temperature resulting from the effects of poor oxygen delivery on tissue metabolism is also a general characteristic of most forms of shock (except that resulting from sepsis; see below). The basic cause of shock is a loss of support of cardiac output. This may arise from direct cardiac dysfunction caused by myocardial

ischemia, infarction, arrhythmias, etc., or from diminished venous filling pressure resulting from loss of whole blood, plasma volume, or venous tone.

Shock is divided into three stages of increasing severity.

Shock can be divided into three, progressively more serious stages, as diagrammed in Figure 17.7. The mildest form of shock is called *nonprogressive* or **compensated shock** because the normal cardiovascular regulatory mechanisms will compensate for the initial decrease in cardiac output and/or arterial pressure. These mechanisms will eventually lead to recovery of the person without the need for clinical intervention. The body's response after an individual donates a unit of blood is a common form of compensated shock.

The compensatory mechanisms in nonprogressive shock are the same as those activated by an acute decrease in blood pressure. These include activation of baroreceptor and other neurogenic pressor reflexes, stimulation of the renin–angiotensin system, and release of AVP. These reflexes and hormones tend to increase blood pressure and cardiac output by increasing heart rate, myocardial contractility, and vascular resistance (especially in skin, splanchnic organs, skeletal muscle, and the kidney) while also promoting renal Na^+ and H_2O retention to increase central venous pressure and stroke volume. In addition, low arterial pressure and increased arterial resistance in the compensated stage of shock reduce

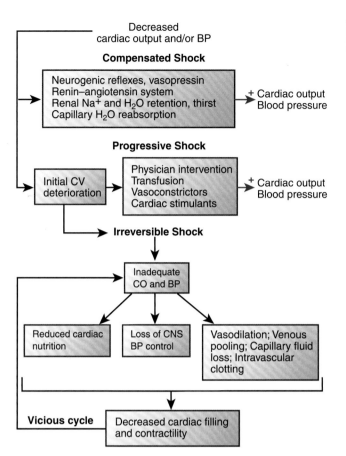

Figure 17.7 Mechanisms of circulatory shock. (See text for details.) BP, blood pressure; CNS, central nervous system; CO, cardiac output; CV, cardiovascular.

capillary hydrostatic pressure. This augments fluid reabsorption from the interstitial fluid, especially in the intestine and kidney. Collectively, these compensatory mechanisms result in the initial clinical presentation of shock, which includes pale/cold skin, rapid pulse, sensation of thirst, hypotension, and reduced urine output.

Cardiovascular Physiology

Progressive shock causes a vicious cycle of cardiac and brain deterioration.

If the initial causes of shock are severe or the body is unable to compensate fully for the malfunctioning cardiovascular system, shock enters into a vicious positive feedback cycle known as **progressive shock**. This occurs when organs maintaining the cardiovascular system, notably the heart and brain, deteriorate as a result of poor blood supply. This usually ensues when, in spite of the efforts of short-term neural/hormonal control mechanisms, blood pressure decreases to a level below the limit of autoregulation in the heart and brain. At this point, because the body's own compensatory mechanisms require proper function of the heart and brain, the body cannot correct the progression of shock. A person in this stage cannot recover without clinical intervention to support the cardiovascular system.

The most important factor involved in the progression of shock is deterioration of the heart itself. During progressive shock, the coronary circulation is compromised and myocardial contractility progressively decreases, which further causes a drop in arterial pressure and further reduction of coronary blood supply. In addition, progressive shock is exacerbated by deterioration of vasomotor centers in the brain, because of impaired cerebral blood flow. These centers are needed to support the heart and increase vascular resistance during hypotension. In addition, during progressive shock, the pre- to postcapillary resistance ratio shifts from that favoring water reabsorption, as in compensated shock, to that favoring water loss through the capillaries. This exacerbates collapse of the circulatory system. Finally, acidotic conditions in peripheral organs, brought on by poor oxygen delivery as well as release of toxins from deteriorating tissues, promote arterial blood clotting, thus exacerbating the already reduced tissue oxygen delivery. Acidosis and toxins also increase capillary permeability, which further enhances fluid loss from the vascular compartment. Without proper intervention, progressive shock will enter into an irreversible

CLINICAL FOCUS | 17.2

Sex Steroids and Cardiovascular Disease in Women

With the exception of reproduction, the physiologic processes presented in most medical physiology textbooks often reflect the basic physiology of the healthy adult male. However, it is widely known that the incidence, morbidity, and mortality associated with cardiovascular disease are much lower in premenopausal women than in men of similar age. After menopause, however, whether surgically induced or from natural processes, women rapidly catch up to men such that by age 60 to 70, their incidence and mortality of cardiovascular disease are as great or greater than it is for men. Currently, cardiovascular disease is the number one source of mortality in women in the United States.

The difference in the prevalence of cardiovascular disease before and after menopause has led to the suggestion that estrogen is cardioprotective in women. Numerous studies and clinical trials have been conducted to investigate this possibility. However, evidence gathered on experimental animals versus that in women in clinical trials, especially in regard to hormone replacement therapy (HRT), has created conflicting insights into the role of estrogens and sex steroids in cardiovascular disease in women. In short, estrogen appears to stimulate the NO system in experimental animals and provide cardiovascular benefits, whereas such benefits appear to be absent in women given HRT. In fact, there is concern that HRT may exacerbate the incidence of morbid

(Continued)

cardiovascular events in postmenopausal women. There is evidence that equine estrogens, other estrogens, or estrogen + progesterone combinations are prothrombotic and precipitate such events.

Estrogen has been shown to be a direct coronary vasodilatory. However, much early evidence to this effect showed that the direct dilatory actions of the steroid occur at dose of 1 μM or more and were thus highly nonphysiologic. Furthermore, other phenol-containing compounds such as plant polyphenolics as well as α-estradiol, the nonactive isomer of physiologically active β-estradiol, are also direct vasodilators at such concentrations. Nevertheless, work conducted over the last decade has uncovered a membrane-bound G-protein–linked estrogen receptor on blood vessels and endothelial cells with some studies revealing that activation of these receptors can lead to vasodilation. More importantly, overnight exposure of coronary arteries to physiologic concentrations of estrogen (1 nM) has been shown to enhance NO-mediated vasodilation or bioavailability and enhance cell mechanisms that inactivate oxygen radicals. Such actions of estrogen would be favorable in reducing the incidence and severity of certain cardiovascular diseases in women.

Nevertheless, past clinical trials testing the effectiveness of hormone replacement therapy on cardiovascular outcomes in postmenopausal women were stopped early because of both a lack of demonstrated efficacy and a slight increase in thrombosis-associated myocardial infarctions associated with postmenopausal women given HRT. Current medical recommendations are that estrogens or HRT not be used in postmenopausal women for potential cardiovascular conditions and instead be reserved for treatment of postmenopausal symptoms or osteoporosis.

An explanation for the extreme difference in suggested benefits of estrogen based on laboratory animal findings and the reality of clinical trials could be rooted in the concept that curatives and preventatives are not the same thing. It is likely that estrogen in premenopausal women protects their cardiovascular system for all the reasons suggested by laboratory studies. However, after menopause, this preventative aspect is lost. Cardiovascular disease then proceeds unabated in a manner similar to that seen in men. Recent laboratory studies support this postulate at least as it relates to the age of an individual receiving estrogen. For example, protective effects of estrogen on inflammatory responses and mitochondrial oxygen radical production are diminished in older versus younger animals. Furthermore, women live decades past menopause, and during that time, many do not take any hormonal replacement therapy. Women in this state likely have accumulated cardiovascular disease that cannot be reversed by the simple addition of estrogen into the woman's system, even though such estrogen prevented the progression of the disease earlier in her life. Indeed, estrogen is slightly prothrombotic, and addition of this steroid to a woman with existing cardiovascular disease, especially coronary artery disease, might be enough to tip the balance into a thrombotic state that causes myocardial infarction. It is interesting that the average age of women was about 65 years in some clinical trials that were stopped because of adverse cardiovascular events associated with estrogen and hormone replacement therapy and included women several years or decades past menopause. Thus, the lack of benefit from estrogen at this stage may have resulted from the buildup of cardiovascular insults prior to taking estrogen rather than from a direct lack of effect of estrogen on the cardiovascular system. Still, additional evidence involving a growing understanding of genomics and metabolomics suggests that not all women may have the genetic/metabolic infrastructure that supports beneficial effects of estradiol on the cardiovascular system. Such women may benefit from more targeted application of specific different types of estrogens, even while their systems are not able to respond favorably to others. In this regard, it has been suggested that estrogen analogs that could target the membrane-bound estrogen receptor, rather than its genomic nuclear receptor, might provide an avenue of providing postmenopausal women with a cardiovascular benefit to estrogen without the other complications of the agent, including potential increased risks of developing cancers. It has also been suggested that SERMs (selective estrogen receptor modulators), such as tamoxifen or raloxifene, which are partial nuclear estrogen receptor agonists, may also provide an avenue for triggering beneficial cardiovascular effects of estrogen while at the same time reducing cancer risks associated with sex steroids. ■

phase. In **irreversible shock**, cardiac and cerebral function are so compromised that no intervention is able to restore normal cardiovascular function. Death is inevitable.

Malfunctions involving the heart, brain, vascular system, or blood volume can cause shock.

Several different conditions can lead to circulatory shock. In addition to blood volume losses resulting from hemorrhage, any loss of plasma volume, or hypovolemia, can reduce cardiac output and induce shock. This can occur from dehydration as a result of severe vomiting, sweating, or diarrhea, as well as from decreased fluid and electrolyte intake, kidney damage, or adrenal cortical destruction (i.e., loss of aldosterone or its effects). Severe burns result in capillary destruction with loss of albumin from the vascular space. This causes transcapillary loss of plasma and a reduction in circulating blood volume. A similar capillary fluid loss is associated with intestinal obstructions that greatly increase intestinal venous pressure. Capillary damage from severe physical injury or trauma also causes transcapillary loss of plasma. An important complication of shock resulting from plasma loss is an increase in blood viscosity, which further impairs the heart's ability to move blood through the peripheral circulation.

Neurogenic shock is a form of circulatory collapse brought about by loss of neurogenic tone to veins and arteries secondary to inhibition or dysfunction of the CNS. This can result from general or spinal anesthesia (i.e., spinal blocks used in childbirth), traumatic brain injury, or depressed vasomotor center function resulting from fever, stress, insomnia, or even severe emotional distress (the last resulting in emotional fainting).

Many people are severely allergic to certain antigens (bee venom, certain foods, etc.). In a serious antigen to antibody reaction, called **anaphylaxis**, tremendous amounts of histamine and other toxins are released into the tissue spaces. This results in destruction of surrounding tissue that increases interstitial osmolality, thus promoting fluid efflux from the capillaries. Histamine also causes arteriolar dilation while increasing the permeability of capillaries and venules to water and proteins. These factors further exacerbate fluid loss from the vascular compartment and result in a decrease in circulating blood volume. A characteristic of anaphylactic shock is that it is not caused by a loss of total body fluid volume but rather from translocation of the volume out of the vascular space into the interstitium.

Septic shock is a form of shock caused by disseminated infection throughout the body. Next to cardiogenic shock, this is the leading cause of death from shock in the United States. It is often brought on by peritonitis secondary to female reproductive organ infection, rupture of the gut during appendicitis, or the introduction of skin bacteria into the bloodstream. Unlike other forms of shock, septic shock is characterized by high body temperature as a result of infection and high cardiac output as a result of intense vasodilation in the infected area. This latter effect is the result of activation of iNOS in vascular smooth muscle in response to endotoxin, LPS, TNF-α, and possibly other cytokines. The excess cardiac output in septic shock represents useless blood flow to infected regions at the expense of perfusion of other organ systems including, potentially, the brain and heart. In addition, tissue destruction in this form of shock produces widespread microvascular blood clots, called *disseminated intervascular coagulation*. This condition greatly impairs O_2 delivery to tissues, thus exacerbating the primary shock condition.

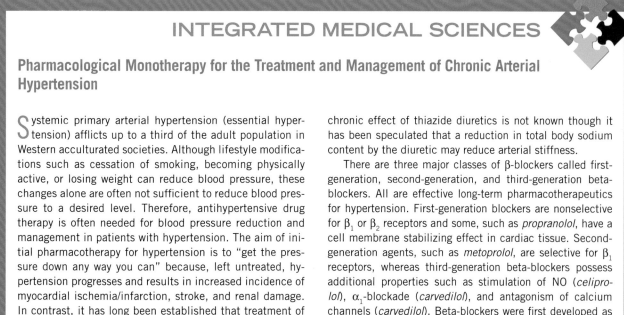

INTEGRATED MEDICAL SCIENCES

Pharmacological Monotherapy for the Treatment and Management of Chronic Arterial Hypertension

Systemic primary arterial hypertension (essential hypertension) afflicts up to a third of the adult population in Western acculturated societies. Although lifestyle modifications such as cessation of smoking, becoming physically active, or losing weight can reduce blood pressure, these changes alone are often not sufficient to reduce blood pressure to a desired level. Therefore, antihypertensive drug therapy is often needed for blood pressure reduction and management in patients with hypertension. The aim of initial pharmacotherapy for hypertension is to "get the pressure down any way you can" because, left untreated, hypertension progresses and results in increased incidence of myocardial ischemia/infarction, stroke, and renal damage. In contrast, it has long been established that treatment of hypertension is of unquestioned benefit.

Currently, there are five types of "first-line" agents recommended as initial monotherapy for the treatment of essential hypertension each of which has a different mechanism of action: thiazide diuretics, β-blockers, ACE inhibitors, angiotensin receptor blockers, and calcium channel antagonists. Thiazide diuretics, such as *chlorothiazide*, produce a mild diuresis and natriuresis. An acute reduction in water volume results along with a decrease in arterial pressure. However, a sustained loss of sodium and water is not compatible with life, and eventually, water and sodium come into balance through activation of the sympathetic nervous system, the RAAS and AVP. This restoration of balance in the presence of chronic diuretic exposure is called diuretic braking. However, even after water and sodium are brought back into balance in the body, blood pressure remains lowered by the diuretic, but at this time, that antihypertensive effect occurs through a reduction in SVR. The mechanism behind this chronic effect of thiazide diuretics is not known though it has been speculated that a reduction in total body sodium content by the diuretic may reduce arterial stiffness.

There are three major classes of β-blockers called first-generation, second-generation, and third-generation beta-blockers. All are effective long-term pharmacotherapeutics for hypertension. First-generation blockers are nonselective for β_1 or β_2 receptors and some, such as *propranolol*, have a cell membrane stabilizing effect in cardiac tissue. Second-generation agents, such as *metoprolol*, are selective for β_1 receptors, whereas third-generation beta-blockers possess additional properties such as stimulation of NO (*celiprolol*), α_1-blockade (*carvedilol*), and antagonism of calcium channels (*carvedilol*). Beta-blockers were first developed as antianginal agents and their antihypertensive action was an unexpected finding revealed in initial clinical trials. The antihypertensive action of these agents is not known completely although inhibition of the RAAS and some central action of the drugs has been implicated. Beta-blockers possess anti-remodeling effects on the heart in CHF and are therefore especially beneficial to patients with the comorbidities of heart failure and hypertension.

Inhibition of the RAAS has an established history as effective pharmacotherapy for chronic hypertension. ACE inhibitors (i.e., *captopril*, *enalapril*, *fosinopril*) have widespread use in patients with hypertension and are effective in patients with either normal or high plasma renin levels. ACE inhibitors are also renal protective and therefore favored in patients with combined diabetes and hypertension. Like beta-blockers, ACE inhibitors prevent ventricular remodeling. This effect, which is added to the ability of ACE inhibitors to reduce elevated plasma volumes, makes ACE inhibitors

(Continued)

a mainstay in treating patients with combined hypertension and CHF. Angiotensin receptor blockers ([ARBs], i.e., *losartan*, *candesartan*, etc.) are newer pharmacological agents that inhibit the effects of the RAAS. They are effective in hypertension, especially in patients with concurrent diabetes. Unlike ACE inhibitors, they do not block inactivation of bradykinin in the lungs and therefore are not associated with the complications of cough and angioedema that can occur with ACE inhibitors. ARBs are potent inhibitors of the RAAS and no amount of stimulation of that system can restore normal tissue responses to angiotensin II.

There are three classes of calcium antagonists currently in use for a variety of cardiovascular diseases including hypertension. The three classes are all effective in treating hypertension but differ in their relative efficacy in inducing vasodilation (an antihypertensive effect) and their cardiac action to reduce myocardial contractility and conduction through the AV node (a property useful in treating myocardial ischemia and supraventricular tachycardia). *Verapamil* and *diltiazem*, respectively, strongly or moderately reduce contractility and conduction, though both are effective antihypertensives. However, members of the 1,4-dihydropyridine class of calcium antagonists (e.g., *nifedipine*, *amlodipine*, *nicardipine*, etc.) are the strongest vasodilators and antihypertensive agents among the calcium antagonists. 1,4-dihydropyridines are therefore the generally favored calcium antagonist for treatment of uncomplicated moderate to severe arterial hypertension. ■

Chapter Summary

- The sympathetic and parasympathetic branches of the autonomic nervous system innervate the heart and act through β-adrenergic and muscarinic cholinergic receptors, respectively.
- Control of heart rate is predominantly by the parasympathetic nervous system, whereas control of contractility of the heart and vascular tone is dominated by the sympathetic nervous system.
- The sympathetic nervous system acts on blood vessels primarily through activation of α-adrenergic receptors.
- Reflex control of the blood pressure involves neurogenic mechanisms that control heart rate, stroke volume, and systemic vascular resistance.
- Behaviors involved in emotion and stress affect cardiovascular responses.

- Baroreceptors and cardiopulmonary receptors are important in the moment-to-moment neural reflex regulation of arterial pressure.
- The renin–angiotensin–aldosterone system, arginine vasopressin, and atrial natriuretic peptide are important in slow hormonal regulation of blood volume and arterial pressure.
- Pressure diuresis in the kidney is the mechanism that ultimately sets the level of mean arterial pressure.
- Circulatory shock occurs in the stages of compensation, progression, and irreversibility.
- Compromised function of the heart and brain as well as abnormalities in the function of blood vessels and blood volume control mechanisms can cause shock.

Chapter Review Questions

1. Which of the following is true with respect to peripheral chemoreceptors?

   A. Activation is important in establishing the set point for mean arterial pressure control.
   B. Activity is increased by increased pH.
   C. They are responsible for moment-to-moment control of mean arterial pressure.
   D. Activation is important in the cardiovascular response to hemorrhagic shock that reduces blood pressure below 80 mm Hg.
   E. They are activated by elevation of the P_{O_2}, of arterial blood.

   The correct answer is D. Carotid chemoreceptors are important in control of blood pressure in emergency situations. They are not involved in setting mean arterial pressure or controlling pressure about that set point during normal activities, such as standing. The chemoreceptors respond to decreased P_{O_2} and pH in the blood.

2. A patient suffers a severe hemorrhagic shock resulting in a lowered mean arterial pressure. Which of the following would be elevated *above normal* levels in the compensatory stage of this shock?

   A. Splanchnic blood flow
   B. Cardiopulmonary receptor activity
   C. Right ventricular end-diastolic volume
   D. Heart rate
   E. Carotid baroreceptor activity

   The correct answer is D. Heart rate would be increased in response to removal of baroreceptor stimulation by the hypotension. The volume loss would reduce ventricular volumes and cardiopulmonary stretch receptor activity in the atria and vena cava. The splanchnic circulation is one of the key components of increased vascular resistance and recruitment of blood in the veins as part of the neural reflex defense against hypotension.

3. Sodium nitroprusside is a powerful arterial vasodilator used in cardiovascular emergencies such as acute heart failure. Which of the following phenomena would be expected to be seen in a patient shortly after being given an IV bolus of sodium nitroprusside?

   A. Tachycardia
   B. Decreased cardiac output
   C. Activation of the vasodepressor area in the medulla
   D. Depressed myocardial contractility with decreased heart rate
   E. Decreased stroke volume with decreased myocardial oxygen demand

   The correct answer is A. A powerful arterial dilator will cause a sudden drop in arterial pressure, which, in turn, will activate the baroreceptor reflex. This reflex will result in activation of sympathetic nerves to the heart and blood vessels and a suppression of vagal activation to the heart. The result, following the initial effect of the nitroprusside, is to increase heart rate, myocardial contractility, and stroke volume. Together, these effects attenuate the initial drop in blood pressure caused by the drug. Depending on the balance between myocardial stimulatory effects on rate and contractility versus the final level of blood pressure, myocardial oxygen consumption may actually increase 1 to 2 minutes after administration of nitroprusside.

4. An older form of treatment for severe chronic hypertension was to use combined pharmacological alpha- and beta-receptor blockade. In this condition, blood pressure regulation in the patient becomes extremely dependent on:

   A. Chemoreceptor reflexes
   B. The central nervous system ischemic response
   C. Factors that alter salt and water balance in the body
   D. Adrenal catecholamines
   E. Vascular resistance

The correct answer is C. Blockade of all adrenergic receptors in the body by combined drug therapy results in a "chemical" sympathectomy. Blood pressure in patients on this therapy becomes exquisitely dependent upon blood volume. Therefore, factors that affect salt and water balance in the body have profound influence over blood pressure (e.g., salt intake, volume depletion, actions of drugs that affect salt and water excretion, and the renin–angiotensin system). The neural arms of the chemoreceptor and central nervous system ischemic reflexes would not function well, if at all, in an individual with chemical sympathectomy. Adrenal or any other catecholamines would have limited effect on the cardiovascular system in which adrenergic receptors were blocked (although the diminution of response would depend on the level of the blockade and the magnitude of catecholamine concentration). The effect of vascular resistance on blood pressure would be markedly blunted by alpha blockade.

Clinical Application Exercises 17.1

MEDICATION-INDUCED POSTURAL HYPOTENSION

A 70-year-old female with a history of angina and mild heart failure is placed on vasodilator therapy with nitroglycerin. She is given this drug as a skin patch, which is to be worn during the day and which releases small amounts of this vasodilator into her circulation continuously through the skin. She is also given nitroglycerin tablets to be taken in case of an anginal attack or as a prophylaxis against angina in anticipation of physical exertion beyond normal daily activities. During one morning, a couple of hours after applying her nitroglycerin skin patch, she takes a nitroglycerin tablet in anticipation of doing some gardening but decides to lie down for a few moments before starting work. Within 10 minutes of lying down, the patient bolts from her supine position to answer her telephone. However, within seconds of rising from the supine to the standing position, she becomes light-headed and dizzy. Her heart rate starts to increase rapidly and she can feel her heart pounding in her chest. Within a couple of steps, her vision narrows and blackens and she collapses onto the floor unconscious. Shortly after this episode, the patient regains consciousness and is able to stand only after slowly arising.

QUESTIONS

1. What cardiovascular phenomenon is most likely responsible for the loss of consciousness in this patient?
2. What changes in blood volume distribution normally occur immediately when one moves from a supine to a standing position? What immediate effects do these have on cardiac output and arterial blood pressure? What reflex mechanisms are brought into play in response to these changes in blood pressure?
3. Nitroglycerin is a rapidly absorbed, quick-acting direct vasodilator that relaxes smooth muscle in veins more than arteries. How does this selective effect of nitroglycerin relate to the responses of the patient suddenly standing?

ANSWERS

1. Loss of consciousness can result from numerous factors, but the most likely related to the cardiovascular system is a sudden severe drop in blood pressure below the autoregulatory limit of the cerebral circulation. This would cause a drop in blood flow and oxygen supply to the brain that would likely induce neurological effects such as darkening vision, dizziness, and fainting.
2. Upon changing from the supine to the standing position, blood tends to fall toward the lower extremities due to the influence of gravity. Blood vessels are compliant with veins being much more compliant and distensible than arteries. Consequently, upon standing, blood tends to pool in the veins of the lower extremities, effectively translocating blood volume away from the central circulation. This causes a sudden drop in ventricular filling and cardiac output along with a drop in arterial pressure. Normally, cardiovascular reflexes activate sympathetic nerves to veins and arteries resulting in venous and arteriolar constriction in response to this initial drop in arterial pressure. Venous constriction helps move blood from the lower extremities into the central circulation and thus supports cardiac output. This output is further augmented by reflex tachycardia upon standing. Stimulation of cardiac output helps prevent any drop in arterial blood pressure when rising to a standing position. Blood pressure is further supported by increased vascular resistance from reflex sympathetic–mediated constriction of systemic arteries in all vascular beds except the heart and brain. This support of blood pressure, along with strong autoregulation of blood flow in the brain, prevents any significant drop in cerebral blood flow that would otherwise occur should the body not be able to compensate for the drop in blood pressure that occurs upon standing against gravity.
3. By directly relaxing smooth muscle in veins and arteries, nitroglycerin antagonizes any constrictor effect on those blood vessels. Because this agent relaxes veins more than arteries, there is a greater reduction in venous than arterial compliance and blood tends to pool more easily in the veins than normal. In addition, the actions of nitroglycerin blunt any reflex sympathetic vasoconstriction of veins in response to standing, preventing those reflexes from counteracting pooling of blood away from the central circulation and into

the lower extremities. The action of nitroglycerin on arteries also antagonizes reflex arterial vasoconstriction during standing. As a result, a patient on nitroglycerin can experience a significant, precipitous drop in arterial pressure when moving suddenly from a supine to a standing position such that blood supply to the brain may become inadequate. If this happens, the patient will likely experience CNS disturbances such as dizziness, loss of vision, and loss of consciousness. Nitroglycerin does not have any direct effect on cardiac or skeletal muscle contraction. Therefore, the drop in arterial pressure seen upon standing while on nitroglycerin will activate neurogenic reflexes that will increase heart rate and myocardial contractility. Therefore, the patient in this study also experienced tachycardia and palpitations upon standing.

thePoint° *Visit* http://thepoint.lww.com/rhoades5e *for additional chapter review Q&A, Clinical Application Exercises, animations, and more!*

18 Ventilation and the Mechanics of Breathing

Active Learning Objectives

Upon mastering the material in this chapter, you should be able to:

- Explain how a pleural pressure is generated.
- Describe how transairway pressure maintains airway patency.
- Explain how changes in alveolar pressure move air in and out of the lungs.
- Explain how spirometry measures lung volumes and airflow in patients.
- Explain why alveolar ventilation measures the amount of fresh air that enters the lung.
- Describe how expired carbon dioxide can be used to measure alveolar ventilation.

- Distinguish between hyperventilation and hyperpnea.
- Explain how lung elastic recoil affects lung compliance.
- Explain how regional lung compliance affects airflow in the lung.
- Explain how surfactant stabilizes alveoli at low lung volumes.
- Explain what keeps the airways from becoming compressed during forced explanation.
- Predict how restrictive and obstructive lung disorders affect the work of breathing.

Breathing is essential to life. The inrush of air at birth sets off a series of events that allows the newborn to progress from a dependent, placental life support system to an independent air breathing system. A breath in and a breath out 12 to 15 times every minute may seem like a simple process to build the entire human respiratory system, which brings in, on an average, 7 L (~1.85 gal) of air per minute into the lungs. At rest, breathing 7 L of air per minute supplies enough oxygen to sustain the metabolism of trillions of cells in the body. However, this simplicity is deceptive because breathing is amazingly responsive to small changes in blood chemistry, mood, level of alertness, and physical activity.

In the previous chapters, you learned that a major function of the cardiovascular system was to distribute blood throughout the body in order to deliver nutrients, oxygen, and other chemicals to the tissues and remove carbon dioxide and other metabolic waste products. In the respiratory chapters, you will learn that the respiratory and cardiovascular systems are closely linked, and the lungs are solely responsible for transferring oxygen from the atmosphere to the blood and removing carbon dioxide from the blood to the atmosphere. You will also learn that the human lungs are so efficiently designed that gas exchange can increase over 20-fold to remove carbon dioxide and to supply oxygen to tissues in order to meet the body's energy demands. The gas exchange process rarely limits body's activity. For example, a marathon runner who staggers across the 26-mile finish line in <3 hours or someone who swims the English Channel in record time is rarely limited by the amount of oxygen

taken up or carbon dioxide removed by the lungs. These examples of human activity not only underscore the functional capacity of the lungs but also illustrate the important role respiration plays in body's extraordinary adaptability to the external environment. Finally, you will learn that respiratory physiology plays a critical role in medicine, because of the impact that many of the respiratory diseases (e.g., **cystic fibrosis**, **asthma**, **chronic obstructive pulmonary disease**, **pulmonary hypertension**, **pulmonary fibrosis**, **pulmonary edema**, **sleep apnea**, and **pneumonia**) have on human health. As a result, respiratory physiology impacts many of the subspecialties including pediatrics, internal medicine, radiology, surgery, ENT, and geriatrics.

Respiration takes place in two stages. The first stage is known as **gas exchange** and the second as **cellular respiration**. Gas exchange involves the transfer of oxygen and carbon dioxide between the atmosphere and the lungs. The second stage, cellular respiration, occurs in two steps. The first involves the exchange of oxygen and carbon dioxide from the cells of the body. The second step is intracellular and involves a series of complex metabolic reactions that burns fuel, releasing carbon dioxide and energy. Recall that oxygen is required in the final step of cellular respiration to serve as an electron acceptor in the process by which cells obtain energy.

The respiratory system can be divided into the following components: ventilation and the mechanics of breathing, gas transfer and transport, pulmonary circulation, and the control of breathing. This chapter discusses ventilation, the mechanics of breathing, and the work of breathing.

Chapter 19 discusses gas uptake and transport. Chapter 20 discusses the pulmonary circulation and the matching of airflow with blood flow in the lungs. Chapter 21 deals with basic breathing rhythms, breathing reflexes, integrated control of breathing, and the control of breathing in unusual environments.

▶ LUNG STRUCTURAL AND FUNCTIONAL RELATIONSHIPS

Although the two lungs are clearly an essential component to the process of breathing, they do not provide the entire picture. The lungs alone cannot bring air in and out of the lungs or exchange oxygen and carbon dioxide from the blood. For example, without respiratory muscles and an airtight chest wall to create a negative pressure within the chest, the lungs would become nonfunctional in gas exchange. Moreover, without blood flow going to the lungs and the matching of airflow with blood flow, there would be essentially no gas exchange.

Airway tree divides repeatedly to increase lung surface area for gas exchange.

The human gas exchange organ consists of two lungs, each divided into several lobes. The lungs comprise two tree-like structures, the vascular tree and the airway tree, which are embedded in highly elastic connective tissue. The vascular tree consists of arteries and veins connected by capillaries (see Chapter 20). The airway tree consists of a series of hollow branching tubes that decrease in diameter at each branching (Fig. 18.1). The main airway, the **trachea**, branches into two

bronchi. Each bronchus enters a lung and branches many times into progressively smaller bronchi, which, in turn, form **bronchioles**.

A functional model of the airway tree is presented in Figure 18.1. The trachea and the first 16 generations of airway branches make up the **conducting zone**. The trachea, bronchi, and bronchioles of the conducting zone have three important functions to (1) warm and humidify inspired air, (2) distribute air evenly to all regions of the lungs, and (3) serve as part of the body's defense system (removal of dust, bacteria, and noxious gases from the lungs). The first four generations of the conducting zone are subjected to changes in negative and positive pressures and contain a considerable amount of cartilage to prevent airway collapse. In the trachea and main bronchi, the cartilage consists of U-shaped rings. Further down, in the lobar and segmental bronchi, the cartilaginous rings give way to small plates of cartilage. In the bronchioles, the cartilage disappears altogether. The smallest airways in the conducting zone are the terminal bronchioles. Bronchioles are suspended by elastic tissue in the lung parenchyma, and the elasticity of the lung tissue helps keep these airways open. The conducting zone has its own separate circulation, the **bronchial circulation**, which originates from the descending aorta and drains into the pulmonary veins. No gas exchange occurs in the conducting zone.

The last seven generations of airways make up the respiratory zone and is the site of gas exchange. The exchange of gases is accomplished in the mosaic of millions of specialized cells that form thin-walled air sacs called **alveoli**. Like the conducting zone, the respiratory zone has its own separate and distinct circulation, the pulmonary circulation.

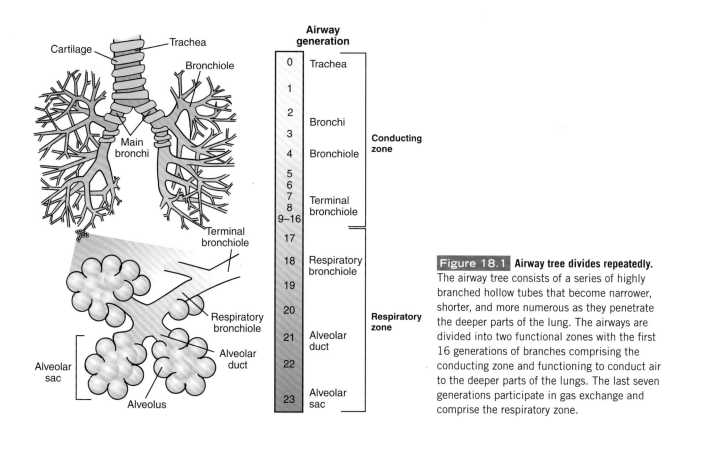

Figure 18.1 **Airway tree divides repeatedly.** The airway tree consists of a series of highly branched hollow tubes that become narrower, shorter, and more numerous as they penetrate the deeper parts of the lung. The airways are divided into two functional zones with the first 16 generations of branches comprising the conducting zone and functioning to conduct air to the deeper parts of the lungs. The last seven generations participate in gas exchange and comprise the respiratory zone.

Respiratory Physiology

The lungs have the most extensive capillary network of any organ in the body. Pulmonary capillaries occupy 70% to 80% of the alveolar surface area. The pulmonary circulation receives all of the cardiac output, and therefore, blood flow is high. One pulmonary arterial branch accompanies each airway and branches with it. Red blood cells can pass through the pulmonary capillaries in <1 second.

The formation of outpockets from the small airways to form alveoli accomplishes an increase in internal surface area (see Fig. 18.1). As mentioned above, a network of capillaries surrounds each alveolus and brings blood into close proximity with air inside the alveolus. Oxygen and carbon dioxide move across the thin-walled alveolus by diffusion. Adult lungs contain 300 to 500 million alveoli, with a combined internal surface area of ~75 m², which is approximately the size of a tennis court. This represents one of the largest biologic membranes in the body. During growth, alveolar surface area increases in two ways: in alveolar number and in diameter, as seen in Table 18.1. However, after adolescence, alveoli only increase in size and, if damaged, have limited ability to repair themselves. Cigarette smoke, for example, can destroy alveoli and lead to concomitant decrease in alveolar surface area for gas exchange.

Vascular and airway trees merge to form a blood–gas interface for gas diffusion.

In the respiratory zone, a group of alveolar ducts and their alveoli merge with pulmonary capillaries to form a terminal respiratory unit; there are ~60,000 of these units in both lungs. The alveolar–capillary membrane of these units forms a blood–gas interface, sometimes referred to as blood–gas barrier that separates the blood in the pulmonary capillaries from the gas in the alveoli (Fig. 18.2). It consists of an alveolar membrane separated by a capillary membrane by interstitial fluid and is essentially a fluid barrier.

The architectural design provides a large surface area for the diffusion of oxygen and carbon dioxide. The alveolar–capillary membrane is exceedingly thin (in some places <0.5 μm) and is composed of alveolar epithelium, interstitial fluid layer, and capillary endothelium. Air is brought to one side of the interface by ventilation—the movement of air to and from the alveoli. Blood is brought to the other side of the interface by the pulmonary circulation. As the blood perfuses the alveolar capillaries, oxygen is taken up and carbon dioxide crosses the blood–gas interface by diffusion.

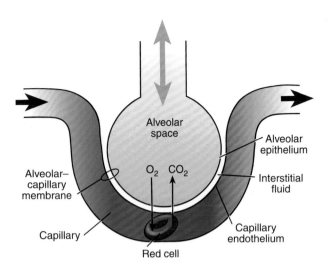

Figure 18.2 Blood–gas interface is the site of gas exchange. The pulmonary capillaries and alveoli form a blood–gas interface. *Thick arrows* indicate direction of blood flow and ventilation, and *thin arrows* indicate the diffusion paths for O_2 and CO_2. The alveolar–capillary membrane is thin (~0.5 mm), and therefore, the diffusion distance for gases is small.

▶ PULMONARY PRESSURES AND AIRFLOW DURING BREATHING

The movement of air in and out of the lungs requires an airtight chest and a set of respiratory muscles. The respiratory muscles include the **diaphragm, intercostal muscles, scalene muscles** of the neck, and the **sternocleidomastoids**, which are inserted into the top of the sternum. The lungs are housed in an airtight thoracic cavity and are separated from the abdomen by the large dome-shaped skeletal muscle, the diaphragm (Fig. 18.3). The thoracic cavity is made up of 12 pairs of ribs, the sternum, and internal and external intercostal muscles, which lie between the ribs. The rib cage is hinged to the vertebral column, allowing it to be raised and lowered during breathing. The space between the lungs and chest wall is the **pleural space**, which contains a thin layer of fluid (~10 μm thick), that functions, in part, as a lubricant so the lungs can slide against the chest wall.

The diaphragm, the main muscle of breathing, expands the thoracic cavity.

Breathing is largely driven by the muscular diaphragm at the bottom of the thorax. Contraction of the diaphragm enlarges the airtight chest cavity with a concomitant, increase in chest volume. When the volume increases in the airtight chest, the pressure decreases causing air to flow into the lungs (and alveoli) (see Fig. 18.3). This enlargement of the airtight thoracic cavity is accomplished in two ways. First, contraction of the diaphragm (which is attached to the lower ribs and sternum) pushes the abdominal contents downward, enlarging the thoracic cavity in the vertical plane. Second, the external intercostal muscles raise the rib cage up and outward, further enlarging the thoracic cavity. The effectiveness of bringing air into the lungs is related to the strength of the contraction of the diaphragm and the intercostal muscles. The more the intercostals

		TABLE 18.1	Age-Related Changes in Alveolar Number and Surface Area in the Human Lung

Age	Number of Alveoli (10^6)	Alveolar Surface Area (m²)	Skin Surface Area (m²)
Birth	24	2.8	0.2
8 y	300	32.0	0.9
Adult	300	75.0	1.8

Inspiration

Expiration

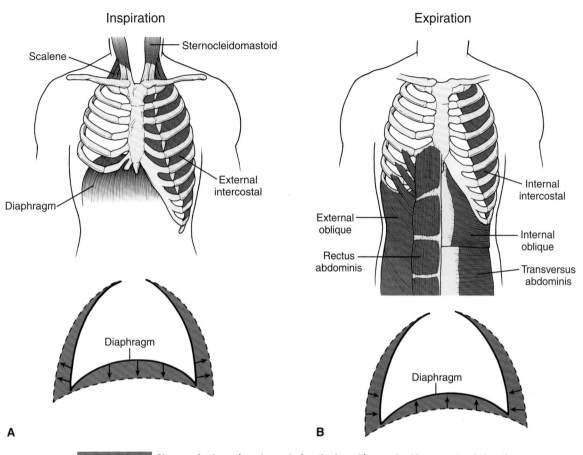

Figure 18.3 **Changes in thoracic volume during the breathing cycle.** Movements of the diaphragm and rib cage change thoracic volume, which allows the lungs to inflate during inspiration and deflate during expiration. **(A)** At rest, during inspiration, the diaphragm contracts and pushes the abdominal contents downward. The downward movement also pushes the rib cage outward. With deep and heavy breathing, the accessory muscles (the external intercostals and sternocleidomastoids) also contract and pull the rib cage upward and outward. **(B)** Expiration is passive during resting conditions. The diaphragm relaxes and returns to its dome shape, and the rib cage is lowered. During forced expiration, however, the internal intercostal muscles contract and pull the rib cage downward and inward. The abdominal muscles also contract and help pull the rib cage downward, compressing thoracic volume.

and diaphragm contract the more the lungs expand. Obesity, pregnancy, and tight clothing around the abdominal wall can impede the effectiveness of the diaphragm in enlarging the thoracic cavity. Damage to the phrenic nerves (the diaphragm is innervated by two phrenic nerves, one to each lateral half) can lead to paralysis of the diaphragm. When a phrenic nerve is damaged, that portion of the diaphragm moves up rather than down during inspiration.

During forced inspiration, in which a large volume of air is taken in, additional accessory muscles are also used. These include the scalene muscles and the sternocleidomastoids, and contraction of these muscles further elevates the upper rib cage and increases the thoracic volume.

The procedure of breathing air out of the lungs is much simpler. During expiration, the process is passive, and the respiratory muscles relax and the lung volume decreases causing pressure in the lungs (and alveoli) to increase. However, with exercise or forced expiration, the expiratory muscles do become active. These muscles include not only

the diaphragm, but also those of the abdominal wall and the internal intercostal muscles (see Fig. 18.3B). Contraction of the abdominal wall pushes the diaphragm upward into the chest, and the internal intercostal muscles pull the rib cage down, reducing thoracic volume. These accessory respiratory muscles are also necessary for such functions as coughing, straining, vomiting, and defecating. The expiratory muscles are extremely important in endurance running and are one of the reasons competitive long-distance runners, as part of their training program, often do exercises to strengthen their abdominal and chest muscles.

Partial pressure drives the diffusion of oxygen and carbon dioxide.

At this point, ventilation has been examined at the systems level. We have reviewed the components, some properties of lung and the chest wall, and the interaction between these structures during inspiration and expiration. A more complete understanding of how the lungs are inflated/deflated and air is inhaled/exhaled, however, requires knowledge of the pressure changes at the biophysical level. A brief review of the gas laws is in order before explaining how changes in pressures within the thoracic cavity and lungs cause changes in lung volumes and air movement.

The atmosphere, the air we breathe and live in, exerts a pressure (P) known as **barometric pressure** (PB). At sea level, PB is equal to 760 mm Hg. The relationship between the total pressure exerted by a mixture of gases and the pressure of individual gases is governed by the **Dalton law**, which states that the total barometric pressure (PB) is equal to the sum of the **partial pressures** of the individual gases and can be written as follows:

$$P_B = P_{N_2} + P_{O_2} + P_{H_2O} + P_{CO_2} \tag{1}$$

where P_{N_2} equals partial pressure of nitrogen, P_{O_2} equals partial pressure of oxygen, P_{H_2O} equals partial pressure of water vapor, and P_{CO_2} equals partial pressure of carbon dioxide. Partial pressure is the individual pressure exerted independently by a particular gas within an air mixture. The air we breathe is a mixture of gases: primarily nitrogen, oxygen, and carbon dioxide. For example, the air you blow into a balloon creates a pressure that causes the balloon to expand. This pressure that is generated is due to all of the molecules of nitrogen, oxygen, and carbon dioxide that collide with the walls of the balloon. However, the total pressure generated inside the balloon is the sum of the individual partial pressure of oxygen, nitrogen, and carbon dioxide. A gas' partial pressure, therefore, is the pressure that the gas exerts if each gas were present alone. Partial pressure is also used to measure how much of that gas is present. To calculate the gas's partial pressure requires rearranging the Dalton law. For example, the partial pressure of oxygen (P_{O_2}), according to the Dalton law, is determined as $P_{O_2} = P_B \times F_{O_2}$, where F_{O_2} is the fractional concentration of oxygen. Because 21% of air is made up of oxygen, the partial pressure (P_{O_2}) exerted by oxygen is 160 mm Hg (760 × 0.21) at sea level. If all of the other gases in the balloon were removed, the remaining oxygen would still exert a pressure of 160 mm Hg. Partial pressure of a gas is often referred to as **gas tension**, and partial pressure and gas tension are used synonymously.

Why is the partial pressure of oxygen inside the airways less than the P_{O_2} in the atmospheric air at sea level? When air is inspired, it is warmed and humidified. The inspired air becomes saturated with water vapor at 37°C. The water vapor exerts a partial pressure that is a function of body temperature, not barometric pressure. At 37°C, water vapor exerts a partial pressure (P_{H_2O}) of 47 mm Hg. Water vapor pressure does not change the percentage of oxygen or nitrogen in a dry gas mixture; however, water vapor does lower the partial pressure of oxygen inside the lungs. The partial pressures of gases in the lungs are calculated on the basis of a dry gas pressure; therefore, water vapor pressure is subtracted when the partial pressure of a gas is determined. The dry gas pressure in the trachea is 760 − 47 = 713 mm Hg, and the individual partial pressures of O_2 and N_2 are:

$$P_{O_2} = 0.21 \times (760 - 47) = 150 \text{ mm Hg} \tag{2}$$

and

$$P_{N_2} = 0.79 \times (760 - 47) = 563 \text{ mm Hg} \tag{3}$$

Table 18.2 lists normal partial pressures of respiratory gases in different locations in the body. When calculating gas tensions in the lung, a good way to remember is as soon as the air hits the nose, always subtract water vapor pressure when converting gas fraction to partial pressure. In Figure 18.2, the schematic depicts mixed venous blood (i.e., blood from organs) coming into the alveolar capillary as blue and the blood leaving the alveolar capillary as red. The color change is due to the partial pressure adding oxygen to hemoglobin. As seen from Table 18.2, these pressures that occur during alveolar gas exchange can be summarized under resting conditions:

Entering Alveolar Capillaries	Alveoli	Leaving Alveolar Capillaries
P_{O_2} = 40 mm Hg	P_{O_2} = 102 mm Hg	P_{O_2} = 95 mm Hg
P_{CO_2} = 46 mm Hg	P_{CO_2} = 40 mm Hg	P_{CO_2} = 40 mm Hg

TABLE 18.2 Partial Pressures and Percentages of Respiratory Gases at Sea Level (PB = 760 mm Hg)

Gas	Ambient Dry Air		Moist Tracheal Air		Alveolar Air		Systemic Arterial Blood	Mixed Venous Blood
	mm Hg	%	mm Hg	%	mm Hg	%	mm Hg	mm Hg
O_2	160	21	150	20	102	14	95	40
CO_2	0	0	0	0	40	5	40	46
Water vapor	0	0	47	6	47	6	47	47
N_2	600	79	563	74	571	75*	571	571
Total	760	100	760	100	760	100	760	704[†]

*Alveolar P_{N_2} increased by 1% because R < 1.
[†]Total pressure in venous blood is reduced because P_{O_2} decreases more than P_{CO_2} increases.

In respiratory physiology, pressures are measured both in mm Hg and centimeters of water (cm H_2O). Gas tensions are measured in mm Hg, but airflow and lung pressures are so small that they are measured in cm H_2O. A pressure of 1 cm H_2O is equal to 0.74 mm Hg (or 1 mm Hg = 1.36 cm H_2O). Changes in lung pressures during breathing are often expressed as *relative pressure*, a pressure relative to atmospheric pressure. For example, the pressure inside the alveoli can be −2 cm H_2O during inspiration. The minus sign indicates that the pressure is subatmospheric, that is, −2 cm below P_B. Conversely, during expiration, the pressure inside the alveoli can be +3 cm H_2O. This means that the pressure is 3 cm H_2O above P_B. A positive or negative pressure indicates that the pressure is relative to atmospheric pressure and is, respectively, above or below P_B. When relative pressures are used, it is important to remember that P_B is set at zero. For example, if airway pressure is zero, the pressure inside the airway equals atmospheric pressure. Unless otherwise specified, the pressures of breathing are relative pressures and the unit is cm H_2O. A list of symbols and abbreviations used in respiratory physiology is shown in Table 18.3.

Pleural pressure is critical for lung inflation and deflation.

Because the thoracic cavity is airtight, an increase in thoracic volume causes the **pleural pressure** (P_{pl}), the pressure in the pleural fluid between the lung and chest wall, to fall. A decrease in P_{pl} causes the lungs to expand and fill with air. This key pressure–volume relationship in breathing is based on two gas laws. The **Boyle law** states that, at a constant

TABLE 18.3   Symbols and Terminology Used in Respiratory Physiology

Symbol	Term
Primary	
C	Compliance
D	Diffusion
F	Fractional concentration of a gas
f	Frequency
P	Pressure or partial pressure
$\dot{Q}$	Volume of blood per unit time (blood flow or perfusion)
R	Resistance
S	Saturation
T	Time
V	Gas volume
$\dot{V}$	Volume of gas per unit time (airflow)
Secondary	
A	Alveolar

Symbol	Term
a	Arterial
Aw	Airway
B	Barometric
D	Dead space
E	Expiratory
I	Inspiratory
L	Lung
c'	Pulmonary end capillary
pl	Pleural
pw	Pulmonary wedge
s	Shunt
T	Tidal
Tp	Transpulmonary
v	Venous
Examples of combinations	
C_L	Lung compliance
D_{LCO}	Lung-diffusing capacity for carbon monoxide
F_{IO_2}	Fractional concentration of inspired O_2
P_B	Barometric pressure
P_{CO_2}	Partial pressure of carbon dioxide
P_A	Alveolar pressure
P_{O_2}	Partial pressure of oxygen
Pa_{CO_2}	Partial pressure of carbon dioxide in arterial blood
P_{ACO_2}	Partial pressure of carbon dioxide in alveoli
P_{IO_2}	Partial pressure of inspired O_2
P_{ECO_2}	Partial pressure of CO_2 in expired gas
$P_{AO_2} - Pa_{O_2}$	Alveolar–arterial difference in partial pressure of O_2
P_{pl}	Pleural pressure
R_{aw}	Airway resistance
Sa_{O_2}	Saturation of hemoglobin with oxygen in O_2 in arterial blood
T_I	Inspiratory time
T_E	Expiratory time
$\dot{V}_A$	Alveolar ventilation
$\dot{V}_A\dot{Q}$	Alveolar ventilation–perfusion ratio
V_D	Dead space volume
$\dot{V}_D$	Dead space ventilation
$\dot{V}_E$	Expired minute ventilation
$\dot{V}_{O_2}$	Oxygen consumption per minute

Note: A dot above a primary symbol denotes flow per unit time.

temperature, the pressure (P) of the gas varies inversely with the volume (V) of gas, or P = 1/V. If either pressure or volume changes and if temperature remains constant, the product of pressure and volume remains constant:

$$P_1 V_1 = P_2 V_2 \qquad (4)$$

The **Charles law** states that if pressure is constant, the volume of a gas and its temperature vary proportionately (V ≈ T). If either temperature or volume changes and pressure remains constant, then:

$$V_1 / T_1 = V_2 / T_2 \qquad (5)$$

These two gas laws can be combined into the **general gas law**:

$$P_1 V_1 / T_1 = P_2 V_2 / T_2 \qquad (6)$$

From the general gas law, at constant temperature, an increase in thoracic volume leads to a decrease in pleural pressure.

Transpulmonary and transairway pressures prevent lung and airway collapse.

In addition to pleural pressure, several other pressures are associated with breathing and airflow (Fig. 18.4). **Alveolar pressure (PA)** is the pressure inside the alveoli. **Transmural pressure** (P_{tm}) is the pressure difference across a wall. In respiration, transmural pressure is the pressure across the airway or across the lung wall or the alveolar wall.

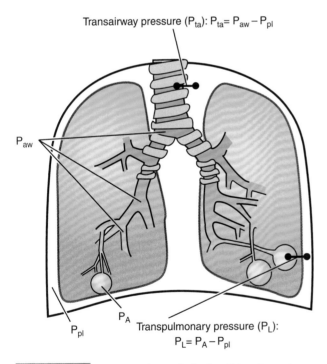

Transairway pressure (P_{ta}): $P_{ta} = P_{aw} - P_{pl}$

P_{aw}

P_A

P_{pl}

Transpulmonary pressure (P_L):
$P_L = P_A - P_{pl}$

Figure 18.4 **Pressures change during breathing.** Several important pressures are involved in breathing: airway pressure (P_{aw}), alveolar pressure (PA), pleural pressure (P_{pl}), transpulmonary pressure (PL), and transairway pressure (P_{ta}). Both transpulmonary pressure and transairway pressure can be defined as the pressure inside minus the pressure outside. In both cases, the pressure outside is pleural pressure (P_{pl}).

Two major transmural pressures are involved in breathing. First, is the pressure difference across the lung wall termed **transpulmonary pressure (PL)**. Transpulmonary pressure describes the difference between the alveolar pressure and the pleural pressure in the lungs and is measured by subtracting pleural pressure from alveolar pressure (PL = PA − P_{pl}). At rest, pleural pressure is −5 cm H_2O (P_{pl} = −5 cm H_2O) and alveolar pressure is zero (PA = 0 cm H_2O). This means, for example, that transpulmonary pressure at rest is 5 cm H_2O [PL = 0 − (−5) = 5 cm H_2O]. It is important to remember that transpulmonary pressure is the pressure that keeps the lungs inflated and prevents the lungs from collapsing. The more positive it becomes, the more the lungs are distended or inflated. In the example above, at rest, the PL is 5 cm H_2O. An increase in PL is responsible for inflating the lungs above the resting volume. The second transmural pressure is **transairway pressure** (P_{ta}), the pressure difference across the airways ($P_{ta} = P_{aw} - P_{pl}$), where P_{aw} is the pressure inside the airway. Transairway pressure is important in keeping the airways open during forced expiration. One way to remember in calculating transairway or transpulmonary pressure is "in minus out," where P_{pl} is always the pressure outside the lung or airway.

Why is pleural pressure negative or subatmospheric? The reason is because of the elastic recoil of the lungs and chest wall. Elastic recoil is analogous to a spring in which the lungs and chest wall when stretched recoils back to their unstretched configuration. At the end of a normal inspiration, the lungs and chest wall are stretched in equal but opposite directions (Fig. 18.5). The stretched lungs have the potential to recoil inwardly, and the stretched chest wall has the potential to recoil outwardly. These two equal but opposing forces cause the pleural pressure to decrease below atmospheric pressure. Pleural pressure is negative or subatmospheric during quiet breathing and becomes more negative with deep inspiration. Only during forced expiration does pleural pressure become positive or rise above atmospheric pressure.

The importance of pleural pressure is seen when the chest wall is punctured (see Fig. 18.5B) and air enters into the pleural space. The stretched lung collapses immediately (recoils inwardly), and the rib cage simultaneously expands outwardly (recoils outwardly). Because the normal pleural pressure is subatmospheric, air will rush into the pleural space any time the chest wall or lung is punctured, and the pleural pressure will become equal to atmospheric pressure because air moves from regions of high to low pressure. In this situation, transpulmonary pressure is zero (PL = 0) because the pressure difference across the lung is eliminated. This condition, in which air or gas accumulates in the pleural space and the lung collapses, is known as **pneumothorax** (see Fig. 18.5B, *right side*). A pneumothorax occurs with a knife or gunshot wound in which the chest wall is punctured or when the lung ruptures from an abscess or severe coughing. In the treatment of some lung disorders (e.g., tuberculosis), a pneumothorax is purposely created by inserting a sterile needle between the ribs and injecting nitrogen into the pleural fluid to rest the diseased lung. It is important to note that the mediastinal membrane keeps the other lung from collapsing.

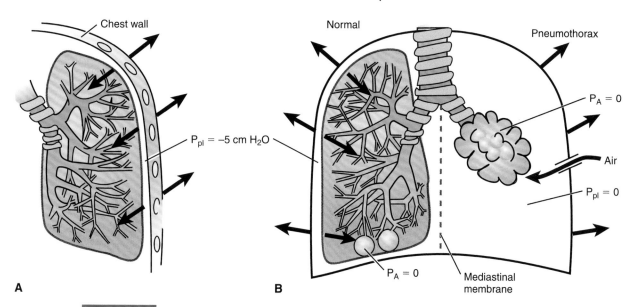

Chest wall

P_{pl} = −5 cm H_2O

A

Normal

Pneumothorax

P_A = 0

Air

P_{pl} = 0

P_A = 0

Mediastinal membrane

B

Figure 18.5 **A negative pleural pressure results from the elastic recoil of the lungs and chest wall pulling in opposite directions. (A)** The stretched lung (at the end of a normal inspiration) tends to recoil inwardly, and the chest wall tends to recoil outwardly, but in equal and opposite directions. Consequently, pleural pressure (P_{pl}) becomes negative (i.e., less than atmospheric pressure). **(B)** Rupture or puncture of the lung or chest wall results in a pneumothorax, during which the transpulmonary pressure becomes zero and elastic recoil causes the lung to collapse. The mediastinal membrane prevents the other lung from collapsing. P_A, alveolar pressure.

Changes in alveolar pressure move air in and out of the lungs.

Pressure changes during a normal breathing cycle are illustrated in Figure 18.6. At the end of expiration, the respiratory muscles are relaxed and there is no airflow. At this point, alveolar pressure is zero (equal to atmospheric pressure or P_B). Pleural pressure is −5 cm H_2O, and transpulmonary pressure is, therefore, 5 cm H_2O [P_{pl} = 0 − (−5 cm H_2O) = 5 cm H_2O].

Inflation of the lungs is initiated by contraction of the diaphragm. If inspiration is started from the end of a maximal expiration, the chest wall can be felt to expand during inhalation. At no time, as our lungs fill, do we feel the need to close our epiglottis to keep the air in. This is because only a slight pressure difference holds air in our lungs. In the example shown in Figure 18.6, pleural pressure goes from −5 to −8 cm H_2O. One of the basic characteristics of gases, such as air, is that the pressures between two regions tend to equilibrate. Therefore, when pleural pressure decreases, transpulmonary pressure increases, and the lungs inflate. Inflation of the lungs causes the alveolar diameter to increase and alveolar pressure to decrease below atmospheric pressure (see Fig. 18.6). This produces a pressure difference between the mouth and alveoli, which causes air to rush into the alveoli. Airflow stops at the end of inspiration because alveolar pressure again equals atmospheric pressure (see Fig. 18.6). The sequence of events is summarized in Figure 18.7.

During expiration, the inspiratory muscles relax, the rib cage drops, pleural pressure becomes less negative, transpulmonary pressure decreases, and the stretched lungs deflate. When alveolar diameter decreases during deflation, alveolar pressure becomes greater than atmospheric pressure and pushes air out of the lungs. Airflow out of the lungs occurs until alveolar pressure equals atmospheric pressure.

▶ SPIROMETRY AND LUNG VOLUMES

Spirometry is the most common of the pulmonary function tests and is used to assess lung function. Specifically, spirometry measures the lung volumes and airflow during inhalation and exhalation. A spirometer displays a *volume–time curve*, showing volume (liters) along the *Y*-axis and time (seconds) along the *X*-axis (Fig. 18.8). The recording from the spirometer is called a **spirogram**.

Spirometry measures specific lung volumes.

The volume of air leaving the lungs during a single breath is called **tidal volume (V_T)**. In the adult lung, under resting conditions, V_T is ~500 mL and represents only a fraction of the air in the lungs. The maximum amount of air in the lungs at the end of a maximal inhalation is the **total lung capacity (TLC)** and is ~6 L in an adult man. Another important spirometry measurement is the **functional residual capacity (FRC)**, the volume of air remaining in the lungs at the end of a normal tidal volume (end of expiration). Note the use of "volume" in the first term and "capacity" in the next two. *Volume* is used when only one volume is involved, and *capacity* is used when a volume can be broken down into two or more smaller volumes; for example, FRC equals **expiratory reserve volume (ERV)** plus **residual volume (RV)**. The various lung volumes and capacities are summarized in Table 18.4.

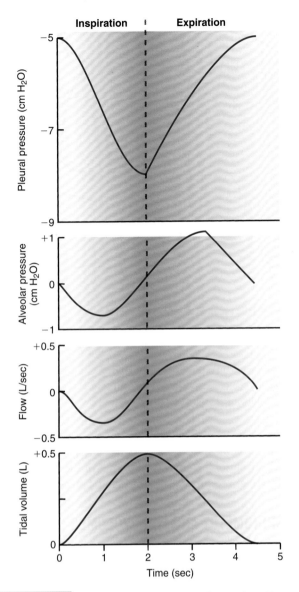

Figure 18.6 **Airflow resistance affects expiratory time.** The inspiratory time (Ti) is 2 seconds and is less than the expiratory time (Te) of 3 seconds. This difference is a result of, in part, a higher airflow resistance during expiration, as is reflected by a higher alveolar pressure (PA) change during expiration (1.2 cm H_2O) than during inspiration (0.8 cm H_2O). An increase in airway resistance will decrease the Ti/Te ratio. Only a small pressure change between the mouth and alveoli is required for a normal tidal volume.

Forced vital capacity is one the most important test in assessing lung function.

The maximum volume of air that can be exhaled after a maximum inspiration is **vital capacity** (**VC**). When expiration is performed as rapidly and as forcibly as possible into a spirometer, this volume is called **forced vital capacity** (**FVC**) and is about 5 L in an adult man (see Fig. 18.8). VC and FVC are the same volume. VC is determined by the sum of ERV, tidal volume, and inspiratory reserve volume. FVC is a direct volume measurement from spirometry and is one of the most useful measurements to assess ventilatory function of the lungs. To measure FVC,

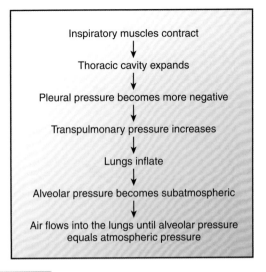

Figure 18.7 **Airflow parallels alveolar pressure during breathing.** The sequence during inspiration results in a fall in alveolar pressure causing air to flow into the lungs.

the person inspires maximally and then exhales into the spirometer as forcefully, rapidly, and completely as possible.

Two additional measurements can be obtained from the FVC spirogram (Fig. 18.9). One is **forced expiratory volume of air exhaled in 1 second** (**FEV$_1$**). This volume has the least variability of the measurements obtained from a forced expiratory maneuver and is considered one of the most reliable spirometry measurements. Another useful way of expressing FEV$_1$ is as a percentage of FVC (i.e., FEV$_1$/FVC × 100), which corrects for differences in lung size. Normally, the FEV$_1$/FVC ratio is 0.8, which means that 80% of a person's FVC can be exhaled in the first second of FVC. This is rather remarkable because small pressure changes are involved to move this volume of air. As you will see later in this chapter FVC and FEV$_1$ are important measurements in the diagnosis of certain types of lung diseases.

A second measurement obtained from the FVC spirogram is forced expiratory flow (FEF$_{25-75}$); it has the greatest sensitivity in terms of detecting early airflow obstruction (see Fig. 18.9). This measurement represents the expiratory flow rate over the middle half of the FVC (between 25% and 75%). FEF$_{25-75}$ is obtained by identifying the 25% and 75% volume points of the FVC and then measuring the volume and time between these two points. The calculated flow rate is expressed in liters per second.

Residual lung volume cannot be measured directly by spirometry.

Because the lungs cannot be emptied completely following forced expiration, neither RV nor FRC can be measured directly by simple spirometry. Instead, they are measured indirectly using a dilution technique involving helium, an inert and relatively insoluble gas that is not readily taken up by blood in the lungs. The subject is connected to a

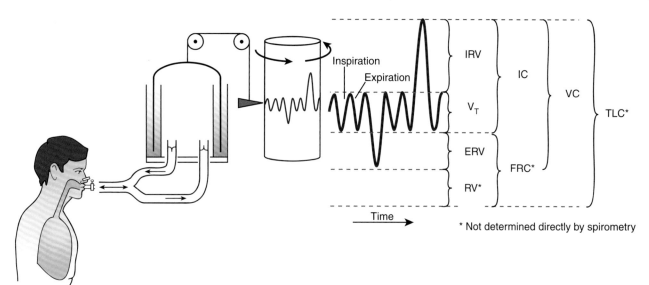

Figure 18.8 **Spirometry measures lung volume.** With expiration, the marker records a downward deflection. Note that residual volume (RV), functional residual capacity (FRC), and total lung capacity (TLC) cannot be measured directly by spirometry. IRV, inspiratory reserve volume; V_T, tidal volume; ERV, expiratory reserve volume; IC, inspiratory capacity; VC, vital capacity.

TABLE 18.4   Abbreviations and Definitions Used in Pulmonary Function

Abbreviation	Term	Definition	Normal Value*
EPP	Equal pressure point	The point at which the pressure inside the airway equals the pressure outside of the airway (i.e., P_{pl})	
ERV	Expiratory reserve volume	The maximum volume of air exhaled at the end of the tidal volume	1.2 L
FEF_{25-75}	Forced expiratory flow	The maximum midexpiratory flow rate, measured by drawing a line between points representing 25% and 75% of the forced vital capacity	5 L/s
FEV_1	Forced expiratory volume	The maximum volume of air forcibly exhaled in 1 s	4.0 L
$FEV_1/FVC\%$	Forced expired volume/ forced vital capacity ratio	The percentage of FVC forcibly exhaled in 1 s	80%
FRC	Functional residual capacity	The volume of air remaining in the lungs at the end of a normal tidal volume	2.4 L
FVC	Forced vital capacity	The maximum volume of air forcibly exhaled after a maximum inhalation	4.8 L
IC	Inspiratory capacity	The maximum volume of air inhaled after a normal expiration	3.6 L
IRV	Inspiratory reserve volume	The maximum volume of air inhaled at the end of a normal inspiration	3.1 L
PEF	Peak expiratory flow	The maximal expiratory flow during an FVC maneuver	7.5 L/s
RV	Residual volume	The volume of air remaining in the lungs after maximum expiration	1.2 L
RV/TLC	Residual volume/total lung capacity ratio	The percentage of total lung capacity made up of residual volume	20%
TLC	Total lung capacity	The volume of air in the lungs at the end of a maximum inspiration	6.0 L
VC	Vital capacity	The maximum volume of air that can be exhaled. (Note that the values for FVC and VC are the same.) VC is calculated from static lung volumes (VC = ERV + VT + IRV). FVC is determined from direct spirometry.	4.8 L
V_D/V_T	Dead space–tidal volume ratio	The fraction of tidal volume made up of dead space	30%
V_T	Tidal volume	The volume of air inhaled or exhaled with each breath	0.5%

*Values are for an average, healthy, young man.

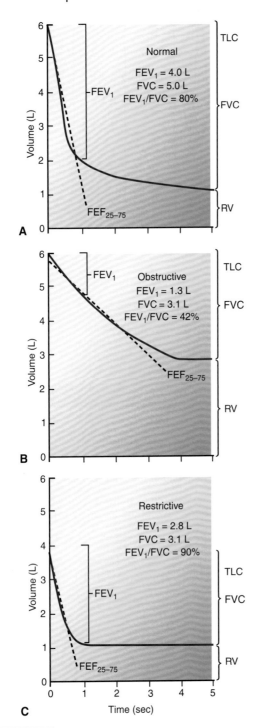

Figure 18.9 Forced vital capacity (FVC) is a useful measurement in assessing ventilatory function. **(A)** A healthy subject inspires maximally to total lung capacity and then exhales as forcefully and completely as possible into the spirometer. Two other measurements can be obtained from this maneuver: the forced expiratory volume in 1 second (FEV_1) and the flow rate over the middle half of the forced vital capacity (FEF_{25-75}). Measurements of FVC, FEV_1, FEV_1/FVC ratio, and FEF_{25-75} are used to detect obstructive and restrictive disorders. **(B)** In an obstructive disorder, expiratory flow rate is significantly decreased, and the FEV_1/FVC ratio is low. **(C)** In a restrictive disorder, lung inflation is decreased, resulting in reduced residual volume (RV) and total lung capacity (TLC). Although FVC is decreased, it is important to note that the FEV_1/FVC ratio is normal or increased in a restrictive disorder.

spirometer filled with 10% helium in oxygen (Fig. 18.10). The lungs initially contain no helium. After the subject rebreathes the helium–oxygen mixture and equilibrates with the spirometer, the helium concentration in the lungs will become the same as in the spirometer. From the conservation of mass principle, we can write:

$$C_1 \times V_1 = C_2(V_1 + V_2) \qquad (7)$$

where C_1 equals the initial concentration of helium in the spirometer, V_1 equals the initial volume of helium–oxygen mixture in the spirometer, C_2 equals helium concentration after equilibration, and V_2 equals unknown volume in the lungs.

$$V_2 = \frac{V_1(C_1 - C_2)}{C_2} \qquad (8)$$

Starting the test at precisely the right time is important. If the test begins at the end of a normal tidal volume (end of expiration), the volume of air remaining in the lungs represents FRC. If the test begins at the end of an FVC, then the test will measure RV. Similarly, if the test starts after a maximal inspiration, then V_2 would equal TLC. In practice, carbon dioxide is absorbed and oxygen is added to the spirometer to make up for the oxygen consumed by the person during the test. Although the helium dilution technique is an excellent test for the measurement of FRC and RV in healthy people, it has a major limitation in patients whose lungs are poorly ventilated because of plugged airways or high airway resistance. In these diseased lungs, helium gives a falsely low FRC value.

▶ MINUTE VENTILATION

So far, ventilation has been described in terms of measurements of static lung volumes and forced lung volumes. Breathing is a dynamic process involving how much air is brought in and out of the lungs in a minute. If 500 mL of air is inspired with each breath (V_T) and the breathing rate (f) is 14 times a minute, then the total **minute ventilation** ($\dot{V}_A$) is the amount of air that enters the lungs each minute ($500 \times 14 = 7,000$ mL/min or 7 L/min). Expired minute ventilation ($\dot{V}_E$) is calculated from the amount of expired air per minute and can be represented by the equation:

$$\dot{V}_E = V_T \times f \qquad (9)$$

Minute ventilation and expired minute ventilation are the same, based on the assumption that the volume of air inhaled equals the volume exhaled. This is not quite true because more oxygen is consumed than carbon dioxide is produced. This difference, for all practical purposes, is ignored.

Not all of the inspired air reaches the alveoli and becomes wasted air.

The tidal volume is distributed between the conducting airways and alveoli. Because gas exchange occurs only in the alveoli and not in the conducting airways, a fraction of the minute ventilation is wasted air. For each 500 mL of air inhaled, ~150 mL remains in the conducting airways and

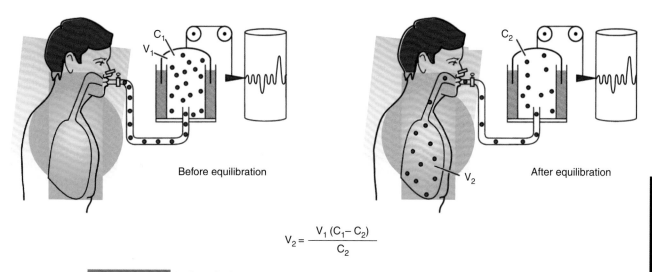

$$V_2 = \frac{V_1(C_1 - C_2)}{C_2}$$

Figure 18.10 **Helium dilution is used to measure residual volume (RV).** *Dots* represent helium before and after equilibration. C, concentration; V, volume.

is not involved in gas exchange (Fig. 18.11). This volume of wasted air is known as **dead space volume (V$_D$)**. Because V$_D$ is a result of the anatomy of the airways, the volume is often referred to as *anatomic V$_D$*.

Picture what occurs during a normal breathing cycle. A normal tidal volume of 500 mL is expired. During the next inspiration, another 500 mL is taken in, but the first 150 mL of air entering the alveoli is V$_D$ (old alveolar gas left behind). Thus, only 350 mL of fresh air reaches the alveoli and 150 mL is left in the conducting airways. The normal ratio of dead space volume to tidal volume (V$_D$/V$_T$) is in the range of 0.25 to 0.35. In this example, the ratio (150:500) is 0.30, which

means that 30% of the tidal volume or 30% of the minute ventilation does not participate in gas exchange and constitutes dead or wasted air.

Dead space air is not confined to the conducting airways alone. Any time gases in the alveoli do not participate in gas exchange; these gases also become part of the wasted air. For example, if inspired air is distributed to alveoli that have no blood flow, this constitutes dead space and is referred to as **alveolar dead space volume** (Fig. 18.12A). Alveolar dead space volume is not confined to alveoli without blood flow. Alveoli that have reduced blood flow exchange less inspired air than normal (see Fig. 18.12B); any portion of alveolar air in excess of that needed to maintain normal gas exchange constitutes alveolar dead space volume. Therefore, dead space volume (V$_D$) may be either anatomic or alveolar in nature. The sum of the two types of dead space is **physiologic dead space volume**. Therefore,

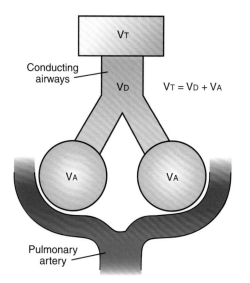

Figure 18.11 **Some of the air in the tidal volume does not participate in gas exchange.** Tidal volume (V$_T$) is represented by the *rectangle* and is the volume of air that will be drawn in during inspiration. V$_T$ will be distributed between the conducting airways and alveoli. The volume of air in the conducting airways does not participate in gas exchange and constitutes dead space volume (V$_D$). V$_A$ is the volume of fresh air added to the alveoli.

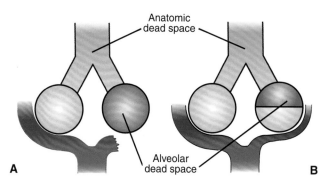

Figure 18.12 **Total wasted air in the lungs is computed from the physiologic dead space volume.** Dead space volume occurs in the conducting airways and in alveoli with poor capillary circulation. **(A)** There is no blood flow to an alveolar region. **(B)** There is reduced blood flow. In both cases, a portion of alveolar air does not participate in gas exchange and constitutes alveolar dead space volume. Note that physiologic dead space is the sum of alveolar dead space plus anatomic dead space.

physiologic V_D = anatomic V_D + alveolar V_D. Physiologic dead space can be measured using the following equation:

$$\frac{V_D}{V_T} = \frac{F_{ACO_2} - F_{ECO_2}}{F_{ACO_2}} \tag{10}$$

Normal values for the V_D/V_T ratio are 0.2 to 0.35. Once the ratio is determined, physiologic dead space can be computed. For example, if a patient has a tidal volume of 500 mL and the V_D/V_T ratio is 0.35, then the physiologic dead space is 175 mL (physiologic V_D = 500 × 0.35). Basically, anatomic V_D, alveolar V_D, and physiologic V_D are terms that denote the volume of inspired gas that does not participate in gas exchange. In one case, there is a fraction of V_T that does not reach the alveoli (anatomic dead space volume). In another, a fraction of the V_T reaches the alveoli, but there is reduced or no blood flow, leading to alveolar dead space volume. Physiologic dead space volume represents the sum of anatomic and alveolar dead space volumes. In healthy people, physiologic V_D is approximately the same as anatomic dead space.

Alveolar ventilation is the amount of fresh air that participates in alveolar gas exchange.

The volume of fresh air per minute actually reaching the alveoli is known as **alveolar ventilation** ($\dot{V}_A$). To determine how much fresh air reaches the alveoli per minute, dead space volume is subtracted from the tidal volume and the result is multiplied by breathing frequency, f, and can be represented by:

$$\dot{V}_A = (V_T - V_D) \times f \tag{11}$$

where V_T equals tidal volume, V_D equals dead space volume, and f equals breathing frequency. For example, if a person has a breathing rate of 14 breaths/min, a V_T of 500 mL, and a V_D of 150 mL, then the volume of air entering the alveoli is 4.9 L/min [(500 − 150 mL) × 14 = 4,900 mL/min]. Only alveolar ventilation represents the amount of fresh air reaching the alveoli, and it is the only air that participates in gas exchange. For instance, a swimmer using a snorkel breathes through a tube that increases dead space volume. Similarly, a patient connected to a mechanical ventilator also has increased dead space volume. Indeed, if minute ventilation is held constant, then alveolar ventilation is decreased with snorkel breathing or with mechanical ventilation.

The significance of dead space volume, minute ventilation, and alveolar ventilation is shown in Table 18.5. Subject C's breathing is slow and deep, B's breathing is normal, and A's

breathing is rapid and shallow. Note that each subject has the same minute ventilation (i.e., the total amount of expired air per minute), but each has marked differences in alveolar ventilation. Subject A has no alveolar ventilation and would die in a matter of minutes, whereas subject C has alveolar ventilation greater than normal. The important lesson from the examples presented in Table 18.5 is that increasing the depth of breathing is far more effective in elevating alveolar ventilation than is increasing the frequency or rate of breathing. A good example is exercise because in most exercise situations, increased alveolar ventilation is accomplished by increases in the depth of breathing rather than in the rate. A well-trained athlete can often increase alveolar ventilation during moderate exercise with little or no increase in breathing frequency (Clinical Focus 18.1).

Alveolar ventilation is determined by measuring the patient's volume of expired carbon dioxide.

Alveolar ventilation is easy to calculate if dead space volume is known. However, dead space volume is not easily determined in a human subject or a patient. Often, dead space is approximated for a seated subject by assuming that dead space (in milliliters) is equal to the subject's weight in pounds (e.g., a subject who weighs 170 lb would have a dead space volume of 170 mL). This assumption is fairly reliable for healthy people but is not in patients with respiratory problems. Alveolar ventilation is calculated in the pulmonary function laboratory from the volume of expired carbon dioxide per minute and fractional concentration of carbon dioxide in the alveolar gas (Fig. 18.13).

This pulmonary function test is based on the concept that (1) no gas exchange occurs in the conducting airways, (2) the inspired air contains essentially no carbon dioxide, and (3) all of the expired carbon dioxide originates from alveoli. Therefore:

$$\dot{V}_{ECO_2} = \dot{V}_A \times F_{ACO_2} \tag{12}$$

where $\dot{V}_{ECO_2}$ equals the volume of carbon dioxide expired per minute and F_{ACO_2} equals the fractional concentration of carbon dioxide in alveolar gas. Rearranging yields the alveolar ventilation equation:

$$\dot{V}_A = \dot{V}_{ECO_2} / F_{ACO_2} \tag{13}$$

The carbon dioxide concentration in the alveoli can be obtained by sampling the last portion of the tidal volume during expiration (end-tidal volume), which contains alveolar gas. Alveolar ventilation can also be determined

Subject	Tidal Volume	×	Frequency	=	Minute Ventilation	−	Dead Space Ventilation*	=	Alveolar Ventilation
	(mL)		(breaths/min)		(mL/min)		(mL/min)		(mL/min)
A	150	×	40	=	6,000	−	150 × 40	=	0
B	500	×	12	=	6,000	−	150 × 12	=	4,200
C	1,000	×	6	=	6,000	−	150 × 6	=	5,100

TABLE 18.5 Effect of Breathing Patterns on Alveolar Ventilation

*Dead space volume is 150 mL in all three subjects.

CLINICAL FOCUS | 18.1

Chronic Obstructive Pulmonary Disease

Chronic bronchitis, emphysema, and asthma are major pathophysiologic disorders of the lungs. Bronchitis is an inflammatory condition that affects one or more bronchi. Emphysema stems from loss of lung elastic recoil and over distention of alveoli. Asthma is marked by spasmodic contractions of smooth muscle in the bronchi. Although the pathophysiology and etiology of chronic bronchitis, emphysema, and asthma are different, they all are classified collectively as **chronic obstructive pulmonary disease (COPD)**. The hallmark of COPD is decreased airflow during forced expiration, which leads to a decrease in **forced vital capacity (FVC)** and **forced expiratory volume in 1 second (FEV$_1$)**.

COPD is characterized by chronic obstruction of the small airways. Airflow can become obstructed in three ways: excessive mucus production (as in bronchitis), airway narrowing caused by bronchial spasms (as in asthma), and airway collapse during expiration (as in emphysema). In the last case, airway collapse stems from abnormally high compliance and concomitant loss of lung elastic recoil. In severe COPD, air is trapped in the lungs during forced expiration, leading to abnormally high residual volume.

Chronic bronchitis and emphysema often coexist. Bronchitis leads to excessive mucus production, which plugs the small airways. A cough is produced to clear the excess mucus from the airways; on repeated exposure to a bronchial irritant, such as tobacco smoke, a persistent cough develops. As a result, the alveoli become over distended and can rupture from excessive pressure, especially in plugged airways.

Late in COPD, patients often experience hypoxemia (low blood oxygen) and hypercapnia (high blood carbon dioxide). These two conditions, especially hypoxemia, cause pulmonary arteries to constrict. Hypoxia-induced pulmonary vasoconstriction leads to further complication by causing pulmonary hypertension (high blood pressure in the pulmonary circulation). If pulmonary hypertension becomes sever, it will cause right heart failure. Based on changes in blood oxygen levels, patients with COPD can be categorized into two types. Those exhibiting predominantly emphysema are referred to as "pink puffers" because their oxygen levels are usually satisfactory and their skin remains pink. They develop a puffing style of breathing to help exhaust the air out of the lungs. Those manifesting predominantly chronic bronchitis are called "blue bloaters" because low oxygen levels give their skin a blue cast and fluid retention from heart failure gives them a bloated appearance.

COPD is by far the most common chronic lung disease in the United States. More than 10 million Americans suffer from COPD, and it is the fifth leading cause of death nationwide. The single most common cause of chronic bronchitis and emphysema is tobacco smoke. ■

from the partial pressure of carbon dioxide in the alveoli (P$_{ACO_2}$) based on the fact that P$_{ACO_2}$ is equal to F$_{ACO_2}$ times total alveolar gas pressure. Equation 13 can be written as:

$$\dot{V}_A\,(L/min) = \frac{\dot{V}_{ECO_2}\,(mL/min) \times 0.863}{P_{ACO_2}\,(mm\,Hg)} \qquad (14)$$

where 0.863 is a constant. Carbon dioxide diffuses readily between the blood and alveoli, and as a result, P$_{ACO_2}$ is in equilibrium with the partial pressure of carbon dioxide in the arterial blood (P$_{aCO_2}$). Recall that a capital "A" signifies alveolar gas and the lower case "a" denotes arterial gas. Because carbon dioxide is in equilibrium between the alveoli and the

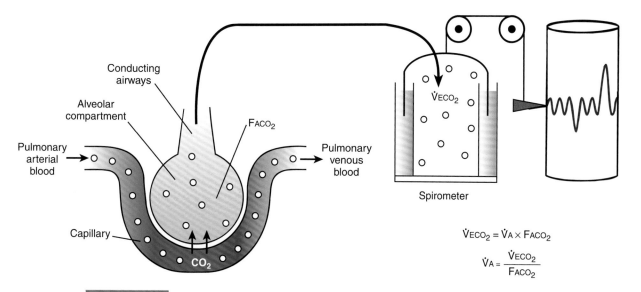

Figure 18.13 **The expired carbon dioxide is used to calculate alveolar ventilation.** Because all of the CO$_2$ (represented by *dots*) in the expired air originates from the alveoli, the fractional concentration of CO$_2$ in alveolar gas (F$_{ACO_2}$) can be obtained at the end of a tidal volume (often referred to as *end-tidal CO$_2$*). $\dot{V}_A$, alveolar ventilation; $\dot{V}_{ECO_2}$, volume of carbon dioxide expired per minute.

arterial blood, arterial carbon dioxide tension (P_{ACO_2}) can be used to calculate alveolar ventilation as follows:

$$\dot{V}_A = \frac{\dot{V}_{ECO_2} \times 0.863}{P_{ACO_2}} \quad (15)$$

From the above equation, it is important to recognize the inverse relationship between $\dot{V}_A$ and P_{ACO_2} (Fig. 18.14). If alveolar ventilation is halved, alveolar P_{CO_2} will double (assuming a steady-state and constant carbon dioxide production). This decrease in alveolar ventilation below normal is called **hypoventilation**. Conversely, increased ventilation, referred to as **hyperventilation**, leads to a fall in blood P_{ACO_2}. Clinically, the adequacy of alveolar ventilation is usually evaluated in terms of arterial P_{CO_2}. An increase in blood P_{ACO_2} reflects hypoventilation, and a decrease in blood P_{ACO_2} reflects hyperventilation. It is important to make the distinction between *hyperventilation* and *hyperpnea*. Remember, hyperventilation is increased alveolar ventilation with a concomitant decrease in arterial P_{CO_2}. **Hyperpnea** is increased alveolar ventilation with *no change* in arterial P_{CO_2}. The reason for this physiologic response is that alveolar ventilation increases proportionally to carbon dioxide production. An example of hyperpnea is seen with exercise. With exercise, the production of CO_2 increases and offsets the increased CO_2 blown off with increased alveolar ventilation.

What happens to P_{AO_2} when alveolar ventilation increases? When alveolar ventilation increases, P_{AO_2} will also increase. However, doubling alveolar ventilation will not lead to a doubling of P_{AO_2}. The quantitative relationship between alveolar ventilation and P_{AO_2} is more complex than that for P_{ACO_2}, for two reasons. First, the inspired P_{O_2} is obviously not zero. Second, the respiratory exchange ratio (R), defined as the ratio of the volume of carbon dioxide exhaled to the volume of oxygen taken up, ($\dot{V}_{CO_2}/\dot{V}_{O_2}$) is usually <1, which means more oxygen is removed from the alveolar gas per

unit time than carbon dioxide is added. The alveolar partial pressure of oxygen (P_{AO_2}) can be calculated by using the alveolar gas equation:

$$P_{AO_2} = P_{IO_2} - P_{ACO_2}[F_{IO_2} + (1 - F_{IO_2})/R] \quad (16)$$

where P_{IO_2} equals the partial pressure of inspired oxygen (moist tracheal air), P_{ACO_2} equals the partial pressure of carbon dioxide in the alveoli, F_{IO_2} equals the fractional concentration of oxygen in the inspired air, and R equals the **respiratory exchange ratio**. When R is 1, the complex term in the brackets equals 1. Also, if a subject breathes 100% oxygen ($F_{IO_2} = 1.0$) for a brief period, the correction factor in the brackets reduces to 1.

In a normal resting person breathing air at sea level with an R of 0.82, a P_{ACO_2} of 40 mm Hg, and a P_{IO_2} of 150 mm Hg, P_{AO_2} is calculated as follows:

$$P_{AO_2} = 150 \text{ mm Hg} - 40 \text{ mm Hg}[0.21 + (1 - 0.21)/0.82]$$
$$P_{AO_2} = 150 \text{ mm Hg} - 40 \text{ mm Hg} \times [1.2] + 102 \text{ mm Hg} \quad (17)$$

Because P_{IO_2} and F_{IO_2} stay fairly constant and P_{ACO_2} equals P_{ACO_2}, the alveolar gas equation can be simplified to:

$$P_{AO_2} = 150 \times 1.2(P_{ACO_2}) \quad (18)$$

The alveolar gas equation for calculating P_{AO_2} is essential to understand any P_{AO_2} value as well as assessing if the lungs are properly transferring O_2 into the alveolar capillary blood.

► ELASTIC PROPERTIES OF LUNG AND CHEST WALL

The lungs, airway tree, and vascular tree are embedded in elastic tissue. When the lungs are inflated, this elastic component is stretched. The degree of lung expansion at any given time in the breathing cycle is proportional to transpulmonary pressure. How well a lung inflates and deflates with a change in transpulmonary pressure depends on its elastic properties. An important feature of an elastic material is that, once stretched, it will recoil to its unstretched position. Therefore, the lung is like a spring; it recoils when stretched.

Elastic recoil of the lungs directly affects inflation and deflation.

When discussing the elastic properties of the lung, there are three basic components that are involved with respiration. These include elastic recoil, stiffness, and lung distensibility. **Distensibility** is the term applied to the ease with which the lungs can be stretched or inflated. Stiffness is defined as resistance to stretch or to inflation. **Elastic recoil** is defined as the ability of a stretched or inflated lung to return to its resting volume (FRC). Elastic recoil of the lung is directly related to lung stiffness, that is, the stiffer the lung, the greater the elastic recoil. An analogy of this relationship is a coiled spring—the more difficult it is to stretch the spring (greater stiffness), the greater the ability to snap back (greater elastic recoil); similarly, a lung that is stiff is more difficult to stretch (inflate), but the inflated lung has a greater ability to recoil back. Elastic recoil plays a key role during expiration in forcing air out of the lungs. Distensibility and elastic recoil, however, are

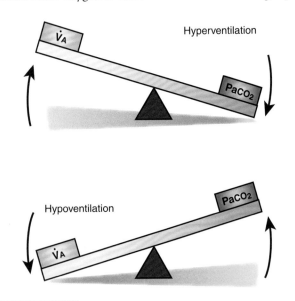

Figure 18.14 Alveolar ventilation and P_{ACO_2} are inversely related. Alveolar ventilation is a key determinant of P_{ACO_2}. In the above figure, the inverse relationship is illustrated. For example, if alveolar ventilation is halved, P_{ACO_2} will double, and conversely, if alveolar ventilation is doubled, P_{ACO_2} will be halved.

inversely related to each other. A lung that is easily inflated has less elastic recoil. Overstretching causes the lungs to lose elastic recoil. Lung distensibility and elastic recoil arise from the elastin and collagen fibers enmeshed around the alveolar walls, adjacent bronchioles, and small blood vessels. Elastin fibers are highly distensible and can be stretched to almost double their resting length. Collagen fibers, however, resist stretch and limit lung expansion at high lung volumes. As the lungs expand during inflation, the fiber network around alveoli, small blood vessels, and small airways unfolds and rearranges—similar to stretching a nylon stocking. When the stocking is stretched, there is not much change in individual fiber length, but the unfolding and rearrangement of the nylon mesh allows the stocking to be easily stretched out to fit the contour of the legs. However, if the nylon stocking is overstretched, it loses its elastic recoil, no longer fits the contour of the legs, and sags or becomes "baggy." In the same way, lungs that lose their elastic recoil also become "baggy." In other words, "baggy lungs" are easy to inflate (stretched) but are difficult to deflate because of their inability to recoil inwardly and force air out of the lungs.

Lung compliance measures distensibility.

Lung distensibility and elastic recoil can be determined from a pressure–volume curve. A simple analogy is the inflation of a balloon with a syringe (Fig. 18.15). For each change in pressure, the balloon inflates to a new volume. The slope of the line of the pressure–volume curve is known as **lung compliance** (C_L). Compliance is a measure of distensibility and is represented by:

$$C_L = \Delta\text{volume}/\Delta\text{pressure} \qquad (19)$$

where Δvolume equals change in volume and Δpressure equals change in pressure (see Chapter 11).

A similar pressure–volume curve can be generated for the human lungs by simultaneously measuring changes in lung volume with a spirometer and changes in pleural pressure with a pressure gauge (Fig. 18.16). Because the esophagus passes through the thorax, changes in pleural pressure can be obtained indirectly from the pressure changes in the esophageal pressure by using a balloon catheter. In practice, a pressure–volume curve is obtained by having the person first inspire maximally to TLC and then expire slowly. During the slow expiration, airflow is periodically interrupted (so that alveolar pressure is zero), and lung volume and pleural pressure are measured. Under these conditions, in which no airflow is occurring, the volume change per unit pressure change ($\Delta V/\Delta P$) is called *static compliance*. Lung compliance is affected by lung volume, lung size, and surface tension inside the alveoli. Because the pressure–volume curve of the lung is nonlinear, compliance is not the same at all lung volumes; it is high at low lung volumes and low at high lung volumes. In the midrange of the pressure–volume curve, lung compliance is about 0.2 L/cm H_2O in adult humans. Lung size affects lung compliance. A mouse lung, for example, has a different compliance than an elephant lung. To allow comparisons among lungs of different sizes, lung compliance is normalized by dividing it by FRC to give a specific compliance.

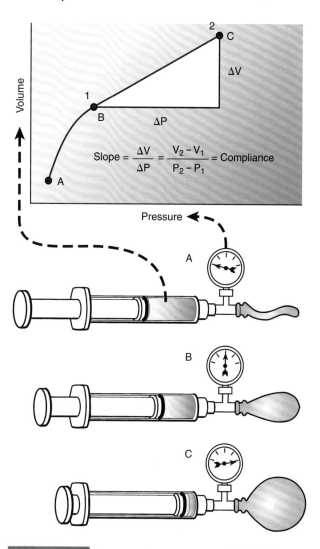

Figure 18.15 **Pressure–volume measurements of a balloon illustrate its elastic properties.** For each change in pressure (shown by movement of the *arrow* on the manometer dial), the balloon inflates to a new volume (plotted on the graph at points A, B, and C). The slope of the line determined by $\Delta V/\Delta P$ between any two points on a pressure–volume curve is compliance. Compliance and elastic recoil are inversely related (C_L = 1/elastic recoil). For example, the lower the compliance, the greater the elastic recoil. P, pressure; V, volume.

What is the significance of an abnormally low or abnormally high compliance? Remember lung compliance is inversely related to lung stiffness (C_L = 1/stiffness). For example, abnormally low lung compliance indicates a stiff lung, which means more work is required to inflate the lungs to bring in a normal tidal volume. An example of abnormally low compliance is seen in a restrictive disorder, a disease state that is characterized by restriction of lung inflation (stiff lungs). The compliance of these lungs is decreased compared with that of a normal lung (Fig. 18.17). Restrictive disorders can be caused by infection or by toxic environmental insults in which the elastic properties are altered. As a result, these lungs become stiffer and lung compliance is decreased. An abnormally high compliance is just as detrimental as a low compliance. Consider the disease **emphysema**, a disease in

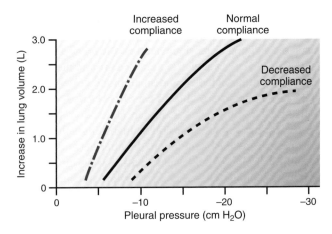

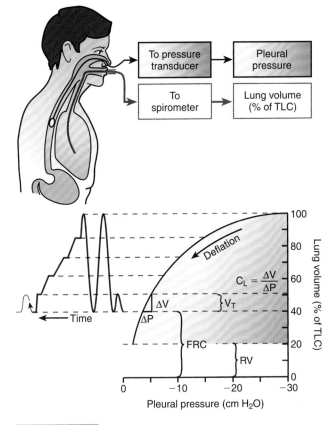

Figure 18.16 Lung compliance is measured from a pressure–volume curve. The subject first inspires maximally to total lung capacity (TLC) and then expires slowly, while airflow is periodically stopped to simultaneously measure pleural pressure and lung volume. Lung compliance (C_L) is measured in L/cm H_2O. Note that pleural pressure is determined by measuring esophageal pressure. FRC, functional residual capacity; P, pressure; RV, residual volume; V_T, tidal volume.

$$C_L = \frac{\Delta V}{\Delta P}$$

Figure 18.17 Obstructive and restrictive diseases alter lung compliance. Patients with a chronic obstructive lung disease, such as emphysema, have abnormally high lung compliance. Patients with restrictive diseases, such as respiratory distress syndrome, have abnormally low lung compliance (see Clinical Focus 18.2).

which lungs have been overstretched from chronic coughing and congested airways. Lungs of patients with emphysema (which is strongly linked to smoking) have a high compliance (high distensibility) and are extremely easy to inflate. However, getting air out again is another matter. Lungs with abnormally high compliance have low elastic recoil, and additional effort is required to force air out of the lungs. Emphysema is part of a broader category termed **chronic obstructive pulmonary disease (COPD)**, which is a diseased state defined as an obstruction of airflow out of the lungs. It is important to remember that lungs with increased compliance have high static lung volumes (increased TLC, FRC, and RV) and low airflow. Therefore, patients with COPD have high lung volumes and retain an abnormally high RV of air.

Elastic recoil of the chest wall affects lung volumes.

Just as the lungs have elastic properties, so does the chest wall. The outward elastic recoil of the chest wall aids lung expansion, whereas the inward elastic recoil of the lungs pulls in the chest wall. The elastic recoil of the chest wall is such that if the chest were unopposed by the recoil of the lung, it would expand to about 70% of TLC. This volume represents the resting position of the chest wall unopposed by the lung.

If the chest wall is mechanically expanded beyond its resting position, it recoils inward. At volumes <70% of TLC, the recoil of the chest wall is directed outward and is opposite to the elastic recoil of the lung. Therefore, the outward elastic recoil of the chest wall is greatest at RV, whereas the inward elastic recoil of the lung is greatest at TLC.

The lung volume at which lung and chest are at equilibrium (i.e., equal elastic recoil but in opposite directions) is represented by FRC. Since the lungs and chest walls are recoiling equally but in opposite directions at FRC, FRC is often referred to as the resting volume of the lungs. A change in the elastic properties of either the lungs or chest wall has a significant effect on FRC. For example, if the elastic recoil of the lungs is increased (i.e., lower C_L), a new equilibrium is established between the lungs and chest wall, resulting in a decreased FRC. Conversely, if the elastic recoil of the chest wall is increased, FRC is higher than normal. The elastic recoil of the chest wall at low lung volumes is a major determinant of RV in young people.

Differences in regional lung compliance cause uneven ventilation.

Lung compliance also causes inspired air to be unevenly distributed in the lungs. In a healthy upright person, compliance at the top part of the lungs is less than that at the base. This difference in compliance between the apex and base, known as **regional compliance**, is caused by gravity (Fig. 18.18). The gravitational effect occurs because lung tissue is ~80% water and gravity exerts a "downward pull," resulting in a lower pleural pressure (i.e., more negative) at the apex than at the base of the lungs. As a result, there is a greater transpulmonary pressure at the apex (see Fig. 18.18A). The higher transpulmonary pressure at the apex causes the alveoli to be more expanded and leads to regional differences in compliance. At any given volume from the FRC and above, the apex of the lung is less distensible (i.e., lower regional compliance) than at the base. This means the base of the lung has both a

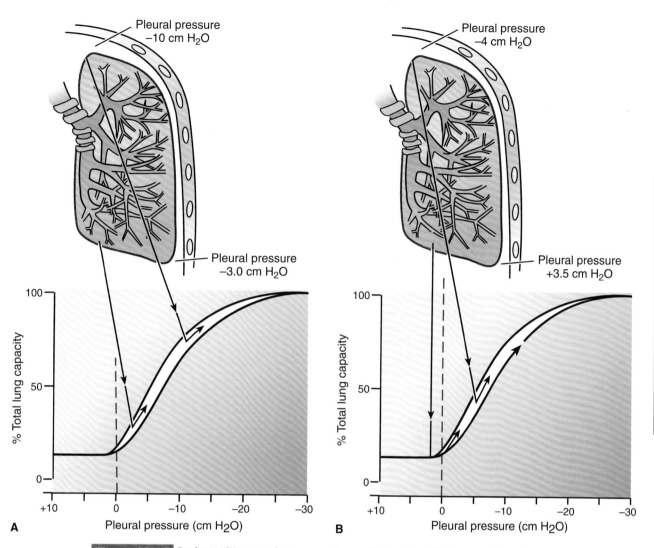

Figure 18.18 **Regional differences in lung compliance affect airflow.** **(A)** Because of gravity, pleural pressure in the upright person is more negative at the apex than at the base of the lungs. Consequently, the base of the lungs is more compliant at functional residual capacity. Because the base of the lung is more distensible, proportionally more of the tidal volume will go to the base with inspiration. **(B)** At low lung volumes (e.g., residual volume), pleural pressure at the base becomes positive. This results in more air going to the top part of the lung during a tidal volume because the apical region of the lung is more compliant.

larger change in volume for the same pressure change and a smaller resting volume than at the apex. In other words, as one takes a breath in from FRC, a greater portion of the tidal volume will go to the base of the lungs, resulting in greater ventilation at the base.

At low lung volumes, alveoli at the apex of the lung are more compliant than that at the base (see Fig. 18.18B). At lung volumes approaching RV, the pleural pressure at the base of the lungs actually exceeds the pressure inside the airways, leading to airway closure. At RV, the base is compressed, and ventilation in the base is impossible until pleural pressure falls below atmospheric pressure. By contrast, the apex of the lung is in a more favorable portion of the compliance curve. Consequently, the first portion of the breath taken in from RV enters the alveoli in the apex, and the distribution of ventilation is inverted at low lung volumes (i.e., the apex is better ventilated than the base).

Alveolar surface tension affects lung compliance.

Another property that significantly affects lung compliance is surface tension, which occurs at the air–liquid interface of the alveoli. The surface of the alveolar membrane is moist, contains water molecules, and is in contact with air, producing a large air–liquid interface. Surface tension increases as water molecules come closer together, which is what happens when the lungs are deflated and the alveoli become smaller (like air leaving a balloon). Potentially, surface tension can cause alveoli to collapse, which would create two serious problems. First, no gas exchange would occur in collapsed alveoli. Second, collapsed alveoli would lead to labored breathing because it is difficult to re-expand the collapsed alveoli during inflation. Fortunately, alveoli don't collapse and inhalation is relatively easy because the lung produces a surface-reducing agent that coats the alveolar lining called **pulmonary surfactant**.

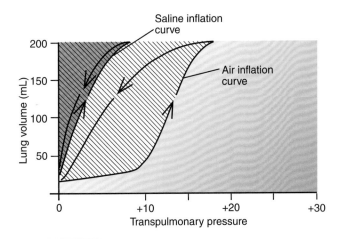

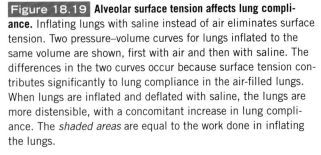

Figure 18.19 **Alveolar surface tension affects lung compliance.** Inflating lungs with saline instead of air eliminates surface tension. Two pressure–volume curves for lungs inflated to the same volume are shown, first with air and then with saline. The differences in the two curves occur because surface tension contributes significantly to lung compliance in the air-filled lungs. When lungs are inflated and deflated with saline, the lungs are more distensible, with a concomitant increase in lung compliance. The *shaded areas* are equal to the work done in inflating the lungs.

Surface tension of the lungs can be studied by examining a pressure–volume curve. Figure 18.19 shows results from lungs that have been removed from the chest (excised lungs) and inflated and deflated in a stepwise fashion, first with air and then with saline. With air-filled lungs, the gas–liquid interface creates surface tension. However, with saline-filled lungs, the air–liquid interface is eliminated and the surface tension is eliminated. In comparing these two pressure–volume curves, two important observations can be made. First, the slope of the deflation limb of the saline curve is much steeper than that of the air curve. This means that when surface tension is eliminated, the lung is far more compliant (more distensible). Second, the different areas to the left of the saline and air inflation curves show that surface tension significantly contributes to the work required to inflate the lungs. Because the area to the left of each curve is equal to work, which can be defined as force (change in pressure) times distance (change in volume), the elastic forces and surface tension can be separated. The area to the left of the saline inflation curve is the work required to overcome the elastic recoil of the lung tissue. The area to the left of the air inflation curve is the work required to overcome both elastic tissue recoil and surface tension. Subtracting the area to the left of the saline curve from the area to the left of the air curve shows that approximately two thirds of the work required to inflate the lungs is needed to overcome surface tension. Lung distensibility and the work of breathing are significantly affected by surface tension.

Surface tension has important ramifications for maintaining alveolar stability. In a sphere such as an alveolus, surface tension produces a force that pulls inwardly and creates an internal pressure. The relationship between surface

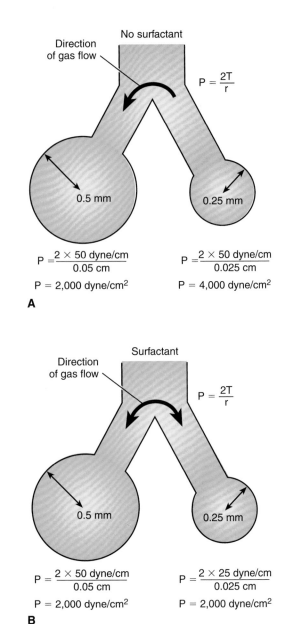

Figure 18.20 **Surfactant promotes alveolar stability at low lung volumes. (A)** If surface tension remains constant (50 dyne/cm), alveoli that are interconnected but differ in diameter become unstable and cannot coexist. Pressure in the smaller alveolus is greater than that in the larger alveolus, which causes air from the smaller alveolus to empty into the larger alveolus. At low lung volumes, the smaller alveoli tend to collapse, a phenomenon known as atelectasis. **(B)** Surfactant lowers surface tension proportionately more in the smaller alveolus. As a result, pressures in the two alveoli are equal, and alveoli of different diameters can coexist. P, pressure; r, radius; T, tension.

tension and pressure inside a sphere is shown in Figure 18.20. The pressure developed in an alveolus is given by **the Law of Laplace** (see Chapter 11, "Overview of the Cardiovascular System and Hemodynamics"):

$$P = 2T/r \qquad (20)$$

where T equals surface tension and r equals radius. Because alveoli are interconnected and vary in diameter, the

CLINICAL FOCUS | 18.2

Acute Respiratory Distress Syndrome

Scot was a healthy 20-year-old man who was starting his 2nd year of college. He was in good condition and worked out regularly by lifting weights and swimming 1 mile each day. However, one fall afternoon, he began to feel ill. He went to the student health center with complaints of a severe headache and pain in the neck and lower abdomen. He was told that he was coming down with the flu and was advised to go back to the dorm and rest. The next day his symptoms, which included fever, cold sweats, skin rash, and nausea, became worse. His roommate took him back to the health center, and Scot was placed in one of the student beds for closer observation. When his condition started deteriorating, he was rushed to the emergency department at the university medical center. On arrival, Scot was in shock with a high fever and a breathing rate of 36 (normal is 12 to 15/min). Because of the danger of respiratory failure, the attending physician immediately had Scot intubated and placed on assisted ventilation. The decision was made to transfer him to the intensive care unit (ICU). Upon arrival at the ICU, Scot's chest radiograph showed patchy infiltrate and pulmonary edema. He was diagnosed as having **acute respiratory distress syndrome** (**ARDS**), with a secondary widespread infection of unknown origin.

Scot's case is similar to those of the other 200,000 to 250,000 persons who develop ARDS each year. Despite Scot's youth and previous history, his anticipated mortality for ARDS was ~80%. Scot was lucky and was able to return to his classes after 12 weeks in the hospital and a bill of >$200,000.

ARDS is the name given to diffuse lung injury of various causes. Few patients have a history of previous lung

disorders. The causes include trauma from chest injury (e.g., car accidents), long bone injury, and pelvic injury. During the Vietnam era, the trauma that caused ARDS was coined "Da Nang lung." Two other major causes include sepsis (presence of pathogen or toxin in the blood) and aspiration of gastric contents. The latter occurs with gastric reflux and usually occurs at night during sleep. Other causes include cardiopulmonary bypass, smoke inhalation, high altitude, and exposure to irritant gases. Exposure to irritant gases was the cause of the widespread ARDS cases that occurred at the disaster in Bhopal, India.

Although the etiology for ARDS is varied in different cases, the pathophysiology is nearly identical. ARDS is characterized by decreased lung compliance, pulmonary edema, focal atelectasis, and hypoxemia (low partial pressure of O_2 in the arterial blood), with an inflammatory reaction that leads to an infiltration of neutrophils into the lung. The increase in lung stiffness (i.e., decreased compliance) is a result of loss of surfactant, which is unrelated to edema.

Neutrophil aggregation is a key underlying mechanism of ARDS. Aggregation of neutrophils causes capillary endothelial damage by releasing a number of toxic products. These include oxygen free radicals, proteolytic enzymes, arachidonic acid metabolites (leukotrienes, thromboxane, prostaglandin), and platelet-activating factor.

New approaches to therapy involve ways to reduce neutrophil chemoattraction and aggregation in the pulmonary capillaries and ways to reduce the amount of toxic substances released by neutrophils. ◼

Laplace equation assumes functional importance in the lung. In the example shown in Figure 18.20, a surface tension is constant at 50 dyne/cm, and an unstable condition results because pressure is greater in the smaller alveolus than in the larger one. Consequently, smaller alveoli tend to collapse, especially at low lung volumes, a phenomenon known as **atelectasis**. Larger alveoli become overdistended when atelectasis occurs. Therefore, two questions arise: How do alveoli of different sizes coexist in the intact lung when interconnected? How does the normal lung prevent atelectasis at low lung volumes? The answers lie, in part, in the fact that the alveolar surface tension is not constant at 50 dyne/cm, as in other biologic fluids (Clinical Focus 18.2).

Surfactant lowers surface tension and stabilizes alveoli at low lung volumes.

The alveolar lining is coated with a special surface-active agent, pulmonary surfactant, which not only lowers surface tension at the gas–liquid interface but also changes surface tension with changes in alveolar diameter (see Fig. 18.20B). Pulmonary surfactant is a lipoprotein rich in phospholipid. The principal agent responsible for its surface tension–reducing properties is **dipalmitoylphosphatidylcholine** (**DPPC**).

The functional importance of surfactant can be demonstrated by using a surface tension balance (Fig. 18.21). Surface tension is measured by placing a platinum flag connected to a force transducer into a trough of liquid. Surface tension creates a meniscus on each side of the platinum flag, and the greater the contact angles of the meniscus, the greater the surface tension. The surface is repeatedly expanded and compressed by a movable barrier that skims the surface of the liquid (simulating lung inflation and deflation). Surface tension of pure water is 72 dyne/cm, a value that is independent of the surface area of water in the balance. Therefore, when the surface is expanded and compressed, surface tension does not change. When a detergent is added, surface tension is reduced but again is independent of surface area. However, when a **lung lavage** (a saline flush from the lungs) is added that contains pulmonary surfactant, surface tension not only is reduced but also changes in a nonlinear fashion with surface area. Therefore, pulmonary surfactant makes it possible for alveoli of different diameters that are connected in parallel to coexist and be stable at low lung volumes, by lowering surface tension proportionately more in the smaller alveoli (see Fig. 18.20B).

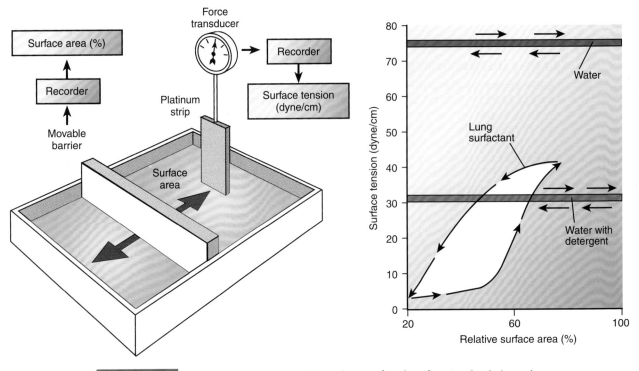

Figure 18.21 **Lung surfactant lowers alveolar surface tension.** A surface tension balance is used to measure surface tension at the air–liquid interface. When distilled water is placed in the balance, surface tension is independent of surface area (a constant 72 dynes/cm). The addition of a detergent reduces surface tension, but it is still independent of surface area. When lung surfactant (obtained from a lung lavage—a volume of saline flushed into the airways to rinse the alveoli) is placed in the balance, surface tension not only is decreased but also changes with surface area.

A schematic showing how pulmonary surfactant works is depicted in Figure 18.22. At low lung volumes, when the molecules are tightly compressed, some surfactant is squeezed out of the surface and forms micelles. On expansion (reinflation), new surfactant is required to form a new film that is spread on the alveolar surface lining. When surface area remains fairly constant during quiet or shallow breathing, the spreading of surfactant is often impaired. A deep sigh or yawn causes the lungs to inflate to a larger volume and new surfactant molecules spread onto the gas–liquid interface. Patients recovering from anesthesia are often encouraged to breathe deeply to enhance the spreading of surfactant and prevent alveolar collapse, a phenomenon known as **atelectasis.** Patients who have undergone abdominal or thoracic surgery often find it too painful to breathe deeply; poor surfactant spreading results, causing part of their lungs to become atelectatic.

Where is pulmonary surfactant produced? The alveolar epithelium consists basically of two cell types: alveolar type I and type II cells (Fig. 18.23). Alveolar type II cells are often referred to as **type II pneumocytes**. The ratio of these cells in the epithelial lining is about 1:1, but type I cells occupy approximately two thirds of the surface area. Type II cells seem to aggregate around the alveolar septa. Surfactant is synthesized in the alveolar type II cell. Compared with type I cells, they are rich in mitochondria and are metabolically very active. Electron-dense **lamellar inclusion bodies** are one of the distinguishing features of the type II cell. These lamellar inclusion bodies, rich in surfactant, are thought to be the storage sites for surfactant.

The process of surfactant synthesis is shown in Figure 18.24. Basic substrates, such as glucose, palmitate, and choline, are taken up from the circulating blood and synthesized into DPPC by alveolar type II cells. Stored surfactant from the lamellar inclusion bodies is discharged onto the alveolar surface. The turnover of surfactant is high because of the continual renewal of surfactant at the alveolar surface during each expansion of the lung. The high rate of replacement of surfactant probably accounts for the active lipid synthesis that occurs in the lung. Because the lungs are among the last organs to develop, the synthesis of surfactant appears rather late in gestation. In humans, surfactant appears at about week 34 (a full-term pregnancy is 40 weeks). Regardless of the total duration of gestation in any mammalian species, the process of lung maturation seems to be "triggered" about the time gestation is 85% to 90% complete. Clearly, the fetal lung is endowed with a special regulatory mechanism to control the timing and appearance of surfactant. Failure of proper lung maturation during the perinatal period is still a major cause of death in newborns. Anatomically, the lung may be structurally intact but functionally immature due to the inadequate amounts of surfactant available to reduce surface tension and stabilize surface forces during breathing.

Respiratory Physiology

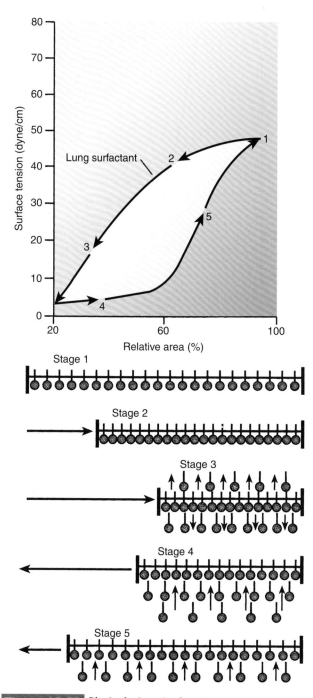

Figure 18.22 **Biophysical mechanism of lung surfactant for lowering alveolar surface tension.** Surfactant molecules are compressed during lung deflation. At stage 3, surfactant molecules form micelles and are removed from the surface. On lung inflation, new surfactant is spread onto the surface film (stage 4). Turnover of lung surfactant is high because of continual replacement of surfactant during lung expansion.

Premature birth and certain hormonal disturbances (such as those seen in diabetic pregnancies) interfere with the normal control and timing of lung maturation. These infants have immature lungs at birth, which often leads to **neonatal respiratory distress syndrome**. Breathing is extremely labored because surface tension is high, making it difficult to inflate the lungs. Because of the high

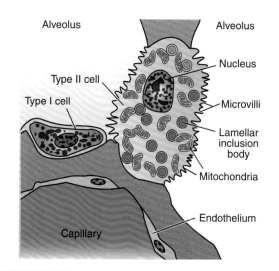

Figure 18.23 **Alveolar type II cells synthesize lung surfactant.** Alveolar type I cells occupy most of the alveolar surface, and alveolar type II cells are located in the corners between two adjacent alveoli. Alveolar type II cells produce lung surfactant. Also shown are endothelial cells that line the pulmonary capillaries.

surface tension, these infants often develop regional atelectasis and pulmonary edema. These infants are at high risk until the lungs become mature enough to secrete surfactant. In addition to lowering alveolar surface tension and promoting alveolar stability, surfactant helps to prevent edema in the lung. The inwardly contracting force that tends to collapse alveoli also tends to lower interstitial pressure, which "pulls" fluid from the capillaries. Pulmonary surfactant reduces this tendency by lowering surface forces. Some pulmonary physiologists think that keeping the lungs dry may be the major role of surfactant, especially in adults.

Alveolar interdependence promotes alveolar stability.

Another mechanism that plays a role in promoting alveolar stability and preventing atelectasis and edema is interdependence or mutual support among adjacent alveoli. Because alveoli (except those next to the pleural surface) are interconnected with surrounding alveoli, they support each other. Studies have shown that this type of structural arrangement, with many connecting links, prevents the collapse of adjacent alveoli. For example, if alveoli tend to collapse, surrounding alveoli would develop large expanding forces. Therefore, interdependence can play a role in preventing atelectasis as well as in opening up lungs that have collapsed. Alveolar interdependence seems to be more important in adults than in newborns because newborns have fewer interconnecting links.

▶ AIRWAY RESISTANCE AND THE WORK OF BREATHING

In addition to the elastic recoil of the lungs and chest wall, resistance to airflow in the respiratory tree is a key factor in contributing to the movement of air in and out of the lungs. Total resistance to airflow in the lungs has two components:

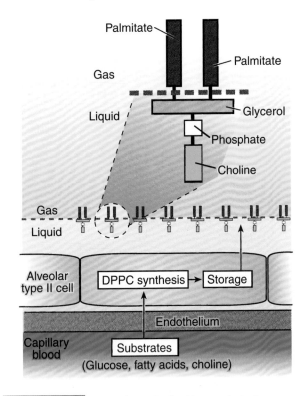

Figure 18.24 Metabolic synthesis of lung surfactant.
Substrates for lung surfactant synthesis are taken up by alveolar type II cells from the pulmonary capillary blood. Surfactant (dipalmitoylphosphatidylcholine or DPPC) is stored in lamellar inclusion bodies and subsequently discharged onto the alveolar surface. Surfactant is oriented perpendicularly to the gas–liquid interface at the alveolar surface; the polar end is immersed in the liquid phase, whereas the nonpolar portion is in the gas phase.

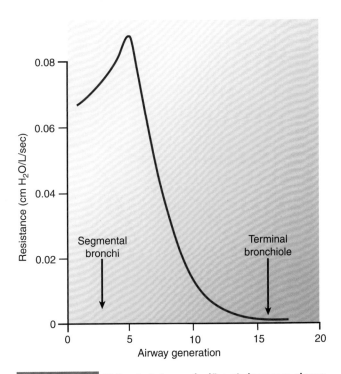

Figure 18.25 Airflow turbulence significantly increases airway resistance. The major sites of resistance are the lobar and segmental bronchi, where airway turbulence is the greatest, down to about the seventh generation of airway branches.

(1) tissue resistance of the lungs and chest wall and (2) airway resistance. Tissue resistance is encountered when the lungs and chest wall expand and contributes about 20% of the total resistance. Airway resistance, on the other hand, is the major factor opposing the flow of air in and out of the lungs and constitutes about 80% of the total resistance. In general, resistance is defined as the ratio of driving pressure (ΔP) to airflow ($\dot{V}$). For total airway resistance (R_{aw}), the driving pressure is the pressure difference between the mouth (P_{mouth}) and the alveoli (P_A). The equation can be written as:

$$R_{aw} = \frac{P_{mouth} - P_A}{\dot{V}} \qquad (21)$$

The unit for resistance is $H_2O/L/sec$. Note that resistance is only meaningful when airflow is occurring. Total airway resistance is small during normal breathing. For example, peak flow is about 2 cm $H_2O/L/s$ ($R = P_{mouth} - P_A/\dot{V} = 1.0$ cm $H_2O/0.5$ L/sec) during quiet inspiration. The low resistance is due to the uniqueness of the airway architecture.

Airway resistance decreases airflow in the lung.

The major site of airway resistance is the medium bronchi (lobar and segmental) and bronchi down to about the

seventh generation (Fig. 18.25). One would expect the major site of resistance, based on **Poiseuille's law** (see Chapter 11), to be located in the narrow airways (the bronchioles), which have the smallest radius. However, measurements show that only 10% to 20% of total airway resistance can be attributed to the small airways (those <2 mm in diameter). This apparent paradox results because so many small airways are arranged in parallel, and their resistances are added as reciprocals. Resistance of each individual bronchiole is relatively high, but the great number of them results in a large total cross-sectional area, causing their total combined resistance to be low.

Many airway diseases often begin in the small airways. Diagnosing a disease in the small airways is difficult, however, because the small airways account for such a low percentage of the total R_{aw}. Early detection is difficult because changes in airway resistance are not noticeable until the disease becomes severe.

Airways are compressed at low lung volumes, causing an increase in airway resistance.

One of the major factors affecting airway diameter, especially the bronchioles, is lung volume. The smaller airways are capable of being distended or compressed. Bronchi and smaller airways are embedded in lung parenchyma and are connected by "guy wires" to surrounding tissue. As the lung enlarges, airway diameter increases, which results in a concomitant decrease in airway resistance during inspiration (Fig. 18.26). Conversely, at low lung volumes, airways are compressed, and airway resistance rises. Note that

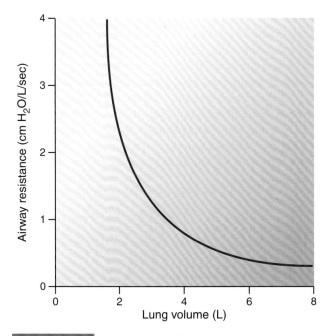

Figure 18.26 Low lung volumes increase airway resistance. Airways are connected to lung parenchyma. During inflation, airway diameter is increased. During deflation, airways are compressed resulting in a concomitant increase in airway resistance.

the inverse relationship between lung volume and airway resistance is nonlinear. At low lung volumes, airway resistance rises sharply.

Airway patency is affected by changes in smooth muscle tone.

Bronchial smooth muscle tone also affects airway diameter with a concomitant change in airway resistance. A change in smooth muscle tone will change airway diameter. The smooth muscles in the airway, from the trachea down to the terminal bronchioles, are under autonomic control. Stimulation of parasympathetic cholinergic postganglionic fibers causes bronchial constriction as well as increased mucus secretion. Stimulation of sympathetic adrenergic fibers causes dilation of bronchial and bronchiolar airways and inhibition of glandular secretion. Drugs, such as isoproterenol and epinephrine, which stimulate β_2-adrenergic receptors in the airways, cause dilation. These drugs alleviate bronchial constriction and are often used to treat asthmatic attacks. Environmental insults, such as breathing chemical irritants, dust, or smoke particles, can cause reflex constriction of the airways. Increased P_{CO_2} in the conducting airways can cause a local dilation. More important, a decrease in P_{CO_2} causes airway smooth muscle to contract.

Unusual gaseous environments alter airway resistance.

In unusual environments in which gas density is changed, airway resistance is altered. The effect of gas density on airway resistance is seen most dramatically in deep-sea diving, in which the diver breathes air the density of which may be greatly increased because of increased barometric pressure. The barometric pressure increases 1 atm for each 10 m or

33 ft underwater (e.g., at 10 m below the surface, P_B = 2 atm or 1,520 mm Hg). As a result of increased resistance, a large pressure gradient is required just to move a normal tidal volume. Instead of breathing air during diving, the diver uses a helium–oxygen mixture because helium is less dense than air and, consequently, makes breathing easier. The fact that density has a marked effect on airway resistance again indicates that the medium airways are the main site of resistance and that airflow here is primarily turbulent (see Chapter 11).

Forced expiration causes airway compression and increases airway resistance.

Airway resistance does not change much during normal quiet breathing. It is significantly increased, however, during forced expiration, such as in vigorous exercise. The marked change during forced expiration is a result of airway compression. This effect of airway compression on airway resistance can be demonstrated with a **flow–volume curve**. A flow–volume curve shows the relationship between airflow and lung volume during a forced inspiratory and expiratory effort (Fig. 18.27). The curve is generated by taking the spirometer tracings of forced inspiratory and expiratory flow and then closing them by connecting the end of maximum expiration back to the beginning of maximum inspiration.

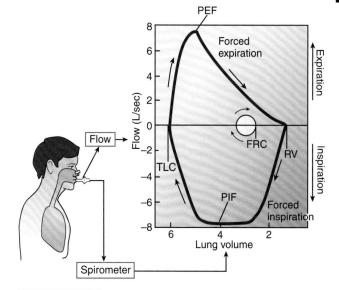

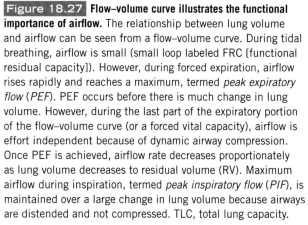

Figure 18.27 Flow–volume curve illustrates the functional importance of airflow. The relationship between lung volume and airflow can be seen from a flow–volume curve. During tidal breathing, airflow is small (small loop labeled FRC [functional residual capacity]). However, during forced expiration, airflow rises rapidly and reaches a maximum, termed *peak expiratory flow (PEF)*. PEF occurs before there is much change in lung volume. However, during the last part of the expiratory portion of the flow–volume curve (or a forced vital capacity), airflow is effort independent because of dynamic airway compression. Once PEF is achieved, airflow rate decreases proportionately as lung volume decreases to residual volume (RV). Maximum airflow during inspiration, termed *peak inspiratory flow (PIF)*, is maintained over a large change in lung volume because airways are distended and not compressed. TLC, total lung capacity.

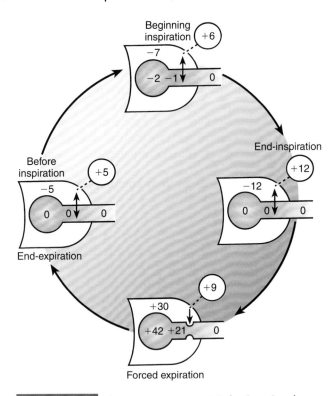

Figure 18.28 Airways are compressed during forced expiration. Transairway pressure is +5 cm H_2O before inspiration and reaches +12 cm H_2O at the end of inspiration. During forced expiration, transairway pressure becomes negative and the small airways are compressed.

The small loop in Figure 18.28 is the flow–volume curve for a normal tidal volume. The large loop shows the maximal flow–volume curve. During forced inspiration, inspiratory flow is limited only by effort—that is, how hard the person tries. During forced expiration (FVC), flow rises rapidly to a maximum value, **peak expiratory flow (PEF)**, and then decreases linearly over most of expiration as lung volume decreases.

The first part of the FEF_{25-75}–volume curve is dependent on effort. Once PEF is achieved, flow is independent of effort in the last part of the flow–volume curve. Actually, over most of the expiratory flow–volume curve, flow is virtually independent of effort. The FEF_{25-75} limitation illustrates the importance of dynamic airway compression. Dynamic airway compression increases airway resistance, which effectively limits the FEF_{25-75}.

How does dynamic airway compression occur during forced expiration? The mechanism is related to changes in transairway pressure (i.e., pressure across the airway) and the compressibility of the airways. For example, in Figure 18.28, pleural pressure (P_{pl}) is ~5 cm H_2O and airway pressure (P_{aw}) is zero before inspiration (no airflow). Transairway pressure ($P_{aw} - P_{pl}$) is 5 cm H_2O [0 − (−5) = +5] and holds the airways open. At the start of maximum inspiration, pleural pressure decreases to ~7 cm H_2O and alveolar pressure falls to ~2 cm H_2O. The difference between alveolar pressure and pleural pressure is still 5 cm H_2O [−2 − (−7) = +5]. However, there is a pressure drop from the mouth to the alveoli because

of resistance to airflow, and the transairway pressure will change along the airway. At the end of maximum inspiration, pleural pressure decreases further, to ~12 cm H_2O, and airway pressure is again zero because of no airflow. During maximum inspiration, airway resistance actually decreases because transairway pressure increases, which enlarges the diameter of the airways, especially the small airways.

On forced expiration, pleural pressure is no longer negative but rises above atmospheric pressure and can increase up to +30 cm H_2O. The added pressure in the alveoli is a result of the elastic recoil of the lungs, is termed *recoil pressure (P_{recoil})*, and can be written as:

$$P_{recoil} = P_A - P_{pl} \qquad (22)$$

where P_A equals alveolar pressure and P_{pl} equals pleural pressure.

The recoil pressure in Figure 18.28 is 12 cm H_2O, because, at the beginning of expiration, lung volume has not appreciably decreased. Note that at the beginning of FVC, a pressure drop occurs along the airway because of airway resistance. Airway pressure falls progressively from the alveolar region to the airway opening (the mouth). The transairway pressure gradient along the airways reverses and tends to compress the airways. For example, at a point inside the airway where the pressure is 21 cm H_2O, the transairway pressure would be 9 cm H_2O, which would tend to close the airway.

At some point along the airway, the airway pressure equals pleural pressure and transairway pressure is zero (Fig. 18.29). This is the **equal pressure point (EPP)**. Theoretically, the EPP divides the airways into an upstream segment (from alveoli to EPP) and a downstream segment (EPP to the mouth). In the downstream segment, the airway pressure is below pleural pressure and the transairway pressure becomes negative. As a result, airways in the downstream segment are compressed or collapsed. The large airways (the trachea and bronchi) are protected from collapse because they are supported by cartilage. However, small airways without this structural support are easily compressed and can collapse.

Lung compliance affects where the equal pressure point is established in airways.

The EPP is established after the PEF is achieved (see Fig. 18.27). As the forced expiratory effort continues, the EPP moves down the airways from larger to smaller airways because the recoil pressure decreases. A greater length of the downstream segment collapses. The driving pressure for airflow, once the EPP is established, is no longer the difference between alveolar pressure and mouth pressure but is alveolar pressure minus pleural pressure (see Fig. 18.29).

Two basic conclusions follow. First, regardless of the forcefulness of the expiratory effort, airflow cannot be increased because pleural pressure increases, causing more airway compression. This explains why the last part of the FVC is independent of effort.

Second, elastic recoil of the lung determines maximum flow rates because it is the elastic recoil pressure that generates the alveolar–pleural pressure difference. As lung volume

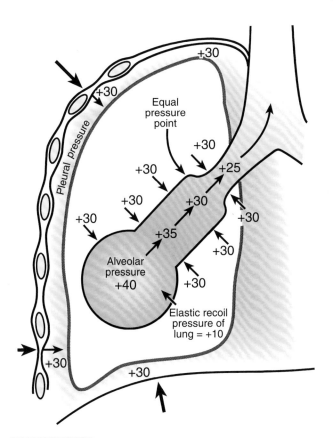

Figure 18.29 **An equal pressure point establishes the driving pressure in the airways.** An equal pressure point (EPP) divides the airways into downstream and upstream segments. The EPP is established at peak expiratory flow (PEF) and occurs when the pressure inside the airway equals the pressure outside the airway (pleural pressure). The upstream segment is represented from the alveoli to the EPP, and the downstream segment is represented from the EPP to the mouth. The driving pressure for airflow is now alveolar pressure minus the pleural pressure. Airways are subjected to compression during forced expiration from the EPP to the trachea.

decreases, so does elastic recoil. The decrease in elastic recoil is the main reason maximum flow falls so rapidly at low lung volumes.

The effect of elastic recoil on expiratory airflow can be demonstrated by comparing a normal lung with an emphysematous lung, in which the compliance is abnormally high. As seen in Figure 18.30, in both instances, pleural pressure rises to +30 cm H_2O with forced expiration. In the normal lung, the recoil pressure of 10 cm H_2O is added to produce an alveolar pressure of +40 cm H_2O. With forced expiration, there is a progressive fall in airway pressure. Because of the elastic recoil, the normal lung has "added" pressure to the small airways that keeps their pressures above pleural pressure, and less collapse occurs. With emphysema, however, the lungs have diminished elastic recoil, resulting in less elastic recoil pressure. In the example shown in Figure 18.31, elastic recoil adds only 2 cm H_2O to alveolar pressure, resulting in an alveolar pressure of +32 cm H_2O. With expiratory effort, flow proceeds along the pressure gradient. But in the emphysematous lungs, the pressure inside the small airways

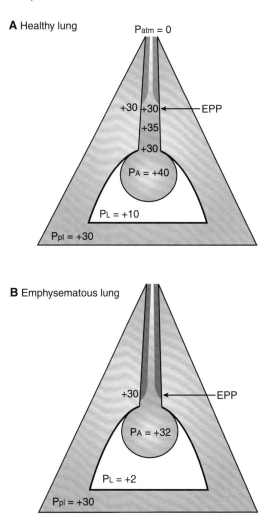

Figure 18.30 **The equal pressure point is influenced by the lung elastic recoil. (A)** In healthy lungs, elastic recoil adds 10 cm H_2O pressure, to produce an alveolar pressure of 40 cm H_2O at the beginning of a forced expiration. As a result, the equal pressure point (EPP) is established in the larger airways. Airway collapse is minimal in the large airways because cartilage supports the airways. **(B)** A loss of elastic recoil causes the EPP to shift downward and to be established in the small airways. In emphysematous lungs, the elastic recoil pressure is low, and little recoil pressure is added to the alveolus. As a result, the EPP is shifted downward and is established in the smaller airways. The small airways are more easily compressed because they are thin and lack cartilaginous support. P_A, alveolar pressure; P_{atm}, atmospheric pressure; P_L, transpulmonary pressure; P_{pl}, pleural pressure.

falls below pleural pressure well before the large airways, resulting in more airway compression and collapsed airways at low lung volumes. As a result, flow stops temporarily in these collapsed airways. Airway pressure upstream to the collapsed segment then rises to equal alveolar pressure, causing the airways to open again. This process repeats itself continually, leading to the "wheeze" that is often heard in emphysematous patients.

In lungs with abnormally high compliance, the position of the EPP with forced expiration is established further down in the small airways, where there is no cartilage to keep them

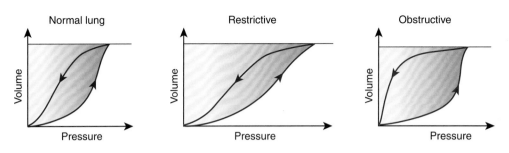

Figure 18.31 Work of breathing is measured from a pressure–volume curve. Pressure is plotted on the *x*-axis, and volume is plotted on the *y*-axis. Work equals force times distance and is represented by the *blue area* above the pressure–volume curve. This area represents the inspiratory work of breathing. Note that with a restrictive disorder, the inspiratory work of breathing is increased. Airway resistance is increased in an obstructive disorder, which requires more energy to force air out of the lungs during expiration.

distended. Therefore, these patients are much more vulnerable to compression and airway collapse. The greatest problem for patients with emphysema is not getting air into the lungs but getting it out. Consequently, they tend to breathe at higher lung volumes, thereby increasing elastic recoil, which reduces airway resistance and facilitates expiration.

Inspiratory muscles inflate the lungs and overcome airway resistance.

During inspiration, muscular work is involved in expanding the thoracic cavity, inflating the lungs, and overcoming airway resistance. Because work can be measured as force x distance, the amount of work involved in breathing can be expressed as a change in lung volume (distance) multiplied by the change in transpulmonary pressure (force). Thus, work (W) is equal to the product of pressure (P) and volume (V). With a volume change, the work involved in taking a breath is defined by this equation:

$$W = P \times \Delta V \qquad (23)$$

where P equals transpulmonary pressure and ΔV equals change in lung volume. During work, energy is expended with muscular contraction to create a force (transpulmonary pressure) to inflate the lungs. When a greater transpulmonary pressure is required to bring more air into the lungs, more muscular work and, hence, greater energy are required.

Figure 18.31 shows how a pressure–volume curve can be used to determine the work required for breathing. The *shaded area* in blue represents the inspiratory work of breathing. In healthy people at rest, the energy needed for breathing represents ~5% of the body's total energy expenditure. During heavy exercise, about 20% of the total energy expenditure is involved in breathing. Breathing is efficient and is most economical when elastic and resistive forces yield the lowest work. Note that the total inspiratory work of breathing in a restrictive lung disorder, compared with the normal lung, is increased and is a result of a greater inspiratory effort required. It is important to remember that lungs with a marked decrease in lung compliance (i.e., restrictive lung disorder) require more work to overcome increase in elastic recoil and surface forces. Patients with a restrictive disorder economize their ventilation by taking rapid and shallow breaths. In contrast, patients with severe airway obstruction tend to do the opposite; they take deeper breaths and breathe more slowly, to reduce their work resulting from the increased airway resistance. Despite this tendency, patients with obstructive disease still expend a considerable portion of their basal energy for breathing. The reason for this is that the expiratory muscles must do additional work to overcome the increased airway resistance. These different breathing patterns help minimize the amount of work required for breathing.

INTEGRATED MEDICAL SCIENCES

The Case of Labored Breathing: Asthma

Jack is a 10-year-old boy who was admitted to the emergency at the university's Children's Hospital with difficulty breathing. His mother reported that Jack had an upper respiratory tract infection for two days and had been using an inhaler more frequently. Today, he has received treatment every three hours, but still complains of shortness of breath and frequent episodes of coughing. His inhaler medication is Flovent (a prescription inhaled corticosteroid medicine for the long-term treatment of asthma in people aged 4 years and older). A call to the pharmacy verifies the correct dosage, and the prescription has been refilled at appropriate interviews. Upon further assessment by the attending resident, Jack's heart rate was 150; respiratory rate 45; blood pressure 95/65; temperature 37°C; and

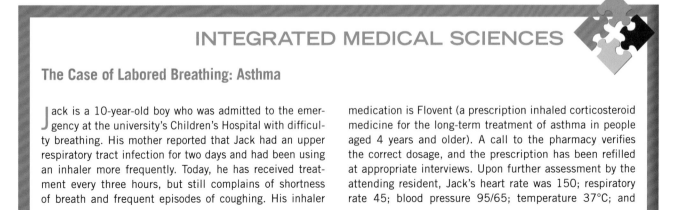

breathing sounds faint expiratory wheezes throughout all lung fields. Physical exam revealed some confusion, sweating, and tightness in the chest.

This case is an example of what many patients with asthma experience. The follow discussion will cover the immunology, pathophysiology, pharmacology, etiology, and the treatment of asthma. Asthma is defined as a chronic inflammatory disease of the airways characterized by reversible bronchospasm. There are three main characteristics of asthma: (1) inflammation, (2) hyperresponsiveness, and (3) airway obstruction. In inflammation, the lining of the airways becomes swollen and produces mucus. In asthma patients, the inflammation is induced by an overreaction to triggers (allergens, pollutants, animal dander, etc.). When the airways (bronchi) become hyperresponsive, they have an exaggerated response that leads to bronchospasms. As a result of the inflammation and hyperresponsiveness, asthmatic patient have severely constricted airways. Airway constriction interferes with gas exchange (CO_2 builds up in the blood, and less O_2 is transferred to the blood). In response to low oxygen, heart and respiratory rate are increased as compensatory mechanism.

The cellular mechanism for induced bronchospasms is mast cells degranulation and the release a variety of mediators (histamine, prostaglandins, thromboxane, and bradykinin). All cause smooth muscle contraction, swelling, and inflammation. The treatment of asthma involves bronchodilators and long-term anti-inflammatory medications. Bronchodilators dilate the bronchi and bronchioles, decrease airway resistance, and increase airflow to the lungs. Bronchodilators are either short acting or long lasting. Short-term dilators provide quick relief from acute bronchoconstriction. Long-term dilators help to control and prevent the symptoms. The bronchodilators are available in inhaled, tablet, liquid, and injection forms, but the preferred method is by inhalation. There three types of bronchodilators: beta agonists (short acting and long acting), anticholinergics (short acting), and theophylline (long acting). Beta agonists and anticholinergic act to increase cellular cAMP. The elevated cAMP, in turn, triggers a cascade of events within the muscle that leads to smooth muscle relaxation.

Long-term anti-inflammatory medicines are prescribed to prevent asthma attacks and reduce inflammation, swelling, and mucus production. These anti-inflammatory asthma inhalers include (1) steroids (e.g., Aerospan and Flovent) and (2) mast cell stabilizers (e.g., Cromolyn). Like most medications, inhaled corticosteroids can have side effects. One common side effect from inhaled corticosteroids is *thrush* (a mouth infection). If taken for long periods, these medications can increase the risk of cataracts and osteoporosis. ■

Chapter Summary

- The primary function of the lungs is gas exchange, which involves five components (ventilation, gas uptake, blood flow, matching airflow and blood flow, and gas transport).
- The alveolar–capillary membrane forms a large blood–gas interface for diffusion of oxygen and carbon dioxide.
- Airflow parallels alveolar pressure.
- A negative alveolar pressure is created during inspiration to bring air into the lungs.
- A positive alveolar pressure is created during expiration to exhaust air out of the lungs.
- Forced vital capacity is one of the most useful spirometry tests to assess lung function.
- Helium dilution is an indirect method for determining residual lung volume.
- Minute ventilation ($\dot{V}$) is the volume of air expired per minute and is equal to expired minute ventilation ($\dot{V}_E$).
- Alveolar ventilation ($\dot{V}_A$) is the amount of fresh air that reaches alveoli and regulates carbon dioxide levels in the blood.

- Physiologic dead space volume is the portion of tidal volume that is wasted air and does not participate in gas exchange. It is the sum of anatomic dead space volume minus the volume of air in the conducting zone alveolar dead space volume minus volume of air in alveoli that does not participate in gas uptake.
- Compliance is a measure of lung distensibility.
- Surfactant and alveolar interdependence maintain alveolar stability.
- Total airway resistance consists of two components: (1) tissue resistance of lungs and chest wall and (2) resistance to airflow in the airways.
- Lung compliance affects airway compression during forced expiration.
- The work of breathing is required to expand the lungs and overcome airway resistance.
- Major lung diseases are categorized into two groups: obstructive and restrictive. Chronic obstructive pulmonary disease obstructs airflow out of the lungs. Restrictive disease restricts lung inflation.

Chapter Review Questions

1. Asthma is a disease that affects primarily the peripheral airways. Which of the following would best characterize pulmonary function in a patient diagnosed with asthma compared to normal predicted values?

   A. FEV_1/FVC ratio is increased.
   B. Residual volume is increased.
   C. Total lung capacity is decreased.
   D. Resistance to airflow is decreased.
   E. Both airway resistance and total lung capacity are decreased.

The correct answer is B. Asthma is one of the diseases that falls under chronic obstructive pulmonary disease (COPD). COPD is characterized by high static lung volumes and low airflow. This means total lung capacity and residual volumes are increased. Airflow is restricted due to the increase in airway resistance. FEV_1, FVC, and FEV_1/FVC ratio are also decreased because of the increase in airway resistance.

2. A 48-year-old patient undergoes a pulmonary function test. At which lung volume would her pleural pressure be most negative?

   A. Residual volume
   B. Functional residual capacity
   C. End of tidal volume
   D. Total lung capacity
   E. Middle of forced vital capacity

The correct answer is D. During inspiration, pleural pressure becomes more negative. Pleural pressure would be the most negative at maximal inspiration (total lung capacity). Pleural pressure is the least negative at residual volume.

During forced vital capacity, pleural pressure is positive. FRC and end-tidal volume are distracters.

3. If alveolar ventilation is doubled while breathing room air, and if CO_2 production remains unchanged, then the most likely effect (increase, decrease, or no change) on alveolar CO_2 tension (P_{ACO_2}), alveolar O_2 tension (P_{AO_2}), arterial O_2 tension (P_{aO_2}), and arterial CO_2 tension (P_{aCO_2}) will be that:

	P_{ACO_2}	P_{AO_2}	P_{aO_2}	P_{aCO_2}
A.	↓	↑	↔	↓
B.	↓	↔	↔	↓
C.	↑	↑	↔	↓
D.	↑	↑	↑	↑
E.	↓	↑	↔	↔

The correct answer is A. If carbon production remains unchanged, one hyperventilates for 2 minutes, both alveolar and arterial carbon tensions decrease. Hyperventilation increases oxygen tension in the alveolar but has little effect on arterial oxygen tension.

4. The following measurements were made on a normal subject:

   Arterial P_{CO_2} = 36 mm Hg
   Arterial P_{O_2} = 100 mm Hg
   Minute ventilation = 9 L/minute
   Alveolar ventilation = 6 L/minute
   Frequency = 15 breaths/minute
   Which one of the following breathing patterns would result in the highest alveolar ventilation?

	Tidal volume, mL	frequency, breaths/minute
A.	400	36
B.	500	30
C.	800	7.5
D.	800	10
E.	900	10

A. A 10% increase in lung compliance and a 10% decrease in chest wall compliance

B. A 10% decrease in lung compliance and a 10% decrease in chest wall compliance

C. A 10% decrease in lung compliance and a 10% increase in chest wall compliance

D. A 10% increase in lung compliance and a 10% increase in chest wall compliance

E. A 5% increase in lung compliance and a 5% increase in chest wall compliance

The correct answer is B. The key to measuring alveolar ventilation is to determine the dead space volume. Dead space ventilation = minute ventilation – alveolar ventilation (9 L/minute – 6 L/minute = 3 L/minute). Dead space volume equals dead space ventilation/breathing frequency (3 L/minute/15 breaths/minute = 200 mL).

Alveolar ventilation = tidal volume – dead space volume × frequency.

5. In a healthy individual, which of the following combination of changes from normal values in chest wall and lung compliance would result in the greatest decrease in functional residual capacity?

The correct answer is C. Since the lung elastic recoil and chest wall elastic recoil are in equal but opposite directions, an increase in lung elastic recoil (decreased lung compliance) and a decrease in chest wall elastic recoil will favor the greatest decrease in functional residual capacity (FRC). Options A and E would result in an increase in FRC. Options B and D would essentially result in no change in FRC.

Clinical Application Exercises 18.1

A DEADLY LUNG DISORDER: EMPHYSEMA

Recovering from a heart attack and fighting pneumonia, a 65-year-old Vietnam veteran was transferred from the VA hospital to the university hospital to be evaluated for surgery to repair his blocked arteries. Wheezing and coughing badly, he learned that he had another deadly ailment that would require treatment first. Upon further evaluation, he had a history of shortness of breath and difficulty breathing on exertion. He also complained of a cough productive of green sputum. He appeared pale and said he felt feverish at home, but denied any shaking chills, sore throat, nausea, vomiting, or diarrhea. He said he was a 2-pack-a-day cigarette smoker who had given up smoking 10 years ago. He had not been previously hospitalized. He is a retired cab driver and lives with his wife; they have no pets. Although he has had dyspnea upon exertion for the past 2 years, he continues to maintain an active lifestyle. Prior to his heart attack, he still mowed his lawn without much difficulty and continued to walk 1 to 2 miles on a flat surface at a moderate pace. The patient said he rarely drinks alcohol.

He denied having had any other significant past medical problems, including childhood asthma or any allergies. He did state that his father, also a heavy smoker, died of emphysema at age 55.

An initial exam shows that the patient is thin but has a large chest. He is in moderate respiratory distress. His blood pressure is 130/80 mm Hg; respiratory rate, 28 to 32 breaths/minute; heart rate, 92/minute; and oral temperature, 37.9°C. His trachea is midline, and his chest expands symmetrically. He has decreased but audible breath sounds in both lung fields, with expiratory wheezing and a prolonged expiratory phase. Head, eyes, ears, nose, and throat findings are unremarkable. A pulse oximetry reading reveals his blood hemoglobin oxygen saturation is 91% when breathing room air.

Pulmonary function tests reveal severe limitation of airflow rates, particularly expiratory airflow. The patient is diagnosed with pulmonary emphysema.

QUESTIONS

1. What are the common spirometry findings associated with emphysema?
2. What are the mechanisms of airflow limitation in emphysema?
3. What is the most commonly held theory explaining the development of emphysema?

ANSWERS

1. The hallmark of emphysema is the limitation of airflow out of the lungs. In emphysema, expiratory flow rates (FVC, FEV_1, and FEV_1/FVC ratio) are significantly decreased. However, some lung volumes (TLC, FRC, and RV) are increased, and the increase is a result of the loss of lung elastic recoil (increased compliance).

2. The mechanisms that limit expiratory airflow in emphysema include hypersensitivity of airway smooth muscle, mucus hypersecretion, and bronchial wall inflammation and increased dynamic airway compression as a result of increased compliance.

3. The patient is one of the estimated 24 million Americans with emphysema, half of whom remain undiagnosed. Many of the pathophysiological changes in emphysema are a result of the loss of lung elastic recoil and destruction of the alveolar–capillary membrane. This is thought to be a result of an imbalance between the proteases and antiproteases (α_1 antitrypsin) in the lower respiratory tree. Normally, proteolytic enzyme activity is inactivated by antiproteases. In emphysema, excess proteolytic activity destroys elastin and collagen, the major extracellular matrix proteins responsible for maintaining the integrity of the alveolar–capillary membrane and the elasticity of the lung. Cigarette smoke increases proteolytic activity, which may arise through an increase in protease levels, a decrease in antiprotease activity, or a combination of the two.

thePoint® *Visit* http://thepoint.lww.com/rhoades5e *for additional chapter review Q&A, Clinical Application Exercises, animations, and more!*

19 Gas Transfer and Transport

Active Learning Objectives

Upon mastering the material in this chapter, you should be able to:

- Describe how alveolar surface area and membrane thickness affect gas diffusion.
- Describe how pulmonary blood flow and blood hematocrit affect lung diffusion capacity.
- Describe how oxygen binds to hemoglobin.
- Describe the concept P_{50} and its effect on hemoglobin's affinity for oxygen.

- Predict how changes in blood pH, blood Pco_2, body temperature, and carbon monoxide will affect the oxyhemoglobin–equilibrium curve.
- Predict the effect of hematocrit on arterial Po_2 and O_2 content.
- Describe the mechanisms by which CO_2 is transported by the blood.
- Describe the physiologic causes of hypoxemia.
- Describe the relationship between venous admixture and the alveolar–arterial oxygen gradient.
- Describe how a low ventilation/perfusion ratio ($\dot{V}A/\dot{Q}$) differs from an anatomic shunt as a cause of hypoxemia.

▶ GAS DIFFUSION AND UPTAKE

In the previous chapter, you learned the unique architecture of the lungs allows for a large alveolar–capillary membrane that forms an interface for the exchange of oxygen and carbon dioxide. You also learned how small pressure changes in the lung can bring large volumes of air into the lungs for gas exchange. You will learn in this chapter that gas exchange is closely linked to pulmonary circulation, and they can increase 20-fold in order to meet the body's energy demands.

Oxygen and carbon dioxide move across the alveolar–capillary membrane by diffusion.

Gas exchange can be defined as the delivery of oxygen from the lungs to the pulmonary capillaries and the removal of carbon dioxide from the capillaries to the lungs. The movement of gases in the alveoli and across the alveolar–capillary membrane is by diffusion; partly due to partial pressure gradients of individual gases. Recall from Chapter 18 that partial pressure or gas tension can be determined by measuring barometric pressure and the fractional concentration (F) of the gas (Dalton law). At sea level, Po_2 is 160 mm Hg (760 mm Hg × 0.21). Fo_2 does not change with altitude, which means that the percentage of O_2 in the atmosphere is essentially the same at 30,000 ft (about 9,000 m) as it is at sea level. Therefore, the difficulty in breathing at high altitudes is due to a decrease in Po_2 rather than a decrease in Fo_2 (Fig. 19.1).

Oxygen is taken up by blood in the lungs and is transported to the tissues. The transfer of oxygen from the alveolar spaces, across the alveolar–capillary membrane to the blood, is referred to as oxygen uptake. Oxygen uptake is determined by three factors: the diffusion properties of the alveolar–capillary membrane, the partial pressure gradient of oxygen, and pulmonary capillary blood flow.

The diffusion of gases is a function of the partial pressure difference of the individual gases. For example, oxygen diffuses across the alveolar–capillary membrane because of the difference in Po_2 between the alveoli and pulmonary

capillaries (Fig. 19.2). The partial pressure difference for oxygen is referred to as the **oxygen diffusion gradient**. In the normal lung, the initial oxygen diffusion gradient, alveolar Po_2 (102 mm Hg) minus venous Po_2 (40 mm Hg), is 62 mm Hg. The initial diffusion gradient across the alveolar–capillary membrane for carbon dioxide (venous Pco_2 [$Pvco_2$] minus alveolar Pco_2 [$Paco_2$]) is about 6 mm Hg, which is much smaller than that of oxygen.

When gases are exposed to a liquid such as plasma, gas molecules diffuse into the liquid and exist in a dissolved state. The dissolved gases also exert a partial pressure. A gas will continue to dissolve in the liquid until the partial pressure of the dissolved gas equals the partial pressure above the liquid. The **Henry law** states that at equilibrium, the amount of gas dissolved in a liquid at a given temperature is directly proportional to the partial pressure and the solubility of the gas. The Henry law accounts only for the gas that is physically dissolved and not for chemically combined gases (e.g., oxygen bound to hemoglobin).

Gas diffusion in the lungs can be described by **Fick's law**, which states that the volume of gas diffusing per minute ($\dot{V}_{gas}$) across a membrane is directly proportional to the membrane surface area (A_s), the diffusion coefficient of the gas (D), and the partial pressure difference (ΔP) of the gas and is inversely proportional to the membrane thickness (T) (Fig. 19.3):

$$\dot{V}_{gas} = \frac{A_s \times D \times \Delta P}{T} \qquad (1)$$

The diffusion coefficient of a gas is directly proportional to its solubility and inversely related to the square root of its molecular weight (MW):

$$D = \frac{\text{solubility}}{(MW)^{1/2}} \qquad (2)$$

Therefore, a small molecule or one that is very soluble will diffuse at a fast rate; for example, the diffusion coefficient

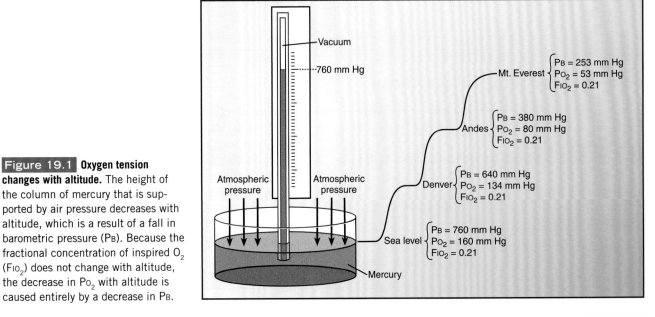

Figure 19.1 **Oxygen tension changes with altitude.** The height of the column of mercury that is supported by air pressure decreases with altitude, which is a result of a fall in barometric pressure (PB). Because the fractional concentration of inspired O_2 (F_{IO_2}) does not change with altitude, the decrease in P_{O_2} with altitude is caused entirely by a decrease in PB.

of carbon dioxide in aqueous solutions is about 20 times greater than that of oxygen because of its higher solubility, even though it is a larger molecule than O_2.

Therefore, Fick's law states that the rate of gas diffusion is inversely related to membrane thickness. This means that the diffusion of a gas will be halved if membrane thickness is doubled. Fick's law also states that the rate of diffusion is directly proportional to surface area (A_s). If two lungs have the same oxygen diffusion gradient and membrane thickness but one has twice the alveolar–capillary surface area, the rate of diffusion will differ by twofold.

Under steady-state conditions, ~250 mL of oxygen per minute are transferred to the pulmonary circulation $\dot{V}_{O_2}$, whereas 200 mL of carbon dioxide per minute are removed $\dot{V}_{CO_2}$. The ratio $\dot{V}_{CO_2}/\dot{V}_{O_2}$ is the **respiratory exchange ratio (R)**, and it is 0.8 in this case.

Pulmonary capillary blood flow is a major determinant in transferring oxygen from the alveoli to the blood.

Pulmonary capillary blood flow has a significant influence on oxygen uptake, and its effect is illustrated in Figure 19.4. The time required for the red cells to move through the capillary, referred to as **transit time**, is ~0.75 seconds. During transit time, the gas tension in the blood equilibrates with the alveolar gas tension. Transit time can change dramatically with cardiac output. For example, when cardiac output increases,

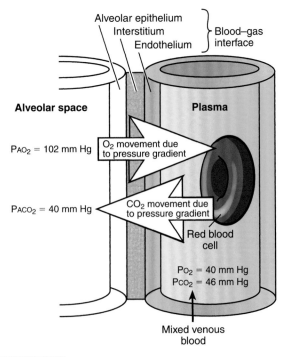

Figure 19.3 **Movement of O_2 and CO_2 across the alveolar–capillary membrane is by diffusion.** Gases move across the blood–gas interface (alveolar–capillary membrane) by diffusion, following Fick's law. P_{AO_2}, partial pressure of alveolar oxygen; P_{ACO_2}, partial pressure of alveolar carbon dioxide.

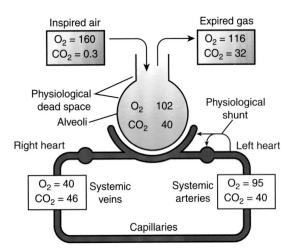

Figure 19.2 **Carbon and oxygen tensions vary between the systemic and pulmonary circulation.** Partial pressure of oxygen (P_{O_2}) is highest when it leaves the lungs, and that of carbon dioxide (P_{CO_2}) is highest when it enters the lungs.

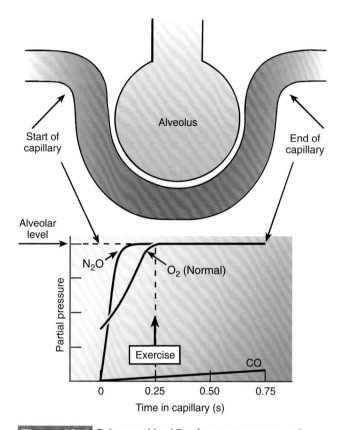

Figure 19.4 **Pulmonary blood flow increases oxygen uptake.** Gas transfer across the alveolar–capillary membrane is affected by pulmonary capillary blood flow. The horizontal axis shows time in the capillary. The average transit time it takes blood to pass through the pulmonary capillaries is 0.75 seconds. The vertical axis indicates gas tension in the pulmonary capillary blood, and the top of the vertical axis indicates gas tension in the alveoli. Individual curves indicate the time it takes for the partial pressure of a specific gas in the pulmonary capillaries to equal the partial pressure in the alveoli. Nitrous oxide (N_2O) is used to illustrate how gas transfer is limited by blood flow; carbon monoxide (CO) illustrates how gas transfer is limited by diffusion. The profile for oxygen is more like that of N_2O, which means oxygen transfer is limited primarily by blood flow. Pulmonary capillary Po_2 equilibrates with the alveolar Po_2 in about 0.25 seconds (*arrow*).

blood flow through the pulmonary capillaries increases, but transit time decreases (i.e., the time blood is in capillaries is less). Figure 19.4 illustrates the effect of blood flow on the uptake of three test gases. In the first case, a trace amount of nitrous oxide (laughing gas), a common dental anesthetic, is breathed. Nitrous oxide (N_2O) is chosen because it diffuses across the alveolar–capillary membrane and dissolves in the blood but does not combine with hemoglobin. The partial pressure in the blood rises rapidly and virtually reaches equilibrium with the partial pressure of N_2O in the alveoli by the time the blood has spent one tenth of the time in the capillary. At this point, the diffusion gradient for N_2O is zero. Once the pressure gradient becomes zero, no additional N_2O is transferred. The only way the transfer of N_2O can be increased is by increasing blood flow. The amount of N_2O

that can be taken up is entirely limited by blood flow, not by diffusion of the gas. Therefore, the net transfer or uptake of N_2O is **perfusion limited**.

When a trace amount of carbon monoxide (CO) is breathed, the transfer shows a different pattern (see Fig. 19.4). CO readily diffuses across the alveolar–capillary membrane, but, unlike N_2O, CO has a strong affinity for hemoglobin. As the red cell moves through the pulmonary capillary, CO rapidly diffuses across the alveolar–capillary membrane into the blood and binds to hemoglobin. When a trace amount of CO is breathed, most is chemically bound in the blood, resulting in low partial pressure (Pco) in the plasma. Consequently, equilibrium for CO across the alveolar–capillary membrane is never reached, and the transfer of CO to the blood is, therefore, **diffusion limited** and not limited by the blood flow.

Figure 19.4 shows that the equilibration curve for oxygen lies between the curves for N_2O and CO. Oxygen combines with hemoglobin but not as readily as CO because it has a lower binding affinity. As blood moves along the pulmonary capillary, the rise in Po_2 is much greater than the rise in Pco because of differences in binding affinity. Under resting conditions, the capillary Po_2 equilibrates with alveolar Po_2 when the blood has spent about one third of its time in the capillary. Beyond this point, there is no additional transfer of oxygen. Under normal conditions, oxygen transfer is more like that of N_2O and is limited primarily by blood flow in the capillary (perfusion limited). Hence, an increase in cardiac output will increase oxygen uptake. Not only does cardiac output increase capillary blood flow, but also it increases capillary hydrostatic pressure. The latter increases the surface area for diffusion by opening up more capillary beds by recruitment, which will be discussed later in the chapter.

The transit time at rest is normally about 0.75 seconds, during which capillary oxygen tension equilibrates with alveolar oxygen tension. Ordinarily, this equilibration takes only about one third of the available time, leaving a wide safety margin to ensure that the end-capillary Po_2 is equilibrated with alveolar Po_2. With vigorous exercise, the transit time may be reduced to one third of a second (see Fig. 19.4). Therefore, with vigorous exercise, there is still time to fully oxygenate the blood. Pulmonary end-capillary Po_2 still equals alveolar Po_2 and rarely falls with vigorous exercise. In abnormal situations, in which there is a thickening of the alveolar–capillary membrane (e.g., edema) so that oxygen diffusion is impaired, end-capillary Po_2 may not reach equilibrium with alveolar Po_2. In this case, there is measurable difference between alveolar and end-capillary Po_2.

▶ DIFFUSING CAPACITY

In practice, direct measurements of A_s, T, and D in intact lungs are impossible to make. To circumvent this problem, Fick's law can be rewritten as shown in Figure 19.5, in which the three terms are combined as lung **diffusing capacity** (**DL**) and is a measure of the lung's ability to transfer gases.

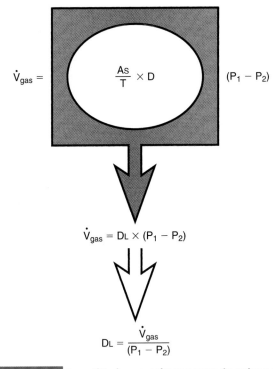

$$\dot{V}_{gas} = \boxed{\frac{As}{T} \times D} (P_1 - P_2)$$

$$\dot{V}_{gas} = D_L \times (P_1 - P_2)$$

$$D_L = \frac{\dot{V}_{gas}}{(P_1 - P_2)}$$

Figure 19.5 **Lung diffusing capacity measures the volume of O_2 taken up per minute.** Membrane surface area (As), the gas diffusion coefficient (D), and membrane thickness (T) affect gas diffusion in the lungs. These properties are combined into one term, *lung diffusion capacity* (DL). The patients DL is equal to the volume of gas transferred per minute ($\dot{V}_{gas}$) divided by the mean partial pressure gradient ($P_1 - P_2$).

Diffusing capacity measures the rate of oxygen transfer across the alveolar–capillary membrane.

The diffusing capacity provides a measure of the rate of gas transfer in the lungs per partial pressure gradient. For example, if 250 mL of O_2 per minute are taken up and the average alveolar–capillary P_{O_2} difference during a normal transit time is 14 mm Hg, then the DL for oxygen is 18 mL/min/mm Hg. Because the initial alveolar–capillary difference for oxygen cannot be measured and can only be estimated, CO is used to determine the lung diffusing capacity in patients. CO offers several advantages for measuring DL:

- Its uptake is limited by diffusion and not by blood flow.
- There is essentially no CO in the venous blood.
- The affinity of CO for hemoglobin is 210 times greater than that of oxygen, which causes the partial pressure of CO to remain essentially zero in the pulmonary capillaries.

To measure the diffusing capacity in a patient with CO, the equation is

$$DL = \frac{\dot{V}_{CO}}{P_{ACO}} \qquad (3)$$

where $\dot{V}_{CO}$ equals CO uptake in milliliters per minute and P_{ACO} equals alveolar partial pressure of CO.

The most common technique for making this measurement is called the **single-breath test**. The patient inhales a single breath of a dilute mixture of CO and holds his or her breath for about 10 seconds. By determining the percentage of CO in the alveolar gas at the beginning and the end of 10 seconds and by measuring lung volume, one can calculate $\dot{V}_{CO}$. The normal resting value for DLCO depends on age, sex, and body size. DLCO ranges from 20 to 30 mL/min/mm Hg and decreases with pulmonary edema or a loss of alveolar membrane (e.g., emphysema).

Blood hematocrit and pulmonary capillary blood volume affect lung diffusing capacity for oxygen.

Diffusing capacity does not depend solely on the diffusion properties of the lungs; it is also affected by blood hematocrit and pulmonary capillary blood volume. Both the hematocrit and capillary blood volume affect DL in the same direction (i.e., a decrease in either the hematocrit or capillary blood volume will lower the diffusing capacity in otherwise normal lungs). For example, if two people have the same pulmonary diffusion properties but one is anemic (reduced hematocrit), the anemic person will have a decreased lung diffusing capacity. An abnormally low cardiac output lowers the pulmonary capillary blood volume, which decreases the alveolar capillary surface area and will, in turn, decrease the diffusing capacity in otherwise normal lungs.

▶ GAS TRANSPORT BY THE BLOOD

The transport of O_2 and CO_2 by the blood, often referred to as *gas transport*, is an important step in the overall gas exchange process and is one of the functions of the systemic circulation.

Most of the oxygen in the blood is transported by hemoglobin.

Oxygen is transported to the tissues in two forms: combined with **hemoglobin (Hb)** in the red cell or physically dissolved in the blood. Approximately 98% of the oxygen is carried by hemoglobin, and the remaining 2% is carried in the physically dissolved form. The amount of physically dissolved oxygen in the blood can be calculated from the following equation:

$$\text{Dissolved } O_2 \,(\text{mL/dL}) = 0.003(\text{mL/dL/mm Hg}) \\ \times P_{AO_2}\,(\text{mm Hg}) \qquad (4)$$

If arterial oxygen tension (P_{AO_2}) equals 100 mm Hg, then dissolved O_2 equals 0.3 mL/dL.

The hemoglobin molecule consists of four oxygen-binding heme sites and a globular protein chain. When hemoglobin binds with oxygen, it is called **oxyhemoglobin (Hb_{O_2})**. The hemoglobin that does not bind with O_2 is called **deoxyhemoglobin (Hb)**. Each gram of hemoglobin can bind with 1.34 mL of oxygen. Oxygen binds rapidly and reversibly to hemoglobin: $O_2 + Hb \leftrightarrow Hb_{O_2}$. The amount of oxyhemoglobin is a function of the partial pressure of oxygen in the blood. In the pulmonary capillaries, in which P_{O_2} is high, the reaction is shifted to the right to form oxyhemoglobin. In tissue capillaries, in which P_{O_2} is low, the reaction is shifted to the left; oxygen is unloaded from hemoglobin and becomes available to the cells. The maximum amount of oxygen that can be carried by hemoglobin is called the **oxygen-carrying** capacity—about 20 mL O_2/dL blood in a healthy young adult. This value is calculated assuming a normal hemoglobin concentration of 15 g Hb/dL of blood (1.34 mL O_2/g Hb × 15 g Hb/dL blood = 20.1 mL O_2/dL blood).

Oxygen content is the amount of oxygen actually bound to hemoglobin (whereas capacity is the amount that can potentially be bound). **Oxygen saturation** is the percentage saturation of hemoglobin with oxygen (So_2) and is calculated from the ratio of oxyhemoglobin content over capacity:

$$So_2 = \frac{Hbo_2 \text{ content}}{Hbo_2 \text{ capacity}} \times 100 \quad (5)$$

Therefore, oxygen saturation is the ratio of the quantity of oxygen *actually bound* to the quantity that can be *potentially bound*. For example, if oxygen content is 16 mL O_2/dL blood and oxygen capacity is 20 mL O_2/dL blood, then the blood is 80% saturated. Arterial blood saturation of hemoglobin with oxygen (Sao_2) is normally about 98%.

Oxyhemoglobin–equilibrium curve illustrates the effect that plasma Po_2 has on the loading and unloading of oxygen from hemoglobin.

Blood Po_2, O_2 saturation, and oxygen content are three closely related indices of oxygen transport. The relationship between these three entities is illustrated by the **oxyhemoglobin–equilibrium curve** (Fig. 19.6). The S shape of the curve results because the hemoglobin affinity for oxygen increases progressively as blood Po_2 increases.

The shape of the oxyhemoglobin–equilibrium (O_2–Hb) curve also reflects several important physiologic advantages. The *plateau region* of the curve is the **loading phase**, in which oxygen is loaded onto hemoglobin to form oxyhemoglobin in the pulmonary capillaries. In particular, the plateau region

illustrates how oxygen saturation and content remain fairly constant despite wide fluctuations in alveolar Po_2. For example, if Pao_2 were to rise from 100 to 120 mm Hg, hemoglobin would become only slightly more saturated (from 97% to 98%). For this reason, oxygen content cannot be raised appreciably by hyperventilation. The *steep region* is the **unloading phase** of the curve and illustrates that large quantities of oxygen are released or unloaded from hemoglobin. The unloading phase occurs in the tissue capillaries where low capillary Po_2 prevails. Thus, the functional importance of the S-shaped oxyhemoglobin–equilibrium curve enables oxygen to saturate hemoglobin under high partial pressures in the lungs and to give up large amounts of oxygen with small changes in Po_2 at the tissue level.

A change in the binding affinity of hemoglobin for O_2 shifts the oxyhemoglobin–equilibrium curve to the right or left of normal (Fig. 19.7). A functional way to assess the binding affinity of

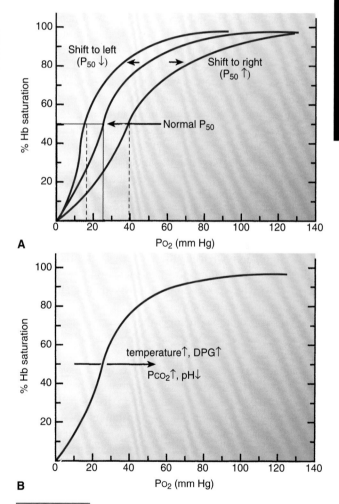

A

B

Figure 19.7 Oxygen has a strong binding affinity for hemoglobin (Hb). **(A)** P_{50} is a measure of Hb's affinity to bind with oxygen. **(B)** An increase in temperature, [H^+], or arterial Pco_2 causes a rightward shift of the oxyhemoglobin equilibrium curve. An increase in P_{50} indicates a lower binding affinity for oxygen, which favors the unloading of O_2 from Hb at the tissue level. An increase in red cell levels of 2,3-diphosphoglycerate (DPG) will also shift the curve to the right. The increase in DPG occurs with hypoxemic conditions. P_{50}, partial pressure of O_2 required to saturate 50% of the hemoglobin with oxygen.

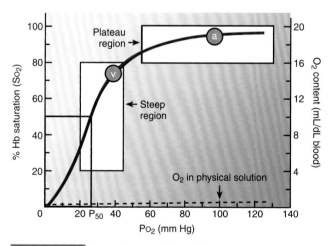

Figure 19.6 The oxyhemoglobin equilibrium curve is nonlinear. The oxygen saturation (left vertical axis) or oxygen content (right vertical axis) is plotted against the partial pressure of oxygen (horizontal axis) to generate an oxyhemoglobin equilibrium curve. The curve is S shaped and can be divided into a *plateau region* and a *steep region*. The dashed line indicates the amount of oxygen dissolved in the plasma. a, arterial; v, venous; Hb, hemoglobin; So_2, oxygen saturation; P_{50}, partial pressure of O_2 required to saturate 50% of the hemoglobin with oxygen.

Hb for O_2 is the P_{50}—the Po_2 at which 50% of the hemoglobin is saturated. As seen in Figure 19.6, the normal P_{50} for arterial blood is 26 to 28 mm Hg. A high P_{50} signifies a decrease in hemoglobin's affinity for oxygen and results in a rightward shift in the oxyhemoglobin–equilibrium curve, whereas a low P_{50} signifies the opposite and shifts the curve to the left. A shift in the P_{50} in either direction has the greatest effect on the steep phase and only a small effect on the loading of oxygen in the normal lung, because loading occurs at the plateau.

Blood pH, body temperature, and arterial Pco_2 significantly alter the P_{50}.

Several factors affect the binding affinity of hemoglobin for O_2, including blood temperature, arterial carbon dioxide tension, and arterial pH. A rise in Pco_2, a fall in pH, and a rise in temperature all shift the O_2–Hb curve to the right (see Fig. 19.7). The effect of carbon dioxide and hydrogen ions on hemoglobin's affinity for oxygen is known as the **Bohr effect**. A shift of the O_2–Hb to the right is physiologically advantageous at the tissue level because the affinity is lowered (increased P_{50}). A rightward shift enhances the unloading of oxygen for a given Po_2 in the tissue, and a leftward shift increases the affinity of hemoglobin for oxygen, thereby lowering the ability to release oxygen to the tissues. A simple way to remember the functional importance of these shifts is that an exercising muscle is hot and acidic and produces large amounts of carbon dioxide (high Pco_2), all of which favor the unloading of more oxygen to metabolically active muscles.

Red blood cells contain 2,3-diphosphoglycerate (2,3-DPG), an organic phosphate compound that can also affect the affinity of hemoglobin for oxygen. In red cells, 2,3-DPG levels are much higher than in other cells because erythrocytes lack mitochondria. An increase in 2,3-DPG facilitates unloading of oxygen from the red cell at the tissue level (shifts the curve to the right). An increase in red cell 2,3-DPG occurs with exercise and with hypoxia (e.g., high-altitude, chronic lung disease).

It is important to remember that oxygen content, rather than Po_2 or Sao_2, is what keeps us alive and serves as a better gauge for oxygenation. For example, a person can have a normal arterial Po_2 and Sao_2 but reduced oxygen content. This situation is seen in patients who have **anemia** (a decreased number of circulating red cells). A patient with anemia who has a hemoglobin concentration half of normal (7.5 g/dL instead of 15 g/dL) will have a normal arterial Po_2 and Sao_2, but oxygen content will be reduced to half of normal. A patient with anemia has a normal Sao_2 because that content and capacity are proportionally reduced. The usual oxyhemoglobin–equilibrium curve does not show changes in blood oxygen content, because the vertical axis is saturation. As a result, the vertical axis is often changed to oxygen content (mL O_2/dL blood) and then changes in content are seen (Fig. 19.8). The shape of the oxyhemoglobin–equilibrium curve does not change, but the curve moves down to reflect the reduction in oxygen content. A good analogy for comparing an anemic patient with a healthy patient is comparing a bicycle tire with a truck tire: Both can have the same air pressure, but the amount of air each tire holds is different.

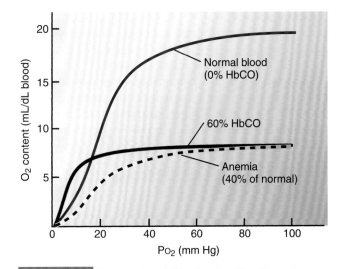

Figure 19.8 Hematocrit and CO poisoning affect the oxyhemoglobin equilibrium curve. Severe anemia can lower the O_2 content to 40% of normal. The blood O_2 content of a person exposed to CO is shown for comparison. When the blood is 60% saturated with carbon monoxide (HbCO), O_2 content is reduced to about 8 mL/dL of blood. Note the leftward shift of the oxyhemoglobin equilibrium curve when CO binds with hemoglobin.

Carbon monoxide has a greater binding affinity for hemoglobin than that of oxygen.

CO interferes with oxygen transport by competing for the same binding sites on hemoglobin as O_2. CO binds to hemoglobin to form **carboxyhemoglobin (HbCO)**. The reaction (Hb + CO ↔ HbCO) is reversible and is a function of Pco. This means that breathing higher concentrations of CO will favor the reaction to the right and form more HbCO. Breathing fresh air will favor the reaction to the left, which will cause CO to be released from the hemoglobin. A striking feature of CO is a binding affinity about 210 times that of oxygen. Consequently, CO will bind with the same amount of hemoglobin as oxygen at a partial pressure 210 times lower than that of oxygen. For example, breathing normal air (21% O_2) contaminated with 0.1% CO would cause half of the hemoglobin to be saturated with CO and half with O_2. With the high affinity of hemoglobin for CO, breathing a small amount of CO can result in the formation of large amounts of HbCO. Arterial Po_2 in the plasma will still be normal because the oxygen diffusion gradient has not changed. However, oxygen content will be greatly reduced because oxygen cannot bind to hemoglobin. This is seen in Figure 19.8, which shows the effect of CO on the oxyhemoglobin–equilibrium curve. When the blood is 60% saturated with CO (carboxyhemoglobin), the oxygen content is reduced to <10 mL/dL. The presence of CO also shifts the curve to the left, making it more difficult to unload or release oxygen to the tissues. CO is dangerous for several reasons:

- It has a strong binding affinity for hemoglobin.

- As an odorless, colorless, and nonirritating gas, it is virtually undetectable.

- Pao_2 is normal, and there is no feedback mechanism to indicate that oxygen content is low.

- There are no physical signs of hypoxemia (i.e., lack of **cyanosis** or bluish color around the lips and fingers) because the blood is bright cherry red when CO binds with hemoglobin.

Therefore, a person can be exposed to CO and have oxygen content reduced to a level that becomes lethal, by causing tissue anoxia, without the person being aware of the danger. The brain is one of the first organs affected by lack of oxygen. CO can alter reaction time and cause blurred vision and, if severe enough, unconsciousness.

The best treatment for CO poisoning is breathing 100% oxygen. Because O_2 and CO compete for the same binding site on the hemoglobin molecule, breathing a high oxygen concentration will drive off the CO and favor the formation of oxyhemoglobin. The addition of 5% carbon dioxide to the inspired gas stimulates ventilation, which lowers the CO and enhances the release of CO from hemoglobin. The loading and unloading of CO from hemoglobin are functions of P_{CO}. Oxygen is not always beneficial—oxygen metabolism can produce harmful products that injure tissues.

Most of the carbon dioxide in the blood is transported as bicarbonate.

Figure 19.9 illustrates the processes involved in carbon dioxide transport. Carbon dioxide is carried in the blood in three forms:

- Physically dissolved in the plasma (10%)
- As bicarbonate ions in the plasma and in the red cells (60%)
- As carbamino proteins (30%)

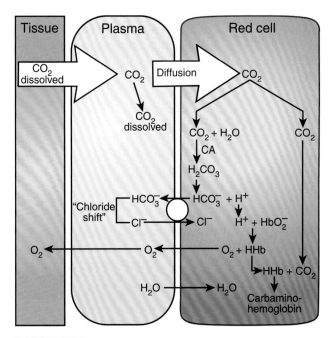

Figure 19.9 Bicarbonate is the main transporter of carbon dioxide. CO_2 is transported in the blood in three forms: physically dissolved, as HCO_3^-, and as carbaminohemoglobin in the red cell (see text for details). The uptake of CO_2 favors the release of O_2. A major portion of the CO_2 is carried in the form of bicarbonate. CA, carbonic anhydrase; Hb, hemoglobin.

The high P_{CO_2} in the interstitial fluid drives carbon dioxide from the tissue into the blood, but only a small amount stays as dissolved CO_2 in the plasma. The bulk of the carbon dioxide diffuses into the red cell, in which it forms either carbonic acid (H_2CO_3) or **carbaminohemoglobin**. In the red cell, carbonic acid is formed in the following reaction:

$$CO_2 + H_2O \xleftarrow{\quad CA \quad} H_2CO_3 \leftrightarrow H^+ + HCO_3^- \qquad (6)$$

The hydration of CO_2 would take place slowly if it were not accelerated about 1,000 times in red cells by the enzyme **carbonic anhydrase (CA)**. This enzyme is also found in renal tubular cells, gastrointestinal mucosa, muscle, and other tissues, but its activity is highest in red blood cells.

Carbonic acid readily dissociates in red blood cells to form bicarbonate (HCO_3^-) and H^+. HCO_3^- leaves the red blood cells, and chloride diffuses in from the plasma to maintain electrical neutrality (see Fig. 19.9). The chloride movement is known as the *chloride shift* and is facilitated by a chloride–bicarbonate exchanger (anion exchanger) in the red blood cell membrane. The H^+ cannot readily move out because of the low permeability of the membrane to H^+. Most of the H^+ is buffered by hemoglobin: $H^+ + HbO_2^- \leftrightarrow HHb + O_2$. As H^+ binds to hemoglobin, it decreases oxygen binding and shifts the oxyhemoglobin–equilibrium curve to the right. This promotes the unloading of oxygen from hemoglobin in the tissues and favors the carrying of carbon dioxide. In the pulmonary capillaries, the oxygenation of hemoglobin favors the unloading of carbon dioxide.

Carbaminohemoglobin is formed in red cells from the reaction of carbon dioxide with free amine groups (NH_2) on the hemoglobin molecule:

$$CO_2 + HbNH_2 \leftrightarrow HbNHCOOH \qquad (7)$$

Deoxygenated hemoglobin can bind much more CO_2 in this way than oxygenated hemoglobin. Although major reactions related to CO_2 transport occur in the red cells, the bulk of the CO_2 is actually carried in the plasma in the form of bicarbonate.

A **carbon dioxide equilibrium curve** can be constructed in a fashion similar to that for oxygen (Fig. 19.10). The carbon dioxide equilibrium curve is nearly a straight-line function of P_{CO_2} in the normal arterial CO_2 range. Note that a higher P_{O_2} will shift the curve downward and to the right. This is known as the **Haldane effect**, and its advantage is that it allows the blood to load more CO_2 in the tissues and unload more CO_2 in the lungs.

Important differences are observed between the carbon dioxide and oxygen equilibrium curves (see Figs. 19.7 and 19.10). First, 1 L of blood can hold much more carbon dioxide than oxygen. Second, the CO_2 equilibrium curve is steeper and more linear, and because of its shape, large amounts of CO_2 can be loaded and unloaded from the blood with a small change in P_{CO_2}. This is important not only in gas exchange and transport but also in the regulation of acid–base balance.

▶ RESPIRATORY CAUSES OF HYPOXEMIA

Under normal conditions, hemoglobin is 100% saturated with oxygen when the blood leaves the pulmonary capillaries, and the P_{O_2} at the end of the capillary (end-capillary P_{O_2})

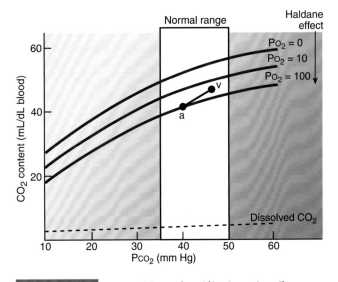

Figure 19.10 Increased O₂ tension shifts the carbon dioxide equilibrium curve. The carbon dioxide equilibrium curve is relatively linear. An increase in P_{O_2} tension causes a rightward and downward shift of the curve. The P_{O_2} effect on the CO_2 equilibrium curve is known as the *Haldane effect*. The *dashed line* indicates the amount dissolved in plasma. a, arterial CO_2 content; v, CO_2 content in mixed venous blood.

equals alveolar P_{O_2}. However, the blood that leaves the lungs (via the pulmonary veins) and returns to the left side of the heart has a lower P_{O_2} than does the pulmonary end-capillary blood. As a result, the systemic arterial blood has an average oxygen tension (Pa_{O_2}) of about 95 mm Hg and is only 98% saturated.

Alveolar–arterial oxygen difference originates because bronchial circulation mixes with oxygenated blood.

The difference between alveolar oxygen tension (Pa_{O_2}) and arterial oxygen tension (Pa_{O_2}) is the **alveolar–arterial oxygen gradient** or A–aO_2 gradient (Fig. 19.11). Because alveolar P_{O_2} is normally 100 to 102 mm Hg and arterial P_{O_2} is 85 to 95 mm Hg, a normal A–aO_2 gradient is between 5 and 15 mm Hg. The A–aO_2 gradient is obtained from blood gas measurements and the alveolar gas equation to determine Pa_{O_2}. Recall from Chapter 18 that the simplified equation is $Pa_{O_2} = F_{IO_2} \times (P_B - 47) - 1.2 \times Pa_{CO_2}$.

The A–aO_2 gradient arises in the healthy person because a fraction of venous blood mixes with oxygenated blood. This mixing of unoxygenated and oxygenated blood is known as **venous admixture**. The venous admixture is the result of two unique feature of the pulmonary circulation. One is due to a small anatomical shunt (e.g., bronchial circulation) that dumps venous blood back into the pulmonary oxygenated venous blood. The second cause is due to regional variations in the ventilation/perfusion ($\dot{V}_A/\dot{Q}$) ratio. The ($\dot{V}_A/\dot{Q}$) ratio is simply the ratio of minute alveolar ventilation to pulmonary blood flow in any unit of the lung. Proper oxygenation of blood in the pulmonary circulation leaving any region of the lung (regional alveoli and their blood supply) occurs when ventilation and perfusion are equally matched in that region. Oxygenation of blood leaving a region of the lung decreases any time there is too little ventilation per minute for the

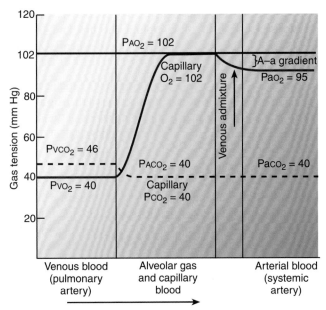

Figure 19.11 An oxygen gradient is established between the alveoli and arterial blood. The diagram shows O_2 and CO_2 tensions in blood in the pulmonary artery, pulmonary capillaries, and systemic arterial blood. The P_{O_2} leaving the pulmonary capillary has equilibrated with alveolar P_{O_2}. However, systemic arterial P_{O_2} is below alveolar P_{O_2}. Venous admixture results in the alveolar–arterial oxygen gradient (A–aO_2). Pa_{CO_2}, partial pressure of alveolar carbon dioxide; Pa_{CO_2}, partial pressure of arterial carbon dioxide; Pa_{O_2}, partial pressure of alveolar oxygen; Pa_{O_2}, partial pressure of arterial oxygen; Pv_{CO_2}, partial pressure of venous carbon dioxide; Pv_{O_2}, partial pressure of oxygen.

amount of blood perfusing that region per minute. A more extensive evaluation of changes in the ($\dot{V}_A/\dot{Q}$) ratio is provided in Chapter 20. Approximately half of the normal A–aO_2 gradient is caused by the bronchial circulation and half is caused by regional variations of the ($\dot{V}_A/\dot{Q}$) ratio. In some pathophysiologic disorders, the A–aO_2 gradient can be greatly increased. A value of >15 mm Hg is considered abnormal and leads to low oxygen in the blood or **hypoxemia**. The normal ranges of blood gases are shown in Table 19.1. Values for Pa_{O_2} below 85 mm Hg indicate hypoxemia. A Pa_{CO_2} < 35 mm Hg is called **hypocapnia**, and a Pa_{CO_2} > 48 mm Hg is called **hypercapnia**. A pH value for arterial blood of <7.35 or >7.45 is called **acidemia** or **alkalemia**, respectively.

TABLE 19.1	Arterial Blood Gases
Parameter	**Normal Range***
Pa_{O_2}	85–95 mm Hg
Pa_{CO_2}	35–48 mm Hg
Sa_{O_2}.	94%–98%
pH	7.35–7.45
HCO_3^-	23–28 mEq/L

*Normal range at sea level.
Pa_{O_2}, partial pressure of arterial oxygen; Pa_{CO_2}, partial pressure of carbon dioxide; Sa_{O_2}, saturation of arterial oxygen.

Free Radical–Induced Lung Injury

An "oxygen paradox" has long been recognized in biology, but only recently has it been well understood: Oxygen is essential for life, but too much oxygen can be harmful to both cells and the organism. The synthesis of **adenosine triphosphate (ATP)** involves reactions in which molecular oxygen is reduced to form water. This reduction is accomplished by the addition of four electrons by the mitochondrial electron transport system. About 98% of the oxygen consumed is reduced to water in the mitochondria. "Leaks" in the mitochondrial electron transport system, however, allow oxygen to accept fewer than four electrons, forming a **free radical**.

A free radical is any atom, molecule, or group of molecules with an unpaired electron in its outermost orbit. Free radicals include the **superoxide ion ($O_2 \bullet^-$)** and the **hydroxyl radical ($\bullet OH$)**. The single unpaired electron in the free radical is denoted by a dot. The $\bullet OH$ radical is the most reactive and most damaging to cells. **Hydrogen peroxide (H_2O_2)**, although not a free radical, is also reactive to tissues and has the potential to generate the hydroxyl radical ($\bullet OH$). These three substances are collectively called **reactive oxygen species (ROS)**. In addition to free radicals produced by leaks in the mitochondrial transport system, ROS can also be formed by cytochrome P450, in the production of nicotinamide adenine dinucleotide phosphate and in arachidonic acid metabolism. A superoxide ion in the presence of nitric oxide will form peroxynitrite, another free radical that is also extremely toxic to cells. Under normal conditions, ROS are neutralized by the protective enzymes superoxide dismutase, catalase, and peroxidases, and no damage occurs. However, when ROS are greatly increased, they overwhelm the protective enzyme systems and damage cells by oxidizing membrane lipids, cellular proteins, and DNA.

The lungs are a major target organ for free radical injury, and the pulmonary vessels are most susceptible as the primary site of injury. Damage to the pulmonary capillaries by free radicals causes the capillaries to become leaky, leading to pulmonary edema. In addition to intracellular production, ROS are produced during inflammation and episodes of oxidant exposure (i.e., oxygen therapy or breathing ozone and nitrogen dioxide from polluted air). During the inflammatory response, neutrophils become sequestered and activated; they undergo a respiratory burst (which produces free radicals) and release catalytic enzymes. This release of free radicals and catalytic enzymes functions to kill bacteria, but endothelial cells can become damaged in the process.

Paraquat, an agricultural herbicide, is another source of free radical–induced injury to the lungs. Crop dusters and migrant workers are particularly at risk because of exposure to paraquat through the lungs and skin. Tobacco or marijuana that has been sprayed with paraquat and subsequently smoked can also produce lung injury from ROS.

Ischemia reperfusion is another cause of free radical–induced injury in organs. In the lungs, ischemia–reperfusion injury results from a blood clot that gets lodged in the pulmonary circulation. Tissues beyond the clot (or embolus) become ischemic, cellular ATP decreases, and hypoxanthine accumulates. When the clot dissolves, blood flow is reestablished. During the reperfusion phase, hypoxanthine, in the presence of oxygen, is converted to xanthine and then to urate. These reactions are catalyzed by the enzyme xanthine oxidase on the pulmonary endothelium, resulting in the production of superoxide ions. Neutrophils also become sequestered and activated in these vessels during the reperfusion phase. Therefore, the pulmonary vasculature and surrounding lung parenchyma become damaged from a double hit of free radicals—those produced from the oxidation of hypoxanthine and those from activated neutrophils. ◼

Regional hypoventilation is the major cause of hypoxemia.

The causes of hypoxemia are classified as respiratory or nonrespiratory (Table 19.2). The respiratory causes of hypoxemia are by far the most common, and they are listed in order of frequency in Table 19.2. As seen from Table 19.2, **regional hypoventilation** is the major cause of hypoxemia (about 90% of cases) and reflects a local ($\dot{V}_A/\dot{Q}$) ratio imbalance. The matching of airflow and blood flow is best examined by considering the **ventilation/perfusion ratio**, which compares alveolar ventilation with blood flow in lung regions. Because resting healthy people have an alveolar ventilation ($\dot{V}_A$) of 4 L/min and a cardiac output ($\dot{Q}$) of 5 L/min, the ideal alveolar ventilation/perfusion ratio ($\dot{V}_A/\dot{Q}$) should be 0.8 (there are no units; this is a ratio). When a partially obstructed airway occurs, a fraction of the blood that passes through the capillary bed of the obstructed airway does not get fully oxygenated, resulting in an increase in venous admixture. Only a small amount of venous admixture is required to lower systemic arterial P_{O_2} as a result of the nature of the oxyhemoglobin–equilibrium curve. This can be seen from Figure 19.12,

TABLE 19.2 Pathophysiologic Causes of Hypoxemia

Causes	Effect on A–aO_2 Gradient
Respiratory	
Regional low ($\dot{V}_A/\dot{Q}$) ratio	Increased
Anatomic shunt	Increased
Generalized hypoventilation	Normal
Diffusion block	Increased
Nonrespiratory	
Intracardiac right-to-left shunt	Increased
Decreased P_{IO_2}, low P_B, low F_{IO_2}	Normal
Reduced oxygen content (anemia and CO poisoning)	Normal

A–aO_2, alveolar–arterial oxygen gradient; F_{IO_2}, fractional concentration of inspired oxygen; P_B, barometric pressure; P_{IO_2}, partial pressure of inspired oxygen; ($\dot{V}_A/\dot{Q}$) ratio, ventilation/perfusion ratio.

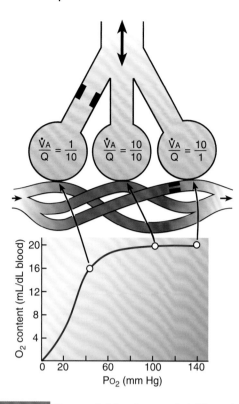

Figure 19.12 **Venous admixture lowers arterial O$_2$ content.** Because of the S-shaped oxyhemoglobin equilibrium curve, a high ventilation/perfusion ratio (ratio) has little effect on arterial O$_2$ content. However, mixing with blood from a region with a low ($\dot{V}_A/\dot{Q}$) ratio can dramatically lower P$_{O_2}$ in blood leaving the lungs.

which depicts oxygen content from three groups of alveoli with low, normal, and high ($\dot{V}_A/\dot{Q}$) ratios. The oxygen content of the blood leaving these alveoli is 16.0, 19.5, and 20.0 mL/dL of blood, respectively. As Figure 19.12 shows, a low ($\dot{V}_A/\dot{Q}$) ratio is far more serious because it has the greatest effect on lowering both the P$_{O_2}$ and the O$_2$ content because of the non-linear shape of the oxyhemoglobin equilibrium curve. Patients who have an abnormally low ($\dot{V}_A/\dot{Q}$) ratio have a high A–aO$_2$ gradient, low P$_{O_2}$, and low O$_2$ content, but usually a normal or slightly elevated Paco$_2$. Paco$_2$ does not change much because the CO$_2$ equilibrium curve is nearly linear, which allows excess CO$_2$ to be removed from the blood by the lungs.

Another cause for a regionally low ($\dot{V}_A/\dot{Q}$) ratio is a large blood clot that occludes a major artery in the lungs. When a major pulmonary artery becomes occluded, a greater portion of the cardiac output is redirected to another part of the lungs, resulting in overperfusion with respect to alveolar ventilation. This causes a regionally low ($\dot{V}_A/\dot{Q}$) and leads to an increase in venous admixture.

The next most common cause of hypoxemia is a **shunt**, either an intracardiac right-to-left anatomic shunt or an intrapulmonary shunt. The latter occurs when an airway is totally obstructed by a foreign object (such as a peanut) or by a lung tumor. Patients with hypoxemia stemming from a shunt also have a high A–aO$_2$ gradient, low P$_{O_2}$, low O$_2$ content, and a normal or slightly elevated Paco$_2$. A test that is often used to distinguish between an abnormally low ($\dot{V}_A/\dot{Q}$) ratio and a shunt is to have the patient breathe 100% O$_2$ for 15 minutes. If the Pao$_2$

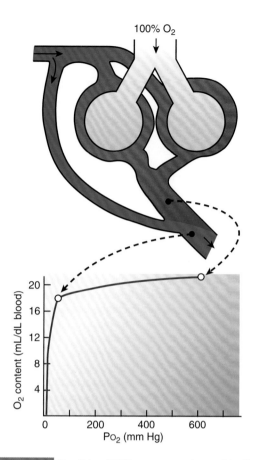

Figure 19.13 **Breathing 100% oxygen can be used to diagnose a shunt.** A shunt can be diagnosed by having the subject breathe 100% O$_2$ for 15 minutes. P$_{O_2}$ in systemic arterial blood in a patient with a shunt does not increase above 150 mm Hg during the 15-minute period. The shunted blood is not exposed to 100% O$_2$, and the venous admixture reduces arterial P$_{O_2}$.

is >150 mm Hg, the cause is a low ($\dot{V}_A/\dot{Q}$) ratio. If the patient's Pao$_2$ is <150 mm Hg, the cause of hypoxemia is a shunt. The principle for using 100% O$_2$ is illustrated in Figure 19.13. The patient with regional hypoventilation who breathes 100% O$_2$ compensates for the low ($\dot{V}_A/\dot{Q}$) ratio, and because all of the blood leaving the pulmonary capillaries is now fully saturated, the venous admixture is eliminated. However, the low arterial P$_{O_2}$ does not get corrected by breathing 100% O$_2$ in a patient with a shunt because the enriched oxygen mixture never comes into contact with the shunted blood.

Generalized hypoventilation, the third most common cause of hypoxemia, occurs when alveolar ventilation is abnormally decreased. This situation can arise from a chronic obstructive pulmonary disorder (such as emphysema) or depressed respiration (as a result of a head injury or a drug overdose, for example). Because alveolar ventilation is depressed, there is also a significant increase in arterial Pco$_2$, with a concomitant decrease in arterial pH. In generalized hypoventilation, total ventilation is insufficient to maintain normal systemic arterial P$_{O_2}$ and Pco$_2$. A feature that distinguishes generalized hypoventilation from the other causes of hypoxemia is a normal A–aO$_2$ gradient, as a result of the alveolar and arterial P$_{O_2}$ being lowered equally. If a patient has a low Pao$_2$ and a normal A–aO$_2$ gradient, the cause of

CLINICAL FOCUS | 19.2

Anemia

Anemia, an abnormally low hematocrit or hemoglobin concentration, is by far the most common disorder affecting erythrocytes. The different causes of anemia can be grouped into three categories: decreased erythropoiesis by bone marrow, blood loss, and increased rate of red cell destruction (hemolytic anemia).

Several mechanisms lead to decreased production of red cells by the bone marrow, including aplastic anemia, malignant neoplasms, chronic renal disease, defective DNA synthesis, defective hemoglobin synthesis, and chronic liver disease. **Aplastic anemia** is the result of stem cell destruction in the bone marrow, which leads to decreased production of white cells, platelets, and erythrocytes. Malignant neoplasms (e.g., leukemia) cause an overproduction of immature red cells. Patients with chronic renal disease have a decreased production of erythropoietin, with a concomitant decrease in red cell production.

Patients with defective DNA synthesis have **megaloblastic anemia**, a condition in which red cell maturation in the bone marrow is abnormal; this may result from vitamin B_{12} or folic acid deficiency. These cofactors are essential for DNA synthesis. Vitamin B_{12} is present in high concentration in liver and, to some degree, in most meat, but it is absent in plants. Vitamin B_{12} deficiency is rare except in strict vegetarians. Folic acid is widely distributed in leafy vegetables; folic acid deficiency commonly occurs when malnutrition is prevalent. **Pernicious anemia** is a form of megaloblastic anemia resulting from vitamin B_{12} deficiency. Most common in adults over 60, it results not from deficient dietary intake but from a decreased vitamin B_{12} absorption by the small intestine. Pernicious anemia is linked to an autoimmune disease in which there is immunologic destruction of the intestinal mucosa, particularly the gastric mucosa.

Iron deficiency anemia is the most common cause of anemia worldwide. Although it occurs in both developed and undeveloped countries, the causes are different. In developed countries, the cause is usually a result of pregnancy or chronic blood loss resulting from gastrointestinal ulcers or neoplasms. In undeveloped countries, hookworm infections account for most cases of iron deficiency anemia.

Acute or chronic blood loss is another cause of anemia. With hemorrhage, red cells are lost and the hypovolemia causes the kidneys to retain water and electrolytes as compensation. Retention of water and electrolytes restores the blood volume, but the concomitant dilution of the blood causes a further decrease in the red cell count, hemoglobin concentration, and hematocrit. Chronic bleeding is compensated by erythroid hyperplasia, which eventually depletes iron stores. Therefore, chronic blood loss results in iron deficiency anemia.

The last category, increased rate of red cell destruction, includes the Rh factor and sickle cell anemia. The Rhesus (Rh) blood group antigens are involved in maintaining erythrocyte structure. Patients who lack Rh antigens (Rh null) have severe deformation of the red cells.

Sickle cell anemia, associated with the abnormal hemoglobin HbS gene, is common in Africa, India, and among African Americans but is rare in the Caucasian and Asian populations. In the sickle cell trait, which occurs in about 9% of African Americans, one abnormal gene is present. A single point mutation occurs in the hemoglobin molecule, causing the normal glutamic acid at position 6 of the beta chain to be replaced with valine, resulting in HbS. The amino acid substitution is on the surface, resulting in a tendency for the hemoglobin molecule to crystallize with anoxia. However, heterozygous people have no symptoms, and oxygen transport by fetal (HbF) and adult hemoglobin (HbA) is normal. The sickle cell trait (i.e., heterozygous people) offers protection against malaria, and this selective advantage is thought to have favored the persistence of the HbS gene, especially in regions in which malaria is common. Sickle cell disease represents the homozygous condition (S/S) and occurs in about 0.2% of African Americans. The onset of sickle cell anemia occurs in infancy as HbS replaces HbF; death often occurs early in adult life. Patients with sickle cell anemia have >80% HbS in their blood with a decrease or an absence of normal HbA.

Whatever the cause of anemia, the pathophysiologic effect is the same—hypoxemia. Symptoms include pallor of the lips and skin, weakness, fatigue, lethargy, dizziness, and fainting. If the anemia is severe, myocardial hypoxia can lead to angina pain. ■

---

hypoxemia is entirely a result of generalized hypoventilation. The best corrective measure for generalized hypoventilation is to place the patient on a mechanical ventilator, breathing room air. This treatment will return both arterial Po_2 and Pco_2 to normal. Administering supplemental oxygen to a patient with generalized hypoventilation will correct hypoxemia but not hypercapnia because ventilation is still depressed.

The least common cause of hypoxemia is a **diffusion block**. This condition occurs when the diffusion distance across the alveolar–capillary membrane is increased or the permeability of the alveolar–capillary membrane is decreased. It is characterized by a low Pao_2, a high A–aO_2

gradient, and a high $Paco_2$. Pulmonary edema is one of the major causes of diffusion block.

In summary, there are four basic respiratory disturbances that cause hypoxemia. Examining the A–aO_2 gradient or $Paco_2$ and/or breathing 100% oxygen distinguishes the four types. For example, if a patient has a low Pao_2, high $Paco_2$, and normal A–aO_2 gradient, the cause of hypoxemia is generalized hypoventilation. If the Pao_2 is low and the A–aO_2 gradient is high, then the cause can be a shunt, a regional low ($\dot{V}_A/\dot{Q}$) ratio, or a diffusion block. Breathing 100% O_2 will distinguish between a low ($\dot{V}_A/\dot{Q}$) ratio and a shunt. Diffusion impairment is the least likely cause and can be deduced if the other three causes have been eliminated.

INTEGRATED MEDICAL SCIENCES

Lethargy and Shortness of Breath

A 28-year-old Hispanic American woman had felt well and worked full time with a landscaping company until she became short of breath whenever she exerted herself. She went to the university hospital with progressive shortness of breath and generalized lethargy. She had a dry cough and some leg swelling. She had no reported fever or chest discomfort. She stated that she lost 15 lb over the last 3 months and had a total loss of energy, which was devastating because she was so active. Following her admission, hypoxia persisted despite the administration of oxygen. She was transferred to the intensive care unit for further observation and monitoring.

The patient was married, was a nonsmoker, and had no history of intravenous drug use. She had no prior history of pulmonary or cardiac conditions and was not taking any prescription or herbal medications. Her older sibling had no significant medical or surgical history. The clinical examination revealed oxygen saturation of 87% breathing ambient air, tachypnea, tachycardia, bilateral pulmonary rales in basal fields, and a low-intensity systolic murmur. The remainder of the examination was normal. The patient received packed red blood cell transfusions to correct for her low O_2 saturation and her low red cell count.

However, her symptoms did not improve and was readmitted to the university hospital. Further blood tests and a transthoracic echocardiogram were performed. The echocardiogram revealed an enlarged heart, tricuspid regurgitation, and moderate-to-severe pulmonary hypertension. A noncontrast computerized tomography (CT) of her thorax was performed and revealed no pulmonary emboli. Her O_2–Hb saturation showed a persistent low 85% breathing room air. Due to multisystem manifestations including moderately severe pulmonary hypertension and profound anemia, human immunodeficiency virus (HIV) testing was performed. She had generalized elevated white cell count and tested positive for HIV. Antiviral therapy was initiated in addition to presumptive antibiotics for possible *Pneumocystis* (*carinii*) *jiroveci* infection. *Pneumocystis jiroveci* is a fungal infection of the lungs. The disease used to be called *Pneumocystis carinii* and is often present in HIV patients. The HIV was traced back to the time patient had a blood transfusion 3 years ago following a work-related accident in which she received a serve cut on the leg with a significant loss of blood.

She did not demonstrate clinical improvement, however, and her cardiopulmonary decline was attributed to HIV-associated pulmonary hypertension. She required transfer to a specialized center for further management.

This case is a common presentation of pulmonary hypertension: a previously health young woman develops a life-threatening disease with no outward manifestation. Because these individuals look normal at rest, family, friends, and coworkers have a difficult time accepting they are sick or have a serious disease. Often, they are initially misdiagnosed when seen by a physician.

Pulmonary arterial hypertension (PAH) is defined as an increased mean pulmonary artery pressure > 25 mm Hg at rest or >30 mm Hg during exercise and a pulmonary capillary wedge pressure of <15 mm Hg. Whatever the initial cause, this devastating disease involves the narrowing of pulmonary blood vessels that increase pulmonary vascular resistance, which makes it harder for the heart to pump blood through the lungs. Over time, the affected blood vessels become stiffer and thicker known as *fibrosis.* This further increases the blood pressure within the lungs and impairs blood flow. As a result, PAH increases the workload of the heart, causing hypertrophy of the right ventricle, ultimately causing right heart failure. Symptoms of PAH are usually nonspecific and include dyspnea (shortness of breath), dizziness, leg swelling, and fatigue.

The molecular mechanism of PAH is not fully known, but it is believed to be linked to endothelial dysfunction, which results in a decreased synthesis of endothelium-derived vasodilators such as nitric oxide and prostacyclin. Moreover, there is increased stimulation of vasoconstrictors such as thromboxane and vascular endothelial growth factor (VEGF). They cause severe vasoconstriction and smooth muscle hypertrophy. Under normal conditions, in the presence of oxygen, the nitric oxide synthase produces nitric oxide from L-arginine. Adenylate cyclase and guanylate cyclase are activated in the presence of nitric oxide, and these enzymes produce cAMP and cGMP, respectively.

In the vascular endothelium, cGMP activates cGMP kinase, which activates the potassium smooth muscle channels subsequently inhibiting the calcium channels. Thus, activation of cGMP leads to a reduction of intracellular calcium and vasodilation. Phosphodiesterase type V (PDE-5) is abundant in the pulmonary tissue. Activation of PDE-5 inhibits cGMP concentration and vasodilation is stopped. The molecular pathway in patients with PAH becomes dysfunctional. They produce less NO, and other vasodilators, and produce more vasoconstrictors.

The etiology of the development of PAH varies. Multiple causes include hypoxia associated with pulmonary and nonpulmonary conditions, thromboembolism, left ventricular muscle or valve diseases, connective tissue disorders, congenital heart diseases, and human immunodeficiency virus (HIV) infection and idiopathic PAH. PAH is a well-established complication of HIV infection and HIV-infected individuals are at higher risk for development of PAH compared to the general population. This suggests that either the virus itself or the consequences of infection may directly be linked with development of PAH. Despite the strong association of PAH with HIV, the underlying pathogenesis of this association remains unclear. A number of HIV viral proteins have been proposed to promote the development of PAH during the course of HIV infection, the most likely of which is the "Nef" protein (negative factor). The Nef protein is an N-myristoylated protein, originally identified as a negative regulatory factor for HIV replication and later recognized as an important protein for the maintenance of high viral loads during the course of HIV infection. The Nef protein is abundantly expressed during early viral infection and appears to promote the initiation and persistence of HIV infection and enhancement of HIV infectivity. The Nef protein also plays an important role in the alteration in pulmonary vascular cells by decreasing endothelin-dependent vasorelaxation as well as NOS (nitrous oxide systems) expression and induces oxidative stress in an experimental model in porcine pulmonary arteries.

Various treatment options are available for PAH. These include phoshoidesterase-5 inhibitors, such as sildenafil; prostanoids, such as epoprostenol; endothelin receptor antagonists, such as bosentan; calcium channel blocker, such as diltiazem; diuretics to reduce swelling in ankles and feet (e.g., hydrochlorothiazide); blood thinner to prevent clots forming (e.g., Coumadin); and digoxin, which helps cardiac contraction.

In conclusion, the advent of newer drugs has doubled the survival of PAH, and many patients are now living well beyond a decade with reasonable function and satisfaction. ▪

Respiratory Physiology

Chapter Summary

- Oxygen uptake across the alveolar–capillary membrane is determined by the O_2 diffusion gradient, pulmonary capillary blood flow, capillary blood volume, and blood hematocrit.
- Under normal conditions, oxygen uptake in the pulmonary capillaries is limited primarily by blood flow.
- Lung diffusion capacity is a measure of the total amount of oxygen transferred. Under resting conditions, ~250 mL of oxygen per minute is transferred to the pulmonary circulation. Total amount of oxygen uptake is measured by lung diffusion capacity.
- Oxygen is transported by the blood in two forms: oxyhemoglobin and dissolved O_2 in the plasma.
- P_{50} is a measure of the binding affinity of hemoglobin (Hb) to oxygen. When Hb-binding affinity to O_2 increases, the P_{50} changes in the opposite direction.

- Changes in blood pH, $Paco_2$, and temperature alter the oxyhemoglobin–equilibrium curve.
- Oxygen content is a better determinant for adequate tissue oxygenation than arterial Po_2 or percentage O_2 saturation. Hematocrit, $Paco_2$, and CO poisoning can affect oxygen content.
- Carbon dioxide is transported in three forms: dissolved, bicarbonate, and hemoglobin.
- An alveolar–arterial oxygen gradient is caused by venous admixture.
- A normal alveolar–arterial oxygen gradient is present because alveolar ventilation and capillary blood flow are not evenly matched in regions of the lung and because bronchial circulation mixes with oxygenated blood.
- Regional hypoventilation (low ventilation/perfusion ratio) is the major cause of hypoxemia.

Chapter Review Questions

1. One of the main causes of a normal $A–aO_2$ gradient in a healthy individual is:

   A. low diffusing capacity for oxygen compared with that for carbon dioxide.
   B. a high ($\dot{V}_A/\dot{Q}$) ratio in the apex of the lungs.
   C. overventilation in the base of the lung.
   D. a small shunt from bronchial circulation.
   E. a right-to-left shunt in the heart.

The correct answer is D. The $A–aO_2$ gradient in a healthy person is due to both a low ($\dot{V}_A/\dot{Q}$) ratio at the base of the lungs and a small shunt from the bronchial circulation.

2. Which of the following will *not* cause a low lung diffusing capacity (D_L)?

   A. Decreased diffusion distance
   B. Decreased capillary blood volume
   C. Decreased surface area
   D. Decreased cardiac output
   E. Decreased hemoglobin concentration in the blood

The correct answer is A. A decrease in the diffusion distance will lead to an increase in D_L. A decrease in capillary blood volume, surface area, cardiac output, and blood hemoglobin concentration will decrease D_L.

3. Which of the following parameters would best reflect adequate oxygenation to the tissues?

   A. Arterial oxygen tension
   B. Arterial oxygen saturation
   C. Arterial oxygen tension required to make the blood 50% saturated (P50)
   D. Arterial–venous O_2 tension difference
   E. Arterial O_2 content

The correct answer is E. Arterial oxygen content provides the best index to tissue oxygenation. Arterial oxygen tension and arterial oxygen saturation can be normal in situations like anemia. Arterial–venous oxygen tension difference can also be unchanged in anemic patients. The P50 reflects the ability of the hemoglobin-binding affinity for oxygen and provides little information about tissue oxygenation.

4. A 40-year-old patient had normal blood gas values with an arterial oxygen tension (Pao_2) = 95 mm Hg, O_2 content = 19 mL/dL, O_2 capacity = 20 mL/dL, and arterial oxygen saturation (Sao_2) = 95%. Three weeks later, she developed a severe case of anemia. Which of the following would best describe her blood gases following anemia compared to her normal values?

	Pao_2	O_2 Content	O_2 Capacity	Sao_2
A.	No change	Decreases	No change	Decreases
B.	Decreases	Decreases	Decreases	No change
C.	Decreases	No change	No change	No change
D.	No change	Decreases	Decreases	No change
E.	Decreases	No change	Decreases	Increases

The correct answer is D. Anemia lowers the hemoglobin concentration in the blood. Consequently, oxygen content and capacity are both decreased. However, the percent oxygen saturation is unchanged. Arterial oxygen tension is unchanged with anemia.

5. An ideal lung unit has a ventilation/perfusion ratio ($\dot{V}_A/\dot{Q}$) of 0.8, with an alveolar oxygen tension (Pao_2) = 100 mm Hg and an alveolar carbon dioxide tension ($Paco_2$) = 40 mm Hg. Which of the following gas tensions would most likely be predicted in a lung unit with a ($\dot{V}_A/\dot{Q}$) ratio of 2?

	P_{AO_2}, mm Hg	P_{ACO_2}, mm Hg
A.	130	30
B.	104	40
C.	95	40
D.	86	46
E.	112	46

The correct answer is A. A lung unit with a high ($\dot{V}_A/\dot{Q}$) results in that region being overventilated with respect to blood flow. This is essentially equivalent to more fresh air reaching these alveoli, which results in a higher oxygen tension and a lower carbon dioxide tension.

Respiratory Physiology

Clinical Application Exercises 19.1

CHEST PAIN

A 27-year-old accountant recently drove cross-country to start a new job in Denver, Colorado. A week after her move, she started to experience chest pains. She drove to the emergency department after experiencing 24 hours of right-sided chest pain, which was worse with inspiration. She also experienced shortness of breath and stated that she felt warm. She denied any sputum production, hemoptysis, coughing, or wheezing. She is active and walks daily and never has experienced any swelling in her legs. She has never been treated for any respiratory problems and has never undergone any surgical procedures. Her medical history is negative, and she has no known drug allergies. Oral contraceptives are her only medication. She smokes a pack of cigarettes a day and consumes wine occasionally. She does not use intravenous drugs and has no other risk factors for HIV disease. Her family history is negative for asthma and any cardiovascular diseases.

Physical examination reveals a mildly obese woman in moderate respiratory distress. Her respiratory rate is 24 breaths/min and her pulse is 115 beats/min. Her blood pressure is 140/80 mm Hg, and no jugular vein distention is observed. Heart rate and rhythm are regular, with normal heart sounds and no murmurs. Her chest is clear, and her temperature is 38°C. Her extremities show signs of cyanosis, but no clubbing or edema is detected. Blood gases, obtained while she was breathing room air, reveal a P_{O_2} of 60 mm Hg and a P_{CO_2} of 32 mm Hg; her arterial blood pH is 7.49. Her A–aO_2 gradient is 40 mm Hg. A Gram-stain sputum specimen exhibited a normal flora. A chest x-ray study reveals a normal heart shadow and clear lung fields, except for a small peripheral infiltrate in the left lower lobe. A lung scan reveals an embolus in the left lower lobe.

QUESTIONS

1. What is the cause of a widened alveolar–arterial gradient in patients with pulmonary embolism?

2. What causes the decreased arterial P_{CO_2} and elevated arterial pH?
3. Why do oral contraceptives induce hypercoagulability?

ANSWERS

1. A normal A–aO_2 gradient is 5 to 15 mm Hg. A pulmonary embolus will cause blood flow to be shunted to another region of the lung. Because cardiac output is unchanged, the shunting of blood causes overperfusion, which causes an abnormally low ($\dot{V}_A/\dot{Q}$) ratio in another region of the lungs. Thus, blood leaving the lungs has a low P_{O_2}, resulting in hypoxemia (a low arterial P_{O_2}). The decrease in arterial P_{O_2} accounts in part for the increase in the A–aO_2 gradient. However, ventilation is also stimulated as a compensatory mechanism to hypoxemia, which leads to hyperventilation with a concomitant increase in alveolar P_{O_2}. The A–aO_2 gradient is, therefore, further increased because of the increased alveolar P_{O_2} caused by hyperventilation.

2. The decreased P_{CO_2} and increased pH are the result of hyperventilation as a result of the hypoxic drive (low P_{O_2}) that stimulates ventilation.
3. The mechanisms by which oral contraceptives increase the risk of thrombus formation are not completely understood. The risk appears to be correlated best with the estrogen content of the pills. Hypotheses include increased endothelial cell proliferation, decreased rates of venous blood flow, and increased coagulability secondary to changes in platelets, coagulation factors, and the fibrinolytic system. Furthermore, there are changes in serum lipoprotein levels with an increase in LDL and VLDL and a variable effect on HDL. Driving cross-country, with long sedentary periods, may have exacerbated the patient's condition.

thePoint® *Visit* http://thepoint.lww.com/rhoades5e *for additional chapter review Q&A, Clinical Application Exercises, animations, and more!*

20 Pulmonary Circulation and Ventilation/Perfusion

Active Learning Objectives

Upon mastering the material in this chapter, you should be able to:
- Predict how changes in cardiac output will affect pulmonary vascular resistance.
- Describe the relationship between capillary recruitment and vascular resistance.
- Describe how low oxygen affects pulmonary vascular resistance.
- Explain the difference between regional and generalized hypoxia and their effect on pulmonary arterial pressure.

- Explain how changes in surface tension will affect interstitial fluid pressure in the lungs.
- Describe how gravity alters blood flow in the base and apex of the lungs.
- Describe how regional ventilation and regional blood flow are matched in the lungs.
- Describe how an anatomic shunt affects the regional ventilation/perfusion ratio.

▶ FUNCTIONAL ORGANIZATION

In the previous chapter, you learned that gas exchange was determined by the partial pressure gradient across the alveolar–capillary membrane. You also learned that uptake of O_2 and the unloading of CO_2 were affected primarily by blood flow. Also, the previous chapter showed that most of the oxygen transported in the systemic circulation was carried in the form of oxyhemoglobin (Hb-O_2), and the amount carried was influenced by changes in blood chemistry (pH, $Paco_2$, and temperature). You also learned that blood flow in the lung was not ideal and some venous admixture occurred that resulted in alveolar–arterial gradient, causing blood leaving the lung to have a lower Po_2 and O_2 content.

In this chapter, you will learn that the pulmonary circulation is a high-flow, low-flow, low-resistance, and a low-pressure system. You also will learn that air flow and blood flow in the lungs are not matched and is the major cause of hypoxemia.

The heart drives two separate and distinct circulatory systems in the body: the pulmonary circulation and the systemic circulation. The pulmonary circulation carries venous blood from the heart to the lungs and arterialized blood in pulmonary veins from the lungs back to the heart. Pulmonary circulation is analogous to the entire systemic circulation because the pulmonary circulation receives all of the cardiac output. Therefore, the pulmonary circulation is not like regional circulation such as the renal, hepatic, or coronary circulations. A change in pulmonary vascular resistance has the same implications for the right ventricle as a change in systemic vascular resistance has for the left ventricle.

The pulmonary arteries branch in the same treelike manner as the airways. Each time an airway branches, the arterial tree branches so that the two parallel each other (Fig. 20.1). Blood in the pulmonary blood vessels comprises >40% of lung weight. The total blood volume of the pulmonary circulation (main pulmonary artery to left atrium) is ~500 mL or 10% of the total circulating blood volume (5,000 mL). The pulmonary veins contain more blood (~270 mL) than do the arteries (~150 mL). The blood volume in the

pulmonary capillaries is approximately equal to the stroke volume of the right ventricle (~80 mL), under most physiologic conditions.

Pulmonary circulation has many secondary functions.

The primary function of the pulmonary circulation is to bring venous blood from the superior and inferior vena cava into contact with alveoli for gas exchange. In addition to gas exchange, the pulmonary circulation has three secondary functions: it serves as a filter, a metabolic organ, and a blood reservoir.

Pulmonary vessels protect the body against **thrombi** (blood clots) and **emboli** (fat globules or air bubbles), preventing them from entering important vessels in other organs. Thrombi and emboli often occur with a sedentary lifestyle, after surgery or injury, and enter the systemic venous blood. Small pulmonary arterial vessels and capillaries trap the thrombi and emboli and prevent them from obstructing the vital coronary, cerebral, and renal vessels. Endothelial cells lining the pulmonary vessels release fibrinolytic substances that help dissolve thrombi. Emboli, especially air emboli, are absorbed through the pulmonary capillary walls. If a large thrombus occludes a large pulmonary vessel, gas exchange can be severely impaired and can cause death. A similar situation occurs if emboli are extremely numerous and lodge all over the pulmonary arterial tree.

Vasoactive hormones are metabolized in the pulmonary circulation. One such hormone is angiotensin I (AI), which is activated and converted to angiotensin II (AII) in the lungs by **angiotensin-converting enzyme** (**ACE**) located on the surface of the pulmonary capillary endothelial cells. Activation is rapid; 80% of AI can be converted to AII during a single passage through the pulmonary circulation. In addition to being a potent vasoconstrictor, AII has other important actions in the body (see Chapter 23). Metabolism of vasoactive hormones by the pulmonary circulation appears to be selective. Pulmonary endothelial cells inactivate bradykinin, serotonin, and the prostaglandins E_1, E_2, and $F_{2\alpha}$. Other prostaglandins, such as PGA_1 and PGA_2,

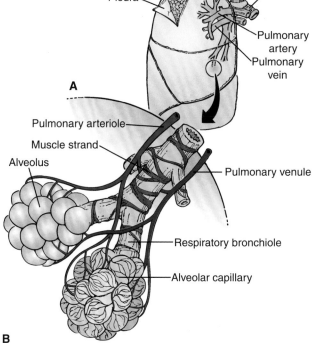

A

B

Figure 20.1 **Pulmonary vessels branch in a treelike fashion similar to the airways. (A)** Systemic venous blood flows through the pulmonary arteries into the alveolar capillaries and back to the heart via the pulmonary veins, to be pumped into the systemic circulation. **(B)** A mesh of capillaries surrounds each alveolus. As the blood passes through the capillaries, it gives up carbon dioxide and takes up oxygen.

pass through the lungs unaltered. Norepinephrine is inactivated, but epinephrine, histamine, and arginine vasopressin pass through the pulmonary circulation unchanged. With acute lung injury (e.g., oxygen toxicity, fat emboli), the lungs can release histamine, prostaglandins, and leukotrienes, which can cause vasoconstriction of pulmonary arteries and pulmonary endothelial damage.

The lungs serve as a blood reservoir. Approximately 500 mL or 10% of the total circulating blood volume is in the pulmonary circulation. During hemorrhagic shock, some of this blood can be mobilized to improve the cardiac output.

The conducting airways have their own separate circulation.

The **bronchial circulation** is distinct and separate from the pulmonary circulation. The primary function of the bronchial circulation is to nourish the walls of the conducting airways and surrounding tissues by distributing blood to the supporting structures of the lungs. Under normal conditions, the bronchial circulation does not supply blood to the terminal respiratory units (respiratory bronchioles, alveolar ducts, and alveoli); they receive their blood from the pulmonary circulation. Venous return from the bronchial circulation is by two routes: bronchial veins and pulmonary veins. About

half of the bronchial blood flow returns to the right atrium by way of the bronchial veins, which empty into the azygos vein. The remainder returns through small bronchopulmonary anastomoses into the pulmonary veins.

Bronchial arterial pressure is approximately the same as aortic pressure, and bronchial vascular resistance is much higher than resistance in the pulmonary circulation. Bronchial blood flow is ~1% to 2% of cardiac output, but in certain inflammatory disorders of the airways (e.g., chronic bronchitis), it can be as high as 10% of cardiac output.

The bronchial circulation is the only portion of the circulation in the adult lung that is capable of undergoing angiogenesis, the formation of new vessels. This is extremely important in providing collateral circulation to the lung parenchyma, especially when the pulmonary circulation is compromised. When a clot or embolus obstructs pulmonary blood flow, the adjacent parenchyma is kept alive by the development of new blood vessels.

▶ HEMODYNAMIC FEATURES

In contrast to the systemic circulation, the pulmonary circulation is a high-flow, low-pressure, low-resistance system. The pulmonary artery and its branches have much thinner walls than the aorta and are more compliant. The pulmonary artery also contains less elastin and smooth muscle in its walls. The pulmonary arterioles are thin walled and contain little smooth muscle and, consequently, have less ability to constrict than the thick-walled, highly muscular systemic arterioles. The pulmonary veins are also highly compliant and contain little smooth muscle compared with their counterparts in the systemic circulation.

The pulmonary capillary bed is also different. Unlike the systemic capillaries, which are often arranged as a network of tubular vessels with some interconnections, the pulmonary capillaries mesh together in the alveolar wall so that blood flows as a thin sheet. It is, therefore, misleading to refer to pulmonary capillaries as a capillary network; they comprise a dense capillary bed. The walls of the capillary bed are exceedingly thin, and a whole capillary bed can collapse if local alveolar pressure exceeds capillary pressure.

The systemic and pulmonary circulations differ strikingly in their pressure profiles (Fig. 20.2). Mean pulmonary arterial pressure is 15 mm Hg, compared with 93 mm Hg in the aorta. The driving pressure (10 mm Hg) for pulmonary flow is the difference between the mean pressure in the pulmonary artery (15 mm Hg) and the pressure in the left atrium (5 mm Hg). Changes in pulmonary venous and left atrial pressures have a profound effect on gas exchange, and pulmonary wedge pressure provides an indirect measure of these important pressures.

The right ventricle pumps mixed venous blood through the pulmonary arterial tree, the alveolar capillaries (where oxygen is taken up and carbon dioxide is removed), the pulmonary veins, and then onto the left atrium. All of the cardiac output is pumped through the pulmonary circulation at a much lower pressure than through the systemic circulation. As shown in Figure 20.2, the 10 mm Hg pressure gradient across the pulmonary circulation drives the same

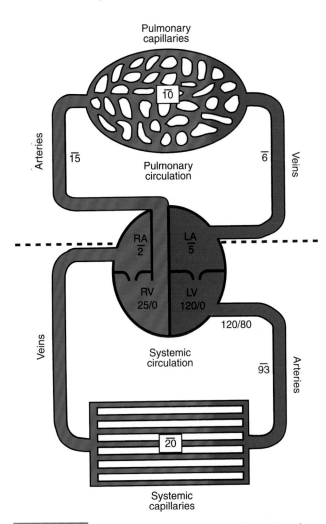

Figure 20.2 Pulmonary circulation has unique hemodynamic features. Unlike the systemic circulation, the pulmonary circulation is a low-pressure and low-resistance system. Pulmonary circulation is characterized as normally dilated, whereas the systemic circulation is characterized as normally constricted. Pressures are given in mm Hg; a bar over the number indicates mean pressure. LA, left atrium; LV, left ventricle; RA, right atrium; RV, right ventricle.

blood flow (5 L/min) as in the systemic circulation, where the pressure gradient is almost 100 mm Hg. Remember that vascular resistance (R) is equal to the pressure gradient (ΔP) divided by blood flow (see Chapter 12):

$$R = \Delta P / \dot{Q} \qquad (1)$$

Pulmonary vascular resistance is extremely low, about 1/10 that of systemic vascular resistance. The difference in resistances is a result, in part, of the enormous number of small pulmonary resistance vessels that are dilated. By contrast, systemic arterioles and precapillary sphincters are partially constricted.

Pulmonary vascular resistance falls with increased cardiac output.

Another unique feature of the pulmonary circulation is its ability to decrease resistance when pulmonary arterial pressure rises. When cardiac output increases, pulmonary

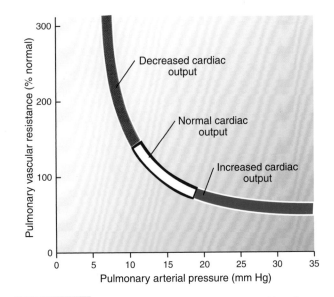

Figure 20.3 Pulmonary vascular resistance falls with a rise in cardiac output. Unlike in the systemic circulation, vascular resistance decreases when perfusion pressure rises (pulmonary arterial pressure). Note that if cardiac output increases, there is a rise in pulmonary arterial pressure and a concomitant fall in pulmonary vascular resistance.

pressure rises, resulting in a marked decrease in pulmonary vascular resistance (Fig. 20.3). Similarly, increasing pulmonary venous pressure causes pulmonary vascular resistance to fall. These responses are different from those of the systemic circulation, where an increase in perfusion pressure increases vascular resistance. Two local mechanisms in the pulmonary circulation are responsible (Fig. 20.4). The first mechanism is known as **capillary recruitment**. Under normal conditions, some capillaries are partially or completely closed, particularly in the top part of the lungs, because of the low perfusion pressure. As blood flow increases, the pressure rises, and the closed vessels are opened, lowering the overall resistance. Capillary recruitment is the primary mechanism for the fall in pulmonary vascular resistance when cardiac output increases. The second mechanism is **capillary distention** or widening of capillary segments.

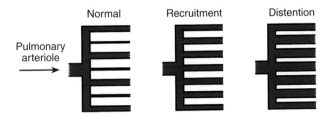

Figure 20.4 Capillary recruitment and capillary distention decrease pulmonary vascular resistance. In the normal condition, not all capillaries are perfused. Capillary recruitment (the opening up of previously closed vessels) results in the perfusion of an increased number of vessels with a concomitant decrease in resistance. Capillary distention (an increase in the caliber of vessels) resulting from high vessel compliance also results in a lower resistance and higher blood flow.

This occurs because the pulmonary arterioles and capillaries are exceedingly thin and highly compliant.

The fall in pulmonary vascular resistance with increased cardiac output has two beneficial effects. It opposes the tendency of blood velocity to speed up with increased flow rate, maintaining adequate time for pulmonary capillary blood to take up oxygen and dispose of carbon dioxide. It also results in an increase in capillary surface area, which enhances the diffusion of oxygen into and carbon dioxide out of the pulmonary capillary blood.

Capillary recruitment and distention also have a protective function. High capillary pressure can cause **pulmonary edema** and is a major threat to lung dysfunction. Pulmonary edema is abnormal accumulation of fluid, which can flood the alveoli and impair gas exchange. When cardiac output increases from a resting level of 5 to 25 L/min with vigorous exercise, the decrease in pulmonary vascular resistance not only minimizes the load on the right heart but also keeps the capillary pressure low and prevents excess fluid from leaking out of the pulmonary capillaries.

Pulmonary vascular resistance increases at high and low lung volumes.

Pulmonary vascular resistance is also significantly affected by lung volume. Because pulmonary arterioles, capillaries, and venules have little structural support, they can be easily distended or collapsed, depending on the pressure surrounding them. It is the change in transmural pressure (pressure across the capillaries) that influences vessel diameter. From a functional point of view, pulmonary vessels can be classified into two types: **extra-alveolar vessels** (pulmonary arteries and veins) and alveolar vessels (arterioles, capillaries, and venules). The *extra-alveolar vessels* are subjected to pleural pressure—any change in pleural pressure affects pulmonary vascular resistance in these vessels by changing the transmural pressure. *Alveolar vessels*, however, are subjected primarily to alveolar pressure.

At high lung volumes, the pleural pressure is more negative. The transmural pressure in the extra-alveolar vessels increases, and they become distended (Fig. 20.5A). However, alveolar diameter increases at high lung volumes, causing the

---

CLINICAL FOCUS | 20.1

Pulmonary Embolism

Pulmonary embolism is clearly one of the more important disorders affecting the pulmonary circulation. The incidence of pulmonary embolism exceeds 500,000 per year, with a mortality rate of ~10%. Pulmonary embolism is often misdiagnosed and, if improperly diagnosed, the mortality rate can exceed 30%.

The term *pulmonary embolism* refers to the movement of a blood clot or other plug from the systemic veins through the right heart and into the pulmonary circulation, where it lodges in one or more branches of the pulmonary artery. Although most pulmonary emboli originate from thrombosis in the leg veins, they can originate from the upper extremities as well. A thrombus is the major source of pulmonary emboli; however, air bubbles introduced during intravenous injections, hemodialysis, or the placement of central catheters can also cause emboli. Other sources of pulmonary emboli include fat emboli (a result of multiple long-bone fractures), tumor cells, amniotic fluid (secondary to strong uterine contractions), parasites, and various foreign materials in intravenous drug abusers.

The etiology of pulmonary emboli focuses on three factors that potentially contribute to the genesis of venous thrombosis: (1) hypercoagulability (e.g., a deficiency of antithrombin III, malignancies, the use of oral contraceptives, the presence of lupus anticoagulant), (2) endothelial damage (e.g., caused by atherosclerosis), and (3) stagnant blood flow (e.g., varicose veins). Risk factors for thrombi include immobilization (e.g., prolonged bed rest, prolonged sitting during travel, or immobilization of an extremity after a fracture), congestive heart failure, obesity, underlying carcinoma, and chronic venous insufficiency.

When a thrombus migrates into the pulmonary circulation and lodges in pulmonary vessels, several pathophysiologic consequences ensue. When a vessel is occluded, blood flow stops, perfusion to pulmonary capillaries ceases, and the ventilation/perfusion ratio in that lung unit becomes high because ventilation is wasted. As a result, there is a significant increase in physiologic dead space. Besides the direct mechanical effects of vessel occlusion, thrombi release vasoactive mediators that cause bronchoconstriction of small airways, which leads to hypoxemia. These vasoactive mediators also cause endothelial damage that leads to edema and atelectasis. If the pulmonary embolus is large and occludes a major pulmonary vessel, an additional complication occurs in the lung parenchyma distal to the site of the occlusion. The distal lung tissue becomes anoxic because it does not receive oxygen (either from airways or from the bronchial circulation). Oxygen deprivation leads to necrosis of lung parenchyma (pulmonary infarction). The parenchyma will subsequently contract and form a permanent scar.

Pulmonary emboli are difficult to diagnose because they do not manifest any specific symptoms. The most common clinical features include dyspnea and sometimes pleuritic chest pains. If the embolism is severe enough, a decreased arterial P_{O_2}, decreased P_{CO_2}, and increased pH result. The major screening test for pulmonary embolism is the perfusion scan, which involves the injection of aggregates of human serum albumin labeled with a radionuclide into a peripheral vein. These albumin aggregates (~10 to 50 μm wide) travel through the right side of the heart, enter the pulmonary vasculature, and lodge in small pulmonary vessels. Only lung areas receiving blood flow will manifest an uptake of the tracer; the nonperfused region will not show any uptake of the tagged albumin. The aggregates fragment and are removed from the lungs in about a day. ◼

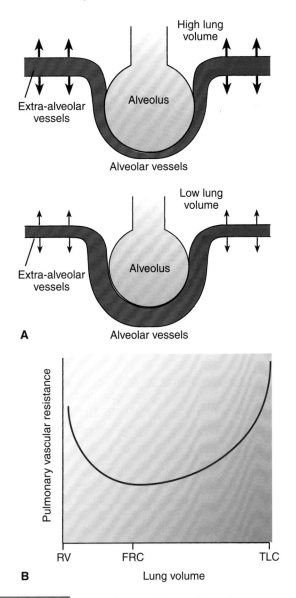

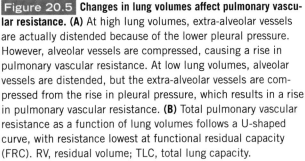

Figure 20.5 **Changes in lung volumes affect pulmonary vascular resistance. (A)** At high lung volumes, extra-alveolar vessels are actually distended because of the lower pleural pressure. However, alveolar vessels are compressed, causing a rise in pulmonary vascular resistance. At low lung volumes, alveolar vessels are distended, but the extra-alveolar vessels are compressed from the rise in pleural pressure, which results in a rise in pulmonary vascular resistance. **(B)** Total pulmonary vascular resistance as a function of lung volumes follows a U-shaped curve, with resistance lowest at functional residual capacity (FRC). RV, residual volume; TLC, total lung capacity.

transmural pressure in alveolar vessels to decrease. As the alveolar vessels become compressed, pulmonary vascular resistance increases. At low lung volumes, pulmonary vascular resistance also increases, as a result of more positive pleural pressure, which compresses the extra-alveolar vessels. Because alveolar and extra-alveolar vessels can be viewed as two groups of resistance vessels connected in series, their resistances are additive at any lung volume. Pulmonary vascular resistance is lowest at **functional residual capacity (FRC)** and increases at both higher and lower lung volumes (Fig. 20.5B).

Because smooth muscle plays a key role in determining the caliber of extra-alveolar vessels, drugs can also cause a change in resistance. Serotonin, norepinephrine, histamine, thromboxane A$_2$, and leukotrienes are potent vasoconstrictors, particularly at low lung volumes when the vessel walls are already compressed. Drugs that relax smooth muscle in the pulmonary circulation include adenosine, acetylcholine, prostacyclin (prostaglandin I$_2$), and isoproterenol. Although the pulmonary circulation is richly innervated with sympathetic nerves, surprisingly, pulmonary vascular resistance is virtually unaffected by autonomic nerves under normal conditions.

Low oxygen tension in the lung causes pulmonary vasoconstriction.

Although changes in pulmonary vascular resistance are accomplished mainly by passive mechanisms, resistance can be increased by low oxygen in the alveoli, alveolar **hypoxia**, and low oxygen in the blood, **hypoxemia**. Recall from Chapter 17 that hypoxemia causes vasodilation in systemic vessels, but in pulmonary vessels, hypoxemia or alveolar hypoxia causes vasoconstriction of small pulmonary arteries. This unique phenomenon is called **hypoxia-induced pulmonary vasoconstriction** and is accentuated by high carbon dioxide and low blood pH. The exact mechanism is not known, but hypoxia can directly stimulate pulmonary vascular smooth muscle cells, independent of any agonist or neurotransmitter released by hypoxia.

Two types of alveolar hypoxia (**regional hypoxia** and **generalized hypoxia**) are encountered in altered lung function, with different implications for pulmonary vascular resistance. In regional hypoxia, pulmonary vasoconstriction is localized to a specific region of the lungs and diverts blood away from a poorly ventilated region, which minimizes the effect on gas exchange (Fig. 20.6A). Regional hypoxia is often caused by bronchial obstruction. Regional hypoxia has little effect on pulmonary arterial pressure, or resistance, and when alveolar hypoxia no longer exists, the vessels dilate and blood flow is restored. Generalized hypoxia, on the other hand, causes vasoconstriction throughout both lungs, leading to a significant rise in resistance and pulmonary artery pressure (Fig. 20.6B). Generalized hypoxia occurs when the partial pressure of alveolar oxygen (Pa$_{O_2}$) is decreased with high altitude or with the chronic hypoxia seen in certain types of respiratory diseases (e.g., asthma, emphysema, and cystic fibrosis). Generalized hypoxia can lead to pulmonary hypertension (high pulmonary arterial pressure), which leads to pathophysiologic changes (hypertrophy and proliferation of smooth muscle cells, narrowing of arterial lumens, and a change in contractile function). Pulmonary hypertension causes a substantial increase in workload of the right heart, often leading to right heart hypertrophy.

Generalized hypoxia plays an important nonpathophysiologic role before birth. In the fetus, pulmonary vascular resistance is extremely high as a result of generalized hypoxia: <15% of the cardiac output goes to the lungs, and the remainder is diverted to the left side of the heart via the foramen ovale and to the aorta via the ductus arteriosus. When alveoli are oxygenated on the newborn's first breath,

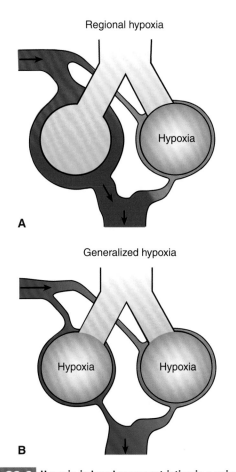

Figure 20.6 Hypoxia-induced vasoconstriction is a unique feature to the pulmonary circulation. Low oxygen tension in the alveoli (alveolar hypoxia) is the major mechanism regulating blood flow within normal lungs. **(A)** With regional hypoxia, precapillary constriction diverts blood flow from poorly ventilated regions, with little change in pulmonary arterial pressure. **(B)** In generalized hypoxia, which can occur with high altitude or with certain lung diseases, precapillary constriction occurs throughout the lungs and there is a marked increase in pulmonary arterial pressure.

pulmonary vascular smooth muscle relaxes, the vessels dilate, and vascular resistance falls dramatically. The foramen ovale and the ductus arteriosus close and pulmonary blood flow increases enormously.

▶ FLUID EXCHANGE IN PULMONARY CAPILLARIES

The forces that govern the exchange of fluid across capillary walls in the systemic circulation also operate in the pulmonary capillaries (see **Starling forces** in Chapter 16). Net fluid transfer across the pulmonary capillaries depends on the difference between hydrostatic and colloid osmotic pressures inside and outside the capillaries. In the pulmonary circulation, two additional forces play a role in fluid transfer—surface tension and alveolar pressure. The force of alveolar surface tension (see Chapter 19) pulls inwardly, which tends to lower the interstitial pressure and draw fluid into the interstitial space. By contrast, the alveolar pressure tends to compress the interstitial space and the interstitial pressure is increased (Fig. 20.7).

Surface tension affects fluid exchange.

Mean pulmonary capillary hydrostatic pressure is normally 8 to 10 mm Hg, which is lower than the plasma colloid osmotic pressure (25 mm Hg). This is functionally important because the low hydrostatic pressure in the pulmonary capillaries favors the net absorption of fluid. Alveolar surface tension tends to offset this advantage and results in a net force that still favors a small continuous flux of fluid out of the capillaries and into the interstitial space. This excess fluid travels through the interstitium to the perivascular and peribronchial spaces in the lungs, where it then passes into the lymphatic channels (see Fig. 20.7). The lungs have a more extensive lymphatic system than most organs. The lymphatic vessels are not found in the alveolar–capillary area but are strategically located near the terminal bronchioles to drain off excess fluid. Lymphatic channels, like small pulmonary blood vessels, are held open by tethers from surrounding connective tissue. Total lung lymph flow is about 0.5 mL/min, and the lymph is propelled by smooth muscle in the lymphatic walls and by ventilatory movements of the lungs.

Pulmonary edema occurs when excess fluid accumulates in the lung interstitial spaces and alveoli and usually results when capillary filtration exceeds fluid removal. Pulmonary edema can be classified as *cardiogenic* pulmonary edema (due to heart dysfunction) or *noncardiogenic* pulmonary edema (due to lung injury). Cardiogenic pulmonary edema is caused by an increase in capillary hydrostatic pressure or by a decrease in plasma colloidal osmotic pressure. The latter occurs when plasma protein concentration is reduced (e.g., starvation).

Increased capillary hydrostatic pressure is the most frequent cause of pulmonary edema and is often the result of an abnormally high pulmonary venous pressure (e.g., with mitral stenosis, left heart failure, or heart attack). The second major cause of pulmonary edema is *noncardiogenic* and is due to increased alveolar surface tension and/or increased

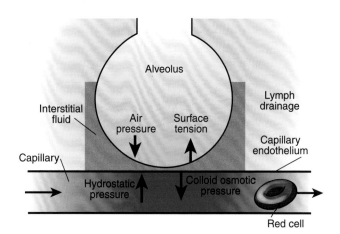

Figure 20.7 Alveolar surface tension and alveolar pressure affect fluid exchange in pulmonary capillaries. Fluid movement in and out of capillaries depends on the net difference between hydrostatic and colloidal osmotic pressures. In the lung, two additional factors (alveolar surface tension and pressure) are involved in fluid exchange. Alveolar surface tension enhances filtration, whereas alveolar pressure opposes filtration. The relatively low pulmonary capillary hydrostatic pressure helps keep the alveoli "dry" and prevents pulmonary edema.

permeability of the alveolar–capillary membrane. Both types of noncardiogenic result in excess fluid and plasma proteins flooding the interstitial spaces and alveoli. Protein leakage makes pulmonary edema more severe because additional water is pulled from the capillaries to the alveoli when plasma proteins enter the interstitial spaces and alveoli. Increased capillary permeability occurs with pulmonary vascular injury, usually from oxidant damage (e.g., oxygen therapy, ozone toxicity), an inflammatory reaction (endotoxins), or neurogenic shock (e.g., head injury). Loss of surfactant leads to high surface tension, which lowers the interstitial hydrostatic pressure with a concomitant increase in capillary fluid entering the interstitial space. Pulmonary edema is a hallmark of **acute respiratory distress syndrome (ARDS)** and is often associated with abnormally high surface tension. Pulmonary edema is a serious problem because excess fluid enters alveoli and hinders gas exchange, causing arterial P_{O_2} to fall below normal (i.e., Pa_{O_2} <85 mm Hg) and arterial P_{CO_2} to rise above normal (Pa_{CO_2} >45 mm Hg). As mentioned earlier, abnormally low arterial P_{O_2} produces hypoxemia and abnormally high arterial P_{CO_2} produces hypercapnia. Pulmonary edema can also flood small airways, thereby obstructing airflow and increasing airway resistance. Lung compliance is decreased with pulmonary edema because of interstitial swelling and the increase in alveolar surface tension. Decreased lung compliance, together with airway obstruction, greatly increases the work of breathing. The treatment of pulmonary edema is directed toward reducing pulmonary capillary hydrostatic pressure. This is accomplished by decreasing blood volume with a diuretic drug, increasing left ventricular function with digitalis, and administering a drug that causes vasodilation in systemic blood vessels.

Although freshwater drowning is often associated with aspiration of water into the lungs, the cause of death is not pulmonary edema but ventricular fibrillation. The low capillary pressure that normally keeps the alveolar–capillary membrane free of excess fluid becomes a severe disadvantage when freshwater accidentally enters the lungs. The aspirated water is rapidly pulled into the pulmonary capillary circulation via the alveoli because of the low capillary hydrostatic pressure and high colloidal osmotic pressure. Consequently, the plasma is diluted and the hypotonic environment causes red cells to burst (hemolysis). The resulting elevation of plasma K^+ level and depression of Na^+ level alter the electrical activity of the heart. Ventricular fibrillation often occurs as a result of the combined effects of these electrolyte changes and hypoxemia. In saltwater drowning, the aspirated seawater is hypertonic, which leads to increased plasma Na^+ and pulmonary edema. The cause of death in this case is asphyxia.

▶ BLOOD FLOW DISTRIBUTION IN THE LUNGS

As mentioned previously, blood in the pulmonary vessels accounts for approximately half the weight of the lungs. The gravitational effect on pulmonary blood flow is dramatic and results in an uneven distribution of blood in the lungs. In an upright person, the gravitational pull on the blood is downward. Because the vessels are highly compliant, gravity causes the blood volume and flow to be greater at the bottom of the lung (the base) than at the top (the apex). Pulmonary vessels can be compared with a continuous column of fluid. The difference in arterial pressure between the apex and the base of the lungs is about 30 cm H_2O. Because the heart is

CLINICAL FOCUS | 20.2

Hypoxia-Induced Pulmonary Hypertension

Hypoxia has opposite effects on the pulmonary and systemic circulations. Hypoxia relaxes vascular smooth muscle in systemic vessels and elicits vasoconstriction in the pulmonary vasculature. Hypoxic pulmonary vasoconstriction is the major mechanism regulating the matching of regional blood flow to regional ventilation in the lungs. With regional hypoxia, the matching mechanism automatically adjusts regional pulmonary capillary blood flow in response to alveolar hypoxia and prevents blood from perfusing poorly ventilated regions in the lungs. Regional hypoxic vasoconstriction occurs without any change in pulmonary arterial pressure. However, when hypoxia affects all parts of the lung (generalized hypoxia), it causes pulmonary hypertension because all of the pulmonary vessels constrict. Hypoxia-induced pulmonary hypertension affects people who live at a high altitude (8,000 to 12,000 ft) and those with chronic obstructive pulmonary disease (COPD), especially patients with emphysema.

With chronic *hypoxia-induced pulmonary hypertension*, the pulmonary artery undergoes major remodeling during several days. An increase in wall thickness results from hypertrophy and hyperplasia of vascular smooth muscle and

an increase in connective tissue. These structural changes occur in both large and small arteries. Also, there is abnormal extension of smooth muscle into peripheral pulmonary vessels, where muscularization is not normally present; this is especially pronounced in precapillary segments. These changes lead to a marked increase in pulmonary vascular resistance. With severe, chronic hypoxia-induced pulmonary hypertension, the obliteration of small pulmonary arteries and arterioles as well as pulmonary edema eventually occurs. The latter is caused, in part, by the hypoxia-induced vasoconstriction of pulmonary veins, which results in a significant increase in pulmonary capillary hydrostatic pressure.

A striking feature of the vascular remodeling is that both the pulmonary artery and the pulmonary vein constrict with hypoxia; however, only the arterial side undergoes major remodeling. The postcapillary segments and veins are spared the structural changes seen with hypoxia. Because of the hypoxia-induced vasoconstriction and vascular remodeling, pulmonary arterial pressure increases. Pulmonary hypertension eventually causes right heart hypertrophy and failure, the major cause of death in patients with COPD. ■

situated midway between the top and the bottom of the lungs, the arterial pressure is about 11 mm Hg less (15 cm H_2O ÷ 1.36 cm H_2O per mm Hg = 11 mm Hg) at the lungs' apex (15 cm above the heart) and about 11 mm Hg more than the mean pressure in the middle of the lungs at the lungs' base (15 cm below the heart). As a result, the low pulmonary arterial pressure results in reduced blood flow in the capillaries at the lungs' apex, whereas capillaries at the base are distended because of increased pressure and blood flow is augmented.

Gravity causes lungs to be underperfused at the apex and overperfused at the base.

In an upright person, pulmonary blood flow decreases almost linearly from the base to the apex (Fig. 20.8). Blood flow distribution is affected by gravity and can be altered by changes in body positions. For example, when a person is lying down, blood flow is distributed relatively evenly from the base to the apex. The measurement of blood flow in a subject suspended upside down would reveal an apical blood flow exceeding basal flow in the lungs. Exercise tends to offset the gravitational effects in an upright person. As cardiac output increases with exercise, the increased pulmonary arterial pressure leads to capillary recruitment and distention in the lungs' apex, resulting in increased blood flow and minimizing regional differences in blood flow in the lungs.

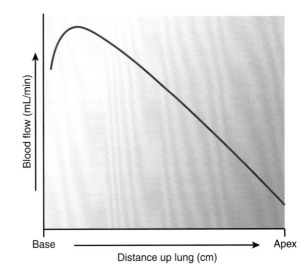

Figure 20.8 **Gravity causes uneven pulmonary blood flow in the upright individual.** The downward pull of gravity causes a lower blood pressure at the apex of the lungs. Consequently, pulmonary blood flow is low at the apex. Toward the base of the lungs, gravity has an added effect on blood pressure, causing an increase in blood flow.

Because gravity causes capillary beds to be underperfused in the apex and overperfused in the base, the lungs are often divided into zones to describe the effect of gravity on pulmonary capillary blood flow (Fig. 20.9). Zone 1 occurs

Respiratory Physiology

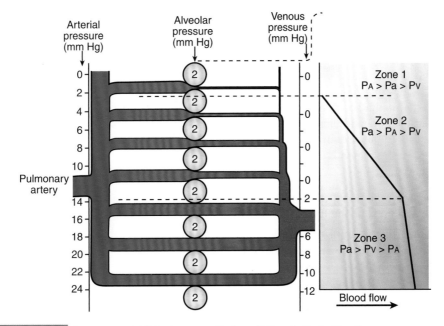

Figure 20.9 **Zones are established as a result of gravitational effects.** The three zones are established in an upright person and are dependent on the relationship between pulmonary arterial pressure (Pa), pulmonary venous pressure (Pv), and alveolar pressure (PA). A zone 1 is established when alveolar pressure exceeds arterial pressure and there is no blood flow. Zone 1 occurs toward the apex of the lung and occurs only in abnormal conditions in which alveolar pressure is increased (e.g., positive pressure ventilation) or when arterial pressure is decreased below normal (e.g., the gravitational pull while standing at attention or during the launching of a spacecraft). A zone 2 is established when arterial pressure exceeds alveolar pressure, and blood flow depends on the difference between arterial and alveolar pressures. Blood flow is greater at the bottom than at the top of this zone. In zone 3, both arterial and venous pressures exceed alveolar pressure, and blood flow depends on the normal arterial–venous pressure difference. Note that arterial pressure increases down each zone, vessel transmural pressure also becomes greater, capillaries become more distended, and pulmonary vascular resistance falls.

when alveolar pressure is greater than pulmonary arterial pressure; pulmonary capillaries collapse and there is no blood flow. Pulmonary arterial pressure (Pa) is still greater than pulmonary venous pressure (Pv); hence, the pressure gradient in zone 1 is represented as PA > Pa > Pv. Zone 1 is usually small or nonexistent in healthy people because the pulsatile pulmonary arterial pressure is sufficient to keep the capillaries partially open at the apex. However, when zone 1 does occur, alveolar dead space is increased in the lungs. This occurs because, in zone 1, that region is still being ventilated but not perfused (no gas exchange). Zone 1 may easily be created by conditions that elevate alveolar pressure or decrease pulmonary arterial pressure. For example, a zone 1 condition can be created when a patient is placed on a mechanical ventilator, which results in an increase in alveolar pressure with positive ventilation pressures. Hemorrhage or low blood pressure can create a zone 1 condition by lowering pulmonary arterial pressure. A zone 1 condition can also be created in the lungs of astronauts during a spacecraft launching. The rocket acceleration makes the gravitational pull even greater, causing arterial pressure in the top part of the lung to fall. To prevent or minimize a zone 1 condition from occurring, astronauts are placed in a supine position during blastoff.

A zone 2 condition occurs in the middle of the lungs, where pulmonary arterial pressure, caused by the increased hydrostatic effect, is greater than alveolar pressure (see Fig. 20.9). Venous pressure is less than alveolar pressure. As a result, blood flow in a zone 2 condition is determined not by the arterial–venous pressure difference but by the difference between arterial pressure and alveolar pressure. The pressure gradient in zone 2 is represented as Pa > PA > Pv. The functional importance of this is that venous pressure in zone 2 has no effect on flow (i.e., lowering venous pressure will not increase capillary blood flow in this zone). In zone 3, venous pressure exceeds alveolar pressure and blood flow is determined by the usual arterial–venous pressure difference. The increase in blood flow down this region is primarily a result of capillary distention.

Gravity causes a mismatch of regional ventilation and blood flow at the base and apex of the lungs.

Thus far, we have assumed that if ventilation and cardiac output are normal, gas exchange will also be normal because ventilation and blood flow are matched. Unfortunately, this is not the case. Even though total ventilation and total blood flow (i.e., cardiac output) may be normal, there are regions in the lung where ventilation and blood flow are not matched and so a certain fraction of the cardiac output is not fully oxygenated.

The matching of airflow and blood flow is best examined by considering the **ventilation/perfusion ratio** ($\dot{V}_A/\dot{Q}$ ratio), which compares alveolar ventilation with blood flow in lung regions. Because resting healthy people have an alveolar ventilation ($\dot{V}_A$) of 4 L/min and a cardiac output ($\dot{Q}$) of 5 L/min, the ideal alveolar ventilation/perfusion ratio ($\dot{V}_A/\dot{Q}$) should be 0.8 (there are no units, because this is a ratio). We have already seen that gravity can cause regional differences in blood flow and alveolar ventilation. In an upright person, the

base of the lungs is better ventilated and better perfused than the apex. Regional alveolar ventilation and blood flow are illustrated in Figure 20.10. Three points are apparent from this figure:

- Ventilation and blood flow are both gravity dependent with airflow and blood flow increasing down the lung.
- There is a fivefold difference in blood flow between the top and the bottom of the lung, whereas ventilation shows about a twofold difference. This causes gravity-dependent regional variations in the $\dot{V}_A/\dot{Q}$ ratio, which range from 0.6 at the base to 3 or higher at the apex.
- Blood flow is proportionately greater than ventilation at the base, and ventilation is proportionately greater than blood flow at the apex.

The functional importance of lung ventilation/perfusion ratios is that the crucial factor in gas exchange is the

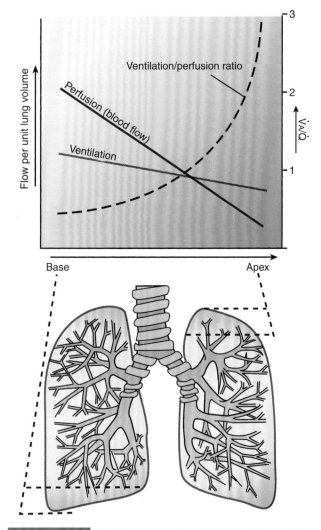

Figure 20.10 Gravity causes a mismatch of blood flow and ventilation in the lung. Both ventilation and perfusion are gravity dependent. At the base of the lungs, alveolar–capillary blood flow exceeds alveolar ventilation, resulting in a low ventilation/perfusion ($\dot{V}_A/\dot{Q}$) ratio. At the apex, the opposite occurs; alveolar ventilation is greater than capillary blood flow, resulting in a high ventilation/perfusion ($\dot{V}_A/\dot{Q}$) ratio.

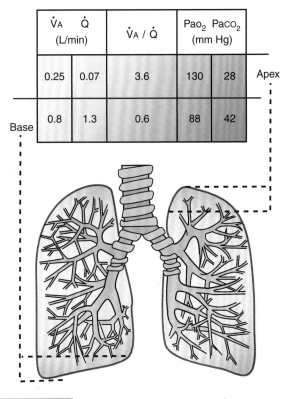

$\dot{V}_A$	$\dot{Q}$	$\dot{V}_A / \dot{Q}$	Pao_2	$Paco_2$	
(L/min)			(mm Hg)		
0.25	0.07	3.6	130	28	Apex
0.8	1.3	0.6	88	42	

Base

Figure 20.11 **Ventilation/perfusion ratios ($\dot{V}_A/\dot{Q}$) affect capillary blood gas tension.** A high ($\dot{V}_A/\dot{Q}$) ratio leads to a high Pao_2 and a low $Paco_2$ in the blood leaving the apex of the lung as a result of overventilation with respect to blood flow. Blood leaving the base of the lung has a low Pao_2 and a high $Paco_2$ as a result of a low ($\dot{V}_A/\dot{Q}$) ratio. $Paco_2$, arterial partial pressure of carbon dioxide; Pao_2, arterial partial pressure of oxygen; $\dot{Q}$, perfusion; $\dot{V}_A$, alveolar ventilation.

matching of regional ventilation and blood flow, as opposed to total alveolar ventilation and total pulmonary blood flow. At the apical region, where the $\dot{V}_A/\dot{Q}$ ratio is >0.8, there is overventilation relative to blood flow. At the base, where the ratio is <0.8, the opposite occurs (i.e., overperfusion relative to ventilation). In the latter case in which blood flow exceeds alveolar ventilation, a fraction of the blood passes through the pulmonary capillaries at the base of the lungs without becoming fully oxygenated.

The effect of regional $\dot{V}_A/\dot{Q}$ ratio on blood gases is shown in Figure 20.11. At the apex, where the lungs are overventilated relative to blood flow, the Pao_2 is high, but the $Paco_2$ is low in this region of the lungs. Oxygen tension (Po_2) in the blood leaving pulmonary capillaries at the base of the lungs is low because the blood is not fully oxygenated as a result of underventilation relative to blood flow. Regional differences in $\dot{V}_A/\dot{Q}$ ratios tend to localize some diseases to the top or bottom parts of the lungs. For example, tuberculosis tends to be localized in the apex because of a more favorable environment (i.e., higher oxygen levels for *Mycobacterium tuberculosis*).

▶ SHUNTS AND VENOUS ADMIXTURE

As a result of the mismatch of air and blood flow, there is "wasted air" in one side of the alveolar–capillary membrane (i.e., physiologic dead space), and on the other side, there is "wasted blood" (Fig. 20.12). Wasted blood refers to any fraction of the venous blood that does not get fully oxygenated. The mixing of unoxygenated blood with oxygenated blood is known as **venous admixture**. There are two causes for venous admixture: a **shunt** and a low $\dot{V}_A/\dot{Q}$ ratio.

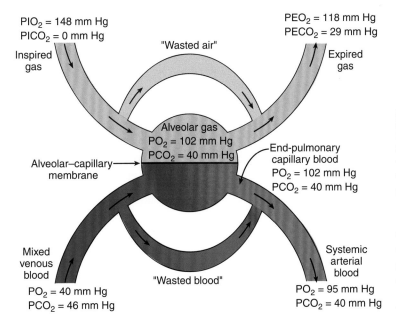

$PIO_2 = 148$ mm Hg
$PICO_2 = 0$ mm Hg

Inspired gas

"Wasted air"

$PEO_2 = 118$ mm Hg
$PECO_2 = 29$ mm Hg

Expired gas

Alveolar gas
$PO_2 = 102$ mm Hg
$PCO_2 = 40$ mm Hg

Alveolar–capillary membrane

End-pulmonary capillary blood
$PO_2 = 102$ mm Hg
$PCO_2 = 40$ mm Hg

Mixed venous blood
$PO_2 = 40$ mm Hg
$PCO_2 = 46$ mm Hg

"Wasted blood"

Systemic arterial blood
$PO_2 = 95$ mm Hg
$PCO_2 = 40$ mm Hg

Figure 20.12 **On one side of the alveolar–capillary membrane, there is some "wasted air" and on the other "wasted blood."** The plumbing on both sides of the alveolar–capillary membrane is imperfect. All of the inspired air does not participate in gas exchange, resulting in some "wasted air." All of the blood entering the lung is not fully oxygenated, leading to some "wasted blood." The total amount of wasted air constitutes physiologic dead space, and the total amount of wasted blood (venous admixture) constitutes physiologic shunt. $PECO_2$, partial pressure of expired carbon dioxide; PEO_2, partial pressure of expired oxygen; $PICO_2$, partial pressure of inspired carbon dioxide; PIO_2, partial pressure of inspired oxygen.

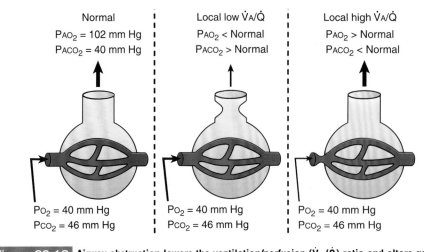

Figure 20.13 **Airway obstruction lowers the ventilation/perfusion ($\dot{V}_A/\dot{Q}$) ratio and alters gas tensions.** Airway obstruction (*middle panel*) causes a low regional ventilation/perfusion ($\dot{V}_A/\dot{Q}$) ratio. A partially blocked airway causes this region to be underventilated relative to blood flow. Note the alveolar gas composition. A low regional $\dot{V}_A/\dot{Q}$ ratio causes venous admixture and will increase the physiologic shunt. A partially obstructed pulmonary arteriole (*right panel*) will cause an abnormally high ($\dot{V}_A/\dot{Q}$) ratio in a lung region. Restricted blood flow causes this region to be overventilated relative to blood flow, which leads to an increase in physiologic dead space. P_{ACO_2}, alveolar partial pressure of carbon dioxide; P_{AO_2}, alveolar partial pressure of oxygen.

An *anatomic shunt* has a structural basis and occurs when blood bypasses alveoli through a channel, such as from the right to left heart through an atrial or ventricular septal defect or from a branch of the pulmonary artery connecting directly to the pulmonary vein. An anatomic shunt is often called a *right-to-left shunt*. The bronchial circulation also constitutes shunted blood because bronchial venous blood (deoxygenated blood) drains directly into the pulmonary veins, which are carrying oxygenated blood.

The second cause for venous admixture is a low regional $\dot{V}_A/\dot{Q}$ ratio. This occurs when a portion of the cardiac output goes through the regular pulmonary capillaries, but there is insufficient alveolar ventilation to fully oxygenate all of the blood. In a healthy person, a low $\dot{V}_A/\dot{Q}$ ratio occurs at the base of the lung (i.e., gravity dependent). A low regional $\dot{V}_A/\dot{Q}$ ratio can also occur with a partially obstructed airway (Fig. 20.13), in which lung regions are underventilated with respect to blood flow, resulting in regional hypoventilation. A fraction of the blood passing through a hypoventilated region is not fully oxygenated, resulting in an increase in venous admixture.

The total amount of venous admixture as a result of anatomic shunt and a low $\dot{V}_A/\dot{Q}$ ratio equals physiologic shunt and represents the total amount of wasted blood that does not get fully oxygenated. *Physiological shunt* is analogous to physiologic dead space; the two are compared in Table 20.1, in which one represents wasted blood flow and the other represents wasted air. It is important to remember that in healthy people, there is some degree of physiologic dead space as well as physiologic shunt in the lungs.

In summary, venous admixture results from anatomic shunt and a low regional $\dot{V}_A/\dot{Q}$ ratio. In healthy people, ~50% of the venous admixture comes from an anatomic shunt (e.g., bronchial circulation) and 50% from a low $\dot{V}_A/\dot{Q}$ ratio at the base of the lungs as a result of gravity. Physiologic shunt (i.e., total venous admixture) represents about 1% to 2% of cardiac output in healthy people. This amount can increase up to 15% of cardiac output with some bronchial diseases, and, in certain congenital disorders, a right-to-left anatomic shunt can account for up to 50% of cardiac output. It is important to remember that any deviation of $\dot{V}_A/\dot{Q}$ ratio from the ideal condition (0.8) impairs gas exchange and lowers oxygen tension in the arterial blood. A good way to remember the importance of a shunt is that it always leads to venous admixture and reduces the amount of oxygen carried in the systemic blood.

TABLE 20.1	Shunts and Dead Spaces Compared	
Shunt	**Dead Space**
Anatomic	Anatomic
+	+
Low ($\dot{V}_A/\dot{Q}$) ratio	Alveolar
=	=
Physiological shunt (calculated total "wasted blood")	Physiologic dead space (calculated total "wasted air")

INTEGRATED MEDICAL SCIENCES

Wheezing and Shortness of Breath

Jack was born June 15th, 2005. At 6½ lb, he was lighter than his two brothers. Jack had difficulty putting on weight after he was born. In fact, during the first month, he dropped to 6 lb. After a spell in the hospital in January 2006, when he was suffering from wheezing and a persistent cough, he was diagnosed with asthma. However, Jack's condition did not improve even with the medications he was taking for bronchitis and asthma. Jack also developed GI complications. After another spell in the hospital, Jack was given a sweat test for cystic fibrosis. The test confirmed that Jack had cystic fibrosis, although there was no history of cystic fibrosis on either side of the family.

This is a typical pattern with children who have been diagnosed with cystic fibrosis (CF) and is the second most common inherited disorder occurring in childhood in the United States, behind sickle cell anemia. More than 10 million Americans carry the defective CF gene without knowing it. About 1,000 new cases of CF are diagnosed each year.

Cystic fibrosis is a genetic disorder that changes a protein that regulates the movement of sodium chloride in and out cells. The specific defect is linked to the chloride ion channel. The result is thick, sticky mucus in the respiratory, digestive, and reproductive systems, as well as increased salt in sweat. The primary target organ is the lungs with long-term difficulties leading to shortness of breath, coughing, frequent lung infections, and hypoxemia. Other symptoms include sinus infections, poor growth, fatty stools, infertility, and clubbing of fingers. The latter is due to hypoxemia. Different patients may have different degrees of symptoms.

Digestive symptoms interfere with the pancreatic function. The thick mucus blocks the pancreatic ducts that carry the digestive enzymes from the pancreas to the small intestine. Without these digestive enzymes, the intestine is unable to absorb nutrients. This often results in foul smelling and greasy stools, poor weight gain and growth, and severe constipation. The blocked pancreatic ducts also lead to painful inflammation (pancreatitis). Also, because of malabsorption, there is poor uptake of vitamin D from the diet. Vitamin D deficiency can lead to osteoporosis in CF patients.

As a result of pancreatic damage, patients often develop diabetes. The diabetes induced by diabetes share characteristics that can be found in both type 1 and type 2 diabetics. Another complication is infertility. Approximately 97% of men with CF are infertile, but not sterile. The main cause of infertility in men is linked to the *vas deferens*, the tubes that connect the prostate gland to the ejaculatory ducts of the penis. With CF, the vas deferens is either blocked or missing entirely. Other mechanisms include poor sperm motility due to a defective chloride channel. About 20% of women with CF have fertility difficulties due to thickened cervical mucus and malnutrition. Malnutrition disrupts ovulation and also causes amenorrhea.

The pathophysiology is an autosomal recessive disorder for the protein, *cystic fibrosis transmembrane conductance regulator* (*CFTR*), which involves the production of sweat, mucus, and digestive fluids. When the CFTR is not functioning, the thin secretions become thick and sticky. The CFTR protein is anchored to the outer membrane of cells in the sweat glands, lungs, pancreas, and all other exocrine glands in the body. The protein acts as a channel connecting the cytoplasm inside the cell to the extracellular fluid. This channel is primarily responsible for controlling NaCl from inside to outside the cell. When the CFTR becomes dysfunctional, the chloride is trapped inside the ducts of the sweat glands and is pumped to the skin. Because chloride is negatively charged, this creates an electrical potential difference across the cell membrane causing cations to move into the cell. Sodium is the most common cation in the extracellular fluid. The excess chloride within the sweat gland prevents sodium reabsorption by epithelial sodium channels and the combination of sodium and chloride creates salt, which is lost in high amounts in the sweat CF patients. The high salt content is the basis for the sweat test. In other organs, most of the damage in CF is due to blockage of narrow airways and ducts with thick secretions. The failure of CFTR protein leads to abnormal chloride movement that causes dehydration of mucus, pancreatic secretions, and biliary secretions.

There is no cure for cystic fibrosis. Pulmonary complications account for ~80% of the deaths. A number of treatment options are available to help reduce the progression of the disease. Airway clearance techniques are used to remove lung mucus, and bronchodilators/anti-inflammatory medicines are used to improve breathing. One technique used to clear mucus from the lungs is called *postural drainage and percussion*. The CF patient sits, stands, or lies in a position to help free up the mucus. The chest and back is then pounded to loosen the mucus. In the hospital setting, chest physiotherapy is utilized. A respiratory therapist percusses the patient's chest with his or her hands several times a day to loosen the mucus. Often, anti-inflammatory medicine, such as ibuprofen, is given to help reduce airway inflammation. Bronchodilators and inhaled hypertonic saline mist can also be beneficial. The dilators help to open the airways, and the hypertonic saline draws water into the small airways and thins the mucus.

Antibiotics for treating lung infections may be given orally, intravenously, or inhaled. Azithromycin is often used long term. Also, Pulmozyme, a mucus-thinning drug, an FDA-approved pharmaceutic agent, is prescribed to reduce lung infection and improve lung function. Digestive supplements include pancreatic enzyme replacement and fat-soluble vitamins can also be prescribed, especially young adults.

Lung transplantation often becomes necessary when lung function and exercise tolerance (e.g., walking up a set of stairs) decline in individuals with CF. A lung transplant is only recommended when the patient's lung function declines to the point that they have difficulty breathing on their own. Although single lung transplantation is advisable with certain lung diseases, individuals with CF must have both lungs replaced because of cross-bacteria contamination to the transplanted lung. In severe cases, a pancreatic and/or liver transplant may be performed at the same time in order to alleviate diabetes and severe GI complications. ◼

Chapter Summary

- Pulmonary circulation is a high-flow, low-resistance, and low-pressure system.
- Capillary recruitment is the primary cause for pulmonary vascular resistance to decrease with increased cardiac output.
- Alveolar hypoxia is the major stimulus for pulmonary vasoconstriction.
- Hypoxia-induced pulmonary vasoconstriction shunts blood away from poorly ventilated regions of the lung.

- Gravity causes regional differences in ventilation/perfusion ($\dot{V}_A/\dot{Q}$) ratios in the lungs.
- Venous admixture occurs when venous blood mixes with oxygenated arterial blood.
- An abnormally low ventilation/perfusion ratio exists when a lung region is overperfused relative to airflow, resulting in an increase in venous admixture.
- Physiologic shunt (wasted blood) is analogous to physiologic dead space (wasted air).

Chapter Review Questions

1. Which of the following best characterizes the pulmonary circulation?

	Flow	Pressure	Resistance	Compliance
A.	Low	High	Low	High
B.	High	Low	High	Low
C.	Low	Low	Low	High
D.	High	Low	High	High
E.	High	Low	Low	High

The correct answer is E. Since the lung receives all of the cardiac output, it is characterized as a high-flow system. Vessel compliance is high, which allows for a low-resistance and a low-pressure system.

2. During moderate to heavy exercise, pulmonary vascular resistance in a healthy individual will:

   A. Increase because cardiac output increases
   B. Decrease because pulmonary arterial pressure decreases
   C. Increase because Pao_2 increases
   D. Decrease because of pulmonary capillary recruitment
   E. Not change

The correct answer is D. With increase exercise, cardiac output increases. The increase flow results in capillary recruitment, which results in a decrease in pulmonary vascular resistance.

3. Which of the following changes would favor the formation of pulmonary edema?

   A. Decreased hydrostatic interstitial pressure around the alveoli
   B. Decreased pulmonary arterial pressure
   C. Decreased pulmonary wedge pressure
   D. Increased oncotic pressure in the pulmonary capillaries
   E. Increased pulmonary blood flow

The correct answer is A. A decrease in pulmonary pressure, wedge pressure, or an increase in oncotic pressure would "pull" fluid from the interstitial space and would not cause edema.

4. A 65-year-old patient has a lung region in which the alveoli are well ventilated but are poorly perfused. This situation will cause alveoli in this region to have:

   A. High Po_2 and a low Pco_2 with increased O_2 uptake
   B. High Po_2 and a low Pco_2 with decreased O_2 uptake
   C. Low Po_2 and a high Pco_2 with decreased O_2 uptake
   D. High Po_2 and a high Pco_2 with increased O_2 uptake
   E. Low Po_2 and a low Pco_2 with decreased O_2 uptake

The correct answer is B. A lung region that is overventilated with respect to blood flow will have an alveolar gas tension in which the Po_2 is greater than normal and the Pco_2 is less than normal. However, the oxygen uptake is less even though the Po_2 is higher because oxygen uptake is most dependent on blood flow. If blood flow is less than normal, then the amount of oxygen taken up (mL/min) is less by the blood passing through this unit.

5. Which of the following best characterizes alveolar ventilation and blood flow at the base, compared with the apex, of the lungs of a healthy standing person?

	Ventilation	Blood Flow	Ventilation/ Perfusion Ratio
A.	Higher	Higher	Normal
B.	Lower	Higher	Lower
C.	Lower	Lower	Lower
D.	Higher	Lower	Higher
E.	Lower	Lower	Higher

The correct answer is B. At the base, the lung is being overperfused and underventilated, which leads to a low ventilation/perfusion ratio.

Clinical Application Exercises 20.1

WATER ON THE LUNGS

A 64-year-old CEO took early retirement and decided to travel. After returning from a monthlong trip, he complained of fatigue and loss of appetite. He saw his internist, and following a physical examination, he was given a CAT scan. The results showed malignant growth on one lobe of his liver. After further x-ray and additional CAT scans, he was diagnosed with primarily metastatic liver cancer in the upper lobe of his liver. The lobe was removed surgically, and the patient was dismissed after being followed by a surgeon and oncologist. Eight months following surgery, he felt tired and began to have difficulty breathing. Upon exertion, he became dyspneic. He was referred to pulmonary specialist and given a chest x-ray. The results confirmed that the patient had pleural effusion.

QUESTIONS

1. What is the pathophysiology of pleural effusion?

2. How is pleural effusion treated?

ANSWERS

1. Pleural effusion is an excess of fluid that accumulates in the pleural cavity, the fluid-filled space that surrounds the lungs. In healthy individuals, the fluid in each pleural space is <1 mL. Under normal conditions, fluid enters the pleural space from the capillaries in the parietal pleura and is removed via the lymphatics situated in the parietal pleura. Fluid can also enter from the interstitial spaces of the lung via the visceral pleura or from the peritoneal cavity through small holes in the diaphragm. The lungs have the capacity to absorb 20 times more fluid than is normally formed. When this capacity is compromised, either through excess formation or decreased lymphatic absorption, pleural effusion develops. The excess fluid accumulations cause labored breathing and dyspnea upon exertion. More than 1 million cases of pleural effusion occur annually in the United States. Pleural effusion is classified into two types: The first type is classified as transudate due to systemic factors (left ventricular failure, pulmonary embolism, and liver cirrhosis). The second type is classified as exudative pleural effusion and is caused by alterations in local factors that influence the formation and absorption of pleural fluid (e.g., bacterial pneumonia, cancer, and viral infection). Pleural effusion is diagnosed based on family history and a physical exam and confirmed by chest x-ray. The type of pleural effusion (transudate or exudate) is based on fluid cytology. Pleural is drawn out of the pleural space in a process called thoracentesis. A needle is inserted through the back of the chest wall into the pleural space to remove fluid for the cytology.

Malignant pleural effusion, secondary to metastatic disease, is the second most common exudative pleural effusion. Over 75% of malignant pleural effusions are caused by lung carcinoma (30%), breast cancer (25%), and tumors of the lymphoma group (20%).

2. Treatment depends on the underlying cause of the pleural effusion. Therapeutic aspiration may be sufficient. However, larger effusions may require insertion of an intercostal drain. Repeated episodes of effusions may require a chemical (talc, bleomycin, or tetracycline/doxycycline) or surgical pleurodesis, in which the two pleural surfaces are fused to each other so that no fluid can accumulate between them.

thePoint® *Visit* http://thepoint.lww.com/rhoades5e *for additional chapter review Q&A, Clinical Application Exercises, animations, and more!*

Respiratory Physiology

Control of Ventilation

Active Learning Objectives

Upon mastering the material in this chapter, you should be able to:
- Describe the major cell groups in the medulla controlling breathing.
- Describe how depth and rate of breathing are regulated.
- Describe how alveolar ventilation is matched to the metabolic needs of the body.
- Predict how changes in blood pH and arterial P_{CO_2} (Pa_{CO_2}) affect the control of breathing.

- Describe how the respiratory system responds to hypoxic-induced stimulation.
- Describe how breathing patterns change during sleep.
- Describe the ventilatory response to high-altitude acclimatization.
- Describe the sequence of events causing underwater blackout.

In the previous chapter, you learned that pulmonary circulation was a high-flow, low-resistance, and a low-pressure system. In addition, you learned the pulmonary circulation had two unique features that were different from systemic circulation. First, vascular resistance decreases with an increased cardiac output. Secondly, hypoxia was the major mechanism for pulmonary constriction. You also learned that pulmonary blood flow and alveolar ventilation were not matched throughout the lungs. The mismatch caused poor gas exchange because of a high ventilation/perfusion ratio at the apex and a low ventilation/perfusion ratio at the base of the lungs. You'll learn in this chapter that the control of breathing is centered primarily on maintaining adequate gas exchange to meet metabolic demands, and Pa_{CO_2} is more a powerful stimulus in the regulation of breathing than Pa_{O_2}. This chapter examines how our basic rhythm and breathing pattern are modified to meet changes in body functions.

▶ NEURAL AND VOLUNTARY CONTROL OF BREATHING

Under resting conditions, ventilation is controlled primarily by the autonomic nervous system and occurs without any conscious effort while we are awake, asleep, or under anesthesia. However, the automatic control can be overridden and conscious effort exerted over ventilation by voluntarily changing the rate and depth of breathing. We can also voluntarily stop breathing for a short period of time until carbon dioxide builds up in the blood, which then will stimulate breathing regardless of how hard we try to hold our breath. The basic pattern of breathing generated in the medulla is extensively modified by several control mechanisms. Figure 21.1 illustrates the overall control of breathing, with signals from the blood, higher cortical regions, and stretch receptors from within and outside the lung modulating our breathing. Minute ventilation is tightly controlled and determined primarily by blood gas levels of Pa_{CO_2} and metabolic rate. Increased metabolism leads to higher levels of carbon dioxide, but also to a lower pH (increased H^+ concentration from lactic acid production and the dissociation

$H_2CO_3 \rightarrow H^+ + CO_3^-$). During strenuous exercise, minute ventilation increases proportionately more because glycolysis facilitates the release of excess lactic acid.

The control of breathing is critical for understanding respiratory responses to activity, changes in the environment, and lung diseases. Our breathing depends on the cyclic excitation of many muscles that influences the volume of the thorax. Control of that excitation is the result of multiple neuronal interactions involving all levels of the nervous system. Furthermore, the muscles used for breathing must often be used for other purposes, as well. For example, talking while walking requires some muscles to simultaneously attend to the tasks of posturing, walking, phonation, and breathing. Because it is impossible to study extensively the subtleties of such a complex system in humans, much of what is known about the control of breathing has been obtained from the study using animal models. Much, however, remains unexplained.

The control of upper and lower airway muscles that affect airway tone is integrated with control of the muscles that start tidal air movements. During quiet breathing, inspiration is brought about by a progressive increase in electrical activation of the inspiratory muscles, most importantly the diaphragm, until tidal volume has been reached. At the end of inspiration, the inspiratory muscles relax, and expiration occurs passively due to the elastic recoil of the lungs and chest wall. However, as more ventilation is required—for example, during exercise—other inspiratory muscles (external intercostals, cervical muscles) are recruited. In addition, when ventilation is elevated, expiration becomes an active process through the use, most notably, of the muscles of the abdominal wall.

The respiratory centers are located within the pons and medulla.

The neural basis of the breathing patterns depends on the generation and subsequent tailoring of cyclic changes in the activity of cells primarily located in the medulla oblongata in the brain. Although the mechanism of generating ventilatory patterns is still not completely understood, but the integration on neural signals by respiratory control centers involve the

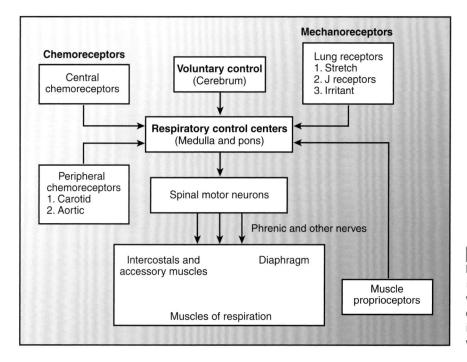

Figure 21.1 **Breathing is regulated by various input signals.** The schematic illustrates the overall control of breathing with various mechanoreceptors, proprioceptors, and chemoreceptors that are involved in adjusting breathing to meet various metabolic demands.

medulla and pons. The central pattern for the basic breathing rhythm has been localized to fairly discrete areas within the medulla oblongata that discharge action potentials in a phasic pattern with respiration. Cells in the medulla oblongata associated with breathing have been identified by noting the correlation between their activity and mechanical events of the breathing cycle. Two different regions in the medulla control the breathing rhythm, and their anatomic locations are shown in Figure 21.2. The **dorsal respiratory**

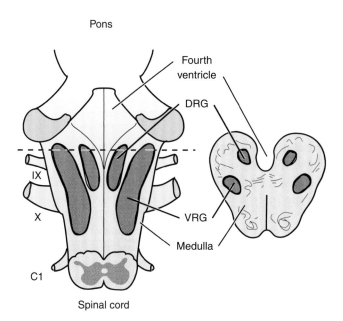

Figure 21.2 **The dorsal respiratory group (DRG) and the ventral respiratory group (VRG) are located in the medulla.** These drawings show the dorsal aspect of the medulla oblongata and the cross-section of the region of the fourth ventricle. C1, first cervical nerve; IX, glossopharyngeal nerve; X, vagus nerve.

group (DRG), named for its dorsal location in the region of the nucleus *tractus solitarii*, predominantly contains cells that are active during inspiration. The **ventral respiratory group (VRG)** is a column of cells in the general region of the *nucleus ambiguus* that extends caudally nearly to the *bulbospinal* border and cranially nearly to the *bulbopontine junction*. The VRG contains both inspiration- and expiration-related neurons. Both groups contain cells projecting ultimately to the bulbospinal motor neuron pools. The DRG and VRG are bilaterally paired, allowing for cross-communication. As a result, they can respond in synchrony allowing respiratory movements to be rhythmic. The DRG controls primarily inspiratory movements and rhythm. The VRG is primarily responsible for controlling voluntary forced inspiration and exhalation. The neural networks forming the central pattern generator for breathing are contained within the DRG/VRG complex. In terms of breathing, inspiratory motor neuron must be activated before expiratory motor neurons.

Central pattern generation probably does not arise from a single pacemaker or by reciprocal inhibition of two pools of cells, one having inspiratory-related and the other having expiratory-related activity. Instead, the progressive rise and abrupt fall of inspiratory motor activity associated with each breath can be modeled by the starting, stopping, and resetting of an integrator of background ventilatory drive. A *rhythmic generator* is responsible for the normal breathing pattern. The rhythmic generator consists of a network of interneurons that communicates with another to effectively produce a repetitive motor pattern that produces a respiratory rhythm of 12 breaths/min. An integrator-based theoretical model, as described below, is suitable for a first understanding of respiratory pattern generation. Many different signals (e.g., anxiety, musculoskeletal movements, pain, chemosensory activity, and hypothalamic temperature) provide a background ventilatory drive to the medulla.

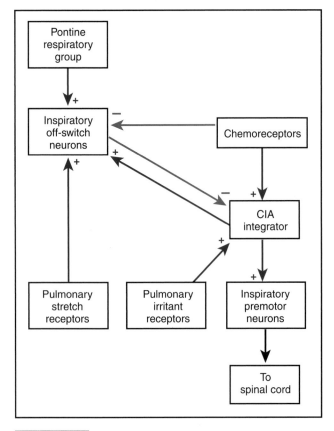

Figure 21.3 Inspiration is initiated by the medullary inspiratory generator. Inspiration begins by the abrupt release from inhibition of a group of cells called the central inspiratory activity (CIA) integrator.

Inspiration begins by the abrupt release from inhibition of a group of cells, **central inspiratory activity (CIA) integrator** neurons, located within the medullary reticular formation that integrate this background drive (Fig. 21.3). Integration results in a progressive rise in the output of the integrator neurons, which, in turn, excites a similar rise in activity of inspiratory premotor neurons of the DRG/VRG complex. The rate of rising activity of inspiratory neurons and, therefore, the rate of inspiration itself can be influenced by changing the characteristics of the CIA integrator. Inspiration is ended by abruptly switching off the rising excitation of inspiratory neurons. The CIA integrator is reset before the beginning of each inspiration, so that activity of the inspiratory neurons begins each breath from a low level.

Inspiratory activity is switched off to initiate expiration.

Two groups of neurons located within the VRG serve as an inspiratory off switch (see Fig. 21.3). Switching occurs abruptly when the sum of excitatory inputs to the off switch reaches a threshold. Adjustment of the threshold level is one of the ways in which depth of breathing can be varied. Two important excitatory inputs to the off switch are a progressively increasing activity from the CIA integrator's rising output and an input from lung stretch receptors, whose afferent activity increases progressively with rising lung volume. (The first of these is what allows the medulla to generate a breathing

pattern on its own; the second is one of many reflexes that influence breathing.) Once the critical threshold is reached, off-switch neurons apply a powerful inhibition to the CIA integrator. The CIA integrator is thus reset by its own rising activity. Other inputs, both excitatory and inhibitory, act on the off switch and change its threshold. For example, chemical stimuli, such as hypoxemia and hypercapnia, are inhibitory, raising the threshold and causing larger tidal volumes.

An important excitatory input to the off switch comes from a group of spatially dispersed neurons in the rostral pons called the **pontine respiratory group**. Electrical stimulation in this region causes variable effects on breathing, dependent not only on the site of stimulation but also on the phase of the respiratory cycle in which the stimulus is applied. It is believed that the pontine respiratory group may serve to integrate many different autonomic functions in addition to breathing.

Expiration comprises two phases.

Shortly after the abrupt termination of inspiration, some activity of inspiratory muscles resumes. This activity serves to control expiratory airflow. This effect is greatest early in expiration and recedes as lung volume falls. Inspiratory muscle activity is essentially absent in the second phase of expiration, which includes continued passive recoil during quiet breathing and activation of expiratory muscles if more than quiet breathing is required.

The duration of expiration is determined by the intensity of inhibition of activity of inspiratory-related cells of the DRG/VRG complex. Inhibition is greatest at the start of expiration and falls progressively until it is insufficient to prevent the onset of inspiration. The progressive fall of inhibition amounts to a decline of threshold for initiating the switch from expiration to inspiration. The rate of decline of inhibition and the occurrence of events that trigger the onset of inspiration are subject to several influences. The duration of expiration can be controlled not only by neural information arriving during expiration but also in response to the pattern of the preceding inspiration. How the details of the preceding inspiration are stored and later recovered is unresolved.

Muscles of the upper airways are also under phasic control.

The same rhythm generator that controls the chest wall muscles also controls muscles of the nose, pharynx, and larynx. However, unlike the inspiratory ramplike rise of the stimulation of chest wall muscles, the excitation of upper airway muscles quickly reaches a plateau and is sustained until inspiration is ended. Flattening of the expected ramp excitation waveform probably results from progressive inhibition by the rising afferent activity of airway stretch reflexes as lung volume increases. Excitation during inspiration causes contractions of upper airway muscles, airway widening, and reduced resistance from the nostrils to the larynx.

During the first phase of quiet expiration, when expiration is slowed by renewed inspiratory muscle activity, there is also expiratory braking caused by active adduction of the vocal cords. However, during exercise-induced **hyperpnea** (increased depth and rate of breathing), the cords are separated during expiration and expiratory resistance is reduced.

Increasing metabolic demands require various control mechanisms to match alveolar ventilation.

Multiple controls provide a greater capability for regulating breathing under a larger number of conditions. Their interactions modify each other and provide for backup in case of failure. The set of strategies for controlling a given variable, such as minute ventilation, typically includes individual schemes that differ in several respects, including choices of sensors and effectors, magnitudes of effects, speeds of action, and optimum operating points. The use of multiple control mechanisms in breathing can be illustrated by considering some of the ways breathing changes in response to exercise. Perhaps the simplest strategies are feedforward mechanisms, in which breathing responds to some component of exercise but without recognition of how well the response meets the demand. One such mechanism would be for the central nervous system (CNS) to vary the activity of the medullary pattern generator in parallel with, and in proportion to, the excitation of the muscles used during exercise. Another prospective feedforward scheme involves sensing the magnitude of the carbon dioxide load delivered to the lungs by systemic venous return and then driving ventilation in response to the magnitude of that load. Experimental evidence supports this mechanism, but the identity of the required intrapulmonary sensor remains uncertain. Still another recognized feedforward mechanism is the enhancement of breathing in response to increased receptor activity in skeletal joints as joint motion increases with exercise.

▶ NEURAL REFLEXES IN THE CONTROL OF BREATHING

Neural reflexes provide feedback for fine tuning, which adjusts frequency and tidal volume to minimize the work of breathing. Reflexes from the upper airways and lungs also act as defensive reflexes, protecting the lungs from injury and environmental insults. Receptors for these reflexes are located centrally and peripheral and consist of *chemoreceptors* and *mechanoreceptors*. Central chemoreceptors of the central nervous system, located in the medulla, are sensitive to changes in $[H^+]$. Peripheral chemoreceptors detect changes in Pao_2, $Paco_2$, and arterial $[H^+]$. Mechanoreceptors located in airways, lung parenchyma, chest wall, and leg muscle/tendons also provide neural reflexes that are involved in the control of breathing (see Fig. 21.1).

Mechanoreceptors mediate reflexes that protect the lung.

Pulmonary mechanoreceptors are located in the airways, and lung parenchyma and can be divided into three groups: *pulmonary stretch*, *irritant*, and *J receptors* (see Fig. 21.1). Afferent fibers of all three types lie predominantly in the vagus nerves.

Pulmonary stretch receptors are sensory terminals of myelinated afferent fibers that lie within the smooth muscle layer of conducting airways. Stretch receptors fire in proportion to applied airway transmural pressure, and their usual role is to sense lung volume. When stimulated, an increased firing rate is sustained as long as stretch is imposed; that is, they adapt slowly and are also called *slowly adapting receptors*. Stimulation of these receptors causes an excitation of the inspiratory off switch and a prolongation of expiration. Because of these two effects, inflating the lungs with a sustained pressure at the mouth terminates an inspiration in progress and prolongs the time before a subsequent inspiration occurs.

This sequence is known as the *Hering-Breuer reflex or lung inflation reflex*. The Hering-Breuer reflex probably plays a more important role in infants than in adults. In adults, particularly in the awake state, this reflex may be overwhelmed by more prominent central control. Because increasing lung volume stimulates stretch receptors, which then excite the inspiratory off switch, it is easy to see how they could be responsible for a feedback signal that results in cyclic breathing. However, as already mentioned, feedback from vagal afferents is not necessary for cyclic breathing to occur. Instead, feedback modifies a basic pattern established in the medulla. The effect may be to shorten inspiration when tidal volume is larger than normal. The most important role of slowly adapting receptors is probably their participation in regulating expiratory time, expiratory muscle activation, and functional residual capacity. Stimulation of the stretch receptors also relaxes airway smooth muscle, reduces systemic vasomotor tone, increases heart rate, and, as previously noted, influences laryngeal muscle activity.

The *irritant receptors* are sensory terminals of myelinated afferent fibers that are found in the larger conducting airways. Also called **rapid adapting receptors**, they have nerve endings, which lie in the airway epithelium, and respond to irritation of the airways by touch or by noxious substances, such as smoke and dust. Irritant receptors are stimulated by histamine, serotonin, and prostaglandins released locally in response to allergy and inflammation. They are also stimulated by lung inflation and deflation, but their firing rate rapidly declines when a volume change is sustained. Because of this rapid adaptation, bursts of activity occur that are in proportion to the change of volume and the rate at which that change occurs. Acute congestion of the pulmonary vascular bed also stimulates these receptors, but, unlike the effect of inflation, their activity may be sustained when congestion is maintained.

Background activity of the irritant receptors is inversely related to lung compliance, and they are thought to serve as sensors of compliance change. These receptors are probably nearly inactive in normal quiet breathing. Based on what stimulates them, their role would seem to be to sense the onset of pathologic events. In spite of considerable information about what stimulates them, the effect of their stimulation remains controversial. As a general rule, stimulation causes excitatory responses such as coughing, gasping, and prolonged inspiration time.

Juxtapulmonary capillary receptors are sensory nerve endings located adjacent to the alveolar wall in juxtaposition to the pulmonary capillaries, hence the name. The shorten term for these receptors are called *J receptors*. These receptors are unmyelinated and also innervated by fibers from the vagus nerve. The J receptors are classified as two populations in the lungs. One group is located adjacent to the alveoli and is accessible from the pulmonary circulation and sometimes referred to as *pulmonary C fibers*. A second group, *bronchial C fibers*, is accessible from the bronchial circulation and, consequently, is located in airways. Both groups play a protective role. The pulmonary C fibers are stimulated by physical engorgement of the pulmonary capillaries, such as pulmonary edema, pulmonary emboli, or congestive heart failure. When stimulated, they cause a reflex of increased in breathing rate. The bronchial C fibers are stimulated by hyperinflation of the lungs and to products of inflammation.

Proprioceptors provide information about the body's activity and muscle tension.

Proprioceptors are located in tendons and skeletal muscle and can play a role in breathing, particularly when more than quiet breathing is called for or when breathing efforts are opposed by increased airway resistance or reduced lung compliance. One type of proprioceptors, *muscle spindle receptors*, provides information about changes in muscle length. A considerable number of muscle spindles are located in the intercostal muscles, but not in the diaphragm. Another type, *Golgi tendon organ*, is located in the tendons that attach muscle to bone and provide information muscle tension. When both types of proprioceptors are stimulated, they cause a reflex of increased ventilation.

Central and peripheral chemoreceptors respond to changes in arterial blood gases and hydrogen ion concentration.

Another set of receptors that profoundly affect breathing are the **chemoreceptors** (see Fig. 21.1). These receptors are stimulated by the hydrogen ion concentration $[H^+]$ and respiratory gas composition of the arterial blood. The general rule is that breathing activity is inversely related to arterial blood P_{O_2}, but directly related to P_{CO_2} and $[H^+]$. Figures 21.4 and 21.5 show the ventilatory responses of a typical person when alveolar P_{CO_2} ($P_{A_{CO_2}}$) and P_{O_2} are individually varied by controlling the composition of inspired gas. Responses to carbon dioxide and, to a lesser extent, blood pH depend on sensors in the brainstem, carotid arteries, and aorta. In contrast, responses to hypoxia are brought about only by the stimulation of arterial receptors.

Cerebrospinal fluid [H⁺] stimulates neuronal cells in the medulla.

Ventilatory drive is exquisitely sensitive to P_{CO_2} of blood perfusing the brain. The source of this chemosensitivity has been localized to bilaterally paired groups of cells just below the surface of the ventrolateral medulla immediately

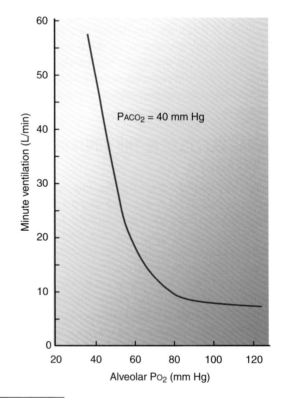

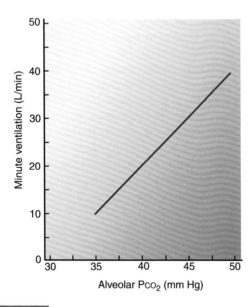

Figure 21.4 **CO₂ is a powerful stimulus to ventilation.** Ventilatory responses to increasing alveolar CO₂ tension are shown with the line, representing the response when alveolar P_{O_2} ($P_{A_{O_2}}$) is held at ≥100 mm Hg to essentially eliminate O_2-dependent activity of the chemoreceptors.

Figure 21.5 **Ventilatory responses to hypoxia are mediated via the peripheral chemoreceptors.** Inspired oxygen is lowered, while alveolar P_{CO_2} ($P_{A_{CO_2}}$) is held at 40 mm Hg by adding CO₂ to the inspired air in order to eliminate the chemoresponse to CO₂-dependent activity. Note that when alveolar carbon dioxide is held constant at 43 mm Hg, minute ventilation does not significantly change until alveolar P_{O_2} is reduced to 60 mm Hg.

caudal to the pontomedullary junction. Each side contains a rostral and a caudal chemosensitive zone, separated by an intermediate zone in which the activities of the caudal and rostral groups converge and are integrated together with other autonomic functions. Exactly which cells exhibit chemosensitivity is unknown, but they are not the same as those of the DRG/VRG complex. Although specific cells have not been identified, the chemosensitive neurons that respond to the [H+] of the surrounding interstitial fluid are referred to as **central chemoreceptors**. The H+ concentration in the interstitial fluid is a function of PCO_2 in the cerebral arterial blood and the bicarbonate concentration of cerebrospinal fluid (CSF).

Cerebrospinal fluid has a weak buffering system and is sensitive to changes in carbon dioxide partial pressure.

CSF is an ultrafiltrate of plasma and is formed mainly by the **choroid plexuses** of the ventricular cavities of the brain. The epithelium of the choroid plexus provides a barrier between blood and CSF that severely limits the passive movement of large molecules, charged molecules, and inorganic ions. However, choroidal epithelium actively transports several substances, including ions, and this active transport participates in determining the composition of CSF. This selective barrier is termed the **blood–brain barrier** and is illustrated in Figure 21.6. CSF formed by the choroid plexuses is exposed

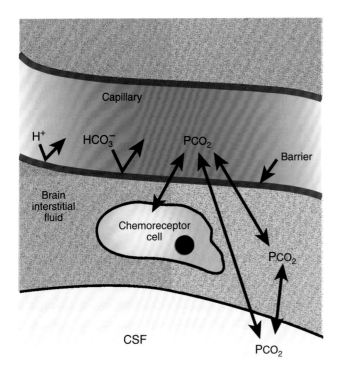

Figure 21.6 **The blood–brain barrier is impermeable to blood H+ and HCO₃⁻.** Movement of H+, HCO₃⁻, and molecular CO_2 is illustrated between capillary blood, brain interstitial fluid, and cerebrospinal fluid (CSF). Because the blood–brain barrier is permeable to CO_2 and not to H+ and HCO₃⁻, the acid–base status of the chemoreceptors can be quickly changed only by changing $PaCO_2$.

to brain interstitial fluid across the surface of the brain and spinal cord, with the result that the composition of CSF away from the choroid plexuses is closer to that of interstitial fluid than it is to CSF as first formed. Brain interstitial fluid is also separated from blood by the blood–brain barrier (capillary endothelium), which has its own transport capability.

Because of the properties of the limiting membranes, CSF is essentially free of protein, but it is not just a simple ultrafiltrate of plasma. CSF differs most notably from an ultrafiltrate by its lower bicarbonate and higher sodium and chloride ion concentrations. Potassium, magnesium, and calcium ion concentrations also differ somewhat from plasma; moreover, they change little in response to marked changes in plasma concentrations of these cations. Most proposed regulatory mechanisms invoke the active transport of one or more ionic species by the epithelial and endothelial membranes. Because of the relative impermeabilities of the choroidal epithelium and capillary endothelium to H+, changes in H+ concentration of blood are poorly reflected in CSF. By contrast, molecular carbon dioxide diffuses readily; therefore, blood PCO_2 can influence the pH of CSF. The pH of CSF is primarily determined by its bicarbonate concentration and PCO_2. The relative ease of movement of molecular carbon dioxide in contrast to hydrogen ions and bicarbonate is depicted in Figure 21.6.

In healthy people, the PCO_2 of CSF is about 6 mm Hg higher than that of arterial blood, approximating that of brain tissue. The pH of CSF, normally slightly below that of blood, is held within narrow limits. CSF pH changes little in states of metabolic acid–base disturbances (see Chapter 24)—about 10% of that in plasma. In respiratory acid–base disturbances, however, the change in pH of the CSF may exceed that of blood. During chronic acid–base disturbances, the bicarbonate concentration of CSF changes in the same direction as in blood, but the changes may be unequal. In metabolic disturbances, the CSF bicarbonate changes are about 40% of those in blood, but with respiratory disturbances, CSF and blood bicarbonate changes are essentially the same. When acute acid–base disturbances are imposed, CSF bicarbonate changes more slowly than blood bicarbonate and may not reach a new steady state for hours or days. As already noted, the mechanism of bicarbonate regulation is unsettled. Irrespective of how it occurs, the bicarbonate regulation that occurs with acid–base disturbances is important because, by changing buffering, it influences the response to a given PCO_2.

Peripheral chemoreceptors respond to PO_2 and PCO_2 and [H+] in the arterial blood.

Peripheral chemoreceptors are located in the carotid and aortic bodies. They both detect changes in arterial blood plasma PO_2, PCO_2, and pH (see Fig. 21.1). **Carotid bodies** are small (~2 mm wide) sensory organs located bilaterally near the bifurcations of the common carotid arteries near the base of the skull. Afferent nerves travel to the CNS from the carotid bodies in the glossopharyngeal nerves. **Aortic bodies** are located along the arch of the aorta and are innervated by vagal afferents.

As with the medullary chemoreceptors, increasing Pa_{CO_2} stimulates peripheral receptors. H^+ formed from H_2CO_3 within the peripheral chemoreceptors (glomus cells) is the stimulus and not molecular CO_2. About 40% of the effect of Pa_{CO_2} on ventilation is brought about by peripheral chemoreceptors, whereas central chemoreceptors bring about the rest. Unlike the central sensor, peripheral chemoreceptors are sensitive to rising arterial blood H^+ and falling Po_2. They alone cause the stimulation of breathing by low oxygen; hypoxia in the brain has little effect on breathing unless it is severe, at which point breathing is depressed.

Carotid chemoreceptors play a more prominent role than do aortic chemoreceptors; because of this and their greater accessibility, they have been studied in greater detail. The discharge rate of carotid chemoreceptors (and the resulting minute ventilation) is approximately linearly related to Pa_{CO_2}. The linear behavior of the receptor is reflected in the linear ventilatory response to carbon dioxide illustrated in Figure 21.4. When expressed using pH, the response curve is no longer linear but shows a progressively increasing effect as pH falls below normal. This occurs because pH is a logarithmic function of $[H^+]$, so the absolute change in $[H^+]$ per unit change in pH is greater when brought about at a lower pH.

The response of peripheral chemoreceptors to oxygen depends on arterial Po_2 (Pa_{O_2}) and not oxygen content. Therefore, anemia or carbon monoxide poisoning, two conditions that exhibit reduced oxygen content but have normal Pa_{O_2}, have little effect on the response curve. The shape of the response curve is not linear; instead, hypoxia is of increasing effectiveness as Po_2 falls below about 90 mm Hg. The behavior of the receptors is reflected in the ventilatory response to hypoxia illustrated in Figure 21.5. The shape of the curve relating ventilatory response to Po_2 resembles that of the oxyhemoglobin equilibrium curve when plotted upside down (see Chapter 20). As a result, the ventilatory response is inversely related in an approximately linear fashion to arterial blood oxygen saturation.

The nonlinearities of the ventilatory responses to Po_2 and pH and the relatively low sensitivity across the normal ranges of these variables cause ventilatory changes to be apparent only when Po_2 and pH deviate significantly from the normal range, especially toward hypoxemia or acidemia. By contrast, ventilation is sensitive to Pco_2 within the normal range, and carbon dioxide is normally the dominant chemical regulator of breathing through the use of both central and peripheral chemoreceptors (compare Figs. 21.4 and 21.5).

There is a strong interaction among stimuli, which causes the slope of the carbon dioxide response curve to increase if determined under hypoxic conditions (see Fig. 21.4), causing the response to hypoxia to be directly related to the prevailing Pco_2 and pH (see Fig. 21.5). As discussed in the next section, these interactions and interaction with the effects of the central carbon dioxide sensor profoundly influence the integrated chemoresponses to a primary change in arterial blood composition.

Carotid and aortic bodies can also be strongly stimulated by certain chemicals, particularly cyanide ion and other poisons of the metabolic respiratory chain. Changes

in blood pressure have only a small effect on chemoreceptor activity, but responses can be stimulated if arterial pressure falls below about 60 mm Hg. This effect is more prominent in aortic bodies than in carotid bodies. Afferent activity of peripheral chemoreceptors is under some degree of efferent control capable of influencing responses by mechanisms that are not clear. Afferent activity from the chemoreceptors is also centrally modified in its effects by interactions with other reflexes, such as the lung stretch reflex and the systemic arterial baroreflex (see Chapter 17). Although the breathing interactions are not well understood in humans, they serve as examples of the complex interactions of cardiorespiratory regulation. Interactions among chemoreflexes, however, are easily demonstrated.

Figure 21.7 is a schematic summarizing the chemical control of ventilation by changes in arterial Po_2, Pco_2, and $[H^+]$.

▶ PHYSIOLOGIC RESPONSES TO ALTERED OXYGEN AND CARBON DIOXIDE

Before the interactions of the chemoresponses to oxygen and carbon dioxide are discussed, a brief review of terminology is order. Hypoxia leads to **hypoxemia** (a condition in which the Pa_{O_2} is below normal). Hypoxemia stimulates respiration, causing **hyperventilation** (increased alveolar ventilation greater than required to meet metabolic needs). Hyperventilation results in excess carbon dioxide being blown off, which, in turn, leads to **hypocapnia** (a condition in which Pa_{CO_2} is below normal). A low Pa_{CO_2} decreases blood $[H^+]$, leading to **alkalosis** (a condition in which arterial blood pH is above normal). Because alkalosis in this case is caused by hyperventilation, the condition is termed **respiratory alkalosis**. When the lungs are underventilated, alveolar ventilation is less than required to meet metabolic needs (**hypoventilation**) leading to a concomitant decrease in Pa_{O_2} and an elevated in Pa_{CO_2} (**hypercapnia**). As a result of hypercapnia, there is an increase in arterial blood $[H^+]$ that leads to **acidosis** (a condition in which blood pH is below normal). Again, because the acidotic condition is caused by a respiratory dysfunction, the condition is referred to as **respiratory acidosis**. Another condition caused by metabolic accumulation of nonvolatile acids, such as lactic acid, leads to **metabolic acidosis**. Respiratory acidosis (accumulation of carbon dioxide) and metabolic acidosis (accumulation of lactic acid) both cause low arterial blood pH; the only difference between them is the cause of the acidotic condition; one is metabolic in nature and the other respiratory.

A distinction needs to be made between hyperventilation and **hyperpnea** (increased alveolar ventilation in proportion to elevated metabolism, e.g., exercise). Remember that *hyperventilation* is characterized by an increase in alveolar ventilation with a concomitant decrease in arterial Pco_2, whereas *hyperpnea* is an increase in alveolar ventilation relative to metabolic carbon dioxide production. The distinguishing feature of hyperpnea is increased ventilation with no change in arterial blood Pco_2, because alveolar ventilation increases in proportion to carbon dioxide production.

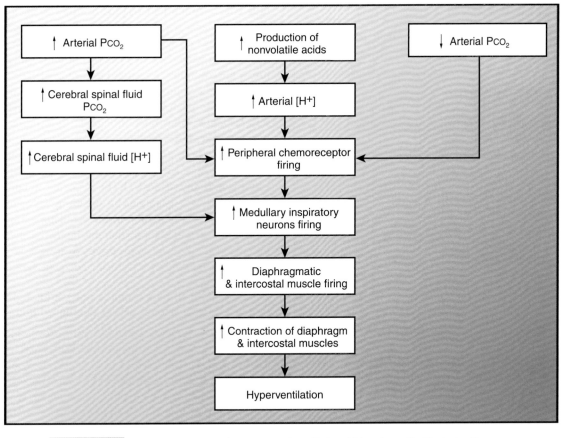

Figure 21.7 **Summary of the chemical control of ventilation.** This schematic illustrates the major chemical inputs that stimulate ventilation. An increase in arterial P_{CO_2} and $[H^+]$ and a decrease in arterial P_{O_2} stimulate ventilation. Conversely, when P_{O_2} is elevated and arterial P_{CO_2} and $[H^+]$ are decreased, ventilation is inhibited.

Responses to altered blood oxygen and carbon dioxide are interdependent.

The effect of P_{O_2} on the response to carbon dioxide and the effect of carbon dioxide on the response to P_{O_2} have already been noted. By virtue of this interdependence, the subsequent increased ventilation is blunted in a hyperventilatory response, unless P_{ACO_2} is somehow maintained, because P_{ACO_2} ordinarily falls as ventilation is stimulated (see Fig. 21.5). The central chemoreceptors, which respond more potently than the peripheral receptors to low P_{ACO_2}, are mainly responsible for blunting the stimulating effect of hyperventilation. The sequence of events in the response to hypoxia (e.g., ascent to high altitude) exemplifies interactions among chemoresponses. For example, if 100% oxygen is given to a person just arriving at high altitude, ventilation is quickly restored to its sea-level value. During the next few days, ventilation in the absence of supplemental oxygen progressively rises further, but it is no longer restored to sea-level value by breathing oxygen. Rising ventilation while acclimatizing to altitude could be explained by a reduction of blood and CSF bicarbonate concentrations. This would reduce the initial increase in pH created by the increased ventilation and allow the hypoxic stimulation to be less strongly opposed. However, this mechanism is not the full explanation of altitude acclimatization. CSF pH is not fully restored to normal, and the increasing ventilation raises P_{aO_2} while further lowering P_{ACO_2}, changes that should inhibit the stimulus to breathe. In spite of much inquiry, the reason for persistent hyperventilation in altitude-acclimatized subjects, the full explanation for altitude acclimatization, and the explanation for the failure of increased ventilation in acclimatized subjects to be relieved promptly by restoring a normal P_{aO_2} are still unknown.

In *metabolic acidosis*, the increase in arterial blood $[H^+]$ initiates and sustains hyperventilation by stimulating the peripheral chemoreceptors (Fig. 21.8). Because of the restricted movement of H^+ into CSF, the fall in blood pH cannot directly stimulate the central chemoreceptors. The central effect of the hyperventilation, brought about by decreased pH via the peripheral chemoreceptors, results in a paradoxical rise of CSF pH (i.e., an alkalosis as a result of reduced P_{ACO_2}) that actually restrains the hyperventilation. With time, CSF bicarbonate concentration is adjusted downward, although it changes less than that of blood, and the pH of CSF remains somewhat higher than pH of blood. Ultimately, ventilation increases more than it did initially as the paradoxical CSF alkalosis is removed.

Respiratory acidosis is rarely a result of elevated environmental CO_2, although this can occur in submarine mishaps, in wet limestone caves, and in physiology laboratories where

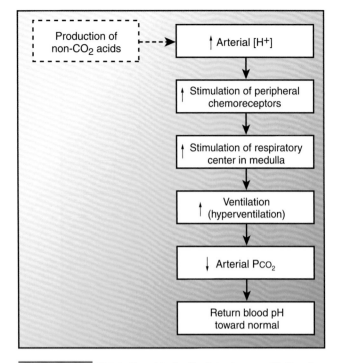

Figure 21.8 **Metabolic acidosis stimulates hyperventilation via the peripheral chemoreceptors.** The accumulation of non-CO_2 acids increases the blood [H^+] and stimulates breathing. The sustained hyperventilation occurs by stimulating the respiratory center in the medulla via the peripheral chemoreceptors. The sustained hyperventilation blows off alveolar CO_2, which lowers $Paco_2$ and compensates by returning blood pH back toward normal.

responses to carbon dioxide are measured. Under these conditions, the response is a vigorous increase in minute ventilation proportional to the level of $Paco_2$; Pao_2 actually rises slightly and arterial pH falls slightly, but these have relatively

little effect. If mild hypercapnia can be sustained for a few days, the intense hyperventilation subsides, probably as CSF bicarbonate is raised. More commonly, respiratory acidosis occurs from generalized hypoventilation and results from failure of the controller to respond to carbon dioxide (e.g., during anesthesia, following brain injury, and in some patients with chronic obstructive lung disease). The hypoventilation response is illustrated in Figure 21.9. Another cause of respiratory acidosis is patients with obstructive lung disease. When these subjects breathe room air, hypercapnia caused by reduced alveolar ventilation is accompanied by significant hypoxia and acidosis. If the hypoxic component alone is corrected—for example, by breathing oxygen-enriched air—a significant reduction in the ventilatory stimulus may promote underventilation, thereby increasing hypercapnia and acidosis. A more appropriate treatment is providing mechanical assistance for restoring adequate ventilation.

Exercise-induced hyperpnea is not stimulated by increased Pco_2, arterial pH, or decreased Pao_2.

With light-to-moderate exercise, Pao_2, Pco_2, and pH are essentially unchanged from normal values. Yet ventilation can be substantially increased in these exercise regimes. Even with heavy exercise, a person with a normal cardiovascular system does not become hypoxemic or acutely hypercapnic. What controls ventilation during exercise if chemoresponses to changes in Pao_2, Pco_2, and atrial pH are not involved? The answer to this question involves various control systems, many of which are not well defined. The control of breathing with exercise occurs in three phases and includes the *neurological phase (phase I)*, the *metabolic phase (phase II)*, and the *compensatory phase (phase III)*. In phase I, ventilation increases almost instantly with the onset of exercise and involves a neurologic response insofar as the increase occurs

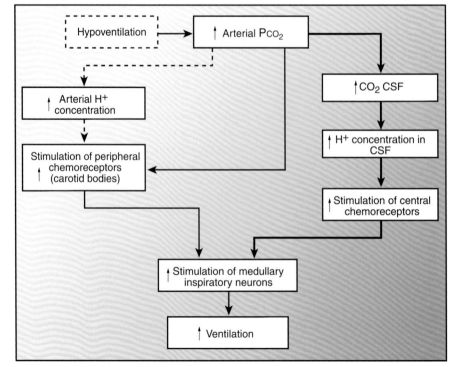

Figure 21.9 **Generalized hypoventilation is the primary cause of respiratory acidosis.** When the lungs are underventilated, arterial blood CO_2 rises (hypercapnia), leading to respiratory acidosis. The common cause for respiratory acidosis is failure of the controller to respond to carbon dioxide (e.g., anesthesia, head injury, and patients with severe chronic obstructive pulmonary disease). CSF, cerebrospinal fluid.

in the absence of any changes in blood gases and metabolism. The neurologic phase involves the medullary generator (see Fig. 21.4) and a feedforward mechanism. During this phase, depth and rate of breathing change in response to the activity of the medullary generator, which increases in parallel with the excitation of the muscles used during exercise. Another recognized feedforward mechanism is stimulation of ventilation in response to increased receptor activity to skeletal joints as joint motion increases with exercise.

In phase II, as exercise continues, ventilation increases linearly with the increase in carbon dioxide production. Exercise-induced hyperpnea occurs in this phase, because the increase in alveolar ventilation is proportional to carbon dioxide production, and as a result, there is no change in $Paco_2$. Remember, not only $Paco_2$ does not change with exercise-induced hyperpnea, but there is no change in Pao_2 and arterial pH as well. Although the increased ventilation is tightly coupled to carbon dioxide production, it is still not known how the medullary controller monitors carbon dioxide production. One hypothesis is that a feedforward scheme senses the magnitude of the carbon dioxide load delivered to the lungs by systemic venous blood, which in return drives ventilation. Experimental evidence supports this mechanism, but specific intrapulmonary sensors have not been identified.

In phase III, the intensity of exercise increases (i.e., exhaustive exercise), and the energy needs outstrip the ability of the cardiovascular system to supply sufficient oxygen for aerobic metabolism. Muscle cells shift to anaerobic metabolism with increased blood lactic acid as a by-product. Sustained ventilation is compensation for the development to metabolic acidosis, hence the term *compensatory phase*. During this phase, the sustained ventilation is driven by the acidotic condition.

▶ CONTROL OF BREATHING DURING SLEEP

We spend about one third of our lives asleep. Sleep disorders and disordered breathing during sleep are common and often have physiologic consequences. Chapter 7 describes the two different neurophysiologic sleep states: **rapid eye movement (REM) sleep** and **slow-wave sleep**. Sleep is a condition that results from withdrawal of the wakefulness stimulus that arises from the brainstem reticular formation. This wakefulness stimulus is one component of the tonic excitation of brainstem respiratory neurons, and one would predict correctly that sleep results in a general depression of breathing. There are, however, other changes, and the effects of REM and slow-wave sleep on breathing differ.

Sleep changes the breathing frequency and inspiratory flow rate.

During *slow-wave sleep*, breathing frequency and inspiratory flow rate are reduced, and minute ventilation falls. These responses partially reflect the reduced physical activity that accompanies sleep. However, because of the small rise in $Paco_2$ (about 3 mm Hg), there must also be a change in either the sensitivity or the set point of the carbon dioxide

controller. In the deepest stage of slow-wave sleep (stage 4), breathing is slow, deep, and regular. But in stages 1 and 2, the depth of breathing sometimes varies periodically. The explanation is that during light sleep, withdrawal of the wakefulness stimulus varies over time in a periodic fashion. When the stimulus is removed, sleep is deepened and breathing is depressed; when returned, breathing is excited not only by the wakefulness stimulus but also by the carbon dioxide retained during the interval of sleep. This periodic pattern of breathing is known as **Cheyne-Stokes breathing** (Fig. 21.10).

In *REM sleep*, breathing frequency varies erratically, whereas tidal volume varies little. The net effect on alveolar ventilation is probably a slight reduction, but this is achieved by averaging intervals of frank **tachypnea** (excessively rapid breathing) with intervals of apnea. Unlike those in slow-wave sleep, the variations during REM sleep do not reflect a changing wakefulness stimulus but instead represent responses to increased CNS activity of behavioral, rather than autonomic or metabolic, control systems (Clinical Focus 21.1).

Sleep blunts the respiratory sensitivity to carbon dioxide.

Responsiveness to carbon dioxide is reduced during sleep. In slow-wave sleep, the reduction in sensitivity seems to be secondary to a reduction in the wakefulness stimulus and its tonic excitation of the brainstem rather than to a suppression of the chemosensory mechanisms. It is important to note that breathing remains responsive to carbon dioxide during slow-wave sleep, although at a less sensitive level, and that carbon dioxide stimulus may provide the major background

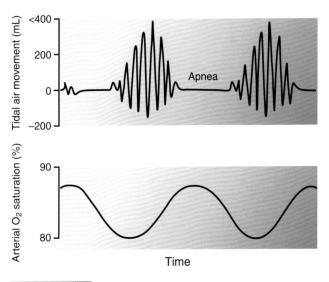

Figure 21.10 **Cheyne-Stokes breathing lowers arterial O_2 saturation.** Cheyne-Stokes breathing occurs frequently during sleep, especially in subjects at high altitude, as in this example. In the presence of preexisting hypoxemia secondary to high altitude or other causes, the periods of apnea may result in further decreases of O_2 saturation to dangerous levels. Falling Po_2 and rising Pco_2 during the apnea intervals ultimately induce a response, and breathing returns, reducing the stimuli and leading to a new period of apnea.

Sleep Apnea Syndrome

The analysis of multiple physiologic variables recorded during sleep, known as polysomnography, is an important method for research into the control of breathing and has been increasingly used in clinical evaluations of sleep disturbances. In normal sleep, reduced dilatory upper airway muscle tone may be accompanied by brief intervals of no breathing movements. Some people, typically overweight men, exhibit more severe disruption of breathing, referred to as **sleep apnea syndrome**. Sleep apnea is classified in two broad groups: obstructive and central.

In *central sleep apnea*, breathing movements cease for a longer-than-normal interval. In obstructive sleep apnea, the fault seems to lie in a failure of the pharyngeal muscles to open the airway during inspiration. This may be the result of decreased muscle activity, but the obstruction is worsened by an excessive amount of neck fat with which the muscles must contend. With *obstructive sleep apnea*, progressively

larger inspiratory efforts eventually overcome the obstruction, and airflow is temporarily resumed, usually accompanied by loud snoring.

Some patients exhibit both central and obstructive sleep apneas. In both types, hypoxemia and hypercapnia develop progressively during the apnea intervals. Frequent episodes of repeated hypoxia may lead to pulmonary and systemic hypertension and to myocardial distress; the accompanying hypercapnia is thought to be a cause of the morning headache these patients often experience. There may be partial arousal at the end of the periods of apnea, leading to disrupted sleep and resulting in drowsiness during the day. Daytime sleepiness, often leading to dangerous situations, is probably the most common and most debilitating symptom. The cause of this disorder is multivariate and often obscure, but mechanically assisted ventilation during sleep often results in significant symptomatic improvement. ■

brainstem excitation in the absence of the wakefulness stimulus or behavioral excitation. Hence, pathologic alterations in the carbon dioxide chemosensory system may profoundly depress breathing during slow-wave sleep.

During intervals of REM sleep in which there is little sign of increased activity, the breathing response to carbon dioxide is slightly reduced, as in slow-wave sleep. However, during intervals of increased activity, responses to carbon dioxide during REM sleep are significantly reduced, and breathing seems to be regulated by the brain's behavioral control system. It is interesting that regulation of breathing during REM sleep by the behavioral control system, rather than by carbon dioxide, is similar to the way breathing is controlled during speech.

Ventilatory responses to hypoxia are probably reduced during both slow-wave and REM sleep, especially in people who have high sensitivity to hypoxia while awake. There does not seem to be a difference between the effects of slow-wave and REM sleep on hypoxic responsiveness, and the irregular breathing of REM sleep is unaffected by hypoxia.

Both slow-wave and REM sleep cause an important change in responses to airway irritation. Specifically, a stimulus that causes cough, tachypnea, and airway constriction during wakefulness will cause apnea and airway dilation during sleep unless the stimulus is sufficiently intense to cause arousal. The lung stretch reflex appears to be unchanged or somewhat enhanced during arousal from sleep, but the effect of stretch receptors on upper airways during sleep may be important.

Arousal mechanisms protect the sleeper.

Several stimuli cause arousal from sleep; less intense stimuli cause a shift to a lighter sleep stage without frank arousal. In general, arousal from REM sleep is more difficult than

from slow-wave sleep. In humans, hypercapnia is a more potent arousal stimulus than does hypoxia, the former requiring a Pa_{CO_2} of about 55 mm Hg and the latter requiring a Pa_{O_2} <40 mm Hg. Airway irritation and airway occlusion induce arousal readily in slow-wave sleep but much less readily during REM sleep.

All of these arousal mechanisms probably operate through the activation of a reticular arousal mechanism similar to the wakefulness stimulus. They play an important role in protecting the sleeper from airway obstruction, alveolar hypoventilation of any cause, and the entrance into the airways of irritating substances. Recall that coughing depends on the aroused state and without arousal airway irritation leads to apnea. Obviously, wakefulness altered by other-than-natural sleep—such as during drug-induced sleep, brain injury, or anesthesia—leaves the person exposed to risk because arousal from those states is impaired or blocked. From a teleological point of view, the most important role of sensors of the respiratory system may be to cause arousal from sleep.

Upper airway tone may be compromised during REM sleep.

A prominent feature during REM sleep is a general reduction in skeletal muscle tone. Muscles of the larynx, pharynx, and tongue share in this relaxation, which can lead to obstruction of the upper airways. Airway muscle relaxation may be enhanced by the increased effectiveness of the lung inflation reflex.

A common consequence of airway narrowing during sleep is snoring. In many people, usually men, the degree of obstruction may at times be sufficient to cause essentially complete occlusion. In these people, an intact arousal mechanism prevents suffocation, and this sequence is not in itself unusual or abnormal. In some people, obstruction is more complete and more frequent, and the arousal threshold may be raised. Repeated obstruction leads to significant hyper-capnia and hypoxemia, and repeated arousals cause sleep deprivation that leads to excessive daytime sleepiness, often interfering with normal daily activities.

► CONTROL OF BREATHING IN UNUSUAL ENVIRONMENTS

Changes in activity and the environment initiate integrated ventilatory responses that involve changes in the cardiopulmonary system. Examples include the response to exercise (see Chapter 29) and the response to altered gas tensions. The importance of understanding integrated ventilatory responses is that similar interactions occur under pathophysiologic conditions in patients with respiratory illnesses.

Breathing with low inspired oxygen tension (P_{IO_2}) leads to low P_{AO_2} (**hypoxia**) to low P_{aO_2} (**hypoxemia**) and is frequently encountered during ascent to high altitudes. How the body responds to high altitude has fascinated physiologists for centuries. The French physiologist Paul Bert first recognized that the harmful effects of high altitude are caused by low-oxygen tension. Recall from Chapter 20 that the percentage of oxygen does not change at high altitude but P_{O_2} decreases due to a drop in barometric pressure (see Fig. 20.1). Thus, the hypoxic response at high altitude is caused by a decrease in P_{IO_2} resulting from a decrease in P_B and not from a change in fraction of inspired oxygen (F_{IO_2}).

In high-altitude physiology and medicine, three altitude regions are recognized that reflect the lowered P_{O_2} in the atmosphere: *high altitude* (1,500 to 3,500 m or 4,900 to 11,500 ft), *very high altitude* (3,500 to 5,500 m or 11,500 to 18,000 ft), and *extreme altitude* (above 5,500 m, 18,000 ft).

When the P_{IO_2} decreases and oxygen supply in the body is threatened, several compensations are made in an effort to deliver normal amounts of oxygen to the tissues. Somewhat surprising is the fact that hypoxia-induced hyperventilation is not significantly increased until the alveolar P_{O_2} decreases to <60 mm Hg (see Fig. 21.5). In a healthy adult, a drop in alveolar P_{O_2} to 60 mm Hg occurs at an altitude of ~4,500 m (14,000 ft). As altitude increases, the body makes noticeable efforts to deliver more oxygen to the tissues. Chief among these responses to altitude is induced hyperventilation—that is, deeper breaths taken more rapidly in which alveolar ventilation is increased. The sequence of events for hypoxia-induced hyperventilation is as follows: (1) a fall in P_{IO_2}, (2) decreased alveolar tension (P_{AO_2}), (3) decreased arterial oxygen tension (P_{aO_2}), (4) increased firing of peripheral chemoreceptors (the carotid bodies), (5) hyperventilation, and (6) increased alveolar and arterial oxygen tension. As mentioned previously, the stimulus for ventilation during hypoxia is a decrease in P_{aO_2} rather than O_2 content or percent O_2 saturation. It is important to realize that there is also an immediate rise in cardiac output to match the increase in ventilation. The increase in cardiac output leads to an increase in pulmonary circulation.

Remember, an increase in pulmonary blood flow reduces both the capillary transit time and pulmonary vascular resistance. The latter is due to capillary recruitment,

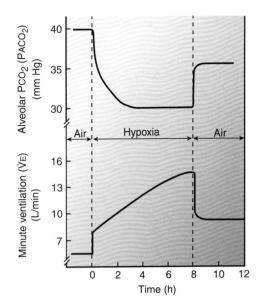

Figure 21.11 **Hypoxia stimulates minute ventilation and causes hypocapnia.** Hypoxia was induced by having a healthy subject breathe 12% O_2 for 8 hours. With hypoxia-induced hyperventilation, excess CO_2 is blown off, resulting in a decrease in alveolar P_{CO_2} and arterial P_{CO_2} with a concomitant rise in blood pH. Minute ventilation remains elevated for awhile after the subject returns to room air.

which occurs primarily in the top part of the lungs, resulting in more even blood flow throughout the lungs. As a result, regional blood flow and airflow in the lungs are better matched (i.e., improved ventilation/perfusion ratio). Thus, the increased cardiac output indirectly increases gas exchange by decreasing transit time and improving the overall ventilation/perfusion ratio in the lungs. Figure 21.11 shows how ventilation and P_{aCO_2} change with hypoxia. The hypoxia-induced hyperventilation appears in two stages. First, there is an immediate increase in ventilation, which is primarily a result of hypoxia-induced stimulation via the carotid bodies. However, the increase in ventilation seen in the first stage is small compared with that of the second stage, in which ventilation continues to rise slowly over the next 8 hours. After 8 hours of hypoxia, minute ventilation is sustained. The reason for the small rise in ventilation seen in the first stage is that the hypoxic stimulation is strongly opposed by the decrease in P_{aCO_2} as a result of excess carbon dioxide blown off with altitude-induced hyperventilation. Remember that hyperventilation elevates P_{aO_2} but significantly lowers P_{aCO_2}, resulting in a concomitant increase in arterial pH. The hypocapnia (due to the decrease in arterial P_{CO_2}) and the rise in blood pH work in concert to strongly blunt the hypoxic drive.

Acclimatization to altitude leads to a sustained increase in ventilation.

The sustained increased ventilation seen in the second stage is referred to as *ventilatory acclimatization*. Acclimatization occurs during prolonged exposure to hypoxia and is a physiologic response, as opposed to a genetic or evolutionary change over generations leading to a permanent **adaptation**.

CLINICAL FOCUS | 21.2

Acute Mountain Sickness

Acute mountain sickness (AMS) is a pathophysiologic condition that is caused by acute exposure to high altitude. AMS is also known as *altitude sickness* and is caused by hypoxia resulting from a decrease in oxygen tension (P_{O_2}) and not from a decrease in the percentage of oxygen in the inspired air.

AMS commonly occurs above 2,400 m (~8,000 ft) and dates back to 30 BCE as indicated by the mention of the "Great Headache Mountain" in Chinese documents. Symptoms are often manifested 6 to 10 hours after ascent and generally subside in 1 to 2 days. AMS symptoms include headache, dizziness, nausea, and sleep disturbance. In most cases, the symptoms are only temporary and usually abate with time as altitude acclimation occurs. Exertion aggravates the situation, and in extreme cases, pulmonary edema and cerebral edema can occur, which can be fatal. The rate of ascent, the altitude, the degree of physical activity, and individual susceptibility are contributing factors to the incidence and severity of AMS.

Many AMS symptoms are caused by hypoxia-induced cerebral blood vessel dilation, which puts pressure on adjacent neuronal tissue. The hypoxia-induced cerebral dilation accounts for the headache, dizziness, nausea, and sleep disturbance. Dehydration can exacerbate these symptoms. The most serious symptoms of AMS are a result of edema (abnormal fluid accumulation) in the lungs and brain. Severe hypoxia causes endothelial cells to become leaky. The edema that is the most life threatening is pulmonary edema. The problem arises because of the oxygen paradox (i.e., hypoxia causes blood vessels to dilate in systemic circulation but causes vasoconstriction in the pulmonary circulation). The hypoxia-induced pulmonary vasoconstriction increases pulmonary vascular resistance. The heart trying to pump blood through "water-soaked lungs" further complicates the situation. As the lungs fill up with water, victims become increasingly short of breath and often spit up foamy blood. Severe pulmonary edema causes death as a result of heart failure.

AMS can be avoided with care. One precaution is to avoid caffeine and alcohol, which tend to exacerbate AMS. A second precaution is to ascend the mountain slowly. Most AMS occurs following a rapid ascent. The third precaution of high-altitude climbing is to follow the climbers' "golden rule": climb high and sleep low, meaning that a climber should stay a few days at base camp to acclimatize, then climb slowly to a higher camp, and return to base camp for the night. This procedure is repeated a few times, each time extending the time spent at higher altitudes to let the body acclimatize to the lower oxygen tension. The general rule is not to ascend >300 m (1,000 ft) per day to sleep. This means a climber can climb from 3,000 m (10,000 ft) to 4,500 m (15,000 ft) in 1 day, but then descend back to 3,300 m (11,000 ft) to sleep. This explains why climbers need to spend days and/or weeks at a time to acclimatize before they can reach a high peak.

Acetazolamide may help speed up the acclimatization process and can treat mild cases of altitude sickness. It reduces fluid formation and helps reduce the severity of cerebral edema. The drug also forces the kidneys to excrete bicarbonate, the base form of carbon dioxide, thus counteracting the effects of hyperventilation that occurs at altitude. ■

Ventilatory acclimatization is defined as a time-dependent increase in ventilation that occurs over hours to days of continuous exposure to hypoxia. After 2 weeks, the hypoxia-induced hyperventilation reaches a stable plateau.

Although the physiologic mechanisms responsible for ventilatory acclimatization are not completely understood, it is clear that two mechanisms are involved. One involves the chemoreceptors, and the second involves the kidneys. When ventilation is stimulated by hypoxia, CSF pH becomes more alkaline. The elevated CSF pH is brought closer to normal by the movement of bicarbonate out of the CSF. Also, during prolonged hypoxia, the carotid bodies increase their sensitivity to P_{aO_2}. These changes result in a further increase in ventilation.

The second mechanism responsible for ventilatory acclimatization involves the kidneys. The alkaline blood pH resulting from the hypoxia-induced hyperventilation is antagonistic to the hypoxic drive. Blood pH is regulated by both the lungs and the kidneys (see Chapter 24), which is illustrated in the following equation:

$$pH = pK_a + \frac{\log[HCO_3^-]}{[H_2CO_3]} \qquad (1)$$

Because pK_a stays constant and H_2CO_3 dissociates into CO_2 and H_2O, the equation can be represented by:

$$pH = \frac{[HCO_3^-]}{Pa_{CO_2}} \frac{20}{1} \frac{Kidney}{Lung} \qquad (2)$$

Equation 2 also illustrates that the blood $[HCO_3^-]$ is regulated primarily by the kidney and that the Pa_{CO_2} is regulated primarily by the lungs. A normal pH of 7.4 is maintained by a ratio of 20:1. The functional importance of the second equation illustrates that it is not the absolute amount of $[HCO_3^-]$ and Pa_{CO_2} that is important in maintaining a normal pH but the 20:1 ratio. During hypoxic-induced hyperventilation, excess CO_2 is blown off, lowering Pa_{CO_2}, which, in turn, increases the ratio and makes the blood pH more alkaline. The kidneys compensate by excreting more bicarbonate, which brings the ratio back to a normal 20:1 and a blood pH toward 7.4. This process occurs over a 2- to 3-day period, and the antagonistic effect resulting from the hypoxia-induced alkaline pH is minimized, allowing the hypoxic drive to increase minute ventilation further. Thus, it is important to remember that the regulation of pH by ventilatory Pa_{CO_2} is fast (seconds to minutes), whereas the regulation of pH by the kidney to adjust the sodium bicarbonate concentration is slow (hours to days).

Cardiovascular acclimatization improves the delivery of oxygen to the tissues.

In addition to ventilatory acclimatization, the body undergoes other physiologic changes to acclimatize to low-oxygen levels. These include increased pulmonary blood flow, increased red cell production, and improved oxygen and carbon dioxide transport. There is an increase in cardiac output at high altitude, resulting in increased blood flow to the lungs and other organs of the body. As mentioned previously, the increase in pulmonary blood flow reduces capillary transit time and improves the overall ventilation/perfusion ratio in the lungs, both of which result in an increase in oxygen uptake by the lungs. Low P_{O_2} causes vasodilation in the systemic circulation. The increase in blood flow resulting from the combined increased vasodilation and increased cardiac output sustains oxygen delivery to the tissues at high altitude.

Erythropoiesis is also increased at high altitude, which improves oxygen delivery to the tissues. Hypoxia stimulates the kidneys to produce and releases a hormone into the circulation (**erythropoietin**), which stimulates the bone marrow to increase the rate production of erythrocytes. The increased hematocrit resulting from the hypoxia-induced **polycythemia** enables the blood to carry more oxygen to the tissues. However, the increased viscosity, as a result of the elevated hematocrit, tends to increase the workload on the heart. In some cases, the polycythemia becomes so severe (hematocrit >70%) at high altitude that blood has to be withdrawn periodically to permit the heart to pump effectively. Oxygen delivery to the cells is also favored by an increased concentration of 2,3-diphosphoglycerate in the red cells. Recall from Chapter 19 that 2,3-diphosphoglycerate shifts the oxyhemoglobin equilibrium curve to the right and favors the unloading of oxygen in the tissues. Although the body undergoes many physiological changes that favor acclimatization to high altitude, there are some undesirable effects. One of these is pulmonary hypertension. There is a hypoxia paradox in circulation. Hypoxia causes vasodilation in the systemic circulation and vasoconstriction in the pulmonary circulation, which results in an increase in pulmonary arterial pressure. Remember that regional hypoxia redirects blood away from poorly ventilated regions in the lung without any change in pulmonary pressure. However, with generalized hypoxia, pulmonary pressure rises because all of the prealveolar vessels constrict. In addition, prolonged hypoxia causes vascular remodeling in which pulmonary arterial smooth muscle cells undergo hypertrophy and hyperplasia. The vascular remodeling results in narrowing of the small pulmonary arteries and increases pulmonary vascular resistance, leading to a further significant increase in pulmonary vascular hypertension. With severe hypoxia, the pulmonary veins are also constricted. The increase in venous pressure elevates the filtration pressure in the alveolar capillary beds and under severe conditions causes **high-altitude pulmonary edema**. Pulmonary hypertension also increases the workload of the right heart, causing right heart hypertrophy, and under severe conditions often leads to death. Extreme altitude also causes **high-altitude cerebral edema**.

An altitude above certain point where the P_{O_2} is insufficient to sustain human life is referred to as the *death zone*. The death zone in mountaineering is generally tagged at 8,000 m (26,000 ft). For example, the summit of Mount Everest is in the death zone. Many deaths occur either directly (loss of vital functions) or indirectly (mistakes made under stress). In the death zone, the human body can no longer acclimate resulting in the loss of consciousness and, ultimately, death.

Diving reflex involves a number of cardiopulmonary responses.

Another respiratory movement is breath-holding, in which normal breathing patterns are voluntarily suspended. Breath-holding can occur until the Pa_{CO_2} rises and overrides the conscious, voluntary effort. Breath-holding is a frequent maneuver with diving and is usually harmless. However, breath-holding becomes dangerous if hyperventilation precedes the event. For example, if a diver hyperventilates before going under water, the diver can hold his or her breath longer, because of the strong inhibitory effect of blowing off excess CO_2. The problem created with hyperventilation is that Pa_{CO_2} is drastically decreased, with no appreciable change in oxygen content. As a result of the low arterial CO_2 tension, the respiratory centers become depressed. At the same time, the exercising muscles rapidly take up oxygen, which lowers both the arterial oxygen content and P_{O_2}. However, the hypoxia-induced stimulation to the carotid chemoreceptors is overridden by the strong inhibitory effect of the low P_{CO_2}. As a result, the brain becomes hypoxic, causing the swimmer to black out under water and drown, a condition commonly referred to as **shallow water blackout**, which is illustrated in Figure 21.12.

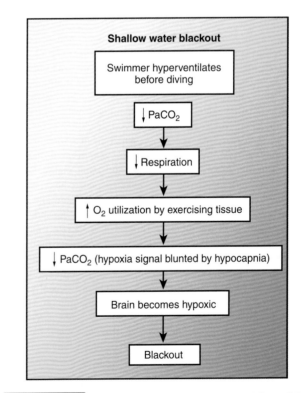

Figure 21.12 **Shallow water blackout is a result of failure of the controller to respond to low oxygen.** When hyperventilation precedes underwater diving, hypocapnia occurs. The hypocapnia blunts the hypoxia-induced ventilation, causing severe hypoxia in the brain.

Another breathing pattern associated with underwater breathing is the **diving reflex**. The diving reflex is most pronounced in newborns and young children and can be elicited when the face is submerged in water. The diving reflex is a protective reflex that encompass a number of cardiopulmonary responses. The reflex causes the person to gasp; breathing stops (apnea), heart rate decreases (bradycardia), and blood flow to the peripheral tissue also decreases (peripheral vasoconstriction). Diverting blood away from the peripheral tissues allows for a heart–brain perfusion. The gasp component of the diving reflex stops breathing and prevents water from entering the lung. The reflex component that induces bradycardia reduces energy requirements for cardiac tissue, and the peripheral vasoconstriction conserves oxygen to

the brain. A number of cases have been reported in which young children have broken through ice on a frozen pond or have fallen into a lake and survived under water for 20 to 30 minutes. Often, the diving reflex by itself is not enough to save victims who remain under water for longer periods of time than 10 to 15 minutes. Lowered body temperature caused by the cold water is equally important. A lower body temperature means a lower metabolic rate, which means less oxygen is required. The reason children survive better than adults is twofold. First, children have a more pronounced diving reflex. Second, their body temperature drops at a faster rate because of their high surface area to body weight ratio. For further details about the relationship between body temperature and surface area, see Chapter 28.

INTEGRATED MEDICAL SCIENCES

Traumatic Head Injury

A professional quarterback, in an attempt to recover a fumble, was hit twice, once in the head and once in the back, and was knocked unconscious. He was taken to the hospital, and upon transport to the hospital, he regained consciousness, but then lost consciousness again. In the emergency room, he displayed the *Cushing triad*, a slow heart rate with high blood pressure and respiratory depression. He was transferred to intensive care and given supplemental oxygen, a sedative, and IV injection containing mannitol. He regained consciousness again and was examined by a neurosurgeon. The patient complained of headache, nausea, and blurred vision and stated his breathing was becoming more difficult. A chest X-ray confirmed pulmonary contusion. The patient was diagnosed with mild to moderate head injury compounded with pulmonary contusion.

This case is a common presentation of traumatic head injury and is the major cause of death and disability worldwide, especially in children and adult males. Causes include car accidents, falls, violence, and contact sports. The brain trauma increases intracranial pressure and alters cerebral blood flow. Sign and symptoms are dependent on the severity of the injury. With mild to moderate traumatic brain injury (TBI), patients often remain conscious, but may lose consciousness for several minutes. Other symptoms include headache, vomiting, nausea, dizziness, difficulty with balance, and blurred vision. Patients frequently complain of ringing in the ears and often exhibit slurred speech and *aphasia* (difficulty finding words). A large percentage of patients that are killed by brain trauma do not die right away. Rather than improving after being in intensive care, over 40% of these patients deteriorate as a result of cardiopulmonary complications and is the underlying cause of morbidity and mortality. Examples include hypertension, arrhythmias, ventricular dysfunction, pulmonary edema, shock, and sudden death.

In the present case, the patient exhibited the Cushing triad, which is a reflex that is responsible for increased systolic and pulse pressure, bradycardia, and depressed respiration. The three-part effect was named after an American neurosurgeon, Harvey Cushing. The Cushing triad is caused by the increased intracranial pressure that compresses cerebral vessels, especially the arterioles, causing cerebral ischemia. In addition, TBI causes a massive sympathetic stimulation to the cardiopulmonary system, which further exacerbates the problem by causing pulmonary edema and elevated systemic blood pressure.

In addition to TBI, this patient also suffered pulmonary contusion. A pulmonary contusion usually occurs from a blow to the front of chest. In this particular case, the patient was hit by a blow to his back. A blow to the back can cause contusion to the front of the lungs because a shock wave travels through the chest, which hits the front of the chest wall. This reflects the energy onto to the front of the lungs. A similar mechanism occurs when a blow occurs to the front of the chest causing a contusion in the back of the lungs. A pulmonary contusion further compounds the problem of a TBI by causing bleeding and fluid leakage into lung tissue and often leads to pulmonary edema. Patients with a pulmonary contusion have difficulty breathing due to a loss of lung elasticity and a decrease in lung compliance (stiff lungs). Contusion also lowers the ventilation–perfusion ratio because the fluid-filled alveoli cannot be filled with air. As a result, the hypoxemia becomes more acute due to pulmonary edema and ventilation–perfusion mismatch.

The pathogenesis can be divided into two phases: primary brain injury and secondary brain injury. The primary injury occurs at the moment of trauma when intracranial pressure increases causing cerebral ischemia and hypoxemia. Secondary injury occurs hours and days following the initial trauma and is caused by a biochemical cascade that leads to inflammation, free radical production, and the release of neurotransmitters. These cellular events are responsible for a series of pathophysiologic events that dramatically alters brain and other organ function. The pathophysiology includes damage to the blood–brain barrier, influx of calcium, and sodium into neuronal tissue, terminal membrane depolarization, and dysfunctional mitochondria. Together, these events lead to loss of cerebral vascular integrity, membrane degradation of neuronal tissue, and *apoptosis* (cell death).

Treatment usually begins in an intensive care unit. Treatment during transport and in the hospital centers on ensuring proper oxygen supply, maintaining adequate blood flow to the brain, and controlling the elevated intracranial pressure. Treatment of the increased intracranial pressure may be done by initially tilting the patient's bed and straightening the head to promote blood flow from the brain through the neck veins. Mannitol, and osmotic diuretic, is often given to reduce the cerebral pressure. Once in intensive care, patients with moderate to severe injuries can have a catheter placed in the ventricle of the brain to drain the cerebrospinal fluid releasing the intracranial pressure. If blood oxygenation is low, endotracheal intubation and mechanical ventilation are used to improve gas exchange.

Approximately 30% to 50% of patients who survive posttraumatic brain injury (TBI) demonstrate endocrine complications. Diabetes insipidus following mild to moderate TBI indicates a posterior pituitary lesion. Anterior hypopituitarism is also associated with moderate to severe TBI that leads to a deficiency in growth hormone (GH), adrenocorticotropic hormone (ACTH), and thyroid-stimulating hormone (TSH). Growth hormone deficiency leads to a decrease in muscle mass, central obesity (increase body fat around waist), and impaired attention and memory. Deficiency in ACTH leads to adrenal insufficiency, a lack of production of glucocorticoids such as cortisol. Chronic symptoms include fatigue, weight loss, and hypoglycemia and anemia. TSH deficiency leads to hypothyroidism (decreased production of thyroxine [T4] and triiodothyronine [T3]). Symptoms include tiredness, intolerance to cold, weight gain, hair loss, and low blood pressure. Finally, post-TBI often includes speech and cognitive disorders. Speech and language therapy, cognitive rehabilitation therapy, and occupational therapy are required for rehabilitation. ■

Respiratory Physiology

Chapter Summary

- Ventilation involves both automatic and voluntary control, in which both require negative and positive feedback systems.
- Automatic control is concerned with the exchange of oxygen and carbon dioxide as well as acid–base balance.
- Voluntary control is concerned with coordinated activities.
- Normal arterial blood gases are maintained, and the work of breathing is minimized despite changes in activity, the environment, and lung function.
- The neural structure responsible for automatic control resides primarily in the medulla.
- The basic breathing rhythm (minute-to-minute breathing) is generated by medullary neurons in the brainstem.
- The rate and depth of breathing are finely regulated by vagal nerve endings that are sensitive to lung stretch.
- The ventral respiratory group (VRG) controls forced exhalation and acts to increase the force of inspiration.
- The dorsal respiratory group (DRG) controls mostly inspiratory movements and their timing.
- Ventilatory rate (minute volume) is tightly controlled and determined primarily by blood levels of carbon dioxide as determined by metabolic rate.

- The autonomic nerves and vagal sensory nerves maintain local control of airway function.
- Mechanical or chemical irritation of the airways and lungs induces coughing, bronchoconstriction, shallow breathing, and excess mucus production.
- Arterial Pco_2 is more important than arterial Po_2 and $[H^+]$ in determining the ventilatory drive in healthy individuals at rest.
- Peripheral chemoreceptors detect changes in arterial Po_2, Pco_2, and pH, whereas central chemoreceptors detect changes only in arterial Pco_2.
- Control of breathing during exercise involves three phases (neurogenic, metabolic, and compensatory).
- Sleep is induced by the withdrawal of a wakefulness stimulus arising from the brainstem reticular formation and results in a general depression of breathing.
- The hypoxia-induced stimulation of ventilation is not significantly increased until the arterial Po_2 drops to <60 mm Hg.
- Chronic hypoxemia causes respiratory acclimatization that result in increased sustained breathing.
- The diving reflex invokes several cardiopulmonary responses.
- Shallow water blackout occurs when the diving reflex is overridden by hypoxia.

Chapter Review Questions

1. Generation of the basic cyclic pattern of breathing in the CNS requires participation of:

   A. the pontine respiratory group.
   B. vagal afferent input to the pons.
   C. vagal afferent input to the medulla.
   D. an inhibitory loop in the medulla.
   E. an intact spinal cord.

The correct answer is D. The basic rhythm exists in the absence of the pontine respiratory group, afferent vagal input to the pons and medulla, or an intact spinal cord. These can modify the rhythm of breathing but are not required.

2. Quiet expiration is associated with:

   A. a brief early burst by inspiratory neurons.
   B. active abduction of the vocal cords.
   C. an early burst of activity by expiratory muscles.
   D. reciprocal inhibition of inspiratory and expiratory centers.
   E. increased activity of slowly adapting receptors.

The correct answer is A. A brief early burst by the inspiratory neurons occurs with expiration.

3. The ventilatory response to hypoxia:

   A. is independent of $Paco_2$.
   B. is more dependent on aortic than carotid chemoreceptors.
   C. is exaggerated by hypoxia of the medullary chemoreceptors.

   D. bears an inverse linear relationship to arterial oxygen content.
   E. is a sensitive mechanism for controlling breathing in the normal range of blood gases.

The correct answer is D. An inverse relationship exists between hypoxia-induced hyperventilation and oxygen content. Hypoxia-induced hyperventilation is dependent on $Paco_2$ and more on carotid than on aortic chemoreceptors.

4. Which of the following is not a consequence of stimulation of lung C fiber endings?

   A. Bronchoconstriction
   B. Apnea
   C. Rapid shallow breathing
   D. Systemic vasoconstriction
   E. Skeletal muscle relaxation

The correct answer is D. Stimulation of lung C fibers will cause bronchoconstriction, apnea, rapid shallow breathing, and skeletal muscle relaxation.

5. Which of the following is true about cerebrospinal fluid?

   A. Its protein concentration is equal to that of plasma.
   B. Its Pco_2 equals that of systemic arterial blood.
   C. It is freely accessible to blood hydrogen ions.
   D. Its composition is essentially that of a plasma ultrafiltrate.
   E. Its pH is a function of $Paco_2$.

The correct answer is E. CSF and plasma differ in protein concentration, Pco_2, and electrolyte composition (including the $[H^+]$).

Clinical Application Exercises 21.1

PICKWICKIAN SYNDROME

A 45-year-old man is referred to the pulmonary function laboratory because of polycythemia (hematocrit of 57%). At the time of referral, he weighs 142 kg (312 lb), and his height is 175 cm (5 ft, 9 in). A brief history reveals that he frequently falls asleep during the day. His blood gas values are Pao_2, 69 mm Hg; Sao_2, 94%; Pco_2, 35 mm Hg; and pH, 7.44. A few days later, he is admitted as an outpatient in the hospital's sleep center. He is connected to an ear oximeter and to a portable heart monitor. Within 30 minutes, the patient falls asleep, and within another 30 minutes, his Sao_2 decreases from 92% to 47%, and his heart rate increases from 92 to 108 beats/min, with two premature ventricular contractions. During this time, his chest wall continues to move, but airflow at the mouth and nose is not detected.

QUESTIONS

1. How would this patient's test results be interpreted?
2. What is the cause of polycythemia?
3. How does hypoxia accelerate heart rate?

ANSWERS

1. This patient is suffering from what has been known as Pickwickian syndrome, a disorder that occurs with severely obese individuals because of their excessive weight. The Pickwickian syndrome was named after Joe, the overweight boy who was always falling asleep in Charles Dickens' novel *The Pickwick Papers*. Pickwickian patients suffer from hypoventilation and often suffer from sleep apnea as well. Pickwickian syndrome is no longer an appropriate name because it does not indicate what type of sleep disorder is involved. About 80% of sleep apnea patients are obese, and 20% are of relatively normal weight.

2. Polycythemia is the result of chronic hypoxemia from hypoventilation, as well as from sleep apnea.

3. An increase in sympathetic discharge is often associated with sleep apnea and is responsible for the accelerated heart rate.

thePoint® *Visit* http://thepoint.lww.com/rhoades5e *for additional chapter review Q&A, Clinical Application Exercises, animations, and more!*

22 Kidney Function

Active Learning Objectives

Upon mastering the material in this chapter, you should be able to:

- Explain the mechanism by which the kidney removes or "clears" substances from plasma.
- Explain how glomerular filtration rate is measured and the factors that affect filtration rate.
- Describe how renal blood flow can be determined from the clearance of p-aminohippurate and the hematocrit and discuss the factors that influence renal blood flow.
- Determine whether there is net reabsorption or secretion of a plasma substance by the kidneys and quantify its value.
- Explain what is meant by tubular transport maximum, threshold, and splay for glucose.
- Explain and contrast the mechanisms of solute and water reabsorption in the proximal convoluted tubule, loop of Henle, and distal nephron.

- Predict the consequences of impairment of any transport process along the nephron.
- Explain how water and ion transport in any section of the nephron is altered by a change in sodium reabsorption in that section.
- Explain mechanisms responsible for secretion of organic anions, cations, acids, and bases by the nephron and how this secretion is altered by urine pH.
- Explain how arginine vasopressin increases collecting duct water permeability.
- Explain how countercurrent mechanisms create and maintain the vertical peritubular osmotic gradient surrounding the renal tubules in cortical and juxtamedullary nephrons.
- Explain how the interplay between the vertical peritubular osmotic gradient and the effect of ADH on the collecting duct allows the kidney to vary its net water reabsorption and the osmolality of the urine it forms.

The function and survival of cells depend on maintaining normal concentrations of NaCl, acids, and other electrolytes in the internal environment. The functional importance of maintaining a normal cellular and extracellular environment was explained in neurophysiology (Chapter 3), muscles (Chapter 8), and cardiovascular function (Chapter 12). The maintenance of normal hydration in cell function was also illuminated in Chapter 2. In the respiratory chapter (Chapter 19), the importance of maintaining a normal acid–base balance was elucidated. In the next three chapters, you will learn how the kidney regulates body fluids, maintains electrolyte balance, eliminates organic waste, and maintains the long-term regulation of acid–base conditions in the body. As you read about kidney function and regulation, you will encounter numerous examples of physiologic principles that have been presented in the Cellular, Neuromuscular, Cardiovascular, and Respiratory Physiology sections. The ability of the kidney to selectively remove waste and at the same time retain essential electrolytes, water, and nutrients is a hallmark of a general physiological principle of renal function.

Accordingly, these paired organs serve as the body's main filtering system for the blood and extracellular fluid (ECF). They maintain normal concentrations of many plasma constituents, especially electrolytes. They maintain normal water volume and osmolarity (solute concentration) in the ECF while removing such metabolic waste products as urea and uric acid. They are also responsible for the reabsorption of filtered water, glucose, and amino acids. Finally, the kidneys produce hormones, including calcitriol (the activated form of vitamin D), and erythropoietin as well as the enzyme renin. This chapter considers the basic renal processes that determine the excretion of various substances.

▶ OVERVIEW OF RENAL FUNCTION

The architectural design of the urinary system consists of the urine-forming organs—the kidneys, the urinary bladder, ureter, and urethra. Peristaltic movements propel the urine down the ureters to the urinary bladder, which stores the urine until the bladder is emptied through the urethra. Each kidney is located in the back of the abdominal wall. The adult kidney weighs about 150 g and is roughly the size of one's fist. The kidneys are highly innervated and have a rich blood supply. Despite their relatively small size, they receive ~20% of the cardiac output. Each kidney is typically supplied by a single renal artery, which branches into anterior and posterior divisions. In turn, these two divisions give rise to a total of five segmental arteries (Fig. 22.1).

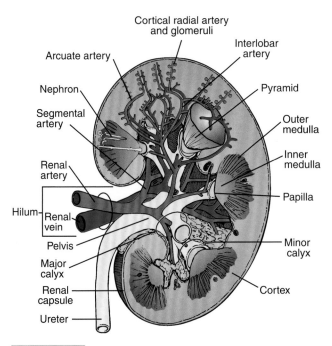

Figure 22.1 The human kidney, sectioned vertically.

Kidneys are highly innervated.

The kidneys are richly innervated by sympathetic nerve fibers, which travel to the kidneys mainly via the thoracic spinal nerves X, XI, and XII and lumbar spinal nerve I. Stimulation of sympathetic fibers causes constriction of renal blood vessels and a decrease in renal blood flow (RBF). Sympathetic nerve fibers also innervate tubular cells and may cause an increase in Na^+ reabsorption by a direct action on these cells. In addition, stimulation of sympathetic nerves increases the release of renin by the kidneys. Afferent (sensory) renal nerves are stimulated by mechanical stretch or by various chemicals in the renal parenchyma.

Kidneys perform a number of key functions.

The important functions performed by the kidneys include the following:

- They regulate the osmolarity of the body fluids by altering renal water reabsorption to excrete osmotically dilute or concentrated urine.
- They regulate the plasma concentrations of Na^+, K^+, Ca^{2+}, Mg^{2+}, Cl^-, HCO_3^-, phosphate, and sulfate.
- They play an essential role in acid–base balance by altering renal H^+ excretion and HCO_3^- reabsorption.
- They regulate the volume of the extracellular fluid by controlling Na^+ and water excretion.
- They help regulate arterial blood pressure by adjusting Na^+ excretion and producing various substances (e.g., renin) that can affect blood pressure.
- They eliminate the waste products of metabolism, including urea (the main nitrogen-containing end product of protein metabolism in humans), uric acid (an end product of purine metabolism), and creatinine (an end product of muscle metabolism).

- They remove many drugs (e.g., penicillin), drug metabolites, and foreign or toxic compounds.
- They are the major sites of production of certain hormones, including erythropoietin (see Chapter 9) and 1,25-dihydroxy vitamin D_3 (see Chapter 35).
- They degrade several polypeptide hormones, including insulin, glucagon, and parathyroid hormone.
- They synthesize ammonia, which plays a role in acid–base homeostasis (see Chapter 24).
- They synthesize substances that affect RBF and Na^+ excretion, including arachidonic acid derivatives (prostaglandins and thromboxane A_2) and kallikrein (a proteolytic enzyme that results in the production of kinins).

When the kidneys fail, a host of problems ensue. Dialysis and kidney transplantation are commonly used treatments for advanced (end-stage) renal failure.

▶ NEPHRON: FUNCTIONAL UNIT OF THE KIDNEY

Each human kidney contains about 1 million microscopic units known as **nephrons** (Fig. 22.2), which consist of a *renal corpuscle* and a *renal tubule*. Each renal corpuscle is derived from a muscular *afferent arteriole* that forms a tuft of capillaries, the **glomerulus**, which is surrounded by the **Bowman capsule**. The glomerulus filters water and solutes from the blood plasma. The Bowman capsule, which cups around the glomerulus, collects the fluid filtered from the glomerulus. The blood leaving the glomerulus that was not filtered is carried via the *efferent arteriole*.

Renal tubules contain unique features.

The renal tubule is divided into several segments (see Fig. 22.2). The part of the tubule nearest the glomerulus is the proximal tubule. This is subdivided into a *proximal convoluted tubule* followed by the *proximal straight tubule*. The straight portion heads toward the medulla, away from the surface of the kidney, and into the loop of Henle. The **loop of Henle** forms a U-shaped loop that dips toward the renal medulla. In nephrons contained within the renal cortex, the loop of Henle includes the proximal straight tubule, a *thin descending limb*, and a *thick ascending limb*. In juxtamedullary nephrons, which are a subgroup of nephrons whose loops of Henle descend deep into the renal medulla, a *thin ascending limb* is interposed between the thin descending and thick ascending limbs. The thick ascending limb of both types of nephrons leads into the next segment of the nephron called the *distal convoluted tubule*. The distal convoluted tubule extends from the **macula densa** (see below) to a connecting tubule, which is connected to the collecting duct system. Each duct in the system receives distal tubular fluid from several nephrons.

Juxtaglomerular apparatus is the site of renin production.

Approximately 12% of the total nephron population contain specialized sets of cells located between the renal corpuscle and the distal convoluted tubule. This specialized

CLINICAL FOCUS | 22.1

Dialysis and Transplantation

Chronic kidney disease is usually progressive and may lead to renal failure. Common causes include diabetes mellitus, hypertension, inflammation of the glomeruli (glomerulonephritis), urinary reflux and infections (pyelonephritis), and polycystic kidney disease. Renal damage may occur over many years and may be undetected until a considerable loss of functioning nephrons has occurred. When glomerular filtration rate (GFR) has declined to 5% of normal or less, the internal environment becomes so disturbed that patients usually die within weeks or months if they are not dialyzed or provided with a functioning kidney transplant.

Most of the signs and symptoms of renal failure can be relieved by **dialysis**, the separation of smaller molecules from larger molecules in solution by diffusion of the small molecules through a selectively permeable membrane. Two methods of dialysis are commonly used to treat patients with severe, irreversible (end-stage) renal failure.

In **continuous ambulatory peritoneal dialysis**, the peritoneal membrane, which lines the abdominal cavity, acts as a dialyzing membrane. About 1 to 2 L of a sterile glucose/salt solution are introduced into the abdominal cavity, and small molecules (e.g., K^+ and urea) diffuse into the introduced solution, which is then drained and discarded. The procedure is usually done several times every day.

Hemodialysis is more efficient in terms of rapidly removing wastes. The patient's blood is pumped through an artificial kidney machine. The blood is separated from a balanced salt solution by a cellophane-like membrane, and small molecules can diffuse across this membrane. Excess fluid can be removed by applying pressure to the blood and filtering it. Hemodialysis is usually done three times a week (4 to 6 hours per session) in a medical facility or at home.

Dialysis can enable patients with otherwise fatal renal disease to live useful and productive lives. However, many physiologic and psychologic problems persist, including alteration of body calcium levels and bone disease, disorders of nerve function, hypertension, atherosclerotic vascular disease, sudden cardiac death due to plasma K^+ disturbances, and disturbances of sexual function. There is a constant risk of infection and, with hemodialysis, clotting and hemorrhage. Dialysis does not maintain normal growth and development in children. Anemia (primarily resulting from deficient erythropoietin production by damaged kidneys) was once a problem but can now be treated with recombinant human erythropoietin.

Renal transplantation is the only real cure for patients with end-stage renal failure. Patients who have had a kidney transplant have more energy, can enjoy a less restricted diet, have fewer complications, and live much longer than if they had stayed on dialysis. In 2015, over 100,000 patients were awaiting kidney transplants. In 2014, about 17,000 kidney transplantation operations were performed in the United States with about two thirds of the donated kidneys coming from living donors and the remaining third from deceased donors. At present, about 95% of kidneys grafted from a living donor related to the recipient function for 1 year; about 90% of kidneys from cadaver donors function for 1 year.

Several problems complicate kidney transplantation. The immunologic rejection of the kidney graft is a major challenge. The powerful drugs used to inhibit graft rejection compromise immune defensive mechanisms so that unusual and difficult-to-treat infections often develop. The limited supply of donor organs is also a major, unsolved problem; there are many more patients who would benefit from kidney transplantation than there are donors. In 2014, over 4,200 patients died awaiting kidney transplants with another 3,600 becoming too ill to receive a kidney transplant. The median waiting time for a kidney transplant is currently about 1,100 to 1,200 days. Finally, the cost of transplantation (or dialysis) is high. Fortunately, for people in the United States, Medicare covers the cost of dialysis and transplantation, but these lifesaving therapies are beyond the reach of most people in developing countries. ■

---

grouping is the **juxtaglomerular apparatus**, so named for its proximity to the glomerulus (Fig. 22.3). The juxtaglomerular apparatus consists of three cellular components: the macula densa of the distal convoluted tubule, **extraglomerular mesangial cells**, and **granular cells** (also known as *juxtaglomerular cells*). The macula densa (dense spot) consists of densely crowded tubular epithelial cells on the side of the thick ascending limb that faces the glomerular tuft; these cells monitor the composition of the fluid in the tubule lumen. The extraglomerular mesangial cells are continuous with mesangial cells of the glomerulus; they may transmit information from macula densa cells to the granular cells. The granular cells are modified vascular smooth muscle cells with an epithelioid appearance, located mainly in the afferent arterioles close to the glomerulus. These cells synthesize and release **renin**, an enzyme that hydrolyses angiotensinogen to form angiotensin I and thus participates in the body's **renin–angiotensin–aldosterone system**. This system contributes to the regulation of sodium and potassium concentration in the ECF (i.e., in blood plasma, lymph, and interstitial fluid), as well as the water volume contained within that fluid compartment. It also contributes to the regulation of arterial blood pressure (see Chapters 17 and 23). In addition to these regulatory roles, the juxtaglomerular apparatus is critical to regulating RBF, and the glomerular filtration rate (see below).

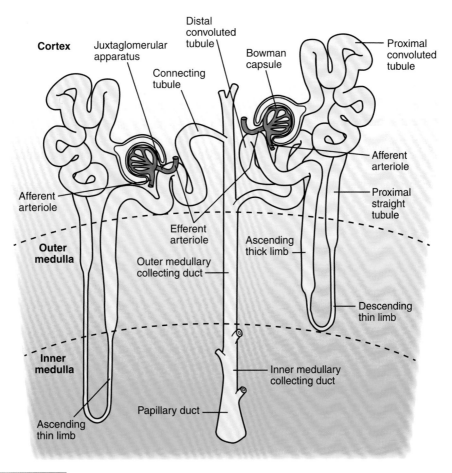

Figure 22.2 **Components of the nephron and the collecting duct system.** On the left is a long-looped juxtamedullary nephron; on the right is a superficial cortical nephron.

Anatomical features of nephrons give rise to different regions in the kidney.

The arrangement of the nephrons within the kidney gives rise to two distinct regions: an outer part, called the **cortex**, and an inner part, called the **medulla** (see Fig. 22.1). The cortex

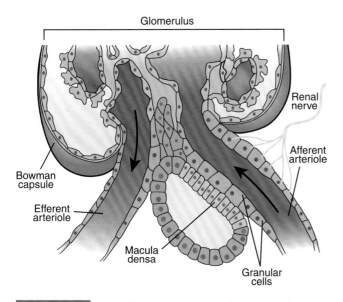

Figure 22.3 **Histologic appearance of the juxtaglomerular apparatus.** A cross-section through a thick ascending limb is below, and part of a glomerulus is on top.

is typically reddish brown and has a granulated appearance. All of the glomeruli, convoluted tubules, and cortical collecting ducts are located in the cortex. The medulla is lighter in color and has a striated appearance that results from the parallel arrangement of the loops of Henle, medullary collecting ducts, and blood vessels of the medulla. Nephrons are often classified into two main types: *cortical* and *juxtamedullary*. All nephrons originate in the cortex, but the glomeruli of the cortical nephrons lie in the outer layer of the cortex, whereas the glomeruli of the juxtamedullary nephrons lie in the inner layer of the cortex next to the medulla. Juxtamedullary nephrons differ in several ways from the other nephron types: they have a longer loop of Henle, longer thin limb (both descending and ascending portions), and different tubular permeability and transport properties.

Urine formation involves three basic processes.

The processes involved in forming urine are glomerular filtration, tubular reabsorption, and tubular secretion (Fig. 22.4). **Glomerular filtration** is the first step and involves filtration of the plasma across the glomerular capillaries. About 20% of the plasma flowing into the glomerulus is filtered into the Bowman capsule, and the remaining 80% flows through the efferent arteriole and into the peritubular capillaries. The filtrate enters the urinary space of the Bowman capsule and then flows downstream through the tubule lumen. The tubule epithelial cells decrease the volume and change the composition

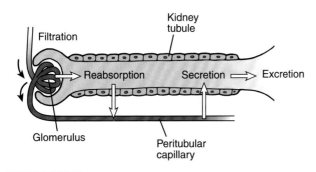

Figure 22.4 **Processes involved in urine formation.** This highly simplified diagram depicts the processes of filtration, reabsorption, secretion, and excretion.

of the fluid in the tubule lumen. Approximately 180 L (~48 gallons) of plasma is filtered through the kidneys each day. The average plasma volume in the adult is about 2.75 L. This means the kidneys filters the body's plasma volume about 65 times per day. If everything filtered appeared as urine, the body's total plasma volume would be urinated in <30 minutes. This doesn't happen because of **tubular reabsorption.** Of the 180 L of plasma filtered per day, ~178.5 L is reabsorbed. The remaining 1.5 L is excreted as urine. In addition to reabsorption of water, tubular reabsorption also involves selective reabsorption of essential substances the body needs. These materials are reabsorbed by diffusion and cell membrane carrier–mediated transport. Reabsorption of material involves the movement of substances out of tubular urine back into the capillary blood, which surrounds the kidney tubules. Reabsorbed substances include many important ions (e.g., Na^+, K^+, Ca^{2+}, Mg^{2+}, Cl^-, HCO_3^-, and phosphate), water, important metabolites (e.g., glucose and amino acids), and even some waste products (e.g., urea and uric acid). The third renal process is **tubular secretion,** which also involves passive diffusion and cell membrane carrier–mediated transport of selective substances from the peritubular capillaries into the tubular urine. Tubular secretion provides a second route for a substance to enter the renal tubules from the bloodstream to be excreted, the first being by glomerular filtration. Tubular secretion provides a mechanism for selectively eliminating substance from the plasma. For example, many organic anions and cations are taken up by the tubular epithelium from the blood surrounding the tubules and added to the tubular urine. Some substances (e.g., H^+ and ammonia) are produced in the tubular cells and secreted into the tubular urine.

The terms *reabsorption* and *secretion* indicate movement out of and into tubular urine, respectively. Tubular transport (reabsorption and secretion) may be either active or passive, depending on the particular substance and other conditions. **Excretion** refers to elimination via the urine. In general, the amount excreted is expressed by the following equation:

$$\text{Excreted} = \text{Filtered} - \text{Reabsorbed} + \text{Secreted} \qquad (1)$$

The functional state of these processes in urine formation can be evaluated using several tests based on the renal clearance concept (see below).

▶ RENAL BLOOD FLOW

Since the kidneys are the body's main filtering system for blood and ECF, they are highly dependent on blood flow. The kidneys have a high blood flow, which allows them to filter the blood plasma at a high rate. Like many other organs, RBF is affected by hormones, extrinsic neural stimulation, and local regulatory factors.

Optimal renal blood flow is maintained by autoregulation.

Autoregulation of blood flow is the mechanism that allows organs to maintain a constant blood flow during changes in mean arterial pressure (see Chapter 15). Although most organs show some degree of autoregulation, the mechanism is most clearly observed in the kidney, heart, and brain. Perfusion of these organs is essential for life and autoregulation allows for a continuous flow of blood despite wide fluctuations of arterial pressure. Autoregulation is an intrinsic property of the kidneys and is observed in an isolated perfused kidney devoid of external nerves. RBF in these isolated perfused kidneys can be kept relatively constant when perfusion pressure is varied from 80 to 180 mm Hg (Fig. 22.5). When the perfusion pressure is raised or lowered, the renal arterioles constrict or dilate, respectively, thereby maintaining a constant blood flow and capillary pressure.

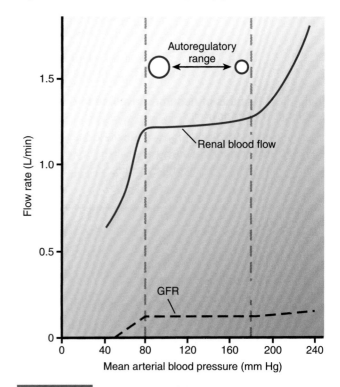

Figure 22.5 **Renal autoregulation, based on measurements in isolated, denervated, perfused kidneys.** In the autoregulatory range, renal blood flow and glomerular filtration rate (GFR) stay relatively constant despite changes in arterial blood pressure. This is accomplished by changes in the resistance (caliber) of preglomerular blood vessels. The circles indicate that vessel radius (r) is smaller when blood pressure is high and larger when blood pressure is low. Because resistance to blood flow varies as r^4, changes in vessel caliber are greatly exaggerated in this figure.

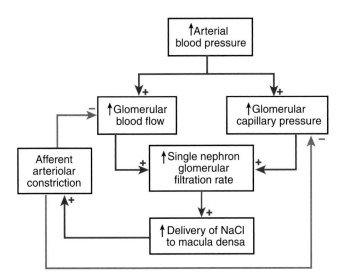

Renal Physiology and Body Fluids

Figure 22.6 **The tubuloglomerular feedback mechanism.** When single-nephron glomerular filtration rate (GFR) is increased—for example, because of an increase in arterial blood pressure—more NaCl is delivered to and reabsorbed by the macula densa, leading to constriction of the nearby afferent arteriole. This negative feedback system plays a role in autoregulation of renal blood flow and GFR.

Two mechanisms account for renal autoregulation: the **myogenic mechanism** and the **tubuloglomerular feedback mechanism**. In the myogenic mechanism, an increase in pressure stretches the afferent arteriolar walls, which then activate stretch-activated cation channels in the arteriolar smooth muscle cells. This causes intracellular Ca^{2+} to rise, resulting in smooth muscle contraction, a reduction in vessel lumen diameter, and an increase in vascular resistance. This counteracts the effect of high perfusion pressure, which would otherwise have increased RBF. Decreased blood pressure causes the opposite changes. In the **tubuloglomerular feedback mechanism** (Fig. 22.6), the transient increase in GFR resulting from an increase in blood pressure leads to increased NaCl delivery to the macula densa. This increases NaCl reabsorption and adenosine triphosphate (ATP) release from macula densa cells. ATP is metabolized to adenosine diphosphate (ADP), adenosine monophosphate (AMP), and adenosine in the juxtaglomerular interstitium. Adenosine combines with receptors in the afferent arteriole and causes vasoconstriction, and thus, blood flow and GFR are lowered to a more normal value. Sensitivity of the tubuloglomerular feedback mechanism is altered by changes in local renin activity, but adenosine, not angiotensin II, is the vasoconstrictor agent. The tubuloglomerular feedback mechanism is a negative feedback system that stabilizes RBF and GFR.

If NaCl delivery to the macula densa is increased experimentally by perfusing the lumen of the loop of Henle, filtration rate in the perfused nephron decreases. This suggests that the purpose of tubuloglomerular feedback may be to control the amount of Na^+ presented to distal nephron segments. These segments have a limited capacity to reabsorb Na^+; if they are overwhelmed, excessive urinary excretion of Na^+ might ensue.

Renal autoregulation is effected through altering the resistance of the afferent, not efferent, arteriole. In this manner, it minimizes the impact of changes in arterial blood pressure on RBF, GFR, and renal-filtered Na^+ load simultaneously. Without renal autoregulation, increases in arterial blood pressure would lead to dramatic increases in GFR and potentially serious losses of NaCl and water from the ECF.

Renal blood flow is altered by sympathetic nerve stimulation and hormones.

The stimulation of renal sympathetic nerves or the release of various hormones may change RBF. Sympathetic nerve stimulation causes vasoconstriction of the afferent and efferent arterioles and a consequent decrease in RBF. Renal sympathetic nerves are activated under stressful conditions, including cold temperatures, deep anesthesia, fearful situations, hemorrhage, pain, and strenuous exercise. In these conditions, renal vasoconstriction may be viewed as an emergency mechanism that helps increase total peripheral vascular resistance, raise arterial blood pressure, and allow more of the cardiac output to perfuse other vital organs, such as the brain and heart, which are more important for short-term survival.

Many substances cause vasoconstriction in the kidneys, including adenosine, angiotensin II, endothelin, epinephrine, norepinephrine, thromboxane A_2, and vasopressin. Other substances cause vasodilation in the kidneys, including atrial natriuretic peptide, dopamine, histamine, kinins, nitric oxide, and prostaglandins E_2 and I_2. Some of these substances (e.g., prostaglandins E_2 and I_2) are produced locally in the kidneys.

A sustained increase in sympathetic nerve activity or plasma angiotensin II concentration stimulates the production of renal vasodilator prostaglandins. These prostaglandins oppose the pure constrictor effect of chronic sympathetic nerve stimulation or angiotensin II, such as that seen in chronic heart failure. They can be viewed as a renal protective mechanism that prevents too severe a reduction in RBF in chronic pathological conditions that could cause renal damage. Antagonism of renal prostaglandin synthesis in these chronic conditions (e.g., through the use of NSAIDs) can cause acute renal failure by negating the renal protective effects of the renal prostaglandins.

▶ GLOMERULAR FILTRATION

The filtration of plasma in the glomerulus is a process of bulk flow where water and low molecular weight substances move from the glomerular capillary across the filtration barrier and into the Bowman capsule. What is filtered is an **ultrafiltrate**. The term ultrafiltrate indicates that glomerular filtration barrier functions as a molecular sieve that allows filtration of small molecules but restricts the passage of macromolecules. The ultrafiltrate includes low molecular weight substances that are freely dissolved in plasma and includes various polar organic molecules such as glucose, amino acids, ions peptides, drugs, and waste products (e.g., creatinine and urea). Since filtration involves bulk flow, the concentration of the substance in the ultrafiltrate is the same as its concentration in the plasma. What are not filtered are blood cells, large proteins, and any small weight molecules that are bound to the large proteins.

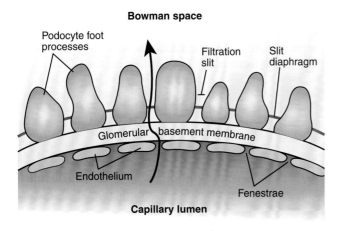

Bowman space

Podocyte foot
processes

Filtration
slit

Slit
diaphragm

Glomerular basement membrane

Endothelium

Fenestrae

Capillary lumen

Figure 22.7 Schematic of the three layers of the glomerular filtration barrier: endothelium, basement membrane, and podocytes. The pathway for filtration is indicated by the *arrow*.

The glomerular filtration barrier is more permeable than the capillaries.

The **glomerular filtration barrier** (Fig. 22.7) consists of three layers: two epithelial layers and basement membrane that lies between them. The first, the capillary **endothelium,** is perforated with large pores called the *lamina fenestra*. At about 50 to 100 nm in diameter, these pores are too large to restrict the passage of the smaller plasma proteins. The second layer, the **basement membrane**, consists of a meshwork of fine fibrils embedded in a gellike matrix. The basement membrane is negatively charged and repels plasma proteins, such as albumin, which are also negatively charged. The third layer is composed of **podocytes**, which constitute the visceral layer of the Bowman capsule. Podocytes ("foot cells") are epithelial cells with extensions that terminate in foot processes that rest on the outer layer of the basement membrane (see Fig. 22.7). The space between adjacent foot processes, called a **filtration slit**, and is about 40 nm wide. A structure called the **slit diaphragm**, is made up of various proteins that are synthesized by the podocytes and extends across the filtration slits. These proteins interconnect with adjacent podocytes to form a meshlike barrier that excludes proteins based on size. The slit diaphragm is an important barrier to the filtration of small proteins.

Thus, the filtered route that substance takes across the glomerular filtration barrier is primarily a physical process and does not involve intracellular pumps. The first step is through the capillary pores, then through the basement membrane, and finally through the filtration slits. The ease in which material is filtered is dependent on molecular size, with large molecules such as albumin completely excluded.

CLINICAL FOCUS | 22.2

Glomerular Disease

The kidney glomeruli may be injured by many immunologic, toxic, hemodynamic, and metabolic disorders. Glomerular injury impairs filtration barrier function and consequently increases the filtration and excretion of plasma proteins (proteinuria). Red cells may appear in the urine, and sometimes, the glomerular filtration rate (GFR) is reduced. Three general syndromes are encountered: nephritic diseases, nephrotic diseases (nephrotic syndrome), and chronic glomerulonephritis.

In the nephritic diseases, the urine contains red blood cells, red cell casts, and mild-to-modest amounts of protein. A red cell cast is a mold of the tubule lumen formed when red cells and proteins clump together; the presence of such casts in the final urine indicates that bleeding had occurred in the kidneys (usually in the glomeruli), not in the lower urinary tract. Nephritic diseases are usually associated with a fall in GFR, accumulation of nitrogenous wastes (urea, creatinine) in the blood, and hypervolemia (hypertension and edema). Most nephritic diseases are a result of immunologic damage. The glomerular capillaries may be injured by antibodies directed against the glomerular basement membrane, by deposition of circulating immune complexes along the endothelium or in the mesangium, or by cell-mediated injury (infiltration with lymphocytes and macrophages). A renal biopsy and tissue examination by light and electron microscopy and immunostaining are often helpful in determining the nature and severity of the disease and in predicting its most likely course.

Poststreptococcal glomerulonephritis is an example of a nephritic condition that may follow a sore throat caused by certain strains of streptococci. Immune complexes of antibody and bacterial antigen are deposited in the glomeruli, complement is activated, and polymorphonuclear leukocytes and macrophages infiltrate the glomeruli. Endothelial cell damage, accumulation of leukocytes, and the release of vasoconstrictor substances reduce the glomerular surface area and fluid permeability and lower glomerular blood flow, causing a fall in GFR.

Nephrotic syndrome is a clinical state that can develop as a consequence of many different diseases causing glomerular injury. It is characterized by heavy proteinuria (>3.5 g/d per 1.73 m² body surface area), hypoalbuminemia (<3 g/dL), generalized edema, and hyperlipidemia. Abnormal glomerular leakiness to plasma proteins leads to increased proximal tubular reabsorption and catabolism of filtered proteins and increased protein excretion in the urine. The resulting loss of protein (mainly serum albumin) leads to a fall in plasma protein concentration (and colloid osmotic pressure). The edema results from the hypoalbuminemia and renal Na⁺ retention. Also, a generalized increase in capillary permeability to proteins (not just in the glomeruli) may lead to a decrease in the effective colloid osmotic pressure of the plasma proteins and may contribute to the edema. The hyperlipidemia (elevated serum cholesterol and elevated triglycerides in severe cases) is probably a result of increased hepatic synthesis of lipoproteins and decreased lipoprotein catabolism. Most often, nephrotic syndrome in young children cannot be ascribed to a specific cause; this is called idiopathic nephrotic syndrome. Nephrotic syndrome in children or adults can be caused by infectious diseases, neoplasia, certain drugs, various autoimmune disorders (such as lupus), allergic reactions, metabolic disease (such as diabetes mellitus), or congenital disorders.

The distinctions between nephritic and nephrotic diseases are sometimes blurred, and both may result in chronic **glomerulonephritis**. This disease is characterized by proteinuria and/or hematuria (blood in the urine), hypertension, and renal insufficiency that progresses over years. Renal biopsy shows glomerular scarring and increased numbers of cells in the glomeruli and scarring and inflammation in the interstitial space. The disease is accompanied by a progressive loss of functioning nephrons and proceeds relentlessly even though the initiating insult may no longer be present. The exact reasons for disease progression are not known, but an important factor may be that surviving nephrons hypertrophy when nephrons are lost. This leads to an increase in blood flow and pressure in the remaining nephrons, a situation that further injures the glomeruli. Also, increased filtration of proteins causes increased tubular reabsorption of proteins, and the latter results in production of vasoactive and inflammatory substances that cause ischemia, interstitial inflammation, and renal scarring. Dietary manipulations (such as a reduced protein intake) or antihypertensive drugs (such as angiotensin-converting enzyme inhibitors) may slow the progression of chronic glomerulonephritis. Glomerulonephritis in its various forms is the major cause of renal failure in people. ■

▶ GLOMERULAR HEMODYNAMIC FORCES

Glomerular filtration is accomplished by hemodynamic forces that drive part of the plasma in the glomerulus through the filtration barrier. No active transport mechanisms are involved in moving the fluid from the plasma across the glomerular filtration barrier into the lumen of the Bowman capsule. Since the glomerulus is a tuft of capillaries, the same hemodynamic forces apply here that causes ultrafiltration across other capillaries (see Chapter 15). Accordingly, **glomerular filtration rate (GFR)** depends on the Starling forces—the balance of hydrostatic and colloid osmotic pressures (COPs) acting across the glomerular filtration barrier.

Glomerular hemodynamics are characterized by high capillary pressure and low vascular resistance.

Figure 22.8 shows how pressures change along the length of a glomerular capillary, in contrast to those seen in a capillary in other vascular beds (in this case, skeletal muscle). Note that average capillary hydrostatic pressure in the glomerulus is much higher than in a skeletal muscle capillary (55 vs. 25 mm Hg). Also, capillary hydrostatic pressure declines little (perhaps 1 to 2 mm Hg) along the length of the glomerular capillary, because the glomerulus contains many (30 to 50) capillary loops in parallel, thereby making the resistance to blood flow in the glomerulus low. In the skeletal muscle capillary, there is a much higher resistance to blood flow, resulting in an appreciable fall in capillary hydrostatic pressure with distance. Finally, note that in the glomerulus, the COP increases substantially along the length of the capillary, because a large volume of filtrate (about 20% of the entering plasma flow) is pushed out of the capillary, and the proteins remain in the circulation. The increase in COP opposes the outward movement of fluid.

Glomerular capillary pressure is the major force that determines glomerular filtration.

In the glomerulus, the driving force for fluid filtration is the glomerular capillary hydrostatic pressure (P_{GC}). This pressure ultimately depends on the blood pressure on the arterial side of the circulation. Filtration is opposed by the hydrostatic pressure in the space of the Bowman capsule (P_{BS}) and by the COP exerted by plasma proteins in glomerular capillary blood. Because the glomerular filtrate is virtually protein free, we neglect the COP of fluid in the Bowman capsule. The *net ultrafiltration pressure gradient* (UP) is equal to the difference between the pressures favoring and opposing filtration:

$$GFR = K_f \times UP = K_f \times (P_{GC} - P_{BS} - COP) \qquad (2)$$

where K_f is the glomerular ultrafiltration coefficient. Estimates of average, normal values for pressures in the human kidney are as follows: P_{GC}, 55 mm Hg; P_{BS}, 15 mm Hg; and COP, 30 mm Hg. From these values, we calculate a net UP of +10 mm Hg (Table 22.1). As seen in the above equation, GFR varies with changes in K_f, hydrostatic pressures in the glomerular capillaries and the Bowman capsule, and the glomerular capillary COP.

Changes in glomerular capillary hydrostatic pressure profoundly affect the GFR.

Glomerular capillary hydrostatic pressure (P_{GC}) is the driving force for filtration; it depends on the arterial blood pressure and the resistances of upstream and downstream blood vessels. Because of autoregulation, P_{GC} and GFR are maintained at relatively constant values when arterial blood pressure is varied from 80 to 180 mm Hg. Below a pressure of 80 mm Hg, however, P_{GC} and GFR decrease, and GFR ceases at a blood pressure of about 40 to 50 mm Hg. One of the classic signs of hemorrhagic or cardiogenic shock is an

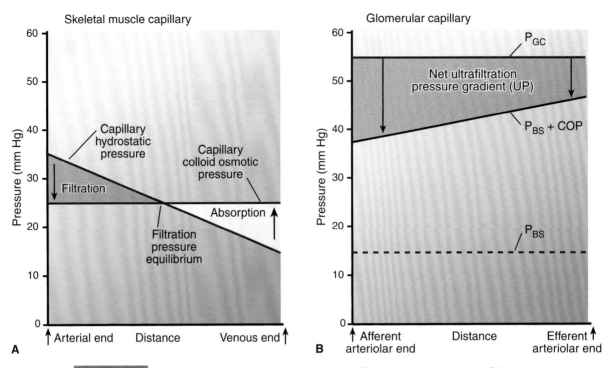

Figure 22.8 **Pressure profiles along a skeletal muscle capillary and a glomerular capillary.** **(A)** In the "typical" skeletal muscle capillary, filtration occurs at the arterial end and absorption at the venous end of the capillary. Interstitial fluid hydrostatic and colloid osmotic pressures (COPs) are neglected here because they are roughly equal and counterbalance each other. **(B)** In the glomerular capillary, glomerular hydrostatic pressure (P_{GC}) (*top line*) is high and declines only slightly with distance. The *bottom (dashed) line* represents the hydrostatic pressure in the Bowman capsule (P_{BS}). The *middle line* is the sum of P_{BS} and the glomerular capillary COP. The difference between P_{GC} and P_{BS} + COP is equal to the net ultrafiltration pressure gradient (UP). In the normal human glomerulus, filtration probably occurs along the entire capillary. Assuming that K_f is uniform along the length of the capillary, filtration rate would be highest at the afferent arteriolar end and lowest at the efferent arteriolar end of the glomerulus.

absence of urine output, which is a result of an inadequate arterial blood pressure, P_{GC}, and GFR.

The caliber of afferent and efferent arterioles can be altered by a variety of hormones and by sympathetic nerve stimulation, leading to changes in P_{GC}, glomerular blood flow, and GFR. Some hormones act preferentially on afferent or efferent arterioles. Afferent arteriolar dilation increases glomerular blood flow and P_{GC} and therefore produces an increase in GFR. Afferent arteriolar constriction produces the exact opposite effects. Efferent arteriolar dilation increases glomerular blood flow but leads to a fall in GFR because P_{GC} is decreased. Constriction of efferent arterioles increases P_{GC} and decreases glomerular blood flow. With modest efferent arteriolar constriction, GFR increases because of the increased P_{GC}. With extreme efferent arteriolar constriction, however, GFR decreases because of the marked decrease in glomerular blood flow.

Normally, about 20% of the plasma flowing through the kidneys is filtered in the glomeruli. This percentage (or fraction) is called the **filtration fraction**. Changes in filtration fraction will result from constriction or dilation of afferent or efferent arterioles. For example, afferent arteriolar dilation or efferent arteriolar constriction leads to an increase in filtration fraction. An increase in filtration fraction will increase the protein concentration of the blood exiting the glomerulus and hence will increase the COP in the peritubular capillaries. Such a change will increase sodium reabsorption in the kidneys (see Chapter 23).

TABLE 22.1	Summary Forces Involved in Glomerular Filtration	
Force	**Magnitude (mm Hg)**	**Effect**
Glomerular capillary pressure	55	Favors filtration
Plasma colloidal osmotic pressure	30	Opposes filtration
Bowman capsule hydrostatic pressure	15	Opposes filtration
Net filtration pressure	10	Favors filtration
	55 − (30 + 15) = 10	

Capillary osmotic pressure and hydrostatic pressure in the Bowman capsule oppose glomerular filtration.

Hydrostatic pressure in the Bowman capsule (P_{BS}) depends on the input of glomerular filtrate and the rate of removal of this fluid by the tubule. This pressure opposes filtration. It also provides the driving force for fluid movement down the length of the tubule. If there is obstruction anywhere along the urinary tract—for example, because of stones, ureteral obstruction, or prostate enlargement—then pressure upstream to the block is increased and GFR consequently falls. If tubular reabsorption of water is inhibited, pressure in the tubular system is increased because an increased pressure head is needed to force a large volume flow through the loops of Henle and collecting ducts. Consequently, a large increase in urine output caused by a diuretic drug may be associated with a tendency for GFR to fall.

The COP opposes glomerular filtration. Dilution of the plasma proteins (e.g., by intravenous infusion of a large volume of isotonic saline) lowers the plasma COP and leads to an increase in GFR. Part of the reason glomerular blood flow has important effects on GFR is that it changes the COP profile along the length of a glomerular capillary. Consider, for example, what would happen if glomerular blood flow were low. Filtering a small volume of fluid out of the glomerular capillary would lead to a sharp rise in COP early along the length of the glomerulus. As a consequence, filtration would soon cease and GFR would be low. On the other hand, a high blood flow would allow a high rate of filtrate formation with a minimal rise in COP. In general, then, RBF and GFR change hand in hand, but the exact relationship between GFR and RBF depends on the magnitude of the other factors that affect GFR.

Glomerular ultrafiltration coefficient depends on the properties of the glomerular filtration barrier.

The **glomerular ultrafiltration coefficient (K_f)** is the glomerular equivalent of the capillary filtration coefficient encountered in Chapter 15. Accordingly, K_f depends on both membrane permeability and surface area of the glomerular filtration barrier. Compared to typical systemic capillaries, the normal K_f for the glomerulus is very high. In chronic renal disease, functioning glomeruli are lost, leading to a reduction in surface area available for filtration and a fall in GFR. Acutely, a variety of drugs and hormones appear to change glomerular K_f and thus alter GFR, but the mechanisms are not completely understood.

In summary, the high GFR (180 L/d) in the human kidney is the result of several factors and far exceeds that in all other capillary beds for several reasons:

- The filtration coefficient is unusually high in the glomeruli. Compared with most other capillaries, the glomerular capillaries behave as though they have more pores per unit surface area; consequently, they have an unusually high hydraulic conductivity. The total glomerular filtration barrier area, about 2 m^2, is large.

- Capillary hydrostatic pressure is higher in the glomeruli than in any other capillaries.

- The high rate of RBF helps sustain a high GFR by limiting the rise in COP, thereby favoring filtration along the entire length of the glomerular capillaries.

In conclusion, glomerular filtration is high because the glomerular capillary blood is exposed to a large, porous surface and there is a high transmural pressure gradient favoring filtration.

Proteinuria is the hallmark of glomerular filtration barrier disorder.

When the filtration membrane is not functioning properly, it ceases to be an effective barrier. What then occurs is **proteinuria**, an abnormal accumulation of protein in the urine. A very small amount of protein gets filtered normally (~5 to 20 mg/d), but this is almost completely reabsorbed in the proximal tubule by endocytosis. If protein is detected in the urine, it usually means that there is a breakdown in the filtration barrier.

Proteinuria is a hallmark of glomerular disease. Proteinuria not only is a sign of kidney disease but also results in tubular and interstitial damage and contributes to the progression of chronic renal disease. This is due to the fact that high levels of proteins in the filtrate have a pathogenic effect on the renal tubules. Increased endocytosis of proteins by the renal tubular cells ultimately stimulates inflammation and fibrosis and leads to the loss of nephrons.

The major disorders that disrupt the glomerular filtration barrier are *diabetes mellitus*, *hypertension*, and *glomerulonephritis*. In diabetes mellitus, hyperglycemia and reduced insulin signaling trigger a set of changes in the filtration membrane that cause a loss of selectivity and result in proteinuria. Hypertension is damaging because the high pressure in the glomerular capillaries damages the filtration barrier. In glomerulonephritis, there is inflammatory damage to the filtration membrane due to an infection.

Severe proteinuria (>3.5 g/d) is called **nephrotic syndrome**. Excessive proteinuria manifests a number of symptoms. It causes low level of protein in the plasma, and this in combination with sodium retention leads to edema. Other symptoms include hyperlipidemia and hypertension.

▶ TUBULAR REABSORPTION

All of the plasma constituents are filtered in the glomerulus with the exception of the larger proteins. Over 70% of the filtered solutes and water are reabsorbed along the proximal convoluted tubule. The reabsorption process involves selecting and retrieving nutrients, electrolytes, and other substances the body needs to function and survive. At the same time, waste products and other toxic substances are not reabsorbed and are eliminated in the urine.

The proximal tubule is responsible for reabsorbing all of the filtered glucose and amino acids. In addition, the proximal tubule reabsorbs the largest fraction of the filtered Na$^+$, K$^+$, Ca^{2+}, Cl$^-$, HCO$_3^-$, and water and secretes various organic anions and organic cations.

TABLE 22.2 Selective Reabsorption of Various Substances Filtered by the Glomerulus

Substances	Percentage of Filtered Substances Reabsorbed	Percentage of Filtered Substances Excreted
Water	99	1
Sodium	99.5	0.5
Glucose	100	0
Urea (waste product from amino acids)	50	50
Phenol (metabolic waste product)	0	100

Tubular reabsorption involves diffusion and active transport.

Tubular reabsorption is highly selective and involves two types of tubular transport—*passive reabsorption and active reabsorption*. The steps in passive reabsorption involve no energy. The net movement of substances in the transepithelial transport from the tubular lumen to the plasma follows an electrochemical gradient. Active reabsorption involves transepithelial transport in the *basolateral membrane* of the proximal tubules. In general, these two reabsorption mechanisms allow the kidneys to have a high reabsorption capacity for substances needed by the body and a low reabsorptive capacity for waste products and toxic substances to be eliminated (Table 22.2).

Water reabsorption occurs in all regions of the renal tubule with the exception of the ascending limb of the loop of Henle and the distal convoluted tubule. The thin and thick ascending limbs are water impermeable. The reason the ascending limb is impervious to water is not due to "tight junctions," but because of the lack of **aquaporins**, which are water channel proteins. These water channels do not exist in the cells of the ascending loop of Henle.

Urea reabsorption is dependent on the reabsorption of water in the proximal tubule.

The reabsorption of urea by the proximal tubules provides an example of reabsorption by passive diffusion. Because urea is freely permeable across the filtration barrier, urea concentration is the same within the Bowman capsule space as in the peritubular capillary and the adjacent interstitial fluid. As the filtered fluid flows through the proximal tubule, water is reabsorbed, and the reabsorption of water increases urea concentration within the tubular fluid. As a result, urea is higher than the surrounding interstitial fluid and peritubular capillaries. The proximal tubule is freely permeable to urea. Accordingly, urea diffuses down the concentration gradient from the tubular lumen to the peritubular capillaries. This is generally thought to occur by both transcellular (across cell membranes) and paracellular (between the cells) routes. Thus, urea reabsorption in the proximal tubule is highly dependent upon the reabsorption of water.

Active Na⁺/K⁺-ATPase pump is essential for sodium reabsorption.

In terms of energy expenditure, over 75% of the total energy used by the kidneys is for the active reabsorption of Na⁺. Unlike other filtered solutes, Na⁺ is reabsorbed throughout the entire tubule, with the exception of the descending loop of Henle. As seen from Table 22.2, 99.5% is reabsorbed. Of the Na⁺ reabsorbed, most occurs in the proximal tubule (~65%), with 25% in the loop of Henle, and the remaining in the distal tubule and collecting ducts.

With the exception of the thin ascending limb of the loop of Henle, Na⁺ reabsorption requires active transport throughout the renal tubule. Figure 22.9 is a model of a proximal tubule cell. Na⁺ enters the cell from the lumen across the apical cell membrane and is pumped out across the basolateral cell membrane by the Na⁺/K⁺-ATPase. Anions, primarily Cl^- and HCO_3^-, accompany this sodium transport across the basolateral membrane. At the luminal cell membrane (brush border) of the proximal tubule cell, Na⁺ enters the cell down combined electrical and chemical potential gradients. The inside of the cell is about −70 mV compared with tubular fluid, and intracellular Na⁺ is about 30 to 40 mEq/L compared with a tubular fluid concentration of about 140 mEq/L. Na⁺ entry into the cell occurs via a number of cotransport

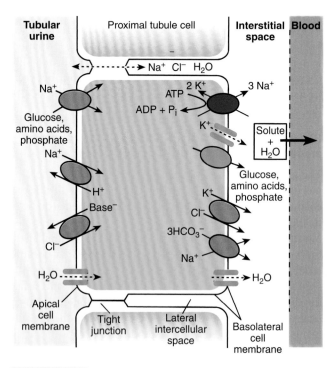

Figure 22.9 Cell model for transport in the proximal tubule. The luminal (apical) cell membrane in this nephron segment has a large surface area for transport because of the numerous microvilli that form the brush border (*not shown*). Glucose, amino acids, phosphate, and numerous other substances are transported by separate carriers. Sodium reabsorption across the apical membrane is accomplished through Na to solute cotransporters, H⁺-driven sodium reabsorption, and paracellular Cl⁻-driven sodium transport. All sodium transport is dependent on the sodium ATPase pumps in the basolateral membrane. ADP, adenosine diphosphate; ATP, adenosine triphosphate.

and antiport mechanisms. These mechanisms rely on the electrochemical gradient produced in part by the Na^+/K^+-ATPase and are therefore considered secondary active transport mechanisms. Na^+ is reabsorbed across the apical membrane together with glucose, amino acids, phosphate, and other solutes by way of separate, specific cotransporters. The downhill (energetically speaking) movement of Na^+ into the cell drives the uphill transport of these solutes. The sodium electrochemical energy available for this process can increase intracellular concentration of the cotransported substances 100 to 1,000 times greater than that in the tubule fluid. Glucose, amino acids, and phosphate, so accumulated in the cell, exit across the basolateral cell membrane by way of separate, Na^+-independent facilitated diffusion mechanisms. The sodium to solute cotransport mechanism in the proximal tubule accounts for reabsorption of all of glucose and amino acids filtered at the glomerulus along with about 7% of the filtered sodium.

Na^+ is also reabsorbed across the luminal cell membrane in exchange for H^+. The Na^+/H^+ exchanger, an antiporter, is a secondary active transport mechanism; the downhill movement of Na^+ into the cell energizes the uphill secretion of H^+ into the lumen. The H^+ in this exchange is derived primarily from two sources. One source is derived from carbonic acid production inside the tubule cells. Proximal tubule cells are richly endowed with carbonic anhydrase, which catalyzes the conversion of water and carbon dioxide (considered a cell "waste product") to carbonic acid (H_2CO_3). Intracellular carbonic acid so formed then quickly dissociates into H^+ and HCO_3^-. HCO_3^- accumulation in the cell is favored by removal of H^+ from the intracellular fluid by the Na^+/H^+ exchanger. A Na^+-HCO_3^- cotransporter uses accumulated bicarbonate to drive sodium from inside the cell into the peritubular fluid. The apical membrane Na^+/H^+ exchanger also removes H^+ formed from the dissociation of formic or oxalic acid within the tubule cell. The anions remaining from these acids are then used to bring Cl^- into the cell across the apical membrane via a Cl^-/base exchanger. Once in the tubular lumen, formate reassociates with any H^+ there. The newly reformed formic acid then diffuses back across the apical membrane where it dissociates and thus allows formate and acid to again exchange for Na^+ and Cl^-. The formic acid to formate anion mechanism is a key means by which Na^+ and Cl^- enter across the apical membrane. Cl^- may leave the cell across the basolateral membrane by way of an electrically neutral K–Cl cotransporter. In total, H^+-driven sodium reabsorption accounts for reabsorption of about 50% of the filtered sodium load at the glomerulus. (This mechanism is also important in the acidification of urine during acid–base homeostasis; see Chapter 24.)

The "tight junctions" between proximal tubule cells are not absolute and are sometimes called "leaky," or loose, tight junctions. Sodium, chloride, and water can pass through these junctions. In the latter part of the proximal tubule, the chloride concentration exceeds that of the peritubular fluid. Chloride is able to diffuse through the tight junctions, down its concentration gradient, and, in so doing, drags sodium with it. This is called chloride-driven sodium transport and accounts for the reabsorption of about 10% of the filtered sodium load.

The reabsorption of Na^+ with its accompanying anions and solutes establishes an osmotic gradient across the proximal tubule epithelium that is the driving force for water reabsorption. Because the water permeability of the proximal tubule epithelium is extremely high, only a small gradient (a few mOsm/kg H_2O) is needed to account for the observed rate of water reabsorption. Water crosses the proximal tubule epithelium through the cells (via water channel proteins—aquaporins—in the cell membranes) and between the cells (leaky tight junctions and lateral intercellular spaces). Upon reabsorption from the proximal tubule, the blood surrounding the tubules then takes up the Na^+, accompanying anions, and water. Filtered Na^+ salts and water are thus returned to the circulation.

In total, the proximal tubule reabsorbs about two third of its filtered load of sodium and water. Because the tubule is freely permeable to water, the osmolarity of tubular fluid is the same as plasma with the sodium concentration being identical in both fluids. However, chloride concentration in the fluid exceeds that in plasma (~132 vs. 110 mM) whereas that for bicarbonate is lower (~8 vs. 24 mM). No amino acids or glucose exist in the tubular fluid exiting the proximal tubule but the reabsorption of water relative to the transport of urea in the tubule renders the tubule urea concentration higher than plasma (~20 vs. 6 mM).

An important feature of sodium reabsorption in the proximal tubule and, indeed, throughout the nephron is that its rate of transport is *load dependent*. In the proximal tubule, an increase in the filtered load of Na^+ from the glomerulus stimulates an increase in sodium reabsorption by the tubule such that the *percent* of sodium reabsorbed remains the same. This phenomenon is called **glomerulotubular balance**, although it is also observed in the distal nephron when sodium load entering the distal nephron from upstream segments changes as well. This intrinsic renal tubular transport property has profound effects on salt and water balance in the body and contributes also to major side effects of renal excretion of ions other than sodium whenever sodium transport in the nephron is altered by disease or clinical drugs such as certain diuretics (see below and Chapter 23).

Transport of NaCl, urea, and water in the loop of Henle is determined by passive and active transport processes in individual sections of the tubule.

The transport of NaCl, urea, and water by the loop of Henle is complex due to differing epithelial permeabilities and transport properties as well as the unusual composition of the peritubular fluid surrounding the loop. The loops of Henle are surrounded by a progressively hyperosmotic **vertical osmotic gradient** in the medullary peritubular fluid (Fig. 22.10). In cortical or superficial nephrons, the loops descend from the cortex to the boundary of the outer and inner medulla. The peritubular fluid in the cortex is the same as plasma (~290 mOsm), but this osmolarity increases progressively to about 600 mOsm at the inner/outer boundary. In juxtamedullary nephrons, the loops are much longer and descend into the inner medullar where peritubular fluid

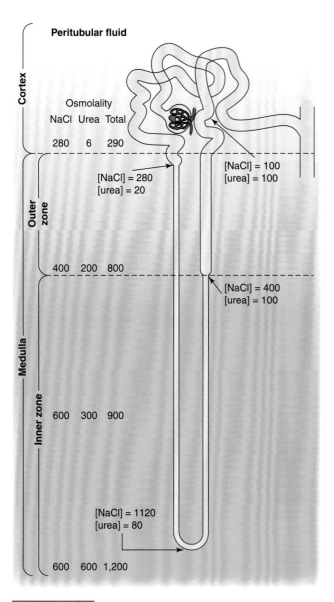

Figure 22.10 **Peritubular and tubular fluid osmolalities in the loop of Henle of a juxtamedullary nephron.** A vertical gradient of increasing osmolality exists around the loop from the cortex to the renal medullary interstitium. The contribution to total osmolality from NaCl and urea are shown. Mechanisms responsible for differences in osmolality due to NaCl and urea within the lumen of the loop and in the peritubular fluids are explained in the text.

reaches 1,200 to 1,400 mOsm. Medullary peritubular fluid osmolality is due to about 50% NaCl and 50% urea, in stark contrast to the osmolality of the end proximal tubular fluid, which is predominantly due to Na⁺ and Cl⁻. (The mechanisms responsible for creating and maintaining the medullary osmotic gradient in the kidney are explained later in this chapter.)

The thin descending limb contains no membrane active transport systems and is freely permeable to water, but it is impermeable to NaCl and urea. Therefore, as fluid from the proximal tubule travels down the thin descending limb, water is osmotically, passively, reabsorbed until its osmolarity is identical to the peritubular fluid surrounding the descending limb. In juxtamedullary nephrons, this equals

about 1,200 to 1,400 mOsm or roughly a fourfold increase in osmolarity over fluid entering from the proximal tubule. This means about three quarters of the water entering the thin limb was osmotically reabsorbed by the time tubular fluid reaches the tip of the loop. This reabsorption of water concentrates NaCl, other solutes, and urea fourfold as well (about 1,120 mOsm due to NaCl and 80 mOsm due to urea).

The thin ascending limb of the loop of Henle is impermeable to water, but permeable to NaCl and somewhat permeable to urea, due to the presence of facilitated urea transporters (UT-A2) in the tubule epithelium. Consequently, as fluid in the loop ascends the limb, NaCl encounters progressively lower NaCl concentrations in the surrounding peritubular fluid and is passively reabsorbed. Urea, however, is slightly, passively secreted down its concentration gradient into the lumen as the tubular fluid ascends the limb. At the end of the ascending thin limb, its concentration reaches ~100 mM. Because this part of the tubule is impermeable to water, concentration changes for NaCl and urea are due solely to their passive transport. However, the reabsorption of NaCl exceeds the secretion of urea into the tubule so that by the end of the thin ascending limb, tubular fluid is hypotonic *relative to peritubular fluid* (about 500 vs. 600 total mOsm) but still hypertonic to plasma (~290 mOsm).

Sodium enters the luminal cell of the thick ascending limb by a Na–K–2Cl cotransporter.

The thick ascending limb of the loop of Henle is characterized by being impermeable to water, NaCl, and urea. However, cells of the thick limb are capable of actively transporting sodium out of the tubular fluid against a steep electrochemical gradient. Figure 22.11 is a model of a thick ascending limb cell. Na⁺ enters the cell across the luminal

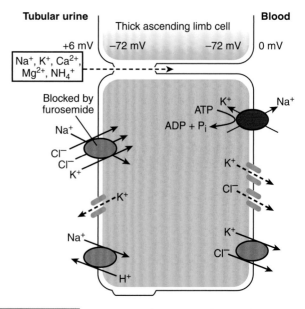

Figure 22.11 **Cell model for ion transport in the thick ascending limb.** Transport of sodium is critically dependent on uptake from the lumen by the Na–K–2Cl cotransporter in the apical membrane. This transporter is inhibited by loop diuretics. ADP, adenosine diphosphate; ATP, adenosine triphosphate.

cell membrane by an electrically neutral Na–K–2Cl cotransporter. This transporter, in particular Cl⁻ binding to this protein, is essential to sodium reabsorption in the thick ascending limb. It is specifically inhibited by "loop" diuretic drugs, such as bumetanide and furosemide, which then, through many steps, inhibits the ability of the kidney to reabsorb NaCl and water (see later sections in this chapter). The downhill movement of Na^+ into the cell results in secondary active transport of one K^+ and two Cl^-. Na^+ is pumped out of the basolateral cell membrane by a vigorous Na^+/K^+-ATPase. K^+ recycles back into the lumen via a luminal cell membrane K^+ channel. Cl^- leaves through the basolateral side by a K–Cl cotransporter or Cl^- channel. The luminal cell membrane is predominantly permeable to K^+, and the basolateral cell membrane is predominantly permeable to Cl^-. Diffusion of these ions out of the cell produces a transepithelial potential difference, with the lumen about +6 mV compared with the interstitial space around the tubules. This potential difference drives reabsorption of small cations (Na^+, K^+, Ca^{2+}, Mg^{2+}, and NH_4^+) out of the lumen, between the cells. The potential difference can be increased to about +9 mV by anything that stimulates sodium reabsorption in the thick ascending limb and decreased to about +3 when sodium transport is impaired. Thus, factors that affect sodium reabsorption in the thick ascending limb indirectly affect reabsorption of other small cations as well. By the end of the thick ascending limb, tubular fluid volume has been reduced by about 75% of that entering the loop and is hypotonic relative to peritubular fluid *and* plasma (about 100 mOsm due equally to NaCl and urea).

The final step in the overall reabsorption of solutes and water in the thin ascending limb is uptake by the peritubular capillaries. This mechanism involves the usual Starling forces that operate across capillary walls. Recall that blood in the peritubular capillaries was previously filtered in the glomeruli. Because a protein-free filtrate was filtered out of the glomeruli, the protein concentration (hence, COP) of blood in the peritubular capillaries is high, thereby providing an important driving force for the uptake of reabsorbed fluid. The hydrostatic pressure in the peritubular capillaries (a pressure that opposes the capillary uptake of fluid) is low, because the blood has passed through upstream resistance vessels. The balance of pressures acting across peritubular capillaries favors the uptake of reabsorbed fluid from the interstitial spaces surrounding the tubules.

The distal nephron handles tubular reabsorption.

The so-called **distal nephron** includes several distinct segments: the distal convoluted tubule; the connecting tubule; and the cortical, outer medullary, and inner medullary collecting ducts (see Fig. 22.3). Note that the distal nephron includes the collecting duct system, which, strictly speaking, is not part of the nephron but, from a functional perspective, this is justified. Transport in the distal nephron differs from that in the proximal tubule in several ways:

- The distal nephron reabsorbs much smaller amounts of salt and water. Typically, the distal nephron reabsorbs 9%

of the filtered Na^+ and 19% of the filtered water, compared with 70% for both substances in the proximal convoluted tubule.

- The distal nephron can establish steep gradients for salt and water. For example, the concentration of Na^+ in the final urine may be as low as 1 mEq/L (vs. 140 mEq/L in plasma) and the urine osmolality can be almost one tenth that of plasma. By contrast, the proximal tubule reabsorbs Na^+ and water along small gradients, and the Na^+ concentration and osmolality of its tubule fluid are normally close to those of plasma.

- The distal nephron has a "tight" epithelium, whereas the proximal tubule has a "leaky" epithelium (see Chapter 2). This explains why the distal nephron can establish steep gradients for small ions and water, whereas the proximal tubule cannot.

- Na^+ and water reabsorption in the proximal tubule are normally closely coupled, because epithelial water permeability is always high. By contrast, Na^+ and water reabsorption can be uncoupled in the distal nephron, because water permeability may be low and variable.

Proximal reabsorption overall can be characterized as a coarse operation that reclaims large quantities of salt and water along small gradients. By contrast, distal reabsorption is a finer process involved more in regulation of water and electrolyte balance in the body.

Sodium and chloride are actively reabsorbed through the luminal cell membrane of the distal convoluted tubule by a Na–Cl cotransporter.

The distal convoluted tubule has functionally "tight" junctions between epithelial cells. It is essentially impermeable to water and urea but can actively reabsorb NaCl against steep electrochemical gradients. Figure 22.12 is a simple model of sodium reabsorption in a distal convoluted tubule cell. In this nephron segment, Na^+ and Cl^- are transported from the lumen into the cell by a Na–Cl cotransporter. (This transporter can be inhibited clinically by thiazide diuretics.)

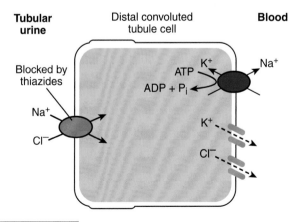

Figure 22.12 Cell model for ion transport in the distal convoluted tubule. ADP, transport of sodium is critically dependent on uptake from the lumen by the Na–Cl cotransporter in the apical membrane. This transporter is inhibited by thiazide diuretics. ADP, adenosine diphosphate; ATP, adenosine triphosphate.

Na⁺ is pumped out the basolateral side by the Na⁺/K⁺-ATPase and chloride is thought to follow via a basolateral membrane channel. Sodium reabsorption coupled with the low water and urea permeability of the tubular epithelium further reduces the osmolality of the tubular fluid without changing its volume or urea concentration. Urea concentration is higher in distal tubules from juxtamedullary compared to cortical nephrons (about 100 vs. 40 mM) because the later nephrons are exposed to lower osmotic and urea concentrations surrounding their loops of Henle. Regardless of the type of nephron, by the end of the distal convoluted tubule, so little of the original filtered NaCl remains that the osmotic contribution of solutes other than NaCl is just as important as that from NaCl. Therefore, at this point in the tubule, osmolality is divided into that due to urea and that due to total **nonurea solutes** (about 50 mOsm).

Aldosterone stimulates Na⁺ reabsorption in the collecting duct.

The collecting ducts are at the end of the nephron system, and what happens there determines the final excretion of Na⁺, K⁺, H⁺, and water. Na⁺ entry into the collecting duct cell is by diffusion through a Na⁺ channel in collecting duct principal cells (Fig. 22.13). This channel has been cloned and sequenced and is known as **ENaC**, for **epithelial sodium (Na) channel**. The entry of Na⁺ through this channel is rate-limiting for overall Na⁺ reabsorption of the cells. The channel can be inhibited by amiloride and similar diuretics, which are known as potassium-sparing diuretics because they do not cause the potassium loss seen with other diuretic classes. Principal cell sodium reabsorption is quite variable due to the load dependency effect and creates a variable lumen-negative transepithelial potential across the tubule, which facilitates secretion of K⁺ and H⁺ by the collecting duct (see next section for details).

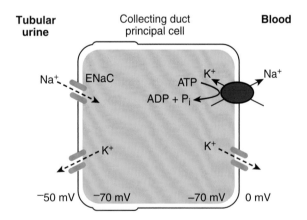

Figure 22.13 Model for Na⁺ and K⁺ transport by a collecting duct principal cell. Transport of sodium is critically dependent on uptake from the lumen by the ENaCl channel, which is inhibited by amiloride-type potassium-sparing diuretics. Sodium transport here is stimulated by the hormone aldosterone and linked to secretion of K⁺ and H⁺ by the collecting duct. Stimulation of sodium transport creates a negative transepithelial electrical potential. ADP, adenosine diphosphate; ATP, adenosine triphosphate.

Ionic transport in the collecting ducts is finely tuned by hormones. Specifically, aldosterone increases Na⁺ reabsorption as well as K⁺ and H⁺ secretion with the secretory effects primarily occurring in the connecting ducts and cortical region of the collecting ducts. Another hormone, arginine vasopressin, also called antidiuretic hormone (ADH) increases water permeability in the collecting ducts. Intercalated cells are scattered among collecting duct principal cells; they are important in acid–base transport (see Chapter 24). A H⁺/K⁺-ATPase is present in the luminal cell membrane of α-intercalated cells and contributes to renal K⁺ conservation when dietary intake of K⁺ is deficient.

▶ TUBULE SECRETION

Tubular secretion moves substance from the peritubular capillaries into the tubule lumen. Similar to glomerular filtration, tubular secretion establishes a pathway from the blood into the tubule. The most important constituents secreted are H⁺, K⁺, and Cl⁻. For the most part, the amount excreted in the urine depends in large measure on the magnitude of tubular transport. Transport of various solutes and water differs in the various nephron segments. Here, we describe transport along the nephron and collecting duct system, starting with the proximal convoluted tubule.

Proximal tubule secretion eliminates many toxins and drugs from the blood.

Recall the proximal convoluted tubule is the first part of the renal tubule and comes right after the Bowman capsule. The proximal convoluted tubule makes up the first 60% of the length of the proximal tubule. Because the proximal straight tubule is inaccessible to study *in vivo*, most quantitative information about function in the living animal is confined to the convoluted portion. Studies on isolated tubules *in vitro* indicate that the two segments of the proximal tubule are functionally similar.

The proximal tubule, both convoluted and straight portions, secretes a large variety of organic anions and organic cations. Many of these substances are endogenous compounds, drugs, or toxins. The organic anions are mainly carboxylates and sulfonates (carboxylic and sulfonic acids in their protonated forms). A negative charge on the molecule appears to be important for secretion of these compounds. Examples of organic anions actively secreted in the proximal tubule include penicillin and PAH. Organic anion transport becomes saturated at high plasma organic anion concentrations (Fig. 22.14), and the organic anions compete with each other for secretion.

Figure 22.15 shows a cell model for active secretion. Proximal tubule cells actively take up PAH from the blood side by exchange for cell α-ketoglutarate. This exchange is mediated by an organic anion transporter (OAT1). The cells accumulate α-ketoglutarate from metabolism and because of a cell membrane Na⁺-dependent dicarboxylate transporters. PAH accumulates in the cells at a high concentration and then moves downhill into the tubular urine in an electrically

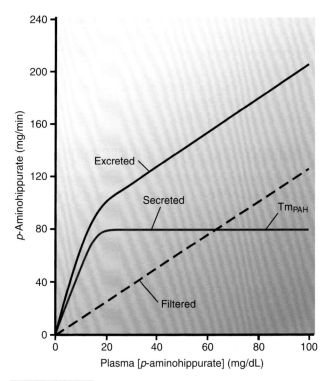

Figure 22.14 **Rates of excretion, filtration, and secretion of** ***p*-aminohippurate (PAH) as a function of plasma PAH.** More PAH is excreted than is filtered; the difference represents secreted PAH. Tm_{PAH}, tubular transport maximum for PAH.

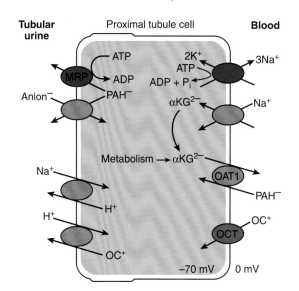

Figure 22.15 **Cell model for the secretion of organic anions (*p*-aminohippurate [PAH]) and organic cations in the proximal tubule.** *Upward slanting arrows* indicate transport against an electrochemical gradient (energetically uphill transport) and *downward slanting arrows* indicate downhill transport. PAH accumulation in the cell is mediated by a basolateral membrane organic anion transporter (OAT) that exchanges PAH for α-ketoglutarate (α-KG^{2-}). The α-KG^{2-} level in the cell is higher than in the blood because of metabolic production and Na^+-dependent uptake of α-KG^{2-}. PAH exits the cell passively via a luminal membrane PAH/anion exchanger or can be actively pumped into the lumen by a multidrug resistance–associated protein (MRP) that consumes adenosine triphosphate (ATP). Organic cations (OC^+) enter the cell down the electrical gradient, a process mediated by a basolateral membrane organic cation transporter (OCT), and are transported uphill into the lumen by an organic cation/H^+ exchanger. ADP, adenosine diphosphate; ATP, adenosine triphosphate.

neutral fashion, by exchanging for an inorganic anion (e.g., Cl^-) or an organic anion via another OAT. Organic anions may also be actively pumped into the tubular urine via a multidrug resistance–associated protein, which is an ATPase.

The organic cations are mainly amine and ammonium compounds and are secreted by other transporters. Entry into the cell across the basolateral membrane is favored by the inside negative membrane potential and occurs via facilitated diffusion, mediated by an organic cation transporter. The exit of organic cations across the luminal membrane is accomplished by an organic cation/H^+ antiporter (exchanger) and is driven by the lumen-to-cell H^+ concentration gradient established by Na^+/H^+ exchange. The transporters for organic anions and organic cations show broad substrate specificity and accomplish the secretion of a large variety of chemically diverse compounds.

In addition to being actively secreted, some organic compounds passively diffuse across the tubular epithelium (Fig. 22.16). Organic anions can accept H^+ and organic cations can release H^+, so their charge is influenced by pH. The nonionized (uncharged) form, if it is lipid-soluble, can diffuse through the lipid bilayer of cell membranes down concentration gradients. The ionized (charged) form passively penetrates cell membranes with difficulty.

Consider, for example, the carboxylic acid probenecid ($pK_a = 3.4$). This compound is filtered by the glomeruli and secreted by the proximal tubule. When H^+ is secreted into the tubular urine (see Chapter 24), the anionic form (A^-) is converted to the nonionized acid (HA, in Fig. 22.15). The concentration of nonionized acid is also increased

because of water reabsorption. A concentration gradient for passive reabsorption across the tubule wall is created, and appreciable quantities of probenecid are passively reabsorbed. This occurs in most parts of the nephron but particularly in those where pH gradients are largest and

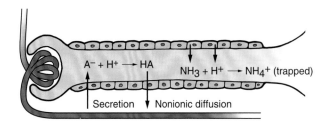

Figure 22.16 **Nonionic diffusion of lipid-soluble weak organic acids and bases.** Small nonionized forms of organic acids and bases can cross the tubular epithelium by simple diffusion whereas their ionized forms cannot. Acidification of the urine converts the organic anion A^- to the undissociated (nonionized) acid HA, which is reabsorbed by diffusion. NH_3, a lipid-soluble base, diffuses into the tubular urine, where it is converted to NH_4^+, thereby trapping the ammonia in the acidic urine.

where water reabsorption has resulted in the greatest concentration (i.e., the collecting ducts; see next section). The excretion of probenecid is enhanced by making the urine more alkaline (e.g., by administering $NaHCO_3$) and by increasing urine flow rate (e.g., by drinking water). In the case of a lipid-soluble base, such as ammonia (NH_3), excretion is favored by making the urine more acidic and enhancing the urine flow.

Finally, a few organic anions and cations are also actively reabsorbed. For example, uric acid is both secreted and reabsorbed in the proximal tubule. Normally, the amount of uric acid excreted is equal to about 10% of the filtered uric acid, so reabsorption predominates. In **gout**, plasma levels of uric acid are increased. One treatment for gout is to promote urinary excretion of uric acid by administering drugs that inhibit its tubular reabsorption.

Cortical collecting duct is the primary site for potassium secretion.

Under normal circumstances, the cortical collecting ducts secrete most of the excreted K^+. With great K^+ excess (e.g., a high-K^+ diet), the cortical collecting ducts may secrete so much K^+ that more K^+ is excreted than was filtered. With severe K^+ depletion, the cortical collecting ducts reabsorb K^+.

K^+ secretion appears to be a function primarily of the collecting duct principal cell (see Fig. 22.13). K^+ secretion involves active uptake by a Na^+/K^+-ATPase in the basolateral cell membrane, followed by diffusion of K^+ through luminal membrane K^+ channels. Outward diffusion of K^+ from the cell is favored by concentration gradients and opposed by electrical gradients. However, the net electrochemical gradient for potassium across the luminal membrane favors its transport into the tubular lumen (i.e., its secretion). This is because the combined effects of epithelial tight junctions and the ENaC channel tend to separate sodium from tubular fluid anions rendering the tubular lumen negatively charged relative to the peritubular fluid. This creates a **negative transepithelial potential** across the tubule epithelium of about −50 mV. This potential favors secretion of K^+ and H^+ and is highly dependent on sodium reabsorption by the principal cells, being as low as −5 mV when sodium reabsorption is impaired to −70 mV when it is stimulated.

The magnitude of K^+ secretion is affected by several factors (see Fig. 22.13):

- The lumen-negative transepithelial electrical potential promotes K^+ and H^+ secretion.
- The activity of the basolateral membrane Na^+/K^+-ATPase is a key factor affecting secretion: the greater the pump activity, the higher the rate of secretion.
- A high plasma K^+ concentration promotes K^+ secretion.
- Increased amounts of Na^+ in the collecting duct lumen (e.g., as a result of inhibition of Na^+ reabsorption by a loop diuretic drug) result in increased entry of Na^+ into principal cells, increased activity of the Na^+/K^+-ATPase, a more negative transepithelial potential, and increased K^+ secretion.

- An increase in permeability of the luminal cell membrane to K^+ favors secretion. This is another effect of aldosterone, which plays a key role in K^+ homeostasis (see Chapter 23).
- A high fluid flow rate through the collecting duct lumen maintains the cell-to-lumen concentration gradient, which favors K^+ secretion.

▶ URINARY CONCENTRATION MECHANISMS

The ability of the human kidney to concentrate and produce hyperosmotic urine is a major determinant in survival. This ability is the effect, or a reflection of, the kidney's ability to maximize reabsorption or "reclamation" of water and defend against dehydration. The human kidneys can reabsorb water to the extent of being able to form a maximal urine concentration of 1,200 to 1,400 mOsm/L, which is up to five times that of plasma (~290 mOsm/L). In this section, we describe the mechanisms by which the hyperosmolarity of the urine is accomplished.

The kidney's ability to concentrate urine osmotically is an important adaptation for survival.

The kidneys have the task of concentrating the urine to save water and at the same time get rid of excess solutes (e.g., urea and various salts), which requires the excretion of water. Suppose, for example, the kidneys were only capable of excreting urine that is isosmotic to plasma (300 mOsm/L), then we would need to excrete remaining solutes in 2.0 L H_2O/d. If we can excrete the solutes in urine that is four times more concentrated than plasma (1,200 mOsm/L), then only 0.5 L H_2O/d would be required. By excreting solutes in osmotically concentrated urine, the kidneys in effect save 1.5 L H_2O (2.0 to 0.5 L H_2O) for the body. The ability to concentrate the urine decreases the amount of water we are obliged to find and drink each day.

Urine-concentrating ability can be looked at in two ways. We can determine what the urine osmolality (or specific gravity) is compared to the plasma, or the U_{osm}/P_{osm} ratio. In people, a maximal value is about 4 to 5, a value that might be observed in a dehydrated, otherwise healthy, individual. Or we can calculate how much solute-free water per unit time the kidneys save or eliminate in the urine; this quantity is called the *free water clearance* (or *free water production*), abbreviated C_{H_2O}. C_{H_2O} is calculated from the following equation:

$$C_{H_2O} = \dot{V} - C_{osm} \tag{3}$$

where $\dot{V}$ is the urine flow rate and C_{osm} (the osmolal clearance) is defined as $U_{osm} \times \dot{V}/P_{osm}$. If we factor out $\dot{V}$ in equation 3, we get:

$$C_{H_2O} = \dot{V}(1 - U_{osm}/P_{osm}) \tag{4}$$

From Equation 12, we see that if the U_{osm}/P_{osm} ratio is >1 (osmotically concentrated urine), C_{H_2O} is negative; if $U_{osm}/P_{osm} = 1$ (urine isosmotic to plasma), then C_{H_2O} is zero; and if U_{osm}/P_{osm} is <1 (osmotically dilute urine), then C_{H_2O} is positive.

Arginine vasopressin is one mechanism that produces osmotically concentrated urine.

Changes in urine osmolality are normally brought about largely by changes in plasma levels of **arginine vasopressin** (**AVP**), also known as **antidiuretic hormone**, or **ADH** (see Chapter 31). In the absence of AVP, the kidney collecting ducts are relatively water impermeable. Continual active reabsorption of NaCl across a water-impermeable epithelium leads to further reduction of the already hypotonic tubular fluid and results in production of osmotically dilute urine (~70 mOsm) with a volume of almost 15% of the original filtered water at the glomerulus. In the presence of AVP, collecting duct water permeability is increased. Because the medullary interstitial fluid is hyperosmotic, water reabsorption in the medullary collecting ducts can lead to the production of osmotically concentrated urine (~1,200 mOsm in a volume of ~0.5% of the original filtered water).

A model for the action of AVP on cells of the collecting duct is shown in Figure 22.17. When plasma osmolality is increased, plasma AVP levels increase. The hormone binds to a specific vasopressin (V_2) receptor in the basolateral cell membrane of principal cells. By way of a guanine nucleotide stimulatory protein (G_s), the membrane-bound enzyme adenylyl cyclase is activated. This enzyme catalyzes the formation of cyclic AMP (cAMP) from ATP. cAMP then activates a cAMP-dependent protein kinase (protein kinase A, or *PKA*) that phosphorylates other proteins. This leads to the insertion, by exocytosis, of intracellular vesicles that contain water channels (aquaporin-2) into the luminal cell membrane. The resulting increase in number of luminal membrane water channels leads to an increase in water permeability. Water leaves the lumen and then exits the cells via aquaporin-3 and aquaporin-4 in the basolateral cell membrane.

The solutes in the collecting duct lumen become concentrated as water leaves. This response to AVP occurs in minutes. AVP also has delayed effects on collecting ducts; it increases the transcription of aquaporin-2 genes and increases the total number of aquaporin-2 molecules per cell.

Countercurrent multiplication in the loops of Henle is the underlying mechanism for urine concentration.

The vertical osmotic gradient in the kidney medulla is created and maintained by the *countercurrent mechanism*. Two countercurrent processes occur in the kidney medulla—**countercurrent multiplication** and **countercurrent exchange**. The term *countercurrent* indicates a flow of fluid in opposite directions in adjacent structures (Fig. 22.18). The loops of Henle are *countercurrent multipliers*. Fluid flows toward the tip of the papilla along the descending limb of the loop and toward the cortex along the ascending limb of the loop. The loops of Henle create the osmotic gradient in the medulla (see below). The vasa recta are *countercurrent exchangers*. Blood flows in opposite directions along juxtaposed descending (arterial) and ascending (venous) vasa recta, and solutes and water are exchanged passively between these capillary blood vessels. The vasa recta help maintain the gradient in the medulla. The collecting ducts act as *osmotic equilibrating devices*; depending on the plasma level of AVP,

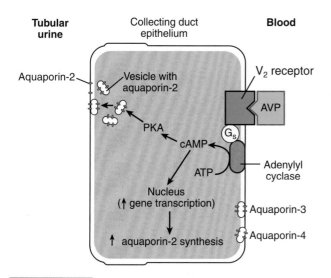

Figure 22.17 **Model for the action of arginine vasopressin (AVP) on the epithelium of the collecting duct.** The second messenger for AVP is cyclic adenosine monophosphate (cAMP). AVP has both prompt effects on luminal membrane water permeability (the movement of aquaporin-2–containing vesicles to the luminal cell membrane) and delayed effects (increased aquaporin-2 synthesis). ATP, adenosine triphosphate; G_s, guanine nucleotide stimulatory protein; PKA, protein kinase A; V_2, type 2 vasopressin receptor.

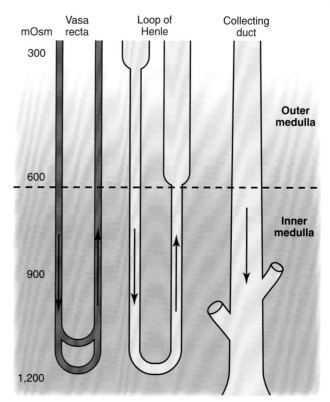

Figure 22.18 **Elements of the urinary concentrating mechanism.** The vasa recta are countercurrent exchangers, the loops of Henle are countercurrent multipliers, and the collecting ducts are osmotic equilibrating devices. Most loops of Henle and vasa recta do not reach the tip of the papilla but turn at higher levels in the outer and inner medulla. There are no thick ascending limbs in the inner medulla.

the collecting duct urine is allowed to equilibrate more or less with the hyperosmotic medullary interstitial fluid.

Countercurrent multiplication is the process whereby a modest gradient established at any level of the loop of Henle is increased (multiplied) into a much larger gradient along the axis of the loop. A simplified model for countercurrent multiplication in the loop of Henle from a cortical nephron shows how this works (Fig. 22.19). Initially, the loop is filled with fluid isosmotic to plasma (~300 mOsm for this example; Fig. 22.19A). Next, we assume that at any level of the loop, the thick ascending limb of the loop can establish an osmotic gradient of 200 mOsm between the tubular and peritubular fluid (Fig. 22.19B). This so-called single effect occurs by active transport of solute (salt) out of the water-impermeable ascending limb and deposition of the salt in the tiny interstitial space. This osmotic gradient causes water to leave the water-permeable descending limb, which equilibrates osmotically with the interstitial space. Next, we add new fluid to the loop and push the fluid in the loop around the bend (Fig. 22.19C). We repeat the single effect (Fig. 22.19D) and continue the process (Fig. 22.19E–H). The final result (see Fig. 22.19H) is a much larger gradient (~400 mOsm/kg H_2O) along the axis of the loop. The interstitial space shares this axial gradient.

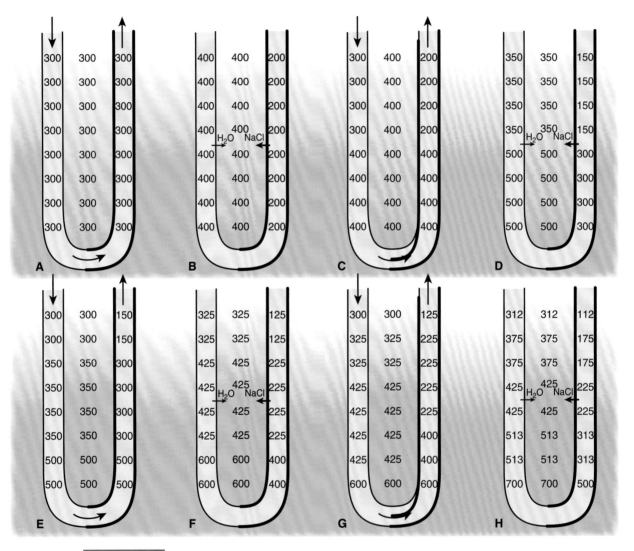

Figure 22.19 **Model for countercurrent multiplication.** The numbers represent osmolality (mOsm/kg H_2O) of tubule fluid and interstitium in a cortical nephron. The establishment of an osmotic gradient along the vertical axis of the loop (or with increasing depth in the medulla) is viewed as the resultant of two successive processes: (1) a shift of fluid within the loop (**A, C, E,** and **G**) and (2) development of an osmotic gradient of 200 mOsm/kg H_2O at any horizontal level of the loop, the so-called single effect (**B, D, F,** and **H**). The single effect involves active transport of solute (mostly NaCl) out of the ascending limb across a water-impermeable barrier (the latter is indicated by heavy outlining along the ascending limb of the loop of Henle) into a tiny interstitial space and osmotic withdrawal of water from the water-permeable descending limb. The single effect of 200 mOsm/kg H_2O is multiplied (magnified) into a larger (~400 mOsm/kg H_2O) gradient along the length of the loop by a stepwise shift of fluid, countercurrent flow, and repetition of the single effect. The interstitium of the medulla shares in the increased osmolality.

The countercurrent multiplication mechanism in juxtamedullary nephrons requires urea.

In the kidney, the highest osmolalities are reached at the bends of the longest loops of Henle belonging to juxtamedullary nephrons (i.e., deep within the inner medulla at the tips of the renal papillae). The countercurrent mechanism in these loops is slightly different than that just described for cortical nephrons. In juxtamedullary nephrons, sodium reabsorption in the thin ascending limb is passive, not active. Passive reabsorption of NaCl in the thin ascending limb occurs down concentration gradients between the tubular and peritubular fluid as the tubular fluid ascends through lower and lower peritubular NaCl concentrations. However, the very high tubular NaCl concentration at the start of the thin ascending limb gets that way by passive reabsorption of water from the thin descending limb, which in turn needs the vertical osmotic gradient in the medullary interstitium to draw water out of it in order to concentrate NaCl within the tubule. Thus, it would appear, paradoxically, that the countercurrent exchange mechanism in this portion of the juxtamedullary nephron, needed to set up the vertical osmotic gradient deep in the medulla, needs the vertical osmotic gradient in the medulla to start the exchange! This problem is solved by the addition of large quantities of urea into the deep medullary interstitium by the following mechanism. In the presence of ADH, water is reabsorbed by the collecting ducts, as described above. The collecting ducts above the level of the papillary medulla are impermeable to urea. Thus, water reabsorption above the papillary medulla concentrates urea within the duct. This level approaches 600 mM. ADH, however, increases the permeability of the papillary collecting duct to urea by stimulating facilitated transporters for urea there (UT-A1). Thus, initially, urea flows out of the duct at that location and into the papillary medullary interstitium, greatly increasing peritubular fluid osmolality there. It is this initial increase in osmolality that draw water out of the thin descending limb, concentrating NaCl there, which then powers the passive "reabsorption" of NaCl up the thin ascending limb. Thus, urea transport into the papillary medullary interstitium "kick starts" countercurrent exchange in juxtamedullary nephrons.

Countercurrent multiplication in all loops of Henle is ultimately dependent on energy derived from ATP, which powers active pumping of Na^+ by the Na^+/K^+-ATPase in the thick ascending limb. However, energy needed to power the single effect there is much less than that which would be needed to pump sodium directly against transtubular gradients seen at the tips of the Loop. Thus, the countercurrent exchanger builds very large osmotic concentrations deep in the kidney at a fraction of the energy that would otherwise be required by a direct active transtubular transport system there.

The extent to which countercurrent multiplication can establish a large axial gradient in a model of the kidney depends on several factors, including the magnitude of the single effect, the rate of fluid flow, and the length of the loop. The larger the single effect, the larger the axial gradient. If flow rate through the loop is too high, not enough time is allowed for establishing a significant single effect, and consequently, the axial gradient is reduced. Finally, if the loops are long, there is more opportunity for multiplication, and a larger axial gradient can be established.

Countercurrent exchange and the vasa recta maintain the vertical osmotic gradient in the renal medullary interstitium.

Countercurrent exchange is a common process in the vascular system. In many vascular beds, arterial and venous vessels lie close to each other, and exchanges of heat or materials can occur between these vessels. For example, because of the countercurrent exchange of heat between blood flowing toward and away from its feet, a penguin can stand on ice and yet maintain a warm body (core) temperature. Countercurrent exchange between descending and ascending vasa recta in the kidney reduces dissipation of the solute gradient in the medulla (see Fig. 22.18). The descending vasa recta tend to give up water to the more concentrated interstitial fluid. The ascending vasa recta, which come from more concentrated regions of the medulla, take up this water. In effect, then, much of the water in the blood short-circuits across the tops of the vasa recta and does not flow deep into the medulla, where it would tend to dilute the accumulated solute. The ascending vasa recta tend to give up solute as the blood moves toward the cortex. Solute enters the descending vasa recta and therefore tends to be trapped in the medulla. Countercurrent exchange is a purely passive process; it helps maintain a gradient established by some other means.

The generation of urine hyperosmolarity requires the integrated functioning of the loops of Henle, vasa recta, and collecting ducts.

Figure 22.20 summarizes the mechanisms involved in producing osmotically concentrated urine. A maximally concentrated urine, with an osmolality of 1,200 mOsm/kg H_2O and a low urine volume (0.5% of the original filtered water), is being excreted.

About 70% of filtered water is reabsorbed along the proximal convoluted tubule, and so, 30% of the original filtered volume enters the loop of Henle. As discussed earlier, proximal reabsorption of water is essentially an isosmotic process, so fluid entering the loop is isosmotic to plasma. As the fluid moves along the descending limb of the loop of Henle in the medulla, it becomes increasingly concentrated. The removal of water along the descending limb leads to a rise in NaCl concentration in the loop fluid to a value higher than in the interstitial fluid. In the thin ascending limb, NaCl is passively reabsorbed across a water-impermeable epithelium, and the NaCl is deposited in the medullary interstitial fluid. In the thick ascending limb, Na^+ transport is active and is powered by a vigorous Na^+/K^+-ATPase. The net addition of solute to the medulla by the loops is essential for the subsequent osmotic concentration of urine in the collecting ducts.

Fluid entering the distal convoluted tubule is hypoosmotic compared with plasma (see Fig. 22.20) because of the removal of solute without water along the ascending limb. In the presence of AVP (ADH), the cortical collecting ducts

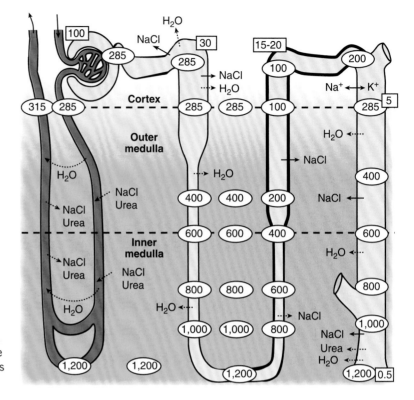

Figure 22.20 **Formation of osmotically con-centrated urine.** This diagram summarizes movements of ions, urea, and water in the kidney during production of maximally concentrated urine (1,200 mOsm/kg H₂O). Numbers in *ovals* represent osmolality in mOsm/kg H₂O. Numbers in *boxes* represent relative amounts of water present at each level of the nephron. *Solid arrows* indicate active transport; *dashed arrows* indicate passive transport. The *heavy outlining* along the ascending limb of the loop of Henle and distal convoluted tubule indicates relative water impermeability.

become water permeable and water is passively reabsorbed into the cortical interstitial fluid. The high blood flow to the cortex rapidly carries away this water, and so, there is no detectable lowering of cortical tissue osmolality. Before the tubular fluid reenters the medulla, it is isosmotic, and its volume is only about 5% of the original filtered volume. The reabsorption of water in the cortical collecting ducts is important for the overall operation of the urinary concentrating mechanism. If this water were not reabsorbed in the cortex, an excessive amount would enter the medulla. It would tend to wash out the gradient in the medulla, leading to an impaired ability to concentrate the urine maximally.

All nephrons drain into collecting ducts that pass through the medulla. In the presence of AVP, the medullary collecting ducts are permeable to water. Water moves out of the collecting ducts into the more concentrated interstitial fluid. At high levels of AVP, the fluid equilibrates with the interstitial fluid, and the final urine becomes as concentrated as the tissue fluid at the tips of the papillae.

Many different models for the countercurrent mechanism have been proposed; each must take into account the principle of conservation of matter (mass balance). In the steady state, the inputs of water and every nonmetabolized solute must equal their respective outputs. This principle must be obeyed at every level of the medulla. Figure 22.21 presents a simplified scheme that applies the mass balance principle to the medulla as a whole. It provides some additional insights into the countercurrent mechanism. Notice that fluids entering the medulla (from the proximal tubule, descending vasa recta, and at some point in the cortical collecting ducts) are isosmotic; they all have an osmolality of about 285 mOsm/kg H₂O. Fluid leaving the medulla in the urine is hyperosmotic. It follows from mass balance

considerations that somewhere a hypo-osmotic fluid has to leave the medulla; this occurs in the ascending limb of the loop of Henle.

The input of water into the medulla must equal its output. Because water is added to the medulla along the descending limbs of the loops of Henle and the collecting ducts, this water must be removed at an equal rate. The ascending limbs of the loops of Henle cannot remove the added water because they are water impermeable. The water is removed by the vasa recta; this is why blood flow in the ascending vasa recta exceeds blood flow in the descending vasa recta (see Fig. 22.21). Blood leaving the medulla is hyperosmotic because it drains a region of high osmolality.

A low-protein diet is associated with a decreased urea concentration in the kidney medulla. It has been known for many years that animals or humans on such a diet have an impaired ability to concentrate the urine maximally, demonstrating a key role for this solute in the urine-concentrating ability of the kidney.

Figure 22.22 shows how urea is handled along the nephron, expressed as a percent of the filtered load of urea. The proximal convoluted tubule is fairly permeable to urea and reabsorbs about 50% of the filtered urea. Fluid collected from the distal convoluted tubule, however, has as much urea as the amount filtered. Therefore, urea is secreted in the loop of Henle as described in previous sections.

The thick ascending limb, distal convoluted tubule, connecting tubule, cortical collecting duct, and outer medullary collecting duct are relatively impermeable to urea. As described earlier in the chapter, the urea concentration rises as water is reabsorbed along cortical and outer medullary collecting ducts. AVP (ADH) stimulates UT-A1 transporters in the deep medullary collecting duct allowing urea diffusion

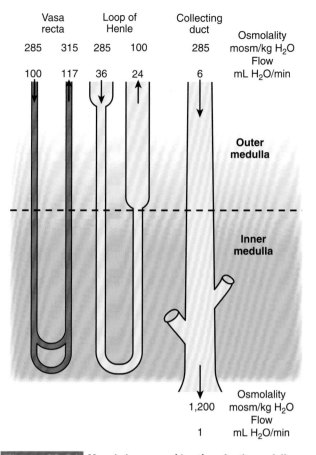

Figure 22.21 Mass balance considerations for the medulla as a whole. In the steady state, the inputs of water and solutes must equal their respective outputs. Water input into the medulla from the cortex (100 + 36 + 6 = 142 mL/min) equals water output from the medulla (117 + 24 + 1 = 142 mL/min). Solute input (28.5 + 10.3 + 1.7 = 40.5 mOsm/min) is likewise equal to solute output (36.9 + 2.4 + 1.2 = 40.5 mOsm/min).

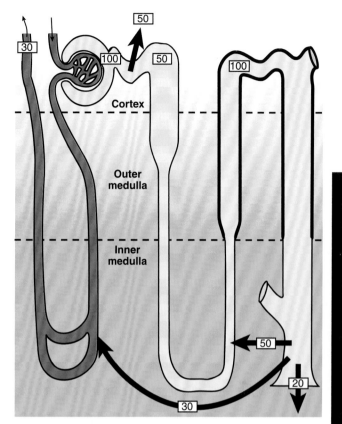

Figure 22.22 Movements of urea along the nephron. The numbers indicate relative amounts (100 = filtered urea), not concentrations. The heavy outline from the thick ascending limb to the outer medullary collecting duct indicates relatively urea-impermeable segments. Urea is added to the inner medulla by its collecting ducts through UT-A1 channels, which are stimulated by AVP; most of this urea reenters the loop of Henle, and the vasa recta remove some.

into the interstitial fluid of the inner medulla. The result is the delivery to the inner medulla of a concentrated urea solution. Urea may re-enter the loop of Henle and be recycled (Fig. 22.22), thereby building up its concentration in the inner medulla. Urea is also added to the inner medulla by diffusion from the urine surrounding the papillae (calyceal urine). Urea accounts for about half of the osmolality in the inner medulla. The urea in the interstitial fluid of the inner medulla counterbalances urea in the collecting duct urine, allowing the other solutes (e.g., NaCl) in the interstitial fluid to counterbalance osmotically the other solutes (e.g., creatinine and various salts) that need to be concentrated in the urine. This enhances the urinary concentrating ability and allows urea to be excreted with less water.

Dilute urine is excreted when plasma AVP levels are low.

Figure 22.23 depicts osmolalities during excretion of dilute urine, as occurs when plasma AVP levels are low. Tubular fluid is diluted along the ascending limb and becomes even more dilute as solute is reabsorbed across the relatively water-impermeable distal portions of the distal convoluted

tubules and collecting ducts. Because as much as 15% of filtered water is not reabsorbed, high urine flow rate results. In these circumstances, the osmotic gradient in the medulla is reduced but not abolished. The decreased gradient results from several factors. First, medullary blood flow is increased during a diuresis, which tends to wash out the osmotic gradient. Second, less urea is added to the inner medulla interstitium because the urea in the collecting duct urine is less concentrated than usual and there is less of a concentration gradient for passive reabsorption of urea. Furthermore, when AVP levels are low, the urea permeability of the inner medulla collecting ducts are low but not zero, and thus, some urea is able to be passively reabsorbed by the medullary collecting ducts. Finally, because of diminished water reabsorption in the cortical collecting ducts, too much water may enter and be reabsorbed in the medulla, lowering its osmotic gradient.

▶ RENAL CLEARANCE AND ASSESSING GLOMERULAR FUNCTION

One of the key functions of the kidneys is their ability to remove or "clear" substances from plasma as a step in their elimination by excretion in the urine. **Renal clearance** of a

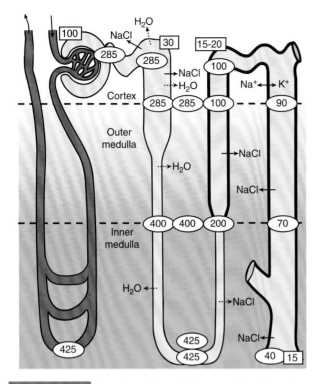

Figure 22.23 **Osmotic gradients during excretion of osmotically dilute urine.** The collecting ducts are relatively water impermeable (heavy outlining) because arginine vasopressin (AVP) is absent. The medulla is still hyperosmotic, but less so than in a kidney producing osmotically concentrated urine.

substance is a key clinical measurement in pharmacotherapy and toxicology where quantification of such processes help the physician understand changes in drug concentrations in the plasma over time and the effectiveness of therapies used to remove toxins from the body. Renal clearance can also be used as an indirect way to measure glomerular function. These tests measure the rates of glomerular filtration, RBF, and tubular reabsorption or secretion of various substances. Measurement of glomerular filtration rate (GFR), in particular, is routinely used to evaluate kidney function.

However, clearance is an unusual measurement and often leads to confusion for students. Clearance is based on the concept that when a substance is excreted in the urine, it is as if a certain volume of plasma was completely cleared of the substance. Thus, renal clearance is defined as the volume of plasma from which a substance is completely removed (cleared) by the kidney per unit time (usually a minute), even though for many substances, the kidney does not actually remove *all* of that substance from the plasma (e.g., one can calculate the renal clearance of sodium, urea, drugs, etc., even though over the period of measurement, the kidney doesn't eliminate them completely from the plasma). For example, the clearance of urea is 65 mL/min. This means that the kidney removed urea from the plasma at a rate equivalent to that which would occur if all of the urea in 65 mL of plasma was removed in 1 minute. In the previous section, urea was shown to play a unique role in concentrating urine and urea clearance assumes physiological importance if the urea clearance is too high or too low.

The formula for renal clearance is:

$$C_X = \frac{U_X \times \dot{V}}{P_X} \qquad (5)$$

where X is the substance of interest, C_X is the clearance of substance X, U_X is the urine concentration of substance X, P_X is the plasma concentration of substance X, and $\dot{V}$ is the urine flow rate. The product of U_X times $\dot{V}$ equals the excretion rate and has dimensions of amount per unit time (e.g., mg/min or mEq/d). Thus, clearance equals the urinary excretion rate divided by plasma concentration. The clearance of a substance can easily be determined by measuring the concentrations of a substance in urine and plasma and the urine flow rate (urine volume/time of collection) and substituting these values into the clearance formula.

Inulin clearance equals the glomerular filtration rate.

An important measurement in the evaluation of kidney function is the **glomerular filtration rate (GFR)**, the rate at which plasma is filtered by the kidney glomeruli. GFR is equal to the sum of the filtration rates of all functional nephrons and is used as index of renal function. A decrease in GFR generally indicates that kidney function is impaired. Thus, GFR is an important diagnostic tool to evaluate kidney disease.

If we had a substance that was cleared from the plasma only by glomerular filtration, it could be used to measure GFR. The ideal substance to measure GFR is **inulin**, a fructose polymer with a molecular weight of about 5,000. Inulin (IN) is suitable for measuring GFR for the following reasons: (1) IN is not bound to plasma proteins and is thus freely filtered by the glomeruli; (2) IN is not reabsorbed or secreted by the kidney tubules; (3) IN is not synthesized, destroyed, or stored in the kidneys; (4) IN is not toxic; and (5) IN concentration in plasma and urine can be determined by simple analysis.

The principle behind the use of IN is illustrated in Figure 22.24. The amount of IN filtered per unit time, the **filtered load**, is equal to the product of the plasma concentration of IN (P_{IN}) times GFR. The rate of IN excretion is equal to the urine IN concentration (U_{IN}) times $\dot{V}$, the

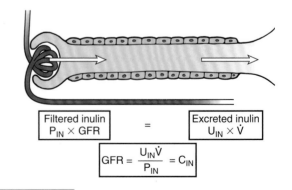

Figure 22.24 **The principle behind the measurement of glomerular filtration rate (GFR).** P_{IN}, plasma inulin concentration; U_{IN}, urine inulin concentration; $\dot{V}$, urine flow rate; C_{IN}, inulin clearance.

urine flow rate. Because IN is not reabsorbed, secreted, synthesized, destroyed, or stored by the kidney tubules, the filtered IN load equals the rate of IN excretion. The equation can be rearranged by dividing by the plasma IN concentration. The expression $U_{IN}\dot{V}/P_{IN}$ is defined as the **inulin clearance**. Therefore, IN clearance equals GFR. Normal values for IN clearance or GFR (corrected to a body surface area of 1.73 m^2) are 110 ± 15 (SD) mL/min for young adult women and 125 ± 15 mL/min for young adult men. In newborns, even when corrected for body surface area, GFR is low, about 20 mL/min per 1.73 m^2 body surface area. Adult values (when corrected for body surface area) are attained by the end of the first year of life. If GFR is 125 mL plasma/min, then the volume of plasma filtered in a day is 180 L (125 mL/min × 1,440 min/d). After the age of 45 to 50, GFR declines, and it is typically reduced by 30% to 40% by age 80.

Plasma creatinine clearance is used clinically to estimate GFR.

IN clearance is the gold standard for measuring GFR and is used whenever highly accurate measurements of GFR are desired. It would be simpler, however, to use an endogenous substance (i.e., one native to the body) that is only filtered, is excreted in the urine, and normally has a stable plasma value that can be accurately measured. There is no such known substance, but creatinine comes close.

Creatinine is an end product of muscle metabolism, a derivative of muscle creatine phosphate. It is produced continuously in the body and is excreted in the urine. Long urine collection periods (e.g., a few hours) can be used, because creatinine concentrations in the plasma are normally stable and creatinine does not have to be infused; consequently, there is no need to catheterize the bladder. Plasma and urine concentrations can be measured using a simple colorimetric method. The **endogenous creatinine clearance** is calculated from the formula

$$C_{CREATININE} = \frac{U_{CREATININE} \times \dot{V}}{P_{CREATININE}} \quad (6)$$

or creatinine clearance equals the ratio of urine to plasma creatinine concentration times the urine flow rate. There are two potential drawbacks of using creatinine to measure GFR. First, creatinine is not only filtered but also secreted by the human kidney. This elevates urinary excretion of creatinine, normally causing a 20% increase in the numerator of the clearance formula. The second drawback is related to errors of measuring creatinine concentration in the plasma. The colorimetric method usually used also measures other plasma substances, such as glucose, leading to a 20% increase in the denominator of the clearance formula. Because both numerator and denominator are 20% too high, the two errors cancel, and so, the endogenous creatinine clearance fortuitously affords a good approximation of GFR when it is about normal. However, when GFR in an adult has been reduced to about 20 mL/min because of renal disease, the endogenous creatinine clearance may overestimate the GFR by as much as 50%. This results from

higher plasma creatinine levels and increased tubular secretion of creatinine. Drugs that inhibit tubular secretion of creatinine or elevated plasma concentrations of chromogenic (color-producing) substances other than creatinine may cause the endogenous creatinine clearance to underestimate GFR.

Plasma creatinine concentration can be used to estimate GFR.

Because the kidneys continuously clear creatinine from the plasma by excreting it in the urine, the GFR and plasma creatinine concentration are inversely related. Figure 22.25 shows the steady-state relationship between these variables—that is, when creatinine production and excretion are equal. If the GFR is halved from a normal value of 180 to 90 L/d, it would result in a doubling of plasma creatinine concentration from a normal value of 1 to 2 mg/dL after a few days. A reduction in GFR from 90 to 45 L/d results in a greater increase in plasma creatinine, from 2 to 4 mg/dL. Figure 22.25 shows that with low GFR values, small absolute changes in GFR lead to much greater changes in plasma creatinine concentration than at high GFR values.

The inverse relationship between GFR and plasma creatinine concentration allows the use of plasma or serum creatinine to estimate GFR, provided that these three cautions are kept in mind:

1. It takes a certain amount of time for changes in GFR to produce detectable changes in plasma creatinine concentration.

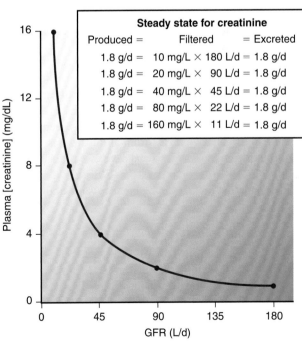

Steady state for creatinine		
Produced =	Filtered	= Excreted
1.8 g/d =	10 mg/L × 180 L/d	= 1.8 g/d
1.8 g/d =	20 mg/L × 90 L/d	= 1.8 g/d
1.8 g/d =	40 mg/L × 45 L/d	= 1.8 g/d
1.8 g/d =	80 mg/L × 22 L/d	= 1.8 g/d
1.8 g/d =	160 mg/L × 11 L/d	= 1.8 g/d

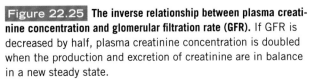

Figure 22.25 **The inverse relationship between plasma creatinine concentration and glomerular filtration rate (GFR).** If GFR is decreased by half, plasma creatinine concentration is doubled when the production and excretion of creatinine are in balance in a new steady state.

2. Plasma creatinine concentration is also influenced by muscle mass. A young, muscular man will have a higher plasma creatinine concentration than an older woman with reduced muscle mass.

3. Some drugs inhibit tubular secretion of creatinine, leading to a raised plasma creatinine, even though GFR may be unchanged.

The relationship between plasma creatinine and GFR is one example of how a substance's plasma concentration can depend on GFR. The same relationship is observed for several other substances whose excretion depends on GFR. For example, the plasma urea concentration or blood urea nitrogen rises when GFR falls. The plasma level of a 13-kDa protein molecule called *cystatin C* also rises when GFR falls, and it has been suggested that serum cystatin C levels can be used to estimate GFR.

Several empirical equations have been developed that allow physicians to estimate GFR from serum creatinine concentration. These equations often take into consideration such factors as age, gender, race, and body size. The equation for adults, GFR (in mL/min per 1.73 m²) = 186 × (serum creatinine in mg/dL)$^{-1.154}$ × (age in years)$^{-0.203}$ × 0.742 (if the subject is female) or × 1.212 (if the subject is black), is recommended by the National Kidney Disease Education Program, which provides GFR calculators on its Web site: www.nih.nkdep.gov. Staging of chronic kidney disease is usually based on estimated GFR measurements; a value <60 mL/min may indicate renal disease.

Para-aminohippurate clearance estimates renal plasma flow.

Renal blood flow (**RBF**) can be determined from measurements of **renal plasma flow** (**RPF**) and blood hematocrit, using the following equation:

$$RBF = RPF/(1 - Hematocrit) \qquad (7)$$

The hematocrit is easily determined by centrifuging a blood sample. RPF is estimated by measuring the clearance of the organic anion *p*-aminohippurate (PAH), infused intravenously. PAH is filtered and so vigorously secreted that it is nearly completely cleared from all of the plasma flowing through the kidneys. The renal clearance of PAH, at low plasma PAH levels, approximates the RPF.

The equation for calculating the true value of the RPF is

$$RPF = C_{PAH}/E_{PAH} \qquad (8)$$

where C_{PAH} is the PAH clearance and E_{PAH} is the extraction ratio (see Chapter 15) for PAH—the difference between the arterial and renal venous plasma PAH concentrations $(P_{PAH}^a - P_{PAH}^{rv})$ divided by the arterial plasma PAH concentration (P_{PAH}^a). The equation is derived as follows. In the steady state, the amounts of PAH per unit time entering and leaving the kidneys are equal. The PAH is supplied to the kidneys in the arterial plasma and leaves the kidneys in urine and renal venous plasma, or:

$$PAH \text{ entering kidneys} = PAH \text{ leaving kidneys}$$
$$RPF \times P_{PAH}^a = U_{PAH} \times \dot{V} + RPF \times P_{PAH}^{rv} \qquad (9)$$

Rearranging, we get:

$$RPF = U_{PAH} \times \dot{V}/(P_{PAH}^a - P_{PAH}^{rv}) \qquad (10)$$

If we divide the numerator and denominator of the right side of the equation by P_{PAH}^a, the numerator becomes C_{PAH} and the denominator becomes E_{PAH}.

If we assume extraction of PAH is 100% ($E_{PAH} = 1.00$), then the RPF equals the PAH clearance. When this assumption is made, the RPF is usually called the *effective RPF*, and the blood flow calculated is called the *effective RBF*. The extraction of PAH by healthy kidneys at low plasma PAH concentrations, however, is not 100% but averages about 91%, so the assumption of 100% extraction results in about a 10% underestimation of the true RPF. To calculate the true RPF or blood flow, it is necessary to sample renal venous blood to measure its plasma PAH concentration, a procedure not often done.

Net tubular reabsorption or secretion of a substance can be calculated from renal clearance.

The rate at which the kidney tubules reabsorb a substance can be calculated if we know how much is filtered and how much is excreted per unit time. If the filtered load of a substance exceeds the rate of excretion, the kidney tubules must have reabsorbed the substance. The equation is

$$T_{reabsorbed} = P_X \times GFR - U_X \times \dot{V} \qquad (11)$$

where T is the tubular transport rate.

The rate at which the kidney tubules secrete a substance is calculated from this equation:

$$T_{secreted} = U_X \times \dot{V} - P_X \times GFR \qquad (12)$$

Note that the quantity excreted exceeds the filtered load, because the tubules secrete X.

In equations 11 and 12, we assume that substance X is freely filterable. If, however, substance X is bound to the plasma proteins, which are not filtered, then it is necessary to correct the filtered load for this binding. For example, about 40% of plasma Ca^{2+} is bound to plasma proteins, and so, 60% of plasma Ca^{2+} is freely filterable.

Equations 11 and 12, which quantify tubular transport rates, yield the *net* rate of reabsorption or secretion of a substance. It is possible for a single substance to be both reabsorbed and secreted; the equations do not give unidirectional reabsorptive and secretory movements, only the net transport.

Renal clearance can estimate glucose reabsorption.

Insights into the nature of glucose handling by the kidneys can be derived from a glucose titration study (Fig. 22.26). The plasma glucose concentration is elevated to increasingly higher levels by the infusion of glucose-containing solutions. IN is infused to permit measurement of GFR and calculation of the filtered glucose load (plasma glucose concentration × GFR). The rate of glucose reabsorption is determined from the difference between the filtered load and the rate of excretion. At normal plasma glucose levels (about 100 mg/dL),

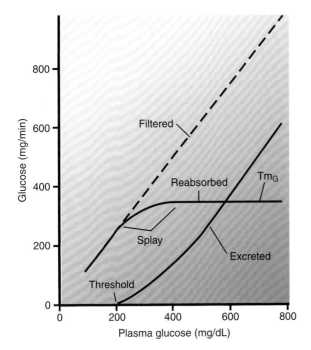

Figure 22.26 **Glucose titration study in a healthy man.** The plasma glucose concentration was elevated by infusing glucose-containing solutions. The amount of glucose filtered per unit time (*top dashed line*) is determined from the product of the plasma glucose concentration and glomerular filtration rate (measured with inulin). Excreted glucose (*bottom line*) is determined by measuring concentration of glucose in the urine and urine flow rate. Reabsorbed glucose is calculated from the difference between filtered and excreted glucose. Tm$_G$, tubular transport maximum for glucose.

all of the filtered glucose is reabsorbed and none is excreted. When the plasma glucose concentration exceeds a certain value (about 200 mg/dL in Fig. 22.26), significant quantities of glucose appear in the urine; this plasma concentration is called the **glucose threshold**. Further elevations in plasma glucose lead to progressively more excreted glucose. Glucose appears in the urine because the filtered amount of glucose exceeds the capacity of the tubules to reabsorb it. At high filtered glucose loads, the rate of glucose reabsorption reaches a constant maximal value, called the **tubular transport maximum (Tm)** for glucose (G). At Tm$_G$, the tubule glucose carriers are all saturated and transport glucose at the maximal rate.

The glucose threshold is not a fixed plasma concentration but depends on three factors: GFR, Tm$_G$, and amount of splay. A low GFR leads to an elevated threshold, because the filtered glucose load is reduced and the kidney tubules can reabsorb all the filtered glucose despite an elevated plasma glucose concentration. A reduced Tm$_G$ lowers the threshold, because the tubules have a diminished capacity to reabsorb glucose.

Splay is the rounding of the glucose reabsorption curve. Figure 22.26 shows that tubular glucose reabsorption does not abruptly attain Tm$_G$ when plasma glucose is progressively elevated. One reason for splay is that not all nephrons have the same filtering and reabsorbing capacities. Thus,

nephrons with relatively high filtration rates and low glucose reabsorptive rates excrete glucose at a lower plasma concentration than nephrons with relatively low filtration rates and high reabsorptive rates. A second reason for splay is the fact that the glucose carrier does not have an infinitely high affinity for glucose, so glucose escapes in the urine even before the carrier is fully saturated. An increase in splay causes a decrease in glucose threshold.

In uncontrolled **diabetes mellitus**, plasma glucose levels are abnormally elevated, so more glucose is filtered than can be reabsorbed. Urinary excretion of glucose, **glucosuria**, produces an osmotic diuresis or an increase in urine output caused by abnormal effective osmoles in the tubular fluid of the nephron. In osmotic diuresis, the increased urine flow results from the excretion of osmotically active solute. *Diabetes* (from the Greek for "siphon") gets its name from this increased urine output and secondary increased water intake.

Tm for PAH provides a measure for functional proximal secretion.

PAH is secreted only by proximal tubules in the kidneys. At low plasma PAH concentrations, the rate of secretion increases linearly with the plasma concentration of PAH. At high plasma PAH concentrations, the secretory carriers are saturated and the rate of PAH secretion stabilizes at a constant maximal value, called the *tubular transport maximum* for PAH (Tm$_{PAH}$). The Tm$_{PAH}$ is directly related to the number of functioning proximal tubules and therefore provides a measure of the mass of proximal secretory tissue. Figure 22.14 illustrates the pattern of filtration, secretion, and excretion of PAH observed when the plasma PAH concentration is progressively elevated by intravenous infusion.

▶ MICTURITION

Recall that the urinary system includes two kidneys, two ureters, the bladder, the urethra, and two sphincter muscles. The kidneys form urine continuously, and urine flows through the ureters to the bladder. The movement of urine to the bladder is aided by the contraction of the smooth muscles in the ureter wall. The bladder is a balloon-like structure with walls containing smooth muscle that stores the urine. The bladder fills with urine at a low pressure but then empties during urination, or **micturition**. In this section, the function of the urinary tract is discussed.

Urinary tract provides the pathway for transporting, storing, and eliminating urine.

As mentioned previously, the **ureters** are muscular tubes that propel the urine from the pelvis of each kidney to the urinary bladder. Peristaltic movements originate in the region of the calices, which contain specialized smooth muscle cells that generate spontaneous pacemaker potentials. These pacemaker potentials trigger action potentials and contractions in the muscular regions of the renal pelvis that propagate distally to the ureter. Peristaltic waves sweep down the ureters at a frequency of one every 10 seconds to one every 2 to 3 minutes.

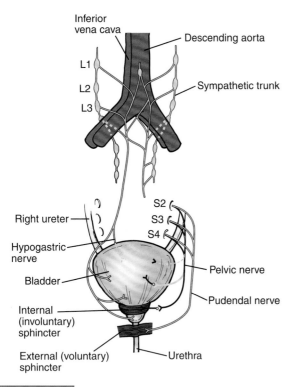

Figure 22.27 **The innervation of the urinary bladder.** The parasympathetic pelvic nerves arise from the S2 to S4 segments of the spinal cord and supply motor fibers to the bladder musculature and internal (involuntary) sphincter. Sympathetic motor fibers supply the bladder via the hypogastric nerves, which arise from lumbar segments of the spinal cord. The pudendal nerves supply somatic motor innervation to the external (voluntary) sphincter. Sensory afferents (*yellow dashed lines*) from the bladder travel mainly in the pelvic nerves but also to some extent in the hypogastric nerves.

The ureters are innervated by sympathetic and parasympathetic nerve fibers. Sensory fibers mediate the intense pain that is felt when a stone distends or blocks a ureter.

The **urinary bladder** is a distensible hollow vessel containing smooth muscle in its wall (Fig. 22.27). The muscle is called the *detrusor*, from Latin for "that which pushes down." The neck of the bladder, the involuntary internal sphincter, also contains smooth muscle. The parasympathetic pelvic nerves and sympathetic hypogastric nerves innervate the body of the bladder and bladder neck. The external sphincter is composed of skeletal muscle and is innervated by somatic nerve fibers that travel in the pudendal nerves. Pelvic, hypogastric, and pudendal nerves contain both motor and sensory fibers.

The bladder has two functions: to serve as a distensible reservoir for urine and to empty its contents at appropriate intervals. When the bladder fills, it adjusts its tone to its content, so that minimal increases in bladder pressure occur. The external sphincter is kept closed by discharges along the pudendal nerves. The first sensation of bladder filling is experienced at a volume of 100 to 150 mL in an adult, and the first desire to void is elicited when the bladder contains about 150 to 250 mL of urine. A person becomes uncomfortably aware of a full bladder when the volume is 350 to 400 mL; at this volume, hydrostatic pressure in the bladder is about 10 cm H_2O. With further volume increases, bladder pressure rises steeply, in part because of reflex contractions of the detrusor. An increase in volume to 700 mL creates pain and, often, loss of control. The sensations of bladder filling, of conscious desire to void, and of painful distention are mediated by afferents in the pelvic nerves.

The **urethra** is a tube that connects the urinary bladder to the exterior of the body. In males, the urethra travels through the penis and carries both urine and semen. In females, the urethra is shorter and emerges in front of the vaginal opening.

Micturition (urination), the periodic emptying of the bladder, is a complex act involving both autonomic and somatic nerve pathways and several reflexes that can be either inhibited or facilitated by higher centers in the brain. The basic reflexes occur at the level of the sacral spinal cord and are modified by centers in the midbrain and cerebral cortex. Distention of the bladder is sensed by stretch receptors in the bladder wall; these induce reflex contraction of the detrusor and relaxation of the internal and external sphincters. This reflex is released by removing inhibitory influences from the cerebral cortex. Fluid flow through the urethra reflexively causes further contraction of the detrusor and relaxation of the external sphincter. Increased parasympathetic nerve activity stimulates contraction of the detrusor and relaxation of the internal sphincter. Sympathetic innervation is not essential for micturition. During micturition, the perineal and levator ani muscles relax, thereby shortening the urethra and decreasing urethral resistance. Descent of the diaphragm and contraction of abdominal muscles raise intra-abdominal pressure and aid in the expulsion of urine from the bladder.

Micturition is, fortunately, under voluntary control in healthy adults. In the young child, however, it is purely reflex and occurs whenever the bladder is sufficiently distended. At about age 2.5 years, it begins to come under cortical control, and, in most children, complete control is achieved by age 3 years.

Damage to the nerves that supply the bladder and its sphincters can produce abnormalities of micturition and incontinence. An increased resistance of the upper urethra commonly occurs in older men and is a result of enlargement of the surrounding prostate gland. This condition is called *benign prostatic hyperplasia* (also known as *BPH*), and it results in decreased urine stream, overdistention of the bladder as a result of incomplete emptying, and increased urgency and frequency of urination.

Urinary incontinence is loss of bladder control.

Urinary **incontinence** is a common and often an embarrassing problem. The severity ranges from an occasional leaking with a sneeze to having an urge to urinate that's so sudden and strong that you lose voluntary control. Incontinence is more common in women. The most common types include *stress incontinence* that results in urine leaks when pressure is exerted on the bladder by coughing, laughing, or lifting a heavy weight. The second type, *urge incontinence*, results from a sudden, intense urge to urinate, followed by an involuntary loss of urine. Urge incontinence often leads to frequent urination, especially throughout the night. The third common type is *overflow incontinence*. This form is characterized by frequent dribbling of urine due to a bladder that doesn't empty completely. Incontinence is aggravated with drinks that contain alcohol or caffeine and certain medications (blood pressure medications, sedatives, and muscle relaxants) because they can act as diuretics.

Often, the cause of incontinence is unknown in individual patients. However, urinary incontinence can be caused by urinary infection, which irritates the bladder, causing strong urges to urinate. Incontinence can also be caused by

physical problems such as pregnancy, childbirth, hysterectomy, and enlarged prostate. During pregnancy, hormonal changes and increased weight of the uterus can lead to stress incontinence. With childbirth, vaginal delivery can weaken muscular control of the bladder and also damage the bladder nerves and supportive tissue. Hysterectomy can lead to damage to muscles supporting the bladder, which can also lead to incontinence. An enlarged prostate (BPH) often causes incontinence in older men.

Finally, neurological disorders, such as multiple sclerosis, Parkinson's disease, stroke, and spinal injury, interfere with neurological control of the bladder, causing urinary incontinence.

INTEGRATED MEDICAL SCIENCES

Glomerulonephritis

A 42-year-old man was admitted to emergency with puffiness of his face, eyes, and trunk. He previously had a 3-month history of intermittent swelling of his ankles and puffiness of his face. Upon exertion, he became short of breath (dyspnea). On examination, he was hypertensive (BP 170/75). His urine showed proteinuria and hematuria with a red cell cast. He was anemic (Hb 10.7 g/dL). His serum albumin was low (2.6 g/dL), and his creatinine clearance was 80 mL/min (NR male: 97 to 137 mL/min.). His serum immunoglobulin IgM and IgA, C3, and C4 levels were normal, but his IgG was low at 5.1 g/L (NR 7.2 to 19.0). Hepatitis B surface antigen was not detected.

These findings are typical of glomerulonephritis, which is an inflammation of the glomeruli. The first indication of glomerular dysfunction is altered fluid retention (edema in abdomen and lower extremities), fatigue from anemia, and altered urine. The dyspnea is the result of pulmonary edema due to fluid retention and high blood pressure. Urine is foamy due to excess protein, and the pink color is from red cells in urine.

Glomerulonephritis can be acute (a sudden inflammatory attack) or chronic that comes on gradually. Nearly all forms of acute glomerulonephritis tend to progress to the chronic stage. The condition is characterized by irreversible and progressive glomerular and tubulointerstitial fibrosis. The condition is triggered by an immunologic reaction causing inflammation and proliferation of glomerular tissue that damages the base membrane and capillary endothelium. The inflammation ultimately leads to a reduction in

the glomerular filtration rate (GFR) and retention of uremic toxins. The low GFR was confirmed with the creatinine clearance test, which provides a good estimate of the GFR. If the progression of the disease is not halted, it leads to chronic kidney disease with severe cardiovascular complications.

The cause of glomerulonephritis is due either to an infection or to an immune disorder. The types of infection include poststreptococcal, bacterial endocarditis, or viral infection. Glomerulonephritis can occur a week or two after recovery from strep throat infection. To fight the streptococcal infection, the body produces extra antibodies that often settle in the glomeruli. If glomerulonephritis occurs on its own, it is known as primary glomerulonephritis. If another disease, such as lupus or diabetes, is the cause, it is called secondary glomerulonephritis. Poststreptococcal renal complications are more likely to occur in children. Bacterial endocarditis is the result of a bacterial infection that spreads through the bloodstream and lodging in the heart, causing infection in one or more of the heart valves. Although there is a connection between endocarditis and glomerular dysfunction, the exact mechanism is not clear. Viral infections, such as the human immunodeficiency virus (HIV) and hepatitis B and C, can trigger glomerulonephritis. Immunological diseases such as lupus and Goodpasture syndrome can trigger glomerulonephritis. Lupus is a chronic inflammatory disease that affects many organs, such as the skin, joints, kidneys, heart, lungs, and blood. Goodpasture syndrome is a rare immunological lung disorder that can mimic pneumonia and causes bleeding in the lungs and glomeruli. ■

Chapter Summary

- The formation of urine involves glomerular filtration, tubular reabsorption, and tubular secretion.
- The kidneys, especially the cortex, have a high blood flow.
- Kidney blood flow is autoregulated; it is also profoundly influenced by nerves and hormones.
- The glomerular filtrate is an ultrafiltrate of plasma.
- Glomerular filtration rate is determined by the glomerular ultrafiltration coefficient, glomerular capillary hydrostatic pressure, hydrostatic pressure in the space of the Bowman capsule, and glomerular capillary colloid osmotic pressure.
- The proximal convoluted tubule reabsorbs about 70% of filtered Na^+, K^+, and water and nearly all of the filtered glucose and amino acids. It also secretes many organic anions and organic cations.
- The transport of water and most solutes across tubular epithelia is dependent on active reabsorption of Na^+.
- The thick ascending limb is a water-impermeable segment that reabsorbs Na^+ via a Na–K–2Cl cotransporter in the apical cell membrane and a vigorous Na^+/K^+-ATPase in the basolateral cell membrane.
- The distal convoluted tubule epithelium is water impermeable and reabsorbs Na^+ via a thiazide-sensitive apical membrane Na–Cl cotransporter.
- Cortical collecting duct principal cells reabsorb Na^+ and secrete K^+.
- The kidneys save water for the body by producing urine with a total solute concentration (i.e., osmolality) greater than that of plasma.
- The loops of Henle are countercurrent multipliers; they set up an osmotic gradient in the kidney medulla. Vasa recta are countercurrent exchangers; they passively help maintain the medullary gradient. Collecting ducts are osmotic equilibrating devices; they have a low water permeability, which is increased by arginine vasopressin.
- The renal clearance of a substance is equal to its rate of excretion divided by its plasma concentration.
- Inulin clearance provides the most accurate measure of glomerular filtration rate.
- The clearance of *p*-aminohippurate is equal to the effective renal plasma flow.
- The rate of net tubular reabsorption of a substance is equal to its filtered load minus its excretion rate. The rate of net tubular secretion of a substance is equal to its excretion rate minus its filtered load.
- The urinary bladder stores urine until it can be conveniently emptied. Micturition is a complex act involving both autonomic and somatic nerves.

Chapter Review Questions

1. An elderly, diabetic woman arrives at the hospital in a severely dehydrated condition and is breathing rapidly. Blood plasma (glucose) is 500 mg/dL (normal is ~100 mg/dL) and the urine (glucose) is zero (dipstick test). What is the most likely explanation for the absence of glucose in her urine?

    A. The amount of splay in the glucose reabsorption curve is abnormally increased.
    B. GFR is abnormally low.
    C. The glucose Tm is abnormally high.
    D. The glucose Tm is abnormally low.
    E. The renal plasma glucose threshold is abnormally low.

The correct answer is B. The patient is elderly and severely dehydrated, so one can expect that the GFR is low. As a consequence, the proximal tubules may be able to reabsorb all of the filtered glucose (because the filtered load is reduced), even though the plasma (glucose) is elevated. If splay is increased, glucose Tm is low, or threshold is low, then glucose should be present (not absent) from the urine. An abnormally high glucose Tm would reduce glucose excretion, but in the scenario presented, this is not a likely cause of the absence of glucose in the urine.

2. In a suicide attempt, a nurse took an overdose of the sedative phenobarbital. This substance is a weak, lipid-soluble organic acid that is reabsorbed by nonionic diffusion in the kidneys. Which of the following would be the best way to promote urinary excretion of this substance?

    A. Abstain from all fluids.
    B. Acidify the urine by ingesting NH_4Cl tablets.
    C. Administer a drug that inhibits tubular secretion of organic anions.
    D. Alkalinize the urine by infusing a $NaHCO_3$ solution intravenously.
    E. Increase the GFR.

The correct answer is D. Excretion of phenobarbital is promoted by increasing urine output and by making the urine more alkaline. The latter would keep phenobarbital in its anionic form, which is not reabsorbed by the kidney tubules. Increasing the GFR would increase the filtration and excretion of phenobarbital but is much less effective than is reducing tubular reabsorption of this compound.

3. A man has progressive, chronic kidney disease. Which of the following indicates the greatest absolute decrease in GFR?

    A. Fall in plasma creatinine from 4 to 2 mg/dL
    B. Fall in plasma creatinine from 2 to 1 mg/dL
    C. Rise in plasma creatinine from 1 to 2 mg/dL
    D. Rise in plasma creatinine from 2 to 4 mg/dL
    E. Rise in plasma creatinine from 4 to 8 mg/dL

The correct answer is C. There is an inverse hyperbolic relationship between plasma (creatinine) and GFR and, therefore, a rise in plasma (creatinine) is associated with a fall in GFR (see Fig. 22.7). The greatest absolute change in GFR occurs when plasma (creatinine) doubles starting from a normal GFR and plasma (creatinine).

Clinical Application Exercises 22.1

NEPHROTIC SYNDROME

A 6-year-old boy is brought to the pediatrician by his mother because of a puffy face and lethargy. A few weeks before, he had an upper respiratory tract infection, probably caused by a virus. Body temperature is 36.8°C, blood pressure 95/65 mm Hg, and heart rate 90 beats/min. Puffiness around the eyes, abdominal swelling, and pitting edema in the legs are observed. A urine sample (dipstick) is negative for glucose but reveals 3+ protein. Microscopic examination of the urine reveals no cellular elements or casts. Plasma $[Na^+]$ is 140 mEq/L, blood urea nitrogen concentration (BUN) 10 mg/dL, glucose 100 mg/dL, creatinine 0.8 mg/dL, serum albumin 2.3 g/dL (normal 3.0 to 4.5 g/dL), and

cholesterol 330 mg/dL. A 24-hour urine sample has a volume of 1.10 L and contains 10 mEq/L Na^+, 60 mg/dL creatinine, and 0.8 g/dL protein.

The child is treated with the corticosteroid prednisone, and the edema and proteinuria disappear in 2 weeks. Puffiness and proteinuria recur 4 months later, and a renal biopsy is done. Glomeruli are normal by light microscopy, but effacement (obliteration) of podocyte foot processes and loss of filtration slits are seen with the electron microscope. No immune deposits or complement are seen after immunostaining. The biopsy indicates minimal change glomerulopathy.

QUESTIONS

1. What features in this case would lead you to suspect the presence of nephrotic syndrome?
2. What is the explanation for the proteinuria?
3. Why does the abnormally high rate of urinary protein excretion underestimate the rate of renal protein loss?
4. What is the endogenous creatinine clearance, and is it normal? (The boy's body surface area is 0.86 m^5.)
5. What is the explanation for the edema?

ANSWERS

1. The child has the classical feature of nephrotic syndrome: heavy proteinuria (8.8 g/d), hypoalbuminemia (<3 g/dL), generalized edema, and hyperlipidemia (plasma cholesterol 330 mg/dL).
2. Proteinuria is a consequence of an abnormally high permeability of the glomerular filtration barrier to the plasma proteins. This condition might be due to an increased physical size of "holes" or pores in the basement membrane and filtration slit diaphragms or a loss of fixed negative charges from the glomerular filtration barrier.
3. Proteins that have leaked across the glomerular filtration barrier are not only excreted in the urine but are reabsorbed by proximal tubules. The endocytosed proteins are digested in lysosomes to amino acids, which are returned to the circulation. Both increased renal catabolism by tubule cells and increased excretion of serum albumin in the urine contribute to the hypoalbuminemia. The liver, which synthesizes serum albumin, cannot keep up with the renal losses.
4. The endogenous creatinine (CR) clearance (an estimate of GFR) equals $(U_{CR} \times V)/P_{CR} = (60 \times 1.10)/0.8 = 82$ L/d. Normalized to a standard body surface area of 1.73 m^5, C_{CR} is 166 L/d per 1.73 m^5, which falls within the normal range (150 to 210 L/d per 1.73 m^5). Note that the permeability of

the glomerular filtration barrier to macromolecules (plasma proteins) was abnormally high, but permeability to fluid was not increased. In some patients, a loss of filtration slits may be marked and may lead to a reduced fluid permeability and GFR.

5. The edema is due to altered capillary Starling forces and renal retention of salt and water. The decline in plasma (protein) lowers the plasma colloid osmotic pressure, favoring fluid movement out of the capillaries into the interstitial compartment. The edema is particularly noticeable in the soft skin around the eyes (periorbital edema). The abdominal distention (in the absence of organ enlargement) suggests ascites (an abnormal accumulation of fluid in the abdominal cavity). The kidneys avidly conserve Na^+ (note the low urine $[Na^+]$) despite an expanded extracellular fluid (ECF) volume. Although the exact reasons for renal Na^+ retention are controversial, a decrease in the effective arterial blood volume may be an important stimulus (see Chapter 23). This leads to activation of the renin–angiotensin–aldosterone system and stimulation of the sympathetic nervous system, both of which favor renal Na^+ conservation. In addition, distal segments of the nephron reabsorb more Na^+ than usual because of an intrinsic change in the kidneys.

thePoint® *Visit* http://thepoint.lww.com/rhoades5e *for additional chapter review Q&A, Clinical Application Exercises, animations, and more!*

23 Regulation of Fluid and Electrolyte Balance

Active Learning Objectives

Upon mastering the material in this chapter, you should be able to:

- Discuss the interrelationships among the major fluid compartments of the body and contrast their relative volumes for an average young adult man and woman.
- Explain the indicator–dilution principle and its application to the measurement of body fluid volumes.
- Describe and contrast the ionic composition of extracellular and intracellular fluids.
- Explain how arginine vasopressin and thirst regulate water balance.
- Compare the amounts of sodium filtered and excreted in a day. Compare the percentage of filtered sodium reabsorbed by the proximal convoluted tubule, loop of Henle, distal convoluted tubule, and collecting ducts.

- Explain how the following affect renal sodium excretion: glomerular filtration rate, the renin–angiotensin–aldosterone system, intrarenal pressure, natriuretic hormones and factors, renal sympathetic nerve stimulation, estrogens, glucocorticoids, osmotic diuretics, poorly reabsorbed anions, and diuretic drugs.
- Explain the relation between sodium chloride, extracellular fluid volume (or effective arterial blood volume), and blood pressure.
- Discuss the importance, amount, and distribution of potassium in the body.
- Explain the mechanisms that affect renal excretion of potassium.
- Explain how the kidneys contribute to calcium, magnesium, and phosphate balance.
- Explain the effects of parathyroid hormone on renal tubular reabsorption of calcium and phosphate and on renal synthesis of $1,25(OH)_2$ vitamin D_3.

The kidneys play a critical role in homeostasis through their ability to regulate the volume, osmolality, electrolyte concentration, and acid–base status of the extracellular fluid (ECF). The kidney accomplishes these homeostatic functions both independently and in concert with other organ systems, especially the neuroendocrine system. Endocrine hormones involved in coordinating these functions include angiotensin II, aldosterone, antidiuretic hormone (ADH), and atrial natriuretic peptide (ANP).

The kidney's ability to regulate the volume and composition of the ECF starts with its prodigious capacity to access plasma through the process of filtration. On a daily basis, the kidneys generate ~180 L of filtrate; the equivalent of filtering an adult's entire plasma volume about 60 times each day. Although a large percentage of the filtrate is reabsorbed by the kidney and retained in the body, about 2 L/d is excreted as urine. The kidney plays a critical role in the regulation of extracellular fluid volume, osmolality, and electrolyte balance. Maintaining the right balance of electrolytes such as Na^+, Ca^{2+}, K^+, and Cl^- is essential for cellular functions, such as neural transmission and muscle contraction. Furthermore, regulating electrolyte concentration is essential for maintaining osmotic balance in the body, which can vary due to alterations in the amount of water in the ECF, that is, due to dehydration or overhydration. These variations can happen with sweating, diarrhea, vomiting, medications, renal disease, or excessive water intake.

This chapter discusses the location, size, and composition of fluid compartments in the body as well as renal function related to sodium, potassium, calcium, magnesium, and phosphate balance.

▶ FLUID COMPARTMENTS OF THE BODY

There are two main fluid compartments in the human body: **intracellular fluid (ICF)** and **extracellular fluid (ECF)**. The ICF compartment makes up ~60% to 67% of body water, and the ECF compartment makes up the other 33% to 40%. Water is the major solvent of all body fluid compartments. **Total body water** averages about 60% of body weight in young adult men and about 55% of body weight in young adult women (Table 23.1). The amount of adipose tissue (fat) a person has affects the percentage of body weight that water occupies. Adipose tissue contains a lower percentage of water (about 10%) than other tissues (e.g., muscle is about 75% water). Thus, an obese person has a lower percentage of body weight that is water than a lean person. Nonobese adult women have relatively less total body water than men, because, on average, they have more subcutaneous fat and less muscle mass. As people age, they tend to lose muscle and add adipose tissue; hence, water content declines with age. In contrast, newborns have a high percentage of body weight as water because of a relatively large ECF volume and little fat.

Total body water is distributed between intracellular fluid and extracellular fluid compartments.

Total body water can be divided into ICF and ECF. ICF is found inside the bilayered cell plasma membrane. It is the medium in which cellular organelles are suspended and where many chemical reactions take place. Under normal circumstances, ICF is in osmotic equilibrium with the ECF.

The volume of fluid in the ICF is slightly greater than that in the ECF (see Table 23.1). However, these two fluids differ

TABLE 23.1	Average Total Body Water and Its Distribution as a Percentage of Body Weight*	
Total Body Water	**Distribution of Total Body Water**	
Adult female = 55%	Intracellular = 30%; extracellular = 25%	
Adult male = 60%	Intracellular = 33%; extracellular = 27%	

*Based on an average young adult weighing 70 kg. Extracellular = fast + slow compartments.

strikingly in terms of their electrolyte composition. Their total solute concentrations and thus osmolalities, however, are normally equal, because of the high water permeability of most cell membranes. Consequently, a momentary osmotic difference between cells and the ECF rapidly disappears.

The ECF is composed of fluid outside of the cells and consists of three subdivisions: the *interstitial compartment*, the *plasma compartment*, and the *third space,* or *transcellular, compartment.* The *interstitial compartment* is the fluid space that surrounds the cells of a given tissue. It is filled with interstitial fluid and comprises about 45% by volume of the ECF. Interstitial fluid provides the microenvironment that allows for movement of ions, proteins, and nutrients across the cell membrane. The fluid is not static but is continuously being turned over and recollected by the lymphatic channels. When excess fluid accumulates in the interstitial compartment, **edema** develops. The *intravascular plasma* is found within the vascular system and comprises ~18% by volume of the ECF (about 3.5 L for the average 70-kg male). The *third space* is part of the ECF compartment (~38%) where fluids are not easily exchangeable with other body water compartments. These fluids are sometimes called **transcellular fluids**. Major examples of such fluids include peritoneal fluid, pleural fluid, cerebrospinal fluid, digestive tract fluid, synovial fluid, renal tubular fluid, and aqueous humor. The collective volume of all transcellular fluids is only about 7 L, but these fluids are physiologically important (i.e., it functions as a lubricant in the pleural space and the fluid surrounding joints). There is a constant turnover of transcellular fluids; they are continuously formed and absorbed or removed. Impaired formation, abnormal loss from the body, or blockage of fluid removal can have serious consequences (e.g., hydrocephalus from excess fluid in the cerebral ventricles).

Indicator–dilution method measures fluid compartment size.

The **indicator–dilution method** can be used to determine the size of body fluid compartments. A known amount of a substance (the indicator), which should be confined to the compartment of interest, is administered. After allowing sufficient time for uniform distribution of the indicator throughout the compartment (e.g., total body water, ECF, or plasma), a plasma sample is collected. At equilibrium, the concentration of the indicator will be the same in the entire compartment, including the plasma. The plasma

concentration is measured, and the distribution volume is calculated from this formula:

$$\text{Volume} = \frac{\text{Amount of indicator}}{\text{Concentration of indicator}} \quad (1)$$

If indicator was lost from the fluid compartment, the amount lost is subtracted from the amount administered.

To measure total body water, heavy water (deuterium oxide [D_2O]), tritiated water (HTO), or antipyrine (a drug that distributes throughout all of the body water) is used as an indicator. For example, suppose we want to measure total body water in a 60-kg woman, we inject 30 mL of D_2O as an isotonic saline solution into an arm vein. After a 2-hour equilibration period, a blood sample is withdrawn, and the plasma is separated and analyzed for D_2O. A concentration of 0.001 mL D_2O/mL plasma water is found. Suppose that, during the equilibration period, urinary, respiratory, and cutaneous losses of D_2O are 0.12 mL. Substituting these values into the indicator–dilution equation, we get: total body water = (30 to 0.12 mL D_2O) ÷ 0.001 mL D_2O/mL water = 29,880 mL or 30 L. Therefore, total body water as a percentage of body weight equals 50% in this woman.

To measure extracellular water volume, the ideal indicator should distribute rapidly and uniformly outside the cells and should not enter the cell compartment. Unfortunately, there is no such ideal indicator, and so the exact volume of the ECF cannot be measured. A reasonable estimate, however, can be obtained using two different classes of substances: impermeant ions and inert sugars. ECF volume has been determined from the volume of distribution of these ions: radioactive Na^+, radioactive Cl^-, radioactive sulfate, thiocyanate (SCN^-), and thiosulfate ($S_2O_3^{2-}$) of which radioactive sulfate ($^{35}SO_4^{2-}$) is probably the most accurate. However, ions are not completely impermeant; they slowly enter the cell compartment, and so measurements tend to lead to an overestimate of ECF volume. Measurements with inert sugars (such as mannitol, sucrose, and inulin) tend to lead to an underestimate of ECF volume, because they are excluded from some of the extracellular water—for example, the water in dense connective tissue and cartilage. In addition, special techniques are required when using these sugars because the kidneys rapidly filter and excrete them after their intravenous injection.

Cellular water cannot be determined directly with any indicator. It can, however, be calculated from the difference between measurements of total body water and extracellular water. The plasma water of the ECF is determined by using Evans blue dye, which avidly binds serum albumin, or radioiodinated serum albumin, and by collecting and analyzing a blood plasma sample. In effect, the plasma volume is measured from the distribution volume of serum albumin. The assumption is that serum albumin is completely confined to the vascular compartment, but this is not entirely true. Indeed, serum albumin is slowly (3% to 4% per hour) lost from the blood by diffusive and convective transport through capillary walls. To correct for this loss, repeated blood samples can be collected at timed intervals, and the concentration of albumin at time zero (the time at which no loss would have occurred) can be determined by extrapolation. Alternatively, the plasma concentration of indicator 10 minutes after injection can be used; this value is usually close to the extrapolated value. If plasma volume and hematocrit are known, total circulating blood volume can be calculated. Unlike plasma volume, the other major component of the ECF, interstitial fluid and lymph volume, cannot be determined directly but can be calculated from the difference between ECF and plasma volumes.

Plasma and intracellular fluid composition are markedly different.

Body fluids contain many uncharged molecules (e.g., glucose and urea), but quantitatively speaking, **electrolytes** (ionized substances) contribute most to the total solute concentration and the effective osmolality of body fluids. Effective osmolality (i.e., the osmolality due to solutes that do not easily passively cross the cell membrane) is of prime importance in determining the distribution of water between ICF and ECF compartments.

The importance of ions (particularly Na^+) in determining the total plasma osmolality (P_{osm}) is exemplified by an equation that is of value in the clinic:

$$P_{osm} = 2 \times [Na^+] + \frac{[glucose] \text{ in } mg/dL}{18} + \frac{[blood \text{ urea nitrogen}] \text{ in } mg/dL}{2.8} \quad (2)$$

If the plasma Na^+ level is 140 mmol/L, blood glucose level is 100 mg/dL (5.6 mmol/L), and blood urea nitrogen level is 10 mg/dL (3.6 mmol/L), then the calculated osmolality is 289 mOsm/kg H_2O. The equation indicates that Na^+ and its accompanying anions (mainly Cl^- and HCO_3^-) normally account for more than 95% of the plasma osmolality. In some special circumstances (e.g., alcohol intoxication), plasma osmolality calculated from the above equation may be much lower than the measured osmolality because of the presence of solutes other than sodium salts, glucose, and urea (e.g., ethanol).

The concentrations of various electrolytes in plasma and ICF are summarized in Table 23.2. The ICF values are based on determinations made in skeletal muscle cells. These cells account for about two thirds of the cell mass in the human body. Concentrations are expressed in terms of milliequivalents (mEq) per kilogram H_2O.

An **equivalent** contains one mole of univalent ions, and a **milliequivalent (mEq)** is 1/1,000th of an equivalent. Equivalents are calculated as the product of moles times valence and represent the concentration of charged species. For singly charged (univalent) ions, such as Na^+, K^+, Cl^-, or HCO_3^-, 1 mmol is equal to 1 mEq. For doubly charged (divalent) ions, such as Ca^{2+}, Mg^{2+}, or SO_4^{2-}, 1 mmol is equal to 2 mEq. Some electrolytes, such as proteins, are polyvalent, and so there are several mEq per mmol. Sodium is the major cation in plasma, and Cl^- and HCO_3^- are the major anions. The plasma proteins (mainly serum albumin) bear net negative charges at physiologic pH. The electrolytes are actually

TABLE 23.2	Electrolyte Composition of Body Fluids	
Electrolyte	Plasma Concentration (mEq/kg H_2O)	Intracellular Fluid* (mEq/kg H_2O)
Cations		
Na^+	153	10
K^+	4.2	159
Ca^{2+}	5.4	1
Mg^{2+}	2.2	40
Total	165	210
Anions		
Cl^-	111	3
HCO_3^-	27	7
Proteins	18	45
Others	9	155
Total	165	210

*Based on skeletal muscle cells.

dissolved in the plasma water, and so the values in Table 23.2 are expressed as concentrations per kilogram H_2O. The water content of plasma is usually about 93%; about 7% of plasma volume is occupied by solutes, mainly the plasma proteins. Interstitial fluid is an ultrafiltrate of plasma. It contains all of the small electrolytes in essentially the same concentration as in plasma, but little protein. The proteins are largely confined to the plasma because of their large molecular size.

ICF composition (see Table 23.2) is different from plasma. The cells have higher K^+, Mg^{2+}, and protein concentrations than plasma. The intracellular Na^+, Ca^{2+}, Cl^-, and HCO_3^- levels are lower. The anions in skeletal muscle cells labeled "Others" are mainly organic phosphate compounds important in cell energy metabolism, such as creatine phosphate, adenosine triphosphate (ATP), and adenosine diphosphate (ADP). As described in Chapter 2, the high intracellular K^+ level and low intracellular Na^+ level are a consequence of plasma membrane Na^+/K^+-ATPase activity; this enzyme extrudes Na^+ from the cell and takes up K^+. The low intracellular Cl^- and HCO_3^- in skeletal muscle cells are primarily a consequence of the inside negative membrane potential (−90 mV), which favors the outward movement of these small negatively charged ions. The intracellular Mg^{2+} is high; most is not free but is bound to cell proteins. Intracellular Ca^{2+} is low; as discussed in Chapter 1, the cytosolic Ca^{2+} in resting cells is about 10^{-7} M (0.0002 mEq/L). Most of the cell Ca^{2+} is sequestered in organelles, such as the sarcoplasmic reticulum in skeletal muscle.

Changes in extracellular fluid osmolality cause changes in intracellular volume.

Despite the different compositions of ICF and plasma, the total solute concentration and osmolality of these two fluid compartments is normally the same. ICF and plasma are in osmotic equilibrium because the high water permeability of

Renal Physiology and Body Fluids

cell membranes prevents a sustained difference in osmolality. If the osmolality changes in one compartment, water moves so as to restore a new osmotic equilibrium (see Chapter 2).

The volumes of ICF and ECF depend primarily on the volume of water present in these compartments. But the latter depends on the amount of solute present and its osmolality. This fact follows from the definition of the term *concentration*: concentration = amount/volume; hence, volume = amount/concentration. The main osmotically active solute in cells is K$^+$; therefore, a loss of cell K$^+$ will cause cells to lose water and shrink (see Chapter 2). The main osmotically active solute in the ECF is Na$^+$; therefore, a gain or loss of Na$^+$ from the body will cause the ECF volume to swell or shrink, respectively.

The distribution of water between intracellular and extracellular compartments changes in various circumstances. However, it is important to recognize that, when such changes occur, the transcellular fluid compartments come into osmotic and volume equilibrium markedly slower than those comprising the intracellular fluid, plasma, and interstitial fluid. The transcellular fluid makes up about 17.5% of total body water by volume, and because of its slow equilibrating properties, it is sometimes called the *slow extracellular fluid compartment*. The remaining extracellular fluid compartments (plasma and interstitial fluid) can quickly equilibrate with the ICF. They are often called collectively the *fast extracellular fluid compartment* and comprise about 27.5% of the total body water. For practical homeostatic purposes, 82.5% of the total body water (55% for the ICF and 27.5% for the fast ECF) participates in rapid moment-to-moment

changes in volume and osmolality between the ICF and ECF. The division of water between the ICF and the fast ECF is thus roughly two third for the ICF and one third for the ECF. Figure 23.1 provides some examples of changes in the ICF and fast ECF. For this illustration, total body water is divided into the two major rapidly equilibrating compartments, ICF and the combined plasma and interstitial compartments (simply called ECF for this illustration). The *y*-axis represents total solute concentration, and the *x*-axis represents the volume; the area of a box (concentration times volume) gives the amount of solute present in a compartment. Note that the height of the boxes is always equal because osmotic equilibrium maintained between these two divisions of total body water.

In the normal situation (shown in Fig. 23.1A), two thirds (28 L for a 70-kg man) of rapidly equilibrating total body water is in the ICF and one third (14 L) is in the ECF. The osmolality of both fluids is 285 mOsm/kg H$_2$O. Hence, the cell compartment contains 7,980 mOsm and the ECF contains 3,990 mOsm.

In Figure 23.1B, 2 L of pure water was added to the ECF (e.g., by drinking water). Plasma osmolality is thereby lowered initially, and water moves into the cell compartment along the osmotic gradient. The entry of water into the cells causes them to swell, and intracellular osmolality falls until a new equilibrium (*solid lines*) is achieved. Because 2 L of water was added to an original total body water volume of 42 L, the new total body water volume is 44 L. No solute was added, and so the new osmolality at equilibrium is (7,980 + 3,990 mOsm)/44 kg

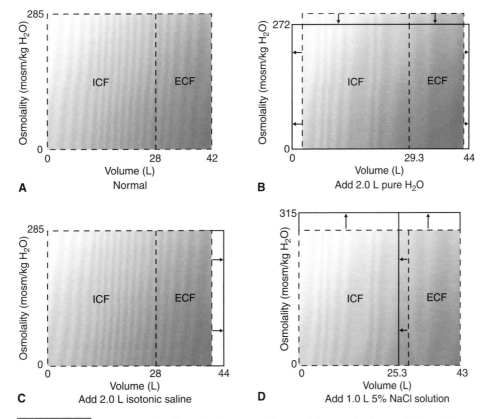

Figure 23.1 **The effects of various disturbances on the osmolalities and volumes of intracellular fluid (ICF) and extracellular fluid (ECF).** The *dashed lines* indicate the normal condition, and the *solid lines*, the situation after a new osmotic equilibrium has been attained. (See text for details.)

= 272 mOsm/kg H_2O. The volume of the ICF at equilibrium is calculated by solving the following equation: 272 mOsm/kg H_2O × volume = 7,980 mOsm, or 29.3 L. The volume of the ECF at equilibrium is 14.7 L. From these calculations, we conclude that two thirds of the added water ends up in the cell compartment and one third stays in the ECF. Stated another way, addition of pure water to the total system distributed in proportion to the relative volume of the two compartments involved (two third of the 2 L to the ICF and one third to the ECF). Note that this description of events is somewhat artificial because in reality the kidneys would excrete the added water over the course of a few hours, thereby minimizing the fall in plasma osmolality and cell swelling.

In Figure 23.1C, 2 L of isotonic saline (0.9% NaCl solution) was added to the ECF. Isotonic saline is isosmotic to plasma or ECF and, by definition, can cause no change in cell volume. Furthermore, the sodium and chloride in the solution are effectively excluded from the intravascular compartment by the sodium/potassium ATPase pumps in the cell membranes and the low passive permeability of the membrane to sodium. Therefore, all of the added isotonic saline is retained in the ECF, and there is no change in osmolality.

Figure 23.1D shows the effect of intravenously infusing 1 L of a 5% NaCl solution (osmolality about 1,580 mOsm/kg H_2O). All the salt stays in the ECF. This exposes the cells to a hypertonic environment and thus water leaves the cells; the solutes left behind become more concentrated as water leaves. A new equilibrium will be established, with the final osmolality higher than normal but equal inside and outside of the cells. The final osmolality can be calculated from the amount of solute present (7,980 + 3,990 + 1,580 mOsm) divided by the final volume (28 + 14 + 1 L), which equals 315 mOsm/kg H_2O. The final volume of the ICF equals 7,980 mOsm divided by 315 mOsm/kg H_2O, or 25.3 L, which is 2.7 L less than the initial volume. The final volume of the ECF is 17.7 L, which is 3.7 L more than its initial value. Thus, the addition of hypertonic saline to the ECF led to its considerable expansion, in large part because of loss of water from the cell compartment.

▶ FLUID BALANCE

Homeostatic fluid balance is dependent on maintaining a balance between input and output. Regulation of fluid balance involves both the control of ECF volume and ECF osmolality. Osmolality is the measure of solute particle concentration and is defined as the number of osmoles (Osm) of solute per kilogram of water (Osm/kg), whereas *osmolarity* is a measure of the osmoles of solute per liter of solution (Osm/L). These values are roughly equivalent in our body. The kidneys control ECF volume by regulating salt balance and control ECF osmolality by regulating water balance. As a result of these homeostatic mechanisms, body fluid volumes and plasma osmolality are kept remarkably constant.

Maintenance of extracellular fluid osmolality requires water input to equal its output.

A balance chart for water for an average 70-kg man is presented in Table 23.3. The person is in a stable balance (or steady state) when the total input and total output

TABLE 23.3	Typical Daily Water Inputs and Outputs (70-kg Male) and Common Abnormal Conditions Affecting Water Balance
Normal Inputs	**Normal Outputs**
Liquid consumption 1–2 L (up to 20 L)	Transpiration through skin, loss through exhalation 0.8–1 L
Water in food 0.8–1 L	Sweat 0.2 L
Oxidative metabolism 0.3–0.4 L	Fecal loss 0.2 L
	Renal output 1–2 L (up to 20 L)
Abnormal Water Gains	**Abnormal Water Losses**
Congestive heart failure	Diarrhea (can be as high as 5 L): vomiting; severe sweating (~10 L)
Renal failure	
High sodium intake	Hemorrhage
Cirrhosis of the liver	Diuretics
Overinfusion of intravenous fluids	Excessive urination (e.g., diabetes, excessive alcohol consumption).

of water from the body are equal (about 2,000 mL/d). On the input side, water is found in the beverages we drink and in the foods we eat. Solid foods, which consist of animal or vegetable matter are, like our own bodies, mostly water. Water of oxidation is produced during metabolism; for example, when 1 mol of glucose is oxidized, 6 mol of water are produced. In a hospital setting, the input of water resulting from intravenous infusions would also need to be considered. On the output side, losses of water occur via the skin, lungs, gastrointestinal (GI) tract, and kidneys. We always lose water by simple evaporation from the skin and lungs; this is called **insensible water loss**.

Appreciable water loss from the skin, in the form of sweat, occurs at high temperatures or with heavy exercise. As much as 4 L of water per hour can be lost in sweat. Sweat, is a hypo-osmotic fluid that contains NaCl; excessive sweating can lead to significant losses of salt. GI losses of water are normally small (see Table 23.3), but with diarrhea, vomiting, or drainage of GI secretions, massive quantities of water and electrolytes may be lost from the body.

The kidneys are the sites of adjustment of water output from the body. Renal water excretion changes to maintain balance. If there is a water deficiency, the kidneys diminish the excretion of water and urine output falls. If there is water excess, the kidneys increase water excretion and urine flow, to remove the extra water. The renal excretion of water is controlled by **arginine vasopressin (AVP)**, which is also called **antidiuretic hormone** or **ADH**.

The water needs of an infant or young child, per kilogram body weight, are several times higher than those of an adult. Children have, for their body weight, a larger body

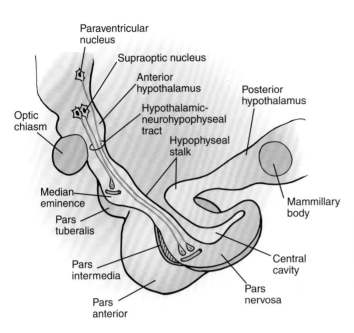

Figure 23.2 **The pituitary and hypothalamus.** Arginine vasopressin (AVP) is synthesized primarily in the supra-optic nucleus and to a lesser extent in the paraventricular nuclei in the anterior hypothalamus. It is then transported down the hypothalamic–neurohypophyseal tract and stored in vesicles in the median eminence and posterior pituitary (pars nervosa), where it can be released into the blood.

surface area and higher metabolic rate. As such they are much more susceptible to volume depletion.

AVP control of water balance is critical in regulating extracellular fluid osmolality.

AVP is a nonapeptide that is synthesized in hypothalamus, then transported via axons, and released from the posterior pituitary into the blood (see Chapter 31). As shown in Figure 23.2, the hormone travels by axoplasmic flow down the hypothalamic–neurohypophyseal tract and is stored in vesicles in nerve terminals in the median eminence and, mostly, the posterior pituitary. Action potentials reaching these nerve terminals release AVP, which then diffuses into nearby capillaries. The hormone is then carried by the bloodstream to the collecting ducts of the kidneys, where it increases water reabsorption (see Chapter 22).

Many factors increase the release of AVP, including pain, trauma, emotional stress, nausea, fainting, most anesthetics, nicotine, morphine, and angiotensin II. These conditions or agents therefore produce a decline in urine output and more concentrated urine. Ethanol and ANP decrease AVP release, leading to the excretion of a large volume of dilute urine.

The main mechanism controlling AVP release under ordinary circumstances is a change in plasma osmolality. Figure 23.3 shows how plasma AVP concentrations vary as a function of plasma osmolality. When plasma osmolality rises, neurons called **osmoreceptor cells**, located in the anterior hypothalamus, shrink. This stimulates the nearby neurons in the paraventricular and supraoptic nuclei to release AVP, and plasma AVP concentration rises. The result is the formation of osmotically concentrated urine by reabsorbing water from the collecting ducts back into the ECF. This reabsorption of solute-free water helps reduce the originally elevated plasma osmolality. It is important for students to understand that not all solutes are equally effective in stimulating the osmo-receptor cells. For example, cell membranes are permeable to

urea, and urea is thus an ineffective osmole at hypothalamic osmoreceptors. Urea can enter these cells and therefore does not cause the osmotic withdrawal of water needed to shrink the cells and stimulate AVP release. Only solutes that cannot or do not rapidly cross the osmoreceptor cell membrane are able to change the volume of the osmoreceptor cells and thus affect AVP release. These solutes are called *effective* osmoles. Extracellular Na^+ and Cl^- are effective osmoles and thus stimulate AVP release.

When plasma osmolality falls in response to the addition of excess water, osmoreceptor cells swell, AVP release is inhibited, and plasma AVP levels fall. In this situation, the collecting ducts express their intrinsically low water permeability, less water is reabsorbed, a dilute urine is excreted (i.e., less water is reabsorbed by the collecting ducts back into the body) and plasma osmolality can be restored to normal by elimination of the excess water. Figure 23.4 shows that the entire range of urine osmolalities, from dilute to concentrated urine, is a linear function of plasma AVP in healthy people.

A second mechanism controlling AVP release is the blood volume—more precisely, the **effective arterial blood volume (EABV)**. An increased blood volume inhibits AVP release, whereas a decreased blood volume (**hypovolemia**) stimulates AVP release. Intuitively, this makes sense, because with excess volume, a low plasma AVP level would promote the excretion of water by the kidneys. With hypovolemia, a high plasma AVP level would promote reabsorption or conservation of water by the kidneys.

The receptors for blood volume include stretch receptors in the right atrium of the heart and in the pulmonary veins within the pericardium. More stretch results in more impulses transmitted to the brain via vagal afferents, which inhibit AVP release. The common experiences of producing a large volume of dilute urine, a **water diuresis**—when lying down in bed at night, when exposed to cold weather, or when immersed in a pool during the summer—may be

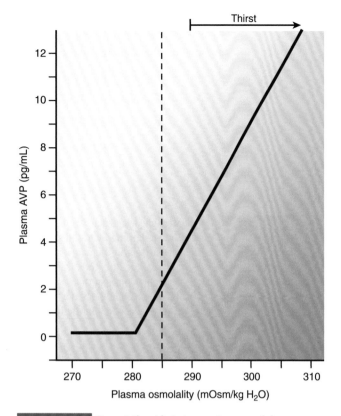

Figure 23.3 The relationship between plasma arginine vasopressin (AVP) level and plasma osmolality in healthy people. Decreases in plasma osmolality were produced by drinking water and increases by fluid restriction. Plasma AVP levels were measured by radioimmunoassay. The plasma AVP level is essentially zero at plasma osmolalities <280 mOsm/kg H_2O. Above this plasma osmolality (threshold), plasma AVP increases linearly with plasma osmolality. Normal plasma osmolality averages about 285 mOsm/kg H_2O (*dashed vertical line*) and so we live above the threshold for AVP release. The thirst threshold is attained at a plasma osmolality of 290 mOsm/kg H_2O; the thirst mechanism is activated only when there is an appreciable water deficit. Changes in plasma AVP and consequent changes in renal water excretion are normally capable of maintaining a normal plasma osmolality below the thirst threshold.

related to the activation of this pathway. In all these situations, an increased central blood volume stretches the atria. Arterial baroreceptors in the carotid sinuses and aortic arch also reflexly change AVP release; however, a fall in pressure at these sites stimulates AVP release. Finally, a decrease in renal blood flow stimulates renin release, which leads to increased angiotensin II production. Angiotensin II stimulates AVP release by acting on the brain.

Relatively large blood losses (more than 10% of blood volume) are required to increase AVP release significantly (Fig. 23.5). With a loss of 15% to 20% of total blood volume, however, very large increases in plasma AVP are observed. Plasma levels of AVP may rise to levels much higher (e.g., 50 pg/mL) than are needed to concentrate the urine maximally (e.g., 5 pg/mL) (compare Figs. 23.4 and 23.5.) With severe hemorrhage, this high circulating level of AVP exerts a significant vasoconstrictor effect, which helps compensate by raising the blood pressure.

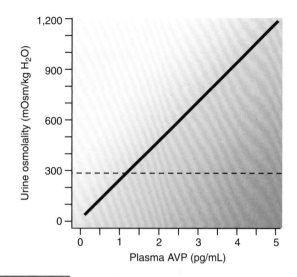

Figure 23.4 The relationship between urine osmolality and plasma arginine vasopressin (AVP) levels. The *dashed horizontal line* reports the normal plasma osmolality (285 mOsm/kg H_2O). With low plasma AVP levels, urine hypo-osmotic to plasma is excreted, and with high plasma AVP levels, hyperosmotic urine is excreted. Maximally concentrated urine (1,200 mOsm/kg H_2O) is produced when the plasma AVP level is about 5 pg/mL.

Plasma osmolality and blood volume work in concert to regulate AVP release.

The two stimuli, plasma osmolality and blood volume, most often work in concert to increase or decrease AVP release. For example, a great excess of water intake in a healthy person will inhibit AVP release because of both the fall in plasma osmolality and increase in blood volume. Conversely, AVP release will be stimulated with excess water loss (e.g., sweating

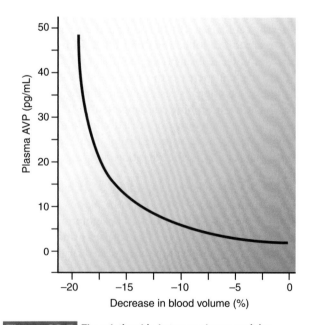

Figure 23.5 The relationship between plasma arginine vasopressin (AVP) and blood volume depletion in the rat. Severe hemorrhage (loss of 20% of total blood volume) causes a striking increase in plasma AVP. In this situation, the vasoconstrictor effect of AVP becomes significant and helps counteract the low blood pressure.

during heavy exercise) because of the concomitant rise in plasma osmolality and decrease in blood volume. However, in certain important circumstances, there is a conflict between these two inputs. In the context of regulating AVP release, the osmoreceptors in the brain are more sensitive to changes in osmolality than are the pressure receptors that indirectly sense change in volume. However, the pressure receptors can elicit a stronger overall response. This relationship has been termed the *Volume Overrides Tonicity Principle*: that is, *AVP release (and thirst) will be stimulated, even if plasma is hypotonic, providing plasma volume is significantly decreased.* This usually occurs when plasma volume is decreased more than 10% to 5%. In this case, the need of the body to support the cardiovascular system, by correcting for volume loss and resulting reduced filling of the heart, takes precedence over using AVP and thirst suppression to try to correct for hypotonicity in the plasma. In general, the principle of volume changes overriding tonicity changes in regard to affecting AVP secretion and thirst is only seen in relatively severe conditions in the body such as hemorrhage, severe dehydration, and edema resulting from loss of fluid out of the CV system (i.e., exclusion of water from the central circulation).

Finally, certain clinical conditions can create a disconnect between plasma volume and osmolality. For example, severe congestive heart failure is characterized by a decrease in the EABV, even though total blood volume is greater than normal. This condition results because the heart does not pump sufficient blood into the arterial system to maintain adequate tissue perfusion. The arterial baroreceptors signal less volume, and AVP release is stimulated. The patient will produce osmotically concentrated urine and will also be thirsty from the decreased EABV, with consequent increased water intake. The combination of decreased renal water excretion and increased water intake leads to hypo-osmolality of the body fluids, which is reflected in a low plasma Na^+, or **hyponatremia**. Despite the hypo-osmolality, plasma AVP levels remain elevated and thirst persists. It appears that maintaining an EABV is of overriding importance, and so osmolality may be sacrificed in this condition. However, the hypo-osmolality can create new problems, such as the swelling of brain cells.

▶ DISTURBANCES IN FLUID–ELECTROLYTE BALANCE

The maintenance of normal volume and normal electrolyte composition of the extracellular fluid is vital for life. In particular, electrolytes play a key role in maintaining homeostasis within the cells, tissues, and organs of the body. They help to regulate cellular secretions, as well as muscle and neurological functions. Fluid–electrolyte disturbances can occur with disease and illness. Examples include excessive ingestion, retention, or elimination of electrolytes. The most common cause of fluid–electrolyte disturbances is renal failure. Examples of causes for fluid loss and gain are shown in Table 23.3.

For a quick review, it is important to remember that: (1) although increased amounts of water and solutes are lost through sweating and breathing during exercise, the loss of excess water or excess solutes depends mainly on regulating

excretion in the urine; (2) the extent of urinary secretion of electrolytes (e.g., Na^+ and Cl^- loss) is the main determinant of extracellular fluid volume; and (3) urinary water loss is the main determinant of extracellular fluid osmolarity.

Thirst center governs water intake.

One of the main ways the body regulates extracellular fluid volume is by adjusting the volume of water intake, mainly by drinking more or less fluid. The thirst center in the hypothalamus governs the urge to drink.

Thirst, a conscious desire to drink water, is basic instinct that arises from a lack of fluids or an increase in certain plasma electrolytes. Continuous dehydration can lead to renal problems and neurological problems, such as seizures. If the body's water volume falls below a certain threshold, or the effective osmolality of the plasma becomes too high, signals from the **thirst center** of the brain are activated. The thirst center is located in the anterior hypothalamus, close to the neurons that produce and control AVP release. This center relays impulses to the cerebral cortex, so that thirst becomes a conscious sensation.

Several factors affect the thirst sensation (Fig. 23.6). The major stimulus is an increase in effective osmolality of the blood plasma, which is detected by osmoreceptor cells in the hypothalamus. These cells are distinct from those that affect AVP release. An increase in effective plasma osmolality of 1% to 2% (i.e., about 3 to 6 mOsm/kg H_2O) is needed to reach the thirst threshold. NaCl is an effective stimulus. In contrast, ethanol and urea are not effective stimuli for the osmoreceptors because they readily penetrate these cells and therefore do not cause them to shrink.

Hypovolemia or a decrease in the effective arterial blood volume (EABV) stimulates thirst. Blood volume loss must be considerable for the thirst threshold to be reached; most blood donors do not become thirsty after donating 500 mL of blood (10% of blood volume). A larger blood loss (15% to 20% of blood volume), however, evokes *intense* thirst. A decrease in EABV resulting from severe diarrhea or vomiting or congestive heart failure may also provoke thirst.

The receptors for blood volume that stimulate thirst include the arterial baroreceptors in the carotid sinuses and

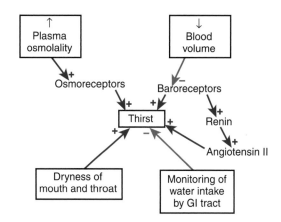

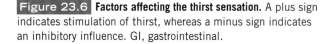
Figure 23.6 Factors affecting the thirst sensation. A plus sign indicates stimulation of thirst, whereas a minus sign indicates an inhibitory influence. GI, gastrointestinal.

aortic arch and stretch receptors in the cardiac atria and great veins in the thorax. The kidneys may also act as volume receptors. When blood volume is decreased, the kidneys release renin into the circulation. This results in production of angiotensin II, which acts on neurons near the third ventricle of the brain to stimulate thirst.

The thirst sensation is reinforced by dryness of the mouth and throat, which is caused by a reflex decrease in secretion by salivary and buccal glands in a water-deprived person. The GI tract also monitors water intake. Moistening of the mouth or distention of the stomach, for example, inhibits thirst, thereby preventing excessive water intake. For example, if a dog is deprived of water for some time and is then presented with water, it will commence drinking but will stop before all of the ingested water has been absorbed by the small intestine. Monitoring of water intake by the mouth and stomach in this situation limits water intake, thereby preventing a dip in plasma osmolality below normal.

Polydipsia is excessive thirst and causes the excretion of large volumes of urine. The condition is most commonly found in diabetics. Polydipsia also occurs when the thirst center, situated in the hypothalamus, gets damaged. Polydipsia can also be caused by mental illness, such as schizophrenia and is termed psychogenic polydipsia. Diabetes insipidus is another disorder that can lead to extreme thirst and polyuria. (See the Integrated Medical Sciences box.)

▶ SODIUM BALANCE

As seen from Table 23.2, Na^+ is the most abundant cation in the plasma and, along with its accompanying anions Cl^- and HCO_3^-, largely determines the osmolality of the ECF. Because AVP, the kidneys, and thirst closely regulate the osmolality of the ECF, it follows that the Na^+ content of the ECF compartment determines the amount of water in it (and hence its volume). The kidneys are primarily involved

---

CLINICAL FOCUS | 23.1

Hyponatremia

Hyponatremia, defined as a plasma Na^+ level of <135 mEq/L, is the most common disorder of body fluid and electrolyte balance in hospitalized patients. Most often, it reflects a problem of too much water, not too little Na^+, in the plasma. Because Na^+ is the major solute in the plasma, it is not surprising that hyponatremia is usually associated with hypo-osmolality. Hyponatremia, however, may also occur with a normal or even elevated plasma osmolality.

Drinking large quantities of water (20 L/d) if not ingested too rapidly rarely causes frank hyponatremia because of the large capacity of the kidneys to excrete dilute urine. If, however, plasma arginine vasopressin (AVP) is not decreased when plasma osmolality is decreased or if the ability of the kidneys to dilute the urine is impaired, hyponatremia may develop even with a normal water intake.

Hyponatremia with hypo-osmolality can occur in the presence of a decreased, normal, or even increased total body Na^+. Hyponatremia and decreased body Na^+ content may be seen with increased Na^+ loss, such as with vomiting, diarrhea, and diuretic therapy. In these instances, the decrease in extracellular fluid (ECF) volume stimulates thirst and AVP release. More water is ingested, but the kidneys form osmotically concentrated urine and so plasma hypo-osmolality and hyponatremia result. Hyponatremia and normal body Na^+ content are seen in hypothyroidism, cortisol deficiency, and the syndrome of inappropriate secretion of antidiuretic hormone (SIADH). SIADH occurs with neurologic disease, severe pain, certain drugs (such as hypoglycemic agents), and some tumors. For example, a bronchogenic tumor may secrete AVP without control by plasma osmolality. The result is renal conservation of water that dilutes plasma sodium. Hyponatremia and increased total body Na^+ content are seen in edematous states, such as congestive heart failure, hepatic cirrhosis, and nephrotic syndrome. The decrease in effective arterial

blood volume stimulates thirst and AVP release. Excretion of dilute urine may also be impaired because of decreased delivery of fluid to diluting sites along the nephron and collecting ducts. Although both Na^+ and water are retained by the kidneys in the edematous states, relatively more water is conserved, leading to a dilutional hyponatremia.

Hyponatremia and hypo-osmolality can cause a variety of symptoms, including muscle cramps, lethargy, fatigue, disorientation, headache, anorexia, nausea, agitation, hypothermia, seizures, and coma. These symptoms, mainly neurologic, are a consequence of the swelling of brain cells as plasma osmolality falls. Excessive brain swelling may be fatal or may cause permanent damage. Treatment requires identifying and then treating the underlying cause. If Na^+ loss is responsible for the hyponatremia, isotonic or hypertonic saline or NaCl by mouth is usually given. If the blood volume is normal or the patient is edematous, water restriction is recommended. Hyponatremia should be corrected slowly and with constant monitoring, because too rapid correction can be harmful.

Hyponatremia in the presence of increased plasma osmolality is seen in hyperglycemic patients with uncontrolled diabetes mellitus. In this condition, the high plasma glucose causes the osmotic withdrawal of water from cells, and the extra water in the ECF space leads to hyponatremia. Plasma Na^+ level falls by 1.6 mEq/L for each 100-mg/dL rise in plasma glucose.

Hyponatremia and a normal plasma osmolality are seen with so-called **pseudohyponatremia**. This occurs when plasma lipids or proteins are greatly elevated. These molecules do not significantly elevate plasma osmolality. They do, however, occupy a significant volume of the plasma, and because the Na^+ is dissolved only in the plasma water, the Na^+ measured in the entire plasma is low. ■

TABLE 23.4	Magnitudes of Daily Filtration, Reabsorption, and Excretion of Ions and Water in a Healthy Young Man on a Typical American Diet					
Substance	Plasma (mEq/L)	GFR (L/d)	Filtered (mEq/d)	Excreted (mEq/d)	Reabsorbed (mEq/d)	% Reabsorbed
Sodium	140	180	25,200	100	25,100	99.6
Chloride	105	180	18,900	100	18,800	99.5
Bicarbonate	24	180	4,320	2	4,318	99.9
Potassium	4	180	720	100	620	86.1
Water	0.93*	180	167 L/d	1.5 L/d	165.5 L/d	99.1

*Plasma contains about 0.93 L H_2O/L.
GFR, glomerular filtration rate.

Renal Physiology and Body Fluids

in the regulation of Na^+ balance. We consider first the renal mechanisms involved in Na^+ excretion and then overall Na^+ balance.

Kidneys excrete only a small percentage of the filtered sodium load.

Table 23.4 shows the magnitude of filtration, reabsorption, and excretion of ions and water for a healthy adult man on an average American diet. The amount of Na^+ filtered (the Na^+ filtered load) was calculated from the product of the plasma Na^+ concentration and the glomerular filtration rate (GFR) because sodium is freely filtered at the glomerulus. The quantity of Na^+ reabsorbed was calculated from the difference between the filtered and excreted amounts. Note that 99.6% (25,100 ÷ 25,200 mEq/d) of the filtered Na^+ was reabsorbed or, in other words, percentage excretion of Na^+ was only 0.4% of the filtered load. In terms of overall Na^+ balance for the body, the quantity of Na^+ excreted by the kidneys is of key importance, because ordinarily about 95% of the Na^+ that we consume is excreted by way of the kidneys. Tubular reabsorption of Na^+ must be finely regulated to keep us in Na^+ balance.

Figure 23.7 shows the percentage of *filtered* Na^+ reabsorbed in different parts of the nephron. Seventy percent of filtered Na^+ is reabsorbed in the proximal convoluted tubule. Because the proximal tubule is highly permeable to water, this percent of sodium reabsorption (along with its attendant anions) causes the tubule to reabsorb the same percentage of filtered water. The loop of Henle reabsorbs about 20% of filtered Na^+, but only 10% of filtered water because the ascending limbs are not permeable to water. The distal convoluted tubule reabsorbs about 6% of filtered Na^+ (and no water), and the collecting ducts reabsorb about 3% of the filtered Na^+ (and 19% of the filtered water, though this varies with plasma AVP levels). Only about 1% of the filtered Na^+ and water is usually excreted. The distal nephron (distal convoluted tubule, connecting tubule, and collecting duct) has a lower capacity for Na^+ transport than more proximal segments and can be overwhelmed if too much Na^+ fails to be reabsorbed in proximal segments. The distal nephron is of critical importance in determining the final excretion of Na^+.

Multiple factors affect renal Na^+ excretion; these are discussed in the subsections that follow. A factor may promote

Na^+ excretion by increasing the amount of Na^+ filtered by the glomeruli, by decreasing the amount of Na^+ reabsorbed by the kidney tubules, or, in some cases, by affecting both processes.

Glomerulotubular balance prevents massive changes in Na excretion.

Theoretically, small changes in GFR, by increasing filtered sodium load, could potentially lead to massive changes in Na^+ excretion if the nephron could not adjust the amount of the filtered load it reabsorbed. If sodium reabsorption was

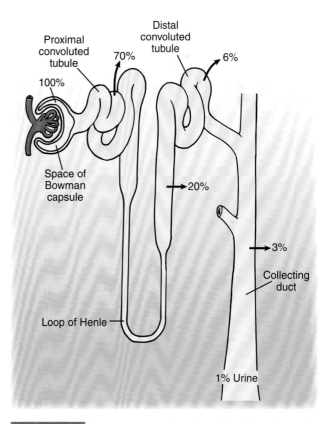

Figure 23.7 **The percentage of the filtered load of sodium reabsorbed along the nephron.** About 1% of the filtered Na^+ is usually excreted.

TABLE 23.5	Glomerulotubular Balance*		
Period	Filtered (mEq/min)	– Reabsorbed (mEq/min)	= Excreted (mEq/min)
1	6.00	5.95	0.05
Increase GFR by one third			
2	8.00	7.90	0.10

*Results from an experiment performed on a 10-kg dog. In response to an increase in glomerular filtration rate (GFR) (produced by infusing a drug that dilated afferent arterioles), tubular reabsorption of Na+ also increased, and so only a modest increase in Na+ excretion occurred. If there had been no glomerulotubular balance and if tubular Na+ reabsorption had stayed at 5.95 mEq/min, the kidneys would have excreted 2.05 mEq/min in period 2. If we assume that the extracellular fluid (ECF) volume in the dog is 2 L (20% of body weight) and if plasma Na+ is 140 mEq/L, an excretion rate of 2.05 mEq/min would result in excretion of the entire ECF Na+ (280 mEq) in a little over 2 h. The dog would have been dead long before this could happen, which underscores the importance of glomerulotubular balance.

constant, even a small change in GFR and filtered sodium load would lead to a large excretion, or loss, of sodium over time. However, Na+ excretion tends to change in the same direction as GFR because of a phenomenon called **glomerulotubular balance** (Table 23.5). Proximal convoluted tubules and loops of Henle reabsorb an essentially *constant fraction, or percent,* of the filtered sodium load, not a constant amount. In other words, the tubules increase the rate of Na+ reabsorption when GFR (and thus filtered sodium load) is increased and decrease the rate of Na+ reabsorption when GFR is decreased (i.e., reabsorbing the same percent of a larger or smaller than normal filtered load yields, respectively, a greater or lesser absolute amount reabsorbed). Glomerulotubular balance reduces the impact of changes in GFR on Na+ excretion and can be thought of a way of shielding potential alterations in sodium balance caused by changes in the cardiovascular system that alter GFR.

Elevated renal capillary hydrostatic pressure increases Na+ excretion.

The increased sodium and water excretion produced by an increase in intravascular pressure in the kidneys is called a *pressure* **natriuresis** and *pressure diuresis,* respectively. The term "natriuresis" means an increase in Na+ excretion. Multiple factors appear to be involved in the mechanism of pressure diuresis. First, an increase in the hydrostatic pressure or a decrease in the colloid osmotic pressure (the so-called **Starling forces**) in peritubular capillaries results in reduced fluid uptake by the capillaries. Such changes would occur, for example, after intravenous infusion of a large volume of isotonic saline. The resulting accumulation of the reabsorbed fluid in the kidney interstitial spaces widens the tight junctions between proximal tubule cells, and the epithelium becomes even more leaky than normal. The result is increased back-leak

of salt and water into the tubule lumen, an overall reduction in net reabsorption of those components of tubular fluid and a resulting increase in salt and water excretion. In addition, an increase in blood pressure rapidly causes Na+/H+ exchangers to be removed from the apical cell membrane of proximal tubule cells and internalized. Basolateral cell membrane Na+/K+-ATPase activity is also decreased. These two changes result in diminished tubular sodium reabsorption and enhanced sodium excretion. This transcellular pathway is another way the kidneys can dispose of excess sodium when arterial or intrarenal pressure is elevated.

Extracellular Na+ is regulated by the renin–angiotensin–aldosterone system and atrial natriuretic peptide.

The **renin–angiotensin–aldosterone system** is a hormone system that regulates blood pressure and water balance. **Renin** is a proteolytic enzyme produced by granular cells, which are located in afferent arterioles in the kidneys (see Chapters 17 and 22). In addition, renin, angiotensinogen and ACE are present in some organs (e.g., the kidneys and brain), so that angiotensin II may also be formed and act locally.

To review, there are three main stimuli for renin release:

1. A decrease in pressure in the afferent arteriole (with the granular cells responding to stretch and functioning as an intrarenal **baroreceptor**)
2. Stimulation of sympathetic nerve fibers to the kidneys to activate β_2-adrenergic receptors on the granular cells
3. A decrease in luminal sodium chloride concentration at the macula densa region of the nephron, resulting, for example, from a decrease in GFR

All three of these pathways are activated and reinforce each other when there is a decrease in the EABV—for example, following hemorrhage, transudation of fluid out of the vascular system, diarrhea, severe sweating, or a low salt intake. Conversely, an increase in the EABV inhibits renin release. In addition, long-term stimulation causes vascular smooth muscle cells in the afferent arteriole to differentiate into granular cells and leads to further increases in renin supply.

Although the renin–angiotensin–aldosterone system is involved in control of blood pressure (see Chapter 17), this system is essentially a salt-conserving system (Fig. 23.8). Angiotensin II has several actions related to Na+ and water balance:

- It stimulates the production and secretion of aldosterone from the zona glomerulosa of the adrenal cortex (see Chapter 33). This mineralocorticoid hormone then acts on the distal nephron to increase Na+ reabsorption.
- Angiotensin II directly stimulates tubular Na+ reabsorption by proximal tubules.
- Angiotensin II stimulates thirst and the release of AVP by the posterior pituitary.

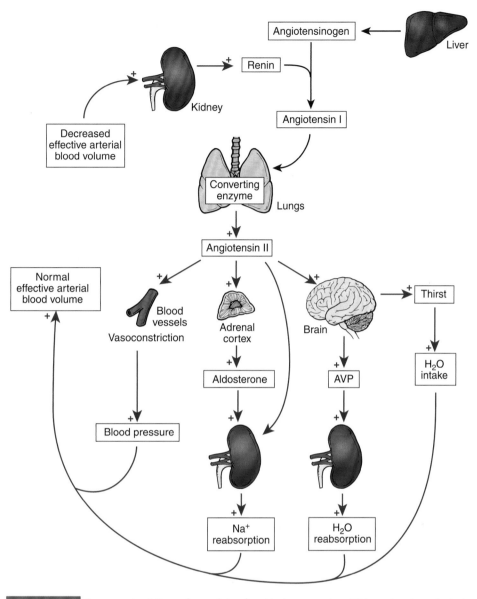

Figure 23.8 **Components of the renin–angiotensin–aldosterone system.** This system is activated by a decrease in the effective arterial blood volume (e.g., following hemorrhage) and results in compensatory changes that help restore arterial blood pressure and blood volume to normal. AVP, arginine vasopressin.

Angiotensin II is also a potent vasoconstrictor of both resistance and capacitance vessels; increased plasma levels following hemorrhage, for example, help sustain blood pressure. Inhibiting angiotensin II production, by giving an ACE inhibitor or inhibiting the binding of angiotensin II to its receptor by using an angiotensin receptor blocker, lowers blood pressure and is used in the treatment of hypertension. Recently, an orally active nonpeptide renin inhibitor called *aliskiren* has been developed; this drug is an effective antihypertensive that may also slow the progression of chronic renal disease by inhibiting intrarenal renin activity.

The renin–angiotensin–aldosterone system plays an important role in the day-to-day control of Na^+ excretion. It favors Na^+ conservation by the kidneys when there is an Na^+ or volume deficit in the body. When there is an excess of Na^+ or fluid volume, diminished activity of the renin–angiotensin–aldosterone system permits enhanced Na^+ excretion.

In the absence of aldosterone (e.g., in an adrenalectomized person) or in a person with adrenal cortical insufficiency, excessive amounts of Na^+ are lost in the urine; percentage reabsorption of Na^+ may decrease from a normal value of about 99.6% to a value of 98%. This change (1.6% of the filtered Na^+ load) may not seem like much, but if the kidneys filter 25,200 mEq/d (see Table 23.4) and excrete an extra $0.016 \times 25,200 = 403$ mEq/d, this is the amount of Na^+ in almost 3 L of ECF (assuming a Na^+ of 140 mEq/L). Such a loss of Na^+ would lead to a decrease in plasma and blood volume, circulatory collapse, and even death.

When there is an extra need for Na^+, people and many animals display a **sodium appetite**, an urge for salt intake, which can be viewed as a brain mechanism, much like thirst, that helps compensate for a deficit. Patients with primary adrenal cortical insufficiency (**Addison disease**) often show a well-developed sodium appetite, which helps keep them alive.

In the case of mineralocorticoid excess, large doses of a potent mineralocorticoid will cause a person to initially retain about 200 to 300 mEq Na^+ (equivalent to about 1.4 to 2 L of ECF) though no obvious edema is seen after retaining this much fluid. However, this sodium retention is not sustained indefinitely, even if dosing with mineralocorticoids is continued, because a sustained salt and water imbalance is not compatible with life. The escape from the salt-retaining action of the mineralocorticoid is called **mineralocorticoid escape**. The fact that the person will not continue to accumulate Na^+ and water is a result of numerous factors that promote renal Na^+ excretion when ECF volume is expanded. These factors overpower the salt-retaining action of mineralocorticoid hormones and include increases in GFR, changes in intrarenal pressures, and release of natriuretic factors (see below).

Atrial natriuretic peptide (ANP) is a 28–amino acid polypeptide synthesized and stored in myocytes of the cardiac atria (Fig. 23.9). It is released by stretch of the atria—for example, following volume expansion. This hormone has several actions that increase Na^+ excretion. ANP acts on the kidneys to increase glomerular blood flow and filtration rate. It also inhibits Na^+ reabsorption by the inner medullary collecting ducts. The second messenger for ANP in the collecting duct is cyclic guanosine monophosphate (cGMP). ANP directly inhibits aldosterone secretion by the adrenal cortex and indirectly inhibits aldosterone secretion by diminishing renal renin release. ANP is a vasodilator and therefore lowers blood pressure. Some evidence

suggests that ANP inhibits AVP secretion. The actions of ANP are in many respects just the opposite of those of the renin–angiotensin–aldosterone system; ANP promotes salt and water loss by the kidneys and lowers blood pressure, whereas activation of the renin–angiotensin–aldosterone system results in salt and water conservation and a higher blood pressure.

Several other natriuretic hormones and factors have been described. **Urodilatin** (kidney natriuretic peptide) is a 32–amino acid polypeptide derived from the same prohormone as ANP. It is synthesized primarily by intercalated cells in the cortical collecting duct where it is secreted into the tubule lumen. It inhibits Na^+ reabsorption by inner medullary collecting ducts via cGMP. **Brain natriuretic peptide (BNP)** was first isolated from the brain but is also produced by myocytes in the cardiac ventricles. Increased plasma levels serve as a marker of cardiac injury and more recently as a measure of the severity of congestive heart failure. Recombinant BNP is used to promote renal sodium excretion in patients with decompensated congestive heart failure. **Guanylin** and **uroguanylin** are polypeptide hormones produced by the small intestine in response to salt ingestion. They activate guanylyl cyclase and produce cGMP as a second messenger, as their names suggest, and induce a natriuresis. **Bradykinin**, a tachykinin involved in inflammation, is produced locally in the kidneys and inhibits Na^+ reabsorption thus acting as a natriuretic agent.

Prostaglandins E_2 and I_2 (prostacyclin) increase Na^+ excretion by the kidneys. These locally produced hormones are formed from arachidonic acid, which is liberated from phospholipids in cell membranes by the enzyme phospholipase A_2. A **cyclooxygenase (COX)** enzyme that has two isoforms, COX-1 and COX-2, mediates further processing. In most tissues, COX-1 is constitutively expressed, whereas COX-2 is generally induced by inflammation. In the kidney, COX-1 and COX-2 are both constitutively expressed in the cortex and medulla. In the cortex, COX-2 may be involved in macula densa–mediated renin release. COX-1 and COX-2 are present in high amounts in the renal medulla, where the main role of the prostaglandins is to inhibit Na^+ reabsorption. The inhibition of Na^+ reabsorption occurs via direct effects on the tubules and collecting ducts and via hemodynamic effects, because the prostaglandins (PGE_2, PGI_2) are vasodilators (see Chapter 22).

Synthesis of these prostaglandins is enhanced in chronic heart failure where they are thought to help support adequate renal blood flow in the face of continual exposure to the effects of activation of sympathetic nerves and the RAAS in that condition. Inhibition of the formation of prostaglandins in this situation, with common nonsteroidal anti-inflammatory drugs (NSAIDs) such as aspirin, may lead to a fall in renal blood flow or Na^+ retention and even renal failure.

In addition to the RAAS and atrial natriuretic peptide, there are other hormones that affect Na^+ balance. For example, estrogens decrease Na^+ excretion, probably by the direct stimulation of tubular Na^+ reabsorption. Most women tend to retain salt and water during pregnancy, which may be related, in part, to the high plasma estrogen levels during this time.

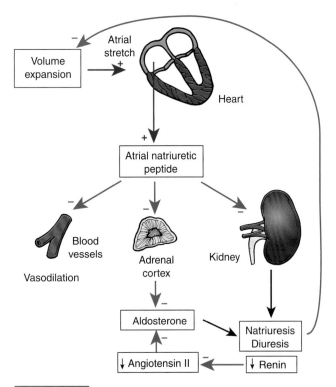

Figure 23.9 Atrial natriuretic peptide (ANP) and its actions. ANP release from the cardiac atria is stimulated by blood volume expansion, which stretches the atria. ANP produces effects that bring blood volume back toward normal, such as increased Na^+ excretion. *Red + arrows* indicate a stimulatory effect by an increase in the variable at the arrow tail.

Glucocorticoids, such as cortisol (see Chapter 33), increase tubular Na⁺ reabsorption but also cause an increase in GFR, which may mask the tubular effect. Usually, a decrease in Na⁺ excretion is seen. Glucocorticoids circulate in the blood at much higher free concentrations than does aldosterone, and they can bind to and activate mineralocorticoid receptors in the kidney but their binding relative to aldosterone is weak. Furthermore, binding and actions in distal nephron cells are minimized by conversion, catalyzed by the enzyme 11β-hydroxysteroid dehydrogenase, to metabolites that do not bind the mineralocorticoid receptor. In general, glucocorticoids cause sodium retention and reduced excretion only at exceptionally high concentrations associated with pathological conditions.

Renal sympathetic nerve stimulation decreases Na⁺ secretion.

Stimulation of renal sympathetic nerves reduces renal Na⁺ excretion in at least three ways:

1. It produces a decline in GFR and renal blood flow, leading to a decreased filtered Na⁺ load and peritubular capillary hydrostatic pressure, both of which favor sodium reabsorption and thus diminished Na⁺ excretion.
2. It has a direct stimulatory effect on Na⁺ reabsorption by the renal tubules.
3. It causes renin release, which results in increased plasma angiotensin II and aldosterone levels, both of which increase tubular Na⁺ reabsorption.

Activation of the sympathetic nervous system occurs in a number of stressful circumstances (such as hemorrhage) in which the conservation of salt and water by the kidneys is of clear benefit.

Diuretics promote Na⁺ secretion by the kidney.

Osmotic diuretics are solutes that are excreted in the urine and increase urinary excretion of Na⁺ and K⁺ salts and water. Examples are urea, glucose (when the reabsorptive capacity of the tubules for glucose has been exceeded), and mannitol (a six-carbon sugar alcohol used in the clinic to promote Na⁺ excretion or cell shrinkage). Osmotic diuretics decrease the reabsorption of Na⁺ in the proximal tubule. This response results from the development of a Na⁺ concentration gradient (lumen Na⁺ < plasma Na⁺) across the proximal tubular epithelium when there is a high concentration of unreabsorbed solute in the tubule lumen. This gradient leads to significant back-leak of Na⁺ into the tubule lumen and consequently decreased net Na⁺ reabsorption. Because the proximal tubule is the place where most of the filtered Na⁺ is normally reabsorbed, osmotic diuretics, by interfering with this process, can potentially cause the excretion of large amounts of Na⁺. Osmotic diuretics may also increase Na⁺ excretion by inhibiting distal Na⁺ reabsorption (similar to the proximal inhibition) and by increasing medullary blood flow.

Most of the diuretic drugs used today are specific Na⁺ transport inhibitors. For example, the loop diuretic drugs (furosemide, bumetanide) inhibit the Na–K–2Cl cotransporter in the thick ascending limb, the thiazide diuretics inhibit the Na–Cl cotransporter in the distal convoluted tubule, and amiloride blocks the epithelial Na⁺ channel in the collecting ducts (see Chapter 22). Spironolactone promotes Na⁺ excretion by competitively inhibiting the binding of aldosterone to the mineralocorticoid receptor. The diuretic drugs are really natriuretic drugs; they produce an increased urine output (diuresis) because water reabsorption is diminished whenever Na⁺ reabsorption is decreased. The loop diuretic drugs produce an especially large increase in Na⁺ excretion, because normally 20% of filtered Na⁺ is reabsorbed in the loop of Henle. More importantly, however, is that by inhibiting reabsorption of NaCl in the thick ascending limb, loop diuretics reduce the medullary vertical osmotic gradient and thereby reduce the ability of the kidney to osmotically reabsorb water from the collecting ducts. This diminished osmotic gradient in the kidney medulla may result in a striking increase in urine output. Diuretics commonly are prescribed for treating hypertension, though the powerful loop diuretics are more often employed to alleviate severe edema.

The principle of glomerular tubular balance is responsible for notable side effects associated with osmotic, loop, and thiazide-type diuretics. Glomerular tubular balance works on loads within adjacent sections of the renal tubule as well as with the kidney as a whole. For example, if sodium reabsorption is inhibited in the thick ascending limb of the loop of Henle by a loop diuretic, that section of the nephron will reabsorb less sodium than normal and thus pass on a larger than normal sodium load to the distal convoluted tubule. Because of glomerular tubular balance, the distal convoluted tubule will then reabsorb more sodium than normal. This effect of glomerular tubular balance has far reaching consequences. Secretion of K⁺ and H⁺ is enhanced, and reabsorption of calcium reduced, when sodium reabsorption is enhanced in the distal nephron because sodium reabsorption there creates a negative transepithelial electrical potential; the larger the sodium reabsorption, the more negative the potential becomes. For this reason, the overall effect of inhibiting sodium reabsorption at the loop of Henle is to enhance excretion of K⁺, H⁺, and Ca²⁺. This loss of other electrolytes with loop diuretics is one of the classic untoward effects of using those agents for treatment of conditions requiring diuresis. This effect is also seen with osmotic and thiazide-type diuretics as well but not with distal tubule sodium channel blockers (e.g., amiloride) or aldosterone antagonists (e.g., spironolactone). These latter diuretics are thus often called *potassium-sparing diuretics* because they cause a diuresis without concurrent enhanced excretion of potassium by the kidney.

Integrated control of sodium balance is by the kidney.

Figure 23.10 summarizes Na⁺ balance in a healthy adult man. Dietary intake of Na⁺ varies and in a typical American diet amounts to about 100 to 300 mEq/d, mostly in the form of NaCl. Ingested Na⁺ is mainly absorbed in the small intestine and is added to the ECF, where it is the major determinant of the osmolality and the amount of water in (or volume of) this fluid compartment. About 50% of the body's Na⁺ is in the ECF, about 40% in bone, and 10% within cells.

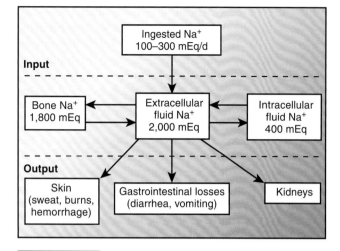

Figure 23.10 **Sodium balance.** Most of the Na^+ that we consume in our diet is excreted by the kidneys.

Losses of Na^+ occur via the skin, GI tract, and kidneys. Skin losses are usually small but can be considerable with sweating or burns. Likewise, GI losses are also usually small, but they can be large and serious with vomiting, diarrhea, or iatrogenic suction or drainage of GI secretions. The kidneys are ordinarily the major route of Na^+ loss from the body, excreting about 95% of the ingested Na^+ in a healthy person. Thus, the kidneys play a dominant role in the control of Na^+ balance. The kidneys can adjust Na^+ excretion over a wide range, reducing it to low levels when there is a Na^+ deficit and excreting more Na^+ when there is Na^+ excess in the body. Adjustments in Na^+ excretion occur by engaging many of the factors discussed above.

In a healthy person, one can think of the ECF volume as the regulated variable in a negative feedback control system (Fig. 23.11). The kidneys are the effectors, and they change Na^+ excretion in an appropriate manner. An increase in ECF

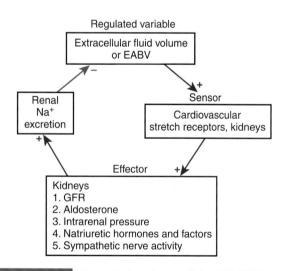

Figure 23.11 **The regulation of extracellular fluid (ECF) volume or effective arterial blood volume (EABV) by a negative feedback control system.** Arterial baroreceptors and the kidneys sense the degree of fullness of the arterial system. The kidneys are the effectors, and they change Na^+ excretion to restore EABV to normal. GFR, glomerular filtration rate.

volume inhibits renal sodium reabsorption and promotes renal Na^+ loss, which restores a normal volume. A decrease in ECF volume leads to enhanced renal sodium reabsorption and decreased renal Na^+ excretion. With continued dietary Na^+ intake, this Na^+ retention leads to the restoration of a normal ECF volume. Closer examination of this idea, particularly when considering pathophysiologic states, however, suggests that it is of limited usefulness. A more considered view suggests that the EABV is actually the regulated variable. In a healthy person, ECF volume and EABV usually change together in the same direction. In an abnormal condition, such as congestive heart failure, however, EABV is low when the ECF volume is abnormally increased. In this condition, there is a potent stimulus for renal Na^+ retention that clearly cannot be the ECF volume.

When EABV is diminished, the degree of fullness of the arterial system is less than normal and tissue blood flow is inadequate. Arterial baroreceptors in the carotid sinuses and aortic arch sense the decreased arterial stretch. This will produce reflex activation of sympathetic nerve fibers to the kidneys, with consequent decreased GFR and renal blood flow and increased renin release. These changes favor renal Na^+ retention. Reduced EABV is also "sensed" in the kidneys in three ways:

1. A low pressure at the level of the afferent arteriole stimulates renin release via the intrarenal baroreceptor mechanism.
2. Decreases in renal perfusion pressure lead to a reduced GFR and hence diminished Na^+ excretion.
3. Decreases in renal perfusion pressure also reduce peritubular capillary hydrostatic pressure, thereby increasing the uptake of reabsorbed fluid and diminishing Na^+ excretion.

When kidney perfusion is threatened, the kidneys retain salt and water, a response that tends to improve their perfusion.

In a number of important diseases, including heart, liver, and some kidney diseases, abnormal renal retention of Na^+ contributes to the development of **generalized edema**, a widespread accumulation of salt and water in the interstitial spaces of the body. The condition is often not clinically evident until a person has accumulated more than 2.5 to 3 L of ECF in the interstitial space. Expansion of the interstitial space has two components: (1) an altered balance of Starling forces exerted across capillaries and (2) the retention of extra salt and water by the kidneys. Total plasma volume is only about 3.5 L; if edema fluid were derived solely from the plasma, hemoconcentration and circulatory shock would ensue. Conservation of salt and water by the kidneys is clearly an important part of the development of generalized edema.

Patients with congestive heart failure may accumulate many liters of edema fluid, which is easily detected as weight gain (because 1 L of fluid weighs 1 kg). Because of the effect of gravity, the ankles become swollen and pitting edema develops. As a result of heart failure, venous pressure is elevated, causing fluid to leak out of the capillaries because of their elevated hydrostatic pressure. Inadequate pumping of blood by the heart leads to a decrease in EABV, and so the kidneys retain salt and water. Alterations in many of the factors discussed above—decreased GFR, increased renin–angiotensin–aldosterone activity, changes in intra-renal pressures, and increased sympathetic nervous system activity—contribute to the renal salt and water retention. To minimize the accumulation of edema fluid, patients are often placed on a reduced Na^+ intake and given diuretic drugs.

▶ POTASSIUM BALANCE

Potassium (K^+) is one of the major electrolytes in the body and the most abundant ion in the ICF compartment. It plays an important role in the electrophysiology of all nerve and muscle tissues and can affect acid–base balance in the body as well. Plasma K^+ concentration is closely regulated by the kidney. Because K^+ is the major osmotically active solute in cells, the amount of cellular K^+ is the major determinant of the amount of water in (and therefore the volume of) the ICF compartment, in the same way that extracellular Na^+ is a major determinant of ECF volume. When cells lose K^+ (and accompanying anions), they also lose water and shrink; the converse is also true.

The distribution of K^+ across cell membranes—that is, the ratio of intracellular to extracellular K^+ concentra-tions—is the major determinant of the resting membrane potential of cells and hence their excitability (see Chapter 3). Disturbances of K^+ balance often produce altered excitability of nerves and muscles. A low plasma K^+ level leads to mem-brane hyperpolarization and reduced excitability; muscle weakness is a common symptom. Excessive plasma K^+ levels lead to membrane depolarization and increased excitability. High plasma K^+ levels can cause cardiac arrhythmias and ventricular fibrillation, which is usually a lethal event.

K^+ balance is linked to acid–base balance in complex ways (see Chapter 24). K^+ depletion, for example, can lead to metabolic alkalosis, and K^+ excess can lead to metabolic aci-dosis. A primary disturbance in acid–base balance can also lead to abnormal K^+ balance. K^+ also affects the activity of enzymes involved in carbohydrate metabolism and electron transport. K^+ is needed for tissue growth and repair, whereas tissue breakdown or increased protein catabolism results in a loss of K^+ from cells.

Distribution of potassium between intracellular and extracellular fluid is tightly regulated.

Total body content of K^+ in a healthy, young adult, 70-kg man is about 3,700 mEq. About 2% of this, roughly 60 mEq, is in the functional ECF (blood plasma, interstitial fluid, and lymph). About 8% of the body's K^+ is in bone, dense con-nective tissue, and cartilage, and another 1% is in transcel-lular fluids. Ninety percent of the body's K^+ is in the cell compartment.

A normal plasma K^+ level ranges from 3.5 to 5.0 mEq/L. By definition, a plasma K^+ level below 3.5 mEq/L is **hypoka-lemia** and a plasma K^+ level above 5.0 mEq/L is **hyperkale-mia**. The K^+ level in skeletal muscle cells is about 150 mEq/L

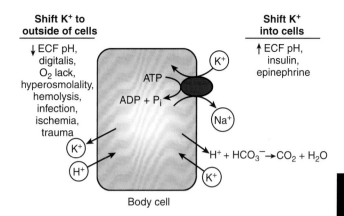

Figure 23.12 Factors influencing the distribution of potassium between intracellular and extracellular fluids. ADP, adenosine diphosphate; ATP, adenosine triphosphate; ECF, extracellular fluid; P_i, inorganic phosphate. K^+ and H^+ passively exchange for each other across the cell membrane.

of cell water. Skeletal muscle cells constitute the largest frac-tion of the cell mass in the human body and contain about two thirds of the body's K^+. One can easily appreciate that abnormal leakage of K^+ from muscle cells, as a result of trauma for example, may lead to dangerous hyperkalemia.

A variety of factors influence the distribution of K^+ between cells and ECF (Fig. 23.12):

- A key factor is the Na^+/K^+-ATPase, which pumps K^+ into cells. If this enzyme is inhibited—as a result of an inad-equate tissue oxygen supply or digitalis overdose, for example—then hyperkalemia may result.

- A decrease in ECF pH (i.e., an increase in ECF H^+) tends to produce a rise in ECF K^+. This results from a passive exchange of extracellular H^+ for intracellular K^+ across the cell membrane. When a mineral acid such as HCl is added to the ECF, a fall of 0.1 unit in blood pH leads to roughly a 0.6-mEq/L rise in plasma K^+ level. When an organic acid (which can penetrate cell membranes) is added, the rise in plasma K^+ level for a given fall in blood pH is considerably less. The fact that blood pH influences plasma K^+ level is sometimes used in the emergency treatment of hyperka-lemia; intravenous infusion of a $NaHCO_3$ solution (which makes the blood more alkaline) causes H^+ to move out of cells in exchange for K^+, which moves into cells.

- Insulin promotes the uptake of K^+ by skeletal muscle and liver cells. This effect appears to be a result of stimula-tion of cell membrane Na^+/K^+-ATPase pumps. Insulin (administered with glucose) is also used in the emergency treatment of hyperkalemia.

- Epinephrine increases K^+ uptake by cells, an effect medi-ated by β_2 receptors.

- Hyperosmolality (e.g., that resulting from hyperglycemia) tends to raise plasma K^+ level; hyperosmolality causes cells to shrink and raises intracellular K^+ level, which then favors outward diffusion of K^+ into the ECF.

- Tissue trauma, infection, ischemia, hemolysis, and severe exercise release K^+ from cells and can cause significant hyperkalemia. An artifactual increase in plasma K⁺ level, **pseudohyperkalemia**, results if blood has been mishandled and red cells have been injured or lysed causing them to leak K⁺.

The plasma K⁺ level is sometimes taken as a rough guide to total body K⁺ stores. For example, if a condition is known to produce an excessive loss of K⁺ (such as taking a diuretic drug), a decrease of 1 mEq/L in plasma K⁺ level may correspond to a loss of 200 to 300 mEq K⁺. Clearly, however, many factors affect the distribution of K⁺ between cells and ECF, and so in many circumstances, the plasma K⁺ is not a good index of the amount of K⁺ in the body.

Abnormal renal K⁺ excretion is the major cause of potassium imbalance.

Figure 23.13 depicts K⁺ balance for a healthy adult man. Most of the foods that we eat contain K⁺. K⁺ intake (50 to 150 mEq/d) and absorption by the small intestine are unregulated. On the output side, GI losses are normally small, but they can be large, especially with diarrhea. Diarrheal fluid may contain as much as 80 mEq K⁺/L. K⁺ loss in sweat is clinically unimportant. Normally, the kidneys excrete 90% of the ingested K⁺.

The kidneys are the major site of control of K⁺ balance; they increase K⁺ excretion when there is too much K⁺ in the body and conserve K⁺ when there is too little. The major cause of K⁺ imbalances is abnormal renal K⁺ excretion. If the kidneys excrete too little K⁺ and the dietary intake of K⁺ continues, hyperkalemia can result. For example, in Addison disease, a low plasma aldosterone level leads to deficient K⁺ excretion because aldosterone normally stimulates potassium secretion directly in the distal nephron and indirectly through stimulating sodium reabsorption, which increases the negative transepithelial electrical potential. Inadequate renal K⁺ excretion also occurs with acute renal failure. Hyperkalemia resulting from inadequate renal excretion is often compounded by tissue trauma, infection, and acidosis,

all of which raise plasma K⁺ level. In chronic renal failure, hyperkalemia usually does not develop until the GFR falls below 15 to 20 mL/min, because of the remarkable ability of the kidney collecting ducts to adapt and increase K⁺ secretion.

Excessive loss of K⁺ by the kidneys leads to hypokalemia. The major cause of renal K⁺ wasting is iatrogenic, an unwanted side effect of diuretic drug therapy. Hyperaldosteronism also causes excessive K⁺ excretion. In uncontrolled diabetes mellitus, K⁺ loss is increased because of the osmotic diuresis resulting from the glucosuria and an elevated rate of fluid flow in the cortical collecting ducts. Several rare inherited defects in tubular transport, including Bartter, Gitelman, and Liddle syndromes, also lead to excessive renal K⁺ excretion and hypokalemia.

Changes in dietary potassium intake change renal potassium excretion in an appropriate direction.

As was discussed in Chapter 22, K⁺ is filtered, reabsorbed, and secreted in the kidneys. Most of the filtered K⁺ is reabsorbed in the proximal convoluted tubule (70%) and loop of Henle (25%), and most K⁺ excreted in the urine is usually the result of secretion by cortical collecting duct principal cells. The percentage of filtered K⁺ excreted in the urine is typically about 15% (Fig. 23.14). With prolonged K⁺ depletion,

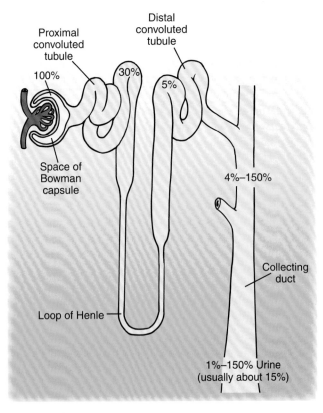

Figure 23.14 **The percentage of the filtered load of potassium remaining in tubular fluid as it flows down the nephron.** K⁺ is usually secreted in the cortical collecting duct. With K⁺ loading, this secretion is so vigorous that the amount of K⁺ excreted may actually exceed the filtered load. With K⁺ depletion, K⁺ is reabsorbed by the collecting ducts.

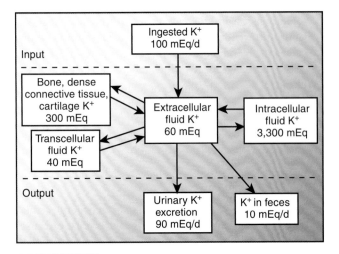

Figure 23.13 **Potassium balance for a healthy adult.** Most of the body's K⁺ is in the cell compartment. Renal K⁺ excretion is normally adjusted to keep a person in balance.

the kidneys may excrete only 1% of the filtered load. On the other hand, excessive K⁺ intake may result in the excretion of an amount of K⁺ that exceeds the amount filtered; in this case, there is greatly increased K⁺ secretion by cortical collecting ducts.

When the dietary intake of K⁺ is changed, renal excretion changes in the same direction. An important site for this adaptive change is the cortical collecting duct. Figure 23.15 shows the response to an increase in dietary K⁺ intake. Two pathways are involved. First, an elevated plasma K⁺ level leads to increased K⁺ uptake by the basolateral cell membrane Na⁺/K⁺-ATPase in collecting duct principal cells, resulting in increased intracellular K⁺, K⁺ secretion, and then K⁺ excretion. Second, an elevated plasma K⁺ level has a direct effect (i.e., not mediated by renin and angiotensin) on the adrenal cortex to stimulate the synthesis and release of aldosterone. Aldosterone acts on collecting duct principal cells to (1) increase the Na⁺ permeability of the luminal cell membrane, (2) increase the number and activity of basolateral cell membrane Na⁺/K⁺-ATPase pumps, (3) increase the luminal cell membrane K⁺ permeability, and (4) increase cell metabolism. These changes increase the transepithelial negative potential as well as enhance K⁺ conductance through the luminal membrane, which collectively result in increased K⁺ secretion.

In cases of decreased dietary K⁺ intake or K⁺ depletion, the activity of the luminal cell membrane H⁺/K⁺-ATPase found in alpha-intercalated cells is increased. This promotes K⁺ reabsorption by the collecting ducts. The collecting ducts

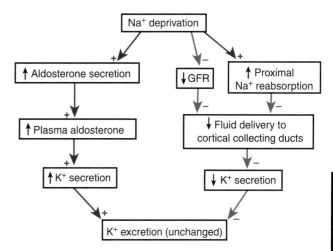

Figure 23.16 **Why sodium depletion does not lead to enhanced potassium excretion.** GFR, glomerular filtration rate.

can greatly diminish K⁺ excretion, but it takes a couple of weeks for K⁺ loss to reach minimal levels.

Net renal potassium excretion may be determined by counterbalancing effects.

Sodium deprivation does not lead to potassium loss via the kidneys. Considering the fact that aldosterone stimulates both Na⁺ reabsorption and K⁺ secretion, why then does Na⁺ deprivation, a stimulus that raises plasma aldosterone levels, not lead to enhanced K⁺ excretion? The explanation is related to the fact that Na⁺ deprivation tends to lower the GFR and increase proximal Na⁺ reabsorption (Fig. 23.16). This response leads to a fall in Na⁺ delivery and decreased fluid flow rate in the cortical collecting ducts, which diminishes K⁺ secretion (see Chapter 22), counterbalancing the stimulatory effect of aldosterone. Consequently, K⁺ excretion is unaltered.

Another puzzling question is why doesn't K⁺ excretion increase during a water diuresis? In Chapter 22, we mentioned that an increase in fluid flow through the cortical collecting ducts increases K⁺ secretion. AVP, in addition to its effects on water permeability, stimulates K⁺ secretion by increasing the activity of luminal membrane K⁺ channels in cortical collecting duct principal cells. Because plasma AVP levels are low during a water diuresis, this will reduce K⁺ secretion, opposing the effects of increased flow, with the result that K⁺ excretion hardly changes.

▶ CALCIUM BALANCE

The kidneys play an important role in the maintenance of Ca²⁺ balance, but it is not the only factor involved. Endocrine effects in the GI tract and on bone metabolism also play key roles in overall calcium balance. Ca²⁺ intake is about 1,000 mg/d and mainly comes from dairy products in the diet. About 300 mg/d is absorbed by the small intestine, a process controlled by 1,25(OH)₂ vitamin D₃. About 150 mg/d of Ca²⁺ is secreted into the GI tract (via saliva, gastric juice, pancreatic juice, bile, and

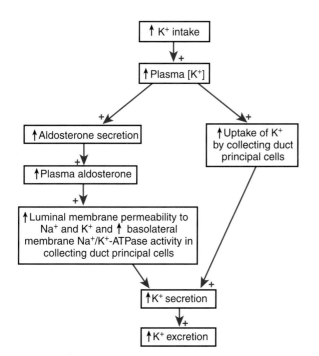

Figure 23.15 **The effect of increased dietary potassium intake on potassium excretion.** K⁺ directly stimulates aldosterone secretion and leads to an increase in cell K⁺ in collecting duct principal cells. Both effects lead to enhanced secretion and, hence, excretion of K⁺.

Renal Physiology and Body Fluids

intestinal secretions), so that net absorption is only about 150 mg/d. Fecal Ca^{2+} excretion is about 850 mg/d, and urinary excretion about 150 mg/d.

Most of the body's Ca^{2+} is in bone (99%), which constantly turns over. In a healthy adult, the rate of release of Ca^{2+} from old bone exactly matches the rate of deposition of Ca^{2+} in newly formed bone (500 mg/d). The kidney can only access calcium that is free in plasma. A normal plasma Ca^{2+} level is about 10 mg/dL, which is equal to 2.5 mmol/L or 5 mEq/L. About 40% of plasma Ca^{2+} is bound to plasma proteins (mainly serum albumin), which are not filtered at the glomerulus, 10% is bound to small diffusible anions (such as citrate, bicarbonate, phosphate, and sulfate), and 50% is free or ionized. It is the ionized Ca^{2+} in the blood that is physiologically important and closely regulated (see Chapter 35).

Ca^{2+} that is not bound to plasma proteins (i.e., 60% of the plasma Ca^{2+}) is freely filterable in the glomeruli. About 60% of the filtered Ca^{2+} is reabsorbed in the proximal convoluted tubule (Fig. 23.17). Two thirds is reabsorbed via a paracellular route in response to solvent drag and the small lumen-positive potential found in the late proximal convoluted tubule. One third is reabsorbed via a transcellular route that includes Ca^{2+} channels in the apical cell membrane and a primary Ca^{2+}-ATPase or 3 Na^+/1 Ca^{2+} exchanger in the basolateral cell membrane. About 30% of filtered Ca^{2+} is reabsorbed along the loop of Henle. Most of the Ca^{2+} reabsorbed in the thick ascending limb is reabsorbed by passive transport through the tight junctions, propelled by the lumen-positive potential there.

Reabsorption continues along the distal convoluted tubule. Reabsorption here is increased by thiazide diuretics. Thiazides inhibit the luminal membrane Na–Cl cotransporter in distal convoluted tubule cells, which leads to a fall in intracellular Na^+ level. In turn, this promotes Na^+/Ca^{2+} exchange and increased basolateral extrusion of Ca^{2+} and, hence, increased Ca^{2+} reabsorption. For this reason, thiazide diuretics may be prescribed in cases of excess Ca^{2+} in the urine (**hypercalciuria**) and **nephrolithiasis** (kidney stone disease). Ca^{2+} reabsorption continues in the connecting tubules and collecting ducts.

The parathyroid hormone (PTH) is the primary hormonal regulator of Ca^{2+} excretion. PTH increases Ca^{2+} reabsorption in the thick ascending limb, distal convoluted tubule, and connecting tubule. Only about 0.5% to 2% of the filtered Ca^{2+} is usually excreted. (Chapter 35 discusses Ca^{2+} balance and its control by several hormones in more detail.)

▶ MAGNESIUM BALANCE

An adult body contains about 2,000 mEq of Mg^{2+}, of which about 60% is present in bone, 39% is in cells, and 1% is in the ECF. Mg^{2+} is the second most abundant cation in cells, after K^+ (see Table 23.2). The bulk of intracellular Mg^{2+} is not free but is bound to a variety of organic compounds such as ATP. Mg^{2+} is present in the plasma at a concentration of about 1 mmol/L (2 mEq/L). About 20% of plasma Mg^{2+} is bound to plasma proteins, 20% is complexed with various anions, and 60% is free or ionized.

About 25% of the Mg^{2+} filtered by the glomeruli is reabsorbed in the proximal convoluted tubule (Fig. 23.18); this is a lower percentage than for Na^+, K^+, Ca^{2+}, or water. The proximal tubule epithelium is rather impermeable to Mg^{2+} under normal conditions, and so there is little passive Mg^{2+} reabsorption. The major site of Mg^{2+} reabsorption is the loop of Henle (mainly the thick ascending limb), which reabsorbs about 65% of filtered Mg^{2+}. Reabsorption here is mainly passive and occurs through the tight junctions, driven by the lumen-positive potential there. Recent studies have identified a tight junction protein that is a channel that facilitates Mg^{2+} movement. Changes in Mg^{2+} excretion result mainly from changes in loop transport. More distal portions of the nephron reabsorb only a small fraction of filtered Mg^{2+} and, under normal circumstances, appear to play a minor role in controlling Mg^{2+} excretion.

An abnormally low plasma Mg^{2+} level is characterized by neuromuscular and central nervous system hyperirritability and cardiac arrhythmias. Abnormally high plasma Mg^{2+} levels have a sedative effect and may cause cardiac arrest. Dietary intake of Mg^{2+} is usually 20 to 50 mEq/d; two thirds is excreted in the feces and one third in the urine. The kidneys are mainly responsible for regulating the plasma

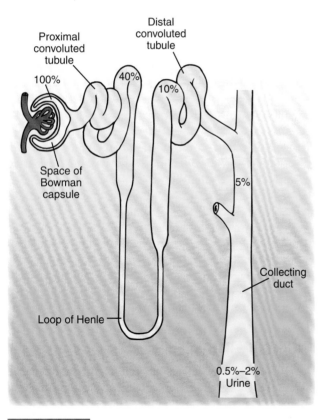

Figure 23.17 **The percentage of the filtered load of calcium remaining in tubular fluid as it flows down the nephron.** The kidneys filter about 10,800 mg/d (0.6 × 100 mg/L × 180 L/d) and reabsorb about 60% of the filtered load in the proximal convoluted tubule and another 30% in the loop of Henle. About 0.5% to 2% of the filtered load (about 50 to 200 mg Ca^{2+}/d) is excreted.

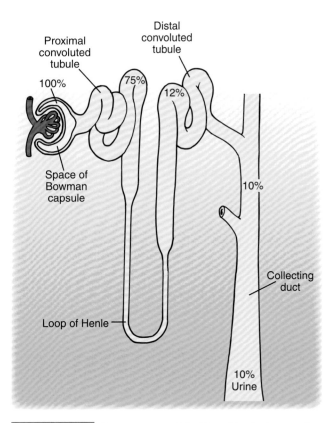

Figure 23.18 **The percentage of the filtered load of magnesium remaining in tubular fluid as it flows down the nephron.** The loop of Henle, specifically the thick ascending limb, is the major site of reabsorption of filtered Mg^{2+}.

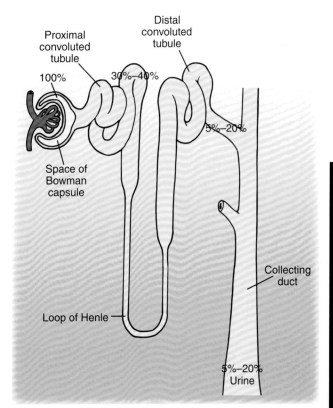

Figure 23.19 **The percentage of the filtered load of phosphate remaining in tubular fluid as it flows down the nephron.** The proximal tubule is the major site of phosphate reabsorption, and downstream nephron segments reabsorb little, if any, phosphate.

Mg^{2+} level. The kidneys rapidly excrete excess amounts of Mg^{2+}. In Mg^{2+}-deficient states, Mg^{2+} virtually disappears from the urine.

▶ PHOSPHATE BALANCE

A normal plasma concentration of inorganic phosphate is about 1 mmol/L. At a normal blood pH of 7.4, 80% of the phosphate is present as HPO_4^{2-} and 20% is present as $H_2PO_4^-$. Phosphate plays a variety of roles in the body. It is an important constituent of bone; it plays a critical role in cell metabolism, structure, and regulation (as organic phosphates); and it is a pH buffer.

Phosphate is mainly unbound in the plasma and is therefore freely filtered by the glomeruli. The proximal tubule is the major site of phosphate reabsorption. About 60% to 70% of filtered phosphate is actively reabsorbed in the proximal convoluted tubule, and another 15% is reabsorbed by the proximal straight tubule via a Na^+–phosphate cotransporter in the luminal cell membrane (Fig. 23.19). The remaining portions of the nephron and collecting ducts reabsorb little, if any, phosphate. Only about 5% to 20% of filtered phosphate is usually excreted. Phosphate in the urine is an important pH buffer and contributes to titratable acid excretion (see Chapter 24).

Phosphate reabsorption is tubular transport maximum (Tm)–limited (see Chapter 22), and the amounts of

phosphate filtered usually exceed the maximum reabsorptive capacity of the tubules for phosphate. This is different from the situation for glucose, in which normally less glucose is filtered than can be reabsorbed. If more phosphate is ingested and absorbed by the intestine, then plasma phosphate rises, more phosphate is filtered, and the filtered load exceeds the Tm more than usual, so the extra phosphate is excreted. Thus, the kidneys participate in regulating the plasma phosphate by an "overflow" type of mechanism. When there is an excess of phosphate in the body, they automatically increase phosphate excretion. In cases of phosphate depletion, the kidneys filter less phosphate and the tubules reabsorb a larger percentage of the filtered phosphate.

Phosphate reabsorption in the proximal tubule is controlled by a variety of factors. PTH is of particular importance; it decreases the phosphate Tm, thereby increasing phosphate excretion. The response to PTH is rapid (minutes) and involves endocytosis of Na^+–phosphate cotransporters from the apical cell membrane and subsequent degradation of the cotransporters in lysosomes. **Fibroblast growth factor 23 (FGF23)** is another protein hormone that inhibits tubular phosphate reabsorption; elevated plasma levels cause hypophosphatemia and, consequently, rickets or osteomalacia.

Patients with chronic renal disease often have an elevated plasma phosphate (**hyperphosphatemia**) level, depending on the severity of the disease. When the GFR falls, the filtered phosphate load is diminished and the tubules reabsorb phosphate more completely. Phosphate excretion is inadequate in the face of continued intake of phosphate in the diet. Hyperphosphatemia is dangerous because of the precipitation of calcium phosphate in soft tissues. For example, when calcium phosphate precipitates in the walls of blood vessels, blood flow is impaired. Hyperphosphatemia can lead to myocardial failure and pulmonary insufficiency.

When plasma phosphate level rises, the plasma ionized Ca^{2+} level tends to fall, for two reasons. First, phosphate forms a complex with Ca^{2+}. Second, hyperphosphatemia decreases production of 1, 25(OH)$_2$ vitamin D$_3$ in the kidneys by inhibiting the 1α-hydroxylase enzyme that forms this hormone. This vitamin, which is a steroid that codes for a calcium-binding protein in intestinal enterocytes, stimulates uptake of calcium by the small intestine. With decreased plasma levels of 1, 25(OH)$_2$ vitamin D$_3$, there is less Ca^{2+} absorption by the small intestine and hence a tendency for hypocalcemia.

A low plasma ionized Ca^{2+} level stimulates hyperplasia of the parathyroid glands and increased secretion of PTH.

A high plasma phosphate level also stimulates PTH secretion directly. PTH then inhibits phosphate reabsorption by the proximal tubules, promotes phosphate excretion, and helps return plasma phosphate levels back to normal.

A primary elevation of PTH levels causes mobilization of both Ca^{2+} and phosphate from bone. Increased bone resorption results, and the bone minerals are replaced with fibrous tissue that renders the bone more susceptible to fracture. The effect of PTH on phosphate excretion in this case can be thought of a means of preventing precipitation of calcium phosphate when calcium and phosphate are coliberated from bone by PTH.

Patients with advanced chronic renal failure are often advised to restrict phosphate intake and consume substances (such as Ca^{2+} salts) that bind phosphate in the intestines, so as to avoid the many problems caused by hyperphosphatemia. Administration of synthetic 1,25(OH)$_2$ vitamin D$_3$ may compensate for deficient renal production of this hormone. This hormone opposes hypocalcemia and inhibits PTH synthesis and secretion. Parathyroidectomy is sometimes necessary in patients with advanced chronic renal failure (Clinical Focus 23.2).

CLINICAL FOCUS | 23.2

Kidney Stone Disease (Nephrolithiasis)

A kidney stone is a hard mass that forms in the urinary tract. About 5% of American women and 12% of men develop a kidney stone some time in their life. A stone lodged in the ureter causes bleeding and intense pain. **Nephrolithiasis** causes considerable suffering and loss of time from work, and it may lead to kidney damage. Once a stone forms in a person, stone formation often recurs.

Stones form when poorly soluble substances in the urine precipitate out of solution, causing crystals to form, aggregate, and grow. About 80% of kidney stones are composed of insoluble Ca^{2+} salts of oxalate and phosphate. There may be excessive amounts of Ca^{2+} or oxalate in the urine as a result of diet, a genetic defect, or unknown causes. Stones may also form from precipitated ammonium magnesium phosphate (struvite), uric acid, and cystine. Struvite stones (about 10% of all stones) are the result of infection with bacteria, usually *Proteus* species. Uric acid stones (9% of all stones) may form in patients with excessive uric acid production and excretion, as occurs in some patients with gout. Defective tubular reabsorption of cystine (in patients with **cystinuria**) leads to cystine stones (1% of stones). The rather insoluble amino acid cystine was first isolated from a urinary bladder stone by Wollaston in 1810, hence its name. Since a low urine flow rate raises the concentration of all poorly soluble substances in the urine, thereby favoring precipitation, a key to prevention of kidney stones is to drink plenty of water and maintain a high urine output day and night.

Fortunately, most stones are small enough to be passed down the urinary tract and spontaneously eliminated.

Microscopic and chemical examination of the eliminated stones is used to determine the nature of the stone and helps guide treatment. Sometimes, a change in diet is recommended to reduce the amount of potential stone-forming material (e.g., Ca^{2+}, oxalate, or uric acid). Thiazide diuretics are useful in reducing Ca^{2+} excretion if excessive urinary Ca^{2+} excretion (hypercalciuria) is the problem. Potassium citrate is useful in treating most stone disease, because citrate complexes Ca^{2+} in the urine and inhibits the crystallization of Ca^{2+} salts. It also makes the urine more alkaline (since citrate is oxidized to HCO_3^- in the body). This treatment is also helpful in reducing the risk of uric acid stones, because urates (favored in alkaline urine) are more soluble than uric acid (the form favored in acidic urine). Administering an inhibitor of uric acid synthesis, such as allopurinol, can help reduce the amount of uric acid in the urine.

If the stone is not passed spontaneously, extracorporeal shock wave lithotripsy, using a lithotripter, is widely used. The patient is placed in a tub of water, and the stone is localized by x-ray imaging. Sound waves are generated in the water by high-voltage electric discharges and are focused on the stone through the body wall. The shock waves break the stone into small pieces that can pass down the urinary tract and be eliminated. Some renal injury is produced by this procedure, and so it may not be entirely innocuous. Other procedures include passing a tube with an ultrasound transducer through the skin into the renal pelvis; stone fragments can be removed directly. A ureteroscope with a laser can also be used to break up stones. ■

INTEGRATED MEDICAL SCIENCES

Pathological Causes of Excessive Urination

Polyuria is a condition that can be defined as excessive urine output of more than 3 L/d although it can often exceed this value considerably. This level of urine production results in frequent urination. There are multiple causes of polyuria, the simplest of which is excessive overconsumption of water well beyond that needed to maintain water balance in the body. Such consumption is sometimes seen as a compulsive drinking behavior associated with psychiatric disorders or, more recently, as part of dangerous fraternity hazing or party games involving "water chugging." The alimentary tract can absorb prodigious amounts of water, and although such addition of water to the ECF will result in a significant polyuria, it also risks rapidly diluting the ECF causing osmotic flux of water into all cells including those in the brain. Ensuing swelling of the brain can result in brain damage or death.

Consumptions of liquids containing ethanol can cause polyuria even if the volume consumption of the liquid containing alcohol is modest. Ethanol inhibits release of ADH by the posterior pituitary thereby leading to less water reabsorption in the renal collecting ducts and increased urine output.

Medicinal diuretics used in the treatment and management of hypertension, CHF, and edematous states (pulmonary or systemic) are common causes of polyuria although in this case the excess urine output is a desired effect of these agents. All diuretic classes produce a diuresis and natriuresis and thus produce increased urine output. Loop diuretics such as furosemide are especially potent diuretics because they reduce the vertical osmotic gradient in the renal medullary interstitium that is exploited to draw water from the collecting ducts. Overdose of diuretics or abuse of these agents can cause severe, rapid diuresis and polyuria. Diuretic abuse has become part of illegal drug use in sporting events as both a performance enhancer and as a masking agent in illegal drug blood doping schemes. Loop diuretics in particular have been used to create rapid weight loss to allow athletes to meet weight requirements in certain events. Diuretics are also employed to create large urine volumes to reduce the concentration of other performance enhancing drugs in the urine. Diuretics have been banned for use in athletes for competition since 1988. Diabetes mellitus is a disorder of glucose and lipid metabolism in which glucose uptake by cells is impaired, either because of reduced or absent pancreatic secretion of insulin (type I) or because tissues such as the liver and skeletal muscle are unresponsive to insulin (type II). In either case, plasma glucose concentration rises to abnormally high levels. This result in a high filtered load of glucose presented to the proximal tubules of the nephron. Normal physiological filtered loads of glucose are totally reabsorbed by the proximal tubule. However, the loads in diabetes mellitus exceed the T_m for glucose resulting in a significant amount of glucose remaining in the tubular fluid. This acts in an osmotic diuretic as well as a natriuretic, which eventually leads to polyuria.

Diabetes insipidus leads to polyuria. Diabetes insipidus does not involve disorders of glucose like diabetes mellitus, but it is characterized by intense thirst, despite ingestion of excess amounts of fluid (polydipsia) and the excretion of large quantities of urine (polyuria). Diabetes insipidus is classified as three types: (1) central diabetes insipidus, (2) nephrogenic diabetes insipidus, and (3) gestational diabetes insipidus. The cause of *central diabetes insipidus* in adults is due to damage of the pituitary gland or to the hypothalamus. The damage is most commonly due to surgery, a tumor, inflammation (e.g., meningitis), head trauma, and injury. In children, the cause is often an inherited genetic disorder. The injury or disorder disrupts the normal synthesis, storage, and release of ADH such that ADH deficiency or absence results. In *nephrogenic diabetes insipidus*, ADH is made and available, but there is a defect in the kidney tubules, the site of water reabsorption, such that they are unresponsive to ADH. This subclass of diabetes insipidus results from the destruction of aquaporin channels. Nephrogenic diabetes insipidus can be due to a genetic disorder or renal infection but may also be caused by lithium toxicity (used to treat psychiatric disorders), the antibiotics amphotericin and tetracycline, hypercalcemia, or hypokalemia. *Gestational diabetes insipidus* occurs only during pregnancy when an enzyme made by the placenta destroys the mother's ADH. Regardless of its etiology, diabetes insipidus is characterized by intense thirst, despite drinking large amounts of fluid (polydipsia), the excretion of large volume of urine (polyuria), and the need to empty the bladder more frequently than the usual four to six times a day.

Tests used to diagnose a patient with polyuria include plasma osmolality, blood glucose, electrolyte levels, and urinalysis. Polyuria due to uncontrolled diabetes mellitus is easily identified by measuring glucose in the urine and the plasma glucose levels. A patient with diabetes insipidus tends to have an elevated plasma osmolality due to the excretion of excessive amounts of dilute urine. A *fluid deprivation test* is the common way to distinguish between the neurogenic and nephrogenic diabetes insipidus. The test involves depriving the patient of water for 8 to 12 hours and a measurement of urine osmolality. The patient is then given 5 units of vasopressin (ADH) subcutaneously, and 2 hours later, urine osmolality is measured. If the polyuria is due to neurogenic diabetes insipidus, the patient's urine osmolality that was persistently diluted (<200 mOsm/kg H_2O) during dehydration will significantly increase in response to the injected vasopressin. If the polyuria is nephrogenic, then the patient's urine osmolality stays diluted in the presence of both dehydration and the injected vasopressin.

Patients with neurogenic diabetes insipidus can be treated with an ADH analog called desmopressin. This agent, however, is useless in nephrogenic diabetes insipidus. Patients with this condition must be on low-salt diets to help reduce excessive urine formation. In addition, thiazide-type diuretics, paradoxically, have been shown to be effective in reducing polyuria in nephrogenic diabetes insipidus. ■

Renal Physiology and Body Fluids

Chapter Summary

- Total body water is distributed in two major compartments: intracellular water and extracellular water. In an average young adult man, total body water, intracellular water, and extracellular water amount to 60%, 40%, and 20% of body weight, respectively. The corresponding figures for an average young adult woman are 50%, 30%, and 20% of body weight.

- The volumes of body fluid compartments are determined by using the indicator–dilution method and the following equation: volume = amount of indicator/concentration of indicator at equilibrium.

- Sodium (Na^+) is the major osmotically active solute in the extracellular fluid (ECF) compartment, and potassium (K^+) has the same role in the intracellular fluid compartment. Cells are typically in osmotic equilibrium with their external environment. The amount of water in (and hence volume of) cells depends on the amount of K^+ they contain, and similarly, the amount of water in (and hence volume of) the ECF is determined by its Na^+ content.

- Plasma osmolality is closely regulated by arginine vasopressin, which governs renal excretion of water, and by both habit and thirst, which govern water intake.

- Arginine vasopressin (AVP) synthesized in the hypothalamus, is released from the posterior pituitary gland and acts on the collecting ducts of the kidney to increase their water permeability.

- The major stimuli for the release of AVP are an increase in effective plasma osmolality (detected by osmoreceptors in the anterior hypothalamus) and a decrease in blood volume (detected by stretch receptors in the atria, carotid sinuses, and aortic arch).

- The kidneys are the primary site of control of Na^+ excretion. Only a small percentage of the filtered Na^+ is usually excreted in the urine, but this amount is of critical importance in overall Na^+ balance.

- Multiple factors affect Na^+ excretion, including glomerular filtration rate, angiotensin II, aldosterone, intrarenal pressures, natriuretic hormones, and renal sympathetic nerve activity. Changes in these factors may account for altered Na^+ excretion in response to excess Na^+ or Na^+ depletion.

- Estrogens, glucocorticoids, osmotic diuretics, poorly reabsorbed anions in the urine, and diuretic drugs also affect renal Na^+ excretion.

- A decrease in EABV leads to Na^+ retention by the kidneys and contributes to the development of generalized edema in pathophysiologic conditions such as congestive heart failure.

- The kidneys play a major role in the control of K^+ balance. K^+ is reabsorbed by the proximal convoluted tubule and loop of Henle and is secreted by cortical collecting duct principal cells. Inadequate renal K^+ excretion produces hyperkalemia, and excessive K^+ excretion produces hypokalemia.

- Calcium balance is regulated on both input and output sides. The absorption of Ca^{2+} from the small intestine is controlled by $1,25(OH)_2$ vitamin D_3, and the excretion of Ca^{2+} by the kidneys is controlled by parathyroid hormone.

- Magnesium is an important intracellular ion and the kidneys regulate the plasma Mg^{2+}.

- Filtered phosphate usually exceeds the maximal reabsorptive capacity of the kidney tubules for phosphate (Tm_{PO4}), and about 5% to 20% of filtered phosphate is usually excreted.

- Phosphate is an important pH buffer in the urine.

- Phosphate reabsorption occurs mainly in the proximal tubules and is inhibited by parathyroid hormone and fibroblast growth factor-23.

- Hyperphosphatemia is a significant problem in chronic renal failure.

Chapter Review Questions

1. A 60-kg woman is given 10 microcuries (mCi) (370 kilobecquerels) of radioiodinated serum albumin (RISA) intravenously. Ten minutes later, a venous blood sample is collected, and the plasma RISA activity is found to be 4 mCi/L. Her hematocrit ratio is 0.4. What is her blood volume?

   A. 417 mL
   B. 625 mL
   C. 2.5 L
   D. 4.17 L
   E. 6.25 L

   The correct answer is D. From the indicator–dilution method, the plasma volume = (10 mCi) 4 mCi/L = 2.5 L. If the hematocrit ratio is 0.4, then the blood volume = (2.5 L plasma) 0.6 L plasma/L blood = 4.17 L.

2. Which of the following leads to decreased Na^+ reabsorption by the kidneys?

   A. An increase in central blood volume
   B. An increase in colloid osmotic pressure in the peritubular capillaries
   C. An increase in GFR
   D. An increase in plasma aldosterone level
   E. An increase in renal sympathetic nerve activity

   The correct answer is A. An increase in central blood volume will stretch the atria, cause the release of atrial natriuretic peptide, and result in decreased Na^+ reabsorption. An increase in colloid osmotic pressure in the peritubular capillaries, an increase in GFR, an increase in plasma aldosterone level, or an increase in renal sympathetic nerve activity increases tubular Na^+ reabsorption.

3. A girl was trapped in the rubble of a collapsed building for 24 hours and both of her legs were crushed. Her plasma potassium concentration is 8 mEq/L, and her ECG does not look normal. She is not producing any urine. Which of the following treatments would be best?

   A. Administer insulin without glucose.
   B. Give an adrenergic blocker to block the action of epinephrine.
   C. Infuse a sodium bicarbonate solution intravenously.
   D. Infuse a solution of 0.1 M HCl into a central vein.
   E. Increase her plasma osmolality be infusing a hypertonic mannitol solution.

The correct answer is C. Infusion of a sodium bicarbonate solution intravenously would lower the plasma [K$^+$] by shifting K$^+$ into the cell compartment. Insulin administration would have the same effect, but giving insulin without glucose might produce hypoglycemia. Making the ECF more acidic, blocking the action of epinephrine, or raising plasma osmolality would all tend to increase, not decrease, the plasma [K$^+$].

4. A dehydrated, hospitalized patient with uncontrolled diabetes mellitus has a plasma [K$^+$] of 4.5 mEq/L (normal 3.5 to 5.0 mEq/L), a plasma (glucose) of 500 mg/dL, and an arterial blood pH of 7 (normal 7.35 to 7.45). These data suggest that the patient has:

   A. a decreased total body store of K$^+$.
   B. a normal total body store of K$^+$.
   C. an increased total body store of K$^+$.
   D. hypokalemia.
   E. hyperkalemia.

The correct answer is A. The low blood pH and hyperglycemia (or hyperosmolality) would tend to raise plasma [K$^+$], yet the plasma [K$^+$] is normal. These findings suggest that the total body store of K$^+$ is reduced. Remember that most of the body's K$^+$ is within cells. In uncontrolled diabetes mellitus, the osmotic diuresis (increased Na$^+$ and water delivery to the cortical collecting ducts), increased renal excretion of poorly reabsorbed anions (ketone body acids), and elevated plasma aldosterone level (secondary to volume depletion) would all favor enhanced excretion of K$^+$ by the kidneys. The subject has normokalemia, not hypo- or hyperkalemia.

5. A 60-year-old woman is always thirsty and wakes up several times during the night to empty her bladder. Plasma osmolality is 295 mOsm/kg H$_2$O (normal range 281 to 297 mOsm/kg H$_2$O), urine osmolality is 100 mOsm/kg H$_2$O, and plasma AVP levels are higher than normal. The urine is negative for glucose. The most likely diagnosis is:

   A. diabetes mellitus.
   B. diuretic drug abuse.
   C. nephrogenic diabetes insipidus.
   D. neurogenic diabetes insipidus.
   E. primary polydipsia.

The correct answer is C. Nephrogenic diabetes insipidus is characterized by increased output of dilute urine. Plasma AVP is elevated because of the volume depletion. Plasma osmolality is on the high side of the normal range, because of the loss of dilute fluid in the urine. The increased urine output is not due to diabetes mellitus because there is no glucose in the urine and the urine is very dilute. Diuretic drug abuse should not produce a very dilute urine, since Na$^+$ reabsorption is inhibited. Neurogenic diabetes insipidus is unlikely because the plasma AVP level is reduced in this case. Primary polydipsia produces output of a large volume of dilute urine, but plasma osmolality and AVP levels are decreased.

Clinical Application Exercises 23.1

A 60-year-old woman with a long history of mental illness was institutionalized after a violent argument with her son. She refused to eat anything since admission, but after maintaining a good fluid intake, she started to compulsively, repeatedly, drink water. She then experienced visual and auditory hallucinations. On one occasion, she ran naked through the ward screaming. On the fifth hospital day, she complained of a slight headache and nausea and had three episodes of vomiting. Later in the day, she was found on the floor in a semiconscious state, confused, and disoriented. She was pale and had cool extremities. Her pulse rate was 70/minute and blood pressure was 150/100 mm Hg. She was transferred to a general hospital, and during transfer, she had three grand mal seizures and arrived in a semiconscious, uncooperative state. A blood sample revealed a plasma [Na$^+$] of 103 mEq/L. Urine osmolality was 362 mOsm/kg H$_2$O, and urine [Na$^+$] was 57 mEq/L. She was given an intravenous infusion of hypertonic saline (1.8% NaCl) and placed on water restriction.

QUESTIONS

1. What is the likely cause of the severe hyponatremia?
2. How much of an increase in plasma [Na$^+$] would an infusion of 1 L of 1.8% NaCl (308 mEq Na$^+$/L) produce? Assume that her total body water is 25 L (50% of her body weight). Why is the total body water used as the volume of distribution of Na$^+$, even though the administered Na$^+$ is limited to the ECF compartment?
3. Why is the brain so profoundly affected by hypo-osmolality? Why should the hypertonic saline be administered slowly?

ANSWERS

1. Hyponatremia is low sodium concentration in the blood (<135 mEq/L) and is considered severe when the serum sodium gets below 125 mEq/L. Low serum sodium can be caused by heart, liver, and kidney failure. However, hyponatremia can also be caused by overhydration, which occurred with the patient in this case. Psychogenic problems started with ingestion of excessive amounts of water. Compulsive water drinking is a common problem in psychotic patients. The increased water intake, combined with an impaired ability to dilute the urine (note the inappropriately high urine osmolality), led to severe hyponatremia and water intoxication.

2. Addition of 1 L of 308 mEq Na^+/L to 25 L produces an increase in plasma [Na^+] of 12 mEq/L. The total body water is used in this calculation because hypertonic NaCl added to the ECF causes movement of water out of the cell compartment, which dilutes the extracellular Na^+.

3. The brain is enclosed in the nondistensible cranium, so when water moves into brain cells and causes them to swell, intracranial pressure can rise to very high values. This can damage nervous tissue directly or indirectly by impairing cerebral blood flow. The neurological symptoms seen in this patient (headache, semiconsciousness, grand mal seizures) are consequences of brain swelling. The increased blood pressure and cool and pale skin may be a consequence of sympathetic nervous system discharge resulting from increased intracranial pressure. Too rapid restoration of a normal plasma [Na^+] can produce serious damage to the brain (central pontine myelinolysis).

thePoint® *Visit* http://thepoint.lww.com/rhoades5e *for additional chapter review Q&A, Clinical Application Exercises, animations, and more!*

Active Learning Objectives

Upon mastering the material in this chapter, you should be able to:

- Define acid, base, acid dissociation constant, weak and strong acids, pKa, pH, and buffer.
- Describe the metabolic processes that produce and consume hydrogen ions, and explain why the body is threatened by net acid gain on a mixed diet.
- Explain the three mechanisms that defend blood pH.
- Explain the role and efficacy of the main chemical buffers present in extracellular fluid, intracellular fluid, and bone.
- Predict the blood pH, given values for plasma HCO_3^- and partial pressure of carbon dioxide (P_{CO_2}).
- Explain why the bicarbonate/CO_2 buffer pair, with a less than an ideal pKa, is the most important pH buffering system in the body.
- Explain the isohydric principle and its significance in control of pH in systems containing multiple buffers.
- Explain the interrelationship between arterial blood pH, P_{CO_2}, and ventilation as it relates to the role of ventilation in the control of arterial pH.

- Calculate renal net acid excretion from excretion of titratable acid, NH_4^+, and bicarbonate.
- Explain the roles of renal reabsorption of filtered bicarbonate, formation of titratable acid, excretion of NH_4^+, and urinary acidification in the kidney's ability to reclaim filtered bicarbonate and generate new bicarbonate to replace that lost titrating fixed acids in the body.
- Discuss the factors that influence renal secretion and excretion of hydrogen ions in the context of the kidney's role in maintaining acid–base balance.
- Explain how cells maintain stability of intracellular pH.
- List the four simple acid–base disturbances, and describe for each the primary defect, changes in arterial blood chemistry (pH, P_{CO_2}, and plasma HCO_3^-), common causes, chemical buffering processes, and degree of respiratory and renal compensations.
- Identify the type of acid–base disturbance present from blood acid–base data and a patient history.
- Calculate the anion gap from plasma electrolyte concentrations, and interpret its meaning.

M etabolic and neural functions are highly sensitive to changes in H^+ concentrations. Outside the acceptable range of pH, protein conformation changes alter enzyme activity, transport proteins, and structural integrity of tissue. Nerve and cardiac function can be markedly impaired as well, which can ultimately result in death. Accordingly, it is not surprising that the H^+ concentration of extracellular fluid is tightly regulated. Acid–base regulation can be viewed in the same way as any other chemical balance, matching gains versus losses. In the case of arterial plasma pH, when loss exceeds gains, H^+ concentration decreases and pH increases above 7.4. When this occurs, it is termed **alkalosis**. Conversely, when gain exceeds loss, plasma H^+ concentration increases and the pH falls below 7.4. This is termed **acidosis**.

Most of this chapter discusses the regulation of H^+ in extracellular fluid (ECF), because ECF is easier to analyze than intracellular fluid (ICF) and it is the fluid used for clinical evaluation of acid–base chemistry. In practice, systemic *arterial* blood is used as the reference sample for this purpose. Measurements on whole blood with a pH meter give values for the H^+ of plasma and, therefore, provide an ECF pH measurement.

▶ BASIC PRINCIPLES OF ACID–BASE INTERACTION

Understanding the physiologic importance of acid–base regulation requires appreciation of some basic laws of chemical reactions. In this section, we briefly review some

basic principles of acid–base chemistry. H^+ ion assumes special significance because of the narrow range that must be maintained for optimal enzymatic and metabolic function. Physiologically important acids include carbonic acid (H_2CO_3), phosphoric acid (H_3PO_4), pyruvic acid ($C_3H_4O_3$), and lactic acid ($C_3H_6O_3$). Physiologically important bases include bicarbonate (HCO_3^-) and biphosphate (HPO_4^{2-}). Acid–base regulation in the body involves primarily two ions, hydrogen (H^+) and bicarbonate (HCO_3^-).

Acids dissociate to release hydrogen ions in solution.

An **acid** is a substance that can release, or donate, H^+; a **base** is a substance that can combine with, or accept, H^+. When an acid (generically written as HA) is added to water, it dissociates reversibly according to the reaction: $HA \rightleftharpoons H^+ + A^-$. The species A^- is a base, because it can combine with an H^+ to form HA. In other words, when an acid dissociates, it yields a free H^+ and its conjugate base. (*Conjugate* means "joined in a pair.")

At equilibrium, the rate of dissociation of an acid to form $H^+ + A^-$ and the rate of association of H^+ and base A^- to form HA are equal. The equilibrium constant (K_a), which is also called the *ionization constant* or *acid dissociation constant*, is given by the expression:

$$K_a = \frac{[H^+] \times [A^-]}{[HA]} \tag{1}$$

The higher the acid dissociation constant, the more an acid is ionized and the greater is its strength. Hydrochloric acid (HCl), for example, is a **strong acid**. It has a high K_a and is almost completely ionized in aqueous solutions. Other strong acids include sulfuric acid (H_2SO_4), phosphoric acid (H_3PO_4), and nitric acid (HNO_3).

An acid with a low K_a is a **weak acid**. For example, in a 0.1 M solution of acetic acid ($K_a = 1.8 \times 10^{-5}$) in water, most (99%) of the acid is nonionized, so that little (1%) is present as acetate$^-$ and H^+. The acidity (concentration of free H^+) of this solution is low. Other weak acids are lactic acid, carbonic acid (H_2CO_3), ammonium ion (NH_4^+), and dihydrogen phosphate ($H_2PO_4^-$).

Acid dissociation constants vary widely and often are small numbers. It is convenient to convert K_a to a logarithmic form, defining pK_a as:

$$pK_a = \log_{10}(1/K_a) = -\log_{10}K_a \qquad (2)$$

In aqueous solution, each acid has a characteristic pK_a, which varies slightly with temperature and the ionic strength of the solution. Note that pK_a is *inversely* proportional to acid strength. A strong acid has a high K_a and a low pK_a. A weak acid has a low K_a and a high pK_a.

pH is inversely related to hydrogen ion concentration.

pH expresses H^+ concentration in aqueous solutions and is often expressed in pH units. The following equation defines **pH**:

$$pH = \log_{10}(1/[H^+]) = -\log_{10}[H^+] \qquad (3)$$

where H^+ concentration is in mol/L. Note that pH is *inversely* related to H^+ concentration. Each whole number on the pH scale represents a 10-fold (logarithmic) change in acidity. A solution with a pH of 5 has 10 times the H^+ of a solution with a pH of 6.

For a solution containing an acid and its conjugate base, we can rearrange the equilibrium expression (equation 1) as:

$$[H^+] = \frac{K_a \times [HA]}{[A^-]} \qquad (4)$$

If we take the negative logarithms of both sides:

$$-\log[H^+] = -\log K_a + \log\frac{[A^-]}{[HA]} \qquad (5)$$

Substituting pH for $-\log[H^+]$ and pK_a for $-\log K_a$, we get:

$$pH = pK_a + \log\frac{[A^-]}{[HA]} \qquad (6)$$

This equation is known as the **Henderson-Hasselbalch equation**. It shows that the pH of a solution is determined by the pK_a of the acid and the ratio of the concentrations of conjugate base to acid.

Buffers protect the stability of the blood pH.

The stability of pH is protected by the action of buffers. A *pH buffer* is defined as an agent that *minimizes* the change in pH produced when an acid or base is added. Note that a buffer *does not prevent* a pH change. A **chemical pH buffer** is a mixture of a weak acid and its conjugate base (or a weak base and its conjugate acid). Following are examples of buffers:

Weak Acid		Conjugate Base	
H_2CO_3 (carbonic acid)	$\rightleftharpoons$	HCO_3^- + H^+ (bicarbonate)	(7)
$H_2PO_4^-$ (dihydrogen phosphate)	$\rightleftharpoons$	HPO_4^{2-} + H^+ (monohydrogen phosphate)	(8)
NH_4^+ (ammonium ion)	$\rightleftharpoons$	NH_3 + H^+ (ammonia)	(9)

Generally speaking, the equilibrium expression for a buffer pair can be written in terms of the Henderson-Hasselbalch equation:

$$pH = pK_a + \log\frac{[\text{conjugate base}]}{[\text{acid}]} \qquad (10)$$

For example, for $H_2PO_4^-/HPO_4^{2-}$:

$$pH = 6.8 + \log\frac{[HPO_4^{2-}]}{[H_2PO_4^-]} \qquad (11)$$

The effectiveness of a buffer—how well it minimizes pH changes when an acid or base is added—depends on its concentration and its pK_a. A good buffer is present in high concentrations and has a pK_a close to the desired pH.

Figure 24.1 shows a titration curve for the phosphate buffer system. As a strong acid or strong base is progressively added to the solution (shown on the *x*-axis), the resulting pH is recorded (shown on the *y*-axis). Going from right to left, as strong acid is added, H^+ combines with the basic form of phosphate: $H^+ + HPO_4^{2-} \rightleftharpoons H_2PO_4^-$. Going from left to right, as strong base is added, OH^- combines with H^+ released from the acid form of the phosphate buffer: $OH^- + H_2PO_4^- \rightleftharpoons HPO_4^{2-} + H_2O$. These reactions lessen the fall or rise in pH.

At the pK_a of the phosphate buffer, the ratio $HPO_4^{2-}/H_2PO_4^-$ is 1, and the titration curve is flattest (the change in pH for a given amount of an added acid or base is at a minimum). In most cases, pH buffering is effective when the solution pH is within ±1 pH unit of the buffer pK_a. Beyond that range, the pH shift that a given amount of acid or base produces may be quite large, so the buffer becomes relatively ineffective.

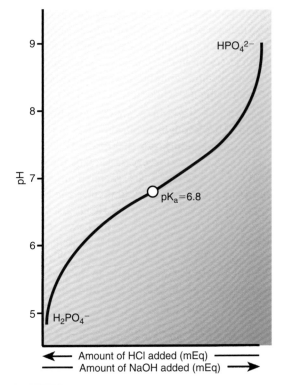

Figure 24.1 **Titration curve for a phosphate buffer.** The pK$_a$ for H$_2$PO$_4^-$ is 6.8. A strong acid (HCl) (*right to left*) or strong base (NaOH) (*left to right*) was added and the resulting solution pH recorded (*y*-axis). Notice that buffering is best (i.e., the change in pH on the addition of a given amount of acid or base is least) when the solution pH is equal to the pK$_a$ of the buffer.

▶ METABOLIC PRODUCTION OF ACIDS

Acids are continuously produced in the body and threaten the normal pH range of the ECF and ICF. Physiologically, acids fall into two groups: (1) H$_2$CO$_3$ (carbonic acid) and (2) all other acids (noncarbonic; also called "nonvolatile" or "fixed" acids). The distinction between these groups arises because H$_2$CO$_3$ is in equilibrium with the volatile gas CO$_2$, which can leave the body via the lungs. The concentration of H$_2$CO$_3$ in arterial blood is, therefore, affected by respiratory activity. By contrast, noncarbonic acids in the body are not directly affected by breathing. Noncarbonic acids are buffered in the body and are then excreted by the kidneys.

Cellular oxidation provides a constant source of carbon dioxide.

A normal adult produces about 300 L of CO$_2$ daily from metabolism. CO$_2$ from the tissues enters the capillary blood, where it reacts with water to form H$_2$CO$_3$, which dissociates instantly to yield H$^+$ and HCO$_3^-$: CO$_2$ + H$_2$O ⇌ H$_2$CO$_3$ ⇌ H$^+$ + HCO$_3^-$. Blood pH would rapidly fall to lethal levels if the H$_2$CO$_3$ formed from CO$_2$ was allowed to accumulate in the body.

Fortunately, H$_2$CO$_3$ is converted to CO$_2$ and water in the pulmonary capillaries and the CO$_2$ is expired (see Chapter 19). In the lungs, the reactions reverse:

$$H^+ + HCO_3^- \rightleftharpoons H_2CO_3 \rightleftharpoons H_2O + CO_2 \qquad (12)$$

As long as CO$_2$ is expired as fast as it is produced, then arterial blood CO$_2$ tension, H$_2$CO$_3$ concentration, and pH do not change.

Nonvolatile acids originate from incomplete metabolism of carbohydrates and fats.

Normally, carbohydrates and fats are completely oxidized to CO$_2$ and water. If carbohydrates and fats are *incompletely* oxidized, nonvolatile acids are produced. Incomplete oxidation of carbohydrates occurs when the tissues do not receive enough oxygen, as during strenuous exercise or hemorrhagic or cardiogenic shock. In such states, glucose metabolism yields lactic acid (pK$_a$ = 3.9), which dissociates into lactate$^-$ and H$^+$, lowering the blood pH. Incomplete fatty acid oxidation occurs in uncontrolled diabetes mellitus, starvation, and alcoholism and produces ketone body acids (acetoacetic and β-hydroxybutyric acids). These acids have pK$_a$ values around 4 to 5. At blood pH, they mostly dissociate into their anions and H$^+$, making the blood more acidic.

Diet of meat and vegetables produces a net acid gain that threatens acid–base balance.

The metabolism of dietary proteins is a major source of H$^+$. The oxidation of proteins and amino acids produces strong acids, such as H$_2$SO$_4$, HCl, and H$_3$PO$_4$. The oxidation of sulfur-containing amino acids (methionine, cysteine, and cystine) produces H$_2$SO$_4$, and the oxidation of cationic amino acids (arginine, lysine, and some histidine residues) produces HCl. The oxidation of phosphorus-containing proteins and nucleic acids produces H$_3$PO$_4$.

A diet containing both meat and vegetables results in a net production of acids, largely from protein oxidation. To some extent, acid-consuming metabolic reactions balance H$^+$ production. Food also contains basic anions, such as citrate, lactate, and acetate. When these are oxidized to CO$_2$ and water, hydrogen ions are consumed (or what amounts to the same thing, HCO$_3^-$ is produced). The average adult who eats a mixed diet results in net production of acid, equivalent to about 1 mEq H$^+$/kg body weight per day. Vegetarians generally have less of a dietary acid burden and a more alkaline urine pH than nonvegetarians, because most fruits and vegetables contain large amounts of organic anions that are metabolized to HCO$_3^-$. The body generally has to dispose of more or less nonvolatile acid, a function performed by the kidneys.

Whether a particular food has an acidifying or an alkalinizing effect depends on whether and how its constituents are metabolized. Cranberry juice has an acidifying effect because of its content of benzoic acid, an acid that cannot be broken down in the body. Orange juice has an alkalinizing effect, despite its acidic pH of about 3.7, because it contains citrate, which is metabolized to HCO$_3^-$. The citric acid in orange juice is converted to CO$_2$ and water, has only a transient effect on blood pH, and has no effect on urine pH.

► INTEGRATION OF THE BODY'S BUFFERING SYSTEMS

The body contains several different buffers that reversibly bind H$^+$ and blunt any change in pH. These buffers include bicarbonate, protein, phosphate, and others.

The bicarbonate buffering system is especially important, as carbon dioxide (CO_2) can be shifted through carbonic acid (H_2CO_3) to hydrogen ions and bicarbonate (HCO_3^-), as shown previously in equation 12. Figure 24.2 shows key buffering mechanisms that tightly regulate pH despite the daily net acid gain. Buffering is accomplished by chemical buffers, the lungs, and the kidneys.

- *Chemical buffering.* Chemical buffers in ECF and ICF and in bone are the first line of defense of blood pH. Chemical buffering minimizes a change in pH but does not remove acid or base from the body.

- *Respiratory response.* The respiratory system is the second line of defense of blood pH. Normally, breathing removes CO_2 as fast as it is produced. Large loads of acid stimulate breathing (respiratory compensation), which removes CO_2 from the body and thus lowers the H_2CO_3 in arterial blood, reducing the acidic shift in blood pH.

- *Renal response.* The kidneys are the third line of defense of blood pH. Although chemical buffers in the body can bind H$^+$ and the lungs can change the H_2CO_3 of blood, the burden of removing excess H$^+$ falls directly on the kidneys. Hydrogen ions are excreted in combination with urinary buffers. At the same time, the kidneys add new HCO_3^- to the ECF to replace HCO_3^- used up in buffering strong acids. The kidneys also excrete the anions (phosphate, chloride, and sulfate) that are liberated from strong acids. The kidneys affect blood pH more slowly than other buffering mechanisms in the body; full renal compensation for an acid–base disturbance may take 1 to 3 days.

Phosphate, protein, and bicarbonate are the body's main buffers.

The body contains many conjugate acid–base pairs that act as chemical buffers (Table 24.1). In the ECF, the main chemical buffer pair is HCO_3^-/CO_2. Plasma proteins and inorganic phosphate are also ECF buffers. Cells have large buffer stores, particularly proteins and organic phosphate compounds. HCO_3^- is present in cells, although at a lower concentration than in ECF. Bone contains large buffer stores, specifically salts of phosphate and carbonate.

When an acid or base is added to the body, the buffers just mentioned bind or release H$^+$, thereby minimizing the change in pH. Buffering in ECF occurs rapidly, in minutes. Acids or bases also enter cells and bone, but this generally occurs more slowly, over hours, allowing cell buffers and bone to share in buffering.

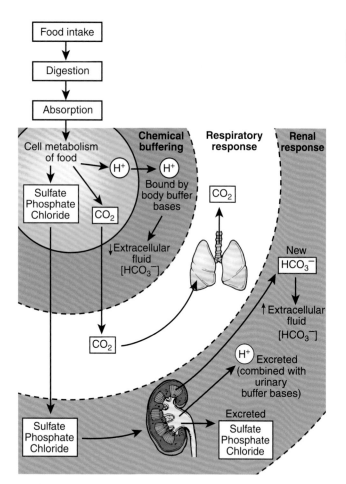

Figure 24.2 **The maintenance of normal blood pH by chemical buffers, the respiratory system, and the kidneys.** On a mixed diet, pH is threatened by the production of strong acids (sulfuric, hydrochloric, and phosphoric) mainly as a result of protein metabolism. These strong acids are buffered in the body by chemical buffer bases, such as extracellular fluid (ECF) HCO_3^-. The kidneys eliminate hydrogen ions (combined with urinary buffers) and anions in the urine. At the same time, they add new HCO_3^- to the ECF to replace the HCO_3^- consumed in buffering strong acids. The respiratory system disposes of CO_2.

TABLE 24.1	Major Chemical pH Buffers in the Body
Buffer	**Reaction**
Extracellular fluid	
Bicarbonate/CO$_2$	$CO_2 + H_2O \rightleftharpoons H_2CO_3 \rightleftharpoons H^+ + HCO_3^-$
Inorganic phosphate	$H_2PO_4^- \rightleftharpoons H^+ + HPO_4^{2-}$
Plasma proteins (Pr)	$HPr \rightleftharpoons H^+ + Pr^-$
Intracellular fluid	
Cell proteins (e.g., hemoglobin [Hb])	$HHb \rightleftharpoons H^+ + Hb^-$
Organic phosphates	Organic $HPO_4^- \rightleftharpoons H^+ + $ organic PO_4^{2-}
Bicarbonate/CO$_2$	$CO_2 + H_2O \rightleftharpoons H_2CO_3 \rightleftharpoons H^+ + HCO_3^-$
Bone	
Mineral phosphates	$H_2PO_4^- \rightleftharpoons H^+ + HPO_4^{2-}$
Mineral carbonates	$HCO_3^- \rightleftharpoons H^+ + CO_3^{2-}$

Phosphate is present as inorganic phosphate in the ECF. The pK_a for phosphate, $H_2PO_4^- \rightleftharpoons H^+ + HPO_4^{2-}$, is 6.8, close to the desired blood pH of 7.4, so, chemically speaking, it would be a good ECF buffer. However, its concentration in the ECF is low (about 1 mmol/L), so it plays a minor role in extracellular buffering.

Phosphate, however, is an important intracellular buffer for two reasons. First, cells contain large amounts of phosphate in such organic compounds as adenosine triphosphate (ATP), adenosine diphosphate (ADP), and creatine phosphate. Although these compounds primarily function in energy metabolism, they also act as pH buffers. Second, intracellular pH is generally lower than the pH of ECF and is closer to the pK_a of phosphate. (The cytosol of skeletal muscle, e.g., has a pH of 6.9.) Phosphate is thus more effective in this environment than in one with a pH of 7.4. Bone has large phosphate salt stores, which also help in buffering.

Proteins are the largest buffer pool in the body and are excellent buffers. Proteins can function as both acids and bases, so they are **amphoteric**. They contain many ionizable groups, which can release or bind H^+. Serum albumin and plasma globulins are the major extracellular protein buffers, present mainly in the blood plasma. Cells also have large protein stores. Recall that the buffering properties of hemoglobin play an important role in the transport of CO_2 and O_2 by the blood (see Chapter 19).

Bicarbonate–carbon dioxide system is a key physiologic buffer.

For several reasons, the HCO_3^-/CO_2 buffer pair is especially important in acid–base physiology:

- Its components are abundant; the concentration of HCO_3^- in plasma or ECF normally averages 24 mmol/L. Although the concentration of dissolved CO_2 is lower (1.2 mmol/L), metabolism provides a nearly limitless supply.

- Despite a pK of 6.10, a little far from the desired plasma pH of 7.40, it is effective because the system is "open" (i.e., its components can be added to or removed from the body at controlled rates).

- Each of its components can be independently controlled, CO_2 by the lungs and HCO_3^- by the kidneys.

CO_2 exists in the body in several different forms: as gaseous CO_2 in the lung alveoli and as dissolved CO_2, H_2CO_3, HCO_3^-, carbonate (CO_3^{2-}), and carbamino compounds in the body fluids. CO_3^{2-} is present at appreciable concentrations only in alkaline solutions, and so we will ignore it. We will also ignore any CO_2 that is bound to proteins in the carbamino form. The most important forms are gaseous CO_2, dissolved CO_2, H_2CO_3, and HCO_3^-.

Dissolved CO_2 in pulmonary capillary blood equilibrates with gaseous CO_2 in the lung alveoli. Consequently, the partial pressures of CO_2 (P_{CO_2}) in alveolar air and systemic arterial blood are normally identical. The concentration of dissolved CO_2 ($CO_{2(d)}$) is related to the P_{CO_2} by the Henry law (see Chapter 19). The solubility coefficient for CO_2 in plasma at 37°C is 0.03 mmol CO_2/L per mm Hg P_{CO_2}. Therefore, $CO_{2(d)}$ = 0.03 × P_{CO_2}. If P_{CO_2} is 40 mm Hg, then $CO_{2(d)}$ is 1.2 mmol/L.

In aqueous solutions, $CO_{2(d)}$ reacts with water to form H_2CO_3: $CO_{2(d)} + H_2O \rightleftharpoons H_2CO_3$. The reaction to the right is called the **hydration reaction** and the reaction to the left is called the **dehydration reaction**. In many cells and tissues, such as the kidneys, pancreas, stomach, and red blood cells, the reactions are catalyzed by **carbonic anhydrase**. At equilibrium, $CO_{2(d)}$ is greatly favored; at body temperature, the ratio of $CO_{2(d)}$ to H_2CO_3 is about 400:1. If $CO_{2(d)}$ is 1.2 mmol/L, then H_2CO_3 equals 3 μmol/L. H_2CO_3 dissociates instantaneously into H^+ and HCO_3^-: $H_2CO_3 \rightleftharpoons H^+ + HCO_3^-$. The Henderson-Hasselbalch expression for this reaction is as follows:

$$pH = 3.5 + \log \frac{[HCO_3^-]}{[H_2CO_3]} \tag{13}$$

Note that H_2CO_3 is a fairly strong acid (pK_a = 3.5). Its low concentration in body fluids lessens its impact on acidity.

Because plasma H_2CO_3 is so low and hard to measure and because $H_2CO_3 = CO_{2(d)}/400$, we can use $CO_{2(d)}$ to represent the acid in the Henderson-Hasselbalch equation:

$$pH = 3.5 + \log \frac{[HCO_3^-]}{[CO_{2(d)}]/400} = 3.5 + \log 400 + \log \frac{[HCO_3^-]}{[CO_{2(d)}]} \tag{14}$$
$$= 6.1 + \log \frac{[HCO_3^-]}{[CO_{2(d)}]}$$

We can also use 0.03 × P_{CO_2} in place of $CO_{2(d)}$:

$$pH = 6.1 + \log \frac{[HCO_3^-]}{0.03\, P_{CO_2}} \tag{15}$$

This form of the Henderson-Hasselbalch equation is useful in understanding acid–base problems. Note that the "acid" in this equation appears to be $CO_{2(d)}$ but is really H_2CO_3 "represented" by CO_2. Therefore, this equation is valid only if $CO_{2(d)}$ and H_2CO_3 are in equilibrium with each other, which is usually (but not always) the case.

Many clinicians prefer to work with H^+ rather than pH. The following expression results if we take antilogarithms of the Henderson-Hasselbalch equation:

$$[H^+] = 24 \times P_{CO_2} / [HCO_3^-] \tag{16}$$

In this equation, H^+ is expressed in nmol/L, HCO_3^- in mmol/L or mEq/L, and P_{CO_2} in mm Hg. If the P_{CO_2} is 40 mm Hg and plasma HCO_3^- is 24 mmol/L, H^+ is 40 nmol/L.

As noted previously, the pK of the HCO_3^-/CO_2 system (6.10) is far from 7.40, the normal pH of arterial blood. From this, one might view this buffer pair as rather poor. On the contrary, it is remarkably effective because it operates in an "open" system; that is, the two buffer components can be added to or removed from the body at controlled rates.

The HCO_3^-/CO_2 system is open in several ways:

- Metabolism provides an endless source of CO_2, which can replace any H_2CO_3 consumed by a base added to the body.

- The respiratory system can change the amount of CO_2 in body fluids by hyperventilation or hypoventilation.

- The kidneys can change the amount of HCO_3^- in the ECF by forming new HCO_3^- when excess acid has been added to the body or by excreting HCO_3^- when excess base has been added.

How the kidneys and respiratory system influence ECF pH by operating on the HCO_3^-/CO_2 system is described below. For now, the advantages of an open buffer system are best explained by an example (Fig. 24.3). Suppose we have 1 L of ECF containing 24 mmol of HCO_3^- and 1.2 mmol of dissolved $CO_{2(d)}$ (P_{CO_2} = 40 mm Hg). Using the special form of the Henderson-Hasselbalch equation described above, we find that the ECF pH is 7.40:

$$pH = 6.10 + \log \frac{[HCO_3^-]}{0.03 P_{CO_2}} = 6.10 + \log \frac{[24]}{[1.2]} = 7.40 \quad (17)$$

Suppose we now add 10 mmol of HCl, a strong acid. HCO_3^- is the major buffer base in the ECF (we will neglect the contributions of other buffers). We predict that the HCO_3^- level will fall by nearly 10 mmol, and from the reaction $H^+ + HCO_3^- \rightleftharpoons H_2CO_3 \rightleftharpoons H_2O + CO_2$, we predict that nearly 10 mmol of $CO_{2(d)}$ will form. If the system was closed and no CO_2 could escape, the new pH would be:

$$pH = 6.10 + \log \frac{[24-10]}{[1.2+10]} = 6.20 \quad (18)$$

This is an intolerably low—indeed a fatal—pH.

If the system is open, then CO_2 can escape into the air. If all of the extra CO_2 is expired by the lungs and the $CO_{2(d)}$ is kept at 1.2 mmol/L, then the pH would be:

$$pH = 6.10 + \log \frac{[24-10]}{[1.2]} = 7.17 \quad (19)$$

Although this pH is low, it is compatible with life.

Still, another mechanism promotes the escape of CO_2 from the body. An acidic blood pH stimulates breathing, which can make the P_{CO_2} lower than 40 mm Hg. If P_{CO_2} falls to 30 mm Hg ($CO_{2(d)}$ = 0.9 mmol/L), the pH would be:

$$pH = 6.10 + \log \frac{[24-10]}{[0.9]} = 7.29 \quad (20)$$

This pH is much closer to normal, demonstrating the effectiveness of an open system in buffering acid. The system is also open at the kidneys, and new HCO_3^- can be added to the plasma to correct the HCO_3^-. Once the pH of the blood is normal, the stimulus for hyperventilation disappears.

Blood pH is protected by a negative feedback of endogenous acids.

Another way in which blood pH may be protected is by changes in endogenous acid production (Fig. 24.4). An increase in blood pH caused by the addition of base to the body stimulates production of lactic acid and ketone body acids, which then reduces the alkaline shift in pH. A decrease

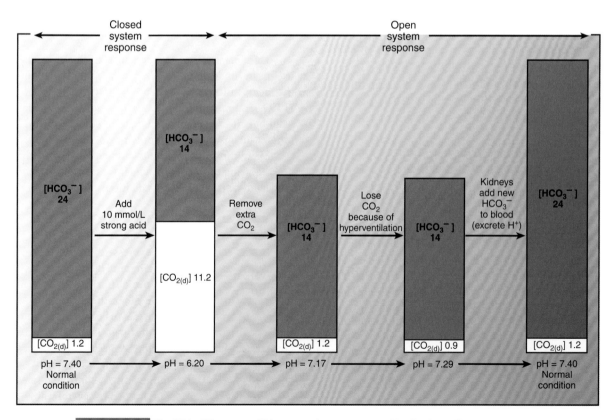

Figure 24.3 **The HCO_3^-/CO_2 system.** This system is remarkably effective in buffering added strong acid in the body, because it is open. HCO_3^- and $CO_{2(d)}$ are in mmol/L.

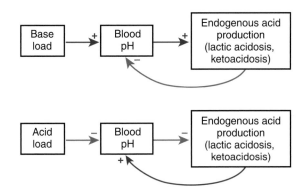

Figure 24.4 Negative feedback control of endogenous acid production. A base load, by raising pH, stimulates the endogenous production of acids. The addition of an exogenous acid load or increased endogenous acid production results in a fall in pH, which in turn inhibits the production of ketone body acids and lactic acid. These negative feedback effects attenuate changes in blood pH.

in blood pH inhibits the production of lactic acid and ketone body acids, which diminishes the acidic shift in pH.

This scenario is especially important when the endogenous production of these acids is high, as occurs during strenuous exercise or other conditions of circulatory inadequacy (lactic acidosis) or during ketosis resulting from uncontrolled diabetes mellitus, starvation, or alcoholism. These effects of pH on endogenous acid production result from changes in enzyme activities brought about by the pH changes, and they are part of a negative feedback mechanism regulating blood pH.

We have discussed the various buffers separately, but in the body, they all work together. In a solution containing multiple buffers, all are in equilibrium with the same H^+ ions. This idea is known as the **isohydric principle** (*isohydric* means "same H^+"). For plasma, for example, we can write:

$$pH = 6.80 + \log\frac{[HPO_4^{2-}]}{[H_2PO_4^-]} = 6.10 + \log\frac{[HCO_3^-]}{0.03PCO_2}$$
$$= pK_{protein} + \log\frac{[proteinate^-]}{H - protein} \quad (21)$$

If an acid or a base is added to such a complex mixture of buffers, all buffers take part in buffering and shift from one form (base or acid) to the other. The relative importance of each buffer depends on its amount, pK, and availability.

The isohydric principle underscores the fact that it is the *concentration ratio* for any buffer pair, along with its pK, that sets the pH. We can focus on the concentration ratio for one buffer pair, and all other buffers will automatically adjust their ratios according to the pH and their pK values.

The rest of this chapter emphasizes the role of the HCO_3^-/CO_2 buffer pair in setting the blood pH. Other buffers, however, are present and participate in buffering. The HCO_3^-/CO_2 system is emphasized because physiologic mechanisms (lungs and kidneys) regulate pH by acting on components of this buffer system.

Lungs regulate blood pH by changing arterial CO_2 tension ($PaCO_2$).

The respiratory system can rapidly and profoundly affect blood pH. Reflex changes in ventilation help to defend blood pH by changing the arterial PCO_2 and hence H_2CO_3 of the blood. As discussed in Chapter 21, a fall in blood pH stimulates ventilation. An elevated arterial blood PCO_2 is a powerful stimulus to increase ventilation; it acts on both peripheral and central chemoreceptors but primarily on the latter. CO_2 diffuses into brain interstitial and cerebrospinal fluids, where it causes a fall in pH, which stimulates chemoreceptors in the medulla oblongata. When ventilation is stimulated, the lungs blow off more CO_2, thereby making the blood less acidic. Conversely, a rise in blood pH inhibits ventilation; the consequent rise in blood H_2CO_3 reduces the alkaline shift in blood pH. Respiratory responses to disturbed blood pH begin within minutes and are maximal in about 12 to 24 hours. However, respiratory responses to such disturbances are self-limiting because in the process of using ventilation to alter pH, the potent stimulant for respiration, PCO_2, is altered such that the change in ventilation needed to return pH toward normal is blunted. For this reason, the respiratory system alone can only bring a deviation of arterial pH to within 50% to 75% of normal arterial pH. Other systems, namely, those in the kidney, are needed to further help the body compensate for an acid–base disturbance.

Kidneys play a crucial role in maintaining the body's acid–base homeostasis.

The kidneys play a critical role in maintaining acid–base balance. First, the kidneys must reclaim HCO_3^- filtered at the glomerulus. Recall from the Henderson-Hasselbalch equation that a decrease in HCO_3^- decreases the pH of the ECF; removal of HCO_3^- is the equivalent of adding acid to the system. Thus, if the HCO_3^- filtered at the glomerulus were to be excreted in the urine, a systemic acidosis will result. Second, the kidneys must remove H^+ if there is excess acid in the body or HCO_3^- if there is excess base. The usual challenge is to remove excess strong acids formed by normal metabolism. As we have learned, body buffer bases, particularly HCO_3^-, first buffer the strong acids produced by metabolism. Thus, ECF HCO_3^- is lost by reacting with strong acids produced in the body. The kidneys then must restore the depleted HCO_3^- by creating "new" HCO_3^- and eliminate H^+ in the urine.

Little of the H^+ excreted in the urine is present as free H^+. For example, if the urine has its lowest pH value (pH = 4.5), the H^+ is only 0.03 mEq/L. With a typical daily urine output of 1 to 2 L, the amount of acid the body must dispose of daily (roughly 70 mEq) obviously is not excreted in the free form. Most of the H^+ combines with urinary buffers to be excreted as **titratable acid** and as NH_4^+. Titratable acid is measured in the clinical laboratory by titrating the urine with a strong base (NaOH) and measuring the number of milliequivalents of OH^- needed to bring the urine pH back to the pH of the blood (usually 7.40). It represents the amount of hydrogen ions that are excreted combined with urinary buffers, such as phosphate, creatinine, and other bases. The largest component of titratable acid is normally phosphate, that is, $H_2PO_4^-$.

Hydrogen ions secreted by the renal tubules also combine with the free base NH_3 and are excreted as NH_4^+. Ammonia (a term that includes both NH_3 and NH_4^+) is produced by the kidney tubule cells and is secreted into the urine. Because the pK_a for NH_4^+ is high (9.0), most of the ammonia in the urine is present as NH_4^+. For this reason, too, NH_4^+ is not appreciably titrated when titratable acid is measured. Urinary ammonia is measured by a separate, colorimetric or enzymatic, method.

Kidneys excrete excess acid to maintain acid–base balance.

In stable acid–base balance, *net acid excretion* by the kidneys equals the net rate of H^+ addition to the body by metabolism or other processes, assuming that other routes of loss of acid or base (e.g., gastrointestinal losses) are small and can be neglected. This is normally the case. The net loss of H^+ in the urine can be calculated from the following equation, which shows typical values in the parentheses:

$$\begin{aligned}
\text{Renal net acid excretion (70 mEq/d)} \\
= \text{urinary titratable acid (24 mEq/d)} \\
+ \text{urinary ammonia (48 mEq/d)} \\
- \text{urinary } HCO_3^- \text{ (2 mEq/d)}
\end{aligned} \tag{22}$$

Urinary ammonia (as NH_4^+) ordinarily accounts for about two thirds of the excreted H^+ and titratable acid for about one third. Excretion of HCO_3^- in the urine represents a loss of base from the body. Therefore, it must be subtracted in the calculation of net acid excretion. If the urine contains significant amounts of organic anions such as citrate that potentially could have yielded HCO_3^- in the body, these should also be subtracted. Because the amount of free H^+ excreted is negligible, this is omitted from the equation

As the urine flows along the tubule, from the Bowman capsule on through the collecting ducts, three processes occur: filtered HCO_3^- is reabsorbed, titratable acid is formed, and ammonia is added to the tubular urine. All three processes involve H^+ secretion by the tubular epithelium that leads to increase urinary acidification. The nature and magnitude of these processes vary in different nephron segments. Figure 24.5 summarizes measurements of tubular fluid pH along the nephron and shows ammonia movements in various nephron segments.

The first step in the urinary acidification process starts in the proximal convoluted tubule. The pH of the glomerular ultrafiltrate is identical to that of the plasma from which it is derived (7.4). Hydrogen ions are secreted by the proximal tubule epithelium into the tubule lumen; about two thirds of this is accomplished by a Na^+/H^+ exchanger and about one third by an H^+-ATPase in the brush border membrane.

In spite of significant addition of H^+ into the tubular fluid, tubular fluid pH falls only to a value of about 6.7 by the end of the proximal convoluted tubule (see Fig. 24.5). The

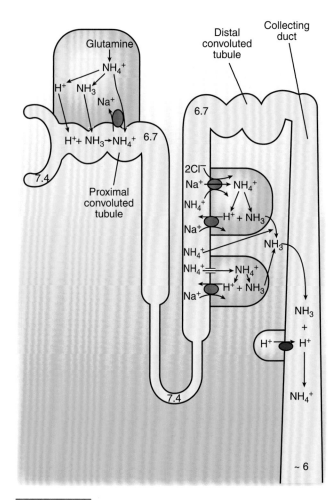

Figure 24.5 Acidification along the nephron. The pH of tubular urine decreases along the proximal convoluted tubule, rises along the descending limb of the Henle loop, falls along the ascending limb, and reaches its lowest values in the collecting ducts. Ammonia ($NH_3 + NH_4^+$) is chiefly produced in proximal tubule cells and is secreted into the tubular urine. NH_4^+ is reabsorbed in the thick ascending limb and accumulates in the kidney medulla. NH_3 diffuses into acidic collecting duct urine, where it is trapped as NH_4^+.

drop in pH is modest for two reasons: buffering of secreted H^+ and the high permeability of the proximal tubule epithelium to H^+. The glomerular filtrate and tubule fluid contain abundant buffer bases, especially HCO_3^-, which soak up secreted H^+, thereby minimizing a fall in pH. The proximal tubule epithelium is also rather "leaky" to H^+, so that any gradient from urine to blood, established by H^+ secretion, is soon limited by the diffusion of H^+ out of the tubule lumen into the blood surrounding the tubules.

Most of the hydrogen ions secreted by the nephron that are secreted in the proximal convoluted tubule are used to bring about the reabsorption of filtered HCO_3^-. Secreted hydrogen ions are also buffered by filtered phosphate to form titratable acid. Proximal tubule cells produce ammonia, mainly from the amino acid glutamine. Ammonia is secreted into the tubular urine by the diffusion of NH_3, which then combines with a secreted H^+ to form NH_4^+, or via the brush border membrane Na^+/H^+ exchanger, which can operate in a Na^+/NH_4^+ exchange mode.

The second step in urinary acidification involves the loop of Henle. Along the descending limb of the loop of Henle, the pH of tubular fluid rises (from 6.7 to 7.4). This rise is explained by an increase in intraluminal HCO_3^- caused by water reabsorption. Ammonia is secreted along the descending limb.

The tubular fluid is acidified by the secretion of H^+ along the ascending limb via a Na^+/H^+ exchanger. Along the thin ascending limb, ammonia is passively reabsorbed. Along the thick ascending limb, the Na–K–2Cl cotransporter in the luminal cell membrane actively reabsorbs NH_4^+ (NH_4^+ substitutes for K^+). Some NH_4^+ can be reabsorbed via a luminal cell membrane K^+ channel. Also, some NH_4^+ can be passively reabsorbed between cells in this segment; the driving force is the lumen-positive transepithelial electrical potential difference (see Chapter 22). Ammonia may undergo countercurrent multiplication in the loop of Henle, leading to an ammonia concentration gradient in the kidney medulla. The highest concentrations are at the tip of the papilla.

The final step in urinary acidification occurs in the distal nephrons. The distal nephron (distal convoluted tubule, connecting tubule, and collecting duct) differs from the proximal portion of the nephron in its H^+ transport properties. It secretes far fewer hydrogen ions, and they are secreted primarily via an electrogenic H^+-ATPase or an electroneutral H^+/K^+-ATPase. The distal nephron is also lined by "tight" epithelia, so little secreted H^+ diffuses out of the tubule lumen, making steep urine-to-blood pH gradients possible (see Fig. 24.5). Final urine pH is typically around 6 but may be as low as 4.5.

The distal nephron usually almost completely reabsorbs the small quantities of HCO_3^- that were not reabsorbed by more proximal nephron segments. Considerable amounts of titratable acid are formed as the urine is acidified. Ammonia that was reabsorbed by the ascending limb of the loop of Henle and accumulated in the medullary interstitial space diffuses as lipid-soluble NH_3 into collecting duct urine and combines with secreted H^+ to form NH_4^+. The collecting duct epithelium is impermeable to the lipid-insoluble NH_4^+, so ammonia is trapped in an acidic urine and is excreted as NH_4^+ (see Fig. 24.5).

The **intercalated cells** of the collecting duct are involved in acid–base transport and are of two major types: an acid-secreting *α-intercalated cell* and a bicarbonate-secreting *β-intercalated cell*. The α-intercalated cell has a vacuolar type of H^+-ATPase (the same kind as is found in lysosomes, endosomes, and secretory vesicles) and a H^+/K^+-ATPase (similar to that found in stomach and colon epithelial cells) in the luminal cell membrane and a Cl^-/HCO_3^- exchanger in the basolateral cell membrane (Fig. 24.6). The β-intercalated cell has the opposite polarity.

A more acidic blood pH results in the insertion of cytoplasmic H^+ pumps into the luminal cell membrane of α-intercalated cells and enhanced H^+ secretion. If the blood is made alkaline, HCO_3^- secretion by β-intercalated cells is increased. Because the amounts of HCO_3^- secreted

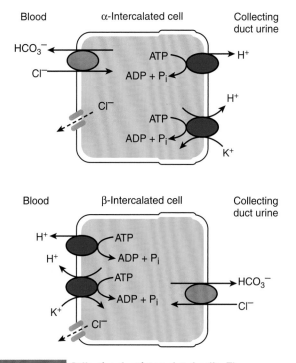

Figure 24.6 Collecting duct intercalated cells. The α-intercalated cell secretes H^+ via an electrogenic vacuolar H^+-ATPase and electroneutral H^+/K^+-ATPase and adds HCO_3^- to the blood via a basolateral cell membrane Cl^-/HCO_3^- exchanger. The β-intercalated cell, which is located in cortical collecting ducts, has the opposite polarity and secretes HCO_3^-. ADP, adenosine diphosphate; ATP, adenosine triphosphate; P_i, inorganic phosphate.

are ordinarily small compared with the amounts filtered and reabsorbed, HCO_3^- secretion will be omitted from the remaining discussion.

Kidneys regulate blood pH by reabsorbing filtered bicarbonate.

The kidney glomeruli filter about 4,320 mEq of HCO_3^- per day (180 L/d × 24 mEq/L). Recall from the Henderson-Hasselbalch equation that a reduction of HCO_3^- in the ECF will result in a reduction of pH; losing HCO_3^- therefore has the same effect as adding acid to the ECF. Urinary loss of even a small portion of this HCO_3^- will lead to an acidic blood and impair the body's ability to buffer its daily load of metabolically produced H^+. The kidney tubules have the important task of recovering the filtered HCO_3^- and returning it to the blood.

At the cellular level (Fig. 24.7), filtered HCO_3^- is not reabsorbed directly across the tubule's luminal cell membrane as is, for example, glucose. Instead, filtered HCO_3^- is reabsorbed indirectly via H^+ secretion, in the following way. About 90% of the filtered HCO_3^- is reabsorbed in the proximal convoluted tubule, and so we will emphasize events at this site. H^+ is secreted into the tubule lumen mainly via the Na^+/H^+ exchanger in the luminal membrane. It combines with filtered HCO_3^- to form H_2CO_3. Carbonic anhydrase (CA) in the luminal membrane (brush border) of the proximal tubule catalyzes the dehydration of H_2CO_3 to CO_2 and water in the lumen. The CO_2 diffuses back into the cell.

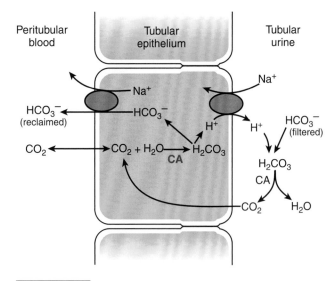

Figure 24.7 **A cell model for HCO$_3^-$ reabsorption.** Filtered HCO$_3^-$ combines with secreted H$^+$ and is reabsorbed indirectly. Carbonic anhydrase (CA) is present in the cells and, in the proximal tubule, on the brush border.

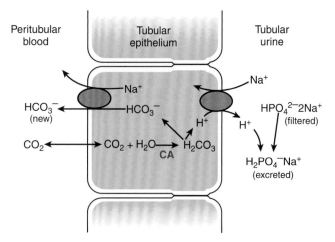

Figure 24.8 **A cell model for the formation of titratable acid.** Titratable acid (e.g., H$_2$PO$_4^-$) is formed when secreted H$^+$ is bound to a buffer base (e.g., HPO$_4^{2-}$) in the tubular urine. For each milliequivalent of titratable acid excreted, a milliequivalent of new HCO$_3^-$ is added to the peritubular capillary blood. CA, carbonic anhydrase.

Inside the cell, the hydration of CO$_2$ (catalyzed by intracellular CA) yields H$_2$CO$_3$, which instantaneously forms H$^+$ and HCO$_3^-$. The H$^+$ is secreted into the lumen, and the HCO$_3^-$ moves into the blood surrounding the tubules. In proximal tubule cells, this movement is favored by the inside negative membrane potential of the cell and by an electrogenic cotransporter in the basolateral membrane that simultaneously transports three HCO$_3^-$ and one Na$^+$. In addition, HCO$_3^-$ can leave the cell via a Cl$^-$/HCO$_3^-$ exchanger in the basolateral cell membrane (not shown in Fig. 24.7).

The reabsorption of filtered HCO$_3^-$ does not result in H$^+$ excretion or the formation of any "new" HCO$_3^-$. The secreted H$^+$ is not excreted because it combines with filtered HCO$_3^-$ that is, indirectly, reabsorbed. There is no net addition of HCO$_3^-$ to the body in this operation. *It is simply a recovery or reclamation process.* In contrast, when H$^+$ is excreted as titratable acid and ammonia, new HCO$_3^-$ is formed and is added to the blood. New HCO$_3^-$ replaces the HCO$_3^-$ used to buffer the strong acids produced by metabolism (see Fig. 24.2).

The formation of new HCO$_3^-$ and the excretion of H$^+$ are like two sides of the same coin. This fact is apparent if we assume that H$_2$CO$_3$ is the source of H$^+$:

$$CO_2 + H_2O \rightleftharpoons H_2CO_3 \Big\langle {}^{\nearrow \ H^+ \ \text{(urine)}}_{\searrow \ HCO_3^- \ \text{(blood)}} \qquad (23)$$

A loss of H$^+$ in the urine is equivalent to adding new HCO$_3^-$ to the blood. The same is true if H$^+$ is lost from the body via another route such as by vomiting of acidic gastric juice. This process leads to a rise in plasma HCO$_3^-$. *Conversely, a loss of HCO$_3^-$ from the body is equivalent to adding H$^+$ to the blood.*

Figure 24.8 shows a cell model for the formation of titratable acid. In this figure, H$_2$PO$_4^-$ is the titratable acid formed. H$^+$ and HCO$_3^-$ are produced in the cell from H$_2$CO$_3$.

The secreted H$^+$ combines with the basic form of the phosphate (HPO$_4^{2-}$) to form the acid phosphate (H$_2$PO$_4^-$). The secreted H$^+$ replaces one of the Na$^+$ ions accompanying the basic phosphate. The new HCO$_3^-$ generated in the cell moves into the blood, together with Na$^+$. For each milliequivalent of H$^+$ excreted in the urine as titratable acid, a milliequivalent of new HCO$_3^-$ is added to the blood. This process gets rid of H$^+$ in the urine, replaces ECF HCO$_3^-$, and helps to restore a normal blood pH.

The amount of titratable acid excreted depends on two factors: the pH of the urine and the availability of buffer. If the urine pH is lowered, more titratable acid can form. However, the supply of phosphate and other buffers in the urine is usually limited. Therefore, to excrete large amounts of acid, the kidneys must rely on increased ammonia excretion.

Figure 24.9 shows a cell model for the excretion of ammonia. The majority of ammonia is synthesized in proximal tubule cells by deamidation and deamination of glutamine:

$$\text{Glutamine} \xrightarrow[\text{Glutaminase}]{NH_4^+} \text{Glutamate}^- \xrightarrow[\text{Glutamate dehydrogenase}]{NH_4^+} \alpha\text{-Ketoglutarate}^{2-} \qquad (24)$$

As discussed earlier, ammonia is secreted into the urine by two mechanisms. As NH$_3$, it diffuses into the tubular urine; as NH$_4^+$, it substitutes for H$^+$ on the Na$^+$/H$^+$ exchanger. In the lumen, NH$_3$ combines with secreted H$^+$ to form NH$_4^+$, which is excreted.

For each milliequivalent of H$^+$ excreted as NH$_4^+$, one milliequivalent of new HCO$_3^-$ is added to the blood. The hydration of CO$_2$ in the tubule cell produces H$^+$ and HCO$_3^-$, as described earlier. Two H$^+$s are consumed when the anion α-ketoglutarate^{2-} is converted into CO$_2$ and water or into glucose in the cell. The new HCO$_3^-$ returns to the blood along with Na$^+$.

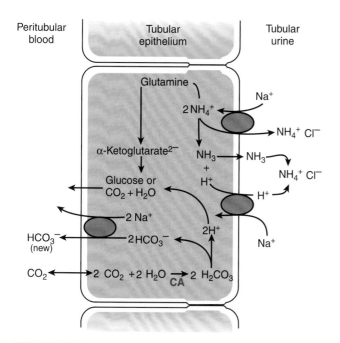

Figure 24.9 **A cell model for renal synthesis and excretion of ammonia.** Ammonium ions are formed from glutamine in the cell and are secreted into the tubular urine (**top**). H$^+$ from H$_2$CO$_3$ (**bottom**) is consumed when α-ketoglutarate is converted into glucose or CO$_2$ and H$_2$O. New HCO$_3^-$ is added to the peritubular capillary blood—1 milliequivalent for each milliequivalent of NH$_4^+$ excreted in the urine. CA, carbonic anhydrase.

If excess acid is added to the body, urinary ammonia excretion increases for two reasons. First, more acidic urine traps more ammonia (as NH$_4^+$) in the urine. Second, glutamine increases renal ammonia synthesis over a period of several days. Enhanced renal ammonia synthesis and excretion is a lifesaving adaptation, because it allows the kidneys to remove large H$^+$ excesses and add more new HCO$_3^-$ to the blood. Also, the excreted NH$_4^+$ can substitute in the urine for Na$^+$ and K$^+$, thereby diminishing the loss of these cations. With severe metabolic acidosis, ammonia excretion may increase almost 10-fold.

Metabolic dynamics influence renal acid secretion.

Many metabolic factors influence the renal excretion of H$^+$ including (1) intracellular pH, (2) arterial blood Pco$_2$, (3) CA activity, (4) Na$^+$ reabsorption, (5) plasma K$^+$, and (6) aldosterone (Fig. 24.10).

The pH in kidney tubule cells is a key factor influencing the secretion and, therefore, the excretion of H$^+$. A fall in pH (increased H$^+$) enhances H$^+$ secretion. A rise in pH (decreased H$^+$) lowers H$^+$ secretion. A decreased pH increases the activity of Na$^+$/H$^+$ exchangers in the luminal cell membrane. The activity of these exchangers is increased by the increased supply of H$^+$ and also by changes in exchanger conformation induced by binding of H$^+$. A low pH increases the recruitment of H$^+$-ATPases from intracellular vesicles and their insertion into the luminal membrane. A fall in intracellular pH also stimulates renal ammonia synthesis, which allows the kidneys to excrete more H$^+$ as NH$_4^+$.

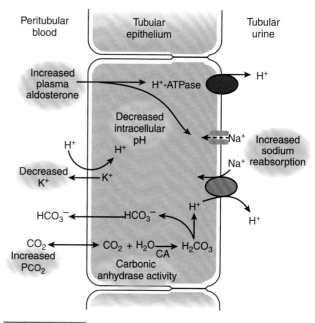

Figure 24.10 **Factors leading to increased H$^+$ secretion by the kidney tubule epithelium.** CA, carbonic anhydrase. (See text for details.)

An increase in arterial Pco$_2$ increases the formation of H$^+$ from H$_2$CO$_3$, leading to enhanced renal H$^+$ secretion and excretion—a useful compensation for any condition in which the blood contains too much H$_2$CO$_3$. (We will discuss this later when we consider respiratory acidosis.) A decrease in arterial Pco$_2$ results in lowered H$^+$ secretion and, consequently, less complete reabsorption of filtered HCO$_3^-$ and a loss of base in the urine (a useful compensation for respiratory alkalosis, also discussed later).

The enzyme carbonic anhydrase catalyzes two key reactions in urinary acidification:

a. The hydration of CO$_2$ in the cells, forming H$_2$CO$_3$ and yielding H$^+$ for secretion

b. The dehydration of H$_2$CO$_3$ to H$_2$O and CO$_2$ in the proximal tubule lumen, an important step in the reabsorption of filtered HCO$_3^-$

Large amounts of filtered HCO$_3^-$ may escape reabsorption if carbonic anhydrase is inhibited (usually by a drug). This situation leads to a fall in blood pH.

Another metabolic factor that is closely linked to acid secretion is Na$^+$ reabsorption. These two ions are directly linked, both being transported by the Na$^+$/H$^+$ exchanger in the luminal cell membrane. The relation is less direct in the collecting ducts. Enhanced Na$^+$ reabsorption in the ducts leads to a more negative intraluminal electrical potential, which favors H$^+$ secretion by its electrogenic H$^+$-ATPase. The avid renal reabsorption of Na$^+$ seen in states of volume depletion is accompanied by a parallel rise in urinary H$^+$ excretion, which can cause systemic alkalosis.

The fifth metabolic factor influencing renal excretion of H$^+$ is a change in plasma K$^+$ concentration. A fall in plasma K$^+$ favors the movement of K$^+$ from body cells into interstitial fluid (or blood plasma) and a reciprocal movement of H$^+$

Renal Physiology and Body Fluids

into cells. In the kidney tubule cells, these movements lower intracellular pH and increase H^+ secretion. K^+ depletion also stimulates ammonia synthesis by the kidneys (probably by lowering the intracellular pH); this increases NH_4^+ excretion. Finally, low plasma K^+ results in increased expression and activity of the H^+/K^+-ATPase in α-intercalated cells; although this allows the kidneys to conserve K^+, it also enhances H^+ secretion. The result is the complete reabsorption of filtered HCO_3^- and the enhanced generation of new HCO_3^- as more titratable acid and ammonia are excreted. Consequently, hypokalemia (or a decrease in body K^+ stores) leads to increased plasma HCO_3^- (*metabolic alkalosis*).

Hyperkalemia (or excess K^+ in the body) results in the opposite changes: an increase in intracellular pH, a decreased in H^+ secretion, an incomplete reabsorption of filtered HCO_3^-, and a fall in plasma HCO_3^- (*metabolic acidosis*).

Finally, the sixth metabolic factor affecting acid secretion by the kidney is aldosterone. Aldosterone stimulates the collecting ducts to secrete H^+ by three actions:

1. It directly stimulates the H^+-ATPase in collecting duct α-intercalated cells.
2. It enhances collecting duct Na^+ reabsorption, which leads to a more negative intraluminal potential and consequently promotes H^+ secretion by the electrogenic H^+-ATPase.
3. It promotes K^+ secretion. This response leads to hypokalemia, which increases renal H^+ secretion.

Hyperaldosteronism results in enhanced renal H^+ excretion and an alkaline blood pH (metabolic alkalosis); the opposite occurs with **hypoaldosteronism**.

▶ REGULATION OF INTRACELLULAR pH

The intracellular and ECF pH are linked by exchanges across cell membranes of H^+, HCO_3^-, various acids and bases, and CO_2. By stabilizing ECF pH, intracellular pH is maintained.

If hydrogen ions were passively distributed across cell membranes, intracellular pH would be lower than what is seen in most body cells. In skeletal muscle cells, for example, we can calculate from the Nernst equation (see Chapter 2) and a membrane potential of -90 mV that cytosolic pH should be 5.9 if ECF pH is 7.4; actual measurements, however, indicate a pH of 6.9. From this discrepancy, two conclusions are clear: hydrogen ions are not at equilibrium across the cell membrane, and the cell must use active mechanisms to extrude H^+.

Cellular pH is maintained by extruding hydrogen ions.

Cells are typically threatened by acidic metabolic end products and by the tendency for H^+ to diffuse into the cell down the electrical gradient (Fig. 24.11). H^+ is extruded by Na^+/H^+ exchangers, which are present in nearly all body cells. Eight different isoforms of these exchangers (designated NHE1, NHE2, etc.), with different tissue distributions, have been identified. These transporters exchange one H^+ for one Na^+ and therefore function in an electrically neutral fashion. Active extrusion of H^+ keeps the internal pH within narrow limits.

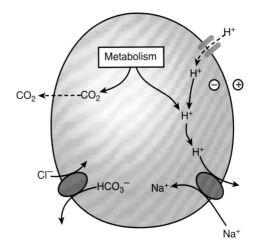

Figure 24.11 **Cell acid–base balance.** Body cells usually maintain a constant intracellular pH. The cell is acidified by the production of H^+ from metabolism and the influx of H^+ from the extracellular fluid (favored by the inside negative cell membrane potential). To maintain a stable intracellular pH, the cell must extrude hydrogen ions at a rate matching their input. Many cells also possess various HCO_3^- transporters (one example is shown) that defend against excess acid or base.

The activity of the Na^+/H^+ exchanger is regulated by intracellular pH and a variety of hormones and growth factors. Not surprisingly, an increase in intracellular H^+ stimulates the exchanger, but not only because of more substrate (H^+) for the exchanger. H^+ also stimulates the exchanger by protonating an activator site on the cytoplasmic side of the exchanger, thereby making the exchanger more effective in dealing with the threat of intracellular acidosis. Many hormones and growth factors, via intracellular second messengers, activate various protein kinases that stimulate or inhibit the Na^+/H^+ exchanger. In this way, they produce changes in intracellular pH, which may lead to changes in cell activity.

Besides extruding H^+, the cell can deal with acids and bases in other ways. First, in some cells, various HCO_3^- transporters (e.g., Na^+-dependent and Na^+-independent Cl^-/HCO_3^- exchangers, electrogenic Na^+/HCO_3^- cotransporters) may be present in plasma membranes. These exchangers may be activated by changes in intracellular pH. Second, cells have large stores of protein and organic phosphate buffers, which can bind or release H^+. Third, various chemical reactions in cells can also use up or release H^+. For example, the conversion of lactic acid to CO_2 and water or to glucose effectively disposes of acid. Fourth, various cell organelles may sequester H^+. For example, H^+-ATPases in endosomes and lysosomes pump H^+ out of the cytosol into these organelles. In summary, ion transport, buffering mechanisms, and metabolic reactions all ensure a relatively stable intracellular pH.

▶ PHYSIOLOGIC DISTURBANCES OF ACID–BASE BALANCE

Table 24.2 lists the normal values for the pH (or H^+), P_{CO_2}, and HCO_3^- of arterial blood plasma. A blood pH below 7.35 (H^+ concentration >45 nmol/L) indicates **acidemia**. A blood

TABLE 24.2 Normal Arterial Blood Plasma Acid–Base Values

Measure	Mean	Range*
pH	7.40	7.35–7.45
[H⁺] (nmol/L)	40	45–35
P_{CO_2} (mm Hg)	40	35–45
[HCO_3^-] (mEq/L)	24	22–26

*The range extends from two standard deviations below to two standard deviations above the mean and encompasses 95% of the population of healthy people.
P_{CO_2}, partial pressure of carbon dioxide. Brackets denote concentration.

pH above 7.45 (H⁺ concentration <35 nmol/L) indicates **alkalemia**. The range of pH values compatible with life is ~6.8 to 7.8 (H⁺ concentration = 160 to 16 nmol/L).

Four **simple acid–base disturbances** may lead to an abnormal blood pH: respiratory acidosis, respiratory alkalosis, metabolic acidosis, and metabolic alkalosis. The word "simple" indicates a single primary cause for the disturbance. **Acidosis** is an abnormal process that tends to produce acidemia. **Alkalosis** is an abnormal process that tends to produce alkalemia. If there is too much or too little CO_2, then a *respiratory disturbance* is present. If the problem is too much or too little HCO_3^-, then a *metabolic (or nonrespiratory) disturbance* of acid–base balance is present. Table 24.3 summarizes the changes in blood pH (or H⁺ concentration), plasma HCO_3^- concentration, and P_{CO_2} that occur in each of the four simple acid–base disturbances.

In considering acid–base disturbances, it is helpful to recall the Henderson-Hasselbalch equation for HCO_3^-/CO_2:

$$pH = 6.10 + \log \frac{[HCO_3^-]}{0.03P_{CO_2}} \qquad (25)$$

If the primary problem is a change in HCO_3^- concentration or P_{CO_2}, the pH can be brought closer to normal by changing the other member of the buffer pair *in the same direction*. For example, if P_{CO_2} is primarily decreased, a decrease in plasma HCO_3^- concentration will minimize the change in pH. In various acid–base disturbances, the lungs adjust the blood P_{CO_2}, and the kidneys adjust the plasma HCO_3^- concentration to reduce departures of pH from normal; these adjustments are called *compensations* (Table 24.3). Compensations attenuate, but do not correct the underlying disorder and often do not bring about a normal blood pH.

Respiratory acidosis and alkalosis are caused by altered levels of $PaCO_2$.

Hypoventilation is the condition when the level of breathing does not keep up with the metabolic production of carbon dioxide. As a result, **respiratory acidosis** occurs and is defined as an abnormal accumulation of CO_2 in the arterial blood. Recall from the control of breathing (Chapter 21) that the elevated arterial P_{CO_2} is referred to as **hypercapnia** that leads to increase in blood [H⁺]. The CO_2 buildup pushes the following reactions to the right:

$$CO_2 + H_2O \rightleftharpoons H_2CO_3 \rightleftharpoons H^+ + HCO_3^- \qquad (26)$$

Blood H_2CO_3 concentration increases, leading to an increase in H⁺ concentration or a fall in pH. Respiratory acidosis is caused by a decrease in overall alveolar ventilation. Hypoventilation occurs with certain lung diseases, with a suppression of ventilation (e.g., excessive central anesthesia, drug overdose, etc.) or as a mismatch between ventilation and perfusion. Respiratory acidosis also occurs if a person breathes CO_2-enriched air (e.g., deep-sea diving or in a submarine).

Hyperventilation results in a decrease in $PaCO_2$. A low $PaCO_2$ (hypocapnia) decreases blood [H⁺] leading to alkalosis. When the alkalosis is caused by the lungs blowing off too much carbon dioxide, the condition is called respiratory alkalosis. This loss causes blood H_2CO_3 concentration and, thus, H⁺ concentration to fall (pH rises). Loss of CO_2 from hyperventilation lowers alveolar and arterial blood P_{CO_2}. Respiratory alkalosis can be caused by voluntary effort, anxiety, direct stimulation of the medullary respiratory center by some abnormality (e.g., meningitis, fever, and aspirin intoxication), or hypoxia resulting from severe anemia or

TABLE 24.3 Directional Changes in Arterial Blood Plasma Values in the Four Simple Acid–Base Disturbances

Disturbance	Arterial Plasma				Compensatory Response
	pH	[H⁺]	[HCO_3^-]	P_{CO_2}	
Respiratory acidosis	↓	↑	↑	↑↑	Kidneys increase H⁺ excretion
Respiratory alkalosis	↑	↓	↓	↓↓	Kidneys increase HCO_3^- excretion
Metabolic acidosis	↓	↑	↓↓	↓	Alveolar hyperventilation; kidneys increase H⁺ excretion
Metabolic alkalosis	↑	↓	↑↑	↑	Alveolar hypoventilation; kidneys increase HCO_3^- excretion

Double arrows indicate the main effect. Brackets denote concentration.
P_{CO_2}, partial pressure of carbon dioxide.

high altitude. Although hyperventilation causes respiratory alkalosis, it also causes changes that inhibit ventilation (i.e., a fall in P_{CO_2} and a rise in blood pH) and, therefore, limit the extent of hyperventilation.

Respiratory acidosis and alkalosis are buffered primarily within cells.

With both respiratory acidosis and alkalosis, more than 95% of the chemical buffering occurs within cells. In the case of respiratory acidosis, the cells contain many proteins and organic phosphates that can bind H^+. For example, hemoglobin (Hb) in red blood cells combines with H^+ from H_2CO_3, thereby minimizing the increase in free H^+. Recall from Chapter 19 the buffering reaction:

$$H_2CO_3 + HbO_2^- \rightleftharpoons HHb + O_2 + HCO_3^- \qquad (27)$$

This reaction raises the plasma HCO_3^-. In *acute* respiratory acidosis, such chemical buffering processes in the body lead to an increase in plasma HCO_3^- concentration of about 1 mEq/L for each 10 mm Hg increase in P_{CO_2} (Table 24.4). Bicarbonate is not a buffer for H_2CO_3 because the reaction

$$H_2CO_3 + HCO_3^- \rightleftharpoons HCO_3^- + H_2CO_3 \qquad (28)$$

is simply an exchange reaction and does not affect the pH.

An example illustrates how chemical buffering reduces a fall in pH during respiratory acidosis. Suppose P_{CO_2}

increased from a normal value of 40 to 70 mm Hg ($CO_{2(d)}$ = 2.1 mmol/L). If there were no body buffer bases that could accept H^+ from H_2CO_3 (i.e., if there were no measurable increase in HCO_3^- concentration), the resulting pH would be 7.16:

$$pH = 6.10 + \log\frac{[24]}{[2.1]} = 7.16 \qquad (29)$$

In acute respiratory acidosis, a 3 mEq/L increase in plasma HCO_3^- concentration occurs with a 30 mm Hg rise in P_{CO_2} (see Table 24.4). Therefore, the pH is 7.21:

$$pH = 6.10 + \log\frac{[24+3]}{[2.1]} = 7.21 \qquad (30)$$

The pH of 7.21 is closer to a normal pH because body buffer bases (mainly intracellular buffers) such as proteins and phosphates combine with H^+ liberated from H_2CO_3.

As with respiratory acidosis, during respiratory alkalosis, most of the buffering occurs within cells. Cell proteins and organic phosphates liberate hydrogen ions, which are added to the ECF and lower the plasma HCO_3^- concentration, thereby reducing the alkaline shift in pH.

With *acute* respiratory alkalosis, plasma HCO_3^- concentration falls by about 2 mEq/L for each 10 mm Hg drop in P_{CO_2} (see Table 24.4). For example, if P_{CO_2} drops from 40 to 20 mm Hg ($CO_{2(d)}$ = 0.6 mmol/L), plasma HCO_3^- concentration falls by 4 mEq/L and the pH will be 7.62:

$$pH = 6.10 + \log\frac{[24-4]}{[0.6]} = 7.62 \qquad (31)$$

If plasma HCO_3^- had not changed, the pH would have been 7.70:

$$pH = 6.10 + \log\frac{[24]}{[0.6]} = 7.70 \qquad (32)$$

Lungs and kidneys compensate for respiratory acidosis and alkalosis.

An initial rise in arterial P_{CO_2} leads to acidosis and a fall in pH. The lungs then compensate for the elevated arterial P_{CO_2} by increasing breathing (see Chapter 21), thereby diminishing the severity of the acidosis.

In addition, the kidneys compensate for respiratory acidosis by adding more H^+ to the urine and adding new HCO_3^- to the blood. The increased P_{CO_2} stimulates renal H^+ secretion, which allows the reabsorption of all filtered HCO_3^-. Excess H^+ is excreted as titratable acid and NH_4^+; these processes add new HCO_3^- to the blood, causing plasma HCO_3^- concentration to rise. This compensation takes several days to develop fully.

With *chronic* respiratory acidosis, plasma HCO_3^- increases, on the average, by 4 mEq/L for each 10 mm Hg rise in P_{CO_2} (see Table 24.4). This rise exceeds that seen

TABLE 24.4	Compensatory Responses in Acid–Base Disturbances[*]
Type	**Response**
Respiratory acidosis	
Acute	1 mEq/L increase in plasma [HCO_3^-] for each 10 mm Hg increase in P_{CO_2}[†]
Chronic	4 mEq/L increase in plasma [HCO_3^-] for each 10 mm Hg increase in P_{CO_2}[‡]
Respiratory alkalosis	
Acute	2 mEq/L decrease in plasma [HCO_3^-] for each 10 mm Hg decrease in P_{CO_2}[†]
Chronic	4 mEq/L decrease in plasma [HCO_3^-] for each 10 mm Hg decrease in P_{CO_2}[‡]
Metabolic acidosis	1.3 mm Hg decrease in P_{CO_2} for each 1 mEq/L decrease in plasma [HCO_3^-][**]
Metabolic alkalosis	0.7 mm Hg increase in P_{CO_2} for each 1 mEq/L increase in plasma [HCO_3^-][**]

[*]Empirically determined average changes measured in people with simple acid–base disorders.
[†]This change is primarily a result of chemical buffering.
[‡]This change is primarily a result of renal compensation.
[**]This change is a result of respiratory compensation.
P_{CO_2}, partial pressure of carbon dioxide. Brackets denote concentration.

with acute respiratory acidosis because of the renal addition of HCO_3^- to the blood. One would expect a person with chronic respiratory acidosis and a Pco_2 of 70 mm Hg to have an increase in plasma HCO_3^- of 12 mEq/L. The blood pH would be 7.33:

$$pH = 6.10 + \log\frac{[24+12]}{[2.1]} = 7.33 \qquad (33)$$

With chronic respiratory acidosis, time for renal compensation is allowed, so blood pH (in this example, 7.33) is much closer to normal than that is observed during acute respiratory acidosis (pH 7.21).

The kidneys compensate for respiratory alkalosis by excreting HCO_3^- in the urine. A reduced Pco_2 reduces H^+ secretion by the kidney tubule epithelium. As a result, some of the filtered HCO_3^- is not reabsorbed. When the urine becomes more alkaline, titratable acid excretion vanishes, and little ammonia is excreted. The enhanced output of HCO_3^- causes plasma HCO_3^- concentration to fall.

Chronic respiratory alkalosis is accompanied by a 4 mEq/L fall in plasma HCO_3^- concentration for each 10 mm Hg drop in Pco_2 (see Table 24.4). For example, in a person with chronic hyperventilation and a Pco_2 of 20 mm Hg, the blood pH is:

$$pH = 6.10 + \log\frac{[24-8]}{[0.6]} = 7.53 \qquad (34)$$

This pH is closer to normal than the pH of 7.62 of acute respiratory alkalosis. The difference between the two situations is largely a result of renal compensation.

Metabolic acidosis is a condition in which tissue and blood pH is abnormally low due to an increase in nonvolatile acids.

An increase in nonvolatile acids, such as lactic acid, leads to **metabolic acidosis**. This is an abnormal process characterized by a gain of acid (other than H_2CO_3) or a loss of HCO_3^-. With metabolic acidosis, both plasma HCO_3^- concentration and pH decrease. If a strong acid increases in the body, the reactions

$$H^+ + HCO_3^- \rightleftharpoons H_2CO_3 \rightleftharpoons H_2O + CO_2 \qquad (35)$$

are pushed to the right. The added H^+ consumes HCO_3^-. If much acid is infused rapidly, Pco_2 rises, as the equation predicts. This increase occurs only transiently, however, because the body is an open system, and the lungs exhaust CO_2 as it is generated. Pco_2 actually falls below normal because an acidic blood pH stimulates ventilation (see Fig. 24.3).

Many conditions can produce metabolic acidosis, including renal failure, uncontrolled diabetes mellitus, lactic acidosis, the ingestion of acidifying agents such as NH_4Cl, excessive renal excretion of HCO_3^-, and diarrhea.

In renal failure, the kidneys cannot excrete H^+ fast enough to keep up with metabolic acid production, and in uncontrolled diabetes mellitus, the production of ketone body acids increases. Lactic acidosis results from tissue hypoxia. Ingested NH_4Cl is converted into urea and a strong acid, HCl, in the liver. Diarrhea causes a loss of alkaline intestinal fluids.

Metabolic acidosis is buffered by cellular fluid, bone, lungs, and kidneys.

Excess acid is chemically buffered in ECF and ICF and bone. In metabolic acidosis, roughly half the buffering occurs in cells and bone. HCO_3^- is the principal buffer in the ECF.

The acidic blood pH stimulates the respiratory system to lower blood Pco_2 by hyperventilation. This action lowers blood H_2CO_3 concentration and, thereby, tends to alkalinize the blood, opposing the acidic shift in pH. Metabolic acidosis is accompanied on average by a 1.3 mm Hg fall in Pco_2 for each 1 mEq/L drop in plasma HCO_3^- concentration (see Table 24.4). Suppose, for example, the infusion of a strong acid causes the plasma HCO_3^- concentration to drop from 24 to 12 mEq/L. If there were no respiratory compensation and the Pco_2 did not change from its normal value of 40 mm Hg, the pH would be 7.10:

$$pH = 6.10 + \log\frac{[12]}{[1.2]} = 7.10 \qquad (36)$$

With respiratory compensation, the Pco_2 falls by 16 mm Hg (12×1.3) to 24 mm Hg ($CO_{2(d)} = 0.72$ mmol/L) and pH is 7.32:

$$pH = 6.10 + \log\frac{[12]}{[0.72]} = 7.32 \qquad (37)$$

This value is closer to normal than a pH of 7.10. The respiratory response develops promptly (within minutes) and is maximal after 12 to 24 hours.

The kidneys respond to metabolic acidosis, as well, by adding more H^+ to the urine. Because the plasma HCO_3^- concentration is primarily lowered by the metabolic acidosis, the filtered load of HCO_3^- drops, and the kidneys can accomplish the complete reabsorption of filtered HCO_3^- (see Fig. 24.7). In addition, more H^+ is excreted as titratable acid and NH_4^+. With chronic metabolic acidosis, the kidneys make more ammonia. The kidneys can therefore add more new HCO_3^- to the blood, to replace lost HCO_3^-. If the underlying cause of metabolic acidosis is corrected, then healthy kidneys can correct the blood pH in a few days.

Plasma anion gap is used to determine the etiology of metabolic acidosis.

The **anion gap** is a useful concept, especially when trying to determine the possible cause of metabolic acidosis. Plasma anion gap is calculated from sodium, chloride, and bicarbonate concentrations. In any body fluid, the sum of the cations and the sum of the anions are equal because solutions are electrically neutral. For blood plasma, we can write:

$$\sum \text{cations} = \sum \text{anions} \quad (38)$$

or

$$[Na^+] + [\text{unmeasured cations}]$$
$$= [Cl^-] + [HCO_3^-] + [\text{unmeasured anions}] \quad (39)$$

The unmeasured cations include K^+, Ca^{2+}, and Mg^{2+} ions. Because these are present at relatively low concentrations (compared with Na^+) and are usually fairly constant, we choose to neglect them. The unmeasured anions include plasma proteins, sulfate, phosphate, citrate, lactate, and other organic anions. If we rearrange the above equation, we get:

$$[\text{unmeasured anions}] \text{ or "anion gap"}$$
$$= [Na^+] - [Cl^-] - [HCO_3^-] \quad (40)$$

In a healthy person, the anion gap falls in the range of 8 to 14 mEq/L. For example, if plasma Na^+ concentration is 140 mEq/L, Cl^- concentration is 105 mEq/L, and if HCO_3^- concentration is 24 mEq/L, the anion gap is 11 mEq/L. If an acid such as lactic acid is added to plasma, the reaction of lactic acid + $HCO_3^- \rightleftharpoons$ lactate$^-$ + H_2O + CO_2 will be pushed to the right. Consequently, the plasma HCO_3^- will be decreased, and because the Cl^- is not changed, the anion gap will be increased. The unmeasured anion in this case is lactate$^-$.

In several types of metabolic acidosis, the low blood pH is accompanied by a high anion gap (Table 24.5). (These can be remembered from the mnemonic MULEPAKS, formed from the first letters of this list.) In other types of metabolic acidosis, the low blood pH is accompanied by a normal anion gap (see Table 24.5). For example, with diarrhea and a loss of alkaline intestinal fluid, plasma HCO_3^- concentration falls, but plasma Cl^- concentration rises, and the two changes counterbalance each other so the anion gap is unchanged. Again, the chief value of the anion gap concept is that it allows a clinician to narrow down possible explanations for metabolic acidosis in a patient.

CLINICAL FOCUS | 24.1

Metabolic Acidosis in Diabetes Mellitus

Diabetes mellitus is a common disorder characterized by an insufficient secretion of insulin (type I) or insulin resistance by the major target tissues (skeletal muscle, liver, and adipocytes; type II). A severe **metabolic acidosis** may develop in uncontrolled diabetes mellitus.

Acidosis occurs because insulin deficiency leads to decreased glucose use, a diversion of metabolism toward the use of fatty acids, and an overproduction of ketone body acids (acetoacetic acid and β-hydroxybutyric acids). Ketone body acids are fairly strong acids (pK_a 4 to 5); they are neutralized in the body by HCO_3^- and other buffers. Increased production of these acids leads to a fall in plasma HCO_3^- concentration, an increase in plasma anion gap, and a fall in blood pH (acidemia).

Severe acidemia, whatever its cause, has many adverse effects on the body. It impairs myocardial contractility, resulting in a decrease in cardiac output. It causes arteriolar dilation, which leads to a fall in arterial blood pressure. Hepatic and renal blood flows are decreased. Re-entrant arrhythmias and a decreased threshold for ventricular fibrillation can occur. The respiratory muscles show decreased strength and fatigue easily. Metabolic demands are increased, as a result in part of activation of the sympathetic nervous system, but at the same time, anaerobic glycolysis and ATP synthesis are reduced by acidemia. Hyperkalemia is favored and protein catabolism is enhanced. Severe acidemia causes impaired brain metabolism and cell volume regulation, leading to progressive obtundation and coma.

An increased acidity of the blood stimulates pulmonary ventilation, resulting in a compensatory lowering of alveolar and arterial blood partial pressure of carbon dioxide (PCO_2). The consequent reduction in blood H_2CO_3 concentration acts to move the blood pH back toward normal. The labored, deep breathing that accompanies severe uncontrolled diabetes is called **Kussmaul respiration**.

The kidneys compensate for metabolic acidosis by reabsorbing all the filtered HCO_3^-. They also increase the excretion of titratable acid, part of which is composed of ketone body acids. These acids can only be partially titrated to their acid form in the urine, because the urine pH cannot go below 4.5. Thus, ketone body acids are excreted mostly in their anionic form; because of the requirement of electroneutrality in solutions, increased urinary excretion of Na^+ and K^+ results.

An important compensation for the acidosis is increased renal synthesis and excretion of ammonia. This adaptive response takes several days to develop fully, but it allows the kidneys to dispose of large amounts of H^+ in the form of NH_4^+. The NH_4^+ in the urine can replace Na^+ and K^+ ions, thereby resulting in conservation of these valuable cations.

The severe acidemia, electrolyte disturbances, and volume depletion that accompany uncontrolled diabetes mellitus may be fatal. Correction of the acid–base disturbance is best achieved by addressing the underlying cause, rather than just treating the symptoms. Therefore, the administration of a suitable dose of insulin is usually the key element of therapy. In some patients with marked acidemia (pH < 7.10), NaHCO$_3$ solutions may be infused intravenously to speed recovery, but this does not correct the underlying metabolic problem. Losses of Na^+, K^+, and water should be replaced. ▪

TABLE 24.5 High and Normal Anion Gap Metabolic Acidosis

Condition	Explanation
High anion gap metabolic acidosis	
Methanol intoxication	Methanol metabolized to formic acid
Uremia	Sulfuric, phosphoric, uric, and hippuric acids retained as a result of renal failure
Lactic acid	Lactic acid buffered by HCO_3^- and accumulated as lactate
Ethylene glycol intoxication	Ethylene glycol metabolized to glyoxylic, glycolic, and oxalic acids
p-Aldehyde intoxication	*p*-Aldehyde metabolized to acetic and chloroacetic acids
Ketoacidosis	Production of β-hydroxybutyric and acetoacetic acids
Salicylate intoxication	Impaired metabolism leading to production of lactic acid and ketone body acids; accumulation of salicylate
Normal anion gap metabolic acidosis	
Diarrhea	Loss of HCO_3^- in stool; kidneys conserve Cl^-
Renal tubular acidosis	Loss of HCO_3^- in urine or inadequate excretion of H^+; kidneys conserve Cl^-
Ammonium chloride ingestion	NH_4^+ is converted to urea in the liver, a process that consumes HCO_3^-; excess Cl^- is ingested

Metabolic alkalosis is a condition in which blood pH is abnormally high due to nonvolatile acid loss.

Metabolic alkalosis is characterized by a gain of a strong base or HCO_3^- or a loss of a nonvolatile acid (other than carbonic acid). Plasma HCO_3^- concentration and pH rise; PCO_2 rises because of respiratory compensation. These changes are opposite of those seen in metabolic acidosis (see Table 24.3). A variety of situations can produce metabolic alkalosis, including the ingestion of excessive amounts of antacids, vomiting of acidic gastric juice, and enhanced renal H^+ loss (e.g., resulting from hyperaldosteronism or hypokalemia).

Metabolic alkalosis is buffered primarily by lungs and kidneys.

Chemical buffers in the body limit the alkaline shift in blood pH by releasing H^+ as they are titrated in the alkaline direction. As a result, the buffering for metabolic alkalosis that occurs in cells is much smaller.

Both the lung and kidney are the major compensatory sites for metabolic acidosis. The respiratory compensation is hypoventilation. An alkaline blood pH inhibits ventilation, which in turn raises the blood PCO_2 and H_2CO_3 concentration. Hypoventilation thereby reduces the alkaline shift in pH. A 1 mEq/L rise in plasma HCO_3^- concentration caused by metabolic alkalosis is accompanied by a 0.7 mm Hg rise in PCO_2 (see Table 24.4). If, for example, the plasma HCO_3^- concentration rose to 40 mEq/L, what would the plasma pH be with and without respiratory compensation? With respiratory compensation, the PCO_2 should rise by 11.2 mm Hg (0.7 × 16) to 51.2 mm Hg ($CO_{2(d)}$ = 1.54 mmol/L). The pH is 7.51:

$$pH = 6.10 + \log\frac{[40]}{[1.54]} = 7.51 \qquad (41)$$

Without respiratory compensation, the pH would be 7.62:

$$pH = 6.10 + \log\frac{[40]}{[1.2]} = 7.62 \qquad (42)$$

Respiratory compensation for metabolic alkalosis is limited because hypoventilation leads to hypoxia and CO_2 retention, and both, in turn, increase breathing.

The kidneys respond to metabolic alkalosis by lowering the plasma HCO_3^- concentration. The plasma HCO_3^- concentration is primarily raised, so more HCO_3^- is filtered than can be reabsorbed (see Fig. 24.7); in addition, HCO_3^- is secreted in the collecting ducts. Both of these changes lead to increased urinary HCO_3^- excretion. If the cause of the metabolic alkalosis is corrected, the kidneys can often restore the plasma HCO_3^- concentration and pH to normal in a day or two. However, metabolic alkalosis accompanied by loss of fluid volume (e.g., severe vomiting) creates a unique complication to the body's compensatory mechanisms for that type of metabolic alkalosis. With a loss of fluid volume in the body, aldosterone will be produced to reabsorb sodium in the distal nephron. This though will also stimulate acid secretion in that part of the nephron that will tend to extend or exacerbate the alkalosis in the body. Therefore, for the body to be able to best compensate for a metabolic alkalosis with a volume loss, the volume loss must be corrected first before compensatory mechanisms for alkalosis can be fully effective.

pH–bicarbonate diagram is a useful way to determine acid–base disorder etiology.

Acid–base data should always be interpreted in the context of other information about a patient. For example, a low blood pH indicates acidosis; a high blood pH indicates alkalosis. If acidosis is present, for example, it could be either respiratory

Vomiting and Metabolic Alkalosis

Vomiting of acidic gastric juice results in **metabolic alkalosis,** along with fluid and electrolyte disturbances. Gastric juice contains about 0.1 M HCl. The acid is secreted by stomach parietal cells; these cells have an H^+/K^+-ATPase in their luminal cell membrane and a Cl^-/HCO_3^- exchanger in their basolateral cell membrane. When HCl is secreted into the stomach lumen and lost to the outside, there is a net gain of HCO_3^- in the blood plasma and no change in the anion gap. The HCO_3^-, in effect, replaces lost plasma Cl^-.

Ventilation is inhibited by the alkaline blood pH, resulting in a rise in P_{CO_2}. This respiratory compensation for the metabolic alkalosis, however, is limited, because hypoventilation leads to a rise in partial pressure of carbon dioxide (P_{CO_2}) and a fall in partial pressure of oxygen (P_{O_2}), both of which stimulate breathing.

The logical renal compensation for metabolic alkalosis is enhanced excretion of HCO_3^-. In people with persistent vomiting, however, the urine is sometimes acidic and renal HCO_3^- reabsorption is enhanced, thereby maintaining the elevated plasma HCO_3^-. This situation arises because vomiting is accompanied by losses of extracellular fluid (ECF) and K^+. Fluid loss leads to a decrease in effective arterial blood volume and engagement of mechanisms that reduce Na^+ excretion, such as decreased glomerular filtration rate and increased plasma renin, angiotensin, and aldosterone levels (see Chapter 23). Aldosterone stimulates H^+ secretion by collecting duct α-intercalated cells. Renal tubular Na^+/H^+ exchange is stimulated by volume depletion because the tubules reabsorb Na^+ more avidly than usual. With more H^+ secretion, more new HCO_3^- is added to the blood. The kidneys reabsorb filtered HCO_3^- completely, even though the plasma HCO_3^- level is elevated and maintain the metabolic alkalosis.

Vomiting results in K^+ depletion because of a loss of K^+ in the vomitus, decreased food intake, and, most important quantitatively, enhanced renal K^+ excretion. Extracellular alkalosis results in a shift of K^+ into cells (including renal cells) and thereby promotes K^+ secretion and excretion. Elevated plasma aldosterone levels also favor K^+ loss in the urine.

Treatment for the metabolic alkalosis primarily depends on eliminating the cause of vomiting. Correction of the alkalosis by administering an organic acid, such as lactic acid, does not make sense, because this acid would simply be converted to CO_2 and H_2O; this approach also does not address the Cl^- deficit. The ECF volume depletion and the Cl^- and K^+ deficits can be corrected by administering isotonic saline and appropriate amounts of KCl. Because replacement of Cl^- is a key component of therapy, this type of metabolic alkalosis is said to be "chloride responsive." After Na^+, Cl^-, water, and K^+ deficits have been replaced, excess HCO_3^- (accompanied by surplus Na^+) will be excreted in the urine, and the kidneys will return blood pH to normal. ■

---

or metabolic. A low blood pH and elevated P_{CO_2} point to respiratory acidosis; a low pH and low plasma HCO_3^- indicate metabolic acidosis. If alkalosis is present, it could be either respiratory or metabolic. A high blood pH and low plasma P_{CO_2} indicate respiratory alkalosis; a high blood pH and high plasma HCO_3^- indicate metabolic alkalosis.

The pH–bicarbonate diagram (Fig. 24.12) is a useful way to look at arterial blood data and determine what type

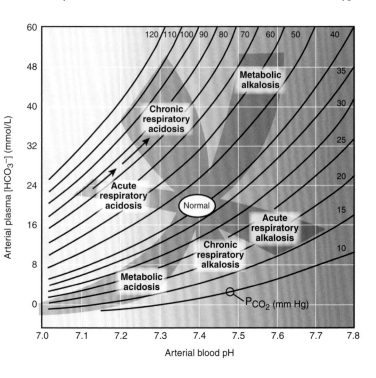

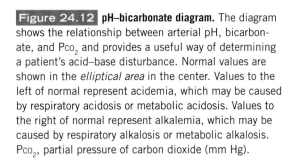

Figure 24.12 pH–bicarbonate diagram. The diagram shows the relationship between arterial pH, bicarbonate, and P_{CO_2} and provides a useful way of determining a patient's acid–base disturbance. Normal values are shown in the *elliptical area* in the center. Values to the left of normal represent acidemia, which may be caused by respiratory acidosis or metabolic acidosis. Values to the right of normal represent alkalemia, which may be caused by respiratory alkalosis or metabolic alkalosis. P_{CO_2}, partial pressure of carbon dioxide (mm Hg).

of acid–base disturbance may be present in a patient. The ellipse in the center of this diagram shows the normal range of values for arterial blood pH, P_{CO_2}, and plasma bicarbonate concentration. Values on the left side of the diagram (acidemia, or a low blood pH) are caused by respiratory acidosis or metabolic acidosis. Values on the right side of the diagram (alkalemia, or an elevated blood pH) are caused by respiratory alkalosis or metabolic alkalosis. The curved lines that slope upward and to the right are P_{CO_2} isobars; all along each line, the P_{CO_2} is the same (*iso* = "same," *bar* = "pressure"). Each point on the diagram must satisfy the Henderson-Hasselbalch equation, so that, for example, if pH and HCO_3^- concentration are known, the P_{CO_2} is automatically defined. The shaded areas include 95% of people with the designated simple acid–base disturbance. Note that a distinction is made between acute and chronic *respiratory* disturbances of acid–base balance but not between acute and chronic *metabolic* disturbances of acid–base balance. This is because the renal compensation for a respiratory disturbance may take days, whereas the respiratory compensation for a

metabolic disturbance is prompt (minutes to hours). Note that compensations tend to return the blood pH closer to normal. For example, if the P_{CO_2} is acutely raised (move to a higher P_{CO_2} isobar), then the pH will fall, but with chronic respiratory acidosis, the pH is closer to normal at the same P_{CO_2} (because of renal addition of bicarbonate to the blood). Mixed acid–base disturbances often, but do not always, fall outside of the shaded areas. For example, values for a patient with chronic respiratory acidosis (e.g., as a result of pulmonary disease) and metabolic acidosis (e.g., resulting from shock) might just happen to fall in the area that indicates acute respiratory acidosis. Values for a patient with a simple disturbance could fall outside the shaded area if insufficient time has elapsed, especially for renal compensation. For example, as illustrated in Figure 24.12 (*arrows on left*), data from a person might have been collected during the period of transition from acute to chronic respiratory acidosis. A complete history and physical examination provide important clues in deciding what acid–base disturbances may be present in a patient.

INTEGRATED MEDICAL SCIENCES

Type 1 (Distal) Renal Tubular Acidosis

A 13-year-old girl, who was adopted at age 5, was evaluated for failure to grow and recurrent pathologic bone fractures. Her height was 1.13 m, and she weighed 21.5 kg. She had suffered several broken bones since the age of 5. Her legs were deformed, and at age 12, she passed the first of several urinary stones. Radiographs of the femurs revealed bowing (characteristic of rickets), a healed fracture, and extreme decalcification of the skeleton. An abdominal radiograph at the level of the kidneys showed dense calcium deposits in the medullas of both kidneys (nephrocalcinosis).

Her plasma $[Na^+]$ was 139 mEq/L, $[K^+]$ 3.2 mEq/L, $[Ca^{2+}]$ 9.0 mg/dL, Cl^- 114 mEq/L, inorganic phosphate 1.9 mg/dL, BUN 10 mg/dL, and total (protein) 6.5 g/dL. Urinalysis showed a trace of protein, no glucose, specific gravity of 1.010, pH=7, no bacteria in cultured urine, and 25 white blood cells per high-power field.

Additional tests indicated an arterial blood pH of 7.30, PaO_2 106 mm Hg, $PaCO_2$ 27 mm Hg, bicarbonate 16 mmol/L, and SaO_2 96% confirming metabolic acidosis. Her estimated glomerular filtration rate was 56 mL/min/1.73 m² body surface area, a value below the normal range but not low enough to account for the abnormalities observed.

The urine pH appeared to be unduly high considering the presence of a metabolic acidosis. To test her ability to acidify the urine, she was given 0.1 g NH_4Cl/kg body weight orally, and urine pH was measured in samples collected over the next 6 hours. The minimum urine pH was 6. Failure to acidify her urine, the diagnosis of distal renal tubular acidosis (RTA) was made. The history of recurrent renal stones,

metabolic acidosis, and alkaline urine was further in keeping with a diagnosis of type (distal) RTA.

The patient was started on a regimen of two glasses of milk, five drops of a vitamin D preparation, and 20 mL Shohl's solution, three times daily. When she was seen 7 months later, the changes were almost miraculous. She had completely recalcified her skeleton, and all radiographic evidence of active rickets had disappeared. The serum phosphate had risen to 5.3 mg/dL. From a frail, bed-ridden child, she had developed into a sturdy, but still undersized, young woman. Her weight increased by 9.5 kg and she had grown 8 cm in height. Her breasts had developed and she had her first menstrual period.

This patient illustrates the classical case for renal tubular acidosis. In RTA, the kidney tubules malfunction, resulting in excess levels of acids in the blood. The disorder is the failure of the distal nephrons to lower urinary pH, due to either a back diffusion of hydrogen ions from the lumen to the blood or inadequate transport of H^+. Electrolyte balance is also affected with RTA that often leads to low blood potassium levels, calcium deposits in the kidneys, dehydration, and painful softening/bending of bones. Muscle weakness and diminished reflexes, due to the electrolyte imbalance, often occur when the disorder has been present for a long time.

There are four types of renal tubular acidosis and are classified as type 1 through 4.

They are distinguished by the specific tubular abnormality that causes the acidosis. Type 3 is extremely rare. The diagnosis for type 1 and 2 RTA is based on the following symptoms: (1) muscle weakness and diminished reflexes,

(Continued)

(2) high levels of acids, and (3) low levels of bicarbonate and potassium in the blood. The low blood potassium is the cause of muscle weakness and diminished reflexes. In type 1, kidney stones can develop causing renal damage. In type 4, renal tubular acidosis is associated with high potassium levels accompanied by high acid levels and low bicarbonate levels in the blood. The exact cause of the RTA in the patient is not known. Type 1 RTA may be the result of an inherited defect, autoimmune disease, treatment with lithium or the antibiotic amphotericin B, or diseases of the kidney medulla. In patients with inherited type 1 RTA, genetic defects in Cl^-/HCO_3^- exchangers and in H^+-ATPases have been identified.

Bone contains large amounts of carbonate and phosphate salts that can be released when the blood becomes too acidic. These buffers help to minimize the fall in pH during metabolic acidosis but has a detrimental effect in demineralization of bone. Respiratory acidosis does not cause calcium to be release from bone, which suggests that the important underlying mechanism is the low plasma bicarbonate, as occurs in metabolic but not respiratory acidosis. Release of calcium from bone leads to increased urinary calcium excretion (hypercalciuria). Chronic metabolic acidosis also inhibits tubular reabsorption of calcium and reduces renal excretion of citrate. Citrate normally inhibits stone formation by complexing calcium and by inhibiting crystallization of calcium salts. A high rate of calcium excretion, low citrate level, and alkaline urine favors precipitation of calcium phosphate (hence the nephrocalcinosis and kidney stones). The deposition of calcium salts in the kidney medulla causes interstitial inflammation and explains the presence of white blood cells in the patient's urine. Blockage of tubules and collecting ducts would reduce the GFR, and damage to the renal medulla and hypokalemia result in an impaired ability to concentrate the urine (urine specific gravity was 1.010).

Mineralization and bone growth depend on multiple factors that can be adversely affected by acidosis: calcium and phosphate concentrations of the extracellular flu osteoblasts and osteoclasts, intestinal absorption of calcium and phosphate, renal excretion of calcium and phosphate, and activity of several hormones (parathyroid hormone, calcitonin, and 1,25-dihydroxy vitamin D3). Human growth hormone release may be trapping in the urine and less formation of titratable acid occurs, so that less new bicarbonate will be added to the blood to replenish this important ECF buffer base.

The treatment of types 1 and 2 includes correction of hypokalemia and alkali replacement.

The hypokalemia should be corrected first, as alkali replacement can worsen the hypokalemia with dangerous consequences. Correcting hypokalemia improves musculoskeletal symptoms, and early treatment also prevents recurrence of kidney stones and the progression of renal failure. The patient was given Shohl's solution, which is a mixture of sodium citrate and citric acid, which is taken orally. Citrate yields three bicarbonate ions when oxidized completely in the body, and this counteracts the acidosis. Citrate is also excreted in the urine and reduces calcium stone formation. Because the patient was also hypokalemic, potassium citrate could have been prescribed. The milk and vitamin D were prescribed to aid in mineralization of the patient's skeleton. ■

Chapter Summary

- The body is constantly threatened by acid resulting from diet and metabolism. The stability of blood pH is maintained by the concerted action of chemical buffers, the lungs, and the kidneys.
- Numerous chemical buffers (e.g., HCO_3^-/CO_2, phosphates, and proteins) work together to minimize pH changes in the body. The concentration ratio (base–acid) of any buffer pair, together with the pK of the acid, automatically defines the pH.
- The bicarbonate/CO_2 buffer system is an effective physiologic buffer in the body because its components are present in large amounts and the system is open.
- The respiratory system influences plasma pH by regulating the arterial P_{CO_2} through changes in alveolar ventilation. The kidneys influence plasma pH by excreting acid or base in the urine.
- Renal acidification involves three processes: reabsorption of filtered HCO_3^-, excretion of titratable acid, and excretion of ammonia. New HCO_3^- is added to the plasma and replenishes depleted HCO_3^- when titratable acid (normally mainly $H_2PO_4^-$) and ammonia (as NH_4^+) are excreted.
- The stability of intracellular pH is ensured by membrane transport of H^+ and HCO_3^-, by intracellular buffers (mainly proteins and organic phosphates), and by metabolic reactions.
- Respiratory acidosis is an abnormal process characterized by an accumulation of CO_2 and a fall in arterial blood pH. The kidneys compensate by increasing the excretion of H^+ in the urine and adding new HCO_3^- to the blood, thereby diminishing the severity of the acidemia.
- Respiratory alkalosis is an abnormal process characterized by an excessive loss of CO_2 and a rise in pH. The kidneys compensate by increasing the excretion of filtered HCO_3^-, thereby diminishing the alkalemia.
- Metabolic acidosis is an abnormal process characterized by a gain of acid (other than H_2CO_3) or a loss of HCO_3^-. Respiratory compensation is hyperventilation, and renal compensation is an increased excretion of H^+ bound to urinary buffers (ammonia and phosphate) and generation of new bicarbonate.
- Metabolic alkalosis is an abnormal process characterized by a gain of strong base or HCO_3^- or a loss of acid (other than H_2CO_3). Respiratory compensation is hypoventilation, and renal compensation is increased excretion of HCO_3^-.
- The plasma anion gap is equal to the plasma $Na^+ - Cl^- - HCO_3^-$ concentration and is most useful in narrowing down possible causes of metabolic acidosis.
- The pH–bicarbonate diagram is used clinically to determine the patient's acid–base disturbance.

Chapter Review Questions

1. A 50-kg woman excreted 20 mEq/d titratable acid, 45 mEq/d NH4+, and 5 mEq/d bicarbonate in her urine. Assuming that she is in acid–base balance, what was the net production of nonvolatile acids in her body?

   A. 20 mEq/d
   B. 45 mEq/d
   C. 60 mEq/d
   D. 65 mEq/d
   E. 70 mEq/d

The correct answer is C. Net acid excretion is calculated from urinary titratable acid + urinary NH4+ – urinary HCO_3^- excretion = 20 + 45 − 5 = 60 mEq/d, in this case. The net acid excretion by the kidneys will be equal to the net renal production of nonvolatile acids in a person in acid–base balance, assuming that nonrenal losses from the body are negligible.

2. Which of the following causes increased tubular secretion of hydrogen ions?

   A. A decrease in arterial P_{CO_2}
   B. Adrenal cortical insufficiency
   C. Administration of a carbonic anhydrase inhibitor
   D. An increase in intracellular pH
   E. An increase in tubular sodium reabsorption

The correct answer is E. When Na^+ reabsorption is stimulated, Na^+/H^+ exchange is increased, resulting in greater H^+ secretion in the proximal tubule and loop of Henle. Additionally, increased Na^+ reabsorption in the collecting ducts renders the duct lumen more negative, which favors H^+ secretion. All of the other factors result in decreased H^+ secretion.

3. The hospital laboratory reports the following measurements on an acutely ill patient:

	Patient	Normal Range
Arterial blood pH	7.25	7.35–7.45
P_{CO_2}, mm Hg	25	35–45
Plasma [HCO_3^-], mEq/L	11	22–26
Plasma [Na^+], mEq/L	140	136–145 mEq/L
Plasma [Cl^-], mEq/L	118	95–105 mEq/L
Anion gap		8–14 mEq/L

What is the most likely cause of the acid–base disturbance?

   A. Cardiogenic shock
   B. Methanol intoxication
   C. Severe diarrhea
   D. Uncontrolled diabetes mellitus
   E. Vomiting of gastric juice

The correct answer is C. The data indicate a normal anion gap (or hyperchloremic) metabolic acidosis (anion gap = 140 − 118 − 11 = 11 mEq/L), with respiratory compensation. This could result from loss of bicarbonate (base) due to diarrhea. Cardiogenic shock (lactic acidosis), methanol intoxication, and uncontrolled diabetes mellitus (ketone body acid production) produce a high anion gap metabolic acidosis. Vomiting of acidic gastric juice produces a metabolic alkalosis.

4. Which of the following arterial blood values might be expected in a mountain climber who has been residing at a high-altitude base camp below the summit of Mt. Everest for 1 week?

	pH	PO_2, mm Hg	Pco_2, mm Hg Plasma	$[HCO_3^-]$, mEq/L
A.	7.18	95	24	9
B.	7.35	50	60	32
C.	7.47	40	20	14
D.	7.53	95	50	40
E.	7.62	40	20	20

The correct answer is C. At a high altitude, because of the low ambient barometric pressure and oxygen tension, hypoxia develops. Therefore, we can immediately rule out the options with a normal arterial blood oxygen tension (options A and D). Option B is a subject with hypoxia that resulted from inadequate ventilation; this subject has CO_2 retention and respiratory acidosis, but the response to high altitude is hyperventilation and consequently a low Pco_2 and respiratory alkalosis. Option E shows values for an acute respiratory alkalosis; the plasma $[HCO_3^-]$ has been lowered by 4 mEq/L, corresponding to the 20 mm Hg decrease below normal in Pco_2 (see Table 24.4). Option C shows typical values for a chronic (1 week) respiratory alkalosis; the kidneys have further lowered the plasma $[HCO_3^-]$ and have, therefore, reduced the severity of the alkalemia. The diagnosis of chronic respiratory

alkalosis can be confirmed by plotting the values on the pH–bicarbonate diagram (see Fig. 24.12).

5. A 24-year-old nurse is brought to the emergency room shortly before midnight. Though somewhat drowsy, she was able to relate that a few hours before she had attempted to kill herself by swallowing the contents of a bottle of aspirin tablets. Which of the following set of arterial blood values is expected?

	pH	PO_2, mm Hg	Pco_2, mm Hg	HCO_3^-, mEq/L
A.	7.24	95	19	8
B.	7.29	55	60	28
C.	7.40	95	40	24
D.	7.59	95	15	14
E.	7.68	95	15	17

The correct answer is D. Aspirin (salicylate) intoxication produces a mixed acid–base disturbance—respiratory alkalosis (due to stimulation of the respiratory center) and metabolic acidosis (due to inhibition of oxidative metabolism and accumulation of lactic and ketone body acids). The respiratory alkalosis predominates during the first several hours in adults; metabolic acidosis occurs at the same time and becomes overwhelming late in the course of the intoxication. Option D shows the predominant respiratory alkalosis. Accumulation of organic acids in the blood contributes to the lowering of plasma $[HCO_3^-]$. Option A represents metabolic acidosis with normal respiratory compensation. Option B represents respiratory acidosis due to alveolar hypoventilation or a mismatch between alveolar ventilation and pulmonary capillary blood flow; note the abnormally low PO_2. Option C represents the normal condition for arterial blood. Option E represents a simple acute respiratory alkalosis. Confirm the correct answer by plotting the point on the pH–bicarbonate diagram (see Fig. 24.12).

Clinical Application Exercises 24.1

A 16-year-old girl was referred to an endocrinology unit with a provisional diagnosis of primary hyperaldosteronism. She had an elevated blood pressure (180/120 mm Hg), hypokalemia (serum K^+ of 2.6 mEq/L), and mild alkalosis. Despite her hypokalemia, she excreted approximately as much K^+ as she was ingesting (80 mEq/d). Her aldosterone secretion rate was found to be abnormally low, even when on a low-sodium diet. The same

constellation of abnormalities was found in several of her 10 siblings, independent of gender. Both her mother and maternal grandmother had severe hypertension and died prematurely. The patient many years later developed renal failure, and at age 45, she received a kidney transplant. The transplant lowered her blood pressure and corrected her hypokalemia.

QUESTIONS

1. Why was primary aldosteronism first suspected as the problem in the patient?
2. What measurement is clearly inconsistent with primary hyperaldosteronism?
3. How can you explain the abnormally low aldosterone secretion rate?

4. Why does the patient have a metabolic alkalosis?
5. What is the genetic disorder in the patient?
6. How might one treat this disorder?
7. What strongly suggests that the patient's problem was an intrinsic renal defect?

ANSWERS

1. Primary aldosteronism typically results in hypertension (due to salt and water retention) and hypokalemia (due to excessive K^+ secretion by cortical collecting duct principal cells), features which were seen in the patient.

2. The aldosterone secretion rate was abnormally low.

3. The abnormally low aldosterone secretion rate was likely a secondary consequence of the electrolyte abnormalities; salt and water retention (volume expansion) would lead to suppression of the renin–angiotensin–aldosterone system, and low plasma $[K^+]$ would inhibit aldosterone secretion by a direct effect on the adrenal cortex.

4. Hypokalemia produces metabolic alkalosis. This may be explained by reciprocal movements of K^+ out of body cells into the ECF and movement of ECF H^+ into cells. The increased $[H^+]$ in kidney tubule cells would promote H^+ secretion and thereby lead to addition of new bicarbonate to the blood (see Chapter 24).

5. The patient has Liddle syndrome, a rare autosomal dominant disorder. The specific defect is a mutation in genes that encode the beta or gamma subunits of the epithelial sodium channel (ENaC), which is located in the luminal cell membrane of collecting duct principal cells. This defect reduces binding to an intracellular ubiquitin ligase that normally removes the sodium channel from the luminal cell membrane. The result is an increased number of ENaC in the luminal cell membrane and consequent excessive reabsorption of Na^+ and excessive secretion of K^+.

6. A low-sodium diet and the potassium-sparing diuretic drugs triamterene or amiloride, both of which block ENaC, are useful in treating this disorder.

7. The hypertension was improved and the patient had a normal plasma $[K^+]$ after she received a new kidney.

thePoint® *Visit* http://thepoint.lww.com/rhoades5e *for additional chapter review Q&A, Clinical Application Exercises, animations, and more!*

25 Gastrointestinal System Functions

Active Learning Objectives

Upon mastering the material in this chapter, you should be able to:

- Describe how salivary secretion is regulated.
- Explain the mechanism by which the stomach secretes hydrochloric acid.
- Describe the phases of acid secretion associated with digestion.
- Explain the phasic secretion of the pancreatic enzymes.
- Explain the role that bile salts play in the absorption of intestinal lipids.
- Describe the difference between primary and secondary bile acids.

- Explain why gallstones are usually formed in the gallbladder.
- Explain how enterocytes digest carbohydrates and transport the digestion products.
- Explain how enterocytes digest proteins and transport the digestion products.
- Explain how the enterocytes digest and transport triglycerides.
- Explain the function of the fat-soluble and the water-soluble vitamins.
- Explain the mechanism of iron absorption by the small intestine.

The digestive system is responsible for the absorption of ingested nutrients, water, minerals, vitamins, and other essential compounds. The digestive system is central to the regulation of the metabolic processes in all of the organs. Normal function of the digestive system is critical for growth, development, maintenance of individual organs, and survival. Moreover, the GI and the cardiovascular systems work in concert to provide and deliver nutrients to the millions of cells in the body. For example, during the process of digestion, blood flow is augmented to the gut by shunting blood from other organs.

Chapter 23 explained how the kidneys regulate water and electrolytes and how electrolyte balance is maintained by renal excretion. These same electrolytes as well as water, and nutrients, enter the body through absorption via the gut. The enteric nervous system first introduced in Chapter 3 interacts with other parts of the nervous and endocrine systems to send information to and from the brain to provide local control of the digestive system. For example, the pH of the stomach's contents is raised and lowered by hormones from the digestive system and neural signals from the peripheral nervous system. Also, the digestion of foods is dependent on basic chemical processes that follow the laws of chemistry. Finally, the interrelationship between structure and function in the unique architectural design of the small intestine creates an enormous surface area for the absorption of nutrients.

▶ FUNCTIONAL OVERVIEW OF DIGESTIVE SYSTEM

The digestive system includes the oral cavity, the gastrointestinal (GI) tract, and associated accessory organs, glands, and specialized structures. The GI tract does not include the mouth or pharynx. It is the hollow, muscular tube starting where the pharynx connects to the upper esophagus and terminating at the anus. The digestive system is physically and functionally divided into regions specialized for digestion, secretion, absorption, motility, and compaction. The teeth, tongue, salivary glands, liver, gallbladder, and pancreas are accessory structures that contribute to the physical, chemical, and enzymatic breakdown processes that occur within the digestive system. Hormones (e.g., gastrin, cholecystokinin [CCK], and secretin) regulate digestive enzyme release, absorption, secretion, and smooth muscle contraction/relaxation along the GI tract. Nervous input (i.e., central nervous system [CNS] and enteric nervous system [ENS]) affects similar digestive system processes (see Chapter 27).

Small intestine is the primary site for digestion and absorption processes.

Enzymatic digestion is initiated in the mouth (i.e., carbohydrates) and absorption begins in the stomach (e.g., medium-chain fatty acids and some drugs). However, most digestion and absorption processes occur in the small intestine. Initial digestion and trituration in the stomach prepare the chyme for final digestion and absorption of nutrients in the small intestine. Specialized small intestine epithelial cells (**enterocytes**) absorb nutrients for transport into either the portal circulation or the lymphatic system. Villi and the brush border membrane of the enterocytes increase the total surface area for digestion and absorption from the lumen. Secretions from the salivary glands, stomach, pancreas, and liver facilitate the processes of digestion and absorption. Secretory glands serve two primary functions. First, digestive enzymes are secreted from salivary glands in the mouth, parietal cells in the gastric pits of the stomach, and the acini in the pancreas that empty into the small intestine. However, complete nutrient digestion only requires the action of pancreatic enzymes. Second, glands secrete mucus for lubrication and protection. Because the oral cavity and the surface of the GI tract are literally exposed to the external environment, the secreted mucus also plays a prominent role in immunity by preventing pathogens from entering the bloodstream and lymph (see Chapter 10 for review). This chapter discusses the basic structure, function, and regulation of digestive system secretion and the role the GI tract plays in the digestion and absorption of carbohydrate, fat, and protein and the absorption of fat-soluble and water-soluble vitamins, electrolytes, bile salts, and water.

▶ SALIVARY SECRETION

Adults add ~7 to 10 L of gastric secretions, including about 1L of saliva in addition to gastric juice, pancreatic juice, and bile, to the GI tract daily for the digestion and absorption of nutrients.

Salivary glands are a heterogeneous group of **exocrine glands** with different proportions of **serous cells** and **mucous cells**. *Serous cells* produce a secretion containing *α-amylase*, an enzyme that breaks down starch, which is stored in **zymogen granules**. *Mucous cells* synthesize, store, and secrete **mucin**, which is composed of glycoproteins and important for lubrication and protection. Three pairs of *major salivary glands* (*extrinsic glands*), the (1) **parotid**, (2) **submandibular**, and (3) **sublingual glands**, connect to the oral cavity via secretory ducts (Fig. 25.1). *Minor salivary glands* (*intrinsic glands*), located in the oral submucosa, secrete mainly mucus and contribute < 10% to the total saliva volume. Salivary glands located on the tongue secrete lingual lipase. This enzyme initiates the digestion of triglycerides. Secretion of saliva is primarily regulated by parasympathetic and sympathetic nerves, but hormones, dry mouth,

the temperature of drinks, and the presence of food in the mouth are examples of other stimuli that impact salivation.

The secretions of the major glands differ significantly. *Parotid glands* secrete primarily a serous secretion that is rich in water and electrolytes, whereas the *submandibular* and *sublingual glands* produce both serous and mucous secretions. Salivary glands are endowed with a significant blood supply, and both the **parasympathetic** (**PNS**) and **sympathetic** (**SNS**) divisions of the **autonomic nervous system** (**ANS**) innervate them. ANS activity affects the quantity and composition of saliva. Although aldosterone is known to modify the ionic composition of saliva (i.e., Na^+ and K^+), it is generally believed that salivary secretion is mainly under ANS and not hormonal control.

Salivon is the functional unit of the salivary gland.

The functional unit of the salivary gland, the **salivon** (Fig. 25.2), consists of the **acinus**, the **intercalated duct**, and the **striated duct**. The specialized *serous cells* and *mucous cells* are organized as a single layer surrounding the saclike *acinus*. The serous cells contain an abundance of rough endoplasmic reticulum (RER), which reflects active protein synthesis, and numerous zymogen granules. Contractile cells (**myoepithelial cells**) are strategically organized around the acini and intercalated ducts. Contraction of these cells facilitates movement of the serous and mucus cell secretions into the ducts and expression of the duct contents in the oral cavity.

Salivary ducts

The intercalated ducts connect to the striated duct, which empties into the excretory duct. The *striated duct* and the large *excretory ducts* are lined with columnar cells and are involved in modifying the ionic composition of saliva. Although the acinar cells synthesize and secrete most salivary proteins, cells of the *intercalated ducts* contain secretory granules and synthesize several proteins, such as epidermal growth factor, ribonuclease, α-amylase, and proteases.

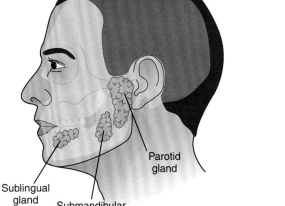

Figure 25.1 **Salivary glands.** The major salivary glands, which are paired, are shown in the above figure.

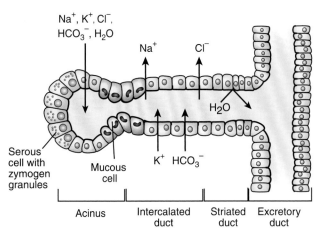

Figure 25.2 **The salivon is the basic unit of the salivary gland.** The acinus and associated ductal system form the salivon.

Saliva electrolyte composition is modified when the striated ducts selectively reabsorb sodium and selectively secrete potassium and bicarbonate.

The primary secretion produced by the acinar cells and intercalated ducts resembles that of plasma. However, samples collected from the striated and excretory (collecting) ducts are hypotonic to plasma, indicating the primary secretion is modified in the striated ducts. Saliva has less Na^+ and Cl^- and more K^+ and HCO_3^- than does plasma, because the cells in the striated duct actively absorb Na^+ and actively secrete K^+ and HCO_3^- into the lumen (see Fig. 25.2). Cl^- ions leave the lumen either in exchange for HCO_3^- ions or by passive diffusion along the electrochemical gradient created by Na^+ absorption.

The electrolyte composition of saliva depends on the rate of secretion. As the secretion rate increases, the electrolyte composition of saliva approaches the ionic composition of plasma, but at low flow rates, it differs significantly. At low secretion rates, the ductal epithelium has more time to modify and, thus, reduce the osmolality of the primary secretion, so the saliva has a much lower osmolality than does plasma. The opposite is true at high secretion rates.

Although the absorption and secretion of ions may explain changes in the electrolyte composition of saliva, these processes do not explain why the osmolality of saliva is lower than that of the primary secretion of the acinar cells. Saliva is hypotonic to plasma because of a net absorption of ions by the ductal epithelium, a result of the action of an Na^+/K^+-ATPase in the basolateral cell membrane. The Na^+/K^+-ATPase transports three Na^+ ions out of the cell in exchange for two K^+ ions transported into the cell. The ductal epithelium is not permeable to H_2O, so H_2O does not follow the absorbed salt. This results in a net absorption of ions.

Salivary proteins

The two major proteins present in saliva are *α-amylase* (*ptyalin*) and *mucin*. Salivary α-amylase is produced predominantly by the parotid glands, and mucin is produced mainly by the sublingual and submandibular glands. Amylase is involved in the digestion of starch and catalyzes the hydrolysis of polysaccharides with α-1,4- glycosidic linkages. It is synthesized in the RER of acinar serous cells and transferred to the Golgi apparatus, where it is packaged into zymogen granules. The zymogen granules are stored at the apical region of the acinar cells and released with appropriate stimuli. Because some time usually passes before acids in the stomach can inactivate the amylase, a substantial amount of the ingested carbohydrate can be digested before reaching the duodenum. (Amylase activity is described later in this chapter.)

Mucin is the most abundant protein in saliva. The term describes a family of glycoproteins, each associated with varying amounts of different sugars. Mucin is responsible for most of saliva's viscosity. It coats the surface of the mouth and teeth. The tooth coating actually aids in the prevention of caries.

Saliva has unique qualities.

Saliva performs various functions. It facilitates chewing and swallowing by lubricating food and the mucosal surface.

The lubrication reduces the frictional damage foods may cause, and it facilitates food movement within the oral cavity and down the esophagus. It helps small food particles stick together to form a bolus, making them easier to swallow. Moistening of the oral cavity facilitates speech. Saliva carries immunoglobulins that combat pathogens and lysozyme that kills bacteria. Salivary enzymes are involved in carbohydrate and fat digestion. Saliva plays an important role in water intake; the sensation of dryness of the mouth as a result of low salivary secretion urges a person to drink. Saliva can dissolve flavorful substances, stimulating the different taste buds located on the tongue. Interestingly, saliva is also involved in the maintenance of taste buds.

Saliva performs immunologic functions by lysing bacteria and killing HIV-infected leukocytes.

Saliva plays an important role in the general hygiene of the oral cavity. The pH of saliva ranges from pH 6.0 to almost neutral (pH 7.0). It contains HCO_3^- to neutralize acidic substances entering the oral cavity, including regurgitated gastric acid. Saliva also contains small amounts of **muramidase** (lysozyme) that breaks down peptidoglycans found in the cell walls of certain bacteria (e.g., *Staphylococcus*); **lactoferrin**, a protein that binds iron, depriving microorganisms of a source of iron for growth; **epidermal growth factor**, which stimulates gastric mucosal growth; *immunoglobulins* (mainly IgA); ABO blood group substances; and proteins with *antiviral* and *antifungal* activity (e.g., lysozyme, peroxidase, salivary agglutinin).

Research has indicated that the hypotonic property of saliva protects against certain infections. Hypotonic saliva has been shown to kill the human immunodeficiency virus (HIV)-infected mononuclear leukocytes and blocking further transmission of the virus. This helps to explain why HIV is rarely transmitted via kissing or normal oral contact. However, the beneficial property of hypotonic saliva is lost when the virus is introduced into the oral cavity via isotonic HIV-infected fluids (e.g., seminal fluid or breast milk).

Acidic food is a potent stimulus of salivary secretion.

In the resting state, salivary secretion is low, ~30 mL/h. The submandibular glands contribute about two thirds of resting salivary secretion, the parotid glands about one fourth, and the sublingual glands the remainder. Stimulation increases the rate of salivary secretion, most notably in the parotid glands, up to 400 mL/h. The most potent stimuli of salivary secretion are acidic-tasting substances (e.g., citric acid). Other stimuli include the smell of food, chewing, and what's in the mouth (i.e., smooth objects like round, hard candies stimulate salivation). Inhibitors of secretion include anxiety, fear, dehydration, medications (e.g., antihistamines), and rough objects.

Saliva secretion is under autonomic control.

As discussed, salivary secretion is predominantly under ANS control. Parasympathetic stimulation of the salivary glands increases activity of the acinar and ductal cells, leading to increased salivation. The PNS control centers are located in the medulla oblongata. Preganglionic fibers from the inferior salivatory nucleus are contained in cranial nerve IX and synapse in the otic ganglion. They send postganglionic fibers to the parotid glands. Preganglionic fibers from the superior salivatory nucleus course with cranial nerve VII and synapse in the submandibular ganglion. They send postganglionic fibers to the submandibular and sublingual glands.

Blood flow is low in resting salivary glands and can increase as much as 10-fold when salivary secretion is stimulated. This increase in blood flow is also under PNS control. Parasympathetic stimulation induces acinar cells to release the serine protease **kallikrein**. Kallikrein acts on the plasma globulin **kininogen** to release **lysyl-bradykinin** (**kallidin**). Kallidin stimulates dilation of the blood vessels supplying the salivary glands. Atropine, an anticholinergic agent, is a potent inhibitor of salivary secretion. Acetylcholinesterase inhibitors (e.g., pilocarpine) enhance salivary secretion. Some parasympathetic stimulation also increases blood flow to the salivary glands directly, apparently via the release of the neurotransmitter **vasoactive intestinal peptide** (VIP).

The SNS also innervates the salivary glands. Sympathetic fibers arise in the upper thoracic segments of the spinal cord and synapse in the **superior cervical ganglion**. Postganglionic fibers leave the superior cervical ganglion and innervate the acini, ducts, and blood vessels. Sympathetic stimulation often results in a short-lived and much smaller increase in salivary secretion than does parasympathetic stimulation. The increase in salivary secretion observed during sympathetic stimulation is mainly via β-adrenergic receptors, which are more involved in stimulating the contraction of myoepithelial cells to increase salivary flow. Although both sympathetic and parasympathetic stimulation increase salivary secretion, the products are different. Parasympathetic stimulation produces a secretion rich in electrolytes and salivary amylase. In contrast, sympathetic stimulation produces a secretion rich in mucus, making the saliva much more viscous.

Hormones affect saliva secretion.

Mineralocorticoid administration reduces the Na^+ concentration of saliva with a corresponding rise in K^+ concentration. Mineralocorticoids act mainly on the striated and excretory ducts. Arginine vasopressin (vasopressin) reduces the Na^+ concentration in saliva by increasing ductal Na^+ reabsorption. Some GI hormones (e.g., VIP and substance P) have been experimentally demonstrated to evoke salivary secretion.

▶ GASTRIC SECRETION

The stomach is a muscular, hollow, dilated part of the GI tract with multiple functions. The stomach transiently stores food, secretes HCl and proteolytic enzymes involved in the gastric digestion of foods, absorbs water-soluble and lipid-soluble substances (e.g., alcohol and some drugs), and prepares the **chyme** for digestion in the small intestine. Chyme, a partially digested semifluid of smaller particles, is produced through a combination of peristaltic movements of the stomach muscles and the chemical and enzymatic digestion initiated by the gastric secretions. A combination of the squirting of

antral content into the duodenum, the grinding action of the antrum, and retropulsion provides much of the mechanical action necessary for the emulsification of dietary fat, which plays an important role in fat digestion. The propulsive, grinding, and retropulsive movements associated with antral peristalsis are discussed in Chapter 27.

Mucus and bicarbonate secreted by epithelial cells protect the stomach from the acidic condition in the lumen.

The stomach wall consists of four distinct layers (from inside to outside): (1) **gastric mucosa**, (2) **submucosa**, (3) **muscularis externa**, and (4) **serosa**. The mucosa layer consists of the epithelium that houses the secretory glands. The submucosa consists of fibrous connective tissue that separates the mucosa from the muscularis externa. The smooth muscle layer of the muscularis externa differs from other portions of the GI tract in that it contains three layers of smooth muscle instead of two. The serosa consists of connective tissue and is continuous with the peritoneum.

Epithelial mucosa

The gastric mucosa epithelium forms deep pits and contains two main types of glands: **pyloric** and **oxyntic** (**gastric**) (Fig. 25.3). *Pyloric glands* secrete gastrin (a hormone) and mucus for protection. They are located in the distal 20% of the stomach, the antral portion. *Pyloric glands* contain cells

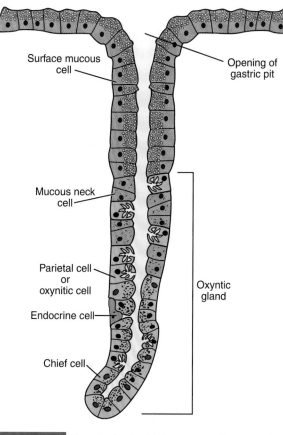

Stomach lumen

Surface mucous cell

Opening of gastric pit

Mucous neck cell

Parietal cell or oxyntic cell

Endocrine cell

Chief cell

Oxyntic gland

Figure 25.3 | **The oxyntic gland in the corpus of a mammalian stomach.** Several cell types, including parietal (oxyntic) cells, open into a common pit and contribute to gastric secretion.

similar to mucous neck cells of oxyntic glands, but the presence of many *gastrin-producing cells* (*G cells*) makes them different. The *oxyntic glands* are the most abundant gastric glands. They are located in the fundus and the corpus, which occupies 80% of the stomach.

Parietal cells of the oxyntic glands secrete hydrochloric acid.

Oxyntic glands contain **parietal (oxyntic) cells**, **chief cells**, **mucous neck cells**, and some *endocrine cells* (see Fig. 25.3). Most of the mucous cells are located in the neck region, and the base of the oxyntic gland contains mostly chief cells, along with some parietal and endocrine cells. Parietal cells secrete mainly **hydrochloric acid (HCl)** and **intrinsic factor** in addition to leptin and growth factors. Chief cells secrete primarily **pepsinogen** along with leptin. (Intrinsic factor and pepsinogen are discussed later in this chapter.) Mucous neck cells secrete mucus.

HCl is secreted across the parietal cell microvillar membrane and flows out of the intracellular canaliculi into the oxyntic gland lumen (Fig. 25.4). The amount of HCl secreted is proportionate to parietal cell number. Surface mucous cells lining the entire surface of the gastric mucosa and the openings of the gastric glands secrete mucus and HCO_3^- to protect the gastric epithelium from the acidic environment. A surface mucous cell contains numerous mucus granules at its apex. The number of granules varies depending on synthesis and secretion. The mucous neck cells of the oxyntic glands resemble surface mucous cells.

Chief cells are morphologically distinguished primarily by the presence of apical zymogen granules and an extensive RER. The zymogen granules contain pepsinogen that is released into the stomach and cleaved in the lumen to pepsin, the active enzyme, by HCl.

Also present in the stomach are various neuroendocrine cells, such as G cells, located predominantly in the antrum. G cells produce the hormone **gastrin**, which stimulates

HCl secretion by gastric parietal cells. An overabundance of gastrin secretion (*Zollinger-Ellison syndrome*) results in gastric hypersecretion and peptic ulceration. In most cases, the tumors arise within the pancreas and/or the duodenum. **D cells**, also present in the antrum, produce **somatostatin**, another important GI hormone.

Mucous gel layer protects the gastric epithelium.

The pH of the stomach lumen becomes quite acidic with pH values reaching 0.7 to 3.8. This raises a question: How is the gastric mucosal surface protected from intrinsic damage? The answer lies in the mucus and HCO_3^- containing secretion produced by the surface mucous cells. The mucus forms a **mucous gel layer** that covers the mucosal surface, and the HCO_3^- trapped in the mucous gel layer neutralizes the HCl.

Hydrochloric acid is formed when the parietal cell pumps hydrogen out in exchange for potassium.

The mechanism of HCl production and secretion by parietal cells located in the corpus and fundus is depicted in Figure 25.4. An **H^+/K^+-ATPase** in the apical (luminal) cell membrane of the parietal cell actively pumps H^+ out of the cell in exchange for K^+ entering the cell. Omeprazole inhibits the H^+/K^+-ATPase. Omeprazole, an acid-activated prodrug that is converted in the stomach to the active drug, binds to two cysteines on the ATPase, resulting in an irreversible inactivation. Omeprazole is used to treat ulcers, gastroesophageal reflux disease (GERD) (often associated with heartburn), and other conditions resulting from HCl production by the stomach. Although the secreted H^+ is often depicted as being derived from carbonic acid (see Fig. 25.4), the source of H^+ is most likely the dissociation of H_2O that also occurs inside the parietal cells. Carbon dioxide (CO_2) and H_2O join to form carbonic acid (H_2CO_3) in a reaction catalyzed by carbonic anhydrase. Acetazolamide inhibits carbonic anhydrase. Metabolic sources inside the cell and diffusion from the blood provide the CO_2.

For the H^+/K^+-ATPase to work, an adequate supply of K^+ ions must exist outside the cell. Although the mechanism is still unclear, there is an increase in K^+ conductance (through K^+ channels) in the apical membrane of the parietal cells simultaneous with active HCl secretion. The H^+/K^+-ATPase recycles K^+ ions back into the cell in exchange for H^+ ions. As shown in Figure 25.4, the basolateral cell membrane has an electroneutral Cl^-/HCO_3^- exchanger that balances the entry of Cl^- into the cell with an equal amount of HCO_3^- entering the bloodstream. The Cl^- inside the cell then leaks down an electrochemical gradient into the lumen through Cl^- channels. Consequently, HCl is secreted into the lumen

The secretion of HCl by parietal cells is balanced by an equal amount of HCO_3^- added to the bloodstream. Therefore, the blood coming from the stomach during periods of active HCl secretion contains much HCO_3^-, a phenomenon called the **alkaline tide**. The osmotic gradient created by the HCl concentration in the gland lumen drives H_2O passively into the lumen, thereby maintaining the iso-osmolality of the gastric secretion.

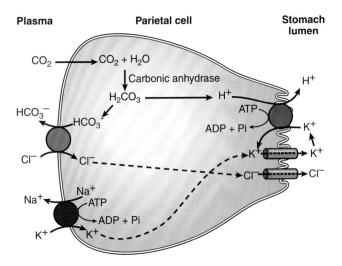

Figure 25.4 Parietal cells secrete gastric acid. The mechanism of HCl secretion is depicted in the figure above. ATPase in the apical cell membrane pumps H^+ out of the cell and into the lumen in exchange for K^+. ADP, adenosine diphosphate; ATP, adenosine triphosphate; Pi, inorganic phosphate.

Gastric secretion occurs in three phases.

Gastric secretion occurs in three phases: (1) **cephalic**, (2) **gastric**, and (3) **intestinal phase** (Table 25.1). The phases are named accordingly to where the stimuli originate. The *cephalic phase* involves the CNS and accounts for about 40% of total acid secretion. Smelling, chewing, and swallowing food (or merely the thought of food) send impulses along parasympathetic fibers in the vagus nerves to enteric neurons in the stomach wall. Nerve input and hormones stimulate parietal cells to release HCl.

The *gastric phase* involves events occurring in the stomach post food intake and accounts for about 50% of total HCl secretion. It is initiated by gastric distention and chemical agents such as digested proteins. Mechanoreceptors located in the stomach wall detect the distention that accompanies the presence of food in the stomach lumen. They respond by stimulating the parietal cells directly through short local (enteric) reflexes and by long **vagovagal reflexes**. Afferent and efferent impulses traveling in the vagus nerve mediate vagovagal reflexes. The products of protein digestion are potent stimulators of HCl secretion, an effect mediated through gastrin release. Several other chemicals (e.g., alcohol and caffeine) stimulate HCl secretion through mechanisms that are not well understood. The stimulation of HCl secretion by alcohol does appear to be related to its alcohol content—beverages lower in alcohol content, such as beer, tend to stimulate acid secretion more effectively than do beverages with higher alcohol content.

The *intestinal phase*, initiated when circulating amino acids, produced during protein digestion in the duodenum, stimulate parietal cells, accounts for about 10% of the total gastric acid secretion. Distention of the small intestine, probably via the release of the hormone **entero-oxyntin** from intestinal endocrine cells, also stimulates HCl secretion.

Acid production in the stomach parallels rates of gastric secretion.

The electrolyte composition of gastric juice changes with secretion rate. At a low secretion rate, Na^+ and Cl^- concentrations are high and K^+ and H^+ concentrations are low. As the secretion rate increases, the concentration of Na^+ decreases whereas that of H^+ and Cl^- increases significantly (Fig. 25.5).

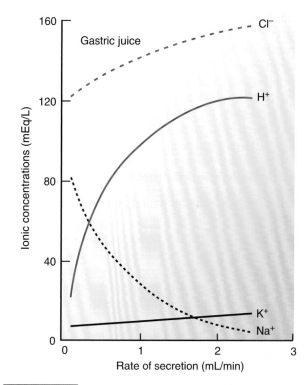

Figure 25.5 **Rate of secretion affects the composition of the gastric juice.** The concentration of electrolytes in the gastric juice is depicted in a healthy young adult male. The K^+ and H^+ ion concentration parallel the secretion rates.

The changes in electrolyte composition that accompany different secretion rates reflect the combined contributions of nonparietal cells and parietal cells to gastric juice. Secretion from nonparietal cells is probably constant; therefore, it is parietal cell secretion of HCl that significantly contributes to the changes in electrolyte composition at higher secretion rates.

Pepsin is the main gastric enzyme in protein digestion.

The enzymes secreted in the stomach are called *gastric enzymes* and include **pepsin**, **gastric amylase**, **gastric lipase**, and **intrinsic factor**. *Pepsin*, an **endopeptidase**, is the main gastric

TABLE 25.1	The Three Phases of Stimulation of Acid Secretion after Ingesting a Meal		
Phase	**Stimulus**	**Pathway**	**Stimulus to Parietal Cell**
Cephalic	Thought of food, smell, taste, chewing, and swallowing	Vagus nerve to:	ACh
		Parietal cells	
		G cells	Gastrin
Gastric	Stomach distention	Local (enteric) reflexes and vagovagal reflexes to:	ACh
		Parietal cells	
		G cells	Gastrin
Intestinal	Protein digestion products in duodenum	Amino acids in blood	Amino acids
		Intestinal endocrine cell	Entero-oxyntin
	Distention		

ACh, acetylcholine.

enzyme that cleaves proteins into smaller peptides. The optimal pH for pepsin activity is 1.8 to 3.5; therefore, it is extremely active in the highly acidic medium of gastric juice. Pepsinogen is formed when HCl in the gastric lumen cleaves pepsinogen. Pepsin also catalyzes its own formation from pepsinogen.

Gastric amylase degrades starch, and gastric lipase acts almost exclusively on butter fat. Intrinsic factor, produced by the parietal cells, is necessary for the absorption of vitamin B$_{12}$ in the terminal ileum. *Gastric lipase*, like lingual and pancreatic lipases, digests fats. *Intrinsic factor* is involved in vitamin B$_{12}$ absorption in the intestine.

Gastric secretion is under neural and hormonal control.

Gastric secretion is mediated through neural and hormonal pathways. Vagus nerve stimulation is the neural effector; histamine and gastrin are the hormonal effectors (Fig. 25.6). Parasympathetic pathways in the vagus nerve connect to the stomach's enteric system. The enteric neurons have stimulatory connections with three cell types in the stomach: the *parietal cells,* the *G cells,* and the **enterochromaffin-like cells (ECL cells)**, a type of neuroendocrine cells located mostly in the acid-secreting regions of the stomach. Enteric nerve endings release several different neurotransmitters including **acetylcholine (ACh)** and **gastrin-releasing peptide (GRP)** that stimulate secretion of HCl and **vasoactive intestinal polypeptide (VIP)** that inhibits HCl secretion. ACh directly stimulates HCl secretion by binding to muscarinic receptors (M$_3$) on the parietal cells, which elevates intracellular Ca^{2+}. GRP stimulates G cells to release gastrin.

Several hormones including **histamine** and gastrin are effectors of HCl secretion (see Fig. 25.6). Parietal cells possess special histamine receptors, **H$_2$ receptors**, whose stimulation increases HCl secretion. ECLs are believed to be the source of this histamine, but the mechanisms that stimulate ECLs

to release histamine are poorly understood. The effectiveness of cimetidine, an H$_2$ blocker, in reducing HCl secretion has indirectly demonstrated the importance of histamine as an effector of HCl secretion. H$_2$ blockers are commonly used for the treatment of peptic ulcer disease or GERD. Ranitidine, a longer-acting H$_2$-receptor antagonist with fewer side effects, has largely replaced cimetidine. Gastrin binds a G protein–coupled cholecystokinin (CCK$_2$) receptor on parietal cells, which elevates intracellular Ca^{2+} that in turn stimulates HCl release. Gastrin also stimulates histamine release by ECLs.

The effects of each of these three stimulants (ACh, gastrin, and histamine) augment those of the others, a phenomenon known as **potentiation**. Potentiation occurs when the effect of two stimulants is greater than the effect of either stimulant alone. For example, the interaction of gastrin and ACh molecules with their respective receptors results in an increase in intracellular Ca^{2+} concentration, and the interaction of histamine with its receptor results in an increase in cellular **cyclic adenosine monophosphate (cAMP)** production. The increased intracellular Ca^{2+} and cAMP interact in numerous ways to stimulate the gastric H$^+$/K$^+$-ATPase, which increases HCl secretion (see Fig. 25.6). Exactly how the increase in intracellular Ca^{2+} and cAMP enhances the effect of the other in stimulating HCl secretion is not well understood.

Gastric hormones inhibit acid secretion.

The inhibition of HCl secretion is physiologically important for two reasons. First, HCl secretion is important only during the digestion of food. Second, excess HCl can damage the gastric and the duodenal mucosal surfaces, causing ulcerative conditions. The body has an elaborate system for regulating HCl secretion by the stomach. Gastric luminal pH is a sensitive regulator of HCl secretion. Proteins in food provide buffering in the lumen; consequently, the gastric luminal pH is usually above 3 after a meal. However, if the buffering capacity of protein is exceeded or if the stomach is empty, the pH of the gastric lumen will fall below 3. When this happens, the endocrine cells (**D cells**) in the antrum secrete *somatostatin*, which inhibits the release of gastrin and, thus, HCl secretion.

Another mechanism for inhibiting HCl secretion is acidification of the duodenal lumen. Acidification stimulates the release of **secretin**, which inhibits the release of gastrin and several peptides, collectively known as **enterogastrones**. Enterogastrones are released by intestinal endocrine cells. Acid, fatty acids, and hyperosmolar solutions in the duodenum stimulate the release of enterogastrones. Two such hormones, **gastric inhibitory peptide (GIP)** and **cholecystokinin (CCK)**, inhibit HCl secretion. CCK is released by "**I**" **cells** stimulating the release of *gastric somatostatin*, which then inhibits HCl.

▶ PANCREATIC SECRETION

The pancreas is located adjacent to the duodenum and functions as both an endocrine and an exocrine gland (Fig. 25.7). The **endocrine** gland secretes several important hormones including insulin, glucagon, and somatostatin. (The endocrine function is further discussed in Chapter 34.) The **exocrine** gland secretes digestive enzymes that pass along the pancreatic duct and into

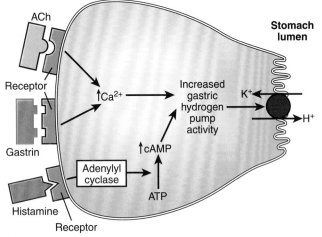

Plasma

Figure 25.6 Acid secretion by the parietal cell is under neural and hormonal control. Vagus nerve stimulation results in the release of Ach. Histamine and gastrin are hormones. Acetylcholine (ACh), histamine, and gastrin stimulate acid secretion. cAMP, cyclic adenosine monophosphate; ATP, adenosine triphosphate.

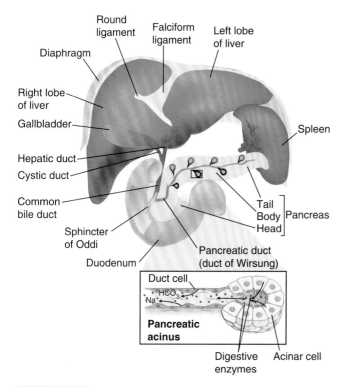

Figure 25.7 The pancreas performs both exocrine and endocrine functions. The exocrine cells of the pancreas secrete digestive juices into the duodenal lumen. Digestive enzymes are secreted by acinar cells, and an aqueous $NaHCO_3$ solution is secreted by duct cells (**inset**).

TABLE 25.2	The Pancreas is the Main Digestive Gland of the Body
Enzyme Secreted	**Hydrolytic Action**
Trypsin	Protease that breaks down proteins at the basic amino acids
Chymotrypsin	Protease that breaks down proteins at the aromatic amino acids
Lipase	Degrades triglycerides into fatty acids and glycerol
Carboxypeptidase	Protease that takes off the terminal acid group from a protein
Elastases	Degrade the protein elastin and some other proteins
Nucleases	Degrade nucleic acids, like DNAse and RNAse
Pancreatic amylase	Besides starch and glycogen, degrades most other carbohydrates; humans lack the enzyme to digest the carbohydrate cellulose

the small intestine. These enzymes break down components present in the human diet including carbohydrates, proteins, fats, DNA, and RNA. The pancreatic enzymes responsible for protein digestion are **trypsin** (the most abundant), **chymotrypsin**, and *carboxypolypeptidase*. *Trypsin* and *chymotrypsin* cleave proteins into polypeptides, and *carboxypolypeptidase* cleaves the polypeptides into amino acids. **Pancreatic amylase** digests carbohydrates, and **pancreatic lipase** is the main enzyme for fat digestion (Table 25.2). Some pancreatic enzymes are secreted as proenzymes, which are activated in the duodenal lumen to

form the active enzymes. (The digestion of nutrients by these enzymes is discussed later in this chapter.).

The chyme leaving the stomach is acidic. It needs to be neutralized to prevent damage to the duodenal epithelium and to bring the luminal contents to a near neutral pH, favorable for enzymes arriving from the pancreas. Pancreatic secretions contain HCO_3^- that raises the duodenal lumen pH.

Pancreatic secretion occurs in three phases.

Pancreatic secretion, like gastric secretion, occurs in three phases: (1) cephalic, (2) gastric, and (3) intestinal phase. Table 25.3 summarizes the stimulation of pancreatic secretion and secretory regulation by various hormonal and neural factors. The cephalic phase, stimulation of pancreatic secretion, is mainly mediated by direct efferent impulses sent by vagal centers in the brain to the pancreas and, to a minor

TABLE 25.3	Factors Regulating Pancreatic Secretion after a Meal		
Phase	**Stimulus**	**Mediators**	**Response**
Cephalic	Thought of food, smell, taste chewing, and swallowing	Release of ACh and gastrin by vagal stimulation	Increased secretion, with greater effect on enzyme output
Gastric	Protein in food	Gastrin	Increased secretion, with greater effect on enzyme output
	Gastric distention	ACh release by vagal stimulation	Increased secretion, with a greater effect on enzyme output
Intestinal	Acid in chyme	Secretin	Increased H_2O and HCO_3^- secretion
	Long-chain fatty acids	CCK and vagovagal reflex	Increased secretion, with greater effect on enzyme output
	Amino acids and peptides	CCK and vagovagal reflex	Increased secretion, with greater effect on enzyme output

Ach, acetylcholine; CCK cholecystokinin.

extent, by the indirect effect of parasympathetic stimulation of gastrin release. The gastric phase is initiated when food enters the stomach and distends it. The vagovagal reflex then stimulates pancreatic secretion. Gastrin may also be involved in this phase. The most important phase, the intestinal phase, is initiated by the entry of acidic chyme from the stomach into the small intestine, which stimulates the intestinal endocrine cells. (See "Pancreatic enzyme secretion is under neural and hormonal control.")

Pancreatic enzymes are produced in the acinar cells of the exocrine pancreas.

Like salivary glands, the exocrine pancreas is composed of numerous acini. Unlike the previously discussed salivary glands that contain two types of acinar cells, the pancreatic acini are composed of a single layer of pyramidal **acinar cells** (see Fig. 25.7). Both salivary and pancreatic acinar cells secrete an ion-rich fluid along with proteins and have elaborate RER and Golgi as well as apical zymogen granules. A few **centroacinar cells** line the lumen of the pancreatic acinus and modify the electrolyte composition of the secretion. Because the processes involved in the secretion or uptake of ions are active, centroacinar cells have numerous mitochondria.

The acini empty their secretions into intercalated ducts, which join to form intralobular and then interlobular ducts. The interlobular ducts empty into two pancreatic ducts: a major duct, the duct of Wirsung, and a minor duct, the duct of Santorini. The duct of Santorini enters the duodenum more proximally than does the duct of Wirsung, which enters the duodenum usually together with the common bile duct. A ring of smooth muscle, the **sphincter of Oddi**, surrounds the opening of the Wirsung and bile ducts at the point of entry into the duodenum and regulates the flow of bile and pancreatic juice into the duodenum. The sphincter also prevents the reflux of intestinal contents into the ducts (Clinical Focus 25.1).

Pancreatic secretions are rich in bicarbonate ions.

Pancreatic juice contains HCO_3^- and H_2O in addition to digestive enzymes (Fig. 25.8). The pancreas secretes about 1 L/day of HCO_3^--rich fluid to neutralize the acidic chyme that enters the duodenum. Unlike salivary fluid, the osmolality of pancreatic fluid equals that of plasma at all secretion rates. When compared to plasma, pancreatic juice has similar Na^+ and K^+ concentrations, is enriched with HCO_3^-, and has a relatively low Cl^- concentration. When the pancreas is stimulated, HCO_3^- concentration can increase to about five times that of plasma, reaching a maximal concentration

CLINICAL FOCUS | 25.1

Peptic Ulcers: When Bacteria Break the Barrier

Peptic ulcers, also known as **peptic ulcer disease**, are erosions of the mucosal lining of the stomach. Ulcerative lesions occur when the mucus barrier is disrupted, and pepsin and hydrochloric acid attack the stomach lining instead of digesting the food. Peptic ulcers affect ~4 million patients in the United States, where duodenal ulcers predominate. In Japan, however, gastric ulcers are more prevalent.

Symptoms include bloating, fullness, and abdominal pain. The symptoms are exacerbated with the ingestion of food (intensifies 3 hours after eating). In contrast, duodenal ulcers are classically relieved by food. If gastric ulcers go untreated, symptoms can escalate to include nausea and vomiting of blood that is accompanied by loss of appetite and weight loss. Many patients confuse heartburn with gastric ulcers. Heartburn, a burning sensation in the chest, is not associated with gastric ulcers. Heartburn is associated with regurgitation of gastric acid (gastric reflux), which is a major symptom of gastroesophageal reflux disease. Ulcers can also be caused or worsened by drugs such as aspirin or other nonsteroidal anti-inflammatory drugs (NSAIDs) that inhibit cyclooxygenase and most glucocorticoids (e.g., dexamethasone and prednisolone) for the treatment of arthritis. NSAIDs interfere with prostaglandin-mediated defense mechanisms against gastric acidity, mucus production, and bicarbonate secretion. Gastrointestinal bleeding is the most common complication. Sudden large bleeding

can be life threatening and occurs when an ulcer erodes a blood vessel.

In the 1990s, a surprising discovery was made in the field of peptic ulcer disease linking *Helicobacter pylori* infection to the cause of gastric and duodenal ulcers. As many as 70% to 90% of peptic ulcers are linked to *H. pylori*, a spiral-shaped bacterium that lives in the acidic environment of the stomach. *H. pylori* appears to be protected by producing large quantities of urease, which hydrolyzes urea to produce ammonia. The ammonia neutralizes acid in the gastric lumen, thus protecting the bacteria from the injurious effects of HCl. The underling mechanism that causes ulcerations appears to be the release of bacterial toxins that leads to chronic inflammation of the gastric mucosal lining. The immune system appears to be unable to clear the infection, despite the appearance of antibodies.

The investigator, Dr. Barry Marshall, who discovered this link, was awarded the Nobel Prize for Medicine and Physiology in 2005 for his pioneering discovery.

Recently, it has been demonstrated that H_2-receptor antagonists (i.e., cimetidine and ranitidine) have no effect on *H. pylori* infection. In contrast, omeprazole (an inhibitor of the H^+/K^+-ATPase) appears to be bacteriostatic. A combined therapy using omeprazole and the antibiotic amoxicillin appears to be quite effective in the eradication of *H. pylori* in 50% to 80% of patients with peptic ulcer disease, resulting in a significant reduction of duodenal ulcer recurrence. ■

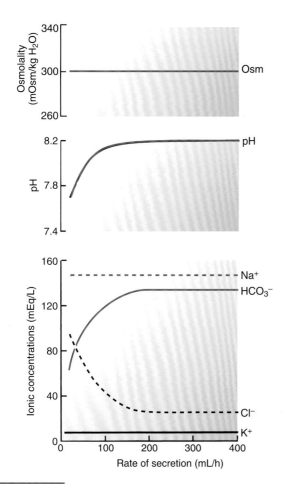

Figure 25.8 | Pancreatic fluid pH and electrolytes are altered with flow. Pancreatic fluid is rich in bicarbonate ions. The HCO_3^- concentration and, hence, the pH parallel the increased rates of secretion. *Note*: Na^+, K^+, and pancreatic osmolality are independent of flow.

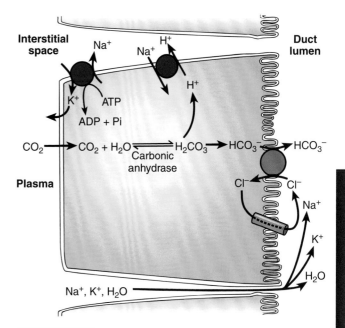

Figure 25.9 | Pancreatic duct cells secrete electrolytes and water into pancreatic fluid. The above model depicts the mechanism for electrolyte secretion by pancreatic duct cells. The luminal membrane Cl^- channel is CFTR (cystic fibrosis transmembrane conductance regulator). ADP, adenosine diphosphate; ATP, adenosine triphosphate; Pi, inorganic phosphate.

Gastrointestinal Physiology

close to 140 mEq/L and a pH of 8.2. A reciprocal relationship exists between the Cl^- and HCO_3^- concentrations in pancreatic juice. As the HCO_3^- concentration increases with secretion rate, the Cl^- concentration falls accordingly, resulting in a combined total anion concentration that remains relatively constant (150 mEq/L) regardless of the pancreatic secretion rate.

Two separate mechanisms have been proposed to explain the secretion of an HCO_3^--rich juice by the pancreas and the changes in HCO_3^- concentration. The first proposes that some cells, probably the acinar cells, secrete a plasmalike fluid containing predominantly Na^+ and Cl^-, whereas other cells, probably the centroacinar and duct cells, secrete an HCO_3^--rich solution when stimulated. Depending on the different rates of cellular secretion, the pancreatic juice can be rich in either HCO_3^- or Cl^-. The second depicts the primary secretion as rich in HCO_3^-. As the HCO_3^- solution moves down the ductal system, HCO_3^- ions are exchanged for Cl^- ions. When the flow is fast, there is little time for ion exchange, so the concentration of HCO_3^- is high. The opposite is true when the flow is slow.

The secretion of electrolytes by pancreatic duct cells is depicted in Figure 25.9. An Na^+/H^+ exchanger is located in the basolateral cell membrane. The energy required to drive the exchanger is provided by the Na^+/K^+-ATPase–generated Na^+ gradient. CO_2 diffuses into the cell and combines with H_2O to form H_2CO_3, a reaction catalyzed by carbonic anhydrase. H_2CO_3 dissociates to H^+ and HCO_3^-. The Na^+/H^+ exchanger extrudes the H^+, and HCO_3^- is exchanged for luminal Cl^- via a Cl^-/HCO_3^- exchanger. Also located in the luminal cell membrane is a protein called **cystic fibrosis transmembrane conductance regulator** (**CFTR**), an ion channel belonging to the ABC (**adenosine triphosphate [ATP]-binding cassette**) family of proteins. Regulated by ATP, its major function is to secrete Cl^- ions into the lumen. These ions are used by the Cl^-/HCO_3^- exchanger. The Na^+/K^+-ATPase removes intracellular Na^+ entering through the Na^+/H^+ antiporter. Na^+ from the interstitial space follows secreted HCO_3^- by diffusing through a paracellular path (between the cells). Movement of H_2O into the duct lumen is passive and driven by the osmotic gradient. The net result of pancreatic HCO_3^- secretion is the release of H^+ into the plasma; thus, pancreatic secretion is associated with an **acid tide** in the plasma.

Pancreatic enzyme secretion is under neural and hormonal control.

Pancreatic secretion is mediated through neural and hormonal pathways that use Ach, CCK, and secretin as mediators. Neural stimulation occurs when **ACh**-releasing parasympathetic fibers in the vagus nerve stimulate pancreatic secretion, which results predominantly in an increase in enzyme secretion—fluid and HCO_3^- secretion are marginally stimulated or unchanged. Sympathetic nerve fibers mainly innervate the blood vessels supplying the pancreas and cause vasoconstriction. Stimulation of the sympathetic nerves neither stimulates nor inhibits pancreatic secretion, most likely due to the reduction in blood flow.

Circulating GI hormones, particularly **secretin** and **CCK**, greatly influence pancreatic secretion of electrolytes and enzymes. Both hormones are produced by the small intestine and bind receptors in the pancreas. *Secretin* is released during the intestinal phase by **S cells** (endocrine cells in the intestinal mucosa). The *secretin* concentration in the plasma increases when the luminal pH in the duodenum decreases. *Secretin* tends to stimulate a HCO_3^--rich secretion. Exposure of the intestinal mucosa to long-chain fatty acids (lipid digestion products) and free amino acids stimulates the release of CCK by the "**I**" cells (endocrine cells in the intestinal mucosa). Circulating CCK and parasympathetic stimulation increase the secretion of pancreatic enzymes through a vagovagal reflex.

Hormones that are structurally similar to secretin and CCK also have similar effects. For example, VIP, structurally similar to secretin, stimulates the secretion of HCO_3^- and H_2O. However, VIP produces a weaker pancreatic response when given together with secretin than when secretin is given alone. Gastrin, structurally similar to CCK, is only a weak agonist for pancreatic enzyme secretion.

Potentiation, as previously described for gastric secretion, also exists in the pancreas. Its effect in pancreatic secretion is a result of the different receptors binding ACh, CCK, and secretin. Secretin binding triggers an increase in adenylyl cyclase activity, which, in turn, stimulates the formation of cAMP (Fig. 25.10). ACh, CCK, and the neuropeptides gastrin-releasing peptide (GRP) and substance P bind to

their respective receptors and trigger the release of Ca^{2+} from intracellular stores. The increase in intracellular Ca^{2+} and cAMP formation results in an increase in pancreatic enzyme secretion. The mechanism by which this occurs is not well understood.

▶ BILIARY SECRETION

Bile, a dark green to yellowish brown fluid produced by the **hepatocytes** in the liver, drains through the bile ducts penetrating the liver (see Chapter 26). Bile facilitates the digestion of lipids in the small intestine by emulsifying fat and forming aggregates around fat droplets called **micelles**. The dispersion of fat droplets into micelles greatly increases the surface area, enhancing the action of pancreatic lipase. Bile is stored in the gallbladder and, under nervous and hormonal control, is discharged into the duodenum. Depending on body size, the human liver can produce close to 1 L of bile a day. Bile contains bile salts, bile pigments (e.g., bilirubin), cholesterol, phospholipids, and proteins. Bile performs several important functions. For example, bile salts play an important role in the intestinal absorption of lipid. Bile salts are derived from cholesterol and, therefore, constitute a path for its excretion. Biliary secretion is an important route for the excretion of bilirubin from the body.

Bile secretion is under neural and hormonal control.

Parasympathetic and sympathetic nerves supply the biliary system. Parasympathetic (vagal) stimulation results in contraction of the gallbladder and relaxation of the sphincter of Oddi as well as increased bile formation. Bilateral vagotomy results in reduced bile secretion after a meal, suggesting that the PNS plays a role in mediating bile secretion. Stimulation of the SNS results in reduced bile secretion and relaxation of the gallbladder.

Gastrin stimulates bile secretion directly by affecting the liver and indirectly by stimulating increased HCl production that results in increased secretin release. When the mucosa of the small intestine is exposed to acidic chyme, it releases secretin. Secretin stimulates HCO_3^- secretion by the cells lining the bile ducts. As a result, bile contributes to the neutralization of acid in the duodenum.

Steroid hormones (e.g., estrogen and some androgens) inhibit bile secretion, and reduced bile secretion is a side effect associated with the therapeutic use of these hormones (e.g., administration of oral contraceptives or postmenopausal hormone replacement therapy). During pregnancy, high circulating levels of estrogen can reduce bile acid secretion.

Cholesterol metabolism in the liver involves the formation of bile acids.

Bile acids are formed in the liver by the cytochrome P450 oxidation of cholesterol. During the conversion, hydroxyl groups (OH^-) and a carboxyl group (COOH) are added to the steroid nucleus. Bile acids are classified as primary or secondary. The hepatocytes synthesize the *primary bile acids*, which include **cholic acid** and **chenodeoxycholic acid**.

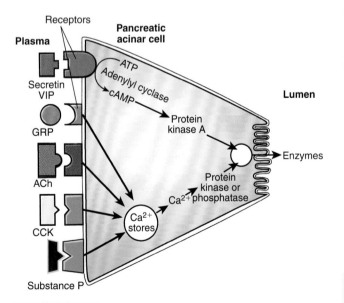

Figure 25.10 Calcium release from intracellular stores stimulates pancreatic enzyme secretion. This diagram illustrates the mechanisms involved in pancreatic secretion by hormones and neurotransmitters. ACh, acetylcholine; ATP, adenosine triphosphate; cAMP, cyclic adenosine monophosphate; CCK, cholecystokinin; GRP, gastrin-releasing peptide; VIP, vasoactive intestinal peptide.

When bile enters the GI tract, bacteria present in the lumen convert primary bile acids to *secondary bile acids* by dehydroxylation. Cholic acid is converted to **deoxycholic acid** and chenodeoxycholic acid to **lithocholic acid**.

At a neutral pH, bile acids are mostly ionized and are referred to as **bile salts**. Conjugated bile acids ionize more readily than do unconjugated bile acids and, thus, usually exist as salts of various cations (e.g., sodium glycocholate). Bile salts are more polar than bile acids and have greater difficulty penetrating cell membranes. Consequently, the small intestine absorbs bile salts much more poorly than it does bile acids. This property of bile salts contributes significantly to their integral role in the intestinal absorption of lipid. Therefore, it is important that the small intestine absorb bile salts only after lipid absorption is completed.

The major lipids in bile are phospholipids and cholesterol. The predominant phospholipid species is phosphatidylcholine (lecithin). The phospholipid and cholesterol concentrations of hepatic bile are 0.3 to 11 mmol/L and 1.6 to 8.3 mmol/L, respectively. Because the gallbladder absorbs a significant amount of H_2O and electrolytes, the concentrations of these lipids in the gallbladder bile are even higher. Cholesterol in bile is responsible for the formation of cholesterol gallstones.

Total bile flow is composed of the ductular secretion and the canalicular bile flow. Cells lining the bile ducts actively secrete HCO_3^- into the lumen, which results in the movement of H_2O into the lumen. A cAMP-dependent Cl^- channel that secretes Cl^- into the ductule lumen may also contribute to ductular secretion of fluid. The electrolyte composition of human bile collected from the hepatic ducts is similar to that of blood plasma, except that the HCO_3^- concentration may be higher, resulting in an alkaline pH.

Cholecystokinin stimulates bile secretion from the gallbladder.

Bile **canaliculi** are fine tubular canals running between the hepatocytes. Bile flows from hepatocytes through the canaliculi to the bile ducts, which drain into the gallbladder. Between meals, the sphincter of Oddi is closed, thereby preventing bile from draining into the intestine and instead to flow into the gallbladder, where it is stored and concentrated from 5 to 10 times its original concentration.

When the products of protein and lipid digestion build up in the small intestine, "I" cells in the duodenum and jejunum are stimulated to release CCK. CCK causes contraction of the gallbladder, which, in turn, causes increased pressure in the bile ducts. As the bile duct pressure rises, the sphincter of Oddi relaxes (another effect of CCK), and the concentrated bile is delivered into the duodenal lumen to complete the digestion of fat. Canalicular bile flow can be conceptually divided into *bile acid–dependent* and *bile acid–independent secretion*.

Canalicular bile acid–dependent flow

Hepatocyte uptake of free and conjugated bile salts is Na^+-dependent and mediated by **bile salt–Na^+ symport** (Fig. 25.11). The transmembrane Na^+ gradient generated by

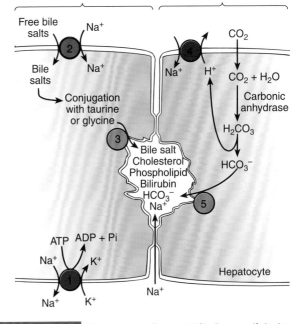

Figure 25.11 Bile salt secretions and bile flow are linked to Na^+/K^+-ATPase activity. This schematic illustrates the mechanism of bile acid secretion and flow. A marked reduction in flow and secretion occurs when Na^+/K^+-ATPase activity is inhibited. (1) Na^+/K^+-ATPase. (2) Bile salt-sodium symport. (3) Canalicular bile acid carrier. (4) Na^+/H^+ exchanger. (5) HCO_3^- transport system. ADP, adenosine diphosphate; ATP, adenosine triphosphate; Pi, inorganic phosphate.

the Na^+/K^+-ATPase, which provides the necessary energy, is a type of secondary active transport because the energy required for the active uptake of bile acid, or its conjugate, is not directly provided by ATP but by an ionic gradient. The free bile acids are reconjugated with taurine or glycine before secretion. Hepatocytes also make new bile acids from cholesterol and secrete bile salts by a carrier located at the canalicular membrane. This secretion is not Na^+-dependent; instead, it is driven by the electrical potential difference between the hepatocyte and the canaliculus lumen.

Other major components of bile, such as phospholipid and cholesterol, are secreted in concert with bile salts. Hepatocytes secrete bilirubin via an active process involving a transporter. Although the secretion of cholesterol and phospholipid is not well understood, it is closely coupled to bile salt secretion. The osmotic pressure generated as a result of the secretion of bile salts draws H_2O into the canaliculus lumen through the paracellular pathway.

Canalicular bile acid–independent flow

As the name implies, this component of canalicular flow is not dependent on the secretion of bile acids (see Fig. 25.11). The Na^+/K^+-ATPase plays an important role in bile acid–independent bile flow, a role that is clearly demonstrated by the marked reduction in bile flow when an inhibitor of this enzyme is applied. Another mechanism responsible for bile acid–independent flow is canalicular HCO_3^- secretion.

Bile acids are potentially toxic to cells, and their concentrations are tightly regulated.

Because bile acids act like a detergent, their concentration is tightly regulated in order to prevent damage to the gut lining. The major determinant of bile acid synthesis and secretion by hepatocytes is the bile acid concentration in hepatic portal blood, which exerts a negative-feedback effect on the synthesis of bile acids from cholesterol. The concentration of bile acids in portal blood also determines bile acid–dependent secretion. Between meals, the portal blood concentration of bile salts is usually extremely low, resulting in increased bile acid synthesis but reduced bile acid–dependent flow. After a meal, there is increased delivery of bile salts in the portal blood, which not only inhibits bile acid synthesis but also stimulates bile acid–dependent secretion.

Gallbladder bile differs from hepatic bile.

Gallbladder bile is more highly concentrated than is hepatic bile. H_2O absorption by the gallbladder epithelium is the major mechanism involved in concentrating the stored hepatic bile. The process is passive and is secondary to active Na^+ transport via an Na^+/K^+-ATPase in the basolateral membrane of the epithelial cells lining the gallbladder. As a result of isotonic fluid absorption from the gallbladder bile, the concentration of the various unabsorbed components of hepatic bile increases as much as 20-fold.

Bile salts are recycled between the small intestine and the liver.

The **enterohepatic circulation** recycles bile acids between the small intestine, the liver, and bile. The **total bile acid pool** is the entire amount of bile acids in the body (primary or secondary, conjugated or free) at any given time. In healthy people, the bile acid pool ranges from 2 to 4 g. By cycling bile acids in this pool through the enterohepatic circulation several times during a meal, a relatively small bile acid pool can provide the body with sufficient amounts of bile salts to promote lipid absorption. In a light eater, the bile acid pool may circulate 3 to 5 times daily, whereas in a heavy eater, it may circulate 14 to 16 times a day.

Bile salts in the intestinal lumen are absorbed via four pathways (Fig. 25.12). First, a small fraction of the total amount of bile salts is absorbed throughout the entire small intestine by passive diffusion. Second, and most important, bile salts are absorbed in the terminal ileum by an extremely efficient, active carrier-mediated process that usually results in < 5% of the bile salts escaping into the colon. Third, bacteria in the terminal ileum and colon deconjugate the bile salts to form bile acids, which are much more lipophilic than are bile salts and, thus, can be absorbed passively. Fourth, these same bacteria are responsible for transforming the primary bile acids to secondary bile acids (deoxycholic and lithocholic acids) by dehydroxylation. Deoxycholic acid may be absorbed, but lithocholic acid is poorly absorbed.

Although bile salt and bile acid absorption is extremely efficient, inflammation of the ileum can lead to their malabsorption and result in the loss of large quantities of bile salts

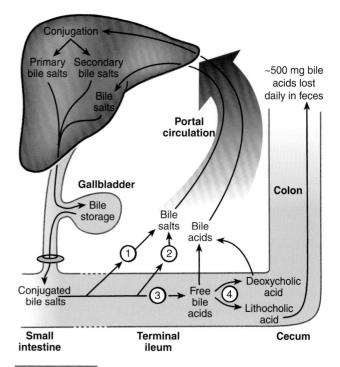

Figure 25.12 **The enterohepatic recycling mechanisms for bile salt.** Approximately 95% of the bile salts secreted are reabsorbed in the terminal ileum and recycled by four mechanisms: (**1**) passive diffusion along the small intestine (playing a relatively minor role), (**2**) carrier-mediated active absorption in the terminal ileum (the most important absorption route), (**3**) deconjugation to primary bile acids before being absorbed either passively or actively, and (**4**) conversion of primary bile acids to secondary bile acids with subsequent absorption of deoxycholic acid.

in the feces. Depending on the severity of illness, malabsorption of fat may result. Regardless, some salts and acids are nonetheless lost with every cycle of the enterohepatic circulation. About 500 mg of bile acids are lost daily and replenished by the synthesis of new bile acids from cholesterol. The loss of bile acids in feces is, therefore, an efficient way to excrete cholesterol.

Absorbed bile salts are transported in the portal blood bound to albumin or **high-density lipoproteins (HDLs)**. In just one pass through the liver, more than 80% of the bile salts in the portal blood is removed by the hepatocytes and then resecreted in bile. The uptake of bile salts is a primary determinant of bile salt secretion by the liver.

Bilirubin is the major component of bile pigments.

The major pigment present in bile is the orange compound **bilirubin**, an end product of hemoglobin degradation in the monocyte–macrophage system in the spleen, bone marrow, and liver (Fig. 25.13). Hemoglobin is first converted to biliverdin with the release of iron and globin. Biliverdin is then converted into bilirubin, which is transported in blood bound to albumin. The liver rapidly removes bilirubin from the circulation and conjugates it with glucuronic acid. The glucuronide is secreted into the bile canaliculi through an active carrier-mediated process.

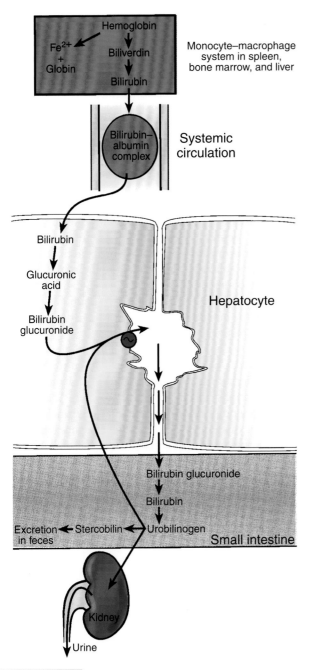

Figure 25.13 **Bilirubin is the end product of hemoglobin degradation.** The liver removes the bile pigment, bilirubin, from the circulation and conjugates it with glucuronic acid.

In the small intestine, bilirubin glucuronide is poorly absorbed. However, bacteria in the colon deconjugate it, and part of the bilirubin released is converted to the highly soluble, colorless compound called urobilinogen. Urobilinogen can be oxidized in the intestine to stercobilin or absorbed by the small intestine. It is excreted in either urine or bile. Stercobilin is responsible for the brown color of the stool.

Gallstones form when cholesterol becomes supersaturated.

Bile salts and lecithin in the bile help solubilize cholesterol. However, when bile contains too much cholesterol and not enough bile salts, it starts to crystallize and forms **gallstones**.

Gallstones can occur anywhere within the biliary tree, including the gallbladder and the common bile duct. Obstruction of the biliary tree can cause jaundice, and obstruction of the outlet of the pancreatic exocrine system can cause pancreatitis. Gallstones can easily be detected on an x-ray image because Ca^{2+} deposits formed in the stones increase their opacity. Gallstones are more prevalent in women than in men.

▶ INTESTINAL SECRETION

The small intestine is where the vast majority of digestion and absorption of food takes place. Duodenal cells secrete mucus as well as 2 to 3 L/d of isotonic alkaline fluid to aid in the digestive process. *Brunner glands* secrete HCO_3^- containing mucus and are located mainly in the proximal region of the duodenum between the entrance from the stomach and the sphincter of Oddi. These secretions aid in protecting the epithelium from HCl degradation as chyme enters the intestinal lumen. The **crypts of Lieberkühn**, tubular glands that dip down into the mucosal surface between the villi (Fig. 25.14), secrete a watery fluid. The intestinal crypts do not secrete digestive enzymes, but do secrete mucus, electrolytes, and H_2O. In addition, the *crypts of Lieberkühn* function to replace intestinal epithelial cells that are continually being sloughed off. Replacement occurs at a rapid rate due to the high mitotic index of epithelial stem cells in the crypts. The new epithelial cells migrate up the villi and replace older epithelial cells at the tips of the villi. The epithelial lining of the small intestine is replaced approximately every 3 days. Due

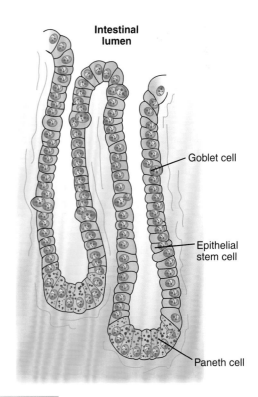

Figure 25.14 **Crypt of Lieberkühn.** Intestinal crypts secrete intestinal fluid and are found in all regions of the small intestine sandwiched in between villi.

to a high rate of cell division, the intestinal crypts are very sensitive to damage caused by radiation and to anticancer drugs used in chemotherapy. This is particularly problematic with patients who have been treated for colorectal cancer because their treatment is specially designed to target and kill tumors in the epithelial lining.

Intestinal secretions provide lubrication and protective functions.

The **Paneth cells** (see Fig. 25.14) are involved in the host defense system in the small intestine. They function like neutrophils and produce antimicrobial substances that provide a protective barrier. The *Paneth cells* are particularly important in protecting the stem cells from damage. **Goblet cells** secrete various mucins (mucoproteins) into the intestinal lumen. Mucins are extremely diverse in structure and are usually large molecules. Mucins are glycoproteins high in carbohydrate and they form gels in solution. The mucus lubricates the mucosal surface and protects it from mechanical damage by solid food particles. In the small intestine, it may also provide a physical barrier against the entry of microorganisms across the epithelial mucosa.

Intestinal secretions probably help maintain the fluidity of the chyme and may also play a role in diluting noxious agents and washing away infectious microorganisms. The HCO_3^- in intestinal secretions protects the intestinal mucosa by neutralizing any H^+ present in the lumen. While especially important in the duodenum, it is also important in the ileum, where bacteria degrade certain foods, which produces acids (e.g., dietary fibers to short-chain fatty acids).

Intestinal fluid hypersecretion is stimulated by toxins and other luminal stimuli.

The small intestine and colon usually absorb the fluid and electrolytes from intestinal secretions, but if secretion surpasses absorption (e.g., in cholera), watery diarrhea may result. If uncontrolled, this can lead to the loss of large quantities of fluid and electrolytes, which can result in dehydration and electrolyte imbalances and, ultimately, death. Several noxious agents, such as bacterial toxins (e.g., cholera toxin), can induce intestinal hypersecretion. Cholera toxin binds to the brush border membrane of crypt cells and increases intracellular adenylyl cyclase activity. Adenylyl cyclase synthesizes cAMP from ATP. The result is a dramatic increase in intracellular cAMP that stimulates active Cl^- and HCO_3^- secretion into the lumen. H_2O follows the osmotic gradient and enters the lumen.

It is well documented that tactile stimulation, or an increase in intraluminal pressure, stimulates intestinal secretion. Other potent stimuli are certain noxious agents and the toxins produced by microorganisms. With the exception of toxin-induced secretion, our understanding of the normal control of intestinal secretion is meager. VIP is known to be a potent stimulator of intestinal secretion. This is demonstrated by a form of endocrine tumor of the pancreas that results in the secretion of large amounts of VIP into the circulation. In this condition, intestinal secretion rates are high.

► CARBOHYDRATE DIGESTION AND ABSORPTION

Carbohydrates are one of the main types of nutrients and the most important source of energy. The digestive system converts carbohydrates into glucose (blood sugar). Sucrose and lactose are two major sources of carbohydrate in the human diet. Sucrose (present in cane sugar and honey) is a disaccharide composed of glucose and fructose. Lactose (the main sugar in milk) is a disaccharide composed of galactose and glucose.

Starches, large polysaccharides, are the third major source of dietary carbohydrates. Starch is produced by all green plants and is the most important carbohydrate in the human diet. Sources include such staple foods as potatoes and most types of grain (e.g., wheat, corn, and rice). All three major dietary sources (sucrose, lactose, and starch) must be hydrolyzed to monosaccharides before they can be absorbed in the small intestine. Although some carbohydrates are digested in the mouth and stomach, most are digested in the small intestine by pancreatic enzymes.

To ensure the optimal absorption of carbohydrates and other nutrients, the GI tract has several unique features. For instance, after a meal, the small intestine undergoes rhythmic contractions called *segmentations* (see Chapter 27), which ensure proper mixing of the small intestinal contents, exposure of the contents to digestive enzymes, and maximum exposure of digestion products to the small intestinal mucosa. The rhythmic segmentation has a gradient along the small intestine, with the highest frequency in the duodenum and the lowest in the ileum. This gradient ensures slow but forward movement of intestinal contents toward the colon.

Another unique feature of the small intestine is its architecture. Spiral or circular concentric folds increase the surface area of the intestine by about three times (Fig. 25.15). Finger-like projections of the mucosal surface called **villi** further increase the surface area of the small intestine to 30 times. To amplify the absorptive surface further, numerous closely packed **microvilli** cover each epithelial cell, or **enterocyte**. The total surface area is increased to 600 times. The GI tract absorbs the various nutrients, vitamins, bile salts, and H_2O by passive, facilitated, or active transport. (The site and mechanism of absorption are discussed below.) The GI tract has a large reserve for the digestion and absorption of various nutrients and vitamins. Malabsorption of nutrients is usually not detected unless a large portion of the small intestine has been lost or damaged because of disease or surgical manipulation (e.g., certain bariatric surgery).

The duodenum and jejunum absorb most nutrients and vitamins, but because bile salts are involved in the intestinal absorption of lipids, the small intestine has adapted to absorb the bile salts later in the terminal ileum through a bile salt transporter. The enterocytes along the villus that are involved in the absorption of nutrients are replaced every 2 to 3 days.

The digestion and absorption of dietary carbohydrates take place in the small intestine. These are extremely efficient processes, in that essentially all of the carbohydrates

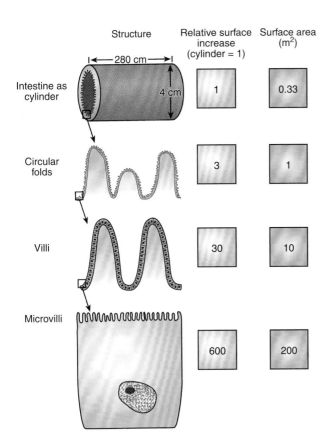

Figure 25.15 **The unique architecture of the small intestine greatly enhances the surface area.** Circular folds, villi, and microvilli are three functional features that greatly increase the surface area of the small intestine for the digestion and absorption of food.

consumed are absorbed. Carbohydrates constitute about 45% to 50% of the typical Western diet and provide the greatest and least expensive source of energy. Carbohydrates must be digested to monosaccharides before absorption by the enterocytes.

Dietary fiber benefits gastrointestinal function.

Dietary fiber, sometimes called *roughage*, consists of non-starch polysaccharides that arise from plants that the GI system cannot digest or absorb. Fiber is classified as soluble or insoluble. Soluble fiber dissolves in water to form a gel-like material and is found in oats, peas, beans, apples, carrots, and citrus fruits. The gel-like material that is formed absorbs sugars and certain fats and is beneficial in lowering blood cholesterol and glucose levels by excreting them in the feces. Insoluble fiber doesn't dissolve in water and is the fiber that increases bulk, softens stools, and shortens transit time in the intestinal tract, all of which facilitates motility and prevents constipation.

By shortening GI transit time, dietary fiber has also been shown to reduce the incidence of colon cancer. A shortened transit time inhibits the formation of carcinogenic bile acids, such as lithocholic acid, as well as reducing the contact time of ingested carcinogens to act on tissue. Some common sources include whole wheat flour, wheat bran, nuts, beans, and vegetables such as green beans.

Carbohydrates are defined by their chemical structure.

Dietary carbohydrates are called monosaccharides, disaccharides, oligosaccharides, and polysaccharides depending on their chemical structure. The monosaccharides are mainly hexoses (six-carbon sugars), and glucose is the most abundant of these. Glucose is obtained directly from the diet or from the digestion of disaccharides, oligosaccharides, or polysaccharides. The next most common monosaccharides are galactose, fructose, and sorbitol. Galactose is present in the diet only as milk lactose. Fructose is abundant in fruit and honey and is usually present as disaccharides or polysaccharides. Sorbitol is derived from glucose and is almost as sweet as glucose, but it is absorbed much more slowly and, thus, maintains a high blood sugar level for a longer period when similar amounts are ingested. It has been used as a weight reduction aid to delay the onset of hunger sensations and is often used in diet foods and diet drinks.

The digestible polysaccharides are starch, dextrins, and glycogen. Starch, by far the most abundant carbohydrate in the human diet, is made of amylose and amylopectin. Amylose is composed of a straight chain of glucose units; amylopectin is composed of branched glucose units. Dextrins, formed from heating (e.g., toasting bread) or the action of the enzyme amylase, are intermediate products of starch digestion. Glycogen is a highly branched polysaccharide that stores carbohydrates in the body. Normally, about 300 to 400 g of glycogen is stored in the liver and muscle, with more stored in muscle than in the liver. Muscle glycogen is used exclusively by muscle, and liver glycogen is used to provide blood glucose during fasting.

Carbohydrate digestion begins with salivary amylase.

The digestion of carbohydrates starts when food is mixed with saliva during chewing (Fig. 25.16). The enzyme salivary amylase acts on starches to release the disaccharide maltose. Because salivary amylase works best at neutral pH, its digestive action terminates rapidly after the bolus mixes with acid in the stomach. If the food is thoroughly mixed with amylase during chewing, a substantial amount of complex carbohy-

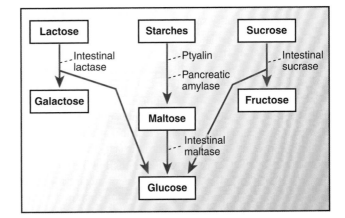

Figure 25.16 **Salivary and pancreatic enzymes break down carbohydrates.** Starch, sucrose, and lactose are the three main dietary sources of carbohydrates. All three are hydrolyzed to monosaccharides for absorption.

CLINICAL FOCUS | 25.2

Celiac Sprue (Gluten-Sensitive Enteropathy)

Celiac sprue, also called **gluten-sensitive enteropathy**, is an autoimmune disorder of the small intestine that occurs in genetically predisposed individuals of all ages. The disease is caused by a reaction to a protein (gluten) found in grain (wheat, barley, oats, and rye). Upon exposure to gluten, the immune system cross-reacts, causing chronic inflammation and primary lesions of the intestinal mucosa. This disorder can result in the malabsorption of all nutrients as a result of the shortening or total loss of intestinal villi, which reduces the mucosal enzymes for nutrient digestion and the mucosal surface for absorption. For example, as the small intestine becomes more damaged, a degree of lactose intolerance may develop. Symptoms include diarrhea, weight loss, failure to gain weight (in young children), and anemia. The latter is due to iron malabsorption. Celiac sprue occurs in about 1 to 6 of 10,000 people in the Western world. The highest incidence is in Western Ireland, where the prevalence is as high as 3 of 1,000 people. People of African,

Chinese, and Japanese descent are rarely diagnosed. Although the disease may occur at any age, it is more common during the first few years and the third to fifth decades of life. Celiac disease has been present since humans started harvesting grains in an organized fashion (about 5,000 BCE); however, the link to wheat was not discovered until the 1940s, by Dutch pediatrician, Dr. Wilhelm-Karel Dicke. He observed during World War II that when flour and bread were scarce, the death rate among children due to celiac sprue dropped to zero. The specific link to gluten was made later in 1952 by a team of physicians in England. The elimination of dietary gluten is a standard treatment for patients with celiac sprue. Occasionally, intestinal absorptive function and intestinal mucosal morphology of patients with celiac sprue are improved with glucocorticoid therapy. Presumably, such treatment is beneficial because of the immunosuppressive and anti-inflammatory actions of these hormones. ∎

---

drates is digested before enzyme inactivation. **Pancreatic amylase** continues the digestion of the remaining carbohydrates. However, pancreatic secretions must first neutralize the chyme because pancreatic amylase works best at neutral pH. Pancreatic amylase digestion of polysaccharides also yields maltose, which is then converted to glucose by maltase (Clinical Focus 25.2).

Enterocytes hydrolyze disaccharides and absorb monosaccharides.

Enterocytes are simple columnar cells lining the villi of the small intestine. Enzymes located at the brush border membrane (microvilli) of the enterocytes digest the disaccharides maltose, sucrose, and lactose (see Fig. 25.16). The final products glucose, fructose, and galactose are then absorbed. The intestinal enterocytes (absorptive cells) absorb monosaccharides using both active and facilitated transport processes. A sodium–glucose symporter (SGLT) (see Chapter 2), located on the brush boarder membrane of the enterocytes, transports two Na$^+$ for every molecule of monosaccharide and absorbs glucose and galactose via secondary active transport (Fig. 25.17). Since glucose and galactose share a common transporter, they compete with each other during absorption. The movement of Na$^+$ into the cell, down concentration and electrical gradients, affects the uphill movement of glucose into the cell. The basolateral membrane Na$^+$/K$^+$-ATPase maintains the low intracellular Na$^+$ concentration. The osmotic effects of sugars increase the Na$^+$/K$^+$-ATPase activity and the K$^+$ conductance of the basolateral membrane. Sugars accumulate in the cell at a higher concentration than in plasma and leave the cell by Na$^+$-independent facilitated transport via a glucose transporter (GLUT-2) or passive diffusion through the basolateral cell membrane.

Fructose is taken up by facilitated transport. This transporter, also in the family of glucose transporters, is named GLUT-5. Although facilitated transport is carrier-mediated, it is not an active process (see Chapter 2). Fructose absorption is much slower than glucose and galactose absorption and is not Na$^+$-dependent. In some animal species, both galactose and fructose can be converted to glucose in enterocytes, but this mechanism is probably not important in humans.

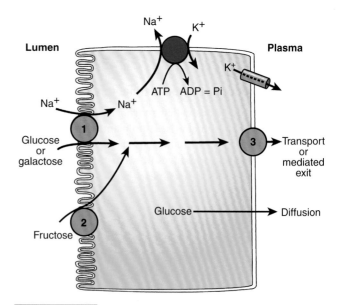

Figure 25.17 **Sugar absorption depends on an Na$^+$-dependent carrier system.** Sodium moves into the cell down a concentration and electrical gradient, which favors the uphill movement of glucose and galactose into the enterocytes. ADP, adenosine diphosphate; ATP, adenosine triphosphate; Pi, inorganic phosphate. Transporters labeled 1, 2, and 3 are members of a family of hexose transporters. 1, SGLUT-1 ; 2, GLUT5; 3, GLUT 2.

The portal blood transports the absorbed sugars to the liver, where they are converted to glycogen or remain in the blood. After a meal, the level of blood glucose rises rapidly, usually peaking at 30 to 60 minutes. Blood glucose levels can be as high as 150 mg/dL and even higher in diabetic patients. Although enterocytes can use glucose for fuel, glutamine is preferred. Both galactose and glucose are used in the glycosylation of proteins in the Golgi apparatus of the enterocytes.

Salivary or pancreatic dysfunction impairs carbohydrate absorption.

Impaired carbohydrate absorption caused by the absence of salivary or pancreatic amylase almost never occurs because these enzymes are usually present in great excess. However, impaired absorption as a result of a deficiency in membrane disaccharidases is common. Such deficiencies can be either genetic or acquired. Among congenital deficiencies, lactase deficiency is, by far, the most common. Affected people suffer from **lactose intolerance**, due to an inability to cleave lactose. Since disaccharides cannot be absorbed by the small intestine, the absence of lactase allows the ingested, uncleaved dairy products to pass into the colon. Bacteria in the colon quickly switch over to lactose metabolism, resulting in *in vivo* fermentation that produces copious amounts of gas. This, in turn, causes a number of abdominal symptoms including stomach cramps, bloating, and flatulence. Fermentation products also raise the osmotic pressure of the colon contents, resulting in excess fluid accumulation and diarrhea (Fig. 25.18).

Congenital sucrase deficiency results in symptoms similar to those of lactase deficiency. Sucrase deficiency can be inherited or acquired through disorders of the small intestine, such as tropical sprue or Crohn's disease.

▶ LIPID DIGESTION AND ABSORPTION

Lipids comprise several classes of compounds that are insoluble in H_2O but soluble in organic solvents. Lipids are a group of naturally occurring dietary molecules that include fats, sterols, monoglycerides, diglycerides, phospholipids, and fat-soluble vitamins (A, D, E, and K). The physiologic function of lipids includes energy storage, structural components for cell membranes, and cell signaling molecules. Although the term "lipid" is sometimes used synonymously with fats, fats are a subgroup of lipids called **triglycerides**. Triglycerides in adipose tissue are the major form of concentrated energy storage in animals. Triglycerides provide 30% to 40% of the daily caloric intake and more than 90% of the daily dietary lipid intake in the Western diet. The human body is capable of synthesizing most of the lipids it requires, with the exception of the **essential fatty acids**. Essential fatty acids must be ingested for normal cell function, because they cannot be synthesized from other foods. Recent studies suggest that the human body may also require dietary omega-3 fatty acids during development. In humans, the essential fatty acids are linolenic acid, linoleic acid, and arachidonic acid. Linolenic acid is an omega-3 fatty acid, and linoleic and arachidonic both omega-6 fatty acids. These essential fatty acids are abundant in seafood, flaxseed (linseed), soya oil, canola oil, sunflower seeds, leafy vegetables, and walnuts. Researchers have recently provided convincing evidence that **eicosapentaenoic acid** (C 20:5) and **docosahexaenoic acid** (C 22:6) are also essential for the normal development of vision in newborns. Docosahexaenoic acid is an important fatty acid present in the retina and parts of the brain.

The other lipids in the human diet are cholesterol and phospholipids. **Cholesterol** is a sterol derived exclusively from animal fat. Humans also ingest a small amount of plant sterols, notably β-sitosterol and campesterol. The phospholipid molecule is similar to a triglyceride, with fatty acids occupying the first two positions of the glycerol backbone (Fig. 25.19). However, the third position is occupied by a phosphate group coupled to a nitrogenous base (e.g., choline or ethanolamine), for which each type of phospholipid molecule is named. For example, phosphatidylcholine (lecithin) is an important phospholipid in cell membranes and lung surfactant.

Bile serves as an endogenous source of cholesterol and phospholipids. Bile contributes about 12 g/d of phospholipid to the intestinal lumen, mostly in the form of phosphatidylcholine, whereas dietary sources contribute 2 to 3 g/d. Another important endogenous source of lipid is desquamated intestinal villus epithelial cells.

Lipases are water-soluble enzymes that break down fats into monoglycerides and free fatty acids.

Fats are digested by lipases (acyl hydrolases). Since lipases are most active at the oil–water interface, the rate of lipolysis depends on the surface area of the interface, which is increased when oil droplets are formed (an emulsion).

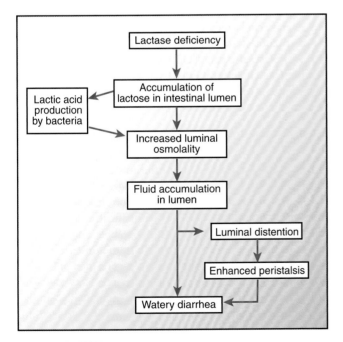

Figure 25.18 Lactase deficiency causes diarrhea. Undigested lactose increases the osmolality of the gut, which results in an increase in osmolality. The high osmolality leads to a concomitant increase in a net water secretion, causing diarrhea.

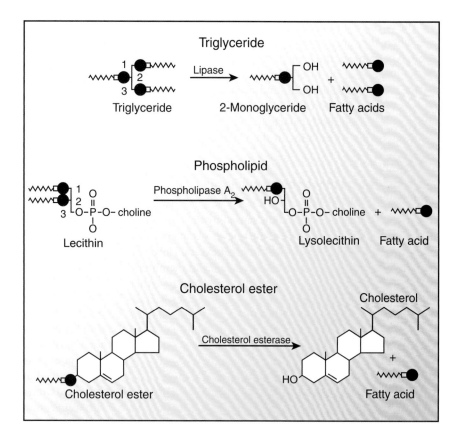

Figure 25.19 Pancreatic lipases digest dietary lipids. Different pancreatic lipases carry out lipid hydrolysis in the gut. *Solid circles* represent oxygen atoms.

Emulsions are formed in the stomach, when churning action mixes proteins and the products of fat digestion, and in the intestine, when components present in bile mix with fats. Lipid digestion starts in the stomach when **gastric lipase** (an acidic lipase) comes in contact with fats. Gastric lipase digests 10% to 30% of the ingested lipids. It has optimal activity in a pH range of 3 to 7 and, therefore, function quite well at normal stomach pH values. Gastric lipase prefers triglycerides containing medium-chain fatty acids (8 to 10 carbons in length), which are abundant in milk. In infants, gastric lipase plays a very important role in the digestion of fat because infants do not have sufficient pancreatic lipase to digest milk fat, which is their major source of fat intake. In adults, gastric lipase is also involved in fat digestion, and the digestion products aid in the emulsification of fat, which greatly facilitates fat digestion by **pancreatic lipase (colipase-dependent lipase)** in the intestinal lumen where the majority of lipid digestion is. Humans secrete an overabundance of pancreatic lipase. Pancreatic lipase is optimally active at a pH of 7 to 8, and it works well in the intestinal lumen after the acidic contents entering from the stomach have been neutralized by pancreatic HCO_3^- secretion. Optimal pancreatic lipase activity requires the presence of the peptide **colipase** also present in pancreatic juice. Colipase binds lipase at a molar ratio of 1:1, thereby allowing the lipase to bind to the oil–water interface where it hydrolyzes the triglyceride molecule yielding a 2-monoglyceride and two fatty acids (see Fig. 25.19). Colipase also counteracts the inhibition of lipolysis by bile salts, which, despite their importance in intestinal fat absorption, prevent the attachment of pancreatic lipase to the oil–water interface.

Phospholipase A₂ and carboxyl ester hydrolase contribute to lipolytic activity.

Phospholipase A$_2$ is the major pancreatic enzyme for digesting phospholipids and forming lysophospholipids and fatty acids. For instance, phosphatidylcholine (lecithin) is hydrolyzed to form lysophosphatidylcholine (lysolecithin) and fatty acid (see Fig. 25.19).

Dietary cholesterol is present as the free sterol or as a sterol ester (cholesterol ester). The pancreatic enzyme *carboxyl ester hydrolase*, also called **cholesterol esterase**, catalyzes the hydrolysis of cholesterol ester (see Fig. 25.19). The digestion of cholesterol ester is important because cholesterol can be absorbed only as the free sterol.

Emulsification by bile acids is the first step in fat digestion.

A layer of poorly stirred fluid, the **unstirred water layer**, coats the surface of the intestinal villi (Fig. 25.20A). This layer reduces the absorption of lipid digestion products because they are H_2O insoluble. **Micellar emulsification** in the small intestinal lumen renders them H_2O soluble and enhances the concentration of these products in the unstirred water layer (Fig. 25.20B). Enterocytes then absorb the lipid digestion products, mainly by passive diffusion. Fatty acid and monoglyceride molecules are taken up individually. Similar mechanisms seem to operate for cholesterol and lysolecithin. However, recent evidence supports a transporter-mediated uptake of cholesterol. Ezetimibe, an antihyperlipidemic medication to lower blood cholesterol, is extremely effective in decreasing intestinal cholesterol absorption. The mechanism appears to involve the binding of this drug to

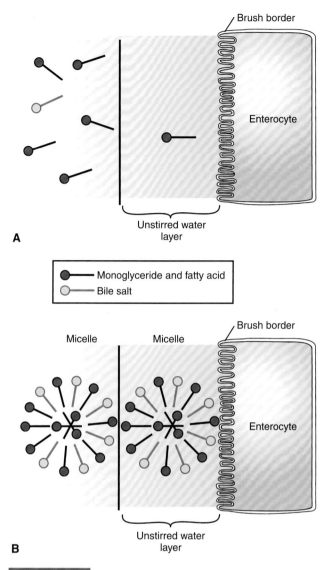

Monoglyceride and fatty acid
Bile salt

Figure 25.20 **Bile salts emulsify lipids for absorption.** Bile acts as a detergent to emulsify fat. Micelles are formed from emulsified lipid, and the micellar solubilization enhances the delivery and absorption of lipid to the brush border membrane. (**A**) The absence of micelles when bile salt is absent. (**B**) The presence of micelles when bile salt is present.

Gastrointestinal Physiology

products water-soluble. Mixed micelles diffuse across the unstirred water layer and deliver lipid digestion products to the enterocytes for absorption. This absorption pathway for lipolytic products is much more efficient than direct absorption from emulsions.

Enterocytes secrete chylomicrons and very low–density lipoproteins.

After entering the enterocytes, the fatty acids and monoglycerides migrate to the smooth endoplasmic reticulum (SER). A fatty acid–binding protein may be involved in the intracellular transport of fatty acids, but whether a protein carrier is involved in the intracellular transport of monoglycerides is still unclear. In the SER, monoglycerides and fatty acids are rapidly reconstituted to form triglycerides (Fig. 25.21). Fatty acids are first activated to form acyl-coenzyme A (acyl-CoA), which is then used to esterify monoglyceride to form diglyceride, which is then transformed into triglyceride. The lysolecithin absorbed by the enterocytes can be re-esterified in the SER to form lecithin.

Cholesterol can be transported out of the enterocytes as free cholesterol or as esterified cholesterol. The enzyme responsible for the esterification of cholesterol to form cholesterol ester is **acyl-CoA cholesterol acyltransferase**.

The reassembled triglycerides, lecithin, cholesterol, and cholesterol esters are then packaged into **lipoproteins** and exported from the enterocytes. The intestine produces two major classes of lipoproteins: **chylomicrons** and **very low–density lipoproteins (VLDLs)**. Both are triglyceride-rich lipoproteins with densities <1.006 g/mL. The small intestine exclusively makes chylomicrons, and their primary function is to transport the large amount of dietary fat absorbed by the small intestine from the enterocytes to the lymph. Chylomicrons are large, spherical lipoproteins with diameters of 80 to 500 nm. They contain less protein and phospholipid than do VLDLs and are, therefore, less dense than are VLDLs. The small intestine continuously makes VLDLs during both fasting and feeding, although the liver contributes significantly more VLDLs to the circulation. VLDL is converted in the bloodstream to **low-density lipoprotein (LDL)**. Since cholesterol is carried by LDL, cholesterol content can be measured on LDL particles. High levels of LDL particles can promote medical problems and they are often referred to as *bad cholesterol* particles (as opposed to HDL, which is often referred to as *good cholesterol*). One of the functions of LDL particles is the transport of cholesterol into the arterial wall where it is retained by arterial proteoglycans. **Macrophages** are attracted to these sites and function to engulf the LDL particles. The process of engulfing LDL particles by macrophages initiates plaque formation in the arterial wall. Increased levels of plaque formation are linked to atherosclerosis. Over time, plaques become vulnerable to rupturing, which, in turn, activates blood clotting and leads to arterial stenosis.

Apoproteins—Apo A-I, Apo A-IV, and Apo B—are among the major proteins associated with the production of chylomicrons and VLDLs. Apo B is the only protein that seems to be necessary for the normal formation of intesti-

the cholesterol transporter, thereby preventing the uptake of cholesterol.

Bile salts are derived from cholesterol, but unlike cholesterol, they are water-soluble. Bile salts have both hydrophilic and hydrophobic constituents, which makes them polar molecules with detergent-like properties. Polar molecules penetrate cell membranes poorly. This property ensures minimal absorption of bile salts by the jejunum, where most fat absorption takes place. At or above a certain concentration of bile salts, the **critical micellar concentration**, the salts aggregate to form micelles; the concentration of luminal bile salts is usually well above the critical micellar concentration. When bile salts alone are present in the micelle, it is called a **simple micelle**. Simple micelles incorporate the lipid digestion products—monoglyceride and fatty acids—to form **mixed micelles**, a process that renders the lipolytic

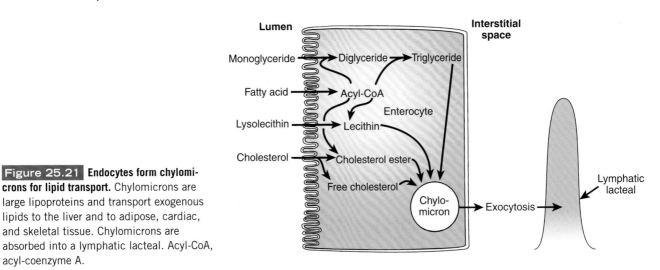

Figure 25.21 **Endocytes form chylomicrons for lipid transport.** Chylomicrons are large lipoproteins and transport exogenous lipids to the liver and to adipose, cardiac, and skeletal tissue. Chylomicrons are absorbed into a lymphatic lacteal. Acyl-CoA, acyl-coenzyme A.

nal chylomicrons and VLDLs. This protein is made in the small intestine. It has a molecular weight of 250,000 and is extremely hydrophobic. Apo A-I, also formed by intestinal cells, is involved in a reaction catalyzed by the plasma enzyme **lecithin cholesterol acyltransferase** (**LCAT**). Plasma LCAT is responsible for the esterification of cholesterol in the plasma to form cholesterol ester with the fatty acid derived from the 2-position of lecithin. After the chylomicrons and VLDLs enter the plasma, Apo A-I is rapidly transferred from chylomicrons and VLDLs to HDLs. Apo A-I is the major protein present in plasma HDLs. The small intestine and the liver make Apo A-IV. Recently, it was shown that Apo A-IV, secreted by the small intestine, may be an important factor contributing to anorexia after fat feeding.

Newly synthesized lipoproteins in the SER are transferred to the Golgi apparatus, where they are packaged into vesicles. Chylomicrons and VLDLs are released into the intercellular space by exocytosis. From there, they are transferred to the central lacteals (the beginnings of lymphatic vessels) by a process (transcellular or paracelluar route) that is not well understood. Intestinal lipid absorption is associated with a marked increase in lymph flow called the **lymphagogic effect** of fat feeding. This increase in lymph flow plays an important role in the transfer of lipoproteins from the intercellular spaces to the central lacteal.

Fatty acids can also travel in the blood bound to albumin. Whereas most of the long-chain fatty acids are transported from the small intestine as triglycerides packaged in chylomicrons and VLDLs, some are transported in the portal blood bound to serum albumin. Most of the medium-chain (8 to 12 carbons) and all of the short-chain fatty acids are transported via the hepatic portal route.

Altered lipase and bile secretion significantly impair lipid digestion and absorption.

In several clinical conditions, lipid digestion and absorption are impaired, resulting in the malabsorption of lipids and other nutrients and fatty stools. Abnormal lipid absorption can result in numerous problems because the body requires certain essential fatty acids (see previous discussion).

Pancreatic deficiency significantly reduces the ability of the exocrine pancreas to produce digestive enzymes. Because the pancreas normally produces an excess of digestive enzymes, enzyme production has to be reduced to about 10% of normal before symptoms of malabsorption develop. One characteristic of pancreatic deficiency is **steatorrhea** (fatty stool), resulting from the poor digestion of fat by pancreatic lipase. Normally, about 5 g/d of fat is excreted in human stool. With steatorrhea, as much as 50 g/d of fat can be excreted.

Fat absorption subsequent to the action of pancreatic lipase requires solubilization by bile salt micelles. Acute or chronic liver disease can cause defective biliary secretion, resulting in bile salt concentrations lower than that necessary for micelle formation. The normal absorption of fat is thereby inhibited.

Abetalipoproteinemia, an autosomal recessive disorder, is characterized by a complete lack of Apo B in the circulation, which is required for the formation and secretion of chylomicrons and VLDLs. Apo B–containing lipoproteins in the circulation, including chylomicrons, VLDLs, and LDLs, are absent. Plasma LDLs are absent because they are derived mainly from the metabolism of VLDLs. Because people with abetalipoproteinemia do not produce any chylomicrons or VLDLs in the small intestine, they are unable to transport absorbed fat, resulting in an accumulation of lipid droplets in the cytoplasm of enterocytes. They also suffer from a deficiency of fat-soluble vitamins (A, D, E, and K). Using molecular biologic techniques, it has been determined that Apo B deficiency is caused by a mutation of a transfer protein called microsomal triglyceride transfer protein. This protein facilitates the transfer of triglyceride formed in the cytoplasm to the ER.

▶ PROTEIN DIGESTION AND ABSORPTION

Dietary proteins are basically long-chain amino acids that are joined together by peptide bonds between the carboxyl and amino groups. Proteins are fundamental structural components of cells and participate in every aspect of cell function. Aside from their role in cellular protein synthesis, amino acids are also an important nutritional source of nitrogen.

Proteins are primarily used in cell function and structure rather than as a source of energy. Most proteins are found in muscle, with the remainder in other cells, blood, body fluids, and body secretions. Enzymes and many hormones are proteins. Examples of their use in cellular function are shown in Table 25.4. Proteins are composed of amino acids and have molecular weights of a few thousand to a few hundred thousand. More than 20 common amino acids form the building blocks for proteins (Table 25.5). Of these, nine are considered essential and must be provided by the diet. Although the **nonessential amino acids** are also required for normal protein synthesis, the body can synthesize them from other amino acids.

Complete proteins are those that can supply all of the essential amino acids in amounts sufficient to support normal growth and body maintenance. Examples are eggs, poultry, and fish. The proteins in most vegetables and grains are called **incomplete proteins** because they do not provide all of the essential amino acids in amounts sufficient to sustain normal growth and body maintenance. Vegetarians need to eat a variety of vegetables and soy protein to avoid amino acid deficiencies.

Some dietary proteins are often the cause of allergies and/or allergic reactions. These reactions occur because the structure of each form of protein is slightly different. Some may trigger an immune response, whereas other structures are perfectly safe. For example, many individuals are allergic

TABLE 25.5 Amino Acids Found in Proteins

Essential	Nonessential
Histidine	Alanine
Isoleucine	Arginine
Leucine	Aspartic acid
Lysine	Citrulline
Methionine	Glutamic acid
Phenylalanine	Glycine
Threonine	Hydroxyglutamic acid
Tryptophan	Hydroxyproline
Valine	Norleucine
	Proline
	Serine
	Tyrosine

to gluten (the protein in wheat and other grains), to casein (the protein in milk), and to particular proteins found in peanuts. Some individuals are also allergic to shellfish and to specific proteins found in other types of seafood. Certain food additives and dyes (food coloring) may also stimulate an immune response.

Nondietary proteins derive from endogenous sources.

The average American adult takes in 70 to 110 g/d of protein. The minimum daily protein requirement for adults is about 0.8 g/kg body weight (e.g., 56 g for a 70-kg person). Pregnant and lactating women require 20 to 30 g of protein above the recommended daily allowance. A lactating woman can lose as much as 12 to 15 g of protein per day as milk protein. Children need more protein for body growth; the recommended daily allowance for infants is about 2 g/kg body weight, and young children may require as much as 4 g/kg body weight.

Although most of the protein entering the GI tract is dietary protein, there are also proteins derived from endogenous sources such as pancreatic, biliary, and intestinal secretions and the cells shed from the intestinal villi. About 20 to 30 g/d of protein enters the intestinal lumen in pancreatic juice and about 10 g/d in bile. Enterocytes of the intestinal villi are continuously shed into the intestinal lumen, and as much as 50 g/d of enterocyte proteins enter the intestinal lumen. An average of 150 to 180 g/d of total protein is presented to the small intestine, of which more than 90% is absorbed.

Digestion of proteins starts in the stomach and continues in the small intestine.

Most of the protein in the intestinal lumen is completely digested into amino acids, dipeptides, or tripeptides before it is taken up by the enterocytes (Fig. 25.22). Protein digestion

TABLE 25.4 Proteins Have a Large Range of Cellular Functions in the Body

Function	Description	Example
Antibody	An antibody neutralizes pathogens such as viruses, bacteria, fungus, and parasites.	IgA, IgE, IgM
Enzyme	Enzymes are biological catalysts that accelerate over thousands of chemical reaction in cells.	Carbonic anhydrase, catalase, lactase
Messenger	Messenger proteins are usually hormones that transmit signals to different parts of the body to coordinate cellular function between different tissues and organs.	Growth hormone, oxytocin, vasopressin
Structural component	Structural proteins are fibrous proteins that provide structure and support for cells, tissues, and organs. On a larger scale, they also allow the body to move.	Actin, myosin, keratins, collagen
Transport proteins	These proteins function to move other molecules throughout the body. There are two types: those that carry molecules across impermeable membranes and those that carry molecules to distant locations in the body.	Na^+/K^+-ATPase, Ca^{2+} ATPase, serum albumin, hemoglobin

Gastrointestinal Physiology

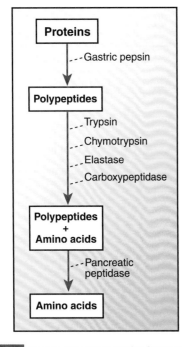

Figure 25.22 Protein absorption begins in the stomach. The first step involves the hydrolysis of proteins by gastric pepsin to form smaller polypeptides. Polypeptides are further acted upon by intestinal enzymes and proteases from both the small intestine and pancreas.

typically begins in the stomach when **pepsinogen** is cleaved by HCl to yield **pepsin**. Pepsin hydrolyzes protein to form smaller polypeptides. This phase of protein digestion is normally not important except in people suffering from pancreatic exocrine deficiency.

Most of the digestion of proteins and polypeptides takes place in the small intestinal lumen by **trypsin** and **chymotrypsin**. Most proteases are secreted in the pancreatic juice as inactive proenzymes. After pancreatic juice enters the duodenum, **enteropeptidase** (**enterokinase**), an enzyme found on the luminal surface of enterocytes, cleaves trypsinogen to **trypsin**. Trypsin then cleaves the other proenzymes to yield active enzymes.

The pancreatic proteases are classified as endopeptidases or exopeptidases (see Table 25.2). **Endopeptidases** hydrolyze certain internal peptide bonds of proteins or polypeptides to release the smaller peptides. The three endopeptidases present in pancreatic juice are trypsin, chymotrypsin, and elastase. **Trypsin** splits off basic amino acids from the C-terminus of a protein, **chymotrypsin** attacks peptide bonds with an aromatic carboxyl terminal, and **elastase** attacks peptide bonds with a neutral aliphatic C-terminus. The **exopeptidases** in pancreatic juice are carboxypeptidase A and carboxypeptidase B. Like the endopeptidases, the exopeptidases are specific in their action. **Carboxypeptidase A** attacks polypeptides with a neutral aliphatic or aromatic C-terminus. **Carboxypeptidase B** attacks polypeptides with a basic C-terminus. The final products of protein digestion are amino acids and small peptides.

Intestinal enterocytes contain specific transporters to take up amino acids and peptides.

Enterocytes take up amino acids in the small intestine via secondary active transport. Six major amino acid carriers in the small intestine have been identified; they transport related groups of amino acids. The amino acid transporters favor the L-form over the D-form. As in the uptake of glucose, the uptake of amino acids depends on an Na^+ concentration gradient across the enterocyte brush border membrane.

The absorption of peptides by enterocytes was once thought to be less efficient than amino acid absorption. However, subsequent studies in humans clearly demonstrated that dipeptide and tripeptide uptake is significantly more efficient than the uptake of amino acids. Dipeptides and tripeptides use different transporters than those used by amino acids. The peptide transporter prefers dipeptides and tripeptides with either glycine or lysine residues. Furthermore, the peptide transporter only poorly transports tetrapeptides and more complex peptides. The **peptidases** (exopeptidases) located on the brush border of the enterocytes can further break down these peptides to dipeptides and tripeptides (see Fig. 25.22). Dipeptides and tripeptides are given to people suffering from malabsorption because they are absorbed more efficiently and are more palatable than free amino acids. Another advantage of peptides over amino acids is the smaller osmotic stress created as a result of delivering them.

In adults, a negligible amount of protein is absorbed as undigested protein. In some people, however, intact or partially digested proteins are absorbed, resulting in anaphylactic or hypersensitivity reactions. The pulmonary and cardiovascular systems are the major organs involved in anaphylactic reactions. For the first few weeks after birth, the newborn's small intestine absorbs considerable amounts of intact proteins. This is possible because of low proteolytic activity in the stomach, low pancreatic secretion of peptidases, and poor development of intracellular protein degradation by lysosomal proteases.

The absorption of immunoglobulins (predominantly IgG) plays an important role in the transmission of **passive immunity** from the mother's milk to the newborn in several animal species (e.g., ruminants and rodents). In humans, the absorption of intact immunoglobulins does not appear to be an important mode of transmission of antibodies for two reasons. First, passive immunity in humans is derived almost entirely from the intrauterine transport of maternal antibodies. Second, human **colostrum**, the thin, yellowish, milky fluid secreted by the mammary glands a few days before or after parturition, contains mainly IgA, which is poorly absorbed by the small intestine. The ability to absorb intact proteins is rapidly lost as the gut matures—a process called **closure**. Colostrum contains a factor that promotes the closure of the small intestine.

After the enterocytes take up dipeptides and tripeptides, peptidases in the cytoplasm break them down to amino acids, which are transported in the portal blood. The small amount of protein that is taken up by the adult intestine is largely degraded by lysosomal proteases, although some proteins escape degradation.

Genetic disorders lead to impaired protein absorption.

Although pancreatic deficiency has the potential to affect protein digestion, it does so only in severe cases and seems to affect lipid digestion more than protein digestion. There are several extremely rare genetic disorders of amino acid carriers. In **Hartnup disease**, the membrane carrier for neutral amino acids (e.g., tryptophan) is defective. **Cystinuria** involves the carrier for basic amino acids (e.g., lysine and arginine) and the sulfur-containing amino acids (e.g., cystine). Cystinuria was once thought to involve only the kidneys because of the excretion of amino acids such as cystine in urine, but the small intestine is involved as well.

Because the peptide transport system remains unaffected, disorders of some amino acid transporters can be treated with supplemental dipeptides containing these amino acids. However, this treatment alone is not effective if the kidney transporter is also involved, as in the case of cystinuria.

▶ VITAMIN ABSORPTION

Vitamins are organic compounds required as a nutrient in small quantities by an animal. Thus, an organic compound is called a *vitamin* when it cannot be synthesized by the body and must be obtained from the diet. By convention, the term "vitamin" excludes dietary minerals, which are inorganic. Currently, 13 vitamins are universally recognized (Table 25.6).

Vitamins are classified in many ways, but, in terms of absorption, they are classified as either water-soluble or fat-soluble. In humans, there are 13 vitamins: 4 fat-soluble vitamins (A, D, E, and K) and 9 water-soluble vitamins (8 B vitamins and vitamin C). Because water-soluble vitamins are easily dissolved in H_2O, they are readily excreted from the body. Thus, urinary output is a strong predictor of B vitamin and C vitamin consumption.

Vitamins have many diverse cell functions, and many are essential for metabolic reactions.

Vitamins act as both catalysts and substrates in chemical reactions of the body. When acting as a catalyst, vitamins are bound to enzymes and are called cofactors (e.g., vitamin K forms part of the proteases involved in blood clotting). Vitamins can also act as coenzymes to carry chemical groups between enzymes (e.g., folic acid carries a carbon group methylene in the cell).

Fruits and vegetables are good sources of vitamins. Until the early 1900s, vitamins were obtained solely through food intake. Therefore, changes in diet often led to vitamin deficiencies and deadly diseases (e.g., lack of citrus fruit led to scurvy, a particularly deadly disease in which collagen is not properly formed and causes poor wound healing, bleeding gums, and severe pain). However, today, vitamins can be produced commercially and made widely available as inexpensive supplementation of dietary intake.

Fat-soluble vitamins: A, D, E, and K

The principal form of **vitamin A** is retinol; the aldehyde (retinal) and the acid (retinoic acid) are also active forms of vitamin A. Retinol can be derived directly from animal sources

TABLE 25.6	Vitamins: Their Solubility and Dietary Sources	
Vitamin	**Solubility**	**Food Source**
Vitamin A (retinol)	Fat	Cod liver oil, carrots
Vitamin B$_1$ (thiamine)	Water	Rice bran
Vitamin C (ascorbic acid)	Water	Citrus, most fresh foods
Vitamin D (calciferol)	Fat	Cod liver oil
Vitamin B$_2$ (riboflavin)	Water	Meat, eggs
Vitamin E (tocopherol)	Fat	Wheat germ oil, unrefined vegetable oils
Vitamin B$_{12}$ (cobalamins)	Water	Liver, eggs, animal products
Vitamin K (phylloquinone/phytol naphthoquinone)	Fat	Leafy green vegetables
Vitamin B$_5$ (pantothenic acid)	Water	Meats, whole grains, and many foods
Vitamin B$_7$ (biotin)	Water	Meats, dairy products, eggs
Vitamin B$_6$ (pyridoxine)	Water	Meat, dairy products
Vitamin B$_3$ (niacin)	Water	Meat, eggs, grains
Vitamin B$_9$ (folic acid)	Water	Leafy green vegetables

or through conversion from β-carotene (found abundantly in carrots) in the small intestine. Micellar solubilization renders vitamin A water-soluble, and the small intestine absorbs it passively. The unesterified retinol is then complexed with retinol-binding protein type 2, and the complex serves as a substrate for the re-esterification of the retinol by the enzyme lecithin: retinol acyltransferase. The retinyl ester is incorporated in chylomicrons and taken up by the liver. Vitamin A is stored in the liver and released to the circulation bound to retinol-binding protein only when needed. Vitamin A is important in the production and regeneration of rhodopsin of the retina and in the normal growth of the skin. People with vitamin A deficiency develop night blindness and skin lesions.

Vitamin D is a group of fat-soluble compounds collectively known as the *calciferols*. The human body has two main sources of vitamin D$_3$ (also called *cholecalciferol* or *activated dehydrocholesterol*)—the skin and the diet. The skin contains a rich source of 7-dehydrocholesterol, which is rapidly converted to cholecalciferol when exposed to ultraviolet light. Like vitamin A, vitamin D$_3$ is absorbed by the small intestine passively and is incorporated into chylomicrons. During the metabolism of chylomicrons, vitamin D$_3$ is transferred to a binding protein in plasma called the **vitamin D–binding protein**.

Unlike vitamin A, vitamin D is not stored in the liver but is distributed among the various organs depending on their lipid content. In the liver, vitamin D_3 is converted to 25-hydroxycholecalciferol, which is subsequently converted to the active hormone **1,25-dihydroxycholecalciferol** in the kidneys. The latter enhances Ca^{2+} and phosphate absorption by the small intestine and mobilizes Ca^{2+} and phosphate from bones.

Vitamin D is essential for normal development and growth and the formation of bones and teeth. Vitamin D deficiency can result in **rickets**, a disorder of normal bone ossification manifested by distorted bone movements during muscular action.

The major dietary **vitamin E** is α-tocopherol. Vegetable oils are rich in vitamin E. It is absorbed by the small intestine by passive diffusion and incorporated into chylomicrons. Unlike vitamins A and D, vitamin E is transported in the circulation associated with lipoproteins and erythrocytes.

Vitamin E is a potent antioxidant and therefore prevents lipid peroxidation. Tocopherol deficiency is associated with increased red cell susceptibility to lipid peroxidation, which may explain why the red cells are more fragile in people with vitamin E deficiency than in healthy people.

Vitamin K can be derived from green vegetables in the diet or the gut flora. The vitamin K derived from green vegetables is in the form of **phylloquinones**. Vitamin K derived from bacteria in the small intestine is in **menaquinones**. The small intestine takes up phylloquinones via an energy-dependent process from the proximal small intestine. In contrast, menaquinones are absorbed from the small intestine passively, depending only on the micellar solubilization of these compounds by bile salts. Vitamin K is incorporated into chylomicrons. It is rapidly taken up by the liver and secreted together with VLDLs. No carrier protein for vitamin K has been identified.

Vitamin K is essential for the synthesis of various clotting factors by the liver. Vitamin K deficiency is associated with bleeding disorders.

Water-soluble vitamins: C, B_1, B_2, B_6, B_{12}, niacin, biotin, and folic acid

Most of the water-soluble vitamins are absorbed by the small intestine by both passive and active processes. The water-soluble vitamins are summarized in Table 25.6.

The major source of **vitamin C (ascorbic acid)** is green vegetables and fruits. It plays an important role in many oxidative processes by acting as a coenzyme or cofactor. It is absorbed mainly by active transport through the transporters in the ileum. The uptake process is sodium-dependent. Vitamin C deficiency is associated with **scurvy**, a disorder characterized by weakness, fatigue, anemia, and bleeding gums.

Vitamin B_1 (thiamine) plays an important role in carbohydrate metabolism. The jejunum absorbs thiamine passively as well as by an active, carrier-mediated process. Thiamine deficiency results in **beriberi**, characterized by anorexia and disorders of the nervous system and heart.

Vitamin B_2 (riboflavin) is a component of the two groups of flavoproteins—flavin adenine dinucleotide and flavin mononucleotide. Riboflavin plays an important role in metabolism. It is absorbed by a specific, saturable, active transport system located in the proximal small intestine. Its deficiency is associated with anorexia, impaired growth, impaired use of food, and nervous disorders.

Niacin plays an important role as a component of the coenzymes NAD(H) and NADP(H), which participate in a wide variety of oxidation–reduction reactions involving H^+ transfer.

At low concentrations, the small intestine absorbs niacin by Na^+-dependent, carrier-mediated facilitated transport. At high concentrations, it is absorbed by passive diffusion. Niacin has been used to treat hypercholesterolemia, for the prevention of coronary artery disease. It decreases plasma total cholesterol and LDL cholesterol yet increases plasma HDL cholesterol.

Niacin deficiency is characterized by many clinical symptoms including anorexia, indigestion, muscle weakness, and skin eruptions. Severe deficiency leads to **pellagra**, a disease characterized by dermatitis, dementia, and diarrhea.

Vitamin B_6 (pyridoxine) is involved in amino acid and carbohydrate metabolism. Vitamin B_6 is absorbed throughout the small intestine by simple diffusion. A deficiency of this vitamin is often associated with anemia and central nervous system disorders.

Biotin acts as a coenzyme for carboxylase, transcarboxylase, and decarboxylase enzymes, which play an important role in the metabolism of lipids, glucose, and amino acids. At low luminal concentrations, biotin is absorbed by the small intestine by Na^+-dependent active transport. At high concentrations, biotin is absorbed by simple diffusion. Biotin is so common in food that deficiency is rarely observed. However, biotin deficiency occurs frequently when parenteral nutrition (nutrition administered intravenously) is administered for a long period.

Folic acid is usually found in the diet as polyglutamyl conjugates (pteroylpolyglutamates). It is required for the formation of nucleic acids, the maturation of red blood cells, and growth. An enzyme on the brush border degrades pteroylpolyglutamates to yield a monoglutamylfolate, which is taken up by enterocytes by facilitated transport. Once inside enterocytes, the monoglutamylfolate is released directly into the bloodstream or converted to 5-methyltetrahydrofolate before exiting the cell. A folate-binding protein binds the free and methylated forms of folic acid in plasma. Folic acid deficiency causes a fall in plasma and red cell folic acid content and, in its most severe form, the development of megaloblastic anemia, dermatologic lesions, and poor growth. A link between folic acid deficiency and spina bifida (opening in the spine) has been demonstrated. Consequently, the U.S. Public Health Service recommends that women of childbearing age should take 0.4 mg of folic acid daily during the first 3 months of pregnancy.

The discovery of **vitamin B_{12} (cobalamin)** followed from the observation that patients with **pernicious anemia**

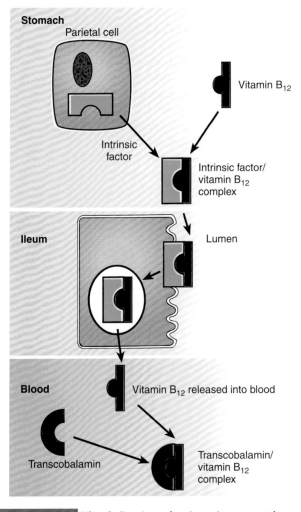

Figure 25.23 **Vitamin B$_{12}$ absorption depends on a gastric intrinsic factor.** Vitamin B$_{12}$ is one of the eight water-soluble B vitamins. The intestinal absorption of B$_{12}$ depends on an intrinsic factor secreted from the stomach parietal cell.

who ate large quantities of raw liver recovered from the disease. Subsequent analysis of liver components isolated the cobalt-containing vitamin, which plays an important role in the production of red blood cells. A glycoprotein called intrinsic *factor* is secreted by the parietal cells in the stomach and binds strongly with vitamin B$_{12}$ to form a complex that is then absorbed in the terminal ileum through a receptor-mediated process (Fig. 25.23). Vitamin B$_{12}$ is transported in the portal blood bound to the protein **transcobalamin**. People who lack the intrinsic factor fail to absorb vitamin B$_{12}$ and develop pernicious anemia.

▶ ELECTROLYTE AND MINERAL ABSORPTION

Nearly all of the dietary nutrients and ~95% to 98% of the H$_2$O and electrolytes that enter the upper small intestine are absorbed. The absorption of electrolytes and minerals involves both passive and active processes, resulting in the movement of electrolytes, H$_2$O, and metabolic substrates into the blood for distribution and use throughout the body.

Minerals are chemical elements required for physiologic function.

Minerals are chemical elements required by the body for normal function. They can be either bulk minerals (required in relatively large amounts) or trace minerals (required in minute amounts). Minerals can be acquired naturally in food or added to the diet (e.g., calcium carbonate or sodium chloride). Bulk minerals include calcium, magnesium, phosphorus, potassium, sodium, and sulfur. Trace minerals include chromium, cobalt, copper, fluorine, iodine, iron, manganese, molybdenum, selenium, and zinc.

Sodium

The GI system is well equipped to handle the large amount of Na$^+$ entering the GI lumen daily—on average, about 25 to 35 g of Na$^+$ every day. Around 5 to 8 g is derived from the diet and the rest from salivary, gastric, biliary, pancreatic, and small intestinal secretions. The GI tract is extremely efficient in conserving Na$^+$: only 0.5% of intestinal Na$^+$ is lost in the feces. The jejunum absorbs more than half of the total Na$^+$, and the ileum and colon absorb the remainder. The small intestine absorbs the bulk of the Na$^+$ presented to it, but the colon is most efficient in conserving Na$^+$.

Several different mechanisms operating at varying degrees in different parts of the GI tract absorb Na$^+$. When a meal that is hypotonic to plasma is ingested, considerable absorption of H$_2$O from the lumen to the blood takes place, predominantly through tight junctions and intercellular spaces between the enterocytes, resulting in the absorption of small solutes such as Na$^+$ and Cl$^-$ ions. This mode of absorption, called **solvent drag**, is responsible for a significant amount of the Na$^+$ absorption by the duodenum and jejunum, but it probably plays a minor role in Na$^+$ absorption by the ileum and colon because more distal regions of the intestine are lined by a "tight" epithelium (see Chapter 2).

In the jejunum, an Na$^+$/K$^+$-ATPase actively pumps Na$^+$ out of the basolateral surface of enterocytes (Fig. 25.24A). The result is low intracellular Na$^+$ concentration, and the luminal Na$^+$ enters enterocytes down the electrochemical gradient, providing energy for the extrusion of H$^+$ into the lumen (via an Na$^+$/H$^+$ exchanger). The H$^+$ then reacts with HCO$_3^-$ in bile and pancreatic secretions in the intestinal lumen to form H$_2$CO$_3$. Carbonic acid dissociates to form CO$_2$ and H$_2$O. The CO$_2$ readily diffuses across the small intestine into the blood. Another mode of Na$^+$ uptake is via a carrier located in the enterocyte brush border membrane, which transports Na$^+$ together with a monosaccharide (e.g., glucose) or an amino acid molecule (a symport type of transport).

In the ileum, the Na$^+$/K$^+$-ATPase in the basolateral membrane also creates a low intracellular Na$^+$ concentration, and luminal Na$^+$ enters enterocytes down the electrochemical gradient. Na$^+$ absorption by Na$^+$-coupled symporters is not as great as in the jejunum because the small intestine has already absorbed most of the monosaccharides and amino acids (see Fig. 25.24B). NaCl is

Lumen

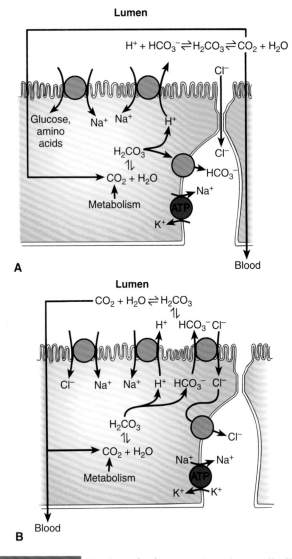

$$H^+ + HCO_3^- \rightleftharpoons H_2CO_3 \rightleftharpoons CO_2 + H_2O$$

Glucose, amino acids

$CO_2 + H_2O$

Metabolism

A

Blood

Lumen

$$CO_2 + H_2O \rightleftharpoons H_2CO_3$$

$CO_2 + H_2O$

Metabolism

Blood

B

Figure 25.24 **Na⁺ absorption in the gut depends on an Na⁺/K⁺-ATPase pump.** Approximately 98% of the electrolytes in the gut are reabsorbed. An Na⁺/K⁺-ATPase actively pumps Na⁺ across the basal lateral surface to be absorbed into the blood. (**A**) Na⁺ absorption by the jejunum. (**B**) Na⁺ absorption by the ileum. ATP, adenosine triphosphate.

transported via two exchangers located at the brush border membrane. One is a Cl^-/HCO_3^- exchanger, and the other is an Na^+/H^+ exchanger. The downhill movement of Na^+ into the cell provides the energy required for the uphill movement of the H^+ from the cell to the lumen. Similarly, the downhill movement of HCO_3^- out of the cell provides the energy for the uphill entry of Cl^- into the enterocytes. The Cl^- then leaves the cell through facilitated transport. This mode of Na^+ uptake is called Na^+/H^+–Cl^-/HCO_3^- countertransport.

In the colon, the mechanisms for Na^+ absorption are mostly similar to those described for the ileum. There is no sugar-coupled or amino acid–coupled Na^+ transport because most sugars and amino acids have already been absorbed. Na^+ is also absorbed here via Na^+-selective ion channels in the apical cell membrane (electrogenic Na^+ absorption).

Potassium

The average daily intake of K^+ is about 4 g. Absorption takes place throughout the intestine by passive diffusion through the tight junctions and lateral intercellular spaces of the enterocytes. The driving force for K^+ absorption is the difference between luminal and blood K^+ concentration. The absorption of H_2O results in an increase in luminal K^+ concentration, resulting in K^+ absorption by the intestine. In the colon, K^+ can be absorbed or secreted depending on the luminal K^+ concentration. With diarrhea, considerable K^+ can be lost. Prolonged diarrhea can be life threatening, because the dramatic fall in extracellular K^+ concentration can cause complications such as cardiac arrhythmias.

Chloride

Most of the Cl^- ions added to the GI tract from the diet and from the intestinal secretions are absorbed. Intestinal Cl^- absorption involves both passive and active processes. In the jejunum, active Na^+ absorption generates a potential difference across the small intestinal mucosa, with the serosal side more positive than the lumen. Cl^- follows this potential difference and enters the bloodstream via the tight junctions and lateral intercellular spaces. In the ileum and colon, enterocytes actively take up Cl^- via Cl^-/HCO_3^- exchange, as discussed above. The presence of other halides inhibits this absorption of Cl^-.

Bicarbonate

Bicarbonate ions are absorbed in the jejunum together with Na^+. In humans, the absorption of HCO_3^- by the jejunum stimulates the absorption of Na^+ and H_2O (Fig. 25.25A). Through an Na^+/H^+ exchanger, H^+ is secreted into the intestinal lumen, where H^+ and HCO_3^- react to form H_2CO_3, which then dissociates to form CO_2 and H_2O. The CO_2 diffuses into the enterocytes, where it reacts with H_2O to form H_2CO_3 (catalyzed by carbonic anhydrase). H_2CO_3 dissociates into HCO_3^- and H^+, and the HCO_3^- then diffuses into the blood.

In the ileum and colon, HCO_3^- is actively secreted into the lumen in exchange for Cl^-. This secretion of HCO_3^- is important in buffering the decrease in pH resulting from the short-chain fatty acids produced by bacteria in the distal ileum and colon.

Calcium

The amount of Ca^{2+} entering the GI tract is about 1 g/day, approximately half of which is derived from the diet. Most dietary Ca^{2+} is derived from meat and dairy products. Of the Ca^{2+} entering the GI tract, about 40% is absorbed. Several factors affect Ca^{2+} absorption. For instance, the presence of fatty acid can retard Ca^{2+} absorption by the formation of Ca^{2+} soap. In contrast, bile salts form complexes with Ca^{2+} ions, which facilitate Ca^{2+} absorption.

Calcium absorption takes place predominantly in the duodenum and jejunum, is mainly active, and involves three steps: (1) Enterocytes take up Ca^{2+} by passive diffusion through a Ca^{2+} channel, because there is a large Ca^{2+} concentration gradient; the luminal Ca^{2+} is about 5 to 10 mM,

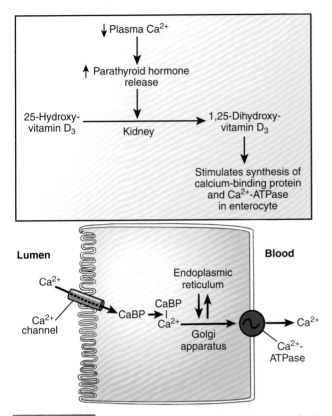

Figure 25.25 **Parathyroid hormone and vitamin D are required for calcium absorption by enterocytes.** Parathyroid hormone stimulates the conversion of vitamin D_3 in the kidney to its active metabolite 1,25-dihydroxycholecalciferol (1,25-diOH-vitamin D_3), which stimulates Ca^{2+} uptake via the Ca^{2+} channels. Additionally, it stimulates the synthesis of both Ca^{2+}-binding protein (CaBP) and the Ca^{2+}-ATPase.

whereas free intracellular Ca^{2+} is about 100 nM. (2) Once inside the cell, Ca^{2+} is complexed with **Ca^{2+}-binding protein, calbindin D (CaBP)**. (3) At the basolateral membrane, Ca^{2+} is extruded from the enterocytes via the Ca^{2+}-ATPase pump. Calcium uptake by enterocytes, the level of CaBP in the cells, and transport by Ca^{2+}-ATPase pumps are increased by 1,25-dihydroxyvitamin D_3. Once inside the cell, the Ca^{2+} ions are sequestered in the ER and Golgi by binding to the CaBP in these organelles.

Calcium absorption by the small intestine is regulated by the circulating plasma Ca^{2+} concentration. Lowering of the Ca^{2+} concentration stimulates the release of parathyroid hormone, which stimulates the conversion of vitamin D to its active metabolite, 1,25-dihydroxyvitamin D_3, in the kidney. This, in turn, stimulates the synthesis of CaBP and the Ca^{2+}-ATPase by the enterocytes (see Fig. 25.25). Because protein synthesis is involved in the stimulation of Ca^{2+} uptake by parathyroid hormone, a lapse of a few hours usually occurs between the release of parathyroid hormone and the increase in Ca^{2+} absorption by the enterocytes.

Magnesium

Humans ingest about 0.4 to 0.5 g/d of Mg^{2+}. The absorption of Mg^{2+} seems to take place along the entire small intestine, and the mechanism involved seems to be passive.

Zinc

The average daily zinc intake is 10 to 15 mg, about half of which is absorbed, primarily in the ileum. A carrier located in the brush border membrane actively transports zinc from the lumen into the cell, where it can be stored or transferred into the bloodstream. Zinc plays an important role in several metabolic activities. For example, a group of metalloenzymes (e.g., alkaline phosphatase, carbonic anhydrase, and lactic dehydrogenase) require zinc to function.

Iron

Iron plays an important role not only as a component of heme but also as a participant in many enzymatic reactions. About 12 to 15 mg/day of iron enters the GI tract and is absorbed mainly in the duodenum and upper jejunum (Fig. 25.26). There are two forms of dietary iron: heme and nonheme. Enterocytes absorb the heme iron intact. Nonheme iron absorption depends on both pH and concentration. Ferric (Fe^{3+}) salts are not soluble at pH 7, whereas ferrous (Fe^{2+}) salts are. Consequently, in the duodenum and upper jejunum, unless Fe^{3+} ion is chelated, it forms a precipitate. Several compounds such as tannic acid in tea and phytates in vegetables form insoluble complexes with iron, preventing absorption. Iron is absorbed by an active process via a carrier(s) located in the brush border membrane. One such transporter, the divalent metal transporter, is expressed abundantly in the duodenum.

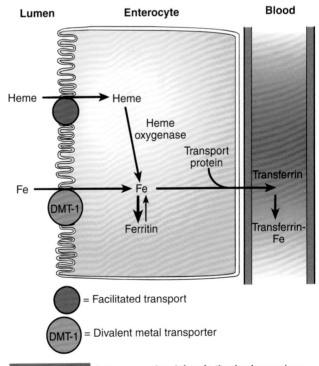

Figure 25.26 **Enterocytes absorb iron in the duodenum.** Iron must be in ferrous form (Fe^{2+}) to be absorbed. A ferric reductase enzyme on the enterocyte brush border reduces ferric iron Fe^{3+} to Fe^{2+}. A protein metal transporter (DMT-1) transports the iron across the enterocyte's cell membrane and into the cell. DMT-1, divalent metal transporter.

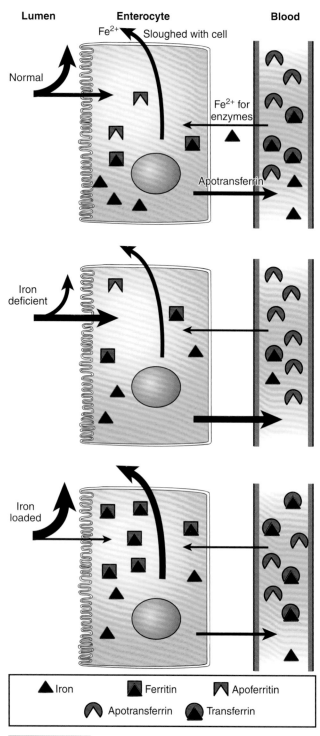

Lumen   **Enterocyte**   **Blood**

Fe²⁺ — Sloughed with cell

Normal

Fe²⁺ for enzymes

Apotransferrin

Iron deficient

Iron loaded

▲ Iron   ◼ Ferritin   ◹ Apoferritin
◠ Apotransferrin   ◗ Transferrin

Figure 25.27 Endocyte pathways regulate the body's iron levels. In healthy subjects, the intestinal lining cells can store iron as ferritin (in which case the iron leaves the body when the cell dies and is sloughed off into feces), or the cell can move it into the blood using a protein transporter and carry it as transferrin to the rest of the body. The body regulates iron levels by regulating these pathways. For example, in patients with iron deficiency (middle figure), cells produce more ferrous reductase, DMT-1, and the protein transporter, which results in a net increase in the amount of iron going to transferrin that can be taken up by red cells. In the bottom figure, iron overload results in more iron converted to ferritin and less converted to transferrin.

Once inside the cell, heme iron is released by the action of heme oxygenase and mixed with the intracellular free iron pool. Iron is either stored in the enterocyte cytoplasm bound to the storage protein **apoferritin** to form **ferritin** or transported across the cell bound to transport proteins, which carry the iron across the cytoplasm and release it into the intercellular space. **Transferrin**, a β-globulin synthesized by the liver, binds and transports iron in the blood.

Iron absorption is closely regulated by iron storage in enterocytes and iron concentration in the plasma. Enterocytes are continuously shed into the lumen, and the ferritin contained within is lost. Normally, iron in enterocytes is derived from the lumen and the blood (Fig. 25.27). The amount of iron stored in enterocytes regulates the amount of iron absorbed. In iron deficiency, the circulating plasma iron concentration is low, which stimulates the absorption of iron from the lumen and the transport of iron into the blood. Moreover, in a deficient state, less iron is stored as ferritin in the enterocytes, so the loss of iron through this means is significantly reduced. In iron-loaded patients, there is less absorption of iron because of the large amount of mucosal iron storage, which increases iron loss as a result of enterocyte shedding. Furthermore, because of the high level of circulating plasma iron, the transfer of iron from enterocytes to the blood is reduced. Through a combination of various mechanisms, body iron homeostasis is maintained.

▶ WATER ABSORPTION

In human adults, the average daily intake of H_2O is about 2 L. As shown in Table 25.7, secretions from the salivary glands, pancreas, liver, and GI tract make up most of the fluid entering the GI tract (about 7 L). Despite this large volume of fluid, only 100 mL is lost in the feces. Therefore, the GI tract

TABLE 25.7 Water Intake, Absorption, and Excretion by the Gastrointestinal (GI) Tract

Source	Amount (mL)
Water added to GI tract	
Food and beverages	2,000
Salivary secretion	1,000
Biliary secretion	1,000
Gastric secretion	2,000
Pancreatic secretion	1,000
Intestinal mucosal secretion	2,000
Water absorbed in feces	
Duodenum and jejunum	4,000
Ileum	3,500
Colon	1,400
Water lost in feces	100

is extremely efficient in absorbing H_2O. Water absorption by the GI tract is passive. The rate of absorption depends on both the region of the intestinal tract and the luminal osmolality. The duodenum, jejunum, and ileum absorb the bulk of the H_2O that enters the GI tract. The colon normally absorbs about 1.4 L of H_2O and excretes about 100 mL. It is capable of absorbing considerably more H_2O (about 4.5 L), however, and watery diarrhea occurs only if this capacity is exceeded.

Water is absorbed in the gut by osmosis.

Because H_2O absorption is determined by the osmolality difference of the lumen and the blood, H_2O can move both ways in the intestinal tract (i.e., secretion and absorption). The osmolality of blood is about 300 mOsm/kg H_2O. The ingestion of a hypertonic meal (e.g., 600 mOsm/kg H_2O) initially leads to net H_2O movement from blood to lumen; however, as the small intestine absorbs the various nutrients and electrolytes, the luminal osmolality falls, resulting in the net H_2O movement from lumen to blood. The water of a hypertonic meal is therefore absorbed mainly in the ileum and colon. In contrast, if a hypotonic meal is ingested (e.g., 200 mOsm/kg H_2O), net H_2O movement is immediately from the lumen to the blood, resulting in the absorption of most of the H_2O in the duodenum and jejunum.

INTEGRATED MEDICAL SCIENCES

Gut Microbiota: A Relationship You Cannot Live Without

A GI tract that is functioning normally has a healthy well-established colonizing microbiota its mucosa and lumen, which is a major contributor to the maintenance of whole body homeostasis. These two communities are not the same and the genera of microbes living in the mucosa vary from those residing in the lumen. Compartmentalization within the GI tract aids in the isolation of genera of microbes to different regions of the tract with the majority of microbes living in the colon. It is well established that the species composition and relative abundance of the gut microbiota are impacted by the diet, lifestyle, and overall health of an individual. Humans, like many other animals, have developed a commensalistic relationship with the gut microbiota. Over time, this relationship has evolved to become a mutualistic and interdependent one, in which the physiological activity of the microbiota has a significant impact on the host and the activity of the host impacts the genera comprising the microbiota. In support of life, gut microbial metabolism supplies the host with short-chain fatty acids, essential vitamins (i.e., vitamins B and K), and contributes to the synthesis and absorption of essential amino acids.

Within the context of this chapter, it is important to note that the digestive, absorptive, and secretory functions of the GI tract are readily affected by the gut's microbiota. The interactions do not end with digestion as microbial chemical signals influence other systems such as the activity of the host's immune cells and immunologic homeostasis. In a normal, healthy individual, the interactions between the host and microbiota are primarily beneficial. However, dysbiosis can lead to, or be a contributing factor in, a number of maladies including inflammatory bowel disease (e.g., Crohn's disease, ulcerative colitis), altered metabolism of bile acids, intestinal malignancy, obesity, and metabolic syndrome.

Clues to the development of the intestinal microbiota have emerged from many different studies. It has long been held that fetuses develop in a sterile environment within the amniotic sac. However, recent studies have shown that certain bacteria are able to cross the placental barrier. When investigators orally administered a genetically labeled species of bacteria to pregnant mice, they noted the appearance of the bacteria in both the amniotic fluid and the meconium of the pups. Others have shown that the oral microbiota is also capable of reaching, and becoming established at, the level of the placenta. Regardless, as indicated by microbial analysis of meconium, infants are born with relatively few microbes colonizing the intestinal lumen.

All indications are that the major microbial colonization of the gut occurs following birth. These colonizers will generally reflect the microbial diversity present in the environment to which the infant is first exposed. For example, a vaginal birth versus one that occurs by cesarean section exposes the infant to very different microbes. Other factors such as the source of nutrition (i.e., breast milk vs. commercial preparations), gestational age at delivery, and treatment with antibiotics significantly alter the genera of gut microbes. Microbial recovery back to homeostasis after antibiotic treatment may take up to several months. Given the many factors that have an impact on the microbial community, there is a wide range of potential microbiota that can ultimately colonize an infant's gut. This results in much variability between individuals. Immediately after birth, the environment in a newborn's gut is favorable for supporting the growth of facultative anaerobic microbes. Later, the environment favors growth of anaerobes, particularly in the distal regions of the gut with aerobes favoring the more proximal regions. By 5 years of age, the child's gut microbial community has changed to be similar to the adult gut. The gut microbiota does have an influence on the early post birth physical changes that take place in the GI tract including changes in crypt size, mucosal thickness, and surface changes in the intestinal epithelium.

The health of the host and the composition of the gut microbiota are intimately linked through the interactions that occur between the gut microbiome and the host's innate and adaptive immune systems. These interactions have a large impact on the development of the immune system early in life and on the possibility of the host developing

(Continued)

Gastrointestinal Physiology

certain diseases in later life. Examples include local inflammatory responses in the gut (mentioned previously) and systemic autoimmune disorders such as rheumatoid arthritis, multiple sclerosis, and type 1 diabetes. Beyond the immune system, the intestinal microbiota is felt to play a role in systemic events outside the digestive tract (e.g., cardiovascular disease) because some bacterial species are able to cross the intestinal mucosa and translocate to other regions of the body. Lymphatic tissue (e.g., Peyer patches, lymphoid follicles, immune cells localized to the lamina propria) provides an important line of immunologic defense against such occurrences happening more frequently. Additionally, the mucus secreted onto the luminal surface helps to prevent microbes from breaching the GI tract. Secreted antimicrobial proteins and IgA play a role in maintaining the effectiveness of that barrier.

Microbes have been implicated in other interactions that occur between the body systems. Such interactions generally result in a healthy individual with a normal body mass index. However, it is known that obese individuals do not have the same genera in their gut microbiota as that of more lean individuals. These differences lead to noticeable imbalances in energy metabolism and fat storage along with insulin resistance and have led to such extreme treatments as fecal transplants to treat obesity. However, the cause and effect may be hard to fully ascertain as the composition of the diet also influences the mixture of genera that comprise the gut microbiota. Recent data on patients who have undergone Roux-en-Y gastric bypass surgery indicate that following the procedure, the patient's gut microbiota changed significantly enough that it contributed to the weight loss observed in these patients, most likely by way of altered fatty acid metabolism and fat storage.

This essay has provided only a glimpse into the role and impact that microbes occupying the GI tract can have. Students are encouraged to consult the literature to learn more about other aspects of the host–microbiota interaction in health and disease. ■

Chapter Summary

- Saliva assists in the swallowing of food, carbohydrate digestion, and the transport of immunoglobulins that combat pathogens.
- The major function of the gastrointestinal tract is storage, mixing, digestion, and absorption of nutrients.
- The stomach prepares chyme to aid in the digestion of food in the small intestine.
- Parietal cells secrete hydrochloric acid and intrinsic factor, and chief cells secrete pepsinogen.
- Gastrin plays an important role in stimulating gastric acid secretion.
- The acidity of gastric secretion provides a barrier to microbial invasion of the gastrointestinal tract.
- Gastric secretion is under neural and hormonal control and consists of three phases: cephalic, gastric, and intestinal.
- Gastric inhibitory peptide, secreted by K cells, is a potent inhibitor of gastric acid secretion and enhances insulin release in elevated concentrations of plasma glucose.
- Pancreatic secretion neutralizes the acids in chyme and contains enzymes involved in the digestion of carbohydrates, fat, and protein.
- Secretin stimulates the pancreas to secrete a bicarbonate-rich fluid, thereby neutralizing acidic chyme.

- Cholecystokinin stimulates the pancreas to secrete an enzyme-rich secretion.
- Pancreatic secretion is under neural and hormonal control and consists of three phases: cephalic, gastric, and intestinal.
- Bile salts play an important role in the intestinal absorption of lipids.
- Proteins are digested to form amino acids, dipeptides, and tripeptides before being taken up by enterocytes and transported in the blood.
- The gastrointestinal tract absorbs water-soluble vitamins and ions by different mechanisms.
- Most of the salt and water entering the intestinal tract, whether in the diet or in gastrointestinal secretions, is absorbed in the small intestine.
- Lipids absorbed by enterocytes are packaged and secreted as chylomicrons into the lymph.
- Carbohydrates, when digested, form maltose, maltotriose, and α-limit dextrins, which are cleaved to brush border enzymes to monosaccharides and taken up by enterocytes.
- Calcium-binding protein is involved in calcium absorption.
- Heme and nonheme iron are absorbed in the small intestine by different mechanisms.
- Minerals are classified as bulk minerals and trace minerals.

Chapter Review Questions

1. Parasympathetic stimulation induces salivary acinar cells to release the protease:

   A. bradykinin.
   B. kallikrein.
   C. kininogen.
   D. kinin.
   E. aminopeptidase.

   The correct answer is B. Parasympathetic stimulation induces the release of kallikrein by the salivary acinar cells, which then converts kininogen to form lysyl-bradykinin (a potent vasodilator). Bradykinin is a vasoactive peptide. Kininogen is the precursor for kinin. Kinins are peptides that are related in amino acid sequence and physiological activity to bradykinin. Aminopeptidase releases amino acids from the amino end of peptides and is found in the brush border membrane and cytoplasm of enterocytes.

2. Which of the following best describes enterokinase, an enzyme necessary for protein digestion?

   A. It is produced by the pancreas and directly converts ribonuclease into its active form.
   B. It is produced by the pancreas and directly converts trypsinogen into trypsin.
   C. It is produced by the pancreas and prevents autodigestion of the pancreas by pancreatic enzymes.

   D. It is produced by the duodenum and directly inactivates trypsin inhibitor.
   E. It is produced by the duodenum and directly converts trypsinogen into trypsin.

   The correct answer is E. Enterokinase is an enzyme produced by the cells of the duodenum (crypt of Lieberkühn) that converts trypsinogen to its active form, trypsin.

3. Hartnup disease is an inherited autosomal recessive disorder involving the malabsorption of amino acids, particularly tryptophan by the small intestine. Feeding di- and tripeptides containing tryptophan to patients with this disease improves their clinical condition because:

   A. di- and tripeptides, unlike free amino acids, can be taken up passively by enterocytes in the small intestine.
   B. peptides, unlike free amino acids, can be taken up by defective amino acid transporters.
   C. di- and tripeptides use transporters that are different from the defective amino acid transporters.
   D. the presence of di- and tripeptides in the intestinal lumen enhances the uptake of amino acids by the transporters.
   E. di- and tripeptides, unlike amino acids, can be taken up passively by the colon.

   The correct answer is C. Amino acids as well as di- and tripeptides use different brush border transporters for their uptake. Di- and tripeptides are not taken up passively by any part of the GI tract.

4. Dietary triglyceride is a major source of nutrient for the human body. It is digested mostly in the intestinal lumen by pancreatic lipase to release:

   A. lysophosphatidylcholine and fatty acids.
   B. glycerol and fatty acids.
   C. diglyceride and fatty acids.
   D. 2-monoglyceride and fatty acids.
   E. lysophosphatidylcholine and diglyceride.

The correct answer is D. Pancreatic lipase hydrolyzes triglyceride to form 2-monoglyceride and fatty acids. Only hydrolysis of phosphatidylcholine results in the formation of lysophosphatidylcholine, so the hydrolysis of triglyceride does not result in the formation of lysophosphatidylcholine. Acid lipase works only in the gastric lumen, not the intestinal lumen. Although diglyceride is an intermediate in the hydrolysis of triglyceride by pancreatic lipase, the hydrolysis continues until 2-monoglyceride and fatty acids are formed. Pancreatic lipase does not hydrolyze triglyceride totally to form glycerol and fatty acids.

Clinical Application Exercises 25.1

LACTOSE INTOLERANCE

A 9-year-old Chinese American boy regularly complains of abdominal cramps, abdominal distention, and diarrhea after drinking milk. A gastroenterologist administers 50 g of lactose by mouth to the child and measures an increase in the boy's expired hydrogen gas.

QUESTIONS

1. Based upon how sugars are normally digested and absorbed by the GI tract, why is this boy suffering from these symptoms?

2. From a physician's standpoint, what recommendations and advice can be given to this patient to address his problem with lactose intolerance?

ANSWERS

1. Lactose is hydrolyzed by a brush border enzyme called lactase to glucose and galactose. The monosaccharides are then absorbed by sodium-dependent secondary active transport. If the lactase enzyme is deficient, lactose will not be broken down and will remain in the intestinal lumen. The osmotic activity of the lactose draws water into the intestinal lumen and results in a watery diarrhea. In the colon, bacteria metabolize the lactose to lactic acid, carbon dioxide, and hydrogen gas. The extra fluid and gas in the intestine result in distention and increased motility (cramps).

2. Avoiding foods that contain lactose (milk, dairy products) is recommended for lactose-intolerant people, but calcium and caloric intake should not be compromised. Milk can be pretreated with an enzyme obtained from bacteria or yeasts that digests lactose, or lactase pills can be taken with meals.

thePoint® *Visit* http://thepoint.lww.com/rhoades5e *for additional chapter review Q&A, Clinical Application Exercises, animations, and more!*

26 Liver Functions and Immune Surveillance

Active Learning Objectives

Upon mastering the material in this chapter, you should be able to:

- Describe how the liver lobule is functionally organized.
- Describe how the architectural arrangement of the hepatocytes allows for the rapid exchange of molecules.
- Describe how the liver is oxygenated.
- Describe the significance of tissue regeneration in the liver.
- Explain the difference between phase I and phase II reactions in drug metabolism of the liver.

- Explain the mechanism by which the liver regulates blood glucose concentration.
- Explain how the liver is involved in the uptake and metabolism of lipids.
- Explain how vitamin A is stored in the liver and transported to other parts of the body when needed.
- Describe the importance and risks of iron storage in the liver.
- Describe how the liver is involved in establishing immunity.

The liver is the largest internal organ in the body, constituting about 2.5% of an adult's body weight. It is composed of several distinct cell types including Kupffer cells, sinusoidal epithelial cells, cholangiocytes, and hepatocytes, but the vast majority, 70% to 80%, is hepatocytes. Given its wide range of functions, the liver is an extremely vital organ for maintaining whole-body homeostasis. Liver functions include glycogen storage, protein synthesis, detoxification, red cell destruction, hormone production, bile production for lipid emulsification, maintenance of blood glucose levels, and regulation of blood lipids by the amount of very-low-density lipoproteins (VLDLs) it secretes. A healthy liver also contributes to both the innate and adaptive immunological defense mechanisms of the body.

The liver is a unique organ because it has a dual blood supply and it is the only internal human organ capable of regenerating lost tissue. With as little as 25% of the liver remaining, it can regenerate into an entire organ, which is possible because hepatocytes are capable of reentering the cell cycle and undergoing mitosis.

This chapter summarizes the liver's many functions including (1) detoxification of hormones, drugs, and waste products; (2) metabolism of carbohydrates, proteins, and fats; (3) storage of iron and vitamins; (4) hormone production; and (5) innate and adaptive immunity.

▶ LIVER STRUCTURE AND FUNCTION

Normal liver function is essential for maintaining the optimal function of other organs. In so doing, the liver interacts with many systems (e.g., cardiovascular, renal, and immune) to maintain homeostasis. Given the liver's wide range of functions, there is surprisingly little specialization among liver cells. The hexagonally shaped lobule is the functional unit of the liver and is built around a central vein (Fig. 26.1). Individual lobules are made up of many cellular plates radiating from the central vein like spokes in a bicycle wheel. Each plate consists of specialized hepatocytes that contain discrete granules and perform a wide variety of metabolic functions and secretory tasks including the manufacture of specific proteins, detoxification of xenobiotics, and the production and secretion of bile.

Cell types other than hepatocytes also have specific functions. **Kupffer cells** (**KCs**) play a role in the production of bilirubin and, along with **natural killer cells** (**NKCs**), are involved in innate immunity. **Cholangiocytes** modulate bile flow through the bile ducts. Sinusoidal epithelial cell (SEC) functions are discussed later in this chapter.

Architectural arrangement of hepatocytes in the liver lobule enhances the rapid exchange of material.

One specialized function of hepatocytes is the formation of bile. The bile canaliculus is formed by the intercellular space located between neighboring hepatocytes. Impermeable tight junctions separate the canaliculus from the pericellular space and prevent the mixing of contents between the two regions (see Fig. 26.1). Bile originating in the bile canaliculus drains into a series of ducts that eventually join the pancreatic duct near where it enters the duodenum. The **sphincter of Oddi**, located at the duodenal connection between the bile duct and the pancreatic duct, regulates drainage of bile and pancreatic juice into the duodenum. The pericellular space between two hepatocytes is continuous with the perisinusoidal space (see Fig. 26.1). A layer of SECs separates the **perisinusoidal space** from the sinusoid. Finger-like projections of the hepatocytes extend into the perisinusoidal space, which greatly increases the contact area between hepatocytes and the perisinusoidal fluid.

Liver cells carry out specialized functions.

Liver endothelial cells, unlike those in other parts of the cardiovascular system, lack a basement membrane. Furthermore, they have sievelike plates that permit the ready exchange of materials, even particles, as big as chylomicrons (80 to 500 nm wide), between the perisinusoidal space and the sinusoid. Despite the barrier's permeability, it does have

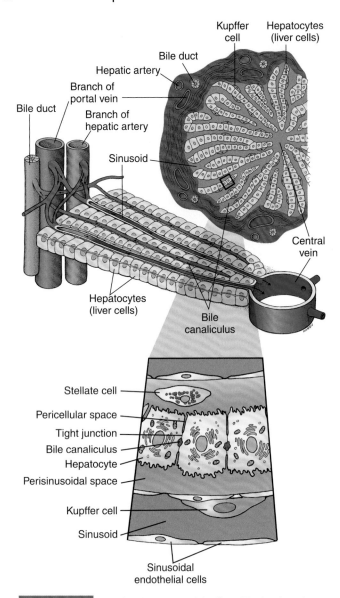

Figure 26.1 **Functional anatomy of the liver.** The basic unit of the liver is the lobule. The lobule is arranged in a hexagonal fashion and is delineated by vascular and bile channels. The lobule contains specialized cells, such as hepatocytes, sinusoidal cells, and Kupffer cells.

some sieving properties. For example, the protein concentration of hepatic lymph, assumed to derive from the perisinusoidal space, is about 10% lower than that of plasma.

While they do have many functions, hepatocytes are not phagocytic. Phagocytosis is accomplished by resident macrophages (KCs) that line the hepatic sinusoids and are part of the reticuloendothelial system. KCs remove unwanted material (e.g., bacteria, virus particles, fibrin–fibrinogen complexes, damaged erythrocytes, and immune complexes) from the circulation via **endocytosis**.

Perisinusoidal cells called **stellate cells** or **Ito cells** store fat in distinct cytoplasmic lipid droplets that contain vitamin A. Through complex and typically inflammatory processes, stellate cells become transformed to myofibroblasts that are capable of secreting collagen and extracellular matrix into the perisinusoidal space.

Hepatic portal vein is the main blood supply to the liver.

The hepatic portal vein and hepatic arteries supply the liver with blood. During rest, their combined blood flow provides the liver with 25% of the cardiac output. Supplying ~75% of the liver's blood supply, the hepatic portal vein carries venous blood drained from the spleen and the gastrointestinal (GI) tract and its associated organs. Unlike blood flowing in the hepatic artery, hepatic portal vein blood is poorly oxygenated. Despite this difference, the liver's O_2 demand is met almost equally by blood flowing from these vessels. What accounts for the high O_2 uptake from the portal vein in which the O_2 tension is so low? Recall from Chapter 19 that O_2 uptake is highly dependent on blood flow. The large flow in the hepatic portal vein offsets the low O_2 tension and therefore significantly contributes to O_2 taken up by the liver. The portal vein branches repeatedly to form smaller venules, and the hepatic artery branches to form arterioles and then capillaries. These venules and capillaries drain into the liver sinusoids, which are considered to be specialized capillaries. The sinusoids empty into the central veins that join to form the hepatic vein flowing into the inferior vena cava.

Liver as a blood reservoir

The vasculature in the liver is characterized as a high-flow, high-compliance, and low-resistance system. High compliance and low resistance allow large quantities of blood to be stored in the liver's vascular tree. Under normal conditions, it contains about 10% of the body's total blood volume. This storage function is especially important in pathophysiologic situations. For example, during hemorrhaging, the liver is capable of supplying extra blood to help maintain central circulating blood volume.

Hepatic blood volume and flow vary inversely with GI activity. Blood flow increases after eating, whereas blood volume decreases. Conversely, during sleep, flow decreases but blood volume slightly increases. The splanchnic arterioles predominantly regulate blood flow to the intestines and spleen and, in turn, in the portal vein. In this way, eating results in increased blood flow to the intestines followed by increased flow to the liver. Portal vein pressure is normally low. Increased resistance to portal blood flow results in **portal hypertension**, which is the most common complication of chronic liver disease and accounts for a large percentage of the morbidity and mortality associated with this disease (Clinical Focus 26.1).

Liver permeability results in large quantities of lymph formation with high-flow rates.

The hepatic sinusoids are very porous and allow a large flux of fluid and protein to move into the perisinusoidal space. Given their high permeability, sinusoidal epithelia form large quantities of lymph, which under resting conditions makes up half of the body's formed lymph. To accommodate both the large quantity and high lymph flow, the hepatic lymphatic system is extensive and is organized around three main areas: adjacent to the central veins, adjacent to the portal veins, and coursing along the hepatic artery. As in other organs, these

Viral Hepatitis

Viral hepatitis refers to the inflammation of the liver caused by the invasion of a number of specific viruses. Many hepatitis viruses have been identified; they are named types A, B, C, D, E, F (not confirmed), and G. Undoubtedly, the known number of viruses that attack the liver will grow as our knowledge of the subject expands. Because of the limitation of space, discussion will focus on the three most common types of hepatitis, A, B, and C.

Viral hepatitis A is a common type of acute liver inflammation caused by the infection of the hepatitis A virus. It is acute but never becomes chronic. It is highly contagious and can spread from person to person like other viral infections. The spread of viral hepatitis A can be rapid and widespread because it is usually caused by unsanitary conditions that allow water or food to become contaminated by human waste containing the hepatitis A virus. Hepatitis A can also spread from person to person through close contact, such as through the passage of oral secretions via intimate kissing and other sexual practices or contact with feces resulting from improper or poor handwashing. Interestingly, some people can harbor the virus with no apparent symptoms; others, however, exhibit symptoms that mimic severe flu. Patients who are infected with hepatitis A usually recover on their own and do not develop chronic hepatitis or cirrhosis.

Hepatitis B, unlike hepatitis A, is a chronic inflammatory liver disease. It is prevalent in the United States, with 200,000 to 300,000 new cases of viral hepatitis B virus infection reported each year. Hepatitis B is spread through sexual contact and the transfer of blood or serum through shared needles in drug abusers, accidental needle sticks with needles contaminated with infected blood, blood transfusions, hemodialysis and by infected mothers to their newborns. The hepatitis B virus can remain in a person for years without manifesting any symptoms. About 6% to 10% of patients with hepatitis B viruses develop chronic liver inflammation, lasting between 6 months and years or decades. Hepatitis B infections can potentially develop into more advanced stages of liver disease associated with cirrhosis of the liver. Cirrhosis is characterized by abnormal structure and function of the liver. The diseases that lead to cirrhosis do so because they injure and kill liver cells and because the inflammation and repair that are associated with the dying liver cells cause scar tissue to form.

The other common virus that attacks the liver is *hepatitis C*. The spread of hepatic C virus is similar to that of hepatitis B. However, the transmission of the hepatitis C virus through sexual contact is not as prevalent as hepatitis B. Patients with chronic hepatitis C infection can continue to infect others. Patients with chronic hepatitis C infection are at risk for developing cirrhosis, liver failure, and liver cancer. ∎

---

channels drain fluid and proteins, but the protein concentration is highest in liver lymphatic fluid.

The perisinusoidal space is the largest space drained by the liver's lymphatic system. Disturbances in the balance of filtration and drainage are the primary causes of **ascites**, the accumulation of serous fluid in the peritoneal cavity. Ascites is another common cause of morbidity in patients with chronic liver disease.

Tissue regeneration in the adult liver is a unique feature to maintain optimal metabolic function.

As mentioned previously, of all of the solid organs in the body, the adult liver is the only internal organ capable of regenerating. There appears to be a critical ratio between functioning liver mass and body mass. Deviations in this ratio trigger a modulation of either hepatocyte proliferation (liver growth) or apoptosis (liver cell death) to maintain the liver's optimal size for metabolic function. Peptide growth factors (e.g., transforming growth factor-α, hepatocyte growth factor, and epidermal growth factor) have been the best-studied stimuli of hepatocyte DNA synthesis. After these peptides bind to their receptors on the living hepatocytes and work their way through a myriad of transcription factors, gene transcription is accelerated, resulting in increased cell number and increased liver mass.

Conversely, enhanced hepatocyte apoptosis rates achieve a decrease in liver volume. Apoptosis is inhibited from occurring prematurely by specific molecules, such as growth factors, that act as antiapoptotic molecules. Apoptosis is therefore a carefully programmed process by which cells kill themselves (cell suicide) and ultimately fragment into apoptotic bodies that are phagocytized by macrophages or other phagocytic cells. Apoptosis can occur via factors that stimulate either extrinsic or intrinsic specific pathways leading to mitochondrial permeabilization. Two types of liver cell death have been described. Hepatic apoptosis describes the occurrence of apoptosis involving multiple cell types within the liver. Hepatocyte apoptosis occurs specifically in the liver hepatocytes. In contrast to hepatic apoptosis, cell death that results from necroinflammatory processes is characterized by a loss of cell membrane integrity (plasma membrane becomes permeable) and the activation of inflammatory reactions that occurs following the cell's rupture.

The events associated with apoptosis can actually contribute to the liver's ability to regenerate. Regeneration is actually a "double-edged sword." Under normal circumstances, the liver's ability to rapidly regenerate tissue is beneficial. However, this capability can become a liability with certain diseases. For example, patients diagnosed with primary hepatic carcinoma often undergo a partial hepatectomy to remove malignant tumors. The hepatectomy not only stimulates tissue regeneration but also activates dormant cancer cells, causing the malignancy to further spread in the liver (Clinical Focus 26.2).

Esophageal Varices, a Common Manifestation of Portal Hypertension

Chronic liver injury can lead to a sequence of changes that terminates with fatal bleeding from esophageal blood vessels. In most forms of chronic liver injury, stellate cells are transformed into collagen-secreting myofibroblasts. These cells are responsible for the deposition of collagen into the sinusoids, thereby interfering with the exchange of compounds between the blood and hepatocytes and increasing resistance to portal venous flow. The resistance appears to be further increased when stellate cells contract. The combination of increased resistance and reduced clearance of vasoconstrictors that have affected arteriolar beds results in increased portal pressure and decreased liver blood flow. This disorder is seen in ~80% of patients with **cirrhosis**. In a compensatory effort, new channels are formed or dormant venous tributaries are expanded, resulting in the formation of varicose veins in the abdomen. Although varicose veins

develop in many areas, portal pressure increases are least opposed in the esophagus because of the limited connective tissue support at the base of the esophagus. This structured condition, along with the negative intrathoracic pressure, favors the formation and rupture of **esophageal varices**. Approximately 30% of those who develop an esophageal variceal hemorrhage die during the episode of bleeding, making it one of the most lethal medical illnesses.

Currently, there are no well-recognized treatments to reverse cirrhosis, but numerous strategies are used to reduce portal hypertension and bleeding. Chief among these is the use of nonselective β-blockers, which have a vasodilator effect. Additionally, bleeding varices are frequently treated by endoscopic ligation of the varices and radiologically or surgically placed portal venous shunts to reduce the portal pressure. ■

▶ DRUG METABOLISM IN THE LIVER

The liver is well suited for its ability to metabolize and detoxify many drugs, including penicillin, ampicillin, erythromycin, and sulfonamides. In a similar manner, many hormones (e.g., thyroxin, aldosterone, estrogen, and cortisol) are either metabolized and/or excreted by the liver. The liver also regulates Ca^{2+} and is the major route of Ca^{2+} secretion via the bile acids that pass into the gut and are excreted in feces.

Hepatocytes are involved in the metabolism of **xenobiotics**, which are compounds not produced by the body and are, therefore, foreign to it. Some xenobiotics (e.g., certain drugs) can be beneficial, but others are toxic. Most environmental xenobiotics are introduced into the body with food, absorbed through the skin, or inhaled during breathing and absorbed into the blood. The kidneys ultimately dispose of these substances, but for effective elimination, they must be rendered hydrophilic (polar and water soluble). Because reabsorption of a substance by the renal tubules depends on its hydrophobicity, the more hydrophobic (nonpolar and lipid soluble) a substance is, the more likely it will be reabsorbed. Many drugs and metabolites are hydrophobic and are converted into hydrophilic compounds by the liver, contributing to their elimination by the kidneys.

Liver metabolism and elimination of drugs and xenobiotics occur in three phases.

Two reactions (phase I and phase II), catalyzed by different enzyme systems, are involved in the metabolism and conversion of xenobiotics and drugs into hydrophilic compounds. In **phase I reactions**, the introduction of one or more polar groups biotransforms the parent compound into a more polar compound. The common polar groups are hydroxyl (OH) and carboxyl (COOH). Most phase I reactions involve oxidation of the parent compound. The majority of the phase I enzymes are located in the smooth endoplasmic

reticulum (SER), but some enzymes are cytoplasmic. For example, alcohol dehydrogenase is located in the cytoplasm of hepatocytes and catalyzes the rapid conversion of alcohol to acetaldehyde. It may also play a role in the dehydrogenation of steroids. Metabolites from phase I reactions are often inactive. However, in some instances, their activity is only modified.

The enzymes involved in phase I reactions (i.e., drug biotransformation) are present as an enzyme complex composed of the **NADPH–cytochrome P-450 reductase** and a series of hemoproteins called **cytochrome P-450** (Fig. 26.2). The drug combines with the oxidized cytochrome P-450³⁺ to

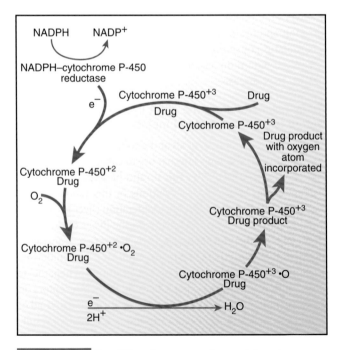

Figure 26.2 Two different enzyme systems catalyze drug metabolism by the liver. The phase I reaction is illustrated here.

form the cytochrome P-450^{3+}–drug complex. The enzyme NADPH–cytochrome P-450 reductase catalyzes the reduction of the cytochrome P-450^{3+}–drug complex to the cytochrome P-450^{2+}–drug complex, which then combines with molecular O_2 to form an oxygenated intermediate. One atom of the molecular O_2 combines with two H^+ atoms and an electron to form H_2O. The other O_2 atom remains bound to the cytochrome P-450^{3+}–drug complex and is transferred from the cytochrome P-450^{3+} to the drug molecule. The drug product with an O_2 atom incorporated is released from the complex. The cytochrome P-450^{3+} released can then be recycled for the oxidation of other drug molecules.

In **phase II reactions**, the phase I reaction products undergo conjugation with several compounds (e.g., glucuronic acid, glycine, taurine, and sulphates) rendering them more hydrophilic resulting in enhanced excretion via the urine in particular but also via the bile. Drug conjugation was once believed to represent the terminal inactivation event, thus rendering the drug nontoxic. However, we know that certain conjugation reactions, for example, *N*-acetylation of isoniazid, may lead to the formation of reactive species responsible for the hepatotoxicity of the drug. Isoniazid is used alone or in conjunction with other drugs to treat tuberculosis.

Phase III reactions involve various transporters (e.g., P-glycoprotein and multidrug resistance–associated protein) that work to eliminate the products of phase II reactions. In doing so, these transporters play a significant role in the absorption, distribution, and excretion of drugs.

Age, nutrition, genetics, sex, and hormones significantly affect drug metabolism and transport in the liver.

The activity of phase I and II enzyme systems is age dependent. These systems are poorly developed in human newborns. It is well known that newborns have a lesser ability to metabolize drugs when compared to adults. At the other end of the spectrum, older adults have a lower capacity than young adults to metabolize drugs. P-450–mediated metabolism may be reduced as much as 35% in older adults, and the clearance of many drugs (e.g., acetaminophen, lidocaine, erythromycin, and triazolam) declines with age. The observed changes in drug metabolism are most likely a reflection of the decrease in liver mass (25% to 35%) and the decrease in blood flow (up to 40%) that accompany aging. The age-related decrease in renal function also impacts drug metabolism.

Nutritional factors can affect the enzymes involved in phase I and II reactions. Insufficient protein intake results in the production of fewer enzyme molecules available to participate in drug metabolism. In fact, low protein intake can result in a decrease of up to 40% in the clearance of certain drugs (i.e., theophylline and phenazone). In some cases, increasing protein intake can return enzyme activity to more normal values.

It is well known that certain factors (e.g., polycyclic aromatic hydrocarbons) can induce drug-metabolizing enzymes. As a result of inhaling polycyclic aromatic hydrocarbons, the metabolism of certain drugs such as caffeine will be increased in cigarette smokers.

The role of genetics in the regulation of hepatic drug metabolism has been investigated to varying degrees in both humans and laboratory animals for at least 60 years, but the impact of genetics on drug metabolisms is still not fully understood. Briefly, some studies have demonstrated loss-of-function polymorphisms and others have shown gain-of-function variants. Both of these changes affect drug clearance and therefore its plasma concentration but in opposite directions. Loss of function results in reduced clearance accompanied by increased plasma concentrations, and gain of function increases clearance and decreases plasma concentrations. Genetic variability combined with the induction or inhibition of P450 enzymes by other drugs or compounds can have a profound effect on what is a safe and effective dose.

Variability in drug metabolism has been observed between individuals and has much to do with the variation in expression and hence the level of activity of the enzymes (e.g., cytochrome P-450s, glutathione transferases, and sulfotransferases) involved in drug metabolism and transport. Studies investigating sex-based differences in drug metabolism have shown that males and females often metabolize the same drug at different rates and that can have a significant impact on the drug's affect(s). Interestingly, adverse drug reactions tend to occur more frequently in women than men and women account for almost 75% of the drug-induced acute liver failure patients seen in the United States. Differences in the expression of growth hormone and sex hormones between males and females are known to affect the metabolism and clearance of drugs by the liver.

▶ ENERGY METABOLISM IN THE LIVER

The liver is a pivotal organ in regulating the metabolism of carbohydrates, lipids, and proteins. The metabolism and interconversion of these compounds by liver cells is highly impacted by hormones and an individual's nutritional state. One hormone, glucagon has a significant impact on liver cells. Its effect on cells of the liver is especially important in maintaining blood glucose homeostasis by stimulating **glycogenolysis** (conversion of glycogen to glucose) and **gluconeogenesis** (conversion of other substances, such as amino acids, into glucose) during periods of low blood glucose. In contrast, when the circulating level of glucose rises (e.g., after meals), the liver will actively store excess glucose as glycogen by a process called **glycogenesis**.

Hepatic portal vein transports nutrients, water-soluble vitamins, and minerals to the liver.

Most water-soluble nutrients and water-soluble vitamins and minerals absorbed from the small intestine are transported via the portal blood to the liver. The water-soluble nutrients will include amino acids, monosaccharides, and fatty acids that are predominantly of short-chain and medium-chain length. Short-chain fatty acids are largely derived from the fermentation of dietary fibers by bacteria in the colon.

Some dietary fibers, such as pectin, are almost completely digested to form short-chain fatty acids (or volatile fatty acids), whereas cellulose is not well digested by the bacteria. The portal blood transports only a small amount of long-chain fatty acids, bound to albumin; most are transported in intestinal lymph as triglyceride-rich lipoproteins (chylomicrons). Chylomicrons serve as a vehicle for delivering fatty acids to muscles and adipose tissue, after which the cholesterol in the chylomicron remnant is transported to and metabolized by the liver.

Liver plays an important role in blood glucose buffering.

The liver is especially important in maintaining normal blood glucose levels. The ability to store glycogen allows the liver to remove excess glucose from the blood and then return it when blood glucose levels begin to fall. The combined result of these pathways is a process called **glucose buffering** (Fig. 26.3) that can be summarized in the following manner. After the ingestion of a meal (usually within 1 to 2 hours), blood glucose concentration increases to a level of 120 to 150 mg/dL (6.6 to 8.3 mmol/L). The hepatocytes remove glucose using a facilitated, carrier-mediated process involving GLUT2. GLUT2 is located in the hepatocyte membrane and its activity is not affected by insulin. Once inside the hepatocytes, glucose is converted to **glucose 6-phosphate (G6P)** and subsequently **uridine diphosphate glucose (UDP-glucose)**. Once phosphorylated, glucose is no longer available and the UDP-glucose is used for glycogenesis. It is generally believed that blood glucose is the major precursor of glycogen. However, recent evidence seems to indicate that the lactate in blood (from the peripheral metabolism of glucose) is also a major precursor of glycogen. Additionally, amino acids (e.g., alanine) can supply pyruvate to synthesize glycogen.

Glycogen, the main carbohydrate stored in the liver, comprises as much as 7% to 10% of the weight of a normal, healthy liver. During fasting, glycogenolysis breaks down glycogen to yield glucose. The enzyme **glycogen phosphorylase** catalyzes the cleavage of glycogen into glucose 1-phosphate that is then converted to G6P by the enzyme phosphoglucomutase. The enzyme **glucose-6-phosphatase**, present in the liver but not in muscle or the brain, converts G6P to glucose that can then be released into the circulation. G6P is an important intermediate in carbohydrate metabolism because it can be used to provide blood glucose or to form glycogen.

Both glycogenolysis and glycogenesis are hormonally regulated. The pancreas secretes insulin into the portal blood. Therefore, the liver, which is hormone sensitive, is the first organ to respond to changes in plasma **insulin** levels. Almost half the insulin in portal blood is removed as it passes through the liver. Insulin tends to lower blood glucose by stimulating glycogenesis and suppressing glycogenolysis and gluconeogenesis. A doubling of portal insulin concentration completely shuts down hepatic glucose production. In contrast to insulin, **glucagon** stimulates glycogenolysis and gluconeogenesis, raising blood sugar levels. **Epinephrine** stimulates glycogenolysis. Regardless, the ultimate balance between hepatic glycogenesis and glycogenolysis involves hormonal input as well as input from glucose sensors in the portal vein.

The closely regulated balance between glycogen synthesis and glucose release by the liver is key to maintaining normal glucose concentrations within a narrow limit of 70 to 100 mg/dL. Although patients with liver disease might be expected to have difficulty regulating blood glucose, this is usually not the case due to the relatively large reserve of hepatic function. However, in the case of chronic liver disease, reduced glycogen synthesis and reduced gluconeogenesis may occur. Some patients with advanced liver disease develop portal hypertension, which induces the formation of portosystemic shunting, resulting in elevated arterial blood levels of insulin and glucagon as a result of suppressed liver removal of these hormones.

Kinases in the liver are involved in monosaccharide metabolism.

Kinases are enzymes that add a phosphate group to a substrate. A **hexokinase** specifically phosphorylates a six-carbon sugar. In the liver (but not in the muscle), the isoform **glucokinase** phosphorylates glucose to form G6P. Depending on the energy requirement, G6P is channeled to glycogen synthesis or used for energy production by the glycolytic pathway.

The liver takes up fructose, and fructokinase phosphorylates fructose to form fructose 1-phosphate, which is either isomerized to form G6P or metabolized by the glycolytic pathway. The glycolytic pathway uses fructose 1-phosphate more efficiently than G6P.

Galactose is used to provide energy and in the biosynthesis of glycoproteins and glycolipids. Galactose taken up by the liver is phosphorylated by galactokinase to form

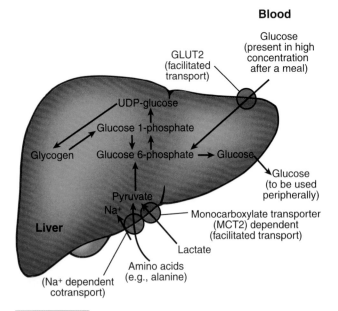

Figure 26.3 **The liver is a key organ in regulating carbohydrate metabolism.** Glycogen is the main carbohydrate stored by the liver. Uridine diphosphate glucose (UDP-glucose) is a precursor to glycogen.

galactose 1-phosphate, which then reacts with UDP-glucose to form UDP-galactose and glucose 1-phosphate. UDP-galactose can be used for glycoprotein and glycolipid biosynthesis or converted to UDP-glucose, which can then be recycled.

Gluconeogenesis in the liver helps maintain blood glucose level during fasting.

Gluconeogenesis is the production of glucose from non-carbohydrate sources such as fat, amino acids, and lactate. The required energy is derived predominantly from the β-oxidation of fatty acids. Pyruvate is the initial substrate, and it can be derived from lactate or the metabolism of glucogenic amino acids (amino acids that can contribute to the formation of glucose). Gluconeogenesis occurs predominately in the liver and the kidneys. However, due to its size, the liver contributes much more to the process.

Gluconeogenesis is important in maintaining blood glucose homeostasis, especially during fasting. The red blood cells and renal medulla totally depend on blood glucose for energy, and glucose is the preferred substrate for the brain. Most amino acids can contribute carbon atoms to glucose synthesis, with alanine from muscle being the most important. The rate-limiting factor in gluconeogenesis is the availability of substrate molecules and not the amount of enzyme. Epinephrine and glucagon stimulate gluconeogenesis, but insulin greatly suppresses it. Thus, in type I diabetic patients, gluconeogenesis is greatly stimulated, contributing to the hyperglycemia observed in these patients (see Chapter 15).

Liver plays a vital role in lipid metabolism.

Most cells in the body can metabolize fatty acids to yield energy The majority of fatty acids however are removed from the blood almost immediately and stored as triglycerides by either fat cells or liver cells until they are needed by other cells for energy (Fig. 26.4). As the primary site for metabolism of triglycerides and the synthesis of lipoproteins, cholesterol, and phospholipids, the liver plays a central and vital role in lipid metabolism. Enzymatic pathways in the liver are also able to synthesize fats using proteins and carbohydrates as substrates. Additionally, the body's tissues benefit from the desaturation of fatty acids in the liver, which they then use in the synthesis of membranes and other cellular structures.

Fatty acid metabolism

After a meal, most of the ingested fat processed by the small intestine is packaged into **chylomicrons** that initially enter the lymph and ultimately the blood. Chylomicrons are comprised almost exclusively of fatty acids (i.e., 80% to 90% of the chylomicron structure), and after they enter the circulation, **lipoprotein lipase** on the endothelial cells of blood vessels hydrolyzes them to liberate fatty acids and glycerol from the triglycerides. Some cells (e.g., heart cells and muscle cells) rely heavily on fatty acids for energy. Adipose cells store them as triglycerides. As metabolism progresses, the chylomicrons shrink and are converted to **chylomicron remnants** that the liver rapidly takes up via chylomicron

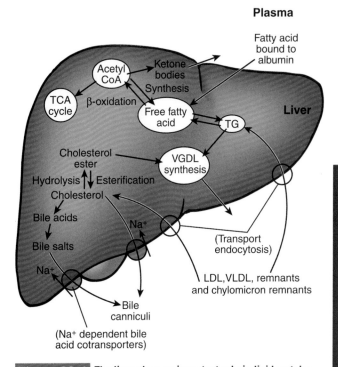

Plasma

Figure 26.4 **The liver plays an important role in lipid metabolism.** The liver is involved in the oxidation of fatty acids, the synthesis of very-low-density lipoproteins (VLDLs), and the regulation of blood triglycerides, low-density lipoproteins (LDLs), and circulating high-density lipoproteins (HDLs). Acetyl-CoA, acetyl-coenzyme A; TCA cycle, tricarboxylic acid cycle; TG, triglyceride.

remnant receptors. The liver then uses the remaining fatty acids to form **very-low-density lipoprotein** (**VLDL**) and for energy production via mitochondrial β-oxidation.

Fatty acids, either absorbed by the digestive tract or released from adipose tissue, are metabolized primarily in the liver. Hepatocytes degrade fatty acids within in the mitochondria yielding acetyl-coenzyme A (acetyl-CoA) in the process. Some acetyl-CoA will enter the citric acid cycle and provide energy to the liver cells. The liver is uniquely able to form acetoacetate, a ketone body, through a condensation reaction between two acetyl-CoAs. Other cells (e.g., muscle, brain, and kidney) absorb the acetoacetate from the blood and convert it back to acetyl-CoA to be used in energy production. In addition to entering the citric acid cycle, acetyl-CoA can be used in the tricarboxylic acid cycle to produce adenosine triphosphate, to synthesize other fatty acids, and to form ketone bodies. Because fatty acids are synthesized from acetyl-CoA, any substances that contribute to the production of acetyl-CoA, such as carbohydrates and proteins, enhance fatty acid synthesis.

The relative amount of fatty acids channeled into the various lipid metabolism pathways of the liver is largely dependent on the person's nutritional and hormonal status. More fatty acid is channeled to ketogenesis or β-oxidation when the supply of carbohydrate is short (during fasting) or under conditions of high circulating glucagon or low circulating insulin (diabetes mellitus). In contrast, more of the fatty acid is used for synthesis of triglyceride for lipoprotein export when the supply of carbohydrate is abundant (during feeding) or under conditions of low circulating glucagon or high circulating insulin.

The liver is one of the main organs involved in fatty acid synthesis. Palmitic acid is synthesized in the hepatocellular cytosol; the other fatty acids synthesized in the body are derived by shortening, elongating, or desaturating the palmitic acid molecule.

Lipoprotein metabolism

A major contribution of the liver to lipid metabolism is the synthesis of lipoproteins, which are the major lipid remaining in the circulation after the metabolism of chylomicrons. The five major classes of circulating plasma lipoproteins are chylomicrons, VLDLs, **low-density lipoproteins (LDLs)**, **intermediate-density lipoproteins (IDLs)**, and **high-density lipoproteins (HDLs)** (Table 26.1). These lipoproteins differ in chemical composition and are usually isolated from plasma using differential centrifugation. Lipoproteins loose buoyancy as the relative amount of protein in the particle increases. Solubility is maintained because the amphipathic phospholipids and proteins surround the otherwise water-insoluble lipid core. This structural arrangement allows lipoproteins to be transported in an aqueous medium (i.e., plasma).

Chylomicrons have a density <0.95 g/mL, which makes them the lightest of the five lipoprotein classes They are synthesized only by the small intestine and are produced in large quantities following fat ingestion. Chylomicrons are large, lipid-rich particles that cannot enter the intestinal capillary beds, but rather are absorbed into the lymphatic system from which they reach the blood via the subclavian vein connection to the thoracic duct.

VLDLs have a density <1.006 g/mL and are smaller than chylomicrons. The liver synthesizes about 10 times more circulating VLDLs than the small intestine. Like chylomicrons, VLDLs are triglyceride rich and carry most of the triglyceride from the liver to the other organs. VLDLs provide cholesterol to organs that need it for the synthesis of steroid hormones (e.g., the adrenal glands, ovaries, and testes). Lipoprotein lipase (LPL) breaks down the triglyceride of VLDLs to yield fatty acids, which can be metabolized to provide energy. When VLDLs are degraded, IDLs are formed. IDLs have a density ranging from 1.006 to 1.019. Only small amounts of IDLs are present in the blood, and they are often overlooked as a class of lipoproteins.

The human liver normally has a considerable capacity to produce VLDLs, but in acute or chronic liver disorders, this ability is significantly compromised. Liver VLDLs are associated with an important class of proteins, **apo B proteins**. The two forms of circulating apo B are B-48 and B-100. The human liver makes only apo B-100 (M.W. ~500,000), which is important for the hepatic secretion of VLDL. In **abetalipoproteinemia**, apo B synthesis is decreased and the secretion of VLDLs is blocked resulting in the appearance of large lipid droplets in the cytoplasm of the hepatocytes of abetalipoproteinemia patients. This lack of VLDL secretion is not entirely related to the decrease in apo B synthesis. A genetic mutation that affects the gene coding for triglyceride transfer protein is also involved. The other major apolipoproteins associated with VLDLs are apo CI, II, and III and apo E.

Although considerable amounts of LDLs (density range 1.019 to 1.063) and HDLs (density range 1.063 to 1.210) are produced in the plasma, the liver also produces a small amount of these two cholesterol-rich lipoproteins. LDLs transport cholesterol ester from the liver to other organs, and HDLs are believed to remove cholesterol from peripheral tissues and transport it to the liver.

Precursors and hormones, such as estrogen and thyroid hormones, regulate the formation and secretion of lipoproteins. For instance, during fasting, the fatty acids in VLDLs are derived mainly from fatty acids mobilized from adipose tissue. In contrast, during fat feeding, fatty acids in VLDLs produced by the liver are largely derived from chylomicrons.

Familial hypercholesterolemia, a disorder in which the liver fails to produce the LDL receptor, exemplifies the

TABLE 26.1 Plasma Lipoproteins

Lipoprotein	Source	Density (g/mL)	Function
Chylomicron	Intestine	<0.95	Composition: 1% protein and 99% lipid; chylomicron is responsible for transporting lipids from the gut to liver, heart, and skeletal muscle
VLDL	Liver	0.95–1.006	Composition: 7%–10% protein, 90%–93% lipid; VLDL is converted to LDL in the liver and functions as the body's internal transport system for lipids (triglycerides, phospholipids, and cholesterol)
IDL	VLDL	1.006–1.019	Composition: 10%–12% protein, 88%–90% lipid; formed when triglycerides present in VLDLs are hydrolyzed by lipoprotein lipase in the capillary beds
LDL	Chylomicron/VLDL	1.019–1.063	Composition: 20% protein, 80% lipid; transports cholesterol into cells
HDL	Chylomicron/VLDL	1.063–1.21	Composition: 35%–60% protein, 40%–65% lipid; smallest in size of all the lipoproteins and functions in the transport of cholesterol and triglycerides; HDL carries about 30% of blood cholesterol and is able to remove cholesterol from atheroma from blood vessels and carry it back to the liver; sometimes called *good cholesterol*

VLDL, very-low-density lipoprotein; IDL, intermediate-density lipoprotein; LDL, low-density lipoprotein; HDL, high-density lipoprotein.

importance of the liver in lipoprotein metabolism. When LDL binds its receptor, it is internalized and catabolized in the hepatocyte. Consequently, the LDL receptor is crucial for the removal of LDL from the plasma. People suffering from familial hypercholesterolemia usually have high plasma LDLs, predisposing them to early coronary heart disease. Often, the only effective treatment is a liver transplantation.

Ketone body production

The formation of ketone bodies as a product of hepatic fatty acid metabolism is both normal and physiologically important. The human liver can produce half of its equivalent weight in ketone bodies per day, but it lacks ketoacid-CoA transferase the enzyme necessary for the metabolism of ketone bodies. Most other organs can use ketone bodies as an energy source. For example, during prolonged fasting, the brain shifts to use ketone bodies (acetoacetate and β-hydroxybutyrate) for energy, although glucose is the preferred substrate. During fasting, a rapid depletion of the liver's glycogen stores occurs, resulting in a shortage of substrates (e.g., oxaloacetate) for the citric acid cycle, and there is a rapid mobilization of fatty acids from adipose tissues to the liver. Under these circumstances, the acetyl-CoA formed from β-oxidation is channeled to ketone bodies. Ketogenesis is also pronounced in diabetes when insulin levels fall precipitously toward zero.

The level of ketone bodies circulating in the blood is usually low, but during prolonged starvation and in diabetes mellitus, it is highly elevated, a condition known as **ketosis**. In patients with diabetes, large amounts of β-hydroxybutyric acid can make the blood pH acidic, a state called **ketoacidosis**.

Blood cholesterol homeostasis

Cholesterol is essential for life, and the liver plays an important role in cholesterol homeostasis. Liver cholesterol is derived from both *de novo* synthesis and the lipoproteins taken up by the liver. Hepatic cholesterol can be used in the formation of bile acids, biliary cholesterol secretion, the synthesis of VLDLs, and the synthesis of liver membranes. Because the absorption of biliary cholesterol and bile acids by the GI tract is incomplete, this method of eliminating cholesterol from the body is essential and efficient. However, patients with high plasma cholesterol levels might be given additional drugs (e.g., statins) to lower their plasma cholesterol levels. Statins act by inhibiting enzymes that play an essential role in cholesterol synthesis.

▶ PROTEIN AND AMINO ACID METABOLISM IN THE LIVER

The liver is one of the major organs involved in synthesizing nonessential amino acids from the essential amino acids. The body can synthesize all but nine of the amino acids necessary for protein synthesis.

The liver synthesizes important plasma proteins.

Of the many circulating plasma proteins synthesized by the liver, albumin may well be the most important (Fig. 26.5).

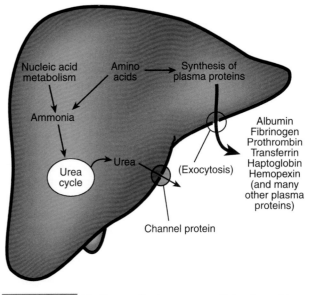

Plasma

Figure 26.5 **The liver synthesizes nonessential amino acids and plasma proteins.** The liver is the body's only source of nonessential amino acids, and it produces most of the circulating plasma proteins.

The liver synthesizes about 3 g of albumin a day. Albumin helps to preserve plasma volume and tissue fluid balance by maintaining the colloid osmotic pressure of plasma. The importance of plasma proteins is illustrated by the fact that both liver disease and long-term starvation result in generalized edema and ascites. Plasma albumin plays a pivotal role in the transport of many substances in blood, such as free fatty acids and certain drugs, including penicillin and salicylate.

The other major plasma proteins synthesized by the liver are components of the **complement system**, components of the **coagulation cascade** (i.e., fibrinogen and prothrombin), globulins, and iron transport proteins (i.e., transferrin, haptoglobin, and hemopexin).

Urea production

Ammonia, derived from protein and nucleic acid catabolism, plays a pivotal role in nitrogen metabolism and is needed in the biosynthesis of nonessential amino acids and nucleic acids. The liver metabolizes ammonia and its ammonia level is 10 times higher than plasma. High circulating levels of ammonia are highly neurotoxic, and hyperammonemia causes irritability, vomiting, cerebral edema, and finally coma leading to death.

The liver synthesizes most of the body's urea by way of the ornithine cycle. Protein intake regulates the enzymes involved in the urea cycle. In humans, starvation stimulates these enzymes causing increased urea production and, therefore, an increase in the circulating urea level. The urea produced by the liver is excreted by the kidneys.

Nonessential amino acids

The **essential amino acids** (see Chapter 25) must be supplied in the diet. The liver can form nonessential amino acids from the essential amino acids. For instance, tyrosine can be synthesized from phenylalanine, and cysteine can be synthesized from methionine.

Glutamic acid and glutamine are important in the biosynthesis of certain amino acids in the liver. Glutamic acid is derived from the amination of α-ketoglutarate by ammonia. This reaction is important because ammonia is used directly in the formation of the α-amino group and constitutes a mechanism for shunting nitrogen from wasteful urea-forming products. Glutamic acid can be used in the amination of other α-keto acids to form the corresponding amino acids, and it can be converted to glutamine by coupling with ammonia, a reaction catalyzed by glutamine synthetase. Following urea, glutamine is the second most important metabolite of ammonia in the liver. It plays an important role in the storage and transport of ammonia in the blood. Through the action of various transaminases, glutamine can be used to aminate various keto acids to their corresponding amino acids. It also acts as an important oxidative substrate, and in the small intestine, it is the primary substrate for providing energy.

▶ LIVER AS A NUTRIENT STORAGE ORGAN

The liver stores and metabolizes fat-soluble vitamins and iron as well as some water-soluble vitamins, particularly vitamin B_{12}. These stored vitamins are released into the circulation as needed throughout the body.

Liver is the site for uptake, storage, and release of fat-soluble vitamins.

Vitamin A comprises a family of compounds related to retinol. Vitamin A is important in vision, growth, the maintenance of epithelia, and reproduction. The liver plays a pivotal role in the uptake, storage, and maintenance of circulating plasma vitamin A levels. **Retinol** (alcohol form of vitamin A) is transported in chylomicrons mainly as an ester of long-chain fatty acids. The retinyl ester is not affected by the metabolism of chylomicrons, and the liver eventually takes up and degrades the chylomicron remnants while storing the retinyl ester.

When the vitamin A level in blood falls, the liver hydrolyzes stored retinyl ester releasing retinol that is then bound to **retinol-binding protein** (**RBP**) that has been synthesized by the liver; the RBP–retinol complex is secreted into the blood. The amount of RBP secreted into the blood depends on vitamin A status. Vitamin A deficiency significantly inhibits the release of RBP, whereas vitamin A loading stimulates its release. RBP binds retinol in a 1:1 stoichiometry. The binding results in the solubilization of retinol and protects it from oxidation.

Hypervitaminosis A develops when massive quantities of vitamin A are consumed. Since the liver stores vitamin A, hepatotoxicity is often associated with hypervitaminosis A. The continued ingestion of excessive amounts of vitamin A eventually leads to portal hypertension and cirrhosis. Most cases of vitamin A toxicity and reported overdose were found in arctic explorers who consumed the livers of polar bears and seals, both of which contain high levels of vitamin A. β-Carotene belongs to a family of natural compounds called *carotenes* or *carotenoids*. It is abundant in plants and gives fruits and vegetables their rich orange and yellow colors. The body converts β-carotene to vitamin A (retinol). Although excessive amounts of vitamin A can be toxic, the body converts only as much vitamin A from β-carotene as needed, thereby making β-carotene a safe source of vitamin A.

Vitamin D is thought to be stored mainly in skeletal muscle and adipose tissue. However, the liver activates vitamin D by converting vitamin D_3 to 25-hydroxyvitamin D_3. Additionally, the liver synthesizes vitamin D–binding protein.

Vitamin K is a fat-soluble vitamin important in the hepatic synthesis of prothrombin. Prothrombin is converted to thrombin, an enzyme necessary for the formation of the fibrin clot. Vitamin K deficiency, therefore, leads to impaired blood clotting.

The dietary vitamin K requirement is small and is adequately supplied by the average Western diet. Bacteria in the GI tract also provide vitamin K. This appears to be an important source of vitamin K because prolonged administration of wide-spectrum antibiotics sometimes results in hypoprothrombinemia. Because vitamin K absorption depends on normal fat absorption, any prolonged malabsorption of lipid can result in its deficiency. The vitamin K store in the liver is relatively limited, and therefore, **hypoprothrombinemia** can develop within a few weeks. Administration of vitamin K usually provides a cure. The largest vitamin K store is in skeletal muscle, but the physiologic significance of this and other body stores is unknown.

Liver plays a key role in the transport, storage, and metabolism of iron.

Iron is essential for life, which makes the regulation of iron metabolism an important factor in many aspects of human health and disease. To safely move iron around the body, it must be bound to a protein, because iron-mediated free radical formation can have dire consequences to tissues and organs (e.g., heart and pancreas). The liver synthesizes several proteins involved in iron transport and metabolism. One of these proteins, **transferrin**, plays a critical role in the transport and homeostasis of iron in the blood. The circulating plasma transferrin level is inversely proportional to the iron load of the body. During iron deficiency, liver synthesis of transferrin is significantly stimulated, which enhances the intestinal absorption of iron. When the amount of iron stored in the liver is high, the rate of transferrin synthesis is lower. Iron is stored intracellularly bound to the intracellular-binding protein **ferritin**. This iron–ferritin complex is soluble in the cytoplasm and can functionally act as a buffering mechanism by regulating the amount of iron available for metabolic processes throughout the body. Liver hepatocytes contain the largest amount of ferritin-bound iron, but measurable amounts also exist in the bone marrow and the spleen. The liver synthesizes two transport proteins that play a vital role in iron conservation: **haptoglobin** that binds free hemoglobin and **hemopexin** that binds free heme when they are present in the plasma. The liver then rapidly removes the hemoglobin–haptoglobin and the heme–hemopexin complexes from the blood conserving the iron for later use in the body.

Iron is an essential component of many molecules, and iron-containing proteins, the heme molecules, are involved in an intracellular energy–yielding process that occurs in every cell's mitochondria. Iron is located in the center of heme molecules, and heme-carrier proteins carry out the redox reactions and electron transport required for oxidative phosphorylation related to energy production. Iron–sulfur proteins are another important group of iron-containing proteins, and some are an essential component in oxidative phosphorylation. Iron is at the center of the heme groups that are part of the hemoglobin molecules found in red blood cells and the myoglobin molecules found in cardiac and skeletal muscle. Iron contributes to the ability of these proteins to transport O_2.

Most of the body's iron is contained in red blood cells, and as a result, hemoglobin deficiency anemia and **iron-deficiency anemia** are the two most common types of diseases associated with iron deficiency.

The spleen is the organ that removes red blood cells that are slightly altered. Kupffer cells of the liver also remove damaged red blood cells, especially those that are moderately damaged (Fig. 26.6). Secondary lysosomes rapidly digest the red cells taken up by Kupffer cells to release heme. Microsomal **heme oxygenase** releases iron from the heme, which then enters the free iron pool and is stored as ferritin or released into the bloodstream, where it binds to apotransferrin. Once iron is bound, the molecule is now

called ferritin. Some of the ferritin iron may be converted to **hemosiderin granules**. It is unclear whether the iron from the hemosiderin granules is exchangeable with the free iron pool.

It was long believed that Kupffer cells were the only cells involved in iron storage, but recent studies suggest that hepatocytes are the major sites of long-term iron storage. Transferrin binds to receptors on the surface of hepatocytes, and the entire transferrin–receptor complex is internalized and processed (Fig. 26.7). The apotransferrin (not containing iron) is recycled back to the plasma, and the released iron enters a labile iron pool. The iron from transferrin is probably the major source of iron for the hepatocytes, but they also derive iron from haptoglobin–hemoglobin and hemopexin–heme complexes. When hemoglobin is released inside the hepatocytes, it is degraded in the secondary lysosomes releasing the heme. Heme is processed in the SER, and the released free iron enters the labile iron pool. A significant portion of the cytosolic free iron most likely combines rapidly with apoferritin to form ferritin. Like Kupffer cells, hepatocytes may transfer some of the iron in ferritin to hemosiderin.

The human body tightly regulates iron absorption and recycling. However, an "iron paradox" exists in physiology; that is, iron is an essential element for life, but the human body has no physiologic regulatory mechanism for excreting iron. Iron overload is prevented solely by the regulation of iron absorption; therefore, those who cannot regulate absorption get disorders of iron overload. With iron overload, iron becomes toxic to cells. Iron toxicity occurs

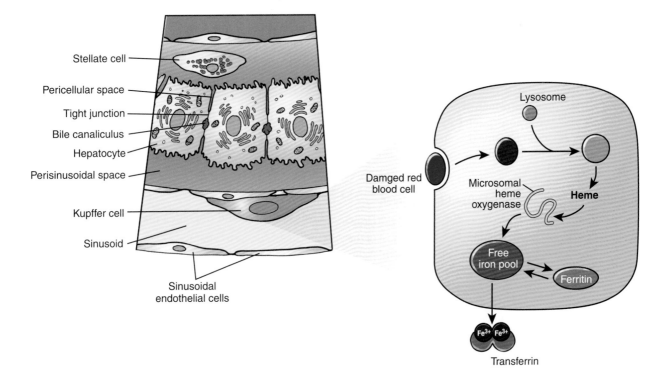

Figure 26.6 Kupffer cells remove damaged red cells from the circulation. Kupffer cells rapidly phagocytize damaged red cells. The iron released from the heme molecule during phagocytosis becomes part of the free iron pool.

Gastrointestinal Physiology

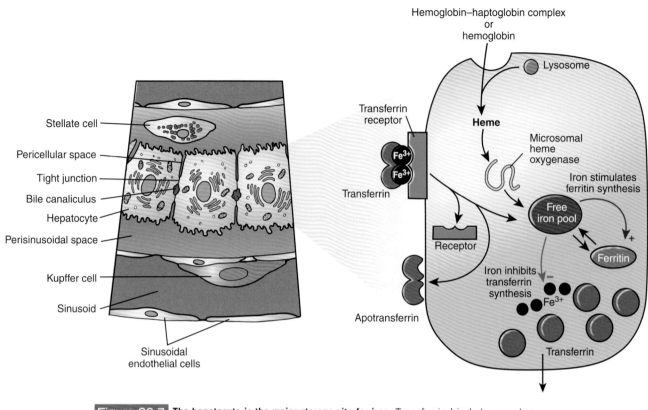

Figure 26.7 **The hepatocyte is the major storage site for iron.** Transferrin binds to receptors on hepatocytes, and as seen in this figure, the entire transferrin–receptor complex is internalized and processed.

when the amount of circulating iron exceeds the amount of transferrin available to bind it. Toxicity arises when intracellular free iron catalyzes the conversion of hydrogen peroxide into free radicals. These free radicals cause extensive cell damage to the heart, liver, and kidney as well as other metabolic organs. Iron toxicity is usually the result of iron supplement overdose (rare), repeated transfusions, or a genetic disease. A classic example of genetic iron overload is **hereditary hemochromatosis (HH)**. The defect in HH causes excessive iron absorption from food that in turn causes the hepatocytes to become defective and fail to perform normal functions. The cellular dysfunction is due to free radical damage.

Coagulation proteins are metabolized by the liver.

Liver cells are involved in both the production and the clearance of coagulation proteins. Hepatocytes secrete most of the known clotting factors and inhibitors, some of them exclusively. Additionally, several coagulation and anticoagulation proteins require a vitamin K–dependent modification following synthesis, specifically factors II, VII, IX, and X and proteins C and S, to make them effective.

The monocyte–macrophage system of the liver, predominantly Kupffer cells, is an important system for clearing clotting factors and factor–inhibitor complexes. Disturbances in liver perfusion or a compromise in liver function can result in the ineffective clearance of activated coagulation proteins,

so patients with advanced liver failure may be predisposed to developing **disseminated intravascular coagulation (DIC)**. DIC is the pathologic activation of blood-clotting mechanisms, leading to the formation of small blood clots inside the blood vessels throughout the body, thereby disrupting normal coagulation and causing abnormal bleeding and possible organ malfunction.

▶ ENDOCRINE FUNCTIONS OF THE LIVER

The liver is important in regulating the endocrine functions of hormones. It can amplify the action of some hormones, and it is the major organ involved in the removal of peptide hormones from the circulation.

Liver activates and degrades many hormones.

The liver converts vitamin D_3 to 25-hydroxyvitamin D3, an essential step before conversion to the active hormone 1,25-dihydroxyvitamin D_3 in the kidneys. The liver is also a major site for the conversion of the thyroid hormone thyroxine (T_4) to the biologically more potent hormone triiodothyronine (T_3). The regulation of the hepatic T_4 to T_3 conversion occurs at both uptake and conversion. Given the liver's capacity to convert T_4 to T_3, hypothyroidism is uncommon in patients with liver disease. However, in advanced chronic liver disease, signs of hypothyroidism can be more evident.

The liver modifies the function of growth hormone secreted by the pituitary gland. **Insulin-like growth factors** made by the liver mediate some growth hormone actions (see Chapter 31).

The liver also helps to remove and degrade many circulating hormones. Although insulin is degraded in many organs, the liver and kidneys are by far the most important. The presence of insulin receptors on the surface of hepatocytes suggests that the binding of insulin to these receptors results in degradation of some insulin molecules. In contrast, the degradation of insulin by hepatocyte proteases does not involve an insulin receptor.

Glucagon and growth hormone are degraded mainly by the liver and the kidneys. Consequently, patients with cirrhotic livers often have high circulating levels of hormones such as glucagon. Additionally, the liver degrades various GI hormones (e.g., gastrin), but the kidneys and other organs probably contribute more significantly to inactivating these hormones.

▶ LIVER AND IMMUNE RESPONSES

Like other organs in the body (e.g., kidney and intestine), the liver has multiple roles, with some considered to be primary functions. For example, the liver's role in processing digestive products, the interconversion of nutrients, and the regulation of circulating glucose levels are well known and often discussed, but the liver also carries out other functions. Discussions on the immune system are generally focused on the importance of bone marrow, the thymus, the spleen, and Peyer patches, but the liver is omitted despite the significant contributions it makes to both innate and acquired immune responses by the body.

Liver architecture contributes to its role in immunity.

The liver receives about 30% of the body's blood flow each minute, which makes it an ideal site for the clearance of blood-borne antigens and pathogens. Over 75% of the blood entering the liver arrives from the digestive system via the portal vein. This direct flow from the GI tact to the liver provides a pathway for the liver's immune cells to clear or internalize antigens and pathogens that were able to escape from the intestinal lumen by crossing the protective GI epithelial barrier. A significant role in immunological clearance function is possible due to the liver's unique internal anatomy (e.g., sinusoidal structure) and the cells types located within the organ (e.g., hepatocytes, lymphocytes, Kupffer cells, and hepatic stellate cells). The sinusoidal structure was discussed previously in this chapter. Briefly, the capillary structure (i.e., decrease in diameter) results in a large drop in both blood pressure and blood flow that is unique to the liver. When combined, these changes increase the time period in which the **antigen-presenting cells** are able to come into contact with nearby hepatic lymphocytes.

Additionally, other physical features of sinusoids such as the lack of a basement membrane and the cellular extensions from lymphocytes and hepatocytes that enter into the space of Disse also increase the likelihood that blood-borne particles and solutes will come into contact with immune-active cells.

Liver cells have immune functions.

Hepatocytes (parenchymal cells) make up about 80% and nonparenchymal cells about 5% of liver volume. The rest is extracellular space and the cells residing there. The nonparenchymal cells include SECs, KCs, lymphocytes, biliary cells, and hepatic stellate cells. Among the many functions that **hepatocytes** carry out within the liver is the detection of pathogens. It is clear that hepatocytes play a role in innate immunity and specifically in an antiviral response. *In vitro*, they have been shown to present antigens to T cells, but the significance of hepatocytes in this role *in vivo* is not entirely clear. Hepatocytes are the source of acute phase proteins and complement components involved in innate immunity. Hepatocytes, KCs, and SECs are all involved in T-cell apoptosis.

It has been well established that SECs are able to remove and process antigens from the circulation and present them to T cells. Both CD4+ and CD8+ T cells are activated by SEC antigen presentation. This activation contributes to the establishment of immune tolerance. Tolerance is important so that individuals do not regularly mount an immune response to antigens derived from the breakdown and absorption of nutrients from the GI tract or against the naturally occurring gut microbial organisms often present in portal blood.

Kupffer cells are the macrophages of the liver, but they remain stationary and do not move through the vasculature. They are either directly responsible for the phagocytosis of bacteria or the recruitment of immune cells to the liver that will then clear the bacteria from the circulation. KCs are capable of presenting antigen (acquired immunity), and they are a source of cytokines and chemokines involved in innate immunity.

Most of the liver's lymphatic cells are NKCs involved in innate immunity and the destruction of infected liver cells that occurs following the release of perforin and granzyme containing granules from the NKCs. NKCs also defend against tumor cells. Other lymphocytes, NK T cells, circulate in the vasculature and destroy pathogens.

Hepatic stellate cells (Ito cells) are located in the space of Disse. This small group of cells, like others in the liver, plays a role in innate immunity. Ito cells have toll-like receptors on their plasma membrane, and once activated, these cells release interferons in addition to cytokines and chemokines. If liver injury occurs, stellate cells are also capable of transdifferentiating and becoming fibrogenic cells.

INTEGRATED MEDICAL SCIENCES

Hepatotoxicity of Nonsteroidal Anti-inflammatory Drugs

Nonsteroidal anti-inflammatory drugs (NSAIDs) are taken by millions of individuals every day, and the majority of them purchase NSAIDs as part of a self-directed treatment regimen as opposed to following a physician-directed program. Tracking the hepatotoxicity or other possible negative impacts of these drugs on human health has historically not been easy. However, testing of hepatic enzyme activity has now become more routine and has provided a more direct means of tracking the impact of drugs on liver function. Some NSAIDs have been found to cause significant liver injury (hepatotoxins) resulting in their withdrawal from the consumer market or an end to their further development (e.g., bromfenac, phenylbutazone, and isoxicam).

Hepatotoxicity does not appear to be related to a patient's age, but certain individual factors do increase a patient's susceptibility to negative effects related to ingesting NSAIDs. A prime example is the association between the development of Reye syndrome in children who have or are recovering from a viral infection and proceed to take aspirin to relieve their symptoms. Investigators have noted that the vast majority of hepatotoxicity is due to idiosyncratic susceptibility (e.g., immunological and genetic idiosyncrasies). Environmental factors cannot be ruled out as contributors or, at the very least, being related to the potential for development of hepatotoxicity.

Certain NSAIDs (e.g., aspirin) can have adverse effects even when taken at normal, therapeutic doses. In contrast, all NSAIDs are going to be toxic at higher exposure levels even though the drug is considered to be safe (i.e., acetaminophen). These situations occur in addition to those that accompany either an unexpected or intentional drug overdose. At least 900 drugs, which include the NSAIDs, have been found to either cause or contribute to hepatotoxicity. The challenge continues to be that drugs like NSAIDs often appear to be safe when first introduced onto the market and any associated hepatotoxicity may take longer to develop and not become apparent until years later. In the last 25 years, however, better testing has led to the detection of hepatic injury much sooner. Regardless, it is believed that liver damage associated with taking NSAIDs is highly underestimated and that the number of cases may be as much as 20 times more frequent in occurrence.

Despite much being known about the action of NSAIDs, the exact causes of hepatotoxicity associated with using these drugs and their specific effect on the liver are not completely understood. This is not surprising given the lipophilic nature of many NSAIDs. However, investigations have provided strong *in vitro* evidence that a common NSAID component, diphenylamine, contributes to the drug-induced damage to liver cells due to an uncoupling of mitochondrial oxidative phosphorylation. As a result of enzymatic bioactivation, acidic NSAIDs are converted to reactive metabolites capable of binding to and altering the activity of liver proteins. This interaction may result in an immune response being mounted against the modified protein. Intrahepatic cholestasis may also be a consequence of drug-induced liver injury that occurs as a result of the NSAID itself or a bioactive drug metabolite. In many situations, a direct correlation between the formation of reactive metabolites and an exact pathway leading to liver injury remains elusive. The ability to fully make a connection and to understand drug–liver interactions is compounded by the fact that a drug can reach the hepatocytes in either the free or protein-bound form, since both can enter the space of Disse. Regardless, NSAID-induced severe or fatal liver injury cases remain rare overall. The majority of cases continue to be either asymptomatic or mild in severity, which is most likely due to the short half-life of many drugs.

Two of the most commonly used NSAIDs are aspirin and Tylenol (acetaminophen or paracetamol). Additionally, Tylenol is often prescribed in combination with or as an active component of other drugs. Both aspirin and Tylenol have clear dose-dependent effects and toxicity. Ingestion of aspirin and Tylenol simultaneously or in combination is extremely dangerous as the negative effects of both are additive. Hepatic injury due to aspirin consumption is generally at the hepatocellular level. While it is difficult to give a precise dosage at which a drug becomes toxic, it has been reported by researchers in London that plasma concentrations in the range of 300 to 500 mg/L produce mild aspirin poisoning, poisoning in the 500 to 700 mg/L range is moderate, and poisoning at levels >750 mg/L was either severe or fatal. These values were determined ~6 hours after ingestion of the drug. Several methods have been reported to treat salicylate poisoning such as rehydration, gastric lavage, activated charcoal, and urinary alkalization. In cases of mild toxicity, it is typical that an aspirin-induced hepatic injury will self-resolve in about 2 weeks after cessation.

Almost half of the acute liver failure cases in the United States are the result of drug-induced liver injury, and the majority of those cases represent an overdose of acetaminophen. Acetaminophen toxicity has four recognized phases of progressing symptoms. In phase I (the first 24 hours post ingestion), symptoms may include vomiting, nausea, or abdominal pain. Once symptoms have progressed to phase IV (96 hours to 3 weeks post ingestion), the patient may be experiencing hepatic failure, coma, or possibly spontaneous recovery. Acetaminophen toxicity may lead to metabolic acidosis. Given the more general nature of phase I symptoms, acetaminophen poisoning may initially be overlooked. To minimize the potential of drug-induced toxicity, it is generally felt that a daily dose for children should remain under 90 mg/kg and the combined adult daily intake should remain <4 g/d. Fortunately, there is an antidote for acetaminophen toxicity, but it must be given as soon as possible after the drug has been consumed. *N*-acetylcysteine is the treatment of choice, and the treatment protocol is determined by the time of administration post drug ingestion. Unfortunately, it cannot reverse any liver damage that has already occurred.

This essay has provided only a glimpse into the potential hepatotoxicity of NSAIDs. Students are encouraged to consult the literature to learn more about the benefits, possible side effects, and potential physiological, pharmacological, and biochemical impact of this group of drugs on individual organs and systems, as well as how to treat possible overdoses. ∎

Chapter Summary

- The basic functional unit of the liver is the liver lobule that contains specialized cells, the hepatocytes.
- The liver plays an important role in maintaining blood glucose levels and in metabolizing drugs and toxic substances.
- The liver is the only organ in the adult that can be regenerated.
- Gluconeogenesis regulates blood glucose during fasting.
- The liver is the first organ to be affected by and respond to changes in plasma insulin levels.

- The liver is one of the main organs involved in fatty acid synthesis.
- The liver aids in the elimination of cholesterol from the body.
- The liver plays an important role in iron metabolism.
- The liver modifies the action of hormones released by other organs.
- The liver plays an important role in establishing immunity.

Chapter Review Questions

1. A 46-year-old patient was not feeling well and was admitted to the emergency department. Upon examination, his liver enzymes were elevated, and he was diagnosed as being jaundiced. Which function would not be affected by his illness?

   A. Albumin synthesis
   B. Urea production
   C. Intestinal glucose absorption
   D. VLDL synthesis
   E. The conjugation of bilirubin with glucuronic acid

   The correct answer is C. Intestinal glucose absorption is not affected by liver dysfunction. All of the other options are correct because they are altered by function of the liver.

2. Both the liver and muscle contain glycogen. However, the liver is the organ capable of contributing glucose to the circulation, because muscle lacks the:

   A. glucose-6-phosphatase enzyme.
   B. glycogen-converting enzyme.
   C. glucose-1-phosphatase enzyme.
   D. ability to convert UDP-D-glucose to glucose.
   E. ability to carry out gluconeogensis.

   The correct answer is A. The liver has the enzyme glucose-6-phosphatase, but muscle does not. Consequently, muscle is incapable of releasing glucose from glucose-6-phosphatase. Option B is incorrect because muscle cannot supply glucose to the circulation. Option C is incorrect because glucose-1-phosphatase is not the right enzyme to hydrolyze glucose-6-phosphatase. Option D is incorrect because UDP-D-glucose is involved in the formation of UDP-galactose, and the enzyme involved is an epimerase. The synthesis of glucose, called gluconeogenesis, is carried out mostly in the liver and to some extent in the kidneys. Option E is incorrect because the synthesis of glucose, called gluconeogenesis, is carried out mostly in the liver and to some extent in the kidneys.

3. The small intestine secretes various triglyceride-rich lipoproteins, but the liver secretes only:

   A. chylomicrons.
   B. VLDLs.
   C. LDLs.
   D. HDLs.
   E. chylomicron remnants.

The correct answer is B. Although both chylomicrons and VLDLs are triglyceride-rich lipoproteins, the liver, unlike the small intestine, produces only VLDLs. LDLs and HDLs are not triglyceride-rich lipoproteins. Chylomicron remnants are generated in the circulation by the metabolism of chylomicrons.

4. A 55-year-old patient was diagnosed with a portacaval shunt (connection between the portal vein and vena cava). Circulating glucagon level in this individual would be expected to be extremely high because the:

   A. pancreas produces more glucagon in these patients.
   B. kidney is less efficient in removing the circulating glucagon in these patients.
   C. liver normally is the major site for the removal of glucagon.
   D. small intestine produces more glucagon in these patients.
   E. circulation to the small intestine is compromised.

The correct answer is C. The liver is one of the major sites for the removal of hormones, including glucagon. Consequently, patients with a portacaval shunt have high levels of circulating glucagon and other hormones because portal blood bypasses the liver. Option A is incorrect because the pancreas does not produce more glucagon in portacaval shunt patients. Option B is incorrect because the kidneys are capable of removing glucagon in these patients. However, the kidneys are not nearly as important as the liver in removing glucagon in healthy individuals. Option D is incorrect because the small intestine does not produce glucagon. Option E is incorrect because blood flow to the small intestine is not compromised in portacaval shunt patients.

Clinical Application Exercises 26.1

AUTOIMMUNE HEPATITIS

A 60-year-old male was previously diagnosed as suffering from rheumatoid arthritis and celiac disease. His current complaint is that he has pain over the liver area, has experienced flulike symptoms, is passing a dark yellow stream of urine, and seems to be having more than the usual joint pain. He has had several tattoos over the past 2 years. An office examination discovered leg bruises, which he could not remember receiving, and evidence of bleeding gums. The remainder of the exam was uneventful. It was concluded that the patient had all the classic signs of acute viral hepatitis but an additional workup was needed.

A liver biopsy was taken and blood was drawn. Renal function was also evaluated, but it was normal. The liver biopsy showed evidence of cirrhosis. The blood draw revealed hypergammaglobulinemia along with elevated aspartate aminotransferase (AST), alanine aminotransferase (ALT), and autoantibody levels including antinuclear antibodies (ANA), antismooth muscle antibodies (ASMA), and antiliver kidney microsomal antibodies (LKM-1). Alkaline phosphatase levels are normal. He was started on prednisone and azathioprine. He will be closely monitored for signs of pancreatitis, which can lead to diabetes.

QUESTIONS

1. Since the patient is now on a drug regimen, why would it be important to be certain that he is suffering from autoimmune hepatitis and not viral hepatitis?

2. After a month, one more autoantibody test was requested. The results came back positive for pANCA (perinuclear neutrophil antibodies). Why would this additional test have been conducted?

ANSWERS

1. Corticosteroids such as prednisone can actually enhance viral replication perhaps by reducing the normal expression of antiviral genes and interfering with innate immunity pathways. It is also possible that corticosteroids may stimulate viral replication rather than stimulate viral elimination. It if the patient has viral hepatitis the treatment choice would be administration of antivirals and interferon. Autoimmune hepatitis was identified around 1950 and is more common in females than males. It can occur at any age; younger patients make up the majority of cases. While it may remain patient dependent, history has shown that if the disease has progressed far enough, up to 20% of patients may not respond to drug treatment (i.e., prednisone). Autoimmune hepatitis is an unresolving, chronic inflammatory response in the liver. Some symptoms such as fatigue, joint pain, and itching are not unique to the disease and can even overlap with those of viral hepatitis. A biopsy is often needed to confirm the

   diagnosis. Viral hepatitis is commonly caused by either the hepatitis A, hepatitis B, or hepatitis C virus, but there are also hepatitis D and E viruses. Hepatitis A, B, and E infections are symptomatic, while hepatitis C–infected individuals may remain asymptomatic. Any of these viral infections can lead to acute hepatitis. The B and C viruses can become chronic hepatitis. Due to the way the viruses replicate, a hepatitis B infection will precede the development of a hepatitis D infection. At this time, hepatitis D is not common in the United States. Symptoms of viral hepatitis can include fatigue, abdominal pain, vomiting, and diarrhea.

2. pANCA autoantibodies are relatively common in autoimmune hepatitis. They are much rarer in viral hepatitis. Both types of hepatitis share many of the same diagnostic abnormalities. Earlier speculation that hepatitis C infection can lead to autoimmune disease has not been definitively proven.

thePoint® *Visit* http://thepoint.lww.com/rhoades5e *for additional chapter review Q&A, Clinical Application Exercises, animations, and more!*

27 Motility and Gastrointestinal Regulation

Active Learning Objectives

Upon mastering the material in this chapter, you should be able to:

- Distinguish the differences between electromechanical and pharmacomechanical coupling in initiation of contractions in gastrointestinal muscles.
- Explain the concept of a "functional electrical syncytium," as applied to the gastrointestinal smooth musculature.
- Understand the forms of electrical activity in the gastrointestinal musculature.
- Understand how the sympathetic, parasympathetic, and enteric divisions of the autonomic nervous system innervate the gut and explain why the enteric nervous system is called a "mini-brain."
- Apply the locations of postsynaptic receptors, presynaptic receptors, and presynaptic autoreceptors on enteric neurons to their physiologic significance.
- Distinguish the role of presynaptic inhibition from that of presynaptic facilitation in the function of enteric neural microcircuits.
- Explain why inhibitory musculomotor neurons are necessary for control of contractile behavior of the intestinal musculature by the enteric nervous system.

- Relate the mechanisms involved in the production of physiologic as opposed to pathophysiological ileus in the intestine.
- Explain how inhibitory neuronal motor control of gastrointestinal sphincters differs from inhibitory control of the longitudinal and circular muscle coats of the small and large intestine.
- Explain how function of the gastric reservoir differs from that of the antral pump in the determination of rate of gastric emptying and the role of the vagovagal reflex.
- Describe three types of smooth muscle relaxation responses that occur in the gastric reservoir.
- Explain why symptoms of early satiety, bloating, and vomiting are experienced by patients after disruption of transmission in vagal nerves.
- Explain how the motor behavior of the stomach and small intestine in the interdigestive state differs from that in the digestive state.
- Describe how the mechanisms of neural control of the internal anal sphincter differ from those of the external anal sphincter and puborectalis muscle.
- Explain how neural deficits lead to gastrointestinal problems.

In the last two chapters, we saw that the gastrointestinal (GI) system was responsible for the absorption of ingested nutrients, H_2O, minerals, and vitamins and the removal of toxic materials from the blood. These processes are essential for the cellular and integrative function of the body. This chapter demonstrates how gut motility and its regulation play a central role in these processes. The GI tract or digestive tract is a long, continuous tube with an oral (i.e., mouth) and an anal opening. The digestive system is physically and functionally divided into regions specialized for digestion, secretion, absorption, motility, and compaction. Skeletal and smooth muscle contractions contribute to the physical breakdown of food and the movement of luminal contents toward the anus. The teeth and tongue and secretions from the salivary glands, liver, gallbladder, and pancreas contribute to the physical, chemical, and enzymatic breakdown processes that occur as ingested materials are mixed and move along the GI tract. Sphincter muscles located at key points along the tube and at critical openings regulate the passage of materials between regions of the tract, the entrance of bile and digestive enzymes into the tract, and the discharge of fecal material from the tract.

Digestive system motility and digestive processes are regulated in a coordinated manner by the sympathetic, parasympathetic, and enteric divisions of the autonomic nervous system, with some conscious control (e.g., the oral and external anal sphincters), along with specific hormones and enzymes. An individual's physiological and psychological

state, as well as immune system activity, can have a significant impact on digestive system functions including salivary secretions, processing and absorption disorders, diarrhea, constipation, abdominal pain, the feeling of satiation, and the species of bacteria that colonize the intestines (see Chapter 25).

Gastroenterology focuses on understanding normal and abnormal functions, diseases, and structural problems of the GI tract and the organs and structures connected to it. **Neurogastroenterology** is a subspecialty of clinical gastroenterology and digestive science. As such, it encompasses understanding nervous system regulation and involvement in GI tract functions. This chapter discusses the organization and motility of the GI tract and the role of skeletal and smooth muscle, the nervous system, and hormones in the process.

▶ ORGANIZATION OF THE DIGESTIVE SYSTEM

The digestive system is divided into the **upper GI tract** (i.e., esophagus, stomach, and duodenum) and the **lower GI tract** (i.e., small and large intestine). The system is structurally and functionally organized to efficiently take in and process significant quantities of food in various forms from large pieces of solids to liquids and purees. Striated muscle is present in the mouth, pharynx, upper esophagus, external anal sphincter, and pelvic floor, while visceral-type smooth muscle is found elsewhere. The contraction and relaxation pattern of the smooth musculature is organized to produce the propulsive forces (peristalsis) responsible for the net movement of luminal contents toward the anus, to mix ingested foodstuffs with digestive enzymes, and to bring nutrients into contact with the mucosa for final digestion and absorption. Muscle contraction also contributes to **trituration**, which decreases particle size. This increases the total surface area available for attack by digestive enzymes, particularly in the small intestine (see Chapter 25). The **enteric nervous system** (**ENS**), together with input from the **central nervous system** (**CNS**), organizes motility into patterns of efficient behavior suited to differing digestive states (e.g., fasting and processing of a meal) as well as abnormal patterns (e.g., during emesis). Functionally, the ENS consists of the myenteric plexus and the submucosal plexus. Output from the myenteric complex increases GI peristalsis and submucosal plexus output regulates secretion and absorption.

Gastrointestinal tract wall contains smooth muscle.

The **muscularis externa** is one of four tissue layers making up the wall of the GI tract and is composed of two distinct sheets of smooth muscle—one circular and one longitudinal (Fig. 27.1). Motility in the stomach and intestines is specifically driven by waves of contraction spreading through the surrounding sheets of smooth muscle. One difference is that the stomach's muscularis externa consists of three sheets of smooth muscle with the long axis of the fibers oriented in a longitudinal, circular, or oblique direction.

Circular and longitudinal muscle layers of the intestine differ in both structure and function.

The long axis of the intestinal **circular muscle fibers** is oriented in the circumferential direction, and contraction in these fibers results in a reduction in diameter that is accompanied by an increase in length of an intestinal segment.

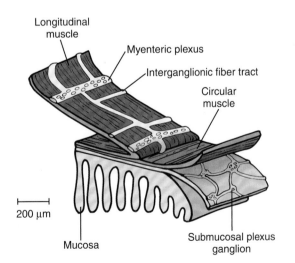

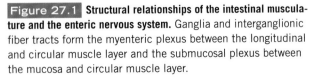

Figure 27.1 **Structural relationships of the intestinal musculature and the enteric nervous system.** Ganglia and interganglionic fiber tracts form the myenteric plexus between the longitudinal and circular muscle layer and the submucosal plexus between the mucosa and circular muscle layer.

The long axis of the **longitudinal muscle fibers** is oriented along the length of the intestine, and contraction in these fibers produces the reverse effect. These concurrent changes occur because the intestine is a cylinder with a constant surface area, and therefore, a change in diameter must be accompanied by an opposite change in length. The circular muscle layer is thicker than the longitudinal layer and is thus capable of exerting more powerful contractile forces onto the lumen.

Sphincters prevent reflux between specialized compartments of the GI tract.

Six strategically placed sphincters regulate movement between compartments of the GI tract. A seventh is located at the junction of the common bile duct with the duodenum. All but the upper esophageal sphincter consist of rings of smooth muscle that remain in a continuous (tonic) state of contraction (positive resting pressure) that occludes the lumen in the region separating two specialized compartments. With the exception of the internal anal sphincter, sphincters function to prevent the backward reflux of intraluminal contents. In this regard, intraluminal pressure is normally highest on the oral side of the sphincter, which will drive net movement in the aboral direction. If the aboral side develops a higher pressure than the oral side, overall movement will be in a retrograde fashion toward the oral cavity (i.e., emesis).

Myogenic contractile activity (i.e., contractile activity initiated within the muscle that does not require any nervous system input) is responsible for maintaining the tonic state of contraction of the smooth muscle sphincters. Specific signals (e.g., distention, pressure, and nervous input) cause the sphincters to transiently relax and allow movement between neighboring compartments and into reservoirs of the stomach and large intestine. Inhibitory sphincteric motor neurons are normally inactive and are switched on to coordinate the opening of the sphincter with physiologic events in adjacent regions (Fig. 27.2). When this occurs, the released inhibitory neurotransmitter causes relaxation in the sphincteric muscle and prevents excitation and contraction in the adjacent muscle from spreading into and closing the sphincter. **Achalasia** is a pathologic state in which smooth muscle sphincters fail to relax and is most often caused by the loss of the ENS and its complement of inhibitory musculomotor innervation of the sphincters.

Upper esophageal sphincter

The **upper esophageal sphincter** (**UES**) is skeletal muscle, and, when compared to the other GI sphincters, it has the highest resting pressure. The UES is located at the junction of the laryngopharynx and esophagus and has two functions. It contracts to close off the esophagus thereby preventing air from entering the digestive tract, and it diverts air into the glottis and trachea during breathing. It relaxes during swallowing to allow food to pass down the esophagus en route to the stomach. During this stage, the epiglottis closes off the glottis. The swallowing center located in the medulla stimulates the UES to relax during swallowing.

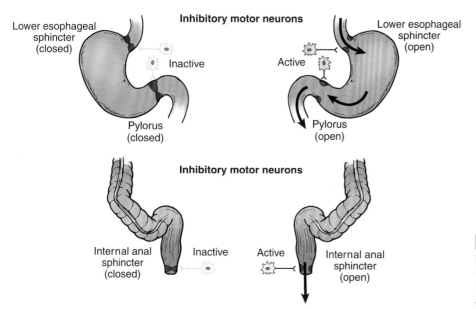

Figure 27.2 Smooth muscle sphincters are closed when their inhibitory innervation is inactive. When inhibitory musculomotor neurons are active, the sphincters are opened.

Lower esophageal sphincter

The **lower esophageal sphincter** (**LES**) is located at the junction of the esophagus and stomach. It prevents the reflux of gastric contents, including HCl, into the esophagus. Incompetence of the LES permits chronic exposure of the esophageal mucosa to gastric acid, which can lead to heartburn, also known as **pyrosis** or **acid indigestion**. Persistent chronic gastric reflux can lead to inflammation and damage to the esophageal mucosa, a major symptom of **gastroesophageal reflux disease** (**GERD**). GERD patients are at risk for **Barrett esophagus**, a condition in which damaged esophageal cells are transformed into metaplastic cells, which are a precursor to carcinoma. Although the risk of cancer is present, the progression from precancerous metaplasia to dysplasia is <20% in Barrett patients.

Pyloric sphincter

The **pyloric sphincter** (**PS**) is located at the gastroduodenal junction. It prevents reflux of duodenal contents into the stomach. Incompetence of the PS can result in the reflux of bile acids and digestive enzymes from the duodenum into the stomach. Bile acids damage the protective barrier of the gastric mucosa, and prolonged exposure can lead to gastritis, formation of ulcers, and risk of perforation.

Ileocecal sphincter

The **ileocecal** (**ileocolonic**) **sphincter** (**IS**) or **ileocecal valve** is located at the junction of the ileum and cecum. It prevents the reflux of colonic contents into the ileum. Incompetence of the IS can allow bacteria from the colon to enter the ileum, which can result in bacterial overgrowth in the small intestine. Bacterial counts are normally low in the small intestine, and bacterial overgrowth has been associated with symptoms of bloating and abdominal pain in a subgroup of patients diagnosed with the irritable bowel syndrome.

Anal sphincters

The **internal anal sphincter** (**IAS**) and the **external anal sphincter** (**EAS**) surround the anal canal. The IAS regulates passage from the rectum into the anal canal and helps to control the movement of flatus and feces out of the anus. The EAS is located at the distal end of the canal and is the last structure separating the GI tract from the external environment. The IAS is under involuntary control only, but the EAS is controlled both involuntarily (i.e., **rectosphincteric reflex**) and voluntarily. Failure of the IAS to relax underlies the fecal retention that occurs in children with **Hirschsprung disease** and adults with **Ogilvie syndrome**. Fecal incontinence is associated with incompetent anal sphincters.

Sphincter of Oddi

The **sphincter of Oddi** surrounds the opening of the bile duct where it enters the duodenum. It prevents the reflux of intestinal contents into the ducts leading from the liver, gallbladder, and pancreas. Failure of this sphincter to open leads to distention in the biliary tree and the associated biliary tract pain that can be felt in the epigastric region and right upper abdominal quadrant. An incompetent sphincter of Oddi is associated with a risk of reflux-induced pancreatitis.

Enteric nervous system controls activity along the GI tract.

The **enteric nervous system** (**ENS**) is a subdivision of the **autonomic nervous system** (**ANS**) and is embedded in the walls of the GI tract where it serves as a "mini-brain" due to its ability to control GI functions without any direct input from the CNS. However, its activity can be modified by transmissions from either the sympathetic (**SNS**) or parasympathetic (**PNS**) subdivisions of the ANS. This allows the ENS to respond to environmental factors such as stress or time of day. The ENS has as many neurons as the spinal cord and is positioned close to the effector systems it controls within the GI tract (i.e., musculature, secretory glands, and blood vessels) (Fig. 27.3).

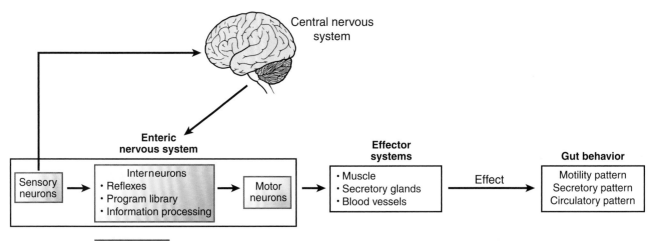

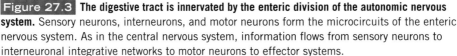

Figure 27.3 **The digestive tract is innervated by the enteric division of the autonomic nervous system.** Sensory neurons, interneurons, and motor neurons form the microcircuits of the enteric nervous system. As in the central nervous system, information flows from sensory neurons to interneuronal integrative networks to motor neurons to effector systems.

The nerve cell bodies of the ENS are clustered in ganglia that are interconnected by fiber tracts resulting in the formation of two plexuses that play a major role in controlling local blood flow, endocrine cell activity, and gut motility, secretion, and absorption. The **myenteric plexus** or **Auerbach plexus** extends the length of the GI tract and runs between the longitudinal and circular muscle layers (see Fig. 27.1). This arrangement makes it ideal for regulating both the strength and frequency of smooth muscle contraction necessary to move materials along the tract. The plexus also controls sphincter activity. **Interstitial cells of Cajal** (see Chapter 25) are located in the myenteric plexus. The **submucosal plexus** or **Meissner plexus** is situated in the submucosal region between the circular muscle and the mucosa (see Fig. 27.1). This plexus is most prominent as a ganglionated network in the small and large intestine, but it is better developed in the small intestine. It does not exist as a ganglionated plexus in the esophagus and is sparse in the submucosal space of the stomach. The submucosal plexus is involved in regulating secretion (i.e., Brunner glands in the duodenum and the crypts of Lieberkühn in the small and large intestine), absorption, blood flow, and endocrine cell activity in addition to regulating the contraction of the submucosal muscle fibers. Contraction and relaxation of these fibers alters the amount of mucosal folding. Although physically separated, the two plexuses are not functionally isolated from one another. Specific neurons link the two networks into a functionally integrated nervous system.

▶ GASTROINTESTINAL SYSTEM MOTILITY

Digestive system motility relies on the integrated function of multiple muscles, nerves, and sometimes endocrine cells for the generation of various motility patterns that are organized to fulfill the specialized function of individual organs and to move nutrients along the digestive tract.

The ENS, together with input from the CNS, organizes muscular activity and motility into functional patterns of wall behavior needed for normal GI function and differing

digestive states (e.g., fasting and processing of a meal) and occasionally into abnormal patterns (e.g., during emesis). Motility patterns differ depending on factors such as awake or sleeping state, time after a meal, and the presence of disease. Nervous system activity is influenced by chemical signals released from **enterochromaffin cells**, **enteroendocrine cells**, and cells associated with the enteric immune system (e.g., **mast cells** and **polymorphonuclear leukocytes**). Specialized pacemaker cells, called **interstitial cells of Cajal** (**ICCs**), are associated with the smooth musculature (see Chapter 25).

Peristalsis and propulsion are the major types of GI motility.

GI motility is divided into two distinct patterns, peristalsis and segmentation. **Peristalsis** (i.e., wavelike contractions in muscle) produces **propulsion**. Propulsion is a basic ENS peristaltic reflex circuit that results in the controlled, progressive, forward movement of food down the esophagus and along the length of the GI tract. It eventually leads to elimination of waste through defecation. **Segmentation** (i.e., the localized division of the intestine into smaller regions lengthwise) occurs when circular muscle contracts. Segmentation aides in the physical breakdown of ingested foods, the mixing of the luminal contents with enzymes for more efficient processing, and the elimination of waste. The **mixing motility pattern** blends pancreatic, biliary, and intestinal secretions with chyme arriving from the stomach and brings digestive products into contact with the absorptive surfaces of the mucosa.

Peristalsis occurs during and shortly after a meal.

Peristalsis is one of the contractions that occur during and shortly after a meal. The contractions occur in wave patterns traveling down short lengths of the GI tract from one section to the next.

The muscle layers of the intestine behave in a characteristic pattern during peristaltic propulsion (Fig. 27.4). Neural networks of the ENS determine this two-component pattern. The contractions occur in waves directly behind a bolus of food, forcing it into the next relaxed section of the GI tract.

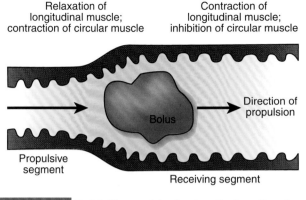

Relaxation of longitudinal muscle; contraction of circular muscle

Contraction of longitudinal muscle; inhibition of circular muscle

Direction of propulsion

Bolus

Propulsive segment

Receiving segment

Figure 27.4 **Peristaltic propulsion involves the formation of a propulsive and a receiving segment, mediated by reflex control of the intestinal musculature.**

As part of the first component of peristaltic propulsion, the longitudinal muscle in the segment ahead of the advancing intraluminal contents contracts at the same time the circular muscle layer relaxes, which results in expansion of the lumen and prepares a *receiving segment* for the forward-moving intraluminal contents.

The second component is contraction of the circular muscle in the segment behind the advancing intraluminal contents with a simultaneous contraction of the longitudinal muscle layer, which converts this region to a *propulsive segment* that propels the luminal contents forward into the receiving segment. The process of lengthening in the longitudinal axis in the propulsive segment is unclear.

Peristalsis is a polysynaptic neural reflex.

The **peristaltic reflex** (i.e., the formation of propulsive and receiving segments) is organized like somatic nervous system–controlled reflexes in skeletal muscle, where reflex circuits with fixed connections in the spinal cord automatically reproduce a stereotypic pattern of behavior each time the circuit is activated (e.g., Achilles and patellar tendon reflexes; see Chapter 5). Connections for these reflexes remain, irrespective of the destruction of adjacent regions of the spinal cord. The peristaltic reflex circuit is similar, with the exception that the basic circuit is repeated continuously within the ENS along and around the entire length of the intestine. In the same way that the monosynaptic reflex circuit of the spinal cord is the terminal circuit for the production of almost all skeletal muscle movements (see Chapter 5), the basic peristaltic reflex circuit of synaptic connections underlies all patterns of propulsive motility. Blocks of the same basic circuit are connected in series up and down the intestine (Fig. 27.5). Distances over which peristaltic propulsion travels are determined by the number of blocks recruited in sequence along the bowel. Synaptic gates between the blocks determine whether recruitment occurs for the next circuit in the sequence. When the gates are opened, neural signals pass between successive blocks of the basic circuit, resulting in propagation of the peristaltic event over extended distances. Long-distance propulsion is prevented when all gates are closed.

Presynaptic inhibition and/or facilitation are involved in gating the transfer of signals between sequential blocks.

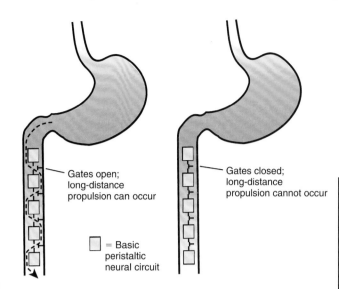

Gates open; long-distance propulsion can occur

Gates closed; long-distance propulsion cannot occur

□ = Basic peristaltic neural circuit

Figure 27.5 **Excitatory synapses connect successive blocks of the basic peristaltic reflex circuit and behave like gates between blocks of circuitry.** Transmission at the synapses opens the gates between successive blocks of the basic peristaltic reflex circuit and accounts for propagation of peristaltic propulsion over long distances. Long-distance propulsion is prevented when all gates are closed.

Synapses between the neurons that carry excitatory signals to the next block are gating points for control of the distance over which peristaltic propulsion travels (Fig. 27.6). Inhibitors of transmitter release at excitatory synapses close the gates to the transfer of information, thereby determining the distance of propagation. Prokinetic drugs that facilitate the release of neurotransmitters at the excitatory synapses (e.g., cisapride and tegaserod) have therapeutic application because they are capable of enhancing propulsive motility.

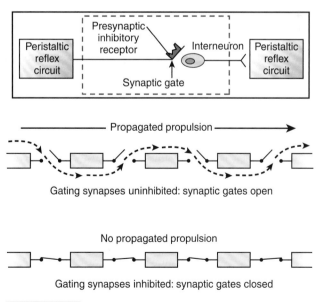

Peristaltic reflex circuit

Presynaptic inhibitory receptor

Interneuron

Peristaltic reflex circuit

Synaptic gate

Propagated propulsion

Gating synapses uninhibited: synaptic gates open

No propagated propulsion

Gating synapses inhibited: synaptic gates closed

Figure 27.6 **Presynaptic inhibitory receptors determine the open and closed states of synaptic gates between blocks of the peristaltic reflex circuit.** In the absence of presynaptic inhibition, propagation can proceed in the direction in which the gates are open. Activation of presynaptic inhibitory receptors closes the gates and no propagated propulsion can occur.

Gastrointestinal Physiology

Emesis

Vomiting (emesis) is the forceful expulsion of the stomach contents out through the mouth and may result from several causes, ranging from gastritis, food poisoning, binge drinking, brain tumors, and brain trauma. Although the terms "vomiting" and "regurgitation" are often used interchangeably, they are different. **Regurgitation** is the return of undigested food back up the esophagus without the forced expulsion associated with vomiting and occurs most commonly in neonates.

Circuits of the ENS are programmed to evoke peristaltic propulsion in either direction along the intestine. If forward passage of the intraluminal contents is impeded in the large intestine, reverse propulsion propels the bolus over a variable distance away from the obstructed segment. **Retropulsion** then stops and forward propulsion moves the bolus again in the direction of the obstruction. During the act of vomiting, retropulsion occurs in the small intestine. In both cases, the musculature for propulsion is stimulated to contract in the reverse direction.

► ESOPHAGEAL AND GASTRIC MOTILITY

The esophagus is divided into three functionally distinct regions: the UES, the esophageal body, and the LES. Esophageal motility involves contraction of striated muscle in the upper esophagus and smooth muscle in the lower third of the esophagus. The esophagus transports ingested solids and liquids downward from the pharynx to the stomach during a swallow and upward during expulsion of gastric contents (i.e., emesis). Swallowing involves voluntary and involuntary muscle contractions. Voluntary muscle contraction in the oral cavity moves the bolus posteriorly and into the pharynx. The swallowing center located in the CNS then initiates involuntary muscle contraction and peristalsis begins. After swallowing, the transit time through the pharynx and esophagus is about 6 to 10 seconds (Clinical Focus 27.1).

Esophageal peristalsis occurs as either a primary or secondary event. **Primary peristalsis** is initiated by the act of swallowing. **Secondary peristalsis** is a subconscious process

---

CLINICAL FOCUS | 27.1

Esophageal Motor Disorders

Dysphagia is defined as difficulty in swallowing. It can result from the failure of mechanisms involving the skeletal muscles of the pharynx, failure of peristalsis in the esophageal body, or failure of the LES to relax. A combination of liquid and solid food dysphagia is a reliable sign of disordered motility in the esophageal body and/or failure of the LES to relax. Dysphagia limited to solid food is most often a symptom of a mechanical obstruction (e.g., malignancies or strictures).

Some dysphagic patients have pressure waves of abnormally high amplitude as peristalsis propagates past the recording ports during diagnostic assessment with manometric catheters. This condition, which is called nutcracker esophagus, is often associated with angina-like chest pain. A manometric diagnosis of nutcracker esophagus is usually made subsequent to confirmation that the patient's chest pain does not reflect blockages in the coronary arteries.

In the absence of chest pain, liquid and solid food dysphagia is likely to be a reflection of LES achalasia. A diagnosis of achalasia is made when manometric recording of esophageal motility fails to show the typical relaxation and reduction of intraluminal pressure in the sphincter coincident with the onset of a swallow (see Fig. 25.24). Successful therapy involves mechanical dilation of the sphincter either by pneumatic inflation of a balloon placed in the sphincter or by passage of a solid bougie dilator. A class of smooth muscle relaxant drugs, known as phosphodiesterase-5 inhibitors (e.g., sildenafil), have been used recently as an alternative treatment for achalasia.

Achalasia, in the majority of cases, reflects the loss of the enteric inhibitory motor innervation of the sphincter. Loss or malfunction of inhibitory motor neurons, which in most cases is an inflammatory ENS neuropathy, is the pathophysiologic starting point for achalasia. Achalasia associated with paraneoplastic syndrome, Chagas disease, and idiopathic ENS degenerative disease are recognizable forms of inflammatory ENS neuropathy. In paraneoplastic syndrome, antigens expressed by a carcinoma (usually small cell carcinoma) in the lung have sufficient similarity to antigenic epitopes expressed by enteric neurons that the immune system simultaneously attacks the tumor and the patient's enteric neurons. Most patients with a diagnosis of achalasia, in combination with small cell lung carcinoma, have circulating immunoglobulin-G autoantibodies that react with and destroy their enteric neurons. The detection of anti-enteric neuronal antibodies is a supplemental means to a specific diagnosis of achalasia. The association of enteric neuronal loss and symptoms of achalasia in Chagas disease also reflects autoimmune attack on ENS neurons. *Trypanosoma cruzi*, the blood-borne parasite that causes Chagas disease, has antigenic epitopes similar to enteric neuronal antigens. This antigenic commonality activates the immune system to assault the ENS coincident with its attack on the parasite.

When chest pain is associated with liquid and solid food dysphagia in the absence of nutcracker-like contractions, the problem is most likely a form of failure of peristaltic propulsion, called diffuse esophageal spasm. A diagnosis of diffuse spasm is made when manometric recording of esophageal motility demonstrates that the act of swallowing results in simultaneous contractions all along the length of the smooth muscle region of the esophageal body. A barium study in patients with diffuse spasm shows the morphometric correlate of diffuse spasm to be a contorted esophageal body, which has been described as a "corkscrew esophagus." Diffuse spasm and LES achalasia are often seen as concurrent pathophysiologies, both of which probably reflect a disorder of enteric inhibitory motor neurons. There are no fully satisfactory treatments for diffuse spasm other than attempts to treat with smooth muscle relaxant drugs. ■

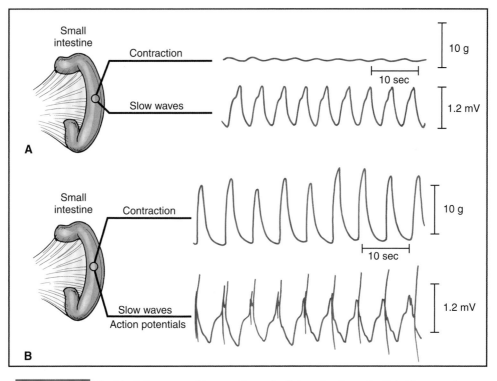

Figure 27.7 **Electrical slow waves trigger action potentials and action potentials trigger contractions.** Electrical slow waves are always present in the small intestine. However, action potentials are not always associated with a slow wave. **(A)** No action potentials appear at the crests of the slow waves, and the muscle contractions associated with each slow wave are small. **(B)** Muscle action potentials appear as sharp upward–downward deflections at the crests of the slow waves. Large-amplitude muscle contractions are associated with each slow wave when action potentials are present.

that is triggered by distention associated with failed downward transport (i.e., food lodging in the esophagus) or by the presence of acid due to gastric reflux. Secondary peristalsis accompanied by the stimulation of salivary secretion is the normal mechanism for "dislodging" the bolus as well as removing acid irritation.

When not involved in swallowing, the muscles of the esophageal body are relaxed and the LES is tonically contracted. In contrast to the intestine, the relaxed state of the esophageal body is not the result of inhibitory motor neuron activity. Rather, the excitability of the muscle is low, and there are no **electrical slow waves** (**ESWs**) to trigger contractions (Fig. 27.7). Activation of excitatory motor neurons, rather than myogenic mechanisms, accounts for the coordinated contractions of the circular and longitudinal muscles during peristaltic propulsion.

Gastric motility is involved in both the filling and emptying of the stomach.

The functions of the stomach include (1) being a reservoir for the storage of ingested food, (2) mixing of food with digestive juices forming chyme, (3) grinding of larger food particles into smaller ones, and (4) emptying. The stomach itself is anatomically divided into four sections, (1) **cardia**, (2) **fundus**, (3) **corpus** (body), and (4) **pylorus** (Fig. 27.8). Each section has different cells and functions. The *cardia* is

where the esophageal contents enter the stomach. The *fundus* forms the upper curvature and the *corpus* forms the main or central region of the stomach. The *pylorus* is the lower section of the stomach that facilitates gastric emptying and the passage of chyme into the duodenum of the small intestine.

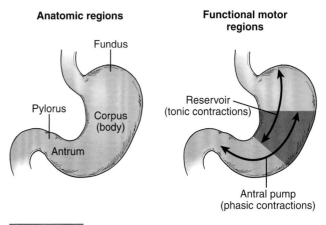

Figure 27.8 **The stomach is divided into three anatomical and two functional regions.** The gastric reservoir is specialized for receiving and storing a meal. The musculature in the region of the antral pump exhibits phasic contractions that function in the mixing and trituration of the gastric contents. No distinctly identifiable boundary exists between the reservoir and the antral pump.

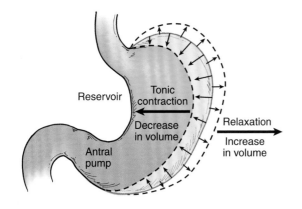

Figure 27.9 **Tonic contraction of the musculature of the gastric reservoir decreases the volume and exerts compressive force on the contents.** Tonic relaxation of the musculature expands the volume of the gastric reservoir. Neural mechanisms of feedback control determine intramural contractile tone in the reservoir (see Fig. 27.29).

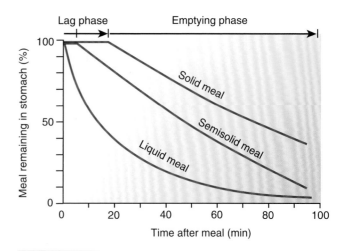

Figure 27.10 **The rate of gastric emptying varies with the composition of the meal.** Solid meals empty more slowly than semisolid or liquid meals. The emptying of a solid or semisolid meal is preceded by a lag phase, which is the time required for particles to be reduced to a sufficiently small size for emptying.

The stomach is divided functionally into two regions based upon their distinct differences in motility. The proximal **gastric reservoir** consists of the fundus and approximately one third of the corpus. The distal **antral pump** consists of the caudal two thirds of the corpus, the antrum, and the pylorus (see Fig. 27.8).

As food enters the esophagus, the stomach muscles are stimulated to relax, **receptive relaxation**, allowing the stomach to accommodate a change in volume. Muscles in the gastric reservoir maintain a state of tonic contraction and do not contract phasically as does the smooth muscle of the antrum (Fig. 27.9). The transition from tonic to phasic contraction occurs in the body of the stomach. The spread of phasic contractions in the region of the antral pump propels the gastric contents toward the gastroduodenal junction. Strong propulsive waves do not occur in the proximal stomach.

Differences in motility between the reservoir and antral pump reflect adaptations for different functions and different states of satiation (i.e., fasting and after eating). Gastric muscle contractions during the fasting state are known as the **migrating motor complex** or **migrating myoelectric complex** (**MMC**) as opposed to that occurs during the digestive phase that follows food intake.

Rate of gastric emptying is determined primarily by the stomach's volume.

Together with storage in the reservoir and mixing and grinding by the antral pump, another important function of gastric motility is the delivery of gastric chyme to the duodenum at a rate that does not overload the digestive and absorptive functions of the small intestine (Clinical Focus 27.2). Neural control mechanisms adjust the rate of gastric emptying to compensate for variations in the volume, composition, and physical state of the gastric contents.

The volume of liquid in the stomach is a major determinant of gastric emptying, but the osmolality, acidity, and caloric content of the gastric chyme are also important. The rate of emptying of isotonic, noncaloric liquids (e.g., 0.9% NaCl)

is proportional to the initial volume in the reservoir: the larger the initial volume, the more rapid the emptying. Isotonic liquids empty faster than hypotonic and hypertonic liquids. The rate of gastric emptying decreases as the acidity of the gastric contents increases. Meals with a low caloric content empty faster than those with a high caloric content. The neurophysiologic control mechanisms for gastric emptying keep the caloric delivery rate to the small intestine within a narrow range, regardless of the source (i.e., carbohydrate, protein, fat, or a mixed meal). Fat is emptied at the slowest rate and is the most potent inhibitor of gastric emptying. This inhibition involves the enteroendocrine release of **cholecystokinin** in the upper small intestine.

With a meal of mixed composition, liquids empty faster than solids (Fig. 27.10). For digestible particles a **lag phase** is required for the grinding action of the antral pump to reduce the particle size to a sufficiently small size for emptying. However, not all particles are released from the stomach at the same time and the smaller particles are selectively released first (**sieving action** of the stomach). Sieving occurs in the antral region. Eventually, the retained larger particles (generally >7 mm) empty at the start of the first MMC as the digestive tract converts motility from the digestive state to the interdigestive state.

▶ SMALL INTESTINAL MOTILITY

The small intestine is divided into three functional regions (i.e., duodenum, jejunum, and ileum) and is the site where most food is enzymatically digested and absorbed. Contractions of the duodenum and hormonal signals arising from there significantly influence gastric emptying. Transit of material through the small intestine is influenced by three fundamental patterns of motility: (1) the **interdigestive pattern**, (2) the **digestive pattern**, and (3) **power propulsion**. Motility is one of the gut behaviors controlled by the ENS (see Fig. 27.3).

CLINICAL FOCUS | 27.2

Disorders of Gastric Motility

Patients with disordered gastric motility fall either into the category of delayed gastric emptying or into a category in which emptying is too rapid. The symptoms associated with accelerated gastric emptying and delayed emptying overlap.

Delayed Gastric Emptying (Gastric Retention)

Delayed gastric emptying occurs in 20% to 30% of patients with diabetes mellitus and is related to vagal neuropathy as part of a spectrum of diabetic autonomic neuropathy. Conduction in vagal afferent and efferent nerve fibers is impaired in diabetic neuropathy. This compromises the vagovagal reflexes, which underlie receptive, adaptive, and feedback relaxation of the gastric reservoir, during intake of a meal (see Figs. 25.28 and 25.29). Vagal nerves are sometimes damaged during laparoscopic surgery for repair of a hiatal hernia or fundoplication as a treatment for GERD. Iatrogenic vagotomy results in a rapid emptying of liquids and a delayed emptying of solids. Truncal vagotomy impairs adaptive relaxation and results in increased contractile tone in the gastric reservoir (see Fig. 25.29). Elevated contractile tone increases pressure in the gastric reservoir, which ejects liquids more forcefully through the antral pump into the duodenum.

Paralysis with a loss of propulsive motility and, consequently, the trituration of solids by the antral pump occur in diabetic autonomic neuropathy and after a vagotomy. The result is **gastroparesis**, which can account for the delayed emptying of solids after a vagotomy.

Delayed gastric emptying with no demonstrable underlying condition is common. Up to 80% of patients with anorexia nervosa have delayed gastric emptying of solids. Another such condition is idiopathic gastric stasis, in which no evidence of an underlying condition can be found. These kinds of disorders for which no physiologic or biochemical explanation for the patient's symptoms can be found are called **functional gastrointestinal motility disorders.** Motility-stimulating drugs (e.g., cisapride, erythromycin, or domperidone) are used successfully in treating patients for both diabetic gastroparesis and the idiopathic form of gastroparesis.

In children, hypertrophic pyloric stenosis impedes gastric emptying. This is a thickening of the muscles of the pyloric canal, which is associated with a loss of the ENS including inhibitory motor neurons to the musculature. The absence of inhibitory motor neurons and achalasia of the circular muscle in the pyloric canal are factors that account for the obstructive stenosis.

Rapid Gastric Emptying

Resection of the distal stomach might be done as a treatment for cancer or peptic ulcer disease. Surgical pyloroplasty used to be done together with vagotomy for the treatment of peptic ulcer disease and might still be done at times in patients with idiopathic gastroparesis. Both resection and pyloroplasty compromise mixing and renders the stomach incontinent for solids. Premature and rapid gastric emptying ("dumping") of solids and liquids into the duodenum in these cases causes hyperglycemia. Vagotomy, done at the same time as pyloroplasty, impairs receptive relaxation and accommodation in the gastric reservoir and exacerbates the dumping symptoms of anxiety, sweating, strong hunger, dizziness, weakness, and palpitations. The **dumping syndrome** is managed by restricting the patient to small meals of complex carbohydrates ingested together with small volumes of liquids. ■

Gastric contents entering the upper GI tract stimulate the enteric nervous system.

Signals transmitted by vagal efferent nerves to the ENS interrupt the MMC and initiate the mixing pattern of motility during ingestion of a meal. The mixing movements are sometimes called segmenting movements or segmentation. The segmentation appearance is the result of peristaltic contractions, which propagate for only short distances and occur simultaneously at multiple sites along the bowel. Receiving segments with an expanded lumen separate circular muscle contractions that form short propulsive segments on either end of the receiving segment (Fig. 27.11). Each propulsive segment jets the contents in both directions into the opened receiving segments, where stirring and mixing occur. This happens continuously at closely spaced sites along the entire length of the small intestine.

► LARGE INTESTINAL MOTILITY

The large intestine consists of the cecum and the colon. It is divided into functionally distinct compartments, which correspond approximately to the ascending colon, transverse colon, descending colon, rectosigmoid region, and IAS (Fig. 27.12). Digestion of food no longer occurs in the large intestine, but absorption of H_2O, minerals, and vitamins and fecal compaction do occur. Contractile activity occurs continuously in the normally functioning large intestine. Whereas the contents of the small intestine move through sequentially with no mixing of individual meals, the large bowel contains a mixture of the remnants of several meals ingested over 3 to 4 days. The arrival of undigested residue from the ileum does not predict the time of its elimination. Normal transit time through the large intestine is about 36 to 48 hours.

Power propulsion is unique to the large intestine.

Between meals the colon is generally quiescent. However, following a meal, motility significantly increases due to ENS electrical activity. In humans, the primary stimulus is the presence of fat from the small intestine. Distention of the colon can also act as a strong stimulus.

Power propulsion is controlled by the ENS. Power propulsion occurs in the transverse and descending colon and

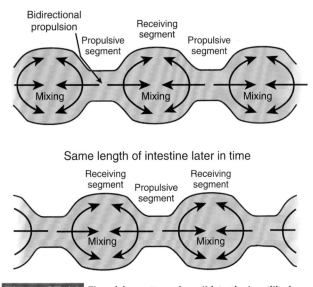

Figure 27.11 **The mixing pattern of small intestinal motility is characteristic of the digestive state.** Propulsive segments separated by receiving segments occur randomly at multiple sites along the small intestine. Mixing of the luminal contents occurs in the receiving segments. Receiving segments convert to propulsive segments, whereas propulsive segments become receiving segments.

fits the general pattern of neurally coordinated peristaltic propulsion. It accomplishes the mass movement of feces over long distances. Increased delivery of ileal contents into the ascending colon, as occurs following a meal, often triggers

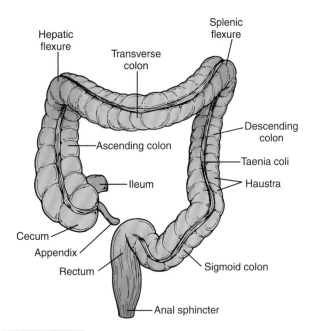

Figure 27.12 **The main anatomical regions of the large intestine are the ascending, transverse, descending, and sigmoid colon and the rectum.** The hepatic flexure is the boundary between ascending and transverse colon; the splenic flexure between transverse and descending colon. The sigmoid colon is well defined by its shape. The rectum is the most distal region. The cecum is the blind ending of the colon at the ileocecal junction. The appendix is an evolutionary vestige. Internal and external anal sphincters close the terminus of the large intestine. The longitudinal smooth muscle layer in humans is restricted to bundles of fibers called *taeniae coli*.

mass movements into the colon. The increased incidence of mass movements and generalized increase in segmental movements following a meal is called the **gastrocolic reflex**. Power propulsion can be initiated by irritant laxatives (e.g., castor oil and senna) and threatening agents, which might be present in the colonic lumen (e.g., parasites, enterotoxins, and food allergens).

Power propulsion in the healthy bowel usually starts in the middle of the transverse colon and is preceded by relaxation of the circular muscle and the downstream disappearance of haustral contractions. An extended length (e.g., 300 cm) of the colon might be emptied as the contents are propelled as far as the rectosigmoid region at rates up to 5 cm/min. Haustration returns after the power contractions pass.

Ascending colon receives intestinal content from the terminal ileum.

Chemoreceptors and mechanoreceptors in the cecum and ascending colon provide feedback for controlled delivery of the ileum contents into the ascending colon, analogous to gastric emptying into the small intestine. Neuromuscular mechanisms, analogous to adaptive relaxation in the gastric reservoir, permit filling of the ascending colon to occur without large increases in intraluminal pressure.

Dwell time in the ascending colon is long when compared to the small intestine but short when compared to the transverse colon. This suggests that the ascending colon is not the primary site for the storage, mixing, and removal of H_2O from the feces.

The ENS-generated motor patterns of the ascending colon consist of peristaltic propulsion that travels at certain times in the orthograde direction and at others in the retrograde direction. The significance of retrograde propulsion in this region is uncertain; it may be a mechanism for temporary retention of the contents in the ascending colon. Forward propulsion is probably controlled by feedback signals related to the fullness of the transverse colon.

Motility of the transverse colon is specialized for storage and removal of water from the feces.

The contents of the transverse colon are retained for about 24 hours, suggesting that this is the primary location for the removal of H_2O and electrolytes and the storage of solid feces. The ENS-driven segmental pattern of motility is responsible for the ultraslow forward movement and compaction of feces in the transverse colon. In this pattern, ringlike contractions of the circular muscle divide the colon into pockets called **haustra** (Fig. 27.13). **Haustration** is reminiscent of the mixing (segmentation) movements in the small intestine (see Fig. 27.11). Haustral formation differs from small intestinal segmentation in that the contracting and receiving segments on either side remain in their respective states for extended periods of time.

Haustrations are dynamic in that they form and reform at different sites. The most common fasting pattern is for the contracting segment to propel the contents in both directions into receiving segments (i.e., haustra). This

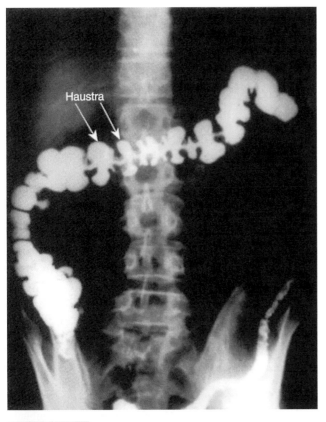

Figure 27.13 **X-ray shows haustral contractions in the ascending and transverse colon.** Between the haustral pockets are segments of contracted circular muscle. Ongoing activity of inhibitory motor neurons maintains the relaxed state of the circular muscle in the pockets. Inactivity of inhibitory motor neurons permits the contractions between the pockets.

mechanism mixes and compresses the semiliquid feces in the haustral pockets and probably facilitates the absorption of H_2O without any net forward propulsion as happens when sequential migration of the haustra occurs along the length of the bowel.

Power propulsion in the descending colon is responsible for mass movements of feces into the sigmoid colon and rectum.

The descending colon is a conduit from transverse to sigmoid colon and feces are not retained for very long. Luminal contents begin to accumulate in the sigmoid colon and rectum about 24 hours after entering the cecum. This suggests that the transverse colon is the main fecal storage reservoir, whereas the descending colon serves as a conduit without long-term fecal retention. Power propulsion in the descending colon drives mass movements of feces into the sigmoid colon and rectum.

Rectosigmoid, anal canal, and pelvic floor musculature preserve fecal continence.

The sigmoid colon and rectum are reservoirs with a capacity of up to 500 mL. Distensibility in this region is an adaptation for temporarily accommodating the mass movements

of feces downstream. The rectum begins at the level of the third sacral vertebra and follows the curvature of the sacrum and coccyx for its entire length. It connects to the anal canal, which is surrounded by the IAS and EAS. Overlapping sheets of striated muscle called the **levator ani** form the pelvic floor. This muscle group, which includes the **puborectalis** and the EAS, is a functional unit that maintains fecal continence. These skeletal muscles behave in many respects like the somatic muscles that maintain body posture (see Chapter 5).

The pelvic floor musculature can be imagined as an inverted funnel consisting of the levator ani and EAS muscles in a continuous sheet from the bottom margins of the pelvis to the anal verge, which is the transition zone between mucosal epithelium and stratified squamous epithelium of the skin. The pelvic floor descends during defecation. When defecation is complete, the levator ani contracts to restore the perineum to its normal position. Fibers of the puborectalis muscle join behind the anorectum and pass around it on both sides to insert on the pubis. This forms a U-shaped sling that pulls the anorectal tube anteriorly such that the long axis of the anal canal lies at nearly a right angle to the rectum (Fig. 27.14). Tonic contraction of the puborectalis narrows the anorectal tube from side to side at the bend of the angle, resulting in a physiologic valve that is important in preventing leakage of feces and flatus.

The puborectalis sling and the upper margins of the IAS and EAS form the anorectal ring, marking the boundary of the anal canal and rectum. When contracted, the EAS compresses the anus into a slit, thereby closing the orifice. The IAS contracts tonically to sustain closure of the anal canal.

Anal canal is innervated by somatosensory nerves.

Mechanoreceptors in the rectum detect distention and supply the ENS with the sensory information needed for feedback control of this region. Unlike the rectum, the anal

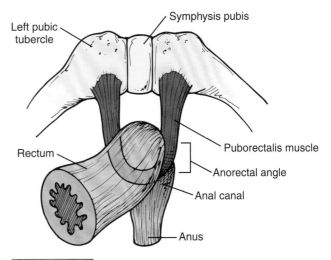

Figure 27.14 **The puborectalis muscle contributes to the maintenance of fecal continence.** One end of the puborectalis muscle inserts on the left pubic tubercle and the other on the right pubic tubercle, forming a loop around the junction of the rectum and anal canal. Thereby, contraction of the puborectalis muscle forms the anorectal angle, which blocks passage of feces.

canal in the region of skin at the anal verge is innervated by somatosensory nerves that transmit signals to the spinal cord and onto higher processing centers in the brain. This region has sensory receptors that detect and transmit information on touch, pain, and temperature with high sensitivity. Processing of information from these receptors allows the person to discriminate consciously between the presence of gas, liquid, and solids in the anal canal. In addition, stretch receptors in the muscle of the pelvic floor detect changes in the orientation of the anorectum as feces are propelled into the region.

Contraction of the IAS and the puborectalis muscle blocks the passage of feces and maintains continence with small volumes in the rectum. When the rectum is distended, the **rectoanal reflex** pathway is activated to relax contractile tone in the IAS. Rectal distention also results in the sensation of rectal fullness.

IAS relaxation allows the rectal contents to come into contact with the sensory receptors in the lining of the anal canal. Signals from these receptors reach conscious perception and provide the person with an early warning of the possibility of a breakdown of continence. When this occurs, voluntary contraction of the EAS and puborectalis muscle maintains continence. The EAS closes the anal canal and the puborectalis sharpens the anorectal angle (see Fig. 27.14). An increase in the anorectal angle works in concert with increases in intra-abdominal pressure to create a "**flap valve.**" This valve is formed by the collapse of the anterior rectal wall onto the upper end of the anal canal, thereby occluding the lumen.

Whereas the ENS mediates the rectoanal reflex, synaptic circuits for the neural reflexes of the EAS and other pelvic floor muscles reside in the sacral portion of the spinal cord. Sensory input from the anorectum and pelvic floor is transmitted over dorsal spinal roots into the sacral cord, and motor outflow to these areas is in ventral sacral root motor nerve fibers. The spinal circuits mediate reflex increases in contraction of the EAS and pelvic floor muscles, which can occur during behaviors that raise intra-abdominal pressure (e.g., coughing, sneezing, lifting weights) and threaten a breakdown in continence.

Neural control of defecation involves the central and enteric nervous systems.

Distention of the rectum evokes an urge to defecate or release flatus. Local processing in the ENS activates the motor program for relaxation of the IAS. At this stage of rectal distention, voluntary contraction of the EAS and the puborectalis muscle prevents leakage and the decision to defecate is voluntary, which results in commands from the brain to the sacral spinal cord to stop excitatory input to the EAS and levator ani muscles. Additional skeletal motor commands contract the abdominal muscles and diaphragm to increase intra-abdominal pressure. Coordination of the skeletal muscle components of defecation results in a straightening of the anorectal angle, descent of the pelvic floor, and opening of the anus.

During defecation, the longitudinal muscle layer in the sigmoid colon and rectum shortens and is followed by strong contraction of the circular muscle layer propelling the luminal contents to the outside (see Fig. 27.4).

A voluntary decision to resist the urge to defecate is eventually accompanied by relaxation of the circular muscle of the rectum, which accommodates the increased volume in the rectum. As wall tension relaxes, the urge to defecate subsides. Receptive relaxation of the rectum is accompanied by a return of contractile tension in the IAS, relaxation in the EAS, and increased pull by the puborectalis muscle sling. When this occurs, the feces remain in the rectum until the next mass movement further increases the rectal volume and again activates mechanoreceptors associated with stimulating defecation.

Motility disorders.

Motility-associated disorders (**dysmotility**) are not uncommon. Without a normal pattern of motility constantly moving along the intestinal tract in an aboral direction, the luminal contents move much slower than normal or they may even stop moving entirely. Motility disorders can occur anywhere along the GI tract from the esophagus through the intestines. The effects of motility disorders will depend upon the region affected. GERD results from a motility problem in the esophagus. Further down in the intestines, problems can range from the less severe (e.g., bloating) to more severe such as nutritional malabsorption leading to malnutrition. The causes of abnormal motility generally fall into two broad categories—myopathy and neuropathy. Chronic intestinal pseudoobstruction can be caused by changes in the regulation of muscle contraction leading to unsynchronized contractions (a **neuropathy**) or contractions may occur but they are weak (a **myopathy**). Irritable bowel syndrome affects large intestine motility.

Physiological ileus is the transient absence of motility (quiescence) in the intestinal tract and is a fundamental behavioral state of the intestine programmed by the ENS. Quiescence of the intestinal circular muscle occurs when inhibitory musculomotor neurons are continuously active and suppress the responsiveness of the circular muscle to the ever-present ESWs. Physiological ileus is a normal state that remains in effect for varying periods of time in different intestinal regions and is affected by factors such as the time after a meal. Ileus disappears after ablation of the ENS. When enteric neural functions are destroyed by pathologic processes, disorganized and nonpropulsive contractile behavior occurs continuously due to myogenic contraction.

The normal state of motor quiescence becomes pathologic when the gates of the peristaltic reflex circuit for the particular motor patterns are rendered inoperative for abnormally long periods. In this state of **pathological ileus** or **paralytic ileus**, the basic circuits are locked in an inoperable state, while unremitting activity of the inhibitory motor neurons continuously suppresses myogenic activity. Paralytic ileus may occur after abdominal surgery.

▶ SMOOTH MUSCLE CONTRACTION

The electrical stimuli for GI smooth muscle contraction may be **myogenic** (signal for contraction originates spontaneously within the smooth muscle itself) or **neurogenic** (contraction is stimulated by the arrival of a nerve impulse). Smooth muscles may also initiate contraction in responses to certain hormones and drugs.

Smooth muscles of the stomach and intestine contract spontaneously in the absence of neural or endocrine influence.

Smooth muscles are classified based on their behavioral properties and associations with nerves. The gastric and intestinal smooth muscle layer behaves like **unitary-type smooth muscle** and contraction is myogenic or a response to stretch. Neuromuscular junctions in the GI tract unitary-type smooth muscle are simpler than the well-structured motor endplates of skeletal muscle (see Chapter 8). Most motor axons in the unitary-type smooth muscle do not release neurotransmitters from terminals as such, but instead, release is from multiple varicosities, which are spaced out along the axon. The neurotransmitter released when an action potential depolarizes the membrane potential of the varicosity diffuses over relatively long distances before reaching the muscle and/or ICCs. This structural organization is an adaptation for the simultaneous application of a neurotransmitter to multiple muscle fibers from a small number of motor axons. Because the muscle fibers are connected by gap junctions, the membrane potential in expanded regions of the musculature can be depolarized to action potential threshold or hyperpolarized below spike threshold by neurotransmitter release from a reduced number of motor axons. The smooth muscle of the esophagus and gallbladder behaves like **multiunit-type smooth muscle**. These muscles contract only in response to nervous input. The nerve runs between the muscle cells and neurotransmitter released from varicosities along the nerve stimulates individual cells. These cells are not connected by gap junctions.

Electromechanical and pharmacomechanical coupling trigger contractions in GI muscles.

GI smooth muscle differs from skeletal muscle in having two mechanisms that initiate the processes leading to contractile shortening and development of tension. In both types of muscle cells, depolarization of the membrane electrical potential leads to the opening of voltage-gated Ca^{2+} channels, followed by the elevation of cytosolic Ca^{2+} and Ca^{2+} activation of contractile proteins. This mechanism is called **electromechanical coupling**. The second mechanism, **pharmacomechanical coupling**, does not occur in skeletal muscle. In this mechanism, the binding of ligands to G protein–coupled receptors on the smooth muscle cell membrane leads to the opening of Ca^{2+} channels and the elevation of cytosolic Ca^{2+} without any change in the membrane electrical potential. The ligands may be chemical substances released from nerves (**neurocrine signaling**), from nonneural cells in close proximity to the muscle (**paracrine signaling**), or from hormones from endocrine cells (see Chapter 1).

GI and esophageal smooth muscles have properties of a functional electrical syncytium.

Neighboring smooth muscle fibers are electrically connected by **gap junctions**, which are permeable to ions and thereby transmit electrical current from one muscle fiber to the next. Ionic connectivity, without cytoplasmic continuity from fiber to fiber, accounts for the **electrical syncytial** properties of smooth muscle just as it does in cardiac muscle (see Chapter 12). Electrical activity and the associated contractions spread from a point of initiation (e.g., a pacemaker region) in three dimensions throughout the bulk of the muscle. The ENS controls the distance and direction the electrical activity spreads in the electrical syncytium. A failure of nervous control can lead to disordered motility that includes spasm and associated cramping abdominal pain.

Inhibitory musculomotor activity determines the distance and direction a contraction travels within the functional electrical syncytium.

Inhibitory musculomotor neurons determine the length of a contracting segment by controlling the distance the action potentials spread within the muscular syncytium (Fig. 27.15). This occurs coincidently with control of contractile force. Contractions can only occur in the circular muscle layer of segments where ongoing inhibition has been inactivated. Contractions are prevented in adjacent segments where the

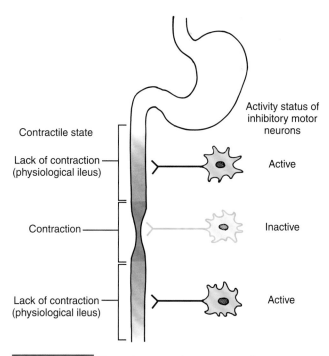

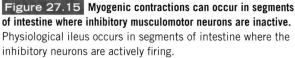

Figure 27.15 **Myogenic contractions can occur in segments of intestine where inhibitory musculomotor neurons are inactive.** Physiological ileus occurs in segments of intestine where the inhibitory neurons are actively firing.

inhibitory innervation is active. The oral and aboral circumferential boundaries of a contracted segment reflect the transition zone from inactive to active inhibitory musculomotor neurons. In this way, the ENS generates short contractile segments during the digestive (mixing) pattern of small intestinal motility and longer contractile segments during propulsive motor patterns, such as power propulsion that travels over extended distances along the intestine.

The directional sequence in which inhibitory musculomotor neurons are inactivated determines whether contractions propagate in the oral or aboral direction (Fig. 27.16). Normally, the neurons are inactivated sequentially in the aboral direction, and thus, contractile activity propagates and moves the intraluminal contents distally. During vomiting, the ENS inactivates inhibitory musculomotor neurons in a reverse sequence, allowing small intestinal propulsion to travel in the oral direction from the midjejunum and propel the contents toward the stomach.

Electrical activity in GI muscles consists of slow waves and action potentials.

ESWs are spontaneous, cyclic changes in the GI smooth muscle membrane potential and are initiated by ICCs. The waves are always present, and if the wave amplitude reaches the electrical threshold of the muscle (this does not occur with every wave), an action potential will be triggered. While ESWs trigger action potentials in some regions, in other regions (e.g., the gastric antrum and large intestinal circular muscle), they represent the only form of electrical activity (Fig. 27.17). In the small intestine, ESWs decrease in frequency along a gradient from the duodenum to the

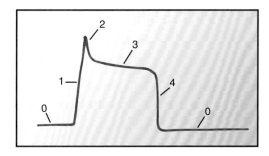

Figure 27.17 Electrical slow waves in gastrointestinal muscles occur in four phases determined by specific ionic mechanisms. Phase 0: Resting membrane potential; outward potassium current. Phase 1: Rising phase (upstroke depolarization); activation of voltage-gated Ca^{2+} channels and activation of voltage-gated K^+ channels. Phase 3: Plateau phase; balance of inward Ca^{2+} current and outward K^+ current. Phase 4: Falling phase (repolarization); inactivation of voltage-gated Ca^{2+} channels and activation of calcium-gated K^+ channels.

ileum. In the gastric antrum, the terms *slow wave* and *action potential* are used interchangeably for the same electrical event. When action potentials are associated with ESWs, they occur during the plateau phase of the slow wave.

Action potentials in GI tract smooth muscle are mediated by changes in conductance in Ca^{2+} and K^+ channels.

The depolarization phase of the action potential is produced by the inward movement of Ca^{2+} through voltage-gated L-type Ca^{2+} channels also called calcium–sodium channels. Repolarization occurs when K^+ channels open and K^+ moves outward as the Ca^{2+} channels are closing at or near the peak of the action potential (see Chapter 3). The L-type Ca^{2+} channels in GI smooth muscle are essentially the same as those found in cardiac and vascular smooth muscle. Therefore, disordered GI motility may be a side effect of treating cardiovascular disease with 1, 4-dihydropyridine drugs that block L-type Ca^{2+} channels.

Electrical slow waves may occur with or without action potentials in the small intestine.

As a general rule, ESWs in the circular muscle of the small intestine trigger action potentials that in turn initiate contractions. However, ESWs are not always accompanied by action potentials, and therefore, contractions will not occur (see Fig. 27.7A). With this method of recording, the size of an action potential appears larger when greater numbers of the total population of muscle fibers are depolarized to action potential threshold by each slow wave. The amplitude of phasic contractions associated with each ESW increases in direct relation to the number of muscle fibers recruited to firing threshold by each slow-wave cycle (see Fig. 27.7B).

Interstitial cells of Cajal generate electrical slow waves.

ICCs are present along the length of the GI tract and generate the ESWs observed in the stomach and intestinal tract. Gap junctions electrically interconnect the ICCs into

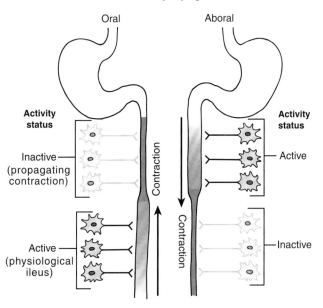

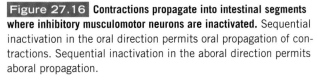

Figure 27.16 Contractions propagate into intestinal segments where inhibitory musculomotor neurons are inactivated. Sequential inactivation in the oral direction permits oral propagation of contractions. Sequential inactivation in the aboral direction permits aboral propagation.

networks, the ICCs to the circular muscle, and the smooth muscle fibers to one another. Gap junctions allow electrical current to flow from the ICC network to the circular muscle leading to the generation of an action potential followed by a contraction.

Pacemaker networks of ICCs surround the circular muscle of the small intestine at the border with the longitudinal muscle layer (myenteric border) and at the border with the submucosa. Traveling through the gap junctions, ESWs generated by the ICC network at the submucosal border spread passively into the bulk of the circular muscle, and those at the myenteric border spread passively into both longitudinal and circular muscle. The slow-wave electrical current then spreads from muscle fiber to muscle fiber.

Electrical slow-wave frequencies differ in the stomach, small intestine, and colon.

ESWs with similar waveforms occur at a frequency ranging from about 3 waves/min in the antrum, 11 to 12 waves/min in the duodenum, and 2 to 13 waves/min in the colon. Since each slow wave does not result in the generation of an action potential and muscle contraction, the contractile frequency of the muscle may be less than but never greater than the frequency of the ESWs. The ENS determines the nature of the contractile response during each slow wave.

▶ NEURAL CONTROL OF GUT MOTILITY AND DIGESTIVE FUNCTION

The digestive system is under ANS and CNS control. Sensory and motor neuron innervation of the digestive tract control muscle contraction, epithelial secretion and absorption, and blood flow and distribution inside the walls of the esophagus, stomach, intestines, and gallbladder. Sensory nerves transmit information from the gut to the brain accounting for sensations that are localized to the digestive tract (e.g., sensations of discomfort, abdominal pain, and chest pain or heartburn). Motor neurons are able to activate and inhibit muscle contraction. Secretion of H_2O, electrolytes, and mucus into the lumen and absorption from the lumen is affected by **secretomotor neurons** of the ENS.

CNS control of digestive functions involves two way communications with the GI system. Sensory neurons carry information from the gut to the brain and spinal cord and motor neurons carry information from the brain and spinal cord to the gut. Outflow may originate in higher processing centers of the brain (the frontal cortex) and accounts for the projection of an individual's emotional state (psychogenic stress) to the gut. This kind of brain–gut interaction underlies the symptoms of diarrhea and cramping lower abdominal pain sometimes reported by individuals during anticipation of a stressful life event (e.g., a difficult exam or job interview).

Neural integrative centers control the moment-to-moment motor activity of the gut.

The SNS, PNS, and ENS innervate the digestive tract. This comprehensive autonomic innervation of the digestive tract consists of interactive interconnections between the brain,

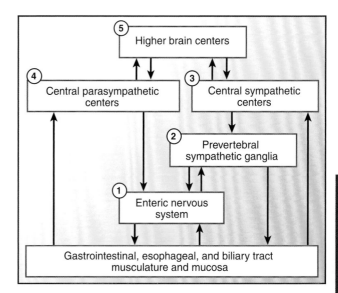

Figure 27.18 **A hierarchy of five levels of neural organization determines the moment-to-moment motor behavior of the digestive tract.** (See text for details.)

the spinal cord, and the ENS. Sympathetic and parasympathetic pathways carry autonomic signals to the gut from the brain and spinal cord and are the extrinsic component of innervation. Neurons of the ENS provide the intrinsic component of innervation.

Neural control of the gut is hierarchical, with five basic levels of integrative organization (Fig. 27.18). Level 1 is the ENS, which behaves like an independent integrative nervous system ("mini-brain") inside the walls of the gut. Level 2 consists of the prevertebral ganglia of the sympathetic nervous system. Levels 3, 4, and 5 are within the CNS. Sympathetic and parasympathetic signals to the digestive tract originate at levels 3 and 4 (central sympathetic and parasympathetic centers) in the medulla oblongata and represent the final common pathways for the outflow of information from the brain to the gut. Level 5 includes higher brain centers that provide input for integrative functions at levels 3 and 4.

Parasympathetic neurons innervate the gut from the medulla oblongata and sacral spinal cord.

PNS stimulation activates the entire ENS and also increases overall blood flow to the gut and general gut activity including increased secretion. The proximal half of the ENS is innervated from the cranial parasympathetic nerve fibers via the vagal nerve. The distal half is innervated via the sacral parasympathetic nerve fibers. The latter gives rise to a rich nerve supply to the sigmoid colon, rectum, and anus. The sacral parasympathetic pathway plays an important role in defecation. Figure 27.19 shows details of the parasympathetic innervation of the gut.

Efferent vagal nerves transmit signals to the ENS to control digestive processes both in anticipation of food intake and following a meal. These ENS circuits regulate gastric smooth muscle contraction and relaxation. Parasympathetic efferents to the wall of the small and large intestine are predominantly stimulatory and they excite musculomotor neurons.

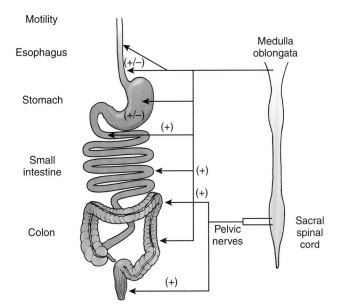

Figure 27.19 **The digestive tract is innervated by the parasympathetic division of the autonomic nervous system.** Signals from parasympathetic centers in the central nervous system are transmitted to the enteric nervous system by the vagus and pelvic nerves. These signals result in either contraction (+) or relaxation (−) of the GI tract musculature.

Dorsal vagal complex in the medulla controls the upper GI tract.

The **dorsal vagal complex** (vagal integrative center in the CNS) consists of the **dorsal motor nucleus** of the vagus, **nucleus tractus solitarius**, **area postrema**, and **nucleus ambiguus**. The center is more directly involved in controlling the specialized functions of the esophagus, stomach, duodenum, gallbladder, and pancreas than those of the distal small intestine and large intestine. The neural networks in the dorsal vagal complex and their interactions with higher centers are responsible for the rapid and precise control necessary for adjustments to rapidly changing conditions in the upper GI tract during anticipation, ingestion, and digestion of meals of varied composition.

Vagovagal reflex controls contractions of GI muscle layers in response to food stimuli.

The **vagovagal reflex** is constituted by afferent and efferent fibers of the vagus nerve that coordinate a response to gut stimuli. One function of the vagovagal reflex is to control GI tract smooth muscle contraction in response to distention of the tract by food; the reflex is activated to relax the stomach muscles in response to swallowing food. This reflex also allows for the accommodation of large amounts of food in the GI tract.

The sensory side of the reflex arc consists of vagal afferent neurons connected with a variety of sensory receptors specialized for the detection and signaling of mechanical parameters (e.g., muscle tension and mucosal brushing) or luminal chemical parameters (e.g., pH, osmolality, and glucose concentration). Afferent neural pathways connect the brain to neural networks in the ENS that innervate and control behavior of the musculature and secretory glands. Cell bodies of the vagal afferents are in **nodose ganglia**.

Efferent vagal fibers form synapses with ENS to activate outflow of signals from motor neurons to the effector systems. When the effector system is the musculature, its innervation consists of both inhibitory and excitatory musculomotor neurons that participate in reciprocal control. If the effector systems are gastric or digestive glands, the secretomotor neurons are excitatory and stimulate secretion.

The circuits for CNS control of the upper GI tract are organized much like those that control skeletal muscle movements and involve spinal reflex circuits (see Chapters 5 and 7). Inputs to spinal reflex circuits organize skeletal muscle contractile activity into functional motor behavior. The basic connections of the vagovagal reflex circuit are like somatic motor reflexes, in being "fine-tuned" from moment to moment by input from higher integrative centers in the brain.

Sympathetic nerves innervate gut blood vessels, mucosa, and muscularis and suppress gut activity when stimulated.

Sympathetic fibers exiting from the thoracic and upper lumbar regions of the spinal cord innervate the gut (Fig. 27.20).

Figure 27.20 **The sympathetic division of the autonomic nervous system innervates the digestive tract from the thoracic and lumbar segments of the spinal cord.** Preganglionic neurons of the sympathetic division of the autonomic nervous system project to the gut from thoracic and upper lumbar segments of the spinal cord. (See text for details.)

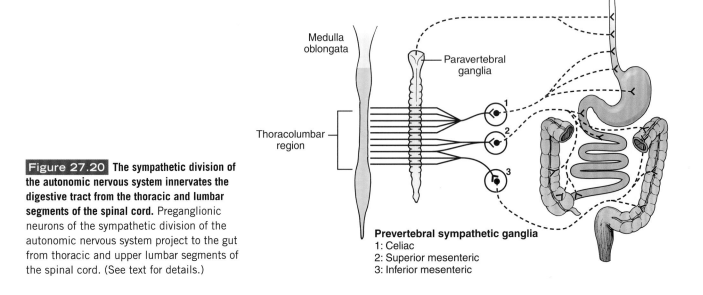

Prevertebral sympathetic ganglia
1: Celiac
2: Superior mesenteric
3: Inferior mesenteric

Efferent sympathetic fibers leave the spinal cord in the ventral roots to make their first synaptic connections with neurons in **prevertebral sympathetic ganglia** located in the abdomen. The prevertebral ganglia are the **celiac, superior mesenteric**, and **inferior mesenteric ganglia**. Cell bodies in the prevertebral ganglia project to the digestive tract, where they synapse with neurons of the ENS in addition to innervating the blood vessels, mucosa, and specialized regions of the musculature. The sympathetic neurons leaving the spinal cord are called **preganglionic sympathetic neurons**; the neurons leaving prevertebral ganglia to the gut are **postganglionic sympathetic neurons**.

CNS activation of sympathetic input to the GI tract shunts blood from the splanchnic to the systemic circulation during exercise and stressful environmental change. Sympathetic suppression of digestive functions, including motility and secretion, occurs as a coincident adaptation for reduced blood flow. **Norepinephrine** (NE) released from sympathetic postganglionic neurons is the main mediator of these effects and acts directly on the LES and IAS to increase tension and keep the sphincter closed. Presynaptic inhibitory action of NE at synapses in the ENS suppresses gastric and intestinal motility. Postsynaptic inhibitory action of NE at secretomotor neurons that innervate the intestinal secretory glands prevents secretion (see Fig. 27.3). Aside from smooth muscle sphincters, most sympathetic innervation goes to the ENS, not the musculature.

Paired splanchnic nerves innervate the gut and carry sensory information from and efferent sympathetic signals to the GI tract.

Splanchnic nerves are mixed nerves (i.e., have both sympathetic efferent and sensory afferent fibers) in the mesentery. They are not part of the SNS, and use of the term *sympathetic afferent* is incorrect. The cell bodies of the sympathetic innervation of the GI tract are positioned in the intermediolateral cell columns of the spinal cord descending from the first thoracic to the third lumbar segment. Efferent sympathetic fibers leave the spinal cord in the ventral roots to make synaptic connections with neurons in prevertebral sympathetic ganglia located in the abdomen (see Fig. 27.20).

Sensory afferent fibers in the splanchnic nerves have their cell bodies in dorsal root spinal ganglia (see Chapter 4). They transmit information from the GI tract and gallbladder to the CNS. These fibers bifurcate within the gut wall and transmit information to the ENS. The afferent fibers branch and send collaterals to form synapses with neurons in prevertebral sympathetic ganglia before projecting onto the spinal cord.

The gut sensory receptors include mechanoreceptors, chemoreceptors, thermoreceptors, and most likely nociceptors (pain receptors). Mechanoreceptors sense mechanical events in the mucosa, musculature, serosal surface, and mesentery. They supply both the ENS and the CNS with information on stretch-related tension and muscle length in the wall and on the movement of luminal contents as they come in contact with the mucosal surface. Mesenteric receptors sense gross movements of the organ. Chemoreceptors

monitor nutrient concentration, nutrient type (e.g., lipids), osmolality, and luminal pH. Most receptors have been found to be multimodal (i.e., respond to both mechanical and chemical stimuli). Thermoreceptors are located along the GI tract. Gastric reflexes are initiated in response to either warm or cold stimuli and intestinal reflexes have been shown to be stimulated by warm stimuli. The presence of GI nociceptors is likely but not unequivocally confirmed, except in the gallbladder. The sensitivity of splanchnic afferents, including nociceptors, may be elevated in cases of intestinal or gallbladder inflammation.

Sensory neurons, interneurons, and motor neurons form the microcircuits that integrate the ENS.

As they are in the CNS, the sensory neurons, interneurons, and motor neurons in the ENS are connected synaptically for the directed flow of information (see Fig. 27.3). Bidirectional communication occurs between the CNS and ENS, but the ENS organizes and coordinates the activity of each effector system. **Musculomotor neurons** control the musculature, and **secretomotor neurons** control the mucosal secretory glands. Interneurons form information-processing networks.

ENS afterhyperpolarization and synaptic-type neurons are distinguished by their electrophysiologic and synaptic behavior.

Two primary types of ENS neurons (i.e., **afterhyperpolarization** (AH)-type and **synaptic** (S)-type) are distinguished by their electrophysiological behavior and morphology.

AH-type neurons

AH-type neurons have multiple long processes, any of which might be an axon or a dendrite, and a long-lasting hyperpolarizing potential (i.e., an AH) that occurs following an action potential (Fig. 27.21). AH-type neurons fulfill the role of interneurons in the ENS. They comprise the largest proportion of myenteric plexus neurons and the smallest proportion of submucosal plexus neurons.

S-type neurons

S-type neurons have a single long axon with multiple short dendrites. S-type enteric neurons can be distinguished from AH-type neurons by the absence of an AH, lower resting membrane conductance, and greater excitability. Additionally, unlike AH-type neurons, application of the Na^+ channel blocker tetrodotoxin abolishes action potentials in most S-type neurons. S-type neurons are so named because nicotinic fast excitatory postsynaptic potentials (EPSPs) were found in virtually all S-type enteric neurons, whereas only restricted subpopulations of AH-type neurons express nicotinic receptors and fast EPSP-like depolarizing responses to the application of ACh.

All motor neurons in the myenteric and submucosal plexuses (i.e., musculomotor, secretomotor, and vasculomotor) are S-type neurons. Subpopulations of interneurons can also exhibit the properties of S-type neurons. Most neurons in the submucosal plexus are S-type neurons.

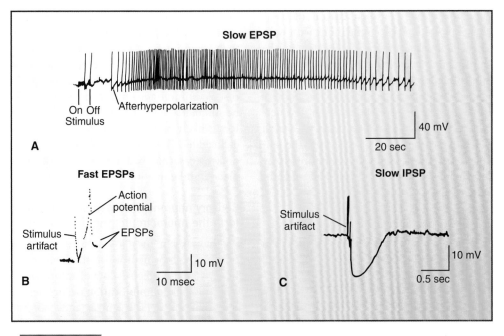

Figure 27.21 Fast excitatory postsynaptic potentials (EPSPs), slow excitatory postsynaptic potentials (EPSPs), and slow inhibitory postsynaptic potentials (IPSPs) are synaptic events in enteric neurons. **(A)** The slow EPSP was evoked by repetitive electrical stimulation of the synaptic input to the neuron. Slowly activating depolarization of the membrane potential continues for almost 2 minutes after the termination of the stimulus. During the slow EPSP, repetitive discharge of action potentials reflects enhanced neuronal excitability. **(B)** The fast EPSPs were also evoked by single electrical shocks applied to the axon that synapsed with the recorded neuron. Two fast EPSPs were evoked by successive stimuli and are shown as superimposed records. Only one of the EPSPs reached the threshold for the discharge of an action potential. The time course of the EPSPs is in the millisecond range. **(C)** The slow IPSP was also evoked by the stimulation of an inhibitory input to the neuron. This hyperpolarizing synaptic potential will suppress excitability (i.e., decrease the probability of action potential discharge), compared with enhanced excitability during the slow EPSP.

▶ SYNAPTIC TRANSMISSION IN THE ENTERIC NERVOUS SYSTEM

Multiple forms of synaptic transmission occur in the ENS, with both "fast" synaptic potentials with durations <50 ms and slow synaptic potentials lasting several seconds occurring. These synaptic events may be EPSPs or inhibitory postsynaptic potentials (IPSPs). Three forms of synaptic transmission can occur in enteric neurons (see Fig. 27.21).

Metabotropic receptors mediate enteric slow excitatory postsynaptic potentials.

Slow EPSPs can last from seconds to minutes. They are mediated by multiple chemical messengers acting at many different G protein–coupled **metabotropic receptors**. Metabotropic receptors characteristically activate a specific intracellular effector (e.g., an enzyme), which in the case of neurons leads to a change in the transmembrane potential. Different kinds of receptors, each of which mediates slow EPSP-like responses, are found in varied combinations on each neuron. Many neurotransmitter-binding metabotropic receptors activate second messenger pathways by activating intracellular cascades. Cyclic AMP (cAMP), diacylglycerol (DAG), inositol triphosphate (IP_3), and Ca^{2+} ions are common

second messengers. The most frequent mode of signal transduction for slow EPSPs in AH-type neurons involves receptor activation of the enzyme adenylyl cyclase, the generation of cAMP, and the phosphorylation of protein kinases and/or channel-associated proteins. Postreceptor signal transduction for slow EPSPs in S-type neurons involves activation of the enzyme phospholipase C, synthesis of IP_3 and DAG, and the elevation of free intraneuronal Ca^{2+}. This pathway also leads to the phosphorylation of channel-associated proteins. In some parts of the ENS, G protein–coupled receptors can directly activate ion channels by way of the G protein that dissociates from the receptors following binding of neurotransmitter. Serotonin, substance P, and ACh are examples of enteric neurotransmitters that evoke slow EPSPs.

In **paracrine signaling**, mediators released from nonneural cells evoke slow EPSP-like responses when released in the vicinity of the ENS. Histamine, for example, is released from mast cells during hypersensitivity reactions (e.g., food allergies and parasitic infections) and acts at the histamine H_2 receptor subtype to evoke slow EPSP-like responses in ENS neurons in rodents. Subpopulations of ENS neurons in specialized regions of the gut (e.g., the upper duodenum) have receptors for hormones such as gastrin and cholecystokinin, which also evoke slow EPSP-like responses.

Slow excitatory postsynaptic potentials prolong neural excitation or inhibition of GI effector systems.

The long-lasting discharge of action potentials during the slow EPSP can result in either prolonged excitation or inhibition at neuronal synapses and neuroeffector junctions.

Muscle contractile responses and glandular secretory responses are slow events and take place over several seconds. Figure 27.22 illustrates how the occurrence of slow EPSPs might evoke prolonged muscle contraction or prolonged secretion. Slow EPSPs in inhibitory musculomotor neurons result in prolonged inhibition of contractile force.

Ionotropic receptors mediate enteric fast excitatory postsynaptic potentials.

Fast EPSPs (see Fig. 27.21) are transient depolarizations of the membrane potential lasting <50 ms. They occur in the ENS throughout the digestive tract. Most fast EPSPs are mediated by ACh acting at **ionotropic receptors**. Ionotropic receptors are coupled directly to ion channels.

Multiple kinds of receptors mediate enteric slow inhibitory postsynaptic potentials.

Peptidergic, purinergic, noradrenergic, somatostatinergic, and cholinergic chemical messengers are examples of substances that evoke slow IPSP-like responses. Adenosine, ATP, or other purinergic analogues experimentally applied to myenteric plexus neurons also mimic slow IPSPs. Opioid peptides and opiate analgesics (morphine) are slow IPSP mimetics. This action is limited to subpopulations of neurons and is blocked by the antagonist naloxone. Addiction to

morphine may be seen in enteric neurons, and withdrawal is observed as a high-frequency spike discharge on the addition of naloxone during chronic morphine exposure.

NE released from sympathetic postganglionic neurons acts at α_2-adrenoreceptors to mimic slow IPSPs in S-type neurons in the submucosal plexus and primarily in secretomotor neurons. Slow IPSPs in submucosal neurons reflect a mechanism by which sympathetic innervation suppresses intestinal secretion during physical exercise, when the blood is shunted from the splanchnic to systemic circulation. Somatostatin released from ENS neurons in the submucosal plexus at synapses on secretomotor neurons mimics the inhibitory action of NE.

Presynaptic inhibition acts like a neurocrine transmitter in gut function.

Presynaptic inhibition (Fig. 27.23) plays a significant role at fast nicotinic synapses, at slow excitatory synapses, and at sympathetic inhibitory synapses in the ENS and at excitatory neuromuscular junctions. Presynaptic inhibition functions like a specialized form of neurocrine transmission. Presynaptic inhibition resulting from paracrine or endocrine

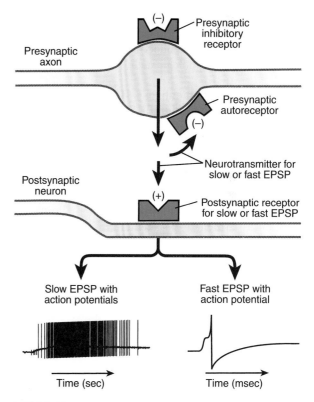

Figure 27.23 **Presynaptic inhibitory receptors are found on axons at neurotransmitter release sites for both slow and fast excitatory postsynaptic potentials (EPSPs).** Different neurotransmitters act at presynaptic inhibitory receptors to suppress release of the transmitters for slow and fast EPSPs. Presynaptic autoreceptors are involved in a special form of presynaptic inhibition whereby the transmitter for slow or fast EPSPs accumulates at the synapse and acts on the autoreceptor to suppress further release of the neurotransmitter. (+), excitatory receptor; (−), inhibitory receptor.

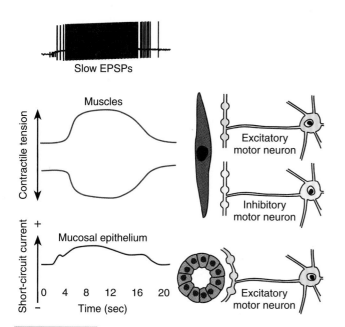
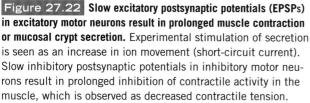

Figure 27.22 **Slow excitatory postsynaptic potentials (EPSPs) in excitatory motor neurons result in prolonged muscle contraction or mucosal crypt secretion.** Experimental stimulation of secretion is seen as an increase in ion movement (short-circuit current). Slow inhibitory postsynaptic potentials in inhibitory motor neurons result in prolonged inhibition of contractile activity in the muscle, which is observed as decreased contractile tension.

mediators binding receptors at presynaptic release sites is an alternative to axoaxonal synapses as a mechanism for enhancing or suppressing transmission at a synapse or neuroeffector junction.

Presynaptic inhibition in the ENS is mediated by multiple chemical mediators and their receptors. Among the known mediators are peptides (cholecystokinin), purinergic substances (ATP), amines (histamine), and ACh. After its release from sympathetic nerve terminals, NE acts at α_2-presynaptic adrenoreceptors to suppress fast EPSPs at nicotinic synapses and slow EPSPs and cholinergic transmission at neuromuscular junctions. A major component of shutdown of gut function by the SNS involves the presynaptic inhibitory action of NE at fast nicotinic synapses. Serotonin suppresses both fast and slow EPSPs in the myenteric plexus. Opiates or opioid peptides suppress some fast EPSPs in the intestinal myenteric plexus.

ACh acts at muscarinic presynaptic receptors to suppress fast EPSPs in the myenteric plexus. ACh released at nicotinic synapses acts in an autoinhibitory way and suppresses ACh release by presynaptic neurons in negative feedback fashion (see Fig. 27.23). Histamine is an example of paracrine-mediated presynaptic suppression in the ENS. After its release from enteric mast cells, histamine acts at histamine H_3 presynaptic receptors to suppress fast EPSPs in the ENS microcircuits of rodents. Presynaptic inhibition mediated by paracrine or endocrine release of mediators is significant in pathophysiologic states (e.g., inflammation).

Presynaptic facilitation enhances the amounts of neurotransmitter released at ENS axon sites.

Presynaptic facilitation refers to an enhancement of synaptic transmission (Fig. 27.24). The phenomenon enhances the amplitude of fast EPSPs in the myenteric plexus of the small intestine and gastric antrum and at noradrenergic inhibitory synapses in the submucosal plexus. Cholecystokinin affects the ENS of the gallbladder in the same way. Presynaptic facilitation is evident experimentally as an increase in amplitude of fast EPSPs at nicotinic synapses and reflects enhanced release of ACh. At noradrenergic inhibitory synapses in the submucosal plexus, the elevation of cAMP in the postganglionic sympathetic fiber appears as an enhancement of slow IPSPs.

Presynaptic facilitation is the mechanism of action of some **prokinetic drugs** (e.g., tegaserod and cisapride), which therapeutically enhance propulsive motility in the GI tract by increasing the amplitude of fast nicotinic EPSPs associated with propulsive motor functions. In the stomach and the intestine, increases in EPSP amplitudes and rates of rise decrease the probability of synaptic transmission failure, thereby increasing the speed of information transfer and enhancing propulsive motility (i.e., gastric emptying and intestinal transit).

▶ ENTERIC MOTOR NEURONS

Both excitatory and inhibitory musculomotor neurons innervate the muscles of the digestive tract and, like spinal motor neurons, are the final pathways for signal transmission from the ENS to the musculature (see Figs. 27.3 and 27.25). Neurotransmitters released from motor neurons act at the neuromuscular junction with smooth muscle fibers and ICCs.

Excitatory musculomotor neurons activate smooth muscle contraction in the gut.

Excitatory musculomotor neurons release neurotransmitters that evoke contraction and increased contractile force in GI smooth muscles. ACh and **substance P** are the main excitatory neurotransmitters released from ENS musculomotor neurons.

Two mechanisms of excitation–contraction coupling are involved in the neural initiation of muscle contraction in the GI tract. Transmitters released from excitatory musculomotor axons can trigger muscle contraction by depolarizing the muscle fiber membrane to the threshold for the discharge of action potentials. Neurally evoked depolarizations of the muscle membrane potential, called **excitatory junction potentials** (**EJPs**), can be distinguished from EPSPs and IPSPs (see Fig. 27.25). In the second mechanism, the activation of G protein–coupled receptors is linked to the direct release of Ca^{2+} from stores inside the muscle fiber (i.e., pharmacomechanical coupling) that triggers contractions independent of any changes in membrane electrical activity (i.e., EJPs or action potentials).

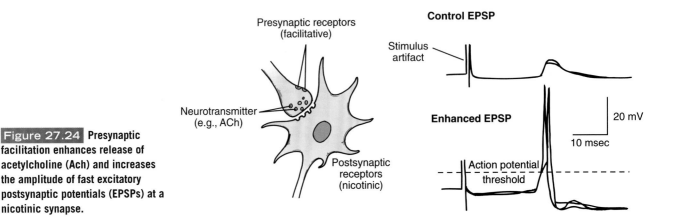

Figure 27.24 Presynaptic facilitation enhances release of acetylcholine (Ach) and increases the amplitude of fast excitatory postsynaptic potentials (EPSPs) at a nicotinic synapse.

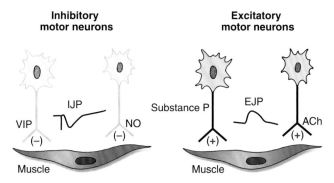

Figure 27.25 Enteric musculomotor neurons are final pathways from the enteric nervous system (ENS) to the gastrointestinal musculature. The motor neuron pool of the ENS consists of both excitatory and inhibitory neurons. Release of vasoactive intestinal peptide (VIP) or nitric oxide (NO) from inhibitory motor neurons evokes inhibitory junction potentials (IJPs). Release of acetylcholine (ACh) or substance P from excitatory motor neurons evokes excitatory junction potentials (EJPs).

The cell bodies of excitatory musculomotor neurons are generally found in the myenteric plexus. In the small and large intestine, they project their axons over relatively short distances to innervate the longitudinal muscle and for longer distances away from the cell body in the oral direction to innervate the circular muscle.

Excitatory motor neurons, also called secretomotor neurons, innervate mucosal secretory glands (i.e., the crypts of Lieberkühn and Brunner glands) and stimulate secretion of NaCl, HCO_3^-, and H_2O in addition to mucus from goblet cells. ACh and vasoactive intestinal peptide (VIP) are the principal excitatory neurotransmitters released by secretomotor neurons. Hyperactivity of these neurons, like that evoked by histamine release from enteric mast cells during allergic responses, can lead to **neurogenic secretory diarrhea**. Suppression of secretomotor neuronal excitability (e.g., by morphine or other opioid analgesics) can result in constipation.

Inhibitory junction potentials diminish gut smooth muscle excitability.

Neurotransmitters released from inhibitory musculomotor neurons activate receptors on the muscle cell membranes to generate **inhibitory junction potentials (IJPs)** (see Fig. 27.25). The transmembrane hyperpolarization that occurs during IJPs prevents depolarization to the action potential threshold by the ever-present ESWs and suppresses propagation of action potentials.

Early evidence supported a purine nucleotide, possibly ATP, as the transmitter released by ENS inhibitory musculomotor neurons, which led to the term **purinergic neuron**. Evidence now exists that VIP, pituitary adenylate cyclase–activating peptide, and nitric oxide (NO) also act as ENS inhibitory transmitters. Inhibitory musculomotor neurons that express VIP and/or NO synthase (NOS) innervate the circular muscle of the stomach, intestine, gallbladder, and various sphincters.

The longitudinal muscle layer of the small intestine appears not to have a significant inhibitory musculomotor

innervation, but there is significant inhibitory musculomotor innervation of the taeniae coli of the large intestine in humans. In contrast to the longitudinal muscle, inhibitory innervation of intestinal circular muscle is essential for ENS programming of intestinal motility.

An absolute requirement for inhibitory neural control of the circular muscle layer is a demand that emerges from the specialized physiology of the musculature (see earlier discussion on electrical syncytial properties). A nonneural pacemaker system of ESWs (i.e., ICCs) accounts for the self-excitable characteristic of the electrical syncytium and the spontaneous contractile activity that characterizes unitary-type smooth muscle. In the integrated system, the ESWs are a separate extrinsic factor to which the circular muscle responds in addition to its motor innervation.

Why does the circular muscle not always respond with action potentials and contract during every slow-wave cycle? Why do action potentials and contractions not spread in the syncytium throughout the entire length and circumference of the intestine each time they occur? The short answer is that inhibitory neuronal motor activity determines when a slow wave can evoke a contraction and determines the distance and direction over which a contraction spreads within the muscular syncytium.

Inhibitory neurotransmitters block phasic contractions in the gut.

Figure 27.26A shows the spontaneous discharge of action potentials occurring in bursts along a neuron in the myenteric plexus of the small intestine of a cat. This kind of continuous discharge of action potentials by subsets of intestinal inhibitory musculomotor neurons occurs in all mammals. The outcome is continuous inhibition of myogenic activity, because in intestinal segments in which neuronal discharge in the myenteric plexus is prevalent, slow wave–associated muscle action potentials and contractile activity are absent or the contractions occur with reduced force with each ESW. Continuous release of VIP and NO has been shown to occur in these kinds of intestinal states. When the inhibitory neuronal discharge is blocked experimentally with tetrodotoxin, every ESW cycle triggers an intense discharge of muscle action potentials associated with a forceful phasic contraction superimposed on a tonic contracture (i.e., a sustained increase in baseline tension). Figure 27.26B shows how phasic contractions, occurring at slow-wave frequency, progressively increase to maximal amplitude during progressive blockade of inhibitory neural activity after the application of tetrodotoxin in the small intestine. This response coincides with a progressive increase in baseline tension.

Tetrodotoxin selectively blocks neural activity without affecting the muscle, which makes it a valuable tool for demonstrating ongoing inhibition of contractile activity. Tetrodotoxin acts by selectively blocking neuronal Na^+ channels. The rising phase of circular muscle action potentials reflects opening of Ca^{2+} channels and inward current that is unaffected by tetrodotoxin (see Chapters 3 and 8).

As a general rule, any treatment or condition that removes or inactivates inhibitory musculomotor neurons

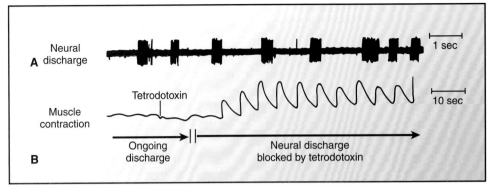

Figure 27.26 Ongoing firing of a subpopulation of inhibitory musculomotor neurons to the intestinal circular muscle prevents electrical slow waves from triggering the action potentials that trigger contractions. When the inhibitory neural discharge is blocked with tetrodotoxin, every cycle of the electrical slow wave triggers discharge of action potentials and large-amplitude contractions. **(A)** Electrical record of ongoing burst-like firing of an inhibitory motor neuron. **(B)** Record of muscle contractile activity before and after application of tetrodotoxin. Application of tetrodotoxin-blocked neuronal firing.

results in tonic contracture and continuous, uncoordinated contraction of the circular musculature. Absence of inhibitory musculoneuronal activity is associated with conversion from a hypoirritable to a hyperirritable state of the circular muscle as occurs, for example, following the application of local anesthetics, hypoxia from restricted blood flow to an intestinal segment, an autoimmune attack that destroys enteric neurons, congenital absence in Hirschsprung disease, treatment with opiate drugs, and inhibition of the synthesis of NO.

Force of gut smooth muscle contraction is directly related to the activation of the inhibitory neurons.

The force of a circular muscle contraction evoked by a slow-wave cycle is a function of the number of inhibitory musculomotor neurons in an active state. The circular muscle in an intestinal segment can respond to ESWs only when the inhibitory musculomotor neurons are turned off by inhibitory synaptic input from other neurons in the control circuits. This means that inhibitory musculomotor neurons determine when the continuously running slow waves initiate a contraction as well as the force of the contraction, which is determined by the number of muscle fibers responding. With maximum inhibition, no contractions are permitted during a slow-wave cycle (see Fig. 27.7A); with no inhibition, contractions of maximum strength occur since all of the muscle fibers in a segment can be activated (see Fig. 27.7B). Intermediate strength contractions are graded in force according to the number of inhibitory musculomotor neurons that are inactivated during each slow-wave cycle.

Motor behavior of the antral pump consists of leading and trailing contractile components triggered by gastric action potentials.

Gastric action potentials are initiated by a dominant pacemaker (believed to be gastric ICCs) located in the corpus distal to the midregion and determine the duration and strength of the phasic contractions of the antral pump. Once initiated,

the action potentials propagate rapidly and spread through the gastric electrical syncytium, traveling around the gastric circumference and triggering a ringlike contraction, which then travels more slowly toward the gastroduodenal junction. The pacemaker region generates action potentials and associated antral contractions at a frequency of 3 contractions/min. The gastric action potential lasts about 5 seconds and has a rising phase (depolarization), a plateau phase, and a falling phase (repolarization) (see Fig. 27.17).

The propulsive contractile behavior in the antral pump has two components; an antral *leading contraction*, of relatively constant amplitude, is associated with the rising phase of the action potential, and an antral *trailing contraction*, of variable amplitude, is associated with the plateau phase (see Fig. 27.27). The gastric pacemaker continuously generates action potentials. Nevertheless, trailing contractions only appear when the plateau phase is at or above threshold, and they increase in strength in direct relation to increases in the amplitude of the plateau potential above threshold.

The leading contractions produced by the rising phase of the gastric action potential have negligible amplitude as they propagate to the pylorus. As the rising phase reaches the terminal antrum and spreads into the pylorus, contraction of the pyloric muscle closes the orifice between the stomach and duodenum. As the trailing contraction, which follows the leading contraction by only a few seconds, approaches the closed pylorus, the gastric contents are forced into an antral compartment of ever-decreasing volume and progressively increasing pressure. This results in jetlike retropulsion through the orifice formed by the trailing contraction (Fig. 27.28). Trituration and reduction in particle size occur as the material is forcibly retropulsed through the advancing orifice and back into the gastric reservoir to wait for the next propulsive cycle. Repetition, at 3 cycles/min, reduces particle size to the 1- to 7-mm range that is necessary before a particle can be emptied into the duodenum during the digestive phase of gastric motility.

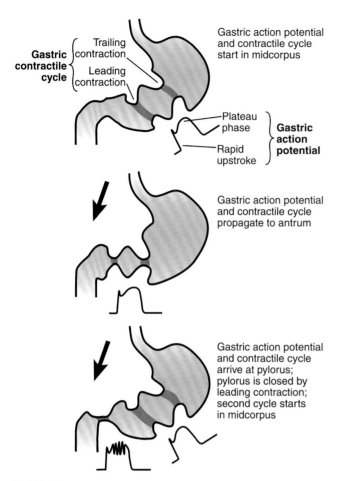

Figure 27.27 **A pacemaker in the antral pump generates gastric action potentials that evoke ringlike contractions as they propagate to the gastroduodenal junction.** Gastric action potentials are characterized by an initial rapidly rising upstroke followed by a plateau phase and then a falling phase back to the baseline membrane potential (see Fig. 27.17). The rising phase of the gastric action potential accounts for a leading contraction that propagates toward the pylorus during one propulsive cycle. The plateau phase accounts for the trailing contraction of the cycle. The strength of the leading contraction is relatively constant; the strength of the trailing contraction is variable and increases in direct relation to neurally evoked increases in amplitude of the plateau phase of the action potential.

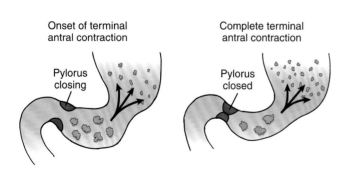

Figure 27.28 **Jetlike retropulsion through the orifice of the antral contraction triturates solid particles in the stomach.** The force for retropulsion is increased pressure in the terminal antrum as the trailing antral contraction approaches the closed pylorus.

Action potentials in the antral pump are modulated by musculomotor neurons in the gastric ENS.

Action potentials in the antral pump are **myogenic** and modulated by gastric ENS musculomotor neurons. Neurotransmitters, including ACh, released from excitatory musculomotor neurons increase the amplitude of both the plateau phase and the strength of the contraction initiated by the plateau. Inhibitory neurotransmitters including NE and VIP have the opposite effect.

The magnitude of the effects is directly related to the concentration of neurotransmitter present in the neuromuscular junction; an increase in the frequency of action potentials causes a corresponding increase in the amount of neurotransmitter. In this way, the firing of musculomotor neurons determines whether or not a trailing antral contraction occurs. With sufficient release of transmitter, the plateau amplitude grows and starts a contraction as it crosses the depolarization threshold. Beyond threshold, the strength of contraction is determined by the amount of neurotransmitter released and present at receptors on the muscle, which in itself determines the extent of membrane depolarization beyond threshold.

The action potentials in the terminal antrum and pylorus differ somewhat in configuration from those in the more proximal regions. The principal difference is the occurrence of spike potentials on the plateau phase (see Fig. 27.27), which trigger short-duration phasic contractions superimposed on the phasic contraction associated with the plateau. These may contribute to the sphincteric function of the pylorus in preventing the reflux of duodenal contents back into the stomach.

Motor behavior of the gastric reservoir differs from the gastric antrum.

The gastric reservoir has two primary functions. One is to accommodate the arrival of a meal, without a significant increase in intragastric pressure and distention inside the reservoir. Failure of this mechanism leads to the sensations of bloating, epigastric pain, and nausea or *dyspepsia*. The second is to maintain constant compressive forces on the reservoir contents, which act to push the contents into the 3-cycles/min motor activity of the antral pump. Agents that relax the musculature of the gastric reservoir (e.g., insulin) neutralize this function and thereby suppress gastric emptying.

The musculature of the gastric reservoir is innervated by both excitatory and inhibitory musculomotor neurons in the ENS (see Figs. 27.3 and 27.25), and their activity is regulated by efferent vagal nerves. Input from these nerves adjusts the volume and pressure of the reservoir to the amount of solid and/or liquid present while sustaining constant compressive forces on the contents. Continuous adjustments are required during both the ingestion and the emptying of a meal.

An increase in activity of excitatory musculomotor neurons, in concert with decreased activity of inhibitory musculomotor neurons, results in increased contractile tone in the reservoir, a decrease in its volume, and an increase in intraluminal pressure (see Fig. 27.9). An increase in activity of inhibitory musculomotor neurons in concert with decreased activity of excitatory musculomotor neurons produces the opposite set of effects.

Three types of relaxation responses occur in the gastric reservoir.

Neurally mediated decreases in muscle tonic contracture are responsible for initiating gastric reservoir relaxation (i.e., increased volume). The act of swallowing initiates **receptive relaxation**, which is a reflex triggered by stimulation of mechanoreceptors in the pharynx. Distention of the gastric reservoir initiates **adaptive relaxation**, which is a vagovagal reflex that is triggered by stimulation of stretch receptors in the gastric wall (Fig. 27.29). The presence of nutrients in the small intestine initiates **feedback relaxation**. It can involve both local reflex connections between receptors in the small intestine and the gastric ENS or hormones that are released from endocrine cells in the small intestinal mucosa.

Adaptive relaxation is lost in patients who have undergone a vagotomy or it can occur as an iatrogenic result of fundoplication surgery for the treatment of an incompetent LES and acid reflux disease. Following a vagotomy, increased tone in the reservoir musculature decreases the wall compliance, which in turn changes the response of gastric stretch receptors to distention of the reservoir (Fig. 27.30). The loss of adaptive relaxation after a vagotomy is associated with a lowered threshold for sensations of epigastric fullness and pain. These effects of vagotomy may explain disordered gastric sensations in diseases with a component of vagal nerve pathology (e.g., autonomic neuropathy of diabetes mellitus).

Motor patterns in the stomach and small intestine reflect the presence and absence of intraluminal nutrients.

The small intestinal ENS "runs" the **digestive state** motor program when nutrients are present and digestive processes are ongoing. Conversion to the **interdigestive state** motor

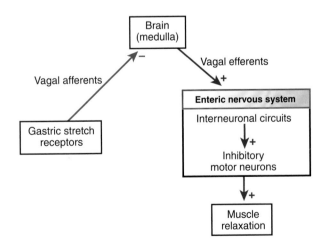

Figure 27.29 Adaptive relaxation in the gastric reservoir is a vagovagal reflex in which information from gastric stretch receptors is the afferent arm of the reflex and outflow from the medullary region of the brain is the efferent component. Vagal efferents transmit to the enteric nervous system, which controls the activity of inhibitory motor neurons that relaxes contractile tone in the reservoir.

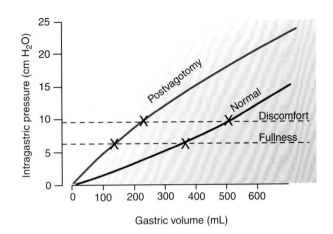

Figure 27.30 Adaptive relaxation of the gastric reservoir is lost following a vagotomy. Loss of adaptive relaxation in the gastric reservoir is associated with a lowered threshold for sensations of fullness and epigastric pain.

program starts when digestion and absorption of nutrients are complete, 2 to 3 hours after a meal. The motility pattern of the interdigestive state in the small intestine is called the **migrating motor complex** (**MMC**). The MMC starts as large forceful contractions at 3 contractions/min in the antral pump. Elevated contractile force in the LES coincides with the onset of the MMC in the stomach, which then migrates into the duodenum and down the small intestine to the ileum (Fig. 27.31).

At a single recording site in the small intestine, the MMC consists of three consecutive phases:

- Phase I: motor silence without contractile activity; corresponds to physiologic ileus
- Phase II: irregularly occurring contractions
- Phase III: regularly occurring contractions

Phase I returns after phase III, and the cycle repeats at the single recording site after 80 to 120 minutes (see Fig. 27.31). With multiple sensors positioned along the intestine, slow propagation of the phase II and phase III activity down the intestine becomes evident.

At a given time, the MMC occupies a limited length of intestine called the **activity front**, which has an oral and aboral boundary (Figs. 27.31 and 27.32). The activity front slowly advances along the intestine at a rate that progressively slows as it approaches the ileum. Peristaltic propulsion of luminal contents occurs in the aboral direction between the oral and aboral boundaries of the activity front. The frequency of the peristaltic propulsive waves within the activity front is the same as the frequency of ESWs in that intestinal segment. Each peristaltic wave consists of propulsive and receiving segments, as described earlier (see Fig. 27.4). Successive peristaltic waves start on average slightly farther in the aboral direction and propagate on average slightly beyond the boundary where the previous one stopped. Therefore, the entire activity front slowly advances down the intestine, sweeping the lumen clean as it goes.

"Phase II" and "phase III" are commonly used terms that have minimal value for understanding the MMC.

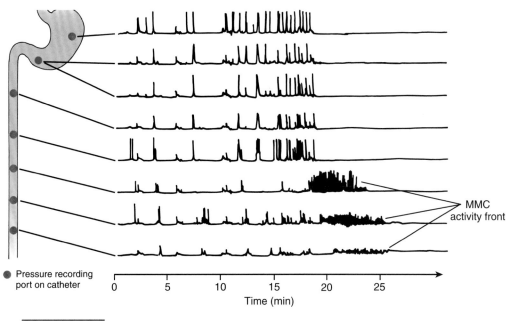

Figure 27.31 **The small intestinal migrating motor complex (MMC) consists of an activity front that starts in the gastric antrum and slowly migrates through the small intestine to the ileum.** Repetitive contractions, which reflect peristaltic propulsion, occur within the activity front.

Contractile activity described as phase II or phase III occurs because of the irregularity of the arrival of peristaltic waves at the aboral boundary of the activity front. On average, each consecutive peristaltic wave within the activity front propagates farther in the aboral direction than the previous wave. Nevertheless, at the lower boundary of the activity front, some waves terminate early and others travel farther

(see Fig. 27.32). Therefore, as the lower boundary of the front passes the recording point, only the waves that reach the sensor are recorded, giving the appearance of irregular contractions. As propagation continues and the midpoint of the activity front reaches the recording point, the propulsive segment of every peristaltic wave is detected. Because the peristaltic waves occur with the same rhythmicity as the

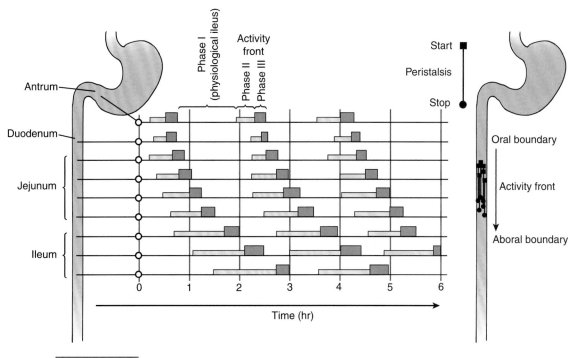

Figure 27.32 **The migrating motor complex (MMC) consists of three phases.** Phase I of physiological ileus refers to the intestinal region with no activity. Phases II and III reflect the migrating activity front, which has an aboral boundary where propulsive contractions stop and an oral boundary where propulsive contractions start. (See text for details.)

ESWs, the contractions are described as being "regular." The regular contractions, which are seen when the central region of the front passes a single recording site, last for 8 to 15 minutes. This time is shortest in the duodenum and progressively increases as the MMC migrates toward the ileum. The MMC occurs in conscious states and during sleep. It starts in the antrum of the stomach as an increase in the strength of the regularly occurring antral contractile complexes and accomplishes the emptying of indigestible particles (e.g., pills and capsules) larger than 7 mm. Eighty to one-hundred and twenty minutes is required for the activity front of the MMC to travel from the antrum to the ileum. As one activity front terminates in the ileum, another begins in the antrum. The time between cycles is longer during the day than that at night. The activity front travels at about 3 to 6 cm/min in the duodenum and progressively slows to about 1 to 2 cm/min in the ileum. It is important not to confuse the speed of travel of the activity front of the MMC with that of the ESWs, action potentials, and peristaltic waves within the activity front. ESWs with associated action potentials and associated contractions of circular muscle travel about 10 times faster.

Cycling of the MMC continues until it is ended by the physical presence of a meal in the upper digestive tract. A sufficient nutrient load terminates the MMC. Intravenous feeding does not stop the MMC. The speed with which the MMC is terminated, wherever it is along the intestine, suggests a neural or hormonal mechanism. Gastrin and cholecystokinin, both of which are released during a meal, terminate the MMC in the stomach and upper small intestine, but not in the ileum, when injected intravenously.

The ENS organizes the MMC (see Fig. 27.3). It continues in the small intestine after a vagotomy or sympathectomy but stops when it encounters a region of the intestine where the ENS has been interrupted. Presumably, command signals to the ENS are necessary for initiating the MMC, but whether the commands are neural, hormonal, or both remains unknown. Although levels of the hormone **motilin** increase in the blood at the onset of the MMC, it is unclear whether motilin is the trigger or is released as a consequence of MMC occurrence.

MMMC acts as "housekeeper" for the small intestine.

Gallbladder contraction and delivery of bile to the duodenum are coordinated with the movement of the MMC into the antroduodenal region. The activity front of the MMC propels the bile from the duodenum through the jejunum to the terminal ileum, where it is reabsorbed into the hepatic portal circulation. This mechanism minimizes the accumulation of concentrated bile in the gallbladder and increases the movement of bile acids in the enterohepatic circulation during the interdigestive state (see Chapter 25).

The MMC also appears to be a mechanism for clearing indigestible debris from the intestinal lumen during the fasting state. Large indigestible particles are emptied from the stomach only during the MMC.

Bacterial overgrowth in the small intestine is associated with an absence of the MMC. This condition suggests that the MMC might play a "housekeeper" role in preventing the overgrowth that would occur if the intestinal contents were allowed to stagnate in the lumen.

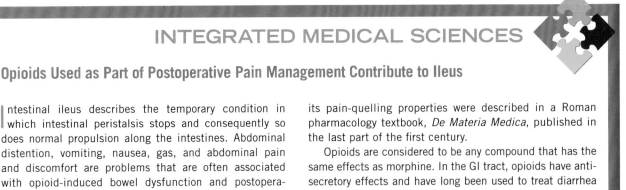

INTEGRATED MEDICAL SCIENCES

Opioids Used as Part of Postoperative Pain Management Contribute to Ileus

Intestinal ileus describes the temporary condition in which intestinal peristalsis stops and consequently so does normal propulsion along the intestines. Abdominal distention, vomiting, nausea, gas, and abdominal pain and discomfort are problems that are often associated with opioid-induced bowel dysfunction and postoperative ileus. The occurrence of postoperative ileus does not appear to be specifically linked to the type of surgery. Ileus is a frequent postoperative problem in patients who have had bowel resection or other procedures that involve abdominal surgery, but ileus can also be observed in patients who have undergone nonabdominal surgeries such as cardiac or orthopedic procedures. This indicates that ileus is not linked to the direct manipulation of the intestine. However, surgical patients do share a common thread and that is the pharmacological management of postoperative pain. This management involves the use of opioids such as morphine and codeine. In fact, opium was used to manage pain by the early Greeks and Romans, and

its pain-quelling properties were described in a Roman pharmacology textbook, *De Materia Medica*, published in the last part of the first century.

Opioids are considered to be any compound that has the same effects as morphine. In the GI tract, opioids have antisecretory effects and have long been used to treat diarrhea (e.g., paregoric, codeine). Opioid receptors are in a family of G protein–coupled receptors. There are three main types of opioid receptors, mu (μ), delta (κ), and kappa (δ). *In vivo*, these receptors normally bind enkephalins, endorphins, and dynorphins, which are naturally occurring opioid peptides. This is one mechanism for *in vivo* modulation of nociception (i.e., pain pathways). Endogenous opioids suppress intestinal motility in certain situations such as during an inflammatory response and a stress response. Exogenously delivered opioids provide analgesic modulation of pain. The μ receptors are the main group to which postoperative prescribed opioids bind. This binding initiates the pathway leading to the manifestation of the clinical effects of these drugs.

Within the CNS, opioids contribute to analgesia, sedation, and the alteration of mood. In the CNS, opioids affect the sympathetic connections, both excitatory and inhibitory motor pathways, to the ENS. This input from the CNS modifies the normal ENS regulation of GI. In the GI tract, opioids act directly in the ENS, which has μ receptors on cells located in the myenteric and submucosal plexuses. The combined end result of activating nervous system opioid receptors is an inhibition of secretion, a decrease in aboral motility, and an inhibition of peristaltic reflexes. This leads to opioid-induced constipation.

Naloxone acts as a μ receptor antagonist to reverse the effects of opioids by overcoming the blocking action opioids have on the peristaltic reflex. However, the drug does not appear to equally impact transit times along the length of the intestine. While it does accelerate colonic transit time, it has no similar affect in the small intestine. Naloxone does effectively restore peristaltic muscle contraction and thus propulsion along the intestinal tract. A drawback to naloxone administration is that it can decrease the analgesic effects of opioids. In clinical investigations, methylnaltrexone was investigated as a possible alternative to naloxone. Methylnaltrexone helps to return gut function while not blocking the analgesic benefit provided by opioid administration. Another alternative, alvimopan, has been shown to have similar postoperative benefits, particularly in patients who have undergone bowel resection. Alvimopan is administered prior to surgery to block opiate binding. The therapeutic benefits of alvimopan show a dose-dependent response. The use of stool softeners and laxatives is generally ineffective at combating opioid-associated constipation. Mild osmotic agents have also been tried.

Opioid receptors are not limited to the ENS and are also present on the surface of intestinal smooth muscle cells. Opioids can affect motility along the GI tract by stimulating a change in the muscle contraction pattern from the normal electrical slow wave–generated contractions driving movement of the luminal contents aborally to a contraction pattern that results in nonpropulsive motility and stagnation. The absence of peristaltic propulsion increases transit time through the intestine. In addition to effecting longitudinal muscle activity, the drugs also affect circular muscle and intestinal segmentation, overall smooth muscle tone, and therefore contraction in intestinal smooth muscle sphincters. Tonus in the pyloric and anal sphincters is increased and relaxation of the LES is inhibited. While the antegrade movement of the luminal contents is delayed, there is a concurrent dehydration of the lumen as a result of increased fluid absorption that is coupled with decreased secretion of both water and electrolytes.

Postoperative ileus is an example of a condition that has no one specific cause, but rather, it is a physiological condition that occurs when several contributing factors interact. The nervous system and various hormones (e.g., vasoactive intestinal peptide and substance P) as well as localized inflammatory responses and mediators such as prostaglandins are thought to play a role in postoperative ileus. In general, GI function and coordinated electrical activity do not return uniformly to the GI tract following postoperative ileus. Initially, during the recovery period, the sympathetic nervous system is highly active and the electrical activity along the tract is disorganized and erratic. Signals from the sympathetic nervous system inhibit activity in the ENS. In the absence of normal electrical patterns, regular peristalsis and propulsion do not occur. It usually takes 3 to 4 days for stomach and small intestine to return to normal activity. The colon, which is the region of the GI tract most affected by anesthetics due to the absence of gap junctions, is often the last section to return to normal. Total recovery of the GI tract from ileus can take as long as 4 to 5 days. ■

Chapter Summary

- The musculature of the digestive tract is mainly smooth muscle.
- Electrical slow waves and action potentials are the primary forms of electrical activity in the gastrointestinal (GI) musculature.
- GI smooth muscles have properties of a functional electrical syncytium.
- A hierarchy of neural integrative centers in the brain, spinal cord, and periphery determines moment-to-moment behavior of the digestive tract.
- The digestive tract is innervated by the sympathetic, parasympathetic, and enteric divisions of the autonomic nervous system.
- Vagus nerves transmit afferent sensory information to the brain and parasympathetic autonomic efferent signals to the digestive tract.
- Splanchnic nerves transmit sensory information to the spinal cord and sympathetic autonomic efferent signals to the digestive tract.
- The enteric nervous system functions as an independent mini-brain in the gut.
- Fast and slow excitatory postsynaptic potentials, slow inhibitory postsynaptic potentials, presynaptic inhibition, and presynaptic facilitation are key synaptic events in the enteric nervous systems.
- Enteric musculomotor neurons may be excitatory or inhibitory.
- Enteric inhibitory musculomotor neurons to the intestinal circular muscle are continuously active and transiently inactivated to permit muscle contraction.

- Enteric inhibitory musculomotor neurons to the musculature of sphincters are inactive and transiently activated for the timed opening and passage of luminal contents.
- A polysynaptic reflex circuit determines the behavior of the intestinal musculature during peristaltic propulsion.
- Physiological ileus is the normal absence of contractile activity in the intestinal musculature.
- Peristaltic propulsion and relaxation of the lower esophageal sphincter are the main motility events in the esophagus.
- The gastric reservoir and antral pump have different kinds of functional motor behavior.
- Vagovagal reflexes are important in the control of gastric motor functions.
- Feedback signals from the duodenum determine the rate of gastric emptying.
- The migrating motor complex is the small intestinal motility pattern of the interdigestive state.
- Mixing movements are the small intestinal motility pattern of the digestive state.
- Intestinal power propulsion is a protective response to harmful agents.
- Motor functions of the large intestine are specialized for storage and dehydration of feces.
- Physiologic functions of the rectosigmoid region, anal canal, and pelvic floor musculature are responsible preserving fecal continence.

Chapter Review Questions

1. A mouse with a new genetic mutation is discovered not to have electrical slow waves in the small intestine. Identify which of the following cell types was most likely affected by the mutation.

   A. Enteric neurons
   B. Inhibitory motor neurons
   C. Mechanoreceptor
   D. Interstitial cells of Cajal
   E. Enteroendocrine cells

   The correct answer is D. Interstitial cells of Cajal are the pacemaker cells that generate electrical slow waves. Enteric neurons, inhibitory motor neurons, mechanoreceptors, and enteroendocrine cells do not generate electrical slow waves.

2. Examination of the properties of a normal sphincter in the digestive tract will show that:

   A. primary flow across the sphincter is unidirectional.
   B. the lower esophageal sphincter is relaxed at the onset of a migrating motor complex in the stomach.
   C. blockade of the sphincteric innervation by a local anesthetic causes the sphincter to relax.

   D. the manometric pressure in the lumen of the sphincter is less than the pressure detected in the lumen on either side of the sphincter.
   E. the inhibitory motor neurons to the sphincter muscle stop firing during a swallow.

   The correct answer is A. Sphincters function to prevent reflux; therefore, flow across a sphincter is generally unidirectional. Tone in the lower esophageal sphincter is increased during the MMC in the stomach. The sphincter cannot be relaxed after blockade of the inhibitory innervation by a local anesthetic. Pressure in the sphincter is higher than in the two compartments it separates. Inhibitory neurons fire to relax the sphincter during a swallow.

3. An 86-year-old female complains of a compromised lifestyle because of fecal incontinence. Examination of this patient will most likely reveal the underlying cause of the incontinence to be:

   A. absence of the rectoanal reflex.
   B. elevated sensitivity to the presence of feces in the rectum.
   C. loss of the enteric nervous system in the distal large intestine (adult Hirschsprung disease)

D. weakness in the puborectalis and external anal sphincter muscles.

E. a myopathic form of chronic pseudoobstruction in the large intestine.

The correct answer is D. Examination of elderly patients often reveals weakness in the pelvic floor musculature. Weakness in the puborectalis muscle allows the anorectal angle to straighten and lose its barrier function to the passage of feces into the anorectum. The rectoanal reflex (i.e., relaxation of the internal anal sphincter in response to distention of the rectum) does not weaken significantly in the elderly. A deficit in sensory detection, not elevated sensitivity, can be a factor in fecal incontinence. Adult Hirschsprung disease results in constipation, not incontinence. The myopathic form of pseudoobstruction is not associated with fecal incontinence because propulsive motility is absent due to weakening of the intestinal smooth muscle.

4. On a return visit after receiving a diagnosis of functional dyspepsia, a 35-year-old female reports sensations of early satiety and discomfort in the epigastric region after a meal. These symptoms are most likely due to:

A. malfunction of adaptive relaxation in the gastric reservoir.

B. elevated frequency of contractions in the antral pump.

C. an incompetent lower esophageal sphincter.

D. premature onset of the interdigestive phase of gastric motility.

E. bile reflux from the duodenum.

The correct answer is A. As the gastric reservoir fills during ingestion of a meal, mechanoreceptors signal reservoir volume to the brain via vagal afferents. When the limits of adaptive relaxation in the reservoir are reached, signals from the stretch receptors in the reservoir's walls account for the sensations of fullness and satiety. Overdistention is perceived as discomfort. Adaptive relaxation malfunctions in the forms of functional dyspepsia characterized by the symptoms described in this question. When adaptive relaxation is compromised (e.g., by an enteric neuropathy), mechanoreceptors are activated at lower distending volumes and the brain wrongly interprets the signals as if the gastric reservoir were full. Elevated frequency of contractions in the antral pump, an incompetent lower esophageal sphincter, premature onset of the interdigestive phase of gastric motility, or bile reflux from the duodenum would not be expected to activate mechanosensory signaling of the state of fullness of the gastric reservoir.

Clinical Application Exercises 27.1

HEARTBURN

A 67-year-old woman experienced heartburns after eating. The heartburn was most severe after a large meal or eating a meal with a high-fat diet. Recently, she started having heartburns after her morning coffee, after taking her vitamins, or in the evening after drinking a glass of wine. At nights, she would wake up coughing with a burning sensation in her throat. She often experienced wheezing after these episodes. At her next visit with her internist, she mentioned her symptoms. She informed her internist that the burning sensation had been going on for the last 5 years but seemed manageable. She said things improved after she started

sleeping on her side instead of her stomach. However, in the last year, the symptoms have gotten worse. After further examination, she was referred to a gastroenterologist for further tests. After a detailed history, a diagnostic endoscopic procedure was performed. The patient was sedated and a thin scope was inserted through the mouth and throat into the esophagus and stomach in order to assess the internal surface of the esophagus, stomach, and duodenum. The endoscopic examination revealed that the patient was suffering from gastroesophageal reflux disease (GERD) with early stages of Barrett esophagus.

QUESTIONS

1. What is the pathophysiology of GERD and what are the contributing factors?
2. What is contributing to the woman's wheezing?
3. What is the relationship between GERD and Barrett esophagus?

ANSWERS

1. Gastroesophageal reflux disease is defined as chronic mucosal damage produced by the abnormal reflux of gastric contents into the esophagus. In healthy patients, the angle of His, the angle at which the esophagus enters the stomach, is intact creating a valve that prevents duodenal bile, enzymes, and gastric acid from traveling back into the esophagus where it can cause burning and inflammation of the sensitive

esophageal tissue. In GERD patients, acid reflux is caused by a failure of the antireflux barrier and its primary component, the gastroesophageal valve.

There are a number of factors that contribute to GERD. Certain type of foods and life styles are considered to promote gastroesophageal reflux. Coffee, alcohol, and large doses of vitamin C supplements are stimulants to gastric acid secretion. Taking these before bedtime compounds the problem of acid reflux. Chocolate and eating spicy foods also contribute to acid reflux. Lifestyles also contribute to gastric acid reflux. Being overweight, eating large meals, foods high in fat, and smoking reduce lower esophageal sphincter competence. Fat also delays the emptying of the stomach. Eating shortly before bedtime also contributes acid reflux.

2. Wheezing usually occurs at night but can occur during the day. Muscles around the esophagus go into spasm with certain postures (slouched or sleeping on ones stomach). As a result, there is no straight path between the stomach and the esophagus. This causes gas and acid to get blocked in the spasm causing coughing and asthma-like symptoms (e.g., wheezing). In severe cases, contents of the acid reflux get into the airway and irritate the lining that induces coughing and wheezing.

3. About 10% of patients who have GERD are diagnosed with Barrett esophagus. Barrett esophagus is a condition in which cells lining the lower esophagus become metastatic. Barrett esophagus is considered to be a premalignant condition and is associated with high risk of esophageal cancer. At present, there is no clinical test to determine which patient with Barrett esophagus become cancerous.

thePoint® *Visit* http://thepoint.lww.com/rhoades5e *for additional chapter review Q&A, Clinical Application Exercises, animations, and more!*

28 Regulation of Body Temperature

Active Learning Objectives

Upon mastering the material in this chapter, you should be able to:

- Explain how the body produces heat and exchanges energy with the environment.
- Explain how sweating, skin blood flow, and metabolic heat production help regulate body temperature.
- Explain how the thermoregulatory set point varies cyclically and is elevated during fever.

- Explain the reflex control of physiologic thermoregulatory responses.
- Explain temperature alterations during exercise in various environments.
- Explain the process of acclimatization to hot and cold environments.
- Explain the adverse effects of excessive heat stress, heat injury, and excessive cold stress.

Humans are warm-blooded animals or **homeotherms**. Like other mammals, humans are able to regulate their internal body temperatures within a narrow range near 37°C, despite wide variations in environmental temperature (Fig. 28.1). In contrast, internal body temperatures of **poikilotherms**, or cold-blooded animals, are governed by environmental temperature. The range of temperatures that living cells and tissues can tolerate without harm extends from just above freezing to nearly 45°C—far wider than the limits within which homeotherms regulate body temperature. What biologic advantage do homeotherms gain by maintaining a stable body temperature? As we shall see, tissue temperature is important for two reasons.

First, temperature extremes injure tissue directly. High temperatures alter the three-dimensional structure of protein molecules, even though the sequence of amino acids remains unchanged. Such alteration of protein structure is called **denaturation**. Because the biological activity of a protein molecule depends on its configuration and charge distribution, denaturation inactivates a cell's proteins and injures or kills the cell. Injury occurs at tissue temperatures higher than about 45°C, which is also the point at which heating the skin becomes painful. The severity of injury depends on the temperature to which the tissue is heated and its duration.

Excessive cold can also injure tissues. As a water-based solution freezes, ice crystals consisting of pure water form and all dissolved substances in the solution are left in the unfrozen liquid. Therefore, as more ice forms, the remaining liquid becomes more and more concentrated. Freezing damages cells through two mechanisms. First, ice crystals mechanically injure the cell. The increase in solute concentration of the cytoplasm as ice forms denatures the proteins by removing their water of hydration, increasing the ionic strength of the cytoplasm and causing other changes in the physicochemical environment in the cytoplasm.

Second, temperature changes profoundly alter biologic function through specific effects on such specialized functions as electrochemical properties and fluidity of cell membranes and through a general effect on most chemical reaction rates. In the physiologic temperature range, most reaction rates vary approximately as an exponential function of temperature (T); increasing T by 10°C increases the reaction rate by a factor of two to three. For any particular reaction, the ratio of the rates at two temperatures 10°C apart is called the Q_{10} for that reaction, and the effect of temperature on reaction rate is called the Q_{10} effect. The notion of Q_{10} may be generalized to apply to a group of reactions that have some measurable overall effect (such as O_2 consumption) in common and, thus, are thought of as comprising a physiologic process. The Q_{10} effect is clinically important in managing patients who have high fevers and are receiving fluid and nutrition intravenously. A commonly used rule is that a patient's fluid and caloric needs are increased 10% to 13% above normal for each 1°C of fever.

The sluggishness of a reptile that comes out of its burrow in the morning chill and becomes active only after being warmed by the sun illustrates the profound effect of temperature on biochemical reaction rates. Homeotherms avoid such a dependence of metabolic rate on environmental temperature by regulating their internal body temperatures within a narrow range. A disadvantage of homeothermy is that, in most homeotherms, certain vital processes cannot function at low levels of body temperature that poikilotherms tolerate easily.

▶ BODY TEMPERATURE AND HEAT TRANSFER

The body is divided into a warm internal core and a cooler outer shell (Fig. 28.2). Because the environment greatly influences the shell temperature, its temperature is not regulated

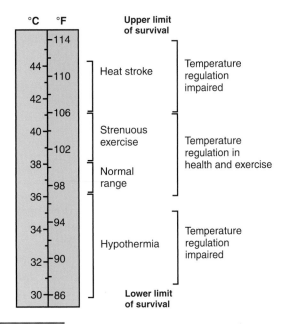

Figure 28.1 Survival ranges of body temperatures in humans.

within narrow limits as is the internal core. This is true despite the thermoregulatory responses that strongly affect the shell, especially its outermost layer, the skin. The thickness of the shell depends on the thermal environment and the body's need to conserve heat. In a warm environment, the shell may be <1 cm thick, but in a subject conserving heat in a cold environment, it may extend several centimeters below the skin. The regulated internal body temperature is the temperature of the vital organs inside the head and

trunk, which, together with a variable amount of other tissue, comprise the warm internal core.

Heat is produced in all tissues of the body but is lost to the environment only from tissues in contact with the environment—predominantly from the skin and, to a lesser degree, from the respiratory tract. Therefore, we need to consider heat transfer within the body, especially heat transfer (1) from major sites of heat production to the rest of the body and (2) from the core to the skin. Heat is transported within the body by two means: **conduction** through the tissues and **convection** by the blood, whereby flowing blood carries heat from warmer tissues to cooler tissues.

Heat flow by conduction varies directly with the thermal conductivity of the tissues, the change in temperature over the distance the heat travels, and the area (perpendicular to the direction of heat flow) through which the heat flows. It varies inversely with the distance the heat must travel. As Table 28.1 shows, the tissues are rather poor heat conductors.

Heat flow by convection depends on the rate of blood flow and the temperature difference between the perfused tissue and its blood supply. Because the vessels of the microvasculature have thin walls and, collectively, a large total surface area, the blood temperature equilibrates with that of the surrounding tissue before it reaches the capillaries. Changes in skin blood flow in a cool environment change the thickness of the shell. When skin blood flow is reduced in a cold environment, the affected skin becomes cooler and the underlying tissues—which in the cold may include most of the limbs and the more superficial muscles of the neck and trunk—become cooler as they lose heat by conduction to the cool overlying skin and, ultimately, to the environment. In this way, these underlying tissues, which in a hot environment were part of the body core, now become part of the shell.

Because the shell lies between the core and the environment, all heat leaving the body core, except that which is lost through the respiratory tract, must pass through the shell before being lost to the environment. Thus, the shell insulates the core from the environment. In a cool subject,

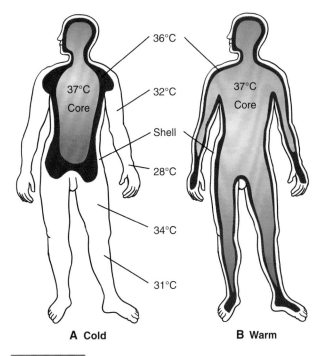

Figure 28.2 Distribution of temperatures in the body's core and shell. **(A)** During exposure to cold. **(B)** In a warm environment. Because the temperatures of the surface and the thickness of the shell depend on environmental temperature, the shell is thicker in the cold and thinner in the heat.

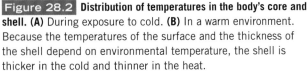

Material	Conductivity (kcal/s·m·°C)	kcal/h	W
Copper	0.09	33,120	38,474
Epidermis	0.00005	18	21
Dermis	0.00009	32	38
Fat	0.00004	14	17
Muscle	0.0001	40	46
Oak (across grain)	0.00004	14	17
Fiberglass insulation	0.00001	3.6	4.2

*Values are calculated for slabs 1 m² in area and 1 cm thick, with a 1°C temperature difference between the two faces of the slab.

skin blood flow is low, so conduction dominates core-to-skin heat transfer; the shell is also thicker, providing more insulation to the core, because heat flow by conduction varies inversely with the distance the heat must travel. Changes in skin blood flow, which directly affect core-to-skin heat transfer by convection, also indirectly affect conductive core-to-skin heat transfer by changing the thickness of the shell. In a cool subject, the subcutaneous fat layer contributes to the insulation value of the shell because the fat layer increases the thickness of the shell and because fat has conductivity about 0.4 times that of dermis or muscle (see Table 28.1). Thus, fat is a correspondingly better insulator. In a warm subject, however, the shell is relatively thin and provides little insulation. Furthermore, a warm subject's skin blood flow is high, and so convection dominates heat flow from the core to the skin. In these circumstances, the subcutaneous fat layer, which affects conduction but not convection, has little effect on heat flow from the core to the skin.

Core temperature approximates central blood temperature.

Core temperature varies slightly from one site to another, depending on such local factors as metabolic rate, blood supply, and the temperatures of adjacent tissues. However, temperatures at different places in the core are all similar to the temperature of the central blood and tend to change together. The notion of a single, uniform core temperature, though not strictly correct, is a useful approximation. The value of 98.6°F, often given as the "normal" body temperature, may give the misleading impression that body temperature is regulated so precisely that it deviates less than a few tenths of a degree. In fact, 98.6°F is simply the Fahrenheit equivalent of 37°C, and body temperature does vary somewhat (see Fig. 28.1). The effects of heavy exercise and fever are familiar; variation among people and such factors as time of day and method of measurement (Fig. 28.3), phase of the menstrual cycle, and acclimatization to heat can also cause differences of up to about 1°C in resting core temperature.

To maintain core temperature within a narrow range (36.5°C to 37.5°C), the thermoregulatory system needs

continuous information about the level of core temperature. Temperature-sensitive neurons and nerve endings in the abdominal viscera, great veins, spinal cord, and especially, the brain provide this information. We consider how the thermoregulatory system processes and responds to this information later in the chapter.

Core temperature should be determined at a site where the measurement is not biased by environmental temperature. Clinically used sites include the rectum, the mouth, and, occasionally, the axilla. The rectum is well insulated from the environment; its temperature is independent of environmental temperature and is a few tenths of 1°C warmer than arterial blood and other core sites. The tongue is richly supplied with blood; oral temperature under the tongue is usually close to blood temperature (and 0.4°C to 0.5°C below rectal temperature), but cooling the face, neck, or mouth can make oral temperature misleadingly low. If a patient holds his or her upper arm firmly against the chest to close the axilla, axillary temperature will eventually come reasonably close to core temperature. However, because this may take 30 minutes or more, axillary temperature is used infrequently. Infrared ear (aural) thermometers are convenient and widely used in the clinic, but temperatures of the tympanum and external auditory meatus are loosely related to more accepted indices of core temperature, and ear temperature in collapsed hyperthermic runners may be 3°C to 6°C below rectal temperature.

Skin is an important organ in heat exchange and thermoregulatory control.

Most heat is exchanged between the body and the environment at the skin surface. Consequently, skin temperature is much more variable than core temperature; thermoregulatory responses such as skin blood flow and sweat secretion, the temperatures of underlying tissues, and environmental factors, such as air temperature, air movement, and thermal radiation, affect it. Skin temperature is one of the major factors in heat exchange with the environment. For these reasons, it provides the thermoregulatory system with important information about the need to conserve or dissipate heat.

Because skin temperature usually is not uniform over the body surface, mean skin temperature ($\overline{T}_{sk}$) is frequently calculated from temperatures at several skin sites, weighting each temperature according to the fraction of body surface area it represents. $\overline{T}_{sk}$ is used to summarize the input to the central nervous system (CNS) from temperature-sensitive nerve endings in the skin. $\overline{T}_{sk}$ is also commonly used, along with core temperature, to calculate a mean body temperature and to estimate the quantity of heat stored in the body because the direct measurement of shell temperature would be difficult and invasive.

Peripheral thermoreceptors in the skin code absolute and relative temperatures.

Cutaneous thermosensation is mediated by various primary afferent nerve fibers that transduce, encode, and transmit thermal information. These peripheral **thermoreceptors** are located in the skin and in the oral and urogenital mucosa

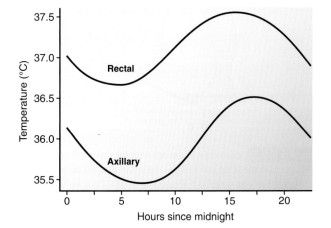

Figure 28.3 Core and axillary temperatures are influenced by circadian rhythm.

and include warm-sensitive and cold-sensitive populations. The most numerous and most superficially situated thermosensors are cold-sensitive. Skin cold sensors are located in or immediately beneath the epidermis, and their afferent input is conveyed by thin myelinated Aδ-fibers. The less common warm-sensitive receptors (10-fold fewer than cold sensors) are located slightly deeper in the dermis, and their input travels via unmyelinated C fibers. In contrast to central thermosensitive neurons, peripheral thermosensors are not pacemakers. The mechanisms of peripheral thermosensitivity involve changes in the resting membrane potential. Importantly, the response of most peripheral thermosensors shows a powerful dynamic (phasic) component. These neurons are very active when the temperature changes but quickly adapt when the temperature stabilizes, enabling rapid reaction to thermal environmental changes.

The mechanism of thermal transduction is believed to involve the action of **transient receptor potential (TRP) ion channels** whose activity depends on the temperature of their environment. Of these, members of three families, the vanilloid TRP channels, the melastatin (or long) TRP channels, and the ankyrin transmembrane protein channels are of particular interest as thermoreceptors. Each of these receptors operates over a specific temperature range, thereby providing a potential molecular basis for peripheral thermosensation. These specialized thermal receptors are embedded in the membranes of afferent fiber terminals as free nerve endings in the skin. Activation of all thermo TRP channels results in an inward, nonselective cationic current and a consequent depolarization of the resting membrane potential. Although individual thermo TRP channels are activated within a relatively narrow temperature range, collectively, their range is quite broad, from noxious cold to noxious heat (see Fig. 28.4).

Thermotransduction in these nerve endings by TRP-independent means is also likely. Experimental evidence indicates cold sensing through cold-induced inhibition of K^+ conductances that are normally active at resting membrane potential, leading to membrane depolarization and action

potential generation, whereas warm sensing may involve the otherwise mechanosensitive K_2P channels and other nociceptive C fibers.

▶ BALANCE BETWEEN HEAT PRODUCTION AND HEAT LOSS

All animals exchange energy with the environment. Some energy is exchanged as mechanical work, but most is exchanged as heat (Fig. 28.5). Heat is exchanged by conduction, convection, and radiation and as latent heat through evaporation or (rarely) condensation of water. If the sum of energy production and energy gain from the environment does not equal energy loss, the extra heat is "stored" in, or lost from, the body. This relationship is summarized in the following heat balance equation:

$$M = E + R + C + K + W + S \qquad (1)$$

where M is metabolic rate; E is the rate of heat loss by evaporation; R and C are rates of heat loss by radiation and

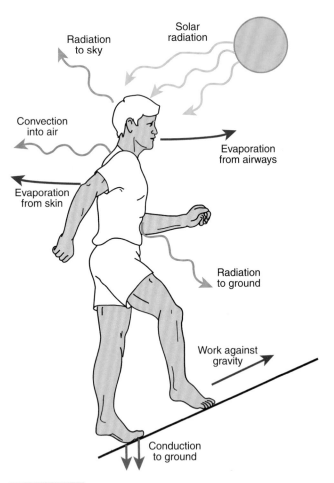

Figure 28.5 Exchange of energy with the environment. This hiker gains heat from the sun by radiation and loses heat by conduction to the ground through the soles of his feet, convection into the air, radiation to the ground and sky, and evaporation of water from his skin and respiratory passages. In addition, some of the energy released by his metabolic processes is converted into mechanical work, rather than heat, because he is walking uphill.

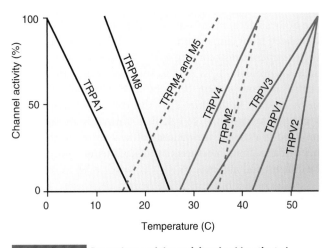

Figure 28.4 Dependence of the activity of cold-activated (*black*) and heat-activated (*green*) thermoTRP channels on temperature. TRP, transient receptor potential.

convection, respectively; K is the rate of heat loss by conduction; W is the rate of energy loss as mechanical work; and S is the rate of heat storage in the body, manifested as changes in tissue temperatures.

M is always positive, but the terms on the right side of equation 1 represent energy exchange with the environment and storage and may be either positive or negative. E, R, C, K, and W are positive if they represent energy losses from the body and negative if they represent energy gains. When S = 0, the body is in heat balance and body temperature neither rises nor falls. When the body is not in heat balance, its mean tissue temperature increases if S is positive and decreases if S is negative. This situation commonly lasts only until the body's responses to the temperature changes are sufficient to restore balance. However, if the thermal stress is too great for the thermoregulatory system to restore balance, the body will continue to gain or lose heat until either the stress diminishes sufficiently or death occurs.

The traditional units for measuring heat are a potential source of confusion because the word calorie refers to two units differing by a 1,000-fold. The *calorie* used in chemistry and physics is the quantity of heat that will raise the temperature of 1 g of pure water by 1°C; it is also called the *small calorie* or *gram calorie*. The *Calorie* (capital C) used in physiology and nutrition is the quantity of heat that will raise the temperature of 1 kg of pure water by 1°C; it is also called the *large calorie, kilogram calorie*, or (the usual practice in thermal physiology) the **kilocalorie** (kcal). Because heat is a form of energy, it is now often measured in joules, the unit of work (1 kcal = 4,186 J), and rate of heat production or heat flow in watts, the unit of power (1 W = 1 J/s). This practice avoids confusing calories and Calories. However, kilocalories are still used widely enough that it is necessary to be familiar with them, and there is a certain advantage to a unit based on water because the body itself is mostly water.

▶ METABOLIC RATE AND HEAT PRODUCTION AT REST

Metabolic energy is used for active transport via membrane pumps, for energy-requiring chemical reactions, such as the formation of glycogen from glucose and proteins from amino acids, and for muscular work. Most of the metabolic energy used in these processes is converted into heat within the body. Other energy is converted to heat only after a delay, as when the energy used in forming glycogen or protein is released as heat when the glycogen is converted back into glucose or the protein is converted back into amino acids.

Among subjects of different body sizes, metabolic rate at rest varies approximately in proportion to body surface area. In a resting and fasting young adult man, it is about 45 W/m² (81 W or 70 kcal/h for 1.8 m² body surface area), corresponding to an O_2 consumption of about 240 mL/min. About 70% of energy production at rest occurs in the body core—trunk viscera and the brain—even though it comprises only about 36% of the body mass (Table 28.2). As a by-product of their metabolic processes, these organs produce most of the heat needed to maintain heat balance at

TABLE 28.2	Relative Masses and Metabolic Heat Production Rates during Rest and Heavy Exercise		
		Percentage of Heat Production	
Region	**Percentage of Body Mass**	**Rest**	**Exercise**
Brain	2	16	1
Trunk viscera	34	56	8
Muscle and skin	56	18	90
Other	8	10	1

comfortable environmental temperatures; only in the cold must heat produced expressly for thermoregulation supplement such by-product heat.

Factors other than body size that affect metabolism at rest include age, sex (Fig. 28.6), hormones, and digestion. The ratio of metabolic rate to surface area is highest in infancy and declines with age, most rapidly in childhood and adolescence and more slowly thereafter. Children have high metabolic rates in relation to surface area because of the energy used to synthesize the fats, proteins, and other tissue components needed to sustain growth. Similarly, a woman's metabolic rate increases during pregnancy to supply the energy needed for the growth of the fetus. However, a nonpregnant woman's metabolic rate is 5% to 10% lower than that of a man of the same age and surface area probably because a higher proportion of the female body is composed of fat, a tissue with low metabolism.

The catecholamines and thyroxine are the hormones that have the greatest effect on metabolic rate. Catecholamines cause glycogen to break down into glucose and stimulate many enzyme systems, increasing cellular metabolism. Hypermetabolism is a clinical feature of some cases of pheochromocytoma, a catecholamine-secreting tumor

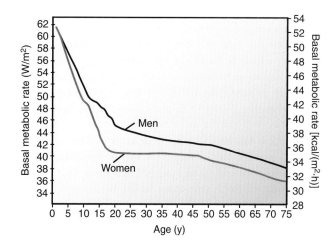

Figure 28.6 Effects of age and sex on the basal metabolic rate of healthy subjects. Metabolic rate here is expressed as the ratio of energy consumption to body surface area.

of the adrenal medulla. Thyroxine magnifies the metabolic response to catecholamines, increases protein synthesis, and stimulates oxidation by the mitochondria. The metabolic rate is typically 45% above normal in hyperthyroidism (but up to 100% above normal in severe cases) and 25% below normal in hypothyroidism (but 45% below normal with complete lack of thyroid hormone). Other hormones have relatively minor effects on metabolic rate.

A resting person's metabolic rate increases 10% to 20% after a meal. This effect of food, called the **thermic effect of food** (formerly known as *specific dynamic action*), lasts several hours. The effect is greatest after consuming protein and less after consuming carbohydrate and fat; it appears to be associated with processing the products of digestion in the liver.

Metabolic rate is the amount of energy expended under standard conditions.

Because so many factors affect metabolism at rest, metabolic rate is often measured under a set of standard conditions to compare it with established norms. Metabolic rate measured under these conditions is called **basal metabolic rate (BMR)**. The commonly accepted conditions for measuring BMR are that the person must have fasted for 12 hours; the measurement must be made in the morning after a good night's sleep, beginning after the person has rested quietly for at least 30 minutes; and the air temperature must be comfortable, about 25°C (77°F). BMR is "basal" only during wakefulness because the metabolic rate during sleep is somewhat lower than the BMR.

Heat exchange with the environment can be measured directly by using a human calorimeter. In this insulated chamber, heat can exit either in the air ventilating the chamber or in water flowing through a heat exchanger in the chamber. By measuring the flow of air and water and their temperatures as they enter and leave the chamber, one can determine the subject's heat loss by conduction, convection, and radiation, and by measuring the moisture content of air entering and leaving the chamber, one can determine heat loss by evaporation. This technique is called **direct calorimetry**, and though conceptually simple, it is cumbersome and costly.

Metabolic rate is often estimated by **indirect calorimetry**, which is based on measuring a person's rate of O_2 consumption because virtually all energy available to the body depends ultimately on reactions that consume O_2. Consuming 1 L of O_2 is associated with releasing 21.1 kJ (5.05 kcal) if the fuel is carbohydrate, 19.8 kJ (4.74 kcal) if the fuel is fat, and 18.6 kJ (4.46 kcal) if the fuel is protein. An average value often used for the metabolism of a mixed diet is 20.2 kJ (4.83 kcal) per liter of O_2. The ratio of CO_2 produced to O_2 consumed in the tissues is called the **respiratory quotient (RQ)**. The RQ is 1.0 for the oxidation of carbohydrate, 0.71 for the oxidation of fat, and 0.80 for the oxidation of protein. In a steady state in which CO_2 is exhaled from the lungs at the same rate it is produced in the tissues, RQ is equal to the respiratory exchange ratio, R (see Chapter 19). One can improve the accuracy of indirect calorimetry by also determining R and either estimating the amount of protein oxidized—which usually is small compared with fat and carbohydrate—or calculating it from urinary nitrogen excretion.

Skeletal muscle is the main source of heat during external work.

Even during mild exercise, the muscles are the principal source of metabolic heat, and during intense exercise, they may account for up to 90%. Moderately intense exercise by a healthy, but sedentary, young man may require a metabolic rate of 600 W (in contrast to about 80 W at rest) and intense activity by a trained athlete 1,400 W or more. Because of their high metabolic rate, exercising muscles may be almost 1°C warmer than the core. The blood perfusing these muscles is warmed and, in turn, warms the rest of the body, raising the core temperature.

Muscles convert most of the energy in the fuels they consume into heat rather than mechanical work. During phosphorylation of **adenosine diphosphate (ADP)** to form **adenosine triphosphate (ATP)**, 58% of the energy released from the fuel is converted into heat, and only about 42% is captured in the ATP that is formed in the process. When a muscle contracts, some of the energy in the ATP that was hydrolyzed is converted into heat rather than mechanical work. The efficiency at this stage varies enormously; it is zero in isometric muscle contraction, in which a muscle's length does not change while it develops tension so that no work is done even though metabolic energy is required. Finally, friction converts some of the mechanical work produced into heat within the body (e.g., the fate of all the mechanical work done by the heart in pumping blood). At best, no more than 25% of the metabolic energy released during exercise is converted into mechanical work outside the body, and the other 75% or more is converted into heat within the body.

Sweating is the primary means of heat loss in humans.

There are four avenues of heat loss: *convection, conduction, radiation*, and *evaporation* (Fig. 28.5). Convection is the transfer of heat resulting from the movement of a fluid, either liquid or gas. In thermal physiology, the fluid is usually air or water in the environment; in the case of heat transfer inside the body, the fluid is blood. To illustrate, consider an object immersed in a fluid that is cooler than the object. Heat passes from the object to the immediately adjacent fluid by conduction. If the fluid is stationary, conduction is the only means by which heat can pass through the fluid, and over time, the rate of heat flow from the body to the fluid will diminish as the fluid nearest the object approaches the temperature of the object. In practice, however, fluids are rarely stationary. If the fluid is moving, heat will still be carried from the object into the fluid by conduction, but once the heat has entered the fluid, the movement of the fluid will carry it—by convection. The same fluid movement that carries heat away from the surface of the object constantly brings fresh cool fluid to the surface, and so the object gives up heat to the fluid much more rapidly than if the fluid were stationary.

Although conduction plays a role in this process, convection so dominates the overall heat transfer that we refer to the heat transfer as if it were entirely convection. Therefore, the conduction term (K) in the heat balance equation is restricted to heat flow between the body and other solid objects, and it usually represents only a small part of the total heat exchange with the environment.

When 1 g of water is converted into vapor at 30°C, it absorbs 2,425 J (0.58 kcal), the **latent heat of evaporation**, in the process. Evaporation of water is, thus, an efficient way of losing heat, and it is the body's only means of losing heat when the environment is hotter than the skin, as it usually is when the environment is warmer than 36°C. Evaporation must then dissipate both the heat produced by metabolic processes and any heat gained from the environment by convection and radiation. Most of the water evaporated in hot environments comes from sweat, but even in cold environments, the skin loses some water by the evaporation of **insensible perspiration**, water that diffuses through the skin rather than being secreted. In equation 1, E is nearly always positive, representing heat loss from the body. However, E is negative in the rare circumstances in which water vapor gives up heat to the body by condensing on the skin (as in a wet sauna).

Heat exchange is proportional to surface area and obeys thermodynamic principles.

Heat exchange always occurs from a higher to a cooler temperature and, thus, follows the second law of thermodynamics. Animals exchange heat with their environment through both the skin and the respiratory passages, but only the skin exchanges heat by radiation. In panting animals, respiratory heat loss may be large and may be an important means of achieving heat balance. In humans, however, respiratory heat exchange is usually relatively small and (though hyperthermic subjects may hyperventilate) is not predominantly under thermoregulatory control. Therefore, we do not consider it further here.

Convective heat exchange between the skin and the environment is proportional to the difference between skin and ambient air temperatures, as expressed by the following equation:

$$C = h_c \times A \times (\overline{T}_{sk} - T_a) \tag{2}$$

where A is the body surface area; $\overline{T}_{sk}$ and T_a are mean skin and ambient temperatures, respectively; and h_c is the convective heat transfer coefficient.

The value of h_c includes the effects of the factors other than temperature and surface area that influence convective heat exchange. For the whole body, air movement is the most important of these factors and convective heat exchange (and, thus, h_c) varies approximately as the square root of the air speed, except when air movement is slight (Fig. 28.7). Other factors that affect h_c include the direction of air movement and the curvature of the skin surface. As the radius of curvature decreases, h_c increases, and so the hands and fingers are effective in convective heat exchange disproportionately to their surface area.

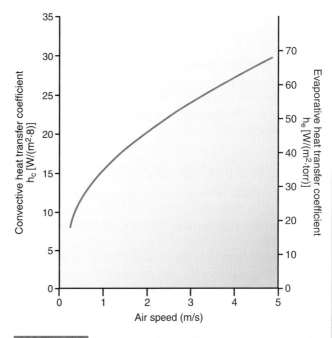

Figure 28.7 Dependence of convection and evaporation on air movement. This figure shows the convective heat transfer coefficient, h_c **(left)**, and the evaporative heat transfer coefficient, h_e **(right)**, for a standing human as a function of air speed. The convective and evaporative heat transfer coefficients are related by the equation $h_e = h_c \times 2.2$°C/torr. The horizontal axis can be converted into English units by using the relation 5 m/s = 16.4 ft/s = 11.2 miles/h.

Radiative heat exchange is proportional to the difference between the fourth powers of the absolute temperatures of the skin and of the radiant environment (T_r) and to the emissivity of the skin (e_{sk}): $R \propto e_{sk} \times (\overline{T}_{sk}^4 - T_r^4)$. However, if T_r is close enough to $\overline{T}_{sk}$ that $\overline{T}_k - T_r$ is much smaller than the absolute temperature of the skin, R is nearly proportional to $e_{sk} \times (\overline{T}_{sk} - T_r)$. Some parts of the body surface (e.g., the inner surfaces of the thighs and arms) exchange heat by radiation with other parts of the body surface, and so the body exchanges heat with the environment as if it had an area smaller than its actual surface area. This smaller area, called the effective radiating surface area (A_r), depends on the body's posture and is closest to the actual surface area in a spread-eagle position and least in a curled-up position. Radiative heat exchange can be represented by the following equation:

$$R = h_r \times e_{sk} \times A_r \times (\overline{T}_{sk} - T_r) \tag{3}$$

where h_r is the radiant heat transfer coefficient, 6.43 W/(m²·°C) at 28°C.

Evaporative heat loss from the skin to the environment is proportional to the difference between the water vapor pressure at the skin surface and the water vapor pressure in the ambient air. These relationships are summarized as follows:

$$E = h_e \times A \times (P_{sk} - P_a) \tag{4}$$

where P_{sk} is the water vapor pressure at the skin surface, P_a is the ambient water vapor pressure, and h_e is the evaporative heat transfer coefficient.

Moving air carries away water vapor, like heat, and so geometric factors and air movement affect E and h_e in the same way that they affect C and h_c. If the skin is completely wet, the water vapor pressure at the skin surface is the saturation water vapor pressure at the temperature of the skin and evaporative heat loss is E_{max}, the maximum possible for the prevailing skin temperature and environmental conditions. This condition is described as follows:

$$E_{max} = h_e \times A \times (P_{sk,sat} - P_a) \qquad (5)$$

where $P_{sk,sat}$ is the saturation water vapor pressure at skin temperature. When the skin is not completely wet, it is impractical to measure P_{sk}, the actual average water vapor pressure at the skin surface. Therefore, a coefficient called skin *wettedness* (w) is defined as the ratio E/E_{max}, with $0 < w <1$. Skin wettedness depends on the hydration of the epidermis and the fraction of the skin surface that is wet. We can now rewrite equation 4 as follows:

$$E = h_e \times A \times w \times (P_{sk,sat} - P_a) \qquad (6)$$

Wettedness depends on the balance between secretion and evaporation of sweat. If secretion exceeds evaporation, sweat accumulates on the skin and spreads out to wet more of the space between neighboring sweat glands, increasing wettedness and E; if evaporation exceeds secretion, the reverse occurs. If sweat rate exceeds E_{max}, once wettedness becomes 1, the excess sweat drips from the body because it cannot evaporate.

Note that P_a, on which evaporation from the skin directly depends, is proportional to the actual moisture content in the air. In contrast, the more familiar quantity **relative humidity** (rh) is the ratio between the actual moisture content in the air and the maximum moisture content possible at the temperature of the air. It is important to recognize that rh is only indirectly related to evaporation from the skin. For example, in a cold environment, P_a will be low enough that sweat can easily evaporate from the skin even if rh equals 100% because the skin is warm and $P_{sk,sat}$, which depends on the temperature of the skin, will be much greater than P_a.

Heat storage is a balance between net heat production and net heat loss.

The rate of **heat storage** is an indicator of a change in heat content of the body and is the difference between heat production and net heat loss (equation 1). (In unusual circumstances in which there is a net heat gain from the environment, such as during immersion in a hot bath, storage is the sum of heat production and net heat gain.) It can be determined experimentally from simultaneous measurements of metabolism by indirect calorimetry and heat gain or loss by direct calorimetry. Storage of heat in the tissues changes their temperature, and the amount of heat stored is the product of body mass, the body's mean specific heat, and a suitable mean body temperature (T_b). The body's mean specific heat depends on its composition, especially the proportion of fat, and is about 3.55 kJ/(kg·°C) [0.85 kcal/(kg·°C)]. Empirical relations of T_b to core temperature (T_c) and T_{sk}, determined in calorimetric studies, depend

on ambient temperature, with T_b varying from $0.65 \times T_c + 0.35 \times T_{sk}$ in the cold to $0.9 \times T_c + 0.1 \times T_{sk}$ in the heat. The shift from cold to heat in the relative weighting of T_c and T_{sk} reflects the accompanying change in the thickness of the shell (see Fig. 28.2).

▶ HEAT DISSIPATION

Figure 28.8 shows rectal and mean skin temperatures, heat losses, and calculated core-to-skin (shell) conductances for nude resting men and women at the end of 2-hour exposures in a calorimeter to ambient temperatures of 23°C to 36°C. Shell conductance represents the sum of heat transfer by two parallel modes: conduction through the tissues of the shell and convection by the blood. It is calculated by dividing heat flow through the skin (HF_{sk}) (i.e., total heat loss from the

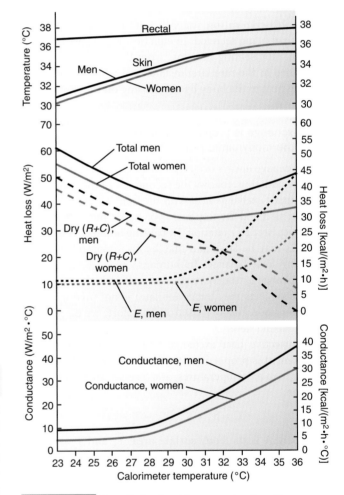

Figure 28.8 Heat dissipation. These graphs show the average values of rectal and mean skin temperatures, heat loss, and core-to-skin thermal conductance for nude resting men and women near steady state after 2 hours at different environmental temperatures in a calorimeter. (All energy exchange quantities in this figure have been divided by body surface area to remove the effect of individual body size.) Total heat loss is the sum of dry heat loss, by radiation (*R*) and convection (*C*) and evaporative heat loss (*E*). Dry heat loss is proportional to the difference between skin temperature and calorimeter temperature and decreases with increasing calorimeter temperature.

body minus heat loss through the respiratory tract) by the difference between core and mean skin temperatures:

$$C = HF_{sk}/(T_c - \overline{T}_{sk}) \qquad (7)$$

where C is shell conductance and T_c and $\overline{T}_{sk}$ are core and mean skin temperatures.

From 23°C to 28°C, conductance is minimal because the skin is vasoconstricted and its blood flow is low. The minimal level of conductance attainable depends largely on the thickness of the subcutaneous fat layer, and women's thicker layer allows them to attain a lower conductance than men. At about 28°C, conductance begins to increase, and above 30°C, conductance continues to increase and sweating begins.

For these subjects, 28°C to 30°C is the zone of **thermoneutrality**, or thermoneutral zone, the range of comfortable environmental temperatures in which thermal balance is maintained without either shivering or sweating. In this zone, controlling conductance and $\overline{T}_{sk}$ and, thus, R and C entirely maintains heat balance. As equations 2 to 4 show, C, R, and E all depend on skin temperature, which, in turn, depends partly on skin blood flow. E depends also, through skin wettedness, on sweat secretion. Therefore, all these modes of heat exchange are partly under physiologic control.

Sweat evaporation can dissipate large amounts of heat.

In Figure 28.8, evaporative heat loss is nearly independent of ambient temperature below 30°C and is 9 to 10 W/m², corresponding to evaporation of about 13 to 15 g/(m²·h), of which about half is moisture lost in breathing and half is insensible perspiration. This evaporation occurs independent of thermoregulatory control. As the ambient temperature increases, the body depends more and more on the evaporation of sweat to achieve heat balance.

The two histologic types of sweat glands are **eccrine** and **apocrine**. Eccrine sweat glands, the dominant type in all human populations, are more important in human thermoregulation and number about 2,500,000. Eccrine secretion is controlled through postganglionic sympathetic C fibers that release acetylcholine (ACh) rather than norepinephrine. Cholinergic stimulation of the sweat gland elicits the secretion of the so-called *precursor fluid*, the composition of which resembles that of plasma, except that it does not

contain the plasma proteins. Therefore, sodium and chloride are the primary electrolytes in sweat, with potassium, calcium, and magnesium present in smaller amounts. As the fluid moves through the duct portion of the sweat gland, its composition is modified by active reabsorption of sodium and chloride. As a result, sweat is hyposmotic compared to plasma, and sodium concentration typically ranges from 10 to 70 mmol/L, whereas chloride ranges from 5 to 60 mmol/L, depending on diet, sweat rate, and degree of heat acclimatization. Eccrine sweat also includes lactate, urea, ammonia, serine, ornithine, citrulline, aspartic acid, heavy metals, organic compounds, and proteolytic enzymes. In contrast, apocrine sweat consists mainly of sialomucin. Although initially odorless, as apocrine sweat comes in contact with normal bacterial flora on the surface of the skin, an odor develops. Apocrine sweat is more viscous and produced in much smaller amounts than eccrine sweat. The exact function of apocrine glands is unclear, though they are thought to represent scent glands. During puberty, **apoeccrine** glands appear. These are hybrid sweat glands that are found in the axilla and may play a role in axillary hyperhidrosis. Their secretory structures have both a small-diameter portion similar to those found in eccrine glands and a large-diameter portion that resembles an apocrine gland. Functionally similar to eccrine glands, they respond mainly to cholinergic stimuli, and their ducts are long and open directly onto the skin surface. Interestingly, apoeccrine glands secrete nearly 10 times as much sweat as eccrine glands do.

A healthy man unacclimatized to heat can secrete up to 1.5 L/h of sweat. Although the number of functional sweat glands is fixed before the age of 3, the secretion capacity of the individual glands can change, especially with endurance exercise training and heat acclimatization. Men well acclimatized to heat can attain peak sweat rates of 1.0 to 2.5 L/h, with sweat rates of more than 2.5 L/h possible when the ambient temperature is high. However, such rates cannot be maintained as the maximum daily sweat output is probably about 15 L (Clinical Focus 28.1).

Increased vasodilation of the skin augments heat loss.

Heat produced in the body must be delivered to the skin surface to be eliminated. When skin blood flow is minimal, shell conductance is typically 5 to 9 W/°C/m² of body surface.

CLINICAL FOCUS | 28.1

Water and Salt Depletion as a Result of Sweating

Changes in fluid and electrolyte balance are probably the most frequent physiologic disturbances associated with sustained exercise and heat stress. Water loss via the sweat glands can exceed 1 L/h for many hours. Salt loss in the sweat is variable; however, because sweat is more dilute than plasma, sweating always results in an increase in the osmolality of the fluid remaining in the body and

increased plasma [Na⁺] and [Cl⁻], as long as the lost water is not replaced. Because people who secrete large volumes of sweat usually replace at least some of their losses by drinking water or electrolyte solutions, the final effect on body fluids may vary. In Table 28.3, the second and third conditions (subject A) represent the effects on body fluids of sweat losses alone and combined with replacement by an

(Continued)

equal volume of plain water, respectively, for someone producing sweat, with a [Na$^+$] and [Cl$^-$] in the upper part of the normal range. By contrast, the fourth and fifth conditions (subject B) represent the corresponding effects for a heat-acclimatized person secreting dilute sweat. Comparing the effects on these two people, we note (1) the more dilute the sweat that is secreted, the greater the increase in osmolality and plasma [Na$^+$] if no fluid is replaced; (2) extracellular fluid volume, a major determinant of plasma volume (see Chapter 15), is greater in subject B (secreting dilute sweat) than in subject A (secreting saltier sweat), whether water is replaced; and (3) drinking plain water allowed subject B to maintain plasma sodium and extracellular fluid volume almost unchanged while secreting 5 L of sweat. In subject A, however, drinking the same amount of water reduced plasma [Na$^+$] by 8 mmol/L and failed to prevent a decrease of almost 10% in extracellular fluid volume. In 5 L of sweat, subject A lost 17.5 g of salt, somewhat more than the daily salt intake in a normal Western diet, and is becoming salt depleted.

Increased osmolality of the extracellular fluid and decreased plasma volume via a reduction in the activity of the cardiovascular stretch receptors stimulate thirst (see Chapter 23). When sweating is profuse, however, thirst usually does not elicit enough drinking to replace fluid as rapidly as it is lost so that people exercising in the heat tend to become progressively dehydrated—in some cases losing as much as 7% to 8% of body weight—and restore normal fluid balance only during long periods of rest or at meals. Depending on how much of his fluid losses he replaces, subject B may either be hypernatremic and dehydrated or be in essentially normal fluid and electrolyte balance. (If he drinks fluid well in excess of his losses, he may become overhydrated and hyponatremic, but this is an unlikely occurrence.) However, subject A, who is somewhat salt depleted, may be dehydrated and hypernatremic, normally hydrated but hyponatremic, or somewhat dehydrated with plasma [Na$^+$] anywhere in between these two extremes. Once subject A replaces all the water lost as sweat, his extracellular fluid volume will be about 10% below its initial value. If he responds to the accompanying reduction in plasma volume by continuing to drink water, he will become even more hyponatremic than shown in Table 28.3.

The disturbances shown in Table 28.3, although physiologically significant and useful for illustration, are not likely to require clinical attention. Greater disturbances, with correspondingly more severe clinical effects, may occur. The consequences of the various possible disturbances of salt and water balance can be grouped as effects of decreased plasma volume secondary to decreased extracellular fluid volume, effects of hypernatremia, and effects of hyponatremia. The circulatory effects of decreased volume are nearly identical to the effects of peripheral pooling of blood (see Fig. 28.12), and the combined effects of peripheral pooling and decreased volume will be greater than the effects of either alone. These effects include impairment of cardiac filling and cardiac output and compensatory reflex reductions in renal, splanchnic, and skin blood flow. Impaired cardiac output leads to fatigue during exertion and decreased exercise tolerance. If skin blood flow is reduced, heat dissipation will be impaired. Exertional rhabdomyolysis, the injury of skeletal muscle fibers, is a frequent result of unaccustomed intense exercise. Myoglobin released from injured skeletal muscle cells appears in the plasma, rapidly enters the glomerular filtrate, and is excreted in the urine, producing myoglobinuria and staining the urine brown if enough myoglobin is present. This process may be harmless to the kidneys if urine flow is adequate. However, a reduction in renal blood flow reduces urine flow, increasing the likelihood that the myoglobin will cause renal tubular injury.

Hypernatremic dehydration is believed to predispose to heatstroke. Both hypernatremia and reduced plasma volume often accompany dehydration. Hypernatremia impairs the heat loss responses (sweating and increased skin blood flow) independently of any accompanying reduction in plasma volume and elevates the thermoregulatory set point. Hypernatremic dehydration promotes the development of high core temperature in multiple ways through the combination of hypernatremia and reduced plasma volume.

Even in the absence of sodium loss, overdrinking that exceeds the kidneys' ability to compensate dilutes all the body's fluid compartments, producing dilutional hyponatremia, which is also called water intoxication if it causes symptoms. The development of water intoxication requires either massive overdrinking or a condition, such as the inappropriate secretion of arginine vasopressin, that impairs the excretion of free water by the kidneys. Overdrinking sufficient to cause hyponatremia may occur in patients with psychiatric disorders or disturbance of the thirst mechanism or may be done with a mistaken intention of preventing or treating dehydration. However, people who secrete copious amounts of sweat with a high sodium concentration, such as subject A or people with cystic fibrosis, may easily lose enough salt to become hyponatremic because of sodium loss. Some healthy young adults who come to medical attention for salt depletion after profuse sweating are found to have genetic variants of cystic fibrosis, which cause these people to have salty sweat without producing the characteristic digestive and pulmonary manifestations of cystic fibrosis.

As sodium concentration and osmolality of the extracellular space decrease, water moves from the extracellular space into the cells to maintain osmotic balance across the cell membranes. The resulting swelling of the brain cells causes most of the manifestations of hyponatremia. Mild hyponatremia is characterized by nonspecific symptoms such as fatigue, confusion, nausea, and headache and may be mistaken for heat exhaustion. Severe hyponatremia can be a life-threatening medical emergency and may include seizures, coma, herniation of the brainstem (which occurs if the brain swells enough to exceed the capacity of the cranium), and death. In the setting of prolonged exertion in the heat, symptomatic hyponatremia is far less common than heat exhaustion but potentially far more dangerous. Therefore, it is important not to treat a presumed case of heat exhaustion with large amounts of low-sodium fluids without first ruling out hyponatremia. ■

For a lean resting subject with a surface area of 1.8 m², minimal whole body conductance of 16 W/°C (i.e., 8.9 W/[°C·m²] × 1.8 m²), and a metabolic heat production of 80 W, the temperature difference between the core and the skin must be 5°C (i.e., 80 W ÷ 16 W/°C) for the heat produced to be conducted to the surface. In a cool environment, $\overline{T}_{sk}$ may easily be low enough for this to occur. However, in an ambient temperature of 33°C, $\overline{T}_{sk}$ is typically about 35°C, and without an increase in conductance, core temperature would have to rise to 40°C—a high, although not yet dangerous, level—for the heat to be conducted to the skin. If the rate of heat production was increased to 480 W by moderate exercise, the temperature difference between core and skin would have to rise to 30°C—and core temperature to well beyond lethal levels—to allow all the heat produced to be conducted to the skin. In the latter circumstances, the conductance of the shell must increase greatly for the body to reestablish thermal balance and continue to regulate its temperature. This is accomplished by increasing the skin blood flow.

Effectiveness of skin blood flow in heat transfer

Assuming that blood on its way to the skin remains at core temperature until it reaches the skin, reaches skin temperature as it passes through the skin, and then stays at skin temperature until it returns to the core, we can compute the rate of heat flow (HF_b) as a result of convection by the blood as follows:

$$HF_b = SkBF \times (T_c - \overline{T}_{sk}) \times 3.85 \text{ kJ/(L} \cdot °C) \qquad (8)$$

where SkBF is the rate of skin blood flow, expressed in liter per second rather than the usual liter per minute, to simplify computing HF in W (i.e., J/s), and 3.85 kJ/(L·°C) (0.92 kcal/[L·°C]) is the volume-specific heat of blood. Conductance as a result of convection by the blood (C_b) is calculated as follows:

$$C_b = HF_b/(T_c - \overline{T}_{sk}) = SkBF \times 3.85 \text{ kJ/(L} \cdot °C) \qquad (9)$$

Of course, heat continues to flow by conduction through the tissues of the shell, and so total conductance is the sum of conductance as a result of convection by the blood and that resulting from conduction through the tissues. Total heat flow is given by the following equation:

$$HF = (C_b + C_0) \times (T_c - \overline{T}_{sk}) \qquad (10)$$

where C_0 is thermal conductance of the tissues when skin blood flow is minimal and, thus, is predominantly a result of conduction through the tissues.

The assumptions made in deriving equation 8 are somewhat artificial and represent the conditions for maximum efficiency of heat transfer by the blood. In practice, blood exchanges heat also with the tissues through which it passes on its way to and from the skin. Heat exchange with these other tissues is greatest when skin blood flow is low; in such cases, heat flow to the skin may be much less than predicted by equation 8, as discussed further below. However, equation 8 is a reasonable approximation in a warm subject with moderate-to-high skin blood flow. Although measuring whole-body SkBF directly is not possible, it is believed to reach several liters per minute during heavy exercise in the heat.

The maximum obtainable is estimated to be nearly 8 L/min. If SkBF = 1.89 L/min (0.0315 L/s), according to equation 9, skin blood flow contributes about 121 W/°C to the conductance of the shell. If conduction through the tissues contributes 16 W/°C, total shell conductance is 137 W/°C, and if $T_c = 38.5°C$ and $T_{sk} = 35°C$, this will produce a core-to-skin heat transfer of 480 W, the heat production in our earlier example of moderate exercise. Therefore, even a moderate rate of skin blood flow can have a dramatic effect on heat transfer.

When a person is not sweating, raising skin blood flow brings skin temperature nearer to blood temperature, and lowering skin blood flow brings skin temperature nearer to ambient temperature. Under such conditions, the body can control dry (convective and radiative) heat loss by varying skin blood flow and, thus, skin temperature. Once sweating begins, skin blood flow continues to increase as the person becomes warmer. In these conditions, however, the tendency of an increase in sweating to cool the skin approximately balances the tendency of an increase in skin blood flow to warm the skin. Therefore, after sweating has begun, further increases in skin blood flow usually cause little change in skin temperature or dry heat exchange and serve primarily to deliver to the skin the heat that is being removed by the evaporation of sweat. Skin blood flow and sweating work in tandem to dissipate heat under such conditions.

Sympathetic control of skin circulation

Because the nutritional needs of the skin are low, the skin blood flow is controlled mainly by reflexes or changes in temperature. In most of the skin, the vasodilation that occurs during heat exposure depends on sympathetic nerve signals that cause the blood vessels to dilate, and regional nerve block can prevent or reverse this vasodilation. Because it depends on the action of neural signals, such vasodilation is sometimes referred to as *active vasodilation*. Active vasodilation occurs in almost all the skin, except in so-called acral regions—hands, feet, lips, ears, and nose. In skin areas where active vasodilation occurs, vasoconstrictor activity is minimal at thermoneutral temperatures, and active vasodilation during heat exposure does not begin until close to the onset of sweating. Therefore, small temperature changes within the thermoneutral range do not much affect skin blood flow in these areas.

Reflex vasodilation is complex, with potentially many layers of redundancy. With whole-body heating, sympathetic adrenergic tone is withdrawn and a sympathetic active vasodilator system is activated, most likely mediated by a cholinergic neurotransmitter system. Neuronal **nitric oxide (NO)** synthase (nNOS)-mediated NO is directly required for 30% to 40% of this response, may also contribute by inhibiting adrenergic function (see below), and appears to have no specific role in response to local skin warming. Other mediators of this response include **vasoactive intestinal peptide (VIP)**, histamine, prostanoids, endothelial-derived hyperpolarizing factor, and **substance P**.

Increased local skin temperature (T_{loc}) also results in cutaneous vasodilation, although through an independent mechanism. Local vasodilatory mechanisms involved in local skin warming can increase SkBF to maximal levels, a biphasic response involving sensory nerves that mediate an initial transient vasodilatory "peak" followed by a prolonged vasodilatory "plateau" that is mediated primarily by endothelial NOS generation of NO.

Active vasodilation operates in tandem with sweating in the heat and is impaired or absent in **anhidrotic ectodermal dysplasia**, a congenital disorder in which sweat glands are sparse or absent. For these reasons, the existence of a mechanism linking active vasodilation to the sweat glands has long been suspected but never established. Recent studies have demonstrated an attenuated sweat rate after NOS inhibition, suggesting that NO augments sweat secretion. Additionally, NOS inhibition attenuates the sweat response to exercise in a warm environment, suggesting that NO has a role in modifying thermoregulatory sweating, specifically by changing sweat gland output. While the overall importance of NO's role remains unknown, the magnitude of reduction in local sweat rate by NOS inhibition is as large as that induced by dehydration. Accordingly, NO appears to play an important role in thermoregulatory control of sweating during exercise in warm environments.

Reflex vasoconstriction, occurring in response to cold and as part of certain nonthermal reflexes such as baroreflexes, is mediated primarily through adrenergic sympathetic fibers distributed widely over most of the skin. The process is initiated when the mean whole-body skin temperature decreases below thermoneutral levels of around 33°C and is mediated by **norepinephrine (NE)** and neuropeptide Y released from vasoconstrictor nerves. Although evidence suggests that endogenous NO can inhibit or modulate this response, these two neurotransmitters appear to be all that is necessary. Accordingly, reducing the flow of impulses in these nerves allows the blood vessels to dilate. In the acral regions and superficial veins (whose role in heat transfer is discussed below), vasoconstrictor fibers are the predominant vasomotor innervation, and the vasodilation that occurs during heat exposure is largely a result of the withdrawal of vasoconstrictor activity. Blood flow in these skin regions is sensitive to small temperature changes even in the thermoneutral range and may be responsible for "fine-tuning" heat loss to maintain heat balance in this range.

There also exists a local vasoconstrictive response to skin cooling that has two components. An adrenergic element includes reduced secretion of norepinephrine that is more than overcome by an increase in α_{2C}-receptor sensitivity via Rho-kinase–mediated signaling. The second element involves a withdrawal of tonic NO production through inhibition of NOS and subsequent steps in the NO system.

▶ THERMOREGULATORY CONTROL

In discussions of control systems, the words "regulation" and "regulate" have meanings distinct from those of the word "control" (see Chapter 1). The variable that a control system acts to maintain within narrow limits (e.g., temperature) is called the *regulated* variable, and the quantities it controls to accomplish this (e.g., sweating rate, skin blood flow,

metabolic rate, and thermoregulatory behavior) are called *controlled* variables.

Humans have two distinct subsystems for regulating body temperature: behavioral thermoregulation and physiologic thermoregulation. Behavioral thermoregulation—through the use of shelter, space heating, air conditioning, and clothing—enables humans to live in the most extreme climates in the world, but it does not provide fine control of body heat balance. In contrast, physiologic thermoregulation is capable of fairly precise adjustments of heat balance but is effective only within a relatively narrow range of environmental temperatures.

Temperature is perceived through thermal sensation.

Sensory information about body temperatures is an essential part of both behavioral and physiologic thermoregulation. The distinguishing feature of behavioral thermoregulation is the involvement of consciously directed efforts to regulate body temperature. Thermal discomfort provides the necessary motivation for thermoregulatory behavior, and behavioral thermoregulation acts to reduce both the discomfort and the physiologic strain imposed by a stressful thermal environment. For this reason, both thermal comfort and the absence of shivering and sweating characterize the zone of thermoneutrality.

Warmth and cold on the skin are felt as either comfortable or uncomfortable, depending on whether they decrease or increase the physiologic strain—a shower temperature that feels pleasant after strenuous exercise may be uncomfortably chilly on a cold winter morning. The processing of thermal information in behavioral thermoregulation is not as well understood as it is in physiologic thermoregulation. However, perceptions of thermal sensation and comfort respond much more quickly than core temperature or physiologic thermoregulatory responses to changes in environmental temperature and, thus, appear to anticipate changes in the body's thermal state. Such an anticipatory feature would be advantageous because it would reduce the need for frequent small behavioral adjustments.

Thermal defense mechanisms operate through changes in heat production and heat loss.

Familiar inanimate control systems, such as most refrigerators and heating and air-conditioning systems, operate at only two levels: on and off. In a typical home-heating system, for example, when the indoor temperature falls below the desired level, the thermostat detects the change and turns on the burner and blower in the furnace; when the temperature is restored to the desired level, the thermostat turns the furnace off. Rather than operating at only two levels, most physiologic control systems produce a graded response according to the size of the disturbance in the regulated variable. In many instances, changes in the controlled variables are proportional to displacements of the regulated variable from some threshold value.

The control of heat-dissipating responses is an example of a proportional control system. Figure 28.9 shows how reflex control of two heat-dissipating responses, sweating and skin blood flow, depends on body core temperature and

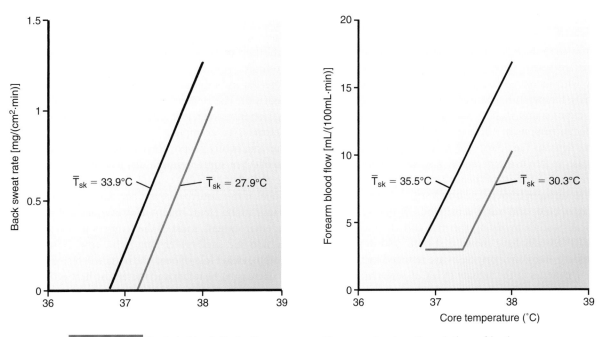

Figure 28.9 **Control of heat-dissipating responses.** These graphs show the relations of back (scapular) sweat rate **(left)** and forearm blood flow **(right)** to core temperature and mean skin temperatures ($\overline{T}_{sk}$). In these experiments, exercise increased core temperature.

mean skin temperature. Each response has a core temperature threshold—a temperature at which the response starts to increase—and this threshold depends on mean skin temperature. At any given skin temperature, the change in each response is proportional to the change in core temperature, and increasing the skin temperature lowers the threshold level of core temperature and increases the response at any given core temperature. In humans, a change of 1°C in core temperature elicits about nine times as great a thermoregulatory response as a 1°C change in mean skin temperature. (Besides its effect on the reflex signals, skin temperature has a local effect that modifies the response of the blood vessels and sweat glands to the reflex signal, discussed later.)

Cold stress elicits increases in metabolic heat production through shivering and nonshivering thermogenesis. **Shivering** is a rhythmic oscillating tremor of skeletal muscles. On a local level, movement is uncoordinated as reflected by rapid (i.e., 250 Hz) electromyographic activity. However, there is an overall 4 to 8 cycles/min "waxing-and-waning" activity that is clinically apparent during intense shivering. The CNS network generating the cold-evoked bursts of α-motor neuron activity is not well understood, but it includes transmission of cutaneous cold afferent signals through the lateral parabrachial nucleus, integration of thermoregulatory signals in the preoptic area (POA), lifting of its inhibition, and activation of fusimotor neurons (β- and γ-motor neurons), which is dependent on descending neurons in the rostral ventromedial medulla. Although these efferent impulses are not rhythmic, their overall effect is to increase muscle tone, thereby increasing metabolic rate. Once the tone exceeds a critical level, the contraction of one group of muscle fibers stretches the muscle spindles in other fiber groups in series with it, eliciting contractions from those groups of fibers via

the stretch reflex, and so on; thus, the rhythmic oscillations that characterize frank shivering begin.

Shivering occurs in bursts, and signals from the cerebral cortex inhibit the "shivering pathway," so that voluntary muscular activity and attention can suppress shivering. Because the limbs are part of the shell in the cold, trunk and neck muscles are preferentially recruited for shivering—the centralization of shivering—to help retain the heat produced during shivering within the body core; and the familiar experience of teeth chattering is one of the earliest signs of shivering. As with heat-dissipating responses, the control of shivering depends on both core and skin temperatures, but the details of its control are not precisely understood.

Hypothalamus integrates thermal information from the core and the skin.

Temperature receptors in the body core and skin transmit information about their temperatures through afferent nerves to the brainstem and, especially, the hypothalamus, where much of the integration of temperature information occurs. The sensitivity of the thermoregulatory system to core temperature enables it to adjust heat production and heat loss to resist disturbances in core temperature. Sensitivity to mean skin temperature lets the system respond appropriately to mild heat or cold exposure with little change in body core temperature, so that changes in body heat as a result of changes in environmental temperature take place almost entirely in the peripheral tissues (see Fig. 28.2). For example, the skin temperature of someone who enters a hot environment may rise and elicit sweating even if there is no change in core temperature. On the other hand, an increase in heat production within the body, as during exercise, elicits the appropriate heat-dissipating responses through a rise in core temperature.

Core temperature receptors involved in controlling thermoregulatory responses are unevenly distributed and are concentrated in the POA of the anterior hypothalamus. Most of these are warm sensitive, meaning they increase their activity with increased brain temperature. In experimental mammals, temperature changes of only a few tenths of 1°C in the hypothalamus elicit changes in the thermoregulatory effector responses. Although much less common, cold-sensitive neurons (i.e., those that increase their activity with a decrease in brain temperature) also exist. However, the cold sensitivity of most of them seems to be due to inhibitory synaptic input from the nearby warm-sensitive neurons. The roles of cold- and warm-sensitive POA neurons are not reciprocal. Studies involving thermal and chemical stimulation of POA cells have shown that both cold-defense and heat-defense autonomic responses are initiated by the corresponding changes in the activity of warm-sensitive neurons; increased activity of warm-sensitive POA neurons triggers heat-defense responses, whereas decreased activity triggers cold-defense responses. Because warm-sensitive POA neurons display spontaneous membrane depolarization, these cells are considered pacemakers; their thermosensitivity is due to currents that determine the rate of spontaneous depolarization between successive action potentials. Furthermore, morphologic identification of thermosensitive neurons in the POA has shown their orientation to be ideally suited to receiving afferent projections from peripheral thermoreceptors via the medial forebrain bundle. In addition to the POA cold- and warm-sensitive neurons, there are peripheral deep-body sensors that respond to changes in core body temperature. They are located in the esophagus, stomach, large intra-abdominal veins, and other organs.

Consider what happens when some disturbance—say, an increase in metabolic heat production resulting from exercise—upsets the thermal balance. Additional heat is stored in the body, and core temperature rises. The central thermoregulatory controller receives information about these changes from the thermal receptors and elicits appropriate heat-dissipating responses. Core temperature continues to rise, and these responses continue to increase until they are sufficient to dissipate heat as fast as it is being produced, restoring heat balance and preventing further increases in body temperatures. In the language of control theory, the rise in core temperature that elicits heat-dissipating responses sufficient to reestablish thermal balance during exercise is an example of a load error. A load error is characteristic of any proportional control system that is resisting the effect of some imposed disturbance or "load." Although the disturbance in this example is exercise, the same principle applies if the disturbance is a decrease in metabolic rate or a change in the environment. However, if the disturbance is in the environment, most of the temperature change will be in the skin and shell rather than in the core. If the disturbance produces a net loss of heat, the body will restore heat balance by decreasing heat loss and increasing heat production.

Set point is the body's "thermostat."

The primary autonomic defenses against heat are sweating and active cutaneous vasodilation, whereas the primary autonomic defenses against cold are cutaneous vasoconstriction and shivering. The core temperature, or reference temperature, triggering each response defines its activation threshold. Body temperature is normally maintained between the sweating and vasoconstriction thresholds, temperatures that define the *interthreshold range* (also referred to as the "null zone"). Onset of sweating, thus, usually defines the upper boundary of normal body temperature, and cutaneous vasoconstriction defines the lower boundary. Typically, the interthreshold range is only a few tenths of a degree, and it is thus likely that temperatures within this range are accurately sensed but do not generate an error signal and trigger thermoregulatory responses. Under normal circumstances, thermoregulatory control can thus be roughly modeled as a discrete yet adjustable "**set point**" (T_{set}) (Fig. 28.10). That thermoregulatory responses are likely to have unique thresholds within the thermal integrative network is exemplified by the fact that the threshold for

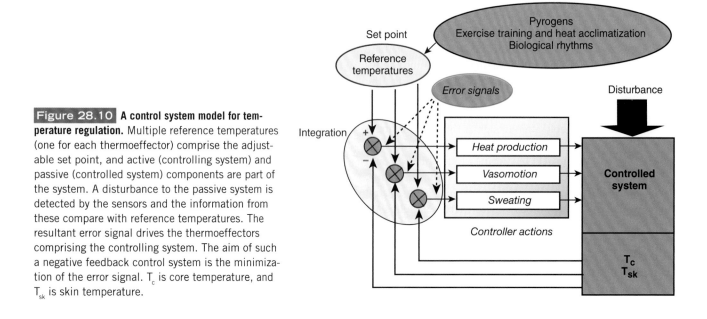

Figure 28.10 **A control system model for temperature regulation.** Multiple reference temperatures (one for each thermoeffector) comprise the adjustable set point, and active (controlling system) and passive (controlled system) components are part of the system. A disturbance to the passive system is detected by the sensors and the information from these compare with reference temperatures. The resultant error signal drives the thermoeffectors comprising the controlling system. The aim of such a negative feedback control system is the minimization of the error signal. T_c is core temperature, and T_{sk} is skin temperature.

shivering is typically a degree less than the vasoconstriction threshold. Shivering is thus a "last resort" response and appropriately so because it is metabolically inefficient.

Effect of nonthermal inputs on thermoregulatory responses

Each thermoregulatory response may be affected by inputs other than body temperatures and factors that influence the set point. We have already noted that voluntary activity affects shivering and certain hormones affect metabolic heat production. In addition, nonthermal factors may produce a burst of sweating at the beginning of exercise, and emotional effects on sweating and skin blood flow are matters of common experience. Skin blood flow is the thermoregulatory response most influenced by nonthermal factors because of its potential involvement in reflexes that function to maintain cardiac output, blood pressure, and tissue O_2 delivery under a variety of disturbances, including heat stress, postural changes, hemorrhage, exercise, hydration state, sleep, and narcosis.

Biologic rhythms and altered physiologic states can change the thermoregulatory set point.

Fever elevates core temperature at rest, heat acclimatization decreases it, and time of day and (in women) the phase of the menstrual cycle change it in a cyclic fashion. Core temperature at rest varies in an approximately sinusoidal fashion with time of day. The minimum temperature occurs at night, several hours before awaking, and the maximum, which is 0.5°C to 1°C higher, occurs in the late afternoon or evening (see Fig. 28.3). This pattern coincides with patterns of activity and eating but does not depend on them, and it occurs even during bed rest in fasting subjects. This pattern is an example of a **circadian rhythm**, a rhythmic pattern in a physiologic function with a period of about 1 day. During the menstrual cycle, core temperature is at its lowest point just before ovulation. During the next few days, it rises from 0.5°C to 1°C to a plateau that persists through most of the luteal phase. Each of these factors—fever, heat acclimatization, the circadian rhythm, and the menstrual cycle—changes the core temperature at rest by changing the thermoregulatory set point, producing corresponding changes in the thresholds for all of the thermoregulatory responses.

Skin temperature alters cutaneous blood flow and sweat gland responses.

The skin is the organ most directly affected by environmental temperature. Skin temperature influences heat loss responses not only through reflex actions (see Fig. 28.10) but also through direct effects on the skin blood vessels and sweat glands.

Local temperature changes act on skin blood vessels in at least two ways. First, local cooling potentiates (and heating weakens) the constriction of blood vessels in response to nerve signals and vasoconstrictor substances. (At low temperatures, however, cold-induced vasodilation [CIVD] increases skin blood flow, as discussed later.) Second, in skin regions where active vasodilation occurs, local heating causes vasodilation (and local cooling causes vasoconstriction) through a direct action on the vessels, independent of

nerve signals. The local vasodilator effect of skin temperature is especially strong above 35°C; and, when the skin is warmer than the blood, increased blood flow helps cool the skin and protect it from heat injury, unless this response is impaired by vascular disease. Local thermal effects on sweat glands parallel those on blood vessels, and so local heating potentiates (and local cooling diminishes) the local sweat gland response to reflex stimulation or ACh, and intense local heating elicits sweating directly, even in skin whose sympathetic innervation has been interrupted surgically.

Sweat gland "fatigue" alters thermoregulation.

During prolonged heat exposure (lasting several hours) with high sweat output, sweating rates gradually decline and the response of sweat glands to local cholinergic drugs is reduced. This reduction of sweat gland responsiveness is sometimes called sweat gland "fatigue." Wetting the skin makes the stratum corneum swell, mechanically obstructing the sweat gland ducts and causing a reduction in sweat secretion, an effect called **hidromeiosis**. The glands' responsiveness can be at least partly restored if air movement increases or humidity is reduced, allowing some of the sweat on the skin to evaporate. Sweat gland fatigue may involve processes besides hidromeiosis because prolonged sweating also causes histologic changes, including the depletion of glycogen, in the sweat glands.

▶ THERMOREGULATORY RESPONSES DURING EXERCISE

During intense physical exercise, metabolic heat production can increase 10- to 20-fold, with <30% of the heat generated converted into mechanical energy. Consequently, more than 70% of metabolic heat generated must be transported to the skin to be dissipated to the environment. Heat starts to accumulate in the body when the heat-dissipating mechanisms are unable to cope with metabolic heat production, leading to an increase in body temperature. Although hot environments also elicit heat-dissipating responses, exercise ordinarily is responsible for the greatest demands on the thermoregulatory system for heat dissipation. Exercise provides an important example of how the thermoregulatory system responds to a disturbance in heat balance. In addition, exercise and thermoregulation impose competing demands on the circulatory system because exercise requires large increases in blood flow to exercising muscle, whereas the thermoregulatory responses to exercise require increases in skin blood flow. Muscle blood flow during exercise is several times as great as skin blood flow, but the increase in skin blood flow is responsible for disproportionately large demands on the cardiovascular system, as discussed below. Finally, if the water and electrolytes lost through sweating are not replaced, the resulting reduction in plasma volume will eventually create a further challenge to cardiovascular homeostasis.

Core temperature rises during exercise, triggering heat loss responses.

As previously mentioned, the increased metabolic heat production during exercise causes an increase in core temperature, which in turn elicits heat loss responses.

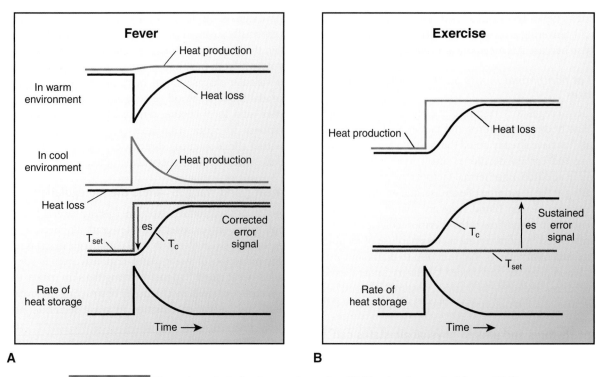

Figure 28.11 **Thermal events during fever and exercise. (A)** The development of fever. **(B)** The increase in T_c during exercise. The es (error signal or load error) is the difference between the T_c and the T_{set}. At the start of a fever, T_{set} has risen, so that T_{set} is higher than T_c and es is negative. At steady state, T_c has risen to equal the new level of T_{set} and es is corrected (i.e., it returns to zero.) At the start of exercise, $T_c = T_{set}$ so that es = 0. At steady state, T_{set} has not changed but T_c has increased and is greater than T_{set}, producing a sustained es. (es is here represented with an *arrow* pointing downward for $T_c < T_{set}$ and with an *arrow* pointing upward for $T_c > T_{set}$.)

Core temperature continues to rise until heat loss has increased enough to match heat production, and core temperature and the heat loss responses reach new steady-state levels. Because the heat loss responses are proportional to the increase in core temperature, the increase in core temperature at steady state is proportional to the rate of heat production and, thus, to the metabolic rate.

A change in ambient temperature causes changes in the levels of sweating and skin blood flow necessary to maintain any given level of heat dissipation. However, the change in ambient temperature also elicits, via direct and reflex effects of the accompanying skin temperature changes, altered responses in the right direction. For any given rate of heat production, there is a certain range of environmental conditions within which an ambient temperature change elicits the necessary changes in heat-dissipating responses almost entirely through the effects of skin temperature changes, with virtually no effect on core temperature. (The limits of this range of environmental conditions depend on the rate of heat production and such individual factors as skin surface area and state of heat acclimatization.) Within this range, the core temperature reached during exercise is nearly independent of ambient temperature. For this reason, it was once believed that the increase in core temperature during exercise is caused by an increase in the thermoregulatory set point, as during fever. As noted, however, the increase in core temperature with exercise is an example of a load error rather than an increase in set point.

This difference between fever and exercise is shown in Figure 28.11. Note that, although heat production may increase substantially (through shivering), when core temperature is rising early during fever, it need not stay high to maintain the fever. In fact, it returns nearly to prefebrile levels once the fever is established. During exercise, however, an increase in heat production not only causes the elevation in core temperature but is necessary to sustain it. Also, although the core temperature is rising during fever, the rate of heat loss is, if anything, lower than it was before the fever began. During exercise, however, the heat-dissipating responses and the rate of heat loss start to increase early and continue to increase as the core temperature rises.

Exercise in the heat can impair cardiac filling.

The rise in core temperature during exercise increases the temperature difference between the core and the skin somewhat but not nearly enough to match the increase in metabolic heat production. Therefore, as we saw earlier, skin blood flow must increase to carry all of the heat that is produced to the skin. In a warm environment, where the temperature difference between core and skin is relatively small, the necessary increase in skin blood flow may be several liters per minute.

The work of providing the skin blood flow required for thermoregulation in the heat may impose a heavy burden on a diseased heart, but in healthy people, the major cardiovascular burden of heat stress results from impaired venous return. As skin blood flow increases, the dilated vascular bed of the skin

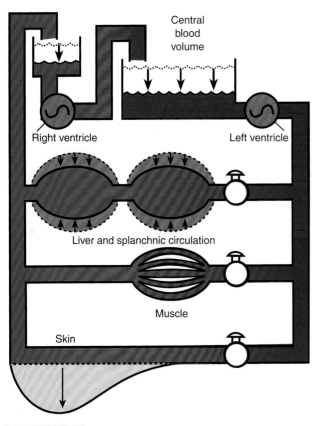

Central blood volume

Right ventricle

Left ventricle

Liver and splanchnic circulation

Muscle

Skin

Figure 28.12 **Cardiovascular strain and compensatory responses during heat stress.** This figure first shows the effects of skin vasodilation on peripheral pooling of blood and the thoracic reservoirs from which the ventricles are filled and, second, the effects of compensatory vasomotor adjustments in the splanchnic circulation. The valves on the *right* represent the resistance vessels that control blood flow through the liver/splanchnic, muscle, and skin vascular beds. *Arrows* show the direction of the changes during heat stress.

becomes engorged with large volumes of blood, reducing central blood volume and cardiac filling (Fig. 28.12). Stroke volume is decreased, and a higher heart rate is required to maintain cardiac output. A decrease in plasma volume, if the large amounts of salt and water lost in the sweat are not replaced, aggravates these effects. Because the main cation in sweat is sodium, a disproportionately large amount of the body water lost in sweat is at the expense of extracellular fluid, including plasma, although this effect is mitigated if the sweat is dilute.

Compensatory responses during exercise in the heat
Several reflex adjustments help maintain cardiac filling, cardiac output, and arterial pressure during exercise and heat stress. The most important of these is constriction of the renal and splanchnic vascular beds. A reduction in blood flow through these beds allows a corresponding diversion of cardiac output to the skin and the exercising muscles. In addition, because the splanchnic vascular beds are compliant, a decrease in their blood flow reduces the amount of blood pooled in them (see Fig. 28.12), helping compensate for decreases in central blood volume caused by reduced plasma volume and blood pooling in the skin.

The degree of vasoconstriction is graded according to the levels of heat stress and exercise intensity. During strenuous exercise in the heat, renal and splanchnic blood flows may fall to 20% of their values in a cool resting subject. Such intense splanchnic vasoconstriction may produce mild ischemic injury to the gut, helping explain the intestinal symptoms that some athletes experience after endurance events. The cutaneous veins constrict during exercise; because most of the vascular volume is in the veins, constriction makes the cutaneous vascular bed less easily distensible and reduces peripheral pooling. Because of the essential role of skin blood flow in thermoregulation during exercise and heat stress, the body preferentially compromises splanchnic and renal flow for the sake of cardiovascular homeostasis. Above a certain level of cardiovascular strain, however, skin blood flow, too, is compromised.

▶ HEAT ACCLIMATIZATION

As in other mammals, thermoregulation is an important aspect of human homeostasis. Humans have been able to adapt to a great diversity of climates, including hot humid and hot arid. High temperatures pose serious stresses for the human body, placing it in great danger of injury and even death. For humans, adjusting to varying climatic conditions includes both physiologic mechanisms as a by-product of evolution and the conscious development of cultural adaptations.

Prolonged or repeated exposure to stressful environmental conditions elicits significant physiologic changes, called **acclimatization**, the process of an individual organism adjusting to a change in its environment. Acclimatization occurs in a short time (days or weeks) and can be reversed once removed from the stress, unlike **adaptation**, which is permanent and passed on to the next generation. Some degree of heat *acclimatization* occurs either by heat exposure alone or by regular strenuous exercise, which raises core temperature and provokes heat loss responses. Indeed, the first summer heat wave produces enough heat acclimatization that most people notice an improvement in their level of energy and general feeling of well-being after a few days. However, the acclimatization response is greater if heat exposure and exercise are combined, causing a greater rise of internal temperature and more profuse sweating. Evidence of acclimatization appears in the first few days of combined exercise and heat exposure, and most of the improvement in heat tolerance occurs within 10 days. The effect of heat acclimatization on performance can be dramatic, and acclimatized subjects can easily complete exercise in the heat that earlier was difficult or impossible.

Changes in basal metabolism, heart rate, cutaneous blood flow, and sweat rate occur with heat acclimatization.

Studies of seasonal variation in basal metabolism indicate an inverse relationship between BMR and ambient temperature. A linear correlation between basal metabolism and mean outdoor temperature within a range of 10°C to 25°C exists, so that a rise of 10°C in monthly mean ambient temperature is accompanied by a fall of 2.5 to 3.0 kcal/m²/h in basal metabolism, a change linked to lower thyroxine levels.

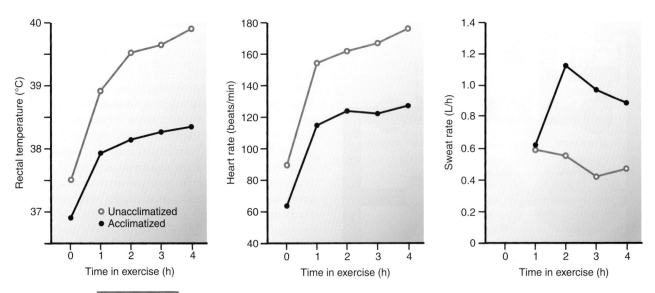

Figure 28.13 Heat acclimatization. These graphs show rectal temperatures, heart rates, and sweat rates during 4-hour exercise (bench stepping, 35 W mechanical power) in humid heat (33.9°C dry bulb, 89% relative humidity, 35 torr ambient vapor pressure) on the first and last days of a 2-week program of acclimatization to humid heat.

During exposure to hot environments, cardiovascular adaptations that reduce the heart rate required to sustain a given level of activity in the heat appear quickly and reach nearly their full development within 1 week. Changes in sweating develop more slowly. After acclimatization, sweating begins earlier and at a lower core temperature (i.e., the core temperature threshold for sweating is reduced). The sweat glands become more sensitive to cholinergic stimulation, and a given elevation in core temperature elicits a higher sweat rate. In addition, the glands become resistant to hidromeiosis and fatigue, and so higher sweat rates can be sustained. These changes reduce the levels of core and skin temperatures reached during a period of exercise in the heat, increase the sweat rate, and enable one to exercise longer. The threshold for cutaneous vasodilation is reduced along with the threshold for sweating, and so the heat transfer from the core to the skin is maintained. The lower heart rate, lower core temperature, and higher sweat rate are the three classic signs of heat acclimatization (Fig. 28.13).

Heat acclimatization modifies fluid and electrolyte balance.

During the first week, total body water and, especially, plasma volume increase. These changes likely contribute to the cardiovascular adaptations. Later, the fluid changes seem to diminish or disappear, although the cardiovascular adaptations persist. In an unacclimatized person, sweating occurs mostly on the chest and back, but during acclimatization, especially in humid heat, the fraction of sweat secreted on the limbs increases to make better use of the skin surface for evaporation. An unacclimatized person who is sweating profusely can lose large amounts of sodium. With acclimatization, the sweat glands become able to conserve sodium by secreting sweat with a sodium concentration as low as 5 mmol/L. This effect is mediated through aldosterone, which is secreted in response to sodium depletion and to exercise and heat exposure. The sweat glands respond to aldosterone

more slowly than the kidneys, requiring several days. Unlike the kidneys, the sweat glands do not escape the influence of aldosterone when sodium balance has been restored but continue to conserve sodium for as long as acclimatization persists. The cell membranes are freely permeable to water so that the movement of water across the cell membranes rapidly corrects any osmotic imbalance between the intracellular and extracellular compartments (see Chapter 2). One important consequence of the salt-conserving response of the sweat glands is that the loss of a given volume of sweat causes a smaller decrease in the volume of the extracellular space than if the sodium concentration of the sweat is high (Table 28.3).

Heat acclimatization is transient, disappearing in a few weeks if not maintained by repeated heat exposure. The components of heat acclimatization are lost in the order in which they were acquired. The cardiovascular changes decay more quickly than the reduction in exercise core temperature and sweating changes.

▶ RESPONSES TO COLD

When the body is exposed to cold temperatures, heat loss is minimized in several ways: (1) sweating stops; (2) minute muscles under the skin contract (**piloerection**), causing "gooseflesh" or "goose bumps"; (3) cutaneous arterioles constrict, thereby rerouting the blood away from the skin; (4) shivering, which increases heat production in skeletal muscles; and (5) fat is converted into energy by mitochondria. Brown fat is specialized for this purpose and is most abundant in newborns.

Arteriole vasoconstriction reroutes blood away from the skin to conserve heat.

Reducing **conductance** is the chief physiologic means of heat conservation during cold exposure. The constriction of cutaneous arterioles reduces skin blood flow and shell

TABLE 28.3 Effect of Sweat Secretion on Body Fluid Compartments and Plasma Sodium Concentration*

| Subject | Condition | Extracellular Space | | Intracellular Space | | Total Body Water | | Osmolality (mOsm/kg) | Plasma [Na⁺] (mmol/L) |
		Volume (L)	Osmotic Content (mOsm)	Volume (L)	Osmotic Content (mOsm)	Volume (L)	Osmotic Content (mOsm)		
A	Initial	15.0	4,350	25.0	7,250	40	11,600	290	140
	Loss of 5 L of sweat, 120 mOsm/L, 60 mmol Na⁺/L	11.9	3,750	23.1	7,250	35	11,000	314	151
	Above condition accompanied by intake of 5 L water	13.6	3,750	26.4	7,250	40	11,000	275	132
B	Loss of 5 L of sweat, 20 mOsm/L, 10 mmol Na⁺/L; accompanied by intake of 5 L water	12.9	4,250	22.1	7,250	35	11,500	329	159
	Above condition accompanied by intake of 5 L water	14.8	4,250	25.2	7,250	40	11,500	288	139

*Each subject has total body water of 40 L. The sweat of subject A has a relatively high [Na⁺] of 60 mmol/L, whereas that of subject B has a relatively low [Na⁺] of 10 mmol/L. Volumes of the extracellular and intracellular spaces are calculated assuming that water moves between the two spaces as needed to maintain osmotic balance.

conductance. However, in extremely cold conditions, excessive vasoconstriction leads to numbness and pale skin. Frostbite only occurs when water within the cells begins to freeze; this destroys the cell, causing the characteristic damage.

Constriction of the superficial limb veins further improves heat conservation by diverting venous blood to the deep limb veins, which lie close to the major arteries of the limbs and do not constrict in the cold. (Many penetrating veins connect the superficial veins to the deep veins so that venous blood from anywhere in the limb potentially can return to the heart via either superficial or deep veins.) In the deep veins, cool venous blood returning to the core can take up heat from the warm blood in the adjacent deep limb arteries. Therefore, some of the heat contained in the arterial blood as it enters the limbs takes a "short circuit" back to the core. When the arterial blood reaches the skin, it is already cooler than the core, and so it loses less heat to the skin than it otherwise would. (When the superficial veins dilate in the heat, most venous blood returns via superficial veins so as to maximize core-to-skin heat flow.) The transfer of heat from arteries to veins by this short circuit is called **countercurrent heat exchange**. This mechanism can cool the blood in the radial artery of a cool but comfortable subject to as low as 30°C by the time it reaches the wrist.

As we saw earlier, the shell's insulating properties increase in the cold as its blood vessels constrict and its thickness increases. Furthermore, the shell includes a fair amount of skeletal muscle in the cold, and although muscle blood flow is believed not to be affected by thermoregulatory reflexes, direct cooling reduces it. In a cool subject, the resulting reduction in muscle blood flow adds to the shell's insulating properties. As the blood vessels in the shell constrict,

blood is shifted to the central blood reservoir in the thorax. This shift produces many of the same effects as an increase in blood volume, including the so-called **cold diuresis** as the kidneys respond to the increased central blood volume.

Shivering is a heat-producing mechanism in response to hypothermia.

Shivering is a physiologic response to early hypothermia in mammals. When core body temperature drops, the shivering reflex, mediated through the hypothalamus, is triggered to maintain homeostasis. Muscle groups around vital organs begin to contract in an attempt to generate heat by expending energy. There are two types of shivering: low intensity and high intensity. During *low intensity*, shivering occurs constantly at low levels over long periods during cold conditions. During *high intensity*, shivering occurs violently over a short period. Both processes consume energy. Low intensity burns fat, whereas glucose is burned with high-intensity shivering.

Once skin blood flow is near minimal, metabolic heat production increases—almost entirely through shivering in human adults. Shivering may increase metabolism at rest more than fourfold—that is, to 350 to 400 W. Although it is often stated that shivering diminishes substantially after several hours and is impaired by exhaustive exercise, such effects are not well understood. In most laboratory mammals, chronic cold exposure also causes **nonshivering thermogenesis**, an increase in metabolic rate that is not a result of muscle activity. Nonshivering thermogenesis appears to be elicited through sympathetic stimulation and circulating catecholamines. It occurs in many tissues, especially in the liver and in **brown adipose tissue**, also called *brown fat,* specialized for nonshivering thermogenesis whose color is imparted by high concentrations of iron-containing respiratory enzymes. Brown adipose tissue is found in human infants, and nonshivering thermogenesis is important for their thermoregulation. The existence of brown adipose tissue and nonshivering thermogenesis in human adults is controversial, but recent evidence strongly suggests the presence of functioning brown adipose deposits in a substantial fraction of adult humans. These are located symmetrically in the supraclavicular and the neck regions with some additional paravertebral, mediastinal, para-aortic, and suprarenal (but no interscapular) localizations and respond with increased activity to sympathetic stimulation and exposure to cold.

Cold acclimatization is an important characteristic in thermal regulation and maintaining homeostasis.

The pattern of human cold acclimatization depends on the nature of the cold exposure. It is partly for this reason that the occurrence of cold acclimatization in humans was controversial for a long time. Our knowledge of human cold acclimatization comes from both laboratory studies and studies of populations whose occupation or way of life exposes them repeatedly to cold temperatures.

Metabolic changes in cold acclimatization

At one time, it was believed that humans must acclimatize to cold as laboratory mammals do—by increasing their metabolic rate. There are a few reports of increased BMR and, sometimes, thyroid activity in the winter. More often, however, increased metabolic rate has not been observed in studies of human cold acclimatization. In fact, several reports indicate the opposite response, consisting of a lower core temperature threshold for shivering, with a greater fall in core temperature and a smaller metabolic response during cold exposure. Such a response would spare metabolic energy and might be advantageous in an environment that is not so cold that a blunted metabolic response would allow core temperature to fall to dangerous levels.

Increased tissue insulation in cold acclimatization

A lower core-to-skin conductance (i.e., increased insulation by the shell) has often been reported in studies of cold acclimatization in which a reduction in the metabolic response to cold occurred. This increased insulation is not a result of subcutaneous fat (in fact, it has been observed in lean subjects) but apparently results from lower blood flow in the limbs or improved countercurrent heat exchange in the acclimatized subjects. In general, the cold stresses that elicit a lower shell conductance after acclimatization involve either cold water immersion or exposure to air that is chilly but not so cold as to risk freezing the vasoconstricted extremities.

▶ CLINICAL ASPECTS OF THERMOREGULATION

Temperature is important clinically because of the presence of fever in many diseases, the effects of many factors on tolerance to heat or cold stress, and the effects of heat or cold stress in causing or aggravating certain disorders.

Fever is one of the body's immunologic responses to a bacterial or viral infection.

Fever is due to an increase in the body temperature regulatory set point. This increase in set point increases muscle tone and shivering. Infection or noninfectious conditions (e.g., inflammatory processes such as collagen vascular diseases, trauma, neoplasms, acute hemolysis, and immunologically mediated disorders) may cause fever. **Pyrogens** are substances that cause fever and may be either exogenous or endogenous. *Exogenous pyrogens* are derived from outside the body; most are microbial products, microbial toxins, or whole microorganisms. The best studied of these is the lipopolysaccharide endotoxin of gram-negative bacteria. Exogenous pyrogens stimulate a variety of cells, especially monocytes and macrophages, to release *endogenous pyrogens,* polypeptides that cause the thermoreceptors in the hypothalamus (and perhaps elsewhere in the brain) to alter their firing rate and input to the central thermoregulatory controller, raising the thermoregulatory set point. The local synthesis and release of **prostaglandin E$_2$** mediate this effect of endogenous pyrogens. Aspirin and other drugs that inhibit the synthesis of prostaglandins also reduce fever.

Fever accompanies disease so frequently and is such a reliable indicator of the presence of disease that body temperature is probably the most commonly measured clinical index. A group of polypeptides called **cytokines** elicits many of the body's defenses against infection and cancer; the endogenous pyrogen is usually a member of this group, **interleukin-1**. However, other cytokines, particularly **tumor necrosis factor**, **interleukin-6**, and the **interferons**, are also pyrogenic in certain circumstances. Elevated body temperature enhances the development of these defenses. How cytokines interact with the hypothalamus remains unclear. However, they are believed to cross the blood–brain barrier by facilitated transport mechanisms, diffuse into areas of the brain where there is no blood–brain barrier, or interact with peripheral neural components of the immune system to signal the hypothalamus to increase the thermal set point. There is evidence to support each mechanism, and it seems likely that each contributes to some extent in various circumstances. If laboratory animals are prevented from developing a fever during experimentally induced infection, survival rates may be dramatically reduced. (Although, in this chapter, fever specifically means an elevation in core temperature resulting from pyrogens, some authors use the term more generally to mean any significant elevation of core temperature.)

Tolerance to heat and cold is dependent on factors affecting thermoregulatory responses.

Regular physical exercise and heat acclimatization increase heat tolerance and the sensitivity of the sweating response. Aging has the opposite effect; in healthy 65-year-old men, the sensitivity of the sweating response is half of that in 25-year-old men. Many drugs inhibit sweating, most obviously those used for their anticholinergic effects, such as atropine and scopolamine. In addition, some drugs used for other purposes, such as glutethimide (a sleep-inducing drug), tricyclic antidepressants, phenothiazines (tranquilizers and antipsychotic drugs), and antihistamines, have some anticholinergic action. All of these and several others have been associated with heatstroke. Congestive heart failure and certain skin diseases (e.g., ichthyosis and anhidrotic ectodermal dysplasia) impair sweating, and in patients with these diseases, heat exposure and especially exercise in the heat may raise body temperature to dangerous levels. Lesions that affect the thermoregulatory structures in the brainstem can also alter thermoregulation. Such lesions can produce **hypothermia** (abnormally low core temperature) if they impair heat-conserving responses. However, **hyperthermia** (abnormally high core temperature) is a more usual result of brainstem lesions and is typically characterized by a loss of both sweating and the circadian rhythm of core temperature.

Certain drugs, such as barbiturates, alcohol, and phenothiazines, and certain diseases, such as hypothyroidism, hypopituitarism, congestive heart failure, and septicemia, may impair the defense against cold. (Hypothermia, instead of the usual febrile response to infection, may accompany septicemia, especially in debilitated patients.) Furthermore, newborns and many healthy older adults are less able than older children and younger adults to maintain adequate body temperature in the cold. This failing appears to be a result of an impaired ability to conserve body heat by reducing heat loss and to increase metabolic heat production in the cold.

Heat stress can lead to pathophysiologic disorders.

The harmful effects of heat stress are exerted through cardiovascular strain, fluid and electrolyte loss, and, especially in heatstroke, tissue injury whose mechanism is uncertain. In a patient suspected of having hyperthermia secondary to heat stress, temperature should be measured in the rectum because hyperventilation may render oral temperature spuriously low.

Heat syncope

Heat syncope is circulatory failure resulting from a pooling of blood in the peripheral veins, with a consequent decrease in venous return and diastolic filling of the heart, resulting in decreased cardiac output and a fall of arterial pressure. Symptoms range from light-headedness and giddiness to loss of consciousness. Thermoregulatory responses are intact, and so core temperature typically is not substantially elevated, and the skin is wet and cool. The large thermoregulatory increase in skin blood flow in the heat is probably the primary cause of the peripheral pooling. Heat syncope affects mostly those who are not acclimatized to heat presumably because the plasma–volume expansion that accompanies acclimatization compensates for the peripheral pooling of blood. Treatment consists in laying the patient down out of the heat to reduce the peripheral pooling of blood and improve the diastolic filling of the heart.

Heat exhaustion

Heat exhaustion, also called *heat collapse*, is probably the most common heat disorder and represents a failure of cardiovascular homeostasis in a hot environment. Collapse may occur either at rest or during exercise and may be preceded by weakness or faintness, confusion, anxiety, ataxia, vertigo, headache, and nausea or vomiting. The patient has dilated pupils and usually sweats profusely. As in heat syncope, reduced diastolic filling of the heart appears to have a primary role in the pathogenesis of heat exhaustion. Although blood pressure may be low during the acute phase of heat exhaustion, the baroreflex responses are usually sufficient to maintain consciousness and may be manifested in nausea, vomiting, pallor, cool or even clammy skin, and rapid pulse. Patients with heat exhaustion usually respond well to rest in a cool environment and oral fluid replacement. In more severe cases, however, intravenous replacement of fluid and salt may be required. Core temperature may be normal or only mildly elevated in heat exhaustion. However, heat exhaustion accompanied by hyperthermia and dehydration may lead to heatstroke. Therefore, patients should be actively cooled if rectal temperature is 40.6°C (105°F) or higher.

The reasons underlying the reduced diastolic filling in heat exhaustion are not fully understood. Hypovolemia contributes if the patient is dehydrated, but heat exhaustion often occurs without significant dehydration. In rats heated to the point of collapse, compensatory splanchnic vasoconstriction develops during the early part of heating but is reversed shortly before the maintenance of blood pressure fails. A similar process may occur in heat exhaustion.

Heatstroke

High core temperature and the development of serious neurologic disturbances with a loss of consciousness and, frequently, convulsions characterize the most severe and dangerous heat disorder. **Heatstroke** occurs in two forms, classical and exertional. In the classical form, the primary factor is environmental heat stress that overwhelms an impaired thermoregulatory system, and most patients have preexisting chronic disease. In exertional heatstroke, the primary factor is high metabolic heat production. Patients with exertional heatstroke tend to be younger and more physically fit (typically, soldiers and athletes) than patients with the classical form. Rhabdomyolysis, hepatic and renal injury, and disturbances of blood clotting are frequent accompaniments of exertional heatstroke. The traditional diagnostic criteria of heatstroke—coma, hot dry skin, and rectal temperature above 41.3°C (106°F)—are characteristic of the classical form. However, patients with exertional heatstroke may have somewhat lower rectal temperatures and often sweat profusely. Heatstroke is a medical emergency, and prompt appropriate treatment is critically important to reducing morbidity and mortality. The rapid lowering of core temperature is the cornerstone of treatment, and immersion in cold water most effectively accomplishes it. With prompt cooling, vigorous hydration, maintenance of a proper airway, avoidance of aspiration, and appropriate treatment of complications, most patients will survive, especially if they were previously healthy.

The pathogenesis of heatstroke is not well understood, but it seems clear that factors other than hyperthermia are involved, even if the action of these other factors partly depends on the hyperthermia. Exercise may contribute more to the pathogenesis than simply metabolic heat production. Elevated plasma levels of several inflammatory cytokines have been reported in patients presenting with heatstroke, suggesting a systemic inflammatory component. No trigger for such an inflammatory process has been established, although several possible candidates exist. One possible trigger is some product(s) of the bacterial flora in the gut, perhaps including lipopolysaccharide endotoxins. Several lines of evidence suggest that sustained splanchnic vasoconstriction may produce a degree of intestinal ischemia sufficient to allow these products to "leak" into the circulation and activate inflammatory responses. Other causes relate to reticuloendothelial system deactivation and heat-induced changes in proinflammatory and anti-inflammatory cytokines.

The preceding diagnostic categories are traditional. However, they are not entirely satisfactory for heat illness associated with exercise because many patients have laboratory evidence of tissue and cellular injury, but they are classified as having heat exhaustion because they do not have the serious neurologic disturbances that characterize heatstroke. Some more recent literature uses the term **exertional heat injury** for such cases. The boundaries of exertional heat injury, with heat exhaustion on one hand and heatstroke on the other, are not clearly and consistently defined, and these categories probably represent parts of a continuum.

Malignant hyperthermia, a rare process triggered by depolarizing neuromuscular-blocking agents or certain inhalational anesthetics, was once thought to be a form of heatstroke but is now known to be a distinct disorder that occurs in people with a mutation of the ryanodine receptor (type 1) on the sarcoplasmic reticulum. In 90% of susceptible people, biopsied skeletal muscle tissue contracts on exposure to caffeine or halothane in concentrations having little effect on normal muscle. Susceptibility may be associated with any of several myopathies, but most susceptible people have no other clinical manifestations. The control of free (unbound) calcium ion concentration in skeletal muscle cytoplasm is severely impaired in susceptible people, and when an attack is triggered, calcium concentration rises abnormally, activating myosin ATPase and leading to an uncontrolled hypermetabolic process that rapidly increases core temperature. Treatment with dantrolene sodium, which appears to act by reducing the release of calcium ions from the sarcoplasmic reticulum, has dramatically reduced the mortality rate of this disorder.

Aggravation of disease states by heat exposure

Other than producing specific disorders, heat exposure aggravates several other diseases. Epidemiologic studies show that during unusually hot weather, mortality may be two to three times that normally expected for the months in which heat waves occur. Deaths ascribed to specific heat disorders account for only a small fraction of the excess mortality (i.e., the increase above the expected mortality). Deaths from diabetes, various diseases of the cardiovascular system, and diseases of the blood-forming organs account for most of the excess mortality.

Drop in core temperature results in hypothermia.

Core temperature drops when the body is exposed to cold and the internal thermoregulatory mechanisms are unable to replenish the heat that is being lost. Hypothermia exhibits characteristic symptoms as body temperature decreases, such as shivering and mental confusion.

Much of our knowledge about the physiologic effects of hypothermia comes from observations of surgical patients.

During the initial phases of cooling, stimulation of shivering through thermoregulatory reflexes overwhelms the Q_{10} effect. Metabolic rate, therefore, increases, reaching a peak at a core temperature of 30°C to 33°C. At lower core temperatures, however, the Q_{10} effect dominates metabolic rate, and thermoregulation is lost. A vicious circle develops, wherein a fall in core temperature depresses metabolism and allows core temperature to fall further, so that at 17°C, the O_2 consumption is about 15% and cardiac output 10% of precooling values.

Hypothermia that is not induced for therapeutic purposes is called **accidental hypothermia**. It occurs in people whose defenses are impaired by drugs (especially ethanol, in the United States), disease, or other physical conditions and in healthy people who are immersed in cold water or become exhausted working or playing in the cold.

Hypothermia is also often induced during surgical procedures, based on the idea that hypothermia reduces metabolic rate via the Q_{10} effect and prolongs the time that tissues can safely tolerate a loss of blood flow. Because the brain is damaged by ischemia soon after circulatory arrest, *controlled hypothermia* is often used to protect the brain during surgical procedures in which its circulation is occluded or the heart is stopped (Clinical Focus 28.2).

CLINICAL FOCUS | 28.2

Hypothermia

Hypothermia is classified according to the patient's core temperature as mild (32°C to 35°C), moderate (28°C to 32°C), or severe (below 28°C). Shivering is usually prominent in mild hypothermia but diminishes in moderate hypothermia and is absent in severe hypothermia. The pathophysiology is characterized chiefly by the depressant effect of cold (via the Q_{10} effect) on multiple physiologic processes and differences in the degree of depression of each process. Other than shivering, the most prominent features of mild and moderate hypothermia are a result of depression of the central nervous system. Beginning with mood changes (commonly, apathy, withdrawal, and irritability), they progress to confusion and lethargy, followed by ataxia and speech and gait disturbances, which may mimic a cerebrovascular accident (stroke). In severe hypothermia, voluntary movement, reflexes, and consciousness are lost and muscular rigidity appears. Cardiac output and respiration decrease as core temperature falls. Myocardial irritability increases in severe hypothermia, causing a substantial danger of ventricular fibrillation, with the risk increasing as cardiac temperature falls. The primary mechanism presumably is that cold depresses conduction velocity in Purkinje fibers more than in ventricular muscle, favoring the development of circus-movement propagation of action potentials. Myocardial hypoxia also contributes. In more profound hypothermia, cardiac sounds become inaudible and pulse and blood pressure are unobtainable because of circulatory depression; the electrical activity of the heart and brain becomes unmeasurable; and extensive muscular rigidity may mimic rigor mortis. The patient may appear clinically dead, but patients have been revived from core temperatures as low as 17°C, and so "no one is dead until warm and dead." The usual causes of death during hypothermia are respiratory cessation and the failure of cardiac pumping because of either ventricular fibrillation or direct depression of cardiac contraction.

Depression of renal tubular metabolism by cold impairs the reabsorption of sodium, causing a diuresis and leading to dehydration and hypovolemia. Acid–base disturbances in hypothermia are complex. Respiration and cardiac output typically are depressed more than metabolic rate, and a mixed respiratory and metabolic acidosis results because of CO_2 retention and lactic acid accumulation and the cold-induced shift of the hemoglobin–O_2 dissociation curve to the left. Acidosis aggravates the susceptibility to ventricular fibrillation.

Treatment consists of preventing further cooling and restoring fluid, acid–base, and electrolyte balance. Patients in mild-to-moderate hypothermia may be warmed solely by providing abundant insulation to promote the retention of metabolically produced heat; those who are more severely affected require active rewarming. The most serious complication associated with treating hypothermia is the development of ventricular fibrillation. Vigorous handling of the patient may trigger this process, but an increase in the patient's circulation (e.g., associated with warming or skeletal muscle activity) may itself increase the susceptibility to such an occurrence, as follows. Peripheral tissues of a hypothermic patient are, in general, even cooler than the core, including the heart, and acid products of anaerobic metabolism will have accumulated in underperfused tissues while the circulation was most depressed. As the circulation increases, a large increase in blood flow through cold, acidotic peripheral tissue may return enough cold, acidic blood to the heart to cause a transient drop in the temperature and pH of the heart, increasing its susceptibility to ventricular fibrillation.

The diagnosis of hypothermia is usually straightforward in a patient rescued from the cold but may be far less clear in a patient in whom hypothermia is the result of a serious impairment of physiologic and behavioral defenses against cold. A typical example is the older person, living alone, who is discovered at home cool and obtunded or unconscious. The setting may not particularly suggest hypothermia, and when the patient comes to medical attention, the diagnosis may easily be missed because standard clinical thermometers are not graduated low enough (usually only to 34.4°C) to detect hypothermia and, in any case, do not register temperatures below the level to which the mercury has been shaken. Because of the depressant effect of hypothermia on the brain, the patient's condition may be misdiagnosed as cerebrovascular accident or other primary neurologic disease. Recognition of this condition depends on the physician's considering it when examining a cool patient whose mental status is impaired and obtaining a true core temperature with a low-reading glass thermometer or other device. ■

INTEGRATED MEDICAL SCIENCES

Therapeutic Hypothermia and Targeted Temperature Management

Cardiac arrest has a grim prognosis with >50% of patients with out-of-hospital cardiac arrest dying at the scene or within a few hours after return of spontaneous circulation (ROSC). The mortality rate after cardiac arrest of patients admitted to the intensive care unit (ICU) is determined largely by (1) their neurological outcomes and (2) the extent of cardiac tissue injury related to an increase in tissue oxidative stress caused by ischemia/reperfusion (I/R) observed within minutes ROSC. In the past decade, this mortality rate has decreased significantly, largely due to increased attention to the post–cardiopulmonary resuscitation (CPR) phase, particularly when therapeutic hypothermia (TH) is included.

Studies have shown that significant levels of oxidative stress in the heart following cardiac arrest stem from the mitochondria within minutes of reperfusion. In isolated cardiomyocytes, simulated I/R induces mitochondrial reactive oxygen species generation, contractile dysfunction, mitochondrial release of cytochrome c, caspase activation, opening of the mitochondrial permeability transition pore, and ultimately cardiomyocyte death. Targeted temperature management (TTM) via TH is the only treatment currently

(Continued)

known to improve survival in clinical post–cardiac arrest patients. This type of cooling is now being induced by emergency personnel during CPR, prior to ROSC, in an attempt to improve cardiovascular function and outcome. In some cardiac arrest survivors, cooling is induced after ROSC, while they are transported to a cardiac catheterization laboratory for possible percutaneous coronary intervention of acute myocardial infarction (AMI). Although intraischemic cooling prior to reperfusion may be highly cardioprotective, the general use of cooling for AMI (non–cardiac arrest) patients remains unproven. Recent studies in mouse cardiomyocytes suggest that cardioprotective hypothermia (5°C drop in temperature) does more than simply slow reperfusion ROS generation. Cardiomyocyte cooling increases Akt phosphorylation resulting from relative inhibition of regulatory phosphatases. Akt is a survival kinase that mediates mitochondrial protection by modulating ROS and nitric oxide (NO) generation. In addition, mild cooling can alter the strength of weak bonds that determine the conformation and association of individual proteins/enzymes that increase protein function—temperature-mediated alterations in mRNA translational efficiency have also been observed. Mildly decreased temperature, like that used in TTM, seem to attenuate proapoptotic signals, including cytochrome c release, Fas and Bax up-regulation, and caspase activation and it activates antiapoptotic mechanisms such as the Erk1/2 pathway and the Akt pathway. The expression of p53 is enhanced by hypothermia, promoting repair after focal ischemia. The levels of neuron-specific enolase, a marker of neuron death, are also reduced in patients treated with

hypothermia following CPR. Cardioprotection is also associated with several kinase signaling pathways including the G-coupled receptors, the reperfusion injury salvage kinase (RISK), and the JAK STAT pathway. The enhanced Akt-NO signaling with reduction of ROS independent of cGMP suggests that TTM is associated with RISK pathway signaling. This is of particular clinical relevance because it suggests that the beneficial effects of TTM soon may be replaced or enhanced by suitable pharmacological agents that target these protein pathways (Vaity et al. *Critical Care* 2015;19:103).

The most beneficial effect of TTM is a decreased metabolic rate (by 6% to 10% per degree of hypothermia), which leads to reduced oxygen consumption. Additionally, neuronal excitotoxicity—where increased calcium influx in the ischemic neurons and elevated glutamate release combine to induce brain damage—can be prevented or diminished by TTM. Furthermore, activation of the physiological immune response by tissue damage, along with leukocyte and macrophage recruitment and complement activation, provokes additional damage in ischemic conditions through the release of free radicals. TTM modulates these immune-related responses and decreases brain edema. These destructive cascades that are the consequence of ischemia result in elevated brain temperature. Both global and localized brain edema impede heat dissipation by venous and lymphatic flow, thereby raising the relative ischemic brain tissue temperature up to 2°C, in contrast to core body temperature. This concept, known as cerebral thermo-pooling, is a critical factor directly affected by TTM. ∎

Chapter Summary

- Humans are homeotherms and regulate temperature within a narrow range.
- Physiologic and physical processes operate continuously to balance heat production and heat loss and to maintain constant body temperature.
- The body can exchange heat energy with the environment via conduction, convection, radiation, and evaporation.
- Internal conduction and circulatory convection transfer heat from the body's core to the shell.
- Most of the metabolic energy used in cell processes is converted into heat within the body, and metabolic rate is the quantity of food energy converted to heat per unit of time.
- Basal metabolic rate is a measure of the metabolic cost of living measured under standard conditions.
- Metabolic rate is estimated by indirect calorimetry, based on measuring a person's rate of oxygen consumption.
- Heat exchange is proportional to surface area and obeys biophysical principles.
- Sweating is essential for heat loss and involves the glandular secretion of a dilute electrolyte solution onto the skin surface that rapidly evaporates in hot environments, cooling the body.
- Raising skin blood flow brings skin temperature nearer to blood temperature, and lowering skin blood flow brings skin temperature nearer to ambient temperature, enabling convective and radiative heat loss.

- Humans have two distinct subsystems for regulating body temperature: behavioral and physiologic thermoregulation; the latter operates through graded control of heat production and heat loss responses.
- Temperature receptors in the body core and skin transmit information about their temperatures through afferent nerves to the brainstem and, especially, the hypothalamus, where much of the integration of temperature information takes place.
- Signals for both heat-dissipating and heat-generating mechanism recruitment are based on the information about core and skin temperatures as well as nonthermal information received by the central nervous system and on the thermoregulatory set point.
- Fever elevates core temperature at rest, heat acclimatization decreases it, and time of day changes it in a cyclic fashion by changing the thermoregulatory set point.
- Intense exercise may increase heat production within the body 10-fold or more, requiring large increases in skin blood flow and sweating to reestablish the body's heat balance.
- Prolonged or repeated exposure to stressful environmental conditions elicits significant physiologic changes that reduce the resulting strain.
- The body maintains core temperature in the cold by minimizing heat loss and, when this response is insufficient, by increasing heat production.
- Fever enhances defense mechanisms, whereas heat and cold stress can cause significant harm.

Chapter Review Questions

1. An unconscious subject is pulled from an icy river and is found to have a mean skin temperature (T_{sk}) of 8°C. Core temperature (T_c), measured rectally, is 30°C. What is this person's mean body temperature (T_b)?

   A. 37°C
   B. 26.9°C
   C. 21.7°C
   D. 17.8°C
   E. 10.6°C

The correct answer is B. As shown in Figure 28.2, in very cold environments, thickness of the shell depends on the thermal environment and the body's need to conserve heat. In the icy river, the shell expands greatly to keep core temperature as high as possible, thus contributing disproportionately to T_b. Using the formula $T_b = 0.65 \times T_c + 0.35 \times T_{sk}$ (in the cold), 19.5 + 2.8 = 22.3°C.

2. A patient is suspected of suffering from mild hyperthyroidism. Her estimated body surface area is 1.6 m², and during a basal metabolic rate test, she consumes 7.5 L of O_2 in 30 minutes. Assuming she is on a mixed diet (energy equivalent of 4.8 Cal/L O_2), what is her metabolic rate?

   A. 18 Cal/m²/h
   B. 24 Cal/m²/h
   C. 36 Cal/m²/h
   D. 45 Cal/m²/h
   E. 54 Cal/m²/h

The correct answer is D. Because BMR is indirectly determined by O_2 consumption at near basal conditions and, in this case, the energy equivalent of that consumption is 4.8 Cal/L O_2, the subject consumed 36 L of O_2 during the 30-minute measurement, corresponding to 36 Cal. Correcting for her body surface area, 1.6 m², and the half hour duration, her metabolic rate is 45 Cal/m²/h. Because the typical female BMR is roughly 36 Cal/m²/h, the suspicion of hyperthyroidism is accurate.

3. What effect on thermoregulatory processes would you expect after infusing N(G)-nitro-L-arginine methyl ester (L-NAME), a competitive inhibitor of nitric oxide synthase (NOS), at 1 mg/kg/h into an exercising subject (treadmill running at room temperature)?

   A. Increased sweating; increased cutaneous vasodilation
   B. Reduced sweating rate; increased cutaneous vasodilation
   C. Reduced sweating rate; decreased cutaneous vasodilation
   D. Increased preoptic area hypothalamic activation
   E. Decreased preoptic area hypothalamic activation

The correct answer is C. As described in the chapter, neuronal nitric oxide (NO) synthase (nNOS)-mediated NO is directly implicated in a significant portion of the withdrawal of sympathetic adrenergic tone and activation of the sympathetic active vasodilator system as well as modifying thermoregulatory sweating, specifically by increasing sweat gland output during whole-body heating. Inhibition of these actions during exercise-induced hyperthermia by the NOS inhibitor L-NAME would be expected to attenuate the response in both thermoregulatory cutaneous vasodilation and sweating.

4. The following are characteristic of which heat stress-related disorder: rapid increase in core temperature due to an uncontrolled hypermetabolic process related to mutation of the ryanodine receptor in susceptible individuals.

   A. Heatstroke
   B. Malignant hyperthermia
   C. Heat exhaustion
   D. Heat syncope
   E. Thermal dystrophy

The correct answer is B. Malignant hyperthermia is a rare process triggered by depolarizing neuromuscular-blocking agents or certain inhalational anesthetics. It was once thought to be a form of heatstroke but is now known to be a distinct disorder that occurs in people with a mutation of the ryanodine receptor (type 1) in the sarcoplasmic reticulum. The other disorders listed are the consequences of either exertional hyperthermia or an inability of thermoregulatory mechanisms to continue to cope with an extremely hot environment.

5. When a subject exercises vigorously in a hot environment, what is the primary thermoregulatory-related cause underlying impaired cardiac filling?

   A. Cutaneous vasodilation
   B. Na^+ loss via sweating
   C. H_2O loss from sweating
   D. Pulmonary hypertension
   E. Reduced ventricular contractility

The correct answer is A. In the heat, skin blood flow increases very significantly (Fig. 28.13) and the dilated vascular bed of the skin becomes engorged with large volumes of blood, reducing central blood volume and venous return. The increased loss of Na^+ and water may aggravate this effect but are secondary. Neither pulmonary hypertension nor reduced ventricular contractility occurs under these conditions. In fact, due to decreased venous return, EDV declines and may thus actually lower contractility.

Clinical Application Exercises 28.1

HEATSTROKE

It was mid-August in San Antonio, another week of consecutive >90°F days with high humidity. Sarah and Katie were using their lunch break to stop by their elderly landlord's house to drop off the rent. When they got there, the landlord was prostrate in his flower garden, semiconscious, his rototiller still idling. While Sarah called 911 on her cell phone, Katie attempted to stir him. He responded weakly, and in a confused state, no sweat evident on his pale dry scalp and arms. EMT personnel advised Sarah to get the victim out of the sun, and, if possible, sprinkle his skin with water,

apply a water-soaked cloth to his forehead, and fan him. Arriving within 10 minutes, the EMT techs found the old man conscious, confused, headachy, and nauseous. At the emergency room, Sarah and Katie learned that their landlord had suffered heatstroke. The physician congratulated them on their quick action, saying their intervention probably saved the old man's life. He assured the two that after intravenous fluid and electrolyte administration and an overnight observation in the hospital, Mr. Leslie would fully recover.

QUESTIONS

1. Define heatstroke and explain how it differs from heat exhaustion.
2. Explain why elderly individuals would have a greater risk of suffering or heatstroke.
3. Explain why sprinkling water on the skin with fanning is useful in this case.
4. When attempting to lower body temperature in response to hyperthermia, one should avoid treatments that induce shivering or vasoconstriction. Why?

ANSWERS

1. Dehydration can lead to heat exhaustion, which is characterized by high heart rate, dizziness, headache, loss of endurance/skill/confusion, and nausea. The skin may still be cool/sweating, but there are clear signs of developing vasoconstriction, for example, pale color. Heat-exhausted individuals have little or no urine production, and that which is produced is highly concentrated. Cramps may be associated with dehydration and rectal temperature may be up to 40°C and the victim may collapse on stopping activity. Heatstroke is similar to heat exhaustion, but with a dry skin. Heatstroke may arise in an individual who has not been identified as suffering from heat exhaustion and has persisted in further activity in the heat.

2. Heat tolerance and the sensitivity of the sweating response are significantly reduced in the elderly. Normal mechanisms for temperature regulation function by causing vasodilation of cutaneous blood vessels when the body temperature rises above the thermoregulatory set point. This increases blood flow near the surface of the body and allows the heat from the core of the body that is carried by the blood to leave the body via radiation, convection, and evaporation.

3. Body heat is transferred to the water on the skin by conduction. The air movement that results from the fan increases the rate of evaporation. As the water on the skin evaporates, it removes heat by endothermic process of evaporative heat loss. These are responses no longer at work in the subject, thus impairing his ability to deal with work in the hot environment.

4. Shivering is an effective homeostatic mechanism that generates heat and raises the body temperature. Peripheral vasoconstriction also occurs as a result of this mechanism and functions to decrease heat loss across the skin thereby keeping warm blood in the core of the body. Both of these responses can occur if treatment for hyperthermia is too intense, and both would thwart body temperature reduction.

thePoint® *Visit* http://thepoint.lww.com/rhoades5e *for additional chapter review Q&A, Clinical Application Exercises, animations, and more!*

Temperature Regulation and Exercise Physiology

29 Exercise Physiology

Active Learning Objectives

Upon mastering the material in this chapter, you should be able to:

- Explain the concept of exercise to predict acute or chronic physiologic responses.
- Explain how maximal oxygen uptake predicts work performance.
- Explain the mechanism whereby substantial regional blood flow shifts occur during dynamic and isometric exercise.
- Explain training effects on both myocardial muscle and the coronary circulation.

- Explain how the respiratory system responds predictably to increased O_2 consumption and CO_2 production with exercise.
- Explain how, in healthy people, muscle fatigue during exercise is mediated.
- Explain how chronic physical activity affects insulin sensitivity and glucose entry into cells.
- Explain how exercise affects age-related functional changes.

Temperature Regulation and Exercise Physiology

Exercise is bodily activity that enhances or maintains physical fitness and overall health or wellness. It is performed for various reasons, including strengthening muscles and the cardiovascular system, honing athletic skills, weight loss, and maintenance, and for enjoyment. Frequent and regular physical exercise boosts the immune system and helps prevent the "diseases of affluence," such as heart disease, cardiovascular disease, type 2 diabetes, and obesity. It also improves mental health, helps prevent depression, helps to promote or maintain positive self-esteem, and can even augment an individual's sex appeal or body image. Exercise, in its many forms, is so common that true physiologic "rest" is rarely observed. During exercise, defined ultimately in terms of skeletal muscle contraction, the body must initiate rapidly integrated adjustments in every organ system in a coordinated response to increased muscular energy demands. So too must the body adjust integratively by adaptive responses in these same systems to days, weeks, and months of repeated bouts of activity. Exercise physiology is thus the branch of physiology that studies the physical and biochemical events that enable physical movement and underlie cellular and systemic adaptations to repeated or chronic activity.

▶ OXYGEN UPTAKE AND EXERCISE

Exercise physiology is the study of acute and chronic changes in response to a wide range of exercise conditions. In addition, many experiments are carried out to study the effect of exercise on pathophysiology and the mechanisms by which exercise can reduce or reverse disease progression. Oxygen uptake is used to quantitate energy expenditure and exercise intensity during these experiments.

Exercise is as varied as it is ubiquitous. A single episode of exercise, or "acute" exercise, may provoke responses different from the adaptations seen when activity is chronic—that is, during *training*. The forms of exercise vary as well. The amount of muscle mass at work (one finger, one arm, both legs), the intensity of the effort, its duration, and the type of muscle contraction (isometric, rhythmic) all influence the body's responses and adaptations.

These many aspects of exercise imply that its interaction with disease is multifaceted. There is no simple answer as to whether or how exercise promotes health. In fact, physical activity can be healthful, harmful, or irrelevant, depending on the patient, the disease, and the specific exercise in question.

Maximal oxygen uptake is used to quantitate energy expenditure during dynamic exercise.

Dynamic exercise is defined as skeletal muscle contractions at changing lengths and with rhythmic episodes of relaxation. Fundamental to any discussion of dynamic exercise is a description of its intensity. Because dynamically exercising muscle primarily generates energy from oxidative metabolism, a traditional standard is to measure, by mouth, the oxygen uptake ($\dot{V}_{O_2}$) of an exercising subject. This measurement is limited to dynamic exercise and usually to the steady state, when exercise intensity and oxygen consumption are stable and no net energy is provided from nonoxidative sources. Three implications of the original oxygen consumption measurements deserve mention. First, the centrality of oxygen usage to work output gave rise to the term "aerobic" exercise. Second, the apparent excess in oxygen consumption during the first minutes of recovery has been termed the **oxygen debt** (Fig. 29.1).

The "excess" oxygen consumption of recovery results from a multitude of physiologic processes, and little usable information is obtained from its measurement. Third, and more useful, during dynamic exercise that uses a large muscle mass, each person has a **maximal oxygen uptake** ($\dot{V}_{O_2}$ max), a ceiling up to 20 times basal consumption that cannot be exceeded, although it can be increased by appropriate training. $\dot{V}_{O_2}$ max is defined as the highest rate at which oxygen can be taken up and used by the body during severe exercise. It is one of the main variables in the field of exercise physiology and is commonly used to indicate cardiorespiratory fitness. $\dot{V}_{O_2}$ max is decreased, all else being equal, by age, bed rest, or increased body fat.

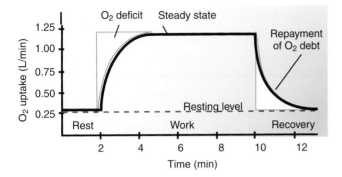

Figure 29.1 Oxygen uptake during rest, exercise, and recovery.

$\dot{V}o_2$ max is also used to express relative work capacity. Obviously, a world-class marathon runner has a greater capacity to consume oxygen than a novice. However, when both are exercising at intensities requiring two thirds of their respective $\dot{V}o_2$ max (the world-class athlete is moving much faster in doing this, as a result of higher capacity), both become exhausted at roughly the same time and for the same physiologic reasons (Fig. 29.2). In the discussion that follows, relative as well as absolute (expressed as L/min of oxygen uptake) work intensities are used to explain physiologic responses. The energy costs and relative demands of some familiar activities are listed in Table 29.1.

What is it that causes oxygen uptake to reach an upper limit? Historically, many arguments claim primacy for either cardiac output (oxygen delivery) or muscle metabolic capacity (oxygen use) limitations. However, it may be that every link in the chain taking oxygen from the atmosphere to the mitochondrion reaches its capacity at about the same time. In practical terms, this means that any lung, heart, vascular,

		% Maximal Oxygen Uptake	
Activity	**Energy Cost (kcal/min)**	**Sedentary 22-Year-Old**	**Sedentary 70-Year-Old**
Sleeping	1	6	8
Sitting	2	12	17
Standing	3	19	25
Dressing, undressing	3	19	25
Walking (3 mi/h)	4	25	33
Making a bed	5	31	42
Dancing	7	44	58
Gardening/ shoveling	8	50	67
Climbing stairs	11	69	92
Crawl swimming (50 m/min)	16	100	
Running (8 mi/h)	16	100	

TABLE 29.1 Absolute and Relative Costs of Daily Activities

or musculoskeletal illness that reduces oxygen flow capacity will diminish a patient's functional capacity. It has been shown that in untrained subjects, the oxygen consumption dominates $\dot{V}o_2$ max, whereas in endurance-trained athletes, oxygen supply is the main limiting factor.

In **isometric exercise**, force is generated at constant muscle length and without rhythmic episodes of relaxation. Isometric work intensity is usually described as a percentage of the **maximal voluntary contraction (MVC)**, the peak isometric force that can be briefly generated for that specific exercise. Analogous to work levels relative to $\dot{V}o_2$ max, the ability to endure isometric effort and many physiologic responses to that effort are predictable when the percentage of MVC among individuals is held constant.

▶ CARDIOVASCULAR RESPONSES TO EXERCISE

Increased energy expenditure with exercise demands more energy production. For prolonged work, the oxidation of foodstuff supplies this energy, with the cardiovascular system carrying the oxygen to working muscles.

Blood flow is preferentially directed to working skeletal muscle during exercise.

Local control of blood flow ensures that only working muscles with increased metabolic demands receive increased blood and oxygen delivery. If the legs alone are active, leg muscle blood flow should increase, whereas arm muscle blood flow remains unchanged or is reduced. At rest, skeletal muscle receives only a small fraction of the cardiac output.

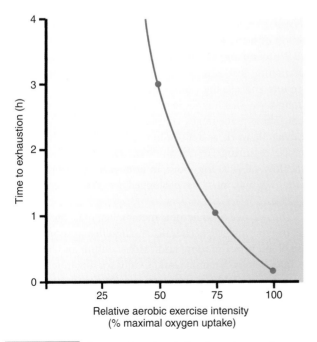

Figure 29.2 Time to exhaustion during dynamic exercise. Exhaustion is predictable on the basis of relative demand on the maximal oxygen uptake.

> **TABLE 29.2** Blood Flow Distribution during Rest and Heavy Exercise in an Athlete

Area	Rest mL/min	Rest %	Heavy Exercise mL/min	Heavy Exercise %
Splanchnic	1,400	24	300	1
Renal	1,100	19	900	4
Brain	750	13	750	3
Coronary	250	4	1,000	4
Skeletal muscle	1,200	21	22,000	86
Skin	500	9	600	2
Other	600	10	100	0.5
Total cardiac output	5,800	100	25,650	100

> **TABLE 29.3** Cardiac Output, Mean Arterial Pressure, and Systemic Vascular Resistance Changes with Exercise

Measure	Rest	Strenuous Dynamic Exercise
Cardiac output (L/min)	6	21
Mean arterial pressure (mm Hg)	90	105
Systemic vascular resistance [(mm Hg • min)/L]	15	5

In dynamic exercise, both total cardiac output and relative and absolute output directed to working skeletal muscle increase dramatically (Table 29.2).

Cardiovascular control during exercise involves systemic regulation (cardiovascular centers in the brain, with their autonomic nervous output to the heart and systemic resistance vessels) in tandem with local control. For millennia, our ancestors successfully used exercise both to escape being eaten and to catch food; therefore, it is no surprise that cardiovascular control in exercise is complex and unique. It is as if a brain software program titled "Exercise" was inserted into the brain as work begins. Initially, the motor cortex is activated; the total neural activity is roughly proportional to the muscle mass and its work intensity. This neural activity communicates with the cardiovascular control centers, reducing vagal tone on the heart (which raises heart rate) and resetting the arterial baroreceptors to a higher level. As work rate is increased further, lactic acid is formed in actively contracting muscles, which stimulates muscle afferent nerves to send information to the cardiovascular center that increases sympathetic outflow to the heart and systemic resistance vessels. However, despite this **muscle chemoreflex** activity, within these same working muscles, low Po_2, increased nitric oxide, vasodilator prostanoids, and associated local vasoactive factors dilate arterioles despite rising sympathetic vasoconstrictor tone. Increased sympathetic drive does elevate heart rate and cardiac contractility, resulting in increased cardiac output; local factors in the coronary vessels mediate coronary vasodilation. Increased sympathetic vasoconstrictor tone in the renal and splanchnic vascular beds and in inactive muscle reduces blood flow to these tissues. Blood flow to these inactive regions can fall 75% if exercise is strenuous. Increased vascular resistance and decreased blood volume in these tissues help maintain blood pressure during dynamic exercise. In contrast to blood flow reductions in the viscera and in inactive muscle, the brain autoregulates blood flow at constant levels independent of exercise. The skin remains vasoconstricted only if thermoregulatory demands are absent. Table 29.3 shows how a profound fall in systemic

vascular resistance matches the enormous rise in cardiac output during dynamic exercise.

Dynamic exercise, at its most intense level, forces the body to choose between maximum muscle vascular dilation and defense of blood pressure. Blood pressure is, in fact, maintained. During strenuous exercise, sympathetic drive can begin to limit vasodilation in active muscle. When exercise is prolonged in the heat, increased skin blood flow and sweating-induced reduction in plasma volume both contribute to the risk of hyperthermia and hypotension (heat exhaustion). Although chronic exercise provides some heat acclimatization, even highly trained athletes are at risk of hyperthermia and hypotension if work is prolonged and water is withheld in demanding environmental conditions.

Isometric exercise causes a somewhat different cardiovascular response. Muscle blood flow increases relative to the resting condition, as does cardiac output, but the higher mean intramuscular pressure limits these flow increases much more than when exercise is rhythmic. Because the blood flow increase is blunted inside a statically contracting muscle, the consequences of hard work with too little oxygen appear quickly: a shift to anaerobic metabolism, the production of lactic acid, a rise in the **adenosine diphosphate (ADP)/adenosine triphosphate (ATP)** ratio, and fatigue. Maintaining just 50% of the MVC is agonizing after about 1 minute and usually cannot be continued after 2 minutes. A long-term sustainable level is only about 20% of maximum. These percentages are much less than the equivalent for dynamic work, as defined in terms of $\dot{V}o_2$ max. Rhythmic exercise requiring 70% of the $\dot{V}o_2$ max can be maintained in healthy subjects for about an hour, whereas work at 50% of the $\dot{V}o_2$ max may be prolonged for several hours (see Fig. 29.2).

The reliance on anaerobic metabolism in isometric exercise triggers muscle ischemic chemoreflex responses that raise blood pressure more and cardiac output and heart rate less than that in dynamic work. Oddly, for dynamic exercise, the elevation of blood pressure is most pronounced when a medium muscle mass is working. This response results from the combination of a small, dilated active muscle mass with powerful central sympathetic vasoconstrictor drive. Typically, the arms exemplify a medium muscle mass; shoveling snow is a good example of primarily arm and heavily isometric exercise. Shoveling snow can be risky for those

TABLE 29.4 Acute Cardiac Response to Graded Exercise in a 30-Year-Old Untrained Woman

Exercise Intensity	Oxygen Uptake (L/min)	Heart Rate (beats/min)	Stroke Volume (mL/beat)	Cardiac Output (L/min)
Rest	0.25	72	70	5
Walking	1.0	110	90	10
Jogging	1.8	150	100	15
Running fast	2.5	190	100	19

in danger of stroke or heart attack because it substantially raises systemic arterial pressure. The elevated pressure places already-compromised cerebral arteries at risk and presents an ischemic or failing heart with a greatly increased afterload.

Cardiovascular responses differ with acute and chronic exercise.

In acute dynamic exercise, vagal withdrawal and increases in sympathetic outflow elevate heart rate and contractility in proportion to exercise intensity (Table 29.4). Factors enhancing venous return also aid cardiac output in dynamic exercise. These include the "muscle pump," which compresses veins as muscles rhythmically contract, and the "respiratory pump," which increases breath-by-breath oscillations in intrathoracic pressure (see Chapter 17). The importance of these factors is clear in patients with heart transplants who lack extrinsic cardiac innervation. Stroke volume rises in cardiac transplant patients with increasing exercise intensity as a result of increased venous return that enhances cardiac preload. In addition, circulating epinephrine and norepinephrine from the adrenal medulla and norepinephrine from sympathetic nerve spillover augment heart rate and contractility.

Maximal dynamic exercise yields a maximal heart rate: further vagal blockade (e.g., via pharmacologic means) cannot elevate heart rate further. Stroke volume, in contrast, reaches a plateau in moderate work and is unchanged as exercise reaches its maximum intensity (see Table 29.4). This plateau occurs in the face of ever-shortening ventricular filling time, testimony to the increasing effectiveness of the mechanisms that enhance venous return and those that promote cardiac contractility. Sympathetic stimulation decreases left ventricular volume and pressure at the onset of cardiac relaxation (as a result of increased ejection fraction), leading to more rapid ventricular filling early in diastole. This helps maintain stroke volume as diastole shortens. Even in untrained subjects, the ejection fraction (stroke volume as a percentage of end-diastolic volume) reaches 80% in strenuous exercise.

The increased blood pressure, heart rate, stroke volume, and cardiac contractility seen in exercise all increase myocardial oxygen demands. A linear increase in coronary blood flow during exercise that can reach a value five times the

basal level meets these demands. Local, metabolically linked factors (nitric oxide, adenosine, and the activation of ATP-sensitive K^+ channels) acting on coronary resistance vessels in defiance of sympathetic vasoconstrictor tone drive this increase in flow. Coronary oxygen extraction, high at rest, increases further with exercise (up to 80% of delivered oxygen). In healthy subjects, there is no evidence of myocardial ischemia under any exercise condition, and there may be a coronary vasodilator reserve in even the most intense exercise (Clinical Focus 29.1).

The heart adapts to chronic exercise overload much as it does to high-demand pathologic states: by increasing left ventricular volume when exercise requires high blood flow and by left ventricular hypertrophy when exercise creates high systemic arterial pressure (high afterload). Consequently, the hearts of subjects adapted to prolonged, rhythmic exercise that involves relatively low arterial pressure exhibit large left ventricular volumes with normal wall thickness, whereas wall thickness is increased at normal volume in those adapted to activities involving isometric contraction and greatly elevated arterial pressure such as lifting weights.

The larger left ventricular volume in subjects chronically active in dynamic exercise leads directly to larger resting and exercise stroke volume. A simultaneous increase in vagal tone and decrease in β-adrenergic sensitivity enhance the resting and exercise bradycardia seen after training, so that in effect the trained heart operates further up the ascending limb of its length–tension relationship. Nevertheless, resting bradycardia is a poor indicator of endurance fitness because genetic factors explain a much larger proportion of the individual variation in resting heart rate than does training.

The effects of endurance training on coronary blood flow are partly mediated through changes in myocardial oxygen uptake. Because myocardial oxygen consumption is roughly proportional to the rate–pressure product (heart rate × mean arterial pressure) and because heart rate falls after training at any absolute exercise intensity, coronary flow at a fixed submaximal workload is reduced in parallel. Training, however, increases the peak coronary blood flow, just as it increases cardiac muscle capillary density and peak capillary exchange capacity. Training also improves endothelium-mediated regulation, responsiveness to adenosine, and control of intracellular free calcium ions within coronary vessels. Preserving endothelial vasodilator function may be the primary benefit of chronic physical activity on the coronary circulation.

Exercise increases levels of "good" cholesterol.

Chronic, dynamic exercise is associated with increased circulating levels of **high-density lipoproteins (HDLs)** and reduced **low-density lipoproteins (LDLs)**, such that the ratio of HDL to total cholesterol is increased. These changes in cholesterol fractions occur at any age if exercise is regular. Weight loss and increased insulin sensitivity, which typically accompany increased chronic physical activity in sedentary

CLINICAL FOCUS | 29.1

Stress Testing

To detect coronary artery disease, physicians often record an electrocardiogram (ECG), but at rest, many disease sufferers have a normal ECG. To increase demands on the heart and coronary circulation, an ECG is performed while the patient walks on a treadmill or rides a stationary bicycle. It is sometimes called a stress test. Exercise increases the heart rate and the systemic arterial blood pressure. These changes increase cardiac work and the demand for coronary blood flow. In many patients, coronary blood flow is adequate at rest but, because of coronary arterial blockage, cannot rise sufficiently to meet the increased demands of exercise. During a stress test, specific ECG changes can indicate that cardiac muscle is not receiving sufficient blood flow and oxygen delivery.

As heart rate increases during exercise, the length of any portion of the ECG (e.g., the R wave) becomes shorter (see Figure).

In patients suffering from ischemic heart disease, however, other changes occur. Most common is an abnormal depression between the S and T waves, known as *ST-segment depression* (see Figure). Depression of the ST segment arises from changes in cardiac muscle electrical activity secondary to lack of blood flow and oxygen delivery.

During the stress test, the ECG is continuously analyzed for changes while blood pressure and arterial blood oxygen saturation are monitored. At the start of the test, the exercise load is mild. The load is increased at regular intervals, and the test ends when the patient becomes exhausted, the heart rate safely reaches a maximum, significant pain occurs, or abnormal ECG changes are noted. With proper supervision, the stress test is a safe method for detecting coronary artery disease. Because the exercise load is gradually increased, the test can be stopped at the first sign of problems. ■

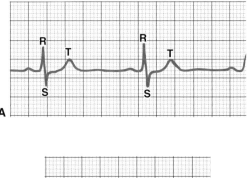

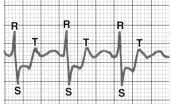

Effect of exercise on the electrocardiogram (ECG) in a patient with ischemic heart disease. (A) The ECG is normal at rest. **(B)** During exercise, the interval between R waves is reduced, and the ECG segment between the S and T waves is depressed.

---

people, undoubtedly contribute to these changes in plasma lipoproteins. Nonetheless, in subjects with lipoprotein levels that place them at high risk for coronary heart disease, exercise appears to be an essential adjunct to dietary restriction and weight loss for lowering LDL cholesterol levels. Because exercise acutely and chronically enhances fat metabolism and cellular metabolic capacities for β-oxidation of free fatty acids, it is not surprising that regular activity increases both muscle and adipose tissue lipoprotein lipase activity. Changes in lipoprotein lipase activity, in concert with increased lecithin–cholesterol acyltransferase activity and apo A-I synthesis, enhance the levels of circulating HDLs. Evidence from interventional studies repeatedly demonstrates the effectiveness of proper diet and exercise in the prevention of metabolic syndrome and type 2 diabetes in a broad range of subjects.

Exercise prevents cardiovascular diseases and reduces mortality.

There is a direct relationship between physical inactivity and cardiovascular mortality, and physical inactivity is an independent risk factor for the development of coronary artery disease. Changes in the ratio of HDL to total cholesterol that take place with regular physical activity reduce the risk of atherogenesis and coronary artery disease in active people, as compared with those who are sedentary. A lack of exercise is established as a risk factor for coronary heart disease similar in magnitude to hypercholesterolemia, hypertension, and smoking. A reduced risk grows out of the changes in lipid profiles noted above, reduced insulin requirements and increased insulin sensitivity, and reduced cardiac β-adrenergic responsiveness and increased vagal tone. When coronary ischemia does occur, increased vagal tone may reduce the risk of fibrillation.

Regular exercise often, but not always, reduces resting blood pressure. Why some subjects respond to chronic activity with a resting blood pressure decline and others do not remain unknown. Responders typically show diminished resting sympathetic tone, so that systemic vascular resistance falls. In obesity-linked hypertension, declining insulin secretion and increasing insulin sensitivity with exercise may explain the salutary effects of combining training with weight loss. Nonetheless, because some obese people who exercise and lose weight show no blood pressure changes, exercise remains adjunctive therapy for hypertension.

Cardiovascular response to pregnancy is similar to chronic exercise.

The physiologic demands and adaptations of pregnancy in some ways are similar to those of chronic exercise. Both increase blood volume, cardiac output, skin blood flow, and caloric expenditure. Exercise clearly has the potential to be deleterious to the fetus. Acutely, it increases body core temperature, causes splanchnic (hence, uterine and umbilical) vasoconstriction, and alters the endocrinologic milieu; chronically, it increases caloric requirements. This last demand may be devastating if food shortages exist: the superimposed caloric demands of successful pregnancy and lactation are estimated at 80,000 kcal. Given adequate nutritional resources, however, there is little evidence of other damaging effects of maternal exercise on fetal development. The failure of exercise to harm well-nourished pregnant women may relate in part to the increased maternal and fetal mass and blood volume, which reduce specific heat loads, moderate vasoconstriction in the uterine and umbilical circulations, and diminish the maternal exercise capacity.

At least in previously active women, even the most intense concurrent exercise regimen (unless associated with excessive weight loss) does not alter fertility, implantation, or embryogenesis, although the combined effects of exercise on insulin sensitivity and central obesity can restore ovulation in anovulatory obese women suffering from polycystic ovary disease. Regular exercise may reduce the risk of spontaneous abortion of a chromosomally normal fetus. Continued exercise throughout pregnancy characteristically results in normal-term infants after relatively brief labor. These infants are usually normal in length and lean body mass but reduced in fat. Maternal exercise through improved glucose tolerance reduces the risk of large infant size for gestational age, which is increased in diabetic mothers. The incidences of umbilical cord entanglement, abnormal fetal heart rate during labor, stained amniotic fluid, and low fetal responsiveness scores may all be reduced in women who are active throughout pregnancy. Further, when examined 5 days after birth, newborns of exercising women perform better in their ability to orient to environmental stimuli and their ability to quiet themselves after sound and light stimuli than weight-matched children of nonexercising mothers.

▶ RESPIRATORY RESPONSES TO EXERCISE

Increased breathing is perhaps the single most obvious physiologic response to acute dynamic exercise. Figure 29.3 shows that minute ventilation (the product of breathing frequency and tidal volume) initially increases linearly with work intensity and then supralinearly beyond that point. Ventilation of the lungs in exercise is linked to the twin goals of oxygen uptake and carbon dioxide removal.

Ventilation matches metabolic demands during a wide range of exercise conditions, but the control mechanisms are poorly understood.

Exercise increases oxygen consumption and carbon dioxide production by working muscles, and the pulmonary response is precisely calibrated to maintain homeostasis

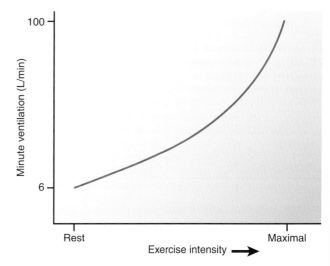

Figure 29.3 **Minute ventilation is dependent on the intensity of dynamic exercise.** Ventilation rises linearly with intensity until exercise nears maximal levels.

of these gases in arterial blood. In mild or moderate work, arterial P_{O_2} (and, hence, oxygen content), P_{CO_2}, and pH all remain unchanged from resting levels (Table 29.5). The respiratory muscles accomplish this severalfold increase in ventilation primarily by increasing tidal volume, without provoking sensations of dyspnea.

More intense exercise presents the lungs with tougher challenges. Near the halfway point from rest to maximal dynamic work, lactic acid formed in working muscles begins to appear in the circulation. This occurs when it is produced faster than it can be removed (metabolized). This point, which depends on the type of work involved and the person's training status, is called the **anaerobic threshold** or lactate threshold. Lactate concentration gradually rises with work intensity, as more and more muscle fibers must rely on anaerobic metabolism. Almost fully dissociated, lactic acid causes metabolic acidosis. During exercise, healthy lungs respond to lactic acidosis by further increasing ventilation, lowering the arterial P_{CO_2}, and maintaining arterial blood pH at normal levels. It is the response to acidosis that spurs the supralinear ventilation rise seen in strenuous exercise (see Fig. 29.3). Through a range of exercise levels, the respiratory system fully compensates the pH effects of lactic acid. However, eventually in the hardest work—near exhaustion—ventilatory compensation becomes only partial, and both pH and arterial P_{CO_2} may fall well below resting values (see Table 29.5). Tidal volume continues to increase until pulmonary stretch receptors limit it, typically at or near half of the vital capacity. Frequency increases at high tidal volume produce the remainder of the ventilatory volume increases.

Hypercapnia relative to carbon dioxide production in heavy exercise helps maintain arterial oxygenation. The blood returned to the lungs during exercise is more thoroughly depleted of oxygen because active muscles with high oxygen extraction receive most of the cardiac output. Because the pulmonary arterial P_{O_2} is reduced in exercise, blood shunted past ventilated areas can profoundly depress systemic arterial oxygen content. Other than having a diminished oxygen content, pulmonary arterial blood flow (cardiac output) rises during

> **TABLE 29.5** Acute Respiratory Response to Graded Dynamic Exercise in a 30-Year-Old Untrained Woman

Exercise Intensity	Ventilation (L/min)	Ventilation/Perfusion Ratio	Alveolar P_{O_2} (mm Hg)	Arterial P_{O_2} (mm Hg)	Arterial P_{CO_2} (mm Hg)	Arterial pH
Rest	5	1	103	100	40	7.4
Walking	20	2	103	100	40	7.4
Jogging	45	3	106	100	36	7.4
Running fast	75	4	110	100	25	7.3

exercise. In compensation, ventilation rises faster than cardiac output: the ventilation/perfusion ratio of the lung rises from near 1 at rest to >4 with strenuous exercise (see Table 29.5). Healthy subjects maintain nearly constant arterial P_{O_2} with acute exercise, although the alveolar-to-arterial P_{O_2} gradient does rise. This increase shows that despite the increase in the ventilation/perfusion ratio, areas of relative pulmonary underventilation and, possibly, some mild diffusion limitation exist even in highly trained, healthy individuals.

The ventilatory control mechanisms in exercise remain undefined. Where there are stimuli—such as in mixed venous blood, which is hypercapnic and hypoxic in proportion to exercise intensity—there are seemingly no receptors. Conversely, where there are receptors—the carotid bodies, the lung parenchyma or airways, the brainstem bathed by cerebrospinal fluid—no stimulus proportional to the exercise demand exists. Paradoxically, the central chemoreceptor is immersed in increasing alkalinity as exercise intensifies a consequence of blood–brain barrier permeability to CO_2 but not hydrogen ions. Perhaps, exercise respiratory control parallels cardiovascular control, with a central command proportional to muscle activity directly stimulating the respiratory center and feedback modulation from the lung, respiratory muscles, chest wall mechanoreceptors, and carotid body chemoreceptors.

Training does not significantly alter the respiratory system.

The effects of training on the healthy pulmonary system are minimal. Lung diffusing capacity, lung mechanics, and even lung volumes change little, if at all, with training. The widespread assumption that training improves vital capacity is false; even exercise designed specifically to increase inspiratory muscle strength elevates vital capacity by only 3%. The demands placed on respiratory muscles increase their endurance, an adaptation that may reduce the sensation of dyspnea in exercise. Nevertheless, the primary respiratory changes with training are secondary to lower lactate production that reduces ventilatory demands at previously heavy absolute work levels. Nonetheless, supervised exercise training that includes whole-body exercise such as cycling and walking is one of the main components of pulmonary rehabilitation in respiratory diseases such as chronic obstructive pulmonary disease (COPD) where training can improve exercise capacity and health-related quality of life (Clinical Focus 29.2).

CLINICAL FOCUS | 29.2

Exercise in Patients with Emphysema

Normally, the respiratory system does not limit exercise tolerance. In healthy people, arterial blood saturation with oxygen, which averages 98% at rest, is maintained at or near 98% in even the most strenuous dynamic or isometric exercise. The healthy response includes the ability to augment ventilation more than cardiac output; the resulting rise in the ventilation/perfusion ratio counterbalances the falling oxygen content of mixed venous blood.

In patients with emphysema, ventilatory limitations to exercise occur long before either skeletal muscle oxidative capacity or the ability of the cardiovascular system to deliver oxygen to exercising muscle imposes ceilings. These limitations are manifest during a stress test on the basis of three primary measurements. First, patients with ventilatory limitations typically cease exercise at a relatively low heart rate, indicating that exhaustion is a result of factors unrelated to

cardiovascular limitations. Second, their primary complaint is usually shortness of breath or dyspnea. In fact, patients with chronic obstructive pulmonary disease often first seek medical evaluation because of dyspnea experienced during such routine activities as climbing a flight of stairs. In healthy people, exhaustion is rarely associated solely with dyspnea. In emphysematous patients, exercise-induced dyspnea results, in part, from respiratory muscle fatigue exacerbated by diaphragmatic flattening brought on by loss of lung elastic recoil. Third, in emphysematous patients, arterial oxygen saturation characteristically falls steeply and progressively with increasing exercise, sometimes reaching dangerously low levels. In emphysema, the inability to fully oxygenate blood at rest is compounded during exercise by increased pulmonary blood flow and by increased exercise oxygen extraction that more fully desaturates blood returning to the lungs. ■

▶ SKELETAL MUSCLE AND BONE RESPONSES TO EXERCISE

Events within exercising skeletal muscle are a primary factor in fatigue. These same events, when repeated during training, lead to adaptations that increase exercise capacity and retard fatigue during similar work. Skeletal muscle contraction also increases stresses placed on bone, leading to specific bone adaptations.

Muscle fatigue is independent of lactic acid.

Muscle fatigue is defined as a loss of muscle power that results from a decline in both force and velocity. Muscle fatigue differs from *muscle weakness* or *damage* in that the loss of power with fatigue is reversible by rest. Although the exact causes of muscle fatigue and the relative importance of particular factors remain controversial, it is clear that an individual's state of fitness, dietary status, fiber-type composition, and intensity and duration of the exercise all affect the process. Historically, an increase in intracellular H^+ (decrease in cell pH) was thought to contribute to muscle fatigue by direct inhibition of the actin–myosin cross bridge, leading to a reduction in contractile force. Although strenuous exercise can reduce intramuscular pH to values as low as 6.8 (arterial blood pH may fall to 7.2), evidence suggests that elevation in $[H^+]$, although a substantial factor in declining force, is not the sole cause of fatigue. The best correlate of fatigue in healthy subjects is ADP accumulation in the face of normal or slightly reduced ATP, such that the ADP/ATP ratio is high. ADP is implicated in slowing cross-bridge cycling. During periods of high-energy demand, the ATP concentration initially remains almost constant, whereas creatine phosphate (CrP) breaks down to Cr and inorganic phosphate (Pi). Although Cr has little effect on contractile function, Pi may cause a marked decrease of myofibrillar force production and Ca^{2+} sensitivity as well as sarcoplasmic reticular Ca^{2+} release. Accordingly, increased Pi is considered to be a major cause of fatigue. There is widespread use of extra Cr intake among athletes, not only in elite athletes but also among people exercising on a recreational level. Cr enters muscle cells via a Na^+-dependent transporter in the sarcolemma. Inside the cell, Cr is phosphorylated by creatine kinase, and the [PCr]/[Cr] ratio basically depends on the energy state of the cell. Extra Cr intake increases the total muscle Cr content by up to ~20%. Cr supplementation has a modest positive effect on muscle performance during bouts of short-term (~10 seconds), high-power exercise, whereas it does not improve function during more long-lasting types of muscle activity. This corresponds to the fact that PCr breakdown contributes to a relatively large fraction of the ATP supply during the first seconds of high-intensity muscle activity, whereas the PCr contribution is minimal during prolonged exercise. Because the complete oxidation of glucose, glycogen, or free fatty acids to carbon dioxide and water is the major source of energy in prolonged work, patients with defects in glycolysis or electron transport exhibit a reduced ability to sustain exercise.

These metabolic defects are distinct from another group of disorders exemplified by the various muscular dystrophies.

In these illnesses, the loss of active muscle mass as a result of fat infiltration, cellular necrosis, or atrophy reduces exercise tolerance despite normal capacities (in healthy fibers) for ATP production. For this reason, Cr supplementation is also used in an attempt to improve muscle function in these diseases. Other potential factors in the etiology of fatigue may occur centrally (pain from fatigued muscle may feed back to the brain to lower motivation and, possibly, to reduce motor cortical output) or at the level of the motor neuron or the neuromuscular junction. Also, skeletal muscles produce ammonia during exercise. This ammonia is taken up by the brain in proportion to its cerebral arterial concentration. Ammonia accumulation is currently thought to be a factor in the sensation of fatigue.

Endurance training enhances muscle oxidative capacity.

Within skeletal muscle, adaptations to training are specific to the form of muscle contraction. Increased activity with low loads results in increased oxidative metabolic capacity without hypertrophy; increased activity with high loads produces muscle hypertrophy. Increased activity without overload increases capillary and mitochondrial density, myoglobin concentration, and virtually the entire enzymatic machinery for energy production from oxygen (Table 29.6). Coordination of energy-producing and energy-using systems in muscle ensures that even after atrophy, the remaining contractile proteins are adequately supported metabolically. In fact, the easy fatigability of atrophied muscle is a result of the requirement that more motor units be recruited for identical external force; the fatigability per unit cross-sectional area is normal. Factors outside the muscle limit the magnitude of the skeletal muscle endurance training response, because cross-innervation or chronic stimulation of muscles in animals can produce adaptations five times larger than those created by the most intense and prolonged exercise.

Local adaptations of skeletal muscle to endurance activity reduce reliance on carbohydrate as a fuel and allow more metabolism of fat, prolonging endurance and decreasing lactic acid accumulation. Decreased circulating lactate, in turn, reduces the ventilatory demands of heavier work. Because metabolites accumulate less rapidly inside trained muscle, there is reduced chemosensory feedback to the central nervous system at any absolute workload. This reduces sympathetic outflow to the heart and blood vessels, reducing cardiac oxygen demands at a fixed exercise level.

Isometric contraction stimulates muscle hypertrophy.

It is common knowledge that walking downhill is easier than walking uphill, but the mechanisms underlying this phenomenon are complex. Forces generated by muscles are identical in the two situations. However, moving the body uphill against gravity involves muscle shortening or **concentric contractions**. In contrast, walking downhill primarily involves muscle tension development that resists muscle lengthening or **eccentric contractions**. All routine forms of physical activity, in fact, involve combinations of concentric, eccentric, and isometric contractions. Because less ATP is required

TABLE 29.6	Effects of Training and Immobilization on the Human Biceps Brachii Muscle in a 22-Year-Old Woman			
Measurement	After Strength Training	After 4 Months' Immobilization	Sedentary	After Endurance Training
Total number of cells	300,000	300,000	300,000	300,000
Total cross-sectional area (cm^2)	10	10	13	6
Isometric strength (% control)	100	100	200	60
Fast-twitch fibers (% by number)	50	50	50	50
Fast-twitch fibers, average area ($\mu m^2 \times 10^2$)	67	67	87	40
Capillaries/fiber	0.8	1.3	0.8	0.6
Succinate dehydrogenase activity/unit area (% control)	100	150	77	100

for force development during a contraction when external forces lengthen the muscle, the number of active motor units is reduced and energy demands are less for eccentric work. However, perhaps because the force per active motor unit is greater in eccentric exercise, eccentric contractions can readily cause muscle damage. This damage includes weakness (apparent the first day), soreness and edema (delayed 1 to 3 days in peak magnitude), and elevated plasma levels of intramuscular enzymes (delayed 2 to 6 days). Histologic evidence of damage may persist for 2 weeks. Damage is accompanied by an acute-phase reaction that includes complement activation, increases in circulating cytokines, neutrophil mobilization, and increased monocyte cell adhesion capacity. Training adaptation to the eccentric components of exercise is efficient; soreness after a second episode is minimal if it occurs within 2 weeks of a first episode.

Eccentric contraction–induced muscle damage and its subsequent response may be the essential stimulus for muscle hypertrophy. Although standard resistance exercise involves a mixture of contraction types, careful studies show that when one limb works purely concentrically and the other purely eccentrically at equivalent force, only the eccentric limb hypertrophies. The immediate changes in actin and myosin production that lead to hypertrophy are mediated at the posttranslational level; after a week of loading, mRNA for these proteins is altered.

Despite our understanding that feeding and resistance exercise affect muscle protein synthesis and growth, researchers have now begun to study the signaling pathways that are activated with either stimulus in humans. It is known that proteins in the Akt/protein kinase B–rapamycin (mTOR)–S6 kinase pathway are involved in stimulating ribosomal assembly, biogenesis, and global protein synthesis, but the extent of signal to response is far from clear. Upstream of mTOR is the tuberous sclerosis complex (TSC), which is a heterodimeric complex of the TSC1 and TSC2 gene products, hamartin and tuberin, respectively. Studies have shown that the TSC1/TSC2 heterodimer regulates cell growth and cell proliferation as a downstream component of the PI3K (phosphoinositide 3-kinase)–Akt signaling, which modulates signal transduction through mTOR. The cellular mechanisms for hypertrophy are

also known to include the induction of insulin-like growth factor-1 and up-regulation of several members of the fibroblast growth factor family, although skeletal muscle hypertrophy can occur in the absence of circulating pituitary hormones, insulin, or local growth factors in response to increased mechanical loading. Although the actual biochemical events that link the mechanical loading of a muscle or muscle fiber to mTOR signaling remain unclear, it is known that mechanical loading or stretch acts independently of amino acids and growth factors to activate mTOR signaling in skeletal muscle. Passive mechanical loading is sufficient to activate mTOR signaling in skeletal muscle, and studies have shown that there is a cytoskeletal requirement for communication of the signaling pathway from mechanical stretch to mTOR. It is likely that this mechanical pathway acts synergistically with changes in amino acid uptake and growth factor availability to contribute to the prolonged activation of mTOR signaling following high-resistance contractions or mechanical overload.

Exercise plays an important role in calcium homeostasis.

Skeletal muscle contraction acts across joints and thus applies force to bone. Because the architecture of bone remodeling involves osteoblast and osteoclast activation in response to loading and unloading, physical activity is a major site-specific influence on bone mineral density and geometry. Repetitive physical activity can create excessive strain, leading to inefficiency in bone remodeling and stress fracture; however, extreme inactivity allows osteoclast dominance and bone loss.

The forces applied to bone during exercise are related both to the weight borne by the bone during activity and to the strength of the involved muscles. Consequently, bone strength and density appear to be closely related to applied gravitational forces and to muscle strength. This suggests that exercise programs to prevent or treat **osteoporosis** should emphasize weight-bearing activities and strength as well as endurance training. Adequate dietary calcium is essential for any exercise effect: weight-bearing activity enhances spinal bone mineral density in postmenopausal women only when calcium intake exceeds 1 g/d. Because exercise may

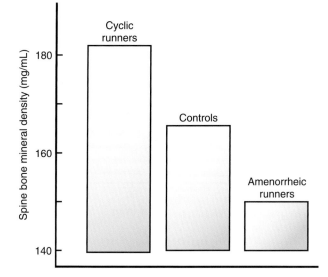

Figure 29.4 **Exercise stimulates bone density.** This graph shows spine bone density in young adult women who are nonathletes (controls), distance runners with regular menstrual cycles (cyclic runners), and distance runners with amenorrhea (amenorrheic runners). Differences from controls indicate the roles that exercise and estrogen play in determination of bone mineral density.

also improve gait, balance, coordination, proprioception, and reaction time, even in older and frail persons, chronic activity reduces the risks of falls and osteoporosis. In fact, the incidence of hip fracture is reduced nearly 50% when older adults are involved in regular physical activity. However, even when activity is optimal, genetic contributions to bone mass are greater than those of exercise. Perhaps, 75% of the population variance is genetic, and 25% is a result of different levels of activity. In addition, the predominant contribution of estrogen to homeostasis of bone in young women is apparent when amenorrhea occurs secondary to chronic heavy exercise. These exceptionally active women are typically thin and exhibit low levels of circulating estrogens, low trabecular bone mass, and a high fracture risk (Fig. 29.4).

Exercise also plays a role in the treatment of **osteoarthritis**. Controlled clinical trials find that appropriate, regular exercise decreases joint pain and degree of disability, although it fails to influence the requirement for anti-inflammatory drug treatment. In **rheumatoid arthritis (RA)**, exercise also increases muscle strength and functional capacity without increasing pain or medication requirements. Whether exercise alters disease progression in either RA or osteoarthritis is not known.

▶ OBESITY, AGING, AND IMMUNE RESPONSES TO EXERCISE

In many cases, obesity and type 2 diabetes can be regarded as diseases of physical inactivity. Regular exercise protects against type 2 diabetes through its positive effects on weight management and metabolic pathways involved in glycemic control that are independent of body weight. Increased physical activity is an effective strategy both for the prevention of type 2 diabetes and its management.

It is also well known that aging lowers maximal dynamic and isometric exercise capacities. There is overwhelming evidence, however, that training can substantially mitigate declines in strength and endurance with advancing age. Changes in functional capacity, as well as protection against heart disease and diabetes, do increase longevity in active persons. However, it remains controversial whether chronic exercise enhances lifespan or whether exercise boosts the immune system, prevents insomnia, or enhances mood.

In obese patients, chronic exercise preferentially increases caloric expenditures over increased appetite.

Obesity is common in sedentary societies. Obesity increases the risk for hypertension, heart disease, and diabetes and is characterized, at a descriptive level, as an excess of caloric intake over energy expenditure. Because exercise enhances energy expenditure, increasing physical activity is a mainstay of treatment for obesity.

The metabolic cost of exercise averages 100 kcal/mi. walked. For exceptionally active people, exercise expenditure can exceed 3,000 kcal/d, added to the basal energy expenditure, which for a 55-kg woman averages about 1,400 kcal/d. At high levels of activity, appetite and food intake match caloric expenditure, though the biologic factors that allow this precise balance have never been defined. In obese patients, modest increases in physical activity increase energy expenditure more than food intake, so progressive weight loss can be instituted if exercise can be regularized. This method of weight control is superior to dieting alone, because substantial caloric restriction (>500 kcal/d) results in both a lowered basal metabolic rate and a substantial loss of fat-free body mass.

Exercise has other, more subtle, positive effects on the energy balance equation as well. A single exercise episode may increase basal energy expenditure for several hours and may increase the thermal effect of feeding. The greatest practical problem remains compliance with even the most precise exercise "prescription"; patient dropout rates from even short-term programs typically exceed 50%.

Exercise is a potent tool in regulating blood glucose in diabetic patients.

Although skeletal muscle is omnivorous, its work intensity and duration, training status, inherent metabolic capacities, and substrate availability determine its energy sources. For short-term exercise, stored phosphagens (ATP and CrP) are sufficient for cross-bridge interaction between actin and myosin. Even maximal efforts lasting 5 to 10 seconds require little or no glycolytic or oxidative energy production. When work to exhaustion is paced to be somewhat longer in duration, high intramuscular ADP concentrations drive glycolysis (particularly in fast glycolytic fibers), and this form of anaerobic metabolism, with its by-product lactic acid, is the major energy source. The carbohydrate provided to glycolysis comes from stored, intramuscular glycogen or blood-borne glucose. Exhaustion from work in this intensity range (50% to 90% of the $\dot{V}_{O_2}$ max) is associated with carbohydrate depletion. Accordingly, factors that increase carbohydrate availability improve fatigue resistance. These include prior high dietary carbohydrate, cellular training adaptations that increase the enzymatic potential for fatty acid oxidation (thereby sparing carbohydrate stores), and oral carbohydrate intake during exercise. Hypoglycemia rarely occurs during even the most prolonged or intense physical activity. When it does, it is usually in association with the depletion of muscle and hepatic stores and a failure to supplement carbohydrate orally.

Exercise suppresses insulin secretion by increasing sympathetic tone at the pancreatic islets. Despite acutely falling levels of circulating insulin, both non–insulin-dependent and insulin-dependent muscle glucose uptake, increase during exercise. Exercise recruits glucose transporters from their intracellular storage sites to the plasma membrane of active skeletal muscle cells. Because exercise increases insulin sensitivity, patients with **type 1 diabetes** (insulin dependent) require less insulin when activity increases. However, this positive result can be treacherous because exercise can accelerate hypoglycemia and increase the risk of insulin coma. Chronic exercise, through its reduction of insulin requirements, up-regulates insulin receptors. This effect appears to be a result less of training than simply of a repeated acute stimulus. The effect is full-blown after 2 to 3 days of regular physical activity and can be lost as quickly. Consequently, healthy active people show strikingly greater insulin sensitivity than do their sedentary counterparts (Fig. 29.5). In addition, up-regulation of insulin receptors and reduced insulin release after chronic exercise are ideal therapy in **type 2 diabetes** (non–insulin-dependent), a disease characterized by high insulin secretion and low receptor sensitivity. In those with type 2 diabetes, a single episode of exercise results in substantial glucose transporter translocation to the plasma membrane in skeletal muscle.

Exercise affects aging more profoundly than longevity.

The influence of exercise on strength and endurance at any age is dramatic. Although the ceiling for oxygen uptake during work gradually falls with age, the ability to train toward an age-appropriate ceiling is as intact at age 70 years as it is at age 20 years (Fig. 29.6). In fact, a highly active 70-year-old person, otherwise healthy, will typically display an absolute exercise capacity greater than that of a sedentary 20-year-old person. Aging affects all the links in the chain of oxygen transport and use, so aging-induced declines in lung elasticity, lung diffusing capacity, cardiac output, and muscle metabolic potential take place in concert. Consequently, the physiologic mechanisms underlying fatigue are similar at all ages.

Regular dynamic exercise, compared with inactivity, increases longevity in rats and humans. In descriptive terms, the effects of exercise are modest; all-cause mortality is reduced but only in amounts sufficient to increase longevity by 1 to 2 years. These facts leave open the possibility that exercise might alter biologic aging. Although physical activity increases cellular oxidative stress, it simultaneously increases antioxidant capacity. Food-restricted rats experience increased lifespan and exhibit elevated spontaneous activity levels, but the role exercise may play in the apparent delay of aging in these animals remains unclear.

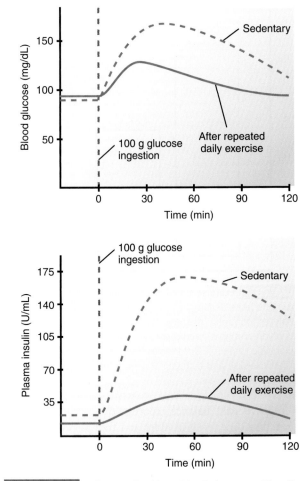

Figure 29.5 **Daily exercise blunts blood glucose and insulin responses to glucose ingestion.** Repeated exercise significantly decreases both responses, demonstrating increased insulin sensitivity.

Exercise has modest effect on immune function.

In protein–calorie malnutrition, the catabolism of protein for energy lowers immunoglobulin levels and compromises the body's resistance to infection. Clearly, in this circumstance, exercise merely speeds the starvation process by increasing daily caloric expenditure and would be expected to diminish the immune response further. Nazi labor camps of the early

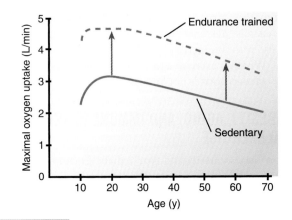

Figure 29.6 **Maximal oxygen uptake is enhanced with endurance training.** Endurance-trained subjects possess greater maximal oxygen uptake than sedentary subjects, regardless of age.

1940s became death camps, partly by severe food restrictions and incessant demands for physical work—a combination guaranteed to cause starvation.

If nutrition is adequate, it is less clear whether adopting an active versus a sedentary lifestyle alters immune responsivity. In healthy subjects, an acute episode of exercise briefly increases blood leukocyte concentration and transiently enhances neutrophil production of microbicidal reactive oxygen species and natural killer cell activity. However, it remains unproven that regular exercise over time can lower the frequency or reduce the intensity of, for example, upper respiratory tract infections (URTIs). In human immunodeficiency virus (HIV)–positive men and in men with AIDS and advanced muscle wasting, strength training and endurance training yield normal gains. There is also incomplete evidence that training may slow progression to AIDS in HIV-positive men, with a corresponding increase in CD4 lymphocytes. In contrast to moderate or intermittent physical activity, prolonged and very intense exertions are associated with numerous changes in the immune system that reflect physiologic stress and suppression. Athletes subjected to regular strenuous exercise may be at increased risk of URTI during periods of heavy exercise and for a several weeks following race events. It has been suggested that the relationship between exercise and URTIs follows a "J curve," with moderate and regular exercise improving the ability to resist infections and heavy acute or chronic exercise decreasing it. Exercise-induced immunodepression has a multifactorial origin, depending on mechanisms related to neuro–immune–endocrine systems. Prolonged periods of intense training may lead to high numbers of neutrophil and low numbers of lymphocytes, impaired phagocytic and neutrophilic function, and decreases in oxidative burst activity, natural killer cell cytolytic activity, and mucosal immunoglobulin levels. Highly intense training may also enhance release of injury-related proinflammatory cytokines, such as tumor necrosis factor-α, interleukin (IL)-1β, and IL-6. Whether these observations have relevance to clinically immune deficiency remains to be determined.

Temperature Regulation and Exercise Physiology

INTEGRATED MEDICAL SCIENCES

Mobilizing Postoperative Patients: Benefits of Exercise through Early Ambulation

Due to a decline in mortality resulting from critical illness in recent years, a new focus on patient outcomes after hospital discharge has emerged. Notable considerations are (1) enhanced recovery from total hip and knee arthroplasty and (2) lengthy stays in the intensive care unit (ICU). Total hip arthroplasty (THA) and total knee arthroplasty (TKA) elicit physiological stressors accompanied by a proinflammatory cascade that leads to postoperative complications and prolonged rehabilitation and recovery, a multifactorial healing process that includes biological, psychological, and social factors. One simple postoperative intervention that has a positive effect on all these factors and reduces recovery time is early ambulation. Patient mobilization (within 24 hours) for THA and TKA is a positive predictor for a shorter hospital stay, fewer postoperative complications, and reduced costs and is associated with lower rates of deep vein thrombosis and pulmonary embolism. Additionally, early (postoperative day 1 or 2) progressive strength training is safe and improves knee extension strength and maximal walking speed. It is well known that a single bout of exercise alters the DNA-binding activity of a variety of transcription factors, including MEF2, HDACs, and NRFs, and transient DNA hypomethylation of gene-specific promoter regions precedes increases in mRNA expression in response to acute exercise (Egan, Zierath. *Cell Metabolism* 2013;17:162–184). These immediate adaptations counter atrophic processes that would otherwise ensue and result in muscle wasting and weakness. Prolonged stays in the intensive care unit (ICU) are commonly associated with the development of persistent and often severe neuromuscular weakness. Prominent in the development of ICU-acquired weakness is immobility related to prolonged bed rest. Studies in unrelated, clinically distinct patient populations have demonstrated that moderate exercise is beneficial in counteracting the inflammatory conditions and atrophy that are associated with immobility and in improving muscle strength and physical function. Prolonged bed rest associated with critical illness typically leads to decreased muscle protein synthesis, increased urinary nitrogen excretion (indicating muscle catabolism), and decreased muscle mass, especially in the lower extremities. This type of immobility leads to a proinflammatory state caused by increased systemic inflammation due to proinflammatory cytokines such as IL-1β, IL-2, and interferon-γ proinflammatory cytokines. This potentiates the production of reactive oxygen species (ROS) with a decrease in antioxidative defenses and activation of both nuclear factor-κB and FOXO signaling pathways, resulting in protein loss. Recent studies in the ICU have demonstrated improved clinical outcomes with early mobility that included ambulation and physical therapy. Early mobilization and decreased sedation were found to be synergistically beneficial in mechanically ventilated patients. Their application was associated with decreased lengths of stay in the ICU and hospital, with a trend toward decreased duration of mechanical ventilation, resulting in overall improvements in physical function and quality of life for ICU survivors. Finally, fully sedated patients admitted to the ICU experience the most extensive muscle wasting. Neuromuscular electrical stimulation (NMES) prevents muscle fiber atrophy in these critically ill, comatose patients during lengthy stays and possibly improves survival and subsequent rehabilitation (Dirks, et al. *Clinical Science.* 2015;128:357–365). ■

Chapter Summary

- Exercise is the performance of some activity that involves muscle contraction and joint flexion or extension and places unique stresses on the body's organ systems.
- Measuring maximal oxygen uptake is the most common method of quantifying dynamic exercise.
- Excess in oxygen consumption during the first minutes of recovery is called the oxygen debt.
- Blood flow is preferentially directed to working skeletal muscle during exercise.
- During exercise, blood pressure, heart rate, stroke volume, and cardiac contractility are all increased.
- Hearts of people adapted to prolonged, rhythmic exercise involving low arterial pressure exhibit large left ventricular volumes with normal wall thickness; in those adapted to activities involving isometric contraction and elevated arterial pressure, wall thickness is increased at normal volume.
- Chronic, dynamic exercise is associated with increased circulating levels of high-density lipoproteins (HDLs) and reduced low-density lipoproteins, such that the ratio of HDL to total cholesterol is increased.

- Exercise has a role in preventing and recovering from several cardiovascular diseases.
- Pregnancy shares many cardiovascular characteristics with the trained state.
- Pulmonary ventilation increases during exercise in proportion to O_2 demand and the need for CO_2 removal.
- In lung disease, respiratory limitations appear as shortness of breath or decreased oxygen content of arterial blood and become more apparent during exercise than at rest.
- Muscle fatigue is defined as an exercise-induced reduction in the maximal force capacity of the muscle and is independent of lactic acid.
- Endurance activity at low loads enhances muscle oxidative capacity without hypertrophy, whereas increased activity at high loads produces muscle hypertrophy.
- Although exercise effects on gastrointestinal function remain poorly understood, chronic physical activity plays a major role in the control of obesity and type 2 diabetes.
- As people age, the effects of exercise on functional capacity are more profound than its effect on longevity.

Chapter Review Questions

1. In an effort to strengthen selected muscles after surgery and immobilization has led to muscle atrophy, isometric exercise is recommended. The intensity of isometric exercise is best quantified:

   A. relative to the maximal oxygen uptake.
   B. as mild, moderate, or strenuous.
   C. as a percentage of the maximum voluntary contraction.
   D. in terms of anaerobic metabolism.
   E. on the basis of the total muscle mass involved.

The correct answer is C. A maximal voluntary contraction involving the identical muscles in an identical form of contraction provides the most readily quantified and accurate basis for normalization of isometric exercise intensity. Choice A is incorrect because the basis for comparison involves rhythmic, dynamic exercise. The other choices also contradict the principle that exercise can only be compared with other exercise involving the same muscles and the same types of muscle contraction.

2. Two people, one highly trained and one not, each exercising at 75% of the maximal oxygen uptake, become fatigued:

   A. for similar physiological reasons.
   B. very slowly.
   C. at different times.
   D. while performing equally well for at least a short period of time.
   E. despite much higher circulating lactic acid levels in the trained person.

The correct answer is A. The physiological responses to dynamic exercise are predictable when healthy individuals differing in endurance exercise capacity are compared at matched levels of relative oxygen transport demand. Exercise at 75% of the maximal oxygen uptake will lead to exhaustion in typically 1 to 2 hours, rendering choice B incorrect. The more highly trained person will show increased work output despite fatiguing at about the same time as the person with lower capacity, rendering choices C and D incorrect. Training lowers lactic acid production at any matched fractional use of the maximal oxygen uptake, making choice E incorrect.

3. A patient completes a graded, dynamic exercise test on a treadmill while showing a modest rise (25%) in mean arterial blood pressure. In contrast, during the highest level of exercise at the end of the test, an indirect method shows that cardiac output has risen 300% from rest. These results indicate that during graded, dynamic exercise to exhaustion, systemic vascular resistance:

   A. is constant.
   B. rises slightly.
   C. falls only if work is prolonged.
   D. falls dramatically.
   E. cannot be measured.

The correct answer is A. Active muscle vasodilation during dynamic exercise is quantitatively much greater than the net vasoconstriction in the gut, skin, kidneys, and inactive muscle. Choices B, C, and D contradict this answer. Total systemic vascular resistance can be measured, albeit indirectly, from measurements of systemic arterial pressure and cardiac output.

4. A patient with inflammatory bowel disease and compromised kidney function asks if exercise will alter blood flow to either the gastrointestinal tract or the kidneys. The answer is that vasoconstriction in both the renal and splanchnic vascular beds during exercise:

   A. rarely occurs.
   B. occurs only after prolonged training.
   C. helps maintain arterial blood pressure.
   D. allows renal and splanchnic flows to parallel cerebral blood flow.
   E. will be balanced by local dilation in these vascular beds.

The correct answer is C. This answer presumes that vasoconstriction occurs in these vascular beds and that its effect is to help balance vasodilation in active skeletal muscles and prevent exercise-induced systemic hypotension. This effect is ubiquitous across all individuals during all forms of dynamic exercise, making choices B, C, and E incorrect. Cerebral blood flow is held constant during all forms of exercise, unlike renal or splanchnic blood flow.

5. A young, healthy, highly trained individual enters a marathon (40 km) run on a warm, humid day (32°C, 70% humidity). The best medical advice for this individual is to be concerned about the possibility for:

   A. heat exhaustion.
   B. coronary ischemia.
   C. renal ischemia and anoxia.
   D. hypertension.
   E. gastric mucosal ischemia and increased risk for gastric ulceration.

The correct answer is A. Even highly trained and heat-acclimatized individuals are at risk for heat-related illness if exercise is sufficiently prolonged and if environmental conditions are sufficiently severe. In healthy persons during exercise, coronary vasodilatory capacity is adequate, renal blood flow reductions in health are entirely safe, and gastric mucosal blood flow reductions are easily tolerated. In long-term exercise in a warm environment, hypotension, not hypertension, is the possible cardiovascular risk.

Clinical Application Exercises 29.1

A PATIENT WITH DYSPNEA DURING EXERCISE

A 56-year-old man complained of shortness of breath and chest pain when climbing stairs or mowing the lawn. He is subjected to a stress test, with noninvasive monitoring of heart rate, blood pressure, arterial blood oxygen saturation, and cardiac electrical activity. His resting heart rate is 73 beats/min; blood pressure, 118/75 mm Hg; arterial blood oxygen saturation, 96%; and the ECG, normal.

After 3.5 minutes of increasingly intense exercise, the test is terminated because of the subject's severe dyspnea. His heart rate is 119 beats/min (his age and sex-adjusted predicted maximal heart rate is 168 beats/min), blood pressure is 146/76 mm Hg, arterial blood oxygen saturation is 88%, and the ECG is normal.

QUESTIONS

1. What are three lines of evidence for ventilatory limitation to this subject's exercise?
2. Why did arterial blood oxygen saturation fall during exercise?
3. Why did exhaustion occur before maximal heart rate was reached?
4. Why did the pulse pressure rise in exercise?
5. Why would endurance exercise training likely increase this individual's exercise capacity?

ANSWERS

1. Ventilatory limitation is evidenced by severe dyspnea as a primary symptom in exercise, falling arterial blood oxygenation, and exercise termination at relatively low heart rate.
2. Arterial blood oxygen saturation fell during exercise because increased cardiac output (increased pulmonary blood flow) and decreased pulmonary arterial blood oxygen content (a result of increased skeletal muscle oxygen extraction) increase demands for oxygenation in the lungs with inadequate diffusing capacity.
3. Exhaustion occurred before a maximal heart rate was reached because lung disease creates severe dyspnea even in mild exercise.
4. The pulse pressure rose during exercise because sympathetic stimulation and enhanced venous return increase the stroke volume at constant arterial compliance.
5. Endurance exercise training would have little effect on any aspect of lung function. However, training would cause adaptations within exercising muscle that would increase muscle oxidative capacity and reduce lactic acid production. By reducing the ventilatory demands of exercise, these changes would increase exercise capacity in this individual.

thePoint® *Visit* http://thepoint.lww.com/rhoades5e *for additional chapter review Q&A, Clinical Application Exercises, animations, and more!*

30 Endocrine Control Mechanisms

Active Learning Objectives

Upon mastering the material in this chapter, you should be able to:

- Explain how, in addition to specific endocrine disease states, other prominent diseases, including atherosclerosis, cancer, and even psychiatric disorders, likely have an underlying endocrine component.
- Explain how hormone receptors, restricted distribution of hormones, and hormone activation processes determine target tissues for a specific hormone.
- Explain how normal feedback relationships permit clinical diagnoses of functional and dysfunctional endocrine systems.

- Explain how secretagogues are used clinically to provide meaningful diagnostic information.
- Explain how chemical differences in hormones influence their endocrine functionality.
- Explain how altered target tissue hormonal responses reflected by dose–response curves can provide useful clinical information regarding the underlying cause of a particular disease.

Endocrinology is the branch of physiology concerned with the description and characterization of processes involved in the regulation and integration of cells and organ systems by a group of specialized chemical substances called **hormones**. The diagnosis and treatment of many endocrine disorders are important aspects of any general medical practice. Certain endocrine disease states, such as diabetes mellitus, thyroid disorders, and reproductive disorders, are fairly common in the general population; therefore, it is likely that they will be encountered repeatedly in the practice of medicine.

In addition, because hormones either directly or indirectly affect virtually every cell or tissue in the body, some other prominent diseases not primarily classified as endocrine diseases may have an important endocrine component. Atherosclerosis, certain forms of cancer, and even certain psychiatric disorders are examples of conditions in which an endocrine disturbance may contribute to the progression or severity of disease.

▶ GENERAL ENDOCRINE CONCEPTS

The word "hormone" is derived from the Greek *hormaein*, which means to "excite" or to "stir up." Glands that comprise the endocrine system produce hormones. The major morphologic feature of endocrine glands is that they are ductless; that is, they release their secretory products directly into the bloodstream and not into a duct system. Figure 30.1 illustrates classical endocrine glands. Many other "nonclassical" hormone-producing glands also exist. These include the central nervous system, kidneys, stomach, small intestine, skin, heart, lung, and placenta. Recent advances in cellular and molecular biology continually broaden our views of

the endocrine system. For example, obesity-related research discovered leptin, a hormone formed in adipocytes that signals to the central nervous system to regulate appetite and energy expenditure. Here, general themes and principles that underpin the functionality of the endocrine system as a whole will be presented.

Hormones are chemicals released by a cell that regulate many functions.

Hormones function as homeostatic blood-borne chemicals to regulate and coordinate various biologic functions. They are highly potent, specialized, organic molecules produced by endocrine cells in response to specific stimuli and exert their actions on specific target cells. These target cells are equipped with receptors that bind hormones with high affinity and specificity; when bound, they initiate characteristic biologic responses by the target cells.

Although the effects of hormones are many and varied, their actions are involved in (1) regulating ion and water balance; (2) responding to adverse conditions, such as infection, trauma, and emotional stress; (3) sequentially integrating features of growth and development; (4) contributing to basic processes of reproduction, including gamete production, fertilization, nourishment of the embryo and fetus, delivery, and nourishment of the newborn; and (5) digesting, using, and storing nutrients.

In the past, definitions or descriptions of hormones usually included a phrase indicating that these substances were secreted into the bloodstream and carried by the blood to a distant target tissue. Although many hormones travel by this mechanism, we now realize that there are many hormones or hormone-like substances that play important roles in cell-to-cell communication that are not secreted directly into the

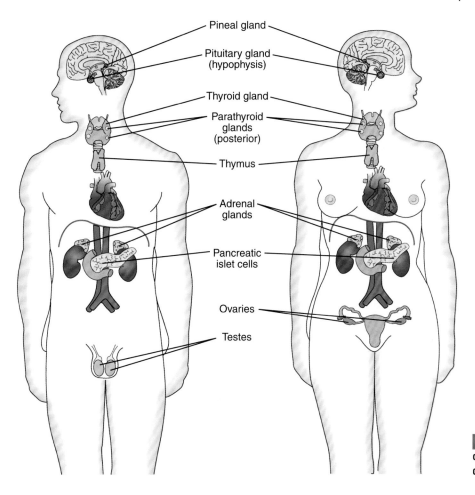

Pineal gland

Pituitary gland
(hypophysis)

Thyroid gland

Parathyroid
glands
(posterior)

Thymus

Adrenal
glands

Pancreatic
islet cells

Ovaries

Testes

Figure 30.1 **Location of many endocrine glands, organs containing endocrine tissue, and associated structures.**

Endocrine Physiology

bloodstream. Instead, these substances reach their target cells by diffusion through the interstitial fluid. Recall the discussion of autocrine and paracrine mechanisms in Chapter 2.

Hormones initiate a cell response by binding to specific receptors.

In the endocrine system, a hormone molecule secreted into the blood is free to circulate and contact almost any cell in the body. However, only **target cells**, those cells that possess specific receptors for the hormone, will respond to that hormone. As presented in Chapter 2, a **hormone receptor** is the molecular entity (usually a protein or glycoprotein) either outside or within a cell that recognizes and binds a particular hormone. When a hormone binds to its receptor, characteristic biologic effects of that hormone are initiated. Therefore, in the endocrine system, the basis for specificity in cell-to-cell communication rests at the level of the receptor. Similar concepts apply to autocrine and paracrine mechanisms of communication.

The restricted distribution of some hormones ensures a certain degree of specificity. For example, several hormones produced by the hypothalamus regulate hormone secretion by the anterior pituitary. These hormones are carried via small blood vessels directly from the hypothalamus to the anterior pituitary, prior to entering the general systemic circulation. The anterior pituitary is, therefore, exposed to considerably higher concentrations of these hypothalamic hormones than the rest of the body; as a

result, the actions of these hormones focus on cells of the anterior pituitary.

Another mechanism that restricts the distribution of active hormone is the local transformation of a hormone within its target tissue from a less active to a more active form. An example is the formation of dihydrotestosterone from testosterone, occurring in such androgen target tissues as the prostate gland. Dihydrotestosterone is a much more potent androgen than testosterone. Because the enzyme that catalyzes this conversion is found only in certain locations, its cell or tissue distribution partly localizes the actions of the androgens to these sites. Therefore, although receptor distribution is the primary factor in determining the target tissues for a specific hormone, other factors may also focus the actions of a hormone on a particular tissue.

Hormone regulation occurs through feedback control.

Feedback mechanisms regulate the endocrine system, just as in many other physiologic systems. The mechanism is usually negative feedback, although a few positive feedback mechanisms are known. Both types of feedback control occur because the endocrine cell, in addition to synthesizing and secreting its own hormone product, has the ability to sense the biologic consequences of secretion of that hormone. This enables the endocrine cell to adjust its rate of hormone secretion to produce the desired level of effect, ensuring the maintenance of homeostasis.

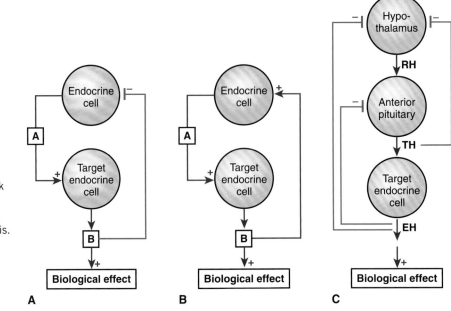

Figure 30.2 Negative, positive, and complex feedback mechanisms in the endocrine system. **(A)** Negative feedback loop. **(B)** Positive feedback loop. **(C)** A complex, multilevel feedback loop: the hypothalamic–pituitary–target gland axis. Red lines indicate stimulatory effects; blue lines indicate inhibitory, negative feedback effects. EH, endocrine cell hormone; RH, releasing hormone; TH, trophic hormone.

In the simplest form, negative feedback is a closed loop in which hormone A stimulates the production of hormone B, which in turn acts on the cells producing A to decrease its rate of secretion (Fig. 30.2A). In the less common positive feedback system, hormone B further stimulates the production of A instead of diminishing it (Fig. 30.2B). Many examples of negative feedback regulation exist and are discussed throughout the endocrine chapters. A typical example of a positive feedback loop is that which exists between luteinizing hormone (LH) and estradiol. As detailed in Chapter 37, positive feedback by estradiol is required to generate an abrupt increase in LH secretion during the menstrual cycle. Endocrine control systems also have feedforward loops, which can be negative or positive and which direct the flow of hormonal information. Unlike negative feedback, which promotes stability, feedforward control anticipates change. Recall discussion of the physiologic regulation in Chapter 2.

More commonly, feedback regulation in the endocrine system is complex, involving second-order or third-order feedback loops. For example, multiple levels of feedback regulation may be involved in regulating hormone production by various endocrine glands under the control of the anterior pituitary (Fig. 30.2C). The regulation of the secretion of target gland hormones, such as adrenal steroids or thyroid hormones, begins with the production of a **releasing hormone** by the hypothalamus. The releasing hormone stimulates the production of a **trophic hormone** by the anterior pituitary, which in turn stimulates the production of the target endocrine cell hormone by the target gland. As indicated by the blue lines in Figure 30.2C, the target endocrine cell hormone may have negative feedback effects to inhibit secretion of both the trophic hormone from the anterior pituitary and the releasing hormone from the hypothalamus. In addition, the trophic hormone may inhibit releasing hormone secretion from the hypothalamus, and in some cases, the releasing hormone may inhibit its own secretion by the hypothalamus.

The more complex multilevel form of regulation appears to provide certain advantages compared with the simpler system. Theoretically, it permits a greater degree of fine-tuning of hormone secretion, and the multiplicity of regulatory steps minimizes changes in hormone secretion in the event that one component of the system is not functioning normally.

Normal feedback relationships that control the secretion of each individual hormone are discussed in the chapters that follow. Clinical diagnoses are often made based on the evaluation of hormone–effector pairs relative to normal feedback relationships. For example, in the case of anterior pituitary hormones, measuring both the trophic hormone and the target gland hormone concentration provides important information to help determine whether a defect in hormone production exists at the level of the pituitary or at the level of the target gland. Furthermore, most dynamic tests of endocrine function performed clinically are based on our knowledge of these feedback relationships. Dynamic tests involve a prescribed perturbation of the feedback relationship(s). The range of response in a healthy person is well established, whereas a response outside the normal range is indicative of abnormal function at some level and greatly enhances information gained from static measurements of hormone concentrations.

Signal amplification is part of the overall mechanism of hormone action.

Another important feature of the endocrine system is signal amplification, a mechanism that increases the amplitude of the signal. For example, blood concentrations of hormones are exceedingly low, generally 10^{-9} to 10^{-12} mol/L. Even at the higher concentration of 10^{-9} mol/L, only one hormone molecule is present for roughly every 50 billion water molecules. Therefore, for hormones to be effective regulators of biologic processes, amplification must be part of the overall mechanism of hormone action.

As presented in Chapter 2, amplification generally results from the activation of a series of enzymatic steps involved in hormone action. At each step, many times more signal molecules are generated than were present at the prior step, leading to a cascade of ever-increasing numbers of signal molecules. The self-multiplying nature of the hormone action pathways provides the molecular basis for amplification in the endocrine system.

Hormones can have multiple, and share some, actions with other hormones.

Most hormones have multiple actions in their target tissues and are, therefore, said to have **pleiotropic** effects. This phenomenon occurs when a single hormone regulates several functions in a target tissue. For example, in skeletal muscle, insulin stimulates glucose uptake, stimulates glycolysis, stimulates glycogenesis, inhibits glycogenolysis, stimulates amino acid uptake, stimulates protein synthesis, and inhibits protein degradation.

In addition, some hormones are known to have different effects in several different target tissues. For example, testosterone, the male sex steroid, promotes normal sperm formation in the testes, stimulates growth of the accessory sex glands, such as the prostate and seminal vesicles, and promotes the development of several secondary sex characteristics, such as beard growth and deepening of the voice.

The ability of several different hormones to regulate a single biological function is referred to as multiplicity of regulation. The input of information from several sources allows a highly integrated response, which is of ultimate benefit to the whole animal. For example, several different hormones, including insulin, glucagon, epinephrine, thyroxine, and cortisol, regulate liver glycogen metabolism.

Hormones are often secreted in rhythmic patterns.

Many hormones are secreted in a defined, rhythmic pattern. These rhythms can take several forms. For example, they may be pulsatile, episodic spikes in secretion lasting just a few minutes, or they may follow a daily, monthly, or seasonal change in overall pattern. Pulsatile secretion may occur in addition to other longer secretory patterns.

For these reasons, a single randomly drawn blood sample for determining a certain hormone concentration may be of little or no diagnostic value. A dynamic test of endocrine function in which a known agent specifically stimulates hormone secretion often provides much more meaningful information (Clinical Focus 30.1).

▶ CHEMICAL NATURE OF HORMONES

Hormones can be categorized by several criteria. Grouping them by chemical structure is convenient, because in many cases hormones with similar structures also use similar mechanisms to produce their biologic effects. In addition, tissues with similar embryonic origins usually produce hormones with similar chemical structures. The three chemical classes include (1) amine-derived, (2) polypeptide-derived, and (3) cholesterol-derived hormones.

Amine-derived hormones consist of one or two modified amino acids.

Hormones derived from one or two amino acids are small in size and often hydrophilic. These hormones are formed by conversion from a commonly occurring amino acid; **norepinephrine** and **thyroxine**, for example, are derived from tyrosine (Fig. 30.3A). Each of these hormones is synthesized

CLINICAL FOCUS | 30.1

Growth Hormone and Pulsatile Hormone Secretion

Growth hormone (**GH**) is a 191-amino acid protein hormone that is synthesized and secreted by somatotrophs of the anterior lobe of the pituitary gland. As described in Chapter 31, GH plays a role in regulating bone growth and energy metabolism in skeletal muscle and adipose tissue. A deficiency in GH production during adolescence results in dwarfism, and overproduction results in gigantism. Measurements of circulating growth hormone levels are, therefore, desirable in children whose growth rate is not appropriate for their age. However, as with many other hormones, the pulsatile nature of GH secretion needs to be considered.

Most healthy children experience episodes or "bursts" of GH secretion throughout the day, most prominently within the first several hours of sleep. To obtain reliable information about growth hormone secretion, insertion of a continuous withdrawal pump or patent indwelling catheter with unrestricted food intake and physical activity is required. Increasing sampling frequencies from every 20-minute to

5-minute or 30-second sampling intervals enhances the threshold for detecting more pulses per hour. Although cumbersome and expensive, this method eliminates the error of isolated peak or trough measurements that might otherwise be obtained by single or multiple random GH samplings.

A less arduous approach with similar discriminating power in diagnosing GH deficiency entails measuring GH levels in the blood shortly after appropriate pituitary stimulation. Pharmacologic agents that stimulate GH secretion include L-dopa, clonidine, propanolol, glucagon, insulin, and arginine. By perturbing the system in a well-prescribed fashion, the endocrinologist is able to gain important information about growth hormone secretion that would not be possible if a random blood sample were used. As further presented in Clinical Focus 31.2, an alternative means of testing for GH deficiency is to measure the levels of insulin-like growth factor-1 and IGF-binding protein-3 in the blood because both predict the presence of GH. ■

by a particular sequence of enzymes that are primarily localized in the endocrine gland involved in its production. Many environmental or pharmacologic agents can influence the synthesis of amino acid–derived hormones in a relatively specific fashion. The steps involved in the synthesis of these hormones are discussed in detail in later chapters.

Polypeptide-derived hormones are diverse in size and complexity.

Hormones in the polypeptide group may be as small as the tripeptide **thyrotropin-releasing hormone** (**TRH**) or as large as **human chorionic gonadotropin**, which is composed of separate α and β subunits, has a molecular weight of ~34 kDa, and is a glycoprotein consisting of 16% carbohydrate by weight. Figure 30.3B provides examples of this second category of polypeptide and protein hormones.

Within the polypeptide class of hormones are many families of hormones, some of which are listed in Table 30.1. Hormones can be grouped into these families as a result of considerable homology with regard to amino acid sequence and structure. Presumably, the similarity of structure in these families resulted from the evolution of a single ancestral hormone into each of the separate and distinct hormones. In many cases, there is also considerable homology among receptors for the hormones within a family.

Polypeptide hormones are synthesized and stored in advance of need. Like other proteins destined for secretion, polypeptide hormones are synthesized with a *prepeptide* or *signal* peptide at their amino terminal end that directs the growing peptide chain into the cisternae of the rough endoplasmic reticulum (RER). Moreover, most, if not all, polypeptide hormones are synthesized as part of an even larger precursor or **preprohormone**. The prepeptide is cleaved off on entry of the preprohormone into the RER to form the **prohormone**. Because the prohormone is processed through the Golgi apparatus and packaged into secretory vesicles, it is proteolytically cleaved at one or more sites to yield active hormone. In many cases, preprohormones may contain the sequences for several different biologically active molecules. Inactive spacer segments of peptide, in some cases, may separate these active elements. Specific mechanisms of polypeptide hormone synthesis, storage, and processing are discussed in later chapters.

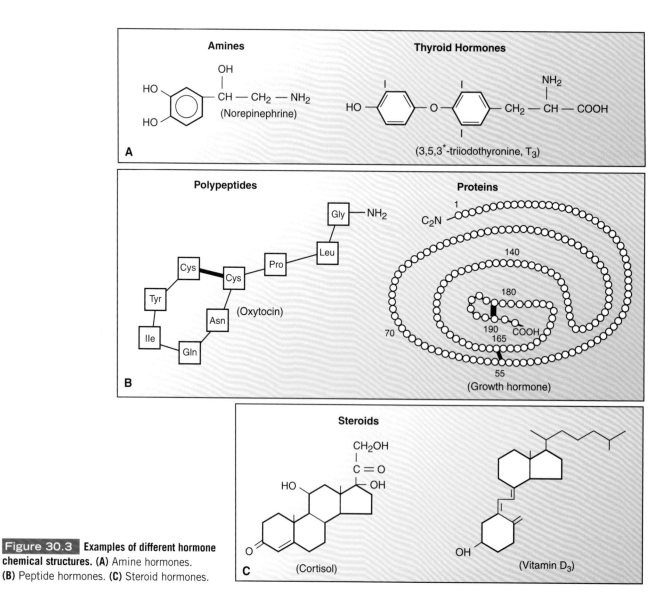

Figure 30.3 Examples of different hormone chemical structures. (**A**) Amine hormones. (**B**) Peptide hormones. (**C**) Steroid hormones.

TABLE 30.1 Examples of Peptide Hormone Families

Insulin Family	Glycoprotein Family	Growth Hormone Family	Secretin Family
Insulin	Luteinizing hormone	Growth hormone	Secretin
Insulin-like growth factor-1	Follicle-stimulating hormone	Prolactin	Vasoactive intestinal peptide
Insulin-like growth factor-2	Thyroid-stimulating hormone	Human placental lactogen	Glucagon
Relaxin	Human chorionic gonadotropin		Gastric inhibitory peptide

Steroid hormones are derived from cholesterol.

Steroids are lipid-soluble, hydrophobic molecules synthesized from cholesterol. Examples include **aldosterone**, **cortisol**, and **androgen**, secreted by the cortex (outer zone) of the adrenal glands; **testosterone**, secreted by the testes; and **estrogen** and **progesterone**, secreted by the ovaries. These adrenal and gonadal steroid hormones contain an intact steroid nucleus, as illustrated for cortisol in Figure 30.3C. In contrast, other hormone derivatives of cholesterol, such as vitamin D and its metabolites, have a broken steroid nucleus (the B ring) (see Fig. 30.3C). Steroid hormones are synthesized and secreted on demand (specific mechanisms are discussed in later chapters).

▶ MEASUREMENT OF CIRCULATING HORMONES

The amounts of different hormones are regulated by varied mechanisms in the blood and target tissues. For example, in the bloodstream, the binding of a plasma protein to a hormone has many effects including inactivation. In the periphery, some hormones are made more active by various transformations.

Hormones can circulate either free or bound to carrier proteins.

Most amine- and polypeptide-derived hormones dissolve readily in the plasma, and thus, no special mechanisms are required for their transport. Steroid and thyroid hormones are relatively insoluble in plasma. Mechanisms are present to promote their solubility in the aqueous phase of the blood and their ultimate delivery to a target cell.

In most cases, 90% or more of steroid and thyroid hormones in the blood are bound to plasma proteins. Some of the plasma proteins that bind hormones are specialized, in that they have a considerably higher affinity for one hormone over another, whereas others, such as serum albumin, bind many hydrophobic hormones. The extent to which a hormone is protein bound and the extent to which it binds to specific versus nonspecific transport proteins vary from one hormone to another. The principal binding proteins involved in specific and nonspecific transport of steroid and thyroid hormones are listed in Table 30.2. The liver synthesizes and secretes these proteins, and changes in various nutritional and endocrine factors influence their production.

Typically, for hormones that bind to carrier proteins, only 1% to 10% of the total hormone present in the plasma exists free in solution. However, only this free hormone is biologically active. Bound hormone cannot directly interact with its receptor and, thus, is part of a temporarily inactive pool. However, free hormone and carrier-bound hormone are in a dynamic equilibrium with each other (Fig. 30.4). The size of the free hormone pool and, therefore, the amount available to receptors are influenced not only by changes in the rate of secretion of the hormone but also by the amount of carrier protein available for hormone binding and the rate of degradation or removal of the hormone from the plasma.

CLINICAL FOCUS | 30.2

Pancreatic β-Cell Function and C-Peptide

β cells of the human pancreas produce and secrete **insulin**. The product of the insulin gene is a peptide known as preproinsulin. As with other secretory peptides, the prepeptide or signal peptide is cleaved off early in the biosynthetic process, yielding proinsulin. Proinsulin is an 86-amino acid protein that is subsequently cleaved at two sites to yield insulin and a 31-amino acid peptide known as **C-peptide**. Insulin and C-peptide are, therefore, localized within the same secretory vesicle and are cosecreted into the bloodstream.

For these reasons, measurements of circulating C-peptide levels can provide a valuable indirect assessment of β-cell insulin secretory capacity. In diabetic patients who are receiving exogenous insulin injections, the measurement of circulating insulin levels would not provide any useful information about their own pancreatic function because it would primarily be the injected insulin that would be measured. However, an evaluation of C-peptide levels in such patients would provide an indirect measure of how well the β cells were functioning with regard to insulin production and secretion. ■

TABLE 30.2	Circulating Transport Proteins
Transport Protein	**Principal Hormone(s) Transported**
Specific	
Corticosteroid-binding globulin (transcortin)	Cortisol, aldosterone
Thyroxine-binding globulin	Thyroxine, triiodothyronine
Sex hormone–binding globulin	Testosterone, estrogen
Nonspecific	
Serum albumin	Most steroids, thyroxine, triiodothyronine
Transthyretin (prealbumin)	Thyroxine, some steroids

In addition to increasing the total amount of hormone that can be carried in plasma, transport proteins also provide a relatively large reservoir of hormone that buffers rapid changes in free hormone concentrations. As unbound hormone leaves the circulation and enters cells, additional hormone dissociates from transport proteins and replaces free hormone that is lost from the free pool. Similarly, following a rapid increase in hormone secretion or the therapeutic administration of a large dose of hormone, most newly appearing hormone is bound to transport proteins because under most conditions these are present in considerable excess. Protein binding greatly slows the rate of clearance of hormones from plasma. It not only slows the entry of hormones into cells, slowing the rate of hormone degradation, but also prevents loss by filtration in the kidneys.

From a diagnostic standpoint, it is important to recognize that most hormone assays are reported in terms of total concentration (i.e., the sum of free and bound hormone), not just free hormone concentration. The amount of transport protein and the total plasma hormone content are known to change under certain physiologic or pathologic conditions, whereas the free hormone concentration may remain relatively normal. For example, increased concentrations of binding proteins are seen during pregnancy, and decreased concentrations are seen with certain forms of liver or kidney disease. Assays of total hormone concentration might be misleading, because free hormone concentrations may be in the normal range. In such cases, it is helpful to determine the extent of protein binding, so that free hormone concentrations can be estimated.

The proportions of a hormone that are free, bound to a specific transport protein, and bound to albumin vary depending on the hormone's solubility, its relative affinity for the two classes of transport proteins, and the relative abundance of the transport proteins. For example, the affinity of cortisol for **corticosteroid-binding globulin** (**CBG**) is more than 1,000 times greater than its affinity for albumin, but albumin is present in much higher concentrations than CBG. Therefore, about 70% of plasma cortisol is bound to CBG, 20% is bound to albumin, and the remaining 10% is free in solution. Aldosterone also binds to CBG, but with a much lower affinity, such that only 17% is bound to CBG, 47% associates with albumin, and 36% is free in solution.

As this example indicates, more than one hormone may be capable of binding to a specific transport protein. When several such hormones are present simultaneously, they compete for a limited number of binding sites on these transport proteins. For example, cortisol and aldosterone compete for CBG binding sites. Increases in plasma cortisol result in displacement of aldosterone from CBG, raising the unbound (active) concentration of aldosterone in the plasma. Similarly, prednisone, a widely used synthetic corticosteroid, can displace about 35% of the cortisol normally bound to CBG. As a result, with prednisone treatment, the free cortisol concentration is higher than might be predicted from measured concentrations of total cortisol and CBG.

Peripheral tissues transform, degrade, and excrete hormones.

As a general rule, hormones are produced in their active form by their endocrine gland or tissue of origin. However, for a few notable exceptions, hormone activity is enhanced or requires peripheral transformation. Well-known transformations are the conversion of testosterone to dihydrotestosterone, as mentioned above and detailed in Chapter 36, and the conversion of thyroxine to triiodothyronine (see Chapter 32). Other examples are the formation of the octapeptide angiotensin II from its precursor, angiotensinogen (see Chapter 33), and the formation of 1,25-dihydroxycholecalciferol from cholecalciferol (see Chapter 35). Specific hormone transformations may be impaired because of a congenital enzyme deficiency or drug-induced inhibition of enzyme activity, resulting in endocrine abnormalities.

As in any regulatory control system, the hormonal signal must dissipate or disappear once appropriate information

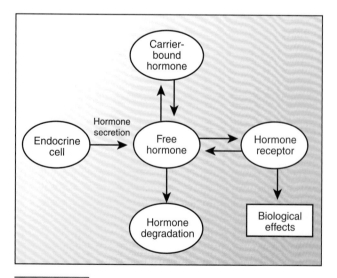

Figure 30.4 The relationship between hormone secretion, carrier protein binding, and hormone degradation. This relationship determines the amount of free hormone available for receptor binding and the production of biologic effects.

has been transferred and the need for further stimulus has ceased. As described earlier, not only the rate of secretion but also the rate of degradation determine steady-state plasma concentrations of hormone. Thus, any factor that significantly alters the degradation of a hormone can potentially alter its circulating concentration. Commonly, however, secretory mechanisms can compensate for altered degradation such that plasma hormone concentrations remain within the normal range. Processes of hormone degradation show little, if any, regulation; alterations in the rates of hormone synthesis or secretion in most cases provide the primary mechanism for altering circulating hormone concentrations.

For most hormones, the liver is quantitatively the most important site of degradation; for a few others, the kidneys play a significant role as well. Diseases of the liver and kidneys may, therefore, indirectly influence endocrine status as a result of altering the rates at which hormones are removed from the circulation. Various drugs also alter normal rates of hormone degradation; thus, the possibility of indirect drug-induced endocrine abnormalities also exists. In addition to the liver and kidneys, target tissues may take up and degrade quantitatively smaller amounts of hormone. In the case of amine and polypeptide hormones, this occurs via receptor-mediated endocytosis.

The nature of specific structural modification(s) involved in hormone inactivation and degradation differs for each hormone class. As a general rule, however, specific enzyme-catalyzed reactions are involved. Inactivation and degradation may involve complete metabolism of the hormone to entirely different products, or it may be limited to a simpler process involving one or two steps, such as a covalent modification to inactivate the hormone. Urine is the primary route of excretion of hormone degradation products, but small amounts of intact hormone may also appear in the urine. In some cases, measuring the urinary content of a hormone or hormone metabolite provides a useful, indirect, noninvasive means of assessing endocrine function.

The degradation of amine and polypeptide hormones has been studied only in a few cases. However, it appears that proteolytic attack inactivates these hormones in many tissues. The first step appears to involve attack by specific peptidases, resulting in the formation of several distinct hormone fragments. Several nonspecific peptidases then metabolize these fragments to yield the constituent amino acids, which can be reused.

The metabolism and degradation of steroid hormones have been studied in much more detail. The primary organ involved is the liver, although some metabolism also takes place in the kidneys. Complete steroid metabolism generally involves a combination of one or more of five general classes of reactions: reduction, hydroxylation, side-chain cleavage, oxidation, and esterification. Reduction reactions are the principal reactions involved in the conversion of biologically active steroids to forms that possess little or no activity. Esterification (or conjugation) reactions are also particularly important. Groups added in esterification reactions are primarily glucuronate and sulfate. The addition of such charged moieties enhances the water solubility of the metabolites,

facilitating their excretion. Steroid metabolites are eliminated from the body primarily via the urine, although smaller amounts also enter the bile and leave the body in the feces.

At times, quantitative information concerning the rate of hormone metabolism is clinically useful. One index of the rate at which a hormone is removed from the blood is the **metabolic clearance rate** (**MCR**). The metabolic clearance of a hormone is analogous to that of renal clearance (see Chapter 22). The MCR is the volume of plasma cleared of the hormone in question per unit time, expressed in milliliter plasma/minute. It is calculated from the equation:

$$MCR = \frac{\text{Hormone removed per unit time (mg/min)}}{\text{Plasma concentration (mg/mL)}} \quad (1)$$

One approach to measuring MCR involves injecting a small amount of radioactive hormone into the subject and then collecting a series of timed blood samples to determine the amount of radioactive hormone remaining. Based on the rate of disappearance of hormone from the blood, its half-life and MCR can be calculated. The MCR and half-life are inversely related—the shorter the half-life, the greater the MCR. The half-lives of different hormones vary considerably, from 5 minutes or less for some to several hours for others. The circulating concentration of hormones with short half-lives can vary dramatically over a short period of time. This is typical of hormones that regulate processes on an acute minute-to-minute basis, such as many of those involved in regulating blood glucose. Hormones for which rapid changes in concentration are not required, such as those with seasonal variations and those that regulate the menstrual cycle, typically have longer half-lives.

Radioimmunoassays, and several adapted methodologies, are popular tools used to measure hormone concentrations.

The concentration of hormone present in a biologic fluid is often measured to make a clinical diagnosis of a suspected endocrine disease or to study basic endocrine physiology. Most hormones are measured in blood or urine, but alternate testing sources, such as saliva and transdermal membrane monitors, are emerging as new tools in endocrinology. Possibly, the greatest contribution to recent advances in the field of endocrinology has been derived from immunochemical methods.

Even before hormones were chemically characterized, they were quantitated in terms of biologic responses they produced. Thus, early assays for measuring hormones were bioassays that depended on a hormone's ability to produce a characteristic biologic response. As a result, hormones came to be quantitated in terms of units, defined as an amount sufficient to produce a response of specified magnitude under a defined set of conditions. A unit of hormone is, thus, arbitrarily determined. Although bioassays are rarely used today for diagnostic purposes, many hormones are still standardized in terms of **biological activity units**. For example, commercial insulin is still sold and dispensed based on the number of units in a particular preparation, rather than by the weight or the number of moles of insulin.

Bioassays in general suffer from a number of shortcomings, including a relative lack of specificity and a lack of sensitivity. In many cases, they are slow and cumbersome to perform, and often, they are expensive, because biologic variability often requires the inclusion of many animals in the assay.

Development of the **radioimmunoassay (RIA)** in the late 1950s and early 1960s was a major step forward in clinical and research endocrinology. Much of our current knowledge of endocrinology is based on this method. An RIA is now available for virtually every known hormone. In addition, RIAs have been developed to measure circulating concentrations of many other biologically relevant proteins, drugs, and vitamins.

The RIA is a prototype for a larger group of assays termed **competitive binding assays**. These are modifications and adaptations of the original RIA, relying to a large degree on the principle of competitive binding on which the RIA is based. It is beyond the scope of this text to describe in detail the competitive binding assays currently used to measure hormone concentrations, but the principles are the same as those for the RIA.

The two key components of an RIA are a specific antibody (Ab) that has been raised against the hormone in question and a radioactively labeled hormone (H*). If the hormone being measured is a peptide or protein, the molecule is commonly labeled with a radioactive iodine atom (^{125}I or ^{131}I) that can be readily attached to tyrosine residues of the peptide chain. For substances lacking tyrosine residues, such as steroids, labeling may be accomplished by incorporating radioactive carbon (^{14}C) or hydrogen (^{3}H). In either case, the use of the radioactive hormone permits detection and quantification of small amounts of the substance.

The RIA is performed *in vitro* using a series of test tubes. Fixed amounts of Ab and H* are added to all tubes (Fig. 30.5A). Samples (plasma, urine, cerebrospinal fluid, etc.) to be measured are added to individual tubes. Varying known concentrations of unlabeled hormone (the standards) are added to a series of identical tubes. The principle of the RIA, as indicated in Figure 30.5B, is that labeled and unlabeled hormones compete for a few antibody-binding sites. The amount of each hormone that is bound to antibody is a proportion of that present in solution. In a sample containing a high concentration of hormone, less radioactive hormone is able to bind to the antibody, and in a sample containing a low concentration of hormone, more radioactive hormone is able to bind to the antibody. In each case, the amount of radioactivity present as antibody-bound H* is determined. The response produced by the standards is used to generate a **standard curve** (Fig. 30.6). Responses produced by the unknown samples are then compared with the standard curve to determine the amount of hormone present in the unknowns (see *dashed lines* in Fig. 30.6).

One major limitation of RIAs is that they measure immunoreactivity rather than biologic activity. The presence of an immunologically related but different hormone or the presence of heterogeneous forms of the same hormone can complicate the interpretation of the results. For example, POMC, the precursor of **adrenocorticotropic hormone**

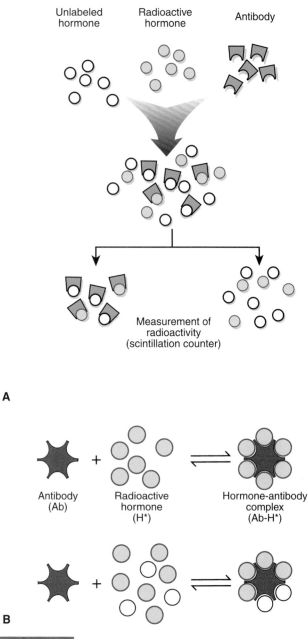

A

B

Figure 30.5 **The principles of radioimmunoassay (RIA).**
(A) When unlabeled hormone (*open circles*) is also introduced into the system, less radioactive hormone binds to the antibody.
(B) Specific antibodies (Ab) bind with radioactive hormone (H*) to form hormone–antibody complexes (Ab-H*).

(**ACTH**), is often present in high concentrations in the plasma of patients with bronchogenic carcinoma. Antibodies for ACTH may cross-react with POMC. The results of an RIA for ACTH in which such an antibody is used may suggest high concentrations of ACTH when actually POMC is being detected. Because POMC has of the biologic potency of ACTH, there may be little clinical evidence of significantly elevated ACTH. If appropriate measures are taken, however, such possible pitfalls can be overcome in most cases and reliable results from the RIA can be obtained.

One important modification of the RIA is the **radioreceptor assay**, which uses specific hormone receptors rather than

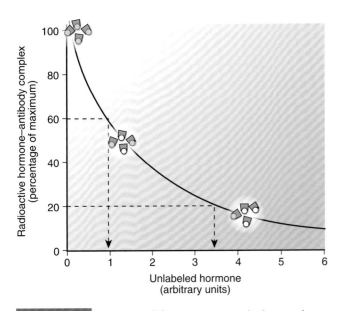

Figure 30.6 **A typical radioimmunoassay standard curve.** As indicated by the *dashed lines*, the hormone concentration in unknown samples can be deduced from the standard curve.

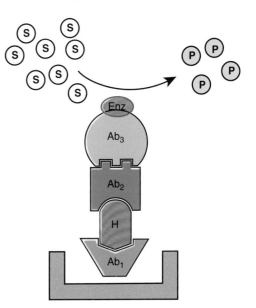

Figure 30.7 **The basic components of an enzyme-linked immunosorbent assay (ELISA).** A typical ELISA is performed in a 3 × 5-in plastic plate containing 96 small wells. Each well is precoated with an antibody (Ab_1) that is specific for the hormone (H) being measured. Unknown samples or standards are introduced into the wells, followed by a second hormone-specific antibody (Ab_2). A third antibody (Ab_3), which recognizes Ab_2, is then added. Ab_3 is coupled to an enzyme (Enz) that will convert an appropriate substrate (S) into a colored or fluorescent product (P). The amount of product formed can be determined using optical methods. After the addition of each antibody or sample to the wells, the plates are incubated for an appropriate period of time to allow antibodies and hormones to bind. Any unbound material is washed out of the well before the addition of the next reagent. The amount of colored product formed is directly proportional to the amount of hormone present in the standard or unknown sample. Concentrations are determined using a standard curve. For simplicity, only one Ab_1 molecule is shown in the bottom of the well when, in fact, there is an excess of Ab_1 relative to the amount of hormone to be measured.

antibodies as the hormone-binding reagent. In theory, this method measures biologically active hormone because receptor binding rather than antibody recognition is assessed. However, the need to purify hormone receptors and the somewhat more complex nature of this assay limit its usefulness for routine clinical measurements. It is more likely to be used in a research setting.

The **enzyme-linked immunosorbent assay** (**ELISA**) is a solid-phase, enzyme-based assay whose use and application have increased considerably over the past two decades. A typical ELISA is a colorimetric or fluorometric assay, and therefore, the ELISA, unlike the RIA, does not produce radioactive waste, which is an advantage, considering environmental concerns and the rapidly increasing cost of radioactive waste disposal. In addition, because it is a solid-phase assay, the ELISA can be automated to a large degree, which reduces costs. Figure 30.7 shows a relatively simple version of an ELISA. More complex assays using similar principles have been developed to overcome a variety of technical problems, but the basic principle remains the same.

▶ MECHANISMS OF HORMONE ACTION

The binding of a hormone to its receptor with subsequent activation of the receptor is the first step in hormone action and also the point at which specificity is determined within the endocrine system. The second step in hormone action entails the receptor-mediated transduction of the extracellular message into an intracellular signal or second messenger. Many intracellular signal cascades exist (recall the discussion of cellular signaling in Chapter 2). Abnormal interactions of hormones with their receptors and/or disturbances in cell signaling are involved in the pathogenesis of many endocrine disease states, and therefore, considerable attention has been paid to these aspects of hormone action. Here, discussion is confined to key principles of hormone–receptor interaction.

Biologic response is partly determined by hormone–receptor binding kinetics.

The probability that a hormone–receptor interaction will occur is related to both the abundance of cellular receptors and the receptor's affinity for the hormone relative to the ambient hormone concentration. The more receptors available to interact with a given amount of hormone, the greater the likelihood there is of a response. Similarly, the higher the affinity of a receptor is for a hormone, the greater the likelihood that a hormone–receptor interaction will occur. The circulating hormone concentration is, of course, a function of the rate of hormone secretion relative to hormone degradation.

The association of a hormone with its receptor generally behaves as if it were a simple, reversible chemical reaction that can be described by the following kinetic equation:

$$[H]+[R]\rightleftharpoons[HR] \qquad (2)$$

where [H] is the free hormone concentration, [R] is the unoccupied receptor concentration, and [HR] is the *hormone–receptor complex* (also referred to as *bound hormone or occupied receptor*). Assuming a simple chemical equilibrium, it follows that:

$$K_a = [HR]/[H] \times [R] \qquad (3)$$

where K_a is the association constant. If R_0 is defined as the total receptor number (i.e., [R] + [HR]), then after substituting and rearranging, we obtain the following relationship:

$$[HR]/[H] = -K_a[HR] + K_aR_0 \qquad (4)$$

Literally translated, this equation states that:

$$\frac{\text{Bound hormone}}{\text{Free hormone}} = -K_a \times \text{Bound hormone} \\ + K_a \times \text{Total receptor number} \qquad (5)$$

Notice that equations 4 and 5 have the general form of an equation for a straight line: $y = mx + b$.

To obtain information regarding a particular hormone–receptor system, a fixed number of cells (and, therefore, a fixed number of receptors) are incubated *in vitro* in a series of test tubes with increasing amounts of hormone. At each higher hormone concentration, the amount of receptor-bound hormone is increased until hormone occupies all receptors. Receptor number and affinity can be obtained by using the relationships given in equation 5 above and plotting the results as the ratio of receptor-bound hormone to free hormone ([HR]/[H]) as a function of the amount of bound hormone ([HR]). This type of analysis is known as a **Scatchard plot** (Fig. 30.8). In theory, a Scatchard plot of simple, reversible equilibrium binding is a straight line (see Fig. 30.8A), with the slope of the line being equal to the negative of the association constant ($-K_a$) and the x-intercept being equal to the total receptor number (R_0). Other equally valid mathematical and graphic methods can be used to analyze hormone–receptor interactions, but the Scatchard plot is probably the most widely used.

In practice, Scatchard plots are not always straight lines but instead can be curvilinear (see Fig. 30.8B). Insulin is a classic example of a hormone that gives curved Scatchard plots. One interpretation of this result is that cells contain two separate and distinct classes of receptors, each with a different binding affinity. Typically, one receptor population has a higher affinity but is fewer in number compared with the second population. Therefore, as indicated in Figure 30.8B, $K_{a1} > K_{a2}$ but $R_{02} > K_{01}$. Computer analysis is often required to fit curvilinear Scatchard plots accurately to a two-site model.

Another explanation for curvilinear Scatchard plots is that occupied receptors influence the affinity of adjacent, unoccupied receptors by **negative cooperativity**. According to this theory, when one hormone molecule binds to its receptor, it causes a decrease in the affinity of nearby unoccupied receptors, making it more difficult for additional hormone molecules to bind. The greater the amount of hormone bound, the lower the affinity of unoccupied receptors. Therefore, as shown in Figure 30.8B, as bound hormone increases, the

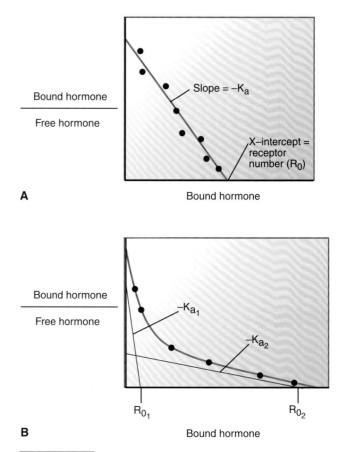

Figure 30.8 **Scatchard plots of hormone–receptor binding data.** **(A)** A straight-line plot typical of hormone binding to a single class of receptors. **(B)** A curvilinear Scatchard plot typical of some hormones. Several models have been proposed to account for nonlinearity of Scatchard plots. K_a, association constant; R_0, total receptor number. (See text for details.)

affinity (slope) steadily decreases. Whether curvilinear Scatchard plots in fact result from two-site receptor systems or from negative cooperativity between receptors is unknown.

Dose–response curves determine changes in responsiveness and sensitivity.

Hormone effects are generally not all-or-none phenomena—that is, they generally do not switch from totally off to totally on and then back again. Instead, target cells exhibit graded responses proportional to the concentration of free hormone present.

The dose–response relationship for a hormone generally exhibits a sigmoid shape when plotted as the biologic response on the *y* axis versus the log of the hormone concentration on the *x* axis (Fig. 30.9).

Regardless of the biologic pathway or process being considered, cells typically exhibit an intrinsic *basal level* of activity in the absence of added hormone, even well after any previous exposure to hormone. As the hormone concentration surrounding the cells increases, a minimal **threshold concentration** must be present before any measurable increase in the cellular response can be produced. At higher

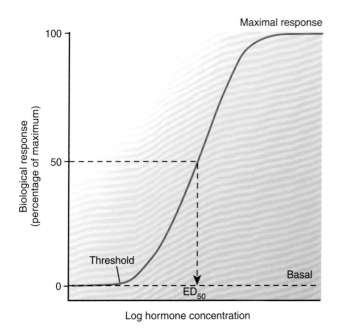

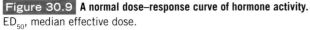

Figure 30.9 **A normal dose–response curve of hormone activity.** ED_{50}, median effective dose.

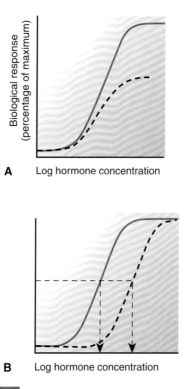

Figure 30.10 **Altered target tissue responses reflected by dose–response curves.** **(A)** Decreased target tissue responsiveness. **(B)** Decreased target tissue sensitivity.

hormone concentrations, a **maximal response** by the target cell is produced, and increasing the hormone concentration cannot elicit greater response. The concentration of hormone required to produce a response half-way between the maximal and basal responses, the **median effective dose (ED_{50})**, is a useful index of the sensitivity of the target cell for that particular hormone (see Fig. 30.9).

For some peptide hormones, the maximal response may occur when only a small percentage (5% to 10%) of the total receptor population is occupied by hormone. The remaining 90% to 95% of the receptors are called spare receptors, because on initial inspection, they do not appear necessary to produce a maximal response. This term is unfortunate, because the receptors are not "spare" in the sense of being unused. Although at any one point in time only 5% to 10% of the receptors may be occupied, hormone–receptor interactions are an equilibrium process and hormones continually dissociate and reassociate with their receptors. Therefore, from one point in time to the next, different subsets of the total population of receptors may be occupied, but presumably all receptors participate equally in producing the biologic response.

Physiologic or pathophysiologic alterations in target tissue responses to hormones can take one of two general forms, as indicated by changes in the hormones' dose–response curves (Fig. 30.10). Although changes in dose–response curves are not routinely assessed in the clinical setting, they can serve to distinguish between a receptor defect and a postreceptor defect in hormone action, providing useful information regarding the underlying cause of a particular disease state. Changes in responsiveness are indicated by an increase or decrease in the maximal response of the target tissue and may be the result of one or more factors (see Fig. 30.10A). Altered responsiveness can be caused by a change in the number of functional target cells in a tissue, by

a change in the number of receptors per cell for the hormone in question, or, if the receptor function itself is not rate limiting for the action of the hormone, by a change in the specific rate-limiting postreceptor step in the hormone action pathway.

A change in sensitivity is reflected as a right or left shift in the dose–response curve and, thus, a change in the ED_{50}; a right shift indicates decreased sensitivity and a left shift indicates increased sensitivity for that hormone (see Fig. 30.10B). Changes in sensitivity reflect (1) an alteration in receptor affinity or, if submaximal concentrations of hormone are present, (2) a change in receptor number. Dose–response curves may also reflect combinations of changes in responsiveness and sensitivity in which there is both a right or left shift of the curve (a sensitivity change) and a change in maximal biologic response to a lower or higher level (a change in responsiveness).

Cells can regulate their receptor number and/or function in several ways. Exposing cells to an excess of hormone for a sustained period of time typically results in a decreased number of receptors for that hormone per cell. This phenomenon is referred to as **down-regulation**. In the case of peptide hormones, which have receptors on cell surfaces, a redistribution of receptors from the cell surface to intracellular sites usually occurs as part of the process of down-regulation. Therefore, there may be fewer total receptors per cell, and a smaller percentage may be available for hormone binding on the cell surface. Although somewhat less prevalent than down-regulation, **up-regulation** may occur when certain conditions or treatments cause an increase in receptor number compared with normal. Changes in rates of receptor synthesis may also contribute to long-term down-regulation or up-regulation.

In addition to changing receptor number, many target cells can regulate receptor function. Chronic exposure of cells to a hormone may cause the cells to become less responsive to subsequent exposure to the hormone by a process termed **desensitization**. If the exposure of cells to a hormone has a desensitizing effect on further action by that same hormone, the effect is termed **homologous desensitization**. If the exposure of cells to one hormone has a desensitizing effect with regard to the action of a different hormone, the effect is termed **heterologous desensitization**.

INTEGRATED MEDICAL SCIENCES

Artificial Pancreas

Of great importance to the practicing physician is the increased number of patients who have **diabetes mellitus**. As described in Chapter 34, diabetes mellitus is an endocrine disorder of metabolic dysregulation, most notably a dysregulation of glucose metabolism. In patients with type 1 diabetes, control of blood glucose may prevent long-term complications. Unfortunately, achieving near-normal glucose control is difficult, and the incidence of hypoglycemia increases when glucose control approaches normal glucose levels. Over the years, biomedical engineers have been trying to develop an "artificial pancreas." Now, with recent advances in *insulin pump* and *continuous glucose monitoring* technologies, the road has been paved for the development of an artificial pancreas system.

This mechanical closed-loop system would determine glucose levels on a continuous basis, via subcutaneous sensors, and transmit data to an insulin pump. Future artificial pancreas systems may have physiologic sensors that determine the onset of a meal, determine the level and duration of physical activity, and measure the concentration of insulin in the blood. These additional measures will advance the computer algorithm contained in the pump to provide the appropriate rate of insulin delivery at the right time.

Moreover, new artificial pancreas systems under development include multiple pancreatic peptide hormones, because the healthy endocrine pancreas controls blood glucose through the balanced and coordinated release of insulin, glucagon, amylin, somatostatin, and other pancreatic hormones (see Chapter 34). Along these lines, development of a closed-loop system that delivers both insulin and glucagon as directed by continuous blood glucose readings has been shown to provide a feasible way to control blood glucose and avoid treatment-requiring hypoglycemia. Because amylin is normally cosecreted with insulin and functionally enhances insulin action and continuous delivery of somatostatin lowers the insulin requirement for diabetic subjects, coadministration of amylin and/or somatostatin with reduced insulin dosing might also be used to decrease hypoglycemic risk. ■

Chapter Summary

- The endocrine system integrates organ function via hormones that are secreted from classical endocrine glands and also from organs whose primary function is not endocrine.
- Hormones may signal to the cells that produced them (autocrine) or neighboring cells (paracrine) but are classically defined as blood-borne chemicals that signal to distant target tissues.
- Target cell recognition of hormones depends on specific, high-affinity receptors, which may be located on the cell surface, within the cytoplasm, or in the target cell's nucleus.
- Hormonal signals are organized in a hierarchy of feedback systems, millionfold amplification cascades, and often definable secretion patterns.

- Most hormones display pleiotropic effects and share their ability to control vital physiologic parameters with other hormones.
- Chemically, hormones may be metabolites of single amino acids, peptides, or metabolites of cholesterol and, depending on their solubility, are either transported in the bloodstream free (amine and peptide hormones) or bound to transport proteins (steroid and thyroid hormones).
- Radioimmunoassay and enzyme-linked immunosorbent assay have provided major advancements in the field of endocrinology, but each type of assay has limitations.
- Disturbances in cell signaling and abnormal interactions of hormones with their receptors may lead to endocrine abnormalities.

Chapter Review Questions

1. A shift to the right in the biological activity dose–response curve for a hormone with no accompanying change in the maximal response indicates:

   A. decreased responsiveness *and* decreased sensitivity.
   B. increased responsiveness.
   C. decreased sensitivity.
   D. increased sensitivity *and* decreased responsiveness.
   E. increased sensitivity.

   The correct answer is C. Right or left shifts in dose–response curves indicate changes in sensitivity. Changes in maximal biological response indicate changes in responsiveness. Because there is no change in maximal response, the correct answer must relate to a change in sensitivity only. A right shift indicates decreased sensitivity.

2. Within the endocrine system, specificity of communication is determined by:

   A. the chemical nature of the hormone.
   B. the distance between the endocrine cell and its target cell(s).
   C. the presence of specific receptors on target cells.

   D. anatomic connections between the endocrine and target cells.
   E. the affinity of binding between the hormone and its receptor.

   The correct answer is C. Hormones produce their effects on target cells by interacting with specific receptors. Hormone binding to its receptor generally initiates a cascade of events that lead to biological effects in the target cells.

3. The ability of hormones to be effective regulators of biological function despite circulating at very low concentrations results from:

   A. the multiplicity of their effects.
   B. transport proteins.
   C. pleiotropic effects.
   D. signal amplification.
   E. competitive binding.

   The correct answer is D. Hormones generally circulate at concentrations from 10^{-9} to 10^{-12} M. They produce much larger changes in a variety of biological parameters as a result of signal amplification, in which the rather weak hormonal signal is amplified into a larger biological response.

Clinical Application Exercises 30.1

A study sought to determine the effect of age and sex on GH secretion. A 24-hour secretory profile of GH was generated by 20-minute sampling in 10 young women (age 15 to 30 years) and 10 young men (age 15 to 30 years). The integrated GH concentration was significantly greater in women than in men and greater in the young than in the old. The mean pulse amplitude, duration, and fraction of GH secreted in pulses were each greater in the young, but not significantly different between the sexes. The mean pulse frequency was not affected by sex or age.

QUESTIONS

1. Why was GH secretion measured by a continuous withdrawal pump?

2. What would be a less arduous approach in diagnosing GH deficiency?

ANSWERS

1. Most healthy children experience episodes or "bursts" of GH secretion throughout the day, most prominently within the first several hours of sleep. To obtain reliable information about growth hormone secretion, insertion of a continuous withdrawal pump or patent indwelling catheter with unrestricted food intake and physical activity is required. Increasing sampling frequencies from every 20-minute to 5-minute or 30-second sampling intervals enhances the threshold for detecting more pulses per hour. Although cumbersome and expensive, this method eliminates the error of isolated peak or trough measurements that might otherwise be obtained by single or multiple random GH samplings.

2. A less arduous approach with similar discriminating power in diagnosing GH deficiency entails measuring GH levels in the blood shortly after appropriate pituitary stimulation. Pharmacologic agents that stimulate GH secretion include L-dopa, clonidine, propanolol, glucagon, insulin, and arginine. By perturbing the system in a well-prescribed fashion, the endocrinologist is able to gain important information about growth hormone secretion that would not be possible if a random blood sample were used.

thePoint® *Visit* http://thepoint.lww.com/rhoades5e *for additional chapter review Q&A, Clinical Application Exercises, animations, and more!*

31

Hypothalamus and the Pituitary Gland

Active Learning Objectives

Upon mastering the material in this chapter, you should be able to:

- Compare and contrast how the functional anatomy of the hypothalamic–pituitary axis results in different mechanisms of synthesis for posterior pituitary hormones (arginine vasopressin, oxytocin) versus anterior pituitary hormones (adrenocorticotropic hormone, thyroid-stimulating hormone, growth hormone, follicle-stimulating hormone, luteinizing hormone, and prolactin).
- Describe how arginine vasopressin (AVP) acts on water reabsorption in the kidneys and predict the AVP secretory response to a rise and fall in blood osmolality.
- Explain the smooth muscle response to oxytocin in the breast and uterus and predict the effect of suckling or cervical dilation on oxytocin secretion.
- Describe the response of pituitary corticotrophs to corticotropin-releasing hormone and predict the effect of adrenocorticotropic hormone to regulate glucocorticoid release from the adrenal cortex in response to physical or emotional stress, arginine vasopressin,

the sleep–wake cycle, and an increase in glucocorticoid in the blood.
- Explain the response of pituitary thyrotrophs to thyrotropin-releasing hormone and predict the effect of thyroid-stimulating hormone to regulate triiodothyronine and thyroxine release from the thyroid follicles in response to cold temperatures, the sleep–wake cycle, and an increase in thyroid hormone in the blood.
- Outline the hypothalamic–pituitary–growth hormone axis, explain the response of pituitary somatotrophs to growth hormone–releasing hormone and somatotropin release–inhibiting factor, and predict the effect of growth hormone (GH), insulin-like growth factor, aging, deep sleep, stress, exercise, and hypoglycemia on GH secretion.
- Outline the hypothalamic–pituitary–gonadal axis and explain the response of pituitary gonadotrophs to luteinizing hormone–releasing hormone.
- Explain the effect of estrogen and dopamine on pituitary lactotrophs and the effect of prolactin on alveolar cells in the mammary gland.

The **pituitary gland** or **hypophysis** is an endocrine gland that protrudes off of the bottom of the hypothalamus at the base of the brain. The pituitary gland is composed of two lobes: the anterior pituitary and the posterior pituitary, with both lobes functionally linked to the hypothalamus by the pituitary stalk. Thus, the hypothalamus controls pituitary gland hormone secretion.

The pituitary gland secretes an array of peptide hormones that regulate almost every aspect of body function. Some pituitary hormones influence key cellular processes to preserve the volume and composition of body fluids. Others bring about changes in body function, which enable the person to grow, reproduce, and respond appropriately to stress and trauma. The pituitary hormones can act directly on their target cells or stimulate other endocrine glands to secrete hormones, which, in turn, bring about changes in body function.

▶ HYPOTHALAMIC–PITUITARY AXIS

The pituitary gland is located at the base of the brain and is connected to the hypothalamus by a stalk. It sits in a depression in the sphenoid bone of the skull called the **sella turcica**. The two morphologically and functionally distinct glands comprising the human pituitary are the adenohypophysis and the neurohypophysis as shown in Figure 31.1. The **adenohypophysis**, or *anterior pituitary*, comprises the **pars tuberalis**, which forms the outer covering of the pituitary stalk, and the **pars distalis** or **anterior lobe**. The **neurohypophysis**, or *posterior pituitary*, comprises the **infundibular**

stem, which forms the inner part of the stalk, the **infundibular process** or **posterior lobe**, and the **median eminence** of the hypothalamus (though some do not consider this a part of the pituitary gland). In most vertebrates, the pituitary contains a third anatomically distinct lobe, the **pars intermedia** or **intermediate lobe**. In adult humans, only a vestige of the intermediate lobe is found, consisting of a thin, diffuse region of cells between the anterior and posterior lobes.

The adenohypophysis and neurohypophysis have different embryologic origins. The adenohypophysis is formed from an evagination of the oral ectoderm called the **Rathke pouch**. The neurohypophysis forms as an extension of the developing hypothalamus, which fuses with the Rathke pouch as development proceeds. The posterior lobe is, therefore, composed of neural tissue and is a functional part of the hypothalamus.

Hypothalamic neurons whose axons terminate in the posterior lobe synthesize posterior pituitary hormones.

The infundibular stem of the pituitary gland contains bundles of nonmyelinated nerve fibers, which terminate on the capillary bed in the posterior lobe. These fibers are the axons of neurons that originate in the **supraoptic nuclei** and **paraventricular nuclei** of the hypothalamus. The cell bodies of these neurons are large compared with those of other hypothalamic neurons; hence, they are called **magnocellular neurons**. The hormones **arginine vasopressin** (**AVP**) and **oxytocin** are synthesized as parts of larger precursor prohormones in the cell bodies of these neurons, which are then packaged into granules and enzymatically processed to

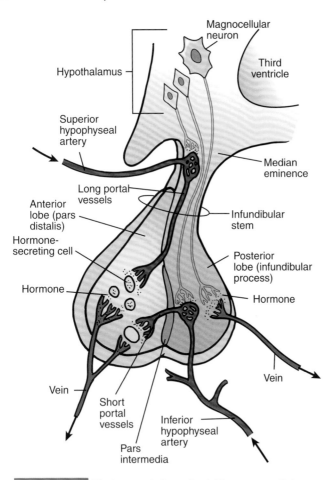

Figure 31.1 **The human pituitary gland.** The magnocellular neuron (large yellow cell body) releases arginine vasopressin or oxytocin at its axon terminals into capillaries that give rise to the venous drainage of the posterior lobe. The neurons with smaller cell bodies are secreting releasing factors into capillary networks that give rise to the long and short hypophyseal portal vessels, respectively. Releasing hormones are shown reaching the hormone-secreting cells of the anterior lobe via the portal vessels.

produce AVP and oxytocin. Axoplasmic flow transports the granules down the axons to accumulate at the axon terminals in the posterior lobe.

Events occurring within or outside the body may generate stimuli for the secretion of posterior lobe hormones. The central nervous system (CNS) processes these stimuli, and the signal for secretion of AVP or oxytocin is then transmitted to neurosecretory neurons in the hypothalamus. Secretory granules containing the hormone are then released into the nearby capillary circulation, from which the hormone is carried into the systemic circulation.

Anterior pituitary hormones are synthesized and secreted in response to hypothalamic-releasing hormones carried in the hypophyseal portal circulation.

The anterior lobe contains clusters of histologically distinct types of cells closely associated with blood sinusoids that drain into the venous circulation. These cells produce anterior pituitary hormones and secrete them into the blood sinusoids. Distinctly different types of cells

produce the six well-known anterior pituitary hormones. **Adrenocorticotropic hormone (ACTH)**, also known as **corticotropin**, is secreted by corticotrophs, **thyroid-stimulating hormone (TSH)** by **thyrotrophs, growth hormone (GH)** by **somatotrophs,** prolactin **(PRL)** by **lactotrophs,** and **follicle-stimulating hormone (FSH)** and **luteinizing hormone (LH)** by **gonadotrophs**.

The cells that produce anterior pituitary hormones are not innervated and, therefore, are not under direct neural control. Rather, **releasing hormones** called **hypophysiotropic hormones**, synthesized by neural cell bodies in the hypothalamus, regulate their secretory activity. Granules containing releasing hormones are stored in the axon terminals of these neurons, located in capillary networks in the median eminence of the hypothalamus and lower infundibular stem. These capillary networks give rise to the principal blood supply to the anterior lobe of the pituitary.

The blood supply to the anterior pituitary is shown in Figure 31.1. The superior and inferior hypophyseal arteries bring blood to the hypothalamic–pituitary region. The **superior hypophyseal arteries** give rise to a rich capillary network in the median eminence. The capillaries converge into long veins that run down the pituitary stalk and empty into the blood sinusoids in the anterior lobe. They are considered to be portal veins because they deliver blood to the anterior pituitary rather than joining the venous circulation that carries blood back to the heart; therefore, they are called **long hypophyseal portal vessels**. The **inferior hypophyseal arteries** provide arterial blood to the posterior lobe. They also penetrate into the lower infundibular stem, where they form another important capillary network. The capillaries of this network converge into **short hypophyseal portal vessels**, which also deliver blood into the sinusoids of the anterior pituitary. The special blood supply to the anterior lobe of the pituitary gland is known as the **hypophyseal portal circulation**.

When a neurosecretory neuron is stimulated to secrete, the releasing hormone is discharged into the hypophyseal portal circulation (see Fig. 31.1). Releasing hormones travel only a short distance before they come in contact with their target cells in the anterior lobe. Neurosecretory neurons only deliver the amount of releasing hormone needed to control anterior pituitary hormone secretion. Consequently, releasing hormones are almost undetectable in systemic blood.

A releasing hormone either stimulates or inhibits the synthesis and secretion of a particular anterior pituitary hormone. **Corticotropin-releasing hormone (CRH), thyrotropin-releasing hormone (TRH),** and **GH-releasing hormone (GHRH)** stimulate the synthesis and secretion of ACTH, TSH, and GH, respectively (Table 31.1). **LH-releasing hormone (LHRH)**, also known as **gonadotropin-releasing hormone (GnRH)**, stimulates the synthesis and release of FSH and LH. In contrast, **somatostatin**, also called **somatotropin release–inhibiting factor (SRIF)**, inhibits GH secretion. All of the releasing hormones are peptides, with the exception of **dopamine**, which is a catecholamine that inhibits the synthesis and secretion of PRL. Releasing hormones can be produced synthetically and used in the diagnosis and treatment

TABLE 31.1	Hypothalamic-Releasing Hormones	
Hormone	**Chemistry**	**Actions on Anterior Pituitary**
Corticotropin-releasing hormone (CRH)	Single chain of 41 amino acids	Stimulates ACTH secretion by corticotrophs; stimulates expression of POMC gene in corticotrophs
Thyrotropin-releasing hormone (TRH)	Peptide of three amino acids	Stimulates TSH secretion by thyrotrophs; stimulates expression of genes for α and β subunits of TSH in thyrotrophs; stimulates PRL synthesis by lactotrophs
Growth hormone–releasing hormone (GHRH)	Two forms in human: single chain of 44 amino acids and single chain of 40 amino acids	Stimulates GH secretion by somatotrophs; stimulates expression of GH gene in somatotrophs
Luteinizing hormone–releasing hormone (LHRH)	Single chain of 10 amino acids	Stimulates FSH and LH secretion by gonadotropin-releasing hormone (GnRH) gonadotrophs
Somatostatin and somatotropin release–inhibiting factor (SRIF)	Single chain of 14 amino acids	Inhibits GH secretion by somatotrophs; inhibits TSH secretion by thyrotrophs
Dopamine	Catecholamine	Inhibits PRL synthesis and secretion by lactotrophs

ACTH, adrenocorticotropic hormone; FSH, follicle-stimulating hormone; GH, growth hormone; LH, luteinizing hormone; POMC, proopiomelanocortin; PRL, prolactin; TSH, thyroid-stimulating hormone.

of diseases of the endocrine system. For example, synthetic GnRH is now used for treating infertility in women.

External events or changes occurring within the body itself generate signals within the CNS that result in releasing hormone secretion. For example, sensory nerve excitation, emotional or physical stress, biologic rhythms, changes in sleep patterns or in the sleep–wake cycle, and changes in circulating levels of certain hormones or metabolites all affect the secretion of particular anterior pituitary hormones. Signals generated in the CNS by such events are transmitted to the neurosecretory neurons in the hypothalamus. Depending on the nature of the event and the signal generated, the secretion of a particular releasing hormone may be either stimulated or inhibited. In turn, this response affects the rate of secretion of the appropriate anterior pituitary hormone. The neural pathways involved in transmitting signals to the neurosecretory neurons in the hypothalamus continue to be defined.

▶ POSTERIOR PITUITARY HORMONES

Magnocellular neurons in the supraoptic and paraventricular nuclei of the hypothalamus produce AVP (also known historically as *antidiuretic hormone*) and oxytocin. Individual neurons make either AVP or oxytocin, but not both. The axons of these neurons form the infundibular stem and terminate on the capillary network in the posterior lobe, where they discharge AVP and oxytocin into the systemic circulation.

AVP and oxytocin are closely related small peptides, each consisting of nine amino acid residues. Although AVP and oxytocin differ by only two amino acid residues, the structural differences are sufficient to give these two molecules different hormonal activities. They are similar enough, however, for AVP to have slight oxytocic activity and for oxytocin to have slight antidiuretic activity.

The genes for AVP and oxytocin are located near one another on chromosome 20. They code for much larger prohormones that contain the amino acid sequences for AVP or oxytocin and for a 93 amino acid peptides called **neurophysin**. Mutations in the neurophysin portion of the AVP gene are associated with **hypothalamic** (or **central**) **diabetes insipidus**, a condition in which AVP secretion is impaired. As described earlier, prohormones for AVP and oxytocin are synthesized in the cell bodies of magnocellular neurons and transported in secretory granules to axon terminals in the posterior lobe. During the passage of the granules from the Golgi apparatus to axon terminals, proteolytic enzymes cleave prohormones to produce AVP or oxytocin and their associated neurophysins.

The hypothalamus and posterior pituitary act as a functional unit in the secretion of AVP and oxytocin. When magnocellular neurons receive neural signals for AVP or oxytocin secretion, action potentials are generated in these cells, triggering the release of AVP or oxytocin and neurophysin from the axon terminals. These substances diffuse into nearby capillaries and then enter the systemic circulation.

Arginine vasopressin increases the reabsorption of water by the kidneys.

Two physiologic signals, a rise in the osmolality of the blood and a decrease in blood volume, generate the CNS stimulus for AVP secretion. Chemical mediators of AVP release include catecholamines, angiotensin II, and **atrial natriuretic peptide** (**ANP**). The main physiologic action of AVP is to increase water reabsorption by the collecting ducts of the kidneys. The result is decreased water excretion and the formation of osmotically concentrated urine (see Chapter 22). This action of AVP works to counteract the conditions that stimulate its secretion. For example, reducing water loss in the urine limits a further rise in the osmolality of the blood and conserves blood volume. Low blood AVP levels lead to diabetes insipidus and the excessive production of dilute urine (see Chapter 23).

Oxytocin stimulates the contraction of smooth muscle in the mammary glands and uterus.

Two physiologic signals stimulate the secretion of oxytocin by hypothalamic magnocellular neurons. Breast-feeding stimulates sensory nerves in the nipple. Afferent nerve impulses enter the CNS to stimulate oxytocin-secreting magnocellular neurons. These neurons fire in synchrony and release a bolus of oxytocin into the bloodstream. Oxytocin stimulates the contraction of **myoepithelial cells**, which surround the milk-laden alveoli in the lactating mammary gland, aiding in milk ejection.

Neural input from the female reproductive tract during childbirth also stimulates oxytocin secretion. Cervical dilation before the beginning of labor stimulates stretch receptors in the cervix. Afferent nerve impulses pass through the CNS to oxytocin-secreting neurons. Oxytocin release stimulates the contraction of smooth muscle cells in the uterus during labor, aiding in the delivery of the newborn and placenta. The actions of oxytocin on the mammary glands and the female reproductive tract are discussed further in Chapter 38.

▶ ANTERIOR PITUITARY HORMONES

The anterior lobe contains clusters of histologically distinct types of cells closely associated with blood sinusoids that drain into the venous circulation. These cells produce anterior pituitary hormones and secrete them into the blood sinusoids.

The anterior pituitary secretes six protein hormones, all of which are small, ranging in molecular size from 4.5 to 29 kDa (Table 31.2). Four of the anterior pituitary hormones have effects on the morphology and secretory activity of other endocrine glands; they are called *tropic* (Greek, meaning "to turn to") or *trophic* ("to nourish") hormones. For example, ACTH maintains the size of certain cells in the adrenal cortex and stimulates these cells to synthesize and secrete the **glucocorticoid hormones, cortisol,** and **corticosterone**. Similarly, TSH maintains the size of the cells of the thyroid follicles and stimulates these cells to produce and secrete the thyroid hormones **thyroxine (T_4)** and **triiodothyronine (T_3)**. The two other tropic hormones, FSH and LH, are called **gonadotropins** because both act on the ovaries and testes. FSH stimulates the development of follicles in the ovaries and regulates the process of **spermatogenesis** in the testes. LH causes **ovulation** and **luteinization** of the ovulated follicle in the ovary of the human female and stimulates the production of the female sex hormones **estrogen** and **progesterone** by the ovary. In the male, LH stimulates the **Leydig cells** of the testis to produce and secrete the male sex hormone, testosterone.

The two remaining anterior pituitary hormones, GH and PRL, are not usually thought of as tropic hormones because their main target organs are not other endocrine glands. However, these two hormones have certain effects that can be regarded as "tropic." The main physiologic action of GH is its stimulatory effect on the growth of the body during childhood. In humans, PRL is essential for the synthesis of milk by the mammary glands during **lactation**. Regulation of the secretion of the gonadotropins and PRL and descriptions of their actions are covered in greater detail in Chapters 36 to 38 (Clinical Focus 31.1).

▷ TABLE 31.2	Hormones of the Anterior Pituitary	
Hormone	**Chemistry**	**Physiologic Actions**
Adrenocorticotropic hormone (ACTH, corticotropin)	Single chain of 39 amino acids; 4.5 kDa	Stimulates production of glucocorticoids and androgens by adrenal cortex; maintains size of zona fasciculata and zona reticularis of cortex
Thyroid-stimulating hormone (TSH, thyrotropin)	Glycoprotein having two subunits, α and β; 28 kDa	Stimulates production of thyroid hormones, T_4 and T_3, by thyroid follicular cells; maintains size of follicular cells
Growth hormone (GH)	Single chain of 191 amino acids; 22 kDa	Stimulates postnatal body growth; stimulates triglyceride lipolysis; inhibits insulin action on carbohydrate and lipid metabolism
Follicle-stimulating hormone (FSH)	Glycoprotein having two subunits, α and β; 28–29 kDa	Stimulates development of ovarian follicles; regulates spermatogenesis in testes
Luteinizing hormone (LH)	Glycoprotein having two subunits, α and β; 28–29 kDa	Causes ovulation and formation of corpus luteum in ovaries; stimulates production of estrogen and progesterone by ovaries; stimulates testosterone production by testes
Prolactin (PRL)	Single chain of 199 amino acids	Essential for milk production by lactating mammary glands

T_3, triiodothyronine; T_4, thyroxine.

CLINICAL FOCUS | 31.1

Pituitary Tumors

Pituitary tumors arise from hormone-secreting adenohypophyseal cells, and the secretory product of the tumor depends on the cell of origin. Generally, tumors of the corticotroph, somatotroph, lactotroph, and thyrotroph hypersecrete their respective hormones. In contrast, gonadotrophic tumors usually do not secrete hormones efficiently. Pituitary adenomas with no clinical symptoms have been reported in about 11% of autopsies. Use of sensitive imaging techniques such as magnetic resonance imaging (MRI) for nonpituitary indications has increased the detection of pituitary abnormalities such that microadenoma is detected in 10% of the normal adult population undergoing MRI of the head. Suspected pituitary tumors are usually examined directly with MRI to assess the effect of tumor mass on surrounding soft tissue structures such as the optic chiasm, which can become compressed by larger tumors, resulting in partial loss of vision.

Testing for aberrant hormone levels in the blood can be used to diagnose the type of pituitary adenoma that is present. For Cushing disease, an elevated 24-hour urinary free cortisol level can be an effective screen for disease. An elevation in insulin-like growth factor 1 (IGF-1) level, compared with age-matched and gender-matched controls, would indicate a growth hormone (GH)–secreting adenoma. Following diagnosis, three modes of therapy are available: surgical, radiotherapeutic, and medical. The goals of therapy are to alleviate compressive mass effects and suppress hormone hypersecretion while maintaining intact function of the unaffected portions of the pituitary.

Pituitary surgery is generally recommended to alleviate pressure effects resulting from tumor mass and to reduce hormone hypersecretion. The transsphenoidal surgical approach is associated with the lowest mortality. Radiotherapy, the delivery of high-energy ionizing radiation, is generally indicated for persistent hormone hypersecretion or residual mass effects after surgery or when surgery is contraindicated. Medical therapy is useful in several cases because pituitary tumors often express receptors for hypothalamic control of hormone secretion. Pharmaceutical ligands that activate the somatotropin release–inhibitory factor (SRIF) receptor on tumors arising from somatotrophs can suppress prolactin (PRL) or GH hypersecretion, block tumor growth, and shrink tumor size. Thus, GH-secreting or PRL-secreting tumors usually respond well to medical therapy. ∎

Adrenocorticotropic hormone regulates the function of the adrenal cortex.

The adrenal glands sit on top of the kidneys but do not play a role in renal function. Each adrenal gland is separated into two distinct functional units, the adrenal cortex and medulla. The *adrenal cortex* produces the glucocorticoid hormones, cortisol and corticosterone, in the cells of its two inner zones, the **zona fasciculata** and the **zona reticularis**. These cells also synthesize **androgens**, or male sex hormones. The main androgen synthesized is **dehydroepiandrosterone**.

Glucocorticoids alter gene transcription on many different target cells. Glucocorticoids permit metabolic adaptations during fasting, which prevent the development of **hypoglycemia** or low blood glucose level. They also play an essential role in the response of the body to physical and emotional stress. Other actions of glucocorticoids include inhibition of inflammation, suppression of the immune system, and regulation of vascular responsiveness to norepinephrine.

The cells of the outer zone of the cortex, the **zona glomerulosa**, produce **aldosterone**, the other physiologically important hormone made by the adrenal cortex. It acts to stimulate sodium reabsorption by the kidneys.

ACTH is the physiologic regulator of the synthesis and secretion of glucocorticoids by the zona fasciculata and zona reticularis. The smallest of the six anterior pituitary hormones, ACTH, stimulates the synthesis of these steroid hormones and promotes the expression of the genes for various enzymes involved in steroidogenesis. It also maintains the size and functional integrity of the cells of the zona

fasciculata and zona reticularis. ACTH is not an important regulator of aldosterone synthesis and secretion.

The actions of ACTH on glucocorticoid synthesis and secretion and details about the physiologic effects of glucocorticoids are described in Chapter 33.

Adrenocorticotropic hormone synthesis

ACTH is synthesized in corticotrophs as part of a larger 30-kDa prohormone, **proopiomelanocortin (POMC)**. Enzymatic cleavage of POMC by **prohormone convertase 1** in the anterior pituitary results in ACTH, an amino terminal protein, and β-**lipotropin** (Fig. 31.2). β-Lipotropin has effects on lipid metabolism, but its physiologic function in humans has not been established. Although POMC can be cleaved into other peptides, such as β-endorphin, only ACTH and β-lipotropin are produced from POMC in the human corticotroph. Proteolytic processing of POMC occurs after it is packaged into secretory granules. Therefore, when the corticotroph receives a signal to secrete, ACTH and β-lipotropin are released into the bloodstream in a 1:1 molar ratio.

Cells of the intermediate lobe of the pituitary gland and neurons in the hypothalamus also synthesize POMC. In the intermediate lobe, the ACTH sequence of POMC is cleaved by **prohormone convertase 2** to release a small peptide, α-melanocyte–stimulating hormone (α-**MSH**), and, therefore, little ACTH is produced. α-MSH acts in lower vertebrates to produce temporary changes in skin color by causing the dispersion of melanin granules in pigment cells. As noted earlier, the adult human has only a vestigial intermediate

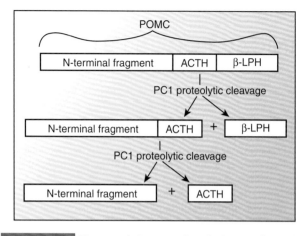

Figure 31.2 The proteolytic processing of adrenocorticotropic hormone (ACTH) from its prohormone proopiomelanocortin (POMC) in the human corticotroph. PC1, prohormone convertase 1; β-LPH, β-lipotropin.

lobe and does not produce and secrete significant amounts of α-MSH or other hormones derived from POMC. However, because ACTH contains the α-MSH amino acid sequence at its N-terminal end, it has melanocyte-stimulating activity when present in the blood at high concentrations. Humans who have high blood levels of ACTH, as a result of **Addison disease** or an ACTH-secreting tumor, are often hyperpigmented. In the hypothalamus, α-MSH is important in the regulation of feeding behavior.

A group of neurons with small cell bodies, called **parvicellular neurons**, synthesize CRH in the paraventricular nuclei of the hypothalamus. The axons of parvicellular neurons terminate on capillary networks that give rise to hypophyseal portal vessels. Secretory granules containing CRH are stored in the axon terminals of these cells. On receiving the appropriate stimulus, these cells secrete CRH into the capillary network; CRH enters the hypophyseal portal circulation and is delivered to the anterior pituitary gland.

CRH binds to receptors on the plasma membranes of corticotrophs, increasing the activity of adenylyl cyclase, which catalyzes the formation of **cyclic adenosine monophosphate (cAMP)** from **adenosine triphosphate (ATP)** (Fig. 31.3). The rise in cAMP in the corticotroph activates **protein kinase A (PKA)**, which then phosphorylates cellular proteins. CRH stimulates the secretion of ACTH and β-lipotropin through PKA-mediated phosphorylation of voltage-dependent calcium channels, although the exact mechanism is still not completely understood.

CRH-stimulated cAMP production in the corticotroph also increases POMC gene expression in these cells (see Fig. 31.3). Thus, CRH both stimulates ACTH secretion and also maintains the capacity of the corticotroph to synthesize the precursor for ACTH.

Glucocorticoids inhibit ACTH secretion

A rise in glucocorticoid concentration in the blood resulting ACH stimulation of the adrenal cortex inhibits the secretion of ACTH. Thus, glucocorticoids have a negative feedback

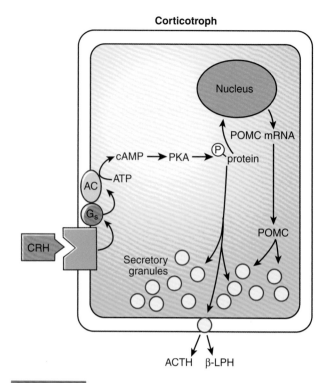

Figure 31.3 The action of corticotropin-releasing hormone (CRH) on a corticotroph. CRH binds to membrane receptors coupled to adenylyl cyclase (AC) by stimulatory G proteins (G$_s$). Cyclic adenosine monophosphate (cAMP) rises in the cell, activating protein kinase A (PKA), which then phosphorylates proteins (P-proteins) that stimulate adrenocorticotropic hormone (ACTH) secretion and proopiomelanocortin (POMC) gene expression. β-LPH, β-lipotropin.

effect on ACTH secretion, which, in turn, reduces the rate of glucocorticoid secretion by the adrenal cortex. If the blood glucocorticoid level begins to fall for some reason, this negative feedback effect is reduced, stimulating ACTH secretion and restoring the blood glucocorticoid concentration. This interactive relationship is called the **hypothalamic–pituitary–adrenal axis** (Fig. 31.4). This control loop ensures that the level of glucocorticoids in the blood remains relatively stable in the resting state, although there is a diurnal variation in glucocorticoid secretion.

The negative feedback effect of glucocorticoids on ACTH secretion results from actions on both the hypothalamus and the corticotroph (see Fig. 31.4). When the concentration of glucocorticoids rises in the blood, CRH secretion from the hypothalamus is inhibited. As a result, the stimulatory effect of CRH on the corticotroph is reduced and the rate of ACTH secretion falls. Glucocorticoids act directly on parvicellular neurons to inhibit CRH release, and indirectly through neurons in the hippocampus that project to the hypothalamus, to affect the activity of parvicellular neurons. At the corticotroph, glucocorticoids inhibit the actions of CRH to stimulate ACTH secretion.

If the blood concentration of glucocorticoids remains high for a long period of time, POMC gene expression is inhibited, POMC mRNA in the corticotroph decreases, and the production of ACTH and the other POMC peptides

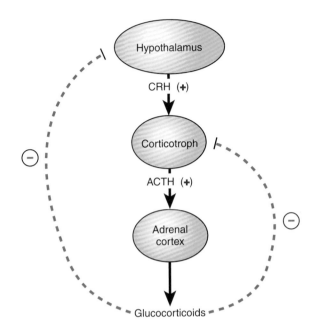

Figure 31.4 The hypothalamic–pituitary–adrenal axis. *Blue dashed lines* indicate the negative feedback actions of glucocorticoids on the corticotroph and the hypothalamus. ACTH, adrenocorticotropic hormone; CRH, corticotropin-releasing hormone.

declines. Glucocorticoids inhibit POMC gene expression in part by suppressing CRH secretion and also by acting directly in the corticotroph itself.

The negative feedback actions of glucocorticoids are essential for the normal operation of the hypothalamic–pituitary–adrenal axis. The disturbances that occur when disease or glucocorticoid administration drastically changes blood glucocorticoid levels vividly illustrate this relationship. For example, if a person's adrenal glands have been surgically removed or damaged by disease (e.g., Addison disease), the resulting lack of glucocorticoids allows corticotrophs to secrete large amounts of ACTH. People with glucocorticoid deficiency caused by inherited genetic defects affecting enzymes involved in steroid hormone synthesis by the adrenal cortex also have high blood ACTH levels resulting from the absence of the negative feedback effects of glucocorticoids on ACTH secretion. High blood concentrations of ACTH cause hypertrophy of the adrenal glands; thus, these genetic diseases are collectively called **congenital adrenal hyperplasia** (see Chapter 33). In contrast, the adrenal cortex atrophies in people treated chronically with large doses of glucocorticoids because the high level of glucocorticoids in the blood inhibits ACTH secretion, resulting in the loss of its trophic influence on the adrenal cortex.

Stress-induced ACTH secretion

Stress greatly influences the hypothalamic–pituitary–adrenal axis. When a person experiences physical or emotional stress, ACTH secretion is increased and the blood level of glucocorticoids rises rapidly. Regardless of the blood glucocorticoid concentration, stress stimulates the hypothalamic–pituitary–adrenal axis because stress-induced neural activity generated at higher CNS levels stimulates parvicellular neurons in the

paraventricular nuclei to secrete CRH at a greater rate. Thus, stress can override the normal operation of the hypothalamic–pituitary–adrenal axis. If the stress persists, the blood glucocorticoid level remains high because the glucocorticoid negative feedback mechanism functions at a higher set point.

Arginine vasopressin

Glucocorticoid deficiency and certain types of stress increase the concentration of AVP in hypophyseal portal blood. The physiologic significance is that AVP, like CRH, can stimulate corticotrophs to secrete ACTH. Acting along with CRH, AVP amplifies the stimulatory effect of CRH on ACTH secretion.

AVP interacts with a specific receptor on the plasma membrane of the corticotroph that is coupled via G proteins to the enzyme **phospholipase C** (**PLC**). The interaction of AVP with its receptor activates PLC, which hydrolyzes plasma membrane **phosphatidylinositol 4,5-bisphosphate** (**PIP**$_2$) to generate the intracellular second messengers **inositol trisphosphate** (**IP**$_3$) and **diacylglycerol** (**DAG**). IP$_3$ mobilizes intracellular calcium stores, and DAG activates the phospholipid-dependent and calcium-dependent **protein kinase C** (**PKC**) to mediate the stimulatory effect of AVP on ACTH secretion.

As noted earlier, magnocellular neurons of the supraoptic and paraventricular nuclei of the hypothalamus produce AVP and oxytocin. These neurons terminate in the posterior lobe, where they secrete AVP and oxytocin into capillaries that feed into the systemic circulation. However, parvicellular neurons in the paraventricular nuclei also produce AVP, which they secrete into hypophyseal portal blood. It appears that much of the AVP secreted by parvicellular neurons is made in the same cells that produce CRH. It is assumed that the AVP in hypophyseal portal blood comes from these cells and from a small number of AVP-producing magnocellular neurons whose axons pass through the median eminence of the hypothalamus on their way to the posterior lobe.

Diurnal variation

Under normal circumstances, the hypothalamic–pituitary–adrenal axis in humans functions in a pulsatile manner, resulting in several bursts of secretory activity over a 24-hour period. This pattern is the result of rhythmic activity in the CNS, which causes bursts of CRH secretion and in turn bursts of ACTH and glucocorticoid secretion (Fig. 31.5). The diurnal oscillation in secretory activity of the axis is thought to result from changes in the sensitivity of CRH-producing neurons to the negative feedback action of glucocorticoids, altering their rate of CRH secretion. As a result, there is a diurnal oscillation in the rate of ACTH and glucocorticoid secretion. This **circadian rhythm** is reflected in the daily pattern of glucocorticoid secretion. In people who are awake during the day and sleep at night, the blood glucocorticoid level begins to rise during the early morning hours, reaches a peak sometime before noon, and then falls gradually to a low level around midnight (see Fig. 31.5). This pattern is reversed in people who sleep during the day and are awake at night. This inherent biologic rhythm is superimposed on the normal operation of the hypothalamic–pituitary–adrenal axis.

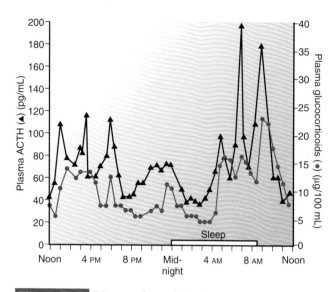

Figure 31.5 **Adrenocorticotropic hormone (ACTH) secretion and the sleep–wake cycle.** Pulsatile changes in the concentrations of ACTH and glucocorticoids in the blood of a young individual over a 24-hour period. Note that the amplitude of the pulses in ACTH and glucocorticoids is lower during the evening hours and increases greatly during the early morning hours illustrating the diurnal oscillation of the hypothalamic–pituitary–adrenal axis.

Thyroid-stimulating hormone regulates thyroid gland function.

The thyroid gland consists of two lobes and is composed of aggregates of follicle cells, which are formed from a single layer of cells. The follicular cells produce and secrete T_4 and T_3, thyroid hormones that are iodinated derivatives of the amino acid tyrosine. The thyroid hormones control the body's basal metabolic rate, heat production, protein synthesis, growth, and CNS development. They act on many cells, changing gene expression and the production of particular proteins in the cell.

TSH is the physiologic regulator of T_4 and T_3 synthesis and secretion by the thyroid gland. TSH also promotes nucleic acid and protein synthesis in the cells of the thyroid follicles, maintaining their size and functional integrity. The actions of TSH on thyroid hormone synthesis and secretion and the physiologic effects of the thyroid hormones are described in detail in Chapter 32.

Thyroid-stimulating hormone synthesis

The α subunit of TSH is a single peptide chain of 92 amino acid residues with two carbohydrate chains linked to its structure. The β subunit is a single peptide chain of 112 amino acid residues, to which a single carbohydrate chain is linked. Noncovalent bonds hold the α and β subunits together. The two subunits combined give the TSH molecule a molecular weight of about 28,000. Neither subunit has significant TSH activity by itself. The two subunits must be combined in a 1:1 ratio to form an active hormone. The gonadotropins FSH and LH are also composed of two noncovalently combined subunits. The α subunits of TSH, FSH, and LH are derived from the same gene and are identical, but the β subunit gives each hormone its particular set of physiologic activities.

Thyrotrophs synthesize the peptide chains of the α and β subunits of TSH from separate mRNA molecules, which are transcribed from two different genes. The peptide chains of the α and β subunits are combined and undergo glycosylation in the rough endoplasmic reticulum (RER). These processes are completed as TSH molecules pass through the Golgi apparatus and are packaged into secretory granules. Normally, thyrotrophs make more α subunits than β subunits. As a result, secretory granules contain excess α subunits. When a thyrotroph is stimulated to secrete TSH, it releases both TSH and free α subunits into the bloodstream. In contrast, little free TSH β subunit is in the blood.

Thyrotropin-releasing hormone regulation of thyroid-stimulating hormone

TRH is a small peptide consisting of three amino acid residues produced by neurons in the hypothalamus. These neurons terminate on the capillary networks that give rise to the hypophyseal portal vessels and secrete TRH into the hypophyseal portal circulation at a constant or tonic rate. Therefore, the thyrotrophs are continuously exposed to TRH.

TRH binds to receptors on the plasma membranes of thyrotrophs that are coupled via G proteins to PLC (Fig. 31.6). TRH binding with its receptor activates PLC,

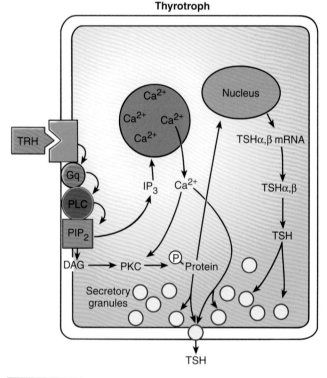

Figure 31.6 **The action of thyrotropin-releasing hormone (TRH) on a thyrotroph.** TRH binds to membrane receptors coupled to phospholipase C (PLC) by G proteins (G_q). PLC hydrolyzes plasma membrane phosphatidylinositol 4,5-bisphosphate (PIP_2), generating inositol trisphosphate (IP_3) and diacylglycerol (DAG). IP_3 mobilizes intracellular stores of Ca^{2+}. The rise in Ca^{2+} stimulates thyroid-stimulating hormone (TSH) secretion. Ca^{2+} and DAG activate protein kinase C (PKC), which phosphorylates proteins (P-proteins) that stimulate TSH secretion and gene expression for the α and β subunits of TSH.

causing the hydrolysis of membrane PIP_2 and the release of the intracellular messengers IP_3 and DAG. IP_3 increases the concentration of Ca^{2+} in the cytosol, which stimulates the secretion of TSH into the blood. The rise in cytosolic Ca^{2+} and the increase in DAG activate PKC, which phosphorylates proteins to promote TSH secretion. TRH also stimulates the expression of the genes for the α and β subunits of TSH in the thyrotroph to maintain the production of TSH fairly constant (see Fig. 31.6).

The exposure of certain animals to a cold environment stimulates TSH secretion. This makes sense from a physiologic perspective because the thyroid hormones are important in regulating body heat production (see Chapter 32). Brief exposure of experimental animals to a cold environment stimulates the secretion of TSH, presumably a result of enhanced TRH secretion. Newborn humans behave much the same way, in that they respond to brief cold exposure with an increase in TSH secretion. This response to cold does not occur in adult humans.

Thyroid hormone regulation of thyroid-stimulating hormone secretion

The thyroid hormones exert a direct negative feedback effect on TSH secretion. For example, when the blood concentration of thyroid hormones is high, the rate of TSH secretion falls. In turn, the stimulatory effect of TSH on the follicular cells of the thyroid is reduced, resulting in a decrease in T_4 and T_3 secretion. When the circulating levels of T_4 and T_3 are low, their negative feedback effect on TSH release is reduced and more TSH is secreted from thyrotrophs, increasing the rate of thyroid hormone secretion. This control system is part of the hypothalamic–pituitary–thyroid axis (Fig. 31.7).

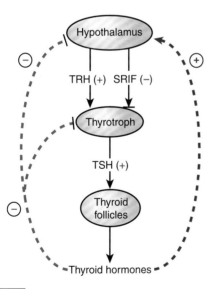

Figure 31.7 **The hypothalamic–pituitary–thyroid axis.** Thyrotropin-releasing hormone (TRH) stimulates and somatostatin (SRIF) inhibits thyroid-stimulating hormone (TSH) release by acting directly on the thyrotroph. The negative feedback loops shown in *blue* inhibit TRH secretion and action on the thyrotroph, causing a decrease in TSH secretion. The feedback loops, shown in *red*, stimulate somatostatin secretion, causing a decrease in TRH secretion. SRIF, somatostatin or somatotropin release–inhibiting factor.

The thyroid hormones exert negative feedback effects on both the hypothalamus and the pituitary. In the hypothalamic TRH-secreting neurons, thyroid hormones reduce TRH mRNA and TRH prohormone to decrease TRH release. The thyroid hormones also increase the release of somatostatin from the hypothalamus. Somatostatin (SRIF) inhibits the release of TSH from the thyrotroph (see Fig. 31.7). In the pituitary, thyroid hormones reduce the sensitivity of the thyrotroph to TRH and inhibit TSH synthesis.

The negative feedback effects of the thyroid hormones on thyrotrophs are produced primarily through the actions of T_3. Both T_4 and T_3 circulate in the blood bound to plasma proteins, with only a small percentage (<1%) unbound or free (see Chapter 32). Thyrotrophs take up the free T_4 and T_3 molecules, and the enzymatic removal of one iodine atom converts T_4 to T_3. The newly formed T_3 molecules and those taken up directly from the blood enter the nucleus, where they bind to thyroid hormone receptors in the chromatin, changing gene expression to decrease the synthesis and secretion of TSH. T_3 also decreases thyrotroph sensitivity to TRH, in part through a reduction in the number of TRH receptors in thyrotroph plasma membrane.

The hypothalamic-pituitary-thyroid axis diurnal rhythm

In humans, peak TSH secretion occurs in the early morning, and a low point is reached in the evening. Physical and emotional stress can alter TSH secretion, but the effects of stress on the hypothalamic–pituitary–thyroid axis are not as pronounced as on the hypothalamic–pituitary–adrenal axis.

Growth hormone promotes growth during childhood and remains important throughout life.

GH is a peptide hormone secreted that promotes the growth of the human body. It does not stimulate fetal growth, and it is not an important growth factor during the first few months after birth. Thereafter, it is essential for the normal rate of body growth during childhood and adolescence.

The anterior pituitary secretes GH throughout life, and GH remains physiologically important even after growth has stopped. In addition to its growth-promoting action, GH has effects on many aspects of carbohydrate, lipid, and protein metabolism. For example, GH is thought to be one of the physiologic factors that counteract and, thus, modulate some of the actions of insulin on the liver and peripheral tissues.

Growth hormone synthesis

GH is a globular 22-kDa protein consisting of a single chain of 191 amino acid residues with two intrachain disulfide bridges, produced in somatotrophs of the anterior pituitary. It is synthesized in the RER as a larger preprohormone that is processed to a prohormone consisting of an N-terminal signal peptide and the 191 amino acid hormones. The signal peptide is then cleaved from the prohormone as it traverses the Golgi apparatus, and the GH is packaged in secretory granules. Human GH has considerable structural similarity to human PRL and placental lactogen.

GHRH regulates the production of GH by stimulating the GH gene expression in somatotrophs. Thyroid hormones also stimulate GH gene expression. As a result, the normal rate of GH production depends on these hormones. For example, a thyroid hormone–deficient person is also GH deficient. This important action of thyroid hormones is discussed further in Chapter 32.

Growth hormone secretion

Two opposing hypothalamic-releasing hormones regulate the secretion of GH. GHRH stimulates GH secretion, and somatostatin inhibits GH secretion by inhibiting the action of GHRH. The net effect of these counteracting hormones on somatotrophs determines the rate of GH secretion. When GHRH predominates, GH secretion is stimulated. When somatostatin predominates, GH secretion is inhibited.

Human GHRH is a single-chain peptide of 44 amino acid residues. A slightly smaller version of GHRH consisting of 40 amino acid residues is also present in humans. The two forms of GH result from posttranslational modification of a larger prohormone. GHRH is synthesized in the cell bodies of neurons in the **arcuate nuclei** and **ventromedial nuclei** of the hypothalamus. The axons of these cells project to the capillary networks giving rise to the portal vessels. When these neurons receive a stimulus for GHRH secretion, they discharge GHRH from their axon terminals into the hypophyseal portal circulation.

GHRH binds to receptors in the plasma membranes of somatotrophs coupled via stimulatory G_s proteins to adenylyl cyclase (Fig. 31.8). GHRH binding its receptor activates adenylyl cyclase to increase cAMP, activating PKA, which, in turn, phosphorylates proteins that stimulate GH secretion and GH gene expression. GHRH binding to its receptor also increases intracellular Ca^{2+}, which stimulates GH secretion. GHRH also stimulates PLC, causing the hydrolysis of membrane PIP_2 in the somatotroph. The importance of this phospholipid pathway for the stimulation of GH secretion by GHRH is not established.

Somatostatin is a small peptide consisting of 14 amino acid residues. Somatostatin neurons are especially abundant in the **anterior periventricular region** (close to the third ventricle), although it is also made by neurons in other parts of the hypothalamus. The axons of these cells terminate on the capillary networks giving rise to the hypophyseal portal circulation, where they release somatostatin into the blood.

Somatostatin binds to receptors in the plasma membranes of somatotrophs that are coupled to adenylyl cyclase by an inhibitory G protein (see Fig. 31.8). Somatostatin binding its receptor decreases adenylyl cyclase activity, reducing intracellular cAMP. Somatostatin binding to its receptor also lowers intracellular Ca^{2+}, reducing GH secretion. When the somatotroph is exposed to both somatostatin and GHRH, the effects of somatostatin are dominant and intracellular cAMP and Ca^{2+} are reduced. Thus, somatostatin has a negative modulating influence on the action of GHRH.

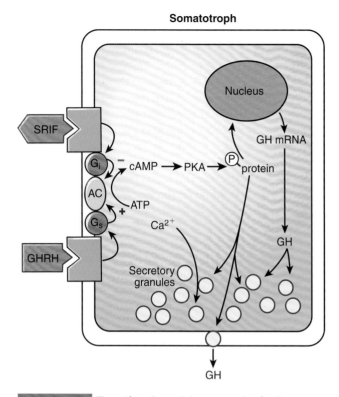

Figure 31.8 **The action of growth hormone–releasing hormone (GHRH) and somatostatin (SRIF) on a somatotroph.** GHRH binds to membrane receptors coupled by stimulatory G proteins (G_s) to adenylyl cyclase (AC). Cyclic adenosine monophosphate (cAMP) rises in the cell and activates protein kinase A (PKA), which then phosphorylates proteins (P-proteins) that stimulate growth hormone (GH) secretion and GH gene expression. Ca^{2+} also facilitates GH secretion. The possible involvement of the phosphatidylinositol pathway in GHRH action is not shown. SRIF binds to membrane receptors coupled to adenylyl cyclase by inhibitory G proteins (G_i), which inhibit GHRH stimulation of adenylyl cyclase.

Insulin-like growth factor 1

GH is not a traditional trophic hormone that directly stimulates the growth of its target tissue. Rather, GH stimulates the production of a trophic hormone called **insulin-like growth factor 1** (**IGF-1** or **IGF-I**), a potent **mitogenic** agent that mediates the growth-promoting action of GH. IGF-1 was originally called *somatomedin C* or *somatotropin-mediating hormone* because of its role in promoting growth. Somatomedin C was renamed IGF-1 because of its structural similarity to proinsulin. Although most effects of GH are mediated by IGF-1, it is important to note that some effects of GH, such as stimulation of lipolysis in adipose tissue (see below) and stimulation of amino acid transport in muscle, occur independently of IGF-1 action.

Insulin-like growth factor 2 (**IGF-2** or **IGF-II**) is structurally similar to IGF-1 and has many of the same metabolic and mitogenic actions. However, the synthesis and release of IGF-2 depends less on GH. Thus, IGF-1 is the more important mediator of GH action.

IGF-1 is a 7.5-kDa protein consisting of a single chain of 70 amino acids. Because of its structural similarity to proinsulin, IGF-1 can produce some of the effects of insulin.

Many cells of the body produce IGF-1; however, the liver is the main source of IGF-1 in the blood. Most IGF-I in the blood is bound to specific IGF-1–binding proteins; only a small amount circulates in the free form. The bound form of circulating IGF-1 has little insulin-like activity, so it does not play a physiologic role in the regulation of blood glucose level.

GH increases IGF-1 gene expression in various tissues and organs, such as the liver, and stimulates the production and release of IGF-1. Excessive GH secretion results in a greater-than-normal amount of IGF-1 in the blood. People with GH deficiency have lower-than-normal levels of IGF-1, but there is still some present, because many hormones and factors, in addition to GH, regulate the production of IGF-1 by cells.

Inhibition of growth hormone secretion

An increase in blood GH concentration has direct negative feedback effects on its own secretion, independent of the production of IGF-1. GH inhibits GHRH secretion and stimulates somatostatin secretion by hypothalamic neurons (Fig. 31.9). GH circulating in the blood can enter the interstitial spaces of the median eminence of the hypothalamus because the blood–brain barrier is porous in this area.

IGF-1 also has a negative feedback effect on the secretion of GH (see Fig. 31.9). It acts directly on the somatotrophs to inhibit the stimulatory action of GHRH on GH secretion. IGF-1 also inhibits GHRH secretion and stimulates the secretion of somatostatin by neurons in the hypothalamus. By stimulating IGF-1 production, GH inhibits its own secretion. This mechanism is analogous to the way ACTH and TSH regulate their own secretion through the respective negative feedback

effects of the glucocorticoid and thyroid hormones. This interactive relationship involving GHRH, somatostatin, GH, and IGF-1 comprises the hypothalamic–pituitary–GH axis.

Growth hormone pulsatile secretory pattern

In humans, GH is secreted in periodic bursts, which produce large but short-lived peaks in GH concentration in the blood. Between these episodes of high GH secretion, somatotrophs release little GH and the blood concentration falls to low levels. It is thought that an increase in the rate of GHRH secretion and a fall in the rate of somatostatin secretion cause these periodic bursts of GH secretion. Increased somatostatin secretion is thought to cause the intervals between bursts, when GH secretion is suppressed. These changes in GHRH and somatostatin secretion result from neural activity generated at higher levels of the CNS, which affects the secretory activity of GHRH and somatostatin-producing neurons in the hypothalamus.

Bursts of GH secretion occur during both the awake and sleep periods of the day; however, GH secretion is maximal at night. The bursts of GH secretion during sleep usually occur within the first hour after the onset of deep sleep (stages 3 and 4 of slow-wave sleep). Mean GH levels in the blood are highest during adolescence (peaking in late puberty) and decline in adults. The reduction in blood GH with aging is mainly a result of a decrease in the size of the GH secretory burst but not the number of pulses (Fig. 31.10).

Many factors affect the rate of GH secretion in humans by changing the hypothalamic neuronal secretion of GHRH and somatostatin. Emotional or physical stress greatly increases the rate of GH secretion, as does vigorous exercise. Obesity results in reduced GH secretion. Changes in the circulating levels of metabolites also affect GH secretion. A decrease in blood glucose stimulates GH secretion, whereas hyperglycemia inhibits it. An increase in the blood concentration of the amino acids arginine and leucine also stimulates GH secretion.

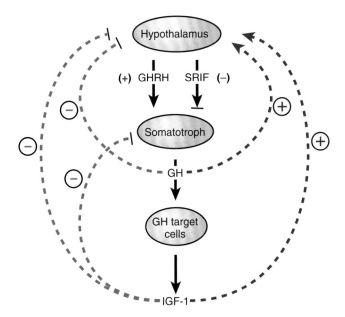

Figure 31.9 **The hypothalamic–pituitary–GH axis.** Growth hormone–releasing hormone (GHRH) stimulates and somatostatin (SRIF) inhibits growth hormone (GH) secretion by acting directly on the somatotroph. The negative feedback loops shown in *blue* inhibit GHRH secretion and action on the somatotroph, causing a decrease in GH secretion. The positive feedback loops shown in *red* stimulate somatostatin secretion, causing a decrease in GH secretion. IGF-1, insulin-like growth factor 1.

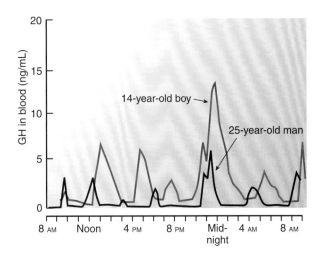

Figure 31.10 **Pulsatile growth hormone (GH) secretion in an adolescent boy and in an adult.** In the adult, GH levels are reduced as a result of smaller pulse width and amplitude rather than a decrease in the number of pulses.

Growth hormone regulation of growth and metabolism

The cells of many tissues and organs of the body have receptors for GH in their plasma membranes through which GH produces its growth-promoting and metabolic effects. The binding of GH to its receptor activates an associated tyrosine kinase (JAK2), which is not an integral part of the GH receptor. Activation of JAK2 initiates changes in the phosphorylation pattern of cytoplasmic and nuclear proteins, ultimately stimulating the transcription of specific genes, including that for IGF-1.

IGF-1 mediates many of the mitogenic effects of GH; however, evidence indicates that GH has direct growth-promoting actions on progenitor or stem cells, such as **prechondrocytes** in the growth plates of bone and **satellite cells** of skeletal muscle. GH stimulates such progenitor cells to differentiate into cells with the capacity to undergo cell division. An important action of GH on the differentiation of progenitor cells is stimulation of IGF-1 gene expression. These cells then produce and release IGF-1, which exerts an autocrine mitogenic action on the cell that produced it or a paracrine action on neighboring cells. In response to IGF-1, these cells undergo division, causing the tissue to grow through cell replication.

GH deficiency in childhood causes a decrease in the rate of body growth. If left untreated, the deficiency results in **pituitary dwarfism**. People with this condition may be deficient in GH only, or they may have multiple anterior pituitary hormone deficiencies. A defect in the mechanisms that control GH secretion or the production of GH by somatotrophs can cause GH deficiency. In some people, the target cells for GH fail to respond normally to the hormone because of several different mutations in the GH receptor.

The excessive secretion of GH during childhood, caused by a defect in the mechanisms regulating GH secretion or a GH-secreting tumor, results in **gigantism**. Affected people may grow to a height of 7 to 8 ft (2.1 to 2.4 m). When excessive GH secretion occurs in an adult, further linear growth does not occur because the growth plates of the long bones have calcified. Instead, it causes the bones of the face, hands, and feet to become thicker and certain organs, such as the liver, to undergo hypertrophy. This condition, known as **acromegaly**, can also result from the chronic administration of excessive amounts of GH to adults.

The main physiologic action of GH is on body growth; however, it also has important effects on certain aspects of fat and carbohydrate metabolism. Its main action on fat metabolism is to stimulate the mobilization of triglycerides from the fat depots of the body. This process, known as **lipolysis**, involves the hydrolysis of triglycerides to fatty acids and glycerol by a number of lipases such as **hormone-sensitive lipase**. The fatty acids and glycerol are released from adipocytes and enter the bloodstream. GH has been shown to directly stimulate lipolysis in some experiments, but most evidence suggests that it causes adipocytes to be more responsive to other lipolytic stimuli, such as fasting and catecholamines.

GH also functions as one of the counterregulatory hormones that limit the actions of insulin on muscle, adipose tissue, and the liver. For example, GH inhibits glucose use by muscle and adipose tissue and increases glucose production by the liver. These effects are opposite those of insulin. GH also makes muscle and fat cells resistant to insulin signaling. Thus, GH normally has a tonic inhibitory effect on insulin action, much like the glucocorticoid hormones (see Chapter 34).

The insulin-opposing actions of GH can produce serious metabolic disturbances in people who secrete excessive amounts of GH (people with acromegaly) or who are given large amounts of GH for an extended time. They may develop insulin resistance and elevated blood insulin levels. They may also exhibit hyperglycemia caused by the inability of insulin to promote glucose utilization by muscle and adipose tissue, coupled with overproduction of glucose by the liver. These disturbances are much like those in people with non–insulin-dependent (type 2) diabetes mellitus. For this reason, this metabolic response to excess GH is called its **diabetogenic action**.

In GH-deficient people, GH has a transitory insulin-like action. For example, intravenous injection of GH in a person who is GH deficient produces hypoglycemia. The hypoglycemia is caused by the ability of GH to stimulate the uptake and use of glucose by muscle and adipose tissue and to inhibit glucose production by the liver. After about 1 hour, the blood glucose level returns to normal. If this person is given a second injection of GH, hypoglycemia does not occur because the person has become insensitive or refractory to the insulin-like action of GH and remains so for some hours. Healthy people do not respond to the insulin-like action of GH, presumably because they are always refractory from being exposed to their own endogenous GH (Clinical Focus 31.2).

Gonadotropins are anterior pituitary hormones that regulate sexual development and reproductive function.

The testes and ovaries have two essential functions in human reproduction. The first is to produce sperm cells and ova (egg cells), respectively. The second is to produce an array of steroid and peptide hormones, which influence virtually every aspect of the reproductive process. The gonadotropic hormones FSH and LH regulate both of these functions. The hypothalamic-releasing hormone LHRH and the hormones produced by the testes and ovaries in response to gonadotropic stimulation, in turn, regulate the production and secretion of the gonadotropins by the anterior pituitary. The regulation of human reproduction by this hypothalamic–pituitary–gonad axis is discussed in Chapters 36 and 37. Here, we describe the chemistry and formation of the gonadotropins.

Like TSH, human FSH and LH are composed of two structurally different glycoprotein subunits, called α and β, which are held together by noncovalent bonds. The β subunit of human FSH consists of a peptide chain of 111 amino acid residues, to which two chains of carbohydrate are attached. The β subunit of human LH is a peptide of 121 amino acid residues. It is also glycosylated with two carbohydrate chains. The combined α and β subunits of FSH and LH give these hormones a molecular size of about 28 to 29 kDa.

CLINICAL FOCUS | 31.2

Growth Hormone Deficiency and Recombinant Human Growth Hormone

Growth hormone (GH) is species specific, and humans do not respond to GH derived from animals. In the past, the only human GH available for treating children who were GH deficient was a limited amount made from human pituitaries obtained at autopsy, but there was never enough to meet the need. Use of pituitary-derived GH was then halted because of concern of a causal relationship between therapy and development of **Creutzfeldt–Jakob disease** (**CJD**), a rare and fatal spongiform encephalopathy. The problems of short supply and potential disease transmission were solved when the gene for human GH was cloned and the production of large amounts of recombinant human GH, with all the activities of the natural substance, became possible. During the 1980s, careful clinical trials established that recombinant human GH was safe to use in GH-deficient children to promote growth. The hormone was approved for clinical use and is now produced and sold worldwide.

Despite the availability of recombinant GH, the diagnosis of GH deficiency has remained controversial. As presented in Figure 31.10, GH release from the pituitary occurs episodically; thus, a random measure of GH in the blood is not useful for diagnosing GH deficiency. However, a random blood sample may be used to detect GH resistance, a syndrome in which the patient exhibits symptoms of GH deficiency but presents with high GH levels in the blood.

Measurement of insulin-like growth factor 1 (IGF-1) and IGF-binding protein 3 (IGFBP3) levels in the blood is a method to assess GH deficiency. IGF-1 mediates many of the mitogenic effects of GH on tissues in the body, and IGF-1 binds to IGFBP3 in the blood. IGFBP3 extends the half-life of the IGF-1, transports it to target cells, and facilitates its interaction with the IGF receptor. GH stimulates the production of IGF-1 and IGFBP3, which are present in the blood at fairly constant, readily detectable levels in healthy people. In children with GH deficiency, the concentrations of IGF-1 and IGFBP3 are low. Treatment with recombinant GH will increase IGF-1 and IGFBP3 in the blood, which will result in increased long bone growth. The epiphyseal growth plate in the bone becomes less responsive to GH and IGF-1 several years after puberty, and long bone growth stops in adulthood (see Chapter 35). GH therapy may be continued after cessation of long bone growth because GH has important metabolic effects, including support of normal gonadal function and attainment of normal adult bone density. ■

As with TSH, the individual subunits of the gonadotropins have no hormonal activity. They must be combined with each other in a 1:1 ratio to have activity. Again, it is the β subunit that gives the gonadotropin molecule either FSH or LH activity because the α subunits are identical.

The same gonadotrophs in the anterior pituitary produce FSH and LH. There are separate genes for the α and β subunits in the gonadotroph; hence, the peptide chains of these subunits are translated from separate mRNA molecules. Glycosylation of these chains begins as they are synthesized and before they are released from the ribosome. The folding of the subunit peptides into their final three-dimensional structure, the combination of an α subunit and a β subunit, and the completion of glycosylation all occur as these molecules pass through the Golgi apparatus and are packaged into secretory granules. As with the thyrotroph, the gonadotroph produces an excess of α subunits over FSH and LH β subunits. Therefore, the rate of β subunit production is considered to be the rate-limiting step in gonadotropin synthesis.

The hormones of the hypothalamic–pituitary–gonad axis regulate the synthesis of FSH and LH. For example, LHRH stimulates gonadotropin production. The steroid and peptide hormones produced by the gonads in response to stimulation by the gonadotropins also affect gonadotropin production. Such hormonally regulated changes in gonadotropin production are caused mainly by changes in the expression of the genes for the gonadotropin subunits. More information about the regulation of gonadotropin synthesis and secretion is found in Chapters 36 and 37.

Prolactin is a peptide hormone that regulates milk secretion from the mammary gland.

Lactation is the final phase of the process of human reproduction. During pregnancy, alveolar cells of the mammary glands develop the capacity to synthesize milk in response to stimulation by many steroid and peptide hormones. Milk synthesis by these cells begins shortly after childbirth. To continue to synthesize milk, PRL must periodically stimulate these cells, and this is thought to be the main physiologic function of PRL in the human female. A role for PRL in the human male is unclear. It is known to have some supportive effect on the action of androgenic hormones on the male reproductive tract, but whether this is an important physiologic function of PRL is not established.

Human PRL is a globular protein consisting of a single peptide chain of 199 amino acid residues with three intrachain disulfide bridges. Its molecular size is about 23 kDa. Human PRL has considerable structural similarity to human GH and to a PRL-like hormone produced by the human placenta called *placental lactogen (hPL)*. It is thought that these hormones are structurally related because their genes evolved from a common ancestral gene during the course of vertebrate evolution. Because of its structural similarity to human PRL, human GH has substantial PRL-like or lactogenic activity. However, PRL and hPL have little GH-like activity. Human placental lactogen is discussed further in Chapter 38.

PRL is synthesized and secreted by lactotrophs in the anterior pituitary. Synthesized in the RER as a larger peptide, the N-terminal signal peptide sequence of PRL is removed and the 199-amino acid protein passes through the Golgi apparatus and is packaged into secretory granules.

Estrogens and other hormones, such as TRH, increase PRL gene expression and stimulate the synthesis and secretion of PRL. However, dopamine inhibits the synthesis of PRL. Dopamine produced by hypothalamic neurons plays a major role in the regulation of PRL synthesis and secretion by the hypothalamic–pituitary axis. The regulation of the synthesis and secretion of PRL and its physiologic actions are discussed in Chapter 38.

INTEGRATED MEDICAL SCIENCES

Hypothalamic–Pituitary Regulation of Energy Intake and Energy Expenditure

Body weight in adults is fairly stable over many years. This stability in body weight occurs despite large fluctuations in caloric intake, demonstrating that energy intake and energy expenditure are precisely matched. The central nervous system (CNS), primarily the hypothalamus, is responsible for matching energy intake and energy expenditure, receiving information relevant to energy balance through metabolic, neural, and hormonal signals. Some signals regulate energy intake over short time periods, for example, acting to terminate a feeding episode, whereas others are active in the long-term regulation of energy intake, ensuring the maintenance of adequate energy stores. Leptin is a hormone that provides information to the hypothalamus on the amount of energy stored in the body as adipose tissue.

Leptin is a 16-kDa protein product of the *LEP* gene (originally termed *ob* gene), which is most highly expressed in adipose tissue but which is also detectable in other tissues including muscle and placenta. Serum leptin increases as adipose tissue mass increases; thus, leptin is significantly greater in obese subjects than in lean subjects.

The receptor for leptin is detectable in several areas of the CNS but is highly expressed in the arcuate nucleus of the hypothalamus (see Figure). Activation of leptin receptors reduces the expression of neuropeptides that stimulate food intake (neuropeptide Y and agouti-related protein) and increases expression of neuropeptides that reduce feeding (α-melanocyte–stimulating hormone). Administration of exogenous leptin to increase serum levels has been tested as a therapy for weight loss in humans, but leptin treatment had only modest effects on appetite and body weight.

In addition to regulating food intake, leptin signaling in the hypothalamus also alters anterior pituitary hormone secretion to influence energy expenditure. Reduced caloric intake and starvation initiate a complex series of biochemical and behavioral adaptations to promote survival, one of which is to reduce whole-body energy expenditure. Thyroid hormones stimulate metabolism and increase energy use (discussed in detail in Chapter 32); thus, it is adaptive to reduce thyroid hormone levels during periods of insufficient food intake. Growth and the ability to reproduce are both energy-intensive processes that are also curtailed during starvation. Although leptin circulates in the blood in proportion to the amount of body fat, serum leptin falls rapidly with restriction of food intake, providing a signal to the

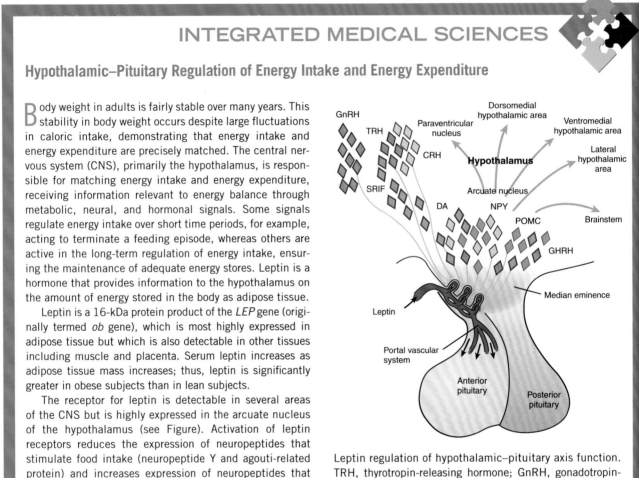

Leptin regulation of hypothalamic–pituitary axis function. TRH, thyrotropin-releasing hormone; GnRH, gonadotropin-releasing hormone; CRH, corticotropin-releasing hormone; SRIF, somatotropin release–inhibitory factor; DA, dopamine; NPY, neuropeptide Y; POMC, proopiomelanocortin; GHRH, growth hormone–releasing hormone.

hypothalamus to conserve body energy stores. Evidence that leptin coordinates the hypothalamic–pituitary response to starvation was originally derived through replacement experiments in rodents. Preventing the starvation-induced fall in leptin by infusion of recombinant protein blunted the reduction in gonadal, adrenal, and thyroid hormones that would normally occur in starved mice. Similar types of replacement experiments have demonstrated that leptin can regulate anterior pituitary hormone secretion in humans. Thus, leptin may prove useful in treatment of disease resulting from reduced hypothalamic–pituitary function secondary to reductions in adipose tissue mass caused by lipodystrophy or excess energy expenditure such as that in highly trained, female athletes. ∎

Chapter Summary

- The hypothalamic–pituitary axis is composed of the hypothalamus, infundibular stalk, posterior pituitary, and anterior pituitary.
- Arginine vasopressin and oxytocin are synthesized in hypothalamic neurons whose axons terminate in the posterior pituitary.
- Arginine vasopressin increases water reabsorption by the kidneys in response to a rise in blood osmolality or a fall in blood volume.
- Oxytocin stimulates milk letdown in the breast in response to suckling and muscle contraction in the uterus in response to cervical dilation during labor.
- Adrenocorticotropic hormone, thyroid-stimulating hormone, growth hormone, follicle-stimulating hormone, luteinizing hormone, and prolactin are synthesized in the anterior pituitary and secreted in response to hypothalamic-releasing hormones carried in the hypophyseal portal circulation.
- Hypothalamic corticotropin-releasing hormone stimulates adrenocorticotropic hormone release from corticotrophs, which, in turn, stimulates glucocorticoid release from the adrenal cortex, to comprise the hypothalamic–pituitary–adrenal axis.
- Glucocorticoids, physical and emotional stress, arginine vasopressin, and the sleep–wake cycle regulate adrenocorticotropic hormone secretion.

- Hypothalamic thyrotropin-releasing hormone stimulates thyroid-stimulating hormone release from thyrotrophs, which, in turn, stimulates triiodothyronine and thyroxine release from the thyroid follicles, to comprise the hypothalamic–pituitary–thyroid axis.
- The thyroid hormones, cold temperatures, and the sleep–wake cycle regulate thyroid-stimulating hormone secretion.
- Hypothalamic growth hormone–releasing hormone increases and hypothalamic somatostatin decreases growth hormone secretion from somatotrophs, which, in turn, stimulate the release of insulin-like growth factor I from liver and other target cells, to comprise the hypothalamic–pituitary–growth hormone axis.
- Growth hormone (GH) itself, insulin-like growth factor I, aging, deep sleep, stress, exercise, and hypoglycemia regulate GH secretion.
- Luteinizing hormone–releasing hormone stimulates the secretion of follicle-stimulating hormone and luteinizing hormone from the anterior pituitary, which, in turn, affects functions of the ovaries and testes, to comprise the hypothalamic–pituitary–reproductive axis.
- Hypothalamic dopamine inhibits prolactin release from lactotrophs in the anterior pituitary.
- Prolactin regulates milk production in the breast.

Endocrine Physiology

Chapter Review Questions

1. Which of the following conditions is consistent with a decreased rate of ACTH secretion?

   A. Hyperosmolality of the blood
   B. Low serum glucocorticoid
   C. Loss of hypothalamic neurons
   D. Primary adrenal insufficiency
   E. Stress as a result of emotional trauma
   F. Increased PKA activity in corticotrophs

The correct answer is C. Destruction of the neurons in the paraventricular nuclei of the hypothalamus decreases CRH release, which, in turn, causes decreased synthesis and secretion of ACTH. Hyperosmolality of the blood would lead to an increase in portal blood AVP, which increases ACTH secretion from corticotrophs. Physical or emotional stress increases ACTH release. Glucocorticoids feedback to the hypothalamus and anterior pituitary to decrease ACTH synthesis and secretion. Primary adrenal insufficiency is characterized by a lack of glucocorticoids in the blood, resulting in an increase in ACTH synthesis and secretion. Increased PKA in corticotrophs increases ACTH synthesis and secretion.

2. Which of the following statements most accurately describes the feedback effects of thyroid hormones?

   A. They increase the sensitivity of thyrotrophs to TRH.
   B. They stimulate transcription of the α and β subunits of TSH in thyrotrophs.
   C. They increase the secretion of TSH by thyrotrophs.
   D. They stimulate the expression of the GH gene in somatotrophs.
   E. They increase IP_3 in thyrotrophs.
   F. They increase ACTH release.

The correct answer is D. Thyroid hormones stimulate the expression of the GH gene in somatotrophs. Thyroid hormones exert a negative feedback signal on the hypothalamic–pituitary–thyroid axis to inhibit their own synthesis and secretion. Therefore, thyroid hormones *decrease* the sensitivity of thyrotrophs to TRH, decrease the formation of IP_3 in thyrotrophs, *inhibit* the expression of the genes for the α and β subunits of TSH in thyrotrophs, and *decrease* the secretion of TSH by thyrotrophs. Thyroid hormones have no effect on ACTH release.

3. A 30-year-old woman completed a routine pregnancy with the uncomplicated delivery of a normal-sized baby girl 6 months ago. The woman is currently experiencing galactorrhea (persistent discharge of milklike secretions from the breast) and has not yet resumed regular menstrual periods. The baby had been bottle-fed since birth. What is the most likely explanation of the galactorrhea?

A. Normal postpartum response
B. Excess PRL secretion
C. Insufficient TSH secretion
D. Reduced GH secretion
E. Increased dopamine synthesis in the hypothalamus

The correct answer is C. Galactorrhea is commonly associated with pituitary tumors secreting excess PRL. Prolactin is important in maintaining milk production in the breast after birth. Insufficient LHRH secretion would result in low levels of PRL, which would not stimulate galactorrhea. Galactorrhea is diagnosed if present longer than 6 months postpartum in a nonnursing mother. The combination of both galactorrhea and amenorrhea is diagnostic of PRL-secreting pituitary tumor. TSH generally has little effect on PRL secretion. GH has lactogenic activity when high not low. Hypothalamic dopamine is an inhibitor of PRL release.

4. A decrease in blood volume would result in an increase in the secretion of:

A. Neurophysin
B. Oxytocin
C. β-lipotropin
D. Somatostatin
E. ACTH
F. POMC

The correct answer is A. Neurophysin is the other product generated when the prohormone for AVP or oxytocin is cleaved. A decrease in blood volume would result in the release of AVP and neurophysin from magnocellular neurons. The hormones oxytocin, β-lipotropin, ACTH, and somatostatin are not involved in the regulation of blood volume.

5. A 50-year-old man complains of decreased muscle strength, libido, and exercise intolerance. Examination reveals a 10% reduction in lean body mass and an increase in body fat, primarily localized to the abdominal region. Thyroid hormone levels are normal. Which diagnosis is most consistent with these symptoms?

A. Glucocorticoid deficiency
B. Addison disease
C. GH deficiency
D. PRL deficiency
E. Acromegaly

The correct answer is C. Growth hormone deficiency in adults is characterized by decreased muscle strength and exercise intolerance, and a reduced sense of well-being (including effects on libido). Lean body mass (muscle) is lost, and excess body fat deposition occurs in the abdominal region. GH replacement can reverse these effects. Thyroid dysfunction is ruled out by the normal thyroid hormone levels. Glucocorticoid deficiency usually results from primary adrenal insufficiency, as in Addison disease. Clinical symptoms include a decreased sense of well-being, gastrointestinal disturbances, and abnormal glucose metabolism. Primary adrenal insufficiency is also characterized by high blood levels of ACTH, which can result in hyperpigmentation due to the melanocyte-stimulating activity of the amino terminal portion of ACTH. Adrenal insufficiency is not usually associated with a redistribution of body fat to central stores. Prolactin does not appear to have a major physiologic effect in human males. Hypoglycemia is a stimulus for GH secretion. Acromegaly results from excessive GH secretion in an adult; the symptoms are not consistent with acromegaly.

Clinical Application Exercises 31.1

GROWTH HORMONE DEFICIENCY

A 6-year-old boy is brought to the endocrine clinic to be evaluated for GH deficiency. The boy's height is between 2 and 3 standard deviations below the average height for his age. Initial physical examination rules out head trauma, chronic illness, and malnutrition. Family history does not suggest similar short stature in immediate relatives. Thyroid hormones are normal.

QUESTIONS

1. Why would you order a blood test for levels of IGFBP3 and IGF-1?
2. The levels of IGFBP3 and IGF-1 are below the normal range in your patient. What does this finding suggest?
3. What is GH resistance and what test would you order to rule out that the patient does not have this syndrome?
4. Why is it important to treat GH deficiency and short stature prior to the onset of puberty?
5. Why is resistance to insulin action a potential side effect of giving extremely high pharmacologic doses of GH for long periods of time?

ANSWERS

1. GH is released in a pulsatile manner, and between pulses of GH secretion, the concentration in the blood may be undetectable in normal individuals. GH induces the synthesis and secretion of IGF-1 and IGFBP3, both of which are easily detectable in the serum. Low IGF-1 and IGFBP3 would indicate insufficient GH secretion.

2. GH stimulates the synthesis and secretion of IGF-1 and IGFBP3. In most cases, low levels of these molecules in the blood would indicate insufficient GH release. However, low levels of IGF-1 and IGFBP3 could also be due to a defect in the GH receptor resulting in GH resistance.

3. GH resistance is characterized by impaired growth due to very low levels of IGF-1 and IGFBP3 in the blood. However, the concentration of GH in the blood is very high. Defects in the GH receptor, which prevent GH from stimulating the production of IGF-1 and IGFBP3, are a common cause of GH resistance. Measurement of GH in the blood should detect the extremely high levels of the hormone to confirm diagnosis.

4. GH and IGF-1 stimulate the epiphyseal growth plate of the long bones to grow. The epiphyseal plate fuses several years after puberty at which time GH and IGF-1 can no longer stimulate the growth of the bone. Therefore, the earlier GH therapy is initiated, the greater chance of achieving normal adult height before long bone growth stops.

5. GH has diabetogenic actions that oppose the actions of insulin. Thus, chronic, high doses of GH can impair the actions of insulin. Insulin resistance is a condition in which tissues in the body do not respond very well to insulin.

thePoint® *Visit* http://thepoint.lww.com/rhoades5e *for additional chapter review Q&A, Clinical Application Exercises, animations, and more!*

32

Thyroid Gland

Active Learning Objectives

Upon mastering the material in this chapter, you should be able to:

- Describe the function of the thyroid follicles and explain the iodination and coupling reactions catalyzed by the enzyme thyroid peroxidase in the synthesis of the thyroid hormones thyroxine and triiodothyronine.
- Outline how thyroid hormones are released from the thyroid gland.
- Explain the function of thyroid-stimulating hormone in regulating the synthesis and release of thyroid hormones.
- Predict the effect of changes in the concentration of thyroid hormones in the circulation on thyroid-stimulating hormone release from the anterior pituitary.

- Explain the function of the peripheral tissue deiodinases in synthesizing the physiologically active thyroid hormone triiodothyronine.
- Describe how triiodothyronine interacts with its receptor and activates transcription of target genes.
- Outline the effects of thyroid hormones on central nervous system development, growth hormone release, and target tissues such as bone.
- Explain how thyroid hormone regulates basal metabolic rate and intermediary metabolism.
- Predict the effects of excess thyroid hormone and thyroid hormone deficiency on metabolic rate, mental status, and body weight.

At the cellular level, the rate of metabolic processes is tightly regulated. The cell not only meets its basic metabolic "housekeeping" needs but also remains poised to do its own special work in the body, such as conducting nerve impulses and contracting, absorbing, and secreting. During its life span, the cell continues to make the enzymatic and structural proteins that ensure the maintenance of an appropriate rate of metabolism.

The thyroid hormones, **thyroxine (T_4)** and **triiodothyronine (T_3)**, play key roles in the regulation of body development and govern the rate at which metabolism occurs in individual cells. These hormones are not essential for life, but without them, cellular housekeeping moves at a slower pace, eventually influencing the ability of individual cells to carry out their physiologic functions. The thyroid hormones exert their regulatory functions by influencing gene expression and affecting the developmental program and amount of cellular constituents needed for the normal rate of metabolism.

▶ THYROID HORMONE SYNTHESIS, SECRETION, AND METABOLISM

The human thyroid gland consists of two lobes attached to either side of the trachea by connective tissue. A band of thyroid tissue, the isthmus, lies just below the cricoid cartilage and connects the two lobes. The thyroid gland is one of the largest endocrine glands in the body, weighing about 20 g in a healthy adult.

Each lobe of the thyroid receives its arterial blood supply from a superior and an inferior thyroid artery, which arise from the external carotid and subclavian arteries, respectively. Blood leaves the lobes of the thyroid by a series of thyroid veins that drain into the external jugular and innominate veins. This circulation provides a rich blood supply to

the thyroid gland, giving it a higher rate of blood flow per gram than even that of the kidneys.

The thyroid gland receives adrenergic innervation from the cervical ganglia and cholinergic innervation from the vagus nerves. This innervation regulates vasomotor function to increase the delivery of **thyroid-stimulating hormone (TSH)**, iodide, and metabolic substrates to the thyroid gland. The adrenergic system can also affect thyroid function by direct effects on the cells.

Follicles within the thyroid gland synthesize and secrete thyroxine and triiodothyronine.

The lobes of the thyroid gland consist of aggregates of many spherical **follicles**, lined by a single layer of epithelial cells (Fig. 32.1). Microvilli cover the apical membranes of the follicular cells, which face the lumen. Pseudopods formed from the apical membrane extend into the lumen. Tight junctions, which provide a seal for the contents of the lumen, connect the lateral membranes of the follicular cells. The basal membranes of the follicular cells are close to the rich capillary network that penetrates the stroma between the follicles.

The lumen of the follicle contains a high viscosity, proteinaceous gel-like substance called **colloid** (see Fig. 32.1). The colloid is a solution composed primarily of **thyroglobulin (Tg)**, a large protein that is a storage form of the thyroid hormones.

The thyroid follicle produces and secretes two thyroid hormones, T_4 and T_3. The molecular structures of these hormones are shown in Figure 32.3. T_4 and T_3 are iodinated derivatives of the amino acid tyrosine, formed by an ether linkage between the phenyl rings of two iodinated tyrosine molecules. The resulting structure is called an **iodothyronine**.

T_4 contains four iodine atoms on the 3, 5, 3′, and 5′ positions of the thyronine ring structure, whereas T_3 has

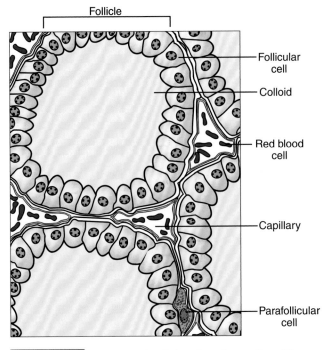

Figure 32.1 A cross-sectional view through a portion of the human thyroid gland. (See text for details.)

only three iodine atoms, at ring positions 3, 5, and 3′ (see Fig. 32.3). Thus, thyroxine is usually abbreviated as T_4 and triiodothyronine as T_3.

Parafollicular cells in the wall of the thyroid follicle synthesize calcitonin.

In addition to the epithelial cells that secrete T_4 and T_3, the wall of the thyroid follicle contains a few **parafollicular cells** (see Fig. 32.1). The parafollicular cell is usually embedded

in the wall of the follicle, inside the basal lamina surrounding the follicle, and its plasma membrane does not form part of the wall of the lumen. Parafollicular cells produce and secrete the hormone calcitonin. The effects of calcitonin on calcium metabolism are discussed in Chapter 35.

Follicular cells synthesize and store iodinated thyroglobulin within the follicle.

The thyroid follicle does not directly synthesize T_4 and T_3 in their final form. Rather, T_4 and T_3 are formed by chemical modification of tyrosine residues in the peptide structure of Tg as the follicular cells secrete it into the follicle lumen. Thus, the T_4 and T_3 formed by this chemical modification are actually part of the amino acid sequence of Tg.

The high concentration of Tg in the colloid provides a large reservoir of stored thyroid hormones for later processing and secretion by the follicle. The synthesis of T_4 and T_3 is completed when Tg is retrieved through **pinocytosis** of the colloid by the follicular cells. Lysosomal enzymes then hydrolyze Tg to its constituent amino acids, releasing T_4 and T_3 molecules from their peptide linkage for secretion into the blood.

The steps involved in the synthesis of iodinated Tg are shown in Figure 32.2. This process involves the synthesis of a thyroglobulin precursor, the uptake of iodide, and the formation of iodothyronine residues.

Thyroglobulin precursor protein

The first step in the formation of T_4 and T_3 is the synthesis of the protein precursor for Tg. This 660-kDa glycoprotein is composed of two similar 330-kDa subunits held together by disulfide bridges. Ribosomes synthesize the subunits on the rough endoplasmic reticulum (ER), which then undergo

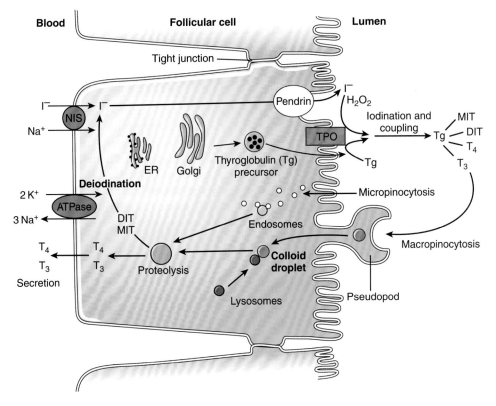

Figure 32.2 Thyroid hormone synthesis and secretion. (See text for details.) ATPase, adenosine triphosphatase; DIT, diiodotyrosine; MIT, monoiodotyrosine; T_3, triiodothyronine; T_4, thyroxine; NIS, sodium–iodide symporter; TPO, thyroid peroxidase.

Endocrine Physiology

dimerization and glycosylation in the smooth ER. The Golgi apparatus packages the completed glycoprotein into vesicles. These vesicles migrate to, and fuse with, the apical membrane of the follicular cell. Thyroglobulin precursor protein is then extruded onto the apical surface of the cell, where iodination takes place.

Iodide uptake and release

The iodide used for iodination of the thyroglobulin precursor protein comes from the blood perfusing the thyroid gland. The **sodium–iodide symporter** (NIS) is located on the basal plasma membrane of the follicular cell, near the capillaries that supply the follicle (see Fig. 32.2). The NIS uses a sodium gradient across the basal membrane to transport one iodide atom along with two Na^+ ions into the cytosol of the follicular cell. Although NIS does not require ATP to transport iodide, the Na^+/K^+-ATPase is required to pump Na^+ back out of the cell and to maintain the Na^+ gradient. The NIS also transports other anions, such as bromide, thiocyanate, and perchlorate, across the basal membrane, and these can be used to reduce iodide uptake by the gland. Through the activity of NIS, the follicular cell concentrates iodide many times higher than the concentration of iodide in the blood; thus, follicular cells are efficient extractors of the small amount of iodide circulating in the blood. Once inside follicular cells, the iodide ions diffuse rapidly to the apical membrane, where **pendrin** transports it into the follicle to be used for iodination of the thyroglobulin precursor.

Thyroid peroxidase

The addition of one or two iodine atoms to certain tyrosine residues in the thyroglobulin precursor protein, a process termed **organification**, is the next step in Tg synthesis. The precursor protein contains 134 tyrosine residues, but only a small fraction of these become iodinated. A typical Tg molecule contains only 20 to 30 atoms of iodine.

The enzyme **thyroid peroxidase** (TPO), which is bound to the apical membrane of the follicular cells, catalyzes the iodination of Tg. TPO binds an iodide ion and a tyrosine residue in the thyroglobulin precursor, bringing them in close proximity. The enzyme oxidizes the iodide ion and the tyrosine residue to short-lived free radicals, using hydrogen peroxide that has been generated within the mitochondria of follicular cells. The free radicals then undergo addition to form a **monoiodotyrosine** (MIT) residue, which remains in peptide linkage in the thyroglobulin structure. This same enzymatic process may add a second iodine atom to a MIT residue, forming a **diiodotyrosine** (DIT) residue.

Iodinated tyrosine residues that are close together in the thyroglobulin precursor molecule undergo a **coupling reaction**, which forms the iodothyronine structure. TPO, the same enzyme that initially oxidizes iodine, is believed to catalyze the coupling reaction through the oxidation of neighboring iodinated tyrosine residues to short-lived free radicals. These free radicals undergo addition to produce an iodothyronine residue and a **dehydroalanine residue**, both of which remain in peptide linkage in the Tg structure. For example, when two neighboring DIT residues couple by this mechanism, T_4 is formed. After being iodinated, the Tg molecule is stored as part of the colloid in the lumen of the follicle (see Fig. 32.2).

Only about 20% to 25% of the DIT and MIT residues in Tg become coupled to form iodothyronines. Thus, a typical Tg molecule contains five to six uncoupled residues of DIT and two to three residues of T_4. However, T_3 is formed in only about one of five Tg molecules. As a result, the thyroid secretes substantially more T_4 than T_3.

Follicular cells phagocytose and hydrolyze thyroglobulin to secrete thyroid hormones.

When the thyroid gland is stimulated to secrete thyroid hormones, vigorous pinocytosis occurs at the apical membrane of follicular cells. Pseudopods from the apical membrane reach into the lumen of the follicle, engulfing bits of the colloid (see Fig. 32.2). Endocytotic vesicles formed by this pinocytotic activity migrate toward the basal region of the follicular cell. Lysosomes mainly located in the basal region of resting follicular cells migrate toward the apical region of the stimulated cells. The lysosomes fuse with the Tg-containing droplet and hydrolyze the Tg to its constituent amino acids. As a result, T_4, T_3, and the other iodinated amino acids are released into the cytosol. The T_4 and T_3 formed from the hydrolysis of Tg are released from the follicular cell and enter the nearby capillary circulation. The DIT and MIT generated by the hydrolysis of Tg are rapidly deiodinated, and the released iodide is reused for the continued iodination of Tg (see Fig. 32.2).

Most of the T_4 and T_3 molecules that enter the bloodstream (~70% and 80%, respectively) become noncovalently bound to **thyroxine-binding globulin** (TBG), a 54-kDa glycoprotein that is synthesized and secreted by the liver. Each molecule of TBG has a single binding site for a thyroid hormone molecule. The remaining T_4 and T_3 in the blood are bound to **transthyretin** or to albumin. Less than 1% of the blood T_4 and T_3 is in the free form, which is in equilibrium with the large protein-bound fraction. It is this small amount of free thyroid hormone that interacts with target cells.

The protein-bound T_4 and T_3 represent a large reservoir of preformed hormone that can replenish the small amount of circulating free hormone as it is cleared from the blood. This reservoir provides the body with a buffer against drastic changes in circulating thyroid hormone levels as a result of sudden changes in the rate of T_4 and T_3 secretion. The protein-bound T_4 and T_3 molecules are also protected from metabolic inactivation and excretion in the urine; thus, thyroid hormones have long half-lives in the bloodstream. The half-life of T_4 is about 7 days; the half-life of T_3 is about 1 day.

Peripheral tissues metabolize thyroid hormones.

Deiodination reactions in the peripheral tissues both activate and inactivate thyroid hormones. The enzymes that catalyze the various deiodination reactions are regulated, resulting in different thyroid hormone concentrations in various tissues under different physiologic and pathophysiologic conditions.

Endocrine Physiology

Thyroxine conversion to triiodothyronine

As noted earlier, T$_4$ is the major secretory product of the thyroid gland and is the predominant thyroid hormone in the blood. About 40% of the T$_4$ secreted by the thyroid gland is converted to T$_3$ by enzymatic removal of the iodine atom at position 5' of the thyronine ring structure in a reaction termed **outer-ring deiodination** (Fig. 32.3). **Deiodinase type 1 (D1)** located in the liver, kidneys, and thyroid gland catalyzes this reaction. The T$_3$ formed by this deiodination and that secreted by the thyroid reacts with **thyroid hormone receptors (TRs)** in target cells; therefore, T$_3$ is the physiologically active form of the thyroid hormones. Skeletal muscle, the central nervous system (CNS), the pituitary gland, and the placenta contain **deiodinase type 2 (D2)**. D2 is believed to function primarily to maintain intracellular T$_3$ in target tissues, but it may also contribute to the generation of circulating T$_3$. A third deiodinase **(type 3 or D3)** catalyzes **inner-ring deiodination** reactions during degradation of thyroid hormones as discussed below. All of the deiodinases contain **selenocysteine** in the active center. This rare amino acid has properties that make it ideal for catalysis of oxidoreductive reactions.

Thyroxine and triiodothyronine inactivation

Whereas the 5'-deiodination of T$_4$ to produce T$_3$ can be viewed as a metabolic activation process, both T$_4$ and T$_3$ undergo enzymatic deiodinations, particularly in the liver and kidneys, which inactivate them (see Fig. 32.3). About 40% of the T$_4$ secreted by the human thyroid gland is deiodinated at the fifth position on the thyronine ring structure by D3 to produce **reverse T$_3$ (rT$_3$)**, which has little or no thyroid hormone activity. This deiodination reaction is the major pathway for T$_4$ metabolic inactivation and disposal. T$_3$ and

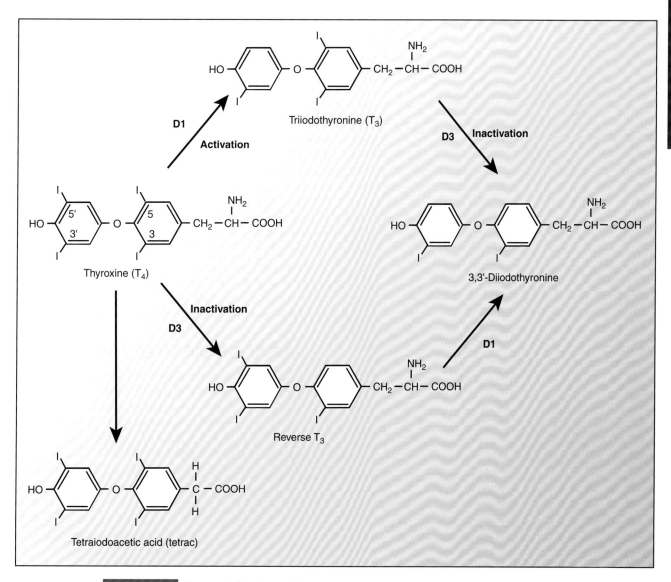

Figure 32.3 **The metabolism of thyroxine.** Deiodinase type 1 (D1) deiodinates thyroxine (T$_4$) at the 5' position to form triiodothyronine (T$_3$), the physiologically active thyroid hormone. Deiodinase type 3 (D3) also enzymatically deiodinates some T$_4$ at the 5' position to form the inactive metabolite, reverse T$_3$. T$_3$ and reverse T$_3$ undergo additional deiodinations to 3,3'-diiodothyronine before being excreted. A small amount of T$_4$ is also decarboxylated and deaminated to form the metabolite, tetraiodoacetic acid (tetrac), which may then be deiodinated before being excreted.

rT$_3$ also undergo deiodination to yield 3,3′-diiodothyronine, which may be further deiodinated before being excreted.

Both physiologic and pathologic factors influence the 5′-deiodination reaction, changing the relative amounts of T$_3$ and rT$_3$ produced from T$_4$. For example, a human fetus produces less T$_3$ from T$_4$ than does a child or adult because the 5′-deiodination reaction is less active in the fetus. During fasting, 5′-deiodination is inhibited, particularly in response to carbohydrate restriction, but it is restored to normal when the person is fed again. Trauma and acute and chronic illnesses also suppress the 5′-deiodination reaction. Under all of these circumstances, the amount of T$_3$ produced from T$_4$ is reduced and its blood concentration falls. The amount of rT$_3$ rises in the circulation, not because its conversion from T$_4$ is increased, as originally believed, but rather because its clearance from the blood is reduced.

Note that during fasting or in the disease states mentioned above, the secretion of T$_4$ is usually not increased, despite the decrease of T$_3$ in the circulation. This response is likely a result of the fact that D2 in the CNS and pituitary is not affected under these conditions and continues to produce T$_3$ in quantities sufficient to maintain relatively normal hypothalamic–pituitary–thyroid axis function.

Thyroxine glucuronidation, decarboxylation, and deamination

T$_4$ and to a lesser extent T$_3$ are catabolized by conjugation with glucuronic acid in the liver. The conjugated hormones are secreted into the bile and eliminated in the feces. Many tissues also catabolize thyroid hormones by modifying the three-carbon side chain of the iodothyronine structure via decarboxylation and deamination. The derivatives formed from T$_4$, such as **tetraiodoacetic acid** (**tetrac**), may also undergo deiodination before being excreted (see Fig. 32.3).

Thyroid hormone synthesis and secretion are controlled by the anterior pituitary.

When the concentrations of free T$_4$ and T$_3$ fall in the blood, the anterior pituitary gland is stimulated to secrete TSH, raising the concentration of TSH in the blood. This results in increased binding of TSH to its receptors on thyroid follicular cells.

The TSH receptor is a seven-transmembrane glycoprotein located on the basal plasma membrane of the follicular cell. G$_s$ proteins couple these receptors to the adenylyl cyclase–cyclic adenosine monophosphate (cAMP)–protein kinase A pathway; however, there is also evidence for effects via phospholipase C (PLC), inositol trisphosphate, and diacylglycerol (see Chapter 2). The physiologic importance of TSH-stimulated phospholipid metabolism in human follicular cells is unclear, because high concentrations of TSH are needed to activate PLC.

Thyroid-stimulating hormone promotion of thyroid hormone secretion

By increasing cAMP, TSH stimulates synthesis of NIS and uptake of iodide by follicular cells, the iodination of tyrosine residues in the thyroglobulin precursor, and the coupling of iodinated tyrosines to form iodothyronines. Moreover, TSH promotes the pinocytosis of colloid by the apical membranes, resulting in a large increase in endocytosis and hydrolysis of thyroglobulin. The overall result of these effects of TSH is an increased release of T$_4$ and T$_3$ into the blood. In addition, TSH rapidly increases energy metabolism in the thyroid follicular cell.

Thyroid-stimulating hormone regulation of thyroid gland growth

Over the long term, TSH promotes protein synthesis in thyroid follicular cells to maintain their size and structural integrity. Evidence of this trophic effect of TSH is seen in the hypophysectomized patient, in whom the thyroid gland atrophies as a result of a reduction in the height of the follicular cells. In contrast, chronic exposure to excessive amounts of TSH causes an increase in follicular cell height and number, resulting in an enlarged thyroid gland called **goiter**. The trophic and proliferative effects of TSH on the thyroid are mediated by cAMP.

Dietary iodide is essential for the synthesis of thyroid hormones.

A continual supply of iodide is required for the synthesis of T$_4$ and T$_3$. If the diet is severely deficient in iodide, as in some parts of the world, the amount of iodide available to the thyroid gland limits T$_4$ and T$_3$ synthesis. As a result, the concentrations of T$_4$ and T$_3$ in the blood fall, causing a chronic stimulation of TSH secretion, which, in turn, produces a goiter. Enlargement of the thyroid gland increases its capacity to accumulate iodide from the blood and to synthesize T$_4$ and T$_3$. However, the degree to which the enlarged gland can produce thyroid hormones to compensate for their deficiency in the blood depends on the severity of the deficiency of iodide in the diet. To prevent iodide deficiency and consequent goiter formation, iodide is added to table salt (iodized salt) in most developed countries. In the United States, iodine intake is decreasing as salt intake is reduced.

▶ THYROID HORMONE EFFECTS ON THE BODY

Most cells of the body are targets for the action of thyroid hormones. The sensitivity or responsiveness of a particular cell to thyroid hormones correlates to some degree with the number of receptors for these hormones. The cells of the CNS appear to be an exception. As is discussed later, the thyroid hormones play an important role in CNS development during fetal and neonatal life, and developing nerve cells in the brain are important targets for thyroid hormones. In the adult, however, brain cells show little responsiveness to the metabolic regulatory action of thyroid hormones, although they have numerous receptors for these hormones. The reason for this discrepancy is unclear.

Triiodothyronine activates a nuclear receptor.

Thyroid hormone receptors (TRs) are located in the nuclei of target cells bound to **thyroid hormone response elements** (**TREs**) in the DNA. TRs are ~50-kDa proteins that are structurally similar to the nuclear receptors for steroid hormones and vitamin D. In humans, two TR genes (a and b) are found on two separate chromosomes and encode several different receptor proteins (TRa1, TRb1, TRb2, and TRb3). The TR receptors are expressed in a tissue-specific manner to impart different effects of thyroid hormone on various tissues. Thyroid receptors bound to the TRE in the absence of T_3 generally act to repress gene expression.

Target cells take up the free forms of T_3 and T_4 from the blood through a carrier-mediated process that requires ATP. Once inside the cell, T_4 is deiodinated to T_3, which enters the nucleus of the cell and binds to its receptor in the chromatin. The TR with bound T_3 forms a complex (called a *heterodimer*) with another nuclear receptor called the **retinoid X receptor** (**RXR**). TR can also complex with other TRs to form homodimers and activate transcription. Transcription coactivators and corepressors also complex with the TR heterodimer or homodimer, resulting in increased or decreased production of certain mRNAs (Fig. 32.4). T_3 can influence cell differentiation by regulating the kinds of proteins produced by its target cells and can influence growth and metabolism by changing the amounts of structural and enzymatic proteins present in the cells (Clinical Focus 32.1).

The transcriptional response to T_3 is slow; thus when T_3 is given to an animal or human, several hours elapse before its physiologic effects can be detected. This delayed action reflects the time required for transcriptional changes and consequent synthesis of key proteins to occur. When T_4 is

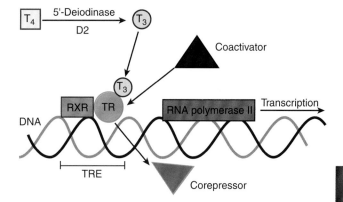

Figure 32.4 **The activation of transcription by thyroid hormone.** Thyroxine (T_4) is taken up by the cell and deiodinated to triiodothyronine (T_3), which then binds to the thyroid hormone receptor (TR). The activated TR heterodimerizes with a second transcription factor, 9-*cis* retinoic acid receptor (RXR) and binds to the thyroid hormone response element (TRE). The binding of TR/RXR to the TRE displaces repressors of transcription and recruits additional coactivators. The final result is the activation of RNA polymerase II and the transcription of the target gene.

administered, its course of action is usually slower than that of T_3 because of the additional time required for the body to convert T_4 to T_3.

Thyroid hormones also have effects on cells that occur much faster and do not appear to be mediated by nuclear TR

CLINICAL FOCUS | 32.1

Resistance to Thyroid Hormone

Mutations in one allele of the thyroid receptor β (TRb) gene result in **resistance to thyroid hormone (RTH)**. These mutations tend to cluster in areas of the thyroid hormone receptor that have important interactions with the hydrophobic ligand-binding domain; thus, the mutant TRb does not bind T_3 with high affinity. The mutations do not interfere with the DNA-binding domain, the corepressor-binding domain, or the region of the receptor that heterodimerizes with 9-*cis* retinoic acid receptor (RXR).

It is thought that RTH is produced when a mutant TRb heterodimerizes with RXR, a normal TRb, or TRa. Receptor dimers containing mutant TRb compete with wild-type (normal) TRb-containing dimers for binding to thyroid hormone response elements (TREs) of thyroid hormone–regulated genes (see Fig. 32.4). The mutant TRb heterodimers also bind corepressor molecules that cannot be released because of the defect in triiodothyronine (T_3) binding; therefore, genes become more repressed than they normally would be.

Over 400 families with RTH have been identified, and it is expected that many more cases likely exist. The frequency of mutation has been estimated at 1:50,000. The mutation effect is characterized as dominant negative inhibition because TRb interferes with the function of the three normal

TR-expressing genes. Patients with RTH may have features of hypothyroidism if the resistance is severe and affects all tissues. Such people are referred to as having *generalized resistance to thyroid hormone*. Alternatively, patients may present with hyperthyroidism if the hypothalamic–pituitary axis is more severely affected—a condition termed *pituitary resistance to thyroid hormone*. The hypothalamic–pituitary axis is primarily regulated by TRb and thus is more sensitive to mutations in the receptor than tissues such as the heart, in which TRa is the dominant thyroid hormone receptor. Although patients may present with symptoms of hyperthyroidism, thyroid-stimulating hormone (TSH) level is normal or only slightly elevated, a finding important in differentiating between a diagnosis of RTH and TSH-secreting pituitary tumor, which results in high TSH levels.

Many patients may present with a mixture of hypothyroid and hyperthyroid conditions. Treatment of patients with RTH is therefore difficult because thyroid hormone analogs designed to suppress the hypothalamic–pituitary axis and reduce the release of T_4 can worsen effects on the heart, which has functional TRa receptors. β-Blockers may be of use in such situations. Development of thyroid hormone analogs that preferentially bind TRb or TRa may prove useful for treatment of RTH. ■

receptors, including effects on signal transduction pathways that alter cellular respiration, cell morphology, vascular tone, and ion homeostasis.

Thyroid hormones are crucial for brain maturation during fetal development.

The human brain undergoes its most active phase of growth during the last 6 months of fetal life and the first 6 months of postnatal life. During the second trimester of pregnancy, the multiplication of neuroblasts in the fetal brain reaches a peak and then declines. As pregnancy progresses and the rate of neuroblast division drops, neuroblasts differentiate into neurons and begin the process of synapse formation that extends into postnatal life.

Thyroid hormones first appear in the fetal blood during the second trimester of pregnancy, and levels continue to rise during the remaining months of fetal life. TRs increase about 10-fold in the fetal brain at about the time the concentrations of T_4 and T_3 begin to rise in the blood. These events are critical for normal brain development because thyroid hormones are essential for timing the decline in nerve cell division and the initiation of differentiation and maturation of these cells.

If thyroid hormones are deficient during these prenatal and postnatal periods of differentiation and maturation of the brain, mental retardation occurs. The cause is thought to be inadequate development of the neuronal circuitry of the CNS. Thyroid hormone therapy must be given to a thyroid hormone–deficient child during the first few months of postnatal life to prevent mental retardation. Starting thyroid hormone therapy after behavioral deficits have occurred cannot reverse the mental retardation (i.e., thyroid hormone must be present when differentiation normally occurs). Thyroid hormone deficiency during infancy causes both mental retardation and growth impairment, as discussed below. Fortunately, this rarely occurs today because thyroid hormone deficiency is usually detected in newborn infants and hormone therapy is given at the proper time.

The exact mechanism by which thyroid hormones influence differentiation of the CNS is unknown. Animal studies have demonstrated that thyroid hormones inhibit nerve cell replication in the brain and stimulate the growth of nerve cell bodies, the branching of dendrites, and the rate of myelinization of axons. These effects of thyroid hormones are presumably a result of their ability to regulate the expression of genes involved in nerve cell replication and differentiation.

Body growth and development require thyroid hormone.

Thyroid hormones are important factors regulating the growth of the entire body. For example, a person who is deficient in thyroid hormones and who does not receive thyroid hormone therapy during childhood will not grow to a normal adult height.

Thyroid hormones promote normal body growth by stimulating growth hormone (GH) gene expression in the somatotrophs of the anterior pituitary gland. In a thyroid hormone–deficient person, GH synthesis by the somatotrophs is greatly reduced, and GH secretion is impaired;

thus, a thyroid hormone–deficient person will also be GH deficient. If this condition occurs in a child, it will cause growth retardation (see Chapter 31).

In tissues such as skeletal muscle, heart, and liver, thyroid hormones have direct effects on the synthesis of many structural and enzymatic proteins. For example, they stimulate the synthesis of structural proteins of mitochondria, as well as the formation of many enzymes involved in intermediary metabolism and oxidative phosphorylation.

Thyroid hormones also promote the growth and calcification of bones in the skeleton. Deficiency in the thyroid hormone early in life leads to delayed, abnormal development of bone and can lead to dwarfism. The effect of thyroid hormones to regulate bone growth is, in part, mediated through its effects on GH and insulin-like growth factor 1 (IGF-1) synthesis.

Thyroid hormones are a major determinant of basal metabolic rate.

When the body is at rest, about half of the ATP produced by its cells is used to drive energy-requiring membrane transport processes. The remainder is used in involuntary muscular activity, such as respiratory movements, peristalsis, and contraction of the heart, and in metabolic reactions requiring ATP such as protein synthesis. The energy required to do this work is eventually released as body heat.

Mitochondria produce ATP by the oxidative phosphorylation of **adenosine diphosphate** (**ADP**). The rate of oxidative phosphorylation depends on the supply of ADP for electron transport. The ADP supply is, in turn, a function of the amount of ATP used to do work. For example, when more work is done per unit time, more ATP is used and more ADP is generated, increasing the rate of oxidative phosphorylation. The rate at which oxidative phosphorylation occurs is reflected in the amount of oxygen consumed by the body because oxygen is the final electron acceptor at the end of the electron transport chain.

Activities that occur when the body is not at rest such as voluntary movements use additional ATP for the work involved. The total amount of oxygen consumed and body heat produced over a 24-hour period is the combination of that needed at rest and that needed during activity.

Thyroid hormones regulate the basal rate at which oxidative phosphorylation takes place in cells. As a result, they set the basal rates of body heat production and of oxygen consumed by the body. This is called the **thermogenic action** of thyroid hormones.

Thyroid hormone levels in the blood must be within normal limits for basal metabolism to proceed at the rate needed for a balanced energy economy of the body. If thyroid hormones are present in excess, oxidative phosphorylation is accelerated and body heat production and oxygen consumption are abnormally high. The converse occurs when the blood concentrations of T_4 and T_3 are lower than normal. The fact that thyroid hormones affect the amount of oxygen consumed by the body has been used clinically to assess the status of thyroid function. The amount of oxygen consumed by the body under resting conditions is called the **basal metabolic rate** (**BMR**) and assumes normal thyroid function. A BMR higher than that expected for a person with normal thyroid function would indicate that concentrations of T_4 and T_3 are higher than normal.

Not all tissues are sensitive to the thermogenic action of thyroid hormones. Tissues and organs that do respond include skeletal muscle, heart, liver, and the kidneys, in which TRs are abundant. The adult brain, skin, lymphoid organs, and gonads show little thermogenic response to thyroid hormones, and with the exception of the adult brain, these tissues contain few TRs.

The thermogenic action of the thyroid hormones is poorly understood at the molecular level. The thermogenic effect takes many hours to appear after the administration of thyroid hormones to a human or animal, probably because of the time required for changes in gene expression. T_3 is known to stimulate the synthesis of cytochromes, cytochrome oxidase, and Na^+/K^+-ATPase in certain cells, suggesting that T_3 may regulate the number of respiratory units in these cells and affecting their capacity to carry out oxidative phosphorylation. A greater rate of oxidative phosphorylation results in greater heat production.

Thyroid hormone also stimulates the synthesis of **uncoupling protein 1** (UCP-1) in brown adipose tissue. ATP synthase produces ATP in the mitochondria when protons flow down their electrochemical gradient. UCP-1 acts as a channel in the mitochondrial membrane to dissipate the ion gradient without making ATP. As the protons move down their electrochemical gradient *uncoupled* from ATP synthesis, energy is released as heat. Adult humans do not appear to have enough brown adipose tissue for UCP-1 to make a significant contribution to nutrient oxidation or body heat production. However, the uncoupling proteins UCP-2 and UCP-3 are found in other tissues, and thyroid hormones regulate their expression. These novel uncoupling proteins may have a role in the thermogenic action of thyroid hormones.

Carbohydrate, fat, and protein metabolism are regulated by thyroid hormone.

In addition to regulating the rate of basal energy metabolism, thyroid hormones influence the rate at which most of the pathways of intermediary metabolism operate in their target cells. When thyroid hormones are deficient, carbohydrate, lipid, and protein metabolism are slowed, and the response of these pathways to other regulatory hormones is decreased. In contrast, these metabolic pathways run at an abnormally high rate when thyroid hormones are present in excess. Thus, thyroid hormones can be viewed as amplifiers of cellular metabolic activity. The amplifying effect of thyroid hormones on intermediary metabolism is mediated through the transcription of gene-encoding enzymes involved in these metabolic pathways.

Thyroid hormones provide negative feedback to control the hypothalamic–pituitary axis.

An important action of the thyroid hormones is the regulation of their own secretion. As discussed in Chapter 31, T_3 inhibits TSH secretion by thyrotrophs in the anterior pituitary gland by decreasing thyrotroph sensitivity to **thyrotropin-releasing hormone (TRH)**. When the circulating

Category	Specific Action
Development of central nervous system	Inhibit nerve cell replication
	Stimulate growth of nerve cell bodies
	Stimulate branching of dendrites
	Stimulate rate of axon myelinization
Body growth	Stimulate growth hormone gene expression in somatotrophs
	Stimulate synthesis of many structural and enzymatic proteins
	Promote calcification of bones
Basal energy expenditure	Regulate basal rates of oxidative phosphorylation, body heat production, and oxygen consumption
Intermediary metabolism	Stimulate synthetic and degradative pathways of carbohydrate, lipid, and protein metabolism
Thyroid-stimulating hormone (TSH) secretion	Inhibit TSH secretion by decreasing sensitivity of thyrotrophs to thyrotropin-releasing hormone

TABLE 32.1 Physiologic Actions of Thyroid Hormones

concentration of free thyroid hormone is high, thyrotrophs are relatively insensitive to TRH and the rate of TSH secretion decreases. The resulting fall in TSH levels in the blood reduces the rate of thyroid hormone release from the follicular cells in the thyroid. When the free thyroid hormone level falls in the blood, however, the negative feedback effect of T_3 on thyrotrophs is reduced and the rate of TSH secretion increases. The rise in TSH in the blood stimulates the thyroid gland to secrete thyroid hormones at a greater rate. This action of T_3 on thyrotrophs is thought to be a result of changes in gene expression in these cells.

The physiologic actions of the thyroid hormones described above are summarized in Table 32.1.

▶ ABNORMALITIES OF THYROID FUNCTION IN ADULTS

A deficiency or excess of thyroid hormone produces characteristic changes in the body. These changes result from dysregulation of nervous system function and altered metabolism.

Hyperthyroidism increases energy expenditure and causes weight loss.

The most common cause of excessive thyroid hormone production in humans is **Graves disease**, an autoimmune disorder caused by antibodies directed against the TSH receptor in the plasma membrane of thyroid follicular cells. These antibodies bind to the TSH receptor, increasing adenylyl cyclase activity, raising cAMP levels, and producing effects similar to those of TSH. The thyroid gland enlarges to form a **diffuse toxic goiter**, which synthesizes and secretes thyroid hormones at an accelerated rate, causing thyroid hormones to be chronically elevated in the blood. Feedback inhibition of thyroid hormone production by the thyroid hormones is also lost.

Less common conditions that cause chronic elevations in circulating thyroid hormones include adenomas of the thyroid gland that secrete thyroid hormones and excessive TSH secretion caused by malfunctions of the hypothalamic–pituitary–thyroid axis. Many changes in body function characterize the disease state that develops in response to excessive thyroid hormone secretion, called **hyperthyroidism** or **thyrotoxicosis**.

People with hyperthyroidism are nervous and emotionally irritable, with a compulsion to constant movement. However, they also experience physical weakness and fatigue. BMR is increased resulting in greater body heat production. Vasodilation in the skin and sweating occur as compensatory mechanisms to dissipate excessive body heat. Heart rate and cardiac output are increased. Energy metabolism increases, as does appetite. However, despite the increase in food intake, a net degradation of protein and lipid stores occurs, resulting in weight loss. Reducing the rate of thyroid hormone secretion with drugs or removal of the thyroid gland by radioactive ablation or surgery can reverse all of these changes (Clinical Focus 32.2).

Hypothyroidism decreases metabolism and causes weight gain.

Thyroid hormone deficiency in humans has many causes. For example, iodide deficiency may result in a reduction in thyroid hormone production. Autoimmune diseases such as **Hashimoto disease**, also known as *Hashimoto thyroiditis*, impair thyroid hormone synthesis. Other causes of thyroid hormone deficiency include heritable defects that alter the biosynthesis of thyroid hormones and hypothalamic or pituitary diseases that interfere with TRH or TSH secretion. Obviously, radioiodine ablation or surgical removal of the thyroid gland also causes thyroid hormone deficiency. **Hypothyroidism** is the disease state that results from thyroid hormone deficiency, and the effects on the body are the opposite of those caused by thyroid hormone excess.

Thyroid hormone deficiency impairs the function of most tissues in the body. As described earlier, thyroid hormone deficiency at birth that is not treated during the first few months of postnatal life causes irreversible mental retardation. Thyroid hormone deficiency later in life also influences the function of the nervous system. For example, all cognitive functions, including speech and memory, are slowed and body movements may be clumsy.

Metabolism is also reduced in thyroid hormone–deficient people. BMR is reduced, resulting in impaired body heat production. Vasoconstriction occurs in the skin as a

CLINICAL FOCUS | 32.2

Autoimmune Thyroid Disease: Postpartum Thyroiditis

Certain diseases affecting the function of the thyroid gland occur when a person's immune system fails to recognize particular thyroid proteins as "self" and reacts to the proteins as if they were foreign. This usually triggers both humoral and cellular immune responses. As a result, antibodies to these proteins are generated, which then alter thyroid function. Two common autoimmune diseases with opposite effects on thyroid function are Hashimoto disease and Graves disease. In Hashimoto disease, the thyroid gland is infiltrated by lymphocytes, and elevated levels of antibodies against several components of thyroid tissue (e.g., antithyroid peroxidase and antithyroglobulin antibodies) are found in the serum. The thyroid gland is destroyed, resulting in hypothyroidism. In Graves disease, stimulatory antibodies to the thyroid-stimulating hormone (TSH) receptor activate thyroid hormone synthesis, resulting in hyperthyroidism.

A third, fairly common autoimmune disease is **postpartum thyroiditis**, which usually occurs within 3 to 12 months after delivery. The disease is characterized by a transient destruction-induced thyrotoxicosis (hyperthyroidism), often followed by a period of hypothyroidism lasting several months. Many patients eventually return to the euthyroid state. Often, only the hypothyroid phase of the disease may be observed, occurring in more than 30% of women with antibodies to

thyroid peroxidase detectable before conception. The disease is also observed in patients known to have Graves disease. The postpartum occurrence of the disorder is likely a result of increased immune system function following the suppression of its activity during pregnancy.

It has been estimated that 5% to 10% of women develop postpartum thyroiditis. The prevalence of the disease has prompted the clinical recommendation that thyroid function (serum thyroxin, triiodothyronine, and TSH levels) be surveyed postpartum at 2, 4, 6, and 12 months in all women with thyroid peroxidase antibodies or symptoms suggestive of thyroid dysfunction. Patients who have experienced one episode of postpartum thyroiditis should also be considered at risk for recurrence after pregnancy.

Treatment for postpartum thyroiditis depends on the time of diagnosis. If symptoms during the period of transient thyrotoxicosis are severe, a β-blocker such as propranolol can be given. Thioamides—drugs that inhibit the oxidation and organic binding of thyroid iodide—are generally not effective for this transient thyrotoxicosis because the release of excess thyroid hormone is not a result of hormone synthesis but rather of destruction of the gland. Thyroid hormone replacement is required to treat the hypothyroidism that occurs later in the progression of the disease. ■

compensatory mechanism to conserve body heat. Heart rate and cardiac output are reduced. The synthetic and degradative processes of intermediary metabolism are slowed and weight gain occurs, despite the fact that food intake is reduced. In severe hypothyroidism, a substance consisting of hyaluronic acid and chondroitin sulfate complexed with protein is deposited in the extracellular spaces of the skin, causing water to accumulate osmotically. This effect gives a puffy appearance to the face, hands, and feet, called **myxedema**. All of the above disorders in adults can be normalized with thyroid hormone therapy. See the Integrated Medical Sciences box for a discussion of euthyroid sick syndrome, in which T_3 levels are reduced to nearly undetectable levels in critically ill patients.

INTEGRATED MEDICAL SCIENCES

Euthyroid Sick Syndrome

Severe caloric restriction and starvation result in reduced T_3 levels, with normal or slightly reduced T_4 and normal TSH levels. This is a result of the decreased release of thyroid hormone and possibly reduced D1 and D2 deiodinase activity in peripheral tissues. Reductions in T_3 result in lower basal oxygen consumption, slower heart rate, and attenuation of nitrogen loss—changes all considered beneficial in adapting to reduced caloric intake. In a patient who is severely ill, T_3 levels can be reduced up to 90% and reverse T_3 (rT_3) increased severalfold, with only slight reductions in T_4. In the most severe cases, the release of TSH and other anterior pituitary hormones is reduced because of the loss of hypothalamic input resulting from endogenous causes and, in some instances, exacerbated by therapy such as glucocorticoids that is given to ill patients. This condition is called **euthyroid sick syndrome** or *low T_3 syndrome*.

The mechanisms resulting in almost undetectable serum T_3 in patients with euthyroid sick syndrome are not entirely clear. It appears that the reduction in serum T_3 is a result of the decreased release of T_3 from the thyroid gland and reduced T_4 outer-ring deiodination. Lower transport of T_4 into tissues such as the liver and kidney, which express D1 deiodinase, and more rapid turnover of D2 deiodinase protein both contribute to reduced T_4 deiodination and lower T_3 levels. The increase in rT_3 in euthyroid sick patients was originally thought to indicate that the specificity of D1 deiodinase was changed from $5'$-T_4 to 5-T_4 deiodination, but it has since been shown that increased rT_3 is a result of reduced clearance, not increased synthesis.

Therapy for euthyroid sick syndrome remains controversial. Despite the extremely low T_3 levels, it is still not clear that T_4 or T_3 supplementation should be done because most controlled studies have not shown a beneficial effect of increasing thyroid hormones alone. A promising alternative strategy that is currently under evaluation is the infusion of a combination of growth hormone (GH)-releasing peptides and thyrotropin-releasing hormone. The administration of these factors corrects the impaired pulsatile secretion of GH, thyrotropin, and gonadotropin, resulting in increased circulating concentrations of IGF-1, GH-dependent IGF-binding proteins, T_4, T_3, and testosterone. Although hypothalamic–pituitary function is clearly improved and catabolism in the severely ill patient reversed, a positive effect of this treatment on survival has still not been definitively established. Normalization of glycemia with exogenous insulin has recently been shown to have a positive impact on survival. Thus, studies to understand the neuroendocrine response to critical illness and the interaction of the hypothalamic–pituitary–adrenal axis with peripheral hormonal and metabolic alterations are ongoing in an attempt to improve survival of the critically ill patient. ■

Chapter Summary

- The thyroid gland consists of two lobes attached to either side of the trachea; its primary function is the synthesis and secretion of thyroxine and triiodothyronine.
- The thyroid gland also produces calcitonin, which plays a key role in calcium homeostasis.
- Thyroid hormones are synthesized by iodination and coupling of tyrosines in reactions catalyzed by the enzyme thyroid peroxidase.
- The proteolysis of thyroglobulin within the follicular cells releases thyroid hormones from the thyroid gland.
- Thyroid-stimulating hormone regulates the synthesis and release of thyroid hormones through activation of adenylyl cyclase and generation of cAMP.
- Thyroid hormones circulating in the blood are bound to transport proteins. Only a small fraction of the circulating hormones are free (unbound) and biologically active.
- The concentration of thyroid hormones in the circulation regulates thyroid-stimulating hormone release from the anterior pituitary.
- In peripheral tissues, 5′-deiodinase deiodinates thyroxine to the physiologically active hormone triiodothyronine.

- In target tissues, triiodothyronine binds to the thyroid hormone receptor (TR), which then associates with retinoic acid receptor or a second TR to regulate transcription.
- TR regulates transcription by binding to specific thyroid hormone response elements in target genes and interacting with specific coactivators and corepressors of transcription.
- Brain cells are a major target for triiodothyronine and thyroxine, which play a crucial role in neuronal maturation during fetal development.
- Thyroid hormones stimulate growth by regulating growth hormone release from the pituitary and by direct actions on target tissues such as bone.
- Thyroid hormones regulate basal metabolic rate and intermediary metabolism through effects on mitochondrial ATP synthesis and the expression of genes encoding metabolic enzymes.
- Excess thyroid hormone (hyperthyroidism) results in nervousness, increased metabolic rate, and weight loss.
- Thyroid hormone deficiency (hypothyroidism) results in decreased metabolic rate and weight gain.

Chapter Review Questions

1. The effects of TSH on thyroid follicular cells include:

   A. stimulation of endocytosis of thyroglobulin stored in the colloid.

   B. release of a large pool of T_4 and T_3 stored in secretory vesicles in the cell.

   C. stimulation of the uptake of iodide from the thyroglobulin stored in the colloid.

   D. Increase in perfusion by the blood.

   E. stimulation of the binding of T_4 and T_3 to thyroxine-binding globulin.

   F. increased cAMP hydrolysis.

The correct answer is A. TSH stimulates the endocytosis of colloid by the apical membrane of the follicular cell. Thyroglobulin in the colloid is then hydrolyzed in the lysosomal vesicles to release thyroid hormones. T_3 and T_4 are stored in thyroglobulin in the colloid, not in secretory vesicles in the follicle cell. TSH stimulates the uptake of iodide from the blood, not the colloid. It has no effect on blood flow to the thyroid gland and no direct effect on the binding of T_3 and T_4 to thyroxine-binding globulin. TSH stimulates an increase in cAMP not an increase in the hydrolysis of this second messenger.

2. A child is born with a rare disorder in which the thyroid gland does not respond to TSH. What would be the predicted effects on mental ability, body growth rate, and thyroid gland size when the child reaches 6 years of age?

   A. Mental ability would be impaired, body growth rate would be slowed, and thyroid gland size would be larger than normal.

   B. Mental ability would be unaffected, body growth rate would be slowed, and thyroid gland size would be smaller than normal.

   C. Mental ability would be impaired, body growth rate would be slowed, and thyroid gland size would be smaller than normal.

   D. Mental ability would be unaffected, body growth rate would be unaffected, and thyroid gland size would be smaller than normal.

   E. Mental ability would be impaired, body growth rate would be slowed, and thyroid gland size would be normal.

   F. Mental ability would be unaffected, body growth rate would be unaffected, and thyroid gland size would be unaffected.

The correct answer is C. Thyroid hormones are important for normal development of the CNS and for body growth. TSH stimulates the synthesis and release of thyroid hormones as well as the growth of the thyroid gland. In a disorder in which the thyroid gland does not respond to TSH, thyroid hormone production would be decreased, resulting in poor development of the CNS and poor body growth. TSH would also not be able to stimulate the growth of the thyroid, resulting in a small gland.

3. If the 6-year-old child described in the previous question is now treated with thyroid hormones, how would mental ability, body growth rate, and thyroid gland size be affected?

   A. Mental ability would remain impaired, body growth rate would be improved, and thyroid gland size would be smaller than normal.

   B. Mental ability would be improved, body growth rate would be improved, and thyroid gland size would be normal.

   C. Mental ability would remain impaired, body growth rate would be improved, and thyroid gland size would be normal.

   D. Mental ability would remain impaired, body growth rate would be improved, and thyroid gland size would be larger than normal.

   E. Mental ability would be improved, body growth rate would remain slowed, and thyroid gland size would be normal.

   F. Mental ability would be improved, body growth rate would remain slowed, and thyroid gland size would be larger than normal.

The correct answer is A. Giving thyroid hormones to the child would improve body growth but not mental ability, because thyroid hormones are most important for CNS development *in utero*. Therefore, giving thyroid hormones after birth would be too late. Thyroid gland size would remain smaller than normal because thyroid hormones have no trophic effect on the gland; only TSH does.

4. Uncoupling proteins:

   A. use the proton gradient across the mitochondrial membrane to facilitate ATP synthesis.

   B. are decreased by thyroid hormones

   C. dissipate the proton gradient across the mitochondrial membrane to generate heat.

   D. are present exclusively in brown fat.

   E. uncouple fatty acid oxidation from glucose oxidation in mitochondria.

   F. are essential for maintaining body temperature in mammals.

The correct answer is C. Uncoupling proteins allow protons to flow down their concentration gradient across the mitochondrial membrane, uncoupled from the synthesis of ATP. The resulting energy generated is released as heat, and ATP is not synthesized. Uncoupling proteins are increased by thyroid hormones. The novel uncoupling proteins are found in many tissues, including muscle and adipose tissue. Fatty acid and glucose oxidation are not coupled in mitochondria, and the uncoupling proteins are not the switch between oxidation of these two substrates. Uncoupling proteins have not been demonstrated to be absolutely essential to the maintenance of body temperature in mammals. However, UCP-1 is important in the ability of small mammals, such as rodents, to tolerate cold temperatures.

5. Triiodothyronine (T_3):

   A. is produced in greater amounts by the thyroid gland than T_4.

   B. is bound by the thyroid receptor present in the cytosol of target cells.

   C. is formed from T_4 through the action of the 5-deiodinase D3.

   D. has a half-life of a few minutes in the bloodstream.

   E. is released from thyroglobulin through the action of thyroid peroxidase.

   F. can be produced by the deiodination of T_4 in pituitary thyrotrophs.

The correct answer is F. T_3 is produced from T_4 by 5'-deiodinase D2 in the anterior pituitary. The major thyroid hormone product of the thyroid gland is T_4. The thyroid hormone receptor (TR) is located in the nucleus. The 5-deiodinase D3 acts on T_4, to make reverse T_3. The half-life of T_3 in the bloodstream is about 1 day because of the protective actions of the thyroid hormone–binding proteins. Thyroid peroxidase catalyzes the iodination of thyroglobulin to form MIT and DIT, the precursor molecules for T_3. Release of T_3 from thyroglobulin is catalyzed by proteases in the lysosomal vesicles.

Clinical Application Exercises 32.1

THYROTOXICOSIS

A 35-year-old woman is seen in the endocrine clinic for evaluation of thyroid disease. The patient complains of weight loss, irritability, and restlessness. Physical examination reveals enlargement of the thyroid gland, weakness in maintaining the leg in an extended position, warm moist skin, and tachycardia. Family history identifies the patient's mother as exhibiting hypothyroidism after the birth of the patient's brother and an aunt with Hashimoto disease.

QUESTIONS

1. Based on the history and physical examination, what would be a reasonable initial diagnosis?

2. From a blood sample what hormone concentrations would you ask the laboratory to measure and what would you expect the results to be?

3. What antibody titers would you request the laboratory to determine? Which antibody titer is the most useful in the diagnosis of Graves disease?

4. What would an increase in antibodies to TPO indicate?

5. The antibody titers indicate that the patient has Graves disease. What treatment would be appropriate for this patient?

ANSWERS

1. The physical findings including the presence of goiter suggest that the patient may be hyperthyroid. However, goiter can also occur in hypothyroidism. As autoimmune thyroid disease "runs in families," the family history of postpartum thyroiditis suggests that the thyroiditis might be due to an autoimmune response. The patient may have Graves disease.

2. The laboratory should determine the levels of thyroid hormone (T_4 and T_3) and TSH. Thyroid hormones should be increased. TSH may be increased if it is early in the progression of Hashimoto disease or decreased if the patient has Graves disease.

3. Antibodies to TSH receptor, thyroid peroxidase, and thyroglobulin can all be elevated in Graves disease. However, the presence of TSH receptor antibodies is diagnostic.

4. Antibodies to thyroid peroxidase (TPO) are elevated to the greatest extent in Hashimoto disease.

5. A thionamide compound should first be used to inhibit thyroid hormone synthesis. This treatment will relieve the symptoms of hyperthyroidism and may result in a reduction in immune response. The drug may be withdrawn after several months of treatment to determine if the disease is in remission. If thyroid hormones increase with cessation of the drug, surgical intervention to remove the thyroid would be indicated.

thePoint® *Visit* http://thepoint.lww.com/rhoades5e *for additional chapter review Q&A, Clinical Application Exercises, animations, and more!*

33 Adrenal Gland

Active Learning Objectives

Upon mastering the material in this chapter, you should be able to:

- Describe the anatomy of the adrenal gland and including which histologically distinct zone synthesizes and secretes glucocorticoids, aldosterone, and adrenal androgens.
- Outline the steps through which cells in the adrenal cortex obtain cholesterol to synthesize steroids, identifying that which is common to steroid synthesis in all three zones of the adrenal cortex.
- Explain the role of the liver in catabolism of adrenal steroids.
- Outline the intracellular signaling steps through which adrenocorticotropic hormone increases glucocorticoid and androgen synthesis in adrenal cortical cells.
- Compare and contrast the mechanism(s) through which angiotensin II and angiotensin III stimulate aldosterone synthesis in the cells of the zona glomerulosa.
- Outline the steps through which glucocorticoids alter gene transcription in target cells and describe the role of glucocorticoids in the adaptation of the body to fasting, injury, and stress.
- Outline the synthesis of epinephrine and norepinephrine and explain how the cellular effects of catecholamines are achieved.
- Compare and contrast how stimuli such as injury, anger, pain, cold, strenuous exercise, and hypoglycemia promote catecholamine release.
- List the effects of catecholamines on liver, muscle, and adipose tissue that regulate blood sugar levels.

To remain alive, the organs and tissues of the human body must have a finely regulated extracellular environment. This environment must contain the correct concentrations of ions to maintain body fluid volume and enable excitable cells to function, as well as an adequate supply of metabolic substrates for cells to generate adenosine triphosphate (ATP). The continual loss of salts, water, and other organic substances from the body through perspiration, respiration, and excretion are normally replenished by the intake of food and liquids. However, a person can survive for weeks on little else but water because the adrenal glands play a key role in adjusting the functions of organs and tissues to preserve body fluid volume and composition. This is readily apparent in an adrenalectomized animal, which cannot survive prolonged fasting. Its blood glucose supply diminishes, ATP generation by the cells becomes inadequate to support life, and the animal eventually dies. Even if fed a normal diet, an adrenalectomized animal typically loses body sodium and water over time and eventually dies of circulatory collapse. Its death is caused by a lack of certain steroid hormones produced by the cortex of the adrenal gland.

▶ ADRENAL CORTEX SYNTHESIZES AND SECRETES STEROID HORMONES

The human adrenal glands are paired, pyramid-shaped organs located on the upper poles of each kidney. The adrenal gland is actually a composite of two separate endocrine organs, one inside the other, each secreting separate hormones and each regulated by different mechanisms. The outer portion or **cortex** of the adrenal gland completely surrounds the inner portion or **medulla** and makes up most of the gland. During embryonic development, the cortex forms from mesoderm; the medulla arises from neural ectoderm.

Adrenal cortex comprises three zones that produce and secrete distinct hormones.

In the adult human, the adrenal cortex consists of three histologically distinct zones or layers (Fig. 33.1). The outer zone, which lies immediately under the capsule of the gland, is called the **zona glomerulosa** and consists of small clumps of cells that produce the mineralocorticoid aldosterone. The **zona fasciculata** is the middle and thickest layer of the cortex and consists of cords of cells oriented radially to the center of the gland. The inner layer, called the **zona reticularis**, is composed of interlaced strands of cells. The zona fasciculata and zona reticularis both produce the physiologically important glucocorticoids cortisol and corticosterone. These layers of the cortex also produce the androgen **dehydroepiandrosterone** (**DHEA**), which is related chemically to the male sex hormone **testosterone**.

Like all endocrine organs, the adrenal cortex is highly vascularized. Many small arteries branch from the aorta and renal arteries and enter the cortex. These vessels give rise to capillaries that course radially through the cortex and terminate in venous sinuses in the zona reticularis and adrenal medulla; therefore, the hormones produced by the cells of the cortex have ready access to the circulation.

Only small amounts of glucocorticoid, aldosterone, and adrenal androgens are found in adrenal cortical cells at a given time because the cells produce and secrete these hormones on demand, rather than storing them. Humans secrete about 10 times more cortisol than corticosterone during an average day (20 vs. 2 mg); thus, cortisol is considered the physiologically important glucocorticoid in humans. Compared with the glucocorticoids, a much smaller amount of aldosterone (0.1 mg) is secreted each day (Clinical Focus 33.1).

Glucocorticoid Excess: Cushing Disease

Cushing syndrome is the name given to the signs and symptoms associated with prolonged exposure to inappropriately elevated glucocorticoids. The use of glucocorticoid in the definition covers both endogenous elevations in cortisol and increased glucocorticoid resulting from endogenous sources (e.g., prednisolone or dexamethasone treatment). Endogenous Cushing syndrome may be adrenocorticotropic hormone (ACTH) dependent or ACTH independent. The term *Cushing disease* is reserved for ACTH-dependent conditions, usually caused by a corticotroph adenoma that secretes excessive ACTH and stimulates the adrenal cortex to produce large amounts of cortisol. ACTH-independent Cushing syndrome is generally the result of an adrenocortical adenoma that secretes large amounts of cortisol.

Weight gain and obesity are the most common clinical signs of Cushing syndrome. In adults, weight gain occurs centrally with excess deposition in the visceral adipose depot (deep fat surrounding the organs), and this can be used to initially distinguish Cushing syndrome from generalized obesity. Patients also develop fat depots over the thoracocervical spine (buffalo hump) and over the cheeks, resulting in a moonlike face. Bright red–purple **striae** >1 cm wide are frequently found on the abdomen. In children, glucocorticoid excess often results in a phenotype lacking a central fat deposition but resulting in an overall generalized obesity. Thus, it is necessary to rule out Cushing syndrome in obese children before beginning treatment for general obesity.

Prolonged exposure of the body to large amounts of glucocorticoids causes the breakdown of skeletal muscle protein, increased glucose production by the liver, and mobilization of lipid from the fat depots. The increased mobilization of lipid provides abundant fatty acids for metabolism, and the increased oxidation of fatty acids by tissues reduces their ability to use glucose. The underuse of glucose by skeletal muscle, coupled with increased glucose production by the liver, results in hyperglycemia, which, in turn, stimulates the pancreas to secrete insulin. However, the rise in insulin is not effective in reducing the blood glucose concentration because glucose uptake and use are decreased in the skeletal muscle and adipose tissue. The net result is that the person becomes insensitive or resistant to the action of insulin and little glucose is removed from the blood, despite the high level of circulating insulin. The persisting hyperglycemia continually stimulates the pancreas to secrete insulin, resulting in a form of "diabetes" similar to type 2 diabetes mellitus (see Chapter 34).

The dexamethasone suppression test is useful for the diagnosis of Cushing syndrome of any cause. In normal subjects, the administration of supraphysiologic doses of glucocorticoid results in the suppression of ACTH and cortisol secretion. There is no suppression of ACTH and cortisol in patients with Cushing syndrome.

Treatment of Cushing syndrome depends on the cause. The removal of adrenal adenomas has a 100% cure rate. Transsphenoidal surgery for pituitary adenoma is also reasonably successful, depending on the type (micro or macro) of adenoma. Features of Cushing syndrome disappear after 2 to 12 months. ■

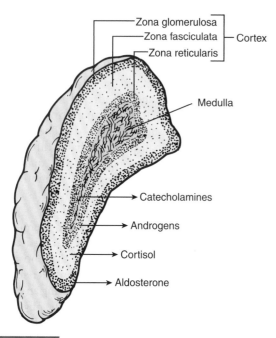

Figure 33.1 **The three zones of the adrenal cortex and corresponding hormones secreted.** Approximately 80% to 90% of the gland is the cortex, with ~10% to 20% comprising the medulla.

Glucocorticoids and aldosterone have overlapping actions due to similarities in their structure. For example, cortisol and corticosterone have some mineralocorticoid activity; conversely, aldosterone has some glucocorticoid activity. However, given the amounts of these hormones secreted under normal circumstances and their relative activities, glucocorticoids are not physiologically important mineralocorticoids, nor does aldosterone function physiologically as a glucocorticoid.

The adrenal cortex also synthesizes and secretes substantial amounts of androgenic steroids. DHEA, in both the free and sulfated form (DHEAS), is the main androgen secreted by the adrenal cortex of both men and women, although lesser amounts of other androgens are also produced. The adrenal cortex is the main source of androgens in women. In men, both the testes and adrenal cortex contribute to androgens in the blood. In both girls and boys, adrenal androgens normally have little physiologic effect other than a role in development before the start of puberty, because the male sex hormone activity of the adrenal androgens is weak. Exceptions occur in people who produce inappropriately large amounts of certain adrenal androgens as a result of diseases affecting the pathways of steroid biosynthesis in the adrenal cortex.

Adrenal steroid hormones are synthesized from cholesterol.

Cholesterol, the starting material for the synthesis of steroid hormones, consists of four interconnected rings of carbon atoms (A through D) and a side chain of eight carbon atoms extending from one ring. In all, there are 27 carbon atoms in cholesterol, numbered as shown at the top of Figure 33.3.

The immediate source of cholesterol used in steroid hormone biosynthesis is the abundant lipid droplets within the adrenal cortical cells, which contain cholesterol esters (single molecules of cholesterol esterified to single fatty acid molecules). **Cholesterol esterase** (**cholesterol ester hydrolase [CEH]**) hydrolyzes the ester bond, generating free cholesterol for steroid biosynthesis within mitochondria located near the lipid droplet. The process of steroid hormone synthesis from cholesterol is initiated when cholesterol enters the mitochondria.

The cholesterol that has been removed from the lipid droplets for steroid hormone biosynthesis is replenished in two ways; uptake from the blood and *de novo* synthesis from acetate (Fig. 33.2). Most of the cholesterol converted to steroid hormones by the human adrenal gland comes from **low-density lipoprotein (LDL)** particles circulating in the blood. LDL particles consist of a core of cholesterol esters surrounded by a coat of cholesterol and phospholipids. **Apolipoprotein B-100 (ApoB-100)**, a 400-kDa protein present on the surface of the LDL particle, is recognized by LDL receptors localized to coated pits on the plasma membrane of adrenal cortical cells. The LDL receptors bind the ApoB-100 and the cell takes up the LDL particle through endocytosis. The endocytic vesicle containing the LDL particles fuses with a lysosome, and the particle is degraded. CEH then hydrolyzes the cholesterol esters in the core of the particle to free cholesterol for steroid synthesis and fatty acid. The enzyme **acyl-CoA: cholesterol acyltransferase (ACAT)** converts any cholesterol not immediately used by the cell to cholesterol ester. The esters are then stored in the lipid droplets of the cell to be used later. When steroid biosynthesis is proceeding at a high rate, cholesterol delivered to the adrenal cell may be used directly for steroid production rather than being esterified and stored. There is evidence that **high-density lipoprotein (HDL)** cholesterol may also be used as a substrate for adrenal steroidogenesis.

In humans, cholesterol that has been synthesized *de novo* from acetate by the adrenal glands is a significant but minor source of cholesterol. The enzyme **3-hydroxy-3-methylglutaryl-CoA reductase (HMG-CoA reductase)** catalyzes the rate-limiting step in this process. The newly synthesized cholesterol is then incorporated into cellular structures such as membranes or converted to cholesterol esters through the action of ACAT and stored in lipid droplets (see Fig. 33.2).

Steroid biosynthesis is catalyzed by P450 enzymes in the mitochondria and endoplasmic reticulum.

Cytochrome P450 enzymes (CYPs) are a large family of oxidative enzymes with a 450-nm absorbance maximum when complexed with carbon monoxide. The adrenal CYPs are often referred to by their trivial names, which denote their function in steroid biosynthesis (Table 33.1).

The conversion of cholesterol into steroid hormones begins with the formation of free cholesterol from the cholesterol esters stored in intracellular lipid droplets. The **steroidogenic acute regulatory protein (StAR)** then facilitates the flow of free cholesterol into mitochondria. Once inside a mitochondrion, single cholesterol molecules bind to the **cholesterol side chain cleavage enzyme (CYP11A1)** embedded in the inner mitochondrial membrane. This enzyme catalyzes the first and rate-limiting reaction in steroidogenesis remodeling the cholesterol molecule into the 21-carbon steroid intermediate **pregnenolone** and releasing isocaproic acid.

Once formed, pregnenolone molecules dissociate from CYP11A1, leave the mitochondrion, and enter the smooth endoplasmic reticulum (ER). The further remodeling of pregnenolone into steroid hormones, summarized in Figure 33.3, depends on whether the process occurs in the zona fasciculata and zona reticularis or the zona glomerulosa.

In cells of the zona fasciculata and zona reticularis, most of the pregnenolone is converted to cortisol and the main adrenal androgen, DHEA. Pregnenolone molecules bind to the enzyme **17α-hydroxylase (CYP17)**, embedded in the ER membrane, which hydroxylates pregnenolone at carbon 17. The product formed by this reaction is **17α-hydroxypregnenolone** (see Fig. 33.3).

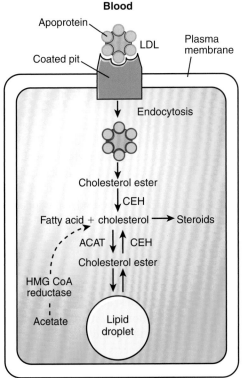

Blood

Apoprotein
LDL
Coated pit
Plasma membrane

Endocytosis

Cholesterol ester
↓ CEH
Fatty acid + cholesterol → Steroids
ACAT ↓↑ CEH
Cholesterol ester

HMG CoA reductase

Acetate

Lipid droplet

Adrenal cortical cell

Figure 33.2 Sources of cholesterol for steroid biosynthesis by the adrenal cortex. Most cholesterol comes from low-density lipoprotein (LDL) particles in the blood, which bind to receptors in the plasma membrane and are taken up by endocytosis. The cholesterol in the LDL particle is used directly for steroidogenesis or stored in lipid droplets for later use. Some cholesterol is synthesized directly from acetate. ACAT, acyl-CoA: cholesterol acyltransferase; CEH, cholesterol ester hydrolase; CoA, coenzyme A; HMG, 3-hydroxy-3-methylglutaryl.

TABLE 33.1 Nomenclature for the Steroidogenic Enzymes			
Common Name	**Previous Form**	**Current Form**	**Gene**
Cholesterol side chain cleavage enzyme	$P450_{scc}$	CYP11A1	$CYP_{11}A_1$
3β-Hydroxysteroid dehydrogenase	3β-HSD	3β-HSD II	HSD_3B_2
17α-Hydroxylase	$P450_{C17}$	CYP17	CYP_{17}
21-Hydroxylase	$P450_{C21}$	CYP21A2	$CYP_{21}A_2$
11β-Hydroxylase	$P450_{C11}$	CYP11B1	$CYP_{11}B_1$
Aldosterone synthase	$P450_{C11AS}$	CYP11B2	$CYP_{11}B_2$

The CYP17 has an additional enzymatic action that becomes important at this step in the steroidogenic process. Once the enzyme has hydroxylated carbon 17 of pregnenolone to form 17α-hydroxypregnenolone, it can lyse or cleave the carbon 20–21 side chain from the steroid structure. Some molecules of 17α-hydroxypregnenolone undergo this reaction and are converted to the 19-carbon steroid DHEA. This action of CYP17 is essential for the formation of androgens (19-carbon steroids) and estrogens (18-carbon steroids), which lack the carbon 20–21 side chain. The lyase activity of CYP17 is important in the gonads, where androgens and estrogens are primarily made. CYP17 does not exert significant lyase activity in children before the age of 7 or 8 years. As a result, young boys and girls do not secrete significant amounts of adrenal androgens. The appearance of significant adrenal androgen secretion in children of both sexes is termed **adrenarche**, which normally occurs before the activation of the hypothalamic–pituitary–gonadal axis, which initiates puberty. The adrenal androgens produced as a result of adrenarche stimulate the growth of pubic and axillary hair.

Those molecules of 17α-hydroxypregnenolone that dissociate from CYP17 bind next to another ER enzyme, **3β-hydroxysteroid dehydrogenase (3β-HSD II)**. This enzyme acts on 17α-hydroxypregnenolone to isomerize the double bond in ring B to ring A and to dehydrogenate the 3β-hydroxy group, forming a 3-keto group. The product formed is **17α-hydroxyprogesterone** (see Fig. 33.3). This intermediate then binds to another enzyme, **21-hydroxylase (CYP21A2)**, which hydroxylates it at carbon 21. The mechanism of this hydroxylation is similar to that performed by the CYP17. The product formed is **11-deoxycortisol**, which is the immediate precursor for cortisol.

To be converted to cortisol, 11-deoxycortisol molecules must be transferred back into the mitochondrion to be acted on by **11β-hydroxylase (CYP11B1)** embedded in the inner mitochondrial membrane. This enzyme hydroxylates 11-deoxycortisol on carbon 11, converting it into cortisol. The 11β-hydroxyl group is the molecular feature that confers glucocorticoid activity on the steroid. Cortisol is then secreted into the bloodstream.

Some of the pregnenolone molecules generated in cells of the zona fasciculata and zona reticularis first bind to 3β-HSD II when they enter the ER. As a result, they are converted to **progesterone**. CYP21A2 hydroxylates some of these progesterone molecules to form the mineralocorticoid

11-deoxycorticosterone (DOC) (see Fig. 33.3). The DOC formed may be either secreted or transferred back into the mitochondrion, where CYP11B1 acts on it to form corticosterone, which is then secreted into the circulation. Progesterone may also undergo 17α-hydroxylation in the zona fasciculata and zona reticularis. It is then converted to either cortisol or the adrenal androgen **androstenedione**.

CYP17 is not present in cells of the zona glomerulosa; therefore, pregnenolone does not undergo 17α-hydroxylation in these cells, and these cells do not form cortisol and adrenal androgens. Instead, the enzymatic pathway for aldosterone synthesis is followed (see Fig. 33.3). Enzymes in the ER convert pregnenolone to progesterone and DOC. The latter compound then moves into the mitochondrion, where it is converted to aldosterone. This conversion involves three steps: the hydroxylation of carbon 11 to form corticosterone, the hydroxylation of carbon 18 to form 18-hydroxycorticosterone, and the oxidation of the 18-hydroxymethyl group to form aldosterone. In humans, these three reactions are catalyzed by a single enzyme, **aldosterone synthase (CYP11B2)**, an isozyme of CYP11B1, expressed only in glomerulosa cells. The CYP11B1 enzyme, which is expressed in the zona fasciculata and zona reticularis, although closely related to CYP11B2, cannot catalyze all three reactions involved in the conversion of DOC to aldosterone; therefore, aldosterone is not synthesized in the zona fasciculata and zona reticularis of the adrenal cortex.

Genetic defects in CYP enzymes alter steroidogenesis.

Inherited genetic defects can cause relative or absolute deficiencies in the enzymes involved in steroid hormone biosynthetic pathways. The consequences of such defects are changes in the type and amount of steroid hormones secreted by the adrenal cortex, resulting in disease.

Most of the genetic defects affecting the steroidogenic enzymes impair the formation of cortisol. A drop in cortisol in the blood stimulates the secretion of **adrenocorticotropic hormone (ACTH)** by the anterior pituitary (see Chapter 31). The consequent rise in ACTH in the blood exerts a trophic (growth-promoting) effect on the adrenal cortex, resulting in adrenal hypertrophy. People with genetic defects affecting adrenal steroidogenesis usually have hypertrophied adrenal glands. These diseases are collectively called **congenital adrenal hyperplasia**.

In humans, inherited genetic defects occur that affect CYP11A1, CYP17, 3β-HSD II, CYP21A2, CYP11B1, and CYP11B2. The most common defect involves mutations in the

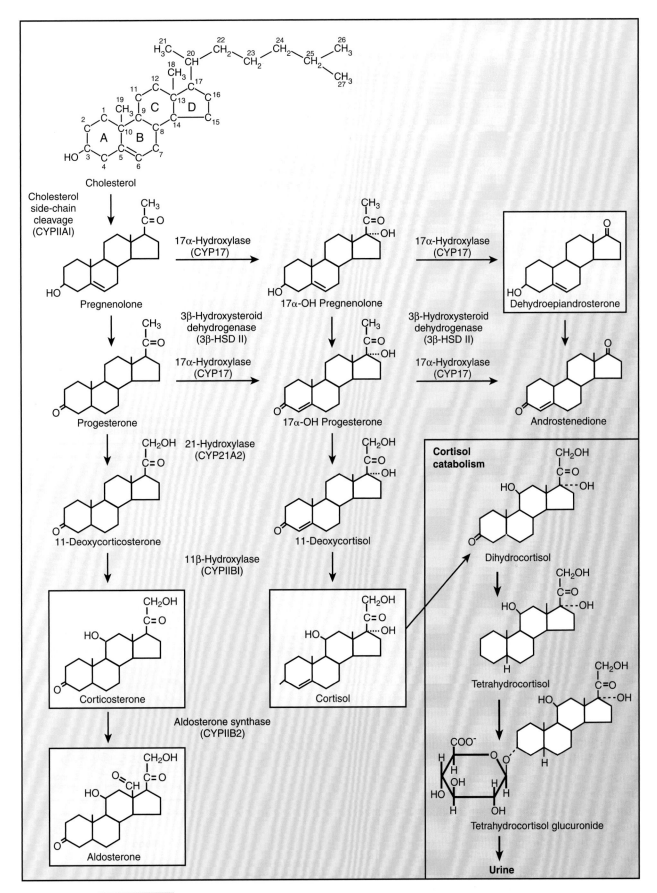

Figure 33.3 **The synthesis of steroids in the adrenal cortex and catabolism of cortisol in the liver.**
The enzymes catalyzing each step are shown, and the major steroidogenic products are boxed.
Note the chemical structure of cholesterol in the upper left, how the four rings are lettered
(*A–D*), and how the carbons are numbered. The hydrogen atoms on the carbons composing the
rings are omitted from the figure. *Inset box lower right*: Cortisol is reduced and conjugated to
glucuronide in the liver and then is excreted in the urine.

gene for CYP21A2 and occurs in 1 of 7,000 people. The gene for CYP21A2 may be deleted entirely, or mutant genes may code for forms of CYP21A2 with impaired enzyme activity. The consequent reduction in the amount of active CYP21A2 blocks the 21-hydroxylation reaction and interferes with the formation of cortisol, corticosterone, and aldosterone. The reduction of cortisol (and corticosterone) secretion in affected individuals promotes ACTH release, which in turn, causes adrenal gland hypertrophy and inappropriate steroidogenesis. With the 21-hydroxylation reaction impaired, pregnenolone is converted to adrenal androgens in inappropriately high amounts. Thus, women afflicted with CYP21A2 deficiency exhibit **virilization** from the masculinizing effects of excessive adrenal androgen secretion. In severe cases, the deficiency in aldosterone production can lead to sodium depletion, dehydration, vascular collapse, and death, if appropriate hormone therapy is not given.

Adrenal steroids are bound to plasma proteins.

Steroid hormones are not stored to any extent by cells of the adrenal cortex but are continually synthesized and secreted. The rate of secretion may change dramatically depending on the stimuli received by the adrenal cortical cells. The accumulation of the final products of the steroidogenic pathways creates a concentration gradient for steroid hormone diffusion through the plasma membranes and into the circulation.

A large fraction of the adrenal steroids entering the bloodstream become noncovalently bound to plasma proteins. **Corticosteroid-binding globulin (CBG)**, a glycoprotein produced by the liver, binds both glucocorticoids and aldosterone, but has a greater affinity for the glucocorticoids. Serum albumin is a high capacity, weak-binding protein for steroids. The binding of a steroid hormone to a circulating protein molecule prevents it from being taken up by cells or being excreted in the urine.

Circulating steroid hormone molecules not bound to plasma proteins are free to interact with receptors within cells and are therefore cleared from the blood. As this occurs, bound hormone dissociates from its binding protein and replenishes the circulating pool of free hormone. Adrenal steroid hormones have long half-lives in the body, ranging from many minutes to hours, because of binding to plasma proteins.

Adrenal steroids are degraded and eliminated from the body in the urine.

Adrenal steroid hormones are structurally modified to destroy their hormone activity and increase their water solubility, primarily in the liver. The most common structural modifications made to adrenal steroids involve reduction of the double bond in ring A and conjugation of the resultant hydroxyl group formed on carbon 3 with glucuronic acid. Figure 33.3 (lower right) shows cortisol modified in this manner to produce a major excretable metabolite, **tetrahydrocortisol glucuronide**. Cortisol and other 21-carbon steroids with a 17α-hydroxyl group and a 20-keto group may undergo lysis of the carbon 20–21 side chain as well. The resultant metabolite, with a keto group on carbon 17, appears as one of the **17-ketosteroids** in the urine. Adrenal androgens are also 17-ketosteroids. They are usually conjugated with sulfuric acid

or glucuronic acid before being excreted and normally comprise the bulk of the 17-ketosteroids in the urine. Before the development of specific methods to measure androgens and 17α-hydroxycorticosteroids in body fluids, the amount of 17-ketosteroids in urine was used clinically as a crude indicator of the production of these substances by the adrenal gland.

Adrenocorticotropic hormone regulates the synthesis of adrenal steroids.

ACTH is the physiologic regulator of the synthesis and secretion of glucocorticoids and androgens by the zona fasciculata and zona reticularis. It has a rapid stimulatory effect on steroidogenesis in these cells, which can raise blood glucocorticoids within seconds. It also exerts several long-term trophic effects on these cells, all directed toward maintaining the cellular machinery necessary to carry out steroidogenesis at a high, sustained rate.

ACTH binds to the melanocortin-2 receptor on the adrenocortical cell, which is coupled by stimulatory guanine nucleotide-binding proteins (G_s proteins) to adenylyl cyclase (Figure 33.4). ACTH increases cAMP and activates protein

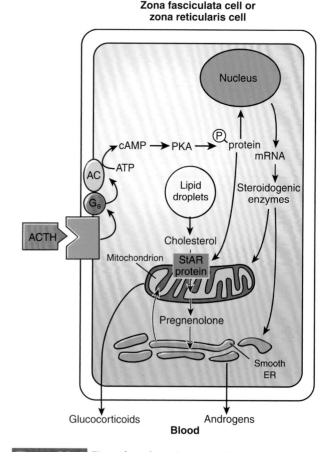

Figure 33.4 The main actions of adrenocorticotropic hormone (ACTH) on steroidogenesis. ACTH binds to the melanocortin-2 receptor, which is coupled to adenylyl cyclase (AC) by stimulatory G proteins (G_s). The intracellular rise in cyclic adenosine monophosphate (cAMP) activates protein kinase A (PKA), which then phosphorylates proteins (P-proteins) that stimulate the expression of genes for steroidogenic enzymes. PKA also phosphorylates steroidogenic acute regulatory (StAR) protein, which mediates the transfer of cholesterol into the mitochondria for steroidogenesis.

kinase A (PKA), which then phosphorylates proteins that regulate steroidogenesis. StAR protein that is already present in the cell is phosphorylated, and the rapid synthesis of new StAR protein is initiated. These actions provide abundant cholesterol for side chain cleavage enzyme, which carries out the rate-limiting step in steroidogenesis.

Steroidogenic enzyme expression

ACTH regulates transcription of genes for many of the steroidogenic enzymes in the zona fasciculata and zona reticularis. Transcription of the genes for side chain cleavage enzyme, CYP17, CYP21A2, and CYP11B1, is increased several hours after ACTH stimulation of adrenal cortical cells. Healthy people are continually exposed to episodes of ACTH secretion (see Fig. 31.5); thus, mRNA for the steroidogenic enzymes is maintained fairly constant.

The importance of ACTH-induced gene expression is evident in hypophysectomized animals or humans chronically treated with large doses of cortisol or related steroids, which suppress ACTH secretion by the anterior pituitary. The chronic lack of ACTH decreases steroidogenic gene transcription, causing a deficiency in steroidogenic enzymes in the adrenals. As a result, the administration of ACTH to such a person does not cause a marked increase in glucocorticoid secretion. Chronic exposure to ACTH is required to restore mRNA levels for the steroidogenic enzymes and, hence, the enzymes themselves to obtain normal steroidogenic responses to ACTH. A patient receiving long-term treatment with glucocorticoid may suffer serious glucocorticoid deficiency if hormone therapy is halted abruptly; withdrawing glucocorticoid therapy gradually allows time for endogenous ACTH to restore steroidogenic enzyme levels to normal.

Cholesterol availability is increased by ACTH.

There are several long-term effects of ACTH on cholesterol metabolism that support steroidogenesis. ACTH increases the abundance of LDL receptors and the activity of HMG-CoA reductase in cells of the zona fasciculata and zona reticularis, increasing the availability of cholesterol for steroidogenesis. However, it is not clear if ACTH exerts these effects directly. The abundance of LDL receptors in the plasma membrane and the activity of HMG-CoA reductase in most cells are inversely related to the amount of cellular cholesterol. By stimulating steroidogenesis, ACTH reduces the amount of cholesterol in adrenal cells; therefore, the increased abundance of LDL receptors and high HMG-CoA reductase activity in ACTH-stimulated cells may merely result from the normal compensatory mechanisms that function to maintain cell cholesterol levels.

ACTH also stimulates cholesterol esterase activity, promoting hydrolysis of the cholesterol esters stored in lipid droplets within adrenal cells, making free cholesterol available for steroidogenesis. The cholesterol esterase in the adrenal cortex is activated when phosphorylated by PKA.

Maintenance of adrenocortical cell size

Cells in the zona fasciculata and zona reticularis, but not the zona glomerulosa, are dependent on ACTH to maintain their size and structure. The trophic effect of ACTH is clearly evident in states of ACTH deficiency or excess. In hypophysectomized or ACTH-deficient people, the cells of the two inner zones atrophy. Chronic stimulation of these cells with ACTH causes them to hypertrophy.

Aldosterone regulates sodium and potassium to maintain fluid homeostasis.

Aldosterone is synthesized in the cells of the zona glomerulosa. ACTH binds to melanocortin-2 receptors on cells in this region of the adrenal cortex, increasing cAMP to stimulate aldosterone secretion to a small extent. However, angiotensin II is a more important physiologic regulator of aldosterone secretion than ACTH. An increase in serum potassium can also stimulate aldosterone secretion as discussed below.

The principal site of aldosterone action on the kidney is in the distal and collecting tubules, where the hormone promotes Na^+ retention and enhances K^+ elimination during urine formation. A secondary effect of Na^+ retention is the osmotic retention of water in the extracellular fluid compartment. The osmotic effect of aldosterone is important in regulating blood pressure (Clinical Focus 33.2).

Angiotensin II formation

Angiotensin II is a short peptide consisting of eight amino acid residues. It is formed in the bloodstream by the proteolysis of **angiotensinogen**, which is secreted by the liver. The formation of angiotensin II occurs in two stages (Fig. 33.5). First, the circulating protease **renin**, produced by juxtaglomerular cells in the kidney, cleaves angiotensinogen at its amino-terminal end, releasing the inactive decapeptide **angiotensin I**. The protease **angiotensin-converting enzyme**, present on the endothelial cells lining the vasculature, then removes a dipeptide from the carboxy-terminal end of angiotensin I, producing angiotensin II. This step usually occurs as angiotensin I molecules traverse the pulmonary circulation. The rate-limiting factor for the formation of angiotensin II is the renin concentration of the blood.

The N-terminal aspartate of angiotensin II is cleaved to form angiotensin III, which circulates at a concentration equivalent to 20% that of angiotensin II. Angiotensin III is as potent a stimulator of aldosterone secretion as angiotensin II.

Angiotensin II action

Angiotensin II stimulates movement of cholesterol into the inner mitochondrial membrane and its conversion to pregnenolone to increase aldosterone synthesis. The primary mechanism is shown in Figure 33.6.

Aldosterone synthesis is initiated when angiotensin II binds to its receptors on the plasma membranes of zona glomerulosa cells, which are linked by a G protein to phospholipase C (PLC). Activated PLC hydrolyzes phosphatidylinositol 4,5-bisphosphate (PIP_2) in the plasma membrane, producing the intracellular second messengers inositol trisphosphate (IP_3) and diacylglycerol (DAG). The IP_3 stimulates calcium release from intracellular structures, increasing the cytosolic calcium concentration. The increase in intracellular calcium and DAG activates protein kinase C (PKC). The rise

CLINICAL FOCUS | 33.2

Primary Adrenal Insufficiency: Addison Disease

Adrenal insufficiency may be caused by the destruction of the adrenal cortex (primary adrenal insufficiency), low pituitary adrenocorticotropic hormone secretion (secondary adrenal insufficiency), or deficient hypothalamic release of corticotropin-releasing hormone (tertiary adrenal insufficiency). Addison disease (primary adrenal insufficiency) results from the destruction of the adrenal gland by microorganisms or autoimmune disease. When Addison first described primary adrenal insufficiency in the mid-1800s, bilateral adrenal destruction by tuberculosis was the most common cause of the disease. Today, autoimmune destruction accounts for 70% to 90% of all cases in developed countries, with the remainder resulting from infection, cancer, or adrenal hemorrhage. The prevalence of primary adrenal insufficiency in the developed world is about 40 to 110 cases per 1 million adults.

In primary adrenal insufficiency, all three zones of the adrenal cortex are usually involved. The result is inadequate secretion of glucocorticoids, mineralocorticoids, and androgens. Major symptoms are not usually detected until 90% of the gland has been destroyed. The initial symptoms generally have a gradual onset, with only a partial glucocorticoid deficiency resulting in inadequate cortisol increase in response to stress. Mineralocorticoid deficiency may only appear as a mild postural hypotension. Progression to complete glucocorticoid deficiency results in a decreased sense of well-being and abnormal glucose metabolism. Lack of mineralocorticoid leads to decreased renal potassium secretion and reduced sodium retention, the loss of which results in hypotension and dehydration. The combined lack of glu-

cocorticoid and mineralocorticoid can lead to vascular collapse, shock, and death. Adrenal androgen deficiency is observed in women only (men derive most of their androgen from the testes) as decreased pubic and axillary hair and decreased libido.

Antibodies that react with all three zones of the adrenal cortex have been identified in autoimmune adrenalitis and are more common in women than in men. The presence of antibodies appears to precede the development of adrenal insufficiency by several years. Antiadrenal antibodies are mainly directed to the steroidogenic enzymes: cholesterol side chain cleavage enzyme (CYP11A1), 17α-hydroxylase (CYP17), and 21-hydroxylase (CYP21A2), although antibodies to other steroidogenic enzymes may also be present. In the initial stages of the disease, the adrenal glands may be enlarged with extensive lymphocyte infiltration. Genetic susceptibility to autoimmune adrenal insufficiency is strongly linked with the HLA-DR3 and HLA-DR4 alleles of human leukocyte antigen.

The onset of adrenal insufficiency is often insidious, and diagnosis is not made until an acute crisis occurs during another illness. Crisis is a medical emergency consisting of hypotension and acute circulatory failure. Treatment for acute adrenal insufficiency should be directed at the reversal of the hypotension and electrolyte abnormalities. Large volumes of 0.9% saline or 5% dextrose in saline should be infused as quickly as possible. Dexamethasone or a soluble form of injectable cortisol should also be given. Daily glucocorticoid and mineralocorticoid replacement allows the patient to lead a normal, active life. ■

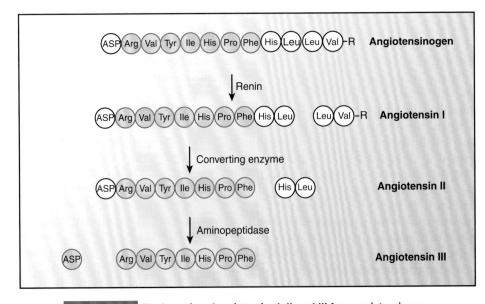

Figure 33.5 The formation of angiotensins I, II, and III from angiotensinogen.

Zona glomerulosa cell

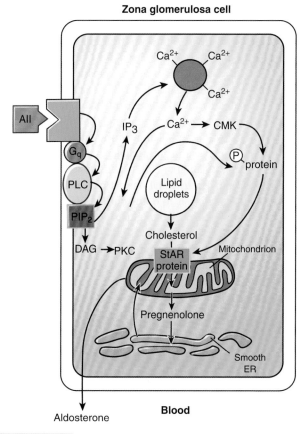

Figure 33.6 **The action of angiotensin II on aldosterone synthesis.** Angiotensin II binds its receptor on the plasma membrane of zona glomerulosa cells, activating phospholipase C (PLC), which is coupled to the angiotensin II (AII) receptor by G proteins (G_q). PLC hydrolyzes phosphatidylinositol 4,5-bisphosphate (PIP_2) in the plasma membrane, producing inositol trisphosphate (IP_3) and diacylglycerol (DAG). IP_3 mobilizes intracellularly bound Ca^{2+}. The rise in Ca^{2+} and DAG activates protein kinase C (PKC) and calmodulin-dependent protein kinase (CMK). These enzymes phosphorylate proteins (P-proteins) that increase the transcription of steroidogenic enzymes and mediate the transfer of cholesterol into the mitochondria (StAR protein). StAR, steroidogenic acute regulatory protein; ER, endoplasmic reticulum.

in intracellular calcium also activates calmodulin-dependent protein kinase (CMK). These enzymes phosphorylate proteins, which initiate steroidogenesis.

Angiotensin II is the final mediator in the physiologic regulation of aldosterone secretion; however, its formation from angiotensinogen depends on the secretion of renin by the kidneys. Thus, the rate of renin secretion ultimately determines the rate of aldosterone secretion. The granular cells in the walls of the afferent arterioles of renal glomeruli secrete renin. These cells are stimulated to secrete renin by three signals that indicate a possible loss of body fluid: a fall in blood pressure in the afferent arterioles of the glomeruli, a drop in sodium chloride concentration in renal tubular fluid at the macula densa, and an increase in renal sympathetic nerve activity (see Chapters 22 and 23). Increased renin secretion results in an increase in angiotensin II formation in the blood, stimulating aldosterone secretion, which

conserves body fluid volume because aldosterone stimulates sodium reabsorption by the kidneys.

Aldosterone synthesis

Cells in the zona glomerulosa are sensitive to extracellular potassium concentrations and increase their rate of aldosterone secretion in response to small increases in blood and interstitial fluid potassium concentration. This signal for aldosterone secretion is appropriate from a physiologic point of view because aldosterone promotes the renal excretion of potassium (see Chapter 23).

A rise in extracellular potassium depolarizes glomerulosa cell membranes, activating voltage-dependent calcium channels in the membranes. The consequent rise in cytosolic calcium is thought to stimulate aldosterone synthesis by the same mechanism described above for angiotensin II.

Glucocorticoids regulate many cellular processes through transcription.

Most cells have receptors for glucocorticoids and are potential targets for their actions; thus, glucocorticoids have been used extensively as therapeutic agents, and much is known about their pharmacologic effects.

Unlike many other hormones, glucocorticoids influence physiologic processes slowly, sometimes taking hours to produce their effects. Glucocorticoids that are free in the blood diffuse through the plasma membranes of target cells to bind tightly but noncovalently to receptor proteins present in the cytoplasm, producing an activated glucocorticoid–receptor complex, which translocates into the nucleus. These complexes then bind to specific regions of DNA called **glucocorticoid response elements** (**GREs**) to either stimulate or inhibit transcription of target genes, changing the abundance of proteins in the cell. The apparent slowness of glucocorticoid action is a result of the time required to change the protein composition of a target cell.

Metabolic adaptation to fasting

During the fasting periods between food consumption, metabolic adaptations occur to prevent hypoglycemia and maintain blood glucose for utilization by the brain. Many of the adaptations that prevent hypoglycemia are not fully expressed in the course of daily life because the person eats before they fully develop. Complete expression of these changes is seen only after many days to weeks of fasting.

At the onset of a prolonged fast, there is a gradual decline in the concentration of glucose in the blood. Within 1 to 2 days, the blood glucose level stabilizes at a concentration of 60 to 70 mg/dL, where it remains even if the fast is prolonged for many days (Fig. 33.7). The production of glucose by the body and the restriction of its use by tissues other than the brain stabilize the blood glucose level. A limited supply of glucose is available from glycogen stored in the liver; the more important source of blood glucose during the first days of a fast is gluconeogenesis in the liver and, to some extent, in the kidneys.

Gluconeogenesis begins several hours after the start of a fast using amino acids derived from tissue protein are the

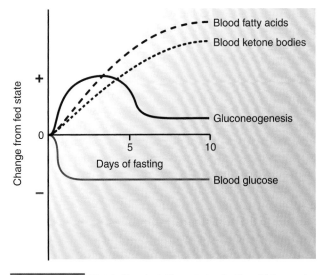

Figure 33.7 **Metabolic adaptations during fasting.** This graph shows the changes in the concentrations of blood glucose, fatty acids, and ketone bodies, and the rate of gluconeogenesis during the course of a prolonged fast. Only the *direction* of change over time, not the magnitude of change, is indicated: increase (+) or decrease (−).

main substrate. Fasting results in protein breakdown in the skeletal muscle and accelerated release of amino acids into the bloodstream. Insulin and glucocorticoids regulate protein breakdown and protein accretion in adult humans in an opposing fashion. During fasting, insulin secretion is suppressed, and the inhibitory effect of insulin on protein breakdown is lost. As proteins are broken down, glucocorticoids inhibit the reuse of amino acids derived from tissue proteins for new protein synthesis, promoting the release of these amino acids from the muscle. The liver and kidneys extract the amino acids from the blood at an accelerated rate and metabolize them to synthesize glucose, which is then delivered to the bloodstream.

The glucocorticoids are essential for the acceleration of gluconeogenesis during fasting, maintaining gene expression and, therefore, the intracellular concentrations of many gluconeogenic enzymes in the liver and the kidneys. Glucocorticoids maintain the amounts of transaminases, pyruvate carboxylase, phosphoenolpyruvate carboxykinase, fructose-1, 6-diphosphatase, fructose-6-phosphatase, and glucose-6-phosphatase needed to carry out gluconeogenesis at an accelerated rate. In an untreated, glucocorticoid-deficient person, the amounts of these enzymes in the liver are greatly reduced. As a consequence, the person cannot respond to fasting with accelerated gluconeogenesis and will die from hypoglycemia.

The other important metabolic adaptation that occurs during fasting is the mobilization and use of stored fat. Within the first few hours of the start of a fast, the concentration of free fatty acids rises in the blood (see Fig. 33.7) due to lipolysis in the adipose tissue. **Adipose triglyceride lipase (ATGL)** and **hormone-sensitive lipase (HSL)** hydrolyze the stored triglyceride to free fatty acids and glycerol, which are released into the blood. ATGL catalyzes the

removal of the first fatty acid from the triglyceride, and HSL acts on the resulting DAG to cleave the second fatty acid. Monoacylglycerol lipase removes the remaining fatty acid from the glycerol backbone.

ATGL transcription and protein expression are increased by glucocorticoids during fasting. HSL is activated when it is phosphorylated by cAMP-dependent PKA. As the level of insulin falls in the blood during fasting, the inhibitory effect of insulin on cAMP accumulation in the fat cell diminishes, resulting in an increase in intracellular cAMP, activation of PKA and phosphorylation of HSL to activate it. The glucocorticoids are essential for maintaining fat cells in an enzymatic state that permits lipolysis to occur during a fast. Accelerated lipolysis does not occur when a glucocorticoid-deficient person fasts.

Many tissues take up the abundant fatty acids produced by lipolysis. The fatty acids enter mitochondria, undergo β-oxidation to acetyl-CoA, and become a substrate for ATP synthesis. The enhanced use of fatty acids for energy metabolism spares the blood glucose supply. There is also significant hepatic gluconeogenesis from the glycerol released from triglycerides during lipolysis. In prolonged fasting, when the rate of glucose production from body protein has declined, a significant fraction of blood glucose is derived from triglyceride glycerol.

Within a few hours of the start of a fast, the increased delivery and oxidation of fatty acids in the liver results in the production of ketone bodies. A gradual rise in ketone bodies occurs in the blood as a fast continues over many days (see Fig. 33.7). Ketone bodies become the principal energy source used by the central nervous system (CNS) during the later stages of fasting.

The increased use of fatty acids for energy metabolism by skeletal muscle results in less use of glucose in this tissue. Acetyl-CoA and citrate, two products resulting from the breakdown of fatty acids, inhibit glycolysis. As a result, the uptake and use of glucose from the blood are reduced.

In summary, the metabolic adaptation to fasting provides the body with glucose produced primarily from protein until the ketone bodies become abundant enough in the blood to be a principal source of energy for the brain. From that point on, the body uses mainly fat for energy metabolism, and it can survive until the fat depots are exhausted. Glucocorticoids do not trigger the metabolic adaptations to fasting but only provide the metabolic machinery necessary for the adaptations to occur. However, when present in excessive amounts, glucocorticoids can trigger many of the metabolic adaptations to the fasting state, resulting in disease.

Glucocorticoid suppression of inflammation and immune system activity

Tissue injury triggers a complex mechanism called **inflammation**, which precedes the actual repair of damaged tissue. Neighboring cells, adjacent vasculature, and phagocytic cells that migrate to the damaged site release a host of chemical mediators into the damaged area including **prostaglandins, leukotrienes, kinins, histamine, serotonin,** and **lymphokines.** These substances exert a multitude of actions at the site of injury and directly or indirectly promote the local vasodilation, increased capillary permeability, and edema formation that characterize the inflammatory response (see Chapters 9 and 10).

Glucocorticoids inhibit the inflammatory response to injury; thus, they are extensively used therapeutically as anti-inflammatory agents. One mechanism through which glucocorticoids reduce inflammation is by inhibiting the production of prostaglandins and leukotrienes, which play a major role in mediating the inflammatory reaction. Prostaglandins and leukotrienes are synthesized from the unsaturated fatty acid arachidonic acid, which is released from plasma membrane phospholipids by the hydrolytic action of **phospholipase A$_2$**. Glucocorticoids stimulate the synthesis of a family of proteins called **lipocortins**, which inhibit the activity of phospholipase A$_2$, reducing the amount of arachidonic acid available for conversion to prostaglandins and leukotrienes.

Glucocorticoids have little influence on the human immune system under normal physiologic conditions. When administered in large doses over a prolonged period, however, they can suppress antibody formation and interfere with cell-mediated immunity. Glucocorticoid therapy is therefore used to suppress the rejection of surgically transplanted organs and tissues.

Exposure to high concentrations of glucocorticoids can kill immature T cells in the thymus and immature B cells and T cells in lymph nodes, decreasing the number of circulating lymphocytes. The destruction of immature T and B cells by glucocorticoids also causes some reduction in the size of the thymus and lymph nodes.

Glucocorticoids and stress

Both physical and psychological stresses stimulate the secretion of ACTH, which increases the secretion of glucocorticoids by the adrenal cortex. Physical stress including fever, surgery, burn injury, hypoglycemia, hypotension, and exercise raise cortisol levels, as does social/emotional stress caused by interpersonal and societal interactions and psychological trauma. The increase in glucocorticoid secretion during stress is a required counterregulatory response. It is well established that glucocorticoid-deficient people require an increase in their replacement therapy to maintain their well-being during periods of stress.

The brain plays a central role in response to both physical and psychological stress. Glucocorticoids along with excitatory neurotransmitters alter neuronal architecture by extending or retracting dendrites and increasing or decreasing synaptic density in different areas of the brain. Chronic stress can lead to epigenetic changes in the brain that can become maladaptive and require therapeutic intervention.

Negative feedback regulation of glucocorticoid release

An important physiologic action of glucocorticoids is the ability to regulate their own secretion. This effect is achieved by a negative feedback mechanism of glucocorticoids on the secretion of **corticotropin-releasing hormone** (**CRH**) and

ACTH and on proopiomelanocortin (POMC) gene expression (see Chapter 31).

▶ ADRENAL MEDULLA CATECHOLAMINES

The catecholamines epinephrine and norepinephrine (NE) are the two hormones synthesized by the chromaffin cells of the adrenal medulla. The human adrenal medulla produces and secretes about four times more epinephrine than NE. Postganglionic sympathetic neurons also produce and secrete NE but do not produce epinephrine.

Adrenal medulla is a modified sympathetic ganglion.

The medulla consists of clumps and strands of **chromaffin cells** interspersed with venous sinuses. Chromaffin cells produce epinephrine and NE, which are stored in granules in the chromaffin cells and discharged into then venous sinuses of the medulla when the adrenal branches of splanchnic nerves are stimulated. The catecholamines then diffuse into capillaries and are transported in the bloodstream. Stimuli such as injury, anger, anxiety, pain, cold, strenuous exercise, and hypoglycemia generate impulses in the cholinergic preganglionic fibers, causing a rapid discharge of the catecholamines into the bloodstream.

Catecholamines mediate the "fight-or-flight" response.

Most cells of the body have receptors for, and are responsive to, catecholamines. There are four structurally related forms of catecholamine receptors, all of which are transmembrane proteins: α_1, α_2, β_1, and β_2. All bind epinephrine or NE to varying extents (see Chapter 3).

Epinephrine and NE produce widespread effects on the cardiovascular system, muscular system, and carbohydrate and lipid metabolism in the liver, muscle, and adipose tissues. In response to a sudden rise of catecholamines in the blood, heart rate accelerates, coronary blood vessels dilate, and blood flow to the skeletal muscles is increased as a result of vasodilation, with vasoconstriction occurring in the skin. Smooth muscles in the airways of the lungs, the gastrointestinal tract, and the urinary bladder relax. Muscles in the hair follicles contract, causing piloerection. Blood glucose levels rise. This overall reaction to the sudden release of catecholamines is known as the "fight-or-flight" response (see Chapter 6).

Catecholamine release defends against hypoglycemia.

Catecholamines secreted by the adrenal medulla, and NE released from sympathetic postganglionic nerve terminals, defend the body against hypoglycemia. Catecholamine release generally starts when the blood glucose concentration falls to the low end of the physiologic range (60 to 70 mg/dL), with further declines producing marked catecholamine release. CNS receptors monitoring blood glucose are activated by hypoglycemia, stimulating the neural fibers innervating the chromaffin cells and sympathetic postganglionic nerve terminals.

Hypoglycemia can result from many situations such as insulin overdosing, catecholamine antagonists, or drugs that block fatty acid oxidation. Hypoglycemia is always a dangerous condition because the CNS will die of ATP deprivation in extended cases. The length of time profound hypoglycemia can be tolerated depends on its severity and the person's sensitivity to the hypoglycemia.

Catecholamines stimulate glucose production in the liver through activation of glycogen phosphorylase, resulting in the hydrolysis of stored glycogen, and stimulation of gluconeogenesis from lactate and amino acids. Catecholamines binding to β receptors on skeletal muscle and adipocytes also activate glycogen phosphorylase by stimulating adenylyl cyclase and increasing cAMP. The glucose 6-phosphate generated in these cells is metabolized intracellularly rather than released into the blood, because the cells lack glucose-6-phosphatase. Glycolysis converts the glucose 6-phosphate in muscle to lactate, much of which is released into the blood. The lactate taken up by the liver is then converted to glucose via gluconeogenesis and returned to the blood.

In adipocytes, the rise in cAMP produced by catecholamines activates HSL, increasing the hydrolysis of triglycerides and the release of fatty acids and glycerol into the bloodstream. These fatty acids provide an alternative substrate for energy metabolism in other tissues, primarily skeletal muscle, and block the phosphorylation and metabolism of glucose.

During profound hypoglycemia, the rapid rise in blood catecholamine levels triggers some of the same metabolic

TABLE 33.2	Catecholamine-Mediated Responses to Hypoglycemia
Target Tissue	**Action**
Liver	Stimulation of glycogenolysis
	Stimulation of gluconeogenesis
Skeletal muscle	Stimulation of glycogenolysis
Adipose tissue	Simulation of glycogenolysis
	Stimulation of triglyceride lipolysis
Pancreatic islets	Inhibition of insulin secretion by β cells
	Stimulation of glucagon secretion by α cells

adjustments that occur more slowly during fasting. During fasting, these adjustments are initiated primarily in response to the gradual rise in the ratio of glucagon to insulin in the blood. The glucagon-to-insulin ratio also rises during profound hypoglycemia, reinforcing the actions of the catecholamines on glycogenolysis, gluconeogenesis, and lipolysis. The catecholamines released during hypoglycemia are thought to be partly responsible for the rise in the glucagon-to-insulin ratio by directly stimulating the secretion of these hormones by the pancreas. Catecholamines stimulate the secretion of glucagon by the alpha cells and inhibit the secretion of insulin by β cells (see Chapter 34). These catecholamine-mediated responses of the major metabolic tissues to hypoglycemia are summarized in Table 33.2.

INTEGRATED MEDICAL SCIENCES

Obesity and 11β-Hydroxysteroid Dehydrogenase

Glucocorticoids regulate multiple metabolic processes, including glucose homeostasis, insulin sensitivity, and lipid metabolism. In addition, glucocorticoids are important in the differentiation of preadipocytes to adipocytes, a process termed adipogenesis. In patients with Cushing syndrome, elevated blood glucocorticoid levels cause excess fat deposition in the abdomen surrounding the organs (visceral obesity), insulin resistance, dyslipidemia, and hypertension. A similar collection of metabolic abnormalities is present in obese people, which has prompted investigation into the possibility that general obesity could be attributed to glucocorticoid excess. Although subtle alterations in the hypothalamic–pituitary–adrenal axis have been observed in obesity, no clear role for increased circulating glucocorticoid has been established.

Until recently, tissue glucocorticoid concentrations were thought to be determined by the level of glucocorticoid in the blood and tissue responses controlled by the availability of glucocorticoid receptors. It is now recognized that an enzyme, 11β-hydroxysteroid dehydrogenase (11βHSD), has a central role in regulating the concentration of glucocorticoid in specific tissues. There are two 11βHSD isozymes. The type 2 enzyme is an NAD-dependent dehydrogenase, converting active cortisol to inactive cortisone (see figure on the next page). Its main role is in aldosterone-sensitive target tissues such as the kidney, colon, salivary glands, and placenta. Cortisol circulates in the blood at concentrations severalfold higher than those of aldosterone, but both hormones have an affinity for the nonselective mineralocorticoid receptor. By inactivating cortisol, 11βHSD2 prevents the flooding of mineralocorticoid receptors by cortisol, leaving free access for aldosterone. The second isozyme, 11βHSD1, is an NADPH-dependent reductase that converts inactive cortisone to active cortisol. 11βHSD1 is expressed primarily in glucocorticoid target tissues, such as the liver, adipose tissue, and the CNS, where it amplifies local glucocorticoid action.

Several lines of evidence point to an important role for 11βHSD1 in the pathogenesis of obesity. The expression and activity of 11βHSD1 are increased in subcutaneous fat of obese humans and in the adipose tissue of obese rodent models. Adipose tissue–specific overexpression of 11βHSD1 in transgenic mice results in a phenotype of central obesity, insulin resistance, and dyslipidemia. Mice with whole-body knockout of the 11βHSD1 allele are protected from the adverse metabolic consequences of obesity such as hyperglycemia that result from high-fat feeding. Furthermore, in mice prone to developing obesity, 11βHSD1 knockout attenuates the effect of a high-fat diet to induce weight gain. Taken together, these observations suggest that 11βHSD1 has a central role in the pathophysiology of obesity

and that there could be significant benefit to 11βHSD1 inhibition.

Carbenoxolone, a licorice-based compound previously used as an antiulcer medication, inhibits 11βHSD1 in the liver and has been shown to lower blood glucose in subjects with type 2 diabetes. However, carbenoxolone is poorly selective for the two 11βHSD isoforms, and an effect in adipose tissue has not been demonstrated. A class of selective 11βHSD1 inhibitors called *arylsulfonamidothiazoles* shows promise and is currently in development. Careful evaluation of potential side effects will be necessary during the development of 11βHSD1 inhibitors because the enzyme is found in many tissues. Despite this, 11βHSD1 remains a useful therapeutic target for the treatment of obesity and its associated comorbidities. ■

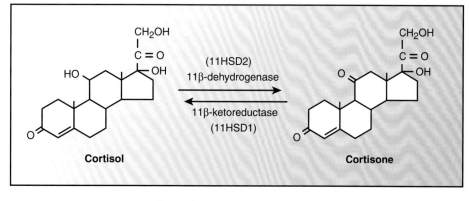

Target tissue cortisol metabolism.

Chapter Summary

- The adrenal gland comprises an outer cortex surrounding an inner medulla. The cortex contains three histologically distinct zones (from outside to inside): the zona glomerulosa, zona fasciculata, and zona reticularis.
- Hormones secreted by the adrenal cortex include glucocorticoids, aldosterone, and adrenal androgens.
- The glucocorticoids cortisol and corticosterone are synthesized in the zona fasciculata and zona reticularis of the adrenal cortex.
- The mineralocorticoid aldosterone is synthesized in the zona glomerulosa of the adrenal cortex.
- Cholesterol, used in the synthesis of the adrenal cortical hormones, comes from cholesterol esters stored in the cells. Stored cholesterol is derived mainly from low-density lipoprotein particles circulating in the blood, but it can also be synthesized *de novo* from acetate within the adrenal gland.
- The conversion of cholesterol to pregnenolone in mitochondria is the common first step in the synthesis of all adrenal steroids and occurs in all three zones of the cortex.
- The liver is the main site for the metabolism of adrenal steroids, which are conjugated to glucuronic acid and excreted in the urine.
- Adrenocorticotropic hormone (ACTH) increases glucocorticoid and androgen synthesis in adrenal cortical cells in the zona fasciculata and zona reticularis by increasing intracellular cAMP. ACTH also has a trophic effect on these cells.
- Angiotensin II and angiotensin III stimulate aldosterone synthesis in the cells of the zona glomerulosa by increasing cytosolic calcium and activating protein kinase C.
- Glucocorticoids bind to glucocorticoid receptors in the cytosol of target cells. The glucocorticoid-bound receptor translocates to the nucleus and then binds to glucocorticoid response elements in the DNA to increase or decrease the transcription of specific genes.
- Glucocorticoids are essential to the adaptation of the body to fasting, injury, and stress.
- The chromaffin cells of the adrenal medulla synthesize and secrete the catecholamines epinephrine and norepinephrine.
- Catecholamines interact with four adrenergic receptors (α_1, α_2, β_1, and β_2) that mediate the cellular effects of the hormones.
- Stimuli such as injury, anger, pain, cold, strenuous exercise, hypoglycemia, and psychological stress generate impulses in the cholinergic preganglionic fibers innervating the chromaffin cells, resulting in the secretion of catecholamines.
- To counteract hypoglycemia, catecholamines stimulate glucose production in the liver, lactate release from muscle, and lipolysis in adipose tissue.

Chapter Review Questions

1. Which of the following sources of cholesterol is most important for sustaining adrenal steroidogenesis when it occurs at a high rate for a long time?

   A. *De novo* synthesis of cholesterol from acetate
   B. Cholesterol in LDL particles
   C. Cholesterol in the plasma membrane
   D. Cholesterol in lipid droplets within adrenal cortical cells
   E. Cholesterol from the endoplasmic reticulum
   F. Cholesterol in lipid droplets within adrenal medullary cells

   The correct answer is B. Cholesterol esters in LDL sources are the most important source of cholesterol for sustaining adrenal steroidogenesis when it occurs at a high rate over a long period of time. This cholesterol can be used directly after release from LDL and not stored. *De novo* synthesis of cholesterol from acetate is a minor source of cholesterol in humans. Cholesterol from the plasma membrane or endoplasmic reticulum is not used for steroidogenesis. Cholesterol esters in lipid droplets within adrenal cortical cells would be used first and depleted during periods of high adrenal steroid hormone synthesis.

2. A 7-year-old boy comes to the pediatric endocrine unit for evaluation of excess body weight. Review of his growth charts indicates substantial weight gain over the previous 3 years but little increase in height. To differentiate between the development of obesity and Cushing syndrome, blood and urine samples are taken. Which of the following would be most diagnostic of Cushing disease?

   A. Increased serum ACTH, decreased serum cortisol, and increased urinary free cortisol
   B. Decreased serum ACTH, increased serum cortisol, and increased serum insulin
   C. Increased serum ACTH, increased serum cortisol, and increased serum insulin
   D. Increased serum ACTH, decreased serum cortisol, and decreased serum insulin
   E. Increased serum ACTH, decreased serum cortisol, and decreased urinary free cortisol
   F. Decreased serum ACTH, decreased serum cortisol, and increased serum insulin

   The correct answer is C. The increase in body weight with little linear growth suggests that the patient has Cushing disease rather than general obesity since linear growth usually continues in obesity syndromes. Laboratory findings in Cushing syndrome include elevated ACTH, serum cortisol, urinary cortisol, and serum insulin (due to the cortisol-induced resistance to insulin action in the peripheral tissues).

3. Congenital adrenal hyperplasia is most likely a result of:

   A. defects in adrenal steroidogenic enzymes.
   B. Addison disease.
   C. defects in ACTH secretion.
   D. defects in corticosteroid-binding globulin.
   E. Cushing disease.
   F. defects in aldosterone synthase.

The correct answer is A. Congenital adrenal hyperplasia is the result of genetic defects that affect adrenal steroidogenic enzymes resulting in impaired formation of cortisol. Low serum cortisol is a stimulus for ACTH release from the hypothalamus. The increase in ACTH has a proliferative effect on the adrenal gland resulting in hyperplasia. Addison disease is the result of pathologic destruction of the adrenal glands by microorganisms or autoimmune disease and would therefore not result in adrenal hyperplasia. ACTH stimulates the growth of the adrenal gland. A reduction in ACTH in the blood would result in atrophy of the adrenal gland. Corticosteroid-binding globulin noncovalently binds steroid hormones in plasma; defects in this protein are not associated with adrenal hyperplasia. Cushing disease results from a pituitary ACTH-secreting tumor; adrenal hyperplasia is secondary, not congenital, in this disease. Aldosterone synthesis is regulated by the renin–angiotensin system. Defective aldosterone synthesis would, therefore, not lead to increased ACTH and adrenal hyperplasia.

4. What is the mechanism through which catecholamines stabilize blood glucose concentration in response to hypoglycemia?

   A. Catecholamines stimulate glycogen phosphorylase to release glucose from muscle.
   B. Catecholamines inhibit glycogenolysis in the liver.
   C. Catecholamines stimulate the release of insulin from the pancreas.
   D. Catecholamines inhibit the release of fatty acids from adipose tissue.
   E. Catecholamines stimulate gluconeogenesis in the liver.
   F. Catecholamines inhibit the release of lactate from muscle.

The correct answer is E. Catecholamines stimulate glycogenolysis and gluconeogenesis in the liver, causing glucose to be synthesized and released into the blood. Catecholamines stimulate glycogen phosphorylase in muscle to free glucose for use by the muscle. Muscle cannot release glucose to the circulation because it lacks glucose-6-phosphatase. However, the muscle can release lactate, which can be used in gluconeogenesis by the liver. Catecholamines inhibit the release of insulin from the pancreas. Insulin would be counterproductive to attempt to increase blood glucose. Catecholamines increase the release of fatty acids from the adipose tissue, to be used in gluconeogenesis by the liver.

5. A patient receiving long-term glucocorticoid therapy plans to undergo hip replacement surgery. What would the physician recommend prior to surgery and why?

   A. Glucocorticoids should be decreased to prevent serious hypoglycemia during recovery.
   B. Glucocorticoids should be increased to stimulate immune function and prevent possible infection.
   C. Glucocorticoids should be decreased to minimize potential interactions with anesthetics.
   D. Glucocorticoids should be increased to stimulate ACTH secretion during surgery to promote wound healing.
   E. Glucocorticoids should be decreased to prevent inadequate vascular response to catecholamines during recovery.
   F. Glucocorticoids should be increased to compensate for the increased stress associated with surgery.

The correct answer is F. Patients on long-term glucocorticoid therapy should have the dose increased prior to undergoing surgery to minimize the effects of surgical stress. These patients cannot mount their own stress response because of the lack of adrenal cortisol release. Glucocorticoid-induced hypoglycemia or interactions with anesthetics are not likely, and these concerns would be secondary to stimulating the response to surgical stress. Glucocorticoids inhibit ACTH release and the immune response. Glucocorticoids increase the response of the vasculature to catecholamines.

Clinical Application Exercises 33.1

CONGENITAL ADRENAL HYPERPLASIA

The pediatric endocrinologist is called in to consult on the case of a 1-week-old female neonate. The baby was born at home and is now being seen in the emergency room because she appears listless and has not nursed at all during the last 24 hours. The parents report that the baby has become increasing listless and less willing to nurse since birth. Upon physical examination, the baby exhibits signs of virilization (growth of pubic hair) and hyponatremia (low plasma sodium), hyperkalemia (high plasma potassium), and volume depletion.

QUESTIONS

1. Based on the physical examination, history, and what you know about adrenal steroidogenesis, what would be a reasonable initial hypothesis?
2. What are the two most likely congenital defects in adrenal steroidogenic enzymes that could explain the findings in this child?
3. From a blood sample, what hormones and metabolites would you ask the laboratory to measure and what would you expect the results to be?
4. From the hormone and metabolite analysis, how would you distinguish between the two most likely causes for this case of congenital adrenal hyperplasia?
5. A genetic screen utilizing DNA from the baby's white cells identifies an inactivating mutation in the gene (*CYP21A2*) for 21-hydroxylase. What would be appropriate hormone replacement for this patient?

ANSWERS

1. A reasonable initial hypothesis is that the baby has a form of congenital adrenal hyperplasia. The virilization (appearance of pubic hair) suggests the presence of excess androgen production by the adrenal gland. The hyponatremia, hyperkalemia, and volume depletion suggest a "salt-wasting" syndrome.

2. Mutations in *CYP21A2*, which encodes 21-hydroxylase, accounts for more than 90% of all cases of adrenal hyperplasia associated with excess androgen production. Mutations in *CYP11B1*, which encodes 11β-hydroxylase, would also result in excess adrenal androgen production.

3. Adrenal androgens would be significantly elevated in patients with virilizing forms of congenital adrenal hyperplasia. Adrenal hyperplasia is usually due to defects in cortisol production. Therefore, the serum concentrations of precursors of cortisol biosynthesis such as progesterone, 17α-hydroxyprogesterone, and 11-deoxycortisol could be elevated. In addition, serum ACTH would be elevated due to the lack of negative feedback from the absent cortisol.

4. Genetic defects in the gene for 11β-hydroxylase resulting in a reduction in the activity of this enzyme would result in increased 11-deoxycortisol. Defects in the gene for 21-hydroxylase, which impair the activity of the enzyme, would not lead to the production of 11-deoxycortisol. Since 11-deoxycortisol has significant mineralocorticoid activity, excess production of this steroid is usually associated with hypertension, rather than volume depletion and hypotension as is observed in this patient.

5. Treatment would be directed toward replacement of glucocorticoids and mineralocorticoids. Glucocorticoids would replace the missing cortisol and also suppress ACTH secretion. With less ACTH stimulation of steroid production from the adrenal gland, the hyperandrogenism should be reduced. Mineralocorticoids are given to treat the "salt wasting" that occurs in the absence of aldosterone.

the**Point** *Visit* http://thepoint.lww.com/rhoades5e *for additional chapter review Q&A, Clinical Application Exercises, animations, and more!*

34 Endocrine Pancreas

Active Learning Objectives

Upon mastering the material in this chapter, you should be able to:
- Explain how cells of the exocrine and endocrine pancreas coordinate to direct many processes related to the digestion, uptake, and use of metabolic fuels.
- Explain how islet cell junctions, vascular supply, and innervation influence islet cell functionality.
- Explain how a key mechanism of insulin secretion led to the development of sulfonylureas.
- Explain how amylin and pancreatic polypeptide work in conjunction with insulin to control fuel metabolism during periods of feeding.

- Explain the molecular aspects of proinsulin biosynthesis that are of clinical significance.
- Explain how mechanistic aspects of the incretin effect have led to a new class of antidiabetic drugs.
- Explain how the etiology of diabetes development is different between types of diabetes.
- Explain how both genetic and nongenetic factors can impair insulin responsiveness.
- Explain how in addition to promoting diabetic complications a series of related variables in metabolic syndrome apparently worsen insulin sensitivity.

The development of mechanisms for the storage of large amounts of metabolic fuel was an important adaptation in the evolution of complex organisms. The processes involved in the digestion, storage, and use of fuels require a high degree of regulation and coordination. Exocrine and endocrine portions of the pancreas play a vital role in these processes. Cells of the **exocrine pancreas** produce and secrete digestive enzymes and fluids into the upper part of the small intestine. Cells of the **endocrine pancreas**, an anatomically small portion of the pancreas (1% to 2% of the total mass), produce hormones involved in regulating fuel storage and use. This chapter focuses primarily on hormones of the endocrine pancreas that coordinate and direct many processes related to the digestion, uptake, and use of metabolic fuels.

▶ ISLETS OF LANGERHANS

The endocrine pancreas consists of numerous discrete clusters of cells, known as the **islets of Langerhans**, which are located throughout the pancreas but are most abundant in the tail region. The human pancreas contains, on average, about 1 million islets, which vary in size from 50 to 300 μm wide. A connective tissue sheath separates each islet from the surrounding acinar tissue. Islets are composed of five major hormone-producing cell types: glucagon-secreting **α cells**, insulin-secreting and amylin-secreting **β cells**, somatostatin-secreting **δ cells**, **pancreatic polypeptide** (**PP**)-secreting **F cells**, and **ghrelin**-secreting **ε cells**.

Cells of the islets of Langerhans display a highly organized arrangement.

In the early 1900s, M. A. Lane established a histochemical method by which two kinds of islet cells could be distinguished. He found that alcohol-based fixatives dissolved the secretory granules in most of the islet cells but preserved

them in a few cells. Water-based fixatives had the opposite effect. He named cells containing alcohol-insoluble granules α cells and those containing alcohol-soluble granules β cells. Many years later, immunofluorescence techniques demonstrated that α cells produce glucagon and β cells produce insulin. Moreover, immunofluorescent staining has shown that α-, β-, δ-, F-, and ε-cell types are arranged in each islet in a pattern suggesting a highly organized cellular community, in which paracrine influences may play an important role in determining hormone secretion rates. As illustrated in Figure 34.1, α cells are generally located near the periphery of the islet, where they form a cortex of cells surrounding the more centrally located β cells. The insulin-producing β cells are the most numerous cell type of the islet, comprising 70% to 90% of the islet. The pancreatic islet δ cells that produce somatostatin are typically located in the periphery, often between β cells and the surrounding mantle of α cells. F cells are the least abundant of the hormone-secreting cells of islets, representing only about 1% of the total cell population. The distribution of F cells is generally similar to that of δ cells. Studies performed mainly in rodents suggest that ε cells are a developmentally regulated islet cell with the cells appearing early in development and then dissipating as β cells appear. Interestingly, the number of ε cells is increased in mouse models of β-cell deficiency. In human and rodent pancreases of neonates and adults, a few ε cells remain visible in marginal areas of the islets. Although studies continue to investigate the exact role that ghrelin-secreting ε cells may have in the pancreas, blockade of ghrelin or ghrelin action in pancreatic islets markedly enhances glucose-induced insulin release.

Cell, vascular, and neural connections likely contribute to islet cell functionality.

Islet cells have both gap junctions and tight junctions, suggesting that cell-to-cell communication within the islet may play a role in regulating hormone secretion. Gap junctions link different

Islet of Langerhans

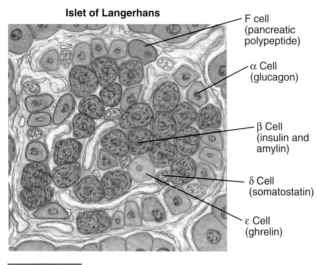

F cell
(pancreatic
polypeptide)

α Cell
(glucagon)

β Cell
(insulin and
amylin)

δ Cell
(somatostatin)

ε Cell
(ghrelin)

Figure 34.1 **Major cell types in a typical islet of Langerhans.**
Note the distinct anatomic arrangement of the various cell types.

cell types in the islet and potentially provide a means for the transfer of ions, nucleotides, or electrical current between cells. The presence of tight junctions between outer membrane leaflets of contiguous cells could result in the formation of microdomains in the interstitial space, which may also be important for paracrine communication. Although the existence of gap junctions and tight junctions in pancreatic islets is well documented, their exact function has not been fully defined.

The arrangement of the vascular supply to islets is also consistent with paracrine involvement in regulating islet secretion. Afferent blood vessels penetrate nearly to the center of the islet before branching out and returning to the surface of the islet. The innermost cells of the islet, therefore, receive arterial blood, whereas those cells nearer the surface receive blood containing secretions from inner cells. Because there is a definite anatomic arrangement of cells in the islet (see Fig. 34.1), one cell type could affect the secretion of others. In general, the effluent from smaller islets passes through neighboring pancreatic acinar tissue before entering into the hepatic portal venous system. By contrast, the effluent from larger islets passes directly into the venous system without first perfusing adjacent acinar tissue. Therefore, islet hormones arrive in high concentrations in some areas of the exocrine pancreas before reaching peripheral tissues. However, the exact physiologic significance of these arrangements is unknown.

Neural inputs also influence islet cell hormone secretion. Islet cells receive sympathetic and parasympathetic innervation. Responses to neural input occur as a result of activation of various adrenergic and cholinergic receptors (described below). Neuropeptides released together with the neurotransmitters may also be involved in regulating hormone secretion.

▶ MECHANISMS OF ISLET HORMONE SYNTHESIS AND SECRETION

Understanding the mechanisms by which islet hormones are synthesized and secreted is of fundamental importance and clinical significance for metabolic health.

Glucose is a major physiological factor regulating insulin synthesis and secretion.

Insulin secretion and synthesis are stimulated when islets are exposed to glucose. Apparently, the metabolism of glucose by the β cell triggers both of these events. Whereas molecular steps by which glucose metabolites regulate insulin secretion have been defined, the basis of glucose-regulated insulin synthesis is not entirely clear. However, studies document that within minutes after raising plasma glucose levels, the rate of proinsulin synthesis increases 5- to 10-fold. At the molecular level, an enhanced efficiency of initiation of translation of proinsulin RNA is observed in the β cell following glucose exposure.

The gene for insulin is located on the short arm of chromosome 11 in humans. Like other hormones and secretory proteins, insulin is first synthesized by ribosomes of the rough endoplasmic reticulum (RER) as a larger precursor peptide that is then converted to the mature hormone prior to secretion (see Chapter 30). The insulin gene product is a 110-amino acid peptide, preproinsulin. Proinsulin consists of 86 amino acids (Fig. 34.2); residues 1 to 30 constitute what will form the B chain of insulin, residues 31 to 65 form the connecting peptide, and residues 66 to 86 constitute the A chain. (Note that "connecting peptide" should not be confused with "C-peptide.") In the process of converting proinsulin to insulin, two pairs of basic amino acid residues are clipped out of the proinsulin molecule, resulting in the formation of insulin and **C-peptide**, which are ultimately secreted from the β cell in equimolar amounts. In addition, small amounts of intact proinsulin and proinsulin conversion intermediates are released. Proinsulin and its related conversion intermediates can be detected in the circulation, where they constitute 20% of total circulating insulin-like immunoreactivity. *In vivo*, proinsulin has a biologic potency that is only about 10% of that of insulin, and the potency of split proinsulin intermediates is between those of proinsulin and insulin.

It is of clinical significance that insulin and C-peptide are cosecreted in equal amounts. The measurement of peripheral insulin concentrations by radioimmunoassay is limited by the fact that 50% to 60% of the insulin produced by the pancreas is extracted by the liver without ever reaching the systemic circulation. In contrast, the liver does not extract C-peptide. Because C-peptide is secreted in equimolar concentrations with insulin and is not extracted by the liver, β-cell secretion rates can be calculated. In addition, as discussed in Chapter 30, another advantage of measuring C-peptide is that the standard insulin radioimmunoassay does not distinguish between endogenous and exogenous insulin, making it an ineffective measure of endogenous β-cell function in an insulin-treated diabetic patient (see Clinical Focus 30.2).

Glucose-stimulated insulin secretion is dose related. Dose-dependent increases in insulin release have been observed after oral and intravenous glucose loads. The insulin secretory response is greater after oral than after intravenous glucose administration. Known as the **incretin effect**, this enhanced response to oral glucose has been interpreted as an indication that absorption of glucose by way of the GI tract stimulates the release of hormones and other

membrane transport (see Chapter 2) that the movement of glucose into a cell requires a transport system. The glucose transporter **GLUT2** mediates movement of glucose into the β cell. In the cytosol, glycolysis converts glucose to pyruvate in a series of enzymatic steps; the first rate-limiting step in this process is the phosphorylation of glucose to glucose-6-phosphate by the enzyme **glucokinase**. There is considerable evidence that glucokinase, by determining the rate of **glycolysis**, functions as the glucose sensor of the β cell and that this is the primary mechanism whereby the rate of insulin secretion adapts to changes in blood glucose. **Adenosine triphosphate (ATP)** is synthesized during this process, and even more is produced during the subsequent oxidation of pyruvate within the mitochondrion. The increase in cytosolic ATP is a key signal that initiates insulin secretion by causing blockade of the **ATP-dependent K$^+$ channel (K$_{ATP}$)** on the β-cell membrane. Blockade of this channel induces membrane depolarization, which activates voltage-gated Ca^{2+} channels. The associated Ca^{2+} influx increases cytosolic Ca^{2+} and additionally boosts cytosolic Ca^{2+} levels via triggering Ca^{2+}-induced Ca^{2+} release. The increase in cytosolic Ca^{2+} is the main spark for exocytosis, the process by which insulin-containing secretory granules fuse with the plasma membrane, leading to the release of insulin into the circulation.

Interestingly, in addition to the initial engagement of exocytosis by ATP (i.e., ATP-mediated K$_{ATP}$ closure), ATP also serves as a major permissive factor for movement of insulin granules and for the priming of exocytosis. The essential role of K$_{ATP}$ channels in insulin secretion is the basis for **sulfonylureas**, a class of drugs used orally in the treatment of **type 2 diabetes**. K$_{ATP}$ channels comprise sulfonylurea receptors, and sulfonylureas are a class of drugs used orally as an insulin **secretagogue** in the treatment of T2D.

In addition to glucose, several other factors such as amino acids, hormones, and neural communication serve as important regulators of insulin secretion (Table 34.1). Among the amino acids, leucine, arginine, and lysine are potent secretagogues. The effects of arginine and lysine on the β cell appear

to be more potent than that of leucine. Mechanistically, the amino acids appear to trigger insulin secretion via ATP production generated via their metabolism. Various lipids and their metabolites also impact insulin secretion. Acutely, lipids and their metabolites enhance glucose-stimulated insulin secretion. However, long-term exposure of islets to lipid impairs glucose-stimulated insulin secretion and biosynthesis, potentially contributing to β-cell failure.

Many GI peptide hormones, including **glucose-dependent insulinotropic peptide, cholecystokinin**, and **glucagon-like peptide 1 (GLP1)**, facilitate the release of insulin from the β cell after a meal. These hormones are released from small intestinal cells following a meal and travel in the blood to reach β cells, where they act through second messengers to increase the sensitivity of these islet cells to glucose. In general, these hormones are not themselves secretagogues, and their effects are evident only in the presence of hyperglycemia. Note that the total amount of insulin that is secreted is greater after oral as opposed to intravenous glucose administration (see Fig. 34.3).

Other intestinal peptide hormones, including VIP, secretin, and gastrin, may also influence the postprandial insulin secretory response, but the exact role of these hormones is currently unclear. The hormones of the neighboring α and δ pancreatic islet cells also modulate β-cell insulin release. Whereas glucagon has a stimulatory effect via a G$_{\alpha s}$-stimulation of adenylyl cyclase, somatostatin suppresses this effect by engaging G$_{\alpha i}$ (see Fig. 34.4). Other hormones reported to exert a stimulatory effect on insulin secretion include growth hormone, glucocorticoids, prolactin, placental lactogen, and the sex steroids.

Sympathetic and parasympathetic divisions of the autonomic nervous system richly innervate islets. Activation of α-adrenoceptors inhibits insulin secretion, whereas β-adrenoceptor stimulation augments insulin secretion. Norepinephrine released by the sympathetic neurons of the pancreas stimulates α-adrenoceptors more than β-adrenoceptors. This response would appear appropriate because during periods of stress and high catecholamine secretion, the desired response is mobilization of glucose and other nutrient stores. Direct infusion of acetylcholine into the pancreatic circulation stimulates insulin secretion, reflecting the role of parasympathetic innervation in regulating insulin secretion.

Hypoglycemia stimulates glucagon secretion.

Similar to insulin, glucagon is first synthesized as part of a larger precursor protein. Many of the factors that regulate insulin secretion also regulate glucagon secretion. In most cases, however, these factors have the opposite effect on glucagon secretion.

Glucagon is a simple 29-amino acid peptide. The initial gene product for glucagon, preproglucagon, is a much larger peptide. As with other peptide hormones, the "pre" piece is removed in the RER, and the prohormone is converted into a mature hormone as it is packaged and processed in secretory granules. Within α cells of the pancreas, proglucagon is cleaved at two positions to yield three peptides, illustrated in Figure 34.5 (left). In contrast, in other cells of the

TABLE 34.1	Factors Regulating Insulin and Glucagon Secretion	
	Insulin	**Glucagon**
Stimulatory agents or conditions	Hyperglycemia	Hypoglycemia
	Amino acids	Amino acids
	Fatty acids (especially long chain)	Acetylcholine
		Norepinephrine
	Gastrointestinal hormones (e.g., GLP1)	Epinephrine
	Acetylcholine	
	Sulfonylureas	
Inhibitory agents or conditions	Somatostatin	Fatty acids
	Norepinephrine	Somatostatin
	Epinephrine	Insulin

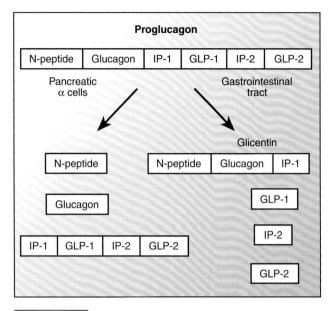

Figure 34.5 **The differential processing of proglucagon.** In α cells of the pancreas **(left)**, the major bioactive product formed from proglucagon is glucagon itself. It is not currently known whether the other peptides are processed to produce biologically active molecules. In intestinal cells **(right)**, proglucagon is cleaved to produce the four peptides shown. Glicentin is the major glucagon-containing peptide in the intestine. GLP1, glucagon-like peptide 1; GLP2, glucagon-like peptide 2; IP-1, intervening peptide-1; IP-2, intervening peptide-2.

gastrointestinal tract in which proglucagon is also produced, the molecule is cleaved at three different positions such that **glicentin, glucagon-like peptide 1**, and **glucagon-like peptide 2** are produced (see Fig. 34.5, right).

The principal factors that influence glucagon secretion are listed in Table 34.1. The primary regulator of glucagon secretion is blood glucose; specifically, when blood glucose falls below about 72 mg/dL, α cells become active and deliver glucagon to the blood. Amino acids are also major secretagogues; however, the concentrations of amino acids required to provoke secretion of glucagon *in vitro* are higher than those generated *in vivo*. This observation suggests that other neural or humoral factors amplify the response *in vivo*, analogous to the effects of incretins on insulin secretion. Somatostatin inhibits glucagon secretion, as it does insulin secretion.

Hyperglycemia and glucagon stimulate somatostatin secretion.

As described earlier, somatostatin is first synthesized as a larger peptide precursor, preprosomatostatin. The hypothalamus also produces this protein, but the regulation of somatostatin secretion from the hypothalamus is independent of that from the pancreatic δ cells. On insertion of preprosomatostatin into the RER, it is initially cleaved and converted to prosomatostatin. The prohormone is converted into active hormone during packaging and processing in the Golgi apparatus. Factors that stimulate pancreatic somatostatin secretion include hyperglycemia, glucagon, and amino acids.

Glucose and glucagon are generally considered the most important regulators of somatostatin secretion.

The exact role of somatostatin in regulating islet secretions has not been fully established. Somatostatin clearly inhibits both glucagon and insulin secretion from α and β cells of the pancreas, respectively, when it is given exogenously. The anatomic and vascular relationships of δ cells to α and β cells further suggest that somatostatin may play a role in regulating both glucagon and insulin secretion. Although many of these data are circumstantial, it is generally accepted that somatostatin plays a paracrine role in regulating insulin and glucagon secretion from the pancreas.

Pancreatic polypeptide secretion and action are regulated by several factors.

Nutrients, hormones, neurotransmitters, gastric distention, insulin-induced hypoglycemia, and direct vagal nerve stimulation regulate PP secretion, whereas hyperglycemia, bombesin, and somatostatin inhibit PP secretion. Furthermore, we now know that a receptor termed Y4, a G-protein–coupled receptor that inhibits **cyclic adenosine monophosphate (cAMP)** accumulation, mediates the actions of PP. The Y4 receptor is expressed in the stomach, small intestine, colon, pancreas, prostate, and enteric nervous system and in select central nervous system neurons. Evidence suggests that PP reduces gastric acid secretion and increases intestinal transit times by reducing gastric emptying and upper intestinal motility. PP also appears to inhibit postprandial exocrine pancreas secretion via a vagal-dependent pathway. Transgenic mice that overexpress PP exhibit reduced weight gain and rate of gastric emptying and decreased fat mass. These findings certainly merit further PP-related research.

▶ INSULIN AND GLUCAGON ACTION

The two main hormones from the endocrine pancreas that direct the storage and use of fuels during times of nutrient abundance (fed state) and nutrient deficiency (fasting) are insulin and glucagon. Insulin is secreted in the fed state and is called the "hormone of nutrient abundance." By contrast, glucagon is secreted in response to an overall deficit in nutrient supply. It should be noted, as discussed above, that the other islet hormones play very important ancillary roles.

Insulin receptor signaling is complex and controls several biological responses.

The insulin receptor (IR) is a heterotetramer, consisting of a pair of α-/β-subunit complexes held together by disulfide bonds (Fig. 34.6). The IR's α subunits are extracellular and contain several insulin-binding sites. The β subunits of the IR span the plasma membrane and function to couple the extracellular event of insulin binding to its intracellular actions. A key intracellular action stimulated by insulin binding is β-subunit autophosphorylation, involving the phosphorylation of a few selected tyrosine residues in the intracellular portion. This event further activates the tyrosine kinase portion of the β subunit, leading to tyrosine phosphorylation of a host of proteins, including members of the insulin receptor

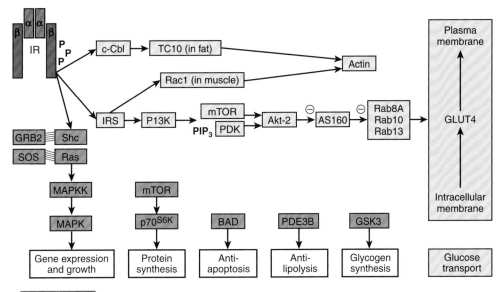

Figure 34.6 **Overview of the complexity of the insulin-signaling network and several biologic responses.** Shown in *black* and *yellow* are intracellular mediators that are involved in GLUT4-regulated glucose transport in muscle and adipose tissue.

substrate (IRS) family (IRS-1, IRS-2, IRS-3, IRS-4, IRS-5, and IRS-6), Gab-1, Shc, p62[dok], c-Cbl, SIRP (signal regulatory protein) family members, and APS (adapter protein containing a pleckstrin homology and Src homology 2 domain). A cascade of phosphorylation and dephosphorylation events, second messenger generation, and protein–protein interactions result, leading to several cellular activities, including gene expression and growth, protein synthesis, antiapoptosis, antilipolysis, glycogen synthesis, and glucose transport, to name a few. Note that IR internalization or dephosphorylation by protein tyrosine phosphatases (PTPs) terminates the activity of the IR. Protein tyrosine phosphatase 1B (PTP1B) and leukocyte antigen-related (LAR) phosphatases are two PTPs shown to attenuate IR activity. A later section on diabetes mellitus highlights that several forms of diabetes mellitus (prediabetes, type 2 diabetes, and gestational diabetes) are complicated, multifactorial diseases in which interactions between multiple genes and environmental factors ultimately result in resistance to the action of insulin. In that regard, both PTP1B and LAR have been reported to be elevated in patients displaying **insulin resistance**. In addition to dampening IR function by tyrosine phosphatase activity, serine–threonine phosphorylation of the β subunit of the IR also has been suggested to decrease the ability of the receptor to undergo autophosphorylation. Many protein kinase C isoforms that catalyze the serine or threonine phosphorylation of the IR are elevated in insulin-resistant tissue from animals and humans.

Insulin stimulates the transport, storage, and metabolism of glucose.

In muscle and adipose tissues, insulin promotes the removal of excess circulatory glucose by regulating the subcellular trafficking of the glucose transporter **GLUT4** that resides in intracellular membrane vesicles (see Fig. 34.6). Under normal insulin responsiveness, insulin promotes the removal of

excess glucose from the circulation by stimulating the exocytic recruitment of the intracellular vesicles containing GLUT4 to the plasma membrane. This stimulated redistribution of these vesicles results in plasma membrane GLUT4 accumulation that facilitates cellular glucose uptake. Activation of GLUT4 vesicle translocation by insulin requires a phosphatidylinositol 3-kinase (PI3K) signal involving the upstream IR and insulin receptor substrate (IRS) activators and the downstream Akt target enzyme. PI3K generates the phospholipid PI3,4,5-P_3 (PIP$_3$) that engages both PI-dependent kinase (PDK) and mTORC2 to phosphorylate/activate the serine/threonine kinase Akt (particularly isoform 2). As shown in Figure 34.6, Akt2 has numerous targets, but Akt2-mediated phosphorylation of Akt substrate of 160 kDa (AS160), a GTPase-activating protein (GAP) toward small G-proteins, inhibits the GAP activity allowing the activation of Rab G-proteins. Studies suggest that Rab8A and Rab13 in muscle cells, whereas Rab10 in adipocytes regulates GLUT4 translocation. In parallel, in both muscle and fat cells, insulin triggers distinct signals that regulate the actin cytoskeleton that is essential for GLUT4 translocation (see Fig. 34.6).

Insulin resistance is associated with insufficient recruitment of GLUT4 to the plasma membrane. This finding emphasizes the importance of understanding the molecular mechanisms of insulin-regulated GLUT4 translocation and glucose uptake so that cellular perturbations associated with insulin resistance can be elucidated and effective treatment and preventative strategies can be designed to bypass or ameliorate insulin resistance.

Besides promoting glucose uptake into muscle and fat cells, insulin promotes its storage. Glucose carbon is stored in the body in two primary forms: as **glycogen** and (by metabolic conversion) as triglycerides. Glycogen is a short-term storage form that plays an important role in maintaining normal blood glucose levels. The primary glycogen storage sites are the liver and skeletal muscle; other tissues,

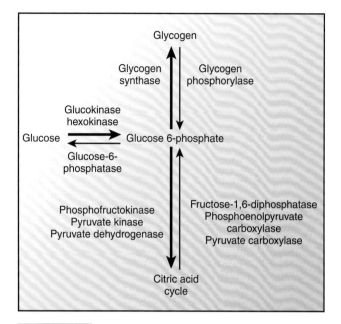

Figure 34.7 Insulin stimulation of glycogen synthesis and glucose metabolism. Insulin promotes glucose uptake into target tissues, stimulates glycogen synthesis, and inhibits glycogenolysis. In addition, it promotes glycolysis in its target tissues. *Bold arrows* indicate processes stimulated by insulin; *light arrows* indicate processes inhibited by insulin.

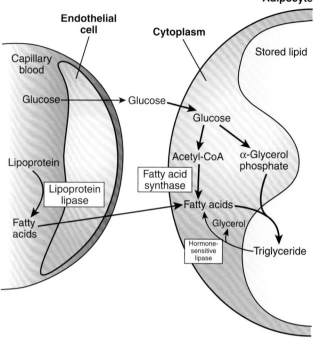

Figure 34.8 Effects of insulin on lipid metabolism in adipocytes. Insulin promotes the accumulation of lipid (triglycerides) in adipocytes by stimulating the processes shown by the *bold arrows* and inhibiting the processes shown by the *light arrows*. Similar stimulatory and inhibitory effects occur in liver cells.

such as adipose tissue, also store glycogen but in quantitatively small amounts. Insulin promotes glycogen storage primarily through two enzymes (Fig. 34.7). It activates **glycogen synthase** by promoting its dephosphorylation and concomitantly inactivates **glycogen phosphorylase**, also by promoting its dephosphorylation. The result is that glycogen synthesis is promoted and glycogen breakdown is inhibited.

In addition to increasing glucose uptake and providing a mass action stimulus for glycolysis, insulin activates the enzymes **glucokinase** (liver), **hexokinase** (muscle), **phosphofructokinase, pyruvate kinase**, and **pyruvate dehydrogenase** of the glycolytic pathway (see Fig. 34.7). Also shown is the inhibition of gluconeogenesis by insulin inhibition of **fructose-1,6-diphosphatase, phosphoenolpyruvate carboxylase**, and **pyruvate carboxylase**.

Insulin has important lipogenic and antilipolytic effects.

In adipose tissue and the liver, insulin promotes **lipogenesis** and inhibits **lipolysis** (Fig. 34.8). Insulin has similar actions in muscle, but because muscle is not a major site of lipid storage, the discussion here focuses on actions in adipose tissue and the liver. By promoting the flow of intermediates through glycolysis, insulin promotes the formation of α-glycerol phosphate and fatty acids necessary for triglyceride formation. In addition, it stimulates **fatty acid synthase**, leading directly to increased fatty acid synthesis. Insulin inhibits the breakdown of triglycerides by inhibiting **hormone-sensitive lipase**, which is activated by many counterregulatory hormones, such as epinephrine and adrenal glucocorticoids. By

inhibiting this enzyme, insulin promotes the accumulation of triglycerides in adipose tissue.

In addition to promoting *de novo* fatty acid synthesis in adipose tissue, insulin increases the activity of **lipoprotein lipase**, which plays a role in the uptake of fatty acids from the blood into adipose tissue. As a result, adipose tissue takes up lipoproteins synthesized in the liver, and fatty acids are ultimately stored as triglycerides.

Insulin enhances the synthesis and suppresses the degradation of proteins.

Insulin promotes protein accumulation in its primary target tissues—liver, adipose tissue, and muscle—in three specific ways. First, it stimulates amino acid uptake. Second, it increases the activity of several factors involved in protein synthesis. For example, it increases the activity of protein synthesis initiation factors, promoting the start of translation and increasing the efficiency of protein synthesis. Insulin also increases the amount of protein synthesis machinery in cells by promoting ribosome synthesis. Third, insulin inhibits protein degradation by reducing lysosome activity and possibly other mechanisms as well.

Glucagon primarily exerts its metabolic actions, via cAMP signaling, in the liver.

Numerous effects of glucagon have been documented in several tissues, primarily adipose tissue, when the hormone has been added at high, nonphysiologic concentrations in experimental situations. Although these effects may play a

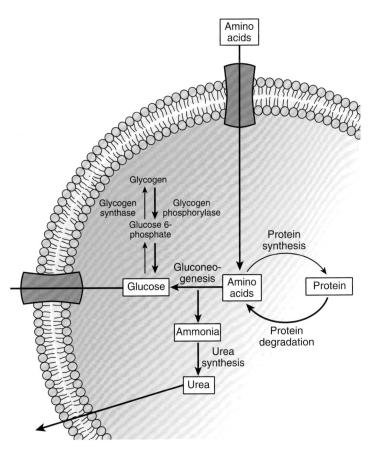

Figure 34.9 **The role of glucagon in glucose production liver cells.** *Bold arrows* indicate processes stimulated by glucagon; the *light arrows* indicate a process inhibited by glucagon.

role in certain abnormal situations, the normal daily effects of glucagon occur primarily in the liver. Glucagon initiates its biologic effects by interacting with one or more types of cell membrane receptors. Glucagon receptors are coupled to G-proteins and promote increased intracellular cAMP, via the activation of adenylyl cyclase or elevated cytosolic calcium as a result of phospholipid breakdown to form inositol triphosphate.

Glucagon promotes hepatic glucose production, coupled with ammonia disposal.

Glucagon is an important regulator of hepatic glucose production. It produces a net effect of glycogen breakdown (**glycogenolysis**) by increasing intracellular cAMP levels, initiating a cascade of phosphorylation events that ultimately results in the phosphorylation and activation of glycogen phosphorylase (Fig. 34.9). Similarly, glucagon promotes the net breakdown of glycogen by inactivating glycogen synthase. In addition, glucagon stimulates hepatic **gluconeogenesis**. It does this principally by increasing the transcription of messenger RNA coding for **phosphoenolpyruvate carboxykinase** (**PEPCK**), a key rate-limiting enzyme in gluconeogenesis. Glucagon also stimulates amino acid transport into liver cells and the degradation of hepatic proteins, helping provide substrates for gluconeogenesis (see Fig. 34.9). The glucagon-enhanced conversion of amino acids into glucose leads to increased formation of ammonia that glucagon alleviates by increasing the activity of the urea cycle enzymes and **ureagenesis**.

Glucagon promotes the oxidation of fats and ketogenesis in the liver.

Glucagon also plays a regulatory role in hepatic lipid metabolism (Fig. 34.10). Mechanistically, glucagon inhibits the activity of acetyl-CoA carboxylase. This inhibition decreases the levels of **malonyl-CoA**, a metabolite that inhibits the activity of the carnitine acyltransferase (CAT) system, which mediates the transfer of fatty acids across the mitochondrial membrane. Therefore, glucagon indirectly increases the delivery of fatty acids into the mitochondria where these become oxidized to produce energy for the liver in the form of ATP. If the rate of fatty acid transport into the mitochondria exceeds the energy needs of the liver, the fatty acids are used in the production of **ketones**. Ketones are an important source of fuel for muscle and heart cells during times of starvation, sparing blood glucose for other tissues that are obligated glucose users, such as the central nervous system. During prolonged starvation, the brain adapts its metabolism to use ketones as a fuel source, lessening the overall need for hepatic glucose production.

Insulin/glucagon ratio determines metabolic status.

In most instances, insulin and glucagon produce opposing effects. Therefore, the relative level of both hormones in the blood plasma, the **insulin/glucagon ratio** (**I/G ratio**), determines the net physiologic response. The I/G ratio is highest in the fed state and is the lowest during fasting. The I/G ratio can vary 100-fold or more because the plasma concentration of each hormone can vary considerably in different

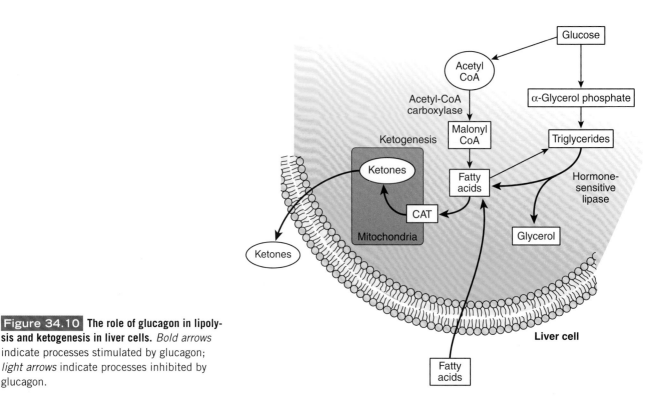

Figure 34.10 **The role of glucagon in lipolysis and ketogenesis in liver cells.** *Bold arrows* indicate processes stimulated by glucagon; *light arrows* indicate processes inhibited by glucagon.

nutritional states. In the fed state, the molar I/G ratio is ~30. After an overnight fast, it may fall to about two, and with prolonged fasting, it may fall to as low as 0.5.

A good example of the profound influence of the I/G ratio on metabolic status is in T1D (see below). Insulin levels are low, so pathways that insulin stimulates operate at a reduced level. However, insulin is also necessary for α cells to sense blood glucose appropriately; in the absence of insulin, the secretion of glucagon is inappropriately elevated. The result is an imbalance in the I/G ratio and an accentuation of glucagon effects well above what would be seen in normal states of low insulin, such as in fasting.

▶ DIABETES MELLITUS

Diabetes mellitus is a systemic disease in which the body does not produce and/or use insulin and is characterized most notably by **hyperglycemia**. In addition to abnormal glucose metabolism, disorders in the metabolism of lipid and protein intensify the seriousness of the disease. Superimposed on the disorders of carbohydrate, fat, and protein metabolism are diabetic-specific microvascular lesions in the retinas, renal glomeruli, and peripheral nerves. Diabetes, if untreated, leads to renal failure, erectile dysfunction, blindness, coronary arterial disease, and increased risk of cancer. Of these, cardiovascular disease (CVD) is the most prominent. For example, more than 65% of people with diabetes die from heart disease. In fact, adults with diabetes have heart disease death rates about two to four times higher than adults without diabetes. Also, stroke accounts for ~20% of diabetes-related deaths, and the risk for stroke is also two to four times higher among people with diabetes. The cause of diabetes continues to be a mystery, although both genetics

and environmental factors such as obesity and lack of exercise appear to play significant roles.

Data from the National Diabetes Statistical Report released in 2014 reported that there are 29.1 million children and adults in the United States, or 9.3% of the population, who have diabetes. Diabetes has several clinical forms, each of which has a distinct etiology, clinical presentation, and course. Diagnosing diabetes mellitus is not difficult to do. Symptoms usually include frequent urination, increased thirst, increased food consumption, and weight loss. Diabetes mellitus is a heterogeneous disorder. The causes, symptoms, and general medical outcomes are variable. There are four main type of diabetes: (1) Type 1 diabetes (T1D) results from the body failure to produce insulin, (2) prediabetes results from insulin resistance, (3) Type 2 diabetes (T2D) results from insulin resistance and β-cell failure, and (4) **gestational diabetes** results from insulin resistance during pregnancy.

Autoimmune disorder underlies type 1 diabetes.

T1D is one of the most intensively studied autoimmune disorders. Although most cases of T1D are a consequence of an inappropriate autoimmune destruction of the β cell, an autoimmune-independent subtype of T1D has recently been described, and recommendations have been put forth to divide T1D into type 1A (immune mediated) and type 1B (other forms of diabetes with severe insulin deficiency). **Type 1B diabetes** appears to be a rare form of T1D in which histologic examination of pancreatic sections demonstrates inflammation but no anti-islet autoantibodies.

Whereas studies to gain deeper insight into type 1B diabetes are currently underway, it is known that **type 1A**

CLINICAL FOCUS | 34.1

Beyond Type 1 and Type 2 Diabetes

It is becoming increasingly apparent that the main forms of diabetes, type 1 diabetes (T1D) and type 2 diabetes (T2D), have subtypes that require the correct diagnosis. A subtype of new concern is termed *latent autoimmune diabetes in adults* ([*LADA*] though some prefer *type 1.5*). This subtype was first revealed in the 1970s while testing a method to identify autoantibodies in the blood of people with T1D. Scientists found that T1D-related autoantibodies were virtually absent in the general population, but they showed up in about 10% of people diagnosed with T2D, which is not an autoimmune disease. Although much debate exists as to whether LADA is distinct from T1D, researchers today are working on a set of criteria for its diagnosis: (1) the presence of autoantibodies in the blood, (2) adult age at onset, and (3) lack of need for insulin treatment in the first 6 months after diagnosis. The first criterion distinguishes LADA from

T2D because of autoantibodies in the blood, and the third criterion differs from T1D because T1D patients need to start insulin immediately. We now know that those with T1D have higher levels and more types of autoantibodies than do those with LADA, which may be the reason β cells are destroyed faster in T1D than in LADA.

Importantly, a recent Japanese study found that early insulin treatment in adults with LADA may help them avoid total dependence on insulin. This study also compared the use of insulin to that of sulfonylurea treatment (a common treatment for people with T2D that enhances β-cell insulin production) and found that, unlike those treated with insulin, patients treated with sulfonylurea lost β-cell function faster and progressed rapidly to insulin dependency. Thus, it is coming to light that a better treatment for this subtype of diabetes may be insulin, not sulfonylureas. ■

diabetes results from a selective destruction of the β cells within the islets. A subtype of type 1A diabetes that has a latent onset in adults also exists, and recent findings suggest that its correct diagnosis and tailored insulin therapy may delay the progression and worsening of the disease (see Clinical Focus 34.1). The development of type 1A diabetes is usually divided into a series of stages, beginning with genetic susceptibility and ending with essentially complete β-cell destruction. Central to the pathology is **insulitis** β-cell injury, involving a lymphocytic attack on β cells.

Studies of identical twins have provided important information regarding the genetic basis of T1D. If one twin develops T1D, the odds that the second will develop the disease are much higher than for any random person in the population, even when the twins are raised apart under different socioeconomic conditions. In addition, people with certain cell-surface human leukocyte antigens bear a higher risk for the disease than others.

Environmental factors are involved as well because the development of T1D in one twin predicts only a 50% or less chance that the second will develop the disease. The specific environmental factors have not been identified, although much evidence implicates viruses. Therefore, it appears that a combination of genetics and environment is a strong contributing factor to the development of T1D.

Because the primary defect in T1D is the inability of β cells to secrete adequate amounts of insulin, these patients must be treated with injections of insulin. In an attempt to match insulin concentrations in the blood with the metabolic requirements of the person, various formulations of insulin with different durations of action have been developed. Patients inject an appropriate amount of these different insulin forms to match their

dietary and lifestyle requirements. New technologies such as insulin pumps and continuous glucose monitors have not only aided treatment but also are paving the road to an artificial pancreas (see Chapter 30 Integrated Medical Sciences Box).

The long-term control of T1D depends on maintaining a balance between several factors including insulin, diet, and exercise, as well as possibly including ancillary hormones such as amylin (see Clinical Focus 34.2). To strictly control their blood glucose, patients are advised to monitor their diet and level of physical activity, as well as their insulin dosage. Exercise *per se*, much like insulin, increases glucose uptake by muscle. Diabetic patients must take this into account and make appropriate adjustments in diet and/or insulin whenever general exercise levels change dramatically.

Insulin resistance is an underlying aspect of prediabetes, type 2 diabetes, and gestational diabetes.

T2D is the predominant form of diabetes worldwide. An epidemic of T2D is underway in both developed and developing countries. Even more alarmingly, it is estimated that twice as many people have prediabetes, a condition that occurs when a person's blood glucose levels are higher than normal but not high enough for the diagnosis of T2D. Recent research has shown that some long-term damage to the body, especially the heart and circulatory system, may already be occurring during prediabetes. An individual is diagnosed with prediabetes if their fasting plasma glucose (FPG) is 100 to 125 mg/dL, if an oral glucose tolerance test (OGTT), which measures blood glucose levels before and 2 hours after drinking a solution containing 75 to 100 g of glucose, is 140 to 199 mg/dL, or if their A1C (sometimes

Amylin

Amylin is a 37-amino acid peptide that is almost exclusively expressed within pancreatic β cells, where it is copackaged with insulin in secretory granules. Consequently, amylin is normally cosecreted with insulin, and the plasma concentrations of the two hormones display a similar diurnal pattern of low fasting levels and rapid and robust increases in response to meals.

As might be expected, because of the colocalization of both hormones within β cells, patients with T1D have an absolute deficiency of both insulin and amylin, whereas patients with T2D have a relative deficiency of both hormones, including a significantly impaired amylin and insulin response to meals. These findings have led to questions of whether amylin deficiency contributes to the metabolic derangements in diabetic patients and, if so, whether amylin replacement might convey clinical benefit when used in conjunction with insulin replacement.

Extensive studies with amylin and amylin antagonists in rodents and with pramlintide (a synthetic analogue of the human amylin hormone) in diabetic patients have provided an understanding of the role of amylin in glucose homeostasis. Preclinical data indicate that amylin acts as a neuroendocrine hormone that complements the actions of insulin in postprandial glucose homeostasis via several mechanisms. These include a suppression of postprandial glucagon secretion and a slowing of the rate at which nutrients are delivered from the stomach to the small intestine for absorption. The net effect of these actions is to mitigate the influx of endogenous (liver-derived) and exogenous (meal-derived) glucose into the circulation and thus to better match the rate of insulin-mediated glucose clearance from the circulation. ■

---

called hemoglobin A1C, HbA1C, or glycohemoglobin), which measures the percent of glucose attached to hemoglobin, is 5.7% to 6.4%. Because red blood cells are constantly forming and dying, but typically they live for about 3 months, the A1C reflects the average of a person's blood glucose levels over the past 3 months. The higher the A1C percentage, the higher a person's blood glucose levels have been. A normal A1C level is below 5.7%. Individuals with A1Cs >6.5% are diagnosed with T2D, and this equates to an FPG of 126 mg/dL or higher or an OGTT measure that is 200 mg/dL or higher.

Central to prediabetes and T2D is insulin resistance, a term that indicates the presence of an impaired biologic response to either exogenously administered or endogenously secreted insulin. Insulin resistance is manifested by decreased insulin-stimulated glucose transport and metabolism in adipose tissue and skeletal muscle and by impaired suppression of hepatic glucose output.

It is clear that the molecular etiology of insulin resistance involves multiple genetic and nongenetic mechanisms contributing to the final phenotype. Furthermore, the resultant hyperglycemia of the syndrome and the compensatory rise in insulin in turn exacerbate insulin resistance, contributing significantly to the pathogenesis of the disease. Superimposed on that backdrop is the negative contribution of the obese state. New lessons learned from microbiology and immunology suggest an apparent role of intestinal microbiota in the etiology of obesity, insulin resistance, and T2D (see Integrated Medical Science Box). Although pinpointing the site of abnormality induced by these conditions is still an intensive area of current research, evidence strongly suggests that the defects induced by each occur at different loci in the insulin-signaling network.

Mutations in the IR are associated with rare forms of insulin resistance. These mutations affect IR number, splicing,

trafficking, binding, and phosphorylation. The affected patients demonstrate severe insulin resistance, including type A syndrome, Donohue syndrome, Rabson–Mendenhall syndrome, and lipoatrophic diabetes.

In some cases, insulin resistance results from the secretion of counterregulatory hormones in a normal (e.g., pregnant) or pathophysiologic (e.g., Cushing syndrome) state. During pregnancy, many women become insulin resistant and develop gestational diabetes that usually disappears after delivery. Women who have had gestational diabetes, however, have an increased risk of developing T2D later in life.

Obesity is closely linked to insulin resistance.

The marked increase in the prevalence of obesity has played an important role in the prevalence of diabetes. According to data from the National Health and Nutrition Examination Survey, two thirds of the adult men and women in the United States with a diagnosis of T2D had a **body mass index** (**BMI**) of 27 kg/m² or greater. This index is calculated by dividing weight (in kilograms) by height (in meters squared). There is a strong curvilinear relation between BMI and relative body fat mass. Men and women with a BMI of 25.0 to 29.9 kg/m² are considered overweight, and those with a BMI of 30 kg/m² or greater are considered obese.

Lifestyle changes and multiple classes of drugs provide therapeutic intervention.

Early on, or in mild cases, diet, weight loss, and exercise can be extremely effective in diabetes therapy and may be the only treatment necessary. Commonly, however, lifestyle intervention is supplemented by treatment with one or more oral agents. Multiple classes of drugs independently address different pathophysiologic features that contribute to the development of T2D. The available oral antidiabetic agents can be divided by mechanism of action into insulin sensitizers with primary action in the liver (e.g., **biguanides**), insulin sensitizers with primary action in peripheral tissues (e.g., **glitazones**), insulin secretagogues (e.g., sulfonylureas), agents that slow the absorption of carbohydrates (e.g., **α-glucosidase inhibitors**), long-acting analogues of incretin hormones (e.g., GLP1 agonists), agents that increase incretin hormone levels (e.g., dipeptidyl peptidase IV inhibitors), and drugs that block the reabsorption glucose in the kidney (e.g., sodium glucose cotransporter inhibitors). In some cases, persons with T2D may be treated with insulin, just like the patient with T1D. Although insulin therapy apparently provides some benefit to type 2 diabetics, this approach is limited in controlling elevated glucose levels or in controlling obesity that predisposes to this disease.

Diabetes mellitus leads to organ dysfunction and tissue damage.

If left untreated or if glycemic control is poor, diabetes leads to acute complications that may prove fatal. However, even with reasonably good control of blood glucose, over a period of years, most diabetics develop secondary complications of the disease that result in tissue damage, primarily involving the cardiovascular and nervous systems.

The nature of acute complications that develop in type 1 and type 2 diabetics differs. Persons with poorly controlled T1D often exhibit hyperglycemia, **glucosuria**, dehydration, and **diabetic ketoacidosis**. As blood glucose becomes elevated above the renal plasma threshold, glucose appears in the urine. As a result of osmotic effects, water follows glucose, leading to **polyuria**, excessive loss of fluid from the body, and dehydration. With fluid loss, the circulating blood volume is reduced, compromising cardiovascular function, which may lead to circulatory failure.

Excessive ketone formation leads to acidosis and electrolyte imbalances in persons with T1D. If uncontrolled, ketones may be elevated in the blood to such an extent that the odor of **acetone** (one of the ketones) is noticeable on the breath. Production of the primary ketones, **β-hydroxybutyric acid** and **acetoacetic acid**, results in the generation of excess hydrogen ions and a metabolic acidosis. Ketones may accumulate in the blood to such a degree that they exceed renal transport capacities and appear in the urine. As a result of osmotic effects, water is also lost in the urine. In addition, the pK of ketones is such that even with the most acidic urine a normal kidney can produce about half of the excreted ketones are in the salt (or base) form. To ensure electrical neutrality, these must be accompanied by a cation, usually either sodium or potassium. The loss of ketones in the urine, therefore, also results in a loss of important electrolytes. Excessive ketone production in T1D results in acidosis, a loss of cations, and a loss of fluids. Emergency department procedures are directed toward immediate correction of these acute problems and usually involve the administration of base, fluids, and insulin.

The complex sequence of events that can result from uncontrolled T1D is shown in Figure 34.11. If left unchecked, many of these complications can have an additive effect to further the severity of the disease state.

Persons with T2D are generally not ketotic and do not develop acidosis or the electrolyte imbalances characteristic of T1D. Hyperglycemia leads to fluid loss and dehydration. Severe cases may result in hyperosmolar coma as a result of excessive fluid loss. The initial objective of treatment in these people is the administration of fluids to restore fluid volumes to normal and eliminate the hyperosmolar state.

With good control of their disease, most persons with diabetes can avoid the acute complications described above; however, it is rare that they will not suffer from some of the chronic secondary complications of the disease. In most instances, such complications will ultimately lead to reduced life expectancy.

Most lesions occur in the circulatory system, although the nervous system is also often affected. Both small and large blood vessels systemic circulation are damaged, producing what are known, respectively, as microvascular and macrovascular complications in both T1D and T2D. The common finding in affected vessels is a thickening of the basement membrane. This condition leads to impaired delivery of nutrients and hormones to the tissues and inadequate removal of waste products, resulting in irreparable tissue damage. Interestingly, micro- and macrovascular complications do not occur in the pulmonary circulation. Although the pulmonary circulation receives all of the cardiac output

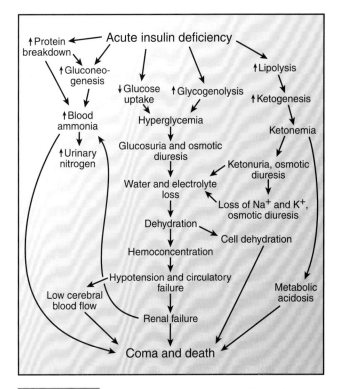

Figure 34.11 **Events resulting from acute insulin deficiency in type 1 diabetes mellitus.** If left untreated, insulin deficiency may lead to several complications, which may have additive or confounding effects that may ultimately result in death.

as well as the same concentration of insulin and glucose as the systemic circulation, it is refractory to diabetes-induced vascular damage.

"Microvascular complications" is a term defining the effects of diabetes on small blood vessels, which include the following: deterioration of blood flow to the retina of the eye, causing **retinopathy** and blindness; deterioration of blood flow to the extremities, causing, in some cases, the need for foot or leg amputation; and deterioration of glomerular filtration in the kidneys, leading to renal failure.

"Macrovascular complications" is a term defining the effect of diabetes on large blood vessels. Diabetes is a major cause of myocardial infarction, stroke, and CVD. Recognized risk factors for CVD (i.e., hypertension, high triglycerides, low high-density lipoprotein cholesterol) are recognized as common in people with diabetes, especially T2D. Patients displaying a constellation of these derangements have been designated as having **metabolic syndrome**.

Another common complication of long-standing diabetes is **diabetic neuropathy**, a disorder affecting the peripheral sensory nerves also known as *peripheral neuropathy*. Diabetic neuropathy is thought to be a microvascular injury involving small blood vessels that supply the nerves. Diabetic neuropathy affects all peripheral nerves (pain fibers and motor neurons) and autonomic nerves.

Many patients with diabetes experience diminished sensation in the extremities, especially in the feet and legs, which compounds the problem of diminished blood flow to these areas. Often, impaired sensory nerve function results in lack of awareness of severe ulcerations of the feet caused by reduced blood flow. Men may develop erectile dysfunction leading to impotence, and both men and women may have impaired bladder and bowel function. Unfortunately, cancer patients (especially colon cancer) suffer from peripheral neuropathy as a side effect of radiation/chemotherapy and manifest the same symptoms.

INTEGRATED MEDICAL SCIENCES

Gut Microbiome

An estimated 100 trillion microbiota colonize the gastrointestinal tract and are considered an essential organ system, with properties that are integrated into the endocrinology, as well as other, systems of the host. Microbiologists refer to these bacteria as the microbiome, which is defined as the genetic material within a microbial community. While the study of the gut microbiome is a relatively new field, evidence suggests that the intestinal microbiota may play an important role in the onset of obesity, insulin resistance, and T2D.

For example, with regard to obesity that sets the stage for insulin resistance and T2D, conventionally raised mice have been reported to have 42% more total body fat than germ-free mice (raised in the absence of any microorganisms), even if their daily caloric intake was 29% less than the germ-free mice. Strikingly, transplanting fecal microbiota from conventionally raised mice into germ-free mice resulted in a 57% increase in the germ-free mice total body fat. Furthermore, germ-free mice transplanted with fecal microbiota from obese donors had a significantly greater increase in total body fat than those colonized with microbiota from lean donors.

Also layered into the integrated science of microbiology and endocrinology is the science of immunology, in that intestinal microbiota are enriched with molecules such as lipopolysaccharide (LPS) and peptidoglycan implicated in the development of inflammation and metabolic diseases. Mice fed a high-fat diet display a significantly altered intestinal microbiota composition and elevated plasma LPS. Normal insulin receptor signaling through IRS-1 is impaired by c-Jun N-terminal kinase (JNK) that is elevated by LPS. This detrimental proinflammatory activity of intestinal microbiota on insulin action is defined as metabolic endotoxemia. In addition to the insulin-signaling dysfunction resulting from increased endotoxemia, metabolic endotoxemia is associated with increased gut permeability. With high-fat feeding, changes in intestinal microbiota reduce tight junction proteins of gut epithelial cells, and the resulting increase in gut permeability sets in motion metabolic endotoxemia and insulin resistance.

Being explored now are therapeutic reversal opportunities to combat the apparent role intestinal microbiota have in the etiology of obesity, insulin resistance, and T2D. Three approaches being investigated are the use of probiotics, antibiotics, and fecal microbiota transplantation. Probiotics are nonpathogenic live microorganisms that have been found to significantly delay the onset of glucose intolerance, hyperglycemia, and hyperinsulinemia in diabetic rats and humans. Antibiotic treatment has also been shown to improve insulin sensitivity with associated lowering of plasma LPS and intestinal expression of inflammatory molecules. Finally, patients with metabolic syndrome who received small intestinal infusions of fecal microbiota from allogenic lean donors for 6 weeks showed an improvement in peripheral and hepatic insulin sensitivity. In closing, this new integrated medical science study of the gut microbiome in metabolic health and disease may be able to boost the development of new effective antidiabetic therapies. ■

Chapter Summary

- The relative distribution of α, β, δ, F, and ε cells within each islet of Langerhans shows a distinctive pattern and suggests that there may be some paracrine regulation of secretion.
- Plasma glucose is the primary physiologic regulator of insulin and glucagon secretion, but amino acids, fatty acids, and some gastrointestinal hormones also play a role.
- Insulin has anabolic effects on carbohydrate, lipid, and protein metabolism in its target tissues, where it promotes the storage of nutrients.

- Effects of glucagon on carbohydrate, lipid, and protein metabolism occur primarily in the liver and are catabolic in nature.
- Type 1 diabetes mellitus results from the destruction of β cells, whereas prediabetes, type 2 diabetes, and gestational diabetes result from a lack of responsiveness to circulating insulin.
- Diabetes mellitus may produce both acute complications, such as ketoacidosis, and chronic secondary complications, such as microvascular disease, macrovascular disease, neuropathy, and nephropathy.

Chapter Review Questions

1. Which of the following stimulate the secretion of both insulin and glucagon from the pancreas?

   A. Acetylcholine
   B. Amino acids
   C. Epinephrine
   D. Sulfonylureas
   E. Both acetylcholine and amino acids

The correct answer is E. Acetylcholine and amino acids both stimulate insulin and glucagon secretion. Epinephrine stimulates glucagon secretion but inhibits insulin secretion. Sulfonylureas stimulate insulin secretion.

2. The effects of insulin include:

   A. inhibition of amino acid uptake into skeletal muscle.
   B. stimulation of glucose uptake into all tissues in the body.
   C. inhibition of protein degradation in skeletal muscle.
   D. stimulation of hormone-sensitive lipase in adipose tissue.
   E. stimulation of glycogenolysis.

The correct answer is C. Insulin stimulates amino acid uptake into skeletal muscle, stimulates glucose uptake specifically into adipose tissue and skeletal muscle, inhibits protein degradation, inhibits hormone-sensitive lipase in adipose tissue, and inhibits glycogenolysis.

3. A 55-year-old man was diagnosed with type 1A diabetes at the age of 8. Which would be most characteristic of his form of the disease?

   A. Absence of anti-islet autoantibodies
   B. Insulin resistance
   C. Sulfonylurea treatment
   D. Treatment with exogenous insulin
   E. Virtual absence of secondary complications

The correct answer is D. Persons with type 1A diabetes have islet autoantibodies, whereas persons with type 1B diabetes do not have islet autoantibodies. Both type 1A and type 1B diabetics are insulin deficient, not insulin resistant and are treated with exogenous insulin. Persons with type 2 diabetes are treated with sulfonylureas. Secondary complications are difficult to avoid in any form of diabetes.

Clinical Application Exercises 34.1

TRANSIENT NEONATAL DIABETES

A 21-day-old male infant was referred to a specialized children's hospital for management of hyperglycemia. Hyperglycemia had been noted on day one of life with an initial blood glucose value of 16 mmol/L. At diagnosis, both the plasma immunoreactive insulin (IRI) and immunoreactive glucagon (IRG) were below the reference range for normal neonates. The IRG, but not the IRI, normalized within the first month. A regimen of daily insulin was begun. Interestingly, from day 41 to day 47, the infant did not receive insulin and a crude control of the blood glucose was demonstrated. Peak levels of blood glucose in excess of 20 mmol/L were followed by drops to levels <2 mmol/L without insulin administration. However, because of unsatisfactory glucose levels, administration of daily insulin was reintroduced. The infant was discharged from the hospital on day 50, and administration of insulin was discontinued uneventfully at 9 months. At 1 year, glycemic control was normal.

QUESTIONS

1. What might be the basis for the infant's reversible bout of hyperglycemia?
2. In addition to hyperglycemia, the patient probably presented with what?

3. Do poorly controlled type 1 diabetic individuals display inappropriate IRG?

ANSWERS

1. The findings indicate an immature function of both beta and alpha cells at the diagnosis with the alpha cells maturing within the first month and the beta-cell recovery later. The exact mechanism to explain this entity is not known, but there are several postulates including transient hypothalamic imbalance, infection, adrenocortical disturbance, insulin resistance, and immaturity of hepatic gluconeogenetic enzymes. However, the most accepted ones are hypoinsulinism with β-cell hypoplasia and temporary delay in β-cell maturity.

2. The infant (even though was full term) might be expected to be small for gestational age with features of dehydration and

wasting in presence of normal intake and absence of vomiting and loose motions.

3. Type 1 diabetes is a consequence of an inappropriate autoimmune (type 1A) or, to a lesser extent, nonimmune (type 1B) destruction of the beta cell. As insulin is also necessary for alpha cells to sense blood glucose appropriately; in the absence of insulin or an insufficient insulin therapy regime, the secretion of glucagon is inappropriately elevated. The result is an imbalance in the I/G ratio and an accentuation of glucagon effects well above what would be seen in normal states of low insulin, such as in fasting.

thePoint® *Visit* http://thepoint.lww.com/rhoades5e *for additional chapter review Q&A, Clinical Application Exercises, animations, and more!*

Endocrine Regulation of Calcium, Phosphate, and Bone Homeostasis

Active Learning Objectives

Upon mastering the material in this chapter, you should be able to:
- Explain how typical daily changes of both calcium and phosphorus between different tissue compartments in a healthy adult are controlled.
- Explain how chondrocytes, osteoblasts, osteocytes, and osteoclasts regulate bone formation and bone reabsorption.

- Explain how parathyroid hormone, calcitonin, and 1,25-dihydroxycholecalciferol affect calcium and phosphate metabolism.
- Explain the underlying mechanisms of osteogenesis imperfecta, osteoporosis, osteomalacia, rickets, and Paget disease.

The plasma calcium concentration is among the most closely regulated of all physiologic parameters in the body. Typically, it varies by only 1% to 2% daily or even weekly. Such stringent regulation in a biologic system usually implies that the parameter plays an important role in one or more critical processes. Phosphate also plays many important roles in the body, although its concentration is not as tightly regulated as that of calcium. Many of the factors involved in regulating calcium also affect phosphate.

▶ OVERVIEW OF CALCIUM AND PHOSPHATE IN THE BODY

Calcium is the most abundant mineral in the body and plays a key role in many physiologically important processes. The average adult contains ~1 kg of calcium in the body, with 99% contained in bone in the form of calcium phosphate. A significant decrease in plasma calcium can rapidly lead to death. A chronic increase in plasma calcium can lead to soft tissue calcification and formation of stones. Phosphate also plays important roles in the body, and because it is a principal component of bone with calcium, phosphate homeostasis is tightly and hormonally coupled to that of calcium.

Calcium and phosphate are major constituents of bone and key cellular functions.

Calcium affects nerve and muscle excitability, neurotransmitter release from axon terminals, and excitation–contraction coupling in muscle cells. It serves as a second or third messenger in several intracellular signal transduction pathways. Some enzymes use calcium as a cofactor, including some in the blood-clotting cascade. Finally, calcium is a major constituent of bone. Of all these roles, the one that demands the most careful regulation of plasma calcium is the effect of calcium on nerve excitability. Calcium affects the sodium permeability of nerve membranes, which influences the ease with which action potentials are triggered. Low plasma calcium (**hypocalcemia**) can lead to the generation of spontaneous action potentials in nerves. When motor neurons are affected, tetany of the muscles of the motor unit may

occur; this condition is called **hypocalcemic tetany**. Certain diagnostically important signs may reveal latent tetany. The **Trousseau sign** is a characteristic spasm of the muscles of the forearm that causes flexion of the wrist and thumb and extension of the fingers. It may occur spontaneously or be elicited by inflation of a blood pressure cuff placed on the upper arm. The **Chvostek sign** is a unilateral spasm of the facial muscles that can be elicited by tapping the facial nerve at the point at which it crosses the angle of the jaw. Although rare, hypocalcemia can also influence myocardial function by changing QT intervals or QRS and ST segments. These changes may mimic acute myocardial infarction or conduction abnormalities. A considerable reduction in serum calcium may lead to **grand mal seizures** or **laryngospasm**.

Phosphorus (usually as phosphate) is also a key constituent of bone. In addition, phosphate serves as an important component of intracellular pH buffering and cellular energy metabolism as part of the adenosine triphosphate molecule. Phosphate is a constituent of many other macromolecules, such as phospholipids and phosphoproteins.

Distributions of calcium and phosphorus differ in bone and cells.

Table 35.1 shows the relative distributions of calcium and phosphate in a healthy adult. The average adult body contains ~1 to 2 kg of calcium, roughly 99% of it is in bones. Despite its critical role in excitation–contraction coupling, only about 0.3% of total body calcium is located in muscles. About 0.1% of total calcium is in extracellular fluid. Of the roughly 600 g of phosphorus in the body, most is in bones (86%). Compared with calcium, a much larger percentage of phosphorus is located in cells (14%). The amount of phosphorus in extracellular fluid is low (0.08% of body content).

Bones also contain a relatively high percentage of the total body content of several other inorganic substances (Table 35.2). About 80% of the total carbonate in the body is located in bones. This carbonate can be mobilized into the blood to combat acidosis; thus, bone participates in pH buffering in the body. Long-standing uncorrected acidosis can result in considerable loss of bone mineral. Significant

TABLE 35.1	Body Content and Tissue Distribution of Calcium and Phosphorus in a Healthy Adult		
Region		**Calcium**	**Phosphorus**
Total body content (g)		1,300	600
Relative tissue distribution (% of total body content)			
Bones and teeth		99	86
Extracellular fluid		0.1	0.08
Intracellular fluid		1.0	14

percentages of the body's magnesium and sodium and nearly 10% of its total water content are in bones.

Calcium and phosphorus exist in the plasma in several forms.

In humans, the normal plasma calcium concentration is 9.0 to 10.5 mg/dL. Plasma calcium exists in three forms: ionized or free calcium (50% of the total), protein-bound calcium (40%), and calcium bound to small diffusible anions, such as citrate, phosphate, and bicarbonate (10%). The association of calcium with plasma proteins is pH dependent. At an alkaline pH, more calcium is bound; the opposite is true at an acidic pH.

Plasma phosphorus concentrations may fluctuate significantly during the course of a day (±50% of the average value) for any particular person. In adults, the normal range of plasma concentrations is 3.0 to 4.5 mg/dL. Phosphorus circulates in the plasma primarily as inorganic **orthophosphate** (**PO$_4$**). At a normal blood pH of 7.4, 80% of the phosphate is in the HPO_4^{2-} form and 20% is in the $H_2PO_4^-$ form. Nearly all plasma inorganic phosphate is ultrafilterable. In addition to free orthophosphate, phosphate is present in small amounts in the plasma in organic form, such as in hexose or in lipid phosphates.

Calcium and phosphorous regulation involves the GI tract, kidneys, and bone.

Both calcium and phosphate are obtained from the diet. The approximate tissue distribution and average daily flux of calcium among tissues in a healthy adult are shown in

TABLE 35.2	Inorganic Constituents of Bone
Constituent	**Percentage of Total Body Content Present in Bone**
Calcium	99
Phosphate	86
Carbonate	80
Magnesium	50
Sodium	35
Water	9

Figure 35.1. Dietary intakes may vary widely, but an "average" diet contains ~1,000 mg/d of calcium. Intakes up to twice that amount are usually well tolerated, but excessive calcium intake can result in soft tissue calcification or kidney stones. Only about one third of ingested calcium is actually absorbed from the GI tract; the remainder is excreted in the feces. The efficiency of calcium uptake from the GI tract varies with the person's physiologic status. The percentage uptake of calcium may be increased in young growing children and pregnant or nursing women; often, it is reduced in older adults.

Figure 35.1 also indicates that ~150 mg/d of calcium actually enters the GI tract from the body. This component of the calcium flux partly results from sloughing of mucosal cells that line the GI tract and also from calcium that accompanies various secretions into the GI tract. This component of calcium metabolism is relatively constant, and so the primary determinant of *net* calcium uptake from the GI tract is calcium absorption. Intestinal absorption is important in regulating calcium homeostasis.

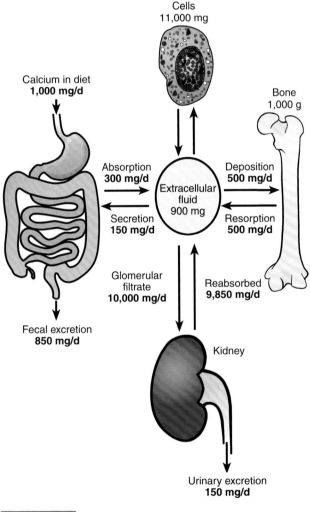

Figure 35.1 Typical daily exchanges of calcium between different tissue compartments in a healthy adult. Fluxes of calcium (mg/d) are shown in *bold font*. Total calcium content in each compartment is shown in *regular font*. Note that most ingested calcium is eliminated from the body via the feces.

Bone in an average person contains ~1,000 g of calcium. Bone mineral is constantly reabsorbed and deposited in the remodeling process. As much as 500 mg/d of calcium may flow in and out of the bones. Because bone calcium serves as a reservoir, both bone resorption and bone formation are important in regulating plasma calcium concentration.

In overall calcium balance, the net uptake of calcium from the GI tract presents a daily load of calcium that will eventually require elimination. The primary route of elimination is via the urine, and therefore, the kidneys play an important role in regulating calcium homeostasis. The 150 mg/d of calcium excreted in the urine represents only about 1% of the calcium initially filtered by the kidneys; the remaining 99% is reabsorbed and returned to the blood. Therefore, small changes in the amount of calcium reabsorbed by the kidneys can have a dramatic impact on calcium homeostasis.

Figure 35.2 shows the overall daily flux of phosphate in the body. A typical adult ingests ~1,400 mg/d of phosphorus. In marked contrast to calcium, most (1,300 mg/d) of this phosphorus is absorbed from the GI tract, typically as inorganic phosphate. There is an obligatory contribution of phosphorus from the body to the contents of the GI tract (about 200 mg/d), much like that for calcium, resulting in a net uptake of phosphorus of 1,100 mg/d and excretion of 300 mg/d via the feces. Thus, most ingested phosphate is absorbed from the GI tract and little passes through to the feces. Because most ingested phosphate is absorbed and circulating phosphate is readily filtered in the kidneys, tubular phosphate reabsorption is a major process regulating phosphate homeostasis.

▶ CALCIUM AND PHOSPHATE METABOLISM

Homeostatic mechanisms are responsible for bringing calcium and phosphate into the body, maintaining normal blood levels of these minerals, transporting them into the cell, and transporting them out again when needed.

Calcium and phosphate are absorbed primarily by the small intestine.

Calcium absorption in the small intestine occurs by both active transport and diffusion. The relative contribution of each process varies with the region and with total calcium intake. Uptake of calcium by active transport predominates in the duodenum and jejunum; in the ileum, passive diffusion predominates. The relative importance of active transport in the duodenum and jejunum versus passive diffusion in the ileum depends on several factors. At high levels of calcium intake, active transport processes are saturated and most of the uptake occurs in the ileum, partly because of its greater length, compared with other intestinal segments. With moderate or low calcium intake, however, active transport predominates because the gradient for diffusion is low. Active transport is the regulated variable in controlling calcium uptake from the small intestine. Metabolites of vitamin D provide a regulatory signal to increase intestinal calcium absorption. Under the influence of 1,25-dihydroxycholecalciferol, calcium-binding proteins in intestinal mucosal cells increase in number, enhancing the capacity of these cells to transport calcium actively.

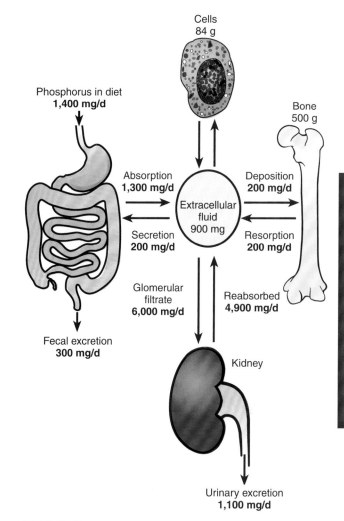

Figure 35.2 Typical daily exchanges of phosphorus between different tissue compartments in a healthy adult. Fluxes of phosphorus (mg/d) are shown in *bold font*. Total phosphorus content in each compartment is shown in *regular font*. Note that most ingested phosphorus is absorbed and eventually eliminated from the body via the urine.

The small intestine is also a primary site for phosphate absorption. Uptake occurs by active transport and passive diffusion, but active transport is the primary mechanism. As indicated in Figure 35.2, phosphate is efficiently absorbed from the small intestine; typically, 80% or more of ingested phosphate is absorbed. However, phosphate absorption from the small intestine is regulated to a minor extent. Active transport of phosphate is coupled to calcium transport, and thus, when active transport of calcium is low, as with vitamin D deficiency, phosphate absorption is also low.

Kidneys play an important role in plasma calcium and phosphate regulation.

Filterable calcium comprises about 60% of the total calcium in the plasma. It consists of free calcium ions and calcium bound to small diffusible anions. The remaining 40% of the total calcium is bound to plasma proteins and is not filterable by the glomeruli. Ordinarily, 99% of the filtered calcium is reabsorbed by the kidney tubules and returned to the plasma. Reabsorption occurs both in the proximal and distal tubules and in the loop of Henle. Approximately 60% of filtered calcium is reabsorbed in the proximal tubule, 30% in the loop of Henle, and 9% in the distal tubule; the remaining 1% is excreted in the urine. Renal calcium excretion is controlled primarily in the late distal tubule; parathyroid hormone (PTH) stimulates calcium reabsorption here, promoting calcium retention and lowering urinary calcium. As detailed later, PTH is an important regulator of plasma calcium concentration.

Most ingested phosphate is absorbed from the GI tract, and the primary route of excretion of this phosphate is via the urine. Ordinarily, about 85% of filtered phosphate is reabsorbed and 15% is excreted in the urine. Phosphate reabsorption occurs via active transport, mainly in the proximal tubule, where 65% to 80% of filtered phosphate is reabsorbed. PTH inhibits phosphate reabsorption in the proximal tubule and has a major regulatory effect on phosphate homeostasis. It increases urinary phosphate excretion, leading to the condition of **phosphaturia**, with an accompanying decrease in the plasma phosphate concentration.

Calcium and phosphate in bone are in continuous flux.

Although bone may be considered as being a relatively inert material, it is active metabolically. Considerable amounts of calcium and phosphate enter and exit the bone each day, and these processes are hormonally controlled.

Mature bone can be simply described as inorganic mineral deposited on an organic framework. The mineral portion of bone is composed largely of calcium phosphate in the form of **hydroxyapatite crystals**, which have the general chemical formula $Ca_{10}(PO_4)_6(OH)_2$. The mineral portion of bone typically comprises about 25% of its volume, but because of its high density, the mineral fraction is responsible for approximately half the weight of bone. Bone contains considerable amounts of the body's content of carbonate, magnesium, and sodium, in addition to calcium and phosphate (see Table 35.2).

The organic matrix of bone on which the bone mineral is deposited is called **osteoid**. **Type I collagen** is the primary constituent of osteoid, comprising 95% or more. Collagen in bone is similar to that of skin and tendons, but bone collagen exhibits some biochemical differences that impart increased mechanical strength. The remaining noncollagen portion (5%) of organic matter is referred to as **ground substance** that consists of various **proteoglycans**. These proteoglycans are high molecular weight compounds consisting of different types of polysaccharides. Typically, they are 95% or more carbohydrate.

Electron microscopic study of bone reveals needle-like hydroxyapatite crystals lying alongside collagen fibers. This orderly association of hydroxyapatite crystals with the collagen fibers is responsible for the strength and hardness characteristic of bone. A loss of either bone mineral or organic matrix greatly affects the mechanical properties of bone. Complete demineralization of bone leaves a flexible collagen framework, and the complete removal of organic matrix leaves a bone with its original shape, but extremely brittle. **Osteogenesis imperfecta**, or brittle bone disease, is a genetic disorder characterized by bones that break easily as a result of defects in the production of the fibrillar collagen matrix.

Osteoblasts, osteocytes, and osteoclasts constitute the major cell types in bone.

Osteoblasts are located on the bone surface and are responsible for osteoid synthesis (Fig. 35.3). Like many cells that actively synthesize proteins for export, osteoblasts have an abundant rough endoplasmic reticulum and Golgi apparatus. Cells actively engaged in osteoid synthesis are cuboidal, whereas those less active are more flattened. Numerous cytoplasmic processes connect adjacent osteoblasts on the bone surface and connect osteoblasts with osteocytes deeper in the bone. Osteoid produced by osteoblasts is secreted into the space adjacent to the bone. Eventually, new osteoid becomes mineralized, and in the process, mineralized bone surrounds osteoblasts.

As osteoblasts are progressively incorporated into mineralized bone, they lose much of their bone-forming ability and become quiescent. At this point, they are called **osteocytes**. Many of the cytoplasmic connections in the osteoblast stage are maintained into the osteocyte stage. These connections become visible channels or **canaliculi** that provide direct contact for osteocytes deep in bone with other osteocytes and with the bone surface. It is generally believed that these canaliculi provide a mechanism for the transfer of nutrients, hormones, and waste products between the bone surface and its interior.

Osteoclasts are cells responsible for bone resorption. They are large, multinucleated cells located on bone surfaces. Osteoclasts promote bone resorption by secreting

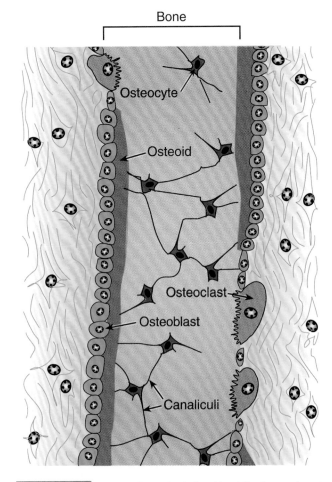

Figure 35.3 The location and relationship of the three primary cell types involved in bone metabolism.

acid and proteolytic enzymes into the space adjacent to the bone surface. Surfaces of osteoclasts facing bone are ruffled to increase their surface area and promote bone resorption. Bone resorption is a two-step process. First, osteoclasts create a local acidic environment that increases the solubility of surface bone mineral. Second, proteolytic enzymes secreted by osteoclasts degrade the organic matrix of bone.

Bone formation starts with neonatal development and continues throughout life.

Early in fetal development, the skeleton consists of little more than a cartilaginous model of what will later form the bony skeleton. The process of replacing this cartilaginous model with mature, mineralized bone begins in the center of the cartilage and progresses toward the two ends of what will later form the bone. As mineralization progresses, the bone increases in thickness and in length.

The **epiphyseal plate** is a region of growing bone of particular interest because it is here that the elongation and growth of bones occur after birth. Histologically, the epiphyseal plate shows considerable differences between its leading and trailing edges. The leading edge consists primarily of **chondrocytes**, which are actively engaged in the synthesis of cartilage of the epiphyseal plate. These cells gradually become engulfed in their own cartilage and are replaced by new cells on the cartilage surface, allowing the process to continue. The cartilage gradually becomes calcified, and the embedded chondrocytes die. The calcified cartilage begins to erode, and osteoblasts migrate into the area. Osteoblasts secrete osteoid, which eventually becomes mineralized, and a new mature bone is formed. In the epiphyseal plate, therefore, the continuing processes of cartilage synthesis, calcification, erosion, and osteoblast invasion result in a zone of active bone formation that moves away from the middle or center of the bone toward its end.

Hormones control chondrocytes of epiphyseal plates. Insulin-like growth factor I, primarily produced by the liver in response to growth hormone, serves as a primary stimulator of chondrocyte activity and, ultimately, of bone growth. Insulin and thyroid hormones provide an additional stimulus for chondrocyte activity.

Beginning a few years after puberty, the epiphyseal plates in long bones (as in the legs and arms) gradually become less responsive to hormonal stimuli and, eventually, are totally unresponsive. This phenomenon is referred to as closure of the epiphyses. In most people, epiphyseal closure is complete by about age 20; adult height is reached at this point, because further linear growth is impossible. Not all bones undergo closure. For example, those in the fingers, feet, skull, and jaw remain responsive, which accounts for the skeletal changes seen in acromegaly, the condition of growth hormone overproduction (see Chapter 31).

The flux of calcium and phosphate into and out of bone each day reflects a turnover of bone mineral and changes in bone structure generally referred to as **bone remodeling**. Bone remodeling occurs along most of the outer surface of the bone, making it either thinner or thicker, as required. In long bones, remodeling can also occur along the inner surface of the bone shaft, next to the marrow cavity. Remodeling is an adaptive process that allows bone to be reshaped to meet changing mechanical demands placed on the skeleton. It also allows the body to store or mobilize calcium rapidly.

▶ PLASMA CALCIUM AND PHOSPHATE REGULATION

Regulatory mechanisms for calcium include rapid nonhormonal mechanisms with limited capacity and somewhat slower hormonally regulated mechanisms with much greater capacity. Similar mechanisms are also involved in regulating phosphate concentrations.

Plasma free calcium can be rapidly buffered by nonhormonal mechanisms.

The calcium bound to plasma proteins and a small fraction of that in bone mineral can help prevent a rapid decrease in the plasma calcium concentration. The association of calcium with proteins is a simple, reversible, chemical equilibrium process. Protein-bound calcium, therefore, has the capacity to serve as a buffer of free plasma calcium concentrations. This effect is rapid and does not require complex signaling pathways; however, the capacity is limited, and the mechanism cannot serve a long-term role in calcium homeostasis.

Recall that ~99% of total body calcium is present in bones, and a healthy adult body has about 1 to 2 kg of calcium. Most of the calcium in bones exists as mature, hardened bone mineral that is not readily exchangeable but can be moved into the plasma via hormonal mechanisms (described below). However, ~1% (or 10 g) of the calcium in bones is in a simple chemical equilibrium with plasma calcium. This readily exchangeable calcium source is primarily located on the surface of newly formed bones. Any change in free calcium in the plasma or extracellular fluid results in a shift of calcium either into or out of the bone mineral until a new equilibrium is reached. Although this mechanism, like that described above, provides for a rapid defense against changes in free calcium concentrations, it is limited in capacity and can provide for only short-term adjustments.

Long-term, hormonal regulation of plasma calcium and phosphate is under the control of parathyroid hormone, calcitonin, and 1,25-dihydroxycholecalciferol.

Hormonal mechanisms have a large capacity and the ability to make long-term adjustments in calcium and phosphate fluxes, but they do not respond instantaneously. It may take several minutes or hours for the response to occur and adjustments to be made. Nevertheless, these are the principal mechanisms that regulate plasma calcium and phosphate concentrations.

One of the primary regulators of plasma calcium concentrations is **parathyroid hormone** (**PTH**). PTH is an 84-amino acid polypeptide produced by the parathyroid glands. Synthetic peptides containing the first 34 amino terminal residues appear to be as active as the native hormone. There are two pairs of parathyroid glands, located on the dorsal surface of the left and right lobes of the thyroid gland. Because of this close proximity, a common cause of chronic hypocalcemia is postsurgical removal of all parathyroid tissue, either during a thyroidectomy and removal of neck-associated malignancies or after inadvertent interruption of blood supply to the parathyroid glands during head and neck surgery. Hypoparathyroidism is a rare complication of radioactive iodine ablation of the thyroid gland in **Graves disease**.

The primary physiologic stimulus for PTH secretion is a decrease in plasma calcium. Figure 35.4 shows the relationship between serum PTH concentration and total plasma calcium concentration. A decrease in the ionized calcium concentration actually triggers an increase in PTH secretion. The net effect of PTH is to increase the flow of calcium into plasma and return the plasma calcium concentration toward normal.

Parafollicular cells of the thyroid gland produce a 32-amino acid polypeptide termed **calcitonin (CT)**, also known as thyrocalcitonin (see Fig. 32.1). Unlike PTH activity, for which only the initial amino terminal segment is required, CT activity requires the full polypeptide. Salmon CT differs from human CT in 9 of the 32 amino acid residues and is 10 times more potent than human CT in its hypocalcemic effect. The higher potency may be a result of a greater affinity for receptors and slower degradation by peripheral tissues. CT is often used clinically as a synthetic peptide matching the sequence of salmon CT.

In contrast to PTH, CT secretion is stimulated by an increase in plasma calcium (see Fig. 35.4). Hormones of the GI tract, especially gastrin, also promote CT secretion. Because the net effect of CT is to promote calcium deposition in bone, the stimulation of CT secretion by GI hormones provides an additional mechanism for facilitating the uptake of calcium into bone after the ingestion of a meal.

The third key hormone shown in Figure 35.5 involved in regulating plasma calcium is **vitamin D$_3$** (cholecalciferol). More precisely, a metabolite of vitamin D$_3$, 1,25-dihydroxycholecalciferol, serves as a hormone in calcium homeostasis. The D vitamins, a group of lipid-soluble compounds derived from cholesterol, have long been known to be effective in the prevention of certain bone diseases such as rickets (discussed below). Research during the past 30 years indicates that vitamin D exerts its effects through a hormonal mechanism. A related compound **vitamin D$_2$** (also known as ergocalciferol) is the form mainly found in plants and yeasts and is commonly used to supplement human foods because of its relative availability and low cost (see Fig. 35.5). Although it is less potent on a mole-per-mole basis, vitamin D$_2$ undergoes the same metabolic conversion steps and, ultimately, produces the same biologic effects as vitamin D$_3$. The physiologic actions of vitamin D$_3$ also apply to vitamin D$_2$.

Vitamin D$_3$ can be provided by the diet or formed in the skin by the action of ultraviolet light on a precursor, **7-dehydrocholesterol**, derived from cholesterol (see Fig. 35.5). In many countries where food is not systematically

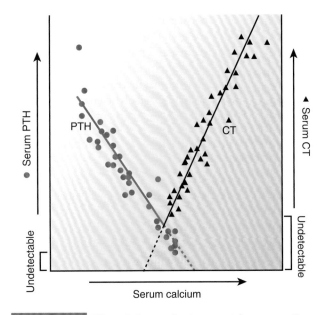

Figure 35.4 Effect of changes in plasma calcium on parathyroid hormone (PTH) and calcitonin (CT) secretion.

supplemented with vitamin D, this pathway provides the major source of vitamin D. Because of the number of variables involved, it is difficult to specify a minimum exposure time. However, exposure to moderately bright sunlight for 30 to 120 min/d usually provides enough vitamin D to satisfy the body's needs without any dietary supplementation.

Vitamins D$_3$ and D$_2$ are by themselves relatively inactive. However, they undergo a series of transformations in the liver and kidneys that convert them into powerful calcium regulatory hormones (Fig. 35.6). The first step occurs in the liver and involves the addition of a hydroxyl group to carbon 25 to form 25-hydroxycholecalciferol (25-OH D$_3$). This reaction is largely unregulated, although certain drugs and liver diseases may affect this step. Next, 25-OH D$_3$ is released into the blood and undergoes a second hydroxylation reaction on carbon 1 in the kidney. The product is **1,25-dihydroxycholecalciferol**, also known as 1,25-dihydroxyvitamin D$_3$ or calcitriol, the principal hormonally active form of the vitamin. The biologic activity of 1,25-dihydroxycholecalciferol is ~100 to 500 times greater than that of 25-OH D$_3$. The enzyme 1α-hydroxylase, which is located in tubule cells, catalyzes the reaction in the kidney.

The final step in 1,25-dihydroxycholecalciferol formation is highly regulated. The activity of 1α-hydroxylase is regulated primarily by PTH, which stimulates its activity. Therefore, if plasma calcium levels fall, PTH secretion increases; in turn, PTH promotes the formation of 1,25-dihydroxycholecalciferol. In addition, enzyme activity increases in response to a decrease in plasma phosphate. This does not appear to involve any intermediate hormonal signals but apparently involves direct activation of either the enzyme or cells in which the enzyme is located. Both a decrease in plasma calcium, which triggers PTH secretion, and a decrease in circulating phosphate result in the activation of 1α-hydroxylase and an increase in 1,25-dihydroxycholecalciferol synthesis.

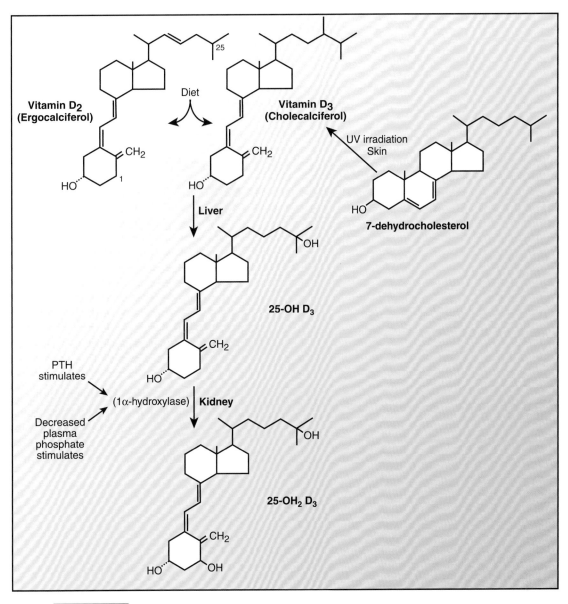

Figure 35.5 The structures of vitamin D$_3$ and vitamin D$_2$ and the conversion pathway of vitamin D$_3$ into 1,25-dihydroxycholecalciferol [1,25-(OH)$_2$ D$_3$]. Note that they differ only by a double bond between carbons 22 and 23 and a methyl group at position 24. 25-OH D$_3$, 25-hydroxycholecalciferol; PTH, parathyroid hormone.

PTH, CT, and 1,25-(OH)$_2$D$_3$ together regulate plasma calcium and phosphate.

PTH is essential for life. The complete absence of PTH causes death from hypocalcemic tetany within just a few days. The condition can be avoided with hormone replacement therapy. The net effects of PTH on plasma calcium and phosphate and its sites of action are shown in Figure 35.6. PTH causes an increase in plasma calcium while decreasing plasma phosphate. This decrease in phosphate is important with regard to calcium homeostasis. At normal plasma concentrations, calcium and phosphate are at or near chemical saturation levels. If PTH were to increase both calcium and phosphate levels, these substances would simply crystallize in bone or soft tissues as calcium phosphate, and the necessary increase in plasma calcium would not occur. Thus, the effect of PTH

to lower plasma phosphate is an important aspect of its role in regulating plasma calcium.

In the kidneys, PTH stimulates calcium reabsorption in the thick ascending limb and late distal tubule, decreasing urine calcium and increasing plasma calcium (see Fig. 35.6). It also inhibits phosphate reabsorption in the proximal tubule, leading to increased urinary phosphate excretion and a decrease in plasma phosphate. Another important effect of PTH is to increase the activity of 1α-hydroxylase in the kidney, which is involved in forming 1,25-(OH)$_2$D$_3$.

In bone, PTH activates osteoclasts to increase bone resorption and the delivery of calcium from bone into plasma (see Fig. 35.6). In addition to stimulating active osteoclasts, PTH stimulates the maturation of immature osteoclasts into mature, active osteoclasts. PTH also inhibits collagen

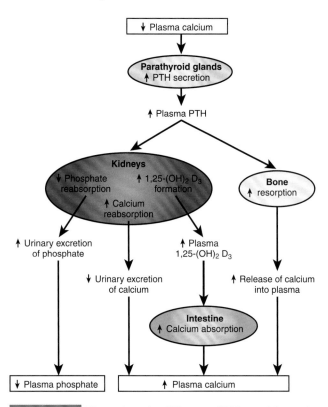

Figure 35.6 Effects of parathyroid hormone (PTH) on calcium and phosphate metabolism. 1,25-(OH)$_2$ D$_3$, 1,25-dihydroxycholecalciferol.

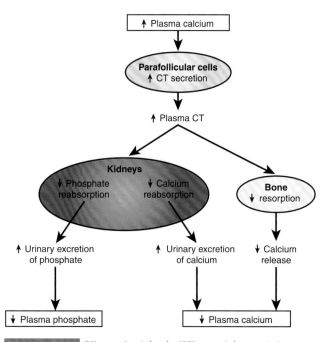

Figure 35.7 Effects of calcitonin (CT) on calcium and phosphate metabolism.

synthesis by osteoblasts, resulting in decreased bone matrix formation and decreased flow of calcium from plasma into bone mineral. All of these actions of PTH promoting bone resorption are augmented by 1,25-(OH)$_2$D$_3$.

PTH does not appear to have any major direct effects on the GI tract. However, because it increases active vitamin D formation, it ultimately increases the absorption of both calcium and phosphate from the GI tract (see Fig. 35.6).

CT is important in several lower vertebrates, but despite its many demonstrated biologic effects in humans, it appears to play only a minor role in calcium homeostasis. This conclusion mostly stems from two lines of evidence. First, CT loss following surgical removal of the thyroid gland (and, therefore, removal of CT-secreting parafollicular cells) does not lead to overt clinical abnormalities of calcium homeostasis. Second, CT hypersecretion such as from thyroid tumors involving parafollicular cells does not cause any overt problems. On a daily basis, CT probably only fine-tunes the calcium regulatory system.

The overall action of CT is to decrease both calcium and phosphate concentrations in plasma (Fig. 35.7). The primary target of CT is bone, although some lesser effects also occur in the kidneys. In the kidneys, CT decreases the tubular reabsorption of calcium and phosphate. This leads to an increase in urinary excretion of both calcium and phosphate and, ultimately, to decreased levels of both ions in the plasma. In bones, CT opposes the action of PTH on osteoclasts by inhibiting their activity. This leads to decreased bone resorption and an overall net transfer of calcium from plasma into bone. CT has little or no direct effect on the GI tract.

The net effect of 1,25-(OH)$_2$D$_3$ is to increase both calcium and phosphate concentrations in plasma (Fig. 35.8).

The activated form of vitamin D primarily influences the GI tract, although it has actions in the kidneys and bones as well. In the kidneys, 1,25-(OH)$_2$D$_3$ increases the tubular reabsorption of calcium and phosphate, promoting the retention of both ions in the body. However, this is a weak and probably

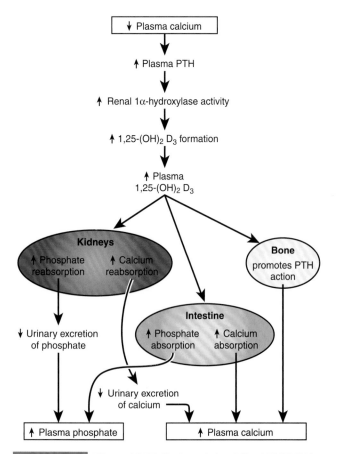

Figure 35.8 Effects of 1,25-dihydroxycholecalciferol [1,25-(OH)$_2$ D$_3$] on calcium and phosphate metabolism. PTH, parathyroid hormone.

only minor effect of the hormone. In bones, as mentioned above, 1,25-$(OH)_2D_3$ promotes actions of PTH on osteoclasts, increasing bone resorption (see Fig. 35.8). In the GI tract, 1,25-$(OH)_2D_3$ stimulates calcium and phosphate absorption by the small intestine, increasing plasma concentrations of both ions. This effect is mediated by increased production of calcium transport proteins resulting from gene transcription events and usually requires several hours to appear.

▶ BONE DYSFUNCTION

There are several **metabolic bone diseases**, all typified by ongoing disruption of the normal processes of either bone formation or bone resorption. There are also genetic bone diseases such as osteogenesis imperfecta (see Clinical Focus 35.1). The conditions most frequently encountered clinically are osteoporosis, osteomalacia, and Paget disease.

Osteoporosis leads to decreased bone density and increased risk of fracture.

Osteoporosis is a major health problem, particularly because older adults are more prone to this disorder and the average age of the population is increasing (Clinical Focus 35.2). Osteoporosis involves a reduction in total bone mass, with an equal loss of both bone mineral and organic matrix. Several factors are known to contribute directly to osteoporosis. Long-term dietary calcium deficiency can lead to osteoporosis because bone mineral is mobilized to maintain plasma calcium levels. Vitamin C deficiency can also result in a net loss of bone because vitamin C is required for normal

collagen synthesis to occur. A defect in matrix production and the inability to produce new bone eventually result in a net loss of bones. For reasons that are not entirely understood, a reduction in the mechanical stress placed on bone can lead to bone loss. Immobilization or disuse of a limb such as with a cast or paralysis can result in localized osteoporosis of the affected limb. Space flight can produce a type of disuse osteoporosis resulting from the condition of weightlessness.

Most commonly, osteoporosis is associated with advancing age in both men and women, and it cannot be assigned to any specific definable cause (see Integrated Medical Sciences Box). For several reasons, postmenopausal women are more prone to develop the disease than are men. Figure 35.9 shows the average bone calcium content for men and women versus age. Until about the time of puberty, males and females have similar bone calcium. However, at puberty, males begin to acquire bone calcium at a greater rate; peak bone mass in men may be ~20% greater than that of women. Maximum bone mass is attained between 30 and 40 years of age and then tends to decrease in both sexes. Initially, this occurs at an approximately equivalent rate, but women begin to experience a more rapid bone calcium loss at the time of menopause (about age 45 to 50 years). This loss appears to result from the decline in estrogen secretion that occurs at menopause. Low-dose estrogen supplementation of postmenopausal women is usually effective in retarding bone loss without causing adverse effects. This condition of increased bone loss in women after menopause is called **postmenopausal osteoporosis**. Along with endocrine mechanisms, it is now appreciated that components of the immune system also have an impact on bone loss in postmenopausal osteoporosis.

CLINICAL FOCUS | 35.1

Osteogenesis Imperfecta

Osteogenesis imperfecta is a genetic disorder characterized by bones that break easily, often from little or no apparent cause. Most patients with osteogenesis imperfecta have defects in the genes for type I collagen. Bones, ligaments, skin, sclera, and teeth are affected. Although the number of people affected with osteogenesis imperfecta in the United States is unknown, the best estimate suggests a minimum of 20,000 and possibly as many as 50,000. There are at least four recognized forms of the disorder (types I to IV), representing a range of severities.

Type I osteogenesis imperfecta is the most common and mildest form, resulting from decreased collagen production but normal molecular structure. The more severe forms, types II and III, involve mutations in the collagen molecule that prevent normal assembly. Type II osteogenesis imperfecta is frequently lethal at or shortly after birth, often as a result of respiratory problems. These patients have numerous fractures, severe bone deformity, and small stature. Type III patients feature progressively deforming bones, usually with moderate deformity at birth. Type IV osteogenesis imperfecta falls between types I and III in severity. Microscopic analyses revealed that some

people who are clinically within the type IV group had a distinct pattern to their bones yet did not have evidence of mutations in the type I collagen genes. These people have been designated as types V and VI osteogenesis imperfecta.

Treatment is directed toward preventing or controlling the symptoms, maximizing independent mobility, and developing optimal bone mass and muscle strength. Care of fractures, extensive surgical and dental procedures, and physical therapy are often recommended for people with osteogenesis imperfecta. Use of wheelchairs, braces, and other mobility aids is common, particularly (although not exclusively) among people with more severe types of osteogenesis imperfecta. A surgical procedure called "rodding" is frequently considered. This treatment involves inserting metal rods through the length of the long bones to strengthen them and prevent and/or correct deformities. Several medications and other treatments are being explored for their potential use to treat osteogenesis imperfecta. For example, antiresorptive therapy with intravenous pamidronate has been shown to decrease fractures in children with severe osteogenesis imperfecta, even before age 3 years. ■

The Toll of Osteoporosis

Osteoporosis is often called the "silent disease" because bone loss initially occurs without symptoms. People may not know that they have significant bone loss until their bones become so weak that a sudden strain, bump, or fall causes a fracture. Osteoporosis is a major public health threat for an estimated 44 million Americans, including 55% of people 50 years of age and older. In the United States, 10 million people are estimated to have the disease and almost 34 million more are estimated to have low bone mass, placing them at increased risk for osteoporosis. Approximately 80% of those affected by osteoporosis are women. Although osteoporosis is often thought of as an older person's disease, it can strike at any age. The seeds of osteoporosis are often sown in childhood, and it takes a lifetime of effort to prevent the disease. A frighteningly large number of children do not get sufficient exercise, vitamin D, and calcium to ensure protection from developing the disease later in life.

Osteoporosis is responsible for more than 1.5 million fractures annually, including 300,000 hip fractures, 700,000 vertebral fractures, 250,000 wrist fractures, and 300,000 fractures at other sites. The estimated national direct care expenditures (including hospitals, nursing homes, and outpatient services) for osteoporotic fractures are >18 billion dollars per year. Additional losses in wages and productivity greatly increase this number. Nearly one third of people who have hip fractures end up in nursing homes within a year; nearly 20% die within a year.

What causes osteoporosis, and what can be done to prevent or treat the disease? Although it is known that a diet low in calcium or vitamin D; certain medications, such as glucocorticoids and anticonvulsants; and excessive ingestion of aluminum-containing antacids can cause osteoporosis, in most cases, the exact cause is unknown. However, several identified risk factors associated with the disease include the following: being a woman (especially a postmenopausal woman), being Caucasian or Asian, being of advanced age, having a family history of the disease, having low testosterone levels (in men), having an inactive lifestyle, cigarette smoking, and an excessive use of alcohol.

A comprehensive program to help prevent osteoporosis includes a balanced diet rich in calcium and vitamin D, regular weight-bearing exercise, a healthy lifestyle with no smoking or excessive alcohol use, and bone density testing and medication when appropriate. Although at present there is no cure for osteoporosis, there are several U.S. Food and Drug Administration (FDA)-approved medications to either prevent or treat the disease in postmenopausal women: estrogens; estrogens and progestins; PTH; alendronate, ibandronate, and risedronate (bisphosphonates); calcitonin; and raloxifene, a selective estrogen receptor modulator. Although 20% of all osteoporosis cases occur in men, only alendronate and risedronate are currently FDA approved for use in men and only for cases of corticosteroid-induced osteoporosis. Also, PTH is approved for the treatment of osteoporosis in men who are at high risk of fracture. Testosterone replacement therapy is often helpful in a man with a low testosterone level. ■

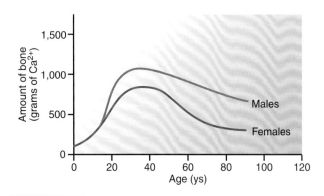

Figure 35.9 Changes in bone calcium content as a function of age in males and females. These changes can be roughly extrapolated into changes in bone mass and bone strength.

Rickets and osteomalacia are disorders of defective bone mineralization.

Rickets is characterized by the inadequate mineralization of new bone matrix, such that the ratio of bone mineral to matrix is reduced. As a result, bones may have reduced strength and are subject to distortion in response to mechanical loads. When the disease occurs in adults, it is called **osteomalacia**; when it occurs in children, it is called rickets. In children, the condition often produces a bowing of the long bones in the legs. In adults, it is often associated with severe bone pain. The primary cause of osteomalacia and rickets is a deficiency in vitamin D activity. Vitamin D may be deficient in the diet; the small intestine may not adequately absorb it; it may not be converted into its hormonally active form; or target tissues may not adequately respond to the active hormone (Table 35.3). Dietary deficiency is generally not a problem in the United States, where vitamin D is added to many foods; however, it is a major health problem in other parts of the world. Because the liver and kidneys are involved in converting vitamin D_3 into $1,25\text{-}(OH)_2D_3$, primary disease of either of these organs may result in vitamin D deficiency. Impaired vitamin D actions are somewhat rare but can be produced by certain drugs. In particular, some anticonvulsants used in the treatment of epilepsy may produce osteomalacia or rickets.

TABLE 35.3	Causes of Osteomalacia and Rickets
Type of Defect	**Cause**
Inadequate availability of vitamin D	Dietary deficiency or lack of exposure to sunlight
	Fat-soluble vitamin malabsorption
Defects in metabolic activation of vitamin D	25-Hydroxylation (liver)
	Liver disease
	Certain anticonvulsants, such as phenobarbital
	1-Hydroxylation (kidney)
	Renal failure
	Hypoparathyroidism
Impaired action of 1,25-dihydroxycholecalciferol on target tissues	Certain anticonvulsants
	1,25-Dihydroxycholecalciferol receptor defects
	Uremia

Paget disease is a chronic disorder leading to enlarged and deformed bones.

Paget disease is rarely diagnosed in patients younger than 40 years. Men are more commonly affected than women, and prevalence of Paget disease in people older than age 40 years is about 1%. It is typified by disordered bone formation and resorption (remodeling) and may occur at a single local site or at multiple sites in the body. Radiographs of the affected bone often exhibit increased density, but the abnormal structure makes the bone weaker than normal. Often, those with Paget disease experience considerable pain, and in severe cases, crippling deformities may lead to serious neurologic complications. The cause of the disease is not well understood. Both genetic and environmental factors (probably viral) appear to be important. Several therapies are available for treating the disease, including treatment with CT, but these typically offer only temporary relief from pain and complications.

INTEGRATED MEDICAL SCIENCES

New Anabolic Therapies for Osteoporosis

Osteoporosis is a major health problem with a significant consequence being fragility fractures. Mainstay therapies for this disease are agents that reduce *bone resorption*. These antiresorptive agents stabilize bone architecture and the incidence of fractures in osteoporosis. Although antiresorptive therapy has been useful in the management of osteoporosis, this therapy cannot restore the bone structure that has been lost because of increased remodeling. New insight into the mechanisms of bone formation may provide new *anabolic therapies* for osteoporosis. For example, **bone morphogenetic proteins** (**BMPs**) and Wnt have been identified and found to induce the differentiation of mesenchymal cells toward osteoblasts. Also insulin-like growth factor 1 (IGF-1) has been shown to enhance the function of mature osteoblasts. In addition to regulated synthesis and receptor binding, studies have found that the activities of BMPs, Wnt, and IGF-1 are tightly controlled by specific extracellular and intracellular regulatory proteins. Therefore, enhancing the synthesis and/or activity of BMPs, Wnt, or IGF-1 or targeting specific extracellular and intracellular proteins that regulate BMP, Wnt, or IGF-1 may provide a new therapeutic strategy to enhance bone formation in osteoporosis.

One interesting example is a protein named *sclerostin* that has BMP and Wnt antagonistic properties. It is expressed by osteoblasts and osteocytes. Interestingly sclerostin synthesis is suppressed by parathyroid hormone (PTH), and the study suggests that it is the enhanced Wnt signaling that contributes to the anabolic effect of PTH in bone. Along with these observations, clinical findings in *high-bone–mass syndrome* indicate that inactivation or neutralization of sclerostin could be used as an approach to enhance Wnt signaling and obtain an anabolic response in bone. Also, humanized monoclonal antibodies to sclerostin cause enhanced Wnt signaling and an increase in bone mass in animal studies. Of potential therapeutic significance, clinical trials have demonstrated that antisclerostin antibodies can increase bone mass density and biochemical markers of bone formation in humans. With regard to IGF-1, the decline in its secretion with aging may play a role in the pathogenesis of osteoporosis. New studies suggest that at low doses, IGF-1 increases bone formation without an effect on bone resorption. As the later effect is seen with higher doses of IGF-1, low-dose IGF-1 therapy may provide a means to increase osteoblast function without a concomitant effect on bone resorption. ■

Chapter Summary

- When plasma calcium levels fall below normal, spontaneous action potentials can be generated in nerves, leading to tetany of muscles, which, if severe, can result in death.
- About half of the circulating calcium is in the free or ionized form, about 10% is bound to small anions, and about 40% is bound to plasma proteins. Most of the phosphorus circulates free as orthophosphate.
- Most ingested calcium is not absorbed by the gastrointestinal (GI) tract and leaves the body via the feces; by contrast, phosphate is almost completely absorbed by the GI tract and leaves the body mostly via the urine.
- A decrease in plasma ionized calcium stimulates the secretion of parathyroid hormone (PTH), a polypeptide hormone produced

by the parathyroid glands. PTH plays a vital role in calcium and phosphate homeostasis and acts on bones, kidneys, and intestine to raise the plasma calcium concentration and lower the plasma phosphate concentration.
- Sequential hydroxylation reactions in the liver and kidneys convert vitamin D to the active hormone 1,25-dihydroxycholecalciferol. This hormone stimulates intestinal calcium absorption and, thereby, raises the plasma calcium concentration.
- Calcitonin, a polypeptide hormone produced by the thyroid glands, tends to lower plasma calcium, but its physiologic importance in humans has been questioned.
- Osteoporosis, rickets and osteomalacia, and Paget disease are the most common forms of metabolic bone disease.

Chapter Review Questions

1. As part of a routine physical exam, a patient's serum electrolyte levels were measured. Among the measurements, it was determined that total plasma calcium concentration was 10.2 mg/dL. What percentage of total plasma calcium is normally present as the free Ca^{2+} ion?

   A. 25%
   B. 50%
   C. 60%
   D. 75%
   E. 100%

   The correct answer is B. Half (50%) of the total plasma calcium is free or ionized.

2. The major route by which ingested phosphate leaves the body is via:

   A. bile.
   B. feces.
   C. sweat.
   D. urine.
   E. both bile and feces.

   The correct answer is D. The majority of ingested phosphate is absorbed by the GI tract and leaves the body via the urine.

3. 1,25-Dihydroxycholecalciferol can be formed in the body by metabolism of cholesterol. Which of the following is not either directly or indirectly involved in formation of 1,25-dihydroxycholecalciferol?

   A. Bone
   B. Skin
   C. Kidney
   D. Liver
   E. Both kidney and liver

   The correct answer is A. Skin, kidney, and liver can all be involved in forming the active metabolite of vitamin D, 1,25-dihydroxycholecalciferol. Bone does not form this hormone but is a target for its actions.

Clinical Application Exercises 35.1

Two siblings presented with a history of back pain. Radiographs showed decreased lumbar bone density and multiple compression fractures throughout the thoracic and lumbar spines of both patients. One child has moderate short stature and mild neurosensory hearing loss. Protein studies demonstrated electrophoretically abnormal type I collagen in samples from both children. Enzymatic cleavage of RNA:RNA hybrids identified a mismatch

in type I collagen alpha2 (COL1A2) mRNA. DNA sequencing of COL1A2 cDNA subclones defined the mismatch as a single-base mutation (1715G → A) in both children. This mutation predicts the substitution of arginine for glycine at position 436 (G436R) in the helical domain of the alpha2(I) chain. Analysis of genomic DNA identified the mutation in the asymptomatic father, who is presumably a germline mosaic carrier.

QUESTIONS

1. What most likely is the disorder?

2. How might these siblings be treated?

ANSWERS

1. Mutations in the type I collagen genes have been identified as the cause of all four types of osteogenesis imperfecta (OI). Type I OI is the most common and mildest form resulting from decreased collagen production but normal molecular structure. The more severe forms, type II and type III, involve mutations in the collagen molecule that prevent normal assembly. Type II OI is frequently lethal at or shortly after birth, often due to respiratory problems. These patients have numerous fractures, severe bone deformity, and small stature. Type III patients feature progressively deforming bones, usually with moderate deformity at birth. Type IV OI falls between type I and type III in severity. Microscopic analyses revealed that some people who are clinically within the type IV group had a distinct pattern to their bone; yet do not have evidence of having mutations in the type I collagen genes. These individuals have been designated as types V and VI OI. Based on the clinical, biochemical, and molecular findings, these siblings might have type III OI or a new variation with only spine manifestations and variable short stature into adolescence.

2. Treatment is directed toward preventing or controlling the symptoms, maximizing independent mobility, and developing optimal bone mass and muscle strength. Care of fractures, extensive surgical and dental procedures, and physical therapy are often recommended for people with OI. Use of wheelchairs, braces, and other mobility aids is common, particularly (although not exclusively) among people with more severe types of OI. A surgical procedure called "rodding" is frequently considered. This treatment involves inserting metal rods through the length of the long bones to strengthen them and prevent and/or correct deformities. Several medications and other treatments are being explored for their potential use to treat OI. For example, antiresorptive therapy with IV pamidronate has been shown to decrease fractures in children with severe OI, even before 3 years of age.

thePoint® *Visit* http://thepoint.lww.com/rhoades5e *for additional chapter review Q&A, Clinical Application Exercises, animations, and more!*

Endocrine Physiology

36 Male Reproductive System

Active Learning Objectives

Upon mastering the material in this chapter, you should be able to:
- Explain how the hypothalamus, anterior pituitary, and testes interactively regulate testicular function.
- Explain how castration causes a large increase in luteinizing hormone levels and why testosterone replacement does not completely correct follicle-stimulating hormone levels.
- Explain how the pampiniform plexus and cremasteric muscle maintain testes at a cooler temperature.

- Explain how mitosis, meiosis, and spermiogenesis mechanistically coordinate the continuous production of fresh spermatozoa.
- Explain how the spermatozoon's head, middle piece, and tail mechanistically function to reach, recognize, and fertilize an egg.
- Explain how testosterone is synthesized and functions.
- Explain how normal spermatogenesis almost never occurs with defective steroidogenesis, but normal steroidogenesis can be present with defective spermatogenesis.

The reproductive system is designed solely for the purpose of survival of the species. Although the reproductive system does not contribute to homeostasis and individual survival, it still plays an important role in physiology and behavior. The reproductive system includes the gonads, reproductive tract, and accessory glands. The overall function of male reproduction is the production of sperm and delivery of the sperm to the female. The testes produce mature male gametes, known as **spermatozoa**, which function to fertilize the female counterpart, the **oocyte**, during conception. This fertilization, in turn, produces a single-celled individual known as a **zygote**. These processes are the cornerstone of sexual reproduction. This chapter focuses on functional aspects of the male reproductive system.

▶ ENDOCRINE GLANDS OF THE MALE REPRODUCTIVE SYSTEM

A diagram of reproduction regulation in the male is presented in Figure 36.1. The system is divided into factors affecting male reproductive function: brain centers, which control pituitary release of hormones and sexual behavior; gonadal structures, which produce sperm and hormones; a ductal system, which stores and transports sperm; and accessory glands, which support sperm viability. The endocrine glands of the male reproductive system include the (1) hypothalamus, (2) anterior pituitary, and (3) testes. The hypothalamus processes information obtained from the external and internal environment, using neurotransmitters that regulate the secretion of **gonadotropin-releasing hormone (GnRH)**. GnRH moves down the hypothalamic–pituitary portal system and stimulates the secretion of **luteinizing hormone (LH)** and **follicle-stimulating hormone (FSH)**

by the gonadotrophs of the anterior pituitary. In the testes, LH binds to receptors on the **Leydig cells**, and FSH binds to receptors on the **Sertoli cells**. Leydig cells reside in the interstitium of the testes, between seminiferous tubules, and produce **testosterone**. Sertoli cells are located within the seminiferous tubules, support spermatogenesis, contain FSH and testosterone receptors, and produce estradiol, albeit at low levels. Sertoli cells also produce glycoprotein hormones—activin, follistatin, and inhibin—that regulate the secretion of FSH.

▶ TESTICULAR FUNCTION AND REGULATION

LH and FSH regulate testicular function. LH regulates the secretion of testosterone by the Leydig cells, and FSH, in synergy with testosterone, regulates the production of spermatozoa.

Hypothalamic GnRH regulates both LF and FSH secretion.

Hypothalamic neurons produce GnRH also called LH-releasing hormone [LHRH]), a decapeptide, which regulates the secretion of LH and FSH. GnRH originates from a large precursor molecule called preproGnRH (Fig. 36.2). PreproGnRH consists of a signal peptide, native GnRH, and a GnRH-associated peptide (GAP). The signal peptide (or leader sequence) allows the protein to cross the membrane of the rough endoplasmic reticulum (RER). However, both the signal peptide and GAP are enzymatically cleaved at the RER prior to GnRH secretion. Although neurons that produce GnRH can be located in various areas of the brain, their highest concentration is in the medial basal hypothalamus, in the region of the infundibulum and arcuate nucleus.

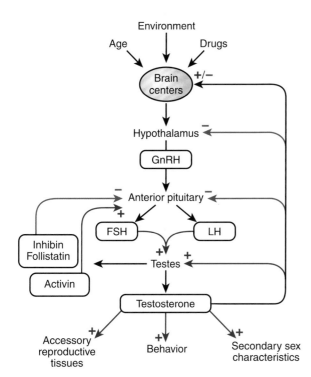

Figure 36.1 Regulation of reproduction in the male. The main reproductive hormones are shown in *boxes. Plus* and *minus* signs depict positive and negative regulations, respectively. FSH, follicle-stimulating hormone; GnRH, gonadotropin-releasing hormone; LH, luteinizing hormone.

GnRH enters the hypothalamic–pituitary portal system and binds to receptors on the plasma membranes of pituitary cells, resulting in the synthesis and release of LH and FSH. Three distinct pituitary LH- and FSH-secreting cells have been identified: (1) LF-secreting **gonadotrophs**, (2) FSH-secreting gonadotrophs, and (3) gonadotrophs that secrete both LH and FSH. GnRH can induce the secretion of both hormones simultaneously because GnRH receptors are present on all of these cell types.

Many external cues and internal signals influence the secretion of GnRH, LH, and FSH. For example, the amounts of GnRH, FSH, and LH secreted change with age, stress levels, and hormonal state. In addition, various disease states lead to hyposecretion of GnRH. Little, if any, secretion of hypothalamic GnRH occurs in patients with prepubertal

hypopituitarism, resulting in a failure of the development of the testes, primarily a result of a lack of LH, FSH, and testosterone.

Male patients with **Kallmann syndrome** are hypogonadal from a deficiency in LH and FSH secretion because of a failure of GnRH neurons to migrate from the olfactory bulbs, their embryologic site of origin. This failure results in an insufficient hypothalamic source of GnRH to maintain secretion of LH and FSH, and the testes fail to undergo significant development.

LH and FSH regulate testosterone secretion and sperm production.

LH and FSH each contain two polypeptide subunits, referred to as α and β chains, that are about 15 kDa in size. These hormones contain the same α subunit but different β subunits. Each hormone is glycosylated prior to release into the general circulation. Glycosylation regulates the half-life, protein folding for receptor recognition, and biologic activity.

LH and FSH regulate testicular function. They bind G-protein–coupled receptors on Leydig and Sertoli cells, respectively, that increase cAMP. For the most part, cAMP can account for all of the actions of LH and FSH on testicular cells via PKA-mediated activation of transcription factors such as steroidogenic factor-1 (SF-1) and cAMP response element–binding protein (CREB). These factors activate the promoter region of the genes of steroidogenic enzymes that control testosterone production by Leydig cells. Similar signaling occurs in Sertoli cells that regulate the production of estradiol. The testis converts testosterone and some other androgens to estradiol by the process of aromatization, although estradiol production is low in males.

Another major function of the testis is the production of mature sperm and of several regulatory proteins (activin, androgen-binding protein, follistatin, and inhibin), characteristics of which are detailed in later sections.

Low-frequency GnRH pulses lead to FSH release, whereas high-frequency GnRH pulses stimulate LH release.

GnRH is secreted in a pulsatile manner. GnRH pulsatility is ultimately necessary for proper functioning of the testes because it regulates the secretion of FSH and LH, which are

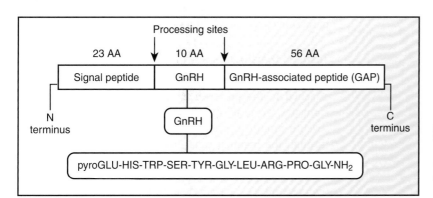

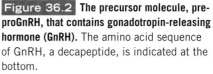

Figure 36.2 The precursor molecule, pre-proGnRH, that contains gonadotropin-releasing hormone (GnRH). The amino acid sequence of GnRH, a decapeptide, is indicated at the bottom.

Reproductive Physiology

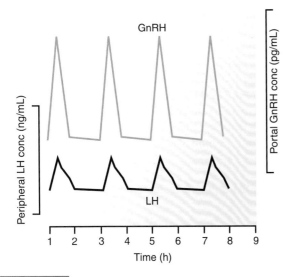

also released in a pulsatile fashion (Fig. 36.3). Continuous exposure of gonadotrophs to GnRH results in desensitization of GnRH receptors, leading to a decrease in LH and FSH release. Therefore, the pulsatile pattern of GnRH release serves an important physiologic function. The administration of GnRH at an improper frequency results in a decrease in circulating concentrations of LH and FSH.

Most evidence for GnRH pulses has come from animal studies because GnRH must be measured in hypothalamic–hypophyseal portal blood, an extremely difficult area from which to obtain blood samples in humans. Because distinct pulses of LH follow discrete pulses of GnRH, measurements of LH in serum indirectly indicate that GnRH pulses have occurred. Unlike LH secretion, FSH secretion is not always pulsatile, and even when it is pulsatile, there is only partial concordance between LH and FSH pulses. Numerous human studies measuring LH and FSH secretion in peripheral blood at various times have provided much of the information regarding the role of LH and FSH in regulating testicular development and function. However, the exact relationship between endogenous GnRH pulses and LH and FSH secretion in humans is unknown.

Hypogonadal **eunuchoid** men exhibit low levels of LH in serum and do not exhibit pulsatile secretion of LH. Pulsatile injections of GnRH restore LH and FSH secretion and increase sperm counts. FSH pulses tend to be smaller in amplitude than LH pulses, mostly because FSH has a longer half-life than LH in the circulation.

While it remains unclear what generates GnRH pulsatility, the activity of the pulse generator is under negative feedback control by gonadal steroids at the level of the hypothalamus and the pituitary. For example, castration causes a large increase in basal LH levels in serum, as evidenced by an increase in frequency and amplitude of LH pulses, indicating that negative feedback of gonadal steroids controls the GnRH pulse generator. Both testosterone and estradiol provide negative feedback. Studies indicate that the negative

feedback of gonadal steroids controls LH at the hypothalamic level. Evidence also supports the concept that testosterone has negative feedback control on LH secretion directly at the pituitary level. In addition to estradiol, dihydrotestosterone, another active metabolite of testosterone, exerts negative feedback control on LH secretion, suggesting that testosterone does not need to be converted to estradiol.

Steroids and polypeptides from the testis inhibit both LH and FSH secretion.

Testosterone, estradiol, inhibin, activin, and follistatin are major testicular hormones that regulate the release of LH and FSH. Testosterone inhibits LH release by decreasing the secretion of GnRH and, to a lesser extent, by reducing gonadotroph sensitivity to GnRH. Estradiol formed from testosterone by aromatase also has an inhibitory effect on GnRH secretion. Acute testosterone treatment does not alter pituitary responsiveness to GnRH, but prolonged exposure significantly reduces the secretory response to GnRH.

Removal of the testes results in increased circulating levels of LH and FSH. Replacement therapy with physiologic doses of testosterone restores LH to precastration levels but does not completely correct FSH levels. This observation led to a search for a gonadal factor that specifically inhibits FSH release. The polypeptide hormone **inhibin** was eventually isolated from seminal fluid. Inhibin is produced by Sertoli cells. Inhibin is composed of two dissimilar subunits, α and β, which are held together by disulfide bonds. There are two β-subunit forms, called A and B. Inhibin B consists of the α subunit bound by a disulfide bridge to the βB subunit and is the physiologically important form of inhibin in the human male. Inhibin acts directly on the anterior pituitary and inhibits the secretion of FSH, but not LH.

Activin is produced by Sertoli cells, stimulates the secretion of FSH, has an approximate molecular weight of 30 kDa, and has multiple forms based on the βA and βB subunits of inhibin. The multiple forms of activin are called activin A (two βA subunits linked by a disulfide bridge), activin B (two βB subunits), and activin AB (one βA subunit and one βB subunit). The major form of activin in the male is currently unknown, although both Sertoli and Leydig cells have been implicated in its secretion.

Follistatin is a single-chain protein hormone, with several isoforms, that binds and deactivates activin. Thus, the deactivation of activin by binding to follistatin reduces FSH secretion. Follistatin is produced by Sertoli cells and acts as a paracrine factor on the developing spermatogenic cells.

Testis is the site of sperm and seminal fluid formation.

During embryonic stages of development, the testes lie attached to the posterior abdominal wall. As the embryo elongates, the testes move to the inguinal ring. Between the seventh month of pregnancy and birth, the testes descend through the inguinal canal into the scrotum. The location of the testes in the scrotum is important for sperm production, which is optimal at 2°C to 3°C lower than the core body temperature. Two systems help maintain the testes at a cooler temperature. One is the **pampiniform plexus** of blood vessels, which serves as a countercurrent heat exchanger between warm arterial blood reaching the testes and cooler venous blood leaving the testes. The second is the **cremasteric muscle**, which responds to changes in temperature by moving the testes closer or farther away from the body. Prolonged exposure of the testes to elevated temperature, fever, or thermoregulatory dysfunction can lead to temporary or permanent sterility as a result of a failure of spermatogenesis, whereas steroidogenesis is unaltered.

A thick, fibrous connective tissue layer, the tunica albuginea, encapsulates the testes. Each human testis contains hundreds of tightly packed **seminiferous tubules**, ranging from 150 to 250 µm wide and from 30 to 70 cm long. The tubules are arranged in lobules, separated by extensions of the tunica albuginea, and open on both ends into the rete testis. Examination of a cross-section of a testis reveals distinct morphologic compartmentalization. Sperm production is carried out in the avascular seminiferous tubules. Each seminiferous tubule is composed of two somatic cell types (myoid cells and Sertoli cells) and germ cells. A basement membrane (basal lamina) with myoid cells on its perimeter, which define its outer limit, surrounds the seminiferous tubule. On the inside of the basement membrane are large, irregularly shaped Sertoli cells, which extend from the basement membrane to the lumen (Fig. 36.4). Tight junctions attach Sertoli cells to one another near their base (Fig. 36.5). The

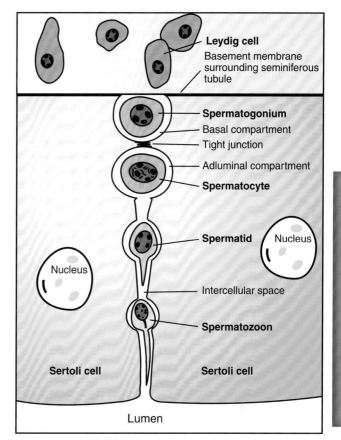

Figure 36.5 **Sertoli cells.** Sertoli cells are connected by tight junctions, which divide the intercellular space into a basal compartment and an adluminal compartment. Spermatogonia are located in the basal compartment and maturing sperm in the adluminal compartment. Spermatocytes are formed from the spermatogonia and cross the tight junctions into the adluminal compartment; they mature into spermatozoa.

tight junctions divide each tubule into a basal compartment, whose constituents are exposed to circulating agents, and an adluminal compartment, which is isolated from blood-borne elements. The tight junctions limit the transport of fluid and macromolecules from the interstitial space into the tubular lumen, forming the blood–testis barrier.

Located between the nonproliferating Sertoli cells are germ cells at various stages of division and differentiation. Mitosis of the **spermatogonia** (diploid progenitors of spermatozoa) occurs in the basal compartment of the seminiferous tubule (see Fig. 36.5). The early meiotic cells (primary spermatocytes) move across the junctional complexes into the adluminal compartment, in which they mature into spermatozoa or **gametes** after meiosis. The adluminal compartment protects spermatozoa from the immune system that does not recognize spermatozoa that develop in the adluminal compartment as "self." If the adluminal compartment is compromised, sperm is allowed to mingle with immune cells from the circulation and males can develop antibodies against their own sperm, resulting in infertility. Sperm antibodies are often present after vasectomy or testicular injury and in some autoimmune diseases in which the adluminal compartment is ruptured.

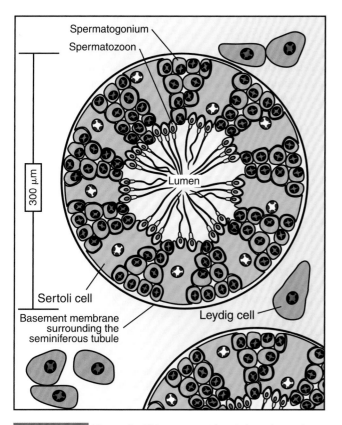

Figure 36.4 **The testis.** This cross-sectional view shows the anatomic relationship of the Leydig cells, basement membrane, seminiferous tubules, Sertoli cells, spermatogonia, and spermatozoa.

Sertoli cells nurture the development of sperm cells through spermatogenesis.

Sertoli cells are critical to germ cell development, as indicated by their close contact. As many as 6 to 12 spermatids may be attached to a Sertoli cell. Sertoli cells phagocytose residual bodies (excess cytoplasm resulting from the transformation of spermatids to spermatozoa) and damaged germ cells, provide structural support and nutrition for germ cells, secrete fluids, and assist in **spermiation**—the final detachment of mature spermatozoa from the Sertoli cell into the lumen. Spermiation may involve **plasminogen activator**, which converts plasminogen to plasmin, a proteolytic enzyme that assists in the release of the mature sperm into the lumen. Sertoli cells also synthesize large amounts of transferrin, an iron transport protein important for sperm development.

During the fetal period, Sertoli cells and gonocytes form the seminiferous tubules as Sertoli cells undergo numerous rounds of cell divisions. Shortly after birth, Sertoli cells cease proliferating, and throughout life, the number of sperm produced is directly related to the number of Sertoli cells. At puberty, the capacity of Sertoli cells to bind FSH and testosterone increases. Through cAMP and PKA signaling mechanisms mentioned above, FSH exerts multiple effects on the Sertoli cell (Fig. 36.6). FSH stimulates the production of androgen-binding protein and plasminogen activator, increases secretion of inhibin, and induces aromatase activity for the conversion of androgens to estrogens. The testosterone receptor is within the nucleus of the Sertoli cell.

Androgen-binding protein (**ABP**) is a 90-kDa protein, made of a heavy and a light chain that has a high binding affinity for dihydrotestosterone and testosterone. It is similar in function, with some homology in structure, to another binding protein, **sex hormone–binding globulin** (**SHBG**), synthesized in the liver. ABP is found at high concentrations in the human testis and epididymis. It serves as a carrier of testosterone in Sertoli cells, as a storage protein for

androgens in the seminiferous tubules, and as a carrier of testosterone from the testis to the epididymis. Recall that other products of the Sertoli cell include inhibin, follistatin, and activin. Whereas activin stimulates the secretion of FSH, inhibin and follistatin decrease the levels of FSH by suppressing FSH release from the pituitary gonadotrophs and reducing FSH secretion induced by activin, respectively.

Luteinizing hormone induces Leydig cells to produce testosterone.

Leydig cells are large polyhedral cells that are often found in clusters near blood vessels in the interstitium between seminiferous tubules. They are equipped to produce steroids because they have numerous mitochondria, a prominent smooth endoplasmic reticulum (SER), and conspicuous lipid droplets. Leydig cells undergo significant changes in quantity and activity throughout life. In the human fetus, the period from week 8 to 18 is marked by active steroidogenesis, which is obligatory for differentiation of the male genital ducts. Leydig cells at this time are prominent and active, reaching their maximal steroidogenic activity at about 14 weeks, when they constitute more than 50% of the testicular volume. Because the fetal hypothalamic–pituitary axis is still underdeveloped, **human chorionic gonadotropin** (**hCG**) from the placenta, rather than LH from the fetal pituitary, controls steroidogenesis (see Chapter 38); LH and hCG bind the same receptor. After this period, Leydig cells slowly regress. At about 2 to 3 months of postnatal life, male infants have a significant rise in LH and testosterone production (infantile testosterone surge). The regulation and function of this short-lived surge remain uncertain, yet it is clearly independent of sex steroids. Leydig cells remain quiescent throughout childhood but increase in number and activity at the onset of puberty.

Leydig cells do not have FSH receptors, but FSH can increase the number of developing Leydig cells by stimulating

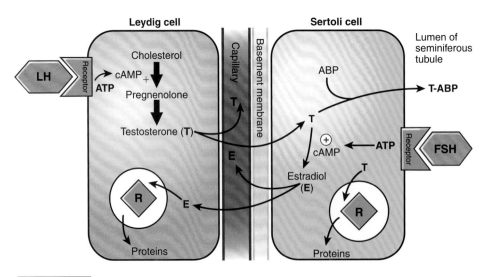

Figure 36.6 **Regulation of hormonal products and interactions between Leydig and Sertoli cells.** ABP, androgen-binding protein; ATP, adenosine triphosphate; cAMP, cyclic adenosine monophosphate; E, estradiol; FSH, follicle-stimulating hormone; LH, luteinizing hormone; R, receptor; T, testosterone.

the production of growth stimulators from Sertoli cells, which subsequently enhance the growth of Leydig cells. In addition, androgens stimulate the proliferation of developing Leydig cells. Estrogen receptors are present on Leydig cells and reduce the proliferation and activity of these cells. Leydig cells have LH receptors, and the major effect of LH is to stimulate androgen secretion via a cAMP-dependent mechanism (see Fig. 36.6). The main product of Leydig cells is testosterone, but two other androgens of less biologic activity, dehydroepiandrosterone (DHEA) and androstenedione, are also produced.

There are bidirectional interactions between Sertoli and Leydig cells (see Fig. 36.6). The Sertoli cell is incapable of producing testosterone but contains testosterone receptors as well as FSH-dependent aromatase. The Leydig cell does not produce estradiol but contains receptors for it, and estradiol can suppress the response of the Leydig cell to LH. Testosterone diffuses from the Leydig cells, crosses the basement membrane, enters the Sertoli cell, and binds to ABP. As a result, androgen levels can reach high local concentrations in the seminiferous tubules. Testosterone is obligatory for spermatogenesis and the proper functioning of Sertoli cells. In Sertoli cells, testosterone also serves as a precursor for estradiol production. The daily role of estradiol in the functioning of Leydig cells is unclear, but it may modulate responses to LH. For example, exposure of Leydig cells to LH or hCG leads to an LH or hCG desensitization. Studies show that estradiol is elevated within 30 minutes after hCG administration and that estrogen antagonists can block the hCG-induced down-regulation phenomenon.

Duct system functions in sperm maturation, storage, and transport sites.

After formation in the seminiferous tubules, spermatozoa travel to the rete testes and, from there, through the efferent ductules to the **epididymis**. Ciliary movement in the efferent ductules, muscle contraction, and the flow of fluid accomplish this movement of sperm.

Figure 36.7A shows a sagittal section of the male pelvis. The two major elements of the male sexual anatomy are the gonads (i.e., testes) and the sex accessories (i.e., the epididymis, vas deferens, seminal vesicles, ejaculatory duct, prostate, bulbourethral or Cowper glands, urethra, and penis). The epididymis is a single, tightly coiled, 4- to 5-m–long duct. It is composed of a head (caput), a body (corpus), and a tail (cauda) (Fig. 36.7B). The functions of the epididymis are storage, protection, transport, and maturation of sperm cells. Maturation at this point includes a change in functional capacity as sperm make their way through the epididymis. The sperm become capable of forward mobility during migration through the body of the epididymis. A significant portion of sperm maturation is carried out in the caput, whereas the sperm are stored in the cauda.

Frequent ejaculations result in reduced sperm numbers and increased numbers of immotile sperm in the ejaculate. The cauda connects to the vas deferens, which forms a dilated tube, the ampulla, prior to entering the prostate. The ampulla also serves as a storage site for sperm. Vasectomy—the cutting

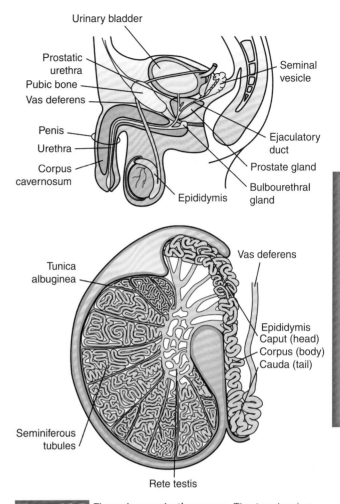

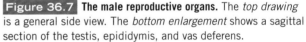

Figure 36.7 **The male reproductive organs.** The *top drawing* is a general side view. The *bottom enlargement* shows a sagittal section of the testis, epididymis, and vas deferens.

and ligation of the vas deferens—is an effective method of male contraception. Because the sperm are stored in the ampulla, men remain fertile for 4 to 5 weeks after vasectomy.

Erection and ejaculation are under neural control.

Erection is associated with sexual arousal emanating from sexually related psychic and/or physical stimuli. During sexual arousal, impulses from the genitalia, together with nerve signals originating in the limbic system, elicit motor impulses in the spinal cord. The parasympathetic nerves in the sacral region of the spinal cord carry these neuronal impulses via the cavernous nerve branches of the prostatic plexus that enter the penis. Those signals cause smooth muscle vasodilation of the arterioles and corpora cavernosa. An important biochemical regulator of arterial and cavernosal relaxations is **nitric oxide (NO)**, which is derived from the nerve terminals innervating the corpora cavernosa, the endothelial lining of penile arteries, and cavernosal sinuses. The actions of NO on the cavernosal smooth muscle and arterial blood flow are mediated through the activation of guanyl cyclase and production of **cyclic guanosine monophosphate (cGMP)** that causes smooth muscle relaxation by lowering intracellular calcium. The relaxation of the smooth muscles in those structures dilates blood vessels, which then begin to engorge with blood. The swelling of the blood-filled arterioles and corpora cavernosa compresses the thin-walled veins, restricting blood flow. The result is a reduction in the outflow of blood from the penis, and blood is trapped in the surrounding erectile tissue, leading to engorgement, rigidity, and elongation of the penis in an erect position.

Erectile dysfunction (**ED**), formerly called impotence, is the repeated inability to get or keep an erection firm enough for sexual intercourse. The word "impotence" may also be used to describe other problems that interfere with sexual intercourse and reproduction, such as lack of sexual desire and problems with ejaculation or orgasm. Using the term erectile dysfunction makes it clear that those other problems are not involved.

Semen, consisting of sperm and associated fluids, is expelled by a neuromuscular reflex that is divided into two sequential phases: (1) emission and (2) ejaculation. Seminal emission moves the sperm and associated fluids from the cauda epididymis and vas deferens into the urethra. The latter process involves efferent stimuli originating in the lumbar areas (L1 and L2) of the spinal cord and is mediated by adrenergic sympathetic (hypogastric) nerves that induce contraction of smooth muscles of the epididymis and vas deferens. This action propels sperm through the ejaculatory ducts and into the urethra. Sympathetic discharge also closes the internal urethral sphincter, which prevents retrograde ejaculation into the urinary bladder. **Ejaculation** is the expulsion of the semen from the penile urethra; it is initiated after emission. The filling of the urethra with sperm initiates sensory signals via the pudendal nerves that travel to the sacrospinal region of the cord. A spinal reflex mechanism that induces rhythmic contractions of the striated bulbospongiosus muscles surrounding the urethra results in propelling the semen out of the tip of the penis.

The secretions of the accessory glands promote sperm survival and fertility. The accessory glands that contribute to the secretions are the seminal vesicles, prostate gland, and bulbourethral glands. The semen contains only 10% sperm by volume, with the remainder consisting of the combined secretions of the accessory glands. The normal volume of semen is 3 mL, with 20 to 50 million sperm per milliliter; normal is considered more than 20 million sperm per milliliter. The seminal vesicles contribute about 75% of the semen volume. Their secretion contains fructose (the principal substrate for glycolysis of ejaculated sperm), ascorbic acid, and prostaglandins. In fact, prostaglandin concentrations are high and were first discovered in semen but were mistakenly considered to be the product of the prostate. Seminal vesicle secretions are also responsible for coagulation of the semen seconds after ejaculation. Prostate gland secretions (~0.5 mL) include fibrinolysin, which is responsible for liquefaction of the coagulated semen 15 to 30 minutes after ejaculation, releasing sperm.

▶ SPERMATOGENESIS

Spermatogenesis is the process of producing mature male gametes (sperm) and functions to fertilize the female counterpart, the oocyte. In humans, it occurs in the male testes

and epididymis in a stepwise fashion in ~64 days. It is a continual process involving mitosis of the male germ cells, which undergo extensive morphologic changes in cell shape and, ultimately, meiosis to produce the haploid spermatozoa.

Spermatogenesis produces an abundance of highly specialized, mobile sperm.

Although sperm production continues throughout life, beginning at puberty, it declines with age. Spermatogenesis is the process of transformation of male germ cells into spermatozoa. This process can be divided into three distinct phases. The phases include cellular proliferation by **mitosis**, two reduction divisions by **meiosis** to produce haploid spermatids, and cell differentiation by a process called **spermiogenesis**, in which the spermatids differentiate into spermatozoa (Fig. 36.8). Spermatogenesis is initiated shortly before puberty, under the influence of the rising levels of gonadotropins and testosterone, and continues throughout life, with a slight decline during old age.

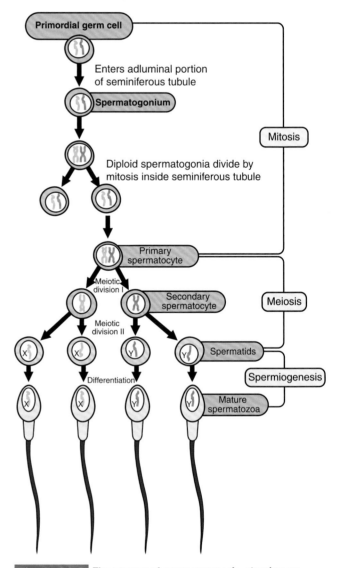

Figure 36.8 The process of spermatogenesis, showing successive cell divisions and remodeling leading to the formation of haploid spermatozoa.

The time required to produce mature spermatozoa from the earliest stage of spermatogonia is 65 to 70 days. Several developmental stages of spermatogenic cells occur during this time frame, and the stages are collectively known as the spermatogenic cycle. There is synchronized development of spermatozoa within the seminiferous tubules, and each stage is morphologically distinct. A spermatogonium becomes a mature spermatozoon after going through several rounds of mitotic divisions, a couple of meiotic divisions, and a few weeks of differentiation. Hormones can alter the number of spermatozoa, but they generally do not affect the cycle duration. Spermatogenesis occurs along the length of each seminiferous tubule in successive cycles. New cycles are initiated at regular time intervals (every 2 to 3 weeks) before the previous ones are completed. As a result, cells at different stages of development are spaced along each tubule in a "spermatogenic wave." This succession ensures continuous production of fresh spermatozoa. Approximately 200 million spermatozoa are produced daily in the adult human testes, which is about the same number of sperm present in a normal ejaculate.

Because sperm cells are rapidly dividing and undergoing meiosis, they are sensitive to external agents that alter cell division. Chemical carcinogens, chemotherapeutic agents, certain drugs, environmental toxins, irradiation, and extreme temperatures are factors that can reduce the number of replicating germ cells or cause chromosomal abnormalities in individual cells. While the immune system normally detects and destroys defective somatic cells, the blood–testis barrier isolates advanced germ cells from immune surveillance.

If physical injury or infection ruptures the blood–testis barrier and sperm cells within the barrier are exposed to circulating immune cells, antibodies can develop to the sperm cells. In the past, it was thought that the development of antisperm antibodies could lead to male infertility. It appears that men with high levels of antisperm antibodies may exhibit some infertility problems. However, men who have developed low or moderate levels of antisperm antibodies after vasectomy and who have had their vas deferens reconnected have normal fertility if the vasectomy was for a relatively short time, according to studies. Vasectomy does not appear to change hormone or sperm production by the testes. Nevertheless, in some cases, a high level of antisperm antibodies in men and women leads to infertility.

Spermatogonia divide mitotically to produce primary spermatocytes, which, in turn, undergo meiosis to become spermatids.

Spermatogonia undergo several rounds of mitotic division prior to entering the meiotic phase (see Fig. 36.8). The spermatogonia remain in contact with the Sertoli cells, migrate away from the basal compartment near the walls of the seminiferous tubules, and cross into the adluminal compartment of the tubule (see Fig. 36.5). After crossing into the adluminal compartment, the cells differentiate into spermatocytes prior to undergoing meiosis. The first meiotic division of **primary spermatocytes** gives rise to diploid (2*n* chromosomes) **secondary spermatocytes**.

The second meiotic division produces haploid (one set of chromosomes) cells called **spermatids**. Of every four spermatids emanating from a primary spermatocyte, two contain X chromosomes and two have Y chromosomes (see Fig. 36.8). Because of the numerous mitotic divisions and two rounds of meiosis, each spermatogonium committed to meiosis should yield 256 spermatids if all cells survive.

There are numerous developmental disorders of spermatogenesis. The most frequent is **Klinefelter syndrome**, which causes hypogonadism and infertility in men. Patients with this disorder have an accessory X chromosome caused by meiotic nondisjunction. The typical karyotype is 47 XXY, but there are other chromosomal mosaics. Testicular volume is reduced >75%, and ejaculates contain few, if any, spermatozoa. Spermatogenic cell differentiation beyond the primary spermatocyte stage is rare.

Formation of a mature spermatozoon requires extensive cell remodeling.

Spermatids are small, round, and nondistinctive cells. During the second half of the spermatogenic cycle, they undergo considerable restructuring to form mature spermatozoa. Notable changes include alterations in the nucleus, the formation of a tail, and a massive loss of cytoplasm. The nucleus becomes eccentric and decreases in size, and the chromatin becomes condensed. The **acrosome**, a lysosome-like structure unique to spermatozoa, buds from the Golgi apparatus, flattens, and covers most of the nucleus. The centrioles, located near the Golgi apparatus, migrate to the caudal pole and form a long axial filament made of nine peripheral doublet microtubules surrounding a central pair (called the "9 + 2 arrangement"). This becomes the **axoneme** or major portion of the tail. Throughout this reshaping process, the cytoplasmic content is redistributed and discarded. During spermiation, most of the remaining cytoplasm is shed in the form of residual bodies.

The reasons for this lengthy and metabolically costly process become apparent when the unique functions of this cell are considered. Unlike other cells, the spermatozoon serves no apparent purpose in the organism. Its only function is to reach, recognize, and fertilize an egg; hence, it must fulfill several prerequisites: it should possess an energy supply and means of locomotion, it should be able to withstand a foreign and even hostile environment, it should be able to recognize and penetrate an egg, and it must carry all the genetic information necessary to create a new person.

The mature spermatozoon exhibits a remarkable degree of structural and functional specialization well adapted to carry out these functions. The cell is small, compact, and streamlined; it is about 1 to 2 μm wide and can exceed 50 μm in length in humans. It is packed with specialized organelles and long axial fibers but contains only a few of the normal cytoplasmic constituents, such as ribosomes, ER, and Golgi apparatus. It has a prominent nucleus, a flexible tail, numerous mitochondria, and an assortment of proteolytic enzymes.

The spermatozoon consists of three main parts: a head, a middle piece, and a tail. The two major components in the head are the condensed chromatin and the acrosome. The haploid chromatin is transcriptionally inactive throughout the life of the sperm until fertilization, when the nucleus decondenses and becomes a pronucleus. The acrosome contains proteolytic enzymes, such as hyaluronidase, acrosin, neuraminidase, phospholipase A, and esterases. They are inactive until the **acrosome reaction** occurs on contact of the sperm head with the egg (see Chapter 38). Their proteolytic action enables the sperm to penetrate through the egg membranes. The middle piece contains spiral sheaths of mitochondria that supply energy for sperm metabolism and locomotion. The tail is composed of a 9 + 2 arrangement of microtubules, which is typical of cilia and flagella, and is surrounded by a fibrous sheath that provides some rigidity. The tail propels the sperm by a twisting motion, involving interactions between tubulin fibers and dynein side arms and requiring ATP and magnesium.

Testosterone is essential for sperm production and maturation.

Spermatogenesis requires high intratesticular levels of testosterone, secreted from the LH-stimulated Leydig cells. The testosterone diffuses across the basement membrane of the seminiferous tubule, crosses the blood–testis barrier, and complexes with ABP. Sertoli cells, but not spermatogenic cells, contain receptors for testosterone. Sertoli cells also contain FSH receptors. FSH function in male subjects is not readily apparent but probably mediates spermatozoa development from spermatids in concert with testosterone, especially as failed spermatogenesis leads to elevated FSH levels.

The actions of FSH and testosterone at each point of sperm cell production are unknown. On entering meiosis, spermatogenesis appears to depend on the availability of FSH and testosterone. In human males, FSH is thought to be required for the initiation of spermatogenesis before puberty. When adequate sperm production has been achieved, LH alone (through stimulation of testosterone production) or testosterone alone is sufficient to maintain spermatogenesis.

▶ ENDOCRINE FUNCTION OF THE TESTIS

A secondary function of the testis is steroidogenesis, the production of the steroid hormones. Testosterone is the main hormone produced by steroidogenesis and is primarily secreted by the testes of males and the ovaries of females. In males, testosterone production is about 10 times greater than that in females and plays a key role in the development of the testes and prostate as well as promoting secondary sexual characteristics, such as increased hair growth, muscle, and bone mass. It is also essential for health and well-being as well as the prevention of osteoporosis. Testosterone is converted to **dihydrotestosterone** (**DHT**), the most biologically active androgen, and to **estradiol**, the most biologically active estrogen.

Testosterone is primarily synthesized in Leydig cells and is derived from cholesterol, involving many enzymatic steps.

The adrenal cortex, ovaries, testes, and placenta produce steroid hormones from cholesterol. Cholesterol, a 27-carbon (C27) steroid, can be obtained from the diet or synthesized

within the body from acetate. Each organ uses a similar steroid biosynthetic pathway, but the relative amount of the final products depends on the particular subset of enzymes expressed in that tissue and the trophic hormones (LH, FSH, and adrenocorticotropic hormone) stimulating specific cells within the organ. The major steroid produced by the testis is testosterone, but other androgens, such as androstenediol, androstenedione, and DHEA, as well as a small amount of estradiol, are also produced.

Cholesterol from lipoproteins is released in Leydig cells and transported from the outer mitochondrial membrane to the inner mitochondrial membrane, a process regulated by **steroidogenic acute regulatory protein**. Under the influence of LH, with cAMP as a second messenger, cholesterol side-chain cleavage enzyme (CYP11A1), which removes six carbons attached to the 21 position, converts cholesterol to **pregnenolone** (C21). Pregnenolone is a key intermediate for all steroid hormones in various steroidogenic organs (Fig. 36.9; see also Fig. 33.5). Specific transport proteins transport pregnenolone out of mitochondria. The pregnenolone then moves by diffusion to the SER, where the remainder of sex hormone biosynthesis takes place.

Pregnenolone can be converted to testosterone via two pathways, the δ5 pathway and the δ4 pathway. In the δ5 pathway, the double bond is in ring B; in the δ4 pathway, the double bond is in ring A (see Fig. 36.9). The δ5 intermediates include 17α-hydroxypregnenolone, DHEA, and androstenediol, whereas the δ4 intermediates are progesterone, 17α-hydroxyprogesterone, and androstenedione.

The conversion of C21 steroids (the progestins) to androgens (C19 steroids) proceeds in two steps: first, 17α-hydroxylation of pregnenolone (to form 17α-hydroxypregnenolone) and second, C17,20 cleavage; therefore, two carbons are removed to form DHEA. A single enzyme, 17α-hydroxylase or 17,20-lyase (CYP17), accomplishes this hydroxylation and cleavage. Another two-step enzymatic reaction converts DHEA to androstenedione: dehydrogenation in position 3 (catalyzed by 3β-hydroxysteroid dehydrogenase [3β-HSD]) and shifting of the double bond from ring B to ring A (catalyzed by δ4,5-ketosteroid isomerase); these two may be the same enzyme. 17-Ketosteroid reductase (17β-hydroxysteroid dehydrogenase) carries out the final reaction yielding testosterone by substituting the keto group in position 17 with a hydroxyl group. Unlike all the earlier reactions, this is a reversible step but tends to favor testosterone.

Although estrogens are only minor products of testicular steroidogenesis, they are normally found in low concentrations in men. The action of the enzyme complex aromatase (CYP19) converts androgens (C19) to estrogens (C18). Aromatization involves the removal of the methyl group in position 19 and the rearrangement of ring A into an unsaturated aromatic ring. The products of aromatization of testosterone and androstenedione are estradiol and estrone, respectively (see Fig. 36.9). In the testis, the Sertoli cell is the main site of aromatization, which is stimulated by FSH; however, aromatization may also occur in peripheral tissues that lack FSH receptors (e.g., adipose tissue).

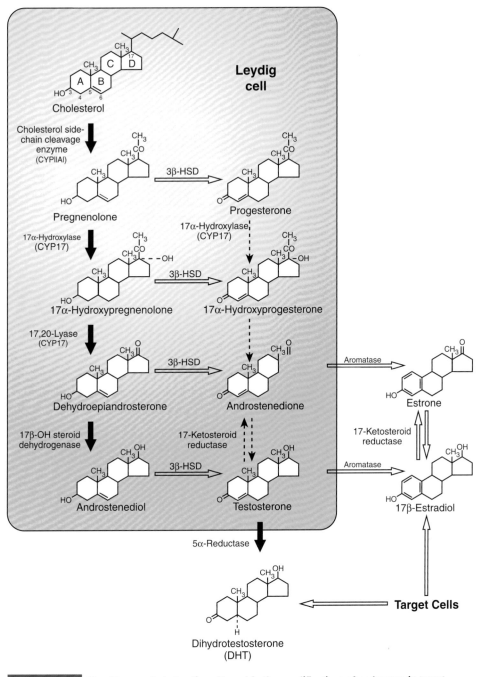

Figure 36.9 Steroidogenesis in Leydig cells and further modifications of androgens in target cells. *Solid arrows* represent the δ5 pathway. *Dashed arrows* represent the δ4 pathway. 3β-HSD, 3β-hydroxysteroid dehydrogenase.

cAMP regulates LH that, in turn, regulates the number of Leydig cells.

The action of LH on Leydig cells is mediated through specific LH receptors on the plasma membrane. A Leydig cell has about 15,000 LH receptors, and occupancy of <5% of these is sufficient for maximal steroidogenesis. This is an example of "spare receptors" (see Chapter 30). Excess receptors increase target cell sensitivity to low circulating hormones by increasing the probability that sufficient receptors will be occupied to induce a response. After exposure to a high LH concentration, the number of LH receptors and testosterone decrease. However, in response to the initial high LH, testosterone

will increase and then decrease. Thereafter, subsequent challenges with LH lead to no response or decreased responses. This so-called *desensitization* involves a loss of surface LH receptors as a result of internalization and receptor modification by phosphorylation.

The LH receptor is a single 93-kDa G protein–coupled receptor. LH receptor binding results in the activation of G_s, increased adenylyl cyclase activity, the production of cAMP, and the activation of PKA (Fig. 36.10). Low doses of LH can stimulate testosterone production without detectable changes in total cell cAMP concentration. However, the amount of cAMP bound to the regulatory subunit of PKA increases in

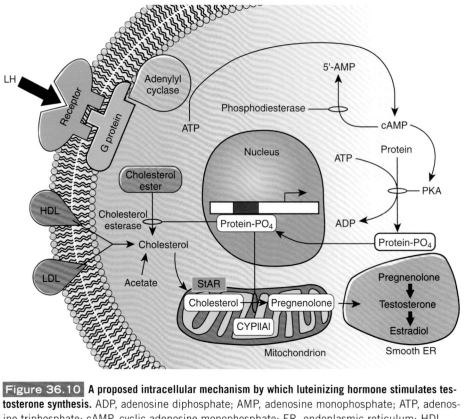

Figure 36.10 **A proposed intracellular mechanism by which luteinizing hormone stimulates testosterone synthesis.** ADP, adenosine diphosphate; AMP, adenosine monophosphate; ATP, adenosine triphosphate; cAMP, cyclic adenosine monophosphate; ER, endoplasmic reticulum; HDL, high-density lipoprotein; LDL, low-density lipoprotein; LH, luteinizing hormone; PKA, protein kinase A; StAR, steroidogenic acute regulatory protein.

response to such low doses of LH. This response emphasizes the importance of compartmentalization for both enzymes and substrates in mediating hormonal action. Other mediators, such as the phosphatidylinositol or calcium, have roles in regulating Leydig cell steroidogenesis, but it appears that the PKA pathway may predominate.

The proteins phosphorylated by PKA are specific for each cell type. Some of these, such as CREB, which functions as a DNA-binding protein, regulate the transcription of cholesterol side-chain cleavage enzyme (CYP11A1), the rate-limiting enzyme in the conversion of cholesterol to pregnenolone. **Phosphodiesterase** inactivates cAMP to AMP. This enzyme plays a major role in regulating LH (and, possibly, FSH) responses because gonadotropin stimulation activates phosphodiesterase. The increase in phosphodiesterase reduces the response to LH (and FSH). Certain drugs can inhibit phosphodiesterase; gonadotropin hormone responses will increase dramatically in the presence of those drugs.

LH stimulates steroidogenesis by two principal activations. One is the phosphorylation of cholesterol esterase, which releases cholesterol from its intracellular stores. The other is the activation of CYP11A1.

Leydig cells also contain receptors for **prolactin (PRL)**. Hyperprolactinemia in men with pituitary tumors, usually microadenomas, is associated with decreased testosterone levels. This condition is a result of a direct effect of elevated circulating levels of PRL on Leydig cells, reducing the number of LH receptors or inhibiting downstream signaling events. In addition, hyperprolactinemia may decrease LH secretion by reducing the pulsatile nature of its release. Under nonpathologic conditions, however, PRL may synergize with LH to stimulate testosterone production by increasing the number of LH receptors.

▶ ANDROGEN ACTION AND MALE DEVELOPMENT

DHT enhances development of the male reproductive tract, accompanying accessory ducts and glands, and male sex characteristics, including behavior. A lack of androgen secretion or action causes feminization.

Testosterone is not stored but circulated and metabolized by peripheral tissue.

Testosterone diffuses into the blood immediately after being synthesized in Leydig cells. An adult male produces 6 to 7 mg of testosterone per day. This amount slowly declines after age 50 and reaches about 4 mg/d in the seventh decade of life. Therefore, men do not undergo a sudden cessation of sex steroid production on aging as women do during their postmenopausal period when the ova are completely depleted.

Testosterone circulates bound to plasma proteins, with only 2% to 3% present as the free hormone. About 30% to 40% is bound to albumin and the remainder to SHBG that binds both estradiol and testosterone with a higher binding affinity for testosterone. Because its production is increased by estrogens and decreased by androgens, plasma SHBG concentration is higher in women than in men. SHBG serves as a reservoir for testosterone, and therefore, a sudden decline in newly formed testosterone may not be evident because of the large pool bound to proteins. SHBG, in effect, deactivates testosterone because only the unbound hormone can enter the cell. It also prolongs the half-life of circulating testosterone because testosterone is cleared from the circulation much more slowly if bound to a protein. Any type of liver damage or disease will generally reduce SHBG production. The latter can upset the hormonal balance between LH and testosterone. For example, if SHBG declines acutely, then free testosterone may increase, while the total amount of circulating testosterone would decrease. In response to the increase in free testosterone, LH levels would decline in a homeostatic attempt to reduce testosterone production.

Once testosterone is released into the circulation, its fate is variable. In most target tissues, testosterone functions as a prohormone and is converted to the biologically active derivatives DHT by **5α-reductase** or estradiol by aromatase (Fig. 36.11). Skin, hair follicles, and most of the male reproductive tract contain an active 5α-reductase. The enzyme irreversibly catalyzes the reduction of the double bond in ring A and generates DHT (see Fig. 36.9). DHT has a high binding affinity for the androgen receptor and is two to three times more potent than testosterone.

Congenital deficiency of 5α-reductase in males results in ambiguous genitalia containing female and male characteristics, because DHT is critical for directing the normal development of male external genitalia during embryonic life (see Chapter 38). Without DHT, the female pathway may predominate, even though the genetic sex is male and small, undescended testes are present in the inguinal region. DHT is nonaromatizable and cannot be converted to estrogens.

Drugs that inhibit 5α-reductase are currently used to reduce prostatic hypertrophy because DHT induces hyperplasia of prostatic epithelial cells. In addition, analogs of GnRH, as either agonists or antagonists, can be given to patients to reduce the secretion of androgen in androgen-dependent neoplasia or cancer. In the case of the GnRH antagonist, this analog blocks the secretion of LH. In contrast, GnRH agonists given in large quantities initially induce the secretion of LH (and androgen). However, GnRH receptor down-regulation on the pituitary gonadotrophs ensues, and this results in a decline in circulating LH and androgen.

Aromatization of some androgens to estrogens occurs in fat, liver, skin, and brain cells. Circulating levels of total estrogens (estradiol plus estrone) in men can approach those in women in their early follicular phase. Men are protected from feminization as long as production of and tissue responsiveness to androgens are normal. The treatment of hypogonadal male patients with high doses of aromatizable testosterone analogs (or testosterone), the use of anabolic steroids by athletes, abnormal reductions in testosterone secretion, estrogen-producing testicular tumors, and tissue insensitivity to androgens can lead to **gynecomastia** or breast enlargement. All of these conditions are characterized by a decrease in the testosterone/estradiol ratio.

Androgens are metabolized in the liver to biologically inactive water-soluble derivatives suitable for excretion by the kidneys. The major products of testosterone metabolism are two 17-ketosteroids, androsterone and etiocholanolone. These, as well as native testosterone, are conjugated in position 3 to form sulfates and glucuronides, which are water soluble, and excreted into the urine (see Fig. 36.11).

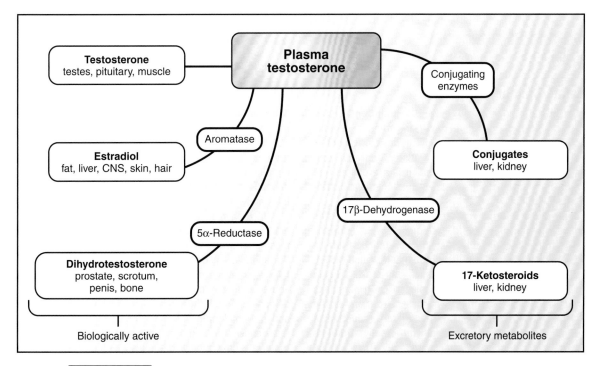

Figure 36.11 Conversion of testosterone to different products in extratesticular sites. CNS, central nervous system.

Androgens target both reproductive and nonreproductive tissues.

An **androgen** is a substance that stimulates the growth of the male reproductive tract and the development of secondary sex characteristics. Androgens have effects on almost every tissue, including alteration of the primary sex structures (i.e., the testes and genital tract), stimulation of the secondary sex structures (i.e., accessory glands), and development of secondary sex characteristics responsible for masculine phenotypic expression. Androgens also affect both sexual and nonsexual behavior. The relative potency ranking of androgens is as follows: DHT > testosterone > androstenedione > DHEA. The action of sex steroid hormones on somatic tissue, such as muscle, is referred to as "anabolic" because the end result is increased muscle size. The same molecular mechanisms that result in virilization mediate this action.

Between 8 and 18 weeks of fetal life, androgens mediate differentiation of the male genitalia. Testosterone, which reaches these target tissues by diffusion rather than by a systemic route, directly influences the organogenesis of the wolffian (mesonephric) ducts into the epididymis, vas deferens, and seminal vesicles. The differentiation of the urogenital sinus and the genital tubercle into the penis, scrotum, and prostate gland depends on testosterone being converted to DHT. Toward the end of fetal life, testosterone and insulin-like hormones from Leydig cells promote the descent of the testes into the scrotum (see Chapter 38).

Enhanced androgenic activity marks the onset of puberty. Androgens promote the growth of the penis and scrotum, stimulate the growth and secretory activity of the epididymis and accessory glands, and increase the pigmentation of the genitalia. Enlargement of the testes occurs under the influence of LH and FSH. Spermatogenesis depends on adequate amounts of testosterone. Throughout adulthood, androgens are responsible for maintaining the structural and functional integrity of all reproductive tissues. Castration of adult men results in regression of the reproductive tract and involution of the accessory glands (Clinical Focus 36.1).

Androgens are responsible for secondary sex characteristics and masculinity.

Androgens effect changes in hair distribution, skin texture, pitch of the voice, bone growth, and muscle development. Hair is classified by its sensitivity to androgens into the following: nonsexual (eyebrows and extremities); ambisexual (axilla), which is responsive to low levels of androgens; and sexual (face, chest, and upper pubic triangle), which is responsive only to high androgen levels. Hair follicles metabolize testosterone to DHT or androstenedione. Androgens stimulate the growth of facial, chest, and axillary hair; however, along with genetic factors, they also promote temporal hair recession and loss. Normal axillary and pubic hair growth in women is also under androgenic control, whereas excess androgen production in women causes the excessive growth of sexual hair (**hirsutism**).

The growth and secretory activity of the sebaceous glands on the face, upper back, and chest are stimulated by androgens, primarily DHT, and inhibited by estrogens. Increased sensitivity of target cells to androgenic action, especially during puberty, is the cause of **acne vulgaris** in both males and females. Skin derived from the urogenital ridge (e.g., the prepuce, scrotum, clitoris, and labia majora) remains sensitive to androgens throughout life and contains an active 5α-reductase. Growth of the larynx and thickening

CLINICAL FOCUS | 36.1

Prostate Cancer

Some prostate cancers highly depend on androgens for cellular proliferation; therefore, physicians attempt to totally ablate the secretion of androgens by the testes. Generally, two options for those patients are surgical castration and chemical castration. Surgical castration is irreversible and requires the removal of the testes, whereas chemical castration is reversible.

One option for chemical treatment of these patients is the use of analogs of gonadotropin-releasing hormone (GnRH), the hormone that regulates the secretion of luteinizing hormone (LH) and follicle-stimulating hormone (FSH). Long-acting GnRH agonists or antagonists reduce LH and FSH secretion by different mechanisms. GnRH agonists reduce gonadotropin secretion by desensitization of the pituitary gonadotrophs to GnRH, leading to a reduction of LH and FSH secretion. GnRH agonists initially stimulate GnRH receptors on pituitary cells but ultimately reduce their numbers. GnRH antagonists bind to GnRH receptors on the pituitary cells, prevent endogenous GnRH from binding to those receptors, and subsequently reduce LH and FSH secretion. Shortly after treatment, testicular concentrations of androgens decline because of the low levels of circulating LH and FSH. The expectation is that androgen-dependent cancer cells will cease or slow proliferation and, ultimately, die.

GnRH agonists (leuprolide acetate [trade name, Lupron]) are usually used in combination with other drugs to block most effectively androgenic activity. For example, one of the androgen-blocking drugs includes 5α-reductase inhibitors that prevent the conversion of testosterone to the highly active androgen dihydrotestosterone. In addition, antiandrogens, such as flutamide, bind to the androgen receptor and prevent binding of endogenous androgen. Some prostate cancers are androgen independent, and the treatment requires nonhormonal therapies, including chemotherapy and radiation. ■

of the vocal cords are also androgen dependent. Eunuchs maintain the high-pitched voice typical of prepubertal boys because they were castrated prior to puberty.

A complex interplay between androgens, GH, nutrition, and genetic factors influences the growth spurt of adolescent males. The growth spurt includes growth of the vertebrae, long bones, and shoulders. The mechanism by which androgens (likely DHT) alter bone metabolism is unclear. Androgens accelerate closure of the epiphyses in the long bones, eventually limiting further growth. Because of the latter, precocious puberty is associated with a final short adult stature, whereas delayed puberty or eunuchoidism usually results in tall stature. Androgens have multiple effects on skeletal and cardiac muscle. Because 5α-reductase activity in muscle cells is low, the androgenic action is a result of testosterone. Testosterone stimulates muscle hypertrophy, increasing muscle mass; however, it has minimal or no effect on muscle hyperplasia. Testosterone, in synergy with GH, causes a net increase in muscle protein.

Androgens affect, directly or indirectly, other nonreproductive organs and systems, including the liver, kidneys, adipose tissue, and hematopoietic and immune systems. The kidneys are larger in males, and androgens induce some renal enzymes (e.g., β-glucuronidase and ornithine decarboxylase). HDL levels are lower and triglyceride concentrations higher in men, compared with premenopausal women, a fact that may explain the higher prevalence of atherosclerosis in men. Androgens increase red blood cell mass (and, hence, hemoglobin levels) by stimulating erythropoietin production and by increasing stem cell proliferation in the bone marrow.

Androgen is involved in sexual differentiation of the brain.

Many sites in the brain contain androgen receptors, with the highest density in the hypothalamus, preoptic area, septum, and amygdala. Most of those areas also contain aromatase, and many of the androgenic actions in the brain result from the aromatization of androgens to estrogens. The pituitary also has abundant androgen receptors, but no aromatase. The enzyme 5α-reductase is widely distributed in the brain, but its activity is generally higher during the prenatal period than in adults. Sexual dimorphism in the size, number, and arborization of neurons in the preoptic area, amygdala, and superior cervical ganglia has been recently recognized in humans.

Unlike most species, which mate only to produce offspring, in humans, sexual activity and procreation are not tightly linked. Superimposed on the basic reproductive mechanisms dictated by hormones are numerous psychological and societal factors. In normal men, no correlation is found between circulating testosterone levels and sexual drive, frequency of intercourse, or sexual fantasies. Similarly, there is no correlation between testosterone levels and impotence or homosexuality. Castration of adult men results in a slow decline in, but not a complete elimination of, sexual interest and activity.

▶ MALE REPRODUCTIVE DISORDERS

Male reproductive dysfunctions may be caused by endocrine disruption, morphologic alterations in the reproductive tract, neuropathology, and genetic mutations. Several medical tests, including serum hormone levels, physical examination of the reproductive organs, and sperm count, are important in ascertaining causes of reproductive dysfunctions.

Hypogonadism leads to decreased functional activity of the gonads.

Male **hypogonadism** may result from defects in spermatogenesis, steroidogenesis, or both. It may be a primary defect in the testes or secondary to hypothalamic–pituitary dysfunction. Determining whether the onset of gonadal failure occurred before or after puberty is important in establishing the cause. However, several factors must be considered. First, normal spermatogenesis almost never occurs with defective steroidogenesis, but normal steroidogenesis can be present with defective spermatogenesis. Second, primary testicular failure removes feedback inhibition from the hypothalamic–pituitary axis, resulting in elevated plasma gonadotropins. In contrast, decreased gonadotropin and steroid levels and reduced testicular size almost always accompany hypothalamic and/or pituitary failure. Third, gonadal failure before puberty results in the absence of secondary sex characteristics, creating a distinctive clinical presentation called **eunuchoidism**. In contrast, men with a postpubertal testicular failure retain masculine features but exhibit low sperm counts or a reduced ability to produce functional sperm.

To establish the cause(s) of reproductive dysfunction, physical examination and medical history, semen analysis, hormone determinations, hormone stimulation tests, and genetic analysis are performed. Physical examination should establish whether eunuchoidal features (i.e., infantile appearance of external genitalia and poor or absent development of secondary sex characteristics) are present. In men with adult-onset reproductive dysfunction, physical examination can uncover problems such as *cryptorchidism* (nondescendent testes), testicular injury, *varicocele* (an abnormality of the spermatic vasculature), testicular tumors, prostatic inflammation, or gynecomastia. Medical and family history help determine delayed puberty, anosmia (an inability to smell, often associated with GnRH dysfunction), previous fertility, changes in sexual performance, ejaculatory disturbances, or impotence.

One step in the evaluation of fertility is semen analysis. Semen is analyzed on specimens collected after 3 to 5 days of sexual abstinence, as the number of sperm ejaculated remains low for a couple of days after ejaculation. Initial examination includes determination of viscosity, liquefaction, and semen volume. The sperm are then counted, and the percentage of sperm showing forward motility is scored. The spermatozoa are evaluated morphologically, with attention to abnormal head configuration and defective tails. Chemical analysis can provide information on the secretory activity of the accessory glands, which is considered abnormal if semen volume is too low or sperm motility is impaired. Fructose and

Reproductive Physiology

prostaglandin levels are determined to assess the function of the seminal vesicles and levels of zinc, magnesium, and acid phosphatase to evaluate the prostate. Terms used in evaluating fertility include aspermia (no semen), hypospermia and hyperspermia (too small or too large semen volume), azoospermia (no spermatozoa), and oligozoospermia (reduced spermatozoa).

Serum testosterone, estradiol, LH, and FSH analyses are performed using radioimmunoassays. Free and total testosterone levels should be measured; because of the pulsatile nature of LH release, several consecutive blood samples are needed. Dynamic hormone stimulation tests are most valuable for establishing the site of abnormality. A failure to increase LH release on treatment with **clomiphene**, an antiestrogen, likely indicates a hypothalamic abnormality. Clomiphene blocks the inhibitory effects of estrogen and testosterone on endogenous GnRH release. An absence of or blunted rise in testosterone after hCG injection suggests a primary testicular defect. Genetic analysis is used when congenital defects are suspected. Karyotyping of cultured peripheral lymphocytes or direct detection of specific Y antigens on cell surfaces can reveal the presence of the Y chromosome.

Hypogonadism or hypergonadism account for most male reproductive disorders.

Endocrine factors are responsible for ~50% of hypogonadal or infertility cases. Those remaining are of unknown etiology or the result of injury, deformities, and environmental factors. Endocrine-related hypogonadism can be classified as hypothalamic–pituitary defects (hypogonadotropic because of the lack of LH and/or FSH), primary gonadal defects (hypergonadotropic because gonadotropins are high as a result of a lack of negative feedback from the testes), and defective androgen action (usually the result of absence of androgen receptors or 5α-reductase). Each of these is further subdivided into several categories, but only a few examples are discussed here.

Hypogonadotropic hypogonadism can be congenital, idiopathic, or acquired. The most common congenital form is Kallmann syndrome, which results from decreased or absent GnRH secretion, as mentioned earlier. It is often associated with anosmia or hyposmia and is transmitted as an autosomal dominant trait. Patients do not undergo pubertal development and have eunuchoidal features. Plasma LH, FSH, and testosterone levels are low, and the testes are immature and have no sperm. There is no response to clomiphene, but intermittent treatment with GnRH can produce sexual maturation and full spermatogenesis.

Another category of hypogonadotropic hypogonadism, **panhypopituitarism** or pituitary failure, can occur before or after puberty and is usually accompanied by a deficiency of other pituitary hormones. **Hyperprolactinemia**, whether from hypothalamic disturbance or pituitary adenoma, often results in decreased GnRH production, hypogonadotropic state, impotence, and decreased libido. It can be treated with dopaminergic agonists (e.g., bromocriptine), which suppress PRL release (see Chapter 37). Excess androgens can also result in suppression of the hypothalamic–pituitary axis, resulting in lower LH levels and impaired testicular function. This condition often results from **congenital adrenal hyperplasia** and increased adrenal androgen production from 21-hydroxylase (CYP21A2) deficiency (see Chapter 33).

Hypergonadotropic hypogonadism usually results from impaired testosterone production, which can be congenital or acquired. The most common disorder is Klinefelter syndrome, as discussed earlier (Clinical Focus 36.2).

Pseudohermaphroditism is a reproductive paradox that results from insensitivity to androgens.

Pseudohermaphroditism produces gonads of one sex and the genitalia of the other in the same person. The term male pseudohermaphrodite is used when a testis is present, and the term female pseudohermaphrodite is used when an ovary is present. The term hermaphrodite is reserved

---

CLINICAL FOCUS | 36.2

Effects of Testosterone Administration

Although testosterone has a role in stimulating spermatogenesis, infertile men with a low sperm count do not benefit from testosterone treatment. Unless given at supraphysiologic doses, exogenous testosterone cannot achieve the required local high concentration in the testis. One function of androgen-binding protein in the testis is to sequester testosterone, which significantly increases its local concentration.

Exogenous testosterone given to men would normally inhibit endogenous luteinizing hormone (LH) release through a negative feedback effect on the hypothalamic–pituitary axis and lead to a suppression of testosterone production by Leydig cells and a further decrease in testicular testosterone concentrations. Ultimately, because LH levels decrease when exogenous testosterone is administered, testicular size decreases, as has been reported for men who abuse androgens.

High androgens have an anabolic effect on muscle tissue, leading to increased muscle mass, strength, and performance, a desired result for bodybuilders and athletes. Androgen abuse has been associated with abnormally aggressive behavior and the potential for increased incidence of liver and brain tumors. ■

for the very rare cases in which both ovarian and testicular tissues are present. One of the most interesting causes of male reproductive abnormalities is the insensitivity of the end organ to androgens. The best characterized syndrome is testicular feminization, an X-linked recessive disorder caused by a defect in the testosterone receptor. In the classical form, patients are male pseudohermaphrodites with a female phenotype and an XY male genotype. They have abdominal testes that secrete testosterone but no other internal genitalia of either sex (see Chapter 38). They commonly have female external genitalia but with a short vagina ending in a blind pouch. Breast development is typical of a female (as a result of peripheral aromatization of testosterone), but axillary and pubic hair, which are androgen dependent, are scarce or absent. Testosterone levels are normal or elevated, estradiol levels are above the normal male range, and circulating gonadotropin levels are high. The inguinally located testes usually have to be removed because of an increased risk of cancer. After orchiectomy, patients are treated with estradiol to maintain a female phenotype.

INTEGRATED MEDICAL SCIENCES

Male Infertility and a Potential Ring of Concern

An active area of bench research is aimed at better understanding the cause of male infertility with an underlying goal of improving semen quality and chances of pregnancy. It is estimated that infertility affects one in every six couples who are trying to conceive, with about half of the cases resulting from the inability of the male to initiate pregnancy. Causes of male infertility include abnormal sperm production or function, impaired delivery of sperm, general health and lifestyle issues, and exposure to certain environmental factors. Of new concern are findings that a modern day environmental factor, the cell phone, commonly carried close to the testicles (i.e., in a trouser pocket or clipped to a waist belt) can reduce a male's fertilizing potential.

Recent studies have highlighted that spermatozoa of infertile men have higher levels of DNA damage, which adversely affects the ability of spermatozoa to reach, recognize, and fertilize an egg. Mechanistically, a key contributing factor of DNA damage in spermatozoa appears to be oxidative stress. Oxidative stress results from an imbalance of **reactive oxygen species** (**ROS**) and antioxidants. Immature spermatozoa with residual cytoplasm produce ROS by having excess nicotinamide adenine diphosphate hydrogen generated via glucose-6-phosphate dehydrogenase. Countering the harmful effect of excess ROS production are antioxidants that are present in normal, healthy seminal ejaculates.

Findings suggesting that electromagnetic waves (EMWs) emitted from cell phones may negatively impact the ROS/antioxidant balance in semen represent a new, modern-day concern. In a large prospective study, conducted by researchers at the Cleveland Clinic, the effects of EMWs emitted by cell phones (900 to 1,900 MHz) on various markers of sperm quality, such as count, motility, morphology, and viability, were assessed. These studies revealed that exposure of ejaculated semen samples to commercially available cell phones for 1 hour caused a significant decrease in sperm motility and viability, increased ROS levels, and decreased ROS–total antioxidant capacity score when compared with neat semen from a nonexposed group.

As importantly noted by this research team, many men carry their cell phones in a trouser pocket (or clipped to a waist belt) exposing testes to high-power cell phone density. As the effects of frequency, distance of the phone from the source, and the duration of talk time on spermatozoa are not known, current investigations by these researchers are now employing a two-dimensional anatomical model of the tissue to extrapolate the effects seen in *in vitro* condition to real-life conditions. ■

Chapter Summary

- In the testes, LH controls the synthesis of testosterone by Leydig cells, and FSH increases the production of androgen-binding protein, inhibin, and estrogen by Sertoli cells.
- Spermatozoa are produced within the seminiferous tubules of both testes. Sperm develop from spermatogonia through a series of developmental stages that include spermatocytes and spermatids.
- The sperm mature and are stored in the epididymis. At the time of ejaculation, muscular contractions of the epididymis and vas deferens move sperm through the ejaculatory ducts into the prostatic urethra. The sperm are finally moved out of the body through the urethra in the penis.

- GnRH controls LH and FSH secretion by the anterior pituitary.
- Testosterone mainly reduces LH secretion, whereas inhibin reduces the secretion of FSH. The testicular hormones complete a negative feedback loop with the hypothalamic–pituitary axis.
- Androgens have several target organs and have roles in regulating the development of secondary sex characteristics, the libido, and sexual behavior.
- The most potent natural androgen is DHT that 5α-reductase produces from testosterone.
- Male reproductive dysfunction is often a result of a lack of LH and FSH secretion or abnormal testicular morphology.

Chapter Review Questions

1. A major causal factor in some cases of hypogonadism is:

   A. reduced secretion of gonadotropin-releasing hormone (GnRH).
   B. hypersecretion of pituitary LH and FSH as the result of increased GnRH.
   C. excess secretion of testicular activin by Sertoli cells.
   D. failure of the hypothalamus to respond to testosterone.
   E. increased number of FSH receptors in the testis.

   The correct answer is A. Reduced secretion of GnRH will result in extremely low levels of circulating LH and FSH, causing testicular atrophy, as in Kallmann syndrome. Hypersecretion of LH and FSH, increased activin, and an increased number of FSH receptors all lead to hyperfunction of the testis, not hypofunction. A failure of the hypothalamus to respond to testosterone increases LH, leading to increased Leydig cell androgens and testicular hypertrophy.

2. The major function of follistatin is to:

   A. bind FSH and increase FSH secretion.
   B. inhibit the production of seminal fluid.
   C. reduce testosterone secretion by Leydig cells.

   D. stimulate the production of spermatogonia.
   E. bind activin and thus decrease FSH secretion.

   The correct answer is E. Follistatin is a binding protein for activin. Activin cannot increase FSH secretion when follistatin is bound to it, so FSH secretion decreases. Follistatin does not bind FSH, does not inhibit seminal fluid production and Leydig cell testosterone secretion, and does not stimulate the production of spermatogonia.

3. A major function of the epididymis is:

   A. storage and transport of mature sperm.
   B. initiating the development of spermatozoa.
   C. secretion of estrogens.
   D. production of inhibin.
   E. secretion of fluids that contribute to semen.

   The correct answer is A. The epididymis and vas deferens are major storage sites of spermatozoa. Spermatozoa develop in the seminiferous tubules. Sertoli cells, not the epididymis, secrete estrogens and inhibin. The prostate gland, seminal vesicles, and bulbourethral glands secrete the seminal fluids.

Clinical Application Exercises 36.1

SEXUAL DEVELOPMENT

A 20-year-old male patient presented with lack of sexual development. On examination, he was eunuchoidal and hypogonadal, and olfactory function testing showed he was anosmic. Biochemical investigations proved he was hypogonadotropic. His appearance was very different from his alleged identical twin who had undergone a normal puberty and had normal plasma testosterone and

gonadotropin levels. However, the twin was hyposmic. Genetic fingerprinting confirmed the twins were identical. Why the syndrome was incompletely expressed in one of them is unexplained. The parents and a normally menstruating sister had normal olfactory function.

QUESTIONS

1. What is the syndrome afflicting the one twin brother?

2. What is the basis for the low gonadotropin levels?

ANSWERS

1. Kallmann syndrome
2. This syndrome results from decreased or absent GnRH secretion. It is often associated with anosmia or hyposmia and is transmitted as an autosomal dominant trait. Patients do not undergo pubertal development and have eunuchoidal features. Plasma LH, FSH, and testosterone levels are low, and the testes are immature and have no sperm.

the Point® *Visit* http://thepoint.lww.com/rhoades5e *for additional chapter review Q&A, Clinical Application Exercises, animations, and more!*

37 Female Reproductive System

Active Learning Objectives

Upon mastering the material in this chapter, you should be able to:

- Explain the mechanism through which pulses of hypothalamic gonadotropin-releasing hormone regulate secretion of luteinizing hormone and follicle-stimulating hormone and the effect of these two gonadotropins on follicular development, steroidogenesis, ovulation, and formation of the corpus luteum.
- Describe how luteinizing hormone and follicle-stimulating hormone, in coordination with ovarian theca and granulosa cells, regulate the secretion of follicular estradiol.
- Explain how positive feedback of follicular estradiol on the hypothalamic–pituitary axis induces luteinizing hormone and follicle-stimulating hormone surges and causes ovulation.
- Outline the sequence of distinct steps in follicular development including how appropriately timed luteinizing hormone and follicle-

stimulating hormone surges, which induce inflammatory reactions in the graafian follicle, lead to follicular rupture and ovulation.
- Predict the effect of gonadotropin suppression on follicular development.
- List the hormones required for the formation of a functional corpus luteum.
- Describe the effects of estradiol and progesterone on the endometrial cycle.
- Describe the changes in hypothalamic secretory function at the onset of puberty that increase luteinizing hormone and follicle-stimulating hormone secretion, enhance ovarian function, and lead to the first ovulation.
- Explain the mechanism that results in menopause and describe the medical complications of this state for older women.

The fertility of a women is cyclic, with the release of a mature ovum approximately once per month. Pulsatile secretion of **luteinizing hormone (LH)** and **follicle-stimulating hormone (FSH)** from the pituitary gland controls the production of ovarian steroid hormones and the release of an ovum during the menstrual cycle. The cyclic changes in steroid hormone secretion from the ovary cause significant changes in the structure and function of the uterus, preparing it for the reception of a fertilized ovum. At different stages of the menstrual cycle, progesterone and estradiol exert negative and positive feedback effects on the hypothalamus and pituitary gonadotrophs, generating the cyclic pattern of LH and FSH release characteristic of the female reproductive system. Hormonal events during the menstrual cycle are delicately synchronized; thus, stress and environmental, psychologic, and social factors can readily affect the menstrual cycle.

▶ HORMONAL REGULATION OF THE FEMALE REPRODUCTIVE SYSTEM

Unlike for men, in whom there is a continuous production of sperm and a constant secretion of testosterone, women experience cyclic swings in the menstrual cycle, ovulation, and secretion of sex hormones. Under normal conditions, these changes follow a 28-day cycle resulting from the interaction of hormones released by the anterior pituitary and ovary. As shown in Figure 37.1, the hypothalamus produces **gonadotropin-releasing hormone (GnRH**; also called **luteinizing hormone-releasing hormone [LHRH]**), which stimulates the secretion of LH and FSH from the anterior pituitary. Release of GnRH from the hypothalamus is regulated by brain areas that monitor various internal and external signals. LH and FSH regulate ovarian

steroidogenesis, which produces **estradiol**, **progesterone**, and **androgen**. The ovarian steroids have positive effects on the reproductive tract and secondary sex characteristics. Ovarian steroids have a negative feedback effect on the ovary to reduce their synthesis and on the anterior pituitary to inhibit LH and FSH secretion. An important exception to the negative feedback effects of ovarian steroids on the anterior pituitary occurs at midcycle, when estradiol has a positive feedback effect on the hypothalamic–pituitary axis to induce significant increases in the secretion of GnRH, LH, and FSH. The ovary also produces three polypeptide hormones that regulate anterior pituitary hormone release. **Inhibin** and **follistatin** (an activin-binding protein) suppress the secretion of FSH, while **activin** (an inhibin-binding protein) increases the secretion of FSH.

The sex steroids in women, estrogen, progesterone, and androgen, are present in and removed from the plasma through the same mechanisms as other steroid hormones (see Chapter 33). Of the three estrogens, estradiol, **estrone**, and **estriol**, estradiol is the most abundant and is many times more potent than estrone and estriol. Estrogens and androgens in the blood are bound to **sex hormone–binding globulin (SHBG)** or with lower affinity to **albumin**. Circulating progesterone binds to the plasma protein **corticosteroid-binding protein (transcortin)** and also to albumin. Binding proteins prolong the lifetime of the sex steroids, which are eventually cleared from the circulation by the liver, undergoing conversion to glucuronides or sulfates, and then excreted in the urine.

Neurons in the hypothalamus release GnRH in a pulsatile fashion.

GnRH is a decapeptide, produced by neurons in the hypothalamic arcuate nucleus (and preoptic area under specific conditions), which is released in a pulsatile manner into the

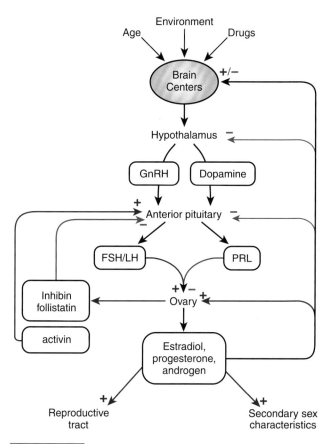

Figure 37.1 **Hormonal regulation of reproduction in the female.** The main reproductive hormones are shown in boxes. Positive regulation is depicted by a plus sign. Negative regulation is depicted by a minus sign. FSH, follicle-stimulating hormone; GnRH, gonadotropin-releasing hormone; LH, luteinizing hormone; PRL, prolactin.

hypophyseal portal circulation to regulate the secretion of LH and FSH (see Chapter 31). Bursts of GnRH occur about once per hour, and the half-life of the peptide in blood is 2 to 4 minutes; thus, there are clear oscillations in the portal blood GnRH concentration. Other brain areas regulate the pulsatile release of hypothalamic GnRH. Neurotransmitters such as epinephrine and norepinephrine stimulate the secretion of GnRH, whereas dopamine and serotonin inhibit the secretion of GnRH. Ovarian steroids and peptides also regulate the secretion of GnRH.

GnRH stimulates the pituitary gonadotrophs to secrete LH and FSH in a pulsatile manner by binding to high-affinity receptors on the anterior pituitary gonadotrophs to activate phosphoinositide–protein kinase C intracellular signaling. LH release throughout the female life span is depicted in Figure 37.2. During the neonatal period, LH is released at low and steady rates without pulsatility; this period coincides with lack of mature ovarian follicles and low ovarian estradiol secretion. Pulsatile LH release begins with the onset of puberty in response to the start of pulsatile GnRH release. For several years, pulsatile GnRH and LH secretion occurs only during sleep; this period coincides with increased, but asynchronous, ovarian follicular development and increased secretion of estradiol. Upon the establishment of regular functional menstrual cycles that include ovulation, LH pulses occur throughout the 24-hour period, changing in a monthly cyclic manner. Early in the follicular phase of the menstrual cycle (discussed in detail below), pituitary gonadotrophs are not very sensitive to GnRH; thus, LH pulses are small. At midcycle, gonadotrophs become more responsive to GnRH and LH pulses are much larger. In postmenopausal women whose ovaries lack sustained follicular development and exhibit low ovarian estradiol secretion, mean circulating LH rises to levels higher than that during the midcycle surge.

▶ FEMALE REPRODUCTIVE ORGANS

The female reproductive tract has two major components: the ovaries, which produce the mature ovum and secrete progestins, androgens, and estrogens, and the ductal system, which transports the ovum, is the place of the union of the sperm and egg, and maintains the developing conceptus until delivery. The morphology and function of these structures change in a cyclic manner under the influence of the reproductive hormones.

The ovaries are in the pelvic portion of the abdominal cavity on both sides of the uterus and are anchored by ligaments (Fig. 37.3). An adult ovary weighs 8 to 12 g and consists of an outer cortex and an inner medulla, without a sharp demarcation. The cortex contains oocytes enclosed in **follicles** of various sizes, **corpora lutea**, **corpora albicantia**, and stromal cells. The medulla contains stromal cells, connective

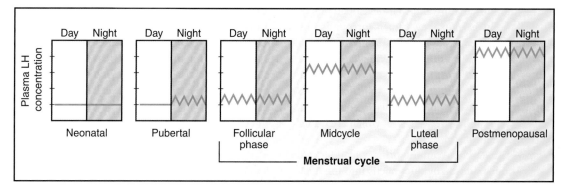

Figure 37.2 **Relative levels of luteinizing hormone (LH) release in human females throughout life.** Note that pulsatile LH release first begins at night during puberty.

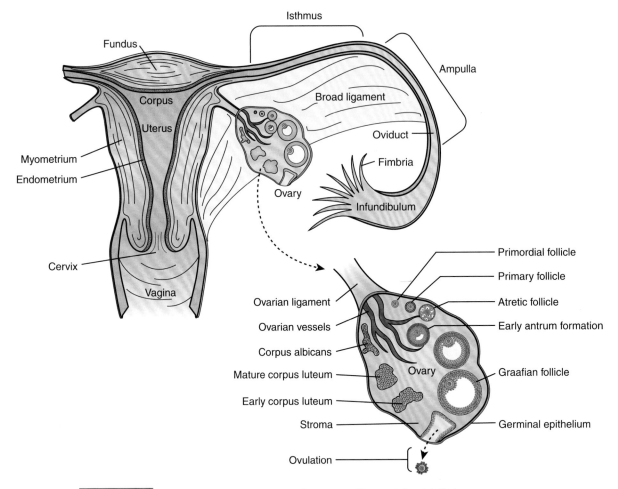

Figure 37.3 **Anatomy of the female reproductive organs.** (See text for details.)

and interstitial tissues. Blood vessels, lymphatics, and nerves enter the medulla of the ovary through the hilus.

The **oviduct** (fallopian tube) receives the ovum immediately after ovulation and provides an environment for fertilization and development of the early embryo. The oviducts are 10 to 15 cm long and composed of sequential regions called the **infundibulum, ampulla**, and **isthmus**. The infundibulum opens into the peritoneal cavity adjacent to the ovary and with finger-like projections called **fimbria**, grasps the ovum at the time of follicular rupture. Densely ciliated projections cover the walls of infundibulum, which facilitate ovum uptake and movement through this region. Fertilization occurs in the ampulla with its thin musculature and well-developed mucosal surface. The isthmus has a narrow lumen surrounded by smooth muscle that provides sphincter-like properties, which can block the passage of germ cells. The oviducts transport the germ cells in two directions: Sperm ascend toward the ampulla and the zygote descends toward the uterus. This requires coordination between smooth muscle contraction, ciliary movement, and fluid secretion, all of which are under hormonal and neuronal control.

The **uterus** is situated between the urinary bladder and rectum. On each upper side, an oviduct opens into the uterine lumen, and on the lower side, the uterus connects to the vagina. The uterus is composed of two types of tissue: the outer part is the **myometrium**, containing multiple layers of smooth muscle. The inner part lining the lumen of the uterus is the **endometrium**, which contains a deep **stromal layer** next to the myometrium and a superficial epithelial layer. The stroma is permeated by spiral arteries and contains much connective tissue. Uterine glands present in the stroma and the epithelial layer produce a viscous secretion under the regulation of estrogen and progesterone. The uterus provides an environment for the developing fetus, and eventually, the myometrium will generate rhythmic contractions that assist in expelling the fetus at delivery.

The **cervix** (neck) is a narrow muscular canal that connects the vagina and the body (corpus) of the uterus. It dilates in response to hormones to allow the expulsion of the fetus. The cervix has numerous glands that produce mucus under the control of estradiol. As greater amounts of estradiol are produced during the follicular phase of the cycle, the cervical mucus changes from a scanty viscous material to a profuse, watery, highly elastic substance called **spinnbarkeit**. The viscosity of the spinnbarkeit can be tested by touching it with a piece of paper and lifting vertically. The mucus can form a thread up to 6 cm under the influence of elevated estradiol. If a drop of the cervical mucus is placed on a slide and allowed to dry, it will form a typical **ferning** pattern when under the influence of estradiol.

The **vagina** is well innervated and has a rich blood supply. It is lined by several layers of epithelium that change histologically during the menstrual cycle. When estradiol levels are low, as during the prepubertal and postmenopausal periods, the vaginal epithelium is thin and the secretions are scanty, resulting in a dry and infection-susceptible area. Estradiol induces proliferation and **cornification** (**keratinization**) of the vaginal epithelium, whereas progesterone opposes those actions and induces the influx of polymorphonuclear leukocytes into the vaginal fluids. Estradiol also activates vaginal glands that produce lubricating fluid during coitus.

▶ OVARIAN CYCLE

Each month, a single **oocyte** or mature egg develops and is released from one of a woman's ovaries. A follicle is the functional structure in which this oocyte develops. Most follicles in the ovary undergo atresia; however, some develop into mature follicles, produce steroids, and ovulate. As follicles mature, oocytes also mature by entering meiosis, which produces the proper number of chromosomes in preparation for fertilization. After rupturing, the follicle becomes a **corpus luteum**.

Oogonia produce an oocyte, which is arrested in meiosis.

Female germ cells develop in the embryonic yolk sac and migrate to the genital ridge, where they participate in the development of the ovary. Without germ cells, the ovary does not develop. The germs cells, called **oogonia**, actively divide by mitosis only during the prenatal period (Table 37.1), so

that by birth, the ovaries contain a finite number of oocytes, estimated to be about 1 million. Most oocytes degenerate and die, a process called **atresia**. By puberty, only 200,000 oocytes remain; by age 30, only 26,000 remain; and by the time of menopause, the ovaries are essentially devoid of oocytes.

When oogonia cease the process of mitosis, they are called oocytes. At that time, they enter the meiotic cycle to prepare for the production of a haploid ovum, become arrested in the prophase of the first meiotic division, and remain arrested in that phase until they either die or grow into mature oocytes at the time of ovulation. The immature oocyte begins development in a **primordial follicle** that is 20 μm wide and which, may or may not be surrounded by a single layer of flattened (squamous) **pregranulosa cells**. When pregranulosa cells eventually grow to surround the oocyte, a basement membrane develops, separating the granulosa cells from the ovarian stroma (see Table 37.1).

Folliculogenesis results in a mature graafian follicle.

The process by which follicles develop and mature is termed **folliculogenesis** (Fig. 37.4). Follicles are in one of the following physiologic states: resting, growing, degenerating, or ready to ovulate. During each menstrual cycle, the ovaries produce a group of growing follicles, most of which fail to grow to maturity and undergo follicular atresia (death) at some stage of development. However, one **dominant follicle** generally emerges from the cohort of developing follicles and ovulates, releasing a mature haploid ovum.

TABLE 37.1	Different Stages in the Development of an Ovum and Follicle		
Stage	**Process**	**Ovum**	**Follicle**
Fetal life	Migration	Primordial germ cells	
	Mitosis	Oogonia	Primordial follicle
	First meiotic division begins	Primary oocyte	Primary follicle
Birth	Arrest in prophase		
	Growth of oocyte and follicle		
Puberty	Follicular maturation		Secondary follicle
Cycle			Antral follicle
Ovulation	Resumption of meiosis	Secondary oocyte	Graafian follicle
	Emission of first polar body		
	Arrest in metaphase		
			Corpus luteum
Fertilization	Second meiotic division complete	Zygote	
	Emission of second polar body		
Implantation	Mitotic divisions	Embryo	
	Blastocyst		
Parturition	Body patterning	Fetus	Corpus albicans

Reproductive Physiology

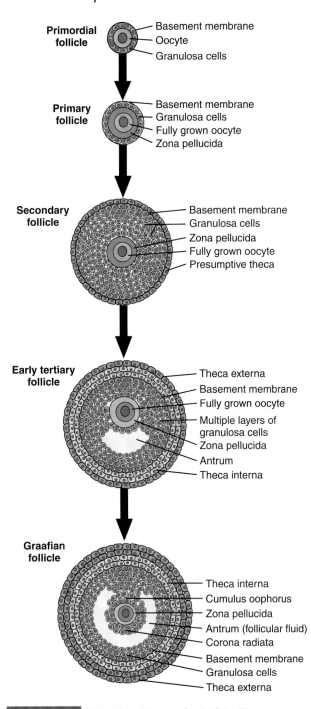

Figure 37.4 **Maturation of the ovarian follicle.** The progression from primordial to primary follicle is relatively constant and independent of gonadotropins. Growth of the primary follicle to a mature graafian follicle requires LH and FSH.

Primordial follicles are considered the nongrowing resting pool of follicles, which gets progressively depleted such that by the time of menopause, the ovaries are essentially devoid of all follicles. Primordial follicles are located in the ovarian cortex or peripheral regions of the ovary.

Progression from the primordial to the next stage of follicular development, the primary stage, occurs at a relatively constant rate throughout fetal, juvenile, prepubertal, and adult life. Once primary follicles leave the resting pool, they are committed to further development or atresia. Studies

in genetically modified mouse models support the concept that the conversion from primordial to primary follicles is independent of pituitary gonadotropins. It is thought that signaling between the oocyte and surrounding granulosa cells, resulting in dramatic oocyte growth, initiates primordial follicle development. There is also evidence that primordial follicles are maintained under a constant inhibitory influence that must be removed for follicle development to occur.

The first sign that a primordial follicle is entering the growth phase is a morphologic change of the flattened pregranulosa cells into cuboidal granulosa cells. The cuboidal granulosa cells proliferate to form a single continuous layer of cells surrounding the oocyte, which has enlarged from 20 μm in the primordial stage to 140 μm wide. At this stage, a glassy membrane, the **zona pellucida**, surrounds the oocyte and serves as the means of attachment through which the granulosa cells communicate with the oocyte. This is the **primary follicular stage** of development, consisting of one layer of cuboidal granulosa cells and a basement membrane (see Fig. 37.4).

The mature graafian follicle contains the oocyte surrounded by granulosa cells and the internal and external theca cell layers.

The follicle continues to grow, mainly through proliferation of its granulosa cells, so that several layers of granulosa cells exist in the **secondary follicular stage** of development (see Fig. 37.4). As the secondary follicle moves deeper into the ovarian cortex, stromal cells near the basement membrane begin to differentiate into cell layers called the **theca interna** and **theca externa**, and a blood supply with lymphatics and nerves forms within the thecal component. The granulosa layer remains avascular.

Theca interna cells eventually become flattened, epithelioid, and steroidogenic. The granulosa cells of secondary follicles acquire receptors for FSH and start producing small amounts of estrogen. The theca externa remains fibroblastic and provides structural support to the developing follicle.

Development beyond the primary follicle is gonadotropin dependent, begins at puberty, and continues in a cyclic manner throughout the reproductive years. As the follicle continues to grow, theca layers expand and **antra** (fluid-filled spaces) begin to develop around the granulosa cells. This early antral stage of follicle development is referred to as the **tertiary follicular stage** (see Fig. 37.4). FSH is the critical hormone responsible for progression from the preantral to the antral stage, stimulating mitosis of the granulosa cells. As the number of granulosa cells increases, the production of estrogens, the binding capacity for FSH, the size of the follicle, and the volume of the follicular fluid all increase significantly.

As the antra increase in size, a single, large, coalesced antrum develops, pushing the oocyte to the periphery of the follicle and forming a large 2- to 2.5-cm-wide **graafian follicle** (also called preovulatory follicle) (see Fig. 37.4). Three distinct granulosa cell compartments are evident in the graafian follicle. **Cumulus granulosa cells** surround the oocyte, with those cells lining the antral cavity termed **antral granulosa cells**. Those cells attached to the basement membrane

TABLE 37.2	**Different Parameters of Follicles During the First Half of the Menstrual Cycle**									
Cycle (d)	Diameter (mm)	Volume (mL)	Granulosa Cells (×10⁶)	FSH	LH	PRL	A	E	P
1	4	0.05	2	2.5	—	60	800	100	—
4	7	0.15	5	2.5	—	40	800	500	100
7	12	0.50	15	3.6	2.8	20	800	1,000	300
12	20	0.50	50	3.6	2.8	5	800	2,000	2,000

Hormone concentrations in mIU or ng/mL.

FSH, follicle-stimulating hormone; LH, luteinizing hormone; PRL, prolactin; A, androstenedione; E, estradiol; P, progesterone.

are called **mural granulosa cells**. Mural and antral granulosa cells produce steroids at a greater rate than cumulus cells.

In addition to exposure to blood-borne hormones, antral follicles have a unique microenvironment in that the follicular fluid contains pituitary hormones, steroids, peptides, and growth factors, some of which are present at a concentration 100 to 1,000 times higher than that in the circulation. As the follicle matures, the intrafollicular concentration of FSH does not change much, whereas that of LH increases and that of PRL declines. The concentrations of estradiol and progesterone increase 20-fold, whereas androgen levels remain unchanged (Table 37.2). The follicular fluid also contains inhibin, activin, GnRH-like peptide, growth factors, opioid peptides, oxytocin, and plasminogen activator.

Granulosa and theca cells cooperate to synthesize estradiol.

The main physiologically active hormone produced by the follicle is **estradiol**, an 18-carbon steroid. Steroidogenesis depends on the availability of cholesterol, which is the

precursor to all steroid synthesis. Low-density lipoprotein (LDL) is the primary source of cholesterol for ovarian steroidogenesis.

The conversion of cholesterol to **pregnenolone** by **cholesterol side-chain cleavage enzyme**, the rate-limiting step in estrogen synthesis, occurs on the inner membrane of mitochondria (Fig. 37.5). LH binds to specific membrane receptors on theca cells, activating adenylyl cyclase through a G protein to increase the production of cyclic adenosine monophosphate (cAMP), which activates protein kinase A (PKA) to phosphorylate intracellular proteins. LH binding increases LDL receptor mRNA expression, LDL cholesterol uptake, and cholesterol transport from the outer to inner mitochondrial membrane, facilitated by the **steroidogenic acute regulatory (StAR) protein**. Once formed, pregnenolone diffuses out of the mitochondria and enters the endoplasmic reticulum (ER), the site of subsequent steroidogenesis. The initiation of steroidogenesis by LH in the theca cell is similar to that for ACTH-induced steroidogenesis in adrenocortical cells (see Figure 33.4 for a detailed illustration).

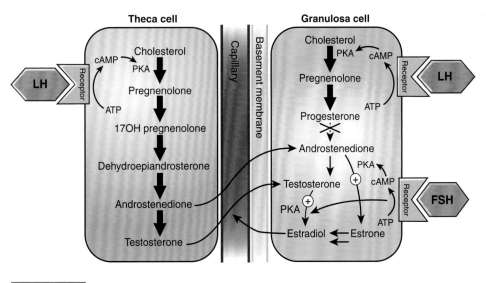

Figure 37.5 **The two-cell, two-gonadotropin hypothesis.** Follicular theca cells under the control of luteinizing hormone (LH) produce androgens that diffuse to the follicular granulosa cells, where they are converted to estrogens via a follicle-stimulating hormone (FSH)-supported aromatization reaction. Granulosa cells lack of the enzyme 17α-hydroxylase and cannot convert progesterone to androstenedione (indicated by the X).

Two pathways are available for steroidogenesis from pregnenolone (see Fig. 36.9). In theca cells, the **delta 5 pathway** (δ5) is predominant, with the **delta 4 pathway** (δ4) utilized by granulosa cells and the corpus luteum. In the δ5 pathway, pregnenolone is sequentially converted to **17α-hydroxypregnenolone, to dehydroepiandrosterone (DHEA), to androstenedione,** and finally to **testosterone**. Testosterone and androstenedione are 19-carbon steroids that diffuse from the thecal cell across the basement membrane to enter the granulosa cells.

In the granulosa cell, FSH increases cAMP to activate PKA-induced protein phosphorylation and the expression of **aromatase**, which converts testosterone and androstenedione to estradiol and **estrone**, respectively. **Aromatase** modifies the A ring of testosterone and androstenedione to remove one carbon, yielding the 18-carbon estrogen hormones (see Fig. 36.9). Estrone is then converted to estradiol by 17β-hydroxysteroid dehydrogenase present in the granulosa cell.

Estradiol synthesis and secretion from the follicle require cooperation between granulosa and theca cells and coordination between FSH and LH. This **two-cell, two-gonadotropin hypothesis** is based on the observation that only granulosa cells possess FSH receptors and respond to the hormone. FSH acts on granulosa cells to produce a broad range of activities including cell proliferation, the induction of aromatase, and increased inhibin and progesterone synthesis. As the follicle matures, FSH stimulates the expression of both LH and FSH receptors on the granulosa cell.

In contrast to the limited expression of FSH receptors, LH receptors are present on theca, granulosa, stromal (interstitial) cells, and the corpus luteum (discussed in detail later). The expression of LH receptors is time dependent as theca cells acquire LH receptors at a relatively early stage, whereas FSH induces LH receptors on granulosa cells in the later stages of the maturing follicle.

In addition to differences in gonadotropin receptor expression, the steroidogenic enzymes are differentially expressed in theca and granulosa cells. Aromatase is expressed only in granulosa cells, and FSH regulates its induction and activation. Granulosa cells are deficient in 17α-hydroxylase and cannot process the C-21 progestins to generate C-19 androgenic compounds (see Fig. 37.5). Consequently, estrogen production by granulosa cells depends on an adequate supply of exogenous aromatizable androgens, provided by theca cells when stimulated by LH.

In the follicle, theca and granulosa cells are exposed to different microenvironments that also influence steroidogenesis. Vascularization is restricted to the theca layer because blood vessels do not penetrate the basement membrane. Thus, theca cells have better access to circulating LDL cholesterol. Granulosa cells must synthesize cholesterol from acetate, a less efficient process.

Androgens are produced by theca and stromal cells as precursors for estrogen synthesis; however, androgens also have a distinct local action at low concentrations to enhance aromatase activity, promoting estrogen production. In contrast, at high concentrations, androgens are converted by 5α-reductase to the more potent androgen dihydrotestosterone. When androgens overwhelm follicles, the intrafollicular androgenic environment antagonizes granulosa cell proliferation and leads to apoptosis of the granulosa cells and subsequent follicular atresia.

Dominant follicle is protected from atresia during FSH suppression.

At the beginning of the menstrual cycle, several follicles begin to grow and develop but only one, the dominant follicle, will go on to ovulate. The remaining follicles will undergo atresia, or destruction of the oocyte and granulosa cells. Atresia is a continuous process and can occur at any stage of follicular development; however, a follicle generally becomes dominant at the preantral stage, with the remaining preantral follicles undergoing atresia. Two main factors contribute to atresia in the nonselected follicles. One is the suppression of plasma FSH by the negative feedback effects of estradiol secreted by the dominant follicle. The decline in FSH support decreases aromatase activity to impair estradiol synthesis and increases production of DHT. Loss of estrogen and expression of the atretogenic DHT interrupts granulosa cell proliferation in the nondominant follicles. The dominant follicle is protected from a fall in circulating FSH levels because it has a healthy blood supply, an accumulation of FSH in the follicular fluid, and an increased density of granulosa cell FSH receptors.

As the dominant follicle grows, vascularization of the theca layer also increases. On day 9 or 10 of the cycle, the vascularity of the dominant follicle is twice that of the other antral follicles, permitting a more efficient delivery of cholesterol to theca cells and better exposure to circulating gonadotropins. At this time, the main source of circulating estradiol in the body is the dominant follicle. Estradiol is the primary regulator of LH and FSH secretion by positive and negative feedback; thus, the dominant follicle ultimately determines its own fate.

Estradiol secretion by the dominant follicle triggers the midcycle LH surge leading to maturation of the oocyte and ovulation.

The midcycle LH surge occurs as a result of rising levels of circulating estradiol that provides a positive feedback effect on the hypothalamic–pituitary axis. Multiple changes in the dominant follicle occur within a relatively short time as a result of the LH surge. These include the resumption of meiosis in the oocyte; granulosa cell differentiation and transformation into luteal cells; the activation of proteolytic enzymes that degrade the follicle wall and surrounding tissues; increased production of prostaglandins, histamine, and other local factors that cause hyperemia; and an increase in progesterone secretion. The midcycle FSH surge is not essential for ovulation; however, only follicles that have been adequately primed with FSH ovulate because they contain sufficient numbers of LH receptors for ovulation and subsequent luteinization.

Periovulatory period

All healthy oocytes in the ovary are arrested in the prophase of the first meiosis, which occurred during fetal life. With the LH surge, the oocyte resumes meiosis and completes its first meiotic division several hours before ovulation (see Table 37.1).

Primary oocytes arrested in meiotic prophase 1 duplicate their centrioles and DNA so that each chromosome has two identical **chromatids**. Crossing over and chromatid exchange occur during this phase to produce genetic diversity. With the resumption of meiosis in metaphase 1, the nuclear membrane disappears, and the chromosomes condense and align at the equator of the spindle. At meiotic anaphase 1, the homologous chromosomes move in opposite directions under the influence of the retracting meiotic spindle at the cellular periphery. At meiotic telophase 1, an unequal division of the cell cytoplasm yields a large **secondary oocyte** and a small, nonfunctional **first polar body**. Both the secondary oocyte and polar body, like secondary spermatocytes (see Fig. 36.8), have a haploid number of duplicated chromosomes (22 somatic and 1 X chromosome). The first polar body degenerates or divides to form nonfunctional cells.

The secondary oocyte begins its second meiotic division and proceeds through a short prophase to become arrested in metaphase, where it remains until fertilization begins with the penetration of a spermatozoon. The second meiotic division resumes and is rapidly completed at the start of fertilization. A second unequal cell division produces a small **second polar body** and a large fertilized egg, the **zygote**, containing a diploid complement of DNA resulting from the haploid contribution of the mother and the father. If fertilization does not occur, the secondary oocyte begins to degenerate within 24 to 48 hours.

Follicle maturation and rupture

The earliest responses of the ovary to the midcycle LH surge are the release of vasodilatory substances including histamine, bradykinin, and prostaglandins, which mediate increased ovarian and follicular blood flow. The highly vascularized dominant follicle becomes hyperemic and edematous, swelling to a size of at least 20 to 25 mm wide. There is also an increased production of follicular fluid, disaggregation of granulosa cells, and detachment of the oocyte–cumulus complex from the follicular wall, moving it to the central portion of the follicle. The basement membrane separating theca cells from granulosa cells begins to disintegrate, granulosa cells begin to undergo luteinization, and blood vessels begin to penetrate the granulosa cell compartment.

At a specific site called the **stigma**, the wall of the follicle thins and bulges due to cellular deterioration; it is at this site on the follicle that rupture will occur. As ovulation approaches, the follicle enlarges and protrudes from the surface of the ovary at the stigma. In response to LH, **plasminogen activator** is produced by theca and granulosa cells, which converts plasminogen to **plasmin**. The proteolytic plasmin enzyme acts directly on the follicular wall to stimulate the production of **collagenase**, an enzyme that digests the connective tissue matrix. The thinning and increased distensibility of the wall facilitate the rupture of the follicle. Smooth muscle contraction extrudes the oocyte–cumulus complex and follicular fluid.

Within a couple of hours after the initiation of the LH surge, the production of progesterone, androgens, and estrogens by the follicle begins to increase. Progesterone, acting through its receptor in granulosa cells, promotes ovulation by releasing mediators that increase the distensibility of the follicular wall and enhance the activity of proteolytic enzymes. As LH levels reach their peak, plasma estradiol levels plunge because of down-regulation of LH and FSH receptors on granulosa cells and the inhibition of granulosa cell aromatase. Eventually, LH receptors on granulosa cells escape the down-regulation, and progesterone production increases.

Corpus luteum forms from the postovulatory follicle.

After ovulation, the wall of the graafian follicle collapses and becomes convoluted, blood vessels course through the granulosa and theca cell layers, and the antral cavity fills with blood. The granulosa cells cease proliferation, undergo hypertrophy, and begin to produce progesterone as their main secretory product. The ruptured follicle forms a solid structure called the **corpus luteum** (Latin for "yellow body"). The numerous biochemical and morphologic changes that result in the corpus luteum occur in response to LH and FSH stimulation and are referred to collectively as **luteinization**. The granulosa cells and theca cells in the corpus luteum are called **granulosa lutein cells** and **theca lutein cells**, respectively.

The corpus luteum is a transient endocrine structure that serves as the main source of circulating steroids during the luteal (postovulatory) phase of the menstrual cycle and is essential for maintaining pregnancy during the first trimester. Continued LH stimulation is needed to ensure healthy luteal cells and progesterone secretion. If pregnancy does not occur, the corpus luteum regresses as luteal cells undergo apoptosis and necrosis, a process termed **luteolysis** or **luteal regression**. Fibrous tissue replaces the luteinized cells, creating a nonfunctional structure called the **corpus albicans**.

Luteinization begins before ovulation occurs. After acquiring a high concentration of LH receptors, granulosa cells enlarge (hypertrophy) and develop smooth ER and lipid inclusions. In contrast to the nonvascular granulosa cells in the follicle, luteal granulosa cells have a rich blood supply. Capillary invasion starts immediately after the LH surge, facilitated by the dissolution of the basement membrane between theca and granulosa cells, and reaches a peak at 7 to 8 days after ovulation. Differentiated theca and stroma cells, as well as granulosa cells, are incorporated into the corpus luteum, and all three classes of steroids, androgens, estrogens, and progestins, are synthesized. Although some progesterone is secreted before ovulation, peak progesterone production is reached 6 to 8 days after the LH surge.

Regression of the corpus luteum occurs about 13 days after ovulation if fertilization does not occur. If fertilization occurs, the embryonic trophoblast produces human chorionic gonadotropin (hCG), an LH-like hormone that rescues the corpus luteum from degeneration (see Chapter 38). hCG binds LH receptors on the corpus luteum to increase cAMP, stimulating progesterone secretion.

Luteal regression in the absence of fertilization and hCG production may occur in two possible ways. The first mechanism is loss of LH stimulation resulting from the fall in circulating LH levels. The second mechanism is inhibition of LH action by locally produced agents such as estrogen, oxytocin, prostaglandins, and GnRH (Clinical Focus 37.1).

CLINICAL FOCUS | 37.1

Luteal Insufficiency

Occasionally, the corpus luteum does not produce sufficient progesterone to maintain pregnancy during its early stages. Initial signs of early spontaneous termination of pregnancy include pelvic cramping and vaginal bleeding, similar to indications of menstruation. If the corpus luteum is truly deficient, then fertilization may occur around the idealized day 14 (ovulation), pregnancy terminates during the deficient luteal phase, and menses will start on schedule. Without measuring levels of human chorionic gonadotropin (hCG), the woman would not know that she is pregnant because of the continuation of regular menstrual cycles.

Analysis of the regulation of progesterone secretion by the corpus luteum provides insights into this clinical problem. There are several reasons for luteal insufficiency. First, the number of luteinized granulosa cells in the corpus luteum may be insufficient because of the ovulation of a small follicle or the premature ovulation of a follicle that was not fully developed. Second, the number of luteinizing hormone (LH) receptors on the luteinized granulosa cells in the graafian follicle and developing corpus luteum may be insufficient. LH receptors mediate the action of LH, which stimulates

progesterone secretion. An insufficient number of LH receptors could be a result of insufficient priming of the developing follicle with follicle-stimulating hormone (FSH). It is well known that FSH increases the number of LH receptors in the follicle. Third, the LH surge could have been inadequate in inducing full luteinization of the corpus luteum, yet there was sufficient LH to induce ovulation. It has been estimated that only 10% of the LH surge is required for ovulation, but the amount required for full luteinization and adequate progesterone secretion to maintain pregnancy is not known.

If progesterone values are low in consecutive cycles at the midluteal phase and do not match endometrial biopsies, exogenous progesterone may be administered to prevent early pregnancy termination during a fertile cycle. Other options include the induction of follicular development and ovulation with clomiphene and hCG. This treatment would likely produce a large, healthy, estrogen-secreting graafian follicle with sufficient LH receptors for luteinization. The exogenous hCG is given to supplement the endogenous LH surge and to ensure full stimulation of the graafian follicle, ovulation, adequate progesterone, and luteinization of the developing corpus luteum. ■

▶ MENSTRUAL CYCLE

The **menstrual cycle** describes the coordinated monthly changes in the ovary and endometrial lining of the uterus in women of reproductive age. The cycle is noted to begin with the onset of menstruation, the flow of blood from the uterus through the vagina, when the lining of the uterus is shed. A woman's first menstruation, termed **menarche**, occurs around age 12 years. The end of a woman's reproductive phase, called **menopause**, commonly occurs between ages 45 and 55 years.

The average menstrual cycle length in adult women is 28 days (see Fig. 37.6), with a range of 25 to 35 days. The interval from ovulation to the onset of menstruation is relatively constant, on average 14 days in most women, dictated by the fixed life span of the corpus luteum. In contrast, the interval from the onset of menses to ovulation (the follicular phase) is more variable and accounts for differences in cycle lengths among ovulating women. Sexual intercourse may occur at any time during the cycle, but fertilization occurs only during the postovulatory period. Once pregnancy occurs, ovulation ceases until after parturition. Lactation can provide continued inhibition of ovulation, but this effect is not absolute.

Menstrual cycles become irregular as menopause approaches at around 50 years of age, and cycles cease thereafter. During the reproductive years, the timing of the menstrual cycle is modulated by physiological, psychological, and social factors.

Puberty marks the start of cyclic reproductive function.

During the prepubertal period, the hypothalamic–pituitary–ovarian axis becomes activated, an event termed **gonadarche**, increasing gonadotropins in the circulation that stimulate ovarian estrogen secretion. Pulsatile release of GnRH from the hypothalamus is the initiating event for gonadarche. Factors that stimulate the secretion of GnRH include glutamate, norepinephrine, and neuropeptide Y emanating from synaptic inputs to GnRH-producing neurons. A decrease in γ-aminobutyric acid (GABA) inhibition of GnRH secretion also occurs at this time. At the time of **puberty**, the pituitary also becomes more responsive to GnRH.

The increase in estradiol release from the ovary induces the expression of **secondary sex characteristics**, including breast development and increased fat deposition on the hips and buttocks. The initiation of breast development under the influence of estrogen is known as **thelarche**. Estrogens also regulate the pubertal growth spurt, induce closure of the epiphyses, have a positive effect to maintain bone formation, and can antagonize the degrading actions of parathyroid hormone (PTH) on bone.

The first few menstrual cycles in puberty are usually irregular and anovulatory, due to delayed maturation of the positive feedback by estradiol on a hypothalamus that thus fails to secrete significant GnRH. During puberty, pulsatile LH secretion occurs more during periods of sleep than during periods of wakefulness, resulting in a diurnal cycle.

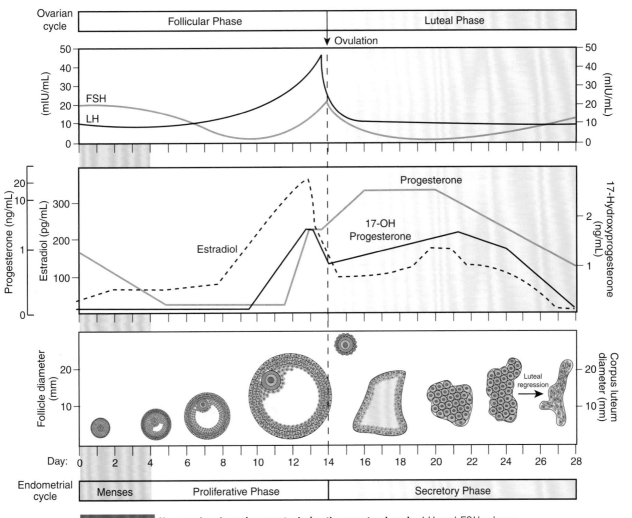

Figure 37.6 **Hormonal and ovarian events during the menstrual cycle.** LH and FSH release from the anterior pituitary **(top panel)** promotes follicular development **(bottom panel)** and follicle steroidogenesis **(middle panel)**. During the follicular phase, estradiol provides positive feedback to induce the LH surge, which is required for ovulation.

At puberty, adrenal androgens promote the development of axillary and pubic hair, a process known as **pubarche**. The adrenals begin to produce significant amounts of androgens such as DHEA and androstenedione 4 to 5 years prior to menarche. The start of adrenal androgen production is called **adrenarche**, and this process is independent of gonadarche (see Chapter 38) (Clinical Focus 37.2).

Hormonal regulation of the ovarian cycle requires synchrony between the ovary, brain, and pituitary.

The ovarian cycle includes the follicular (days 0 to 13), ovulatory (days 13 to 14), and luteal phase (days 14 to 28). The ovarian cycle overlaps with the endometrial cycle, which is comprised of the menstrual (days 0 to 4), proliferative (days 4 to 14), and secretory phases (days 14 to 28). The endometrial cycle is discussed in detail in the next section of this chapter.

The menstrual cycle requires several coordinated elements: hypothalamic control of pituitary function, follicular and luteal changes in the ovary, and positive and negative feedback of ovarian hormones at the hypothalamic–pituitary

axis. We have discussed separately the mechanisms that regulate the synthesis and release of the reproductive hormones; we now put them together in terms of sequence and interaction.

Follicular phase

The follicular phase begins with the start of menstruation, which is characterized by low estrogen, progesterone, and inhibin levels resulting from the luteal regression that has just occurred, and low estrogen synthesis by immature follicles. Plasma FSH levels are high due to the removal of negative feedback by estrogen, progesterone, and inhibin. The elevated levels of FSH act on a cohort of follicles recruited 20 to 25 days earlier from the resting pool. These 4- to 6-mm-wide follicles are stimulated to grow into the preantral stage between days 3 and 5. FSH promotes granulosa cell proliferation and increased aromatase activity, resulting in plasma estradiol levels rising slightly between days 3 and 7. The designated dominant follicle is selected between days 5 and 7 and increases in size and steroidogenic activity. Between days

Precocious Puberty

Signs of secondary sexual maturation at an age 2 standard deviations or more below the mean for pubertal onset is referred to as *precocious puberty*. The average age of pubertal onset has gradually declined as a result in part to better nutrition and general health; thus, present data indicate that the lower limit of puberty in normal boys is age 9 years, but that the lower limit for white girls is age 7 years and for black girls is age 6 years. Thus, signs of secondary sexual maturation earlier than these lower limits should be evaluated.

Two types of sexual precocity are recognized. If there is evidence of premature reactivation of the hypothalamic LHRH pulse generator, the condition is referred to as *complete isosexual precocity*, *true precocious puberty*, or *central precocious puberty*. In this condition, there is pulsatile release of LH that mimics that seen in puberty, and LH increases in response to administration of LHRH. If extrapituitary gonadotropin secretion or gonadal steroid secretion occurs independently of pulsatile LHRH generation, the condition is termed *incomplete isosexual precocity*, *pseudoprecocious puberty*, or *LHRH-independent precocious puberty*. In this condition, there is no pulsatile release of LH as seen in puberty, and LH does not increase in response to administration of LHRH. In both forms of sexual precocity, increased gonadal steroid secretion results in an increased rate of linear growth, somatic development, and rate of skeletal maturation.

Hamartomas, congenital malformations composed of a heterotropic mass of nerve tissue usually located on the floor of the third ventricle or attached to the tuber cinereum, appear to be a major cause of central precocious puberty. Hamartomas are not neoplastic and do not progress or enlarge. LHRH-releasing cells have been found in hamartomas and appear to release LHRH in a pulsatile manner unrestrained by the central nervous system (CNS) mechanisms that inhibit the normal LHRH pulse generator. It has also been suggested that hamartomas release factors such as transforming growth factor α (TGF-α) that drive LHRH secretion from normal neurons in the hypothalamus. Central precocious puberty can also result from gliomas, CNS infection, CNS radiation, and head trauma.

LHRH agonists are the treatment of choice for central precocious puberty after the possibility of an expanding intracranial lesion has been eliminated. Chronic administration of LHRH analogs results in suppression of pulsatile LH and FSH release, gonadal steroidogenesis, and gametogenesis. This occurs because the constant presence of LHRH results in down-regulation of its receptors and uncoupling of the receptor from the intracellular signal transduction mechanism.

LHRH-independent precocious puberty results from inappropriate gonadal or adrenal steroid secretion. Congenital adrenal hyperplasia (see Chapter 33) leads to elevated androgens and masculinization of boys and girls. McCune-Albright syndrome, resulting from gain of function mutations in activating G proteins coupled to gonadotropin receptors, results in increased ovarian estrogen synthesis in girls and increased testicular testosterone synthesis in boys. In boys, human chorionic gonadotropin (hCG)-secreting tumors can result in precocious puberty because hCG has LH-like activity (serum LH and FSH are not elevated). hCG-secreting tumors do not result in signs of precocity in girls, unless the tumor also secretes estrogen. Estrogen-secreting follicular cysts are the most common ovarian cause of precocity in girls.

Treatment of LHRH-independent precocious puberty centers on inhibition of steroidogenesis. Medroxyprogesterone acetate, which has a direct suppressive effect on gonadal steroidogenesis, is commonly prescribed for both boys and girls.

---

8 and 10 plasma estradiol levels rise sharply, reaching peak levels above 200 pg/mL on day 12, 1 day before the LH surge.

During the early follicular phase, LH pulsatility is of low amplitude and frequency coinciding with once per hour pulses of GnRH. As estradiol levels rise, GnRH pulse frequency increases as do LH pulses. The resulting increase in plasma LH further supports follicular steroidogenesis because FSH has increased the number of LH receptors on growing follicles.

During the midfollicular to late follicular phase, rising estradiol and inhibin from the dominant follicle suppress FSH release. The decline in FSH, together with an accumulation of androgens not aromatized to estrogens, induces atresia in the nonselected follicles. The dominant follicle is saved by virtue of its high density of FSH receptors, the accumulation of FSH in its follicular fluid (see Table 37.2), and the acquisition of LH receptors by the granulosa cells.

Ovulatory phase

The midcycle surge of LH is short lived (24 to 36 hours) and results from positive feedback by estradiol. For the LH surge to occur, estradiol must be maintained at a critical concentration (about 200 pg/mL) for a sufficient duration (36 to 48 hours) prior to the surge. A reduction in the critical concentration, or a rise that is too small or too short eliminates, reduces the LH surge.

The mechanism(s) that transforms estradiol from a negative to a positive regulator of LH to induce the LH surge is incompletely understood. It is known that the high estradiol concentrations increase the number of GnRH receptors on the gonadotrophs, making the pituitary more responsive to GnRH. Estradiol may also promote the conversion of a storage pool of LH, perhaps within a subpopulation of gonadotrophs, to a more readily releasable pool.

There is also evidence that estradiol may induce greater GnRH release to induce the LH surge, an effect on GnRH neurons in the preoptic area of the hypothalamus. It is thought that these neurons do not participate in the daily pulsatile release of GnRH generated by neurons within the arcuate nucleus of the hypothalamus. The high estradiol concentration thus recruits these neurons to secrete additional GnRH to promote the surge. It should be noted, however, that estradiol has been shown to induce an LH surge in monkeys in which an effect to increase GnRH could not occur due to ablation of GnRH neurons in the hypothalamus.

A small but distinct rise in progesterone occurs before the LH surge, which is important for augmenting the LH surge. In contrast, if progesterone levels become too high, it prevents estradiol from inducing the LH surge. The small rise in progesterone together with estradiol promotes a concomitant surge in FSH. There are indications that the midcycle FSH surge is important for inducing enough LH receptors on granulosa cells for luteinization, stimulating plasminogen activator for follicular rupture and oocyte release, and activating a cohort of follicles destined to develop in the next cycle.

The LH surge reduces the concentration of 17α-hydroxylase and subsequently decreases androstenedione production by the dominant follicle. Estradiol levels decline, 17-hydroxyprogesterone increases, and progesterone levels rise. The prolonged exposure to high LH levels during the surge down-regulates the ovarian LH receptors, accounting for the immediate postovulatory suppression of estradiol.

Luteal phase

As the corpus luteum matures, it increases progesterone production and reinitiates estradiol secretion, to achieve high plasma concentrations of these hormones on days 20 to 23, about 1 week after ovulation. The elevated steroids suppress circulating FSH levels. The LH pulse frequency is reduced during the early luteal phase, but the amplitude is higher than that during the follicular phase. LH is important at this time for maintaining corpus luteum function and sustaining steroid production. In the late luteal phase, a progesterone-dependent, opioid-mediated suppression of the GnRH pulse generator reduces both LH pulse frequency and amplitude.

After the demise of the corpus luteum on days 24 to 26, estradiol and progesterone levels plunge, causing the withdrawal of support of the uterine endometrium, culminating within 2 to 3 days in menstruation. The reduction in ovarian steroids acts centrally to remove feedback inhibition. The FSH level begins to rise and a new cycle is initiated.

Ovarian steroids estradiol and progesterone regulate the endometrial cycle.

The endometrial cycle is composed of the menstrual (days 0 to 4), proliferative (days 4 to 14), and secretory phases (days 14 to 28), which are under the influence of ovarian steroids. The endometrium (also called *uterine mucosa*) lining the uterus is composed of a superficial layer of epithelial cells and an underlying stromal layer. The epithelial layer contains glands that penetrate the stromal layer. A secretory columnar epithelium lines the glands. Significant changes in the function and histology of the uterine endometrium, the composition of cervical mucus, and the cytology of the vagina occur across the endometrial cycle (Fig. 37.7).

Proliferative phase

The **proliferative phase** coincides with the mid to late follicular phase of the ovarian cycle. Under the influence of rising plasma estradiol concentrations, the stromal and epithelial layers of the uterine endometrium undergo hyperplasia and hypertrophy, increasing in size and thickness. The endometrial glands elongate and are lined with columnar epithelium. The endometrium becomes vascularized with more spiral arteries developing to provide a rich blood supply to this region. Estradiol also induces the expression of progesterone receptors and increases myometrial excitability and contractility.

Secretory phase

The **secretory phase** begins on the day of ovulation and coincides with the early to midluteal phase of the ovarian cycle. Under the combined action of progesterone and estrogen, the endometrial glands become coiled, store glycogen, and secrete large amounts of carbohydrate-rich mucus. The stroma increases in vascularity and becomes edematous, and the spiral arteries become tortuous (see Fig. 37.7). Peak secretory activity, edema formation, and overall thickness of the

Reproductive Physiology

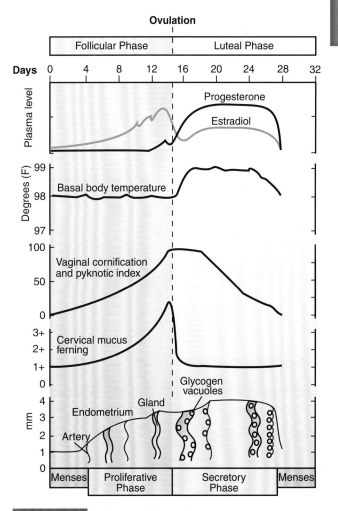

Figure 37.7 Uterine, cervical, vaginal, and body temperature change in relation to ovarian steroids and ovulation during the menstrual cycle.

endometrium are reached on days 6 to 8 after ovulation in preparation for implantation of the blastocyst. Progesterone antagonizes the effect of estrogen on the myometrium to reduce spontaneous myometrial contractions.

Menstrual phase

Desquamation and sloughing of the entire functional layer of the endometrium occur during the menstrual phase (*menses*). This process begins with ischemia in the endometrial tissue (sometimes referred to as the ischemic phase), initiated by the declining levels of progesterone and estradiol that result from regression of the corpus luteum. Necrotic changes and abundant apoptosis occur in the secretory epithelium as it collapses. The arteries constrict, reducing the blood supply to the superficial endometrium. Leukocytes and macrophages invade the stroma and begin to phagocytose the ischemic tissue. Leukocytes persist in large numbers throughout menstruation, providing resistance against infection to the denuded endometrial surface.

At the cellular level, the reduction in steroids destabilizes lysosomal membranes in the endometrial cells, resulting in the liberation of proteolytic enzymes and increased production of vasoconstrictor prostaglandins such as prostaglandin F2. The prostaglandins induce vasospasm of the spiral arteries, and the proteolytic enzymes digest the tissue. Eventually, the blood vessels rupture and blood is released, together with cellular debris.

The endometrial tissue is expelled through the cervix and vagina, along with blood from the ruptured arteries. The menstrual flow lasts 4 to 5 days and averages 30 to 50 mL in volume. The menstrual blood flow does not clot due to the presence of fibrinolysin released from the necrotic endometrial tissue.

Estradiol and progesterone prepare the ductal system and signal ovulation.

In addition to regulating the cyclic changes in the endometrium, the ovarian steroids have effects on the ductal system (see Fig. 37.7). Estrogen maintains the ciliated nature of the epithelium and increases the motility of the oviducts. This is important for uptake of the oocyte following ovulation, as well as movement of sperm and the zygote. Lack of estrogen following ovariectomy causes a loss of the cilia from ductal cells. Very high estrogen, as would be achieved with exogenous administration, may increase oviduct contractility to the extent that a fertilized egg becomes trapped, resulting in an ectopic pregnancy. Progesterone opposes the effects of estrogen on smooth muscle contractility of the ductal system.

The cervical mucus undergoes cyclic changes in composition and volume that are important for sperm transport. During the follicular phase, estrogen increases the quantity, alkalinity, viscosity, and elasticity of the cervical mucus. By the time of ovulation, elasticity of the mucus or spinnbarkeit is greatest, and sperm can readily pass through this estrogen-dominated mucus. As progesterone rises after ovulation, during pregnancy, or with low-dose administration of progestogen during the cycle, it opposes these effects of estrogen. Thus, the quantity and elasticity of the cervical mucus declines making it difficult for sperm to pass through, but also providing better protection against infection.

The vaginal epithelium proliferates under the influence of estrogen early in the follicular phase. The columnar epithelium then becomes cornified (keratinized) and reaches its peak in the periovulatory period. During the postovulatory period, progesterone induces the formation of thick mucus, the epithelium becomes infiltrated with leukocytes, and cornification decreases.

At the time of ovulation, there is also a small but detectable rise in basal body temperature, caused by progesterone. The rise in body temperature can be utilized to monitor ovulation in attempts to either maximize or minimize the chances of fertilization and pregnancy.

Menopause is the cessation of ovarian function and reproductive cycles.

A woman is formerly defined to be in menopause 12 months following the completion of her final menstrual period. As a woman approaches her final menses, the cycles and bleeding become irregular, with shorter cycles due to lack of complete follicular development. The ovaries atrophy and are characterized by the presence of few to no healthy follicles.

The decline in ovarian function is associated with a decrease in estrogen secretion and a concomitant increase in LH and FSH, which can be used as a diagnostic tool (Table 37.3). The elevated LH stimulates ovarian stroma cells

TABLE 37.3	Serum Gonadotropin and Steroid Levels in Premenopausal and Postmenopausal Women				
		Menstrual Cycle			
Hormone	**Units**	**Follicular**	**Preovulatory**	**Luteal**	**Postmenopausal**
LH	mIU/mL	2.5–15	15–100	2.5–15	20–100
FSH	mIU/mL	2–10	10–30	2–6	20–140
Estradiol	pg/mL	70–200	200–500	75–300	10–60
Progesterone	ng/mL	≤0.5	≤1.5	4–20	≤0.5

LH, luteinizing hormone; FSH, follicle-stimulating hormone.

to continue producing androstenedione. Estrone, derived almost entirely from the peripheral conversion of adrenal and ovarian androstenedione, becomes the dominant estrogen (see Fig. 36.9). As androgens rise, some women exhibit hirsutism. The lack of estrogen causes atrophic changes in the breasts and reproductive tract, accompanied by vaginal dryness, which often causes pain and irritation during intercourse.

Hot flashes, resulting from the loss of vasomotor tone, osteoporosis, and an increased risk of cardiovascular disease are important medical conditions for women entering and in menopause. Hot flashes occur primarily during the menopausal transition and are characterized by episodic increases in upper body and skin temperatures, peripheral vasodilation, and sweating. Hot flashes occur concurrently with LH pulses, but are not caused by the gonadotropins because they are evident in hypophysectomized women. It is thought that hot flashes reflect temporary disturbances in the hypothalamic thermoregulatory centers, which are linked to the GnRH pulse generator in some manner.

Osteoporosis increases the risk of hip fractures, and estrogen replacement therapy reduces the risk. Estrogen antagonizes PTH-induced calcium mobilization from bone but enhances PTH-stimulated calcium retention by the kidney. Estrogen also promotes the intestinal absorption of calcium through 1,25-dihydroxyvitamin D_3. See the Integrated Medical Sciences box for a detailed discussion of hormone replacement therapy (HRT) in menopause.

▶ INFERTILITY

Infertility is arbitrarily defined as the inability to conceive within 1 year of unprotected intercourse and should be considered as a medical problem for both partners. We focus here on abnormalities in women that can cause infertility (see Chapter 36 for a discussion of infertility in men). Environmental factors, disorders of the CNS and pituitary, and ovarian abnormalities can interfere with follicular development and/or ovulation. If a normal ovulation occurs, structural, pathologic, and/or endocrine problems associated with the oviduct and/or uterus can prevent fertilization, impede the transport or implantation of the embryo, and, ultimately, interfere with the establishment or maintenance of pregnancy.

Amenorrhea is the most easily recognized and common sign of ovulatory dysfunction and is defined as the absence of menarche by 16 years of age, or no menses for more than three cycles in a woman who previously had cyclic menses. The absence of menses is a result of anovulation, the etiology of which can be broadly characterized as due to hypothalamic dysfunction, hyperprolactinemia, androgen excess or premature ovarian insufficiency.

Hypothalamic amenorrhea encompasses a broad range of conditions, each of which results in insufficient gonadotropin release. In the absence of gonadotropins, follicular development is impaired and downstream estradiol production low. Genetic or anatomic disorders such as tumors or head trauma cause amenorrhea by interfering with GnRH release. However, the most commonly observed causes are emotional distress, excessive exercise, or rapid weight loss, resulting in the condition being termed **functional hypothalamic amenorrhea** because it is assumed that the neuroendocrine pathways are intact. Highly trained athletes and dancers often exhibit energy intakes that are insufficient to match their high energy expenditure. Eating disorders such as **anorexia nervosa** or **bulimia** also result in insufficient energy intake leading to weight loss. In both cases, low levels of the adipose tissue hormone leptin signal the hypothalamus that energy stores are low, resulting in reduced GnRH release. Prolonged daily stress due to work or social issues can also impair GnRH release, resulting in amenorrhea. Therapy can include administration of GnRH or lifestyle modification, depending on the cause of the problem.

Hyperprolactinemia is most commonly due to pituitary prolactinoma and is often associated with galactorrhea, a persistent milklike discharge from the nipple in nonlactating women and men. Other causes of elevated prolactin levels are hypothalamic disorders, trauma to the pituitary stalk, and medications, all of which result in reduced hypothalamic dopamine release. Prolactin suppresses ovarian function by reducing GnRH release. Hyperprolactinemia is treated by surgical removal of the pituitary adenoma or pharmacologic intervention with bromocriptine, a dopaminergic agonist that reduces the size and number of the pituitary lactotrophs.

Polycystic ovarian syndrome is the most common form of androgen excess. This is a heterogeneous disorder associated with elevated LH with normal FSH, polycystic ovaries, hirsutism, insulin resistance, and obesity. Elevated insulin in response to insulin resistance can stimulate LH gene transcription and androgen synthesis. Weight loss and interventions to reduce insulin resistance can be of some therapeutic value.

Premature ovarian insufficiency is defined as the early depletion of follicles prior to 40 years of age and is a state of hypogonadotropic hypogonadism. Patients go through puberty with variable subsequent menstrual cycling. Clinical findings include low estrogen levels and high FSH, with symptoms similar to those of menopause, including hot flashes and an increased risk of osteoporosis. The etiology is variable, including chromosomal abnormalities; lesions resulting from irradiation, chemotherapy, or viral infections; and autoimmune conditions.

INTEGRATED MEDICAL SCIENCES

Hormone Replacement Therapy and the Role of Selective Estrogen Receptor Modulator in Menopause

Clinical features of menopause can be separated into those more closely associated with the transition to menopause, such as hot flashes and night sweats, vaginal dryness, and irritation, versus long-term complications such as bone loss and cardiovascular disease. HRT, consisting of estrogen, or the combination of estrogen plus progestin, has been shown in large clinical trials to reduce hot flashes and vaginal discomfort in most women. Progestin is given with estrogen to protect the endometrium from estrogen-induced hyperplasia or carcinoma. HRT reduces spine and hip fractures resulting from osteoporosis, especially if given during the first 3 years of menopause. Recognized risks of HRT include a small increased incidence of breast cancer, an established greater incidence of endometrial cancer, and a modest increased incidence of deep vein thrombosis.

It has long been thought that estrogen protects against atherosclerosis and cardiovascular disease in women. In premenopausal women, total and LDL cholesterol are lower, HDL cholesterol is higher, and myocardial infarction is six times less common than in men of the same age. During menopause, total and LDL cholesterol levels rise, as does cardiovascular disease risk. Estrogen therapy in postmenopausal women has a favorable impact on lipid profile, lowering total and LDL cholesterol. This finding supported the conclusion that estrogen use in menopause would protect against cardiovascular disease. However, the Women's Health Initiative, a large randomized study of estrogen therapy, was halted early because of a small, but significant, increased incidence of cardiovascular disease and breast cancer. Analyses of data from that trial and others suggest that HRT may increase cardiovascular risk in older women starting therapy many years after menopause, or in women on therapy for many years, but not in women starting therapy to relieve symptoms at the onset of menopause. Thus, current recommendations are that HRT is used for symptoms of menopausal transition but in the long term should be weighed against other therapeutic interventions that are more effective in reducing cardiovascular disease.

Selective estrogen receptor modulators (SERMs) are synthetic nonsteroidal agents with varying estrogen agonist and antagonist activities in different tissues. They are activators of estrogen receptor (ER) in some tissues (bone, liver, and the cardiovascular system), inhibitors in other tissues (brain and breast), and mixed activator/inhibitors in the uterus. The goal of therapy with SERMs is to mimic the beneficial effects of estrogen on the bones and heart, but to inhibit estrogen action in the breast and uterus, minimizing harmful effects in these tissues. FDA approved SERMs include tamoxifen to prevent or treat breast cancer, raloxifene for the treatment of osteoporosis, and ospemifene for postmenopausal vaginal atrophy.

Two distinct ERs exist (ERα and ERβ) that differ in the N-terminal region, which is important in initiation of transcription. The receptors also exhibit different expression patterns, with ERα found predominantly in the breast, uterus, and vagina, and ERβ expressed in tissues of the CNS, cardiovascular system, immune system, gastrointestinal system, kidney, lungs, and bone.

Several mechanisms have been proposed to explain the ability of SERMs to be estrogenic in some tissues and antiestrogenic in others. SERMs can block estrogen binding to its receptor to inhibit transcription or can promote the interaction of the N-terminal domain of the ER with other coactivators resulting in transcriptional activation. SERMs may also recruit different coactivators and coinhibitors in different tissues than estrogen usually does. Finally, the actions of SERMs can differ depending on which ER isoform they bind. SERMs function as pure antagonists when acting through ERβ on genes containing ER-binding elements, but can function as a partial agonist when acting on them through ERα. Continuing research should lead to a better understanding of the mechanism(s) of action of SERMs and development of new molecules designed for specific and beneficial actions on estrogen-responsive tissues. ■

Chapter Summary

- Pulses of hypothalamic GnRH regulate the secretion of LH and FSH, which regulate follicular development, steroidogenesis, ovulation, and formation of the corpus luteum.
- LH and FSH in coordination with ovarian theca and granulosa cells regulate the secretion of follicular estradiol.
- Ovulation occurs as the result of a positive feedback of follicular estradiol on the hypothalamic–pituitary axis, which induces surges in LH and FSH.
- Follicular development occurs in distinct steps: primordial, primary, secondary, tertiary, and graafian follicle stages.
- Follicular rupture (ovulation) requires the coordination of appropriately timed LH and FSH surges that induce inflammatory reactions in the graafian follicle, leading to dissolution of the follicle wall by ovarian enzymes.
- Follicular atresia results from the withdrawal of gonadotropin support.

- The formation of a functional corpus luteum requires the presence of the LH surge, adequate numbers of LH receptors, sufficient granulosa cells, and significant progesterone secretion.
- Estradiol and progesterone regulate the endometrial cycle, with estradiol inducing proliferation of the endometrium, and progesterone inducing the endometrial glands to secrete carbohydrate-rich mucus.
- During puberty, the hypothalamus begins to secrete increasing quantities of GnRH, which increases LH and FSH secretion, activates ovarian function, and leads to the first ovulation.
- Menopause is characterized by a lack of oocytes in the ovary and the subsequent failure of follicular development and estradiol secretion. LH and FSH levels rise due to the lack of negative feedback by estradiol.

Chapter Review Questions

1. Estradiol synthesis in the graafian follicle involves:

    A. activation of LH-stimulated granulosa production of androgen.
    B. stimulation of aromatase in the granulosa cell by FSH.
    C. decreased secretion of progesterone from the corpus luteum, resulting in increased LH.
    D. inhibition of the LH surge during the preovulatory period.
    E. synergy between FSH and progesterone.

The correct answer is B. Aromatase is present only in granulosa cells and is regulated mainly by FSH. Although LH may stimulate aromatase in granulosa cells, granulosa cells do not produce androgens. Estradiol synthesis in the graafian follicle is unrelated to progesterone synthesis in the corpus luteum and does not increase LH during this phase. Estradiol increases LH secretion during the LH surge. There is no evidence for synergy between FSH and progesterone in regulating estradiol secretion by the graafian follicle.

2. Granulosa cells do not produce estradiol from cholesterol because they do not have an active:

    A. 17α-hydroxylase.
    B. aromatase.
    C. 5α-reductase.
    D. sulfatase.
    E. steroidogenic acute regulatory protein.

The correct answer is A. Granulosa cells do not have the enzyme called 17α-hydroxylase, which converts progesterone to 17α-hydroxyprogesterone. Aromatase is the enzyme that converts androgens to estrogens. 5α-Reductase converts testosterone to dihydrotestosterone. Sulfatase is an enzyme that conjugates steroids with sulfate for subsequent excretion in the urine. Steroidogenic acute regulatory protein transports cholesterol from the outer to the inner mitochondrial membrane.

3. A clinical sign indicating the onset of the menopause is:

    A. the onset of menses near age 50.
    B. an increase in plasma FSH levels.
    C. an excessive presence of corpora lutea.
    D. an increased number of cornified cells in the vagina.
    E. regular menstrual cycles.

The correct answer is B. One of the first clinical measures for menopause is an increase in the serum concentration of FSH (and LH), indicative of the lack of ovarian function. Menses starts at age 12, not age 50, and its onset at this time would not indicate menopause. Excessive corpora lutea would likely indicate multiple ovulations or a failure of luteal regression. Increased vaginal cornification is an indicator of estrogen secretion, which does not occur in menopause. Menstrual cycles become irregular at menopause.

4. Increased progesterone during the postovulatory period is associated with:

    A. proliferation of the uterine endometrium.
    B. enhanced development of graafian follicles.
    C. luteal regression.
    D. an increase in basal body temperature by 0.5°C to 1.0°C.
    E. increased secretion of FSH.

The correct answer is D. Progesterone has a thermogenic effect on the hypothalamus, increasing the basal body temperature for a few days after ovulation. Women who, because of ovulatory problems, are having trouble getting pregnant are sometimes asked to record their daily oral temperatures and look for the increase in basal body temperature, indicating an increase in progesterone (which indicates ovulation). Progesterone induces a secretory type of endometrium, whereas estrogens induce a proliferative type. During the luteal phase, when progesterone is increasing, graafian follicles are not present. Progesterone levels decrease during luteal regression. FSH decreases when progesterone is rising.

5. The theca interna cells of the graafian follicle are distinguished by:

   A. their capacity to produce androgens from cholesterol.
   B. the lack of cholesterol side-chain cleavage enzyme.
   C. aromatization of testosterone to estradiol.
   D. the lack of a blood supply.
   E. the production of inhibin.

The correct answer is A. Theca interna cells produce androgens under the influence of LH, whereas granulosa cells do not produce androgens. Theca interna cells do contain cholesterol side-chain cleavage enzyme, which converts cholesterol to pregnenolone. Because theca cells do not express aromatase, they cannot convert testosterone to estradiol. The theca interna has a rich blood supply. Granulosa cells produce inhibin.

Clinical Application Exercises 37.1

EARLY SPONTANEOUS PREGNANCY TERMINATION

A 38-year-old woman visited her obstetrician gynecologist and complained that she was unable to get pregnant. Upon taking a medical history, the physician notes that the patient had regular 28- to 30-day cycles during the past year and engaged in unprotected intercourse. She does not smoke and does not use caffeine, drugs, or alcohol. She appears to be in good health. Her ovaries and uterus appear normal in size for her age. Laboratory tests indicate that her preovulatory (late follicular phase) estradiol is 40 pg/mL (normal, 200 to 500 pg/mL) and midluteal phase progesterone is 3 ng/mL (normal, 4 to 20 ng/mL). Her husband's sperm count is 30 million/mL.

QUESTIONS

1. What are the clinical indications of a fertility problem with this patient?
2. Based on the clinical signs, what basic physiological principles provide insight into the infertility?
3. What are some theoretical treatment options for this patient?

ANSWERS

1. There are two laboratory tests that indicate a problem, the low late follicular phase plasma estradiol concentration and the low midluteal phase plasma progesterone concentration.
2. The low estradiol could be due to the development of a small dominant graafian follicle with insufficient numbers of granulosa cells. The reduced number of granulosa cells would not contain sufficient aromatase to synthesize the high levels of estradiol required during the late follicular phase. In addition, low estradiol could be due to inadequate FSH receptors on the granulosa cells or inadequate FSH secretion. The low estradiol could also be explained by a lack of LH stimulation of thecal androgen production from the small dominant follicle, possibly the result of inadequate LH receptors on theca cells or low LH levels. The low progesterone during the luteal phase might be due to the ovulation of a small follicle or premature ovulation of a follicle that was not fully developed. The number of LH receptors on the luteinized granulosa cells in the graafian follicle and developing corpus luteum may be insufficient, or LH secretion may be deficient. LH receptors mediate the action of LH, which stimulates progesterone secretion. An insufficient number of LH receptors could be due to insufficient priming of the developing follicle with FSH. Finally, the LH surge may be insufficient for maximal progesterone secretion.

3. Theoretical treatment options for the patient include exogenous progesterone during the luteal phase, which would raise the overall circulating progesterone to levels compatible with maintaining pregnancy, allowing implantation of the embryo. Another option would be to use exogenous FSH to stimulate follicular development to produce larger follicle(s) with sufficient estradiol secretion and LH receptors. Follicles with adequate LH receptors would respond to an LH surge with increased progesterone in the normal range. Another option is the administration of hCG during the periovulatory period for inducing ovulation and full luteinization. The latter would overcome any deficiency in the endogenous LH surge. Finally, the use of clomiphene, an antiestrogen, would increase FSH (and LH) secretion in the follicular phase and, subsequently, induce follicular development with sufficient estradiol to induce a full LH surge. Exogenous hCG could be given during the ovulatory phase to ensure full luteinization of the corpus luteum with sufficient progesterone to maintain pregnancy.

thePoint® *Visit* http://thepoint.lww.com/rhoades5e *for additional chapter review Q&A, Clinical Application Exercises, animations, and more!*

Active Learning Objectives

Upon mastering the material in this chapter, you should be able to:

- Outline the steps required to unite the sperm and egg to achieve fertilization, and explain the roles of progesterone and estrogen in the preparation of the oviduct and uterus for receiving the developing embryo.
- List the steps and timing of blastocyst implantation into the uterine wall.
- Describe the synthesis of human chorionic gonadotropin and its role in the function of the corpus luteum early during pregnancy.
- Outline the steps required in the development of the placenta and explain the importance of this organ for steroid production during pregnancy.
- List the major hormones produced by the cooperation between the maternal compartment and fetoplacental unit, and explain the role of each during the course of pregnancy.

- Describe the cause, and explain the consequences, of maternal insulin resistance during the latter half of pregnancy.
- Outline the mechanism(s) that result in parturition, including the function of cortisol, oxytocin, estrogen, and prostaglandins.
- Explain the roles of prolactin, insulin, glucocorticoids, and oxytocin in lactogenesis.
- Describe the mechanism through which lactation results in anovulation and suppression of menstrual cycles.
- Explain the processes that activate the hypothalamic–pituitary axis to initiate puberty and the effect of this activation on steroid output by the gonads.
- Outline examples of chromosomal and hormonal alterations that result in disorders of sexual development in males and females.

ertilization, the successful union of a sperm and an egg, is the start of pregnancy in a female. The fertilized egg, or zygote, begins to divide immediately after fertilization, and a new life begins. Cell division eventually produces a **blastocyst,** which enters the uterus and implants itself into the uterine endometrium. **Implantation** can only occur in a uterus that has been primed by gonadal steroids and is receptive to the blastocyst. At the time of implantation, the trophoblast cells of the early embryonic placenta begin to produce **human chorionic gonadotropin** (**hCG**), a hormone that signals the ovary to continue to produce progesterone, the major hormone required for the maintenance of pregnancy. As a signal from the embryo to the mother to extend the life of the corpus luteum, hCG prevents the onset of the next menstruation and ovulatory cycle. The placenta, an organ produced by the mother and fetus, exists only during pregnancy to regulate the supply of oxygen and energy, and the removal of wastes, for the fetus. The placenta also produces protein and steroid hormones that are important for fetal development and maintenance of pregnancy. The fetal endocrine glands regulate important functions *in utero* including sexual differentiation.

    Parturition, the expulsion of the fully formed fetus from the uterus, is the final stage of gestation. The onset of parturition is triggered by signals from both the fetus and the mother and involves biochemical and mechanical changes in the uterine myometrium and cervix. After delivery, the mammary glands of the mother must be fully developed and secrete milk to provide nutrition to the newborn baby. The act of suckling, through neurohormonal signals, stimulates milk production and prevents new ovulatory cycles. Ovulatory cycles return at the end of the breast-feeding

period. Sexual maturity is attained during puberty, which occurs at approximately 11 to 12 years of age. The onset of puberty requires changes in the sensitivity, activity, and function of several endocrine organs, including those of the hypothalamic–pituitary–gonadal axis (Clinical Focus 38.1).

▶ FERTILIZATION AND IMPLANTATION

Fertilization (conception) is the fusion of the male and female gametes (sperm with an ovum), which leads to the formation of an embryo and the development of a new organism. The oviduct facilitates the movement of the gametes toward each other, and after fertilization provides nourishment for the developing embryo while transporting it to the uterus.

Cilia and smooth muscle transport the gametes toward each other within the female genital tract.

During intercourse, a fertile man will deposit 2 to 6 mL of **semen**, the ejaculatory fluid containing 20 to 30 million sperms per milliliter, in the vagina of a woman at the time of ejaculation. Only about 50 to 100 sperms will reach the ampulla of the fallopian tube for fertilization. Major losses of sperm occur in the vagina due to the acidic pH, in the uterus due to phagocytosis by leukocytes, and at the uterotubal junction. Spermatozoa that survive reach the ampulla within 5 to 10 minutes after coitus. Sperm motility; muscular contractions of the vagina, cervix, and uterus; ciliary movement; peristaltic activity; and fluid flow in the oviducts assist transport of the sperms to the ampulla. Sperms remain motile for up to 4 days in the female reproductive tract; however, the capacity for fertilization is limited to 1 to 2 days.

In Vitro Fertilization

Candidates for *in vitro* fertilization (IVF) are women with disease of the oviducts, unexplained infertility, or endometriosis (occurrence of endometrial tissue outside the endometrial cavity, a condition that reduces fertility) and those whose male partners are infertile (e.g., low sperm count). Follicular development is induced with one or a combination of GnRH analogs, clomiphene, FSH, and menopausal gonadotropins (a combination of LH and FSH). Follicular growth is monitored by measuring serum estradiol concentration and by ultrasound imaging of the developing follicles. When the leading follicle is 16 to 17 mm wide and/or the estradiol level is >300 pg/mL, hCG is injected to mimic an LH surge and induce final follicular maturation, including maturation of the oocyte. Approximately 34 to 36 hours later, oocytes are retrieved from the larger follicles by aspiration using laparoscopy or a transvaginal approach. Oocyte maturity is judged from the morphology of the granulosa cells and the presence of the germinal vesicle and first polar body. The mature oocytes are then placed in culture media.

The donor's sperms are prepared by washing, centrifuging, and collecting those that are most motile. About 100,000 spermatozoa are added for each oocyte. After 24 hours, the eggs are examined for the presence of two pronuclei (male and female). Embryos are grown to the four-cell to eight-cell stage, about 60 to 70 hours after their retrieval from the follicles. Approximately three embryos are often deposited in the uterine lumen to increase the chance for a successful pregnancy. To ensure a receptive endometrium, daily progesterone administrations begin on the day of retrieval. The rate of success for IVF techniques depends on several factors, including the age of the female partner and normal semen analysis. ■

Sperms undergo final maturation within the oviduct in a process called **capacitation**. As sperms develop in the epididymis of a male, surface glycoproteins are acquired that act as stabilizing factors, but which prevent the interaction of sperm and egg. During capacitation, surface glycoproteins are removed, membrane cholesterol is depleted increasing plasma membrane fluidity, and sperm motility increases. Capacitation can occur anywhere in the female reproductive tract where pH is elevated, as shown when sperms are deposited in the intraperitoneal cavity near the fimbria during assisted fertilization procedures. Sperms can also undergo capacitation in chemically defined medium during *in vitro* fertilization.

A secondary oocyte arrested in metaphase of the second meiosis is released from the ovary approximately every 28 days in response to the LH surge (see Table 37.1). Ciliated fingerlike projections of the oviducts, called **fimbria**, grasp the ovum and propel it into the oviduct. The ciliary movement and muscle contractions of the oviduct are stimulated by the estrogen synthesized by the ovary during the periovulatory period. The human ovum is most viable for fertilization for about 24 hours, with fertilization usually occurring within the 2 days following ovulation. There is a space between the fimbria and ovary that is open to the peritoneal cavity (see Fig. 37.3); if the oviducts do not actively pick up the egg when released, it can enter the abdominal cavity. An **ectopic pregnancy** may result if an ovum is fertilized in the abdomen.

The use of chemotaxic signals by sperm to find the egg is well established for marine invertebrates such as sea urchin and starfish. There is suggestive evidence that human spermatozoa are sensitive to progesterone, which is secreted by the cumulus cells surrounding the ovulated oocyte and could thus act as a chemoattractant. However, some sperms that arrive in the ampulla within the vicinity of the egg exit into the abdominal cavity, suggesting that contact between the sperm and egg may be random.

Fertilization begins as the sperm attaches to the zona pellucida and undergoes the acrosomal reaction.

The zona pellucida, or jelly coat, surrounding the ovum contains specific glycoproteins that serve as sperm receptors. When a sperm contacts one of these receptors, an increase in intracellular calcium is triggered, initiating the **acrosomal reaction** in the sperm (Fig. 38.1). The membrane surrounding the acrosome, a large secretory vesicle containing hydrolyzing enzymes, fuses with the plasma membrane of the sperm head. The proteolytic enzymes are then released to dissolve the zona pellucida, and the sperm uses the propulsive force of its tail to move toward the oocyte membrane in a process that may take up to 30 minutes.

After entering the **perivitelline space**, the sperm head becomes anchored to the **oolemma** (plasma membrane) of the egg through a second set of sperm receptors, with microvilli protruding from the membrane surrounding and eventually engulfing the sperm, incorporating the head and tail into the **ooplasm**.

As the sperm enters the egg, it triggers a rise in intracellular calcium, causing the **cortical reaction.** Lysosome-like organelles located just underneath the oolemma begin to fuse with the membrane, beginning at the point of sperm attachment and propagating over the entire egg surface. The cortical granules contain enzymes that are released into the perivitelline space and diffuse into the zona pellucida. These enzymes act on glycoproteins in the zona pellucida, causing them to harden in a process termed the **zona reaction**. The hardened zona pellucida prevents other sperms from gaining access to the egg, preventing polyspermia (see Fig. 38.1).

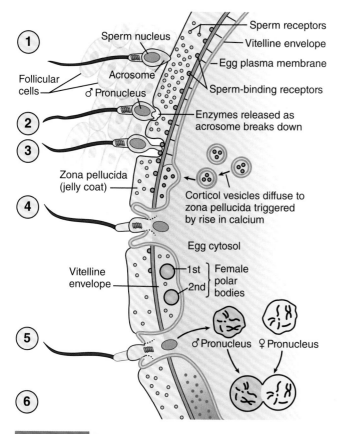

Figure 38.1 **The process of fertilization.** (*1*) The sperm cell weaves past follicular cells to sperm receptors in the zona pellucida. (*2*) Intracellular calcium rises in the sperm initiating the acrosomal reaction. (*3*) Proteolytic enzymes released from the acrosome digest the zona pellucida. (*4*) The sperm penetrates the oolemma, stimulating a rise in intracellular calcium in the egg, which initiates the cortical reaction. (*5*) The rise in intracellular calcium initiates the completion of the second meiotic division and generation of the female pronucleus and second polar body. (*6*) The sperm head enlarges to become the male pronucleus, which fuses with the female pronucleus.

The increase in intracellular calcium initiated by entry of the sperm into the ooplasm also triggers completion of the second meiotic division. The chromosomes of the egg separate, and half of the chromatin is extruded as a small second polar body. The remaining haploid nucleus with its 23 chromosomes transforms into the **female pronucleus** (see Fig. 38.1). As the nuclear envelope of the sperm disintegrates, the **male pronucleus** forms and increases four to five times in size. Contractions of microtubules and microfilaments visible 2 to 3 hours after the entry of the sperm into the egg move the two pronuclei to the center of the cell. Replication of the haploid chromosomes begins in both pronuclei. Pores form in the nuclear membranes of the two pronuclei, which then fuse. Successful fertilization thus restores the full complement of 46 chromosomes, initiating the development of an embryo.

The first mitotic division occurs 24 to 36 hours following fertilization and yields two cells called **blastomeres**. Successive divisions produce four- and eight-blastomere structures, at 48 and 72 hours later, respectively. The morula,

a solid ball of at least 12 blastomeres, is formed by 96 hours postfertilization. Cell division occurs without growth; thus, cells in the developing embryo become progressively smaller. The developing embryo also remains enclosed in the zona pellucida, which protects it from damage, adhesion to the uterine wall, and immunologic rejection by the mother. Development up to the morula stage occurs within the oviduct because smooth muscle contractions of the isthmus prevent advancement of the conceptus into the uterus while the endometrium is prepared for implantation.

Process of implantation proceeds through stages: apposition, adhesion, and invasion.

The morula enters the uterus 4 to 6 days after fertilization. It remains suspended in the uterine cavity for 2 to 3 days, continuing to undergo cell division, nourished by constituents of the uterine fluid. At the 20- to 30-cell stage, a fluid-filled cavity (blastocoele) appears and enlarges until the embryo becomes a hollow sphere, called a **blastocyst**. The cells of the blastocyst undergo significant differentiation. A single outer layer consists of extraembryonic ectodermal cells called the **trophoblast**, which facilitates implantation, forms the embryonic contribution to the placenta and embryonic membranes, produces hCG, and provides nutrition to the embryo. A cluster of smaller, centrally located cells constitutes the **embryoblast** or inner cell mass and gives rise to the fetus.

As the blastocyst approaches the uterine endometrium, the surrounding zona pellucida is ruptured as the blastocyst expands, aided by proteases released from the endometrium. A loose association of the blastocyst with the endometrial wall, called **apposition**, then occurs (Fig. 38.2). The blastocyst secretes cytokines and hormones including interleukin-1α and interleukin-1β that promote endometrial receptivity. The endometrium responds by secreting leukemia inhibitory factor, colony-stimulating factor-1, and other signals that increase trophoblast protease production. The paracrine interaction between the blastocyst and endometrium results in **adhesion** between the two structures. Cellular adhesion molecules including integrins are expressed by both the trophoblast cells and the endometrium. Expression of the integrins αVβ3 and α4β1 by the endometrium, induced by luteal phase estrogen and progesterone production, is considered a receptivity marker for the blastocyst.

Differentiation of the trophoblast to syncytiotrophoblasts

Following adhesion, the trophoblast differentiates into large polyhedral **cytotrophoblasts,** with well-defined cell borders that surround the blastocyst, and multinucleated **syncytiotrophoblasts** that lack distinct cell boundaries (see Fig. 38.2). The syncytiotrophoblast extends long protrusions between the uterine epithelial cells, secreting proteases that digest the basement membrane to eventually reach the uterine stroma and initiate the **decidual reaction**, resulting in a highly specialized endometrium of pregnancy. The endometrial stromal cells hypertrophy to contain large amounts of glycogen and lipid and modify their plasma membrane, creating a specialized group of cells called the **decidua**. Decidualization

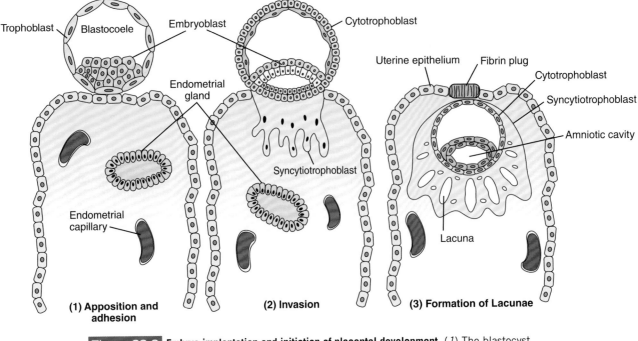

Trophoblast   Blastocoele   Embryoblast

Cytotrophoblast

Uterine epithelium   Fibrin plug
Cytotrophoblast
Syncytiotrophoblast
Amniotic cavity

Endometrial gland

Syncytiotrophoblast

Endometrial capillary

Lacuna

(1) Apposition and adhesion   **(2) Invasion**   **(3) Formation of Lacunae**

Figure 38.2 **Embryo implantation and initiation of placental development.** (*1*) The blastocyst loosely adheres to the endometrium (apposition) and then adheres by binding between cellular adhesion molecules. (*2*) The trophoblast differentiates into the syncytiotrophoblast and cytotrophoblast. The syncytiotrophoblast invades the epithelium to reach the uterine stroma. (*3*) The blastocyst is fully embedded in the uterine wall. Lacunae form within the syncytiotrophoblast and fill with maternal blood. Eventually, the lacunae will coalesce into the intervillous space, and the chorionic villi will form, creating the fully developed placenta.

is initiated by estrogen and progesterone, with the cellular transformation completed by factors released by the invading blastocyst. The decidua is classified into three parts based on anatomic location. Decidua directly beneath the implanting embryo, which is modified by the invading syncytiotrophoblast, becomes the **decidua basalis**. The **decidua capsularis** overlies the enlarging blastocyst and separates the conceptus from the rest of the uterine cavity. The **decidua parietalis** covers the remainder of the uterine wall.

As the syncytiotrophoblast continues to invade the endometrium, fluid-filled holes called lacunae begin to form. These lacunae eventually fill with maternal blood as the syncytiotrophoblast permeates first small veins and then small arteries within the endometrium. Mesenchymal cells from the embryonic mesoderm protrude into the lacunae and form fetal blood vessels, establishing a functional communication between the developing embryonic vascular system and the maternal blood (see Fig. 16.6).

Embryoblast differentiation

By day 10 to 12 following ovulation, the blastocyst is totally encased with the endometrium. The inner cell mass differentiates into an ectoderm, endoderm, and mesoderm. The ectoderm will eventually form the epidermis, appendages (nails and hair), and the entire nervous system. The endoderm gives rise to the epithelial lining of the digestive tract and associated structures, and the mesoderm will form the bulk of the body, including connective tissue, muscle, bone, blood, and lymph.

▶ PLACENTAL NUTRIENT UPTAKE, WASTE ELIMINATION, AND GAS EXCHANGE

Chorionic villi are the functional units of the placenta (Fig. 38.3). As mentioned above, these structures begin to form on days 12 to 15, as cytotrophoblast cells extend into the syncytiotrophoblast to the lacunae containing endometrial blood. Mature chorionic villi form when mesenchymal cells from the embryonic mesoderm migrate into the lacunae and form fetal blood vessels. Chorionic villi initially spread over the entire surface of the **chorionic sac**, the outmost membrane surrounding the developing embryo. However, by the third month of development, the chorionic villi are confined to the **chorionic plate**, an area of the chorion directly opposite of the decidua basalis. The chorionic villi aggregate into groups known as **cotyledons** that protrude into the **intervillous space**, a large contiguous area formed as the maternal blood–filled lacunae expand and merge with one another. The decidua basalis and chorionic plate together constitute the placenta proper (see Fig. 38.3).

As the placenta forms, the decidua capsularis surrounding the fetus and the decidua parietalis fuse to occlude the uterine cavity, and the extraembryonic membranes develop. From the fourth month onward, the fetus is fully enclosed within the **amnion** and **chorion**, connected to the placenta by the **umbilical cord**. Two umbilical arteries approach the placenta and branch into capillaries within the chorionic villi. These arteries carry deoxygenated blood to the placenta, which returns to the fetus with oxygen and nutrients through

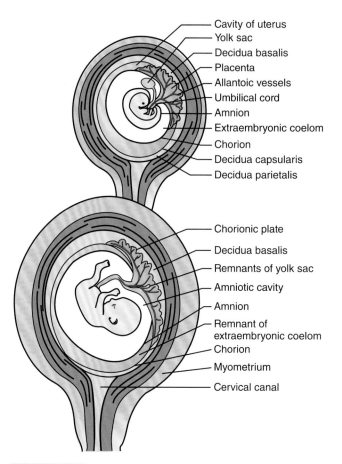

Figure 38.3 Two stages in the development of the placenta, showing the origin of the membranes around the fetus. Note the three decidual layers and their relationship to the developing fetus. Chorionic villi develop early in placental development.

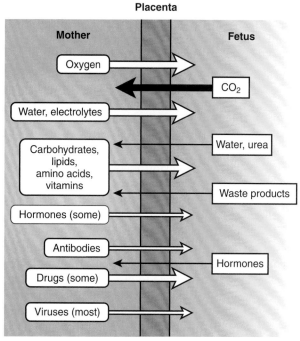

Figure 38.4 Role of the placenta in exchanges between the fetal and maternal compartments. The size of the *arrows* indicates the amount of exchange between the compartments.

a single umbilical vein. The fetal and maternal circulations remain separated by the chorionic villi, where the maternal blood surrounds the fetal endothelium. Thus, the human placenta is classified as hemochorial, a type of placenta in which maternal blood is in direct contact with chorionic villi.

Oxygen diffuses from maternal to fetal blood down an initial gradient of 60 to 70 mm Hg. **Fetal hemoglobin** has a high affinity for oxygen and enhances the oxygen-transporting capacity of fetal blood. The P_{CO_2} of fetal arterial blood is 2 to 3 mm Hg higher than that of maternal blood, allowing the diffusion of carbon dioxide toward the maternal compartment. Other compounds, such as glucose, amino acids, free fatty acids, electrolytes, vitamins, and some hormones, are transported by diffusion, facilitated diffusion, or pinocytosis. The lipid-soluble steroid hormones readily pass through the placenta. Some immunoglobulins, viruses, and drugs can also transfer across the blood–placental barrier from the mother to the fetus (Fig. 38.4).

Waste products, such as urea and creatinine, diffuse down their concentration gradients from the fetus into the amniotic fluid. The fetal kidneys mature at 10 to 12 weeks, after which they produce 75% of the amniotic fluid. The fetal gastrointestinal tract, amnion, and lungs remove the amniotic fluid, sending the waste products to the material compartment.

HORMONES REQUIRED FOR A SUCCESSFUL PREGNANCY

The placenta produces **progesterone** and **estrogens**, which are essential for the continuance of pregnancy. The placenta also secretes peptide hormones including **human placental lactogen (hPL)** and human chorionic gonadotropin (hCG). Corticotropin-releasing hormone (CRH), gonadotropin-releasing hormone (GnRH), and insulin-like growth factors synthesized by the placenta function as paracrine factors.

During a nonconception menstrual cycle, the corpus luteum regresses and circulating progesterone and estrogen decline. With the loss of ovarian steroidal support, the superficial endometrial layer of the uterus degrades, resulting in menstruation. If the egg is fertilized, the developing embryo signals its presence by producing hCG, which extends the life of the corpus luteum, a process called the **maternal recognition of pregnancy**. Syncytiotrophoblast cells produce hCG, which enters the maternal circulation 6 to 8 days after fertilization. hCG can be detected in the urine of a pregnant woman, which is the basis of colorimetric pregnancy tests.

Human chorionic gonadotropin maintains corpus luteum steroidogenesis.

hCG is a glycoprotein structurally similar to LH, FSH, and TSH, with which it shares an identical 92–amino acid α subunit. The β subunit of hCG resembles the β subunit of LH, but with a 24–amino acid extension at the C-terminal end, and includes six N-linked and O-linked oligosaccharide units. Due to this extensive glycosylation, the half-life of hCG in the circulation is longer than that of LH. The major function of hCG in early pregnancy is the stimulation of luteal steroidogenesis. hCG binds to LH receptors on the corpus luteum to activate the rate-limiting step in steroidogenesis, the synthesis of pregnenolone from cholesterol.

Maternal plasma hCG levels double every 2 to 3 days during early pregnancy and reach peak levels at about 10 to 15 weeks of gestation. At week 25, hCG is reduced about 75% from peak and remains at that lower level until term (Fig. 38.5). Fetal concentrations of hCG follow a similar pattern. Maternal plasma hCG levels are higher in pregnancies with multiple fetuses. During the first trimester, GnRH produced by the cytotrophoblasts regulates hCG production through paracrine signaling. The suppression of hCG release during the second half of pregnancy results from negative feedback by placental progesterone and other steroids produced by the fetus. Corpus luteum progesterone secretion is maximal at 4 to 5 weeks after conception but declines despite the rising hCG levels. Corpus luteum refractoriness to hCG results from receptor desensitization and the rising levels of placental estrogens. From weeks 7 to 10 of gestation, steroid production by the placenta gradually replaces steroid production by the corpus luteum. Surgical removal of the corpus luteum after week 10 does not terminate pregnancy. Activin, inhibin, and transforming growth factors α and β are additional placenta-derived growth factors that regulate hCG production.

Other important targets of hCG are the fetal adrenal gland and testes. hCG is important in the sexual differentiation of the male fetus, which depends on testosterone production by the fetal testes. Peak testosterone production occurs 11 to 17 weeks after conception, coinciding with peak hCG production and predating the functional maturity of the fetal hypothalamic–pituitary axis (thus, fetal LH levels are low). Fetal Leydig cell proliferation and testosterone biosynthesis are regulated by hCG through LH/hCG receptors that appear early in fetal testes development. LH/hCG receptors are not present on fetal ovaries; thus, a role for hCG in fetal ovarian development is less clear.

Human placental lactogen regulates fetal fuel metabolism.

Human placental lactogen (hPL) is a single-chain polypeptide hormone of 191 amino acids with two disulfide bridges. Its structure and function resemble those of prolactin and growth hormone. As a result, it is also called chorionic somatomammotropin and chorionic growth hormone. Three genes encode chorionic somatomammotropin, but only two are transcribed. hPL is synthesized by syncytiotrophoblasts in proportion to placental mass and secreted into the maternal circulation where its levels gradually rise from the third week of pregnancy until term.

The main function of hPL is to regulate fuel availability for the fetus by antagonizing maternal glucose consumption and enhancing maternal fat mobilization. The effects of hPL on carbohydrate, protein, and fat metabolism are similar to those of GH. Despite its important metabolic effects, lack of hPL does not impair pregnancy.

Maternal and fetoplacental units cooperate to produce steroids during pregnancy.

The corpus luteum is the primary source of progesterone and estrogen during the first ~8 weeks of pregnancy. As the placenta develops, the trophoblast cells gradually become the major site of progesterone and estrogen production, although the corpus luteum continues to secrete progesterone. Blood progesterone levels gradually rise during early pregnancy and plateau during the transition period from corpus luteal to placental production (see Fig. 38.5). Thereafter, plasma progesterone levels rise to almost 150 ng/mL near the end of pregnancy. Estradiol and estriol rise gradually during the first half of pregnancy, but increase steeply in the latter half to more than 25 ng/mL near term.

Progesterone and estrogen have numerous functions throughout gestation. Estrogens increase the size of the uterus and uterine blood flow, are critical to the timing of implantation of the embryo into the uterine wall, induce the formation of uterine receptors for progesterone and oxytocin, enhance fetal organ development, stimulate maternal hepatic protein production, and increase the mass of the pregnant woman's breast and adipose tissues. Progesterone is essential for maintaining the uterus and early embryo, inhibits myometrial contractions, and suppresses maternal immunologic responses to fetal antigens. Progesterone also serves as a precursor for steroid production by the fetal adrenal glands and plays a role in the onset of parturition.

The placenta and fetus lack certain steroidogenic enzymes and therefore cannot synthesize all steroids required during pregnancy and fetal development. Thus, the placenta, fetus, and mother share precursors and intermediates to facilitate steroidogenesis.

The placenta obtains cholesterol for steroidogenesis from LDL particles in the maternal blood. The trophoblast cells express **LDL receptors**, which bind and internalize LDL particles, from which cholesterol ester hydrolase produces free cholesterol. The cholesterol side-chain cleavage enzyme then synthesizes pregnenolone from the free cholesterol followed by the conversion of pregnenolone to progesterone by

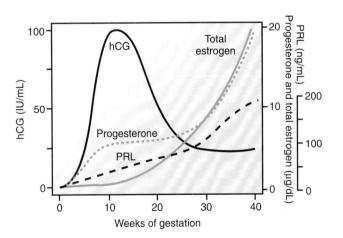

Figure 38.5 Human chorionic gonadotropin (hCG), progesterone, total estrogens, and prolactin (PRL) levels in the maternal blood throughout gestation. Total estrogen represents estradiol and estriol.

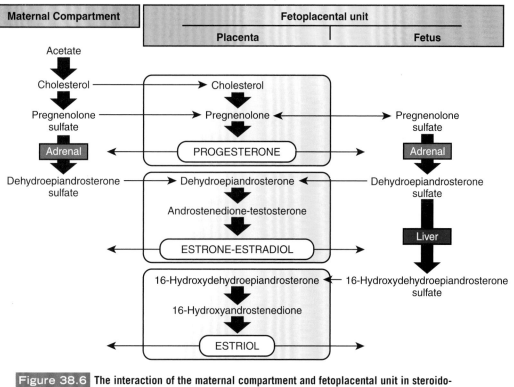

Figure 38.6 **The interaction of the maternal compartment and fetoplacental unit in steroido-genesis.** Estriol is the major estrogen produced by the placenta and is the product of reactions occurring in the fetal adrenal, fetal liver, and placenta.

3β-hydroxysteroid dehydrogenase. Progesterone produced by the placenta then diffuses to the maternal and fetal circulations (Fig. 38.6).

The placenta lacks 17α-hydroxylase, which is required to convert progesterone to 17α-progesterone and thus cannot make androgens. Of importance, 17α-hydroxyprogesterone released by the corpus luteum is detectable in maternal blood during the first trimester and serves as a marker of corpus luteum function. To produce estrogens, the placenta uses androgenic substrates derived from both the fetus and the mother (see Fig. 38.6). The primary androgenic precursor is **dehydroepiandrosterone sulfate** (**DHEAS**), which is produced by the maternal and fetal adrenal gland. The fetal adrenal gland is extremely active in the production of steroid hormones, but it lacks 3β-hydroxysteroid dehydrogenase and cannot make progesterone. Therefore, the fetal adrenals use progesterone from the placenta to synthesize and sulfate dehydroepiandrosterone. The conjugation of androgenic precursors to sulfate ensures greater water solubility, aids in transport, and reduces biologic activity while in the fetal circulation. DHEAS diffuses into the placenta and is cleaved by a sulfatase to yield dehydroepiandrosterone, which is converted to androstenedione and then testosterone. The placenta has an active aromatase that converts testosterone to estradiol and estrone, which enter the maternal and fetal circulation.

The main estrogen produced by the placenta is estriol, which has weak estrogenic activity. Estriol is produced by cooperation between the placenta and fetus (see Fig. 38.6). Within the fetal liver, 16α-hydroxylase converts DHEAS produced by the fetal adrenal gland to 16-hydroxydehydro

epiandrosterone sulfate (16-OH-DHEAS), which diffuses to the placenta where the sulfate group is removed. Placental 3β-hydroxysteroid dehydrogenase converts 16-OH-DHEAS to 16-hydroxyandrostenedione, which is subsequently aromatized to estriol. The maternal adrenal can synthesize 16-OH-DHEAS from maternal DHEAS but does not do so to any significant extent; thus, it is estimated that 90% of estriol is derived from the fetal 16-OH-DHEAS. The levels of estriol in plasma, amniotic fluid, or urine can be used as an index of fetal well-being. Low levels of estriol would indicate potential fetal distress, although rare inherited sulfatase deficiencies can also lead to low estriol.

Maternal metabolism and endocrine function change to support gestation.

The pregnant woman undergoes significant adjustments in her pulmonary, cardiovascular, renal, metabolic, and endocrine systems to support her growing fetus. Notable changes include hyperventilation, reduced arterial blood PCO_2 and osmolality, increased blood volume and cardiac output, increased renal blood flow and glomerular filtration rate, and substantial weight gain. These changes are brought about by the rising levels of estrogens, progesterone, hPL, and other placental hormones and by mechanical factors such as the expanding size of the uterus and the development of uterine and placental circulations.

There are a number of adaptations in the maternal endocrine system. The high levels of sex steroids suppress the hypothalamic–pituitary–ovarian axis; thus, circulating gonadotropins are low, and ovulation does not occur during pregnancy. In contrast, the rising levels of estrogens stimulate

PRL release, which begins to rise during the first trimester, increasing gradually to reach a level 10 times higher near term (see Fig. 38.5). Estrogen stimulates hyperplasia and hypertrophy of the pituitary lactotrophs that synthesize PRL, resulting in enlargement of the pituitary gland. In contrast, the somatotrophs that produce GH are reduced in size, and GH levels are low throughout pregnancy.

The thyroid gland enlarges during pregnancy, but TSH levels remain in the normal range. T_3 and T_4 increase, but estrogen also increases thyroxine-binding globulin synthesis; thus, the pregnant woman remains euthyroid. The parathyroid glands enlarge, and PTH secretion increases during the third trimester. PTH stimulates calcium mobilization from maternal bone stores to meet the growing fetal demand for calcium. Mineralocorticoid and glucocorticoid release from the maternal adrenal increase. Plasma-free cortisol is higher because it is displaced from the cortisol-binding globulin transcortin by the high levels of progesterone, but hypercortisolism does not develop. Maternal ACTH levels rise throughout pregnancy to levels fivefold higher than that prior to pregnancy. ACTH falls 50% just before parturition and then increases 15-fold during the stress of delivery. Both the maternal pituitary and the placenta contribute to the increase in ACTH.

Maternal metabolism also responds to the increasing nutritional demands of the fetus. The major net weight gain of the mother occurs during the first half of gestation, resulting primarily from effects of progesterone to increase appetite and divert glucose into fat synthesis and storage. When the metabolic requirements of the fetus peak later in pregnancy, and also during periods of starvation, the extra fat stores are used as an energy source. Toward the second half of gestation, the mother develops insulin resistance, brought about by the combined effects of GH, PRL, hPL, glucagon, and cortisol, hormones that antagonize insulin action. As a result, maternal glucose usage declines and gluconeogenesis increases, maximizing glucose availability for the fetus.

Fetal endocrine system develops early to regulate fetal homeostasis.

The protective intrauterine environment postpones the initiation of some physiologic functions that are essential for life after birth. For example, the fetal lungs and kidneys do not act as organs of gas exchange and excretion because the placenta carries out these functions. Constant isothermal surroundings alleviate the need to expend calories to maintain body temperature. The gastrointestinal tract does not carry out digestion, and fetal bones and muscles do not support weight or locomotion. Exposure to limited levels of external stimuli results in the slow development of the fetal nervous and immune systems. In contrast, the fetal endocrine system plays a vital role in fetal growth, development, and homeostasis.

The blood–placental barrier excludes most protein and polypeptide hormones from the fetus; thus, the maternal endocrine system has limited direct effects on the fetus. Rather, the fetus is fairly self-sufficient in its hormonal requirements, with the exception of the steroid hormones produced by the fetoplacental unit, which cross easily between the different compartments to carry out integrated functions in both the fetus and the mother. In general, fetal hormones perform the same functions for the fetus as in the adult, but they also subserve unique processes such as sexual differentiation and the initiation of labor.

Hypothalamic–pituitary axis development

The fetal hypothalamic nuclei develop and express releasing hormones such as TRH and GnRH by ~12 weeks of gestation. The anterior pituitary, which forms from the Rathke pouch (an ectodermal evagination of the roof of the fetal mouth) beginning at gestation week 4, is present by gestation week 8, and most anterior pituitary hormones can be identified. The posterior pituitary, which forms by evagination of the floor of the primitive hypothalamus, is present by week 14 and arginine vasopressin (AVP) and oxytocin can be detected. Overall, the hypothalamic–pituitary axis is functional by 12 to 17 weeks of gestation, although the maturation of the portal vascular system continues until 30 to 35 weeks' gestation. The release of pituitary hormones occurs prior to the full maturation of the portal system, suggesting that the hypothalamic-releasing hormones diffuse down to the pituitary from hypothalamic sites of synthesis.

Adrenal gland production of steroidogenic precursors

The fetal adrenal glands are present as distinct organs by 8 weeks' gestation and are composed of three functional zones. The outer definitive zone synthesizes glucocorticoids and mineral corticoids, with the transitional zone producing cortisol. The inner fetal zone, which is much larger than the other two zones, synthesizes significant amounts of androgens such as DHEAS, which are used by the placenta to make estrogens (see Fig. 38.6). The fetal adrenal glands grow rapidly during gestation and at birth are 10- to 20-fold larger than adult adrenals. Postnatally, the adrenal gland undergoes rapid involution due to regression of the inner fetal zone, which is absent by 6 months of age. ACTH is potent stimulus for fetal adrenal steroidogenesis and is required for adrenal growth, as evidenced by the observation that anencephalic fetuses with low ACTH experience fetal zone involution at 15 weeks' gestation. The adrenal medulla develops by about week 10 and is capable of producing epinephrine and norepinephrine. Cortisol has multiple functions in the fetus such as stimulating pancreas and lung maturation, inducing expression of liver enzymes, and promoting intestinal tract cytodifferentiation.

PTH and calcitonin promotion of calcium uptake and bone growth

Calcium is in large demand because of the rapid growth and large amount of bone formation of the fetus. The placenta has a specialized calcium pump that transfers calcium to the fetus, resulting in sustained increases in calcium and phosphate throughout pregnancy. PTH and calcitonin are evident in the fetus near week 12 of gestation and are important in promoting placental uptake and calcium and bone anabolism, respectively. Fetal PTH also acts on the fetal kidney to promote synthesis of 1,25-dihydroxyvitamin D, which stimulates cartilage and bone growth. At the end of gestation, calcium and phosphate levels in the fetus are higher than those in the mother. However, after delivery, neonatal calcium levels decrease and PTH levels increase to raise serum calcium levels.

Insulin and insulin-like growth factors are required for fetal growth.

The rate of fetal growth increases significantly during the last trimester and includes significant adipose tissue deposition. Surprisingly, GH of maternal, placental, or fetal origin has little effect on fetal growth, as judged by the normal weight of hypopituitary dwarfs or anencephalic fetuses. However, the insulin-like growth factors (IGF-1 and IGF-2) have been shown in mouse studies to be required for placental and fetal growth. Absence of these hormones or their receptors results in significant growth retardation with small placentas. Expression of IGF-1 and IGF-2 is not regulated by GH, which increases in the fetus up to midgestation and then decline when fetal weight is increasing significantly. Rather, nutrition is the major factor modulating IGF production in the fetus.

Insulin is produced by the fetal pancreas by week 10 of gestation at a relatively constant rate, increasing only slightly in response to a rapid rise in blood glucose levels. Glucose is the main metabolic fuel for the fetus, but insulin is not needed for fetal tissue glucose uptake as a relatively constant glucose supply is provided by the mother. However, near-term fetal insulin can increase glucose uptake and lipolysis in fetal tissues. Insulin is considered a fetal growth factor because babies born to women with **diabetes mellitus** and hyperinsulinemia are larger with more adipose tissue deposition. Babies with congenital hyperinsulinemia are also born larger. In contrast, insulin receptor mutations in humans lead to severe intrauterine growth retardation. In women who develop **gestational diabetes**, which results in chronically elevated blood glucose, the fetal pancreas becomes enlarged and fetal insulin levels increase. Consequently, fetal growth is accelerated, and infants of women with uncontrolled diabetes are also born large for gestational age.

There are a large number of additional factors that are involved in fetal growth and development including epidermal growth factor, nerve growth factor, platelet-derived growth factors, fibroblast growth factors, angiogenic growth factors, and members of the transforming growth factor β family. Expression of loss of function and gain of function mutations in these factors in either the fetus or placenta can result in developmental abnormalities.

Placental and fetal factors induce parturition

Parturition is the culmination of a pregnancy, occurring on average 270 ± 14 days from the time of fertilization. Uncoordinated uterine contractions start about 1 month before the end of gestation. Strong rhythmic contractions (*labor*) that may last several hours and eventually generate enough force to expel the fetus terminate the pregnancy. There are a number of mechanisms through which parturition is achieved.

In humans, placental CRH levels in the maternal circulation increase exponentially throughout gestation and may be a coordinating signal for the fetus and uterus to initiate labor. Maturation of the fetal lung is required for survival of the fetus following delivery. Placenta CRH can stimulate fetal pituitary ACTH release and cortisol synthesis, promoting lung maturation. In the mother, placental CRH also increases cortisol levels, which results in maternal androgen synthesis and greater estriol production from the placenta. The rise in estriol acts on the myometrium to induce contractile proteins. Late in pregnancy, CRH also appears to alter the expression of progesterone receptors in the myometrium.

Throughout most of pregnancy, progesterone keeps the uterus quiescent by reducing expression of contractile proteins and hyperpolarizing myometrial cells, a process termed **progesterone block**. It also prevents the release of phospholipase A_2, the rate-limiting enzyme in prostaglandin synthesis (see below). In many species, a sharp decline in progesterone at term induces uterine contractions; however, in humans, progesterone does not significantly fall before parturition. Rather, expression of an alternative progesterone receptor is induced by the elevated CRH, and this receptor functions as an inhibitor of progesterone action on target genes, resulting in withdrawal of the progesterone block.

Prostaglandins produced by uterine decidual cells play a key role in the initiation of labor. Prostaglandin F_{2a} and E_2 stimulate uterine contractions in a paracrine manner and potentiate oxytocin-induced contractions by promoting formation of gap junctions between uterine smooth muscle cells. The prostaglandins also cause softening, thinning, and dilation of the cervix. The myometrium, decidua, and chorion all produce prostaglandins, and shortly before the onset of parturition, the prostaglandin concentration in amniotic fluid rises abruptly. Aspirin and indomethacin block prostaglandin synthesis, inhibiting labor and prolonging gestation.

Cessation of uterine growth toward the end of pregnancy may also contribute to the initiation of parturition. Stretch, which would be induced by the increasing tension on the uterine wall, usually leads to contraction of smooth muscle.

Maternal **oxytocin** is not involved in initiating labor but has an important role in maintaining uterine contractions once labor has been initiated. Rising estrogen levels significantly increase oxytocin receptors on the myometrium and placental decidual tissue during the last few weeks of gestation. **Relaxin**, a large polypeptide hormone produced by the corpus luteum and the decidua, also increases oxytocin receptors. Distension of the cervix initiates bursts of oxytocin release from the maternal posterior pituitary, which stimulates myometrial contractions. Oxytocin is used clinically to facilitate labor and reduce postpartum bleeding after delivery (Clinical Focus 38.2).

Reproductive Physiology

Pharmacologic Induction and Augmentation of Labor

Several drugs are currently used to assist in the therapeutic induction and augmentation of labor. Therapeutic induction implies that the use of a drug initiates labor. Augmentation indicates that labor has started and that a therapeutic agent further stimulates the process.

Oxytocin, the natural hormone produced from the posterior pituitary, is widely used to induce and augment labor. Several synthetic forms of oxytocin can be used by intravenous routes. Prostaglandins (F_{2a} and E_2) have also been used to induce and augment labor and cervical ripening. They promote dilation and effacement of the cervix and can be used intravaginally, intravenously, or intra-amniotically. ■

► POSTPARTUM LACTATION

Lactation, the production and removal of milk from the breast, occurs at the final phase of the reproductive process. Prolactin (PRL) promotes mammary growth, initiation of milk secretion, and maintenance of milk production (**galactopoiesis**) once it has been established. Oxytocin activates **milk ejection** or **milk letdown**, the process by which stored milk is released from the mammary glands.

Mammogenesis occurs at three distinct developmental stages.

The growth and differentiation of the mammary glands occur *in utero*, at puberty, and during pregnancy. The **mammary glands** begin to differentiate in the pectoral region as an ectodermal thickening on the epidermal ridge at gestation weeks 7 to 8. The prospective mammary glands lie along bilateral mammary ridges or milk lines extending from the axilla to groin on the ventral side of the fetus. In humans, most of the ridge disintegrates except in the axillary region. Mammary buds derive from the surface epithelium, which invade the underlying mesenchyme. During the fifth gestation month, the buds elongate, branch, and sprout, eventually forming the **lactiferous ducts** (primary milk ducts), which continue to branch and grow throughout life. The ducts unite, grow, and extend to the site of the future nipple. The primary buds give rise to secondary buds, which are separated into lobules by connective tissue and become surrounded by **myoepithelial cells**. The **nipple** and **areola** form during the eighth month of gestation.

The mammary glands of male and female infants are identical and, although underdeveloped, can secrete small amounts of milk ("witch's milk") at birth. Fetal mammary tissue is responsive to the lactogenic hormones of pregnancy and the withdrawal of placental steroids at birth, resulting in milk production in some infants. Sexual dimorphism in breast development begins at the onset of puberty. The male breast is fully developed at about 20 years of age and is similar to the female breast at an early stage of puberty.

Estrogen exerts a major influence on breast growth at puberty in females. The first response is an increase in size and pigmentation of the areola and accelerated deposition of adipose and connective tissues. In association with menstrual cycles, estrogen stimulates the growth and branching of the ducts, whereas progesterone acts primarily on the alveolar components. The action of both hormones, however, requires synergism with PRL, GH, insulin, cortisol, and thyroxine.

The mammary glands undergo significant changes during pregnancy. The ducts become elaborate during the first trimester, and new lobules and alveoli are formed in the second trimester. The alveolar cells differentiate into secretory cells, replacing most of the connective tissue. The development of the secretory capability requires estrogen, progesterone, PRL, and placental lactogen, as well as insulin, cortisol, and several growth factors. Lactogenesis begins during the fifth month of gestation, but only **colostrum** (initial milk) is produced. Elevated progesterone levels antagonize the action of PRL to prevent full lactation during pregnancy. The ovarian steroids synergize with PRL in stimulating mammary growth but antagonize its actions in promoting milk secretion.

Lactogenesis is fully expressed after parturition, on the withdrawal of placental steroids. Lactating women produce up to 600 mL of milk each day, increasing to 800 to 1,100 mL/d by the sixth postpartum month. Milk is isosmotic with plasma, and its main constituents include proteins (casein and lactalbumin), lipids, and lactose. Colostrum, produced in small quantities during the first postpartum days, is higher in protein, sodium, and chloride content and lower in lactose and potassium than normal milk. It also contains immunoglobulin A, macrophages, and lymphocytes, which provide passive immunity to the infant. During the first 2 to 3 weeks, the protein content of milk decreases, whereas that of lipids, lactose, and water-soluble vitamins increases.

Within the mammary gland, the milk-secreting **alveolar cells** form a single layer of epithelial cells around a lumen into which milk components are deposited (Fig. 38.7). The alveolar cells have numerous microvilli on the luminal surface and are surrounded on their basolateral side by contractile myoepithelial cells. Alveolar cells have a well-developed endoplasmic reticulum and Golgi apparatus, numerous mitochondria and lipid droplets, and plasma membrane PRL receptors. In synergism with insulin and glucocorticoids, PRL is critical for lactogenesis, promoting mammary cell division and differentiation and increasing the synthesis of milk constituents.

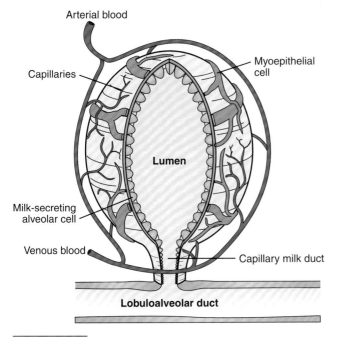

Figure 38.7 **The structure of a mammary alveolus.** A meshwork of contractile myoepithelial cells surrounds milk-producing cells.

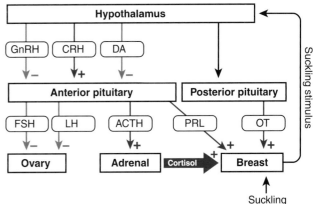

Figure 38.8 **Effect of suckling on hypothalamic, pituitary, and adrenal hormones.** Neural signals initiated by suckling stimulate hormone release from the anterior pituitary to promote milk secretion and removal. Neural signals also inhibit GnRH release to prevent return of the normal menstrual cycle during the lactation period.

Suckling reflex maintains lactation and inhibits ovulation.

The suckling reflex coordinates the release of PRL and oxytocin to maintain lactation and delay the onset of ovulation. Lactation involves two components: milk secretion, which is a continuous process, and milk removal, which is intermittent. PRL is the major regulator of milk secretion, the process through which alveolar cells synthesize the milk constituents and release the formed milk into the alveolar lumen (see Fig. 38.7). Oxytocin is responsible for milk removal by activating milk ejection.

The stimulation of sensory nerves in the breast by the infant initiates the **suckling reflex**. In contrast to neural–neural reflexes, the afferent arc of the suckling reflex is neural and the efferent arc is hormonal. The suckling reflex increases the release of PRL, oxytocin, and ACTH and inhibits the secretion of gonadotropins (Fig. 38.8). Sensory receptors in the nipple initiate nerve impulses in response to breast stimulation that travel via ascending fibers in the spinal cord to the midbrain. These impulses eventually reach the supraoptic and paraventricular nuclei of the hypothalamus to trigger the release of oxytocin from the posterior pituitary into the general circulation (see Chapter 31). Oxytocin induces the contraction of myoepithelial cells in the mammary gland, increasing intramammary pressure and forcing milk into the main collecting ducts. The milk ejection reflex can be conditioned, occurring in anticipation of nursing or in response to a baby's cry.

Blood PRL levels are elevated by the end of gestation and decline by 50% within the first postpartum week, further decreasing to near pregestation levels by 6 months. Suckling elicits a rapid and significant rise in plasma PRL in proportion to the intensity and duration of nipple stimulation.

Dopamine release from the hypothalamus into the portal circulation normally inhibits PRL release from lactotrophs in the anterior pituitary. Suckling inhibits dopamine release from the arcuate nucleus, relieving inhibition of the lactotrophs to allow a burst of PRL release. Dopaminergic agonists that reduce PRL or the discontinuation of suckling terminates lactation.

Lactation is associated with the suppression of cyclicity and **anovulation.** The contraceptive effect of lactation is moderate in humans. In non–breast-feeding women, the menstrual cycle may return within 1 month after delivery, whereas lactating women have a period of several months of **lactational amenorrhea**, with the first few menstrual cycles being anovulatory. The cessation of cyclicity results from the combined effects of the act of suckling and elevated PRL levels. Neural signals from suckling directly inhibit pulsatile GnRH release from the hypothalamus (see Fig. 38.8). PRL may also inhibit pulsatile GnRH release and suppress pituitary responsiveness to GnRH. Although lactation reduces fertility, numerous other methods of contraception are more reliable.

▶ PUBERTY ONSET

The onset of puberty depends on a sequence of maturational processes that begin during fetal life. *In utero*, the GnRH pulse generator is operative, and gonadotropins are released in a pulsatile manner, with LH and FSH levels in fetal blood elevated to near-adult values at midgestation. As placental steroids increase, negative feedback reduces GnRH release, slightly lowering LH and FSH levels prior to parturition.

At birth, the newborn is abruptly deprived of maternal and placental steroids, reducing negative feedback on the hypothalamus and pituitary, resulting in increased GnRH, LH, and FSH release. The rise in gonadotropins stimulates the gonads, resulting in increased serum testosterone and estradiol in male and female infants, respectively. At ~3 months of age, the levels of both gonadotropins and gonadal steroids remain in the low-to-normal adult range. Circulating gonadotropins then decline to low levels by 6 to 7 months and remain suppressed until the onset of puberty.

The prepubertal restraint of gonadotropin secretion is the result of two processes. First, the GnRH pulse-generating neurons are extremely sensitive to negative feedback by gonadal steroids. This permits the very low levels of gonadal steroids in the circulation to inhibit GnRH release. The second mechanism is inhibition of the GnRH pulse generator by the central nervous system (CNS). γ-Aminobutyric acid (GABA) and endogenous opioid peptides in the CNS are major inhibitors of the GnRH pulse generator. Thus, the combination of extreme sensitivity to steroid negative feedback and CNS inhibition suppresses the amplitude and frequency of GnRH pulses, resulting in diminished secretion of LH, FSH, and gonadal steroids. Throughout this period of quiescence, the pituitary and the gonads can respond to exogenous GnRH and gonadotropins, but at a relatively low sensitivity.

The hypothalamic–pituitary–gonadal axis becomes reactivated to initiate puberty. This response involves a decrease in hypothalamic sensitivity to sex steroids, so that higher levels are needed to inhibit GnRH release, and a reduction in CNS inhibition of the GnRH pulse generator. The mechanisms underlying the reduction in central inhibition are not fully understood, but involve stimulation of GnRH neurons by kisspeptin, excitatory amino acids, neuropeptide Y, and prostaglandins. The adipose tissue hormone leptin, released in proportion to body energy stores, is also required as a permissive signal that the body has sufficient energy stores for reproduction to take place. As a result of disinhibition, the frequency and amplitude of GnRH pulses increase. Initially, pulsatility is most prominent at night, entrained by deep sleep; later, it becomes established throughout the 24-hour period (see Fig. 37.2). GnRH acts on the gonadotrophs of the anterior pituitary to increase the number of GnRH receptors and augment the synthesis, storage, and secretion of LH and FSH. The increased responsiveness of FSH to GnRH in females occurs earlier than that of LH, accounting for a higher FSH/LH ratio at the onset of puberty than during late puberty and adulthood. A reversal of the ratio is seen again at menopause.

The increased pulsatile release of GnRH and gonadotropins stimulates gonadal steroid synthesis, inducing the progressive development of the secondary sex characteristics and the establishment of the adult pattern of negative feedback on the hypothalamic–pituitary axis. Activation of the positive feedback mechanism in females that results in an estrogen-induced LH surge is a later event that occurs in mid to late puberty.

Genetic, nutritional, climatic, and geographic factors determine the onset of puberty.

The onset of puberty in humans begins on average at age 10 to 12 years (see Clinical Focus Box 37.2). Lasting 3 to 5 years, the process involves the development of secondary sex characteristics, a growth spurt, and the acquisition of fertility. Over the last 150 years, the age of puberty has declined by 2 to 3 months per decade due to improvements in nutrition and general health. Leptin is a hormone produced by adipose tissue that interacts with receptors in the hypothalamus to regulate satiety and energy metabolism. Leptin has a permissive effect on the initiation of puberty, providing a signal to the CNS that there are sufficient energy stores to support reproduction. People with leptin deficiency resulting from mutations in the gene-encoding leptin do not progress through puberty, but will do so if treated with exogenous hormone. See Chapter 31 Integrated Medical Sciences Box for an in-depth discussion of the role of leptin in regulation of the hypothalamic–pituitary axes.

The first physical signs of puberty in girls are breast budding, **thelarche**, and the appearance of pubic hair. Axillary hair growth and peak height spurt occur within 1 to 2 years. **Menarche**, the beginning of menstrual cycles, occurs at a median age of 12.8 years in Caucasian girls in the United States and at 12.2 years in African American girls. The first few cycles are usually anovulatory. The first sign of puberty in boys is enlargement of the testes, followed by the appearance of pubic hair and enlargement of the penis. The peak growth spurt and appearance of axillary hair in boys usually occur 2 years later than in girls. The growth of facial hair, deepening of the voice, and

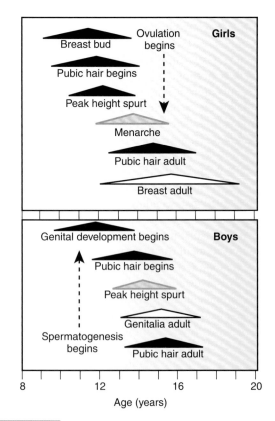

Figure 38.9 Pubertal maturation of secondary sex characteristics in girls and boys.

broadening of the shoulders are late events in male pubertal maturation (Fig. 38.9).

The adrenal androgens DHEA and DHEAS are primarily responsible for the development of pubic and axillary hair. Adrenal maturation (or **adrenarche**) precedes gonadal maturation (or **gonadarche**) by 2 years. The pubertal growth spurt requires the action of sex steroids, GH and IGF-1. Plasma IGF-1 concentrations increase significantly during puberty, with peak levels observed earlier in girls than in boys. The gonadal steroids act primarily by augmenting pituitary GH release, which stimulates the production of IGF-1 in the liver and other tissues.

Defects in hypothalamic–pituitary function can alter the timing of pubertal onset.

Precocious **puberty** has been traditionally defined as physical signs of sexual maturation before the age of 8 years in girls and 9 years in boys. However, given nutritional improvements as discussed above, age alone cannot be used in diagnosis. True precocious puberty results from premature activation of the hypothalamic–pituitary–gonadal axis, leading to the development of secondary sex characteristics as well as gametogenesis. The most frequent causes are CNS lesions or infections, hypothalamic disease, and hypothyroidism. **Pseudoprecocious puberty** is the early development of secondary sex characteristics without gametogenesis. It can result from the abnormal exposure of immature boys to androgens and of immature girls to estrogens. Augmented steroid production can be of gonadal or adrenal origin. See Clinical Focus Box 37.2 for a detailed discussion of precocious puberty.

The definition of **delayed puberty** is the absence of physical signs by age 13 years in girls and 14 years in boys. This condition can result from constitutional delay of growth in which children are below the third percentile for height and have delayed skeletal maturation. It can also result from lack of gonadal development and steroid production. **Hypothalamic–pituitary hypogonadism** refers to conditions where hypothalamic GnRH or pituitary gonadotropins, or both, are deficient or inactive. Single-gene defects have been identified in ~30% of cases. Kallmann syndrome is an example of GnRH deficiency resulting from a defect in the *KAL* gene, which directs GnRH neuronal migration during embryogenesis.

▶ SEXUAL DEVELOPMENT

Normal sexual development consists of three sequential processes. **Chromosomal sex** or **genetic sex** is determined at the time of fertilization by the random unification of an X-bearing egg with either an X-bearing or Y-bearing spermatozoon. The second step takes place during early embryonic life with the development of male or female gonads. Finally, gonadal hormones act at critical times to control the differentiation of the internal and external genitalia. The process of sexual development is incomplete at birth, with the secondary sex characteristics and functional reproductive system fully activated at puberty.

Sex chromosomes dictate the development of the fetal gonads.

Human somatic cells have 44 autosomes and 2 sex chromosomes. The female is **homogametic** (having two X chromosomes) and produces similar X-bearing ova. The male is **heterogametic** (having one X and one Y chromosome) and generates two populations of spermatozoa, one with X chromosomes and the other with Y chromosomes. The X chromosome is large, containing 80 to 90 genes responsible for many vital functions. The Y chromosome is much smaller, carrying only a few genes responsible for testicular development and normal spermatogenesis. Genetic mutations on an X chromosome result in the transmission of X-linked traits, such as hemophilia and color blindness, to male offspring, cannot compensate with an unaffected allele.

Theoretically, by having two X chromosomes, the female has an advantage over the male, who has only one. However, because one of the X chromosomes is inactivated at the morula stage, there is no advantage. Each cell randomly inactivates either the paternally or the maternally derived X chromosome, and this continues throughout the cell's progeny. The inactivated X chromosome is recognized cytologically as the **Barr body**. In males with more than one X chromosome or in females with more than two, extra X chromosomes are inactivated and only one remains functional. This does not apply to the germ cells. The single active X chromosome of the spermatogonium becomes inactivated during meiosis, and a functional X chromosome is not necessary for the formation of fertile sperm. The oogonium, however, reactivates its second X chromosome, and both are functional in oocytes and important for normal oocyte development.

Testicular development requires a Y chromosome and occurs even in the presence of two or more X chromosomes. The testis-determining gene *SRY* (sex-determining region, Y chromosome) regulates **gonadal sex** determination. Located on the short arm of the Y chromosome, *SRY* encodes an HMG-box transcription factor. The presence or absence of *SRY* in the genome determines whether male or female gonadal differentiation takes place. Thus, in 46,XX fetuses, ovaries develop.

Whether possessing the 46,XX or 46,XY karyotype, every embryo goes through an initial **ambisexual stage** and has the potential to acquire either masculine or feminine characteristics. A 4- to 6-week-old human embryo possesses indifferent gonads and undifferentiated pituitary, hypothalamus, and higher brain centers. The indifferent gonad consists of a **genital ridge**, derived from coelomic epithelium and underlying mesenchyme, and primordial germ cells, which migrate from the yolk sac to the genital ridges. Depending on genetic programming, the inner **medullary tissue** becomes the testicular component, and the outer **cortical tissue** develops into an ovary. The primordial germ cells become oogonia or spermatogonia. In a 46,XY fetus, the testes differentiate between weeks 6 and 8 of gestation as the cortex regresses, the medulla enlarges, and the seminiferous tubules become distinguishable. Sertoli cells line the basement membrane of the tubules, and Leydig cells undergo rapid proliferation. Development of the ovary does not begin until gestation weeks 9 to 10. Primordial follicles, composed of oocytes surrounded by a single layer of granulosa cells, are discernible in the cortex between weeks 11 and 12 and reach maximal development by weeks 20 to 25.

Hormones from the fetal testes regulate differentiation of the internal and external genitalia.

During the indifferent stage, the primordial genital ducts are the paired **mesonephric (wolffian) ducts** and the paired **paramesonephric (müllerian) ducts**. In the genetic male fetus, the wolffian ducts give rise to the epididymis, vas deferens, seminal vesicles, and ejaculatory ducts, whereas the müllerian ducts become vestigial. In the genetic female fetus, the müllerian ducts fuse at the midline and develop into the oviducts, uterus, cervix, and upper portion of the vagina, whereas the wolffian ducts regress (Fig. 38.10). The mesonephros present at the indifferent stage becomes the embryonic kidney.

Between weeks 6 and 8 of gestation, Leydig cells, either autonomously or under regulation by hCG, start producing testosterone. Sertoli cells also become active and produce two nonsteroidal compounds. One is **anti-müllerian hormone** (AMH), a large glycoprotein with a sequence homologous to inhibin and transforming growth factor β, which inhibits cell division and induces apoptosis, resulting in involution of the müllerian ducts. The second is **androgen-binding protein (ABP)**, which binds testosterone. Peak production of these compounds occurs between weeks 9 and 12, coinciding with the time of differentiation of the internal genitalia along the male line. The ovary does not produce hormones that regulate differentiation of the internal genitalia.

The primordial external genitalia include the genital tubercle, genital swellings, urethral folds, and urogenital sinus. Differentiation of the external genitalia occurs between gestation weeks 8 and 12, is determined by the presence or absence of male sex hormones, and requires active **5α-reductase**, which converts testosterone to dihydrotestosterone (DHT). Without DHT, regardless of the genetic, gonadal, or hormonal sex, the external genitalia develop along the female pattern. The development of male and female external genitalia from primordial structures is illustrated in Figure 38.11, and a summary of sexual differentiation during fetal life is shown in Figure 38.12. Androgen-dependent differentiation occurs only during fetal life and is thereafter irreversible. However, the exposure of females to high androgens either before or after birth can cause clitoral hypertrophy. Testicular descent into the scrotum, which occurs during the third trimester, is also controlled by androgens.

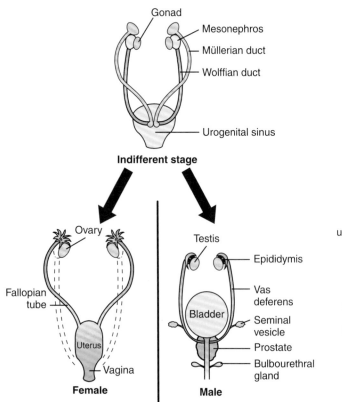

Figure 38.10 Differentiation of the internal genitalia from the sexually indifferent duct system.

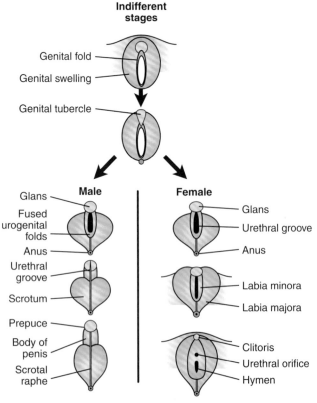

Figure 38.11 Differentiation of the external genitalia from the indifferent primordial structures.

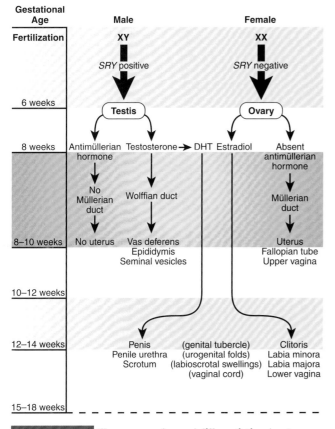

Figure 38.12 The process of sexual differentiation *in utero*.

Disorders of sex development occur when chromosomes, gonads, or development of reproductive anatomy is atypical.

The term **disorders of sex development** (**DSD**) describes congenital conditions in which chromosomal, gonadal, or genital sex is not coincident. Classification of DSD is complex, but these conditions can be broadly defined in three categories: sex chromosome DSD, disorders of testes development and androgenization (46,XY DSD), and disorders of ovary development and androgen excess (46,XX DSD). The nomenclature for DSD was adopted at a consensus conference in 2005, which determined that older terms such as hermaphroditism and intersex were not appropriate given our increasing understanding of the molecular mechanisms regulating sex differentiation.

Sex chromosome DSDs result when the number of sex chromosomes differs from the normal XX or XY distribution, resulting from errors in the first or second meiotic division such as chromosomal nondisjunction, translocation, rearrangement, or deletion. The two most common disorders are Klinefelter syndrome (47,XXY) and Turner syndrome (45,X). Individuals with a 47,XXY karyotype have normal testicular function *in utero* in terms of testosterone and AMH production and no ambiguity of the genitalia at birth. However, the extra X chromosome interferes with the development of the seminiferous tubules and Leydig cell proliferation. Such males have small testes, are azoospermic,

and often exhibit some eunuchoidal features. In Turner syndrome (45,X), there is no gonadal development during fetal life, and at birth, the external genitalia present as female. Given the absence of ovarian follicles, such patients have low levels of estrogens and primary amenorrhea and do not undergo normal pubertal development, which is when the syndrome may be diagnosed. The failure of ovarian development highlights the importance of two X chromosomes in this process.

46,XY DSD can result from defects in testes development or disorders of androgen synthesis and action. The external phenotype can range from complete absence of male genitalia to inadequate androgenization of the external genitalia (e.g., a micropenis). Infertility is often also present. There are a number of genetic defects that can impair testes development with subsequent loss of androgen production. Defects anywhere in the androgen synthetic pathway also impair production of the hormone. 5α-reductase deficiency, an autosomal-recessive disorder results in the inability to convert testosterone to DHT. Infants with 5α-reductase deficiency have ambiguous or female external genitalia and normal male internal genitalia. Affected individuals are often raised as females but undergo a complete or partial testosterone-dependent puberty, including enlargement of the clitoris-like small penis, testicular descent, and the development of male psychosexual behavior. Gender role changes may occur at puberty.

Insensitivity to androgens results from defects in the androgen receptor. Complete androgen insensitivity syndrome results in a complete female phenotype in a 46,XY individual with normally formed testes and age-appropriate testosterone production. Partial or minimal androgen insensitivity can result in partial development of the male external genitalia.

46,XX DSD can result from defects in ovarian development, disorders in androgen synthesis, and other conditions that affect sex development. Defects in ovarian development can result from a number of conditions but is most frequently seen in Turner syndrome (46,X) as discussed above. **Congenital adrenal hyperplasia** (**CAH**) is the most common cause of androgen excess in females. This is an inherited abnormality in adrenal steroid biosynthesis resulting most often from 21-hydroxylase (CYP21) deficiency but also from lack of 11β-hydroxylase (see Chapter 33). A female with CAH will have normal ovaries and internal genitalia, but with increasing degrees of androgenization of the external genitalia up to the full male phenotype without palpable testes, resulting from exposure to excessive androgens *in utero*. In CAH, cortisol production is low, resulting in increased ACTH production by the activated hypothalamic–pituitary axis (Fig. 38.13). The elevated ACTH induces adrenal hyperplasia and an abnormal production of androgens and corticosteroid precursors. When aldosterone levels are also affected, a life-threatening salt-wasting disease results.

Reproductive Physiology

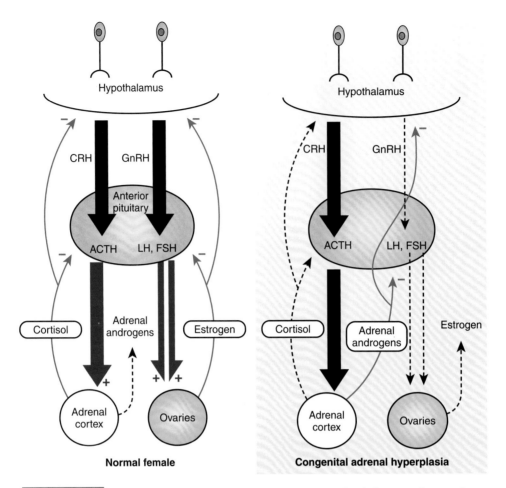

Figure 38.13 The hypothalamic–pituitary–adrenal and hypothalamic–pituitary–ovarian axes in normal female development and with congenital adrenal hyperplasia. *Dashed arrows* indicate low production of the hormone. *Bold arrows* indicate increased hormone production. *Plus* and *minus* signs indicate positive and negative effects.

INTEGRATED MEDICAL SCIENCES

Contraceptive Methods

Interfering with the association between the sperm and ovum, preventing ovulation or implantation, or terminating an early pregnancy can regulate fertility. Most current methods regulate fertility in women, with only a few contraceptives available for men.

Prevention of Egg and Sperm Contact

Methods to prevent contact between the germ cells include coitus interruptus (withdrawal before ejaculation), the rhythm method (no intercourse at times of the menstrual cycle when an ovum is present in the oviduct), and barriers. Rhythm methods require meticulous monitoring of menstrual cycles and body temperature or cervical mucus characteristics, but can still fail as a result of the unpredictable duration of the follicular phase of the menstrual cycle. Both coitus interruptus and the rhythm method require cooperation and discipline in both partners. Barrier methods include condoms,

diaphragms, and intrauterine devices (IUD). The most effective barrier method is the IUD, which reduces the viability of sperm and eggs and creates a local inflammation in the endometrium that prevents implantation. Condoms are the most widely used reversible contraceptives for men but have a high failure rate if not used properly. Condoms also provide protection against the transmission of venereal diseases and AIDS. Diaphragms and cervical caps seal off the opening of the cervix. Postcoital douching with spermicides is not an effective contraceptive method because sperms can rapidly enter the uterus and oviduct.

Surgical methods of contraception are generally considered nonreversible. Vasectomy, cutting of the two vasa deferentia, prevents sperm from passing into the ejaculate. Tubal ligation is the closure or ligation of the oviducts. Surgery for the reversal of a tubal ligation and a vasectomy can be performed; success is variable.

Hormonal Methods (Oral Contraceptives)

Oral contraceptive steroids (birth control pills) prevent ovulation by reducing LH and FSH secretion through negative feedback. The reduction in LH and FSH retards follicular development and prevents ovulation. Birth control pills also adversely affect the environment within the reproductive tract, making it unlikely for pregnancy to result even if fertilization were to occur. Exogenous estrogen and progesterone alter normal endometrial development, and progesterone thickens cervical mucus and reduces oviductal peristalsis, impeding gamete transport.

Noncontraceptive benefits of oral contraception include a reduction in excessive menstrual bleeding, alleviation of premenstrual syndrome, and some protection against pelvic inflammatory disease. In addition, oral contraception results in a significant decrease in incidence of endometrial and ovarian carcinoma. The most commonly reported adverse effects of oral contraception include nausea, headache, breast tenderness, water retention, and weight gain, some of which disappear after prolonged use. Breakthrough bleeding can also be expected in up to 30% of women during the first 3 months of use. Birth control pills are contraindicated in women with a history of thromboembolic disease, atherosclerosis, or breast carcinoma because of the effects of estrogen on these systems.

Emergency Contraception

Women can use emergency contraception to prevent pregnancy after known or suspected failure of birth control or after unprotected intercourse. Two prescription formulations are specifically available for emergency contraception, but over-the-counter preparations are also available. The mechanism of action of emergency contraception is similar to that for standard oral contraceptive use: prevention of ovulation, insufficient corpus luteum function, and induction of an adverse environment in the female reproductive tract. Emergency contraception is most effective when started within the first 72 hours after unprotected intercourse. Thus, it has been recommended that physicians provide an advance prescription in states where the treatment is not available over the counter. ■

Reproductive Physiology

Chapter Summary

- Fertilization of the ovum occurs in the ampulla of the oviduct.
- Sperms undergo capacitation to fully mature and the acrosome reaction to penetrate the zona pellucida surrounding the oocyte.
- The cortical reaction occurs after a sperm penetrates the egg to prevent polyspermy.
- The blastocyst travels the oviduct, enters the uterus, ruptures the zona pellucida, and implants into the uterine wall on about day 7 of gestation. Progesterone and estrogen released from the ovary prepare the uterus to receive the blastocyst.
- Trophoblast cells of the developing placenta produce human chorionic gonadotropin, which stimulates the corpus luteum to continue producing progesterone to maintain pregnancy.
- Following implantation, the placenta develops from embryonic and maternal cells and becomes the major steroid-secreting organ during pregnancy.
- Major hormones produced by the fetoplacental unit are progesterone, estradiol, estrone, estriol, human chorionic gonadotropin, and human placental lactogen. Elevated estriol levels indicate fetal well-being, whereas low levels can indicate fetal stress. Human placental lactogen regulates fetal fuel metabolism.
- Maternal physiology and endocrine function change to support the developing fetus.
- The pregnant woman becomes insulin resistant during the latter half of pregnancy to reduce maternal glucose consumption, making glucose available for the developing fetus.

- The fetal endocrine system develops early in gestation to regulate fetal homeostasis.
- Placental and fetal factors induce parturition.
- Oxytocin maintains strong uterine contractions once initiated. Estrogens, relaxin, and prostaglandins are involved in softening and dilating the uterine cervix so that the fetus may exit.
- Lactogenesis (milk production) requires prolactin (PRL), insulin, and glucocorticoids. Galactopoiesis is the maintenance of an established lactation and requires PRL and other hormones.
- Milk ejection is the process by which stored milk is released. Oxytocin released by the suckling reflex contracts the myoepithelial cells surrounding the milk-producing alveolar cells to eject milk into the alveolar ducts.
- Lactation suppresses the menstrual cycle via a neural signal that reduces GnRH release.
- The GnRH pulse generator in the central nervous system becomes activated at the start of puberty. Increased frequency and amplitude of GnRH pulses initiate increased LH and FSH secretion and increased steroid output by the gonads.
- Sexual development depends on genetic and hormonal factors for differentiation of the internal and external genitalia.
- Disorders of sex development occur when chromosomes, gonads, or development of reproductive anatomy is atypical.

Chapter Review Questions

1. The suckling reflex:

    A. has afferent hormonal and efferent neuronal components.
    B. increases placental lactogen secretion.
    C. increases the release of dopamine from the arcuate nucleus.
    D. triggers the release of oxytocin by stimulating the supraoptic nuclei.
    E. reduces PRL secretion from the pituitary.

The correct answer is D. Suckling involves hormonal and neuronal components, but the hormonal component is efferent and the neuronal component is afferent. When the baby suckles, neural signals from the nipple travel via nerves to the spinal cord and up to the brain (afferent component), which triggers the release of oxytocin from the supraoptic nuclei (efferent component). Oxytocin enters the circulation, enters the breasts, and causes contraction of the myoepithelial cells. Placental lactogen is no longer present after parturition; it is a placental hormone. Dopamine release is decreased by suckling, and as a result, PRL secretion is increased.

2. Implantation occurs:

    A. on day 4 after fertilization.
    B. after the endometrium undergoes a decidual reaction.
    C. when the embryo is at the morula stage.
    D. only after priming of the uterine endometrium by progesterone and estrogen.
    E. on the first day after entry of the embryo into the uterus.

The correct answer is D. Under normal circumstances, the uterus must be primed with both progesterone and estrogen for successful implantation. Implantation occurs on day 7 after fertilization. The decidual reaction occurs as the result of the implanting blastocyst. The embryo is in the blastocyst stage of development at the time of implantation. A morula does not implant. The developing embryo enters the uterus on day 3 or 4, it remains suspended in the uterus for 3 or 4 more days, and implantation occurs on day 7.

3. Upon contact between the sperm head and the zona pellucida, penetration of the sperm into the egg is achieved because of:

   A. the acrosome reaction.
   B. the zona reaction.
   C. the perivitelline space.
   D. pronuclei formation.
   E. cumulus expansion.

The correct answer is A. The acrosome reaction causes a fusion of the plasma membrane and the acrosomal membrane of the sperm, with subsequent release of proteolytic enzymes that help the sperm enter the ovum. The zona reaction and pronuclei formation occur after the sperm has entered the ovum. Sperms enter the perivitelline space after penetration; there is no evidence that this space has any role in penetration. Cumulus expansion assists in movement of the sperm through the mass of granulosa cells for the sperm to get to the surface of the zona pellucida. However, the cumulus cells do not assist in actual penetration of the zona.

4. The next ovulatory cycle after implantation is postponed because of:

   A. high levels of PRL.
   B. the production of hCG by trophoblast cells.
   C. the production of prostaglandins by the corpus luteum.
   D. the depletion of oocytes in the ovary.
   E. low levels of progesterone.

The correct answer is B. The production of hCG by trophoblast cells stimulates the corpus luteum to continue to produce progesterone so that luteal regression does not occur at the end of the anticipated cycle. Although PRL levels increase throughout pregnancy, PRL is not responsible for maintenance of the corpus luteum of pregnancy. Prostaglandins are generally luteolytic, causing regression of the corpus luteum, termination of the luteal phase, and return to the next cycle; they do not prolong the cycle or postpone it. Oocytes are not depleted after implantation. In fact, pregnancy tends to preserve oocytes, as ovulation ceases during pregnancy. Plasma progesterone levels are high during pregnancy as a result of activation of the corpus luteum and placental production of progesterone. Elevated progesterone blocks follicular development and the ensuing LH surge; low levels of progesterone would permit a return to cyclicity.

5. Polyspermy block occurs as a result of the:

   A. cortical reaction.
   B. enzyme reaction.
   C. acrosome reaction.
   D. decidual reaction.
   E. inflammatory reaction.

The correct answer is A. Fertilization by more than one sperm is prevented by the cortical reaction. Cortical granules containing proteolytic enzymes fuse just beneath the entire surface of the oolemma. The proteolytic enzymes are released to the perivitelline space, destroy the sperm receptors, and harden the zona, preventing other sperms from penetrating the fertilized ovum. Enzyme reaction is a nonspecific term with little meaning for polyspermy. The acrosome reaction allows the sperm to penetrate the zona. The decidual reaction is an inflammatory-like reaction that occurs simultaneously with implantation of the blastocyst into the uterine endometrium.

Clinical Application Exercises 38.1

FEMALE INFERTILITY

A 25-year-old woman and her 29-year-old husband have been trying to have a baby for 1 year. She has regular menstrual cycles of 24 to 26 days in length. The couple has intercourse three times a week, with no physical problems, and they try to time intercourse around the time of her ovulation. Physical examination and history for the wife are normal. The husband's physical examination is normal. The husband's semen analysis reveals a semen volume of 4 mL, pH 7.5, sperm count of 30 million/mL, and normal morphology and motility of the sperm. Because of the short cycles (24 to 26 days vs. 28 days), the wife's plasma progesterone level during the midluteal phase is assessed and determined to be 10 ng/mL, which is considered normal (4 to 20 ng/mL).

QUESTIONS

1. What hormones can be measured in the blood to determine why the woman is not able to get pregnant?

2. Based on the hormone measurements, what treatment would likely result in a successful pregnancy?

ANSWERS

1. Estradiol should be measured at the end of the anticipated follicular phase. High serum levels of estradiol would indicate that a dominant follicle has been recruited and is active; low levels would indicate a subnormal dominant follicle or lack of a dominant follicle, which could be verified by ultrasound of the ovaries. Serum LH should be measured during the anticipated preovulatory period. High serum levels of LH would indicate that the dominant follicle is receiving the

signal to ovulate. Low levels of LH may lead to an unruptured dominant follicle that fails to ovulate but luteinizes, leading to progesterone levels in the normal range for the luteal phase. Plasma concentrations of hCG could be determined during the midluteal to late luteal phase to determine whether she was pregnant.

2. Low estradiol indicates a lack of development of a dominant follicle. Therapies such as gonadotropins and clomiphene would be appropriate to stimulate follicular development, estradiol secretion, and ovulation. If a dominant follicle is present, then hCG can be given to induce follicular rupture. hCG binds to the LH receptor and is preferred over LH and GnRH for induction of ovulation because of its longer half-life.

the Point® *Visit* http://thepoint.lww.com/rhoades5e *for additional chapter review Q&A, Clinical Application Exercises, animations, and more!*

Common Abbreviations in Physiology

A	amount; area
$_A$	alveolar
a	arterial; ambient
ABP	androgen-binding protein
AC	adenylyl cyclase
ACE	angiotensin-converting enzyme
ACh	acetylcholine
AChE	acetylcholinesterase
ACTH	adrenocorticotropic hormone (corticotropin)
ADH	antidiuretic hormone
ADP	adenosine diphosphate
AE	anion exchanger
AMH	anti-müllerian hormone (müllerian-inhibiting substance)
AMP	adenosine monophosphate
ANP	atrial natriuretic peptide
ANS	autonomic nervous system
AQP	aquaporin
A_r	effective radiating surface area
ARDS	adult respiratory distress syndrome
ATP	adenosine triphosphate
AV	atrioventricular
A-V	arteriovenous
AVP	arginine vasopressin (antidiuretic hormone or ADH)
aw	airway
$_B$	barometric
b	body; blood
BF	blood flow
BMR	basal metabolic rate
$_{BS}$	urinary space of Bowman capsule
BSC	bumetanide-sensitive (Na^+–K^+–$2Cl^-$) cotransporter
BUN	blood urea nitrogen
C	concentration; compliance; capacitance; capacity; clearance; kilocalorie; conductance
C	heat loss by convection
c	core
CA	carbonic anhydrase
CaBP	calcium-binding protein (calbindin)
CaM	calmodulin
CAM	cell adhesion molecule
cAMP	cyclic AMP (cyclic adenosine 3′,5′-monophosphate)
CBG	corticosteroid-binding globulin
CCK	cholecystokinin
CFC	capillary filtration coefficient
CFTR	cystic fibrosis transmembrane conductance regulator
cGMP	cyclic GMP (cyclic guanosine monophosphate [guanosine 3′,5′-monophosphate])
CGRP	calcitonin gene–related peptide
CIA	central inspiratory activity
CMK	calmodulin-dependent protein kinase
CNS	central nervous system
CO	cardiac output; carbon monoxide
COP	colloid osmotic pressure (oncotic pressure)
COPD	chronic obstructive pulmonary disease
COX	cyclooxygenase
CRH	corticotropin-releasing hormone
CSF	cerebrospinal fluid
CT	computed tomography; calcitonin (thyrocalcitonin)
CYP	cytochrome P450 enzyme
D	diffusion coefficient; diffusing capacity
$_D$	dead space
DAG	diacylglycerol
DHEAS	dehydroepiandrosterone sulfate
DHT	dihydrotestosterone
DIT	diiodotyrosine
D_L	CO lung diffusion capacity
DMT	divalent metal transporter
DNA	deoxyribonucleic acid
DPG	diphosphoglycerate
DPPC	dipalmitoylphosphatidylcholine
E	extraction (extraction ratio)
E	evaporative heat loss
$_E$	expiratory
e	emissivity
EABV	effective arterial blood volume
ECaC	epithelial calcium channel
ECF	extracellular fluid
ECG	electrocardiogram
ECL	cell enterochromaffin–like cell
ED_{50}	median effective dose
EDRF	endothelium-derived relaxing factor (NO)
EDV	end-diastolic volume
EEG	electroencephalogram
EF	ejection fraction
EGF	epidermal growth factor
E_{ion}	equilibrium potential for an ionic species
EJP	excitatory junction potential
ELISA	enzyme-linked immunosorbent assay
E_m	membrane potential
ENaC	epithelial sodium channel
ENS	enteric nervous system
EPP	equal pressure point
EPSP	excitatory postsynaptic potential
ER	endoplasmic reticulum
ERV	expiratory reserve volume
ESR	erythrocyte sedimentation rate
ESV	end-systolic volume
F	fractional concentration of gas; farad; Faraday constant
f	frequency
F_ECO_2	fractional concentration of expired carbon dioxide
FEF	forced expiratory flow
FEV	forced expiratory volume
FEV_1	forced expiratory volume in 1 second
FGF	fibroblast growth factor
F_I	fractional composition of any gas in a mixture
FL	focal length
FM	frequency modulation
F_{max}	maximal force
FRC	functional residual capacity
FSH	follicle-stimulating hormone
FVC	forced vital capacity
g	ionic conductance
G	protein guanine nucleotide–binding protein
GABA	γ-aminobutyric acid
GAP	GnRH-associated peptide
$_{GC}$	glomerular capillary
GDP	guanosine diphosphate
GFR	glomerular filtration rate
GH	growth hormone (somatotropin)
GHRH	growth hormone–releasing hormone
GI	gastrointestinal
GIP	gastric inhibitory peptide (glucose-dependent insulinotropic peptide)

GLP	glucagonlike peptide
GLUT	facilitated glucose transporter
GnRH	gonadotropin-releasing hormone (LHRH)
GPCR	G-protein–coupled receptor
GRE	glucocorticoid response element
GRP	gastrin-releasing peptide
GTO	Golgi tendon organ
GTP	guanosine triphosphate
Hb	hemoglobin
h_c	convective heat transfer coefficient
hCG	human chorionic gonadotropin
HDL	high-density lipoprotein
h_e	evaporative heat transfer coefficient
HF	heat flow
HGF	hepatocyte growth factor
hPL	human placental lactogen (human chorionic somatomammotropin)
HR	heart rate
HRE	hormone response element
HSL	hormone-sensitive lipase
5-HT	5-hydroxytryptamine (serotonin)
5-HTP	5-hydroxytryptophan
I	inspiratory
IC	inspiratory capacity
ICC	interstitial cell of Cajal
ICF	intracellular fluid
IGF-I	insulin-like growth factor I (somatomedin C)
I/G	ratio insulin/glucagon ratio
I_{ion}	ionic current
IJP	inhibitory junction potential
IL	interleukin
IP_3	inositol 1,4,5-trisphosphate
IPSP	inhibitory postsynaptic potential
IRDS	infant respiratory distress syndrome
IRV	inspiratory reserve volume
J	flow (or flux) of a solute or water; joule
K	heat loss by conduction
K_f	ultrafiltration coefficient
K_h	hydraulic conductivity
L	lung
LDL	low-density lipoprotein
L-DOPA	L-3,4-dihydroxyphenylalanine
LH	luteinizing hormone
LHRH	luteinizing hormone–releasing hormone
L_o	optimum length
M	metabolic rate
MAP	microtubule-associated protein
MAP	kinase mitogen–activated protein kinase
MCH	mean cell (corpuscular) hemoglobin
MCHC	mean cell (corpuscular) hemoglobin concentration
MCR	metabolic clearance rate
M-CSF	macrophage–colony-stimulating factor
MCV	mean cell (corpuscular) volume
MIT	monoiodotyrosine
MLCK	myosin light chain kinase
MLCP	myosin light chain phosphatase
MMC	migrating motor complex
MSH	melanocyte-stimulating hormone
MVC	maximal voluntary contraction
NANC	nonadrenergic noncholinergic
NaPi	sodium-coupled phosphate transporter
NE	norepinephrine
NHE	Na^+/H^+ exchanger
NK	natural killer
NMDA	N-methyl-D-aspartate
NO	nitric oxide
NOS	nitric oxide synthase
OAT	organic anion transporter
OCT	organic cation transporter

P	pressure; partial pressure; permeability; permeability coefficient; plasma; plasma concentration
P_{50}	Po_2 at which 50% of hemoglobin is saturated
$Paco_2$	partial pressure of carbon dioxide in alveoli
PAH	p-aminohippurate
Pao_2	partial pressure of oxygen in arterial blood
Pao_2	partial pressure of oxygen in alveoli
$Pao_2–Pao_2$	alveolar arterial difference in partial pressure of oxygen
Pc'	pulmonary end-capillary
Pco_2	partial pressure of carbon dioxide
PCr	phosphocreatine (creatine phosphate)
PDE	phosphodiesterase
PEF	peak expiratory flow
PG	prostaglandin
PI	phosphatidylinositol
P_i	inorganic phosphate
PIF	peak inspiratory flow
PIP_2	phosphatidylinositol 4,5-bisphosphate
PKA	protein kinase A
PKC	protein kinase C
PKG	cGMP-dependent protein kinase
pl	pleural
PLC	phospholipase C
PNS	peripheral nervous system
Po_2	partial pressure of oxygen
POMC	proopiomelanocortin
PRL	prolactin
PRU	peripheral resistance unit
PT	prothrombin time
PTH	parathyroid hormone (parathormone)
PTT	partial thromboplastin time
PVC	premature ventricular complex
$Pvco_2$	partial pressure of carbon dioxide in venous blood
pw	pulmonary wedge
$\dot{Q}$	generic symbol for flow of any kind
R	respiratory exchange ratio; resistance; universal gas constant
R	heat loss by radiation
r	radius; radiant environment
RAAS	renin–angiotensin–aldosterone system
RBF	renal blood flow
RBP	retinol-binding protein
RDA	recommended daily allowance
Re	Reynolds number
REM	rapid eye movement
rh	relative humidity
RIA	radioimmunoassay
RISA	radioiodinated serum albumin
ROS	reactive oxygen species
RPF	renal plasma flow
RQ	respiratory quotient
RV	residual volume
RVD	regulatory volume decrease
RVI	regulatory volume increase
RV/TLC	residual volume to tidal volume ration
S	saturation; siemens
S	rate of heat storage in the body
s	shunt
SA	sinoatrial
Sao_2	saturation of hemoglobin with oxygen in arterial blood
SF-1	steroidogenic factor 1
SGLT	Na^+–glucose cotransporter
SH2	src homology domain
SHBG	sex hormone–binding globulin
SIADH	syndrome of inappropriate ADH
sk	skin
SN	single nephron
SPCA	serum prothrombin conversion accelerator

SRIF	somatotropin release–inhibiting factor (somatostatin)	T	tubule transverse tubule
SRY	sex-determining region, Y chromosome	U	urine concentration
SSRI	selective serotonin reuptake inhibitor	UCP	uncoupling protein
StAR	steroidogenic acute regulatory protein	UDP	uridine diphosphate
SV	stroke volume	UP	ultrafiltration pressure gradient
SVR	systemic vascular resistance (total peripheral resistance)	V	volume; volt; vasopressin; vacuolar
		v	velocity; venous
SW	stroke work	$\dot{V}$	gas volume per unit time (airflow); minute ventilation; urine flow rate
T	tension; temperature; time		
T	tidal	$\dot{V}_A$	alveolar ventilation
t	time	$\dot{V}_A/\dot{Q}$	alveolar ventilation/perfusion ratio
T_3	triiodothyronine	VC	vital capacity
T_4	thyroxine	V_D	dead space volume in the lung (i.e., volume of air that does not participate in gas exchange)
ta	transairway		
TBG	thyroxine-binding globulin	V_D/V_T	ratio of dead space volume to tidal volume
TBW	total body water	$\dot{V}_E$	expired minute ventilation (note: minute ventilation and expired minute ventilation are the same)
TEA	tetraethylammonium		
TF	tubular fluid		
TGF	transforming growth factor	VEGF	vascular endothelial growth factor
TLC	total lung capacity	VIP	vasoactive intestinal peptide
tm	transmural	VLDL	very-low-density lipoprotein
Tm	tubular transport maximum	V_{max}	maximum theoretical velocity of muscle contraction
Tn-C	calcium-binding troponin	$\dot{V}_{O_2}$	oxygen consumption per minute
TNF	tumor necrosis factor	*W*	heat loss as mechanical work
Tn-I	troponin that inhibits actin–myosin interactions	w	wettedness
Tn-T	tropomyosin-binding troponin	z	valence of an ion
tp	transpulmonary	φ	osmotic coefficient
TPA	tissue plasminogen activator	η	viscosity
TR	thyroid hormone receptor	λ	length (space) constant; wavelength
TRE	thyroid hormone response element	μ	electrochemical potential
TRH	thyrotropin-releasing hormone	π	osmotic pressure; 3.14 (pi)
TSC	thiazide-sensitive (Na+–Cl-) cotransporter	ρ	density
TSH	thyroid-stimulating hormone	σ	reflection coefficient
		τ	time constant

Normal Blood, Plasma, or Serum Values

Internationale	Fluid*	Reference Range	System Reference Range
Alanine aminotransferase	S	8–20 U/L	0.13–0.33 mkat/L
Aldosterone, with normal sodium intake of 100–200 mEq/d	S	Supine: <16 ng/dL	<444 pmol/L
Aldosterone, with normal sodium intake of 100–200 mEq/d		Upright: 4–31 ng/dL	111–860 pmol/L
Arginine vasopressin	P	1.0–13.3 pg/mL	1.0–13.3 ng/L
Aspartate aminotransferase	S	9–40 U/L	0.15–0.67 mkat/L
Bicarbonate, arterial	P	22–26 mEq/L	22–26 mmol/L
Bilirubin, total	S	0.1–1.0 mg/dL	2–17 µmol/L
Calcium	S	8.4–10.2 mg/dL	2.1–2.6 mmol/L
Carbon dioxide (P_{CO_2}), arterial	B	35–45 mm Hg	4.7–6.0 kPa
Chloride	P, S	95–105 mEq/L	95–105 mmol/L
Cholesterol, recommended	S	<200 mg/dL	<5.2 mmol/L
Cortisol	S	08:00 h: 5–23 µg/dL	138–635 nmol/L
		16:00 h: 3–15 µg/dL	82–413 nmol/L
		20:00 h: <50% of 08:00 h value	<0.50 of 08:00 h value
Creatinine	S	0.6–1.2 mg/dL	53–106 µmol/L
Creatine kinase	S	Female: 10–70 U/L	0.17–1.17 µkat/L
		Male: 25–90 U/L	0.42–1.50 µkat/L
1,25-Dihydroxyvitamin D	S	16–42 pg/mL	38–101 pmol/L
Estradiol	P, S	Female:	
		Prepubertal: <20 pg/mL	<73 pmol/L
		Premenopausal adult: 23–361 pg/mL	84–1,325 pmol/L
		Postmenopausal adult: <30 pg/mL	<110 pmol/L
		Male: <50 pg/mL	<184 pmol/L
Ferritin	S	Male: 15–200 ng/mL	15–200 µg/L
		Female: 12–150 ng/mL	12–150 µg/L
Follicle-stimulating hormone	S	Male: 4–25 mIU/mL	4–25 U/L
		Female:	
		Premenopause: 4–30 mIU/mL	4–30 U/L
		Midcycle peak: 10–90 mIU/mL	10–90 U/L
		Postmenopause: 40–250 mIU/mL	40–250 U/L
Glucagon	P	50–200 pg/mL	50–200 ng/L
Glucose	S	Fasting: 70–110 mg/dL	3.9–6.1 mmol/L
		2-h postprandial: <140 mg/dL	<7.8 mmol/L
Growth hormone	P	2.0–6.0 ng/mL	2.0–6.0 mg/L

Internationale	Fluid*	Reference Range	System Reference Range
HDL cholesterol, as major risk factor	S	<35 mg/dL	<0.91 mmol/L
25-Hydroxyvitamin D	S	8–80 ng/mL	20–200 nmol/L
Insulin	S	0–29 µU/mL	0–208 pmol/L
Iron	S	50–170 µg/dL	9–30 µmol/L
Lactate dehydrogenase	S	100–190 U/L	1.7–3.2 µkat/L
LDL cholesterol, desirable	S	<130 mg/dL	<3.36 mmol/L
Luteinizing hormone	S	Male: 6–23 mIU/mL	6–23 U/L
		Female:	
		Follicular phase: 5–30 mIU/mL	5–30 U/L
		Midcycle: 75–150 mIU/mL	75–150 U/L
		Postmenopause: 30–200 mIU/mL	30–200 U/L
Magnesium	S	1.5–2.0 mEq/L	0.75–1.0 mmol/L
Osmolality	P, S	275–295 mOsm/kg H_2O	275–295 mOsm/kg H_2O
Oxygen content, arterial	B	17–21 mL/dL	0.17–0.21 L/L
Oxygen saturation, arterial	B	96%–100%	0.96–1.00 mol/mol
Oxygen tension (Po_2), arterial	B	75–100 mm Hg	10.0–13.3 kPa
Parathyroid hormone	P	10–60 pg/mL	10–60 ng/L
pH, arterial	B	7.35–7.45 [H⁺]	35–45 nmol/L
Phosphatase, alkaline	S	20–70 U/L	0.33–1.17 µkat/L
Phosphorus, inorganic	S	3.0–4.5 mg/dL	1.0–1.5 mmol/L
Potassium	S	3.5–5.0 mEq/L	3.5–5.0 mmol/L
Prolactin	S	Female:	
		Nonpregnant: 3–30 ng/mL	3–30 µg/L
		Pregnant: 10–209 ng/mL	10–209 µg/L
		Postmenopausal: 2–20 ng/mL	2–20 µg/L
		Male: 2–18 ng/mL	2–18 µg/L
Proteins			
Total	S	6.0–7.8 g/dL	60–78 g/L
Albumin	S	3.5–5.5 g/dL	35–55 g/L
Globulin	S	2.3–3.5 g/dL	23–35 g/L
Sodium	S	136–145 mEq/L	136–145 mmol/L
Testosterone, total (morning)	P	Male: 300–1,100 ng/dL	10.4–38.1 nmol/L
		Female: 20–90 ng/dL	0.7–3.1 nmol/L
Thyroid-stimulating hormone	S	0.5–5.0 µU/mL	0.5–5.0 mU/L
Thyroxine (T4), total	S	5–12 µg/dL	64–155 nmol/L
Triglycerides, fasting	S	35–160 mg/dL	0.4–1.8 mmol/L
Triiodothyronine (T3)	S	115–190 ng/dL	1.8–2.9 nmol/L
Urea nitrogen (BUN)	S	7–18 mg/dL	2.5–6.4 mmol urea/L
Uric acid	S	3.0–8.2 mg/dL	0.18–0.49 mmol/L

Internationale	Fluid*	Reference Range	System Reference Range
Hematologic Values			
Blood volume		Male: averages 69 mL/kg body weight	
		Female: averages 65 mL/kg body weight	
Plasma volume		Male: 25–43 mL/kg body weight	0.025–0.043 L/kg body weight
		Female: 28–45 mL/kg body weight	0.028–0.045 L/kg body weight
Erythrocyte count		Male: $4.3–5.9 \times 10^6/mm^3$	$4.3–5.9 \times 10^{12}/L$
		Female: $3.5–5.5 \times 10^6/mm^3$	$3.5–5.5 \times 10^{12}/L$
Erythrocyte sedimentation rate		Male: 0–15 mm/h	0–15 mm/h
		Female: 0–20 mm/h	0–20 mm/h
Hematocrit		Male: 41%–53%	0.41–0.53
		Female: 36%–46%	0.36–0.46
Hemoglobin, blood		Male: 13.5%–17.5 g/dL	2.09–2.71 mmol/L
		Female: 12.0–16.0 g/dL	1.86–2.48 mmol/L
Hemoglobin A_{1c}		≤6% of total Hb	≤0.06 of total Hb
Leukocyte count and differential			
Leukocyte count		4,500–11,000/mm³	$4.5–11.0 \times 10^9/L$
Segmented neutrophils		45%–74%	0.45–0.74
Bands		0%–4%	0.00–0.04
Eosinophils		0%–7%	0.00–0.07
Basophils		0%–2%	0.00–0.02
Lymphocytes		16%–45%	0.16–0.45
Monocytes		4%–10%	0.04–0.10
Mean corpuscular hemoglobin		25.4–34.6 pg/cell	0.39–0.54 fmol/cell
Mean corpuscular hemoglobin concentration		31–36 g/dL	4.81–5.58 mmol Hb/L
Mean corpuscular volume		80–100 mm³	80–100 fL
Partial thromboplastin time, activated		25–40 s	25–40 s
Platelet count		150,000–400,000/mm³	$150–400 \times 10^9/L$
Prothrombin time		11–15 s	11–15 s
Red cell distribution width		11.5%–14.5%	0.115–0.145
Reticulocyte count		0.5%–1.5% of red cells	0.005–0.015

*B, blood; P, plasma; S, serum.

GLOSSARY

A band band of muscle myofilament that appears dark under a microscope and contributes to muscles' striated appearance; contains myosin and overlapping projections of actin; *anisotropic*

ABCA1 ATP-binding cassette transporter member 1 is encoded by the ABCA1 gene. This transporter is a major regulator of cellular cholesterol and phospholipid homeostasis

abdominal wall muscles along the abdomen that are involved in forced expiration

abduction movement away from the midline

abetalipoproteinemia condition in which apo B synthesis cannot occur, thereby blocking production and secretion of very-low-density lipoproteins by the liver; condition is marked by large accumulated lipid droplets in the cytoplasm of hepatocytes

absence seizure characterized by the lack of jerking or twitching movements of the eyes; the person seems to stare off into space; formerly known as *petit mal seizures*

absolute refractory period window of time following one action potential during which an axon cannot generate an action potential no matter how strong the stimulus

absolute threshold of hearing intensity level below which no sound can be heard; for the human ear, it has a value of 0.0002 dyne/cm^2

absorptive hyperemia increase in intestinal blood flow that accompanies digestion; local response to elevated tissue metabolism

absorptivity property of the body surface that depends on the temperature of the body and the wavelength of the incident radiation; dimensionless value measured as the fraction of incident radiation that is absorbed by the body

accessory muscles muscles that are brought into play during forced inspiration and expiration

accessory structures structures that either enhance the specific sensitivity of a sensory receptor or exclude unwanted stimuli such as the lens of the eye

accidental hypothermia drop of central temperature below 35°C not induced for therapeutic purposes; occurs in people impaired by drugs or ethanol, disease, and other physical conditions and in healthy people who are immersed in cold water or become exhausted in the cold

acclimation functional change in response to an environmental stress; changes are not permanent and can be reversed when the environmental stress is removed

acclimatization process of an organism adjusting to changes in its environment, often involving temperature or climate; usually occurs in a short time and within one organism's lifetime

accommodation (1) process of focusing the eye to adjust to varying object distances; (2) decrease in sensory action potential frequency with a constant generator potential

acetoacetic acid unstable acid found in abnormal amounts in the blood and urine in some cases of impaired metabolism (such as diabetes mellitus or starvation)

acetone chemical formed in the blood when the body uses fat instead of glucose (sugar) for energy

acetylcholine (ACh) one of the two major neurotransmitters used by the autonomic nervous system; the transmitter molecule present in cholinergic synapses and the myoneural junction; the neurotransmitter present in cholinergic synapses in the brain; major neurotransmitter in cognitive function, learning, and memory

acetylcholinesterase (AChE) enzyme that hydrolyzes acetylcholine and terminates its action as a transmitter substance

achalasia failure of any sphincter to relax; in clinical practice, *achalasia* most often refers to a specific condition in which the lower esophageal sphincter fails to relax

acid substance that can release or donate a hydrogen ion

acid dissociation constant (K_a) ionization constant for an acid

acid indigestion indigestion resulting from hypersecretion of acid in the stomach, a synonym for pyrosis

acid tide condition in which the blood from the pancreas is acidic as a result of the bicarbonate secretion in pancreatic juice

acidemia condition in which arterial blood pH is < 7.35

acidosis condition in which blood pH is below 7.35; also known as *acidemia*

acinar cells major cells found in the exocrine pancreas actively involved in the production of enzymes

acinus blind sac containing mainly pyramidal cells found in the salivary glands of humans and animals that aids in the secretion of saliva

acne vulgaris disease characterized by comedones, pustules, papules, nodules, and cysts; most common in teenagers and young adults

acoustic reflex automatic contractions of the stapedius and tensor tympani muscles that protect the ear from excessively loud noises

acromegaly thickening of the bones of the face, hands, and feet and liver hypertrophy resulting from excess growth hormone in an adult

acrosome caplike structure that surrounds the sperm head and contains enzymes that help penetration of the egg

acrosome reaction fusion of the outer acrosomal membrane of the sperm head with the sperm plasma membrane, forming a composite membrane with openings through which acrosomal enzymes necessary for sperm penetration of the egg pass

actin globular protein that, in its polymeric filamentous form, makes up the bulk of thin myofilaments in muscle and other contractile systems

α-actinin filamentous muscle protein that anchors thin filaments to the structures of Z lines

actin-linked regulation muscle control system based on controlled alteration of the constituents of the actin filaments

action potential short-lasting, all-or-none change in membrane potential of a nerve or muscle cell that arises when a graded membrane depolarization passes a threshold for opening voltage-gated sodium channels; usually becomes a propagated nerve or muscle cell action potential; also known as *spike potential*

action tremor tremor present during movement

activated partial thromboplastin time (aPTT) test to monitor the intrinsic and common coagulation pathways

activation gate molecular gate in voltage-gated ion channels that opens in response to a polarization change

active hyperemia blood flow increase in response to increased tissue metabolism or reduced oxygen content in the blood

active immunity immunity resulting from activation of lymphocytes (natural) or vaccination (artificial)

active state the state of the voltage-gated sodium channel in which the channel is open and allowing the flow of sodium ions

active tension component of muscle force that arises from crossbridge interactions

active transport membrane transport of solute or ions against the prevailing chemical and/or electrical gradient and driven by input of metabolic energy usually in the form of adenosine triphosphate (ATP); when solute movement is directly coupled to ATP hydrolysis, the process is termed *primary active transport*, whereas *secondary active transport* of solute is driven by an ion gradient (usually Na$^+$) that is maintained by a separate primary active transport process (usually Na$^+$/K$^+$-ATPase)

active zones specialized regions of presynaptic terminals where the synaptic vesicles dock, fuse, and release their transmitter

activin nonsteroidal regulator, composed of two covalently linked β subunits, that is synthesized in the pituitary gland and gonads and stimulates the secretion of follicle-stimulating hormone; the actions of activins are the opposite of those of inhibins

activity front length of the intestine with an oral and aboral boundary occupied by the migrating mot or complex that advances (migrates) along the intestine at a rate that progressively slows as it approaches the ileum

actomyosin ATPase enzymic entity, formed by the interaction of actin and myosin, that hydrolyzes adenosine triphosphate and converts its chemical energy into a mechanical change

acute kidney injury (AKI) serious condition characterized by a rapid decline in glomerular filtration rate, retention of nitrogenous waste products, and disturbances of extracellular fluid volume and electrolyte and acid–base balance; also known as *acute renal failure*

acute mountain sickness (AMS) pathophysiologic condition caused by acute exposure to high altitude (from hypoxia resulting from a decrease in oxygen tension, not from a decrease in the percentage of oxygen in the inspired air); also known as *altitude sickness*

acute-phase blood proteins serum proteins elevated during acute inflammation

acute respiratory distress syndrome (ARDS) diffuse lung injury that is characterized by abnormally low compliance, pulmonary edema, and focal atelectasis

acyl-CoA: cholesterol acyltransferase (ACAT) enzyme responsible for the esterification of cholesterol to form cholesterol ester

adaptation genetic change that enhances the ability of the body to adjust to the environment; changes (e.g., sweat glands) are passed on to the next generation

adaptive immune system antigen-specific interacting components; also called *acquired immune system*

adaptive relaxation vagovagal reflex triggered by stretch receptors in the gastric wall, transmission over vagal afferents to the dorsal vagal complex, and efferent vagal fibers back to inhibitory musculomotor neurons in the gastric enteric nervous system

Addison disease disorder characterized by hypotension, weight loss, anorexia, weakness, and sometimes a bronze-like melanotic hyperpigmentation of the skin; results from primary deficiency of adrenal cortex function and inadequate secretion of aldosterone and cortisol

adduction movement toward the midline

adenohypophysis anterior pituitary

adenosine inhibitory neurotransmitter believed to play a role in promoting sleep and suppressing arousal, with levels increasing with each hour an organism is awake

adenosine diphosphate (ADP) nucleotide released upon platelet activation, causing platelet aggregation; the product of adenosine triphosphate (ATP) dephosphorylation by ATPases; ADP is converted back to ATP by ATP synthases

adenosine triphosphate (ATP) nucleotide important in transport of energy within cells for metabolism

adenylyl cyclase (AC) enzyme that catalyzes the synthesis of cyclic adenosine monophosphate from adenosine triphosphate

adequate stimulus biologically correct stimulus for a given sensory receptor

adipose triglyceride lipase (ATGL) triglyceride lipase that catalyses the removal of the first fatty acid from the glycerol backbone during triglyceride hydrolysis

adrenal cortex outer portion of the adrenal gland

adrenal medulla inner portion of the adrenal gland

adrenarche adrenal maturation during pubertal development in boys and girls

adrenergic characterized by nerves releasing or tissues responding to the catecholamines norepinephrine or epinephrine

adrenocorticotropic hormone (ACTH) hormone product of corticotrophs located in the anterior pituitary; also known as *corticotropin*

adventitia outer elastic layer of arteries and veins that gives some structural integrity to blood vessels but more so in arteries than in veins

aerobic pertaining to a biochemical or physiologic process that requires O$_2$ for its function

affect psychological impression of the degree of "pleasantness" of a stimulus

affective disorders illnesses marked by abnormal mood regulation and associated signs and symptoms; also known as *mood disorders*

afferent carrying toward

afferent nerve (afferents) nerve fiber (usually sensory) carrying impulses from an organ or tissue toward the central nervous system or the information-processing centers of the enteric nervous system; examples are vagal and spinal afferents

afterhyperpolarization transient period of increased membrane potential that follows the repolarization phase of the action potential

afterload force a muscle experiences (exerts) after it has begun to shorten

aganglionosis congenital absence of ganglion cells in the enteric nervous system in a segment of bowel; peristaltic propulsion is impossible in the absence of the enteric nervous system and is the pathophysiologic basis for Hirschsprung disease

age-related macular degeneration (AMD) decreased sharpness of central vision, especially in bright light; progressive condition whose causes are obscure and for which no cure is available

agonist (1) synthetic drug molecule that selectively binds to a specific receptor and triggers a signaling response in the cells; agonists mimic the action of endogenous biochemical molecules, such as hormones or neurotransmitters that bind to the same receptor; (2) muscle producing the desired movement

agouti-related protein (AgRP) a neuropeptide synthesized in neuropeptide Y containing neurons in the arcuate nucleus of the hypothalamus; increases appetite

agranulocyte mature cell, such as a monocyte or lymphocyte; also known as *monomorphonuclear leukocyte*

AH-type enteric neuron multipolar neuron with multiple long processes so named because a characteristic long-lasting hyperpolarizing potential (i.e., an *afterhyperpolarization* or *AH*) occurs after discharge of an action potential; fulfills the role of interneurons in the enteric neural networks; previously thought to be intrinsic primary afferent neurons

AIDS acquired immunodeficiency syndrome; disease of the human immunodeficiency virus

airway resistance resistance of air movement in airways resulting from turbulent airflow

airway tree airway branching, which occurs in a treelike fashion in the lung

albumin main plasma protein and key regulator of oncotic pressure of blood

albuminuria presence of serum albumin in the urine

aldosterone mineralocorticoid hormone produced by the zona glomerulosa of the adrenal cortex that stimulates tubular reabsorption of sodium and secretion of potassium and hydrogen ions in the kidneys

aldosterone synthase (CYP11B2) substance that synthesizes aldosterone from pregnenolone

alkalemia condition in which arterial blood pH is >7.45

alkaline tide phenomenon that occurs when blood coming from the stomach during active acid secretion contains an overabundance of bicarbonate

alkalosis condition in which blood pH is above 7.45; also known as *alkalemia*

all-or-none phenomenon once threshold is reached, a full action potential is generated

allergy type I immediate hypersensitivity disorder

allo Greek for "other"

allograft transplantation between genetically different people

alpha cells see *α cell*

alpha motor neuron see *α motor neuron*

alpha waves electroencephalogram wave pattern with a rhythm ranging from 8 to 13 Hz, observed when the person is awake but relaxed with the eyes closed

alveolar arterial oxygen gradient pressure difference between the partial pressure of alveolar oxygen and arterial oxygen

alveolar cells milk-secreting cells in the breast

alveolar dead space volume fraction of air in the alveoli that does not participate in gas exchange; occurs because regional blood flow and airflow are not evenly matched (ventilation/perfusion inequalities)

alveolar gas equation formula used to calculate partial pressure of gases in the alveoli

alveolar pressure pressure inside the alveoli

alveolar ventilation amount of fresh air (L/min) brought into the alveoli

alveoli small air sacs in which oxygen and carbon dioxide are exchanged between the atmosphere and the pulmonary capillary blood

Alzheimer's disease (AD) most common form of dementia characterized by plaques and neurofibrillary tangles; symptoms include confusion, long-term memory loss, irritability and aggression, mood swings, language breakdown, and sensory decline; disease is incurable, degenerative, and terminal

amacrine cells interneurons in the retina that are responsible for 70% of input to retinal ganglion cells and that regulate bipolar cells, which are responsible for the other 30% of input to retinal ganglia

ambisexual stage time in development when the embryo has the potential to acquire either masculine or feminine characteristics

amenorrhea absence of menstruation

γ-aminobutyric acid (GABA) inhibitory neurotransmitter

amnion extraembryonic membrane that forms a closed fluid-filled sac around the embryo and fetus

amniotic sac structure that forms the amnion around the developing fetus

amphipathic condition in which both hydrophilic and hydrophobic regions are present in the same molecule, for example, phospholipids

amphoteric capable of acting as either an acid or a base

ampulla woolen end of the oviduct, which receives the egg after it is passed into the duct by the fimbria

amygdala region of the cerebrum involved in visceral components of emotional responses

amylase important digestive enzyme secreted by both the salivary glands and the pancreas that is involved in the digestion of carbohydrates

amylin hormone formed by β cells in the pancreas that regulates the timing of glucose release into the bloodstream after eating by slowing the emptying of the stomach; also known as *islet amyloid-associated peptide*

anaerobic pertaining to a biochemical or physiologic process that does not require O_2 for its function

anaerobic threshold exercise intensity at which lactate (lactic acid) starts to accumulate in the blood, which happens when lactate is produced faster than it can be removed (metabolized)

anamnestic response increased immune response to repeated antigen exposure

anaphylaxis condition resulting from severe allergic reaction and characterized by intense local and sometimes systemic vasodilation that results in shock; histamine released during the reaction dilates arterioles and increases capillary permeability; fluid loss from transcapillary filtration contributes to plasma volume loss that exacerbates the shock caused by arteriolar dilation

anatomic dead space volume amount of air found in the lungs' airways that does not participle in gas exchange

anatomic shunt blood that bypasses alveoli and does not get oxygenated as a result of a structural defect in the circulation

androgen type of hormone that promotes the development and maintenance of male sex characteristics

androgen-binding protein (ABP) protein produced by Sertoli cells that binds testosterone

androstenedione adrenal androgen made by 17α-hydroxylation of progesterone; serves as a precursor for estrogen synthesis in the ovaries

anemia reduction in erythrocytes, which leads to a decrease in the oxygen-carrying capacity of the blood

anemia of inflammation long-term condition characterized by a lower-than-normal number of healthy red blood cells; also called *anemia of chronic disease*

aneurysm weakening of the arterial wall that causes it to bulge outward, leak blood, and eventually rupture

angiogenesis formation of new blood vessels

angiotensin I inactive decapeptide formed during the cleavage of angiotensinogen

angiotensin II peptide hormone formed from the proteolysis of angiotensin I by converting enzyme in the lungs; potent vasoconstrictor; enhances the effectiveness of sympathetic nerve stimulation to blood vessels and promotes salt and water retention in the body primarily by stimulating the release of aldosterone from the adrenal cortex

angiotensin-converting enzyme (ACE) enzyme located in the capillary endothelial cells of the lung and kidney that converts angiotensin I to angiotensin II

angiotensinogen $α_2$-globulin produced by the liver that is the substrate for renin

anhidrotic ectodermal dysplasia hereditary condition characterized by abnormal development of the skin, hair, nails, teeth, and sweat glands and in which sweat glands are sparse or absent; affected adults are unable to tolerate a warm environment and require special measures to maintain normal body temperature

anion negatively charged ion, especially the ion that migrates to an anode in electrolysis

anion exchange protein (AE1) transports protein responsible for mediating the electroneutral exchange of chloride (Cl^-) for bicarbonate (HCO_3^-) across a plasma membrane

anion gap difference (in mEq/L) between plasma [Na$^+$] and plasma ([Cl$^-$] + [HCO$_3^-$])

anisocytosis large variation in the size of erythrocytes

anorectal angle (approximately) 90° angle between the rectum and the anal canal; becomes more obtuse during defecation and more acute when holding back stool; formed by the contraction of the puborectalis muscle

anorexia nervosa severe behavioral disorder associated with the lack of food intake, extreme malnutrition, and endocrine changes secondary to psychological and nutritional disturbances

anorexigenic causing a loss of appetite

anovulation period in which no ovulation occurs

antagonist muscle opposing the desired movement

antagonistic pair pair of muscles arranged so that the shortening of one lengthens the other, and vice versa

anterior cavity forward section of the eyeball that contains the iris and the lens

anterior periventricular region portion of the hypothalamus close to the third ventricle

anterograde direction of axonal transport from the cell body toward the axon terminal

anterograde amnesia loss of memory going forward from a trauma

antibody secreted protein that can bind antigen

anterograde transport transport of cellular components from the soma to the nerve process

anticoagulant substance that prevents coagulation

antidiuretic hormone (ADH) peptide hormone synthesized in the anterior hypothalamus and released from the posterior pituitary, whose primary action is to increase the water permeability of the kidney collecting ducts; now known as *arginine vasopressin (AVP)*

antidromically conduction of nerve action potentials in the direction opposite to that which is usual

antigen substance that binds to antigen receptors and antibody

antigen-presenting cell (APC) cell displaying antigen on its surface together with major histocompatibility complex class II

antigen recognition molecules T-cell receptor, B-cell receptor, and major histocompatibility complex proteins

antioxidant molecule capable of inhibiting the oxidation of another molecule, for example, the enzyme catalase reduces superoxide to oxygen and water

antiport membrane transport process in which the driver ion (usually Na$^+$) and the solute move in opposite directions

antra fluid-filled cavities in a tertiary ovarian follicle

antral dysrhythmia abnormality of electrical rhythm in the gastric antrum (bradygastria/tachygastria) analogous to cardiac arrhythmia

antral granulosa cells cells that line the antral cavity of a graafian follicle

antral pump caudal two thirds of the gastric corpus, the antrum, and the pylorus

aorta largest artery in the body; sole arterial outlet from the left ventricle

aortic bodies specialized innervated nodules located in the aorta that sense low Po$_2$ and pH as well as high Pco$_2$; send impulses to the medullary area of the brain to elicit reflex activation of the sympathetic nervous system to increase blood pressure and do not respond to low arterial pressure unless the pressure is low enough to affect oxygen delivery to the bodies (<80 mm Hg)

aortic baroreceptors spray nerve endings in the wall of the aortic arch that sense stretch and, therefore, serve as sensory receptors for changes in arterial pressure

aortic valve thin, flexible, trileaflet structure that separates the aorta from the left ventricle; opens into the aorta and closes off the aorta when the left ventricle relaxes

aphasia loss of the ability to use or understand language, usually resulting from injury or stroke

apical that side of a cell or plasma membrane facing the lumen

apical membrane the surface of the plasma membrane of a polarized cell that faces inward to the lumen

aplastic anemia stem cell destruction in the bone marrow that leads to the decreased production of white blood cells, platelets, and red blood cells

apnea temporary cessation of breathing

Apo B proteins class of apolipoproteins that are a component of liver-synthesized very-low-density lipoprotein particles; Apo B48 produced by the small intestine and apo B100 produced exclusively by the liver are the circulating forms of the apo B proteins

apocrine type of exocrine gland from which the apical portion of the secreting cell is released along with the secretory products, for example, sweat glands

apoeccrine hybrid sweat glands found in the axilla with secretory structures similar to those found in both apocrine and eccrine glands but are functionally similar to eccrine glands; they secrete nearly 10 times as much sweat as eccrine glands

apoferritin storage protein found in the enterocyte cytoplasm that binds and stores iron

apoprotein B100 400-kDa protein present on the surface of the low-density lipoprotein (LDL) particle and recognized by LDL receptors

apoproteins major proteins associated with chylomicrons and very-low-density lipoproteins

apoptosis deliberate type of programmed cell death that is mediated by an intracellular signaling pathway; a type of *programmed cell death*; contrast with *necrosis*

aquaporin (AQP) small (30-kDa) integral membrane protein that functions as a water channel and explains the rapid movement of polar water molecules across the lipid bilayer of plasma membranes

aqueous humor thin, clear liquid that fills the anterior chamber of the eyeball

arachidonic acid fatty acid released from phospholipids in the membrane when ligand-gated receptors activate phospholipase A2

arcuate nuclei (ARC) hypothalamic nuclei that synthesize growth hormone–releasing hormone and regulate feeding

area postrema area in the lateral wall of the fourth ventricle, which is one of the few locations in the brain where the blood–brain barrier is open; it contains neural networks that trigger vomiting

areflexia absence of muscle stretch reflexes

areola pigmented region of skin surrounding the nipple

arginine vasopressin (AVP) substance produced by magnocellular neurons in the supraoptic and paraventricular nuclei of the hypothalamus that regulates water resorption by the kidneys

aromatase an enzyme that modifies the A ring of testosterone and androstenedione to remove one carbon, yielding the 18-carbon estrogen hormones

arrhythmia abnormality in heart rate; irregularity in the rhythm of the heart or pattern

arterioles small arteries of ~10 to 1,000 μm in diameter that join large arteries to organ capillary networks; they have the same wall structure as larger arteries but generally have thicker walls and smaller lumen diameters and are the primary site of changes in vascular resistance and blood flow distribution within the cardiovascular system; they are also involved with the control of intracapillary hydrostatic pressure

arteriosclerosis a condition of reduced compliance of large arteries, usually caused by breakdown of the organization of elastic fibers in the adventitia as a natural result of the aging process

arteriovenous anastomoses generally, a direct connection between the lumen of an artery and vein that bypasses the capillary bed of an organ; in the skin, arteriovenous anastomoses that are highly innervated by sympathetic nerves link arterioles to subcutaneous venous plexi that serve to conduct heat to or restrict loss of heat from the skin surface; these anastomoses play a key role, therefore, in temperature regulation in the body

artery blood vessel that carries blood away from the heart and composed of three layers: an outermost stiffly elastic adventitia, a middle zone composed of circular layers of smooth muscle, and a single-cell lining of epithelial cells called the *endothelium*; has structural rigidity in that it retains circular shape even without any positive transmural pressure

ascending reticular activating system system of neurons in the brainstem that project to the thalamus and are involved in arousal and motivation

ascites sites of edema in the organs of the abdominal cavity

aspartate amino acid used as an excitatory neurotransmitter

association cortex cortical region involved in higher-order processing of information

asthma spasmodic contraction of smooth muscle in the bronchi leading to airway obstruction; symptoms include labored breathing accompanied especially by wheezing and coughing and by a sense of constriction in the chest; it is triggered by hyperreactivity to various stimuli (as allergens or rapid change in air temperature)

astigmatism vision defect caused by an eyeball with differing radii of curvature at differing orientations; causes vertical and horizontal lines to have different focal distances

astrocytes glial cells in the CNS that help form the blood–brain barrier and support neuronal functions

ataxia inability to coordinate muscle activity during voluntary movement; most often due to disorders of the cerebellum or the posterior columns of the spinal cord; may involve the limbs, head, or trunk

atelectasis lack of gas exchange in alveoli due to alveolar collapse or fluid consolidation

atherosclerosis thickening of arterial walls from cholesterol plaque buildup, possibly leading to blockage

atopy genetic predisposition to allergens

ATP synthase see *F-type ATPase*

ATPase enzyme or membrane transport system that hydrolyses adenosine triphosphate (ATP) to adenosine diphosphate and free phosphate; *P-type ATPases* are ATPases that are phosphorylated during the hydrolysis of ATP and are found on the plasma membrane; *V-type ATPases* are found on intracellular vacuolar structures; *F-type ATPases* are located on the inner mitochondrial membrane and synthesize ATP rather than hydrolyze it

ATP-binding cassette (ABC) transporters superfamily of adenosine triphosphate–binding cassette transporters composed of two transmembrane domains and two cytosolic nucleotide-binding domains; the transmembrane domains recognize specific solutes and transport them across the membrane by using a number of different mechanisms, including conformational change

ATP-dependent K+ channel (K_{ATP}) class of K channels inhibited by the binding of intracellular adenosine triphosphate

atresia degeneration of follicles in the ovary

atrial fibrillation random, chaotic electrical activation of atrial muscle cells, often resulting in irregular pacing of the ventricles

atrial natriuretic peptide (ANP) peptide hormone that is released from the right atrium when it is stretched, usually from an increased blood volume; promotes the loss of sodium and, indirectly, water from the kidney, thereby helping to restore normal blood volume

atrioventricular (AV) node specialized conduction tissue in the septum of the heart between the atria and ventricles and the only normal path by which electrical activation of the atria spreads into the ventricles; also delays activation of the ventricles during normal cardiac activation, thereby increasing the time the ventricles have to fill with blood during diastole

attention ability to concentrate on a single stimulus

atypical antipsychotics drugs that treat both the positive and negative symptoms of schizophrenia

auditory cortex region of the brain responsible for the higher-level processing of auditory signals

Auerbach plexus ganglionated plexus of the enteric nervous system situated between the longitudinal and circular muscle coats of the muscularis externa of the digestive tract; also known as the *myenteric plexus*

augmented leads system of three unipolar frontal plane leads in the standard clinical recording of an electrocardiogram (ECG); leads are oriented toward the right shoulder, left shoulder, and feet and are designated aVR, aVL, and aVF, respectively; ECG recordings in these leads are amplified or "augmented" relative to the standard bipolar frontal plane ECG leads

auto Greek for "self"

autocrine signaling form of signaling in which a cell secretes a chemical messenger that signals the same cell

autograft transplanted tissue grafted into a new position on the same person

autoimmunity failure of the body to recognize its own tissue as self, allowing attacks against its own cells and tissues

automaticity property of all cardiac cells in which they have the ability to generate their own action potentials

autonomic nervous system (ANS) part of the nervous system that regulates involuntary functions

autoreceptor receptor found on the presynaptic membrane that responds to the transmitter released by the neuron; usually involved in feedback regulation of transmitter synthesis or release

autoregulation in the cardiovascular system, the ability of an organ to maintain near-normal blood flow in the face of changes in perfusion pressure

A wave venous pulse wave created by backflow into the superior vena cava following contraction of the right atrium; amplified with stenosis of the right atrioventricular valve

axial streaming tendency of blood cells and elements to compact into the center of a blood vessel, where flow velocity is highest, when blood flows through the vessel; results in the phenomenon of plasma skimming, whereby the cellular elements of blood are pushed away from the arterial wall, leaving the wall exposed to essentially cell-free plasma

axon hillock initial segment of an axon adjacent to the neuronal cell body; except in pseudounipolar cells, the region of the axon that initiates the action potential; also known as *initial segment* or *trigger zone*

axon terminal region of an axon most distal from the cell body and containing synaptic boutons that signal to the target cell

axoneme main core of a flagellum, consisting of two central microtubules surrounded by nine double microtubules

axons fibers that carry impulses away from the perikaryon of a nerve cell

axoplasmic transport molecular motor-driven mechanism to transport proteins, lipids, synaptic vesicles, and other organelles from the cell body to the axon and its terminals; also used to transport lysosomes and trophic factors from the axon terminal to the cell body

band cell cell undergoing granulopoiesis to become a mature granulocyte

barometric pressure atmospheric pressure that surrounds us; at sea level, the barometric pressure is 760 mm Hg

baroreceptor stretch receptor sensitive to blood pressure and to the rate of change of blood pressure

baroreceptors the name given to neural sensors in the carotid sinus and aortic arch that are activated by stretch and thereby secondarily activated by an increase in arterial pressure. The term is also applied to similar neurosensors in the atria, pulmonary arteries, and atriavenojunction that respond to stretch secondary to a change in blood volume (sometimes called "low-pressure receptors").

baroreceptor reflex broad term used to describe the marshaling of the autonomic nervous system to the heart, arteries, veins, adrenal medulla, and kidney (renin release) in order to counteract changes in arterial pressure; also sometimes called *buffer reflex* in that it prevents large deviations in arterial blood pressure brought about by moment-to-moment changes in the internal or external environment of the body

Barr body condensed mass of chromatin in female cells resulting from inactivation of one X chromosome

Barrett esophagus precancerous metaplasia associated with abnormal exposure of the esophageal mucosa to reflux of gastric contents

Bartter syndrome rare, inherited condition of the thick ascending limb of the kidney thought to be caused by a defect in the kidney's ability to reabsorb sodium, causing a rise in aldosterone and, consequently, potassium wasting and hypokalemic alkalosis

basal forebrain nuclei subcortical groups of cholinergic neurons located at the base of the forebrain that are involved in cognition

basal ganglia subcortical groups of neurons involved in motor control

basal metabolic rate (BMR) oxygen consumption of the body measured under resting conditions

base substance that can bind or accept a hydrogen ion

basement membrane second layer of the glomerular membrane of the kidney, consisting of a meshwork of fine fibrils embedded in a gellike matrix

basic metabolic panel (BMP) group of blood chemistry tests that includes the analysis of glucose, Ca^{2+}, Na^+, K^+, CO_2 (or HCO_3^-), Cl^-, blood urea nitrogen, and creatinine

basilar membrane vibrating membrane in the cochlea along which different frequencies of sound are detected

basolateral that side of a cell or plasma membrane facing the blood supply

basolateral membrane the surface of the plasma membrane of a polarized cell that forms its basal and lateral surfaces, it faces outwards towards the interstitium and away from the lumen

basophil granulocyte of the myeloid series that plays a role in inflammation and immediate allergic reactions

B cell type of lymphocyte with immunoglobulin on the cell membrane surface

B-cell receptor (BCR) antigen recognition molecule of B cells

Becker muscular dystrophy (BMD) inherited disorder that involves slowly worsening muscle weakness of the legs and pelvis; less common than Duchenne muscular dystrophy, affecting 3 to 6 out of every 100,000 males

benign paroxysmal positional vertigo (BPPV) severe vertigo, with incidence increasing with age; thought to be a result of the presence of canaliths, debris in the lumen of one of the semicircular canals

benzodiazepines drugs used to treat anxiety that act at γ-aminobutyric acid receptors

beriberi disorder of the nervous system and heart characterized by anorexia caused by a deficiency in thiamine

Bernoulli principle principle derived from the fact that total energy in a system cannot be created or destroyed but only transferred between different forms. In the cardiovascular system, it is the principle that describes the transfer of energy between kinetic (from flow velocity) and potential (from lateral pressure) energy. The principle predicts that as the velocity of flow in a fluid in a tube increases, the lateral pressure on the wall of the tube decreases

beta cells see β *cell*

beta waves electroencephalogram wave pattern with a rhythm ranging from 13 to 30 Hz, observed when the person is awake but relaxed with the eyes open

biguanides class of drugs used in the treatment of type 2 diabetes that lower glucose via effects on peripheral tissues, primarily the liver

bile acids one of the major components of bile, they are formed in the liver from cholesterol. There are two types of bile acid: primary and secondary

bile salts cholesterol derivatives that play an important role in the intestinal absorption of lipids. They are water soluble and are essentially detergents. They are polar molecules that penetrate cell membranes poorly, an important property because it ensures their minimal absorption of fat

bile salt–sodium symport mechanism of bile acid secretion and bile flow that involves hepatocyte uptake of free and conjugated bile salts and is calcium dependent and calcium mediated. Bile salts serve the important function of absorbing fat

bilirubin major pigment present in bile; an orange compound that is an end product of hemoglobin degradation in the monocyte–macrophage system in the spleen, bone marrow, and liver

biliverdin green metabolite product of hemoglobin catabolism

biological activity unit quantitated amount of a hormone defined as an amount sufficient to produce a response of specified magnitude under a defined set of conditions

biologic clock circadian pacemaker located in the hypothalamus that regulates sleep and wakefulness and metabolic pathways

biotin coenzyme for carboxylase, transcarboxylase, and decarboxylase enzymes, which play an important role in the metabolism of lipids, glucose, and amino acids

bipolar cell as part of the retina, these cells exist between photoreceptors (rod cells and cone cells) and ganglion cells and act, directly or indirectly, to transmit signals from the photoreceptors to the ganglion cells

bipolar disorder psychiatric mood disorder in which periods of profound depression are followed by periods of mania, in a cyclic pattern; formerly known as *manic–depressive disorder*

blastocyst stage in embryonic development after the morula stage that consists of an inner cell mass, an outer trophoblast layer, and a blastocele

blastomeres cells produced by mitotic division of a zygote

bleeding time medical test of hemostasis

bloating sensation of abdominal distention that may or may not be visible

blood indices indices including mean cell (corpuscular) hemoglobin concentration, mean cell (corpuscular) hemoglobin, mean cell (corpuscular) volume, and blood oxygen-carrying capacity

blood lipid profile blood tests including total cholesterol, low-density lipoprotein (LDL), high-density lipoprotein (HDL), LDL/HDL ratio, and total triglycerides

blood smear spread of a blood drop for microscopic analysis

blood–brain barrier system of anatomical and transport barriers that controls the entry and rates of transport of substances into the brain extracellular space from the capillary blood

blood–gas interface interface between the atmosphere and blood, comprising the alveolar–capillary membrane; the site for gas exchange in the lung

blood–placental barrier barrier that regulates the transmission of substances from the maternal circulation to the fetus

body mass index (BMI) number, derived by using height and weight measurements, that gives a general indication of whether weight falls within a healthy range

body plethysmograph airtight box in which patient sits to measure the functional residual capacity and total lung capacity of the lungs; also called *pulmonary plethysmograph* and *body box*

Bohr effect rightward shift of the oxygen equilibrium curve resulting from a rise in blood carbon dioxide and hydrogen ions

bone conduction means of hearing in which sound vibrations are carried to the inner ear by the bones of the skull

bone morphogenetic proteins (BMPs) growth factors found to induce the differentiation of mesenchymal cells toward osteoblasts as well as orchestrating tissue architecture throughout the body

bone remodeling turnover of bone mineral and changes in bone structure from daily flux of calcium and phosphate into and out of bone; generally occurs along most of the outer surface of the bone, making it either thinner or thicker, as required

botulinum toxin toxin produced by the bacterium *Clostridium botulinum*, which, in small concentrations, prevents presynaptic transmitter release at the myoneural junction

bouton the specialized axonal nerve ending from which neurotransmitters are released

Boyle law gas law that states that at a constant temperature, the pressure of a gas varies inversely with gas volume

bradykinin nonapeptide hormone that causes a slowly developing contraction of intestinal smooth muscle, vasodilation, and increased renal sodium excretion

brain natriuretic peptide (BNP) peptide hormone produced in the brain and cardiac ventricles that increases renal sodium excretion

brainstem region of the central nervous system between the spinal cord and diencephalon comprising the midbrain, pons, and medulla

brain–gut axis bidirectional nervous connections between the brain (central nervous system [CNS]) and gut that serve various physiologic functions; visceral afferent fibers project to somatotypic, emotional, and cognitive centers of the CNS, producing many interpretations to the stimuli based on prior learning and one's cognitive and emotional state; in turn, the CNS can inhibit or facilitate afferent nociceptive signals, motility, secretory function, or inflammation

breast cancer resistance protein (BRCP) member of the ABC transporter family of multidrug resistance proteins

Broca area area of the frontal lobe of the brain involved in motor control of speech, usually on the left hemisphere

bronchi bifurcation of the airway or the two primary divisions of the trachea that lead respectively into the right and the left lung

bronchial circulation circulation that supplies oxygenated blood to the conducting zone (i.e., the first 17 generations of the lung airways), which originates from the descending aorta and drains into the pulmonary vein

bronchioles small, thin airways that lack cartilage and are subjected to lung pressures; they are branches of the bronchi and terminate by entering the circular sacs called *alveoli*; also called *bronchioli*

bronchioli see *bronchioles*

bronchitis inflammation of the bronchi

brown adipose tissue one of the two types of adipose tissue (the other being white adipose tissue) that are present in many newborn or hibernating mammals; in contrast to white adipocytes (fat cells), which contain a single, large fat vacuole, brown adipocytes contain several smaller vacuoles, a much higher number of mitochondria, and more capillaries because it has a greater need for oxygen than most tissues; also called *brown fat*

buffer agent that minimizes the change in pH produced when an acid or base is added

buffy coat fraction of an anticoagulated blood sample after centrifugation, containing most of the white blood cells and platelets

bulimia a behavioral disorder involving distortion of body image and an obsessive desire to lose weight, in which bouts of extreme overeating are followed by depression and self-induced vomiting, purging, or fasting

bundle of His specialized conduction tissue that receives impulses from the atrioventricular node. It branches into right and left structures, which in turn give rise to the Purkinje fiber system

C wave venous pulse wave resulting from bulging of the right atrioventricular valve into the superior vena cava during systole followed by descent of the heart. It is amplified when the valve is incompetent (will not close)

Ca²⁺-binding protein, calbindin D (CaBP) protein that binds with calcium in the duodenum and jejunum that aids in the absorption of calcium

calcitonin (CT) hormone secreted by the thyroid that lowers blood calcium; also known as *thyrocalcitonin*

calcium pump enzyme that uses the energy generated by the hydrolysis of adenosine triphosphate to move calcium ions out of the cytoplasm or, in the case of muscle cells, from the cytoplasm to the sarcoplasmic reticulum; also known as *Ca²⁺ ATPase*

calcium release channel membrane channel in the sarcoplasmic reticulum through which calcium ions exit into the myofilament space or cytoplasm

calcium sparks small localized releases of calcium intracellularly following the activation of one coupled voltage-gated calcium channel:sarcoplasmic reticular–calcium release channel pair

calcium-induced calcium release (CICR) release of calcium from the sarcoplasmic reticulum–calcium release channel in response to a calcium entry signal through the plasma membrane voltage-gated calcium channel

calmodulin (CaM) small calcium-binding protein that confers calcium sensitivity to an enzyme by binding to it and makes its activity subject to calcium control; for example, calmodulin complexed with calcium binds to myosin light-chain kinase to promote its activation

calor heat associated with acute inflammation

cAMP-dependent protein kinase enzyme whose activity (such as regulation of glycogen, sugar, and lipid metabolism) is dependent on cyclic adenosine monophosphate; also known as *protein kinase A*

canaliculi tiny cytoplasmic connections between osteocytes, where much of calcium homeostasis takes place

canaliths debris in the lumen of one of the semicircular canals often causing vertigo

cancer immunotherapy therapeutic use of the body's immune system to fight cancer

capacitance two conductors separated by an insulator with the ability to store an electrical charge. The biologic capacitor is the lipid bilayer of the plasma membrane, which separates two conductive regions, the extracellular and intracellular fluids

capacitation maturation of sperm in the female genital tract, which involves a loss of surface glycoprotein, a loss of lipid, and an increase in motility

capillaries smallest elements of the vascular system, composed of a single layer of endothelial cells. They are the primary site of transport between the bloodstream and the extracellular fluid surrounding all tissues. Capillaries can be joined tightly or exist with intermittent openings in their walls, depending on the organ from which they come

capillary distention dilation of a bed of highly compliant small pulmonary arterioles with increased perfusion pressure

capillary filtration coefficient (CFC) constant relating Starling forces to movement of water across the capillaries. It is a measure of how much water can flow per unit of time per net positive pressure moving fluid out of the capillaries

capillary recruitment opening of a bed of pulmonary capillaries with increased perfusion pressure

carbamino proteins proteins that bind with carbon dioxide

carbaminohemoglobin compound of hemoglobin and carbon dioxide, which is one of the forms in which carbon dioxide is transported by the blood

carbon dioxide equilibrium curve curve that reflects the relationship between carbon dioxide content and partial pressure in the blood

carbonic anhydrase (CA) enzyme that catalyzes the reversible reaction $CO_2 + H_2O \Rightarrow H_2CO_3$ or, more correctly, $CO_2 + OH^- \Rightarrow HCO_3^-$

carboxyhemoglobin (Hbco) carbon monoxide bound to hemoglobin

carboxypeptidase A exopeptidase pancreatic protease found in pancreatic juice that is specific in its action to attack polypeptides with a neutral aliphatic or aromatic carboxyl terminal to produce the final products of protein digestion, amino acids, and small peptides

carboxypeptidase B exopeptidase pancreatic protease found in pancreatic juice that is specific in its action to attack polypeptides with a basic carboxyl terminal to produce the final products of protein digestion, amino acids, and small peptides

carboxypolypeptidase pancreatic enzyme that converts peptides into amino acid

cardia one of four stomach sections, where the esophageal contents are emptied

cardiac function curve relationship depicting how cardiac output varies as a function of central venous pressure

cardiac gland one of three main types of glands found in the mucosal lining of the stomach, specifically in a small area adjacent to the esophagus and lined by mucus-producing columnar cells that secrete mucous and bicarbonate ions that protect the stomach from the acid in the lumen

cardiac glycoside drug used in the treatment of congestive heart failure and cardiac arrhythmia; see also *digoxin*

cardiac index cardiac output normalized for body surface area

cardiac muscle striated muscle tissue that constitutes the contractile pumping elements of the heart in the form of the right and left atria and ventricles. It is an excitable tissue that can contract and generate force when activated intrinsically by an initiating action potential from the SA node or, in abnormal conditions, from any other myocardial cell. It does not require nerves for activation, but its contraction is modified by both branches of the autonomic nervous system. Histologically, it is a functional syncytium whereby all cardiac muscle cells are connected to one another electrically unlike the motor unit construction of skeletal muscle.

cardiac output (CO) total flow output of the heart, often expressed as L/min or mL/min. It is typically between 4 and 5 L/min in healthy adults. It can also be thought of as the sum of all the individual organ blood flows in the body

cardiac tamponade pericardial constriction of the ventricles resulting from bleeding beneath the pericardium. It reduces passive stretch of the ventricle during filling

cardiopulmonary baroreceptors receptors located on the right side of the circulation that sense changes in circulating blood volume or central blood volume through the effects of volume on the stretch of the receptors. They elicit mechanisms to stimulate salt and water loss whenever blood or central volume is too high; also known as "*low pressure*" or "*volume*" receptors

carotid bodies specialized innervated nodules located in the bifurcation of the internal and external carotid arteries that sense

low Po_2 and pH as well as high Pco_2. They send impulses to the medullary area of the brain to elicit reflex activation of the sympathetic nervous system to increase blood pressure. They do not respond to low arterial pressure unless the pressure is low enough to affect oxygen delivery to the bodies (<80 mm Hg)

carotid sinus slight dilation of the common carotid artery at its bifurcation into external and internal carotids; it contains baroreceptors that, when stimulated, cause slowing of the heart, vasodilation, and a fall in blood pressure and is innervated primarily by the glossopharyngeal nerve

carotid sinus baroreceptors spray nerve endings in the wall of the carotid sinus that sense stretch, and therefore, serve as sensory receptors for changes in arterial pressure

carrier-mediated diffusion see *passive transport*

caspases family of cysteine–aspartic proteases involved in apoptosis, cell necrosis, and cell death

catalase enzyme that protects cells by catalyzing hydrogen peroxide

cataract increase in the opacity of the lens of the eye

catechol-*O*-methyltransferase (COMT) enzyme that degrades the catecholamines by methylating the 3-OH group on the catechol ring

cation ion or group of ions having a positive charge and characteristically moving toward the negative electrode in electrolysis

caudate nucleus one of the basal ganglia; receives input from the cerebral cortex

caveolae microdomain or lipid raft that contains the protein caveolin and forms an invagination of the plasma membrane

caveolin one of a family of integral membrane proteins present in lipid rafts (termed caveolae) and involved in receptor-independent endocytosis

CCK cholecystokinin; satiety signal produced by endocrine cells in the gut wall mediated via the vagus

CD cluster of differentiation; distinct plasma membrane molecules used as cell markers

celiac ganglion prevertebral sympathetic ganglion involved in gastrointestinal control

celiac sprue (gluten-sensitive enteropathy) common disease involving a primary lesion of the intestinal mucosa caused by the sensitivity of the small intestine to gluten that results in the malabsorption of all nutrients as a result of the shortening or a total loss of intestinal villi, which reduces the mucosal enzymes for nutrient digestion and the mucosal surface for absorption

α cell one of five types of cells found in the islets of Langerhans of the endocrine pancreas; this one produces the hormone glucagon

β cell one of five types of cells found in the islets of Langerhans of the endocrine pancreas; this one produces the hormones insulin and amylin

δ cell one of five types of cells found in the islets of Langerhans of the endocrine pancreas; this one produces the hormone somatostatin

ε cell one of five types of cells found in the islets of Langerhans of the endocrine pancreas; this one produces ghrelin

cell adhesion molecules plasma membrane glycoproteins that promote cell adhesion

cellular immunity immunity mediated by activated T cells

cellular respiration final stage of respiration in which food molecules are broken down in the cell, releasing carbon dioxide and oxygen

central autonomic network neurons and axon pathways in the central nervous system that control autonomic function

central blood volume theoretical volume of blood contained in the veins of the thoracic cavity. It is usually measured as pressure in the right atrium

central chemoreceptors chemosensitive neurons located adjacent to the dorsal and ventral respiratory group complex that respond to H⁺ of the surrounding interstitial fluid

central inspiratory activity (CIA) integrator group of cells in the medulla that activates inspiration

central integrative system brain area involved in the coordinated activities of human behavior

central nervous system (CNS) brain and the spinal cord

central pattern generator (CPG) group of nerve cells producing rhythmic action potentials important in movements

central sulcus a deep fissure on the lateral surface of the cerebral cortex that separates the frontal lobe from the parietal lobe

central venous pressure venous pressure in the right atrium. It is representative of the effect of changes in central blood volume

centroacinar cells cells lining the lumen of the acinus that modify the electrolyte composition of pancreatic secretion

cephalic phase one of three phases that stimulate acid secretion during the ingestion of a meal; involves the central nervous system and is induced by the simple process of smelling, chewing, swallowing, and the thought of food. The cephalic phase accounts for about 40% of total acid secretion

cerebellum brainstem structure important in movement coordination

cerebral cortex (1) sheet of neural tissue outermost to the cerebrum of the mammalian brain that plays a key role in memory, attention, perceptual awareness, thought, language, and consciousness; (2) the outer shell of the brain, containing the cell bodies of neurons

cerebral edema edema formation in the brain. It causes an increase in intracranial pressure that results in restriction of cerebral blood flow

cerebral peduncles pathway in the upper brainstem for corticospinal tract axons

cerebrocerebellum region of the cerebellar cortex that receives input from cerebral cortex, active in control of limb muscles

cervical mucus viscous substance on the surface of the cervix, made up of mucopolysaccharide and water

cervix narrow muscular canal that connects the vagina and the body (corpus) of the uterus

C-fiber endings unmyelinated nerve endings that are located adjacent to alveoli and small bronchioles, that are sensitive to stretch, and that play a protective role in the lung

cGMP-dependent protein kinase enzyme that requires cyclic guanosine monophosphate to phosphorylate potential target substrates, such as calcium pumps in the sarcoplasmic reticulum or sarcolemma, leading to reduced cytoplasmic levels of calcium and, in turn, muscle relaxation; also known as *protein kinase G*

Chagas disease neuropathic degeneration of autonomic neurons resulting from autoimmune attack in patients infected with the blood-borne parasite *Trypanosoma cruzi*

channel gating regulation of the opening or closing of ion channels by changes in transmembrane voltage, mechanosensitivity, or ligand binding

channelopathies diseases caused by dysfunction of ion channels

Charcot-Marie-Tooth disease (CMT) group of inherited disorders that affect the peripheral nerves, specifically, involving damage to the covering (myelin sheath) around nerve fibers (either destruction of the myelin sheath or erosion of the central [axon] portion of the nerve cell); nerves that stimulate movement (the motor nerves) are most severely affected, and the nerves in the legs are affected first and most severely

Charles law gas law that states that at a constant pressure, gas volume varies proportionately with temperature

chemical pH buffer mixture of a weak acid and its conjugate base (or a weak base and its conjugate acid)

chemical potential gradient created by different concentrations of an uncharged solute on opposite sides of a membrane

chemical synapse site of communication between a neuron and another neuron or other target cell mediated by the release of a chemical neurotransmitter

chemoattractants cytokines that attract white blood cells

chemokine cytokine that attracts and activates leukocytes

chemoreceptors neurosensitive endings that respond to changes in blood gases (oxygen, carbon dioxide, and pH)

chemosensitive nerve endings efferent arm of the chemoreflex that increases cardiac output and improved blood flow to the ischemic skeletal muscle

chenodeoxycholic acid primary bile acid

Cheyne-Stokes breathing abnormal breathing pattern characterized by periods of breathing with gradually increasing and decreasing tidal volume interspersed with periods of apnea

chief cells cells that reside in the oxyntic glands in the stomach and secrete pepsinogen

chloride bicarbonate exchanger exchange mechanism that allows the red blood cell to maintain electrical neutrality by exchanging Cl⁻ for HCO₃⁻ as it diffuses out of the cell

chloride shift movement of chloride ions into the red blood cell coupled with movement of HCO₃⁻ out to maintain electrical neutrality of the cell

cholangiocytes epithelial cells lining the bile duct

cholecystokinin (CCK) hormone produced by the small intestine that is responsible for gallbladder contraction as well as the secretion of enzyme-rich pancreatic juice

cholesterol sterol derived exclusively from animal fat

cholesterol ester hydrolase (CEH) substance that hydrolyzes the ester bond of esterified cholesterol to generate free cholesterol

cholesterol esterase pancreatic enzyme that hydrolyzes cholesterol ester

cholesterol esters single molecules of cholesterol esterified to single fatty acid molecules

cholesterol side-chain cleavage enzyme (CYP11A1) rate-limiting enzyme in steroidogenesis

cholic acid primary bile acid

choline acetyltransferase enzyme that catalyzes synthesis of acetylcholine from acetyl-CoA and choline

cholinergic neurons releasing acetylcholine as a neurotransmitter

cholinergic receptor receptors with which acetylcholine interacts

chondrocyte nondividing cartilage cell; occupies a lacuna within the cartilage matrix

chorion extraembryonic membrane that forms the fetal portion of the placenta

chorionic plate membrane forming the fetal portion of the placenta

chorionic villi finger-like projections of the chorion that contact maternal blood

choroid plexuses region of the ventricular cavities of the brain that forms the cerebrospinal fluid

choroidal neovascularization (CNV) creation of new blood vessels in the choroid layer of the eye; common symptom of age-related macular degeneration

chromaffin cells cells in the adrenal medulla that synthesize and release epinephrine

chromatid one strand of a chromosome

chromophore part of a molecule responsible for its color

chronic granulomatous disease caused by the lack of phagocytic NADPH oxidase and characterized by recurrent life-threatening bacterial and fungal infections

chronic kidney disease (CKD) slowly progressive disorder that results in an abnormal internal environment, and eventual loss of kidney function is typically a slowly progressive disorder that results in an abnormal internal environment

chronic myeloid leukemia (CML) cancer of the white blood cells characterized by increased and unregulated clonal proliferation of myeloid cells in the bone marrow; results from an inherited chromosomal abnormality that involves a reciprocal translocation or exchange of genetic material between chromosomes 9 and 22 and was the first malignancy to be linked to a genetic abnormality

chronic obstructive pulmonary disease (COPD) classification of lung diseases (e.g., asthma, bronchitis, and emphysema) that obstruct airflow out of the lung

chronotropic related to effects on heart rate

Chvostek sign spasm of the facial muscles elicited by tapping the facial nerve in the region of the parotid gland; seen in tetany

chylomicron remnants remains of chylomicron particles after being broken down during metabolism. These remnants can be rapidly cleared from circulation by the liver via chylomicron remnant receptors

chylomicrons least dense (<0.95 g/mL) of the four lipoprotein classes; produced by the small intestine; their production increases during ingestion of a fat-containing meal. Chylomicrons function in the transport of lipids to the bloodstream

chyme semifluid material produced by the gastric digestion of food that results partly from the conversion of large solid particles into smaller particles via the combined peristaltic movements of the stomach and contraction of the pyloric sphincter

chymotrypsin one of three endopeptidases present in pancreatic juice that attack peptide bonds with an aromatic carboxyl terminal

ciliary ganglion location of synapses between parasympathetic preganglionic and postganglionic neurons of the eye

cingulate gyrus area of the cerebral cortex involved in the limbic system

circadian rhythm roughly 24-hour cycle in organismal physiologic processes. Circadian rhythms are endogenously generated and modulated by external cues such as sunlight and temperature

circumventricular organs regions of the central nervous system that do not have a blood–brain barrier and thus can sense signals in the blood

cirrhosis chronic liver disease characterized by the replacement of normal liver tissue with fibrous tissue as a result of the transformation of stellate cells into collagen-secreting myofibroblasts. Stellate cells deposit collagen into the sinusoids, thereby leading to loss of functional hepatocytes

11-*cis*-retinal aldehyde form of vitamin A_1 (retinal); chromophore that starts the process of photoisomerization of retinal for vision

clearance volume of plasma per unit time from which a substance is completely removed (cleared)

climbing fibers afferent axons from the inferior olive of the medulla to the cerebellum that synapse on Purkinje cells

clomiphene fertility drug that stimulates the production of one or more follicles and, therefore, increases the chances of pregnancy

clonal selection collection of specific lymphocytes grouped by stimulation or inhibition of growth (positive or negative selection)

clonus repetitive, self-sustaining muscle stretch reflexes

closure inability to absorb intact proteins as a result of the maturation of the gut

clot retraction phenomenon that may occur within minutes or hours after clot formation in which the clot draws together, pulling the torn edges of the vessel closer together and reducing residual bleeding and stabilizing the injury site; requires the action of platelets, which contain actin and myosin; reduces the size of the injured area, making it easier for fibroblasts, smooth muscle cells, and endothelial cells to start wound healing

coagulation cascade pair of pathways (*intrinsic* and *extrinsic coagulation pathways*) in which a series of reactions ultimately result in a stable fibrin clot

coagulation factors series of 13 proteins (I to XIII) synthesized in the liver that circulate in the plasma in an inactive state

coated pit specialized regions of the cell membrane composed of pits made of the protein clathrin

cochlea bony structure of the inner ear, containing the organ of Corti, which is the organized transducer of sound vibrations into nerve impulses

cochlear microphonic generator potential of the inner ear

cognitive science interdisciplinary study of how information (e.g., concerning perception, language, speech, reasoning, and emotion) is represented and transformed in the brain

cold diuresis increased urine production on exposure to cold as a result of peripheral vasoconstriction leading to a transient increase in central blood volume, and consequently, an increase in glomerular filtration rate

cold pressor response reflex increase in heart rate, cardiac output, vascular resistance, and mean arterial pressure following pain associated with sudden exposure, usually of a hand or extremity, to extreme cold. It results from simultaneous activation of the sympathetic nervous system and reduced activity of the parasympathetics to the heart

cold-induced vasodilation (CIVD) acute increase in peripheral blood flow observed during cold exposures

colipase peptide in pancreatic juice necessary for the normal digestion of fat by pancreatic lipase

collagenase enzyme that digests the connective tissue matrix

collateral ganglia collective name for the three prevertebral sympathetic ganglia of the abdominal cavity

collateral vessels vascular connections between adjacent arteries such that when one artery is obstructed, the area it normally supplies with blood is supplied through the collaterals by the neighboring artery

colloid thick, proteinaceous gel-like substance in the thyroid follicle

colloid osmotic pressure osmotic pressure resulting from proteins in water

colostrum thin, yellowish, milky fluid secreted by the mammary glands a few days before or after parturition

commissures bundles of axons that connect neurons in the two sides of the brain

compensated shock stage of shock in which normal mechanisms within the body designed to elevate blood pressure and restore plasma volume correct the shock without the need for external intervention

compensatory pause pause following an extrasystole, when the pause is long enough to compensate for the prematurity of the extrasystole; the short cycle ending with the extrasystole plus the pause following the extrasystole together equal two of the regular cycles

competitive binding assay assay in which a biologically specific binding agent competes for radioactively labeled or unlabeled compounds, used especially to measure the concentration of hormone receptors in a sample by introducing a radioactively labeled hormone

complement serum proteins that can be activated and may lead to cell lysis, formation of opsonins, and regulation of inflammation

complete atrioventricular block any condition in which no action potentials from the atria reach the ventricles. It usually arises from blockage at the level of the atrioventricular node and results in atria and ventricles that are paced at two different rates

complete blood count (CBC) blood tests including white blood cell count, red blood cell count, platelet count, hemoglobin, hematocrit, mean cell (corpuscular) volume, mean cell (corpuscular) hemoglobin, mean cell (corpuscular) hemoglobin concentration, and platelet indices; also known as *full blood count (FBC)*

complete metabolic panel (CMP) includes the tests of the basic metabolic panel (see also *basic metabolic panel*) and the following additional tests: albumin, total protein, alkaline phosphatase, alanine transaminase, aspartate transaminase, and bilirubin

complete proteins proteins that supply all of the essential amino acids in amounts sufficient to support normal growth and body maintenance, such as eggs, poultry, and fish

compliance (1) change in volume in a segment of blood vessel or vessels per unit of change in transmural pressure, or $\Delta V / \Delta P$; (2) capability of a region of the gut to adapt to an increased intraluminal volume

compression first step in process of encoding electrical signals to be sent to the central nervous system, involving increasing the density of a medium; opposite of *rarefaction*

concave lens lens that causes light rays to diverge

concentric contractions muscular contractions that permit the muscle to shorten, because the load on the muscle is less than the maximum tetanic tension the muscle is capable of generating

conductance (1) ease with which ions flow across the membrane through their channels; (2) property of a material that indicates its ability to conduct heat

conducting zone first 16 generations of airways (trachea down to the terminal bronchioles), with its own separate circulation (bronchial circulation). It has a major function of conducting air to the deeper parts of the lung

conduction transfer of energy through matter from particle to particle; the transfer and distribution of heat energy from atom to atom within a substance

conduction velocity rate of impulse conduction in a peripheral nerve or its various component fibers, generally expressed in meters per second

conductive deafness hearing impairment due to sound vibrations being interrupted in the outer or middle ear and not reaching the inner ear and its nerve endings

cone cell cell in the retina responsible for color (photopic, chromatic) vision. There are separate cone cells for red, blue, and green wavelengths of light

congenital adrenal hyperplasia inherited condition that affects the adrenal glands; these glands are located on top of the kidneys and produce three types of hormones, called *cortisol, aldosterone*, and *androgens*; females with classical congenital adrenal hyperplasia are born with masculine-appearing external genitals but with female internal sex organs; males with classical congenital adrenal hyperplasia appear normal at birth; males and females with classical congenital adrenal hyperplasia are likely to have trouble retaining salt, a condition that can be life threatening

congenital nephrotic syndrome rare, inherited condition characterized by excessive filtration of plasma proteins due to a mutation of nephrin

connective tissue matrix complex ground substance between and among cells that provides mechanical integrity to a tissue; also called *stroma*

connexin four-pass transmembrane proteins that in a group of six form a hemichannel called a *connexon*

connexon assembly of six proteins called *connexins* that forms a bridge called a *gap junction* between the cytoplasms of two adjacent cells

constitutive exocytosis form of exocytosis in which the vesicles are continuously filled and released into the extracellular fluid at a constant rate

continuous ambulatory peritoneal dialysis (CAPD) procedure in which peritoneal membrane, which lines the abdominal cavity, acts as a dialyzing membrane: about 1 to 2 L of a sterile glucose/salt solution are introduced into the abdominal cavity, and small molecules (e.g., K^+ and urea) diffuse into the introduced solution, which is then drained and discarded; the procedure is usually done several times every day

contractile activity muscular activity of the gut wall, either of short duration (phasic contractions) or more sustained activity (tonic contractions)

contractility synonym for the inotropic state of the heart

convection transfer of heat by the actual movement of warmed (or cooled) matter. Convection is the transfer of heat energy in a gas or liquid by movement of currents

convex lens converging lens that brings light rays to a focus; it can form a real image

cornea outer, transparent, covering of the front of the eyeball

corneal reflex involuntary blinking of the eye elicited by stimulation of the cornea, bright light, or very loud sounds; also known as the *blink reflex*

cornification (keratinization) formation of a group of intermediate filaments in epithelial cells

coronary reserve amount of blood flow possible in the coronary circulation above resting values; sometimes used to refer to the dilating capacity of the coronary circulation beyond its resting value

corpora albicantia or corpus albicans connective tissue–filled structure; remnant of the regressed corpus luteum

corpora lutea or corpus luteum yellowish endocrine gland formed from the wall of an ovulated follicle

corpus callosum major axonal fiber bundle in the brain that connects the two hemispheres

corpus of the stomach main or central region of the stomach (body); one of four stomach sections

cortex outer portion of an organ, such as the kidney, as distinguished from the inner, or medullary, portion

cortical granules granules in the cortical vesicles at the edge of the ovum cytoplasm; penetration of sperm causes these granules to release proteases and inhibitors to prevent more sperm from penetrating the egg

cortical tissue part of the indifferent gonad of the embryo; will develop into the ovary

corticospinal tract axons arising from the cortical motor area

corticosteroid-binding globulin (CBG) glycoprotein produced by the liver; binds glucocorticoids and aldosterone in the blood

corticosteroid-binding protein (transcortin) binding protein for progesterone in the blood

corticosterone product of the adrenal cortex; regulates the body's response to fasting, injury, and stress

corticotroph cells located in the anterior pituitary that secrete adrenocorticotropic hormone

corticotropin-releasing hormone (CRH) polypeptide from the hypothalamus; stimulates secretion of adrenocorticotropic hormone

cortisol product of the adrenal cortex; regulates the body's response to fasting, injury, and stress

cotyledons aggregate groups of chorionic villi surrounded by maternal blood

coumarin common oral anticoagulant

countercurrent flowing in an opposite direction

countercurrent exchange passive exchange of solutes, water, or heat between two adjacent streams flowing in opposite directions

countercurrent heat exchange exchange of heat between two separated streams flowing in opposing directions, commonly used to minimize heat loss from the circulatory system by transferring heat from the arteries to the veins

countercurrent multiplication energy-demanding process that sets up a solute (osmotic) gradient along the length of two streams flowing in opposite directions

coupling reaction reaction that forms the iodothyronine structure from iodinated tyrosines

C-peptide peptide made when proinsulin is split into insulin and C-peptide on release from the pancreatic β cell into the blood

cranial nerves 12 pairs of nerves that originate directly from the brain

craniosacral regional locations of the parasympathetic preganglionic neurons

creatine phosphate pool cellular reservoir of high-energy phosphate that can rephosphorylate the adenosine triphosphate consumed in muscle contraction

creatinine anhydride of creatine, being the end product of creatine metabolism, found in muscle and blood and excreted in the urine

creep slow and time-dependent elongation of a viscoelastic substance subjected to an external force

cremasteric muscle muscle with origin from the internal oblique and inguinal ligament, with insertion into the cremasteric fascia and pubic tubercle, with nerve supply from the genitofemoral nerve, and whose action raises the testicle; in male, the muscle envelops the spermatic cord and the testis; in female, it envelops the round ligament of the uterus

Creutzfeldt-Jakob disease (CJD) degenerative brain disorder caused by death of nerve cells leading to progressive dementia, speech impairment, ataxia, and seizures; most common form of transmissible spongiform encephalopathy

critical micellar concentration certain point of bile salt concentration in the intestinal lumen at which bile salts aggregate to form micelles

crossbridge transient structure in muscle formed by the myosin heads projecting from thick filaments and attached to binding sites on the thin filament actin

crossbridge cycle series of adenosine triphosphate–consuming and work-producing reactions between actin and myosin that produces muscle contraction

cross-reactivity antibody that reacts with antigen that is similar to the one that induced its formation

crypts of Lieberkühn tubular glands located at the base of intestinal villi in the small intestine. Cells in the crypts of Lieberkühn secrete isotonic alkaline fluid

cumulus granulosa cells granulosa cells surrounding the oocyte in a graafian follicle

curare drug that is a competitive inhibitor of acetylcholine at the myoneural junction

Cushing reflex sympathetic nerve reflex initiated by an increase in intracranial pressure. It significantly elevates arterial pressure by causing severe, sometimes occlusive, constriction of arterioles in systemic organs. It is a defense mechanism that helps maintain cerebral blood flow in the face of elevated intracranial pressure by forcing blood vessels open

Cushing syndrome name given to the signs and symptoms associated with prolonged exposure to inappropriately elevated glucocorticoids; the disease can result from excess adrenocorticotropic hormone secretion from the pituitary (specifically called *Cushing disease*) stimulating cortisol production from the adrenal gland, or it can result from an adrenocortical adenoma secreting large amounts of cortisol

cyanosis blue coloration of the skin and mucous membranes due to the presence of excess deoxygenated hemoglobin in blood vessels near the skin surface; it is caused by low hemoglobin concentration (>5 g/dL) or from breathing low oxygen (hypoxia)

cyclic adenosine monophosphate (cAMP) second messenger important in many biologic processes; derived from adenosine triphosphate and used for intracellular signal transduction in many different organisms

cyclic guanosine monophosphate (cGMP) intracellular second messenger derived from guanine triphosphate that causes smooth muscle relaxation by lowering intracellular calcium

cyclooxygenase (COX) enzyme that catalyzes the conversion of arachidonic acid to prostaglandins

cystic fibrosis hereditary disease prevalent especially in Caucasian populations that appears usually in early childhood. It is inherited as an autosomal recessive monogenic trait, involves functional disorder of the exocrine glands, and is marked by faulty digestion due to a deficiency of pancreatic enzymes, by difficulty in breathing due to excess mucus secretion and accumulation in airways and by excessive loss of salt in the sweat

cystic fibrosis transmembrane conductance regulator (CFTCR) ion channel belonging to the adenosine triphosphate–binding cassette family of proteins, whose major function is to secrete chloride ions out of the cells, providing chloride in the lumen for the chloride–bicarbonate exchanger to work

cystinuria autosomal recessive disorder associated with a defect in reabsorptive transport of cystine and the dibasic amino acids ornithine, arginine, and lysine by the kidney. Cystinuria also affects the transporter in the small intestine

cytochrome P-450 complex of iron-containing proteins responsible for the oxidation–reduction reactions that transform a nonpolar xenobiotic compound to one that is more polar by introducing one or more polar groups onto the molecule. This particular complex contains the enzyme NADPH–cytochrome P-450 reductase, which is responsible for the reduction of the cytochrome P-450–drug complex being metabolized

cytokines group of protein-signaling compounds that, similar to hormones and neurotransmitters, are used extensively for intercellular communication

cytoskeleton internal meshwork of proteins adjacent and attached to the plasma membrane of the cell that maintains cell shape and other functions

cytosol liquid or matrix found within a living cell

cytotrophoblast cellular component of the trophoblast; gives rise to the chorion

Dalton law gas law that states that total barometric pressure is equal to the sum of the partial pressure of the individual gases

D cells endocrine cells in the antrum that produce somatostatin, a gastrointestinal hormone that inhibits the release of gastrin and, thus, gastric acid secretion

dead space volume air that is inhaled by the lungs but does not take part in gas exchange, because not all the air in each breath is able to be used for the gas exchange and represents the air in the conducting zone; about a third of every resting breath is exhaled exactly as it came into the body (in adults, it is usually in the range of 150 mL)

decerebrate rigidity describes the involuntary extension of the upper extremities in response to external stimuli: the head is arched back, the arms and elbows are extended by the sides, the legs are extended and rotated internally, and the patient is rigid, with the teeth clenched; these signs can be on just one or the other side of the body or on both sides, and it may be just in the arms and may be intermittent

decibel (dB) logarithmic unit that indicates the ratio of a physical quantity relative to a specified or implied reference level; widely known as a measure of sound pressure level

decidua site of implantation and the maternal contribution to the placenta

decidua basalis maternal part of the placenta

decidua capsularis part of the endometrium that grows over and covers the fetus

decidua parietalis pregnant endometrium not contained in the placenta

decidual reaction reaction that occurs in the endometrium and consists of dilation of blood vessels, increasing capillary permeability, edema formation, and increased proliferation of endometrial glandular and epithelial cells

decidualization hypertrophy of endometrial cells that contain large amounts of glycogen and lipid to ready the endometrium for implantation

declarative memory memory of events (episodic) or facts (semantic); also known as *explicit memory*

defensins antibacterial proteins that can enter phagosomes

degranulation release of granules from cells, such as mast cells and basophils

dehydration removal of water

dehydration reaction chemical reaction in which a hydrogen atom is removed from the end of one molecule that was originally a hydroxyl; the remaining O, from the original OH that only the H was taken from, is then bonded with the other monosaccharide; the H and OH are then released as a molecule of water (H_2O)

dehydroalanine residue product of free radical reaction to produce an iodothyronine residue

7-dehydrocholesterol provitamin D; the presence of this compound in human skin enables humans to manufacture vitamin D_3 from ultraviolet rays in the sunlight, via an intermediate isomer previtamin D_3

dehydroepiandrosterone (DHEA) androgenic precursor for estrogen and testosterone synthesis produced by ovaries and testes

dehydroepiandrosterone sulfate (DHEAS) sulfate salt of dehydroepiandrosterone

deiodinase type 1 (D1) substance located in the liver, kidneys, and thyroid gland; catalyzes outer-ring deiodination

deiodinase type 2 (D2) substance believed to function primarily to maintain intracellular T_3 in target tissues

deiodinase type 3 (D3) substance that catalyzes inner-ring deiodination reactions during degradation of thyroid hormones

delayed puberty condition in which physical signs of puberty have not presented by age 13 years in girls and 14 years in boys

delayed afterdepolarizations (DADs) occur late in phase 3 or during phase 4 of the ventricular action potential and are associated with increased intracellular calcium concentration within myocytes, which can thus depolarize the cell membrane to the threshold for an action potential

δ (delta) 4 pathway dominant steroidogenic pathway in granulosa cells and the corpus luteum

δ (delta) 5 pathway dominant steroidogenic pathway in theca cells

delta cells see *δ cell*

delta waves electroencephalogram wave pattern with a rhythm ranging from 0.5 to 4 Hz, observed when the person is in deepest sleep

delusions fixed false beliefs

dementia progressive loss of brain function, including memory, behavior, and cognition

denaturation structural change in macromolecules caused by extreme conditions, for example, protein denaturation by extreme heat

dendrites fibers that receive synaptic inputs and transmit electrical signals toward the perikaryon of a nerve cell

dendritic cell phagocyte; antigen-presenting cell of epithelia and blood

dendritic spines small membrane protrusions from dendrites that receive synaptic input

denervation hypersensitivity heightened reactivity of an organ to a neurotransmitter following denervation of the organ

dense bodies intracellular protein complexes at the margins of smooth muscle cells that serve as attachment points between intracellular actin and extracellular connective tissue

density mass per unit volume

dentate nucleus largest of the four deep cerebellar nuclei, responsible for the planning, initiation, and control of volitional movements; receives its afferents from the premotor cortex and the supplementary motor cortex, and its efferents project via the superior cerebellar peduncle through the red nucleus to the ventrolateral thalamus

deoxycholic acid secondary bile acid

11-deoxycorticosterone (DOC) mineralocorticoid made from progesterone by 21-hydroxylase

11-deoxycortisol immediate precursor for cortisol synthesis

deoxyhemoglobin hemoglobin that does not bind with oxygen

depolarize when the membrane potential becomes more positive (less negative)

depolarization change in electrical potential in the direction of a more positive potential inside cells; in excitable cells, depolarization to a threshold potential triggers an action potential

depolarizing blockers agents that interfere with neuromuscular transmission by causing the postsynaptic membrane to remain depolarized

desensitization process by which effector cells are rendered less reactive or nonreactive to a given stimulus

diabetes insipidus disorder characterized by excretion of a large volume of osmotically dilute urine, resulting from either deficient production or release of arginine vasopressin

diabetes mellitus disease in which plasma glucose control is defective because of insulin deficiency or decreased target cell response to insulin

diabetic ketoacidosis condition in which extremely high blood glucose levels, along with a severe lack of insulin, result in the breakdown of body fat for energy and an accumulation of ketones in the blood and urine

diabetic neuropathy family of nerve disorders caused by diabetes; also known as *peripheral neuropathy*

diabetogenic action effect of growth hormone to oppose the actions of insulin

diacylglycerol (DAG) intracellular second messenger generated by hydrolysis of phosphatidylinositol 4,5-bisphosphate

dialysis separation of smaller molecules from larger molecules in solution by diffusion of the small molecules through a selectively permeable membrane; two methods of dialysis are commonly used to treat patients with severe, irreversible ("end-stage") renal failure; see also *continuous ambulatory peritoneal dialysis (CAPD)* and *hemodialysis*

diapedesis movement of cells from blood to tissue during acute inflammation

diaphragm dome-shaped sheet of internal skeletal muscle that extends across the bottom of the rib cage and separates the thoracic cavity (heart, lungs, and ribs) from the abdominal cavity; it is the main muscle of breathing and is innervated by the phrenic nerves

diastole period of time in which the heart is in its relaxation phase

diastolic pressure blood pressure in the ventricle or arterial system during the relaxation of the heart, or diastole

diencephalon region of the central nervous system that includes the thalamus and hypothalamus

differential white blood cell count proportions of the different types of circulating white blood cells

diffuse esophageal spasm diagnosis of diffuse spasm is made when manometric recording of esophageal motility demonstrates that the act of swallowing results in simultaneous contractions all along the length of the smooth muscle region of the esophageal body

diffuse toxic goiter enlarged thyroid gland; secretes thyroid hormones at an accelerated rate

diffusion random movement of particles, such as ions and solutes in solution, that leads to mixing and elimination of concentration gradients

diffusion block abnormal increase in the diffusion distance across the alveolar–capillary membrane or to a decrease in the alveolar–capillary permeability leading to decrease in oxygen uptake by the lungs (hypoxemia); pulmonary edema is one of the major causes of a diffusion block and is characterized by a low Pa_{O_2}, elevated Pa_{CO_2}, and a high $A–aO_2$ gradient

diffusion capacity see *lung diffusion capacity*

diffusion limited condition in which the uptake of an alveolar gas is limited by the diffusion properties of the alveolar–capillary membrane; see also *Fick's law*

diffusion-limited transport transport of substances across the capillary in which the rate of diffusion is what limits the rate of transport

diffusive membrane transport see *passive transport*

digestive state gastrointestinal motor behavior (mixing/segmentation pattern) in effect when nutrients are present in the upper gut

digoxin binds to the extracellular aspect of the a-subunit of the Na^+/K^+ ATPase and decreases the transport function of the pump; often used to treat atrial fibrillation or atrial flutter; also known as *digitalis*

dihydropyridine receptors (DHRPs) voltage-sensitive protein molecules that respond to changes in the T-tubule membrane of skeletal muscle as an important step in the excitation–contraction coupling process

dihydrotestosterone (DHT) potent androgen derived from testosterone

1,25-dihydroxycholecalciferol substance that enhances calcium and phosphate absorption by the small intestine and mobilizes calcium and phosphate from bones; the active form of vitamin D; also known as *1,25-dihydroxycholecalciferol* or *calcitriol*

diiodotyrosine (DIT) intermediate in the biosynthesis of thyroid hormone

diopter (D) measure of lens-converging power; the inverse of the focal length in meters

dimer chemical or biologic entity consisting of two structurally similar subunits that are joined by bonds, which can be strong or weak

dimerization (1) aggregation of two monomers of a protein (e.g., receptor), usually mediated by binding of a ligand; (2) chemical reaction in which two monomers combine to form a dimer

dipalmitoylphosphatidylcholine (DPPC) surface-reducing material lining the alveoli that is responsible for reducing surface tension and increasing the distensibility of the lung

diplopia double vision, caused by partial failure of the convergence mechanism driven by the extraocular muscles

dipole difference in the electrical polarity between two physical points

direct calorimetry measurement of the metabolic rate of an organism by measuring the amount of heat energy released as heat over a given period

direct pathway basal ganglia pathway for axons passing from the caudate directly to the globus pallidus internus

discomfort (upper abdomen) subjective, unpleasant sensation or feeling that is not interpreted as pain according to the patient and that, if fully assessed, can include nausea, fullness, bloating, and early satiety

disinhibitory motor disease disordered gastrointestinal motility resulting from neuropathic degeneration of enteric inhibitory motor neurons

disorders of sex development (DSD) congenital conditions in which chromosomal, gonadal, or genital sex is not coincident

disseminated intravascular coagulation (DIC) pathologic activation of blood-clotting mechanisms, leading to the formation of small blood clots inside the blood vessels throughout the body, thereby disrupting normal coagulation and causing abnormal bleeding and possible organ malfunction

distal nephron distal convoluted tubule, connecting tubule, and collecting duct, considered as one functional entity

distending pressure pressure required to stretch material (e.g., the pressure required to inflate the lung)

distensibility measure of the force required to stretch material

diuresis increased urine flow rate

diurnal rhythm 24-hour cycle of activity entrained to the day–night cycle

divergence tendency for spread of sympathetic responses from a lesser number of preganglionic neurons to a greater number of postganglionic ones

diving reflex reflex that occurs when the face is submerged under water, which is found in all mammals (including humans, although less pronounced). The reflex puts the body into an oxygen-saving mode to maximize the time that can be spent underwater. The reflex includes bradycardia, peripheral vasoconstriction, and a decreased metabolic rate

diving response cardiovascular reflex to immersion of the body or even the face underwater. It is characterized by intense slowing of heart rate and sympathetic peripheral vasoconstriction. It is a means of preserving oxygen to the brain and heart when submerged underwater

D_1-like receptor dopamine receptor coupled to stimulatory G proteins, which activate adenylyl cyclase

D_2-like receptor dopamine receptor coupled to inhibitory G proteins, which inhibit adenylyl cyclase

DNA vaccine gene that expresses, when incorporated in human cells, immunogenic protein

docosahexaenoic acid omega-3 fatty acid found abundantly in seafood; is present in parts of the brain and the retina and is essential for the normal development of vision in newborns

dolor pain associated with acute inflammation

dominant follicle one follicle from the cohort of developing follicles that will ovulate

dopamine (DA) catecholamine used as a neurotransmitter; important in motor function and limbic pathways; also inhibits the synthesis and secretion of prolactin

dopaminergic neurons relating to nerve cells or fibers that employ dopamine as their neurotransmitter

Doppler ultrasound ultrasound imaging technique that works similar to radar speed detection devices; gives an ultrasound image of the heart but is used primarily for detecting flow velocity abnormalities through the heart valves and in the aorta

dorsal motor nuclei region of the medulla where parasympathetic neurons involved in cardiac, respiratory, and gastrointestinal functions are located

dorsal motor nucleus of vagus location in the medulla oblongata (brainstem) containing cell bodies of vagal efferent (motor) fibers to the digestive tract, excluding parts of the esophagus

dorsal respiratory group (DRG) cluster of neuronal cells located in the dorsal portion of the medulla that is active during inspiration

dorsal vagal complex combined structures of the dorsal vagal motor nucleus, nucleus tractus solitarius, and area postrema in the medulla oblongata (brainstem)

down-regulation process by which a cell decreases the quantity of a cellular component, such as a protein, in response to an external variable

drusen localized deposits of cellular debris whose accumulation distorts the shape of the retina in age-related macular degeneration

Duchenne muscular dystrophy (DMD) rapidly worsening muscle weakness caused by a defective gene for dystrophin (a protein in the muscles); although inherited, it also occurs in those without a family history of the disease

ductus arteriosus fetal blood vessel connection between the pulmonary artery and aorta that shunts blood ejected from the pulmonary artery to the aorta

ductus venosus fetal blood vessel connection between the umbilical veins and the inferior vena cava that shunts oxygenated blood from the veins into the vena cava while bypassing the liver

dumping syndrome syndrome that occurs after a meal due to rapid delivery of gastric chyme into the duodenum, characterized by flushing, sweating, dizziness, weakness, and vasomotor collapse, resulting from rapid passage of large amounts of food into the small intestine, with an osmotic effect removing fluid from plasma and causing hypovolemia

dynamic exercise performance of some activity that involves muscle contraction and joint flexion or extension; in contrast to static exercise, in which joint movement is absent

dynein microtubule-associated protein involved in retrograde axonal transport of organelles and vesicles (from the plus to the minus ends of the microtubules) via the hydrolysis of adenosine triphosphate

dysphagia difficulty swallowing; sensation of abnormal bolus transit through the esophageal body

dyspnea shortness of breath; a perceived difficulty breathing or pain on breathing

dystrophin filamentous muscle protein that lies just inside the sarcolemma and participates in the transfer of force from the contractile system to the outside of the cells via integrins

early afterdepolarizations (EADs) depolarizations of the muscle cells that occur late in phase 2 or early in phase 3 of the ventricular action potential and are more likely to be induced by factors that prolong the action potential duration, such that slow calcium channels have time to recover and, thus, fire an additional low-amplitude action potential

early satiety feeling that the stomach is overfilled soon after starting to eat; the sensation is out of proportion to the size of the meal being eaten, so that the meal cannot be finished

eccentric contraction contraction characterized by involving the lengthening of muscle fibers while they are maintaining contractile tension; related movements frequently involve deceleration and high passive tension

eccrine ordinary, or simple, sweat gland of the merocrine type. These unbranched, coiled, tubular glands are distributed over almost all of the body surface and promote cooling by evaporation of their secretion

echinocyte spiked erythrocyte; also known as *Burr cell*

echocardiography general classification of noninvasive techniques by using ultrasound echoes to image the heart wall, its chambers, and movement of the valves during the cardiac cycle

2D echocardiography (1) two-dimensional echocardiography; used to examine ventricular chambers and movement of valves in the beating heart *in situ*; (2) type of ultrasound cross-sectional image of the heart used to measure ventricular wall thickness, valve motion and abnormalities, and wall motion during the cardiac cycle; it is often used to estimate the ejection fraction of the left ventricle

ectopic foci portion of the myocardial muscle that fires its own action potentials, causing activation of the heart from an area other than the sinoatrial node

ectopic pregnancy implantation of an embryo outside of the uterus

edema clinically apparent increase in interstitial fluid volume

Edinger-Westphal nucleus nucleus region of cranial nerve III where parasympathetic neurons arise

effective arterial blood volume (EABV) degree of fullness of the arterial system, which determines the perfusion of the body's tissues

effector small molecule used to relay signals within a cell. Effectors bind to a protein and alter the activity of that protein

effector cell executing cell of immune responses; in adaptive immunity: plasma cells, T helper cells, and cytotoxic T cells

effector systems musculature, secretory epithelium, and blood–lymphatic vasculature in the digestive tract

efferent carrying away from

efferent nerve (efferents) nerve fibers carrying impulses away from the central nervous system that cause muscles to contract and glands to secrete (inhibitory efferent nerves)

efflux transfer of material from the interior to the exterior of a cell

eicosapentaenoic acid omega-3 fatty acid found in fish, essential for the normal development of vision in newborns

Einthoven triangle theoretical triangle in the frontal plane of the body created from the interconnected electrodes making up the standard bipolar electrocardiogram lead system

ejaculation abrupt discharge of fluid; the expulsion of seminal fluid from the urethra of the penis during orgasm

ejection fraction amount of blood ejected in one systole by the left ventricle expressed as a percentage of the left ventricular residual volume; it is used clinically to detect impaired performance of the left ventricle

elastase one of three endopeptidases present in pancreatic juice that attacks peptide bonds with a neutral aliphatic carboxyl terminal

elastic recoil degree to which stretched material returns to its unstretched position

elastic response proportionate and time-independent force response to the extension of a deformable body such as muscle

elasticity ability of a material when stretched to return to its unstretched position (e.g., a balloon). This property differs from plasticity (the capability of being stretched, but not returning to its unstretched position, e.g., putty)

electrical slow waves omnipresent form of electrical activity (rhythmic depolarization and repolarization) in gastrointestinal muscle cells

electrical synapse site of communication between neurons mediated by movement of ions from one cell to the other via gap junctions

electrical syncytium tissue (e.g., gastrointestinal and cardiac muscle) in which the cells are electrically coupled one to another at cell-to-cell appositions that do not include cytoplasmic continuity; accounts for three-dimensional spread of excitation in excitable tissues

electrochemical potential (gradient) combined effect of electrical and chemical gradients that determines the direction of movement of ions and charged solutes

electroencephalogram (EEG) recording of the electrical activity of the brain

electroencephalography measurement of the electrical activity of the brain

electrogastrography recording of gastric electrical activity from surface electrodes positioned on the abdominal wall

electrogenic system a system in which there is a net movement of charge

electrogenic transport membrane transport system that produces net movement of charge across the membrane, for example, H^+ ATPase, Na^+/glucose cotransport

electrolyte substance that is ionized in solution and thus becomes capable of conducting electricity

electromechanical coupling in gastrointestinal smooth muscle, depolarization of the membrane electrical potential that leads to the opening of voltage-gated calcium channels, followed by the elevation of cytosolic calcium, which, in turn, activates the contractile proteins

electroneutral system a system in which there is no net movement of charge

electroneutral transport membrane transport system that produces no net movement of charge across the membrane, for example, Na^+/H^+ exchange

electroneutrality principle that, in an electrolytic solution, the concentrations of all the ionic species are such that the solution as a whole is neutral

electroretinogram generator potential of the retina, recorded extracellularly and at some distance from its source

electrotonic conduction local spread of electrical activity within a group of cells by ionic current flows that do not involve action potential mechanisms; also known as *passive conduction*

electrotonic potential nonpropagated potential in a nerve or muscle membrane generated by the flow of ions across the membrane

electrotonically way in which electrical activity spreads within a group of cells by ionic current flows that do not involve action potential mechanisms

emboli blood clots that form in the bloodstream

embryoblast cluster of small centrally located cells within the blastocyst that will give rise to the fetus

emesis vomiting

emissivity ratio of energy radiated by a black body at the same temperature. It is a measure of a material's ability to absorb and radiate energy. A true black body would have $\varepsilon = 1$, whereas any real object would have $\varepsilon < 1$

emmetropia normal vision, free from refractive errors

emphysema chronic lung disorder that leads to abnormally high compliance as a result of distention and eventual rupture of the alveoli with progressive loss of pulmonary elasticity; symptoms include shortness of breath with or without cough, which may lead to impaired heart action

end-tidal volume volume of air measured at the end of a tidal volume

end-diastolic volume volume within the ventricle at the end of diastole

endocannabinoids endogenous ligands for the receptors that mediate the effects of the psychoactive component of marijuana, Δ^9-tetrahydrocannabinol

endocrine glands glands that secrete their products (hormones) directly into the bloodstream (ductless glands) or release hormones (paracrines) that affect only target cells nearby the release site

endocrine pancreas portion of the pancreas that secretes hormones such as insulin, glucagon, and somatostatin into the bloodstream

endocrinology branch of physiology concerned with the description and characterization of processes involved in the regulation and integration of cells and organ systems by specialized chemical substances called *hormones*

endocytosis invagination of the plasma membrane to pinch off and internalize substances that are otherwise unable to cross the plasma membrane. Some endocytic mechanisms involve binding of extracellular solute to a specific membrane receptor protein (receptor-mediated endocytosis)

endogenous originating or produced within the body

endogenous creatinine clearance renal clearance of endogenous (i.e., not administered) creatinine; used to estimate the glomerular filtration rate

endolymph fluid that fills the scala media of the cochlea and the semicircular canals of the vestibular apparatus

endometrial cycle life cycle of the endometrial lining of the uterus; consists of four parts

endometrial veins veins that drain the maternal blood–filled sinus of the placenta

endometrium inner membrane lining the lumen of the uterus

endopeptidases one of two classifications of pancreatic proteases; include trypsin, chymotrypsin, and elastase; are present in pancreatic juice; and hydrolyze certain internal peptide bonds of proteins or polypeptides to release the smaller peptides

endorphins endogenous opioid neurotransmitters that can produce a feeling of well-being

endothelial-derived relaxing factor (EDRF) a compound produced and released by the endothelium to promote smooth muscle relaxation. The best-characterized is nitric oxide. Some sources equate EDRF and nitric oxide.

endothelial growth factor (EGF) cytokine involved in wound healing

endothelial vesicles pinocytotic vesicles used to transport substances that are too large to cross the capillary membrane or pores

endothelin extremely potent vasoconstrictor produced by the endothelium, especially in cardiovascular disease states; the most potent vasoconstrictor yet found

endothelium thin epithelial lining of all blood vessels and the inner surface of the chambers of the heart. It contains anticoagulant factors such as heparin and forms a nonthrombotic surface in arteries. It is also a barrier between the bloodstream and the inner vascular tissue. Endothelium is the prime source of nitric oxide and PGI_2, which have vasodilator, antithrombotic, and antiatherogenic properties as well as antimitogenic properties on the underlying vascular smooth muscle. The endothelium also produces growth factors that are thought to be responsible for angiogenesis and intimal repair in blood vessels. The endothelium and its ability to produce nitric oxide are impaired in all known forms of cardiovascular disease, especially those associated with hypertension, atherosclerosis, and diabetes

endothelium-derived relaxing factor (EDRF) original name for *endothelial nitric oxide*

endotoxin toxins of gram-negative bacteria associated with lipopolysaccharide

endplate current postsynaptic ionic current at the myoneural junction whose local flow excites adjacent areas of muscle membrane to produce action potentials

endplate membrane specialized postsynaptic membrane at the myoneural junction, containing the acetylcholine receptors

endplate potential proportionate depolarization of the endplate membrane caused by the permeability to Na^+ and K^+ ions that results from the binding of acetylcholine

end-product inhibition end product of a biosynthetic pathway, which inhibits the enzymes that catalyze the first step in the pathway

end-systolic pressure–volume relationship considered the ideal measure of the inotropic state of the heart, this straight line sets the limit of contraction for the whole heart as it generates pressure and reduces volume during ejection that is formed from all the points of volume and pressure at the end of systole in the whole heart generated from multiple different preloaded and afterloaded contractions of the whole heart. The position, or slope, of this line is independent of the loading conditions on the myocardium but is shifted by positive or negative inotropic influences on the heart

end-systolic volume volume of blood in the ventricle (usually as concerns the left ventricle) and the end of systole; although it is sometimes called the *residual volume*, it actually best represents the volume in the ventricle at the end of ventricular ejection rather than the start of diastole

enkephalins endogenous ligand that binds to opioid receptors to regulate nociception

enteric mini-brain term used in reference to the brainlike functions of the enteric nervous system

enteric nervous system division of the autonomic nervous system situated within the walls of the digestives tract and involved in independent integrative neural control of digestive functions

enteroceptors sensory receptors classified by "vantage point" in the body; detect stimuli from inside the body

enterochromaffin cells enteroendocrine cells located in the mucosal epithelium of the digestive and respiratory tracts that secrete most of the body's serotonin; cells are stimulated by "brushing" of the mucosa to release 5-hydroxytryptamine as a paracrine signal to neurons in the enteric nervous system

enterochromaffin-like (ECL) cells special neuroendocrine cells located mostly in the acid-secreting regions of the stomach that are believed to be the source of histamine

enterocytes specialized epithelial cells in the small intestine that take up the digested products of nutrients for transport into either the portal circulation or the lymphatic system

enteroendocrine cells cells found throughout the digestive tract that produce hormones and paracrine signals to enteric neurons and sensory neurons

enterogastrone hormone released by the small intestine that inhibits gastric motility and secretion

enterohepatic circulation recycling of bile salts between the small intestine and the liver

entero-oxyntin hormone present in intestinal endocrine cells whose release is stimulated by the distention of the intestine; it stimulates acid secretion

enteropeptidase enzyme found on the luminal surface of enterocytes that converts trypsinogen to trypsin when pancreatic juice enters the duodenum; also known as *enterokinase*

enzyme-linked immunosorbent assay (ELISA) biochemical technique used mainly in immunology to detect the presence of an antibody or an antigen in a sample. It uses two antibodies, one of which is specific to the antigen and the other of which is coupled to an enzyme. This second antibody gives the assay its "enzyme-linked" name and will cause a chromogenic or fluorogenic substrate to produce a signal

eosin orange dye used in polychrome stains

eosinophil granulocyte of the myeloid series involved in allergic reactions and defense against parasites

ependymal cells glial cells that help form the blood–CSF barrier

epidermal growth factor (EGF) substance in saliva that stimulates gastric mucosal growth and the growth of epithelial cells in the skin and other organs

epididymis tiny tube where sperm collect after leaving the testis

epigenetics a regulatory mechanism in which genes can be switched off and on without altering their genetic code

epigenome the outer structure of the gene that acts as a switch to turn the cell's DNA off and on

epilepsy neurologic disorder of the brain characterized by repeated spontaneous discharges of electrical activity (seizures)

epinephrine (EPI) catecholamine neurotransmitter released by the adrenal medulla

epinephrine reversal phenomenon whereby the vasoconstrictive actions of epinephrine are blocked resulting in its vasodilatory actions being manifested

epineurium the fibrous coat enveloping PNS axons to form nerves

epiphyseal plate cartilage (hyaline) region (between the epiphysis and diaphysis) in bone that allows lengthwise growth of a bone. In puberty, rapid growth in this region leads to height increase, which then ossifies in the adult

episodic memory memory of events

epithelial sodium channel (ENaC) channel through which Na^+ enters into the collecting duct cell during K^+ regulation

epitope part of the molecule that is recognized by the immune system

epsilon cells see ε *cell*

equal pressure point (EPP) point in the airway at which the pressure inside equals the pressure outside of the airway

equilibrioception physiologic sense of balance

equilibrium stable situation that occurs when the internal processes of a system are in balance and no overall change in the system occurs

equilibrium potential concentration at which there is a balance between the forces due to the concentration gradient for the ion and the electrical gradient across the membrane

equivalent in chemistry, a unit of measure that contains the Avogadro number (or a mole) of positive or negative charges, each the quantity of electricity possessed by a proton or electron

erectile dysfunction (ED) inability to attain and/or maintain an erection suitable for penetration

erythroblast stem cell committed to differentiate into erythrocytes

erythrocyte see *red blood cell*

erythrocyte sedimentation rate (ESR) test to measure the settling of erythrocytes per time

erythromelalgia rare neurovascular peripheral pain disorder in which blood vessels, usually in the lower extremities, are episodically blocked and then become hyperemic and inflamed; there is severe burning pain (in the small fiber sensory nerves) and skin redness

erythropoiesis process by which red blood cells are produced in hematopoietic tissues

erythropoietin hormone produced by the kidney that stimulates the bone marrow to produce red blood cells (such as in hypoxia)

esophageal varices varicose veins occurring at the base of the esophagus that form as a compensatory mechanism to the increased resistance to (and, therefore, dramatically decreased) portal venous flow. Conditions at the base of the esophagus (including minimal connective tissue support and the associated negative intrathoracic pressure of the area) together with the increased demand on these vessels to support an increasing load of blood flow lead to a tendency for these varices to rupture, a condition leading to morbidity in 30% of occurrences

essential amino acids amino acids that cannot be synthesized by the liver and, therefore, supplied through dietary consumption. The essential amino acids include histidine, methionine, threonine, tryptophan, isoleucine, leucine, valine, phenylalanine, and lysine

essential fatty acids fatty acids that humans and other animals must ingest for normal cell function, because they cannot be made from other foods; the three essential fatty acids include linolenic acid, linoleic acid, and arachidonic acid

essential light chains small protein constituents of the myosin molecule that are necessary for its function (but not its regulation)

estradiol steroid with 18 carbons

estriol major estrogen produced by the placenta

estrogen steroid that promotes maturation of the female reproductive organs and secondary sexual characteristics

estrone estrogen secreted by the ovaries, placenta, and fat

eunuchoid sexually deficient person; especially one lacking in sexual differentiation and tending toward the intersexual state

eunuchoidism male hypogonadism; the state of being a eunuch (either because of lacking testicles or because they failed to develop)

euthyroid sick syndrome state of adaptation/dysregulation of feedback control with triiodothyronine and/or thyroxine at low levels but normal thyroid-stimulating hormone levels

excitable cells cells such as neurons and muscle with membranes capable of giving rise to action potentials and propagated impulses

excitation–contraction coupling series of steps that begins with the depolarization of a muscle cell membrane and the mechanical events of contraction

excitatory junction potentials (EJPs) neurally evoked depolarizations of the muscle membrane potential

excitatory neurotransmitters neurotransmitters that result in membrane depolarization

excitatory postsynaptic potentials (EPSPs) depolarization of the neuronal membrane at the postsynaptic side of a synapse as a result of ions flowing through channels that open when neurotransmitter released from the presynaptic terminal binds to its receptor

excretion elimination of a substance from the body in the urine or feces

excretory ducts part of the network of ducts in the salivary glands that are lined with columnar cells and play a role in modifying the ionic composition of saliva

execution orchestration of complex sensorimotor sequences in a seamless path toward a goal

exercise performance of some activity that involves muscle contraction and joint flexion or extension, often to develop or maintain physical fitness and overall health

exertional heat injury syndrome occurring in association with strenuous exercise characterized by hyperthermia with exhaustion or collapse, which is accompanied by evidence of tissue and organ damage

exocrine glands glands that secrete their products (excluding hormones and other chemical messengers) into ducts leading directly into the external environment; contrast with *endocrine glands*

exocrine pancreas portion of the pancreas that secretes digestive enzymes and HCO_3^- into the intestinal lumen

exocytosis fusion of a secretory vesicle with the plasma membrane and expulsion of the vesicle contents into extracellular fluid

exopeptidases substance present in pancreatic secretion that could be a carboxypeptidase or an aminopeptidase, depending on whether it releases amino acid from the carboxyl end or the amino end of the peptides

expiratory reserve volume (ERV) maximum volume of air exhaled at the end of a tidal volume

expired minute ventilation (E) volume of air expired from the lungs in 1 minute

extension movement that increases the angle between two moving parts of the body

extensors skeletal muscles whose action extends a limb or body part

external anal sphincter ring of skeletal muscle surrounding the anal canal that can be voluntarily contracted to postpone defecation

external intercostal muscles muscles located between the ribs that are involved in forced inspiration

exteroceptors sensory receptors classified by "vantage point" in the body; detect stimuli from outside the body

extra-alveolar vessels small pulmonary vessels that are not part of the blood vessels surrounding the alveoli

extracellular fluid (ECF) fluid outside of cells

extraction (E) or extraction ratio amount of a substance removed from the bloodstream from arterioles to venules by the capillaries

extrafusal muscle fibers population of muscle fibers that produce force during contraction

extramedullary hematopoiesis blood cell formation occurring outside bone medulla

extraocular muscles muscles within the orbit but outside of the eyeball, including the four rectus muscles (superior, inferior, medial, and lateral), two oblique muscles (superior and inferior), and the levator of the superior eyelid (levator palpebrae superioris)

extrathoracic blood volume blood volume in the veins of the systemic circulation

extrinsic coagulation pathway pathway in which blood clots in response to tissue injury

exudate tissue fluid that seeps out of blood vessels into inflammatory tissue

facilitated diffusion see *passive transport*

factor III see *tissue thromboplastin*

factor X see *prothrombinase*

factor XIII see *fibrin-stabilizing factor*

familial hypercholesterolemia disorder in which the liver fails to produce the low-density lipoprotein (LDL) receptor, which is essential for LDL clearance from plasma. The condition is characterized by elevated levels of plasma LDLs, which typically leads to a predisposition for early coronary heart disease

farad (F) unit of measure of membrane capacitance

fast axoplasmic transport mechanism for transporting proteins, organelles, and other cellular materials needed for the maintenance of the cell along the length of axons

fast synaptic transmission neurotransmission mediated by activation of ligand-gated receptors

fastigial nucleus deals with antigravity muscle groups and other synergies involved with standing and walking; receives afferent input from the vermis, and most efferent connections travel via the inferior cerebellar peduncle to the vestibular nuclei

fast-twitch fibers skeletal muscle fibers specialized for rapid contractions

fatty acid synthase multienzyme complex required to convert carbohydrates to fatty acids

F cells one of five types of cells found in the islets of Langerhans of the endocrine pancreas; this one produces pancreatic polypeptide

feedback relaxation involves both local reflex connections between receptors in the small intestine and the gastric enteric nervous system (ENS) or hormones that are released from endocrine cells in the small intestinal mucosa and transported by the blood to signal the gastric ENS and stimulate firing in vagal afferent terminals in the stomach

feedforward control kind of system that reacts to changes in the environment, usually to maintain some desired state of the system

female athletic triad condition of disordered eating, amenorrhea, and osteoporosis occurring in female athletes and dancers

ferning crystal pattern formed by dried cervical mucus; indicates high estrogen levels

ferritin intracellular protein that binds and stores iron

ferroportin membrane protein for transport of iron from cell cytoplasm into the interstices

fertilization fusion of the male and female gametes (sperm with an ovum), which leads to the formation of an embryo and the development of a new organism; also known as *conception*

fetal hemoglobin major hemoglobin of the fetus, which exhibits much higher affinity for O_2 than does adult hemoglobin

fibrin fibers that form stable blood clots

fibrin-stabilizing factor (factor XIII) clotting factor that catalyzes the formation of covalent bonds between strands of polymerized fibrin, stabilizing and tightening the blood clot

fibrinolysis process by which a fibrin clot is broken down

fibroblast growth factor-23 (FGF-23) protein hormone produced mainly by bone cells that inhibits tubular reabsorption of phosphate and 1α-hydroxylase activity in the kidneys and secretion of PTH; the result of these actions is a lower plasma phosphate concentration

Fick's law gas law that states that the amount of gas that diffuses across a biologic membrane per minute is directly proportional to the membrane surface area (A_s), the diffusion coefficient of the gas (D), and the partial pressure gradient (ΔP) of the gas and inversely proportional to membrane thickness (T); also known as *Fick law*

Fick's principle principle based on mass conservation that uses oxygen content of the blood and whole-body oxygen consumption to measure cardiac output

fight-or-flight response cardiovascular response to fear or stress that is similar to the baroreceptor reflex except that vasodilation occurs in the skeletal muscle circulation instead of vasoconstriction on initiation of the reflex

filopodia long, thin, transient actin-containing protrusions from growth cones

filtered load quantity of a substance, per unit time, that is filtered by the glomeruli and is therefore presented to the tubules; for a freely filterable substance, this is equal to the product of the plasma concentration and the glomerular filtration rate

filtration fraction fraction of the plasma flowing through the kidneys that is filtered; the ratio of glomerular filtration rate to renal plasma flow

filtration pressure equilibrium condition in which the pressures favoring fluid movement out of a capillary are exactly balanced by pressures favoring fluid movement into the capillary, so that there is no net fluid movement across the capillary wall

filtration slit space between adjacent glomerular podocytes about 40 nm wide and bridged by a diaphragm composed of nephrin

fimbria finger-like projections on the edge of the infundibulum of each oviduct

final common pathway neurons lower motor neurons and their axons that control skeletal muscles

first-degree atrioventricular block slower-than-normal transmission of action potentials from the atria to the ventricles. It is revealed on an electrocardiogram recording as a PR interval >0.2 seconds but with each P wave being followed by a QRS complex

first heart sound series of low-pitched sounds following the start of ventricular contraction created by vibrations in the chordae tendineae and blood in the ventricles

first messengers the messenger that binds to a receptor to initiate a signaling cascade. Examples are hormones, peptides, gases such as nitric oxide or CO, ions such as Ca^{2+}, and adenosine triphosphate. Note that some first messengers, such as Ca^{2+}, can also be second messengers, depending on whether they initiate or amplify the signaling cascade

first polar body small haploid cell produced by unequal first meiotic division of the oocyte

fixed monocyte–macrophage system cells that remain at a strategic fixed location and when encountering pathogenic material, thus being able to differentiate from monocytes to macrophages to ingest and thereby remove the material from the circulation

flap valve term used to describe the effects of contraction of the puborectalis muscle to interfere with the movement of feces or flatus in the direction of the anus; a component of the mechanisms that sustain fecal continence

flexion movement that decreases the angle between two moving parts of the body

flexor withdrawal reflex complex spinal reflex triggered by cutaneous stimulus resulting in withdrawal of the body part from the stimulus

flexors skeletal muscles whose action retracts a limb or body part

flow-limited transport transport of materials across the capillaries that is limited not by diffusion but instead by the rate at which the materials can be delivered by blood flow into the capillary network of an organ

flow-mediated vasodilation dilation of arteries and arterioles following an increase in blood flow. It is caused by increases in shear stress stimulation of nitric oxide release from the arterial endothelium

flow velocity measure of how fast blood moves from one point to the next downstream in the cardiovascular system. It has units of cm/s and is an important determinant of lateral pressure, intimal shear stress, and flow turbulence when blood flows within blood vessels

flow–volume curve curve that reflects airflow during forced expiration and forced inspiration

fluid mosaic model model conceived by S.J. Singer and G. Nicolson in 1972 to describe the structural features of biologic membranes

fluid-phase endocytosis see *exocytosis*

flux net transport of ions or solutes in free solution or via membrane transport

focal adhesions dense bodies involved in cell-to-substrate attachment in smooth muscle

focal length (FL) distance behind a positive (converging) lens at which parallel rays from a distant object are brought to a focus

folia folds of the cerebellar cortex

folic acid water-soluble vitamin essential for the formation of nucleic acids, the maturation of red blood cells, and growth

follicles structures containing the oocyte in the ovary

follicle-stimulating hormone (FSH) in women, the pituitary hormone responsible for stimulating follicular cells in the ovary to grow, triggering egg development and production of the female hormone estrogen; in male, the pituitary hormone that travels through the bloodstream to the testes and helps stimulate them to manufacture sperm

follicular phase part of the menstrual cycle during which follicle development occurs

folliculogenesis follicular development

follistatin single-chain protein hormone, with several isoforms, that binds and deactivates activin

foramen ovale opening between the right and left atria of the fetal heart. It shunts oxygenated blood from the right atrium of the fetus into the left atrium

forced expiratory flow (FEF$_{25-75}$) maximum amount of air (L/s) forced out of the lungs between two given lung volumes (25% and 75%)

forced expiratory volume (FEV$_1$) maximum amount of air (L) forced out of the lungs in 1 second after maximal inhalation

Forced Expiratory Volume in 1 second (FEV$_1$) the volume of air exhaled during the first second of a forced expiratory maneuver started at total lung capacity. FEV$_1$ is the most frequently used test for assessing airway obstruction, bronchoconstriction, or bronchodilation.

forced vital capacity (FVC) maximum volume of air forcibly exhaled after maximal inhalation

force–velocity curve expression of the inverse relationship between muscle force and shortening

fornix fiber bundle in the brain containing axons from hippocampal neurons that project to the hypothalamus and basal forebrain and axons from neurons in those regions that project to the hippocampus

free radical atom and/or molecule with unpaired electrons that is damaging to cell structure and function

fresh frozen plasma (FFP) blood product frozen within 6 hours of collection

frontal lobe region of the brain located most anterior, which controls motor and cognitive functions

fructose 1,6-diphosphatase focal enzyme in gluconeogenesis via its conversion of fructose 1,6-diphosphate (FDP) to fructose 6-phosphate (F-6-P), which permits endogenous glucose production from gluconeogenic amino acids (e.g., alanine and glycine), glycerol, or lactate

F-type ATPase proton pump located in the inner mitochondrial membrane that synthesizes adenosine triphosphate using the energy stored in a gradient of protons that crosses the inner mitochondrial membrane from outside to inside the mitochondria down its electrochemical gradient; also called *ATP synthase*

functio laesa loss of function associated with acute inflammation

functional gastrointestinal motility disorder disorder for which the physical or biochemical cause of the patient's symptoms cannot be determined

functional residual capacity (FRC) volume of air remaining in the lungs at the end of a normal tidal volume

functional syncytium electrical property of the heart muscle created by gap junction connections between cells that allows all heart muscle to be activated following the generation of an action potential in any myocardial cell

fundus of the stomach upper curvature of the organ; one of four stomach sections

fused tetanus series of skeletal muscle twitches occurring in such rapid succession that there is no intervening relaxation

fusimotor system γ motor neurons and the intrafusal muscle fibers within muscle spindles

G cells neuroendocrine cells present in the stomach and located predominantly in the antrum that produce the hormone gastrin, which stimulates acid secretion by the stomach

G protein–coupled receptors (GPCRs) large protein family of transmembrane receptors that sense molecules outside the cell and activate signal transduction pathways and, ultimately, cellular responses, internally; these cell surface receptors are implicated in many diseases and are the target of about 30% of all drug therapies

GABAergic neurons neurons that use γ-aminobutyric acid (GABA) as their neurotransmitter; in most cases, inhibitory neurons

galactopoiesis maintenance of lactation; regulated by prolactin

galactorrhea persistent milklike discharge from the nipple in nonlactating people

gallstone crystalline concretion formed within the gallbladder by accretion of bile components; these calculi are formed in the gallbladder but may pass distally into other parts of the biliary tract such as the cystic duct, common bile duct, pancreatic duct, or the ampulla of Vater

gamete specialized reproductive cell through which sexually reproducing parents pass chromosomes to their offspring; a sperm or an egg

gamma globulin see γ *globulin*

gamma motor neuron see γ *motor neuron*

ganglia grouping of nerve cell bodies situated outside the central nervous system

ganglion cell type of neuron located near the inner surface (the ganglion cell layer) of the retina of the eye that receives visual information from photoreceptors via bipolar and amacrine cells

ganglionated plexus array of ganglia and interganglionic fiber tracts forming parts of the enteric nervous system

gap junction (1) specialized protein channels in the plasma membrane composed of *connexins*, which form a *connexon*; allows the flow of ions (electrical current) from cell to cell; (2) membrane-to-membrane apposition in gastrointestinal smooth muscle, which mediates electrotonic coupling and allows electrical (ionic) current to pass from one muscle fiber to another

gas diffusion process in which gas moves across the alveolar–capillary membrane

gas exchange transfer of oxygen and carbon dioxide between the atmosphere and blood via the alveoli

gas tension partial pressure of an individual gas; used interchangeably with partial pressure of oxygen and carbon dioxide (e.g., oxygen tension = Po_2; carbon dioxide tension = Pco_2)

gastric adaptive relaxation relaxation triggered by distention of the gastric reservoir; a vagovagal reflex triggered by stretch receptors in the gastric wall, transmission over vagal afferents to the dorsal vagal complex, and efferent vagal fibers to inhibitory motor neurons in the gastric enteric nervous system

gastric amylase digests starch that was not digested in the mouth; it is of minor significance in the stomach

gastric feedback relaxation relaxation triggered in the gastric reservoir by the presence of nutrients in the small intestine; it can involve both local reflex connections between receptors in the small intestine and the gastric enteric nervous system or hormones that are released from endocrine cells in the small intestinal mucosa and transported by the blood to signal the gastric enteric nervous system and stimulate firing in vagal afferent terminals in the stomach

gastric glands see *oxyntic glands*

gastric inhibitory peptide (GIP) enterogastrone produced by the small intestinal endocrine cells that inhibits parietal cell acid secretion by the stomach. Recently, it has been named the glucose-dependent insulinotropic peptide

gastric lipase acidic lipase secreted by the gastric chief cells in the fundic mucosa in the stomach with a pH optimum of 3 to 6; acidic lipases do not require bile acid or colipase for optimal enzymatic activity and make up 30% of lipid hydrolysis occurring during digestion in the human adult, with gastric lipase contributing the most of the two acidic lipases

gastric mucosa mucous membrane layer of the stomach wall (one of four layers) containing the glands and the gastric pits; it is ~1 mm thick and its surface is smooth, soft, and velvety; it consists of epithelium, lamina propria, and the muscularis mucosae

gastric phase one of three phases that stimulates acid secretion resulting from the ingestion of food. Acid secretion during the gastric phase is mainly a result of gastric distention and the digested peptides in the gastric lumen. The gastric phase accounts for ~50% of total gastric acid secretion

gastric receptive relaxation relaxation initiated in the gastric reservoir by the act of swallowing; a reflex triggered by stimulation of mechanoreceptors in the pharynx followed by transmission over afferents to the dorsal vagal complex and activation of efferent vagal fibers to inhibitory motor neurons in the gastric enteric nervous system

gastric reservoir reservoir consisting of the fundus and approximately one third of the corpus; the muscles of the gastric reservoir are adapted for maintaining continuous contractile tone (tonic contraction) and do not contract phasically

gastrin hormone that stimulates acid secretion by the stomach

gastrin-releasing peptide (GRP) peptide released by the nerves that stimulates G cells to release gastrin, which stimulates parietal cell acid secretion by the stomach. GRP is atropine-resistant, indicating that it works through a noncholinergic pathway

gastrocolic reflex mass movement of feces in the colon that is generally preceded by a similar movement in the small intestine, which often occurs immediately following the intake of a meal

gastroesophageal reflux retrograde flow of gastric contents into the esophagus

gastroesophageal reflux disease (GERD) chronic symptoms or mucosal damage produced by the abnormal reflux of stomach acid to the esophagus, typically accompanied by pyloris (heartburn)

gastrointestinal motility organized application of forces of muscle contraction that results in physiologically significant movement or nonmovement of intraluminal contents

gastroparesis weakness of phasic propulsive contractions in the antral pump, which results in delayed gastric emptying

general gas law gas equation that combines the Charles and Boyle laws

generalized edema widespread accumulation of salt and water in the interstitial spaces of the body

generalized hypoventilation condition in which the entire lung is underventilated, leading to low alveolar ventilation

generalized hypoxia condition in which the entire lung becomes hypoxic, as opposed to only a local region

generalized seizure widespread seizure that involves both cerebral hemispheres simultaneously

generator potential local electrical activity of a sensory receptor that is proportional to the intensity of a stimulus. The magnitude of the generator potential controls the frequency of action potentials along the sensory nerve

genetic sex gender determined by the chromosomes; also known as *chromosomal sex*

genital ridge embryonic ridge on the dorsal wall of the abdominal cavity; will form the gonads

gestational diabetes form of diabetes that may develop during pregnancy in women who do not otherwise have diabetes

ghrelin signal produced by the epithelia of the stomach and small intestine that stimulates food intake by activating the NPY neurons in the arcuate nucleus of the hypothalamus

gigantism excess growth in height as a result of inappropriately high levels of growth hormone

Gitelman syndrome inherited defect in the distal convoluted tubule of the kidneys, causing the kidneys to pass sodium, magnesium, chloride, and potassium into the urine, rather than allowing them to be reabsorbed into the bloodstream

glandular gastric mucosa stomach's mucosal lining, which contains three main types of glands: cardiac, pyloric, and oxyntic

glaucoma condition that results when drainage of the ocular aqueous humor is impaired, pressure builds up in the anterior chamber, and internal structures are compressed, leading to optical nerve damage that can ultimately cause blindness

glia neuronal support cells

glial cells nonneuronal cells of the nervous system that mediate metabolic functions and myelination

glicentin peptide fragment cleaved from glucagon by prohormone convertase

glitazones antidiabetic sulfonylurea drug that stimulates insulin secretion

γ globulin electrophoretically similar serum proteins, including antibodies

globulins plasma proteins, including antibodies, enzymes, carriers, and transporters

globus pallidus (GP) group of nerve cells in the basal ganglia whose output influences the thalamus and cerebral cortex, divided into an external and internal segment

glomerular filtration process in the kidney glomeruli in which an essentially protein-free filtrate is pushed out of the glomerular capillaries into the urinary space of the Bowman capsule

glomerular filtration barrier structures that separate the blood plasma and the urinary space of the Bowman capsule, namely, capillary endothelium plus glomerular basement membrane plus podocytes

glomerular filtration rate (GFR) rate at which the kidneys filter the plasma

glomerular ultrafiltration coefficient (K_f) product of both the hydraulic conductivity (fluid permeability) and capillary surface area of the glomerular filtration barrier

glomerulonephritis disease characterized by proteinuria and/or hematuria, hypertension, and renal insufficiency that progresses over years

glomerulotubular balance proportionate change in sodium reabsorption by the proximal convoluted tubule and loop of Henle when glomerular filtration rate is changed

glomerulus in the kidney, the tuft of capillaries, surrounded by the Bowman capsule, where the blood is filtered

glucagon hormone produced by the α cells of the pancreas that increases the level of glucose in the blood

glucagon-like peptide-1 (GLP-1) peptide generated from proglucagon that is one of the most potent incretins

glucagon-like peptide-2 (GLP-2) peptide generated from proglucagon that is not an incretin and whose biologic actions are not known

glucocorticoid hormones products of the adrenal cortex

glucocorticoid response elements (GREs) specific regions of DNA that bind glucocorticoid receptor

glucokinase enzyme that facilitates phosphorylation of glucose to glucose 6-phosphate in cells in the liver, pancreas, gut, and brain of humans and most other vertebrates

gluconeogenesis occurring mainly in the liver and kidneys, the energy-dependent production of glucose from noncarbohydrate sources, such as fat, amino acids, and lactate; energy for this process is derived from B oxidation of fatty acids

glucose simple sugar (monosaccharide) and an important carbohydrate used by cells as a source of energy and a metabolic intermediate; one of the main products of photosynthesis that starts cellular respiration

glucose buffering process by which the liver maintains normal blood glucose levels; its ability to store glycogen allows the liver to remove excess glucose from the blood and then return it when blood glucose levels begin to fall

glucose threshold plasma level at which glucose first appears in urine

glucose-6-phosphatase enzyme present exclusively in the liver that catalyzes the breakdown of glycogen to glucose molecules for release into the circulation

glucose 6-phosphate intermediate product in the conversion of glucose to glycogen and in the breakdown of glycogen to glucose

glucose-dependent insulinotropic peptide (GIP) gut hormone secreted by the duodenum that stimulates insulin secretion

glucose–galactose malabsorption (GGM) condition in which the cells lining the intestine cannot transport glucose and galactose, preventing proper absorption of these nutrients

α-glucosidase inhibitors oral medications used to treat type 2 diabetes that decrease the absorption of carbohydrates in the small intestine

glucosuria presence of glucose in the urine

GLUT2 facilitative glucose transporter expressed in the pancreas, liver, intestine, and kidney

GLUT4 insulin-responsive facilitative glucose transporter expressed in muscles and adipose tissues

GLUT5 a fructose transporter expressed on the apical border of enterocytes in the small intestine

glutamate (GLU) amino acid used as an excitatory neurotransmitter

glutamatergic neurons neurons that use glutamate as their neurotransmitter

glutamine precursor for glutamate

glycine (GLY) amino acid used as an inhibitory neurotransmitter

glycinergic neurons neurons that use glycine as their neurotransmitter

glycogen polysaccharide that is the main form of carbohydrate storage in animals and occurs mainly in liver and muscle tissue; readily converted to glucose

glycogen phosphorylase first-acting enzyme in the breakdown of glycogen to glucose, responsible for the production of glucose 1-phosphate from the glycogen substrate, which is subsequently converted to glucose. The enzyme acts on the α-1,4-glycosidic linkage yielding glucose polymers

glycogen synthase enzyme that catalyzes the transfer of glucose from uridine diphosphate–glucose to glycogen

glycogenesis occurring mainly in the liver, the process by which glycogen is synthesized from circulating glucose, lactate, and pyruvate after ingestion of a meal

glycogenolysis process by which glycogen is broken down to glucose 6-phosphate and then to glucose for subsequent release into circulation. The reaction is specific to the liver

glycolipid lipid molecule that contains one or more monosaccharide units

glycolysis adenosine triphosphate–generating metabolic process of most cells in which carbohydrates are converted to pyruvic acid

glycolytic pathway (or glycolysis) series of cytoplasmic chemical reactions that break down glucose molecules and produce adenosine triphosphate for energy-requiring cellular processes

glycoprotein protein that contains one or more monosaccharide units

goblet cells cells that serve to secrete various mucins (mucoproteins) found in intestinal secretions

goiter enlarged thyroid gland

Goldman equation value of the membrane potential when all the permeable ions are accounted for. In its usual form, it shows the relationship between intracellular and extracellular concentrations of Na^+, K^+, and Cl^-, the plasma membrane permeability to these ions, and the membrane potential

Golgi apparatus eukaryotic organelle that processes and packages proteins, lipids, and other molecules after their synthesis

Golgi tendon organs (GTOs) sensory receptors located in tendons that sense the force of muscle contraction

gonadal sex sex of the structures that produce the gametes

gonadarche gonadal maturation during pubertal development in boys and girls

gonadotrophs cells located in the anterior pituitary that secrete luteinizing hormone and follicle-stimulating hormone

gonadotropin-releasing hormone (GnRH) hormone made by the hypothalamus (part of the brain); causes the pituitary gland to make luteinizing hormone and follicle-stimulating hormone, which are involved in reproduction; also known as *luteinizing hormone–releasing hormone (LHRH)*

gonadotropins products of the gonadotrophs in the anterior pituitary; luteinizing hormone and follicle-stimulating hormone

gout disorder in which plasma uric acid is elevated and urate crystals precipitate in the joints, causing pain and inflammation and swelling of the affected joints (often in the big toe)

graafian follicle mature preovulatory follicle

graded response sensory response in approximate proportion to the intensity of the stimulus

graft tissue or organ transplanted from a donor to a recipient

graft versus host disease (GVHD) disease in which T cells of the graft attack tissue of the recipient as foreign

grand mal seizure see *tonic–clonic seizure*

granular cells modified pericytes of glomerular arterioles; also known as *juxtaglomerular cells*

granular layer nuclear layer in the cerebellar cortex rich in interneurons

granulocyte mature cell of the myeloid series, including neutrophils, eosinophils, and basophils; also known as *polymorphonuclear leukocyte*

granulosa lutein cells cells that are found in the corpus luteum and formed from the granulosa cells

granzyme enzyme family present in the granules of cytotoxic T cells and natural killer cells

Graves disease autoimmune disorder caused by antibodies directed against the thyroid-stimulating hormone receptor

gray matter portions of the CNS that histologically appear gray due to cell bodies and nonmyelinated processes

gray ramus nerve branch that carries postganglionic sympathetic axons back to the spinal nerve

ground substance intercellular material in which the cells and fibers of connective tissue are embedded

growth cone leading edge of a growing axon that samples the environment and makes decisions about the direction of growth

growth factors naturally occurring proteins that are capable of stimulating cellular proliferation, differentiation, or other cellular responses

growth hormone (GH) hormone product of somatotrophs located in the anterior pituitary

growth hormone–releasing hormone (GHRH) polypeptide from hypothalamus; stimulates secretion of growth hormone

guanosine diphosphate (GDP) product of guanosine triphosphate phosphorylation

guanosine triphosphate (GTP) nucleotide that provides source of energy for protein synthesis; essential for G protein signal transduction by converting to guanosine diphosphate via GTPases

guanylin polypeptide hormone produced by the small intestine that increases renal salt excretion

guanylyl cyclase (GC) enzyme that catalyzes the conversion of guanosine triphosphate to 3′,5′-cyclic guanosine monophosphate and pyrophosphate

gustducin G protein associated with tasting bitterness in the gustatory system

gynecomastia excessive development of the male breasts

gyrus convolution on the surface of the brain

H^+-ATPases (proton pump) located in the membranes of lysosomes and the Golgi apparatus and pumps protons from the cytosol into the organelles to keep the inside of the organelle more acidic than the cytoplasm; classified as *V-type ATPases* because they were first discovered in intracellular vacuolar structures; now known to also exist in the plasma membranes

H^+/K^+-ATPase mechanism present in the apical (luminal) cell membrane of the parietal cell that is involved in the secretion of hydrochloric acid by the stomach

hair cells mechanosensitive cells within the organ of Corti and the vestibular apparatus

Haldane effect effect of oxygen on the carbon dioxide equilibrium curve. Oxygen causes a downward shift of the carbon dioxide equilibrium curve

hallucination perception in the absence of a stimulus

hamartoma nonneoplastic congenital malformation composed of a heterotropic mass of nerve tissue usually located on the floor of the third ventricle or attached to the tuber cinereum

hapten nonimmunogenic antibody-binding molecule, unless linked to a carrier molecule

haptoglobin large glycoprotein (molecular weight of 100,000 Da) that binds free hemoglobin in blood, forming a hemoglobin–haptoglobin complex, which is rapidly removed from circulation by the liver, thereby conserving iron in the body

Hartnup disease extremely rare genetic disorder of amino acid carriers. The membrane carrier for neutral amino acids (e.g., tryptophan) is defective

Hashimoto disease autoimmune disease resulting in impaired thyroid hormone synthesis; also known as *Hashimoto thyroiditis*

haustration process of formation of a haustrum (haustral sac) in the colon

heart four-chambered, hollow organ composed primarily of striated muscle that is responsible for pumping blood through the lungs and the systemic circulation. It is composed of two upper chambers, a right and left atrium, and two larger, lower chambers called the right and left ventricles. The right atrium receives blood from the peripheral veins and delivers it through the tricuspid valve to the right ventricle. The right ventricle pumps blood through the pulmonic valve into the pulmonary circulation. Blood returning from the pulmonary circulation arrives in the left atrium, where it is pumped through the mitral valve into the left ventricle. This chamber pumps blood through the aortic valve into the aorta and thus into the systemic circulation. In this arrangement, the right and left sides of the heart are actually two pumps arranged in series

heart murmur sound heard over the chest with a stethoscope that results from turbulent flow in the heart, aorta, or pulmonary circulation, usually resulting from stenotic or incompetent heart valves

heart rate number of times the normal heart contracts in 1 minute; in a healthy individual, this rate can range between 50 and 180 contractions per minute

heartburn episodic retrosternal burning

heat exhaustion most common heat-related illness, involving mild to moderate dysfunction of temperature control associated with elevated ambient temperatures and/or strenuous exercise resulting in dehydration and salt depletion. It may rapidly progress to heatstroke when the body's thermoregulatory mechanisms become overwhelmed and fail

heat storage difference between heat production and net heat loss

heat syncope body temperature above 40°C (104°F) with fainting or weakness but without mental confusion; results from circulatory failure as a result of peripheral venous pooling, with a consequent decrease in venous return. Heat syncope is caused by mild overheating with inadequate water or salt

heatstroke extreme hyperthermia, typically above 100.4°F (40°C), associated with a systemic inflammatory response, which leads to end-organ damage with universal involvement of the central nervous system. Heatstroke traditionally is divided into exertional and classic varieties, which are defined by the underlying etiology but are clinically indistinguishable

heel stick procedure in which a newborn baby's heel is pricked and then a small amount of the blood is collected, usually with a narrow-gauge ("capillary") glass tube or a filter paper

hem(at)/o referring to blood

hematocrit (Hct or Ht) percentage of whole blood that is composed of erythrocytes

hematologist blood specialist

hematopoiesis generation of blood cells

hematopoietins subclass of cytokines regulating hematopoiesis

heme oxygenase microsomal enzyme responsible for the release of iron from heme. Iron then enters the cellular free iron pool and can be stored as ferritin or released into the bloodstream

hemochromatosis condition caused by iron overload that is characterized by excessive amounts of hemosiderin in the hepatocytes, rendering them defective and unable to perform many normal, essential functions. The excess iron is stored in the organs, especially the liver, heart, and pancreas, causing damage to these organs and leading to life-threatening illnesses such as cancer, heart complications, and liver disease

hemocytometer chamber for manual blood cell counts

hemodialysis procedure usually done three times a week (4 to 6 hours per session) in a medical facility or at home in which kidney-failure patient's blood is pumped through an artificial kidney machine; the blood is separated from a balanced salt solution by a cellophane-like membrane, and small molecules can diffuse across this membrane; excess fluid can be removed by applying pressure to the blood and filtering it

hemodynamics study of the physics of the containment and movement of blood in the cardiovascular system

hemoglobin (Hb) oxygen-transporting protein of erythrocytes

hemoglobinopathies diseases caused by mutated hemoglobin

hemolysis bursting (*lysis*) of red blood cells leading to the release of hemoglobin

hemolytic anemia anemia caused by destruction of erythrocytes that exceeds their generation

hemopexin protein synthesized by the liver that is involved in the transport of free heme in the blood. It forms a complex with free heme, and the liver rapidly removes the complex

hemophilia hereditary bleeding disorders

hemorrhage severe bleeding

hemosiderin iron-storing protein related to ferritin

hemosiderin granules iron stored as ferritin that is sometimes converted to these bodies

hemostasis cessation of bleeding occurring in four steps: (1) compression and vasoconstriction, (2) formation of a temporary loose platelet plug (also called *primary hemostasis*), (3) formation of stable fibrin clot (also called *secondary hemostasis*), and (4) clot retraction and dissolution

Henderson-Hasselbalch equation logarithmic form of the acid dissociation equation

Henry law gas law that states that the amount of gas dissolved in a liquid at a given temperature is directly proportional to the solubility and the partial pressure of the dissolved gas

heparin injectable anticoagulant

hepatic arterial buffer response phenomenon in which reduction in portal blood supply to the liver is compensated by an increase in supply from the hepatic artery and vice versa

hepatocytes highly specialized cells arranged along the liver sinusoids that facilitate the rapid exchange of molecules, as they possess numerous finger-like projections that extend into the perisinusoidal space, thereby increasing the surface area over which they are in contact with the perisinusoidal fluid

hepcidin hormone produced by the liver that regulates iron homeostasis by inhibiting ferroportin

hereditary hemochromatosis (HH) inherited disorder that increases the amount of iron that the body absorbs from the gut, causing excess iron to be deposited in multiple organs of the body; excess iron in the liver causes cirrhosis (which may develop into liver cancer); iron deposits in the pancreas can result in diabetes; and excess iron stores can also cause cardiomyopathy, pigmentation of the skin, and arthritis

hereditary spherocytosis genetically transmitted (autosomal dominant) form of spherocytosis characterized by the production of red blood cells that are sphere-shaped and, therefore, more prone to hemolysis

Hering-Breuer reflex reflex that involves the pulmonary stretch receptors and that prevents overinflation of the lungs; also known as *lung inflation reflex*

heterogametic having one X and one Y chromosome; a male

heterologous desensitization process by which effector cells are rendered less reactive or nonreactive to a given stimulus by a different stimulus

heteroreceptor presynaptic receptor regulating the release of a different neurotransmitter

hexokinase enzyme that facilitates phosphorylation of glucose to glucose 6-phosphate in cells in yeast and muscle

hidromeiosis condition caused by excessive wetting of the skin and subsequent interference with the free flow of eccrine sweat during exercise in humid environments; may be compounded by fatigue of sweat glands as a result of a lack of neurotransmitters

high-altitude cerebral edema swelling in the brain due to cerebral edema and elevated cerebral blood volume. High-altitude cerebral edema is the result of rapid accent to high altitude (5,280 ft above sea level or higher) and is caused by hypoxemia. Initial symptoms include headaches, insomnia, nausea, dizziness, and loss of coordination.

high-altitude pulmonary edema a noncardiogenic form of pulmonary edema resulting from alveolar capillary membrane leakage. The edema is caused by hypoxic-induced pulmonary arterial vasoconstriction and occurs from short-term exposures to altitudes in excess of 2,000 m above sea level (6,560 ft).

high-density lipoprotein (HDL) particles circulating in the blood that contain cholesterol esters; removes cholesterol from cells and transports it back to the liver for excretion or reutilization; also known as *good cholesterol*

hilum depression or pit at that part of an organ where the blood vessels and nerves enter

hippocampus three-layered cortical structure in the temporal lobe involved in learning and memory

Hirschsprung disease congenital dilation and hypertrophy of the colon resulting from the absence of the enteric nervous system (aganglionosis) of the rectum and a varying but continuous length of gut above the rectum

hirsutism excessive increase of hair growth, sometimes leading to male pattern hair growth in a female

histamine inflammatory mediator, released by mast cells and basophils

HMG-CoA reductase see *3-hydroxy-3-methylglutaryl CoA reductase*

homeostasis property of a living organism to regulate its internal environment to maintain a stable, constant condition, by means of multiple dynamic equilibrium adjustments, controlled by interrelated regulation mechanisms

homeotherms warm-blooded animals who keep their core body temperature at a nearly constant level regardless of the temperature of the surrounding environment

homogametic having two X chromosomes; a female

homologous desensitization loss of sensitivity only to the class of agonist used to desensitize the tissue

hopping reaction limb reflex that promotes automatic forward movement of a weight-bearing limb

horizontal cells laterally interconnecting neurons in the outer plexiform layer of the retina of mammalian eyes that help integrate and regulate the input from multiple photoreceptor cells; responsible for allowing eyes to adjust to see well under both bright and dim light conditions as well as other functions

hormone chemical or protein messenger made by one cell that serves as a signal to a target cell

hormone receptor molecular entity (usually a protein or glycoprotein) either outside or within a cell that recognizes and binds a particular hormone

hormone response element (HRE) short DNA sequence that is usually within the promoter or repressor regions of genes and binds a specific hormone/receptor complex to regulate transcription of the gene

hormone-sensitive lipase enzyme that plays an important role in controlling the balance of energy by breaking down fats

Horner syndrome rare condition resulting from damage to the sympathetic nervous system that affects the nerves to the eye and face and causes decreased sweating on the affected side of the face, ptosis, sinking of the eyeball into the face, and a mall (constricted) pupil

host versus graft response condition in which the recipient's immune system attacks transplanted leukocytes

H₂ receptors special histamine receptors found in the parietal cells whose stimulation results in increased acid secretion by the stomach

human chorionic gonadotropin (hCG) protein secreted by the blastocyst and placenta; similar in biologic action to luteinizing hormone

human leukocyte antigen (HLA) human mean cell (corpuscular) hemoglobin protein

human placental lactogen (hPL) substance synthesized by the placenta; its structure and function resemble those of prolactin and growth hormone

humoral immunity part of adaptive immunity mediated by soluble factors, especially antibodies

Huntington's disease disease of the basal ganglia producing uncontrollable spontaneous movements

hydration combination with water

hydration reaction chemical reaction, usually happening in a strong acidic solution, in which a hydroxyl group and a hydrogen cation (an acidic proton) are added to two carbon atoms

hydraulic conductivity for water measure of hydraulic resistance across a capillary network; high conductivity means that water passes easily across the network. It is a measure of how much water can be moved in bulk across the capillaries per minute for each 1 mm Hg of driving pressure

hydrochloric acid (HCl) important constituent of gastric acid that aids in the digestion of food; secreted by the parietal cells and responsible for the change in the electrolyte composition of gastric juice

hydrogen peroxide (H₂O₂) reactive oxygen species that is not a free radical and is damaging to cell structure and function

hydrophilic highly soluble in water, for example, ions and polar molecules

hydrophobic having low solubility in water, for example, nonpolar molecules

3-hydroxy-3-methylglutaryl CoA reductase (HMG-CoA reductase) substance that catalyzes the rate-limiting step in *de novo* cholesterol synthesis

hydroxyapatite crystals form of calcium phosphate in the mineral portion of bone

β-hydroxybutyric acid β derivative of hydroxybutyric acid that is found in the blood and urine in some cases of impaired metabolism

hydroxyl radical (·OH) reactive oxygen species that is damaging to cell structure and function

11β-hydroxylase (CYP11B1) substance that hydroxylates 11-deoxycortisol to form cortisol

17α-hydroxylase substance that hydroxylates pregnenolone to form 17α-hydroxypregnenolone

21-hydroxylase (CYP21A2) substance that converts 17α-hydroxyprogesterone to 11-deoxycortisol

17α-hydroxypregnenolone intermediate in steroid synthesis

17α-hydroxyprogesterone intermediate in steroid synthesis

3β-hydroxysteroid dehydrogenase (3β-HSD II) substance that acts on 17α-hydroxypregnenolone to produce 17α-hydroxyprogesterone

5-hydroxytryptamine (5-HT) monoamine neurotransmitter; also known as *serotonin*

hyperaldosteronism medical condition characterized by too much aldosterone being produced by the adrenal glands, which can lead to lowered levels of potassium in the blood

hypercalciuria excess calcium in the urine

hypercapnia high partial pressure of blood carbon dioxide (>45 mm Hg) resulting from underventilation of the lung

hyperglycemia presence of an abnormally high concentration of glucose in the blood

hypergonadotropic hypogonadism defective development of ovaries or testes; associated with excess pituitary gonadotropin secretion; results in delayed sexual development and growth delay

hyperkalemia condition in which the plasma K^+ level is > 5.0 mEq/L

hyperkinesis excessive movement

hyperopia (farsightedness) condition caused by the eyeball being physically too short to focus on distant objects

hyperosmotic see *osmotic pressure*

hyperphagia abnormally increased appetite and consumption of food

hyperphosphatemia condition in which plasma phosphate concentration is elevated

hyperpnea increased minute ventilation via increased depth and frequency of breathing (unlike tachypnea, which is rapid, shallow breaths) to meet increased metabolic demand (such as during or following exercise); ventilation increases proportionally to carbon dioxide production, but Pa_{CO_2} remains the same, which is the hallmark feature of hyperpnea (unlike in hyperventilation, in which Pa_{CO_2} is significantly lowered, leading to alkalosis)

hyperpolarization change in electrical potential in the direction of a more negative potential inside cells; makes excitable cells less excitable by moving the membrane potential away from the threshold for discharge of an action potential

hyperpolarize when the membrane potential becomes more negative than the resting membrane potential

hyperprolactinemia elevated prolactin secretion from the pituitary causing amenorrhea

hypersensitivity damaging immune responses to allergens, resulting in inflammation and organ dysfunction

hyperthermia acute condition that occurs when the body produces or absorbs more heat than it can dissipate

hyperthyroidism or thyrotoxicosis excessive thyroid hormone secretion

hypertonic see *tonicity*

hyperventilation state of breathing faster and/or deeper than necessary that causes excess carbon dioxide to be blown off; it can result from a psychological state such as a panic attack, can result from a physiologic condition such as metabolic acidosis, or can be brought about by voluntarily forced ventilation and is characterized by increased minute ventilation with a concomitant decrease in Pa_{CO_2}; also known as *overbreathing*

hypervitaminosis A condition resulting from consumption of massive quantities of vitamin A that is often associated with hepatotoxicity and that eventually leads to portal hypertension and cirrhosis

hypoaldosteronism condition characterized by decreased levels of aldosterone, which may result in hyperkalemia and urinary sodium wasting, leading to volume depletion and hypotension

hypocalcemia deficiency of calcium in the blood

hypocalcemic tetany condition characterized by low blood calcium levels, resulting in convulsions

hypocapnia condition characterized by low partial pressure of blood carbon dioxide (<35 mm Hg) as a result of overventilation of the lung

hypochromic condition in which there is less hemoglobin per cell than normal

hypocretin/orexin neuropeptide found in neurons in the posterolateral hypothalamus that are sensitive to nutritional status, simulate appetite, and project to arousal areas of the brain; a functional deficit in hypocretin causes narcolepsy

hypoglycemia condition characterized by low blood sugar level

hypogonadism condition in which sex organs such as the testes or ovaries are underactive

hypogonadotropic hypogonadism absent or decreased function of the gonads, the male testes or the female ovaries, resulting from the absence of the gonadal stimulating pituitary hormones, follicle-stimulating hormone, and luteinizing hormone

hypokalemia condition in which the plasma K^+ level is <3.5 mEq/L

hypokinesis condition characterized by less than a normal amount of movement

hyponatremia condition in which the plasma Na^+ level is <135 mEq/L

hypophyseal portal circulation special blood supply to the anterior lobe of the pituitary gland

hypophysiotropic hormones releasing hormones synthesized by neural cell bodies in the hypothalamus and that regulate secretion of posterior pituitary hormones

hypopituitarism deficiency of one or more hormones of the pituitary gland

hypoprothrombinemia a deficiency in prothrombin (factor II) resulting in extended blood clotting time

hyposmotic see *osmotic pressure*

hypotension condition characterized by lower-than-normal arterial pressure

hypothalamic amenorrhea a broad range of conditions, each of which results in insufficient gonadotropin release

hypothalamic amenorrhea, functional amenorrhea resulting from emotional distress, excessive exercise or rapid weight loss, but where the neuroendocrine pathways are intact

hypothalamic diabetes insipidus condition in which arginine vasopressin secretion is impaired

hypothalamic hormones hormones released by the hypothalamus to reach the anterior pituitary lobe by a portal system of capillaries; designated as releasing factors because they regulate the secretion of most hormones of the endocrine system

hypothalamic–pituitary axis functional connection between the brain and the pituitary, in which the hypothalamus plays a central role: the brain links the pituitary gland to events occurring within or outside the body, which call for changes in pituitary hormone secretion

hypothalamic–pituitary–adrenal axis interactive relationship regulating glucocorticoid levels in the blood

hypothalamic–pituitary–gonad axis interactive relationship regulating gonadotropin levels in the blood

hypothalamic–pituitary–growth hormone axis interactive relationship regulating growth hormone levels in the blood

hypothalamic–pituitary–thyroid axis interactive relationship regulating thyroid hormone levels in the blood

hypothalamohypophyseal portal blood vessels system of blood vessels in the pituitary stalk into which neurons within specific nuclei of the hypothalamus secrete releasing factors that are then transported to the anterior pituitary, where they stimulate secretion of hormones

hypothalamohypophyseal tract axons within the pituitary stalk derived from magnocellular neurons whose cell bodies are located in the supraoptic and paraventricular hypothalamic nuclei; represents the efferent limbs of neuroendocrine reflexes that lead to the secretion of the hormones vasopressin and oxytocin into the blood within the posterior pituitary

hypothalamus region of the brain that integrates autonomic responses to temperature and hunger

hypothermia drop of central temperature below 35°C

hypothyroidism condition characterized by inadequate production of the thyroid hormone thyroxine and by low circulating levels of thyroxine and other thyroid hormones

hypotonia reduction in the amount of normal resistance to a muscle undergoing passive stretch

hypotonic see *tonicity*

hypoventilation decreased alveolar ventilation

hypovolemia condition characterized by an abnormally low circulating blood volume

hypoxemia condition characterized by an abnormally low oxygen content or low P_{O_2} in the arterial blood

hypoxia condition in which the inspired oxygen is below normal leading to abnormal low oxygen tension in the tissues

hypoxia-induced pulmonary vasoconstriction phenomenon in which small pulmonary vessels constrict with low oxygen, which is just the opposite of that which occurs in the systemic circulation

H zone pale region of A band that contributes to striated appearance in muscle

I band band of actin myofilament in muscle that shows low density under a microscope and contributes to muscles' striated appearance; *isotropic*

I cell type of endocrine cell in the small intestine that releases cholecystokinin

Ig immunoglobulin; a membrane-bound or soluble antigen-binding protein

ileocolonic sphincter prevents reflux of colonic contents into the ileum; incompetence can allow entry of bacteria into the ileum from the colon that may result in bacterial overgrowth

immediate early genes genes that are activated transiently and rapidly in response to a cellular stimuli; the first round of response to stimuli, before any new proteins are synthesized

immune deficiency disease resulting from absent or malfunctioning immune components

immune surveillance theory that the immune system recognizes and destroys tumor cells that are constantly arising during the life of the individual

immune thrombocytopenia purpura (ITP) common autoimmune disorder caused by autoantibodies to platelets formed in the spleen and by the platelet–antibody aggregates being destroyed by the spleen

immune tolerance process by which the immune system does not attack antigens and includes self-tolerance (in which the body does not mount an immune response to self-antigens) and induced tolerance (in which tolerance to external antigens can be created by manipulating the immune system); the three forms are central, peripheral, and acquired tolerance

immunocontraception use of antibodies against the egg, sperm, or zygote or against hormones such as human chorionic gonadotropin or follicle-stimulating hormone to prevent pregnancy (this method of contraception is still being developed for use in humans)

immunogen substance that provokes the immune system

immunologic synapse complex junction between a T cell and an antigen-presenting cell

immunosenescence gradual deterioration of the immune system because of advancing age

implantation attachment of the blastocyst to the surface endometrial cells of the uterine wall, beginning on days 7 to 8 after fertilization

impulse initiation region region of the receptor membrane that is the location of the formation of action potential in the neuron and constitutes an important link in the sensory process

inactivation voltage-gated sodium channel entering the inactive state

inactivation gate molecular gate in voltage-gated sodium channels that closes after the channel opens and remains closed for a finite period of time and until the membrane is repolarized

inactive state the state of the voltage-gated sodium channel in which it is closed and incapable of being opened

incisura small indentation that appears in the aortic pressure waveform when the aortic valve closes

incomplete proteins proteins that do not provide all of the essential amino acids in amounts sufficient to sustain normal growth and body maintenance, such as those found in most vegetables and grains

incretin effect principle stating that a glycemic stimulus derived from enteral nutrients exerts a greater insulinotropic response than a comparable isoglycemic challenge achieved through parenteral glucose administration

indicator–dilution method technique used to determine cardiac output based on the disappearance of concentration of an indicator over time

indirect calorimetry measure of an organism's metabolic rate by calculating oxygen uptake, assuming that burning 1 cal (kilocalorie) requires 208.06 mL of oxygen

indirect pathway basal ganglia pathway in which axons from the caudate synapse in the globus pallidus externus before reaching the globus pallidus externus

indirect sympathomimetic chemical that produces activation of the sympathetic nervous system by increasing the release of norepinephrine

inferior cervical ganglia lowermost of the sympathetic ganglia located in the neck; also known as *stellate ganglion*

inferior hypophyseal arteries arteries that provide arterial blood to the posterior lobe

inferior mesenteric ganglion most caudal of the three prevertebral sympathetic ganglia of the abdomen

inferior salivatory nucleus region of the medulla where neurons involved in the control of submandibular, sublingual, and parotid glands are located

inferior vena cava (IVC) single major vein that collects blood from the lower extremities and abdominal organs and that empties into the right atrium

inflammation pathologic process including cytologic changes, cellular infiltration, and mediator release; also known as *inflammatory response*

influx transfer of material from the exterior to the interior of a cell

infundibular process major portion of the neurohypophysis; also known as *posterior lobe*

infundibular stem inner part of the pituitary stalk

infundibulum end of the oviduct that receives the ovulated egg from the ovary; has fimbria

inhibin substance secreted by the testes and ovaries that regulates follicle-stimulating hormone levels

inhibitors of apoptosis (IAPs) family of functionally and structurally related proteins that serve as endogenous inhibitors of programmed cell death; there are eight human IAPs

inhibitory junction potential (IJP) membrane hyperpolarization leading to decreased excitability in gastrointestinal muscles; evoked by the release and action of inhibitory neurotransmitters from enteric motor neurons

inhibitory motor neuron neuron that releases neurotransmitter, which suppresses contractile or secretory behavior in the gastrointestinal tract

inhibitory neurotransmitters neurotransmitters that result in membrane hyperpolarization

inhibitory postsynaptic potentials hyperpolarization of neuronal membrane at the postsynaptic side of a synapse resulting from ions flowing through channels that open when neurotransmitter released from the presynaptic terminal binds to its receptor

initial velocity earliest steady velocity of a contracting muscle

innate immune system nonspecific defense mechanisms; also called *nonspecific* or *natural immunity*

inner ear portion of the ear containing the cochlea, where the actual sound transduction takes place

inner-ring deiodination enzymatic removal of an iodine atom at the 5 position on the thyronine ring structure

innervated supplied with intact nerves

iNOS inducible nitric oxide synthase; an enzyme that converts L-arginine to L-citrulline and nitric oxide that is not constitutively expressed in tissues. Its appearance in tissues such as vascular smooth muscle and leukocytes occurs after exposure of the tissues to an inducing factor such as endotoxins or lipopolysaccharides

inositol trisphosphate (IP$_3$) intracellular second messenger generated by hydrolysis of phosphatidylinositol 4,5-bisphosphate

inotropic state contractile state of cardiac muscle independent of effects of loading condition; a function of factors that modify the contractile force of cardiac muscle at the level of the cell

insensible perspiration those evaporative losses of water from the moist surfaces of the body (such as the skin and respiratory tree) not resulting from the secretory activity of glands

insensible water loss water loss from the skin and lungs that is not ordinary

insular area region of the temporal cerebral cortex involved in coordination of olfactory and gustatory stimuli for autonomic functions

insulin hormone produced by the β cells of the pancreas that decreases the level of glucose in the blood

insulin resistance inability of a known quantity of exogenous or endogenous insulin to increase glucose uptake and use in an individual as much as it does in a normal population

insulin/glucagon ratio (I/G ratio) net physiologic response determined by the relative levels of insulin and glucagon in the blood plasma

insulin-like growth factor-1 (IGF-1) potent trophic hormone whose secretion is stimulated by growth hormone; also known as *somatomedin C*

insulin-like growth factor-2 (IGF-2) trophic hormone

insulin-like growth factors (IGFs) class of small peptides responsible for mediation of some growth hormone actions that increase the rate of cartilage synthesis by promoting the uptake of sulfate and the synthesis of collagen; also known as *somatomedins*

insulitis penetration by lymphocytes into pancreatic islets of Langerhans that produces an inflammatory or autoimmune response and results in destruction of the β cells of the pancreas

integral (intrinsic) membrane protein protein that is embedded in or completely spans the phospholipid bilayer region of a cell membrane, for example, membrane transport proteins and ion channels

integral proteins a type of membrane protein that is permanently attached to the biological membrane

integration process of assembling parts together to make a whole; a neural network processing resulting in output being a function of input other than one; the organization of behavior of individual digestive effector systems into harmonious function of the whole organ

integrins protein important for platelet adhesion in hemostasis

intention shaping of behavior in accordance with internal motivation and external context

intercalated cells cells in the connecting tubules and collecting ducts of the kidneys that are involved in acid–base transport

intercalated duct one of three ducts that comprise the salivon of the human submandibular gland. The intercalated ducts are connected to the striated duct, which eventually empties into the excretory duct

intercostal muscles muscles between the ribs that are involved in forced expiration

interdigestive state fasting state in the upper gastrointestinal that begins with the absorption of all nutrients, which is characterized by the migrating motor complex behavioral pattern

interferons class of natural proteins produced by the cells of the immune system in response to challenges by foreign agents such as viruses, bacteria, parasites, and tumor cells. Interferons belong to the large class of glycoproteins known as cytokines

interganglionic fiber tracts bundles of nerve fibers connecting adjacent ganglia of the enteric nervous system

interleukin cytokine made by one leukocyte and acting on another

interleukin-1 cytokine secreted by macrophages, monocytes, and dendritic cells as part of the inflammatory response to infection. It enables transmigration of leukocytes to sites of infection and resets the hypothalamic thermoregulatory center, leading to an increased body temperature (fever), thereby acting as an endogenous pyrogen

interleukin-6 proinflammatory cytokine secreted by T cells and macrophages to stimulate immune response to trauma, especially burns or other tissue damage leading to inflammation. IL-6 is one of the most important mediators of fever

intermediate filaments cellular constituents of smooth muscle that form part of the cytoskeleton

intermediate-density lipoprotein (IDL) particles circulating in the blood that contain cholesterol esters; falls between low-density and very-low-density lipoprotein in density

internal anal sphincter smooth muscle sphincter surrounding the anal canal, which provides a passive barrier to the leakage of liquid and gas from the rectum

internal capsule pathway for corticospinal tract neurons within the deep cerebral white matter

international normalized ratio (INR) system established to report the results of blood coagulation (clotting) tests; results are standardized using the international sensitivity index for the particular thromboplastin reagent and instrument combination used to perform the test

interneuron internuncial neuron that is neither sensory nor motor and that connects neurons with neurons

interposed (interpositus) nucleus deep nucleus of the cerebellum composed of the globose nuclei and the emboliform nuclei; receives its afferent supply from the anterior lobe of the cerebellum and sends output via the superior cerebellar peduncle to the red nucleus; modulates muscle stretch reflexes of distal muscle groups

interstitial cells of Cajal (ICCs) specialized cells of mesodermal origin believed to be the pacemaker cells for intestinal electrical slow waves

interstitial fluid fluid outside blood vessels that directly bathes most body cells

intestinal phase one of three phases that stimulate acid secretion resulting from the ingestion of food. During the intestinal phase, protein digestion products in the duodenum stimulate gastric acid secretion through the action of the circulating amino acids on the parietal cells. Distention of the small intestine stimulates acid secretion. The intestinal phase accounts for about 10% of total gastric acid secretion

intracellular fluid (ICF) fluid within cells

intrafusal muscle fibers muscle cells within a muscle spindle whose contraction changes tension on spindle sensory endings

intrinsic coagulation pathway blood-clotting pathway in response to an abnormal vessel wall

intrinsic factor glycoprotein secreted by the parietal cells in the stomach that binds strongly with vitamin B_{12} to form a complex that is then absorbed in the terminal ileum through a receptor-mediated process

inulin fructose polymer whose renal clearance is used to measure glomerular filtration rate

inulin clearance procedure by which the filtering capacity of the kidney glomeruli is determined by measuring the rate at which inulin, the test substance, is cleared from blood plasma

inverse myotatic reflex type of muscle stretch reflex that helps regulate the tension of contracting muscle

inwardly rectifying K^+ channels channels responsible for the hyperpolarizing potassium currents that are the main determinant of the resting membrane potential in myocardial cells. The term "inwardly" is an electrophysiologic convention that refers to the fact that the channels pass potassium currents more readily in the inward than in the outward direction, even though such inward potassium currents never occur in myocardial cells

iodothyronine molecule formed by the coupling of the phenyl rings of two iodinated tyrosine molecules in an ether linkage

ion channel opening through the plasma membrane composed of several polypeptide subunits that span the membrane and contain a gate that determines whether the channel is open or closed; regulates the flow of ions across the membrane in all cells

ion pump integral membrane protein that uses energy derived from adenosine triphosphate hydrolysis to drive ion movement across the membrane against the electrochemical gradient, for example, Na^+/K^+-ATPase

ionotropic type of membrane receptor that is a channel that opens (or closes) when the receptor is bound by ligand

ionotropic receptor receptors that are coupled directly to ion channels; fast excitatory postsynaptic potentials in the enteric nervous system are mediated by ionotropic receptors

iron profile blood tests including serum iron, ferritin, total iron-binding capacity, transferrin, and transferrin saturation

iron deficiency anemia condition characterized by decreased healthy red blood cell count because of a dietary iron deficiency and, hence, functioning hemoglobin

irreversible shock stage of shock in which the function of organs needed to maintain cardiac output and blood pressure (heart and brain) is so compromised that the person cannot survive even with prompt and aggressive medical intervention

irritable bowel syndrome (IBS) group of functional bowel disorders in which abdominal discomfort or pain is associated with defecation or a change in bowel habit and with features of disordered defecation

irritant receptors rapidly adapting receptor that is found in the large conducting airways and that is sensitive to irritation by noxious substances

ischemia reperfusion condition in which blood flow is restricted, followed by normal blood flow being reestablished. This condition leads to the formation of reactive oxygen species, which often leads to tissue damage

ischemic phase (of the endometrial cycle) phase in which necrotic changes and abundant apoptosis occur in the secretory epithelium as it collapses

islets of Langerhans cell clusters in the pancreas that form the endocrine part of that organ and that secrete insulin and other hormones

isocaproic acid product of side-chain cleavage of cholesterol

isoelectric in cardiovascular physiology, the portion of the normal electrocardiogram where the tracing is at zero or near-zero potential, which occurs in the PR and ST intervals

isograft transplantation between genetically identical people

isohydric principle idea that all buffer systems in a solution are in equilibrium with the same hydrogen ion concentration

isometric contraction contraction at constant length that produces force but no muscle shortening

isometric exercise form of exercise involving the contraction of a muscle without the shortening of the angle of the joint

isosmotic see *osmotic pressure*

isotonic denoting a solution having the same osmotic pressure as some other solution, especially one in a cell or a body fluid

isotonic contraction muscle contraction at constant force that produces a change in length

isotonic solution see *tonicity*

isovolumic contraction portion of systole in which all valves are closed, and therefore, blood does not enter or exit the heart

isovolumic relaxation portion of diastole in which all valves are closed, and therefore, blood does not enter or exit the heart

isthmus portion of the oviduct connecting the ampulla to the uterus

Ito cells perisinusoidal cells that contain lipid droplets in the cytoplasm and that function in the storage of fat. During the inflammatory process, these cells have the ability to transform into myofibroblasts, which secrete collagen into the space of Disse and regulate sinusoidal portal pressure by contracting or relaxing. Ito cells may be involved in the pathologic fibrosis of the liver; also known as *stellate cells*

J receptors unmyelinated C-fiber nerves that are located adjacent to alveoli that are stimulated by stretch; also referred to as *juxtapulmonary capillary receptors* or *C-fiber endings*

junctional complex functional unit in muscle comprising both dihydropyridine and ryanodine receptors

juxtaglomerular apparatus structure in the kidney consisting of macula densa, extraglomerular mesangial cells, and granular cells, located where the end of the thick ascending limb touches afferent and efferent arterioles of its parent glomerulus

juxtamedullary nephron nephron with its glomerulus deep in the cortex, next to the medulla, and having long loops of Henle

juxtapulmonary capillary receptors see *J receptors*

kallidin see *lysyl-bradykinin*

kallikrein serine protease released by the acinar cells that acts on endogenous peptides present in body fluid to release lysyl-bradykinin (kallidin), which causes dilation of the blood vessels supplying the salivary glands

Kallmann syndrome syndrome characterized by a congenital absence of gonadotropin-releasing hormone in the hypothalamus (causing, in women, primary amenorrhea and anovulation and, in men, failure of puberty) in combination with a congenitally absent sense of smell

ketoacidosis condition found in diabetics that is characterized by large amounts of β-hydroxybutyric acid in the blood

ketogenesis process by which ketone bodies are produced as a result of fatty acid breakdown

ketones acidic substances produced when the body uses fat, instead of sugar, for energy

ketosis condition observed in diabetic patients and during periods of prolonged starvation that is marked by highly elevated levels of circulating ketone bodies

17-ketosteroids excretable metabolite formed by lysis of the carbon 20 to 21 side chain of 21-carbon steroids

kilocalorie unit of measurement for energy that approximates the heat energy needed to increase the temperature of 1 kg of water by 1°C, equivalent to the dietary "calorie"

kinesin microtubule-associated protein involved in anterograde axonal transport of organelles and vesicles (from the minus to the plus ends of the microtubules) via the hydrolysis of adenosine triphosphate

kininogen endogenous peptide present in body fluid that releases lysyl-bradykinin (kallidin)

kinins mediator of the inflammatory response

kinocilium nonmotile cilium-like structure on each hair cell of the human ear involved in converting mechanical movement of endolymph into electrical signals

Klinefelter syndrome congenital abnormality of men in which there is one X chromosome too many; men with this condition are usually sterile

Korotkoff sounds sounds made by the first spurts of blood escaping from a compressed artery as the compression is slowly released. They are detected by stethoscope in the measurement of arterial pressure by sphygmomanometer and used as the indicator of peak systolic pressure

Krebs cycle series of mitochondrial chemical reactions that produces adenosine triphosphate in the presence of O_2, a process called *oxidative phosphorylation*

Kupffer cells one of the types of cells lining hepatic sinusoids; resident macrophages of the "fixed monocyte–macrophage system" that act in the removal of unwanted material (bacteria, viral particles, etc.) from the circulation through endocytosis

Kussmaul respiration labored, deep breathing that accompanies severe uncontrolled diabetes

labeled line concept that the central nervous system identifies a particular sensation on the basis of the pathway by which it arrives at the brain

lactation secretion of milk

lactational amenorrhea period of time with no menstrual cycles, resulting from lactation

lactic acid end product of glycolysis that accumulates under anaerobic conditions and is oxidized in the Krebs cycle in muscle

lactiferous ducts primary milk ducts

lactoferrin protein that binds iron

lactogenesis milk production by alveolar cells; also known as *lactogenic activity*

lactose intolerance condition in which the ingestion of milk products results in severe osmotic diarrhea

lactotrophs cells located in the anterior pituitary that secrete prolactin

lacunae blood-filled spaces formed in the endometrium

lag phase of gastric emptying time required for the grinding action of the antral pump to reduce the size of larger particles to particles of sufficiently small size for emptying (i.e., < ≈7 mm)

Lambert-Eaton myasthenic syndrome (LEMS) rare presynaptic disorder of neuromuscular transmission in which quantal release of acetylcholine is impaired, causing a unique set of clinical characteristics, which include proximal muscle weakness, depressed tendon reflexes, posttetanic potentiation, and autonomic changes

lamellar inclusion bodies electron-dense bodies found in type II epithelial cells of the alveoli that are the storage sites of lung surfactant

laminar flow relative motion of elements of a fluid along smooth parallel paths, which occurs at lower values of Reynolds number

laminin external filamentous muscle protein that forms a link between integrins and the extracellular matrix

Langerhans cells dendritic epidermal cell present in lymph nodes and other organs

laryngospasm severe constriction of the larynx in response to the introduction of water, allergies, or noxious stimuli

latch state modification of the crossbridge cycle in smooth muscle that results in long contractions with little expenditure of energy

latency period time period between stimulus onset and response occurrence

latent heat of evaporation energy required to overcome the molecular forces of attraction between the particles of a liquid and to bring them to the vapor state, in which such attractions are minimal

lateral inhibition capacity of an excited neuron to reduce the activity of its neighbors

lateral sulcus (sylvian fissure) deep valley that separates the temporal lobe from the frontal and parietal lobes of the brain

LDL receptors low-density lipoprotein receptors; receptors present on the surface of steroid-producing and other cells; bind cholesterol-containing LDL particles in the blood

leak channels channels that are always open, allowing the passage of sodium ions (Na^+) and potassium ions (K^+) across the membrane to maintain the resting membrane potential of −70 mV; also called *passive channels*

lecithin cholesterol acyltransferase (LCAT) plasma enzyme that is responsible for the esterification of cholesterol in the plasma to form cholesterol ester with the fatty acid derived from the 2 position of lecithin

left atrium left upper muscular chamber of the heart, which receives blood from the pulmonary veins and delivers it through the mitral valve into the left ventricle

left bundle-branch block condition in which transmission of action potentials is blocked in the left branch of the bundle of His

left ventricle left lower muscular chamber of the heart, which receives blood from the left atrium and delivers it through the aortic valve into the aorta

left ventricular end-diastolic pressure (LVEDP) blood pressure within the left ventricle at the end of diastole; function of both the volume contained within the ventricle and ventricular compliance

left ventricular end-diastolic volume (LVEDV) volume of blood in the ventricle at the end of diastole and just before the initiation of ventricular contraction that determines the passive stretch on myocardium before contraction and, therefore, is considered the equivalent of muscle preload in the heart *in situ*

left ventricular hypertrophy abnormally increased muscle mass in the left ventricle, occurring in response to any factor that chronically increases wall stress on the ventricle, such as systemic arterial hypertension or a stenotic aortic valve

length–tension curve expression of the isometric force a muscle can produce at different fixed lengths

leptin anorexigenic hormone released by white fat cells (adipocytes); levels increase as fat stores increase

leukemia bone marrow cancer resulting in uncontrolled proliferation of leukocytes

leukocyte see *white blood cell*

leukocytosis condition characterized by a higher-than-normal number of circulating white blood cells

leukopenia condition characterized by a lower-than-normal number of circulating white blood cells

leukotrienes mediator of the inflammatory response

levator ani muscle compound muscle of the pelvic floor formed by pubococcygeus and iliococcygeus muscles that functions to resist prolapsing forces and draws the anus upward following defecation; supports the pelvic viscera

lever system system of bones and joints that confers a mechanical advantage (or disadvantage) on a skeletal muscle

Lewis hunting response response characterized by an intermittent supply of warm blood to cold-exposed extremities via vasodilation that recurs in 5- to 10-minute cycles to provide some protection from the cold

Leydig cells cells within the seminiferous tubules of the testicles that produce testosterone

Liddle syndrome autosomal dominant disorder characterized by hypertension associated with low plasma renin activity, metabolic alkalosis due to hypokalemia, and hypoaldosteronism; one of several conditions with this unusual set of characteristics known collectively as *pseudohyperaldosteronism*

ligand molecule that binds to another, that is, a hormone or neurotransmitter that binds to a receptor

ligand-gated ion channels channels in membrane receptors that open (or close) when a ligand binds the receptor

limbic system interconnected structures in the central nervous system associated with olfaction, emotion, motivation, and behavior; includes amygdala, cingulate gyrus, hippocampus, hypothalamus, olfactory cortex, and thalamus

lipid bilayer structure formed by spontaneous assembly of amphipathic lipid molecules in an aqueous solution; hydrophobic regions or fatty acids are buried in the interior, and hydrophilic head groups are on the exterior, in contact with the solution

lipid rafts highly organized, free-floating "microdomains" within the plasma membrane that are aggregates of sphingolipids and cholesterol; serve to compartmentalize cellular processes by serving as organizing centers for the assembly of signaling molecules, influencing membrane fluidity and membrane protein trafficking, and regulating neurotransmission and receptor trafficking

lipocortin substance that inhibits the activity of phospholipase A_2

lipogenesis conversion of carbohydrates and organic acids to fat

lipolysis mobilization of fatty acids from triglycerides stored in adipose tissue

lipopolysaccharide (LPS) major component of gram-negative bacterial membrane

lipoprotein lipase enzyme responsible for hydrolysis of triacylglycerol molecules packaged and transported in very-low-density lipoprotein particles to fatty acids, which can then in turn be metabolized for energy

lipoproteins complexes or compounds containing lipid and protein. The intestine produces two major classes of lipoproteins: chylomicrons and very-low-density lipoproteins

β-lipotropin substance derived from proopiomelanocortin that may have effects on lipid metabolism

lithium-induced polyuria excretion of large volumes of dilute urine occurring in patients receiving chronic lithium therapy for psychiatric disorders

lithocholic acid carcinogenic secondary bile acid derived from chenodeoxycholic acid

liver lobule basic functional unit of the liver, which is hexagonal in shape and built around a central vein

loading phase flat portion of the oxygen–equilibrium curve that reflects the loading of oxygen onto blood hemoglobin

local axon reflex pathway for propagation of action potentials within a portion of a neuron; does not pass through the cell body

local excitatory current current through an axon membrane created by the generator potential that provides the link between the formation of the generator potential and the excitation of the nerve fiber membrane

locus caeruleus collection of noradrenergic neurons in the rostral pons that innervate widespread areas of the cerebrum, brainstem, and spinal cord

long hypophyseal portal vessels vessels that deliver blood to the anterior pituitary

long-loop reflexes reflexes that require processing by several spinal or higher levels

long-term memory memory that can last as little as a few days or as long as decades; includes *declarative* and *nondeclarative* subtypes

long-term potentiation increased synaptic excitability and altered chemical state on repeated synaptic stimulation with a persistence beyond the cessation of electrical stimulation; thought to underlie learning and memory and be dependent on Ca^{2+} entry through activation of *N*-methyl-D-aspartate receptors

loop of Henle portion of the nephron leading from the proximal straight tubule to the distal convoluted tubule; main function is to create a concentration gradient in the medulla of the kidney: a countercurrent multiplier system using sodium pumps creates an area of high concentration near the collecting duct; water present in the filtrate in the collecting duct flows through aquaporin channels out of the collecting duct, moving passively down its concentration gradient; this process reabsorbs water and creates a concentrated urine for excretion

low-density lipoprotein (LDL) complex containing a single apolipoprotein B100 molecule, fatty acids, and cholesterol that transports cholesterol and fatty acids to cells; also known as *bad cholesterol*

lower esophageal sphincter smooth muscle sphincter separating the esophagus from the stomach

L-type Ca^{2+} channels voltage-gated channels that conduct primarily depolarizing Ca^{2+} currents into myocardial cells and nodal tissue during phase 2 of the action potential. The gates are opened and closed by membrane depolarization in a manner analogous to that seen for voltage-gated sodium channels except that their opening and closing kinetics are slower

lung compliance (C_L) measure of lung distensibility; represented by Δvolume/Δpressure

lung diffusion capacity amount of oxygen taken up by the lung per minute per mm Hg of oxygen partial pressure

lung inflation reflex see *Hering-Breuer reflex*

lung lavage medical procedure in which a bronchoscope is passed through the mouth or nose into the lungs and fluid is squirted into a small part of the lung and then recollected for examination; often referred to as *bronchoalveolar lavage*

luteal insufficiency condition in which the corpus luteum does not produce sufficient progesterone to maintain pregnancy in the early stages

luteal phase part of the menstrual cycle during which progesterone is produced; lasts 14 days

luteinization maturation of the corpus luteum

luteinizing hormone (LH) hormone secreted by the pituitary gland in the brain that stimulates the growth and maturation of eggs in females and sperm in males

luteinizing hormone–releasing hormone (LHRH) polypeptide from the hypothalamus that stimulates secretion of follicle-stimulating hormone and luteinizing hormone; also known as *gonadotropin-releasing hormone (GnRH)*

luteolysis process of regression of the corpus luteum; also known as *luteal regression*

lymph transparent, slightly yellow liquid, found in the lymphatic vessels and derived from tissue fluid

lymph node lymphoid organ filled with leukocytes

lymphagogic effect marked increase in lymph flow that plays an important role in the transfer of lipoproteins from the intercellular spaces to the central lacteal vessel

lymphatic bulbs blind lymphatic end tubes that are the origin of lymphatic vessels in the microcirculation

lymphatic vessels flexible, predominantly endothelial vessels that drain the interstitium of plasma filtrate from the capillaries and send the fluid back into the venous side of the circulation at the level of the right atrium. Larger vessels contain one-way valves directed away from the tissues and toward the heart. Muscle cells in the bulb ends of the vessels help with forward flow of fluid into larger lymphatic vessels

lymphoblast stem cell committed to differentiate into lymphocytes

lymphocyte agranular leukocyte: B cell, T cell, or null cell; involved in body defenses against disease

lymphoid organs organs that include bone marrow and thymus (*primary*) and lymph follicles, lymph nodes, spleen, thymus, and tonsils (*secondary*)

lymphoid series series of morphologically distinguishable cells once thought to represent stages in lymphocyte development; now known to represent various forms of mature lymphocytes

lymphokine cytokine made by lymphocytes

lymphokine-activated killer cell (LAK) cell derived from natural killer cells

lymphoma cancer of the lymphatic system

lysis decomposition or destruction of cells by antibodies called *lysins*

lysozyme bacteriolytic enzyme

lysyl-bradykinin potent vasodilator that causes dilation of the blood vessels that supply the salivary glands; also known as *kallidin*

M line thin line in middle of H zone of muscle

macrocyte larger-than-normal erythrocyte

macrophage phagocytic tissue cell, derived from monocytes

macula sensory structure in each semicircular canal of the vestibular apparatus, used to sense rotation

macula densa closely packed group of densely staining cells in the distal tubular epithelium of a nephron, in direct apposition to the juxtaglomerular cells; they may function either as chemoreceptors or as baroreceptors feeding information to the juxtaglomerular cells

macula lutea area of the retina about 1 mm² in area, specialized for very sharp color vision

magnocellular neurons hypothalamic neurons with large cell bodies that synthesize arginine vasopressin and oxytocin

major depression one of a class of illnesses marked by abnormal mood regulation and associated signs and symptoms called *affective disorders* or *mood disorders*. Major depression can be so profound as to provoke suicide

major histocompatibility complex (MHC) gene-encoding proteins involved in antigen presentation

malignant hyperthermia life-threatening condition resulting from a genetic sensitivity of skeletal muscles to volatile anesthetics and depolarizing neuromuscular-blocking drugs that occurs during or after anesthesia

malonyl-CoA condensation product of malonic acid and coenzyme A and an intermediate in fatty acid synthesis

mammary glands paired structures on the thoracic wall of both sexes; in females, they secrete milk

mammillothalamic tract neuroanatomical tract in the limbic system connecting the neurons in the mammillary bodies of the hypothalamus with the anterior nucleus of the thalamus

mammogenesis differentiation and growth of the mammary glands

manometric related to pressure measurements (in this context, within the gut lumen)

manometric catheter gut manometry is the measurement of intraluminal pressures in a segment of the gut to study its motor function: contractions of muscles within or around the gut wall produce intraluminal pressure waves that are detected by manometry

margination loose adhesion of leukocytes to endothelial cells

marrow stroma bone marrow tissue that is not directly involved in hematopoiesis but provides the necessary microenvironment and factors for it

mast cell connective tissue cell in the intestine that releases paracrine signals, such as histamine, to act at receptors on enteric neurons; functionally resembles basophil

maternal recognition of pregnancy the secretion of human chorionic gonadotropin by a developing embryo to sustain the corpus luteum

maximal oxygen uptake (V$_{O_2}$ max) highest rate at which oxygen can be taken up and used during exercise. The derivation is V = volume per time, O$_2$ = oxygen, and max = maximum; expressed either as an absolute rate (L/min) or as a relative rate (mL/kg/min), the latter expression is often used to compare the performance of endurance sports athletes

maximal response greatest, most complete, or best response

maximal voluntary contraction (MVC) the peak isometric force that can be briefly generated for a specific exercise; used as an index of muscle function and, more specifically, isometric work intensity

McCune-Albright syndrome (MAS) rare multisystem disorder characterized by replacement of normal bone tissue with areas of abnormal fibrous growth (fibrous dysplasia), patches of abnormal skin pigmentation (i.e., areas of light-brown skin [café au lait spots] with jagged borders), and abnormalities in the glands that regulate the body's rate of growth, its sexual development, and certain other metabolic functions (multiple endocrine dysfunction)

mean circulatory filling pressure equilibrium pressure throughout all components of the vascular system when cardiac output is zero

mean corpuscular hemoglobin (MCH) index of average hemoglobin per erythrocyte

mean corpuscular hemoglobin concentration (MCHC) index of average hemoglobin in circulating erythrocytes

mean corpuscular volume (MCV) index of average volume of each erythrocyte

mean electrical QRS axis average vector representing the wave of depolarization through the ventricles

mean platelet volume (MPV) evaluates size of platelets

mechanical obstruction any physical obstacle to the caudal movement of intestinal contents, including tumors, scar tissue, or volvulus (twisting); distinguished from a functional obstruction, which is a result of absent or abnormal contractions of the intestine

mechanoreceptors sensory receptors that respond to physical deformations both externally and internally

medial forebrain bundle fiber bundle interconnecting the brainstem (especially the midbrain tegmentum) with the hypothalamus, nucleus accumbens, and basal forebrain

medial prefrontal area region of frontal lobe cerebral cortex involved in coordination of autonomic responses to olfactory and emotional stimuli

median effective dose (ED$_{50}$) concentration of a substance (e.g., hormone) required to produce a response halfway between the maximal and basal responses

median eminence region in the floor of the hypothalamus that allows release of neurohormones and access to the hypothalamus of blood-borne substances

medulla innermost portion of an organ or structure

medullary reticulospinal tract descending inhibitory brainstem pathway to spinal interneurons that originates in the medulla

medullary tissue part of the indifferent gonad of the embryo, which will become the testicular components

megacolon colon filled with impacted feces above the terminal aganglionic region of large intestine in Hirschsprung disease

megakaryocyte bone marrow precursor cell for platelets

megaloblast nucleated, large, abnormal red blood cell

megaloblastic anemia condition in which red blood cell maturation in the bone marrow is abnormal, possibly resulting from vitamin B$_{12}$ or folic acid deficiency

meiosis process of two consecutive cell divisions in the diploid progenitors of sex cells, resulting in four rather than two daughter cells, each with a haploid set of chromosomes

Meissner plexus named for Georg Meissner (1829 to 1905), a German histologist; see also *submucosal plexus*

melanin pigment that in humans is the primary determinant of skin color; in the pigment epithelium of the eye, melanin helps to sharpen an image by preventing the scattering of stray light

α-melanocyte–stimulating hormone (α-MSH) hormone derived from proopiomelanocortin that regulates feeding in humans

melatonin hormone released by the pineal gland in darkness; the mechanism by which the suprachiasmatic nucleus of the hypothalamus regulates diverse functions in a diurnal fashion

membrane lipid bilayer that acts as a boundary to various cellular structures or the environment

membrane attack complex (MAC) complement proteins that form a channel in the target cell membrane

membrane potential difference between voltage measured inside a cell and the value (0) detected by a reference electrode located in the extracellular fluid

membrane-associated dense bodies dense bodies associated with cell margins that serve as anchors for thin filaments and transmit the force of contraction to adjacent cells; also called *focal adhesions*

memory cell T or B cell that recognizes previously encountered antigens

menaquinones vitamin K derived from bacteria in the small intestine

menarche beginning of menstrual cycles during pubertal development in girls

Ménière disease disorder of the inner ear that can affect hearing and balance to a varying degree and is characterized by episodes of both vertigo and tinnitus and by progressive hearing loss, usually in one ear

menopause final menses at the end of the reproductive cycle in the female

menstrual cycle monthly reproductive cycle in which periodic uterine bleeding occurs

menstrual phase or menses or menstruation bleeding phase of the menstrual cycle or endometrial cycle, which lasts about 5 days

mesangial cell specialized cell found between capillary loops in the glomerulus, which has supportive, phagocytic, and contractile functions

mesolimbic/mesocortical system projection of dopaminergic neurons from the midbrain ventral tegmental area to the limbic system, especially the prefrontal cortex, nucleus accumbens, and hypothalamus

mesonephric (wolffian) ducts ducts that give rise to the epididymis, vas deferens, seminal vesicles, and ejaculatory ducts in male

messenger low molecular weight diffusible molecule used in signal transduction to relay signals within a cell

metabolic acidosis abnormal condition in which the blood becomes too acidic, leading to a decrease in pH. The acidosis is a result of any cause other than elevated carbon dioxide level

metabolic alkalosis abnormal process characterized by a gain of strong base or bicarbonate or a loss of acid (other than carbonic acid), which results in a rise in arterial blood pH

metabolic bone diseases diseases characterized by ongoing disruption of the normal processes of either bone formation or bone resorption

metabolic clearance rate (MCR) volume of plasma cleared of a substance, such as a hormone, in question per unit time

metabolic syndrome cluster of conditions that often occur together, including obesity, high blood glucose, hypertension, and high triglycerides, which can lead to cardiovascular disease

metabotropic characteristic of receptors that do not form an ion channel, instead, causing indirect alterations in current via signal transduction mechanisms or causing only metabolic changes

metabotropic receptor (mGLUR) type of receptor that is linked through G proteins to intracellular production of 1,2-diacylglycerol and inositol 1,4,5-trisphosphate

methemoglobin (metHb) abnormal hemoglobin with oxidized ferric iron

methylene blue component of blood dyes such as Wright stain

MHC class I molecule molecules found in every nucleated cell of the body that bind peptides mainly generated from degradation of cytosolic proteins; see also *major histocompatibility complex*

MHC class II molecule molecules found only on antigen-presenting cells (macrophages, dendritic cells, and B cells); see also *major histocompatibility complex*

micellar emulsification process that enhances the delivery of lipid to the brush border membrane by making it water soluble

micelle aggregate of surfactant molecules dispersed in a liquid colloid

microalbuminuria condition characterized by the excretion of 30 to 300 mg of serum albumin in the urine per day, usually a sign of kidney disease

microcirculation region of the circulation involved with the control of blood pressure, blood flow, and transcapillary transport of substances, composed of terminal arterioles, capillaries, and venules

microfilaments elements of the neuronal cytoskeleton that are composed of actin, are 4 to 5 nm wide and about a micrometer in length (except in growth cones, where they can be much longer), and are concentrated in dendritic spines, growth cones, and presynaptic terminals

microglia glial cells in the CNS serving an immune function similar to macrophages

microspherocyte erythrocyte that is smaller than normal

microtubules elements of the neuronal cytoskeleton responsible for the rapid movement of material in axons and dendrites; composed of a core polymer of 50-kDa tubulin subunits that align to form protofilaments, several of which join laterally to form a hollow tube whose diameter is about 23 nm

microtubule-associated proteins (MAPs) accessory proteins associated with microtubules that bind tubulin and, in neurons, are thought to be required for specific sorting of materials to axons or dendrites

microvilli numerous closely packed villi that cover the enterocytes to amplify further the villi that increase the surface area of the small intestine

micturition emptying of the urinary bladder; also known as *urination*

middle cervical ganglia middle of the three sympathetic ganglia located in the neck

middle ear portion of the ear, containing the ossicles, that couples eardrum movements to the sensory structures of the inner ear

migrating motor complex (MMC) organized, distinct, cyclically recurring and distally propagating sequence of contractile activity occurring in the small intestine during the fasting state; may be associated with similar activities in the stomach and proximal colon

milk ejection release of stored milk from the mammary glands by the action of oxytocin; also known as *milk letdown*

milliequivalent (mEq) one-thousandth of an equivalent; calculated from the product of millimoles times valence

mineralocorticoid steroid hormone produced by the adrenal cortex and involved in the regulation of electrolyte and water balance

mineralocorticoid escape limitation of the salt-retaining effects of a mineralocorticoid hormone (such as aldosterone), expressed by increased renal sodium excretion despite the continued presence of high levels of mineralocorticoid

miosis pupillary constriction

mitochondria membrane-bound organelle that supplies energy in the form of adenosine triphosphate to cells

mitochondrial permeability transition pore (MPTP) protein pore that is formed in the membranes of mitochondria under certain pathologic conditions; induction of MPTP can lead to mitochondrial swelling and cell death

mitochondrial voltage-dependent anion channel (VDAC) channel for metabolite diffusion across the mitochondrial outer membrane

mitogen agent that induces proliferation; many growth factors and hormones are mitogens

mitogenic agent agent that stimulates the growth of target tissues

mitosis replication of a cell to form two daughter cells with identical sets of chromosomes

mitral valve thin, flexible, bileaflet structure that separates the left atrium from the left ventricle. It opens into the left ventricle during diastole but closes off the left atrium when the left ventricle contracts

mitral valve prolapse condition whereby either the mitral valve bulges or the leaflets open backward into the pulmonary vein during systole. In the latter case, blood flows back into the pulmonary circulation from the left ventricle during systole

mittelschmerz (German for "middle pain") lower abdominal and pelvic pain that occurs roughly midway through a woman's menstrual cycle, associated with ovulation; pain can appear suddenly and usually subsides within hours

mixed acid–base disturbance simultaneous presence of two or more primary acid–base disturbances

mixed contractions skeletal muscle contractions that begin isometrically and become isotonic as force becomes sufficient to lift the afterload

mixed micelles substance that aids in rendering lipid digestion products water soluble. Mixed micelles diffuse across the unstirred water layer and deliver lipid digestion products to the enterocytes for absorption

mixing motility pattern motility pattern of the digestive state

M-mode echocardiography ultrasound image of the heart used to examine motion of the ventricular walls, changes in chamber size, and valve motion during the cardiac cycle

modulator peptide released at a synapse that modifies the response of a coreleased transmitter interacting with its own receptor in the synapse

molecular switch intracellular protein that can transition between "on" and "off" states in response to a cellular stimuli (e.g., G protein)

monoamine oxidase (MAO) a mitochondrial enzyme that degrades catecholamines by removing the amine

monoamine oxidase inhibitors (MAOIs) powerful class of antidepressant drugs

monoaminergic type of neuron that uses a catecholamine or indolamine as a neurotransmitter

monoblast stem cell committed to differentiate into monocytes

monocyte agranular phagocytic white blood cell that is a precursor of macrophages

monocytic series succession of developing cells that ultimately culminates in the monocyte

monoiodotyrosine (MIT) iodinated tyrosine in peptide linkage in the thyroglobulin structure

monokine cytokine made by monocytes

mononuclear leukocyte agranulocyte: either lymphocyte or monocyte

mood sustained emotional state

mood disorders see *affective disorders*

morula solid ball of cells that develops from a fertilized egg and enters the uterus at day 3 to 4 following fertilization

mossy fibers afferent axons to the cerebellum from pontine nuclei

motilin signal peptide released from enteroendocrine cells and associated with the onset of the migrating motor complex

motility movements within the intestinal tract, encompassing the phenomena of contractile activity, myoelectrical activity, tone, compliance, wall tension, and transit within the gastrointestinal tract

motor endplate see *myoneural junction*

motor neuron nerve cell that sends an axon to a muscle or, by extension, any effector

α motor neuron innervates the extrafusal muscle fibers, which are responsible for force generation

γ motor neuron controls contraction of the intrafusal muscle cells within a muscle spindle

motor unit alpha motor neuron, its axon, and all the muscle fibers controlled by that motor neuron

motor unit summation see *spatial summation*

mucin one of two major proteins (a glycoprotein high in carbohydrate) present in saliva that form gels in solution; the most abundant and produced mainly by the sublingual and submandibular salivary glands. Mucins lubricate the mucosal surface and protect it from mechanical damage by solid food particles. Mucins may also provide a physical barrier in the small intestine against the entry of microorganisms into the mucosa

mucous cells cells that line the entire surface of the gastric mucosa and the openings of the cardiac, pyloric, and oxyntic glands; secrete mucus and bicarbonate to protect the gastric surface from the acidic environment of the stomach

mucous gel layer layer covering the surface of the gastric mucosa that traps bicarbonate and neutralizes acid, thus preventing damage to the mucosal cells

mucous neck cells one of several types of cells in the oxyntic glands. The neck region in the gland is where mucous cells are most numerous. Mucous neck cells secrete mucus

müllerian-inhibiting substance (MIS) glycoprotein that inhibits cell division of the müllerian ducts; also known as *antimüllerian hormone*

multidrug resistance insensitivity of various tumors to a variety of chemically related anticancer drugs; mediated by a process of inactivating the drug or removing it from the target tumor cells

multidrug resistance–associated protein (MRP) transporters class of ABC transporters that interfere with antibiotic therapy and chemotherapy

multiple myeloma typically incurable cancer of plasma cells

multiple sclerosis demyelinating autoimmune disease

multipotent stem cell cell, such as hematopoietic stem cell, that can give rise to different specialized cells

multiunit-type smooth muscle type of smooth muscle that predominates in the pupil of the eye and urinary bladder; it does not contract spontaneously, is not activated by stretch, and expresses structured neuromuscular junctions; contrast with *unitary-type smooth muscle*

mural granulosa cells cells attached to the basement membrane in a graafian follicle

muramidase lysozyme that can lyse the muramic acid of certain bacteria (e.g., *Staphylococcus*); present in saliva in small amounts

muscarine compound found in poisonous mushrooms that activates receptors for acetylcholine known as *muscarinic receptors*

muscarinic receptor metabotropic-type receptor for acetylcholine

muscle chemoreflex reflex initiated by stimulation of the chemosensitive nerve endings as a result of a mismatch between skeletal muscle blood flow and an increase in metabolites that stimulate

muscle fibers contractile cells of a muscle

muscle stretch reflex contraction of a skeletal muscle provoked by a sudden stretch of its tendon

muscularis externa longitudinal and circular muscle coats of the stomach wall (one of four layers); lies over the submucosa and differs from that of other GI organs in that it has three layers of smooth muscle instead of two

musculomotor neuron neuron that innervates the gastrointestinal musculature

myasthenia gravis autoimmune disease that produces weakness by reducing the number of postsynaptic acetylcholine receptors in skeletal muscle

mydriasis pupillary dilation

myelin lipid structure formed by oligodendrocytes in the central nervous system or Schwann cells in the peripheral nervous system that wraps around neuronal axons and serves to increase the speed with which they conduct action potentials

myelin sheath the lipid surrounding many axonal processes that acts to insulate that portion of the axon

myeloblast stem cell committed to differentiate into cells of the myeloid series

myeloid series various stages of cell development such as those seen in the granulocytic and erythrocytic series

myenteric plexus ganglionated plexus of the enteric nervous system situated between the longitudinal and circular muscle coats of the muscularis externa of the digestive tract; also known as *Auerbach plexus*

myocardial infarction pathologic condition in which myocardial tissue has died. This usually results from extended exposure of the tissue to ischemic conditions

myocardial ischemia any condition in which the oxygen supply to myocardial tissue cannot meet the oxygen demand of that tissue

myoclonus brief, involuntary muscle twitches that commonly occur during light sleep

myoepithelial cells cells that surround milk-secreting alveolar cells and that contract and expel milk from the lactiferous duct in response to oxytocin

myofibrils longitudinal columns of series-connected sarcomeres in a skeletal muscle

myogenic originating in muscle tissue

myogenic mechanism regulatory mechanism whereby arteries or veins contract in response to an increase in transmural pressure. It is related to activation of stress-operated calcium channels in the smooth muscle itself that open with stretch, thereby increasing calcium entry into the cells and causing contraction of vascular smooth muscle. It is one mechanism responsible for the ability of organs to autoregulate their blood flow in the face of changes in perfusion pressure

myoglobin oxygen-binding protein in skeletal muscle that facilitates oxygen diffusion

myokines are endocrine-like peptides that are released during muscle contraction that exert a local effect on muscle metabolism, repair, tissue regeneration, and immunomodulation

myometrium outer part of the uterus, composed of multiple layers of smooth muscle

myoneural junction see *neuromuscular junction*

myopathic pseudoobstruction pathologic failure of propulsion in the gastrointestinal tract related to muscular degeneration and weakened contractility

myopia (nearsightedness) condition caused by the eyeball being physically too long to focus on near objects

myosin fibrous protein dimer that makes up both the structural and chemically active portions of the thick filaments in muscle

myosin light-chain kinase (MLCK) calcium/calmodulin-dependent enzyme that phosphorylates the light chains of smooth muscle myosin and initiates contraction. In skeletal muscle, phosphorylation modulates tension during contraction

myosin light-chain phosphatase (MLCP) enzyme that catalyzes the dephosphorylation of the regulatory light chains of muscle, particularly those of smooth muscle myosin

myosin-linked regulation muscle control system that is based on controlled alteration of the constituents of the myosin filaments

myotatic another name for the muscle stretch reflex

myxedema puffy appearance to the face, hands, and feet as a result of hypothyroidism

Na⁺/K⁺-ATPase plasma membrane protein that uses the energy of adenosine triphosphate hydrolysis to transport three Na^+ ions out of the cell while transporting two K^+ ions into the cell; it is a *P-type ATPase* because the protein is phosphorylated during the transport cycle

NADPH oxidase enzyme-producing superoxide as part of a respiratory burst

NADPH–cytochrome P-450 reductase (NADPH–cytochrome C reductase) the enzyme in the cytochrome P-450 complex that is responsible for the phase I reactions of the drug biotransformation. Specifically, it reduces the drug complexed with cytochrome P-450 so as to make it a more polar compound

narcolepsy neurologic disorder characterized by excessive daytime sleepiness, cataplexy (sudden loss of muscle tone), and vivid hallucinations at sleep onset or on awakening

natriuresis increase in renal sodium excretion

natural killer (NK) cell cytotoxic lymphocyte and major component of the immune system; plays major role in the rejection of tumor and cells infected by virus; kills cells by releasing perforin and granzyme, which cause the target cell to die by apoptosis

nebulin filamentous muscle protein that may play a role in stabilizing thin filament length during muscle development

necrosis unprogrammed cell death, less orderly than apoptosis, results from acute tissue injury; initiates an immune response to remove cell debris (*immunogenic*)

negative cooperativity cooperativity in which the dissociation constant for each successive ligand bound is higher than for the one preceding it, so that the binding affinity is successively decreased

negative feedback type of feedback in which the system responds in an opposite direction to the perturbation

negative selection process by which T cells that recognize self-MHC (major histocompatibility complex) molecules with high affinity are deleted from the repertoire of cells, producing central tolerance and preventing autoimmunity; occurs after positive selection

neonatal respiratory distress syndrome syndrome of premature infants in which developmental insufficiency of surfactant production and structural immaturity of the lungs cause pulmonary edema and atelectasis; also known as *infant respiratory distress syndrome* or *IRDS*

nephrogenic diabetes insipidus condition characterized by lack of response of collecting ducts in the kidney to arginine vasopressin

nephrolithiasis kidney stone disease

nephrin protein necessary for the proper functioning of the renal filtration barrier, which consists of fenestrated endothelial cells, the glomerular basement membrane, and the podocytes of epithelial cells; nephrin is a transmembrane protein involved with the filtration slit diaphragm

nephron basic unit of kidney structure and function, consisting of a renal corpuscle and its attached tubule

nephrotic syndrome kidney disorder characterized by edema, proteinuria, hypoproteinemia, and, usually, elevated blood lipids

Nernst equation chemical equation that allows calculation of the equilibrium potential for any ion

Nernst equilibrium potential value of the electrical potential difference across the plasma membrane that is necessary for a specific ion to be at equilibrium (no net transport)

nerve growth factor protein molecule that enhances nerve cell development and stimulates the growth of axons

neural networks aggregates of interconnected neurons that produce a range of behaviors associated with the central and enteric nervous systems

neural presbycusis progressive loss of the ability to hear high frequency with increasing age as the hair cells "wear out" over time

neuregulin axonally derived epidermal growth factor–like substance that causes glial cells to myelinate the axon

neuroendocrine both neural and endocrine in structure or function

neuroendocrine system coordinates the interaction of the endocrine and nervous systems to provide dual control in regulating homeostasis and many physiologic processes in the human body

neuroendoimmunology study of the interaction between the nervous, endocrine, and immune systems

neurofilaments cytoskeletal intermediate filaments in neurons that are found in both axons and dendrites and are thought to provide structural rigidity. Neurofilaments are composed of a 70-kDa core protein, are about 10 nm wide, and form ropelike filaments that are several micrometers in length

neurogastroenterology subdiscipline of gastroenterology that encompasses all basic and clinical aspects of nervous system involvement in normal and disordered digestive functions and sensations

neurogenic diabetes insipidus condition characterized by deficient production or release of arginine vasopressin; also called *central*, *hypothalamic*, and *pituitary diabetes insipidus*

neurogenic secretory diarrhea watery diarrhea that reflects stimulated activity of secretomotor neurons in the submucosal plexus

neurohormone neurotransmitters with endocrine effects

neurohypophysis neural portion of the hypophysis; also known as *posterior pituitary*

neuroleptics class of drugs used to treat schizophrenia and other psychotic disorders; also known as *typical antipsychotics*

neuromodulator neuropeptides that modulate the postsynaptic actions of a primary neurotransmitter

neuromuscular junction (NMJ) excitatory nicotinic synapse between a motor nerve terminal and a skeletal muscle fiber; also known as *motor endplate* and *myoneural junction*

neurons nerve cells

neuropathic pseudoobstruction pathological failure of propulsion in the gastrointestinal tract related to enteric neuropathy and loss of nervous control mechanisms

neuropeptide Y (NPY) neurotransmitter secreted by the hypothalamus associated with a number of physiologic processes in the brain, including the regulation of energy balance, memory and learning, and epilepsy, but its main effect is increased food intake and decreased physical activity

neurophysin part of the arginine vasopressin gene; important in the processing and secretion of arginine vasopressin

neuropil tangle of fine nerve fibers (dendrites and arborizations of axons) and their endings

neurosecretory (neuroendocrine) cells specialized nerve cells that directly convert a neural signal into a hormonal signal

neurotransmitter chemical substance released from nerve cells at synapses that amounts to a signal from one neuron to another or from motor neurons to effectors

neutropenia hematological disorder characterized by an abnormally low number of neutrophils

neutrophil most common type of granulocyte, which destroys pathogens by phagocytosis and respiratory burst

niacin water-soluble vitamin that is absorbed by the small intestine by calcium-dependent, carrier-mediated facilitated transport at low concentrations and by passive diffusion at high concentrations. Niacin plays an important role in coenzymes that participate in many oxidation–reduction reactions involving hydrogen transfer. It has been used to treat hypercholesterolemia for the prevention of coronary artery disease by decreasing plasma total cholesterol and low-density lipoprotein cholesterol and increasing plasma high-density lipoprotein cholesterol

nicotine compound found in tobacco that activates receptors for acetylcholine known as *nicotinic receptors*

nicotinic acetylcholine (ACh) receptor postsynaptic membrane–based protein assembly in the myoneural junction and at synapses in the central and autonomic nervous systems; its associated channel becomes permeable to Na^+ and K^+ ions when ACh is bound; stimulated selectively by nicotine and blocked by a specific class of receptor-blocking drugs

nicotinic receptor ionotropic-type receptor for acetylcholine

nigrostriatal system projection of dopaminergic neurons from the midbrain substantia nigra to the striatum (caudate and putamen) in the cerebrum and involved in control of movement; loss of these projections results in parkinsonism

nipple conical structure on the breast that is the external opening to the lactiferous ducts

nitrergic neurons neurons that use nitric oxide as a neurotransmitter

nitric oxide (NO) endothelial autocoid (biological factor that acts like a local hormone) that is a potent relaxer of vascular smooth muscle; it is released from the endothelium in response to calcium activation of the enzyme NO synthase (calcium activation can occur in response to activation of various receptors on the endothelial cell by circulating agonists and by shear forces on the endothelial cells); putative inhibitory neurotransmitter released by enteric motor neurons to gastrointestinal muscles; previously called *endothelial-derived relaxing factor (EDRF)*

nitric oxide synthase (NOS) enzyme that converts L-arginine to L-citrulline and nitric oxide

NMDA receptor glutamate receptor named for the synthetic analog that specifically activates it, *N*-methyl-ᴅ-aspartate; an ionotropic receptor that is a nonselective cation channel permeable to Na^+, K^+, and Ca^{2+}

nociception experience of a stimulus as harmful in contrast to a pleasant sensation

nociceptors receptors that sense unpleasant or harmful stimuli, external and internal

nodes of Ranvier apposition between two internodes of myelin along an axon; a high concentration of voltage-gated Na^+ channels are localized here

nodose ganglion location of cell bodies of vagal afferent neurons

nonadrenergic noncholinergic (NANC) autonomic neurons that release neither norepinephrine nor acetylcholine

nonalcoholic steatohepatitis (NASH) presents as fat in the liver, along with inflammation and fibrosis; patients with NASH often feel healthy and are unaware they have a liver problem, but NASH can be severe and can lead to cirrhosis and consequent failure of the liver

nondeclarative or procedural memory memory for new skills and procedures; sometimes referred to as *implicit memory*

nonessential amino acids amino acids that are required for normal protein synthesis and that the body synthesizes from other amino acids; comprise 9 of the 20 total common amino acids that form the building blocks for proteins

nongated ion channels pores in the membrane that allow ions to flow across the membrane according to electrochemical parameters

nonrapid eye movement (NREM) sleep see *REM sleep*

nonshivering thermogenesis mechanism involving the generation of heat in brown adipose tissue as a result of uncoupled mitochondria located there; also known as *diet-induced thermogenesis*

noradrenergic neuron neuron that uses noradrenaline (norepinephrine) as a neurotransmitter

norepinephrine (NE) catecholamine produced by the adrenal medulla, postganglionic sympathetic axons (in the gut), and central neurons, especially those in the locus caeruleus

normoblast nucleated precursor of mature erythrocytes

normochromic erythrocyte with a normal concentration of hemoglobin

nuclear bag fibers type of intrafusal muscle fiber in muscle spindles

nuclear chain fibers type of intrafusal muscle fiber in muscle spindles

nuclear pore complex opening in membrane surrounding the nucleus, formed by a large assembly of proteins, through which molecules from the cytoplasm travel into the nucleus

nucleus accumbens cholinergic cell group in the base of the forebrain involved in reward pathways; part of the "brain's pleasure center"

nucleus ambiguus column of efferent neurons in the ventrolateral medulla oblongata from which its fibers leave with the vagus and glossopharyngeal nerves and innervate the striated muscle fibers of the pharynx, parts of the esophagus, and the vocal cord muscles of the larynx

nucleus tractus solitarius located in the medulla oblongata, containing cell bodies of second-order neurons in vagal afferent sensory pathways

null cell lymphocyte without T-cell or B-cell markers

nutcracker esophagus condition in which propulsive esophageal contractions are found to have abnormally high amplitudes during manometric testing; can be a source of angina-like chest pain

nystagmus reflexive sideways movement of the eyeballs that allows visual fixation during rotation of the head. It occurs in response to information from the vestibular apparatus

obesity excess of caloric intake over energy expenditure

occipital lobe posterior region of the cerebrum; contains the visual cortex

olfaction process of sensing odors

olfactory bulb complex multilayer neural architecture located in the forebrain that serves as a relay station to transmit information from the nose to the brain

oligodendrocytes glial cells in the CNS that produce a myelin sheath around neuronal axons in the CNS

oligomenorrhea infrequent or irregular menstruation

oncogenes genes that turn immune cells into tumors cells

oncotic pressure osmotic pressure exerted by colloids in a solution; also called *colloidal osmotic blood pressure*

oocyte immature ovum

oogonia germ cells

oolemma plasma membrane of the egg

ooplasm cytoplasm of an egg

opioids endogenous peptides that bind to opiate receptors

opsonin molecule that coats pathogens to facilitate their phagocytosis

optic chiasma part of the brain where the optic nerves (cranial nerve II) partially cross, located immediately below the hypothalamus

optic disc region of the retina where the optic nerve leaves the eye; the lack of photoreceptors in this region produces a blind spot

optic nerve transmits visual information from the retina to the brain; also called *cranial nerve II*

optic radiations collections of axons from relay neurons in the lateral geniculate nucleus of the thalamus carrying visual information to the visual cortex along the calcarine fissure; there is one such tract on each side of the brain; also known as *geniculostriate pathways*

optic tract continuation of the optic nerve that runs from the optic chiasm (where half of the information from each eye crosses sides, and half stays on the same side) to the lateral geniculate nucleus

optimum length (L_o) length corresponding to the maximum isometric force of muscle contraction

orbit bony socket for the eyeball

orexigenic causing stimulation of the appetite

organ of Corti complex organized structure of the inner ear that transforms sound vibrations into nerve impulses

organic anion transporting polypeptides (OATPs) highly expressed in liver, kidney, and brain, OATPs generally transport anionic and cationic chemicals, steroid, and peptide backbones into cells

organification addition of one or two iodine atoms to tyrosine residues in the thyroglobulin precursor protein

orthodromically conduction of nerve action potential in the conventional direction, efferently for motor and afferently for sensory

orthophosphate (PO_4) chemistry-based term that refers to an organic phosphate in which the phosphate is attached on the ortho position in a benzene ring

osmolality total solute concentration in a solution expressed as osmoles per kilogram of H_2O

osmolarity total solute concentration, expressed as osmoles per liter of solution

osmole Avogadro number (or a mole) of solute particles

osmoreceptor cells neurons in the anterior hypothalamus stimulated by increased extracellular fluid osmolality; they increase the release of arginine vasopressin and thirst

osmoregulation regulation of water potential in an organism by promoting either water retention or secretion, depending on the environmental situation

osmosis passive transport of water across a membrane in response to a gradient of osmotic pressure. Water always moves from the region of low osmotic pressure (low solute concentration) to the region of high osmotic pressure (high solute concentration). Water crosses cell membranes via water channel proteins (aquaporin)

osmotic diuretic solute excreted in the urine that increases urinary excretion of Na^+, K^+, and water

osmotic pressure property of a solution defined as the pressure necessary to stop the net movement of water across a selectively permeable membrane that separates the solution from pure water: a hyperosmotic solution has a greater solute concentration and exerts a greater pressure than another solution, whereas a hyposmotic solution has a lower solute concentration and exerts a lower pressure than another solution, and an isosmotic solution has the same solute concentration and exerts the same pressure as another solution

osteoarthritis arthritis characterized by erosion of articular cartilage, either primary or secondary to trauma or other conditions, which becomes soft, frayed, and thinned with eburnation of subchondral bone and outgrowths of marginal osteophytes; pain and loss of function result; mainly affects weight-bearing joints; more common in older persons

osteoblast cell that makes bone. It does so by producing a matrix that then becomes mineralized; bone mass is maintained by a balance between the activities of osteoblasts that form bone and other cells called *osteoclasts* that break it down

osteoclast large multinucleate cell that breaks down bone and is responsible for bone resorption

osteocyte branched cell embedded in the matrix of bone tissue

osteogenesis imperfecta hereditary disease characterized by abnormally brittle, easily fractured bones

osteoid uncalcified bone matrix, the product of osteoblasts

osteomalacia softening of the bones as a result of a deficiency of vitamin D; also known as *adult rickets*

osteoporosis disease of bone in which the bone mineral density is reduced, bone microarchitecture is disrupted, and the amount and variety of noncollagenous proteins in bone is changed; osteoporotic bones are more susceptible to fracture

otic ganglion location of pre- to postganglionic synapse of parasympathetic neurons from the inferior salivatory nucleus through cranial nerve IX, which are involved in control of the parotid gland

otolithic organs utricle and saccule in the vestibular apparatus responsible for sensing body position and acceleration

otoliths small particles, composed of a combination of a gelatinous matrix and calcium carbonate, in the viscous fluid of the saccule and utricle of the ear; inertia of these small particles causes them to stimulate hair cells when the head moves

outer-ring deiodination enzymatic removal of the iodine atom at position 5′ of the thyronine ring structure

outwardly rectifying potassium channels myocardial membrane channels that are responsible for carrying hyperpolarizing K^+ currents out of myocardial cells, resulting in cell repolarization during phase 3 of the myocardial cell action potential

overshoot part of the action potential in which the inside of the cell becomes positive with respect to the extracellular compartment

oviduct fallopian tube, which receives the ovum immediately after ovulation

ovulation release of an egg from a follicle in the ovary

ovulatory phase part of the menstrual cycle during which an egg is released; lasts 24 hours

oxidative burst cellular generation and release of antimicrobial reactive oxygen species; also known as *respiratory burst*

oxidative phosphorylation generation of adenosine triphosphate by the reactions of the Krebs cycle in the presence of O_2

oxidative stress imbalance between production and degradation of reactive oxygen species

oxidoreductase enzyme that catalyzes the transfer of electrons from one molecule (the reductant, also called the *hydrogen* or *electron donor*) to another (the oxidant, also called the *hydrogen* or *electron acceptor*)

oxygen-carrying capacity (OCC) total maximum amount of oxygen (mL/dL) carried by 100 mL of blood

oxygen content amount of oxygen (mL/dL) bound to hemoglobin in 100 mL of blood

oxygen debt measurably increased rate of oxygen consumption following strenuous activity. The extra oxygen is used in the processes that restore the body to a resting state and adapt it to the exercise just performed (hormone balancing, replenishment of fuel stores, cellular repair, innervation, and anabolism); also known as *excess postexercise oxygen consumption*

oxygen diffusion gradient partial pressure difference for oxygen across the alveolar–capillary membrane

oxygen saturation ratio of the amount of oxygen bound to hemoglobin to the carrying capacity of oxygen (O_2 content/O_2 capacity)

oxygen uptake transfer of oxygen from the alveoli to the pulmonary capillary blood

oxyhemoglobin (Hbo$_2$) oxygen-saturated form of hemoglobin

oxyhemoglobin equilibrium curve curve that reflects the relationship between oxygen saturation and the partial pressure of oxygen

oxyntic glands one of three types of glands found in the fundus and corpus of the stomach that contribute to gastric secretions; they are the most abundant and contain parietal (oxyntic) cells, chief cells, mucous neck cells, and some endocrine cells; also known as *gastric glands*

oxytocin substance that is produced by magnocellular neurons in the supraoptic and paraventricular nuclei of the hypothalamus and that stimulates smooth muscle contraction in the mammary glands and uterus

Pacinian corpuscle very highly adaptive receptor that is responsible for sensitivity to pressure as well as pain

Paget disease condition in which the existing bone is resorbed, with the simultaneous overgrowth of new, poorly calcified, irregular bone

palmitoyltransferase dimeric complex of the endoplasmic reticulum that catalyzes *S*-palmitoylation, the addition of palmitate (C16:0), or other long-chain fatty acids to proteins at a cysteine residue

pampiniform plexus venous plexus accompanying the testicular artery up to the superficial inguinal ring and thought thereby to be involved in maintenance of the testicles at some 3°C below body temperature. It forms from veins draining the testis and epididymis at the mediastinum of the testis and unites within the inguinal canal to form the left and right testicular veins

pancreatic amylase substance that continues in the digestion of the remaining carbohydrates at the point at which salivary amylase stops. Pancreatic secretions must first neutralize chyme for pancreatic amylase to work best, at a neutral pH level. The products of pancreatic amylase digestion of polysaccharides are maltose, maltotriose, and α-limit dextrins

pancreatic lipase substance that hydrolyzes lipids, specifically the triglyceride molecule to a 2-monoglyceride and two fatty acids, in the intestinal lumen

pancreatic polypeptide (PP) 36-amino acid peptide secreted by islet cells of the pancreas in response to a meal; of uncertain physiologic function

Paneth cells found in the intestinal tract, these cells contain zinc and lysozyme as well as large eosinophilic refractile granules within their apical cytoplasm; their exact function is unknown but due to the presence of lysozyme, it is likely that Paneth cells contribute to host defense and protection of intestinal stem cells; when exposed to bacteria or bacterial antigens, Paneth cells secrete a number of lysozymes into the lumen of the crypt, thereby contributing to maintenance of the gastrointestinal barrier

panhypopituitarism generalized or particularly severe hypopituitarism, which in its complete form leads to absence of gonadal function and insufficiency of thyroid and adrenal cortical function. Dwarfism, regression of secondary sex characteristics, loss of libido, weight loss, fatigability, bradycardia, hypotension, pallor, depression, and many other manifestations may occur; also known as *pituitary failure*

papillary light reflex in controlling the diameter of the pupil, this reflex regulates the intensity of light entering the eye

parabrachial nucleus location of neurons in the upper pons that coordinate cardiac and respiratory functions

paracrine signaling signaling in which the signal substance is released into the local environment and moves to receptors on target cells by diffusion in the extracellular milieu

paradoxical sleep stage of sleep, also known as REM (rapid eye movement), in which the electroencephalogram exhibits unsynchronized, high-frequency, low-amplitude waves similar to the awake state, yet the subject is as difficult to arouse as when in stage 4 slow-wave (deep) sleep

parafollicular cells cells that are embedded in the wall of the thyroid follicle and that produce and secrete calcitonin

parallel fibers axons from cerebellar granule cells to the dendrites of the Purkinje cells

paralytic ileus obstruction of the bowel resulting from paralysis of the bowel wall, usually as a result of open abdominal surgery, peritonitis, or shock; sometimes called *adynamic ileus*

paramesonephric (müllerian) ducts ducts that fuse at the midline and develop into the oviducts, uterus, cervix, and upper portion of the vagina in females

paraneoplastic syndrome neuropathic degeneration of autonomic neurons resulting from autoimmune attack in patients with small cell carcinoma of the lung

paraquat compound found in herbicides that is toxic to lung tissue and skin

parasympathetic division division of the autonomic nervous system, with its outflow from the central nervous system in certain cranial nerves and sacral nerves, respectively, and having its ganglia in or near the innervated viscera

parasympathetic nervous system the division of the autonomic nervous system generally responsible for maintaining normal organ functioning, referred to as the "rest and digest" system

parathyroid hormone (PTH) hormone made by the parathyroid gland that helps the body store and use calcium; also known as *parathormone* or *parathyrin*

paraventricular nuclei portion of the hypothalamus in which magnocellular neuron cell bodies are found

paravertebral sympathetic ganglia collections of sympathetic neurons located parallel to the spinal cord; site of most synapses between preganglionic and postganglionic sympathetic neurons

parietal lobe posterolateral region of the cerebrum

parietal (oxyntic) cells one of two primary sources of gastric secretion; secrete hydrochloric acid and intrinsic factor, a peptide that is crucial for the absorption of vitamin B_{12}

parietal–temporal–occipital association cortex interface between the parietal, temporal, and occipital lobes of the cerebrum, which integrates visual, somatosensory, auditory, and cognitive stimuli

Parkinson's disease (PD) neurological disease characterized by hand tremor and slowed body movements as a result of loss of neurons in the substantia nigra

parotid gland largest salivary gland, found wrapped around the mandibular ramus, that secretes saliva into the oral cavity to facilitate mastication and swallowing

pars distalis major part of the adenohypophysis of the pituitary gland; also known as *anterior lobe*

pars intermedia thin diffuse region of cells between the anterior and posterior lobes of the pituitary in humans; also known as *intermediate lobe*

pars reticulata (SNr) group of neurons in the substantial nigra producing inhibitory output to eye movement control regions of the midbrain

pars tuberalis upward extension of the anterior lobe that wraps around the infundibular stalk; its cells, mostly gonadotropic, are arranged in cords and clusters; it is supplied by the superior hypophyseal arteries and contains the first capillary bed and the venules of a portal system that carries neurosecretory factors from the hypothalamus to a second capillary bed in the adenohypophysis where they regulate the release of hormones

partial pressure independent pressure exerted by an individual gas within a mixture of gases

partial seizures seizure preceded by an isolated disturbance of a cerebral function, such as twitching of a limb

partition coefficient measure of a molecule's solubility in oil compared with its solubility in water. The value will be high for a lipid-soluble molecule and low for a water-soluble molecule

parturition birth

parvicellular neurons neurons located in the paraventricular nuclei of the hypothalamus that synthesize corticotropin-releasing hormone

passive conduction passive spread of electrical signal over short segments of the axon

passive force length-dependent component of muscle force that is produced by stretching an inactivated muscle

passive immunity transmission of antibodies from the mother to the fetus via intrauterine transport and to the infant via the mother's colostrum, the thin, yellowish, milky fluid secreted by the mammary glands a few days before or after parturition. The absorption of immunoglobulins (predominantly IgG) plays an important role in the transmission of passive immunity from the mother's milk to the newborn in several animal species

passive transport membrane transport of a solute or ions down the prevailing electrochemical gradient. Solute movement dissipates the gradient and an equilibrium state is eventually reached at which net movement ceases. Passive transport of lipid-soluble solutes can occur without a carrier protein (diffusive membrane transport). Passive transport of ions occurs through ion channel proteins, and polar hydrophilic solutes must use a carrier protein. These are examples of facilitated diffusion

patch clamp sophisticated laboratory technique for studying the behavior of individual ion channels, based on detecting the small current generated when ions flow through an open channel

pathogen agent, primarily an organism, that causes disease

peak expiratory flow (PEF) maximum flow at the outset of forced expiration, which is reduced in proportion to the severity of airway obstruction, as in asthma

pellagra severe niacin deficiency characterized by dermatitis, dementia, and diarrhea

pendrin substance that transports iodide into the follicle to be used for iodination of the thyroglobulin precursor

pepsin endopeptidase in gastric juice that cleaves protein molecules from the inside, resulting in the formation of smaller peptides. Protein digestion starts in the stomach with the action of pepsin, which is secreted as a proenzyme and activated by acid in the stomach

pepsinogen substance secreted by chief cells that is a precursor of the enzyme pepsin and is found in gastric juice

peptic ulcer disease ulcerative lesions of the gastroduodenal area caused by reduced bicarbonate secretion and, in many circumstances, increased acid secretion resulting from an increased mean number of gastric parietal cells that appear to have increased sensitivity to gastrin. The stomach-emptying rate is greatly increased in patients with duodenal ulcers

peptidase any enzyme capable of hydrolyzing one of the peptide links of a peptide

peptide hormones peptides, such as prolactin, secreted in the bloodstream that serve endocrine functions

perforin protein of natural killer and cytotoxic T cells that forms a pore in target cells

perfusion limited characterized by the limitation of the uptake of an alveolar gas, primarily by pulmonary capillary blood flow

perfusion pressure pressure force driving blood through the circulation of an organ; usually synonymous with the blood pressure in the artery supplying an organ or tissue

periaqueductal gray matter region of the midbrain where integration of afferent and nerve impulses from higher cerebral centers occurs

pericarditis inflammation of the pericardium. It often restricts the ability of the ventricle to fill during diastole

perikaryon cell body of a neuron containing the nucleus; also called *soma*

perilymph fluid that fills the scala vestibuli and scala tympani and is high in sodium and low in potassium

peripheral chemoreceptors neurosensitive endings that respond to changes in blood chemistry and that are located in the carotid artery and the aorta

peripheral nervous system ganglia and nerves located outside of the CNS; the division of the nervous system localized outside of the brain and spinal cord and composed of nerves and ganglia

peripheral protein protein that is attached to but does not penetrate the phospholipid bilayer region of a cell membrane, for example, spectrin in the red blood cell

perisinusoidal space (space of Disse) space that is continuous with the pericellular space and separated from the sinusoid by a layer of sinusoidal endothelial cells

peristalsis coordinated contraction of inner circular muscle layer and longitudinal layer of the gastrointestinal tract that propel a bolus of food through the tract

perivitelline space region between the zona pellucida and the vitelline membrane of an oocyte

permeability coefficient number expressing the ease with which a solute can cross a plasma membrane. It accounts for factors such as the thickness of the membrane and the lipid solubility of the solute

permeability surface area coefficient coefficient that characterizes how much material per unit of time can pass across the capillaries by diffusion for a given concentration gradient across the capillary membrane. It is a function of the capillary surface area over which diffusion occurs as well as the permeability of the capillary membrane to the diffusing substance in question

pernicious anemia chronic, progressive anemia found in older adults as a result of failure of vitamin B_{12} absorption, usually resulting from the lack of secretion of intrinsic factor. Pernicious anemia is characterized by numbness and tingling, weakness, diarrhea, loss of weight, and fever

peroxidases enzymes that neutralize peroxynitrites

peroxynitrite free radical formed in the presence of a superoxide ion and nitrous oxide

pH measure of the acidity or alkalinity of a solution. It is equal to the negative logarithm (to the base 10) of the hydrogen ion concentration (the latter is measured in mol/L)

phagocyte cell capable of engulfing particulate material

phagocytosis engulfment of large extracellular particles and microorganisms by expansion of the plasma membrane around the particle to eventually surround it. A typical example is the destruction of invading bacteria by macrophage cells

phagosome phagocytic vacuole for the killing of enclosed pathogens

pharmacomechanical coupling mechanism in gastrointestinal smooth muscles in which the binding of a ligand to its receptor on the muscle membrane leads to the opening of calcium channels and the elevation of cytosolic calcium without any change in the membrane electrical potential; the ligands may be chemical substances released as signals from nerves (neurocrine), from nonneural cells in close proximity to the muscle (paracrine), or from endocrine cells as hormones delivered to the muscle by the blood

phase I reactions reactions in which a parent xenobiotic compound is biotransformed into a more polar compound by the introduction of one or more polar groups, usually hydroxyl (OH) or carboxyl (COOH). Usually, this is accomplished by the oxidation of the parent compound

phase II reactions reactions in which phase I products conjugate to various possible products (i.e., glucuronic acid, catalyzed by glucuronyl transferases), yielding a more hydrophilic end compound

phasic contraction contraction not long maintained (transitory), in contrast to a tonic contraction

phlebotomist technician trained to draw blood

phorbol ester parent alcohol of the cocarcinogens, which are 12,13(9,9a) diesters of phorbol found in croton oil; the hydrocarbon skeleton is a cyclopropabenzazulene; phorbol esters mimic 1,2-diacylglycerol as activators of protein kinase C

phosphatidylinositol 4,5-bisphosphate (PIP$_2$) membrane lipid found in all eukaryotic cells, which regulates many important cellular processes, including ion channel activity

phosphaturia excess of phosphates in the urine

phosphodiesterase (PDE) enzyme that cleaves phosphodiester bonds to yield a phosphomonoester and a free hydroxyl group; for example, cyclic adenosine monophosphate (cAMP) is inactivated by phosphodiesterase action to 5′ AMP; this cleavage is mediated by a cyclic nucleotide phosphodiesterase enzyme

phosphoenolpyruvate carboxykinase (PEPCK) enzyme in the gluconeogenic pathway catalyzing oxaloacetate conversion to phosphoenolpyruvate

phosphoenolpyruvate carboxylase enzyme in the family of carboxy-lyases that catalyzes the addition of bicarbonate to phosphoenolpyruvate to form the four-carbon compound oxaloacetate

phosphofructokinase enzyme that catalyzes the conversion of fructose 6-phosphate to fructose 1,6-diphosphate

phospholamban regulatory protein that inhibits sarcoplasmic reticulum Ca^{2+} pumps. Its inhibitory effects are removed when it is phosphorylated

phospholipase A$_2$ major pancreatic enzyme for digesting phospholipids, thereby forming lysophospholipids and fatty acids

phospholipase C (PLC) lipase that hydrolyzes phosphatidylinositol 4,5-bisphosphate

phospholipid subclass of lipid molecules based on glycerol 3-phosphate to which are attached two long-chain fatty acids (hydrophobic), for example, palmitic acid, and a specific polar or charged head group (hydrophilic), for example, choline or inositol

phosphorylation transfer of the terminal phosphate of adenosine triphosphate to a protein, under the control of a kinase-type enzyme

photon quantum of the electromagnetic interaction and the basic unit of light and all other forms of electromagnetic radiation

photopic (daytime) vision color vision that is served by the cone cells of the retina

photoreceptors sensory receptors that respond to visual stimuli

phylloquinones vitamin K that is derived from green vegetables

physiological dead space volume total volume of wasted air that does not participate in gas exchange and is the sum of anatomical dead space and alveolar dead space volume

physiological ileus normal absence of motility in a region of the gastrointestinal tract; requires an intact enteric nervous system in the intestine

physiological shunt total amount of venous admixture from anatomical shunts and regions of low ventilation/perfusion ratios

physostigmine (eserine) blocker of the neuromuscular junction that inhibits the action of acetylcholinesterase

piloerection mechanism whereby furred or hairy animals can increase the thickness of their coat and insulating properties by making the hairs stand on end. This response makes a negligible contribution to heat conservation in humans, but manifests itself as gooseflesh

pineal gland gland located superficial to the dorsal midbrain and posterior dorsal thalamus that produces melatonin in response to signals from the suprachiasmatic nucleus of the hypothalamus relayed through the sympathetic nervous system

pinocytosis process by which a cell ingests extracellular fluid and its contents by forming narrow channels through its membrane that pinch off into vesicles and fuse with lysosomes that hydrolyze contents

pituitary dwarfism decrease in the rate of body growth as a result of growth hormone deficiency

pituitary gland gland lying in the sphenoid bone and connected to the hypothalamus; also known as *hypophysis*

place theory theory of hearing that states that our perception of sound depends on where each component frequency produces vibrations along the basilar membrane, such that the pitch of a musical tone is determined by the places where the membrane vibrates, based on frequencies corresponding to the tonotopic organization of the primary auditory neurons

placenta structure composed of maternal and fetal tissues attached to the inner surface of the uterus

placing reaction cortical reflex producing limb movement, provoked in a dependent limb touching a horizontal surface

plaque the name given to an advanced atherosclerotic lesion in the intima of large arteries consisting of a necrotic, fibrotic core of lipid-laden cells, cholesterol crystals, and calcium deposits covered by an unstable fibrous cap. Plaques can rupture, exposing thrombogenic materials to blood, which then forms a local thrombosis that in turn can occlude the artery.

plasma liquid portion of the blood; an extracellular fluid

plasma cell B-cell–derived cell that secretes antibody

plasma membrane outer surface of the cell, which separates the contents of the cell from the extracellular fluid

plasma membrane calcium ATPase (PMCA) transport protein in the plasma membrane of cells that serves to remove calcium from the cells

plasmapheresis therapeutic plasma exchange

plasmin proteolytic enzyme that acts on the follicular wall during rupture

plasminogen inactive proenzyme that is converted to plasmin by tissue plasminogen activator, which is released by activated endothelial cells

plasminogen activator see *tissue plasminogen activator*

plateau in cardiovascular physiology, the extended depolarized plateau in phase 2 of the action potential in myocardial cells. It is the basis of the long refractory period seen in ventricular and atria muscle cells and is the reason tetanic contractions cannot be produced in those cells.

platelet anuclear, irregular disklike fragment of megakaryocytes that is important for hemostasis

platelet count (Plt) used as a starting point in the diagnosis of hemostasis

platelet-derived growth factor (PDGF) cytokine involved in wound healing; stimulates the proliferation of vascular smooth muscle and endothelial cells

plegia complete paralysis of a muscle

pleiotropic producing more than one effect

plethysmograph airtight box designed to measure a patient's residual lung volume

pleural pressure (P_{pl}) pressure between the chest wall and the lungs (i.e., the pleural space)

pleural space fluid-filled space between the chest wall and the lungs

pneumonia disease of the lungs characterized by inflammation and consolidation of lung tissue and accompanied by fever, chills, cough, and difficulty in breathing; it is caused chiefly by infection and is categorized as *bronchopneumonia*, *lobar pneumonia*, or *primary atypical pneumonia*

pneumothorax collapsed lung

podocytes epithelial cells with extensions that terminate in foot processes that rest on the outer layer of the basement membrane of the glomerular membrane of the kidney

poikilocyte irregularly shaped erythrocyte

poikilotherms poikilothermic animal; denoting the so-called cold-blooded animals, such as the reptiles and amphibians

Poiseuille's law law that quantifies the relationship between pressure, flow, and resistance in the cardiovascular system. It states that blood flow is proportional to the pressure difference across any two points along the vascular system and inversely proportional to the resistance to flow. It can also be used to show that pressure anywhere in the cardiovascular system is the direct product of blood flow and vascular resistance. The work of Poiseuille showed that vascular resistance is proportional to vessel length and blood viscosity but that it is inversely proportional to the fourth power of the internal vessel radius

polyclonal antibody antibodies obtained from different B-cell resources; they are a combination of immunoglobulin molecules secreted against a specific antigen, each identifying a different epitope

polycystic kidney disease (PKD) most common life-threatening inherited disorder in which numerous cysts develop in both kidneys, producing massive enlargement of the kidneys and ultimately complete renal failure

polycystic ovarian syndrome heterogeneous group of disorders characterized by amenorrhea or anovulatory bleeding, an elevated luteinizing hormone/follicle-stimulating hormone ratio, high androgen levels, hirsutism, and obesity

polycythemia proportion of blood volume that is occupied by red blood cells increases abnormally due either to an increase in the mass of red blood cells (*absolute polycythemia*) or to a decrease in the volume of plasma (*relative polycythemia*); blood volume proportions can be measured as hematocrit level

polydipsia Extreme thirst in response to excess fluid loss. Common causes of increased fluid loss are from rigorous exercise, diarrhea, vomiting, profuse sweating, hemorrhaging and certain prescription medications. Increase thirst can also occur as a result of high blood sugar and is often one of the first symptoms of both diabetes mellitus and diabetes insipidus.

polymorphonuclear leukocyte granulocyte, with a segmented or lobated nucleus

polyspermy fertilization of an egg by more than one sperm

polyuria condition characterized by excessive excretion of dilute urine, for example, diabetes insipidus

pontine respiratory group group of cells in the medulla responsible for integrating signals and acting as an "off" switch for inspiration and expiration

pontine reticulospinal tract descending brainstem pathway carrying excitatory impulses to spinal interneurons from the pons

pores in the cardiovascular system, the gaps between endothelial cells, which allow the passage of water and solutes between the tissues and capillary blood

portal hypertension high blood pressure in the portal vein and its tributaries; it is often defined as a portal pressure gradient (the difference in pressure between the portal vein and the hepatic veins) of 10 mm Hg or greater

portal vein primary large vein that collects blood exiting the splanchnic organs, such as the stomach, pancreas, small intestine, and portions of the large intestine. It supplies blood into the hepatic circulation, where it is mixed with blood from the hepatic artery

positive feedback feedback that enhances the input and responds in the same direction as the perturbation and that involves the return of some of the output of the system as input, which helps control the process

positive selection process by which cells are recruited with a T-cell receptor able to bind major histocompatibility (MHC) class I and II molecules with at least a weak affinity, thereby eliminating (in "death by neglect") those T cells that would be nonfunctional due to an inability to bind MHC

postcapillary venules first veins in which blood enters from the capillaries

postganglionic axon axon of the postganglionic autonomic neuron

postganglionic sympathetic neurons neurons leaving prevertebral ganglia to the gut

postjunctional folds invaginations in the endplate membrane that increase its surface area

postmenopausal osteoporosis bone loss resulting from the deficiency of estrogen associated with menopause; also known as *type 1 osteoporosis*

postpartum thyroiditis usually occurring within 3 to 12 months after delivery, this disease is characterized by a transient destruction-induced thyrotoxicosis (hyperthyroidism), often followed by a period of hypothyroidism lasting several months; many patients eventually return to the euthyroid state

postprandial after meals

postrotatory nystagmus eyeball movements that follow an episode of body or head rotation, resulting from the continued motion of the endolymph in the semicircular canals

poststreptococcal glomerulonephritis nephritic condition that may follow a sore throat caused by certain strains of streptococci; endothelial cell damage, accumulation of leukocytes, and the release of vasoconstrictor substances reduce the glomerular surface area and fluid permeability and lower glomerular blood flow, causing a fall in the glomerular filtration rate

postsynaptic terminal the dendritic nerve ending specialized to receive information

potentiation interaction between two or more drugs or agents, resulting in a pharmacologic response greater than the sum of individual responses to each drug or agent

power output rate at which muscle does work on the environment

power propulsion intestinal motility pattern associated with defense and pathologic conditions; rapid propulsion of luminal contents over extended lengths of intestine

power stroke conformational change in the myosin molecule (crossbridge) that provides the driving force in muscle contraction

PR interval interval of time on the electrocardiogram between the end of the P wave and the start of the QRS complex. It is an estimate of the time it takes for action potentials to proceed from the atria into the ventricles and is a primary reflection of the delay of conduction through the atrioventricular node

prechondrocyte precursor to chondrocyte in the growth plates of bone

precocious puberty physical signs of sexual maturation before the age of 7 years in girls and 9 years in boys

precordial leads set of six unipolar electrocardiogram leads that are oriented in the horizontal plane of the body at the level of the heart. They are designated V_1 to V_6 and are placed in a semicircle around the left side of the chest, from just right of the sternum to the left midaxial point

prediabetes condition that occurs when a person's blood glucose levels are higher than normal but not high enough for a diagnosis of type 2 diabetes

prefrontal association cortex region of the cerebrum responsible for executive function (making decisions) and, by virtue of its connections with the limbic system, coordinating emotionally motivated behaviors

prefrontal cortex most anterior region of the cerebrum

preganglionic axon axon of the preganglionic autonomic neuron

preganglionic sympathetic neurons neurons in the intermediolateral cell column of the spinal cord that project axons to make synapse with one of the three prevertebral sympathetic ganglia (celiac, superior mesenteric, and inferior mesenteric ganglia)

pregnenolone precursor for steroid hormone synthesis derived from cholesterol

pregranulosa cells cells that surround the oocyte in a primordial follicle

preload passive stretch placed on a muscle prior to an active contraction; initial passive length to which muscle is stretched prior to the start of a contraction

premature atrial contraction (PAC) premature electrical activation of any portion of the atria outside the sinoatrial node or atrial ectopic foci

premature ovarian insufficiency early depletion of follicles prior to 40 years of age—is a state of hypogonadotropic hypogonadism

premature ventricular atrial contraction (PVC) premature electrical activation of any portion of the ventricles or ventricular ectopic foci

premotor area region of the cerebrum just anterior to the primary motor cortex; serves as the association cortex for the motor system responsible for coordinating movements around a joint

premotor cortex motor-related cortical region located on the lateral surface of the frontal lobe anterior to the primary motor cortex

preprohormone precursor form of a protein hormone extended at the amino termini by a hydrophobic amino acid sequence called *leader* or *signal peptide* and in addition containing internal cleavage sites that on enzymatic action yield different bioactive peptides

presbyopia condition characterized by reduced converging power (accommodation) of the eye, caused by increasing stiffness of the crystalline lens resulting from age

pressure technically, the force on a specific surface area and expressed as force per unit area (i.e., pounds per square inch); in physiological systems, it is measured as the force required to raise a column of water or mercury a given distance vertically against gravity and is expressed as mm Hg or cm H_2O

pressure diuresis response of the kidney to increase salt and water excretion whenever arterial pressure is elevated. It has the capacity to continue to increase excretion until blood pressure is returned to a level set by the body for mean arterial pressure. It is believed to be the mechanism whereby the kidney sets the long-term mean arterial pressure set point in the body, around which all cardiovascular reflexes operate

pressure–volume curve curve generated by measuring the amount of pleural pressure required to inflate the lungs to a given volume. The slope of the pressure–volume curve provides a measure of lung compliance

presynaptic facilitation effects of chemical mediators or drugs to increase the release of neurotransmitter at a synapse

presynaptic inhibition receptors on axons suppress the release of neurotransmitters at neural synapses and neuroeffector junctions

presynaptic inhibitory receptors receptors on axons that suppress the release of neurotransmitters at neural synapses and neuroeffector junctions

presynaptic terminal the axonal nerve ending from which neurotransmitters are released

prevertebral sympathetic ganglia sympathetic ganglia (celiac, aorticorenal, and superior and inferior mesenteric) lying in front of the vertebral column, as distinguished from the ganglia of the sympathetic trunk (paravertebral ganglia); these ganglia occur mostly around the origin of the major branches of the abdominal aorta, and all are in the abdominopelvic cavity; the neurons comprising the ganglia send postsynaptic sympathetic fibers to the gut

primary active transport mechanism see *active transport*

primary axons (type Ia) largest myelinated afferent axons from the muscle spindle

primary follicular stage ovarian follicle consisting of one layer of cuboidal granulosa cells and a basement membrane surrounding the oocyte

primary lymphoid organs see *lymphoid organs*

primary motor cortex (M1) region of the cerebral cortex that fits all four parameters of a cortical motor area and is located just anterior to the central sulcus

primary oocyte oocyte in first meiotic arrest

primary peristaltic wave peristaltic propulsion in the esophagus initiated by swallowing

primary somatosensory cortex (S1) region of the cerebral cortex that receives tactile input and is located just posterior to the central sulcus

primary spermatocyte spermatocyte that divides into two secondary spermatocytes

prime mover contracting muscle that contributes most to a movement

primordial follicle ovarian follicle consisting of a primary oocyte surrounded by a layer of pregranulosa cells

progesterone major natural progestogen, promoting the secretory function of the uterus

progesterone block maintenance of uterine quiescence throughout gestation by progesterone to prevent premature delivery

progestin or progestogen substance that promotes secretory function of the uterus

programmed cell death death of a cell mediated by signaling pathways invoked in the cell in response to unhealthy growth conditions (e.g., loss of growth factors, infection by a virus); a deliberate cell suicide; this death is altruistic because the whole organism benefits from the death of a cell, thus it does not initiate an immune response (*nonimmunogenic*); *apoptosis* is one type of programmed cell death; contrast with *necrosis*

progressive shock stage of shock in which the function of organs needed to maintain cardiac output and blood pressure (heart and brain) is compromised but in which the person can survive with prompt and aggressive medical intervention

prohormone precursor form of a protein hormone extended at the amino termini by a hydrophobic amino acid sequence called *leader* or *signal peptide*

prohormone convertase 1 (PC1) enzyme encoded by the *PCSK1* gene that differentially cleaves proopiomelanocortin to adrenocorticotropic hormone and β-lipotropin

prohormone convertase 2 (PC2) enzyme responsible for the first step in the maturation of many neuroendocrine peptides from their precursors, such as the conversion of proinsulin to insulin intermediates

prokinetic drug agent that acts on enteric nerve endings to enhance propulsion of contents through the gut and that may act by direct muscle stimulation, release of motor neurotransmitters, or blockade of inhibitory neurotransmitters

prolactin (PRL) hormone product of lactotrophs located in the anterior pituitary

proliferative phase phase of the endometrial cycle in which stromal and epithelial layers of the uterine endometrium undergo hyperplasia

pronucleus haploid nucleus with 23 chromosomes. One female pronucleus and one male pronucleus present in an egg 2 to 3 hours after sperm penetration and fuse to form the nucleus of a fertilized egg

proopiomelanocortin (POMC) prohormone from which adrenocorticotropic hormone, β-lipotropin, and α-melanocyte–stimulating hormone are derived

propagated potential the property of action potentials to regenerate and move along a nerve processes

proprioceptors sensory receptors that provide information as to body and limb position, muscle force, and similar properties from within the body

propriospinal cells interneurons that convey action potentials between adjacent spinal segments

propulsion controlled movement of ingested foods, liquids, gastrointestinal secretions, and sloughed cells from the mucosa through the digestive tract

propulsive segment intestinal segment identified by contraction of the circular muscle layer behind a receiving segment in which the circular muscle is inhibited and the longitudinal muscle layer contracted to produce a segment with expanded lumen

prostaglandin derivative of arachidonic acid that usually acts as a local hormone to affect many functions, such as blood flow and inflammation

prostaglandin E_2 substance released by blood vessel walls in response to infection or inflammation and acting on the brain to induce fever. The enzyme mPGES-1 is involved in the production of prostaglandin E_2 and is an important "switch" for activating the fever response

protein inclusion bodies protein aggregates that typically represent sites of viral multiplication in bacteria; can also signal genetic diseases such as Parkinson's disease

protein kinase A (PKA) see *cAMP-dependent protein kinase*

protein kinase C (PKC) phospholipid-dependent and calcium-dependent kinase that, when activated, phosphorylates serine or threonine residues in target proteins to regulate their function

protein kinase G (PKG) see *cGMP-dependent protein kinase*

protein S deficiency most common inherited thrombotic disorder

proteinuria condition characterized by the presence of proteins in the urine

proteoglycans mortar-like substances made from protein and sugar that are the building blocks of cartilage

prothrombin time (PT) test to monitor the extrinsic coagulation pathway

prothrombinase (factor X) clotting factor that initiates the common coagulation pathway

protooncogenes normal gene that, when altered by mutation, becomes an oncogene

proton pump see *H⁺-ATPases*

pseudohyperkalemia spurious elevation of the serum concentration of potassium occurring when potassium is released *in vitro* from cells in a blood sample collected for a potassium measurement

pseudohyponatremia artifactual decrease in plasma Na^+ resulting from greatly elevated levels of lipids or proteins in plasma

pseudomaturation process in which the oocyte enters meiosis during atresia of the follicle

pseudoobstruction pathological failure of propulsion in the gastrointestinal tract in the absence of mechanical obstruction

pseudoprecocious puberty early development of secondary sex characteristics in the absence of gametogenesis. Results from exposure of immature boys to androgens or immature girls to estrogen

pterygopalatine ganglion location of pre- to postganglionic synapse of parasympathetic neurons from the superior salivatory nucleus through cranial nerve VII, which are involved in control of the lacrimal and nasal glands

ptosis drooping of the eyelid as a result of loss of sympathetic activation of the Müller muscle of the eyelid

P-type ATPase see *ATPase*

pubarche development of pubic hair during pubertal development in boys and girls

puberty development of physical characteristics of sexual maturation

puborectalis muscle sling muscle that anchors to the symphysis pubis anteriorly and loops around the rectum to form the anorectal angle; this muscle is important to continence for stool

pulmonary artery single artery that transmits all blood pumped from the right ventricle into the lungs. This blood is deoxygenated venous blood from all the peripheral organs

pulmonary circulation lung circulation that perfuses the respiratory zone and participates in alveolar gas exchange

pulmonary edema abnormal collection of fluid in the lung, which can flood the alveoli and impair gas exchange

pulmonary embolism movement of a blood clot or other plug from the systemic veins through the right heart and into the pulmonary circulation, where it lodges in one or more branches of the pulmonary artery; most pulmonary emboli originate from thrombosis in the leg veins but can also originate from the upper extremities; other major sources of pulmonary emboli include air bubbles introduced during intravenous injections, hemodialysis, placement of central catheters, fat emboli (a result of multiple long bone fractures), tumor cells, amniotic fluid (secondary to strong uterine contractions), parasites, and various foreign materials in intravenous drug abusers

pulmonary embolus blood clot that has lodged in the microcirculation or farther upstream in the pulmonary arterial circulation; large pulmonary emboli can impede filling of the left heart to the extent that death results

pulmonary fibrosis (PF) the most common group of diseases known as interstitial diseases. The term fibrosis means scaring, and interstitial means the condition occurs in the area of the lung between alveoli. It is the scarring of the tissue between the alveoli that interferes with a patient's ability to breathe. In most cases, the cause is unknown, and these cases are diagnosed as idiopathic pulmonary fibrosis (IPF).

pulmonary hypertension potentially severe disease characterized by an increase in blood pressure in the lung vasculature including the pulmonary artery, pulmonary vein, or pulmonary capillaries; symptoms include shortness of breath, dizziness, fainting, and others, all of which are exacerbated by exertion; markedly decreased exercise tolerance and heart failure may result

pulmonary stretch receptors slowly adapting receptors located in the smooth muscle of the conducting airways

pulmonary surfactant a lipoprotein that secreted by alveolar epithelial type II cells that coats the inner lining of the alveoli. The major functions are to reduce surface tension at the alveolar air–liquid interface and to maintain alveolar stability at low lung volumes.

pulmonary wedge pressure pressure measured by wedging a Swan-Ganz catheter–tipped balloon up against small pulmonary vessels

pulmonic valve thin, flexible, trileaflet structure that separates the right ventricle from the pulmonary artery. It opens into the pulmonary artery during systole but closes during diastole

pulse pressure difference in systolic and diastolic arterial pressure. It is the change in pressure associated with each cardiac cycle and is a function of arterial compliance and stroke volume

purinergic neuron enteric neurons that use a purine nucleotide (e.g., adenosine triphosphate or adenosine) as a neurotransmitter

Purkinje cells largest neurons of the cerebellum and the source of efferent cerebellar axons

Purkinje fibers specialized set of modified myocardial fibers that line the endocardium and provide rapid spread of electrical excitation to the ventricles

putamen one of the basal ganglia, receiving input from the cerebral cortex

pyloric glands one of three main types of glands in the stomach's mucosal lining. These glands contain mucous cells that secrete mucus and bicarbonate ions, which protect the stomach from the acid in the stomach lumen. The pyloric glands are located in a larger area adjacent to the duodenum and contain cells similar to mucous neck cells. The pyloric glands consist of many gastrin-producing cells, called *G cells*

pyloric sphincter sphincter at the boundary of the gastric pylorus and duodenum

pylorus lower section of four stomach sections that facilitates gastric emptying into the small intestine

pyramidal tract major motor control pathway for axons from the cerebral cortex to the spinal cord

pyrogens exogenous and endogenous substances that induce fever, such as the bacterial substance LPS and interleukin-1, which both increase the thermoregulatory set point in the hypothalamus. The endogenous pyrogens may also come directly from tissue necrosis

pyrosis burning sensation in the chest usually associated with gastroesophageal reflux; also called *heartburn* or *acid indigestion*

pyruvate carboxylase enzyme of the ligase class that catalyzes the irreversible carboxylation of pyruvate to form oxaloacetate

pyruvate dehydrogenase allosteric enzyme that transforms pyruvate into acetyl-CoA, which is then used in the citric acid cycle

pyruvate kinase enzyme that catalyzes the transfer of a phosphoryl group from phosphoenolpyruvate to adenosine diphosphate, yielding a pyruvate molecule

pyruvic acid end product of the glycolytic pathway

P wave initial, small positive deflection on the electrocardiogram resulting from depolarization of the atria

Q₁₀ phenomenon in which the rate of a biological process increases when the temperature is increased by 10°C; this value is obtained using the following formula: $Q_{10} = $ rate at $(T + 10°C)/$ rate at T, where T = starting temperature; for biological systems, the Q_{10} value is generally between 1.5 and 2.5

QRS complex complex, spiked, largely positive deflection on the electrocardiogram resulting from depolarization of the ventricles

QRS interval time interval associated with ventricular depolarization and represented by the time interval of the QRS complex on the electrocardiogram

QT interval interval of time associated with depolarization and repolarization of the ventricles. It measures the interval of time between the initiation of the Q wave and the end of the T wave on the electrocardiogram and is sometimes referred to as *electrical systole*

quantum number of transmitter molecules released by any one exocytosed synaptic vesicle

quantum content total number of quanta (vesicles filled with a certain number of transmitter molecules) released when the synapse is activated

Q wave first downward (negative) deflection of the electrocardiogram following the isoelectric interval after completion of the P wave

radioimmunoassay (RIA) scientific method used to test antigens (e.g., hormone levels in the blood) without the need to use a bioassay. It involves mixing known quantities of radioactive antigen (frequently labeled with γ-radioactive isotopes of iodine attached to tyrosine) with antibody to that antigen and then adding unlabeled or "cold" antigen and measuring the amount of labeled antigen displaced

radioreceptor assay important modification of the radioimmunoassay that uses specific hormone receptors rather than antibodies as the hormone-binding reagent. In theory, this method measures biologically active hormone, because receptor binding rather than antibody recognition is assessed

rapid adaptation process in phasic sensory receptors in which the output to the central nervous system decreases quickly while the applied stimulus is maintained in intensity

rapid adapting receptors sensory terminals of myelinated afferent fibers found in large conducting airways; also referred to as *irritant receptors*

rapid ejection phase portion of the cardiac cycle starting immediately after the opening of the aortic valve, in which blood is ejected into the aorta from left ventricular contraction and continuing up to the peak ejection flow rate and peak aortic systolic pressure; it comprises the first third of the ejection phase of ventricular contraction, during which 70% of the stroke volume is ejected

rapid eye movement (REM) sleep see *REM sleep*

rapid ventricular filling first third of filling of the ventricle during diastole

rarefaction reduction of a medium's density; the opposite of *compression*

ras (RAt sarcoma) protein subfamily of small GTPases that are generally responsible for signal transduction leading

Rathke pouch evagination of the oral ectoderm from which the adenohypophysis is formed

reactive hyperemia response of a tissue to a period of ischemia or no blood flow in which blood flow exceeds the value seen before the ischemia or flow stoppage once the ischemia or stoppage is removed

reactive oxygen species (ROS) collective term to include free radicals (hydroxyl radical, superoxide ion, and hydrogen peroxide)

receptive relaxation reflex triggered by stimulation of mechanoreceptors in the pharynx followed by transmission over afferents to the dorsal vagal complex and activation of efferent vagal fibers to inhibitory musculomotor neurons in the gastric enteric nervous system

receptor protein molecule on the cell surface or located within the cell that receives and responds to a neurotransmitter, hormone, or other substance

receptors signaling systems that reside either in the plasma membrane or within cells and are activated by a variety of extracellular signals or first messengers, including peptides, protein hormones and growth factors, steroids, ions, metabolic products, gases, and various chemical or physical agents (e.g., light); divided into two categories: *cell surface receptors*

(G protein–coupled receptors, ion channel–linked receptors, and enzyme-linked receptors) and *intracellular receptors* (steroid and thyroid hormone receptors)

receptor (membrane) membrane surface proteins with unique conformations for specific binding of chemical signal substances and triggering of changes in the cell's behavior

receptor adaptation general loss in the efficacy of a receptor-mediated response after sustained stimulation of the receptor

receptor-mediated endocytosis see *exocytosis*

receptor potential characteristic electrical record of a sensory receptor; a generator potential specific to a given organ, such as the electroretinogram or the cochlear microphonic

reciprocal inhibition synaptic pattern that causes inhibition of antagonist muscles during contraction of an agonist

rectoanal reflex relaxation of the internal anal sphincter in response to mass movement of feces distention of the rectosigmoid region of the large intestine

red blood cells most common cell in the blood and the principal means of delivering oxygen to cells via the circulatory system; also referred to as *erythrocytes*

red cell distribution width (RDW) measures the degree of the average size dispersion of erythrocytes and thereby indicates the degree of anisocytosis

red marrow part of bone marrow that forms blood cells

red nucleus group of neurons in the midbrain that receive input from the cerebral cortex and cerebellum and produce output to the spinal cord via the rubrospinal tract

reduced ejection phase ejection phase of the cardiac cycle in which the rate of ventricular outflow starts to diminish. It is the last phase of systole

reduced ventricular filling filling phase of the cardiac cycle in which the rate of ventricular filling begins to diminish. It is the last phase of diastole

5α-reductase steroidogenic enzyme that converts testosterone to dihydrotestosterone

reentry tachycardia type of ventricular tachycardia caused by continual recycling of conducted action potentials through an abnormal pathway in the heart

referred pain pain from deep structures perceived as arising from a surface area remote from its actual origin; the area where the pain is appreciated is innervated by the same spinal segment(s) as the deep structure; also known as *reflective pain*

reflection coefficient measure of the relative permeability of a particular membrane to a particular solute; calculated as the ratio of observed osmotic pressure to that calculated from van't Hoff law; also equal to 1 minus the ratio of the effective pore areas available to solute and to solvent

reflex neuronal event occurring beyond volition; in neurophysiology, it is a relatively simple behavioral response of an effector produced by influx of sensory afferent impulses to a neural center and its reflection as efferent impulses back to the periphery to the effector (e.g., muscle); the neural center may consist of interneurons; the simplest reflex circuit consists only of sensory and motor neurons

reflex sympathetic dystrophy (RSD) chronic, painful, and progressive neurological condition that affects the skin, muscles, joints, and bones; usually develops in an injured limb, such as a broken leg, or following surgery; however, many cases of RSD involve only a minor injury, such as a sprain, and in some cases, no precipitating event can be identified; also called *complex regional pain syndrome*

regional compliance pressure–volume relationship at a specific region of the lung (i.e., the apex and base); *compliance* is the ability of the lungs to stretch during a change in volume relative to an applied change in pressure

regional hypoventilation condition in which a region of the lung is being underventilated with respect to an adjacent region of the lung

regional hypoxia hypoxic condition in a regional part of the lung, as contrasted with generalized hypoxia; see *generalized hypoxia*

regulated exocytosis form of exocytosis in which vesicles are filled, stored in the cytoplasm, and their contents released only when a specific extracellular stimulus arrives at the cell membrane

regulatory light chains small protein constituents of the myosin molecule that regulate its interaction with actin

regulatory volume decrease mechanism by which solute and water exit the cell to correct a previous increase in volume resulting from the osmotic entry of water

regulatory volume increase (RVI) mechanism by which solute and water enter the cell to correct a previous decrease in volume resulting from the osmotic exit of water

relative humidity quantity of water vapor that exists in a gaseous mixture of air and water

relative pressure pressure relative to atmospheric pressure

relative refractory period window of time following one action potential during which a greater-than-usual stimulus is required to elicit an action potential

relaxin a polypeptide hormone produced by the corpus luteum and the decidua that assists parturition by softening the cervix and increasing oxytocin receptors

releasing factors see *releasing hormones*

releasing hormones hypothalamic hormones that reach the anterior pituitary lobe by a portal system of capillaries and regulate the secretion of most hormones of the endocrine system; also known as *releasing factors*

REM sleep rapid eye movement sleep; a phase of sleep in which the eye movement becomes rapid and the electroencephalograph becomes desynchronized, with a loss of skeletal muscle tone

renal blood flow (RBF) volume of blood delivered to the kidneys per unit time

renal plasma flow (RPF) volume of blood plasma per unit time flowing through the kidneys

renin proteolytic enzyme produced by granular cells in the kidney that catalyzes the formation of angiotensin I from angiotensinogen

renin–angiotensin–aldosterone system salt-conserving system

renin–angiotensin–converting enzyme enzyme in lung endothelium that converts angiotensin I to angiotensin II

repolarization change in membrane potential that returns the membrane potential to a negative value after the depolarization phase of an action potential has just previously changed the membrane potential to a positive value

residual volume (RV) volume remaining in the ventricle at the end of systole

resistance capacity of a membrane that inhibits flow of ions across the membrane

resistance to thyroid hormone (RTH) rare syndrome in which thyroid hormone levels are elevated but thyroid-stimulating hormone level is not suppressed

respiration process of exchanging oxygen and carbon dioxide from the cells

respiratory acidosis abnormal condition caused by an accumulation of carbon dioxide causing the blood to become too acidic, with a concomitant decrease in pH

respiratory alkalosis abnormal process characterized by the loss of too much CO_2, which leads to a rise in arterial blood pH

respiratory burst rapid cellular release of reactive oxygen species; also known as *oxidative burst*

respiratory exchange ratio (R) ratio of the volume of carbon dioxide exhaled to the volume of oxygen taken up

respiratory quotient (RQ) steady-state ratio of carbon dioxide produced by tissue metabolism to oxygen consumed in the same metabolism; for the whole body, normally about 0.82 under basal conditions; in the steady state, the respiratory quotient is equal to the respiratory exchange ratio

respiratory sinus arrhythmia repetitive increase and decrease in heart rate seen on inspiration and expiration of the lungs

resting force force present in a muscle prior to stimulation

resting length length of an unstimulated muscle; its natural length in the body

resting membrane potential membrane potential of an excitable cell in the unstimulated state. It is negative relative to the extracellular fluid and is a steady-state potential defined by the electrochemical gradients for ions that can pass through leakage channels in the membrane and the contribution of an electrogenic pump

resting state the state of the voltage-gated sodium channel in which it is closed but capable of being opened

reticular formation region in the central portion of the brainstem where collections of nerve cells regulate integrative autonomic and spinal reflex functions and coordinate levels of consciousness and attention

reticulocyte young erythrocyte with residual nuclear material

reticuloendothelial system (RES) phagocytic cells in connective tissue responsible for removing old cells, debris, and pathogens from the bloodstream

retinohypothalamic tract projection of axons of retinal ganglion cells in the eye to the suprachiasmatic nucleus of the hypothalamus

retinoid X receptor (RXR) nuclear hormone receptor that binds to other nuclear receptors to form heterodimers that activate transcription

retinol principal form of vitamin A; plays a primary role in vision as well as growth and differentiation. It is derived directly from animal sources or through conversion from beta-carotene (found abundantly in carrots) in the small intestine

retinol-binding protein (RBP) protein synthesized by the liver that binds with retinol in response to falling levels of vitamin A. RBP secretion depends on vitamin A availability

retinopathy deterioration of the blood vessels of the retina, leading to blindness

retinyl ester reesterification of retinol by the enzyme lecithin: retinol acyltransferase. Retinyl ester is incorporated in chylomicrons and taken up by the liver

retrograde direction of movement of material from the tip of an axon toward the cell body of the neuron

retrograde amnesia loss of memory laid down prior to a certain point in time

retrograde transport transport of cellular components from the nerve process to the soma

retropulsion movement of luminal contents in the oral direction

reversal potential voltage across the endplate membrane determined by the Na^+ and K^+ concentrations inside and out

reverse T$_3$ (rT$_3$) reverse triiodothyronine; a substance generated by inner-ring deiodination of thyroxine

reward system circuit in the brain involved in pleasurable sensation comprising projections from neurons in the midbrain ventral tegmental area to the nucleus accumbens in the cerebrum and other limbic structures

Reynolds number (Re) calculated unit used to predict when turbulence will occur in a fluid flow stream. It takes into account the contribution of fluid flow velocity, fluid viscosity, and the geometry of the tube through which the flow is occurring

rheology study of the fluid properties of blood, which is a suspension and, therefore, a nonhomogeneous fluid

rhesus factor (Rh) protein on the erythrocyte surface of some people

rheumatoid arthritis (RA) traditionally considered a chronic, inflammatory autoimmune disorder that causes the immune system to attack the joints. It is a disabling and painful inflammatory condition, which can lead to substantial loss of mobility as a result of pain and joint destruction

rhodopsin pigment of the retina that is responsible for both the formation of the photoreceptor cells and the first events in the perception of light; also known as *visual purple*

rhythmicity property of all heart cells by which they can generate their own action potentials in a repetitive, uniform, or rhythmic manner. It naturally occurs in the nodal tissue of the healthy heart and is responsible for setting the intrinsic heart rate but can be manifested in ectopic foci and thus cause cardiac arrhythmias

rib cage 12 pairs of ribs that are hinged to the vertebral column

rickets disorder of normal bone ossification manifested by distorted bone movements during muscular action. Rickets is caused by vitamin D deficiency

right atrium upper chamber of the right heart. It receives the entire venous blood supply returning from all the systemic organs. It empties into the right ventricle through the tricuspid valve

right bundle-branch block condition in which transmission of action potentials is blocked in the right branch of the bundle of His

right ventricle lower chamber of the right heart that receives blood from the right atrium through the tricuspid valve and pumps blood across the pulmonic valve into the pulmonary artery

right ventricular hypertrophy abnormally increased muscle mass in the right ventricle. It occurs in response to any factor that chronically increases wall stress on the right ventricle, such as pulmonary hypertension or a stenotic pulmonic valve

right-to-left shunt anatomical shunt in which blood goes directly from the pulmonary artery to the pulmonary vein and bypasses the alveoli. Shunted blood does not get oxygenated

rigor crossbridge myosin crossbridge strongly attached to actin and lacking adenosine triphosphate

rigor mortis postmortem muscle stiffness resulting from rigor crossbridges in the absence of adenosine triphosphate

rising phase the portion of the action potential in which the membrane potential is approaching 0 mV

rolling slow movement of leukocytes in blood vessels along the endothelium

rough endoplasmic reticulum (RER) eukaryotic organelle that forms an interconnected network of tubules, vesicles, and cisternae within cells; functions to synthesize proteins

rouleaux phenomenon stacking of erythrocytes like coins in blood

rubor redness associated with acute inflammation

rubrospinal tract motor control pathway for axons from the red nucleus to the spinal cord

rugae gastricae longitudinal folds in the corpus of the empty stomach

R wave in the normal heart, it is the highest amplitude, sharp, positive deflection of the QRS complex of the electrocardiogram (ECG); it is associated with a positive deflection in all frontal ECG leads except aV_R and is caused by ventricular depolarization during electrical activation of the heart

ryanodine receptor (RyR) controllable channel in the sarcoplasmic reticulum of muscle that allows calcium release

SA node sinoatrial node; the normal "pacemaker" of the heart. It can generate its own action potentials in a repetitive manner and is responsible for setting the intrinsic heart rate. It is the first area of the heart to display electrical activity during the cardiac cycle

saccades rapid sideways movements of the eyeballs that change the direction of vision

safety factor at the myoneural junction, the amount by which the released transmitter substance exceeds that necessary for effective transmission

salivary glands heterogeneous group of exocrine glands that produces saliva; innervated by both the parasympathetic and sympathetic divisions of the autonomic nervous system. Parasympathetic stimulation of the salivary glands results in increased salivary secretion. The parotid, submandibular (submaxillary), and sublingual glands are the major salivary glands. They are drained by individual ducts into the mouth

salivon basic unit of the human submandibular gland, consisting of the acinus, the intercalated duct, the striated duct, and the excretory (collecting) duct—the network of acini and ducts that comprises the salivary glands

saltatory conduction process in a myelinated nerve whereby action potentials are successively generated at neighboring nodes of Ranvier and thus the action potential appears to jump from one node to the next

sarcolemma plasma membrane of a muscle cell, along with a fine fibrillar connective tissue network

sarcomere fundamental organized contractile unit of a skeletal or cardiac muscle; includes the overlapping myofilaments between two Z lines

sarcoplasmic reticulum elaboration of the endoplasmic reticulum of muscle that contains, releases, and takes up calcium during the control of contraction

satellite cell precursor to a muscle cell

satiety sensation of fullness or being well fed

S cells type of endocrine cell in the intestinal mucosa that releases secretin induced by the entry of acidic chyme from the stomach into the small intestine during the intestinal phase of digestion

scalene muscles three paired accessory skeletal muscles in the lateral neck that elevate the rib cage during forced inspiration

Scatchard plot graph used to find the equilibrium constant for a reaction such as $[H] + [R] \rightleftharpoons [HR]$ ($[HR]/[H]$) versus $[HR]$, or any functions proportional to these quantities; the magnitude of the slope of the graph is the equilibrium constant

schistocyte erythrocyte fragment generated as cell flows through damaged vessels

schizophrenia psychiatric illness marked by disordered thoughts, delusions (fixed false beliefs), inappropriate emotional response, auditory hallucinations, social withdrawal, and lethargy

Schwann cell glial cells in the PNS that produce a myelin sheath around neuronal axons in the periphery

scintigraphic techniques use of radioisotopes, either ingested with food or released from swallowed capsules (in this case) to measure transit times through different regions of the gastrointestinal tract

scotopic (nighttime) vision noncolor vision that is served by the rod cells of the retina

scotopsin protein moiety in retinal rods that combines with 11-*cis*-retinal to form rhodopsin

scurvy disorder characterized by weakness, fatigue, anemia, and bleeding gums caused by vitamin C deficiency

second heart sound second normally audible sound of contraction of the heart caused by vibrations set up in the ventricular chamber following abrupt closure of the aortic and pulmonic valves at the onset of ventricular relaxation (diastole); its intensity is proportional to the intensity of valve closure and is increased whenever aortic or pulmonary pressure is abnormally high; thus, it is taken as a clinical indicator of possible systemic or pulmonary hypertension, respectively

second messenger intermediary molecule that is generated as a consequence of the binding of a first messenger to a receptor. These intermediary molecules are generated in large quantities, and this amplifies the signaling initiated by the first messenger. Examples of second messengers are cyclic adenosine monophosphate, cyclic guanosine monophosphate, calcium, inositol 1,4,5-triphosphate, and diacylglycerol

second polar body small haploid cell produced by unequal second meiotic division of an oocyte

secondary active transport mechanisms see *active transport*

secondary axons (type II) afferent axons from nuclear chair fibers of the muscle spindle, smaller in diameter than primary (type I) endings

secondary follicular stage stage during which several layers of granulosa cells and a single theca surround the oocyte

secondary lymphoid organs see *lymphoid organs*

secondary oocyte haploid oocyte produced by completion of the first meiotic division

secondary pacemaker ectopic foci that repetitively activates in the myocardium, thereby pacing the heart at a rate other than that set by the sinoatrial node

secondary peristaltic wave peristaltic propulsion in the esophagus initiated by an initial block of forward movement of a swallowed bolus or by the presence of acid in the esophagus

secondary sex characteristics external sex-typical male and female structures

secondary spermatocytes spermatocyte that divides to form the spermatids

second-degree atrioventricular block any condition in which some, but not all, action potentials from the atria reach the ventricles. It usually arises from blockage at the level of the atrioventricular node and results in an irregular heart rate

secretagogue substance stimulating secretion

secretin hormone secreted by the duodenal and jejunal mucosa when exposed to acid chyme that stimulates pancreatic secretion rich in bicarbonate ions. Secretin inhibits gastrin release

secretomotor neuron motor neurons in the submucous plexus, which innervate and evoke secretion from the intestinal glands

secretory phase phase of the endometrial cycle during which the endometrial glands become coiled, store glycogen, and secrete large amounts of carbohydrate-rich mucus. The stroma increases in vascularity and becomes edematous, and the spiral arteries become tortuous

segmentation specific pattern of small intestinal motility that starts with the ingestion of a meal and accomplishes mixing of the contents in the lumen; also called *digestive motility*

seizure sudden abnormal synchronized discharge of electrical activity in the brain

selective serotonin reuptake inhibitors (SSRIs) class of compounds typically used as antidepressants to treat depression, anxiety disorders, and some personality disorders

selenocysteine rare amino acid with properties that make it ideal for catalysis of oxidoreductive reactions

self-tolerance see *immune tolerance*

sella turcica depression in the sphenoid bone of the skull

semantic memory memory of facts

semen combination of sperm, seminal fluid, and other male reproductive secretions

semicircular canals three orthogonally oriented fluid-filled tubes of the vestibular apparatus

seminal emission movement of sperm and associated fluids from the cauda epididymis and vas deferens into the urethra

seminal plasma fluid portion of semen, secreted by the male sex accessory glands

seminiferous tubules tiny tubes in the testes where sperm cells are produced, grow, and mature

sensorineural deafness hearing impairment due to damage of the inner ear nerve endings, the cochlear portion of the eighth cranial nerve, the vestibulocochlear nerve, or the cortical hearing center

sensory modality specific type of sensory activity; the kind of sensation

sensory neuron neuron that conducts impulses arising in a sense organ or at sensory nerve endings

sensory receptor cell or part of a cell specialized and normally functioning to convert environmental stimuli into nerve impulses or some response, which, in turn, evokes nerve impulses; most sensory receptor cells are nerve cells, but some nonnervous cells can be receptors (e.g., intestinal enteroendocrine cells); one method of classifying receptors is by the form of adequate stimulus: chemoreceptors, osmoreceptors, and mechanoreceptors are normally stimulated by chemicals, osmotic pressure differences, and mechanical events, respectively

sensory transduction process whereby environmental signals (radiation, mechanical forces, or chemical quantities) are transformed into trains of nerve impulses

sepsis bloodstream infection causing excessive inflammation

septal nuclei group of cholinergic neurons located in the basal forebrain area with reciprocal connections with the hippocampus

septic shock decrease in blood pressure associated with overproduction of nitric oxide in infected tissue. It is characterized by high, unnecessary blood flow to the affected organs at the expense of flow to other organs in the body

serosa one of four layers of the stomach wall; lies over the muscularis externa, consisting of layers of connective tissue continuous with the peritoneum

serotonergic neuron type of neuron that uses serotonin as a neurotransmitter

serotonin mediator of the inflammatory response and a neurotransmitter used by central and peripheral neurons; also known as *5-hydroxytryptamine* or *5-HT*

serous cells cells that are found in the salivary glands, contain an abundance of rough endoplasmic reticulum, and secrete digestive enzymes

Sertoli cell cell, found in large numbers lining the semen-producing tubules of the testis, that provides support and nourishment for developing sperm

serum blood plasma without clotting factors

serum iron levels test measuring iron levels in blood as part of the iron profile

serum protein electrophoresis electrophoretic separation test of blood proteins

set point any one of a number of quantities or physiological variables (e.g., body weight, body temperature) that the body tries to keep at a particular value or in a state of dynamic constancy

sex hormone–binding globulin (SHBG) substance that binds sex steroids in the blood

SH2 domain structurally conserved protein domain contained in intracellular signal–transducing proteins that "helps" a protein find its way to another protein by recognizing phosphorylated tyrosine on the other protein

shallow water blackout condition in which a person passes out underwater as a result of hypoxia to the brain

shear stress "rubbing" or frictional effect exerted along the walls of a cylinder as a result of fluid flowing in the cylinder. This force tends to strain or deform the inner portion of the wall in the direction of flow. In arteries, shear stress has an important influence on the function of the endothelium and influences the location of the development of atherosclerotic plaques

shivering bodily function in response to early hypothermia in warm-blooded animals. When the core body temperature drops, the shivering reflex is triggered. Muscle groups around the vital organs begin to shake in small movements in an attempt to create warmth by expending energy. Shivering can also be a response to a fever, as a person may feel cold, although the core temperature is already elevated

short hypophyseal portal vessels vessels that deliver blood into the sinusoids of the anterior pituitary

short-term memory newly acquired information that can be readily recalled for only a few minutes

shunt diversion of blood away from the alveoli

sickle cell anemia hereditary disorder that leads to an abnormal production of erythrocytes

sieving action selective gastric emptying of particles according to the particle size with the smallest particles emptying first

signal cascade series of biochemical reactions in which the products of one reaction are consumed in the next reaction

signal transduction process by which a cell converts one kind of signal or stimulus into another through a series of biochemical reactions that are often linked by second messengers, which serve to amplify the initial stimulus

silent nociceptor unresponsive sensory afferent nerve terminal that becomes responsive to mechanical stimulation during inflammatory states

simple acid–base disturbance one of the four primary acid–base disturbances

simple micelle micelle in which bile salts alone are present

single breath test test to determine oxygen uptake in a patient, using a single breath of a low concentration of carbon monoxide

single effect establishment of a modest osmotic gradient at any level of the loop of Henle. This gradient is developed into a larger gradient along the axis of the loop by countercurrent multiplication

single nucleotide polymorphism (SNP) DNA sequence variation occurring when a single nucleotide in the genome differs between members of a species or paired chromosomes in an individual

sinus bradycardia decreased heart rate caused by normal physiological effects on the sinoatrial node

sinus tachycardia elevated heart rate caused by normal physiological effects on the sinoatrial node

sinusoidal endothelial cells cells that lack a basement membrane and have sievelike plates, thereby permitting the ready exchange of materials between the perisinusoidal space and the sinusoid

size principle pattern of muscle activation in which the smallest motor neurons and associated motor units are activated first and then, as more power is needed, larger size motor neurons are activated

skeletal muscle striated muscle tissue that is associated with the skeleton as well as the diaphragm, parts of the esophagus, and some voluntary alimentary sphincters (e.g., the external anal sphincter). It is an excitable tissue that can contract and generate force when activated electrically by neurotransmitters released from its associated motor neuron.

sleep apnea a serious sleep disorder that occurs when breathing is interrupted during sleep. There are two types of sleep apnea: (1) obstructive sleep apnea and (2) central sleep apnea. Obstructive sleep apnea is by far the more common of the two forms, and is caused by airway blockage. The blockage usually occurs when soft tissue in the back of the throat collapses during sleep. With central sleep apnea the airways are not blocked, but it is due to a dysfunction in the respiratory control center in the brain that fails to signal the respiratory muscles to contract.

sleep apnea syndrome sleep disorder characterized by abnormal pauses in breathing that last a few seconds to minutes

sliding filament theory schema for muscle contraction involving the relative motion of actin and myosin filaments driven by the action of myosin crossbridges

slow adaptation process in static sensory receptors in which the output to the central nervous system is maintained (or falls slowly) while the applied stimulus is maintained in intensity

slow axoplasmic transport mechanism by which structural proteins, such as actin, neurofilaments, and microtubules, are transported from the cell body to the axon; occurs at 1 to 2 mm/d

slow excitatory postsynaptic potential a postsynaptic potential characterized by membrane depolarization and augmented excitability that lasts for extended periods of seconds or minutes

slow wave membrane electrical potential that waxes and wanes in a rhythmic fashion in the gastrointestinal tract; the slow wave determines the timing of a contraction but is not enough to produce a contraction in the absence of a burst of spike potentials triggered by the release of a neurotransmitter

slow-twitch fibers skeletal muscle fibers specialized for long-duration activity, using aerobic metabolism

slow-wave sleep (SWS) four stages of progressively deepening sleep during which the electroencephalogram becomes progressively slower in frequency and higher in amplitude

small-molecule transmitters amino acids and monoamines used as neurotransmitters

smooth muscle muscle tissue that lacks the striated microscopic appearance of skeletal and cardiac muscle. It is the most diverse and complex of all the muscle tissues, being able to generate complex contraction patterns to a multitude of physical, chemical, hormonal, paracrine, and neurotransmitter stimuli. Unlike skeletal and cardiac muscle, smooth muscle can be made to relax actively.

SNARES proteins found on both the synaptic vesicle and plasma membrane that cause the synaptic vesicle to dock and bind to the presynaptic terminal membrane

sodium appetite craving for salt

sodium–iodide symporter (NIS) substance located on the basal plasma membrane of the thyroid follicular cell that transports iodide into the cell

solvent drag one mechanism by which sodium is absorbed by the duodenum and jejunum

soma neuronal cell body; also called *perikaryon*

somatic pain sensation of pain arising from nonvisceral parts of the body

somatomedins class of small peptides responsible for the mediation of some growth hormone actions that increase the rate of cartilage synthesis by promoting the uptake of sulfate and the synthesis of collagen; also known as *insulin-like growth factors*

somatostatin (1) gastrointestinal hormone secreted by endocrine cells in the antrum of the stomach that inhibits the release of gastrin and consequently gastric acid secretion when the pH of the stomach lumen falls below 3; (2) polypeptide from the hypothalamus that inhibits secretion of growth hormone; also known as *somatotropin release–inhibiting factor (SRIF)*

somatotopic maps discrete groupings of motor or sensory neurons associated with functions of specific areas of the body

somatotrophs cells located in the anterior pituitary that secrete growth hormone

somatotropin release–inhibiting factor see *somatostatin*

space constant (λ) length along a membrane required for an electrotonic membrane potential to decay to 37% of its maximal value; also called *length constant*

spasticity feature of altered skeletal muscle performance occurring in disorders of the central nervous system impacting the upper motor neuron in the form of a lesion; when there is a loss of descending inhibition from the brain to the spinal cord, such that muscles become overactive, this loss of inhibitory control can cause an ongoing level of contraction, with decreased ability for the affected individual to volitionally control the muscle contraction, and increased resistance felt on passive stretch

spatial summation summation of two or more membrane potentials generated at different sites; also called *motor unit summation*

special senses (i.e., hearing, sight, smell, and taste) senses that have specialized organs devoted to them

specific compliance lung compliance divided by the functional residual capacity; used to correct for lung size

specific gravity measure of blood density

spectrin cytoskeletal protein that lines the intracellular side of the plasma membrane, forming a scaffold important role in maintenance of plasma membrane integrity and cytoskeletal structure

spermatids any of the four haploid cells formed by meiosis in a male organism that develop into spermatozoa without further division

spermatogenesis process of producing sperm, the male reproductive cells

spermatogonia any of the cells of the gonads in male organisms that are the progenitors of spermatocytes

spermatozoa mature sperm cells

spermiation release of mature spermatozoa from the surface of the Sertoli cell into the lumen of the seminiferous tubule

spermiogenesis process by which spermatids mature into sperm cells

spherocyte abnormal, spherical-shaped erythrocyte

sphincter of Oddi ring of smooth muscle surrounding the opening of bile ducts in the duodenum that regulates the flow of bile and pancreatic juice into the duodenum and prevents the reflux of intestinal contents into the pancreatic ducts

sphygmomanometer external, noninvasive device for measuring arterial pressure producing external pressure on a limb that is measured by a column of mercury or calibrated gauge

spinal shock decrease in blood pressure resulting from physical trauma to the spinal cord

spinnbarkeit highly elastic cervical mucus

spinobulbar muscular atrophy neurodegenerative neuromuscular disease associated with mutation of the androgen receptor rarely affecting females; also known as *Kennedy disease*

spinocerebellar tracts afferent pathways bringing sensory axons from the spinal cord to the cerebellum or associated brainstem nuclei

spinocerebellum portion of the cerebellum concerned with coordination of axial and proximal limb motor function

spiral arteries specialized arteries that supply blood into the maternal placental sinus

spirogram recording from the spirometer, which is an instrument for recording respiratory movements

spirometry measurement of lung volumes by a spirometer

splanchnic nerves mixed nerves in the mesentery that have both sympathetic efferent and sensory afferent fibers

splay renal physiology, the rounding of the corner on the graph relating tubular reabsorption or secretion of a substance to its arterial plasma concentration, resulting in part from the fact that some nephrons reach their tubular maximum before others

splenomegaly enlargement of the spleen usually associated with hyperfunction of the organ SRY sex-determining region gene on the Y chromosome; required to produce male gonadal differentiation

SRY (sex-determining region, Y chromosome) found on the Y chromosome; the gene encodes a transcription factor required for the development of the testes

S-type enteric neurons synaptic-type enteric neurons; neurons with a single long process (axon) and so named because nicotinic fast excitatory postsynaptic potentials occur in virtually all of the S-type enteric neurons

standard curve quantitative research tool and method of plotting assay data that are used to determine the concentration of a substance, particularly proteins and DNA

standard frontal lead system three frontal bipolar lead system developed by Einthoven. It consists of a lead I, in which the right arm has a negative polarity and the left arm is positive; a lead II, in which the right arm has a negative polarity and the left foot is positive; and a lead III, in which the left arm has a negative polarity and the left foot is positive

Starling forces all of the hydrostatic and osmotic forces within the capillary plasma and the interstitium that determine the direction and rate of bulk fluid transfer between the capillary blood and the interstitium

Starling-Landis equation illustrates the role of *hydrostatic* and *oncotic* forces in the movement of fluid across *capillary membranes*, which may occurs from diffusion, filtration, or pinocytosis: $J_v = K_h A [(P_c - P_t) - \sigma(COP_p - COP_t)]$

Starling's law of the heart law that states that stroke volume increases when ventricular filling increases and decreases when ventricular filling decreases; also known as *Starling law of the heart*

static compliance compliance curve generated with no airflow

statin drug used to lower plasma cholesterol level; inhibits HMG-CoA reductase, the rate-limiting enzyme of the mevalonate pathway of cholesterol synthesis

steady state state of a system in which the recently observed behavior of the system will continue into the future

steatorrhea excretion of unabsorbed lipids in the feces, caused by a pancreatic deficiency that significantly reduces the ability of the exocrine pancreas to produce digestive enzymes. Normally about 5 g/d of fat are excreted in human stool, but as much as 50 g/d can be excreted in the case of steatorrhea

stereocilia apical modifications of the cell characterized by their length and their lack of motility; found in the inner ear, the ductus deferens, and the epididymis

sternocleidomastoids paired muscles in the superficial layers of the anterior portion of the neck that, in addition to flexing and rotating the head, function as accessory muscles to elevate the rib cage during forced inspiration, along with the scalene muscles of the neck

steroid hormone lipid hormones generally synthesized from cholesterol in the gonads and adrenal glands that can freely traverse the plasma membrane to enter the cytoplasm as well as membranes of other organelles such as the nucleus they are carried in the blood bound to a specific carrier protein

steroidogenesis process of steroid hormone production

steroidogenic acute regulatory (StAR) protein protein that facilitates transport of cholesterol into the mitochondria

stigma specific site where the follicular wall thins by cellular deterioration and bulges for rupture

stimulus input to a sensory receptor that causes it to produce its characteristic output

stellate cells specialized cells that store vitamin A and are located in the space of Disse in the liver

stress relaxation slow and time-dependent fall in force of a viscoelastic substance held at a stretched length

stretch-activated calcium channels unique calcium channels found in the plasma membrane of vascular smooth muscle cells that open when the cell, or the artery in which it is contained, is stretched.

This results in calcium influx and contraction of the smooth muscle cell/artery. This channel is a separate entity from ligand or voltage-gated calcium channels in smooth muscle, and it believed to be responsible for the myogenic contractile response in arteries when they are exposed to an increase in intra-arterial pressure.

stria vascularis pseudostratified columnar vascularized epithelium containing melanocytes, which help maintain the high potassium concentration of the endolymph that is produced here

striae irregular areas of skin that appear as red or purple bands, stripes, or lines

striated duct duct lined with columnar cells that is part of the network of ducts that comprise the salivary glands, whose major function is to modify the ionic composition of the saliva

striatum caudate nucleus and putamen of the basal ganglia

stroke volume volume of blood ejected from the heart in one systole

stromal layer part of endometrium of the uterus; the closest layer to the myometrium

strong acid acid with a high dissociation constant, which consequently is highly ionized in solution

subcortical structures collections of neuronal cell bodies in the cerebrum that are located deep to the cerebral cortex

sublingual glands salivary glands that empty saliva into the mouth; located anterior to the submandibular gland under the tongue, beneath the mucous membrane of the floor of the mouth

submandibular ganglion location of synapse between pre- and postganglionic parasympathetic neurons from the superior salivatory nucleus through cranial nerve VII, which are involved in control of the submandibular glands

submandibular glands (submaxillary glands) paired salivary glands located beneath the floor of the mouth, accounting for 70% of the salivary volume

submucosa layer of dense irregular connective tissue in the stomach wall (one of four layers) or loose connective tissue that supports the mucosa as well as joins the mucosa to the bulk of underlying smooth muscle (fibers running circularly within layer of longitudinal muscle)

submucosal plexus ganglionated plexus of the enteric nervous system situated between the mucosa and circular muscle coat of the small and large intestine; it consists of an inner ganglionated plexus called the *Meissner plexus* and an outer ganglionated plexus called the *Schabadasch plexus*

substance P neuropeptide used as a neurotransmitter by small dorsal root ganglion cells that signal pain

substantia nigra group of nerve cells in the upper brainstem, where dopaminergic neurons important in Parkinson's disease are located

subthalamic nucleus group of nerve cells in the basal ganglia that receives input from the other basal ganglia nuclei and provides output to the globus pallidus internus

succinylcholine blocker of the myoneural junction that causes maintained depolarization of the endplate membrane

suckling reflex reflex in which the stimulation of sensory nerves in the breast by the infant initiates a neural signal to the hypothalamus, resulting in release of oxytocin to stimulate milk letdown

sudden infant death syndrome (SIDS) sudden death of an infant that is unexpected by history and cannot be explained following a thorough forensic autopsy; also known as *crib death*

sulcus valley on the surface of the brain

sulfonylureas any of a group of hypoglycemic drugs that act on the β cells of the pancreas to increase the secretion of insulin

superior cervical ganglion uppermost of the sympathetic ganglia located in the neck. Innervated by the sympathetic nervous system, sympathetic fibers arise in the upper thoracic segments of the spinal cord and synapse in the superior cervical ganglion.

Postganglionic fibers leave the superior cervical ganglion and innervate the acini, ducts, and blood vessels, resulting in a short-lived and small increase in salivary secretion

superior colliculus layered structure in which the superficial layers are sensory related and receive input from the eyes as well as other sensory systems; the deep layers are motor related, capable of activating eye movements as well as other responses; and the intermediate layers have multisensory cells and motor properties

superior hypophyseal arteries arteries that give rise to a rich capillary network in the median eminence

superior mesenteric ganglion uppermost of the prevertebral sympathetic ganglia located in the abdomen

superior parietal lobe region of the cerebral cortex that integrates complex sensory information

superior salivatory nucleus region of the pons where parasympathetic preganglionic neurons involved in control of the lacrimal and nasal glands are located

superior vena cava (SVC) single major vein that collects blood from the brain, head, and upper extremities and empties into the right atrium

superoxide dismutase (SOD) antioxidant enzyme that neutralizes superoxide ions in cells by catalyzing the breakdown of superoxide to oxygen and hydrogen peroxide (H_2O_2)

superoxide ion ($O_2 \bullet^-$) free radical that is formed during oxygen use in the mitochondria. The free radical is an oxygen molecule with an unpaired electron

supplementary motor area (SMA) cortical region on the medial aspect of the hemispheres that is active in motor control

suprachiasmatic nucleus (SCN) collection of neurons in the hypothalamus responsible for generating the body's circadian rhythms and entraining them to the day–night cycle

supraoptic nuclei portion of the hypothalamus in which magnocellular neuron cells bodies are found

supraventricular tachycardia abnormally high heart rate resulting from an ectopic foci in a region anatomically superior to the ventricles, usually the atria

surfactant surface-reducing material (lipoprotein) lining the alveoli

S wave first downward (negative) deflection of the electrocardiogram following the R wave

sympathetic division division of the autonomic nervous system, with its outflow from the central nervous system in the thoracolumbar segments of the spinal cord and having its ganglia in a pair of chains on either side of the spinal cord and in a grouping of three (celiac and superior and inferior mesenteric) in the abdomen

sympathetic nervous system the division of the autonomic nervous system modulating organ functioning during times of arousal, stress, and danger, referred to as the "fight-or-flight" system

sympathetic tone concept that some level of neuronal activity is present in the sympathetic neurons, even in an apparent resting state

sympathy coordination of bodily activities by the nervous system

symport membrane transport process in which the driver ion (usually Na^+) and the solute move in the same direction

synapse functional (noncytoplasmic) connection between neurons consisting of a presynaptic site of chemical transmitter release and a postsynaptic site of action of the released transmitter

synapsin a class of proteins in the nerve terminal that anchor neurotransmitter-containing vesicles within the cytoplasm

synaptic cleft extracellular space separating the presynaptic and postsynaptic membranes at a chemical synapse

synaptobrevin integral vesicular membrane SNARE protein

synaptotagmin integral vesicular membrane protein that acts as the Ca^{2+} sensor and binds proteins that cause the vesicles to dock and bind to the presynaptic terminal membrane

syncytiotrophoblast syncytial component of the trophoblast that lacks distinct cell boundaries

syndrome of inappropriate antidiuretic hormone (SIADH) disorder characterized by persistent hyponatremia

synergist muscle that acts in concert with the prime mover muscle

synthase enzyme that catalyzes synthesis of neurotransmitters in the enteric nervous system

systemic circulation circulation outside the thoracic cavity, not including the pulmonary circulation

systole period of time in which the heart is in its contraction phase

systolic hypertension a subclassification of chronic elevated arterial pressure that affects the systolic but not diastolic blood pressure. It is usually seen in the elderly as a result of the general decline of arterial compliance associated with the aging process.

systolic pressure blood pressure in the ventricle or arterial system during the contraction of the heart, or systole

tachyphylaxis rapid decrease in the response to a drug after repeated doses over a short period of time; increasing the dose of the drug will not increase the pharmacological response

tachypnea increased ventilation through rapid shallow breathing (i.e., increased breathing rate)

target cell (1) abnormal erythrocyte with central staining and a pale surrounding area; also known as *codocyte* or *leptocyte*; (2) cell that possesses specific receptors for a particular hormone

taste buds organized structures, located on the tongue and pharynx, that serve the gustatory sense

taste pore minute opening of a taste bud on the surface of the oral mucosa through which the gustatory hairs of the specialized neuroepithelial gustatory cells project

T cell type of lymphocyte with specific membrane receptors to recognize peptide antigens

T-cell receptor (TCR) recognition molecule of T cells for antigen–major histocompatibility complex protein complex

telencephalon region of the central nervous system; also called the *cerebrum* or *brain*

temporal lobe inferior and lateral region of the cerebrum

temporal summation time-wise addition of muscle force in response to repeated stimulation; the summation of two or more membrane potentials generated at the same location in quick succession

tension (1) *partial pressure* of dissolved gas in a liquid; (2) force exerted on the wall of a cylindrical structure that would have the tendency to rip the wall as a result of a positive transmural pressure. It is directly proportional to both the transmural pressure and the radius of the cylinder, such that in two cylinders exposed to the same transmural pressure, the cylinder with the larger internal radius experiences a greater wall tension. This principle is applied to blood vessels and other hollow organs, such as the heart and urinary bladder

terminal arterioles smallest arterioles; they lead directly into capillaries

terminal bronchioles smallest airways in the conducting zone

terminal cisternae lateral regions of the sarcoplasmic reticulum of striated muscle, from which calcium is released and where it is stored at rest; also called *lateral sacs*

terminal respiratory unit merging of alveolar ducts and their alveoli with adjacent pulmonary capillaries

tertiary follicular stage stage characterized by a large ovarian follicle containing an antral cavity

testosterone main steroid producing the male secondary characteristics

tetanic fusion frequency stimulation rate beyond which individual muscle twitches are no longer discernible

tetanus maintained muscle contraction caused by restimulating a muscle before mechanical relaxation is complete

tetanus/twitch ratio comparison between the maximal tetanic contraction and that of a single twitch

tetrahydrocortisol glucuronide major excretable metabolite of cortisol degradation

tetraiodoacetic acid (tetrac) derivatives formed by degradation of thyroxine

thalamus collection of neuronal cell bodies deep in the cerebrum adjacent to the midline that relays information to and from the cortex

theca externa external dense vascular connective tissue layer of the theca

theca interna internal dense vascular connective tissue layer of the theca

theca lutein cells cells found in the corpus luteum that are formed from the theca cells

thelarche development of breast buds in girls during puberty

The Law of Laplace mathematical relationship relating tension pulling on the wall of a thin-walled vessel to the transmural pressure across the wall. It is given by $T = P_r$, where T is wall tension and r is the internal vessel radius

therapeutic hypothermia (TH) medical treatment that lowers a patient's body temperature order to help reduce the risk of the ischemic injury to tissue that follows a period of insufficient blood flow; also called *controlled hypothermia*

thermic effect of food increment in energy expenditure above resting metabolic rate resulting from the cost of processing food for storage and use; one of the components of total metabolism, along with the resting metabolic rate and the exercise component; also known as *specific dynamic action*

thermodilution in cardiovascular physiology, a technique used to measure cardiac output by the indicator dilution technique in which temperature is used as the indicator

thermogenic action action of thyroid hormones to regulate the basal rate of body heat production and of oxygen consumed

thermoneutrality range of ambient temperatures at which energy expenditure is lowest and no metabolic heat is required to maintain body temperature; thermal balance is maintained with neither shivering nor sweating

thermoreceptors sensory receptors that respond to variation in environmental temperature; separate populations of receptors serve the sensations of warmth and coldness

theta waves electroencephalogram wave pattern with a rhythm ranging from 4 to 7 Hz, observed during sleep

thirst conscious sensation associated with craving for drink, ordinarily interpreted as a desire for water

thirst center located in the anterior hypothalamus, close to the neurons that produce and control arginine vasopressin release; this center relays impulses to the cerebral cortex, so that thirst becomes a conscious sensation

thoracic cavity cavity that houses the lung and is synonymous with the chest cavity

thoracolumbar division division analogous to the sympathetic division of the autonomic system; also called *sympathetic division*

threshold intensity level below which sensation is not possible. Accessory structures and processes may vary this level

threshold concentration concentration that must be exceeded to begin producing a given effect or result or to elicit a response

threshold potential the membrane potential at which an action potential is generated

thrombi fat globules or air bubbles that form in the bloodstream that act like clots and block vital blood flow

thrombin coagulation protein

thrombocyte cells that play a key role in blood clotting: in mammals, thrombocytes are anucleated cell fragments called *platelets*;

thrombocytes of nonmammalian vertebrates have a nucleus and, thus, resemble B lymphocytes

thrombocytopenia condition characterized by a lower-than-normal number of circulating platelets

thrombopoietin liver/kidney hormone regulating production of thrombocytes

thymus lymphoid organ in which T lymphocytes mature

thyroglobulin (Tg) large protein that is a storage form of the thyroid hormones

thyroid hormone receptor (TR) receptor located in the nuclei of cells that binds thyroid hormones to facilitate their action on transcription

thyroid hormone response elements (TREs) sequence of DNA in a gene that binds the thyroid hormone receptor

thyroid peroxidase (TPO) substance that catalyzes the iodination of thyroglobulin

thyroid-stimulating hormone (TSH) hormone product of thyrotrophs located in the anterior pituitary

thyrotroph cells located in the anterior pituitary that secrete thyroid-stimulating hormone

thyrotropin-releasing hormone (TRH) polypeptide from the hypothalamus that stimulates secretion of thyroid-stimulating hormone

thyroxine (T_4) peptide hormone produced by the thyroid gland that is a precursor to triiodothyronine

thyroxine-binding globulin (TBG) glycoprotein synthesized and secreted by the liver that binds thyroid hormone in the blood

tidal volume volume of air inhaled with each breath

tight junction closely associated areas of two cells whose membranes join together forming a virtually impermeable barrier to fluid

time constant (τ) time required for the membrane potential to decay to 37% of its initial peak value following an influx of ions; a function of membrane resistance and capacitance

tissue growth factors chemical messengers that influence cell division, differentiation, and cell survival; they may exert effects in an autocrine, paracrine, or endocrine fashion

tissue thromboplastin (factor III) factor extrinsic to blood but released from injured tissue during hemostasis; also called *tissue factor*

tissue plasminogen activator (TPA) substance that has the ability to cleave plasminogen and convert it into plasmin, its active form

titin large filamentous muscle protein that may help prevent overextension of sarcomeres and maintain the central location of A bands

titratable acid acid involved in renal physiology; this term is used to explicitly exclude ammonium (NH_4^+) as a source of acid and is part of the calculation for net acid excretion

toll-like receptor transmembrane protein that recognizes pathogens, activates innate immune responses, and regulates adaptive immune responses

tonic contraction degree of tension, firmness, or maintained contraction in a muscle, in contrast to a phasic contraction

tonic–clonic seizures type of generalized seizure affecting the entire brain; convulsion, characterized by generalized muscle spasms and loss of consciousness, most commonly associated with epilepsy; formerly known as *grand mal seizures* or *grand mal seizures*

tonicity quality of a solution related to the concentrations of nonpenetrating solutes, that is, solutes that do not cross the plasma membrane of a cell. Cells placed in a **hypotonic** solution will accumulate water and their volume will increase. Cells placed in a **hypertonic** solution will lose water and their volume will decrease. Cell volume will not change in an **isotonic** solution because there is no net movement of water into or out of the cell

tonotopic organization place-specific frequency discrimination provided over the length of the basilar membrane in the organ of Corti

total bile acid pool total amount of bile acids in the body, primary or secondary, conjugated or free, at any time

total body water volume of water in the body

total iron-binding capacity (TIBC) test analyzing the amount of iron needed to bind to all transferrin

total lung capacity (TLC) total volume of air in the lungs at the end of maximum inspiration

total peripheral vascular resistance (TPR) resistance to flow created by all the systemic vascular segments in the cardiovascular system from the origin of the aorta to the entry of the right atrium. Quantitatively, this value is dominated by the contribution made from the resistance of the arterial side of the circulation, which is much greater than the resistance to flow attributed to the veins. TPR is used in Poiseuille's law to calculate mean arterial pressure or cardiac output in the cardiovascular system as MAP = CO × TPR. MAP, microtubule-associated protein; CO, carbon monoxide

totipotency ability of a cell to divide and produce all the differentiated cells in an organism

toxoid inactivated toxin, used as vaccine against the toxin

trachea main airway of the lung through which respiratory air is brought into the lungs; it is held open by up to 20 C-shaped cartilage rings and prevents airway collapse during forced expiration; commonly known as *windpipe*

tractus solitarius brainstem structure that carries and receives visceral sensation and taste from the facial (VII), glossopharyngeal (IX), and vagus (X) cranial nerves

transairway pressure pressure difference across the airways (airway pressure–pleural pressure)

transcellular fluid specialized extracellular fluid that is separated from the blood plasma not only by a capillary endothelium but by a continuous layer of epithelial cells. It includes aqueous humor, cerebrospinal fluid, digestive secretions, sweat, synovial fluid, renal tubular fluid, and bladder urine

transcellular transport transport of molecules through an epithelial cell, generally from the apical to basolateral side

transcobalamin protein to which vitamin B_{12} binds and by which it is transported in the portal blood

transducer any device that changes one form of energy (e.g., from the environment) to a signal for further signal processing (e.g., by the central nervous system)

transducin heterotrimeric G protein naturally expressed in vertebrate retina rods and cones by activation of the effector cGMP phosphodiesterase, which degrades cGMP to 5′GMP

transferrin molecule that functions in the plasma transport of iron and in the maintenance of circulating iron level homeostasis. Plasma transferrin levels are inversely proportional to the iron load of the body

transferrin saturation ratio of iron serum levels and total iron-binding capacity

transient outward K⁺ channels voltage-gated myocardial cell membrane channels that carry outward, hyperpolarizing K⁺ current shortly after the initiation of a myocardial cell action potential. They are responsible for phase 1 of the action potential

transient receptor potential (TRP) channels family of loosely related ion channels that are relatively nonselectively permeable to cations, including sodium, calcium, and magnesium

transit time time required for red blood cells to move through alveolar capillaries; under normal resting conditions, the transit time is 0.75 seconds

transmigration see *diapedesis*

transmitter-specific transporter integral membrane proteins that remove neurotransmitter from the synaptic cleft by high-affinity reuptake into the presynaptic terminal or perisynaptic astrocytes (in the central nervous system)

transmural pressure literally transmural or "across the wall"; the pressure difference between the inside and outside of a hollow structure in the body

transneuronal transport transport of trophic (or other) substances from one cell to another in either the retrograde or anterograde direction

transpulmonary pressure pressure difference across the lung (i.e., alveolar pressure–pleural pressure)

transthyretin binds thyroxine and triiodothyronine in the blood

transverse tubules (T tubules) membrane-bound invaginations of the striated muscle surface membrane that conduct surface electrical activity inward as part of the excitation–contraction coupling process

triacylglycerols see *triglycerides*

triad structure in skeletal muscles, formed by a transverse (T) tubule surrounded by sarcoplasmic reticulum (including two terminal cisternae)

tricuspid valve a thin, flexible, trileaflet structure that separates the right atrium from the right ventricle. It opens into the right ventricle during diastole but closes off the right atrium when the right ventricle contracts

triglycerides most abundant dietary lipids, consisting of a glycerol backbone esterified in the three positions with fatty acids (more than 90% of the daily dietary lipid intake is in the form of triglycerides); also known as *triacylglycerols*

triiodothyronine (T_3) peptide hormone produced by the thyroid gland that regulates growth, development, and metabolism

trituration process of crushing and grinding ingested food by the gastric pump

trophic hormones hormones produced and secreted by the anterior pituitary that target endocrine glands

trophoblast single outer layer of the blastocyst consisting of extraembryonic ectodermal cells

tropomyosin fibrous protein constituent molecule of the actin filaments that participates in regulating striated muscle contraction

troponin protein complex in striated muscle thin filaments that binds calcium and, acting through tropomyosin, allows interaction between actin and myosin

Trousseau sign indication of latent tetany in which carpal spasm occurs when the upper arm is compressed, as by a tourniquet

trypsin one of three endopeptidases present in pancreatic juice that convert proenzymes to active enzymes. Trypsin splits off basic amino acids from the carboxyl terminal of a protein

tuberoinfundibular system neurons located within the hypothalamus, with cell bodies in the arcuate nucleus and periventricular nuclei and terminals in the median eminence on the ventral surface of the hypothalamus; responsible for the secretion of hypothalamic-releasing factors into a portal system that carries them through the pituitary stalk into the anterior pituitary lobe

tubular reabsorption transport of materials by the kidney tubule epithelium out of the tubular urine

tubular secretion transport of materials by the kidney tubule epithelium into the tubular urine

tubular transport maximum (Tm) maximum rate at which a particular substance is reabsorbed (e.g., glucose) or secreted (e.g., *p*-aminohippurate) by the kidney tubules

tubuloglomerular feedback mechanism blood flow control mechanism operating in the kidneys that limits changes in glomerular filtration rate

tubulovesicular membranes extensive smooth endoplasmic reticulum membrane system in the stomach

tumor swelling associated with acute inflammation

tumor necrosis factor (TNF) important cytokine (TNF-α, cachexin, or cachectin) involved in systemic inflammation and the acute-phase response; released by white blood cells, endothelium, and several other tissues in the course of damage, for example, by infection. Its release stimulates fever

tumor-specific antigens molecules present only on tumor cell surfaces to be recognized by cytotoxic T cells or B cells

turbulence chaotic random motion of particles in a flow stream. It results when the kinetic component of flow exceeds the viscous forces holding fluid elements together. In the cardiovascular system, turbulence dissipates more pressure energy than does laminar flow of similar quantitative magnitude

turbulent flow random movement of fluid in a tube (airway or blood vessel)

Turner syndrome congenital abnormality caused by a nondisjunction of one of the X chromosomes, resulting in a 45 XO chromosomal karyotype

T wave represents repolarization of the ventricles in electrocardiography

twitch single brief contraction of a muscle as the result of a single action potential

two-cell, two-gonadotropin hypothesis steroidogenesis completed by cooperation between granulosa and theca cells

type I collagen type of protein chemical substance that is the main support of skin, tendon, bone, cartilage, and connective tissue

type 1 diabetes (T1D) autoimmune disorder in which the body's immune system attacks the β cells in the islets of Langerhans, resulting in a decreased or complete absence of the production and secretion of insulin; formerly known as *childhood* or *juvenile diabetes* or *insulin-dependent diabetes*

type 1A diabetes subclass of type 1 diabetes characterized as immune-mediated

type 1B diabetes subclass of type 1 diabetes characterized as non–immune-mediated

type 2 diabetes (T2D) metabolic disorder characterized by insulin resistance, relative insulin deficiency, and hyperglycemia; formerly known as *diabetes mellitus type II*, *non–insulin-dependent diabetes (NIDDM)*, *obesity-related diabetes*, or *adult-onset diabetes*

type II pneumocytes granular, cuboidal cells typically found at the alveolar–septal junction that cover a much smaller surface area than *type I cells* (<5%) but are much more numerous; they are responsible for the production and secretion of surfactant (the majority of which are dipalmitoylphosphatidylcholine), a group of phospholipids that reduce the alveolar surface tension and are stored in type II pneumocytes in lamellar bodies (specialized vesicles unique to type II cells)

tyrosine hydroxylase rate-limiting enzyme in the synthesis of catecholamines that catalyzes the conversion of L-tyrosine to L-3,4-dihydroxyphenylalanine

tyrosine kinase receptors family of receptors each having a tyrosine kinase domain (which phosphorylates proteins on tyrosine residues), a hormone-binding domain, and a carboxyl terminal segment with multiple tyrosines for autophosphorylation

UDP glucose uridine diphosphate glucose; a molecule involved in the biosynthesis of glycoproteins or glycolipids

ultrafiltrate filtrate formed by ultrafiltration

ultrafiltration filtration through a selectively permeable membrane that allows passage of small molecules but restricts the passage of large molecules, such as proteins

umami one of five main taste sensations referring to that which is "meaty" or savory

umbilical arteries fetal blood vessels that carry blood containing wastes and CO_2 from the fetus into the fetal placenta

umbilical cord cord that connects the fetal circulation to the placenta

umbilical veins fetal blood vessels that pick up oxygen from the placental sinus and transport oxygenated blood to the fetus

uncoupling protein-1 mitochondrial protein that uncouples the membrane proton gradient from adenosine triphosphate synthesis

unipolar neuronal cells with only one process emanating from the soma

unitary-type smooth muscle predominant kind of smooth muscle in the gastrointestinal tract; it contracts spontaneously, contracts in response to stretch, and does not have structured neuromuscular junctions; contrast to *multiunit-type smooth muscle*

universal donor person with type O blood who can, therefore, donate to people with all blood groups

universal recipient person with AB blood group who can, therefore, receive blood from someone with any other blood type

unloading phase steep portion of the oxygen equilibrium curve that reflects the unloading of oxygen from hemoglobin

unsaturated iron–binding capacity (UIBC) calculated amount of transferrin not occupied by iron; UIBC equals total iron-binding capacity minus iron

unstirred water layer layer of poorly stirred fluid that coats the surface of the intestinal villi and that reduces the absorption of lipid digestion products

up-regulation increase in the number of receptors on the surface of target cells, making the cells more sensitive to hormones or other agents

ureagenesis formation of urea, especially the metabolism of amino acids to urea

ureter tube that conducts the urine from the renal pelvis to the bladder; it consists of an abdominal part and a pelvic part, is lined with transitional epithelium surrounded by smooth muscle (both circular and longitudinal), and is covered externally by a tunica adventitia

urethra tube that carries urine from the urinary bladder to the outside of the body

urine titratable acid amount of acid excreted in the urine combined with buffers such as phosphate, determined by measuring the milliequivalents of strong base (NaOH) needed

urodilatin peptide produced by the kidneys that increases renal sodium excretion

uroguanylin polypeptide hormone produced by the small intestine that increases renal sodium excretion

uterus pear-shaped organ located in the abdomen of a female; the site of pregnancy

V_{max} hypothetical maximum velocity of shortening of an isotonic contraction of muscle; also the theoretical velocity obtained by muscle shortening with zero afterload; this value is essentially constant in a given skeletal muscle but can be altered by physiological and pathophysiological influences in cardiac and smooth muscle

vaccine weakened or killed pathogen or piece of pathogen, administered to stimulate immunity

vagina tube extending from the uterine cervix to the vestibule

vagovagal reflexes pertaining to a process that uses both afferent and efferent vagal fibers: during the gastric phase of acid secretion, distention of the stomach stimulates mechanoreceptors, which stimulate the parietal cells directly through short local (enteric) reflexes and long vagovagal reflexes

varicosities specialized regions along postganglionic axons from which neurotransmitter release occurs

vascular capacitance measure of the volume contained in a segment of blood vessel or vessels per unit of transmural pressure, measured in mL/mm Hg

vascular endothelial growth factor (VEGF) factor responsible for the growth of blood vessels (angiogenesis) at many sites in the body

vascular function curve relationship depicting how central venous pressure varies as a function of cardiac output

vascular tree branching of the pulmonary circulation, which parallels airway branching

vasoactive intestinal peptide (VIP) putative inhibitory neurotransmitter released by enteric musculomotor neurons to gastrointestinal muscles; also a putative excitatory neurotransmitter that is a potent stimulator of intestinal secretion by parasympathetic stimulation to increase blood flow to the salivary glands

vasovagal syncope loss of consciousness from profound fear or emotional stress; accompanying cardiovascular events include marked parasympathetic-induced bradycardia and depressed resting sympathetic vasoconstrictor tone, which cause a dramatic drop in heart rate, cardiac output, systemic vascular resistance, and, thus, mean arterial pressure

vein any blood vessel that carries blood toward the heart. It is composed of three layers: an outermost elastic adventitia, a middle zone composed of circular layers of smooth muscle, and a single-cell lining of epithelial cells called the endothelium. Veins are thinner than arteries of similar outer diameter. They do not have structural rigidity and collapse without any positive transmural pressure in them

venipuncture process of obtaining intravenous access for the purpose of intravenous therapy or obtaining a sample of venous blood

venous admixture mixing of venous blood with oxygenated blood

venous–arteriolar response increase in arteriolar constriction in response to an increase in postcapillary venule pressure. It is a mechanism that helps prevent capillary hydrostatic pressure from becoming too high

ventilation movement of air between the mouth and alveoli

ventilation/perfusion ratio ratio of alveolar ventilation to simultaneous alveolar–capillary blood flow in any part of the lung; because both ventilation and perfusion are expressed per unit volume of tissue and per unit time, which cancel, the units become liters of gas per liter of blood

ventral respiratory group (VRG) group of neuronal cells located in the ventral portion of the medulla that is active during inspiration and expiration

ventricular fibrillation random, chaotic electrical activation of all the ventricular muscle cells in which the heart no longer can effectively contract. It results in death of the person if not abolished by some intervention

ventricular tachycardia abnormally high heart rate resulting from an ectopic foci in any area of the ventricles

ventrolateral medullary area region of the medulla where neurons controlling cardiovascular and respiratory functions are located

ventromedial nuclei hypothalamic nuclei, which synthesize growth hormone–releasing hormone

venules small veins ~10 to 1,000 μm wide that drain capillary networks and pass blood to larger veins in the circulation. They are also involved with the control of intracapillary hydrostatic pressure and have some transport function between the bloodstream and the extracellular fluid surrounding all tissues, especially in the presence of histamine

vertigo type of dizziness in which there is a feeling of motion even while stationary due to a dysfunction of the vestibular system in the inner ear; often associated with nausea and vomiting as well as difficulties standing or walking

very-low-density lipoproteins (VLDLs) one of two major classes of lipoproteins produced by the intestine; triglyceride-rich lipoproteins with densities <1.006 g/mL, made continuously by the small intestine during both fasting and feeding

vesicles small membrane-delimited spheres in nerve terminals filled with neurotransmitters whose contents are released into the synaptic cleft on stimulation of the nerve

vestibular apparatus sensory structures of the middle ear that are responsible for sensing body position and acceleration

vestibular nuclear complex groups of neurons in the pons and medulla that receive afferent axons from the vestibular portion of the inner ear

vestibulocerebellum medial portion of the cerebellum that is active in control of eye motions and equilibrium

vestibuloocular reflex (VOR) reflex eye movement that stabilizes images on the retina during head movement by producing an eye movement in the direction opposite to head movement, thus preserving the image on the center of the visual field

vestibulospinal tract axon pathway from the lateral vestibular nucleus to the spinal cord, which influences postural reflexes in the spinal cord

villi finger-like projections of the mucosal surface that increase the surface area of the small intestine by about 30 times

virilization masculinizing effects of excessive adrenal androgen secretion on females

visceral hyperalgesia gut hypersensitivity; the appreciation, or exaggeration, of gut symptoms with stimuli (such as balloon distention of the gut) that would ordinarily not be noticed or considered noxious

visceral pain sensation of pain arising from visceral parts (i.e., internal organs) of the body; often perceived as arising in somatic structures

viscoelastic materials substances whose mechanical reaction to applied forces is time dependent

viscosity measure of the blood property that offers resistance to flow

vital capacity (VC) maximum volume of air that can be exhaled; the sum of expiratory reserve volume, tidal volume, and inspiratory reserve volume (note that this value is the same as *forced vital capacity*)

vitamin A lipid-soluble vitamin important in vision and in the normal growth of the skin

vitamin B$_1$ (thiamine) water-soluble vitamin that plays an important role in carbohydrate metabolism; absorbed by the jejunum passively and by an active, carrier-mediated process

vitamin B$_2$ (riboflavin) water-soluble vitamin that plays an important role in metabolism. It is absorbed by a specific, saturable, active transport system located in the proximal small intestine

vitamin B$_6$ (pyridoxine) water-soluble vitamin involved in amino acid and carbohydrate metabolism; absorbed throughout the small intestine by simple diffusion

vitamin B$_{12}$ (cobalamin) water-soluble, cobalt-containing vitamin that plays an important role in the production of red blood cells

vitamin C (ascorbic acid) water-soluble vitamin found in green vegetables and fruits that plays an important role in many oxidative processes by acting as a coenzyme or cofactor; absorbed mainly by active transport through the transporters in the ileum; uptake process is sodium dependent

vitamin D group of fat-soluble compounds collectively known as the *calciferols* that plays an important indirect role in the absorption of calcium by the gastrointestinal tract and is essential for normal development and the formation of bones and teeth; derived from the skin, which contains a rich source of 7-dehydrocholesterol that is rapidly converted to cholecalciferol when exposed to ultraviolet light and dietary vitamin D$_3$; absorbed by the small intestine passively and incorporated into chylomicrons

vitamin D$_2$ form of vitamin D generated by the irradiation of ergosterol; also known as *ergocalciferol*

vitamin D$_3$ form of vitamin D normally manufactured in the skin, where ultraviolet light activates the compound 7-dehydrocholesterol

vitamin D–binding protein binding protein in plasma that transfers vitamin D$_3$ during the metabolism of chylomicrons

vitamin E vitamin that is absorbed by the small intestine by passive diffusion and incorporated into chylomicrons and transported in the circulation associated with lipoproteins and erythrocytes and is a potent antioxidant that prevents lipid peroxidation

vitamin K vitamin derived from green vegetables in the diet or the gut flora and incorporated into chylomicrons, where it is rapidly taken up by the liver and secreted together with very-low-density lipoproteins; essential for the synthesis of various clotting factors by the liver

vitreous humor thick clear gel that fills the bulk of the eyeball; provides support for the shape of the eyeball and its retinal structures; also known as *vitreous body*

voltage-dependent anion channel (VDAC) mitochondrial channel for metabolite diffusion across the mitochondrial outer membrane

voltage-gated ion channels ion-selective pores in the membrane of excitable cells that open (or close) in response to a change in membrane potential

von Willebrand disease most common hereditary bleeding disorder; caused by deficiency of von Willebrand factor

von Willebrand factor a glycoprotein important for coagulation

V-type ATPase see *ATPase*

V wave venous pressure wave that gradually increases during systole as blood continues to return to the heart while the atrioventricular valves are closed

wall stress force exerted on the wall of a cylindrical structure, as modified by the thickness of the wall of the structure, that would have the tendency to rip the wall apart as a result of a positive transmural pressure. It is directly proportional to both the transmural pressure and the radius of the cylinder but inversely proportional to the wall thickness. Wall stress is less in thick-walled vessels than in thin-walled vessels of the same internal radius when exposed to the same transmural pressure. This principle is applied to blood vessels and other hollow organs, such as the heart and urinary bladder

water diuresis excretion of a large volume of dilute urine

wavelength spatial period of the wave—the distance over which the wave's shape repeats

weak acid acid with a low dissociation constant, which consequently is poorly ionized in solution

Wernicke area area at the junction of the temporal and parietal lobes of the brain involved in sensory perception of language, usually on the left hemisphere

white blood cell (WBC) cell delivered by the blood to sites of infection or tissue disruption, where it defends the body against infecting organisms and foreign agents in conjunction with antibodies and protein cofactors in blood; five main types are neutrophils, eosinophils, basophils, lymphocytes, and monocytes; also known as *leukocyte*

white blood cell count measures the concentration of leukocytes in blood as an indicator of the body's immunologic stress state

white matter portions of the CNS that histologically appear white due to myelinated processes

white ramus branch of a spinal nerve that carries preganglionic sympathetic axons from the spinal cord to the sympathetic ganglia

whole-gut transit time time required for transit from the mouth to the anus

working memory mechanism whereby new experiences are processed, drawing on skills such as language and mathematical ability, to integrate new inputs in light of previously acquired learning

Wright-Giemsa common dual stain for peripheral blood and bone marrow smears

xanthine oxidase enzyme that catalyzes xanthine to urate. During ischemia–perfusion injury, hypoxanthine builds up and, in the presence of oxygen hypoxanthine, is converted to xanthine. Xanthine oxidase neutralizes xanthine by converting it to urate

xenobiotics hydrophobic compounds foreign to the body, some of which are toxic, that are metabolized by the liver. Xenobiotics are converted into hydrophilic compounds that are then absorbed and excreted by the kidneys

xenograft transplantation in which animal tissue is used to replace human organs

yellow marrow bone marrow of long bones consisting of fat and primitive blood cells

yolk sac extraembryonic membrane that contains a slight amount of yolk

Z line dark structure that crosses the center of the I band in muscle; sometimes termed a *Z disk* to emphasize its three-dimensional nature

zinc finger small globular protein structures that can coordinate zinc ions with cysteine residues to help stabilize their folds; they can be classified into families and typically function as interaction molecules that bind DNA, RNA, proteins, or small molecules

zona fasciculata inner zone of the adrenal cortex, which produces the glucocorticoid hormones: cortisol and corticosterone

zona glomerulosa outer zone of the adrenal cortex, which produces aldosterone

zona pellucida acellular membrane between the oocyte and membrana granulosa of an ovarian follicle, which surrounds the ovulated egg prior to implantation

zona reaction modification of the zona pellucida that blocks polyspermy; enzymes released by cortical granules digest sperm receptor protein so that they can no longer bind sperm

zona reticularis inner zone of the adrenal cortex, which produces the glucocorticoid hormones: cortisol and corticosterone

zones various regions of the lung (e.g., apex, base) through which blood flows, with respect to pulmonary arterial pressure, pulmonary venous pressure, and alveolar pressure

zygote fertilized egg

zymogen granules granules found in the apical region of acinar cells that store salivary amylase

INDEX

Note: Page numbers in *italics* indicate figures; those followed by "t" indicate tables; those followed by "b" indicate boxes.

A

A band, of skeletal muscle, 145
A wave, 264
ABCA1, 15
abduction, defined, 86
abetalipoproteinemia, 528, 548
ABO blood group, 213
abscess, 208
absence seizures, 134
absolute refractory period, 40
absorption
  of carbohydrates, 522–525, *523–524*
  disorders of
    abetailpoproteinemia, 528
    cystinuria, 531
    Hartnup disease, 531
    lactose intolerance, 525, *525*
  of electrolytes, 533–536
  of lipids, 525–528, *526–528*
  of minerals, 533–536
  of proteins, 528–531
  of vitamins, 531–533
  of water, 536–537
absorptive hyperemia, 321
A1C, 699
accessory structures, in sensation, *56*
accidental hypothermia, 608
acclimatization
  cardiovascular, 425
  cold, 606
  heat, 603–604, *604,* 605t
  ventilatory, 423–424
accommodation, in vision, 64
acetoacetic acid, 701
acetone, 701
acetylcholine (ACh), 48, 147, 514
  autonomic postganglionic fibers, 110
  origins and innervations, 106, 108t
  parasympathetic postganglionic neurons, 106
  synthesis and degradation, *109*
acetylcholinesterase (AChE), 147
acetyl-CoA carboxylase, 697
achalasia, 558
acid
  defined, 485
  strong, 486
  weak, 486
acid dissociation constant, 485–486
acid-base balance
  acid production
    alkalinizing effect, 487
    nonvolatile, 487
    protein metabolism, 487
    source of carbon dioxide, 487
  balance disturbance
    compensation, 497
    etiology, 498t, 501–503, *502*
    metabolic acidosis, 499
    metabolic alkalosis, 498t, 501
    plasma anion gap, 499–500, 501t

    respiratory acidosis (*see* respiratory acidosis)
  buffering systems
    bicarbonate, 488, *488*
    chemical buffering (*see* chemical buffering)
    kidney, 491–495
    lungs, 491
    intracellular pH regulation, 496, *496*
   principle
    acid dissociation constant, 485–486
    conjugate base, 485
    Henderson–Hasselbalch equation, 486
    K_a logarithmic expression, 486
    pH stability, 486–487, *487*
acide tide, 517
acidemia, 390
acidosis
  acid-base homeostasis, 497
  metabolic, 499
  respiratory acidosis (*see* respiratory acidosis)
acinar cells, of pancreas, 516
acinus, 509
acne vulgaris, 730
acoustic reflex, 71
acquired immune system. *see* adaptive immune system
acromegaly, 654
acrosome, 725
acrosome reaction, 726
actin, 145
actin filaments, 73, 145
actin-linked regulation, 151
action potentials, 22, 38
activated partial thromboplastin time (aPTT), 189
activation gate, 164
active hyperemia, 305
active immunization, 215
active membrane transport
  calcium pumps, 14
  proton pumps, 14
  sodium-potassium pump, 15
active state, 40
active transport, 9
active zone, 43
activin, 720, 736
activity front, migrating motor complex (MMC), 580
acute mountain sickness (AMS), 424b
acute respiratory acidosis, 498
acute respiratory distress syndrome (ARDS), 371b, 404
acute-phase blood proteins, 210
acute-phase serum amyloid A (A-SAA) proteins, 210
acyl-CoA cholesterol acyltransferase, 527, 675
acyl-coenzyme A (acyl-CoA), 527
adaptation, 59, 423
adaptive immune system, 195, 197
  antibodies of, 205–207
  cellular

    helper T cells, 204
    immunological synapse, 202, *203*
    killer T cells, 204
    major histocompatibility complex, *202,* 202–203, *203*
    memory T cells, 204
    T regulatory cells, 204
    T-cell differentiation, 203
    time delay in, 204
  clonal selection, 201
  defined, 200
  diversity of, 200–201
  humoral
    antibody action, 207, *207*
    antibody classes, 206–207, 206t
    antibody structure, *205,* 205–206
  memory of, 201, *201*
  specificity of, 200
adaptive relaxation, of gastric reservoir, 566
addiction, 47
Addison disease, 471, 680b
adenosine diphosphate (ADP), 592, 666
  oxidative phosphorlation of, 666
  thyroid hormone regulation of, 666
adenosine triphosphate (ATP), 391b, 592, 693
adenylyl cyclase (AC), 24, 168
adequate stimulus, 56
adipocytes, 684
adipose tissue, lipogenesis in, 696
Adipose triglyceride lipase (ATGL), 682
adrenal cortex
  cholesterol esters, 675
  dehydroepiandrosterone (DHEA), 673
  hormones, 673–683
    aldosterone (*see* aldosterone)
    cholesterol (*see* cholesterol)
    glucocorticoids (*see* glucocorticoids)
    steroids (*see* steroids)
  synthesis
    ACTH regulation, *678,* 678–679
    from cholesterol, 675, *675, 677*
  zona glomerulosa, 673
  zona reticularis, 673
adrenal gland
  cortex synthesis, 673–674
  disorders of
    Addison disease, 471
    congenital adrenal hyperplasia, 649, 676, 732, 767
    Cushing syndorme, 684b
  steroid hormones synthesis, *675, 675–676, 677*
  steroidogenic precursors production, 760
adrenal hyperplasia, congenital, 676, 767
adrenal medulla, 110
  chromaffin cells, 683
  hormones
    catecholamines, 683–684, 684t
    epinephrine and nonepinephrine, 683

adrenal medulla, sympathetic innervation of, *110*
adrenalin, 110
adrenarche, 676, 745, 765
adrenergic receptors, in autonomic nervous system, 48
adrenocorticotropic hormone (ACTH), 636, 644, 676
    action on steroidgenesis, *678,* 678–679
    cholesterol availability, 679
    in cholesterol metabolism, 647
    deficiency, 649, 679
    proteolytic processing of, *648*
    in regulation of adrenal cortex, 647
    secretion of, 644
        arginine vasopressin, 649
        corticotropin-releasing hormone, 648, *648*
        diurnal variation, 649–650, *650*
        glucocorticoid inhibition, 648–649, *649*
        proopiomelanocortin, 647–648, *648*
        and sleep–wake cycle, *650*
        stress-induced, 649
        and synthesis, 648, *648*
adventitia, of blood vessels, 223
aerobic metabolism, muscle fibers and, 161
aetylcholinesterase (AChE), 119–120
affect, 131
affective disorders, 137–138
afferent muscle innervation, 88–90
afterhyperpolarization, 40
afterhyperpolarization (AH)-type neurons, 573
afterload
    cardiac muscle contraction, 265, *265*
    defined, 155
age-related macular degeneration (AMD), 83b
aggression, limbic system mediation of, 137
aging, 3, 120
    exercise, 623, 624, *624*
    influences on liver metabolism, 545
agonist muscle, 86
agonists, defined, 26, 117
agoutirelated protein (AgRP), 126
airflow
    airway resistance, 374
    lung volumes, 374–375, *375*
    regional lung compliance, 368–369, *369*
airway
    during forced expiration, *375,* 375–376
    lung compliance, 376–378, *377, 378*
airway resistance
    lung volumes, 374–375, *375*
    major sites, *374*
    overcoming, 378
    smooth muscle tone effect, 375
airway tree, *353,* 353–354
albumin, 175, 549, 736
aldosterone, 633, 647, 674
    angiotensin II action, 679, 681, *681*
    angiotensin II formation, 679, *680*
    synthesis, 681
aldosterone synthase (CYP11B2), 676
alkalemia, 390
alkalemia, acid–base homeostasis, 497
alkaline tide, 512
alkalinizing effect, acid–base homeostasis, 487
alkalosis
    acid–base homeostasis, 497
    defined, 418

metabolic, 501, 501t
    respiratory, 418
allergies, 211b, 213
allografts, 212
α_1-adrenergic receptor blockers, 168
α-amino-3-hydroxy-5-methyl-4-isoxazole propionic acid (AMPA) receptor, 49
alpha cells, 684
alpha (α) globulins, 175
α-glucosidase inhibitors, 701
α-melanocyte-stimulating hormone (α-MSH), 647
α motor neurons, 87, *87*
α_1 receptors, 114
α_2 receptors, 114
5α-reductase, 729, 766
alpha waves, 132
alveolar hypoxia, 402
alveolar pressure, 358
alveolar surface tension, 403, *403*
alveolar ventilation, 415
alveolar vessels, 401
alveolar–arterial oxygen gradient, 390
alveoli
    anatomic considerations, *353*
    defined, 353
    partial pressures, 356t
    stability, 370, *370*
    surface tension, 369–371
Alzheimer's disease (AD), 3, 119, 139
amacrine cells, for vision, 68
ambisexual stage, 765
amenorrhea
    infertility, 749
    lactational, 763
amine, 446
amino acids
    liver metabolism of, *549,* 549–550
    nonessential, 549–550
ammonia, 417, 434
amniotic sac, 537b
amphipathic, defined, 9
ampulla, 738
amygdala, 136
amylase, 509, 510
amylin, 700b
amylopectin, 523
amylose, 523
anaerobic metabolism
    isometric exercise, *615,* 616–617
    muscle blood flow, 323, 324
    muscle fibers, 619
anaerobic pathway, 160
anaerobic threshold, 619
anamnestic response, 201
anaphylaxis, 213, 347
androgen-binding protein (ABP), 722, 766
androgens, 633, 730, 736
androstenedione, 676, 742
anemia, 177, 388, 393b
    aplastic, 393b
    hemolytic, 191b
    iron-deficiency, 183, 393b
    megaloblastic, 393b
    pernicious, 393b, 532
    red blood cells (erythrocytes), 182–183
    sickle cell anemia, 180, 393b
aneurysms, 317

angina, 30b–31b
    exertional, 318b
angiogenesis, 190
angiotensin I, 679
angiotensin II action, 679, 681, *681*
angiotensin II formation, 679, *680*
angiotensin-converting enzyme (ACE), 398, 679
angiotensinogen, 470, 679
anion exchange protein (AE1), 14
anion gap, metabolic acidosis, 499
anions, composition in body fluids, 175
anisocytosis, 180, 181
anorexia nervosa, 749
anorexigenic, 126
anovulation, 763
ANS receptor element, 169
antagonistic pairs, 160
antagonists, 86, 117
anterior and posterior chambers, 63
anterior lobe, 643
anterior periventricular region, 652
anterograde transport, 37
antibodies, 200, 205, *205,* 510, 530, *767*
    classess, 206–207, 206t
    polyclonal, 200
    structure, *205,* 205–206
anticholinergic agents (blockers), 169
anticoagulant, 174
antidiuretic hormone (ADH), 18, 464
antigen-presenting cells (APCs), 196, 553
antigen-recognition molecules, 200
antigens, 196
    epitopes, 200
    immune system, 215
anti-müllerian hormone (AMH), 766
antioxidants, 83b
antiport, 16
antra, 740
antral granulosa cells, 740
aorta, 222, 282–283, *283*
aortic baroreceptors, 336
aortic bodies, 417
aortic compliance, 282–283, *283*
aortic valve, 222
aphasias, 140
apical membrane, 17
aplastic anemia, 393b
apnea, 422
apo B proteins, 548
apoferritin, 536
apolipoprotein B-100, 675
apoproteins, 527–528
apoptosis, 197
apposition, 755
aquaporins, 12, 18, 440
arachidonic acid, 50
arcuate nuclei, 652
arcuate nucleus (ARC), 126
area postrema, 572
areflexia, 93
areola, 762
arginine vasopressin (AVP), 18, 124, 334, 337, 464, 643, 645, 649
aromatase, 742
arrhythmias
    compensatory pause, 253
    complete atrioventricular block, 253
    myocardial ischemia and, 244
    premature atrial contractions (PACs), 253

respiratory sinus, 251, *252*
sinus bradycardia, 252
sinus tachycardia, 252
supraventricular tachycardia, 253
ventricular fibrillation, 253
ventricular tachycardia, 253
arterial baroreceptors, 336
arterial blood gases, 390t
arterial blood pressure
arterial compliance, 282–283, *283*
cardiac output and, 282, *283*
determinants of, 282–284, *283–284*
factors affecting, 285, 286b
heart rate, 283–284, *284*
mean artrial pressure
cardiac output and, 282
defined, *283*
equation for, 282
interactions of, 283–284, *283–284*
measurement of, 284–285, *285*
measurement of, 284–285, *285*
pulse pressure
arterial compliance and, 282–283, *283*
defined, 282
interactions of, 283–284, *283–284*
stroke volume and, 282–283, *283*
systemic vascular resistance and, 284, *284*
arterial compliance
cardiac output and, 288–289
pulse pressure, 282–283, *283*
arteries, 222
blood circulation (*see* hemodynamics)
factors that affect muscular contraction, 223, 224t–225t
inferior hypophyseal, 644
pulmonary, 398, *399*
spiral, 327
umbilical, 326
arterioles, 222
arteriosclerosis, 285b–286b
defined, 285b
systolic hypertension and, 285b–286b
arteriovenous anastomoses, 324, *325*
arthritis
osteoarthritis, 623
rheumatoid arthritis (RA), 213, 623
artificial organs, 212
artificial pancreas, 640b
ascending reticular activating system, 130, *130, 132*
ascites, 543
association cortex, 135
asthma, 210, 211b, 378b–379b
astigmatism, 63
astrocytes, 34
ataxia, 102
atelectasis, 372
atherosclerosis, 176, 292b
ATPases, 14
ATP-binding cassette (ABC) transporters, 15
ATP-dependent K^+ channel (K_{ATP}), 693
atresia, 739
atrial fibrillation, *252*, 253
atrial natriuretic peptide (ANP), 334, 343, 472, 645
atrial septal defects, 329b
atrioventricular block, 253
atrioventricular (AV) node
functions of, 241–242

as secondary pacemaker, 253
as slow action potentials, 238, *239*
attention, 136
atypical antipsychotics, 138
auditory cortex, 74
auditory system
cochlea, 72, *72*
decibel (dB), 70
external ear functions, 70
hearing loss, 75
inner ear, 71–75
middle ear, 70–71, *71*
sound, 70
sound wave formations, *70*
Auerbach plexus, 560
augmented leads, 248
autocrine signaling, 22
autograft, 212
autoimmune disease, 176, 214
thyroiditis, 668b
autoimmune disorders, 214t
autoimmunity, 214–215
autologous, 187b
automaticity, of cardiac cells, 238
autonomic integration
aqueous humor and intraocular pressure, 116, *116*
epinephrine reversal, 117
nitric oxide-induced vascular smooth muscle relaxation, 116, *117*
presynaptic inhibition, 115
pupil light adaptation, 115, *115*
receptors and pathways, 117, 118t–119t
autonomic nervous system (ANS), 2, 509
anatomy, 105–106
autonomic integration, 115–116
heart rate determination, 242
neurotransmitters, 106–110
organ-specific arrangement, *107*
parasympathetic nervous system
cholinergic receptors, 112
ganglia, 106, *107*, 108t
homeostatic functions, 110, 111t–112t
preganglionic neurons, 106, *107*, 108t
vs. somatic motor system, *106*
sympathetic nervous system
adrenergic receptors, 113–114
arousal, stress, and danger, 113, 113t
ganglia, 106, *107*, 108t
organization, 112, *113*
preganglionic neurons, 106, *107*, 108t
treatments, 117, 118t–119t
autonomic regulation
aqueous humor and intraocular pressure, 116, *116*
pupil light adaptation, 115, *115*
autoreceptors, 43
autoregulation, 305, 309–310, 317, 320–321, *434*, 434–435
axial streaming, 232–233
axon hillock, 36
axonal transport, 37, *38*
axoneme, 725

baroreceptor, 470
baroreceptor reflex
arterial pressure, 336–337
blood flow, 338
cardiac output and total peripheral resistance, 336–337
hormonal systems, 337
baroreceptors
aortic, 336
arterial, 336
cardiopulmonary, 336
carotid sinus, 336
Barr body, 765
Bartter syndrome, 476
basal forebrain nuclei, 132
basal ganglia
Huntington's disease, 99
motor control, 98–100, *99*
basal ganglial, 135
basal metabolic rate (BMR), 126, 666–667
base, defined, 485
basement membrane, 436
basic metabolic panel (BMP), 176
basilar membrane, 72
basket cells, 101
basolateral membrane, 17
basophils, 183
B-cell receptors (BCR), 200
behavioral conditioning, cardiovascular responses, 340
benign paroxysmal positional vertigo (BPPV), 79b
benzodiazepines, 114b
beriberi, 532
Bernoulli principle, 229
β-adrenergic agonists, 169
beta-adrenergic receptor antagonists, 256b
beta cells, 311b
beta (β) globulins, 175
β-hydroxybutyric acid, 701
11β-hydroxylase (CYP11B1), 676
3β-hydroxysteroid dehydrogenase (3β-HSD II), 376
11β-hydroxysteroid dehydrogenase (11βHSD), 684b–685b
β-lipotropin, 647, *648*
$β_1$ receptors, 114
$β_2$ receptors, 114
$β_3$ receptors, 114
beta waves, 133
bicarbonate, 389, *389*, 417–420, 534
bicarbonate buffering system, 488, *488*
biguanides, 701
bile acids, 518
bile canaliculi, 519
bile salt–Na^+ symport, 519
bile salts, 519
bilirubin, 183, 520
biliverdin, 183
bioelectrical stimulation, 2
biologic clock, 128–129, *129*
biotin, 532
bipolar cells, 36, 68
bipolar disorders, limbic system, 137
birth control pills, 769b
bitter receptors, 81
bladder, 456
blastocyst, 753, 755
blastomeres, 755

B

B cells, 185
Babinski sign, 98
band cells, 180
barometric pressure, 356

bleeding time, 190
blind spot, 63
blood
  circulation of (*see* hemodynamics)
  clotting of, 187–190, *188* (*see also*
    hemostasis)
  coagulation of
    activated partial thromboplastin time,
      189–190
    coagulation cascade, 188, 189t
    commom pathway in, 189
    disorders of, 188
    prothrombin time, 189
  components of (*see also* blood cells)
    plasma, 174, 174t
  doping, erythropoietin and, 187b
  functions
    hemostasis, 173
    homeostasis, 173–174
    immunity, 174
    transport, 173
  gas transport by, 386–389
  indices for, 178
  laboratory tests for
    blood smears, 180
    complete blood count, 178, 178t–179t, 180
    hematocrit (Hct), 177
    iron profile, 183, 183t
  pH regulation
    bicarbonate buffering, 488
    chemical buffering (*see* chemical
      buffering)
    kidneys, 493–495
    lungs, 491
    negative feedback of endogenous acid,
      490–491, *491*
  viscosity, 177
  volume of (*see* blood volume)
  whole, 174
blood cells, 173, 174, *177*, 180–183
blood clot, 188
blood flow
  anaerobic metabolism, 323, 324
  baroreceptor reflex, 338
  heart, 227–233
  lungs, 404–407
  skin in regulation of, 597
blood group systems, 179b–180b
blood indices, 178
blood lipid profile, 176
blood pressure
  diastole, defined, 222–223
  factors affecting, 285, 286b
  measurement of, 284–285, *285*
  systolic pressure, defined, 225
blood smear, 180
blood vessels
  factors that affect muscular contraction, 223,
    224t–225t
  structure of, 223
blood volume
  central
    cardiac output and, 287–288
    central venous pressure and, 287
    defined, 287
  in circulation, 287
  extrathoracic, 287
  systemic, 287
  vein compliance, 287

    venous pressure in, 287
blood–brain barrier, 34, *35*, 125, 417
body buffer, 489
body fluids, 462t
body mass index (BMI), 700
body movement, 86
body temperature, 587–610
  in cold acclimatization, 606
  in cold response, 604–606
  control of (*see* thermoregulation)
  core
    distribution of, 587, *588*
    factors influencing, 600
    in heat acclimatization, 603–604, *604*
    measurement of, 589, *589*
    normal, 589
    set point and, 600–601
    thermal receptors **lor**, 600
  extreme, effects of, 605
  fever, 601
  in heat acclimatization, 603–604, *604*
  heat balance in, 590–593
  heat dissipation
    skin blood flow, 597
    sweating, 595
  heat exchange
    convection, 558–592, *593*, 597
    evaporation, 592, *593*, 594–595
    radiation, *590*, 590–593
    surface area and, *593*, 593–594
  heat production
    vs. heat loss, *590*, 590–591
    metabolism in, 591, 594
  heat storage, 591, 594
  heat stress
    disorders related to, 607–608
    effects of, 598
  heat transfer in, 587–590, 588t
  hypothermia, 605–606, 608b–610b
  nonthermal effects on, 601
  normal range, *588*, 595b–596b
  shell, 587–589, 594, 595
    in cold reponse, 604–606
    conductance, 594–597, 604–605
  skin in regulation of
    blood flow and, 597
    convection, 592, 593, *594*
    evaporation, 593
    radiation, 592, 593, *594*
    surface area and, *593*, 593–594
  skin temperature
    in heat exchange, 589
    set point and, 600–601
    in thermoregulation, 599
body water
  extracellular fluid (ECF) in, 460
  in fluid compartments, 460–464
  intracellular fluid (ICF) in, 460
  total
    by age, 460
    distribution of, 460–461
    measurement of, 461–462
body weight, 126, 127, 656b
Bohr effect, 388
bone
  disorders of
    osteogenesis imperfecta, 708, 713b
    osteomalacia, 714, 715t
    osteoporosis, 713, *714*, 714b, 715b

    Paget disease, 715
    rickets, 714, 715t
  exercise effects on, 621–623
  remodeling, 709
  rickets, 532
bone marrow transplantation, 212
bone morphogenetic proteins (BMPs), 715b
bony labyrinth, 72
botulinum toxin, 148
Bowman capsule, 431
Boyle law, 357
bradycardia, 252
bradykinin, 472
brain
  functions of, 124–125, 125t, *126*
  split, 141b
brain natriuretic peptide (BNP), 472
brain-derived neurotrophic factor (BDNF), 2
brain's reward system, 137
brainstem, 124
  autonomic reflexes, 117, *117*
  motor control, *93*, 93–94
breathing. *see also* ventilation
  airflow, 354–359
  changes in thoracic volume during, 354, *355*
  neural and voluntary control of, 412–415
  neural reflexes, 415–418
  pons and medulla oblongata, 412–413
  pulmonary pressures and airflow, 354–359
  during sleep
    arousal mechanisms, 422
    inspiratory flow rate, 421
    responsiveness to carbon dioxide, 421–422
    upper airway tone, 422–423
  unusual environments, 423–426
    acclimatization, 423–424
    cardiovascular acclimatization, 425
    diving reflex, 425–426
Broca area, 140, *140*
Brodmann cytoarchitectural map, 95
bronchial circulation, 390, 399
bubble-boy disease, 213
buffer, chemical
  bicarbonate/carbon dioxide buffer
    acid–base physiology, 489
    Henderson–Hasselbalch equation, 489
    open system, 489–490, *490*
  changes, acid production, 490–491, *491*
  isohydric principle, 491
  metabolic acidosis, 499
  metabolic alkalosis, 501
  phosphate buffer, 489
  protein buffer, 489
buffering systems
  bicarbonate, 488, *488*
  chemical buffering (*see* chemical buffering)
  kidney, 491–495
  lungs, 491
buffy coat, 175
bulimia, 749
bulk flow, 435
bundle of His, 241
burr cells, 181

C

C-peptide, 633b, 690
C wave, 264
Ca^{2+}-binding protein, 535
calbindin D (CaBP), 535

calcitonin, 661, *661,* 710
calcium, 534–535
  absorption in small intestine, 707
  in bone
    continuous flux, 708
    neonatal development, 708
    osteoblasts, 708, *708*
    osteoclasts, *708,* 708–709
    osteocytes, 708, *708*
  distribution in bone cells, in healthy adult, 705–706, 706t
  inorganic constituents, 705–706, 706t
  kidneys, 707–708
  nerve and muscle function, 705
  plasma, 706
    calcitonin, 710, 712, *712*
    nonhormonal mechanisms, 709
    parathyroid hormone, 709–710, *710, 712, 712*
    vitamin D$_3$ and 1,25-dihydroxycholecalciferol, 709–713, *711, 712*
  regulations in GI tract, kidneys and bone, *706,* 706–707, *707*
calcium channels
  in cardiac action potential(s), 238
  L-type, 240
calcium homeostasis, 706, *706*
calcium pumps, 14, 151
calcium release channel, 29, 150
calcium sparks, 261
calcium-induced calcium release, 29, 164, 261
calmodulin (CaM), 30, 166
cAMP-dependent protein kinase, 27
canal of Schlemm, 63, 116
canaliculi, 708
canaliths, 79b
cancer immunotherapy, 215
cannabinoids, 50
capacitance, 226
capacitation, 754
capillaries, 222
  capillary distention, 400
  capillary filtration coefficient (CFC), 303
  diffusion, 386
  glomerular filtration
    capillary hydrostatic pressure, 437–438
    capillary pressure, 437
  microcirculation
    anatomic considerations, 297, *297*
    defined, 297
    endothelial vesicles in, *297,* 297–298
    pores in, 297
    porosity of, 298
  recruitment, 400
capillary distention, 400, *400*
capillary filtration coefficient (CFC), 303
capillary recruitment, 400, *400*
carbenoxolone, 685b
carbohydrates
  absorption, 522–525
  digestion, 522–525
  metabolism of, 545–546, *546*
carbon dioxide
  effects on oxyhemoglobin equilibrium curve, 388, *388*
  physiologic responses
    blood oxygen, 419–420
    exercise-induced hyperpnea, 420–421

carbon dioxide equilibrium curve, 389
carbon monoxide (CO), 181
carbonic acid, 389, 485–488, 501, 512, 533
carbonic anhydrase, acid–base homeostasis, 455, 489
carboxyhemoglobin (Hbco), 388
carboxypeptidase A, 530
carboxypeptidase B, 530
carboxypolpeptidase, 515
cardiac action potentials, 238
cardiac and gastrointestinal functions, 2
cardiac arrhythmias, 256b–257b
cardiac cells
  automaticity of, 238
  rhythmicity of, 238
cardiac cycle, *263,* 263–265
  rapid ejection phase, 264
  reduced ejection phase, 264
  venous pressure waveformsl, 264–265
  ventricular diastole, 264
cardiac dipole, 244, *245*
cardiac function curve, 290, *290*
cardiac glycoside, 14
cardiac muscle, 144
  contraction
    action potential and isometric twitch, *261*
    afterload, 265, *265*
    calcium, 261, *262*
    cardiac cycle, 263–265
    ejection fraction, 271
    excitation–contraction coupling, 260–263
    force generation, 261
    force-velocity curves, *267*
    isotonic, 266, *267*
    length-tension curve, *267*
    preload, 265, *265*
    pressure-volume loops, 268, *268*
    stroke volume, 268–270
  electrophysiology of
    cardiac action potential, 238, *238*
    depolarization, 239, *239*
    pathological conditions, 243
    repolarization, 240
    resting membrane potentials, 238–239, *239*
    voltage-gated sodium channels in, 239–240
  vs. skeletal muscle, 237
cardiac output, 228
  arterial compliance and, 288–289
  arterial pressure and, 282, *283*
  blood volume and, 287–288
  cardiac function curve, 290, *290*
  cardiac index, 272
  effects on pulmonary vascular resistance, *400,* 400–401
  equilibriums in, *290,* 290–291
  factors influencing, 272t
  heart rate, 273–274
  imaging technique, 275–277, *276*
  indicator-dilution technique, *274,* 274–275
  mean artrial pressure, 282, *283*
  measurement, 274–275
  stroke volume, 272–273
  venous filling pressure and, 289, *289*
cardiac tamponade, 270
cardiopulmonary baroreceptors, 336
cardiovascular disease in women, 345b–346b
cardiovascular diseases, 3

cardiovascular system
  autonomic innervation, *335*
  behavioral conditioning, 340
  blood flow to organs, 223
  blood vessel structure, 223
  blood volume in, 227–228, *228*
  circulation (*see* circulation)
  disorders of
    arterial hypertension, 230b, 285b, 286b
    atherosclerosis, 232b, 285b
  estrogen benefits, 345b–346b
  functional organization of, 222–223, 224t–225t
  and hemodynamics, 221–234
  nitric oxide (NO), 341b
  output of right and left heart, 223
  overview, 221–222, *222*
  pressure profile across, 228, *229*
carnitine acyltransferase (CAT) system, 697
carotid bodies, 417
carotid sinus baroreceptors, 336–337
carrier-mediated diffusion, 12
cataracts, 64
catecholamines, 591, 683–684, 684t
catechol-*O*-methyltransferase (COMT), 48, 109
cations, 175
caudate nucleus, 98
caveolae, 9, 163
caveolin, 9
celiac, 573
celiac ganglion, *107*
celiac sprue, 524b
cell death, 12
cell model
  HCO$_3$-/CO$_2$ reabsorption, *494*
  renal synthesis, *495*
  titratable acid formation, *494*
cell-derived mediators, 209
cells
  blood (*see* blood cells)
  muscle (*see* muscle fibers)
  nerve (*see* neurons)
cellular signaling
  autocrine signaling, 22
  cell-surface receptors
    G protein-coupled receptors (GPCRs) in, 23–25, *24*
    ion channel-linked receptors in, 25
    types of, 23
    tyrosine-kinase receptors in, *25,* 25–26
  first messengers, 23
  gap junctions in, 8, *8*
  hormone receptor signaling, 26–27, *27*
  hormones in, 22–23
  molecular basis of, 23–27
  paracrine signaling, 22
  second messengers in, 23, 27–30
  signal cascade in, 23
  transducers, 56
central auditory pathways, 74–75
central blood volume, 287
central chemoreceptors, 417
central inspiratory activity (CIA) integrator neurons, 414
central nervous system (CNS), 34, 128b
central pattern generator, 413
central sulcus, 135
central venous pressure, 287
centroacinar cells, 516

cephalic phase, 513, 513t
cerebellar circuitry, 101, *101*
cerebellar peduncles, 101
cerebellum
 cerebellar lesions, 102
 intrinsic circuitry, *101,* 101–102
 structure, 100, *101*
cerebral circulations
 autoregulation, *319*
 blood pressures, 317–318
 cerebral edema, 319–320
 chronic hypertension, 319, *319*
 sympathetic nerve activity, 319
 vasodilation by CO_2 and H^+, 318
cerebral cortex, 135
cerebral edema, 319–320
cerebral hemispheres, *135*
cerebral peduncles, 98
cerebrocerebellum, 100, *101*
cervical ganglion, 511
 superior, 511
cervix, 738
C-fiber endings, 416
cGMP phosphodiesterase (PDE), 25
cGMP-dependent protein kinase (PKG),
 28, 168
Charles law, 358
"cheese effect," 117b
chemical buffering, 488
 bicarbonate/carbon dioxide buffer
 acid–base physiology, 489
 Henderson–Hasselbalch equation, 489
 open system, 489–490, *490*
 changes, acid production, 490–491, *491*
 isohydric principle, 491
 metabolic acidosis, 499
 metabolic alkalosis, 501
 phosphate buffer, 489
 protein buffer, 489
chemical potential, 21
chemical signals, 127
chemical synapses, 43
chemoattractants, 190
chemokines, 209
chemoreceptors, 56, 338–339, 416
chenodeoxycholic acid, 518
chest leads, 248
Cheyne-Stokes breathing, 421
chief cells, 512
chloride, 534
cholecystokinin (CCK), 126, 127, 508, 514, 693
cholesterol, 8, 9, 525
 acyl-CoA, 675
 adernal steroid eliminated, 678
 adernal steroid to plasma proteins, 678
 ApoB-100, 675
 cholesterol acyl-transferase (ACAT), 676
 cholesterol esterase, 675
 cytochrome P450 enzymes
 adrenarche, 676
 aldosterone synthase (CYP11B2), 676
 androstenedione, 676
 cholesterol side-chain cleavage
 enzyme, 675
 11-deoxycorticosterone (DOC), 676
 17α-hydroxylase (CYP17), 675
 21-hydroxylase (CYP21A2), 676
 11β-hydroxylase (CYP11B1), 676
 17α-hydroxypregnenolone, 675, *677*
 17α-hydroxyprogesterone, 676

3β-hydroxysteroid dehydrogenase
 (3β-HSD II), 376
 isocaproic acid, 675
 pregnenolone, 675
 progesterone, 676
 steroidogenic acute regulatory protein
 (StAR), 675
 genetic defects, 676, 678
 high-density lipoprotein (HDL), 675
 3-hydroxy-3-methylglutaryl-CoA
 reductase, 675
 low-density lipoprotein (LDL), 675
 pregnenolone formation, 675, *675*
cholesterol ester hydrolase (CEH), 675
cholesterol esterase, 526, 675
cholesterol side-chain cleavage enzyme, 675, 741
cholic acid, 518
choline, 106
cholinergic receptors, 48, 169
cholinergic system, 106, 131
chondrocytes, 709
chorionic villi, 756
choroid layer, 62
choroid plexuses, 417
choroidal neovascularization, 83b
chromaffin cells, 683
chromatids, 743
chromophore, 67
chromosomal sex, 765
chronic granulomatous disease, 184
chronic myeloid leukemia (CML), 26
chronic obstructive pulmonary disease (COPD),
 186, 368, 404b
chronic respiratory acidosis, 498–499, 503
Chvostek sign, 705
chylomicron remnants, 547
chylomicrons, 176, 527
 characteristics of, 548, 548t
 defined, 547
 functions of, 548
chyme, 511
chymotrypsin, 515, 530
ciliary body, 63
ciliary ganglion, *107*
cilium, 66
cingulate gyrus, 136
circadian rhythm, 128, *129,* 649
circulation
 bronchial, 353
 cerebral, 317–320, *319*
 coronary, 315–317, *317*
 cutaneous, 324–326
 fetal, 326–328
 hepatic, 322–323, *323*
 hormonal control, 340–344
 arginine vasopressin, 343
 atrial natriuretic peptide, 343
 epinephrine, 340–342, *341*
 RAAS, 342–343
 renal hypoxia, 343
 renin–angiotensin–aldosterone system,
 342, *342*
 short-and long-term blood pressure
 control, 343–344, *344*
 neural control of, 334–340
 arterial pressure, 336
 carotid sinus baroreceptors, 336–337
 central blood volume, 338
 cerebral and coronary blood flow, 338
 chemoreceptor, 338–339

 higher-order ANS responses, 339–340
 hormonal systems, 337
 pain and myocardial ischemia, 339
 parasympathetic and sympathetic nerves,
 334–336
 pressure-and time-dependent, 337–338
 placental, 326–328
 shock (*see* circulatory shock)
 skeletal muscle, 323–324
 small intestine, *320,* 320–322
 systemic
 arterial pressures in, 282–285, 282–286,
 283–285
 blood volume, 287–288
 cardiac function in, 288–291
circulatory shock
 compensated, 344–345
 etiology, 344
 irreversible, 346
 neurogenic, 346
 progressive, 345–346
 septic, 347
circumventricular organs, 125
climbing fibers, 102
clomiphene, 732
clonal expansion, 185
clonal selection, 185, 201
clonic phase, 134
closure, 530
clot retraction, 190
coagulation, 188
coagulation cascade, 188, 189t, 549
coagulation factors, 188
coated pits, 10
cocaine, 117b
cochlea, 72, *72*
cochlear implants, 78
cochlear nerve, 73
coding region, 57
cold acclimatization, 606
cold diuresis, 605
cold pain, 62
cold pressor response, 339
cold stress, 599
colipase, 526
colipase-dependent lipase, 526
collagenase, 184, 743
collateral (prevertebral) ganglia, 106
colloid, 660
colloid, in thyroid gland, 660, *661*
colloid osmotic pressure, 302
colloidal osmotic blood pressure, 175
colon, 571
colony-stimulating factors, 184
colostrum, 530
commissures, 135
communication, 124
compensations, acid–base homeostasis, 497
compensatory pause, 253
compensatory response, acid–base
 disturbances, 498t
competitive binding assays, 636
competitive exclusion, 198
complement fixation, 207, *207*
complement proteins activation, 185, 197
complement system, 549
complete blood counts (CBCs), 175
complete heart block, 253
complete metabolic panel (CMP), 176
complete protein, 529

complex sound waves, 70
complex spike, 102
compliance, 226
compression, 59–60
concentric contraction, 86, 621
conductance, 38
conducting zone, 353
conductive deafness, 75
conductive hearing loss, 78
cone cells, 66
cones, 63
congenital adrenal hyperplasia, 649, 676,
        732, 767
conjugate base, 485
connective tissue matrix, 163
connexin, 9
connexon, 9
constrictor muscles, 115
continuous ambulatory peritoneal dialysis, 432
contraceptives, 768b–769b
contractility, 262
convection
        in body heat exchange, 558–592, *593*, 597
        defined, 592
cornea, 63
corneal reflex, 69
cornification (keratinization), 739
coronary artery bypass grafts (CABGs), 329b
coronary artery disease, 618b
        drug-eluting stents, 329b–330b
        exertional angina, 318b
coronary circulation
        autoregulation, 317
        autoregulation of, 317
        blood pressure and, 315–316, *317*
        oxygen demand and, 316
        sympathetic nervous system in, 316–317
coronary reserve, 315
corpus albicans, 743
corpus callosum, 135
corpus luteum, 739, 743
cortex, *431, 433*, 673
cortical granules, 754
cortical motor area, 93
cortical tissue, 765
corticospinal tract, *97*, 97–98
corticosteroid-binding globulin (CBG), 634, 678
corticosteroid-binding protein (transcortin), 736
corticosteroids, 211
corticosterone, 646, 674
corticotropin, 644
corticotropin-releasing hormone (CRH), 644,
        648, 757
        action of, 645t, 648, *648*
        structure of, 645t
cortisol, 633, 646, 674
cotyledons, 756
cough reflex, 2
coumarin, 189
countercurrent heat exchange, 605
coupling reaction, in thyroid hormone
        synthesis, 662
cranial nerves, 106
craniosacral regions, 106
creatine phosphate pool, 1560
creatinine, 453
cremasteric muscle, 721
crib death, 131
crista ampullaris, 76
critical micellar concentration, 527

crossbridge cycle, 146, *167*
crypt of Lieberkühn, 521, *521*
cumulus granulosa cells, 740
curare, 146b, 148
Cushing reflex, 320
Cushing syndrome, 674b
Cushing triad, 426b
cutaneous circulations, 324–326
cyanosis, 181
cyclic adenosine monophosphate (cAMP), 24,
        24, 168, 514, 648, 694
        TSH effects on, 664
cyclic GMP activates protein kinase G, 110
cyclic guanosine monophosphate (cGMP),
        25, 723
cyclooxygenase (COX), 211, 472
cystic fibrosis (CF), 409b
cystic fibrosis transmembrane conductance
        regulator, 409b, 517
cystinuria, 480b, 531
cytochrome P-450 enzyme, 544
cytochrome P-450 reductase, 544
cytokines, 29, 606
cytoskeleton, 9
cytotrophoblasts, 755

D

D cells, 512, 514
dacryocytes, 182
Dalton law, 356, 383
dark adaptation, in vision, 68
dead space volume, 363
deafness, 75
deamination, 664
decarboxylation, 664
decerebrate rigidity, 95
decibel (dB), 70
decidua, 755, 756, 761
decidua basalis, 756
decidua capsularis, 756
decidua parietalis, 756
decidual reaction, 755
decidualization, 755
declarative memory, 139
decoding, in sensory transduction, 59
deep pain, 62, 339
defecation, 560, 567, 568
defensins, 184, 198
degranulation, 185, 188
dehydration reaction, acid-base
        homeostasis, 489
dehydroalanine residue, 662
7-dehydrocholesterol, 710
dehydroepiandrosterone (DHEA), 647, 673, 742
dehydroepiandrosterone sulfate (DHEAS), 759
deiodinase type 1, 663, *663*
deiodinase type 2, 663
deiodination
        5′, 663–664
        inner ring, 663
        outer ring, 663
        in thyroid hormone matabolism, *663, 664*
delayed afterdepolarizations (DADs), 244
delayed gastric emptying, 565b
delayed pain, 62
delayed puberty, 765
delayed-type hypersensitivity, 204, 214t
delta 4 pathway (δ4), 742
delta 5 pathway (δ5), 742
delta waves, 133

delusions, 138
dementia, 3, 139
denaturation, 587
dendrite, 36
dendritic cells, 198, 200, 202
denervation hypersensitivity, 341
dense bodies, 163
dentate, 100
deoxycholic acid, 519
11-deoxycorticosterone (DOC), 676
deoxyhemoglobin (Hb), 181, 386
depolarization blockade, 148
depolarization, in cardiac muscle, 239, *239*
depolarization phase, 40
depolarize, 38
depolarizing blockade, 119
desensitization, 640
1,4 DHP, 168
diabetes insipidus, 468, 481b, 645
diabetes mellitus, 311b, 455, 640b
        autoimmune disorder, 698–699
        gestational, 700
        lifestyle changes and drug classes, 700–701
        obesity and insulin resistance, 699
        organ dysfunction
                diabetic neuropathy, 702
                macrovascular complications, 702
                metabolic and rena complications, 701
                microvascular complications, 702
        prediabetes, 699–700
        type 2 diabetes, 699–700
                treatment, 701
diabetic ketoacidosis, 701
diabetic neuropathy, 702
diabetogenic action, 654
diacylglycerol (DAG), 112, 649
dialysis, 432b
diapedesis, 208
diaphragm in breathing, 354–355
diastole, defined, 222–223
diastolic pressure, 225
dibetes mellitus, 761
diencephalon, 124
differential white blood count, 180
diffuse toxic goiter, 668
diffusing capacity, 386
diffusion, 11
        alveolar-capillary membrane, 383–384
        blood hematocrit, 386
        in capillaries, 386
        pulmonary capillary blood flow, 384–385
diffusion-limited transport, 301
digestion
        carbohydrate, 522–525
        lipids, 525–528
        protein, 528–531, 529t, *530*
digestive system, 508
        circular muscle fibers, 558
        division, 557
        enteric nervous system, 559–560, *560*
        motility and digestive processes, 567
        muscularis externa, 558, *558*
        sphincters
                external anal, 559
                ileocolonic, 559
                internal anal, 559
                lower esophageal, 559
                pyloric, 559
                sphincter of Oddi, 559
                upper esophageal, 558

digoxin, 14
dihydropyridine receptors (DHPRs), 150
dihydrotestosterone (DHT), 726
1,25-dihydroxycholecalciferol, 532, 710
diiodotyrosine (DIT), 662
dilation, 115
dimerization, 26, 27
dimerization domain, 27
dipalmitoylphosphatidylcholine (DPPC), 371
dipeptidyl peptidase IV inhibitors, 701
dipole, 244
direct calorimetry, 592
direct pathway, 99
disorders of sex development (DSD), 767, *768*
    sex chromosome, 767
    46,XX, 767
    46,XY, 767
disseminated intervascular coagulation, 347
disseminated intravascular coagulation, 552
distal nephron
    potassium secretion, *444,* 446
    proximal reabsorption, 443
distensibility, 366
diuresis, cold, 605
diurnal rhythm, 128
diving reflex, 425–426
diving response, cardiovascular responses, 340
DNA, 3
DNA-binding domain, 27
docosahexaenoic acid, 525
dominant follicle, 739
dopamine (DA), 49, 107, 644
dopamine transporter (DAT), 117b
doppler ultrasound, 277
dorsal motor nuclei, 108t
dorsal motor nucleus, 572
dorsal respiratory group (DRG), in respiratory
    centers, 413, *413*
dorsal vagal complex, 572
down-regulation, 639
drepanocytes, 182
drug-eluting stents (DES), 329b–330b
drugs, metabolism of, *544,* 544–545
drusen, 83b
duchenne muscular dystrophy, 197
duct intercalated cells, *493*
ductus arteriosus, 328
ductus venosus, 327
dynamic exercise, 284
dyspnea, 620
dystonias
    treatment of, 149b
    types of, 149b

E

early afterdepolarizations (EADs), 244
eccentric contractions, 86, 621
eccrine sweat glands, 595
echinocytes, 181
echocardiography, 277
ecotopic pregnancy, 754
ectopic foci, 243
edema, 461
    myxedema, 669
effective arterial blood volume (EABV), 465
effector, 6
effector T cells, 201
eicosanoids, 50
eicosapentaenoic acid, 525
Einthoven triangle, 248, *249*

ejaculation, 724
ejection fraction, 271
elastase, 530
elastic recoil, 366
electrical repolarization, 168
electrical synapses, 43
electrical syncytial property, 569
electrocardiogram (ECG), 618b
    cardiac stress testing with, 255b
    dipole recording in, 244, 245, *246,* 247, 248,
        249, 250, 251, *254*
    information from
        drug effects, 254
        mean electrical QRS axis, 251, *251*
        metabolic conditions, 254, *255*
        ventricular hypertrophy, 254, *254*
    leads for, 248–250, *249, 250*
    normal waveforms, 245–246, *246*
electrochemical potential, 21
electroencephalogram (EEG)
    abnormal brain waves, 134
    brainwave patterns on, *132,* 132–133
    defined, 132
    functions of, 132
    sleep stages, 133–134
electroencephalography (EEG), 132
electrogenic system, 17
electrolyte absorption, 533–536, *534–536*
electrolyte panel, 176
electrolytes, 462
electroneutral system, 17
electroneutrality, 175
electrotonic conduction, 38
emboli, 398
emboliform nuclei, 100
embryoblast, 755, 756
embryonic hemoglobin, 181
emergency contraception, 769b
emissivity, 593
emotional stress, cardiovascular responses, 339
emphysema, 186, 620
end-diastolic volume, 264
endocannabinoids, 50, 127
endocrine control, circulation, 340–344
endocrine gland, 514
endocrine muscle, 161
endocrine system
    anatomic considerations, 628, *629*
    feedback regulation in, 629–630, *630*
    pleiotropic effects, 631
    signal amplification, 630–631
endocytosis, 10, *10,* 542
endogenous creatinine clearance, 453
endogenous mediators, 208–209
endolymph, 72
endometrial cycle
    menstrual phase, 748
    proliferative phase, 747
    secretory phase, 747–748
endometrial veins, 327
endometrium, 738
endopeptidase, 513, 530
endorphins, 216
endothelial vesicles, *297,* 298
endothelial-derived relaxing factor (EDRF), 28
endothelin, 307
endothelium, 223, 436
endothelium and endothelial cells
    of liver, 541–542
    sinusoidal, 541–542, *542*

endothelium-derived relaxing factor
    (EDRF), 306
endotoxin, 208
endplate membrane, 147
endplate potential, 147
end-systolic pressure-volume relationship
    (ESPVR), 268
end-systolic volume, 264
energy expenditure, 126
energy intake, 126
energy-saving process, 1
enkephalins, 216
enteric nervous system, 105, 559–560, *560*
    excitatory musculomotor neurons, 576–577
    gastric reservoir, 579–582, *580–581*
    inhibitory junction potentials, 577
    inhibitory musculomotor activity, 577
    inhibitory neurotransmitters, 577–578,
        577–579
    MMC, housekeeper role, 582
    musculomotor neurons, 579
    synaptic transmission
        ionotropic receptors, *574,* 575
        metabotropic receptors, 574
        presynaptic facilitation, 576
        presynaptic inhibition, 575–576
        slow excitatory postsynaptic potentials, 575
enterochromaffin cells, 560
enterochromaffin-like cells, 514
enterocytes, 508, 522
enteroendocrine cells, 560
enterogastrones, 514
enterohepatic circulation, 520
entero-oxyntin, 513
enteropeptidase, 530
enzyme-linked immunosorbent assay (ELISA),
    637, *637*
enzymes
    in drug metabolism, *544,* 545
    of glycolytic pathway, 546
    pancreatic, 516
eosinophils, 184–185, 200
ependymal cells, 36
epidermal growth factor (EGF), 23, 190, 510
epididymis, 723
epigenetic-induced functional changes, 3, 4
epigenetics, 3, 161
epigenome, 3
epilepsy, 132, 133, 134
epileptic seizures, 134
epinephrine (EPI), 48–49, 110, 546, 683
epinephrine reversal, 117
epineurium, 41
"EpiPens," 117
epiphyseal plate, 709
epithelial sodium channel, 80
epitopes, 200
equal pressure point (EPP), 376
equilibrium, 7
equilibrium, fluid compartments in, 462–463
equilibrium potential, 38
erectile dysfunction (ED), 30, 31, 724
erection, 723–724
erythroblasts, 186
erythrocytes, 174
erythropoiesis, 182, 186–187, *187*
erythropoietin, 186, 187b, 425
esophageal motility, 562–563
esophageal motor disorders, 562b
essential amino acids, 549

essential fatty acids, 525
essential light chain, 145
estradiol, 726, 736, 741
estrogens, 633, 646, 656, 737
estrone, 736, 742
eunuchoidism, 731
eustachian tube, 70
euthyroid sick syndrome, 669b
evaporation
    in body heat exchange, 592, *593*, 594–595
    latent heat of, 593
    of sweat, 592
excitation–contraction coupling, 146
excitatory postsynaptic potential (EPSP), 38, *40*
excretion, 434
    of electrolytes, 462
    renal clearance, 434
excretory duct, of salivary gland, 509
excretory (collecting) ducts, 510
execution, 136
exercise, 161
    aging, 623, 624, *624*
    benefits through early ambulation, 625b
    calcium homeostasis, 622–623, *623*
    caloric intake, 623
    cardiovascular response to, 340
        acute and chronic exercise, 617
        blood flow, 615–617, 616t
        cholesterol level, 617–618
    diabetes, 623–624, *624*
    dynamic, 284
        acute respiratory response to, 620t
        cholesterol, 617
    immune response, 624–625
    ischemic heart, ECG, *618*
    isometric, defined, 615
    obesity, 623
    oxygen uptake
        dynamic exercise, 614–615, *615*, 615t
        isometric exercise, 615
        maximal oxygen uptake, 614
        maximal voluntary contraction
            (MVC), 615
        oxygen debt, 614, *615*
    pregnancy and, 619
    respiratory response
        anaerobic threshold, 619
        effects of training, 620
        emphysema, 620b
        hypercapnia, 619–620
        minute ventilation, *619*
    skeletal muscle, 621–623
    thermoregulation during, *602*, 602–603, *603*
    training, 614
        cardiovascular disease, 618
        endurance, 621, 622t
        respiratory system, 620
exercise testing, 255b
exertional angina, 318b
exocrine glands, 509, 514
exocytosis, 10, *10*, 10–11
exogenous mediators, 208
exopeptidases, 530
expiration, 414
expiratory reserve volume (ERV), 359
expired minute ventilation, 362
extension, 86
extension movements, 86
extensors, 160
exteroceptors, 56

extracellular fluid (ECF), 460
    defined, 462
    measurement of, 461
    osmolality of, 462
    osmotic equilibrium and, 462–463
    pH of, 487
    in potassium balance, 475
    potassium distribution, 475–476
extraction (E), 300
extraction ratio, 300
extrafusal muscle fibers, 87
extramedullary hematopoiesis, 186
extraocular muscles, 62
extrathoracic blood volume, 287
extrinsic coagulation pathway, 189
exudate, 208

F

fab region, of antibody, 206
facilitated diffusion, 12
familial hypercholesterolemia, 548–549
farsightedness, 64
"fast" mushroom poisoning, 116
"Fast response" action potentials, 238
fast synaptic transmission, 45
fastigial, interposed, 100
fasting plasma glucose (FPG), 699
fast-twitch fibers, 161
fat
    brown, 604
    dietary, 285, 511, 527
    soluble vitamins, liver storage of, 550
fatty acid synthase, 696
fatty acids, 547–548
fear, cardiovascular responses, 339
fecal continence, 567, *567*
feedforward control, 6, *6*–7
female and male sex hormones, 216
female pronucleus, 755
female reproductive system
    ampulla, 738, *738*
    atresia, 739
    cervix, 738, *738*
    cornification, *738*, 739
    corpora albicantia, 737
    corpora lutea, 737, *738*
    endometrium, 738, *738*
    ferning, 738, *738*
    fimbria, 738, *738*
    follicles, 737
    hormonal regulation
        hypothalamus production, 736
        neurons in GnRH, 736–737, *737*
        sex steroids, 736
    infertility, 749
    infundibulum, 738, *738*
    isthmus., 738, *738*
    menstrual cycle, 744–749
    myometrium, 738, *738*
    oogonia, 739, 739t
    ovarian cycle
        corpus luteum, 743–744
        dominant follicle, 742
        estradiol synthesis, *741*, 741–742
        folliculogenesis, 739–740, *740*
        graafian follicle, *740*, 740–741, 741t
        hormonal regulation, 745–747
        LH surge, 742–743
        oogonia, 739, 739t
    oviduct, 738, *738*

spinnbarkeit, 738, *738*
stromal layer, 738, *738*
uterus, 738, *738*
vagina, *738*, 738–739
ferritin, 183, 536
fertilization
    and implantation
        acrosomal reaction, 754–755, *755*
        gametes transportation, 753–754
        process of, 755–756, *756*
    process of, 754–755, *755*
fetal development
    bone formation, 709
    of central nevous system (CNS), 666
    endocrine system
        growth, 760–761
        hormones, 760–761
    hormones, 766
    parturition, 761
    sex chromosomes, 765–766
    thyroid hormones in, 666
fetal hemoglobin, 181, 327
fetus
    circulation, 327–328
    heart, *326*
    vasculature of, *326*, 327
fever
    pyrogens in, 606
    thermoregulation during, 601, *601*
fibrin, 175
fibrin stabilizing factor (factor XIII), 189
fibrinogens, 175
fibrinolysis, 190
fibroblast growth factor 23 (FGF23), 479
Fick's law, 11, 383
fight-or-flight response, 105, 113, 113t, 117,
    137, 339
filamentous actin (F-actin), 145
filtration fraction, 438
filtration pressure equilibrium, *438*
filtration slit, 436
fimbria, 738, 754
final common path, 87–88
first heart sound, 263–264
first messengers, 23
first polar body, 743
first-degree atrioventricular block, 253
fissures, 135
flecainide, 257b
flexion movement, 86
flexor withdrawal reflex, 92, *92*
flexors, 160
flow velocity, 229
flow-limited transport, 301
flow-mediated vasodilation (FMD), 307, 308b
fluid balance, composition in body fluids
    extracellular fluid osmolality, 464–465, 464t
    hyponatremia, 468b
    plasma osmolality, 466
        AVP disorders, 467
        blood volume depletion, 466–467
    water balance, 465
        arginine vasopressin, 465–466
fluid compartments
    body water in, 461–462, 461t
    extracellular fluid (ECF) in, 462–464, *463*
    intracellular fluid (ICF) in, 460–462, *463*
    size, measurement of, 461–462
fluid-phase endocytosis, 10, *10*
focal adhesions, 163

focal length (FL), 63
focal point, 63
folia, 100
folic acid, 532
follicles, 660
follicle-stimulating hormone (FSH), 644, 718, 736
follicular phase, 745–746
folliculogenesis, 739–740, *740*
follistatin, 720, 736
foramen ovale, 328
fore brain, cerebral cortex, *135*, 135–136
formed elements, 174, 177
fornix, 125
fragment, crystallizable (Fc region), 206
free radical, 391b
free radical–induced lung injury, 391b
fresh frozen plasma (FFP), 175
frontal lobe, 135
fructose-1,6-diphosphatase, 696
F-type ATPases, 15
functio laesa, 208
functional hypothalamic amenorrhea, 749
functional residual capacity (FRC), 402
functional syncytium, heart as, 237
fused tetanus, *152*, 153
fusimotor system, 90

G

G cells, 514
G protein–coupled receptors (GPCRs), 23, 45–46
G proteins
    ras family of monomeric, 26
    trimeric, 23–24
gallstones, 521
γ-aminobutyric acid (GABA), 49
γ motor neurons, 87
gametes, 721
gamma globulins, 175
ganglia, 34
ganglion, 105
ganglion cell layer, 68
ganglion, cervical, 511
ganglionic cell, 105
gap junctions, 9, 43, 163
gas
    diffusing capacity, 385–386
    diffusion and uptake, 383–385
    transport by blood, 386–389
gas tension, 356
gas transport
    alveolar–arterial oxygen gradient, 390, *390*
    bicarbonate, 389, *389*
    carboxyhemoglobin, *388,* 388–389
    of oxygen, 386–388
    oxyhemoglobin-equilibrium curve, 387–388
gastric amylase, 513
gastric inhibitory peptide, 514
gastric lipase, 513, 526
gastric motility, 562–564
    antral pump, 564
    filling and emptying of stomach, *563,* 563–564, *564*
    gastric reservoir, 564
    receptive relaxation, 564
gastric mucosa, 511
gastric phase, 513, 513t
gastric secretions

anatomic considerations, 511
chyme in, 511
endocrine control of, 514, *514*
gastric acid, 512, *512*
hormonal control of, 514, *514*
neural control of, 514, *514*
oxyntic glands in, 512, *512*
parietal cells in, 512, *512*
phases of digestion, 513, *513*
production of, 513, *513*
rate of, 513, *513*
gastrin, 512
gastrin-releasing peptide, 514
gastrointestinal hormones, 127
gastrointestinal motility patterns
    emesis, 562
    peristalsis
        occurrence, 560–561, *561*
        polysynaptic neural reflex, 561, *561*
        and propulsion, 560
gastrointestinal secretions
    biliary, 518–521
    intestinal, *521*, 522
    pancreatic, 514–518
    saliva, 509–511
    of stomach, 513
gastrointestinal (GI) tract, 508, 537b, 538b
general gas law, 358
generalized edema, 474
generalized hypoventilation, 392
generalized hypoxia, 402
generalized seizures, 134
generator potentials, 56–57
genetic considerations, in liver metabolism, 545
genetic sex, 729, 765
genital ridge, 765
gestational diabetes, 700, 761
GH-releasing hormone (GHRH), 644
ghrelin, 126
gigantism, 654
Gitelman syndrome, 476
glaucoma, 63
glial cell, 34, *35*
glicentin, 694
glitazones, 701
globose nuclei, 100
globulins, 175
globus pallidus (GP), 94
glomerular filtration
    disorders
        nephrotic syndrome, 436b–437b
        poststreptococcal glomerulonephritis, 436b
        proteinuria, 439
    hemodynamic forces
        capillary hydrostatic pressure, 437–438
        capillary pressure, 437
        filtration coefficient, 439
        glomerular ultrafiltration coefficient, 439
        layers of, 436, *436*
        osmotic pressure and hydrostatic pressure, 439
        tubular reabsorption, 434
        tubular secretion, 434
        urine formation, 433–434
        vascular resistance, 437, *438*
    layers, 436, *436*
glomerular filtration rate (GFR), 452, 469
    inulin clearance, 452–453

plasma creatinine, 453, *453*
plasma urea concentration, *453*, 454
renal blood flow, 454
serum creatinine concentration, 453, *453*
glomerulonephritis, 437b, 457b
glomerulotubular balance, 470
glomerulus, 431
GLP1 agonists, 701
glucagon, 546
glucagon-like peptide 1, 693, 694
glucagon-like peptide 2, 694
glucocorticoid hormones, 646, 648–649
glucocorticoid response elements (GREs), 681
glucocorticoids, 674
    inflammation and immune system, 682–683
    inhibition, 683
    metabolic adaptations during fasting, 681–682, *682*
    negative feedback regulation of, 683
    and stress, 683
glucokinase, 546, 693, 696
gluconeogenesis, 545, 681–682
glucose, 160
    absorption of, 690
    blood, 547
    in glucagon secretion, 693–694
    handling by kidneys, 454
glucose buffering, 546
glucose metabolism, 696, *696*
glucose 6-phosphate (G6P), 546
glucose reabsorption, 454–455, *455*
glucose threshold, 455
glucose-dependent insulinotropic peptide, 693
glucose-galactose malabsorption (GGM), 17
glucose-stimulated insulin secretion, 690–691, *692*
glucosuria, 455, 701
glucuronidation, thyroxine, 664
GLUT2, 693
GLUT4, 695, *695*
glutamate, 49
glutamic acid decarboxylase (GAD), 49
gluten-sensitive enteropathy, 524b
GLUTS, 17
glycogen, 696
glycogen phosphorylase, 546, 696
glycogen synthesis, 696, *696*
glycogenesis
    defined, 545
    in metabolism of glucose, 545
glycogenolysis, 545, 697
glycolipids, 9
glycolysis, 160, 693
glycolytic pathway, 160
glycoproteins, 9
goblet cells, 10, 522
goiter, diffuse toxic, 668
Goldman equation, 38
Golgi cells, 101
golgi tendon organs (GTO), 88, *88*, 90
gonadal sex, 765
gonadarche, 744, 765
gonadotrophs, 644
gonadotropin-releasing hormone (GnRH), 127, 644, 718, 736, 757
    high frequency, 719–720, *720*
    hypothalamic, 718–719, *719*
    low frequency, 719–720, *720*
    in puberty sexual development, 767
gonadotropins, 646, 654–655

graafian follicle, 740, *740*
graded response, 57
graft-*versus*-host disease (GVHD), 213
grand mal seizures, 134, 705
granular cells, 432
granular layer, 101
granule cells, 101
granulocytes, 184
granulosa cells
    antral, 740
    cumulus, 740
    mural, 741
granulosa lutein cells, 743
granzyme, 200
Graves disease, 71, 215, 667
gray matter, 34
green-sensitive pigment, 66
ground substance, 708
growth factors, 190
growth hormone (GH), 216, 644
    action of, *652,* 654
    disorders of
        Creutzfeldt–Jakob disease, 655b
        deficiency, 654
        excessive secretion, 654
    expression of, thyroid hormones in, 666
    function, 654
    insulin-like growth factor I, 652–653
    insulin-opposing actions of, 654
    pulsatile hormone secretion, 631b
    pulsatile secretory pattern, 653, *653*
    regulation of growth and metabolism,
        651, 654
    secretion, 644
        IGF-1 inhibition of, 653, *653*
        regulation, 652, *652*
    synthesis, 651–652
growth hormone-releasing hormone (GHRH),
    644, 645t
    actions of, 645t
    structure of, 645t
guanosine diphosphate (GDP), 24
guanosine triphosphate (GTP), 23–24
guanylin, 472
guanylyl cyclase (GC), 28
gustationl, 78
gustatory system
    chemoreception
        bitter receptors, 81
        salt receptors, 80
        sour receptors, 80–81
        sweet receptors, 81
        umami receptors, 81
    taste buds, 78–80, *80*
gustducin, 81
gut microbiome, 702b
gut microbiota, 537b
gut motility, neural control of
    dorsal vagal complex, 572
    enteric nervous system, 573
    motor activity, 571
    myenteric and submucosal plexuses, 573
    parasympathetic neurons, 571
    splanchnic nerves, 573
    sympathetic nerves, 572–573
    vagovagal reflex, 572
GVHD, 212
gynecomastia, 729
gyrus, 135

H

H_2 receptors, 514
H zone, of skeletal muscle, 145, *147*
habenula, 136
hair cells, 72–74
    anatomic considerations, 73
    inner, 72, 73
    outer, 72–73
    sound transduction in, 73, *73*
hair-follicle receptors, *61,* 198
Haldane effect, 389
hallucinations, 138
hamartomas, 746b
haptens, 196
haptoglobin, 183, 550
Hartnup disease, 531
Hashimoto disease, 668, 668b
H^+-ATPases, 14
H_2CO_3. *see* carbonic acid
HCO_3. *see* bicarbonate
HCO_3^-/CO_2 system, 489–490, *490*
HDAC inhibitors, 161
hearing, 69, 135, 138
hearing loss, 75
    cochlear implants for, 78b
    conductive, 75
    sensorineural, 75
heart, 222
    blood flow
        autoregulation of, 310, *310*
        baroreceptor reflex, 336, *337*
    electrophysiology of
        cardiac action potential, 238, *238*
        depolarization, 239, *239*
        pathological conditions, 243
        repolarization, 240
        resting membrane potentials, 238–239, *239*
        voltage-gated sodium channels in, 239–240
    as functional syncytium, 237
    output of right and left, 223
heart block, 253, 266
heart failure
    muscle mechanics, 272b
    treatment, 277b
heart rate
    acute exercise effects, 252
    arterial blood pressure, 283–284, *284*
    autonomic nervous system and, 273
    in heat acclimatization, 603–604
    sinoatrial (SA) node, 273
heart sound, 263, 264
heart valves
    anatomic considerations, 223
heat
    as energy exchange, 590, 591
    extreme, effects on body, 605
    measurement of, 594
    as by product of metabolism, 591–594
heat acclimatization, 603–604, *604,* 605t
heat balance, equation for, 590
heat dissipation
    skin blood flow, 597
    sweating, 595
heat exchange
    conduction in, 558–591
    convection in, 588–592, *593,* 597
    countercurrent, 605
    equation for, 593
    evaporation in, 592, *593,* 594–595

heat dissipation
    skin blood flow, 597
    sweating, 595
heat production
    vs heat loss, *590,* 590–591
    at rest, 591–594, 591t, *593*
    radiation, *590,* 590–593
    skin temperature, 589
    surface area and, *593,* 593–594
    total health flow in, 588
heat exhaustion, 607
heat flow, equations for, 597
heat injury, exertional, 608
heat storage, 591, 594
heat stress
    disorders related to, 607–608
    effects of, 598
    exercise and, 602
heat stroke, 607–608
heat syncope, 607
heat transfer. *see* heat exchange
heavy meromyosin (HMM), 145
heel stick, 175
helicotrema, 72, 74
helium dilution technique, for residual
    volume, 362
hematocrit (Hct)
    blood disorder, 177
    blood flow velocity and, 232
    defined, 177
hematologic malignancies, 215
hematologist, 174
hematopoiesis, 186
hematopoietic stem cells, 186, *186*
hematopoietins, *186*
heme oxygenase (HO), 536, 551
hemifacial spasm, 149b
hemochromatosis, 552
hemodialysis, 432b
hemodynamics
    axial streaming in, 232–233
    Bernoulli principle, 229
    defined, 213
    flow velocity in, 230–231, *231*
    fluid dynamic properties and, 231–232, *232*
    physical determinants of, 227–233
    Poiseuille's law, 228–229
    pressure in, 226–227, *227*
        compliance, 226
        effects of gravity, 225, *226*
        profile across cardiovascular system,
            227–228, *228*
        pulmonary arteries, 230–231, *231*
        transmural, 226
        vascular capacitance, 226
    Reynolds number, 231
    shear stress in, 230
    total resistance of multiple vascular
        segments, 229
    turbulence in, 231
    wall stress in, 227
hemoglobin (Hgb or Hb), 173
hemoglobinopathies, 181
hemolysis, 174, 182
hemolytic anemia, 191b
hemopexin, 183, 550
hemophilia, 190
hemorrhage, 173
hemosiderin granules, 551

hemostasis, 173
  activation of, 189
  angiogenesis, 190
  blood coagulation in, 187
  clot retraction in, 190
  defined, 173
  platelet plug formation in, 187
  steps in, 187–190
  vasoconstriction in, 187
Henderson–Hasselbalch equation, 486
Henry law, 383
heparin, 189, 200
heparin therapy, 189
hepatic arterial buffer response, 323
hepatic circulations, 322–323
hepatic portal vein, 542
hepatocytes, 518, 553
hereditary hemochromatosis (HH), 552
Hering–Breuer reflex, 415
hermaphroditism, 767
heterogametic, 765
heterologous desensitization, 640
heteroreceptors, 115
hexokinase, 546, 696
hidromeiosis, 601, 604
high-density lipoproteins (HDLs), 15, 176, 520,
    548, 617
hippocampus, 136
hirsutism, 730
histamine, 50, 185, 200, 514, 682
histaminergic neurons, 131
H+/K+-ATPase, 14, 512
homeostasis, 1, 173–174
  coordinated body activity, 6
  defined, 6
  equilibrium *vs.* steady state, 7–8, *8*
  forward control system in, 6, 6–7
  internal environment and, 6
  intracellular, 6
  negative feedback in, 6, *6*
  positive feedback in, 7
homeotherms, 587
homogametic, 765
homologous desensitization, 640
hopping reaction, 96
horizontal cell, *66*, 68
hormonal mediators, 216
hormone control
  of circulation
    arginine vasopressin (AVP), 343
    atrial natriuretic peptide (ANP), 343
    blood pressure, 343–344
    circulating epinephrine, 340–342
    of gastric secretions, 514, *514*
    renal hypoxia stimulates, 343
    renin–angiotensin–aldosterone system
        (RAAS), *342*, 342–343
hormone replacement therapy, 345b, 518, 668b,
    711, 749, 750b
hormone response element (HRE), 27
hormone-binding domain, 27
hormones
  amine-dervied, 631–632, *632*
  circulating transport proteins, 633–634, 634t
  C-peptide, 633b
  defined, 628–629
  degradation and excretion, 634–635
  effects of, 628
  function, 628

hormone receptor, 629
  measurement of
    bioassay measurements, 635
    enzyme-linked immunosorbent assay,
        637, *637*
    radioimmunoassay, *636*, 636–637, *637*
  mechanisms of
    dose–response relationship, 638–640, *639*
    hormone-receptor binding kinetics,
        637–638, *638*
  mutliple actions with target tissues, 631
  pancreatic β-cell function, 633b
  peptide-derived hormones, 632, *632*
  peripheral transformation, 634–635
  pleiotropic effects, 631
  polypeptide hormone, 632, *632*, 633t
  preprohormones, 632
  prohormones, 632
  receptors for, 629
  regulation through feedback control,
      629–630, *630*
  releasing, 630
  secretion pattern, 631
  smiple amplification, 630–631
  steroid hormones, *632*, 633
  target cells, 629
  tropic, 630
hormone-sensitive lipase (HSL), 654, 682, 696
human chorionic gonadotropin (hCG), 632, 722,
    753, 757
human leukocyte antigen (HLA), 213
human placental lactogen (hPL), 757, 758
humidity, defined, 594
humoral immune system, 204
Huntington's disease, 99
hydration reaction, acid–base homeostasis, 489
hydraulic conductivity for water, 302
hydrochloric acid (HCl), 512
hydrogen, 324, 388, 416, 485
hydrogen ion concentration, 486, 516
hydrogen peroxide (H_2O_2), 184, 391b
hydrophilic, defined, 9
hydrophobic, defined, 9
hydrostatic pressure, 301–302, *302, 303*, 310,
    321, 403, 404
hydroxyapatite crystals, 708
hydroxyl radical (•OH), 391b
17α-hydroxylase (CYP17), 675
21-hydroxylase (CYP21A2), 676
3-hydroxy-3-methylglutaryl-CoA reductase, 675
17α-hydroxypregnenolone, 675, *677*, 742
17α-hydroxyprogesterone, 676
5-hydroxytryptamine (5-HT), 49
hyperaldosteronism, acid–base homeostasis, 496
hypercalciuria, 478
hypercapnia, 390, 497, 619–620
hyperemia
  absorptive, 321
  active, 305, 308b
  reactive, 308b, 309b, 310
hyperglycemia, 701
hypergonadotropic hypogonadism, 732
hyperkalemia, 224t, 239, 243, 254, *255*, 475
hyperkinesis, 99
hyperkinetic disorders, 99
hyperlipoproteinemia, 292b
hypernatremia, 596b
hyperopia, 64
hyperosmotic, defined, 19

hyperphagia, 128b
hyperphosphatemia, 479
hyperpnea, 418
hyperpolarization, 49, 68, 73, 76, 168, 224t–225t,
    240, *241*
hyperpolarization-activated cyclic nucleotide-
    gated channels, 242
hyperpolarizes, 38
hyperprolactinemia, 732, 749
hypersensitivity reactions, 204, 213, 530
hypertension
  arterial, 286b, 329b
  chronic, 329b
  hypoxia-induced pulmonary, 404b
  pharmacological monotherapy, 347b–348b
  portal, 542
hyperthermia
  defined, 607
  due to lesion, 607
  malignant, 608
  therapeutic, 608
hyperthyroidism, 668
hyperventilation, 418
hypervitaminosis A, 550
hypoaldosteronism, acid–base homeostasis, 496
hypocalcemia, 705
hypocalcemic tetany, 705
hypocapnia, 390, 418
hypochromic cells, 181
hypocretin, 129
hypogastric nerves, 724
hypoglycemia, 647, 683–684
hypogonadism, 731
  hypergonadotropic, 732
  hypogonadotropic, 732
hypokalemia, 239, 242, 254, *255*, 475
hypokinesis, 99
hyponatremia, 467, 468b, 596b
hypophyseal arteries
  inferior, 644
  superior, 644
hypophyseal portal circulation, 644
hypophyseal portal vessels, 644, 650
hypophysiotropic hormones, 644
hypophysis. *see* pituitary gland
hypopituitarism, 719
hypoprothrombinemia, 550
hyposmotic, defined, 19
hypotension, 322
hypothalamic amenorrhea, 749
hypothalamic (or central) diabetes insipidus, 645
hypothalamic–pituitary axis, 643–645, *644*, 645t
hypothalamic–pituitary axis development, 760
hypothalamic–pituitary–adrenal axis, 648, *649*
hypothalamic-pituitary-GH axis, 653, *653*
hypothalamic-pituitary-thyroid axis, 651, *651*, 668
hypothalamic-releasing hormones, 124
hypothalamohypophyseal portal blood vessels,
    124, *126*
hypothalamo-hypophyseal tract, 124
hypothalamus, 124
  components of, 124, *125*
  eating behavior, 126
  functions of
    biological clock, 128–129, *129*
    metabolism and eating behavior,
        126–127, *127*
    reticular formation, 129–130
    sexual activity, 137

hormones in, 124
interactions with pituitary gland, 124
limbic system, 125
nuclei of, *126*
paraventricular, 643
suprachiasmatic, 125
supraoptic, 643
structure of, 124, *125*
hypothermia, 605–606, 608b–610b
accident, 608
classification of, 608b
defined, 607, 608
diagnosis of, 609b
due to lesion, 607
etiology of, 609b
signs and symptoms of, 608
treatment of, 609b
hypothyroidism, 668–669
hypotonia, 102
hypotonic, 19, 20
hypoventilation, 391, 392, 393, 418, *420*, 422,
490, 502b
hypovolemia, 465
hypoxemia, 389–392, 402, 418, 423
diffusion block, 393
generalized hypoventilation, 392–393
pathophysiologic causes, 391, 391t
respiratory dysfunction, 389–393
shunt, 392, *392*
venous admixture, 390, *392*
hypoxia, 402, 416, *416*, 418, 422, 423
hypoxia-induced pulmonary hypertension, 404b
hypoxia-induced pulmonary
vasoconstriction, 402

I

I band, 145, *147*
I cells, 518
Ig isotype switching, 201
IgA antibody, 206–207
IgD antibody, 206
IgE antibody, 200, 207
IgG antibody, 199, 206
IgM antibody, 206
ileocolonic sphincter, 559
ileus
paralytic, 568
pathological, 568
immediate hypersensitivity reactions, 205
immune defense system, 195
immune deficiency, 213–214
immune disorders, 214t
immune response, 201
immune surveillance, 215
immune system, *196*
activation of
large complex molecules and proteins,
196–197
severe injury and necrosis, 197
adaptive, 197
antibodies of, 205–207
antigens in, 215
cells of, *196*
cellular, 201–205
clonal selection, 201
defense mechanism
surface barriers, 198
defined, 195, *197*
disorders of

allergies, 211b
asthma, 211b
cancer and, 215
etiology, 213
rheumatoid arthritis, 213
types of, 214t
diversity of, 200–201
functions of, 195
humoral, 205–207
inflammation and
acute inflammation, 208–210
anti-inflammatory drugs, 211
chronic inflammation, 210–211
memory of, 201, *201*
neuroendoimmunology, 215–216
organ transplantation, 212–213
organs, *196*
overview of, *197*
self tolerance in, 198
specificity of, 200
structures of, *205*, 205–206
immune system, *197*
innate, 195, *197*
cellular elements, 199t
characteristics, 199t
components of, 195–196, *196*
defense mechanism, 195
defined, 195
humoral component of, 198
immune therapies, 216
immune thrombocytopenia purpura, 188
immunity
adaptive, 195, *197*
humoral, defined, 201
innate, 195, *197*
immunization. *see also* vaccinations
active, 215
immunodeficiency disorders, 214t
immunogens, 196
immunoglobulins. *see* antibodies
immunologic synapse, 202
immunosenescence, 216
immunosuppressive drugs, 213
immunosuppressive effects, 524b
implantation, 753
implicit memory, 139
impotence, defined, 724, 731–733
impulse initiation region, 57
in vitro fertilization, 754b
inactivation gate, 40, 164
inactive state, 40
incisura, 264
incomplete isosexual precocity, 746b
incomplete proteins, 529
incretin effect, 690
incretin hormones, 694, 701
incus, 71
indicator-dilution technique, *274*, 274–275
indirect calorimetry, 592
indirect pathway, 99
indirect sympathomimetics, 114b–115b
inducible form of nitric oxide (NO) synthase in
septic shock, 347
induction, of labor, 762b
inferior hypophyseal arteries, 644
inferior olive, 102
inferior vena cava (IVC), 222, 240, 287, 328, 329,
329b, 542
infertility

amenorrhea, 749
definition, 749
inflammation
acute
biomarker, 209–210
characterization, 208
inflammatory mediators, 208–209
self-terminates, 210
vasodilation, 208
chronic, characterization of, 210
inflammatory response, or inflammation, 197
infundibular process, 643, *644*
infundibular stem, 643, *644*
infundibulum, 738
inhibin, 720, 736
inhibitory junction potentials (IJPs), 577
inhibitory postsynaptic potential (IPSP), 38, *40*
initial pain, 62
innate immune system, 195, *197*
cellular elements, 199t
characteristics, 199t
components of, 195–196, *196*
defense mechanism, 195
nonphagocytic leukocytes, 200
phagocytic leukocytes, 199–200, 199t
defined, 195
humoral component of, 198
innate immunity, 197
inositol trisphosphate (IP$_3$), 112, 649
inotropic state, 262
insensible perspiration, 593
insensible water loss, 464
inspiration, 414, *414*
insulin, 126, 633b
amino acid stimulation, 697
biologic response, 694–695, *695*
degradation of by liver, 553
GLUT4, 695
glycogen synthesis and glucose
metabolism, *696*
hormone of nutrient abundance, 694
insulin receptor signaling, 694–695, *695*
insulin-signaling network, 694, *695*
lipogenic and antipolytic effects, 696, *696*
protein synthesis, *697*
Insulin receptor (IR), 694
Insulin receptor substrate (IRS) family, 695
insulin resistance, 695
insulin/glucagon ratio (I/G ratio), 697–698
insulin-like growth factor I (IGF-I), 715
growth hormone effects, 652–653
inhibition of GH secretion, 653, *653*
measurement of, 655b
insulin-like growth factor II (IGF-II), 652
insulin-like growth factors, 23, 553
insulitis β-cell injury, 699
integral proteins, 9
intensity, 55
intensity receptors, 59
intention, 136
intercalated cells, acid–base homeostasis, 493
intercalated duct, 509
interferon-γ, 210
interferons, 198
interleukin-1, 29, 209–210
interleukin-6, 2, 210
interleukin-8, 210
interleukins, 209
intermediate filaments, 163

intermediate lobe, 643
intermediate zones, 100, 101
intermediate-density lipoproteins, 176, 548
internal anal sphincter (IAS), 559
internal auditory meatus, 73
internal capsule, 98
interstitial cells of Cajal (ICCs), 560
interstitial fluid, 462t
intestinal phase, 513, 513t
intestinal secretions
    crypts of Lieberkühn, 521, 521
    goblet cells, 522
    mucoproteins, 522
    paneth cells, 522
    protective functions, 522
intracellular fluid (ICF), 460
intrafusal muscle fibers, 87
intralaminar nuclei of the thalamus, 130
intrinsic coagulation pathway, 189
intrinsic factor, 512, 513, 533
inulin, 453
inverse myotatic reflex, 92, 92
iodine, in thyroid hormone synthesis, 664
iodothyronine, 660, 661
ion channels, 25
ion pumps, 14
ionization constant, acid–base homeostasis,
    485–486
iris, 62
iron, 535–536, 551, 552
iron homeostasis, liver in, 551
iron profile, 183, 183t
iron-deficiency anemia, 183, 393b, 551
irritant receptors, 415
ischemia
    myocardial
        arrhythmias, 244
        ectopic foci related to, 243
        reduced refractory period in, 244
    reperfusion, 391b
ischemia-reperfusion injury, 309b
ischemic heart disease, ECG, 618
ischemic phase, 748
islets of langerhans
    diabetes mellitus (see diabetes mellitus)
    functional units
        β cells, 689
        cell types, 689, 690
        α cells, 689
        δ cells, 689
        F cells, 689
        vascular supply, 690
    glucagon
        cAMP-mediated, 694
        gluconeogenesis, 697, 697
        glycogenolysis, 697
        ketogenesis, 697
        lipolysis, 698
        ureagenesis, 697
    hyperglycemia, 694
    hypoglycemia
        glucagon secretion, 693–694
        proglucagon synthesis, 693
    insulin (see insulin)
    insulin/glucagon ratio
        fasting, 697, 698
        type 1 diabetes, 698
    physiologic factor
        β-cell signaling, 692, 693

insulin-secretion regulation, 693, 693t
    proinsulin biosynthesis, 690, 691
isoelectric potential, 245
isograft, 212
isohydric principle, 491
isometric contraction, 86, 154, 154, 155
isometric exercise, 615
isosmotic, defined, 19
isotonic contraction, 86
isotonic, defined, 19
isotonic or concentric contraction, 86
isovolumic contraction, 263
isovolumic relaxation, 264
isthmus, 738
Ito cells, 542, 553

J
J receptors, 416
jejunum. see small intestine
junctional complex, 150
juxtaglomerular apparatus, 431–432, 433
juxtapulmonary capillary receptors, 416

K
kainate (KA) receptor, 49
kallikrein, 511
Kallmann syndrome, 719
ketoacidosis, 491, 501t, 549, 800
ketogenesis, glucagon in, 697
ketone bodies
    for energy, 547, 549
    during fasting, 682
ketone, production of, glucagon in, 549, 682
ketones, in diabetes mellitus complications, 701
ketosis, 491, 549
17-ketosteroids, 678
key inflammatory mediators, 2
kidney, pH regulation
    acid–base balance, 491–492
    bicarbonate generation, 494
    net renal acid excretion, 492–493
    reabsorption of filltered bicarbonate,
        493–495, 494
    renal secretion, 495, 495–496
    titratable acid, 491
    urinary acidification, 492, 492–493, 493
kidney stones, 706
kidney tubule epithelium, 495
kidneys
    in acid–base balance, 460, 475–476
    acid–base homeostasis, 491–495, 492–495
    anatomic considerations, 431, 433
    blood supply to, 232b
    calcium and phosphate regulation, 706,
        706–707, 707
    in calcium homeostasis, 706–707
    collecting duct system, 413, 417, 431
    functions, 430–431
    glomerular filtration (see glomerular
        filtration)
    glomerular function (see glomerular filtration
        rate (GFR))
    glucose titration, 454, 455
    innervated, 431
    micturition, 455–457
    nephron, 431–433, 433
    proximal tubule (see proximal tubule)
    proximal tubule transport, 435–437
    renal blood flow, 434–435

renal clearance, 451–455
structure and function
    glomerulus, 431
        juxtaglomerular apparatus, 431–432, 433
        nephron, 431–434, 433
    transplantation of, 212, 432b
    tubular reabsorption (see tubular
        reabsorption)
    tubular secretion, 444–446
    urinary concentration and dilution (see
        urinary concentration mechanism)
    urine formation, 433–434, 434
kilocalorie, in heat measurement, 591
kininogen, 325, 511
kinins, 682
kinocilium, 73, 76
Klinefelter syndrome, 725, 732
Korotkoff sounds, 284
Krebs cycle, 49, 160
Kupffer cells, 200, 541, 542, 551, 551, 552, 553
Kussmaul respiration, 500b

L
labeled line, 56
labor
    and delivery, 761, 763
    hormones of, 760, 761
    induction of, 762b
laboratory tests, of blood, 177
    blood smear, 180
    complete blood count, 175, 178, 178t
    plasma, 177
lactase, 525, 525
lactate, 160
lactate threshold, during exercise, 619
lactation, 646
    lactogenesis, 762
    milk ejection, 646
    oxytocin in, 646
lactational amenorrhea, 763
lactic acid, 160
lactoferrin, 510
lactogenesis, 762
    defined, 762
    process of, 762
lactose, 522, 523, 523, 524b
lactose deficiency, 525
lactose intolerance, 524b, 525
lactotrophs, 644, 645, 655
lacuna, 756, 756
lamellar inclusion bodies, 372–374
langerhans cells, 200
language, association cortex in, 140
large intestinal motility
    ascending colon, 566
    defecation, 567
    descending colon, 567
    fecal continence, 567
    power propulsion, 565–566
    somatosensory nerves, 567–568
    transverse colon, 566–567
large intestine. see also colon
    motility of, 565–568
    secretions of, 516
laryngospasm, 705
latch state, 166–167, 167
latch state, of smooth muscle, 166–167
latency period, 91
lateral geniculate nucleus (LGN), 68

lateral inhibition, of retina, 68
lateral reticulospinal tract, 93
law of Laplace, 227, 296
LDL receptors, 758
"leak" channels, 163
leak channels, in smooth muscle contraction, 163
learning, 124, 138–139
  cholinergic innervation in, 139
  long-term potentiation in, 139
  studies of, 139
lecithin cholesterol acyltransferase (LCAT), 528
lectin pathway, 207
left atrium, 222
left atrium, anatomical considerations, 222
left axis deviation, 254
left bundle branch block, 254
left ventricle, 222
left ventricle, anatomical considerations, 222
left ventricular end-diastolic volume (LVEDV), 267, *268*
left ventricular hypertrophy, 251, 254
length-tension curve
  of cardiac muscle, 260, *265*, 268
  of skeletal muscle, *156*, 156–157
  of smooth muscle, *167*
leptin, 126, 216, 656b
lethargy, 394b–395b
leukemia, 215
leukocyte antigen-related (LAR)
  phosphatases, 695
leukocytes, 177
leukocytosis, 180
leukopenia, 180
leukotrienes, 682
lever system, 159, *160*
Leydig cells, 646, 718
Liddle syndrome, 476
lidocaine, 257b
ligand, 26
ligand-gated ion channels, 12–13
ligand-gated receptor, 45
light, 63
light adaptation, 68
light chains, 145
limbic system, 125, 135
  aggressive behavior, 137
  cerebral cortex, 135–136
  dopaminergic pathway, 130, *130*
  language and speech, 140, *140*
  learning and memory systems, 138–139
  limbic reward system, 137
  long-term and short-term memory, 139–140
  monoaminergic reticular formation cell groups, 130
  noradrenalin pathways, 131, *131*
  psychiatric disorders, 137–138
  serotonergic neurons, 131
  sexual arousal, 137
  structure of, 135, *135*
lipid bilayer, 8
lipid microdomains, 9
lipid raft, 9, 23
lipids
  absorption, 525–528
  digestion, 525–528
  metabolism, liver, 545–549, *546*
lipocortins, 683

lipogenesis, 696
lipolysis, 654, 696
lipopolysaccharide (LPS), 202
lipoprotein lipase, 547, 696
lipoproteins
  characteristics of, 548, 548t
  high-density, 548, 548t
    characteristics of, 548, 548t
    liver metabolism of, 548
    liver synthesis of, 549
  low density
    characteristics of, 548, 548t
    liver metabolism and removal of, 549
  very-low-density
    characteristics of, 548, 548t
    functions of, 548t
    synthesis of, 548
lithocholic acid, 519
liver, 541–554
  anatomic considerations, *542*
  in blood glucose regulation, 546
  disorders of
    abetalipoproteinemia, 548
    ascites, 543
    blood glucose regulation and, 548
    cirrhosis, 544b
    portal hypertension, 542, 544b
    viral hepatitis, 543b
  endocrine functions of, 552–553
  functions of, 541–542
  hormone degradation of, 552–553
  and immune responses, 553
  in iron homeostasis, 551, *551*
  lymphatic functions of, 542–543
  metabolism in
    of ammonia, 549
    of carbohydrates, *544*, 545
    of cholesterol, 549
    of coagulation proteins, 552
    of drugs and xenobiotics, *544*, 544–545
    of glucose, 545, *546*
    of lipids, 545
    of lipoprotein, *547*, 548–549
    of proteins, 552, *552*
    of urea, 549
  nutrient supply to, 545–546
  plasma proteins synthesized by, 549
  regeneration of tissue of, 543
  storage of fat-soluble vitamins, 550
  synthesis of blood-clotting proteins, 550
loading phase, 387
local excitatory current, 57
localized reactions, 213
locus coeruleus, 131
long hypophyseal portal vessels, 644
long receptor, 57
long-loop reflexes, 94
long-term memory, 139
long-term potentiation, 139
loop of Henle, 431
  ascending limb of, 440
  sodium, 440
loudness, 75
low-density lipoproteins (LDLs), 176, 527, 617, 675
  characteristics of, 548, 548t
  liver metabolism and removal of, 549
lower esophageal sphincter (LES), 559
L-type calcium, 168

lung inflation reflex, 415
lung injury, free radical-induced, 391b
lung volumes
  effect on pulmonary vascular resistance, 401–402, *402*
  forced vital capacity, 360, *360*
  functional residual capacity, 359
  minute, 362–366
  residual lung volume, 360–362, *363*
  residual volume, 359
  spirometry, 359–362
  tidal volume, 359
lungs
  age-related changes, 354, 354t
  airway tree, *353*, 353–354
  alveolar pressure, 359
  blood flow, 404–407
    gravity, *405*, 405–406
    regional ventilation, 406, *406*
    ventilation/perfusion ratios, 406–407, *407*
    zones of, *405*, 405–406
  blood–gas interface, 354, *354*
  diffusing capacity, 386
  elastic properties and chest wall, 366–373, *367–374*
  free radical-induced lung injury, 391b
  reflexes, 415–416
  spirometry, 359–362
  structural and functional relationships, 353–354
  ventilation/perfusion ratio, 391
luteal insufficiency, 744b
luteal phase, 747
luteal regression, 743
luteinization, 646, 743
luteinizing hormone (LH), 644, 718, 736
luteinizing hormone–releasing hormone (LHRH), 644, 736
  actions of, 645t
  structure of, 645t
luteolysis, 743
lymph, 299
lymph nodes, 196
lymphagogic effect, 528
lymphatic bulbs, 298
lymphatic system, 542–543
lymphatic vessels, 196, 298, *299*
lymphocytes, 180
lymphoid leukocytes, 184
lymphokine-activated killer (LAK) cells, 200
lymphokines, 209, 216, 682
lymphomas, 215
lysozyme, 184
lysyl-bradykinin (kallidin), 511

M

M line, of skeletal muscle, 145, *147*
macrocytes, 181
macrophages, 10, *10*, 185, 199–200, 202, 527
macula densa, 431
magnesium, 535
  balance, 462t
magnocellular neurons, 643
major depression, 137
major histocompatibility complex (MHC), 197, 202
malabsorption of hexoses, 17
male infertility, 733
male pronucleus, 755

male reproductive system
  androgen action
    peripheral tissue, 728–729
    puberty, 730
    secondary sex characteristics and
        masculine phenotype, 730–731
    sexual differentiation, brain, 731
  disorders
    hypogonadism, 731–732
    hypogonadotropic hypogonadism, 732
    prostate cancer, 730b
    pseudohermaphroditism, 732–733
  duct system, 723, *723*
  endocrine glands of, 718, *719*
  erection and ejaculation under neural control,
    723–724
  luteinizing hormone, *722,* 722–723
  neural control, 723–724
  sertoli cells, 722
  spermatogenesis, 724–726
  testicular function and regulation
    GnRH, 718–720, *719, 720*
    LH and FSH, 719
    steroids and polypeptides, 720
    testis, 720–721, *721*
malignant hyperthermia, 608
malleus (hammer), 71, *71*
malonyl-CoA, 697
maltose, 523
mammillothalamic tract, 126
mammogenesis, 762
mania, limbic system role in, 137
margination, 208
Marijuana *(Cannabis sativa),* 128b
mast cells, 185, 200
maternal recognition of pregnancy, 757
maximal oxygen uptake, 614
maximal response, 639
maximal voluntary contraction (MVC), 615
McCune-Albright syndrome, 746b
MDR-associated protein transporters, 15
mean artrial pressure
  cardiac output and, 282
  defined, *283*
  equation for, 282
  interactions of, 283–284, *283–284*
  measurement of, 284–285, *285*
mean circulatory filling pressure (MCP), 287
mean corpuscular (or cell) hemoglobin
  (MCH), 178
mean corpuscular (or cell) hemoglobin
  concentration (MCHC), 178
mean corpuscular (or cell) volume (MCV), 178
mean electrical QRS axis, 250
mean platelet volume (MPV), 179t
mechanoreceptors, 56, 62
medial forebrain bundle, 126
medial preoptic area (MPOA), 127–128
medial vestibulospinal tract, 94
median effective dose (ED$_{50}$), 639
median eminence of the hypothalamus, 643, *644*
mediastinal membrane, 358, *359*
medical pharmacotherapy, 168b–169b
medium spiny neuron, 99
medulla, *431, 433,* 673
  adrenal, 110, *110, 113*
  renal
    inner, 441, *447*
    outer, 441
medulla oblongata, 336

in breathing rhythm, 413
  in respiratory centers, 412–413
  ventrolateral, 336
medullary pyramid, 97, 98
medullary reticulospinal tract, 94
medullary tissue, 765
megakaryocyte, 174
megaloblast, 182
megaloblastic anemia, 393b
meiosis, 724
Meissner corpuscles, 61
Meissner plexus, 560
melanocortin system, 126
melatonin, 129, 216
membrane attack complex (MAC), 207
membrane-associated dense bodies, 163
membranous labyrinth, *76*
memory, 124
  declarative, 139
  long-term, 139
  nondeclarative, 139
  short-term, 139
memory B cells, 205
memory cells, in immune system, 201, *201*
memory loss, 139
menaquinones, 532
menarche, 744, 764
menopause, 744, 748–749
  hormone replacement therapy, 749
  osteoporosis and, 713
  selective estrogen receptor modulators in, 750b
menses, 748
menstrual cycle
  estradiol and progesterone
    cervical mucus changes, 748
    ductal system and signal ovulation
      preparation, *747,* 748
    menstrual phase, 748
    proliferative phase, 747
    secretory phase, *747,* 747–748
  menopause, 748t, 749
  ovarian cycle (*see* ovarian cycle)
  puberty, 744–745
menstrual phase, 748
menstruation, 744
Merkel disks, 61
mesenteric ganglia
  inferior, *572,* 573
  superior, *572,* 573
mesenteric receptors, in gastrointestinal motility
  control, 573
mesocortical pathway, 131
mesoderm, 756
mesolimbic pathway, 131
mesolimbic/mesocortical system, 131
mesonephric (wolffian) ducts, 766
metabolic acidosis, 418, 419, 499
  anion gap in, 499
  diabetes mellitus and, 499, 500b
metabolic alkalosis, 501, 501t
  compensation in, 501
  defined, 501
  vomiting in, 502b
metabolic bone diseases, 713
metabolic clearance rate (MCR), 635
metabolic condition, electrocardiogram (ECG)
  in, 254–255, *255, 256*
metabolic energy, 591
metabolic rate
  in hypothermia, 608, 609b

in response cold, 606
metabolic syndrome, 702
metabolism
  of cholesterol, 549
  by liver
    drugs and xenobiotics, *544,* 544–545
    of lipoprotiens, 548
  of xenobiotics, *544,* 544–545
methemoglobin (metHb), 181
methylene blue, 180
MHC class I molecules, 202
MHC class II molecules, 202
micellar emulsification, 526
micelles, 518
microbial infection, 208
microcirculation, 296
  arterial microvasculature, 296–297
  capillaries in
    anatomic considerations, 297, *297*
    defined, 297
    endothelial vesicles in, *297,* 298
    pores in, 297
    porosity of, 298
  defined, 296
  diabetes mellitus, 311b
  lymphatic (*see* lymphatic system)
  overview, 306, *307*
  regulation of
    autoregulation, 309–310, *310*
    endothelium cells, *306,* 306–308, *307,*
      307–308
    myogenic, 304–305
    organ blood flow, 305–306, *306,*
      306–307, *307*
    reactive hyperemia, 310
    sympathetic nervous system (SNS),
      308–309
  solute exchange in
    capillary filtration coefficient, 303
    capillary pressure, 302
    colloid osmotic pressure, 302
    diffusion, 301
    diffusion-limited transport, 301
    extraction ratio, 300
    filtration, 303
    flow-limited transport, 301
    hydrostatic pressure, 302, *302,* 303
    pathological conditions related to, 300
    permeability surface area coefficient, 300
    Starling-Landis equation, 302–303
  structure and function, 296–298, *297*
  vascular resistance and, 296–297
  venules in, 298
microcytes, 181
microfilaments, 37
microglia, 34
microtubules, 37
microvascular disease, 311b
microvilli, 522
micturition
  defined, 455
  urinary incontinence, 456–457
  urine elimination, 455–456
midbrain extrapyramidal area, 99, *99*
migrating motor complex (MMC)
  activity front, *581*
  phase of, *581*
milk ejection, 763
milliequivalent, 462
mineral absorption, 533–536, *534–536*

mineralocorticoid escape, 472
mineralocorticoids. *see also* aldosterone
   action of various, 472
   excess of, 472
minerals, absorption, 533–536
minute ventilation
   alveolar ventilation, 364, 364t
   dead space volume, 363
   expired, 364–366, *365*
miosis, 115
mitochondria, 144
mitochondrial permeability transition pore, 12
mitochondrial voltage-dependent anion channel
   (VDAC), 12
mitogen-activated protein kinase (MAP
   kinase), 25
mitogenic agent, 652
mitogens, 190
mitosis, 724
mitral and tufted cells, 82
mitral valve, 222
mitral valve prolapse, 277
mixed acid–base disturbances, 503
mixed contractions, 155
mixed micelles, 527
M-mode echocardiography, 277
monoamine oxidase (MAO), 48
monoamine oxidase inhibitors (MAOIs),
   117b, 138
monoaminergic neurons, 130
monoamines, 48
monoblasts, *186*
monochromatic vision, 66
monocyte/macrophage, 210
monocyte-macrophage system, fixed, 552
monocytes, 183
monoiodotyrosine (MIT), 662
monokines, 209
mononuclear leukocytes, 184
mood, 137
mood disorders, 137
morula, 755
mossy fibers, 101
motion sickness, 78
motivational systems, 124
motor control
   basal ganglia, 98–100, *99*
   brainstem in, *93*, 93–94
   cerebellum, *101*, 101–102
   cerebral cortex role in, *95*, 95–98, *96*, *97*
   Golgi tendon organs in, 88, *88*, 90
   motor cortex, primary, 95–96, *96*
   motor neurons
      alpha, 87–88
      gamma, 87, *89*, 90
   muscle fibers in, 87–88
   muscle spindles in, *89*, 89–90, *90*
   rubrospinal tract, 93
   sensory receptors in, 89
   spinal cord in
      anatomical consideration, 90–91
      reflexes, *91*, 91–93, *92*
motor endplate, 147, *148*. *see also* neuromuscular
   junctions
motor neurons
   alpha, 87–88
   gamma, 87, *89*, 90
motor reflex, 572
motor unit, 87, 153. *see also* motor neurons
motor unit summation, *153*, 154

mucins, 509
mucous cells, 509, 512
mucous gel layer, 512
mucous neck cells, 512
multidrug resistance (MDR), 15
multiple myeloma, 215
multiple sclerosis, 36, 216
multipolar cells, 36
multipotent stem cells, 186
mural granulosa cells, 741
muramidase, 510
muscarinic receptors, 48, 116, 120
muscle(s)
   blood flow during exercise, 588, 592
   skeletal (*see* skeletal muscle)
   smooth (*see* smooth muscle)
muscle chemoreflex activity, 616
muscle fatigue, defined, 621
muscle fibers, 144
muscle function, 86–87
muscle plasticity, 161
muscle spindles, 59, *89*, 89–90, *90*
muscularis externa, 511, 558
musculomotor neurons, 573, 579
musculoskeletal grafts, 212
myasthenia gravis, 119
mydriasis, 115
myelin sheath, 34, 36
myeloid leukocytes, 184
myenteric plexus, 560
myocardial infarction, 243
myocardial ischemia
   arrhythmias, 244
   defined, 244
   reduced refractory period in, 244
myocarditis, 210
myoclonus, 133
myoepithelial cells, 509, 646
myofibrils, 145
myogenic contraction, 187, 558
myogenic mechanism, 435
myogenic regulation, 304–305
myoglobin, 160
myokines, 2
myometrium, 738
myoneural junction, 147, *148*
myopia, 64
myosin, 144
myosin light-chain kinase (MLCK), 166
myosin light-chain phosphatase (MLCP), 166
myosin-linked regulation, 165
myotatic (muscle stretch) reflex, *91*, 91–92
myxedema, 669

N

NADPH oxidase (NOX), 184
NADPH-cytochrome P-450 reductase, 544
Na^+/K^+-ATPase, 38, 510
narcolepsy, 129
natriuresis, 470
natural immunity, 199
natural killer (NK) cells, 185, 541
necrosis, 197
negative cooperativity, 638
negative feedback, 6, *6*, 112
negative selection, 203
nephritis, 210
nephrolithiasis, 478, 480b
nephron, acidification, *492*, 492–493, *493*
nephrons, 431–434, *433*, *434*

nephrotic syndrome, 436b–438b, 439
Nernst equation, 21, 38
Nernst equilibrium potential, 21
nerve growth factors, 23
nerve transmission, 2
nervous system
   action potentials
      conduction velocity, 41–43, *42*, 42t
      excitatory and inhibitory postsynaptic
         potentials, 38, *40*
      phases, 39–40, *41*
      resting membrane potential, 38, *39*
      sodium and potassium channels, 38, *39*
      unidirectional propagation, 40, *42*
   blood–brain barrier, 34, *35*
   central nervous system, 34
   neuronal transport, 37, *38*
   peripheral nervous system, 34
   primary cell types, 34–36, *35*
   synaptic transmission, 43–45
      autoreceptors, 43
      multistep process, 43, *44*
      neurohormones, 45
      neurons communication, 43
      presynaptic membrane, 43–45
      synapsins, 43
neural mediators, 216
neural network layer, 68
neural reflexes, 415–418
neuroendocrine, 23
neuroendocrine system, 2
neuroendoimmunology, 215–216
neurofilaments, 37
neurogastroenterology, 557
neurogastroenterology and motility, 471
   enteric nervous system (*see* enteric nervous
      system)
   gastrointestinal tract, 558
neurogenic secretory diarrhea, 577
neurohormones, 43, 45
neurohypophysis, 643, *644*
neuroleptic drugs, 131
neurological disorders, 3
neuromuscular blocking agents, 146
neuromuscular junctions (NMJ), 105, *106*,
   147–148, *148*
neuromuscular toxins, 146b
neurons, 34
   morphology and function, 36, *37*
neuropeptide Y (NPY), 126
neuropeptides, 50, 216
neurophysin, 645
neurosecretory cells, 23
neurotransmitters, 9, 22, 43
   acetylcholine, 48
   arachidonic acid, 50
   autonomic nervous system, 106–110
   cannabinoids, 50
   dopamine, 49
   eicosanoids, 50
   endocannabinoids, 50
   epinephrine, 48–49
   excitatory and inhibitory effect, 45, *46*
   G protein–coupled receptors, 45–46
   GABA, 49
   glutamate, 49
   histamine, 50
   ligand-gated, 45
   monoamines, 48
   neuropeptides, 50

neurotransmitters (*Continued*)
    norepinephrine (NE), 48–49
    opioids, 50
    purines, 50
    serotonin, 49
neutralization, 207
neutrophils, 180, 199
niacin, 532
nicotinamide adenine dinucleotide phosphate
    (NADPH), 184
nicotinic ACh receptors, 147
nicotinic receptors, 48
nigrostriatal system, 130, 138
nipple, 762
nitric oxide (NO), 22, 110, 306, 311b, 723
nitric oxide synthase (NOS), 28
nitroglycerin, 168b
N-methyl-d-aspartate (NMDA) receptor, 49
NO synthase, 306
nociceptors, 56, 62
nocturnal rhythms, 128
nodes of Ranvier, 36
nodose ganglia, 572
nonadrenergic noncholinergic (NANC)
    fibers, 110
nondeclarative memory, 139
nonessential amino acids, 529
nongated channels, 12
nonphagocytic leukocytes, 200
non-rapid eye movement (NREM) sleep, 133
nonspecific immunity, 199
nonsteroidal anti-inflammatory drugs, 211
nonvital transplantations, 212
nonvolatile acid-base homeostasis, 487
noradrenaline, 106
noradrenergic neurons, 131, *131*
norepinephrine (NE), 48–49, 106–110, 113, 631,
    632, 683
    autonomic postganglionic fibers, 110
    sympathetic postganglionic neurons, 106
    synthesis and degradation, *109*
norepinephrine transporter (NET), 117b
normoblast, 182
normochromic cells, 181
nuclear bag fibers, 89
nuclear chain fibers, 89
nuclear pore complex, 12
nucleus accumbens, 136
nucleus ambiguus, 572
nucleus tractus solitarius, 572
null cells, *177*
nystagmus, 149b

O

obesity, 3, 684b–685b
occipital lobe, 68, 135
olfactory glomeruli, 82
olfactory system
    mitral cells, 82
    olfactory bulb, 82
    sensory cells, *82*
    sneezing, 82–83
oligodendrocytes, 34
oncogenes, 26, 215
oncotic pressure, 175
oocytes, 718, 739
oogonia, 739
oolemma, 754
ooplasm, 754

open-angle glaucoma, 116
opioid use, 116
opioids, 50
opsonin, 184, 199
opsonization, 207
optic disc, 63
optic nerve, 63
optic radiations, 68
optimum length (L_o), 156
oral contraceptives, 769b
oral contraceptives steroids, 769b
oral glucose tolerance test (OGTT), 699
orexigenic, 126
orexin, 129
organ of Corti, 72, 74–75
organ systems, 1
organic anion transporting polypeptides
    (OATPs), 15
organification, 662
organophosphate poisoning, 119–120
organophosphates, 120
orthophosphate, 706
osmolality, 19, 442
osmoreceptor cells, 465
osmoregulation, 6
osmotic diuretics, 473
osmotic pressure, 18–19
osteoarthritis, 623
osteoblasts, 708, *708*
osteoclasts, 14, *708,* 708–709
osteocytes, 708, *708*
osteogenesis imperfecta, 708, 713b
osteoid, 708
osteomalacia, 714, 715t
osteoporosis, 622, 713, *714, 714b, 715b*
oval window, 71
ovarian cycle
    corpus luteum, 743–744
    dominant follicle, 742
    estradiol synthesis, *741,* 741–742
    folliculogenesis, 739–740, *740*
    graafian follicle, *740,* 740–741, 741t
    hormonal regulation of
        follicular phase, 745–746
        luteal phase, 747
        ovulatory phase, 746–747
    LH surge
        follicle maturation and rupture, 743
        periovulatory period, 742–743
    oogonia, 739, 739t
overshoot phase, 40
oviduct, 738
ovulation, 646
ovulatory phase, 746–747
oxidative burst, 184
oxidative phosphorylation, 160, 666
oxygen
    demands, small intestine, 320–321
    normal range, 356
    partial pressure measurement, 355–356
    at sea level, 356, 356t
    uptake, 383–385, *384*
oxygen carrying capacity (OCC), 180
oxygen content, 386–387
oxygen debt, 614, *615*
oxygen diffusion gradient, 383
oxygen saturation, 387
oxygen uptake, 383
oxygen-carrying capacity, 386–387

oxyhemoglobin (HbO_2), 181, 386
oxyhemoglobin-equilibrium curve, 387–388
oxyntic glands, 511, 512
oxytocin, 124, 643, 761

P

P wave, 245
pacemaker
    for antral pump of stomach, *563,* 564
    in gastrointestinal motility, 560
pacinian corpuscle, 61
Paget disease, 715
pain
    cold, 62
    deep, 62, 339
    delayed, 62
    initial, 62
    referred, 60, 62
    superficial, 62
    visceral, 62
pain sensory receptor, 56
pampiniform plexus, 721
pancreatic amylase, 515, 524
pancreatic deficiency, 531
pancreatic lipase, 515, 526
pancreatic peptide
    F cell production of, 689
    secreation of, 689
pancreatic proteases, 530
Paneth cells, 522
panhypopituitarism, 732
para-aminohippurate (PAH) clearance, 454
paracrine mediators, 574
paracrine signaling, 22
paradigm shift, 3–4
paradoxical cold, 62
paradoxical sleep, 133
parafollicular cells, 71, 661, *661*
parahippocampal gyrus, 136
parallel fibers, 101
paralytic ileus, 568
paramesonephric (müllerian) ducts, 766
parasympathetic activity, 110, 115
parasympathetic nervous system (PNS), 105
    autonomic integration, 115–116
    cholinergic receptors, 112
    cholinergic system, 106
    ganglia, 106, *107,* 108t
    homeostatic functions, 110, 111t–112t
    preganglionic neurons, 106, *107,* 108t
    salivary secretion of, 509
parathyroid hormone (PTH)
    in calcium regulation, 447–478, *478*
    in phosphate balance, *479,* 479–480
paraventricular nuclei, 643
paravermal zone, 100
parietal (oxyntic) cells, 512
parietal lobe, 135
parietal-temporal-occipital association
    cortex, 136
Parkinson's disease (PD), 51
    basal ganglia in, 99
    deep brain stimulation, 100
    stereotactic neurosurgery, 100
parotid glands, anatomic considerations,
    509, *509*
pars distalis, 643, *644*
pars intermedia, 643
pars reticulata (SNr), 99

pars tuberalis, 643, *644*
partial pressure
    of arterial carbon dioxide, 407, *407*
    oxygen
        of gas in liquid, 7
        measurement of, 355–356
        normal range, 356
        in respiratory gas exchange, 356, 356t
        at sea level, 356, 356t
    peripheral chemoreceptors, 417–418
    in respiratory gas exchange, 356, 356t
partial pressures, 356
partial seizures, 134
partition coefficient, 11
parturition, 753
parvicellular neurons, 648
passive conduction, 38
passive force, 156
passive immunity, 530
passive length, 156
passive transport, 9
patches, 163
pathogens
    defined, 196
    transmission of, 196
peak expiratory flow, during forced expiration, 361t, 376
pedunculopontine nucleus, 132
pellagra, 532
pendrin, *661*, 662
penile erection, 110, *111*
pepsin, 198, 513, 530
pepsinogen, 512, 530
peptic ulcers, 516b
peptidases, 530
peptide hormones, 26
peptides
    absorption of, 528–531
    pancreatic, *530*
    in protein digestion, 528–531
    in regulation of insulin, 693
perception, 62
perforin, 200
perfusion pressure, 317
perfusion-limited gas uptake, 385
pericellular space, hepatic, 541, *542*
perilymph, 72
peripheral lymphoid organs, 196
peripheral nervous system (PNS), 34
peripheral neuropathy, 702
peripheral proteins, 9
perisinusoidal space, 541, *542*
perivitelline space, 754
permeability coefficient (P), 11
permeability surface area coefficient, 300
pernicious anemia, 393b, 532
petit mal seizures, 134
pH
    buffer, 486
    cerebrospinal fluid in ventilation, 416–417
    definition, 486
    of saliva, 510
phagocytic leukocytes, 199–200
phagocytosis, 10, 184, *184*, 198
phagosome, 184
pharmacomechanical coupling, 164
phases of digestion
    gastric secretions, 513, *513*
phasic receptors, 59

pheochromocytoma, 118
phorbol esters, 28
phosphate
    absorption in small intestine, 707
    in bone, continuous flux, 708
    distribution in bone cells, in healthy adult, 705–706, 706t
    inorganic constituents, 705–706, 706t
    kidneys, phosphaturia, 708
    pH buffering, 705
    plasma, 706
    regulations in GI tract, kidneys and bone, *706*, 706–707, *707*
phosphate buffer, acid-base homeostasis, 489
phosphatidylinositol 4,5-bisphosphate (PIP2), 649
phosphaturia, 708
phosphodiesterases (PDEs), 28, 728
phosphoenolpyruvate carboxykinase, 682, 697
phosphoenolpyruvate carboxylase, 696
phosphofructokinase, 696
phospholamban, 263
phospholipase A₂, 526, 683
phospholipase C (PLC), 28, 168, 649
phospholipids, 8, 9
phosphorus, daily flux of, 707, *707*
phosphorus homeostasis, 707, *707*
phosphorylation, 166
phosphorylation, oxidative
    in skeletal muscle contraction, 160
    in smooth muscle contraction, 165–166
photons, 63
photopic (daytime) vision, 65
photoreceptors, 56
phylloquinones, 532
physiologic regulation, 5–8
physiology, 2
physostigmine (eserine), 148
pigment epithelium, 63, 65, *66*
pill rolling, 99
piloerection, 604
pineal gland, 1229
pinna, 70, *70*
pinocytosis, 661
pitch discrimination, 75
pituitary dwarfism, 654
pituitary gland, 643
    adenohypophysis, 643, *644*
    anatomic considerations, 643, *644*
    anterior hormones, 646–656
    blood supply to, 644–645
    disorders of
        congenital adrenal hyperplasia, 649
        tumors of, 647b
    neurohypophysis, 643, *644*
    posterior hormones
        arginine vasopressin, 645
        hypothalamic-releasing hormones, 645, 645t
        oxytocin, 646
pituitary resistance to thyroid hormone, 665b
pituitary stalk, 124
pituitary tumors, 647b
place theory, 74
placenta
    circulation of, 327
    development of, 655
    vasculature of, *326*, 327
placental villus, *326*, 327

placing reaction, 96
plasma, 174, 174t, 175
    electrolyte composition of, 462t
    as extracellular fluid (ECF), 460
    fresh frozen, 175
    proteins in (see plasma proteins)
plasma carbohydrate, 176
plasma cells, 185, 201
plasma lipids, 176
plasma membrane, 8
    fluid mosaic model of, 8, *8*
    memebrane transport, 9–18
    structure, *8*, 8–9
        lipid bilayer, 8, *8*
        lipids, 8–9
        proteins, 9
plasma proteins, 175
    in hormone transport, 633, 634t
    synthesized by liver, 549
plasma volume
    in circulatory shock, 346
    effects of sweating on, 605t
    in heat acclimatization, 604
plasma water
    electrolyte composition of, 462t
plasma-derived mediators, 209
plasmapheresis, 175
plasmin, 190, 743
plasminogen, 190
plasminogen activator, 722, 743
plateau region, of action potential, 238
platelet count (Plt), 180
platelet formation, 185
platelet-derived growth factor (PDGF), 23, 188, 190
platelets, 174
plegia, 93
pleural pressure
    effect on pulmonary vascular resistance, *400*, 400–401
    in forced expiration, 355, 358
    pneumothorax, 358
pleural space, 354
pneumothorax, 358
podocytes, 436
poikilocytes, 181
poikilotherms, defined, 587
Poiseuille's law, 228–229, 374
polyclonal antibody, 200
polycyclic aromatic hydrocarbons, 545
polycystic kidney disease, 432b
polycystic ovarian syndrome, 749
polycythemia, 177, 425
polydipsia, 468
polymorphonuclear leukocytes (PMNs), 184
polyuria, 701
pons, in respiratory centers, 412–413
pontine respiratory group, 414
pontine reticulospinal tract, 94
pores, 297
porphyrin, 183
portal hypertension, 542, 544b
portal vein, 223
positive feedback, 7
positive selection, 203
postcapillary venules, 303
posterior lobe, 643
postganglionic axon, 105
postganglionic fibers, 105, 113

postganglionic neuron, 105
postganglionic sympathetic neurons, 573
postmenopausal osteoporosis, 713
postpartum and prepubertal periods
    hypothalamic GnRH
        newborn, 763–764
            physical signs, puberty, 764
        mammogenesis and lactogenesis, 762
        suckling reflex, 763
postpartum thyroiditis, 668b
postrema, 572
poststreptococcal glomerulonephritis, 436b
postsynaptic terminals, 36
potassium, 534
    abnormal levels of
        plasma, in pH regulation, 175
    acid–base balance, 475
    hyperkalemia, 475
    hypokalemia, 475
    metabolism, 475
potassium channels
    inwardly rectifying, 44
    outwardly rectifying, 240
    transient outward, 240
potentiation, 514
power stroke, 151
PR interval, 245
prechondrocytes, 654
precocious puberty, 746b, 765
precordial leads, 248
prefrontal association cortex, 136
prefrontal cortex, 136
preganglionic fiber axon, 105
preganglionic neuron, 105
preganglionic sympathetic axon, 110
preganglionic sympathetic neurons, 573
pregnancy
    hormones required for
        endocrine function changes, 759–760
        estrogens, 757
        fetal endocrine system, 760–761
        hCG, 757–758, *758*
        hPL, 758
        insulin and insulin-like growth factors, 761
        maternal and fetoplacental interactions, 758–759, *759*
        parturition, 761
        progesterone, 757
    postpartum thyroiditis, 668b
pregnenolone, 675, 741
    in testosterone production, 726
pregranulosa cells, 739
preload, 155
premature atrial contractions (PACs), *252*, 253
premature ovarian insufficiency, 749
premature ventricular contractions (PVCs), 244
premotor cortex, 97, 136
presbyopia, 64
pressure, 225
pressure diuresis, 343–344, *344*
presynaptic terminal/bouton, 36
pretectal nuclei, 69
prevertebral sympathetic ganglia, 573
primary active transport mechanism, 14
primary adrenal insufficiency, 680b. *see also* Addison disease
primary follicular stage, 740, *740*
primary (or central) lymphoid organs, 196
primary motor cortex, 95–96, *96*

primary oocytes, 743
primary receptors, 81
primary somatosensory cortex, 97
primary spermatocytes, 725
prime mover, 86
primordial follicle, 739
procedural memory, 139
progesterone, 633, 646, 676, 736
progesterone block, 761
progestin, 737, 750b
proglucagon, 693, *694*
prohormone, 632
prohormone convertase 1, 647
prohormone convertase 2, 647
prolactin (PRL), 216, 644, 655, 728
proliferative phase, 747
proopiomelanocortin (POMC), 126, 647–648, *648*
propagated potential, 40
proprioceptors, 56, 88, 416
propriospinal cells, 91
prostaglandins, 606, 682
prostaglandins E$_2$, 472
prostate cancer, 730b
protein buffer, acid–base homeostasis, 489
protein C, 190
protein inclusion bodies, 182
protein kinase A, 28
protein kinase A (PKA), 648
protein kinase C (PKC), 28, 649
protein kinase G (PKG), 28
protein S deficiency, 190
protein tyrosine phosphatase 1B (PTP1B), 695
proteins
    absorption, 528–531, 529t, *530*
    aminoacids in, 529–530, 529t
    complete, 529
    digestion, 528–531, 529t, *530*
    genetic disorders, 531
    incomplete, 529
    metabolism of, *549,* 549–550
proteoglycans, 708
prothrombin, 550
prothrombin time (PT), 189
prothrombinase (factor X), 189
proton pumps, 14
protooncogenes, 26
proximal tubule
    cell model for transport, 440, *440*
    sodium reabsorption, 441
    Tm for PAH, 455
    toxins and drugs, 444–445, *445*
    water reabsorption, 440
pseudoephedrine, 117b
pseudohermaphrodite, 732–733
pseudohermaphroditism, 732–733
pseudohyperkalemia, 476
pseudohyponatremia, 468b
pseudoprecocious puberty, 765
P-type ATPase, 14
pubarche, 745
puberty, 744
    delayed, 765
    factors, 764–765
    hypothalamic–pituitary function defects, 765
    precocious, 765
    pseudoprecocious, 765
    secondary sex characteristics, *764,* 764–765
pulmonary arterial hypertension (PAH)

    causes, 394b
    definition, 394b
    etiology, 394b–395b
    molecular mechanism, 394b
    treatment, 395b
pulmonary arteries, 222, 398, *399*
pulmonary capillaries
    anatomic considerations, 403
    fluid exchange, 403–404
    gravity effects on, 225
pulmonary circulation, 398–399, *399*
    blood flow in lungs, 425
    gravity effects on, 225
    hemodynamic features, 399–400, *400*
    secondary functions, 398–399
pulmonary edema, 403–404
pulmonary embolism, 215, 401b
pulmonary embolus, 223
pulmonary hypertension, hypoxia-induced, 404b
pulmonary stretch receptors, 415
pulmonary vascular resistance, *400,* 400–401
pulmonary vasoconstriction, 402–403
    hypoxemia, 402
    hypoxia-induced, 402
pulmonary wedge pressure, 399
pulmonic valve, 222
pulse pressure
    arterial compliance and, 282–283, *283*
    defined, 282
    interactions of, 283–284, *283–284*
    stroke volume and, 282–283, *283*
    systemic vascular resistance and, 284, *284*
pupil, 62
pupillary constriction, 115
pupillary light reflex, 69
purinergic neuron, 577
purines, 50
Purkinje cells, 101
Purkinje fibers, 238
putamen, 98
pyloric glands, 511–512
pyloric sphincter (PS), 559
pyramidal tract, 97
pyrogens, role in fever, 606
pyruvate carboxylase, 696
pyruvate dehydrogenase, 696
pyruvate kinase, 696
pyruvic acid or pyruvate, 160

Q

Q$_{10}$ effect, 608b
Q wave, 245
QRS complex, 245
QT interval, 246
quantum, in synaptic transmission, 147

R

R wave, 245
radial bundles, 72
radiation, in body heat exchange, *590,* 590–593
radioimmunoassay (RIA), *636,* 636–637, *637*
radioreceptor assay, 636–637
rapid adapting receptors, 415
rapid ejection phase, 264
rapid eye movement (REM), 133–134, 421
rapid ventricular filling, 264
ras family, 26
Rathke pouch, 643

reactive hyperemia, 310
reactive oxygen species (ROS), 12, 391b
receptor(s), 23, 43
    baroreceptors, 336–338
    in cellular signalling, 23
    chemoreceptors (see chemoreceptors)
    for hormones (see hormones)
    mechanoreceptors (see mechanoreceptors)
    photoreceptors (see photoreceptors)
    sensory (see sensory receptors)
    thermal, 589–590, 600
    thyroid hormone, 663
receptor adaptation, 338
receptor potential, 57
receptor-mediated endocytosis, 10
reciprocal inhibition, 86, 91
red blood cells (erythrocytes), 173, 180–183
    anemia and, 182, 183
    destruction of, 174
    disorders of, 181–182
    formation of, 185–187, 186
    hematocrit (Hct), 177
    hemoglobin in, 181
    iron recycling, 183, 183t
    life span of, 185
    mean corpuscular hemoglobin (MCH),
        178, 178t
    mean corpuscular hemoglobin concentration
        (MCHC), 178, 178t
    mean corpuscular volume (MCV), 178, 178t
    morphology of, 181, 182
    normal ranges for, 177
    in oxygen transport, 180–181, 181
    structure and functions of, 181, 181, 182
Red blood cells (RBCs), 173
red cell distribution width (RDW), 180
red nucleus, 93
red-sensitive pigment, 66
reduced ejection phase, 264
reduced ventricular filling, 264
reentry tachycardia, 243b
referred pain, 60, 62
reflection coefficient, 302
reflective pain, 62
reflex(es), 91
    lung
        irritant receptors, 415
        J receptors, 416
        mechanoreceptors, 415
        proprioceptors, 416
        pulmonary stretch receptors, 415
    neural
        mechanoreceptors, 415
        proprioceptors, 416
regional compliance, 368
regional hypoventilation, 391
regional hypoxia, 402
regulatory light chain, 145
regulatory volume decrease (RVD)
    mechanisms, 19
regulatory volume increase (RVI)
    mechanisms, 20
relative refractory period, 40
relative WBC count, 180
relaxin, 761
releasing factors, 124
releasing hormones, 644
renal blood flow (RBF), 434–435
    autoregulation, 434, 434–435

sympathetic nerve stimulation and
    hormones, 435
renal clearance
    definition, 451–452
    glucose reabsorption, 454–455
    inulin clearance, 452–453
    net tubular reabsorption, 454
    para-aminohippurate clearance, 454
renal compensation
    metabolic acidosis, 499
    metabolic alkalosis, 501
renal hypoxia, 343
renal response, acid–base homeostasis, 488
renin, 432, 470, 679
renin-angiotensin-aldosterone system, 342,
    342–343
repolarization, 40
residual volume (RV), 264
resistance, 38
resistance to thyroid hormone (RTH), 665b
respiration
    cellular, 352
    defined, 352
    gas exchange, 352
respiratory acidosis, 418, 419–420
respiratory acidosis/alkalosis
    altered levels of $PaCO_2$, 497–498
    cell chemical buffering, 498
    definition, 497
    lungs and kidneys compensate, 498–499
respiratory alkalosis, 418
respiratory burst, 184
respiratory compensation
    metabolic acidosis, 499
    metabolic alkalosis, 501
respiratory exchange ratio, 384, 590
respiratory quotient (RQ), 592
respiratory response, acid–base homeostasis, 488
respiratory sinus arrhythmia, 251, 252
respiratory system compensation, metabolic
    alkalosis, 501
rest-and-digest system, 115
resting force, 156
resting length, 155
resting membrane potential
    cardiac, 238–239, 239
    electrochemical potential, 20–21
    Goldman equation, 21
    Nernst equation, 21
    steady state, 20, 20
resting state, 40
reticular formation, 93, 129–130
reticulocytes, 182
reticuloendothelial system (RES), 200, 542
reticulospinal tracts, 94
retina, 63
retinohypothalamic tract, 129
retinoid X receptor (RXR), 665
retinol-binding protein (RBP), 550
retinopathy, 702
retrograde transport, 37
Reynolds number, 231
rhesus factors (Rh), 179b
rheumatoid arthritis (RA), 213, 623
rhodopsin, 66
RhoGTP/Rho kinase system, 166
rhythmicity, of cardiac cells, 238
rickets, 532, 714, 715t
right atrium, 222

right axis deviation, 254
right bundle branch block, 254
right ventricle, 222
right ventricular hypertrophy, 251
rigor crossbridge, 151, 151
rigor mortis, 152
rising phase, 40
rod cell, 66
rods and cones, 70, 63
rolling, 208
rubrospinal tract, 93
ryanodine receptor (RyR), 29, 150

S

S cells, 518
S wave, 245
saccule, 75, 76
safety factor, 148
saliva
    electrolyte composition of, 510
    formation of, 510–511
    in immune response, 206
    osmolality, 510
    protective functions of, 510
    proteins in, 510
salivary glands, 509
    anatomic considerations, 509, 509
    excretory duct of, 509
    primary secretion of, 509
salivary secretions
    acidic food, 510
    anatomic considerations, 509, 509
    electrolyte composition, 510
    hormones affect, 511
    immunologic functions, 510
    parasympathetic nervous system in, 510–511
    saliva, 510
salivon, 509, 509
salt concentration, 6
salt receptors, 80
saltatory conduction, 41
salty taste, 80
sarcolemma, 149
sarcomere, 145, 147
    defined, 145
    of skeletal muscle, 145
    structure of, 145
sarcoplasmic reticulum (SR), 149
satellite cells, 654
satiety, 127
saturation, of transport systems, 12
scala media, 72
scala tympani, 72
scala vestibuli, 72, 72
scalene muscle, 354
Scatchard plot, 638
schistocytes, 181
schizophrenia
    defined, 138
    limbic system, 138
Schwann cells, 34, 147, 148
sclera, 62
SCN, 129
scotopic (nighttime) vision, 65
scotopsin, 67
scratch reflex, 2
scurvy, 532
secodary functions, pulmonary circulation,
    398–399

second heart sound, 264
second messengers, 23, 26, 27–30
    1, 2-diacylglycerol (DAG), 28, *29*
    calcium, 29–30, *30*
    in cellular signaling, 23
    cGMP, 28–29
    CO, 27
    cyclic adenosine monophosphate (cAMP),
        24, 27–28
    inositol triphosphate (IP_3), 28–29, *29*
    lipids, 27
    NO, 28–29
    protein kinase A, 28, *28*
    reactive oxygen species, 27
    types of, 27
second polar body, 743
secondary active transport mechanisms, 16
secondary characteristics
    female, 127
    male, 127
secondary follicular stage, 740, *740*
secondary lymphoid organs, 196
secondary oocyte, 743
secondary pacemaker, atrioventricular (AV)
    node, 242
secondary sex characteristics, 744
secondary spermatocytes, 725
second-degree atrioventricular block, 253
secretagogues, 693, 694
secretin, 514, 518
    gastric secretion of, 514
    in pancreatic secretion, 518
secretomotor neurons, 573
secretory immunoglobulin (sIgA), 206
secretory phase, *747,* 747–748
segmental arteries, 430, *431*
segmentation, of small intestine, 560
seizure, defined, 134
selective estrogen receptor modulators
    (SERMs), 750b
selective serotonin reuptake inhibitors
    (SSRIs), 138
selectively permeable, 18
selenocysteine, 663
self-tolerance, of immune system, 198
sella turcica, 643
semen, 724, 753
semen secretions, 724
seminal emission, 724
seminal vesicles, secretion of, 631
seminiferous tubules, 721
senile plaques, 140b
sensation
    accessory structures for, 56
    adaptation in, 59
    hearing, 55
    sensory modality, 55
    smell sense, 81–83, *82*
    taste sense, 78–80, *80*
sensorineural deafness, 75
sensorineuronal hearing loss, 78
sensory information
    encoding and decoding in, 59–60
    transfer of, 60
sensory modality, 55
sensory presbycusis, 75
sensory receptors
    accessory structures for, 56
    accommodation in, 59

action potential generation in, 56–57, *57*
    adaptation in, 59, *59*
    generator potential in, 56–57, *57*
    hearing, 55
    smell sense, 81–83, *82*
    taste sense, 78–80, *80*
    touch sense, 61
sensory signals, 127
sepsis, 210
septal nuclei, 136
septic shock, 347
septicemia, 210
serosa, 511
serotonergic neurons, 131
serotonin, 49, 682
serotonin transporter (SERT), 117b
serous cells, 509
Sertoli cells, 718
    Leydig cells interactions with, 718
    sperm cell development, 722
    spermatogenesis in, 718
serum, 175
serum iron levels, 183
serum protein electrophoresis, 176
set point, in thermoregulation
    defined, 600
    factors influencing, 601
severe combined immunodeficiency
    (SCID), 2136
sex
    chromosomal, 765
    genetic, 765
    gonadal, 765
    secondary characteristics
        female, 127
        male, 127
    steroids, 345b–346b
sex chromosome DSDs, 767
sex hormone–binding globulin (SHBG),
    722, 736
sexual activity
    hypothalamus role in, 137
    limbic system and, 137
sexual arousal, 137
sexual development
    chromosomal sex, 765
    disorders, 767, *768*
    fetal gonads, 765–766
    genetic sex, 765
    internal and external genitalia differentiation,
        766, *766*
    sexual differentiation, 766, *767*
SH2 domain, 26
shallow-water blackout, 425, *425*
shear stress, 230
shell conductance, 594
shivering
    during cold stress, 599
    control of, 599
    primary motor center, 599
shock
    circulatory, 344–347, *345*
    neurogenic, 346
    septic, 347
    spinal, 92
short hypophyseal portal vessels, 644
short receptors, 57
shortness of breath, 394b–395b, 409b
short-term memory, 139

shunts, 407
    anatomic, 408
    vs. dead spaces, 408, 408t
    hypoxemia related to, 392
    right-to-left, 408
sickle cell anemia, 180, 393b
side-polar, 163
sieving action, of stomach, 564
sigmoid colon, *566*
signal cascade, 23
signal transduction
    G protein-coupled receptors (GPCRs),
        23–25, *24*
    ion channel-linked receptors, 23, 25
    types of, 23
    tyrosine kinase receptors, *25,* 25–26
    visual, 64–68, *66–68*
signaling. *see* cellular signaling
signaling cascade, 26
simple acid–base disturbances, 497
simple micelle, 527
simple reflex, 2
single-breath test, 386
sinoatrial (SA) node
    cardiac output, *238,* 242
    functions of, 240–241
    heart rate and, 242
    as slow action potential, *238, 238*
sinus bradycardia, 252
sinus tachycardia, 252, *252*
sinusoid, hepatic, 541–543, *542*
sinusoidal endothelial cells, 541, *542*
sinusoidal sound waves, 70
size principle, 88
skeletal muscle
    antagonistic pairs of, 160
    blood circulation, 323–324
    vs. cardiac muscle, 237
    contraction of
        adenosine diphosphate (ADP), 151
        adenosine triphosphate, 160
        calcium, *150,* 150–151
        crossbridge cycle, 146
        excitation-contraction coupling, 146–152
        force-velocity curve, 157–158, *158*
        isometric, 154, *154,* 155
        isotonic, 155, *155*
        length–tension relationship, 155–157,
            156, 157
        mixed, 155
        motor unit summation, 153, *153*
        sliding filament theory, 150
        twitch, 152, *152*
    disorders of
        myasthenia gravis, 149
    isotonic and isometric, *154,* 154–155, *155*
    lever system, 159, *160*
    mechanical factors, 154
    metabolic process of, 160
    metabolism and fiber types, 160–161
    neuromuscular junction (NMJ), 147–148, *148*
    red muscle fibers, 161
    sarcomere, 145, *147*
    thick filaments, 144–145, *145*
    thin filaments, 144–145, *145*
    white muscle fibers, 161
    Z lines, 145
skeleton, 86, *87*
skin

in body temperature regulation
    convection, 588, 592, *594*
    evaporation, 593
    radiation, 592, 593, *594*
    shell conductance, 594–595, 599
    surface area and, *593*, 593–594
  peripheral factors influencing, 589–590
  in response cold, 604–606
  set point and, 600–601
  sympathetic nervous system in, 597
  temperature
    in heat exchange, 589
    in thermoregulation, 599
  in thermoregulation, 599–600
  wettedness, 594
sleep
  apnea syndrome, 422b
  arousal from, 422
  debt, 134
  disorders of
    apnea, 422b
    narcolepsy, 129
  non-rapid eye movement (NREM), 133
  rapid eye movement (REM), 133–134
  REM, 422–423
  slow-wave, 133
sliding filament hypothesis, 146
slit diaphragm, 436
slow excitatory postsynaptic potentials, in
    gastrointestinal motility, 574
slow inhibitory postsynaptic potentials, in
    gastrointestinal motility, 575
"slow response" action potentials, 238
slow-twitch fibers, 161
slow-wave sleep (SWS), 133, 421
small intestinal motility
  activity front, 580
  digestive state, 564
  ENS, 565
  interdigestive state, 564
  migrating motor, 580–582
  MMC as "housekeeper" for, 582
  power propulsion, 564
small intestine
  absorption by, 321–322
  anatomic consideration, *323*
  bile salt recycling in, 520
  blood flow
    autoregulation, 320–321
    nutrients absorption, 321–322
    sympathetic nerve activity, 322
    venous-arteriolar response, 321
    during water absorption, 322
  digestion and absorption processes, 508–509
  vasculature of, 320, *320*
smell sense, 81–83, *82*
smooth muscle
  adaptation of, 163
  of bladder, 162
  of blood vessels, 161, 223, 224t–225t
  contraction of
    calcium, 163–164, *164*
    cell-to-cell transmission, 163
    control of, 162
    latch state, 166–167, *167*
    myosin filament proteins, phosphorylation,
      165, 165–166
    relaxation, 168
  crossbridge cycle, 165b
  force–velocity curve, 167b, 168

mechanical activity in, *167*, 167–168
  mechanisms, 168b
  uterus, 738
smooth muscle contraction
  Ca^{2+} and K^+ channels, 570
  electrical slow waves, 570–571, *570–571*
  electrical syncytial properties of, 569–570
  electromechanical and pharmacomechanical
    coupling, 569
  endocrine influence, 569
  gap junctions of, 569
SNARES, 43
sodium, 533–534, *534*
  appetite, 471
  balance, 473–475, *474*
  dietary intake, 473–474, *474*
  diuretic drugs, effect of, 473
  in heat acclimatization, 604
  hormonal and neural factors
    estrogens hormones, effect of, 472
    glomerular filtration rate, effect of,
      469–470, 469t
    hydrostatic pressure, effect of, 470
    natriuretic hormones, effect of, 472, *472*
    osmotic diuretics, effect of, 473
    renin-angiotensin-aldosterone, effect of,
      470–473, *471*
    sympathetic stimulation, effect of, 473
  kidneys excrete, *469*, 469–470, 470t
sodium channel antagonists, 256b
sodium channel blockers, treatment of cardiac
    arrhythmias, 256b–257b
sodium channels, voltage-gated, *239*,
    239–240, *240*
sodium glucose cotransporter inhibitors, 701
sodium-dependent glucose transporter-1
    (SGLT1), 16
sodium-iodide symporter (NIS), 662
solute transport
  active
    calcium pumps, 14
    defined, 9
    proton pumps, 14
    secondary, 15–17, *16*
    sodium-potassium pumps, 14, *15*
  endocytosis in, 10, *10*
  exocytosis in, *10*, 10–11
  passive, 9
  passive diffusion (*see* diffusion)
  phagocytosis in, 10, *10*
  water movement and
    cell volume regulation, 19–20, *20*
    osmotic pressure, 18–19
    water channels, 18
solvent drag, 533
soma, 36
somatosensory system
  nociceptors, 62
  skin receptors, 60–61
  thermoreceptors, 61–62
somatostatin, 512, 644, 645t
somatotopic maps, 95
somatotrophs, 644
somatotropin release-inhibiting factor (SRIF),
    644, 645t
  action of, 645t
  structure of, 645t
sound
  pressure, 69–70, *70*
  transduction of, 70

wavelength, 69, *70*
sour receptors, 80–81
sour taste, 80
spasmodic dysphonia, 149b
spasmodic torticollis, 149b
spasticity, 98
spatial summation, 38, 153, *153*, 154
special senses, 60
spermatogenesis, 646, 718, 724–726
  meiosis, 725
  mitosis, 725
  phases of, *724*, 724–725
  spermatozoon formation, 725–726
  testosterone, 726
spermatogenic cycle, 725
spermatogonia, 721
spermatozoa, 718
spermiation, 722
spermiogenesis, 724
spherocytes, 181
sphincter muscles, 115
sphincter of Oddi, 516, 541
sphingolipids, 9
sphygmomanometer, 284
sphygmomanometry, 284–285, *285*
spinal cord motor neuron pools, 93, *93*
spinal shock, 92, 336
spinnbarkeit, 738, *738*
spinocerebellar tracts, 90
spinocerebellum, 101
spiral arteries, 327
spiral ganglia, 72
spiral ligament, 72
spirogram, 359
spirometry, airflow, 359
splanchnic nerves, 573
splay, 455
spleen, functions of, 551
split brain, 141b
SRY (sex-determining region, Y
    chromosome), 765
standard curve, 636, *637*
standard frontal lead system, 248
stapedius muscle, 71
Starling forces, 303, 403, 470
Starling's law of the heart, 269
Starling-Landis equation, 302–303
steady state, 7
steatorrhea, 528
stellate cells, 101, 542, 553
stereocilia, 73, *73*
steroid hormones, 26
steroidogenesis, 726
  ACTH action on, *678*, 678–679
  genetic defects in CYP enzymes, 676, 678
steroidogenic acute regulatory protein (StAR),
    675, 726, 741
steroids
  ACTH, 678–679
  adrenocortical cell size, 679
  cholesterol availability, 679
  steroidogenic enzyme expression, *678*, 679
stigma, 743
stimuli, 55
stomach
  anatomic considerations, 511
  functions of, 511
  mucosal lining of, 516b
  secretion of, 513
stomatocytes, 182

stress response, 215
stress test, 618b
stretch reflex, 2
stretch-activated calcium channels, 165
stria terminalis, 125
stria vascularis, 72
striae, 674b
striated duct, 509
striatum, 98
stroke volume, 264
    arterial pressure, 283–284, *284*
    pulse pressure, 282–283, *283*
stroma, 163
stromal layer, 738
strong acid, acid-base homeostasis, 486
subcortical structures, 135
sublingual glands, anatomic considerations, 509, *509*
submandibular ganglion, 511
submandibular glands, anatomic considerations, 509, *509*
submucosa, 511
substantia nigra, 98
subthalamic nucleus, 98
sudden infant death syndrome (SIDS), 131
sulcus, 135
sulfonylureas, 693
superior cervical ganglion, 511
superior colliculus, 69
superior hypophyseal arteries, 644
superior mesenteric, 573
superior parietal lobe, 97
superior vena cava (SVC), 222
superoxide ion, 184, 391b
supplementary motor area (SMA), 95–97
suprachiasmatic nucleus (SCN), 69, 125
supraoptic nuclei, 643
supraventricular tachycardia, 253
surface tension, 403–404
surfactant, 198
sweat glands, 109–110
    absence of, 596b, 598
    apocrine, 595
    eccrine, 595
    evaporative heat loss from, 595
sweating
    evaporation and, 595
    in heat acclimatization, 595
    set point and, 596b
    skin temperature and, 601
    skin wettness and, 594
    in thermoregulation, 595, 598
    water loss and, 596b
sweet receptors, 81
sympathetic (SNS), 509
sympathetic "chain," 106
sympathetic nervous system (SNS), 105, 215
    acetylcholine, 106
    adrenergic receptors, 113–114
    arousal, stress, and danger, 113, 113t
    autonomic integration, 115–116
    in circulation, skin, 597–598
    epinephrine, 110
    fight or flight response, 113, 113t
    ganglia, 106, *107*, 108t
    norepinephrine, 106–110
    organization, 112, *113*
    postganglionic axons, 108
    postganglionic fibers, 113
    preganglionic neurons, 106, *107,* 108t

sympathetic tone, 340
symport, 16
synapse, 22, 36
synapsins, 43
synaptic cleft, 36
synaptic (S)-type neurons, 573
syncytiotrophoblasts, *326,* 327, 755–756
syndrome of inappropriate secretion of antidiuretic hormone (SIADH), 468b
synergist, 86
systemic circulation
    arterial pressures in, 282–285, 282–286, *283–285*
    blood volume, 287–288
    cardiac function in, 288–291
systemic inflammatory response syndrome, 210
systemic lupus erythematosus (SLE), 214–215
systemic reactions, 213
systemic vascular resistance
    arterial pressure and, 284, *284*
    interaction of, 283–284, *283–284*
    mean artrial pressure, 282, *283,* 283–284, *284*
systole, defined, 222
systolic hypertension, 286b
systolic pressure, 225

T

T cell–mediated immunity, 199, 201–207
    antigen, *203*
    clonal selection in, 201, *201*
    cluster of differentiation in, 201, *202*
    immunological synapse in, 202
    negative selection in, 203
    positive selection in, 203
T cells, 185
    cytotoxic CD8[+], 202
    double negative, 203
    double positive, 203
    effector, 204t
    helper CD4[+], 204
    memory, 204
    specific antigen, 200
T helper 1 cells, 204
T helper 2 cells, 204
T regulatory cells, 204
T wave, 245
tachycardia
    reentry, 243b
    sinus, 252
    supraventricular, 253, 273t
    ventricular, 243, 253, 257b
tachyphylaxis, 30
tachypnea, 421
target cells, 181, 629
taste buds, 78–80, *80,* 510
taste pores, 78
taste sense, 80
T-cell receptors (TCRs), 200
T1D, 701, *701*
technetium-99 scans, of heart, 276
tectorial membrane, 73, 74
telencephalic, 98
telescoping system, 157
temperature
    body (*see* body temperature)
    environmental
        during exercise, 602
        in hypothermia, 608
temperature sensory receptors, 600
temporal lobe, 74, 135, *135*

temporal summation, 38, *152,* 153
tensions, 7, 227
tensor tympani muscle, 71, *71*
terminal cisternae, 150
terminal respiratory unit, 354, 399
tertiary follicular stage, 740, *740*
testicular feminization, 733
testis, 720–721, *721*
    luteinizing hormone, 727–728, *728*
    steroidogenesis, Leydig cells, 726, *727*
testosterone, 633, 673, 718, 742
tetanic fusion frequency, 153
tetanus, *152,* 153
tetanus/twitch ratio, *152,* 153
tetany, hypocalcemic, 705, 711
tetrahydrocortisol glucuronide, 678
tetraiodoacetic acid, *663,* 664
thalamus, 126
theca externa, 740
theca interna, 740
theca lutein cells, 743
thelarche, 744, 764
thermal receptors, 589–590, 600
thermic effect of food, 592
thermodilution, in cardiac output measurement, 275
thermogenesis
    nonshivering, 599, 605
    thyroid hormone regulation of, 666
thermoneutrality, 595, 598
thermoreceptors, 56
thermoregulation, 598–610. *see also* body temperature
    behavioral, 598
    body temperature, 598–600, 605, 606, 607, 608
    central nervous system in, 589, 599
    central thermoregulatory controller, 606
    clinical aspects of, 606–610
    cold acclimatization, 606
    cold response, 604–606
    cold stress, 606
    during exercise
        cardiovascular system effects, 601, 608
        compensatory responses, 603, *603*
        vs. fever, 606
        in heat, 602–603
    during fever, 606
    heat acclimatization, 603–604, *604,* 605t
    heat stress
        disorders related to, 607–608
        effects of, 598
    hypothermia, 605–606, 608b–610b
    nonthermal effects on, 601
    physiological, 598
    set point in
        defined, 600
        factors influencing, 601
    skin temperature, 599–600, 602
    thermal stress, 591
    thermoneutrality, 595, 598
theta waves, 133
thiamine, 531t, 532
thick filaments, 163
    of skeletal muscle, 151
thin filaments, 163
third-degree atrioventricular block, 253
thirst center, 467
thoracic cavity, components of, 354, 355
thoracolumbar region, 106